Managing
for
Healthy
Ecosystems

Managing *for* Healthy Ecosystems

Edited by

David J. Rapport
William L. Lasley
Dennis E. Rolston
N. Ole Nielsen
Calvin O. Qualset
Ardeshir B. Damania

Associate Editors

Daniel W. Anderson
Darwin Anderson
James R. Carey
Santiago Carrizosa
Daniel P.Y. Chang
Gary N. Cherr
Alexander H. Harcourt

Robert T. Lackey
Jean Lebel
Nicholas W. Lerche
Jonna A.K. Mazet
Albert C. Medvitz
Ganapati P. Patil
Ruth A. Reck

Terrell P. Salmon
Marc B. Schenker
David Waltner-Toews
Bruce A. Wilcox
Barry W. Wilson

LEWIS PUBLISHERS

A CRC Press Company
Boca Raton   London   New York   Washington, D.C.

Cover photo credits:
Background of water and birds: Milton Friend, U.S. Geological Survey, La Quinta, CA, USA.
Woman with basket of food, and the frog: Santiago Carrizosa, Genetic Resources Conservation Program, University of California, Davis, CA, USA.
Rice terraces of Radi, eastern Bhutan: B.R. Lu, E. Hettel, and G.C. Loresto, International Rice Research Institute, The Phillippines.

Library of Congress Cataloging-in-Publication Data

Managing for healthy ecosystems / edited by David J. Rapport ... [et al.].
    p. cm.
    Papers from an international congress held at the University of California, Davis, Aug. 1999.
    Includes bibliographical references.
    ISBN 1-56670-612-2 (alk. paper)
       1. Ecosystem management—Congresses. I. Rapport, David.

QH75 .A1 M37 2002
333.95—dc21   2002070099

Visit the CRC Press Web site at www.crcpress.com

© 2003 by CRC Press LLC
Lewis Publishers is an imprint of CRC Press LLC

No claim to original U.S. Government works
International Standard Book Number 1-56670-612-2
Library of Congress Card Number 2002070099
Printed in the United States of America 1 2 3 4 5 6 7 8 9 0
Printed on acid-free paper

Preface

This collection of papers reflects the diversity of concepts, methods, and case studies that address the challenge of *Managing for Healthy Ecosystems* as holistic environmental management in the context of health, integrity, and sustainability. *Ecosystem health* embodies the capacity of ecosystems to function without impairment, while *management* concerns the assessment of ecosystem conditions relative to social goals and adoption of interventions to achieve designated goals.

An international congress, Managing for Ecosystem Health, an activity of the International Society for Ecosystem Health, was hosted by the University of California, Davis, at Sacramento in August 1999. The congress was designed to cover most aspects of ecosystem management, hence the congress provided a solid framework for this book. The many contributors expressed widely diverging views as to the meanings and relevance of ecosystem health as a goal and practical guide to environmental management. The conventional means of dealing with problems in the environment by technological fixes, by relying solely on reductionist science, is directly confronted. What is sought is a balance between reductionist and holistic approaches to gain a clearer understanding of the complexity of the Earth's ecosystems and the interconnectivity between the condition of ecosystems and human activities and aspirations.

The contributions are grouped into three major parts. The first, Emerging Concepts, explores the diverse meanings of ecosystem health within ecological, socioeconomic, and human health perspectives, and the linkages to related concepts such as ecological integrity, sustainable development, and ecological footprints. In the second part, Issues and Methods, more than 60 chapters introduce methods for assessing and monitoring ecosystem health, including strategies for gaining political and stakeholder input and support for science-based ecosystem management. Both large- and small-scale ecosystem management schemes are evaluated. New representational and statistical approaches to compiling and summarizing complex data derived from the field and remote sensing are introduced and analyzed. The final part, Case Studies, reports experiences of interdisciplinary teams grappling with specific issues in a variety of aquatic and terrestrial ecosystems. Throughout these contributions, one underlying theme prevails: the necessity of forging interdisciplinary solutions, integrating across the natural, social, and health sciences, and of incorporating considerations of ethics, policy, and politics to put human ingenuity to the service of managing change in the world's ecosystems.

Over the past decade great attention has been paid to the fact that degradation of the Earth's ecosystems is curtailing ecosystem services and thus adversely impacting people by undermining the economy, social structures, and public health. Evidence for this abounds: loss of fisheries and forests; impacted local and regional economies; soil fertility harmed by salinity and other factors that adversely affect agricultural communities in developed and developing countries; loss of soil by wind and water erosion, threatening agriculture around the world; negative impacts of coastal pollution and eutrophication on marine and freshwater fisheries; mounting risks to public health; and many other direct and secondary effects of anthropogenic impacts on large- and small-scale ecosystems. It is encouraging that these issues have received wide recognition, and the goal of achieving healthy ecosystems has been articulated by many agencies and incorporated into the basic principles for achieving sustainable development. However, the challenge remains: how to manage for healthy ecosystems — that is, how to go beyond good intent and achieve results.

This book was not designed to be encyclopedic or a how-to manual; rather, it presents snapshots of the condition or health of our physical environment and how researchers and managers are approaching the need for systematic data and rational analysis for repairing damaged ecosystems. The reader will find numerous examples among the chapters in this book from which to draw future work. This snapshot of managing for healthy ecosystems is indeed an indication of the vigorous research and desire for mitigating the trends and directions of impaired environmental conditions throughout the globe. There is by no means consensus on the most appropriate concepts and methods to achieve healthy ecosystems, but the growing body of knowledge relevant to managing for ecosystem health is embedded in a growing number of case studies that demonstrate applications to environmental management with holistic approaches crossing disciplinary boundaries.

The contributions in this volume form an important part of that ongoing inquiry. They will not, and should not, put closure to the issues of how to achieve ecosystem health; rather, they serve to broaden the discussion of useful approaches toward achieving this goal. At present, degradation of the Earth's ecosystems continues virtually unchecked due to human population growth, consumerism, and the dominance of economic growth-oriented policies over environmental concerns. Thus, it is timely to redouble efforts to devise more effective strategies for achieving ecological as well as socioeconomic sustainability. This volume points the way by providing holistic concepts, state-of-the-art methodologies, and leading examples of integrative science that recognize and more effectively tackle these real-world complexities that challenge our common future.

The basic structure of this book was planned by the program committee that designed the content of the international congress. An international steering committee provided topical ideas and assisted in identifying authors. Many of the program committee members served as associate editors to conduct peer review of the manuscripts within sections and provide interpretation of the chapters in overviews that precede each of the 23 sections. The editors are extremely grateful for these contributions to the quality of this book. We especially acknowledge Deborah Rogers, who so ably facilitated the program committee for more than a year of intense activity. Claudette Oriel and Nancy Barker provided excellent logistical support before, during, and after the congress. Angela Moskow coordinated fund raising, Susan Donohue edited the Web site, and Heather Todd coordinated poster presentations and abstracts. A book of abstracts was produced that can be viewed at http://www.grcp.ucdavis.edu/publications/index.htm.

The congress could not have been held, nor could this book have been produced, without financial and in-kind support from many organizations. These are acknowledged with great appreciation elsewhere in this volume. Jennifer Hounsel provided important linkage of the International Society for Ecosystem Health between the congress secretariat at the University of California Genetic Resources Conservation Program throughout the whole activity. The CALFED Bay-Delta Ecosystem Restoration Program provided financial support, allowing us to engage the following for detailed editing of all of the manuscripts: Marti Childs, Jeff March, Linda Sears, Eileen O'Farrell, Penny Walgenbach, Joyce Nettleton, Jill Nicholls, and Susan Schideler. Their expert editing skills helped to smooth out much of the diversity in presentations by the authors. We are pleased that CRC Press/Lewis Publishers had the courage to join in this large effort. CRC's David Packer and staff, especially Randi Gonzalez, Debbie Vescio, and Erika Dery, and Gail Renard and James Yanchak of the Production Department, have worked so diligently and efficiently to produce an excellent book. Among the book editors we must confess to unequal participation by Ardeshir Damania, who handled each of the manuscripts and correspondence with authors persistently and efficiently. Patrick McGuire at the Genetic Resources Conservation Program assembled the electronic version in preparation for publishing.

The Center for Ecological Health Research, largely supported by grants from the U.S. EPA and ably managed by Cheryl Smith, provided logistical and financial support to the Congress.

David J. Rapport
William L. Lasley
Dennis E. Rolston
N. Ole Nielsen
Calvin O. Qualset
Ardeshir B. Damania

International Congress on Ecosystem Health

CONVENORS

David J. Rapport

William L. Lasley

INTERNATIONAL ORGANIZING COMMITTEE

David J. Rapport, *Chair*
Bernardo Aguilar
Val Beasley
György Miklós Böhm
Donald E. Buckingham
J. Baird Callicott
Norman L. Christensen
Robert Costanza
Elizabeth Dowdeswell
William H. Farland
William S. Fyfe
Tee L. Guidotti
Ann-Mari Jansson
Hans de Kruijf
William L. Lasley

Orie Loucks
Thomas E. Lovejoy
Anthony J. McMichael
Norman Myers
N. Ole Nielsen
Ganapati P. Patil
Jonathan Patz
Christian Thorpe
Mostafa Kamal Tolba
Ola Ullsten
David Waltner-Toews
Robert T. Watson
Laura Westra
Walter G. Whitford

PROGRAM DEVELOPMENT COMMITTEE

William L. Lasley, *Chair*
Daniel W. Anderson
Michael Barbour
Kurt Benirschke
James R. Carey
Daniel P.Y. Chang
Dick A. Daniel
Montague W. Demment
Holly D. Doremus
Michael D. Fry
Charles R. Goldman
Alexander H. Harcourt
David E. Hinton
Michael L. Johnson
Robert A. Johnston
Lynn Kimsey
Jerhold Last
Robert T. Lackey
Eugenia Laychak
Jonna Mazet

Mike McCoy
Frederick A. Murphy
N. Ole Nielsen
Richard B. Norgaard
Robert Pearcy
Calvin O. Qualset
John Reuter
Deborah L. Rogers
Dennis E. Rolston
Paul Sabatier
Terrell P. Salmon
Marc B. Schenker
Stephen H. Schneider
James N. Seiber
Cheryl Smith
Daniel Sperling
Ronald S. Tjeerdema
Llewellyn R. Williams
Barry W. Wilson

CONGRESS COORDINATING COMMITTEE

Calvin O. Qualset, *Chair*
William L. Lasley
N. Ole Nielsen

Deborah L. Rogers
Dennis E. Rolston
Cheryl Smith

SECRETARIAT
UNIVERSITY OF CALIFORNIA
GENETIC RESOURCES CONSERVATION PROGRAM

Calvin O. Qualset, *Director*
Nancy L. Barker, *Program Facilitator*
Susan Donahue, *Web Designer*
Claudette Oriol, *Event Coordinator*

Angela Moskow, *Fund Development Specialist*
Deborah L. Rogers, *Program Facilitator*
Heather Todd, *Poster Session Coordinator*

INTERNATIONAL SOCIETY FOR ECOSYSTEM HEALTH

David J. Rapport, *President*

Jennifer Hounsell, *Coordinator*

Supporters of the International Congress on Ecosystem Health

California Department of Water Resources
    CALFED Bay-Delta Program
California State Water Resources Control Board
U.S. Department of the Interior
    Bureau of Reclamation
    National Park Service
U.S. National Aeronautics and Space Administration
U.S. Department of Energy
    National Institute of Global Environmental Change
U.S. National Institute of Environmental Health Sciences
U.S. Department of Commerce
    National Oceanic and Atmospheric Administration
U.S. Department of Agriculture
    Cooperative State Research, Education, and Extension Service
U.S. Environmental Protection Agency
The University of Western Ontario Faculty of Medicine and Dentistry (Canada)
University of California Genetic Resources Conservation Program
University of California, Davis
    College of Agriculture and Environmental Sciences
    Center for Ecological Health Research
    Center for Environmental Health Sciences
    Commission on Environment
    John Muir Institute of the Environment
    NIEHS–UCD Superfund Basic Research Program
    Office of the Vice Chancellor for Research
    School of Medicine
    School of Veterinary Medicine
    Systemwide Ecotoxicology Lead Campus Graduate Program
University of Guelph (Canada)
Encyclopedia of Life Support Systems (U.K.)
International Development Research Center (Canada)
Natural Resources Canada
Health Canada
Medical Research Council (Canada)
The World Bank

Contributors

Pauzi Abdullah
LESTARI
Selangor, Malaysia

Saiful Arif Abdullah
University Kebangsaan Malaysia
Selangor, Malaysia

Muhammad Abu Yusuf
Ministry of Science and Technology
Dhaka, Bangladesh

Heiner Acevedo
Instituto Nacional de Biodiversidad
Santo Domingo de Heredia, Costa Rica

W. Neil Adger
Center for Social and Economic Research on
 the Global Environment
Norwich, Norfolk, England

Miguel A. Altieri
University of California
Berkeley, CA

Ernesto Alvarado
USDA Forest Service
Seattle, WA

Saúl Alvarez-Borrego
Centro de Investigación Científica y de
 Educación Superior de Ensenada
Ensenada, Mexico

Vittorio Amadio
Ministry of Environment
Rome, Italy

Daniel W. Anderson
University of California
Davis, CA

Darwin W. Anderson
University of Saskatchewan
Saskatoon, Saskatchewan, Canada

Erin Anderson
University of California
Davis, CA

Jock R. Anderson
The World Bank
Washington, DC

James K. Andreasen
U.S. Environmental Protection Agency
Washington, DC

Erik Arvin
Technical University of Denmark
Lyngby, Denmark

Lowell Ashbaugh
University of California
Davis, CA

Francis R. Beck
The Pennsylvania State University
University Park, PA

Ken W. Belcher
University of Saskatchewan
Saskatoon, Saskatchewan, Canada

A.L. Bern
University of California
Davis, CA

Paul Bertram
U.S. Environmental Protection Agency
Chicago, IL

M. Boehm
University of Saskatchewan
Saskatoon, Saskatchewan, Canada

André F. Boshoff
University of Port Elizabeth
Port Elizabeth, South Africa

Peter Brang
Swiss Federal Institute for Forest, Snow, and
 Landscape Research
Birmensdorf, Switzerland

Robert P. Brooks
The Pennsylvania State University
University Park, PA

Peter N. Brostrom
California Department of Water Resources
Sacramento, CA

Sharon F. Browder
U.S. Fish and Wildlife Service
Stevensville, MT

Lisa Burley
International Development Research Center
Ottawa, Ontario, Canada

Richard T. Burnett
Health Canada
Ottawa, Ontario, Canada

William C.G. Burns
Pacific Institute for Studies in Development,
 Environment, and Security
Oakland, CA

Thomas A. Cahill
University of California
Davis, CA

John C. Callaway
University of San Francisco
San Francisco, CA

Michael C. Calver
Murdoch University
Murdoch, Perth, Western Australia

Sally Campbell
USDA Forest Service
Portland, OR

Victor Canton
Ministry of Housing, Land Planning, and
 Environment
Montevideo, Uruguay

James R. Carey
University of California
Davis, CA

Maria Carpio-Obeso
California Regional Water Quality Control
Palm Desert, CA

Santiago Carrizosa
University of California
Davis, CA

Omar Carvacho
University of California
Davis, CA

Marco Castro
Instituto Nacional de Biodiversidad
Santo Domingo de Heredia, Costa Rica

Joseph J. Cech, Jr.
University of California
Davis, CA

Allan J. Cessna
Environment Canada
Saskatoon, Saskatchewan, Canada

Daniel P.Y. Chang
University of California
Davis, CA

Gary N. Cherr
University of California
Bodega Marine Lab
Bodega Bay, CA

Monica Heekyoung Choi
Chevron
Richmond, CA

Scott Christiansen
USDA–ARS–OIRP
Beltsville, MD

Steven S. Cliff
University of California
Davis, CA

Dean O. Cliver
University of California
Davis, CA

Michael Cohen
Pacific Institute for Studies in Development,
 Environment, and Security
Oakland, CA

Cecilia Collados
Independent Researcher
Walnut Creek, CA

Barbara Conkling
North Carolina State University
Research Triangle Park, NC

Pedro Cordero
Instituto Nacional de Biodiversidad
Santo Domingo de Heredia, Costa Rica

John Coulston
North Carolina State University
Research Triangle Park, NC

Margaret M. Crossley
University of Saskatchewan
Saskatoon, Saskatchewan, Canada

Lisa L. Dale-Burnett
University of Regina
Regina, Saskatchewan, Canada

Ardeshir B. Damania
University of California
Davis, CA

Hans A.M. de Kruijf
Ecoassistance, UNITAR
Breedenbroek, The Netherlands

Jos N.M. Dekker
Utrecht University
Utrecht, The Netherlands

Diana Deumling
Redefining Progress
Oakland, CA

Ashk Dhawan
University of California
Intermountain Research and Extension Center
Tulelake, CA

Julian Dumanski
Consultant in Sustainable Land Management
Nepean, Ontario, Canada

C.A. Eagles-Smith
University of California
Davis, CA

Curtis M. Edmonds
U.S. Environmental Protection Agency
Las Vegas, NV

Robert G. Eilers
University of Manitoba
Winnipeg, Manitoba, Canada

Keith Endres
U.S. Environmental Protection Agency
Research Triangle Park, NC

Alan Ewert
Indiana University
Bloomington, IN

Exequiel Ezcurra
Instituto Nacional de Ecología
Coyoacan, Mexico

Sally K. Fairfax
University of California
Berkeley, CA

Vincent Favrel
Free University of Brussels
Brussels, Belgium

Linda Fernandez
University of California
Riverside, CA

William S. Fisher
U.S. Environmental Protection Agency
Gulf Breeze, FL

Allan K. Fitzsimmons
Balanced Resource Solutions
Woodbridge, VA

Robert Flocchini
University of California
Davis, CA

Igor V. Florinsky
University of Manitoba
Winnipeg, Manitoba, Canada

John Freemuth
Boise State University
Boise, ID

Milton Friend
U.S. Geological Survey
Madison, WI

Rodney M. Fujita
Environmental Defense
Oakland, CA

William S. Fyfe
The University of Western Ontario
London, Ontario, Canada

Flávio Bertin Gandara
University of Sao Paulo
Piracicaba, Brazil

Jaqueline García
Centro de Investigaciones en Alimentación y
  Desarollo
Guaymas, Sonora, Mexico

Rospidah Ghazali
LESTARI
Selangor, Malaysia

James D. Giattina
U.S. Environmental Protection Agency
Stennis Space Center, MS

Edward P. Glenn
University of Arizona
Tucson, AZ

Robert J. Glennon
University of Arizona
Tucson, AZ

Charles R. Goldman
University of California
Davis, CA

Luis González
Instituto Nacional de Biodiversidad
Santo Domingo de Heredia, Costa Rica

John W. Grandy
The Human Society of the United States
Washington, DC

Melissa M. Grigione
University of South Florida
Tampa, FL

Paul D. Gunderson
Marshfield Clinic
Marshfield, WI

Louise Hagel
University of Saskatchewan
Saskatoon, Saskatchewan, Canada

Richard R. Hamilton
Hamilton Ranch
Rio Vista, CA

Jack Hanson
AgResource Solutions
Susanville, CA

Alexander H. Harcourt
University of California
Davis, CA

Walter Hecq
Free University of Brussels
Brussels, Belgium

Daniel T. Heggem
U.S. Environmental Protection Agency
Las Vegas, NV

Kurt J. Hembree
University of California
Fresno County Cooperative Extension
Fresno, CA

E.V. Herrero
University of California
Yuba–Sutter Counties Cooperative Extension
Yuba City, CA

Paul F. Hessburg
USDA Forest Service
Wenatchee, WA

Alan C. Heyvaert
University of California
Davis, CA

Harry Hirvonen
Canadian Forest Service
Ottawa, Ontario, Canada

Phan Nguyen Hong
Center for Natural Resources and
  Environmental Studies
Hanoi, Vietnam

John D. Hopkins
University of California
Davis, CA

Paul Horvatin
U.S. Environmental Protection Agency
Chicago, IL

Ann G. Houck
University of California
Davis, CA

Lynn Huntsinger
University of California
Berkeley, CA

Shahruddin Idrus
LESTARI
Selangor, Malaysia

D.G. Irvine
University of Saskatchewan
Saskatoon, Saskatchewan, Canada

Laura E. Jackson
U.S. Environmental Protection Agency
Research Triangle Park, NC

Teresa James
University of California
Davis, CA

Alan D. Jassby
University of California
Davis, CA

Barry Jessiman
Health Canada
Ottawa, Ontario, Canada

Steen Joffe
Consultant in Agricultural Economics
Linton, Kent, England

Douglas H. Johnson
U.S. Geological Survey
Jamestown, ND

Glen D. Johnson
New York State Department of Health
Troy, NY

Mike B. Johnson
University of California
Davis, CA

K. Bruce Jones
U.S. Environmental Protection Agency
Las Vegas, NV

Stephen R. Kaffka
University of California
Intermountain Research and Extension Center
Tulelake, CA

Paulo Y. Kageyama
University of Sao Paulo
Piracicaba, Brazil

James R. Kahn
Washington and Lee University
Lexington, VA

Maarten Kappelle
Utrecht University
Utrecht, The Netherlands

Arturo A. Keller
University of California
Santa Barbara, CA

P. Mick Kelly
University of East Anglia
Norwich, Norfolk, England

Graham I.H. Kerley
University of Port Elizabeth
Port Elizabeth, South Africa

Kirke A. King
U.S. Fish and Wildlife Service
Phoenix, AZ

Donald W. Kirby
University of California
Intermountain Research and Extension Center
Tulelake, CA

Michael H. Knight
South African National Parks
Humewood, South Africa

Lada V. Kochtcheeva
University of Oregon
Eugene, OR

Ibrahim Komoo
University Kebangsaan Malaysia
Selangor, Malaysia

R. Sari Kovats
London School of Hygiene and Tropical
  Medicine
London, England

Daniel Krewski
University of Ottawa
Ottawa, Ontario, Canada

Armand M. Kuris
University of California
Santa Barbara, CA

Janis C. Kurtz
U.S. Environmental Protection Agency
Gulf Breeze, FL

Robert T. Lackey
U.S. Environmental Protection Agency
Corvallis, OR

G.W. Lammers
National Institute of Public Health and the
  Environment
Bilthoven, The Netherlands

Wayne G. Landis
Western Washington University
Bellingham, WA

W. Thomas Lanini
University of California
Davis, CA

William L. Lasley
University of California
Davis, CA

Jean Lebel
International Development Research Center
Ottawa, Ontario, Canada

D.L. Ledingham
Public Health Services
Saskatoon, Saskatchewan, Canada

Glenn W. Lelyk
University of Manitoba
Winnipeg, Manitoba, Canada

Nicholas W. Lerche
University of California
Davis, CA

Karen Levy
Environmental Defense
Oakland, CA

Susan Lindstrom
Consulting Archeologist
Truckee, CA

Mary B. MacDonald
University of Saskatchewan
Saskatoon, Saskatchewan, Canada

Camilla Maclean
University of Warwick
Coventry, England

Ronald W. Matheny
U.S. Environmental Protection Agency
Research Triangle Park, NC

Robert T. Matsumura
T2 Systems, Inc.
Yuba City, CA

Csaba Mátyás
University of West Hungary
Sopron, Hungary

Jonna A.K. Mazet
University of California
Davis, CA

Helen H. McDuffie
University of Saskatchewan
Saskatoon, Saskatchewan, Canada

S. C. McHatton
Novozymes Biologicals
Salem, VA

John F. McLaughlin
Western Washington University
Bellingham, WA

Anthony J. McMichael
Australian National University
Canberra, Australia

Albert G. Medvitz
McCormack Sheep and Grain
Rio Vista, CA

Megan H. Mehaffey
U.S. Environmental Protection Agency
Las Vegas, NV

Eric Mellink
Centro de Investigación Cientifíca y de
  Educación Superior de Ensenada
Ensenada, Mexico

Edgar Méndez
Instituto Nacional de Biodiversidad
Santo Domingo de Heredia, Costa Rica

Donna Mergler
Centre for the Study of Biological Interactions
  in Human Health
Montreal, Quebec, Canada

Philip Milner
University of Bath
Bath, England

Jeff P. Mitchell
University of California
Kearney Agricultural Center
Parlier, CA

E.M. Miyao
University of California
Yolo–Solano Counties Cooperative Extension
Woodland, CA

Ahmad Fariz Mohamed
LESTARI
Selangor, Malaysia

Huberth Monge
Instituto Nacional de Biodiversidad
Santo Domingo de Heredia, Costa Rica

Miguel A. Mora
U.S. Geological Survey
College Station, TX

Michael J. Moran
U.S. Geological Survey
Rapid City, SD

Eberhard Morgenroth
University of Illinois
Urbana, IL

Jason Morrison
Pacific Institute for Studies in Development,
  Environment, and Security
Oakland, CA

J.F. Mount
University of California
Davis, CA

W. Mohd Muhiyuddin
LESTARI
Selangor, Malaysia

Judith H. Myers
University of British Columbia
Vancouver, British Columbia, Canada

Wayne L. Myers
The Pennsylvania State University
University Park, PA

Maliha S. Nash
U.S. Environmental Protection Agency
Las Vegas, NV

Anne C. Neale
U.S. Environmental Protection Agency
Las Vegas, NV

Peter Neitlich
USDI National Park Service
Winthrop, WA

D.C. Nelson
University of California
Davis, CA

David Newsome
Murdoch University
Murdoch, Perth, Western Australia

Clara Ines Nicholls
University of California
Berkeley, CA

N. Ole Nielsen
University of Guelph
Guelph, Ontario, Canada

M. Nordin
University Kebangsaan Malaysia
Selangor, Malaysia

Ligia Noronha
Tata Energy Research Institute
Panaji, India

Reed Noss
Conservation Science, Inc.
Corvallis, OR

Timothy J. O'Connell
Penn State Cooperative Wetlands Center
University Park, PA

Kenneth Olden
National Institute of Environmental Health
  Sciences
Research Triangle Park, NC

Michael A. O'Malley
University of California
Davis, CA

Robert V. O'Neill
Oak Ridge National Laboratory
Oak Ridge, TN

Roger D. Ottmar
USDA Forest Service
Seattle, WA

Eduardo Palacios
Centro de Investigación Científica y de
  Educación Superior de Ensenada
Ensenada, Mexico

Ganapati P. Patil
The Pennsylvania State University
University Park, PA

Jonathan A. Patz
Johns Hopkins University
Baltimore, MD

Joy Jacqueline Pereira
University Kebangsaan Malaysia
Selangor, Malaysia

Christian Pieri
The World Bank
Washington, DC

Timothy S. Prather
University of Idaho
Moscow, ID

Jeff Price
American Bird Conservancy
Boulder, CO

Calvin O. Qualset
University of California
Davis, CA

Todd A. Radenbaugh
University of Regina
Regina, Saskatchewan, Canada

Iral Ragenovich
USDA Forest Service
Portland, OR

David J. Rapport
University of Guelph
Guelph, Ontario, Canada

Ruth A. Reck
University of California
Davis, CA

William E. Rees
University of British Columbia
Vancouver, British Columbia, Canada

Rudo Reiling
National Institute of Public Health and the
    Environment
Bilthoven, The Netherlands

Richard J. Reiner
The Nature Conservancy
Chico, CA

William K. Reisen
University of California
Arbovirus Field Station
Bakersfield, CA

Gail Remus
University of Saskatchewan
Saskatoon, Saskatchewan, Canada

John E. Reuter
University of California
Davis, CA

P.J. Richerson
University of California
Davis, CA

Kurt H. Riitters
U.S. Geological Survey
Research Triangle Park, NC

David M. Rizzo
University of California
Davis, CA

Tonie E. Rocke
U.S. Geological Survey
Madison, WI

Dennis E. Rolston
University of California
Davis, CA

Alan M. Rosenberg
University of Saskatchewan
Saskatoon, Saskatchewan, Canada

Orazio Rossi
University of Parma
Parma, Italy

Allen T. Rutberg
The Humane Society of the United States
Washington, DC

Paul Safonov
Russian Academy of Sciences
Moscow, Russia
and
Free University of Brussels
Brussels, Belgium

Raphael D. Sagarin
University of California
Hopkins Marine Station
Pacific Grove, CA

Terrell P. Salmon
University of California
Davis, CA

R. Brion Salter
USDA Forest Service
Wenatchee, WA

Abdullah Samat
LESTARI
Selangor, Malaysia

Marc Schenker
University of California
Davis, CA

G. Schladow
University of California
Davis, CA

K.M. Semchuk
University of Saskatchewan
Saskatoon, Saskatchewan, Canada

A. Senthilselvan
University of Alberta
Edmonton, Alberta, Canada

M. Sexton
University of California
Davis, CA

Abdul Hadi Harmon Shah
LESTARI
Selangor, Malaysia

Harvey Shear
Environment Canada
Downsview, Ontario, Canada

Deborah J. Shields
USDA Forest Service
Fort Collins, CO

Daniel Simberloff
University of Tennessee
Knoxville, TN

Ashbindu Singh
UNEP GRID
Sioux Falls, SD

Nicholas C. Slosser
Conservation Biology Institute
Corvallis, OR

D.G. Slotten
University of California
Davis, CA

K. Shawn Smallwood
Consulting in the Public Interest
Davis, CA

E.M.W. Smeets
National Institute of Public Health and the
 Environment
Bilthoven, The Netherlands

Bradley G. Smith
USDA Forest Service
Bend, OR

William Smith
USDA Forest Service
Research Triangle Park, NC

Rick Soehren
California Department of Water Resources
Sacramento, CA

Slavko V. Šolar
Geological Survey of Slovenia
Ljubljiana, Slovenia

Colin L. Soskolne
University of Alberta
Edmonton, Alberta, Canada

Paul J. Squillace
U.S. Geological Survey
Rapid City, SD

Nancy Stadler-Salt
Environment Canada
Burlington, Ontario, Canada

David Stieb
Health Canada
Ottawa, Ontario, Canada

John J. Streicher
U.S. Environmental Protection Agency
Research Triangle Park, NC

Dean T. Stueland
Marshfield Clinic
Marshfield, WI

Thomas H. Suchanek
University of California
Davis, CA

Rita Schmidt Sudman
Water Education Foundation
Sacramento, CA

Brian J. Swisher
University of Vermont
Burlington, VT

Charles Taillie
The Pennsylvania State University
University Park, PA

Steve R. Temple
University of California
Davis, CA

E.R. Terry
The World Bank
Washington, DC

Lori Ann Thrupp
U.S. Environmental Protection Agency
San Francisco, CA

Susanna T.Y. Tong
U.S. Environmental Protection Agency
Cincinnati, OH

Nguyen Hoang Tri
Center for Natural Resources and
  Environmental Studies
Hanoi, Vietnam

Ola Ullsten
World Commission on Forests and Sustainable
  Development
Burlington, Ontario, Canada

Karma Ura
The Center for Bhutan Studies
Thimphu, Bhutan

Christina Echavarría Usher
Universidad de Antioquia
Medellin, Colombia

Carlos Valdes-Casillas
Centro de Conservación para el
  Aprovechamiento de Recursas Naturales
Guaymas, Sonora, Mexico

C. Martijn van der Heide
Agricultural University
Wageningen, The Netherlands

Ekko C. van Ierland
Agricultural University
Wageningen, The Netherlands

Rick D. van Remortel
Lockheed Martin Environmental Services
Las Vegas, NV

Detlef P. van Vuuren
National Institute of Public Health and the
  Environment
Bilthoven, The Netherlands

Marylou Verder-Carlos
California Environmental Protection Agency
Sacramento, CA

H.J. Verkaar
National Institute of Public Health and the
  Environment
Bilthoven, The Netherlands

Mathis Wackernagel
Redefining Progress
Oakland, CA

Timothy G. Wade
U.S. Environmental Protection Agency
Research Triangle Park, NC

David Waltner-Toews
University of Guelph
Guelph, Ontario, Canada

Alyson Warhurst
University of Warwick
Coventry, England

L.B. Webber
University of California
Davis, CA

Richard Weisskoff
University of Miami
Coral Gables, FL

Laura Westra
York University
Toronto, Ontario, Canada

James D. Wickham
U.S. Environmental Protection Agency
Research Triangle Park, NC

Bruce A. Wilcox
University of Hawaii
Honolulu, HI

Barry W. Wilson
University of California
Davis, CA

Maiken Winter
State University of New York
Syracuse, NY

Nancy B. Young
Marshfield Clinic
Marshfield, WI

Terry F. Young
Environmental Defense
Oakland, CA

Lester Yuan
U.S. Environmental Protection Agency
Washington, DC

Joy B. Zedler
University of Wisconsin
Madison, WI

Minghua Zhang
University of California
Davis, CA

R. Zierenberg
University of California
Davis, CA

John S. Zogorski
U.S. Geological Survey
Rapid City, SD

Giovanni Zurlini
University of Lecce
Lecce, Italy

Table of Contents

PART I

Emerging Concepts

SECTION I.1

Setting the Stage

CHAPTER 1

Regaining Healthy Ecosystems: The Supreme Challenge of Our Age

David J. Rapport

RECOGNIZING THE CONCEPT

The first sign of progress toward any goal is recognition that the goal is worthy of the effort. A decade ago, applying the concept of health beyond the individual and population level was deemed foolhardy. Ecologists in particular were adamant that ecosystems were not organized as organisms. Thus the concept of health, which applied to individuals and, later, populations, was inappropriate for ecosystems. They were partly right; indeed, ecosystems are not organized in the same manner as organisms. Ecosystems are not superorganisms. Ecosystems do, however, have organization, structure, and function. They constitute another level above that of populations — but below that of landscapes and biomes — in the biological hierarchy from cells to the biosphere. Thus, ecologists were wrong to assume that, because ecosystems are not organized in the same manner as organisms, the concept of *health* had no application to this level.

Clearly the concept of health — which at root refers to the capacity of a system (whether biological, social, or mechanical) to perform normal functions — is not restricted to the hierarchical levels in which the particular systems operate. There are healthy and unhealthy cells, tissues, organs, organisms, populations, biotic communities, ecosystems, and landscapes. When it comes to the biosphere, that all-encompassing dimension of life, its state of health also depends on the degree to which its functions are unimpaired.

Today the notion of healthy ecosystems, if not yet mainstream, has more than a toehold in the public domain. Numerous programs within government agencies are targeting the health of ecosystems as a key priority. For example, in a national audit of Australia, ecosystem health is one of the focus areas. In the guidelines for a recently launched project of the National Round Table on the Environment and the Economy (Canada) on indicators of sustainability, ecosystem health indicators with respect to natural resource systems are specified in the design. Major projects in managing forests, estuaries, and lakes around the world now focus on the degree to which the essential functions of these ecosystems have become compromised and how they may be restored. And, while species at risk are still major drivers of public concern about the environment, that concern has widened to consider ecosystems at risk.

The goal of managing for healthy ecosystems recognizes explicitly the human component. No longer is it tenable to consider ecosystems isolated from humans. Of course it has been recognized for decades — even made explicit in the writings of Aldo Leopold in the 1940s — that human activity has compromised ecosystem functions. However, the connections between humans and ecosystems go much deeper. True, humans have compromised ecosystem functions; but it is also true that ecosystem degradation has compromised humans physiologically, socially, economically, culturally, and psychologically.

How could it be otherwise? We are part of ecosystems and fully dependent on the functions of these systems for our very lives. We rely on the food, water, and shelter they provide, on their value for spiritual healing, and on the cultural aspects, particularly in those societies in which whole cultures co-evolved in intimate association with particular ecosystems. Destroy those ecosystems, and inevitably the cultures are destroyed as well.

Clarification of the concept and the role of culture in defining the meaning of healthy ecosystems are central to a number of the themes within this volume. It is particularly prominent in the section on Emerging Concepts (Part I). In a practical vein the CALFED Bay–Delta Program is building a broad consensus among the stakeholders as to what constitutes a healthy environment, bringing together landowners, farmers, natural resource managers, nongovernment organizations (NGOs), and others. At root the issues are political: what steps can be taken to rehabilitate the Bay Delta without undermining the economy or the social/cultural values of the area?

In a more theoretical vein, the issues of ecosystem health, quality of life, and ecological integrity are the topics of Section I.4, Setting Goals and Objectives in Managing for Healthy Ecosystems. Questions arise as to the appropriate use of ecosystem health concepts in public policy, whether the use of the concept should be the basis for federal and land use management, and the degree to which the concept relates to sustainability.

DEVELOPING THE INDICATORS

Management requires information, and information requires both data and understanding. Recognition is increasing about the importance of maintaining ecosystem functions and the high costs to society if these are impaired or lost (in terms of human health vulnerabilities, access to safe food and water, and the ensuing risks of civil disobedience). This increased recognition has led to a growing concerted effort to report on the state of health of major ecosystems: forests, coastal marine systems, great lakes, prairies, etc. Such reporting requires a judicious choice of indicators (Rapport, 1992). Obviously, when major ecosystems such as the San Francisco Bay or Lake Tahoe have become radically transformed as a consequence of chronic cumulative stress from human activities, there will be inevitable changes in nearly all of their biophysical, socioeconomic, human health, and cultural dimensions. What are the most sensitive and meaningful parameters to measure as indicators of the entire process of transformation? To what degree can one validate the indicators as correlating with the larger system change? In other words, which are the most informative parameters to track? Which are the most reliable groups of indicators that provide an overview of changes in the state of the ecosystem over time? This is one of the major challenges within the ecological indicator field.

A framework for addressing this issue arises from the notion of an *ecosystem distress syndrome* (EDS) (Rapport et al., 1985). EDS comprises a group of signs commonly observed in ecosystems under stress, such as the loss of biodiversity, altered primary productivity, altered nutrient cycling, increased dominance of biotic communities by exotic (non-native) species, and the like. One of the common patterns in all of these changes is that the ecosystems under stress appear to have the characteristics of *retrogression* — that is, the appearance of a system moving to an earlier stage of succession. Under these conditions, terrestrial ecosystems lose accumulated nutrients (within the soils), while aquatic systems (lower in the landscape) tend to accumulate nutrients. In both cases, the biotic community tends to become dominated by a few opportunistic species and biodiversity within the ecosystem declines.

However, an important distinction separates a healthy ecosystem at an early stage of primary or secondary succession from an ecosystem that is in retrogression. In the healthy system, these early stages of colonization set the stage for later stages, characterized by increasing biocomplexity. In the unhealthy (stressed) system, retrogression is not often reversed without outside intervention (rehabilitation). Even with such interventions, success in restoring health is not a foregone conclusion (Rapport and Whitford, 1999).

Ecological (biophysical) signs of degrading ecosystems are now well documented in the literature (e.g., Hilden and Rapport, 1993; Vitousek et al., 1997). While a variety of specific mechanisms are involved in producing any one of the indicators (for example, many different ecological pathways may result in a decrease in nutrient availability to the biotic components of the ecosystem), the broad manifestations of ecosystem degradation are highly similar (Rapport and Whitford, 1999).

The question of indicators is, as one might expect, a central underlying theme in almost all of the chapters of this book, for obvious reasons: indicators are the basis for managing for ecosystem health or any other societal goal. Section I.5, Finding Indicators, includes indicators for forest ecosystem health, for the impact of ecological degradation on human health, for terrestrial and aquatic system function, and for ecological and economic sustainability.

The concrete application of indicators is the subject of Part II, Section II.1, in which environmental issues are the focus. These issues include biodiversity, climate change, and invasion of exotic species, among others. Each issue requires hard information upon which to make judgments, calling into play indicators of ecosystem health.

INTEGRATING THE DISCIPLINES

Ecosystem health is now emerging as a societal goal, and in this process it has become clear that the concept of healthy ecosystems extends well beyond the ecological domain. Ecosystem health assessments require an understanding of the social/cultural determinants of human health and of evolving systems of governance, ethics, and environmental management. Each of these dimensions has largely been a world unto itself. Integrative concepts, such as ecosystem health, however, force communication among the silos of disciplinary knowledge and the emergence of new, more effective transdisciplinary approaches to problems.

There are various possible levels of integration within a branch of science. For example, within ecology, integration among subdisciplines may bring together behavioral ecology with population ecology. Understanding changes in biological community structure may require integration across fields within a branch of science such as sociology, anthropology, and economics. Understanding the dynamics of human communities and the factors which interactively alter the health of ecosystems may call for integration of social, biological, and health sciences. There are even further possibilities in transcending disciplines, such as crossing the great divide between the arts and the sciences. Some fields are, by their nature, more naturally predisposed to integration than others. For example, medicine attempts to train physicians to integrate all the elements of medical education (a whole host of fields, e.g., immunology, physiology, anatomy, neurology) as well as ethics, and, to some extent, social factors.

Bringing these heretofore largely isolated silos into contact with one another is very much in evidence in the chapters that comprise this volume. For example, in the CALFED project to restore health to the San Francisco Bay, biotic communities, politics, culture, and tradition are viewed as interactive. In the case studies presented in Part III, particularly those on mining, forest health, and grazing animals/rangelands, long-overdue connections are being forged between the social sciences and natural sciences. People are not only part of (and not *apart from*) the ecosystem, but their views, values, and goals must also be central to any viable path for managing for ecosystem health.

MOVING THE CONCEPT TO PRACTICE

As amply illustrated within the present work, the notion of ecosystem health has now moved far beyond mere philosophical discussion. While still debated in terms of its fundamental meanings, it has entered the world of policy and practice (e.g., Lackey, 2001).

What is evident, even at this comparatively early stage, is that the link between healthy ecosystems and healthy people has begun to resonate within the body politic. As a consequence, ecosystem health is finding new academic homes within the health sciences and the associated professional schools, particularly human and veterinary medicine (e.g., Rapport et al., 2001) and public health (de Gruijl, 2000; Hales et al., 2000; Haines and Kammen, 2000).

One of the major factors providing the impetus to move ecosystem health into practice has been the support from the medical and public health fields. I have long puzzled over why ecosystem health found its initial acceptance within medicine and public health rather than within ecology. One reason, no doubt, is the fact that the terminology initially established for ecosystem health derived from medicine (Rapport et al., 1979; 1989). Description of ecosystem conditions in terms of health, pathology, and distress syndromes has a familiar ring to medical practitioners, but not to ecologists. Indeed, the notion that ecosystems have the property of *health* has been actively resisted by some ecologists. They contend that health implies value judgments (as to the desirable state of nature) and that, while this may be accepted with respect to individuals, it does not extend to higher levels of biological organization. If this argument were to be accepted, it would invalidate the entire field of public health and population medicine.

Perhaps the overriding reason why ecosystem health has more readily found acceptance in the health sciences than in ecology is that the health sciences are, by their nature, caring professions and mission oriented. Their mission of restoring function, alleviating suffering, etc. is not unlike the mission of ecosystem health — restoring functions to degraded landscapes and ecosystems. Thus, ecosystem health and medicine do have a similar orientation. Another connection between these fields is that interventions aimed at restoring ecosystem functions serve directly to enhance human health (e.g., by enhancing the availability of such basic amenities as water and food). Permanent solutions to certain public health problems may demand a different perspective. For instance, New York City recently suffered an increased risk to public health from declining water quality due to land use changes in the drainage system. The solution may be found not by technical interventions (i.e., by building a bigger and better sewage treatment plant), but rather by managing the Catskill watershed (that supplies New York City with its water) in a manner that restores health and the ecological services that healthy watersheds provide.

Since the Stockholm Conference on the Environment (1972) and the United Nations Conference on Environment and Development (Rio Declaration, 1992), it is also recognized that economic development (and more generally the health of economies) is also dependent on maintaining healthy ecosystems. The Rio Declaration on Environment and Development was explicit on this point and committed the parties to develop a new legal framework that integrates resource management, protection of the environment, and future economic growth (Cicin-Sain, 1993). This mandate calls for a comprehensive approach to human futures, integrating economic policies with full recognition of environmental constraints. Principle 7 of the Rio Declaration calls upon nation states, individually and collectively, to take responsibility to "conserve, protect and restore the health and integrity of the Earth's ecosystems" (Rio Declaration, 1992). It is this declaration that has given rise to the so-called Agenda 21, the action plan to achieve the goals of the Rio Declaration. In effect, the action plan details through treaties and accords (such as the Kyoto Protocol) the means of achieving healthy ecosystems.

In Part III of this work, case studies illustrate how the concepts of ecosystem health are applied to practical efforts to restore function to some of the world's most damaged environments. Five studies ranging from the Colorado River delta to the Canadian prairies to the Langat Basin of Malaysia all demonstrate the potential for improving the health of large-scale ecosystems by taking a broad systems approach. These studies also suggest that times are changing, and that today one finds teams of applied ecologists that are adopting an ecosystem health perspective. An additional study also found in this section (on the impacts of a motor fuel additive on human health and the economics of the transport industry) further illustrates the potential applications of an ecosystem health perspective in action.

NEW GOALS FOR ENVIRONMENTAL MANAGEMENT

The history of environmental management dates to times of antiquity, when it was observed that humans impact their surroundings and that some practices are more deleterious than others. The earliest applications were undoubtedly in agriculture, where, particularly in dry or relatively infertile regions, managing the environment has long been practiced and continues to this day.

Over the past half century it has became apparent that many agricultural and industrial activities have cumulatively rendered ecosystems and landscapes less functional. These deleterious changes have often had dramatic and negative impacts on the human community through loss of income, medicinal plants, and culture, and increased threats or risks to human health. It has also become apparent that neighboring ecosystems, particularly aquatic systems, have become degraded as a consequence of certain management practices, such as agriculture and urbanization. As the interconnections of human activity, ecosystem dynamics, human health, international trade, and sustainable economic development have become better articulated, many new concerns are coming to light. Among these are the likely role of antibiotics (used as growth promoters in poultry, swine, and cattle production) in directly and indirectly increasing the prevalence of antibiotic resistance in certain human pathogens, especially those that invade the gut. Climate change increases human health vulnerabilities by spreading the range of vector-borne diseases as well as contributing to a greater frequency of extreme weather events.

These realities are encouraging a rapid evolution in the practice of environmental management, from protection of the environment from various sources of contamination — the focus being on clean air and clean water — to protection of the vitality, organization, and resilience of whole ecosystems, of which humans are an integral part. The new approach to environmental management goes well beyond crisis-driven management. No longer are single factors identified as the likely sole cause; rather, the search is for multiple stresses that singly and interactively degrade major ecosystems over time. Efforts at restoration are increasingly directed toward reestablishing critical ecosystem functions rather than one particular species.

Finally, monitoring for ecosystem health is far broader in scope than simply monitoring biophysical attributes. A watchful eye is also required to identify the socioeconomic, cultural, and human health manifestations of dysfunctional ecosystems (Maffi, 2001). Clearly the great challenge ahead in applying ecosystem health to environmental management is to make use of new methods that allow continuous monitoring (particularly from remote sensing) of biophysical conditions of large-scale ecosystems (Rapport et al., 1995) and to integrate these findings with socioeconomic, cultural, and human health trends. Further, if ecosystem health is to be maintained — and restored where it has been lost — what is needed above all else is the recognition that the politics of the environment inevitably must change. The current stance of *politics as usual* that applies to events that seem of no particular consequence must change to one of *politics of urgency* that applies to extreme threats to the future of humanity. While the insidious erosion of ecosystem health has none of the public appeal aroused by the September 11 crisis, it may well cost ultimately far more in lives. If ecosystem health is to be restored, it is time to back away from politics as usual and play hardball with the environment. Toward this end, the present volume provides abundant evidence not only that ecosystem health is a worthy goal for environmental management, but also that the concept finds ready application in solutions to real-world problems. In the past few years a multitude of new programs in ecosystem health (and ecosystem approaches to human health) have been initiated in the U.S., Canada, Australia, South America, and elsewhere, both within academic institutions and within government programs. Increasingly, there is recognition that healthy ecosystems perform services upon which humans and other biota are dependent. Thus, actions that threaten ecosystem health and threaten the continued supply of ecosystem services are coming under more intense scrutiny. The public and decision makers are becoming aware that, unless the health of the world's ecosystems is restored and maintained, socioeconomic aspirations will become increasingly compromised.

REFERENCES

Cicin-Sain, B., Sustainable development and integrated coastal management, *Ocean Coastal Manage.*, 21:11–43, 1993.

De Gruijl, F.R., Health effects from the sun's ultraviolet radiation and ozone as a stratospheric sunscreen, *Global Change Hum. Health*, 1:26–40, 2000.

Haines, A. and Kammen, D.M., Sustainable energy and health, *Global Change Hum. Health*, 1:78–87, 2000.

Hales, S., Kovats, S., and Woodward, A., What El Niño can tell us about human health and global climate change, *Global Change Hum. Health*, 1:66–77, 2000.

Hildén, M. and Rapport, D.J., Four centuries of cumulative cultural impact on a Finnish river and its estuary: an ecosystem health approach, *J. Aquat. Ecosyst. Health*, 2:261–275, 1993.

Lackey, R.T., Values, policy, and ecosystem health, *BioScience*, 51:437–444, 2001.

Maffi, L., *On Biocultural Diversity: Linking Language, Knowledge and the Environment*, Smithsonian Institution Press, Washington, D.C., 2001.

Rapport, D.J., What constitutes ecosystem health? *Persp. Biol. Med.*, 33: 120–132, 1989.

Rapport, D.J., Evolution of indicators of ecosystem health, in McKenzie, D.H., Hyatt, E.D., and McDonald, J.V. Eds., *Ecological Indicators*, Elsevier, Amsterdam, 1992, pp. 121–134.

Rapport, D.J. and Whitford, W.G., How ecosystems respond to stress: common properties of arid and aquatic systems, *BioScience*, 49:193–203, 1999.

Rapport D.J., Thorpe, C., and Regier, H.A., Ecosystem medicine, *Bull. Ecol. Soc. Am.*, 60: 180–182, 1979.

Rapport, D.J., Regier, H.A., and Hutchinson, T.C., Ecosystem behavior under stress, *Am. Nat.*, 125: 617–640, 1985.

Rapport, D.J., Gaudet, C., and Calow, P., Eds., *Evaluating and Monitoring the Health of Large-Scale Ecosystems*, Springer-Verlag, Heidelberg, 1995.

Rapport, D.J., Howard, J.M., Lannigan, R., Angema, C.M., and McCauly, W., Strange bedfellows: ecosystem health in the medical curriculum, in *Ecosyst. Health* 7(3): 155–162, 2001.

Rio Declaration on Environment and Development, 1992, in Johnson, S., Ed., *The Earth Summit: The United Nations Conference on Environment and Development*, Graham and Troutman/Martinus Nijhoff, London, 1993, 118.

Stuart, L.P., Márcio, A., Balmford, A., Branch, G., Brandon, K., Brooks, T., Bustamante, R., Costanza, R., Cowling, R., Curran, L.M., Dobson, A., Farber, S., da Fonseca, G.A.B., Gascon, C., Kitching, R., McNeely, J., Lovejoy, T., Mittermeier, R.A., Myers, N., Patz, J.A., Raffle, B., Rapport, D., Raven, P., Roberts, C., Rodríguez, J. Paul, Rylands, A., Tucker, C., Samper, C., Stiassny, M.L.J., Safina, C., Supriatna, J., Wall, D.H., and Wilcove, D., Can we defy nature's end, *Science*, 293:2207–2208, 2001.

Vitousek, P.M., Mooney, H.A., Lubchenco, J., and Melillo, J.M., Human domination of earth's ecosystems, *Science*, 277:494–499, 1997.

CHAPTER 2

The Politics of the Environment

Ola Ullsten

People are becoming more aware that we live in the transition from a world with an abundance of natural resources to one of severe limits. The world population is likely to increase with another 90 million per annum and will reach another doubling from today's 6 billion before it levels off. At the same time, access to water, soil, land, fishery, and forest resources is becoming increasingly scarce. Global warming adds to the problem. The space for expansion of the human enterprise is shrinking rapidly, and nothing seems to be stopping this trend. This is a somewhat dismal picture.

On the other hand, we can see ambitious efforts by governments in most countries introducing new legislation for environmental protection. In addition, numerous examples of successful projects of pollution control have given us cleaner air, lakes, and rivers, locally and regionally. A noticeably more proactive stand has been taken on environmental issues within the corporate world. Scientific achievements, emerging popular pro-environmental movements, and a subsequent growing awareness of the public are pushing government and businesses to do their part in protecting the environment.

How do we explain those two very contradictory pictures of the same piece of reality? Does it prove that current attempts to remedy what is wrong with our environment are meaningless? Is it meaningless to plant trees to combat deforestation; to establish protected areas of forest and marine reserves to preserve plant and animal species and to save wildlife; to treat industrial waste to protect groundwater; to prohibit the use of chemicals that are hazardous to the environment; to sort out garbage; and to ecolabel agricultural and manufactured products?

All these efforts are useful in themselves. Great value can be attained by having cleaner air and water. Earnings go down for businesses as a result of a more effective use of raw material, cleaner equipment, less waste, and less energy. However, ecology-minded businesses know they look good on the stock market and to their customers. Destroying forests and fisheries can hardly be justified from any point of view, especially if it happens voluntarily, and other options are available. All of these policies, when properly pursued, are as good for the local environment as they are for the economy of governments, businesses, or individuals.

The problem may be that they are only policies and nothing more. Actions may not reach beyond the improvement in the local environment and are limited to a handful of high-tech, high-income, and generally prosperous countries with a well-educated population. It is part of business as usual in the affluent societies, but with an ecolabel attached.

Technical solutions, upon which environmental policies are usually based, are necessary means for achieving tangible results in a process toward sustainable development. But sustainable development, solemnly embraced by political leaders worldwide, is meant as a new paradigm for

1-56670-612-2/03/$0.00+$1.50

development with an important new message — the health of the ecosystems places restraints on human economic activities. The message asks for something more than technical solutions. It speaks about the need to review the conventional wisdom that relies on the opposite assumption — that the economy and what is technologically possible set the limits.

It could hardly have been expected that the politically bipartisan Brundtland Commission, in which this new eco/economic doctrine was framed, could have given it a clear political interpretation. It is even less likely that the Earth Summit in Rio in 1992 could have approved the doctrine and what it embodies in political terms. The summit, after all, was a highly politicized event representing all possible political persuasions and well-defined national interests.

Sustainable development is a deceptive term. It is logical and easy to accept by most people until they begin dismantling the parts on which it is built. Then it becomes clear how many challenges it offers in its implementation. We endorse sustainable development in principle, but we have only a blurred vision of what it implies in terms of practical politics. It is too politically costly, too challenging for the political cultures of the Northern Hemisphere, and it is beyond the reach of the Southern Hemisphere. The farsighted approach, which is the doctrine's strength, is difficult to fully embrace by people who are fighting day by day, and by whatever means, for their survival.

The term *sustainability* is intrinsically holistic and interdisciplinary, which requires a policy-making process that has proven to be difficult to implement by any national administration, in rich or poor countries, or by international organizations, regardless of the issue. It embodies an extreme degree of complexity, ethically and technically. It makes value judgments about equity, which is a matter at the core of controversy in all domestic politics and an issue that flares up north/south tension at all international gatherings. It is long-term in character, which is inconsistent with the timescale of both elected governments and profit-driven business activities, particularly in developing countries. It demands changes in the way people behave, the true meaning of which repels the conservative instinct that dominates all societies (Dowdeswell, L., personal communication).

Most new development doctrines have an initial phase of popularity that tends to hide the costs, politically and financially, for making them work. Sustainability is no exception. One popular line is that sustainability is a win–win strategy. Sustainable management of the world's renewable resources is a condition for human survival, so what is good for the environment is good for development. Businesses like it, because it sounds like a low-cost investment. Politicians like it because they never like asking the electorates for sacrifices. Nongovernment organization (NGO) activists like it because it makes an easier sell.

There is a catch, though. Sustainability is a doctrine of limits, while conventional politics is caught up in an always-more-never-enough syndrome. That culture will have to change. It is not to be taken for granted that the world's natural resources can sustain a continued growth of the world's population coupled with a constant increase of consumption in already affluent countries. It is indeed an illusion to think that such consumption is possible if we also want to be serious about alleviating poverty in developing countries and elsewhere.

Thus, environmental policies so far have become halfway measures. They are tolerated as long as they help maintain established economic wisdom, which still remains based on the assumption that societal goals are to be expressed in economic terms, and environmental concerns are secondary. Typically, if not explicitly expressed that way, new environmental demands are always viewed as expensive concessions to the environmentalists and as something that compromises the overriding and more important economic goals. It is basically a conservative approach.

As necessary as they are, technical solutions to environmental problems can never be fully effective globally unless they become part of structural reforms with implications for our economic and governmental systems. We will have to accept that today's pressures on all resources limit the choices in management of human affairs. The policy measures this requires should not be dictated by a fear that abiding by the laws of ecology typically hurts the economy, but rather by the greater risk that declining ecosystems eventually may undermine a nation's prosperity and threaten the stability of the biosphere.

Ample examples exist, historically and in the present, of countries and regions that suffer widespread social misery, political instability, and cultural impoverishment due to declining ecosystems. The signs are easily visible for anyone to read about or view firsthand. Historically, the fall of the Mycenaean state and, later the civilization in ancient Greece, paralleled the loss of forests and the impoverishment of the landscape. Today's Haiti, Somalia, Ethiopia, parts of Russia, and Afghanistan (to mention a few) show signs of dysfunctioning landscapes — barren lands that were once majestic forests; deserts of salt due to diversion of rivers; uncovered rock instead of fertile soil; dried-up lakes and rivers; groundwater too deep for poor people to access; shortage of fuel wood; and farming conditions in decline.

Declining ecosystems are also present in more prosperous countries. The difference so far is that they have financial and technical means by which to remedy the damage and make the negative consequences of the decline less visible. But this can come only at some cost. The North American Free Trade Agreement (NAFTA) study (January 2002) concluded that Canada, over a 5-year period, spent 1.9 billion Canadian dollars in unemployment programs due to excessive fishing. And the country is likely to spend another half billion dollars over the next 3 years. The same report also supports the view that something can be done about this, if the desire is strong enough, and that environmental conservation measures can have dual benefits. It was found that the implementation of the U.S. Clean Air Act over 20 years ago has cost the U.S. $524 million but has saved the economy more than $6 trillion.

In a discussion on how to build into the decision-making process the understanding that unhealthy ecosystems also destroy the economy and the well-being of people, focus must be placed on *the politics of the environment* and how leaders in politics and business, and their electorates and shareholders, perceive the task of restoring a declining environment. It is also a matter of how progress is defined and, not least, how we measure progress so that the relationship between economic achievements and ecological costs becomes clear. The NAFTA study mentioned above suggested that Mexico's average GNP between 1985 and 1992 would have to be reduced by as much as one fourth (from 2.2 to 1.3% annually) when the depletion of natural assets was taken into account.

A distinction can be made between *soft politics* with local reach, which is what we basically have today, vs. *hard politics,* yet to be practiced (MacNeil, J., personal communication). Although commendable, as such, what we have seen so far are reforms that have been politically cost-free — with no demands for changes in how we are used to doing things; no claims on changing of lifestyle — basically about actions for which politicians and business leaders easily received approval from voters and stock markets, and safely within the framework of established models for how to achieve economic growth.

Hard politics is likely to be less cozy. It would have to be about abolishing billions of dollars in subsidies that favor special interests and unsustainable practices. It will have to do with gradually moving away from levying taxes on labor to increasing taxes on, for example, energy. It involves ratifying (and implementing) treaties on greenhouse gas reductions even if a hint of inconvenience and a need for thorough adjustments of economic policies are involved. It is about reviewing trade policies, structural adjustments, and privatization by paying as much attention to ecological factors as to economical aspects. Hard politics will mean a change of focus from the treatment of symptoms to challenging the causes of environmental degradation. It implies admitting that solving the big environmental issues carries big political costs. And, again, that the health of ecosystems places restraints on human economic activities.

A need exists for a process of change that is based both on the seriousness of our situation and on a realistic assessment of a practical way of achieving results. One obvious but important observation is that people are not exploiting the world's natural capital beyond its regenerating capacity out of ill will. The declines occur as an unintended result of values, priorities, and behaviors that have become part of political cultures.

We depend on the social and economic circumstances that surround us. The landless shifting cultivators and their families in Brazil, or elsewhere in the tropics, live under extreme poverty. They have no choices other than to survive by clearing a piece of virgin forest on which to grow their crops. Hundreds of thousands of landless farmers live in tropical countries, and the only farming practice available for them is to move on from one woodlot to another as the fertility of the cleared land runs out. So they account for a substantial part of deforestation in the tropical areas.

Governments can change this situation by providing the landless with other types of land, by changing tax and subsidy policies that encourage mining, logging, and other such activities in sensitive forest areas that open up roads so crucial for the encroachment of people into the forests. The reason governments do not act to change these situations, at least on any broad scale, is that it means challenging the status quo. Generally the entrenched interests that are promoting the status quo are far stronger than those of poor farmers and more convincing in the short term than the international communities' pleas for an improved global environment.

It is easy for pro-environmental movements, media, and individuals in the West to place blame on governments in third-world countries for not daring to show more courage in challenging those interests. And, no doubt, their people would benefit if governments would have the means and the will to do that.

The politics become more delicate and complicated when the same reasoning is applied to our own situation. The behavior of ordinary suburbanite families in a typical Western city also has implications for the health of ecosystems and the global environment. Not because they lack choices — they have lots of choices — but typically they choose to live as lavishly as they can and consume as much as their economic situation permits. There is nothing that forces them to do so, but there are lots of incentives created through political decisions or by prevailing economic systems to encourage them. In fact, governments tend to be rather worried if we do not spend enough because it hurts our economy. We all get worried when Christmas shopping is down or other planned spending feasts do not deliver as expected.

It is natural that people, given existing values and available choices, tend to do what is best for them in the time perspective they can foresee. As individuals, we have little chance of knowing what impact our choices may have on the global environment. It is the responsibility of governments to look after the public interest and to set the objectives for a country's environmental policies. Governments, nationally and through international agreements, have a duty to assess how to keep the landscape integral and functioning for humans and other species. The marketplace does not serve that function.

The search is on again for a new order in the world. We are confronted with a deepening of the conflict between growing ecological decline and challenges to established patterns of social behavior on the one hand, and economic expansion on the other. The free market models, the liberalization of trade, globalization, the role of technology, and the degree of virtue of economic growth are targets of considered analyses and populist protests.

We probably cannot do without either globalization or economic growth. So the discussion should be about what kind of globalization and what kind of growth we want to have — a globalization for all, or for the rich only? An economic growth with a social purpose and based on a sustainable use of natural resources, or one for the sake of growth itself? If economic growth is to be what it claims to be, it cannot be based on the use of resources of others or paid for by others (e.g., by future generations). If economic growth has its origin in an increase in efficiency and productivity, it can be sustained. As we know, the most important source of growth is innovations based on acquired new knowledge. Such new knowledge that the capacity of the renewable natural resources to sustain our economic activities sets the limits for growth. Today's economic strategies are based on the opposite order of things and will have to be changed.

Also consider this: a sixth of the world's population receives 78% of the world's income, while *three fifths* of the world's population living in 61 countries receive only 6% of the world's income. Still, equity, within and between generations, is a crucial element in the doctrine of sustainability.

Faced with a system that does not work, it is no surprise that people who care are looking for a system that works. They are eager to see the gaps filled between awareness of the problems and actions to solve them.

CHAPTER 3

Environmental Health Research Challenges

Kenneth Olden

Over the past 50 years, we have dramatically reduced the human health threats posed by the environment. Environmental remediation and pollution reduction have been one of the nation's public health success stories. In fact, we have been so successful in improving the quality of our environment that there are those who argue that the environment no longer represents a serious threat to human health. They argue that the job is done, that the low-level exposures experienced by most Americans pose no significant health threat.

Well, we have cleaned up the "big dirties" of the 1950s and 1960s. Rivers are no longer catching on fire and burning for days. Stacks bellowing smoke have largely disappeared from our urban landscape. And the density of particulates and other pollutants in the air of most American cities has been greatly reduced. For most Americans, high-dose exposure is no longer a reality. Managing today's risks at low-dose exposures will require new science, new technologies, and new ways of conceptualizing and managing the risks. Managing today's risks will also require consideration of biological concepts that were not part of the environmental health science vernacular as recently as 10 years ago. Concepts such as susceptibility, environmental genomics, high-throughput screening, and transgenic technology were not among the priorities of the environmental health research enterprise.

Environmental health decisions of the future will require better information than now exists. The default assumptions and uncertainties used to assess risk at high-dose exposures are even more unreliable at low-dose exposures.

First, I want to make the point that human illnesses are caused by both genetic and environmental factors acting alone or in concert over the course of many years. In fact, we know of only a few diseases that are caused solely by genetic factors or solely by environmental factors. Most diseases are caused by gene–environment interactions. Therefore, to understand the etiology and prevent chronic illnesses, we are going to have to make the investment to understand both the genetic and environmental contributions to disease.

Behavior is also an important contributor to human illness, but behavior is either genetically predisposed or environmentally acquired. So it is captured in my earlier statement about genes and environment, as a function of age or stage of development, being the major determinants of health. However, children may be at increased risk to environmental exposures because of their behaviors in addition to their rapid growth and development. Seniors may have increased risk because of a compromised immune system and an accumulation of genetic lesions.

A quotation in *USA Today* by Judith Stern, Professor of Nutrition and Internal Medicine, University of California, Davis, captures the nature of the interaction between genes and the environment: "Genetics loads the gun but environment pulls the trigger." As you know, a loaded gun by itself causes no harm. It is only when the trigger is pulled that the potential for harm is released or initiated. Likewise, one can inherit a predisposition (a susceptibility) for a devastating disease (e.g., Alzheimer's, Parkinson's, cancer) yet never develop the disease unless exposed to the environmental trigger. A well-known example of this is asthma. People with asthma are not always in respiratory distress (although somewhat compromised with respect to respiratory capacity). It is only when they are exposed to certain environmental triggers (such as tobacco smoke, dust mites, cockroach allergens, acid aerosols, or ozone) that they experience their severe symptoms.

Let me take a moment to define the environment. The environment is the chemical, physical, and biological agents to which humans (as well as plants and other animal species) are exposed during the course of their lives at home or at work. The environment also includes diet, nutrition, behavior, lifestyle choices, and poverty. So the environment is more than synthetic chemicals and industrial by-products. In fact, the thousands of natural chemicals to which humans are exposed are often overlooked as possible causes of human illnesses. But there is no reason to conclude that natural products are harmless.

I am pleased to say that I expanded the definition of the environment to encompass all the factors I have just mentioned when I became director of the National Institute of Environmental Health Sciences (NIEHS) in 1991. Today, the institute supports research dealing with all aspects of the environment with an emphasis on prevention and public health.

Diseases once thought to be genetically determined by one or two genes are now proving to be more complex. For example, in early reports, the contribution of the breast cancer susceptibility gene (BRCA1) to early onset of premenopausal breast cancer was overestimated. When first isolated by Dr. Roger Weisman (scientist at NIEHS) and Dr. Mark Skolnick (University of Utah), it was estimated that 85% of women who inherited BRCA1 would have breast cancer and/or ovarian cancer. Now the estimate is that only about 50 to 60% of women who inherit BRCA1 will have breast cancer. It now appears that other genes (i.e., polygenic inheritance) and/or environmental agents play a role in the development of these diseases. There are numerous examples like this where the penetrance (i.e., the expression) of specific genes is much lower than originally thought. The fact is we now know what the causes are for most chronic diseases.

We know much more about curative treatment than we do about the initiation and progression of disease. As a nation, we are preoccupied with the development of treatment strategies or "magic bullets" to prevent death from end-stage disease, whereas our public health success stories over the past two centuries have been in the arena of prevention through public health measures. Yet we continue to emphasize treatment in our public health policies even though we have been unable to develop effective treatments or cures for the most devastating diseases such as Alzheimer's, Parkinson's, diabetes, and most cancers.

The good news about gene–environment interaction is that the environmental triggers can be manipulated. All the components discussed under the definition of the environment (chemical, physical, and biological agents, lifestyle choices, behavior, and poverty) can be controlled or manipulated. Even though one is genetically predisposed to develop a specific disease, one can prevent initiation and/or progression by manipulating the environmental triggers.

However, to realize the public health potential of environmental health research, several critical investments will be required. The necessity for better science is clear if the public is to have confidence in government decisions.

The first critical investment that we need to make is to develop new high-throughput technologies to assess environmental agents for carcinogenicity and toxicity. Estimates are that 70 to 75% of the high-volume, high-use chemicals in commercial use in the U.S. have not been assessed for human toxicity. While many of these (if not most) may not require testing since they are very similar to chemicals already tested, several do need more testing. In fact, we can never satisfy the testing requirements using current technologies. Without better test systems (ones that are faster and cheaper), we will continue to operate in a state of "toxic ignorance." By exploiting recent advances in human genetics and recombinant DNA technology, we can develop animal models and *in vitro* assay systems to identify carcinogens and toxicants in a matter of days or weeks rather than years. Two promising approaches involve manipulation of the expression of tumor suppressor or proto-oncogenes in animal models and development of cDNA microarray systems for measurement of gene expression.

The second crucial investment is to determine the genetic basis for the wide variation in susceptibility to environmental toxicants. It is well known that most smokers do not develop lung cancer and most women exposed *in utero* to diethylstilbestrol (DES) never develop cancer of the reproductive organs. A common question asked of physicians is, "Why me, Doc?" Well, genetically determined differences in susceptibility are at least in part responsible for individual or population differences in susceptibility.

With rapid advances in cloning and sequencing of the human genome, it is now possible to identify the genetic polymorphism responsible for differences in susceptibility. For example, by identifying genetic alterations that lead to functional changes in critical pathways involved in the metabolism of xenobiotics, one will have a better understanding of differences in susceptibility. The important point here is that the current use of the *one size fits all* approach in environmental health risk assessment is not consistent with our understanding of human biology and surely leaves some individuals or subpopulations unprotected from risk for diseases from environmental exposures.

Also, susceptibility is influenced by the timing (childhood vs. adulthood) of exposure, the gender and behavior of the individual, the nutritional state, and the socioeconomic status. For example, exposure during rapid or critical stages of development of the various organ systems is likely to be an important consideration in the development of disease. Critical stages include embryonic development, adolescence, puberty, and old age. At present, we have little information to guide decision making with respect to these issues because most toxicologic assessments have been done in adults, both animals and humans, under otherwise optimal conditions.

To investigate the genetic basis for differences in susceptibility or predisposition to disease, two related gene resequencing efforts were initiated at the National Institutes of Health (NIH). The first to be initiated is called the Environmental Genome Project, and the second is the Single Nucleotide Polymorphism Project. Both seek to identify genetic changes that increase risk for disease development. The results from these studies will have profound implications for the practice of medicine and environmental health risk assessment.

To address the concern about timing of exposure, we have initiated studies in which animals are exposed corresponding to the various stages of human development (e.g., *in utero,* infancy, adolescence). We have also developed eight Children's Environmental Health and Disease Prevention Research Centers, in collaboration with the Environmental Protection Agency, to focus on basic research, prevention, and treatment of pediatric illnesses. Vice President Gore announced the creation of the Centers Program and recipients of the awards from the White House in August 1998.

The third crucial investment for the NIH is to understand the causes of health disparities in American society. It is well established that the poor and racial and ethnic minorities have higher morbidity and mortality from various diseases. Since members of these groups are more likely to live and work in the most hazardous environments, they experience higher levels of exposure to environmental stressors in terms of both frequency and magnitude. They are also less able to deal with these hazards because of limited knowledge and a higher level of disenfranchisement from the political process. Several important questions need to be answered. What are the causes of the observed disparities? Which of the causes require more research, and which can be addressed by application of current knowledge and technologies? Health disparity is a very complex issue with important social and economic implications. To address this issue we need a well-thought-out strategic plan.

I am very pleased to report that The Office of Minority Health of the Department of Health and Human Services (DHHS) has identified health disparities as a top priority.

Gender differences in response to environmental insults is another area requiring special attention. Because of the important role that hormones play in regulating homeostasis, gender-specific responses can be expected. For example, hundreds of synthetic and natural chemicals in the environment can modulate hormone responses. Of the so-called endocrine disruptors, those that mimic estrogen are the best studied. Environmental chemicals with estrogenic activity are putative risk factors for breast and ovarian cancers, uterine fibroids, various autoimmune diseases, and endometriosis. NIEHS has been very interested in research on the health effects of synthetic estrogen since the early 1970s. In fact, our current understanding of the carcinogenic and reproductive health effects of DES can be largely attributed to the body of research generated with NIEHS support over the past 30 years.

Researchers at the NIEHS and elsewhere are developing transgenic animal models containing the mutated human breast cancer genes (BRCA1 and BRCA2) for use in identifying environmental agents that interact with these genes to promote mammary cancers in rodents.

NIEHS scientists have also developed animal models that lack functional estrogen receptors. Since these receptor molecules are crucial for estrogen action, the so-called knockout animal models are useful in studying the mechanism of estrogen action. Also, we and others have now identified 39 to 40 chemicals in the environment that cause mammary tumors in mice. Chemicals that cause tumors in rodents (mice and rats) have the potential to cause cancer in humans.

Recently, NIEHS joined forces with Phyllis Greenberger and the Society for the Advancement of Women's Health Research (SAWHR) to focus attention on the possible adverse health effects of herbal medicines. The products are used regularly by millions of Americans (mostly women), yet most have not been evaluated for either efficacy or toxicity (safety). The SAWHR brought this issue to our attention. To date, we have co-sponsored (with the SAWHR, the NIH Office of Dietary Supplements, the HHS Office of Disease Prevention and Health Promotion, and the Food and Drug Administration Office of Special Nutrition) a very successful symposium to assess the research needs to ensure that herbal products are safe for human use. We plan to evaluate several of these products for safety over the next 5 years. Exposure assessment is another critical area of investigation. The truth is that we do not know what the American people are exposed to. Exposure is typically estimated using indirect surrogates, based on toxic release and production inventories. Exposure is highly variable for individuals and subpopulations.

Exposure is really a function of individual uptake, metabolism, excretion, and behavior. So the assumption that all individuals (men, women, and children) living in the same geographic area have similar exposures is seriously flawed. We need direct measures of exposure based on tissue analysis or deposition. NIEHS has supported an effort by the Centers for Disease Control and Prevention (CDC) to develop methodologies for direct assessment of human exposure to environmental estrogen-like chemicals.

Exposure to complex mixtures is another area of environmental health research in need of attention. Typically, toxicity and carcinogenicity are assessed in animal models (mice and rats) one chemical at a time. This contrasts with the real world, where we are exposed to multiple agents (chemical, physical, and biological) at any given time in the form of mixtures. Current assumptions are that components of mixtures behave in an additive fashion (i.e., one plus one equals two). However, we are aware of situations in which components of mixtures behave synergistically (i.e., one plus one equals some number significantly higher than two). For example, carcinogens in cigarette smoke interact synergistically with radon to produce lung cancer, and aflatoxin B1 interacts synergistically with the hepatitis B virus to produce liver cancer. I believe that we now have the capacity to develop the technology (e.g., cDNA microarray gene expression systems) to assess the toxicity of mixtures.

Another critical area of investigation is understanding the mechanism by which environmental insults initiate or promote disease development. It is not sufficient to know that chemical A causes cancer; it would be very useful to know *how* chemical A causes cancer in order to develop preventive, diagnostic, and treatment strategies. Also, knowledge of mechanisms is very important in assessing risk to humans based on experimental data generated in animal models. If disease initiation and progression due to a given chemical exposure occur by pathways common to both humans and rodents, then that chemical can be reasonably anticipated to be a human carcinogen based on data from animal studies.

The final critical investment of research that I want to discuss here is population-based studies. Most of the large cohort studies undertaken to identify disease risk factors were not designed to elucidate the role of most environmental agents in the etiology of disease. Nevertheless, epidemiology has been an important tool in identifying environmental causes of diseases. However, most of these studies were conducted using defined populations exposed accidentally or occupationally to high doses of the specific chemicals. To identify environmental risk factors at low-dose chronic exposures will require specially designed epidemiologic studies using large populations.

CHAPTER **4**

Toward Ecoresponsibility:
The Need for New Education, New Technologies,
New Teams, and New Economics

William S. Fyfe

INTRODUCTION

We have lived through the two giant centuries of science. Just imagine a world without the dynamo (Faraday, 1821), a world without antibiotics (Fleming, penicillin, 1928), or even a world without jet engines and atomic bombs. In a Royal Society of London publication, *Population: The Complex Reality* (1994), the past president of the Royal Society, Sir Michael Atiyah, wrote, "Most of the problems we face are ultimately consequences of the progress of science, so we must acknowledge a collective responsibility. Fortunately, science opens up possibilities of alleviating our problems, and we must see that these are pursued." In the same volume, Sir Crispin Tickell, former British ambassador to the UN, wrote, "It would be nice to think that the solutions to some of our present problems could be drawn from past experience, but in this case the past is a poor guide to the future ... our current situation is unique."

Yes, we are living in a new world with a new demography. Imagine the year 2050, when Europe and North America will make up only about 10% of the global population. And the present state of the global society is not sustainable. According to the *Manchester Guardian Weekly*, November 2, 1998, in its article "The Politics of Hunger," in 1960 the richest 20% of world population made 30 times more than the poorest 20%; but by 1995, that ratio had changed to 82 times. They stated that 3 billion people live on less than $2 per day, and at least 2 billion suffer from lack of food. Renner (1999) wrote a report on world conflict pointing out that, in the period 1900 to 1995, 110 million people had been killed. If we cannot accept such realities, admit to our failures, is there hope for a positive, sustainable future for all people who live on our planet?

In Canada (Mungall and McLaren, 1995), a group of us wrote a book for schools on the state of the planet. In this I used a quotation from Aldous Huxley (*Managing the Double Crisis*, 1948):

Industrialism is the systematic exploitation of wasting assets. In all too many cases, the thing we call progress is merely an acceleration in the rate of that exploitation. Such prosperity as we have known up to the present is the consequence of rapidly spending the planet's irreplaceable capital. Sooner or later mankind will be forced by the pressure of circumstances to take concerted action

1-56670-612-2/03/$0.00+$1.50

against its own destructive and suicidal tendencies. The longer such action is postponed, the worse it will be for all concerned. ... Overpopulation and erosion constitute a Martian invasion of the planet. ... Treat Nature aggressively, with greed and violence and incomprehension: wounded Nature will turn and destroy you. ... if presumptuously imagining that we can "conquer" Nature, we continue to live on our planet like a swarm of destructive parasites — we condemn ourselves and our children to misery and deepening squalor and the despair that finds expression in the frenzies of collective violence.

OUR POWER OF OBSERVATION

Considering Huxley's words, there is no longer any excuse today for incomprehension. Modern science has provided our incredible powers of observation. We can watch galaxies, planets, atoms, and molecules, and, today, the details of biosystems at the micro level. And we can watch the actions of humans in every part of the globe. We can see the changes in the chemistry of our atmosphere caused by our technologies. We can watch global soil erosion, deforestation. We can watch water pollution and its dramatic influences on human health as recently reported by Pimentel et al. (1999). The fact is that if our future population of about 10 billion continues to behave like my generation, we will destroy the life support systems for most humans and a host of our biological friends. Our systems of the past century have failed.

NEEDED NEW LIFE-SUPPORT TECHNOLOGIES

The quality of life for billions of people today is related to a number of major components. We all need clean air to breathe; and in many of the growing urban regions of the world, breathing has become a major cause of death (*Ecosystem Health*, 1998). Air in Tehran, Iran, contains 209 micrograms of SO_2 per cubic meter; Rio de Janeiro's air contains 129. (*The Economist Pocket World in Figures,* 1999). For example, the World Resources Institute, (1998) predicted that SO_2 emissions in Asia will rise to 110 million tons by 2020. Bright (1999) reports that in parts of the Smoky Mountains in the Appalachians, the dew can be at pH 2.1.

The quality of our nutrition depends on soil quality, water quality, climate, diversity of foods, and our transport systems (which allow Canadians to eat fresh fruit from Argentina and Chile all winter). But in almost all nations, soil quality and quantity are deteriorating due to catastrophic erosion (Fyfe, 1989). In work we started in Portugal (Fonesca et al., 1998) we showed that often the best soil components collect in hydroelectric reservoirs and other dam systems. And on a recent visit to the Yangtze region, China, I was amazed by recent soil erosion. Around the world, the need for irrigation increases. Few rivers today flow freely to the oceans. More and more groundwater is used for agriculture. In parts of India the water table is dropping 1 to 3 meters per year. If water does not flow off the continents, eventually the result is salinization of soils, a disaster on 15% of irrigated land.

The world climate is changing, and rainfall distribution is changing. As recently reported by the U.K. Department of the Environment (Fisk, 1997), it is predicted that the tropical forest may be transformed into desert in major regions of northern Brazil and the eastern Amazon basin, and it is happening now.

When I was a baby in the 1920s, who used bottled water? Water pollution is now a major cause of disease in the world (Pimentel et al., 1999). They estimated that 90% of all infectious diseases in developing nations are related to water quality. In many nations — Bangladesh is a prime example — arsenic in water is causing vast problems. We are working on this problem, and there are simple techniques to remove arsenic. But where do you put the filtered waste materials?

WASTES

One of the greatest problems today and in the future is that of managing our waste products. According to *The Economist* data of 1999, solid, hazardous waste generation has reached 2731 kg per capita in Belgium and 1059 kg in the U.S., the world leaders. In careful nations like Germany, the number is 82 kg. With industrial wastes and municipal wastes, the U.S. leads the world. I was recently in Hamburg, Germany, and met with city engineers. Hamburg no longer needs landfills. There is a rule: if waste cannot be reused or recycled, then the parent product cannot be used.

Two extreme cases of hazardous waste today involve combustion gases and nuclear wastes. These wastes perfectly illustrate international problems, the need for international systems and controls, and new ethics.

With regard to combustion gases (carbon dioxide, acid rain, nitrogen oxides, etc.), a key question is whether they can be disposed of underground. We have been studying disposal in rocks like the basalts of Hawaii; Parana; Brazil; and Deccan, India. These rocks, with help from microorganisms, can remove them (Fyfe, 1996). It is interesting to note that the Department of Energy (DOE) has recently announced programs in this area. I have found great interest in India and China, which will use coal for decades to come. If all the world produced CO_2 like North America, the planet would not be a greenhouse; it would be Venus.

Nuclear waste is another perfect example of the need for international responsibility. Given that such wastes are dangerous for a million years, who will live where in those future times? The great issue is the location of the best place on our planet where we can guarantee isolation from the biosphere (Fyfe, 1999).

NEEDED NEW TECHNOLOGIES, NEW SYSTEMS, NEW ECONOMICS

As Sir Michael Atiyah of the Royal Society of London wrote, we can solve many of the urgent problems. We must change our energy systems (Fyfe, 1998). With the sun and the deep warm Earth, we have an almost infinite source of clean energy. Slowly, the world is turning to solar (Brown et al., 1999). Many of the poorest nations are rich in potential solar energy. The production of solar, photovoltaic devices is rising. I was interested to read that in the U.K., large turbines using ocean currents may soon produce 19% of the energy and at reasonable cost. In my home country, New Zealand, over 30% of the electrical energy is generated by geothermal systems using the regions of active volcanism. I have seen an experiment in which a small town in Sweden is heated from drill holes using the normal geothermal gradient of about 30°C/km depth.

Recent work in many nations, Japan leading, has shown that when sunlight shines on the surface of certain materials, such as copper oxide (Cu_2O) with a film of water, the water is dissociated into $2H_2 + O_2$, almost the perfect transport fuel (*Chem. Eng. News*, 1998). By using such fuels and electric motors, we can clean our cities.

When one examines these types of problems, it is obvious that solutions can be found by forming teams, integrating the work of experts from a range of disciplines. For example, to produce clean, sustainable solar energy, we need experts in climate and solar radiation, surface physics, and chemistry and engineering. To develop geothermal energy, we need experts from rock mechanics, geophysics, deep-water hydrology, geochemistry, and engineering. New Zealand developed such teams long ago and led the world in such developments.

But to accelerate the desperate, urgent need for the use of such systems, we must involve those from local government particularly from economics. What is the cost of polluted urban air, polluted water, or soil erosion on decade time scales? Recently the *Worldwatch Journal* (1999) had some spectacular, alarming reviews on such topics (Bright, 1999). The costs of environmental deterioration are becoming vast. I was recently at a meeting in Lebanon with earth scientists from more than 20 Arab nations. I met with many students from the region. They are angry. In many of these nations one of their few ultimate resources, fossil fuels, are being wasted by advanced nations. When we plan, how often do our leaders think of the next generations? Here I again quote Sir Crispin Tickell (1993):

> I was recently asked if I was an optimist or a pessimist. The best answer was given by someone else. He said that he had optimism of the intellect but pessimism of the will. In short we have most of the means for coping with the problems we face, but are distinctly short on our readiness to use them. It is never easy to bring the long term into the short term. Our leaders, whether in politics or business, rarely have a time horizon of more than five years.

We need new systems, new teams, new economics; and as *Worldwatch* wrote (Brown et al., 1999), "We need a new moral compass to guide us into the twenty-first century — a compass grounded in the principles of meeting human needs sustainably."

EDUCATION THE KEY

When one examines the state of nations, the state of their environment, and their respect for ecology biodiversity, it is clear that the quality of education is the most fundamental factor. If one examines data on literacy (World Resources Institute, 1996) the message is clear. In most of Europe, literacy for both males and females is nearly 99%. But when we examine data from many of the poor nations, the situation is a disaster (e.g., female literacy in Ethiopia is 21%; in Niger it is 5%; and in India it is 34%). One of the most striking developments in recent times, the social acceptance of limits, is shown by trends in fertility. The World Resources Institute (1998) reports that fertility (the number of children per family) is 7.1 in Niger; 7.0 in Ethiopia; 3.1 in India; 1.2 in Italy; 1.5 in Japan; and 1.8 in China. It is clear that the state of nations is related to female literacy.

It is essential that those with expertise devote more effort to world problems. At the present time I am involved in problems such as water purification, combustion gas disposal, nuclear waste disposal, and soil remediation. But recently, while trying to find financial support, a reviewer of a project wrote that this was not at the cutting edge of science! I often observe that many of my academic colleagues devote their lives to finishing their Ph.D.s. During many of our university programs, the students are rarely exposed to the great world problems. Yes, we must have experts, but experts who can communicate with other experts and society.

If we are to have a positive 21st century, one priority must be to universally educate the populations of the world in literature, math, and the sciences. Those of us in the rich nations must help those less fortunate. It is interesting to examine the nations who give foreign aid generously. The largest donors as a percentage of gross domestic product (GDP) are Denmark (1), Norway (2), and Sweden (3), with the U.S. ranking 23. Denmark gives 1.04% of GDP; in contrast, the U.S. donates only 0.12%.

If our planet is to be a beautiful place by the year 2100, we must increase our foreign aid; and priority number one must be universal education, training teachers, building schools, and providing infrastructure.

Are we *Homo sapiens*, or did Linnaeus (Carl von Linné of Sweden) make a mistake? Are we *Homo horribilis*? As the great French philosopher Montaigne said in 1580, the most universal quality is diversity, and diversity is security! And as Ayres (1999) recently asked, "Why can we not learn from history?"

REFERENCES

Ayres, E., The paradox of surprise, *Worldwatch J.*, 12(3):3–4, 1999.
Bright, C., The nemesis effect, *Worldwatch J.*, 12(3):12–23, 1999.
Brown, L.R., Renner, M., and Halweil, B., *Vital Signs*, Worldwatch Institute, Washington, D.C., 1999.

Brown, L.R., Flavin, C., and French, H., *State of the World, 1999*, W.W. Norton, New York, 1999.

Chem. Eng. News, American Chemical Society, February 16, 1998, p. 26.

The Economist Pocket World in Figures, John Wiley & Sons, New York, 1999.

Ecosystem Health, 4(4), December 1998.

Fisk, D., *Climate Change and Its Impacts: A Global Perspective*, Bracknell, U.K., 1997.

Fonesca, R., Barriga, F.J.A.S., and Fyfe, W.S., Reversing desertification by using dam reservoir sediments as agricultural soils, *Episodes,* 21:218–224, 1998.

Fyfe, W.S., Soil and global change, *Episodes,* 12:249–254, 1989.

Fyfe, W.S., Energy and wastes: from plutonium to carbon dioxide, *Sci. Int. Newsl.*, 62:14–17, 1996.

Fyfe, W.S., Clean energy for 10 billion humans in the 21st century: is it possible? *Int. J. Coal Geol.,* 776, 1998.

Fyfe, W.S., Nuclear waste isolation: an urgent international responsibility, *Eng. Geol.,* 52:159–161, 1999.

Huxley, Aldous, The double crisis, in *World Review*, Chatto and Windus, London, 1948.

Mungall, C. and McLaren, D.J., Eds., *Planet under Stress*, Oxford University Press, Oxford, U.K., 1995.

Pimentel, D., Tort, M., D'Anna, L., Krawic, A., Berger, J., Rossman, J., Mugo, F., Doon, N., Shriberg, M., Howard, E., Lee, S., and Talbot, J., Ecology of increasing disease, *BioScience,* 48:817–826, 1999.

Renner, M., Ending Violent Conflict, Worldwatch Paper 146, Worldwatch Institute, Washington, D.C., 1999.

Tickell, C., The future and its consequences, *Br. Assoc. Lect.,* The Geological Society of London, UK:20–24, 1993.

World Resources Institute, *World Resources, 1998–99*, Oxford University Press, Oxford U.K., 1998.

Worldwatch Journal, Working for a sustainable future, 12:3, 1999.

Perspectives on Ecosystem Health

CHAPTER 5

Overview: Perspectives on Ecosystem Health

David J. Rapport

Ecosystem health is about a lot of things. It is about the state of a community of organisms (including humans as part of that community) that interact with their environment, sometimes in ways that compromise ecological functions, and in the process compromise the health of the entire ecosystem and its components. It is about questions of the relationship between the health of ecosystems and their integrity, i.e., how far they may diverge from a pristine state. It is also about human health — a topic that is receiving increased attention as issues of climate change, ozone depletion, food safety, exotic pathogens, and drinking water come to the fore. And at root it is about sustainability — that is, about the conditions necessary to sustain the functions of the ecosystem, the human community, and the other species that are totally dependent on those functions. The trend continues for more of the world's population to be concentrated in urban areas (more than half the world's population now lives in urban ecosystems, and in developed countries the percentage is substantially higher). Therefore, it is about the health of those systems and their interrelations with other ecosystems (forests, lakes, rivers, grasslands, etc.). Section I.2 provides a kaleidoscope of many of the philosophical, ethical, and practical issues that are part and parcel of the ecosystem health perspective. The chapter by Laura Westra examines the conceptual, historical, ethical, and philosophical dimensions of ecosystem health. Her chapter provides guidance in making a useful distinction between ecological integrity, which is oriented toward the preservation of unfettered nature, and ecosystem health, which is oriented toward preserving the functions of ecosystems, even though the system may be considerably altered as a result of human domination. The ethics of integrity (and ecosystem health) can be boiled down to this: if integrity (or, one might argue, ecosystem health) is basic for life, then the duty "to respect/protect life must come before all other obligations we might have, as no other *good* and no other *right* may be present, if life is absent." Westra's chapter goes on to explore the implications of this ethic, as well as to explore the relative strengths and the interface of ecosystem integrity and health. Both ecological integrity and ecosystem health are gaining in popularity as policy goals for environmental management.

The chapter by William Rees argues that ecosystem integrity and health require movement toward a steady state economy, and that neither goal is likely to be achieved with a conventional economic growth scenario. The reasoning is simple and convincing. Human-dominated ecosystems are sapping the materials and energies (both finite, even allowing for ever-increasing technological efficiencies) needed to sustain life systems. Rees emphasizes that the real issue is not an environmental crisis, as many would have us believe, thereby externalizing the problem, as if it were something outside of our control that has run amok. Rather, the crisis is a human ecology issue, whereby the pressures that humans have brought to bear on ecological systems have become so great as to place these systems in critical condition. Reviewing the fundamentals of human ecology, and drawing upon principles of nonequilibrium thermodynamics and the evidence that human pressures have coincided in many areas of the world with massive prehistoric extinctions, the conclusion seems clear: unless we rein in the size of our ecological footprint — particularly in the developed countries — we will continue to extinguish not only species but whole ecosystems and, finally, the very foundations of our life support.

The next four chapters focus on some specific consequences of degrading environments, namely, the threat posed to human health. This topic is given much attention as the range of threats to human health from the loss of ecosystem health becomes more apparent. Burnett et al. address one of the long-suspected problem areas: the impact of air pollution on human health. They conclude that, although there are many confounding factors, the overall consistency of epidemiological data suggests strongly that premature mortality, cardiorespiratory hospitalization, and emergency department visits for respiratory diseases, asthma attacks, etc., are correlated with air pollution. Placing these findings within a cost–benefit analysis, it appears that billions of health care dollars (and much human misery) would be spared if more stringent air quality standards were adopted.

McMichael and Kovats shift the focus from the local to the regional and global changes that endanger human health. Climate change (climate warming), stratospheric ozone depletion, and invasive species are separately and interactively exacerbating existing threats to human health and giving rise to new ones. While mathematical model-driven estimates of likely impacts from global change have wide ranges owing to underlying uncertainties, it seems clear enough that the impacts of degrading environments on human health are significant. For example, one scenario-based model of the impact of stratospheric ozone destruction indicates that European and North American populations will experience significant increases in skin cancer — on the order of 5 to 10% in excess of present rates by the middle decades of this century. Among the many kinds of ecological disruptions that threaten human health are invasions of exotic species that provide habitat for human pathogens. The introduction of the water hyacinth to East Africa's Lake Victoria from Brazil, for example, has provided fertile breeding grounds for the water snail that transmits schistosomiasis as well as for bacteria causing diarrheal diseases.

Jonathan Patz looks at quantitative methods that might be used to assess public health implications of global climate change. He focuses on six specific threats: (1) mortality related to heat waves in the U.S.; (2) hantavirus, which is transmitted by human contact with rodents and can be fatal; (3) malaria, which is on the rise owing to habitat disturbances that create more breeding grounds for the mosquito vector; (4) dengue fever (another vector-borne disease, also involving mosquitoes); (5) water-borne cryptosporidiosis; and (6) cholera. Patz examines how ecological and climatic factors might be related to specific human health risks. His work provides quantitative evidence that the loss of ecosystem health is associated with increased risks to human health.

William Reisen's chapter focuses more specifically on the potential of correlating meteorological variables to encephalitis risk in California. Both western equine encephalomyelitis and St. Louis encephalitis are public health concerns in the area. They are thus the motivation for mosquito abatement actions. Reisen explores the complexities of the transmission cycles of the pathogens (in this case the viruses that live in their mosquito hosts) and explores how changes in temperature alter both mosquito abundance and the virus transmission rates. Some meteorological variables (snow pack and river runoff) were correlated with mosquito abundance during the following season, while others (winter rainfall) were not. This is illustrative of the potential for *data mining* — associating meteorological variables with ecological change and ultimate consequences for human health.

Mathis Wackernagel returns to the "big picture," examining the demands of people for the limited energy and material resources of the planet. He focuses on cities, and describes the toolkit that Redefining Progress is developing for assisting cities and towns to achieve a better balance between material and energy consumption and the capacity of ecosystems to supply these requirements on a sustained basis. His focus is on assessing current progress toward sustainability and on developing activities that maintain community interest in building a sustainable future.

CHAPTER 6

The Ethics of Ecological Integrity and Ecosystem Health: The Interface

Laura Westra

INTRODUCTION TO INTEGRITY

Despite the bad press that generally followed the most recent occurrence of El Niño on November 1, 1997, the Italian News Channel (RAI) and the *U.S. Sunday Report* showed a marvel engendered by El Niño: the flowering of the Chilean desert. It is important to note this example, because it shows clearly why the insistence on largely unmanipulated (if not intact, pristine, or virgin) systems is so vital to the understanding of integrity and to life on Earth. A seemingly barren desert area in Chile, without a discernible complement of species in recent times, changed dramatically after El Niño. Because the latent biological processes specific to deserts were present there, the unusual rains brought by El Niño produced a wonderland of flowers and grasses, with all the accompanying complementary species of insects, such as bees, ants, butterflies, and an abundance of other species.

This burst of life occurred because anthropogenic stress was largely absent from the history of the desert; that is, that landscape was not subjected to the chemical technological stress that prevails in exploited ecosystems around the world. In essence the desert retained its capacity because its vital state had not been reduced by human disturbance. The main point of this example is to emphasize the difference between a landscape that has been heavily utilized and one that has been left — at least for the most part — in its natural condition, following its own evolutionary trajectory. At one end of the spectrum, the remote desert area retained most of its capacities for development; largely untouched, the desert flowered. At the other end, the petroleum-laced fields where Royal Dutch/Shell Oil carries on its ecologically destructive enterprise in Ogoniland, Nigeria, will not burst into flower under any circumstances. Note that we, the general public, did not know the immense potential for diverse life that was present in that "barren" desert in Chile (although desert ecologists may be familiar with such phenomena). But its integrity guaranteed that — under changed conditions — one of its possible developmental evolutionary trajectories might come to be.

The generic concept of integrity connotes a valuable whole, "the state of being whole, entire, or undiminished," or "a sound, unimpaired, or perfect condition" (*Random House Dictionary of the English Language*, 1967). We begin with the recognition that integrity, in common usage, is an umbrella concept that encompasses a variety of other concepts (Westra, 1994). The example of the blooming desert illustrates a number of the themes associated with ecological integrity.

1. The example is drawn from *wild nature*, i.e., nature that is relatively unimpacted by human presence or activities. Although the concept of integrity may be applied in other contexts, wild nature provides the paradigmatic examples for our reflection and research. Because of the extent of human exploitation of the planet, such examples are most often found in those places that, until recently, have been least hospitable to dense human occupancy and industrial development: deserts, the high arctic, high-altitude mountain ranges, the deep ocean, and the less accessible reaches of forest and jungle. Wild nature is also found in locations whose capacity to evoke human admiration won their protection in natural parks.

2. The rapid bloom of desert organisms illustrates in a dramatic fashion some of the *autopoietic (self-creative) capacities* of life to organize, regenerate, reproduce, sustain, adapt, develop, and evolve.

3. These self-creative capacities of life are *dynamically temporal*. The present display of living forms and processes in the desert gains significance through its past and its future.

   a. Conjoined with its past, the Chilean desert is a part of *nature's legacy*, the product of natural history. Because of the relative absence of anthropogenic impacts, the desert biota is the creature largely of "evolutionary and biogeographical processes at that place" (Angermeier and Karr, 1994). It thus illustrates what nature is and does in the absence of the human design and impacts that dominate the built, modified, and impacted environments in which we live most of our lives.

   b. The events of their past and present demonstrate the capacity of desert life forms to maintain their functions, to respond to changing conditions, and to evolve. If those capacities are not destroyed, we may anticipate their evolving *future realizations*. Indeed, continued evolution provides evidence that these adaptive capacities have not been destroyed.

4. Desert conditions, relieved by rains at rare intervals, are themselves the products of larger regional and global weather patterns. Indeed, both the biological and geoclimatic processes that led to the blooming desert play themselves out on a stage with much larger *spatial scope* (Westra et al., 2000).

5. Note finally that ecological integrity is *valuable* and *valued*. In the case of the Chilean desert, the dramatic transformation of "barren" desert into a vital and diverse biotic community provoked wonder and appreciation. Other ecological communities, such as reefs and rain forests, display their prolific lives in a more continuous, less seasonal or episodic fashion. More generally, the kinds of processes at work in these instances gave rise to the totality of life on Earth, including ourselves, and together maintain the conditions for the continuation of life as we know it. Thus, natural ecosystems are valuable to and in themselves for their continuing support of life on Earth, for their aesthetic features, and for the goods and services they provide to humankind. Ecological integrity is thus essential to the maintenance of ecological sustainability as a foundation for a sustainable society. For these reasons, a growing body of policy and law mandates the protection and restoration of ecological integrity (Pimentel et al., 2000).

THE GLOBAL INTEGRITY PROJECT AND THE ETHICS OF INTEGRITY

This project was initiated in 1992, when I first sought funds to investigate the meaning of the expressions *ecosystem integrity* or *ecological integrity* that appeared in regulations, laws, and mission statements, not only in Canada and the U.S. but all over the world (Westra, 1994). The concept of integrity had been introduced in 1972 in the U.S. Clean Water Act, and its use multiplied and accelerated in the years after that. The Great Lakes Water Quality agreement (1978, ratified 1988) is a clear example of its use.

The purpose of the agreement is to restore and maintain the chemical, physical, and biological integrity of the waters of the Great Lakes Basin Ecosystem, where the latter is defined as: "the interacting components of air, land, water and living organisms, including humans within the drainage basin of the St. Lawrence River" (Westra, 1994).

In most documents that refer to integrity, only the vaguest definition of the concept is available; and most of the time no attempt is made to clarify it. This is surprising, to say the least, in regard to a concept that is intended to represent the goal of legislation. For this reason, my first effort was to secure funding to examine the concept of integrity from many points of view and from the perspective of several disciplines. The funding we received was used to conduct workshops on business, business ethics, agriculture, fisheries, and other enterprises. The questions were, in each case, how do we understand ecosystem integrity and how do we define it?

Following a series of meetings, a team comprised of two complex-systems theorists and ecologists, James Kay and Robert Ulanowicz; an ecologist, Henry Regier; and a physicist and ecologist, Don de Angelis; and I drafted a definition after careful discussion of each word and concept (Westra, 1994). This definition was later shown to conservation biologist Reed Noss and to ecologist and biologist James Karr, both of whom were largely in agreement with most of the language of the definition.

I believed then, as I do now, that if integrity is so vitally important, so fundamental to life on Earth as we jointly believed it to be, then morality must also acknowledge its centrality, and so should public policy. The *Principle of Integrity* (PI) recognized this centrality and proposed an obligation for morality as such, not simply an "environmental ethic."

The reasoning employed is simple: If integrity is basic for life, then our duty to respect and protect life must come *before* all other obligations we might have, as no other *good* and no other *right* may be present if life is absent. In *Perspectives on Ecological Integrity* (Westra, 1995) scientists and philosophers argued about the meaning of integrity from various points of view. In the PI, I set out what seemed to follow from the recognition of the centrality of integrity, and considered the role other *basic* fundamental goods such as *justice* or *happiness* played in providing a starting point and central consideration for moral theories.

If we trace briefly the history of ethics in recent times, say from the 18th to the 20th century, we note that some of the major figures we study today (Bentham, Mill, Kant, and — more recently — Rawls), all depend on at least *one* factual or natural characteristic of human beings and, from that starting point, move to construct the moral principles for which they argue. *Pace* Hume, unless human (and many non-human animals) were capable of feeling pain and pleasure, (a fact), Bentham's "ought" would have no basis. A similar case can be made for the work of John Stuart Mill and again for Immanuel Kant, whose categorical imperatives prescribe universalized and reversible respect for the dignity of autonomous humans. Dignity and autonomy are grounded in the presence of rationality and the capacity to exercise free will, and both abilities are unique to humankind, according to Kant. Finally, Rawls' principles, required to establish fairness and justice in society, are also based on discourse among rational contractors. The capacity to enter into this discourse also requires rationality, a human factual characteristic (Figure 6.1).

All these approaches then depend and are based on a *single* factual characteristic, whose presence is acknowledged and taken as a given, rather than argued for. But when we come to the work of Aldo Leopold, we discover that his main maxim, "a thing is right when it tends to preserve the integrity, stability, and beauty of the biotic community. It is wrong when it tends otherwise" (Leopold, 1949) appeals to a whole as the object of moral action, rather than to a single component of that whole, or even less to a single characteristic of any individual.

One might argue that something underlying Leopold's position is unique and general enough to guarantee the value of individuals and processes as well as the value of the whole: there is the biotic connectedness Leopold emphasizes — that is, the presence of life. I have argued from this point of view because I believe that the reconnection between anthropocentrism and nonanthropo-centrism is possible only at the most basic level, based on the value of life, intrinsic as it is to individuals, aggregates, and wholes — although the latter encompass abiotic components as well (see Figure 6.1).

With the arrival of Aldo Leopold on the scene of ethics and public policy, we have a new approach that is not flawed, as some would argue, by its reliance on the *naturalistic fallacy*. We have instead a radically new approach — one that is based on a wider actual reality and one that recognizes the primacy of natural wholes, in line with the new science of ecology. In addition, the insights of conservation biology, complex systems theory, and even environmental epidemiology are compatible with Leopold's principle, as they indicate clearly that both moral theory and public policy must be integrative rather than reductionist; holistic, rather than individualistic or aggregative; and, most of all, fundamentally nonanthropocentric, as they must recognize the value of *all* life, human and nonhuman.

Table 6.1   Moral Theories, Principles, and Natural Characteristics

Philosopher	Principle	Factual/Natural Characteristic
Jeremy Bentham (1748–1832) Basis of principle: The promotion of pleasure	Principle of Utility An action conforms to the utility principle (or promotes the interests of the individual) when it tends to add to the sum total of his pleasures, or … to diminish the sum total of his pains	Capacity for pleasure/pain in sentient creatures (includes human and nonhuman animals; individualistic)
John Stuart Mill (1806–1873) Basis of principle: The greatest happiness overall	Greatest Happiness Principle Actions are right in proportion as they tend to promote the greatest good for the greatest number	Human capacity for pleasure/pain (anthropocentric individualistic)
Immanuel Kant (1724–1804) Basis of principle: The dignity of human reason and free will	The Categorical Imperative (two formulations) 1. Always act so that you may also will your maxim (for action) to become a universal law 2. Always act so as to treat humanity in your person or that of another, as an end in itself, never merely as means	Human reason and free will (anthropocentric individualistic)
John Rawls Basis of principle: Justice as equality/fairness contractarianism	Original Position: Two Principles of Justice 1. The first principle of justice requires an equal distribution (since it is not reasonable for anyone to expect more than an equal share of social goods) 2. Inequalities are permissible when they maximize or at least contribute to the long-term expectations of the least fortunate group in society	Rational contractors' dialogue (anthropocentric)
Aldo Leopold (1887–1947) Basis of principle: Interdependence of all nature	The Land Ethic A thing is right when it tends to preserve the integrity, stability, and beauty of the biotic community; it is wrong when it tends otherwise	Interconnectedness of all life • holistic • integrative • non-anthropocentric

Based on this point of view, the ethics of integrity proposes a first *principle of integrity* and a series of eight second-order principles (SOPs) intended to clarify, apply, and implement the PI (Westra, 1994; 1998). These principles do not simply employ a deductive argument starting from *what is* to prescribe *what ought to be*. The principles acknowledge that, as *ought* indeed implies *can*, moral prescriptions must incorporate the actual conditions and circumstances that render both first-order principles and second-order, derived prescriptions consonant with reality.

For instance, on utilitarian grounds we might say that because human beings can and do feel pain, it is morally wrong to cause them pain-producing harms. We should add yet another specific natural fact or law — that is, the law of gravity, as applied to dropping heavy objects from above on innocent citizens. We can therefore conclude that because of *two* actual facts (one, the capacity of humans to feel pain; two, the law of gravity as it applies to heavy objects and their earthbound trajectories), it is morally wrong to drop bricks from my window on those who might pass below because it would harm them.

The addition of a holistic perspective to this procedure ensures the inclusion of all life, sentient or not (as the latter also participates in providing natural services in support of all life; see Figure 6.1), as an object of our concern to avoid inflicting harm.

In addition, this approach does justice to the mutualism and the interdependencies present in all nature as facts and laws that will inform our decisions about what kind of interference with

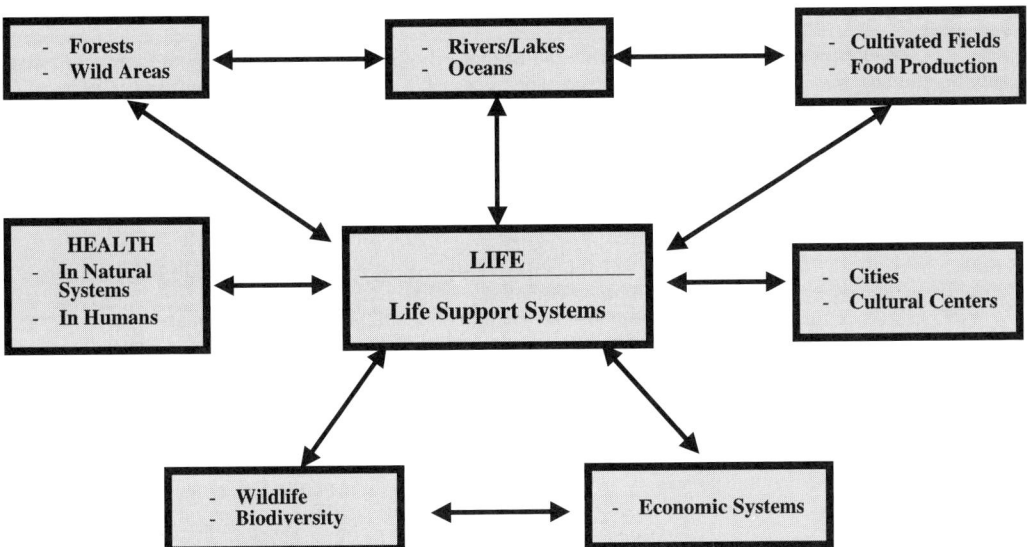

Figure 6.1   Mutual and reciprocal impacts for an integrated ethic (symbolizing the work of the Global Integrity Project, 1996–1999)

natural systems would or would not inflict a harm (such as diminished integrity) upon the wholes under consideration.

In this way, the discourse and the argument are radically different as they necessarily include an ongoing dialogue with scientists, who can research and update the true meaning of *harm* as well as that of *interference.*

Public policy then should not depend on bureaucrats or politicians, let alone the interests and biases of transnational corporations. A dialogue between ethicists and scientists, as envisioned by both postnormal science (Funtowicz and Ravetz, 1995) and by the Precautionary Principle (Brown, 1995) will be required to characterize the true impact of our interference, hence to suggest the limits required for all our activities. Such a dialogue will best help us to understand and explain what constitutes a *harm* and to whom or to what these harms will apply. These procedures will ensure that the principles used will be radically integrative and holistic. As a consequence, the role of the moral philosopher will also be radically altered, as her contribution must also be integrated within scientific discourse in a way that will enrich both science and morality (see Table 6.1).

In fact, there have always been two parallel paths originating from the same starting point in the Global Integrity Project: one tracked the *ethics* of integrity, the other the *science* of integrity. The main point is that the latter is basic to the formulation of the former. This is because, through integrity as a factual condition, we recognize the need for a holistic, integrated approach that the ethics of integrity support a moral stance *quite different* from that of other traditional moral theories (see Table 6.2). With such a radical departure from traditional approaches, the principle I provided was also viewed as giving little or no guidance for specific cases and situations, without fitting into any other established approach.

Following on the path of the *PI*, I saw the necessity to provide second-order principles to render the PI operational. Even for that endeavor I still needed interaction and dialogue with other disciplines so that the SOPs would reflect the joint conclusions of our research. This goal was eventually met as I wrote *Living in Integrity* (1998), a work for which I relied on the critique and advice of the collaborators and the co-investigators in the new grant (1996–1999). I proposed eight SOPs that seemed to capture best the meaning of living out the principle of integrity and its rather sketchy categorical mandates.

Table 6.3   Integration and Synthesis in Ethics: the Ethics of Integrity.

Philosopher	Principle of Integrity (PI)	Factual/Natural Characteristics
Laura Westra (living)	The first moral principle is that nothing can be moral that is in conflict with the physical realities of our existence or cannot be seen to fit within the natural laws of our environment in order to support the primacy of integrity	Interconnectedness of all life; holistic, integrative nonanthropocentric
Basis of principle: Primacy of life and life-support systems		

Scientific collaborators:	Second-order principles (SOPs) (to render *PI* operational)
J. Karr	1) In order to protect and defend ecological integrity, we must start by designing policies that embrace complexity
R. Noss	
O. Loucks	2) Accepting the paradigm change of postnormal science for public policy
D. Pimentel	3) Human activities ought to be limited by the requirements of the Precautionary Principle
R. Goodland	
W. Rees	4) We must accept an *ecological worldview* and thus reject our present *expansionist worldview* and reduce our *ecological footprint*
C. Soskolne	5) We must reject the pressures of *technological maximality* and reduce many of our wasteful human rights
J. Kay	
R. Ulanowicz	6) It is necessary for humanity to learn to live as in a buffer; zoning will impose limits on both the *quality* and the *quantity* of our activities
	7) We must respect micro-integrity (for single organisms)
	8) We need to accept both *risk thesis* and *potency thesis*

1996–1999 The Global Ecological Integrity Project (two recommendations)
No.   9 Reduce consumption in northwest affluent countries at least by 50%
No. 10 Practice contextual vegetarianism and always attempt to eat lower on the food chain

Beyond the efforts to find appropriate prescriptions for an ethic derived from integrity, the primary strand of our investigation is and has been the transdisciplinary research that leads to an understanding of the actual conditions that define a landscape with integrity. This understanding informs all efforts to defend it as an ultimate value in morality and in public policy.

INTEGRITY AND HEALTH: THE INTERFACE

The focus of this publication is on ecosystem health, so it is important to show why concern for ecosystem integrity and for ecosystem health are different but compatible. In fact, they represent two aspects of the same concern. The difference between the two concepts is an important one. The simplest way to distinguish one from the other is to start by saying that ecological integrity includes health (Westra, 1994), but ecosystem health as such does not include integrity. The starting section of this chapter makes clear that biological integrity is intended to refer to the areas that are largely unmanipulated and that are "the product of bioevolutionary processes" (Karr, 1995; 1996). But a healthy ecosystem is, almost by definition, one that *functions* and *produces* well in response to various human needs and wants — that is, one that is used or manipulated in a nondestructive appropriate way. This is the reason integrity is absolutely necessary, even if our main focus might be an interest in ecosystem health. In Costanza et al. (1992), B. Haskell says:

> Since fast-changing human cultures are embedded in large-scale, slow changing ecological systems, we must develop policies that allow human cultures to thrive *without changing the life-support, functions, diversity and complexity of ecological systems.*

The authors do not seem to appreciate the inherent conflict between "human thriving," at least as it is understood in the modern technologically advanced countries in the northwest, and the possibility of maintaining "without change the life-support functions, diversity and complexity of ecological systems." Any natural system that has been substantially altered by human beings has correspondingly lost its ability to provide life support, and this occurs in proportion to the severity of the disruption and modifications that have happened over time. Nor is it sufficient to regulate for ecosystem health, as it does not exclude human use and presence. Unless use and presence are understood in terms a hunter–gatherer society or a native tribe might accept, these are incompatible with the *full* presence of life support, as can be provided by a system that is largely unmanipulated and intentionally left wild.

Some argue that if we have the concept and the goal of ecosystem health, integrity is unnecessary. After all, there are no "untouched" places left on Earth, so why — so the argument goes — should we insist on integrity rather than health as an environmental goal? This argument, like the one in the paragraph cited, is incorrect. It is disingenuous to expect the same services from a cultivated field or a suburban park that one can expect from an old-growth forest. Of course, human beings need various natural products to survive. But if they assume the freedom to co-opt the whole Earth for their purpose, they will eliminate the most important elements needed for their survival, as Costanza et al. indicate, when they speak of the conditions for life support.

Some percentage of wild (if not pristine) nature *must* be preserved, if the goal advocated by Costanza et al. (1992) is to be achieved. Once again, it is not possible to utilize and manage all corners of the Earth and expect — at the same time — that it will provide the services that only the wild can offer. Whether the idea offends modern sensibilities, we do not *belong* anywhere we choose on Earth; because we, unlike nonhuman animals, are not prepared to enter nature as we are, without all the technological aids we take for granted, from computers to chemical toilets, from all-terrain vehicles to televisions. Given this inescapable aspect of modernity, it is absurd to say that we are "animals like other animals" and that the disruptions we cause can be compared to those caused by other animals, beavers, ants, or wolves.

If we were prepared to live in nature like those other animals, fostering the limited, localized disruptions, maybe even extinctions, that animal activities and predations produce, then that line of argument might be valid. As it is, and because we are not, for the most part, prepared to live an existence that might bring us close to the land and its rhythm, the argument that we still are just natural animals, one among many, will not stand. We need only to consider that even nonhuman animals, if they have been technologically altered, wreak havoc in their (formerly) natural habitats. Consider, for instance, farmed salmon when it escapes into wild ecosystems and, in general, the effect of all "alien" plants and creatures in ecosystems (Westra, 1998).

In essence, if we admit the necessity for life support systems, then we need wild areas in some proportions to provide sufficiently for those functions (Noss and Cooperrider, 1994). In addition, it is also true that we must support ecosystem health for all areas we need to utilize. But both our life and health, and that of areas intended for careful human use, *also* require some percentage of wildness and wild areas to ensure continued support.

Thus the main message I want to convey is that there is no conflict between the concern for biological integrity and the concern for human and ecosystem health, as the former supports the latter and the two remain complementary. Because of the way we understand ecosystem health, another difference arises: integrity is a basic value from both an anthropocentric and a nonanthropocentric perspective. Health remains an anthropocentric concept when applied to ecosystems. For integrity, the main point is that "nature's services" provide life support for all biota, within and without the confines of the wild areas, with minimal human impacts, where integrity can be recognized. All that live within those areas need both components and processes for their optimum flourishing as they have evolved naturally in that landscape.

Hence, from a moral point of view, the primacy of integrity mirrors the primacy of nonanthropocentrism. That is, by emphasizing the intrinsic value of all natural things, individuals, aggregates, wholes, and processes, it establishes a clear starting point for moral considerability. Why the necessary connection between integrity and natural services? Common sense indicates that parking lots, city centers, and skyscrapers do not provide the natural services on which we all depend; in contrast, an old-growth forest does a lot better in that regard, although the precise point where natural services are sufficiently provided (or totally lost) is not obvious. Both need to be identified through scientific research and placed somewhere along the continuum from "nothing" to (almost) "everything" that is required for the presence of integrity and the life support upon which it is based.

In addition, in order to understand better the connection between integrity and ecological health, we need to understand that human and nonhuman health, in general, are best protected by the presence of areas of integrity, in some proportions and percentage (somewhere around 30% of the whole Earth), on both land and water (Noss and Cooperrider, 1994; Lubchenco, 1999; Pauly, 2000). But humans, like most other animals, also need to disrupt natural evolutionary development to some extent. When we use earth or sea, whether we extract resources or modify and cultivate, we need to be aware of the necessity to maintain health (hence sustainability) in the areas we modify and to ensure that our interventions do not cause stress to the components or the processes of those other areas, the "core" or "wild" areas of integrity, that we do not use.

During the period from 1992 to 1999, the Global Ecological Integrity Project established a transdisciplinary dialogue among various experts that led to a number of discoveries for each one of us. One of the most important discoveries for me was the extent to which even set-aside protected forests had suffered from the impact of modern human civilization as it proliferated just outside their borders. Orie Loucks' research showed unequivocally a 50% loss of biomass (worms) in the "protected" forests he studied in Ohio (Loucks, 2000).

This discovery emphasizes the main point I have been making: a *healthy* functioning forest is good and necessary, as are sustainable fields, sustainable fisheries, and sustainable plantations of all kinds. However, all of these are not *sufficient*, because unless humankind manages and restrains its activities outside wild protected areas, hence ensuring that enough fully protected areas exist to provide the life support we all need, their sustainability and that of all life is — at best — temporary. Their functions (like ours) need to be supported by areas of integrity, smaller than healthy areas used for human purposes, but stronger in themselves as more "complete," as they approximate as closely as possible the products of natural bioevolution.

Therefore, areas of integrity are fundamental for health and sustainability for all biota but are also required to support a moderate and respectable amount of human industrial enterprise. Neither economy nor trade can subsist without sustainable practices and a healthy natural base, protected by human restraints about the activities they practice and the areas they occupy. The ethics of integrity recommend that humans should live "as in a buffer" in order to ensure that areas of integrity retain the full range of their capabilities, as appropriate to their geographical location. Hence, health and integrity are essentially two aspects of the same sustainable whole.

Those who imagine a conflict between ecosystem health and ecosystem integrity as a goal of protection and regulation are simply mistaken. Those who perceive a conflict between the respect for integrity and that for human health are equally fundamentally in error. The World Health Discussion Document (Soskolne and Bertollini, 1999) makes this fact abundantly clear. The workshop's executive summary says:

> Humans, like other forms of life on Earth, are dependent upon the capability both of local ecosystems and of the global ecosphere for maintaining health. However, in relatively recent times, humans, particularly in industrialized countries, have developed an erroneous perception of being separate from nature's processes.

That is why we believe that the emphasis on "ecosystem health," as the focus of sustainability for our human projects and enterprises, must be balanced by the focus on ecological integrity and, with it, on the sustainability of all life, not only our own. Adopting integrity as the main focus is also best for human health, despite the nonanthropocentric thrust of this approach. The WHO Discussion Document says:

> The Workshop assessed that current world-wide patterns of over-consumption, population growth, and the inappropriate uses of technology, are unsustainable. Assessment methods included the eco-

logical footprints of nations, the Index of Biological Integrity, the Measure of Mean Functional Integrity and the World Wide Fund for Nature (WWF) analyses. The use of each of these indices reveals that environmental degradation has the potential for severe negative human health impact. Infrastructure is needed to establish and maintain data that are adequate for improved scientific assessments of resource depletion and health risks (Soskolne and Bertollini, 1999).

Paradoxically, then, adopting a nonanthropocentric holistic perspective based on Leopold's first normative concern, the promotion of integrity, is more conducive to the protection of human life and health than any other perspective one might adopt. In addition, even ecosystem health is protected if the ethics of integrity are implemented. A perspective based on respect for life embraces both humans and nonhumans, individuals, wholes, processes, and natural functions. Although no one — to my knowledge — has proposed an ethic of ecosystem health, that approach might ultimately be self-defeating. Because it would highlight only anthropocentric values, it would fail in its support of the very value it intended to protect. The ethics of integrity, with its wider reach, can extend protection and support to all life, including all the components that are vitally necessary for it.

REFERENCES

Angermeier, P.L. and Karr, J.R., Biological integrity versus diversity as policy directives, *BioScience*, 44, 690–697, 1994.

Brown, D.A., The role of law in sustainable development and environmental protection decision making, in *Sustainable Development: Science, Ethics and Public Policy*, Kluwer Academic Press, Dordrecht, the Netherlands, 1995, pp. 64–76.

Costanza, R., Norton, B., and Haskell, B., *Ecosystem Health — New Goals for Environmental Management*, Island Press, Washington, D.C., 1992, pp. 3–20.

Daily, G., Ed., Introduction, in *Nature's Services*, Island Press, Washington, D.C., 1997, pp. 3–4.

Funtowicz, S. and Ravetz, J., Science for the post-normal age, in *Perspectives on Ecological Integrity*, Kluwer Academic Publishers, Dordrecht, the Netherlands, 1995, pp. 146–161.

Karr, J.R., Ecological integrity and ecological health are not the same, in *Engineering within Ecological Constraints*, National Academy Press, Washington, D.C., 1996, pp. 100–113.

Karr, J.R., and Chu, E., Ecological integrity: reclaiming lost connections, in *Perspectives on Ecological Integrity*, Westra, L. and Lemons, J., Eds., Kluwer Academic Publishers, Dordrecht, the Netherlands, 1995, pp. 34–48.

Leopold, A., *A Sand County Almanac and Sketches Here and There*, Oxford University Press, New York, 1949.

Loucks, O., Pattern of forest integrity in the eastern U.S. and Canada: measuring loss and recovery, in *Ecological Integrity: Integrating Environment, Conservation and Health*, Pimentel, D., Westra, L., and Noss, R., Eds., Island Press, Washington, D.C., 2000, pp. 177–190.

Lubchenco, J., Thinking Like an Ocean, Plenary presentation for the NATO Advanced Research Workshop, Budapest, June 26, 1999.

Noss, R.F. and Cooperrider, A.Y., *Saving Nature's Legacy*, Island Press, Washington, D.C., 1994.

Pauly, D., Global change, fisheries and the integrity of marine ecosystems: the future has already begun, in *Ecological Integrity: Integrating Environment, Conservation and Health*, Pimentel, D., Westra, L., and Noss, R., Eds., Island Press, Washington, D.C., 2000, pp. 22–23.

Pimentel, D., Westra, L., and Noss, R., Eds., *Ecological Integrity: Integrating Environment, Conservation and Health*, Island Press, Washington, D.C., 2000.

Soskolne, C. and Bertollini, R., Global Ecological Integrity and "Sustainable Development": Cornerstones of Public Health, A Discussion Document of the World Health Organization (WHO) and the European Centre for Environment and Health (ECEH), 1999, World Wide Homepage of WHO/ECEH Rome Division, http://www.who.it.

Westra, L., *An Environmental Proposal for Ethics: The Principle of Integrity*, Rowman Littlefield, Lanham, MD, 1994.

Westra, L., *Perspectives on Ecological Integrity*, Westra, L. and Lemons, J., Eds., Kluwer Academic Publishers, Dordrecht, the Netherlands, 1995.

Westra, L., *Living in Integrity: A Global Ethic to Restore a Fragmented Earth*, Rowman Littlefield, Lanham, MD, 1998.

Westra, L. and Wenz, P., *The Faces of Environmental Racism: the Global Equity Issues*, Rowman Littlefield, Lanham, MD, 1995.

Westra, L. and Werhane, P., *The Business of Consumption — Environmental Ethics and the Global Economy*, Rowman Littlefield, Lanham, MD, 1998.

Westra, L., Miller P., Karr, J., Rees, W., and Ulanowicz, R., Ecological integrity and the aims of the global integrity project, in *Ecological Integrity: Integrating Environment, Conservation and Health*, Island Press, Washington, D.C., 2000, pp. 19–41.

CHAPTER 7

Ecological Integrity and Material Growth: Irreconcilable Conflict?*

William E. Rees

INTRODUCTION AND ANALYTIC FRAMEWORK

Many ecologists, other scientists, and ordinary citizens who simply love nature see the loss of biodiversity and the threat to ecosystems integrity as one of the most pressing problems confronting humankind. Maintaining ecosystems structure and function has become part of even the mainstream sustainable development agenda. For example, the Brundtland (United Nations) World Commission on Environment and Development, the commission that popularized the concept of *sustainable development,* argued that we could protect the environment by using more benign technologies, even as we strive for a five- to tenfold expansion of industrial activity by 2040 (WCED, 1987).

This is a tall order. I argue below that there is an unavoidable conflict between such growth-oriented economic development and maintaining ecological integrity. This conflict is rooted in the very nature of human beings as a highly successful species but one whose activities are now constrained by planetary limits governed ultimately by the second law of thermodynamics.

The problem goes beyond concern for the natural world — some analyses suggest there is virtually no possibility for an industrial society of 6 to 10 billion people using prevailing or anticipated technologies to live sustainably on Earth. Greenhouse gas accumulation, climate change, ozone depletion, fisheries collapse, land degradation, falling water tables, deforestation, toxic contamination, endocrine (hormone) mimicry, accelerating species loss — these and related trends, both local and global, are indicators that the scale of the human enterprise already exceeds the long-term carrying capacity of the ecosphere.

Despite increasing scientific awareness of biophysical limits, human pressure on the planet is relentlessly increasing. Global population reached 6 billion in July, 1999, and was growing by 80 million per year. By the end of the decade it had almost doubled twice in the 20th century. All these people, rich and poor alike, have rising material expectations sustained by an economic system that assumes the latter are insatiable.

Unfortunately, society seems paralyzed by conflicting perceptions of the so-called environmental crisis. The majority, including mainstream policy makers, most economists and techno-optimists,

---

* Parts of this chapter are revised from Rees, W.E., Patch disturbance, eco-footprints, and biological integrity: revisiting the limits to growth, in Pimentel, D., Westra, L., and Noss, R.F., Eds., *Ecological Integrity: Integrating Environment, Conservation and Health,* Island Press, Washington, D.C., 2000, pp. 139–156. An extended earlier version appears in Nemetz, P., Ed., *Bringing Business on Board*, JBA Press, Vancouver, B.C., 2002. It is reprinted here with permission.

---

and many humanists, remain dedicated to growth and consumer ideals. They see freer markets and a new efficiency revolution as the only politically feasible solution to both global ecological decline (wealth can allegedly purchase a "cleaner" environment) and to the social problems caused by persistent poverty and growing material inequity.

On the other hand, a growing minority, mainly ecocentric and community-oriented groups and individuals, see the growth ethic and consumerism themselves as the issues. They empha-size that excess production and consumption are the *causes* of resource depletion (including biodiversity loss) and pollution. Moreover, at least in high-income countries, there may be few compensating benefits. The evidence suggests that beyond a certain income level there is actually an *inverse* relationship between income growth and perceived well-being (Lane, 2000). Further growth may therefore be unnecessary for either ecohealth or social sustainability — the solutions lie more in changing consumer behavior to ease pressures on ecosystems, and in policies to ensure more equitable distribution of the world's present economic output. (Certainly growth-through-globalization is not effective in reducing inequity. In 1970 the richest 10% of the world's citizens earned 19 times as much as the poorest 10%. By 1997, the ratio had increased to 27:1. At that time, the wealthiest 1% of the world's people commanded the same income as the poorest 57% (income ratios reflect purchasing power parity; data from UNDP, 2001).

Environmentalism Is *Not* Human Ecology

Whatever the proposed solutions, almost everyone in the mainstream shares the perception that this is an *environmental crisis* rather than a *human ecological crisis*. The distinction is not a trivial one. The former term literally externalizes the problem, effectively blaming it on an environment gone wrong or on defective resource systems that need to be managed more effectively. This perception reduces the destruction of the ecosphere to mere mechanics, a problem readily amenable to the "technical quick fix" approach favored by industrial society. By contrast, seeing the crisis as a human ecological problem places blame squarely where it belongs — on the nature and behavior of people themselves. It also suggests that it is human wants that must be better controlled. This is probably the least comfortable policy domain for the denizens of modern consumer societies to contemplate.

In this chapter, I start from the premise that the current dilemma is at least partly rooted in this perceptual or psychological tension. The Cartesian dualism that underpins Western techno-scientific culture has created a psychological barrier between humans and the rest of nature that keeps us from truly knowing ourselves. Indeed, it is a deep irony of the human-induced environmental crisis that people have a dismally ill-developed understanding of themselves as ecological beings.

My overall purpose, therefore, is to reinterpret the environmental crisis as a problem of gross *ecological dysfunction* stemming, ironically, from humanity's unquestionable evolutionary success. This requires acceptance of two material facts that are virtually ignored in conventional analysis. First, the human ecological niche is now so extensive that *Homo sapiens* has become a significant participant in most of the world's major ecosystems. Second, human economic activity (like the economic activity of any other species) requires continuous, irreversible energy and material transformations that are ultimately subject to constraints imposed by the second law of thermody-namics (Georgescu-Roegen, 1971; Daly, 1991a,b; Rees, 1999). In this light, we can see that material economic activity is really human ecology and that the economy itself is an inextricably integrated, completely contained, and wholly dependent growing subsystem of a nongrowing ecosphere (Daly, 1992; Rees, 1995).

As noted, this perspective contrasts sharply with the dominant worldview. The latter assumes that the human enterprise is now more or less independent of nature so that the "environment" no longer constrains the economy. It is this ecologically empty vision that drives the prevailing global development paradigm and has created the sustainability conundrum.

FUNDAMENTALS OF HUMAN ECOLOGY

Far-from-Equilibrium Thermodynamics

The second law of thermodynamics is fundamental to all processes involving energy and material transformations. Since ecosystem analysis often begins with energy and material relationships, understanding the second law is essential to understanding both the structure of whole ecosystems and the functional *niches* of individual species, including humans.

In its simplest form, the second law states that any isolated system* will tend toward equilibrium; alternately, the entropy of any isolated system always increases. This means that available energy spontaneously dissipates, concentrations disperse, and gradients disappear. An isolated system thus becomes increasingly unstructured in an inexorable slide toward thermodynamic equilibrium. This is a state of maximum entropy in which "nothing happens or can happen" (Ayres, 1994).

Early formulations of the second law referred strictly to simple isolated systems close to equilibrium. We now recognize, however, that *all* systems, whether isolated or not, near equilibrium or not, are subject to the same forces of entropic decay. Thus *any* differentiated far-from-equilibrium system has a natural tendency to erode and crumble.

But not all such systems unravel as expected. Indeed, for much of this century, people believed that living entities were exempt from the second law since they obviously do not spontaneously run down (they do not "tend toward equilibrium"). On the contrary, biological systems from individual fetuses to the entire ecosphere seem to gain in mass and organizational complexity over extended periods of time — they actually *increase* their distances from equilibrium. It seems that for describing nature, "distance from equilibrium becomes an essential parameter ... much like temperature [is] in equilibrium thermodynamics" (Prigogine, 1997).

This seemingly paradoxical behavior can be reconciled with the second law in light of the hierarchical organization and relationships of thermodynamic systems in nature (Kay, 1991). Biophysical systems exist in loose, nested, and overlapping hierarchies, each component system being contained by the next level up and itself comprising a chain of linked subsystems at lower levels.**

Living systems are thus able to import available energy and material (essergy) from their host environments and use it to maintain their internal integrity and to grow. The second law also dictates that such self-producing systems export or dissipate the resultant waste (entropy) back into their hosts. In short, modern formulations of the second law posit that all highly ordered systems develop and grow (increase their internal order) "at the expense of increasing disorder at higher levels in the systems hierarchy" (Schneider and Kay, 1994). Living systems maintain their distance from equilibrium by "feeding" on their hosts. Because such systems continuously degrade and dissipate available energy and matter, they are called *dissipative structures*.

Self-Production by the Ecosphere...

The highest order dissipative structure on Earth is the ecosphere itself. The ecosphere comprises all the biomes and individual ecosystems on the planet, and all of these subsystems develop and maintain themselves in a far-from-equilibrium steady state by dissipating light from the sun (the next level up in the systems hierarchy). In effect, the ecosphere feeds on the sun. (The existence of a gradient of exogenous energy is a prerequisite for life.)

In nature, green plants are the factories. Through photosynthesis, plants use solar energy and the simplest of low-grade inorganic chemicals (mainly water, carbon dioxide, and a few mineral nutrients) to assemble the high-grade fats, carbohydrates, proteins, and nucleic acids upon which

---

* An isolated system can exchange neither energy nor matter with its environment.
** For example, consider the following nested hierarchy of biological organization (from high to low): ecosystem, population, individual, organ–system, organ, tissue, cell, cellular micro-organelles. Kay et al. (1999) define such complex hierarchic structures as "self-organizing holarchic open (SOHO) systems."

most other life forms and the overall functioning of the ecosphere depend. Because they are essentially self-feeding and use only dispersed (high-entropy) substances for their growth and maintenance, green plants are called primary *producers*.

By contrast, all animals, bacteria, and fungi are primary, secondary, or tertiary *consumers* whose growth and development is achieved through the assimilation and dissipation of organic energy and matter originally assembled by plants. Although dependent on plants, consumer organisms are also essential to the continuing functioning of the ecosystem. Consumers degrade the complex energy-rich chemicals produced by plants, dissipating the energy and releasing simple organic and inorganic chemicals back into their host ecosystems for reuse by the producer plants. (Nature invented material recycling.)

... and Consumption by the Economy

Where do humans and their economies fit in? Both economists and ecologists would agree that human beings are consumer organisms. In fact, in today's increasingly market-based society, people are as likely to be called consumers as they are citizens, even when the context is a noneconomic one.

Ecologists would actually refer to humans as *macro*-consumers. In general, macro-consumers are large organisms, mainly animals, that consume (i.e., dissipate) either green plants or other animals to grow, develop, and satisfy their own basic metabolic needs. Humans are particularly catholic and adaptive consumers — we have wide-ranging omnivorous tastes; and if we cannot consume something directly (such as grass), we domesticate an animal that will and then eat the animal. "Indeed, if one feature sets humans apart from other animals, it is the breadth of the ecological niche we presently occupy" (Flannery, 1994). This, combined with learning and cumulative technology, makes us particularly formidable competitors in the global ecological arena.

Economists and ecologists also both see humans as producers. However, there is a fundamental difference between production by green plants and production by the economy. As noted, green plants are *primary* producers that assimilate simple dispersed materials and sunlight (a relatively low-grade form of energy) to produce the most complex and energy-rich molecules known to science. By contrast, human beings are strictly *secondary* producers. The production and maintenance of our bodies, and all the products of human factories, require enormous inputs of high-grade energy and concentrated material resources from the rest of the ecosphere. That is, all production by the human enterprise, from the increase in population to the accumulation of manufactured capital, requires the consumption and dissipation of a much larger quantity of available energy and material *first produced by nature*.

This relationship implies, on theoretical grounds alone, that the continuous growth of the economy will ultimately generate an ecological crisis. Both the economy and the ecosphere are self-organizing, far-from-equilibrium dissipative structures; but as previously noted, the former is an embedded dependent subsystem of the latter. In structural terms, therefore, the expanding human enterprise is potentially parasitic, positioned to consume the ecosphere from within (Rees, 1999).

HOMO SAPIENS: THE ARCHETYPAL PATCH-DISTURBANCE SPECIES

The consequences of this relationship have been felt at every stage of human evolution. The mere existence of people, even pre-industrial hunter–gatherers, in a given habitat invariably results in significant changes in ecosystem structure and function. This is the inevitable consequence of two simple (but generally unacknowledged) facts of human biology: first, humans are large animals with correspondingly large individual energy and material requirements; and second, we are social beings who universally live in extended groups. The invasion of any previously stable ecosystem by people therefore necessarily produces changes in established energy and material pathways. There will be a reallocation of resources among species in the system to the benefit of some and the detriment of others. To this extent at least, humans invariably perturb the systems of which they are a part.

These facts, together with food productivity data for typical terrestrial ecosystems, are enough to suggest the following hypothesis: *in most of the potential habitats on Earth, groups of human hunter–gatherers will sooner or later overwhelm local ecosystems and be forced to ramble farther afield in search of sustenance.* The productivity of most unaltered ecosystems is inadequate to support more than a handful of people for very long in the immediate vicinity of a temporary camp. In pre-agricultural times, when a group of human foragers had hunted out and picked over a given area, they were forced to move on. This enabled the abandoned site to recover, perhaps to be revisited in a few years or decades. Thus, by moving among favored habitat sites, exploiting one, allowing others to rebound, early humans could exist in an overall dynamic equilibrium with the ecosystems that sustained them, albeit across their total home ranges. Hunting–gathering and closely related swidden (slash-and-burn) agriculture, with their long fallow periods after episodes of intensive use, may well be the most nearly sustainable lifestyles ever adopted by humans (Kleinman et al., 1995, 1996).

Such quasi-steady-state systems, established after long periods of human habitation, would, of course, be quite different from the ecosystems that would exist in the same regions in the absence of people. Perhaps the most dramatic evidence of this is the major ecosystem alterations that occur when humans first invade and settle a new habitat or land mass. Some of these changes are permanent. The recent paleoecological, anthropological, and archaeological literature tells a convincing story of the extinctions of large mammals and birds that accompanied first contact and settlement of their habitats by human beings (Ponting, 1991; Diamond, 1992; Flannery, 1994; Pimm et al., 1995; Tuxill, 1998). "For every area of the world that paleontologists have studied and that humans first reached within the last fifty thousand years, human arrival approximately coincided with massive prehistoric extinctions" (Diamond, 1992).

The species so extirpated include not only those upon which humans preyed, but also various other predators to whom humans proved to be competitively superior (Table 7.1). The resultant changes to ecosystem structure were considerable. In North America, South America, and Australia, about 72, 80, and 86%, respectively, of large mammal genera became extinct after human arrival (Diamond, 1992). Pimm et al. (1995) estimate "that with only Stone Age technology, the Polynesians exterminated >2000 bird species, some ~15 percent of the world total."

Table 7.1 Example of Areas Where Significant Extinction Is Thought to Have Accompanied Human Occupation

Geographic Area	Species Extinguished by Humans (in last 50,000 years)
Africa	Giant buffalo, giant hartebeest, giant horse
Australia	Diprotodonts (marsupial equivalent of cows, and rhinos), giant wombat, giant kangaroo
Crete and Cyprus	Pygmy hippos and giant tortoises, dwarf elephants and deer
Europe	European rhino, cave-bear, long-tusked elephant, hippopotamus, Irish elk, woolly mammoth and woolly rhino
Hawaii	Flightless geese and ibises; 50 species of small birds
Henderson Island	Three large pigeons, one small pigeon, three seabirds
Madagascar	Half-dozen species of giant flightless "elephant" birds, two giant land tortoises, a dozen species of large lemurs, pigmy hippo, large mongoose-like carnivore
New Zealand	Moas; giant duck, goose, coot, raven, and eagle; pelican; swan; numerous small songbirds and mammals; frogs; snails, and giant insects
North and South America, Australia	Numerous large mammals
West Indies	Several monkeys; ground sloths, bear-sized rodent; several owls — normal, colossal, and titanic

Data from Diamond (1992, 1998), Flannery (1994), Ponting (1994).

As noted, prehistoric humans eventually came to live in a more or less stable dynamic equilibrium (or steady state) within their altered ecosystems, perhaps for thousands of years. However, with improvements in tools and weapons such as the shift from stone to metal, these long-term stable relationships ultimately broke down. Human hunter–gatherers seem to have been instrumental in the eventual extinction of various large animals with which they had long coexisted in many parts of the world. These species include everything from giant deer and mammoths in Eurasia, through giant buffalo, antelopes, and horses in Africa, to bears, wolves, and beavers in Britain (Diamond, 1992). All in all, the slow spread of pre-agricultural humanity over the Earth, accelerated by the diffusion of more advanced hunting technologies and the inexorable expansion of human numbers, seems, invariably, to have been accompanied by significant, permanent changes in the structure and function of ecosystems at all spatial scales.

Upping the Impact Ante

The already significant impact of humans on ecosystems escalated dramatically with the shift away from the hunter–gatherer lifestyle. With the dawn of agriculture 10,000 years ago and the much larger populations it could support, people acquired the capacity permanently to alter entire landscapes. For the first time, we were not merely consuming other species; humans were now appropriating entire ecosystems and diverting much of the photosynthetic energy flow to their own use.

Increased food production enabled the establishment of permanent settlements, the division of labor, the evolution of class structure, the development of government including bureaucracies and armies, and other manifestations of civilization. Indeed, as Diamond has eloquently argued, "the adoption of food production exemplifies what is termed an autocatalytic process — one that catalyzes itself in a positive feedback cycle, going faster and faster once it has started" (Diamond, 1998). More food calories made higher population densities possible, enabled large permanent settlements with the specialized skills and inventiveness this implies, and shortened the time spacing between children. This, in turn, enabled the higher populations to produce still more people, which increased both the demand for food and the technical and organizational capacity to produce it. Pressure on the land and ecosystems increased accordingly and, in the process, ended even the possibility of returning to hunting–gathering for the majority of people. The ancient trend has accelerated throughout the scientific–industrial era: for example, 11% of the 4400 mammal species extant today are endangered or critically endangered, and 25% of all mammal species are on a path of decline which, if not halted, is likely to end in extinction (Tuxill, 1998).

All this is to emphasize that humans are, *by nature*, a typical patch-disturbance species, a distinction we share with other large mammals ranging from beavers to elephants (*BioScience*, 1988).*

In simple terms, "large animals, due to their size, longevity, and food and habitat requirements tend to have a substantial impact on ecosystems" (Naiman, 1988). Thus, a patch-disturbance species may be defined as *any organism which, usually by central place foraging, degrades a small "central place" greatly and disturbs a much larger area away from the central core to a lesser extent* (definition revised from Logan, 1996).

Patch Disturbance and Ecological Integrity

The fact that humans are natural patch disturbers is problematic when it comes to understanding the relationships among humankind, biodiversity, and ecological integrity. As noted, there can be little question that the species composition and developmental trajectory of an ecosystem inhabited by even pre-agricultural humans would be different from the trajectory it would follow without them. However, the same statement could be made about many other ecosystems with reference to their respective keystone species. For example, the extirpation of large predators (jaguars,

---

* Predictably, with the exception of a passing reference to modern humans as "primary agents of environmental change," people are not included in this special issue on "How Animals Shape Their Ecosystems."

cougars, and harpy eagles) on Barro Colorado Island in Panama was followed by a ripple effect that saw major increases in the relative abundance of smaller predators and medium-sized seed-eaters which, in turn, led to the local extinction of small antbirds and several similar species, as well as to a large-scale shift in forest tree composition (Diamond, 1992). It seems that the presence of large predators significantly affects the structure of the entire ecosystem, a fact that may not be fully appreciated until their *removal* appears to disrupt the established order of things.

Now, biological integrity is sometimes seen as a property of near-pristine ecosystems, i.e., ecosystems not significantly modified by human intervention. But this begs the question of why ecosystem modification by humans should be treated differently from that induced by other species like jaguars and cougars. After all, the presence of these large cats results in a different climax system from that which develops in their absence; yet few would argue that Central American forests naturally populated by jaguars and cougars lack ecological integrity because of the unseen structural effects of their presence.

In fact, I would argue that there *is* a major difference, that the quality of human predation and patch disturbance should be distinguished from those of other species. Other large mammals may significantly affect the structure of their ecosystems, but the resultant relationship is relatively stable over the very long term. By contrast, *H. sapiens* has unique qualities that continuously increase the pressure on ecosystems and whose cumulative impacts now affect the entire ecosphere.

First, *H. sapiens* displays a uniquely broad and ever-widening food niche that extends from nearly pure carnivory to obligate herbivory and thus affects a wide range of plant and animal species. Second, humans are uniquely adaptive, which enables the species to exploit virtually all the ecosystems and habitats on Earth. Finally, and most important, because of the evolution of language and culture, human knowledge and technology are cumulative. As a result, human patch disturbance has tended over the millennia to intensify both gradually and in irregular spurts and now threatens to diverge permanently from equilibrium with supporting ecosystems. Humanity's drift from long-term steady state with nature has been accelerating since the Neolithic era. It received a major boost with agriculture, and really broke free with the use of fossil fuels* and the industrial revolution.

In summary then, I am arguing that the makings of the ecological crisis are genetically programmed into the ecology and sociobiology of our species. We are naturally a highly successful patch-disturbance species whose capacity to disrupt the ecosphere has been greatly augmented by behavioral plasticity and cumulative learning. Because of our extraordinary energy and material demands, and the "nesting" of the economy within the ecosphere, the growth of human numbers and cultural artifacts (manufactured capital assets) inevitably occurs at the expense of other organisms. (A form of competitive exclusion is operating here — what we get, other species do not.) The loss of ecosystems integrity or health associated with human beings is thus measured by persistent trends, accelerating biodiversity loss, and increasing systems variability.

OUR ECOLOGICAL FOOTPRINT: QUANTIFYING THE MODERN HUMAN "PATCH"

Despite — or because of — the marvels of modern technology, the industrial era has brought little change to the nature of human–ecosystems interaction except to increase the scale of the "patches" we disturb, the intensity of the disturbance, and the risk to our own survival. This can be shown using ecological footprint analysis, a method I have pioneered with my graduate students. In effect, eco-footprinting estimates the area of the modern human patch (Rees, 1992, 1996, 2000; Rees and Wackernagel, 1994; Wackernagel et al., 1999).

Ecological footprint analysis is an adaptation of trophic analysis, applied to humans and extended to include both our biological and industrial metabolisms (Ayres and Simonis, 1994). It

---

* Cheap, plentiful fossil fuels have enabled humans to accelerate the exploitation of everything else.

recognizes that all our toys and tools, factories and infrastructure are "the exosomatic equivalent of organs" (Sterrer, 1993) and, like bodily organs, require continuous flows of energy and material from and to the ecosphere for their production, maintenance, and operation.

We start by constructing what is, in effect, an elaborate food web for a specified human population, showing its material connections to the rest of the ecosphere. This step quantifies the material and energy flows (resources) required to support that population and identifies significant sources and sinks for its wastes.

Eco-footprinting is further based on the fact that many material and energy flows can be converted into land- and water-area equivalents. These are the ecosystem areas effectively appropriated for human use to produce food, fiber, and minerals, and to assimilate selected categories of waste.* Thus, the theoretical ecological footprint of a specified population is *the area of productive land and water ecosystems required on a continuous basis to produce the resources that the population consumes, and to assimilate the wastes that the population generates, wherever on Earth the relevant land/water may be located* (Rees, 2001). It therefore includes both the area appropriated through commodity trade and the area needed for the referent population's share of certain ecosystems-based services of nature (e.g., waste assimilation and nutrient recycling).**

How Big Is Our Ecological Footprint?

Since the beginning of the industrial revolution, the economy has been largely propelled by the constantly expanding use of extrasomatic (outside the body) energy supplies, particularly fossil fuels. By the early 1990s, extrasomatic energy consumption totaled about 407.5×10^{15} Btu annually by a population of about 5.5 billion people. "It is as if every [person] in the world had fifty slaves. In a technological society like the United States, every person has more than 200 such 'ghost slaves' " (at an assumed working level of consumption of 4000 Btu person^{-1} day^{-1}) (Price, 1995).***

As might be expected, much of this energy subsidy has been used to increase humanity's harvest of biomass (photosynthetic energy) and, indeed, of all other resources from groundwater to mineral ores. As a result, humans have become the dominant consumer organism in virtually all the major ecosystems types on Earth. For example, we are certainly the most ecologically significant marine mammal. Fossil energy and modern technology enable the global fishing fleet to appropriate seafood for humans that represents 25 to 35% of net marine primary productivity from shallow coastal shelves and estuaries. This 10% of the world's oceans produces 96% of the catchable fish (Pauly and Christensen, 1995). Similarly, humans are the principal consumer in most of the world's significant terrestrial habitats, diverting from grasslands and forests at least 40% of the products of photosynthesis for direct and indirect human use (Vitousek et al., 1986; Haberl, 1997).

It should therefore be no surprise that the residents of wealthy developed countries now have per-capita ecological footprints of 5 to 10 hectares (Wackernagel et al., 1999; WWF, 2000). This means that the ecosystem area required to support a typical high-income city is two to three orders of magnitude larger than the geographic area it physically occupies. For example, assuming Vancouverites are typical Canadians with an ecological footprint of about 7.7 hectares per capita, the 472,000 residents of Vancouver proper require an area of 3,634,400 ha, or 319 times the political area of their city (11,400 ha) to support their consumer lifestyles. In a particularly comprehensive analysis, Folke et al. (1997) estimate that the 29 largest cities of the Baltic basin in Northern Europe appropriate for their resource consumption and waste assimilation an area of forest, agricultural, marine, and wetland ecosystems 565 to 1130 times larger than the area of the cities themselves.

---

* To facilitate simple international comparisons, we generally use world average land productivities in eco-footprint estimates.

** In practice, only a limited number of wastes (e.g., carbon dioxide and plant nutrients such as nitrates and phosphates) can readily be converted to land-area equivalents with present knowledge. Some contaminants such as ozone-depleting chemicals or endocrine mimics cannot be included at all in eco-footprint analyses.

*** So dependent is industrial society on fossil energy that some authors predict the decline and collapse of civilization as supplies run out in the next few decades (e.g., Price, 1995; Duncan, 1993).

We usually think of cities as the productive engines of national economic growth. However, the strictly biophysical perspective of ecological footprint analysis adds another layer of complexity. Cities and urbanized regions are also intensive nodes of *consumption* whose economic and social activities are sustained almost entirely by bio-production and life-support processes occurring in a vastly larger area outside their political and geographic boundaries (Rees and Wackernagel, 1996; Rees, 1997).

Ecological Disparity and the Fair Earth-Share

Ecological footprinting reveals another insight relevant to future human development and conservation efforts. As noted, the citizens of high-income countries typically have eco-footprints of 5 to 12 hectares. This compares to a hectare or less per capita for most developing countries, such as India and Bangladesh, and a world average eco-footprint of 2.8 hectares (Figure 7.1, 1996 data from WWF, 2000). Because of their large per-capita footprints and high population densities, many first-world countries have an aggregate ecological footprint several times the size of their domestic territories. They are living on trade and imposing heavily on the global commons — i.e., by importing ecosystems goods and services (carrying capacity) from elsewhere. This poses a problem for prevailing international development strategies. Such strategies assume that we can relieve chronic poverty through sheer economic growth, thus avoiding political pressure for income redistribution. Indeed, the Brundtland Commission believed we could bring 8 to 10 billion people up to 1980s Western European material standards by the middle of the 21st century (albeit using more ecologically benign technologies [WCED, 1987]). Regrettably, on a finite planet, not all countries can be net importers of carrying capacity.

This reality begs the following question: What would be the global ecological footprint if the entire world attained a typical late-1990s Western European lifestyle using prevailing fossil fuel–based technologies? At about 6 hectares per capita, the present world population of over 6 billion people would need approximately 36 billion hectares to supply its resources and assimilate its wastes. If the human family grows by an additional 3 billion by, say 2050, the human eco-footprint could reach 54 billion hectares. The problem is that there are only about 12 billion hectares of adequately productive land and water on Earth. If the world's ecologically productive space were evenly divided among the present human population, we would get only about 1.5 ha of terrestrial

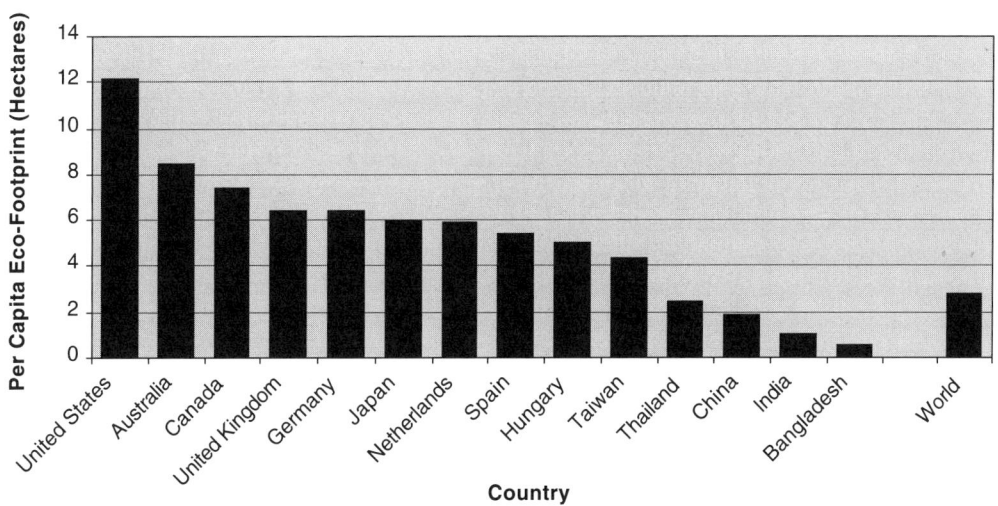

Figure 7.1 Ecological footprints of selected countries.

and 0.5 ha of aquatic ecosystem for a total of 2 hectares per capita.* In other words, the citizens of wealthy countries already inadvertently appropriate three to six times their "fair Earthshare" (Rees, 1996). (Note that these data take no account of the needs of other species.) Since we are unlikely to acquire three or four additional Earth-like planets over the next half century, it seems that the world will have to deal with socioeconomic and ecological inequity by means other than economic growth if it is serious about achieving sustainability.

DISCUSSION AND CONCLUSION

Many technological optimists believe that industrial "man" is no longer bound by ecological constraints. As the late famed growth advocate, Julian Simon, recently wrote: "We have in our hands now ... the technology to feed, clothe, and supply energy to an ever-growing population for the next seven billion years ..." (Simon, 1995). By contrast, this chapter shows that humankind is as dependent on nature as ever. Rather than substitute for nature, the principal ecological effect of technological advance has been to extend the scope and intensity of human exploitation of supporting ecosystems beyond sustainable limits. Humanity's ecological footprint is steadily expanding.

Because it remains a dependent subsystem of the ecosphere, the continuous growth of the human economy inevitably reduces the diversity, biomass, and resilience of exploited ecosystems. Energy and material flows appropriated from the ecosphere for humans are irreversibly unavailable for use by other species or to help maintain life-support functions. Indeed, since the Paleolithic era, humans have invariably expanded their presence on Earth by:

1. Depleting both self-producing and nonrenewable "natural capital" stocks (e.g., other species populations such as flightless birds, large slow mammals, fish; whole ecosystems such as forests; groundwater; soils; and petroleum and other hydrocarbons).
2. Displacing other species from their niches and appropriating their habitats (e.g., various ungulates in Africa, bison in North America, and thousands of species from tropical forests).
3. Eliminating competition for "our" food (e.g., seals in the case of fisheries; wolves in the case of ungulates, wild and domestic; and insects in the case of crops).

Production agriculture has been particularly damaging in its gross simplification of ecosystems to channel the product of photosynthesis to humans through a relatively few crop species. The spread of energy-intensive technologies, and most recently, globalization, has intensified all these processes.

Because of the expanding human ecological footprint, the Earth is now experiencing the greatest extinction episode since the natural catastrophes at the end of the Paleozoic and Mesozoic eras (Wilson, 1988; Pimm et al., 1995). Overharvesting, pollution, and particularly habitat destruction are producing what some analysts are calling the *sixth extinction*. The current extirpation rate is 100 to 1000 times prehuman levels inferred from the fossil and paleontological record (Pimm et al., 1995).

Ominously, evidence is gathering that humans themselves will increasingly bear the consequences of ecological degradation. In 1998, a record number of natural disasters caused massive property damage and drove more people from the land and their homes than war and civil conflict combined. According to the International Red Cross 1999 *World Disasters Report*, singular events such as Hurricane Mitch and the El Niño weather phenomenon, plus declining soil fertility and deforestation, drove a record 25 million people from the countryside into crowded underserviced shantytowns around the developing world's fast-growing cities. This is 58% of the world's total refugees. The *World Disasters Report* predicts that developing countries in particular will continue to be hit by super-disasters driven by human-induced atmospheric and climatic change, ecological degradation, and rising population pressures.

---

* Recall that the world average eco-footprint is already 2.8 ha per capita (Figure 7.1). This discrepancy shows that even at current average levels of consumption, the population of *H. sapiens* already exceeds the long-term carrying capacity of the Earth. We are living, in part, by depleting *our* natural capital by consuming the ecosphere from within (Rees, 1999).

Where does all this leave prospects for maintaining ecosystem health? It seems clear at this stage in global development that there is an irreconcilable conflict between further expansion of the human enterprise and contemporary conservation efforts. Nevertheless, *and this is critical*, the majority of humankind is understandably not satisfied with its present material lot and is determined to improve it. In these circumstances, conservation initiatives based on large set-asides of relatively undisturbed habitat may be absolutely necessary but by no means constitute a sufficient strategy for biodiversity preservation or the maintenance of ecosystems health. Population pressures, ecological collapse elsewhere, economic desperation, and hunger will eventually drive millions of impoverished people to seize ecological reserves, wilderness areas, national parks, and related conservation areas to satisfy their own needs. This is already happening in some developing countries. The illegal landings of unregistered ships overloaded with desperate Asian would-be immigrants on the Pacific Coast of Canada and the U.S. and in Australia in recent years is a striking reminder of the same general problem.

This chapter started from the premise that the "environmental crisis" is really a problem of human evolutionary success. Our competitive superiority has led to widespread human ecological dysfunction as the growing human enterprise breaches the biophysical limits of a finite planet. Those who would work to maintain biodiversity and ecosystem health must first confront this reality. Humanity has outgrown the Earth: we cannot for long safely consume more than nature produces and produce more wastes than nature can assimilate. To ensure local ecosystems health and the long-term integrity of the ecosphere, humanity's efforts must be directed toward the most effective means of reaching a stable steady state in our material relationship with nature.

Achieving a steady state will be no mean task in a world addicted to growth and material consumption — it requires nothing less than an absolute reduction in the total dissipative load imposed on ecosystems by people. The main items on the agenda include (1) lowering human populations everywhere; (2) reducing material consumption in the high-income countries (both through more efficient technologies and by adopting simpler, less materially intensive lifestyles); and (3) improving living conditions for the chronically impoverished. Unfortunately, in the prevailing geopolitical and economic climate, discussion of the first objective is suppressed (one is likely to be labeled racist); most politicians, particularly in North America, dismiss the second objective as too politically volatile (moreover, the available evidence suggests that efficiency gains in market economies result in increased per-capita and gross consumption [WRI, 2000.); and policies to achieve the third objective through sheer economic growth are failing. Chronic poverty prevails in much of the south, and the absolute income gap between high-income OECD countries and developing countries is still increasing. Even the relative income gap is growing except in the case of East Asia (UNDP, 2001).

This state of affairs makes clear that despite the rise in political rhetoric, the goal of an ecologically sustainable, socially equitable, and economically viable global village is eluding us. Society has misunderstood the problem, our solutions have deepened the crisis, and we remain in deep denial of the evidence that the present approach to sustainability is failing on its own terms. In this light, the greatest contribution to ecospheric health and sustainability will come, not from better resource management and nature conservation per se, but rather from a direct focus on prevailing cultural beliefs and values and on fundamental human behaviors. Shifting policy attention from proximate symptoms to deeper causes provides the greatest leverage in moving humanity toward a collective steady state with nature.

It also presents a daunting challenge because it requires a confrontation between human nature and present reality. *Homo sapiens* will become truly human — i.e., different from other species' populations that exceed their carrying capacity and crash — only if we succeed in culturally overriding the same hard-wired behavioral tendencies that have led to our evolutionary success. This will require a reasoned collective international response to the evidence that these tendencies are now maladaptive, that they have taken us too far. It is therefore unfortunate that for the past quarter century global culture has emphasized individual self-interest, sanctified greed, and downplayed the role of governments, all at the expense of community and a sense of shared commitment to the common good.

REFERENCES

Ayres, R.U., *Information, Entropy and Progress: A New Evolutionary Paradigm*, Woodbury, AIP Press, New York, 1994.

Ayres, R.U. and Simonis, U., *Industrial Metabolism: Restructuring for Sustainable Development*, United Nations University Press, New York, 1994.

BioScience, How Animals Shape Their Ecosystems, special issue, 38:11, 1988.

Daly, H.E., The circular flow of exchange value and the linear throughput of matter-energy: a case of misplaced concreteness, in *Steady-State Economics,* 2nd ed., Island Press, Washington, D.C., 1991a, pp. 195–210.

Daly, H.E., The concept of a steady-state economy, in *Steady-State Economics,* 2nd ed., Island Press, Washington, D.C. 1991b, pp. 14-49.

Daly, H.E., Steady-state economics: concepts, questions, policies, *Gaia,* 6:333–338 1992.

Diamond, J., *The Third Chimpanzee*, HarperCollins, New York, 1992.

Diamond, J., *Guns, Germs and Steel: The Fates of Human Societies,* W.W. Norton and Company, New York, 1998.

Duncan, R.C., The life expectancy of industrial civilization: the decline to global equilibrium, *Popul. Environ.,* 14:325–357, 1993.

Flannery, T.F., *The Future Eaters: An Ecological History of the Australasian Lands and Peoples*, Reed Books, Chatsworth, U.K., 1994.

Folke, C., Jansson, A., Larsson, J., and Costanza, R., Ecosystem appropriation by cities, *Ambio,* 26:167–172, 1997.

Georgescu-Roegen, N., *The Entropy Law and the Economic Process*, Harvard University Press, Cambridge, MA, 1971.

Haberl, H., Human appropriation of net primary production as an environmental indicator: implications for sustainable development, *Ambio,* 26:143–146, 1997.

Kay, J.J., A nonequilibrium thermodynamic framework for discussing ecosystem integrity, *Environ. Manage.,* 15:483–495, 1991.

Kay, J.J. and Regier, H.A., Uncertainty, complexity, and ecological integrity, in Crabbé, P., Holland, A., Ryszkowski, L., and Westra, L., Eds., I*mplementing Ecological Integrity: Restoring Regional and Global Environment and Human Health*, NATO Science Series IV: Earth and Environmental Sciences, Vol. 1, Dordrecht, Kluwer Academic Publishers, 2000, pp. 121–156.

Kleinman, P.J.A., Pimentel, D., and Bryant, R.B., The ecological sustainability of slash-and-burn agriculture, *Agric. Ecosyst. Environ.,* 52:235–249, 1995.

Kleinman, P.J.A., Bryant, R.B., and Pimentel, D., Assessing ecological sustainability of slash-and-burn agriculture through soil fertility indicators, *Agron. J.,* 88:122–127, 1996.

Lane, R.E., *The Loss of Happiness in Market Democracies,* Yale University Press, New Haven, 2000.

Logan, J., Patch disturbance and the human niche, Manuscript at http://dieoff.org/page78.htm, 1996.

Naiman, R.J., Animal influences on ecosystem dynamics, *BioScience,* 38:750–752, 1988.

Nemetz, P., Ed., *Bringing Business on Board*, JBA Press, Vancouver, B.C., 2002.

Pauly, D. and Christensen, V., Primary production required to sustain global fisheries, *Nature,* 374:255–257, 1995.

Pimm, S.L., Russell, G.J., Gittleman, J.L., and Brooks, T.M., The future of biodiversity, *Science,* 296:347–350, 1995.

Ponting, C., *A Green History of the World*, Sinclair-Stevenson, London, 1991.

Price, D., Energy and human evolution, *Popul. Environ.,* 16:301–317, 1995.

Prigogine, I., *The End of Certainty: Time, Chaos and the New Laws of Nature*, The Free Press, New York, 1997.

Rees, W.E., The ecology of sustainable development, *Ecologist,* 20:18–23, 1990.

Rees, W.E., Ecological footprints and appropriated carrying capacity: what urban economics leaves out, *Environ. Urban.,* 4:121–130, 1992.

Rees, W.E., Achieving sustainability: reform or transformation? *J. Plann. Lit.,* 9:343–361, 1995.

Rees, W.E., Revisiting carrying capacity: area-based indicators of sustainability, *Popul. Environ.,* 17:195–215, 1996.

Rees, W.E., Is "sustainable city" an oxymoron? *Local Environ.,* 2:303–310, 1997.

Rees, W.E., Consuming the Earth: the biophysics of sustainability. *Ecol. Econ.,* 29:23–27, 1999.

Rees, W.E., Patch disturbance, eco-footprints, and biological integrity: revisiting the limits to growth, in Pimentel, D., Westra, L., and Noss, R.F., Eds., *Ecological Integrity: Integrating Environment, Conservation and Health,* Island Press, Washington, D.C., 2000, pp. 139–156.

Rees, W.E., Ecological footprint, Concept of, *Encyclopaedia of Biodiversity,* Levin, S., Ed. in Chief, Vol. 2, San Diego, Academic Press, 2001, pp. 229–244.

Rees, W.E. and Wackernagel, M., Ecological footprints and appropriated carrying capacity: measuring the natural capital requirements of the human economy, in Jansson, A-M. et al., Eds., *Investing in Natural Capital: The Ecological Economics Approach to Sustainability,* Island Press, Washington, D.C., 1994, pp. 362–390.

Rees, W.E. and Wackernagel, M., Urban ecological footprints: why cities cannot be sustainable and why they are a key to sustainability, *Environ. Impact Assess. Rev.,* 16:223–248, 1996.

Schneider, E. and Kay, J., Life as a manifestation of the second law of thermodynamics, *Math. Comput. Modeling,* 19:6–8; 25–48, 1994.

Simon, J., The state of humanity: steadily improving, in *Cato Policy Report* 17:5. The Cato Institute, Washington, D.C., 1995.

Sterrer, W., Human economics: a non-human perspective, *Ecol. Econ.,* 7:183–202, 1993.

Tuxill, J., Losing strands in the web of life: vertebrate declines and the conservation of biological diversity, Worldwatch Paper 141, The Worldwatch Institute, Washington, D.C., 1998.

UNDP, *Human Development Report 2001,* Oxford University Press, Oxford, U.K., 2001.

Vitousek, P., Ehrlich, P.R., Ehrlich, A.H., and Matson, P., Human appropriation of the products of photosynthesis, *BioScience,* 36:368–374, 1986.

Wackernagel, M. and Rees, W.E., *Our Ecological Footprint: Reducing Human Impact on the Earth,* New Society Publishers, Gabriola Island, B.C., 1996.

Wackernagel, M., Onisto, L., Bello, P., Linares, A.C., Falfán, I.S.L., Garcia, J.M., Guerrero, A.I.S., and Guerrero, M.G.S., National natural capital accounting with the ecological footprint concept, *Ecol. Econ.,* 29:375–390, 1999.

WCED, *Our Common Future,* Oxford University Press, Oxford, U.K., 1987.

Wilson, E.O., The current state of biological diversity, in Wilson, E.O., Ed., *Biodiversity,* National Academy Press, Washington, D.C., 1988, pp. 3–18.

WRI, *Weight of Nations: Material Outflows from Industrial Economies,* World Resources Institute, Washington, D.C., 2000.

WWF, *Living Planet Report 2000,* Loh, J., Ed., Worldwide Fund for Nature, Gland, Switzerland, 2000.

CHAPTER 8

Population Health Issues in the Management of Air Quality*

Richard T. Burnett, Barry Jessiman, David Stieb, and Daniel Krewski

INTRODUCTION

Historical extreme air pollution events, such as those experienced in London in the 1950s and 1960s, clearly demonstrated the potential of ambient air pollution to exacerbate cardiorespiratory disease, as reflected in premature mortality and admission to the hospital. In the intervening years, considerable effort has been made to reduce pollution from the combustion of fossil fuels. Several countries, including Canada and the U.S., have established new stringent guidelines and standards for air pollutants such as sulfur dioxide, nitrogen dioxide, carbon monoxide, ozone, and particulate matter. At present, Canadian National Ambient Air Quality Objectives for these pollutants are rarely violated.

In 1983, David Bates and Ron Sizto published their analyses of the association between the daily variations in ambient summertime ozone concentrations in southwestern Ontario and daily fluctuations in the number of admissions to the hospital for respiratory diseases in 79 acute care hospitals (Bates and Sizto, 1983). The existence of a positive association between ozone and respiratory conditions was somewhat unpredicted in that, although southwestern Ontario experiences the highest levels of ozone in Canada, these concentrations are considered low to moderate for many locations in the U.S. and Europe. In subsequent analyses, Bates and Sizto (1987) suggested that this association could be explained by daily variations in ambient particulate sulfate levels.

The most striking aspect of their work was that ambient air pollutants were apparently related to adverse health effects in the Canadian population at concentrations much lower than previously thought possible. The adequacy of the National Air Quality Objectives in protecting human health was thus brought into question. As summarized below, this pioneering work by Bates and his colleagues spawned an extensive series of epidemiological investigations on the effects of ambient air pollution on population health in Canada. These studies have been of value not only in assessing population health risks within a Canadian context but also in contributing to a growing international literature on the adverse health effects of ambient air pollution.

* This chapter has been published in *Ecosystem Health*, 6:67–78. It is reprinted here with permission of Blackwell Science, Inc.

SUMMARY OF CANADIAN STUDIES

Over the last decade, a series of studies has been published linking daily variations in either deaths or admissions to the hospital for cardiorespiratory diseases and daily fluctuations in a number of ambient air pollutants (Lipfert and Wyzga, 1995). Although the majority of studies have focused on U.S. and European locations with much higher air pollution levels than normally experienced in Canada, some of the most convincing evidence linking air pollution to health has been obtained from data collected in Canada.

Burnett (1990) demonstrated a positive association between decrements in lung function of both asthmatic and nonasthmatic children at a summer camp in central Ontario and the ambient air pollution mix including particulate matter. Similar associations were observed in another summer camp study on the north shore of Lake Erie (Raizenne et al., 1989). Residing in communities experiencing elevated concentrations of sulfates and acid aerosols was associated with declines in the lung function of primary school children in Canada (Stern et al., 1989, 1994) and North America (Raizenne et al., 1996).

Following Bates and Sitzo's previous work, summertime concentrations of both ozone and particulate matter have been linked to respiratory hospitalizations in southern Ontario (Burnett et al., 1994), Toronto, Ontario (Thurston et al., 1994; Burnett et al., 1997) and for 16 of Canada's largest cities (Burnett et al., 1997a). Summertime ozone levels have also been associated with visits to the emergency department in the Saint John Regional Hospital for patients presenting with asthma (Stieb et al., 1996) and to emergency department visits for respiratory diseases in Montreal (Delfino et al., 1996). Elevated ambient levels of carbon monoxide have been linked to hospitalizations for respiratory (Burnett et al., 1997a) and cardiac diseases (Burnett et al., 1997b) in several Canadian cities. Particulate sulfates were related to respiratory hospitalizations in southern Ontario (Burnett et al., 1994, 1995) and the coefficient of haze was linked to increases in respiratory hospitalizations in 16 cities spanning the breadth of the country (Burnett et al., 1997a). Particulate matter and carbon monoxide were also associated with daily mortality in Toronto over the 15-year period from 1980 to 1994 (Burnett et al., 1998).

AMBIENT AIR POLLUTION POLICY INITIATIVES

Based on the epidemiological evidence collected from Canadian studies and investigations demonstrating similar associations between ambient air pollution and adverse health effects worldwide, the Canadian government has embarked on several new initiatives, including redefinition of the National Ambient Air Quality Objectives (NAAQOs) for established criteria pollutants, establishing new PM_{10} and $PM_{2.5}$ air quality objectives, and assessing the public health benefits of reducing the sulfur content in gasoline.

National Ambient Air Quality Objectives (NAAQOs) and Canada-Wide Standards

Until 1998, air quality risk management initiatives were predicated largely on the development of NAAQOs for specific pollutants. The criteria for their development and use must be based on scientifically defensible evidence while incorporating a margin of protection that reflects three factors: variability in the levels of exposure and their associated effects, uncertainty in the degree of both health and environmental risk, and the values of Canadian society.

The legal basis for NAAQOs is the Canadian Environmental Protection Act (CEPA) (Armstrong and Newhook, 1992), originally passed into law in 1988 (a new version came into effect in 1999), which established a comprehensive legal framework to manage toxic substances in Canada. It addresses pollution problems on land, in water, and through all layers of the atmosphere. NAAQOs are developed by the Working Group on Air Quality Objectives and Guidelines (WGAQOG), which reports to the Federal/Provincial Advisory Committee (F/PAC) under CEPA. The Working Group includes representatives of federal, provincial, and territorial departments of environment and health.

In January 1998, an agreement between federal and provincial environment ministers led to the development of a Harmonization Accord on joint actions to address environmental issues of pressing concern. A major subagreement, the Canada Wide Standards (CWS), established a framework for the development of national standards for the highest priority pollutants. This subagreement laid out a process for selecting national targets with associated compliance dates and management plans. At the time the CWS process was agreed upon, NAAQOs for particulate matter ($PM_{2.5}$ and PM_{10}) were in preparation. These proposed NAAQOs for PM were withdrawn and replaced by CWSs in 1999. The framework for the development and implementation of NAAQOs and CWSs is described in Figure 8.1.

The ambient air quality objectives are designed within a two-tier framework, in which the first tier identifies the scientific reference level, defined as "a level above which there are demonstrated adverse effects on human health and/or the environment (Working Group on Air Quality Objectives and Guidelines, 1996). The second tier, called the Air Quality Objective, is a subjective interpretation of the scientific information designed to provide protection to the general population and environment. The formal definition of the objective is: "the air quality management goal for the protection of the general public and the environment in Canada. It is a level selected based upon the consideration of scientific, social, economic and technical factors" (Working Group on Air Quality Objectives and Guidelines, 1996).

The principles by which objectives are developed are embodied in a formal protocol (Working Group on Air Quality Objectives and Guidelines, 1996) that provides a framework for evaluating the scientific basis for ambient air quality objectives. This protocol is designed to ensure consistency in the evaluation approach and the degree of protection and to provide guidelines for the scientific evaluation that can be systematically reviewed with respect to completeness, rationale, and degree of protection.

For the CWSs, the basis for action is developed elsewhere; initiation of the process generally means governments have accepted the need to take some action. The purpose of the CWS process is to set achievable targets based on technological and economic considerations while providing some measure of population health or environmental protection.

For both NAAQOs and CWSs, federal and provincial governments have specific and sometimes overlapping responsibilities. Federal responsibilities rest largely in the international arena (such as the Canada–U.S. Air Quality Agreement) or on national standards for specific issues (such as the sulfur content of gasoline). Provincial responsibility rests largely, but not exclusively, with pollution sources and related processes that are local or regional in nature. Whereas CWSs have considerable government support in advance of their formal deployment, NAAQOs remain recommendations until such time as they are incorporated in legislation or other formal controls.

Canada-Wide Standards for PM_{10} and $PM_{2.5}$

The first application of the CWS process in Canada involved particulate matter and ozone. Previously, the only air quality objective in Canada for particles was the set of Total Suspended Particulate (TSP) guidelines (a desirable concentration of 0 to 60 $\mu g/m^3$; acceptable annual and 24-hour average concentrations of 60 to 70 $\mu g/m^3$ and 0 to 120 $\mu g/m^3$ respectively; and a 24-hour average tolerable level of 120 to 400 $\mu g/m^3$). Ozone has an existing NAAQO of 80 ppb (1-hour average).

The major impetus for a regulatory review of particulate matter in Canada came from the appearance of a large number of well-conducted epidemiological studies linking ambient particulate matter levels with adverse health outcomes over the last several years. These included both short-term (time-series) and long-term (cohort) studies demonstrating associations between particulate matter and a spectrum of adverse cardiorespiratory outcomes, including mortality and hospitalization, increased illness, reduced activity days, and decreased lung function.

It has been noted that there are several important confounders that need to be considered in the interpretation of the above studies. Most studies have used techniques that reduce the effect of these confounders and covarying pollutants, though it is not possible to eliminate this concern completely. Nonetheless, given the wide variety of locales, pollutant mixtures, and exposure levels

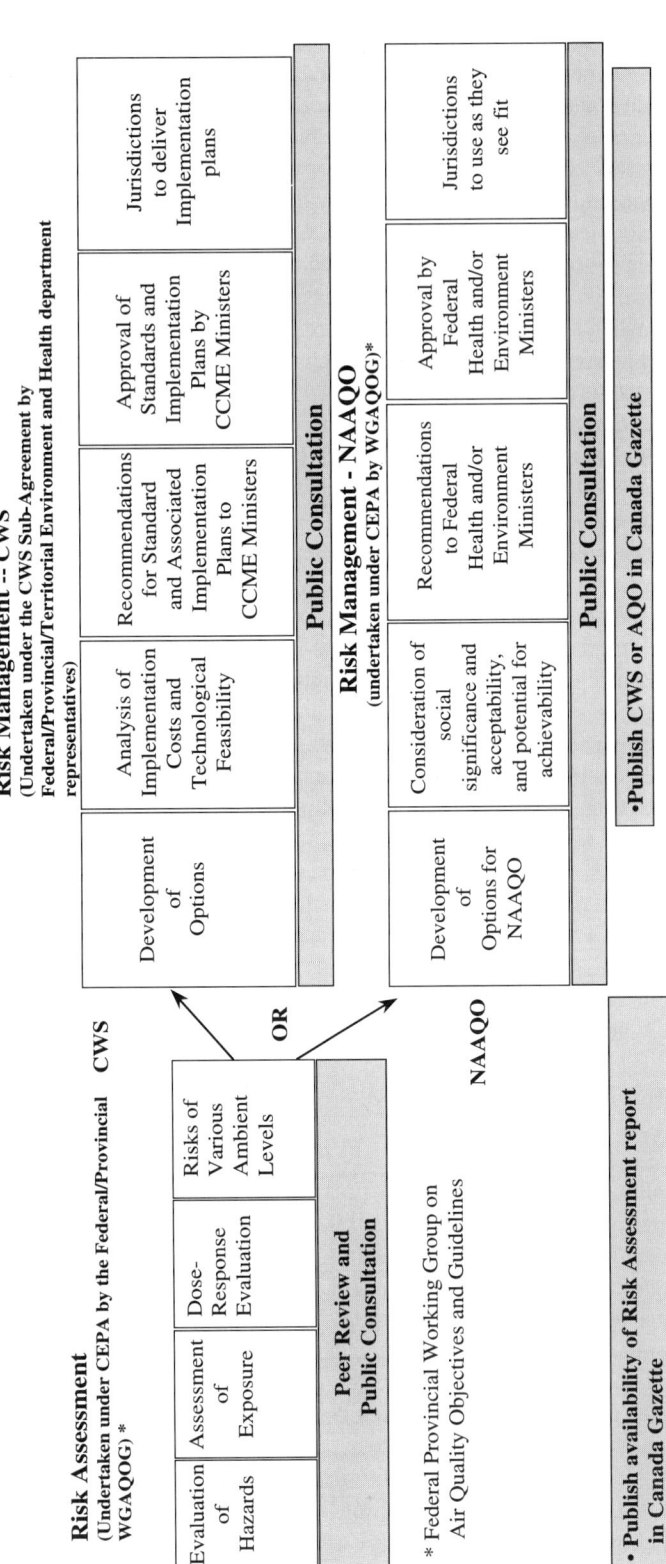

Figure 8.1   Developing and implementing NAAQOs and CWSs.

assessed in these studies and the consistently strong evidence of association, confounding cannot explain all the observed impacts.

Though the mode of action is not clear, and potential causal mechanisms are only now being elucidated, the coherence and consistency of the body of evidence cannot be dismissed. It was judged by several government bodies that it would be prudent to act on the assumption that the association between particulate matter and cardiorespiratory disease is causal.

At this time, the available evidence provides little indication of a threshold, with mortality and morbidity increasing with ambient particle concentrations. While there remains uncertainty as to the most biologically relevant index of particulate matter, there is sufficient evidence to warrant targeting of the fine fraction (<2.5 μm in aerodynamic diameter) along with gaseous co-pollutants in any contemplated control measures. This perspective acknowledges both the covarying nature of the pollutant mix and the precursor nature of many of these gaseous pollutants with respect to fine particulate matter. While control measures should emphasize the importance of the fine fraction, there remains sufficient concern with the coarse fraction of PM_{10} to warrant the development of standards for both PM_{10} and $PM_{2.5}$.

The federal–provincial Working Group (WG) has developed a science assessment document (Working Group on Air Quality Objectives and Guidelines, 1999) that details the scientific basis for conclusions on the health effects of PM. This WG has also developed reference levels for both PM_{10} and $PM_{2.5}$. Options for CWSs for both PM_{10} and $PM_{2.5}$ as well as ground-level ozone have been developed and presented to a multi-stakeholder forum.

Economic Evaluation

Methods for assessing the value for money obtained from alternative public policies or programs can essentially be divided into cost-effectiveness analysis (CEA) and cost–benefit analysis (CBA). CEA compares the cost of introducing a program with its incremental effects, measured in natural units (such as the cost per life saved or quality-adjusted life years — QALYs — gained). CBA, on the other hand, expresses all effects of the program in dollar terms and determines net benefits by subtracting costs (the dollar value of negative effects) from benefits (the dollar value of positive effects). Although CEA avoids measurement issues associated with "monetization" or "valuation," it does not provide guidance whether the positive effects justify the costs (Gold et al., 1996; Drummond et al., 1987). Also, it must be borne in mind that any decision about whether to implement a program to improve health or environmental quality implicitly involves valuation. CBA makes the valuation process systematic and explicit (Mooney, 1986).

Although measures such as avoided health care costs and avoided lost wages as a proxy for increased productivity (essentially the costs of illness) can be used to represent the value of avoided adverse health effects, these are likely to result in underestimates of value because they exclude the value of avoided out-of-pocket expenses to individuals as well as the value of improved quality of life and increased longevity. The basic approach to determine more comprehensively the value of a *good* in a manner consistent with fundamental welfare economics axioms is to determine individuals' willingness to pay (WTP) to obtain the good, or willingness to accept compensation (WTA) to give it up (Gafni, 1991).

There are a variety of empirical methods for measuring WTP or WTA that fall into essentially two categories — those which rely on observed market behavior (*revealed preference methods*) and those which do not (*stated preference methods*) (Gafni, 1991). Some aspects of the value of health and environmental quality can be assessed using market data; examples include analysis of wages in relation to occupational risks to determine the value of avoided mortality (WTA), or of costs of travel or use permits for recreational sites to determine the value of wilderness areas (WTP). In other cases, market data are not available and surveys are used to directly measure people's preferences for various goods in the context of a hypothetical market.

Contingent valuation (CV) studies have in particular been the subject of considerable discussion. The basic concern about CV or stated preference studies in general is whether investigators have the technical skills to measure values using CV surveys and whether survey respondents have the cognitive capabilities to express accurately their own preferences for nonmarket goods.

A number of attempts have been made to synthesize the evidence on the reliability and validity of CV studies (Portney, 1994; Hanemann, 1994; Diamond and Hausman, 1994; Mitchell and Carson, 1989; Cummings et al., 1986; National Oceanic and Atmospheric Administration, 1994). Although some authors have highlighted perceived shortcomings of CV, the prevailing view appears to be that, when carried out meticulously and in the appropriate context, CV or stated preference methods can generate valid and reliable valuations.

A number of activities have been undertaken involving the application of economic evaluation techniques in assessing the health effects of air pollution. In addition, methodological advances have led to the development of a framework for the application of economic evaluation principles devised for health care settings to environmental health risk management programs (Krewski et al, 1989; Torrance and Krewski, 1986; Letourneau et al., 1992). The essence of this framework is the damage function approach. *Damages* refer to negative impacts on human health or the environment. The converse of damages is *benefits*, which refers to the reduction in damages as a consequence of a given policy or program (Freeman, 1993; Freeman et al., 1994).

In the context of air pollution control programs, this approach involves four principal steps: (1) estimating the impact of the proposed policy on emissions of air pollutants; (2) modeling the impact of emission changes on ambient air quality; (3) applying dose–response relationships to predicted changes in air quality, thus estimating the number of avoided health and environmental impacts; and (4) applying monetary values to the avoided impacts (Freeman, 1993; Freeman et al., 1994). Values are then aggregated across the range of impacts; uncertainty analysis is carried out to determine a range of possible results given observed uncertainty in the various steps described above; and, finally, sensitivity analysis is conducted in order to test the sensitivity of results to alternative assumptions at various stages of the assessment. This approach is consistent with the economic theory of environmental impact valuation and is generally preferred to alternative approaches (Freeman, 1993).

This approach, or components of it, has been applied in a number of studies of the benefits and costs of air pollution control programs (Cannon, 1985; 1990). A study by Hall (1989) in the South Coast Air Basin in California estimated that the annual benefits of meeting the federal ozone and PM_{10} standards (120 ppb and 150 $\mu g/m^3$, respectively) ranged from \$4.8 to 20.4 billion, with a central estimate of \$9.4 billion. The annual benefits of meeting the more stringent California standards (90 ppb and 50 $\mu g/m^3$, respectively) ranged from \$7.4 to 31.1 billion, with a central estimate of \$14.3 billion (all estimates in 1988 \$U.S.).

When this approach was applied to the Lower Fraser Valley in British Columbia, it was determined that the annual economic value (1990 \$Canadian) of health impacts due to mobile source particulates was approximately \$600 to 700 million, depending on assumptions about background particulate concentrations (SENES Consultants Limited, 1994). An additional study in this region determined that the Air Quality Management Plan for the Greater Vancouver Regional District (GVRD) generated \$5.4 billion in benefits in the GVRD at a cost of \$3.8 billion (1993 \$Canadian) over the years 1985 to 2020 (ARA Consulting Group Inc. and Bovar-Concord Environmental, 1994). A study by Oxman and Julian (1993) found that the benefits of eliminating exceedances of Canada's national ambient air quality objective for ozone (80 ppb) amounted to \$40 million per year (1990 \$Canadian); however, this study examined only reductions in the occurrence of symptoms for which dose–response functions were available from random-ized–controlled trials, ignoring more severe effects of air pollution for which evidence is derived from nonexperimental studies.

A recent study for the U.S. Environmental Protection Agency estimated that by 2010, U.S. Clean Air Act amendments to reduce SO_2 emissions by 40% from 1980 levels will result in annual health benefits in the eastern U.S. and Canada of $12 to 80 billion, with a central estimate of $40 billion (1994 $U.S.) (Chestnut, 1995).

The study on Environmental and Health Benefits of Cleaner Vehicles and Fuels (Lang et al., 1995) was commissioned by the Canadian Council of Ministers of the Environment (CCME), Task Force on Cleaner Vehicles and Fuels. It was one of a series of studies commissioned by the CCME to provide information to policy makers on economic aspects of proposed policies on vehicle emission standards and the reformulation of transportation fuels. This study utilized the damage function approach to assess the health and environmental benefits of two vehicle emission and fuel reformulation scenarios. Innovative aspects of this study included assessment of the impact of age on the value of avoiding death (this was seen as an important consideration, given that air pollution–associated death is suspected of disproportionately affecting elderly individuals with host conditions favoring susceptibility to the adverse effects of ambient air pollution); and adjusting cost of illness-based valuations upward, based on the ratio between willingness to pay and cost of illness valuation estimates (Rowe and Chestnut, 1986).

This study estimated that 100% use of low-sulfur diesel by 1997, Canada-wide use of gasoline containing reduced levels of sulfur (200 ppm) and benzene (1% by volume) by 1998, and introduction of alternative low-emission vehicles into Canada's fleet in 2001 would result in a wide spectrum of avoided health effects that would translate into $11 to 29 billion in cumulative health benefits (central estimate of $23 billion) between 1997 and 2020 (1994 $Canadian). An accelerated program including a phased approach for implementing various classes of low-emitting vehicles, a minimum requirement for zero-emission vehicles, introduction of lower-sulfur-content gasoline (40 ppm), and the same target for low-sulfur diesel as the previous scenario, would result in $14 to 38 billion in cumulative health benefits (central estimate of $30 billion) between 1997 and 2020. (Benefits estimates exclude British Columbia, which was the subject of a separate study, as well as the Yukon Territory and Northwest Territories, for which vehicle fleet data were not available.)

The approach utilized in the Cleaner Vehicles and Fuels study was subsequently generalized to a spreadsheet model, the Air Quality Valuation Model (AQVM), applicable to a wide variety of air pollution control programs that result in improvements in ambient air quality (Lang et al., 1996). This model includes baseline air quality and population data; dose–response relationships for health and ecological impacts; economic value measures for these impacts; and Monte Carlo simulations for the purpose of uncertainty analysis. It requires as inputs absolute or percent changes in ambient air quality specified at the national, provincial, or census division level and predicts the annual number and value of avoided health and environmental impacts. Although the AQVM is based largely on the approach used in the Cleaner Vehicles and Fuels study, it quantifies the value of environmental as well as health impacts, including material damages and soiling, visibility, recreational fishing, agricultural crops, and climate change. This model ensures consistency in approaches to benefit assessment, while allowing for flexibility to change key parameters (e.g., dose–response functions or valuations) as new epidemiological and economic data become available. The AQVM was utilized to expand on the work carried out in the Cleaner Vehicles and Fuels study, more specifically to examine the benefits of reduced levels of sulfur in gasoline, as described in the next section.

Several new primary valuation studies have also been undertaken. The basis for the Saint John Emergency Department study is an examination of the relationship between daily levels of air pollution and the daily frequency of cardiorespiratory emergency department visits (Stieb et al., 1995). This study expands on the limited amount of data normally available from administrative databases by collecting enhanced data, both at the time of the emergency department visit and in follow-up 2 weeks later. In addition to attempting to answer specific epidemiological questions related to the role of air pollution relative to other triggers in precipitating disease episodes, this study is also attempting to measure the broad quality of life and economic impacts of the health effects of air pollution. Analyses to date confirm that outcomes such as reduced activity days, bed days, additional medication use, lost time from work and school, physician office visits, and hospital admissions commonly accompany emergency department visits and that health care costs associated with these episodes are significant (Stieb et al., 2000; Anis et al., 2000).

To complement these data on the costs of air pollution-related illness, a survey has been developed to measure the value to Canadians (willingness to pay) of improved health and increased longevity in relation to cleaner air (Johnson et al., 1999). This survey uses *conjoint analysis* methodology to measure trade-offs between costs and health status, thus determining willingness-to-pay values. The cardiorespiratory health conditions that form the basis of the trade-off questions in this survey were specifically selected because they have been linked with air pollution in various studies. Each health condition is described in terms of several attributes including symptoms, number of episodes per year, effects on daily activities such as work, physical activity, social activities, and costs. While this is a primary valuation study, the systematic classification of health conditions into a set of several attributes facilitates benefit transfer to a variety of conditions and policy contexts.

The cost of illness and WTP-based results of these new valuation studies will be integrated to reflect a comprehensive societal value for Canada of avoiding air pollution–associated morbidity and incorporated directly into the AQVM for subsequent benefits assessments.

A survey instrument for eliciting information on valuation of mortality risk reductions has also been developed (Krupnick et al., 1999). This instrument will provide alternative valuation estimates to those based on wage premiums in relation to occupational mortality risks. Like the morbidity valuation survey described above, the mortality risk valuation survey is computer administered and makes extensive use of interactive graphics and voice-overs to communicate risk information. After pilot tests in Japan and the U.S., the survey was taken into the field in Canada in the summer of 1999. This instrument will probably also be used in a number of countries around the world, facilitating internal comparisons on the valuation of mortality risk reductions.

Reduced Sulfur in Gasoline: A Case Study in Air Quality Management

Based on concerns over the contribution of vehicles to poor air quality, the Canadian government, in conjunction with the petroleum industry, undertook a detailed analysis of the impact of sulfur levels in gasoline on air quality, resultant health impacts, and the implications for the fuel industry in Canada. This process was a multi-stakeholder one with the direct participation of government, petroleum and motor vehicle manufacturers, and public interest groups. Additionally, a component of the process allowed for the participation of any interested party in the fact-finding portion of the analysis (Figure 8.2).

This process was designed in three stages: process setup, fact-finding, and options and recommendations. The first stage involved government and industry determining the structure of the effort and led to stage two, wherein three expert panels were established to deal with atmospheric modeling (Report of the Atmospheric Science Expert Panel on the Sulphur Content of Gasoline, 1997), health and environmental impact (Report of the Health and Environmental Impact Assessment Expert Panel on the Sulphur Content in Gasoline, 1997), and industry cost and competitiveness issues (Report of the Cost and Competitive Expert Panel on the Sulphur Content of Gasoline, 1997). The third stage took the results of the fact-finding stage and, in an intergovernmental forum, developed regulations for sulfur levels in gasoline for Canada. Sulfur levels in gasoline had not yet been regulated and varied widely across the country according to feedstock quality and refinery capability.

The fact-finding stage focused on a number of scenarios, including six potential regulated levels of sulfur in gasoline ranging from 30 to 350 ppm. The atmospheric modeling panel used these scenarios to project tailpipe emissions changes and resultant effects on air quality. The health and environment panel used these projected changes to estimate impacts on human and environmental health. The most significant changes in air quality were for SO_2 and SO_4. The human health benefits resulting from these changes focused on sulfate and were quantified using the available epidemiological health effects literature. This literature contained sufficient material to estimate benefits

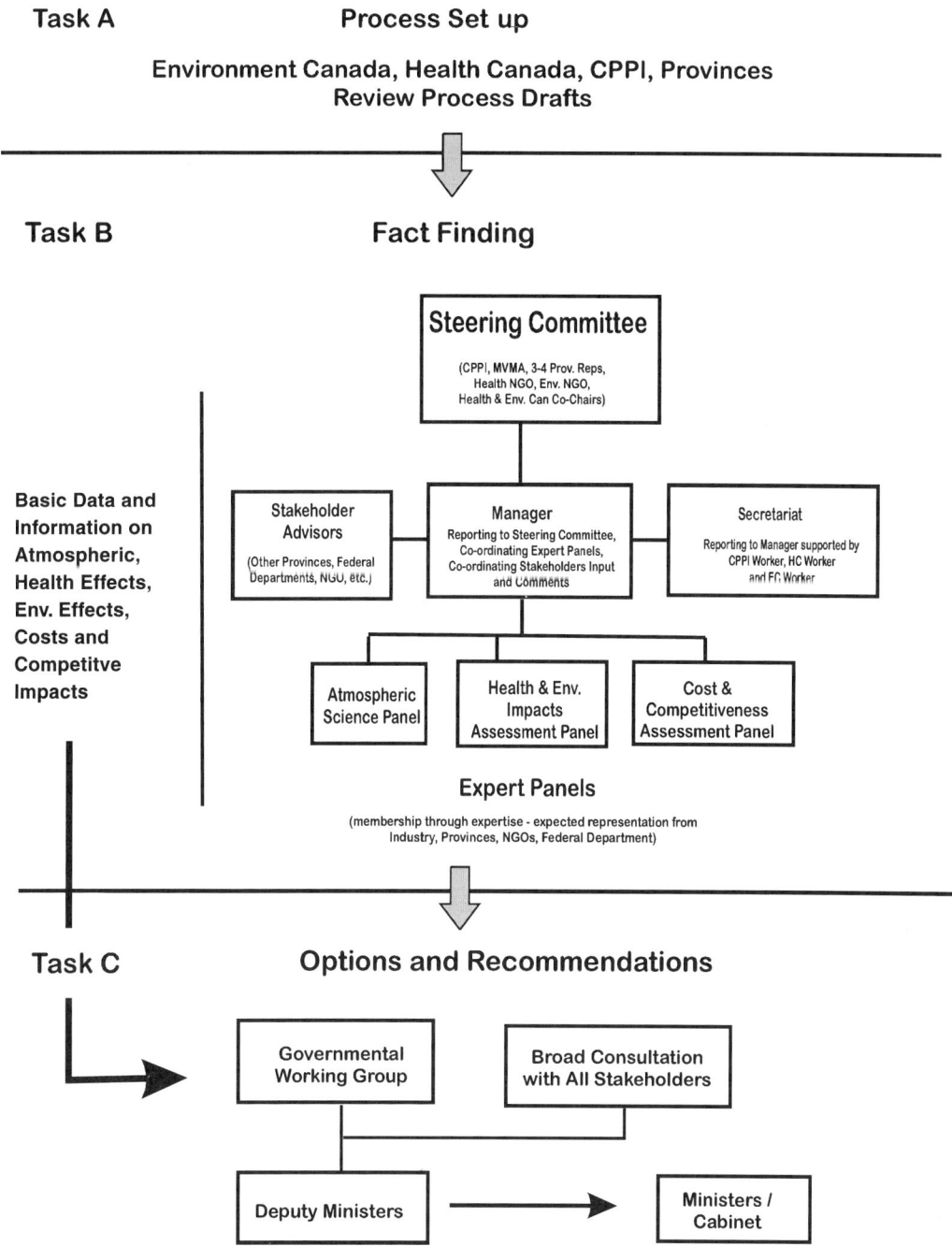

Figure 8.2 Studying air quality management.

of avoided premature mortality, chronic respiratory disease, cardiac and respiratory hospitalization, emergency room visits, asthma symptoms, restricted activity, acute respiratory symptoms, and lower respiratory illness in children. The health panel also performed a monetary valuation of the avoided effects based on the health economics literature. The cost and competitiveness panel examined the actual costs to retrofit refineries to meet the various proposed sulfur standards, and the impact on the competitiveness of the industry, both within Canada and in the international setting.

After considering the findings of the panel process and the recommendations of the intergovernmental working group, the Government of Canada decided on the most stringent option examined, 30 ppm. Key factors were the effects on catalytic converters, direct health benefits, and reasonable costs. A modifying factor in the details of the regulation (a 3-year transition period) was designed to allow independent gasoline marketers to maintain access to supply. The regulation as embodied under CEPA requires a 150 ppm (average) standard to be met starting in July 2002, with the 30 ppm provision following in January 2005. This regulation will bring Canadian sulfur levels into line with current Japanese and California standards, as well as with proposed U.S. and European standards.

FUTURE DIRECTIONS

It has now become apparent that more people visit emergency departments, are admitted to the hospital, and die on days of elevated air pollution compared to those days during which ambient air pollution concentrations are relatively low. There is limited evidence from clinical and toxicological studies to support a causal link with such adverse health effects at the relatively low ambient concentrations observed in these studies. We hypothesized that air pollution may be disrupting homeostasis through a variety of potential mechanisms (Burnett et al., 1995; 1998) in those people already severely compromised with preexisting disease. These mechanisms may include effects on macrophage function leading to increased bacterial infectivity, regulation of metabolic activities in the lung endothelium, and effects on coronary circulation and vasculature.

Three epidemiological studies were conducted to examine this hypothesis. In the first, patients visiting the emergency department for cardiorespiratory diseases of the Saint John, New Brunswick Regional Hospital were interviewed between April 1994 and March 1996 (Stieb et al., 1995). Saint John experiences the highest concentrations of acid aerosols in Canada and has the highest acid-to-particle mass ratio ever recorded anywhere. Use of medication, time activity patterns in the 2 weeks prior to the visit, date of onset of signs or symptoms, smoking habits, and disease history were obtained. The purpose of the study was to examine the influence of these factors on the air pollution–visit association. In the second study, death certificates were linked to physician billing, hospital use, and drug prescription records in Montreal. Death prediction indices were developed based on this co-morbid information to test the hypothesis that those people who were severely ill were at the greatest risk of death from air pollution exposures.

In the final study, the daily number of nonaccidental deaths in metropolitan Toronto over the 15-year period from 1980 to 1994 was related to ambient air pollution measures. A clear positive association was detected (Burnett et al., 1998). It is not clear, however, what percentage of these deaths were due to *mortality displacement*, a process in which air pollution plays a role in setting the specific date of death but accelerates the death process by only a few days. Under this scenario, those people would have likely died shortly after their actual date of death in the absence of exposure to ambient air pollution.

CONCLUSION

Concentrations of ambient air pollution currently experienced by Canadians have been clearly linked to adverse health events in the population, increasing the numbers of daily visits to emergency departments, hospital admissions, and even deaths. This information has been used to assess the impact of air pollution on public health. Further research is being conducted to examine the role that severity of preexisting disease plays in the association between ambient air pollution and health in Canada. Estimating the independent adverse health effects of each pollutant that comprises the atmospheric mix has proven to be a challenge (Moolgavkar and Luebeck, 1996) due to the high colinearity between ambient air pollutants and the varying degree of exposure error. Studies have been undertaken to examine the health effects of ambient air pollutants at selected locations in Canada with diverse populations, climates, and pollution mixes. Further work has focused on determining the relation between concentrations measured at fixed outdoor monitoring sites and personal exposure levels, using time activity patterns and spatial modeling techniques.

Weighing the costs and benefits of public programs is a central feature of Canada's regulatory policy, as administered by the Treasury Board (Treasury Board Secretariat, 1995) and of similar policies in the U.S. (Arrow et al., 1996). It has been suggested that, although cost–benefit analysis should be viewed as neither necessary nor sufficient for making good public policy, it represents a useful analytical tool. At a minimum, it provides a systematic framework for organizing and synthesizing a wide variety of information that is salient to a given policy question (Arrow et al., 1996). In terms of impact, a recent assessment of regulations introduced by the U.S. government revealed that departments (notably the Department of Transportation) that consistently applied cost–benefit analyses had a record of regulations that more effectively maximized benefits and minimized costs (Viscusi, 1996). Agencies such as Health Canada, with a mandate to manage environmental health risks, can play a similar role by ensuring that programs to improve environmental quality represent the best possible investment in population health.

REFERENCES

Anis, A.H., Guh, D., Stieb, D.M., Leon, H., Beveridge, R.C., Burnett, R.T., and Dales, R.E., The costs of cardiorespiratory disease episodes in a study of emergency department use, *Can. J. Public Health,* 91:103–106, 2000.

ARA Consulting Group, Inc. and Bovar-Concord Environmental, Clean Air Benefits and Costs in the GVRD, prepared for Greater Vancouver Regional District, B.C. Ministry of Environment Lands and Parks, Environment Canada, 1994.

Armstrong, V.C. and Newhook, R.C., Assessing the health risks of priority substances under the Canadian Environmental Protection Act, *Regul. Toxicol. Pharmacol.,* 15:111–121, 1992.

Arrow, K.J., Cropper, M.L., Eads, G.C., Hahn, R.W., Lave,L.B., Noll, R.G., Portney, P.R., Russell, M., Schmalensee, R., Smith, K.V., and Stavins, R.N., Is there a role for benefit-cost analysis in environmental health, and safety regulation? *Science,* 272:221–222, 1996.

Bates, D.V. and Sizto, R., Relationship between air pollution levels and hospital admissions in southern Ontario, *Can. J. Public Health,* 74:117–122, 1983.

Bates, D.V. and Sizto, R., Air pollution and hospital admissions in southern Ontario: the acid summer haze effect, *Environ. Res.,* 43:317–331, 1987.

Burnett, R.T., Raizenne, M.E., and Krewski, D., Acute health effects of transported air pollution: a study of children attending a residential summer camp, *Can. J. Stat.,* 18:367–373, 1990.

Burnett, R.T., Dales, R.E., Raizenne, M.E., Krewski, D., Summers, P.W., Roberts, G.R., Raad-Young, M., Dann, T., and Brook, J.R., Effects of low ambient levels of ozone and sulfates on the frequency of respiratory admissions to Ontario hospitals, *Environ. Res.,* 65:172–194, 1994.

Burnett, R.T., Dales, R.E., Krewski, D., Vincent, R., and Brook, J.R., Associations between ambient particulate sulfate and admissions to Ontario hospitals for cardiac and respiratory diseases, *Am. J. Epidemiol.,* 142:15–22, 1995.

Burnett, R.T., Brook, J.R., Yung, W.T., Dales, R.E., and Krewski, D., Association between ozone and hospitalization for respiratory diseases in 16 Canadian cites, *Environ. Res.,* 72:24–31, 1997a.

Burnett, R.T., Cakmak, S., Brook, J.R., and Krewski, D., The role of particulate size and chemistry in the association between summertime ambient air pollution and hospitalization for cardio-respiratory diseases, *Environ. Health Perspect.,* 105:614–620, 1997b.

Burnett, R.T., Dales, R.E., Brook, J.R., Raizenne, M.E., and Krewski, D., Association between ambient carbon monoxide levels and hospitalizations for cardiac diseases in ten Canadian cites, *Epidemiology,* 8:162–167, 1997c.

Burnett, R.T., Brook, J.R., Philps, O., Cakmak, S., Raizenne, M.E., Stieb, D., Vincent, R., Ozkaynak, H., and Krewski, D., The association between ambient concentrations of carbon monoxide and daily mortality in Toronto, Canada, *J. Air Waste Manage. Assoc.,* 48:689–700, 1998.

Cannon, J.S., *The Health Costs of Air Pollution: An Updated Survey of Studies between 1978 and 1983,* American Lung Association, New York, 1985.

Cannon, J.S., *The Health Costs of Air Pollution: A Survey of Studies Published 1984–1989,* American Lung Association, New York, 1990.

Chestnut, L.G., *Human Health Benefits from Sulfate Reductions under Title IV of the 1990 Clean Air Act Amendments,* Prepared for U.S. Environmental Protection Agency, Office of Air and Radiation, EPA Contract No. 68-D3–0005, Work Assignment No. 2F-03 and 3F12, 1995.

Cummings, R.G., Brookshire, D., and Schulze, W.D., Eds., *Valuing Environmental Goods: A State of the Arts Assessment of the Contingent Valuation Method,* Rowman and Allanheld, Totowa, NJ, 1986.

Delfino, R.J., Murphy-Moulton, A.M., Burnett, R.T., Brook, J.R., and Becklake, M.R., Effects of ozone and particulate air pollution on emergency room visits for respiratory illnesses in Montreal, *J. Respir. Crit. Care Med.,* 155:568–576, 1996.

Diamond, P.A. and Hausman, J.A., Contingent valuation: is some number better than no number? *J. Econ. Perspect.,* 8:45–64, 1994.

Drummond, M.F., Stoddart, G.L., and Torrance, G.W., *Methods for the Economic Evaluation of Health Care Programmes,* Oxford University Press, Oxford, 1987.

Freeman, A.M., *The Measurement of Environmental and Resource Values,* Resources for the Future, Washington, D.C., 1993.

Freeman, A.M., Rowe, R.D., and Thayer, M., Economic issues in the application of externalities to electricity resource selection, in *Technical Review of Externalities,* Electric Power Research Institute, EPRI TR-104813, 1994.

Gafni, A., Willingness-to-pay as a measure of benefits: relevant questions in the context of public decision making about health care programs, *Med. Care,* 29:1246–1252, 1991.

Gold, M.R., Siegel, J.E., Russell, L.B., and Weinstein, M.C., *Cost-Effectiveness in Health and Medicine,* Oxford University Press, Oxford, U.K., 1996.

Hall, J.V., Economic Assessment of the Health Benefits from Improvements in Air Quality in the South Coast Air Basin, Final Report to South Coast Air Quality Management District, Contract No. 5685, 1989.

Hanemann, W.M., Valuing the environment through contingent valuation, *J. Econ. Perspect.,* 8:19–43, 1994.

Johnson, F.R., Ruby, M.C., and Desvousges, W.H., Willingness to pay for improved respiratory and cardiovascular health: a multiple-format stated preference approach, *Health Econ.,* 9:295-317, 2000.

Krewski, D., Oxman, A., and Torrance, G.W., A decision-oriented framework for evaluation environmental risk management strategies: a case study of lead in gasoline, in Paustenbach, D.J., Ed., *The Risk Assessment of Environmental and Human Health Hazards,* John Wiley & Sons, New York, 1989, pp. 1047–1062.

Krupnick, A., Cropper, M.L., O'Brien, B., Goeree, R., Simon, N., Kropf, M., Futo, L., and Heintzelman, M., Preliminary Report: Mortality Risk Valuation: Survey Approach, prepared by Resources for the Future under contract for Health Canada and Environment Canada, March 31, 1999.

Lang, G., Yarwood, G., Lalonde, F., and Bloxam, R., Environmental and Health Benefits of Cleaner Vehicles and Fuels, summary report prepared for Canadian Council of Ministers of the Environment, Task Force on Cleaner Vehicles and Fuels, 1995.

Lang, C., Keffe, S., Chestnut, L., and Rowe, R., Air Quality Valuation Model (AQVM), prepared for Environment Canada and Health Canada, 1996.

Letourneau, E.G., Krewski, D., Zielinski, J.M., and McGregor, R.G., Cost effectiveness of radon mitigation in Canada, *Radiat. Prot. Dosimetry,* 45:593–598, 1992.

Lipfert, F.W. and Wyzga, R.E., Air pollution and mortality: issues and uncertainties, *J. Air Waste Manage. Assoc.,* 45:949–966, 1995.

Mitchell, R.C. and Carson, R.T., *Using Surveys to Value Public Goods: The Contingent Valuation Method.,* Resources for the Future, Washington, D.C., 1989.

Moolgavkar, S.H. and Luebeck, E.G., A critical review of the evidence on air pollution and mortality, *Epidemiology,* 7:420–428, 1996.

Mooney, G.H., *Economics, Medicine and Health Care,* Wheatsheaf, Brighton, U.K., 1986.

National Oceanic and Atmospheric Administration, U.S. Department of Commerce, Natural Resource Damage Assessments, *Federal Register,* 59:1062–1190, 1994.

Oxman, A. and Julian, J., The Health Impacts of Ground-Level Ozone in Canada, prepared for the Environmental Health Directorate, Health and Welfare Canada, 1993.

Portney, P.R., The contingent valuation debate: why economists should care, *J. Econ. Perspect.*, 8:3–17, 1994.

Raizenne, M.E., Burnett, R.T., Stern, B., Franklin, C.A., and Spengler, J.D., Acute lung function responses to ambient acid aerosol exposures in children, *Environ. Health Perspect.*, 79:179–185, 1989.

Raizenne, M.E., Neas, L.M., Damokosh, A.I., Dockery, D.W., Spengler, J.D., Koutrakis, P., Ware, J.H., and Speizeret, F.E., Health effects of acid aerosols on North American children: pulmonary function, *Environ. Health Perspect.*, 5:506–514, 1996.

Report of the Atmospheric Science Expert Panel on the Sulphur Content in Gasoline, Oil, Gas and Energy Division, Environmental Protection Service, Environment Canada, Ottawa, March, 1997.

Report of the Health and Environmental Impact Assessment Expert Panel on the Sulphur Content in Gasoline, Oil, Gas and Energy Division, Environmental Protection Service, Environment Canada, Ottawa, Canada, March, 1997.

Report of the Cost and Competitiveness Expert Panel on the Sulphur Content in Gasoline, Oil, Gas and Energy Division, Environmental Protection Service, Environment Canada, Ottawa, Canada, March, 1997.

Rowe, R.D. and Chestnut, L.D., Oxidants and Asthmatics in Los Angeles: A Benefits Analysis, prepared for U.S. Environmental Protection Agency and California Air Resources Board, 1986.

SENES Consultants Limited, Screening Level Valuation of Air Quality Impacts Due to Particulates and Ozone in the Lower Fraser Valley, Prepared for Government of British Columbia, Ministry of Transportation and Highways, Planning and Services Branch, Vancouver, B.C., 1994.

Stern, B.R., Jones, L., Raizenne, M.E., Burnett, R.T., Meranger, J.C., and Franklin, C.A., Respiratory health effects associated with ambient sulphates and ozone in two rural Canadian communities, *Environ. Res.*, 49:20–39, 1989.

Stern, B.R., Raizenne, M.E., Burnett, R.T., Jones, L., Kearney, J., and Franklin, C.A., Air pollution and childhood respiratory health: exposure to sulfate and ozone in 10 Canadian rural communities, *Environ. Res.*, 66:125–142, 1994.

Stieb, D., Beveridge, R.C, Brook, J.R., Burnett, R.T., Anis, A.H., and Dales, R.E., The Saint John Particle Health Effects Study: Measuring Health Effects, Health Costs and Quality of Life Impacts Using Enhanced Administrative Data, in *Particulate Matter: Health and Regulatory Issues (VIP-49)*, 1995, pp. 131–142, Proceedings of an International Specialty Conference, Air and Waste Management Association, Pittsburgh, PA, April 1995.

Stieb, D., Beveridge, R.C., Smith-Doiron, M., Burnett, R.T., Judek, S., Dales, R.E., and Anis, A.H., Beyond administrative data: characterizing cardiorespiratory disease episodes among patients visiting the emergency department, *Can. J. Public Health*, 91:107–112, 2000.

Stieb, D., Burnett, R.T., and Beveridge, R.C., Association between ozone and asthma emergency department visits in Saint John, New Brunswick, Canada, *Environ. Health Perspect.*, 104:1354–1360, 1996.

Thurston, G., Ito, K., and Lippmann, M., Respiratory hospital admissions and summertime haze air pollution in Toronto, Ontario: consideration of the role of acid aerosols, *Environ. Res.*, 65:271–290, 1994.

Torrance, G.W. and Krewski, D., Economic evaluation of toxic chemical control programs, *Toxic Substances J.*, 7:53–71, 1986.

Treasury Board Secretariat, Government of Canada, *Benefit-Cost Analysis Guide for Regulatory Programs*, Minister of Public Works and Government Services Canada, Ottawa, 1995.

Viscusi, W.K., Economic foundations of the current regulatory reform efforts, *J. Econ. Perspect.*, 10:119–134, 1996.

Working Group on Air Quality Guidelines and Objectives, A Protocol for the Development of National Ambient Air Quality Objectives, Science Assessment Document and Derivation of the Reference Level(s), Ottawa, 1996.

Working Group on Air Quality Objectives and Guidelines, National Ambient Air Quality Objectives for Particulate Matter, Part 1, Science Assessment Document, Report of the CEPA/FPAC Working Group on Air Quality Objectives and Guidelines, Bureau of Chemical Hazards, Ottawa, 1999.

CHAPTER 9

Global Environmental Changes and Health: Approaches to Assessing Risks

Anthony J. McMichael and R. Sari Kovats

INTRODUCTION

In the 18th century, Europe's main continuing environmental health hazard — as in many preceding centuries — was malnutrition and famine. After the 1740s this ancient health hazard began to recede in Europe as the modern agricultural revolution began. The extreme urban crowding and working-class poverty due to early industrialization in the 19th century resulted in infectious diseases becoming the dominant environmental health hazard. With the rise of modern large-scale industry and of synthetic organic chemistry in the 20th century, pollution of local environments — air, water, soil, and food — became the major focus of environmental health concern in developed countries.

Thus, for the past two centuries, our environmental health concerns have typically focused on toxicological or microbiological risks to health from factors within the local environment. However, as we enter the 21st century, the scale of environmental health hazards is increasing. As the scale of human impact on the environment has begun to alter global biophysical systems (such as the climate system), a range of larger-scale environmental hazards to human health has emerged. This includes the health risks posed by climate change, stratospheric ozone depletion, loss of biodiversity and its effects on ecological systems, stresses on terrestrial and ocean food-producing systems, changes in hydrological systems and the supplies of fresh water, and the global dissemination of persistent organic pollutants (McMichael, 1999; Watson et al., 1998). We are learning, today, to understand better the determinants of human health at the population level and within a predominantly ecological framework.

This *ecological* perspective recognizes that the foundations of long-term good health in populations reside in the continued stability and functioning of the biosphere's ecological and physical systems — often referred to as *life-support systems*. The deliberate modification of these natural systems by human societies throughout history has conferred many social, economic, and public health benefits. However, disturbance of these systems has also often caused new risks to health. A major subtext to the development of agriculture, beginning ten millennia ago, has been the acquisition and mobilization of infectious agents, the depletion of fresh water, and the emergence of various nutritional deficiencies (McNeill, 1976).

In this chapter we first say a few things about the long human prehistory and history that has brought us to the major environmental crossroads at which we now stand. Then we examine these contemporary global environmental changes in a little more detail. Finally, we consider how to develop appropriate research methods, both for empirical study and for scenario-based health risk assessment.

HUMAN BIOLOGICAL AND CULTURAL EVOLUTION: DEMOGRAPHIC, ENVIRONMENTAL, AND DISEASE TRANSITIONS

The story of the modern human species *Homo sapiens* spans several hundred thousand years. The highlights are these:

- Continuing biological and cultural adaptation, including the refinement of toolmaking, in Africa — our place of origin
- Dispersal of hunter–gatherer humans out of Africa from around 100,000 years ago
- Adaptation and survival, in diverse environments, during the last long glaciation from around 80,000 years ago to 15,000 years ago
- Onset of agriculture and village life, in favored centers, from around 10,000 years ago — and acquisition of many new infectious diseases from domesticated animals
- The rise of city-states, trade, and warfare
- During the past 3000 years, increasing transcontinental and, later, intercontinental exchange of local germs
- Since 1500, the rise of Europe, favored by climate, soils, coastline, and the robust legacy of Middle East farming and animal husbandry
- Onset of industrialization and urban consumerism — and their associated diseases

During the 19th century, the wealth of imperial Europe increased and industrialization spread. Living conditions in urban–industrial environments were predominantly crowded, squalid, and disease-ridden. From around midcentury, conditions became better with improvements in sanitation, food supplies, literacy, and domestic hygiene. Death rates from infectious diseases declined, especially in infancy and childhood. Meanwhile, a new set of diseases of later adulthood emerged. These diseases, characteristic of the *epidemiological transition* and reflecting a radical shift in human ecology, include chronic bowel disorders, heart disease, stroke, diabetes, various cancers, chronic respiratory disorders, arthritis, and dementia. With improved material living conditions, the decline in infectious diseases, and modern medical science to keep us alive longer, life expectancy has continued to edge up.

Meanwhile, the disease profile of low-income countries has also been changing. Rapidly developing countries, especially those of East and Southeast Asia, are acquiring "Western" disease profiles (Shetty and McPherson, 1997). Recently, however, a long shadow has been cast over life expectancy projections in many poorer countries by the spread of HIV/AIDS, along with the resurgence of tuberculosis and the widespread increases in malaria. Average life expectancy in several sub-Saharan African countries has dropped by 10 to 15 years over the past decade (WHO, 1999).

GLOBAL ENVIRONMENTAL CHANGE: A LENGTHENING SHADOW OVER HEALTH?

Another long shadow is now appearing over human population health — a shadow that extends well into the future. It is the consequence of humankind's capacity to deplete and damage the Earth on such a scale that Earth's basic life-supporting processes are being weakened (Daily, 1997; McMichael, 1993). This great aggregate impact of humankind is a function of our increasing numbers, our levels of consumption of energy and materials, and our massive waste generation.

Like other large terrestrial mammal species, we are *patch disturbers*: we reduce the biodiversity and the integrity of ecosystems in our immediate living environment, and we then move on and disturb a nearby patch (Rees, 1999). Unfortunately, modern, energy-intensive, consumer urban society no longer sees the evidence of its patch disturbance. Urban populations require much larger areas of Earth's surface to supply materials and to absorb waste. London, which occupies 160,000 hectares of space, depends on a surface area of 20 million hectares to supply and absorb — i.e., a ratio of 125. This ratio is a measure of our *ecological footprint* (Rees, 1992). There is simply not enough of planet Earth for all of next century's cities to have urban footprints with even half that ratio.

Increasingly, our patch disturbance is regional and global in scale, hence, the recent advent of *global environmental changes* — including stratospheric ozone depletion, climate change, the accelerating loss of biodiversity, the spread of invasive species, the degradation of food-producing ecosystems on land and at sea, and the global dissemination of persistent, biologically active, organic chemical pollutants.

IDENTIFYING AND ESTIMATING RISKS TO HEALTH

These global environmental changes have substantial implications for the future of human population health. These changes therefore present a great and somewhat unfamiliar challenge to epidemiologists and other public health scientists. Let us explore this, first, in relation to global climate change.

Greenhouse Gas Accumulation and Global Climate Change

Global climate change will occur — indeed, it is apparently already occurring (IPCC, 2001) — because of the human augmentation of greenhouse gases in the lower atmosphere. Average world temperatures have been increasing for over two decades. Other changes in physical and biological systems — glaciers, plant migration, animal behaviors, insect distributions — corroborate this apparent change in world climate (Epstein et al., 1998). Questions about the future health consequences of climate change are therefore now firmly on our environmental health agenda (McMichael et al., 1996; McMichael and Haines, 1997).

Epidemiologists face a major challenge in estimating future health impacts. This requires working from projected scenarios of future environmental conditions — e.g., a possible world climate, as modeled by climatologists for the 2020s or the 2050s. To these scenarios we apply our existing knowledge of how gradients in climatic conditions affect some particular health outcome. For example, if we know how death rates in certain cities are affected by heat waves in today's world (e.g., Rooney et al., 1998), then we can estimate how those urban populations would respond to a summer season in a future warmer world in which the frequency of heat waves is, say, tripled.

A more complicated task is to estimate how climatic changes would affect the potential geographic range of transmission of mosquito-borne infectious diseases such as malaria and dengue fever. Considerable effort has recently gone into developing mathematical models for making such projections (Martens, 1998; Patz et al., 1998). The models in current use have well-recognized limitations — but they have provided an important start. For example, the latest malaria models estimate (1) the future geographical distribution of changes in the seasonal transmission of malaria, based on the climate model-generated scenarios; (2) the potential geographical limits to malaria transmission; and (3) changes in future populations at risk of malaria (Martens et al., 1999).

Other modeling studies have estimated the impacts of climate change upon regional and global cereal grain yields (which account for two thirds of world food energy). Integrated assessment is now often used to link models together. For example, the crop yield models are first used to simulate the effects of climate change (with a global climate model scenario) and increased CO_2 (which has a fertilization effect) on the yield of the major cereal crops. An established world food trade model is then used to simulate the economic consequences of yield changes to determine future populations at risk of hunger (Parry et al., 1999).

Stratospheric Ozone Depletion

Meanwhile, higher in the atmosphere, depletion of stratospheric ozone by human-made gases such as chlorofluorocarbons (CFCs) is already occurring. As a result, ambient ground-level levels of ultraviolet irradiation (UVR) are estimated to have increased by up to 10% at middle to high latitudes over the past two decades. Via the Montreal Protocol of 1987, we have managed to curtail the release of many of these gases; but a problem remains with black-market sales and with the rapid increase in production of halons by China and other low-income countries temporarily exempted from the production ban.

Scenario-based modeling, integrating across four submodels (gaseous emissions, stratospheric ozone destruction, UVR flux at Earth's surface, and cancer induction) indicates that European and U.S. populations will experience a 5–10% excess in skin cancer incidence during the middle decades of the coming century (Slaper et al., 1996). With a better information base in future, similar modeling could be carried out for UVR impacts on the human eye (e.g., cataract formation) and immune system activity.

Other Global Environmental Changes and Health

As the human demand for space, materials, and food increases, so populations and species of plants and animals are being extinguished — apparently faster than the great natural extinctions that have occurred in eons past. This has been referred to as the *Sixth Great Extinction* (Leakey and Lewin, 1996). The most important consequence for humans is the weakening and disruption of ecosystems such that *natural goods and services* decline (Daily, 1997; Griffo, 1997). The loss of biodiversity also means that we are losing, prior to their discovery, many of the chemicals and genes available from nature of the kind that have already conferred enormous medical and health improvement benefits upon us. Myers has calculated that the usual hit rate in the bioprospecting of tropical plant extracts for new medicinal drugs is of the order of one in half a million (Myers, 1997). Therefore, as an indicative estimate, he points out that the as-yet-unprospected extracts should yield at least 375 new drugs, compared with the 48 in current use (including curare, quinine, codeine, and the life-saving cancer treatment drugs vincristine and vinblastine).

Meanwhile, invasive species are spreading into new non-natural environments as a result of intensified methods of human food production and long-distance commerce and mobility (McMichael and Bouma, 2000). As with species loss, these changes in the composition of the living world have myriad consequences for human health. Just one example: the choking spread of water hyacinth in eastern Africa's Lake Victoria, introduced from Brazil as a decorative plant, is now a fertile breeding ground for the water snail that transmits the infectious disease schistosomiasis and for bacteria causing diarrheal diseases (Epstein, 1998).

The ever-increasing pressures of agricultural and livestock production are stressing the world's arable lands and pastures. We enter the 21st century with an estimated one-third of the world's previously productive land seriously damaged by erosion, compaction, salinization, waterlogging, and chemicalization that destroys organic content (WRI, 1998). Similar pressures on the world's ocean fisheries have left a majority of them severely depleted or seriously stressed. All these changes to the integrity of soils and waterways and to the productivity of fisheries are compromising the capacity of the world to continue to provide, sustainably, sufficient food to nourish humankind. Unless we find environmentally and socially acceptable applications of genetic engineering to increase food yields, we will face a struggle over the coming decades.

There are three other great worldwide environmental changes that portend risks to future human population health. First, freshwater aquifers in all continents are being depleted of their ancient "fossil water" supplies. Agricultural and industrial demands, amplified by population growth, often greatly exceed the rate of natural recharge. Water-related political and public health crises loom in some regions within decades (Gleick, 2001).

Second, various semivolatile organic chemicals (such as PCBs) are now known to be dissem-inated worldwide via a sequential "distillation" process in the cells of the lower atmosphere, thereby transferring chemicals from their usual origins in low to middle latitudes to high — indeed, polar — latitudes (Watson et al., 1998). Consequently, there is growing evidence of maximal levels occurring in polar mammals and fish and in traditional human groups that depend on those food sources. Chemical pollutants are no longer just an issue of local toxicity.

Third, we have greatly altered the biogeochemical circulation of sulfur and nitrogen (Watson et al., 1998). Since the 1940s there has been a marked upturn in the human "fixation" of nitrogen, converted from the inert form to biologically active nitrate and ammonium ions. Most of the sixfold increase in humankind's contribution has been due to the use of nitrogenous fertilizers. Likewise, we have changed the cycling of sulfur. By affecting the acidity and nutrient balances of the world's soils and waterways, these two chemical changes are already affecting various ecosystems (including the enhancement of algal blooms) and are likely to impair the capacity of food-producing systems.

HEALTH IMPACT ASSESSMENT: METHODOLOGICAL STRATEGIES AND DIFFICULTIES

Scientist and policy maker must both now contend with assessing the health impacts — and the avoidance strategies and costs — of these global environmental changes. The three main challenges for scientists are (1) operating within an enlarged spatial and temporal frame, (2) dealing with complex, dynamic, nonequilibrial natural systems, and (3) handling the irreducible uncertain-ties inherent in such scenario-based modeling. For policy makers the challenges are (1) understand-ing these complex environmental processes, (2) taking a long-term, precautionary, and perhaps electorally unrewarding view, and (3) achieving cooperation and equity with other constituencies with whom the future is to be shared.

Most of the anticipated health effects of global environmental change have yet to occur — that is, they will occur in response to future configurations of environmental change. Meanwhile, some other health effects may now be occurring, but we are not yet able to discern them.

There are three different logical strategies for assessing the health impacts of global environ-mental change:

- By direct inference from existing (past or present) analogue situations that are believed to approx-imate aspects of future environmental change. The analogue situations, which arise from (appar-ently) naturally occurring situations, may entail either temporal or geographic variation in environmental conditions.
- By seeking evidence, in response to observed anthropogenic environmental change, of changes in indicators of health risk or health status that are especially climate sensitive. (While related to the preceding item, this study strategem depends on the existence of identifiable global environmental change or its local manifestation. Further, one must rely on the assessment of the climatologists that the environmental change is anthropogenic and not a manifestation of natural variability.)
- By using existing knowledge and theory to conduct mathematical modeling (or other types of formalized assessment) of likely future health outcomes (McMichael, 1997). The three main types of models are (1) empirical statistical models (based on statistical equations derived from directly relevant sets of observations), (2) process-based models (in which key processes are represented by theoretically formulated relationships, widely accepted in the literature), and (3) integrated assessment modeling (which extends the other two approaches by the inclusion of a much fuller range of interacting and modulating influences, including social, economic, behavioral, and tech-nical influences).

The fact that there is as yet little evidence of changes in population health attributable to the incipient change in global environmental conditions poses a problem for scientists, policy makers, and the public. Science classically operates empirically — via observation, interpretation, replication, prediction, and, as necessary, modification of hypotheses. However, having unintentionally initiated a one-off global experiment, we cannot sensibly plan to wait decades for the accrual of sufficient empirical evidence of health consequences. That would be too great a gamble with an uncertain future. We must therefore conduct risk assessment in relation to future environmental scenarios — i.e., scenario-based health risk assessment. Further, the inherent complexity and dynamic properties of the world's biophysical systems means that we should expect interactions, feedbacks, threshold phenomena, and surprises (Levins, 1995).

The research task of assessing the actual and potential health impacts of climate change has several distinctive characteristics and poses four major challenges to scientists:

1. The anticipated anthropogenic climate change will be a gradual and long-term process. This projected change in mean climatic conditions is likely to be accompanied by regional changes — mostly increases — in the frequency of extreme weather events. Changes in particular health outcomes may already be occurring, or may soon begin to occur, in response to the recent and ongoing change in world climate (whatever the cause of that change). The identification of such health effects will require carefully planned epidemiological studies and the implementation of appropriate monitoring activities (Haines and McMichael, 1997). Meanwhile, much research effort in the near future is likely to continue to go either into studies of the *acute* health impacts of natural climate variability (e.g., El Niño events) or into the scenario-based forecasting of health outcomes as projected by simple extrapolation, mathematical modeling, or expert judgment. There will also probably be increased research into intervention strategies (adaptations) to reduce the health impacts of climate change.

2. In empirical epidemiological studies, there are nearly always difficulties in estimating the role of climate per se as a cause of change in health status. Most diseases have multiple contributory causes. Often, two or more of the established causal factors are present, which makes it difficult to apportion causation between or among them. In other situations pertaining to health outcomes for which the causal factors have been less well established, it is difficult to know which factors have truly influenced the disease occurrence and which are merely "guilty by association." For example, in a particular locale the clearing of forest to plant crops and the extension of the irrigation system may coincide with a rise in regional temperature. Since all three factors could affect mosquito abundance, it is difficult to apportion between them the causation of any change in malaria incidence. This category of the problem is well recognized by epidemiologists as the *confounding* of effects.

3. It is equally important to distinguish coexistent *confounding* factors that can independently influence health outcomes from factors that modify the vulnerability of a particular population to climate stress. The latter factors comprise either endogenous characteristics of the population (such as nutritional or immune status in relation to infectious disease) or contextual circumstances that influence the sensitivity of the population's response to the climate change (such as access to air conditioning during heat waves). This modification of the effect of one factor by the action of another is commonly called *interaction*.

4. The assessment of scenario-based health risks with mathematical models entails three key challenges relating to validity, uncertainty, and contextual realism:
   • The adequate and valid representation of the central set of environmental and biological relationships, and of the interacting ecological and social processes that influence the impact of those upon health, is difficult. A balance must be attained between complexity and simplicity (transparency).
   • The complex configuration of causal and modifying factors results in substantial uncertainties in the model's output projections. In the first instance, there is the uncertainty attached to the input scenarios of climate change (and of associated social, demographic, and economic trends). Subsequently, there are three main types of uncertainties in the modeling process: (1) *normal* statistical variation (reflecting the stochastic processes of the real world); (2) uncertainty about the correct or appropriate values of key parameters in the model; and (3) incomplete knowledge about the structural relationships represented in the model.

- None of the global environmental changes impinges in isolation upon human health. As processes, these changes frequently modify one another, and they can impinge on human population health simultaneously and often interactively (Watson et al., 1998). An obvious example is in relation to agricultural crop yields, where climatic conditions, ultraviolet radiation levels (reflecting stratospheric ozone depletion), soil fertility, freshwater availability, and the ecological balance between pests and their predators all affect yields. Similar configurations of environmental stresses influence patterns of vector-borne infectious diseases.

Human populations, as with individuals, vary in their vulnerability — or susceptibility — to the impact of environmental exposures upon their health status. The vulnerability of a population depends on factors such as population density, level of economic development, food availability, local environmental characteristics, pre-existing health status, and the quality and availability of public health care. In the specific context of climate change, a population's vulnerability has been described as a function of the extent to which a health outcome is sensitive to climate change and of the capacity of the population to adapt to new climate conditions (Parry and Carter, 1998).

CONCLUSION

It is becoming increasingly clear that global environmental changes will have a range of impacts — some of them beneficial, most of them adverse — upon human health. While some of these health impacts would result from direct changes in the physical conditions of living, many of the impacts would result from more complex changes in the biophysical and ecological systems that are the preconditions for good, sustained population health. Epidemiologists and other public health scientists must therefore think about environmental health problems not only within the framework of environment-as-hazard but also within the framework of environment-as-habitat (McMichael, 1999).

There are great and interesting challenges to researchers: in the imaginative use of historical and spatial analogues to enhance understanding and foresight; in the development of mathematical modeling techniques; and in the handling of unusual configurations of uncertainties. Koopman, recognizing this general contemporary challenge, states that "epidemiology is in transition from a science that identifies risk factors for disease to one that analyzes the systems that generate patterns of disease" (Koopman, 1996). We should not, of course, replace the former with the latter — but, substantively and soon, we must *supplement* the former with the latter.

REFERENCES

Daily, G., Ed., *Nature's Services: Societal Dependence on Natural Ecosystems,* Island Press, Washington, D.C., 1997.

Epstein, P.R., Climate and health, *Science,* 285:347–348, 1999

Epstein, P.R., Weeds bring disease to the east African waterways, *Lancet,* 351:577, 1998.

Epstein, P.R., Diaz, H.F., Elias, S., Grabherr, G., Graham, N.E., Martens, W.J., Mosley-Thomson, E., and Susskind, J., Biological and physical signs of climate change: focus on mosquito–borne diseases, *Bull. Am. Meteorol. Soc.,* 78:409–417, 1997.

Gleick, P., *The World's Water 2000–2001*, Island Press, Washington, D.C., 2001.

Griffo, F., *Biodiversity and Health,* Island Press, Washington, D.C., 1997.

Haines, A. and McMichael, A.J., Climate change and human health: implications for research, monitoring and policy, *Br. Med. J.,* 315:870–874, 1997.

Intergovernmental Panel on Climate Change, Climate Change, 2001 — The Science of Climate Change: Contribution of Working Group I to the Second Assessment Report of the Intergovernmental Panel on Climate Change, Houghton, J.T. et al., Eds., Cambridge University Press, Cambridge, 2001

Koopman, J.S., Emerging objectives and methods in epidemiology, *Am. J. Public Health,* 86: 630–632, 1996.

Leakey, R. and Lewin, R., *The Sixth Extinction,* Doubleday, New York, 1996.

Levins, R., Preparing for uncertainty, *Ecosyst. Health,* 1:47–57, 1995.

Martens, W.J.M., *Health and Climate Change: Modelling the Impacts of Global Warming and Ozone Depletion,* EarthScan, London, 1998.

Martens, W.J.M., Kovats, R.S., Nijhof, S., de Vries, P., Livermore, M.T.J., Bradley, D., Cox, J., McMichael, A.J., Climate change and future populations at risk of malaria. *Global Environmental. Change,* 9 Supplement:S89–S107, 1999.

McMichael, A.J., *Planetary Overload: Global Environmental Change and the Health of the Human Species,* Cambridge University Press, Cambridge, 1993.

McMichael, A.J., Integrated assessment of potential health impact of global environmental change, prospects and limitations, *Environ. Modeling Assess.,* 2:129–137, 1997.

McMichael, A.J., From hazard to habitat: rethinking environment and health, *Epidemiology,* 10:460–464, 1999.

McMichael, A.J. and Bouma, M., Global changes, invasive species and human health, in Mooney, H. and Hobbs, R., Eds., *The Impact of Global Changes on Invasive Species,* Island Press, Washington, D.C., 2000.

McMichael, A.J. and Haines, A., Climate change and potential impacts on human health, *Br. Med. J.,* 315: 805–809, 1997.

McMichael, A.J., Haines, A., Slooff, R., and Kovats, R.S., Eds., *Climate Change and Human Health,* WHO, Geneva, 1996.

McNeill, W.H., *Plagues and Peoples,* Penguin, Middlesex, England, 1976.

Myers, N., Biodiversity's genetic library, in Daily, G.C., Ed., *Nature's Services: Societal Dependence on Natural Ecosystems,* Island Press, Washington, D.C., 1997, pp. 255–273.

Parry, M.L. and Carter, T., *Climate Impact and Adaptation Assessment,* EarthScan, London, 1998.

Parry, M.L. et al., Climate change, world cereal grain yields and human food adequacy, *Global Environ. Change,* 9:S51–S67, 1999.

Patz, J.A., Epstein, P.R., Burke, T.A., and Balbus, J.M., Global climate change and emerging infectious diseases, *JAMA,* 275: 217–223, 1996.

Patz, J.A., Martens, W.J.M., Focks, D.A., and Jetten, T.H., Dengue fever epidemic potential as projected by general circulation models of global climate change, *Environ. Health Perspect.,* 106:147–152, 1998.

Rees, W., Ecological footprints and appropriated carrying capacity: what urban economics leaves out, *Environ. Urbanization,* 4(2):121–130, 1992.

Rees, W., A human ecological assessment of economic and population health, in Crabbe, P., Westra, L., and Holland, A., Eds., *Implementing Ecological Integrity,* Kluwer, Dordrecht, The Netherlands, 1999.

Rooney, C., McMichael, A.J., Kovats, R.C., and Coleman, M., Excess mortality in England and Wales during the 1995 heat wave, *J. Epidemiol. Community Health,* 1998.

Shetty, P. and McPherson, K., Eds., *Diet, Nutrition and Chronic Disease. Lessons from Contrasting Worlds,* Wiley, Chichester, U.K., 1997.

Slaper, H., Velders, G.J.M., Daniel, J.S., de Gruijl, F.R., and van der Leun, J.C., Estimates of ozone depletion and skin cancer incidence to examine the Vienna Convention achievements, *Nature,* 384, 256–258, 1996.

Watson, R.T., Dixon, J.A., Hamburg, S.P., et al., Eds., *Protecting Our Planet. Securing our Future. Linkages among Global Environmental Issues and Human Needs,* UNEP/USNASA/World Bank, 1998.

World Health Organization (WHO), *The World Health Report 1999: Making a Difference,* WHO, Geneva, 1999.

World Resources Institute (WRI), *World Resources 1998–1999 Environment and Health,* Oxford University Press, Oxford, 1998.

CHAPTER **10**

Climate Change and Health: New Research Challenges*

Jonathan A. Patz

According to the United Nations Intergovernmental Panel on Climate Change (IPCC), anthropogenic greenhouse gas emissions are significantly accelerating current global surface warming trends and are inconsistent with natural climate variation (Houghton et al., 1996). This century the Earth has warmed by about 0.6°C, and the midrange estimates of future temperature change and sea level rise are 2.8°C and 40 cm by the year 2100, respectively. If such changes occur, diseases that are influenced by weather factors could be expected to respond to an altered climate regime.

The commonly used term *global warming* is inadequate to describe the range of climate effects of greenhouse gas accumulation. In addition to concerns about temperature and sea level rise, climatologists, ecologists, and health and other scientists are concerned that accelerated evaporation will drive a more extreme hydrological cycle, producing more frequent or severe floods and droughts. The regional impacts of these climate changes will vary widely depending on existing population vulnerability.

Extreme weather variability associated with climate change may add an important new stress to developing nations that are already vulnerable as a result of environmental degradation, resource depletion, overpopulation, or location (e.g., low-lying coastal deltas). Persistent poverty, accompanied by inadequate sanitation and public health infrastructure, will also limit many populations' capacities to adapt to this newly recognized environmental risk factor.

The broad spectrum of multisectorial impacts posed by climate change illustrates the need for truly multidisciplinary research and integrated assessments that address the range of regional climate and health effects. This chapter reviews major climate-sensitive health outcomes, many of which are currently being studied by our program and collaborating institutions.

OVERVIEW OF HEALTH IMPACTS OF CLIMATE CHANGE

There are three well-recognized physical consequences of climate change: (1) temperature rise, (2) sea level rise, and (3) extremes in the hydrologic cycle. A recent paper in the journal *Nature* (Shindell et al., 1998) presented a fourth complication, accelerated ozone depletion, which results from the trapping of heat in the troposphere, thereby cooling the stratosphere. These four physical attributes of climate change are expected to increase the frequency of heat waves and potentially air pollution episodes; increase the number of extreme weather events; cause coastal flooding and

* Adapted from Patz et al., 1998, 2000.

1-56670-612-2/03/$0.00+$1.50
© 2003 by CRC Press LLC

77

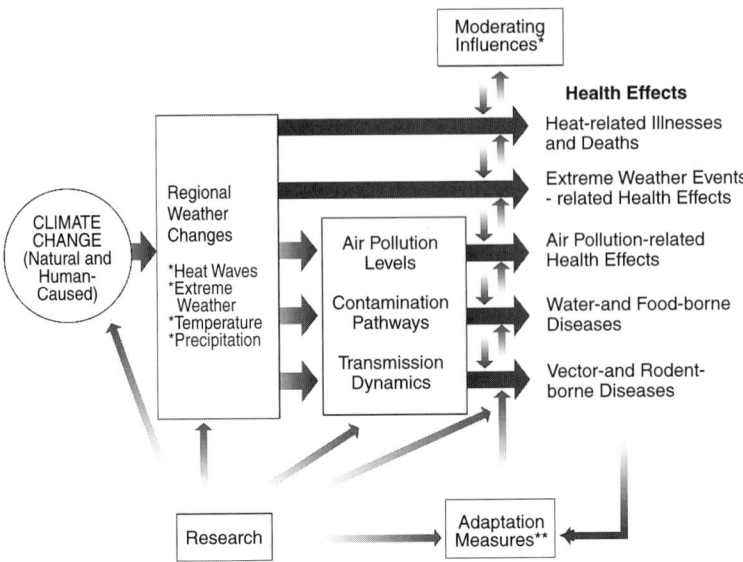

Figure 10.1 Schematic of potential health effects (and moderating factors) of climate variability and change. (From Patz, J.A. et al., *Environ. Health Perspect.,* 108:367–376, 2000.)

salination of freshwater aquifers; and (if ozone depletion is exacerbated) increase ultraviolet radiation exposure. Potential health effects from these consequences are illustrated in Figure 10.1.

Seasonality in disease incidence can often infer an association with weather factors. In sub-Saharan Africa, epidemics of meningococcal meningitis consistently erupt during the hot dry season and subside soon after the onset of the rainy season (Moore, 1992). In the southwestern U.S., rodent-borne hantavirus has been linked to El Niño-driven flooding, leading to an upsurge in mouse populations (Glass et al., 2000). In Peru, prevalence of the diarrheal disease cyclospora peaks in summer and wanes during cooler winter months (Madico et al., 1997). The current extent of our understanding of climate/disease relationships varies considerably, as discussed in more depth for other examples below.

Heat-Related Illness

While the relationship between ambient temperatures and seasonal variability in mortality rates has been extensively studied, this expected direct health consequence of climate change is far from resolved. Cardiovascular mortality in elderly people comprises the largest number of heat-related fatalities; however, physiological responses to both extreme heat and cold are not straightforward. For example, blood viscosity and cholesterol have been found to increase with high temperatures (Keating et al., 1986), whereas blood pressure and fibrinogen levels increase during winter (although outdoor temperature does not seem to determine the seasonal variation of fibrinogen) (van der Bom et al., 1997). In some cases, rainfall and snowfall have been found to influence winter mortality more than temperature, further complicating the health assessment of future winters that may be warmer but wetter.

Mortality curves assume a classic J- or V-shape, with highest mortality occurring at both temperature extremes. Generally, populations in warmer regions tend to be most vulnerable to cold (Eurowinter Group, 1997), and those residing in cold climates are most sensitive to heat. In temperate regions, mortality rates are highest during the winter.

The Chicago heat wave of July 1995 led to over 700 excess deaths in the metropolitan area. Climatologists project a doubling in the frequency of heat waves associated with a rise of 2 to 3°C in average summer temperature. A study of 44 U.S. cities found that, after adjusting for some expected acclimatization, heat-related mortality could increase by 70 to 150% (Kalkstein and Greene, 1997). A meta-analysis of 20 international cities, however, found a reduction in mortality due to less deaths during winter (Martens, 1998). In short, there is a need for more extensive net annual mortality estimates stratified by season, cause of death, and other main confounders.

Air Pollution

Increased ambient temperature and altered wind and air mass patterns can affect atmospheric chemistry and, thereby, air pollution. For example, there is a nonlinear relationship between temperature and the formation of ground-level ozone (photochemical urban smog): above 90°F there is a strong positive relationship with temperature. The relatively high ozone levels in the U.S. during 1988 and 1995 were likely due in part to the hot, dry, stagnant conditions.

With increasing demand for automobiles, ozone is a growing problem in the developing world and continues to be the most pervasive air pollution problem in the U.S., with an estimated 71 million people living in counties that exceeded the National Ambient Air Quality Standards (NAAQS) in 1995 (U.S. EPA, 1996). Ozone is a potent lung irritant and can heighten the sensitivity of asthmatics to allergens. An increased frequency of heat waves could be expected to worsen ozone problems in urban areas. One study that held emissions and other weather factors constant showed that a 4°C warming could (1) increase maximum ozone concentration by about 20% and double the area out of compliance with the ozone NAAQS in the San Francisco Bay area; and (2) nearly triple the areas exceeding national standards in the U.S. Midwest and Southeast (U.S. EPA, 1989).

Malaria and Other Vector-Borne Diseases

Malaria is believed to be the most climate-sensitive vector-borne disease and thus most sensitive to climate change (Figure 10.2) (WHO, 1996). There are 300 to 500 million new cases of malaria

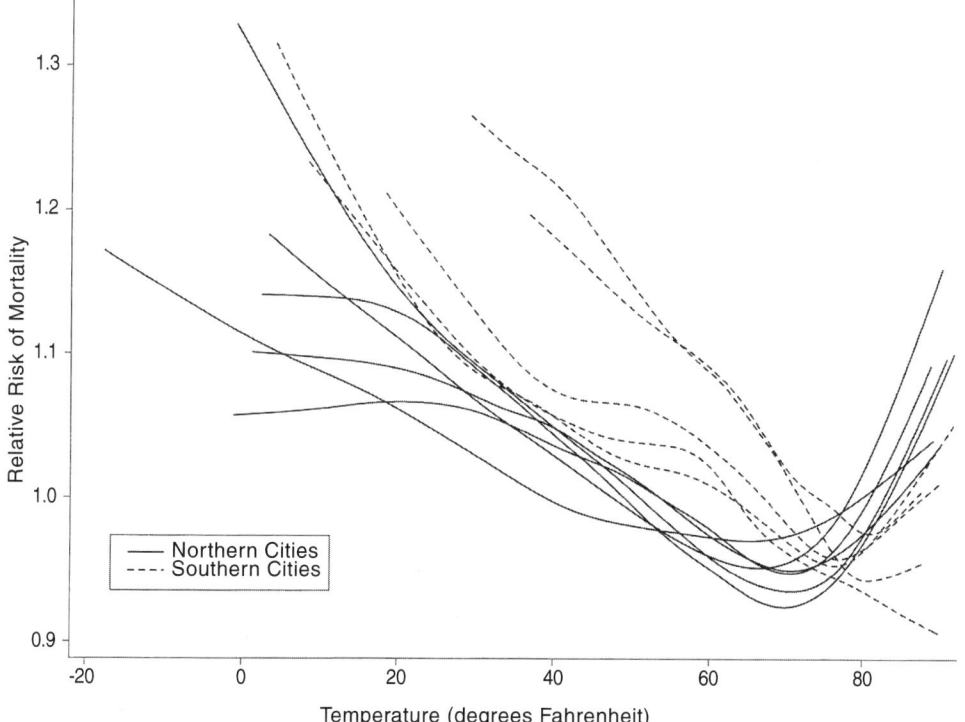

Figure 10.2 Temperature-relative risk functions for 11 U.S. cities, 1973 to 1994. (From Curriero, F.C. et al., *Am. J., Epidemiol.*, 155: 80–87, 2002. With permission.)

annually, with an estimated 2 million annual malaria fatalities, the majority young children. Infectious disease agents that must cycle through cold-blooded insect vectors to complete their development are quite sensitive to subtle climate variations (Patz et al., 1996).

The minimum temperature for parasite development of *Plasmodium falciparum* and *P. vivax* approximates 18 and 15°C, respectively. In the typically nonendemic highlands of Kenya (Garnham, 1948), Rwanda (Loevinsohn, 1994), and Zimbabwe (Freeman and Bradley, 1996), increases in ambient temperature have been linked to malaria epidemics. Also, the incidence and prevalence of malaria is closely associated with altitude (Taylor and Mutambu, 1986), a good proxy for temperature. One scenario-based climate change/malaria modeling study concludes that the global population living within a potential malaria transmission zone could increase from 45% to 60% by the year 2100 (Martens et al., 1995).

While malaria tends to be seasonal, there is substantial interannual heterogeneity of malaria incidence around the globe (Fontenille et al., 1997). Extremes of rainfall (both drought and floods) associated with El Niño events have been linked to variability in malaria incidence in some regions (Bouma et al., 1994, 1997). But the link between El Niño frequency and future climate change is still uncertain.

Other climate-sensitive diseases include mosquito-borne arboviruses, such as encephalitis, dengue fever, and Ross River virus. Field and lab studies on St. Louis encephalitis (Reeves et al., 1994), as well as lab studies on dengue virus, indicate that higher temperatures hasten viral development (or extrinsic incubation period, EIP) inside the mosquito (Watts et al., 1987). Human outbreaks of St. Louis encephalitis are highly correlated with several-day periods when the temperature exceeds 29°C (Monath and Tsai, 1987). Ross River virus, causing epidemic polyarthritis in Australia, shows a positive association with increases in minimum temperatures and rainfall (Tong et al., 1998).

Disease dynamics and insect ecology are complex. For example, higher temperatures also reduce adult mosquito survival. However, since minimum temperatures are expected to rise disproportionately compared to maximum temperatures, pathogen EIP may rise faster than adult survival declines. Climate change modeling studies of dengue fever using a vectorial capacity model (which combines variables including mosquito survival, EIP, biting rates, and others) have found increases in the potential transmission of dengue due to climate change (Jetten and Focks, 1997; Patz et al., 1998). Yet actual risk, as opposed to potential global or regional risk, will depend on site-specific factors; fully parameterized climate-driven models have successfully predicted dengue risk as validated against historical data in some locations (Focks et al., 1995).

Water-Borne Diseases

In 1995, 3.1 million people died from diarrheal diseases, 80% of them children (WHO, 1996). Extremes of the hydrologic cycle could worsen the problem since both water shortages and flooding are associated with diarrheal diseases. In developing countries, water shortages cause disease through poor hygiene. On the other extreme, flooding can contaminate drinking water from watershed runoff or sewage overflow. Since 1900, many regions of the U.S. have experienced an increased intensity of precipitation, e.g., more frequent heavy downpours, and the trend is expected to strengthen due to climate change (Karl et al., 1995).

Cryptosporidiosis, a zoonotic disease associated with domestic livestock, may be affected by altered weather patterns. Floods, storms, heavy rainfall, and snow melt wash material of fecal origin into surface drinking water sources, and the oocyst is resistant to chlorine treatment. The Milwaukee outbreak in 1993, which resulted in 403,000 reported cases, coincided with unusually heavy spring rains and runoff from melting snow. Preliminary research results from our program show that the majority of historical water-borne outbreaks in the U.S. (involving multiple agents) are preceded by heavy rainfall events.

In the marine environment, warm water and nitrogen favor blooms of dinoflagellates that cause red tides, which can cause paralytic shellfish poisoning, diarrheic shellfish poisoning, and amnesiac shellfish poisoning. During the 1987 El Niño, a red tide of *Gymnodinium breve*, previously confined to the Gulf of Mexico, extended northward after warm Gulf Stream water extended far up the East coast, resulting in human neurological shellfish poisonings and substantial fish kills (Tester, 1991).

Similarly, that year an outbreak of amnesiac shellfish poisoning occurred on Prince Edward Island when warm eddies of the Gulf Stream neared the shore and heavy rains increased nutrient-rich runoff.

Copepods (or zooplankton), which feed on algae, can serve as reservoirs for *Vibrio cholerae* and other enteric pathogens. In Bangladesh, cholera follows seasonal warming of sea surface temperature that can enhance plankton blooms (Colwell, 1996). The Peruvian outbreak in 1991 coincided with an El Niño event, and satellite images of sea surface temperature confirm a close temporal relationship between the arrival of warmer waters along the Peruvian coast with the epidemic starting soon afterward. Simultaneous initial coastal outbreaks that occurred hundreds of kilometers apart imply a marine environmental reservoir (likely zooplankton), and this hypothesis is under epidemiologic investigation.

Water Resources and Agriculture

Several factors come into play when predicting the impact of climate change on crop and livestock production. First are the direct effects of temperature, precipitation, CO_2 levels (e.g., the CO_2 fertilization effect), and extreme climate variability and sea level rise (McCarthy et al., 1996). Next are the indirect effects of climate-induced changes in soil quality, incidence of plant diseases, weed and insect populations, and enhanced food spoilage from heat and humidity. Finally, the extent to which adaptive responses are available to farmers must be considered.

Developing countries already struggle with large and growing populations and malnutrition and would be particularly vulnerable to changes in food production. A recent analysis indicates that an additional 40 million to 300 million people (relative to a projected baseline of 640 million people by year 2060) could be at risk from malnutrition due to climate change (Rosenzweig and Parry, 1994).

Extreme Weather Events

Tropical cyclones represent the most destructive form of recurring natural disaster. Historical analysis shows that hurricanes only form in regions where sea surface temperatures are above 26°C (Gray, 1979). A recent modeling study of the western Pacific concluded that a sea surface warming of just over 2°C would intensify hurricane wind speed by 3 to 7 meters per second (or 5 to 12%) (Knudson et al., 1998). Yet hurricane frequency has not increased in the last 50 years, and current climate models lack the spatial resolution to predict a change in cyclone formation.

While hurricane predictions are still elusive, heavy rainfall events are anticipated with climate change. In the U.S., flash floods are currently the leading cause of weather-related mortality (Noji, 1997). In addition to immediate drowning deaths, floodwaters can cause the release of dangerous chemicals from storage and waste disposal sites and precipitate outbreaks of vector- and waterborne diseases. For example, extreme flooding and hurricanes have been responsible for outbreaks of the spirochetal zoonosis leptospirosis in Nicaragua and Barbados (Centers for Disease Control and Prevention, 1995). Mosquito-borne Rift Valley fever occurs in association with El Niño–driven flooding in east Africa (Nicholls, 1993), as demonstrated by the serious outbreak in Kenya during this year's strong El Niño.

Environmental Refugees

Sea level rise, in combination with extremes of the hydrologic cycle, could pose serious repercussions for coastal communities and cause large population displacement. Of the world's 20 current megacities, 13 are at sea level. A 1-meter sea level rise would inundate low-lying areas affecting 18.6, 13.0, 3.5, and 3.3 million people in China, Bangladesh, Egypt, and Indonesia, respectively (Strzepek and Smith, 1995).

Rising seas result in salination of coastal freshwater aquifers and threaten drinking water resources and coastal farmland. Far larger numbers than those threatened by frank inundation may be indirectly impacted in this way. For example, a 1-meter sea level rise could affect 60%

of the population of Bangladesh and 100% of many island nations (Strzepek and Smith, 1995). In places such as Alexandria, Egypt, agricultural depletion of groundwater and reduced siltation from upstream dams and levees are already causing land subsidence, thus decreasing the threshold for impact. Lagos, Nigeria, acquires 60% of its water from a shallow aquifer that is merely 1 meter above current sea level. Finally, sea level rise could disrupt stormwater drainage and sewage disposal, compromising sanitation.

The multisectoral impacts of sea level rise, combined with more severe drought and floods in varying regions, may displace substantial numbers of persons (Myers and Kent, 1995). Environmental refugees could potentially present the most serious health consequences of climate change, considering the associated risks that stem from overcrowding, virtually absent sanitation, scarcity of shelter and natural resources, and heightened tensions leading potentially to war. Environmentally forced population migration may unfold to be the largest challenge beneath the "tip of the iceberg" of climate change health impacts.

Ozone Depletion and Ultraviolet Radiation

By trapping heat at the Earth's surface, greenhouse gases cool the stratosphere, which enhances chlorofluorocarbon (CFC) destruction of the ozone layer (UNEP, 1991). The direct health impacts from increases in UV-B include (1) skin cancer, (2) cataract and other ocular diseases, and (3) immunosuppression. Indirect effects to health may occur primarily through UV-mediated crop damage and by photochemical formation of tropospheric ozone (Figures 10.3 and 10.4).

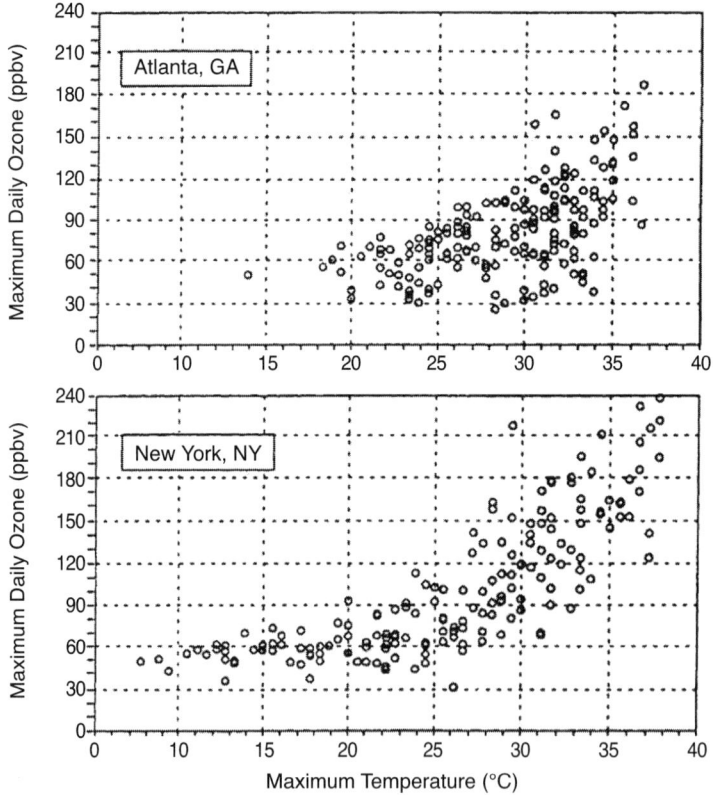

Figure 10.3 Ozone and temperature relationship.

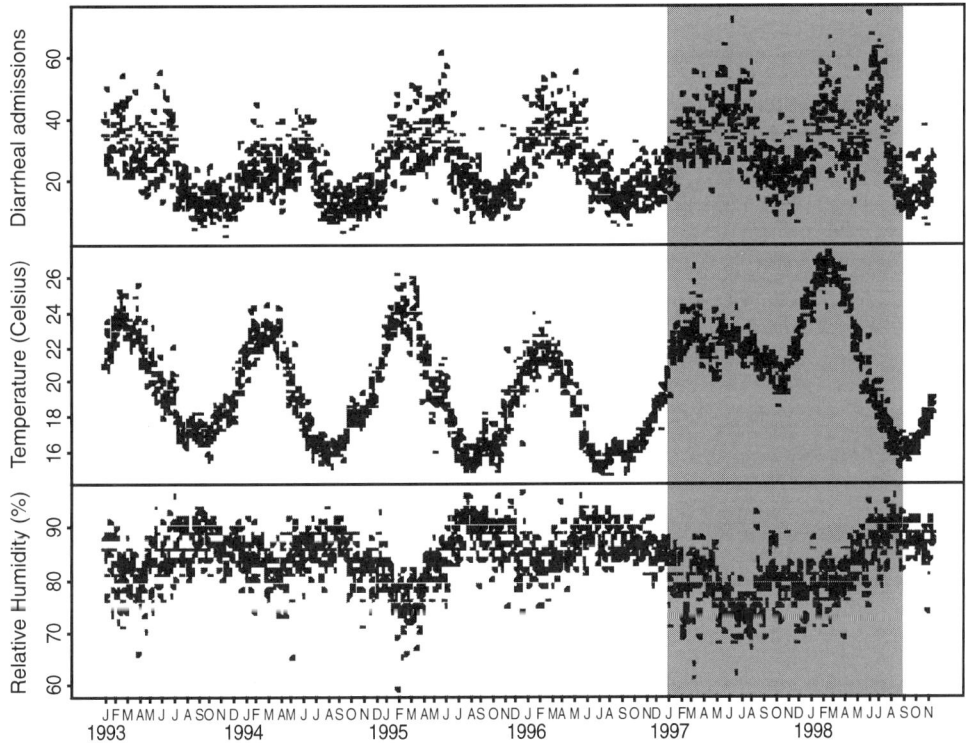

Figure 10.4 Daily time series between January 1, 1993, and November 15, 1998, for admissions for diarrhea, mean ambient temperature, and relative humidity in Lima, Peru. Note that winter temperatures during the 1997–1998 El Niño were approximately 5°C above normal, and admissions for childhood diarrhea doubled during that winter period. (From Checkley, W., et al., Lancet, 355: 442–450, 2000. With permission.)

It is estimated that for a sustained 10% decline in the stratospheric ozone layer, non-melanoma skin cancer cases could rise by 26%, or 300,000 globally per year; melanoma could increase by 20%, or 4500 more cases annually (UNEP, 1991). Cataracts account for half of all blindness in the world. A 10% sustained loss of stratospheric ozone would result in nearly 1.75 million extra cataracts annually (UNEP, 1991).

RESEARCH TOOLS AND STRATEGIES

The following research methods are among those required to address this cross-cutting, ecologically complex, and long-term public health challenge:

1. Time-series and regression analysis of historical data, including analogue situations of extreme climate variability (e.g., El Niño)
2. Geographic analysis of disease incidence based on weather and land use/land cover variables (taking advantage of geographic information systems, or GIS, and satellite remote sensing)
3. Scenario-based mathematical and predictive modeling with uncertainty analysis
4. Generalized integrated assessments that include demographic, social, and economic disruptions

For any of these analytical methods, human health data are the most unreliable due to reporting bias and variability in detection methods. There is a critical need for *capacity building* to improve surveillance and monitoring to detect changes that may be a result of global climate and ecological change. Yet surveillance alone is not sufficient to prevent illness, and continued efforts to develop predictive models should be a priority.

Comment on Modeling: Predictive models are essential to improving proactive preventive health measures. Even though no model can completely simulate real life, models are useful in conceptualizing dynamic processes and their outcomes. Although empirical studies have limits, they are the foundation upon which modeling parameters are determined. While not necessarily more accurate, mathematical models can achieve a better conceptual representation of interrelated systems. Multiple iterations of well-conceptualized models that help identify key knowledge gaps, thus obtaining a "correct" model projection, might not uniformly be the most valuable outcome for scientists.

CONCLUSION

New understanding of linkages between public health and global life-support systems is emerging in the literature (McMichael, 1993). The long-term and complex problems posed by climate change may not be readily discernible over short time spans and therefore demand an expanded effort in scenario-based risk assessment in parallel with historical validation.

The results of the studies reviewed in this article must be viewed in the context of many other environmental and behavioral determinants. Future studies must consider, along with climatological factors, key variables such as poverty, sanitation, land use changes, and public health surveillance and mitigation programs. Studies of potential risk at the level of global climate models, while instructive, can only translate directly into actual risk or vulnerability when these local factors are included in assessments.

Analyzing the role of climate in determining human health outcomes will require interdisciplinary cooperation among health scientists, climatologists, biologists, ecologists, and social scientists. Increased disease surveillance, integrated modeling, and use of geographically based data systems will afford more anticipatory measures by the medical community. Understanding the linkages between climatological and ecological change as determinants of disease will ultimately help in constructing predictive models to guide proactive prevention.

REFERENCES

Bouma, M.J., Sondorp, H.E., and van der Kaay, H.J., Climate change and periodic epidemic malaria, *Lancet,* 343:1440, 1994.

Bouma, M.J., Poveda, G., Rojas, W., et al., Predicting high-risk years for malaria in Colombia using parameters of El Niño southern oscillation, *Trop. Med. Intl. Health,* 2:1122–1127, 1997.

Centers for Disease Control and Prevention, Outbreak of acute febrile illness and pulmonary hemorrhage — Nicaragua, 1995, *Morbidity and Mortality Weekly Report,* 44:841–843, 1995.

Checkley, W., Epstein, L.D., Gilman, R.H., Figueroa, D., Cama, R.I., Patz, J.A., and Black, R.E., Effects of the El Niño and ambient temperature on hospital admissions for diarrhoeal disease in Peruvian children, *Lancet,* 355: 442–450, 2000.

Colwell, R.R., Global climate and infectious disease: the cholera paradigm, *Science,* 274:2025–2031, 1996.

Curriero, F.C., Heiner, K., Zeger, S., Samet, J., and Patz, J.A., Analysis of heat-mortality in 11 cities of the eastern United States, *Am. J., Epidemiol.,* 155: 80–87, 2002.

Eurowinter Group, The, Cold exposure and winter mortality from ischaemic heart disease, cerebrovascular disease, respiratory disease, and all causes in warm and cold regions of Europe, *Lancet,* 349:1341–1346, 1997.

Focks, D.A., Daniels, E., Haile, D.G., and Keesling, J.E., A simulation model of the epidemiology of urban dengue fever: literature analysis, model development, preliminary validation, and samples of simulation results, *Am. J. Trop. Med. Hyg.,* 53:489–506, 1995.

Fontenille, D., Lochouarn, L., Diagne, N., et al., High annual and seasonal variations in malaria transmission by anophelines and vector species composition in Dielmo, a holoendemic area in Senegal, *Am. J. Trop. Med. Hyg.*, 56:247–253, 1997.

Freeman, T. and Bradley, M., Temperature is predictive of severe malaria years in Zimbabwe, *Trans.R. Soc. Trop. Med. Hyg.*, 90:232, 1996.

Garnham, P.C., The incidence of malaria at high altitudes, *J. Natl. Malaria Soc.*, 7:275–284, 1948.

Glass, G.E., Cheek, J., Patz, J.A., et al., Using remotely sensed data to identify areas at risk for hantavirus pulmonary syndrome, *J. Emerg. Infect. Dis.*, 6:238–246, 2000.

Gray, W., Hurricanes: their formation, structure and likely role in the tropical circulation, in Shaw, D.B., Ed., *Meteorology over the Tropical Oceans*, Royal Meteorological Society, London, 1979, pp. 155–218.

Houghton J.T., Ding, Y., Griggs, D.J., Noguer, M., van der Linden, P.J., Dai, X., Maskell, K., and Johnson, C.A., Eds., *Climate Change, 2001: The Scientific Basis*, Intergovernmental Panel on Climate Change, Cambridge University Press, Cambridge, 1996.

Jetten, T.H. and Focks, D.A., Potential changes in the distribution of dengue transmission under climate warming, *Am. J. Trop. Med. Hyg.*, 1997.

Kalkstein, L.S. and Greene, J.S., An evaluation of climate/mortality relationships in large U.S. cities and possible impacts of a climate change, *Environ. Health Perspect.*, 105:2–11, 1997.

Karl, T.R., Knight, R.W., and Plummer, N., Trends in high-frequency climate variability in the twentieth century, *Nature*, 377:217–220, 1995.

Keating, W.R., Coleshaw, S.R., Easton, J.C., Cotter, F., Mattock, M.B., and Chelliah, R., Increased platelet and red cell counts, blood viscosity, and plasma cholesterol levels during heat stress, and mortality from coronary and cerebral thrombosis, *Am. J. Med.*, 81:795–800, 1986.

Knudson, T.R., Tuleya, R.E., and Kurihara, Y., Simulated increase of hurricane intensities in a CO_2-warmed climate, *Science*, 279:1018–1020, 1998.

Loevinsohn, M., Climatic warming and increased malaria incidence in Rwanda, *Lancet*, 343:714–718, 1994.

Madico, G., Checkley, W., and Gilman, R.H., Epidemiology and treatment of *Cyclospora cayetanenis* infection in Peruvian children, *Clin. Infect. Dis.*, 24:977–981, 1997.

Martens, W.J.M., Climate change, thermal stress and mortality changes, *Soc. Sci. Med.*, 46:331–344, 1998.

Martens, W.J.M., Niessen, L.W., Rotmans, J., Jetten, T.H., and McMichael, A.J., Potential impact of global climate change on malaria risk, *Environ. Health Perspect.*, 103:458–464, 1995.

McCarthy, J.J., Canziani, O.F., Leary, N.A., Dokken, D.J., and White, K.S., Eds., *Climate Change 2001 — Impacts, Adaptations and Vulnerability*, Cambridge University Press, Cambridge, 1996.

McMichael, A.J., Global environmental change and human population health: conceptual and scientific challenge for epidemiology, *Int. J. Epidemiol.*, 22:1–8.

Monath, T.P. and Tsai, T.F., St. Louis encephalitis: lessons from the last decade, *Am. J. Trop. Med. Hyg.*, 37:40, 1987.

Moore, P.S., Meningococcal meningitis in sub-Saharan Africa: a model for the epidemic process, *Clin. Infect. Dis.*, 14:515–525, 1992.

Myers, N. and Kent, J., *Environmental Exodus: An Emergent Crisis in the Global Arena*, Climate Institute, Washington, D.C., 1995.

Nicholls, N., El Niño-southern oscillation and vector-borne disease, *Lancet*, 342:1284–5, 1993.

Noji, E.K., The nature of disaster: general characteristics and public health effects, in Noji, E.K., Ed., *The Public Health Consequences of Disasters*, Oxford University Press, Oxford, 1997, pp. 3–20.

Patz, J.A., Climate change and health: new research challenges, *Health Environ.*, 12, 49–53, 1998.

Patz, J.A., Climate change and health: new research challenges, *Ecosyst. Health*, 6:52–58, 2000.

Patz, J.A., Epstein, P.R., Burke, T.A., and Balbus, J.M., Global climate change and emerging infectious diseases, *J. Am. Med. Assoc.*, 275:217–223, 1996.

Patz, J.A., Martens, W.J.M., Focks, D.A., and Jetten, T.H., Dengue fever epidemic potential as projected by general circulation models of global climate change, *Environ. Health Perspect.*, 106:147–153, 1998.

Patz, J.A., McGeehin, M.A., Bernard, S.M., Ebi, K.L., Epstein, P.R., Grambsch, A., Gubler, D.J., Reiter, P., Romieu, I., Rose, J.B., Samet, J.M., and Trtanj, J., The potential health impacts of climate variability and change for the United States: executive summary of the report of the health sector of the U.S. National Assessment, *Environ. Health Perspect.*, 108, 367–376, 2000.

Reeves, W.C., Hardy, J.L., Reisen, W.K., and Milby, M.M., The potential effect of global warming on mosquito-borne arboviruses, *J. Med. Entomol.*, 31:323–332, 1994.

Rosenzweig, C. and Parry, M.L., Potential impact of climate change on world food supply, *Nature,* 367:133–138, 1994.

Shindell, D.T., Rind, D., and Lonergan, P., Increased polar stratospheric ozone losses and delayed eventual recovery owing to increasing greenhouse-gas concentrations, *Nature,* 392:589–592, 1998.

Strzepek, K.M. and Smith, J.B., Eds., *As Climate Changes; International Impacts and Implications,* Cambridge University Press, Cambridge, 1995.

Taylor, P. and Mutambu, S.L., A review of the malaria situation in Zimbabwe with special reference to the period 1972–1981, *Trans. R. Soc. Trop. Med. Hyg.,* 80:12–19, 1986.

Tester, P., Red tide: effects on health and economics, *Health Environ. Dig.,* 5:4–5, 1991.

Tong, S., Bi, P., Parton, K., Hobbs, J., and McMichael, A.J., Climate variability and transmission of epidemic polyarthritis, *Lancet,* 351:1100, 1998.

UNEP, *Environmental Effects of Ozone Depletion: 1991 Update,* UN Environment Programme, Nairobi, 1991.

U.S. EPA, National Air Quality and Emissions Trends Report, 1995, U.S. Environmental Protection Agency, Washington, D.C., 1996.

U.S. EPA, The potential effects of global climate change on the United States, appendix F: air quality, in Smith, J.B. and Tirpak, D.A., Eds., U.S. Environmental Protection Agency, Office of Policy, Planning and Evaluation, Washington, D.C., 1989.

van der Bom, J.G., de Maat, M.P., Bots, M.L., Hofman, A., Kluft, C., and Grobbee, D.E., Seasonal variation in fibrinogen in the Rotterdam study, *Thromb. Haemost.,* 78:1059–1062, 1997.

Watts, D.M., Burke, D.S., Harrison, B.A., Whitmire, R.E., and Nisalak, A., Effect of temperature on the vector efficiency of *Aedes aegypti* for dengue 2 virus, *Am. J. Trop. Hyg.,* 36:143–152, 1987.

WHO, *Climate Change and Human Health,* World Health Organization, Geneva, 1996.

WHO, *World Health Report 1996: Fighting Disease Fostering Development,* World Health Organization, Geneva, 1996.

CHAPTER 11

Use of Meteorological Data to Predict Mosquito-Borne Encephalitis Risk in California: Preliminary Observations in Kern County

William K. Reisen

INTRODUCTION

In 1952 above-normal snowpack in the Sierra Nevada, a cool, wet spring, and a strong earth-quake centered in the Tehachapi Mountains combined to produce conditions that led to the largest epidemic of mosquito-borne encephalitis in California history (Reeves and Hammon, 1962). The elevated snowpack and wet spring led to flooding along the Kern River and the emergence of a large mosquito population — while the earthquake destroyed most of the city of Bakersfield, forcing the population into temporary shelters and thereby increasing human contact with mosquitoes. Morbidity rates for western equine encephalomyelitis virus (WEE) were as high as 100 per 100,000 population in infants (<1 year age group), and many of these patients experienced serious neuro-logical sequelae requiring institutional care (Finley et al., 1955; Longshore et al., 1956). Additional clinical cases were diagnosed with a second virus, St. Louis encephalitis virus (SLE). This epidemic occurred when the size of the human population residing in the southern San Joaquin Valley was relatively small and residents in rural high-risk communities had acquired immunity rates >75% (Reeves and Hammon, 1962). Clearly, an epidemic of this magnitude under current conditions would be devastating and overwhelm the health care system, because population size has increased dramatically and the current population is essentially nonimmune (Reeves and Milby, 1989), with seroprevalence rates <1% (Reisen and Chiles, 1997).

Both arboviruses remain of public health importance in California today. WEE, an alphavirus, is amplified during spring and summer in a primary cycle involving wild birds (mostly house sparrows and house finches) and the mosquito, *Culex tarsalis* (Reisen and Monath, 1989). As summer progresses, *Cx. tarsalis* expands its blood meal host selection pattern to include other birds such as doves, chickens, and blackbirds as well as rabbits. Rabbits develop a viremia (virus concentration in the peripheral bloodstream) of sufficient titer to infect susceptible *Aedes* mosqui-toes, and a secondary rabbit–*Aedes* cycle may become established by mid to late summer (Hardy, 1987). Humans and equines (horses, mules, donkeys) become infected tangentially to these primary or secondary cycles, but fail to develop sufficient viremias to infect mosquitoes and therefore are dead-end hosts for the virus. SLE, a flavivirus, is amplified during spring and summer in a parallel primary cycle involving the same avian hosts and *Cx. tarsalis* (Tsai and Mitchell, 1989). In urban/

suburban environments, a secondary cycle involving other *Culex* mosquitoes may become established during summer. Human involvement again is tangential to these cycles.

Concepts germane to both cycles have emerged. First, despite over 50 years of investigation, the mechanisms responsible for virus persistence between seasons remains unknown (Reeves, 1990). This precludes focusing surveillance on meteorological or other factors responsible for the initiation of transmission. Second, most of the virus life history is spent in the mosquito host. Birds produce a high-titered viremia of short duration (2 to 3 days), whereas mosquitoes become infected for life. Transmission requires that the mosquito host acquire virus infection during one blood meal, amplify and disseminate the infection to its salivary glands, and then transmit to a susceptible host during a subsequent blood meal. Because mosquito body temperature approximates environmental temperature, virus transmission rates are related to ambient temperature (Reisen et al., 1993). Third, WEE and SLE are zoonoses. Therefore, human cases provide a poor indication of virus activity, and extensive avian epizootics may occur without detectable human infection (Reisen et al., 1995). Investigations of the impact of meteorological factors on virus transmission dynamics therefore must focus on components of the primary enzootic cycle.

The threat of a widespread epidemic in California has led to the development of a statewide Encephalitis Virus Surveillance (EVS) program under the direction of the California Department of Health Services to forecast the risk of human infection (Longshore, 1960) and to the establishment of an extensive assemblage of special tax-funded districts to provide intervention through mosquito control. The surveillance program currently monitors information on (1) funding for mosquito and encephalitis virus surveillance and control, (2) weather, (3) mosquito abundance, (4) enzootic virus activity, and (5) human and equine cases. The Center for Vector-borne Disease Research, in collaboration with the California Department of Health Services and the Mosquito and Vector Control Association of California and its corporate members, is conducting research to modernize this system (Eldridge et al., 1999). Developing quantitative forecasting relationships among climate variability, mosquito abundance, and encephalitis risk is one of five objectives of this program and the goal of the preliminary research described in the current chapter.

DATA SOURCES

Weather data are measured by California Irrigation Management Information System and National Oceanographic and Atmospheric Administration stations throughout California, and data are available from their Web site (http://www.dla.water.ca.gov/cgi-bin/cimis/cimis/hq/main.pl). Precipitation, river runoff, and snow water content data are available from the California Department of Water Resources Web page (http://cdec.water.ca.gov). Information on sea surface temperatures and the El Niño–Southern Oscillation (ENSO) anomaly was obtained from the National Oceanographic and Atmospheric Administration (http://www.cpc.ncep.noaa.gov/data/indices/).

Mosquito abundance data were obtained from the Kern and Coachella Valley Mosquito and Vector Control Districts and were collected by two types of traps:

1. New Jersey light traps (Mulhern, 1942) were operated for 2 to 5 consecutive nights per week at 9 to 15 permanent locations from April through October each year. Data from these traps are available for >30 years, but recent information is confounded by increased human populations and associated competing illumination in the form of street and security lighting (Milby and Reeves, 1989).

2. CDC-style traps (Sudia and Chamberlain, 1962) were baited with dry ice and operated without lights on alternate weeks at 25 to 28 permanent locations on the floor of the Coachella and San Joaquin Valleys from April through October each year since 1990. CDC traps collect more mosquitoes per night and therefore are more sensitive to low population abundance levels, rely on a CO_2 chemotaxis as the primary attractant, and are less affected by background illumination (Reisen et al., 1999a).

Enzootic virus activity is monitored by testing pools of ≤50 *Cx. tarsalis* for virus infection and by sequentially bleeding sentinel chickens in 9 to 10 flocks to detect seroconversions (Reisen, 1995a). Chicken sera were submitted for testing to the California Department of Health Services or the University of California at Davis.

ANALYSIS

Temperature and precipitation affect host populations and therefore encephalitis virus transmission in different ways. Temperature alters the rates of biological functions and has been shown to be an important factor in mosquito population modeling attempts (Eisenberg et al., 1995). Warming temperature increases the rate of mosquito immature development, decreases population generation time, and therefore increases the rate of population growth (Reisen, 1995b). In contrast, adult survivorship decreases by about 1% per day for every 1°C increase in temperature (Reeves et al., 1994). The duration of the extrinsic incubation period (time between virus acquisition and transmission) in the mosquito vector decreases as a function of increasing temperature that exceeds the minimal developmental temperature threshold for virus replication, thereby increasing the transmission rate (Reisen et al., 1993). The impact of these temperature effects can be observed in nature by comparing the dynamics of virus transmission between the relatively warm Coachella Valley and cool San Joaquin Valley. The Coachella Valley averages approximately 3 to 5°C warmer than the San Joaquin Valley each month of the year (Reisen et al., 1993). *Culex tarsalis* abundance (measured by number of females collected per CDC trap per night) increases more rapidly during spring in the Coachella Valley, but then declines in midsummer when mean daily temperatures exceed 30°C, producing in a bimodal pattern (Figure 11.1A). In contrast, abundance in the southern San Joaquin Valley (Kern County) increases later in the spring and most typically exhibits a unimodal pattern with a peak during late summer (Figure 11.1A). WEE activity as indicated by seroconversions of sentinel chickens similarly peaks earlier in the summer in Coachella than Kern as shown in Figure 11.1B.

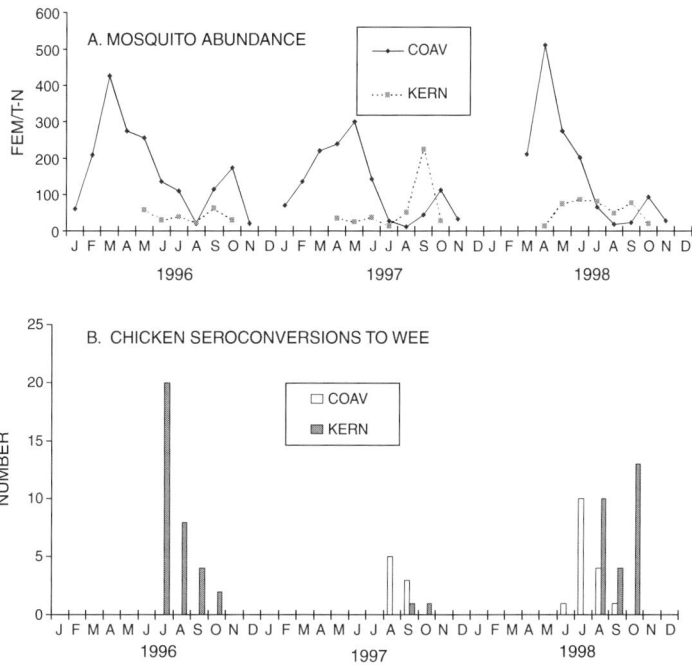

Figure 11.1 A, Mosquito abundance in females per CDC trap-night. B, The number of sentinel chicken seroconversions to WEE in Coachella Valley (COAV) and the southern San Joaquin Valley (KERN) per month, 1996 to 1998.

The effects of water on mosquito abundance were investigated using data from Kern County, because 10 years of CDC trap data were available in electronic format (Wegbreit and Reisen, 1999). Rainfall in California falls primarily during the winter months, whereas mosquitoes are most abundant during summer. Consequently, there was no simple positive correlation between the quantity of winter rain on the valley floor and mosquito abundance the following summer ($R^2 = 0.13$), even when lags of 1 to 4 months were included (Wegbreit and Reisen, 1999). *Culex tarsalis* terminates reproductive diapause by January, immediately blood-feed, and then shortly thereafter must locate a suitable site for oviposition (Bellamy and Reeves, 1963). Therefore, the quantity of winter rain would seem critical in providing surface water habitats for oviposition prior to the initiation of agricultural irrigation or snowmelt. Abundance was well correlated with the impaired runoff of the Kern River, lagged 1 month ($R^2 = 0.67$) (Wegbreit and Reisen, 1999). Impaired runoff was measured at Bakersfield after irrigation water was removed and therefore reflected discharge onto the San Joaquin Valley floor. Presumably, these data also were indicative of the general magnitude of vernal runoff from other smaller drainage systems that frequently inundate considerable areas of the west valley. As expected, vernal runoff was well correlated ($R^2 = 0.88$) with the water content of accumulated snowpack during the previous winter and, therefore, the annual mean abundance of *Cx. tarsalis* during the following summer ($R^2 = 0.70$) (Wegbreit and Reisen, 1999).

Models for long-term weather forecasting based on Pacific Ocean sea surface temperature anomalies recently have been developed, and studies have related these phenomena to climatic variation and the occurrence of mosquito-borne arboviruses outbreaks (Linthicum et al., 1999). Having developed firm associations among winter snowpack, vernal river runoff, and mosquito abundance over a 10-year period, it seemed useful to examine long-term relationships among sea temperature, snowpack, and WEE activity. WEE was extremely active in Kern County during the strong ENSO event of 1982–83 (Reisen, 1984) but then disappeared until 1996–98 (Reisen et al., 1999b), a period also dominated by a strong ENSO event. To examine this relationship further, surface sea temperature anomaly and snowpack in the southern Sierra Nevada were plotted for the past 30 years and years when WEE was detected in the southern San Joaquin Valley indicated (Figure 11.2). No relationship was readily apparent. A simple deviation analysis (Figure 11.3) indicated that changes in sea temperature did not appear to be related directly to snowpack ($R^2 = 0.07$) (Wegbreit and Reisen, 1999).

The long-term prediction of climate variation and its impact on mosquito abundance and encephalitis risk could be of considerable value in planning intervention strategies. However, establishing quantitative relationships among meteorological events, mosquito abundance, enzootic virus activity levels, and the risk of human infection will require additional detailed retrospective analyses of existing weather and surveillance data, a topic we plan to address in future research.

ACKNOWLEDGMENTS

We especially thank the Kern MVCD (Robert Quiring, manager; Richard Takahashi, entomologist) for providing mosquito abundance data. The Kern MVCD surveillance system was established by Dr. R.P. Meyer, the data collected each summer by Boyd R. Hill, and the database maintained by Sherry Sharp. Data from Coachella Valley was collected by Hugh Lothrop, University of California, Davis. This research was funded, in part, by a subcontract of a grant from the Human Health Initiative, NASA, to Jonathan Day, University of Florida. The following MVCDs provided partial financial support: Butte, Coachella Valley, Contra Costa, Greater L.A., Marin–Sonoma, Northwest, Orange, Sacramento–Yolo, San Joaquin, and Sutter–Yuba. The Viral and Rickettsial Disease Laboratory, California Department of Health Services (DHS) and the Arbovirus Research Laboratory of the University of California at Berkeley and now at Davis tested mosquito pools and sentinel chicken specimens. Selected surveillance data summaries were provided by Lucia Hui and Stan Husted of the Vector Control Section, DHS. Bruce Eldridge, University of California, Davis, provided office space, logistical support, and helpful assistance during the protocol and manuscript preparation phases of this project.

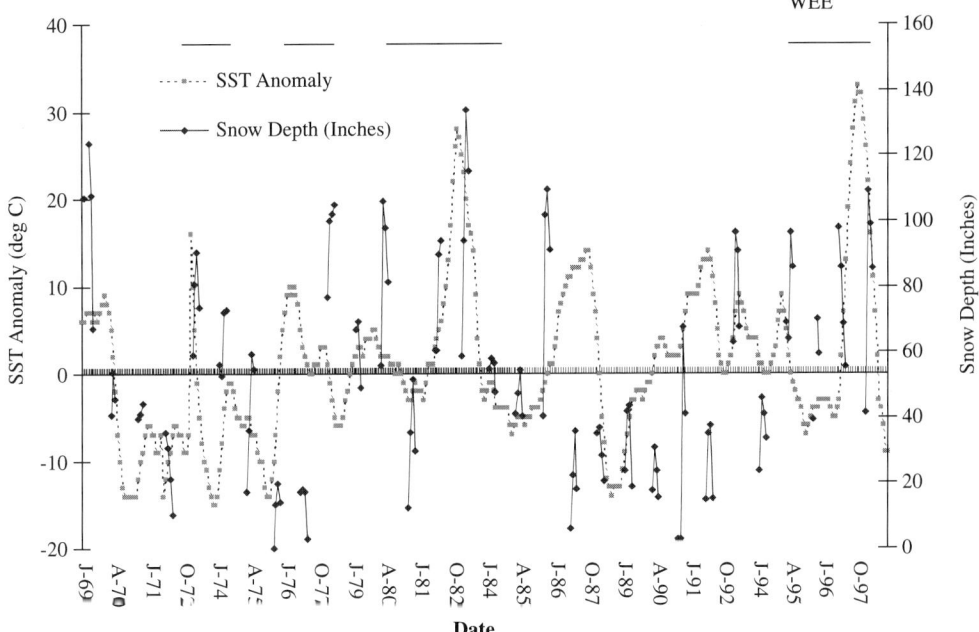

Figure 11.2 Sea surface temperature anomaly (SST in C) and snowpack depth in the southern Sierra Nevada from 1969 to 1998. Bars indicate periods when WEE activity was detected in the southern San Joaquin Valley.

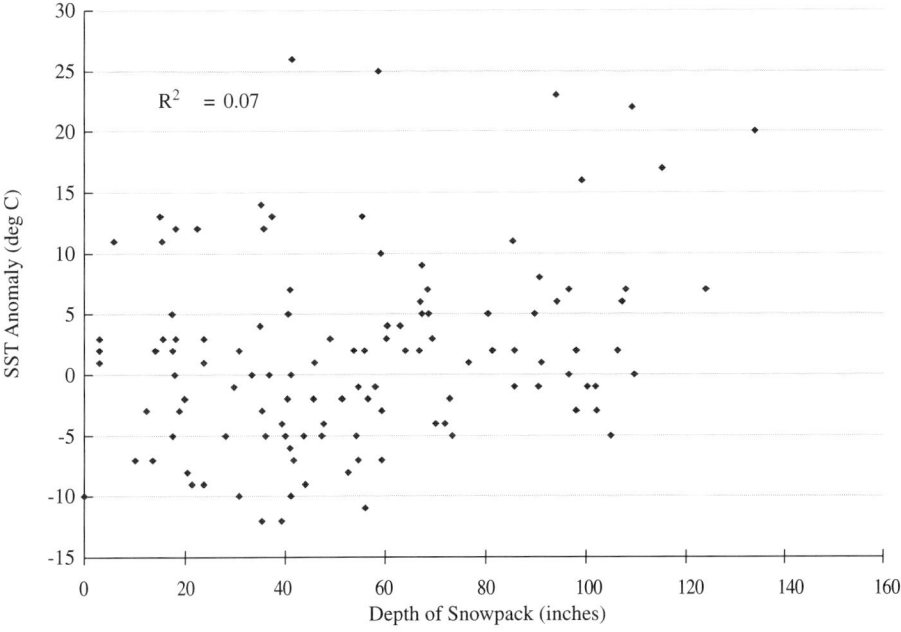

Figure 11.3 Sea surface temperature anomaly (SST in °C) plotted as a function of snowpack depth in the southern Sierra Nevada from 1969 to 1998.

REFERENCES

Bellamy, R.E. and Reeves, W.C., The winter biology of *Culex tarsalis* (Diptera:Culicidae) in Kern County, California, *Ann. Entomol. Soc. Am.*, 56:314–323, 1963.

Eisenberg, J.N., Reisen, W.K., and Spear, R.C., Dynamic model comparing the bionomics of two isolated *Culex tarsalis* (Diptera: Culicidae) populations: model development, *J. Med. Entomol.*, 32:83–97, 1995.

Eldridge, B.F., Reisen, W.K., and Scott, T.W., A model surveillance program for vector-borne diseases in California: 1997–1998, *Proc. Mosq. Vector Control Assoc. Calif.*, 66:48–55, 1999.

Finley, K.H., Longshore, W.A., Palmer, R.J., Cooke, R.E., and Riggs, M.S., Western equine and St. Louis encephalitis, preliminary report of a clinical follow-up study in California, *Neurology*, 5:223–235, 1955.

Hardy, J.L., The ecology of western equine encephalomyelitis virus in the Central Valley of California, 1945–1985, *Am.J. Trop. Med. Hyg.*, 37:18s–32s, 1987.

Linthicum, K.J., Anyamba, A., Tucker, C.J., Kelley, P.W., Myers, M.F., and Peters, C.J., Climate and satellite indicators to forecast Rift Valley fever epidemics in Kenya, *Science*, 285:397–400, 1999.

Longshore, J., California encephalitis surveillance program introduction, *Am. J. Hyg.*, 71:363–367, 1960.

Longshore, J., Stevens, I.M., Hollister Jr., A.C., Gittelsohn, A., and Lennette, E.H., Epidemiologic observations on acute infectious encephalitis in California, with special reference to the 1952 outbreak, *Am. J. Hyg.*, 63:69–86, 1956.

Milby, M.M. and Reeves, W.C., Comparison of New Jersey light traps and CO_2-baited traps in urban and rural areas, *Proc. Calif. Mosq. Vector Control Assoc.*, 57:73–79, 1989.

Mulhern, T.D., The New Jersey mechanical trap for mosquito surveys, *NJ Agric. Exp. Sta. Circ.*, 421:1–8, 1942.

Reeves, W.C., Overwintering of arboviruses, in *Epidemiology and Control of Mosquito-Borne Arboviruses in California, 1943–1987*, Reeves, W.C., Ed., California Mosquito Vector Control Association, Inc., Sacramento, 1990, pp. 357–382.

Reeves, W.C. and Hammon, W.M., Epidemiology of the arthropod-borne viral encephalitides in Kern County, California, 1943–1952, *Univ. Calif. Berkeley Publ. Health*, 4:1–257, 1962.

Reeves, W.C. and Milby, M.M., Changes in transmission patterns of mosquito-borne viruses in the U.S.A., in Service, M.W., Ed., *Demography and Vector-Borne Diseases*, CRC Press, Boca Raton, 1989, pp. 121–141.

Reeves, W.C., Hardy, J.L., Reisen, W.K., and Milby, M.M., Potential effect of global warming on mosquito-borne arboviruses, *J. Med. Entomol.*, 31:323–332, 1994.

Reisen, W.K., Observations on arbovirus ecology in Kern County, California, *Bull. Soc. Vector Ecol.*, 9:6–16, 1984.

Reisen, W.K., Guidelines for surveillance and control of arbovirus encephalitis in California, in Reisen, W.K., Kramer, V.L., and Mian, L., Eds., *Interagency Guidelines for the Surveillance and Control of Selected Vector-Borne Pathogens in California*, Mosquito and Vector Control Association of California, Sacramento, 1995a, pp. 77–90.

Reisen, W.K., Effect of temperature on *Culex tarsalis* (Diptera: Culicidae) from the Coachella and San Joaquin Valleys of California, *J. Med. Entomol.*, 32:636–645, 1995b.

Reisen, W.K. and Chiles, R.E., Prevalence of antibodies to western equine encephalomyelitis and St. Louis encephalitis viruses in residences of California exposed to sporadic and consistent enzootic transmission, *Am. J. Trop. Med. Hyg.*, 57:526–529, 1997.

Reisen, W.K. and Monath, T.P., Western equine encephalomyelitis, in Monath, T.P., Ed., *The Arboviruses: Epidemiology and Ecology*, CRC Press, Boca Raton, FL, 1989, pp. 89–138.

Reisen, W.K. Meyer, R.P., Presser, S.B., and Hardy, J.L., Effect of temperature on the transmission of western equine encephalomyelitis and St. Louis encephalitis viruses by *Culex tarsalis* (Diptera:Culicidae), *J. Med. Entomol.*, 30:151–160, 1993.

Reisen, W.K., Boyce, K., Yoshimura, G., Lemenager, D., and Emmons, R.W., Enzootic transmission of western equine encephalomyelitis virus in the Sacramento Valley of California during 1993 and 1994, *J. Vector Ecol.*, 20:153–163, 1995.

Reisen, W.K., Boyce, K., Cummings, R.F., Delgado, O., Gutierrez, A., Meyer, R.P., and Scott, T.W., Comparative effectiveness of three adult mosquito sampling methods in habitats representative of four different biomes of California, *J. Am. Mosq. Control Assoc.*, 15:24–31, 1999a.

Reisen, W.K., Lundstrom, J.O., Scott, T.W., Eldridge, B.F., Chiles, R.E., Cusack, R., Martinez, V.M., Lothrop, H.D., Gutierrez, D., Hartman, C.A., Wright, S., Boyce, K., and Hill, B.R., Patterns of avian seroprevalence to western equine encephalomyelitis and St. Louis encephalitis viruses in three biomes of California, *J. Med. Entomol.,* 37:507–527, 1999b.

Sudia, W.D. and Chamberlain, R.W., Battery-operated light trap, an improved model, *Mosq. News,* 22:126–9, 1962.

Tsai, T.F. and Mitchell, C.J., St. Louis encephalitis, in Monath, T.P., Ed., *The Arboviruses: Epidemiology and Ecology,* CRC Press, Boca Raton, 1989, pp. 431–458.

Wegbreit, J. and Reisen, W.K., Relationships among weather, mosquito abundance and encephalitis virus activity in California: Kern County 1990–1998, *J. Am. Mosq. Control Assoc.,* 16:22–27, 1999.

CHAPTER 12

Preparing for the Quantum Leap to Sustainability: A Toolkit for Future-Friendly Cities

Mathis Wackernagel and Diana Deumling

WHY A MUNICIPAL TOOLKIT FOR PLANNING A SUSTAINABLE FUTURE?

Why Municipal?

Although the challenge of sustainability has implications from the global to the individual level (WCED, 1987), it is the local level, where all its components visibly interact, that offers a key leverage point for confronting the challenge. Towns and cities that set a sustainability agenda are the pioneers of the sustainable future. We call such communities future-friendly, and they hold the potential to find ways out of the sustainability dilemma.

But local stakeholders have a more powerful motivation than the distant prospect of solving the long-term problems of humanity. More important from their point of view is the fact that future-friendly places also reap many immediate benefits. Because they are ahead of the game, future-friendly communities are more competitive, more livable, and more secure. They are more competitive because they can use resources more efficiently and run lower ecological deficits (Sturm et al., 1999; von Weizsäcker et al., 1997; Hawken et al. 1999). They are more livable, because sustainable communities offer more attractive features (Fodor, 1999; Engwicht, 1993; Roseland, 1998) and they are more secure because they are economically prepared for a more resource-scarce world and tend to build stronger social communities.

Why Indicators?

Without proper feedback, no organism, whether a bacteria, an organization, or a society, can secure its health and vitality. Sensors tell the organism where danger looms and rewards wait. If these sensors do not operate adequately, the organism can get hurt because it does not find out in time that it needs to react. Therefore, health depends on a system of well-functioning sensors that respond quickly and that measure all signals relevant to the organism.

Indicators are society's sensors (Tyler Norris Associates et al., 1997). When they work properly, they furnish us with a feedback system. Once people get appropriate feedback they become well equipped to deal with complex challenges (Hanhart, 1989). Appropriate feedback tells them (1) what they are doing *now*, (2) what they *should do*, and (3) how they can improve the results. These are the conditions for self-improvement and learning; they also lie at the root of the quality control principles widely accepted in Japanese industry (Juran, 1974). Therefore, the first step toward sustainability is building better sensors and feedback systems for society. With a more accurate picture of reality, people can more meaningfully perceive the costs and stresses of unsustainable living and act accordingly. Such sensors are most effective if they are tied to immediate concerns of decision makers in government, business, and households. The idea of the toolkit for future-friendly cities and regions is to develop sensors that link concerns to action.

Why Is the Toolkit the Best Next Step?

A first generation of initiatives on community indicators has focused on building such feedback systems and rapidly spread throughout the country. It has been able to mobilize many communities and get people actively involved in thinking about what matters to them.

What we have learned, however, is that the conventional indicator approach taken by these initiatives has not been sufficiently effective in moving beyond gathering information. Too much emphasis has been placed on the details of finding indicators for every single issue facing the community. Consequently, many participants get overwhelmed with the long list and the process gets bogged down. Common sense erodes, the thrust weakens, and the purpose becomes diffuse. Participants get disenchanted because the effort results in little action (Cobb and Rixford, 1998).

Building on these experiences, we plan to move to the next generation of sustainability indicators. Here, we propose to develop and test an indicators toolkit for future-friendly cities that should be more effective in generating adequate action because it:

- Is based on a clear purpose (here the purpose is building sustainability)
- Uses a simple yet comprehensive framework explaining the challenge and trade-offs (the framework is avoiding the dominant conflict of securing quality of life vs. maintaining the Earth's biocapacity)
- Is imaginative and realistic about possible intervention points (the focus is on generating the will rather than "planning ways")
- Envisions next steps beyond the project before starting it (the project is seen as a stage for a more informed and focused sustainability discussion that will be able to attract meaningful sustainability projects at the local level)

Because it is based on two simple measures, the municipal toolkit approach is simpler, more conceptual, and more engaging than conventional indicators initiatives. It provides tools and processes ideas designed to allow communities to measure how sustainable they are and how fast they are moving toward their goals. Indicators work most effectively when they are as clear and simple as a red traffic light and as tangibly linked with life concerns.

PROJECT APPROACH

We build our project on a simple and straightforward definition of the sustainability dilemma: *Everyone needs to be able to secure a satisfying life, within the means of nature.* Obviously, this shorthand refers to a much fuller agenda: healthy communities, equity (intragenerational, intergenerational, and interspecies), social cohesion, economic vitality, long-term competitiveness, and ecological integrity, to name a few larger items. Although this formulation is short and snappy, it synthesizes the essence of the sustainability agenda and makes "sustainability" accessible.

After all, the competition between these two goals — staying within the capacity of our ultimate means (natural capital) and reaching our ultimate ends (satisfying lives for everybody) — is the core of humanity's current self-destructive path. It is this conflict that becomes the

---

BOX 12.1
SUSTAINABILITY: LIVING WELL WITHIN THE MEANS OF NATURE

Sustainability requires promoting everyone's quality of life within the means of nature. Living beyond our ecological means will lead to the destruction of humanity's only home. Having insufficient natural resources and living in unsatisfactory and inequitable ways will cause conflict and degrade our social fabric.

Sustainability means settling the conflict between two competing goals: the sustenance of human life and the integrity of nature. Indeed, this is the conflict between ultimate ends (a good life for everybody) and ultimate means (the capacity of the biosphere) we need to resolve.

We need to know how much quality we are getting — or losing — in our communities. We need to know the degree to which life quality can be nonmaterial and thus less resource-intensive, and whose quality of life is already eroding due to degradation of natural and social systems. We will explore this question by asking people how satisfied they are with their lives and their communities.

At the same time, we need to start monitoring whether we are living within our ecological means or at what rate humanity is depleting the biosphere. We must ask: "How much nature does humanity, our country, or our household use to sustain itself?" After all, people are part of nature and depend on its steady supply of the basic requirements for life: energy for heat and mobility, wood for housing, furniture, and paper products, fibers for clothes, quality food and water for healthy living, ecological sinks for waste absorption, and many life-support services for maintaining good living conditions on our planet. This use of nature is measured by the ecological footprint and compared to the biocapacity that exists on this planet.

---

hurdle to sustainability. Also, didactically, it is useful to capture sustainability in this format: two conflicting poles make it easy to understand and watch the development. It is like a soccer game. Two teams struggling for the ball is a simple and clear process to follow. However, once three, four, or even more teams play against each other simultaneously, the game becomes unintelligible and confusing. Not only will people lose interest, but they will be confused about how to intervene.

To frame this dilemma and to explain how it becomes the driving force behind today's self-destructive behavior, the toolkit will have two direct outcome measures capturing the ultimate means and the ultimate ends: (1) a *satisfaction barometer* that documents people's levels of satisfaction with their society, community, neighborhood, and personal life, as well as their attitude toward sustainability measures that helps identify what kind of sustainability initiatives people feel most excited about; and (2) the *ecological footprint* that compares people's consumption with the local and global ecological capacity. Each of these measures can then be fleshed out and enriched by subsidiary projects and applications that focus on more specific aspects of or interesting applications for future-friendly town and cities.

BACKGROUND FOR THE SATISFACTION BAROMETER

Why Satisfaction?

A sustainable society cannot be built on martyrdom and suffering. People want fulfilling lives. To make sustainability come true, we must find ways by which people can thrive in all senses, without senselessly overtaxing the ecosystems that support us. In other words, everybody should be able to realize a satisfying life in a way that limits humanity's ecological impact on the regenerative capacity of the biosphere.

BOX 12.2

SUMMARY OF THE ACCOUNTING TOOLS THAT REPRESENT THE TWO RIVALING POLES OF SUSTAINABILITY

"Does Everybody Secure A Satisfying Life

The Satisfaction Barometer

The barometer is a psychological (or subjective) measure. It assesses people's satisfaction with their society, community, and personal lives. In contrast, physical (or objective) measures of well-being oppose sustainability as it is by definition impossible to improve the material situation of a growing population without also increasing the resource throughput of society. Also, psychological outcomes are more meaningful to people. For example, subjective feelings such as depression, love, fear, hope, or loneliness more fundamentally determine people's satisfaction with life than the number of television sets in their homes or even the number of doctors in their municipality.

Selling benefits, not features. A sustainable society cannot be built on martyrdom and suffering. People want fulfilling lives. To make sustainability come true, we must find ways that people can thrive in all senses without senselessly overtaxing the ecosystems that support us. In other words, everybody should be able to realize a satisfying life, and do that within the regenerative capacity of the biosphere.

The dimensions of satisfaction. We believe that nurture, pleasure, empowerment, acceptance, "flow," control, excitement, and the sense of belonging and contributing are drivers of people's satisfaction with life. It is control over one's time and relationships rather than financial wealth that become critical. In fact, satisfaction may not be tied as closely to resource consumption as we seem to believe. Therefore, the barometer will ask such questions as: Do you have time to focus on what is important to you? Do you live in a community that you love and that loves you? Do you feel you can make a contribution to the world? Are you in control of your life, or is your life controlling you? How many times a week do you have to get up using an alarm clock? How important are these issues to you?

... Within the Means of Nature?

The Ecological Footprint

The footprint measures human impact on nature. In order to live, people consume what nature offers. Every one of us has an impact on our planet. This is not bad as long as we don't take more from the Earth than it has to offer. But are we taking more than we should? The ecological footprint measures what we consume from nature. It shows how much productive land and water we occupy to produce all the resources we consume and to absorb our waste.

How big is your footprint? The average American uses 24 acres to support his or her current lifestyle. This corresponds to the size of 24 football fields put together. In comparison, the average Canadian lives on a third of that footprint, and the average Italian on half.

How much can nature provide? Nature provides an average of about 5 acres of bioproductive space for every person in the world. With a global population of 9 billion in the year 2050, the available space will be reduced to over 3 acres, and that is without setting aside area for the millions of other species. Our research shows that humanity's footprint is over one quarter larger than what the planet can regenerate.

What can we do? We can become part of the sustainability movement and make it possible that everyone secure a satisfying life within the means of nature. We can better use resources, for example, by using energy-efficient lamps or by composting. We can consume less by having a smaller population and decreasing our consumption. Buying fewer cars and other disposable products will save us money and grief and give us more spare time.

Are People Satisfied with Their Lives? Measures of Well-Being

To get there, we need measurement tools that can assess whether society is moving in the desired direction of preserving both quality of life and the natural system that sustains it. While the ecological footprint tells us about the natural side of the equation, we need a complementary indicator to tell us about the side that addresses the quality of life, especially one that can identify where needs are.

Various measures have been developed to allow comparisons among the situations of people and societies. Although Gross Domestic Product (GDP), the most widespread and influential economic indicator, has been abandoned as an indicator for human welfare, new ones have emerged, including the Index of Sustainable Economic Welfare, developed by Clifford Cobb, John Cobb and Herman Daly and now further developed by Redefining Progress (RP) as the "Genuine Progress Indicator" (Daly and Cobb, 1989). As these modified national account approaches still focus primarily on economic performance rather than on a wider set of parameters determining social well-being, other measures have been developed, such as the Human Development Index, which is compiled annually by the United Nations Development Program (UNDP).

Subjective Satisfaction Measures

But these material or *objective* measures of well-being undermine the goal of sustainability, since by definition it is impossible to improve the material situation of a growing population without also increasing the resource throughput of society. Therefore, a more constructive way to address this dilemma is to emphasize the *subjective* side of well-being. Subjective well-being is as real as objective well-being, although its manifestation is more complex and, therefore, not directly measurable. In the end, however, it is more meaningful to people. For example, subjective feelings such as depression, love, fear, hope, or loneliness more fundamentally determine people's satisfaction with life than the number of television sets in their homes or even number of doctors in their municipality. Focusing on subjective well-being does not deny the material dimension of human life. Obviously, subjective well-being depends on some material inputs. The point, however, is that material throughput does not determine people's satisfaction with their lives.

The RP Satisfaction Barometer

The RP satisfaction barometer is a survey tool based on existing measures of well-being. It is designed to assess people's satisfaction with their society, their community, and their personal lives in a subjective (nonmaterial-based) measure, doing so eventually at various scales. At the same time, the measure invites people to think about the bountiful possibilities of improving quality of life and to seek ways to develop more fulfilling lives while actually reducing their footprint. It focuses on issues that are central to fulfilled lives, flourishing communities, and a healthy biosphere. In this way it can address the need for positive and engaging outlets for people's creativity, warm and gratifying relationships, and opportunities to connect with the natural world that supports our lives.

Do you have time to focus on what is important to you? Do you live in a community that you love and that loves you? Do you feel you can make a contribution to the world? Are you in control of your life or is your life controlling you? How many times a week do you have to get up using an alarm clock? How important are these issues to you? These are among the driving questions behind our search for exploring people's satisfaction with life.

Apart from analyzing people's satisfaction, we also need to find out what kind of sustainability initiatives people feel most excited about. Without understanding people's attitudes and resistance toward sustainability, and without a clear idea of their perceptions of limitations and possibilities, the concept of sustainability remains abstract. Worse, the emergence of possibly synergistic and widely supported sustainability strategies remains unlikely. To prepare the ground for such strategies, the satisfaction barometer will be refined through discussions with community leaders, extensive focus groups, surveys, and other complementary tools for identifying the sustainability strategies most likely to succeed.

BACKGROUND FOR THE ECOLOGICAL FOOTPRINT

The Ecological Context

Humanity has become the dominant species on the planet — the largest resource consumer and waste producer. Human activities have grown to such an extent that they have begun to exceed the global carrying capacity, resulting in a loss of ecological productivity. In other words, humanity has come to a point where further expansion impoverishes it. What is called *development* has more often become a liquidation of assets, mainly natural capital.

To be sustainable, humanity must not diminish nature's capacity to generate and regenerate its functions and services. Ecological economists call this *maintaining natural capital*. This capital is humanity's ultimate source of wealth without which human life and activities become impossible. Planning for sustainability, therefore, depends on measuring this ecological bottom-line condition. People need to know how much of nature a given population uses compared to how much there is. If a population exceeds ecological limits, it needs to be able to monitor whether it is reducing this deficit. If it cannot gauge this, it will continue to grope unproductively in the dark.

What an Ecological Footprint Analysis Offers

Ecological footprints are accounting tools for natural capital.* Such tools can help people assess their ecological impact and compare this impact to nature's capacity to regenerate. In other words, footprints contrast people's impact with nature's carrying capacity. Such analyses provide a benchmark for present ecological performance compared with potential available natural capital; identify challenges for lightening an economy's ecological load, and document gains as a country, region, or city moves toward sustainability.

Footprint-based assessments of regions start with an inventory of the ecologically productive spaces within that area. Most of these land-use data already exist and, in the case of more precise analyses, can be complemented by local surveys and remote-sensing data from satellites. Such an inventory would provide an estimate of the local ecological capacity — the supply side.

The demand side — that is, the footprint of the local population — can be documented with various degrees of precision. Given an estimated per capita demand for ecosystem services and the productivity of the ecologically productive space categories in a region, an estimate of the human carrying capacity deficit (or remainder) of the region can be obtained.

Applying the Footprint Measure at the Municipal Level

Applying the footprint measure at the municipal level expands the approach described for Santiago de Chile, completed for the International Council on Local Environmental Initiatives (ICLEI) (Wackernagel, 1998). Our ecological footprint analyses of nations and regions are continually improving and becoming more sophisticated (WWF et al., 2002; Wackernagel et al., 1999a,b,c,; van Vuuren et al., 1999). National footprints serve as reference data sets for assessing

---

* In the early 1990s, William Rees and Mathis Wackernagel developed the original Ecological Footprint concept. Their book *Our Ecological Footprint: Reducing Human Impact on the Earth* was published in 1996 by New Society Publishers. Ecological Footprint studies have become popular on all continents. Updates and links to footprint initiatives can be found at http://www.redefiningprogress.org.

subnational footprints. The footprint of a region is extracted from the national footprint by comparing the extent to which consumption patterns in the region differ from the national average and adjusting the national footprint accordingly. This approach to calculating local footprints generates results that are directly comparable to national and global averages. For instance, municipal footprints can be compared to global and national footprint and biocapacity averages as well as to local biocapacity. The latter can be assessed using maps or GIS data and local yield figures. RP has conducted local ecological footprint assessments for Duluth, MN; Marin County, CA; Ojai Valley, CA; Sonoma County, CA; Paris, France.

JOINING THE TOOLS TO BUILD THE TOOL BOX

We have not yet tested this package approach of joining an accounting tool for the ultimate means (natural capital) with one for the ultimate ends (quality of life) in a real community situation. We are discussing the package with potential partners who may join us in testing this concept.

We hope this approach could kick-start the next generation of sustainability indicators. Significant benefits could result from joining the satisfaction measure to the ecological footprint. A well-designed scale of social and individual satisfaction, combined with an effective tool for quantifying (and communicating) human impact, will allow people to document sustainability constraints and current conflicts between social and ecological sustainability imperatives; it will also help us discover ways to resolve these conflicts. Further, this approach will help advance the scientific understanding of the relation between ecological impacts and human well-being. It will identify opportunities for improving people's quality of life on lower ecological footprints, thus providing a tangible and useful base to further the public debate on sustainability and encouraging appropriate action.

This approach moves away from first focusing on indicators without having a framework to embed them. Past approaches have led to much discussion on a few details while obscuring the overarching questions. We hope that by putting the framework first and keeping the focus on the current contradiction between expanding human demands and limited ecological capacity, we will be effective in encouraging more systemic approaches at the community level. After all, sustainability is about reconciling difficult trade-offs.

This does not mean that traditional indicators are not useful. The question is more when and how to use them. The satisfaction barometer and the ecological footprint are not advocated to be the only indicators. We view them as an accounting framework. They summarize the points of tension. Indeed, the many indicators that are used at the community level are the elements that make up these two measures representing the poles of the sustainability tension.

NEXT STEPS FOR THE PROJECT

The first phase of the project began with outreach to communities and planners interested in measuring where they are in terms of sustainability and finding out what they can do about it. With their help, we propose to pilot the toolkit in municipal applications.

After testing and evaluating the indicators, we will develop the final toolkit, a handbook that should enable others to assess a U.S. municipality's present footprint and quality of life and generate scenarios for future paths of development. The simply written and well-illustrated handbook will include:

- Background on the ecological footprint and satisfaction barometer
- Instructions for calculating and interpreting the ecological footprint and for giving and interpreting satisfaction surveys
- Diskettes with calculation forms and reference data to help people perform their own community calculations

- Suggestions and practical strategies for engaging a community in a constructive debate on quality of life
- Suggestions and practical strategies for moving from framing issues to developing action and maintaining commitment to the sustainability agenda

To make the manual user-friendly, we will strongly emphasize evaluation and testing of the material. Therefore, the research will involve not only hands-on workshops with nongovernmental organizations (NGOs) and local government officials but also paid outside evaluation by experienced individuals in academia and the community.

EXPECTED RESULTS OF THE PROJECT

The municipal toolkit is intended to help communities frame sustainability in a more constructive way. It would provide local governments and NGOs with a standard for the tangible realization of their sustainability goals. It could become a platform for municipal planning of housing, energy infrastructure, transportation, education, etc.

The goal is to help communities build consensus for focusing on sustainability and to provide a meaningful context for decision making. More specifically, this kit should enable communities to:

- Define their sustainability goals in more specific and measurable terms
- Assess their ecological impact and compare it to nature's capacity to regenerate
- Evaluate the community's quality of life and identify its weak spots
- Identify areas for reducing an economy's ecological load while improving its performance in providing quality of life
- Assess current progress toward realizing goals
- Engage in constructive debate
- Create strategies for implementing goals
- Develop activities that maintain the community's interest in building a sustainable future
- Motivate policies and changes (such as local incentive programs) in the municipality that bias private and public decision making toward sustainable choices
- Communicate best practices to other communities, once the handbook has been developed in test communities

The concept we offer is new. The handful of communities that have explored parts of it or similar models worldwide show promising results. The Green Smiles Project in Holland is one example of an initiative inspired by ecological footprint work (Redczus, 1999). The Global Living Project in Canada invited researchers in a 6-week program to reduce their footprints while maintaining a high quality of life. For 2 years in a row, they achieved a footprint as low as 0.9 hectares per person, less than 1/10 of the average U.S. footprint (Merkel, 1999). Many local Agenda 21 initiatives have used the ecological footprint as a starting metaphor.

In the short term, we may have to judge the project's success with circumstantial evidence. Increasing demand for the handbook would be a good sign. A better sign would be community groups using the kit as a platform for developing more creative proposals for moving toward sustainability.

Over time, the toolkit for future-friendly cities and regions will have the benefit of being replicated easily in many places and helping to build the local stages for developing sustainability. We will know we are succeeding when a growing network of communities is fruitfully applying the toolkit to increase residents' satisfaction while decreasing footprints. Along the way, we hope that it can help current indicator initiatives to become more effective and inspire a new generation of local policies whose goals are reducing footprints while improving the quality of life of all their residents.

REFERENCES

Cobb, C. and Rixford, C., Lessons learned from the history of social indicators, Report, Redefining Progress, San Francisco, 1998.

Daly, H.E. and Cobb, J., *For the Common Good*, Beacon Press, Boston, 1989.

Engwicht, D., *Reclaiming Our Cities and Towns: Better Living with Less Traffic*, New Society Publishers, Gabriola Island, British Columbia, Canada, 1993.

Fodor, E., *Better Not Bigger: How to Take Control of Urban Growth and Improve Your Community*, New Society Publishers, Gabriola Island, British Columbia, Canada, 1999.

Hanhart, J., Ecofeedback: Feedback as a Tool in Restoring Environmental and Humanitarian Equilibrium, Internal report, Rosmalen, the Netherlands, 1989.

Hawken, P., Lovins, A., and Lovins, H., *Natural Capitalism*, Little Brown, New York, 1999.

Juran, J.M., Gryna, F.M., Jr., and Bingham, R.S., Jr., Eds., *Quality Control Handbook*, 3rd ed., McGraw Hill, New York, 1974.

Merkel, J., Global Living Project, Winlaw, British Columbia, Canada, http://www.globallivingproject.org/, 1999.

Redczus, H., Milieutrouw, Leeuwarden, N.L., http://www.milieutrouw.nl, private communication, 1999.

Roseland, M., *Toward Sustainable Communities: Resources for Citizens and Their Governments*, New Society Publishers, Gabriola Island, British Columbia, Canada, 1998.

Sturm, A., Wackernagel, M., and Müller, K., *The Winners and Losers in Global Competition*, Verlag Rüegger, Chur/Zurich, 1999.

Tyler Norris Associates, *The Community Indicators Handbook*, Redefining Progress, San Francisco, 1997.

United Nations Development Program (UNDP), *Human Development Report*, Oxford University Press, Oxford, 2000.

van Vuuren, D., Smeets, E.M.W., and de Kruijf, H.A.M., The Ecological Footprint of Benin, Bhutan, Costa Rica and The Netherlands, RIVM Report 87005 004, National Institute of Public Health and the Environment (RIVM), Bilthoven, the Netherlands, 1999.

von Weizsäcker, E.U., Lovins, A., and Lovins, H., *Factor Four: Doubling Wealth — Halving Resource Use*, Earthscan, London, 1997.

Wackernagel, M., The ecological footprint of Santiago de Chile, *Local Environ.*, 3(1):7–25, 1998.

Wackernagel, M. and Rees, W.E., *Our Ecological Footprint: Reducing Human Impact on the Earth*, New Society Publishers, Gabriola Island, British Columbia Canada, 1996.

Wackernagel, M., Onisto, L., Bello, P., Callejas Linares, A., López Falfán, I.S., Méndez García, J., Suárez Guerrero, A.I., and Guadalupe Suárez Guerrero, Ma., National natural capital accounting with the Ecological Footprint concept, *Ecol. Econ.*, 29:375–390, 1999a.

Wackernagel, M., Lewan, L., and Hansson, C., Evaluating the use of natural capital with the ecological footprint: applications in Sweden and subregions, *Ambio*, 28:604–612, 1999b.

WCED, World Commission on Environment and Development, *Our Common Future*, Oxford University Press, New York, 1987.

World Wide Fund for Nature International (WWF), UNEP World Conservation Monitoring Centre, Redefining Progress, and Center for Sustainability Studies, *Living Planet Report 2000*, WWF, Gland, Switzerland, 2000.

Building Policies and Linkages

CHAPTER 13

Overview: Building Policies and Linkages

Barry W. Wilson

The cliché "all politics are local" expresses a general characteristic of human decision making, and the generalization "all ecosystems are local" defines a natural property of living communities, from molecules to ecosystems, from mountains to seas. The onrush of industrialization, the plundering of nonrenewable resources, and the pressure of increasing human population have set the stage for negative changes in species survival and the integrity of the soil. Such rapid changes have not occurred perhaps since the Permian era. In the past, shifts in the planet's population were due to natural forces. In contrast, those facing us today are likely due to our own activities. Recognition of the problem is the first step in finding a solution. "We have met the enemy," said comic strip character Pogo, "and he is us!" However, this is not enough. Avoiding disastrous global warming, stopping habitat destruction and overfishing, ending smog and acid rain, and correcting all the insults to which the environment is heir sometimes appear beyond our abilities. To whom shall we turn? Big Business? Big Government? Big Green?

There are no widely accepted rules regarding the formulation and application of environmental conservation and remedial programs. Some public policies are based on tradition, e.g., "We always plow this way," and are unlikely to change. Change does occur as a result of a dedicated few who make heroic efforts. Some of these dedicated people influence policy makers directly, while others appeal to the masses by pointing out what we are losing. "Where have all the birds gone," said Olga Owens Huckins to Rachel Carson (1962), "Spring is without bird song and I cannot bear it." "Stick with it! And stick it to them!" is a motto of environmental groups that have taken a leaf from the stylebook of activist community efforts exemplified by Saul Alinsky's Chicago projects in the 1940s and 1950s (Alinsky, 1946). Many despair that even these extreme actions will not make a dent in the planet's wide destruction and pollution. But others, in spite of the many voices that preach environmental disaster, cling to the notion that there has been, and will continue to be, progress in reversing the destructive trends. Over the centuries, step by step, we learn about the web of life, its controls and limits, and what to do about it. But the level of inertia is high. Once, discussing the impact of humans on wildlife (Wilson, 1980), I said: "The issues lie between the living, the dead and the unborn. Each generation feels the Earth is theirs; each knows that they are the most modern, progressive and important of people. They see themselves as heirs to the ages, entitled to the riches of the Earth. Then they die, passing what is left to their children. The ethic of conservation is one of common sense; it says we are caretakers, holding the land in trust for future generations." To hold the land as sacred requires a change not only of the mind but of the heart.

Words come easy, actions hard. As long as people look upon land as real estate and coastal waters and river dams as the makings of waterfront property, they are not apt to change their minds or the

1-56670-612-2/03/$0.00+$1.50

habits of centuries. There is still much to learn about the determinants of the web and the tapestry of life and how to protect it. We seem to commit the same mistakes over and over again. Ecosystem truisms such as "everything must go someplace" and "everything is connected to everything else" (Wilson, 1998) too often take second place to the "law of unintended consequences." When it comes to ecosystems, it is often not the future, but past mistakes that lie before us. Decision makers in the public and private sectors too often accept the most expedient or profitable process, tout convenient but unconfirmed research findings, or substitute benchmark data for hard-won field studies and modeling for reality. It may take more than facts to convince scientists and permit the lay public to understand the reality. In addition to an appreciation of the processes that underlie life on Earth, we need to understand the processes by which human beings decide what to do about themselves and their environment. The former is the province of ecological science, the latter the realm of politics and social psychology. There is a complex interplay between individual desires and government programs, a complex dance between the wishes of the individual and society.

Three chapters in this section touch upon several issues inherent in understanding the origins of environmental policies. Chapter 16, Attitudes and Their Influence on Nature Valuation and Management in Relation to Sustainable Development, by Hans A.M. de Kruijf et al., discusses some of the factors that determine the way people view policy making and management of nature. Chapter 17, Humane Values as a Basis for Ecosystem Health, by John Grandy and Allen Rutberg, focuses on one attitude, concern for the welfare of animals, and its increasingly important role in environmental decision making. Chapter 18, The Role of the Water Education Foundation in Creating Factual Awareness and Facilitating Consensus in Western Water Issues, by Rita Schmidt Sudman, is an example of the growth and impact of one foundation dedicated to water issues. Its mission is based on the insight that the apportioning of water more than the mining of gold is a determining factor in the history of the West.

Have we already turned an important corner? Only time will tell. Speaking at a ceremony dedicating a monument to the last passenger pigeon, ecologist, wildlife manager, poet, and prophet Aldo Leopold (1949) said that for one species to mourn the extinction of another is "a new thing under the sun." Leopold's formulation of a "Land Ethic" highlights the feelings of many of us: "When we see land as a community to which we belong, we may begin to use it with love and respect. That land is a community is the basic concept of ecology, but that land is to be loved and respected is an extension of ethics." Looking back at the previous century, we see that more people are concerned about and learning about ecosystems and the health of the planet than ever before. Each in their own way, elected representatives work from the top down, forming government agencies like the Environmental Protection Agency (EPA) and writing the Endangered Species Act; individuals work from the bottom up, forming political lobbies like Friends of the Earth and the Sierra Club. These lobbyists work to change attitudes, bringing about a metamorphosis where a swamp is not only a piece of real estate but also a magical home for many creatures. Is time on our side? "Had we but world enough, and time" mourned Elizabethan poet Andrew Marvell (Marvell, 1997) as he tried to convince a coy lady to surrender. What persuasions will we use to convince the contemporary lumberman to put down his axe and the developer his steam shovel? Our failures litter the centuries; let us hope that our successes will provide a better world for our children.

REFERENCES

Alinsky, S.D., *Reveille for Radicals*, University of Chicago Press, Chicago, 1946.
Carson, R., *Silent Spring*, Houghton Mifflin, Cambridge, MA, 1962.
Leopold, A.A., *A Sand County Almanac and Sketches from Here and There*, Oxford University Press, New York, 1949.
Marvell, A., *To His Coy Mistress and Other Poems*, Dover Publications, New York, 1997.
Wilson, B.W., *Birds: Readings from Scientific American*, W.H. Freeman, San Francisco, 1980.
Wilson, B.W., Ecosystem health: some perspectives, in Cech, J.J. Jr., Wilson, B.W., and Crosby, D.G. Eds., *Multiple Stresses in Ecosystems*, Lewis Publishers, Boca Raton, FL, 1998.

CHAPTER 14

The CALFED Bay–Delta Ecosystem Restoration Program: Complexity and Compromise*

Rick Soehren

GEOGRAPHY, HYDROLOGY, AND HISTORY

The great Central Valley of California, ringed by mountains, has only one outlet: the San Francisco Bay/Sacramento–San Joaquin Delta estuary (Bay–Delta). The two largest rivers flowing to the Bay–Delta are the Sacramento and the San Joaquin. Smaller streams include the Mokelumne and Cosumnes Rivers. Together these rivers and smaller tributaries in the system drain an area of 158,000 km^2. Runoff is highly variable and ranges from less than 10 km^3 to over 60 km^3 annually, and has averaged 29.6 km^3 per year (Monroe and Kelly, 1992). The estuary itself is a maze of braided channels, sloughs, and diked islands covering just less than 300,000 ha in five California counties (California Department of Water Resources, 1993). It is the largest estuary on the west coast of North America.

The Bay–Delta is the hub of California's two largest water distribution systems, the Central Valley Project (CVP), operated by the U.S. Bureau of Reclamation, and the State of California's State Water Project (SWP). In addition, at least 7000 other water diverters, large and small, take water from the watershed feeding the estuary. Together, these water developments divert 20 to 70% of the natural flow in the system depending on the amount of runoff available in a given year. Water diverted from the system is used to irrigate about 2.8 million ha of highly productive farmland and provides a source of drinking water for two thirds of California's growing population (CALFED Bay–Delta Program, 2000).

The estuary has a long history of disturbance. During California's gold rush, a technique of hydraulic mining was used to wash away entire hillsides to extract the gold within. Between 1853 and 1884, when the practice was outlawed by federal injunction, silt streamed from the Sierra foothills. This silt raised the bed level of rivers and channels in the system and covered the floor of San Francisco Bay. Mercury was used in the gold extraction process, and much of it escaped with the silt, producing a toxic legacy that remains today.

Starting about 1869, settlers began to dike and drain the land of the estuary. Levees were built to protect parcels of land from surrounding channels, and these parcels came to be called *islands*. Fertile peat soils and an abundant water supply made the region attractive to farmers, and the lands of the estuary continue to be significant contributors to California's agricultural economy. Within

---

* Opinions expressed are those of the author and do not necessarily represent positions of the CALFED Bay–Delta Program or the California Department of Water Resources.

about a generation, the estuary was converted from a sea of tules ringed by riparian forest to one of the garden baskets of California.

Peat soils are fertile but fragile. Once drained, the high proportion of organic matter in the soil is subject to oxidation, wind erosion, and even fire. The soil has been mined in this way for over a century. Today, some islands in the Bay–Delta are more than 5.5 m below sea level, ringed by channels that carry flood flows during most winters (California Department of Water Resources, 1993). Levee maintenance and repair has become a significant expense for those who depend on the integrity of Bay–Delta channels.

This highly disturbed system has become a haven for introduced species, including both intentional and accidental introductions. Nearly every new species survey turns up new organisms in the Bay–Delta, so an accurate count is impossible. A recent thorough review of exotics in the system (Cohen and Carlton, 1995) put the tally at 212 species. The diversity of introduced species is astonishing, and their effects on native biota are likely significant. An early and intentional introduction was the striped bass, *Morone saxatilis*. About 135 juvenile fish from the Navesink River in New Jersey were brought west by rail and planted in the Carquinez Strait at Martinez circa July 1879. Within 10 years there was a thriving commercial fishery. By the 1970s the bass population had begun to decline, and artificial propagation was used to bolster the population. In 1992 this stocking of hatchery fish was temporarily halted because of possible predation by stripers on native fishes (Dill and Cordone, 1997). Several of these native fish species are listed by state and federal governments as endangered, threatened, or species of special concern.

Invertebrate introductions are usually less visible to the lay public, but their roles in altering the food web of the Bay–Delta is at least as significant. The Asian clam; *Potamocorbula amurensis,* apparently introduced about 1986, has been recorded at densities of 48,000 individuals per m^2 of substrate in Suisun Marsh, a large wetland area in the estuary (Peterson, 1996). This filter feeder has reduced food availability for larval fish and increased the clarity of Bay–Delta waters.

Against this backdrop of disturbance, the CVP and SWP began diversions from the Bay–Delta in 1940 and 1967, respectively. Each project has its own massive pumping plant adjacent to channels at the southern edge of the estuary, far enough upstream so that the intrusion of saltwater from San Francisco Bay is usually moderated sufficiently by outflow from the rivers that feed the system. Aqueducts the size of rivers carry water south to local and regional water agencies that have contracted with the CVP or SWP for their water supplies. Water deliveries to farms in the San Joaquin Valley and urban areas of Southern California have allowed these areas to grow and prosper, contributing significantly to California's powerful economy. These water diversions have increased over time to satisfy growth, reaching some of their highest levels in the 1980s or 1990s before conflicts over endangered species protection began to force curtailment of water exports from the system.

The pumping plants of the CVP and SWP exert considerable influence on the hydrodynamics of the Bay–Delta when the pumps are in operation. At times, certain channels in the Bay–Delta reverse flow due to the influence of the pumps, flowing upstream toward the pumps rather than downstream toward the ocean. Throughout the estuary, there is an effect on natural flows as water is drawn from the largest tributary at the north edge of the Bay–Delta to the pumping plants in the south, rather than the generally east–west flows that predominate when the pumps are not in operation.

In addition to the effect of altered hydrodynamics on aquatic species, there is direct mortality of fish and eggs resulting from entrainment in the water diverted from the system. Despite ambitious efforts to screen fish at the pumps and return them to channels of the Bay–Delta, the water projects represent a major stressor on estuarine species. These effects on the biota of the system have placed water operations at odds with endangered species protection. By 1993, two fish species had been listed under the federal Endangered Species Act (ESA), petitions to list other fish had been filed, and the listings had resulted in restrictions on the operations of the projects. There was a significant effect on the amount of water the projects could deliver, jeopardizing the economic well-being of much of the state.

THE CALFED BAY–DELTA PROGRAM

In 1994, the confrontational demeanor that had been the hallmark of California water history yielded to an era of cooperation and compromise. In June 1994, state and federal agencies — those responsible for water deliveries as well as those responsible for environmental protection — signed the Framework Agreement, pledging cooperation and "development of a long-term solution to fish and wildlife, water supply reliability, flood control, and water quality problems in the Bay–Delta Estuary." In December 1994, these agencies, together with representatives of water users and environmental organizations, penned an agreement that extended the level of compromise. These Principles for Agreement, often referred to as the Bay–Delta Accord, called for increased freshwater flows through the Bay–Delta to protect water quality and native fish. To provide greater certainty for water users, the agreement stipulated that any additional environmental water needs due to any new listings under ESA would be met by water purchases financed by the federal government and undertaken on a willing-seller basis, not through regulatory reallocation of water. These principles were intended to be in force for 3 years, reinforcing the need for state and federal agencies to develop a long-term solution (Rieke, 1995).

The resulting collaborative effort between California and the federal government has come to be called the CALFED Bay–Delta Program, or simply CALFED. The CALFED letterhead is a veritable *who's who* of the government players in water and resource management. Current CAL-FED agencies are listed in Table 14.1.

The collaboration has not always come easily, and the process has been marked by occasional internal conflict because each agency maintains its separate responsibilities and authorities. Still, agencies that had sometimes seen themselves as adversaries are now working together. Staffers are forging personal working relationships that are making it easier to compromise and appreciate alternative positions on issues.

Table 14.1 CALFED Bay–Delta Agencies

State Agencies

Resources Agency of California*
   Department of Water Resources
   Department of Fish and Game
   Reclamation Board
   Delta Protection Commission
   Department of Conservation
   San Francisco Bay Conservation and Development Commission
California Environmental Protection Agency
   State Water Resources Control Board
California Department of Food and Agriculture
California Department of Health Services

Federal Agencies

U.S. Department of Interior
   Bureau of Reclamation*
   Fish and Wildlife Service*
   Bureau of Land Management
   Geological Survey
U.S. Army Corps of Engineers*
U.S. Environmental Protection Agency*
U.S. Department of Commerce
   National Marine Fisheries Service*
U.S. Department of Agriculture
   Natural Resources Conservation Service*
   Forest Service
Western Area Power Administration

* Co-lead agencies for Environmental Impact Statement/Environmental Impact Report.

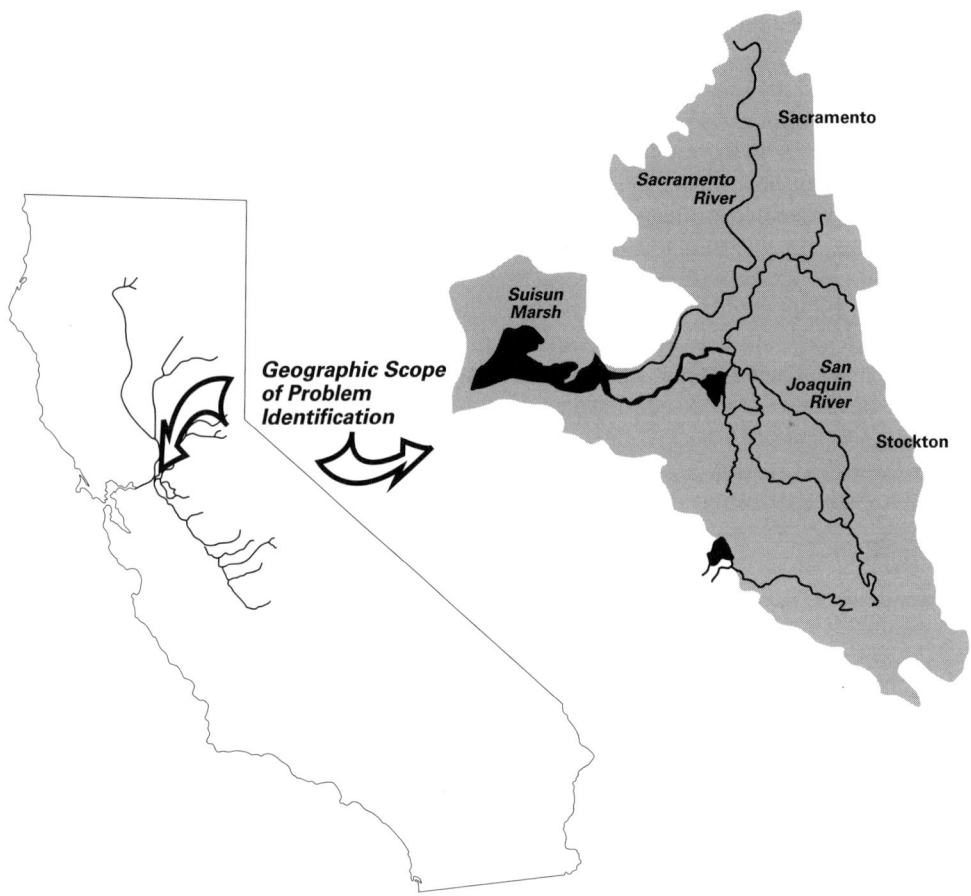

Figure 14.1 CALFED problem area geographic scope.

Initial CALFED planning began in 1995 and envisioned a three-phase program. In Phase I, completed in August 1996, CALFED identified the problems confronting the Bay–Delta, developed a mission statement and guiding principles, and devised three preliminary categories of solutions, or program alternatives. Figure 14.1 shows the geographic scope of the CALFED problem area. Solutions to problems in the Bay–Delta were sought throughout the watershed as well as in areas of Southern California that receive water exported from the system. The CALFED charge was to address four interrelated problem areas: water quality, water supply reliability, levee system integrity, and ecosystem restoration. Rather than solving problems in one of these areas at the expense of another, the goal was to develop a comprehensive solution that offered benefits for all interests.

Following scoping, public comment, and agency review, CALFED concluded that all three of the program alternatives could share many types of proposed actions. These types of proposed actions became central parts of the CALFED program and were called *program elements*. The program elements addressed ecosystem restoration, water use efficiency, levee system integrity, and water quality. Two additional elements (water transfers and watershed management) were added to each alternative because of their value in helping the program meet its multiple objectives. These six program elements have generally been referred to as the *common programs*. The three preliminary alternatives CALFED identified for further analysis in Phase II represented three differing approaches to conveying water through the delta, the uppermost freshwater end of the estuary. The first conveyance configuration relied primarily on existing channels with some minor changes near the project pumping plants at the southern edge of the delta. The second configuration relied on enlarging channels within the delta. The third configuration included in-delta channel modifications and a new constructed conveyance channel that would move some water around the delta. This configuration was a variation on a project rejected by California voters in 1982, which had been known as the peripheral canal due to its intended transport of water in a constructed channel around the periphery of the delta. Each of the three alternatives also included consideration of new groundwater and surface water storage options.

The level of detail of planning was intended to be programmatic. That is, CALFED recognized that it would eventually specify hundreds of actions that together would constitute the CALFED solution. The intent was to outline these actions, refine them, and add more specifics later as the actions were being implemented. Accordingly, the environmental impact statement prepared during Phase II to comply with the National Environmental Policy Act (NEPA) was also programmatic in nature and is intended to be followed by site-specific or project-specific environmental documents. (The same programmatic document also served as the environmental impact report required under the California Environmental Quality Act, CEQA.)

During Phase II, CALFED developed a Preferred Program Alternative, conducted comprehensive programmatic environmental review, and developed an implementation plan focusing on the first 7 years of the anticipated 30 years needed to carry out the entire program. One of three delta conveyance alternatives — construction of a new conveyance channel that would move some water around the delta — was attractive to the CALFED planners for several reasons. It could be screened to protect fish more effectively than existing project pumping plants, it avoided effects on delta hydrodynamics such as reverse flows, and it protected export water quality from agricultural runoff and natural organic carbon in delta channels. But it also would make a huge change in the physical system, and the consequences of such a change would be impossible to predict in the complex and dynamic Bay–Delta environment. This alternative also suffered from the specter of a similar idea that the voters had rejected 18 years earlier. Ultimately, the compromise solution included in the Preferred Program Alternative was to continue the use of existing channels for water transport across the delta, with some modifications to channels and pumping plants.

The CALFED Ecosystem Restoration Program

The plan that CALFED released in July 2000 is one of the most comprehensive and ambitious resource management plans ever issued by government agencies in the U.S. It describes thousands of actions to be implemented during Phase III of the program, an implementation period of more than 30 years. During just the first stage of implementation, a period of 7 years, the estimated cost would be $8.4 billion (CALFED Bay–Delta Program, 2000). The CALFED Bay–Delta Program's Ecosystem Restoration Program Plan (ERPP) is the most detailed program element plan among those that CALFED published. It fills four volumes: Volume I describes ecological attributes of the San Francisco Bay–Delta watershed, and Volume II describes ecological management zone visions. In addition, there is a volume of maps and a strategic plan for ecosystem restoration volume.

Volume I, Ecological Attributes of the San Francisco Bay–Delta Watershed, presents the visions for ecological processes and functions, fish and wildlife habitats, species, and stressors that impair the health of the processes, habitats, and species. The visions presented in this volume are the foundation of the ERP and display how the many ecosystem elements relate to one another and establish a basis for actions, which are presented in Volume II, Ecological Management Zone Visions.

Ecological Management Zone Visions presents the visions for the 14 ecological management zones identified by CALFED and subdivisions of these zones called ecological management units. Each individual ecological management zone vision contains a brief description of the management zone and units, important ecological functions associated with the zone, important habitats, species that use the habitats, and stressors that impair the functioning or use of the processes and habitats. Volume II also contains strategic objectives, targets, programmatic actions, and conservation mea-

Table 14.2 CALFED Bay–Delta Ecosystem Restoration Program Goals

1. Achieve recovery of at-risk native species dependent on the delta and Suisun Bay as the first step toward establishing large, self-sustaining populations of these species; support similar recovery of at-risk native species in the Bay–Delta estuary and the watershed above the estuary; and minimize the need for future endangered species listings by reversing downward population trends of native species that are not listed.
2. Rehabilitate natural processes in the Bay–Delta estuary and its watershed to fully support, with minimal ongoing human intervention, natural aquatic and associated terrestrial biotic communities and habitats in ways that favor native members of those communities.
3. Maintain and/or enhance populations of selected species for sustainable commercial and recreational harvest, consistent with the other ERP strategic goals.
4. Protect and/or restore functional habitat types in the Bay–Delta estuary and its watershed for ecological and public values such as supporting species and biotic communities, ecological processes, recreation, scientific research, and aesthetics.
5. Prevent the establishment of additional nonnative invasive species and reduce the negative ecological and economic impacts of established nonnative species in the Bay–Delta estuary and its watershed.
6. Improve and/or maintain water and sediment quality conditions that fully support healthy and diverse aquatic ecosystems in the Bay–Delta estuary and watershed; and eliminate, to the extent possible, toxic impacts to aquatic organisms, wildlife, and people.

sures that describe the ERP approach. Rationales are also contained in Volume II, which clarify, justify, or support the targets and programmatic actions.

The Strategic Plan for Ecosystem Restoration was not originally envisioned to be part of the ERPP. In 1997, CALFED convened a panel of independent scientists to review an early draft of the three other volumes of the ERPP. A key criticism by the panel was that the 1997 version of the ERPP was a menu of options without a clear strategy for implementation. The panel provided specific recommendations for preparing a concise strategy document. The resulting strategic plan was a collaborative effort of independent scientists and agency staff. It defines an ecosystem-based approach that is comprehensive, flexible, and iterative, designed to respond to changes in the complex Bay–Delta system and changes in the understanding of how this system works. The strategic plan establishes adaptive management as the primary tool for achieving restoration objectives, describes opportunities and constraints to be considered, presents broad goals and more specific objectives, and presents a stepwise procedure for selecting actions to be implemented. The strategic plan also describes how conceptual models should be used in developing restoration programs and defining information needs. Table 14.2 lists the six goals of the Ecosystem Restoration Program as presented in the strategic plan.

Throughout the planning phases of the CALFED Program, every effort was made to keep the program as balanced as possible with respect to all the major interest groups. Urban, agricultural, and environmental interests were consulted regularly and were represented on advisory councils. As mentioned above, the CALFED charge was to address four interrelated problem areas. The package of actions that was eventually developed reflected this careful topical balance. There were actions to protect and improve water quality, increase water supply reliability, maintain and improve levee system integrity, and pursue ecosystem restoration. The goal was to develop a comprehensive solution that offered benefits for all interests, and the CALFED planning documents met this goal.

Early Implementation

Despite this careful balancing, there was from the beginning a recognition that CALFED was spawned out of crisis surrounding the protection of endangered species in the Bay–Delta. This recognition led to consensus on implementation of some ecosystem restoration actions even before the CALFED plan was complete. Local and regional water agencies were some of the strongest supporters of early implementation of ecosystem restoration actions and even funded many of these actions. A bond act passed by California voters in 1996 provided additional funds for this early ecosystem restoration effort. By the time CALFED completed its program plan and environmental documents in 2000, the program had already spent or committed $254 million to fund 240 restoration projects.

The approach to implementation of ecosystem restoration actions that CALFED developed at that time, and continues to follow today, is remarkable for its open and inclusive nature. Rather than following the old model of tasking state or federal agencies with the job of planning and carrying out restoration actions, CALFED awards grants through an annual proposal solicitation process, or PSP, that is open to virtually anyone with an appropriate project. Successful project proponents have included grassroots citizen groups, environmental groups, local agencies, educational institutions, as well as state and federal agencies. Restoration priorities are adjusted from year to year but generally follow the ERPP. The competitive selection process includes peer review for scientific and technical merit. Funded projects are diverse, including control of invasive species, promotion of wildlife-friendly agriculture, physical habitat restoration, environmental education, and water quality improvement.

These restoration projects are not limited to the estuary but are located throughout the watershed of the Bay–Delta system. One example of restoration opportunities far up in the watershed is at Battle Creek, a Sierran tributary of the Sacramento River where CALFED has funded the removal or modification of barriers to fish passage. Battle Creek is a cold, spring-fed stream with exceptionally high flows during the summer dry season, making it the only Sacramento River tributary that is highly resistant to drought. It offers productive habitat similar to that which was lost to the system upon the completion of Shasta Dam, the centerpiece of the CVP. Battle Creek apparently once supported all four races of chinook salmon (*Oncorhynchus tshawytscha*) as well as steelhead (*O. mykiss*). At present, the main population of endangered winter-run chinook salmon spawns in the Sacramento River. During droughts, winter-run chinook suffer severe mortality from high water temperatures. Restoration of Battle Creek will provide an alternative spawning area for this race of chinook. The Battle Creek project will remove five dams and place fish ladders at others. Diversions will be screened and flows improved. These actions will result in improved water quality and access to 71 km of historical anadromous fish habitat (CALFED Bay–Delta Program, 1999).

IMPLEMENTATION CHALLENGES

By 2001, CALFED was poised to implement its plan. The years of careful planning, the unprecedented level of involvement by a spectrum of interest groups, the painstaking balancing to ensure that every interest group gained something from the program, and the collaboration among more than 20 state and federal agencies were all designed to produce an implementable program. The beginning of CALFED implementation has been marked by several challenges. For the most part, these challenges are being successfully overcome. Most of the agencies and organizations involved with CALFED are eager to see the collaborative CALFED approach succeed and want to avoid the controversy and policy stalemate that has marked California water policy development for at least a generation.

The strategy of providing something for everyone so that everyone would support the program was not completely successful. Individuals and groups that had not been directly involved in formulation of the plan questioned the level of balance in the final compromise. Some interests that had been deeply involved decided that they might improve their position through the courts, and several lawsuits were filed against CALFED and its member agencies. Remarkably, the suits were all filed by water interests and agricultural interests. The environmental community, long accused of using the courts to pursue its agenda, filed no suits against CALFED. While the lawsuits slowly work their way through the judicial system, CALFED continues to implement its plan.

What other challenges can be anticipated as the ambitious CALFED ecosystem restoration program is implemented? There will likely be some unanticipated difficulties, but some challenges can already be foreseen. These include difficulties related to implementing new and innovative programs, maintaining a balanced program, adhering to an implementation program rooted in adaptive management, and coping with drastic change in the natural system.

Challenges of New and Innovative Programs

One of the major sources of conflict in the Bay–Delta system prior to CALFED was protection of listed species, particularly delta native fishes, and the impact of this protection on water supply reliability and the ability to export water from the Bay–Delta. The most useful tool available to regulatory agencies had been curtailment of the export of delta water supplies to farms and cities to the south, and these curtailments had a huge economic impact on water customers that received water supplies from the Bay–Delta. Water managers sometimes viewed regulatory water curtailments as being made without any regard for the economic impacts these reductions in water deliveries might have.

CALFED devised an innovative mechanism to maintain protection for aquatic species while ensuring that regulatory agencies would share the perspectives of water managers. This mechanism is called the *Environmental Water Account* (EWA). The EWA is based on the concept that flexible management of water will achieve fishery and ecosystem benefits more efficiently and to a greater degree than a completely prescriptive regulatory approach. The EWA provides assets of water, water storage capacity, money, and rights to use facilities of the CVP and SWP. Three regulatory agencies, the U.S. Fish and Wildlife Service, the National Marine Fisheries Service, and the California Department of Fish and Game, manage the EWA. When they call for a curtailment in water exports to protect fish, they compensate water customers for the loss of that water by providing alternative water supplies from the EWA.

During development of the EWA concept, a number of simulations were conducted to better understand how an EWA might have been operated in real time if it had existed during the 1981 through 1994 water years. This period included a variable hydrologic sequence of wet years and dry years to test the EWA. In each simulation, the EWA had access to a different collection of facilities, contracts, water rights, and income. Based on these simulations, the CALFED agencies agreed to test the EWA concept for 4 years. The EWA would be authorized to acquire, bank, transfer, and borrow water and arrange for its conveyance. Each year, water "deposits" would be made to the account through several complex mechanisms and formulae, totaling roughly 468 million m^3 per year. Operations of the EWA were agreed to in a document of operating principles signed by the involved CALFED agencies and published as part of the CALFED Record of Decision in August 2000.

The agencies acknowledged that the EWA alone would not be sufficient to provide the water needs of species in the Bay–Delta. They envisioned three tiers of environmental water. Tier 1 is the regulatory baseline of water protections, including existing regulation and operating flexibility. Tier 2 is the EWA. Finally, Tier 3 would provide additional water only when necessary; and the CALFED agencies committed to making that water available should it be needed so that there would be no additional uncompensated loss of water by customers south of the Delta who received their supplies from the CVP and SWP.

The EWA represents an entirely new way of protecting listed species in the Bay–Delta, so it is an unfamiliar tool to all that are using it. During the simulations conducted during EWA development, there was recognition that the simulations did not and could not reflect all the variations that EWA management could encounter. The government planners devised a mechanism (Tier 3) to provide protection should EWA assets prove insufficient; but it was clear that finding the funding to acquire any needed Tier 3 water would be challenging, and reliance on Tier 3 would erode the compromise that was reflected in the innovative EWA.

By March 2001 the EWA was in operation and facing its first difficult period. A situation occurred that EWA planners had never anticipated. The outmigrating smolts of a protected species, the endangered winter-run race of chinook salmon, began showing up in fish salvage facilities in the south delta, near the giant export pumps, at unprecedented levels (Greene, 2001). To protect them, EWA managers ordered the curtailment of project pumping and expended EWA assets at a rate that far exceeded expected use of the account. Still, EWA managers allowed some pumping so that they could save some EWA assets to protect other native fish such as delta smelt (*Hypomesus transpacificus*) later in the year. The result was a take of more than 19,000 winter-run salmon at the SWP pumps. The future for the EWA looked perilous. Only 3 months into its existence, with its water supplies already drawn down more than anticipated and most of the first year still ahead, the EWA faced the possibility of running out of assets within its first year.

Fortunately, the EWA was able to end the water year with a slight surplus of its water assets. The surprisingly large take of winter-run salmon fostered constructive discussion among biologists and regulators over the relative merits of various population estimation techniques. Finally, an independent scientific assessment of the EWA found several positive aspects of the EWA in its first year of operation (EWA Review Panel, 2001).

Maintaining a Balanced Program

Another challenge for CALFED will be maintaining a balanced program. Although painstaking CALFED efforts to achieve balance did not produce unanimous initial support, the program did attract an impressive level of support considering the contentious history of California water policy development. The support of various interests can be expected to wax or wane according to how well the program lives up to their expectations. For example, the CALFED program includes the commitment to pursue several new or expanded reservoirs. Any such effort is fraught with challenges — technical, financial, environmental, and regulatory. At the insistence of political leaders, CALFED committed to extremely ambitious work schedules. It is highly likely that planning, permitting, or construction of some of these projects will fall behind schedule.

To guard against unequal implementation, CALFED committed to annual progress reviews by the governor of California and the secretary of interior. If the schedules are not substantially adhered to, the governor and secretary will prepare a revised schedule that ensures achievement of balanced solutions in all program areas. In other words, when the reservoir projects bog down, CALFED could delay work in other areas such as ecosystem restoration. At that point, advocates of ecosystem restoration might be expected to argue that a restored ecosystem takes years or decades to become fully functional, so the ecosystem restoration program is really years behind construction projects. Unless various interest groups are patient and flexible, political leaders will then face a new conflict that could prove difficult to resolve to the satisfaction of all interests.

Implementing Adaptive Management

Additional seeds of conflict are sown in the CALFED program. A prominent feature of the CALFED program is a dedication to *adaptive management*. The ERPP includes a scholarly discussion of the topic, and the term has been used often by staff of CALFED and its agency members as they describe the way the program will be carried out. It is not clear that the parties involved have considered all the possibilities that might be presented if the program adheres to a policy of adaptive management. Continued research into the structure and function of the complex estuary and the reasons for the decline of many of its native species may lead to conclusions that are difficult for some interests to accept. If any interests are later asked to give up more and get less than they did in the original carefully crafted program, support for the program is likely to erode, and the level of conflict is likely to rise.

Coping with Unanticipated Change

Finally, the CALFED program may not be resilient to unanticipated changes in the Bay–Delta system if the type or degree of change is outside the scope of existing planning efforts. Existing data suggest at least two scenarios that could occur, and either one would sharply increase the level of conflict among those with an interest in the Bay–Delta system. These two scenarios are global warming with a concomitant sea level rise and prolonged drought.

Exploration of sea level rise and its effect on the CALFED effort will illustrate the challenge. CALFED documents mention global warming and related sea level rise as an example of the type of situation that could be approached through adaptive management. However, it is likely that the consequences of sea level rise on the economy and environment of the Bay–Delta would overwhelm the collaborative decision-making process that CALFED has crafted. Land subsidence resulting from draining and cultivation of delta peat soils has left much of the delta below sea level. Even today, the coincidence of high flows, high tides, and wind often causes one or more delta levees to fail and the protected islands to flood. A gradual rise in sea level would result in calls for greater spending to maintain and strengthen levees, and this could tip the careful CALFED balance. A rise in sea level might also cause greater saline intrusion up into the channels of the estuary, resulting in higher salinities at the CVP and SWP intakes. There would be a call for a new channel to bring fresh water to the intakes from further upstream where salinity is lower. CALFED rejected this alternative in 2000 when it selected through delta conveyance as a component of the preferred alternative. It is not clear whether CALFED agencies and interest groups would readily accept such a fundamental change in the program, regardless of the technical or biological merits.

In conclusion, the development, adoption, and initial implementation of the CALFED program have been major accomplishments. The level of collaboration among federal, state, and local agencies, public interest groups, and other interested parties has been unprecedented in California water policy development. The well-funded CALFED Ecosystem Restoration Program, guided by science and adaptive management, is a great advance over project-by-project mitigation for environmental impacts. The CALFED future will undoubtedly be a perilous one. The program has its detractors now, and the situation will likely become even more difficult and contentious in the future. But CALFED was formed because conflict had made it impossible for any Bay–Delta interests to achieve their objectives, whether these objectives related to the recovery of endangered species or the delivery of a reliable water supply. The new way of collaboration and comprehensive problem solving is difficult, but the old way had become impossible. Ultimately, the success of CALFED and its ambitious ecosystem restoration program may depend on the ability of Californians to remember this history lesson.

REFERENCES

CALFED Bay–Delta Program, Phase II Report, Final Programmatic EIS/EIR Technical Appendix, CALFED Bay–Delta Program, Sacramento, CA, 2000.

CALFED Bay–Delta Program, Ecosystem Restoration Program Plan Volume I: Ecological Attributes of the San Francisco Bay–Delta Watershed, Final Programmatic EIS/EIR Technical Appendix, CALFED Bay–Delta Program, Sacramento, CA, 2000.

CALFED Bay–Delta Program, Ecosystem Restoration Program Plan Volume II: Ecological Management Zone Visions, Final Programmatic EIS/EIR Technical Appendix, CALFED Bay–Delta Program, Sacramento, CA, 2000.

CALFED Bay–Delta Program, Ecosystem Restoration Program Plan: Strategic Plan for Ecosystem Restoration, Final Programmatic EIS/EIR Technical Appendix, CALFED Bay–Delta Program, Sacramento, CA, 2000.

CALFED Bay–Delta Program, Ecosystem Restoration Program Plan: Maps, Final Programmatic EIS/EIR Technical Appendix, CALFED Bay–Delta Program, Sacramento, CA, 2000.

CALFED Bay–Delta Program, Restoring the Environment, Investing in the Future: 1999 Annual Report, CALFED Bay–Delta Program, Sacramento, CA, 1999.

California Department of Water Resources, California Water Plan Update, California Department of Water Resources, Sacramento, CA, 1998.

California Department of Water Resources, Sacramento San Joaquin Delta Atlas, California Department of Water Resources, Sacramento, CA, 1993.

Cohen, A.N. and Carlton, J.T., Nonindigenous Aquatic Species in a United States Estuary: A Case Study of the Biological Invasions of the San Francisco Bay and Delta, United States Fish and Wildlife Service, Washington, D.C., 1995.

Dill, W.A. and Cordone, A.J., History and Status of Introduced Fishes in California, 1871–1996, Fish Bulletin 178, California Department of Fish and Game, Sacramento, CA, 1997.

EWA Review Panel, First Annual Review of the Environmental Water Account for the CALFED Bay–Delta Program, CALFED Bay–Delta Program, Sacramento, CA, 2001.

Greene, S., Winter run and older juvenile chinook loss at the Delta Fish Facilities, 01 Oct 2000 through 12 Mar 2001, unpublished data, California Department of Water Resources, Sacramento, CA, 2001.

Monroe, M.W. and Kelly, J., State of the Estuary, A report on conditions and problems in the San Francisco Bay/Sacramento-San Joaquin Delta estuary, San Francisco Estuary Project, Oakland, CA, 1992.

Peterson, H., *Potamocorbula amurensis* Spatial Distribution Survey, Interagency Ecological Program Newsletter, Winter:18–19, 1996.

Rieke, B., The Bay–Delta accord: a stride toward sustainability, *Resource Law Notes*, Natural Resource Law Center, University of Colorado, 35:5–11, 1995.

Scoonover, M., The CALFED Bay–Delta Program: Making Sense in the Delta, *Calif. Environ. Law Rep.*, 1999:271–277, 1999.

CHAPTER 15

Understanding the Politics of Ecological Regulation: Appropriate Use of the Concept of Ecosystem Health

John Freemuth

Unless the scientist reads outside her or his field, or takes a sociology class and learns how Kuhnian paradigms as "cultural baggage" lead humans by the nose in all they think and do, one can go through life ignorantly believing that science is detached, objective, factual, unmythic, and withal goal-setting.

Michael Barbour, 1995

Where there is certain knowledge, true science or absolute right, there is no conflict that cannot be resolved by reference to the unity of truth, and thus there is no necessity for politics.

Benjamin Barber, 1994

INTRODUCTION: ECOSYSTEM HEALTH AS POLITICAL ARGUMENT

This chapter works from a premise about our discussions of environmental management. Those who support ecosystem health as a useful way to understand the management of the natural (and to some extent the human) world are making a political argument about the use of a very powerful and appealing idea. Ecosystem health has the same emotional and intuitive appeal as words like justice, equality, and freedom.

Proponents of ecosystem health value certain things at the expense of other things. They are making claims about decision processes and who ought to make decisions. As noted by political scientist Deborah Stone, "Every idea about policy draws boundaries. It tells what or who is included or excluded in a category. These boundaries are more than intellectual — they define people in or out of a conflict or place them on different sides" (Stone, 1997).

SETTING THE STAGE

Two examples from *Ecosystem Health* (Rapport et al., 1998) illustrate this introductory point. In the first we read that society should ensure that "ecosystem health is not further compromised by human activity." Note how a boundary is drawn here: certain human activities compromise ecosystem health, so they should probably not be allowed. My activity is acceptable, yours is not.

This act of drawing boundaries, of giving to and taking from, puts ecosystem health squarely into the middle of political talk and discussion. Someone has to draw and enforce those boundaries.

In the second example, again from *Ecosystem Health*, we learn that:

> [I]ndividual values generally prevail on local scales, community values prevail on regional scales, and societal values dominate on still larger scales. Ecosystems and landscapes are regional in extent and generally involve time frames that span more than one generation. Thus, the community, which has a longer time horizon than the individual, represents the best "match" in terms of relating values to ecosystem outcomes (Rapport et al., 1998).

This is political theory, not scientific fact. It asks the reader to accept the claim that the community (which remains undefined) rather than the nation (the state? the city?) is where we should apply the ecosystem health paradigm. It also unconsciously follows a tradition that has been going on in political theory for a very long time. James Madison once argued that large republics were better than small republics. In *Federalist 10* he said: "Extend the sphere and you take in a greater variety of parties and interests; you make it less probable that a majority of the whole will have a common motive to invade the rights of other citizens" (Rossiter, 1961). Arguments about the proper size of a community, which makes decisions that allocate values, are thus political arguments, not scientific ones.

THE PROBLEM OF POLITICS: ECOSYSTEM HEALTH AND POLITICAL POWER

Those who advocate for ecosystem health need to be fully aware of how our current institutional arrangements affect the success of implementing ecosystem health as a management paradigm. Note, though, that these arrangements are based on assumptions that lead to structuring of political power relationships in a certain way. Ecosystem health proponents do this as well, as will be discussed later.

There is no better voice here than that of James Madison, who explains one of the key assumptions of the authors of the Constitution this way in *Federalist 51*:

> Ambition must be made to counteract ambition …. If men were angels no government would be necessary. If angels were to govern men, neither external nor internal controls on government would be necessary. In framing a government of men over men, the great difficulty lies in this: you must first enable the government to control the governed and in the next place, oblige it to control itself. A dependence on the people is, no doubt, the primary control on the government but experience has taught mankind the necessity of auxiliary precautions (Rossiter, 1961).

The precautions, of course, are the commonly understood checks and balances, separation of powers, federalism, and republicanism. Power is diffused in the U.S. political system. Policy change is often difficult to achieve.

Madison also remarks in *Federalist 10* that one of the most important reasons for checking power is the existence of factions (today's interest groups): "a majority or minority of the whole who are united and actuated by some common impulse of passion, or of interest, adverse to the rights of other citizens, or to the permanent and aggregate interests of the community"; hence the need to check Madison's "mischiefs of faction" by representative government, larger political units, and so forth.

Putting all of the above in more modern terms, there is thus a designed tendency of the political system to gridlock and for policy shifts to happen rarely. But we do know that there are examples where the political system overcame the tendency for political gridlock. One example of particular interest to proponents of ecosystem health is the development of certain policies during the progressive era at the turn of the last century. Scholars and practitioners interested in the implementation of an ecosystem health-based management regime would do well to revisit the early days of the progressive movement for clues as to how to develop and implement a management regime accepted by an entire society.

This era was the time of Gifford Pinchot, Teddy Roosevelt, and the birth of the conservation movement. The progressive era, of course, institutionalized science-based, expert-centered management as a general approach to the growing complexity of society at the time. For example, the federal bureau that best represented the progressive era in land management was the United States Forest Service. Samuel Hays, in his seminal work *Conservation and the Gospel of Efficiency,* noted that:

> Conservationists were led by people who promoted the "rational" use of resources, with a focus on efficiency, planning for future use, and the application of expertise to broad national problems. But they also promoted a system of decision making consistent with that spirit, a process by which the expert would decide in terms of the most efficient dovetailing of all competing resource users according to criteria which were considered to be objective, rational, and above the give-and-take of political conflict (Hays, 1975).

In the case of the Forest Service, the expertise brought to bear on forest management questions came from the science of forestry.

What is most important about that earlier movement, however, may well be how its themes captured the public imagination. Gifford Pinchot discovered that "in the long run, forestry cannot succeed unless the people who live in and near the forest are for it and not against it" (Pinchot, 1947). Pinchot helped lead the effort for professional management of the national forests. But the key to Pinchot's success lay not solely in his advocacy of professionalism and expertise but also in the service of both to a democratic vision.

In the words of Bob Pepperman Taylor, "For Pinchot the conservation of natural resources is of fundamental democratic value because it allows for the possibility of equality of opportunity (access to public resources) for all citizens" (Taylor, 1992). Taylor adds, "If we remove the vision of Progressive democracy from Pinchot's work, we are left merely with the scientific management and control of nature for no other purpose than brute human survival."

It is also true that later foresters "became progressively more narrow in outlook as a result of the kind of specialized education they [Pinchot] encouraged" (Clary, 1986). The vision may have become less successful over time because it lost its ability to speak in nonspecialized terms. The point to remember, though, is that early public land management was successful because of its link to a democratic vision accepted by the majority of society at the time, representing an underlying consensus about how a large amount, but not all, of the federal estate should be managed. The estate should be managed wisely for an ongoing supply of goods and services for the industrial society. It is clear that progressive-era public land management was centered as much on this vision of society as it was on a technical expert-centered land management regime. This, then, is part of the roadmap for proponents of ecosystem health. They should make clear how ecosystem health offers a vision for society that might link up democratic institutions and processes with economic and environmental equality. But one other component needs discussion.

THE PROBLEM OF SCIENCE AND HIDDEN VALUES

Perhaps it would be good to begin this part of the discussion with a brief look at the development of ecosystem management as a relevant case study. There appear to be two general approaches to how ecosystem management was originally conceived: the expert-centered and technically rational, and the democratic. The first approach is illustrated in one definition of the term: "Ecosystem management integrates scientific knowledge of ecological relationships within a complex sociopolitical and values framework toward the general goal of protecting native ecosystem integrity over the long term" (Grumbine, 1994).

This definition appears technically rational and expert-centered for several reasons. One, it seems to imply some sort of expert ecosystem manager who will integrate an established knowledge base to achieve a goal that apparently has already been defined and agreed upon. Such a stance is very similar to the expert-centered decision-making model first developed in the progressive era, without the accompanying underlying consensus that ecosystem management is acceptable to society, if its goal is protecting native ecosystem integrity.

Two, this approach appears to view the people who live within various ecosystems as another component to be managed. These people are to be studied, surveyed, and so on by the same managers who will then take whatever knowledge is discovered and include it as social data in an ecosystem management framework. There does not appear to be much need for a discourse to be established with local people who might happen to live in a defined ecosystem. Nor, even more fundamentally, do these definitions consider that the local communities may wish to, or currently do, disapprove of ecosystem management as they understand it. The people should be for it and not against it, to paraphrase Pinchot.

There are, however, other definitions and ways to look at ecosystem management that clearly take a more democratic approach. Robert Lackey moves in this direction when he asserts that, "Ecosystem management should maintain ecosystems in the appropriate condition to achieve desired social benefits; the desired social benefits are defined by society, not scientists" (Lackey, 1994). Lackey's point can be better understood if we consider his view on biodiversity, often mentioned as a key component of ecosystem management. He asserts: "Ecosystem management may or may not result in emphasis on biological diversity as a social benefit." Finally, Hal Salwasser has observed that "ecosystem management is about people. Other views of ecosystem management have placed goals for nonhuman biological and physical attributes of ecosystems ahead of goals for human attributes" (Salwasser, 1994).

My point in contrasting these various definitions of ecosystem management is to illustrate a fundamental difference in the approach and assumptions of the definitions. That difference centers on whether ecosystem management is a technical process best left to experts who will manage any public component of the process of implementation, or more of a democratic process requiring the development of a new and democratically derived consensus position. It has been my contention that, for ecosystem management to succeed, it must center on that democratic process.

Proponents of ecosystem health are faced with a similar choice. If they adopt the highly rationalized and democratically passive management alternative suggested above, they risk failure, as they have provided no way to show that there is any public acceptance of their new management paradigm. That then runs the risk of appearing to dismiss and disdain the opinions and values of citizens.

A second and closely related component to this analysis is the growing use of what might be termed *advocacy science*. Advocacy science can take two closely related forms. The first clearly mixes up values and science, where what are clear values preferences end up masked as scientific truths. The second works by adopting a certain value preference as a policy goal (logging is harmful or logging is beneficial) and then finding the science that demands a conclusion that turns out to be the prechosen goal (logging harms biodiversity, so we must stop logging; or forests are unhealthy, so we should selectively cut them).

Consider one example from one side of the grazing debate. In the December 1994 issue of *Conservation Biology*, a fascinating editorial was written about the role of conservation biology in range management questions. The opinion piece takes issue with the answer to a question asked and answered by Reed Noss: whether conservation biologists should "link arms with activists in efforts to reform grazing practices" (Brussard et al., 1994). The authors conclude that they should not. Worried that conservation biologists would damage their credibility by openly advocating political positions, the authors instead suggest asking a different question. That question is, "how can livestock grazing be managed to have the fewest impacts on biodiversity and ecosystem integrity?" The authors claim that a special symposium on grazing, which precipitated their editorial, offered no help on this question.

In a powerful conclusion to their editorial they express concerns about:

> ... the inherent flaw of deductive reasoning which asks one simply to accept "that range management must be dramatically reformed" ... how can we hope to advance the Society's mission to preserve biological diversity if our audience of policymakers assumes that we intend to "prove" a presumed conclusion (Brussard et al., 1994).

Lackey also does a good job with his discussion of the values vs. science problem. Writing about the increasingly contentious issue of salmon recovery in the Pacific Northwest, he urges:

> ... those of us who are technocrats, scientists, biological resource managers, or scientific advisors should remain humble in our dealings with the public and elected officials and overcome the tendency to advocate political choices driven by strong personal interest and packaged under the guise of a scientific imperative (Lackey, 1997).

The success of ecosystem health as the leading paradigm for environmental management will depend on its integration with forms of democratic discussion and deliberation. As noted in *Ecosystem Health,* success will depend on whether society adopts ecosystem health as an important goal and supports the allocation of resources toward achieving that goal. A good distance remains to be traveled, however. In that same book is the assertion that ecosystem health "challenges us to identify the societal goals that are compatible with sustainable life systems, to identify and validate indicators of ecosystem function that are essential to its evolution and perpetuation, and to merge goals (societal values) with biophysical realties." This statement is potentially unnerving. It suggests, certainly, a cadre of elites setting goals for society. It is exactly the sort of science –values confusion that Lackey and others caution against. Science cannot set goals for society, but it can inform those goals. As Gregg Cawley and I have noted:

> Indeed, science strives to construct a picture of the physical world based on empirical observation. A management regime, in contrast, must remain continually attentive to the impact human values and interests have on allocating meanings to scientific observations. A rather simple example might help make this point.

> Scientific observation remains more or less content with the conclusion that a sixteen-ounce container has eight ounces of fluid in it. From a management perspective, however, the crucial question may very well be whether the container is "half-full" or "half-empty." It is this question which determines the appropriate course of action — whether or not more fluid is needed, as it were. The problem, of course, is that "half-full/half-empty" are value judgments derived from the interests of people. As such, they are open to negotiation at any given moment, and over time, unless we assume that values and interests remain constant. Scientists and managers, who center decisions solely on science, do not have any *special* position in negotiations over value questions (Cawley and Freemuth, 1993).

There is a recent indication that some of Lackey's advice has begun to be heeded. At a recent meeting of the Idaho chapter of the American Fisheries Society, a vote was taken as to the opinion of the membership regarding the best option for salmon recovery. Some 90% of those in attendance voted that dam breaching had the best option of restoring salmon runs, but the vote was couched in the language of probability, not certainty. More importantly, according to chapter president-elect Ted Koch, "The chapter does not advocate dam breaching. It's up to our elected political leaders to make the hard choice for the region" (Barker, 1999).

This action appears to be close to what is emerging as the proper role for science in difficult public policy debates according to many observers: informing policy debate, not making policy decisions. Again, as Lackey and others argue, "The complete implications of each alternative policy choice should be fully and clearly explained, including the short- and long-term consequences, and especially the level of scientific uncertainty" (Lackey, 1997).

Salmon recovery is a complex issue, but it pales alongside the wholesale adoption of management paradigms for the entire society. It may well be that ecosystem health is an inappropriate paradigm to follow. Some have made that case; others disagree with that assessment. That debate will continue. What this chapter has tried to do is suggest that the strategies and arguments that are employed by ecosystem health advocates are as much about politics and policy as they are about science. They are in large part about the society we ought to envision for ourselves.

ACKNOWLEDGMENTS

The author wishes to thank the two reviewers of an earlier version of this chapter, as well as Gregg Cawley of the Department of Political Science of the University of Wyoming, for their helpful comments.

REFERENCES

Barber, B., *Strong Democracy*, University of California Press, Berkeley, CA, 1984.

Barbour, M.G., Ecological fragmentation in the 1950s, in Cronon, W., Ed., *Uncommon Ground*, W.W. Norton, New York, 1995, pp. 233–255.

Barker, R., Scientists say dams must go to save salmon, *Idaho Statesman*, 1999, p. A1.

Brussard, P., Murphy, D., and Tracy, R.C., Cattle and conservation biology — another view, *Conserv. Biol.*, 8:919–921, 1994.

Cawley, R.M. and Freemuth, J., Tree farms, mother earth and other dilemmas: the politics of ecosystem management in greater Yellowstone, *Soc. Nat. Resour.*, 6:41–53, 1993.

Clary, D., *Timber and the Forest Service*, University Press of Kansas, Lawrence, KS, 1986.

Grumbine, R.E., What is ecosystem management? *Conserv. Biol.*, 8:27–38, 1994.

Hays, S.P., *Conservation and Gospel of Efficiency: The Progressive Conservation Movement 1890–1920*, Atheneum, New York, 1975.

Lackey, R., Seven pillars of ecosystem management, *Landscape Urban Plan.*, 40:21–30, 1994.

Lackey, R., Restoration of Pacific salmon: the role of science and scientists, in Report 92, Water Resources Center, University of California, Berkeley CA, 1997, pp. 35–40.

Pinchot, G., *Breaking New Ground*, Harcourt, Brace, New York, 1947.

Rapport, D., Costanza, R., Epstein, P., Gaudet, C., and Levins, R., *Ecosystem Health*, Blackwell Science, London, 1998.

Rossiter, C., Ed., *The Federalist Papers*, New American Library, New York, 1961.

Salwasser, H., Ecosystem management, *J. For.*, 92:6–10, 1994.

Stone, D., *Policy Paradox*, Norton, New York, 1997.

Taylor, B.P., *Our Limits Transgressed: Environmental Political Thought in America*, University Press of Kansas, Lawrence, KS, 1992.

CHAPTER 16

Attitudes and Their Influence on Nature Valuation and Management in Relation to Sustainable Development

Hans A.M. de Kruijf, Ekko C. van Ierland, C. Martijn van der Heide, and Jos N.M. Dekker

DILEMMAS OF VALUATION AND BIODIVERSITY MANAGEMENT

In this chapter the concept of biodiversity is viewed as analogous to the Convention on Biological Diversity: "Biodiversity is the variability among living organisms from all sources, including *inter alia*, terrestrial, marine and other aquatic ecosystems, and the ecological complexes of which these are a part: this includes diversity within species, between species and of ecosystems" (UNCED, 1992).

It is widely recognized that biodiversity is threatened and that extinction of several species occurs or may occur in the near future. Increasing population pressure and per capita income will have tremendous impact on ecosystems and the species that inhabit it (Heywood and Baste, 1995). Pressures on biodiversity are described in more detail in an OECD report (1996) and summarized in Table 16.1.

In protecting biodiversity, many policy decisions have to be made on how to protect biodiversity: creating reserves; protecting natural ecosystems by direct regulation; reducing pollution that is affecting ecosystems; constructing corridors between ecosystems; or other policy measures (Table 16.2). But first, when dealing with biodiversity, policy makers should make choices about what to protect or sacrifice, where to protect, and how to finance the protection of biodiversity (Table 16.3). Although valuations are not always directly expressed in monetary terms, in practically all cases, implicit value assessments are made in the decisions to protect or sacrifice ecosystems and manage nature.

Answering these questions involves value judgments about biodiversity and nature. Different nations have different value systems and different attitudes toward the protection of biodiversity (Furze et al., 1996; Perrings, 1995).

It would be helpful if a common framework of analysis could be developed to understand and analyze the problems of biodiversity valuation (Bateman, 1993). There are many indications that economic valuation methods are important in making explicit the values that play a role in protection of biodiversity.

Many politicians and economists agree that some of these valuation methods assist in decision making about biodiversity protection. Certainly not all valuation questions related to biodiversity can be solved by monetary valuation methods. Ethical and religious aspects are important in considering the nature of the values; the "public" good character of environmental amenities may not be automatically compatible with value judgments based on private and individual value assessments (Furze et al., 1996).

Table 16.1 Human Actions and Biodiversity

	Cause	Comment
I. Proximate Pressures		
a. Land conversion	Development for agriculture	Losses of ecosystems
b. Exploitation wild species	Over-exploitation, exceeding regenerative capacity	Ecosystem instability
c. Introducing exotic species	Changing ecosystem structures	Loss of endemic species
d. Homogenization	Competing in global agricultural markets by specialization for higher productivity	Relatively few plants account for the world's food production; serious implications for food security
e. Pollution	Pollutants from different sources	Reduce and eliminate populations of sensitive species
f. Global environmental change	Global warming, caused by greenhouse gases	Large-scale extinctions have occurred in the past; further losses in biological diversity may be expected
II. Underlying Pressures		
a. Consumption and production	Increasing and unsustainable consumption of energy and natural resources	Increased impact on resources and habitat conversion and degradation
b. Patterns of population growth and distribution	Greater need for agriculture and industrial produce	Impacts on level of biological diversity
c. Economic failure	Uncontrollable use of natural resources, conversion of land	Loss of precious habitats, valuable species

Adapted from OECD, Saving Biological Diversity: Economic Incentives, OECD Publications, Paris, 1996; Van Ierland, E.C., de Kruijf, H.A.M., and van der Heide, C.M., Attitudes and the Value of Biodiversity, Department of Economics, Wageningen Agricultural University, Wageningen, the Netherlands, 1998.

Table 16.2 How to Deal with Biodiversity

Should species be protected?	Who is responsible for protection?
Should ecosystems be protected?	Where should they be protected?
Which objectives could be formulated for biodiversity?	When should they be protected?
	Who will enjoy benefits of protection?
Which ecosystems should be protected?	Who will bear the costs of protection?
How can biodiversity be protected?	

Data from OECD, 1996; Barrett, 1995; Begon et al., 1990; Miller et al., 1995; Barbault and Sastrapradja, 1995; Stork and Samways, 1995; Turner et al., 1994.

Table 16.3 Options for Protecting Biodiversity

I. At the National Levels	II. At the International level
1. Reduction of pollution	1. Reduction of pollution
2. Reduction of physical impacts	2. International treaties for protection
3. Creation of wildlife parks	3. Debt-for-nature swaps
4. Protection of habitats	4. Global environmental facility
5. Creation of ecological main structures	5. Joint implementation programs
6. Reduction of population pressure	

Data from Keating, 1993; Swanson, 1995; McNeely et al., 1995; Barbault and Sastrapradja, 1995; Turner et al., 1994; Barbier and Rauscher, 1995.

If monetary valuation methods cannot provide a full solution to dealing with biodiversity, which other methods for policy analysis are available and can be applied that take into account monetary and nonmonetary values?

This chapter analyzes basic attitudes toward biodiversity in countries and how policymakers and stakeholders with different attitudes cope with decision making about biodiversity.

ATTITUDES AND DETERMINANTS

Attitudes

Attitudes toward biodiversity are value-loaded perspectives on possible behavior with regard to biodiversity. Valuations of biodiversity are embedded in these attitudes, which connect values to possible actions, and vice versa. Attitudes are steering elements for strategies and policies.

Opinions and attitudes on biodiversity differ among people, countries, religions, political parties, and societies. People have different — even opposite — attitudes, which creates ambivalence. Moreover, attitudes can change in time. The relationship of determinants and attitudes to valuation methods and policy making is illustrated in Figure 16.1. Finally, valuation methods and policy making are needed to assess biodiversity.

Classification of attitudes toward biodiversity assists understanding attitudes and thus encourages communication about strategies and policies for the management of biodiversity. However, classifying can also restrict and close such communication. There are many different classifications; most are one dimensional with man on one side, nature on the other, and mixed attitudes in between. Then one can conceive attitudes with dominating man and rich nature, or powerless man and poor

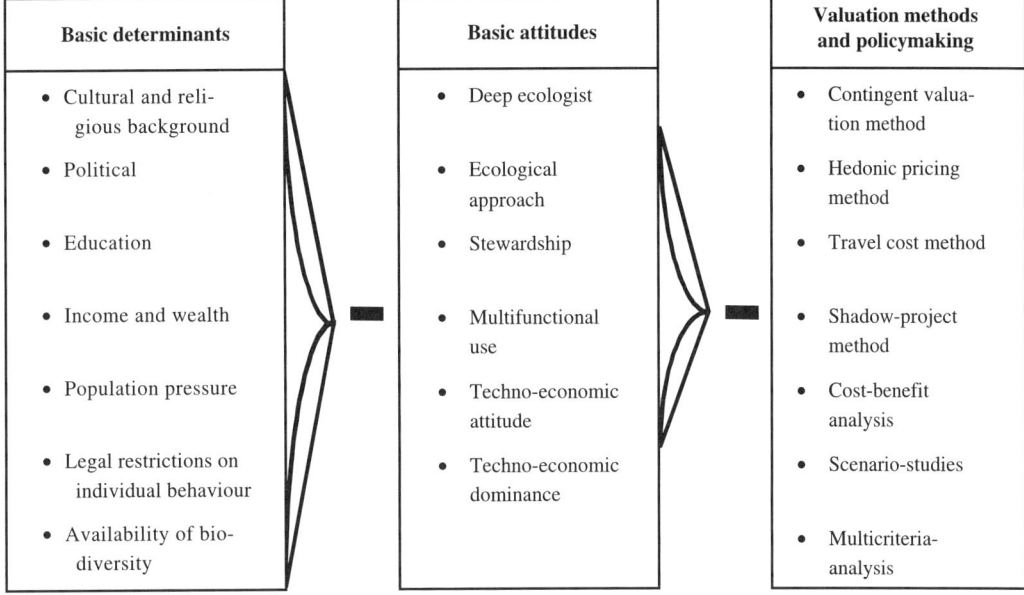

Figure 16.1 Relationship of determinants and attitudes to valuation methods and policy making. Attitudes are determined by various basic determinants. These attitudes determine the valuation of biodiversity and policy choices that can be made about biodiversity protection. Deep ecologist and techno-economic dominance are attitudes that reflect an extreme position. For these attitudes, valuation methods are not necessarily required since choices have already been made. (Adapted from Achterberg, W., *Partners in de Natuur (Partners in Nature)*, Van Arkel, Utrecht, the Netherlands, 1986.)

nature. For example, ecological restoration in the U.S. and nature development in the Netherlands are policies based on the supposition that man can produce nature in a systematic way with modern technology. However, not all people will accept this result as nature. In both developing and developed countries, poverty can result in neglecting nature. Some may appreciate this result as an enrichment of nature, while others will see it as marginalization.

The classification of attitudes (Figure 16.1) ranging from the *deep ecologist* attitude to *techno-economic dominance* will be used (adapted from Achterberg, 1986).

Societies, over time and space, can be placed on a continuum between these opposites as human attitudes change due to internal social dynamics and external influences. However, such attitudes greatly influence how humans use natural resources and, therefore, the impacts they have on biodiversity (McNeely et al., 1995).

- The *deep ecologist* gives first priority to biodiversity. Mankind should live in harmony with nature, without interfering in ecological processes through economic activity; this view limits the options for economic activity and potential for traditional economic growth. Biodiversity protection, nature conservation, and pristine landscapes have top priority.
- The *ecological approach* is also focused on biodiversity, nature conservation, and priority for ecosystems. Some impacts on the environment and the ecological system are necessary for mankind to survive and are accepted; discussion is ongoing to define the economic minimum standards and identify acceptable impacts on biodiversity and ecosystems.
- *Stewardship* can be characterized as the attitude which accepts that humankind can manipulate nature and ecosystems but that responsibility exists to deal very carefully with ecosystems in order to provide opportunities to next generations. The main question is the extent to which nature and ecosystems can be manipulated and the limits to consider.
- The attitude of *multifunctional use* focuses on the combination of economic activity and conservation of nature and biodiversity. It sacrifices large areas of ecosystems in order to provide space for agriculture, infrastructure, housing, and industrial activities. In doing so it tries to protect some areas and exploit ecosystems without destroying them. The main emphasis is on matching economic activity with protection of biodiversity and nature.
- The *techno-economic attitude* is characterized by confidence in technical solutions and confidence in the resilience and robustness of ecosystems. Economic growth is an important aspect. Sometimes technological devices and technologies are applied to maintain biodiversity artificially, e.g., by constructing dikes or manipulating water tables. In this context ecological engineering is considered a useful tool to combine economic activity and protection of biodiversity.
- *Techno-economic dominance* is characterized by full dominance of the drive for economic growth and technological development and is negligent of nature conservation, biodiversity, or ecosystems protection. Only economic issues matter; the more goods and services produced, the better it is. Techno-economic dominance is highly optimistic about self-regulation of nature and the resilience and robustness of ecosystems.

In the two extreme cases, no policy choices have to be made. *Deep ecologists* have, by definition, chosen for ecosystems protection and biodiversity. Everything is to be protected. At the other extreme, *techno-economic dominance* focuses completely on economic growth, sacrificing what is necessary to obtain these goals. All other attitudes have, in one way or another, the challenging task to decide about what, where, and how to protect, and how to finance protection of ecosystems and biodiversity. Attitudes play an important role in the decision-making process and the preparation of biodiversity policies.

A serious problem of most classifications is that they are not operationalized. Identifying concrete attitudes within the classification categories is difficult. For example, a traditional agricultural method of shifting cultivation may be identified by conservationists as an indicator of a techno-economic dominance because it reduces biodiversity. From a broader perspective, it can be seen as an indicator of multifunctional use, stewardship, or even an ecological approach. Thus, trying to classify this method and the culture clearly generates debate and disagreement.

Basic Determinants

Most losses of biodiversity are due to independent decisions of billions of individual users of environmental resources. Underlying causes of the impacts and the decisions include the objectives that motivate decisions, preferences behind the demand for goods and services, property rights that define individual endowments, relative prices that determine the market opportunities associated with those endowments, and attitudes toward biodiversity (Perrings, 1995). These attitudes are determined by a large variety of characteristics and circumstances of basic determinants as described below.

Cultural and Religious Background

Cultural and biological diversity are intimately and inextricably linked (McNeely et al., 1995). Religion is also closely related to attitudes toward biodiversity. Perrings (1995) and others (Furze et al., 1996) demonstrate that some religions (Islam and Christianity) have gone far in setting humans apart from nature; other cultures with religious traditions such as Buddhism, Jainism, and Hinduism did not go as far.

Political Conviction

To some extent the political persuasion of individuals determines the attitudes toward biodiversity. When policy influences local populations, political decision making can change the attitudes of individuals; for example, where central government agencies have established control over forests previously managed by the community, attitudes changed from the ecological approach to the multifunctional approach (Neba, 1998).

Education

The higher the level of education, the greater the awareness of biodiversity. In industrialized countries a tendency toward techno-economic attitude or techno-economic dominance can be observed. In isolated populations, where individuals have never been in school, people know intuitively how to manage biodiversity in a sustainable way. Nevertheless, proper attention to biodiversity issues in education is an important element in developing support for policies to protect biodiversity.

Income and Wealth

Income and wealth are important determinants of basic attitudes. Poverty sets priorities on economic development and exploitation of natural resources.

Population Pressure

In extreme population densities, insufficient opportunities exist to maintain biodiversity. Population pressure tends to set priorities on economic development, exploitation, and conversion of national or rural areas toward agriculture. McNeely et al. (1995) demonstrated that human population density is clearly linked to loss of biodiversity.

Legal Restrictions on Individual Behavior That Prescribe the Range of Admissible Actions

Legal arrangements can be considered an expression of certain attitudes, while at the same time they contribute to the development of attitudes. Societies with an individualistic approach will not be in favor of protective policies. Societies with a more collective attitude may be more inclined to a larger extent toward collective actions for protection.

Valuation Methods

Valuation methods can be used to some extent to establish the benefits of biodiversity protection. The contingent valuation method is an *expressed* preference method; the hedonic pricing method and the travel cost method are referred to as a *revealed* preference method. Another important valuation method, the shadow-project method, can be considered as a *nondemand curve approach.*

The *contingent valuation method* (CVM) approach relies on a direct survey in which respondents are confronted with a hypothetical market situation. Given information on a specific environmental issue (e.g., forest conservation or the presence of certain species in nature), they are asked to make bids. Thus, they reveal their values and indicate their willingness to pay or their willingness to accept. There has been much controversy concerning the philosophy and practice of contingent valuation surveys (Cameron, 1997). Nevertheless, under some conditions it seems to be a helpful method (Bishop et al., 1983; Munasinghe, 1995).

The *hedonic pricing method* (based on the assumption of hedonism as the basis for human behavior) derives a value for some aspects of environmental quality from actual market prices of certain private goods. For example, we can expect real estate prices near a national park to be higher than those in industrial areas. Other things being equal, the difference in housing prices shows the premium paid for the local environment: how the consumer values the availability of nature. Thus, the purchaser's willingness to pay is indicated (Pearce and Markanday, 1987; Hanley, 1992).

The *travel cost method* is one of the earliest methods for valuing recreational functions of the environment. Just like the hedonic pricing method, the travel cost method is based on observed behavior (Clawson and Knetsch, 1966; Knetsch and Davis, 1966; Brown and Mendelsohn, 1984; Howe, 1991). The travel cost method has been particularly useful for assessing recreational values.

The *shadow-project method* is closely linked to the names of Klaassen and Botterweg (1976). The method is based on the view that no further deterioration of natural areas should take place. The shadow-project method determines the price of a natural area by calculating the costs of reproducing such an area at another location or by calculating the costs of an alternative project that avoids the damage.

Finally, valuation of the benefits of biodiversity and nature are far from trivial. The valuation methods can be separated into two basic approaches — those which value biodiversity following a demand curve and those which do not. The former approach consists of two kinds of methods: expressed preference method and revealed preference method.

ANALYSIS

Better understanding of attitudes, values, and biodiversity can only be gained when functions of biodiversity are recognized and valuation methods are appraised.

Functions of Biodiversity

Quality of life (well-being) and human welfare depend directly or indirectly on the availability of environmental goods and services. Early recognition of the functions and the role ecosystems play in maintaining these functions is essential for ensuring the long-term integrity of biodiversity. Four functional categories can be distinguished (Table 16.4; de Groot, 1992).

1. *Regulation:* the capacity of natural and seminatural ecosystems to regulate essential ecological processes and life-support systems that, in turn, contribute to the maintenance of a healthy environment by providing clean air, water, and soil.
2. *Carrier:* natural and seminatural ecosystems provide space and a suitable medium for many human activities, such as habitation, cultivation, and recreation.

Table 16.4 Functions of Natural Environment

Regulation Functions

1. Protection against harmful cosmic influences
2. Regulation of the local and global energy balance
3. Regulation of the chemical composition of the atmosphere
4. Regulation of the chemical composition of the oceans
5. Regulation of the local and global climate (including the hydrological scale)
6. Regulation of runoff and flood prevention (watershed protection)
7. Water catchment and groundwater recharge
8. Prevention of soil erosion and sediment control
9. Formation of topsoil and maintenance of soil fertility
10. Fixation of solar energy and biomass production
11. Storage and recycling of organic matter
12. Storage and recycling of nutrients
13. Storage and recycling of human waste
14. Regulation of biological control mechanisms
15. Maintenance of migration and nursery habitats
16. Maintenance of biological (and genetic) diversity

Carrier Functions

Providing Space and a Suitable Substrate for:
1. Human habitation and (indigenous) settlements
2. Cultivation (crop growing, animal husbandry, aquaculture)
3. Energy conversion
4. Recreation and tourism
5. Nature protection

Production Functions

1. Oxygen
2. Water (for drinking, irrigation, industry, etc.)
3. Food and nutritious drinks
4. Genetic resources
5. Medicinal resources
6. Raw materials for clothing and household fabrics
7. Raw materials for building, construction, and industrial use
8. Biochemicals (other than fuel and medicines)
9. Fuel and energy
10. Fodder and fertilizer
11. Ornamental resources

Information Functions

1. Aesthetic information
2. Spiritual and religious information
3. Historic information (heritage value)
4. Cultural and artistic inspiration
5. Scientific and educational information

Adapted from de Groot, R.S., *Functions of Nature*, Wolters Noordhoff, Groningen, the Netherlands, 1992.

3. *Production:* nature's many resources range from food and raw materials for industrial use to energy resources and genetic material.
4. *Information:* natural ecosystems contribute to the maintenance of mental health by providing opportunities for reflection, spiritual enrichment, cognitive development, and aesthetic experience.

Functions of ecosystems can be mutually exclusive or competitive; one function cannot be fulfilled if another function is dominating (Holling et al., 1995; Mooney et al., 1995). This competition can be qualitative (polluted water reduces the regulation function) or quantitative (habitation requires space that is not available for other purposes).

Also, there are probably many unknown goods and services (functions) with considerable (potential) benefit to human society (De Groot, 1992). Table 16.5 describes 37 separate functions of nature; a complete listing and division of functions may be impossible to produce.

Closely related to the functions of nature are the categories of value distinguished in economics.

Value Categories

The economic taxonomy of the value of biodiversity distinguishes between direct and indirect use values and adds the so-called non-use value (Perrings, 1995). The components of value are illustrated schematically in Figure 16.2. This figure illustrates the use and non-use values provided by a multiattribute environmental asset such as a woodland (Bateman and Turner, 1993). By definition, *use value* refers to the direct value individuals attribute to the services an ecosystem provides, e.g., recreational value or production value relating to hunting, fishing, and collecting plants or fruits. Use value is a measure of willingness to pay for goods and services, backed up by utility and purchasing power (Kadekodi and Ravindranath, 1997). For production value we can usually measure the market price of the goods and services. For values without market prices such as recreational values, we need to use other valuation methods (Clawson and Knetsch, 1966; Douglas and Johnson, 1992).

Option values based on Turner et al. (1994) need further explanation. They are essentially expressions of preference: a willingness to pay for the preservation of environmental systems or components of systems against some probability that the individual will make use of them at a later date. Provided the uncertainty concerning future use relates to uncertainty in the availability or *supply* of the environment, economic theory indicates that this option value is *likely* to be positive (Weisbrod, 1964). A related form of value is *bequest value*, a willingness to pay to preserve the environment for the benefit of one's children and grandchildren. It is not a use value for the current individual value but a potential future use or non-use value for his or her descendants.

Finally, there are *non-use values*. They suggest noninstrumental, anthropocentric values in the real nature of the thing, but not associated with actual use or even the option to use the thing. "Instead such values are taken to be entities that reflect people's preferences, but include concern for, sympathy with and respect for the rights or welfare of non-human beings" (Bateman and Turner, 1993). An example of these non-values is the *existence value* or intrinsic value (Krutilla, 1967; Pearce and Turner, 1990). It refers to the value that the public attributes to the ecosystems just because of their presence, without making direct or indirect use of them. It includes a recognition of the value of the very existence of certain species or whole ecosystems. *Total economic value*, then, is made up of actual use value plus option value plus existence value.

The non-use values of ecosystems are, in general, very difficult to establish. Non-use values typically have a public good character for which no market prices are available.

If biodiversity is important to the functioning of some ecological system, it will not automatically be valuable to society, or vice versa. Economic assessments of biodiversity are also influenced by ethical judgements. They merely make explicit something that is currently implicit. However, there remains a gap between the market price of biodiversity and its value to individuals and societies (Perrings, 1995).

Trade-Offs

Trade-offs are benefits from one decision that are bought at the expense of another. If conserving biodiversity conflicts with economic activity, the question is which view should

Table 16.5 Overview of Characteristics of Attitudes, Valuation, and Nature Management in Six Countries

Country/ Region	Dominant Religious Background	Dominant Attitude	Example of Values	Consequences for Nature Management	Value and Valuation Method
Bali	Balinese culture endowed by the Hindu religion	Humans have use of, not ownership of or dominance over the environment	Great value on continuity of customary belief and community behavior	Few constraints to fell forests, few protected areas	Bequest and existence value CVM[a], MCA[b]
Benin	Urban: Christian, Islamic; Rural: animism	Urban: multifunctional use Rural: ecological approach	Holy forests with religious and cultural functions	To be strongly protected, preferably by reinforcement of traditional laws	Option and existence value CVM, MCA
Bhutan	Buddhism	Compassion and respect to all forms of life, a deep ecologist view	Forest Act (1959 and adapted 1997)	Protection of nearly all forests of Bhutan, strong control	Existence and indirect use value CVM, MCA
Cameroon	Urban: Christian Rural: animism	Ecological approach supported by taboos	Conservation of forests in sud-Cameroon	Proposed restrictions for protection of forests in favor of local communities	Existence and indirect use value CVM, MCA
Costa Rica	Christian	Multifunctional approach; stewardship	Originally forests being converted into land for cattle	Laws for protecting remaining forests and other biodiversity assets	Direct use, option and bequest value CVM, travel cost, market prices
The Netherlands	Christian	Diverse, but dominant: stewardship	Wadden Sea, strong clashes on ecological and nature issues by groups with different attitudes, slight movement toward ecological approach	National: results in the development of ecological main structures Local: management of the Wadden Sea	Direct use, existence value; market price, hedonic pricing and CVM

[a] CVM: Contingent valuation method.
[b] MCA: Multicriteria analysis.

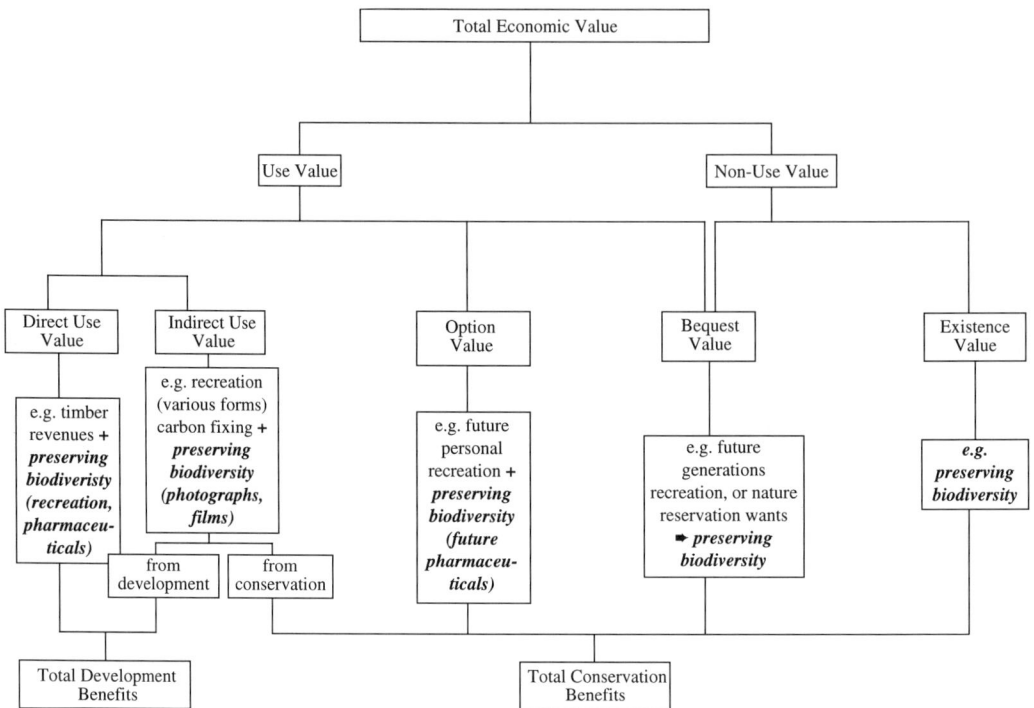

Figure 16.2 The total economic value of biodiversity. (Adapted from Bateman, I.J. and Turner, R.K., *Sustainable Environmental Economics and Management; Principles and Practice*, Belhaven Press, London, 1993, pp. 120–191.)

prevail? Decision making is based on the idea that we do only those things that yield us net gains. In other words, where we have to choose between alternatives, we choose the one that offers the greatest net gain. This is the foundation of the cost–benefit analysis (Turner et al., 1994). If the cost and the benefits of biodiversity protection are known, economic decisions on biodiversity protection can be made.

Local and Global Benefits

The benefits of biodiversity can be separated into local and global benefits. These benefits often have no market; but when they can be estimated, they are often global in nature (Turner et al., 1994). A good example is given by Costanza et al. (1997), who estimated the current economic value of 17 ecosystem services, such as water regulation, soil formation, biological control, genetic resources, cultural assets, etc.

The Costs of Protection and Sharing the Burden

Biodiversity protection implies a cost to society: the opportunities forgone by society by protecting such diversity. Protection of biodiversity also includes direct operating and capital costs. When the benefits of biodiversity protection can be classified as global benefits, it will always be in the interest of any one country to see that all other countries, except itself, pay for such a protection. The global existence value of biodiversity (Figure 16.2) often accrues to rich nations but does not have a cash flow.

Removing Barriers for Protection

One way to encourage protection is for countries possessing biodiversity to appropriate the global benefits they are providing the rest of the world (Turner et al., 1994). Further, mental barriers can be removed by changing the basic attitudes on biodiversity.

Many economic activities are subsidized or encouraged in some way by governments. Whereas conservation is not subsidized (generally), alternative land uses are. Therefore, biodiversity conservation has to face unfair competition. Conservation of biodiversity should be evaluated along with other goods and services provided by environmental resources and should form an integral part of land use and management. The aim of conservation should be to optimize the provision of ecological and other services, and this does not necessarily imply the preservation of the existing resource in all its facets (Perrings, 1995).

For making cost-effective solutions, it is fundamentally important to know what is being traded off against what. The basic rule is that, for conservation to be *economically* justified, the benefit of protection minus the cost of protection should be larger than the benefit of economic activity minus the cost of economic activity. But, as already mentioned, "biodiversity often has no market price tag and a considerable amount of uncertainty can surround their true value and significance" (Turner et al., 1994).

Since not all value categories of biodiversity can be valued and expressed in monetary units, it is essential to develop consistent policy plans for biodiversity protection. These plans are based on scenario studies for biodiversity that describe and analyze past and existing situations and sketch a future situation and how to arrive at it.

To develop scenarios it is essential to describe and analyze options for biodiversity protection and include the economic costs of carrying them out. Next, the policy measures can be ranked according to cost effectiveness, and choices can be made for the priorities in biodiversity protection.

The scenarios can be analyzed by partial cost–benefit analysis, assessment of indicators, and by multicriteria analysis, which rank alternative policy plans on the basis of explicit weights attached to the various criteria. These criteria may include economic costs, quality of landscape, protection of biodiversity (expressed by means of indicators), and impacts on employment.

Once scenarios have been developed, it is possible to analyze their impacts and discuss the distribution of costs and benefits among various stakeholders, including local producers and consumers, national economic agents, and the international community. At the same time, one can analyze how biodiversity protection can be financed from domestic or international sources. The scenario studies can be supported by geographical information systems, decision support systems, and expert systems that document the present situation and make it possible to analyze future developments.

Research along these lines can contribute to developing national and international policy for biodiversity protection that is cost effective and clearly indicative of how well-defined targets for biodiversity protection can be reached in various regions of the countries concerned (de Kruijf and van Vuuren, 1998).

OVERVIEW OF CASE STUDIES

This section gives an overview of the background and attitudes to nature and biodiversity of selected populations and communities. Based on the foregoing discussion, we indicate the values of and consequences for nature management. For Bali, we used an extensive review by Martopo and Mitchell (1995) and some personal observations; for Cameroon, we used a thesis and some personal observations; for the other countries we used different reports and publications (see van Ierland et al., 1998).

Bali

In Balinese society, human ecological relationships are governed by philosophies and cosmologies that place humans in a *middle* world between spirit worlds *above* and *below*. Life revolves around finding and maintaining harmony and balance among gods, humans, and the environment. Nevertheless, although there are small areas of protected forests, there seem to be few constraints on felling forests. Still, the possible presence of spirits slows the rate of attrition. Protection and use of natural assets typically are conflicting activities. In Balinese culture, based on Hinduism, there seems to be a delicate equilibrium between the ecologist view and the multifunctional approach (Boehmer and Wickham, 1995; Whitten, 1995).

Benin: The Holy Forests

Small, isolated forests near villages are set apart by local and regional authorities for religious and cultural rites. Only initiated men are allowed to enter. Values of the forests are considered only from religious perspectives. The forests are still very well protected but face challenges from the population due to conversion to Islam and Christianity. Another problem is the exchange of endangered, but in principle protected species (e.g., the pythons), as these forests are quite isolated. Reinforcement of traditional law is the best option, but it is difficult to achieve. Creation of corridors between the holy forests is not feasible because agricultural areas would have to be taken from the population. A clash exists between traditional — animism — and modern ways of management.

Bhutan: Biodiversity and Nature

In Bhutan 72.5% of the land is covered by forest (26% is protected), and nearly all of it is protected by law and religious conviction (Buddhism). Ethical and aesthetic roles and values of biodiversity are integral components of the culture, as rooted in the Buddhist philosophy of "compassion and respect to all forms of life." What is taken away should be given back. The intrusion of modernity has not, until now, disrupted the ancient equilibrium of a hierarchical, well-ordered, agricultural, and pastoral society, under which evolved a way of life, architecture, and art firmly rooted in the philosophy of Buddhism and exceptionally well related to its natural environment. Bhutan is practicing an ecological approach in the management of its environment nearest to the deep-ecologist attitude. Opening to modern lifestyle will in the long run challenge this view, approach, and decision-making system.

Cameroon

In Southern Cameroon, as in Benin, taboos form the basic drive for local management of forests. This approach determines the way the autochthone population is allowed to use and manage the forests. The forests are multifunctional and, therefore, extremely valuable to the local people. The intention to develop nature reserves in these areas demands an approach that considers these views. Traditional views are in conflict with the technical management views of the government, which tries increasingly to limit the use of forests by the riverine people. Cameroon represents the typical conflict between the original multifunctional approach and the modern ecological approach of nature management (Neba, 1998).

Costa Rica

Christian tradition has governed the use and management of nature in the last few hundred years. Exploitation was governed by the needs of the population and economic demands. This is the typical stewardship approach. Serious changes can be observed in the last decade, partly initiated by the UNCED and other comparable activities; these changes have led to a more multifunctional approach of the management of nature and biodiversity. Judicial systems are not yet sufficiently developed, nor is the population fully aware of these changes of philosophy in managing nature.

The Netherlands: The Wadden Sea

The open connection of the Wadden Sea to the North Sea has an enormous importance for wildlife as a breeding area for North Sea fisheries and as a sanctuary for migratory birds twice a year. The presence of large, economically interesting mineral resources, as well as the commercial exploitation of some areas within the Wadden Sea, has led to heated discussion about the management of the Wadden Sea. Attitudes ranging from ecologist to techno-economic are present in these discussions. Policy measures most related to the ecologist view have been taken to protect the area, but they are constantly challenged on the basis of economic welfare.

OBSERVATIONS AND CONCLUSIONS

Attitudes

Attitudes toward biodiversity and nature differ among policy makers, individuals, and stakeholders and range from deep ecologist to techno-economic dominance. The examples show that attitudes of the more relevant policy-making institutions range from ecological approach to techno-economic attitude. At the local level, decision makers or responsible local authorities tend to the more extreme end of the classification. It is unclear when and how the local attitudes will prevail or influence decision making on biodiversity protection, use, or exploitation because these attitudes may well differ from the national views.

In addition to socioeconomic and political considerations, traditions, religions, and culture are important in defining the values and uses of biodiversity, especially at the local and regional levels. Therefore, these factors were considered in decision-making processes in biodiversity. In spite of strong influences of attitudes and religions in Benin and Bhutan, trade-offs between biodiversity protection and economic development can and are being made under highly specific circumstances. People accept that some negative impacts on biodiversity may occur, if they are sufficiently compensated by economic gains and if extensive cultural explanations are allowed. It was also shown in the case studies that attitudes toward biodiversity and nature conservation can be influenced by education, religion, practical use of the natural goods, and experience.

Classification of Economic Values Related to Biodiversity

The classification of economic value categories (direct use, indirect use, option, bequest, and existence value) is considered to be an anthropocentric, mainly economic classification, acceptable in principle to many decision makers. Besides direct and indirect use values, option, bequest, and existence values make explicit that other value categories (related to ethical, cultural, and religious perspectives) are important in decision making on biodiversity conservation. How to assess option, bequest, and existence values remains a complicated issue. Further studies and practical experience should reveal to what extent actual monetary valuations for these latter values can be made.

Valuation Methods and Their Applicability

Economic valuation methods are considered useful tools to assess the above-mentioned value categories but should not be considered a panacea for all valuation questions related to biodiversity. Different valuation methods can be applied to various functions of biodiversity or value categories (see Table 16.5).

Table 16.5 shows that most methods for monetary valuation can only be used for a limited number of value categories. The CVM method is, in principle, applicable for all value categories; but a scientific debate is going on about its validity, strengths, and drawbacks.

It is recommended that the relevant valuation methods in practice for the appropriate value categories (Table 16.5) be applied to obtain the best information about the monetary value of various functions of ecosystems and biodiversity. Valuation methods make it possible to pay attention explicitly to the various functions of biodiversity and ecosystems. However, not all value categories can be captured by monetary valuation methods.

Monetary valuation methods can only be usefully applied if the policy makers and other stakeholders support their use. In countries with a strong neoclassical economic tradition, such as the Netherlands and Costa Rica, the application of economic valuation methods are more likely to be accepted. In Bali, Benin, and Cameroon, an interest also exists in the application of monetary valuation methods. For Bhutan it remains to be discussed whether and how economic valuation methods can be applied. Valuation is an essential tool to support the policy-making process, e.g., at the local level and in land use planning, agriculture, and infrastructure activities. Monetary valuation is an additional step that may provide relevant information. More experience is necessary to assess to which cases to apply monetary valuation.

Analytical Tools for Biodiversity Protection

If economic valuation methods cannot contribute to proper decision making on biodiversity protection, what other methods are available? Decision making on biodiversity protection can be supported by means of scenario studies, land use planning, and environmental impact assessment that show how various functions of ecosystems will develop over time under different policy scenarios. The scenarios can suggest implications for the international aspects of biodiversity, and national, regional, and local impacts. In this way the global and national benefits of various plans for biodiversity protection can be assessed, at least qualitatively. Policy makers and other stakeholders can then express their preferences for various scenarios, thus implicitly indicating how they value the various characteristics of the scenarios. Ranking the scenarios can be established by multicriteria analysis.

Research is then required to analyze, describe, and sketch future developments. Good indicators are required for describing the developments and the state of biodiversity. Values based on attitudes and well-documented information should be obtained to sketch the various scenarios and implications for biodiversity at the level of ecosystems, species, and genetic material. In addition to valuation, other essential tools to assist in conservation and sustainable use of biodiversity include scenarios, protection plans, environmental impact assessment, and spatial planning. They must also be applied on the basis of data and should consider the different values and attitudes of stakeholders. Knowledge derived from continuous data collection and inventory of biodiversity and its components largely determines the effectiveness of the application of these tools. Therefore, a continuous scientific effort must determine:

- Inventory of ecosystems, species, and genetic information
- Qualitative description of functions
- Quantitative analysis

A thorough decision-making process, based on sustainable land use planning that includes local participation and the application of tools such as valuation, will provide elements for sustainable use and conservation of biological resources.

Economic Benefits from Biodiversity Protection

Biodiversity protection will result in many economic benefits attributable to the various functions of ecosystems, species, or genetic information. Economic benefits include carbon sequestration, watershed protection, water supply, hydropower, and development of genetic potentials for pharmaceuticals, agricultural use, recreation, and tourism. Still, numerous economic benefits are underutilized and need further attention if they are to be taken into account in the economic decision-making process. Attitudes are very important in these processes as shown by the case studies. For further development to be long-lasting, successful, and inclusive of many views, information about attitudes is essential.

Protection of property rights, particularly for benefits sharing, is essential. Well-designed laws and relevant human resources are also essential in this regard.

International Aspects

When applying valuation of biodiversity, it is important to recognize that values may differ among stakeholders. At the international level, attitudes and values may differ from those at the regional and local levels. The difficult issue of when international interests, national interests, regional interests, or local interests receive priority certainly cannot be solved by means of monetary valuation. Harmonization of legislation on transboundary biological resources and adequate enforcement are needed both between countries and between relevant international agreements (e.g., with CBD and CITES). Enforcement of existing regulation should be strengthened.

The international community should support conservation measures that consider peoples' attitudes and values and provide global as well as local benefits through financial means and appropriate technological transfers.

ACKNOWLEDGMENTS

We are grateful to a number of colleagues who participated in workshops trying to get hold of the complicated aspects of values, attitudes, and nature. In particular, we thank Gaetan Agbangla and Fulbert G. Amoussouga from Benin; Edmundo Castro, Luis Gamez, and Erick Varga from Costa Rica; Karma C. Nyedrup and Kunzang Norbu from Bhutan; and Jacob-Jan Bakker, Arthur Eijs, and Chris Enthoven from the Netherlands. Also we acknowledge the help of WWF in Cameroon. The Foundation Ecooperation supported part of this study.

REFERENCES

Achterberg, W., *Partners in de Natuur (Partners in Nature)*, Van Arkel, Utrecht, the Netherlands, 1986.

Barbault, R. and Sastrapradja, S., Generation, maintenance and loss of biodiversity, in Heywood, V.H., Ed., *Global Biodiversity Assessment*, United Nations Environment Program, Cambridge University Press, Cambridge, 1995, pp. 193–274.

Barbier, E.B. and Rauscher, M., Policies to control tropical deforestation: trade intervention versus transfers, in Perrings, C. et al. (Eds.), *Biodiversity Loss; Economic and Ecological Issues*, Cambridge University Press, Cambridge, 1995, pp. 260–282.

Bateman, I.J., Valuation of the environment and techniques: revealed preference methods, in Turner, R.K., Ed., *Sustainable Environmental Economics and Management; Principles and Practice*, Belhaven Press, London, 1993, pp. 192–265.

Bateman, I.J. and Turner, R.K., Valuation of the environment, methods and techniques: the contingent valuation method, in Turner, R.K., Ed., *Sustainable Environmental Economics and Management; Principles and Practice*, Belhaven Press, London, 1993, pp. 120–191.

Bishop, R.C., Heberlein, T., and Kealy, M.J., Contingent valuation of environmental assets: comparisons with a simulated market, *Nat. Resour. J.,* 23:619–633, 1983.

Boehmer, K. and Wickham, T., Linking Bali's past with a sustainable future, in Martopo, S. and Mitchell, B., Eds., *Bali, Balancing Environment, Economy and Culture*, Department of Geography Series Pub. No. 44, University of Waterloo, Kitchener-Waterloo, Ontario, 1995, pp. 437–464.

Brown, G. and Mendelsohn, R., The hedonic travel cost method, *Rev. Econ. Stat.*, 66:427–433, 1984.

Cameron, J.I., Applying socioecological economics: a case study of contingent valuation and integrated catchment management, *Ecolog. Econ.*, 2: 155–165, 1997.

Clawson, M. and Knetsch, J., *Economics of Outdoor Recreation*, Resources for the Future, Washington, D.C., 1966.

Costanza, R., d'Arge, R., de Groot, R., Farber, S., Grasso, M., Hannon, B., Naeem, S., Limburg, K., Paruelo, J., O'Neill, R.V., Raskin, R.G., Sutton, P., and van den Belt, M., The value of the world's ecosystem services and natural capital, *Nature*, 387:253–260, 1997.

de Groot, R.S., *Functions of Nature*, Wolters Noordhoff, Groningen, the Netherlands, 1992.

de Kruijf, H.A.M. and van Vuuren, D.P., Following sustainable development in relation to the North-South Dialogue: ecosystem health and sustainability indicators, *Ecotoxicol. Environ. Saf.*, 40:414, 1998.

Douglas, A.J. and Johnson, R.L., Congestion and recreation site demand: a model of demand-induced quality effects, *J. Environ. Manage.*, 36:201–213, 1992.

Furze, B., deLacy, T., and Birckhead, J., *Culture, Conservation and Biodiversity; The Social Dimension of Linking Local Level Development and Conservation through Protected Areas*, John Wiley & Sons, Brisbane, 1996.

Hanley, N., Are there environmental limits to cost benefit accounting? *Environ. Resour. Econ.*, 1:33–59, 1992.

Heywood, V.H. and Baste, I., Introduction, in Heywood, V.H., Ed., *Global Biodiversity Assessment*, United Nations Environment Program, Cambridge University Press, Cambridge, 1995, pp. 1–19.

Holling, C.S., Schindler, D.W., Walker, B.W., and Roughgarden, J., Biodiversity in the functioning of ecosystems: an ecological synthesis, in Perrings, C. et al. Eds., *Biodiversity Loss; Economic and Ecological Issues*, Cambridge University Press, Cambridge, 1995, pp. 44–83.

Howe, C.W., Taxes versus tradable permits: the views from Europe and the United States, University of Colorado (mimeo), 1991.

Kadekodi, G.K. and Ravindranath, N.H., Macro-economic analysis of forestry options on carbon sequestration in India, *Ecol. Econ.*, 3:201–223, 1997.

Keating, M., *The Earth Summit's Agenda for Change: A Plain Language Version of Agenda 21 and Other Rio Agreements*, Centre for Our Common Future, Geneva, 1993.

Klaassen, L.H. and Botterweg, T.H., Project evaluation and intangible effects: a shadow project approach, in Nijkamp, P., Ed., *Environmental Economics Volume 1: Theories*, Martinus Nijhoff Social Sciences Division, Leiden, the Netherlands, 1076, pp. 33–49.

Knetsch, J. and Davis, R.K., Comparisons of methods for recreation evaluation, in Kneese, A. and Smith, S., Eds., *Water Research*, John Hopkins University Press, Baltimore, 1966, pp. 125–142.

Krutilla, J., Conservation reconsidered, *Am. Econ. Rev.*, 4:777–786, 1967.

Martopo, S. and Mitchell, B., Bali, Balancing Environment, Economy and Culture, Department of Geography Publ., series 44, University of Waterloo, 1995.

McNeely, J.A., Gadgil, M., Leveque, C., Padoch, C., and Redford, K., Human influences on biodiversity, in Heywood, V.H., Ed., *Global Biodiversity Assessment*, United Nations Environment Programme, Cambridge University Press, Cambridge, 1995, pp. 711–821.

Miller, K., Allegretti, M.H., Johnson, N., and Jonsson, B, Measures for conservation and sustainable use of its components, in *Global Biodiversity Assessment*, United Nations Environment Programme, Cambridge University Press, Cambridge, 1995, pp. 915–1061.

Mooney, H.A., Lubchenco, J., Dirzo, R., and Sala, O.E., Biodiversity and ecosystem functioning: basic principles, in Heywood, V.H., Ed., *Global Biodiversity Assessment*, United Nations Environment Programme, Cambridge University Press, Cambridge, 1995, pp. 275–325.

Munasinghe, M., Environmental Economics and Sustainable Development, World Bank Environmental Paper 3, Washington, D.C., 1993.

Neba, G.M., Sur l'attitude au nature des peuples sud-cameronais, Thesis, Yaounde, Cameroon, 1998.

OECD, *Saving Biological Diversity: Economic Incentives*, OECD Publications, Paris, 1996.

Pearce, D. and Markandya, A., *The Benefits of Environmental Policies*, Environment Committee Group of Economic Experts, OECD, Paris, 1987.

Pearce, D.W. and Turner, R.K., *Economics of Natural Resources and the Environment,* Harvester Wheatsheaf, Hemel Hempsted, U.K., 1990.

Perrings, C., The economic value of biodiversity, in Heywood, V.H., Ed., *Global Biodiversity Assessment*, United Nations Environment Program, Cambridge University Press, Cambridge, 1995, pp. 823–914.

Perrings, C., Mäler, K-G., Folke, C., Holling, C.S., and Jansson, B-O., Introduction: framing the problem of biodiversity loss, in Perrings, C. et al., Eds., *Biodiversity Loss; Economic and Ecological Issues*, Cambridge University Press, Cambridge, 1995, pp. 1–17.

Stork, N.E. and Samways, M.J., Inventorying and monitoring, in Heywood, V.H., Ed., *Global Biodiversity Assessment*, United Nations Environment Programme, Cambridge University Press, Cambridge, 1995, pp. 453–543.

Swanson, T., The international regulation of biodiversity decline: optimal policy and evolutionary product, in Perrings, C. et al., Eds., *Biodiversity Loss; Economic and Ecological Issues*, Cambridge University Press, Cambridge, 1995, pp. 225–259,

Turner, R.K., Pearce, D., and Bateman, I., *Environmental Economics; An Elementary Introduction,* Harvester Wheatsheaf, Hemel Hempsted, U.K., 1994.

UNCED, Agenda 21: Programme of Action for Sustainable Development; RIO Declaration on Environment and Development, Statement of Forest Principles; Rio de Janeiro, Chapter 15: Conservation of Biological Diversity, 131–135, 1992.

Van Ierland, E.C., de Kruijf, H.A.M., and van der Heide, C.M., *Attitudes and the Value of Biodiversity,* Department of Economics, Wageningen Agricultural University, Wageningen, the Netherlands, 1998.

Weisbrod, B.A., Collective-consumption services of individual-consumption goods, *Q. J. Econ.*, 58:471–477, 1964.

Whitten, A.J., Natural areas and nature of inland Bali, in Martopo, S. and Mitchell, B., Eds., *Bali, Balancing Environment, Economy and Culture*, Department of Geography Series Pub. No. 44, University of Waterloo, Kitchener-Waterloo, Ontario, Canada, 1995, pp. 237–262.

CHAPTER 17

Humane Values as a Basis for Ecosystem Health

John W. Grandy and Allen T. Rutberg

INTRODUCTION

Humane values have broad public appeal. Care and compassion for our fellow creatures are deeply rooted in our emotional makeup and in our fundamental human abilities to understand and empathize with the experience of other people and animals (Kellert and Wilson, 1993). The emotional connections we form with the dogs or cats in our households, the horses in our pastures, and the many other animals in our lives are real and profound, and spring from the same sources as our loving connections with family and friends.

More and more people now recognize and appreciate the fundamental rightness of humane values and that the quality of life for all of us will be improved if these values guide our actions. As a result, charitable organizations such as The Humane Society of the United States (HSUS), which advances the values of care and compassion for all creatures, have experienced tremendous growth. In the last decade of the millennium, the constituency of the HSUS grew more than fivefold and now exceeds over 7 million concerned citizens. Other organizations dedicated to advancing humane values have also thrived.

This growth in membership and support base, which we believe is underestimated even by these growth levels, represents both a sea change in public support for animal protection and conservation programs and an unparalleled opportunity to make significant progress in meeting our shared objectives for ecosystems, habitats, and animals that are affected by humans.

The lives of wild animals and humans have become increasingly entangled. People have always affected and been affected by wildlife: as prey or predators, as competitors, as codependents, and as biological signposts in our attempts to map our world. All these relationships remain intact today; but now, more than ever, the fate of wild animals will depend on how they are treated by humans. The modification, fragmentation, degradation, and destruction of wildlife habitat by people are now, at least in the short run, the most important forces determining where, how, and whether wild animals live on Earth.

To cement the link between humane values and wildlife protection, and ultimately to ecosystem protection, we focus our efforts on that large segment of the public that lives in cities, towns, and suburbs. That is where the heart of public opinion lies. (According to the 2000 U.S. Census, 80% of the population now lives in metropolitan areas of 50,000 or more.) It is also where most people encounter wildlife most frequently. These people know, and want to know, the beauty of wildlife first-hand. They are not concerned about whether wildlife is deemed feral, native, or nonnative; rather, they understand it as beautiful, majestic, and awe-inspiring. For many, wildlife still sleeps in back yards; feeds in gardens, flower beds, and berry bushes; burrows under the front steps; forages in the vacant lot across the street; and fills the park near the outskirts of town. Even in the heart of Washington, D.C. and in its densely settled suburbs, our yards are home to raccoons, opossums, voles, salamanders, woodpeckers, crows, and chickadees. We and others are richer for each encounter with these wonderful creatures and for the increasing level of interaction with Canada geese and white-tailed deer in the larger community.

ENHANCING PUBLIC APPRECIATION OF WILDLIFE

We believe that if people appreciate and enjoy wildlife in their lives, even vicariously, they will want to help improve the welfare of wildlife, improve and save wildlife habitat, and support local, national, and international programs and policies that benefit wildlife. This is an ascending cycle of understanding, concern, and involvement. Consequently, we have put into place a number of programs that promote appreciation for wildlife, help people solve any conflicts that may arise with their wild neighbors through humane and effective techniques, and simultaneously build concern and encourage involvement in wildlife protection.

Our urban wildlife protection program is the centerpiece of these efforts. The HSUS Urban Wildlife Program started (and continues) with a series of 1- and 2-day workshops, held all over the country, aimed at animal control officers, shelter staff, nuisance wildlife control operators, and others who deal directly with wildlife conflicts experienced by the public. Participants are trained in the latest techniques in humane wildlife conflict resolution and receive information on wildlife diseases and legal constraints. Information presented in each workshop is tailored to the region in which it is conducted.

From these origins, the HSUS Urban Wildlife Program grew to include publication of brochures and two books. The first book was the very successful *Pocket Guide to the Humane Control of Wildlife in Cities and Towns* (Hodge, 1991), which was then expanded into *Wild Neighbors* (Hadidian et al., 1997). HSUS personnel regularly attend and present at wildlife damage control meetings and work closely with government personnel, wildlife rehabilitators, nuisance wildlife control operators, and others to explore new techniques and set standards for handling wildlife.

Our latest urban wildlife program is the HSUS Urban Wildlife Sanctuary Program. Born out of the National Institute of Urban Wildlife's urban sanctuary program, led by Dr. Lowell Adams, our backyard sanctuary program is a powerful educational tool for encouraging appreciation of backyard wildlife, disseminating practical information on the humane resolution of wildlife conflicts, and providing guidance on how to improve backyard habitat for wildlife. Landowners who apply to the HSUS for urban sanctuary designation receive a copy of *Wild Neighbors*, a subscription to the quarterly newsletter *Wild Neighbor News*, and access to a variety of other helpful resources. This enterprise engages not only small private landowners but also neighborhood associations, planners, developers, corporate landowners, and even whole municipalities. We see unlimited potential here to influence the urban landscape and to preserve the opportunity for people to see and appreciate wildlife.

Complementing the Urban Wildlife Sanctuary Program is the HSUS Wildlife Land Trust. Although only a few years old, the land trust now protects 46,296 acres on 35 properties in 17 states as well as one foreign country, and it is growing rapidly. As with many conservation-oriented land trusts, the Wildlife Land Trust uses donations of land, land purchases, and conservation easements to restrict further use and development on properties valuable to wildlife. But the HSUS Wildlife Land Trust is different from others. While there are hundreds of land trusts in the U.S. (the Land Trust Alliance reports that it works with over 1200 such groups), from organizations with a national or international focus, like the Nature Conservancy, to the smallest local trust, the HSUS Wildlife Land Trust is the only national land trust committed to conserving land under principles of *humane stewardship*. For the HSUS, humane stewardship means that the animals living on land trust properties are protected in perpetuity from commercial and recreational trapping and hunting; that wildlife conflicts arising on trust properties are solved humanely; and that management is directed toward the benefit of the animals themselves, not toward human use. Moreover, the HSUS Wildlife Land Trust makes one additional commitment: if it agrees to take a parcel of land, it will protect or insure its protection *in perpetuity*. It will either keep it fully protected under its ownership, or, if the land is donated to another protected use, it will insure that protective covenants are maintained in perpetuity.

Thus, in addition to the direct and immediate benefits to wildlife of protecting their homes forever, the HSUS Wildlife Land Trust serves another function: to send a message to land managers everywhere that wildlife is worth preserving for its own sake, not just for its utility to hunters, trappers, or even bird-watchers. The HSUS Wildlife Land Trust takes a critical step in inserting humane values into the preservation of ecosystem health by demonstrating that conservation and protection of wildlife habitats can be mutually reinforcing of humane values.

BUILDING HUMANE VALUES INTO SCIENCE

Securing the health of ecosystems requires good science, as is demonstrated by the other chapters in this volume. The HSUS maintains, however, that the science of wildlife and land management itself must, for the future, be founded on humane values. After all, the role of science is to describe nature by collecting and organizing observations of the natural world. But the choice of questions that scientists ask about nature, what observations are made, and how those observations and questions are organized are guided by values as well as by the observations themselves (Rutberg, 1997). We believe that science founded on humane values such as kindness, compassion, and reverence for life will better inform and guide wildlife management policies in a progressive and positive direction.

In the 20th century, wildlife management was founded on and steeped in killing — sport hunting, "management hunting," commercial and recreational trapping, predator control, and other lethal forms of animal damage control. To increase deer numbers, state wildlife agencies sold licenses to shoot bucks; they then used the revenue to disrupt the habitats of other creatures to create more deer habitat, pushing up the reproductive rates of the does. To increase waterfowl numbers, state and federal wildlife management agencies inhumanely trapped raccoons, foxes, and other nest predators — again using the revenues generated from the sale of hunting licenses. Until the passage of the Endangered Species Act (ESA) in 1973, nongame species were largely ignored. Even including ESA-driven programs, government expenditures on nongame species were still, at the end of the century, dwarfed by hunting- and trapping-related expenditures (Hagood, 1997). In FY 1995, for example, approximately $30 million were appropriated by Congress to implement the ESA; by contrast, $209 million of federal Pittman–Robertson money were channeled to the state wildlife agencies to support hunting-related activities. Expenditures by the states are even more lopsided in favor of hunting.

But the restoration and preservation of healthy, sustainable ecosystems cannot be based on an ethical system indifferent to the value of life and to the well-being of our fellow creatures. For the next century, we must develop new methods and new technologies that intrude as little as possible into the lives of wildlife, thereby recognizing the value of all life.

For wildlife, the other side of the deteriorating habitat problem is increased conflicts with people where wildlife and people come together. The HSUS believes that tolerance for wildlife and modification of human behavior are the appropriate first responses to wildlife conflicts. However, it has become clear in the last decade or so that, in some locations, densities of wildlife are so high as to cause conflicts intolerable to the majority of people and injurious to the animals themselves (Conover et al., 1995). In the eastern U.S., dense populations of white-tailed deer are involved in many thousands of collisions with vehicles, with fatal consequences to the deer and trauma and sometimes injury to the occupants of the vehicle. In other situations, high densities of deer, wild horses, pigs, and other species may result in harm to habitat, biodiversity, and other wildlife that is deemed unacceptable (DeCalesta, 1997; Mac et al., 1998).

Recognizing that wildlife overabundance can be a genuine concern, but insisting that the traditional method of dealing with wildlife overabundance — killing — is ethically bankrupt, the HSUS has committed itself to leading a major research program in wildlife immunocontraception. By helping to develop a safe, effective, and humane nonlethal method to control wildlife populations, we hope to help ease the conflicts that develop where wildlife and people meet as neighbors, and thereby to help engender public tolerance for wildlife. At the same time, the act of applying immunocontraception to wild animals carries with it a message of concern for the lives of the animals we meet, disturb, and sometimes displace, and acknowledges our responsibility for humane stewardship.

The immunocontraception program is one of the largest carried out by the HSUS. We collaborate in this endeavor with at least half a dozen major universities, including the University of California, Davis; with zoos and aquaria around the world; with the National Park Service, the U.S. Fish and Wildlife Service, the Bureau of Land Management, the Department of Commerce, and Navy; with South Africa National Parks; and numerous state and local agencies. Our research focuses principally on white-tailed deer, wild horses, and African elephants; but it also encompasses dozens of species in zoos and a handful of other species in the field (Kirkpatrick and Rutberg, 2001).

Since our program began in earnest in the early 1990s, we have made extraordinary progress. The National Park Service is using immunocontraception to control wild horse populations on several East Coast barrier island beaches, most notably Assateague Island National Seashore, where immunocontraception has been the management tool of choice since 1995 (National Park Service, 1995). Our research team, in cooperation with the Bureau of Land Management, has treated over one thousand wild horses on public lands with immunocontraceptive vaccines and is testing it for large-scale management use across the west (Turner et al., 2001). We have learned much about the vaccine itself (Thiele, 1999; Naugle et al., 2002; Walter et al., 2002) and are testing prototypes of an effective tool — a one-shot, remotely delivered vaccine — for the management of white-tailed deer. Research we supported has even shown that the vaccine can be effective in wild African elephants when delivered remotely from helicopters in Kruger National Park in South Africa (Fayrer-Hosken et al., 2000). Possible applications appear enormous.

Moreover, immunocontraception has kindled the imagination of the public. Even without an effective management tool, the HSUS annually receives hundreds of requests for access to the technology from constituents, communities, land managers, and the public. They ask us: We have a deer problem; can we use your birth control method to help solve it? Does it work on beavers, on raccoons, on pigs, on goats, on bison? And the interest is confirmed in public opinion surveys. When asked what their preferred method for controlling wildlife populations is, the public consistently places birth control at or near the top.

Immunocontraception is not, of course, the ultimate answer to wildlife overabundance. As we assert above, the fate and welfare of wildlife depends ultimately on changing human behavior. Still, we are heartened and gratified by the outpouring of public demand for its use. It means we are right; the public cares deeply about wildlife and would like to see wild animals managed with kindness and compassion.

CONCLUSION

Ecosystems are slippery things. They are of intense interest to scientists, but the public has trouble understanding what ecosystems are, let alone supporting their protection. Ecosystem health, likewise, is of intense interest to scientists, as evidenced by this volume and participation in the conference upon which it is based. However, from this and other efforts must come the recognition that conservation and humane communities must increasingly come together to support protection for wildlife, its habitat, and the ecosystems upon which it depends. In 1938, the National Council of Parent Teacher Associations urged nationwide programs for the education of children emphasizing four fundamental humane values: kindness, compassion, justice, and reverence for all life. These are the values that drive the HSUS and its constituents today. And these are the values that drive the true conservation and animal protection communities. Clearly, what we must build for the future is a better world based on these core values and the healthy ecosystems they will support.

Healthy ecosystems will need healthy values. Care, compassion, and acknowledgment of the value of all life are, we believe, the right medicine for healthy ecosystems.

REFERENCES

Conover, M.R., Pitt, W.C., Kessler, K.K., DuBow, T.J., and Sanborn, W.A., Review of human injuries, illnesses, and economic losses caused by wildlife in the United States, *Wildl. Soc. Bull.,* 23:407–414, 1995.

DeCalesta, D.S., in McShea, W.B., Underwood, H.B., and Rappole, J.H., Eds., Deer and ecosystem management, in *The Science of Overabundance: Deer Ecology and Population Management,* Smithsonian Press, Washington, D.C., 1997, pp. 267–279.

Fayrer-Hosken, R.A., Grobler, D., Van Altena, J.J., Bertschinger, H.J., and Kirkpatrick, J.F., Immunocontraception of African elephants, *Nature,* 407:149, 2000.

Hadidian, J., Hodge, G.R., and Grandy, J.W., *Wild Neighbors, The Humane Approach to Living with Wildlife,* Fulcrum Publishing, Golden, CO, 1997.

Hagood, S., *State Wildlife Management: The Pervasive Influence of Hunters, Hunting Culture and Money,* The Humane Society of the United States, Washington, D.C., 1997.

Hodge, G.R., *Pocket Guide to the Humane Control of Wildlife in Cities and Towns,* Falcon Press, Helena, MT, 1991.

Kellert, S.R. and Wilson, E.O., Eds., *The Biophilia Hypothesis,* Island Press, Washington, D.C., 1993.

Kirkpatrick, J.F. and Rutberg, A.T., Fertility control in animals, in Salem, D.J. and Rowan, A.N., Eds., *The State of the Animals 2001*, Humane Society Press, Washington, D.C., 2001, pp. 183–198.

Mac, M.J. et al, *Status and Trends of the Nation's Biological Resources,* U.S. Department of the Interior, U.S. Geological Survey, Reston, VA, 1998.

National Park Service, Alternatives for Managing the Size of the Feral Horse Population of Assateague Island National Seashore, Environmental Assessment, U.S. Department of Interior, 1995.

Naugle, R.N., Rutberg, A.T., Underwood, H.B., Turner, Jr., J.W., and Liu, I.K.M., Field testing of immunocontraception on white-tailed deer (*Odocoileus virginianus*) on Fire Island National Seashore, New York, USA, in *Reproduction,* Supplement 60, 143–153, 2002.

Rutberg, A.T., The science of deer management: an animal welfare perspective, in McShea, W.J., Underwood, H.B., and Rappole, J.H., Eds., *The Science of Overabundance: Deer Ecology and Population Management,* Smithsonian Press, Washington, D.C., 1997, pp. 37–54.

Thiele, L.A., A Field Study of Immunocontraception in a White-Tailed Deer Population, M.S. thesis, University of Maryland, College Park, MD, 1999.

Turner, J.W., Liu, I.K.M., Flanagan, Jr., D.R., Rutberg, A.T., and Kirkpatrick, J.F., *J. Wildl. Manage.,* 65:235–241, 2001.

Walter, W.D., Pekins, P.J., Rutberg, A.T., and Kilpatrick, H.J., Evaluation of immunocontraceptive adjuvants, titers, and fecal pregnancy indicators in free-ranging white-tailed deer, *Wildl. Soc. Bull.*, in press.

CHAPTER 18

The Role of the Water Education Foundation in Creating Factual Awareness and Facilitating Consensus in Western Water Issues

Rita Schmidt Sudman

When the Western Water Education Foundation was incorporated as a California nonprofit foundation in 1977, few expected it to become a major force in water education. After all, there were many groups espousing the same goal of water education for Californians. The exact goal of the founding board of directors, a group largely aligned with water agencies in California, was to implement public information programs on water conservation, reclamation, reuse, and development. The founding board of the foundation believed in presenting unbiased information and allowing the major sides of the water debate to air their differences in the foundation's main vehicle for information, *Western Water* magazine. Some of the board members were confident that the public would support more water development if the facts were known. All board members were fair-minded people who believed in open discussion of the issues. This spirit of open discussion formed the basis of the growth of the Water Education Foundation to become a trusted source of information and a major credible organization.

The first issues of the magazine were distributed to several thousand people around California. Today it is distributed to over 20,000 — a list that includes reporters, interest groups, schools, libraries, and public agencies. This nonpartisan discussion of the issues in *Western Water* magazine immediately set the foundation apart from other water interest groups. One of the foundation's first projects was a "white paper" format, the *Layperson's Guide* series. It was designed to explore a particular water issue in more depth. The first guide in the series was Auburn Dam; some things never change. Today the series includes 16 guides. The guides are distributed similarly to *Western Water* magazine, but to a smaller audience.

At the time the foundation began, many groups were representing themselves as unbiased, and the foundation spent the early 1980s establishing credibility from all sides of the water debate. Funding was modest (about $30,000 the first year) and came from many sources, including subscriptions to the fledgling magazine, sales of a newly produced California Water Map, small contributions from those involved in the water business, and even a raffle to fund the water map. *Western Water* was written about a hot water topic, as it is today. In the early years, some pioneering work — such as the new topic of water marketing and agricultural water conservation — was analyzed for the public for the first time in the magazine's pages. Other topics, which gained public attention for the foundation, included the fight over the waters going into Mono Lake and state board hearings on water quality and conservation. A briefing for the business community and other

1-56670-612-2/03/$0.00+$1.50

forums for discussion were established as a regular part of the program. The reporters' briefings and other events to educate journalists about water issues were begun. We also established an award for the best water reporting in small and large newspapers. School education outreach was begun with packets of materials for teachers and students — a program that now reaches over 500 teachers and more than 650,000 students yearly. We began to diversify our board of directors to better reflect the major points of view in the water world — agricultural, urban, environmental. and public interest. And we changed our name to the Water Education Foundation to reflect better our broader view of water issues.

Additionally, we developed the first in a series of field trip tours to major water facilities and rivers. These were established to bring decision makers to see the controversial areas of the state. Typically two buses of water directors, city council members, reporters, legislative aides, and state and federal agency staff would travel on 3-day tours. These tours greatly facilitated discussion among people sufficiently involved in water issues to spend 3 days touring and analyzing the issues. The early tours covered the Bay–Delta and Northern California. Also in the 1980s we developed funding and support for the statewide Water Awareness campaign with state and local agencies. This annual program, carried out in the month of May, today reaches millions through public events, television and radio spots, and school programs.

By 1990 we were positioned to move into more multimedia education to reach a wider audience. During the height of the early 1990s drought, we obtained grant funding from a private foundation, the Hans and Margaret Doe Charitable Trust, to partner with public television to produce a documentary on California water issues, "To Quench a Thirst," with Roger Mudd as host. This important program brought a thoughtful discussion on water and the visualization of the problems into the homes of 3 million Californians. We updated the program several years later, and it is still being shown on cable television and distributed by the Water Education Foundation. Shortly after the program aired, Governor Wilson, who was looking for a forum cosponsor to announce his 1992 statewide water policy framework in San Diego, chose the nonpartisan Water Education Foundation.

Also in the early 1990s, recognizing the connection of California to the other Western states, the foundation began outreach to the other six Western states that share the Colorado River. The goal was to educate a wider group about Western water issues and seek resolution of these problems. We changed the mission of the foundation to reflect that change. No longer were we going to report only on these issues; we were also going to facilitate solutions through stakeholder conferences, public forums, published papers, and leadership training. This newer work would still be done within our continuing role as a nonpartisan organization.

Our first work in Western issues began with the publication of the Colorado River map, a three-state tour of the lower Colorado region, speaking engagements in the west, and a special issue of *Western Water* on Western drought. In the early 1990s we responded to hundreds of inquiries from the press and public on information on the drought. We also began a major involvement with two national information organizations: The National Geographic Society and National Public Radio. We assisted the reporters and photographer to produce the special issue of *National Geographic* magazine (November 1993) on water in California and reviewed the article for factual content. Also we assisted in the production of National Public Radio's "Radio Expeditions" series on Western water issues. Since politics tends to be local, we decided to develop a series of town hall meetings on local water issues. The purpose of the town hall meetings would be to (1) disseminate information and (2) facilitate solutions. During several summers we traveled to five different towns annually to present forums on specific local water issues and worked with the local media to cover those forums.

During this time revenue grew to several hundred thousand dollars, and funding was roughly divided as one third from grants, one third from contributions, and one third from tours, briefings, and publications. By 1992, when the foundation was 15 years old, our work on television documentaries increased. Our ability to bring groups together on issues led to the foundation's selection by the Hewlett Foundation to assist in coordinating and becoming the fiscal agent for the loose coalition of interest groups called the Three Way Group — a group representing agricultural, urban, and environmental groups. This group met at the height of the early 1990s drought, when pressures from the Endangered Species Act were causing pumping restrictions and water shortages among agricultural and urban users and when drought was putting more pressure on the environment. It was the first attempt to have formal and informal meetings among these diverse interest groups. In this age of political consensus, the Three Way Group seems archaic; but at the time it was a revolutionary means of changing policy. It failed because there was no leadership to force a decision or compromise. Also, there was no crisis point until 1994, when the Bay–Delta Accord was agreed upon by all sides under the threat of federal government implementing water quality standards for the state of California. During both the Three Way Process and the Bay–Delta Accord, our main role continued to be to cover the story. Our coverage of the issues, particularly in *Western Water* magazine, helped the decision makers themselves to see points where there may be compromise and today continues to be the basis of factual information for all interested people.

After our involvement with the Three Way process and the Bay–Delta Accord, the Water Education Foundation became involved in a number of consensus-based meetings and discussions. In 1996, the foundation was a member of the planning committee and the conference coordinator of the California Assembly Process on the Bay–Delta issues. Dwight Eisenhower had established the Assembly process when he was president of Columbia University. In this particular process, members of diverse interest groups debated concerns about the new CALFED process. About 300 stakeholders participated in the 3-day process. We assisted the group to develop a final document by compiling the voting results of the general discussions and breakout groups. The results were sent on to CALFED and formed a basis for its early discussions.

For us, the larger outreach for Bay–Delta education came with the development of our public television documentary on the issue. We have convinced a number of well-known actors to host these water documentaries, and their involvement seems to bring us wider attention and more coverage on public television stations. Tim Busfield hosted this 1-hour special on the history and current issues of the Bay–Delta area. In developing the documentary, we involved the interest groups in an advisory committee and therefore continued our involvement in the collaborative process of seeking solutions.

Another consensus process we became involved in was the search by the Sacramento area water interest groups for a way to share local water resources. Although this was a fully facilitated process lasting several years, the foundation cosponsored and moderated panels for the 1996 Water Forum. Our materials, again, were used as a basis for discussion, and we have analyzed the results of the forum process through the years.

Also in the 1990s we made a significant addition to our water education program by emphasizing groundwater education. Groundwater accounts for at least one third of the water used in California. It is a hidden resource but vital to the economy and ecosystem stability of the state. We began our groundwater work with a poster map of the groundwater basins in California. By 1995, we had added groundwater school programs, videos, a groundwater table flow model, and a statewide groundwater tour. We joined forces with the national Groundwater Guardian program to help communities gain recognition for local work to protect the resource. In several communities we staged community forums seeking solutions to overdraft and contamination.

In our continuing search for factual information, we developed a water input study with the University of California to study the amount of water associated with food production. For example, we analyzed how much water it takes to produce a hamburger. This figure takes into account the water to produce the feed for the cow, produce, and package the hamburger. There have been quick studies by groups who released certain numbers on water input in food production, but this joint foundation and University of California study was the first methodical study of its kind. Needless to say, the results were quite controversial. We were threatened with a lawsuit by an agricultural producer and criticized by some environmental interests. We survived the storm and stood by the numbers.

Through the years, we have coordinated a series of exclusive in-depth personal interviews in the pages of *Western Water* and in the television documentaries. Top state and federal officials such as Interior Secretary Bruce Babbitt, Environmental Protection Agency Administrator Carol Browner, California Governor Wilson, and, recently, California Resources Agency Director Mary Nichols were interviewed in *Western Water*. We have also reached out to specific professional audiences. We have even tried to educate attorneys. We began the Continuing Legal Education program in 1994. And in our attempt to raise the level of knowledge of young professionals who may work in water issues, or at least be impacted by them, we began the water leaders mentor program in 1995. There have been four classes of 10 to 20 young professionals each year in our water leaders class. They have set aside some of their time to attend our tours, briefings, read our publications, interview top leaders in the water world, and produce a class report on a specific water issue.

Another group we have worked with in recent years is teachers; we provided high-quality water education materials that fit into established curriculum guidelines. By conducting teacher training seminars, we have increased the acceptance of our materials statewide.

As water issues in California intensified, we were asked to assist water education on an international level. In 1993, we assisted the city water department of Barcelona, Spain, in setting up a public education campaign to convince the Spanish people to value and conserve water. Shortly afterward, we sponsored a major conference on similarities in the problems and solutions of California's Bay–Delta, the Netherlands, and Chesapeake Bay, Maryland. In 1995, we began a series of trips to Sweden, at the request of and with some expenses paid by the Swedish government, to discuss strategies for educating the public and specific audiences. We are now involved with the American advancement of the Junior Swedish Water Prize, a program to educate high school students about water and science. In 1997, we assisted the University of California with the meeting of the Rosenberg International Forum on Water Policy. The Water Education Foundation was the coordinator of the scientific field tour for the visiting professors from the world's arid countries.

In addition to publishing our own material, in 1996, the foundation's chief writer, Sue McClurg, was selected to prepare a study for Congress and the administration of the Sacramento–San Joaquin basin for the Western Water Policy Review Advisory Commission. This was one of several such basin studies developed for the commission. McClurg and the foundation were selected because of the commission's respect for our unbiased writing about Western water issues. Another basin report described the issues around water use of the Truckee River and Pyramid Lake. In an Emmy-nominated public TV program, seen by millions around the country, the struggle over the water of Pyramid Lake for the Paiute Indians, farmers, growing cities, and States of California and Nevada were chronicled in the foundation's documentary, "Healing the Water." Hosted by actor Robert Conrad, the history and recent negotiations were analyzed. The diverse funding and reputation of the Water Education Foundation are part of our success in getting our programs on public TV. Another program, now in production, covers issues in the Klamath basin and involves similar interest groups and the States of California and Oregon. The message of all these documentaries is basically the same: what is at stake in the ecosystem and the economy if people of good will do not come together and try to find compromise and solutions to these Western water problems?

That search for solutions and consensus led us, in 1997, to use the 75th anniversary of the Colorado River Compact, the document that divided the water of the Colorado River between the upper and lower basin states, as a pretext for a major symposium of invited stakeholders and top policy makers. At the symposium, the first draft of a California plan for the state to live within its 4.4 million acre-feet water limitation was formulated. Of the 125 invited stakeholders, 60 were involved and the current interior secretary and a former interior secretary addressed the group. A 200-page proceedings of the symposium was published in the conservation-style format of the conference and has been used by reporters and others interested in river issues. A similar symposium was scheduled in Keystone, Colorado.

One of the groups not included in the original division of the Colorado River was the American Indian tribes living along the river. We are now writing more about tribal water issues. Also, with a grant from the U.S. Environmental Protection Agency, we are developing a tribal workbook to help tribes assess and protect their water quality. The results will be placed on the EPA and our Web pages.

The World Wide Web has taken us around the world. It has allowed us to make briefings on water issues free to public. We also sell our materials on our Web site, and sales generate revenue to continue our work. By the end of the 1990s, our budget grew to $1.4 million, with funds continuing to come from diverse sources. Our tour agendas and upcoming events also are publicized on the Web. We see the huge potential to use the Web in water education. It is certainly a way that our small staff, ten people, can reach millions.

Our latest project is a literature and photography book on California water. It is meant to be a beautiful book, one to inspire the next generation and maybe even a few of today's folks, to care about these issues. Finally, the foundation's History Project brings us full circle. In an attempt to learn from our past, in the last few years we have sponsored a series of exhibitions, historical reenactments, and publications on how other generations made decisions on their water problems and how those decisions affect us today. Yes, we are still following Bay–Delta issues, and we even look at the proposed Auburn Dam when it periodically gets discussed.

As we enter the 21st century, we hope the Water Education Foundation will continue to be a reliable source of unbiased information on water issues. We hope this information and our work in the collaborative process will help Westerners become good stewards of our water systems and encourage care for our ecosystem.

CHAPTER 19

Addressing Threats to the Health of Coastal and Near-Coastal Ecosystems — The Gulf of Mexico*

James D. Giattina

INTRODUCTION

The diversity of the Gulf of Mexico, from the hypersaline waters of the Laguna Madre to the mangrove swamps of south Florida, contributes to the variety and productivity of life and human custom that characterize this sea. When a Spanish fleet sailed into the Gulf waters in 1497, the human diversity and culture of the region already paralleled the physical diversity. From the Colusa of Florida's southwestern coast to the empire of the Aztecs, the depth and complexity of the native civilizations was largely unappreciated by the Europeans. In the 500 years since its discovery by Spain, the Gulf of Mexico has become the Americas' "Mediterranean" — a wellspring of commerce and transportation as well as an area of strategic military importance (Gore, 1992). The Gulf region has flown under many flags, and that diverse political and cultural heritage is still evident in the architecture, linguistics, and political structure of the area.

It is that vastness and diversity that often precludes us from seeing the fundamental relationships between the living and nonliving workings of this magnificent ecological system. As a result, the vastness of the Gulf is both its strength and its weakness. Its size and diversity contribute to its productivity, yet they also lull us into a sense of complacency with regard to our day-to-day actions and their potential impact on the Gulf.

ENVIRONMENTAL AND PUBLIC HEALTH THREATS TO THE GULF OF MEXICO

A closer look at more specific areas and issues within the Gulf ecosystem reveals that many of the effects of human action are threatening the health and well-being of the Gulf. Those effects range from the localized, but ubiquitous, closure of shellfish-growing waters after heavy rains to a larger area of low dissolved oxygen on the Louisiana inner continental shelf, known as Gulf hypoxia. The former problem results from ineffective or inappropriate sewage treatment. The latter results from the cumulative effects of myriad human actions within the Mississippi River basin — a basin that covers over 40% of the continental U.S.

Throughout the Gulf natural processes combine with human actions in ways that are detrimental to the economy as well as to the ecological stability of the Gulf region. For example, today 85%

---

* Portions of this chapter were published in *The Coastal Society Bulletin,* 20(3):8–12.

of our nation's remaining coastal wetlands are in the southeastern U.S., with a significant portion of those in coastal Louisiana. Yet we are continuing to lose portions of coastal Louisiana at a rate of approximately 25 square miles every year. This is due to a combination of factors, including natural subsidence that has been exacerbated by the leveeing of the lower Mississippi River. Levees, while providing vital flood protection, have also cut off the river from its natural floodplain, reducing the essential flow of nutrients and sediments needed to replenish the marshes. This effect, in turn, has been further accentuated by channeling of the coastal marshes to allow for oil and gas exploration and production and improved transportation. The dissection of the coastal marshes, which was essential to the energy policy of the day, both caused physical destruction and allowed the introduction of saltwater into previously freshwater marsh areas, hastening their demise. Finally, we cannot ignore the impact on the coastal marshes of nutria, a nonindigenous species. These voracious, beaver-like rodents are making their own contribution to the loss of Louisiana wetlands because of their propensity for eating marsh grasses — roots and all.

The conversion of land and marsh to open water in coastal Louisiana is resulting in the tragic loss of a unique American place — one that is rich not only biologically but also culturally and historically. Furthermore, these wetlands serve as important protection from tropical storms; thus, this loss also increases the risk to billions of dollars of infrastructure — homes, schools, roads, towns, cities, ports, businesses, industries, and storage facilities — even from storms that are less than Category 5 catastrophes. Coastal marshes are essential to various life stages of over 90% of commercially and recreationally important species. The Louisiana marshes alone account for over 40% of the commercial landings in the U.S. waters of the Gulf, and marsh loss will continue to impact this sector of the economy and a way of life for many people.

As further evidence of the potentially destructive combination of natural and human forces, an unprecedented red tide occurred in the Gulf during the fall of 1996, impacting beaches and shellfish waters from Florida to Texas. Red tides — blooms of tiny organisms, known as dinoflagellates — have occurred in the Gulf for many years. In the past these events were confined to the waters of the western and northwestern Florida continental shelf, the east coast of Texas, and the Bay of Campeche, Mexico. The unusual 1996 event caused large-scale mortalities in fishes and invertebrates and significant mortalities in the endangered Florida manatee. In addition, as these blooms approached shore, the toxic by-products of the bloom organisms contaminated shellfish, causing closures and harvest limitations. In some areas wave action generated toxic aerosols, causing respiratory problems, eye irritations, and allergic reactions for beachgoers. The linkage of the 1996 red tide event and land-based pollution, particularly excessive nutrient enrichment of Gulf coastal waters, is speculative at this time. The hope is future research will shed additional light on the causes of these blooms and those environmental factors that contribute to their persistence and movement.

GULF OF MEXICO PROGRAM — OFFERING LEADERSHIP BEFORE CRISIS

At the 1998 National Ocean Conference, the Honorable Leon Panetta observed that, "in a democracy you get things done either through crisis or leadership" (National Ocean Conference, 1998). The warning signs we can already see reinforce the need for leadership in both protecting and fostering the sustainable use of resources of the Gulf. Leadership requires foresight, and foresight is contingent on our ability to evaluate our current state, assess and predict trends in key indicators (social and economic indicators, as well as ecological ones), and act to prevent those actions that will lead to crisis in both ecological and economic terms. The way to prevent a systematic crisis in the Gulf of Mexico is to redouble our efforts to remediate the problems of today. These problems are harbingers of impacts to come as the underlying stresses of population, coastal development, energy production, and transportation continue to expand on the Gulf — the nation's second fastest growing coastline.

In the 1980s, a variety of factors and events culminated in the formation of a multiagency leadership forum on Gulf environmental and natural resources issues — the Gulf of Mexico Program. Citizens were becoming more vocal in their concerns over the environmental changes they were witnessing on the Gulf coast — habitat losses, endangered species, and toxic pollution. The latter concerns were most vividly expressed in 1983 at a public hearing on ocean incineration of hazardous waste conducted by the U.S. Environmental Protection Agency (USEPA). This single hearing drew approximately 6000 people. Citizen opposition from the Gulf, Atlantic, and Pacific states was massive and adamant and contributed to the cancellation of the ocean incineration program in 1988 (*Rachel's Hazardous Waste News*, 1988).

The ocean incineration issue galvanized public concern for the Gulf and served as a lightning rod for public frustration with what many saw as duplicate and conflicting policies both within and among government agencies. Recognizing the need for coordinated policy development and implementation as well as increased financial and technical assistance, citizens called for the formation of a program capable of addressing those concerns and responsible for taking a Gulf-wide perspective on problems and their solutions. Federal and Gulf state agencies responded and, under the leadership of the USEPA, formed the Gulf of Mexico Program in 1988.

The Gulf Program was modeled after the organization and management concepts embraced by the National Estuary Programs. Founded on the threefold principles of partnership, science-based information, and citizen involvement, the Gulf Program joined the Great Lakes and Chesapeake Bay Programs as flagships of the nation's efforts to apply an adaptive management approach to large coastal freshwater and marine ecosystems.

From its inception, the Gulf Program was envisioned as a multiagency partnership endeavor based on the simple premise that no one agency or institution alone had the technical skills, financial resources, or legislative authority needed to resolve the environmental or natural resource problems confronting an ecological system the size of the Gulf. Lacking a specific legislative mandate, the program was structured around a policy review board and a technical advisory committee composed of representatives from the major federal and Gulf state agencies concerned with environmental protection and natural resource management, as well as representatives from business, industry, the environmental community, and academia. Eight subcommittees were formed to characterize both the effects and the causes of key problems confronting the Gulf and to develop action agendas that could be implemented by the partnering agencies to resolve those problems. The issues included living marine resources, public health, habitat loss, freshwater inflow, shoreline erosion, nutrient enrichment, marine debris, and toxic substances. The program was bolstered by two support committees — one addressing public education and outreach, and the other data and information transfer among agencies.

Also key to the early momentum of the program was a citizens advisory committee appointed by the governors' offices. The committee was composed of five representatives from each state, with each of the five representing a different interest — environmental, fisheries, tourism, business and industry, and agriculture.

Throughout the program's development, the principles of involving a broad constituency, building consensus, and basing decisions on technical information and science have been followed. The Gulf Program was established under the general authority of the Clean Water Act and operates under the guidelines and procedures of the Federal Advisory Committee Act (FACA). Day-to-day administration and facilitation of the various program committees and teams has been the responsibility of the Gulf of Mexico Program Office, which is underwritten by the USEPA.

Over its 10-year history, the Gulf of Mexico Program has assisted in characterizing some of the most difficult environmental issues confronting coastal waters and has implemented a variety of demonstration projects and studies that offer solutions to those problems.

PROGRAM IN TRANSITION — A PARTNERSHIP WITH A PURPOSE

Beginning in 1996 and stimulated by concerns over the program being "federally intrusive" and fostering a regulatory agenda, the Gulf Program's policy review board recommended a series of changes to the program's organization and management to accomplish three objectives: (1) increase the involvement of key nongovernment interests; (2) enhance the opportunities for greater Gulf state leadership; and (3) focus the efforts of the partnership on a narrower range of environmental objectives in order to achieve demonstrable progress and measurable outcomes. Specific changes were instituted, while the fundamental principles of partnership, science-based decision making, and public involvement were reinforced.

The program has streamlined federal agency representation on the policy review board to improve the opportunities for strategic leadership by the Gulf state agencies and the nongovernment interests. At the same time, the management committee has been expanded to embrace a broader constituency so that key organizations and agencies are engaged more effectively in the operational aspects of the program. Three key interest groups have also been brought into leadership positions on the board and are involved in all activities of the program: business and industry, represented by the Gulf of Mexico Business Coalition; production agriculture, represented by the American Farm Bureau Federation's Gulf of Mexico Committee; and environmental and social justice interests, represented by the Gulf Restoration Network.

Finally, the program has targeted four "focus areas" for concerted action: (1) public health (specifically shellfish and recreational water contamination and hazardous algal blooms); (2) excessive nutrient enrichment in numerous bays and estuaries and in the Gulf hypoxic zone on the Louisiana inner continental shelf; (3) habitat loss and degradation, with initial emphasis on emergent coastal wetlands and submerged aquatic vegetation; and (4) nonindigenous species introductions, principally targeting shrimp viruses and ballast water releases from commercial vessels. Focus teams have been formed for each area and charged with establishing measurable goals and objectives, working with the management committee and policy review board to implement those actions, and identifying the key indicators for measuring progress.

The program's focus teams are supported by three operational committees: the two noted earlier — public education and outreach and data and information transfer — and a new modeling, monitoring, and research committee. Finally, a scientific review committee has been established to oversee technical peer reviews of specific program products and plans.

THE CHALLENGE OF SCALE

Kai N. Lee, director of the Center for Environmental Studies at Williams College, noted in his seminal work *Compass and Gyroscope: Integrating Science and Politics for the Environment*:

> Social learning is most urgently needed in large ecosystems: territories with a measure of ecological integrity that are divided among two or more governing jurisdictions. Large ecosystems present some of the most difficult problems of environmental science and policy. They are complex, often badly damaged, riven by deep-rooted rivalries among several jurisdictions, and essential to the well-being of large populations.... Large ecosystems provide opportunities for learning from and about the real world. Their governance presents challenges of science, management, and politics, often entangled in ways that resist simple approaches. But without some degree of simplification there can be no learning and no transfer from one case to others" (Lee, 1993).

Even though the ecological problems confronting the Gulf are complex and enormous on a geographic scale, these sometimes seem to pale in comparison to the institutional barriers fostered by the legal, regulatory, and organizational landscape we have created. Those barriers are founded on the very real challenges associated with human relationships, including our ability to effectively communicate and to develop not only shared goals but also common and complementary approaches to achieving those goals. Governance of the Gulf of Mexico truly encompasses the challenges and opportunities in the areas of science, management, and politics noted by Lee (1993). But however great the challenges, the opportunities for learning and taking positive actions are even greater.

To successfully seize those opportunities and achieve measurable improvements in the quality of life for the Gulf's citizens and in the quality of the Gulf's ecological foundation — its water, land, air, and living resources — the Gulf Program must form effective working partnerships — partnerships with a purpose. The institutions and interests now at the table must be willing and capable of taking decisive action to resolve existing and emerging problems before they lead to systemic crises that cannot be ignored. In essence, the Gulf of Mexico Program must be viewed by all partners as an integral forum for identifying and resolving environmental issues in the Gulf ecosystem. Ultimately, the sphere of influence generated by the current partners will have to be broadened to embrace and invite the leadership of our international counterparts in Mexico, Central America, and the Caribbean.

CONCLUSION

If we can bring to light the recognition that the Gulf is one resource shared by all, then there is very likely the possibility that the experiment in governance we call the Gulf of Mexico Program will provide the leadership needed to avoid crisis. As former Administrator of the USEPA William K. Reilly observed, the challenges and opportunities presented by the Gulf of Mexico "call upon us as never before to look beyond isolated pollution problems to chart a course of action broadly based on the health of the entire ecosystem, a course drawing together the creative energies of all levels of government, academia, business, environmental groups, and private citizens" (Weber et al., 1992). The Gulf program is doing just that in order to create a future Gulf that continues to flourish in all its natural richness and variety while embracing the needs and desires of its people. Both are inextricably linked and essential to the cultural fabric of the region and to the economic well-being of both the region and the nation.

REFERENCES

Gore, R.H., *The Gulf of Mexico*, Pineapple Press, Sarasota, FL, 1992.

Lee, K.N., *Compass and Gyroscope: Integrating Science and Politics for the Environment*, Island Press, Washington, D.C., 1993.

National Ocean Conference — Oceans of Commerce, Oceans of Life, U.S. Department of Commerce, National Oceanic and Atmospheric Administration, Silver Spring, MD, 1998, p. 116.

Rachel's Hazardous Waste News, EPA abandons efforts to license ocean-going incineration ships. Environmental Research Foundation, Annapolis, MD, No. 64, February 15, 1988.

Weber, M., Townsend, R.T., and Bierce, R., *Environmental Quality in the Gulf of Mexico: A Citizen's Guide*, Center for Marine Conservation, Washington, D.C., 1992.

Setting Goals and Objectives in Managing for Healthy Ecosystems

Overview: Setting Goals and Objectives in Managing for Healthy Ecosystems

Robert T. Lackey*

Section I.4 focuses on policy goals and objectives — the explicit policy targets that provide meaning and definition to the platitudes that typically dominate much of the political discourse on ecological policy. To move beyond the realm of policy platitudes (e.g., protect our planet, assure sustainable development, embrace smart growth, implement community-based environmental protection, perpetuate our cherished natural legacy, restore degraded ecosystems, achieve ecosystem health) and toward policy evaluation and implementation requires that society, through its mechanisms of governance, decide which societal values and preferences to adopt. Societal values and preferences are the criteria society uses to select from among opposing policy goals and objectives.

Scientific input is important in selecting policy goals and objectives because not all goals are feasible. Even among the goals and objectives that are ecologically feasible, decision makers, and especially the public, rarely understand the ecological consequences of each option. Ecological policy goals typically conflict, may be mutually exclusive, and have ecological consequences, each of which are known with varying levels of certainty.

Ecological goals and objectives are often cast in terms of ecosystem *restoration*, but exactly what ecological feature does society wish to restore and to what extent? What makes one ecosystem more important to society than another? For example, if society wishes to receive the benefits of a roadway, should the adverse ecological effects of highway construction be mitigated? If so, how should they be mitigated? As Zedler and Callaway illustrate in Chapter 21, not all ecological restoration efforts replace what was lost, nor can even the trajectory for restoration be predicted in advance.

Many discussions about goals and objectives end up enmeshing values and preference within the scientific information essential to evaluate the consequences of policy options. For example, scientists providing technical information in policy discussions are often accused of offering *normative science* by implicitly advocating policy and value judgments under the banner of impartial science. Normative science is science based on implicit policy preferences. An example of normative science is the use of adjectives such as *degraded* or *healthy* in describing the condition of a particular ecosystem. Such terminology under the guise of "science" conveys the message as to which ecological state is (or should be) desired and which is not. Often scientists are unaware that they have moved from science devoid of a policy preference to science that

---

* The views and opinions expressed do not necessarily represent those of any organization with which Dr. Lackey is affiliated.

implies that a particular policy option is preferred. The notion of ecosystem health is often criticized because of its tacitly derived value and preference character. In Chapter 22, I review the characteristics of normative science and propose a proper role for scientists to play when providing information in policy deliberations.

Debates over goals and objectives often become the crux of approaches to addressing ecological policy problems. For example, ecosystem management has burst on the land management agencies in North America as the policy approach for this century. What exactly is ecosystem management and how does it differ from past approaches to implementing ecological policy? Does it only apply to publicly owned lands, or are private lands within its scope? Is ecosystem health sufficiently robust to underlay implementation of ecosystem management? Fitzsimmons provides a critical review of the concept of ecosystem health as a basis for managing lands in North America in Chapter 23.

Most governmental policy favors, even encourages, economic development, but how do such policies relate to concepts of ecosystem health? Do healthy ecosystems imply that human populations are prospering? In some sections of the world, it appears that relatively pristine ecosystems support (by Western standards) a very unhealthy human population. If an organization such as the World Bank has alleviating poverty as one of its central policy goals, how is this goal reconciled with "healthy" ecosystems? Anderson (Chapter 34) explores the often confusing and contradictory worlds of *ecosystem health* and *economic development*.

Food security is of widespread concern and a feature of many governmental goals and objectives, but how does it relate to ecological policy? For many years agriculture operated by reducing biological diversity and channeling photosynthesis through a few plants and animals. Few would argue that biological diversity, at least in a general sense, is important to past and continuing agricultural development, but what should be the relationship between biodiversity and agriculture given that the amount of "natural" ecosystems being converted to farming continues to increase? Thrupp (Chapter 35) evaluates the relationship of biological diversity and agriculture from the perspective of assuring a long-term food supply.

Ecological goals and objectives deal with more than producing food and fiber. How does a person's perception of quality of life relate to ecological policy? There does appear to be, at least for some people, a connection between what are often described as healthy ecosystems and their perceived quality of life. Is this relationship true only under circumstances where people are relatively affluent? In Chapter 24, Ewert explores the connection between perceived quality of life, recreation in natural ecosystems, and individual policy preferences.

Traditionally, economic development has been predicated on the natural resource development model. Early in the development of a country, its economy tends to be extractive. As the economy develops and expands, the economy generally shifts toward manufacturing and possibly toward a "service" economy. Is this the most desirable trajectory? Are concepts of *natural capital* useful in describing more effective approaches to economic sustainability? In Chapter 25, Collados provides a critical look at natural capital and ecological sustainability and their implications for developmental policies of nations.

The chapters in this section attempt to move beyond the platitudes so typical of ecological policy discourse. Each author critically evaluates the nature and character of potential goals and objectives and, in some cases, how such goals might be achieved. Some of the chapters also document what are clearly inappropriate goals because they rely on the values and preferences of scientists, rather than reflecting the values and preferences of society.

CHAPTER 21

Adaptive Restoration: A Strategic Approach for Integrating Research into Restoration Projects

Joy B. Zedler and John C. Callaway

INTRODUCTION

The San Francisco Bay–Delta and the southern California coast are both highly populated regions (over 16 million people in southern California and nearly as many in the Bay area), and both have experienced substantial impacts on natural wetlands. The Bay–Delta, however, had large wetland resources historically, whereas southern California never had large wetlands; and less than 10% of the historical extent has survived coastal development. Thus, efforts to restore lost wetland habitat and ecosystem functions (services) are relatively advanced in southern California, and experiences and recommendations in this rapidly developing region may be useful to efforts in northern California and other coastal areas.

The San Diego Association of Governments predicts that southern California's population will double by 2040. As human pressures continue, opportunities for wetland restoration occur both as mitigation for the filling and dredging of wetlands and as habitat improvement measures supported by resource agencies. In an earlier paper, Zedler (1996) recommended that restoration and mitigation activities be undertaken within a regional context. The Southern California Wetlands Recovery Project has taken a bold step toward implementing that recommendation. In a recent state budget, California targeted approximately $6 million for wetland restoration along the region's coast, following a regional inventory of coastal wetlands (http://www.coastalconservancy.ca.gov/index.html).

Recent experiences with habitat restoration and planning in southern California are relevant to the California–Federal (CALFED) program for the San Francisco Bay–Delta. First, we explore how a 13-year mitigation program at San Diego Bay became a model of adaptive management; next we describe how modular restoration at Tijuana Estuary was planned to incorporate scientific experiments. The experiments were planned to learn how subsequent modules should be restored. That is, there was insufficient knowledge to decide exactly how to restore habitat so that objectives would be achieved quickly and efficiently. For both San Diego Bay and Tijuana Estuary, project managers were amenable to accommodating research at the restoration sites, but the information needs developed at different times. At San Diego Bay, the studies developed along the way — an example of adaptive management, in that needs suggested research and results from the studies led to management recommendations (Table 21.1). At Tijuana Estuary, the questions were clear in advance, and research formed the basis for the restoration design (adaptive restoration). The term

Table 21.1 Research Findings from the San Diego Bay Mitigation Project

Specific Findings	Ref.
Documents that short cordgrass is related to coarse soil	Langis et al. 1991
Clapper rails need a tall canopy for nesting; author suggests height standards for suitable nesting habitat	Zedler 1993
One-time N addition produce tall plants, but effects are short-lived	Gibson et al. 1994
Scale insect outbreak occurred in short cordgrass; adding N reduced plant susceptibility	Boyer and Zedler 1996
Site supported the native invertebrates, but at low densities	Scatolini and Zedler 1996
Remote sensing approach helped assess compliance of constructed habitat with mitigation requirements	Phinn et al. 1996
Endangered plant can be reintroduced, but pollinators limit seed production	Parsons and Zedler 1997
Site is not stable hydrologically; long-term future is hard to predict	Haltiner et al. 1997
N additions increase cordgrass growth in proportion to amount added, but effects are not sustained after additions cease	Boyer and Zedler 1998
N addition at the large scale favors a native annual plant over perennial cordgrass	Boyer and Zedler 1999
Mitigation site did not follow desired trajectory	Zedler and Callaway 1999
Fish use small tidal creeks, e.g., as a killifish nursery	Desmond et al. 2000
Summaries for Broader Audiences	
Summarizes the project midway	Zedler 1991
Provides functional equivalency index	Zedler and Langis 1991
Summary for a general audience	Zedler 1992
Conceptual model of interactions among soil, plants, insects, and clapper rails	Zedler and Powell 1993
Overview of the project	Zedler 1998
Provides broader context for environmental lawyers and planners	Zedler 1997

adaptive restoration has been coined to describe projects designed as large-scale restoration experiments; that is, the site itself is divided into replicate subareas where different treatments are applied.

Compared with the San Francisco Bay–Delta, southern California has few opportunities for restoration and small areas available for experimentation. San Diego Bay's two excavated wetlands, Connector Marsh and Marisma de Nación, total 28 acres (12 ha) and are a significant part of the overall opportunity for restoration at the U.S. Fish and Wildlife Service (FWS) Refuge. At Tijuana Estuary, the Tidal Linkage and Model Marsh are the initial steps of a 500-acre (200-ha) planned tidal restoration program. With large areas at Tijuana Estuary (and even larger for CALFED), it is possible to experiment early on to develop and improve methods for future restoration.

To direct research toward the most pressing issues, an assessment of the ecosystem in question and its potential for restoration is needed. With basic understanding of how reference systems operate and a list of questions about how natural resources can best be restored, restoration experiments can be developed and prioritized.

THE SAN DIEGO BAY ADAPTIVE MANAGEMENT EXPERIENCE

Although San Diego Bay is large, the remaining wetland habitats are tiny. Plans to widen Interstate 5 and construct an interchange for a new highway with a major flood control between its west- and east-bound lanes were, thus, highly controversial. These three federal projects were, nevertheless, approved and construction began in 1984. However, mitigation for phase 1 was not undertaken on schedule, and phase 2 construction was begun in violation of environmental agreements. The Sierra Club and League for Coastal Protection sued three federal agencies and won. The lawsuit reopened the Endangered Species Act's Section 7 Consultation, in which the FWS is required to identify endangered species whose populations are jeopardized by federal projects and to identify suitable mitigation measures. These were determined to be the excavation of about 12 acres (4.8 ha) of fill from a degraded wetland and the excavation of 17.1 acres (6.9 ha) of fill from a dredge spoil deposit (sediments previously removed from shipping channels in San Diego Bay). The sediments removed from both sites were designated for use in constructing highway ramps at the edge of the site.

The lawsuit was settled by a federal judge in 1988; this marked the beginning of a 10-year association of unlikely teammates: the California Department of Transportation and the U.S. Army Corps of Engineers (Caltrans and CoE, the mitigators); the U.S. Fish and Wildlife Service (FWS, the resource agency and regulator); and the Pacific Estuarine Research Laboratory (PERL, the monitoring and research group, headed by J. Zedler). This team worked together to evaluate the results of ecosystem monitoring and to identify shortcomings of habitat designed for three endangered species. The field studies are the subject of numerous scientific papers (Table 21.1). The aim here is to describe the adaptive management process and its outcome.

FWS established a set of criteria that were to be met for Caltrans to be relieved of habitat construction work and monitoring. Some criteria required that specific standards be met over a 3-year period, which was an innovation at the time; previous monitoring requirements were for set 1- to 2-year periods. Other standards required self-sustainability during a 3-year period; in other words, mitigation efforts would need to continue until self-sustainability was demonstrated for 3 years in succession. We knew of no other mitigation requirements that were as open-ended as these.

PERL was hired by Caltrans to begin monitoring the Connector Marsh in 1989. Caltrans selected the attributes that we should evaluate, as compliance was not expected until some time after excavation of Marisma de Nación was completed in 1990. We began by assessing the use of channels by fishes and invertebrates (required as forage for endangered birds) and the growth of cordgrass (*Spartina foliosa*, required as nesting habitat by an endangered bird). We also began sowing seeds of an endangered plant *(Cordylanthus maritimus maritimus),* which was to be reintroduced to the site as part of the mitigation requirement.

In our field studies, it became clear that the standards were not written clearly enough to specify monitoring protocols; that the endangered plant (an annual species) was not setting sufficient seed; and that the cordgrass was not growing as tall as in natural marshes. After conferring with Caltrans, we met with representatives of the management team to clarify both the requirements and the information needs. That meeting established what became our adaptive approach. We agreed to make the standards more clear; i.e., we recorded our perceptions of what was initially intended and achieved concurrence from the management team. For example, the requirement that fish and invertebrate species densities be 75% of those in natural channels was interpreted to mean densities of *native* species, not exotics. We agreed that the designated site for plant reintroduction be changed from an island that had been created by dredging channels around it (and which seemed to be too isolated from pollinators to yield adequate seed) to a remnant natural marsh where the plant had historically occurred. We further agreed that we could begin experimenting with soil amendments in the newly excavated 17-acre (6.5 ha) site to find ways to grow taller cordgrass. Caltrans agreed not to plant the entire site until we had results from the 1990 field experiment. In 1991, Caltrans implemented our recommendation that nitrogen-rich fertilizer be added to holes where plugs of cordgrass were transplanted. Later, when we saw that one-time additions were insufficient to sustain tall cordgrass (Gibson et al., 1994), we were permitted to undertake additional experiments (Boyer and Zedler, 1998) in the Connector Marsh. Independent funding was obtained for much of the research that went beyond routine monitoring. Support came from the California Sea Grant College, the NOAA Coastal Ocean Program, and the FWS Biological Resources Survey (now the USGS Biological Resources Division). The infusion of research money, although difficult to obtain and of short duration, was critical to the adaptive management program.

Gradually, we institutionalized the following process: PERL sampled as directed, with frequent interaction with the Caltrans biologist. Results were provided in a draft report to Caltrans before being distributed to the full team. Included in the report were findings from associated research projects and recommendations for work the following year. PERL then organized an annual meeting of all team members, who had received copies of the draft report in advance. At the annual meeting, PERL presented results relevant to each set of criteria (i.e., for each of the three endangered species deemed jeopardized by the project). At the end of each presentation, PERL called for decisions as needed. For example, when we had evidence suggesting that the criteria for the endangered California least tern (*Sterna albifrons browni*) had been met, we called for a decision from FWS that Caltrans was in compliance for that species. When we proposed to apply soil amendments in large-scale plots (400 m^2), that decision was agreed upon by all team members as an appropriate follow-on study from experiments in 4-m^2 plots. We recommended using remote sensing to assess the area of cordgrass and of high marsh in both excavated wetlands, and suggested that Dr. Doug Stow, an expert in remote sensing, be brought in to apply the technique to restored wetlands. Such decisions were recorded and added to the draft report; the final report was then distributed to team members. The budget for monitoring and assessment was also discussed annually, as duties changed and some approaches were more costly than anticipated by Caltrans. Each year, the budget for monitoring was negotiated, with PERL suggesting what *could* be done, FWS indicating which tasks *should* be done, and Caltrans indicating what constraints there were on funding. Negotiations were facilitated by a sincere interest by all parties to learn what was needed to restore the site to levels that would support endangered species.

As various students undertook studies related to the project, we asked them to incorporate their findings into the annual report and to contribute to the annual presentations. Stuart Phinn was a frequent contributor, because his dissertation involved the remote sensing of different habitat types; and only he could explain where the imagery was most reliable and why digital maps of the sites were not perfectly matched with regular photographs. The oral presentations personalized the information and helped team members see the broader value of the work.

By 1996, criteria for the endangered California least tern and for the endangered plant had been met, but the site was clearly slow to produce tall cordgrass required for the endangered light-footed clapper rail *(Rallus longirostris levipes)*. As much of the research on cordgrass had been paid for by funding agencies that required broad dissemination of results, our finding that the restoration was not achieving its target became widely known. Among the magazines that publicized stories based on our research were *National Geographic, National Wildlife* magazine, *Scientific American, Time* magazine, and *Science*. An April 17, 1998, headline in *Science* read, "Restored Wetlands Flunk Real-World Test." Understandably, Caltrans administrators were not happy with this publicity. Reporters and editors tended to pay more attention to the "flunking" than to the years of sincere effort and the compliances for two out of the three endangered species whose habitat was jeopardized.

Several observations in the field led us to conclude that the site would not achieve compliance for clapper rail nesting habitat in a timely manner. The large-scale experiments with nitrogen addition gave results counter to those of small-scale plots, and it appeared that fertilizing the marsh would aid its conversion to succulents, rather than lead to self-sustaining tall cordgrass (Boyer and Zedler, 1999). We summarized our long-term data on key variables (soil organic matter, soil nitrogen levels, and tall stems), and the quickest to match reference sites was soil nitrogen, which we predicted would match that of the reference site after 40 years (Zedler and Callaway, 1999). The number of tall cordgrass stems was shown to be declining, rather than increasing, as the site gradually accumulated sediments and nutrient status remained inadequate. Our prediction that compliance would be unlikely in the near future was critical to ending the mitigation program. We recommended that the FWS select an alternative penalty and end the mitigation program. Our recommendation was probably made more credible by the fact that implementing it would end our annual monitoring contracts; that is, it was clear that PERL recommendations were not biased to continue funding. The recommendation was accepted, and FWS asked Caltrans to remove sediments from a nearby site to complete the mitigation program. Many details are avoided in this summary, but the process was fairly simple once all players were familiar with the annual routine. Absolutely key to the achievement of consensus on all decisions were the dedicated team members (essentially the same individuals from about 1990 on) and their shared goal of providing habitat for endangered species.

THE TIJUANA ESTUARY RESTORATION PLAN

At Tijuana Estuary, PERL played a major role in designing an overall restoration plan (Entrix et al., 1991) for some 500 acres (200 ha) of habitat improvement that would be implemented in 10 to 20 modules, with each module incorporating scientific experiments in the design. This approach was accepted by numerous participating agencies (FWS, NOAA Sanctuaries and Programs Division, National Marine Fisheries Service at the federal level, California Department of Fish and Game, and California Coastal Commission at the state level), and it passed public review at various hearings during the environmental impact analysis phase. The State Coastal Conservancy paid for the planning effort and guided the environmental impact statement (EIS) through the review process. As at San Diego Bay, the dedication and long-term involvement of a few individuals made this plan a reality.

Key to planning and implementing the two projects at Tijuana Estuary has been the existence of a National Estuarine Research Reserve, a resident research group, long-term research funding, managers who facilitate scientific involvement, and the long-standing assistance of the California State Coastal Conservancy. The outcome will be answers to questions that are needed to proceed with the full 200-ha restoration program at this National Estuarine Research Reserve. An award from Earth Island Institute supported several years of research. The importance of a funding base to fill in between short-term awards cannot be overstated.

Adaptive Restoration at the Tidal Linkage

The first restoration module, just 1.74 acres (0.7 ha), was excavated in 1997. This *Tidal Linkage* site has a channel that supports native fishes, invertebrates, and birds, a northern marsh plain that was planted with salvaged marsh plugs, and a southern marsh plain that was devoted to an ambitious field experiment. Because of our understanding of salt marshes, we knew that some eight halophytes would potentially grow at the site. Previous restoration efforts, however, included few species and mostly those grown from rhizomes. We hypothesized that low species diversity would slow the development of marsh functions. Hence, we designed an experiment that would show how many of the native salt marsh species needed to be planted to achieve desired ecosystem functions.

Because the question (does diversity matter?) was of broad scientific interest, the National Science Foundation funded three of us (J. Zedler, J. Callaway, and G. Sullivan) to evaluate the field experiment. As is often the case, funding was not obtained on first request, and we had to revise our proposal; hence, funds were not in hand in time to support our experimental planting of the site. If PERL had not had an award from Earth Island Institute for improving restoration methods, we would not have been able to produce the 6000-plus seedlings nor to pay helpers to plant them in the 87 field plots. Although the site was small, we designed a large uniform marsh plain to allow planting of 87 plots, each 2×2 m, as both a native plant-dominated salt marsh and a field experiment. Admittedly, it did not look like a natural marsh for the first few years, because the plots and paths between them gave it a patchwork quilt appearance. What might have been a negative feature became an advantage for public interpretation, as a nearby footbridge offered an overlook that allowed visitor center staff to describe our work and its importance to the National Estuarine Research Reserve.

Preliminary results show that species richness does little to affect seedling recruitment — the process is mostly restricted to three opportunistic species, and their seedling distributions are mostly determined by physical factors, such as elevation and open space. However, plots with diverse plantings have more complex canopies (which benefit wildlife) and more biomass belowground. Analyses of nitrogen accumulation rates are under way. We also see species-specific effects. In general, our preliminary conclusion is that species richness, species composition, and abiotic factors act together to influence the development of ecosystem structure and function. This finding supports three sides of a debate that is receiving considerable attention in the ecology literature.

Adaptive Restoration at the Model Marsh

The second module is larger (20 ac; 8 ha) and has a more ambitious experiment. Again, knowledge of how the natural ecosystem functions and observations of constructed or restored salt marshes played a central role in prioritizing the experimental approach. Whereas most natural marshes are riddled with small tidal creeks, marshes were constructed as round or square basins with perhaps a round, mounded island to provide habitat heterogeneity. It seemed reasonable to hypothesize that tidal creeks would affect nearly every aspect of marsh functioning, including plant composition and productivity, algal growth, invertebrate abundance and composition, fish use, and bird use. Hence, the model marsh was designed to compare ecosystem functioning in replicate areas with and without tidal creek networks.

Although designed in 1988, it took 10 years to obtain the engineering designs, provide the necessary supplementary impact analyses, develop sediment-control measures upstream, gain permits, and identify sources for $3.1 million needed to excavate the site. Construction began in late 1999. PERL has added a number of experiments concerning how to prepare the site, and we will evaluate several hypotheses about how tidal creeks affect ecosystem structure and functioning. Again, the Earth Island Institute award made the planning and implementation of these adaptive restoration experiments possible.

LESSONS FOR THE BAY–DELTA PROGRAM

The CALFED program is designed to be adaptive. Although many of the restoration measures concern alteration of flows and other off-site measures, there are many opportunities for restoration site-based experiments, e.g., restoration of tidal emergent vegetation and riparian vegetation, levee setbacks, and channel dredging. Using the terminology introduced here, these efforts could be undertaken as adaptive restoration, that is, with subareas established as replicated treatments in scientific field experiments.

We have several general recommendations that may be helpful in establishing adaptive restoration modules:

- Develop an adaptive management team that meets annually, identifies priority research needs, prioritizes sites where adaptive restoration might take place, reviews research results, and recommends future actions.
- Design the restoration site to facilitate experimentation and to test alternative approaches to restoration. We emphasize the need for careful consideration of experimental design, replication of treatments, data gathering, and statistical analyses.
- Using the general goals that have been established for the Bay–Delta Program, set site-specific goals with appropriate assessment criteria so that projects can be evaluated as part of the adaptive process. While it may be necessary to modify goals during the process, an initial target is desirable and suitable for the adaptive process.
- Establish priorities for research questions based on management concerns and current knowledge; this is an area where there is much opportunity for interaction between researchers and resource managers.

- Carefully consider the scale of experimentation. We recommend that experiments be conducted along a series of scales, from mesocosms to large-scale experiments, so that specific cause–effect relationships can be identified at the small scale, while general and more diverse response patterns can be characterized at the large scale.
- Because the Bay–Delta Program offers large areas for restoration and adaptive management, it is important to phase the work and the research so that information gained in early modules can be used in subsequent modules. Initial experimentation should help improve restoration methods.
- Provide backup funding for research teams that are dedicated to adaptive restoration projects, while encouraging researchers to obtain additional competitive grants. Adding studies reviewed by the National Science Foundation will help the research gain credibility and aid in information dissemination.
- Include scientists in the adaptive management team and reward them for publishing peer-reviewed research as well as for writing guidelines for future restoration implementers and managers.

SUMMARY

At San Diego Bay, a project designed to mitigate damages due to highway and flood channel construction evolved into a model adaptive management program based on restoration. For approximately 10 years, the California Department of Transportation and the U.S. Army Corps of Engineers (mitigators), the U.S. Fish and Wildlife Service (resource agency and regulator), and the Pacific Estuarine Research Laboratory (monitors and researchers) worked together to evaluate the results of ecosystem monitoring and to identify shortcomings of habitat designed for endangered species. Multiple attempts to improve habitat based on field research were followed by the prediction that some mitigation standards would not be achieved for ≥40 years. The finding that habitat for the light-footed clapper rail could not be provided in a timely manner was critical to ending this phase of the mitigation program (an alternative penalty was agreed upon by all parties).

At Tijuana Estuary, an improved adaptive restoration approach has scientific experiments built into the first two modules of a 200-ha restoration plan. The first module (the 0.7-ha Tidal Linkage) has one half of its marsh plain devoted to an experiment that shows which and how many of the native salt marsh species need to be planted to achieve desired ecosystem functions. The second module (an 8-ha model marsh) is designed to compare ecosystem functioning in replicate areas with and without tidal creek networks.

Key to adaptive restoration is the willingness of managers to facilitate scientific involvement, availability of funding for restoration and research, and interest of researchers. An additional helpful condition is the ability to restore large sites as phased modules, with results from early efforts informing the planning and implementation of later modules. Given these attributes and sufficient time, the adaptive restoration approach should be able to answer critical questions about how to implement even the largest and most ambitious restoration projects.

ACKNOWLEDGMENTS

We are grateful for the support of the National Science Foundation (DEB96-19875) and the Earth Island Institute. We thank Dick Daniels for inviting this contribution.

REFERENCES

Boyer, K.E. and Zedler, J.B., Damage to cordgrass by scale insects in a constructed salt marsh: effects of nitrogen additions, *Estuaries,* 19:1–12, 1996.
Boyer, K.E. and Zedler, J.B., Effects of nitrogen additions on the vertical structure of a constructed cordgrass marsh, *Ecol. Appl.,* 8:692–705, 1998.

Boyer, K.E. and Zedler, J.B., Nitrogen addition could shift species composition in a restored California salt marsh, *Restoration Ecol.,* 7:74–85, 1999.

Desmond, J., Williams, G.D., and Zedler, J.B., Fish use of tidal creek habitats in two southern California salt marshes *Ecol. Eng.,* 14:233–252, 2000.

Entrix Inc., PERL, and PWA Ltd., Tijuana Estuary Tidal Restoration Program, Draft Environmental Impact Report/Environmental Impact Statement, Vol. I-III, California Coastal Conservancy and U.S. Fish and Wildlife Service, Lead Agencies, Oakland, CA, 1991.

Gibson, K.D., Zedler, J.B., and Langis, R., Limited response of cordgrass *(Spartina foliosa)* to soil amendments in constructed salt marshes, *Ecol. Appl.,* 4:757–767, 1994.

Haltiner, J., Zedler, J.B., Boyer, K.E., Williams, G.D., and Callaway, J.C., Influence of physical processes on the design, functioning and evolution of restored tidal wetlands in California, *Wetlands Ecol. Manage.,* 4:73–91, 1997.

Langis, R., Zalejko, M., and Zedler, J.B., Nitrogen assessments in a constructed and a natural salt marsh of San Diego Bay, California, *Ecol. Appl.,* 1:40–51, 1991.

Parsons, L. and Zedler, J.B., Factors affecting reestablishment of an endangered annual plant at a California salt marsh, *Ecol. Appl.,* 7:253–267, 1997.

Phinn, S.R., Stow, D.A., and Zedler, J.B., Monitoring wetland habitat restoration in southern California using airborne digital multispectral video data, *Restoration Ecol.,* 4:412–422, 1996.

Scatolini, S.R. and Zedler, J.B., Epibenthic invertebrates of natural and constructed marshes of San Diego Bay, *Wetlands,* 16:24–37, 1996.

Zedler, J.B., The challenge of protecting endangered species habitat along the southern California coast, *Coastal Manage.,* 19:35–53, 1991.

Zedler, J.B., Restoring cordgrass marshes in southern California, in Thayer, G.W., Ed., *Restoring the Nation's Marine Environment,* Maryland Sea Grant College, College Park, MD, 1992, pp. 7–51.

Zedler, J.B., Canopy architecture of natural and planted cordgrass marshes: selecting habitat evaluation criteria, *Ecol. Appl.,* 3:123–138, 1993.

Zedler, J.B., Coastal mitigation in southern California: the need for a regional restoration strategy, *Ecol. Appl.,* 6:84–93, 1996.

Zedler, J.B., Adaptive management of coastal ecosystems designed to support endangered species, *Ecol. Law Q.,* 24:735–743, 1997.

Zedler, J.B., Replacing endangered species habitat: the acid test of wetland ecology, in Fiedler, P.L. and Kareiva, P.M., Eds., *Conservation Biology for the Coming Age,* Chapman & Hall, New York, 1998, pp. 364–379.

Zedler, J.B. and Powell, A., Problems in managing coastal wetlands: complexities, compromises, and concerns, *Oceanus,* 36(2):19–28, 1993.

Zedler, J.B. and Callaway, J.C., Tracking wetland restoration: do mitigation sites follow desired trajectories? *Restoration Ecol.,* 7:69–73, 1999.

Zedler, J.B. and Langis, R., Comparisons of constructed and natural salt marshes of San Diego Bay, *Restoration Manage. Notes,* 9(1):21–25, 1991.

Zedler, J.B., Williams, G.D., and Desmond, J., Wetland mitigation: can fishes distinguish between natural and constructed wetlands? *Fisheries,* 22:26–28, 1997.

CHAPTER 22

Appropriate Use of Ecosystem Health and Normative Science in Ecological Policy

Robert T. Lackey*

INTRODUCTION

Complex, challenging, and important ecological policy issues confront society (National Research Council, 1997; Rapport et al., 1998; Science Advisory Board, 2000). Significant improvements in some aspects of the environment have been realized. Many of the more egregious forms of pollution in North America have been reduced, but the continuing increase in the human population and associated human activities have created a tangled array of ecological policy challenges (e.g., land use alteration, hydrologic modification, climate change, change in biological diversity, introduction of nonnative species [also called exotic or alien species], concern about ecological sustainability, cumulative effects of man-made chemicals, etc.) (U.S. Environmental Protection Agency, 1999). Further, the fact that commerce is increasingly international in scope complicates already befuddled ecological policy issues. Recent treaties, for example, address climate change, biological diversity, waste transport, and environmental equity — and the directives contained in such legally binding agreements must be considered when addressing domestic ecological policy issues.

Traditional approaches to implementing ecological policy typically follow the *command-and-control* (*promulgate-and-police*) paradigm (Carnegie Commission on Science, Technology, and Government, 1990). With the command-and-control approach, a narrow (e.g., water, air, chemical, or effluent), technically based standard is promulgated as a surrogate for a larger, often nebulous ecological or public health policy goal. Adherence to achieving the standard is enforced by a regulatory bureaucracy (Elliott, 1997). In practice, the typical result is a centralization of political power.

> This strategy may be characterized in simple terms as relying on an elaborate system of planning in which a central administration imposes production quotas on different plants and industries through directives specifying the amount of pollution allowed to escape into the air, water, and land (Carnegie Commission on Science, Technology, and Government, 1990).

Command-and-control approaches to implementing ecological policy tend to be reductionist, thereby limiting the kinds of policy problems that can be addressed effectively (Science Advisory

* The views and opinions expressed do not necessarily represent those of any organization with which Dr. Lackey is affiliated.

Board, 1990). Attempts to correct one environmental problem sometimes create or exacerbate others (National Research Council, 1997). The command-and-control approach fits reasonably well for comparatively narrow policy problems (e.g., water quality and air quality), but it does not mesh well with complex policy problems such as the consequences of land-use changes, maintenance of biological diversity, nor the impacts of the introduction of exotic species.

The command-and-control approach to implementing ecological policy is criticized frequently as ineffective or insufficient in addressing the most important ecological concerns. For example, the U.S. Environmental Protection Agency's Science Advisory Board (1990) concluded a decade ago:

> … controlling the end of the pipe where pollutants enter the environment, or remediating problems caused by pollutants after they have entered the environment, is not sufficient.

In some cases, command-and-control approaches have been effective (although perhaps not efficient cost-wise) at ameliorating the most conspicuous forms of pollution, but important ecological concerns today are not easily nor efficiently amenable to end-of-the-pipe and command-and-control approaches (Science Advisory Board, 2000).

Another criticism of the command-and-control approach is its tendency to polarize the public and rouse strong opposition to the proposed policy or regulation (Elliott, 1997). The very nature of the command-and-control approach engenders centralized decision making, top-down policy making, and public resistance (Carnegie Commission on Science, Technology, and Government, 1990). The U.S. Environmental Protection Agency (1998) has concluded:

> In the past, there has been a "command and control" approach to regulation…. As with centralized decision making, the regulations have been made clear, unbending, and applicable nationally.

Lack of public support is understandable, even predictable, because ecological issues and socio-economic issues are intertwined. There are winners and losers in policy choices, so the prospect of authentic win–win solutions is illusory. Even so, many perceive that command-and-control approaches to implementing ecological policy create excessive societal strife.

Another widely voiced perception is that many command-and-control regulations are excessively intrusive, especially when the ecological benefits are not obvious or are of only local concern. Some efforts to comply with the U.S. Endangered Species Act, for example, can be expensive and socially disruptive for little apparent benefit to society or even the species being protected.

Other critics assert that command-and-control approaches do not effectively use new scientific and technical information (Elliott, 1997). Current understanding of the functioning of ecosystems, for example, has moved away from the assumption that the natural or climax condition of an ecosystem is fairly predictable, e.g., the old *balance of nature* idea (De Leo and Levin, 1997). Current thinking is that the state of ecosystems is less circumscribed (e.g., "chaotic" events are often decisive). Although rarely explicitly stated, much of the command-and-control approach to implementing ecological policy has been predicated, in part, on the balance of nature worldview.

Command and control, characteristically narrowly focused (e.g., policy reductionism), often reinforces the proclivity of many scientists to simplify science and research (e.g., science reductionism). That is, many scientists prefer to reduce complex policy problems into small, compartmentalized research pieces that can be addressed in scientifically credible ways. Thus, for scientists working in a command-and-control bureaucratic environment, there is a propensity for *both* scientists and policy makers to fall victim to the reductionist snare. Research reductionism results in excellent science that withstands rigorous scientific scrutiny, but it is not necessarily useful to policy makers in selecting from among policy options. Many scientists tend to eschew research problems that deal *directly* with complex policy problems, because such problems tend to be scientifically intractable: the results of such research would be unlikely to weather the scrutiny of other scientists.

ALTERNATIVE APPROACHES

Because the limitations of the command-and-control approach are widely recognized, many experts contend that effectively resolving complex, divisive ecological policy issues requires a different approach (Carnegie Commission on Science, Technology, and Government, 1990; Science Advisory Board, 1990). Issues in ecological policy are now less focused on relatively isolated questions, such as whether it is safe to license a certain chemical, whether it is good policy to build a particular dam, or whether we ought to spend resources to control exotic species such as the zebra mussel. Alternative and competing approaches are widely discussed in the professional literature (Norton, 1995; Gaudet et al., 1997). For example, as the National Research Council (1997) concluded:

> ... efforts to solve a specific problem must be considered within a broader context. This is particularly true of the growing number of regional- and global-scale problems associated with population growth, industrial development, and the corresponding pressure on limited natural resources.

Specific but disparate examples of popular modifications and permutations of the command-and-control approach are ecosystem management, community-based environmental protection, the Precautionary Principle, bioregional management, watershed management, and imposition of over-arching public policy goals such as ecological sustainability, ecosystem integrity, or ecosystem health. Each alternative is championed, sometimes energetically, by its partisans.

In some alternatives (e.g., ecosystem management and community-based environmental protection), command and control is often viewed as one of several possible policy *tools* to help achieve the overarching policy goal. Other alternatives (e.g., bioregionalism) are the antithesis of the centralized, bureaucratic command-and-control philosophy. Taylor (1991) portrays the political propensity of adherents to bioregionalism as one of devolved decision making:

> Bioregionalism envisions communities of creatures living harmoniously and simply within the boundaries of distinct ecosystems. It criticizes growth-based industrial societies preferring locally self-sufficient and ecologically sustainable economies and decentralized political self-rule.

It is easy to dismiss as scholarly quibbling the arguments about which of the competing approaches for implementing ecological policy or natural resource management ought to be adopted, but that would be a mistake. It is unfortunate that the discussion about the competing concepts has the flavor of "a battle of buzzwords" (Noss, 1995), because the discussion is more than a mere scholarly debate; the future direction of ecological policy will be determined, in part, by which concept wins.

ECOSYSTEM HEALTH

Ecosystem health is the most popular of the emerging modifications of command and control (Gaudet et al., 1997; Belaoussoff and Kevan, 1998; Rapport et al., 1998). Many of the popular alternatives and modifications to command and control (e.g., ecosystem management and ecosystem sustainability) have notions of ecosystem health at their core (Lackey, 1998). Adoption of ecosystem health as a public policy goal could have major, although often unclear, ramifications:

> ... an ecosystem health focus sets the stage for a new environmental ethic — one in which actions may be judged by their contribution to maintaining or enhancing the health of the regional ecosystem (Rapport, 1995).

Ecosystem health enjoys a wide following, especially in the popular press and with some environmental advocacy groups (Gaudet et al., 1997). Part of the appeal is that it appears to be a simple, straightforward concept (Ryder, 1990; De Leo and Levin, 1997). Applying the human health metaphor to ecosystems, it proposes a model of how to view ecological policy questions. By implication, the metaphor also defines what types of scientific information are essential to help decision makers (Shrader–Frechette, 1997).

Ecosystem health, especially in the 1970s and 1980s, was often defined in nebulous terms — definitely not as clearly articulated constructs (Steedman, 1994). It was typically depicted as a broad societal aspiration rather than a precise policy goal or management target. Lacking precise definition, it was difficult to consider the concept as a practical public policy goal. As the concept emerged from semantic ambiguity with more precise definition and description, it became a serious topic for discussion and, predictably, a lightning rod for conflict.

The most alluring feature of the human health metaphor is that most people have an inherent sense of personal health (Ryder, 1990). By extension, many proponents argue that most people almost instinctively envision a "healthy" ecosystem (e.g., a forest, lake, or pastoral landscape) as pristine or at least appearing to be minimally altered by human action.

Most concepts of human health focus on the *individual* human, whereas ecosystem health treats the *ecosystem* as the unit of policy concern, not the individual animal or plant (Schaeffer et al., 1988). Concerns about *individual* animals or plants — the typical focus of animal rights and animal welfare policy — are usually not the level at which *ecological* policy is debated.

There remains considerable variation in the concept conveyed by the words *ecosystem health* (Calow, 1992; De Leo and Levin, 1997). Karr and Chu (1999), for example, reflect a common, but not universal, position that concepts of ecosystem *health* and *integrity* are fundamentally different. They define ecosystem *health* as the *preferred* state of ecosystems modified by human activity (e.g., farmland, urban environments, airports, managed forests). In contrast, ecological *integrity* is defined as an *unimpaired* condition in which ecosystems show little or no influence from human actions. Ecosystems with a high degree of integrity are natural, pristine, and often labeled as the baseline or benchmark condition. *Natural* ecosystems would continue to function in essentially the same way if humans were removed (Anderson, 1991).

Others make no such clear distinction and may describe ecosystem health and integrity as different words for the same general concept. Regier (1993), for example, concludes that:

> ... the notion of ecosystem integrity is rooted in certain ecological concepts combined with certain sets of human values.

and, thus, a desired ecosystem condition:

> ... other than the pristine or naturally whole may be taken to be "good and normal."

Hence, if one accepts that there are multiple (and equally acceptable) benchmarks for ecosystems with integrity, then the terms *ecosystem health* and *ecosystem integrity* would be conceptually the same. However, for the remainder of this chapter, I will use the concepts and definitions of ecosystem health and ecosystem integrity used by Karr and Chu (1999), where the two notions represent different but related intellectual constructs.

The majority of ecological policy debates concern ecosystem *health* rather than ecosystem *integrity* (Westra, 1998). Such an emphasis on health (altered ecosystems) is understandable because the vast majority of ecosystems are not pristine; hence, according to the definitions used here, altered ecosystems lack at least some integrity. Westra (1998) clearly describes the relationship between the two concepts:

> ... an ecosystem can be said to possess integrity when it is wild — that is, free as much as possible from human intervention today, and "unmanaged," although not necessarily pristine. This aspect of integrity is the most significant one; it is the aspect that differentiates the wild from ecosystem health, which allows support and manipulation.

NORMATIVE BASIS

The concept and implementation of ecosystem health are surrounded by controversy (Jamieson 1995; Wicklum and Davies, 1995; Callicott, 1995; Belaoussoff and Kevan, 1998). Addressing questions of ecosystem health might appear to be a fairly scholarly, perhaps even arcane, activity, free from the policy intrigue that dominates much of the science and policy underlying environmental management, but such is not the case. Concepts of ecosystem health are seldom afforded the luxury of dispassionate discussion because, as Wicklum and Davies (1995) observe:

> The phrases ecosystem health and ecosystem integrity are not simply subtle semantic variations on the accepted connotations of the words health and integrity. Health and integrity are not inherent properties of ecosystems.

Wicklum and Davies (1995) realize that the words *health* and *integrity* elicit powerful, positive images even if their meanings are ambiguous. Therefore, they argue, a precise understanding of these words is essential because they are likely to be used, and given a variety of meanings, by policy advocates, politicians, bureaucrats, and the general public. In practice, it may fall to scientists and other technocrats to provide operational clarity to these perplexing, value-laden, normative concepts that appeal on an intuitive level to nearly everyone. Unfortunately, but typically, normative ecological concepts, such as ecosystem health, become general perceptions, perhaps useful in general conversation but impossible to quantify (Ryder, 1990).

Ecosystem health and other normative concepts have become highly charged political terms (Jamieson, 1995), often to the extent that they have become shorthand descriptors for one faction in political debates. Even in the relatively isolated venues of academic and government laboratories, an assertion that ecosystem health and integrity are not intellectually sound concepts may be sufficient to have the perpetrator branded as a political reactionary. Conversely, proponents may be categorized as zealots whose political aspirations have corrupted the sanctity of the scientific method. As Callicott et al. (1999) maintain, "partisans of a single normative concept try to make it cannibalize or vanquish all the rest."

Some (Shrader–Frechette, 1997; Kapustka and Landis, 1998) have counseled against using the concept of ecosystem health to communicate to the public about environmental issues. To be sure, thoughtful discussions about ecosystem health and similar concepts are usually abstract, often contentious, and rarely lead to consensus; but is the use of the health metaphor, even as a heuristic tool, ill-advised? Kapustka and Landis (1998) exhort against the metaphor because it is misleading and based on the chosen values and judgments, not an *independent* scientific reality. Conversely, Callicott (1995) concludes that ecosystem health is intellectually defensible and heuristically valuable; but he concedes that the *value*, thus the calibration, of ecosystem health is subjective. Indeed, Callicott et al. (1999) classify it as an "ill-defined normative concept" that reflects the "occurrence of normal ecosystem processes and functions;" but most discussions rarely explain clearly how current policies would change if attainment of ecosystem health became a public policy goal. Perhaps one way to make progress would be to move discussions beyond policy platitudes and definitional nuances and toward assessments of the specific *implications* for individuals and society of implementing the concept.

Most frustrating to some critics of the health metaphor is the charge that they have rejected a concept but not offered an alternative. Even many supporters of the utility of the notion of ecosystem health concede that it is easy to identify its conceptual limitations (Callicott, 1995). Developing alternatives that overcome the shortcomings has been much more difficult. If critics end up spurning ecosystem health, what do they offer as an alternative? Better alternatives are not obvious.

Regardless of the merit and direction of the scholarly debate, notions of ecosystem health frame important public policy issues (i.e., sustainability of agriculture, overuse of marine ecosystems, scarcity of water for domestic and agricultural use, and ecological consequences of introduced species) (Shrader–Frechette, 1997). Ecological policy issues are not mere abstract intellectual concerns, but matters that affect people's daily lives.

IMPLICIT ASSUMPTIONS

At the core of the debate over ecosystem health is a number of implicit but highly contested assumptions. First and foremost is the long-debated assumption that ecosystems are *real* (Calow, 1992; Callicott, 1995). Kapustka and Landis (1998), however, assert that "no human has ever seen an ecosystem" because it is not a discrete unit like individual birds, trees, or worms, or even populations of organisms. When a science or policy problem is specified (i.e., a "salmon" issue), then the ecological boundaries (i.e., the *ecosystem*) follow intuitively. Thus, ecosystems are context-specific because they cannot be delimited without a science or policy concern or issue; therefore, they may have heuristic and problem-solving value but are not analogous to the patient in medicine (Suter, 1993).

Although rarely stated explicitly, in most formulations of ecosystem health there is a premise that *natural* systems are healthier than human-altered systems (Wicklum and Davies, 1995). For example, consider a defined geographic location and, given the alternatives of a pristine woodland, a housing subdivision, or an industrial complex, which is the healthiest? The subdivision may be necessary, even somewhat aesthetically pleasing, and the industrial complex may serve a worthy purpose; but almost everyone implicitly considers the "unaltered" woodland to be the *healthiest*. Tacitly, the assumption is that pristine, or the less altered, is good and preferred; highly altered ecosystems, in contrast, are less desirable, if not *degraded*. Thus, recognizing the normative basis for ecosystem health, Fairbrother (1998) concludes: "use of the term ecosystem health as a definition of an idealized state is not an appropriate paradigm."

Another common assumption involves the importance of biological diversity to *society*. Biological diversity is certainly an important element in understanding the structure and function of ecosystems, but the key policy assumption revolves around the level of importance *society* assigns to biological diversity or its constituent elements. For example, some argue that biological diversity is such a core (i.e., societal) policy value that scientists should actively lobby for it. As Meffe and Viederman (1995) bluntly recommend:

> Scientists can take a clear stand that biodiversity is good, that functioning and intact ecosystems are good, that continued evolutionary change and adaptation are good, and that diversity and variation in general is good. Scientists cannot and should not remove themselves from these usually unstated value judgments.

Meffe and Viederman (1995) assert that values in science are always present, whether admitted or formally expressed by scientists, and that the policy process merely focuses values more clearly and honestly. Therefore, scientists should drop the facade of policy neutrality and lobby for those policies they deem to be in the best interests of society.

Invariably, concepts of ecosystem health implicitly assume that certain ecosystem features such as biological diversity have an *inherent* policy importance (Schaeffer et al., 1988). Ecosystems are complex, typically in both structure and function, and the diversity of species within an ecosystem is important to determining how that particular ecosystem functions; but biological diversity is *inherently* no more important to ecosystems than are nutrient cycling, carbon storage, or the rate of photosynthesis. As a public policy priority, and apart from its ecological function, society collectively may ascribe high (or low) value to preservation of certain, perhaps all, species based on *human* values and preferences.

Although not universally assumed, a common implicit assumption is that there is a *natural* ecosystem state (i.e., balance of nature) akin to the simple homeostatic dynamics of physiological systems (Anderson, 1991). The existence of such a natural state is appealing because disruption of ecosystem balance — deviation from the natural state — could be used to define and measure health. Unfortunately, this idealized view of ecosystems does not typically exist. Ecosystems may not predictably approach single-point equilibrium but may oscillate over time in a fairly indeterminate manner.

Another assumption concerns the degree to which human activities should be considered *natural*. Many proponents of ecosystem health contend that a fundamental goal of managing ecosystems is to maintain or restore their *natural* structure and function (Hunter, 1996). Outside of ecosystem reserves, some deviation from natural would be tolerated to meet human needs, but the benchmark would be the *natural* state of the ecosystem in question (Anderson, 1991). Even defining the *natural* state of an ecosystem is *de facto* an implicit policy preference when used in policy discussions. For example, in North America, is the natural condition that which existed at the time of initial human arrival (~13,000 to 15,000 years ago) or at the time of European and African arrival (~500 years ago)? To a dissimilar degree, both groups of immigrants and their offspring altered ecosystems (Hunter, 1996). Selecting which of these two benchmarks (or another one) is *natural* is a value-based decision.

NORMATIVE SCIENCE

Few challenge the assertion that societal aspirations drive the environmental management goals inherent in implementing ecosystem health, but the question remains: Which societal aspirations will be selected (Gaudet et al., 1997)? Society is not a monolith, and there are many competing opinions of what is important.

The language and discussion of ecosystem health is value laden (Jamieson, 1995), but how are societal values and preferences to be incorporated when using ecosystem health in public policy? The crux of the policy challenge is deciding which of the diverse set of societal preferences is to be adopted. Resolving policy issues always consists of trade-offs, partially or entirely exclusive alternatives, winners and losers, and plenty of compromises.

Consider any specific ecological policy issue and ask: Who are the stakeholders and how would their input be used to define ecosystem health? The task is relatively easy when policy problems are defined narrowly, such as licensing a particular chemical or authorizing a timber harvest rate for an individual forest, but what about for achieving a broad societal aspiration (like ecosystem health)? For example, are the stakeholders for a national forest local or national? Obviously, local residents are most directly affected by policy decisions, but the land is "owned" by everyone in the nation. The policy preferences of local residents are likely to differ from those with a national perspective.

What role should science and scientists play in defining ecosystem health? Scientific information is important, even essential, but it is only part of what is needed (Gaudet et al., 1997). Most important ecological policy issues involve coarse scales. Unfortunately, most scientific information is finely scaled and narrowly focused, thus not directly relevant to many ecological policy questions. Further, political institutions (legislative and regulatory agencies) must balance competing values and preferences, so scientific information is merely one facet of decision making. For adjudicating conflicts over values and preferences, science offers no moral or ethical guidance (Kapustka and Landis, 1998).

An argument is sometimes advanced that, because ecosystem health shrouds difficult and painful trade-offs under the guise of science, its use inhibits incorporation of societal values and preferences by not forcing an explicit selection from competing policy options. As Suter (1993) observes in evaluating various attempts to implement ecosystem health:

> Use of unreal properties (particularly unreal properties with imposing names) in environmental regulation obscures the bases for decision making; increases the opportunity for arbitrariness; and decreases the opportunity for informed input by the public, regulated parties, or advocacy groups.

Toll (1999) unequivocally concludes that "environmental problems cannot be solved without applying some sort of value system." Shrader-Frechette (1997) charges that the concept of ecosystem health does little, in spite of the volume of rhetoric, to improving decision making because proponents have failed to:

> ... clarify the precise respects in which the term yields additional scientific explanation beyond those provided by assessments of production, biodiversity, and so on.

APPROPRIATE USE

Regardless of the precise notion of ecosystem health asserted, it is important to understand its use in implementing ecological policy. The most redeeming feature is its ability to help clarify complex policy questions (Calow, 1992). The metaphor of health applied to ecosystems is simple whereas ecological policy problems are complex; the decision options are many and sometimes counterintuitive; and the consequences of implementing each option are rarely certain.

Although not essential, concepts of ecosystem health *may* help explain to the public the ecological consequences of policy choices, thus potentially reducing the likelihood (to the public) of *unexpected* consequences. Helping avoid surprises that result from policy decisions is a useful characteristic of any decision-support tool; but unfortunately, surprises are a common trait of ecological policy decisions (National Research Council, 1997). If society decides, for example, to have a dam constructed which causes an *unexpected* (to the public) loss of a migratory population of fish, then society lacked the appropriate scientific understanding of the consequences of the decision. If, on the other hand, loss of the fish population was *expected* (by the public), then the benefits of dam construction were judged by the public to be sufficient to warrant the loss of the fish population.

Another feasible use of ecosystem health is that it potentially allows society to understand more easily complex ecological policy questions (Shrader-Frechette, 1997). If the ecological information is complex, as is often the case, it is difficult to provide helpful, understandable information to decision makers unless there is a relatively simple intellectual organizing framework such as ecosystem health. However, excessive simplification of scientific information has the risk of misleading decision makers. Thus, the complexity of ecological systems should not be overlooked in an attempt to provide helpful information to decision makers (National Research Council, 1997).

There are properties of ecosystem health that make it prone to misuse. Misuses may be intentional and done in an effort to achieve advantage in policy debates, or simply be due to ignorance of the fact that the concept has a normative basis.

The most pervasive misuse of ecosystem health and similar normative notions is insertion of personal values under the guise of *scientific* impartiality. Most concepts of ecosystem health require a benchmark (i.e., a *desired, preferred,* or *reference* condition) of an ecosystem. Often, the implicit assumption is that an undisturbed or natural ecosystem is somehow superior, thus preferable to an altered one (Anderson, 1991). An ecosystem, once altered by human activity, is different from the previous state, but there is nothing *scientific* that compels any ecological state to be considered preferred or better (healthier). Lele and Norgaard (1996) caution those searching for scientifically derived benchmarks for ecosystems: "Naturalness as the benchmark is neither value-free nor logically or practically usable."

Practical expressions of ecosystem health and ecosystem integrity should reflect societal values and preferences (Gaudet et al., 1997). A misuse of the concepts is the situation in which professionals, usually operating from bureaucratic positions, *de facto* determine healthy (i.e., preferred) target ecosystems conditions. Concepts of ecosystem health or ecosystem integrity are normative because someone must decide what ecosystem condition or function is *good* (Sagoff, 1995). Ecosystems have no preferences about their states; thus preferred states or *benchmarks* must come from the individuals doing the evaluation (Jamieson, 1995). One common approach is to arbitrarily select reference sites to serve as the benchmarks for the ecosystems in question. Kapustka and

Landis (1998) conclude that the principal danger for scientists attempting to define healthy eco-systems comes from the incorporation of beliefs, morals, values, and ethics as *properties* of ecological systems.

Another less obvious but disconcerting use of the concepts of ecosystem health and integrity is defining a public policy goal in vague terms that engender broad public support, labeling it ecosystem health or ecosystem integrity, but camouflaging the ramifications of its adoption. Indeed, there is general public support for the idea of maintaining ecosystem health, but few members of society grasp the consequences of such a policy approach. The implications for democratic processes are rarely revealed, much less debated. Westra (1996), for example, candidly stated some far-reaching political consequences:

> … no country's unilateral decision, no matter how representative it might be of its citizens' values, should be permitted to prevail, unless it does not conflict with the global requirements of the ethics of integrity, thus with true sustainability.

Another inappropriate use of the concept of ecosystem health is pejoratively categorizing opposing policy choices. After all, the competing policy choices must, by definition, not be appropriate for achieving ecosystem health. One policy choice then becomes identified as promoting *health,* with the alternatives struggling to avoid being dismissed as arguing for *sickness.* For example, a policy decision to drain a wetland to create a cornfield might legitimately be categorized as appropriate to maintain ecosystem health. Either the wetland or cornfield could be healthy, depending on the societal preferences embraced. Because *health* conveys a positive political con-notation, the common practice in policy debates is to capture the high ground by labeling *your* policy choices as necessary for health and those of your opponents as leading to sickness or ecosystem degradation.

Environmental managers are culpable, often unintentionally, of misusing the concept of eco-system health. Understandably, those responsible for making difficult, controversial policy decisions may be reluctant to define their goals clearly; so they sometimes, perhaps unintentionally, embrace ecosystem health in the belief that it is a scientifically operational term. After evaluating the potential uses of the health metaphor in environmental management, Suter (1993) concludes:

> … environmental managers are active agents, translating the inchoate norms of the current generation and the poorly predicted needs of future generations into specific actions to protect or restore real, valued properties of actual ecosystems…. Hence, the decision to abandon ecosystem health as a goal is not just a matter of semantics.

As Kapustka and Landis (1998) admonish: "If we are to manage the environment, it should be done with the clear knowledge that choices will have to be made, not fueled by misplaced desires or myths."

ALTERNATIVES

Ecological policy issues such as managing the consequences of human land use, reduced biological diversity, or the cumulative effects of chemical use, are real and demand serious attention by society (Science Advisory Board, 2000). Concepts based on normative science can be compel-ling, but even most proponents concede that there are serious conceptual or operational difficulties with such concepts. What are the alternatives, if any?

The most direct alternative to using normative science is to cease using words such as ecosystem health and simply describe what is proposed. More specifically, rather than propose a policy objective of managing a forest for health, express exactly and clearly the management objective.

Another alternative is to demand coherent, clear definitions of the normative concepts of ecosystem health. There are multiple definitions for the same words, so consensus on the exact meaning is essential to focusing policy debate on societal trade-offs, not semantic niceties. The Environmental Protection Agency (1998), for example, defined ecological *integrity* as the "ecosystem structure and function characteristic of a reference condition deemed appropriate for its use by *society*" (emphasis added). Thus, by adopting this definition, the appropriate ecological reference condition — the benchmark for normative evaluation — is decided by society, not by scientists applying their own policy preferences.

Regardless of whether normative concepts are used in ecological policy deliberations, public involvement (even as fractured as the public often appears to be) is essential because *values* drive policy. Public involvement should be at the essence of using normative concepts because of their requirement for *inherent* value judgments. As Rykiel (1998) explains:

> In a simplistic sense, science deals with true and false, whereas society deals with good and bad. Science can delineate the possibilities and describe the system that is likely to result from a policy, but it cannot decide if the resulting system is good or bad.

Thus, policy *decisions* are, by definition, normative because values and preferences were used by the decision maker to select a particular option.

Another alternative to using ecosystem health is to treat ecological policy issues as yet another complex public policy question and not to rely on any metaphor. Other policy issues (e.g., welfare, education, energy, transportation) are also complex and challenging, but overarching, explicit heuristic models or metaphors are not typically used.

CONCLUSION

Ecology has become much more than a scholarly discipline; it has impacts far beyond simply enhancing our understanding of ecosystems. Many uses of ecology have a strong normative flavor. As Worster (1990) observes:

> The science of ecology has had a popular impact unlike that of any other academic field of research. Consider the extraordinary ubiquity of the word itself: it has appeared in the most everyday places and the most astonishing, on day-glo T-shirts, in corporate advertising, and on bridge abutments.

The future role of normative science (and ecosystem health in particular) is uncertain. At the ideological extreme, there are stark opinions. Some argue that normative science is desirable — even essential — for implementing ecological policy. Scientists, they assert, have an obligation to incorporate *policy* value judgments into ecology, even to the point that such *science* concepts as ecosystem health should be adopted as the cornerstone of ecological policy (Callicott, 1995). Some scientific disciplines and professions (e.g., conservation biology, restoration ecology) unapologetically embrace normative science postulates as the core of their trade (e.g., biological diversity is inherently good; extinction of populations and species is inherently bad; ecological complexity is inherently good; evolution is good; biological diversity has intrinsic value) (Soulé, 1985).

Others, however, assert that normative science (e.g., ecosystem health) hides, under a veneer of science, the reality of trade-offs involving competing personal and societal values and preferences (Kapustka and Landis, 1998). The proper role of science is to help lay out options and assess the consequences of various choices, and it is only part of the needed input (Tingey et al., 1990; Shaw et al., 1999).

Scientists and scientific information will continue to play an important role in resolving ecological policy, but the role, in my opinion, should be carefully circumscribed (Lackey, 1999). Often, even among scientists, ecology has been treated more as a belief system than a science. It is easy, even encouraged, for scientists to abuse privileged roles in ecological policy debates by surreptitiously labeling personal values and policy preferences as "science" (Salzman, 1995).

Understanding the values and preferences of society is crucial to appropriately implementing concepts of ecosystem health, but obtaining such understanding credibly is difficult. Political institutions do not provide such understanding or guidance in efficient ways. To assert that concepts of ecosystem health are merely scientific constructs is incorrect. As Russow (1995) concludes, "the claim that scientific descriptions in general or measures of ecosystem health in particular are value neutral is simply false." The likely alternative to public involvement is that the values of scientists and other technocrats will be used as surrogates for societal values and preferences.

Perhaps the term *ecosystem health* has already become a political term, a code word for a particular policy or political position. Even now, invoking ecosystem health often is equated with a green political position. Becoming identified as a political term is unfortunate because the word and concept lose usefulness in serious policy, public, and scientific debate.

A different risk for the future of ecosystem health is that it becomes co-opted and, ultimately, marginalized. For example, if everyone adopts the term and becomes an advocate of ecosystem health, then the term and concept have lost their usefulness. In policy deliberations the terms are now political rhetoric — encompassing a suite of meanings that everyone readily accepts as reflecting each's individual, though divergent, policy positions.

The ecological policy concerns that engender widespread debate over ecosystem health and other normative constructs will not disappear. These concerns need to be addressed because of the increasing demand on limited ecological resources (Salwasser et al., 1997). The resolution of ecological policy is likely to become increasingly challenging because interactions among the planet, its nonhuman inhabitants, and its large and still expanding, human population constitute a dynamic system of increasing complexity (National Research Council, 1997). Whether one finds intellectual sustenance in the notion of ecosystem health, the policy concerns it attempts to confront are genuine.

REFERENCES

Anderson, J.E., A conceptual framework for evaluating and quantifying naturalness, *Conserv. Biol.*, 5:347–352, 1991.

Belaoussoff, S. and Kevan, P.G., Toward an ecological approach for the assessment of ecosystem health, *Ecosyst. Health*, 4:4–8, 1998.

Callicott, J.B., A review of some problems with the concept of ecosystem health, *Ecosyst. Health*, 1:101–112, 1995.

Callicott, J.B., Crowder, L.B., and Mumford, K., Current normative concepts in conservation, *Conserv. Biol.*, 13:22–35, 1999.

Calow, P., Can ecosystems be healthy? Critical consideration of concepts, *J. Aquat. Ecosyst. Health*, 1:1–5, 1992.

Carnegie Commission on Science, Technology, and Government, E^3: *Organizing for Environment, Energy, and the Economy in the Executive Branch of the U.S. Government*, Task Force on Environment and Energy, Carnegie Commission on Science, Technology, and Government, New York, 1990.

De Leo, G.A. and Levin, S., The multifaceted aspects of ecosystem integrity, *Conserv. Ecol.*, 1:3, 1997. (Available online at http://www.consecol.org/vol1/iss1/art3.)

Elliott, E.D., Toward ecological law and policy, in Chertow, R.M. and Daniel, E.C., Eds., *Thinking Ecologically: The Next Generation of Environmental Policy*, Yale University Press, New Haven, CT, 1997, pp. 170–186.

Fairbrother, A., Establishing the health of ecosystems, in Cech, J.J., Wilson, B.W., and Crosby, D.G., Eds., *Multiple Stresses in Ecosystems*, Lewis Publishers, Boca Raton, FL, 1998, pp. 101–107.

Gaudet, C.L., Wong, M.P., Brady, A., and Kent, R., How are we managing? The transition from environmental quality to ecosystem health, *Ecosyst. Health*, 3:3–10, 1997.

Hunter, M., Benchmarks for managing ecosystems: are human activities natural? *Conserv. Biol.*, 10:695–697, 1996.

Jamieson, D., Ecosystem health: some preventative medicine, *Environ. Values*, 4:333–344, 1995.

Kapustka, L.A. and Landis, W.G., Ecology: the science versus the myth, *Hum. Ecol. Risk Assess.*, 4:829–838, 1998.

Karr, J.R. and Chu, E.W., *Restoring Life in Running Waters: Better Biological Monitoring*, Island Press, Washington, D.C., 1999.

Lackey, R.T., Seven pillars of ecosystem management, *Landscape Urban Plann.*, 40:21–30, 1998.

Lackey, R.T., The savvy salmon technocrat: life's little rules, *Environ. Pract.*, 1:156–161, 1999.

Lele, S. and Norgaard, R.B., Sustainability and the scientist's burden, *Conserv. Biol.*, 10:354–365, 1996.

Meffe, G.K. and Viederman, S., Combining science and policy in conservation biology, *Wildl. Soc. Bull.*, 23:327–332, 1995.

National Research Council, *Building a Foundation for Sound Environmental Decisions*, National Academy Press, Washington, D.C., 1997.

Norton, B., Ecological integrity and social values: at what scale? *Ecosyst. Health*, 1:228–241, 1995.

Noss, R.F., Ecological integrity and sustainability: buzzwords in conflict, in Westra, L. and Lemons, J., Eds., *Perspectives on Ecological Integrity*, Kluwer Academic Publishers, Dordrecht, 1995, pp. 60–76.

Rapport, D.J., Ecosystem health: exploring the territory, *Ecosyst. Health*, 1:5–13, 1995.

Rapport, D.J., Defining ecosystem health, in Rapport, D.J. et al., Eds., *Ecosyst. Health*, Blackwell Science, Malden, MA, 1998, pp. 18–33.

Regier, H.A., The notion of natural and cultural integrity, in Woodley, S.J., Kay, J.J., and Francis G., Eds., *Ecological Integrity and the Management of Ecosystems*, St. Lucie Press, Boca Raton, FL, 1993, pp. 3–18.

Russow, Lilly-Marlene, Ecosystem health: an objective evaluation? *Environ. Values*, 4:363–369, 1995.

Ryder, R.A., Ecosystem health, a human perception: definition, detection, and the dichotomous key, *J. Great Lakes Res.*, 16:619–624, 1990.

Rykiel, E.J., Relationships of scale to policy and decision making, in Peterson, D.L. and Parker, V.T., Eds., *Ecological Scale: Theory and Applications*, Columbia University Press, New York, 1998, pp. 485–497.

Sagoff, M., The value of integrity, in Westra, L. and Lemons, J., Eds., *Perspectives on Ecological Integrity*, Kluwer Academic Publishers, Dordrecht, 1995, pp. 162–176.

Salwasser, H., MacCleery, D.W., and Snellgrove, T.A., The Pollyannas vs. the Chicken Littles — enough already! *Conserv. Biol.*, 11:283–286, 1997.

Salzman, L., Scientists and advocacy, *Conserv. Biol.*, 9:709–710, 1995.

Schaeffer, D.J., Herricks, E.E., and Kerster, H.W., Ecosystem health: I. Measuring ecosystem health, *Environ. Manage.*, 12:445–455, 1988.

Science Advisory Board, Reducing Risk: Setting Priorities and Strategies for Environmental Protection, U.S. Environmental Protection Agency, Publication SAB-EC-90–021, 1990.

Science Advisory Board, Toward Integrated Environmental Decision-making, U.S. Environmental Protection Agency, Publication SAB-EC-00–011, 2000.

Shaw, C.G., Everest, F.H., Swanston, D.N., Julin, K.R., and Allen, S.D., Independent scientific review in natural resources management: a recent example from the Tongass Land Management Plan, *Northwest Sci.*, 73:58–62, 1999.

Shrader-Frechette, K., Ecological risk assessment and ecosystem health: fallacies and solutions, *Ecosyst. Health*, 3:73–81, 1997.

Soulé, M.E., What is conservation biology? *BioScience*, 35:727–734, 1985.

Steedman, R.J., Ecosystem health as a management goal, *J. North Am. Benthol. Soc.*, 13:605–610, 1994.

Suter, G.W., A critique of ecosystem health concepts and indexes, *Environ. Toxicol. Chem.*, 12:1533–1539, 1993.

Taylor, B., The religion and politics of Earth First! *The Ecologist*, 21:258–266, 1991.

Tingey, D.T., Hogsett, W.E., and Henderson, S., Definition of adverse effects for the purpose of establishing secondary National Ambient Air Quality Standards, *J. Environ. Qual.*, 19:635–639, 1990.

Toll, J.E., Elements of environmental problem-solving, *Hum. Ecol. Risk Assessment.*, 5:275–280, 1999.

United States Environmental Protection Agency, Ecological Research Strategy, EPA/600/R-98/086, Washington, D.C., 1998.

United States Environmental Protection Agency, EPA's Framework for Community-Based Environmental Protection, EPA 237-K-99–001, Washington, D.C., 1999.

Westra, L., Ecosystem integrity and the "fish wars," *J. Aquat. Ecosys. Health*, 5:275–282, 1996.

Westra, L., The ethics of integrity, in *The Land Ethic: Meeting Human Needs for the Land and Its Resources*, Society of American Foresters, Bethesda, MD, 1998, pp. 31–44.

Wicklum, D. and Davies, R.W., Ecosystem health and integrity? *Can. J. Bot.*, 73:997–1000, 1995.

Worster, D., The ecology of order and chaos, *Environ. Hist. Rev.*, 14:1–18, 1990.

CHAPTER 23

Ecosystem Health: A Flawed Basis for Federal Regulation and Land-Use Management

Allan K. Fitzsimmons

INTRODUCTION

In the U.S. many scientists, politicians, environmentalists, and others see ecosystem management as the path to a correct relationship between people and nature. They argue for the new paradigm of ecosystem management because, wrote Carl Reidel and Jean Richardson (1995), "Put simply the old paradigms are no longer scientifically or politically valid." The traditional anthropocentric approach to natural resource policy that blends multiple use, conservation, preservation, and protection of human health leads to "collapse of life and living as we know it" and must be replaced by "ecocentrism ... and ecosystem management" according to Roderick Nash (1994). In the new paradigm world of ecosystem management, protection of ecosystem health, integrity, and sustainability replaces concern for improvement in human well-being at the core of government environmental and natural resource policies. Government must use regulations, permitting, planning, and other means to control land use on public and private lands to see to it that ecosystems remain healthy or are restored to a state of health. In 1998 former President Clinton's last chief of the U.S. Forest Service (FS), Mike Dombeck, said, "Our first priority is to maintain and restore the health of our ecosystems." Dale Bosworth (2001), the new FS chief under President Bush, echoed Dombeck, "Today, the role of the Forest Service is to restore and maintain healthy ecosystems to meet the needs of present and future generations."

In this chapter, however, I argue that the ecosystem concept itself and the notion of ecosystem health are too arbitrary and ambiguous to provide a viable basis for prudent public policy or land use management.

VAGUENESS OF THE ECOSYSTEM HEALTH CONCEPT

Researchers vigorously debate the validity of the ecosystem health concept and its value for public policy making, but observers of all stripes generally agree that the idea remains imprecise (Shrader-Frechette and McCoy, 1993; Goulden, 1994; Jamieson, 1995; Callicott, 1995; Lackey, 1996; Sagoff, 1997; De Leo and Levin, 1997; Rapport, 1998; Wilkins, 1999; Karr, 2000; Lancaster, 2000; Levin, 2001; Lackey, 2001). David Rapport (1998), a strong advocate of making the protection of ecosystem health "a super societal goal," nonetheless pointed out that, "The

1-56670-612-2/03/$0.00+$1.50

question of what constitutes ecosystem health remains somewhat perplexing and controversial." Benjamin Haskell, Bryan Norton, and Robert Costanza (1992) found that there is no "clear conception of the term" and that it has "never been defined well enough" to make it useful for policy purposes, while Peter Calow (1995) concluded that the search for a definition "has proved somewhat elusive." After reviewing the work of his colleagues, Rapport (1995) observed that there "are a plethora of attempts to define ecosystem health ... [that] range widely from very broad definitions which incorporate biophysical, human, and socioeconomic components to definitions focusing primarily on the biophysical aspects to definitions which focus on a single indicator within the biophysical domain." The public can rightly ask why protection of ecosystem health should become the centerpiece of natural resource policy when its proponents cannot even agree on a definition of the term.

Citizen perplexity will only increase with further investigation of the idea. I will not review the variety of definitions offered for ecosystem health, but a look at some thoughts of notable scholars working in the field is instructive. For Bryan Norton and Robert Ulanowicz (1996), "The capacity for creativity constitutes the crux of what is normally referred to as ecosystem health." But what is *creativity*? In their view it is the ability of an ecosystem to solve problems and that in turn requires the ecosystem to possess "a requisite amount of ordered complexity" and an adequate level of "internal incoherence." And what is *ordered complexity*? For them, an ecosystem must contain enough "apparatus" to respond to events via "a channelized sequence of reactions." And what is *internal incoherence*? That is the presence of "dysfunctional repertoires" not normally used by the ecosystem but that serve as reservoirs of "stochastic, disconnected, inefficient features that constitute the raw building blocks of effective innovation." This sequence of fuzzy ideas led them to accept the Haskell et al. (1992) view: "An ecological system is healthy and free from 'distress syndrome' if it is stable and sustainable; i.e., if it is active and maintains its organization and autonomy over time."

Let me apply these ideas to a specific ecosystem: a whale carcass. (Norman Christensen and his fellow members of the Ecological Society of America's Ad Hoc Committee on Ecosystem Management, 1996, told us, "A dung pile or whale carcass are ecosystems as much as a watershed or lake.") We can all understand how the whale, when it was alive, was active and maintained its organization and autonomy over time. The brain directed purposeful actions to secure the whale's continued existence. Various organs and other parts of the whale performed specific functions that combined to sustain its life. While alive, the whale consciously sought to maintain itself.

A moment after the whale's demise, however, what had been a whale became an ecosystem made up of large amounts of now-dead tissue plus various biota that had found the once-living whale a nourishing habitat. With no brain or central nervous system, what now supervises the effort to keep the carcass-as-ecosystem stable and sustain its presence on the beach over time? How does this supervisory entity do its job? Or, if there is nothing in charge of maintaining the carcass, how do separate players or forces residing within it collectively decide that keeping the carcass together is a good idea? How do they coordinate with each other to achieve their common goal? In short, what allows the carcass to be creative in defense of itself as an organized whole? We know that agents of other ecosystems — crabs, seagulls, a host of other living things, and physical components of the environment — will make short work of the carcass, thus dooming efforts of the new ecosystem to maintain itself. At the very least, therefore, we must conclude that either carcasses-as-ecosystem are inherently unhealthy or that the notions of stability, sustainability, and purposeful action over time by an ecosystem to maintain its organization and autonomy are poor criteria of ecosystem health.

How do ideas in the Norton and Ulanowicz view of ecosystem health and those in the Haskell et al. definition square with other aspects of current ecological understanding? For example, most ecologists renounce the equilibrium model in favor of a nonequilibrium model (Hagen, 1992; Rapport et al., 1995; Christensen et al., 1996; De Leo and Levin, 1997). The ideas of stability and sustainability, however, flow from the rejected equilibrium model; so what are their meanings here? The nonequilibrium model makes it clear that the landscape constantly changes; therefore, how can an ecosystem's health depend on its ability to achieve the impossible goal of maintaining itself over time?

Can ecosystems have properties like creativity, or be capable of innovation, or actively seek to perpetuate themselves? Together these attributes assume that ecosystems exist as real and discrete organizational entities above the species level that have inventoried themselves so that they know what they should look like and what they should be doing. It suggests that ecosystems deliberately gather and evaluate information to determine what combination of factors are harmful to their self-preferred state and subsequently plot a course to avoid or overcome injurious changes. Such attributes presuppose consciousness, the ability to be aware of external happenings and take purposeful action to achieve a particular goal. Yet such abilities are characteristics of living things, which most researchers now steadfastly maintain ecosystems are not. Ecologists widely agree that the discipline abandoned superorganicism in decades past. As Frank Golley (1993) wrote, the ecosystem concept itself is properly criticized "when ecosystem scientists propose that ecosystems are self-regulated superorganisms with the purpose of maintaining stability." The preface to the proceedings of a NATO advanced research workshop on ecosystem health includes the finding, "All who have been involved in this Workshop reject the concepts of ecosystems as superorganisms and as homeostatic systems" (Rapport et al., 1995). One cannot reconcile the abandonment of the equilibrium paradigm and superorganicism with claims that ecosystems are creative, innovative, and seek to maintain themselves over time.

Given the nebulous nature of the ecosystem health idea, one should not be surprised there is little agreement on its measurement. Shrader-Frechette and McCoy (1993) pointed out that people know the norms of human health but comparable norms for ecosystems do not exist. Why the difference? The medical community knows about diseases that attack various components of the human body and their respective threats to the body's ability for continued functioning. When considering human health, a real and living patient is before the doctor. Every other physician who enters the examination room would instantly recognize the exact same patient. They would agree about function, structure, and placement of the patient's component parts as well as generally concur about the relative importance of each. They understand that the individual parts do not have life separate from the whole and that the parts perform their functions to serve the central purpose of maintaining the patient's life. Doctors can differentiate between life and death of the patient.

In contrast, when considering ecosystem health, the ecodoctor's patient is neither real nor alive. Other ecodoctors view the same landscape and may see quite different patients. Even if they can agree on the presence of a particular ecopatient, ecodoctors may or may not concur on the patient's parts or their functioning or their relative importance. They may or may not share the same views regarding what conditions represent the patient's death. Perhaps the only thing they may agree on is that, unlike the medical doctor's patient, separate parts of the ecosystem-as-patient possess life without any regard to the whole.

A landscape in constant flux confronts scholars seeking to give substance to the ecosystem health notion. Biota continuously redistribute themselves in space. Populations wax and wane. Rates of change vary from species to species and place to place and over time. Physical processes likewise vary though space and time at different rates. How can we objectively proclaim some changes as healthy but label others as indicators of disease for public policy purposes? Jill Lancaster (2000) concluded, "Ecological health (and its synonyms) cannot be defined or measured objectively and that claims to the contrary are essentially fraudulent." Vagueness and internal contradiction render the ecosystem health concept inappropriate as a guide to better environmental policy making. But these flaws, while fatal in their own right, do not get to the heart of the matter; and that is the nature of the presumed patient, ecosystems themselves.

NATURE OF ECOSYSTEMS

Ecosystems are mental constructs, heuristic devices, rather than tangible landscape units possessing intrinsic characteristics like health or integrity. Ecosystems represent geographic conveniences fabricated by researchers, not objective reality. They are but portions of the landscape — regions — that researchers arbitrarily label as this or that ecosystem. What some people call the Greater Yellowstone Ecosystem is no more a discrete, organized entity seeking to maintain itself against external or internal threats than is the Great Plains region a physical totality defending itself against the Midwest region to its east or from a growing greater Denver region within its borders.

The study of landscapes and regions is the cornerstone of one our oldest fields of knowledge, geography. More than half a century ago, after a lengthy and spirited debate, geographers abandoned the notion that the regions they studied were actual objects. They recognized that regions are tools that may play an important role as intellectual devices for organizing, analyzing, and presenting information in a spatial context while rejecting the idea that regions are present on the landscape awaiting discovery by suitably diligent and well-informed researchers (Hartshorne, 1959; Hart, 1982). Scholars can say the same, but not more, about ecosystems.

Scientists do not discover ecosystems on the landscape via the application of theory and diligent fieldwork; instead, they arbitrarily label a portion of the landscape as an ecosystem to suit the needs of the project at hand. Simon Levin (2001) observed that ecosystems "are better thought of as idealizations rather than real entities." He goes on to note, "What we call an ecosystem is an operational convenience, a loosely defined and ephemeral assemblage of co-occurring organisms ... [whose] ... boundaries are defined by the investigator for convenience of description." Ecosystems are "temporary conveniences," wrote Norton (1992) "to be discarded and replaced at will." James Agee and Darryll Johnson (1988) found that an ecosystem is "any part of the universe chosen as the area of interest, with the line around that area being the ecosystem boundary." Bruce Hannon (1991) agreed, saying, "The delimitation of the [eco]system is strictly up to the observer; i.e., the system boundaries and the list of internal elements may be chosen at will." Ecologists Lawrence Kapustka and Wayne Landis (1998) pointed out, "No human has ever seen an ecosystem." These observations confirm Robert Ricklefs' (1987) conclusion that ecosystems "have no boundaries in space or time — they are not discrete, identifiable units like organisms." Efforts to identify the Greater Yellowstone Ecosystem (GYE) illustrate these points.

The idea of the GYE emerged in the 1970s. Grizzly bear researcher Frank Craighead, Jr., produced the first published map of the GYE in 1979. His version covers roughly 5 million acres enclosed within a seven-sided polygon, determined by the range of the grizzly bear population roughly centered on Yellowstone National Park. Others latched on to the idea of the GYE but took different approaches to its construction. The first edition of *The Greater Yellowstone: the National Park & Adjacent Wildlands* appeared in 1984. Author Rick Reese, a founder and first president of the Greater Yellowstone Coalition, wrote that the GYE extended over "8 to 10 million acres." In the second edition he estimated the size of the GYE at 18 million acres (Reese, 1991). The fire of 1988 notwithstanding, during the interval between the first and second editions, the landscape underwent no fundamental change; the GYE doubled in size simply because those doing the labeling changed their minds. "How we define the geographic extent of the Greater Yellowstone," Reese said, "depends on which of its characteristics we use as criteria for delineating boundaries." He and his colleagues at the Greater Yellowstone Coalition invoked a bewildering array of criteria to consider when determining the GYE, including climate; actual vegetation; potential natural vegetation; hydrography; topology; geothermal features; and ranges and habitat requirements of grizzlies, elk, deer, bison, and other biota (Glick et al., 1991). Each one of these separate spatial variables has its individual (and often shifting and imperfectly known) distribution. The coalition did not say what these patterns were or tell us how they combined them into their version of the boundary for the GYE. Meanwhile, the Congressional Research Service (1986) prepared a map that largely followed existing administrative boundaries and enclosed some 14 million acres.

Confusion surrounds the GYE. Law professor Robert Keiter (1991) observed that the Greater Yellowstone Ecosystem "cannot be defined by boundaries of any lasting significance." Fellow student of the GYE, Duncan Patton (1991) said that "the GYE has no definite boundaries" while simultaneously claiming it "is bound by its ecological cohesiveness." Tim Clark and Steven Minta (1994) — who, like Keiter and Patton, are advocates for protection of the GYE — wrote "Some definitions in currency are ... clearly at odds with one another ... there is disagreement on the location and extent of the GYE ... [which simultaneously] cannot be 5 million and 19 million acres." No agreement exists on which spatial variables analysts should use in delimiting the GYE or how they should meld those variables into a single boundary. No wonder Jonathon Taylor and Nina Burkhardt (1993) opined, "Clear designation of the boundaries of the greater Yellowstone ecosystem remains somewhat vague."

Over 20 years have passed since the GYE surfaced in print, but its students cannot agree on where it is, how to define it, or what shape it has. They cannot agree on how big it is, how much public land it includes, or how much private land it covers. They cannot tell you where it begins or ends, describe its core characteristics, or classify it with respect to other ecosystems. "Boundary definition," wrote John Freemuth (1997), "is stunningly problematic." He concluded that the debate about boundaries "is no mere quibble, and may lie at the core of problems in both defining and implementing ecosystem management." That is because sound land management policy cannot be built on geographic uncertainty, yet what we see with the GYE is typical of what happens when researchers move the ecosystem concept from the textbook to the landscape.

Mark Sagoff (1997) pointed out that one area of intellectual enterprise vital to the growth of a science is the ability for practitioners to classify the objects they study. The National Research Council (1993) likewise emphasized the importance of classification. "Ecological classifications," they wrote, "are extremely important for the management of biological resources on an ecosystem ... basis." Classifications "are needed for recognizing and mapping" ecosystems "and for communicating about their status, distributions, and trends." Ecosystem maps, in turn, are "a prerequisite to planning for, budgeting, authorizing, and appropriating funds for, and ultimately managing activities on the basis of ecological units" according to the U.S. General Accounting Office (1994) in a report to Congress on ecosystem management. Yet more than 60 years after Sir Arthur Tansley (1935) proposed the ecosystem concept, after decades of dominance in graduate schools, after the publication of thousands of articles, books, and studies, researchers still have not developed a recognized classification system for ecosystems (Sagoff, 1997). As the National Research Council (1993) concluded, "There are currently no broadly accepted classification schemes for ... ecological units above the level of species."

Nebulousness reigns. Scholars cannot agree on what constitutes the core characteristics of ecosystems which, after all, can be anything from a drop of water to a dung pile to a whale carcass to the entire planet (National Research Council, 1993; Christensen et al., 1996). Kurt Jax et al. (1998) observed considerable disagreement about the basic features of ecosystems. They go on to say that no agreement exists on how to define and measure the "spatiotemporal continuity of an [ecosystem] through time, during which the essential characteristics of the [ecosystem] are retained." Sagoff (2000) concurred, writing that researchers cannot agree on ways to define ecosystems and to redefine them through time and change. Put another way, scholars cannot objectively determine if an ecosystem remains itself in space and time or clearly establish when or where it becomes a different ecosystem. Yet such capabilities are indispensable to answering questions about the status and trends of an ecosystem's health, integrity, and sustainability — the presumed attributes of ecosystems which champions of ecosystem-based policies want to make the centerpiece of federal land management.

Essentially researchers, policy makers, land managers, and the public are left with idiosyncrasy from which not even the most knowledgeable and energetic among us can fashion a cogent and verifiable ecosystem classification system or the definitive ecosystem maps necessary to direct the application of laws and regulations intended to restore or maintain ecosystem health, integrity, or sustainability.

Let me review. Scholars cannot classify ecosystems or point to their core characteristics. They cannot determine where ecosystems begin or end in space or time. People can label any portion of the Earth's surface they choose as an ecosystem, using whatever criteria they select. They can draw its boundaries wherever they please using methods of their own choosing so that on the landscape the ecosystem concept represents a geographic free-for-all. Ecosystems are mental constructs, not living entities seeking to sustain themselves either spatially or temporally. How, then, can ecosystems form a rational basis to organize the landscape for the purpose of governance? They cannot, as two recent efforts by the federal government to implement ecosystem-based land use management make plain.

EFFORTS TO IMPLEMENT ECOSYSTEM MANAGEMENT

On November 9, 2000, the U.S. Forest Service (2000) announced a final rule to govern planning for the 191-million-acre national forest system. When the proposal was first announced, then-Agriculture Secretary Dan Glickman (1999) heralded it as a fundamental change in philosophy. It proclaimed, "The first priority for planning to guide management of the National Forest System is to maintain or restore ecological sustainability." That required the Forest Service to "put greater emphasis on maintaining and restoring ecosystem health in order to promote" sustainability. The rule reflects the uncertainty and confusion inherent in the ecosystem concept and in the notion of ecosystem health.

The entire rule rests on the idea that ecosystems are discrete, knowable, and real entities on the landscape, possessing attributes that the government should restore and protect as its first order of business. The rule offers no justification for such contentions and fails to define either ecosystems or ecosystem health. This is especially disturbing in light of the substantial literature arguing that ecosystems are nothing more than *ad hoc* spatial conveniences and that the notion of health cannot be properly applied to ecosystems. The final rule tacitly acknowledges that the landscape contains a limitless number of ecosystems (the proposed rule freely admitted there are an infinite number of ecosystems on the landscape), but it does not specify how managers are to select the relatively few ecosystems whose maintenance and restoration will become the focus of agency efforts. It establishes no procedures for determining ecosystem boundaries in space or time; instead it simply directs managers to choose "appropriate spatial and temporal scales." Spatially, these scales may range from a few acres to millions of acres and, temporally, from decades to centuries, depending on agency opinion. Guesswork abounds.

How people are to judge the success of plans developed according to the rule is equally obtuse. The rule defines ecological sustainability as "the maintenance or restoration of the composition, structure, and processes of ecosystems including … the productive capacity of ecosystems." Productive capacity is said to be "the ability of an ecosystem to maintain primary productivity including its ability to sustain desirable conditions" that encompass the ability for ecosystem self-renewal after disturbance. But as we have seen, the scientific community has no agreed-upon way of judging when an ecosystem system ceases to be itself and becomes a new ecosystem. Further, because the nonequilibrium model makes it plain that the landscape constantly changes, the idea of maintaining composition and structure appears to be a fruitless exercise that does not offer a rational basis for assessing the success of government programs.

When one examines the rule more closely to see how people are to judge plan success, things become even more confusing. Section 219.20 (b) tells "responsible officials," when making decisions "that will affect ecological sustainability," they must consider "ecosystem diversity," "species diversity," and "Federally listed threatened and endangered species." The measure of sustainability, and therefore plan success, rests heavily on the idea of protecting ecosystem diversity. The rule offers no definition of this nebulous idea, describing it in terms of maintaining an ecosystem within its natural range of variation which is "the expected range of variation in ecosystem composition and structure that would be expected under natural disturbance regimes in the current climate period." The agency does not tell us how one is to measure, predict, and evaluate the natural range of variation. More importantly, the FS offers no justification for why deviation from the arbitrary and wholly unscientific standard of *nature knows best* should guide management of public land (Peters, 1991).

In short, the agency wants to make application of undefined standards to protect unclear values found in unspecified places to be determined later by unnamed means the basis for managing the nation's national forests. This is not the stuff of sound land use management.

The Interior Columbia Basin Ecosystem Management Project (ICBEMP) provides a second example of a regulatory effort at making ecosystem protection the centerpiece of federal land management. A combined effort of the FS and the Department of the Interior's Bureau of Land Management (BLM), it aims at making protection and restoration of ecosystems the fundamental management goal for some 75 million acres of public land from Oregon and Washington to Montana. Like the FS planning rule, it reflects all the vagueness embodied in the ecosystem concept and its associated ideas like ecosystem health.

The project began in 1993 with a directive from former President Clinton to develop an ecosystem-based strategy for federal land use management in the region. Its purpose is to "restore and maintain long-term ecosystem health and ecological integrity" and, while human use of the area is permitted, it must be "consistent with maintaining healthy, diverse ecosystems" (U.S. Department of Agriculture, 2000a). Yet, after 7 years of effort and over $40 million in project costs, the agencies still cannot tell us which ecosystems will be protected or restored (U.S. General Accounting Office, 1999). The Draft Environmental Impact Statements (which, with appendices, exceed 2000 pages) contain over 100 maps, but none shows the ecosystems of policy interest (U.S. Department of Agriculture, 1997a,b). The Supplemental Draft Environmental Impact Statement contains 64 maps, and the Final Environmental Impact Statement has three maps; none show the ecosystems whose health and integrity the agencies intend to restore or protect (U.S. Department of Agriculture, 2000a,b). The documents also fail to specify how such ecosystems may be identified in the future. About ecosystems in general, however, they do say that we can never know for certain what a particular ecosystem contains, but we do know they constantly change and have artificial boundaries.

Project officials define ecosystem health as a "condition where the parts and functions of an ecosystem are sustained over time and where the system's capacity for self-repair is maintained, such that goals for uses, values, and services of the ecosystem are met." Questions arise immediately. Because ecosystems constantly change over time and space, the notions of sustainability and a capacity for self-repair have no substantial meaning. Ecosystems simply blend into one another as time passes (or one replaces another after cataclysmic events such as the eruptions on Mount St. Helens); so how can one reasonably say one sort of change is healthful and another pathologic? What do we use as a baseline to compare change against? Today's conditions? How things were in 1900? How about conditions at the time when Europeans arrived? How do we compare different kinds of change for overall health assessment purposes? For the sake of argument, suppose we decide to place carbon sequestration and species abundance on our list of ecosystem health measures. Suppose further that after 20 years we find an ecosystem sequesters 10% more carbon but lost 10% of its species. Is it more or less healthy than it was?

ICBEMP strives to restore and protect ecosystem integrity as well as ecosystem health. In so doing, it wanders farther into a buzzword world. Admitting that "absolute measures of integrity do not exist," project officials defined integrity as "the degree to which all ecological components and their interactions are represented and functioning; the quality of being complete; a sense of wholeness" (U.S. Department of Agriculture, 2000b). Although this view of integrity comes from within the scientific literature (Kay, 1993; Angermeier and Karr, 1994), one is hard pressed to find an objective, verifiable foundation for it. For Karr (2000) a place with maximum integrity is one that supports "a biota that is the product of evolutionary and biogeographic processes with little or no influence from industrial society." Again we are confronted with an arbitrary *nature knows best* standard that flows from individual likes and dislikes rather than from science. The judgment that one landscape is complete or whole, whereas another is not, does not rise above personal opinion. If wild salmon disappeared tomorrow from the ICBEMP area, would any ecosystems therein lose their sense of wholeness? If so, which ones; and were these systems complete or whole before salmon appeared in the region? Suppose, after a species of salmon disappeared from a river, three other fish species joined the community. Would the ecosystem then be more or less complete or whole than before the salmon departed?

Science cannot answer such questions because the notions of completeness and wholeness have no scientific basis. Like the ideas of ecosystem health and integrity, they are nothing more than subjective personal landscape preferences, states of mind rather than conditions of the environment. Levin (2001) noted, "The ecosystem is not an organism and has not been shaped by evolution to perform particular functions" thus undercutting "all efforts to measure ecosystem 'health' and 'integrity.'" Lancaster (2000) was more direct, writing, "Ecological health is a nebulous concept that should be stricken from the vocabulary. Likewise, all synonymous terms [like ecosystem integrity] are ridiculous in a scientific context." The ICBEMP effort, like the FS planning rule, rests on buzzwords.

A FINAL POINT ON ECOSYSTEMS AND GOVERNANCE

Americans require their government to be neither arbitrary nor capricious. Spatially, our government rests on precisely known, universally recognized, and fixed geopolitical units. Consider the states of the U.S. Only one pattern of states blankets that portion of the planet called the U.S.; that is, there are a fixed number of states whose shape, size, and location are known to all. States do not overlap other states, and there are no gaps between states. Boundaries between states are well defined, and they do not change with the seasons or the years. A given portion of the landscape lies in exactly one state. This geopolitical exactness provides citizens a certainty that, for example, officials in state A who are unaccountable to people in neighboring state B cannot dictate to people in state B. Such geopolitical exactness is a necessary condition for stability in a democratic society.

Imagine for a moment that states possessed the same spatial characteristics as ecosystems. Then, for example, people would not really know the location of state boundaries; state boundaries would change through space and time, transforming individual states into geographic amoebas oozing over the landscape; states would overlap so that a given area could, say, simultaneously be in California, Nevada, and Oregon; some states would cease to exist while new ones would appear; and several different patterns of states would simultaneously cover the U.S. No one would seriously propose giving states these characteristics because governing would become arbitrary and capricious if it were possible at all. So why, then, should we use ecosystems as the spatial guides for the application of government rules and regulations?

CONCLUSION

Sagoff (1997) wrote, "Ecological systems are the conceptual constructs of a theoretical ecology, the old equilibrium ecology, that is now defunct. Just as the smile of the Cheshire Cat survived its demise ... the concept of the ecosystem haunts the ecological literature as an apparition without substance." Many legitimate reasons exist for society to concern itself with land use and land management, but protecting or restoring the health of ecosystems is not among them. Nebulousness and uncertainty so surround the concepts of the ecosystem and ecosystem health that they do not provide a cogent foundation for government action.

A hundred years ago Joel Chandler Harris wrote his classic Uncle Remus stories. Through the antics and adventures of his characters, Harris continues to impart wisdom to generations of readers. Chief among his players are Brer Rabbit and his antagonists Brer Fox and Brer Bear. In the story of the "Tar Baby," Brer Fox and Brer Bear fashion a figure from tar and dress it like child so as to ensnare our unsuspecting hero. When Brer Rabbit encounters the Tar Baby along the side of the road, he thinks he has found a new friend and says, "Howdy." The Tar Baby, however, fails to answer and thus does not respond as Brer Rabbit hopes. After repeated "Howdies" do not get the desired response, Brer Rabbit — refusing to accept the now obvious — pounds on the Tar Baby in a fruitless attempt to force it into doing something it cannot do — talk. Eventually our rabbit friend is hopelessly enmeshed in goo, but the Tar Baby still does not speak.

It seems to me that is where we are with respect to ecosystems and public policy. The tar baby of ecosystems attracts many people. But for reasons I note here and elsewhere, defending illusions on the landscape is not a responsible goal for public policy (Fitzsimmons, 1999). Instead of continuing to try and turn the ecosystem tar baby into something it cannot become — a rational foundation for land use management — I suggest we extract ourselves and move on down the road in search of better constructs because, at least in the realm of public policy, the ecosystem concept and its attendant notions of health, integrity, and sustainability represent intellectual and evolutionary dead ends.

REFERENCES

Agee, J.K. and Johnson, D.R., Eds., *Ecosystem Management for Parks and Wilderness*, University of Washington Press, Seattle, WA, 1988.

Angermeier, P.L. and Karr, J.R., Biological integrity versus biological diversity as policy directives, *BioScience,* 44:690–697, 1994.

Bosworth, D., Working together for the health of the land, speech delivered in Sacramento, CA, on September 28, 2001, available online at http://www.fs.fed.us/intro/speech/2001/2001sep28-Sierrafinl.htm.

Callicott, J.B., The value of ecosystem health, *Environ. Values,* 4:345–361, 1995.

Calow, P., Ecosystem health — a critical analysis of concepts, in Rapport, D.J., Gaudet, C.L., and Calow, P., Eds., *Evaluating and Monitoring the Health of Large Scale Ecosystems*, Springer, New York, 1995, pp. 33–41.

Christensen, N.L., Bartuska, A.M., Brown, J.H., Carpenter, S., D'Antonio, C., Frances, R., Franklin, J.F., MacMahon, J.A., Noss, R.F., Parsons, D.J., Peterson, C.H., Turner, M.G., and Woodmansee, R.G., Report of the ecological society of America committee on the scientific basis for ecosystem management, *Ecol. Appl.,* 6:665–691, 1996.

Clark, T. and Minta, S., *Greater Yellowstone's Future*. Homestead Publishing, Moose, WY, 1994.

Congressional Research Service. *Yellowstone: Ecosystem, Resources, and Management*, Corn, M.L. and Gorte, R.W., Library of Congress, 86-1037 ENR, Washington, D.C., Dec. 12, 1986.

Craighead, F., Jr., *Track of the Grizzly*, Sierra Club Books, San Francisco, 1979.

De Leo, G.A. and Levin, S., The multifaceted aspects of ecosystem integrity, *Conserv. Ecol.,* 1, 1997, available online at http://www.consecol.org/vol1/iss1/art3.

Dombeck, M., A gradual unfolding of a national purpose: a natural resource agenda for the 21st century, speech delivered to Forest Service employees in Washington, D.C., March 2, 1998, available online at http://www.fs.fed.us/news/agenda/sp30298.html.

Fitzsimmons, A.K., *Defending Illusions: Federal Protection of Ecosystems*, Rowman & Littlefield, Lanham, MD, 1999.

Freemuth, J., Ecosystem management and its place in the national park service, *Denver Univ. Law Rev.,* 74:697–727, 1997.

Glick, D., Carr, M., and Harting, B., Eds., *An Environmental Profile of the Greater Yellowstone Ecosystem*, The Greater Yellowstone Coalition, Bozeman, MT, 1991.

Glickman, D., Glickman proposes new rules for managing the national forests, U.S. Department of Agriculture press release number 0399.99, September 30, 1999. Available online at http://www.usda.gov/news/releases/1999/09/0389.

Golley, F., *A History of the Ecosystem Concept in Ecology*, Yale University Press, New Haven, CT, 1993.

Goulden, C.E., Ecological comprehensiveness, *Science,* 264:726–727, 1994.

Hagen, J.B., *An Entangled Bank: The Origins of Ecosystem Ecology*, Rutgers University Press, New Brunswick, NJ, 1992.

Hannon, B., Accounting in ecological systems, in Costanza, R., Ed., *Ecological Economics: The Science of Management of Sustainability*, Columbia University Press, New York, 1991, pp. 234–252.

Hart, J.F., Presidential address: the highest form of the geographers art, *Ann. Assoc. Am. Geogr.,* 72:1–29, 1982.

Hartshorne, R., *Perspectives on the Nature of Geography*, Rand McNally, Chicago, IL, 1959.

Haskell, B.D., Norton, B.G., and Costanza, R., Introduction: what is ecosystem health and why should we worry about it? in Haskell, B.D., Norton, B.G., and Costanza, R., Eds., *Ecosystem Health: New Goals for Environmental Management*, Island Press, Washington, D.C., 1992, pp. 3–19.

Jamieson, D., Ecosystem health: some preventative medicine, *Environ. Values,* 4:333–344, 1995.

Jax, K., Jones, C.G., and Pickett, S.T.A., The self-identity of ecological units, *Oikos*, 82:253–264, 1998.

Kapustka, L.A. and Landis, W.G., Ecology: the science versus the myth, *Hum. Ecol. Risk Assess.,* 4:829–837, 1998.

Karr, J.R., Health, integrity, and biological assessment: the importance of measuring the whole thing, in Pimentel, D., Westra, L., and Noss, R., Eds., *Ecological Integrity*, Island Press, Washington, D.C., 2000, pp. 209–226.

Kay, J., On the nature of ecological integrity: some closing comments, in Wooley, S., Kay, J., and Francis, G., Eds., *Ecological Integrity and the Management of Ecosystems*, St. Lucie Press, Del Ray Beach, FL, 1993, pp. 201–212.

Keiter, R.B., An introduction to the ecosystem management debate, in Keiter, R.B. and Boyce, M.S., Eds., *The Greater Yellowstone Ecosystem*, Yale University Press, New Haven, CT, 1991, pp. 1–18.

Lancaster, J., The ridiculous notion of assessing ecological health and identifying the useful concepts underneath, *Hum. Ecol. Risk Assess.,* 6:213–222, 2000.

Lackey, R.T., Pacific salmon, ecological health, and public policy, *Ecosyst. Health,* 2:61–68, 1996.

Lackey, R.T., Values, policy and ecosystem health, *BioScience,* 51:437–443, 2001.

Levin, S., The problem of pattern and scale in ecology, *Ecology,* 73:1943–1967, 1992.

Levin, S., Immune systems and ecosystems, *Conserv. Ecol.,* 5, 2001, available online at http://www.consecol.org/vol5/iss1/art17.

Nash, R., Historical and philosophical considerations of ecosystem management, in Ecosystem Management: Status and Potential, report of a workshop convened by the Congressional Research Service, March 24–25, 1994, pp. 25–32.

National Research Council, *A Biological Survey for the Nation*, National Academy Press, Washington, D.C., 1993.

Norton, B.G., A new paradigm for environmental management, in Costanza, R., Norton, B.G., and Haskell, B.D., Eds., *Ecosystem Health: New Goals for Environmental Management*, Island Press, Washington, D.C., 1992, pp. 23–41.

Norton, B.G. and Ulanowicz, R.E., Scale and biodiversity policy: a hierarchical approach, in *Ecosystem Management*, Samson, F.B. and Knopf, F.L., Eds., Springer, New York, 1996, pp. 424–434.

Patton, D.T., Defining the Greater Yellowstone Ecosystem, in Keiter, R.B. and Boyce, M.S., Eds., *The Greater Yellowstone Ecosystem*, Yale University Press, New Haven, CT, 1991, pp. 19–26.

Peters, R.H., *A Critique for Ecology*, Cambridge University Press, Cambridge, U.K., 1991.

Rapport, D.J., Ecosystem health: an emerging integrative science, in Rapport, D.J., Gaudet, C.L., and Calow, P., Eds., *Evaluating and Monitoring the Health of Large Scale Ecosystems*, Springer, New York, 1995, pp. 5–31.

Rapport, D., Defining ecosystem health, in Rapport, D. et al., Eds., *Ecosystem Health,* Blackwell Science, London, 1998, pp. 18–33.

Rapport, D.J., Gaudet, C.L., and Calow, P., Preface, in Rapport, D.J., Gaudet, C.L., and Calow, P., Eds., *Evaluating and Monitoring the Health of Large Scale Ecosystems*, Springer, New York, 1995.

Reese, R., *Greater Yellowstone: The National Park & Adjacent Wildlands,* 2nd ed., American and World Geographic Publishing, Helena, MT, 1991.

Reidel, C. and Richardson, J., Strategic environmental leadership in a time of change, inaugural Donlon lecture, State University of New York, College of Environmental Science and Forestry, Syracuse, NY, 1995.

Ricklefs, R., Structure in ecology, *Science,* 236:206–207, 1987.

Sagoff, M., Muddle or muddle through? Takings jurisprudence meets the endangered species act, *William Mary Law Rev.,* 38:825–993, 1997.

Sagoff, M., Ecosystem design in historical and philosophical context, in Pimentel, D., Westra, L., and Noss, R., Eds., *Ecological Integrity,* Island Press, Washington, D.C., 2000, pp. 61–77.

Shrader-Frechette, K.S. and McCoy, E.D., *Method in Ecology,* Cambridge University Press, Cambridge, U.K., 1993.

Tansley, A.G., The use and abuse of vegetational concepts and terms, *Ecology,* 16:284–307, 1935.

Taylor, J. and Burkhardt, N., Introduction: the Greater Yellowstone Ecosystem — biosphere reserves and economics, *Soc. Nat. Resour.,* 6:105–108, 1993.

U.S. Department of Agriculture, *Upper Columbia River Basin Draft Environmental Impact Statement,* Vol. 1, Interior Columbia Basin Ecosystem Management Project, Boise, ID, 1997a.

U.S. Department of Agriculture, *Eastside Draft Environmental Impact Statement,* Vol. 1, Interior Columbia Basin Ecosystem Management Project, Walla Walla, WA, 1997b.

U.S. Department of Agriculture, *Interior Columbia Basin Supplemental Draft Environmental Impact Statement,* Vol. 1, Interior Columbia Basin Ecosystem Management Project, Walla Walla, WA, 2000a.

U.S. Department of Agriculture, *Interior Columbia Basin Final Environmental Impact Statement,* Interior Columbia Basin Ecosystem Management Project, Walla Walla, WA, 2000b.

U.S. Forest Service, National forest system land planning and resource management planning final rule, *Fed. Regist.,* 65:67514–67581, 2000.

U.S. General Accounting Office, *Ecosystem Management: Additional Actions Needed to Adequately Test a Promising Approach,* GAO/RCED-94–111, General Accounting Office, Washington, D.C., 1994.

U.S. General Accounting Office, *Ecosystem Planning: Northwest Forest and Interior Columbia Basin Plans Demonstrate Improvements in Land Use Planning,* GAO/RCED-99–64, General Accounting Office, Washington, D.C., 1999.

Wilkins, D.A., Assessing ecosystem health, *Trends Ecol. Evol.,* 14:69, 1999.

CHAPTER 24

Quality of Life, Recreation, and Natural Environments: Exploring the Connection

Alan Ewert

INTRODUCTION

The interface between people and natural environments occurs in a variety of complex ways that affect the individual's economic, cultural, physical, and spiritual well-being. Although historically this interaction has been characterized in economic terms (e.g., commodity production) or cultural aspects (e.g., number of jobs created in a community), there is currently a trend that recognizes the physical and spiritual effects of natural environments upon individuals or communities (Driver et al., 1996; Schroeder, 1996).

Moreover, natural environments appear to play an increasingly important role in an individual's perceived quality of life (QOL). The purpose of this chapter is to discuss the connection between natural environments, outdoor recreation, and QOL issues. To discuss these and other points the following framework will be used:

- A definition and description of the construct of QOL
- A discussion of natural environments and their relationships to human health and QOL
- The connection between outdoor recreation and pro-environment behaviors and attitudes

In this chapter, ecosystem is defined as a community of organisms working together with their environments as integrated units (Salwasser et al., 1993). Natural environments are thought of as areas involving differing scales (e.g., stand, watershed, region, etc.) in which natural processes are essentially predominant and human development is minimized. Moreover, while technically not equivalent, the terms *ecosystem* and *environment* will be used interchangeably throughout this chapter.

Two issues tend to complicate any discussion about QOL. First, individuals who are part of broader social structures, such as neighborhoods or communities, often associate QOL with variables such as accessible education, affordable housing, and just law enforcement systems. Individuals who are members of natural resource management and research professions often equate QOL with living environments rich in biodiversity, sufficient wilderness, and undeveloped landscapes managed under a sustainable practices strategy. In both cases, however, QOL is ultimately associated with the life and work space of the individual and community.

A second, more fundamental issue related to discussing QOL revolves around the operational aspects of QOL. That is, QOL for many is equated with personal income, size of house, or the number of cars owned (Hatcher, 1994). Yet others think of QOL in terms of the amount and type of leisure time and recreational opportunities available to them. For this second group QOL is related to their ability to recreate and enjoy pristine natural environments (Peterson et al., 1988; Loomis and Walsh, 1997). Indeed, while managers and scientists discuss the importance of issues such as sustainability, ecosystem restoration, and exotic invasions, a substantial portion of the public equate QOL with the ability to enjoy outdoor recreation actively on a regular basis and in relatively natural landscapes.

Consequently, a major point of this chapter is that participation in recreational activities located in natural environments may offer an effective vector for educating the public about environment-related issues and provide a direct linkage to the attitude–behavior phenomenon. In other words, if people are able to enjoy natural surroundings and environments, they will be more willing to seek protection for these areas and practice environmentally sustainable behaviors such as reduced energy consumption, reduction of littering, and less materialism. Thus, the premise of this chapter is that public education and awareness will ultimately serve to protect and promote natural eco-systems and environments.

QUALITY OF LIFE

QOL has been defined in a variety of ways using a number of different terms, including a sense of happiness, manifest joy, and some measure of self-actualization. Consistent in many of these definitions have been the terms *well-being* and *wellness*. In this chapter, well-being is considered an internal sense of satisfaction or the fulfillment of real as well as perceived health. In this instance, *health* is defined as optimal social, physical, spiritual, or emotional conditions of the individual. In a similar vein, wellness can be thought of as a function of health, with an individual's sense of wellness at any one time located on a continuum from poor to optimal health. It is worth noting, however, that the construct of health is dynamic and subject to constant reevaluation of what constitutes human health (VanLeeuwen et al., 1999).

In addition to well-being and wellness, QOL can be characterized as falling within two general approaches to treatment and intervention — one associated with the medical sciences and the other closely aligned with the behavioral sciences. Within the medical sciences, QOL is often assessed along parameters such as health status, physical functioning, energy levels, or general perceptions of health. From the behavioral sciences perspective, QOL is often linked to issues such as need satisfaction, sense of independence, competence, and a meaningful use of time. Thus, in a larger sense, QOL acquires the characteristic of a latent variable by subsuming a number of manifest or indicator variables such as a sense of wellness, security, or a sense of personal control, depending on one's perspective and situation. While it is clear that QOL is a construct highly dependent on how and by whom it is defined, there are some common human needs that are thought of as important components for most QOL indices. These common needs involve personal health, economic needs, self-esteem, social support systems, and aesthetic needs (Sirgy and Efraty, 1998).

ECOSYSTEM HEALTH AND THE INDIVIDUAL

Worldwide, significant changes in the attitudes, behaviors, and priorities concerning the manage-ment and use of natural resources among the population have been observed (Costanza et al., 1992; Bengston, 1994; Ewert and Kessler, 1996). Fundamental to these changes has been an awareness that the interface between humans and natural environments produces both benefits and costs that have significant implications for an individual's perception of QOL (Driver et al., 1991; Suzuki, 1997).

Increasingly clear is the linkage between healthy, natural environments and elevated QOL indicators. This linkage may be regarded as bidirectional in that humans not only impact natural environments, but natural environments can also significantly impact humans. In the vernacular, "we affect nature; nature affects us."

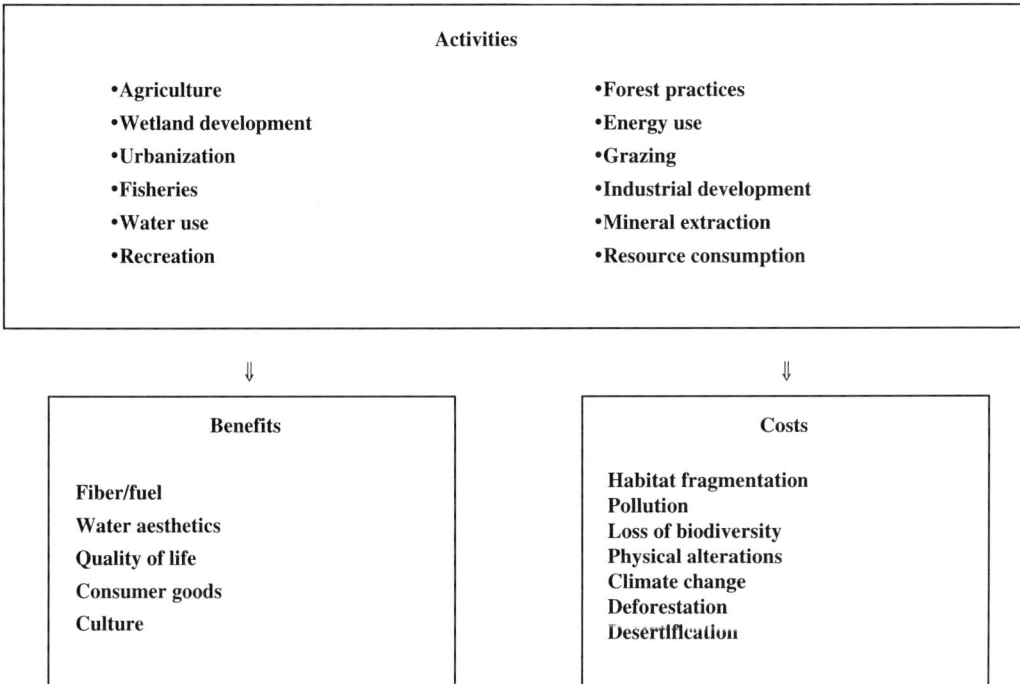

Figure 24.1 Human activities impacting the sustainability of natural ecosystems.

Human Impacts on Natural Environments

From the perspective of human impacts on natural environments, numerous human activities serve as catalysts for creating benefits from, and costs to, natural environments. Benefits, discussed in the next section, imply goods and services. As listed in Figure 24.1, sample costs include habitat fragmentation, introduction of exotic species, levels of nonsustainable consumption, and alteration of physical systems such as the atmosphere. Porter and Brown (1991) suggest that some of the forces driving these negative impacts include increased population, consumption patterns, disparities in economic wealth, patterns of global trade, and national and international politics. As a result, many communities (towns, cities, countries, etc.) often embrace an orientation toward unsustainable growth and development that often leads to unrestrained parameters of consumption, instability of systems, and an increased ability to impact and modify natural environments.

Environmental Impacts on Humans

Environmental impacts on human systems can occur along three different dimensions: direct, indirect, and goods and services. Direct impacts are functionally associated with an interaction between humans and natural environments and include recreational, cathartic, and aesthetic experiences. Other direct impacts to humans are health-related and include pesticide use, water and air quality, atmospheric warming, and land use for development and recreation (Table 24.1). These issues are not only significant by themselves but are also regarded as part of a larger pattern moving from specific issues (e.g., use of dioxin) to broader concerns such as anxiety over the food supply, water quality, or air-related illness. These effects can be pervasive and have now assumed a significant level of concern among the public (Hahn, 1994; Graham and Rhomberg, 1996). Foster (1993) supports this view by suggesting four major determinants to human health: human biology, lifestyle, medical care, and the environment.

**Table 24.1 Major Health-Related Concerns
Related to Ecosystem Management**

Pesticides
Air/water quality
Toxic contaminants
Release of hazardous materials
Atmosphere warming and ozone depletion
Excessive noise
Loss of biodiversity
Psychological stress

Indirect impacts on humans encompass vicariously experienced stress reduction, restorative environments, and *existence valuation* (Krutilla, 1967; Loomis and Walsh, 1997). In this case, a benefit can be described as an improved condition or the prevention of a worse condition (Lee and Driver, 1999). Goods and services can be thought of as one form of benefit. Westman (1977) characterized goods as products derived from the natural ecosystem, while services provide a functional role representing a component or combination of components (e.g., water purification through wetlands) in a natural ecosystem.

THE RELATIONSHIP BETWEEN OUTDOOR RECREATION
AND ENVIRONMENTAL HEALTH

It is in the realm of personal and health-related concerns that outdoor recreation can provide a segue between healthy, natural ecosystems and healthy humans. This role assumes two dimensions. First, the literature strongly supports something already well known: namely, that recreation and leisure can be effective agents toward improving human health in a variety of ways (Driver et al., 1991). Examples include physiological benefits, relaxation, catharsis, and sociological benefits (Figure 24.2).

Recreational activities appear to create positive benefits for humans in a number of ways: personal enjoyment, personal growth, social harmony, and social change (Wankel and Berger, 1990). In addition, two mechanisms useful in coping with life stress and developed through recreational activities involve building a sense of hardiness and internal locus of control through the enhancement of social support networks and heightened feelings of self-determination (Coleman and Iso-Ahola, 1993). Thus, participation in recreational activities in natural environments benefits the individual not only through enhancing one's level of resiliency (i.e., ability to return to baseline) through physical fitness, relaxation, and exposure to pristine environments but also by strengthening the hardiness (i.e., ability to resist negative change) of an individual through the use of social networks and a sense of personal empowerment (Driver et al., 1991).

Participation in a recreational activity in a natural setting can have beneficial effects on individual health and these benefits can have important ramifications in a QOL assessment. In a sense, natural

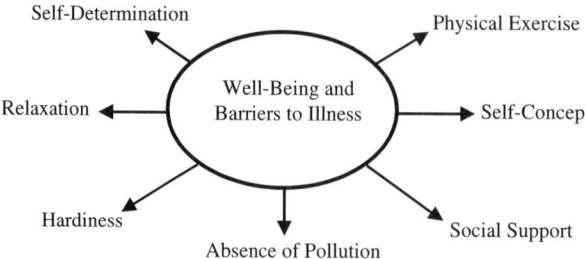

Figure 24.2 Recreation and leisure contributions to human health.

ecosystems become giant health machines through the goods and services they provide and the personal recreation opportunities they offer.

It is the second dimension, however, that is perhaps the most important factor in the long-term interaction between human actions and natural ecosystems. Salwasser (1999) suggests that the practice of ecosystem management will primarily be accomplished through resource management agencies, governments, and large corporations. This could be true for the management of large ecosystems on federal and other public lands, but I believe he misses the point. To paraphrase from Breslow (undated), "It is what you do hour by hour and day by day that largely determines your state of health." The same thing could be said about ecosystem management and preservation. The day-by-day, individual-by-individual things eventually determine whether an ecosystem will be preserved or properly cared for. However, without public backing, government intervention is usually of short-term duration, at best. In a sense, our natural ecosystems are as dependent on individual actions, behaviors, and commitments as they are on large governmental bodies or corporate entities (most of whom take their lead from public trends and aspirations).

And what better way to develop public understanding and support for ecosystem protection than by allowing individuals to form connections and positive attitudes toward the natural environment through direct experience in those environments? Many authors suggest that experience-building can play a critical role in the development of solutions to environmental issues and willingness to become involved in those issues (Kraft and Kielsmeier, 1995; Glover and Deckert, 1998). Moreover, there are a number of complex cognitive schema that can be experienced and learned in an outdoor environment — competence, empowerment, symbolic interaction, and personal redirection.

Using a modification of Ajzen and Fishbein's (1980) Model of Reasoned Behavior as an organizing framework, one possible explanation of how this experiential component through outdoor recreation may function is in the sequence illustrated in Figure 24.3. In this model, recreational experiences in natural environments provide a platform from which the individual becomes more aware of natural environments and develops a greater willingness to protect those environments. Thus, the experience acts as a mediating variable in affecting the development of a particular set of attitudes that, in turn, have direct impacts on an individual's behaviors and actions (Vaske and Donnelly, 1999). In part, this effect may be the result of heightened levels of awareness or knowledge and a sense that individual actions can make a difference in protecting natural environments (Shepard and Speelman, 1986; Ewert, 1996). Keen and Fisher (1992) suggest that the development

Direct experience through outdoor recreational behaviors

▼

Emotive connections and values

▼

Development of a belief system

▼

Alteration of attitudes and behaviors toward environmental stewardship

Figure 24.3 Model of effect of experience and environmental behaviors.

of this awareness, and concomitant attitudes and behaviors, is a necessary condition for advancing a willingness to engage in and promote sustainable practices. Orr (1995) goes so far as to suggest that traditional university curricula worldwide need to change radically in order to address the growing crisis in natural environmental degradation. This chapter has suggested that direct participation in recreational activities in natural environments, besides benefiting the individual, may help meet the challenges of environmental degradation.

In conclusion, this chapter has described the connections among QOL, natural environments, and outdoor recreation. Natural environments provide a host of products and services for human use in addition to recreational location. However, the direct experience with natural environments through recreation increases the likelihood that an individual will become more aware of the threats and impacts to an area and consequently become more willing to practice behaviors that serve to protect those environments. This approach to increasing individual action may ultimately be the true value of outdoor recreational activities because the problems of, and solutions for, natural resource management and ecosystem protection are human-centered.

REFERENCES

Ajzen, I. and Fishbein, M., *Understanding Attitudes and Predicting Behavior*, Prentice-Hall, Englewood Cliffs, NJ, 1980.

Bengston, D., Changing forest values and ecosystem management, *Soc. Nat. Resour.*, 7:515–533, 1994.

Coleman, D. and Iso-Ahola, S.E., Leisure and health: the role of social support and self-determination, *J. Leisure Res.*, 25:111–128, 1993.

Costanza, R., Norton, B., and Haskell, B., *Ecosystem Health: New Goals for Environmental Management*, Island Press, Covelo, CA, 1992.

Driver, B.L., *Nature and the Human Spirit: Toward an Expanded Land Management Ethic*, Venture Publishing, State College, PA, 1996.

Driver, B.L., Brown, P.J., and Peterson, G.L., Eds., *Benefits of Leisure*, Venture Publishing, State College, PA, 1991.

Ewert, A., Experiential education and natural resource management, *J. Experiential Educ.*, 19(1):29–33, 1996.

Ewert, A.W. and Kessler, W.B., Human health and natural ecosystems: impacts and linkages, *Ecosyst. Health*, 2(4):271–278, 1996.

Foster, H.D., Health as a renewable resource: management implications of the health field concept, in Foster, H., Ed., *Advances in Resource Management: Tributes to W.R. Derrick Sewell*, Belhaven Press, London, 1993, pp. 313–333.

Glover, J.M. and Deckert, L., Beyond praying for earthquakes: what works in environmental education, *Parks Recreation*, 33(11):30, 1998.

Graham, J.D. and Rhomberg, L., How risks are identified and assessed, in Kunreuther, H. and Slovic, P., Eds., *The Annals of the American Academy of Political and Social Science: Challenges in Risk Assessment and Risk Management*, Sage Publications, Thousand Oaks, CA, 1996, pp. 15–24.

Hahn, D.B. and Payne, W.A., *Focus on Health*, 2nd ed., Mosby-Year Book, St. Louis, MO, 1994.

Hatcher, R.L., Measuring human ecology: indicators of a sustainable community, Paper presented at the Society for Human Ecology, Seventh International Conference, Michigan State University, East Lansing, MI, April 1994.

Keen, M. and Fisher, F., Environmental and outdoor education: some Australian views on a false distinction, *World Leisure Recreation*, 34(2):37–41, 1992.

Kraft, R. and Kielsmeier, J., *Experiential Learning in Schools and Higher Education*, Kendall/Hunt Publishing, Dubuque, IA, 1995.

Krutilla, J., Conservation reconsidered, *Am. Econ. Rev.*, 57:787–796, 1967.

Lee, M.E. and Driver, B.L., Benefits-based management: a new paradigm for managing amenity resources, in Aley, J. et al., Eds., *Ecosystem Management: Adaptive Strategies for Natural Resources Organizations in the 21st Century*, Taylor & Francis, Philadelphia, PA, 1999, pp. 143–154.

Loomis, J.B. and Walsh, R.G., *Recreation Economic Decisions: Comparing Benefits and Costs*, Venture Publishing, State College, PA, 1997.

Orr, D.W., Educating for the environment: higher education's challenge of the next century, *Change,* May/June:43–45, 1995.

Peterson, G.L., Driver, B.L., and Gregory, R., *Amenity Resource Valuation: Integrating Economics with Other Disciplines,* Venture Publishing, State College, PA, 1988.

Porter, G. and Brown, J.W., *Global Environmental Politics,* Westview Press, Boulder, CO, 1991.

Salwasser, H., Ecosystem management: a new perspective for national forests and grasslands, in Aley, J. et al., Eds., *Ecosystem Management: Adaptive Strategies for Natural Resources Organizations in the 21st Century,* Taylor & Francis, Philadelphia, PA, 1999, pp. 85–96.

Salwasser, H., MacCleery, D.W., and Snellgrove, T.A., An ecosystem perspective on sustainable forestry and new directions for the U.S. National Forest Service, in Aplet, G. et al., Eds., *Defining Sustainable Forestry,* Island Press, Washington, D.C., 1993, pp. 44–89.

Schroeder, H.W., Ecology of the heart: understanding how people experience natural environments, in Ewert, A.W., Ed., *Natural Resource Management: The Human Dimension,* Westview Press, Boulder, CO, 1996, pp. 13–27.

Shepard, C. and Speelman, L., Affecting environmental attitudes through outdoor education, *J. Environ. Educ.,* 17(2):31–40, 1986.

Sirgy, M.J. and Efraty, D., A quality of life model, paper presented at the Second Annual Conference of the International Society for Quality-of-Life Studies, Williamsburg, VA, December 1998.

Suzuki, D.T., *The Sacred Balance: Rediscovering Our Place in Nature*, Greystone Books, Vancouver, B.C., 1997.

VanLeeuwen, J.A., Waltner-Toews, D., Abernathy, T., and Smitt, B., Evolving models of human health toward and ecosystem context, *Ecosyst. Health,* 5:204–219, 1999.

Vaske, J.J. and Donnelly, M.P., A value–attitude–behavior model predicting wildland preservation voting intentions, *Soc. Nat. Resour.,* 12:523–537, 1999.

Wankel, L.M. and Berger, B.G., The psychological and social benefits of sport and physical activity, *J. Leisure Res.,* 22:167–182, 1990.

Westman, W., How much are nature's services worth? *Science,* 197:960–964, 1993.

CHAPTER 25

Natural Capital Differentiation, Sustainability, and Regional Environmental Policy*

Cecilia Collados

INTRODUCTION

Chile is a country that has grown at a fast rate in the last two decades. This economic growth comes mostly from exporting natural resources. This growth process, which includes mining, fishing, industrial agriculture, exploitation of native forests, and the construction of hydroelectric dams, has deteriorated or destroyed important ecosystems. These actions not only have direct negative environmental, social, and economic consequences for the regions but also erode the natural capital basis of new development for the whole country. What explains the Chilean choice of a nonsustainable path of development?

QUALITY OF LIFE AND USE OF NATURAL CAPITAL: THE THEORY

The Neoclassical Model

The Chilean model of development is based on the predominant neoclassical economic paradigm that focuses on efficiency. Neoclassical economics considers the economic system to be separated from the natural and other social systems. Natural capital is only a source of materials that comes into the production process at no cost, except the cost of extraction. Development is defined as increase in gross domestic product (GDP) and can be accomplished with an efficient price system that reflects scarcity and stimulates technical improvement. This technical change counteracts depletion by substitution and by facilitating extraction. Substitution among different kinds of capital (and between capital and labor) is infinite, depending only on technological change (Bator, Cobb–Douglas analyses**). Irreversibility of damage to natural capital is therefore not an issue. The sustainability criteria derived from this view are highly influenced by the assumption of complete substitutability between the factors of production, including natural capital. Economic

---

* This chapter is a summary of part of Collados' dissertation work for the Ph.D. degree in City and Regional Planning, University of California, Berkeley (Collados, 1999). Part of this chapter has been included in Collados and Duane (1999).
** Bator analysis for the determination of the best configuration of inputs, outputs, and commodity distribution for a two-input, two-output, two-person situation uses land and labor as perfectly substitutable inputs (Bator, 1957). The Cobb–Douglas analysis determines value added as a function of capital stock and employment that can be substituted among themselves as long as neither input goes to zero (Heathfield and Wibe, 1987).

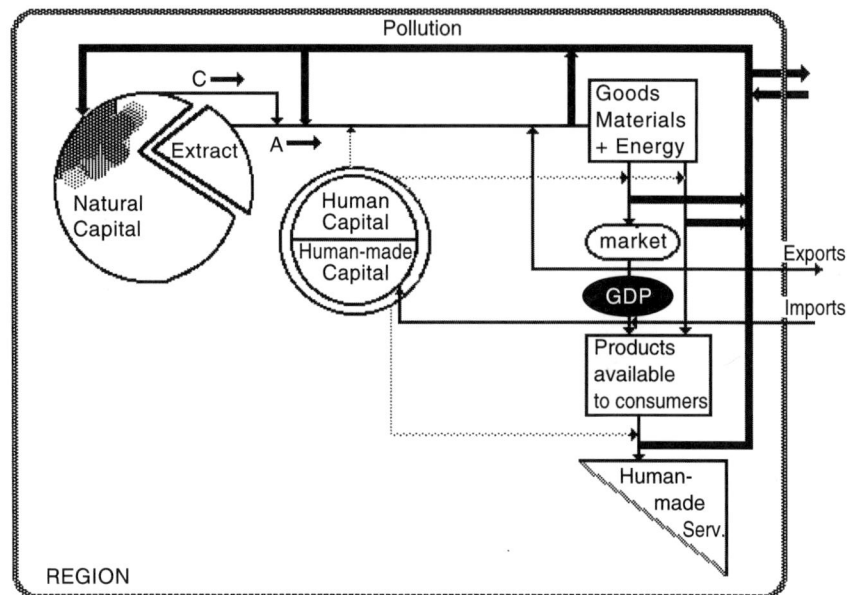

Figure 25.1 Traditional view. *Thin arrows:* processes that contain natural resources: production processes (A and C), distribution, consumption, and import-export. *Thick arrows:* waste. *Dotted lines:* processes without natural resources component. Some processes in this figure have been kept simple: *Distribution:* makes processed materials and energy, including intermediate and capital goods, accessible to consumers and producers (some occurs through the market, some does not). *Consumption:* changes materials and energy into services. *Import–export:* brings materials and energy in and out of the region. (From Collados, C. and Duane, T.P., *Ecol. Econ.*, 30:441–460, 1999. With permission.)

development is sustainable if the overall stock of assets remains constant over time. Any one asset can be reduced as long as another asset is increased to compensate for it. This is what has been called *very weak sustainability* (Turner et al., 1994).

A regional model of these views is summarized in Figure 25.1, showing the relationship of the economic system with natural capital. Natural capital includes all natural assets — everything that is not human-made. It can be altered by humans, and its reproduction can be enhanced by humans, but it cannot be created by humans. The categories of natural and human-made capital are not sharply divided. Natural capital can have human modifications, as will be discussed later. Examples of natural capital are ecosystems (forests, rivers, the water cycle). Human-made capital is always an alteration of natural capital.

According to this model, natural capital generates two kinds of production processes, A and C. Process A starts with extractive activities, while process C obtains natural resources without destroying ecosystem integrity.* Both processes can use recycled waste. After natural resources are introduced into the economic process by A or C, with the intervention of people (human capital), and human-made capital (which includes institutions and other forms of social capital), these resources are transformed in material goods and energy that are distributed, with or without the use of the market, and made available to consumers and producers. These goods and energy are transformed into human-made services through a consumption process.

This paradigm presents some strong assumptions: among others, a narrow definition of development that includes only part of human-made services; the existence of a price system that reflects scarcity accurately; and a high degree of substitution among different kinds of capital. If working

---

* *Extraction* is defined as any kind of destruction of the integrity of the ecosystem. It will be defined more precisely when natural capital categories are explained.

well, it allocates resources to where they are most needed, but takes them from where it is less expensive to extract in traditional cost terms, not in environmental services terms.

Environmental economists propose two variants to the neoclassical approach. The first is the Pigouvian taxes approach, which controls pollution with taxes; the second is the property rights approach, which offers a market solution to externalities. This second approach maintains that, by redefining property rights, pollution problems can be solved by polluters and sufferers through bargaining; no state intervention should be necessary. (The basis for this approach is an interpretation of the Coase theorem, 1960). This solution allocates resources efficiently but does not consider the fact that other assignments of rights can also have efficient allocations, that there are income distribution effects when assigning property rights, or that the transaction costs involved in bargaining can be very high.

Pearce and Turner (1991) use "the main body of economic thought to derive important propositions about the linkages between the economy and the environment.... Rather than looking for some different economics, we are seeking to expand the horizons of economic thought." They do not look for alternative economics, arguing that many of the concerns of those who are motivated to search for alternatives can be accommodated within the central body of modern economics.

Most of the theoretical contributions of environmental economists deal with the effects of pollution on natural capital and the possibility of internalization of these negative externalities into the economic system. This body of theory has been the basis for the *polluter pays* principle and for policies such as tradeable pollution permits.

Another branch of environmental economics is resource economics, which applies welfare economics principles to determine optimal use and depletion of resources (Fisher, 1981). The goal is, then, the maximization of the present value of the resource to society. In the case of renewable resources, efficient use may entail depletion of the resource to extinction, or its sustained use at either higher or lower stocks than currently exist. In the case of exhaustible resources, it means finding an optimal rate to deplete a resource completely. The Hotelling model of efficient resource extraction was developed in 1931 under the assumption that resource allocators are aware of the total stock of resources, the course of technological development, and the level of demand throughout the future. This model indicates that, in the simplest case, with costless extraction, the resource should be extracted in a way that the prices rise at the discount rate (Pearce and Turner, 1991).

Another approach in the resource economics literature tries to deduce whether resources are scarce from costs and price paths. Critics of this empirical model, and in general to all economic indicators of long-run scarcity, argue that such approaches implicitly assume information about quality and total quantity of the resource in question, which is precisely what they are trying to determine (Norgaard, 1990). Other problems with this model are its absence of concern for future generations' interests (Howarth and Norgaard, 1992; Norgaard, 1992) and the implied conviction that technology and substitution will solve the problems after resources are depleted. The policies derived from this discipline recommend a better management of nature respecting the environmental linkages — economic incentives like prices, credit, and exchange rates, and changes in the tenure of land (Pearce and Turner, 1991).

Natural capital is recognized by these authors as an important contributor to human welfare, as a life support system that provides a supply of resources, as a waste receptor, and as a provider of amenities. The opportunity cost of the use of natural capital is emphasized by Krutilla and Fisher (Fisher and Krutilla, 1975). The rule for natural capital conservation is that resource stocks should be kept constant over time. The stock of renewable resources should not decline over time (understanding the waste assimilative capacity as a renewable resource), and depletion of exhaustible resources should be replaced by increases in renewable resources or human-made capital (assuming substitutability). This is what has been called *weak sustainability* (Turner et al., 1994).

These contributions have emphasized the idea that environmental services are important as a component of well-being and have introduced a general criterion of conservation of natural capital; but they do not specify the linkages between the different kinds of natural capital according to its function in producing environmental services and improvements in the quality of life.

Regional Development Theory

Regional development theory treats land and other environmental components as space, and in most cases treats space as an abstraction that does not include social or physical characteristics. The social vacuum in such theory has been analyzed by authors like Charles Gore (Gore, 1984), and many institutional and cultural aspects have been introduced by authors that come from a sociological or geographical background. But environmental degradation is still ignored or treated as a marginal consequence, and natural capital is viewed only as a source for extraction of commodities. It is not recognized as an essential component of an ecological system that has a value per se or as a factor of production of environmental services, culture, and social space.

This view impedes inclusion of sustainability issues in regional theory. The contributions of this body of knowledge to the sustainability debate are limited by its failure to recognize that regional development always involves environmental opportunity costs and that society has to deal with these costs somehow, by exporting them to other societies, by transferring them to future generations, or by paying for them locally now.

Many authors have dealt with sustainability problems from a local point of view, trying to define self-sufficient regions. These efforts have led to very diverse and interesting concepts. Examples of these concepts are bioregions, most commonly associated with water basins. The goal is "to create a society that will maintain its environment as a congenial and pleasant place to live in now and in the future" (Atkinson, 1992). The concept of agroecology is also related to bioregionalism in the context of developing countries. It proposes a comprehensive reframing of rural development strategies, with priority given to empowerment of local communities, poverty reduction, natural resource conservation, secure food supply, and promotion of self-sufficiency (Altieri, 1993).

There is also the argument that environmental quality can be an important driving force of the economy that can bring in people, and employers would follow. This is especially important in areas where extraction activities have been considered the economic engine for years. In these cases the opportunity cost in terms of quality of living environments might be too high today and needs a careful evaluation (Power, 1996). In the case of minerals, this assertion is confirmed by recent estimates of mineral supplies for the next century. Given the lack of pending crises in availability of minerals, mining can no longer presume the *de facto* claim as the best of all possible uses for land (Hodges, 1995).

Industrial ecology is another small-scale approach to environmental problems. It involves the design of industrial processes and products from the dual perspectives of product competitiveness and environmental interactions. It is a system-oriented vision with the premise that industrial design and manufacturing processes are not performed in isolation from their surroundings but rather are influenced by them and, in turn, have influence on them. It seeks to optimize the total materials cycle from virgin material to finished material, to component, to product, to obsolete product, and to ultimate disposal. Factors to be optimized include resources, energy, and capital. As in a biological system, it rejects the concept of waste and considers all possible uses of residues. These objectives can completely change the product development processes, because many variables are added to the production processes and the product definition (Graedel and Allenby, 1995). These ideas can bring about a new rationality for agglomerating industries that can use each other's wastes and by-products.

The main problems in implementing these ideas relate to the boundaries of the regions. Political boundaries could be changed; but often the most relevant physical boundaries do not coincide with cultural differences, and these could be quite problematic. In today's global economy it is almost impossible to refer to regions as geographically discrete places in terms of the land they depend on for their activities (the *ecological footprints**). Cities, for example, appropriate the carrying capacities of distant *elsewheres* (Rees and Wakernagel, 1994).

---

* The opposite idea or *environmental bleeding* could also be useful, especially from the point of view of those regions which export natural resources: determining who are the beneficiaries of the environmental damage on a particular region.

Ecological Economics and Co-Evolutionary Economics

Ecological economics represents a major challenge to neoclassical economics by making the socioeconomic system part of the ecological system and addressing the relationships between them. Within this approach it is generally recognized that environmental problems have a structural character and cannot be approached simply by using the concepts of externalities and substitution.

The materials balance approach introduces entropy limits to the economic analysis; it argues that pollution is not only the result of market failure but an inevitable phenomenon dictated by the laws of thermodynamics.

The ideas of Boulding, who introduced the relevance of thermodynamics to economics in 1966 (expanded by Nicholas Georgescu-Roegen in the early 1970s) are very important to the development of this paradigm. In very simple terms, low entropy is the ultimate means of production and exists for humans as terrestrial stock and solar flow, both limited in practical terms. Humans have almost complete control over the use of stocks (which could yield only a few days of sunlight, and could be used very quickly — 4 days [Ayres, 1950]), but almost no control over the flow of solar radiation intercepted by the Earth (Georgescu-Roegen, 1980).

Based on these concepts of limits, Herman Daly, in 1977, argued for a steady-state economics.

> Steady-state economy is an economy with constant stocks of people and artifacts, maintaining some desired, sufficient levels by low rates of maintenance *throughput* — that is, by the lowest feasible flows of matter and energy from the first stage of production (depletion of low-entropy materials from the environment) to the last stage of consumption (pollution of the environment with high-entropy waste and exotic materials) (Daly, 1991).

To define the optimum stock to be maintained in a steady state, Daly first defines our dependence on the natural world:

> This dependence takes two forms — that of a source of low-entropy inputs and that of a sink for high-entropy waste outputs. Services come from two sources: the stock of artifacts and the natural ecosystems. The stock of artifacts requires throughput for its maintenance, which requires depletion and pollution of the ecosystem (Daly, 1991).

From this perspective he defines an optimum stock as:

> Total service (the sum of services from the economy and the ecosystem) is maximum. This occurs when the addition to service arising from a marginal addition to the stock is equal to the decrement to service arising from impaired ecosystem services that result from the incremental throughput required by the increment in stock (marginal cost or service sacrificed equals marginal benefit or service gained). This is the rule defining the optimum level of stocks to be maintained in a steady state (Daly, 1991).

Coevolutionary theory, developed by Richard Norgaard in the early 1990s, refers to an ongoing process between the social and ecological systems, through feedback loops, that determines the possible development of both. This theory considers economic development as a process of adaptation to a changing environment while being itself a source of change. Coevolution can continue indefinitely, reaching equilibriums and coming out of them as the social and ecological systems evolve. Norgaard (1994) argues that Western science and modernization break this pattern of coevolution by producing new ways of organizing and new technologies that are aimed more to control nature than relate deeply to it.

Ecological economics and coevolutionary approaches emphasize physical limits and environmental processes that are essential for human life and the uncertainty that surrounds these processes. They bring together multidisciplinary fields of study and go back to the questions classic economists tried to answer, in a more ample context that includes social and natural systems. They are also characterized by including a view of how the world should look in the long term. These transdisciplinary fields address three main goals: first, assessing and ensuring that the scale of human activities is ecologically sustainable; second, ensuring that the distribution of resources is fair within the current generation, between future generations, and among species; and third, efficiently allocating marketable and nonmarketable resources under those limits. Natural capital, human capital, and manufactured capital are "interdependent and to a large extent complementary" (Folke et al., 1994). Natural capital consists of two major subtypes: nonrenewable resources, such as oil, coal, and minerals; and renewable resources such as ecosystems. Ecosystems are sources of raw materials, food, and a wide range of environmental services.

These authors emphasize the importance of environmental services, which include maintenance of the composition of the atmosphere, amelioration of climate, operation of the hydrological cycle, waste assimilation, nutrient recycling, soil generation, crop pollination, maintenance of biodiversity and landscape, and aesthetic and amenity services. "These services can hardly be produced by any other form of capital, and consequently are complementary to them. Depletion and pollution are making natural capital a limiting factor for further development" (Daly, 1991).

The sustainability concept for this paradigm is keeping the life-support ecosystems and interrelated socioeconomic systems resilient to change. "It is not sufficient to just protect the overall level of capital, rather natural capital must also be protected, because at least some natural capital is nonsubstitutable" (Turner et al., 1994). What is stressed in this approach is the combination of factors (e.g., irreversibility and uncertainty), not their presence in isolation. Only if this combination of factors is maintained can we have sustainability. This has been called *strong sustainability* by the authors just cited.

Because these theories are large in scope and are not very specific in establishing the links between natural capital attributes and quality of life, it is very difficult to use them as a basis for regional policy.

Table 25.1 presents a summary of how these different theories conceptualize the relationship between the economic and ecological systems, define natural capital, prevent damage to natural capital, cope with scarcity, and define sustainability.

THE PROPOSED MODEL

This model explains how the natural capital of a region contributes to its quality of life. This model is based mainly on ecological and co-evolutionary economics concepts, adapting them to the regional level. It specifically defines the relationship between the different kinds of natural capital and the quality of life of the region in an attempt to make it a framework for policy, development, and evaluation (Figure 25.2).

In Figure 25.2 natural capital is still represented as a circle (as in Figure 25.1) for analytical purposes but dispersed irregularly throughout the region as components of ecosystems. This model adds another production process to the neoclassical view presented in Figure 25.1 — process B, which generates environmental services. These services are consumed by humans directly, becoming part of the quality of life, and also go back to regenerate natural capital. Production process B does not require human participation (human capital, human-made capital, or technology). Examples of these services are pollination, climate control, natural water purification, maintenance of biodiversity, and aesthetic services. These processes do not compromise the integrity of the natural capital permanently because their waste is processed by the natural systems. Some environmental services are indispensable for the regeneration of natural capital and, in this way, critical for any kind of development.

The natural capital regeneration rate and ability to produce environmental services is diminished by the waste produced by all of these processes (thick arrows). It is also diminished by the destruction of its integrity due to the extraction or fragmentation of some of its components (process A).

Table 25.1 Five Views of the Economic System and the Environment

Views	Relationship between Economic and Ecological System	Natural Capital	Prevention and Treatment of Damage to Natural Capital	Solutions to Scarcity	Sustainability
Neoclassic economics	Separated systems	Materials source Exhaustible: (Hotelling model) Renewable: (harvest = yield)	Pollution treated as externality Taxes Market (legal system)	Technical change, recycling Allocative efficiency and substitution	Very weak = sum all assets constant
Environmental economics	Connected systems — looks for links, interactions	Inputs Receiving media Amenities	Keep within bounds of assimilative capacity of environment	Substitute exhaustible resources by renewable ones	Weak = renewable conserved, nonrenewable replaced by renewable
Ecological economics	Economic system is a subsystem of ecological system	Inputs Receiving media Amenities Environmental services Biodiversity Complementary to human-made capital	Carrying capacity is the limit to biophysical throughput	Recognizes limits to substitution	Strong = combination of natural capital and human-made capital constant
Steady-state economics	Same	Same	Same	No substitution	Very strong = sum of services from the economy and ecosystems is maximum
Coevolutionary economics	Coevolving, relationship between both	Same	Same	Substitution is part of the co-evolving process, dynamic, unpredictable	Very strong = natural capital and human-made capital maintain an evolving equilibrium

Collados, C. and Duane, T.P., *Ecol. Econ.*, 30:441–460, 1999. With permission.

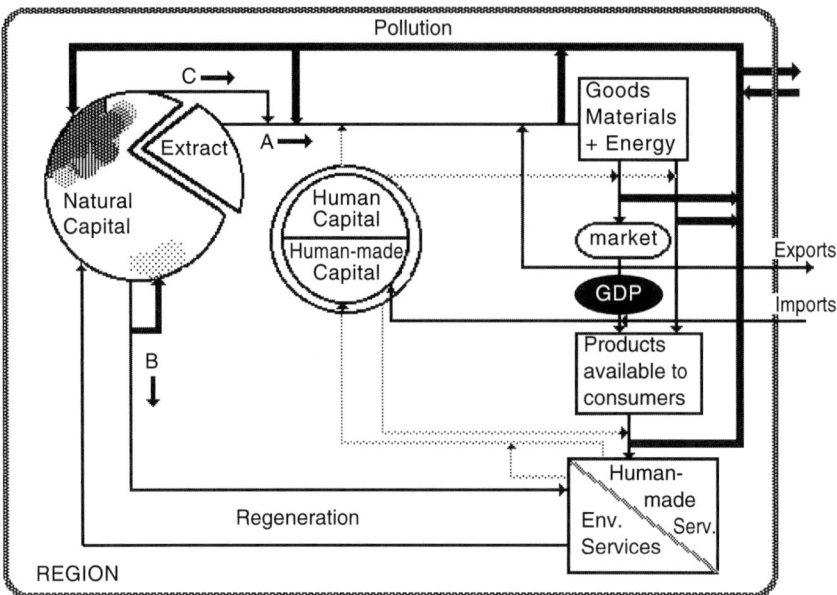

Figure 25.2 The proposed model relating quality of life and natural capital. (From Collados, C. and Duane, T.P., *Ecol. Econ.*, 30:441–460, 1999. With permission.)

From a regional point of view, if the ability of natural capital to produce environmental services is impaired by the damage caused by extraction and pollution beyond its regeneration capacity, both production processes would be affected. Goods and energy can be imported, but environmental services can only be produced locally. The boundaries of the region are permeable to environmental services, but the region does not have control over this permeability.

Natural capital contributes to the quality of life of a region in two complementary ways: first, by directly providing a flow of environmental services that cannot be imported; and second, by supplying the natural resources that, through a human-controlled production process, become valuable to humans.* Environmental services also determine of the ability of natural capital to regenerate itself. Ecosystems and other components of the regional natural capital produce environmental services that provide life-support functions necessary for natural capital reproduction. The economic process also has waste, which alters natural capital and, as a consequence, the quality of life both in the short and long term. Quality of life is determined by the combination of human-made services and environmental services.

Human-Made Services

Human-made services are the services produced directly by people and those derived from goods during the consumption process. These services can be essential — necessary for human life. There are many approaches to define what is essential for a society. These approaches are related to different theories of well-being. These theories interpret well-being as a state of mind, a state of the world, human capability, or as the satisfaction of underlying needs (Dodds, 1997). For the purpose of this study, it is important that what is essential be determined socially. For some regions, weapons or cars may be considered essential, while to others these could be nonessential goods.

---

* The relation between human-made services and environmental services as components of the QOL is a complex one. For a detailed analysis of increases and decreases of human-made services and environmental services, see Collados (1999).

Environmental Services

Environmental services are functions provided by natural systems. They cannot be imported; they become part of the quality of life because of their importance in the surroundings where humans live. They can be classified as:

Life-supporting environmental services are necessary for maintaining all life and for the regeneration of natural systems (e.g., the regeneration of soils, pollination of crops, maintenance of the atmosphere, waste assimilation, or purification of water). These services make ecosystems dynamic self-regulating systems that maintain their essential structures in the face of exogenous shocks (homeostatic quality), that adapt to changes in their input and output environments (adaptiveness), and are conducive to change in the species composition with time (succession) (Dryzek, 1992).*

All other environmental services are part of the quality of life but do not determine the survival or reproduction of the ecological system. They include recreational and aesthetic environmental services not provided with human intervention.

The Stock of Natural Capital and Its Differentiation

Natural capital is all that is not created by humans. It can be destroyed by human actions, and its reproduction can be influenced by human actions, such as in agriculture or plantation forests. It is the basis for the creation of most human-made capital and services. It is also the source of all environmental services, including the necessary life-support mechanisms for humans and other species, that allow for its own regeneration. Natural capital is classified in this model according to its ability to produce services and other attributes in four categories: essential, nonsubstitutable, nonreconstitutable, and life-supporting natural capital. These categories, and how they are linked to the quality of life of the region, are presented in Figure 25.3.

Essential natural capital is used by humans to produce services that are essential for human survival, such as food, shelter, and clothing, as previously defined. This subset of natural capital, determined by the need of essential human-made services, can change because population numbers or expectations change. Technology can also change and allow use of different kinds of natural capital to satisfy the same need. An example could be replacing the construction of hydroelectric plants by producing more efficient electric bulbs. In this case the same service would be obtained with a different kind of natural capital. No number has been associated with essential natural capital in this model because the *range of essentiality* should be determined by each region.

How the region is defined (and the size of it) will make a difference in determining what services are essential and, in turn, the natural capital that is essential. The size of the region will also alter the availability of natural capital to satisfy essential and other needs. When the demand of different social groups contained in a region for essential natural capital overlaps, a social problem may arise. This may be a social — not environmental — problem, but it can become an environmental problem if the conflict is resolved in a way damaging to the environment, e.g., war or destruction of crops.

Life-supporting or critical natural capital (Figure 25.2, Items 1 to 4) produces life-supporting environmental services and are necessary for life reproduction. Examples are riverine ecosystems, wetlands, and forests that provide habitat for many species. This quality of natural capital is determined by nature.

Extraction is defined in this model as any action that destroys the integrity of an ecosystem. This integrity is maintained by what has been previously defined as life-supporting environmental services that are responsible for its self-regulation. Therefore, any use of life-supporting natural capital is an extraction since it will stop the flow of life-supporting environmental services and impair the ability of an ecosystem to adapt and reproduce in the presence of exogenous stress.

---

* This definition of life-supporting environmental services derives from a dynamic conception of ecosystems that are perpetually changing and undergoing disturbance (natural and human). Preserving nature *untainted* or in stable balance can be, in many cases, counterproductive, impairing the self-organizing and self-recreating characteristics of ecosystems.

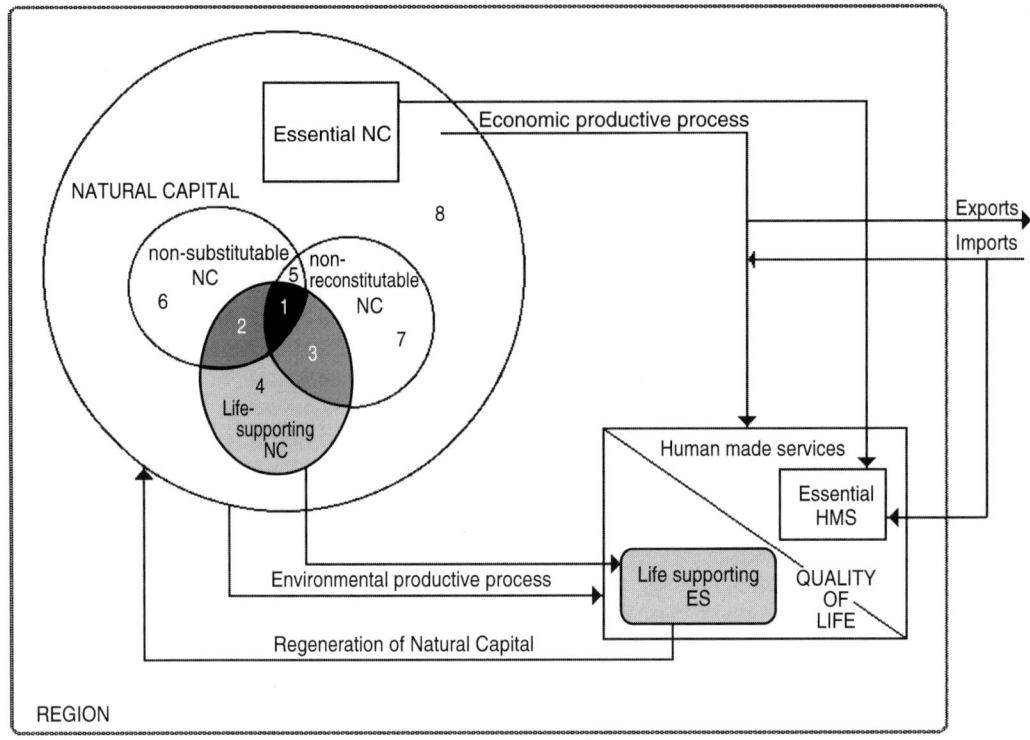

Figure 25.3 Natural capital categories. Examples of natural capital in the different categories: 1. Rivers and riverine ecosystems that cannot be reconstituted and provide nonsubstitutable support to life. 2. Artificial forests that provide environmental services necessary for life that cannot be provided with other kinds of capital but could be reconstructed. 3. Riverine fishing areas that could be substituted by non-life-supporting natural capital or human-made capital. 4. Wetlands as purifiers of water, to some extent reconstitutable, and substitutable by artificial purifiers. 5. Landscape, aesthetical natural capital, recreation areas. 6. Gardens, agricultural land. 7. Minerals. 8. Artificial lakes, pasture land with low biodiversity. (From Collados, C. and Duane, T.P., *Ecol. Econ.*, 30:441-460, 1999. With permission.)

Nonsubstitutable natural capital (Figure 25.2, Items 1, 2, 5, and 6) cannot be substituted by human-made capital. Examples are forests that regenerate soils, birds and insects that pollinate crops, and their habitats.

Nonreconstitutable natural capital (Figure 25.2, Items 1, 3, 5, and 7) does not regenerate once destroyed, or are too costly or impossible to reconstruct artificially. They include what economists call exhaustible resources, e.g., minerals and complex ecosystems like rivers, native forests, or wetlands that are impossible to create or recreate. In the case of complex ecosystems, there is not only the difficulty and cost of putting it back together, but also the lack of information once it has been destroyed. In the case of minerals, possible reconstitution is mostly a matter of cost; the energy needed to restore them in their original form is simply too high.

What is nonreconstitutable and nonsubstitutable natural capital at a given moment is determined by the evolving responses of nature and past human actions. It is discovered by humans through science or other sources of knowledge. The categories presented in Figure 25.3 are not sharply divided, they vary, and they are constantly changed by alterations in the environment and new discoveries. Such changing of categories makes planning the use of natural capital a complex task that requires constant adjustment of these categories in order to maintain a sustainable state.

Categories 1, 2, and 3, which result from the overlap of life-supporting or critical natural capital with nonsubstitutable and nonreconstitutable natural capital, are of special interest for conservation policy. Category 1 includes life-supporting natural capital for which there is no substitute or possible reconstruction. An example is riverine ecosystems. Category 2 includes life-supporting natural capital that cannot be substituted by human-made capital, such as planted forests. Category 3 is natural capital that is life-supporting and cannot be reconstituted but could be substituted. An example could be water purification by wetlands.

Uses of Natural Capital

Figure 25.3 represents an ideal utilization of natural capital. It discriminates the use of natural capital according to its ability to provide critical environmental services and essential human-made services. It would use essential natural capital only for the production of essential human-made services and noncritical, nonessential natural capital for the rest of the human-made services.

In a market economy without environmental planning, the criteria for using natural capital is determined by the cost of extraction or accessibility of the resource in need. Natural capital is extracted in most cases without consideration of its ability to provide environmental services, obtained indiscriminately from all categories of natural capital. This practice is conducive to an erratic use of natural capital that can be unsustainable if it destroys critical natural capital.

Interrelations among critical and essential natural capital define different scenarios and planning problems that require specific approaches.

Essential and critical natural capital do not overlap: By substituting noncritical for critical natural capital to satisfy essential needs, the production of life-supporting environmental services can be increased, improving the quality of life of the region. This is a problem that can be solved with adequate environmental planning institutions that administer natural capital according to its ability to produce environmental and human made services.

Essential and critical natural capital overlap: Where critical natural capital is the basis for satisfying essential needs, there is a problem that cannot always be solved by better administration. An example could be a region that needs wood as fuel and can take it only from native forests that are considered critical for their function as habitat. Confronting these problems could require changes in the scale by which the problem is defined, changes in values, or changes in the objectives of investments and new technologies (creating new opportunities for substitution, thereby reducing or eliminating the competition between satisfying essential needs and using natural capital).

Overlap of these two categories, without possible substitution of natural capital internally, can explain, for example, the cycles of increasing poverty in some regions without access to resources from other regions. This dilemma is illustrated by the overexploitation of forests for fuel or the clearing of land for agriculture beyond their abilities to reproduce. The traditional planning processes used in these situations tend to increase the production of human-made services by increasing productivity of labor, often damaging natural capital at a faster rate. In such cases this model shows that there would be no internal solution to the problem, and a connection with other regions should be sought to solve it. These solutions could include migration or a change in the export–import flow to correct the use of internal critical natural capital. It is often said that poverty is the main enemy of the environment. This model shows that the problem may not be poverty per se, but the isolation of poor regions which have limited access to outside sources of noncritical natural capital.

Conditions for Sustainability

Which kind of natural capital goes into the economic process has a twofold direct impact on the quality of life of the region: it determines the quantity and quality of human-made services and the quantity and quality of the flow of environmental services.

This decision also has an indirect impact on the regional quality of life in the long term: it determines the availability of life-supporting environmental services which, in turn, establish the regeneration capacity of natural capital. Since natural capital is the basic source for maintaining or increasing quality of life, its regeneration is a condition for long-term or sustainable development.

If a region uses the critical natural capital that is most essential for the production of environmental services and for its own regeneration, it will have a development path that diminishes the environmental services component of quality of life in the long term. If, instead, it uses its natural capital so that ecosystems are maintained to function and produce environmental services, it will maintain or increase this component of quality of life in the long term.

In a market economy, where an important part of natural capital is in private hands, there are economic incentives to utilize natural capital to produce human-made services. These incentives do not exist for the production of environmental services, however, because they are considered positive externalities of natural capital. This asymmetry and the goals of economic growth (increases in human-made services) drive the system in one direction: transferring natural capital from producing environmental services to producing human-made services.

Under these conditions the transfer of natural capital from producing environmental services to producing human-made services can only be done in a sustainable way without impairing the reproduction of natural capital and the production of environmental services if the economic system uses natural capital that is not critical. Besides the conditions imposed by strong sustainability criteria, a necessary condition for sustainable development is *not to use critical natural capital.*

To comply with this necessary condition, the function of environmental planning is to balance the use of essential and natural capital and strictly avoid using critical natural capital. This concept resembles co-evolution (Norgaard, 1994), in which social and environmental systems evolve and adapt to each other in a constantly changing balance; it differs, however, by limiting the changes that can occur. These limits are determined by the nonreconstitutability and nonsubstitutability of some life-supporting natural capital, because the destruction of these kinds of natural capital is conducive to irreversible system breakdown, or *distress syndrome* (Haskell et al., 1992).

THE INSTITUTIONAL INTERFACE

Institutional Framework

The social decision to use natural capital is made through an institutional system that acts as an interface between natural capital and the quality of life. How the institutional system comes to play in the decision to bring natural capital to the economic process is illustrated in Figure 25.4.*

The institutional framework, composed of institutions from inside and outside of the region, may be divided in a layer of basic underlying institutions and a layer of organizations that comprise the public sector and those associated with it.

The market, the legal system, the educational system, and the political voting system are the main institutions involved in natural capital use.

The market allows for the expression of demand-and-supply forces through prices of goods, prices of natural resources, and through extraction costs. For any given income distribution and a set of preferences, it allocates resources where they are most needed. In general, the market directs the use of natural capital to where it is least expensive to extract, not where it is least costly in terms of environmental services.

An important effect of the market is that it acts as a filter of information by separating the consumer from the human, natural, and social capital that generated the product to be consumed. For example, in the furniture marketplace, advertising, labels, display announcements, and other information mechanisms communicate to prospective buyers the qualities of products; but they do not indicate whether the products were produced with child labor, native woods, or under labor laws that do not respect basic human health conditions. This filtering action becomes very important

---

* This model shows only the interactions among the components of the institutional system that may be significant for the design of policies affecting the use of natural capital; it is not comprehensive.

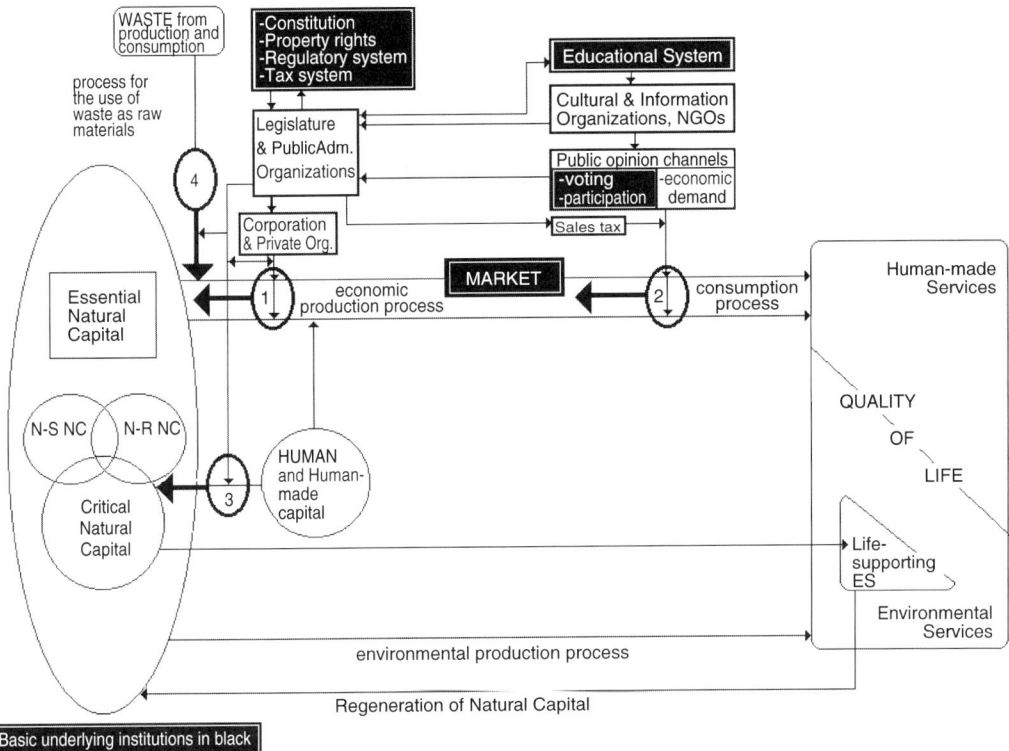

Figure 25.4 Some intersects among institutional and economic systems.

in the complicated economic relations of a global economy and the many elements that must be combined to produce some goods.

The legal system links basic institutions with organizations that implement laws. It also establishes the relationship between natural capital and people through three main mechanisms: property rights that assign natural capital to people and specify its use; the regulatory system that limits private property rights; and the tax system that directly or indirectly encourages or dissuades the use of critical natural capital.

The educational system provides social knowledge which, in turn, influences the public's preferences and determines its economic demand, the way it votes, and how it participates in the decision-making processes.

In a democracy the political voting system elects representatives and provides a mechanism to convey preferences on social issues. For the system to be effective, the issues should be clearly defined and well known by voters. Environmental issues present a particular problem because of complexity and because information about them diffuses slowly; thus, it takes considerable time until they can be acted upon by voters.* There are other forms of public participation, such as lobbying and comments on the environmental impact assessments by nongovernment organizations or the public.

Over this first layer of basic institutions is an organizational setting consisting of government institutions that make decisions and nongovernment organizations that function in association with them. These organizations include the public administrative system, the private sector, international organizations, cultural organizations, and informational organizations.

The organizations and the basic institutions are linked and dependent on each other, thus forming an institutional structure. For example, these organizations may be related in a

* Some political systems, such as those in California, also allow direct votes either on voter initiatives or legislative action.

hierarchical structure, connected in an advisory relationship, or in conflict with each other. The institutional structure and the relative influence of each organization determine how each of them contributes to the planning process and the final decisions over the use of natural capital. Some organizational views dominate and others may be ignored, while some may not even be considered in the final decision.

Interaction of the Institutional and Economic System

There are many significant points for the use of natural capital at which the institutional system and the economic system intersect in a market economy. This chapter identifies four of these *policy points* at which the consideration of the categories defined in the proposed model could improve the design of sustainable development policy.

The first point of intersect (1, Figure 25.4) occurs before the natural resources enter the production process and the market. At this point, besides the accessibility costs, the basic underlying institutions determine the use of natural capital.

Natural capital is assigned to public, private, or joint entities through a property rights regime that allows a range of uses for natural capital (Hanna, 1994). These uses can result in the resource being destroyed or remaining virtually unchanged. In the case of water, for example, it can be taken out of the source (a river, a lake) and consumed, or it can be taken, used, and then returned to the source in the same condition in which it was obtained. The property rights regime is not necessarily related to all attributes of natural capital as defined in this research. Usually, property rights are only associated with aesthetic values, as is the case of most national parks. A possible intervention here is designing property rights that are more restricted according to the scale of priorities suggested by differentiating natural capital in the proposed model.

At this point the regulatory system also influences the decision to bring natural capital into the economy. Laws such as the Endangered Species Act, which limit exploitation rights when endangered species may be affected, are a good example of regulations that restrict the use of natural capital. Environmental impact assessments, which sometimes require reconstruction of affected natural capital, are another good example. These regulations could be guided by the priorities suggested by this research as well.

The property tax subsidy system can also modify behavior. By creating subsidies to owners of critical natural capital who do not extract it, society can foster the production of life-supporting environmental services. In the same way, taxing extraction of the most valuable natural capital can contribute to its conservation.

The second place (2, Figure 25.4), where the institutional system interacts with the economic process, is the point of consumer choice. Consumer decisions, in turn, affect the market for resource extraction, thereby determining the use of natural capital. Consumer demand is culturally determined and can be influenced by education and information.

According to neoclassical economic theory, consumer demand is utility driven and influenced by available information. Consumer decisions are made not only by the direct utility expected from the good to be purchased but also by the overall utility of the consumer. One factor that can diminish this overall utility is the decrease of environmental services from damage to the environment. For instance, buying gasoline to drive a car provides consumers a benefit from which they subtract the benefits of breathing pure air.

Awareness of these indirect effects on the environment is important when consumers decide what to buy and it is not provided by the market. Worse, the market obscures such awareness. At this point education and information systems are the institutions through which planning can influence the use of natural capital. Introducing reforms to school programs, for example, or expanding public access to environmental information will be conducive to a better understanding of production and consumption processes (from cradle to grave, or, with recycling, from cradle to cradle).

Another way to inform consumers is to introduce mandatory labels for products that identify the damages and benefits to the environment. Manufacturers and producers stress positive environmental outcomes in advertising or product labels; but other than health warnings, negative outcomes are never reported.

The third point (3, Figure 25.4) in which institutions influence the use of natural capital is direct intervention on natural capital regeneration to maintain or accelerate its reproduction.

This is an important point of action for environmental policy; it does not interfere directly with the economic process except by subtracting some human and human-made capital for the enhancement of reproducing critical natural capital. Examples of direct intervention include government initiatives such as nurseries for native plants or reproduction of native fish to be reintroduced in areas where they have become extinct or endangered. Currently only a few intervention programs that involve private parties have been implemented in India and the U.S. (Gadgil and Rao, 1995). These programs could improve the welfare of the region, not only by improving the flow of environmental services but also by creating jobs that provide income and access to imports from other regions.

The fourth point of interaction (4, Figure 25.4) is the input of waste to the economic process that replaces the extraction of virgin natural capital. This use of waste requires the development of new technologies and different kinds of investments that need government participation.

To utilize this source of materials and energy, industry (mining, agriculture, forestry, and manufacturing) has to be designed in accordance with its surroundings, following the principles of industrial ecology. The factors to be optimized include resources, energy, and capital (Graedel and Allenby, 1995). Profound changes are needed to design competitive industry that utilizes waste as inputs. An example of this kind of investment is the $600 million recycled newsprint mill in the South Bronx. These kinds of projects face "huge economic and cultural obstacles, especially in urban settings where their environmental and social value could be greatest."*

THE EMPIRICAL WORK

This research was designed to evaluate if the institutional recognition of the natural categories defined by the proposed model contributes to a sustainable path of regional development.

The empirical study examined the new Chilean environmental institutional system at point (1) in Figure 25.4. This is the point at which the decision to bring natural resources to the production process is made. In order to evaluate to what extent basic institutions and environmental organizations have changed in their recognition of the different kinds of natural capital, two cases were selected: the Ralco and Pangue hydroelectric dams in the Bio-Bío River in the Alto Bio-Bío region in southern Chile.

Pangue had been already constructed, and Ralco was projected to start operating in 2002. Besides their actual and potential damage to critical natural capital, the Alto Bio-Bío cases were chosen because they are separated (in time) by the creation of environmental and social institutions designed to implement environmental policy and achieve the goal of sustainable development. The first dam, Pangue, was planned and approved before the existence of any environmental laws (Collados, 1993). The second one, Ralco, has been subject to new legislation that includes an Environmental Law and an Indigenous Peoples Law.

The decisions made in both cases were compared to conclude whether there was a change toward a more sustainable use of natural capital.

---

* Presentation of Dr. Allen Hershkowitz, Senior Scientist, Natural Resource Defense Council (NRDC), University of California at Berkeley, March 4, 1999.

The Chilean Basic Institutions*

The basic institutions gave greater recognition to the categories of natural capital defined by the proposed model after the new environmental law (No. 19,300) was approved in March 1994. The categorization of natural capital by this law is mostly embedded in Articles 34 to 39, which refer to protected areas. Article 30 specifies the need for protection of some areas in order to "assure biological diversity, preserve nature, and conserve the national patrimony," recognizing the value of biological diversity as an important life-support factor and the value of nature as capital. Articles 37 and 38 indicate the need to classify flora and fauna species to protect biodiversity according to the following categories: extinct, in danger of extinction, vulnerable, rare, with insufficient information, and out of danger of extinction. This classification could be associated with the nonreconstitutable and nonsubstitutable categories of the habitats of these species.

The Organizations

Those at the organizations researched who participated in the decision processes differed in their recognition of critical natural capital; however, the general tendency was to recognize the categories established by the theoretical model. Life-supporting characteristics were the most recognized, nonreconstitutability was the least recognized.

The Chilean Environmental Decision Process

The Pangue process occurred before the environmental law was in place. It was subject to environmental review only because Pangue S.A., a subsidiary of ENDESA (Empresa Nacional de Electricidad, a private enterprise) requested a loan from a foreign institution, the International Financial Corporation, to finance part of the project. In turn, the financiers requested an environmental impact assessment (EIA) and some compensation for the affected people. Part of this EIA was carried out before the loan was approved and part was not. The EIA was not publicly distributed, but it started a debate that had great public impact. Eventually the debate contributed to the creation of environmental institutions, as this was the first case to go through an EIA in Chile.

At the time the Chilean institutional framework could only impose conditions on the project when its presenter applied for water rights. Traditionally, hydroelectric projects were requested to leave a 10% minimum flow. In this case, the only requirement was to leave a 10% minimum flow in a tributary that joins the Bio-Bío River 3 km downstream from the dam (Collados, 1993).

The Ralco project was one of the first cases to go through the Sistema de Evaluación de Impacto Ambiental (SEIA) carried out by Comisión Nacional del Medio Ambiente (CONAMA). When this took place, controversies about the Pangue dam were still fresh in the public mind. Points of conflict were much greater in the Ralco case, however, because the project was bigger (seven times the inundated surface of the Pangue dam) and because the affected community was larger, had great cohesion, and had conserved its culture practically intact. The process, occasionally bumpy, was characterized by the confrontation of professional and political views and the reversal of previous decisions. The process included the removal of two political appointees from the two agencies with the most decision-making power, after the appointees changed their views in light of further study and professional opinion. In addition, the final decision was made by the Directive Commission, which is composed of 13 ministers, and this commission can reverse all previous decisions. In this case it reversed most of the environmental conditions that had been imposed by CONAMA. The way the institutions are set, the executive branch has the power to appoint the heads of the agencies involved; if that is not enough, it can impose its view at the end of the process. The SEIA is, then, mostly a political decision. In this case, the decision was dominated by the short-term goals of

---

* For a detailed analysis of these basic institutions, see Collados (1999).

rapid development, which are part of the political agenda. Most dominant in the process were ENDESA and top executives of the government.

Ralco followed the approval process established under the environmental law, and its construction was approved in August 1997. Its compliance with the Indigenous Peoples Law is still under consideration, but construction of roads and other complementary features has already begun despite the complaints of environmentalists and some of the Pehuenche people (40% of the investment had been done by July 1999).

The Ralco case process has fleshed out many weaknesses of the SEIA as a tool for the protection of natural capital — weaknesses that the private sector has detected and used to its advantage. Many of the necessary changes have already been pointed out by the Chilean scientific, legal, and political communities; but the will to change has to be political, and seems to be lacking.

Figure 25.5 graphically presents the Pangue and Ralco processes. The results that affect the use of natural capital differ only in the condition for the Ralco case to preserve an area equivalent to the inundated area of 1700 hectares and the obligation to maintain a minimum flow of 27.1 m^3/s.

Besides the environmental law, the existence of the Indigenous People's Law as part of the approval process of the Ralco project made the procedure of this case very different from Pangue. The dam reservoir will inundate indigenous peoples' land. According to the law, these lands should be protected and cannot be taken away without consent of the owners and the National Corporation of Indigenous Development. Neither consent has been given; thus, the inundation of the land is illegal. The construction of the dam has, nevertheless, begun. The campaign to acquire the land has been long, bitter, and cruel. The economic and political power of ENDESA and the government have been used against the Pehuenche people, one of the poorest groups in the country, while the congress and most of the rest of the country watch in silence.

The Effects of the Alto Bio-Bío Dams

The construction of these dams will continue to create important social problems and considerable destruction of critical natural capital (Figure 25.6).

The environmental consequences of the inundation are severe (Goldsmith and Hildyard, 1984). First, there is a loss of the functions that a forest provides, such as water replenishment, climatic stabilization, soil renewal, air purification, and wildlife habitat. This kind of damage can also be associated with the deforestation necessary for the installation of the transition lines. The loss and alteration of habitats that are part of ecosystems rich in biodiversity are highly significant in the long term and reduce the adaptability of species to change (Jensen at al., 1993).

Second, there is a loss of riverine environment. Rivers and river banks are very rich habitats that are destroyed when a lake is created (Chapel, 1992; FEMAT, 1993; Koski et al., 1984; Erman et al., 1977). Over 75% of the terrestrial flora and fauna in semiarid climates is linked directly or indirectly to rivers (Kondolf et al., 1987). Fish and birds that nest in river banks are highly affected, since the waters are detained and the level of the artificial lake created may vary greatly and rapidly during the production of electricity.

Third, there is an interruption of water flow during dam construction. In general, ecological processes are an important attribute of ecosystems and determinants of biodiversity. They create the initial conditions that allow species to colonize an area or maintain an environment conducive to species' survival. These processes can be large-scale and dramatic, such as severe storms or wildfires, or routine, everyday occurrences, sometimes difficult to discern (Peck, 1998). Dams have downstream and upstream effects derived from the interruption of the water flow process that can be considerable. Once the river is interrupted by a dam, the ecological balance is further altered when the flow of suspended material and dissolved gases change. Fish migration is disrupted, and spawning and rearing conditions are altered. The total river population decreases substantially, and some species disappear completely. Furthermore, in the long term, the release of clear water by the dam with excess stream power erodes the channel bed and the banks

THE ENVIRONMENTAL PROCESS IN CHILE

Before the Environmental Law

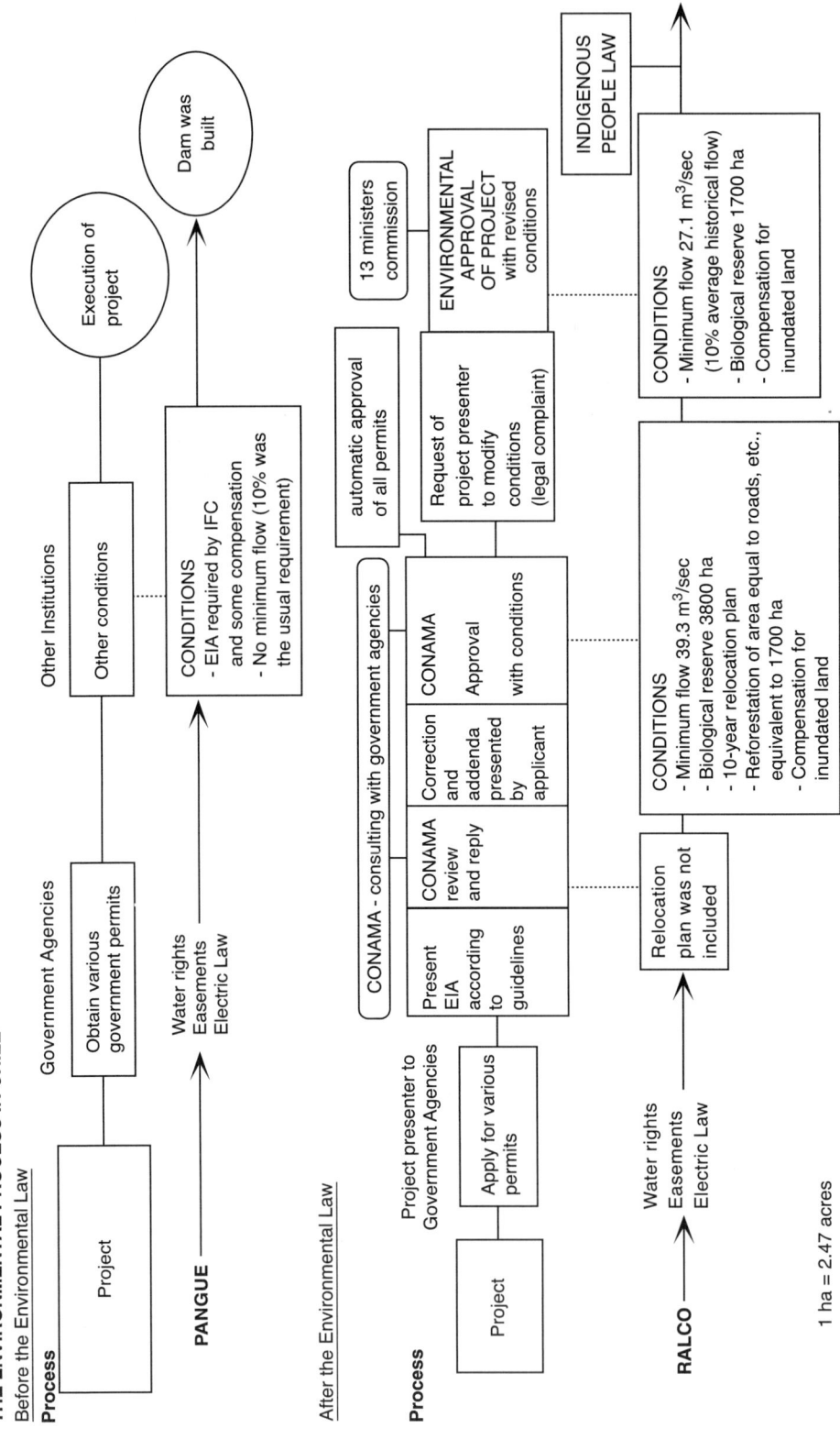

Figure 25.5 The Pangue and Ralco decision process.

downstream, and producing changes in the channel morphology that affect the living communities and riparian vegetation of the banks.

In this area, where seismicity and volcanic activity are present, the risks derived from these events are augmented by the presence of the dam. The undercutting of steep slopes by the rapid variation of the lake level makes landslides and debris flows a more probable occurrence. There is also some evidence that earthquakes could be triggered by the pressure of the immense mass of water in the lake on the geological structure (Keller, 1992). When dams are constructed for hydroelectric or irrigation purposes, it is convenient to keep the highest level of water in the reservoir; hence, the dangers of flooding can be exacerbated rather than diminished.

These factors also contribute to shortening the useful life of the projects, which is already limited because of the heavy sediment content of the rivers. The Pangue dam is estimated to trap 87% of the sediment load, an equivalent of over 4000 tons daily, thus giving an estimated life for the project of 113 years (Ecology and Environment, Inc. and Agrotec, Ltd., 1991).

Results of the Empirical Study

- The case of Ralco, when compared to Pangue, presents some evidence of a more systematic use of natural capital according to its ability to produce environmental services. The changes are minor: 1700 hectares have to be protected in exchange for the destruction of the forests that will be inundated.
- A compulsory maintenance of a minimum flow for dammed rivers was established at 10% of the annual flow average.

In the case of Chile, moderate changes in environmental institutions that allow for the recognition of critical natural capital, under the conditions of constant technology, have resulted in only

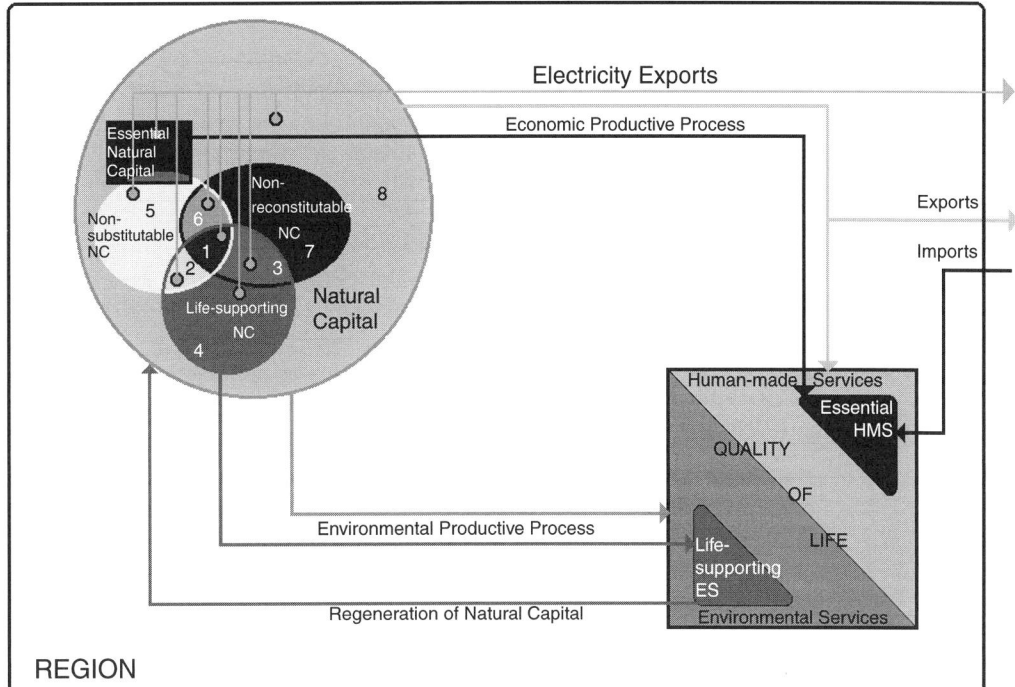

Figure 25.6 Use of natural capital by the Pangue and Ralco dams. *Critical natural capital*: 1. River bed and buffer zones. 2. Planted forests as soil regenerators. 3. Riverine fishing areas. 4. Wild fruit trees and bushes. *Noncritical natural capital*: 5. Landscape, hot springs, rapids for sports. 6. Agricultural land. 8. Pasture lands. *Local essential natural capital*: Overlap with natural capital #6.

minor changes in the use of natural capital. The recognition of these kinds of natural capital is a necessary but not sufficient condition for sustainability.

CONCLUSION

What explains the Chilean choice of a nonsustainable path of development? The analysis of the Ralco approval process shows that the short-term goals of economic development dominate decision making. The new Chilean environmental institutions are designed to allow, at many stages of the process, for the reversal of long-term decisions already made. They tend to imitate the steps and procedures followed in developed countries, and in many cases directly borrow standards and reference terms from foreign legislation. One important difference with other countries is that the process of approval includes, and is determined by, the participation of government organizations whose direction can be changed. The ease in removal of decision makers from the process keeps its control in the hands of the executive branch, making the process a completely political one in which technical considerations may have no impact at all and short-term goals may dominate.

The fact that the new environmental institutions were created in Chile mainly to comply with international demand and commercial agreements (which resulted from the increase of international exchange due to an open economy policy) has had an effect on the way these institutions were designed and the way they operate. In general they have the shape, staff, and presence of those in developed countries, but they serve only as a facade for international trade and domestic appeasement, without contributing to immediate improvements in the local use of natural capital or to a fair distribution of environmental costs. The only positive effect in the long term is creating public awareness of the environmental consequences of infrastructure projects.

In absolute terms, the approval of the Ralco project, as well as Pangue, represents an irreversible loss of critical natural and cultural capital for Chile. The new environmental institutions have had almost no impact in reducing this loss, which is their mandate. The long-term problem of reducing environmental damage has been transformed, through the decision process, into a short-term equity problem: how best to compensate immediate losers. The majority of the environmental costs are still transferred to future generations, perpetuating unsustainable development.

REFERENCES

Altieri, M.A., Sustainable rural development in Latin America: building from the bottom-up, *Ecol. Econ.,* 7:93–121, 1993.

Atkinson, A., The urban bioregion as sustainable development paradigm, *Third World Plann. Rev.,* 14:4, 1992.

Ayres, E., Power from the sun, *Sci. Am.,* August, 1950.

Bator, F.M., The simple analytics of welfare maximization, *Am. Econ. Rev.,* March, 1957.

Boulding, K., *Ecodynamics,* Sage, Beverly Hills, 1977.

Cecilia, C. and Duane, T.P., Natural capital and quality of life: a model for evaluating the sustainability of alternative regional development paths, *Ecol. Econ.,* 30(3):441–460, 1999.

Chapel, M., Recommendations for managing late-seral-forest and riparian habitats on the Tahoe National Forest, USDA Pacific Northwest region report, 1992.

Collados, C., Infrastructure and Environmental Planning for the Construction of the Pangue Project in Chile, thesis for master of city planning degree, Department of City and Regional Planning, University of California, Berkeley, 1993.

Collados, C., A Model of Natural Capital Differentiation for Sustainable Regional Development: The Case of the Alto Bío-Bío in Chile, dissertation, University of California, Berkeley, 1999.

Daly, H.E., *Steady-State Economics,* Island Press, Washington, 1991.

Dodds, S., Towards a "science of sustainability:" improving the way ecological economics understands human well-being, *Ecol. Econ.,* 23(2):95–111, 1997.

Dryzek, J., *Rational Ecology: Environmental and Political Economy*, Blackwell Publishers, Cambridge, MA, 1992.

Ecology and Environment, Inc. and Agrotec Ltda., *Evaluación de Impactos Ambientales Relevantes del Proyecto Pangue*, report prepared for Pangue S.A., Volumes 1 and 2, 1991.

Erman, D.C., Newbold, J.D., and Roby, K.B., *Evaluation of Streamside Buffer Strips for Protecting Aquatic Organisms*, California Water Resources Center, University of California, Davis, CA, 1977.

FEMAT (Forest Ecosystem Management Assessment Team), Supplemental Environmental Impact Statement on Management of Habitat for Late-Successional and Old Growth Forest Related Species within the Range of the Northern Spotted Owl, USDA, Forest Service, U.S. Department of the Interior, Bureau of Land Management, 1993.

Fisher, A.C., *Resource and Environmental Economics*, Cambridge University Press, Cambridge, MA, 1981.

Fisher, A. and Krutilla, J., *The Economics of Natural Environments*, Resources for the Future, Washington, D.C., 1975.

Folke, C., Hammer, M., Costanza, R., and Jansson, A., Investing in natural capital — why, what, and how? in Jansson, A. et al., Eds., *Investing in Natural Capital: The Ecological Economics Approach to Sustainability*, Island Press, Washington, D.C., 1994.

Gadgil, M. and Seshagiri Rao, P.R., Designing incentives to conserve India's biodiversity, in *Property Rights in a Social and Ecological Context: Case Studies and Applications*, Hanna, S. and Munasinghe, M., Eds., The Beijer International Institute of Ecological Economics and The World Bank, 1995.

Georgescu-Rogen, N., The entropy law and the economic problem, in Daly, H.E., Ed., *Economics, Ecology, Ethics: Essays toward a Steady-State Economics*, W.H. Freeman, San Francisco, 1980.

Goldsmith, E. and Hildyard, N., *The Social and Environmental Effects of Large Dams*, Wadebridge Ecological Centre, Camelford, Cornwall, U.K., 1984.

Gore, C., *Regions in Question: Space, Development Theory and Regional Policy*, Methuen, New York, 1984.

Graedel, T.E. and Allenby, B.R., *Industrial Ecology*, Prentice Hall, Englewood Cliffs, NJ, 1995.

Hanna, S.S., Linking Human and Natural Systems through Property Rights Regimes, Department of Agricultural and Resource Economics, Oregon State University, paper presented in Costa Rica International Symposium, Down to Earth: Practical Applications of Ecological Economics, 1994.

Haskell, B.D., Norton, B.G., and Costanza, R., What is ecosystem health and why should we worry about it? In Haskell, B.D., Norton, B.G., and Costanza, R., Eds., *Ecosystem Health, New Goals for Environmental Management*, Island Press, Covelo, CA, 1992.

Heathfield, D.F. and Wibe, S., *An Introduction to Cost and Production Functions*, Humanities Press International, Atlantic Highlands, NJ, 1987.

Hodges, C.A., Mineral resources, environmental issues, and land use, *Science*, 268:1305–1312, 1995.

Howarth, R.B. and Norgaard, R.B., Environmental valuation under sustainable development, *Am. Econ. Rev.*, 80(2):473–477, 1992.

Jensen, D.B., Torn, M.S., and Harte, J., *In Our Hands: A Strategy for Conserving Biodiversity in California*, University of California Press, 1993.

Keller, E.A., *Environmental Geology*, 6th ed., Macmillan, New York, 1992.

Kondolf, G.M., Webb, J.W., Sale, M.J., and Felando, T., Basic hydrologic studies for assessing impacts of flow diversions on riparian vegetation: examples from streams of the Eastern Sierra Nevada, *Calif. Environ. Manage.*, 11:757–769, 1987.

Koski, K.V., Heifezt, J., Johnson, S., Murphy, M., and Thedinga, J., Evaluation of buffer strips for protection of salmonids rearing habitat and implications for enhancement, in Hassier, J., Ed., *Pacific Northwest Stream Habitat Management Workshop*, 1984.

Norgaard, R.B., Economic indicators of resource scarcity: a critical essay, *J. Environ. Econ. Manage.*, 19:1990.

Norgaard, R.B., Sustainability as intergenerational equity: economic theory and environmental planning, *Environ. Impact Assess. Rev.*, 12:85–124, 1992.

Norgaard, R.B., *Development Betrayed: The End of Progress and a Coevolutionary Revisioning of the Future*, Routledge, London, 1994.

Pearce, D.W. and Turner, R.K., *Economics of Natural Resources and the Environment*, Second ed., The Johns Hopkins University Press, Baltimore, 1991.

Peck, S., *Planning for Biodiversity: Issues and Examples*, Island Press, Washington, D.C., 1998.

Power, M.T., *Lost Landscapes and Failed Economies: The Search for a Value of Place*, Island Press, Washington, D.C., 1996.

Rees, W.E. and Wakernagel, M., Ecological footprints and appropriated carrying capacity: measuring the natural capital requirements of the human economy, in Jansson, A. et al., Eds., *Investing in Natural Capital: The Ecological Economics Approach to Sustainability*, Island Press, Washington, D.C., 1994.

Turner, R.K., Doktor, P., and Adger, N., Sea-level rise and coastal wetlands in the U.K.: mitigation strategies for sustainable management, in Jansson, A. et al., Eds., *Investing in Natural Capital: The Ecological Economics Approach to Sustainability*, Island Press, Washington, D.C., 1994.

SECTION I.5

Finding Indicators

Overview: Can We Develop and Utilize Indicators of Ecological Integrity to Manage Ecosystems Successfully?

Gary N. Cherr

Section I.5, Finding Indicators, contains six chapters that provide thought-provoking insight into the approaches and processes used to develop suites of indicators for establishing integrity of a particular ecosystem. Along the broad scale of ecosystem integrity, environmental managers as well as the public can typically differentiate highly degraded ecosystems from those that are more toward the pristine end of the scale. However, the assessment of an ecosystem that is undergoing a slow decline in its integrity (or the often misused word *health*) over many decades, or of one which has not been characterized in its previous, more intact state, offers challenges to scientists and environmental managers. A major goal of environmental management is not only to develop a suite of indicators of ecosystem integrity but also to synthesize these indicators into technically defensible assessments of environmental health. An important part of this goal is to develop indicators of integrity for specific floral and faunal populations within an ecosystem and the health of humans who exist within the ecosystem and utilize its resources. Existing monitoring approaches, such as population assessments, epidemiology, routine collections, toxicity testing, and chemical analyses can and should all be components of an integrated approach to environmental health assessment. However, these approaches are at present not adequate by themselves to definitively address ecosystem integrity; thus, additional research is needed to define measurement endpoints and the integration of those metrics into a conceptual framework that environmental managers can directly utilize for decision making.

Understanding the impact of multiple anthropogenic stressors on ecosystems is an urgent challenge for environmental management. A major proportion of the human population lives on or near watersheds, impacts the quality of these ecosystems, and, in turn, experiences adverse health impacts due to environmental degradation. These impacts are both ecological and economic and can include altered food production and safety, with direct impacts on both wildlife and human populations. While pollution from large industrialized cities and agricultural runoff have increasingly stressed watersheds, other human-related impacts, including poor logging practices, freshwater diversions, invasive species, pathogens, and climatic change have also contributed significantly to declining ecosystem integrity. Detecting the effects of the multiple environmental stressors is particularly difficult in complex ecosystems. For example, environments such as estuaries are particularly difficult to assess given their dynamic natures and wide fluctuations in freshwater input, tidal circulation, and biotic abundance (Breitburg et al., 1999). Developing ecological indicators for complex ecosystems is as challenging as it is essential.

The criteria for selecting and validating indicator endpoints have been discussed by numerous authors (e.g., Kelly and Harwell, 1990; Huggett et al., 1992; Karr, 1993; Peakall and Shugart, 1993; Suter, 1993; Mineau, 1998; Harwell et al., 1999; Jackson et al., 1999; Van Dam et al., 1998), and these vary with the goals to be achieved. Some of the most relevant include early warning of impact, ecological relevance, variability, specificity, selectivity, broad geographic applicability, and feasibility of implementation. For example, some indicators of ecological integrity may have the advantage of being highly relevant to managers because they measure the status of an organism or population of organisms, but lack specificity and often exhibit high temporal and spatial variability. In contrast, specific biomarker responses can have high specificity and can also provide early warning of environmental impact, but need to be applied in conjunction with other tools to be relevant for the population- and ecosystem-level problems faced by managers. Thus, it is critical that combined information from the multiple types of endpoints, expressed at different levels of biological organization and spatial and temporal scales, will provide more meaningful assessments of environmental stress than could be achieved with a less diverse suite of endpoints.

Integration of ecological indicators involves combining measurement endpoints in different ways and interpreting the results as an index of ecosystem integrity. If this can be accomplished, then the specific measurements become useful indicators for managers. As such, one cannot determine, *a priori,* which specific measurement endpoints for a particular ecosystem will be the most useful ecological indicators until they are incorporated in an environmental framework in the appropriate context that utilizes previous studies for guidance. There are several general approaches one can take in establishing an ecological framework for assessing ecosystem integrity. The assessment of a number of endpoints that relate directly to measures of ecosystem integrity needs to be conducted to determine the appropriate statistical combinations of endpoints and evaluate their applicability for different environments. These measurements can then be incorporated into process-based models that permit projections of population and ecosystem change over time in order to predict ecosystem sustainability.

The chapters in this section cover ecosystem indicators that range from aquatic and terrestrial environments to human health. The contributions in this section provide a case study for establishing a conceptual framework for selecting indicators of ecological integrity using the San Francisco Bay–Delta–River system. This estuarine system, the largest in the western U.S., is rapidly becoming one of the most managed ecosystems because of its extreme loss of ecosystem integrity. Chapter 27 discusses the opportunity to employ a conceptual framework that incorporates a suite of comprehensive indicators to provide direct ecosystem health assessment. The second contribution to this section, Chapter 28, establishes specifications of ecological indicators of sustainability. This includes the establishment of five specific requirements for developing useful indicators of sustainability. The next contribution, Chapter 29, looks at the relationship between ecological decline and human health through the power of epidemiological approaches. This is truly a holistic approach to ecosystems, where the human health element is so often given low priority in indicator integration. Chapter 30 focuses on a terrestrial index of ecological integrity with the emphasis on landscape-level indicators of entire watersheds that can be tracked over time. Chapter 31 discusses the U.S. EPA Office of Research and Development technical guidelines for evaluation of ecological indicators for monitoring programs. There are 15 guidelines presented, organized into four evaluation phases. Several examples are provided to enable the reader to see how these guidelines provide a framework for large-scale assessments. Finally, Chapter 32 addresses concepts and models as applied to forest ecosystems. Specifically, it discusses the feasibility of using a Forest Capital Index to manage forest ecosystems on a worldwide scale.

The themes that run through Section I.5 address the ideas mentioned above, including suites of indicators of ecological integrity, development of conceptual frameworks and modeling, and its utilization by managers. However, throughout this section the different contributions consistently go beyond assessment of ecosystem integrity and deal with the issue of sustainability. Although an obvious issue to discuss, sustainability has often been left out of many ecosystem indicator discussions; but this section provides an overarching conceptual link among these excellent contributions.

ACKNOWLEDGMENT

The author would like to acknowledge the U.S. EPA-supported Pacific Estuarine Ecosystem Indicator Research (PEEIR) Consortium, which is part of the Estuarine and Great Lakes Indicators Research Program (EaGLe).

REFERENCES

Breitburg, D.L., Seitzinger, S., and Sanders, J., Eds., The effects of multiple stressors on freshwater and marine ecosystems. *Limnol. Oceanogr.,* 44:24–3590, 1999.

Harwell, M.A., Myers, V., et al., A framework for an ecosystem integrity report card, *BioScience,* 497:543–556, 1999.

Huggett, R.J., Foster, L.M., Melancon, M.J., Shugart, L.R., Hinton, D.E., Weeks, B.A., and Stegeman, J.J., *Biomarkers: Biochemical, Physiological, and Histological Markers of Anthropogenic Stress,* Lewis Publishers, Boca Raton, FL, 1992, p. 347.

Jackson, L., Kurtz, J., et al., *Evaluation Guidelines for Ecological Indicators,* U.S. Environmental Protection Agency, Office of Research and Development, Research Triangle Park, NC, 1999.

Karr, J.R., Measuring biological integrity: lessons from streams, in *Ecological Integrity and the Management of Ecosystems,* Woodley, S., Kay, J., and Francis, G., Eds., CRC Press, Boca Raton, FL, 1993, 83–103.

Kelly, J.R. and Harwell, M.A., Indicators of ecosystem recovery, *Environ. Manage.,* 145:527–546, 1990.

Mineau, P., Are there linkages to ecological effects? in *Effects of Multiple Impacts on Ecosystems,* Cech., J.J., Wilson, B.W., and Crosby, D.C., Eds., Lewis Publishers, Boca Raton, FL, 1998.

Peakall, D.B. and Shugart, L.R., Biomarkers: research and application in the assessment of environmental health, *NATO ASI Series, Series H Cell Biology,* 68:119, 1993.

Suter, G.W., A critique of ecosystem health concepts and indexes, *Environ. Toxicol. Chem.,* 12:1533–1541, 1993.

Van Dam, R.A., Camilleri, C., et al., The potential of rapid assessment techniques as early warning indicators of wetland degradation: a review, *Environ. Toxicol. Water Qual.,* 134:297–312, 1998.

CHAPTER 27

A Conceptual Framework for Choosing Indicators of Ecological Integrity: Case Study of the San Francisco Bay–Delta–River System

Karen Levy, Terry F. Young, and Rodney M. Fujita

Expert opinion — from scientists, managers, and policy makers alike — has begun to converge on ecosystem management as an appropriate method for managing activities that affect natural systems. The federal Interagency Ecosystem Management Task Force, created under the Clinton administration, defines the goal of ecosystem management as "to restore and maintain the health, sustainability, and biological diversity of ecosystems while supporting sustainable economies and communities" (GAO, 1994). Grumbine (1994) states that "ecosystem management integrates scientific knowledge of ecological relationships within a complex sociopolitical and value framework toward the general goal of protecting native ecosystem integrity over the long term."

Ecosystem management programs are currently being designed and implemented across the U.S. and elsewhere (Yaffee et al., 1996; Christensen et al., 1996; GAO, 1994). All four of the primary federal land management agencies (the National Park Service, Forest Service, Fish and Wildlife Service, and Bureau of Land Management), which collectively manage all federal lands (30% of the total surface area of the U.S.), have planned ecosystem-based approaches to managing their lands and natural resources and have formed an interagency group to coordinate these efforts (GAO, 1994; Beattie, 1996; Dombeck, 1996; Thomas, 1996). Many other groups have formulated ecosystem management strategies as well, including the National Oceanic and Atmospheric Administration and even the Department of Defense (Griffis and Kimball, 1996; Goodman, 1996). The U.S. General Accounting Office states that "a consensus has emerged that ecosystem management provides a sounder approach for meeting the federal stewardship mandates of protecting natural resources and sustaining long-term commodity production and other uses on federal lands" (GAO, 1994).

The Ecological Society of America Committee on the Scientific Basis for Ecosystem Management describes ecosystem management as "management driven by explicit goals, executed by policies, protocols, and practices, and made adaptable by monitoring and research based on our best understanding of the ecological interactions and processes necessary to sustain ecosystem structure and function" (Christensen et al., 1996). The committee recognized eight essential components of ecosystem management: (1) long-term sustainability as fundamental value; (2) clear operational goals; (3) sound ecological models and understanding; (4) understanding complexity and interconnectedness; (5) recognition of the dynamic character of ecosystems; (6) attention to context and scale; (7) acknowledgment of humans as ecosystem components; and (8) commitment to adaptability and accountability.

The ecosystem management paradigm is also being applied to large-scale *restoration* programs. For example, the National Research Council (1992) concluded that restoration of an aquatic ecosystem requires coordinated, comprehensive management of all significant ecological elements, often on a watershed or other landscape scale. The Kissimmee River Restoration Project, which aims to restore the historic form, hydrology, and associated ecological resources of 70 km of continuous river channel and 11,000 hectares of floodplain in Florida (Arrington, 1995), has the explicit goal of "restoring the ecological integrity of the Kissimmee River" (Toth, 1993).

While a common vision may be emerging about the prerequisites for sound ecosystem restoration and management, however, a similar convergence of opinion has not yet developed on how to implement this new paradigm. In this chapter, we address the challenge of operationalizing the emerging paradigm of ecosystem management, using a case study of the San Francisco Bay–Delta–River system as an illustration. We propose a methodology to implement part of this challenge — describing a large system in order to evaluate its overall health or sustainable ecological integrity — and report on the experience of our case study. We also discuss issues pertaining to the decision-making processes involved, specifically our attempt to achieve broad stakeholder consensus on scientific ideas.

OPERATIONALIZING THE ELUSIVE CONCEPT OF ECOLOGICAL INTEGRITY

A common understanding of the concepts of sustainability and ecological integrity is still evolving in the scientific literature, largely because ecosystems and landscapes are highly complex entities. *Ecological integrity* and *ecosystem health* have been described in a variety of ways, from ecological persistence to resilience, homeostatic balance, increased probability of survival, and availability of habitat (Scrimgeour and Wicklum, 1996). Karr (1993) defines ecosystem health as the condition in which a system realizes its inherent potential, maintains a stable condition, preserves its capacity for self-repair when perturbed, and needs minimal external support for management.

Although these qualities are intuitively understandable, implementing ecosystem management requires that they be captured with an operational definition (Wrona and Cash, 1996). To do this, specific, measurable indicators of ecological health are needed. To use a human health analogy, we know humans *should* be able to walk and talk and recover when hit by a tennis ball, but we use specific measurements based on a scientific understanding of the human body, such as pulse and temperature, to evaluate whether the patient is in a healthy state overall. Similarly, for ecosystem integrity, it is useful to develop a model of the ecological system, and then define measurable parameters that, when taken together, report on the status of the key properties of the system. Ecological indicators, defined here as attributes of a system that can be measured to provide information about the health or integrity of that system, provide such measurements.

To successfully develop a suite of ecological indicators on a landscape scale, we advocate the use of a formal conceptual framework as an organizing tool. The framework ensures that the suite of indicators is ecologically sound and that the indicators (and the restoration or management plan) are comprehensive — that they address the integrity of the entire landscape as an ecological whole, as well as addressing each of the component parts. An additional benefit of using an explicit framework is to demonstrate to policy makers and the public the importance and relevance of each of the indicators.

CASE STUDY: THE SAN FRANCISCO BAY–DELTA–RIVER SYSTEM

As a result of a confluence of events, a major restoration effort for the San Francisco Bay–Delta–River system is currently being undertaken by a consortium of federal and state agencies, with extensive stakeholder input. The impetus for the restoration effort grew out of more than 20 years of conflict among agricultural, urban, and environmental interests regarding the appropriate apportionment of water resources. In the latest chapter of this saga, a truce was declared in 1994, when the State of California, the federal government, and a variety of stakeholders signed the "Principles for Agreement on Bay–Delta Standards between the State of California and the Federal Government," more commonly known as the *Bay–Delta Accord*. The accord required the parties to develop an environmental restoration program and associated long-term agreements regarding urban and agricultural water use in 3 years, providing about $60 million annually for the 3-year period to support restoration efforts, primarily derived from assessments on water users. It also established the CALFED Bay–Delta program, a consortium of state and federal agencies responsible for developing a long-term plan for managing the Bay–Delta. Additional state funding was secured in November of 1996, when California voters approved Proposition 204, a $995 million general obligation water bond measure containing about $600 million to fund Bay–Delta ecosystem restoration efforts. In addition, federal funding in the amount of $430 million ($143 million per year over 3 years) was authorized by the passage of the "Bay–Delta Act" (California Bay–Delta Environmental Enhancement and Water Security Act, P.L. 104–333, Div. I, Title XI, 1996).

Although the restoration plan for the system is being developed by a consortium of interests not dominated by scientists, the basic premise of the program is remarkably consistent with the tenets gleaned from the scientific literature and listed above. In addition to formal acknowledgment of a whole-system management approach and broad stakeholder input, the program seeks to "take advantage of natural processes, ... restore some of the ecosystem's natural resilience," and "support sustainable populations of diverse and valuable plant and animal species" (CALFED, 1996a).

Few precedents exist for restoring a system this large. The San Francisco Bay–Delta–River system, as defined herein, comprises the watersheds of the Sacramento and San Joaquin Rivers, their delta, San Francisco Bay, and the nearshore ocean off the Golden Gate Bridge (see Figure 27.1). This system is commonly referred to as the San Francisco Bay–Delta; we have added the word "river" to its name to emphasize the need to think of the system as a whole, from the headwater streams in the Sierra Nevada out to the Pacific Ocean. The estuary alone supports 380 wildlife species, 22 of which are listed as threatened or endangered by the state or federal governments (SFEP, 1992).

This watershed comprises one of the most modified systems in the U.S. (Nichols et al., 1986). It provides drinking water to 20 million people — two thirds of the state of California. Water diverted from the watershed irrigates over 1.8 million hectares of farmland (SFEP, 1996) that provide 45% of the nation's produce (CALFED, 1996a). Given the size and complexity of this system, translating the general ecological health-related goal into measurable indicators of success presents a challenge.

It is within this context that we initiated a series of meetings and workshops with CALFED staff, other stakeholders, and outside scientists to develop a suite of ecological indicators for the San Francisco Bay–Delta–River system. The purpose of the suite of indicators is to provide a scientifically valid definition of ecological integrity that can be used to help develop the restoration program and ultimately to determine whether its goal has been met. During the first phase of this effort, Environmental Defense and its partners developed and tested a conceptual framework for indicator development that has proved useful in promoting a comprehensive, whole-systems approach.

PROTOCOL FOR DEVELOPING A SUITE OF INDICATORS

In order to develop indicators that could be broadly accepted in both academic and political arenas, Environmental Defense, the Bay Institute of San Francisco, and the University of California, Berkeley, convened two workshops, in October 1995 and January 1996, whose missions were to develop a scientifically defensible suite of ecological indicators for the San Francisco Bay–Delta–River system. Participants included local experts, outside experts with experience in other systems (e.g., the Rhine, Mississippi, and Kissimmee Rivers), selected policy makers, and

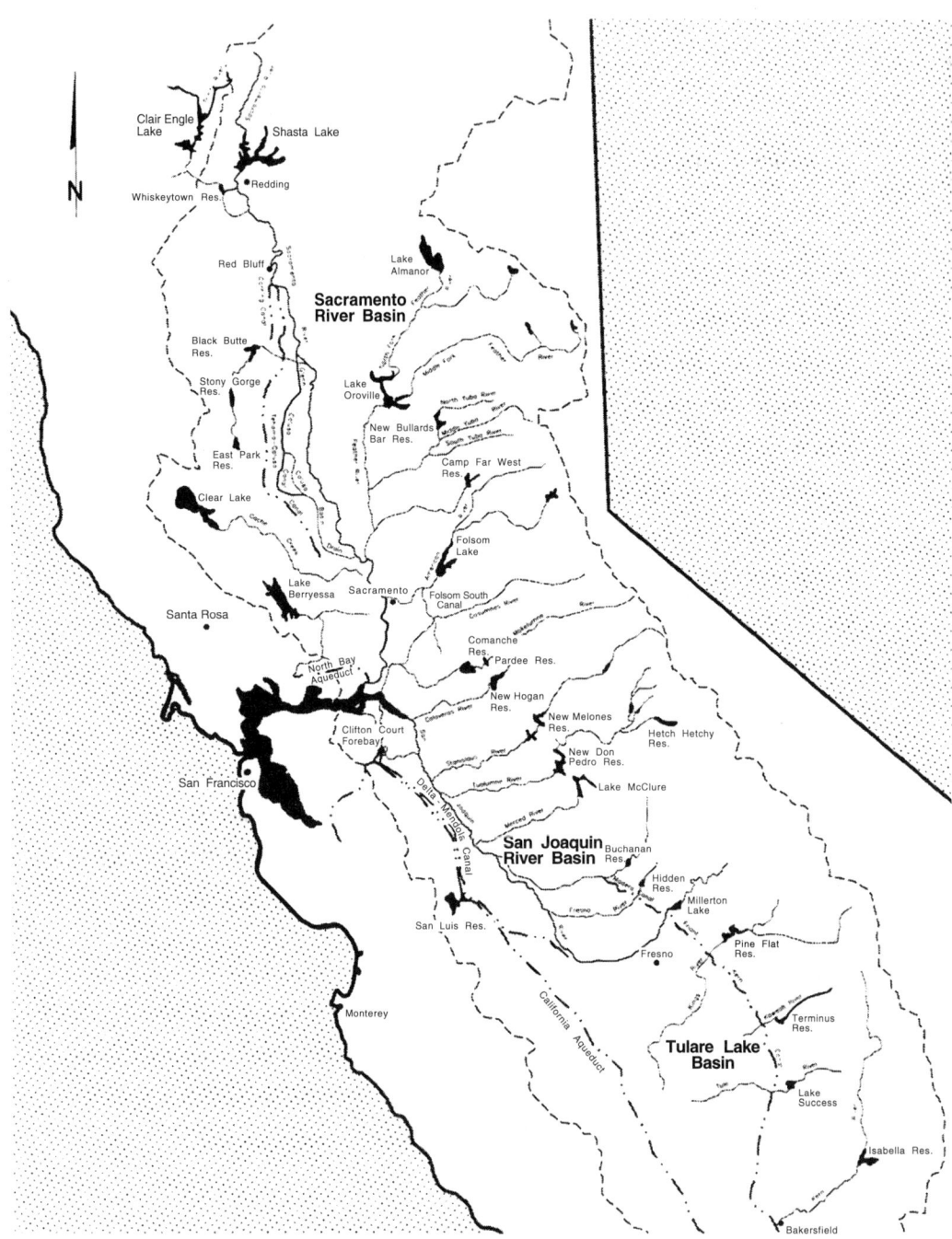

Figure 27.1 Map of the San Francisco Bay–Delta–River system. (Used with permission of the CALFED Bay–Delta Program.)

stakeholder scientists willing to check their stakeholder hats at the door. We provided background materials to the participants, including a strawman proposal for a framework and protocol for indicator development. The first workshop helped refine that framework, and the second workshop developed an initial list of indicators in accordance with the framework.

The basic procedure we undertook for indicator development followed a four-step process suggested by Keddy et al. (1993) for translating the ideas of ecosystem health into practice and establishing ecological indicators: (1) define ecological integrity or health in an operational way; (2) select appropriate indicators of health or integrity; (3) identify target levels of selected indicators; and (4) develop a monitoring system to provide feedback. The first workshop focused on step one and the second on step two; together they laid the groundwork for further refinement of the indicator system as the design of the restoration program evolved.

Step One: Define Ecological Integrity in an Operational Way

Step one — defining ecological integrity in an operational way — may begin with the development of general goals and objectives for the ecosystem management or restoration program. We distinguished between *goals*, which describe the "big picture" overview of what the restoration program is trying to achieve, and *objectives*, which are more precise descriptions of the particular steps necessary to achieve the program goal.

The workshops built on previous goal statements put forth by various government agencies and private organizations. Synthesizing these and the scientific input of workshop participants, we developed the following suggested goal statement for Bay–Delta–River restoration and management:

> Establishment of a healthy system that supports a diversity of habitat types along with their resident communities of plants and animals, supports essential ecological functions, and is self-sustaining (requiring minimal intervention) and resilient to stresses.

This goal statement assumes that the system will continue to accommodate human use of natural resources.

A series of objectives also evolved from workshop discussions (Table 27.1). These objectives provide more specific descriptions of the components and processes that appear necessary to achieve the overall goal for the Bay–Delta–River system and to successfully implement the whole-system management envisioned by current law and policy. They also relate more directly to indicators than does the general goal. Note that these eight objectives are not prioritized. All must be achieved in order to support the system-wide goal stated above.

The derivation of these objectives, as well as the selection of indicators in step two, ideally should be carried out in the context of working models of the ecological system at both the landscape and smaller scales. In this case study, however, ecological models were not explicitly derived at the outset, except to the extent that a habitat typology (see below) provided a partial surrogate.

In order to take the next step — translating these objectives into a suite of indicators — an ecologically sound conceptual framework is required to ensure that the indicators include each of the key components and processes referenced in the objectives (and ideally also identified in ecological models). The framework that we propose is adapted from Noss' (1990) conceptual model of biodiversity at multiple levels of organization. In this model, structural

---

Table 27.1 Restoration Objectives for the San Francisco Bay–Delta–River System

A. Ensure conditions necessary to support and protect native biodiversity
B. Protect and/or restore conditions necessary to increase populations of valuable plant and animal species (in a manner consistent with protecting native biodiversity)
C. Ensure sufficient extent, diversity, quality, connectivity, and range of successional states of natural habitats
D. Protect and/or restore the natural trophic structure of communities
E. More closely approximate the natural patterns of transport of essential materials (water, sediments, nutrients)
F. More closely approximate the natural hydrological regime
G. Protect and/or restore water quality
H. Provide for societal uses (harvest, recreation, aesthetics)

(i.e., "physical organization or pattern of a system"), functional (i.e., "ecological and evolutionary processes"), and compositional (i.e., "identity and variety of elements [e.g., species or communities] in a collection") biodiversity, shown as interconnected spheres, each encompass multiple scales of organization. The scales of interest range from the landscape to the genetic level. Although Noss' model was developed to facilitate selection of indicators that represent the many aspects of biodiversity that warrant attention in environmental monitoring and assessment programs (Noss, 1990; 1995), it appears useful in applications beyond the context of biodiversity assessment.

Accordingly, the framework proposed here includes several elements: first, a classification scheme or typology that defines the components and scales of interest within the ecological hierarchy; second, for each component of this typology, a set of properties and corresponding measurable indicators that represents structural, compositional, and functional attributes (particularly those that relate to the restoration objectives or identified in an ecological model); and third, a set of criteria for eliciting and then screening the indicators.

Define the Components of Ecological Hierarchy: Habitat Typology

Participants at the first workshop suggested that a typology should be developed in order to divide the Bay–Delta–River system into working units for analysis and management. Typology is defined here to mean a hierarchical classification system depicting various major levels of ecological organization of the entire system. The typology we developed characterizes the Bay–Delta–River system at three basic scales: (1) *the entire landscape* (in order to consider the interactions among each of the different components of the system); (2) *ecological zones* (corresponding to major biomes at the landscape scale); and (3) *habitat types* (ecologically distinct areas within each ecological zone) (Figure 27.2). Ultimately, resolution down to the subhabitat level will be necessary to implement a restoration program.

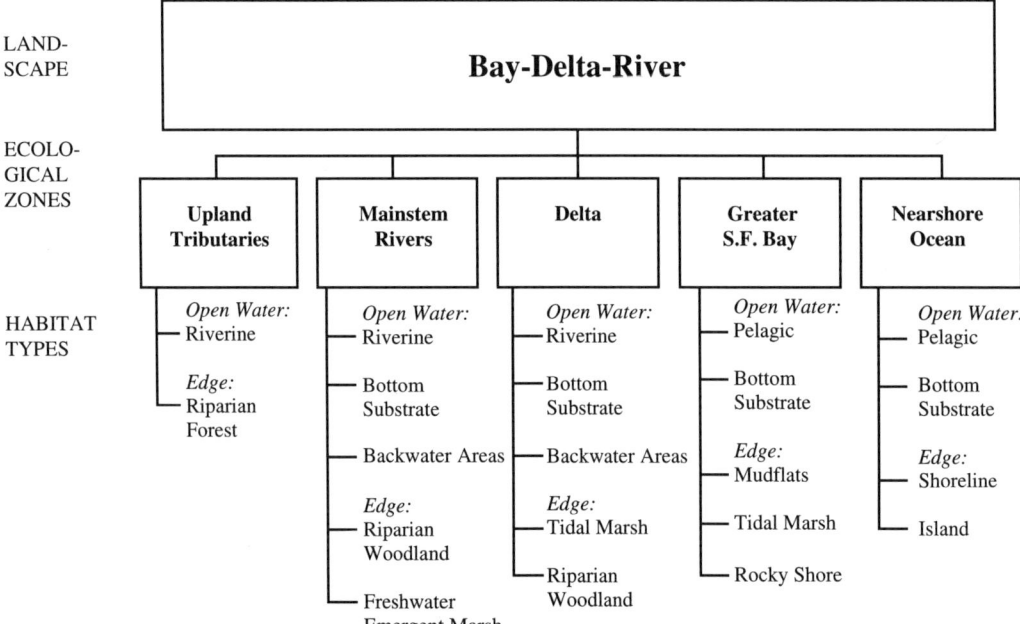

Figure 27.2 Schematic of the habitat typology used for indicator development and organization, indicating the five ecological zones upon which workshop participants came to consensus. The *suite* of indicators for the Bay–Delta–River system will incorporate a *set* of indicators for each component of this typology.

The landscape/seascape level is defined as the entire watershed of California's Central Valley and San Francisco Bay, including freshwater, estuarine, and marine portions (see Figure 27.1). The five major ecological zones are defined as:

- *Upland Tributaries and Watersheds*: streams and rivers of the Sierra and Coast Ranges (fresh water)
- *Lowland Rivers*: the alluvial river–floodplain systems of the California Central Valley, most notably the Sacramento and San Joaquin Rivers above the tidal influence (fresh water)
- *The Delta*: the tidally influenced portion of the Sacramento and San Joaquin Rivers (estuarine)
- *Greater San Francisco Bay*: a series of embayments (Suisun Bay, San Pablo Bay, Central Bay, and South Bay) linking the delta and the ocean (estuarine)
- *Nearshore Ocean*: the corridor extending approximately 40 kilometers north and south of the Golden Gate Bridge and west to the continental shelf break (marine)

A list of habitats (e.g., riparian woodland, riverine, brackish tidal marsh) was also described for each of these zones.

Using this typology, the *suite* of indicators for the Bay–Delta–River system as a whole would incorporate a *set* of indicators for the entire landscape as well as *sets* for each component ecological zone and habitat type.

Define Attributes Relating to Structure, Function, and Composition

Because the typology organizes an extensive system like the Bay–Delta–River into components that relate to an ecological hierarchy, indicators developed for each component of the typology taken together paint a picture of the health of the whole system. We developed a template matrix for indicator development (Figure 27.3), which served as an organizing tool for developing each set of indicators. Following the Noss (1990) model, spaces were provided for both structural and functional indicators at the landscape and zone levels and, for each habitat type, structural and functional indicators at both the community and population levels. Our use of the term *structure* incorporates both the concepts of "structure" and "composition" as used by Noss; we adopted this compromise of nomenclature at the request of the workshop participants. The purpose of the matrix was to ensure completeness by providing a checklist. This device may be particularly useful in systems where reference to existing monitoring programs or a focus on particular species or habitat types might otherwise tend to dominate indicator selection.

For the Landscape and Each Ecological Zone

	Structure	Function
	A e.g., habitat acreage	B e.g., hydrologic connectivity

For each Habitat Type:

	Structure	Function
Community/Ecosystem	C e.g., species diversity/richness	D e.g., primary production
Population/Species	E e.g., population age structure	F e.g., fecundity

Figure 27.3 Matrix used by workshop participants to guide indicator development.

Step Two: Select Indicators According to an Organizing Framework

Based on this proposed organizing framework, participants in the second conference were asked to fill in the matrix checklist with indicators of structure or function relating to the appropriate typology component. The participants were further asked to select only indicators that met suggested criteria.

Many authors have suggested criteria for selecting ecological indicators (see, e.g., Noss, 1990; Keddy et al., 1993; Angermeier and Karr, 1994; Barbour et al., 1995; Cairns et al., 1993). Drawing on these, we developed a consolidated, short list of *essential* criteria, and a longer list of *desirable* criteria. In the former category, indicators *must* be (1) ecologically relevant and (2) scientifically defensible. An ecologically relevant indicator is closely related to or reflective of key ecological characteristics of a system or habitat. Scientifically defensible indicators are quantitative, with sufficient accuracy and precision to allow for ready interpretation. The relationship of an indicator to the property it reflects should be unambiguous and demonstrable. Exceptions to the *essential* criteria may be made if a certain indicator has significant public appeal, high economic significance, or is especially relevant to policy makers for some other reason.

Other criteria define beneficial, but not crucial, qualities of indicators. These *desirable* qualities include (1) ease of measurement; (2) sensitivity (including timely response to stress/perturbation, ability to provide early warning of disturbance); (3) existence of a historical database; (4) environmentally benign to measure (monitoring of the indicator is not damaging to the environment); (5) generality (applicable to different habitat types); and (6) ability to aid in distinguishing between natural processes and anthropogenic effects.

An inconsistency emerged from the workshops in that some participants confined their selection of indicators to measurable attributes, while others listed certain key *properties* of the system that needed to be assessed but were not in themselves directly measurable. For example, the "areal extent of wetlands" and "connectivity" of wetlands to some other habitat may both be considered key structural attributes of the system. However, while the former may be directly measured, the latter is not, but rather must be evaluated through a set of surrogate measures (e.g., amount of common boundary with another habitat; flow rates between other habitats and wetlands). We therefore distinguished between *properties* of the system that workshop participants believed necessary to assess and actual *indicators*.

For examples of properties and indicators developed at the workshops, see Table 27.2. The entire suite is reproduced in Levy et al. (1996). It provides a valuable starting point for the development of an indicator suite that can be used to assess ecosystem condition and restoration progress.

In order to further refine the indicator suite for use as an ecosystem management tool, several additional steps are required. Following the initial development of indicators, a final quality control check should be undertaken to ensure that the candidate suite of indicators meets the criteria laid out above. Also, the framework provides only an aid to ensure completeness, so it may be necessary to analyze the suite to make sure that the stated objectives will be met if the indicators all fall within the "healthy" target range. Part of the completeness check should include a determination of whether increases or decreases in the main *stressors* impacting the system will be detected by one or more indicators. Finally, the smallest number of indicators that will effectively evaluate the system is desirable.

Step Three: Develop Target Values

Once indicators are selected, a range of target values, from tolerable to desirable levels, should be developed for each. This constitutes step three of the Keddy et al. (1993) process. Because determining the target range of indicator values from first principles is difficult, comparisons with reference systems or historic states are sometimes used. In disturbed ecosystems such as the

Table 27.2 Examples of Indicators Developed at the San Francisco Bay–Delta–River Workshops[a]

Scale of Interest	Objective[b]	Sample Property Assessed	Sample Indicator
Landscape	A, B, C, E, H	Connectivity of habitats (Structural)	Total number of barriers to anadromous fish migration
	C, E, F, G	Hydrologic regime (Functional)	Variability in flooding duration and frequency
Rivers	A, B, C	Extent and quality of edge habitat (Structural)	Length of shaded riparian bank
	E	Deviation from natural sediment budget (Functional)	Percentage of pre-dam supply of gravel and sand-sized sediment delivered to the reach
Delta	A, B, C, G, H	Quality of anadromous fish habitat (Structural)	Smolt survival through zone
	C, E, F, G	Deviation from natural hydrograph (Functional)	Percent of monthly inflow not diverted
Bay	A, B, C	Habitat quality (Structural)	Channel density
	A, B, C, D, G	Capacity to support resident fish and wildlife (Functional)	Habitat acreage
Ocean	A, B, C	Habitat extent and diversity (Structural)	Total area and proportionate amount of primary habitat types
	A, B, D	Production (Functional)	Average chlorophyll levels (as determined by landsat imagery)

[a] The complete results of the workshops are given in Levy et al. (1996). Although lowland alluvial rivers differ ecologically from upland tributaries, we consolidated the two zones for ease of discussion.
[b] Refers to objectives listed in Table 27.1.

Bay–Delta–River system, it is clearly unreasonable to strive for the restoration of pristine conditions in all parts of the system, because certain changes in the system (e.g., major dams and urban development) are not reversible within a reasonable time period. However, an analysis of historical conditions and processes can provide insights into realistic target levels (see, e.g., Steedman et al., 1996; The Bay Institute of San Francisco, 1998).

Step Four: Monitoring and Adaptive Management

Step four of the process involves monitoring the indicators to determine whether the management and/or restoration program is having its intended effect. Monitoring provides the foundation for adaptive management, whereby management actions and targets are evaluated and altered according to the results of monitoring data, and carefully designed management experiments are conducted to reduce uncertainty. Once there is consensus on the suite of indicators, monitoring can be taken up by the pertinent government agencies, although continued interaction with scientists is important.

Monitoring also allows for public accountability. It is important to communicate results of restoration projects to the general public to garner support for restoration and management efforts. Indicators that have been selected for their ability to communicate the status of the system as a whole in a way that the public can readily understand are particularly important for maintaining accountability, interest, and support.

LESSONS LEARNED

The process of developing a general goal and then translating the goal into more specific objectives (step one) provided a solid foundation for development and later refinement of the indicators. The experience from this case study highlights the benefits of involving scientists in this step — in contrast to the current trend toward maximizing involvement of stakeholders along with agency managers.

The formal conceptual framework used as an organizing tool for translating the goals and objectives into an indicator suite (step two) aided in the development of indicators that comprehensively reflect the most important attributes of the system. A secondary benefit of this formalized approach was to catalyze experienced scientists to "think outside of the box" rather than simply fall back on the elements covered in existing monitoring programs, or within their own specialties.

The development of the typology provided a tool for integrating existing, smaller-scale ecosystem management and restoration programs in the region into a larger vision for restoration on the landscape scale. The CALFED Bay–Delta Program adopted a similar typology for its ecosystem restoration planning process, in which specific restoration objectives and targets are being developed for the system.

The workshops also generated broad consensus on many indicators. It proved important to achieve this broad buy-in on metrics to describe the health of the system before trying to set the far more controversial *target ranges* for those metrics. In this way, the comprehensiveness and quality of the indicators remained the focus of discussion, rather than stakeholder issues such as economics or equity. CALFED was able to utilize the initial suite of indicators developed at the workshops immediately in the formulation of its ecosystem restoration program plan (D. Daniel, personal communication; CALFED, 1996b), although it remains to be seen whether CALFED will actually use a comprehensive indicator suite to evaluate its program.

Lack of time and failure to conduct each task in an entirely consistent way resulted in a suite of indicators that was unwieldy and somewhat internally inconsistent. In retrospect, it appears that spending additional time and resources to first develop theoretical models of the ecological system, in addition to the typology, may have paid off by providing a mechanism to focus and integrate the indicators derived for different scales. Using the formal conceptual framework for indicator selection *in tandem with* a set of theoretical models of the ecological system may well yield optimum results: the framework and the models together define the most important structural and functional elements within the system; the conceptual models display the relationships among these elements; and the framework for indicator selection provides the assurance that coverage is complete.

CONCLUSION

Increasingly, resource managers across the nation are adopting an ecosystem approach to natural resources management. The universal challenge facing any ecosystem management or large-scale restoration program is to translate the general and often amorphous goal defined by the public — e.g., a healthy system — into a set of *measurable* attributes that, taken as a whole, provide an assessment of overall ecological integrity. In the San Francisco Bay–Delta–River case study described here, we attempted to help accomplish this task by using a reproducible and methodical procedure that draws heavily on scientific literature and expert opinion, and applying this procedure in the context of the current political process aimed at forging a plan for ecosystem restoration and management. This ecologically based conceptual framework helped translate the abstract concepts of *ecosystem health* and *ecological integrity* into usable tools to guide the long-term process of large-scale restoration.

Comprehensive, operational definitions of ecological integrity also serve a broader purpose by helping to provide assurances to taxpayers and stakeholders that restoration programs will actually achieve the goals that have been set out. Because of the amount of uncertainty inherent in the science of ecology, as well as the infancy of the field of restoration ecology, a program of adaptive management will be necessary in most systems to ensure that the restoration actions (and associated dollar expenditures) are having the desired effect. Yet this flexible implementation approach may well exacerbate suspicions among stakeholders that the program will not ultimately be carried out in a way that achieves the original goal. In this context, a transparent, agreed-upon suite of indicators provides an important mechanism to ensure accountability. By providing an explicit set of performance measures, the indicator suite can help to ensure that the adaptive management program remains on course to achieve the original goal.

Ecosystem protection and restoration goals are sometimes limited to capturing the goods and services that the ecosystem may provide (NHI, 1995). The utility of using a stepwise process and a conceptual framework, such as the one described here, is to try to make the suite of indicators — i.e., the explicit statements of the program goal and the measures of success — as ecologically relevant as possible, by capturing functional processes and major components of the system at scales that reflect the system's organizational hierarchy. Doing so will not only maximize the ultimate chances for designing and implementing a successful restoration program but also contribute to continuing public support for the endeavor.

ACKNOWLEDGMENTS

The authors would like to acknowledge Bill Alevizon, Hans Bernhardt, Pete Chadwick, Dick Daniel, Chuck Hanson, Bruce Herbold, Paul Keddy, Matt Kondolf, Peter Moyle, Fred Nichols, Palma Risler, Emery Roe, Charles Simenstad, Lou Toth, and Philip Williams for help with the workshop process and development of ideas. This work was made possible through grants from the United States Environmental Protection Agency and the CALFED Bay–Delta Program, administered through the UC Berkeley Center for Sustainable Resource Development. The Environmental Defense's Western Water Program is generously supported by the David and Lucile Packard, Dean Witter, Wallace Alexander Gerbode, and William and Flora Hewlett Foundations.

REFERENCES

Angermeier, P.L. and Karr, J.R., Biological integrity versus biological diversity as policy directives, *BioScience*, 44:690–697, 1994.

Arrington, D.A., Preface: the restoration process, *Restoration Ecol.*, 3:146, 1995.

Barbour, M.T., Stribling, J.B., and Karr, J.R., Multimetric approach for establishing biocriteria and measuring biological condition, in Davis, W.S. and Simon, T.P., Eds., *Biological Assessment and Criteria: Tools for Water Resource Planning and Decision Making*, Lewis Publishers, Boca Raton, FL, 1995, pp. 63–80.

Bay Institute of San Francisco, From the Sierra to the Sea: Lessons from the Ecological History of the San Francisco Bay–Delta Watershed, The Bay Institute of San Francisco Report, 1998.

Beattie, M., An ecosystem approach to fish and wildlife conservation, *Ecol. Appl.*, 6:696–699, 1996.

Cairns, J., Jr., McCormick, P., and Niederlehner, B., A proposed framework for developing indicators of ecosystem health, *Hydrobiologia*, 263:1–44, 1993.

CALFED Bay–Delta Program, Phase I Final Report, Sacramento, CA, 1996a.

CALFED Bay–Delta Program, Introduction to restoration targets: workshop packet, Sacramento, CA, 1996b.

Christensen, N.L., Bartuska, A.M., Brown, J.H., Carpenter, S., D'Antonio, C., Francis, R., Franklin, J.F., MacMahon, J.A., Noss, R.F., Parsons, D.J., Peterson, C.H., Turner, M.G., and Woodmansee, R.G., The report of the Ecological Society of America Committee on the scientific basis for ecosystem management, *Ecol. Appl.*, 6:665–691, 1996.

Dombeck, M.P., Thinking like a mountain: BLM's approach to ecosystem management, *Ecol. Appl.*, 6:699–702, 1996.

GAO (United States General Accounting Office), Ecosystem management: additional actions needed to adequately test a promising approach, Report to Congressional Requesters, GAO/RCED-94-111, Washington, D.C., 1994.

Goodman, S.W., Ecosystem management at the Department of Defense, *Ecol. Appl.*, 6:706–707, 1996.

Griffis, R.B. and Kimball, K.W., Ecosystem approaches to coastal and ocean stewardship, *Ecol. Appl.*, 6:708–712, 1996.

Grumbine, R.E., What is ecosystem management? *Conserv. Biol.*, 8:27–38, 1994.

Karr, J.R., Measuring biological integrity: lessons from streams, in Woodley, S., Kay, J., and Francis, G., Eds., *Ecological Integrity and the Management of Ecosystems,* CRC Press, Boca Raton, FL, 1993, pp. 83–104.

Keddy, P.A., Lee, H.T., and Wisheu, I.C., Choosing indicators of ecosystem integrity: wetlands as a model system, in Woodley, S., Kay, J., and Francis, G., Eds., *Ecological Integrity and the Management of Ecosystems*, CRC Press, Boca Raton, FL, 1993, pp. 61–79.

Levy, K., Young, T.F., Fujita, R.M., and Alevizon, W., Restoration of the San Francisco Bay–Delta–River System: Choosing Indicators of Ecological Integrity, Report prepared for the CALFED Bay–Delta Program and U.S. Environmental Protection Agency, June 1996.

National Research Council (NRC), *Restoration of Aquatic Ecosystems*, National Academy Press, Washington, D.C., 1992.

Natural Heritage Institute (NHI), Goals for Restoring a Healthy Estuary: Report on Results of a Workshop, San Francisco, CA, 1995.

Nichols, F.H., Cloern, J.E., Luoma, S.N., and Peterson, D.H., The modification of an estuary, *Science,* 231:567–573, 1986.

Noss, R.F., Indicators for monitoring biodiversity: a hierarchical approach, *Conserv. Biol.,* 4:355–364, 1990.

Noss, R.F., Maintaining Ecological Integrity in Representative Reserve Networks, World Wildlife Fund Canada/World Wildlife Fund United States Discussion Paper, 1995.

San Francisco Estuary Project (SFEP), State of the Estuary: A Report on Conditions and Problems in the San Francisco Bay/Sacramento–San Joaquin Delta Estuary, U.S. Environmental Protection Agency and Association of Bay Area Governments, Oakland, CA, 1992.

San Francisco Estuary Project (SFEP), How We Use the Estuary's Water, SFEP, Oakland, CA, 1996.

Scrimgeour, G.J. and Wicklum, D., Aquatic ecosystem health and integrity: problems and potential solutions, *J. North Am. Benthol. Soc.,* 15:254–261, 1996.

Steedman, R.J., Whillans, T.H., Behm, A.P., Bray, K.E., Cullis, K.I., Holland, M.M., Stoddart, S.J., and White, R.J., Use of historical information for conservation and restoration of great lakes aquatic habitat, *Can. J. Fish. Aquat. Sci.,* 53 (Suppl. 1):415–423, 1996.

Thomas, J.W., Forest Service perspective on ecosystem management, *Ecol. Appl.,* 6:703–705, 1996.

Toth, L.A., The ecological basis of the Kissimmee River restoration plan, *Fla. Sci.,* 56:25–51, 1993.

Wrona, F.J. and Cash, K.J., The ecosystem approach to environmental assessment: moving from theory to practice, *J. Aquat. Ecosyst. Health,* 5:89–97, 1996.

Yaffee, S.L., Phillips, A.F., Frentz, I.C., Hardy, P.W., Maleki, S.W., and Thorpe, B.E., *Ecosystem Management in the United States: An Assessment of Current Experience*, Island Press, Washington, D.C., 1996.

CHAPTER 28

Establishing Specifications for Ecological Indicators for the Prediction of Sustainability

Wayne G. Landis and John F. McLaughlin

INTRODUCTION

In Chapter 22, Robert T. Lackey presented in detail the difficulties in using the term *ecological health* for the management of ecological structures. Although this phrase is pervasive in popular print, radio, and television, the lack of a quantitative description of ecosystem health prevents its use in a scientific process. Lackey also states that currently an alternative model for setting the priorities of managing ecological structures does not exist. Allan K. Fitzsimmons (Chapter 23) details effectively the problems in using ecosystem health to make policy and to manage ecological structures. This chapter provides a further critique of the use of ecosystem health as a metaphor, evaluates current definitions of sustainability, and presents an alternative quantitative model, *The Box*. Finally, we present criteria for the development of indicators to facilitate the management of sustainable ecological structures.

Ecological Health and Integrity

Attempts to define ecological health and integrity have foundered. To date, health and integrity have been described vaguely: these words mean different things to different people, and explicit criteria have not been provided to explain how these terms should be applied to environmental decisions and setting policy. Two definitions of ecosystem health illustrate our point. Haskell et al. (1992) offered the following working definition:

> An ecological system is healthy and free from "distress syndrome" if it is stable and sustainable — that is, if it is active and maintains its organization and autonomy over time and is resilient to stress.

In this definition, *distress syndrome* means "the irreversible processes of system breakdown leading to death" (Costanza, 1998). Karr (1992) provided a more explicit definition:

> A biological system, whether individual or ecological, can be considered healthy when its inherent potential is realized, its condition is stable, its capacity for self-repair when perturbed is preserved, and minimal external support for management is needed.

The utility of these definitions is diminished by controversies over the terms *ecosystem stability* and *resilience*" (Pimm, 1984) as well as the use of terms that elude practical use and measurement. How can one measure ecosystem *autonomy, inherent potential*, or *capacity for self-repair* in ways that would seem objective to stakeholders with divergent values?

We have three main concerns about use of the term *ecosystem health*. First, the term *health* suggests that ecosystems are homeostatic, although they are nonequilibrium systems (De Leo and Levin, 1997).

Second, even if *health* were an appropriate term to describe ecosystem status, it would be related indirectly to the system properties of concern. Management decisions are made relative to specific entities and processes; discussion would be clearer if those properties were addressed directly and explicitly. Even in medicine, where health clearly is an appropriate concept, the term rarely is used in practice to make diagnoses, to determine treatments (for preventative or corrective purposes), or to monitor progress. Instead, direct measures of the properties or functions of concern are used. Similar approaches should be adopted for ecosystems; direct references to properties of concern are even more important to managing the status of ecosystems than they are to managing the status of individuals. Just as vague allusions to individual health ignore progress in medicine this century, reverting to ideas of ecosystem health disregards advances in knowledge about ecosystem function won by ecosystem science.

Our greatest criticism of the term *ecosystem health* is that it has very limited practical application to management. Because health is a metaphorical model, it cannot be measured directly. Advocates of the terminology suggest measuring surrogate variables. Instead, we offer an alternative term that more clearly addresses the same concept, while meeting Brooks's (1992) criteria for an operational definition.

Our alternative defines *sustainability* for ecosystems quantitatively using an *assessment hypervolume*. This approach bases assessments on the variables themselves rather than on surrogates. This involves defining the desired ranges for various ecosystem functions and/or structural characteristics and then determining if a given system falls within or exceeds those ranges. Management goals and actions can be directed explicitly at those variables explicitly rather than at vague descriptions of system health. In this way, assessments about system status relate directly to properties of concern.

SUSTAINABILITY DEFINED

Next we present a contrast to the previous descriptions of health and integrity. These paragraphs set the context for listing specifications for parameters that lead to predictive indicators useful for management decisions.

Assessment Hypervolume (The Box)

Landis et al. (1994) originally defined the assessment hypervolume for risk assessment, and this concept provides the basis of our approach to defining sustainability. Kersting (1988) and Johnson (1988a,b) also proposed multidimensional models for assessing impacts to ecological systems. Here we apply the assessment hypervolume quantitative definition for sustainability.

Ecosystem response variables related to stakeholder values define the axes of an assessment space. These values can be derived from surveys, stakeholder meetings, or established management goals as represented by legislation. Examples of such values are presented in Table 28.1. These values were derived from the Willamette Valley Livability Forum, a group established by the Governor of Oregon with a charge of establishing management goals for the ecological services provided by the Willamette River and its tributaries. The process was driven by consensus for the time up to 2050. The first column lists the goals as defined by

Table 28.1 Subset of Values for the Construction of an Assessment Hyperplane for the Middle Willamette and McKenzie Rivers[a]

Stakeholder Values	Assessment Endpoint
Fisheries	
Summer steelhead	Population meets Oregon Department of Fish and Wildlife basin fisheries plan: maintain a potential sport catch of 250 in the mainstem above Willamette Falls[5]
	Maintain an annual catch of 1200 on the McKenzie River[4]
	Maintain a return of 2400 to the McKenzie subbasin[4]
Native Fish Populations	
River sustains thriving populations of native fish	Populations of spring chinook, rainbow trout, cutthroat trout, and winter steelhead meet ODFW basin fishery plans[1–6]
Spring chinook	Increase production to 100,000 fish entering the Columbia River[1]
	Increase the number of wild spring Chinook to the McKenzie River to 10,000[1]
Fish Organismal Health	Less than 5% of the mountain whitefish (*Prosopium williamsoni*) and large-scale suckers (*Catostomus macrocheilus*) show either external or internal anomalies
Potentially Conflicting Values	
Floodplain management for human health and safety	Flow is controlled to prevent damage to human lives or property in urban areas
Water quantities sustain human communities	Crop irrigation and human consumption
Maintaining reservoirs for fishing, boating, and windsurfing	
No loss of recreation including boating	
Fishing	e.g., Summer steelhead

[a] Note the conflicting values.

References:

[1] Oregon Department of Fish and Wildlife, Spring Chinook chapters, Willamette Basin Fish Management Plan, Portland, OR, 1998.

[2] Oregon Department of Fish and Wildlife, Willamette Basin Fish Management Plan, Willamette River Subbasin, Portland, OR, 1991.

[3] Oregon Department of Fish and Wildlife, Willamette Basin Fish Management Plan, Willamette Mainstem, Portland, OR, 1991.

[4] Oregon Department of Fish and Wildlife, Willamette Basin Fish Management Plan, McKenzie River, Portland, OR, 1991.

[5] Oregon Department of Fish and Wildlife, Willamette River Subbasin Salmon and Steelhead Production Plan, Portland, OR, 1990.

[6] Oregon Department of Fish and Wildlife, Willamette Basin Fish Management Plan, Portland, OR, 1988.

Data collected by Matt M. Luxon.

this group. The second column is the quantitative measure that we used to define this goal. In some areas there are conflicts where two desired goals appear incompatible, but our goal is to be as inclusive as possible. This process provides the limits of our hypervolume, which are tied directly to stakeholder values.

Within this space, the tolerable limits to each ecosystem response define the assessment hypervolume (Figure 28.1a) (Landis et al., 1994). Thus, the boundaries of the assessment hyperplane are social constructs, determined by stakeholder values. The stakeholder values can be derived by expressions of limits set by regulatory agencies, public forums, or scientifically based surveys.

The numerical range of criteria of the stakeholder values for a given system determine its position within the assessment space. One can follow the trajectory of a system relative to stakeholder values by plotting the position of the system through time. All judgments regarding the

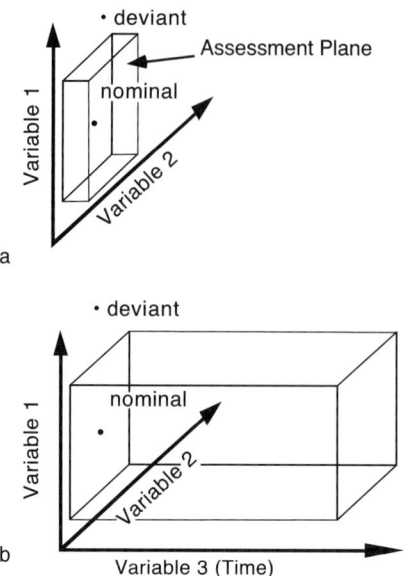

Figure 28.1 Creation of an assessment hypervolume.

success of environmental management or the suitability of management goals should refer to the system trajectory relative to the assessment hypervolume.

The time frame for defining sustainability should be considered as a separate axis (Figure 28.1b). Time horizons need to be specified to make predictions and provide boundaries for management goals. The time horizon must be finite, because no system can be sustainable indefinitely (Chapin et al., 1996; Costanza et al., 1998). Construction of the assessment hypervolume is complete after the time horizon is determined (Figure 28.1b). Then developing operational definitions for sustainability becomes straightforward.

Operational Definitions

Sustainability: Management, development, or use is sustainable if it maintains system responses within the assessment volume during a specified time period. Societal values determine the response variables, their limits, and the time horizon that delimit the assessment volume.

Ecological Nominality: A system is nominal if it is within the assessment hyperpolygon, that is to say, if it is in the Box. (Ecological integrity can be thought of as the state of being within the assessment hyperpolygon.) The system is deviant if it falls outside the hyperpolygon or is out of the box. These states are depicted in Figure 28.2. This definition meets the criteria stated by De Leo and Levin (1997): "Measures of integrity must reflect the ability of ecosystems to maintain services of value to humans" because the assessment hypervolume represents societal values.

Indicators of Sustainability: These are variables that predict whether a system will persist within the assessment volume during the specified time period, whether an initially nominal system will depart from the assessment volume, or whether a deviant system will enter the assessment volume within the assessment time.

These definitions satisfy Brooks's (1992) requirement for an operational definition: the assessment volume provides an objective and measurable set of criteria against which ecosystem responses can be judged. They also suggest how sustainability or its loss can be predicted, subject to the degree that ecosystem stressor–response relationships can be quantified.

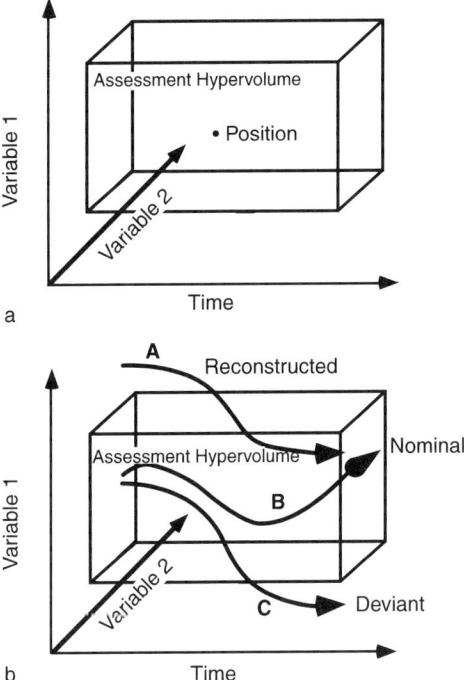

Figure 28.2 Trajectories of a sustainability evaluation.

SPECIFICATIONS FOR INDICATORS OF SUSTAINABILITY

Most monitoring programs are conducted to determine the current status of an ecological system (Usher, 1991). This status, or state, represents a point within the assessment hypervolume (Figure 28.2a). Since trajectories passing through a given point rarely are determined, monitoring programs often ignore system dynamics unless monitoring continues over long periods. Even with long-term monitoring, the focus on current status makes predicting trajectories difficult.

Management for sustainability, however, requires predictions. Three different kinds of predictions are needed, depending on a system's initial position relative to the assessment volume (Figure 28.2b). First, restoration or recovery is predicted (Figure 28.2b, arrow "A") when the system starts outside the assessment volume and the forecast trajectory moves into the volume within the time horizon. The second prediction is sustainability: the system will remain within the assessment volume throughout the specified time (Figure 28.2b, arrow "B"). Anthropogenic or nonanthropogenic factors may affect system dynamics, but these changes do not force the system outside the box. The third prediction is that the system is not sustainable: it will exceed the boundaries of the assessment volume because of widely fluctuating responses or prolonged trends caused by anthropogenic or nonanthropogenic impacts (Figure 28.2b, arrow "C"). Each kind of prediction should be made probabilistically because stochastic factors affect all ecological systems.

The assessment hypervolume approach also has three clear characteristics that are important in making predictions and managing ecological structures. First, the approach is inherently multivariate. Even if the assessment box is reduced to one variable over time, several factors determine the outcome of that parameter. Approaches that do not recognize this inherent property of ecological systems are not accurate models of ecological reality. So-called multimetric approaches do not take into account the sample or intrinsic distribution of ecological parameters. Second, the assessment hypervolume implicitly recognizes the dynamic nature of ecological systems. No assumptions are made about unimpacted control or reference states. Finally, prediction is implicit and central to this strategy. To implement the assessment hypervolume approach successfully, the characteristics of ecological measurements need to be examined.

Measurement of Ecological Parameters for the Prediction of Sustainability

Classical measures of health and sustainability deny the inherent dynamics of ecological systems. Point measures such as species presence or absence, the index of biological integrity and its derivatives, species diversity, and the wide variety of water quality indices (reviewed in Matthews et al., 1998) all fit into this category. Each of these measures may have utility for describing the position of the ecological system within the assessment box, but none is tied implicitly to social or cultural values.

What are the goals of ecological measurements in the context of the assessment hypervolume? The measures need to place the ecological system being managed along the axes and make predictions about the future trajectory. We will now examine the properties of measurement variables and the importance of the stochastic features putting the system in a landscape context. Finally, we introduce the interaction between landscape and the timescale of the prediction.

Measurement Variables

It has been demonstrated that the variables that best differentiate between impacted and non-impacted systems change during the time course of the experiment or field study. This property of variables has been examined in a series of microcosm experiments (Landis et al., 1993a,b; 1994; 1996; 2000; Matthews et al., 1996) and in a field study (Shubert, 1997).

Several clustering methods were used in each report. Cosine and Euclidean metric clustering and nonmetric clustering and association analysis (NCAA) were performed on each sampling date. Patterns were examined to test the hypothesis that the four treatment groups within each experiment could be identified. In other words, could a microcosm treatment replicate be placed properly into its treatment group? In each experiment the variables that were best at correctly identifying the treatment of each replicate changed over the course of the experiment. In back-to-back replicate experiments (Landis et al., 2000), there was little correspondence among important variables on the same dates. Although the algal, daphnid, blue-green, and ostracod dynamics had similar patterns overall, the intrinsic variability of the system precluded an exact correspondence. Of course, microcosms change dramatically over the course of each experiment. At the beginning, systems were dominated by green algae and the daphnids. At the end of these experiments, ostracods were the most plentiful invertebrate and the systems had changed to a detritus economy.

Shubert found a similar variability in several plots in Prince William Sound following the 1989 oil spill. Some sites were oiled, some oiled and steam cleaned, and other sites were never oiled. The best variables for determining treatment group using NCAA changed from year to year.

Upon reflection, these findings are not surprising. Each system, whether microcosm or Prince William Sound, changes dramatically over time. When responding to a disturbance, ecosystem change can occur at an even more rapid rate. Colonization and growth of new organisms have a large stochastic component that has an important role in determining the composition of an ecological community.

These microcosm experiments and the field research demonstrate the persistence of effects. In each case the treatment group of the replicate or study site was correctly identified throughout the duration of the studies. This characteristic has been labeled *community conditioning* (Matthews et al., 1996; Landis et al., 1996) and is a fundamental characteristic of ecological systems. Community conditioning recognizes that ecological systems are historical and complex structures. Information about past events remains within the system for extended periods and perhaps for the temporal extent of the system. Complex systems also have the property of being nonequilibrium systems. Of course, ecological systems have stochastic features and exist within a spatial context.

Stochastic and Deterministic Variability as a Predictor

Variation, due to both deterministic and stochastic processes, is important in determining the future trajectory of an ecological system. The importance of understanding variation can be illustrated using a simple difference equation:

$$Nt + 1 = N[1 + r(1 - N/K)],$$

where

$$N = \text{ population size at time } t$$
$$Nt + 1 = \text{ population size at the next time interval}$$
$$K = \text{ carrying capacity of the environment}$$
$$r = \text{ intrinsic rate of increase over the time interval}$$

In our simulations K is allowed to vary as a random number selected between the specified boundaries. The rate of increase is 2.1, which does not put the system in a chaotic domain but is in a region where a simple bifurcation in the dynamics occurs when the carrying capacity is held constant. The output from such simulations is presented in Figure 28.3. In each instance the figure portrays the outcome of five replicate simulations. The simulation portrayed in the top of the figure is the result of letting K vary from 8000 to 10,000. All five populations remain extant and exist between 11,000 and 3000 as the upper and lower bounds. The lower panel represents a simulation with K varying from 7000 to 10,000. Extinction of all five populations occurs before iteration (time interval) 70. Different distributions of the variation within K will give different probabilities of extinction over the period of the simulation.

In summary, temporal variability is an important measure if prediction is a goal. Variability can be measured and analyzed and forms an important part of a suite of predictive indicators. In the case of ecological systems there is an important spatial component to consider.

Spatial and Temporal Considerations in Landscapes

There is an extensive literature, mostly theoretical, on the distribution of organisms within a landscape. McLaughlin and Landis (2000) have reviewed this literature with regard to the effects of contaminants and other disturbances upon populations. We would like to summarize some of the critical features of how stressors affect spatially dispersed populations.

One of the critical features of spatially dispersed populations is that an impact upon one patch can affect sites with no direct exposure to the stressor. This feature of dispersed populations, "action at a distance" (Spromberg et al., 1998), has been demonstrated by computer simulations and experimentally by L. Macovsky (1999). This research and further investigation (Landis and McLaughlin, 2000; McLaughlin and Landis, 2000) have demonstrated that the impacts of a disturbance, natural or anthropogenic, can be affected by a variety of spatial factors. The size and distribution of the population patches, the relative arrangement of patches to the stressor, and the migration rates between patches affect the magnitude and type of effects upon a dispersed population. The magnitude of indirect effects generally decreases with distance. Because stressor exposure and organismal dispersal have large stochastic components, in certain regions of the initial state space, several distinct outcomes can occur within the population patches. Probabilities can be assigned to achieving carrying capacity or becoming extinct.

One of the other critical features of spatially dispersed populations is the destruction of the reference site construct. Since the effects of a stressor can be transferred throughout a landscape by migration, no site is unaffected. Once the distance is so great that an effect does not occur, then other historical and geological factors will also be different.

The dynamics of the impacts are therefore complicated by spatial structure. Contaminant effects can be transmitted to populations not exposed. Blockage of dispersal pathways can cause extinction

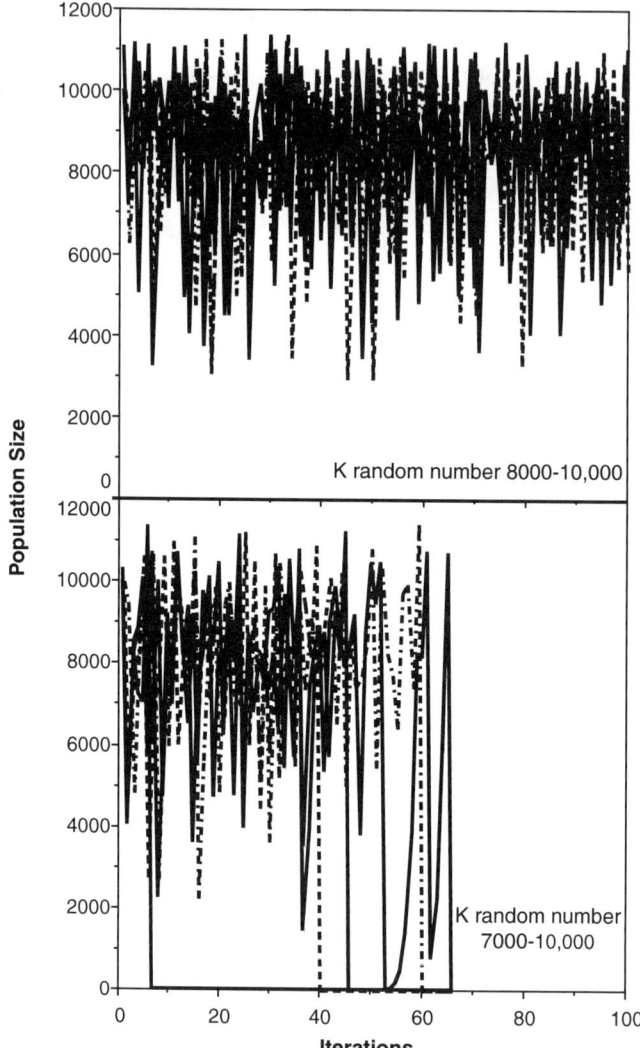

Figure 28.3 The importance of variability in the prediction of dynamics. Five replicate simulations are plotted in each graph. A change in the variability of the carrying capacity (K) results in the extinction of the model population.

of a population within a patch by not allowing migration from surrounding patches to reestablish the population. A mere census of populations within patches cannot reflect the dynamics of spatially structured systems. Populations in all patches may be similar, but one may be increasing due to migration while another is decreasing as outmigration rescues an affected population. Only an understanding of the stochastic and deterministic features of a spatially structured system will allow accurate assessment and prediction.

Other features of a landscape are also critical in understanding the potential impacts of stressor event in an ecological structure. We will use a section of the Willamette and McKenzie Rivers in central western Oregon to illustrate these features. In a regional risk assessment conducted by Matt Luxon and W. Landis, the area has been broken down into specific risk regions depending on characteristics such as the type of stressors, land use, morphology of the rivers, and urbanization. The risk assessment focuses upon impacts to the main channel due to industrialization, effluents, dams, and land use within the study area. One of our findings to date is the non-Euclidean nature of distances within the watershed to the main channel. In several instances potential stressors that

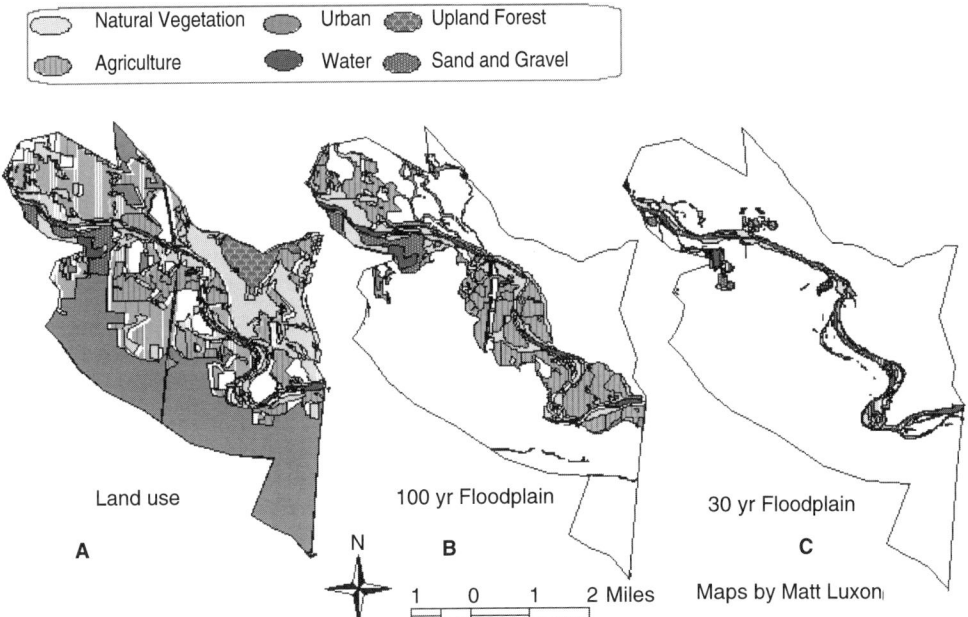

Figure 28.4 The effects of temporal specifications upon the area to be considered in the sustainability of an aquatic resource. The land use types to be considered for a direct impact on the mainstream of the river depends on the specified time frame of the assessment.

are nearly adjacent within the watershed have impacts in the main channel kilometers apart. This contradiction is because the potential sources of stressors are along different tributaries, and these tributaries feed into the main channel at very different sites.

The second critical feature is the importance of time in determining what part of a landscape may have an impact on the main channel of a river. Figure 28.4 illustrates the importance of time frame in assessing impacts. Figure 28.4A depicts the land use types along one reach of the watershed of the Willamette as it passes through Eugene–Springfield, Oregon. However, in judging the risks to the river posed by that landscape because of flooding, a different picture emerges. If the time frame is 100 years (Figure 28.4B), about half of the region will be flooded, with very little of the urban area included. If the time frame is 30 years, only those lands adjacent to the main channel will be flooded (Figure 28.4C). The number of land uses in this area is similarly limited. The time horizon of an assessment must be specified if the outcome is to be accurate and useful.

In review, the measures that best determine the current and future state of an ecological structure change over time. Stochastic and deterministic variability are important features to quantify if prediction is a goal. The distribution within and utilization of the landscape by populations has to be understood to predict the dynamics of change from a stressor event. Finally, the temporal extent of the assessment has to be specified. Each of the measures needs to be incorporated into an assessment of sustainability and with it a recognition of the temporal and spatial dynamics of ecological systems. Unlike a *health* approach, there is an explicit recognition of these features in a dynamic nonequilibrium system.

In summary, there are five specific design requirements for developing a useful set of indicators regardless of the ecological system.

1. The indicators must be placed in the context of the assessment hypervolume and tied to cultural values of the stakeholders.
2. Because projections cause the loss of information, a variety of methods with different assumptions about the nature of the system must be used.

3. The best attributes to measure alteration of the system change over time. To use just one set may miss the dynamic complexity of the system and subsequent impacts.
4. Ecological systems are both deterministic and stochastic. Each property is important to represent in an evaluative or predictive context.
5. Spatial context must be known and use of the landscape by the biota understood.

Comparisons and Damnations

In the introduction to Section I.4, Lackey described the failings of the *ecosystem health* approach to assessing and managing ecological systems. He lamented that no alternative was currently available. The assessment hypervolume is a workable alternative.

In contrast to the ecosystem health model, the assessment hypervolume approach clearly ties stakeholder values to the dynamic processes of ecological structures. The goals of the assessment and ecological management process are specified and quantified in the formulation of the assessment hypervolume. Reference sites are not required. The assessment hypervolume explicitly recognizes the dynamic nature of ecological systems and embraces measurement of dynamics. Recognizing that ecological systems and their management are multivariate discourages the use of indices to estimate risks of impacts. None of these features is an aspect of the ecosystem health model. Also, note that classic features of ecosystem health such as integrity, recovery, resilience, or unimpacted reference are not required by the assessment hypervolume approach.

Furthermore, the assessment hypervolume approach works. We have used the assessment hypervolume and the relative risk model tools successfully in assessing risks in the fjord of Port Valdez (Wiegers et al., 1998; Landis and Wiegers, 1997) and sections of the Willamette and the McKenzie Rivers in Oregon. We have started a similar assessment for Codorus Creek in Pennsylvania. Other research groups have assessments under way in New Zealand (G. Lewis,), Australia (Walker et al., 2001), and a Brazilian rain forest (Moraes et al., in press).

Ecosystem health and related ideas of integrity have been long recognized as constructs with severe problems (Suter, 1993; Lackey, 1996). The idea of ecosystem health is not necessary in the performance of assessments and management plans. In fact, ecosystem health is a misleading, unobservable, and inaccurate model of ecological systems. However, its use as witnessed by this symposium persists.

An argument often employed for using health, integrity, and values such as "good" in describing ecological systems is that these are terms the public readily understands; they can therefore be used to advance the cause of environmental stewardship. Kapustka and Landis (1998) published a response to this practice:

> We contend that this is wrong on both ethical and practical grounds. The ethical component is that misrepresenting scientific issues undermines the foundation of the scientific process. In science the experiment defines reality, not public opinion or fashionable morality. Scientists have an obligation to present reality as best they know it, independent of funding source, home institution, or other cultural constraints.

The outcomes of the assessments and management of ecological systems have deep personal, economic, and social impacts. In our view, this requires us to use the best predictive methodologies available and to communicate effectively with the public. The use of misleading or incomplete descriptions reduces our credibility and the importance of the field. Given the obvious problems with the use of ecosystem health as a model, we must abandon it.

ACKNOWLEDGMENTS

We would like to thank Matt Luxon for the preparation of the maps of the Willamette and for the compilation of stakeholder values for the Willamette Valley. This project was supported in part by a grant from the National Council for Air and Stream Improvement (NCASI).

REFERENCES

Brooks, H., Sustainability and technology, in *Science and Sustainability: Selected Papers on IIASA's 20th Anniversary*, International Institute for Applied Systems Analysis, Laxenburg, Austria, 1992, pp. 29–60.

Chapin, F.S., Torn, M.S., and Tateno, M., Principles of ecosystem sustainability, *Am. Nat.*, 148:1016–1037, 1996.

Costanza, R., Mageau, M., Norton, B., and Patten, B.C., What is sustainability? in Rapport, D. et al., Eds., *Ecosystem Health*, Blackwell Science, Malden, MA, 1998, pp. 231–250.

De Leo, R., Toward an operational definition of ecosystem health, in Costanza, R., Norton, B.G., and Haskell, B.D., Eds., *Ecosystem Health: New Goals for Environmental Management*, Island Press, Washington, D.C., 1992, pp. 239–256.

De Leo, G.A. and Levin, S., The multifaceted aspects of ecosystem integrity, *Conserv. Ecol.*, 1(1):3, 1997. Available online at http://www.consecol.org/vol1/iss1/art3.

Fitzsimmons, A.K., Ecosystem health: a flawed basis for federal regulation and land use management, *Proc. Int. Congr. Ecosystem Health*, Sacramento CA, in review.

Haskell, B.D., Norton, B.G., and Costanza, R., Introduction: What is ecosystem health and why should we worry about it? in Costanza, R., Norton, B.G., and Haskell, B.D., Eds., *Ecosystem Health: New Goals for Environmental Management*, Island Press, Washington, D.C., 1992, pp. 3–20.

Johnson, A.R., Evaluating ecosystem response to toxicant stress: a state space approach, in Adams, W.J., Chapman, G.A., and Landis, W.G., Eds., *Aquatic Toxicology and Hazard Assessment: 10th Volume*, ASTM STP 971, American Society for Testing and Materials, Philadelphia, PA, 1988a, pp. 275–285.

Johnson, A.R., Diagnostic variables as predictors of ecological risk, *Environ. Manage.*, 12:515–523, 1988b.

Kapustka, L.A. and Landis, W.G., Ecology: the science versus the myth, *Hum. Ecol. Risk Assess.*, 4:829–838, 1998.

Karr, J.R., Ecological integrity: protecting Earth's life support systems, in Costanza, R., Norton, B.G., and Haskell, B.D., Eds., *Ecosystem Health: New Goals for Environmental Management*, Island Press, Washington, D.C., 1992, pp. 223–238.

Kersting, K., Normalized ecosystem strain in micro-ecosystems using different sets of state variables, *Verh. Int. Verein. Limnol*, 23:1641–1646, 1988.

Lackey, R.T., Pacific salmon, ecological health, and public policy, *Ecosyst. Health,* 2:61–68, 1996.

Lackey, R.T., Appropriate use of ecosystem health and normative science in ecological policy, *Proc. Int. Congr. Ecosystem Health*, Sacramento, CA, in review.

Landis, W.G. and McLaughlin, J.F., Design criteria and derivation of indicators for ecological position, direction and risk, *Environ. Toxicol. Chem.*, 19:1059–1065, 2000.

Landis, W.G. and Wiegers, J.A., Design considerations and a suggested approach for regional and comparative ecological risk assessment, *Hum. Ecol. Risk Assess.*, 3:287–297, 1997.

Landis, W.G., Markiewicz, A.J., Matthews, R.A., and Matthews, G.B., A test of the community conditioning hypothesis: persistence of effects in model ecological structures dosed with the jet fuel, JP-8, *Environ. Toxicol. Chem.*, 19:327–336, 2000.

Landis, W.G., Matthews, R.A., Markiewicz, A.J., Shough, N.A., and Matthews, G.B., Multivariate analyses of the impacts of the turbine fuel Jet-A using a microcosm toxicity test, J. *Environ. Sci.*, 2:113–130, 1993a.

Landis, W.G., Matthews, R.A., Markiewicz, A.J., and Matthews, G.B., Multivariate analysis of the impacts of the turbine fuel JP-4 in a microcosm toxicity test with implications for the evaluation of ecosystem dynamics and risk assessment, *Ecotoxicology,* 2:271–300, 1993b.

Landis, W.G., Matthews, G.B., Matthews, R.A., and Sergeant, A., Application of multivariate techniques to endpoint determination, selection and evaluation in ecological risk assessment, *Environ. Toxicol. Chem.*, 12:1917–1927, 1994.

Landis, W.G., Matthews, R.A., and Matthews, G.B., The layered and historical nature of ecological systems and the risk assessment of pesticides, *Environ. Toxicol. Chem.*, 15:432–440, 1996.

Macovsky, L.M., The effects of toxicant related mortality upon metapopulation dynamics: a laboratory model, M.S. thesis, Western Washington University, Bellingham, 1998.

Matthews, R.A., Landis, W.G., and Matthews, G.B., Community conditioning: an ecological approach to environmental toxicology, *Environ. Toxicol. Chem.*, 15:597–603, 1996.

Matthews, R.A., Matthews, G.B., and Landis, W.G., Application of community level toxicity testing to environmental risk assessment, in Newman, M.N. and Strojan, C.L., Eds., *Risk Assessment: Logic and Measurement,* Ann Arbor Press, Chelsea, MI, 1998, pp. 225–253.

McLaughlin, J.F. and Landis, W.G., Effects of environmental contaminants in spatially structured environments, in *Environmental Contaminants in Terrestrial Vertebrates: Effects on Populations, Communities, and Ecosystems,* Albers, P.H. et al., Eds., Society of Environmental Toxicology and Chemistry, Pensacola, FL, 2000.

Moraes, R., Landis, W.G., and Molander, S., Regional risk assessment of a Brazilian rain forest reserve, *Hum. Ecol. Risk Assess.,* in press.

Pimm, S.L., The complexity and stability of ecosystems, *Nature,* 307:321–326, 1984.

Shubert, K.F., A non-metric, non-parametric analysis of the low intertidal epibiota fraction of Prince William Sound data collected during the post-*Exxon Valdez* oil spill assessment, master's thesis, Western Washington University, Bellingham, WA, 1997.

Spromberg, J.A., Johns, B.M., and Landis, W.G., Metapopulation dynamics: indirect effects and multiple discrete outcomes in ecological risk assessment, *Environ. Toxicol. Chem.,* 17:1640–1649, 1998.

Suter, G.W. II, A critique of ecosystem health concepts and indexes, *Environ. Toxicol. Chem.,* 12:1521–1531, 1993.

Usher, M.B., Scientific requirements of a monitoring program, in Goldsmith, F.B., Ed., *Monitoring for Conservation and Ecology,* Chapman & Hall, New York, 1991, pp. 15–32.

Walker, R., Landis, W.G., and Brown, P., Developing a regional ecological risk assessment: a case study of a Tasmanian agricultural catchment, *Hum. Ecol. Risk Assess.,* 7:417–439, 2001.

Wiegers, K., Feder, H.M., Mortensen, L.S., Shaw, D.G., Wilson, V.J., and Landis, W.G., A regional multiple stressor rank-based ecological risk assessment for the fjord of Port Valdez, AK, *Hum. Ecol. Risk Assess.,* 4:1125–1173, 1998.

Measuring the Impact of Ecological Disintegrity on Human Health: A Role for Epidemiology

Colin L. Soskolne

INTRODUCTION

For several decades, ecologists and biologists have been warning of the extinction of various species, of declines in both the quality and extent of ecological capital, and of general and specific environmental degradation. More recently, climate change as a consequence of global warming has been added to the list of environmental concerns worldwide. The concerns arise from the view that global ecological integrity — simply interpreted as the ability of nature's life-support systems to withstand perturbations and continue to provide their usual life-sustaining services — is a first-order principle for the sustainability of life on Earth (Soskolne and Bertollini, 1999). Two texts in the early 1990s served to bring these concerns to epidemiologists' attention (World Health Organization, 1992; McMichael, 1993).

Global ecological disintegrity, the converse of *global ecological integrity*, is a function of three factors: over-consumption, population growth, and the inappropriate applications or misuses/abuses of technology. All three of these factors are, in fact, human-induced and are therefore potentially controllable. Thus, human influence in perpetuating the current course of environmental degradation is at the core of this chapter. It is posited that if it could be demonstrated that current socioeconomic policies indeed are associated with negative human health impacts, then alternative policies likely would be introduced that could serve to steer humanity from a course of nonsustainability.

In essence, environmental degradation is impacting those ecological systems upon which human beings depend for their well-being and their survival, including the life-support systems of air, water, soil, forests, fisheries, and climate. If life-support systems indeed are collapsing, the need for any of the sciences to bring the seriousness of the problem to the attention of those responsible (i.e., humans) for causing the negative environmental impacts is paramount. If inherently anthropocentric human beings could be shown scientific evidence that escalating environmental degradation is having, or could have, negative human health consequences of epidemic proportions, then those sciences could play a critical public health role by impacting policies to avert potential disasters from occurring. Thus, this chapter is based on the belief that if negative impacts on human health and well-being could be associated with environmental degradation, then adequate justification would be at the disposal of policy makers to change the course of human development toward one of sustainability. Sustainable development has been defined as development that meets the needs of the present without compromising the ability of future generations to meet their own needs (United Nations World Commission on Environment and Development, 1987).

Because epidemiology is the applied science basic to public health policy formulation, it is proposed that this discipline has a major role to play in bridging those concerns noted by the ecologists, biologists, and climatologists with the appropriate endpoints of human health and well-being. The definition of epidemiology (Last, 2001) is: "The study of the distribution and determinants of health-related states or events in specified populations, and the application of this study to control health problems." As such, epidemiology is the science that is basic to public health. Of all the sciences, epidemiology is the applied science that provides the rational basis for policy formulation. It is the science that informs the policy process. Its goal is to protect the public's health through preventing disease, disability, and premature mortality.

If catastrophic events of epidemic proportions are even potentially preventable through reducing the risk of environmental collapses, then epidemiologists would have a significant role to play in bridging the ecological and biological assessments to human health. Epidemiologists thus must join the ranks of the ecologists and biologists in making recommendations to prevent such collapses. In this way, the public's attention would be directly focused on human concerns related to their health and well-being and, hence, with a potentially more immediate and urgent effect on policy. After all, it is one thing to be concerned about the extinction of flora and even fauna; but the motivation for the present work lies in the public's attention being more directly gained if human life itself is recognized as being demonstrably at risk.

EPIDEMIOLOGIC METHODS IN POPULATION RESEARCH

Epidemiologic researchers have at their disposal a broad array of methods for determining the distribution and determinants of diseases in populations. The methods range from experimental (i.e., randomized controlled trials) to observational studies that fall into the class of either *descriptive studies* (e.g., disease patterns, or the distribution of diseases over time) or *analytical studies* (e.g., etiologic studies, or studies of disease determinants). These include several designs: ecological/correlational studies; cross-sectional studies; case-control studies; and cohort studies (Beaglehole, et al., 1993). An appropriate study design must be selected to best address the health question of concern; the choice will depend on the available resources and on the existing body of knowledge to justify one design over another. The question of concern in traditional etiologic research is one that aims to link a proximate exposure to the development of a particular disease. While this falls into the broad rubric of applied science, its analogy lies in basic science with its reductionist approach of relating a specific cause to a particular outcome.

The question of a health concern is one of applied science, and it usually finds its way to the desk of the epidemiologist. These concerns are identified not so much from the basic sciences through, say, the conduct of toxicological experimentation that then must be tested in human studies, but often through anecdotal evidence that is brought to the epidemiologist's attention. Anecdotal evidence could come to the epidemiologist in the form of an astute observation by a layperson who might observe that a particular disease or type of death is occurring among the person's peer group with unusual frequency. Perhaps it will be presented by an astute professional practitioner who will report in the scientific literature a cluster of events/cases, or a series of events/cases that have come to light through a narrow time frame. Such anecdotal evidence has led to epidemiologists entering the scene to scientifically quantify the problem suggested by the anecdotal reports. Thus, in the more traditional reductionist paradigm and the one that has led to many worthwhile scientific discoveries, anecdotal evidence often has provided the justification to invoke the scientific method.

If environmental degradation were to be negatively impacting human health, then epidemiology, in order to fulfill its traditional role, should be able to demonstrate that, over time, as the environment is degraded, human health and well-being decline (or suffer some loss). The latter would usually be measured in terms of increasing rates of disease, disability, and death. If such a pattern were discernible epidemiologically, then no threshold effect would be operating. On the other hand, if such trends could not be demonstrated, then a threshold effect is the most likely scenario. If epidemiologists were to wait for threshold limits to be exceeded, then their preventive role would be nil; under the threshold model, health indicators would remain stable and even show signs of improvement. But these improvements would continue only until such time as the life-support systems failed, followed by rapidly acting processes with dramatic and massive impacts on human health and well-being. For epidemiology to play a useful preventive role, access to appropriate indicators of a sensitive and specific enough nature will be needed to maximize the detection of health trends well in advance of allowing declines to approach threshold-bearing levels.

UPSTREAM DETERMINANTS OF HEALTH AND WELL-BEING

Because disease prevention is at the core of maintaining public health, there have been appeals to epidemiologists to focus more on the so-called upstream (or distant) determinants of public health (Pearce, 1996; Susser and Susser, 1996; Shy, 1997; Susser, 1998; Porta and Alvarez-Dardet, 1998; Pearce and McKinlay, 1998; Pearce, 1999). This would require that epidemiologists make a paradigm shift in their conception of both the models and methods used in their research. That is, epidemiologists would need to move from reductionist approaches to more systemic or holistic approaches in their research. Accomplishing this shift will require transdisciplinary approaches to health research.

Transdisciplinary approaches to human health are approaches that integrate the natural, social, and health sciences in a humanities context and, in so doing, transcend each of their traditional boundaries. Emergent concepts and methods are the hallmark of the transdisciplinary effort. This approach will be essential if models are to take variables related to overconsumption, population growth, and abuses of technology into account while simultaneously controlling for the potential confounding effects of economic development as measured by, say, gross domestic product (GDP). Epidemiology is a discipline that is accustomed to inter- and multidisciplinary collaboration. The challenge will be to move into the realm of transdisciplinarity.

All of this is not to suggest that funding should be shifted from reductionism to holism; it is rather to say that some funding should be made available to stimulate systemic (or holistic) research. To help make this shift, training programs and continuing education courses will be needed to equip people with skills in complex systems analysis. Understanding the distinction between poverty and wealth and between growth and development will be needed. Integrating moral philosophy into the recommendations arising from epidemiologic research and directed at informing policy will be needed for understanding the relevance of the Precautionary Principle and future generational issues. Skills will be needed for conducting integrative risk assessments that are scenario based. The latter may require working closely with experts in the area of futures studies. Finally, understanding mass behaviors will be critical for designing interventions to promote paradigm shifts in social policy.

PAYING ATTENTION TO ANECDOTAL EVIDENCE

Many of the indicators from research to date point to dramatic declines in numerous life-support systems. Indeed, indicators of environmental decline, independently derived, show as much as a 50% reduction in biodiversity associated with selected living systems over the 15-year period 1982 to 1997 (Soskolne and Bertollini, 1999). While there is an inherent level of uncertainty associated with any scientific assessment or projection, there are some assessments for which minimal uncertainty exists.

The uncertainty associated with climate change, the associated extreme weather events, and the reemergence of some infectious diseases are a reality. The rate of species extinction continues to escalate while biological diversity declines. Fisheries are collapsing and fish stocks are depleting worldwide. Existing freshwater scarcity and increased desertification make the issue of access to fresh water the most serious threat to the security of the planet in the next 10 years. Nonrenewable resources continue to be consumed at increasing rates. The number of ecorefugees exceeded the number of war refugees in 1998, for the first time in recorded history. All indicators, independently derived, point to declines in ecological integrity.

World population continues to grow, having exceeded the 6 billion mark in October 1999. For the first time in human history, some 50% of the world's population now resides in cities. Threats to safety abound through the inappropriate uses of technology. For example, bioterrorism poses ever-greater threats to human security, while nuclear reactors and waste disposal present growing problems of pollution. The transportation of hazardous products and wastes results in accidental spills (e.g., oil, gas blowouts). Transboundary movement of pollutants and the movement of species high on the food chain have resulted in contamination of every part of the planet by some aspect of human activity.

If classical public health epidemiology has successfully paid attention to anecdotal evidence to justify studies that have led to preventive health programs, then, in the presence of the evidence pointing to declines in ecological integrity, surely epidemiology should explore ways to address the issues presented in this book in order to play its part in averting potential catastrophe.

AN INITIAL ATTEMPT TO MEASURE HUMAN HEALTH IMPACTS

As noted earlier, epidemiology is as much an exploratory, descriptive science as one that contributes to the establishment of causality. An exploratory, cross-sectional aggregate data study at the international level has been undertaken (Sieswerda, 1999). In this study, we examined the relationship between some available measures of ecological (dis)integrity and some selected general health indicators using aggregate data from 203 countries. Measures of ecological disintegrity served as *exposure* in the epidemiologic sense; our variables were chosen according to their plausibility as measures of *intactness* or *wildness* of ecosystems. The lack of an integrated ecological model means that we had no objective criteria for the validity of these choices.

Drawing on our experience with the definition of ecological integrity from the Global Ecological Integrity Project (noted in the acknowledgments), we selected the following variables as indicators of ecological disintegrity:

- Percent of land highly disturbed by human activity
- Threatened species (%) (total for mammals, birds, higher plants, reptiles, and amphibians)
- Partially protected areas (International Union for the Conservation of Nature, IUCN, categories IV to V) as a percentage of total area
- Totally protected areas (IUCN categories I to III) as a percentage of total area
- Forest remaining since preagricultural times (%)
- Average annual deforestation (%)

We modeled these indicators of ecological disintegrity against three health outcomes: life expectancy at birth, infant mortality rate, and percent of low birth weight babies.

In addition, to ensure that any potential relationships among the exposure variables and the health outcomes were not attributable to socioeconomic factors, our analyses controlled for several socioeconomic variables:

- Carbon dioxide emissions per capita (a surrogate for industrial activity)
- Percent urbanization

- Population density
- Gross Domestic Product (GDP) per capita (adjusted for cost of living)
- Gini Index coefficient (a measure of income distribution inequality)
- Adult male literacy

Despite the shortcomings of having to deal with population-based averages, variable data quality, uncertain causal mechanisms, and the necessity of using surrogates of ecological disintegrity instead of direct measurements, we found a few associations between some of our ecological disintegrity variables and the three human health outcomes that we considered. In the extremely narrow time frame of this cross-sectional study, the conversion of land to permanent human use has a small but stable association with improving health outcomes. Deforestation appears to have an unstable association with improving health. Neither biodiversity nor land protection had any association with our health indicators. Industrialization was important, but its association varied depending on whether countries were low, middle, or high income. In low- and middle-income countries, industrialization appears to be associated with improvements in the health indicators examined. In high-income countries, further industrialization appears to be associated with negative health impacts.

The main result of this study may provide some answers as to why the environment is often a secondary consideration for policy makers. This study showed that ecological disintegrity appears to be disconnected from human health in that, based on the available metrics, countries that tend to be ecologically impoverished tend to perform best on the health indicators that we used. There may exist a trade-off between improving conditions for human life and depleting or destroying the environment. So far, that trade-off has favored continued development at the expense of the environment. Thus, this study could not demonstrate directly that human health is at risk from ecological degradation. Indirectly, however, it reveals much information that is, at least, strongly suggestive.

In many ways, these findings should not come as a surprise. The interpretation given has arisen based on evidence of our species' ability to survive, and even thrive, in "islands" of ecological disintegrity, such as major cities. This perpetuates the view that humans can survive and achieve high levels of public health, independent of nature. In reality, this level of health is maintained by drawing on healthy or productive ecosystems elsewhere. In other words, trade and technology serve to distance human life from its very life source. Here lies the disconnect. Future epidemiologic studies of this issue will necessarily need to incorporate not only improved methods, data, and an integrated ecological model for human and public health; they must also recognize the realities of global technology and trade in masking the true connectedness of human health and ecological integrity.

The models developed for this study were based on past events and are inadequate for predicting catastrophic events. With the help of an integrated ecological model, our model could perhaps be refined with additional variables. Results from this initial foray into studying these questions should therefore be interpreted with great caution (Soskolne et al., 2000; Sieswerda et al., 2001).

QUESTIONS THAT EMERGE

We have conducted (Sieswerda, 1999) a correlational analysis examining associations between broad measures of ecosystem health and the current health status of human populations involving 203 countries around the world. This study revealed that, likely owing to world trade, the health of human populations (as measured by life expectancy, percent low birth weight, and infant mortality) improves as ecological capital is lost. Two serious questions emerge from such a finding:

1. How long can the health of human societies be sustained while continuing to draw down ecological capital (i.e., exploit nonrenewable resources) from other regions of the world?
2. Are the measures used from among those available the most sensitive and specific to assess human health impacts from declines in ecological integrity?

RECOMMENDATIONS: A STRATEGIC FRAMEWORK FOR PUBLIC HEALTH RESEARCH AND PRACTICE

Government offices tend to plan on the basis of 4- to 5-year horizons, or until the next election. Yet, when considering ecological degradation from human activity and the potential to measure these impacts on human health and well-being, longer time periods are needed. A list of research priorities is proposed. The time frame for these proposed priorities is longer than a 5-year horizon, starting immediately, but extending to a more appropriate long-term horizon of, say, 30 to 60 years. The framework that follows is based on a 10-year time frame.

- Provide incentives for gathering more directly relevant data/variables from which to derive indicators for modeling earlier health impacts (e.g., mental health, endocrine disruption, indicators of personal and societal well-being). Ecoregions (as opposed to geopolitical boundaries) will provide needed information against which to make meaningful comparisons of rates of disease and well-being associated with ecosystem declines, both within and across countries. Distinguish between sensitive and specific indicators of global changes and measurable human health impacts. The routine monitoring, surveillance, and assessment of global change-related variables is needed. Full participation in global observation systems' technologies is encouraged.
- Provide incentives for the development of methods capable of distilling the complex array of multidisciplinary inputs about ecosystem degradation into a risk assessment paradigm that better helps in guiding policy. Standardized metrics are essential for integrating both current and needed data into models to assess the likelihood of collapses in life-support systems and their consequences for human health and well-being. Classical risk assessment is inadequate for handling more than one variable at a time. New methods would need to be scenario based.
- Educate professions in the adaptation (i.e., adjustment strategies and approaches) and mitigation (i.e., preventative) measures needed for future projected environmental scenarios in the presence of global change. In particular, global change warrants public health involvement. Incentives for collaboration with others globally that have considered these matters would be an advantage.
- Enhance infrastructure to cope with demands under the more likely future scenarios consequent to environmental degradation. For instance, emergency and disaster relief services will need to be strengthened.
- In Canada, infrastructure includes organizations such as the Canadian Association of Physicians for the Environment (note their work to date on "Implications for Human Health of Biodiversity") and nongovernmental organizations such as the David Suzuki Foundation. Public discourse will facilitate needed shifts in policy that have yet to be considered. Effective engagement (including communication) of all stakeholders will be essential to effect needed shifts in policy.
- Build local, regional, national, and international partnerships for addressing both mitigation and adaptation measures.
- Integrate the *Precautionary Principle* into economic, social, legal, and other thinking.
- Revisit this framework continually, especially as more data become available.

ACKNOWLEDGMENTS

An international pilot workshop took place at the European Centre for Environment and Health, Rome Division of the World Health Organization, December 3–4, 1998. This event arose from the findings and collaborative efforts of a 3-year Social Sciences and Humanities Research Council of Canada Grant 806-96-0004, entitled: "Global Ecological Integrity: The Relation between the Wild, Health, Sustainability, and Ethics" (1996–1999), with Professor Laura Westra as the principal investigator. This grant included several co-investigators from various disciplines, and I was the epidemiologist on the research team. The product of the workshop is a 74-page discussion document (Soskolne and Bertollini, 1999), the most downloaded document on the WHO Euro Web site to date. In this chapter, I have distilled the conclusions of the workshop from the epidemiological and public health perspectives. These conclusions were reached with the extensive inputs of the workshop participants and, in particular, other co-investigators on the original grant, including William Rees and Orie Loucks. James Karr, and Anthony McMichael also provided substantial input. The content of this chapter flows from the referenced discussion document (Soskolne and Bertollini, 1999).

REFERENCES

Beaglehole, R., Bonita, R., and Kjellstrom, T., *Basic Epidemiology,* World Health Organization, Geneva, 1993, Chapter 3.

Last, J.M., Ed., *A Dictionary of Epidemiology,* 4th ed., Oxford University Press, New York, 2001.

McMichael, A.J., *Planetary Overload: Global Environmental Change and the Health of the Human Species,* Cambridge University Press, Cambridge, U.K., 1993.

Pearce, N., Traditional epidemiology, modern epidemiology, and public health, *Am. J. Public Health,* 86:678–683, 1996.

Pearce, N., Epidemiology as a population science, *Int. J. Epidemiol.,* 28:S1015–S1018, 1999.

Pearce, N. and McKinlay, J., Back to the future in epidemiology and public health, *J. Clin. Epidemiol.,* 51:643–646, 1998.

Porta, M. and Alvarez-Dardet, C., Epidemiology: bridges over (and across) roaring levels (editorial), *J. Epidemiol. Community Health,* 52:605, 1998.

Shy, C.M., The failure of academic epidemiology: witness for the prosecution, *Am. J. Epidemiol.,* 145:479–484; discussion 485–487, 1997.

Sieswerda, L.E., Towards Measuring the Impact of Ecological Disintegrity on Human Health, master's thesis, University of Alberta, Canada, 1999.

Sieswerda, L.E., Soskolne, C.L., Newman, S.C., Schopflocher, D., and Smoyer, K.E., Toward measuring the impact of ecological disintegrity on human health, *Epidemiology,* 12:28–32, 2001.

Soskolne, C.L. and Bertollini, R., Global Ecological Integrity and "Sustainable Development": Cornerstones of Public Health — A Discussion Document, 1999, available online at http://www.euro.who.int/document/gch/ecorep5.pdf.

Soskolne, C.L., Sieswerda, L.E., and Scott, H.M., Epidemiologic methods for assessing the health impact of diminishing ecological integrity, in Pimentel, D., Westra, L., and Noss, R.F., Eds., *Ecological Integrity: Integrating Environment, Conservation, and Health,* Island Press, Covelo, CA, 2000, pp. 261–277.

Susser, M., Does risk factor epidemiology put epidemiology at risk? peering into the future, *J. Epidemiol. Community Health,* 52:608–611, 1998.

Susser, M. and Susser, E., Choosing a future for epidemiology: I. eras and paradigms; II. from black boxes to Chinese boxes, *Am. J. Public Health,* 86:668–678, 1996.

U.N. World Commission on Environment and Development (WCED), Brundtland Commission, *Report: Our Common Future,* United Nations, New York, 1987.

World Health Organization, *Our Planet, Our Health: Report of the WHO Commission on Health and the Environment,* WHO, Geneva, 1992.

CHAPTER **30**

Development of a Terrestrial Index of Ecological Integrity (TIEI), a New Tool for Ecosystem Management

James K. Andreasen, Reed Noss, and Nicholas C. Slosser

INTRODUCTION

Ecological systems are inherently complicated, interacting systems composed of biological and physical components that are so complex that some ecologists believe they cannot be understood (Egler, 1977). However, management and policy decisions require information on the status, condition, and trends of these complex ecosystems and their components. Scientists often become mired in the details about what exactly should be measured, what spatial scale should be studied, how frequently measurements should be taken, and how much data should be accumulated before recommendations are presented to decision makers. Those ecologists who are more oriented to problem solving point out that every detail of an ecosystem does not have to be understood in order to make reasonably intelligent decisions about how to manage and conserve sensitive or valued resources.

The objective of this research is to investigate whether an index, or indices, can be developed that will summarize the condition of ecosystems or watersheds so that changes can be tracked over time and this information utilized as a tool to support environmental decision making. The index might also be used to prioritize management options in watersheds so that those systems in the poorest condition could receive attention first, or that those in more pristine condition could be preserved from further degradation. Indices of integrity can function both in assessments (i.e., snapshots of the condition of a particular site or region) and in monitoring (i.e., periodic measurements used to track changes in condition over time).

There are two research questions to be answered: (1) is it possible to construct an index of ecological integrity for a given region that takes into account ecological and biological phenomena at the relevant levels of organization and spatiotemporal scales, and (2) what are the requirements for such an index?

WHAT IS ECOLOGICAL INTEGRITY?

The first reference to integrity in the environmental literature was Aldo Leopold's (1949) statement in his essay on land ethics: "A thing is right when it tends to preserve the integrity, stability, and beauty of the biotic community. It is wrong when it tends otherwise." Leopold did not explain what he meant by integrity, and it was several decades before the concept appeared again in the environmental literature. Other ecologists consider ecological integrity merely a convenient way to describe and summarize the condition of a biological community or ecosystem. However, integrity is a key concern in several policies of the U.S. and Canada. For instance, the U.S. Water Quality Amendments of 1972 called for the restoration and maintenance of "the chemical, physical, and biological integrity of the Nation's waters."

It was not until Karr and Dudley (1981) offered the following insight that a definition finally gained wide acceptance. Biological integrity describes the:

> ... ability of an aquatic ecosystem to support and maintain a balanced, adaptive community of organisms having a species composition, diversity, and functional organization comparable to that of natural habitats within a region ... the summation of chemical, physical, and biological integrity can be equated with ecological integrity. A system possessing integrity can withstand, and recover from, most perturbations imposed by natural environmental processes, as well as many major disruptions induced by man.

The problem is how to measure ecological integrity, monitor its change through time, and describe the results to people — to the individuals who make decisions about resource management and environmental regulation and to the general public. Ecological integrity is concerned not only with ecological services for humans but also with the condition of the ecosystem for its own sake and for the sake of the other species that are dependent on it.

WHAT ARE THE COMPONENTS OF INTEGRITY?

Stability or Resilience

Ecosystems with high integrity should be relatively resistant to environmental changes and stresses and should be able to recover their original conditions or trajectories relatively quickly after a perturbation. As suggested by Holling (1996), "On the one hand, destabilizing forces are important in maintaining diversity, resilience, and opportunity. On the other hand, stabilizing forces are important in maintaining productivity and biogeochemical cycles, and even when these features are perturbed, they recover rapidly if the stability domain is not exceeded." A resistant system is one that undergoes small changes in a state or flux variable after disturbance, whereas a resilient system is one that returns to the reference state following disturbance (DeAngelis et al., 1989).

Sustainability

As related to integrity, a holistic concept of sustainability focuses on sustaining ecosystems and all their components and processes in a condition such that they continue to provide all the goods and ecological services of which they are capable, or that managers want from that ecosystem.

Naturalness

Natural is one of the most difficult concepts to define, but it signifies something of great aesthetic and spiritual importance to people. Natural areas are generally assumed to have greater ecological integrity than human-modified habitats. Although naturalness (like integrity) is often assumed to be unmeasurable, some researchers have attempted to formalize criteria for its assessment. Bonnickson and Stone (1982), for example, argued that:

A natural system is defined as one that portrays, to the extent feasible, either the same scene that was observed by the first European visitor to the area or the scene that would have existed today, or at some time in the future, if European settlers had not interfered with natural processes.

Angermeier (2000) states the naturalness of most ecosystems or ecosystem alterations can be assessed objectively despite imperfect knowledge if evolutionary limits and natural ranges of variability are carefully considered. Most conservationists value naturally evolved biotic elements such as genomes and communities over artificial elements. This judgment, which is not shared by society at large, is based on intrinsic and instrumental values, including respect for nature and recognition that many amenities stem from natural processes. Although sometimes difficult to assess, naturalness is a more reasonable guide than are other ecosystem features such as diversity, productivity, or evolution.

SOME REQUIREMENTS OF A USEFUL TIEI

Comprehensive and Multiscale

As broadly defined, ecological integrity encompasses ecosystem health, biodiversity, stability, sustainability, naturalness, wildness, and beauty. As more narrowly defined, but more easily measurable, it encompasses chemical, physical, and biological integrity. By most definitions, integrity suggests wholeness, completeness, and intactness, and an adequate index should capture these qualities.

An index of ecological integrity should incorporate all the components and phenomena of interest in the defined ecosystem (Karr, 1993; Fore et al., 1996). The boundaries of the ecosystem must be determined by the questions that are being addressed, but within those boundaries there will be many subsystems. These are often organized as a nested hierarchy (Allen and Starr, 1982; Angermeier and Karr, 1994) but also can be seen as a mosaic of subsystems with varying degrees of interaction and dependence on one another (Norton, 1991). Furthermore, natural communities are influenced by surrounding terrestrial, freshwater, estuarine, and marine ecosystems, by the airshed, and by invasive exotic species. An adequate index of ecological integrity must account for all these communities, their ecotones, and the interactions among them.

Although an index must be multiscale, not every conceivable spatial and temporal scale needs to be covered. Holling (1992) suggested that the behavior of ecosystems can be understood by paying attention to a relatively small number of dominant processes operating at definable scales of time and space. For most assessments of ecological integrity, data from 1-m^2 plots and global carbon models can be ruled out as being directly informative. Nevertheless, the landscape and intermediate temporal scales cover a lot of ground, and a useful index of ecological integrity should cover these thoroughly.

Grounded in Natural History

The importance of grounding an index of ecological integrity in natural history cannot be overstated. The process of selecting appropriate indicators, identifying and measuring them properly, and interpreting what they tell us depends on expert judgment. Expert judgment, in turn, must be based on a thorough understanding of the natural history and autecology of the organisms selected as indicators, their habitat affiliations and interactions with other organisms, and their roles in the broader ecosystem.

Relevant and Helpful

An index of ecological integrity must be applicable for measuring trends of concern to the public and decision makers in relevant landscapes, not just trends interesting to scientists. Indicators that relate to specific trends and inform people about the status of resources they value must be part of any acceptable index. Perhaps the general public does not have to understand every component, but the overall index or suite of indicators must be interpreted and presented in a way that is understandable (Fore et al., 1996).

Able to Integrate Concerns from Aquatic and Terrestrial Ecology

The integrity of aquatic habitats is closely tied to the condition of the entire watershed, which is composed mostly of terrestrial features (Schlosser and Karr, 1981; Hunsaker and Levine, 1995). There is a great need to integrate aquatic and terrestrial ecology in environmental monitoring, assessment, regulatory, and conservation programs. Maintaining and restoring ecological integrity will require that landscapes are viewed and managed as a whole.

Flexible

Ecosystems are in a dynamic, constantly shifting state of change; and the metrics used in the index also should change as needed to properly capture these changes. In addition, the availability, type, amount, and quality of scientific information are constantly changing. An index should be able to incorporate any new and relevant information that may enhance the understanding of ecosystem function and biotic integrity. For example, certain ecosystem components that were not previously considered important, and thus were left out of an index, may, after further research, turn out to be critical factors. In such a case, flexibility would ensure that the index could be altered to reflect this new understanding. Development and application of any index should be responsive to both ecological dynamics over space and time and the generation of new scientific information.

Measurable

To be useful, an index must be measurable. Many authors have provided guidelines for selection of ecological indicators. Among the more common and useful criteria for narrowing the field of potential indicators to a workable, measurable set are (1) a relationship to the phenomenon of interest; (2) convenience, ease, and cost-effectiveness for repeated measurement; (3) ability to provide an early warning of change; (4) distribution over a broad geographic area or otherwise widely applicable; (5) capability of providing a continuous assessment over a wide range of stress; (6) relative independence of sample size; and (7) ability to distinguish changes caused by human activity from natural changes (Cook, 1976; Sheehan, 1984; Munn, 1988, 1993; Noss, 1990; Marshall et al., 1993; Woodley, 1993; Barbour et al., 1995; Fore et al., 1996; de Soyza et al., 1997).

COMPONENTS OF A POSSIBLE TIEI

If indicators of ecological integrity should be comprehensive, multiscale, grounded in natural history, relevant and helpful, able to integrate terrestrial and aquatic environments, flexible, and measurable, then what ecological components fit these criteria that might be selected for inclusion in the TIEI?

Composition (what is there?), structure (how is it arranged?), and function (what does it do?) are equally important attributes of ecosystems, and a list of indicators should include representatives of all these, ideally at several levels of organization. However, the interdependence of composition, structure, and function suggests that indicators in one group might be surrogates for the others. For instance, information on the structure and function of the ecosystem might be provided by monitoring the biota. Conversely, it might be faster and cheaper to measure habitat structure directly, especially at the landscape scale, where remote sensing and geographic information system (GIS) technology permit rapid quantitative measurements.

Composition

The various forms of existing aquatic indices, such as the Index of Biotic Integrity (IBI), are compositional (biological) measures of integrity; that is, they focus on individuals, populations, functional groups, and communities and do not consider habitats (structure) or processes (function) at any scale directly. Possibly the most important step for compositional measures is selecting the biological entities (species, functional groups, etc.) to use as metrics for the index. This step generally has been accomplished by using expert biological knowledge, and some amount of expertise in natural history is necessary for making appropriate judgments (Karr and Chu, 1999).

Because compositional indices for terrestrial ecosystems are mostly in the early stages of development and have not been well tested, the taxonomic groups, functional groups or guilds, and species most appropriate to use as metrics are unknown. Choices are highly context dependent and must vary from region to region and ecosystem to ecosystem. The taxonomic groups should be selected based on the nature of the biotic community, available expertise, ease of sampling, and other ecosystem-specific factors.

Structure

As noted earlier, the IBI in its many forms focuses directly on the biota, not on their habitats. Nevertheless, in many cases indices built on habitat measurements — either on a site or landscape scale — might be more cost-effective than indices based on organisms. Landscape metrics are in common use today, but often they do not have a close tie to the biota.

The major components of landscape spatial structure are (1) amount of habitat in the landscape, (2) mean size of habitat patches, (3) mean interpatch distance, (4) variance in patch sizes, (5) variance in interpatch distances, and (6) landscape connectivity. In a hypothetical landscape of fixed size, the components of habitat amount, mean patch size, and mean interpatch distance cannot be varied independently of one another, a limitation that complicates their use in developing the TIEI.

Habitat fragmentation is one of the best-studied problems in landscape ecology; and effects at population, community, ecosystem, and landscape levels of organization have been documented (Burgess and Sharpe, 1981; Noss, 1983; Harris, 1984; Wilcove et al., 1986; Franklin and Forman, 1987; Saunders et al., 1991; Noss and Csuti, 1997). In real landscapes, fragmentation involves both a reduction in the amount of habitat (habitat loss) and the apportionment of remaining habitat into smaller and more isolated pieces (fragmentation per se). Hence, fragmentation is often confounded with habitat loss and is difficult to study rigorously. Indices of fragmentation might be applicable to ecological integrity in landscapes that have passed some threshold of fragmentation. This threshold will be species specific, as species vary tremendously in their sensitivity to fragmentation.

Function

In the long run nothing is more crucial to the health and integrity of the ecosystem than the continued operation of natural biogeochemical, hydrological, ecological, and evolutionary processes. Key processes are biotic (e.g., predation, herbivory, competition) and abiotic (e.g., weathering, hydroperiod), as well as mixtures of the two (e.g., decomposition, disturbance, succession). The sensitivity of organisms to the rate, magnitude, and other characteristics of natural processes suggests that measuring processes directly without information from the biota may not be helpful for assessing integrity. However, direct measures of processes combined with information on responses of process-sensitive or process-limited species may be worthwhile to incorporate into an index of ecological integrity (Lambeck, 1997).

WHAT CAN BE LEARNED FROM PREVIOUS AND ONGOING RESEARCH?

Defining ecosystem integrity and choosing indicators for it would be a relatively simple task if the science of ecology were able to provide us with simple, rigorous models for describing and predicting these states of ecosystems. Unfortunately, a major problem in modern ecology is that we do not know which state variables are important and which ones are not. Neither do we have simple quantitative models to describe relationships among these state variables (Keddy et al., 1993).

A number of approaches to developing indices of ecological integrity have been suggested, but the only reasonably well accepted index is Karr's index of biotic integrity, the IBI (Karr and Chu, 1999). Application and testing of the IBI, however, has been confined almost entirely to aquatic ecosystems, especially streams. Although the aquatic biota reflects the condition of watersheds, no well-tested and direct index of biotic or ecological integrity exists for terrestrial ecosystems or for entire landscapes comprising terrestrial and aquatic habitats (Slocombe, 1992).

Early attempts to develop an IBI for terrestrial communities were not successful, in large part because of the enormous number of confounding variables in terrestrial systems (Karr, 1999). Applying multimetric concepts to terrestrial environments has so far been limited, and there has been less emphasis on understanding and documenting broader biological responses to human influence. Most relevant studies have examined individual biological attributes rather than a set of metrics.

Although there is no thoroughly tested terrestrial IBI, several recent studies using birds as indicators of ecological integrity are promising (Blair, 1996; Bradford et al., 1998). O'Connell et al. (1998) developed a bird community index (BCI) modeled on the IBIs previously developed for aquatic communities. The BCI is a landscape-scale index based on songbird response guilds, defined as groups of species with similar requirements for habitat, food, or other elements for survival. A high-integrity condition was indicated by a bird community associated with structural, functional, and compositional elements typical of the study region in the absence of human disturbance. It should be noted that the BCI addresses only the dominant natural vegetation of the region — eastern deciduous forest — and not wetlands, grasslands, streams, and other special communities. Hence, it is not a complete landscape IBI.

Another promising attempt to develop a terrestrial, single-community IBI is in progress at the Hanford Nuclear Reservation in Washington. This work, by J. Karr and colleagues at the University of Washington, is examining species, taxa, and ecological groups of invertebrates and plants in the shrub steppe to determine their utility as indicators of varying levels of disturbance.

CRITICISM OF INTEGRITY AND INDICES

Although widely accepted and applied, multimetric indices in general have received criticism. Suter (1993) suggested several potential problems, including ambiguity, eclipsing (i.e., component scores offsetting each other when combined into an index), arbitrary combining functions, arbitrary variance, unreality, *post hoc* justification, unitary response scales, lack of diagnostic results, disconnection from testing and modeling, nonsensical results, and improper analogy to other indices such as economic models. Wicklum and Davies (1995) argued that integrity is an ill-defined concept and "not an objective, quantifiable property of ecosystems." Furthermore, the authors contended that both health and integrity carry with them connotative baggage that does not necessarily apply to ecosystems.

These criticisms have been answered by Simon and Lyons (1995) and Karr and Chu (1999) among others. For example, multimetric indices lose information only if the information on individual metrics is discarded. IBI-type indices retain this information and condense, integrate, and summarize individual metrics into an index that enables quick, practical comparisons among sites (Karr and Chu, 1999).

In evaluating potential structural, functional, and compositional metrics of ecological integrity, it will be important to avoid confounding stressors and responses in the same index. For example, measures of habitat fragmentation and measures of bird species composition should not be combined in a single index, as the latter responds directly to the former. Confounding stressors and responses could lead to a tautological (circular) index that is difficult to interpret (Gerritsen, 1999).

CAUTIONS THAT NEED TO BE CONSIDERED

Ecoregional Characteristics

Ecoregions are defined by climate, geology, physiography, soils, vegetation, and biogeography. Natural processes also vary among ecoregions and among the many natural communities within ecoregions. An index of ecological integrity developed for one ecoregion should not be applied to a dissimilar ecoregion without substantial modification to reflect the particular characteristics of the region. Individual indicators and metrics must correspond to the natural history of each case.

Community Type

Even within an ecoregion, different aquatic and terrestrial communities will require different kinds of metrics. A landscape or regional-scale index that spans many community types must account for the unique characteristics of each system.

Disturbance History

Ecosystems with different disturbance regimes, or examples of the same kind of ecosystem at different times since the last disturbance, can be expected to diverge in many ways. Appropriate metrics for communities in early stages of succession will differ from those suited for later stages. Selection of reference sites to match the successional stages and natural disturbance history of altered sites will be important for addressing the major challenge of separating effects of human perturbations from natural variation in ecosystems.

Impacts of Exotics

Introduced plants and animals can alter the characteristics of a biological community or eco-system radically. Exotics have proven to be a highly reliable metric in aquatic IBIs (Karr and Chu, 1999) and have increased in bird communities along an urbanization gradient (Blair, 1996) and in plant communities at the Hanford site (Hawk, 1999). Many exotic plants have been documented to alter ecological processes (Gordon, 1998), and, in some cases, systems heavily invaded by exotics may shift irreversibly to an entirely different condition.

Spatial and Temporal Scale

Indices of biotic and ecological integrity have been and are being developed at many different spatial scales. The implicit temporal scale of interest also varies widely. Studies at a local or site level will require very different metrics than those at landscape, regional, or continental scales. The "best" scale depends on the particular scientific and management questions being asked.

The terrestrial IBIs under development may turn out to be reliable and informative indices for the particular communities involved. However, it is doubtful that such IBIs could be adapted to other vegetation, habitat types, or regions without substantial modification. Furthermore, IBIs (terrestrial or aquatic) developed at the scale of a given site may not be informative of the condition of the landscape at watershed, regional, or broader scales unless a large sample of site-level IBIs exists. Even then, the emergent properties of ecosystems at higher levels in the ecological hierarchy (Allen and Starr, 1982; O'Neill et al., 1986) will require measurements of their own.

Reference conditions and measurements from areas with minimal human disturbance are critically important for evaluating sites. More research is needed into how to define and choose reference sites or reference conditions for a variety of landscape and ecosystem types.

CONCLUSION

A terrestrial index of ecological integrity will be a useful tool for ecosystem managers and decision makers. The requirements of the TIEI are that it be comprehensive and multiscale, grounded in natural history, relevant and helpful, able to integrate concerns from aquatic and terrestrial ecology, and flexible and measurable. However, much research remains to be completed before a well-functioning index will be available. Deciding which indicators or variables should be included, how they will be measured, combined, and interpreted will not be trivial tasks. After the TIEI is developed, it will have to be thoroughly tested and evaluated in the field.

REFERENCES

Allen, T.F.H. and Starr, T.B., *Hierarchy: Perspectives for Ecological Complexity,* University of Chicago Press, Chicago, IL, 1982.

Angermeier, P.L., The natural imperative for biological conservation, *Conserv. Biol.,* 14:373–381, 2000.

Angermeier, P.L. and Karr, J.R., Biological integrity versus biological diversity as policy directives, *BioScience,* 44:690–697, 1994.

Barbour, M.T., Stribling, J.B., and Karr, J.R., Multimetric approach for establishing biocriteria and measuring biological condition, in Davis, W.S. and Simon, T.P., Eds., *Biological Assessment and Criteria: Tools for Water Resource Planning and Decision Making,* Lewis Publishers, London, 1995, pp. 63–77.

Blair, R.B., Land use and avian species diversity along an urban gradient, *Ecol. Appl.,* 6:506–519, 1996.

Bonnicksen, T.M. and Stone, E.C., Restoring naturalness to national parks, *Environ. Manage.,* 9:479–486, 1982.

Bradford, D.F., Franson, S.E., Miller, G.R., Neale, A.C., Canterbury, G.E., and Heggem, D.T., Bird species assemblages as indicators of biological integrity in Great Basin rangeland, *Environ. Monitoring Assess.,* 49:1–22, 1998.

Burgess, R.L. and Sharpe, D.M., Eds., *Forest Island Dynamics in Man-Dominated Landscapes*, Springer-Verlag, New York, 1981.

Cook, S.E.K., Quest for an index of community structure sensitive to water pollution, *Environ. Pollut.,* 11:269–288, 1976.

DeAngelis, D.L., Mulholland, P.J., Palumbo, A.V., Steinman, A.D., Huston, M.A., and Elwood, J.W., Nutrient dynamics and food web stability, *Annu. Rev. Ecol. Syst.,* 20:71–95, 1989.

de Soyza A.G., Whitford, W.G., and Herrick, J.E., Sensitivity testing of indicators of ecosystem health, *Ecosyst. Health,* 3:44–53, 1997.

Egler, F., *The Nature of Vegetation: Its Management and Mismanagement*, Aton Forest, Norwalk, CT, 1977.

Fore, L.S., Karr, J.R., and Wisseman, R.W., Assessing invertebrate responses to human activities: evaluating alternative approaches, *J. North Am. Benthol. Soc.,* 15:212–231, 1996.

Franklin, J.F. and Forman, R.T.T., Creating landscape patterns by forest cutting: ecological consequences and principles, *Landscape Ecol.,* 1:5–18, 1987.

Gordon, D.R., Effects of invasive, nonindigenous plant species on ecosystem processes: lessons from Florida, *Ecol. Appl.,* 8:975–989, 1998.

Harris, L.D., *The Fragmented Forest: Island Biogeography Theory and the Preservation of Biotic Diversity*, University of Chicago Press, Chicago, IL, 1984.

Holling, C.S., Cross-scale morphology, geometry, and dynamics of ecosystems, *Ecol. Monogr.*, 62:447–502, 1992.

Holling, C.S., Engineering resilience versus ecological resilience, in Schulze, P.C., Ed., *Engineering within Ecological Constraints*, National Academy Press, Washington, D.C., 1996, pp. 31–44.

Hunsaker, C.T. and Levine, D.A., Hierarchical approaches to the study of water quality in rivers, *BioScience*, 45:193–203, 1995.

Karr, J.R., Defining and assessing ecological integrity: beyond water quality, *Environ. Toxicol. Chem.*, 12:1521–1531, 1993.

Karr, J.R. and Chu, E., *Restoring Life in Running Waters: Better Biological Monitoring*, Island Press, Washington, D.C., 1999.

Karr, J.R. and Dudley D.R., Ecological perspective on water quality goals, *Environ. Manage.*, 5:55–68, 1981.

Keddy, P.A., Lee, H.T., and Wisheu, I.C., Choosing indicators of ecosystem integrity: wetlands as a model system, in Woodley, S., Kay, J., and Francis, G., Eds., *Ecological Integrity and the Management of Ecosystems*, Heritage Resources Centre, University of Waterloo and Canadian Parks Service, Ottawa, 1993, pp. 61–79.

Lambeck, R.J., Focal species: a multispecies umbrella for nature conservation, *Conserv. Biol.*, 11:849–856, 1997.

Leopold, A., *A Sand County Almanac*, Oxford University Press, Oxford, 1949.

Marshall, I.B., Hirvonen, H., and Wiken, E., National and regional scale measures of Canada's ecosystem health, in Woodley, S., Kay, J., and Francis, G., Eds., *Ecological Integrity and the Management of Ecosystems*, CRC Press, Boca Raton, FL, 1993, pp. 117–130.

Munn, R.E., The design of integrated monitoring systems to provide early indications of environmental/ecological changes, *Environ. Monitoring Assess.*, 11:203–217, 1988.

Munn, R.E., Monitoring for ecosystem integrity, in Woodley, S., Kay, J., and Francis, G., Eds., *Ecological Integrity and the Management of Ecosystems*, Heritage Resources Centre, University of Waterloo and Canadian Parks Service, Ottawa, 1993, pp. 105–115.

Norton, B.G., *Toward Unity among Environmentalists*, Oxford University Press, Oxford, 1991.

Noss, R.F., A regional landscape approach to maintain diversity, *BioScience*, 33:700–706, 1983.

Noss, R.F., Indicators for monitoring biodiversity: a hierarchical approach, *Conserv. Biol.*, 4:355–364, 1990.

Noss, R.F. and Csuti, B., Habitat fragmentation, in Meffe, G.K. and Carroll, R.C., Eds., *Principles of Conservation Biology*, 2nd ed., Sinauer Associates, Sunderland, MA, 1997, pp. 269–304.

O'Connell, T.J., Jackson, L.E., and Brooks, R.P., A bird community index of biotic integrity for the Mid-Atlantic Highlands, *Environ. Monitoring Assess.*, 51:145–156, 1998.

O'Neill, R.V., DeAngelis, D.L., Waide, J.B., and Allen, T.F.H., *A Hierarchical Concept of Ecosystems*, Princeton University Press, Princeton, NJ, 1986.

Saunders, D.A., Hobbs, R.J., and Margules, C.R., Biological consequences of ecosystem fragmentation: a review, *Conserv. Biol.*, 5:18–32, 1991.

Schlosser, I.J. and Karr, J.R., Riparian vegetation and channel morphology impact on spatial patterns of water quality in agricultural watersheds, *Environ. Manage.*, 5:233–243, 1981.

Sheehan, P.J., Effects on community and ecosystem structure and dynamics, in Sheehan, P.J. et al., Eds., *Effects of Pollutants at the Ecosystem Level*, John Wiley & Sons, New York, 1984, pp. 51–99.

Simon, T.P. and Lyons, J., Application of the index of biotic integrity to evaluate water resource integrity in freshwater ecosystems, in Davis, W.S. and Simon, T.P., Eds., *Biological Assessment and Criteria: Tools for Water Resource Planning and Decision Making*, CRC Press, Boca Raton, FL, 1995, pp. 245–262.

Slocombe, D.S., Environmental monitoring for protected areas: review and prospect, *Environ. Monitoring Assess.*, 21:4978, 1992.

Suter, G.W. II, A critique of ecosystem health concepts and indexes, *Environ. Toxicol. Chem.*, 12:1533–1539, 1993.

Wicklum, D. and Davies, R.W., Ecosystem health and integrity? *Can. J. Bot.*, 73:997–1000, 1995.

Wilcove, D.S., McLellan, C.H., and Dobson, A.P., Habitat fragmentation in the temperate zone, in Soulé, M.E., Ed., *Conservation Biology: The Science of Scarcity and Diversity,* Sinauer Association, Sunderland, MA, 1986, pp. 237–256.

Woodley, S., Monitoring and measuring ecosystem integrity in Canadian National Parks, in Woodley, S., Kay, J., and Francis, G., Eds., *Ecological Integrity and the Management of Ecosystems,* CRC Press, Boca Raton, FL, 1993, pp. 155–176.

CHAPTER 31

U.S. EPA Office of Research and Development Guidelines for Technical Evaluation of Ecological Indicators

William S. Fisher, Laura E. Jackson, and Janis C. Kurtz

INTRODUCTION

Worldwide concern over environmental threats and sustainable development has led to increased efforts to monitor status and trends in environmental condition. In the past, environmental monitoring focused on obvious, discrete sources of stress such as chemical emissions. It became evident that remote and combined stressors, while difficult to measure, can also significantly alter environmental conditions. Consequently, monitoring efforts began to examine valued ecological components that reflected effects of multiple and sometimes unknown stressors. To characterize the condition of these components, national, state, and community-based environmental programs increasingly explored the use of ecological indicators. These are biological, chemical, or physical measurements, indices or models used to characterize or summarize a critical, and usually complex, component of an ecosystem. Indicators are signs or signals that relay a complex message, potentially from numerous sources, in a simplified and useful manner. Indicators are essential when it is implausible to measure and interpret all the factors that comprise an ecological issue.

There are several types of ecological indicators, and they may serve several purposes. Indicators reflect biological, chemical, or physical attributes of ecological condition and may be used to characterize status, to track or predict change, or to identify ecosystem stress. They are used in environmental risk assessments (Simon and Davis, 1992; EPA, 1992; 1998), environmental management actions (EPA, 1990; 1995) and a variety of monitoring programs, such as the U.S. Environmental Protection Agency (EPA) Environmental Monitoring and Assessment Program (EMAP) (EPA, 1990), that are intended to establish status and trends in environmental condition.

Selection of an indicator is driven by the objective of the assessment. Under the EPA Framework for Ecological Risk Assessment (EPA, 1992), indicators must provide information relevant to specific assessment questions, which are developed to focus monitoring data on environmental management issues. The process of identifying environmental values, developing assessment questions, and identifying potentially responsive indicators is a challenge that must be met through close collaboration of scientists, risk assessors, and resource managers (Bardwell, 1991; Cowling, 1992; EPA, 1994a, b). The importance of appropriate assessment questions cannot be overstated; without well-considered assessment questions, an indicator may provide accurate information that is ultimately useless for management decisions. In addition, development of assessment questions can be controversial because of competing interests for environmental resources. Issues may include multiple social values that cross resource types (e.g., forest, stream, estuary) and political boundaries.

Because of the variety of environmental issues and the complexity of environmental data and management decisions, there must be a broad array of available indicators. An indicator may be required to elucidate status, trend, risk, vulnerability, or restoration success. It may be needed to describe past, present, or possible future conditions, or to identify an environmental stressor. It must work at a geographical scale determined by the extent of the resource at risk, whether that is local, watershed, ecoregion, or global. And it must ultimately be applied within a monitoring design (landscape, representative subsample, or index site) selected to best address the issue of concern. Each of these factors enters into the development and selection of a successful ecological indicator.

ORD ECOLOGICAL INDICATORS WORKING GROUP

The Office of Research and Development (ORD) within EPA formed an Ecological Indicators Working Group to guide indicator research and facilitate indicator application, particularly in the EMAP program. The working group consisted of researchers from all of the ORD National Research Laboratories — Health and Environmental Effects, Exposure, and Risk Management — as well as the ORD National Center for Environmental Assessment. Interactions of the working group with industry (Stahl et al., 2000) and the National Research Council (NRC, 2000) identified the need for a common set of criteria to serve as a basis for selection and comparison of ecological indicators. Since the measurable characteristics of an ecosystem are essentially limitless, so are the number of possible indicators. And, whereas many different indicators can be proposed for a variety of purposes, not all will meet the program objectives.

Numerous sources have developed criteria to evaluate environmental indicators (e.g., Macek et al., 1978; Hammons, 1981; Suter, 1989; Kelly and Harwell, 1990; Kerr, 1990; Cairns et al., 1993). The working group assembled those factors most relevant to ORD-affiliated ecological monitoring and assessment programs, drawing heavily from early guidance developed for EMAP (Messer, 1990; EPA, 1991; 1994a). The resulting document, *Evaluation Guidelines for Ecological Indicators* (EPA, 2000), presents 15 guidelines in a format intended to facilitate consistent and technically defensible indicator research and review. Consistency is critical to developing a dynamic and iterative base of knowledge on the strengths and weaknesses of individual indicators; it allows comparisons among indicators and documents progress in indicator development. The evaluation guidelines will be used primarily to examine indicators for their suitability in ecological monitoring and assessment programs, but may also serve technical needs of users who are evaluating ecological indicators for other programs, including regional, state, and community-based initiatives.

The working group elected a phased approach to indicator evaluation, a concept originally suggested by Barber (EPA, 1994a). The phases describe an idealized progression for indicator development that flows from fundamental concepts to methodology, then to examination of data from pilot or monitoring studies, and last to consideration of how the indicator serves the program objectives. The guidelines are presented in this sequence because progression from one phase to the next can represent a large commitment of resources (e.g., conceptual fallacies may be resolved less expensively than issues raised during method development or a large pilot study). However, in practice, application of the guidelines may be iterative and not necessarily sequential. For example, as new information is generated from a pilot study, it may be necessary to revisit conceptual or methodological issues. Or, if an established indicator is modified for a new use, the first step in an evaluation may concern the indicator's feasibility of implementation rather than its well-established conceptual foundation. Each phase in an evaluation process will highlight strengths or weaknesses of an indicator at its current stage of development. Weaknesses may be overcome through further indicator research and modification. Alternatively, weaknesses might be overlooked if an indicator has strengths that are particularly important to program objectives. The evaluation guidelines may be customized to suit the needs and constraints of many different applications. Certain guidelines may be weighted more heavily or reviewed more frequently. The phased approach described here allows interim reviews as well as comprehensive evaluations. Finally, there were no intended restrictions on the types of information (journal articles, data sets, unpublished results, models, etc.) that could be used to support an indicator during evaluation, so long as they were technically and scientifically defensible.

FIFTEEN GUIDELINES FOR EVALUATION OF ECOLOGICAL INDICATORS

Phase 1: Conceptual Relevance

The indicator must provide information that is relevant to societal concerns about ecological condition. The indicator should clearly pertain to one or more identified assessment questions. These, in turn, should be germane to a management decision and clearly relate to ecological components or processes deemed important in ecological condition. Often, the selection of a relevant indicator is obvious from the assessment question and from professional judgment. However, a conceptual model can be helpful to demonstrate and ensure an indicator's ecological relevance, particularly if the indicator measurement is a surrogate for measurement of the valued resource. This phase of indicator evaluation does not require field activities or data analysis. Later in the process, however, information may come to light that necessitates reevaluation of the conceptual relevance, and possibly indicator modification or replacement. Likewise, new information may lead to a refinement of the assessment question.

Guideline 1: Relevance to the Assessment

Early in the evaluation process, it must be demonstrated in concept that the proposed indicator is responsive to an identified assessment question and will provide information useful to a management decision. For indicators requiring multiple measurements (indices or aggregates), the relevance of each measurement to the management objective should be identified. In addition, the indicator should be evaluated for its potential to contribute information as part of a suite of indicators designed to address multiple assessment questions. The ability of the proposed indicator to complement indicators at other scales and levels of biological organization should also be considered. Redundancy with existing indicators may be permissible, particularly if improved performance or some unique and critical information is anticipated from the proposed indicator.

Guideline 2: Relevance to Ecological Function

It must be demonstrated that the proposed indicator is conceptually linked to the ecological function of concern. A straightforward link may require only a brief explanation. If the link is indirect or if the indicator itself is particularly complex, ecological relevance should be clarified with a description or conceptual model. A conceptual model is recommended, for example, if an indicator is comprised of multiple measurements or if it will contribute to a weighted index. In such cases, the relevance of each component to ecological function and to the index should be described. At a minimum, explanations and models should include the principal stressors that are presumed to impact the indicator, as well as the resulting ecological response. This information should be supported by available environmental, ecological, and resource management literature.

Phase 2: Feasibility of Implementation

Adapting an indicator for use in a large or long-term monitoring program must be feasible and practical. Methods, logistics, cost, and other issues of implementation should be evaluated before routine data collection begins. Sampling, processing, and analytical methods should be documented for all measurements that comprise the indicator. The logistics and costs associated with training, travel, equip-ment, and field and laboratory work should be evaluated; and plans for information management and quality assurance should be developed.

Guideline 3: Data Collection Methods

Methods for collecting all indicator measurements should be described. Standard, well-docu-mented methods are preferred. Novel methods should be defended with evidence of effective performance and, if applicable, with comparisons to standard methods. If multiple methods are necessary to accommodate diverse circumstances at different sites, the effects on data comparability across sites must be addressed. Expected sources of error should be evaluated.

Methods should be compatible with the monitoring design of the program for which the indicator is intended. Plot design and measurements should be appropriate for the spatial scale of analysis. Needs for specialized equipment and expertise should be identified.

Sampling activities for indicator measurements should not significantly disturb a site. Evidence should be provided to ensure that measurements made during a single visit do not affect the same measurement at subsequent visits or, in the case of integrated sampling regimes, simultaneous measurements at the site. Also, sampling should not create an adverse impact on protected species, species of special concern, or protected habitats.

Guideline 4: Logistics

The logistical requirements of an indicator can be costly and time-consuming. These requirements must be evaluated to ensure the practicality of indicator implementation and to plan for personnel, equipment, training, and other needs. A logistics plan should be prepared that identifies requirements, as appropriate, for field personnel and vehicles, training, travel, sampling instruments, sample trans-port, analytical equipment, and laboratory facilities and personnel. The length of time required to collect, analyze, and report the data should be estimated and compared with the needs of the program.

Guideline 5: Information Management

Management of information generated by an indicator, particularly in a long-term monitoring program, can become a substantial issue. Requirements should be identified for data processing, analysis, storage, and retrieval, and data documentation standards should be developed. Identified systems and standards must be compatible with those of the program for which the indicator is intended and should meet the interpretive needs of the program. Compatibility with other systems should also be considered, such as the Internet, established federal standards, geographic informa-tion systems, and systems maintained by intended secondary data users.

Guideline 6: Quality Assurance

For accurate interpretation of indicator results, it is necessary to understand their degree of validity. A quality assurance plan should outline the steps in collection and computation of data and should identify the data quality objectives for each step. It is important that means and methods to audit the quality of each step are incorporated into the monitoring design. Standards of quality assurance for an indicator must meet those of the targeted monitoring program.

Guideline 7: Monetary Costs

Cost is often the limiting factor in implementing an indicator. Estimates of all implementation costs should be evaluated. Cost evaluation should incorporate economy of scale, since cost per indicator or cost per sample may be considerably reduced when data are collected for multiple indicators at a given site. Costs of a pilot study or any other indicator development needs should be included if appropriate.

Phase 3: Response Variability

It is essential to understand the components of variability in indicator results to distinguish extraneous factors from a true environmental signal. Total variability includes both measurement error introduced during field and laboratory activities and natural variation, which includes influences of stressors. Natural variability can include temporal (within the field season and across years) and spatial (across sites) components. Depending on the context of the assessment question, some of these sources must be isolated and quantified in order to interpret indicator responses correctly. It may not be necessary or appropriate to address all components of natural variability. Ultimately, an indicator must exhibit significantly different responses at distinct points along a condition gradient. If an indicator is composed of multiple measurements, variability should be evaluated for each measurement as well as for the resulting indicator.

Guideline 8: Estimation of Measurement Error

The process of collecting, transporting, and analyzing ecological data generates errors that can obscure the discriminatory ability of an indicator. Variability introduced by human and instrument performance must be estimated and reported for all indicator measurements. Variability among field crews should also be estimated, if appropriate. If standard methods and equipment are employed, information on measurement error may be available in the literature. Regardless, this information should be derived or validated in dedicated testing or a pilot study.

Guideline 9: Temporal Variability within the Field Season

It is unlikely in a monitoring program that data can be collected simultaneously from a large number of sites. Instead, sampling may require several days, weeks, or months to complete, even though the data are ultimately to be consolidated into a single reporting period. Thus, within-field season variability should be estimated and evaluated. For some monitoring programs, indicators are applied only within a particular season, time of day, or other window of opportunity when their signals are determined to be strong, stable, and reliable, or when stressor influences are expected to be greatest. This optimal time frame, or index period, reduces temporal variability considered irrelevant to program objectives. The use of an index period should be defended, and the variability within the index period should be estimated and evaluated.

Guideline 10: Temporal Variability across Years

Indicator responses may change over time, even when ecological condition remains relatively stable. Observed changes in this case may be attributable to weather, succession, population cycles, or other natural inter-annual variations. Estimates of variability across years should be examined to ensure that the indicator reflects true trends in ecological condition for characteristics that are relevant to the assessment question. To determine inter-annual stability of an indicator, monitoring must proceed for several years at sites known to have remained in the same ecological condition.

Guideline 11: Spatial Variability

Indicator responses to various environmental conditions must be consistent across the monitoring region if that region is treated as a single reporting unit. Locations within the reporting unit that are known to be in similar ecological condition should exhibit similar indicator results.

If spatial variability occurs due to regional differences in physiography or habitat, it may be necessary to normalize the indicator across the region, or to divide the reporting area into more homogeneous units.

Guideline 12: Discriminatory Ability

The ability of the indicator to discriminate differences among sites along a known condition gradient should be critically examined. This analysis should incorporate all error components relevant to the program objectives, and should separate extraneous variability to reveal the true environmental signal in the indicator data.

Phase 4: Interpretation and Utility

A useful ecological indicator must produce results that are clearly understood and accepted by scientists, policy makers, and the public. The statistical limitations of the indicator's performance should be documented. A range of values should be established that defines ecological condition as acceptable, marginal, and unacceptable in relation to indicator results. Finally, the presentation of indicator results should highlight their relevance for specific management decisions and public acceptability.

Guideline 13: Data Quality Objectives

The discriminatory ability of the indicator should be evaluated against program data quality objectives and constraints. It should be demonstrated how sample size, monitoring duration, and other variables affect the precision and confidence levels of reported results, and how these variables may be optimized to attain stated program goals. For example, a program may require that an indicator be able to detect a 20% change in some aspect of ecological condition over a 10-year period with 95% confidence. With magnitude, duration, and confidence level constrained, sample size and extraneous variability must be optimized in order to meet the program's data quality objectives. Statistical power curves are recommended to explore the effects of different optimization strategies on indicator performance.

Guideline 14: Assessment Thresholds

To facilitate interpretation of indicator results by the user community, threshold values or ranges of values should be proposed that delineate acceptable from unacceptable ecological condition. Justification can be based on documented thresholds, regulatory criteria, historical records, experimental studies, or observed responses at reference sites along a condition gradient. Thresholds may also include safety margins or risk considerations. Regardless, the basis for threshold selection must be documented.

Guideline 15: Linkage to Management Action

Ultimately, an indicator is useful only if it can provide information to support a management decision or to quantify the success of past decisions. Policy makers and resource managers must be able to recognize the implications of indicator results for stewardship, regulation, or research. An indicator with practical application should display one or more of the following characteristics: responsiveness to a specific stressor, linkage to policy indicators, utility in cost–benefit assessments, limitations and boundaries of application, and public understanding and acceptance. Detailed consideration of an indicator's management utility may lead to a reexamination of its conceptual relevance and to a refinement of the original assessment question.

APPLICATION OF THE GUIDELINES

The evaluation guidelines were generated both to guide indicator development and to facilitate indicator review. Researchers can use the guidelines informally to find weaknesses or gaps in indicators that may be corrected with further development. Indicator development will also benefit from formal peer reviews, accomplished through a panel or other appropriate means that bring experienced professionals together. It is important to include both technical experts and environmental managers in such a review, since the evaluation guidelines incorporate issues from both arenas. It is recommended that a review address information and data supporting the indicator in the context of the four phases described. The guidelines included in each phase are functionally related and allow the reviewers to focus on four fundamental questions, adapted from Barber (EPA, 1994a) and Fisher (1998):

> *Phase 1 — Conceptual Relevance:* Is the indicator relevant to the assessment question (management concern) and to the ecological resource or function at risk?
>
> *Phase 2 — Feasibility of Implementation:* Are the methods for sampling and measuring the environmental variables technically feasible, appropriate, and efficient for use in a monitoring program?
>
> *Phase 3 — Response Variability:* Are human errors of measurement and natural variability over time and space sufficiently understood and documented?
>
> *Phase 4 — Interpretation and Utility:* Will the indicator convey information on ecological condition that is meaningful to environmental decision making?

It is important to recognize that the evaluation guidelines do not determine indicator applicability or effectiveness. Users must decide the acceptability of an indicator in relation to their specific needs and objectives. The evaluation guidelines document was developed to evaluate indicators for ORD-affiliated monitoring programs, but it should be useful for other programs as well. To increase its potential utility, this document avoids labeling individual guidelines as either essential or optional and does not establish thresholds for acceptable or unacceptable performance. Some users may be willing to accept a weakness in an indicator if it provides vital information. In other cases, the cost may be too high for the information gained. These decisions should be made on a case-by-case basis. The evaluation guidelines document (EPA, 2000) provides three example indicators to illustrate application of the guidelines. The examples include a direct measurement (dissolved oxygen concentration), an index (benthic condition), and a multimetric indicator (stream fish assemblages) of ecological condition. All three examples employ data from EMAP studies, but each varies in the type of information and extent of analysis provided for each guideline as well as the approach and terminology used. The authors of these chapters present their best interpretations of currently available information.

ACKNOWLEDGMENTS

The authors wish to thank the many people in the ORD Ecological Indicators Working Group for the development and evaluation of the ecological indicator evaluation guidelines. This is Gulf Ecology Division contribution 1096.

REFERENCES

Bardwell, L.V., Problem-framing: a perspective on environmental problem-solving, *Environ. Manage.,* 15:603–612, 1991.

Cairns, J., Jr., McCormick, P.V., and Niederlehner, B.R., A proposed framework for developing indicators of ecosystem health, *Hydrobiologia,* 263:1–44, 1993.

Cowling, E.B., The performance and legacy of NAPAP, *Ecol. Appl.,* 2:111–116, 1992.

EPA, Ecological Indicators for the Environmental Monitoring and Assessment Program, Hunsaker, C.T. and Carpenter, D.E., Eds., EPA 600/3–90/060, U.S. Environmental Protection Agency, Office of Research and Development, Research Triangle Park, NC, 1990.

EPA, Indicator Development Strategy for the Environmental Monitoring and Assessment Program, Knapp, C.M., Ed., EPA/600/3–91/023, U.S. Environmental Protection Agency, Office of Research and Development, Corvallis, OR, 1991.

EPA, Framework for Ecological Risk Assessment, EPA/630/R-92/001, U.S. Environmental Protection Agency, Washington, D.C., 1992.

EPA, Environmental Monitoring and Assessment Program Indicator Development Strategy, Barber, M.C., Ed., EPA/620/R-94/022, U.S. Environmental Protection Agency, Washington, D.C., 1994a.

EPA, Environmental Monitoring and Assessment Program: Assessment Framework, Thornton, K.W., Saul G.E., and Hyatt, D.E., Eds., EPA/620/R-94/016, U.S. Environmental Protection Agency, Office of Research and Development, Research Triangle Park, NC, 1994b.

EPA, A Conceptual Framework to Support Development and Use of Environmental Information in Decision-Making, EPA/239-R-95–012, U.S. Environmental Protection Agency, Office of Policy Planning and Evaluation, Washington, D.C., 1995.

EPA, Guidelines for Ecological Risk Assessment, EPA/630/R-95/001F, U.S. Environmental Protection Agency, Washington, D.C., 1998.

EPA, Evaluation Guidelines for Ecological Indicators, Jackson, L., Kurtz, J., and Fisher, W., Eds., EPA/620/R-99/005, U.S. Environmental Protection Agency, Office of Research and Development, Research Triangle Park, NC, 2000.

Fisher, W.S., Development and validation of ecological indicators: an ORD approach, *Environ. Monitoring Assess.,* 51:23–28, 1998.

Hammons, A., *Methods for Ecological Toxicology,* Ann Arbor Science, Ann Arbor, MI, 1981.

Kelly, J.R. and Harwell, M.A., Indicators of ecosystem recovery, *Environ. Manage.,* 14:527–545, 1990.

Kerr, A., Canada's National Environmental Indicators Project: Background Report, Sustainable Development and State of the Environment Reporting Branch, Environment Canada, Ottawa, 1990.

Macek, K., Birge, W., Mayer, F., Buikema, A., Jr., and Maki, A., Discussion session synopsis, in Cairns, J., Jr, Dickson, K., and Maki, A., Eds., *Estimating the Hazard of Chemical Substances to Aquatic Life,* STP 657, American Society for Testing and Materials, Philadelphia, PA, 1978, pp. 27–32.

Messer, J.J., EMAP indicator concepts, in Hunsaker, C.T. and Carpenter, D.E., Eds., Environmental Monitoring and Assessment Program: Ecological Indicators, EPA/600/3–90/060, U.S. Environmental Protection Agency, Office of Research and Development, Research Triangle Park, NC, 1990, pp. 2.1–2.26.

NRC, Ecological Indicators for the Nation, Report from the National Research Council Committee to Evaluate Indicators for Monitoring Aquatic and Terrestrial Environments, Commission on Geosciences, Environment and Resources. National Academy Press, Washington, D.C., 2000.

Simon, T.P. and Davis, W.S., Proc. 1991 Midwest Pollut. Control Biol. Meeting, Environmental Indicators: Measurement and Assessment Endpoints, EPA/905/R-92/003, U.S. Environmental Protection Agency, Region V, Chicago, IL, 1992.

Stahl, R.G., Jr., Orme-Zavaleta, J., Austin, K., Berry, W., Clark, J.R., Cormier, S., Fisher, W., Garber, J., Hoke, R., Jackson, L., Kreamer, G.-L., Muska, C., and Sierszen, M., Ecological indicators in risk assessment: workshop summary, *Hum. Ecol. Risk Assess.,* 6:671–677, 2000.

Suter, G., Ecological endpoints, in Warren-Hicks, W., Parkhurst, B., and Baker, S., Jr., Eds., Ecological Assessment of Hazardous Waste Sites: A Field and Laboratory Reference, EPA 600/3–89–013, Washington, D.C., 1989, pp. 2.1–2.6.

CHAPTER 32

Toward a Forest Capital Index

Bruce A. Wilcox, K. Shawn Smallwood, and James R. Kahn

INTRODUCTION

A significant effort has been under way since the 1992 Earth Summit to develop an international indicator framework for monitoring trends and conditions in forests and forest management. The effort, which has focused on "criteria and indicators for sustainable forest management" (FAO, 1999), has produced a set of remarkably comparable indicator frameworks (Wijewardena et al., 1997). Recently, the World Commission on Forests and Sustainable Development (WCFSD) proposed the development of a "uniform numerical indicator for each nation's current forest capital" (WCFSD, 1999), suggesting that a "Forest Capital Index" (FCI) could be developed that would permit evaluation of progress in sustaining forest capital for each country; serve as a benchmark for assessing whether forest capital is increasing or declining; facilitate a global framework for valuation of forest ecosystem services; and create market mechanisms to compensate countries for ecological services.

It will require nearly another decade before national programs based on the criteria and indicators presently agreed upon are widely implemented. The WCFSD proposal would add another dimension to the international forest indicator challenge. This begs several technical questions beyond those relating to the policy implications of developing another global forest indicator program: What would such an indicator look like? How would we approach its development? And what kinds of data would be required?

Advances in forest ecology, ecological economics, environmental indicator approaches, and in the information technology necessary to gather and analyze forest data (particularly spatial data) remotely suggest at least the outline of an indicator framework that will achieve the above objectives. This framework would draw mainly on concepts of ecosystem health and sustainability, natural capital, the Pressure–State–Response model (P-S-R) for environmental policy indicators, and Geographic Information Systems (GIS).

In this chapter we describe these concepts and models as they apply to forest ecosystems, provide examples of the kinds of indicators and approaches that may be useful for a forest capital index, and discuss the data requirements. We conclude that while much remains to be learned about forest function, forest ecosystem goods and services, the linkage between them, and the effect of detectable forest change, an adequate understanding exists to develop an FCI. Like many of the economic indicators in use, developed despite the lack of a precise understanding of economic systems, it could prove highly useful as both a policy and management device.

BACKGROUND

Since the Rio Earth Summit of 1992 first focused world attention on the importance of forests in achieving sustainable development, more than 150 countries have become actively involved in seven regional or ecoregional initiatives to define national-level "criteria and indicators for forest management" (FAO, 1999). Taking place largely outside the normal U.N. negotiating framework, the general approach is described as follows (Wijewardana et al., 1997):

> "Criteria" define the essential components of sustainable forest management. These include vital forest functions such as biological diversity and forest health, the multiple socioeconomic benefits of forests such as wood production and cultural values and, in most cases, the legal and institutional framework needed to facilitate sustainable forest management. "Indicators" are ways to measure or describe a criterion and over time can demonstrate trends.

> Many indicators are quantitative, such as the percent of a country's forest cover. Others are qualitative or descriptive, such as indicators related to forest planning, public participation, and investment and taxation policies. Both are important to assessing sustainable forest management at the national level.

> Criteria and indicators are tools for assessing national trends in forest conditions and forest management. They provide a common framework for describing, monitoring and evaluating, over time, progress toward sustainable forest management. As such, they help provide a common understanding and implicit definition of what is meant by sustainable forest management.

In view of the diversity of nations involved in the development of criteria and indicators, the degree of similarity or comparability among initiatives is striking. Indicators that Wijewardana et al. refer to as quantitative can be viewed as focusing on forest characteristics. Indicators referred to as qualitative or descriptive can be viewed as focusing on factors affecting forest utilization. The different initiatives across countries incorporate both types of indicators, and all incorporate to some degree the following seven fundamental elements:

1. Extent of forest resources
2. Biological diversity
3. Forest health and vitality
4. Productive functions of forests
5. Protective functions of forests (e.g., soil and water conservation)
6. Socioeconomic benefits and needs
7. Legal, policy, and institutional framework

It is important to have a wide array of indicators, as individual indicators can be misleading. For example, an indicator such as the percent of forest cover would include not only land area covered by pristine forest but also land area covered by second-growth forest or forest plantations. These three different types of forest, however, will be associated with very different levels of ecological services such as biodiversity or watershed protection. Thus the need for additional indicators.

However, the devil is in the details — that is, the scientific and conceptual specifics. The indicator development efforts are in the early stages, and the development of the institutional infrastructure and technical protocols to gather the indicator data is nonexistent in all but a few countries. In particular, as Wijewardana et al. (1997) point out, there is a significant need to improve scientific understanding of a number of indicators, particularly those that go beyond traditional sustained yield factors to capture more recently recognized dimensions of sustainable forest management, including forest fragmentation, nonmarket benefits, and social aspects.

While this international effort has been progressing, three other lines of pursuit dealing with policy-relevant environmental indicators or other measures applicable to ecological systems have been developing in parallel. One is the effort to create national environmental policy indicators that address the environment very broadly, focusing primarily on pollution, waste, and consumption and less on the "life support system" (Adriaanse, 1993; Schulze and Colby, 1994; Hammond et al., 1995). A second is research activity, largely within the ecological health field, addressing the problem of indicators and indices of health or biotic integrity of ecosystems, with the most progress made for aquatic ecosystems (Rapport et al., 1985; Costanza et al., 1992; Davis and Simon, 1994). However, some important steps have been taken more recently toward developing an understanding of ecological patterns and processes and conceptually framing ecosystem health indicators at the landscape scale (Karr, 1994; Rapport et al., 1998). The third has been the development of valuation approaches based on ecosystem functioning, particularly the development of the concept of natural capital in terms of ecosystem goods and services (Costanza and Daly, 1992; Costanza et al., 1992). These ideas, concepts, and methodologies collectively emerging mainly from the field of ecological economics have not yet, as a whole, played a significant part of the international dialogue on sustainable forest criteria and indicators.*

Integrating the above developments within a single indicator framework would be highly desirable, especially given the need to strengthen the scientific understanding of forest indicators. Improved understanding of ideas like natural capital and ecosystem health applied to forests is needed, particularly their role in conceptually framing the indicator challenge and operationalizing indicator methods. This should help accelerate international harmonization of national-level forest indicators, perhaps the implementation of national indicator programs, and, most important, begin reversing many of the current trends in forest degradation globally.** The point of this chapter is to attempt, at least preliminarily, to consider forest indicators in light of the Pressure–State–Response (P-S-R) approach and the natural capital and health of ecosystems, while addressing the technical feasibility of a global Forest Capital Index.

PRESSURE–STATE–RESPONSE INDICATOR APPROACH

The P-S-R model has been adopted by many developed countries as the basis of the conceptual framework for policy-relevant environmental indicators. As shown in Figure 32.1, the model combines the measurement of the pressures (stresses) on the ecosystem, the biophysical criteria that measure the state of an ecosystem, and criteria that address policy actions and their efficacy on mitigating the pressures or improving the state (or health) of an ecosystem. First developed by the Organization of Economic Cooperation and Development, based on work by Canada and subsequently the Netherlands, several variations of the model exist in which an *effect* or *impact* term is added between State and Response (Schulze and Colby, 1994). The model represents both a biophysical cause–effect chain and a sociocultural response.

The P-S-R framework addresses the need for indicators that answer three basic questions relating to environmental change and social responses. First, what is happening to the environmental or a natural resource? Second, why is it happening? Third, what is being done about it? Indicators of state and change address the first question; indicators of pressures or stresses the second; and

---

* An area of resource valuation that probably has had more reach in these forest management policy circles is that dealing with natural resource depletion in terms of national income accounts (e.g., Repetto et al., 1989). However, this work focuses on natural capital that is utilized directly in the production of economic output. Therefore, this discussion does not necessarily involve the ecosystem functions that underlie resource values and have a more indirect (but extremely important) impact on economic output.
** Harmonization does not necessarily imply an identical set of indicators, but it would suggest indicators in the same general area. For example, in temperate forests nutrients are stored in the soil, while in tropical rain forests they are stored in the living biomass of the trees. Although one would want an indicator related to the integrity of the nutrient cycle in both tropical and temperate forests, in each case a different variable must be chosen to measure this characteristic.

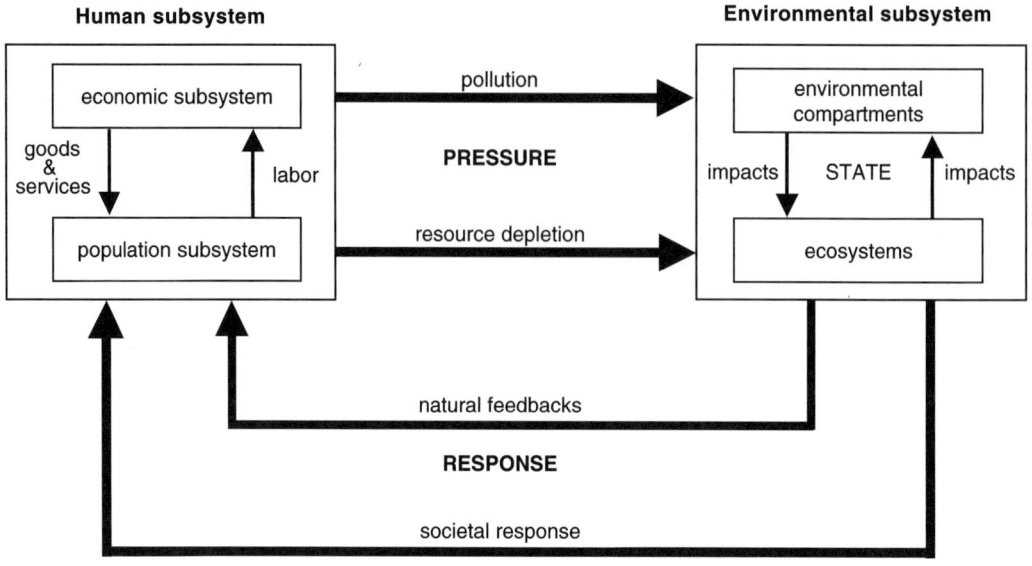

Figure 32.1 P-S-R framework for indicators.

indicators of policy response measure the third (Hammond et al., 1995). The development of national environmental indicators has focused mainly on pollution emissions, or waste generation, and depletion of natural resources. The data employed are national statistics in weight, volumetric, or aerial units. For pollution indicators such as greenhouse gas emissions, for example, emission of carbon dioxide (in units of mass) and the atmospheric concentration of greenhouse gases are recorded as pressure and state indicators. For land quality, the proportion of land having different levels of human-induced soil degradation and soil-climatic zone constraints are proposed for pressure and state indicators. For forests, measures of total forest cover and the rate of loss of forest cover typically serve as pressure and state indicators. A weakness recognized in this existing framework and set of indicators is they do not address the issue of sustainability in terms of the viability of "ecosystems to provide life-supporting goods and services," nor do they provide "adequate measures of the state of ecosystems, such as ecosystem health" (SCOPE, 1994).

DEFINING FOREST CAPITAL

Costanza et al. (1997) refer to natural capital as the ecosystem *infrastructure* necessary to provide the numerous categories of services ranging from the regulation of atmospheric gases, climate, disturbance, and water regimes to the provision of water, food, raw materials, genetic resources, waste treatment, recreational experiences, and others (see Table 1 in Costanza et al., 1997). It should be noted that many economists define natural capital merely as nature's extractable resources, that is, goods — not both goods and services (sometimes referred to as *environmental services*). Franceschi and Kahn (in press) differentiate between *natural capital* and *environmental capital*. Since many neoclassical economics treatments of natural capital view it as consisting solely of extractive resources, Franceschi and Kahn use the term *environmental capital* to refer to our stock of environmental resources that yield a flow of ecological services. The concept of natural capital provides a useful way of framing the issue of ecosystem sustainability and for deriving indicators that address viability in terms of ecosystem goods and services and ecosystem health.

When we utilize natural capital, we withdraw a portion of the stock, whether it is renewable or nonrenewable. However, when we utilize environmental resources, we consume the flow of ecological services such as nutrient cycling, maintenance of global climate, flood protection, or biodiversity. It is interesting to note that some types of resources can be viewed as both natural capital and environmental capital. For example, since trees are a source of wood, they can be viewed as natural capital, but the forest itself is environmental capital that yields ecological services.

Another important distinction between natural and environmental capital has to do with their substitutability with other forms of capital. For example, artificial capital and human capital are good substitutes for natural capital such as oil or coal deposits. If we run out of oil in 30 or 40 years, it won't matter, because substitutes (such as battery-powered cars charged with electricity from solar power) will be developed. However, it is unreasonable to assume that artificial or human capital can provide ecological services on the same level as an ecosystem. This has been demonstrated in the Midwest, where the engineering of flood-control systems of the Mississippi River has not been able to compensate for the flood-control services of the wetlands that have been eliminated.

Another way of distinguishing between the two is related to biophysical and geophysical properties. Extractable resources include nonrenewable resources, such as minerals, and renewable resources such as trees. Though ecosystems include minerals and processes involving their transformation and flux, the capital represented by ecosystems is, by definition, living and therefore renewable. In other words, goods produced by or associated with geophysical processes do not qualify as ecosystem capital, while those of biophysical processes do.

Forest capital can, therefore, be operationally defined in terms of biophysical attributes, the ecosystem infrastructure providing the goods and services derived from forests. Globally, according to the latest U.N. figures, this infrastructure is represented by about 3.5 billion hectares, or one third of the area of terrestrial ecosystems worldwide, and it is declining at an average rate of about 3% annually (FAO, 1999). Costanza et al. (1997) estimate that global forest capital accounts for 38% of all ecosystem goods and services. However, measurement of the decline of natural capital in these terms could be highly misleading. Coverage measures have limited meaning in the absence of further information on how the spatial pattern, location, and other factors affect ecosystem functioning and goods and services.

The forest ecosystem infrastructure providing the basis of these goods and services can be comprised of functional, structural, and compositional elements. Major functional elements are simply those processes corresponding to many of the services enumerated above: regulation of atmospheric chemical composition, global temperature, and hydrological flows (Costanza et al., 1997). Compositional elements are represented by the biophysical components — what generally is thought of as biological diversity. These are the constituents at different hierarchical levels of organization (Wilcox, 1984; Solbrig, 1991) — genes, populations, species, and finally, communities (or local ecosystems). *Structure* is represented by the spatial or geometric patterns or arrangements of the constituents. Later we describe the measurement of these elements in terms of ecosystem health and forest capital.

There are two major parts to the problem of measuring state and change of this infrastructure in terms of capital. First, there is an economic part that is considered goods and services; thus, natural capital provided by a particular ecosystem is socioculturally determined. In particular, the stock of natural capital is related to the uses we generate for the resources, the price of the resources, and the technology that is available to discover, produce, utilize, and consume the resources. For example, if the price of energy rises, or technological innovation in drilling methods occurs, the measured stock of natural resources will increase. Second, there is the ecological problem of determining the relationship of the ecosystem infrastructure to the provision of ecological services. This is a particularly difficult problem, as the relationship between social welfare and the level of ecological services is likely to be a nonlinear function (increasing at a decreasing rate), as depicted

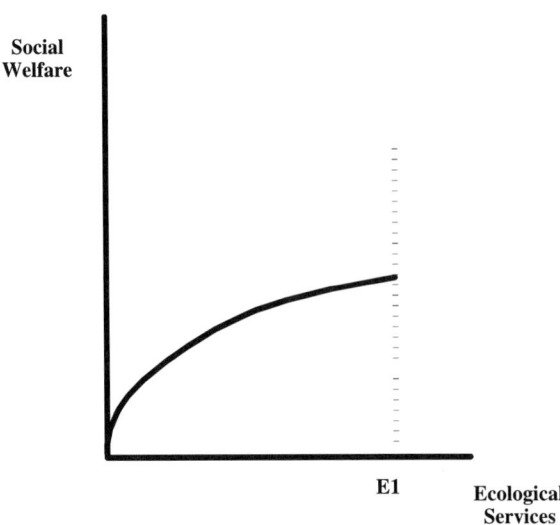

Figure 32.2 Functional relationship between social welfare and ecological services.

in Figure 32.2. This means that the value of a unit of ecological services (or the value of the stock of environmental capital from which the ecological services flow) depends on the current location on the function.

The Relationship of Ecosystem Health and Environmental Capital

How do the concepts of ecosystem health and environmental capital relate? As initially conceived, the ecosystem health model applied the idea of health in holistic medicine and physiology to ecosystems (Rapport et al., 1985). The application was to ecosystems in biophysical terms, with particular regard to state and change and emergent properties such as resilience and physiological cause and effect (i.e., stress and response). Thus, the condition of *health* of an ecosystem can be described as free of distress syndrome, stable, sustainable, maintaining organization and autonomy over time, and resilient to stress (Costanza et al., 1992). The ecosystem health field has increasingly entered the realm of the humanities and social sciences, becoming a transdisciplinary effort involving consideration of the interrelations between human activity, ecological change, ecosystem services, and economic and human health risks (Rapport et al., 1999).

As the disciplinary boundaries of this pursuit become more blurred, discussion of such issues as the role of values in determining condition and trends in the health of ecosystems can be confusing. In fact, what constitutes the condition of health as perceived by both the general public and experts in the fields of medicine or public health is a normative issue. However, the attributes of ecological health described above can, at least potentially, be characterized largely independent of cultural values and based on biophysical attributes that underlie the emergent properties equated with ecosystem health.

One possible solution to this dilemma is to define ecosystem health as a set of physical outcomes related to fundamental health properties such as stability and resilience. The recognition of values can take place in terms of social valuation of the alternative sets of outcomes. In myriad fashions, these outcomes will influence social welfare through the provision of clean water, recreational opportunities, aesthetic values, and so on. Although this approach has potential, one difficulty that will be encountered with the valuation process is that the path by which ecological services influence an individual's welfare is less direct than other services such as recreational opportunities. Thus, valuation processes could bias decision making toward these more direct use values.

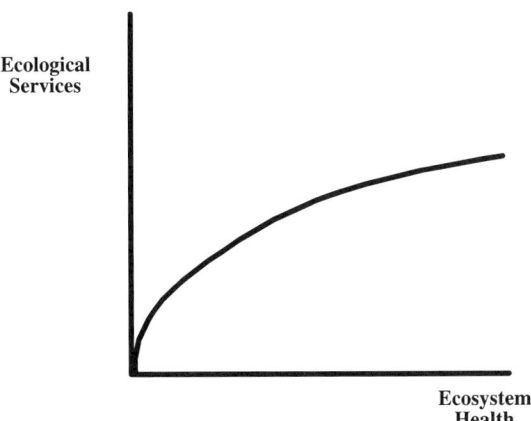

Figure 32.3 Functional relationship between ecological services and ecosystem health.

By contrast to the concept of ecosystem health, the value of the capital represented by ecosystems is a function of society's preferences for alternative states of the environment and for the mix of outputs that are produced. An important determinant of value in this regard is scarcity. As the value of particular ecosystems and ecological services increases, the scarcer these ecosystems and services become. Approaching value in this way suggests that some trade-offs are permissible, even desirable. However, the scarcer a resource, the greater value the alternative must have to justify the trade-off. For example, in the U.S., old-growth forests have become so scarce that the loss of social value from further harvesting is extremely high. On the other hand, eastern second-growth forest is very abundant, so its marginal value is low. The distinction between marginal and total value is extremely important when value is used to inform policy decisions. However, this marginal decision making can conflict with systems of value that do not arise from the neoclassical economics paradigm. For example, the concept of ecological integrity, as articulated by Karr and others (Karr, 1994) is reserved for the notion of *pristine* — that is, the condition of being *intact* with regard to a predisturbed or nondisturbed reference state.

Given what is currently the generally accepted meaning of these concepts, we can reasonably state that the health of the system increases as its ability to provide ecological services increases. As a system becomes less and less healthy, it eventually crosses a threshold and collapses; and the capacity to provide ecological services is lost completely. This functional relationship is approximated graphically in Figure 32.3.

INTEGRATING THE PRESSURE–STATE–RESPONSE AND ECOSYSTEM MODELS

Although the P-S-R model is applied to environmental systems (i.e., countries), it has not been adapted to ecosystems (natural or human) as spatially discrete entities. Here we describe how the P-S-R and ecosystem models can be readily integrated. The P-S-R models and the ecosystem ecologist's view of ecosystem behavior both follow a systems scheme of causal relations in which external factors affect a system, the system responds, and the potential outcome is a new state. It can be useful to distinguish driving variables from controlling variables among the external factors, as shown in Figure 32.4. Driving variables include the size of the human population and the nature of both public and private decision making, including decisions related to land use, waste disposal, pollution emissions, and other impacts called *pressures* in the P-S-R model and *stress* or *stressors* in the ecosystem health model.

Controlling variables are useful to acknowledge because they affect the decision-making process, which leads to a multiplicity of indirect effects, particularly in terms of the kind of land,

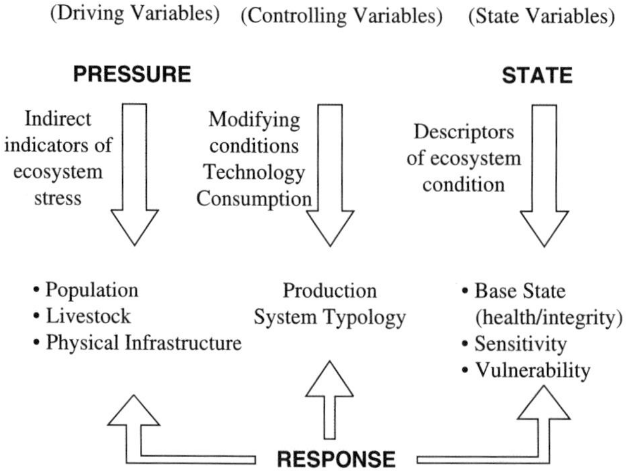

(Driving Variables)   (Controlling Variables)   (State Variables)

PRESSURE   **STATE**

Indirect   Modifying
indicators of   conditions   Descriptors
ecosystem   Technology   of ecosystem
stress   Consumption   condition

• Population   Production   • Base State
• Livestock   System Typology   (health/integrity)
• Physical Infrastructure   • Sensitivity
                                       • Vulnerability

RESPONSE

Figure 32.4 Ecosystem indicators model.

cultural, or economic production system through which the pressure is mediated. For example, the decision to clear forests is related to a set of variables including income (Brown and Pearce, 1994), the definition of property rights (Repetto and Gillis, 1988), the level of technical information (Caviglia–Harris and Kahn, 2001) and the external indebtedness of a nation (Kahn and McDonald, 1995). State variables are the *descriptors* describing the condition of an ecosystem. They include the biophysical attributes that provide measures of health or integrity.

The ecosystem health model can be integrated within this framework by employing the four major categories of stress as pressures and relating each to classes of state variables (or indicators): harvesting of renewable resources, physical restructuring, pollutant discharges, introduction of exotics, and extreme natural events (Rapport et al., 1985). Finally, effect or impact and response variables can be related to these stresses in a complex and nonlinear fashion.

Thus, ecosystem indicators can be seen to represent a special case of environmental indicators in that they reflect attributes characterizing a particular geographic area constituting an ecosystem, whether it is a local ecological community or a region. This provides the opportunity to employ a GIS for the management, analysis, and communication of information, including that on the indicator approach itself. Also, GIS, with remotely sensed data, provides a powerful method for analyzing spatial patterns, particularly fragmentation patterns. Figure 32.5 illustrates how these land-based ecosystem indicators can be implemented using spatial data consisting of zones and boundaries representing ecological and human land use characteristics. These data layers can correspond to the major variable categories in the above ecosystem indicator model.

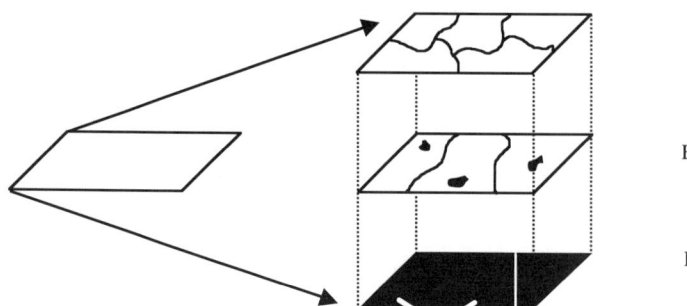

Land units
(administrative, ecoregions,
or arbitrary polygons)

Ecological zones and natural habitat
determinants of land capability/potential

Land uses, human infrastructure,
and population

Figure 32.5 Georeferenced data layers for ecosystem-based indicators.

DERIVING FOREST CAPITAL INDICATORS

The size of the land units to which P-S-R descriptors can be meaningfully applied is, of course, another critical issue in indicator development. Rapport et al. (1998) provide a useful overview of why the landscape is the appropriate scale for ecosystem health indicators. They give three reasons: at this ecosystem scale, values tend to focus on processes that contribute to the resilience of regional ecosystems; biophysical, sociocultural, and economic factors or processes meaningfully integrate; and the concept of the intergenerational community applies.

Thus, they further conclude, assessment of landscape health depends upon identification of societal values and the nature of biophysical processes. Both aspects provide criteria for management of human activity in specific regions. Health in the landscape is dependent on simultaneously meeting two primary goals: (1) providing ecosystem services undiminished in quantity and quality by human activity; and (2) maintaining future management options so as to accommodate changes in societal values. Both are constrained by biophysical limitations. These biophysical processes include, for example, the sequestering, dispersion, and inactivation of toxic substances, conservation and recycling of water, maintenance of soil quality and prevention of erosion, and maintenance of biodiversity.

Implementation of Indicators

Indicators of health, such as forest capital, can be distilled to three aspects of ecosystems: resilience, productivity, and organization (Mageau et al., 1995). These aspects, in turn, can be drawn from measures related to the attributes of structure, function, and composition described earlier. For any such measure, the information ultimately required is the functional relationship of health to the measure. For example, we need to know for a given change in a forest characteristic, such as soil moisture or species richness, how resilience, productivity, and organization are diminished, and, given this loss, what the reduction is in the flow of ecological services. While this answer will be unique to each measure and ecological context, and the data necessary for quantifying these relationships are not generally available, it is possible to postulate very general properties for a system of indicators.

First, we can assume ecological services increase at a decreasing rate for a given ecological characteristic. Thus, the relationship between ecological characteristics and ecosystem services has the functional form shown in Figure 32.6. This form, of course, ignores the nonlinearity that characterizes the complete collapse of a forest ecosystem and its "flip" to a different stability domain represented by another ecosystem type entirely.

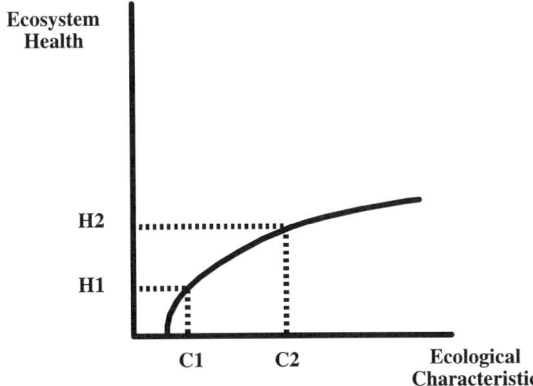

Figure 32.6 Relationship between an ecological characteristic and ecosystem health.

Second, while such dynamics are inherently unpredictable, threshold levels can be roughly approximated, assuming a linear relationship for a particular characteristic and ecological and economic context. While this is far from an exact science, we believe forest ecologists could, for a range of measures (e.g., forest cover, hydrological features, soil parameters, species abundance), reasonably approximate or at least bracket threshold values. These thresholds could be narrowed by considering the pattern of forest loss, type of forest, and various measures of forest quality (e.g., soil condition, age structure, health of individual trees).

In fact, for the ecological characteristic species richness, the functional form and approximate threshold range are well established. The well-known species–area relationship fits a linear proportionality function whose parameter values are known universally to fall within a particular range. Thus, it can be shown in general that significant species losses will begin to occur when the loss of total habitat area reaches a level approximately 70 to 90%. Obviously there will be exceptions, based mainly on the spatial (fragmentation) patterns of remnant forest in relation to the distribution of species richness and the population biology of the species involved. There also is the issue of unequal value of species in terms of ecological services. However, these exceptions potentially can be taken into account.

Spatial scale also is an issue. What may represent a catastrophic collapse of ecological services locally may or may not impact forest capital at the landscape level; and a collapse at this level may or may not have an impact at a regional scale, depending on the mosaic and condition of landscapes. In this regard, it is noteworthy to consider the effect of forest cover loss on ecosystem functioning and the cumulative effect of local habitat loss and fragmentation on the biota at a regional scale. In Amazonia, for example, a large percentage of rainfall originates as water vapor from the forest itself. Thus hydrological and meteorological functions may be so altered that the climate, fire regime, and overall ecological character of the forest ecosystem, including its species composition, is changed (Nepstad et al., 1996).

Ultimately, a Forest Capital Index will require models that incorporate the behavior of ecosystems at local, landscape, and regional scales and integrate processes across scales in terms of the hydrological cycle, species composition, and other factors described by Rapport et al. (1998). Although the state of knowledge about forests in this regard is fragmentary at best, we believe enough is known to begin formulating an FCI.

Fragmentation Measures as Indicators of Ecosystem Health

We believe measures of ecosystem fragmentation — slope, terrain, riparian data, along with simple ground measures to calibrate the remotely sensed data — offer the best way to develop meaningful national level indicators. Broadly defined, fragmentation represents one of the five general classes of stress on ecosystems (Rapport et al., 1985), namely, physical restructuring, and, conveniently, it is highly amenable to mapping and measurement for use as a landscape scale indicator.

Fragmentation is a disruption of the linkages among patches that exchange ecologically important resources, but the critical linkages among patches occur at various scales (Karr, 1994). From the perspective of physical resources such as soil and water, fragmentation occurs when land conversions disrupt flows and storages of materials, energy, or landscape functions that are vital to the maintenance of soil and water qualities at particular places (Dynesius and Nilsson, 1994). Habitat fragmentation can be defined as the process in which the removal or alteration of vegetative cover produces remnant areas of smaller average size and greater isolation than were originally present (Wilcox and Murphy, 1985). In general, fragmentation is the loss of contiguity of accessible landscape from the perspective of the organism or some other ecologically important element.

Three major kinds of forest functions are affected by fragmentation: (1) hydrological and meteorological processes; (2) species' population processes; and (3) disturbance regime. These functions and the changes in these functions associated with fragmentation pattern are described as follows.

Components of the water cycle, including precipitation, runoff, infiltration, transpiration, and evaporation — and associated geomorphological processes including erosion, transport, and deposition of sediment — are strongly influenced by the spatial pattern and character of the vegetation cover. Context, in terms of geology and geomorphology, is critically significant. Climate and fire regimes are, in turn, influenced by the water cycle of a landscape, particularly where precipitation is continental in origin, as is largely the case for the Amazon Basin, for example.

Population processes, including demographics, dispersal, and, ultimately, the persistence of a regional population of a species — and so-called metapopulation dynamics — are largely determined by the spatial pattern of habitat. Some populations, particularly large vertebrate carnivores and herbivores, influence the habitat in terms of its structure, community structure, and species composition. Thus, alteration of the spatial pattern of forest habitat can significantly alter species persistence and, in turn, the integrity and health of a forest landscape.

Patch dynamics, which are driven by disturbance factors like wind, fire, or disease outbreaks, are an integral component of the dynamics and spatial patterning of closed canopy temperate and tropical forests. These disturbance dynamics are key to maintaining ecosystem resilience and biological diversity and, thus, to the integrity and health of forests (Holling et al., 1995). Therefore, fragmentation alters forest health by contracting forest area and creating remnants too small to "support" natural disturbance regimes, and allowing human disturbances larger and more frequent than natural regimes.

Some Candidate Components for Indicator Development

We will discuss two areas of applied ecological research that are particularly promising as key indicator components of an FCI: hydrologic function and vertebrate umbrella species. In both, relatively recent syntheses have substantially increased understanding of their provision of ecological functions or services, their impairment from physical alteration or fragmentation, and their possible detection and measurement using spatial analysis. The disruption and measurement of hydrological function recently has been gaining attention, and we will briefly summarize its indicator possibilities. Less well appreciated, perhaps, is the potential for developing an indicator component based on the population biology of large vertebrates serving as *umbrella* species, to which we give more attention.

Hydrological Indicators

Human activities disrupt natural hydrological functioning in multiple ways (Gleick, 1998). Dams, especially large dams which in some river systems have a combined storage capacity that exceeds annual runoff, alter flow, sediment loads, river chemistry and channel processes. In extreme cases surface flow no longer reaches the sea in normal precipitation years. Land use, including the conversion of forested to agricultural land and the increase in impervious surfaces in urban areas, changes flow pathways and often results in more frequent extreme events, including floods and droughts. Channel alterations, including navigation dams, levees, and dredging to increase shipping, constitute another major class of human alterations to natural hydrologic functioning. Flow regime is increasingly being viewed as a *master variable* that influences virtually all aspects of the physical, chemical, and biological functioning of river ecosystems (Poff et al., 1997). Thus, the integrity of the water flow regime is a key component of a forest landscape health.

We believe it is feasible to develop an index of hydrologic function (or dysfunction) applicable at the national level and scaled to economic benefits. The following characteristics, in particular, are candidates for inclusion in such an index: (1) hydrologic variability, which can be quantified using an existing and tested Index of Hydrologic Alteration (Richter et al., 1996); (2) flooding periodicity for floodplain rivers, which can be assessed using SAR imagery; and (3) change analysis of channel sinuosity and extent of flooding derived from bankfull estimates, which can be assessed using Landsat imagery and aerial photography where available. The development of the hydrological component of a forest capital index would require testing a large suite of candidate indicators to gain as full an understanding as possible of hydrologic function, and then settling on a subset of measures to apply widely.

Umbrella Species

The capacity of a patchwork or forest to support a community of large vertebrates, particularly carnivores, is a reasonable surrogate for forest health. For example, the absence of top predators from neotropical forests has been linked to changes in the abundance of seed-dispersing herbivores — which, in turn, influences the composition of forest canopy trees (Dirzo and Miranda, 1990; Terborgh, 1992). Large-bodied vertebrates are generally more vulnerable to habitat fragmentation and therefore represent good surrogates for the vertebrate biota; their habitat requirements can serve as an *umbrella* for other species. Thus, assuming a careful selection of such species, determination of the degree to which spatial and habitat requirements are adequate to support viable populations should be a reasonable measure of a biota's intactness and resilience and, thus, ecosystem health at the landscape level (Smallwood et al., 1998).

Ecologists can now estimate the spatial areas critical for viability of larger-bodied animal species. Holling (1992) studied body size relationships in different ecological communities to identify robust scale domains within which terrestrial mammalian carnivore populations function. Smallwood and Schonewald (1996, 1998) integrated population estimates of terrestrial mammalian carnivores and found that most of the variation in density was explained by the size of the study area. Moving away from densities and log transformations, Smallwood (1999) identified scale domains in which species' populations likely occur, with an average of 52 individuals per population among species of carnivora. Based on these findings, the effect of habitat fragmentation on the capacity of a landscape to support different levels of vertebrate species richness can be approximated empirically for any ecosystem. Figure 32.7 from Smallwood (unpublished data) shows the fitted relationship of fragmentation to mammal species richness for the Headwaters Forest region in California. This figure essentially represents a surrogate curve relating ecosystem health and

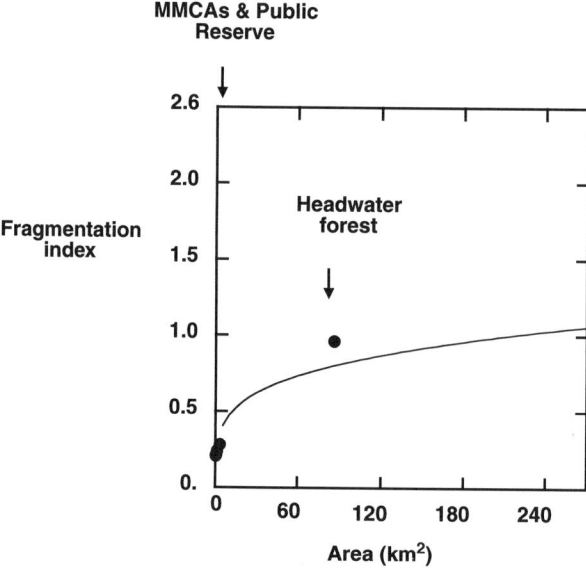

Figure 32.7 Example of an empirically determined relationship between an indicator of forest ecosystem health (Forest Fragmentation Index) and an ecological characteristic (habitat area).

ecological characteristics. Because these curves can be estimated on the basis of body size alone, they can be generated for any mammalian fauna.

CONCLUSION

This chapter suggests an approach for developing indicators of forest capital. It describes how such an indicator approach could be conceptually framed and what ideas, methodologies, tools, and kinds of forest data could be used to develop an indicator system. We suggest that the understanding of the ecological and economic aspects of forest ecosystems has a level now that makes possible an indicator system adequate for decision making at the national level — and for policy prescriptions at regional and international institutions. This understanding results from the relatively recent emergence of concepts like ecosystem health and natural capital; a more comprehensive understanding of the effect of pattern and scale of forest cover and hydrological functioning in relation to the flow of ecological services; and from remote sensing and computer-based information technology for generating and handling geospatial data. Our chapter, however, represents only a preliminary step in designing and implementing a system of indicators of forest capital. Research into the design and implementation of indicators must be complemented by research that focuses on ways to use these indicators to inform the decision-making process.

REFERENCES

Adriaanse, A., Environmental Policy Performance Indicators: A Study on the Development of Indicators for Environmental Policy in the Netherlands, Ministry of Housing, Amsterdam, 1993.

Brown, K. and Pearce, D., Eds., *The Causes of Tropical Deforestation*, University College London Press, London, 1994.

Cavilgia-Harris, J. and Kahn, J.R., Diffusion of sustainable agriculture in the Brazilian tropical rain forest: a discrete choice analysis, *Econ. Dev. Cult. Change*, 49:311–334, 2001.

Costanza, R. and Daly, H.D., Natural capital and sustainable development, *Conserv. Biol.*, 6:37–47, 1992.

Costanza, R.B., Norton, G., and Haskell, B.D., *Ecosystem Health: New Goals for Environmental Management*, Island Press, Washington, D.C., 1992.

Costanza, R., d'Arge, R., de Groot, R., Farber, S., Grasso, M., Hannon, B., Limberg, K., Naeen, S., O'Neill, R.V., Paruelo, J., Raskin, R.G., Sutton, P., and van den Belt, M., The value of the world's ecosystem services and natural capital, *Nature*, 387:253–260, 1997.

Davis, W.S. and Simon, T.P., *Biological Assessment and Criteria: Tools for Water Resource Planning and Decision Making*, Lewis Publishers, Boca Raton, FL, 1994.

Dirzo, R. and Miranda, A., Contemporary neotropical defaunation and forest structure, function, and diversity — a sequel to John Terborgh, *Conserv. Biol.*, 4:444–447, 1990.

Dynesius, M. and Nilsson, C., Fragmentation and the flow regulation of river systems in the northern third of the world, *Science*, 266:753–762, 1994.

Food and Agricultural Organization (FAO), State of the World's Forests, Food and Agriculture Organization of the United Nations, Rome, 1999.

Franceschi, D. and Kahn, J.R., *Performance Bonding as a Continuous Economic Incentive for Environmental Improvement*, Washington and Lee University, Lexington, VA, in press.

Gleick, P.H., *The World's Water*, Island Press, Washington, D.C., 1998.

Hammond, A., Adriaanse, A., Rodenburg, E., Bryant, D., and Woodward, R., Environmental Indicators: A Systematic Approach to Measuring and Reporting on Environmental Policy Performance in the Context of Sustainable Development, World Resources Institute, Washington, D.C., 1995.

Holling, C.S., Cross-scale morphology, geometry, and dynamics of ecosystems, *Ecol. Monogr.*, 62:447–502, 1992.

Holling, C.S., Schindler, D.W., Walker, B.W., and Roughgarden, J., Biodiversity in the functioning of ecosystems: an ecological synthesis, Chapter 2, in Perrings, C. et al., Eds., *Biodiversity Loss: Economic and Ecological Issues*, Cambridge University Press, Cambridge, U.K., 1995.

Kahn, J.R. and McDonald, J.A., Third-World debt and tropical deforestation, *Ecol. Econ.*, 12:107–123, 1995.

Karr, J.R., Landscapes and management for ecological integrity, in Kim, K.C. and Weaver, R.D., Eds., *Biodiversity and Landscape: A Paradox for Humanity*, Cambridge University Press, New York, 1994.

Laurance, W.F. and Bierregaard, R.O., Jr., *Tropical Forest Remnants: Ecology, Management, and Conservation of Fragmented Communities*, University of Chicago Press, Chicago, IL, 1997.

Mageau, M.T., Costanza, R., and Ulanowicz, R.E., The development and initial testing of a quantitative assessment of ecosystem health, *Ecosyst. Health*, 1:201–213, 1995.

Nepstad, D.C., Moutinho, P.R., Uhl, C., Vieira, I.C., and da Silva, J.M.C., The ecological importance of forest remnants in an eastern amazonian frontier landscape Chapter 7, in Schelhas, J. and Greenberg, R., Eds., *Forest Patches and Tropical Landscapes*, Island Press, Washington, D.C., 1996.

Poff, N.L., Allan, J.D., Bain, M.B., Karr, J.R., Prestegaard, K.L., Richter, B.D., and Sparks, R.P., The natural flow regime: a paradigm for river conservation, *BioScience*, 47:69–84, 1997.

Rapport, D.J., Regier, H.A., and Hutchinson, T.H., Ecosystem behavior under stress, *Am. Nat.*, 125:617–614, 1985.

Rapport, D.J., Gaudet, C., Harris, J.R., Baron, J.S., Bohlen, C., Jackson, W., Jones, B., Naiman, R.J., Norton, B., and Pollock, M.M., Evaluating landscape health: integrating societal goals and biophysical processes, *J. Environ. Manage.*, 53:1–15, 1998.

Rapport, D.J., Bohm, G., Buckingham, D., Cairns, Jr., J., Costanza, R., Karr, J.R., de Kruijf, H.A.M., Levins, R., McMichael, A.J., Nielsen, N.O., and Whitford, W.G., Ecosystem health: the concept, the ISEH, and the important tasks ahead, *Ecosyst. Health.* 5:82–90, 1999.

Repetto, R. and Gillis, M., Public Policies and the Misuse of Forest Resources, World Resources Institute, Washington, D.C., 1988.

Repetto, R., Magrath, W., Wells, M., Beer, C., and Rossini, F., Wasting Assets: Natural Resources in the National Income Accounts, World Resources Institute, Washington, D.C., 1989.

Richter, B.D., Baumgartner, J.V., Powell, J., and Braun, D.P., A method for assessing hydrologic alteration within ecosystems, *Conserv. Biol.*, 10:1163–1174, 1996.

Schulze, I. and Colby, M., *A Conceptual Framework to Support the Development and Use of Environmental Information*, Environmental Indicators Team, U.S. EPA, Office of Policy, Planning, and Evaluation, Washington, D.C., 1994.

Scientific Committee on Problems of the Environment (SCOPE), A Systematic Approach to Measuring and Reporting on Environmental Policy Performance in the Context of Sustainable Development, Paris, 1994.

Smallwood, K.S., Scale domains of abundance among species of mammalian Carnivora, *Environ. Conserv.*, 26:102–111, 1999.

Smallwood, K.S. and Schonewald, C., Scaling population density and spatial pattern for terrestrial, mammalian carnivores, *Oecologia*, 105:329–335, 1996.

Smallwood, K.S. and Schonewald, C.M., Study design and interpretation for mammalian carnivore density estimates, *Oecologia*, 113:474–491, 1998.

Smallwood, K.S., Wilcox, B., Leidy, R., and Yarris, K., Environmental auditing: indicators assessment for habitat conservation plan of Yolo County, California, USA, *Environ. Manage.*, 22:947–958, 1998.

Solbrig, O.T., *From Genes to Ecosystems: A Research Agenda for Biodiversity*, The International Union of Biological Sciences, Cambridge, U.K., 1991.

Terborgh, J.T., Maintenance of diversity in tropical forests, *Biotropica*, 24:283–292, 1992.

Wijewardana, D., Caswell, S.J., and Palmberg-Lerche, C., Criteria and indicators for sustainable forest management, in *Proc. XI World Forestry Congr.*, Vol. 6, Antalya, Turkey, 1997.

Wilcox, B.A. and Murphy, D.D., Conservation strategy: the effects of fragmentation on extinction, *Am. Nat.*, 125:879–887, 1985.

Wilcox, B.A., *Concepts in Conservation Biology: Application to the Management of Biological Diversity*, Institute of Ecology, University of Georgia, Athens, GA, 1984.

World Commission on Forests and Sustainable Development (WCFSD), *Our Forests Our Future, Report of the Commission on Forests and Sustainable Development*, Cambridge University Press, Cambridge, U.K., 1999.

Monitoring, Learning, and Adjusting

CHAPTER 33

Overview: Monitoring, Learning, and Adjusting

David Waltner-Toews

Monitoring indicators, learning from changes observed, and adjusting policies and actions based on an assessment of those indicators are the hallmarks of the kinds of learning organizations we must create if we wish to learn our way into a sustainable, healthy future. Furthermore, one of the fundamental insights gained from the successes and failures of the past several decades of managing agroecosystems has been that indicators of health or sustainability cannot be derived or understood without referring to goals, and that goals are necessarily embedded in specific historical socioecological contexts. Thus, it should come as no surprise that the authors of this chapter have identified the characteristics of this process as (1) participatory, grounded in local communities but multiscalar; (2) embracing multiple perspectives among which trade-offs may need to be negotiated; and (3) rooted in a complex understanding of reality in which various components affect each other, often in surprising and sometimes disconcerting ways, through feedback loops across various timescales.

Two major perspectives, both historically and socioecologically rooted, are offered in this chapter on how one can view and assess the health of agroecosystems. The socioeconomic perspective, described by Jock Anderson in Chapter 34, views agricultural activity primarily from the perspective of its role in general economic development and its potential for achieving poverty alleviation. Lori Ann Thrupp, on the other hand, brings to the fore the deep interactions between agricultural activities and biodiversity (Chapter 35). As might be expected, given the different perspectives, they differ on what should be monitored and what might be viewed as success. Nevertheless, both authors clearly see the connections between their views and identify some key contradictions, such as the high water demand of some high-yielding crop varieties (one tool, perhaps, to alleviate poverty) and the severe water constraints faced by many developing nations. They also agree that the way forward lies through integrated, participatory, and multiscalar approaches.

The other two chapters in this section examine two broad, integrative approaches, again from slightly different perspectives and with different scalar foci. In Chapter 36, Dumanski et al., drawing on a definition of "sustainability as opportunity," focus on sustainable land management as a way to capture a complex set of goals, and they point the way toward optimizing trade-offs in evolving "resilient production systems that are well suited to local socioeconomic and physical conditions." In Chapter 37, Van Vuuren et al. take a specific integrative assessment tool — the Ecological Footprint — and apply it to four countries on four continents. In the spirit of this chapter, they use this assessment as an opportunity for learning not only about the performance of these countries but also about the strengths and limitations of the assessment tool and the nature of sustainability itself.

Anderson, echoing an OED assessment, suggests that there is as yet no integrative strategy for understanding rural development. Nevertheless, the characteristics identified, and many of the tools and concepts used by the authors in this section, are all features of what has been called *The Ecosystem Approach*. This approach, recently explored by the UNDP as a way to operationalize its Sustainable Livelihoods Program, is rooted in adaptive environmental management, complex systems theories, and what has been called *postnormal science* — that is, the science of extended peer groups (Ravetz and Funtowicz, 1999). This "new science" is necessary to resolve issues in which the decision-making stakes and the scientific, ethical, and epistomological uncertainties are high (Kay et al., 1999). Clearly, agroecosystem health management, with its central role in producing food and thus making possible all human development, both rural and urban, as well as its implications for socioeconomic justice and environmental sustainability, is such an issue. In practice, the ecosystem approach leads to integrative research and management methods that are iterative, participatory, multiscalar, and systemic — precisely the characteristics sought by the authors of this chapter (Waltner-Toews et al., 2003). For anyone interested in agroecosystem sustainability and health, it is reassuring that the empirical and the theoretical explorations converge in such a remarkable manner.

REFERENCES

Kay, J., Regier, H., Boyle, M., and Francis, G., An ecosystem approach to sustainability: addressing the challenge of complexity, *Futures,* 31:721–742, 1999.
Waltner-Toews, D. et al., Adaptive methodology for ecosystem sustainability and health: an introduction, in Midgley, G. and Ochoa-Arias, A., Eds., *Community Operational Research: Systems Thinking for Community Development*, Plenum Publications/Kluwer Academic, 2003.

CHAPTER 34

Ecosystem Health and Economic Development: Rural Vision to Action

Jock R. Anderson

Globalization and localization are transforming many aspects of the human experience — so many that only a comprehensive, multilayered response of policy and institutional reform will be adequate. If we fail to meet this challenge, we will condemn the world's poor to a cycle of instability, hunger, and despair. By seizing the opportunities presented by the dawn of the 21st century, together we can turn our dream into a reality — a world free of poverty (World Bank, 1999, p. 174).

INTRODUCTION

Rural development must be a central element of economic development in much of the developing world. Of the nearly 1 billion poor identified in some 60 poverty profiles completed by the World Bank, over 70% live in rural areas. Despite rapid urban growth, a majority of the poor will continue to live in the countryside until at least 2015. Without a major sustained effort to reduce rural poverty, global efforts in poverty reduction cannot be successful. Thus, policy makers and development bureaucrats must pay more attention to the poor who live in rural areas.

But there is more to rural development than only alleviating poverty. The rural environment is compromised in many places to the point where the ecosystem can hardly be described as healthy. Connectivity to the world economy is poor in many rural areas distant from urban centers and often effectively disconnected from trade opportunities.

Global interdependence of nations increased phenomenally in the past three decades because of rapid growth in international capital markets, the adoption of flexible exchange rates, and the explosion of information technology. In the realm of trade, successive rounds of the General Agreement on Tariffs and Trade (GATT) negotiations have substantially reduced trade barriers. In the most recent Uruguay Round, agriculture and textiles, long heavily protected sectors, were finally brought under GATT rules. The new World Trade Organization (WTO) should provide for better enforcement of rules and more effective dispute settlement. However, some trade policies still discourage what should be facilitated growth and investment, as well as better environmental management.

These and many other aspects of economic development are dealt with quite comprehensively in the World Bank's *World Development Report 1999/2000* (*WDR99*) (World Bank, 1999). There is little explicitly rural in this report, but there is much discussion of ecosystem maladies, and reference will be made to some of these that link to the rural sector. A companion chapter in this section (Dumanski et al., Chapter 36, this volume) takes up in more detail the global issues of direct relevance to the rural sector, also addressed by the World Bank (1997b).

Hundreds of millions of private farmers, large and small, men and women, are stewards of the vast majority of the globe's renewable natural resources. Most recognize the importance of maintaining and enhancing resource productivity and have shown that they can do so, given proper incentives. But the resources are severely undervalued by inappropriate accounting methods, policies, and institutional frameworks. Of the world's fresh water used directly by people, agriculture uses (or misuses) more than 70% for irrigation. Many water-linked ecosystems are not in good health (Water and Nature, 1999). Unsustainable agricultural practices are major contributors to nonpoint-source pollution. Deforestation remains a critical issue (Sharma, 1992). The global challenges of climate change (Alexandratos, 1995; World Bank, 1997b) and loss of biodiversity (Thrupp, 1998) require major efforts. If production per unit of land is nearly to double in the next 30 years, the policy and institutional failures that cause or contribute to the negative environmental impacts of agriculture must be reversed, and sustainable production systems developed, encouraged, and applied (Pannell and Schilizzi, 1999).

There is growing recognition (*WDR99*; World Bank, 1999) of the diverse contributions that forest resources make to development and to the global commons; forests provide not only wood but also nuts, fruits, soil and water protection, flood control, recreational opportunities, carbon sequestration as a means of offsetting global warming, and perhaps the most complex and extensive source of biodiversity habitat. However, capture of the benefits from forest cover — especially by people living in or near forests — is frequently impeded by tenure and forest-use policies (Anderson, 1987; Sanchez, 1994b; von Amsberg, 1998). The lack of effective mechanisms to transfer compensation from those who benefit from sustainable use and conservation of forests, including the international community as a whole for carbon sequestration and biodiversity preservation (Dumanski et al., Chapter 36, this volume), to those who can gain more from conversion of forest land, compromises sustainability. In far too many countries, forests remain severely undervalued because of the combination of strong rent-seeking behavior by commercial users and forest officials and highly centralized and bureaucratic forest services. This leads to poor forest management and protection and a stifling of local communities' participation in the use and protection of the resources. Regrettably, the measurement of deforestation and forest degradation, and the valuation of biodiversity and carbon sequestration, are matters that have not yet been satisfactorily addressed.

IMPORTANCE OF RURAL DEVELOPMENT IN BROADER ECONOMIC DEVELOPMENT

Early development economists tended to disregard the contribution of the rural economy to development. In the 1950s and later in many countries, the emphasis was on industrial investment to foster economic growth, usually with a strong urban bias. At best, rural areas were seen as merely the source of underemployed labor that could be redeployed in industrial enterprises. The many failures of this model led to an eventual reconsideration of possible agricultural contributions to development, particularly after the successes of the green revolution, and especially so in South Asia.

This extensive literature has been analyzed by many (Hayami and Ruttan, 1984; Hoff et al., 1993; Carruthers and Kydd, 1997). Notwithstanding the lessons of the past (e.g., Vasey, 1992), in most of these and earlier development discussions the environment generally, and rural ecosystems specifically, seldom featured prominently or at all (until the late 1980s, as noted by Staatz and Eicher, 1998), except in those notorious cases of highly obvious land degradation, such as the eroded slopes of Nepal. Ironically, contemporary perspectives on some of these cases is that much of the most dramatic cases of erosion are more geological than anthropogenic (Biot et al., 1995), and thus lead to overstatement of the extent of land degradation (Anderson and Thampapillai, 1990; Leach and Mearns, 1996; Scherr, 1999). Another example of changing perspective is given by *WDR99* in its overview of changed thinking in development economics. The World Bank notes that in the 1950s and 1960s, large dams were seen as synonymous with development, whereas in more recent times, the sharpened appreciation for the human and ecological problems that are associated with some of these schemes have led to much greater emphasis on the adverse effects that tend to go with large dams.

As is outlined in following sections, the new World Bank rural strategy (World Bank, 1997a, hereafter *VtoA*) departed from the earlier tradition of mainstream development economics and gave rural development a central place in economic development of agrarian countries. Rural development is seen not only as an engine of growth but also as a primary focal point of poverty alleviation and responsible (to present and future generations) use of the natural resources underpinning agriculture and the rural landscape (e.g., Crosson and Anderson, 1993; Dumanski et al., Chapter 36, this volume). The new vision of better resource management is now widely shared among key players in rural development such as the Food and Agriculture Organization (FAO), bilateral donors, nongovernmental organizations (NGOs), and the multilateral development banks (Alexandratos, 1995; Lutz et al., 1998).

RETHINKING RURAL DEVELOPMENT STRATEGY AT THE END OF THE CENTURY

As rural development moved up in the development agenda during the 1960s and early 1970s, analysis of problems in the rural sector, and project interventions therein, identified issues of ownership, design, implementation, and supervision, many of which applied also to those of the integrated agricultural or rural development era. The era of Integrated Rural Development Projects (IRDPs) had emerged in the early 1970s. Fortunately, there has been a good deal of reflection on the experiences of this era in the World Bank, donor, NGO, and academic communities.*

Several basic concerns drove the IRDP approach to rural development (Ruttan, 1984). The multiplicity of constraints faced by smallholders were felt to be inadequately addressed in projects with a piecemeal approach to selected weaknesses in local institutions and infrastructure. In preparation for Robert S. McNamara's famous 1973 Nairobi speech, there was a renewed interest in the Bank to strive to deliver project benefits to the poor. Explicitly addressing human development aspects such as health, nutrition, and education along with directly productive aspects of the agricultural environment were (and still are) felt likely to be both socially vital and synergistically productive. Special efforts are needed to reach the weakest and most remote members of society, who seldom benefit from trickle-down effects of interventions not directly targeted to them. Active participation by *all* potential beneficiaries in change processes was desired to move marginalized members of society out of oppressive dependency structures. Designing project interventions to reflect these constraint-removing desiderata proved difficult from the outset, dating from U.S. Agency for International Development (USAID) efforts in North Africa and elsewhere in the 1950s and 1960s (Kumar, 1987). Activity levels and project preparation between the donor community and the bank increased in the 1970s, when enthusiasm for broad-based projects of this type reached a high pitch.

Critical scrutiny of many aspects of implementation grew rapidly in the 1970s (Lele, 1975, 1979; Klitgaard, 1981; Korten and Alfonso, 1981; Moris, 1981; Chambers, 1983), and, more recently, an Operations Evaluation Department** (OED) study in Brazil (Tendler, 1993) was part of a proliferating literature on key design issues. Among many concerns was the consistent tendency

---

* For example, Tendler, 1993; and the Bank's (1978) Rural Development Projects: A Retrospective View of Bank Experience in Sub-Saharan Africa, Report No. 2242.
** The World Bank's Operations Evaluation Department is a relatively independent unit that reviews completed activities.

for underperformance in the planned monitoring and evaluation work designed in the projects, because the same inherent complexity in scope of project activities led any effective monitoring to be similarly complex, difficult, and expensive.

With this history, it was inevitable that the OED would review experience in this area of rural project work, and a study began in 1987. The draft report was discussed and contrasted to the experience of other donors in a Development Assistance Committee of the Organization for Economic Cooperation and Development (OECD) meeting of June 1988. A synthesis was eventually made by Binnendijk (1989).

The OED review of Bank rural development experience was critical of many design elements including setting up enclave projects outside regular administrative structures, excessive reliance on expatriate technical assistance, and a tendency to upscale projects before adequate pilot experience had been gained (OED, 1988).

Linkage to civil society was usually vague and fragile, especially in the early projects, although subsequent project preparation (along with considerable institutional commitment in the 1990s such as the appointment of responsible offices in most resident missions), and implementation of rural development initiatives, has emphasized close working arrangements with diverse NGOs. Also, this has come with a sharpened understanding of the need for more decentralized approaches and administrative structures to engage effectively with communities in both design and implementation of rural development efforts.

The following observations recap (Ridley Nelson, World Bank, personal communication) the problems of this era (Donaldson, 1991):

- The starting point of IRDPs was predominantly households with productive land assets, and design was strongly oriented to soil, crops, and livestock. There was a tacit recognition that such projects could do little directly to help the landless.
- IRDPs were supply-driven, central-planning exercises, although often decentralized to the regional or provincial level. They were designed and implemented as blueprints with limited flexibility.
- IRDPs were only holistic in the sense that they were multisectoral, *Christmas tree* projects. There was limited analysis of the cross-sectoral complementarities *ex ante*, and in evaluations these were not obvious *ex post*.
- There was little institutional analysis in IRDPs; typically, institutions were accepted as found, usually government departments and parastatals. There was little involvement of local governments and limited building of capacity either at the national or local levels. Generally decentralization per se was not an element.
- IRDPs failed partly because of inadequate policy environments. Where policies were recognized as serious constraints, sometimes attempts were made to resolve them, but largely through conditionality rather than strong analysis; usually the extent of single-project policy leverage was limited. Inadequate policies would, of course, also have affected most types of rural projects.
- There was limited local participation or use of NGOs (not widely available in the 1970s) in the design and implementation of IRDPs, although substantial use was made of sometimes massive farm management surveys. These surveys gave those preparing the projects quite good contact with local households and a valuable quantitative analytical base, something that is often missing from present-day Participatory Rural Appraisal techniques. Furthermore, IRDPs typically took place within a governance environment that was largely inimical to local consultation and feedback.
- Partnerships and donor coordination were limited in the IRDP era. Regions or provinces were parceled out, and donors went about their business largely in isolation, although with some attention to comparative advantage.
- Typically, IRDPs were enclaved from the rest of the sector and economy. IRDPs were typically managed by project management units that lay outside the normal government structure. There was usually a presumption that expatriate technical assistance was essential, which, for complex projects with the limited local capacities at the time, was partly correct.
- IRDPs generally did not consider the importance of privatization and cost recovery, although this direction has been followed in Bank projects for some time.

- Typically IRDPs exhibited weak financial sustainability analysis. There was limited focus in IRDPs on environmental sustainability, apart from some soil conservation components, and limited attention to longer term operation and maintenance.
- Generally there was quite strong official local ownership in IRDPs, although in many of the earlier cases the concept was introduced initially by donors.
- Monitoring and evaluation (M&E) was typically a substantial component in IRDPs, with sometimes massive expatriate-staffed M&E units. However, there were serious questions about efficiency, with large household surveys yielding limited results usable by management. This legacy still influences M&E thinking today.

Needless to say, understanding of these intrinsic weaknesses featured strongly in the evaluation of the Bank's new rural vision and was reflected in the 1996–1997 development strategy (*VtoA*) as well as the Bank's emerging fresh framework (the CDF, defined and discussed later in this chapter).

VISION AND ACTION

In 1995, the opportunity arose for a fresh approach to rural development with the arrival of a new president at the World Bank. In a Bankwide consultative process nurtured by the new director of agriculture, a vision of what was needed was put together by distilling the lessons of past endeavors in and outside Bank operations. When the president said: "That's great, but what are you really going to do on my watch?" a further extensive brainstorming took place in the Bank to devise an action plan for a new rural development strategy (*VtoA*) that was debated and accepted by the Bank's board of executive directors. Thus *VtoA* became one of the planks of a new strategic compact for Bank renewal, with special additional funding for a period of 4 years (terminating in fiscal year 2000).

The new rural strategy *From Vision to Action* had four goals:

- Poverty reduction
- Widely shared growth
- Household, national, and global food security
- Sustainable natural resource management

These goals were guided by five key principles (World Bank, 1997a). First, the policy and institutional framework must be supportive of project success. Second, where possible, the private sector must be mobilized to provide investment capital, production, and most services. Third, the state must shift away from heavy intervention in the economy toward enabling macroeconomic, fiscal, and sector policy environments. Fourth, a plurality of institutions must be involved — community-level groups and lower-level governments as well as central ministries. Fifth, where feasible and advantageous, projects and programs are decentralized, designed, and executed with a high degree of influence and participation by community associations and local governments.

In addition, a Strategic Checklist for Rural Development was developed (World Bank, 1997a):

- Macroeconomic and sectoral policies are stable. The foreign exchange, trade, and taxation regimes do not discriminate against agriculture.
- The growth of private agriculture is encouraged by minimizing distortions in input and output markets and by market development for agricultural products.
- Public expenditure programs for economic and social infrastructure, health, nutrition, education, and family planning services do not discriminate against rural populations.
- Large farms and large agroindustrial firms do not receive special privileges and are not able to reduce competition in output, input, land, or credit markets.
- The agrarian structure is dominated by efficient and technologically sophisticated family operators. The rights of women farmers and laborers are explicitly recognized.

- Security of land and water rights is actively promoted. Restricting land rentals hurts the poor, so their access may need protection. Land reform is needed where land distribution is highly unequal. Decentralized, participatory, and market-assisted approaches to land reform are preferred.
- Private and public sectors complement each other in generating and disseminating knowledge and technologies. Public-sector financing is generally important for areas that are probably of limited interest to the private sector, such as strategic research and smallholder extension.
- Rural development programs mobilize the skills, talents, and labor of the rural population through administrative, fiscal, and management systems that are decentralized and participatory, and through private-sector involvement.
- Rural development programs are designed so that the rural poor and other vulnerable groups are fully involved in the identification, design, and implementation.

This list is not comprehensive, especially with respect to natural resource management and rural resource stewardship.

NATURAL RESOURCE MANAGEMENT ISSUES IN *VISION TO ACTION*

The diagnosis in *VtoA* went from the global to the regional as follows. Future increases in food supplies must come primarily from rising biological yields rather than from area expansion and more irrigation, because land and water are becoming increasingly scarce (Crosson and Anderson, 1992; Mitchell et al., 1997; McCalla, 1998). Many new lands brought under cultivation are marginal and ecologically fragile and cannot make up for the land removed from cultivation each year because of urbanization and land degradation. Sources of water that can be developed cost-effectively for irrigation are nearly exhausted, and irrigation water will increasingly need to be reallocated for municipal and industrial use. Thus, expansion of irrigation or increased intensification through irrigation are not viewed as major opportunities. New irrigation projects are increasingly expensive and subject to much stricter environmental standards.

The analysis identified that production on existing rainfed land will need to nearly double to provide a diversified, storable, and transportable food supply to an expanding urban population. The particular problems of livestock production systems and their often unhealthy links to their surrounding ecosystems must be addressed (de Haan et al., 1997).

REGIONAL PRIORITIES FOR DEVELOPMENT

Thus, the four major challenges of reducing rural poverty and hunger, raising economic growth, increasing global food production, and halting natural resource degradation require action on a broad and complex rural development front. It is not only a matter of raising food output but also of ensuring that growth is widely shared and sustainable.

Africa

The specifics of what must be done vary around the world. The long-recognized (FAO, 1986) overwhelming challenge in Africa is to increase food production and raise incomes in rural areas (Lele, 1991; Crosson and Anderson, 1995). Seventy percent of people in sub-Saharan Africa live in rural areas; agriculture accounts for 30% of GDP, 40% of exports, and 70% of employment. Far more so than in any other region, a prosperous agriculture is the engine without which poverty cannot be reduced, natural resources cannot be managed sustainably, and food security cannot be assured (Cleaver and Schrieber, 1994). New lending must focus on a few selected national programs of high impact (Cleaver, 1997) — in research, extension, animal health, and natural resources management — water, soil fertility (Sanchez, 1994a; Pieri, 1999), forests, pastures, and wildlife. The second major challenge is to pursue rural growth rather than only agricultural growth. This is being done by designing a common strategy for development of the rural economy, with investment in infrastructure and social services to be undertaken through nonagricultural projects. The third major challenge is to make national programs work. Sector investment lending is rapidly becoming a privileged instrument; although still not yet proven, it has the potential to generate systemic change in the whole public rural expenditure program. This is particularly important in Africa where, with official development assistance recently at around 10% of GNP* and many donors active in rural areas, donor coordination is essential. Agricultural policy reform would continue to be vigorously pursued through adjustment operations.

Eastern Europe and Central Asia

The main challenge in Eastern Europe and Central Asia is to reform agricultural policies that in the past encouraged inefficient and environmentally unfriendly farming and other practices. The strategy is to base rural lending on a graduated response to thresholds of policy reform including price and trade liberalization, change in agribusiness and farmland ownership, demonopolization and deregulation of marketing, and financial sector reforms. Exceptions are made for projects with long lead times, such as research, or for activities such as land registration that will enable a policy change, when made, to take effect promptly. A second major challenge, unique to the region, is the severity of the needed restructuring and rebuilding of agriculture and the agroindustrial complexes. This is a task without precedent. The Bank will continue to devote a large share of its resources to nonlending services, particularly to analyze land and rural property issues, and the impact of privatization on efficiency and equity, and to assist with determining the necessary legal and regulatory framework for agribusiness, environmental, and rural services.

Middle East and North Africa

The most scarce resource in the Middle East and North Africa is water, and the overwhelming challenge is to increase the efficiency of water use in agriculture. Water charges are merely symbolic in nearly all cases, and operations and maintenance are inadequate because of weak user participation. Intersectoral water transfers are inevitable in the future, and urgent action is sought by the Bank through a series of investment operations in the water sector. A second major challenge is to improve the competitiveness of agriculture, while taking account of most governments' preoccupation with maintaining low urban food prices. In the past, this preoccupation has led to the prevalence of state marketing companies that stifle commerce and the implementation of ineffectively targeted food subsidies that result in enormous fiscal drain. While sector adjustment operations have made significant progress in addressing aspects of competitiveness, much more remains to be done in designing targeted food consumption programs that are effective and politically acceptable. A third major challenge in the Middle East and North Africa is the scarcity of arable land, the severity of soil erosion, the indifference of many governments, and the low levels of investment in watershed management. Many of the region's poorest people live in the upper watersheds, eking out livelihoods from forests, rangelands, and steppes, and degrading watersheds in the process. Given the pervasiveness of urban bias in public expenditures, more persuasive analysis is required to focus attention to investment in watershed management.

South Asia

South Asia has the world's largest concentration of rural poor, particularly in rainfed areas, where the resource base is both more limited and more fragile than in the irrigated plains. Landlessness is far more common in this region than elsewhere. Thus, poverty reduction is the main challenge. Another major challenge is water resources management. This has several dimensions: intersectoral

* One contextual note on this is the observation in *WDR99* that in much of the developing world, environmental damage amounts to something of the order of 5% of GNP.

and intercountry disputes over allocation of scarce water, a crisis in irrigation service delivery, a worsening problem of waterlogging and salinization, and the buildup of pollution of ground and surface water that threatens rural domestic water supply. Innovative projects in irrigation and drainage, flood control, and rural water supply are planned.

Latin America and the Caribbean

A striking characteristic of Latin America and the Caribbean region is the large diversity of socioeconomic conditions and agricultural practices. In many cases, there is extreme income inequality in the rural population as a result of a long history of governments providing discriminatory access to land and capital to the wealthy and powerful. Poverty is deep and prevalent in rural areas, and a high proportion of the rural poor have fled to urban peripheries. Of those remaining, rural poverty is closely connected to issues of natural resources management (Lopez, 1996), and many rural poor earn their livelihoods as laborers or subsistence smallholders in environmentally fragile areas or on the frontiers. At the same time, commercial farmers are beginning to better understand concepts of land conservation and sustainable farming and are seeking assistance in developing and implementing these techniques. Poverty reduction, better use of existing productive capacity, and sustainable natural resource management are the main challenges. Meeting these challenges often requires, however, a redefinition of the roles of government, emphasizing community-based approaches to natural resources management, targeting interventions to reduce rural poverty, protecting the rights of indigenous peoples, increasing efforts to speed the pace of land reform and land allocation, and working with governments to remove policy biases against small farmers (Binswanger and Deininger, 1997).

East Asia and the Pacific

The main characteristics of East Asia and the Pacific are (until recently) generally fast growth, including fast agricultural growth (4% per year since 1980), and good provision of social services nationwide in many countries. The result, however, is that environmental problems abound, especially in watershed management. Unless action is taken soon, these may reach catastrophic proportions in the near future. Another major challenging aspect is the sheer heterogeneity of the region, which ranges from some of the largest economies (China and Indonesia) to some of the smallest (Laos, Mongolia, and the Pacific Islands).

PROGRESS IN IMPLEMENTING THE NEW STRATEGY

While *VtoA* did not set targets for rural lending, there was a clear expectation that the recommended actions would result in a resurgence of lending for agriculture in particular and continued growth in other rural lending. The overall trend has been generally positive, although not as pronounced as had been anticipated:

- Rural lending is expected to increase from about 58 to perhaps 75 projects a year, if present intentions come to pass. The uncertainties, however, are considerable.
- In terms of commitments (Figure 34.1), agricultural projects are expected to increase from $2.6 billion a year in FY98–99 to $4.3 billion in FY00–01, with rural nonagricultural projects increasing from $0.8 billion a year to $1.2 billion. Total rural lending is thus anticipated to increase from $3.4 billion a year to $5.5 billion, or approximately 20% of total bank lending.
- Among subsectors, the largest increase in shares of commitments is expected in a category called *Other Agriculture* that includes community-based rural development and natural resources management. This increase will offset the apparent decline in projects formally classified as natural resource management (NRM).

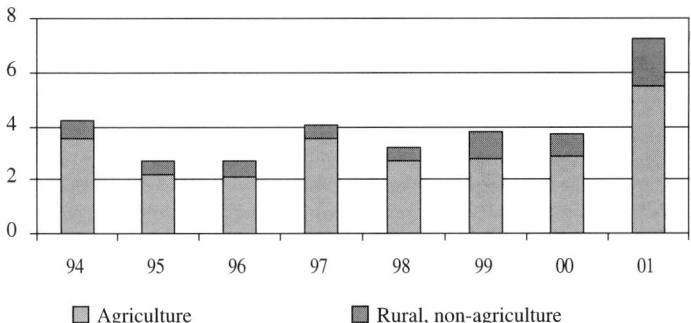

Figure 34.1 Rural approvals: U.S. dollars in billions.

Reviewing Country Assistance Strategies

VtoA is a strategy to ensure that rural development is better reflected in Country Assistance Strategies (CASs) that are the most important investment plans for Bank lending. Now that the country dimension of Bank management has been strengthened through the Bank's latest reorganization, it follows that the CAS process gained in authority. CASs are the vehicles through which key decisions about resource allocation are made.

A recent review by OED showed that the current approach to rural development was judged to be broader than before, but the extent of this breadth is underestimated; and there is no integrated strategy that trades on the large role rural areas play in the Bank's work program. There are several reasons for this. The Bank's knowledge base was judged to serve the cause of rural development poorly. Rural sector work is declining. Poverty assessments rarely contain good estimates of the incidence of rural poverty, even in that majority of countries where most of the poor live in rural areas. Many agree that the off-farm dimension is important, but the dynamics of nonfarm rural enterprise are still poorly understood. There is, unfortunately, no magic bullet for rural development and no consensus about which of the lengthy menu of possible interventions is likely to have the largest payoff in terms of growth and poverty reduction.

Changing Portfolio

Changes by subsector have been significant (Figure 34.2).

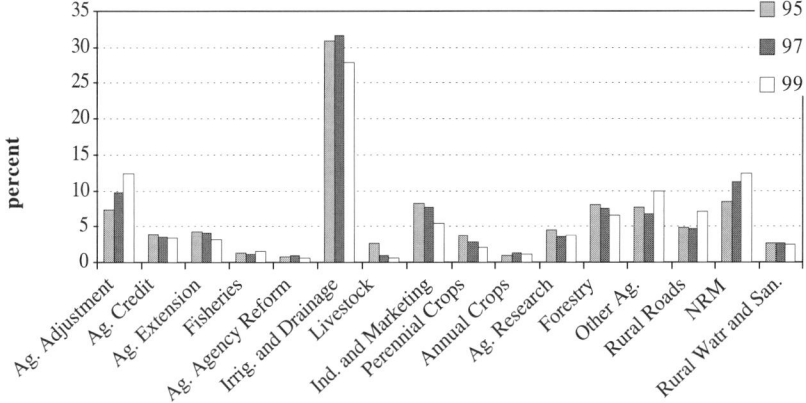

Figure 34.2 Subsectoral share of rural portfolio commitments.

- Irrigation and drainage continues to be, by far, the most important rural subsector. While its share in number of rural projects has been close to 20% for some time, its rather larger share in commitments has now slipped from a high of nearly 33% in FY98 to about 28%.
- Adjustment and natural resources management (and in FY99 also rural roads) are the subsectors that have grown most rapidly, particularly in terms of the volume of commitments.
- Forestry, perennial crops, extension, and livestock have all declined, both by number of projects and commitments (although part of the decline in free-standing extension and livestock projects has been offset by the inclusion of these components in agricultural services projects).

Two aspects with particular relevance to a more comprehensive approach to World Bank forest policy have already become clear. First, the frontiers of the forestry issue have broadened vastly over the past 5 years. To reach any considered position on a bank forest policy, it is now essential to bring into focus the potential impacts on forests of economic adjustment policies, energy policies, agriculture policy, land use legislation, public expenditure strategies, biodiversity, etc. Second, it is increasingly appreciated that not only are many of the poor in rural areas but also rural development is central to combatting poverty, but that many of the poor in rural areas rely at least partly on forest resources, and what is done in forests, and by extension to other natural resources, can be central to rural poverty and global goods, such as containment of atmospheric carbon (Lopez 1996). Of course, while national governments can do good for their citizens through better attention to forest management and policy, they may also be contributing to global goods, some of which may be partially marketable. An example is the deal between Costa Rica and Merck in the work of INbio (*WDR99*; World Bank, 1999).

THE COMPREHENSIVE DEVELOPMENT FRAMEWORK

Further improvements in project quality may require more attention to institutional development, a message that is reiterated in the latest and still emerging bank development thrust: the *Comprehensive Development Framework*. The framework, or CDF, includes the following as one of its five principles: "Sustainable poverty reduction requires holistic, long-term country development strategies with equal weight to the social and structural as to the macro and trade agenda." Some recent analysis of aid effectiveness (World Bank, 1998) reinforces this message. Now that many of the macro and sector policy distortions identified in the 1980s have been sorted out, a good case can be made for paying more attention to institutional capacity, especially resource custody. Fortunately much emphasis is also given to effective partnerships, a concern increasingly mirrored by many potential partners (e.g., Farrington and Bebbington, 1993; FIPA/IFAP, 1996; Thrupp, 1996; Gibbs et al., 1998). Many of the Bank's partners have their own vision of how better to proceed with future rural development work, and there is an active interplay between their ideas and the Bank's own emerging thinking as it seeks to work more effectively with its diverse partners. By way of illustration, one of these deserving mention is that crafted by the U.K. Department for International Development, namely, Sustainable Rural Livelihoods (Carney, 1998; Farrington et al., 1999). In this, the central focus on poverty elimination is set around an appreciation of the several different types of capital (human, financial, physical, social, and natural) available to rural households, their vulnerability to change, and instruments to increase their well-being without undermining the natural resource base. Unfortunately, this strategy is still incomplete with respect to rural landscapes. The logical framework increasingly used in this work makes clear distinctions between outputs, project development objectives, and sector-related goals from the CAS; there is some hope that it will help to ensure clarity in stating objectives and systematically describing project design.

CONCLUSION

This chapter describes the current status and the evolving work programs of the World Bank related to ecosystem health. Progress has been made in rural development, especially in learning from past experience; but there is much that is increasingly urgent to do. In the current global picture, there are still too many poor people, too many in poor health, too many living in poor housing, too many poorly fed and subsisting in poorly managed ecosystems. The discussion has concentrated on project-level interventions, but there are also other entry points for the Bank.

On the global environmental agenda, the Bank is mainstreaming the relevant international conventions including the Convention to Combat Desertification and the Conventions on Climate Change and Bio-diversity, and it is working toward capturing the benefits that sustainable land management provides in conserving biodiversity, sequestering carbon, and reducing global warming. Neglected areas of land management and fisheries are also being tackled — the former described in the companion Bank paper in this session, the latter in a proposal to form a worldwide coalition of multilateral agencies, development banks, bilateral and private donors, civil society, national governments, and the private fishing and processing sectors. This Forum for Sustainable Fisheries will be a coordinating mechanism aimed at better husbandry of marine resources and conservation of the ocean's natural resources.

There is clearly much yet to be done in fostering better ecosystem health. Many of the challenges are in the rural sectors of developing countries. The World Bank has a responsibility to assist its client countries to meet the challenges through provision of analysis, advice, and financial resources. But the tasks are great and many players must be involved. The commitment to work closely with all relevant players is at the center of the new bank approaches, both to the rural sector specifically through *VtoA* and more generally through the emerging CDF.

REFERENCES

Alexandratos, N., Ed., *World Agriculture: Towards 2010: An FAO Study*, Food and Agriculture Organization, Wiley, Chichester, U.K., 1995.

Anderson, J.R. and Thampapillai, J., *Soil Conservation in Developing Countries*: *Project and Policy Intervention*, Policy & Research Series PRS8, World Bank, Washington, D.C., 1990.

Binnendijk, A., Rural Development: Lessons from Experience: Highlights of the Seminar Proceedings, A.I.D. Program Evaluation Discussion Paper No. 25, U.S. Agency for International Development, Washington, D.C., 1989,

Binswanger, H.P. and Deininger, K., Explaining Agricultural and Agrarian Policies in Developing Countries, Policy Research Working Paper 1765, World Bank, Washington, D.C., 1997.

Biot, Y., Blaikie, P.M., Jackson, C., and Palmer-Jones, R., Rethinking Research on Land Degradation in Developing Countries, World Bank Discussion Paper No. 289, World Bank, Washington, D.C., 1995.

Carney, D., Ed., *Sustainable Rural Livelihoods: What Contribution Can We Make?* Department for International Development, London, 1998.

Carruthers, I. and Kydd, J., The development and direction of agricultural development economics: requiem or resurrection? *J. Agric. Econ.,* 48:223–238, 1997.

Chambers, R., *Rural Development: Putting the Last First*, Longman, London, 1983.

Cleaver, K., *Rural Development Strategies for Poverty Reduction and Environmental Protection in Sub-Saharan Africa*, Directions in Development, World Bank, Washington, D.C., 1997.

Cleaver, K.M. and Schrieber, G.A., *Reversing the Spiral: The Population, Agriculture, and Environment Nexus in Sub-Saharan Africa*, Directions in Development, World Bank, Washington, D.C., 1994.

Crosson, P. and Anderson, J.R., Resources and Global Food Prospects: Supply and Demand for Cereals to 2030, World Bank Technical Paper 184, World Bank, Washington, D.C., 1992.

Crosson, P. and Anderson, J.R., Concerns for Sustainability: Integration of Natural Resource and Environmental Issues for the Research Agendas of NARSs, Research Report 4, ISNAR, The Hague, 1993.

Crosson, P. and Anderson, J.R., Achieving a Sustainable Agricultural System in Sub-Saharan Africa, AFTES Post-UNCED Series Paper No. 4, World Bank, Washington, D.C., 1995.

de Haan, C., Steinfeld, H., and Blackburn, H., *Livestock and the Environment: Finding a Balance*, Commission of the European Communities, Brussels, 1997.

Deininger, K., Squire, L., and Basu, S., Does economic analysis improve the quality of foreign assistance, *World Bank Econ. Rev.*, 12:385–418, 1998.

Donaldson, G., Government-sponsored rural development: experience of the World Bank, in Timmer, C.P., Ed., *Agriculture and the State: Growth, Employment, and Poverty in Developing Countries*, Cornell University Press, Ithaca, NY, 1991, pp. 156–190.

FAO, *African Agriculture: The Next 25 Years*, Main Report, FAO, Rome, 1986.

Farrington, J. and Bebbington, A., *Reluctant Partners? Non-Governmental Organizations: The State and Sustainable Agricultural Development*, Routledge, New York, 1993.

Farrington, J., Carney, D., Ashley, C., and Turton, C., Sustainable livelihoods in practice: early applications of concepts in rural areas, *Nat. Resour. Perspect.*, 42, ODI, London, 1999.

FIPA/IFAP, *Farmer's Strategy for Agricultural Development and World Food Security*, International Federation of Agricultural Producers, Paris, 1996.

Gibbs, C., Fumo, C., and Kuby, T., *Nongovernmental Organizations in Bank-Supported Projects: A Review*, Operations Evaluation Department, World Bank, Washington, D.C., 1998.

Hayami, Y. and Ruttan, V.W., Eds., *Agricultural Development: An International Perspective*, Johns Hopkins University Press, Baltimore, MD, 1985.

Hoff, K., Braverman, A., and Stiglitz, J., Eds., *The Economics of Rural Organization: Theory, Practice, and Policy*, Oxford University Press for the World Bank, New York, 1993.

IUCN–UNEP–WWF, *Caring for the Earth: A Strategy for Sustainable Living*, IUCN, Gland, Switzerland, 1991.

Korten, D.C. and Alfonso, F.B., *Bureaucracy and the Poor: Closing the Gap*, McGraw-Hill, Singapore, 1981.

Kumar, K., A.I.D.'s Experience with Integrated Rural Development Projects, A.I.D. Program Evaluation Report 19, U.S. Agency for International Development, Washington, D.C., 1987.

Leach, M. and Mearns, R., Eds., *The Lie of the Land: Challenging Received Wisdom on the African Environment*, International African Institute with James Currey, Oxford, U.K., 1996.

Lele, U., *The Design of Rural Development: Lessons from Africa*, Johns Hopkins University Press for the World Bank, Baltimore, MD, 1975, 1979.

Lele, U., Ed., *Aid to African Agriculture: Lessons from Two Decades of Donors' Experience*, Johns Hopkins University Press for World Bank, Baltimore, MD, 1991.

Lopez, R., *Policy Instruments and Financing Mechanisms for the Sustainable Use of Forests in Latin America*, ENV-106, Inter-American Development Bank, Washington, D.C., 1996.

Lutz, E. et al., Eds., *Agriculture and the Environment: A World Bank Symposium*, World Bank, Washington, D.C., 1998.

McCalla, A.F., Agriculture and food needs to 2025, in Eicher, C.K. and Staatz, J.M., Eds., *International Agricultural Development*, 3rd ed., Johns Hopkins University Press, Baltimore, MD, 1998, pp. 39–54.

Mitchell, D.O., Ingco, M.D., and Duncan, R.C., *The World Food Outlook*, Cambridge University Press, Cambridge, U.K., 1997.

Moris, J., *Managing Induced Rural Development*, International Development Institute, Bloomington, IN, 1981.

OED, *Rural Development: World Bank Experience, 1965–86*, Operations Evaluation Department, World Bank, Washington, D.C., 1988.

Pannell, D.J. and Schilizzi, S., Sustainable agriculture: a matter of ecology, equity, economic efficiency or expedience? *J. Sustainable Agric.*, 13:57–66, 1999.

Pieri, C., Soil Fertility Improvement as a Key Connection between Sustainable Land Management and Rural Well Being, Paper for 16th World Congress of Soil Science, Commission IV, Soil Fertility and Plant Nutrition, 1999.

Ruttan, V.W., Integrated rural development, *World Dev.*, 12:393–401, 1984.

Sanchez, P.A., Tropical soil fertility research: towards the second paradigm, *Trans. 15th World Congr. Soil Sci.*, 1:65–88, 1994a.

Sanchez, P.A., Alternatives to slash and burn: a pragmatic approach for mitigating tropical deforestation, in Anderson, J.R., Ed., *Agricultural Technology: Policy Issues for the International Community*, CAB International, Wallingford, U.K., 1994b, pp. 451–79.

Scherr, S.J., Soil Degradation: A Threat to Developing-Country Food Security? Food, Agriculture, and the Environment Discussion Paper 27, IFPRI, Washington, D.C., 1999.

Sharma, N.P., Ed., *Managing the World's Forests: Looking for Balance between Conservation and Development*, Kendall/Hunt, Dubuque, IA, 1992.

Staatz, J.M. and Eicher, C.K., Agricultural development ideas in historical perspective, in Eicher, C.K. and Staatz, J.M., Eds., *International Agricultural Development,* 3rd ed., Johns Hopkins University Press, Baltimore, MD, 1998, pp. 8–38.

Tendler, J., *New Lessons from Old Projects: The Workings of Rural Development in Northeast Brazil*, Operations Evaluation Study, World Bank, Washington, D.C., 1993.

Thrupp, L.A., Ed., *New Partnerships for Sustainable Agriculture*, World Resources Institute, Washington, D.C., 1996.

Thrupp, L.A., *Cultivating Diversity: Agrobiodiversity and Food Security*, World Resources Institute, Washington, D.C., 1998.

Vasey, D.E., *An Ecological History of Agriculture: 10,000 B.C.–A.D. 10,000,* Iowa State University Press, Ames, IA, 1992.

Von Amsberg, J., Economic parameters of deforestation, *World Bank Econ. Rev.,* 12:133–53, 1998.

Water and Nature, *A Vision for Water and Nature: Environment and Ecosystems*, IUCN with World Water Council, World Bank and GEF, Montreal, 1999, available online at http://www.waterandnature.org.

World Bank, Rural development: from vision to action: a sector strategy, *Environ. Sustainable Dev. Stud. Monogr.*, Series 12, World Bank, Washington, D.C., 1997a.

World Bank, Guidelines for climate change global overlays, *Environment Department Papers, Climate Change*, Series 47, World Bank, Washington, D.C., 1997b.

World Bank, *Assessing Aid: What Works, What Doesn't, and Why*, Oxford University Press for World Bank, New York, 1998.

World Bank, *Entering the 21st Century: World Development Report 1999/2000*, Oxford University Press for World Bank, New York, 1999.

CHAPTER 35

Agricultural Biodiversity: A Key Element of Ecosystem Health and Sustainable Food Security*

Lori Ann Thrupp

INTRODUCTION

There is growing worldwide realization that biodiversity is fundamental to agricultural production and food security as well as a valuable ingredient of ecosystem health and environmental conservation. Yet predominant patterns of agricultural and economic growth have eroded biodiversity in agroecosystems, including plant and animal genetic resources. This erosion has caused economic losses, jeopardizing productivity and food security in some areas and leading to broader social costs. Equally alarming is the loss of biodiversity in natural habitats from the expansion of agricultural production to frontier areas.

The conflicts between agriculture and biodiversity are by no means inevitable. With sustainable farming practices and changes in agricultural policies and institutions, they can be overcome. Historical evidence and current observations show that biodiversity and the maintenance and integration of agricultural systems can have multiple ecological and socioeconomic benefits, including sustaining productivity, ecosystem functions, and food security. Practices that conserve and enhance *agrobiodiversity* are necessary at all levels. The sustainable use of diverse biological resources is critical for food production, nutrition, and for maintaining or improving the health and sustainability of ecosystems.

This chapter summarizes the main conflicts and complementarities between biodiversity and agriculture, discusses the ecosystem services provided by agricultural biodiversity, and highlights principles, policies, and practices that enhance diversity in agroecosystems.

AGROBIODIVERSITY AS A BASIS FOR PRODUCTION AND SURVIVAL

Biodiversity and detailed knowledge about it have allowed farming systems to evolve since agriculture began some 12,000 years ago (Shand, 1997; FAO, 1998; UNDP, 1992). Although sometimes perceived as an enemy of biodiversity, agriculture is actually *based* on richly diverse biological resources. Likewise, agriculture comprises a variety of managed ecosystems — *agroecosystems* — that benefit from resources in natural habitats.

Agrobiodiversity is an important feature of farming systems around the world. It encompasses many types of biological resources tied to agriculture, including:

* This chapter is derived from a longer report published by the World Resources Institute in 1998 titled *Cultivating Diversity: Agrobiodiversity and Food Security*, by Lori Ann Thrupp.

- Genetic resources — the essential living materials of plants and animals
- Edible plants and crops
- Livestock (small and large, lineal breeds, or thoroughbreds) and freshwater fish
- Soil organisms vital to soil fertility, structure, quality, and soil health
- Naturally occurring insects, bacteria, and fungi that control insect pests and diseases of domesticated plants and animals
- Agroecosystem components and types (polycultural/monocultural, small/large scale, rainfed/irrigated, etc.) indispensable for nutrient cycling, stability, and productivity
- *Wild* resources (species and elements) of natural habitats and landscapes that can provide services (for example, pest control and ecosystem stability) to agriculture (Brookfield and Padoch, 1994; Altieri, 1991; FAO, 1998; Shand, 1997; UNDP, 1992)

Agrobiodiversity therefore includes not only a wide variety of species but also the many ways in which farmers can *use* biological diversity to produce and manage crops, land, water, insects, and biota. (Figure 35.1) The concept also includes habitats and species outside of farming systems that benefit agriculture and enhance ecosystem functions, such as host plants that harbor natural enemies and predators of agricultural pests.

As recorded by Egyptian, Mesopotamian, Chinese, and Andean civilizations, ancient agricultural settlements made use of diverse plants, livestock, and agroecosystems. Over many centuries, farmers have employed numerous practices to use, enhance, and conserve this diversity in traditional farming systems. Many such practices continue today: the use of diverse species for pest control and the integration of trees and woody shrubs into farming systems are two examples. Wild plant and animal species in surrounding habitats also provide services and value to the farming system. Such practices are a basis of survival and well-being for millions of people.

Examples of these traditional farming systems that maximize diversity are the polycultural plots, sometimes called home gardens, found in many regions, including Central America, Southeast Asia, sub-Saharan Africa, and even Europe. Others are traditional agroforestry systems, such as the shaded coffee plantations common throughout Central and South America. These traditional agroforestry systems commonly contain well over 100 annual and perennial plant species per field (Altieri, 1991). Farmers often integrate leguminous trees, fruit trees, trees for fuelwood, and types that provide fodder on their coffee farms. The trees also provide habitat for birds and animals that benefit the farms. For example, a shaded coffee plantation in Mexico supports up to 180 species of birds that help control insect pests and disperse seeds (Greenburg, 1994). Ethnobotanical studies show that the Tzeltal Mayans of Mexico can recognize more than 1200 species of plants, while the P'urepechas recognize more than 900 species and Yucatan Mayans some 500 (Altieri, 1991). Such knowledge is used to make production decisions.

Numerous studies show that shifting cultivation systems in traditional forms are agroecologically diverse and contain numerous plant species. These can also be relatively sustainable in certain areas of the world, especially where economic and demographic pressures for growth are low

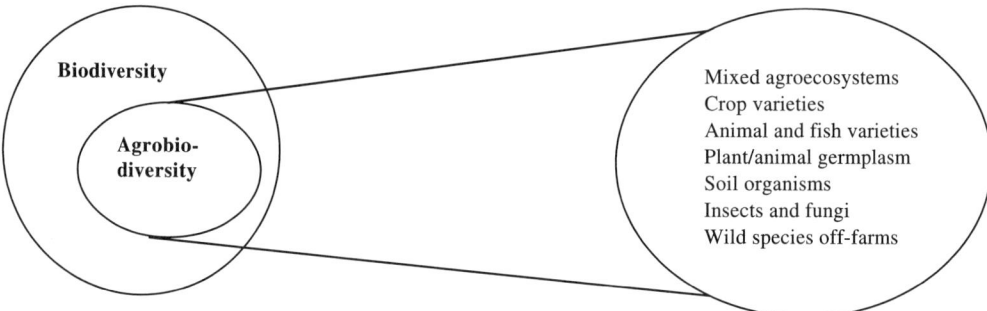

Figure 35.1 Conceptual view of agrobiodiversity.

(Brookfield and Padoch, 1994; Thrupp, 1996). Another important dimension of traditional agro-biodiversity is the use of so-called folk varieties, also known as *landraces*. Defined as "geographically or ecologically distinctive populations [of plants and animals] which are conspicuously diverse in their genetic composition" (Brown, 1978), landraces are products selected by local farmers over time for their various production benefits (Cleveland et al., 1994). In some areas in the Andean region, for example, farmers have also developed complex techniques to select, store, and propagate the seeds of landraces.

Many practices for using and enhancing agrobiodiversity are tied to the rich cultural diversity and local knowledge that are valuable elements for community livelihood. Rural women are particularly knowledgeable about diverse plants and tree species and about their uses for health care, fuel, and fodder, as well as food (Abramowitz and Nichols, 1993; Thrupp, 1984). Many principles, as well as intuitive knowledge, from traditional systems are applied today in both large- and small-scale production. In fact, "traditional multiple cropping systems still provide as much as *20 percent* of the world food supply" (UNDP, 1995).

From ancient times to the present day, plant collecting has also enhanced agrobiodiversity. The earliest record dates back to 1495 B.C., when Egyptians gathered plants for food, medicine, and other purposes from various parts of the world. Throughout the colonial period, the search for and collection of diverse plants and foods was a driving interest of European explorers and played an important role in colonial expansion.

Also significant was the work of N. Vavilov, a renowned Russian botanist who carried out systematic plant collection, pioneering research, and conservation of crop diversity starting in the early 20th century. Vavilov developed a theory of the origin of domesticated crops and launched numerous worldwide expeditions to collect crop germ plasm. He established an immense seed bank in St. Petersburg that still endures, now containing some 380,000 specimens from more than 180 locations in the world (Raeburn, 1995). Vavilov also identified major areas of high concentrations of crop diversity around the world, most of which are in developing countries.

The innovative uses of plant genetic resources has continued to be valuable for scientific advances in plant and livestock breeding and seed improvements up to the present day. These discoveries have been important to all forms of agriculture, from large-scale industrial agribusiness to traditional small-scale farming (Raeburn, 1995). Access to germ plasm is vital for modern agriculture and for the development of medicinal products, fibers, and foods. In the U.S., for example, exotic germ plasm "adds a value of $3.2 billion to the nation's $1 billion annual soybean production and $7 billion to its $18 billion annual maize crop" (Shand, 1997).

In sum, it is clear that the many forms of agrobiodiversity yield an array of benefits (Box 35.1). They reduce risk and contribute to resilience, food security, and income generation. Agricultural biodiversity also provides ecosystem services, such as biological control of pests and improving the health of soils, and benefits nutrition and productivity in many cases.

THE DECLINE OF AGROBIODIVERSITY

The links between agriculture and biodiversity have changed over time and should be viewed in the context of global agricultural development trends. Agricultural production and productivity have increased greatly in the last 30 years, and the international trade and exchange of seeds and food products have also expanded. This growth stems from both the expansion of cultivated area (*extensification*) and the increased output per unit of land (*intensification*) through technological inputs, improved varieties, and the management of biological resources such as soil and water. Agricultural development trends have also failed to overcome problems: hunger and malnutrition persist; food security has declined in many developing countries; food and land are distributed highly inequitably; and biological resources have been degraded. Consequently, the ecosystem services provided by agricultural biodiversity have declined, therefore undermining

---

BOX 35.1
BENEFITS OF AGROBIODIVERSITY

Experience and research have shown that agrobiodiversity can:

- Increase productivity, food security, and economic returns
- Reduce the pressure of agriculture on fragile areas, forests, and endangered species
- Make farming systems more stable, robust, and sustainable
- Contribute to sound insect pest and disease management
- Conserve soil and increase natural soil fertility and health
- Contribute to ecosystem services and ecosystem health
- Diversify products and income opportunities
- Reduce or spread risks to individuals and nations
- Reduce dependency on external inputs
- Improve human nutrition and provide sources of medicines and vitamins

Adapted from UNDP, 1985; Altieri, 1987; Brookfield and Padoch, 1994; Thrupp, 1998.

---

ecosystem health. These problems pose tremendous challenges to meet growing food needs while conserving resources.

These general trends in agriculture and biodiversity have been shaped by demographic pressures, including high population growth rates, the migration of people into frontier areas, and imbalances in population distribution. Additional influential forces are the predominant paradigms of industrial agriculture and the green revolution, beginning in the 1960s. These approaches generally emphasize maximizing yield per unit of land, uniform varieties, reduction of multiple cropping, standardized farming systems (particularly generation and promotion of high-yielding varieties), and the use of agrochemicals. Seed and agrochemical companies have also influenced these trends. High-yielding varieties (HYVs) are now planted on high percentages of agricultural land — estimated as 52% for wheat, 54% for rice, and 51% for maize. Use of HYVs has increased production in many regions and sometimes reduced pressure on habitats by curbing the need to farm new lands.

EROSION OF CROP AND LIVESTOCK DIVERSITY

Although people consume approximately 7000 species of plants, only 150 species are commercially important; and about 103 species account for 90% of the world's food crops. Just three crops — rice, wheat, and maize — account for about 60% of the calories and 56% of the protein people derive from plants. Along with this trend toward uniform monocropping, the dependence on high levels of inputs such as irrigation, fertilizers, and pesticides has increased worldwide. The reduction in diversity often increases vulnerability to climate and other stresses, raises risks for individual farmers, and can undermine the stability of agriculture. In Bangladesh, for example, "promotion of HYV rice monoculture has decreased diversity, including nearly 7000 traditional rice varieties and many fish species. The production of HYV rice per acre in 1986 dropped by 10% from 1972, in spite of a 300% increase in agrochemical use per acre" (Hussein, 1994). In the Philippines, HYVs have displaced more than 300 traditional rice varieties that had been the principal source of food for generations. In India, by 1968, the so-called miracle HYV seed had replaced half of the native varieties, but these seeds were not high-yielding unless cultivated on irrigated land with high rates of fertilizer, which is often unaffordable to poor farmers (Shiva, 1991). Thus, the expected production increases were not realized in many areas.

Table 35.1   Extent of Genetic Uniformity in Selected Crops

Crop	Country	Number of Varieties
Rice	Sri Lanka	From 2000 varieties in 1959 to fewer than 100 today 75% descend from a common stock
Rice	Bangladesh	62% of varieties descend from a common stock
Rice	Indonesia	74% of varieties descend from a common stock
Wheat	U.S.	50% of crop in nine varieties
Potato	U.S.	75% of crop in four varieties
Soybeans	U.S.	50% of crop in six varieties

Data from World Conservation Monitoring Centre, *Global Biodiversity: Status of the Earth's Living Resources*, Groombridge, B., Ed., Chapman & Hall, London, 1992.

In Africa, transfer of the green revolution model has also reduced diversity. In Senegal, for example, a traditional cereal called fonio (*Panicum laetum*) — which is highly nutritious as well as robust in lateritic soils — has been threatened by extinction because of its replacement by modern crop varieties (IFOAM, 1994). In the Sahel, reports also confirm that traditional systems of polyculture are being replaced with monocultures that cause further food instability (Mann, 1994) (Table 35.1).

Homogenization also occurs in high-value export crops. Nearly all the coffee trees in South America, for example, are descended from a single tree in a botanical garden in Holland. *Coffea arabica* was first obtained from forests of southwest Ethiopia that have virtually disappeared (Fowler and Mooney, 1990). Uniform varieties are also common in export crops of bananas, cacao, and cotton, replacing traditional diverse varieties. Such changes have increased productivity, but the risks of narrowing varietal selection have become clear over time. In the north, similar losses in crop diversity are occurring. Many fruit and vegetable varieties listed by the USDA in 1903 are now extinct. Of more than 7000 apple varieties grown in the U.S. between 1804 and 1904, 86% are no longer cultivated, and 88% of 2683 pear varieties are no longer available (Fowler and Mooney, 1990) (Table 35.2). Evidence from Europe shows similar trends — thousands of varieties of flax and wheat vanished after HYVs were introduced (Mooney, 1979). Similarly, varieties of oats and rye are also declining in Europe (Vallve, 1993). In Spain and Portugal, various legumes that had been an important part of the local diet are being replaced by homogeneous crops; and in the Netherlands, four crops are grown on 80% of Dutch farmlands (Vallve, 1993).

Livestock is also suffering genetic erosion; the FAO estimates that somewhere in the world at least one breed of traditional livestock dies out every week. Many traditional breeds have disappeared as farmers focus on new breeds of cattle, pigs, sheep, and chickens (Plunknett and Horne,

Table 35.2 Reduction of Diversity in Fruits and Vegetables, 1903 to 1983 (Varieties in NSSL Collection)

Vegetable	Taxonomic Name	Number in 1903	Number in 1983	Loss (%)
Asparagus	*Asparagus officinalis*	46	1	97.8
Bean	*Phaseolus vulgaris*	578	32	94.5
Beet	*Beta vulgaris*	288	17	94.1
Carrot	*Daucus carota*	287	21	92.7
Leek	*Allium ampeloprasum*	39	5	87.2
Lettuce	*Lactuca sativa*	497	36	92.8
Onion	*Allium cepa*	357	21	94.1
Parsnip	*Pastinaca sativa*	75	5	93.3
Pea	*Pisum sativum*	408	25	93.9
Radish	*Raphanus sativus*	463	27	94.2
Spinach	*Spinacia oleracea*	109	7	93.6
Squash	*Cucurbita* spp.	341	40	88.3
Turnip	*Brassica rapa*	237	24	89.9

Data from Fowler, C. and Mooney, P., *The Threatened Gene — Food, Politics, and the Loss of Genetic Diversity*, The Lutworth Press, Cambridge, 1990.

1992). Of the 3831 breeds of cattle, water buffalo, goats, pigs, sheep, horses, and donkeys believed to have existed in this century, 16% have become extinct, and a further 15% are rare (Hall and Ruanne, 1993). Some "474 of extant [livestock] breeds can be regarded as rare. A further 617 have become extinct since 1892" (Hall and Ruanne, 1993). Over 80 breeds of cattle are found in Africa, and some are being replaced by exotic breeds (Rege, 1994). These losses weaken breeding programs that could improve hardiness of livestock.

As agricultural biodiversity is eroded, food security is often reduced and economic risks increased. Evidence indicates that such changes can decrease sustainability and productivity in farming systems (Tillman et al., 1996). Loss of diversity also reduces the resources available for future adaptation.

INCREASED VULNERABILITY TO INSECT PESTS AND DISEASES

Homogenization of varieties increases vulnerability to insect pests and diseases that can devastate a uniform crop, especially on large plantations. History has shown serious economic losses and suffering from relying on monocultural uniform varieties (Table 35.3). Among renowned examples are the potato famine of Ireland during the 19th century, a winegrape blight that wiped out valuable vines in both France and the U.S., a virulent disease *(Sigatoka)* that damaged extensive banana plantations in Central America in recent decades, and devastating mold that infested hybrid maize in Zambia.

In addition, a serious decline in soil organisms and soil nutrients has occurred. Beneficial insects and fungi also suffer under agriculture that involves heavy pesticide inputs and uniform stock — practices that make crops more susceptible to pest problems. These losses, along with fewer types of agroecosystems, also increase risks and can reduce productivity. In addition, many insects and fungi commonly seen as enemies of food production are actually valuable. Some insects benefit farming — for pollination, contributions to biomass, natural nutrient production and cycling, and as natural enemies to insect pests and crop diseases. Mycorrhizae, the fungi that live in symbiosis with plant roots, are essential for nutrient and water uptake.

The global proliferation of modern agricultural systems has eroded the range of insects and fungi, a trend that lowers productivity. Dependence on agrochemicals, and particularly the heavy use or misuse of pesticides, is largely responsible. Agrochemicals generally kill natural enemies and beneficial insects as well as the target pest. "Pesticides [especially when overused] destroy a wide array of susceptible species in the ecosystem while also changing the normal structure and function of the ecosystem" (Pimentel, 1992).

This disruption in the agroecosystem balance can lead to perpetual resurgence of pests and outbreaks of new pests — as well as provoke resistance to pesticides. This disturbing cycle often

Table 35.3 Past Crop Failures Due to Genetic Uniformity

Date	Location	Crop	Effects
1846	Ireland	Potato	Potato famine
1800s	Sri Lanka	Coffee	Farms destroyed
1940s	U.S.	U.S. crops	Crop loss to insects doubled
1943	India	Rice	Great famine
1960s	U.S.	Wheat	Rust epidemic
1970	U.S.	Maize	$1 billion loss
1970	Philippines, Indonesia	Rice	Tungo virus epidemic
1974	Indonesia	Rice	3 million tons destroyed
1984	U.S. (Florida)	Citrus	18 million trees destroyed

Data from World Conservation Monitoring Centre, *Global Biodiversity: Status of the Earth's Living Resources,* Groombridge, B., Ed., Chapman & Hall, London, 1992.

leads farmers to apply increasing amounts of pesticides or to change products — a strategy that is not only ineffective but also further disrupts the ecosystem functions and elevates costs. This *pesticide treadmill* has occurred in numerous locations throughout the world. Reliance on monocultural species and the decline of natural habitat around farms also cuts beneficial insects out of the agricultural ecosystem.

ADDITIONAL LOSSES — HABITATS, NUTRITION, AND KNOWLEDGE

Agricultural expansion has also reduced the diversity of natural habitats, including tropical forests, grasslands, and wetland areas. Projections of food needs in the coming decades indicate probable further expansion of cropland, which could add to this degradation. Modifying natural systems is necessary to fulfill the food needs of growing populations, but many conventional forms of agricultural development, particularly large-scale conversion of forests or other natural habitats to monocultural farming systems, erode the biodiversity of flora and fauna. Intensive use of pesticides and fertilizers can also disrupt and erode biodiversity in natural habitats and ecosystem services that surround agricultural areas, particularly when these inputs are used inappropriately.

Other direct effects of reduced diversity of crops and varieties include:

- Decline in the variety of foods adversely affects nutrition (IIED, 1995).
- High-protein legumes have often been replaced by less nutritious cereals (Shiva, 1991).
- Local knowledge about diversity is lost as uniform industrial agricultural technologies predominate (Altieri, 1991).
- Institutions and companies in the north have unfair advantages in exploiting the diverse biological resources from the tropics (Shand, 1997).

In sum, the loss of agrobiodiversity adds immediate costs to producers, creates social costs to communities and nations, has long-term effects on agricultural productivity, and jeopardizes food security.

CAUSES OF AGROBIODIVERSITY DECLINE

The causes of agrobiodiversity loss are very complex and they vary under different conditions. In general, however, the proximate explanations are often tied to the use of unsustainable farming methods and degrading land use practices. More deeply, the root causes that underlie the erosion of agricultural biodiversity are generally tied to policies, economic, and institutional factors that largely determine field-level practices and behaviors. Other root causes include demographic pressures, disparities in resource distribution, the prevalent influence of industrial monocultural agricultural models or paradigms that contribute to decline of biodiversity, the depreciation and devaluation of diversity and accumulated local knowledge, and market demands for standardized products (Shand, 1997; Raeburn, 1995; Fowler and Mooney, 1990). Changing these patterns requires reforming land use and technology policies, broader socioeconomic changes that give the rural poor more economic and educational opportunities, and changes in the predominant scientific models or paradigms.

CONSERVATION AND SUSTAINABLE USE OF AGROBIODIVERSITY

Humanity faces a major challenge to prevent and mitigate the losses of agricultural biodiversity and to build strategies to integrate biodiversity and agriculture. The conservation and enhancement of agrobiodiversity have been upheld in major international conventions, particularly the Convention on Biodiversity (CBD, 2001) and the World Food Summit of 1996. These significant global conventions establish a framework for biodiversity conversation for all signatory nations, and they identify specific mandates for implementing agrobiodiversity conservation measures, sustainable use, and benefit sharing of plant genetic resources. The FAO has also recommended strategies and approaches for the conservation and sustainable use of agricultural biodiversity, to implement the Convention on Agrobiodiversity (FAO, 1998, 1999).

DIVERSITY THROUGH SUSTAINABLE AGRICULTURE PRINCIPLES AND PRACTICES

Proven practical experiences provide lessons and opportunities to integrate biodiversity in agriculture. Such experiences and practices must be strongly supported and multiplied. Effective approaches to conserve and enhance agrobiodiversity fit within a general framework of sustainable agriculture. This approach merges the goals of productivity, food security, and social equity with ecological health. A shift to sustainable agriculture requires changes in production methods, models, and policies, as well as the full participation of local people. In this approach, scientific advancements in genetics and improved varieties can have significant roles, but they need to be reoriented toward conserving and using diversity in farming systems.

To achieve such transformations for the conservation and enhancement of agricultural biodiversity, the following strategic principles are critical:

- Application of *agroecological principles* helps conserve, use, and enhance biodiversity on farms and can increase sustainable productivity and intensification that avoids extensification, thereby reducing pressure on off-farm biodiversity.
- *Participation and empowerment of farmers and indigenous peoples,* and protection of their rights, are important means of conserving agrobiodiversity in research and development.
- *Adaptation of methods* to local agroecological and socioeconomic conditions, building upon existing successful methods and local knowledge, is essential to link biodiversity and agriculture and to meet livelihood needs.
- Conservation of plant and animal genetic resources — especially *in situ* efforts — helps protect biodiversity for current livelihood security as well as future needs and ecosystem functions.
- Reforming genetic research and breeding programs to enhance agrobiodiversity is essential and can also benefit production.
- Creating a *supportive policy environment* — including eliminating incentives for uniform varieties and for pesticides and implementing policies for secure tenure and local rights to plant genetic resources — is vital to enhance agricultural biodiversity and food security.

Applying these basic principles can generate considerable public and private benefits. Specific practices that have proven effective for this purpose have been discovered and adapted in many areas of the world; they are listed in Box 35.2. Building on the knowledge of rural people has also proven to be effective in many contexts to make scientific advancements and to help ensure adoption and spread the benefits of agrobiodiversity innovations (Altieri and Merrick, 1988; Brookfield and Padoch, 1994; Michon and de Foresta, 1990). The use of such principles and practices has resulted in production increases in both small- and large-scale farms. Additional advantages include improvement of soil nutrient cycles and soil quality, added economic value, increase in sustainability and stability of systems, and alleviation of pressures on habitats.

Ecologically oriented Integrated Pest Management (IPM) methods illustrate well the use and benefits of biodiversity. IPM approaches usually highlight diversity as a key feature. Examples of best practices that are effective for insect management include:

BOX 35.2
KEY PRINCIPLES AND PRACTICES TO USE AND ENHANCE AGROBIODIVERSITY

1. Management of diverse productive resources
   a. Diversification and diversity enhancement
      • Temporal (crop rotation, sequences)
      • Spatial (polycultures, agroforestry, crop/livestock systems, intercropping)
      • Genetic (multiple species/varieties, multilines, interspecies)
      • Regional (i.e., variation in watersheds, zones)
   b. Recycling and conservation of soil nutrients and organic matter
      • Plant biomass (green manures, crop residues, mulch, for diverse soil nutrients)
      • Animal biomass (manure/dung, urine, etc.)
      • Reuse of nutrients and resources internal and external to the farm (e.g., tree litter)
      • Integrate diverse plants or organisms (vermiculture, cover crops, mainly legumes)
   c. Integrated Pest Management, stressing agroecological approaches
      • Natural biological control (enhancing natural control agents)
      • Imported biological control methods (e.g., add natural enemies, botanical products)
      • Diverse cropping/soil management methods to enhance natural fauna
      • Enhancing use of habitats and species in habitats
2. Conservation and regeneration of resources (stressing diversity aspects)
   a. Germ plasm conservation (plant and animal species, landraces, adapted germ plasm)
   b. Beneficial fauna and flora (multiple use vegetation, pollinators, natural enemies)
   c. Soil health (erosion control, fertility enhancement, see recycling above)
   d. Water (harvesting, conservation, management, irrigation)

Adapted from UNDP, 1995; Altieri, 1991.

- Multiple cropping and crop rotations, used to prevent buildup of pests
- Intercropped plants that house predators of insect pests
- Intercropped plants that act as alternative host plants for pests; for example, in Tlaxcala, Mexico, farmers grow lupinus plants in their corn to attract the scarab beetles, and thus protect corn; in California, vineyards and orchards use cover crops for similar purposes
- Use of certain plants as natural pesticides; for example, in Ecuador, castor leaves that contain a paralyzing agent are used to control the tenebronid beetle
- Weeds that are used to repel insects; for example in Colombia, grassweeds are grown around bean fields to repel leafhoppers; and in Southern Chile, the shrub *Cestrum parqui* is used to repel beetles in potato fields
- Integration of biocontrol agents, including various parasites, animals, and fish that consume insect pests (such as use of ducks and fish in rice paddies in Asia)
- Elimination or reduction of pesticide use to avoid adverse agroecological effects on the insect diversity in agroecosystems

Effective disease management practices using agrobiodiversity include:

- Mixed crop stands that slow the spread of diseases by altering the microenvironment; for example, in Central America, cowpeas grown with maize are less susceptible to the fungus *Ascochyta phaselolorum* and to the cowpea mosaic virus
- Use of nonhost plants as decoy crops to attract fungus (or nematodes) (Altieri, 1987; Thrupp, 1996; UNDP, 1995, 1992; Lee, 1991; NRC, 1989)

Successful IPM programs in Asia illustrate that building agrobiodiversity — particularly using diverse beneficial insects — is a key ingredient of effective pest management in rice production. These initiatives, coordinated by the Food and Agriculture Organization along with governmental and nongovernmental organizations, have resulted in remarkable reductions of pesticide use and increased rice yields. In Bangladesh, for example, thousands of farmers involved in IPM projects have also integrated fish into rice paddies, have adopted agroecological methods to restore the natural balance between insects and other fauna, and have planted vegetables on the dikes around the edges. This approach has increased rice yields, provided new sources of nutrition, and made hazardous chemical use unnecessary. For example, farmers in the pilot IPM program achieved an 11% increase in rice production while eliminating pesticides (Thrupp, 1996).

Practices for improving soil fertility and soil health also make use of agrobiodiversity. Good examples include:

- Compost from crop residues, tree litter, and other plant and organic residues
- Intercropping and cover crops, particularly legumes, which add nutrients, fix nitrogen, and pump nutrients to the soil surface
- Use of mulch and green manures (through collection and spread of crop residues, litter from surrounding areas, and organic materials and under crop)
- Integration of earthworms (vermiculture) or other beneficial organisms and biota into the soil to enhance fertility, organic matter, and nutrient recycling
- Reduction of agrochemicals — especially toxic nematicides — that destroy beneficial diverse soil organisms and organic material (UNDP, 1992; Altieri, 1991, 1987; Lee, 1991)

These kinds of soil management practices have proven effective and profitable in a variety of farming systems. Agroforestry illustrates *best practice* of using agrobiodiversity that also generates multiple benefits (Michon and Foresta, 1990). In many contexts, the integration of trees into farming systems is highly efficient, and the trees have multiple functions, such as providing fuel, fodder, shade, nutrients, timber for construction, and aiding soil conservation and water retention. (In West Sumatra, agroforestry gardens occupy 50% to 85% of the total agricultural land.) Complex forms of agroforestry exhibit forest-like structures as well as a remarkable degree of plant and animal diversity, combining conservation and natural resource use. (In Indonesia, for example, smallholder "jungle rubber" gardens incorporate numerous tree species.) Agroforestry systems in traditional forms also shelter hundreds of plant species, constituting valuable forms of *in situ* conservation (Michon and Foresta, 1990). Many of the practices noted here serve multiple purposes. For example, intercropping provides pest and soil management as well as enhanced income. For example, an estimated 70% to 90% of beans and 60% of maize in South America are intercropped with other crops. Farmers in many other parts of the world have recognized such diversity as valuable sources of soil nutrients, nutrition, and risk reduction — essential for livelihoods as well as other economic values (UNDP, 1995).

A common misperception is that agrobiodiversity enhancement is feasible only in small-scale farms. In fact, experience shows that large production systems also benefit from incorporating these principles and practices. Crop rotations, intercropping, cover crops, IPM, and green manures are the most common methods used profitably in larger commercial systems, both in the north and in the south (Box 35.3).

These situations illustrate *sustainable approaches to intensification.* Examples are found in tea and coffee plantations in the tropics and in vineyards and orchards in temperate zones (UNDP, 1995). In most large-scale settings, the change from monocultural to diverse systems and practices entails transition costs and sometimes trade-offs or profit losses for the first two or three years. However, after the initial transition, producers have found that agroecological changes are profitable as well as ecologically sound for commercial production and that they present new valuable opportunities.

USING PARTICIPATORY APPROACHES

The incorporation of farmers' local knowledge, practices, and experimentation is advantageous in such efforts in agrobiodiversity and sustainable agriculture. Evidence shows that full involvement of local farming practices in agricultural research and development — through participation and

BOX 35.3
DIVERSITY IN LARGE-SCALE FARMING: COFFEE, TEA, AND GRAPES

India

The Singampatti Group of Estates is a large tea plantation (312 hectares) located in Tamil Nadu, India. A subdivision of the Bombay Burmah Trading Corporation Ltd., it emphasizes integrated, organic management. Several practices enhance diversity. Shade trees (e.g., *Grevilla robusta*, *Erythrina lithosperma*, and *Gliricidia sepium*) are cultivated along with commercial bush and tree species like cinchona and cardamom. The shade trees fix nitrogen, recycle nutrients, and prevent nutrient leaching. Open areas are planted with leguminous crops to control erosion and weeds. The tea bushes are planted in trenches, which are filled with compost, prunings, and castor or neem cakes to conserve water, increase nutrients, and reduce erosion. Cattle are a source of dairy products and income for the workers, and their dung is also used to produce gas. Labor conditions are also improved in this estate. The plantation houses many species of endangered animals, including tigers, the lion-tailed macaque, and the Malabar squirrel. The yields on this organic estate are 11% higher than in conventional production, but cultivation costs are about twice that of conventional tea production. The market price for organic tea is about 80% higher than conventional tea. The greatest concern of the plantation is the "loss of … genetic diversity" of seed tea because of the tradition of planting clones. Although the clone has built-in resistance to blister blight, the managers fear that uniformity invites pests and diseases.

Mexico

Started in 1928, Finaca Irlanda, in the state of Chiapas in Mexico, is one of the oldest organic and biodynamic coffee estates in the world. The owner, a naturalist, is committed to maintaining diversity on his farm. The farm area is 320 hectares, and the main crops are Arabica and Robusta varieties of coffee, intercropped with cardamom and cacao. The farm also raises dairy and beef cattle. All waste from the farms' animals and plants are used to improve nutrients. More than 40 varieties of leguminous trees provide both shade and nitrogen. Pest control is practiced by maintaining crop diversity and by using biological agents, such as a wasp introduced in Mexico from Africa to control the fungi *Beauvaria bassiana*. Indigenous wild animals that are threatened with extinction, such as puma, wild boar, pheasants, and toucans, are protected on the plantation. Some of the animals, such as the ocelot and gray fox, are used as natural predators to control pests such as rodents. Even though the gross income on this organic farm is much higher than on a conventional farm, the costs of maintenance are higher on an organic farm since it is more labor-intensive, so the profit margin is narrower. However, the coffee estate serves as a center for learning for other farmers and "serves as a tool for local extension."

Vineyards

Large-scale commercial winegrape plantations, both in the north and the south, are increasingly turning to integrated crop/pest management methods and organic production. Renowned California wine companies (such as Fetzer, Gallo, and Mondavi) are converting from conventional chemical-intensive practices to diverse cover-cropping and other sustainable practices. They generally plant a mixture of cover crops between rows of vines; these include clovers, vetch, native grasses, radishes, and flowering grasses. The cover crops are then mowed and incorporated into the soil several times in a year. The cover crops have multiple purposes:

(Continued)

BOX 35.3 (*CONTINUED*)
DIVERSITY IN LARGE-SCALE FARMING: COFFEE, TEA, AND GRAPES

- Improving soil structure and water penetration, adding organic matter and roots to increase soil aeration, and decreasing tillage requirements
- Preventing soil erosion, partly by spreading and slowing the movement of water
- Improving soil fertility by adding organic material
- Helping control insect pests by harboring beneficial insect predators and parasites

This management method has proven both economical and sustainable. Grape growers in other countries have also begun to adopt such methods.

Cases adapted from UNDP, 1995.

---

leadership of local people — has had beneficial outcomes and needs to be done consistently. In other words, a farmer-friendly approach is essential to develop changes. Efforts to integrate agrobiodiversity in farming can be made more relevant by drawing on farmers' own informal knowledge and methods of experimenting with varieties and new practices. At the same time, the involvement of farmers as partners in research and development helps to ensure adoption of agroecological methods and can help to empower local people (Chambers et al., 1987; Rajesekaran, 1993; Thrupp, 1996; Altieri and Merrick, 1988).

In Mexico, for example, researchers worked with the local people to recreate *Chinampas* — multicropped, species-diverse gardens developed from reclaimed lakes — that were native to the Tabasco region and part of Mexico's pre-Hispanic tradition (Morales, 1984; Altieri, 1991). A similar project conducted in Veracruz also incorporated the traditional Asiatic system of mixed farming — mixing Chinampas with animal husbandry and aquaculture. These gardens also made more productive use of local resources and integrated materials from plant and animal waste as fertilizers. Yields of such systems equaled or surpassed those of conventional systems.

In Burkina Faso, on the other hand, a soil conservation and integrated cropping project in Yatenga province was based largely on an indigenous technology of Dogon farmers in Mali — building rock bunds for preventing water runoff (UNDP, 1995). The project added innovations — bunds along contour lines — and revived an indigenous technique called *zai*, which is adding compost to holes in which seeds of millet, sorghum, and peanut are planted. These crops are in a multicropping system. Animals are incorporated for their manure. In fields using these techniques, yields were consistently higher than in fields using conventional practices, ranging from 12% higher in 1982 to 91% in 1984. Yields in the zai method reached 1000 to 1200 kg/ha, compared to conventional yields of 700 kg/ha. Water management was enhanced and food security of the local people was also improved through this approach. The techniques have been widely adopted, covering 3500 hectares by the end of 1988 (UNDP, 1995).

In these efforts, the active participation of women has significant benefits. As managers of biodiversity in and around farming systems in many areas of the world, women can make important contributions and have a promising role in research, development, and conservation of agrobiodiversity. In Rwanda, for example, in a plant breeding project of the International Center for Tropical Agriculture, scientists worked with women farmers from the early stages of a project on breeding new varieties of beans to suit local peoples' needs (CGIAR, 1994). Together they identified the characteristics desired to improve beans, run experiments, manage and evaluate trials, and make decisions on the trial results. The experiments resulted in stunning outcomes: the varieties selected and tested by women farmers over four seasons "performed better than the scientists' own local mixtures 64% to 89% of the time" (CGIAR, 1994). The women's selections also produced substantially more beans, with average production increases as high as 38%.

The development of participatory approaches requires deliberate measures, training, and time to change the conventional approaches of agricultural R&D (Chambers et al., 1987; Thrupp, 1996). The application of such participatory methods improves the likelihood of adoption and success of agrobiodiversity efforts. Some basic principles of participatory research and development include joint problem-solving among farmers and scientists and responsiveness to local needs, mutual listening and learning between farmers and scientists, flexibility in selecting methods so that they are adjusted to local timing, and interdisciplinary and holistic perspectives (Chambers et al., 1987; Thrupp, 1996; Rajasekaran, 1993).

In sum, the use of these participatory approaches can help planners and communities to identify and develop best practices in sustainable production, meaning practices that are adapted to diverse local conditions, integrating agriculture with biodiversity while meeting socioeconomic needs.

CONFRONTING THE CAUSES OF AGROBIODIVERSITY DECLINE THROUGH POLICY CHANGES

Efforts to conserve and enhance agrobiodiversity must also address the underlying causes that accelerate its loss. Therefore, changes are required in agricultural policies, and commitments are needed by governments and institutions to confront the roots of the problems.

Several policy initiatives and institutions have already been established to address some of these policy issues. For example, several international institutions influence and regulate the use of plant genetic resources. Among the key players are the Consultative Group on International Agricultural Research, the International Plant Genetic Resources Institute, the Food and Agriculture Organization, the Commission on Plant Genetic Resources, and the World Intellectual Properties Organization. Recent important international conventions and agreements, particularly the Convention on Biological Diversity and the General Agreement on Tariffs and Trade, are also influential in setting guidelines that affect agrobiodiversity and use of genetic resources.

Concerns about the control of plant genetic resources have generated intellectual property regulations that govern the activities of public institutions and private companies and affect farmers' legal access to genetic resources. Gene banks conserve a remarkable diversity of plant genetic resources, and increasing numbers of agricultural research institutes have begun *in situ* conservation projects as well. Along with these large formal institutions, many NGOs and local farmer organizations are also increasingly involved in promoting the conservation and equitable distribution of benefits from agrobiodiversity.

Although many institutions are already actively involved, more coordination and work is needed *at all levels* to ensure effective reforms and agrobiodiversity conservation policies that benefit the public. Policy changes that attack the roots of problems and ensure citizens' rights can help achieve these aims. Strategies needing further attention include:

- Ensuring public participation in the development of agricultural and resource use policies
- Reducing or eliminating subsidies and credit policies for HYVs, fertilizers, and pesticides to encourage the use of more diverse seed types and farming methods
- Policy support and incentives for effective agroecological methods that make sustainable intensification possible
- Reform of tenure and property systems that affect the use of biological resources to ensure that local people have rights and access to necessary resources
- Development of markets and business opportunities for diverse organic agricultural products
- Changing marketers' and retailers' demands to favor diversity of food instead of standardized uniform products, and to reflect the actual diversity of preferences in human populations

Integrating biodiversity and agriculture will also require changes in agricultural research and development, land use, and breeding approaches. Such changes are urgently needed to overcome threats from the ongoing erosion of genetic resources and biodiversity. Experience shows that enhancing agrobiodiversity economically can benefit both small- and large-scale farmers, while at the same time serving the broader social interests of ecosystem health and food security.

ACKNOWLEDGMENTS

I am grateful to research assistants Rita Banerji, Dina Matthews, and Nabiha Megateli, and to the following WRI colleagues for sharing their expertise and assistance: Walt Reid, Kenton Miller, Thomas Fox, Robert Blake, Arthur Getz, Nels Johnson, Paul Faeth, and Consuelo Holguin. I appreciate the support and comments from the World Bank Environment and Agriculture Departments, particularly to World Bank staff Jitendra Srivastava, Lars Vidaeus, and John Kellenberg (Kellenberg and Vidaeus, 1997). The comments of Calestous Juma, Miguel Altieri, Kristin Schafer, Daniel Debouck, David Williams, and Frances Seymour are also appreciated. I thank the Swedish International Development Cooperation Agency for support for the project on agrobiodiversity.

REFERENCES

Abramowitz, J. and Nichols, R., Women and agrobiodiversity, *Soc. Int. Dev. J. Dev.*, 1993.
Altieri, M., *Agroecology: The Scientific Basis of Sustainable Agriculture*, Westview Press, Boulder, CO, 1987.
Altieri, M., Traditional farming in Latin America, *Ecologist,* 21(2):93, 1991.
Altieri, M. and Merrick, L., Agroecology and *in situ* conservation of native crop diversity in the third world, *Biodiversity*, National Academy Press, Washington, D.C., 1988.
Brookfield, H. and Padoch, C., Appreciating agrodiversity: a look at the dynamism and diversity of indigenous farming practices, *Environment,* 36(5):7–44, 1994.
Brown, A.H.D., Isozymes, plant population genetic structure, and genetic conservation, *Theor. Appl. Genet.,* 52:145–157, 1978.
CBD, Convention on Biological Diversity, Secretariat for the Convention on Biological Diversity, U.N. Environment Programme, Nairobi, Kenya, 2001, available online at http://www.biodiv.org.
CGIAR, *Partners in Selection*, Consultative Group on International Agricultural Research, Washington, D.C., 1994.
Chambers, R., Pacey, A., and Thrupp, L.A., *Farmer First: Farmer Innovation and Agricultural Research*, IT Publications, London, 1987.
Cleveland, D., Soleri, D., and Smith, S., Do folk crop varieties have a role in sustainable agriculture? *BioScience,* 44(11):740–751, 1994.
FAO, Agricultural biodiversity: assessment of ongoing activities and priorities for a programme of work, UNEP/CMD/SBSTTA/5/10, Food and Agriculture Organization of the U.N., Rome, 1998, available online at http://www.fao.org/biodiversity/.
FAO, Background paper on agricultural biodiversity, Food and Agriculture Organization of the U.N., Rome, 1999, available online at www.fao.org/biodiversity/.
Forno, D. and Smith, N., Biodiversity and Agriculture: Implications for Conservation and Development, World Bank Technical Paper 321, The World Bank, Washington, D.C., 1996.
Fowler, C. and Mooney, P., *Shattering: Food, Politics, and the Loss of Genetic Diversity.* University of Arizona Press, Tucson, AZ, 1990a, p. 104.
Fowler, C. and Mooney, P., *The Threatened Gene — Food, Politics, and the Loss of Genetic Diversity*, The Lutworth Press, Cambridge, 1990b.
GRAIN, Biodiversity in agriculture: some policy issues, *IFOAM Ecol. Farming,* January 14, 1994.
Greenburg, R., Phenomena, comment and notes, *Smithsonian,* 25(8):24–27, 1994.

Hall, S.J.G. and Ruane, J., Livestock breeds and their conservation: a global overview, *Conserv. Biol.,* 7(4):815–825, 1993, cited in Smith, N., The Impact of Land Use Systems on the Use and Conservation of Biodiversity, draft paper, World Bank, Washington, D.C., 1996.

Hussein, M., Regional focus news Bangladesh, *Ecol. Farming: Global Monitor, IFOAM,* January 20, 1994.

IFOAM, Biodiversity: crop resources at risk in Africa, *Ecol. Farming: Global Monitor,* January 5, 1994.

IIED, *Hidden Harvests Project Overview: Sustainable Agriculture Program*, International Institute for Environment and Development, London, 1995.

Kellenberg, J. and Vidaeus, L., Mainstreaming Biodiversity in Agricultural Development: toward Good Practice, World Bank Environment Paper 15, World Bank, Washington, D.C., 1997.

Lee, K.E., The diversity of soil organisms, in Hawksworth, D.K., Ed., *The Biodiversity of Microorganisms and Invertebrates: Its Role in Sustainable Agriculture*, CAB International, London, 1991, pp. 73–87.

Mann, R.D., Time running out: the urgent need for tree planting in Africa, *Ecologist,* 20(2):48–53, 1994.

Michon, G. and de Foresta, H., Complex agroforestry systems and the conservation of biological diversity, in harmony with nature, *Proc. Int. Conf. Trop. Biodiversity*, SEAMEO-BIOTROP, Kuala Lumpur, Malaysia, 1990.

Mooney, P., *Seeds of the Earth: A Private or Public Resource?* Canadian Council for International Cooperation, Ann Arbor, 1979.

Morales, H.L., Chinampas and integrated farms: learning from the rural traditional experience, in De Castri, F., Baker, G., and Hadley, M., Eds., *Ecology and Practice. Volume 1, Ecosystem Management*, Tycooly, Dublin, 1984.

Pimentel, D., Conserving biological diversity in agricultural/forestry systems, *BioScience,* 42(5):360, 1992.

Plucknett, D. and Horne, M.E., Conservation of genetic resources, *Agric. Ecosyst. Environ.,* 42:75–92, 1992, cited in Smith, N., The Impact of Land Use Systems on the Use and Conservation of Biodiversity, draft paper, World Bank, Washington, D.C., 1996.

Raeburn, P., *The Last Harvest: The Genetic Gamble That Threatens to Destroy American Agriculture*, Simon & Schuster, New York, 1995.

Rajasekaran, A framework for incorporating indigenous knowledge systems into agricultural extension, *Indigenous Knowledge Dev. Mon.,* 1(3):21–24, 1993.

Rege, J.E.O., International livestock center preserves Africa's declining wealth of animal biodiversity, *Diversity,* 10(3):21–25, 1994.

Shand, H., *Human Nature: Agricultural Biodiversity and Farm-Based Security*, RAFI, Ottawa, Ontario, 1997.

Shiva, V., The green revolution in the Punjab, *Ecologist,* 21(2):57–60, 1991.

Thrupp, L.A., Women, wood, and work: in Kenya and beyond, *Unasylva,* FAO, Rome, December, 1984.

Thrupp, L.A., Ed., *New Partnerships for Sustainable Agriculture*, World Resources Institute, Washington, D.C., 1996.

Thrupp, L.A., *Cultivating Diversity: Agrobiodiversity and Food Security,* World Resources Institute, Washington, D.C., 1998.

Thrupp, L.A., Hecht, S., and Browder, J., *Diversity and Dynamics of Shifting Cultivation: Myths, Realities, and Human Dimensions*, World Resources Institute, Washington, D.C., 1996.

Tillman, D., Wedline, D., and Knops, J., Productivity and sustainability influenced by biodiversity in grassland ecosystems, *Nature,* 379(22):718–720, 1996.

UNDP, *Benefits of Diversity: An Incentive toward Sustainable Agriculture*, U.N. Development Program, New York, 1992.

UNDP, *Agroecology: Creating the Synergism for a Sustainable Agriculture,* United Nations Development Program, New York, 1995.

UNDP, *Agroecology: Creating the Synergisms for Sustainable Agriculture*, United Nations, New York, 1996.

UNEP, *Global Biodiversity Assessment*, United Nations Environmental Program, Nairobi, Kenya, 1995.

Vallve, R., The decline of diversity in European agriculture, *Ecologist,* 23(2):64–69, 1993.

World Conservation Monitoring Centre, *Global Biodiversity: Status of the Earth's Living Resources*, Groombridge, B., Ed., Chapman & Hall, London, 1992.

CHAPTER **36**

Evolving Opportunities for Management of Agricultural Landscapes and Ecosystem Health: Sustainability by Opportunity

Julian Dumanski, Steen Joffe, E.R. Terry, and Christian Pieri

INTRODUCTION

The intimate connections between economic growth and ecosystem health are becoming increasingly clear. Pursuing these concomitantly makes good business sense and provides for considerable growth opportunities. The only choices are short-term economic gains vs. long-term prosperity. The debate between "jobs or the environment" is becoming a debate of the past.

Agriculture and related biology-based land uses, including forestry and agroforestry, occupy major areas of the Earth's land areas. Vitousek (1994) estimates that currently about one third to one half of the Earth's nonglaciated land areas are regularly managed, and up to 70% receive some degree of human intervention. How these lands are managed and the nature of management systems have a direct impact on the health of local and even global ecosystems and the services derived therefrom. These sectors have both major opportunities and responsibilities to ensure healthy ecosystems, but their current performance leaves much to be desired. Also, even though there are major trends toward global urbanization, the proper husbandry of rural landscapes has huge consequences to the provision of high-quality environmental services, such as clean water, to urban dwellers.

In the years since the 1992 Rio accords, economic growth of $2.4 trillion and population growth of about 400 million have placed continuing pressures on the Earth's natural resources and ecosystems (Brown et al., 1998). Tropical forest cover, wetlands, and other natural habitats have declined by 3.5%, carbon emissions have increased by 4%, while natural carbon sinks in soils and forests have been degraded or lost. As much as 10 million hectares of land are lost annually to severe degradation; we currently consume about half of the available fresh water, and more nitrogen is fixed by humanity than by all natural sources combined. Each year about 1000 new chemicals are released for use without any knowledge of their biological or synergistic effects, but only a few are regularly monitored. About one quarter of the bird species has been driven to extinction, and two thirds of the global fishery is fully exploited or overexploited.

We are transforming the Earth through land clearing, forestry, overgrazing, urbanization, mining, trawling, dredging, and so on. We remove species through overexploitation while adding new variations through biotechnology and other means. In these activities, we alter the major biogeochemical cycles, with impacts on the global climate and, increasingly, on global life-support systems. A statement from the Ecological Society of America (Lubchenko, 1998) asserts that "environmental problems resulting from human activities have begun to threaten the sustainability of earth's life-support system. Among the most critical challenges facing humanity are the conservation, restoration, and wise management of the earth's resources."

For the first time in history, global populations are of such magnitude that how we manage the land has a direct impact on the ecological services that support life on this planet. These services include purification of air and water, mitigation of floods and droughts, detoxification and decomposition of wastes, generation and renewal of soil and soil fertility, pollination of crops and natural vegetation, control of potential agricultural and forestry pests, dispersal of seeds, translocation of nutrients, maintenance of biodiversity (from which humanity derives major agricultural, medicinal, and industrial benefits), protection from harmful ultraviolet rays, stabilization of climate and moderation of climatic extremes, provision of aesthetic beauty, and support of human cultures. The importance of the global environment and ecosystem health is not an issue just for the resource industries. Increasingly, it is viewed also in the context of human health, world trade, social justice, and even national security (Lubchenko, 1998). As a society, we have never been at this point and we are unsure how to proceed.

Many of the problems arise from our penchant to control rather than live within the bounds of nature. In the process we create wastes and other excesses beyond the capacity of global and local ecosystems to absorb. Examples are soil erosion beyond the capacity to form new soil, CO_2 emissions beyond the capacity for carbon sequestration, pollution beyond the capacity of local ecosystems to filter the wastes, crop yields beyond levels of ecological baseline productivity, groundwater withdrawals beyond recharge capacity, and deforestation beyond sustainable yields. In most cases the problems arise with natural resources that are simultaneously viewed as depletable and renewable.

OPPORTUNITIES THROUGH THE INTERNATIONAL CONVENTIONS

Agriculture has historically been a major contributor to environmental degradation, but under improved systems of land management, it could be a major partner in the environmental solution. Land management decisions by individual farmers have implications for many environmental goods and services,* such as effects on *habitats* for fauna and flora, on a variety of *ecological services*, and on *amenity* or aesthetic values (Joffe et al., 2002). The impacts may arise directly on land managed for agriculture and livestock, or indirectly as a consequence of fragmentation and degradation of natural (less managed) habitats such as forests and wetlands. Many of the environmental benefits associated with sustainable land management will accrue locally and nationally. These include productivity effects such as pollination, biological control, nutrient cycling, and soil conservation, as well as off-site effects such as water regulation and supply, disturbance regulation, waste treatment, and flood control. Others are more clearly global, or at least *supranational* in scope, such as climate regulation, conservation of genetic resources with potential value in plant breeding or pharmaceuticals, international tourism, and transboundary water-mediated effects (Figure 36.1). Sometimes it is essential to promote intensified uses in more favored areas to reduce pressures on marginal, fragile environments.

Although the ecological functions are discrete, in practice the boundaries are far from clearcut. The global environmental benefits and costs are those for which the global community, through international agreements and nascent trading frameworks, has expressed a willingness to pay. The rationale is on the grounds that:

- They would normally receive suboptimal attention within a national accounting and planning framework
- They are considered highly valuable or irreplaceable
- There is considered to be an unacceptable economic or humanitarian risk associated with further depletion.

---

* Also referred to in this chapter as *biodiversity goods and services*. The stock of natural resources from which these goods and services flow is also referred to as *natural capital*. An important principle is that sustainably managed land constitutes a form of natural capital from which a variety of key services may be derived in the long term.

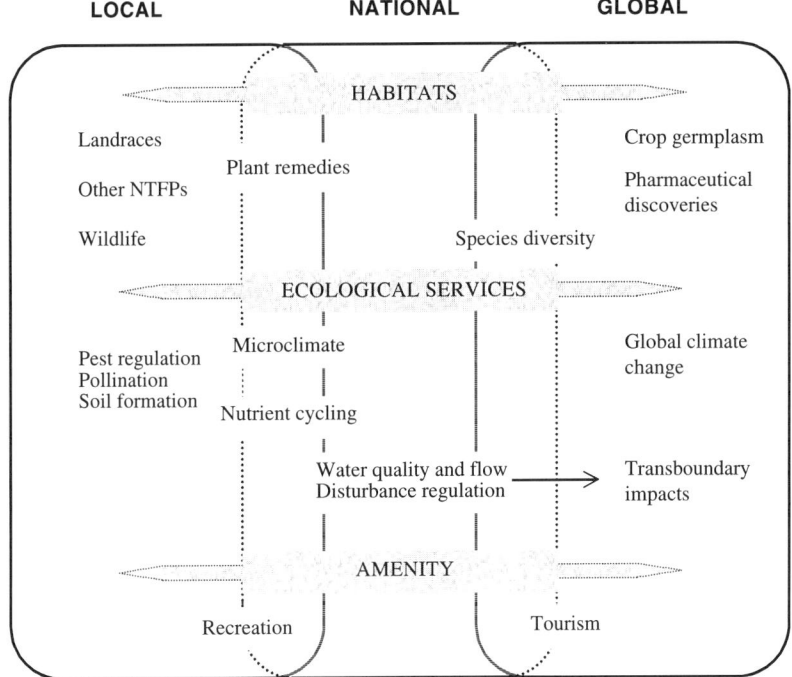

Figure 36.1 Examples of environmental benefits of sustainable land management: local, national, and global layers. (From Joffe, S. et al., *Opportunities to Capture the National and Global Environmental Benefits of Sustainable Land Management*, The World Bank, Geneva, in press.)

The three categories relate to (1) biodiversity (which embraces all goods and services associated with terrestrial ecosystems); (2) climate change (to do with concerns about emissions of greenhouse gases); and (3) international waters (where the negative impacts of depleted water flow or quality have serious transboundary implications). In all cases, there are direct linkages between global environmental change and land management, as well as opportunities under the international agreements to generate funding to achieve these objectives. The main relevant United Nations agreements and financing mechanisms are:

- Convention on Biological Diversity (CBD)
- Framework Convention for Climate Change (UNFCCC)
- Convention to Combat Desertification (CCD)
- Global Environment Facility (GEF)

Biodiversity and Agriculture

Biodiverse* ecosystems have a fundamental role and importance in sustainable development and provide many important benefits. They often contain a variety of economically useful products that can be harvested or serve as inputs for *production* processes. In addition, they provide *habitats* for flora and fauna and many key *ecological services* including those associated with nutrient cycling, disturbance regulation, availability and quality of water for agriculture, industry, or human consumption. Agriculture remains dependent on many biological services, such as provision of

---

* Biological diversity, often shortened to biodiversity, embraces the whole of *life on Earth*; it encompasses the variability among living organisms from all sources, including terrestrial and aquatic ecosystems and their ecological complexes. This includes diversity within species, among species, and of ecosystems. Decline in biodiversity includes all those changes that have to do with reducing or simplifying biological heterogeneity — from individual members of a species to regional ecosystems.

genes for improved varieties and livestock breeds, but also for crop pollination, soil fertility services provided by microorganisms, and pest control services provided by insects and wildlife. Conversely, sustainably managed agricultural landscapes are important to the conservation and enhancement of biodiversity. The term *agrobiodiversity* has been coined to describe the important subset of biodiversity that contributes to agriculture.

Climate Change and Agriculture

The linkages between agricultural land use and greenhouse gases relate to land use dynamics and management of rural landscapes. During the 19th century, rapid agricultural expansion, primarily in temperate regions, led to widespread clearing of land and losses of organic carbon in vegetation and soils. In recent years, deforestation in temperate regions has been reversed, but land conversions in the tropics has greatly expanded. This has become a major source of CO_2 emissions to the atmosphere, in the order of 1.6 Gt C or about 20% of total anthropogenic CO_2 emissions, the continuing net global loss of C from cultivated soils contributes approximately an additional 5% of anthropogenic CO_2 (Paustian et al., 1998). Also, agriculture contributes around 50% of anthropogenic CH_4 emissions globally, primarily from the rumen of livestock and from flooded rice fields, and about 70% of anthropogenic N_2O, largely as a result of nitrogen inputs (synthetic fertilizers, animal wastes, and biological nitrogen fixation) (IPCC, 1996).

International Waters and Agriculture*

More than 200 river basins are shared by two or more countries, accounting for about 60% of the Earth's land area. For example, at least 54 rivers cross or form international boundaries in sub-Saharan Africa (SSA), and 10 river basins have drainage areas greater than 350,000 km², affecting 33 SSA countries and Egypt.

Many of these shared watercourses are subject to alarming rates of environmental degradation that have strong linkages with land and water management. For example:

- Water withdrawals from lakes and reservoirs, water diversions, upstream dams, and lake reclamation for agriculture and aquaculture significantly deplete the size of the water bodies, destroying habitats for plants and animals and sometimes causing very high levels of salinization.
- Deforestation and land degradation in international watersheds such as the Nile, Niger, or Indus affect rainfall patterns, increase the range of local temperatures, and cause major variations in water flow and quality. Soil erosion leads to siltation and sedimentation of lakes and reservoirs, shortens their lifetimes, destroys aquatic environments, reduces the productivity of their ecosystems, and diminishes their flood control capacity.

LINKAGES BETWEEN SUSTAINABLE LAND MANAGEMENT AND THE INTERNATIONAL CONVENTIONS

Any measure of physical and biological sustainability must combine measures of productivity enhancement, measures of natural resource protection, and measures of social acceptability. Thus, it is essential to integrate concepts of those who focus on resource quality with those who emphasize economic productivity measures. Such an integrated approach is being developed as a *Framework for Evaluation of Sustainable Land Management* (FESLM) (Smyth and Dumanski, 1993).

The FESLM was developed through collaboration among international and national institutions as a practical approach to assessing whether farming systems are trending toward or away

---

* Although the other conventions clearly identify global environmental benefits, the benefits under the various agreements on international waters are primarily national and transboundary.

from sustainability. In this context sustainable land management (SLM) combines technologies, policies, and activities aimed at integrating socioeconomic principles with environmental concerns so as to simultaneously:

- Maintain or enhance productivity/services
- Reduce the level of production risk
- Protect the potential of natural resources and prevent degradation of soil and water quality
- Be economically viable
- Be socially acceptable

These factors are referred to as the five pillars* of sustainable land management, and they can be applied for sustainable agriculture. Performance indicators for each pillar are used for assessing the contribution of that pillar to the general objectives of sustainable land management. Thus, for any given agricultural development activity, sustainability can be predicted if the objectives of all five pillars are achieved simultaneously. However, as is the likely case in the majority of situations, only degrees of sustainability can be predicted if only some of the pillars are satisfied, and this results in partial or conditional sustainability. The recognition of partial sustainability, however, provides valuable direction for the interventions necessary to enhance sustainability.

The linkages between the pillars of sustainable land management, the agricultural challenges to achieve food security, and the international conventions are shown in Table 36.1. This shows that agricultural objectives, under food security, still emphasize productivity and economic returns, whereas the international conventions encompass environmental goods and services. This dichotomy, developed under the historic paradigm of economically driven national development and nonlimiting natural resources, is being challenged in many parts of the world. The evolving debate creates opportunities to better harmonize the common ground and to capture the synergy between sustainable land management and food production. However, this requires that agricultural systems recognize that sustainable systems must also ensure ecosystem health. Conversely, it requires that conservationists recognize that improved local, national, and global environmental management will not be realized without first securing economically viable rural economies.

Capturing the mutually reinforcing interactions (synergy) between these two aspects — the biophysical and the social — is the major opportunity for achieving the concomitant objectives of improved food security and ecosystem health. An important premise is that farmers normally are keen to manage their land better if they know how, but their management choices are often constrained by outside influences, including financial, marketing, and other constraints.

Table 36.1 Relationships among Sustainable Land Management, Food Security, and the International Conventions

	The Pillars of Sustainable Land Management				
	Productivity	Risk Management	Soil/Water Conservation	Economic Viability	Social Acceptability
Food security	X			X	X
Biodiversity	X	X	X		X
Climate change		X	X		X
International waters		X	X		X

* The pillars are carefully designed so that they can be aggregated into the physical, economic, and social dimensions of sustainability.

SELECTED EXAMPLES OF CONCOMITANT BENEFITS BETWEEN AGRICULTURE AND ECOSYSTEM HEALTH

Examples of natural resource improvements are becoming increasingly available, and much can be learned from the factors and processes that led to their success. In some cases, e.g., Brazil, millions of hectares have become more stable, productive, and sustainable under improved land management systems, whereas in others, e.g., Kenya, profound improvements in land quality have been made in local areas, but the technologies have not yet spread to large areas. In either case, success is triggered by two key components of true sustainability (El-Swaify et al., 1999):

- Resilience reflected in the inherent capacity of living systems, including soils, to regenerate themselves when suitably managed
- Deployment of the latent skills and enthusiasm of rural people when adequately informed, supported, and empowered

Parana, Southern Brazil

A microwatershed, integrated soil and water management program was adopted in 1987 as the basis for comprehensive statewide rural development. The basis of the program was to help local communities address their land management problems while improving the economic viability of their operations. Local decision making, planning, implementation, and monitoring (participatory approach) was ensured from the beginning. Integrated technology adoption (no magic bullet) was employed, based on conservation tillage and crop residue management, cover crops and crop rotations, and fertilizer and lime applications. Agricultural cooperatives provided efficient marketing and a reliable supply of required inputs. There was strong political support over the life of the program, and funding was provided by the state and complemented by additions from the World Bank and others. Within 5 years the program was working with local communities in 2100 small catchments, covering over 5 million ha and 165,000 farm properties. After 10 years of close monitoring, it was shown that the program was successful in controlling soil erosion, while agricultural incomes and rural economies improved. Also, it was shown that reduced sedimentation of waterways and improved water quality alone more than paid for the cost of the program.

Kakamega, Kenya

This is a program to assist self-help groups of small, resource-poor farmers to improve their livelihoods through soil conservation coupled with a strong marketing component in local and export markets to improve cash flow. The basic philosophy is that smallholder farmers can, with enhanced skills and their own enthusiasm, bring themselves out of poverty by making better use of the natural resources to which they have access, notably rainwater, organic materials, and soils. Products are marketed under predetermined *conservation supreme* standards, and under the *Farmers' Own* label, which enables the group to gain and retain market share. Since the start in 1993, 300 farm families in over 150 groups have joined the process, and participation is expanding rapidly. It is promoted by a small nongovernmental organization, the Association for Better Land Husbandry.

Gestion de terroirs, Sahelian West Africa

This approach evolved from the need to mitigate the serious impacts of land degradation and desertification on local livelihoods. The process introduced involves management of common lands by the community, under sets of locally adapted rules. The necessary steps are:

- Participatory diagnosis by local people to identify resource problems
- Election of a village or community committee responsible for resource management decisions
- Establishment of the committee's jurisdiction
- Preparation of a management plan, zonation, and allowable land uses
- Implementation, monitoring, and evaluation

The approach requires commitment to local decision making, local diagnosis of problems, and creation of clear lines of responsibility. Although not universally successful, the approach deals up front with issues of equity, power-sharing, self-reliance, and community environmental management. The approach has become widely endorsed by national governments and many donors.

Changar, Northwest India

The issues centered on severe land degradation, runoff, and erosion, especially on common land and state forestlands, due to inappropriate management of pastures and forests. The consequences were decreases in productivity and severe water scarcity in summer months. The project worked with local communities (570 villages with 140,000 people) on catchment-based development in the highly fragile sediment hills in the western Himalayan foothills. The project used a combination of interdisciplinary measures focused on better land management, effective soil and water conservation, and community-based forestry and animal husbandry. Through strengthening community organizations in land management, the project contributed to the simultaneous economic development and ecological stabilization of the region.

The Catskill Watershed, New York

New York City's water supply comes from a watershed in the Catskill Mountains. In recent times, poor land management in the upper watershed and contamination from sewage, fertilizers, and pesticides deteriorated the water quality to below EPA standards. The options were to build a super filtration plant at a cost of $6 to 8 billion, plus annual maintenance costs of about $300 million, or restore the ecosystem integrity of the Catskill watershed. The city chose to purchase the land in and around the watershed so that its use could be restricted (agricultural land uses that contributed to pollution were eliminated), and to upgrade existing treatment plants for a total cost of $1 to 1.5 billion. Thus an investment of $1 to 1.5 billion to restore ecosystem functions resulted in a saving of $6 to 8 billion in physical capital. The project is expected to give a rate of return of 90 to 170% in a payback period of 4 to 7 years (Chichilnisky and Heal, 1998).

Summary

These few examples illustrate the opportunities that can be gained when the objectives of economic development and ecosystem maintenance are combined. In all cases, successes were achieved by building on the knowledge and capacity of rural communities to manage their resources better. New technologies were introduced in most cases, but only under conditions of local decision making. The international conventions provided an enabling policy framework and often incremental funding to cover initiation and transaction costs. These are classical win–win illustrations of sustainable rural development.

THE WAY FORWARD: SUSTAINABILITY AS OPPORTUNITY

The challenges for the future are considerable, but so are the opportunities. However, we need to change our perceptions and develop some new ways of thinking if we are to take advantage of the opportunities available. Sustainability will never be achieved by *overcoming constraints*, which has been the driving paradigm of the past. Overcoming constraints promotes the concepts of human dominance over nature and recognizes neither the limits nor the capacity of natural systems.

A more promising emerging paradigm, which is promoted by the World Bank, is *sustainability as opportunity*. This approach is based on the practical reality that sustainable agricultural systems require economic growth while simultaneously protecting natural resources.

A broadly acceptable definition of sustainable agriculture has been proposed by the Technical Advisory Committee (TAC) of the Consultative Group on International Agricultural Research (CGIAR):

> Sustainable agriculture involves the successful management of resources for agriculture to satisfy changing human needs, while maintaining or enhancing the quality of the environment and conserving natural resources.

This is a practical approach to sustainability as it recognizes the legitimate use of natural and anthropogenic resources for satisfaction of human needs, but it cautions against the exploitation of these resources in a manner that would degrade the quality and potential of the resources on which production depends. More important, the definition recognizes that human needs change and, therefore, the systems of production also must change. This definition embodies the concepts of system flexibility and natural resource resilience as primary criteria for achieving sustainability.

A logical outgrowth of this approach to sustainability leads to the concept of sustainability as opportunity.* This can be defined as ensuring that the choices for future production systems are not reduced by decisions made in the present.

That view of sustainability is based on the following:

- Sustainability will not be achieved by overcoming constraints, but rather as a process to concomitantly capture economic and environmental opportunities. For example, investment in new knowledge has allowed many traditional farming systems to become more sustainable in the face of rapid population growth, by shifting from extensification based on expanding land area, to intensified use of existing land area.
- Considerable substitution is possible in agricultural systems (the physical, biological, economic, and social dimensions of sustainability), but the substitution is not perfect. For example, most agricultural production systems allow for a certain amount of input substitution, such as among different sources of crop nutrients, substitution of labor for land, and so forth. Such substitutions may contribute positively to sustainability as long as the impacts of the substitution are reversible and they contribute to more resilient and flexible systems. Substitutions that lock a system into a restricted range of options, e.g., reduced agrobiodiversity, monocropping, etc., will not lead to sustainability.
- Agricultural systems with the capacity to change — that is, the capacity to remain flexible and resilient in response to major influences such as shifts toward a global market, important recent advances in science and technology, and changes in labor availability — will ensure more sustainable systems. Static agricultural systems are not sustainable systems. These continual shifts and adjustments often lead to new opportunities to intensify production, increase productivity, and exploit emerging commodity markets. In other cases, negative market forces and adverse natural conditions beyond the control of producers will lead to transition out of agriculture and a search for off-farm employment.
- Whereas the principles and criteria in sustainable land management systems are universal and transferable, technologies and application must be local. In sustainable agriculture there are no single solutions, shortcuts, or magic bullets. In fact, the magic bullet approach must be consciously avoided.** Blanket recommendations are rarely successful, and innovations found to be successful in one area will likely have to be modified somewhat to be successful in another. For example, zero-till is a successful technology contributing to sustainable land management,

---

* Sustainability as opportunity was first proposed by Serageldin (1995) as a definition for sustainable development: sustainability is to leave future generations as many, if not more, opportunities as we have had ourselves.
** For the extension agent, this means the requirement to work more actively in a participatory approach with farmers and to develop an enabling atmosphere for local farmer innovation regarding what will work and what is not acceptable. However, providing technological backup to the store of local farmer knowledge is often a critical component.

but some local modifications of the technology are almost always required for it to be successful. However, the basic principles of zero-till — that is, minimal disturbance of the soil surface and maintenance of continual soil cover — are universal and therefore transferable.

- The primary agents of change toward sustainable agricultural systems are the rural communities (farmers, pastoralists) who depend on the land for their livelihoods, and the primary emphasis is on community-based or farmer-centered interventions. Rural families make decisions about production practices and land use in line with their objectives, production possibilities, and constraints; but these decisions are part of a wider process to secure and improve the family's food security and livelihood. They are in turn influenced by government policies and market forces.

- Many of the environmental benefits associated with sustainable land management, such as pollination, biological control, nutrient cycling, soil conservation, water quality, waste treatment, etc., accrue locally and nationally. Others are more clearly global, or at least supranational in scope, such as climate regulation, conservation of genetic resources of potential value in plant breeding or pharmaceuticals, international tourism, and transboundary water-mediated effects. Policies and programs designed to capture economic growth as well as environmental benefits are more likely to ensure sustainability.

- A strong local and national commitment toward sustainable land management is required to bring down transaction costs and to capture national and global environmental benefits. In this respect, functioning rural markets for traditional goods and services, an appropriate system of land tenure and property rights, and a broadly enabling policy framework for natural resources conservation are prerequisites to the capture of global flows.

CONCLUSION

The need for agricultural growth strategies that can achieve the required growth and food security, while reversing the historical conflict with natural resources conservation, is now a front-line issue for global sustainable development. It is widely recognized that agriculture and environmental management are inseparably linked and that tackling problems of natural resource degradation must be seen as part of a wider set of actions to revitalize the rural sector as a whole. Promoting rural development strategies that have win–win outcomes for agricultural livelihoods and the environment is mainstream policy for the World Bank and other major development agencies and is considered vital to provide a sustainable basis for future productivity growth and poverty alleviation.

The Bank's Rural Vision to Action (World Bank, 1997) and related policies establish sustainable land management at the heart of such strategies. It is increasingly recognized that well-designed, farmer-centered, sustainable land management interventions have distinct advantages as vehicles for pursuit of joint agriculture–environment objectives. The pillars of sustainable land management (Smyth and Dumanski, 1993) are the application of agroecological principles to farming; an emphasis on human resource development and knowledge-based management techniques; a participatory and decentralized approach; the value placed on natural and social capital enhancements in addition to economic efficiency gains; and the role of strong and self-reliant rural institutions.

Agriculture that is truly sustainable will not be business as usual. It will be a type of agriculture that will provide environmental, economic, and social opportunities for the benefit of present and future generations, while maintaining and enhancing the quality of the resources that support agricultural production. This will not be the agriculture of today or of the recent past, with its emphasis on *maximizing* yields and economic returns, but rather one with the objectives of *optimizing* productivity and conserving the natural resource base. The objective of optimization implies *trade-offs* in the production systems to ensure maintenance of environmental quality and global life-support systems. Experience indicates that these trade-offs will be defined and implemented voluntarily by farmers and other rural land users, or they will be implemented through policies and legislation. Society is beginning to demand that agriculture become more than simply putting food on the table; it is beginning to demand that it also becomes the steward of rural landscapes.

The objective is to evolve sustainable systems in which appropriate technological and policy interventions have created resilient production systems that are well suited to local socioeconomic and physical conditions and that are supported by affordable and reliable policies and support services. However, these systems cannot be static systems but must be carefully designed to be flexible and responsive to change — that is, systems in transition. Sustainability of any system cannot be ensured unless the production technologies and associated management practices continuously evolve to accommodate changes in the agroclimatic, economic, and demographic environment in which agricultural intensification is undertaken. The appropriate level of analysis of sustainability is at the level of the cropping or farming system on a relatively homogeneous agroecological resource base, within which similar choices of crop and livestock management decisions can be made.

Agricultural sustainability is a concept that is continuously evolving, and therefore it is difficult to define and measure. Because it reflects our understanding of systems that in themselves are in continual transition, sustainability should be approached as a concept to strive for, such as social well-being, rather than an objective that can be measured with common analytical techniques. In this sense, the ability to track the performance of core indicators toward this goal is more useful than setting specific targets to be achieved (although setting targets is often useful to identify levels of satisfaction). This is not unlike tracking the performance of national economies, where, in most cases, this is done simply to know how we are doing and in what direction the economies are trending.

REFERENCES

Brown, L.R., Flavin, C., and French, H., *State of the World*, W.W. Norton and Co., New York, 1998.

Chichilnisky, G. and Heal, G., Economic returns from the biosphere, *Nature,* 391:629–630, 1998.

El-Swaify, S.A., with an international group of contributors, *Sustaining the Global Farm — Strategy Issues, Principles, and Approaches*, International Soil Conservation Organization (ISCO) and the Department of Agronomy and Soil Science, University of Hawaii at Manoa, Honolulu, 1999.

IPCC, *Climate Change 1995: Impacts, Adaptations, and Mitigation of Climate Change: Scientific–Technical Analyses*, Watson, R.T. et al., Eds., Cambridge University Press, Cambridge, U.K., 1996.

Joffe, S., Dumanski, J., Pieri, C., and Forno, D., *Opportunities to Capture the National and Global Environmental Benefits of Sustainable Land Management,* The World Bank, Geneva, in press.

Lubchenco, J., Entering the century of the environment: a new social contract for science, *Science,* 279:491–497, 1998.

Paustian, K., Cole, C.V., Sauerbeck, D., and Sampson, N., CO_2 mitigation by agriculture: an overview, *Climate Change,* 40:135–162, 1998.

Serageldin, I., *Sustainability and the Wealth of Nations: First Steps in an Ongoing Journey,* Third Annual World Bank Conference on Environmentally Sustainable Development, World Bank, Washington, D.C., 1995.

Smyth, A.J. and Dumanski, J., *FESLM, An International Framework for Evaluating Sustainable Land Management*, World Soil Resources Report 73, FAO, Rome, 1993.

Vitousek, P.M., Beyond global warming: ecology and global change, *Ecology,* 75:1861–76, 1994.

World Bank, *Rural Development: From Vision to Action: A Sector Strategy,* Environmentally Sustainable Development Studies and Monographs Series 12, World Bank, Washington, D.C., 1997.

CHAPTER 37

The Ecological Footprint as Indicator for Sustainable Development — Results of an International Case Study

Detlef P. van Vuuren, E.M.W. Smeets, and Hans A.M. de Kruijf

INTRODUCTION

The topic of the conference session for which this contribution was originally made was formulated as an objective: *securing everybody's quality of life within the means of nature.* This objective, although sometimes worded differently, is also referred to as *sustainable development.* The concept of sustainable development has become (at least officially) a widely accepted goal of national and international policy making, since it was used by the World Commission on Environment and Development (WCED) to advocate a kind of development that would not only aim for economic objectives per se but also include a fuller sense of human and ecological well-being (WCED, 1987). Since 1987, considerable attention has been paid to developing indicators that could support decision making in this context (e.g., Hammond et al., 1995; Meadows, 1998). The ecological footprint (EF) has frequently been mentioned as one of these. The EF is defined as the amount of the world's environmental resources required to support the consumption of a defined population regardless of where this capacity is used (Wackernagel and Rees, 1996).

Most attention to the EF has come from educational and environmental organizations. They have used the results of the EF to convey the message that at the moment: (1) human consumption requires more natural resources than can be provided in a sustainable way; and (2) many industrialized countries use a disproportionately large share of available resources. The strongest use of the EF so far is as a heuristic device, as its results are readily understandable and can be directly related to personal activities.

However, the question is now raised whether EF information can also be used in decision-making processes (for instance, the question was asked explicitly in the Netherlands by the Minister of the Environment; Pronk, 1998). For this purpose, it should comply with strict criteria with respect to measurability, usefulness/policy relevance, and analytical soundness. In this chapter, we will use the results of an application of the EF to Benin, Bhutan, Costa Rica, and the Netherlands to discuss whether the EF could be a useful sustainable development indicator. The methodology of the case studies and its results have been described earlier (Van Vuuren and Smeets, 2000; Van Vuuren et al., 1999). Following a brief introduction to sustainable development and sustainable development indicators, the main results of the case study are discussed. Finally, we draw attention to both strong and weak points of the EF based on this application and the ongoing discussion in the literature. The focus of this discussion is on the environmental aspects of sustainable development.

SUSTAINABLE DEVELOPMENT AND SUSTAINABLE DEVELOPMENT INDICATORS

Although a large body of literature exists on different definitions and interpretations of sustainable development, no definition has attained general approval and support. Most scholars now seem to agree that sustainable development encompasses economic, ecological, and social objectives. This is elegantly shown by the Sustainable Development Indicator Framework proposed by the Balaton Group (Meadows, 1998) (Figure 37.1). In this framework, well-being is seen as the ultimate end of development supported by various types of intermediate means (infrastructure, social relationships) and ultimate means, which are natural resources. Sustainable development can now be defined as the long-term growth potential of all these types of capital (Serageldin, 1996). In many cases, however, development decisions imply value-laden trade-offs between different types of capital. Obviously, it is clear that any useful set of sustainable development indicators would consist not only of the economic indicators mostly used now but also include indicators capturing the ultimate means and ends of development. Such a set should show us whether imbalances develop between the different types of capital or whether critical limits are passed.

The objective of sustainable development can also be characterized by a growing awareness of and responsibility for the impacts of human actions in space and time. For environmental capital, for instance, sustainable development requires an additional focus on long-term effects and management of global environmental resources, in combination with more traditional policies looking at short-term effects and the direct environment. Obviously, this is a difficult task as everybody will automatically spend the most time thinking about here and now. Indicators that are able to bridge the gap between these different levels of abstraction could be very important in motivating people to action.

These notions have also been the basis for the project Multidisciplinary Indicators for Sustainable Development (Van Vuuren and de Kruijf, 1998). Within this project, institutes in Benin, Bhutan, Costa Rica, and the Netherlands have been collaborating to develop indicators for sustainable development. These countries are all relatively small and, thus, potentially very open to influences from outside their national borders. This is certainly the case for the Netherlands, which is densely

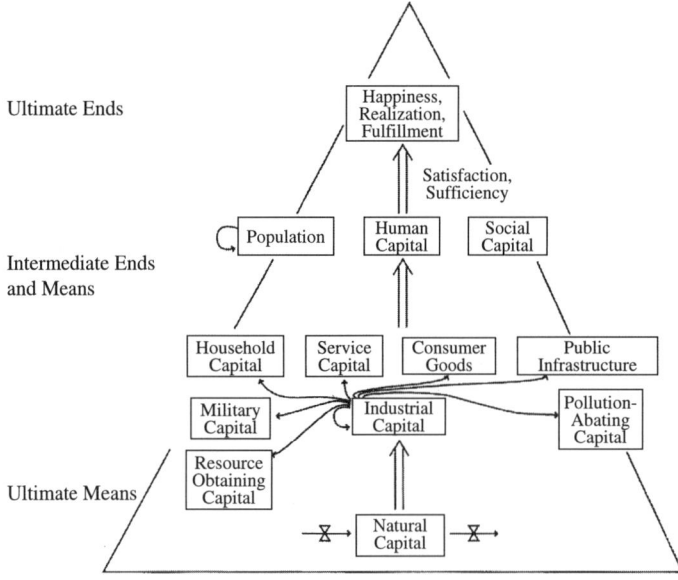

Figure 37.1 Sustainable development indicator framework of the Balaton Group. (From Meadows, D., Indicators and Information for Sustainable Development: A Report to the Balaton Group, The Sustainability Institute, Hartland Four Corners, VT, 1998. With permission.)

populated and highly industrialized. In many other respects, the countries are totally different — which makes the project interesting as a potential case study for indicators meant to be universally applicable. The collaborating institutes decided to apply the EF concept in a case study to consider whether this indicator could be used further within the project.

MAIN RESULTS OF THE EF APPLICATION IN BENIN, BHUTAN, COSTA RICA, AND THE NETHERLANDS

The methodology of the calculations shown here is based on Wackernagel and Rees (1996) and Wackernagel et al. (1997). Land use and carbon dioxide emissions are attributed to consumption in each country on the basis of production, trade, and consumption data. However, we have made two important modifications (Figure 37.2):

- We disaggregated the indicator into separate indicators for land and carbon dioxide emissions, thus avoiding the controversial topic of combining real land use and land equivalents for carbon sequestration.
- We calculated land use on the basis of both local yields and global average yields to show the real land use of countries vis à vis a more comparable land-use measure.

The data were collected from FAO and national data sources (Van Vuuren et al., 1999).

Looking at overall land use, the results of our assessments indicate that total land use in the four countries increased between 1980 and 1994 (Figure 37.3). This is obviously a disturbing trend, because globally land is clearly a limited resource. Comparing land use of each country to its surface area, one can see that, for the Netherlands, more land is used than is domestically available (Figure 37.4). This was expected in view of the high population density. For consumption in the other three countries, less land is required than is available within their borders. This results mainly from relatively large areas that are still undomesticated and indicates the trade-off between the size of the EF in land use and protection of biodiversity by large natural areas. In addition, in Benin and Costa Rica, much land is allocated to export of fruit and coffee (thus needs to be attributed to land use of other countries).

Both differences in consumption per capita and productivity determine the size of the total land use. These differences in productivity complicate comparisons among different countries because productivity is influenced (actively) by human management factors and (passively) by natural circumstances (such as soil and climate). Thus, countries with less favorable conditions for agriculture (savannah, alpine areas) necessarily use more land per unit of consumption and thus have higher EFs, if calculated on the basis of real productivity. In most current EF work, the issue is altogether avoided by calculating *virtual land use* for all types of consumption directly based on global average yields. For national governments, however, land use based on local yields might be much more relevant because the calculated area is in that case equal to a real, touchable area.* The results differ enormously; in fact, on the basis of local yields, per capita land use is smallest for the Netherlands (0.7 ha per capita), while per capita land use of Benin, Bhutan, and Costa Rica are, respectively, 0.7, 0.9, and 1.2 ha. If, instead of local yields, global average yields are used, the EF directly reflects consumption levels: it is smallest for Bhutan (0.6 ha), followed by Benin (0.7 ha), Costa Rica (1.7 ha), and the Netherlands (3.0 ha). The reason for this is that agricultural production per hectare is much higher in the Netherlands than the global average yields.

The EF can also be calculated for carbon dioxide emissions. Here, the EF increased for all four countries both in per capita and absolute terms between 1980 and 1994. For the Netherlands, the EF for 1994 was 8.9 tons of carbon dioxide per capita, two times the global average level. The EF for the Netherlands was, however, 20% lower than the domestic emissions of the country (due to net exports of energy-intensive products). Emissions of the other three countries were below the global average: 2.5 ton per capita and less than 0.1 ton per capita for Benin and Bhutan, respectively.

* Local yields are defined as the real yields for each product in the country of production.

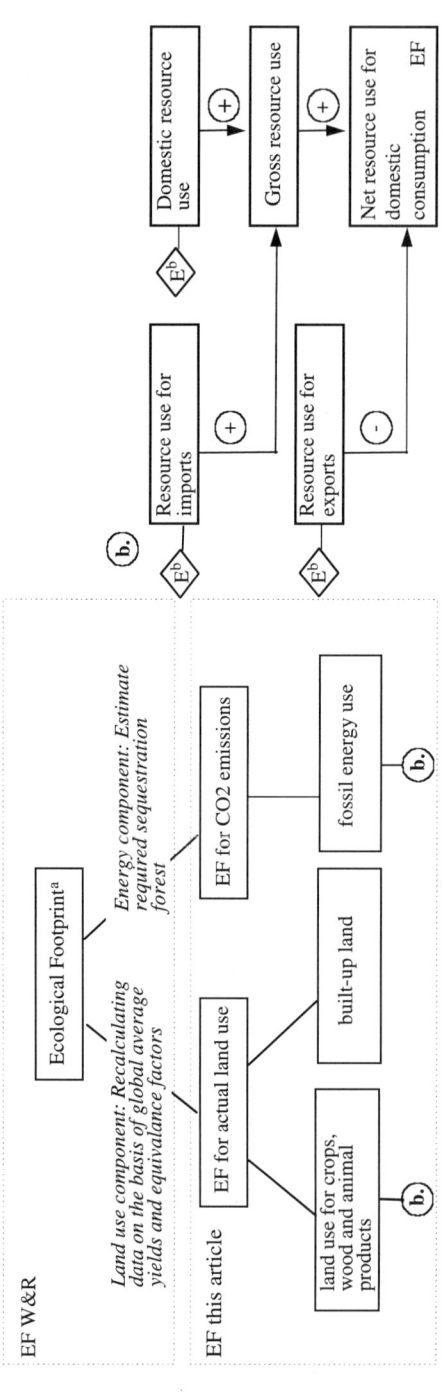

Figure 37.2 Components of the EF indicating differences in focus between Wackernagel and Rees and this chapter. E indicates efficiency of resource use; averages are either local or global. (From Van Vuuren, D.P. and Smeets, E.M.W., Ecological footprints of Benin, Bhutan, Costa Rica and the Netherlands EF, information to stimulate debate or debatable information *Ecol. Econ.*, 34:115–130, 2000. With permission.)

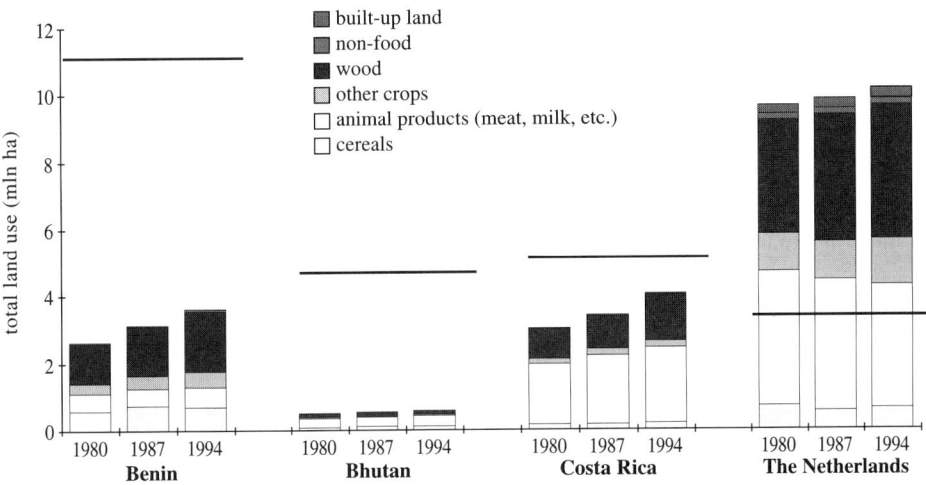

Figure 37.3 Net land use for agricultural products and wood. The horizontal line indicates the actual size of each country; nonfood includes land use for cotton and rubbe. (From Van Vuuren, D.P. and Smeets, E.M.W., Ecological footprints of Benin, Bhutan, Costa Rica and the Netherlands EF, information to stimulate debate or debatable information *Ecol. Econ.*, 34:115–130, 2000. With permission.)

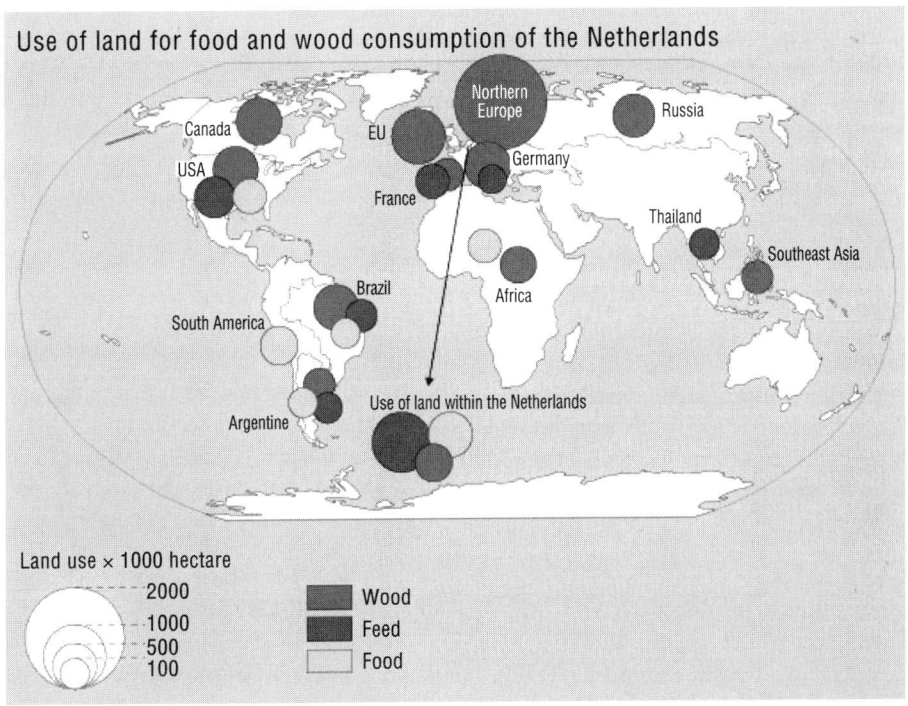

Figure 37.4 Land use for food and wood consumption of the Netherlands. (From RIVM, *Milieubalans 1999 [Environmental Balance]*, National Institute of Public Health and the Environment, Bilthoven, The Netherlands, 1999.)

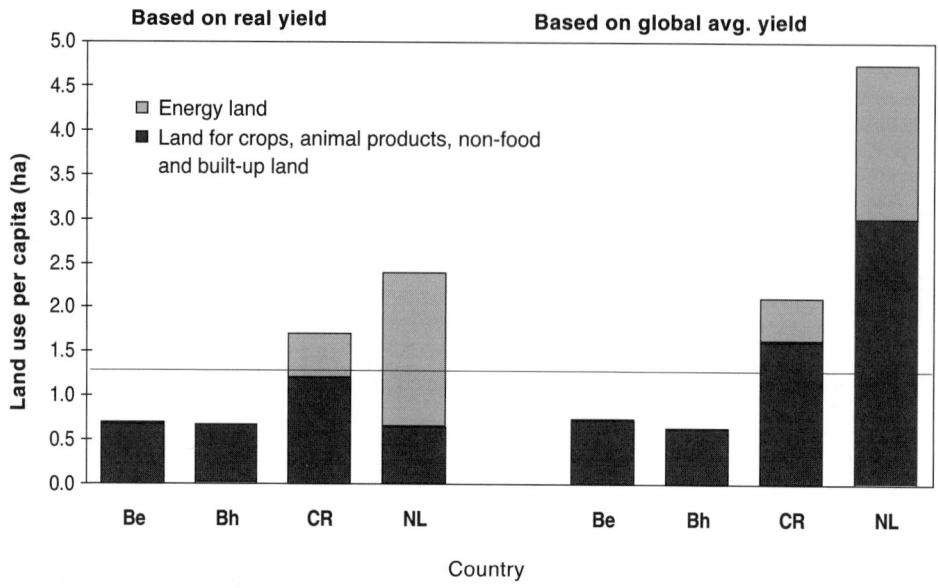

Figure 37.5 Aggregated EF following definition of Wackernagel and Rees, 1996. Land use is included using real land use and global average yield as a base. The horizontal line indicates the level considered to be sustainable on the global scale by Wackernagel et al. (1997). Be = Benin; Bh = Bhutan; CR = Costa Rica; NL = Netherlands. (From Van Vuuren, D.P. and Smeets, E.M.W., Ecological footprints of Benin, Bhutan, Costa Rica and the Netherlands EF, information to stimulate debate or debatable information *Ecol. Econ.*, 34:115–130, 2000. With permission.)

In addition to these EFs for specific resources, the aggregated EF (according to Wackernagel and Rees, 1996) can now be calculated by translating carbon dioxide emissions into the hectare carbon-sink forest that would be required for sequestering these emissions. The resulting figures can be compared to the globally available capacity — that is, the available ecologically productive land (Wackernagel et al., 1997) (Figure 37.5). The following results arose:

- The Netherlands has the largest EF, independent of the question of whether global average yields are used.
- The EF of the Netherlands is made up largely by carbon dioxide emissions — which are indeed considered by many in the Netherlands to be the most important environmental problem. In the other three countries, land use represents the largest pressure — again according to our expectations.
- The EF per capita of the Netherlands is larger than the per capita globally available capacity.
- The EF of Costa Rica is currently near this level. Trend extrapolation indicates that, if unabated, the EF of Costa Rica will rapidly increase. This suggests that industrializing low-income countries such as Costa Rica approach the high consumption levels of high-income countries. Benin and Bhutan are still far below the global available capacity.

DISCUSSION ON THE POTENTIAL USE OF THE EF

In the previous section we have shown that the EF can be actually applied to different countries on the basis of available data. The numbers give interesting information — but the question still remains whether it might also be useful for policy-making purposes. On the basis of our work and current discussions on the EF, particularly in the Netherlands, the following conclusions could be drawn:

1. The Indicator Should Be Part of a Larger Set

EF-type indicators should be part of a larger set of indicators, as they cover only some of the environmental topics and do not contain any information on how biophysical constraints can be linked effectively to economic and social policy or on social aspects of sustainable development.

2. The Use of the EF Is Value Related

In the Netherlands, the Minister of the Environment recently indicated his interest in the EF and asked an important advisory council to look deeper into the issue. A dynamic debate started, part of which could be followed in the international literature (van den Bergh and Verbruggen, 1999). Several nongovernmental organizations (NGOs) and dedicated scientists promoted greater use of the EF for policy development — while, in particular, some environmental economists vigorously rejected it. How could such a lively debate come about? The debate reflected the fact that indicators are not absolute truth machines but function within a network of values, worldviews, and paradigms (see Rotmans and de Vries, 1997).

In this debate, the economists' position includes the assumptions that resources (e.g., the types of capital in Figure 37.1), in principle, can be substituted, and that sustainable development is the optimal allocation of resources. Necessarily, there is less focus on absolute limits — in contrast to most of the EF literature, which has tended to emphasize limits. The implicit weighting method within the EF was regarded by the involved economists as too rigid because it does not reflect changes in scarcity (Verbruggen, 1999). The economists in the debate indicated that the most important criterion for judging the EF is *scientific soundness* (van den Bergh and Verbruggen, 1999); according to them, the EF fails to meet this criterion. Finally, they argued that the EF implicitly promotes autarchy or is against trade. However, there is nothing on how the EF is calculated that substantiates this argument. If international trade is truly based on ecologically comparative advantages (including the effects of transport), it will result in a lower overall EF. Their argument must, therefore, be directed against some of the EF literature, which has put forward self-sufficiency as a mean of restoring the direct link between environmental pressure and consumption.*

In contrast, the NGOs began their argument from the assumption that limits are a crucial aspect of ecology. Therefore, for sustainable development to occur, all resources need to be protected. Moreover, this position is combined with the view that equity is an important aspect of sustainable development. It is clear that the EF does exactly fit into this position (focus on environmental resources, emphasis on personal and local responsibility). But above all, they will argue that the most important criterion to judge the EF as an indicator is its power to motivate (Juffermans, 1999).

Fortunately, all parties seem to disagree less on policies. All agreed about a need to strengthen policies and research with regard to (1) environmental consequences of consumption patterns, (2) environmental impacts of trade, and (3) current state of natural resources. One could say that despite all discussion, the EF stimulated a debate on certain issues that previously had not had a high priority on the policy-making agenda.

The final conclusion reached in the advisory council (in which both parties mentioned above participated) is that the EF is not a policy indicator. The most important reasons were that the EF is too aggregated and at the same time leaves out several issues considered to be important for sustainable development (VromRaad, 1999). Immediately after this conclusion, however, the council also concludes that the issues raised by the EF are very important and the council in fact welcomes the boost to the discussion on sustainable development in the Netherlands given by EF. The council also acknowledges that the EF can play a role in raising awareness by application at the individual level. Finally, it should be noted that in the state-of-the-environment reports and the environmental statistical compendium of the Netherlands, EF-type indicators are now regularly used.

* That some of the proponents of the EF indeed emphasize production and consumption within each range can be seen in Ferguson (2001) and the response by van Vuuren and Smeets (2001).

3. From a Policy Perspective, the EF Is Able to Generate Some Relevant Information; but There Are Obstacles to the Approach

On the basis of our own work, we conclude that EF-type research could be an assessment framework complementing or adding to production or domestic pollution-oriented approaches. EF-type indicators are attractive because they highlight several interrelated topics of sustainable development:

- *Focus on consumption:* Traditional environmental policies have often aimed at reducing emissions on the production side of the economy (industry, business, agriculture). However, state-of-the-environment reports on the national, continental, and global scale (RIVM, 1998; EEA, 1998; UNEP, 1997) indicate that, for several environmental issues, results of such policies are partly or completely offset by increased consumption (for example, global warming). The EF focuses on consumption and is able to quantify the consequences of reallocation of environmental pressures, squandering of resources, and impacts of the size and composition of consumption patterns.
- *Focus on renewable resources:* An attractive attribute of the EF is its focus on resources that are key to sustainable development (land, carbon dioxide emissions). The UNEP Global Environment Outlook, for instance, concluded that land — suffering from degradation and becoming increasingly scarce — is one of the key resources for sustainable development in most of the world's regions (UNEP, 1997).
- *Distribution of available environmental resources:* The EF addresses the current distribution of resource use by calculating the per capita footprint for individuals, cities, or countries and comparing them to a global averages. In the context of ongoing climate change negotiations, this is already an important issue, with proposals to base burden-sharing initiatives in the next century on an equitable distribution of per capita emissions rights.
- *Communication and proactive aspects:* Results of the EF are relatively easy to visualize. Moreover, since the EF can be calculated on a per capita basis, EFs can be compared among countries, with a global average or a potentially more sustainable level. These features allow for powerful communication of the results; this explains its popularity among advocacy groups.
- *Aggregation:* Different authors have argued for the development of aggregated environmental indicators that are able to give a rough overview of environmental pressure or sustainability (for example, Jesinghaus, 1996). At the same time, aggregation is extremely difficult and controversial for complicated systems (such as the environment) since there are no general scientific rules or theories to evaluate different impacts from an ecological perspective.

There also remain some obstacles to overcome with the method. Issues in particular need of attention are:

- *Possibilities and limitation of aggregation of resources:* van den Bergh and Verbruggen (1999) argue that by using land as an aggregated indicator, the EF implies that we are talking of a specific environmental resource (land); actually it represents (at least partly) a hypothetical amount of land. As we have shown in our work, it is possible to avoid the issue of aggregation by focusing not on the aggregated EF but on its components in their own typical units. We think this improves the policy relevance of the indicator.
- *Differences in agricultural productivity in international comparisons:* This chapter has shown the enormous consequences of using local instead of global average productivities.
- *Accounting for unsustainable use of resources in the context of the EF:* In calculating EFs, no distinction is made between sustainable and unsustainable use of land; this limitation might lead to the paradoxical situation that unsustainable agricultural production methods (contributing to soil degradation) decrease EF because of higher yields.

Table 37.1 summarizes some of the attractive attributes and obstacles found. Overall, we conclude that the EF has been a positive force in evoking a debate on important sustainable development issues. As a communication instrument, the EF can be very useful. For policy purposes, the focus is the underlying information, not the aggregated EF; overcoming the obstacles identified will improve its analytical soundness.

Table 37.1 Overview of Attractive Attributes, Limitations, and Obstacles of the Ecological Footprint

Attractive Attributes	Limitations and Obstacles
Vivid metaphor	Aggregation/false concreteness
Focuses on consumption	Agricultural productivity
Focuses on key resources (implicitly land and carbon dioxide)	Multifunctional land use
	Unsustainable land use
Connects environmental problems and distribution of material wealth (built-in references)	Does not and will not include all relevant resources
Easy to communicate	
Aggregates	
Consistent accounting scheme	

REFERENCES

EEA, European Environment Agency, *Europe's Environment, The Second Assessment*, Copenhagen, Denmark, 1998.

Fergusson, A.R.B., Comments on ecofootprinting, *Ecol. Econ.,* 37:1–2, 2001.

Hammond, A., Adriaanse, A., Rodenburg, E., Bryant, D., and Woodward, R., *Environmental Indicators: A Systematic Approach to Measuring and Reporting on Environmental Policy Performance in the Context of Sustainable Development*, World Resources Institute, Washington, D.C., 1995.

Jesinghaus, J., The European pressure index project, in MacGillivray, A., Ed., *Accounting for Change*, New Economics Foundation, London, 1996.

Juffermans, J., Snel te begrijpen (Easy to understand), in Dijksterhuis, Ed., *De Ecologische Voetafdruk (The Ecological Footprint)*, Stromen, July, 9, 1999.

Meadows, D., Indicators and Information for Sustainable Development: A Report to the Balaton Group, The Sustainability Institute, Hartland Four Corners, VT, 1998.

Pronk, J., Advies aanvraag aan de VROM Raad, Ecologische Voetafdruk (Request for advice from the Minister of Public Housing, Spatial Planning and the Environment to the VROM Raad), Letter, September 23, 1998.

RIVM, *Milieubalans 1998 (Environmental Balance)*, National Institute of Public Health and the Environment, Bilthoven, the Netherlands, 1998.

RIVM, *Milieubalans 1999 (Environmental Balance),* National Institute of Public Health and the Environment, Bilthoven, the Netherlands, 1999.

Rotmans, J. and de Vries, B., *Perspectives on Global Change — The TARGETS Approach*, Cambridge University Press, Cambridge, U.K., 1997.

Serageldin, I., *Sustainability and the Wealth of Nations — First Steps in an Ongoing Journey*, World Bank, Washington, D.C., 1996.

UNEP, *Global Environment Outlook*, Oxford University Press, New York, 1997.

van den Bergh, J.C.J.M. and Verbruggen, H., Spatial sustainability, trade and indicators: an evaluation of the "ecological footprint," *Ecol. Econ.*, 29:61–87. 1999.

Van Vuuren, D.P. and De Kruijf, H.A.M., Compendium of Data and Indicators for Sustainable Development in Benin, Bhutan, Costa Rica, and the Netherlands, RIVM report 807005001, National Institute of Public Health and the Environment, Bilthoven, the Netherlands, 1998.

Van Vuuren, D.P. and Smeets, E.M.W., Ecological footprints of Benin, Bhutan, Costa Rica and the Netherlands EF, information to stimulate debate or debatable information, *Ecol. Econ.*, 34:115–130, 2000.

Van Vuuren, D.P. and Smeets, E.M.W., Ecological footprints: reply to A.R.B. Ferguson, *Ecol. Econ.*, 37:2–3, 2001.

Van Vuuren, D.P., Smeets, E.M.W., and De Kruijf, H.A.M., The Ecological Footprint of Benin, Bhutan, Costa Rica, and the Netherlands, RIVM Report 807005004, National Institute of Public Health and the Environment, Bilthoven, the Netherlands.

Verbruggen, H., Grote bezwaren (Strong objections), in Dijksterhuis, Ed., *De Ecologische Voetafdruk (The Ecological Footprint),* Stromen, July 9, 1999.

Vromraad, Global Sustainability and the Ecological Footprint, Advice 016E, Vrom-council, The Hague, the Netherlands, 1999.

Wackernagel, M. and Rees, W.E., *Our Ecological Footprint: Reducing Human Impact on the Earth*, New
    Society Publishers, Gabriala Island, B.C., Canada, 1996.
Wackernagel, M., Onisto, L., Callejas Linares, A., López Falfán, I.S., Méndez Garía, J., Suárez Guerrero, A.I.,
    and Suárez Guerrero, M.G., *Ecological Footprints of Nations: How Much Nature Do They Use? How
    Much Nature Do They Have*? International Council for Local Environmental Initiatives, Toronto,
    Canada, 1997.
WCED, *Our Common Future*, Oxford University Press, Oxford, U.K., 1987.

PART II

Issues and Methods

SECTION II.1

Managing for Biodiversity

CHAPTER 38

Overview: Managing for Biodiversity

Jonna A.K. Mazet

In this age of technology and information, humans are better equipped than ever before to assess the impact that our activities have on the Earth and its ecosystems. Our ability to investigate these changes has rapidly increased our awareness of the speed and scale of the current progression of environmental degradation. Fortunately, that awareness has directed us to use our newfound tools, not only to monitor the damaging processes, but also to evaluate strategies to minimize environmental degradation and initiate restorative processes.

In North America, management methods intended to reduce and resolve environmental problems traditionally have followed a resource path; in other words, agencies were created to protect and regulate the use of land, air, and water resources, as well as to manage specific flora and fauna. The effectiveness of these traditional methods has recently come into question because of our failure to protect our ecosystems as a result of focusing on one piece at a time. Major reorganizations of large bureaucratic machines have begun to change the management focus to the larger reticulated systems in which survival of one species requires the vitality of numerous others. This major management shift, driven by evidence of dwindling biodiversity, is definitely encouraging, but requires such an enormous cultural change in existing institutions that its benefits may not be achieved for decades.

In the meantime, scientists continue to use technological innovations to direct management resources in the most beneficial fashion and to decrease the lag time between investigative processes and changes in environmental policy. Unfortunately, humans are, and will continue to be, the source of the majority of our ecosystem problems. No matter how much we want to reduce our impact on the Earth that sustains us, we cannot diminish the innate drive to propagate and strive for a more comfortable lifestyle. Ironically, these innate forces will no doubt result in the loss of the biodiversity on which we are dependent for healthy and aesthetically pleasing lives. Thus, we must continue to seek solutions to environmental problems that protect the reticulated systems of which we are only a part.

The chapters contained in this section focus on balancing solutions and application of technologies to respond to the rapid loss of biodiversity on four continents. Development of progressive solutions like these must continue and should be encouraged by governing bodies responsible for ecosystem management. As individuals, we must also recognize our impacts on the systems in which we thrive and must teach our children to respect the Earth as a whole rather than only the little piece in which they live.

1-56670-612-2/03/$0.00+$1.50
© 2003 by CRC Press LLC

The Greater Addo National Park, South Africa: Biodiversity Conservation as the Basis for a Healthy Ecosystem and Human Development Opportunities

Graham I. H. Kerley, André F. Boshoff, and Michael H. Knight

INTRODUCTION

The recognition that ecosystem health is strongly linked to human welfare, and that many ecosystems have been heavily degraded under human domination — resulting in reduced capacity to support human populations — is a dominant feature of the environmental debate (e.g., Rapport et al., 1998). This has led to a search for ecosystem management strategies to maintain ecosystem health, ranging from water pollution management to disease control and sustainable resource utilization. To some extent this process has been hampered by the inability to look beyond conventional management strategies in order to recognize and develop new opportunities for extracting resources from ecosystems, while maintaining these systems in a healthy and functional state. This deficit is particularly apparent in rangeland ecosystems that traditionally have been used for domestic herbivore production through pastoralism, despite considerable evidence of the threats to ecosystem health that this strategy imposes (e.g., Fleischner, 1994). We present here the background of ecosystem degradation and loss of ecosystem resources due to pastoralism in the Eastern Cape Province (hereafter "Eastern Cape") in South Africa (Figure 39.1), an area of spectacular biodiversity, and assess the consequences of alternate management strategies. We show how an initiative to address these problems, based on the recognition that biodiversity conservation yields tangible human development opportunities that include the full range of ecosystem services, is developing.

DESERTIFICATION OF THE THICKET BIOME

The Thicket Biome, one of the seven terrestrial biomes in South Africa (Low and Rebelo, 1996), is largely confined to the hot, dry valleys of the Eastern Cape, hence its alternative name of Valley Bushveld (Acocks, 1975). In the pristine state, thicket is characterized by dense, low-growing (3 to 5 m) vegetation that includes small trees, woody and succulent shrubs, grasses, lianas, forbs, and geophytes (Low and Rebelo, 1996). These plants tend to be evergreen; succulence and/or spinescence is common, and they form a virtually impenetrable, spiny thicket (Everard, 1987). The high diversity of growth forms reflects the transitional nature of these thickets, which occur at the interface of four major phytochoria or phytogeographical regions

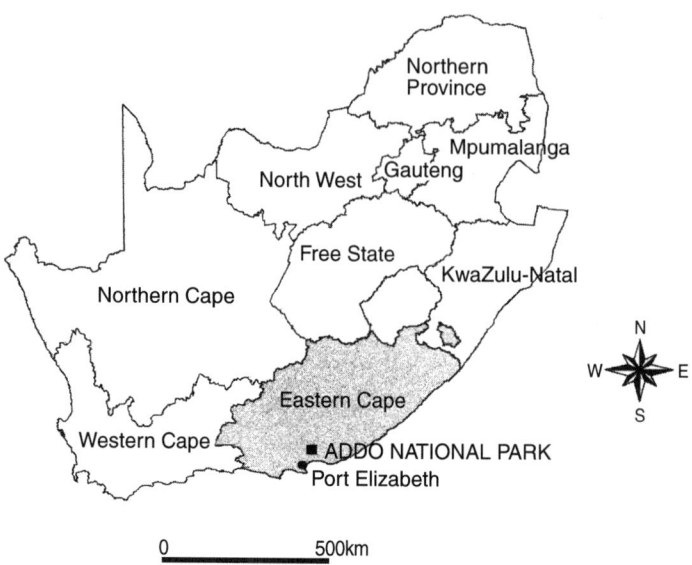

Figure 39.1 The nine provinces of South Africa and the location of the City of Port Elizabeth and the Addo Elephant National Park.

(Cowling, 1984; Lubke et al., 1986). The wealth of growth forms and the transitional nature of the thicket result in very high plant diversity, and these may be the most species-rich formations of woody plants in southern Africa (Hoffman and Everard, 1987). In addition, these thickets form part of the IUCN-recognized Albany center or *hot spot* of biodiversity (Cowling and Hilton-Taylor, 1997). At a regional level these thickets have the highest endemism (30%) within the Eastern Cape (Lubke et al., 1986). This endemism is not symmetrical across the growth forms; it is highest among the succulent forbs and shrubs and geophytes (Hoffman and Cowling, 1991; Moolman and Cowling, 1994).

In addition to the high floristic diversity, the Thicket Biome historically supported a high diversity of vertebrate herbivores, including charismatic species such as the African elephant, *Loxodonta africana,* and black rhinoceros, *Diceros bicornis,* as well as a range of smaller species such as Cape buffalo, *Syncerus caffer;* eland, *Taurotragus oryx;* kudu, *Tragelaphus strepsiceros;* bushbuck, *T. scriptus*; bushpig, *Potamochoerus porcus;* and common *Silvicapra grimmia* and blue duikers, *Philantomba monticola* (Skead, 1987). Associated with these herbivores was the full complement of large predators, including lion, *Panthera leo*; leopard, *P. pardus*; African wild dog, *Lycaon pictus;* brown hyena, *Hyaena brunnea;* and spotted hyena, *Crocuta crocuta.* This high faunal diversity extends also to the birds (more than 300 species), herpetofauna, and probably invertebrates (Kerley and Boshoff, 1997).

Fire plays a minor role in the natural dynamics of the Thicket Biome, and drought, another major form of disturbance, appears to have little impact on the flora. Hence, animal/plant interactions appear to constitute the most important process to generate and maintain the high floristic diversity (Kerley et al., 1995). Animal-driven seed dispersal is a common feature of the reproductive strategies of many of the plants (Cowling, 1984); in addition, the evergreen, nutritious, and spinescent nature of the vegetation supports the hypothesis that herbivory is an important process in structuring the vegetation, with strong co-evolution between the flora and associated herbivores. The megaherbivores (elephant and rhinoceros) are thought to be particularly important in terms of herbivory, patch creation, and seed and nutrient dispersal (Kerley et al., 1995).

Whatever the role of the fauna in structuring the floristic communities, these processes were severely disrupted by the extirpation of many of these species, coincident with the arrival of modern pastoralists of European and African origin (Skead, 1987). Large predators (lion, cheetah, hyenas) were exterminated in an attempt to protect domestic herbivores, and leopards managed to survive in only the most remote areas. Herbivores were actively hunted for meat, leather, and ivory, and suffered the familiar loss of habitat. Elephants were reduced to a population of 11 individuals in 1931 and did not receive full protection until 1955. The black rhinoceros was extirpated by 1858, remaining so until reintroduced in 1961; buffalo persisted under the umbrella of elephant conservation (Kerley and Boshoff, 1997). The medium-sized herbivores (kudu, bushbuck, duiker) managed to persist in the dense thickets.

The dominant land use in the Thicket Biome is now pastoralism, particularly the husbandry of goats, *Capra hircus*. Pastoralism has led to not only the loss of a large proportion of the indigenous fauna but also to a collapse of ecosystem health. These changes can be expressed at a number of different levels. At the landscape level, pastoralism is associated with a loss of landscape heterogeneity, which reflects a loss of landscape elements critical for hosting a range of habitats and other resources for species of plants and animals (Fabricius, 1997). At the floristic level, pastoralism has led to a loss of plant cover and phytomass, a replacement of dominant species by less palatable species, a decrease in the cover of perennial plants, and an increase in the cover of annuals, as well as invasion by alien plant species and extirpation of plant species (Hoffman and Cowling, 1990, 1991; Kerley et al., 1999; Moolman and Cowling, 1994). These localized plant extinctions are not random among growth forms; succulents and geophytes, many of which are endemics, suffer the most (Moolman and Cowling, 1994). Soils also are affected, with pastoralism leading to large-scale erosion; loss of soil organics, N, Mg, Ca, and moisture; and an increase in soil Al (Kerley et al., 1999; La Cock, 1992; Palmer et al., 1988). These observed indicators of ecosystem degradation also are expressed in terms of productivity: degraded thicket habitats can support fewer herbivores. In addition, forage production is more variable in degraded thickets, virtually collapsing during drought (Stuart-Hill and Aucamp, 1993). Finally, these changes are apparently irreversible within normal management time frames, and therefore this loss of ecosystem resources and productivity can be characterized as desertification (Kerley et al., 1995).

The mechanisms of the degradation, through pastoralism, of a system that evolved with high levels of herbivory appear to occur through the management and specific foraging behavior of domestic herbivores, chiefly goats. Goats are stocked at high densities in response to the economic constraints of farmers (Kerley et al., 1995). In addition, goats are able to overcome the physical plant defenses that protect the plants from the indigenous herbivores with which they have co-evolved (Haschick and Kerley, 1999). Goat feeding behavior also exposes the soil to erosion (Kerley et al., 1999) and prevents vegetative regeneration of some plants (Stuart-Hill, 1992). In addition, goats are poor dispersers of seeds compared to the indigenous herbivores (Sigwela, 1999).

The degradation of the Thicket Biome can be expressed in terms of the three indicators of ecosystem health listed by Rapport et al. (1998): vigor (i.e., loss of productivity, soil resources); organization (loss of biodiversity, keystone species); and resilience (shift in herbivores and the increased response of the system to drought). Ecosystem services, although not well documented, also are strongly affected. Loss of phytomass — ~150,000 kg wet plant mass/ha (Penzhorn et al., 1974) — and soil carbon — ~20,000 kg C/ha in the top 10 cm (Kerley et al., 1999) — indicates a reduced ability of the system to counter global warming through carbon sequestration. Similarly, the reduction in productivity has consequences for human resource extraction and leads to an increase in rural poverty and increased urbanization (Geach, 1997).

LAND USE OPTIONS: ECOLOGICAL AND ECONOMIC SUSTAINABILITY

Although pastoralism is the dominant form of land use in the Thicket Biome at present, it is fraught with problems, as discussed earlier. Game ranching and ecotourism/conservation, based on the indigenous fauna (primarily herbivores) are the other major land use options. The desirability of these alternative land use options has to be evaluated in terms of ecological and economic sustainability — i.e., in terms of the consequences of pastoralism, game ranching, and ecotourism/conservation to ecosystem health.

Although pastoralism is apparently financially rewarding in the short to medium term, it cannot be considered to be economically sustainable, given the observed erosion of the natural resource capital upon which it is based (Birch, 1991; Kerley et al., 1995). However, government subsidy policies and artificially inflated land prices have contributed to maintaining pastoralism. Associated with ecosystem degradation has been an observed depopulation of rural areas (urbanization) and increased rural poverty. Ecosystem services (soil stabilization, primary productivity, carbon sequestration, biodiversity, and economic opportunities) suffer under pastoralism (Kerley and Boshoff, 1997).

More and more previously pastoral operations are switching to game ranching, relying on the diversity of medium-sized herbivores (particularly kudu) to produce venison and to attract local and foreign hunters. All available evidence indicates that the indigenous, medium-sized herbivores are ecologically sustainable (Stuart-Hill, 1992). There are some indications that areas from which megaherbivores are excluded do tend to lose some patch diversity (Fabricius, 1997; Stuart-Hill, 1992), although the amount of such patch diversity before the extirpation of the megaherbivores is unknown. Game ranching therefore appears to maintain healthy ecosystems.

In financial terms, game ranching is not as lucrative as pastoralism, generating only about 40% of the income that a comparably sized pastoral operation would generate (Kerley et al., 1995). The critical issue here is that, although game ranching generates less income in the short term, it is ecologically sustainable and hence can continue operating long after a pastoral operation has collapsed due to the loss of the natural resources (Kerley et al., 1995). Attention needs to be paid to the problems of appropriate scale and technology, efficient marketing, and adding value to the products through producer-based processing (Kerley et al., 1995) in order to improve the economic opportunities of game ranching if this form of land use, with its associated ecosystem health benefits, is to thrive. Given the benefits accruing to society through the ecological and economic sustainability of game ranching, consideration should be given to developing incentives (e.g., tax rebates, advice) to encourage and assist landowners in making the transition from pastoralism to game ranching.

The third major land-use option in the Thicket Biome is that of ecotourism/conservation. Under this model, tourists pay to experience the indigenous biodiversity, particularly the megaherbivores, within formal conservation areas. The best developed and studied of these is the Addo Elephant National Park (AENP), northeast of Port Elizabeth (Figure 39.1). Although elephants (78% of vertebrate herbivore biomass in the AENP) do have a significant impact on the vegetation, they are ecologically sustainable in the Thicket Biome at densities of much less than two elephants/km (Stuart-Hill, 1992). In the process of protecting elephants, conservation is also conferred upon the full suite of biodiversity (excluding the large predators at present) as well as maintaining ecosystem services (Kerley et al., 1995, 1999).

Ecotourism is also highly desirable in financial and economic terms: for example, in 1994 the AENP generated a profit and employed twice as many staff members at four times the salary of a comparably sized pastoral operation (Kerley et al., 1995). The benefits also extend well beyond the borders of the AENP. For example, a travel cost study estimated a recreational value of approximately $60 million for the nearly 100,000 tourists who visited the AENP in 1996, which clearly represents significant economic activity (Geach, 1997). Furthermore, it has been estimated (Kerley and Boshoff, 1997) that the visitors to the AENP generate at least 4000 additional employment opportunities in the tourism-related industry. Ecotourism/conservation in the Thicket Biome is therefore a winning option in both ecosystem health and sociopolitical terms.

THE EASTERN CAPE AS A BIODIVERSITY HOT SPOT

Southern African Biodiversity in the World Context

A comparison of the world's 18 biological hot spots (areas with high species richness, a high concentration of endemics, and a high degree of threat) highlights the importance of the temperate Mediterranean climatic areas (Myers, 1990). In particular, southern Africa and its Cape Floristic Region were identified as one of the hottest of all hot spots in terms of plant diversity and threats (Myers, 1990). On a subcontinental scale, southern Africa registers in the upper quarter of the world's 12 megabiodiversity countries (McNeely et al., 1990). Subsequently, an additional hot spot, the Succulent Karoo, has been recognized in South Africa (Mittermeier et al., 1999).

Yet within southern Africa, South Africa is distinct as one of the most biologically diverse countries in the world. Its 23,404 recorded vascular plant species, at an average density of 8.1 species/106 km^2, exceeds species-rich countries such as Brazil, Ecuador, Madagascar, and other countries in southeast Asia, all of which have rich tropical forests (Cowling et al., 1989; Gibbs Russel, 1985). In addition, South Africa has an exceptionally large proportion of endemic vascular plants (80%), exceeded only marginally by New Zealand with 82% endemic species (Cowling and Hilton-Taylor, 1994).

Among the vertebrate fauna of southern Africa, a similar species-rich picture emerges. The region's amphibian, reptilian, avian, and mammalian faunas account for about 2, 6, 7, and 6%, respectively, of the world's total species for each taxon (Siegfried, 1989). On average, the density of these southern African taxa exceeds the African and world species densities by 6.6 and 5.5 times, respectively (Siegfried, 1989). Faunal endemicity in southern Africa shows marginal congruence between taxa with the amphibians and reptiles (44 and 31%, respectively) having relatively high levels of endemic species, while the mammals and birds are relatively less well endowed with 6 and 15%, respectively (Drinkrow and Cherry, 1995; Siegfried, 1989; Skinner and Smithers, 1990).

To a large degree, southern Africa's rich biodiversity and high numbers of endemics can be attributed to contemporary ecological conditions (Lombard, 1995). Southern Africa offers transitional climates between the subtropical summer rainfall and temperate winter rainfall, in combination with complex topography, geology, pedology, biomes, vegetation types, and habitats, all situated at the base of the continent (Branch et al., 1995; Coe and Skinner, 1993; Cowling and Hilton-Taylor, 1994; Gelderblom and Bronner, 1995). The complexity of the region is reflected in the description of five phytochoria (White, 1983), seven biomes (Low and Rebelo, 1996; Rutherford and Westfall, 1986), 68 vegetation types (Low and Rebelo, 1996), and 5.2% of the world's vertebrates, on 0.8% of the land area (Siegfried, 1989).

Yet this rich biodiversity is also under great threat; about 47% of South Africa's natural vegetation already has been transformed (Low and Rebelo, 1996). South Africa has the third highest number of threatened reptile, amphibian, and invertebrate species (IUCN, 1996). Although only 13% of its diverse mammalian fauna is considered threatened, the country is still noted to be among those areas under severe threat of extinctions (IUCN, 1996). Among the vascular plants, southern Africa records the highest number of Red Data Book species (2575 species) per area, exceeding Australia, India, and Mexico (McNeely et al., 1990). Thus, in the global context, South Africa, with its rich biodiversity and increasing levels of threat, is worthy of conservation attention. South Africa contains multiple hot spots, of which the Eastern Cape clearly stands out.

The Eastern Cape Biodiversity Hot Spot

The Eastern Cape of South Africa sits in a transition zone of topographical, geological, pedological, and climatic complexity (Lubke et al., 1986). The wide range of habitats is conducive to supporting a wide diversity of plants and animals. The diversity of abiotic conditions prevalent in the Eastern Cape has made the region the most botanically diverse area of the country. Of the subcontinent's five major phytochoria, the Eastern Cape forms a major transition or tension zone between four of them: the Cape, Afromontane, Karoo-Namib, and Tongaland-Pondoland (Werger, 1978). That tension zone is characterized by the convergence within the Eastern Cape of all (Nama Karoo, Succulent Karoo, Fynbos, Savanna, Grassland, Forest, and Thicket) of the seven recognized biomes in the country (Low and Rebelo, 1996) (Figure 39.2). Furthermore, within these biomes in the Eastern Cape, a total of 27 different vegetation types are represented, more than any of the other eight provinces of South Africa (Low and Rebelo, 1996). Thus, at the level of the vegetation type and biome, biodiversity in the Eastern Cape is the highest in southern Africa.

The Albany plant diversity hot spot in the Eastern Cape has been identified as one of the subcontinent's eight biodiversity hot spots (Cowling and Hilton-Taylor, 1997). Although the approximately 2000 species and endemism (10%) within the Albany hot spot is not particularly high, the species-to-area relationships compare with the other southern African hot spots, which together should be ranked among the world's most conservation-worthy areas. In addition, the transitional

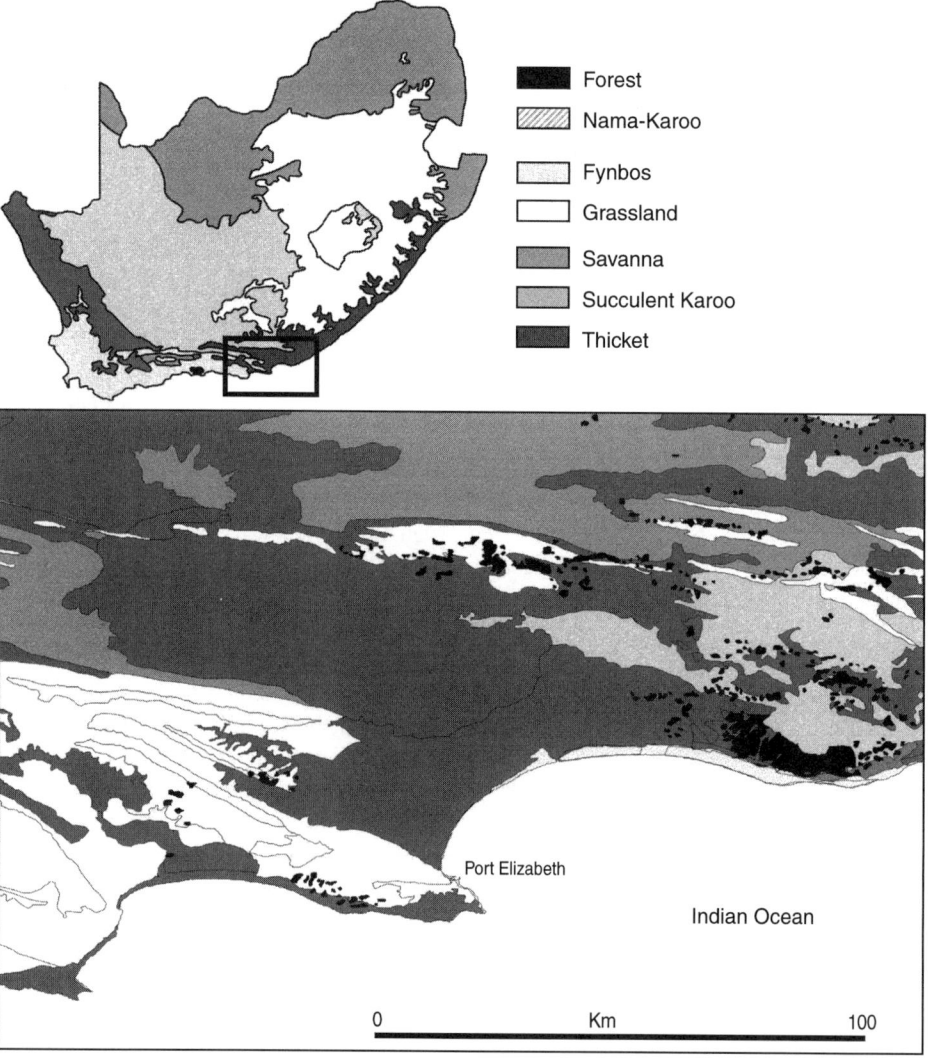

Figure 39.2 The seven South African biomes (bioclimatic regions), with details of the area in the vicinity of the Addo Elephant National Park.

nature of the vegetation types within the Albany area — where many of the species are at their distributional limits — is unique. This phenomenon likely accounts for the relatively low endemicity in the Albany hot spot because most of the species tend to be generalists rather than specialists. Endemics within the Albany hot spot typically are succulents from the succulent thicket vegetation types. For example, 30% of southern Africa's succulent *Euphorbia* species are represented in the Eastern Cape, of which 48% are endemic to the region (Court, 1988). In addition to the diverse terrestrial elements, the Eastern Cape also includes the coastal, marine, and estuarine components, further adding to the biodiversity of the area.

Biodiversity at the Biome Level

Table 39.1 shows that the Eastern Cape is particularly important in terms of the representation, and hence conservation, of the biomes and vegetation types of South Africa. All seven biomes are represented in the province.

Biodiversity at the Landscape Level

The Eastern Cape encapsulates a broad ecological and geomorphic diversity within relatively short distances. This diversity ranges from the grass-covered escarpment to the flat semiarid Karoo plains on the African erosion surface with its previously large herds of springbok; to the folded sandstone Zuurberg and Baviaanskloof Mountains, with their accompanying high rainfall vegetation types on the south-facing slopes; to drier rain-shadow north-facing slopes that are home to the endemic Cape mountain zebra, *Equus zebra zebra*. It also encompasses the more nutrient-rich post-African surface south of the mountains, with its dense succulent vegetation and rich mammalian browsing community; the marine and coastal aeolian sediments on the coast; and the marine province with its continental islands (Kerley and Boshoff, 1997). In the aggregate, these terrestrial and marine landscapes comprise the most biologically diverse area in southern Africa.

The Eastern Cape marine component includes a diversity of sandy and rocky shores and two island groups (Bird Island and St. Croix), adding to the biological diversity of the region. This section of the South African coast falls within the South Coast marine biogeographical province — one of three identified (Hockey and Buxton, 1989) — and has been independently identified as

Table 39.1 The Conservation Status of Biomes and Their Vegetation Types in South Africa and Specifically in the Eastern Cape

Biome	Biomes		Vegetation types		
	Percent Conserved Nationally	Percent Conserved in Eastern Cape	Number and Percent in Eastern Cape	Approximate Percent Transformed	Restricted to the Eastern Cape
Thicket	4.50	4.20	5 (100)	45	Xeric succulent thicket, Mesic succulent thicket
Nama Karoo	0.47	0.01	3 (50)	?	None
Succulent Karoo	2.82	17.40	1 (0.25)	?	None
Fynbos	11.84	21.40	3 (30)	47	None
Forest	17.90	15.80	2 (67)	44	Coastal forest
Savanna	9.40	0.33	4 (16)	46	Eastern thorn bushveld, sub-arid thorn bushveld
Grassland	2.23	0.30	9 (60)	58	Coastal grassland, Southeastern mountain grassland

Data adapted from Low, A.B. and Rebelo, A.G., *Vegetation of South Africa, Lesotho and Swaziland*, Department of Environment Affairs and Tourism, Pretoria, South Africa, 1996.

a region in need of protection (Payne and Crawford, 1989). Much like the terrestrial areas in proximity to Algoa Bay, this section of the coast appears to be a transitional area of marine species from the cool temperate west coast and warm tropical elements, and it is particularly important for its diversity of bivalves, limpets, and endemic fish species. East of Algoa Bay, the proportion of endemic fish species rapidly declines (Hockey and Buxton, 1989).

Biodiversity at the Species Level

Invertebrates

Little is known about the invertebrate fauna of the Eastern Cape, although researchers believe that the conservation of the larger mammalian species as flagship species within conservation areas will benefit the invertebrates (Fabricius, 1997). For example, conservation of the larger mammalian herbivores in Addo has been beneficial to the endemic flightless dung beetle, *Circellium bacchus* (Kerley and Boshoff, 1997).

Vertebrates

Comparisons between the distribution of total species richness, endemics, and rarity across six vertebrate taxa (fish, tortoises, frogs, snakes, birds, and mammals) in South Africa revealed little congruence within and between taxa, indicative of the different environmental conditions necessary for each taxon (Lombard, 1995). However, the general trend to emerge is one of species richness in the northeastern subtropical areas of the country, with centers of endemism in the more remote south and southwestern temperate regions, and the rare species split between the two. The Eastern Cape emerges as a species-rich transitional area. Analysis of each of the six taxa revealed the following:

1. *Fish*: The northeastern subtropical low-lying areas of South Africa have the greatest diversity of freshwater fish species, yet the majority of the country's 33 endemic species are restricted to the southern regions, with the Sundays River area identified as one of 10 national endemic species hot spots (Skelton et al., 1995). The Zuurberg Mountains of the Eastern Cape provide protection to two indigenous species of minnows, *Barbus pallidus* and *B. anoplus*, both of which are considered in need of conservation (Skelton, 1987).

2. *Frogs*: Southern Africa has a particularly rich amphibian fauna; and the Eastern Cape, with 10 to 18 amphibian species, of which 5 to 8 are endemic, is considered to be one of the country's amphibian hot spots (Drinkrow and Cherry, 1995).

3. *Reptiles*: The Eastern Cape harbors a high diversity of reptiles, particularly snakes and lizards, 9% of which are endemic to the Eastern Cape. Such reptiles include the Tasman's girdled lizard, *Cordylus tasmani,* the Cape legless burrowing skink, *Scelotes anguina,* and southern dwarf chameleon, *Bradypodion ventrale ventrale* (Branch, 1988a,b). In this regard the Algoa basin is a minor center of endemicity (Branch, 1988b). The Eastern Cape also has the most diverse tortoise fauna in the world — five species, three endemics (Branch et al., 1995).

4. *Birds*: More than 500 bird species have been recorded in the Eastern Cape, with more than 300 species breeding in the area (Harrison et al., 1997; Skead, 1967), 13 of which are Red Data Book species (Brooke, 1984). The Eastern Cape also has an endemic hot spot located in the coastal areas west of Algoa Bay (Lombard, 1995). Unique features include the presence of four of southern Africa's five large eagle species: namely, the black eagle *Aquila verreauxii*, crowned eagle *Stephanoaetus coronatus*, martial eagle *Polemaetus bellicosus*, and the fish eagle *Haliaeetus vocifer.* Additionally the region hosts three species of large bustard: the Kori *Areotis kori*, Stanley's *Neotis denhami,* and Ludwig's *N. ludwigii* bustards (Kerley and Boshoff, 1997).

5. *Mammals*: Although the Eastern Cape is neither species rich nor an endemic mammalian hot spot, it does register as an important hot spot for endemic insectivore species that are associated with the indigenous forests (Gelderblom and Bronner, 1995; Lombard, 1995). The Eastern Cape is

home to the second-largest population of elephants *Loxodonta africana* in South Africa (Kerley and Boshoff, 1997). In addition, the Eastern Cape, particularly in the Thicket Biome, supports the second-largest population of the threatened black rhinoceros *Diceros bicornis* in South Africa (Hall-Martin and Knight, 1994).

Flora

The juxtaposition of the seven biomes within the Eastern Cape provides for great botanical diversity yet relatively low endemism, led by the grassy fynbos and succulent thicket vegetation types (Cowling, 1983). Within the latter, most of the endemics are geophytes and succulents, and these are predominantly restricted to the understory layer. This makes them particularly prone to grazing pressure mainly from domestic stock, although high densities of elephants also have a deleterious affect (Cowling and Holmes, 1991; Johnson, 1992; Moolman and Cowling, 1994; Stuart-Hill, 1992).

Of the indigenous forest patches, the Alexandria forest is particularly important, given its mix of Cape and Tongaland-Pondoland species. A total of 27 (16%) of its species are found either exclusively in this forest or within a single other forest patch in the area (Phillipson and Russell, 1988). Three species are endemic to this forest — the Cape wing-nut *Atalaya capensis*, the buig-my-nie *Smelophyllum capense,* and the Cape star-chestnut *Sterculia alexandria.* The region harbors seven (25%) of South Africa's 28 species of cycad *Encephalartos* spp. (Giddy, 1974) and the endemic Willowmore cedar *Widdringtonia schwarzii* (Van Wyk et al., 1988). Many of the indigenous trees such as the cycads are now not only under threat by collectors but also through the harvesting of their products for medical purposes (Cunningham, 1988; La Cock and Briers, 1992; Simelane, 1996).

Three main points emerge from the evidence presented in this section. First, the Eastern Cape, and particularly the region encompassing Algoa Bay and its hinterland, possesses an immense diversity of plants, animals, and landscapes. Second, much of this biodiversity is under threat due to desertification and to poorly planned urban and rural development. Third, the region holds great potential for increased conservation action as well as for underpinning a vibrant tourism industry, based on its spectacular natural assets.

THE PROPOSED GREATER ADDO NATIONAL PARK

Emerging from the considerations of the ecological and economic sustainability of land-use options in the Thicket Biome is the recognition that conservation/ecotourism is the best form of land use for ecosystem health and that conservation should be expanded in order to take advantage of the opportunities presented (Kerley et al., 1995). This area boasts spectacular biodiversity from the species to the landscape level, providing further opportunity to combine conservation and tourism. The region is also well placed to exploit the steady growth in the number of tourist arrivals in South Africa since the ground-breaking 1994 democratic elections. In addition, the territory (terrestrial and marine) acquired for national parks in the Eastern Cape has been increased significantly during the last 20 years. This combination of factors led to a proposal to develop a major conservation area that would have explicit human development opportunities (Kerley and Boshoff, 1997).

This vision sees the amalgamation of two large protected areas with one small, existing protected area — the Addo Elephant National Park (60,000 ha), the Woody Cape (24,142 ha), and Tootabie (343 ha) Nature Reserves — to form a core conservation area, The expansion of this core area would create a Greater Addo National Park (GANP) (Figure 39.3), thereby providing an opportunity for a viable regional and national development and conservation initiative. The selection of the footprint of the proposed GANP (Figure 39.3) is based on 11 recognized criteria and on national and international environmental legislation and treaties. These criteria encompass issues related to

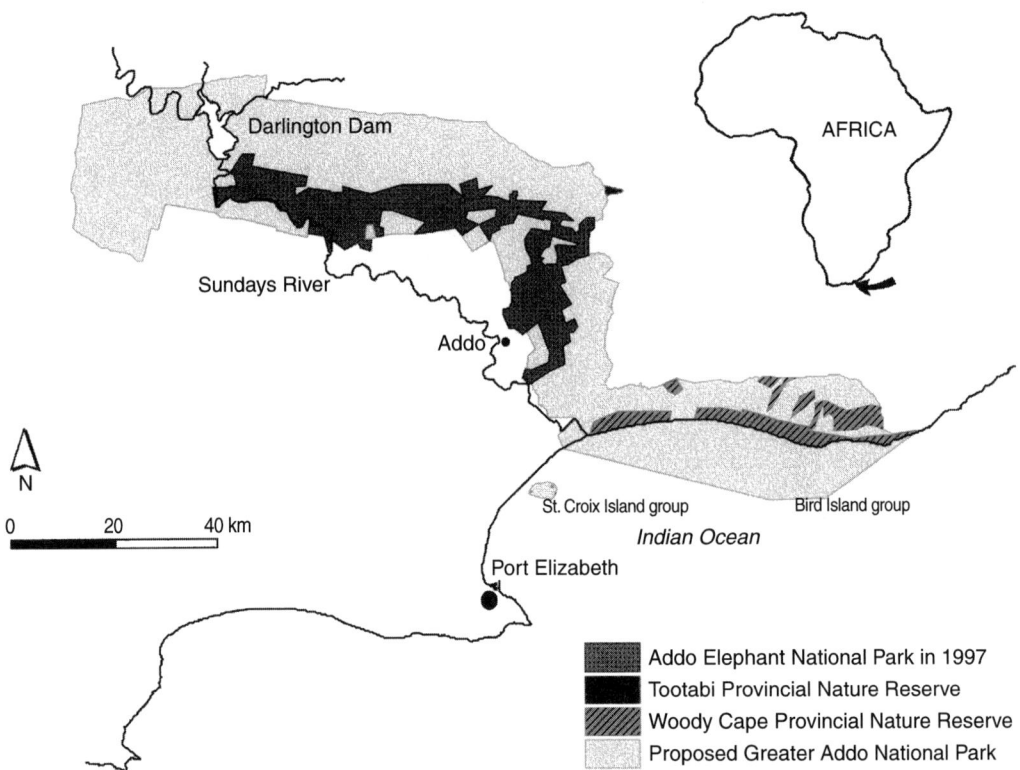

Figure 39.3 The proposed Greater Addo National Park: the core conservation areas in 1997.

biodiversity, spatial complexity and representativeness, ecosystem functioning, naturalness and uniqueness, protection of flagship species, multipurpose (buffer) zones, financial viability, economic potential, ecotourism, and management factors. The vision for the proposed GANP is fully moti-vated on the basis of sound conservation, ecotourism, and economic principles, perspectives, and information. Space precludes the inclusion of a fully referenced account of the GANP proposal here. However, details are available in Kerley and Boshoff (1997) and at the Web site http://www.upe.ac.za. A brief synopsis of the key features and advantages of the proposed GANP follows.

The proposed GANP is some 398,000 ha in size, consisting of a 341,000-ha terrestrial zone and a 57,000-ha marine zone. The former includes the Darlington Dam and almost 90 km of the Sundays River, and the latter includes the Bird and St. Croix island groups in the Indian Ocean (Figure 39.3). The proposed park, which forms a continuous conservation area more than 200 km in length, will be the third largest national park in South Africa. It will be geomorphologically and biotically the most diverse conservation area in South Africa, and probably one of the most diverse in the world. In addition it will create enormous potential for socioeconomic development in the economically depressed Eastern Cape, a province identified by the government as a priority for development.

In terms of its conservation value, the GANP will contribute significantly to South Africa's conservation requirements. It will be unique through the inclusion of examples of six of the seven biomes identified in South Africa (Low and Rebelo, 1996) as well as a diverse marine component. Major landscapes included are the Zuurberg Mountain range, part of the former African land surface, the Alexandria coastal dune field, fossil dune ridges, and karstic landforms. The Alexandria coastal dune field is the largest, most impressive, and least degraded coastal dune field in South Africa, and one of the most spectacular in the world. The GANP also will offer some protection to fragile and threatened river systems, with the Sundays River estuary of particular conservation significance.

The Bird and St. Croix island groups are of immense conservation value; they support, for example, the largest population of the threatened African penguin *Spheniscus demersus* and the largest Cape gannet colony *Sula capensis* in the world, as well as a range of other species of special conservation significance, notably the Cape fur seal *Arctocephalus pusillus*. The marine zone also supports populations of threatened and ecologically and economically important species, e.g., reef and game fish, bottle-nosed *Tursiops truncatus* and humpback *Sousa plumbea* dolphins, southern right *Eubalaena glacialis* and Bryde's *Balaenoptera edeni* whales, and great white sharks *Carcharadon carcharias,* all of which are also important ecotourism resources.

A particular attraction in the proposed GANP will be the megaherbivores (elephant, black rhinoceros) and other charismatic animals, ranging from the large predators such as lion *Panthera leo* and cheetah *Acinonyx jubatus*, to an endemic flightless dung beetle. The GANP will be a true *Big Five* (lion, leopard, elephant, rhinoceros, buffalo) park, a critical feature for tourism success. It will ultimately be able to carry genetically viable populations of most of the large species. With more than 400 species within its boundaries, it will provide habitat for almost half of the bird species recorded in South Africa, and it will play a significant role in conserving the region's reptiles and amphibians.

The proposed GANP also will contribute significantly to the conservation of a range of threatened freshwater and estuarine fish species as well as to the conservation of a number of threatened invertebrates — e.g., rare butterfly species *Aloeides clarki* and *Lepidochrysops bacchus*, an endemic dune grasshopper *Acrotylos hirsutus,* and the flightless dung beetle.

The proposed GANP will conserve an impressive array of plant species, ranging from the desert-adapted succulents in the Karoo to the stately trees of the Alexandria coastal forest. It is characterized by a wide range and high diversity of plant species and by the proximity of several very different and unrelated vegetation types. Part of the Albany hot spot, recognized by the WWF–IUCN as a global center of plant biodiversity, falls within the GANP footprint. The GANP also will provide increased protection for 25% of South Africa's cycad species.

The proposed GANP and its surrounds additionally have an important and interesting paleontological record, consisting of a range of plant and animal fossils; these include dinosaur bones and unique fossil fish deposits. The fact that the GANP includes areas of six biomes ensures that at least some of these bioclimatic regions will persist there in the face of global climate change; in this regard the marked altitudinal variation over a relatively short distance — sea level to 1000 m above sea level over 50 km — within the proposed boundary is noteworthy.

The proposed GANP is located in an area with an extensive and interesting archaeological and historical record. This region includes an important Late Stone Age site; remains of Khoi/San settlements; and sites of conflict between early African and European pastoralists, and between the Boers and the British. The two island groups also have a particularly interesting and valuable history of whaling and shipwrecks.

All the above features and attractions will combine to ensure the success of the proposed GANP as national and international ecotourism destination. Available evidence clearly indicates that ecotourism/conservation is an ecologically sustainable form of land use and that it is successful in terms of wealth generation, economic activity, and job creation. In terms of ecotourism and its economic role, the proposed GANP has enormous potential and would be ideally located to exploit the fast-growing ecotourism market. It will offer a wide and exciting range of attractions, from lions to penguins, from forests to deserts, and from mountains to beaches. The climate of the region is ideally suited to ecotourism, the area has a good tourist safety record, and it is free of malaria. The GANP would be well serviced in terms of tourism infrastructure, such as an airport, transport, surfaced roads, and hospitality facilities. The GANP would provide opportunities for linkages with private conservation areas within the context of the creation of a biosphere reserve, and it is thus obvious that it would be a highly desirable form of land use and development in the region.

The development and operation of a major national park such as the GANP will undoubtedly provide a number of socioeconomic benefits at the local, regional, and national levels. For example, numerous permanent and casual jobs will be created within the GANP and within service and peripheral industries. The potential for the long-term viability of these jobs is considered to be good, and local communities stand to gain most from the economic activity that the GANP will foster. The Mayibuye Ndlovu Crafts Project, involving communities neighboring the present Addo Elephant National Park, has laid a solid foundation for further community relationships with respect to a GANP. A wide range of institutional, social, and management issues were identified and discussed at a GANP Stakeholders Workshop held in February 1999; workshop delegates unanimously endorsed the GANP concept in 1999 (Boshoff and Kerley, 1999).

The potential of the proposed GANP as a national development opportunity has been accepted by the South African government. At a presidential jobs summit organized by the national government in October 1998, one of the three areas in the country singled out was the Algoa Bay region. As a key part of the overall plan for this region, emphasis was laid on "the development of tourism centred on an enlarged Greater Addo [Elephant] National Park" (*Eastern Province Herald,* 3 November 1998). A GANP is now regarded as an integral part of the Fish River Spatial Development Initiative — a development project of the national Department of Trade and Industries, which commissioned an economic viability analysis of the proposal by an independent firm of chartered accountants. The proposed GANP was named as a provincial achievement by Premier of the Eastern Cape Province Government, Mr. M. Stofile, in his speech at the opening of the Provincial Legislature on 12 February 1999; President Thabo Mbeki confirmed the government's commitment to the concept at the opening of the South African Parliament in February 2001.

In November 1998, the Global Environment Facility (GEF) Operational Focal Point, attached to the Department of Environmental Affairs and Tourism in Pretoria, provided formal country endorsement of the proposal in a letter to the World Bank Biodiversity Program in Washington, D.C. On the strength of this, the World Bank, operating as an agent of GEF, invited the submission of a formal project proposal to World Bank/GEF for the establishment of GANP. This will include all aspects of park planning and development except land purchase. The required process has been initiated, and the first step was the previously mentioned inclusive participatory GANP Stakeholder Workshop in February, 1999. The workshop, which attracted more than 150 delegates representing a wide range of stakeholders (Boshoff and Kerley, 1999), was followed by the submission of a proposal to the World Bank for a Project Development Fund grant. This project development phase was initiated in November 2000, with GEF funding, and was completed in 2002. The full project, supported by GEF, will commence in January 2003.

Since the GANP proposal was launched in September 1997, the expansion of the Addo Elephant National Park (AENP) has proceeded apace. By the end of 2001 the AENP had increased in size by some 36,000 ha, with units of land acquired in the south, in the Zuurberg Mountains in the western part, and in the noorsveld to the northwest of the Zuurberg Mountains. Negotiations are under way or have been completed for the use or purchase of a further 30,000 ha of land. In November 2001, the Woody Cape Provincial Nature Reserve (24,142 ha) was handed over to South African National Parks for inclusion into the GANP. The latter acquisition completed the link between the present AENP and the coast and, with this extension, the AENP had effectively doubled in size since the proposal was developed.

It is noteworthy that a substantial proportion of the funds used for the purchase of additional land is in the form of donations from the private sector, in particular from the International Fund for Animal Welfare and the Leslie Hill Succulent Karoo Trust. The Worldwide Fund for Nature–South Africa also has indicated its willingness to purchase land for the GANP as funds become available. Furthermore, the GANP proposal, with its explicit linkages between conservation and socioeconomic development opportunities, has provided the incentive for the South African government to budget funds for the purchase of land for the expansion of the AENP.

THE PROPOSED GANP AND ECOSYSTEM HEALTH

Environmental degradation, as documented earlier, is not restricted to the Thicket Biome; the other biomes that fall within the GANP footprint also exhibit various levels of degradation. Problems include overgrazing, soil erosion, invasion by alien plants, and poorly planned rural and industrial development (Low and Rebelo, 1996). Although the ecosystem health benefits of the proposed GANP will be confined largely to the GANP footprint, the proposal serves as a model and catalyst for other similar initiatives elsewhere in South Africa.

The benefits of a healthy ecosystem, expressed in terms of the contribution of the proposed GANP, are characterized by the provision of clean water and air, the sequestration of carbon, the maintenance of stable soils, enhanced plant productivity, and enhanced biodiversity conservation (species, ecosystems, and landscapes) (Rapport et al., 1998). In addition, the latitudinal and altitudinal variations within the GANP footprint will promote biodiversity persistence in the face of global climate change. We consider that the proposed GANP will presently more than adequately satisfy all three of the indicators of ecosystem health — vigor, organization, and resilience (Rapport et al., 1998).

CONCLUDING STATEMENTS

We contend that the conservation/ecotourism type of land use espoused in the GANP proposal, and based on the tenets of sustainable utilization of natural resources, holds the greatest potential for the meaningful improvement of human communities, especially in the vicinity of the park. There are ultimately only two choices for the region: one that will enable a healthy ecosystem, providing ecosystem services and concomitant opportunities for socioeconomic development, or desertification followed by loss of these services and exacerbated human poverty. Lackey (1998) contends that ecosystem management is a society-driven process. In this regard, the stakeholder endorsement of the GANP concept, and the tangible support for its implementation, indicates the recognition of the benefits of ecosystem health by society.

Our overriding conclusion is that ecosystem health can generate political and financial support, if couched in both conservation and human terms — *conservation for the people*.

ACKNOWLEDGMENTS

This research has been funded by the National Research Foundation (Pretoria), South African National Parks, and the University of Port Elizabeth, South Africa.

REFERENCES

Acocks, J.P.H., Veld types of South Africa, *Mem. Bot. Surv. S. Afr.,* 40:1–128, 1975.

Birch, T., A farmer's perspective of Fish River Valley Bushveld, in Zacharias, P.J.K., Stuart-Hill, C.G., and Midgley, J., Eds., *Proc. First Valley Bushveld/Subtropical Thicket Symp.,* Grassland Society of Southern Africa, Pietermaritzburg, South Africa, 1991, pp. 14–16.

Boshoff, A.F. and Kerley, G.I.H., Eds., *Proc. Greater Addo Natl. Park Stakeholders Workshop,* Terrestrial Ecology Research Unit, Report 25, University of Port Elizabeth, South Africa, 1999.

Branch, W.R., Distribution and diversity of reptiles and amphibians of the Eastern Cape, in Bruton, M.N. and Gess, F.W., Eds., *Towards an Environmental Plan for the Eastern Cape,* Rhodes University, Grahamstown, South Africa, 1988a, pp. 218–245.

Branch, W.R., *South African Red Data Book – Reptiles and Amphibians,* South African National Science Programme Report, 151:1–241, 1988b.

Branch, W.R., Benn, G.A., and Lombard, A.T., The tortoises (*Testudinae*) and terrapins (*Pelomedusidae*) of southern Africa: their diversity, distribution and conservation, *S. Afr. J. Zool.*, 30:91–103, 1995.

Brooke, R.K., South African Red Data Book – Birds, *S. Afr. Natl. Sci. Prog. Rep.*, 97:1–213, 1984.

Coe, M.J, and Skinner, J.D., Connections, disjunctions and endemism in the eastern and southern African mammal faunas, *Trans. R. Soc. S. Afr.*, 48:133–254. 1993.

Court, G.D., The genus *Euphorbia* in the Eastern Cape, in Bruton, M.N. and Gess, F.W., Eds., *Towards an Environmental Plan for the Eastern Cape*, Rhodes University, Grahamstown, South Africa, 1988, pp. 144–148.

Cowling, R.M., Phytochorology and vegetation history in the southeastern Cape, South Africa, *J. Biogeogr.*, 10:393–419, 1983.

Cowling, R.M., A syntaxonomic and synecological study in the Humansdorp region of the Fynbos biome, *Bothalia*, 15:175–227, 1984.

Cowling, R.M. and Hilton-Taylor, C., Patterns of plant diversity and endemism in southern Africa: an overview, in Huntley, B.J., Ed., *Botanical Diversity in Southern Africa*, Strelitzia, 1, National Botanical Institute, Pretoria, South Africa, 1994, pp. 31–52.

Cowling, R.M. and Hilton-Taylor, C., Phytogeography, flora and endemism, in Cowling, R.M., Richardson, D.M., and Pierce, S.M., Eds., *Vegetation of South Africa*, Cambridge University Press, Cambridge, U.K., 1997, pp. 43–61.

Cowling, R.M. and Holmes, P.M., Subtropical thicket in the southeastern Cape: a biogeographical perspective, in Zacharias, P.J.K., Stuart-Hill, G.C., and Midgley, J., Eds., *Proc. First Valley Bushveld/Subtropical Thicket Symposium*, Grassland Society of Southern Africa, Pietermaritzburg, South Africa, 1991, pp. 3–4.

Cowling, R.M., Gibbs-Russel, G.E., Hoffman, M.T., and Hilton-Taylor, C., Patterns of plant diversity in southern Africa, in Huntley, B.J., Ed., *Biotic Diversity in Southern Africa: Concepts and Conservation*, Oxford University Press, Cape Town, South Africa, 1989, pp. 19–50.

Cunningham, T., Overexploitation of medicinal plants in Natal/Kwazulu: root causes, *Veld Flora*, 74:85–87, 1988.

Drinkrow, D.R. and Cherry, M.I., Anuran distribution, diversity and conservation in South Africa, Lesotho and Swaziland, *S. Afr. J. Zool.*, 30:82–90, 1995.

Everard, D.A., A classification of the subtropical transitional thicket in the Eastern Cape, based on syntaxonomic and structural attributes, *S. Afr. J. Bot.*, 53:329–338, 1987.

Fabricius, C., The Impact of Land Use on the Biodiversity in Xeric Succulent Thicket, South Africa, Ph.D. thesis, University of Cape Town, South Africa, 1997.

Fleischner, T.L., Ecological costs of livestock grazing in western North America, *Conserv. Biol.*, 8:629–644, 1994.

Geach, B.G.S., The Addo Elephant National Park as a Model of Sustainable Land Use through Ecotourism, M.Sc. thesis, University of Port Elizabeth, 1997.

Gelderblom, C.M. and Bronner, G.N., Patterns of distribution and current protection status of the endemic mammals of South Africa, *S. Afr. J. Zool.*, 30:127–136, 1995.

Gibbs Russel, G.E., Analysis of the size and composition of the southern African flora, *Bothalia*, 15:613–629, 1985.

Giddy, C., *Cycads of South Africa*, Purnell, Cape Town, South Africa, 1974.

Hall-Martin, A.J. and Knight, M.H., Conservation and management of black rhinoceros in South African national parks, in Penzhorn, B.L. and Kriek, N.P.J., Eds., *Rhinos as Game Ranch Animals*, University of Pretoria, Onderstepoort, South Africa, 1994, pp. 11–19.

Harrison, J.A., Allan, D.G., Underhill, L.G., Herremans, M., Tree, A.J., Parker, V., and Brown, C.J., *The Atlas of Southern African Birds, including Botswana, Lesotho, Namibia and Zimbabwe, Vol. 1, Non-Passerines* and *Vol. 2, Passerines*, BirdLife South Africa, Johannesburg, 1997.

Haschick, S.L. and Kerley, G.I.H., Plant spinescence and biodiversity: responses of indigenous and domestic browsers to plant defenses, in *Proc. VI Intl. Rangeland Congr.*, 1:544–546, 1999.

Hockey, P.A.R. and Buxton, C.D., Conserving biotic diversity on southern Africa's coastline, in Huntley, B.J., Ed., *Biotic Diversity in Southern Africa: Concepts and Conservation*, Oxford University Press, Cape Town, South Africa, 1989, pp. 289–309.

Hoffman, M.T. and Cowling, R.M., Desertification in the lower Sundays River Valley, South Africa, *J. Arid Environ.*, 19:105–117, 1990.

Hoffman, M.T. and Cowling, R.M., Phytochorology and endemism along aridity and grazing gradients in the lower Sundays River Valley: implications for vegetation history, *J. Biogeogr.,* 18:189–201, 1991.

Hoffman, M.T. and Everard, D.A., Neglected and abused — the eastern Cape subtropical thickets, *Veld Flora,* 73:43–45, 1987.

IUCN, *1996 IUCN Red List of Threatened Animals,* IUCN, Gland, Switzerland, 1996.

Johnson, S.D., Plant–animal relationships, in Cowling, R.M., Ed., *Fynbos: Nutrients, Fire and Diversity,* Oxford University Press, Cape Town, South Africa, 1992, pp. 289–309.

Kerley, G.I.H. and Boshoff, A.F., A Proposal for a Greater Addo National Park: A Regional and National Development and Conservation Opportunity, Report 17, Terrestrial Ecology Research Unit, University of Port Elizabeth, 1997.

Kerley, G.I.H., Knight, M.H., and De Kock, M., Desertification of subtropical thicket in the Eastern Cape, South Africa: are there alternatives? *Environ. Monitoring Assessment,* 37:211–230, 1995.

Kerley, G.I.H., Tongway, D., and Ludwig, J., Effects of goat and elephant browsing on soil resources in Succulent Thicket, Eastern Cape, South Africa, *Proc. VI Intl. Rangeland Congr.,* 1:116–117, 1999.

Lackey, R.T., Seven pillars of ecosystem management, *Landscape Urban Plann.,* 40:21–30, 1998.

La Cock, G.D., The Conservation Status of Subtropical Transitional Thicket and Regeneration through Seeding of Shrubs in the Xeric Succulent Thicket of the Eastern Cape, Report, Port Elizabeth, Eastern Cape Nature Conservation, 1992.

La Cock, G.D. and Briers, J.H., Bark collecting at Tootabie Nature Reserve, South Africa: Eastern Cape, *S. Afr. J. Bot.,* 58:505–509, 1992.

Lombard, A.T., The problems with multispecies conservation: do hotspots, ideal reserves and existing reserves coincide? *S. Afr. J. Zool.,* 30:145–163, 1995.

Low, A.B. and Rebelo, A.G., *Vegetation of South Africa, Lesotho and Swaziland,* Department of Environment Affairs and Tourism, Pretoria, South Africa, 1996.

Lubke, R.A., Everard, D.A., and Jackson. S., The biomes of the eastern Cape, with emphasis on their conservation, *Bothalia,* 16:251–261, 1986.

McNeely, J.A., Miller, K.R., Reid, W.V., Mittermeier, R.A., and Werner, T.B., *Conserving the World's Biological Diversity,* IUCN, WRI, CI, WWF-US, WB, Gland, Switzerland, 1990.

Mittermeier, R.A., Myers, N., Gil, P.R., and Mittermeier, C.G., *Hotspots: Earth's Biologically Richest and Most Endangered Terrestrial Ecoregions,* CEMEX, Conservation International, Mexico City, Mexico, 1999.

Moolman, H.J. and Cowling, R.M., The impact of elephant and goat grazing on the endemic flora of South African succulent thicket, *Biol. Conserv.,* 68:53–61, 1994.

Myers, N., The biodiversity challenge: expanded hotspots analysis, *Environmentalist,* 10:243–255, 1990.

Palmer, A.R., Cook, B.J.S., and Lubke, R.A., Aspects of the vegetation and soil relationships in the Andries Vosloo Kudu Reserve, Cape Province, *S. Afr. J. Bot.,* 54:309–314, 1998.

Payne, A.I.L. and Crawford, R.J.M., *Oceans of Southern Africa,* Vlaeberg Publishers, Cape Town, South Africa, 1989.

Penzhorn, B.L., Robbertse, P.J., and Olivier, M.C., The influence of the African elephant on the vegetation of the Addo Elephant National Park, *Koedoe,* 17:137–158, 1974.

Phillipson, P.B. and Russell, S., Phytogeography of the Alexandria Forest (southeastern Cape Province), *Monogr. Syst. Bot. Miss. Bot. Gardens,* 25:661–670, 1988.

Rapport, D.J., Costanza, R., and McMichael, A.J., Assessing ecosystem health, *Trends Ecol. Evol.,* 13:397–402, 1998.

Rutherford, M.C. and Westfall, R.H., Biomes of southern Africa — an objective categorization, *Mem. Bot. Surv. S. Afr.,* No. 54, Department of Agriculture and Water Supply, Pretoria, South Africa, 1986.

Siegfried, W.R., Conservation status of terrestrial ecosystems and their biota, in Huntley, B.J., Ed., *Biotic Diversity in Southern Africa: Concepts and Conservation,* Oxford University Press, Cape Town, South Africa, 1989, pp. 186–201.

Sigwela, A.M., Goats and Kudu in Subtropical Thicket: Dietary Competition and Seed Dispersal Efficiency, M.Sc. thesis, University of Port Elizabeth, 1999.

Simelane, T.S., The Traditional Use of Indigenous Vertebrates, M.Sc. thesis, University of Port Elizabeth, 1996.

Skead, C.J., Ecology of birds in the Eastern Cape Province, *Ostrich,* Suppl. 7:1–103, 1967.

Skead, C.J., *Historical Mammal Incidence in the Cape Province, Vol. 2: The Eastern Half of the Cape Province, including the Ciskei, Transkei and East Griqualand,* Chief Directorate of Nature and Environmental Conservation, Cape Town, South Africa, 1987.

Skelton, P.H., South African Red Data Book — Fishes, *S. Afr. Natl. Sci. Prog. Rep.*, 137:1–199, 1987.

Skelton, P.H., Cambray, J.A., Lombard, A., and Benn, G.A., Patterns of distribution and conservation status of freshwater fishes in South Africa, *S. Afr. J. Zool.*, 30:71–81, 1995.

Skinner, J.D. and Smithers, R.H.N., *The Mammals of the Southern African Subregion,* University of Pretoria, Pretoria, South Africa, 1990.

Stuart-Hill, G.C., The effects of elephants and goats on the Kaffrarian succulent thicket of the Eastern Cape, South Africa, *J. Appl. Ecol.,* 29:699–710, 1992.

Stuart-Hill, G.C. and Aucamp, A.J., Carrying capacity of the succulent bushveld of the Eastern Cape, *Afr. J. Range Forage Sci.,* 10:1–10, 1993.

Van Wyk, B.E., Novellie, P.A., and van Wyk, C.M., Flora of the Zuurberg National Park. 1: characterization of major vegetation units, *Bothalia,* 18:211–220, 1988.

Werger, M.J.A., Biogeographical division of southern Africa, in Werger, M.J.A., Ed., *Biogeography and Ecology of Southern Africa*, Vol 1., W. Junk, The Hague, 1978, pp. 301–462.

White, F., The vegetation of Africa: a descriptive memoir to accompany the UNESCO/AETFAT/UNISO vegetation map of Africa, *Natl. Resour. Res.,* 20, UNESCO, Paris, 1983.

CHAPTER 40

The Role of an Accidentally Introduced Fungus in Degrading the Health of the Stirling Range National Park Ecosystem in Southwestern Australia: Status and Prognosis

David Newsome

INTRODUCTION AND BACKGROUND

The ecological status of national parks and nature reserves is usually considered to be healthy when compared to intensively managed landscapes and land that has been cleared for agriculture. The Stirling Range National Park (Figure 40.1) is a large area (115,600 ha) of original vegetation that was designated national park status in 1913 (Underwood and Burbidge, 1993). It is a zone of exceptional biodiversity and contains 1500 plant species, 87 of which are endemic to the park. Historical records of vertebrate fauna show that the park supported 27 species of terrestrial mammals; however, 12 of these are now extinct. Their decline is largely attributed to predation by introduced carnivores (Friend and Muir, 1993). The avifauna is more diverse with 140 species recorded, 90 of which are known to have bred or breed in the park (Burbidge and Rose, 1993).

The park is now completely surrounded by cleared land comprising agricultural ecosystems, parts of which are showing signs of degradation in the form of land salinization. The impact of this secondary salinity, brought about by clearing natural vegetation and perturbing the water table, is profound and obvious. Because of this, agricultural land is degraded, and conventional forms of agriculture and associated economic activity are rendered useless.

The Stirling Range by contrast appears to comprise a healthy original ecosystem. Significantly, however, it is argued here that the Stirling Range ecosystem is in the process of being degraded. The loss of mammal species could perhaps be perceived as an indicator that ecosystem health has been compromised. A more serious threat to the Stirling Range ecosystem is in the form of an accidentally introduced fungus, *Phytophthora cinnamomi*, which has been shown to kill a wide range of plant species. The disease manifestation is the development of chlorotic tissues, the death of leaves, and eventual plant death. It is referred to as *dieback* and is caused by the fungus preventing water uptake by the roots. Although the Stirling Range is a reserve and has therefore been subject to minimal disturbance, the disease continues to spread throughout the park.

This chapter draws on the work that has been done to identify, assess, and manage the dieback problem in the Stirling Range National Park. The objective of this work is to place the dieback problem within the framework of ecosystem health. The significance of placing dieback within

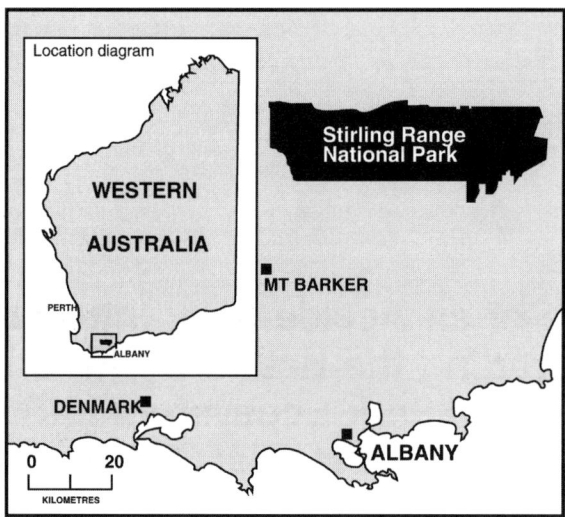

Figure 40.1 Location map.

such a context is in demonstrating the role of exotic organisms in degrading valued ecosystems and in the need to communicate this problem to the public.

ECOLOGICAL SERVICES PROVIDED BY THE STIRLING RANGE NATIONAL PARK

Like many national parks around the world, the Stirling Range provides for tourism, recreation, and the conservation of biodiversity. The mountainous scenery and rich array of wildflowers in spring makes the Stirling Range an increasingly popular tourism destination, with the number of visitors increasing on an annual basis (Figure 40.2). Almost all visitors walk in the park, with a significant percentage undertaking self-guided touring by car. Roads for tourist access were initiated around 1920, and since that time further road improvements and the development of walking tracks have occurred (CALM, 1996). A number of mountain peaks provide for hiking, wildflower viewing, and the appreciation of scenery. In 1992–1993 the Stirling Range National Park had 209,000 visitors, and the highest peak, Bluff Knoll, received 36,000 walkers (Wills and Robinson, 1994).

The Stirling Range National Park contains some of the only mountainous scenery in southwest Western Australia, and this alone is a significant attraction for many visitors. Western Australia is, however, well known for spectacular displays of wildflowers during the period of September to November. This wildflower resource is valued in the region of Au$280 million to $560 million annually (Wills and Robinson, 1994), and wildflower tourism is actively promoted both at national and international levels. The Stirling Range National Park ecosystem is a particularly impressive example of such a wildflower resource. Diverse assemblages of plants, coupled with spectacular flowering (e.g., *Banksia* spp.) and variety of leaf form, constitutes a major part of the tourism experience. A survey in 1992 revealed that 90% of the visitors staying overnight in the nearby Stirling Range Caravan Park had come to view wildflowers (Wills and Robinson, 1994).

The large number of plant species in the Stirling Range provides for a substantial genetic library. Most of these plant species have not been investigated for possible economically significant secondary compounds. In recent years, extracts from a species of *Conospermum* (Proteaceae) were found to contain compounds that could be used in the treatment of human immune deficiency viral infections. Potential royalties from commercial interests were at the time estimated at Au$100 million per annum (Wills and Robinson, 1994). The genetic resources of the park are also of

Figure 40.2 Bluff Knoll track, Stirling Range National Park.

potential significance for the cut wildflower industry. Various species of *Banksia* in the park are significant in this regard.

The Stirling Range is also a significant refuge for fauna, especially birds, and is therefore an important site in relation to conserving terrestrial biota typical of the wider region. The park also provides suitable habitat for reintroduction programs for those species that have become locally extinct.

An ecosystem service that is often not recognized without education and interpretation of the natural environment is one of aesthetic enjoyment per se. Although this service is appreciated by some, for example, ecotourists, much work must be done in order to achieve widespread acknowledgment by the public that the conservation of nature, and therefore ecosystem health, is essential in promoting and maintaining our quality of life.

MEASURES OF ECOSYSTEM HEALTH IN THE STIRLING RANGE NATIONAL PARK

The health of the Stirling Range ecosystem can potentially be measured against the component biodiversity baseline. In simple terms, it answers two fundamental questions — what species should occur in the Stirling Range, and what should the natural communities look like? From an ecological standpoint, the answer to these questions is often very difficult to resolve, mainly as a result of incomplete knowledge about an ecosystem in undisturbed or pristine condition, and the role that humans have played in modifying that ecosystem. If we assess the ecosystem according to the services it provides — that is, attach a value to the services — then measuring and assessing the health of an ecosystem becomes much more straightforward. In the case of the Stirling Range, in which the flora controls many of the ecosystem services, the characteristics of the plant communities provide the measures of ecosystem health.

Pignatti et al. (1993) provide a comprehensive analysis of the plant communities based on extensive phytosociological studies. Eight main communities are described, with Kwongan or heathland as the most widespread community, and rich in flora, with an average of 56.3 species occurring per sampling site. Keighery and Beard (1993) provide a simpler description of the vegetation based on structural characteristics.

Five structural types of plant community are seen to occur in the Stirling Range and consist of the following:

1. *Thicket*: Present on the upper levels of all the major peaks; characterized by a diverse array of genera that belong to the family Proteaceae, e.g., *Banksia*, *Dryandra*, *Adenanthos* and *Isopogon*
2. *Mallee heath*: Variable in species composition and a widespread plant community; contains the main populations of species endemic to the Stirling Range; particularly diverse where it grades into the thicket community
3. *Mallee*: Not well represented in the park but characterized by well-spaced, emergent eucalyptus with a heath understory
4. *Low Woodland*: *Eucalyptus marginata* and *Corymbia calophylla* in low-lying and valley areas; *Banksia attenuata* and *Banksia coccinea* woodland present on deep, sandy soils
5. *Woodlands*: *Eucalyptus marginata* and *Corymbia calophylla* on lateritic soils; *E. wandoo* occurs in association with *C. calophylla*

These five structural types of vegetation contain 90 plant families and 1517 plant species. There is a high degree of endemism (87 species), and the mallee heath (Kwongan) community contains 60 to 120 species/100 m. The Proteaceae, Epacridaceae, Myrtaceae, and Papilionaceae are major plant families. Keighery and Beard (1993) note that the Stirling Range is a zone of species richness for several genera within the Myrtaceae and a center of diversity within the Proteaceae and Epacridaceae.

The diversity of plants, their biology, and variety of vegetation, structure, and habitat, supports 15 species of native mammals and 140 species of birds. The Stirling Range is also an important refuge for reptiles and amphibians due to widespread clearing beyond the park boundaries (Dell and Harold, 1993). The park is also important for invertebrates and provides habitat for a number of relict species (Main, 1993).

The abundance and diversity of plants and animals provides a baseline against which the health of the Stirling Range ecosystem can be measured (Yazvenko and Rapport, 1996). These measures also fit with indicators of ecosystem health identified by Rapport et al. (1998), who use vigor, organization, and resilience to assess ecosystem health. In the Stirling Range any loss of vigor can be measured in terms of declines in species abundance (for example, *Banksia coccinea*). Ecosystem organization can be measured against the baseline level of the morphological and functional diversity of species. Rapport and Whitford (1999) have demonstrated this in three different ecosystems. For example, changes in community structure were used as a measure of ecosystem health in the North American desert grasslands. Because of the ecosystem services provided by vegetation, changes in community structure would be an important measure of health in the Stirling Range.

Resilience is identified by Rapport et al. (1998) as the capacity of an ecosystem to maintain structure and function in the presence of stresses with which it has evolved. In the Stirling Range, the prevailing climate with summer drought and wet winters, combined with nutrient-poor soils, provide for regular stress on the vegetation. Fire causes the largest changes in vegetation structure but is a regular feature in the Australian landscape. The plant communities are well adapted to these prevailing environmental stresses and, although the structure of vegetation may vary depending on the fire regime, under natural conditions the vegetation maintains its original form. The ability of plant communities to recover from stress and disturbance events is therefore a useful measure of ecosystem health.

FACTORS COMPROMISING ECOSYSTEM HEALTH IN THE STIRLING RANGE NATIONAL PARK

The Stirling Range National Park is managed for conservation, recreation, and tourism (CALM, 1996). Management programs are also in place to deal with the fire risk during summer and to control introduced species that pose a threat to the ecological integrity of the park (Gillen and Watson, 1993; Wills and Kinnear, 1993; CALM, 1996). Active management is necessary to monitor visitor activity, maintain tourism facilities, and prevent degradation, such as path erosion. The fire hazard is managed according to a program of prescribed burning and firebreak corridors (Gillen and Watson, 1993). Management action to control weeds and feral animals is expensive, but considerable progress has been made in controlling the introduced European fox *(Vulpes vulpes)* in recent years.

The task of managing these threats to the integrity of the Stirling Range ecosystem requires further funding and additional human effort because the introduced species problem remains unresolved. Degradation of the park continues in the form of edge effects from surrounding agricultural land. This includes increased fire hazard, weed invasion, and the immigration of feral animals from adjacent farmland. The Stirling Range already has lost 12 species of mammals, and any introduction program is likely to be compromised by the continuing presence of introduced carnivores. Fire management and erosion control constitute ongoing management needs.

The single most serious threat to the health of the Stirling Range ecosystem is an introduced fungus, *Phytophthora cinnamomi* (CALM, 1992; Wills, 1992; Gillen and Watson, 1993; Wills and Kinnear, 1993; CALM, 1996). The fungus is considered to have been introduced in 1960, when a number of roads, footpaths, and fire management tracks were constructed (Gillen and Watson, 1993; CALM, 1996). According to Wills and Kinnear (1993), the fungus had spread throughout the park by 1975, before the biology of the disease was properly understood. The disease is originally thought to have come from Southeast Asia and was probably introduced to Western Australia around 1900 (Wills and Kinnear, 1993).

ACTIVITY, SPREAD, AND ECOLOGICAL IMPACT OF *PHYTOPHTHORA CINNAMOMI* (DIEBACK DISEASE) IN THE STIRLING RANGE NATIONAL PARK

Phytophthora cinnamomi is a soil-borne fungus that enters the roots of plants that are susceptible to attack. Vegetative reproduction produces sporangia (Figure 40.3), which release mobile zoospores that infect the roots of host plants. It then spreads by root-to-root contact, through the movement of soil, and is readily transported in water (Shearer, 1994). Widely dispersed infection is the result of soil movement on the wheels of vehicles (Figure 40.4) or on footwear. Activities such as road building and maintenance and the construction of firebreaks would have been the initial mechanism through which the disease spread through the park (Gillen and Watson, 1993; CALM, 1996). The fungus also has spread along recreational hiking tracks as a result of infected soil carried on the boots of walkers. Wills and Kinnear (1993) report that nearly all of the walking tracks in the park are, in part, infected and that some of the infection is of recent occurrence. A significant correlation exists between the distribution of the fungal infection and the more accessible and popular tourist peaks (CALM, 1996). The presence of the disease in upland areas is of particular significance because of the potential for downhill spread of infectious zoospores through surface and subsurface runoff (Gillen and Watson, 1993; CALM, 1996).

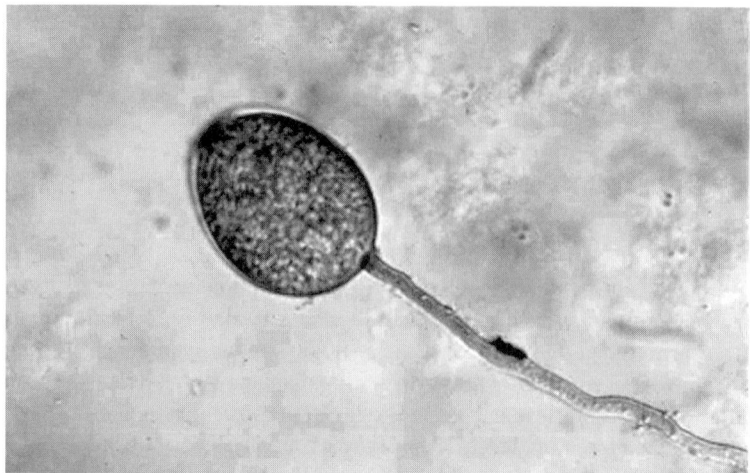

Figure 40.3 *Phytophthora cinnamomi* sporangium containing zoospores.

Figure 40.4 Mud on the wheel of a vehicle.

Figure 40.5 Mass death of native vegetation caused by *P. cinnamomi* infection.

Infected host plants are killed as a result of impaired root function. The disease is characterized by the death of photosynthetic tissues and is known as dieback (Figure 40.5). Research into host susceptibility has found that a wide range of species can become infected. Susceptible species are contained mostly in the plant families Proteaceae, Mytaceae, Papilionaceae, and Epacridaceae (CALM, 1992; Wills, 1992; Shearer, 1994). These plant families comprise a very high proportion of the plants that make up the biotic communities in the Stirling Range.

Wills and Kinnear (1993) estimate that up to 85% of species in the Proteaceae are at risk. In the Stirling Range the entire population of the rare *Banksia brownii* is infected with *Phytophthora cinnamomi* (CALM, 1992). Bridgewater and Edgar (1994) identify the local and uncommon *B. coccinea* as highly susceptible, and Shearer (1994) points out that two species of *Dryandra*, *B. brownii* and *Lambertia orbifolia,* are threatened with extinction as a result of *P. cinnamomi* infection.

More important than individual species, however, the health of the Stirling range ecosystem is under threat from the total effect of *P. cinnamomi* on plant communities. The disease is known to have a major impact on the species-rich understory that is an essential component of mallee heath in Western Australia. Wilson et al. (1994) report that as many as 60% of the component species present can be destroyed following infection by *P. cinnamomi*. Studies carried out in the Stirling Range National Park by Wills (1992) have shown an abundance of susceptible species to be present and that the disease is widespread (Figure 40.6). Moreover, because the Proteaceae contribute the bulk of floristic structure to many plant communities in the park, their loss due to *P. cinnamomi* infection causes a decline in species richness, changes in community structure, and a reduction in plant biomass.

Building on the work of Rapport et al. (1998) in assessing ecosystem health, these changes can be represented diagrammatically (Figure 40.7). The loss of a large number of species from the Proteaceae is a critical factor in ecosystem degradation. The death of species of *Banksia* and *Dryandra* and other members of the Proteaceae cause marked changes in community structure and a decline in species richness (Wills, 1992). The observed and predicted changes to vegetation are illustrated in Figure 40.8. The overall effect on an infected site can be summarized by:

- Change in community structure
- Loss of plant biomass and productivity
- Reduced capacity of a site to support dependent biota (Wills, 1992)

The resultant change in community structure and function degrades the vegetation and its capacity to provide suitable habitats and resources for fauna. Wilson et al. (1994) provide an overview of the likely effects on wildlife. They point out the potential consequences for arboreal mammals and birds that rely on the canopy of dominant species, such as *Banksia* spp., for food and shelter.

Resilience is viewed by Rapport et al. (1998) as an important indicator of ecosystem health. The ecosystem changes described in this chapter are predicted to be long-lasting and even permanent, as has been demonstrated at *P. cinnamomi*–infected sites in eastern Australia (Kennedy and Weste, 1986; Weste and Ashton, 1994). These changes reflect the low resistance of many species to fungal infection. Moreover, recovery following fungus-induced death of vegetation appears to be low, as evidenced by the complete loss of key species (such as *B. coccinea*), reduced seedling regeneration, increased invasion of annual weeds, and colonization by resistant native species such as *Lepidosperma* spp. (Wilson et al., 1994; Wills and Keighery, 1994). Furthermore, as long as the fungus is present, the plant community will not return to its original condition.

Such changes reveal a lack of plant community resilience to *P. cinnamomi* infection. Although at present only pockets of infection causing ecosystem distress syndrome have materialized, up to 70% of the Stirling Range National Park is to some degree infected with *P. cinnamomi*. The implications of this widespread but localized pattern of infection (Figure 40.6) for ecosystem health

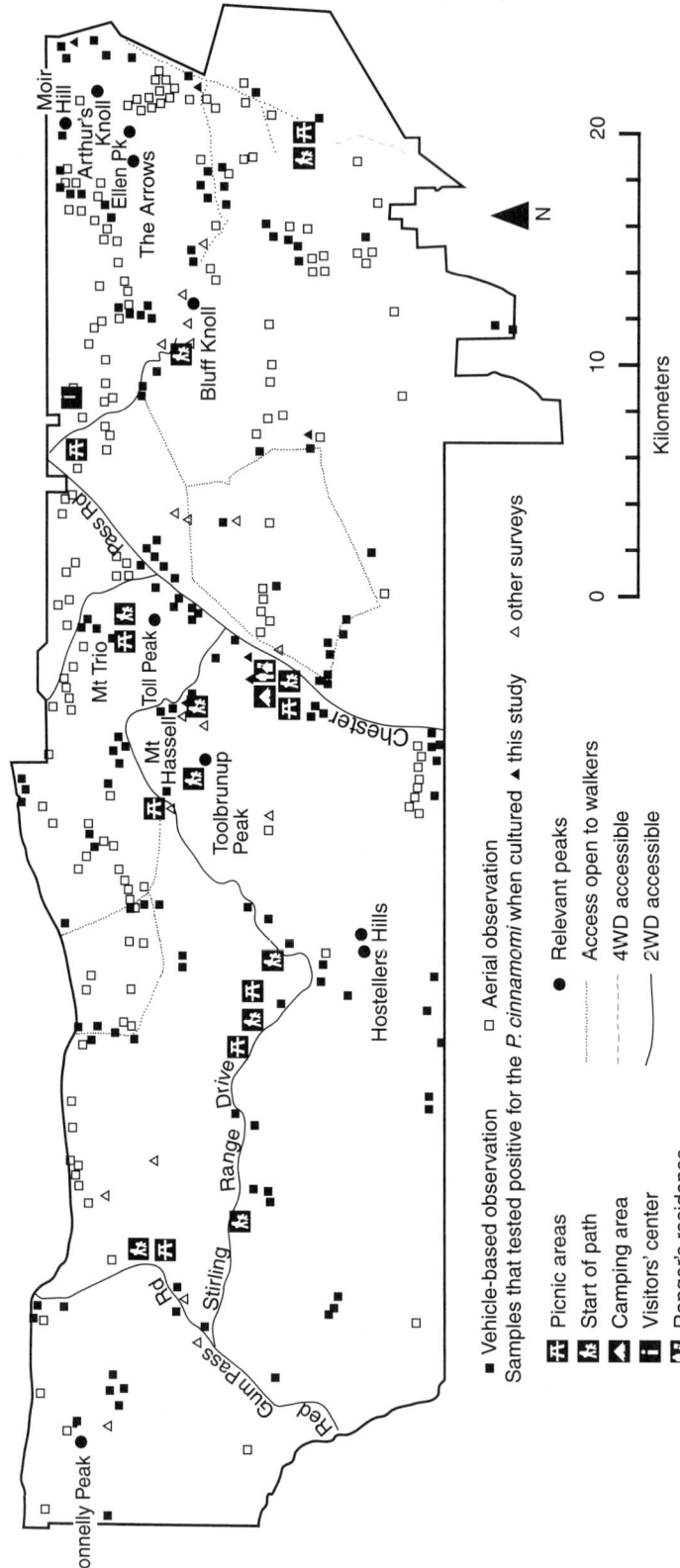

Figure 40.6 Distribution of *Phytophthora cinnamomi* infection in the Stirling Range National Park. (Adapted from Wills, R., Ecological impact of *Phytophthora cinnamomi* in the Stirling Range National Park, *Aust. J. Ecol.*, 17:145–159, 1992, and CALM, Stirling Range and Porongorup National Parks, Draft Management Plan, Department of Conservation and Land Management, Perth, Western Australia, 1996.)

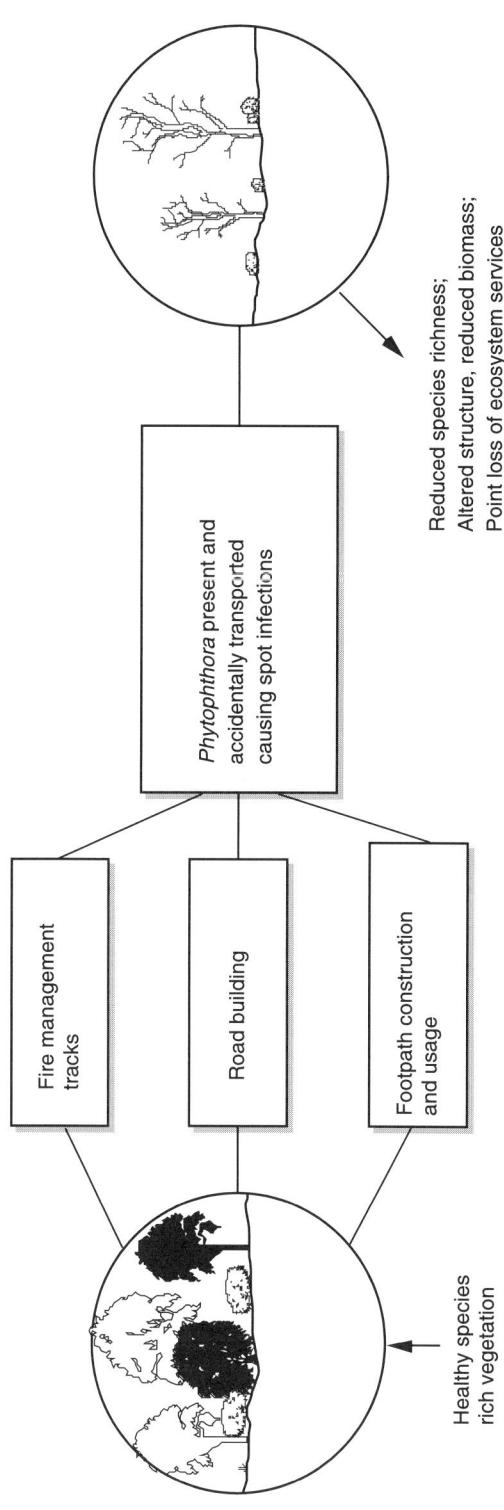

Figure 40.7 *Phytophthora cinnamomi* as an agent in compromising the health of vegetation in the Stirling Range National Park. (Adapted from Newsome, D., Moore, S.A., and Dowling, R.K., *Natural Area Tourism: Ecology, Impacts and Management*, Channel View Publications, Clevedon, U.K., 2002.)

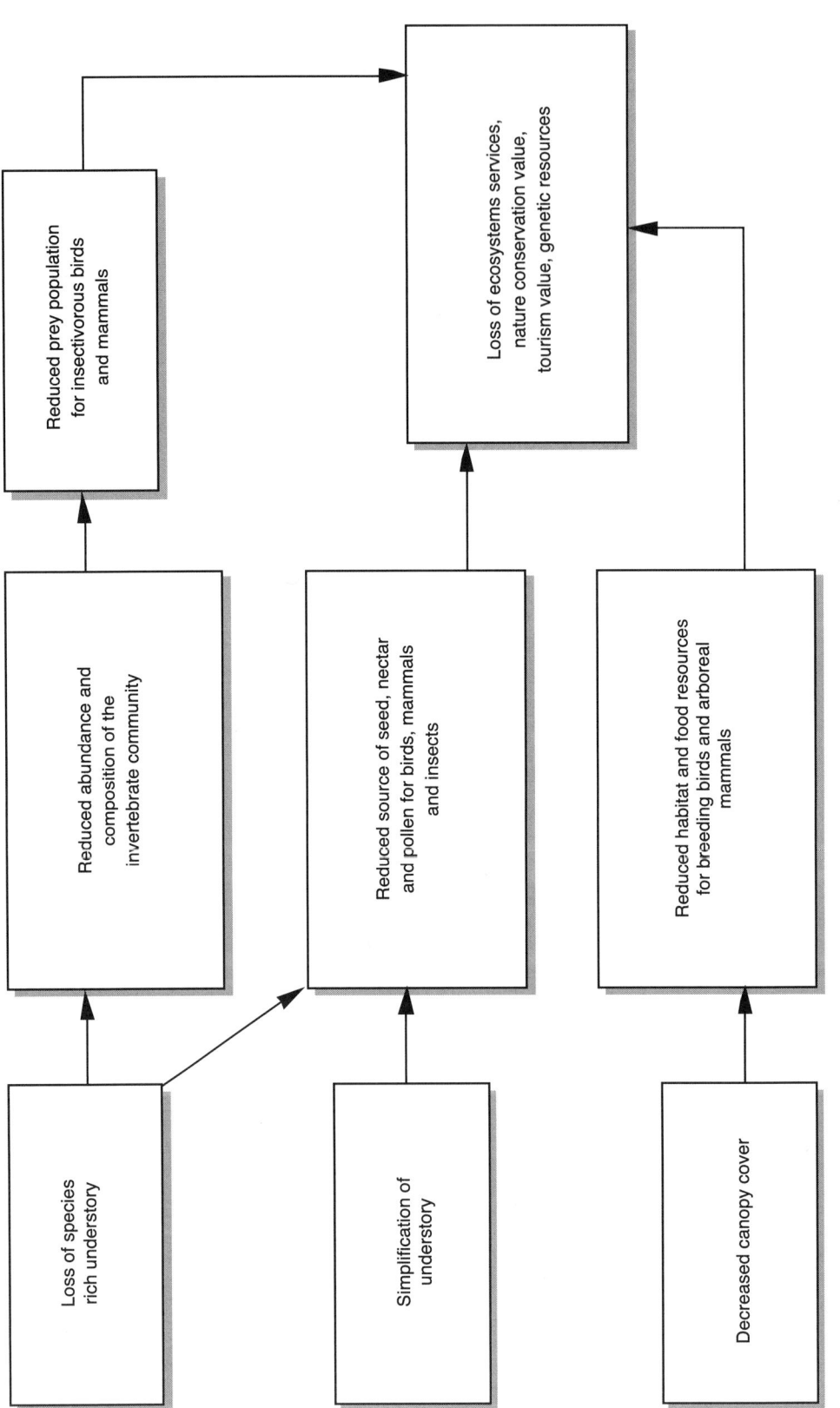

Figure 40.8 Predicted ecosystem damage caused by *Phytophthora cinnamomi* infections. (From Newsome, D., Moore, S.A., and Dowling, R.K., *Natural Area Tourism: Ecology, Impacts and Management*, Channel View Publications, Clevedon, U.K., 2002. With permission.)

are profound, and without some measure of control, the prognosis is for a gradual degradation of vegetation communities and progressive loss of ecosystem services.

THE IMPORTANCE OF APPLYING THE CONCEPT OF ECOSYSTEM HEALTH TO THE *PHYTOPHTHORA CINNAMOMI* PROBLEM IN WESTERN AUSTRALIAN NATIONAL PARKS AND NATURE RESERVES

Widespread infection by *Phytophthora cinnamomi* has the capacity to degrade the conservation and tourism value of the Stirling Range National Park. Although much research and effort is exerted toward controlling the disease, it continues to spread. The Stirling Range National Park presents managers with a significant problem in controlling the spread of the disease because of the need to maintain access for fire management, scenic driving, hiking, picnicking, and wildflower viewing.

Attempts to control the spread of *P. cinnamomi* infection consist of:

- Minimizing spread through quarantine and hygiene practices
- Monitoring its distribution and assessing the risk of spread
- Educating the public and researching methods of control
- The use of a chemical treatment, phosphonate, either as a foliar application or directly injected into woody plants, is showing some promise as a means of controlling the disease (Hardy et al., 1994).

Because controlling spread is critical in managing the disease in the Stirling Range National Park, managers have implemented a number of strategies to reduce spread of the fungus. These consist of designating special conservation zones that are closed to vehicles and walkers, track management to avoid erosion, the closure of uninfected peaks, and public education (CALM, 1996).

Gillen and Watson (1993) emphasize the problems associated with controlling the spread of *P. cinnamomi* from a tourism perspective. More importantly, they point out the lack of appreciation by the public about the significance of the disease and how it can be spread by activities commonly perceived as low impact. For example, many international tourists are accustomed to restriction of access. However, Australians have an attitude and tradition of belief that they can go where they please, and consequently visitors can perceive control of access as a restriction on their recreational experience. Walkers are still seen along the uninfected Mount Monderup track despite its closure and the presence of a public information sign.

Many visitors see the Stirling Range and other national parks as safe from environmental degradation because they are protected areas, because only low-impact activities are permitted, and because a management authority is in attendance. The public is largely unaware that *P. cinnamomi* poses a serious threat to the ecological integrity of the park. Although education and public information programs are in place about the *P. cinnamomi* problem, for a large part the public still does not appreciate natural landscapes for the services they provide. Furthermore, although many people do recognize the salt-damaged land in Western Australia, they cannot see the impact of *P. cinnamomi* infection until mass deaths occur, such as is the case in the eucalyptus forests that lie north and west of the Stirling Range.

This chapter proposes that a recognition and promotion of ecosystem health will make a significant contribution to educating the general community about the importance of natural ecosystems to society. The task needs to be done initially as a scientific dialogue that must be communicated to the public. The intelligent layperson will understand the scientific arguments presented in relation to assessing ecosystem health. For others, however, a more human-centered approach will be required, such as the cancer analogy to describe *P. cinnamomi* as a malignant tumor that is gradually spreading and compromising the health of plant communities that are central to the health of the Stirling Range ecosystem. Recognition of ecosystem health thus becomes another much-needed approach in attempting to reverse the gradual and continuous degradation of many ecosystems in Australia and elsewhere in the world.

ACKNOWLEDGMENTS

I thank G. Hardy and P. Ladd for providing photographs and useful discussion about the *Phytophthora cinnamomi* problem in Western Australia. M. Roeger and C. Ferguson helped with cartographic work.

REFERENCES

Bridgewater, P.B. and Edgar, B., Ecosystem pathogens: a view from the centre, *J. R. Soc. West. Aust.,* 77:109–111, 1994.

Burbidge, A. and Rose, A., Birds, in Thomson, C., Hall, G., and Friend, G., Eds., *Mountains of Mystery: A Natural History of the Stirling Range,* Department of Conservation and Land Management, Perth, Western Australia, 1993, pp. 99–108.

CALM, Biological diversity in Western Australia, in *A Nature Conservation Strategy for Western Australia,* Draft, Department of Conservation and Land Management, Perth, Western Australia, 1992.

CALM, Stirling Range and Porongorup National Parks, Draft Management Plan, Department of Conservation and Land Management, Perth, Western Australia, 1996.

Dell, J. and Harold, G., Frogs and reptiles, in Thomson, C., Hall, G., and Friend, G., Eds., *Mountains of Mystery: A Natural History of the Stirling Range,* Department of Conservation and Land Management, Perth, Western Australia, 1993, pp. 109–116.

Friend, G. and Muir, B., Mammals, in Thomson, C., Hall, G., and Friend, G., Eds., *Mountains of Mystery: A Natural History of the Stirling Range,* Department of Conservation and Land Management, Perth, Western Australia, 1993, pp. 87–97.

Gillen, K. and Watson, J., Controlling *Phytophthora cinnamomi* in the mountains of southwestern Australia, *Aust. Ranger,* 27:18–20, 1993.

Hardy, G.E., O'Brien, P.A., and Shearer, B.L., Control options of plant pathogens in native plant communities in southwestern Australia, *J. R. Soc. West. Aust.,* 77:169–177, 1994.

Keighery, G. and Beard, J., Plant communities, in Thomson, C., Hall, G., and Friend, G., Eds., *Mountains of Mystery: A Natural History of the Stirling Range,* Department of Conservation and Land Management, Perth, Western Australia, 1993, pp. 43–54.

Kennedy, J. and Weste, G., Vegetation changes associated with invasion by *Phytophthora cinnamomni* on monitored sites in the Grampians, Western Victoria, *Aust. J. Bot.,* 34:251–279, 1986.

Main, B., Spiders and other invertebrates, in Thomson, C., Hall, G., and Friend, G., Eds., *Mountains of Mystery: A Natural History of the Stirling Range,* Department of Conservation and Land Management, Perth, Western Australia, 1993, pp. 117–125.

McCaw, L. and Gillen, K., Fire, in Thomson, C., Hall, G., and Friend, G., Eds., *Mountains of Mystery: A Natural History of the Stirling Range,* Department of Conservation and Land Management, Perth, Western Australia, 1993, pp. 143–148.

Newsome, D., Moore, S.A., and Dowling, R.K., *Natural Area Tourism: Ecology, Impacts and Management,* Channel View Publications, Clevedon, U.K., 2002.

Pignatti, E., Pignatti, S., and Lucchese, F., Plant communities of the Stirling Range, Western Australia, *J. Vegetation Sci.,* 4:477–488, 1993.

Rapport, D.J. and Whitford, W.G., How ecosystems respond to stress, *BioScience,* 49:193–203, 1999.

Rapport, D.J., Costanza, R., and McMichael, A.J., Assessing ecosystem health, *Tree,* 13:397–402, 1998.

Shearer, B.L., The major plant pathogens occurring in the native ecosystems of southwestern Australia, *J. R. Soc. West. Aust.,* 77:113–122, 1994.

Underwood, R. and Burbidge, A., Introduction, in Thomson, C., Hall, G., and Friend, G., Eds., *Mountains of Mystery: A Natural History of the Stirling Range,* Department of Conservation and Land Management, Perth, Western Australia, 1993, pp. 3–4.

Weste, G. and Ashton, D.H., Regeneration and survival of indigenous dry sclerophyll species in the Brisbane Ranges, Victoria, after a *Phytophthora cinnamomi* epidemic, *Aust. J. Bot.,* 42:239–253, 1994.

Wills, R., Ecological impact of *Phytophthora cinnamomi* in the Stirling Range National Park, *Aust. J. Ecol.,* 17:145–159, 1992.

Wills, R. and Keighery, G.J., Ecological impact of plant disease on plant communities, *J. R. Soc. West. Aust.,* 77:127–131, 1994.

Wills, R. and Kinnear, J., Threats to the Stirling Range, in Thomson, C., Hall, G., and Friend, G., Eds., *Mountains of Mystery: A Natural History of the Stirling Range*, Department of Conservation and Land Management, Perth, Western Australia, 1993, pp. 135–141.

Wills, R. and Robinson, C.J., Threats to flora-based industries in Western Australia from plant disease, *J. R. Soc. West. Aust.,* 77:159–162, 1994.

Wilson, B.A., Newell, G., Laidlaw, W.S., and Friend, G., Impact of plant diseases on faunal communities, *J. R. Soc. West. Aust.* 77:139–143, 1994.

Yazvenko, S.B. and Rapport, D.J., A framework for assessing forest ecosystem health, *Ecosyst. Health,* 2:40–51, 1996.

CHAPTER 41

Mangrove Conservation and Restoration for Enhanced Resilience

Nguyen Hoang Tri, Phan Nguyen Hong, W. Neil Adger, and P. Mick Kelly

TRENDS IN MANGROVE ECOSYSTEM CONVERSION

Mangrove forests encompass those areas that include some species of mangrove trees and are intertidal tropical and subtropical coastal wetlands. Despite being among the most productive ecosystems in the world, the area of mangrove forests has been globally declining to less than half the former area from the beginning of the 20th century (Field et al., 1998). These trends are due to conversion for agriculture, forestry, and urban uses, and to extraction of timber for fuel and other uses (Farnsworth and Ellison, 1997).

In Vietnam, large areas of mangroves have been converted to agriculture and, in particular, to shrimp aquaculture, causing ecological disturbance and enhancing instability in the coastal physical environment (Hong and San, 1993). The total mangrove area of Vietnam has been in decline through the second half of the 20th century, according to contemporary historical estimates of this total area (see Figure 41.1). Before the two Indochina wars, mangrove areas were distributed extensively in coastal and estuarine areas and were estimated to have an area of around 400,000 ha in 1943 (Maurand, 1943), with over 250,000 ha in the southern parts of the country concentrated in the Ca Mau peninsula (150,000 ha) (Hong, 1994). Conversion occurred throughout the past half century, and an estimate by Rollett (1963) puts the area at less than 300,000 ha in the early 1960s.

From 1962 to 1971 nearly 40% of the mangrove area in southern and central Vietnam, or almost 159,200 ha, was destroyed as a result of chemical warfare, including over half of the mangrove in the Ca Mau peninsula (Hong, 1994). Estimates of the impact on productive areas suggest that more than 4 million m^3 of timber was destroyed during this period, with severe impact on the fauna and flora in these areas.

The legacy of Agent Orange and other dioxins remains until the present day not only in the higher fauna but also in human populations in these areas. But as shown in the estimates of total mangrove area in Figure 41.1, a significant proportion of the area degraded by herbicides was subsequently replanted, such that a mid-1980s estimate of total areas suggests around 250,000 ha remaining (Hong, 1994). Despite this restoration effort, areas all along the coast of Vietnam, including those rehabilitated areas, have now been converted again for use in agriculture, aquaculture, and human settlement. Where agricultural land has been reclaimed, it is often necessary to build significant sea dikes to protect the land from storm surges, where once the land would

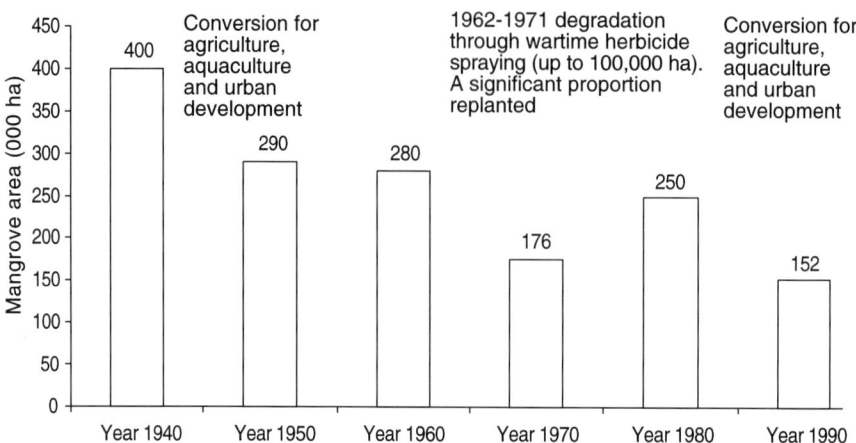

Figure 41.1 Total mangrove area in Vietnam, 1945–1995. (Based on Hong, P.H., Causes and effects of the deterioration in the mangrove resources and environment in Vietnam, in *Reforestation and Afforestation of Mangroves in Vietnam: Proceedings of the National Workshop, Ho Chi Minh City*, Mangrove Ecosystem Research Centre, Hanoi National Pedagogic University, and Action for Mangrove Reforestation, Hanoi, 1994; Maurand, P., *L'Indochine Forestiere*, Institute Recherché Agronomie de Indochine, Paris, 1943; and Rollett, B., *Note sur le Vegetation du Vietnam au Sud du 17e Parallele Nord*, Archives du Forete Recherché Institute, Saigon, 1963.)

Table 41.1 Estimated Timber Loss (m³) from Direct Degradation by Herbicides and Subsequent Productivity Loss for Selected Sites in Southern Vietnam

	Prompt Loss		Loss of Productivity		Total	
Location	Good Timber	Mixed Wood	Good Timber	Mixed Wood	Good Timber	Mixed Wood
Can Gio, HCMC	1,962,998	881,850	1,206,371	705,480	3,169,369	1,587,330
Mekong River mouths	1,258,350	1,258,350	1,721,423	1,208,016	2,979,773	2,466,366
Ca Mau Peninsula	16,016,157	903,898	4,267,656	740,900	20,283,813	1,644,798

From Hong et al., 1997.

have been buffered by the mangrove forests. The best estimate of the remaining area of mangroves is shown by province in Table 41.2, which also shows that nearly one quarter of the coastline of Vietnam is now protected by sea dikes.

Mangrove conversion is an important ecological issue as well as an economic issue. The functions and services provided by mangrove areas are diverse and have been well documented and appraised (Lugo and Snedaker, 1974; Mitsch and Gosselink, 1993; Field et al., 1998). Mangrove ecosystems are diverse in both their species and in terms of their functional diversity, yet little is really known about the role of biodiversity per se in maintaining the services they provide. The environmental determinants of the successful restoration of mangrove habitats may be related to two observations:

- Substrate and physical factors only loosely explain the natural zonation of mangrove trees (Ewel et al., 1998a). In other words, there is some flexibility in the conditions in which mangroves can be planted.
- Disturbed mangrove ecosystems lead to rapid vegetation reestablishment and substrate stabilization (Field et al., 1998).

These two factors suggest that the physical niches in which mangroves can enhance landscape resilience are widespread. This is particularly so given the circumstances in Vietnam, where much planting is simply restoring former mangrove areas.

Table 41.2 Coast and Sea Defense Length and Mangrove Area for the Coastal Provinces of Vietnam

Coastal Provinces	Coastline (km)	Protective Dikes			Mangrove (ha)
		River Dikes (km)	Sea Dike (km)	Total (km)	
North					
Quang Ninh	377	66	64	130	13,294
Hai Phong	110	48	67	115	3382
Thai Binh	50	135	135	270	4200
Nam Dinh	65	15	76	91	4000
Ninh Binh	17	6	24	30	400
Thanh Hoa	83	59	41	100	700
Nghe An	83	120	33	153	600
Ha Tinh	130	304	17	321	645
Central					
Quang Binh	115	92	4	96	100
Quang Tri	65	94	132	226	0
Thua Thien Hue	105	30	126	156	10
Quang Nam Da Nang	189	68	31	99	30
Quang Ngai	122	71	65	136	205
Binh Dinh	206	92	28	120	10
Phu Yen	204	39	6	45	54
Khanh Hoa	422	0	12	12	33
Ninh Thuan	115	0	0	0	9
Binh Thuan	216	0	0	0	9
Ba Ria-Vung Tau	104	0	9.08	8.08	300
South					
Ho Chi Minh	51	0	1.3	1.3	40,000
Tien Giang	40	0	43	43	4000
Kien Giang	154	0	146	146	2532
Ben Tre	103	0	33.7	33.7	6126
Tra Vinh	68	0	48	48	5591
Soc Trang	75	0	13.5	13.5	3058
Bac Lieu - Ca Mau	435	0	239	239	65,779
Total	**3704**	**1239**	**1394.58**	**2632.58**	**155,067**

Economic analysis has recognized that the functions and services provided by mangroves, and wetlands in general, have economic value and that these are often ignored in the ongoing process of mangrove conversion (Barbier, 1993; Ruitenbeek, 1994; Barbier and Strand, 1998). Mangrove wetlands display the features of public goods in that their use is nonexclusive, and they are converted because these functions are undervalued. Often mangrove conversion takes place through overriding traditional common management of the resources (Adger and Luttrell, 1998; Bailey and Pomeroy, 1996). Identification of the functions and services, the incorporation of these into the land allocation process, and the encouragement of appropriate property rights — whether communal or private — are, therefore, necessary first steps in promoting sustainable utilization of such resources.

The economic rationale behind mangrove restoration in the case of three coastal districts of Nam Dinh Province in northern Vietnam is reported. In these areas, mangrove rehabilitation is often subsidized by international development agencies such as the World Food Programme, Oxfam, and Save the Children Fund as part of rural income-generating projects, based largely on an assumed benefit to local communities. This chapter provides some quantification of the economic and environmental parameters of restoration in the context of resilience and adaptability at different scales.

MANGROVE RESTORATION IN NAM DINH PROVINCE

Nam Dinh's Remaining Mangroves

This section describes conditions in Nam Dinh Province, located in the southwest portion of the Red River delta in northern Vietnam. The province includes three coastal administrative units: Xuan Thuy, Hai Hau, and Nghia Hung districts. The region has a sea dike system to protect people, houses, and crops. Freshwater reserves help mitigate against damage from saline intrusion, flood, storm, and seawater incursion. The total area of the three coastal districts is approximately 72,052 ha (Table 41.3).

Within Nam Dinh Province, the impacts of severe storms are generally concentrated in the coastal districts. The total population of the three coastal districts of Xuan Thuy, Hai Hau, and Nghia Hung is almost 0.5 million, with a population density of 1076 per km², which is typical of the densely populated areas of the Red River delta plain. The economy of these districts is primarily dependent on agriculture. Paddy cultivation, aquaculture, and salt making are the major agricultural activities. Each of these activities is susceptible to, and differentially affected by, typhoons.

Given the prevailing circumstances in the coastal districts of Nam Dinh Province and similar regions elsewhere, it is clear that mangrove restoration can have a variety of benefits where the topography of the coastal shelf and other social, physical, and ecological factors are appropriate. In such situations, mangrove restoration can provide income where households are often severely constrained in cash income sources, and restoration can yield environmental benefits through productive assets and reduction of potential damage from coastal storm surges.

Table 41.3 Mangrove Area in Nam Dinh Province

District	Present Mangrove Areas (ha)	Land Estimated to be Available for Planting (ha)
Xuan Thuy	3000	7640
Hai Hau	200	641
Nghia Hung	5200	9826
Total	8400	18107

From Nam Dinh Province data.

Mangrove Restoration: Economic Analysis

Some economic values of the goods and services can be assessed by observation of existing markets, but many functions and services associated with replanting mangroves result in indirect or functional benefits. The crucial aspects of value for local decision making and for the differential impacts of environmental change are determined by whether these benefits stem from direct or indirect use. The major goods and services that accrue from mangroves in this area are outlined in Table 41.4. Some economic benefits of the mangrove resource will increase in value over time, while others will remain constant or decline. For example, as agricultural development intensifies, the potential economic losses from storm surges increase; thus, the value of the coastal protection function of the mangroves will rise accordingly. Exogenous environmental change associated with global climate change may increase the frequency and intensity of storm surges, and hence the value of this function of the mangroves is likely to rise over time.

The first step in this appraisal framework is therefore to delineate the resource issue and potential environmental change to be examined. In this case, the analysis allows examination of the resource efficiency of using land, labor, and capital resources to rehabilitate or restore mangroves in the coastal areas of Vietnam. Table 41.4 highlights an important distinction between the direct benefits of using mangroves, almost always extractive in nature and located within the mangrove areas, and those which are largely off site and indirect. It is often more difficult to quantify these indirect benefits because of the dynamic nature and ecological complexity of the relationship between the productive output

Table 41.4 Location and Nature of the Economic Benefits of Mangrove Restoration and Conservation

Type of Service	Location of Goods and Services	
	On Site	Off Site
Marketed goods and services	Timber, tannin	Honey from beekeeping, fishing
Nonmarketed benefits	Fish nursery function, medicinal plants,* crisis foodstuff, fodder, wildlife habitat	Storm protection function: Avoidance of cost of maintaining sea defenses and dikes Avoidance of impact on agriculture and infrastructure

Note: Goods and services in italics are quantified in this study.

* Medicinal plants utilized are surveyed and classified, but no economic value is estimated (see text).

Adapted from Dixon, J.A. and Lal, P.N., The management of coastal wetlands: economic analysis of combined ecologic–economic systems, in Dasgupta, P. and Mäler, K.G., Eds., The Environment and Emerging Development Issues, Clarendon, Oxford, 1997.

and the mangrove forest. The fish nursery function of mangroves, for example, is well established (Primavera, 1998), yet it is difficult to attribute the value of final commercial or subsistence fish catch to this single function among many of the whole system (see Barbier and Strand, 1998).

The second step in this appraisal is the identification among these services of a set of costs and benefits. The allocation of effects into costs and benefits involves determining the current situation, and focusing in partial analysis on the values of the marginal changes. The economic cost–benefit analysis of mangrove rehabilitation schemes in this case is of the form:

$$\text{NPV} = \sum_{i=1}^{\gamma} \frac{B_t^T + B_t^{NT} + B_t^P - C_t}{(1+r)^t}$$

where
NPV = net present value (VND per ha)

B_t^T = net value of the timber products in year t (VND per ha)

B_t^{NT} = net value of the nontimber products in year t (VND per ha)

B_t^P = value of the protection of the sea defenses in year t (VND per ha)

C_t = costs of planting, maintenance, and thinning of mangrove stand in year t (VND per ha)

r = rate of discount

t = time horizon (20-year rotation).

Estimates of the data sources and methods for carrying out the quantification and valuation of costs and benefits in establishing the rehabilitated mangrove stands are summarized in Table 41.5.

Direct Costs and Benefits of Restoration

The costs of establishing the rehabilitated mangrove stands are estimated primarily based on the cost of labor for the activities described. The survey research was carried out in 1994, when the cost for a workday was typically 2.5 kg of rice or VND 5500. Planting of 1 ha of mangroves required 95 workdays or VND 522,000, as shown in Table 41.5. The estimates are averaged across the three districts, with variations in costs dependent on where the seedlings were obtained. The planting and handling fees for seedlings obtained from forests in the area under rehabilitation are not significant compared to costs for collecting, handling, and transportation for other areas that

Table 41.5 Benefits, Costs, and Valuation of Mangrove Rehabilitation in Vietnam

Impact or Asset Valued	Method and Assumptions for Valuation	Timing of Costs and Benefits
Benefits		
Timber benefits	Market data: thinning (VND 180 per tree); extraction mature trees (VND 5000)	Thinning and extraction from year 6 with 3-year rotation
Fish	Market data: mean price of VND 12,500 per kg; yield 50 kg per ha	Fishing benefits from year 2 after planting
Honey	Market data: potential yield estimated at 0.21 kg per ha	Honey collected from year 5 after planting
Sea dike maintenance costs avoided	Morphological model: costs avoided three-dimensional (stand width, age, mean wavelength)	Benefits rising from year 1
Costs		
Planting, capital, and recurrent costs	Market and labor allocation data: costs of seedlings and capital (VND 440,000 per ha); workdays valued at local wage in rice equivalent (VND 5500 per day)	Planting costs at year 1; thinning from year 6 on 3-year rotation

US$1 = VND 11,000.

increase depending on the distance from the seedling source site to the planting site. The seed mortality rate between time of collection and time of planting adds an additional cost factor.

For some mangrove species, such as *Sonneratia* sp., *Avicennia* sp., *Aegiceras* sp., and others, planting directly onto mud flats is unsuccessful due to the exposure to strong wind and wave forces that wash away the seedlings. The cost of raising such species in a nursery and transplanting them at 8 months old is relatively high, with fees for maintaining the nursery, care, protection, and transportation adding to overall expenditure. The costs of establishing a stand, including planting, gapping, and protection, occur mainly in the first year. Maintenance, beginning with the second year, incurs an estimated annual expenditure of VND 82,500 per hectare. The cost of thinning occurs in years 6, 9, 12, 15, 20, and 25.

The benefits from wood and fuelwood sources from the processes of periodic thinning and extraction are derived from observations in local markets and are shown in Table 41.5. The timber benefits represent wood for poles and fuelwood. The benefits from direct fishing sources were estimated on site. Fishing activities in the three districts are undertaken through the use of simple fishing nets, simple tools, or even by hand. Aquatic products include fish, crab, shrimp, and shellfish. The yield is estimated at approximately 50 kg per hectare within mature mangrove stands annually for all types of aquatic products. The average unit price in 1994 was around VND 12,650 per kg averaged across the products. Some evidence exists indicating that present exploitation of mangrove aquatic products in the Red River delta may lead to declines in fish stocks, although they are considered conservative for the districts surveyed.

Honey from beekeeping is derived from the flowers of a number of mangrove species, though the season spans a limited number of months. The honey from mangroves is obtained during the first flowering season of *Kandelia candel,* from January to March, and from July to September for other mangrove species and the second flowering season of *K. candel.* The potential yield from this bee honey source was estimated to be an annual minimum of 0.21 kg per hectare. Honey production is possible from 5 years after planting, though some species of mangrove can flower after 3 to 4 years, and even after 1.5 years from planting.

In addition to these direct uses, mangroves produce other subsistence and indirect services as shown in Table 41.4. In particular the diverse species within a mangrove forest are used extensively for medicinal purposes. For the established mangrove area of Xuan Thuy, the major utilized flora are shown in Table 41.6 along with other subsistence uses. This information is based on extensive

Table 41.6 Useful Species within the Mangrove Area of Xuan Thuy Protected Ramsar Site

Species	Timber, Fuelwood, and Charcoal	Tannin	Green Manure	Food	Fodder	Beekeeping	Medicinal Plant	Medicinal Use
Acanthius ebracteatus							+	Boiled bark and roots used for cold symptoms, skin allergies; bark in malaria treatment and back pain
Acanthius illicifolius							+	Leaves for relief of swelling, rheumatic pain, neuralgia
Achyranthes aspera							+	Leaves as poison antidote and applied to boils
Acrostichum aureum							+	
Aegiceras corniculatum	+	+				+		
Avicennia marina	+					+	+	Bark used in contraceptives, leaves on abscesses
Azeratum conyzoides			+				+	
Bidens pilosa							+	
Bruguiera gymnorhiza	+	+	+			+	+	
Casuarina equisetfolia	+	+	+					
Chenopodium album			+	+				
Clerodendron inerme					+		+	Leaves used in treatment of jaundice, dried roots for wounds and colds
Crinum asiaticum							+	Tonic, laxative, and expectorant
Cynodon dactylon			+				+	
Cyperus malaccensis					+			
Cyperus tegetiformis								
Cyperus stolonferus								
Datura fastuosa							+	
Derris trifoliata			+				+	Leaves as laxative; roots used in treatment of malnutrition
Eclipta alba								
Eupatorium oderatum			+		+			
Excoecaria agallocha	+	+	+				+	Bark for leprosy; leaves for ulcers and epilepsy
Heliotropium indicum			+				+	
Hibiscus tiliacetus								

(Continued)

Table 41.6 (continued) Useful Species within the Mangrove Area of Xuan Thuy Protected Ramsar Site

Species	Timber, Fuelwood, and Charcoal	Tannin	Green Manure	Food	Fodder	Beekeeping	Medicinal Plant	Medicinal Use
Ipomoea pes-caprae								
Kanedlia candel	+	+	+		+	+		
Lumnitzera racemosa	+	+	+			+		
Pluchea pteropoda							+	Treatment of scalds and burns
Rhizophora stylosa	+	+	+			+	+	Bark as astringent, anti-diarrhea, for scalds and burns
Scripus kimsonensis				+				
Sesumvium portulacastrum			+					
Sonneratra caseolaris	+		+			+		
Sporolobus virginicus					+			
Thespesia populnea	+		+			+	+	Scabies
Wedelia calendulacea							+	

Based on Tri, N.H. et al., Economic Valuation Studies of Mangrove Conservation and Rehabilitation in Nam Ha Province, Red River Delta, Progress report for SARCS/WOTRO/LOICZ, Mangrove Ecosystem Research Centre and CERED, Hanoi, Vietnam, 1997; and Hong, P.N. and San, H.T., *Mangroves of Vietnam*, IUCN, Bangkok, 1993.

surveys in Xuan Thuy District and on the identification of these uses more generally by Hong and San (1993). Leaves, bark, or roots from more than 30 species of plants are commonly used in medicines in the district, with other parts of the plants used for tannin, manure, and directly as food in times of food shortage. These direct consumptive uses are not included in the cost–benefit analysis presented here. Medicinal uses of plants and their associated cultural significance constitute primary reasons for conservation of forest areas, to an extent more significant than can be captured in any economic analysis (Crook and Clapp, 1998; Brown, 1995). The utilization of these areas, particularly for subsistence and nonmarket uses, adds weight to the argument for restoration and conservation.

MANGROVES AS COASTAL PROTECTION

The planting of mangroves on the seaward side of the extensive sea dike system helps reduce costs that would otherwise be expended in maintenance of these defenses. Such maintenance takes place on an annual basis in most of the coastal districts of Vietnam — previously through the obligatory labor of district inhabitants organized by the district committees, and now primarily financed through local land taxes. These commitments draw a heavy burden on labor-scarce households and constitute a source of conflict over the interdistrict allocation of labor contracts (Adger, 1998).

The impact of typhoons on coastal Vietnam is significant. One typhoon in October 1985 was responsible for the loss of almost 900 lives. In addition, 3300 boats were sunk and over half a million people were rendered homeless. While this was a particularly extreme single disaster, over 400,000 hectares of crops were lost in the coastal provinces of Vietnam as a result of tropical cyclone impacts over the 10-year period 1977 to 1986 (Thu, 1991). Protecting vulnerable coastal areas from typhoon impacts is, therefore, of high social and economic importance. Yet in a nation where resources are limited, affording adequate protection can prove difficult even in the present day (Wickramanayake, 1994). Estimates of the magnitude of impacts in Nam Dinh Province from floods and typhoons for the 20 years between 1973 and 1992 show that more than 990 people were injured or killed, and damage totaling VND 470 billion (1993 constant prices) occurred as a result of severe storms.

The evaluation of the role of mangroves in protecting sea dikes is estimated from expenditure on their maintenance and repair in comparison with a case in which no mangroves exist, using a control situation assumed to have similar morphological characteristics. In general terms, the greater the area of mangrove, the greater the benefit in terms of avoided maintenance costs. Establishing a precise set of relationships in order to estimate the benefits is not, however, a straightforward matter because the mechanisms by which mangroves protect the adjacent dike are complex. Mangrove stands provide a physical barrier, resulting in drag effects and dissipation of wave energy. They also help stabilize the seafloor, trapping sediment, and can affect the angle of slope of the sea bottom, resulting in dissipation of wave energy.

Studies in southern China have demonstrated an empirical relationship through which the benefit, in terms of avoided cost (B_t^p), can be expressed as a function of the width of the mangrove stand as a proportion of the average wavelength of the ocean waves to which the stand is exposed, as well as various parameters related to the age of the stand (mangrove size and density) expressed as a buffer factor. The key parameters are illustrated in Figure 41.2. The relationship was developed and tested in mangrove stands in southern China and has been calibrated in Vietnam through simulation (Vinh, 1995). We have used a simplified version of this relationship in estimating indirect use value in this study. Here, the buffer factor, α, is given by:

$$\alpha = \frac{2\pi R^2}{1.73 b^2}$$

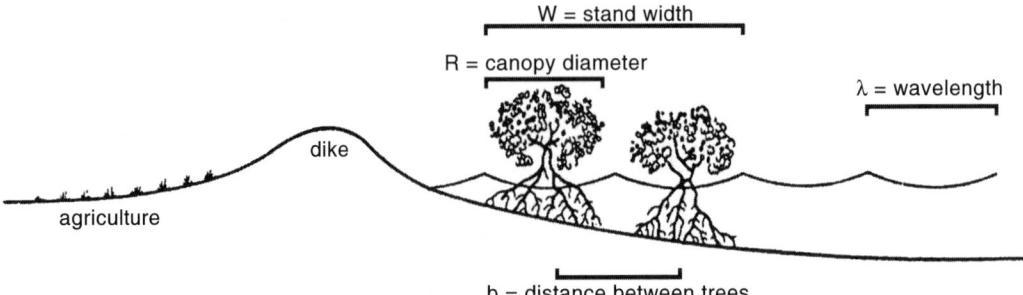

Figure 41.2 Profile of rehabilitated mangrove stands showing parameters of estimation of avoided maintenance cost.

where R is the mean radius of the canopy of an individual tree (m), which increases with age, and b is the typical distance between trees (m), which generally increases with time. As the stand matures, α increases from a minimum of around 0.1 to close to 1.0 as the stand presents a more and more effective obstacle.

Observations indicate that a mature stand will avert 25 to 30% of the costs of dike maintenance, assuming a stand width at least comparable to the characteristic wavelength of the incident waves. Tri et al. (1998) extended this analysis and showed that beyond a certain point, increasing stand width results in decreasing gains in protection. Typical wavelengths would be between 25 and 75 m, suggesting that the stand width should be on the order of 50 to 100 m.

For the Nam Dinh example, the model is calibrated using survey data on the annual costs of maintenance of sea dikes in each of the three coastal districts and data on mangrove productivity (growth in terms of mean annual increment, height, canopy density) for *Rhizophora apiculata* (Aksornkoae, 1993). The model was tested for its sensitivity to various parameters including the costs of maintenance in the districts and the design of the protection schemes in terms of the width of the stand in front of the sea dikes.

The model used here must be regarded as a provisional attempt to estimate the benefits associated with reduced maintenance costs. In particular, we consider the model may be overestimating the benefits when the stand is not fully developed or the width of the stand is much less than the incident wavelength. Nevertheless, uncertainties in this area may not be critical for two reasons. First, as will be seen, the direct benefits from use of the resources are considerably more significant than this indirect use value; and second, the value estimated here is only part of the true storm protection value, which must also include broader damage avoidance benefits and is, therefore, a lower bound figure.

The model of maintenance costs avoided was used for the three coastal districts to derive the indirect benefit of mangrove rehabilitation. The baseline costs of maintenance are incurred by the district committees, which keep detailed records of workdays and expenditure on annual maintenance. Recent estimates of the number of person–days a year spent on dike maintenance were used in the calculations. Because the results represent the average situation, the impact of the most severe storm surges on both the cost of maintenance and repair of dikes is not tabulated. In addition, this model does not account for other damage costs associated with storm occurrence, such as agricultural losses.

COMPARING THE COSTS AND BENEFITS OF RESTORATION

The full results of the cost–benefit analysis are presented in Table 41.7. This cost–benefit analysis is of a partial nature, comparing establishment and extraction costs with the direct benefits from extracted marketable products and with the indirect benefits of avoided maintenance of the sea dike system. We assumed that present-day conditions continue to prevail with respect to storm

Table 41.7 Costs and Benefits Comparison of Direct and Indirect Use Values of Mangrove Restoration

Discount Rate	Direct Benefits (PV million VND per ha)	Indirect Benefits (PV million VND per ha)	Costs (PV million VND per ha)	Overall B/C Ratio
3	18.26	1.40	3.45	5.69
6	12.08	1.04	2.51	5.22
10	7.72	0.75	1.82	4.65

Notes: Stand width = 100 m; incident wavelength = 75 m. US$1 = VND 11,000. B/C ratio = NPV Total Benefits/NPV Costs.

Data from Tri, N.H., Adger, W.N., and Kelly, P.M., Natural resource management in mitigating climate impacts: mangrove restoration in Vietnam, *Global Environ. Change,* 8:49–61, 1998.

frequency. The results show a benefit–cost ratio in the range of four to five for a range of discount rates. The low relative changes in benefit–cost ratios illustrates that most of the costs, as well as the benefits of rehabilitation, occur within a relatively short time frame, with even the reduced maintenance cost beginning to accrue within a few years of initial planting.

Choosing the rate of discount is considered by many economists to be somewhat arbitrary and dependent on whether the project to be appraised is undertaken in the public or private domains. A range of real discount rates from 1 to 20 have been used in many circumstances, but rates at the lower end of this range tend to reflect the time preferences implicitly applied by governments in investments on behalf of society (see Markandya and Pearce, 1991). The results presented in Figure 41.3 appear to be robust in comparison to the discount rates adopted as a sensitivity test.

Figure 41.3 illustrates that the direct benefits from mangrove rehabilitation are more significant in economic terms than the indirect benefits associated with sea dike protection over a range of realistic parameter values. As might be expected, the greater the stand width, the more important the direct benefits in comparison to the avoidance of maintenance costs (Figure 41.3a). Yet even at the lower end of the range of realistic stand widths, offering the greatest return per hectare given suitable conditions, the direct benefits dominate (Figure 41.3b).

As shown in Table 41.4, the sea dike protection estimates do not include the benefits of reduced repair after serious storm damage or the potential losses of agricultural produce when flooding occurs. Flooding associated with severe tropical storms can lead to large economic losses, as well as to loss of life, and a reduced probability of flooding associated with the protection afforded by the mangrove would be an additional indirect benefit. This benefit has not been estimated to date, though the impact of historic storms can be discerned by examining aggregate agricultural production from district archival records. Figure 41.4 shows total rice production in one of the case study districts, Xuan Thuy, from 1981 to 1995.

The radical increase in agricultural production over the period coincides with the liberalization of agricultural production practices and distribution of leaseholds to individual households, beginning with the *output contract* system in the early 1980s. But the major storms crossing the coast close to this district over the period can be seen to have some effect on agricultural production, at least in 1986 and in 1994. However, despite little evidence of the impact of the 1992 storms on agricultural output, we know that the timing of extreme events as well as institutional and other factors directly affect vulnerability in both physical and social senses. The impact of the presence or absence of mangroves in its role as protecting agriculture in coastal areas cannot be directly deduced from these data. Again, the complexity of the relationship between mangroves and economic and social systems is the limiting factor in such analysis.

In any event, it is clear from the results of the economic analysis summarized in Figure 41.3 that the direct benefits from mangrove rehabilitation mean that this activity is economically desirable, as evidenced by the positive net present values at all discount rates considered. The increase in net present value associated with mangrove planting resulting from the inclusion of dike main-

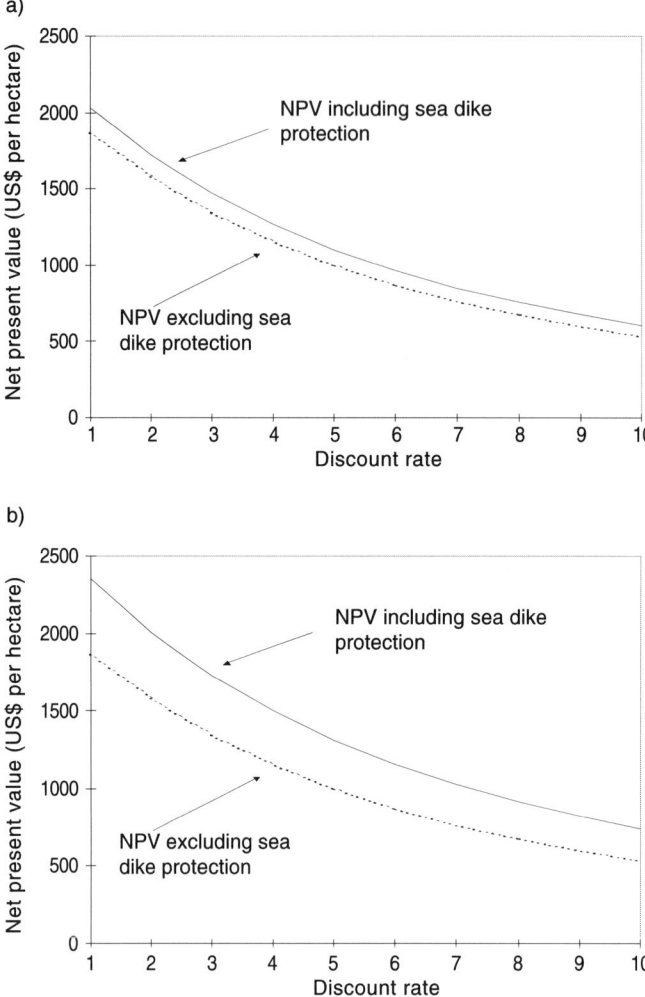

Figure 41.3 Net present value of mangrove rehabilitation, including value of sea dike protection, for two cases: (a) stand width = 100 m; incident wavelength = 75 m, and (b) stand width = 33.3 m; incident wavelength = 25 m.

tenance savings supports the desirability of planting. The results presented in Table 41.4 show that this indirect joint product benefit of mangrove rehabilitation is significant in further strengthening the economic case for such action in these locations.

CONCLUSION

This chapter argues that mangrove restoration has the potential, under suitable social and physical conditions, to provide *win–win* situations, whereby the dichotomy between short- and long-term concerns in adaptation to environmental change is avoided. It can be contrasted with one alternative course of action — building higher sea dikes — which, while possibly necessary if the threat of increased storm impacts materializes, provides limited benefits in the short term.

A broader lesson can be drawn from this analysis regarding approaches to the problematic issue of adaptive responses to long-term environmental change. Decision makers face difficult choices in assigning priorities when faced with an uncertain future in a resource-limited present.

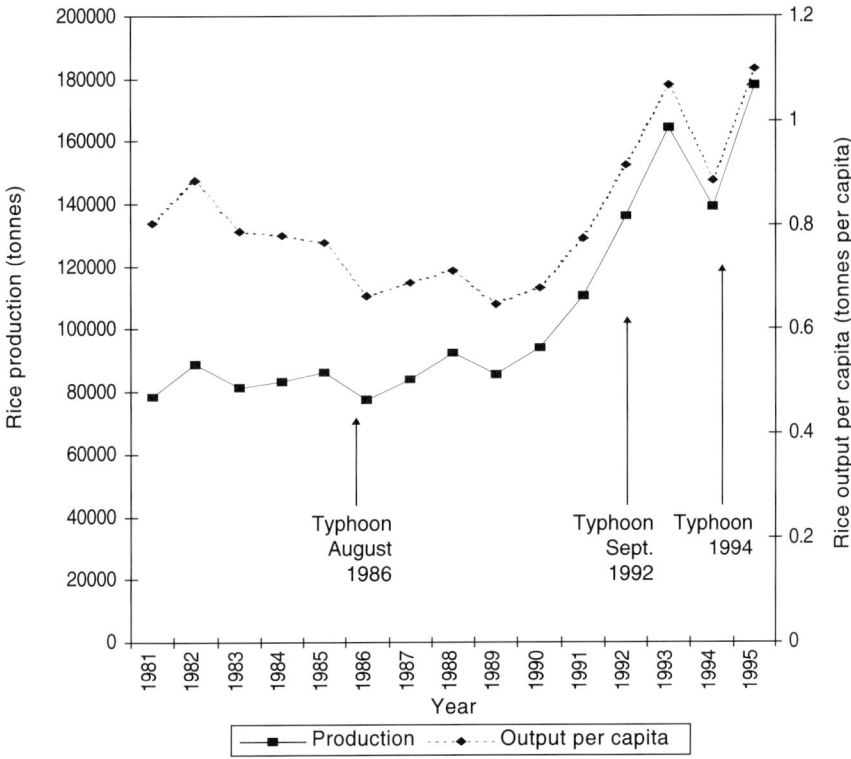

Figure 41.4 Total rice production and production per capita for Xuan Thuy District and timing of major storms, 1981–1985. (Data from Xuan Thuy District archives.)

We would argue that this difficulty can be minimized by adopting a precautionary approach that focuses attention on present-day or near-future benefits that will accrue regardless of the nature and magnitude of the impact of environmental change. In the example presented in this chapter, mangrove restoration — a practice that is widespread and is based on available and appropriate technology — provides immediate economic benefits to residents of the adjacent settlements that are most vulnerable to storm impacts, while reducing the potential for storm damage over the near and long term.

REFERENCES

Adger, W.N., Observing Institutional Adaptation to Global Environmental Change: Theory and Case Study from Vietnam, Global Environmental Change Working Paper 98–21, Centre for Social and Economic Research on the Global Environment, University of East Anglia and University College, London, 1998.

Adger, W.N. and Luttrell, C., Property rights and the utilization of wetlands, in Soderqvist, T., Ed., *Wetlands: Landscape and Institutional Perspectives. Proceedings of the Fourth Global Wetlands Economics Network*, Beijer International Institute of Ecological Economics, Stockholm, 1998.

Aksornkoae, S., *Ecology and Management of Mangroves*, IUCN, Bangkok, 1993.

Bailey, C. and Pomeroy, C., Resource dependency and development options in coastal South East Asia, *Soc. Nat. Resour.,* 9:191–199, 1996.

Barbier, E., Sustainable use of wetlands valuing tropical wetland benefits: economic methodologies and applications, *Geogr. J.,* 159:22–32, 1993.

Barbier, E.B. and Strand, I., Valuing mangrove fishery linkages: a case study of Campeche, Mexico, *Environ. Resour. Econ.,* 12:151–166, 1998.

Brown, K., Medicinal plants, indigenous medicine and conservation of biodiversity in Ghana, in Swanson, T., Ed., *Intellectual Property and Biodiversity Conservation*, Cambridge University Press, Cambridge, 1995.

Crook, C. and Clapp, R.A., Is market-oriented forest conservation a contradiction in terms? *Environ. Conserv.*, 25:131–145, 1998.

Dixon, J.A. and Lal, P.N., The management of coastal wetlands: economic analysis of combined ecologic–economic systems, in Dasgupta, P. and Mäler, K.G., Eds., *The Environment and Emerging Development Issues*, Clarendon, Oxford, 1997.

Ewel, K.C., Bourgeois, J.A., Cole, T.G., and Zheng, S., Variation in environmental characteristics and vegetation in high rainfall mangrove forests, *Global Ecol. Biogeogr. Lett.*, 7:49–56, 1998.

Ewel, K.C., Twilley, R.R., and Ong, J.E., Different kinds of mangrove forests provide different goods and services, *Global Ecol. Biogeogr. Lett.*, 7:83–94, 1998.

Farnsworth, E.J. and Ellison, A.E., The global conservation status of mangroves, *Ambio*, 26:328–334, 1997.

Field, C.B., Osborn, J.G., Hoffman, L.L., Polsenberg, J.F., Ackerley, D.D., Berry, J.A., Bjorkman, O., Held, A., Matson, P.A., and Mooney, H.A., Mangrove biodiversity and ecosystem function, *Global Ecol. Biogeogr. Lett.*, 7:3–14, 1998.

Hong, P.H., Causes and effects of the deterioration in the mangrove resources and environment in Vietnam, in *Reforestation and Afforestation of Mangroves in Vietnam: Proceedings of the National Workshop, Ho Chi Minh City*, Mangrove Ecosystem Research Centre, Hanoi National Pedagogic University, and Action for Mangrove Reforestation, Hanoi, 1994.

Hong, P.N. and San, H.T., *Mangroves of Vietnam*, IUCN, Bangkok, 1993.

Hong, P.N., Ba, Tran van, San, Hoang Thi, Tre, Le Thi, Tri, Nguyen Hoang, Tuan, Mai Sy, and Tuan, Le Xuan, *The Role of Mangroves in Vietnam, Planting Techniques and Maintenance*, Agricultural Publishing House, Hanoi, 1997, 224 pp. (in Vietnamese).

Kelly, P.M., Granich, S.L.V., and Secrett, C.M., Global warming: responding to an uncertain future, *Asia Pac. J. Environ. Dev.*, 1:28–45, 1994.

Lugo, A.E. and Snedaker, S.C., The ecology of mangroves, *Annu. Rev. Ecol. Syst.*, 5:39–64, 1974.

Markandya, A. and Pearce, D.W., Development, the environment and the social rate of discount, *World Bank Res. Observer*, 6:137–152, 1991.

Maurand, P., *L'Indochine Forestiere*, Institute Recherché Agronomie de Indochine, Paris, 1943.

Mitsch, W.J. and Gosselink, J.G., *Wetlands,* 2nd ed., Van Nostrand Reinhold, New York, 1993.

Primavera, H., Mangroves as nurseries: shrimp populations in mangrove and non-mangrove habitats, *Estuarine Coastal Shelf Sci.*, 46:457–464, 1998.

Rollett, B., *Note sur le Vegetation du Vietnam au Sud du 17e Parallele Nord*, Archives du Forete Recherché Institute, Saigon, 1963.

Ruitenbeek, H.J., Modelling economy–ecology linkages in mangroves: economic evidence for promoting conservation in Bintuni Bay, Indonesia, *Ecol. Econ.*, 10:233–247, 1994.

Thu, T.V., Advances in Forecast Dissemination and Community Preparedness Tactics in Vietnam, Paper presented at the Second International Workshop on Tropical Cyclones, Hydrometeorological Service, Hanoi, 1991.

Tri, N.H. et al., Economic Valuation Studies of Mangrove Conservation and Rehabilitation in Nam Ha Province, Red River Delta, Progress report for SARCS/WOTRO/LOICZ, Mangrove Ecosystem Research Centre and CERED, Hanoi, Vietnam, 1997.

Tri, N.H., Adger, W.N., and Kelly, P.M., Natural resource management in mitigating climate impacts: mangrove restoration in Vietnam, *Global Environ. Change,* 8:49–61, 1998.

Vinh, T.T., Tree Planting Measures to Protect Sea Dike Systems in the Central Provinces of Vietnam, Paper presented at the Workshop on Mangrove Plantation for Sea Dike Protection, 24–25 December 1995, Hatinh, Vietnam, 1995.

Wickramanayake, E., Flood mitigation problems in Vietnam, *Disasters,* 18:81–86, 1994.

CHAPTER 42

A Comparison of Landscape
Change Detection Methods

Curtis M. Edmonds, Anne C. Neale, Daniel T. Heggem, James D. Wickham,
and K. Bruce Jones

INTRODUCTION

This chapter compares change detection analysis using Normalized Difference Vegetation Index (NDVI) data derived from satellite images, with the analysis of change based on the same images classified into land cover. The NDVI index has provided the researcher with a reliable method to measure vegetative biomass through the use of satellite imagery (Tucker, 1979) for nearly 20 years. NDVI can be used as a fast and convenient method for measuring temporal changes associated with vegetation (Lyon et al., 1998). This chapter will show that the NDVI change detection process can provide a cost-effective method of accurately targeting areas within a region that have undergone extensive human land use change.

STUDY AREA LOCATION

The data for this evaluation were made available through a study conducted in the Tensas River Basin, located in northeast Louisiana and documented in *An Ecological Assessment of the Louisiana Tensas River Basin* (Heggem et al., 1999b). The study area encompasses approximately 930,000 acres of historical Mississippi River alluvial floodplain (Figure 42.1). The Tensas River is now hydrologically connected to the Atchafalaya River, which is a major distributary of the Mississippi River (Gosselink et al., 1990). Historically, most of the Tensas River Basin was covered with bottomland hardwood forested wetlands. The bottomland hardwoods of the Tensas River Basin have been described as some of the richest ecosystems in the country in diversity and productivity of plant and animal species (Rainer et al., 1994). At the same time, these lands are recognized as among the nation's most productive regions for grain and fiber farming (Townsley, 1996). The result is a conflict between advocates of land use for traditional row crop agriculture and defenders of a healthy, diverse, and stable ecosystem. In years past, the stakeholders of freshwater marshes, stream bank areas, and bottomland swamps of the basin were under strong development pressure to increase agriculture production. As a result, large portions of forest near streams and in backwater swamp areas were converted to agriculture (Heggem et al., 1999a). This loss of forested areas allowed increasing levels of pollution to enter streams, lakes, and estuaries. Loss of wetland forests

Tensas River Basin

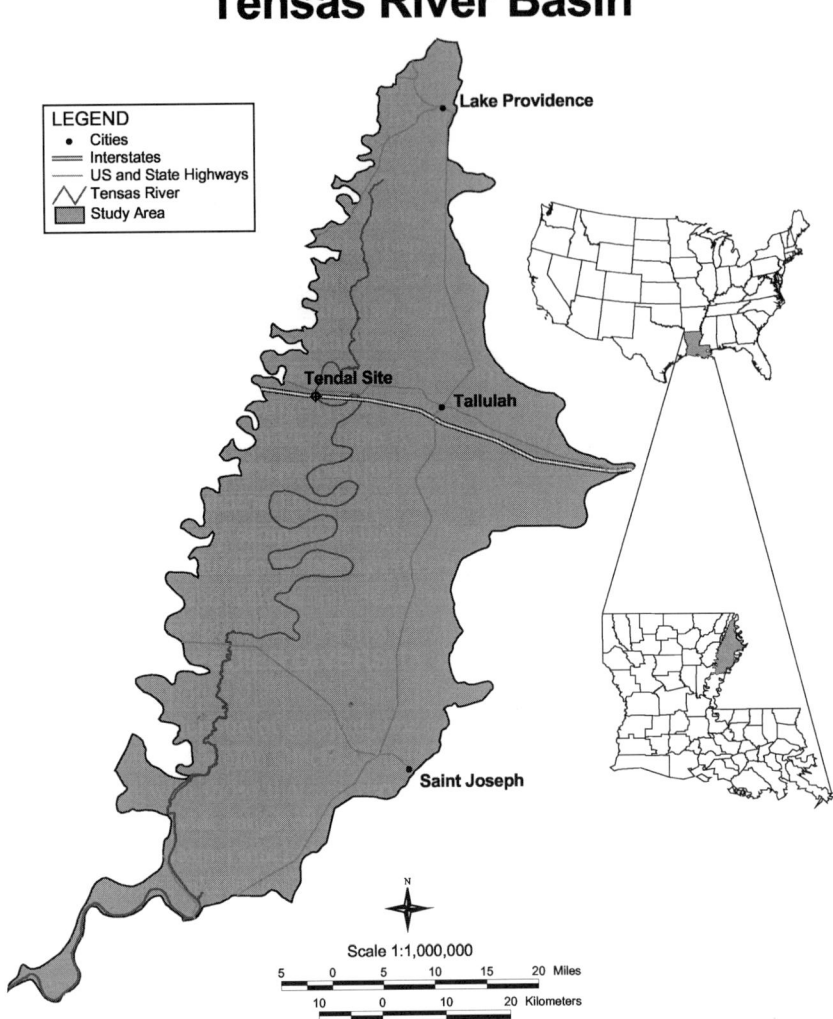

Figure 42.1 Tensas location map.

also impeded flood control efforts because wetlands absorb peak flows during floods, reducing damage to downstream farms and cities. These economic and environmental benefits, coupled with their recreational value, make preserving and restoring the wetland forests an important concern for the people living within the Tensas River Basin.

MATERIALS AND METHODS

Our study area roughly corresponds to the U.S. Geological Survey (USGS) Hydrologic Unit 08050003, which includes the area around the main channel of the Tensas River. During the course of our study, we divided the watershed into topographically relevant subwatersheds, or zones, and examined landscape indicators based on these subwatersheds. These subwatersheds, known as 11-digit hydrological accounting zones, were defined by combining the 8-digit hydrologic unit boundary with the Natural Resources Conservation Service (NRCS) 11-digit boundaries (Figure 42.2). Note that zones 2, 7, and 9 were not defined as 11-digit hydrologic accounting units by NRCS,

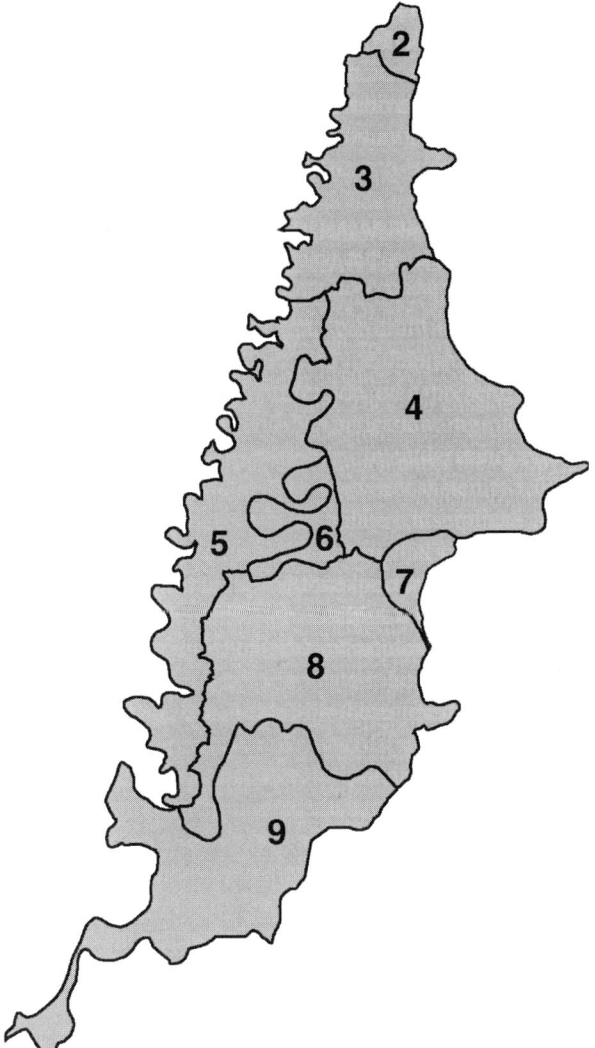

Figure 42.2 The 11- and 8-digit HUCs from Tensas River.

but they do fall within the boundary of the 8-digit USGS hydrologic unit code (HUC) boundary. Subwatersheds 2 and 7 may be parts of other subwatersheds or contain, most likely, bayous in which water could flow in many directions.

The North American Landscape Characterization (NALC) (Lunetta et al., 1993; 1998) data set contains triplicate images from the early 1970s through the early 1990s and covers the lower 48 states and Mexico. These data were derived from multispectral scanner imagery and processed to a resolution of 60 m. Images were selected from this data set to cover the Tensas River Basin. The Tensas River Basin study area is not contained within one path/row scene. Therefore, it was necessary to mosaic (combine) two NALC images to cover the study area for the 1970s (Figure 42.3a) and again for the 1990s (Figure 42.3b). The images that were available for the 1970s were created on the same date in 1972; however, the two images for the 1990s were from different years, 1991 and 1992. The 1970s and 1990s mosaicked images were then adjusted using a multiple-date empirical radiometric normalization process (Jensen, 1996) in order to produce a statistical comparison between the two time periods.

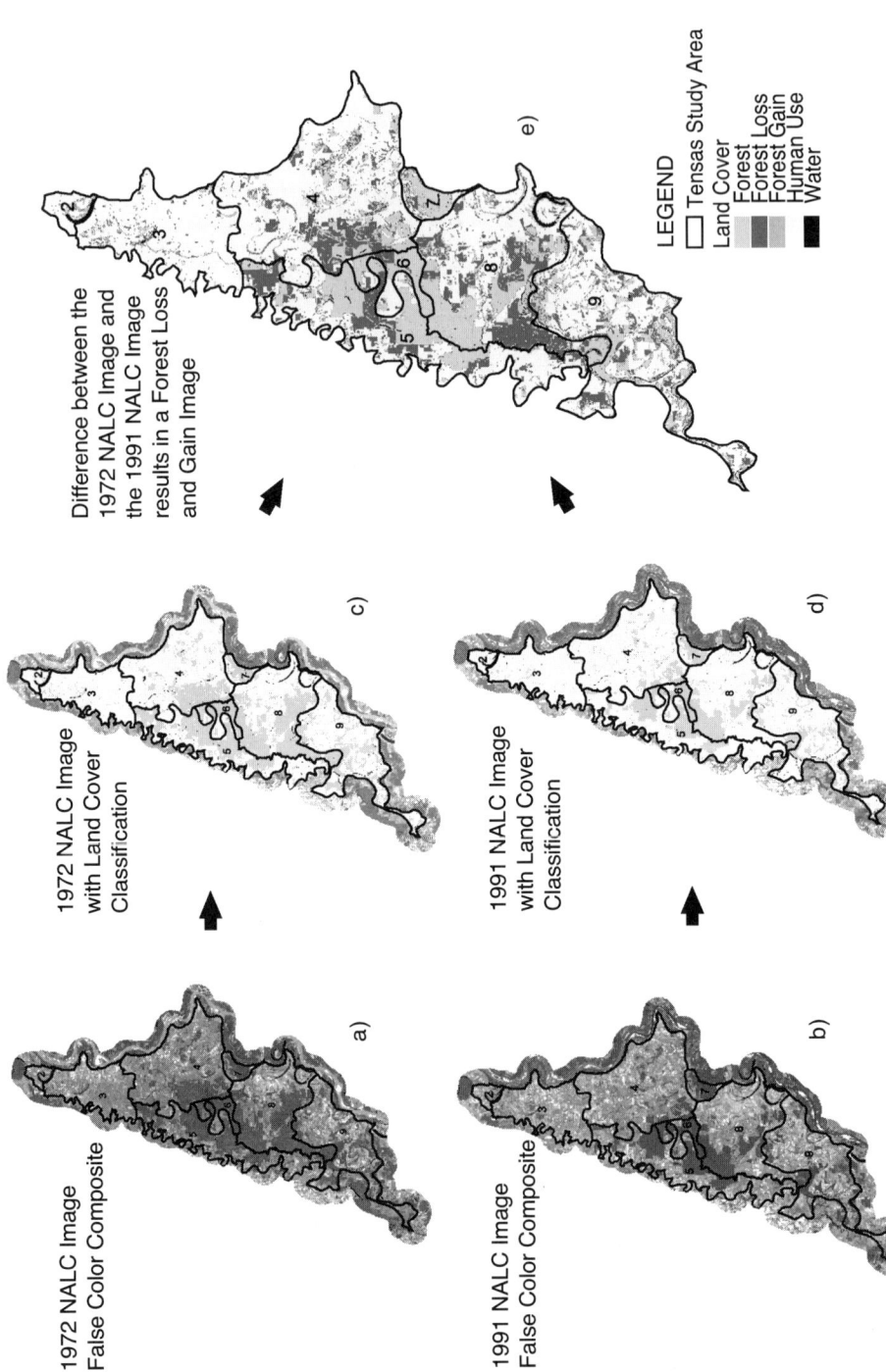

Figure 42.3 a, 1972 mosaic of Tensas River Basin (cut out to show only the study area). b, 1991 and 1992 mosaic of Tensas River Basin (cut out same as a). c, Land cover derived from the 1972 mosaic image. d, Land cover derived from the 1990s mosaic images. e, Difference between the 1972 and 1990s mosaic images, showing gains and losses.

Classification Method

The images were classified into a three-class system of forest, human use (urban and agriculture), and water. Classification of images was conducted using a combination of alternating supervised and unsupervised classifications (Lillesand and Kiefer, 1994) until a high percentage of features on both images (i.e., 1970s and 1990s) were classified. In areas difficult to classify, digital raster graphic maps were used to locate and identify features. In addition, air photos (1:24,000) were used to identify forests and wet agricultural fields. Analyses were performed by zone for each image, and then the number of picture elements (pixels) within each land cover type was calculated. The 1970s image and the 1990s images were compared, and the resultant change image is shown in Figure 42.3e.

Normalized Difference Vegetation Index (NDVI) Method

The same images that were used in the land cover classification method, previously described, also were used to calculate NDVI. Images were then created from the NDVI values. NDVI is calculated from the following equation:

$$NDVI = \frac{(NIR - red)}{(NIR + red)}$$

where NIR represents the band in the near infrared, and red represents the band in the visible red. Band seven (800 to 1100 nm) from the NALC data was used for the NIR band, and band five (600 to 700 nm) was used for the visible red band. Values derived using NDVI range from –1.0 to +1.0, where negative index values represent clouds, water, and snow; index values near zero represent barren soil and rocks; and positive index values are indicators of the vegetation biomass (Lillesand and Kiefer, 1994). A difference image was then generated by subtracting the 1970s NDVI image from the 1990s NDVI image. Positive values represent gains in vegetation, while negative values depict losses in vegetation. The standard deviation was calculated from the difference image across the entire basin. In Figure 42.4, pixels that exceeded 1.5 standard deviations of loss were set to red, and pixels that exceeded 1.5 standard deviations of gain were set to green. This method of differentiating change is described in the appendix of *An Ecological Assessment of the United States Mid-Atlantic Region* (Jones et al., 1997). The large patches of forest loss shown in Figure 42.3e and in Figure 42.4a clearly bear a strong resemblance to the large patches of NDVI loss shown in Figure 42.4b.

RESULTS

The majority of changes in the Tensas River Basin result from the conversion of forested areas to agricultural use. The classification method showed a net loss of 46,234 hectares, or 12.3% of the study area. The NDVI method showed a net loss of 42,807 hectares, or 11.4% of the study area. Loss and gain data are represented in hectares and percentages of area by subwatershed in Table 42.1 for the land cover classification method and in Table 42.2 for the NDVI method. The data from both of these processes are then ranked (greatest to least loss) by subwatershed and are presented in Table 42.3.

The data presented in Table 42.3 reveal that the results ranked from the NDVI change detection are quite similar to those resulting from the land cover change detection method.

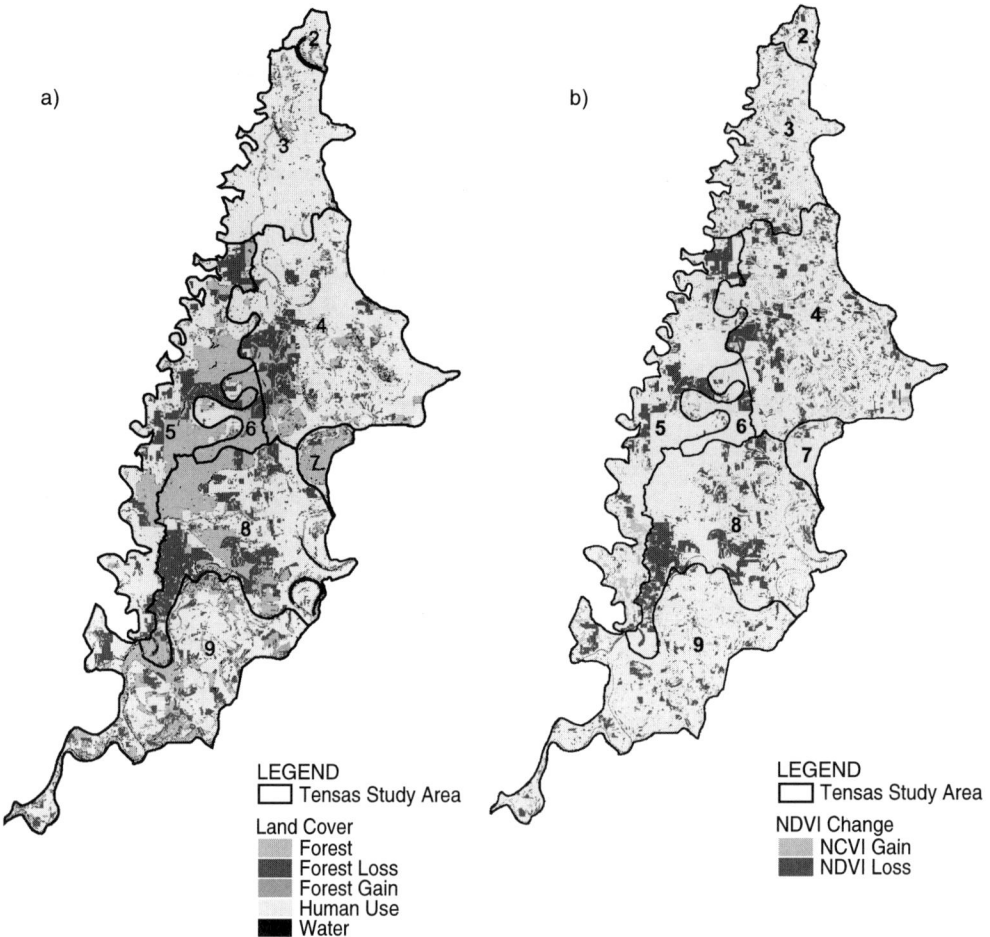

Figure 42.4 NDVI change image with loss set to black and gain set to gray (1.5 standard deviation).

DISCUSSION AND CONCLUSIONS

In this study, the two methods produced very similar results. This is an encouraging finding because the NDVI method is far less time-consuming than the land cover classification method. However, several issues should be analyzed and reconciled before this method of analysis could be applied confidently to a study area.

One issue concerns NDVI changes occurring as a result of agricultural practices. For example, subwatershed 3 had very little forested area in the 1970s image (less than 2%) and, thus, any loss in vegetation as calculated by NDVI change may be the result of agricultural practices, such as crop rotation, or growing season and harvesting differences. Another issue concerns climate differences. Images that are derived, for example, during particularly wet or dry time periods could conceivably cause misleading results when performing change detection. The causes of changes in NDVI may not be as easily identifiable as changes found by the land cover classification method. Losses in NDVI, for example, could be caused by road improvements, new residential developments, urbanization projects, and construction of reservoirs, as well as by increases in agriculture. Gains in NDVI may be caused by additions of golf courses and municipal parks or by increases in forested areas. Obviously, interpretation of NDVI values for actual land cover change requires additional work beyond simply calculating the difference over time; it requires a knowledge of the area studied.

Table 42.1 Land Cover Analysis

Zone	Total No. (pixels)	Total (ha)	Forest (pixels)	Forest %	Forest Gain (pixels)	Forest Gain %	Forest Loss (pixels)	Forest Loss %	Loss (ha)	NetLoss (ha)	Net Forest Loss %
Total	1043907	375806.5	202203	19.4	19409	1.9	147836	14.2	53221.0	-46233.7	-12.3
2	15559	5601.2	503	3.2	473	3.0	209	1.3	75.2	95.0	1.7
3	126076	45387.4	2210	1.8	1419	1.1	3140	2.5	1130.4	-619.6	-1.4
4	256864	92471.0	26983	10.5	3765	1.5	34739	13.5	12506.0	-11150.6	-12.1
5	171505	61741.8	54148	31.6	2487	1.5	30141	17.6	10850.8	-9955.4	-16.1
6	30906	11126.2	18090	58.5	578	1.9	3294	10.7	1185.8	-977.8	-8.8
7	20405	7345.8	10572	51.8	1531	7.5	1731	8.5	623.2	-72.0	-1.0
8	221426	79713.4	56488	25.5	4261	1.9	51239	23.1	18446.0	-16912.1	-21.2
9	200588	72211.7	32586	16.2	6039	3.0	24496	12.2	8818.6	-6644.5	-9.2

Table 42.2 NDVI Change

Zone	Total No. (pixels)	Total (ha)	NDVI (pixels)	NDVI %	NDVI Gain (pixels)	NDVI Gain %	NDVI Loss (pixels)	NDVI Loss %	Loss (ha)	Net Loss (ha)	Net NDVI Loss %
Total	1045254	376291.4	170118	16.8	25605	2.4	144513	13.8	52024.7	-42806.9	-11.4
2	15626	5625.4	1663	10.6	167	1.1	1496	9.6	538.6	-478.4	-8.5
3	126486	45535.0	16243	12.8	3003	2.4	13240	10.5	4766.4	-3685.3	-8.1
4	257020	92527.2	46229	18.0	8442	3.3	37787	14.7	13603.3	-10564.2	-11.4
5	172230	62002.8	31302	18.2	5576	3.2	25726	14.9	9261.4	-7254.0	-11.7
6	30934	11136.2	3149	10.2	730	2.4	2419	7.8	870.8	-608.0	-5.5
7	20420	7351.2	1663	8.1	453	2.2	1210	5.9	435.6	-272.5	-3.7
8	221532	79751.5	46126	20.8	2315	1.0	43811	19.8	15772.0	-14938.6	-18.7
9	201006	72362.2	23743	11.8	4919	2.4	18824	9.4	6776.6	-5005.8	-6.9

Table 42.3 Subwatershed Ranking by Percent; 2 and 7 Not Shown (see text)

Land Cover Forest		NDVI Vegetation	
Zone	Net Change	Zone	Net Change
8	−21.3	8	−18.7
5	−15.9	5	−11.7
4	−12.0	4	−11.4
9	−9.1	9	−6.9
6	−8.9	3	−8.1
3	−1.4	6	−5.5

The availability of a temporal date data set like NALC, coupled with NDVI change detection methods presented in this chapter, gives researchers a fast and consistent method of targeting ecosystems at varied scales. For example, with the need to study the hypoxia problem in the Gulf of Mexico, this method could be used to target high-nutrient runoff areas within the Mississippi Basin requiring further study. This may be a more cost-effective method than analyzing millions of individual farm fields in the basin, as other investigators have suggested (Warrick, 1999).

ACKNOWLEDGMENTS

The authors would like to thank Lee Bice and Karen Lee of Lockheed Martin Corporation for their support during the preparation of this document. Their contribution has been funded by the U.S. Environmental Protection Agency's Region 6 under Contract 68-C5–0065 to Lockheed Martin Corporation. This chapter has been subjected to the agency's review, and no official endorsement should be inferred.

REFERENCES

Gosselink, J.G., Shaffer, G.P., Lee, L.C., Burdick, D.M., Childers, D.L., Leibowitz, N.C., Hamilton, S.C., Boumans, R., Cushman, D., Firlds, S., Koch, M., and Visser, J.M., Landscape conservation in a forested wetland watershed, *BioScience,* 40(8):588–600, 1990.
Heggem, D.T., Edmonds, C.M., Neale, A.C., Bice, L., and Bruce Jones, K., Forested wetland restoration, identifying potential sites in northeast Louisiana, *Geo. Info. Syst.,* 9(5):34–39, 1999a.
Heggem, D.T., Neale, A.C., Edmonds, C.M., Bice, L., and Bruce Jones, K., An Ecological Assessment of the Louisiana Tensas River Basin, EPA 600/R-99/016, U.S. Environmental Protection Agency, Office of Research and Development, Washington, D.C., 1999b.
Jensen, J.R., *Introductory Digital Image Processing,* 2nd ed., Prentice-Hall, Englewood Cliffs, NJ, 1996, p. 116.
Jones, K.B., Riitters, K.H., Wickham, J.D., Tankersley, R.D., Jr., O'Neill, R.V., Chaloud, D.J., Smith, E.R., and Neale, A.C., *An Ecological Assessment of the United States Mid-Atlantic Region,* EPA 600/R-97/130, U.S. Environmental Protection Agency, Office of Research and Development, Washington, D.C., 1997.
Lillesand, T.M. and Kiefer, R.W., *Remote Sensing and Image Interpretation,* 3rd ed., John Wiley & Sons, New York, 1994.
Lunetta, R.S., Lyon, J.G., Sturdevant, J.A., Dwyer, J.L., Elvidge, C.D., Fenstermaker, L.K., Yuan, D., Hoffer, J.R., and Werrackoo, R., North American Landscape Characterization: Research Plan, EPA 600/R-93/135, U.S. Environmental Protection Agency, Office of Research and Development, Washington, D.C., 1993.
Lunetta, R.S., Lyon, J.G., Guindon, B., and Elvidge, C.D., North American Landscape Characterization Dataset Development and Data Fusion Issues, *Photogrammetric Eng. Remote Sensing,* 64(8):821–829, 1998.
Lyon, J.G., Yuan, D., Lunetta, R.S., and Elvidge, C.D., A change detection experiment using vegetation indices, *Photogrammetric Eng. Remote Sensing,* 64(2):143–150, 1998.

Rainer, M., Conti, J., Yantis, B., and Townsley, G., Selecting Sites for Wetland Restoration in the Tensas River Basin, Louisiana: A Case Study of Landscape Analysis Using the Synoptic Assessment Methodology, U.S. Department of Agriculture, Soil Conservation Service, Water Resources Planning Staff, Alexandria, LA, 1994.

Townsley, G., Selecting Sites for Wetland Restoration in the Tensas River Basin, Paper presented at the Louisiana Delta Conference, Memphis, TN, 1996.

Tucker, C.J., Red and Photographic Infrared Linear Combinations for Monitoring Vegetation, *Remote Sensing Environ.,* 8:127–150, 1979.

Warrick, J., Death in the Gulf of Mexico, *Natl. Wildl.,* June/July:48–52, 1999.

CHAPTER 43

Relationships among Environmental Stressors and Fish Community Composition and Health: Case Study of Chesapeake Bay

Susanna T.Y. Tong

INTRODUCTION

Located within the states of Maryland and Virginia, the Chesapeake Bay is the largest estuary in the U.S. Biologically, it is one of the most productive systems in the world. The primary production exceeds 400 g/Cm^{-2}/year (Sellner, 1987). Stretching from the mouth of the Susquehanna River at Havre de Grace in Maryland seaward to Cape Charles and Cape Henry, it is about 322 km long and varies in width from 5 to 48 km. The bay's drainage basin is 166,000 km^2 in extent, and it includes more than 150 rivers, creeks, and branches. About 94% of Maryland's land area drains into the Chesapeake Bay. The Susquehanna, Potomac, and James Rivers are the largest rivers in terms of drainage areas and freshwater input to the bay system.

Chesapeake Bay is a very young estuary, less than 10,000 years old. It was formed during the most recent rise in sea level in the postglacial period. More than 12.7 million people live in the region. Virtually every type of economic activity and land use is found in its watershed. Agriculture, residential, and forest are the predominant land use types. Although it is situated so closely to human residential and industrial activities, the estuary encompasses a variety of aquatic environments that support a rich and diverse assemblage of freshwater and marine resources of flora and fauna. Besides the emergent tidal wetlands, the estuary hosts more than 20 species of submerged vascular plants in seven families (Hershner and Wetzel, 1987) and well over 900 different species of benthic invertebrates (Diaz, 1987). The bay produces over one half of the nation's blue crabs and soft-shelled clams and one fourth of its oyster harvests. Striped bass, menhaden, shellfish, finfish, and rockfish are also commercially harvested (Sellner, 1987).

Since the time of the industrial revolution, sound ecological management practices were neglected in the Chesapeake Bay and water quality has been deteriorating. This decline has accelerated throughout the past three decades. During this period, the bay region has undergone serious losses of farm and forest and has been subject to an increased threat from atmospheric deposition of nitrogen. Most important, the region is faced with the prospect of 1 million additional people moving into the watershed by the next decade. The rapid urbanization process in the watershed, widespread use of fertilizer and pesticides in the farms, and physical modification of the shoreline, combined with overharvesting, have caused substantial degradation of the estuarine habitat, reduction of the reproductive capabilities of many species, and depletion of the renewable natural resources. Numerous chemical and physical pollutants are found in the Chesapeake Bay and its tributaries. Sediments, fertilizers, pesticides, herbicides, heavy metals, sewage, and petroleum products arrive in the bay from both point and nonpoint sources.

In the 1970s, the bay was so deteriorated that the distribution of the submerged and emergent aquatic vegetation, as well as that of the macro-invertebrates and game fish, had dramatically declined. More than 10 species of submerged aquatic vegetation have experienced significant decreases in abundance in Susquehanna Flats, in the lower reaches of the Patuxent, Potomac, Rappahannock, and York Rivers, and in innumerable other places throughout the bay (Price et al., 1985). Fishes collected from the Elizabeth River and Baltimore Harbor have a variety of maladies: lesions of the skin, liver, and gill, fin erosion, and cataracts. Today, probably no pristine, uncontaminated sites remain in the Chesapeake Bay.

In 1976, the USEPA established the multiyear, $27 million Chesapeake Bay Program. The bay was the first ecosystem in the U.S. to be targeted for restoration. In 1983, a comprehensive study of the bay was concluded and the Chesapeake Bay Study Report was released. It has identified four areas of greatest concern: nutrient enrichment and associated algal growth; erosional inputs of sediments; the presence of toxic chemicals (such as agricultural herbicides, atrazine, and linuron); and the loss of submerged aquatic vegetation (DeFur, 1997). Among these, nutrient enrichment — regarded as the most important — is a problem in all watersheds, especially on the eastern shore (Kemp et al., 1983).

In the Chesapeake Bay, the primary production and the carbon dynamics and flux are dominated by the photosynthetic activity of algae rather than vascular plants. The distribution of phytoplankton reflects the concentrations of nutrients and turbidity. Excessive nutrient loading, coupled with a high temperature, often provides a favorable condition for the growth of blue-green algae, even to the detriment of higher aquatic organisms. Massive blooms of harmful blue-green algae and dinoflagellates, such as *Pfiesteria piscidida*, have been observed frequently in recent years. In aquatic ecosystems, nitrogen concentrations of 0.5 to 1.0 mg/L are commonly used as threshold values for eutrophication. However, in estuarine environments, where salinity levels are greater and more sensitive to eutrophication, the threshold level drops to <0.6 mg/L (USDA, 1991). In addition to eutrophication, nitrate pollution poses a direct health threat to humans and other mammals. In cattle, nitrates can cause anemia as well as abortions. Nitrate levels of 40 to 100 mg/L in livestock drinking water are considered risky. High concentrations of nitrate also have been linked to methemoglobinemia in infants. The USEPA has established a maximum contaminant level for nitrate nitrogen in drinking water of 10 mg/L to protect the safety of U.S. drinking water supplies (USEPA, 1985). In many areas of the Delmarva peninsula, groundwater nitrate levels often exceed the 10 mg/L drinking water standard (Hamilton and Shedlock, 1992).

Agriculture is a major nonpoint source of nutrients in the Chesapeake Bay drainage basin (USEPA, 1983). Maryland agricultural statistics document approximately 18,000 farms in the state, totaling 2,650,000 acres. In the Chesapeake Bay watershed alone, approximately 17.9% of the land is cropland with more than 540 farms. Pastures comprise another estimated 16.5% of the land (Macknis et al., 1983). Fertilizers applied to farms contain nitrogen and phosphorus. Agricultural runoff from cropland and pasture is often enriched with nutrients. Consequently, runoff from these fields after rainfall constitutes one of the leading causes for nutrient loading in rural rivers. The USEPA and USGS estimated that annual total nitrogen loading to the bay ranged from 84.0 to 179.5×10^6 kg per year, depending on the dryness or wetness of individual years (USEPA, 1983; Lang, 1982). During the past 30 years, nutrients and chlorophyll *a*, an indicator of phytoplankton biomass, have increased in many areas of the bay. For example, in the upper bay, north of the Patapsco River, annual average concentrations of total nitrogen, phosphorus, and chlorophyll *a* have approximately doubled during this period. On the eastern shore, agriculture was estimated to contribute approximately 87% of the phosphorus and 70% of the nitrogen deposited into the bay. A study of the Upper Potomac River basin found that 28% of the nitrogen load to the watershed came from atmospheric deposition, 12% from biotic fixation and adsorption on surfaces of leaves,

16% from fertilizer, and 44% from animal waste (Jaworski et al., 1992). Another estimate suggested that 40% of the nitrogen load to the Chesapeake Bay watershed came from atmospheric deposition, 33% from livestock waste, and 27% from fertilizer (Fisher and Oppenheimer, 1991).

If we are to restore the ecological integrity of the Chesapeake Bay area and protect its fish resources, the trend of increasing nutrient loading must be reversed. Therefore, it is critical to understand the environmental factors that are controlling the nitrate levels and the health and composition of the fish communities in the bay area. By using a quantitative exploratory approach involving spatial and multivariate analyses, this chapter examines these intrinsic relationships.

METHODS

The data used included the water chemistry, habitat, and fish information from the 1995–1997 Maryland Biological Stream Survey, MBSS (Maryland Department of Natural Resources, 1997a,b). The MBSS is a statistical survey designed to characterize the biological and habitat conditions of the first-, second-, and third-order nontidal streams. Land use data were derived from the 1991–1993 leaves-on summer Landsat TM data. They were then calculated as percentage of land use for each 11-digit hydrological unit code (HUC) using a geographical information system (GIS). A list of the environmental attributes, together with their abbreviations used in this chapter, is shown in Table 43.1.

Table 43.1 Environmental Attributes Used in the Analyses

Environmental Attribute	Abbreviation
Stream Habitat Variables	
Aesthetic rating	AESTH
Bank stability	BNKST
Channel alteration	CHALT
Epifaunal substrate	EPSUB
In-stream habitat	INSTH
Pool/glide/eddy quality	PQUAL
Riffle/run quality	RQUAL
Remoteness	RMOTE
Velocity/depth diversity	VDPTH
Stream order	STORD
Channel flow status	CHFLO
Embeddedness	EMBED
Shading	SHADE
Water Chemistry	
pH	pH
Specific conductance	COND
Acid neutralizing capacity	ANC
Dissolved organic carbon	DOC
Sulfate	SO_4
Nitrate	NO_3
Land Use Types	
Deciduous forest	DECID
Mixed forest	MIXED
Agriculture	AGR
Pasture	PAST
Low-density residential	LRES
High-density residential	HRES
Industrial	INDUS

Fish, a valuable biological resource of food for humans, is important to conserve because of its commercial and sport recreation values. Fish communities are at the top of the food chain. They are highly visible. As one of the most sensitive components of aquatic ecosystems, they respond predictably to changes in both abiotic and biotic factors. Hence, they are useful indicators of biological integrity and ecosystem health. When used in bioassessment, they can indicate the long-term conditions of the aquatic habitat. For these reasons, this research focuses on the composition and health of the fish communities.

Fish IBI is an *Index of Biological Integrity* for biological assessment of stream conditions. It is a synthetic index that combines several metrics, including tolerance, trophic status, native and introduced species, and lithophillic spawners (see Maryland Department of Natural Resources, 1997a,b). The MBSS IBI values range from 1 to 5. A score of 3 or higher represents a stream condition that is comparable to the minimally affected reference sites. Scores of 3 or lower represent conditions that are influenced by human activities and therefore are significantly different from the reference sites. Hence, in certain respects, IBI denotes the general structure, ecology, and health of the fish communities. The percentage anomalies (expressed as the number of anomalies for game and nongame fish divided by the total number of fish examined × 100%) is also used as an indicator of the general health of stream fish. Anomalies refer to hooking injuries, cuts, presence of visible parasites, and pathological abnormalities (such as ocular abnormalities, skin lesions, and skeletal deformities).

The approach to the analyses entailed, first of all, the examination of the relationships of in-stream nitrate levels, fish IBI, and percent of fish anomalies with environmental attributes in a few selected watersheds for which a complete set of environmental and fish data is available. The Maryland Biological Stream Survey sampled 18 subwatersheds. However, a complete set of environmental data is available for only nine of these subwatersheds (see Table 43.2 and Figure 43.1). Moreover, data documenting fish anomalies were lacking in most subwatersheds. Consequently, analyses of fish anomalies were confined to only three river basins.

To identify the watersheds with the highest concentrations of nitrate, the mean nitrate levels in each HUC of each subwatershed were calculated and plotted in a map. Correlation analysis was then conducted to examine the associations of in-stream nitrate levels and environmental attributes. For those watersheds that have elevated levels of nitrate, further regression analyses were conducted to quantify the relationships.

Fish IBI and percentage of anomalies were selected to indicate the general structure and health of the fish communities. They were used in correlation and stepwise multiple regression analyses

Table 43.2 The Mean In-Stream Nitrate Values (mg/L) in the Subwatersheds with Complete Data; the *F*-Value is the Result of ANOVA on Nitrate Values among the Different Subwatersheds

River	Subwatershed Code	Mean*	Standard Deviation	Number of Observations
Choptank	CK	4.227	2.797	39
Chester	CR	3.252	2.817	45
Lower Potomac	LP	0.459	0.291	51
Middle Potomac	MP	3.380	2.103	108
North Branch Potomac	NO	0.714	0.473	64
Nanticoke	NW	4.889	4.401	17
Patapsco	PP	2.965	1.999	122
Upper Potomac	UP	1.595	2.250	64
Youghiogheny	YG	0.505	0.394	55

* *F*-value = 22.01 for differences among means, P < 0.0001.

From Tong, S.T.Y., An integrated approach to examining the relationship of environmental stressors and fish responses, *J. Aquatic Ecosystem Stress Recovery*, 9(1):1, 2001. With permission from Kluwer Academic Publishers, Dordrecht, The Netherlands.)

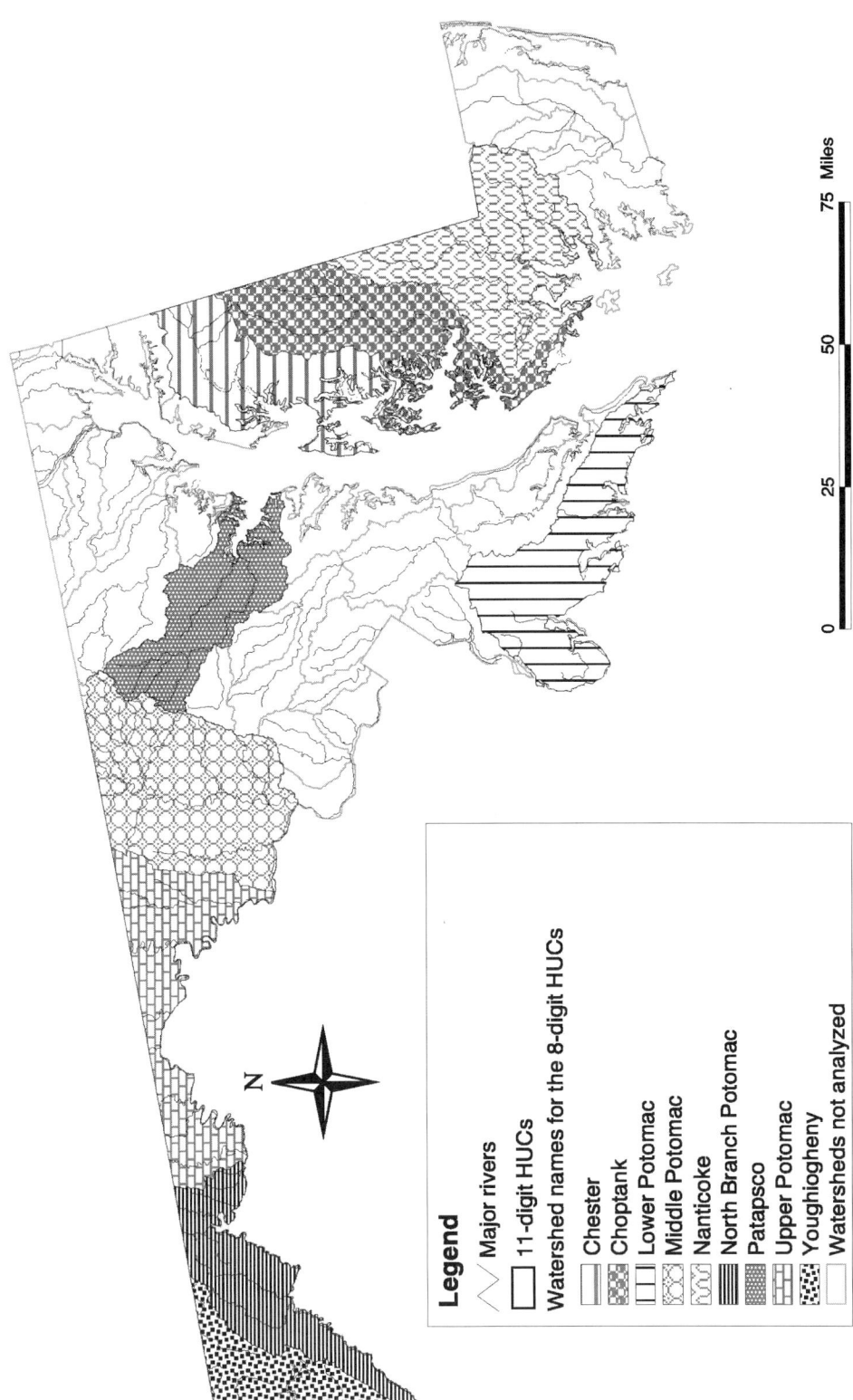

Figure 43.1 A map of the Chesapeake Bay in Maryland with the major river systems. (From Tong, S.T.Y., An integrated approach to examining the relationship of environmental stressors and fish responses, *J. Aquatic Ecosystem Stress Recovery*, 9(1):1, 2001. With permission from Kluwer Academic Publishers, Dordrecht, The Netherlands.)

Legend

/\/ Major rivers

☐ 11-digit HUCs

Watershed names for the 8-digit HUCs

Chester
Choptank
Lower Potomac
Middle Potomac
Nanticoke
North Branch Potomac
Patapsco
Upper Potomac
Youghiogheny
Watersheds not analyzed

N

0   25   50   75   Miles

with environmental factors. However, these analyses cannot provide much information about the species composition and distribution. To ascertain the relationships among environmental factors and fish assemblages, an ordination technique was adopted.

Multivariate ordination techniques are useful exploratory tools to help elucidate latent environmental relationships and to define specific biocriteria. In this chapter, a nonmetric multidimensional scaling ordination technique was employed. This is a multivariate indirect ordination analytical method. The choice of nonmetric multidimensional scaling in this research is based on the fact that the model uses nonparametric mathematical algorithm and theoretical assumption; as such, it does not require metric, ratio, and normally distributed data, nor any linear and metric relationships among the variables. In addition, since it is an indirect ordination technique, a prior knowledge or assumption about the dependent and independent variables is not needed. The environment is holistic. Many factors are interrelated or interdependent to each other. The causal relationships among these factors are often very complex and may not be apparent to investigators. The indirect technique is therefore advantageous. Furthermore, nonmetric multidimensional scaling is proven to be a powerful and robust technique, capable of revealing the latent and meaningful environmental structure (Gauch, 1982; Davidson, 1983; Tong, 1989). Hence, it is a more flexible, unbiased, objective, yet reliable diagnostic approach to exploratory analyses in ecological studies (Minchin, 1987).

Based on the presence and absence values of fish species in each sampled site, the nonmetric multidimensional scaling was used to arrange the sites. According to the species distribution and composition, the pairwise similarities between the sites were calculated and a distance matrix constructed. The sampling sites were arranged in a special manner, the order of which is in accordance with the species distribution, thereby imparting the latent environmental gradients that are controlling the intrinsic structure of the fish species composition. The output of the ordination analyses were in the forms of site coordinates. To characterize the environmental gradient, the site coordinates were correlated and regressed with the environmental attributes, such as land use types and nitrate levels. However, since the ordination axes may not parallel the actual environmental gradients, the ordination axes were rotated in $10°$ steps through $180°$. This generated 18 coordinate sets for each dimension. For each rotation, correlation and stepwise multiple regression analyses were performed using the habitat, land use, and water chemistry variables as the independent variables. The use of stepwise multiple regression was chosen to reduce the multicollinearity between the independent variables. Rotation of $1°$ is not feasible in this case, because doing so would generate 180 coordinate sets and 180 regression equations for each dimension. Environmental variables/factors that were highly related to the site coordinates were graphed against the angle of rotation (x-axis), with the y-axis depicting the level of correlation. Based on the graph, a rotational angle is selected for each dimension that resulted in the maximal level of correlation between the most important variable and the rotated site coordinate. The stepwise regression equation at that rotational angle was then selected for detailed interpretation. The environmental variables/factors that are featured in the graph and the stepwise regression equation were regarded as the latent controls for that environmental gradient.

Using the nitrate map generated earlier, the most contaminated sites with the highest nutrient loading and the worst fish performance (such as IBI) were depicted through GIS overlays. These *hot spots* represented the most deteriorated sites with adverse ecological effects. They are the sites that warrant our immediate attention. Such information would be useful to land use planners, resource managers, and personnel at regulatory agencies for assessing the current conditions and prioritizing restoration efforts.

RESULTS

Table 43.2 and Figure 43.2 show the nitrate content in each watershed. Out of the nine watersheds examined, six have mean in-stream nitrate nitrogen levels above 1.0 mg/L, which is the threshold level for eutrophication. Among these watersheds, Nanticoke has the highest nitrate level. Other streams that

Table 43.3 Correlation Coefficients for In-Stream Nitrate Nitrogen Concentrations with Environmental Attributes in Nine Subwatersheds

Variable	CR	YG	UP	PP	NW	CK	MP	LP	NO	All sites
AESTH				0.34						
EMBED	−0.58		0.37							
EPSUB	0.42									0.12
INSTH	0.42						−0.37			
SHADE										-0.16
RQUAL	0.41									0.13
CHALT			−0.66	0.27						
CHFLO			0.39	0.23						0.19
VDPTH										0.11
BNKST							−0.37			−0.18
DOC	−0.63			−0.41		−0.45				
pH		0.56	0.57			0.43				0.18
ANC		0.55	0.87	−0.50			0.41		0.51	0.22
COND			0.84	−0.32	0.87		0.33	0.50		
SO₄			0.56	−0.50				0.48		−0.10
MIXED					−0.83		−0.44			−0.49
DECID			−0.65	0.29			−0.62		0.34	−0.49
AGR			0.61	0.31		−0.43	0.63	0.40	−0.34	0.53
PAST			0.64	0.29		0.52				0.48
LRES			0.61	−0.32			−0.48			
HRES			0.62	−0.26			−0.49			
INDUS			0.64	−0.26			−0.46			

[a] Table 43.1 contains definitions of all acronyms.
All the correlations shown are significant at $P \leq 0.01$.

From Tong, S.T.Y., An integrated approach to examining the relationship of environmental stressors and fish responses, *J. Aquatic Ecosystem Stress Recovery*, 9(1):1, 2001. With permission from Kluwer Academic Publishers, Dordrecht, The Netherlands.

are enriched by nitrates are the Choptank and Middle Potomac. Stepwise multiple regression of land use variables (independent variables) on in-stream nitrate nitrogen (dependent variable) demonstrates that land use is an important factor in affecting the nitrate levels. For example, in Nanticoke, mixed forest and agricultural land use accounts for 77% of the variation of nitrates. In the Middle Potomac, deciduous forest and pasture account for 46% of the variation in nitrates (Tables 43.3 and 43.4).

The relationships of percent fish anomalies with environmental attributes were performed in only three watersheds: Nanticoke, Middle Potomac, and Choptank (Tables 43.5 and 43.6). In Nanticoke, only DOC is significantly related to fish anomalies. Land use variables (MIXED, DECID, AGR, and PAST) together explain about 40% of the variations of fish anomalies. In the Middle Potomac, VDPTH, EMBED, and RQUAL are apparently important in affecting the percentage of fish anomalies. SHADE, EMBED, and PAST are the variables that are significantly related to the percentage of fish anomalies in the Choptank.

The results of the IBI analyses for each subwatershed are shown in Tables 43.7 to 43.9 and Figures 43.3 and 43.4. The river basins that have high mean IBI values are the Choptank, Chester, and Lower Potomac — meaning that they are the "better" basins, comparable to the reference sites. Based on the IBI values, the state of Maryland (Chesapeake Bay and Watershed Programs 1997, 1998a,d) reported that 52% of the stream miles in the Choptank were in good condition. In Chester, however, the portion in good condition was 17%, and in the Lower Potomac it was 28%. River basins like North Branch Potomac and Upper Potomac have low IBI scores and are more degraded. The relationships of nitrates and fish IBI are unclear because not all of the basins that have high nitrate levels also have low IBI scores, or vice versa. The Nanticoke, Choptank, and Middle Potomac are the river basins that have the

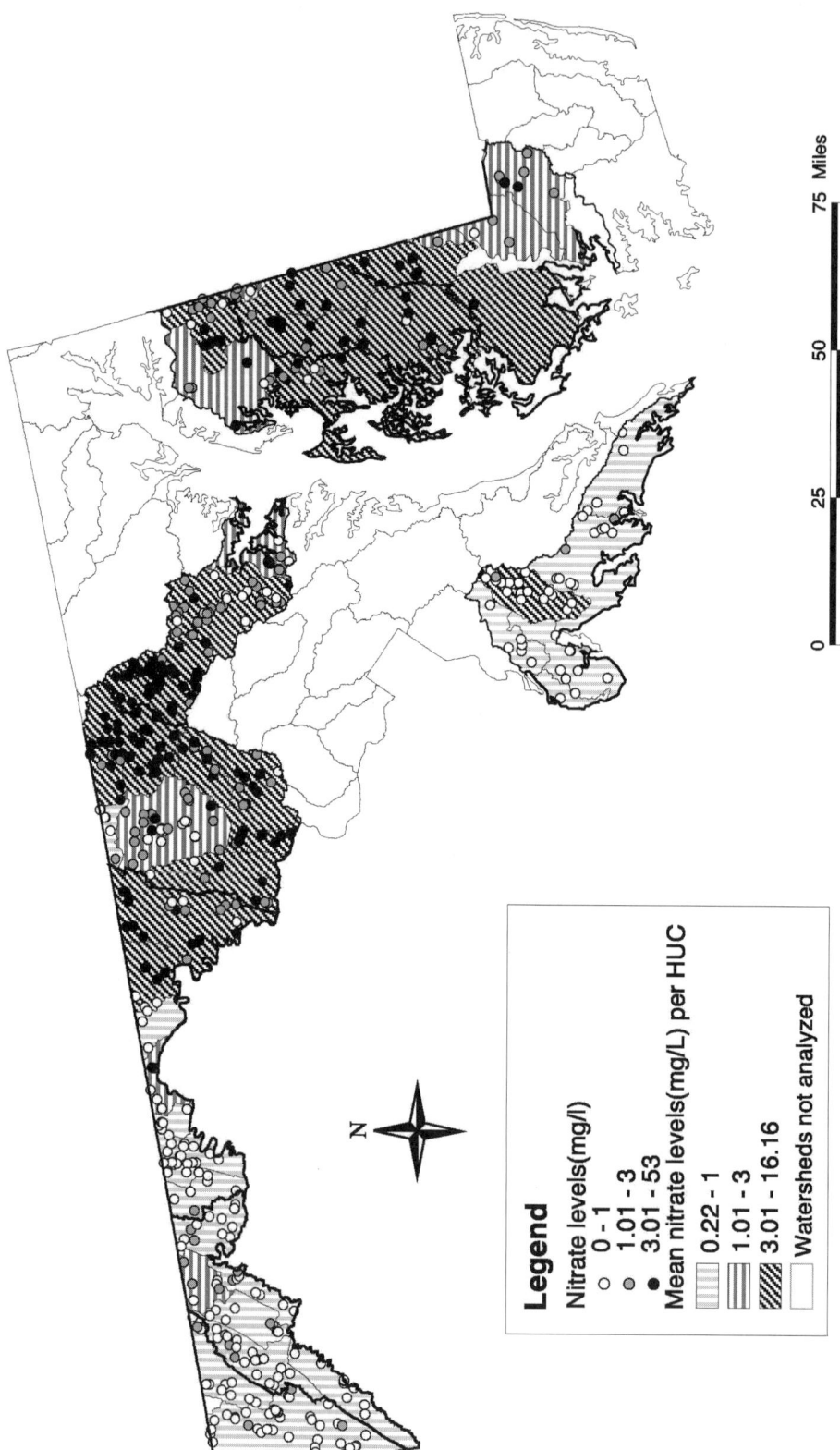

Figure 43.2 The distribution of in-stream nitrate levels in Maryland. (From Tong, S.T.Y., An integrated approach to examining the relationship of environmental stressors and fish responses, *J. Aquatic Ecosystem Stress Recovery*, 9(1):1, 2001. With permission from Kluwer Academic Publishers, Dordrecht, The Netherlands.)

Table 43.4 Relationship of Land-Use Variables with Nitrate Concentration for the Subwatersheds That Were Enriched with In-Stream Nitrate

Subwatershed	Regression Equation	R^2
NW	$NO_3 = 27.31 - 1.93$ MIXED $- 0.26$ AGR	0.77
MP	$NO_3 = 12.46 - 0.22$ DECID $- 0.12$ PAST	0.46
CK	$NO_3 = -0.32 + 0.21$ PAST	0.25

From Tong, S.T.Y., An integrated approach to examining the relationship of environmental stressors and fish responses, *J. Aquatic Ecosystem Stress Recovery*, 9(1):1, 2001. With permission from Kluwer Academic Publishers, Dordrecht, The Netherlands.

Table 43.5 Correlation Results of Percentage Fish Anomalies with Environmental Attributes in Selected Subwatersheds*

	Subwatershed		
Variable	NW	CK	MP
PQUAL			0.29
RQUAL			0.39
VDPTH			0.45
EMBED		−0.57	−0.43
SHADE		−0.76	−0.26
pH			0.37
DOC	−0.42		
MIXED			−0.29
AGR			0.25
PAST		−0.55	

* Only significant relationships at $P \leq 0.01$ are listed.

From Tong, S.T.Y., An integrated approach to examining the relationship of environmental stressors and fish responses, *J. Aquatic Ecosystem Stress Recovery*, 9(1):1, 2001. With permission from Kluwer Academic Publishers, Dordrecht, The Netherlands.

Table 43.6 Relationships of Percentage Fish Anomalies with Environmental Attributes for the Subwatersheds Enriched with In-Stream Nitrate

Subwatershed	Regression Equation	R^2
NW	% anomalies = $2.48 - 0.26$ MIXED $+ 0.52$ DECID $- 0.22$ AGR $+ 0.29$ PAST	0.40
MP	% anomalies = $76.78 - 0.85$ EMBED $- 0.58$ SHADE $+ 6.55$ VDPTH	0.40
CK	% anomalies = $19.91 - 0.12$ EMBED $- 0.16$ SHADE $- 0.03$ ANC $+ 1.74\ NO_3 - 1.28$ CHALT $+ 0.82$ RQUAL $+ 0.82$ RMOTE $- 0.75$ VDPTH	0.95

From Tong, S.T.Y., An integrated approach to examining the relationship of environmental stressors and fish responses, *J. Aquatic Ecosystem Stress Recovery*, 9(1):1, 2001. With permission from Kluwer Academic Publishers, Dordrecht, The Netherlands.

MANAGING FOR HEALTHY ECOSYSTEMS

Table 43.7 The Mean Fish IBI Scores in Different Subwatersheds

Subwatershed	Mean	Standard Deviation	Number of Observations
CK	4.045	0.575	22
CR	3.610	0.910	41
LP	3.572	1.041	45
MP	3.367	1.118	95
NO	2.670	1.234	52
NW	3.235	0.752	17
PP	3.275	1.082	115
UP	2.720	1.115	52
YG	3.175	0.960	39

From Tong, S.T.Y., An integrated approach to examining the relationship of environmental stressors and fish responses, *J. Aquatic Ecosystem Stress Recovery*, 9(1):1, 2001. With permission from Kluwer Academic Publishers, Dordrecht, The Netherlands.

Table 43.8 Correlation of Fish IBI Values with Environmental Attributes for Individual Subwatersheds[a]

	Subwatershed								
Variable	CK	CR	LP	MP	NO	NW	PP	UP	YG
AESTH					0.39		0.70		
CHALT							0.36		
EPSUB			0.42	0.44		0.72	0.28		0.62
INSTH		0.47	0.61	0.45		0.72	0.29	0.42	0.55
PQUAL			0.62	0.44					
RQUAL		0.60	0.43	0.54	0.39		0.34		
RMOTE							0.43		
VDPTH		0.59	0.64	0.52			0.32	0.45	0.59
CHFLO							0.42		
EMBED		−0.49		−0.45			−0.41		−0.52
pH		0.60							
COND				−0.43			−0.55		
ANC				−0.39			−0.63		
DOC		−0.68					−0.50		
SO$_4$							−0.59		
NO$_3$		0.47					0.49		
DECID					−0.59		0.37		
MIXED	−0.80				0.58		−0.41		
AGR					0.40		0.50		
PAST	0.68				−0.39		0.48		
LRES	−0.78				−0.60		−0.52		
HRES	−0.78						−0.41		
INDUS	−0.82				−0.41		−0.37		

[a] Only significant relationships ($P \leq 0.01$) are listed.

From Tong, S.T.Y., An integrated approach to examining the relationship of environmental stressors and fish responses, *J. Aquatic Ecosystem Stress Recovery*, 9(1):1, 2001. With permission from Kluwer Academic Publishers, Dordrecht, The Netherlands.

highest levels of nitrate, yet their IBI scores are all acceptable with a score of 3 and above. Out of the nine basins analyzed, only two — the Lower Potomac and the Youghiogheny — have low mean nitrate levels and high IBI scores (see Tables 43.2 and 43.7 and Figure 43.3). In the Chester and Patapsco, the correlation coefficients of nitrates and IBI are significant; yet they are positively related. It means that as the levels of nitrates increase, the fish IBI values will increase and the stream conditions will be better. Hence, contrary to common belief, in-stream nitrate levels may not be a good diagnostic indicator of fish IBI in these river basins. Instead, other environmental factors are more prominent in

Table 43.9 Regression Equations for Fish IBI with Environmental Attributes in Individual Subwatersheds

Subwatershed	Regression Equation	R^2
CK	IBI = 4.62 − 1.90 INDUS	0.66
	IBI = 4.09 + 0.06 SO_4 − 0.06 DOC − 1.83 INDUS	0.84
CR	IBI = 0.63 − 0.01 SHADE + 0.66 pH − 8.14 COND − 0.06 BNKST − 0.05 CHALT + 0.09 INSTH + 0.08 RQUAL	0.80
LP	IBI = − 8.60 + 0.17 MIXED + 1.43 pH −10.76 COND + 0.08 VDPTH + 0.45 LRES − 3.86 INDUS + 1.02 NO_3	0.72
MP	IBI = 4.62 − 0.01 EMBED − 0.40 pH − 0.02 SO_4 − 0.03 BNKST + 0.07 PQUAL + 0.04 RQUAL + 1.29 INDUS	0.61
NO	IBI = 10.08 + 0.69 MIXED − 0.19 DECID − 0.81 AGR − 0.16 LRES	0.63
	IBI = − 0.74 + 0.23 MIXED − 1.00 PAST + 0.05 CHALT − 0.03 RMOTE − 0.63 LRES + 6.52 INDUS	0.74
NW	IBI = 0.22 + 12.42 COND − 0.09 SO_4 + 0.08 DOC + 0.05 INSTH + 0.06 RMOTE	0.96
PP	IBI = 3.15 − 0.34 COND − 0.09 DOC + 0.06 AESTH − 0.06 HRES	0.69
UP	IBI = − 1.56 + 0.02 PAST + 0.59 pH − 4.30 COND − 0.06 AESTH − 0.06 BNKST + 0.07 VDPTH	0.63
YG	IBI = 11.79 − 0.02 CHFLO − 0.12 DECID − 0.04 SO_4 − 0.05 AESTH + 0.07 EPSUB + 0.13 VDPTH − 0.75 NO_3	0.83

From Tong, S.T.Y., An integrated approach to examining the relationship of environmental stressors and fish responses, *J. Aquatic Ecosystem Stress Recovery*, 9(1):1, 2001. With permission from Kluwer Academic Publishers, Dordrecht, The Netherlands.

influencing IBI. For example, in the Choptank, correlation and regression results show that land use variables are more important. Industrial land use (INDUS) alone explains 66% of the variation of fish IBI in this river basin. Similarly, in North Branch Potomac, many land use variables are significantly related to IBI. LRES, DECID, AGR, MIXED together explain 63% of the IBI variation in the stepwise multiple regression. Once again, the harmful effects of land use are prevalent only in these watersheds. Consequently, land use cannot be regarded as a universal stressor for the region. Figure 43.4 offers an example of the relationships of land use with IBI. The GIS overlay shows that the North Branch Potomac is the only area with low AGR and high IBI values. In the Chester, Lower Potomac, Middle Potomac, Upper Potomac, and Youghiogheny, stream habitat variables like VDPTH, INSTH, EPSUB, and RQUAL are more important in affecting fish IBI. The *r* values for most of these variables with IBIs are above 0.41. Water chemistry, such as DOC, is more influential in the Patapsco and perhaps also in the Chester.

Ordination analyses implicate that different environmental factors are in control of the fish composition in different subwatersheds. However, a few environmental factors are more prominent. They are in-stream nitrate levels, land use, and velocity-depth/remote/riffle quality/shading (Table 43.10). In Nanticoke, Patapsco, and Upper Potomac, NO_3 is more important. The simple correlation coefficients of NO_3 with the rotated site coordinates for these river basins are all significant at a level above 0.45. COND is significant in the Youghiogheny, where the maximum *r* value is 0.74. It is the only variable that is featured prominently in the site coordinates rotation. Land use (DECID, HRES, and MIXED) is a significant environmental factor in the North Branch Potomac and Lower Potomac. Although the ordination results reveal two significant variables (DECID and DOC) in the North Branch Potomac, land use variables (in terms of DECID, AGR) are found in both stepwise regression equations. In the Lower Potomac, land use variables (HRES and MIXED) are featured in both rotated axes. Hence, it seems that land use has an important role to play in influencing the species composition and distribution in these two river basins. In the Middle Potomac, Choptank, and Chester, in-stream habitat characteristics, such as VDPTH, SHADE, RQUAL, and RMOTE, are the major environmental gradients. The maximum *r* value for VDPTH in the Middle Potomac is −0.68. It is the only variable isolated in the rotation. In the Choptank, both regression equations are predominated by in-stream habitat variables (SHADE, RQUAL, and EMBED). The R^2 values for both equations are above 0.75.

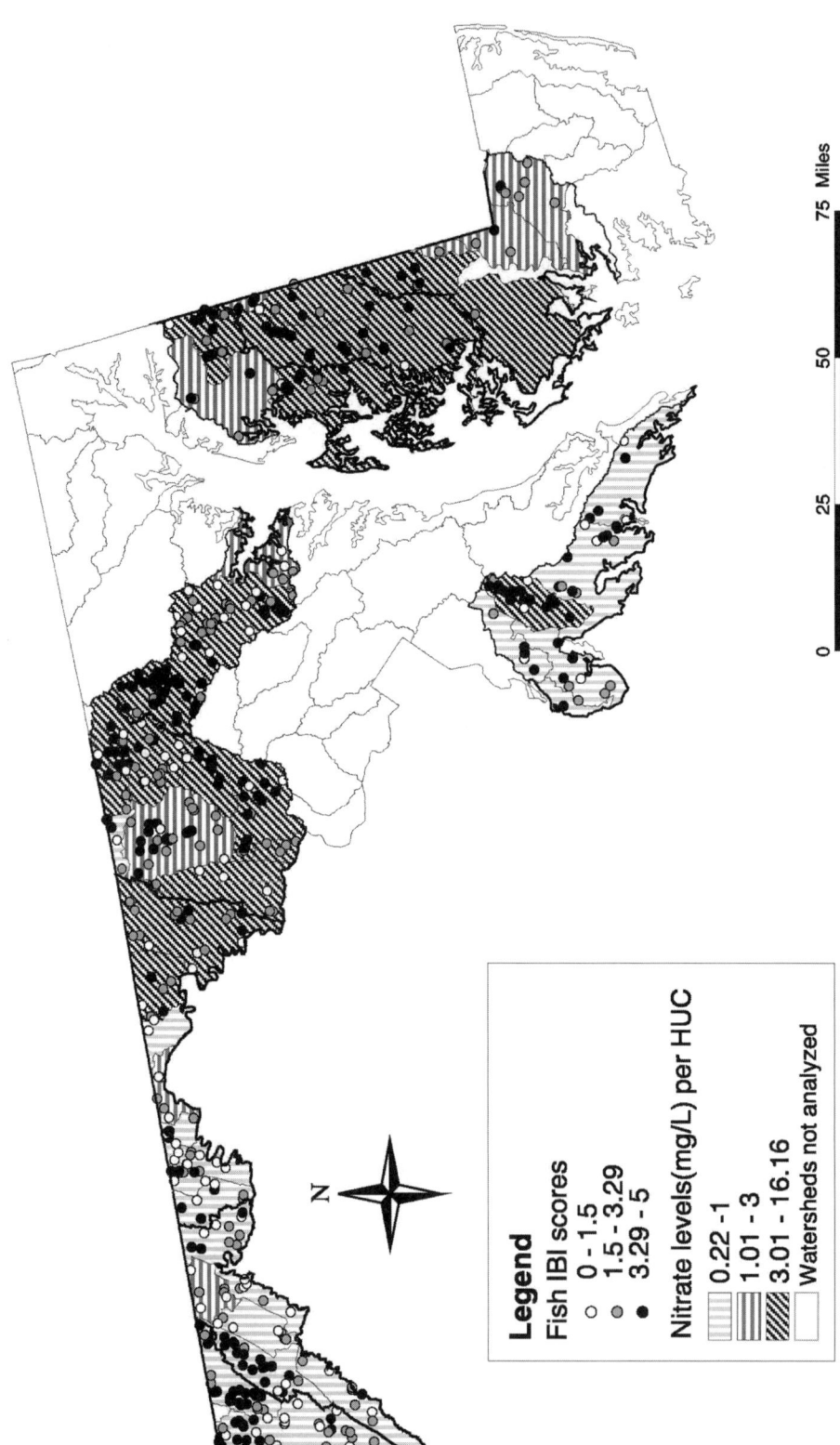

Figure 43.3 The relationship between fish IBI scores and in-stream nitrate levels. (From Tong, S.T.Y., An integrated approach to examining the relationship of environmental stressors and fish responses, *J. Aquatic Ecosystem Stress Recovery*, 9(1):1, 2001. With permission from Kluwer Academic Publishers, Dordrecht, The Netherlands.)

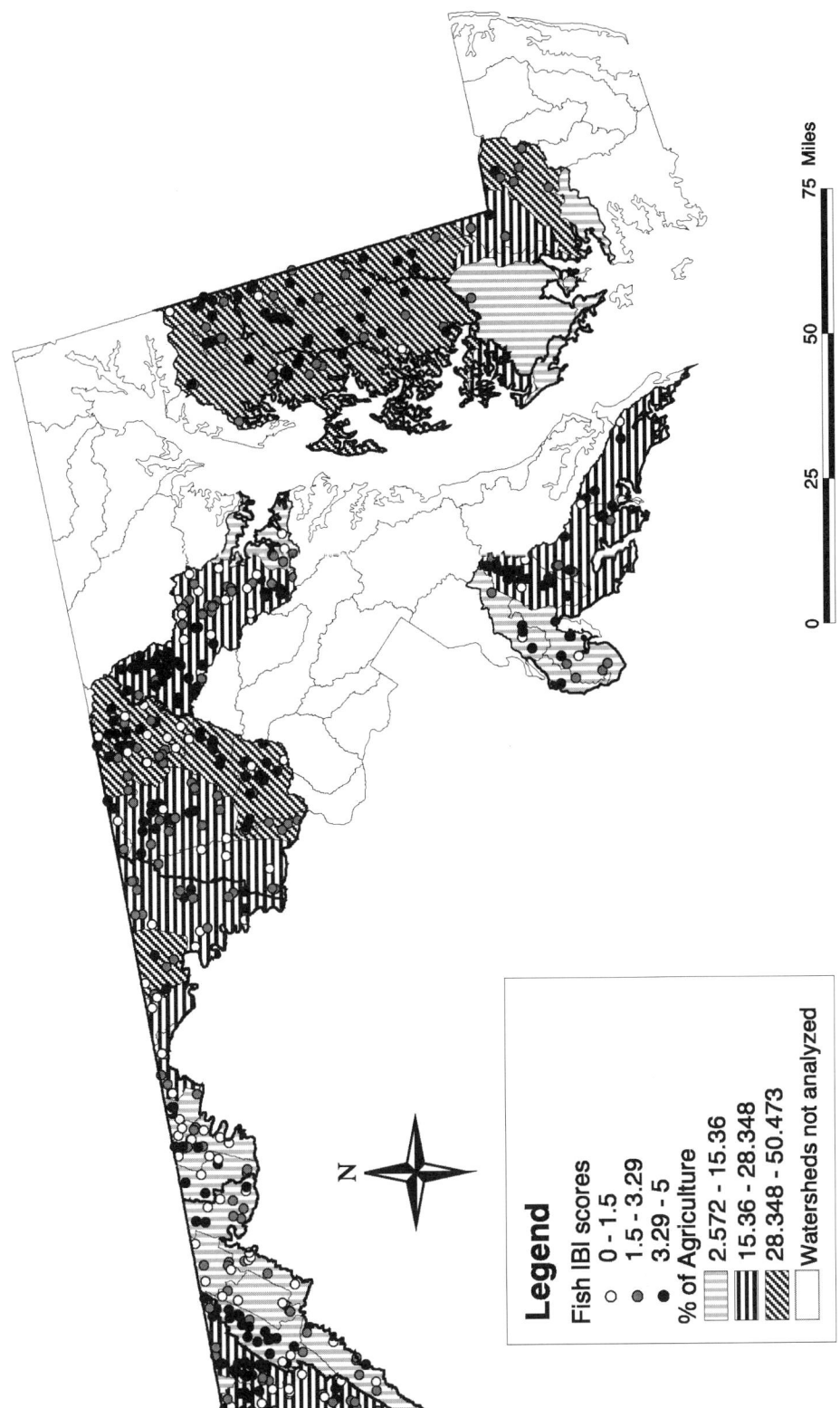

Figure 43.4 The relationship between fish IBI scores and percent agriculture. (From Tong, S.T.Y., An integrated approach to examining the relationship of environmental stressors and fish responses, *J. Aquatic Ecosystem Stress Recovery*, 9(1):1, 2001. With permission from Kluwer Academic Publishers, Dordrecht, The Netherlands.)

Table 43.10 Correlation and Stepwise Multiple Regression Results of the Environmental Variables on the Rotated Site Coordinates from NMDS Ordination

Subwatershed	Variable (max r)	Stepwise Multiple Regression	R^2
NW	DOC (−0.81)	Y = −0.10 + 0.03 SO_4 − 10.14 DOC + 0.03 AESTH +0.03 BNKST − 0.01 RQUAL	0.98
	NO_3 (−0.59)	Y = 1.42 − 20.81 COND + 0.02 SO_4 + 0.03 CHALT − 0.04 RQUAL + 0.25 NO_3	0.93
PP	NO_3 (−0.45)	Y = 180.93 + 37.45 MIXED + 0.08 ANC + 42.52 DOC − 12.82 INSTH	0.32
	VDPTH (0.40)	Y = 1054.33 + 3.17 EMBED + 106.23 VDPTH +18.79 DECID − 0.19 ANC + 31.47 INSTH	0.34
CK	SHADE (0.84)	Y = − 133.16 + 0.44 EMBED + 1.53 SHADE	0.84
	RQUAL (0.77)	Y = 121.71 − 1.02 EMBED + 2.82 PAST − 7.40 INSTH − 5.12 RMOTE + 8.84 RQUAL	0.91
NO	DECID (0.45)	Y = −5515.67 + 66.88 DECID + 73.66 AGR +11.56 AESTH + 19.10 BNKST − 17.19 CHALT	0.47
	DOC (0.45)	Y = 3442.16 − 2.66 SHADE + 100.56 STORD − 42.49 DOC − 64.70 AGR + 94.08 DECID − 31.29 BNKST +16.51 CHALT + 9.02 EPSUB + 101.51 NO_3	0.65
LP	HRES (0.46)	Y = 1.57 + 0.01 SHADE + 0.11 DOC + 0.07 HRES	0.37
	MIXED (0.38)	Y = 0.97 − 0.09 MIXED + 0.07 DOC	0.20
CR*	RMOTE (−0.44)	Y = − 72.62 − 1.33 EMBED + 5.19 AGR + 14.19 AESTH + 4.98 CHALT − 16.16 EPSUB + 8.27 INSTH − 7.77 PQUAL − 13.23 RMOTE	0.76
MP*	VDPTH (−0.68)	Y = 833.14 + 1.39 SHADE − 4.03 CHFLO − 119.91 STORD + 9.85 PAST + 0.10 ANC + 4.15 SO_4 + 11.29 EPSUB − 21.60 PQUAL − 42.86 VDPTH	0.70
YG*	COND (0.75)	Y = −92.41 + 0.02 DECID + 0.11 COND + 0.15 AESTH − 0.02 EPSUB + 0.01 RQUAL	0.54
UP*	NO_3 (−0.54)	Y = 0.54 − 0.12 SO_4 − 0.03 EPSUB − 0.23 NO_3	0.47

*Only one environmental factor predominated in the ordination analyses.

From Tong, S.T.Y., An integrated approach to examining the relationship of environmental stressors and fish responses, *J. Aquatic Ecosystem Stress Recovery*, 9(1):1, 2001. With permission from Kluwer Academic Publishers, Dordrecht, The Netherlands.

DISCUSSION AND CONCLUSION

In the Chesapeake Bay, the Nanticoke, Choptank, and Middle Potomac Rivers are enriched with nitrates. According to the Chesapeake Bay and Watershed Programs (1998a–c), elevated nitrate nitrogen above 1 mg/L occurred at 92% of the stream miles in the Nanticoke, 82% in the Choptank, and 83% in Middle Potomac. Reduction of the nitrate levels in the Chesapeake Bay will require concentration of efforts in these subwatersheds. Results from stepwise multiple regression and correlation analyses implicate that land use variables (percent of cropland and pasture) are important in affecting the levels of nitrate nitrogen in these streams. These results are in agreement with the suggestions of the Chesapeake Bay and Watershed Programs (1998a–c).

Similar results were reported by Lowrance (1983). He found that the movement of nitrate nitrogen from row crops was significantly higher than in pastures and upland pine forest. Likewise, Hall et al. (1996) concluded that nutrients were influenced by agricultural activities. The mean nitrate value and IBI scores for streams in agriculture areas were significantly higher than in forested areas. Correll (1987) has also noted that cropland/riparian forest watersheds discharged five times as much nitrogen and seven times as much phosphorus in runoff as forested watersheds, while pasturelands had intermediate discharges. The nitrate concentrations decreased as the proportion of forested land in the watershed increased. In drainage basins that are more than 85% forested, the nitrate concentrations were below 0.7 mg/L.

In terms of fish IBI values, the North Branch Potomac and Upper Potomac have the lowest scores. The conditions of these river basins are far from pristine, and they warrant our restoration effort if the ecology and structure of the fish communities are our major concerns.

The factors that control the health, ecology, structure, and composition of fish communities are revealed by synthesizing the results from nonmetric multidimensional scaling ordination and statistical analyses. A few environmental factors, including NO₃/DOC/ANC/COND (water chemistry), DECID/HRES/MIXED (land use), and VDPTH/SHADE/RQUAL/RMOTE (stream habitat), appear to be important controls. But the effect of each environmental variable is different in individual watersheds. For example, the predominant environmental controls for the health and distribution of fish assemblages in the Nanticoke appear to be DOC, NO₃, and perhaps land use. The Patapsco is dominated by ANC/COND/NO₃, and possibly land use. Likewise, land use is a major environmental gradient in the North Branch Potomac. SHADE and land use are the key factors in the Choptank. In the Middle Potomac, VDPTH, and perhaps EMBED and RQUAL, seems to be important. Hence, evidence indicates that nitrates, land use, and — in the case of the Middle Potomac — velocity–depth ratio are the latent environmental structures in these river basins.

The research methodology used in this project involved an exploratory approach integrating statistical, ordination, and GIS analyses. The results of the analyses show that this approach is vigorous and reliable. This chapter has identified the watersheds that are enriched with nitrates as well as with low IBI scores. These may be the areas that warrant our restoration efforts. Moreover, this chapter quantified the relationships of nitrates and land use. Most importantly, it examined not only the health, structure, and ecology but also the composition and distribution of fish assemblages, identifying the environmental factors that are significant in controlling the fish communities.

ACKNOWLEDGMENTS

This project was performed when the author was a summer research faculty fellow with the National Exposure Research Laboratory at the USEPA in Cincinnati, OH. The author is thankful to many contractors, especially Sherry Hu, Rona Fan, Jian Liang, and Jill McAfee of SoBran Inc.; Amy Liu of PAI, Inc.; and Keith Adams of OAO, Inc., for their valuable statistical, GIS, graphics, and data management support. Their help has greatly expedited the progress of the project, and their contributions are gratefully acknowledged.

Special thanks are also due to Paul Kazyak and Scott Stranko of the Maryland Department of Natural Resources for the use of their survey data (MBSS data). In addition, Susan Cormier and Brian Hill of the National Exposure Research Laboratory, USEPA, and Gabriel Senay of SoBran, Inc., have provided helpful discussion about the project.

The author is indebted to Kluwer Academic Publishers for their kind permission to use their copyrighted materials, including Tables 43.2 to 43.10 and Figures 43.1 to 43.4, that appear in the journal article: Susanna T.Y. Tong, "An integrated exploratory approach to examining the relationships of environmental stressors and fish responses," *Journal of Aquatic Ecosystem Stress and Recovery*, Vol. 9(1), 1–19, 2001.

REFERENCES

Chesapeake Bay and Watershed Programs, Chester River Basin, Environmental assessment of stream conditions, CBWP-MANTA-EA-96-4, Maryland Department of Natural Resources, 1997.

Chesapeake Bay and Watershed Programs, Choptank River Basin, Environmental assessment of stream conditions, CBWP-MANTA-EA-98-6, Maryland Department of Natural Resources, 1998a.

Chesapeake Bay and Watershed Programs, Nanticoke/Recommit River Basin, Environmental assessment of stream conditions, CBWP-MANTA-EA-96-6, Maryland Department of Natural Resources, 1998b.

Chesapeake Bay and Watershed Programs, Middle Potomac River Basin, Environmental assessment of stream conditions, CBWP-MANTA-EA-98-5, Maryland Department of Natural Resources, 1998c.

Chesapeake Bay and Watershed Programs, Lower Potomac River Basin, Environmental assessment of stream conditions, CBWP-MANTA-EA-96-5, Maryland Department of Natural Resources, 1998d.

Correll, D.L., Nutrients in Chesapeake Bay, in Matchmaker, S.K., Hall, L.W., Jr., and Austin, H.M., Eds., *Contaminant Problems and Management of Living Chesapeake Bay Resources*, The Pennsylvania Academy of Science, 1987, pp. 298–319.

Davidson, M.L., *Multidimensional Scaling*, John Wiley & Sons, New York, 1983.

DeFur, P.L., The Chesapeake Bay Program: an example of ecological assessment, *Am. Zool.*, 37:641–649, 1997.

Diaz, R.J., Benthic resources of the Chesapeake Bay estuarine system, in Majumdar, S.K., Hall, L.W. Jr., and Austin, H.M., Eds., *Contaminant Problems and Management of Living Chesapeake Bay Resources*, The Pennsylvania Academy of Science, 1987, pp. 158–164.

Fisher, D.C. and Oppenheimer, M., Atmospheric nitrogen deposition and the Chesapeake Bay Estuary, *Ambio*, 20:102–108, 1991.

Gauch, H.G., *Multivariate Analysis in Community Ecology*, Cambridge University Press, Cambridge, U.K., 1982.

Hall, L.W., Jr., Scott, M.C., Killen Jr., W.D., and Anderson, R.D., The effects of land use characteristics and acid sensitivity on the ecological status of Maryland Coastal Plain streams, *Environ. Toxicol. Chem.*, 15:384–394, 1996.

Hamilton, P.A. and Shedlock, R.J., Are Fertilizers and Pesticides in the Ground Water? A Case Study of the Delmarva Peninsula, Geological Survey Circular 1080, 1992.

Hershner, C. and Wetzel, R.L., Submerged and emergent aquatic vegetation of the Chesapeake Bay, in Majumdar, S.K., Hall, L.W., Jr., and Austin, H.M., Eds., *Contaminant Problems and Management of Living Chesapeake Bay Resources*, The Pennsylvania Academy of Science, 1987, pp. 116–133.

Jaworski, N.A., Groffman, P.M., Keller, A.A., and Prager, J.C., A watershed nitrogen and phosphorus balance: the Upper Potomac River basin, *Estuaries*, 15:83–95, 1992.

Kemp, W.M., Twilley, R.R., Stevenson, J.C., Boynton, W.R., and Means, J.C., The decline of submerged vascular plants in Upper Chesapeake Bay: summary of results concerning possible causes, *MTS J.*, 17:78–89, 1983.

Lang, D.J., Water Quality of the Three Major Tributaries to the Chesapeake Bay, the Susquehanna, Potomac and James Rivers, January 1979 to April 1981, USGS Water Resource Investigation 82-32, 1982.

Lowrance, R.R., Waterborne nutrient budgets for the riparian zone of an agricultural watershed, *Agric. Ecosyst. Environ.*, 10:371–384, 1983.

Macknis, J., Gillelan, M.E., and Glotfelty, C.E., Land use methodology and data, Appendix B, Section 2, in USEPA Chesapeake Bay Program, *Chesapeake Bay: A Framework for Action*, Washington, D.C., 1983.

Maryland Department of Natural Resources, *Maryland Biological Stream Survey: Ecological Status of Non-Tidal Streams in Six Basins Sampled in 1995*, Chesapeake Bay and Watershed Programs, Monitoring and Non-tidal Assessment, 1997a.

Maryland Department of Natural Resources, *Guide to Using 1995 Maryland Biological Stream Survey Data*, Chesapeake Bay and Watershed Programs, Monitoring and Non-tidal Assessment Division, 1997b.

Minchin, P.R., An evaluation of the relative robustness of techniques for ecological ordination, *Vegetatio*, 69:89–107, 1987.

Price, K.S., Flemer, D.A., Taft, J.L., Mackiernan, G.B., Nehlsen, W., and Biggs, R.B., Nutrient enrichment of Chesapeake Bay and its impact on the habitat of striped bass: a speculative hypotheses, *Trans. Am. Fish. Soc.*, 114:97–106, 1985.

Sellner, K.G., Phytoplankton in Chesapeake Bay: role in carbon, oxygen and nutrient dynamics, in Majumdar, S.K., Hall, L.W., Jr., and Austin, H.M., Eds., *Contaminant Problems and Management of Living Chesapeake Bay Resources*, The Pennsylvania Academy of Science, 1987, pp. 134–157.

Tong, S.T.Y., An integrated approach to examining the relationship of environmental stressors and fish responses, *J. Aquat. Ecosyst. Stress Recovery*, 9(1):1–19, 2001.

Tong, S.T.Y., On nonmetric multidimensional scaling ordination and interpretation of the *matorral* vegetation in lowland Murcia, *Vegetatio*, 79:65–74, 1989.

USEPA, *Chesapeake Bay: A Framework for Action*, USEPA, Washington, D.C., 1983.

USEPA, National primary drinking water regulations: synthetic organic chemicals, inorganic chemicals, and microorganisms, Proposed Rule, *Fed. Regis.*, 50:46935–47022, 1985.

USDA, *Nitrate Occurrence in U.S. Waters*, USDA, Washington, D.C., 1991.

Assessing and Monitoring Biodiversity

CHAPTER 44

Overview: Assessing and Monitoring Biodiversity

Santiago Carrizosa

Since the late 1980s, when the concept of biodiversity was brought to the public domain, scientists, conservationists, and policy makers have increased research activities in order to understand and manage the complexities of our natural world. Since then, biodiversity management objectives have not changed much; they still seek to protect biodiversity, to restore it, and to use it in a sustainable way. In contrast, new biodiversity management strategies and approaches have been undertaken during the past 10 years, as our understanding of biodiversity threats and ecosystem dynamics have increased. These new approaches include *in situ* strategies (e.g., conservation corridors); restoration and rehabilitation strategies (e.g., reintroduction of native species and repair of ecosystem services); major land use strategies (e.g., bioprospecting initiatives and sustainable use practices included in forestry, fisheries, agriculture, wildlife management, and tourism); and policy and institutional strategies (e.g., methods such as access laws, subsidies, and taxes to regulate access to and use of land resources).

In addition, several government and nongovernment organizations have designed national biodiversity action plans and bioregional management programs in order to carry out specific biodiversity management strategies. These strategies have been enacted with different levels of success all over the world. In general, ecologists thus far have not found sound strategies to reconcile the threats to biodiversity with the complex nature of tropical and some temperate ecosystems. The tropics contain the world's least understood ecosystems, which are being depleted at a faster rate than any other habitat. If that damage continues unchecked, we are destined to lose at least 50% of all species found on Earth (Myers, 1988). Discovery of new biodiversity management approaches is imperative. Fortunately, the concept of ecosystem health may have paved the way for a relatively new understanding and management of tropical ecosystems.

In 1992, Haskell et al. defined ecosystem health as a "characteristic of complex ecosystems" and concluded: "An ecological system is healthy and free from 'distress syndrome' if it is stable and sustainable — that is, if it is active and maintains its organization and autonomy over time and is resilient to stress."

Sustainability is certainly a key word in this definition. The ecosystem has to be sustainable from the biological, social, cultural, and economic point of view. It has to maintain its structure, diversity, and biological relationships, and at the same time be able to produce services to society and to recover from external threats. This definition also suggests that the ecosystem health paradigm may be used to assess and understand the ecosystem's overall performance and to restore health to ecological processes at all levels.

Health as a concept has been traditionally associated with humans, but while physicians focus on the health of a single patient, biodiversity managers have to concentrate on the multiple social, economic, and natural levels and scales of the ecosystem (Haskell et al., 1992). The human health paradigm can also be used to develop the following parallel between the approaches followed by physicians and those of biodiversity managers to assess and monitor the health of humans and biodiversity, respectively.

Human Health	Ecosystem Health
Identify symptoms	*Identify "Pressure" indicators;* they measure the factors that have an effect on ecosystem health (e.g., concentration of pollutants such as heavy metals)
Identify and measure vital signs	*Identify "State" indicators* to measure the present condition of biodiversity (e.g., species diversity/richness)
Make a diagnosis and prognosis	*Develop a comprehensive assessment study;* based on the above indicators and additional primary and secondary information, this assessment would describe the many causes of ecosystem stress
Prescribe a treatment	*Design and enact a management plan*
Conduct tests to verify the effectiveness of the treatment	*Develop and implement a monitoring and evaluation plan* (this includes collecting information over time for the *State, Pressure,* and *Response* indicators[a] in order to monitor biodiversity); *Response* indicators are used to measure the activities enacted to obtain a desired biodiversity condition (e.g., national parks declared)

[a] The *State, Pressure,* and *Response* indicators framework has been proposed by organizations such as the OECD (1994), United Nations (1996), and SBSTA (1997) to monitor sustainable development and biodiversity conditions. This system can be adapted to respond to the needs of field scientists and policy makers alike.

The ecosystem health problems described in the chapters in this section indicate the need to bridge the gap between science and action; the State, Pressure, and Response indicators presented above constitute a framework that can be used by both scientists and policy makers for monitoring and evaluation goals. In Chapter 45, Gandara and Kageyama, for example, document scientific evidence about tropical rain forest density patterns, pollination, and seed dispersal patterns that can contribute to development of State, Pressure, and Response indicators for management and monitoring purposes. The authors provide a good description of the methods used to analyze tree demographics, genetic, and ecological relationships, and they establish sound connections between their findings and recommendations for a proper management and conservation plan for rare tree species in Brazilian rain forests. For example, the authors found that rare tree species with low population densities (i.e., 1 tree/ha to 1 tree/50 ha) have a low natural regeneration rate and long-distance pollinators and seed dispersals. Therefore, these species are not recommended for sustainable management programs. In this case, regeneration rate and pollination distance can be used as a State indicator of tropical forest health, and it can be used to guide biodiversity managers and politicians.

Chapter 46 by Grigione and Johnson presents an interesting case study about the effects of roads on mountain lions in California. Although the general influences of roads on wildlife have been documented by numerous authors, the specific consequences of roads (e.g., car collisions and habitat alteration) on mountain lions are not well understood. Roads apparently exert greater influence on populations inhabiting the Coastal Range Mountains than those of the Sierra Nevada. The authors also found differences in mountain lion habitat between the Sierra Nevada and Coastal ranges. They conclude that in the Sierra Nevada factors such as tracking deer may be more important than the presence of roads to the survival of the mountain lion. Furthermore, despite lion mortality on highways, they do not appear to be selecting areas with lower road densities. These types of case studies are particularly useful and pertinent for scientists and policy makers to develop State and Pressure indicators that can be used for monitoring and management purposes.

Recent and rapid developments in remote sensing imagery and geographical information systems also have supported efforts to assess, monitor, and manage biodiversity. Chapter 47 by Kappelle et al. presents an overview of ecosystem mapping methods developed in Costa Rica during the past 30 years and proposes a new method in rapid ecosystem mapping. This method uses techniques of stereoscopic interpretation to identify nine physiognomic classes from major Costa Rican habitats. The proposed physiognomic classes were developed based on the tree, shrub, and herb cover that the researchers were able to distinguish from semi-detailed aerial photographs. This is also a rapid, efficient, effective, and relatively inexpensive method that will allow researchers to repeat their efforts in order to monitor vegetation changes over time. This information will be valuable for Costa Rican ecosystem and wildlife managers who are currently developing a strategic plan to monitor biodiversity based upon the presence of State indicators such as ecosystem area, species richness, and species population viability.

Since 1992, when the Convention on Biological Diversity (CBD) was signed by 157 nations, the term *biodiversity* became part of the lexicon of law and policy makers all over the world. The countries that signed and ratified the CBD committed to develop plans and to enact legislation to help achieve the main objectives and initiate activities proposed by the CBD. Several years ago the Global Environment Facility (GEF) became the financial mechanism of the CBD, and since then it has been supporting countries willing to develop biodiversity national strategies and plans. Uruguay is among these countries. In Chapter 48 Victor Canton, a Uruguayan government official, presents an overview of the participatory process that led to the identification of assessment and monitoring recommendations in the context of the national biodiversity strategy. His chapter offers valuable lessons for countries engaged in the development of national biodiversity strategies and action plans. The chapters presented in this section also demonstrate that current biodiversity assessment and monitoring efforts must be strongly supported in biodiversity-rich countries in which current development trends and lack of local capacity are threatening the survival of many species.

REFERENCES

Haskell, B.D., Norton, B.G., and Costanza, R., Introduction: What is ecosystem health and why should we worry about it? in Costanza, R., Norton, B.G., and Haskell, B.D., Eds., *Ecosystem Health: New Goals for Environmental Management,* Island Press, Washington, D.C., 1992, pp. 3–20.

Myers, N., Tropical forests and their species: going, going...? in Wilson, E.O., Ed., *Biodiversity,* National Academy Press, Washington, D.C., 1988, pp. 28–35.

Organization for Economic Cooperation and Development (OECD), *Environmental Indicators*, OECD Core Set, Paris, 1994.

Subsidiary Body on Scientific, Technical and Technological Advice (SBSTA), Recommendations for a Core Set of Indicators, background paper prepared by the liaison group on indicators of biological diversity, third meeting, Montreal, Canada, September, 1997.

United Nations, *Indicators of Sustainable Development Framework and Methodologies,* UN, New York, 1996.

CHAPTER 45

Management and Conservation of Tropical Forests with Emphasis on Rare Tree Species in Brazil

Flávio Bertin Gandara and Paulo Y. Kageyama

INTRODUCTION

Conservation of tropical forests is one of the most urgent and challenging tasks of foresters, biologists, social scientists, environmental managers, nonprofit organizations, and public administrators in relevant departments. The species richness of tropical forests has been greatly emphasized, and the serious consequences of deforestation and consequent loss of biodiversity have been highlighted. However, of all tropical forests, those considered as lowland tropical rain forests contain the greatest number of species, and their destruction would have the worst and most extensive global effects (UNEP, 1995; Whitmore and Sayer, 1992; Alvarez-Buylla et al., 1997).

The conservation of tropical forests poses a number of biological and practical constraints. The number of species that require attention in a single region is large, running into the thousands. Thus, the resources needed for research and conservation have to be prioritized. An overwhelming majority of plant species in tropical forests is involved in mutualistic interactions for growth and reproduction. Pollination and reproduction of forest species involves a wide variety of insects and animals, both as pollen and seed dispersal agents. The prevalence of mutualistic interactions further justifies the conservation of these resources in their natural habitats.

Many tropical forest trees, particularly in lowland rain forests, occur in extremely low densities. For large trees, more than 50% of the species in any one community may have densities as low as one adult or less per hectare. Because the majority of these species is outcrossed, their interbreeding populations can cover areas hundreds of square kilometers in size. Deforestation leading to smaller areas is likely to cause inbreeding, with possible loss of genetic diversity. Because of their long lives, trees can survive as adults without regeneration. Thus, in many areas, populations can occur as *living dead* (Janzen 1988), conveying the illusion of persistence when, actually, the population is on its way to extinction.

Our ignorance about the basic biology of most tropical forest plants is overwhelming. A large number of species that are found in rapidly disappearing habitats require conservation measures. Therefore, to be effective, conservation of tropical forests needs to be integrated into the overall framework of sustainable forest management. In the past, efforts in conserving protected areas, national parks, and reserves have not given sufficient consideration to the distribution of species and their intraspecific variation, an understanding of both of which is central to ensuring the maintenance of adaptive capacity and production potential to meet present and future needs (FAO, 1993; McNeely, 1993).

The Atlantic Tropical Forest in Brazil, specifically, is considered a hot spot, i.e., an area of high species diversity and a high rate of endemism (McNeely, 1990), and therefore an absolute priority for conservation actions. On the other hand, the rate at which these ecosystems have been destroyed from the beginning of Brazil's colonization to date has resulted in elimination of all but 5% of the original vegetation covering (Consorcio Mata Atlântica, 1992). This destruction clearly illustrates the importance of the remaining fragments as models for identifying alternatives for resource conservation and utilization, as well as for restoration of the existing degraded areas.

Genetic diversity has a fundamental role in evolutionary theory, because natural selection chooses among the variants that occur within populations based on their adaptations to the immediate environment. Therefore, information on the level and distribution of genetic variation is crucial for the conception of scientifically sound strategies and implementation of cost-effective and efficient plans of *in situ* and *ex situ* genetic conservation.

Advances in genetic diversity quantification techniques — initially with quantitative genetic techniques and establishment of progeny and provenance trials (Kageyama, 1990), and later with isozymes and DNA polymorphisms — have resulted in several studies on Atlantic Forest tree species. Empirical results, which have been obtained for genetic variation within species of well-characterized groups, show that model species may represent part of the community of tropical tree species, and studies with samples of these different groups may permit an advance in the genetic understanding of natural populations of such species (Gandara et al., 1997).

TREE SPECIES DIVERSITY OF TROPICAL FORESTS

A conspicuous characteristic of tropical forests is the diversity of types of tree species. High population density trees (common species), such as those with 100 or more adult individuals per hectare, can be found with those that are very rare, such as 1 tree per 50 hectares. Between the two extreme densities, a wide range of species may be found, most of which are rare tree species. This is why so many species per unit of area are found in these tropical forest ecosystems (Kageyama and Gandara, 1994). From 100 to 300 different tree species per hectare have frequently been reported; moreover, the maximum described to date exceeded 400 tree species per hectare in the Atlantic Forest.

On the other hand, sampling rare species is more difficult than sampling common ones in small plots. Foster and Hubbell (1990) showed that of 409 known species of trees and shrubs in BCI–Panama (1500 ha), only 306 were found in the 50-ha plot studied. Do special characteristics keep 103 species from occurring in the plot? It is important to emphasize that 61% (63) of the species not found in the plot are listed in the forest flora as very rare. Also, it is noteworthy that 1-ha plots in this area contained an average of 180 tree species, constituting only 44% of the total number of species in the 1500 ha area.

Kageyama et al. (1991), cited by Kageyama and Gandara (2000), quantified the participation of the most rare and common species in seven phytosociological surveys conducted in the Atlantic Forest of the state of São Paulo. These authors pointed out that 30% of the species proved to be rare or had sparse distribution, with only one individual per plot of slightly less than 1 ha; in the same samples, 28% of the individuals belonged to only three of the most common tree species. From these data we can clearly see that, by withdrawing rare species from the survey, we exclude many species but few individuals.

How did such rare tree species evolve to this low population density in the forest? Gandara (1996), studying *Cedrela fissilis* in the Atlantic Forest, and Lepsch-Cunha (1996), studying *Couratari multiflora* in the Amazon, with eight and ten adults per hectare, respectively, found that the high cross-pollination rates ($t = 0.92$ and 0.95, respectively) associated with pollen flight distances (950 and 1000 m, respectively) show that the species under study had essential reproduction characteristics to attain rarity and may have evolved adaptations to deal with their low densities.

Besides population density, other ecological and demographic traits can distinguish different species. Several papers have hypothesized the existence of groups of species with common characteristics and functions, also called *guilds*, which might also have similar populational and genetic structures (Gandara et al., 1997). We putatively suggest division of species into five groups, based on ecological (successional status) and demographic (commonness or rarity) features and responses to anthropic disturbances. These ecological groups are:

Group 1 species are rare in primary forests and become abundant in secondary forests after human disturbance. They are generally emergent or canopy, and secondary tree species.

Group 2 species are rare in primary forests and absent in secondary forests. They require the primary forest environment and are generally shade-tolerant and understory or canopy tree species.

Group 3 species are specialized in forest environments with restrictive edaphic characteristics, such as the tops of hills, seashores, and shallow soils. They may become very common in secondary forests.

Group 4 includes pioneer species limited to large gaps. They are considered rare in primary forests, and occur in high density in the early stage of succession.

Group 5 are shade-tolerant or climax species that are common under the canopy of primary forests and have variable responses in secondary forests.

The task involves monitoring of demographic parameters of model species (abundance, age distribution, mortality, and growth rates) and genetic (expected heterozigosity, outcrossing rate, and spatial autocorrelation). Because these parameters are highly dependent on interactions with other species (through pollination, seed dispersal, predation, competition, and other means), the stability of demographic and genetic parameters would indicate maintenance of the interactions of a considerable part of the community. Data obtained from these few model species can then be extrapolated to the community.

Genetic Structure and Fragmentation

During the last decade many studies have been conducted on population genetics of tropical tree species with increasing sampling efforts and technology. The amount of data generated during this time has shown some important directions to minimize the environmental impacts in these ecosystems. Ecological and genetic studies on tree species in primary and secondary forests show the effects of anthropic disturbances, which helps to define most adequate genetic parameters to monitor actions in these ecosystems.

Gene Flow in Fragments

Gene flow through pollen has rarely been measured in tropical tree species. Estimations of gene flow through pollen have been made directly or by observing the behavior of different pollinators. Measurements show that gene flow encompasses large areas, proving that many pollinators fly long distances. On the other hand, we lack data that quantify the effects of the absence of pollinators in fragmented landscapes.

Estimations of effective gene flow distance in rare and common tropical tree species can be evaluated and compared through the data of Gandara (1996) and Lepsch-Cunha (1996) — evaluating two rare tree species (*Cedrela fissilis* and *Couratari multiflora*, respectively) — and Reis (1996), studying a common species (*Euterpe edulis*). Data show that *Cedrela fissilis* (one adult tree per 8 ha) in Atlantic Forest and *Couratari multiflora* (one adult tree per 10 ha) in Amazon Forest have long-distance pollen flow, quantified through genetic markers (isozymes) in mapped population of 950 and 1000 m, respectively. *Euterpe edulis* (122 adult trees per ha) in Atlantic Forest registered a distance of 56 m.

These data show that rare tree species probably will have bigger reproduction problems after fragmentation as their demes occupy larger areas. Although gene flow through pollen can reach long distances in natural forests, in a fragmented landscape these distances would be much shorter due to difficulties for dispersers' movement. Besides that, forest fragments generally are further apart than the distance found for pollen flow in natural forests (Viana and Pinheiro, 1998; Kageyama et al., 1998).

Genetic Effects of Fragmentation

Forest fragmentation decreases the number of individuals in a population, leading to loss of genetic variation. These small populations may suffer genetic drift in a short term and/or an increase of inbreeding due to the higher probability of selfing and biparental crosses. Few studies exist showing these aspects in tropical trees. A reduction of genetic variation was observed in *Eucalyptus albens* (Young et al., 1996). Hall et al. (1996) showed lower levels of genetic variation in smaller populations of *Pithecelobium elegans.*

The Escola Superior de Agricultura "Luis de Queiroz"/Universidade de São Paulo (ESALQ/ USP), through the Laboratory of Reproduction and Genetics of Tree Species, conducted two studies about fragmentation and anthropic disturbance in populations of two tropical tree species. Results show that the genetic structure of *Cedrela fissilis* Vell. (Meliaceae) exhibited significant changes comparing primary and secondary forests (Gandara et al., 1997). In the same way, the effects of fragmentation on *Chorisia speciosa* St. Hil. (Bombacaceae) show some adequate genetic parameters to quantify genetic drift in fragmented populations (Souza, 1997).

The population density of *Cedrela fissilis* may increase in secondary forests in comparison to primary ones, but that case holds the possibility of spatial genetic structuring, i.e., neighboring trees that are genetically related. This genetic structure may result, in the future, in an increase of inbreeding due to the preferential crossing among neighbors. The spatial genetic structure was quantified through spatial autocorrelation with Moran's I (Table 45.1). Primary forest population (with a population density of one adult tree per 8 ha) has a Moran's I not significantly different from zero, indicating lack of spatial structure. In secondary forest population, Moran's I reached a medium valium of 0.43, revealing the existence of spatial structure.

Souza (1997) detected that the heterozygosity was not adequate to quantify the genetic changes after fragmentation. On the other hand, by analyzing the allelic frequencies, she discovered loss of alleles and fixation of loci as well as oscillation of allelic frequencies. Comparing a big reserve (around 200 ha) and three small fragments (25, 23, and 50 ha), the loss of alleles varied from 5 to 16%; and the F_{st} was 0.183, indicating high genetic divergence among these close populations (Table 45.2).

Table 45.1 Spatial Autocorrelation (Moran's I) of Alleles among Near Neighbors, Mean Distance among Compared Individuals, and Number of Comparisons in a Primary Forest Population — Intervales State Park (Sete Barras — São Paulo State) — and in a Secondary Forest Population — Matão Reserve (Arapoti — Parana State) of *Cedrela fissilis*

Locus	Allele	Moran's I	
		Primary Forest	Secondary Forest
Pgi-2	1	0.117*	0.334**
Idh	1	−0.194*	0.456**
	2	−0.125*	0.451**
Mdh-3	1	0.173*	0.370**
Mdh-4	1	−0.181*	0.546**
Mean distance (m)		94.0	14.6
Number of comparisons		24	40

* Not significantly different from zero.
** Different from zero at 1% probability level.

Table 45.2 Loss of Alleles in Natural Populations of *Chorisia speciosa* in Forest Fragments in São Paulo State, Brazil

| | Fragment | | | | Metapopulation |
	1	2	3	4	
Area (ha)	200	25	23	50	—
No. of observed alleles	18	16	18	18	19
No. of lost alleles	2	3	2	2	—

FINAL CONSIDERATIONS

Clearly, the huge original area of the Atlantic Forest permitted the occurrence of an explosion of different tree species whose populations are, at present, isolated in large or small fragments of this biome. The study and interpretation of genetic variation between and within populations, based on adequate sampling criteria, undoubtedly represent a major advance in the understanding of such ecosystems and consequently help focus their utilization, conservation, and restoration.

Conservation of tropical tree species in a fragmented landscape can be enhanced through genetic enrichment of secondary forests with the replanting of extinct or genetic eroded species; exchange of seeds or saplings among neighbor fragments; increase of fragment size with the restoration of surrounding areas promoting population growth; and implementation of forest corridors, facilitating gene flow among small populations.

Genetic and reproduction studies of Atlantic Forest tree species, based on the small, directed samples studied to date, disclosed three key findings: the outcrossing rate of tree species in general has been higher than 90%, or they are preferentially cross-pollinated; genetic diversity within populations is invariably much higher than that between populations; and populations of rare tree species are more difficult to handle using management practices and require more efforts to sustain.

The results obtained to date indicate that any action in the Atlantic Forest should be taken only when heeding three considerations: seed collection of tree species may, in principle, be regarded as outcrossed, based on the concept of effective size, to define adequate populations; economic common tree species permit better control of initial and post-exploited populations and are more favorable for sustainable management; and rare species present difficulties in population handling and management, which makes their sustainability difficult.

REFERENCES

Alvarez-Buylla, E.R., García-Barrios, R., Lara-Loreno, C., and Martinez-Ramoz, M., Demographic and genetic models in conservation biology: application and perspectives for tropical rain forest tree species, *Annu. Rev. Ecol. Syst.*, 27:387–421, 1977.

Consórcio Mata Atlântica, *Reserva da Biosfera da Mata Atlântica, Plano de Ação*, Vol. 1: *Referências Básicas*, Consórcio Mata Atlântica/UNICAMP, 1992.

FAO, Conservation of genetic resources in tropical forest management, FAO Forestry Paper 107, Rome, 1993.

Foster, R.B. and Hubbell, S.P., The floristic composition of the Barro Colorado Island forest, in Gentry, A., Ed., *Four Neotropical Rainforests*, Yale University Press, New Haven, CT, 1990, pp. 85–98.

Gandara, F.B., Diversidade genética, taxa de cruzamento e estrutura espacial dos genótipos em uma população de *Cedrela fissilis* Vell. (Meliaceae), M.Sc. thesis, IB/UNICAMP, Campinas, 1996.

Gandara, F.B., Grattapaglia, D., Kageyama, P.Y., Batista, J.L.F., Ciampi, A.Y., Walter, B.M.T., Cavalcanti, T.B., Udry, C., and Abdala, G., Towards the development of genetic and ecological parameters for *in situ* conservation of forest genetic resources, in IBAMA, Ed., *Workshop sobre Monitoramento da Biodiversidade em Unidades de Conservação Federais*, IBAMA/GTZ, Brasilia, Brazil, 1997, pp. 95–111.

Hall, P., Walker, S., and Bawa, W.S., Effect of forest fragmentation on genetic diversity and mating system in tropical tree *Pithecellobium elegans, Conserv. Biol.*, 10:757–768, 1996.

Janzen, D.H., Management of habitat fragments in a tropical dry forests: growth, *Ann. Miss. Bot. Garden*, 75:105–116, 1988.

Kageyama, P.Y., Genetic structure of tropical tree species of Brazil, in Bawa, K.S. and Hadley, M., Eds., *Reproductive Ecology of Tropical Forest Plants*, UNESCO, Paris, 1990, pp. 375–387.

Kageyama, P.Y. and Gandara, F.B., Dinâmica de populações de espécies arbóreas: implicações para o manejo e a conservação, *Proceedings, III Simpósio de Ecossistemas da Costa Brasileira*, ACIEP, 1994, pp. 1–9.

Kageyama, P.Y. and Gandara, F.B., Recuperação de áreas ciliares, in Rodrigues, R.R. and Leitão Filho, H.F., Eds., *Matas Ciliares: Conservação e Recuperação*, EDUSP, São Paulo, Brazil, 2000, pp. 249–269.

Kageyama, P.Y., Gandara, F.B., and Souza, L.M.I., Conseqüências genéticas da fragmentação sobre populações de espécies arbóreas, *Série Técnica IPEF*, 12:65–70, 1998.

Lepsch-Cunha, N.M., Variabilidade genética, fenologia e sistema reprodutivo de *Couratari* spp. na Amazônia, M.Sc. thesis, ESALQ/USP, Piracicaba, 1996.

McNeely, G.A., Conserving the World's Biological Diversity, The World Bank, WRI, IUCN, CI, and WWF, Washington, D.C., 1990.

McNeely, G.A., Lessons from the past: forests and biodiversity, *Proc. Global Forest Conf.*, Bandung, Indonesia, February 1993.

Reis, M.S., Distribuição e dinâmica da variabilidade genética em populações naturais de palmiteiro (*Euterpe edulis* Martius), Ph.D. thesis, ESALQ/USP), Piracicaba, 1996.

Souza, L.M.I., Estrutura genética de populações naturais de *Chorisia speciosa* St. Hill. (Bombacaceae) em fragmentos florestais na região de Bauru (SP), Brasil, M.Sc. thesis, ESALQ/USP, Piracicaba, 1997.

United Nations Environment Program (UNEP), *Global Diversity*, Cambridge University Press, Cambridge, U.K., 1995.

Viana, V.M. and Pinheiro, L.A.F.V., Conservação da biodiversidade em fragmentos florestais, *Série Técnica IPEF*, 12(32):25–42, 1998.

Whitmore, T.C. and Sayer, J.A., *Tropical Deforestation and Species Extinction*, Chapman & Hall, London, 1992.

Young, A., Boyle, T., and Brown, T., The population genetics consequences of habitat fragmentation for plants, *Tree*, 11:413–418, 1996.

CHAPTER 46

The Effects of Roads on Carnivores: A Case Study of Mountain Lions (*Puma concolor*) in California

Melissa M. Grigione and Mike B. Johnson

INTRODUCTION

Vehicular traffic in environmentally sensitive areas poses a serious conservation threat to individuals, most wild animal populations, even entire species, and their respective ecosystems and landscapes. This chapter reviews the general ways in which access to their habitat may affect carnivores and their prey, using specific studies to quantify impacts associated with roads. We will then present results from our research on the effects of roads on mountain lions (*Puma concolor*) in three regions of California.

The specific effects of roads on wildlife are manifold and still not entirely understood. However, some of the most prominent threats include habitat fragmentation, direct mortality, direct habitat loss, displacement and avoidance, and associated development (Forman, 2000; Trombulak and Frissell, 2000; Ruediger, 1998). Habitat fragmentation isolates populations and limits successful dispersal opportunities for individuals, subjecting them to demographic, genetic, and stochastic factors that reduce their persistence. Direct mortality involves animals killed by automobiles as well as the amount of poaching that occurs when people can access wild areas readily. Studies that document direct mortality are abundant (Gunter et al., 1998; Tewes and Blanton, 1998; Calvo et al., 1996; Gilbert, 1996; Lehnert et al., 1996; Ruediger, 1996). In addition to direct mortality, roads can displace species or prevent species from accessing important habitats (Forman and Deblinger, 1998; Van Riper and Ockenfels, 1998; Paquet and Callaghan, 1996). In certain circumstances animals are attracted to roads, to their own detriment (Gibeau and Herrero, 1998).

CARNIVORE VULNERABILITY

Carnivores are particularly vulnerable to roads because of their inherently low population densities, their long lives, their late onset of first reproduction (for example, 8 years or older for grizzly bears), and their low reproductive rates (Ruediger, 1996). Carnivores often roam in very large home areas, and dispersal from these areas can include excursions of several hundred miles. Carnivores require large, interconnected areas for long-term persistence and viability.

CASE STUDY: THE EFFECTS OF ROADS ON MOUNTAIN LIONS IN CALIFORNIA

Our research involves the study of changes in home range patterns of mountain lions as road densities increase from relatively undisturbed areas to more urban environments. The overall goal of our study is to compare the spatial ecology of mountain lions in three study areas of California (Figure 46.1).

Study Areas

Mountain lions in the California coastal ranges, consisting of the Diablo Range and the Santa Ana Mountains, inhabit low-elevation and chaparral–oak woodlands. Differentiation between seasons is subtle, and snowfall is uncommon in these areas. The main prey for mountain lions in these ranges is black-tailed deer (*Odocoileus hemionus*) that are nonmigratory. However, mountain lions in California's central Sierra Nevada inhabit a high-elevation, mixed-conifer ecosystem with two distinct seasons. Snowfall is common, and the main prey for Sierra Nevada mountain lions is black-tailed deer that are migratory. Road densities are highest in the Santa Ana Mountains (0.00196 m/m²), lowest in the Sierra Nevada (0.00084 m/m²), and the Diablo Range densities (0.00119 m/m²) are somewhere in between.

Study Objectives

The specific objectives of this study are to answer two questions. How does mountain lion home range size and home range overlap vary for males and females, during winter and summer

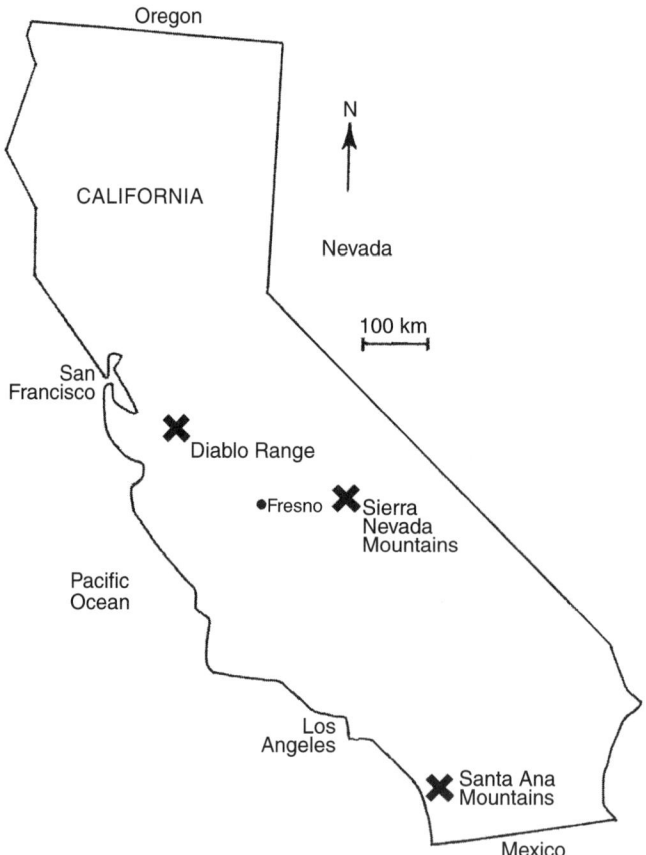

Figure 46.1 Map of three study sites in California.

seasons, within each study area, and among all study areas? How do densities of roads, rivers, and the number of vegetation types differ between exclusive and nonexclusive portions of mountain lion home ranges? For the purposes of this chapter, we focused solely on the effects of roads on mountain lions.

A home range is an area that an animal uses to complete most of its daily tasks. It is in this area that an animal obtains food and shelter in addition to reproducing and raising young. Exclusive home ranges are areas that are normally defended in order to keep other mountain lions from obtaining any of the resources found within the home range. In this study, we are assuming that exclusive home ranges (i.e., home ranges that are not shared with other mountain lions) have limited and valuable resources within them that are worth defending. When high levels of home range overlap occur, we assume that these overlap areas either have a bountiful resource providing enough for all mountain lions, or it may be of so little value that an animal expends little to no energy keeping conspecifics out.

We used theories associated with food distribution for carnivores (which involve the distribution and predictability of a food resource, i.e., black-tailed deer for our study) and carnivore social structure (which for our study involves a polygynous mating system, with no typical breeding season or reproductive peak) to develop hypotheses and generate predictions about home range size, home range overlap, and the effect of roads, rivers, and the number of vegetation types on mountain lion space use. This provided us with a means for determining the relative importance of social structure vs. food resources for mountain lion space use.

We divided home range hypotheses into three groups: hypotheses of home range size within and among study areas; home range overlap within and among study areas; and the effects of roads, rivers, and number of vegetation types in exclusive vs. nonexclusive portions of home range for each mountain lion. We reviewed several studies that investigated the effects of anthropomorphic elements, such as logging activities and roads, on mountain lions. Based on these studies, we predicted that exclusive home ranges should have a lower density of roads than nonexclusive areas (Belden and Hagedorn, 1993; Spreadbury et al., 1996; Smallwood, 1994; Van Dyke et al., 1986).

Examples of the Effects of Roads on Mountain Lions

In British Columbia, over a 2-year period, Spreadbury et al. (1996) studied a population of mountain lions that lived near a coal mine. Four of the 34 known mountain lions were killed by automobiles. The authors believed that mountain lions were changing their peak activity hours to coincide with the absence of people at the coal mine.

In Arizona and Utah, Van Dyke et al. (1986) found that mountain lions did not use timber sale areas in proportion to their occurrence. Rarely were lions found in or near active or inactive timber sale areas until 6 or more years after the logging had ceased. Mountain lions in the immediate vicinity of disturbance shifted their activity peaks from daylight hours to after sunset, concentrated other activities during evening hours, and were inactive rather than active at sunrise. The authors believe that habitat alteration itself was the most important factor behind mountain lion avoidance of timber areas.

In a California study, Smallwood (1994) surveyed roads for mountain lion tracks in 1992 and found a 61% decrease in track sets in areas from which timber was removed since 1986 (the year of his previous survey). Deer densities on these transects remained the same.

In a study in the Santa Ana Mountains, California, Beier (1993) documented that 32% (ten animals) of all known animal mortalities in his study population were due to car collisions. Of these ten animals, four were young males trying to disperse.

Literature is also available for the Florida panther (*Puma concolor coryi*), an endangered subspecies of the mountain lion. In urban southern Florida, 47% of mountain lions were killed on highways (Maehr et al., 1991). In northern Florida, 2 of 19 mountain lions were killed on highways, and 2 of 19 were illegally shot (Belden and McCown, 1995). In the same area, mountain lions established home ranges in areas containing half the density of roads found in the entire study area (Belden and Hagedorn, 1993). For these reasons, we hypothesized that roads were in fact undesirable for mountain lions and not worthy of home range defense. Hence, we predicted that exclusive home ranges in all of our study areas would have a lower density of roads than nonexclusive portions of the home range.

Study Methods

We used Universal Transverse Mercator grid coordinates (UTMs) collected by the primary investigators to construct 95% Minimum Convex Polygons (MCPs). We exported these home ranges into arcinfo that intersected river-, road-, and vegetation-type data. The road data used in our Geographic Information Systems (GIS) analysis were created at 1:100,000 scale. We used one-way analysis of variance (ANOVAS) with the GIS data to understand differences in road density between exclusive and nonexclusive portions of the home range.

Results

In the Sierra Nevada (i.e., the area with the lowest road densities), we found significant differences with respect to road density and home ranges for 2 of 11 seasons (Table 46.1). During the winter of 1985, for male mountain lions, road densities in areas shared with female mountain lions were significantly greater than those in their exclusive home areas, as well as in areas shared with other males. In addition, exclusive home areas for males had significantly greater road densities than those areas shared with other males ($p = 0.001$; adjusted alpha = 0.01). During the summer of 1985, the density of roads was significantly greater in areas females shared with males than in those areas shared with other females; however, these areas were not significantly different from road densities in exclusive portions of the home range ($p = 0.009$; adjusted alpha = 0.01).

We found significant differences with respect to road density and home range in the coastal ranges as well (i.e., the areas having medium and high road densities). When these differences occur, we see a greater density of roads in exclusive portions of the home range than in nonexclusive portions. In the Santa Ana Mountains, significant differences in road densities occurred between exclusive and nonexclusive portions of the home range for 3 of 18 seasons (Table 46.1). For females during the winter and summer of 1986, exclusive home areas contained a greater density of roads than areas shared with males. Areas that females shared with other females had a greater density of roads than areas shared with males ($p = 0.004, 0.0001$; adjusted alpha = 0.0083). However, during the winter of 1991, a greater density of roads existed in areas females shared with males than existed in exclusive areas or in areas shared with females ($p = 0.0001$; adjusted alpha = 0.0083). In the Diablo Range, significant differences occurred in road densities between exclusive and nonexclusive portions of the home range for 5 of 19 seasons (Table 46.1). During four of these seasons, a greater density of roads was shown in exclusive portions of the home range than in nonexclusive areas ($p = 0.001, 0.001, 0.003, 0.006$; adjusted alpha = 0.0083). In contrast, during the summer of 1985, a greater density of roads existed in areas females shared with males than in exclusive female areas or in areas shared with other females ($p = 0.007$; adjusted alpha = 0.0083).

During the study period, no car-related deaths occurred in the Sierra Nevada or the Diablo Range study areas. However, in the Santa Ana Mountains, 10 of 31 mountain lions died car-related deaths (Beier, 1993).

Discussion

Home range overlap is a potentially confounding variable because, while overlapping ranges are easy to detect, it is more difficult to prove that home ranges are exclusive and not shared. Also, two individual animals may have a large area of overlap used little by each or a small area of

Table 46.1 P-Values for One-Way ANOVAS of Road Density — Home Range Data*

Study Area	Year						Adjusted Alpha 0.01
Sierra Nevada	1983	1984	1985	1986	1987		
Male-Winter			0.001* olm-f>ex, olm** ex>olm				
Male-Summer			0.275				
Female-Winter	0.193	0.884	0.42	0.738	0.298		
Female-Summer		0.301	0.009* olf-m>olf	0.439	0.183		
Santa Ana	1987	1988	1989	1990	1991	1992	0.0083
Male-Winter				0.234	0.25	0.009	
Male-Summer				0.719	0.041	0.101	
Female-Winter	0.137	0.004* ex>olf-m; olf>olf-m	0.431	0.248	0.0001* olf-m>olf, ex	0.213	
Female-Summer	0.029	0.0001* ex>olf-m; olf>olf-m	0.02	0.828	0.115	0.074	
Diablo Range	1984	1985	1986	1987	1988	1989	0.0083
Male-Winter	0.065	0.001* ex>olm, olm-f	0.003* ex>olm; olm-f>olm	0.243	0.931		
Male-Summer	0.261	0.443	0.089	0.188	0.056		
Female-Winter	0.173	0.007* ex>olf; olf-m>olf	0.929	0.394	0.086		
Female-Summer	0.016	0.006* olf-m>ex, olf	0.025	0.164	0.474	0.001* ex>olf-m, olf	

* Ex = Exclusive home range areas; OL = Home range overlap areas; F = Female; M = Male.

**Reads: For males during winter 1985, there is a greater road density in areas shared with females than in exclusive areas or areas shared with males; and exclusive home areas have a greater density of roads than areas shared with males.

overlap used intensively by each. An animal's home range, however, is where it spends most of its life and the area in which most of its interactions with other organisms and its environment occur.

The importance of road densities varied between our two systems, with Coastal Range studies having higher road densities in exclusive portions of the home range than in nonexclusive portions. In the Santa Ana Mountains, a large proportion of the radio-collared population was killed by cars, in comparison to the Diablo Range and Sierra Nevada, where none of the study animals was killed by cars. It is beyond the scope of this analysis, however, to assess whether mountain lions, in any of the study locations, were actively avoiding roads.

The Diablo Range, with its intermediate road densities, is where we see the greatest difference in road densities between exclusive and nonexclusive portions of the home range, with exclusive areas having greater road densities. This is contrary to our original predictions. Could it be that mountain lions in this area are temporally avoiding roads by adjusting their activity patterns (Van Dyke, 1986)? It appears that mountain lions in this area and in the Santa Ana Mountains have activity peaks before sunrise and after sunset (Beier and Barret, 1993; Padley, 1990; Hopkins, 1989). This activity cycle resembles that of mountain lions in undisturbed areas (Hansen, 1992). In the Diablo Range these activity peaks match those of black-tailed deer. Are the deer (and, hence, mountain lions) drawn to roads because of favorable changes in the adjacent vegetative community? Although our information about deer distribution is limited, this could be a possibility since areas adjacent to roads in the Diablo Range are relatively intact as compared to many Southern California locations.

The GIS road density data for this analysis were created at the 1:100,000 scale. This scale simplifies landscape features and as a result does not represent the true sinuosity of roads, hence decreasing the *true* density of roads within an animal's home range. In addition, our analysis did not differentiate between the road types. By treating all roads equally, the analysis could be missing key avoidance areas. Although comparing differences between exclusive and nonexclusive portions of home ranges is revealing, depending on the research question, the scale of GIS-related data (e.g., roads, rivers) may have to be reduced in order to detect important relationships at the level of the mountain lion home range. Despite the large scale at which our data were created, we were able to detect differences in mountain lion habitat use between our two regions (Sierra Nevada and Coastal Ranges).

In the Sierra Nevada, where deer are migratory, few differences occurred between exclusive and nonexclusive portions of mountain lion home ranges for any of our study variables (e.g., road densities, river densities, vegetation type). For mountain lions in this ecosystem, tracking deer both temporally and spatially is crucial to their survival. Tracking deer may not involve intimate familiarity with one home area as much as an ability to move through home areas efficiently and effectively (Grigione et al., in preparation).

In the Coastal Ranges, where ecological variables differed more between exclusive and nonexclusive portions of mountain lion home ranges, deer are nonmigratory (Grigione et al., in preparation; Beier et al., 1995; Neal, 1990; Padley, 1990; Hopkins, 1989). In this system, mountain lions access deer by positioning themselves in a home area that will provide the most deer over time. This "positioning" allows mountain lions to use their immediate home areas in more subtle ways (Grigione et al., in preparation). This could in turn accentuate the importance of ecological variables between exclusive and nonexclusive portions of the home range and possibly negate the effect of roads on mountain lion home ranges.

In conclusion, it is important to examine several variables when exploring questions at the ecological scale. In the Sierra Nevada, we cannot say that roads are unimportant to mountain lions — only that other factors, such as tracking deer, may be more important to their survival. For all of our study areas, all direct mortalities occurred in the Santa Ana Mountains, the area with the highest road densities. However, mountain lions in this system do not appear to be selecting areas of low road density — perhaps they cannot in such a road-dense place. We see a similar pattern in the Sierra Nevada — mountain lions do not appear to be selecting areas with lower road density, albeit for different reasons (e.g., roads are in low density to begin with and deer are migratory).

It is in the Diablo Range, where road densities are intermediate in comparison to our other areas, that we are seeing the most significant differences in road densities between exclusive and nonexclusive portions of the home range. While these differences do not support our original hypothesis, their occurrence is frequent enough (approximately 25%) to warrant additional analysis. We are going to investigate this relationship further by performing a "nearest neighbor analysis" to determine if single point locations (UTMs) of individual home ranges are significantly different in distance to roads than randomly generated points within the same study area.

ACKNOWLEDGMENTS

This project would not have been possible without the collaboration and generous support associated with the mountain lion research of Paul Beier, Rick Hopkins, Don Neal, and Doug Padley, and the GIS data and analyses of Joshua Viers and Jim Quinn of the University of California, Davis, Division of Environmental Science and Policy.

REFERENCES

Beier, P. and Barrett, R.H., The Cougar in the Santa Ana Mountain Range, California, Final Report, Orange County Cooperative Mountain Lion Study, University of California, Berkeley, 1993.

Beier, P., Choate, D., and Barrett, R.H., Movement patterns of mountain lions during different behaviors, *J. Mammal.*, 76:1056–1070, 1995.

Belden, R.C. and Hagedorn, B.W., Feasibility of translocating panthers into northern Florida, *J. Wildl. Manage.*, 57:388–397, 1993.

Belden, R.C. and McCown, J.W., Florida Panther Reintroduction Feasibility Study, Florida Game and Freshwater Fish Commission, Final Report Study 7507, Tallahassee, FL, 1995.

Calvo, R. and Silvy, N., Key deer mortality, U.S. 1 in the Florida Keys, in Evink, G.L., Garrett, P., Zeigler, D., and Berry, J., Eds., *Trends in Addressing Transportation-Related Wildlife Mortality, Proc. Transp. Related Wildl. Mortality Semin.*, FL-ER-58-96, Florida Department of Transportation, Tallahassee, FL, 1996.

Forman, R.T.T., Estimate of the area affected ecologically by the road system in the United States, *Conserv. Biol.*, 14(1):31–35, 2000.

Forman, R.T.T. and Debliger, R.D., The ecological road-effect zone for transportation planning and Massachusetts highway example, in Evink, G.L., Garrett, P., Zeigler, D., and Berry, J., Eds., *Proc. Intl. Conf. Wildl. Ecol. Transp.*, FL-ER-69-98, Florida Department of Transportation, Tallahassee, FL, 1998, pp. 78–96.

Gibeau, M., Effects of transportation corridors on large carnivores in the Bow River Valley, Alberta, in Evink, G.L., Garrett, P., Zeigler, D., and Berry, J., Eds., *Trends in Addressing Transportation-Related Wildlife Mortality, Proc. Transp. Related Wildl. Mortality Semin.*, FL-ER-58-96, Florida Department of Transportation, Tallahassee, FL, 1996.

Gibeau, M. and Herrero, S., Roads, rails and grizzly bears in the Bow River Valley, Alberta, in Evink, G.L., Garrett, P., Zeigler, D., and Berry, J., Eds., *Proc. Intl. Conf. Wildl. Ecol. Transp.*, FL-ER-69-98, Florida Department of Transportation, Tallahassee, FL, 1998, pp. 104–108.

Gilbert, T., An overview of black bear roadkill in Florida 1976–1995, in Evink, G.L., Garrett, P., Zeigler, D., and Berry, J., Eds., *Trends in Addressing Transportation-Related Wildlife Mortality: Proc. Transp. Related Wildl. Mortality Semin.*, FL-ER-58-96, Florida Department of Transportation, Tallahassee, FL, 1996.

Grigione, M.M., Viers, J.H., Quinn, J.F., Beier, P., Padley, W.D., Hopkins, R.A., Neal, D., and Johnson, M.L., *The Effect of Biological and Anthropogenic Features on Mountain Lion* Puma concolor *Home Range Size and Home Range Overlap in California*, in preparation.

Gunter, K.A., Biel, M.K.J., and Robison, H.L., Factors influencing the frequency of road-killed wildlife in Yellowstone National Park, in Evink, G.L., Garrett, P., Zeigler, D., and Berry, J., Eds., *Proc. Intl. Conf. Wildl. Ecol. Trans.*, FL-ER-69-98, Florida Department of Transportation, Tallahassee, FL, 1998, pp. 32–42.

Hansen, K., *Cougar: The American Lion,* Northland Publishing, Flagstaff, Arizona, 1992.

Hopkins, R.A., Ecology of the Puma in the Diablo Range, California, Ph.D. dissertation, University of California, Berkeley, 1989.

Lehnert, M. and Bissonette, J., Mule deer highway mortality in northeastern Utah: causes, patterns and a new mitigative technique, in Evink, G.L., Garrett, P., Zeigler, D., and Berry, J., Eds., *Trends in Addressing Transportation Related Wildlife Mortality, Proc. Transp. Related Wildl. Mortality Semin.*, FL-ER-58-96, Florida Department of Transportation, Tallahassee, FL, 1996.

Maehr, D.S., Lang, E.D., and Roelke, M.E., Mortality patterns in panthers in southwest Florida, *Proc. Annu. Conf. Southeastern Assoc. Fish Wildl. Agencies*, 45:201–207, 1991.

Neal, D.L., The effect of predation on deer in the Central Sierra Nevada, in Giusti, G.A., Timm, R.M., and Schmidt, R.H., Eds., Predator Management in North Coastal California, Proceedings of a Workshop Held in Ukiah and Hopland, California, University of California, Hopland Field Station Publication 101, University of California, 1990, pp. 53–61.

Padley, W.D., Home Ranges and Social Interactions of Mountain Lions (*Felis concolor*) in the Santa Ana Mountains, California, Master's thesis, California State Polytechnic University, Pomona, CA, 1990.

Paquet, P. and Callahan, C., Effects of linear developments on winter movements of gray wolves in the Bow River Valley of Banff National Park, Alberta, in Evink, G.L., Garrett, P., Zeigler, D., and Berry, J., Eds., *Trends in Addressing Transportation-Related Wildlife Mortality, Proc. Transp. Related Wildl. Mortality Semin.*, FL-ER-58-96, Florida Department of Transportation, Tallahassee, FL, 1996.

Ruediger, B., The relationship between rare carnivores and highways, in Evink, G.L., Garrett, P., Zeigler, D., and Berry, J., Eds., *Trends in Addressing Transportation-Related Wildlife Mortality, Proc. Transp. Related Wildl. Mortality Semin.*, FL-ER-58-96, Florida Department of Transportation, Tallahassee, FL, 1996.

Ruediger, B., Rare carnivores and highways: moving into the 21st century, in Evink, G.L., Garrett, P., Zeigler, D., and Berry, J., Eds., *Proc. Intl. Conf. Wildl. Ecol. Transp.*, FL-ER-69-98, Florida Department of Transportation, Tallahassee, FL, 1998, pp. 10–16.

Smallwood, K.S., Trends in mountain lion populations, *Southwest. Nat.*, 39:67–72, 1994.

Spreadbury, B.R., Musil, K., Musil, J., Kaisner, C., and Kovak, J., Cougar population characteristics in southeastern British Columbia, *J. Wildl. Manage.*, 60:962–969, 1996.

Tewes, M.E. and Blanton, D.R., Potential impacts of international bridges on ocelots and jaguarundis along the Rio Grande wildlife corridor, in Evink, G.L., Garrett, P., Zeigler, D., and Berry, J., Eds., *Proc. Intl. Conf. Wildl. Ecol. Transp.*, FL-ER-69-98, Florida Department of Transportation, Tallahassee, FL, 1998, pp. 135–139.

Trombulak, S.C. and Frissell, C.A., Review of ecological effects of roads on terrestrial and aquatic communities, *Conserv. Biol.*, 14(1):18–30, 2000.

Van Dyke, F.G., Brocke, R.H., Shaw, H.G., Ackerman, B.B., Hemker, T.P., and Lindzey, F.G., Reactions of mountain lions to logging and human activity, *J. Wildl. Manage.*, 50:95–102, 1986.

Van Riper, C. III, and Ockenfels, R., The influence of transportation corridors on the movement of pronghorn antelope over a fragmented landscape in Northern Arizona, in Evink, G.L., Garrett, P., Zeigler, D., and Berry, J., Eds., *Proc. Intl. Conf. Wildl. Ecol. Transp.*, FL-ER-69-98, Florida Department of Transportation, Tallahassee, FL, 1998, pp. 241–248.

CHAPTER 47

A Rapid Method in Ecosystem Mapping and Monitoring as a Tool for Managing Costa Rican Ecosystem Health

Maarten Kappelle, Marco Castro, Heiner Acevedo, Pedro Cordero, Luis González, Edgar Méndez, and Huberth Monge

BRIEF HISTORY OF ECOLOGICAL MAPPING IN COSTA RICA

Since the late 1940s Costa Rica has suffered from a severe change in its dense tropical forest vegetation cover and agroecosystem distribution (Sader and Joyce, 1988; Sánchez-Azofeifa, 1997; Van Omme et al., 1998). Over the past decades the original area covered by tropical premontane rain forest, or lower montane moist forest, has been reduced significantly due to forest conversion for agricultural land. Since the late 1960s and particularly throughout the 1980s, degraded or unproductive agricultural lands such as lowland and montane pastures have been abandoned and have given way to successional processes leading to a large area at present covered by secondary forests (Holl and Kappelle, 1999). Simultaneously, the area covered by traditional crops, such as banana, in some regions was reduced due to economic recessions by the mid 1980s. Other areas have lost coffee plantations as a result of the increased extension of urban areas such as Costa Rica's central valley. In order to understand the processes related to these intense land cover changes and the consequences for biodiversity conservation, we recognize a need to map the current vegetation types present in the country.

Citing Dirzo (in press) "the most synthetic and revealing way of describing Central American ecosystems is through the description of the vegetation. A premise for this is that the vegetation constitutes the most obvious descriptor of the ecosystem, constitutes the basis of the food chain, and provides the structural matrix which most communities and populations of animals live or indirectly depend upon." This statement justifies the use of vegetation mapping as an indirect way of ecosystem mapping. Therefore, we have focused on vegetation mapping as a tool in understanding ecosystem makeup and distribution and managing ecosystem health.

In Costa Rica, vegetation mapping has received major attention since the establishment of the Holdridge World Life Zone Classification System (Holdridge, 1967), which describes and predicts the distribution of tropical ecosystems. The Life Zone System formed the basis for Tosi's (1969) ecological map of Costa Rica, which shows potential distribution of the 12 Life Zones (climate-based ecological zones defined by specific temperature, humidity, and evapotranspiration regimes). However, growing empirical evidence has called into question the predictive power of the Holdridge model (Cornell, 1996). Indeed, throughout the world, criticism has been raised about the Holdridge Life Zone Classification System, which does not take into account parameters related to, for instance, vegetation physiognomy (e.g., evergreen, broad-leaved forest vs. deciduous, thorny scrub). Although environmental aspects such as rainfall distribution (length of the dry season) or edaphic characteristics (e.g., alluvial, inundated vegetation on sandy soils vs. peaty paramo bogs) are included at a second level of organization (associations) within the Holdridge System (Bolaños and Watson, 1993), very few sites in the tropics are classified at this level. The third level of organization, which concerns successional phases and land use types (Bolaños and Watson, 1993), has received even less attention from scientists. Indeed, the Holdridge Life Zone System in Costa Rica is supported by fewer than 50 sample points throughout the country (Holdridge et al., 1971). These limitations contribute to the fact that the Life Zone approach has rarely been applied outside Costa Rica and Puerto Rico — countries in which Holdridge and his collaborators have conducted the majority of their studies (Cornell, 1996).

As a reply to the Holdridge classification system, Gómez (1986) proposed his Vegetation Macrotypes System, which is based on climatic and physiognomic, floristic, and edaphic characteristics rather than primarily on climatic aspects on which the Holdridge system is based. This approach coincides with vegetation classification systems as proposed by the Anglo-American School with scholars such as Beard (1944), Grubb et al. (1963), Whitmore (1990), and, more recently, Ross et al. (1992). Recently, the Gómez Vegetation Macrotypes System has proved its usefulness in a gap analysis directed at defining threatened biodiverse areas (local hot spots) located in critical zones outside Costa Rican protected wildlife areas in need of major conservation efforts (The GRUAS Project; García, 1997). Even more recently, Herrera and Gómez (1993) presented their Biotic Units Map integrating both Holdridge's climatic classification concepts (Holdridge, 1967). It was adapted on the basis of more detailed and recent meteorological data and analysis (Herrera, 1986) and the vegetational and edaphic concepts published in Gómez (1986). However, these three ecological maps show Costa Rica's potential vegetation at scales of 1:750,000 to 1:200,000. The maps indicate the presence of vegetation types void of any human influence in Costa Rica, therefore limiting their usefulness for decision making in actual biodiversity conservation and day-to-day protected area management on the ground.

At a subnational scale, several efforts have been made to map vegetation in Costa Rican protected areas (e.g., Kappelle, 1991; Vargas, 1992). Most of these studies use plant sociological approaches as proposed by the French-Swiss School of Phytosociology (Braun-Blanquet, 1965). These detailed scale maps have been useful in answering specific plant ecological research questions related to floristic composition and plant association distribution but have not been consulted for planning in conservation and development due to limited dissemination or popularization outside the academic scene.

A few years ago, Savitsky and Tarbox (1995) mapped (scale 1:500,000) the current distribution of Costa Rica's vegetation, superimposing deforestation patterns on the distribution of Holdridge's Life Zones. Besides Holdridge's Life Zones (e.g., the deforested area covering more than 60% of the country's territory), areas included eight land cover types: paramos, mangrove forests, pastures, croplands, urban areas, bodies of water, barren land, and unknown cover types that are not in accordance with the Holdridge Life Zone System as represented by the polygons corresponding to remaining primary forest fragments.

Next to these ecological maps, a series of maps shows Costa Rica's dense forest cover at present and over the past few decades (Hartshorn, 1982; Fundación Neotrópica–Conservation International, 1988). The authors of these maps, however, do not clearly indicate what is meant by use of the term *dense forest*. Does *dense forest* refer only to undisturbed primary forest? To what extent are these forests really undisturbed, if we know that the presence of indigenous groups in Costa Rica dates back to more than 10,000 B.C.? And are those forests that have been selectively logged over the years included? How *dense* are all these so-called *dense forests*?

POLITICAL AND SCIENTIFIC NEEDS FOR ECOSYSTEM AND LAND COVER MAPS

Urgent political and scientific needs exist today for development of a series of ecological vegetation and land cover maps (Castro, 1998) to (1) show Costa Rica's present (actual) — not potential — ecosystem and vegetation distribution; (2) display natural as well as seminatural and (agri-)cultural vegetation types; (3) show at a semidetailed scale of 1:50,000, which corresponds to the scale of the country's intensively used topographic sheets; (4) help in defining sites for conducting intensive biodiversity inventories including different taxonomic groups (one of the main activities of Costa Rica's National Biodiversity Institute, INBio; Gámez, 1996); (5) serve as a reference for future monitoring of land use and land cover changes over time; (6) include both digital and printed formats; and, most important, (7) serve as a decision-support tool for local individuals in charge of managing ecosystem health.

These needs are emphasized at the decision-making units of the regional offices of Costa Rica's Sistema Nacional de Areas de Conservación (SINAC). This system is part of the Ministry of Environment and Energy in charge of the administration of the country's approximately 150 protected wildlife areas, which cover some 24% of the nation's territory and are distributed throughout 11 environmental administrative regions. Over the last decade, SINAC and its precursor, the National Park Service, have been considered a model for international conservation (Gámez and Ugalde, 1988).

In order to respond to these needs, INBio and SINAC recently set up the Netherlands-funded ECOMAPAS Project (http://www.inbio.ac.cr/en/mi/ecomapas.html), whose goal is to map land cover (principally vegetation) within five conservation areas (Osa, Amistad–Pacific Region, Amistad–Caribbean Region, Arenal, and Tempisque) over a 4-year period (1998–2001). In accordance with INBio and SINAC's mission of saving, knowing, and sustainably using tropical biodiversity (Gámez 1991, 1996), and as a follow-up of the 1998-approved national Biodiversity Law, the ECOMAPAS Project contributes in an advanced and innovative manner to the development of knowledge and wise use of Costa Rica's ecosystems.

A NEW METHOD IN RAPID ECOLOGICAL VEGETATION MAPPING

We have developed a new methodology in rapid ecological vegetation mapping that emphasizes analysis of spatiotemporal dynamics in land cover and vegetation distribution as a decision-support tool in biodiversity conservation planning and ecosystem health management. This innovative methodology, now widely implemented by the ECOMAPAS Project, is based upon principles formulated by Eiten (1968), the UNESCO (1973), Van Gils and Van Wijngaarden (1984), and approaches implemented by the U.S. Geological Service and U.S. National Park Service. This methodology departs from the characteristics that can be distinguished from semidetailed aerial photographs (1:40,000) through techniques of stereoscopic interpretation (Zonneveld, 1995). These photo characteristics include form, size, structure, texture, color, and shade of polygons that can be identified during aerial photo interpretation (API) procedures (Van Gils and van Wijngaarden, 1984).

Based on these characteristics, a classification system was developed using the proportion of tree, shrub, and herb cover in a polygon as a vegetation structure parameter (Figure 47.1). The following nine physiognomic classes could be distinguished as follows: (1) dense shrubland (Spanish: *matorral denso*), (2) shrubby herbland (*herbazal arbustivo*), (3) sparse shrubland (*matorral ralo*), (4) dense, woody shrubland (*matorral denso arbolado*), (5) dense herbland (*herbazal denso*), (6) sparse herbland (*herbazal ralo*), (7) woody herbland (*herbazal arbolado*), (8) sparse forest (*bosque ralo*), and (9) dense forest (*bosque denso*). The spatial distribution of trees, shrubs, and herbs for each of these physiognomic classes is illustrated in Figure 47.2. In addition to these nine classes, five additional cover classes were added in order to map entire areas that also harbor

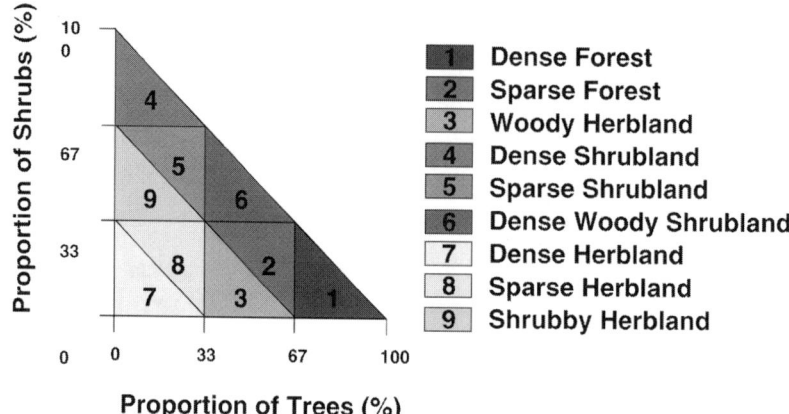

Figure 47.1 Structural physiognomic vegetation typification for aerial photo-interpretation of areas in which different horizontal layers (strata) can be observed. Nine physiognomic vegetation classes have been distinguished, as occurring along two axes representing proportions of trees and shrubs as observed in mapped polygons.

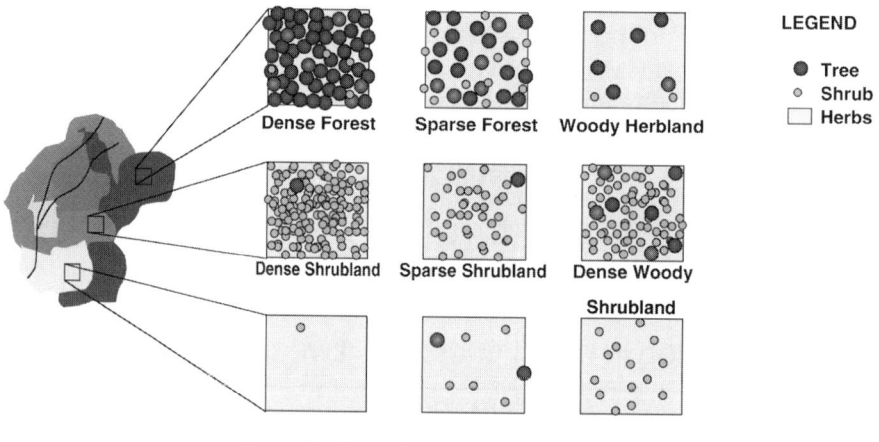

Figure 47.2 Example of an area mapped on the basis of aerial photo-interpretation, showing different physiognomic vegetation classes. The structure and texture of the nine physiognomic classes as defined in Figure 47.1 and seen from the birds-eye perspective are presented. The minimum mappable area is about 3 ha.

nonvegetated and unidentifiable land cover types: (1) bare soil and rock outcrops, (2) water bodies, (3) infrastructures, (4) shades, and (5) clouds.

IMPLEMENTING THE RAPID ECOLOGICAL VEGETATION MAPPING METHOD

In an attempt to put this methodology into practice, the vegetation of the Osa Conservation Area in Southern Pacific Costa Rica was mapped at a scale of 1:50,000 using 1:40,000 color aerial photographs taken in 1995 and 1996 by the German Hansa–Luftbild Company within the scope of the INBio-GEF Project. In late 1998 and early 1999, approximately 325 aerial photos were stereoscopically interpreted, using the aforementioned land cover classification and vegetation typification system. Preliminary map data were scanned and digitized into a geographical information system (GIS) using the Arc/View and Arc/Info software programs. Subsequently, ground-truthing was conducted by comparing preliminarily mapped (1995–1996) and current (1999) land cover (both vegetated and nonvegetated patches).

More than 150 sample points were established at which cartographic, geographic, and ecological data were collected. The sample points were selected on the basis of representativeness, in a stratified manner, covering all physiognomic land cover classes. Sample points were well distributed over the full geographic extension of the Osa Conservation Area. More sample points were established in those physiognomic units that covered a greater extension in order to maintain equal sample point densities. Sample points were selected in areas with high to medium levels of accessibility (e.g., near roads, trails, and rivers). In several cases, week-long expeditions (hiking, horseback riding) were organized. Replicates were included in order to generate reliable information on each land cover class.

Measured parameters included Global Positioning System–based altitude, latitude, and longitude, information on physiognomic vegetation class (the nine aforementioned types), vegetation height (in m), leaf phenology (evergreen vs. [semi]-deciduous, or annual vs. perennial in the case of herbaceous vegetation), dominant leaf type (broad-leaved vs. coniferous vs. graminoid), hydrological regime (palustrine, lacustrine, riverine, marine, according to Cowardin et al., 1979), flooding and tidal influences (e.g., in the case of a mangrove forest), current land use (forestry, cropping, grazing), state (natural and seminatural vs. cultural), and grade of human intervention. Additionally, at each sample point the observed presence of vascular plant species was recorded as an indicator of the present floristic composition. Data on species abundance were not collected due to time and budget constraints. However, the establishment of permanent plots in each vegetation type is planned for a second phase dedicated to the monitoring of land cover and vegetation changes — a step particularly important in developing ecosystem health management plans.

Subsequently, different thematic layers will be added to the land cover and vegetation map, including air temperature and humidity regimes, length of the dry season, soil orders, and geomorphological types (landforms). Climatic layers will be based upon cartographic information provided by Herrera and Gómez (1993), who modified the Holdridge System on the basis of much more detailed meteorological information available today (Herrera, 1986). In this way, the up-to-date biotic land cover and vegetation layer generated by ECOMAPAS will be integrated with nonbiotic environmental layers into a flexible decision-support GIS.

As a next step, preliminary map data were digitally adjusted on the basis of ground-truthed field information. The quality of the GIS-based data was controlled to guarantee information reliability and fidelity. Finally, the process of preparing digital maps was initiated and directed to different kinds of users, such as the protected area managers based at SINAC's central and regional offices.

BRIEF DESCRIPTION OF FLORISTIC COMPOSITION OF PHYSIOGNOMIC CLASSES

As a result of the botanical fieldwork at the sample points, the nine physiognomic land cover classes can be typified at the floristic level. More than 700 vascular plant species were recorded. A large number of plant specimens were collected for identification and storage in the herbarium at INBio. The most important plant species, in terms of dominance and exclusiveness, are listed in Table 47.1 for each of the nine physiognomic land cover classes.

EFFECTIVENESS AND APPLICATION

The currently developed methodology has proven to be most useful and effective in the area of project resource limitations and operational capacity restrictions (budget, time, staff, field area accessibility). During the course of about 6 to 8 months, the land cover of an entire conservation

Table 47.1 Most Important Plant Species, in Terms of Dominance and Exclusiveness, Listed for Nine Physiognomic Land Cover Classes

Trees	Shrubs	Herbs
1. Dense Shrubland		
Psidium guajava L.	*Trema micrantha* (L.) Blume	*Lantana camara* L
Luehea seemannii Triana & Planch.	*Palicourea guianensis* Aubl.	*Hyptis capitata* Jacq.
Ochroma pyramidale (Cav. ex) Lam. Urb.	*Vismia baccifera* (L.) Triana & Planch.	*Desmodium incanum* DC.
Spondias mombin L.	*Palicourea guianensis* Aubl.	
Cecropia obtusifolia Bertol.	*Byrsonima crassifolia* (L.) Kunth	
Attalea butyracea (Mutis ex) L. f. Wess. Boer.		
Trichospermum galeottii (Turcz.) Kosterm.		
Guazuma ulmifolia Lam.		
Miconia argentea (Sw.) DC.		
Jacaranda caucana Pittier.		
Cecropia peltata L.		
Diphysa americana (Mill.) M. Sousa		
Apeiba tibourbou Aubl.		
2. Shrubby Herbland		
Cecropia peltata L.	*Vismia baccifera* (L.) Triana & Planch.	*Cyperus* sp.
Senna papillosa (Britt. & Rose) H.S.Irwin & Barn.		*Ludwigia* sp.
Senna reticulata (Willd.) H. S. Irwin & Barneby.		*Phyllanthus urinaria* L.
Guazuma ulmifolia Lam.		*Sida rhombifolia* L.
Ochroma pyramidale (Cav. ex) Lam. Urb.		*Paspalum* sp.
Cecropia obtusifolia Bertol.		*Desmodium incanum* DC.
Luehea seemannii Triana & Planch.		*Lantana camara* L
		Triumfetta lappula L.
		Mimosa pudica L.
		Blechum pyramidatum (Lam.) Urb.
3. Sparse Shrubland		
Attalea butyracea (Mutis ex) L. f. Wess. Boer.	*Senna reticulata* (Willd.) H. S. Irwin & Barn.	*Lantana camara* L.
Psidium guajava L.	*Hamelia magnifolia* Wernham.	*Phyllanthus urinaria* L.
Luehea seemannii Triana & Planch.	*Palicourea guianensis* Aubl.	*Desmodium incanum* DC.
Miconia argentea (Sw.) DC.	*Vismia baccifera* (L.) Triana & Planch.	*Sida rhombifolia* L.
Croton schiedeanus Schltdl.	*Trema micrantha* (L.) Blume	*Heliconia latispatha* Benth.
Castilla tunu Hemsl.	*Myriocarpa longipes* Liebm.	
Lacistema aggregatum (Bergius) Rusby.		
Virola koschnyi Warb.		
Cecropia obtusifolia Bertol.		
Spondias mombin L.		
Apeiba tibourbou Aubl.		
4. Dense Woody Shrubland		
Cecropia peltata L.	*Myriocarpa longipes* Liebm.	*Sida rhombifolia* L.
Luehea seemannii Triana & Planch.	*Vismia baccifera* (L.) Triana & Planch.	*Ludwigia* sp.
Anacardium excelsum (Bert. & Balb.ex K.) Skeels.		*Hyptis capitata* Jacq.
Trichospermum galeottii (Turcz.) Kosterm.		*Calathea lutea* (Aubl.) Schult.
		Desmodium incanum DC.
		Cyclanthus bipartitus Poit.
		Lantana camara L.

Table 47.1 *(continued)*   Most Important Plant Species, in Terms of Dominance and Exclusiveness, Listed for Nine Physiognomic Land Cover Classes

Trees	Shrubs	Herbs
Psidium guajava L.		
Ochroma pyramidale (Cav. ex) Lam. Urb.		
Spondias mombin L.		
Miconia argentea (Sw.) DC.		
Ficus crassivenosa W. C. Burger.		
Apeiba tibourbou Aubl.		
Jacaranda caucana Pittier.		
Lacistema aggregatum (Bergius) Rusby.		

5. Dense Herbland

Trees	Shrubs	Herbs
Guazuma ulmifolia Lam.		*Paspalum* sp.
Psidium guajava L.		*Cyperus* sp.
Diphysa americana (Mill.) M. Sousa		*Ludwigia* sp.
		Blechum pyramidatum (Lam.) Urb.
		Hyptis capitata Jacq.
		Desmodium incanum DC.
		Lantana camara L.
		Phyllanthus urinaria L.
		Sida rhombifolia L.
		Mimosa pudica L.
		Hyptis vilis Kunth & Bouche.

6. Sparse Herbland

Trees	Shrubs	Herbs
Psidium guajava L.		*Lantana camara* L.
Gliricidia sepium (Jacq.) Kunth ex Walp. ex		*Paspalum* sp.
Cecropia obtusifolia Bertol.		*Cyperus* sp.
Diphysa americana (Mill.) M. Sousa		*Cyathula achyranthoides* (Kunth) Moq.
Ochroma pyramidale (Cav. ex) Lam. Urb.		*Hyptis capitata* Jacq.
Attalea butyracea (Mutis ex) L. f. Wess. Boer.		*Caperonia palustris* (L.) A. St.-Hil.
Apeiba tibourbou Aubl.		*Priva lappulacea* (L.) Pers.
		Iresine diffusa HBK. ex Willd.
		Mimosa pudica L.
		Blechum pyramidatum (Lam.) Urb.
		Phyllanthus urinaria L.
		Ludwigia sp.
		Desmodium incanum DC.

7. Woody Herbland

Trees	Shrubs	Herbs
Cecropia peltata L.		*Caperonia* sp.
Simarouba amara Aubl.		*Ludwigia* sp.
Ficus crassivenosa W. C. Burger.		*Calathea lutea* (Aubl.) Schult.
Attalea butyracea (Mutis ex) L. f. Wess. Boer.		*Phyllanthus urinaria* L.
Luehea seemannii Triana & Planch.		*Caperonia palustris* (L.) A. St.-Hil.
Ochroma pyramidale (Cav. ex) Lam. Urb.		*Desmodium incanum* DC.
		Hyptis capitata Jacq.
		Triumfetta lappula L.
		Cyathula achyranthoides (Kunth) Moq.
		Sida rhombifolia L.
		Lantana camara L.
		Paspalum sp.

(Continued)

Table 47.1 (continued)   Most Important Plant Species, in Terms of Dominance and Exclusiveness, Listed for Nine Physiognomic Land Cover Classes

Trees	Shrubs	Herbs
8. Sparse Forest		
Brosimum utile (Kunth) Oken.	*Vismia baccifera* (L.) Triana & Planch.	*Carludovica drudei* Mast.
Miconia argentea (Sw.) DC.		*Piper peltatum* L.
Lacistema aggregatum (Bergius) Rusby.		
Alchornea costaricensis Pax & K. Hoffm.		
Luehea seemannii Triana & Planch.		
Apeiba tibourbou Aubl.		
Spondias mombin L.		
Jacaranda caucana Pittier.		
Cecropia obtusifolia Bertol.		
Trichospermum galeottii (Turcz.) Kosterm.		
Croton schiedeanus Schltdl.		
Inga sapindoides Willd.		
9. Dense Forest		
Virola koschnyi Warb.	*Myriocarpa longipes* Liebm.	*Heliconia latispatha* Benth.
Castilla tunu Hemsl.	*Palicourea guianensis* Aubl.	*Piper reticulatum* L.
Virola sebifera Aubl.	*Vismia baccifera* (L.) Triana & Planch.	*Carludovica drudei* Mast.
Ficus crassivenosa W. C. Burger.		*Calathea lutea* (Aubl.) Schult.
Luehea seemannii Triana & Planch.		
Apeiba tibourbou Aubl.		
Brosimum alicastrum Sw.		
Carapa guianensis Aubl.		
Brosimum utile (Kunth) Oken.		
Clarisia biflora Ruiz & Pav.		
Cecropia obtusifolia Bertol.		
Hyeronima alchorneoides Allemao.		
Alchornea latifolia Sw.		
Trichospermum galeottii (Turcz.) Kosterm.		
Croton schiedeanus Schltdl.		
Jacaranda caucana Pittier.		
Symphonia globulifera L. f.		
Caryocar costaricense Donn. Sm.		
Calophyllum brasiliense Cambess.		

area (Osa, in southern Pacific Costa Rica, covering 422,008 ha) was mapped at the semidetailed scale of 1:50,000, including an extensive ground-truthing effort. Further repeated field research at permanent plots in selected vegetation/ecosystem patches belonging to different physiognomic classes will help understand land cover and vegetation changes over time. Their monitoring effort should include the inventory of other taxonomic groups than the mere group of vascular plant species. Currently, INBio and SINAC are developing a strategic plan to monitor biodiversity at the ecosystem level over time. Such a monitoring plan will be based on the presence and abundance of key indicator species (e.g., selected plants) that are diagnostic for different levels of ecosystem health. Specific indicators such as ecosystem area, species richness, and species population viability may also be included in the monitoring system (Carrizosa, S., personal communication).

One of the specific uses for this type of map is ecosystem health management, including restoration. Patches representing seminatural vegetation types such as dense, woody shrublands or sparse shrublands are experiencing processes of vegetation recovery following clearing of tropical rain forest, subsequent cattle-ranching, and final abandonment. Restoration efforts to recover biodiversity and manage ecosystem health (to be stimulated by the Costa Rican Government through SINAC), depend heavily on technically processed information on the monitored presence and distribution of different stages of secondary vegetation as expressed in different kinds of shrublands (Castro et al., 1999). This is of particular interest in those areas in which the establishment of biological corridors is at stake.

The ECOMAPAS Project will contribute significantly to the knowledge base developed on biodiversity at the ecosystem level for governmental and nongovernmental decision-making organizations such as SINAC and INBio, which direct their efforts toward the integration of biodiversity conservation and the sustainable use of natural resources *in situ* for the benefit of Costa Rica's society and the world community at large (Lovejoy and Gámez, 1995).

ACKNOWLEDGMENTS

We thank the staff at SINAC, particularly Luis Rojas, Vera Violeta Montero, Luis Barquero, Sara Araya, Jairo Mora, Celso Alvarado, and Jose Miguel Valverde, for technical cooperation and facilitation of field operations. We are very grateful to Rodrigo Gámez, Alfio Piva, Erick Mata, Jesús Ugalde, Randall García, Nelson Zamora, and Gabriela Córdoba of INBio. We thank Reynaldo Aguilar and José González for their botanical assistance. This study was financed by the Netherlands Ministry for Development Cooperation, represented by the Netherlands Embassy in San José.

REFERENCES

Beard, J.S., Climax vegetation in Tropical America, *Ecology,* 25:127–158, 1944.

Bolaños, R.A. and Watson, V., *Mapa Ecológico de Costa Rica*, Tropical Science Center and Instituto Costarricense de Electricidad, San José, 1993.

Braun-Blanquet, J.J., *Plant Sociology: The Study of Plant Communities*, Hafner, London, 1965.

Castro, M.V., Memoria del Taller Técnico para la Definición de Necesidades Cartográficas–Ecológicas dentro del Proyecto "Desarrollo del Conocimiento del Biodiversidad y Uso Sostenible en Costa Rica," Instituto Nacional de Biodiversidad (INBio)–Sistema Nacional de Áreas de Conservación (SINAC), Santo Domingo de Heredia, Costa Rica, 1998.

Castro, M.V., Kappelle, M., and Montero, V.V., Mapping and monitoring degraded and recovering forests for restoration in Costa Rican conservation areas, in *Abstracts of the Symposium "Tropical Forest Regeneration in Abandoned Agricultural Lands: Implications for Restoration Ecology,"* as part of the International Conference "Tropical Restoration for the new Millennium," San Juan, Puerto Rico, May 1999, p. 41.

Cornell, J.D., The status of the Holdridge lifezone model on its 50th anniversary in Abstracts of the Annual Meeting of the Association for Tropical Biology, Baltimore, Maryland, *Biotropica* (Suppl.) 30(2), 1996.

Cowardin, L.M., Carter, V.V., Golet, F.C., and LaRoe, E.T., *Classification of Wetlands and Deepwater Habitat of the United States,* Publication FWS/OBS-79/31, Biological Service Program, U.S. Fish and Wildlife Service, Washington, D.C., 1979.

Dirzo, R., Ecosystems of Central America, in Panissidi, N., Ed., *Encyclopedia of Biodiversity*, Academic Press, London, in press.

Eiten, G., Vegetation forms, *Bol. Inst. Bot.,* 4:18–29, 1968.

Fundación Neotrópica–Conservation International, *Assessment of the Conservation of Biological Resources*, Fundación Neotrópica–Conservation International, San José, 1988.

Gámez, R., Biodiversity conservation through the facilitation of its sustainable use, *Trends Ecol. Evol.,* 6, 377–378, 1991.

Gámez, R., Inventories: preparing biodiversity for non-damaging use, in di Castri, F. and Younés, T., Eds., *Biodiversity, Science and Development: Towards a New Partnership*, CAB International, Oxon, U.K., 1996, pp. 180–183.

Gámez, R. and Ugalde, A., Costa Rica's National Park System and the preservation of biological diversity: linking conservation with socioeconomic development, in Almeda, F. and Pringle, C.M., Eds., *Tropical Rainforests: Diversity and Conservation*, California Academy of Sciences, San Francisco, 1988, pp. 131–142.

García, R., *Biología de la Conservación y Áreas Silvestres Protegidas: Situación Actual y Perspectivas en Costa Rica*, INBio, Santo Domingo de Heredia, 1997.

Gómez, L.D., Vegetación de Costa Rica, Vol.1., in Gómez, L.D., Ed., *Vegetación y Clima de Costa Rica*, EUNED, San José, 1986.

Grubb, P.J., Lloyd, J.R., Pennington, T.D., and Whitmore, T.C., A comparison of montane and lowland rain forest in Ecuador: the forest structure, physiognomy, and floristics, *J. Ecol.*, 51:567–601, 1963.

Hartshorn, G.S., Ed., *Costa Rica: Perfil Ambiental*, Tropical Science Center, USAID, San José, 1982.

Herrera, W., Clima de Costa Rica, Vol. 2., in Gómez, L.D., Ed., *Vegetación y Clima de Costa Rica*, EUNED, San José, 1986.

Herrera, W. and Gómez, L.D., *Mapa de Unidades Bióticas de Costa Rica*, U.S. Fish and Wildlife Service–TNC–INCAFO–CBCCR–INBio–Fundación Gómez-Dueñas, San José, 1993.

Holdridge, L.R., *Life Zone Ecology*, Tropical Science Center, San José, 1967.

Holdridge, L.R., Grenke, W.C., Hatheway, W.H., Liang, T., and Tosi, J.A., *Forest Environments in Tropical Life Zones: A Pilot Study*, Pergamon Press, Oxford, U.K., 1971.

Holl, K.D. and Kappelle, M., Tropical forest recovery and restoration, *Trends Ecol. Evol.*, 14: 378–379, 1999.

Kappelle, M., Distribución altitudinal de la vegetación del Parque Nacional Chirripó, Costa Rica, *Brenesia*, 36:1–14, 1991.

Lovejoy, A. and Gámez, R., INBio as a pilot project: a new approach to the management of biodiversity, in *Proc. Intl. Conf. Biosphere Reserves*, Seville, Spain, 1995.

Ross, M.S., O'Brien, J.J., and Flyn, L.J., Ecological site classification of Florida Keys terrestrial habitats, *Biotropica*, 24(4):488–502, 1992.

Sader, S.A. and Joyce, A.T., Deforestation rates and trends in Costa Rica, 1940 to 1983, *Biotropica*, 20(1):11–19, 1988.

Sánchez–Azofeifa, G.A., Assessing Land Use/Cover Change in Costa Rica, Ph.D. thesis, University of New Hampshire, Durham, 1997.

Savitsky, B. and Tarbox, D., *Habitats de Costa Rica*, Universidad Nacional de Costa Rica, Heredia, 1995.

Tosi, J.A., *Mapa Ecológico de Costa Rica, Basado en la Clasificación Vegetal Mundial de L.R. Holdridge*, Tropical Science Center, San José, 1969.

UNESCO, *International Classification and Mapping of Vegetation*, UNESCO, Paris, 1973.

Van Gils, H.A.M.J. and van Wijngaarden, W., Vegetation structure in reconnaissance and semi-detailed vegetation surveys, *ITC J.*, 3:213–218, 1984.

Van Omme, E., Kappelle, M., and Juárez, M.E., Land Use/Land Cover Changes and Deforestation Trends over 55 years (1941–1996) in a Costa Rican Cloud Forest Watershed Area, on CD-ROM with the *Proc. Intl. Conf. Geo-Inf. Sustainable Land Manage.*, August 1997, ITC, Enschede, the Netherlands, 1998.

Vargas, G., *Cartografía Fitogeográfica de la Reserva Biológica Carara*, Universidad de Costa Rica, San Pedro, 1992.

Whitmore, T.C., *An Introduction to Tropical Rain Forests*, Clarendon, Oxford, U.K., 1990.

Zonneveld, I.S., *Land Ecology*, SPB Academic Publishing, Amsterdam, 1995.

CHAPTER 48

Identification and Monitoring in the Context of the National Biodiversity Strategy in Uruguay

Victor Canton

INTRODUCTION

The Convention on Biological Diversity (CBD) was one of the most important results from the United Nations Conference on Environment and Development (UNCED), also called the Earth Summit, that was held in Rio de Janeiro, Brazil, in 1992. The CBD is a legally binding international instrument with three main objectives: the conservation of biological diversity, the sustainable use of its components, and the fair and equitable sharing of benefits arising from the use of the genetic resources.

Today almost 170 countries worldwide have ratified the CBD, and most of them are working to the full implementation of the CBD commitments — actually, a very hard task, if we take into account the complexity of the issue and the different kinds of problems that must be resolved to achieve the three CBD objectives.

The CBD refers to different levels and scales about the biodiversity problems: species, ecosystems, and genes. During the last negotiations of the conference of the parties of the CBD, the Subsidiary Body of Scientific, Technical and Technological Advice (SBSTTA) recommended the use of the ecosystem approach to effectively deal with the issues related to the management of biological resources. Article 6 of the CBD asks to develop national strategies and action plans for conservation and sustainable use of biodiversity in areas within the limits of its national jurisdiction, and integrate this concept into relevant sectoral or cross-sectoral plans, programs, and policies.

Uruguay, as a party of the CBD, started in early 1998 a multisectorial and participative process to develop its national biodiversity strategy and action plan with the financial support of the GEF (Global Environment Facility), the participation of UNDP (United Nations Development Program) as implementing agency, and the National Environmental Division (DINAMA) of the Ministry of Housing, Land Planning and Environment (MVOTMA) as the coordinating agency.

STATUS AND TRENDS OF URUGUAYAN BIODIVERSITY

Uruguay, located in the southern cone of South America between Brazil and Argentina, has a temperate climate and is a transition zone from the subtropical ecosystems of southern Brazil and the flat pampas of Argentina. The country is included within the broader biological entity called the biogeographical Pampas Province, characterized by the dominance of grasslands.

Uruguay is located downstream of the Parana River Watershed and the Uruguay River Watershed, both included in the major Cuenca del Plata Watershed; thus, terrestrial, aquatic, and marine biodiversity are influenced by this factor.

With grasslands as a predominant natural ecosystem, Uruguay has important areas of wetlands, gallery forest, ridge forest, palm groves, and coastal and lagoon ecosystems. Uruguay shares important areas of estuarine and marine ecosystems with Argentina (Rio de la Plata and the Atlantic continental shelf) that have a rich marine biological diversity.

The existing animal species correspond to the available habitats, comprising some 1226 species of vertebrates that have been identified in Uruguay. These species include 600 fish, 111 mammals, 62 reptiles, 41 amphibians, and 412 birds. The modifications suffered by the natural ecosystems with the development of livestock and later agriculture (early main economic activities) generated a loss of habitats for native biodiversity. In the case of animals, various species have disappeared, particularly the large carnivores (puma and jaguar), the peccary, and the swamp deer.

Today, other activities are arising as a potential menace to biodiversity (mining, tourism, exotic forestation planting), but DINAMA, in combination with the public and private sectors, is working to develop strategies and actions to achieve conservation and sustenance of Uruguayan biodiversity.

THE STRATEGY PROCESS: FIRST STEPS

One of the first steps in the Uruguayan biodiversity strategy process was gathering existing information about Uruguayan biodiversity and, as far as possible, identifying the activities that threaten the species and ecosystems.

It was necessary to determine where information was available, and who produced any information; in other words, the situation was to locate information that could be useful to develop a strategy and identify gaps of knowledge. A small team of biologists, a marine biologist, an agronomist, a veterinarian, and a geographer spent several months on research. Article 7 of the CBD requires identification of components that are critical to conservation of national biodiversity. Parties are to identify activities or processes that have adverse impacts on biodiversity and organize that data to support future activities to implement the national strategy and action plan. Influenced by DINAMA, we started the task with an orientation to the policy-making and management contexts in order to work with the best knowledge available and to take actions as soon as possible. One of the constraints in this process in developing countries is the lack of baseline information (several times dispersed) and a short time frame to take advantage of protective measures from the government.

CONTINUING THE PROCESS: PUBLIC INVOLVEMENT
AND MULTISECTORIAL APPROACH

DINAMA started to organize workshops with broad participation from representatives of the public sector (ministries, agencies, local governments, etc.) and the private sectors (universities, educational institutions, and NGOs). About 125 people from the 58 different institutions attended the workshops and contributed their points of view in strategy discussions.

The main goal in each workshop was to reach a consensus about the main guidelines to implement the CBD objectives in Uruguay. This part of the process took about 1 year. The issues of the workshops concentrated on how to implement the articles of the CBD at the national level. The workshops addressed Article 7 (identification and monitoring), Article 8 (*in situ* conservation), Article 9 (*ex situ* conservation), Article 10 (sustainable use of the biodiversity components), Article

11 (incentive measures), Article 12 (research and training), Article 13 (public education and awareness), Article 14 (impact assessment and minimizing adverse impacts), and Article 15 (access to genetic resources). In every workshop the concern about the topic of identification and monitoring arose among the participants and was included as a priority issue.

This participative way to resolve the biodiversity issues is strongly recommended from the well-respected organizations involved (UNEP, UICN, WRI) and is believed to be the best way to move forward and take advantage of concrete actions; gain a minimum consensus between the various sectors necessary to make progress; and include the general populace, which must be part of the process of construction of the national strategy on biodiversity.

Biodiversity is a part of the environment; it is not possible to separate biodiversity as an isolated topic. The strategy to protect Uruguay's biodiversity must be part of the more comprehensive strategy to protect the whole environment (ecosystems health).

In December 1998, Uruguay sent to the CBD Secretariat its Advanced Draft of the National Biodiversity Strategy. The document was published by MVOTMA in late 1999.

RECOMMENDATIONS TO IDENTIFICATION AND MONITORING

The experience of the Uruguayan biodiversity effort leads to some comments and recommendations about identification and monitoring on the conservation process.

1. The monitoring system must be in accordance with the policy-making schedule to take action (e.g., ecological rapid assessments). In developing countries there is a lack of experience in long-term planning processes. In most cases, the problem is that politicians need concrete deadlines to take biodiversity conservation measures.
2. The spatial perspective (Maps/GIS/Remote Sensing) is an important tool for the monitoring system; species and ecosystems have a place in the territory, and it is necessary to be familiar with them to appropriately manage; otherwise, developing a set of regulations to protect biodiversity can lose effectiveness if the decision makers do not know where the problems are located in the territory.
3. Indicators of biological diversity must be developed at the national and local levels. These indicators must draw from the experience of the developed countries but should be adapted to the biogeographical conditions of each country.
4. It is very important to identify those activities that can threaten the biological diversity, such as mining, deforestation, large-scale forestation with exotic species, unsustainable agriculture, spread of invasive species, major infrastructure constructions, and so on.
5. It is equally important to identify other components such as sectorial policies, international agreements, market changes, and other factors that can influence biodiversity.
6. Environmental Impact Assessments and Environmental Strategy Assessments could be excellent tools to prevent or minimize the loss of biodiversity and also should be considered sources of information about status and trends.
7. More research is needed about native species, ecosystems, and genetic resources as well as investments in capacity building and transfer of technology.
8. More education and awareness about the significance of biological diversity is necessary, not only in the educational system but also for civil society and policy makers. At times the public confuses the concept of biodiversity with the megabiodiversity that is often found in tropical countries. However, flora and fauna of each country result from the climate, the soils, and other conditions such as altitude, latitude, water resources, etc. This point is a constraint in a temperate country such as Uruguay; it is important to explain the need for protecting the native biological diversity to the civil society.
9. It is also necessary to give an economic value to biodiversity; biological resources are a raw material for food, medicine, etc., as a direct value and provide us with indirect values such as ecological services, leisure, aesthetics, etc. It is a priority to determine how to transmit this concept to policy makers and economists to get their support to take actions for conservation.

10. Public participation in identification and monitoring is very important. The people who are living close to the biodiversity problems can understand better and support actions for conservation and sustainable use. They can also constitute an early warning system against activities that lead to the loss of biodiversity.

11. Conservation efforts are hampered by *lack of financial resources*, *lack of coordination* among the involved institutions (public, private, academia, etc.), and *gaps in trained people* to perform biological inventories. Additional *requirements of cooperation* must be undertaken soon to attain success in the conservation and sustainable use of Uruguayan biodiversity. Greater cooperation among participating organizations is necessary before real progress is made.

REFERENCES

Global Biodiversity Assessment — Summary for Policy Makers UNEP, Cambridge University Press, Cambridge, U.K., 1995.

National Environmental Study of Uruguay/OAS/BID/OPP, 1992.

Planeación Nacional de la Biodiversidad WRI/PNUMA/UICN, 1995.

Propuesta de Estrategia Nacional para la Conservación y Uso Sostenible de la Diversidad Biológica en Uruguay/MVOTMA/PNUD, text of the United Nations Convention on Biological Diversity, 1999.

Uruguay National Report to UNCED/MVOTMA, 1992.

Climate Change and Ecosystem Health

CHAPTER 49

Overview: Climate Change and Ecosystem Health

Ruth A. Reck

Given that an ecosystem is defined by a dynamic arrangement of competing and evolving populations controlled by common nutrient cycles and energy pathways, consideration of how climate change can alter the stability of these cycles and pathways may be useful. Change in climatic variables — either in terms of the mean, the seasonal cycle, or day-to-day variability — can affect all constituents of ecosystems, flora, and fauna, including mammals and birds as well as herbivorous insects, parasitic plants, and fungi. A change in any constituent's essential role can change the details of the ecosystem's structure and function. Depending upon the scale of a disturbance, climate change can alter the sustainability of the ecosystem and can upset the hierarchy and the composition of biotic species.

RELATIONSHIP OF ECOSYSTEM HEALTH TO CLIMATE CHANGE

Many persons have some idea about what is important or unimportant if ecosystem health is affected by climate change, but in general the concept appears to be difficult to define. Ecological variables may be chosen to relate to climate change, but for these to be useful they must be measurable, affordable, and reliable. And for any database to be useful for detecting changes in ecosystem health, it must be capable of describing both structural and functional characteristics. Some researchers have suggested that an ecosystem's condition can be characterized by key elements: rate of plant growth and primary productivity; the physical conditions necessary for support of life; available nutrients, particularly nitrogen and phosphorus; abundance of chemical contaminants and how they are altered by climate change; the biological community condition; condition of native species and whether the group might be imperiled by climate change; and the effects of biological invasions, outbreaks, and diseases (see *The State of the Nation's Ecosystems*, 2000). A significant change in each variable must be identified as either having values within normal behavior or as having values outside the normal range, which then could be termed pathological. Developing these concepts is a tall order when the climate is expected to change over decades to centuries. Alas, following the impacts of climate change on ecosystems will be a very challenging but important area for study in the future.

The certain common physical factors that can help to define the role of climate impact include changes in water availability leading to drought or flooding, changes in the effects of insects and pathogens, changes in nutrient cycles, and shifts in climate zones. Both succession and competition can interact, but in different ways, depending upon whether changes are the least or greatest benefit to a given member of the population. Alteration in the availability of solar radiation from changes in the amount or type of cloudiness, or in the abundance and optical characteristics of aerosols (Ramanathan et al., 1987), can immediately alter plant growth and primary productivity. Shifts in seasonality — either the length of the season or changes in weather variables such as temperature and wind convection, or in nighttime minimum temperature and related humidity — can greatly alter the physiology of one species compared with another and can lead to changes in the biological community condition.

EFFECTS OF ECOSYSTEMS ON CLIMATE

Scientists accept the fact that ecosystems affect the weather on regional and continental scales. Large tracts of forest, for example, tend to help regulate rainfall. Researchers have estimated that 50 to 80% of the moisture in the Amazonia region is held by the forest system. The water is constantly recycled between the growing plants and the atmosphere, released to the atmosphere to again condense and form storm systems and precipitation.

Almost three quarters of the Earth's surface is water covered, and most of the remainder (26.34%) contains the terrestrial biosphere. The sun is the only source of energy to fuel the Earth–atmosphere system. Incident solar radiation interacts with the Earth's surface and, depending upon the optical properties of the surface — i.e., the color of the surface, meaning its albedo (or reflectivity) — greater or lesser amounts of sunlight are reflected back outward to space. Hence, the color of the surface features in a given region greatly influences the climate every bit as much as the abundance of greenhouse gases.

Figure 49.1 shows the effects of seasonality in a section of Wisconsin. The figure shows that large shifts in surface albedo result from snowfall, which changes the local albedo from around 12 to 15% to as high as 70%. Drought can change the color of the surface characteristics in much the same way as snow, from dark green (which can absorb as much as 80% or more of the incoming radiation) to sand-colored (which can be much more reflecting, changing the absorption from 80% to 20 to 30%. So the ecological well-being not only influences the water cycle but also has a most profound effect on the amount of solar energy available to run the Earth–atmosphere weather system. Changes in the ecosystem's optical characteristics and moisture possibly could alter where and how storm patterns can form, and in a most intimate way can influence the weather it experiences. Table 49.1 shows the surface albedos for different types of ecosystem surface cover, and Figure 49.2 shows how it is distributed across the globe on an annual basis. From the values in Table 49.1 and the distribution in Figure 49.2, the reader can perform mental experiments to identify some of the most critical areas of

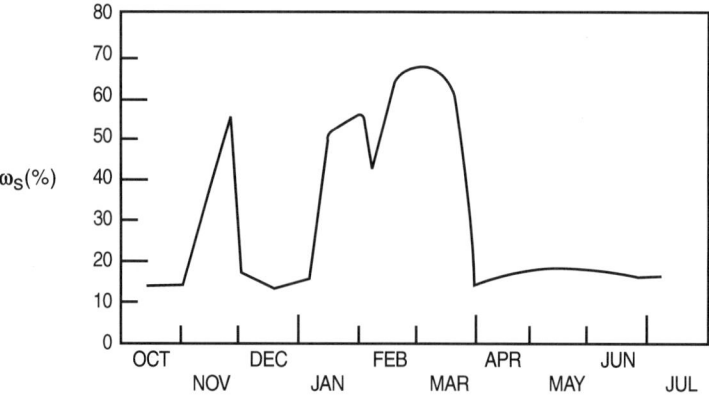

Figure 49.1 Measured values of the surface albedo for each month of a typical year in a small area of Wisconsin. (From Bauer, K.G., and Dutton, J.A., Albedo variation from an airplane over several types of surfaces, *J. Geophys. Res.*, 67, 2367–2376, 1962. With permission.)

Table 49.1 Ecological Surface Area Types across the Globe, Percentage of Surface Area Covered, and the Seasonal Values of the Surface Albedo (in the Northern Hemisphere)

Surface Category	Surface Area (km²)	Percent of Total Surface Area	Albedo (%)			
			January-March	April-June	July-September	October-December
Arable land with intensive farming						
Snow in the winter	5.237×10^6	1.026	50.0	16.0	15.0	27.0
No winter snow	1.598×10^7	3.129	14.0	16.0	15.0	14.0
Grazing and marginal farming lands						
Snow in the winter	5.488×10^6	1.075	50.0	18.0	16.0	20.0
Heavy snow in winter	3.804×10^5	0.074	50.0	38.6	16.0	20.0
No winter snow	2.520×10^7	4.935	20.0	16.0	18.0	20.0
Rice lands or regions where paddies dominate	2.633×10^6	0.516	12.0	12.0	12.0	12.0
Other irrigated land in dry areas where paddies do not dominate						
Snow in winter and fall	1.687×10^5	0.033	50.0	39.0	17.0	32.0
Snow in winter	2.008×10^5	0.039	50.0	17.0	17.0	26.0
No winter snow	1.246×10^6	0.244	20.0	20.0	20.0	20.0
Coniferous forests						
Heavy snow areas	1.019×10^7	1.995	47.0	47.0	14.0	47.0
Snow in winter	2.625×10^6	0.514	36.0	16.0	14.0	11.0
No winter snow	4.197×10^5	0.082	16.0	16.0	14.0	11.0
Deciduous forest						
Snow in winter	3.508×10^5	0.069	33.0	14.0	14.0	19.0
No winter snow	4.654×10^6	0.911	19.0	14.0	14.0	19.0
Mixed coniferous and deciduous forests						
Snow in winter	2.025×10^6	0.397	34.5	15.0	15.0	19.0
No winter snow	2.117×10^6	0.415	15.0	15.0	15.0	15.0
Tropical woodlands and grasslands	6.641×10^6	1.300	16.0	16.0	16.0	16.0
Equatorial (rain) forests	1.513×10^7	2.962	7.0	7.0	7.0	7.0
Deserts						
Shrubland with winter snows	7.591×10^4	0.015	36.0	22.0	22.0	22.0
Shrubland with no winter snow	1.447×10^7	2.833	22.0	22.0	22.0	22.0
Sand	4.769×10^6	0.934	42.0	42.0	42.0	42.0
Marshes						
Snow in winter	8.494×10^5	0.166	55.0	34.5	14.0	55.0
No winter snow	1.951×10^6	0.382	10.0	10.0	10.0	10.0
Tundra	1.172×10^7	2.295	82.0	82.0	17.0	82.0
Total	1.345×10^8	26.34				

From Hummel, J.R. and Reck, R.A., A global surface albedo model, *J. Appl. Meteorol.*, 18:239–253, 1979. With permission.

the globe that could influence the surface albedo in the future. To compare with the well-recognized surface heating effect from carbon dioxide, a 5% change in the rain forest area from a surface albedo of 7% to a value of 42% (an extreme case) — an albedo value characteristic of sand — would be equivalent to decreasing the atmospheric concentration of carbon dioxide by ~70%. The previous estimate assumes the sensitivity of a 1% change in global surface albedo and is equivalent to a 50%

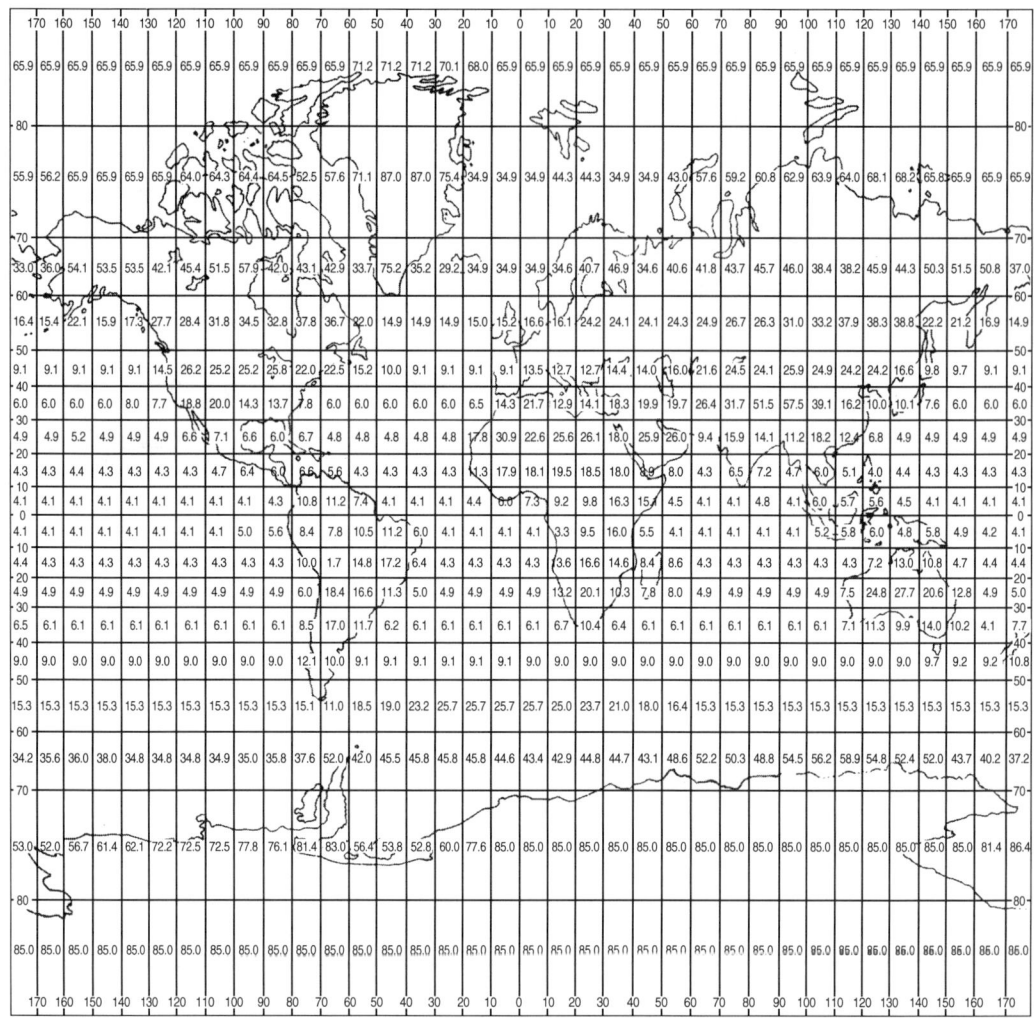

Figure 49.2 Annual values of the surface albedo at each square degree area across the globe (the plot is a modified Gall projection). (From Hummel, J.R and Reck, R.A., A global surface albedo model, *J. Appl. Meteorol.*, 18:239–253, 1979. With permission.)

change in atmospheric CO_2 — surface sensitivity values obtained many years ago with a one-dimensional radiative-convective model (Reck, 1985).

INDIRECT CLIMATE REGULATION BY ECOSYSTEMS

It is well accepted by this time that ecosystems are important regulators of the greenhouse gas, carbon dioxide. Plants, forests, and agriculture take up huge amounts of carbon as carbon dioxide. Ecosystems serve as "sinks" for a portion of the carbon dioxide emitted from burning fossil fuels. Likewise, deforestation and slash-and-burn farming release the carbon previously sequestered. Today's estimates indicate that about one half of the forest carbon is located in boreal forests, and one third in tropical forests (Matthews and Hammond, 1999). If the deforestation numbers are approximately correct, tropical forests possibly could be releasing as much as 1.6 billion tons of carbon per year, equivalent to 20% of the annual rate of human emissions. In contrast, researchers have estimated that forest expansion in the temperate regions sequesters on the order of 0.7 billion tons of carbon per year, making the net effects of forests a carbon source, a factor *contributing* to global climate change (Matthews and Hammond, 1999). That result is surprising to many people who think of forests as beneficial in the global balance. The Intergovernmental Panel on Climate Change has suggested that changes in the management of forests through conservation and replanting could help to offset the emissions of carbon dioxide by as much as 12 to 15% up to the year 2050.

FUTURE RESEARCH NEEDS — A SUMMARY OF THE DISCUSSIONS BY CONFERENCE PARTICIPANTS

Climate models were identified as having higher credibility for larger scales but lacked credibility for smaller scales. Understanding the kinds of mechanisms that would be operative under climate change conditions would require identification of a present-day base case with typical responses, and combination of information with an understanding of how nature would respond under future stresses. Participants noted the difficulty in understanding larger scale ecosystems using an experimental approach (see "Relationship of Ecosystem to Climate Change" section above). The need for a common language among related disciplines — for example, a set of glossaries for related fields — was also clearly identified.

Participants described the need for identifying threshold values of ecosystems. Some suggested that stakeholders should specify which combination of variables matter to a given ecosystem. With the identification of these ecosystem threshold values in place, ecologists could then ask climate modelers to focus on the threshold values of ecosystems. For the concept to be useful, of course, the variability of the constituents would need to be identified for current-day conditions. For example, one suggestion urged ecologists to look for critical community properties.

Other participants advocated encouraging interdisciplinary scientists to approach the future with no preconceptions, to start their studies anew. In particular, participants noted that studies of carbon dioxide exchange and balance were difficult to scale up from a local level, and they discussed the concept of strategic cyclical scaling.

More questions than answers developed in the course of the session. Discussions regarding management strategies did not clearly identify what the management efforts should target. Does it in fact make any sense to try to revert a changed ecosystem back to its past state? Should not the future have the priority? How should a generation make decisions for the future? And, more specifically, what determines the regional vulnerability for a region of the globe that is of concern? Should ecosystems be managed with public opinion in mind, and should cost–benefit analysis play a role? Ideas included a suggestion that perhaps the solution should involve ecosystem flexibility so those ecosystems can fill multiple niches. In that regard, scientists can use conventional tools for helping policy makers while they, in turn, allow for imperfect foresight.

Conference participants agreed that lands should be managed rather than holding vegetation fixed in its place. Perhaps a compensation program could be enacted for those whose interests are not served by management practices. One participant suggested that local people could play a supporting role by means of incentives to shape the future of ecosystem management and to ultimately connect this regional-scale idea to broader, global-scale thinking for all land management.

With regard to forests, when should action be taken? Should we not by policy preserve habitats as diverse as possible, recognizing the importance of primary forests? Participants recognized that, for the purpose of taking action and giving credit to owners, *a forest baseline value must be established throughout the globe.* Once an adequate baseline is calculated, determination of how quickly forest areas are cut can be made, and considerations can be given, comparing local vs. global extinction.

One issue that arose in the course of the discussion was the role of the scientist. Should scientists serve as informing voices, or should they advocate a particular policy? At the very least, the scientist should be able to predict what could possibly happen as well as the odds of potential occurrences.

Advocacy promotion could then be left to the value judgments of other individuals. For the sake of good decision making, however, the advocacy should be kept separate from the facts. And in that regard, one participant was asked, "How would it be possible to introduce the best information on global climate change and the potential effects on ecosystems throughout our education system?"

As a final topic of the discussion, scientists at the conference recognized that management would have to continue in the face of uncertainty. Explanation of the levels of uncertainty and integration of those concepts into policy decisions would be necessary to diminish difficulties in identifying proper management approaches. Scientists need to specify ecosystem values in terms of ranges and to identify the precision of their measurements or models. They need to explain their methods with due care and to identify the likelihood of structural possibilities. Scientists need to consider what happens when ecosystems aggregate with consideration for equity and fairness. And when disaggregation is mandated by policy, it should be necessary to give a traceable account of just how the dis-aggregation is to be accomplished.

REFERENCES

Bauer, K.G., and Dutton, J.A., Albedo variation from an airplane over several types of surfaces, *J. Geophys. Res.*, 67, 2367–2376, 1962.

Hummel, J.R and Reck, R.A., A global surface albedo model, *J. Appl. Meteorol.*, 18:239–253, 1979.

Matthews, E. and Hammond, A., *Critical Consumption Trends and Implications: Degrading Earth's Ecosystems*, World Resources Institute, 1999.

Ramanathan, V., Callis, L., Cess, R., Hansen, J., Isaksen, I., Kuhn, W., Lacis, A., Luther, F., Mahlman, J., Reck, R., and Schlesinger, M., Climate-chemical interactions and effects of changing atmospheric trace gases, *Rev. Geophys.*, 25:1441–1482, 1987.

Reck, R.A., CO_2 and climate: effect of natural variations in atmospheric parameters, in *Ninth Conference on Probability and Statistics in Atmospheric Sciences,* American Meteorological Society, Boston, MA, October 1985.

The State of the Nation's Ecosystems, Summary and Findings, The H. John Heinz III Center for Science, Economics and the Environment, available online at http://www.us-ecosystems.org/findings.html, September 7, 2000.

CHAPTER 50

Climate Change, Birds, and Ecosystems — Why Should We Care?

Jeff Price

INTRODUCTION

A functioning ecosystem is made up of animals and plants, in a matrix of soil and surrounding abiotic conditions, interacting in myriad different ways. However, much of the published work on the potential impacts of climate change on ecosystems has dealt primarily with vegetative responses. While the amount of work on climate change and birds has recently increased, it is still relatively scant. Much of this research has focused on potential impacts on single organisms, or, at most, suites of organisms taken together. Little work has been done on how changes to organisms will, in turn, affect the ecosystems of which they are a part. A given ecosystem might continue to function if species were removed (or added). However, how well it would function would depend on the role of the lost species. While there may be redundancies in the roles of many species in maintaining ecosystem health, as more species are removed the probability increases that the system will break down.

Much of the climate change impacts research has focused on factors known to feed back on climate, such as albedo and vegetation. The strength of the feedbacks caused by birds, from even potentially large changes, has been assumed to be relatively low. Consequently, very little research has been conducted investigating ways in which climatic changes may affect birds. Yet birds are necessary or valued for a number of reasons, including economic benefits and, especially, the ecosystem services (and subsequent influences on climate) they exert.

One important concern of researchers examining the potential impacts of climate change on ecosystems is whether ecosystems will become decoupled. That is, will the species within an ecosystem respond differentially to a changing climate? The ecosystem services birds perform contribute significantly to a healthy ecosystem. Indeed, some ecosystems might break down if a number of birds and their accompanying services disappear.

OVERVIEW OF METHODS

In order to determine how climate change might affect bird distributions, investigators should first verify whether bird distributions are associated with climate at all. If they are (and most are), then models based on current climate can be projected into the future to see how the climatic distributions — that area having a climate suitable for the occurrence of a species — might change under various climate scenarios. To date, models depicting how the suitable summer climatic distributions of more than 150 species of North American birds might change with a changing climate have been developed using the techniques outlined by Price (1995). Logistic regression was used to develop models of the association between bird distributions based on Breeding Bird Survey (BBS) data and 18 climate variables. The time period used for developing these models was 1985 to 1989. The short time period was chosen in order to minimize changes from anthropogenic influences or random noise. The models were then statistically validated against a separate data set. Maps of the model distribution were created following the techniques of Price et al. (1995). The maps were used to determine whether the model of the distribution was similar to the distribution developed from the BBS data (Price et al., 1995).

Output from the $2xCO_2$ equilibrium run of the Canadian Climate Centre's General Circulation Model 2 (GCM2) was used to project how the climate of each of the BBS routes might change. These projected climate data were then processed through the logistic regression equations to determine the probability of a species occurrence on each BBS route. Maps were then generated from these data, and the maps were compared with the current distribution of the species. In short, these maps depict the area where the climate may be suitable for the species in the future based on information about their current distributions.

DISCUSSION

Birds and other wildlife perform a number of valuable services for the ecosystems of which they are a part. These services include seed dispersal, pollination, decomposition of organic matter, primary productivity, and predation on insect pests. This investigation examined how avian predation of insect pests benefits boreal forest, grassland, and agriculture ecosystems, and how these systems might break down as the climate changes.

Case Study 1: Birds, Spruce Budworms, and the Southern Boreal Forest

The eastern spruce budworm (*Choristoneura fumiferana*) is a common, native phytophagous (plant-eating) insect of boreal forests in eastern North America (Fleming and Volney, 1995). Its preferred foods include balsam fir (*Abies balsamea*) and several spruce species (*Picea* spp.). Budworms are estimated to annually defoliate approximately 2.3 million hectares in the U.S. (Haack and Byler, 1993) and affect 51 million m^3 of forests in Canada (Fleming and Volney, 1995). Losses in Canada attributed to budworms can be 1.5 times that due to wildfire and can constitute a notable proportion of the annual harvest volume. Spruce budworms are usually present at low densities (<100,000 larvae/ha), but periodic outbreaks occur throughout the species range with densities reaching 22 million larvae/ha (Crawford and Jennings, 1989). These outbreaks can extend over more than 70 million hectares and last for up to 15 years, causing an annual loss totaling greater than 80% of foliage on fir and spruce trees. This foliage loss kills almost all of the trees in mature stands of balsam fir (Crawford and Jennings, 1989; Fleming and Volney, 1995). Trees not killed by defoliation are often at risk from other phytophagous insects and pathogenic diseases, and standing dead trees increase the risk of fire.

While the exact mechanisms behind the beginning and end of outbreaks are unknown, weather is thought to play a role in at least a portion of the budworm's range. Outbreaks in some areas frequently follow droughts (Mattson and Haack, 1987), and outbreaks in central Canada tend to start in stands with high concentrations of mature firs flowering after hot, dry summers (Fleming and Volney, 1995). Drought stresses host trees, leading to increases in concentrations of sugar and sugar alcohols but to decreases in complex carbohydrates. This frequently leads to a reduction in plant defense mechanisms. Drought also changes the microhabitat around affected plants. Drought-stressed plants average 2 to 4°C warmer (maximum 15°C warmer) than abundantly watered plants

(Mattson and Haack, 1987). Average temperature also plays a role in regulating the number of eggs laid by spruce budworms. The number of eggs laid at 25°C is 50% greater than the number laid at 15°C (Jardine, 1994). Drought and higher temperatures may also shift the timing of reproduction in budworms sufficiently that they may no longer be affected by some of their natural parasitoid predators (Mattson and Haack, 1987). Weather may also play a role in stopping outbreaks. Many outbreaks in central Canada are thought to have been halted by late spring frosts that killed new growth on trees, depriving budworms of their food source.

The control of some endemic populations of eastern spruce budworm may be aided by avian predators, especially some of the wood warblers (Crawford and Jennings, 1989). Several warbler species, including Cape May warbler (*Dendroica tigrina*), bay-breasted warbler (*D. castanea*), Blackburnian warbler (*D. fusca*), Tennessee warbler (*Vermivora peregrina*), and Nashville warbler (*V. ruficapilla*) are important predators of spruce budworms. Many of these species, but especially bay-breasted warbler, show functional responses to increases in budworms — individuals moving into the area and increasing their reproductive output in response to increases in the insect's population (Crawford and Jennings, 1989). Birds consume up to 84% of budworm larvae and pupae when budworm populations are low (approximately 100,000/ha) and up to 22% when populations reach approximately 500,000/ha. However, bird predation cannot effectively neutralize budworm populations in concentrations exceeding 1 million larvae/ha (Crawford and Jennings, 1989). These warblers, in concert with other budworm predators (mostly predatory insects), are thought to be able to control some populations of budworms under normal conditions.

Climate change may influence almost every component of this system, both individually and in species interactions. For example, the spruce budworm's northern range may increase with rising temperatures. Currently, this species' distribution is thought to be tied to the completion of larval development before autumn freezes begin (Jardine, 1994). Increasing temperatures and a greater frequency of drought could lead to more frequent and possibly more severe outbreaks. This change would occur because of the effects of both drought and temperature on host plants and insects (Mattson and Haack, 1987; Fleming and Volney, 1995). Increasing temperatures may also reduce the frequency of late spring frosts in southern boreal forests, possibly increasing the duration of budworm outbreaks in those areas.

A changing climate might also decouple budworm population cycles from those of some of its parasitoid and avian predators (Mattson and Haack, 1987). Distributions of most of the warblers that feed on spruce budworms could shift significantly farther north. One set of models projects that three of the most significant predators — Tennessee, Cape May, and bay-breasted warblers — may all be extirpated from below 50°N latitude with a doubling of CO_2 (Price, 2001). Indeed, the average latitude of occurrence of both Cape May and bay-breasted warblers has already shifted significantly farther north in the last 24 years.

This decoupling of predator and prey may exert the greatest change on southern boreal forests. Replacing biological control mechanisms with chemical control mechanisms (e.g., pesticides) could ultimately yield a different set of problems. Besides the economic and social issues relating to large-scale pesticide application, some pesticides may not work as well as expected. Some insects are known to have reduced susceptibility to pyrethrins, DDT, and some carbamate pesticides with increased temperatures (Mattson and Haack, 1987).

While there may be decoupling of some budworm populations from their control mechanisms, the range and timing of reproduction of the species would probably stay synchronized with their host species. A large outbreak area could contain as many as 7.2×10^{15} insects. Even assuming a budworm genetic mutation rate on the order of 1×10^{-5} per generation, billions of rare genotypes would still be present in the population (Fleming and Volney, 1995). These genotypes could help budworms adapt to remain in synchrony with their host species.

One result of these changes may be an increase in the rate of conversion from boreal forests to other habitat types (especially in southern boreal forests). Rising temperatures, drought conditions, and insect damage may increase the likelihood of major fires, particularly in combination.

Indeed, fire frequency in Canada, the western U.S., Alaska, and Russia has increased during the last several decades (Jardine, 1994), although several likely causes may be involved. Scientists need to confront two key questions: how quickly might southern boreal forests be converted to a different habitat type, and what might that habitat be? As fire frequency increases, the successional pathway changes from fir/spruce to aspen/birch to grasslands (Jardine, 1994). If the climate changes quickly, more southerly species may not have the time to migrate into areas as fast as boreal species are lost. That could mean a shift to grassland or grass/shrubland in areas currently dominated by southern boreal forest. That, in turn, may affect regional economies and distributions of species using southern boreal forests.

Case Study 2: Birds, Grasshoppers, and Grasslands

In pastureland and grassland ecosystems, birds are important predators of grasshoppers. Bioenergetic models estimate that a single pair of Savannah sparrows (*Passerculus sandwichensis*) raising their young consumes approximately 149,000 grasshoppers over a breeding season. Considering typical bird densities in these same areas, this implies that approximately 218,000 grasshoppers/ha are consumed each season (Kirk et al.,1996). In many areas, the economic threshold for spraying insecticides occurs when densities reach approximately 50,000 grasshoppers/ha (McEwen, 1987). Thus, birds may help keep grasshopper populations at lower levels, avoiding pesticide spraying. However, models project that the distribution of Savannah sparrows could shift north with climate change (Price, 1995). This shift may then decouple this predator from its prey in much of the U.S. Many other grassland birds are also projected to undergo similar shifts in their distributions (Price, 1995). Unless all components (grasslands, grasshoppers, and birds) of this ecosystem were to change at the same rate — an unlikely prospect — the decoupling of the predators from their prey could lead to an increase in grasshopper outbreaks in the future.

Case Study 3: Birds in Agricultural Ecosystems

Birds may also be effective predators of insect pests in agricultural settings. Numerous studies have documented the efficiency of bird predation on overwintering codling moth (*Cydia pomonella*) larvae in orchards in different parts of the world, with predation levels reaching 90% or more (Neff, 1942; MacLellan, 1959; Hagley, 1970; Solomon et al., 1976; Solomon and Glen, 1979; Stairs, 1985). In southeastern U.S. pecan orchards, a single tufted titmouse (*Parus bicolor*) feeding on pecan nut casebearers (*Acrobasis nuxvorella*) is estimated to save nut growers 52,000 pecans annually (Whitcomb, 1971). While models have yet to be developed for those birds preying on codling moths, the models for tufted titmouse project that the range of that species does not shift out of the southeastern U.S. (J.T. Price, personal communicaton, 1999).

CONCLUSION

While changes in bird distributions due to incipient modifications to climate may not exert strong feedbacks on climate (except possibly in the southern boreal forest), they could be significant because of disruptions to ecosystem services. Species within ecosystems have evolved together over time and depend on each other to various degrees for services such as food, protection, or reproduction. Rapidly separating, or decoupling, these species may cause detrimental changes. The examples presented here are just a few of the myriad interactions taking place in ecosystems on a daily basis. Furthermore, climate change is just one of the stressors that species in these systems face.

Humans may be able to intervene to help prevent disruption of many ecosystem services, but the costs of such intervention could be large. For example, loss of predators may lead to the need to apply more pesticides. This will not only have direct monetary costs but also indirect costs such as other ecosystem damages or possible human health problems. Over time, new predators might migrate into the ecosystem, or the prey might become pesticide resistant. Problems may be caused not necessarily by the change itself, but rather by the rate of change. Unless all components of the ecosystem change at the same rate, the systems will decouple, and the new systems may bear little resemblance to the ecosystems to which managers are accustomed. More research is needed to determine not only how more of the pieces of the ecosystem work together but also how they might move and at what rates relative to each other.

REFERENCES

Crawford, H.S. and Jennings, D.T., Predation by birds on spruce budworm *Choristoneura fumiferana*: functional, numerical and total responses, *Ecology,* 70:152–63, 1989.

Fleming, R.A. and Volney, W.J.A., Effects of climate change on insect defoliator population processes in Canada's boreal forest: some plausible scenarios, *Water Air Soil Pollut.,* 82:445–454, 1995.

Haack, R.A. and Byler, J.W., Insects and pathogens: regulators of forest ecosystems, *J. For.,* September:32–37, 1993.

Hagley, E.A.C., The distribution and survival of overwintering codling moth larvae in southern Ontario, *Proc. Entomol. Soc. Ontario,* 100:40–47, 1970.

Jardine, K., Finger on the carbon pulse: climate change and the boreal forests, *Ecologist,* 24:220–224, 1994.

Kirk, D.A., Evenden, M.D., and Mineau, P., Past and current attempts to evaluate the role of birds as predators of insect pests in temperate agriculture, in Nolan, V., Jr. and Ketterson, E., Eds., *Current Ornithology* Vol. 13, Plenum Press, New York, 1996, pp. 175–269.

MacLellan, C.R., Woodpeckers as predators of codling moth in Nova Scotia, *Can. Entomol.,* 91:673–80, 1959.

Mattson, W.J. and Haack, R.A., The role of drought in outbreaks of plant-eating insects, *BioScience,* 37:110–118, 1987.

McEwen, L.C., Function of insectivorous birds in a shortgrass IPM system, in Capinera, J.L., Ed., *Integrated Pest Management on Rangeland: A Shortgrass Prairie Perspective,* Westview Press, Boulder, CO, 1987, pp. 324–333.

Morris, R.F., Cheshire, W.F., Miller, C.A., and Mott, D.G., The numerical response of avian and mammalian predators during a gradation of the spruce budworm, *Ecology,* 39:487–494, 1958.

Neff, J.A., Comments on birds and codling moth control in the Ozarks, *Wilson Bull.,* 54:21–24, 1942.

Price, J.T., Potential impacts of global climate change on the summer distributions of some North American grassland birds, Ph.D. dissertation, Wayne State University, Detroit, MI, 1995.

Price, J.T., Climate change, warblers and spruce budworms, in Green, R.E. et al., Eds., *Impacts of Climate Change on Wildlife*, Royal Society for the Preservation of Birds, U.K., 2001, pp. 60–63.

Price, J.T., Droege, S., and Price, A., *The Summer Atlas of North American birds*, Academic Press, San Diego, CA, 1995.

Solomon, M.E. and Glen, D.M., Prey density and rates of predation by tits (*Parus* spp.) on larvae of codling moth (*Cydia pomonella*) under bark, *J. Appl. Ecol.,* 16:49–59, 1979.

Solomon, M.E., Glen, D.M., Kendall, D.A., and Milsom, N.F., Predation of overwintering larvae of codling moth (*Cydia pomonella* L.) by birds, *J. Appl. Ecol.,* 13:341–52, 1976.

Stairs, G.R., Predation on overwintering codling moth populations by birds, *Ornis Scand.,* 16:323–324, 1985.

Whitcomb, W.H., The tufted titmouse, *Parus bicolor,* as a predator of the pecan nut casebearer, *Acrobasis caryae,* in Komarek, E.V., Ed., *Proc. Tall Timbers Conf. Ecol. Anim. Control Habitat Manage.*, Vol. 2, Tall Timbers Research Station, Tallahasee, FL, 1971, pp. 305–308.

CHAPTER 51

A Checklist for Historical Studies of Species' Responses to Climate Change

Raphael D. Sagarin

INTRODUCTION

Biologists have been widely involved in our understanding of global climate change since the early 1980s. The vast majority of work includes studies that attempt to predict future biological changes in response to future climatic change. These studies range from speculation of range shifts and changes in species interactions based on assumptions that species' distributions and behaviors are broadly controlled by temperature, to highly detailed bioclimatic mapping models that identify the climatic variables most important in determining species' ranges and then map future species' distributions based on expected changes in these variables. Such studies cover a broad spectrum of biological systems, including terrestrial plants (Woodward, 1992; Beerling, 1993), infectious disease (Patz et al., 1996), insects (Rubenstein, 1992), other invertebrates (Bhaud et al., 1995), birds (Root, 1993), freshwater fishes (Scott and Poynter, 1991), algae (Breeman, 1990), marine fishes (Frank et al., 1990), coastal marine communities (Fields et al., 1993; Lubchenco et al., 1993), and native human populations (Langdon, 1995). The vast majority of these studies consider climate change solely in terms of its future signal as predicted by large-scale global circulation models. The significant warming trend seen during the 20th century (Intergovernmental Panel on Climate Change, 2001) and its effects on natural populations is largely ignored by these treatments.

Recently, however, a growing number of studies have shown long-term biological changes during this century that are argued to be correlated with warming trends over the same time period. These studies, like their more speculative antecedents, cover a wide range of taxa including butterflies (Parmesan, 1996; Parmesan et al., 1999), birds (Thomas and Lennon, 1999), reptiles and amphibians (Pounds et al., 1999), fish (Holbrook et al., 1997), and marine invertebrates (Barry et al., 1995; Sagarin et al., 1999). These studies are largely examples of "historical science" (Francis and Hare, 1994) rather than strict experimental procedures. While experimental studies of species' responses to climate change have an essential role in determining important mechanisms that will drive species' responses (Chapin et al., 1995) or reveal unexpected indirect effects (Davis et al., 1998; Sanford, 1999), they are largely inadequate or impractical for tracking species' responses to climate change in the present (Sagarin, 2001).

1-56670-612-2/03/$0.00+$1.50
© 2003 by CRC Press LLC

HISTORICAL SCIENCE DEFINED

In Francis and Hare's (1994) definition, historical-descriptive science begins with observations of a system. At this stage, the observations may be anecdotal (e.g., fishermen noticing unusual species in their catches during warm-weather periods) or tangential to a main line of research (e.g., an investigator noticing changes in species composition over many years at a study site). Whatever the source, these observations are intriguing enough that an investigator uses them as a launching point for a more in-depth study of the system.

Simple models, or narratives, are then developed that suggest hypotheses of how the various observations fit together. The many studies that have posited predictions of population or community changes under expected warming climate are examples of these narratives. For example, a large number of these studies suggest that species' ranges should shift poleward or toward higher altitudes with warming climate.

In the final step, historical observations from a variety of sources are examined for patterns that support or do not support the model. The physical data used typically include long-term time-series data for one or more climatic variables, such as daily temperature records or annual tallies of frost-free days. Historical data on biotic systems might include presence or absence data, population counts, or measures of community composition and diversity through time. These data may take the form of snapshot data taken at the same location at separate points in time, or preferably, as time-series data collected for many years, such as catch records for fisheries or Christmas bird counts. The analysis of these data may be as simple as noting a gradual increase in temperature alongside gradual changes in species' abundances (Southward et al., 1995), or more rigorous attempts to correlate physical and biological variables using time-series analysis (Francis and Hare, 1994; McGowan et al., 1996).

The aim of historical science is not to prove a single hypothesis, or to decide in favor of one of two competing hypotheses, but to reject multiple alternative hypotheses, leaving a single hypothesis as the best remaining explanation of the data. Whereas experimental studies are carefully controlled and, in the ideal situation, all components of the system are accounted for, a historical study may be plagued by a number of uncontrolled variables and methodological uncertainties. For these reasons, it is imperative that the investigator thoroughly research all available historical information.

EXAMPLES OF HISTORICAL SCIENCE IN CLIMATE CHANGE STUDIES

Research that I have conducted with Sarah Gilman (University of California, Davis), James Barry (Monterey Bay Aquarium Research Institute), and Charles Baxter (Stanford University, retired) (Sagarin et al., 1999) fits the historical science model well. Our efforts were initiated by anecdotal observations by Charles Baxter that changes were occurring to the intertidal flora and fauna of Stanford University's Hopkins Marine Station (HMS) in Monterey, CA. In order to quantify these changes, we replicated an extensive survey of intertidal invertebrates conducted by W.G. Hewatt at HMS in the early 1930s (Hewatt, 1937). We recorded abundance of more than 100 invertebrate species in 57 0.84-m^2 plots placed in the precise location of Hewatt's original plots. A quantitative comparison of 62 species for which reliable records could be compared confirm the initial observations (Figure 51.1); 46 species showed a significant change in abundance (paired t-test, $P < 0.05$), but no overall direction of change occurred, with 24 species increasing and 22 decreasing. In fact, no pattern could be discerned from the data until species' changes were separated by their geographic ranges relative to HMS. In this case it becomes clear that southern species primarily declined, northern species increased, and cosmopolitan (wide-ranging) species were split between those that increased and those that decreased (Figure 52.2). We argued that these range-related changes are related to long-term warming trends of nearly 1°C on average since 1930 in nearshore sea temperatures taken daily at HMS.

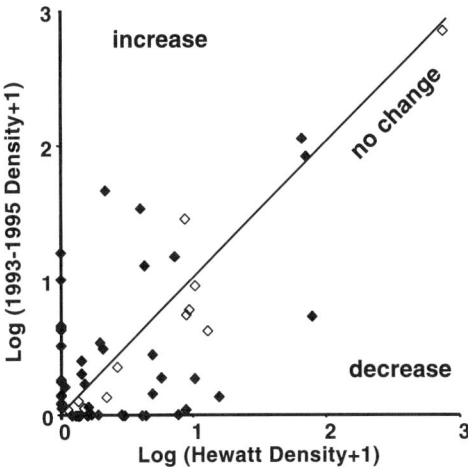

Figure 51.1 Plot of density in 1995 vs. density in Hewatt's study in 57 paired plots for 62 species. Solid symbols are species that showed a significant change in abundance between the two studies (paired t-test, $P < 0.05$).

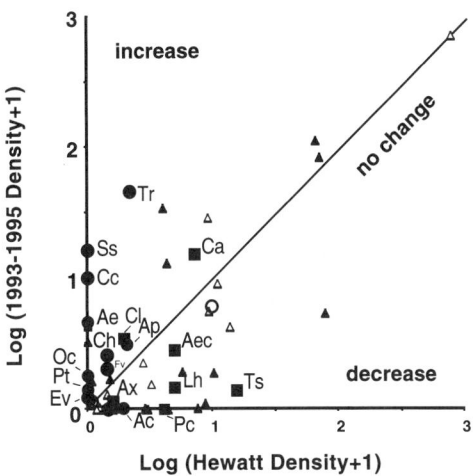

Figure 51.2 Plot of density in 1995 vs. density in Hewatt's study in 57 paired plots for 62 species. Solid symbols are species that showed a significant change in abundance between the two studies (paired t-test, $p < 0.05$). Open symbols are non-significant changes. Southern species: circles. Northern species: squares. Cosmopolitan species: triangles. Abbreviations are as follows: *Acanthina punctulata* (Ap), *Alpheus clamator* (Ac), *Anthopleura elegantissima* (Ae), *Anthopleura elegantissima* (clonal form) (Aec), *Anthopleura xanthogrammica* (Ax), *Calliostoma ligatum* (Cl), *Corynactis californica* (Cc), *Crepidula adunca* (Ca), *Lepidochitona* (*Cyanoplax*) *hartwegii* (Ch), *Erato vitellina* (Ev), *Fissurella volcano* (Fv), *Leptasterias hexactis* (Lh), *Ocenebra circumtexta* (Oc), *Petrolisthes cinctipes* (Pc), *Pseudomelatoma torosa* (Pt), *Serpulorbis squamigerus* (Ss), *Tectura scutum* (Ts), *Tetraclita rubescens* (Tr). (From Sagarin, R.D. et al., *Ecol. Monogr.*, 69, 465–490, 1999. With permission.)

Because of the large uncertainty in comparing communities in two widely disparate points in time, we considered many alternative hypotheses to explain the range-related pattern of change we observed. These alternatives included anthropogenic effects, habitat changes, seismic effects, the return of sea otters and oyster catchers, shifts in tidal heights by species, sea level changes, upwelling effects, and El Niño effects. Through analysis of our data as well as many relevant contributions from other scientists and historians, we were able to weigh the importance of these alternatives to varying degrees. In all cases we found that even when the alternative hypotheses could explain some of the changes, they could not account for the overall range-related pattern of change we observed (Sagarin et al., 1999).

A CHECKLIST FOR HISTORICAL STUDIES

Through analysis of this historical research, as well as several other historical studies of species' responses to climate change, I have compiled a checklist of the eight most important features of such study programs. No study to date has covered all these features and I am doubtful that any one study will do so in the future. In fact, many convincing studies have a weakness or are altogether lacking in some of the checklist areas. The purpose of the checklist, then, is not to reject outright studies that don't measure up to its standards, but to highlight potential weaknesses of studies. This checklist also can be used to prescreen existing historical data sets to determine if they are fit for retrospective study. Finally, the list can be used as a prescriptive of important areas to consider when initiating new long-term study programs. Each feature on the list is discussed in the context of studies of species' responses to climate change, and positive and negative examples of its use are given from the literature.

1. Well-marked study site — In most biological systems spatial location and spatial heterogeneity play an important role in species' distribution and abundance. In the rocky intertidal, for instance, rock type, slope, tidal height, exposure to waves, aspect relative to the sun, and biological cover all influence the distribution and abundance of a given organism, which in turn affect the populations of the species that interact with that organism. Thus, any attempt to compare populations or locations of intertidal species from separate time periods is potentially confounded by differences in any of these factors. While experimental studies attempt to overcome this spatial heterogeneity by using fully randomized designs (Underwood, 1997), the many uncontrolled variables in historical studies, the limited sample size of the available data, and the simple fact that many older sampling programs were not designed with statistical considerations in mind, often make this impossible or impractical. Therefore, the best way to rule out the effects of spatial heterogeneity is to resample in the precise location of the original sample size.

Alterations of the physical habitat also contribute to the importance of well-marked study sites. For example, seismic activity can lead to massive habitat changes in the intertidal (Castilla, 1988). In our study, we were able to rule out seismic effects because we were able to compare the tidal height of the study transect relative to sea level and confirm that this height was the same as reported in the original study. As another example, erosion of rock surfaces would be expected to have effects on intertidal populations. We were able to reject this explanation of changes in our system because we had maps drawn by the original investigator that showed details and contours of rock surfaces that are still clearly visible today.

Many investigators have been thwarted in their attempts to assess long-term changes in their systems simply because they could not adequately relocate baseline studies. Even cases in which the original investigators are still alive and able to provide information about the initial study, it is not always possible to relocate the exact study site (Zavaleta, E., personal communication).

2. Protected study area — Historically based research programs have suffered from lack of protected study sites. In a major study of changes to European butterflies, Camille Parmesan et al. (1999) had to eliminate the entire central portion of most species' ranges from consideration because habitat transformation would have confounded interpretations of climate-related species' shifts.

If sites have been protected by law, it is important to recognize that many anthropogenic effects are not mitigated by politically designated boundaries. Airborne or waterborne pollutants originating outside the site, as well as the heat-island effect of cities are examples of such uncontainable effects. Furthermore, in most cases, humans have already significantly altered study systems, and we may never know the true baseline conditions of a system (Dayton et al., 1998). Nonetheless, conducting historical studies in areas protected from human tampering in the years between surveys allows us to rule out the role of direct influence of some anthropogenic effects on the study system. In the case of our research, a village at the current site of HMS collected large numbers of intertidal invertebrates for human consumption until at least 1907 (Lydon, S., personal communication), making it highly probable that the site was not pristine when the first investigations were conducted there. However, the subtidal and intertidal area at HMS has been formally protected from collecting and disturbance as part of the state-designated Hopkins Marine Life Refuge since 1931, almost the same time that the original surveys were conducted. Thus, we know that human collections have not played a significant role in the abundance changes we observed.

3. Knowledge of data — One of the most important issues in any historical research is the reliability of historical sources of information. In historical scientific research, this takes on added weight because the data are to be compared with recently collected data that are (presumably) held to a high standard of scientific merit. Obviously, this problem is eliminated if the same investigator has performed both the earlier as well as the later research. Even if the methodology has changed, the investigator would be aware of this and could account for it in the data analysis. The second best option is to have direct contact with the original investigator. This is especially important for filling in gaps in materials and methods. In many cases, however, the original investigators have long departed, and the retrospective analysis must rely completely on written records. In this case, the modern investigators must be very clear on whether data collection was conducted exactly as specified by the original author, inferred from the author's writing, or with no knowledge of the original author's methods.

Ultimately, the determining factor in judging a historical study that must rely on methodological uncertainty is whether or not a bias is introduced into the study by assuming that the modern methods approximate the historical methods. For example, an observed trend toward earlier budding in plants is biased if earlier investigators recorded first budding as plants with 25% of their stems containing buds while later investigators recorded the very first bud on a plant. Likewise, observed range shifts can be biased if earlier records ignored solitary sightings of accidentals in their range maps and later records included accidentals in their maps.

4. Quantitative data — Many ecological surveys record species as present or absent, or categorically as rare, common, or abundant. Except in the most extreme cases, or with a tremendous amount of data, such data will be inadequate to determine species' responses to climate change. Most predictive papers focus on species' range shifts as responses to climate change, yet responses at sites may first be seen in abundance changes that will not be picked up in comparisons of presence or absence data, and cannot be quantified with categorical data. This is clearly the case in our investigations. When we observe our data as if only presence or absence of species was recorded, no range-related trend is apparent (Figure 51.3). While we did find some local arrivals of southern

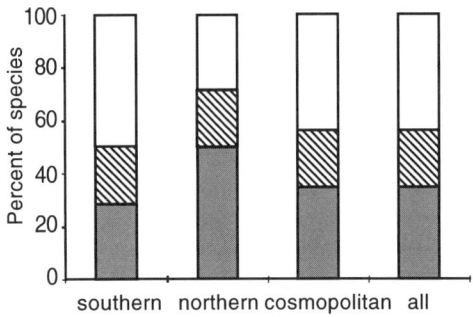

Figure 51.3 Changes in presence or absence of all species recorded by either Hewatt (1937) or Sagarin et al. (1999) in the 57 resurveyed plots. "Additions" (unfilled areas) are species recorded by Sagarin et al., but not Hewatt. "Deletions" (hatched areas) are species recorded by Hewatt, but not Sagarin et al. "Both studies" (filled areas) are species recorded by both Hewatt and Sagarin et al.

species and local extinctions of northern species, and these may be related to actual range shifts, the strong range-related pattern of change is primarily driven by abundance changes.

In fact, in some cases, abundance data are not enough to reveal long-term trends. For example, Kullman's (1998) long-term studies in Sweden show shifts in altitudinal tree limits that are associated with the condition of trees, rather than tree abundance. Likewise, the many phenological studies that have been published recently (Beebee, 1995; Sparks and Yates, 1997; McCleery and Perrins, 1998; Brown et al., 1999; Menzel and Fabian, 1999) show a strong response to warming trends in timing of biological events, rather than in population numbers.

5. Relevant physical data —When a pattern of biological change is discovered that is consistent with predicted responses to climate change, it is tempting for researchers to attribute the biological signal to the climatic warming seen in average temperatures during this century. It is important to remember, however, that even the coarsest global circulation models do not predict uniform warming throughout the globe (Intergovernmental Panel on Climate Change, 2001), and many areas have experienced both warm and cool periods during this century. In addition to the spatial variability of climate, the temporal component is also important to biological systems. For example, Welch (1998) found that sharp limits in salmon distributions appeared when comparing distributions with monthly temperature data, but these distinctive limits disappear when seasonally or annually averaged temperatures were used. Likewise, Pounds et al. (1999) argued that daily records of mist frequency must be used in order to see an apparent correlation between species' declines and cloud canopy height in the Monteverde Cloud Forest Preserve in Costa Rica.

Thus, climatic information that is relevant to the biological study system should be incorporated into any historical study program. Individual investigators should first determine if the data available are appropriate for their site. For our studies of the intertidal, it was important for us to have long-term records of sea temperatures taken at the shores of HMS, because large local variations occur in sea surface temperature along the California coast (MacCall, 1996) that would make interpreting the role of temperature changes recorded elsewhere on the coast problematic. Nevertheless, it is unlikely that the most relevant physical factor data will be available and researchers will have to use proxies. In our studies, for instance, local sea surface temperature is used as an approximation of the incredibly complex and poorly understood microhabitat temperatures (Helmuth, 1998) that are expected to have a more direct effect on species.

6. Taxonomic resolution — Climate changes would not be expected to target individual species or groups in the same manner as disease vectors, harvesting, or predation. Thus, gathering data from many different species, with a wide variety of life-history strategies, and from multiple trophic levels is an excellent strategy for eliminating alternate hypotheses to the climate change hypothesis. In our study, we compared 62 species quantitatively. This included a group representing several phyla, predators and prey; herbivores and filter feeders; solitary and colonial; mobile and sessile; cryptic and noncryptic; high to low intertidal; and animals with dispersive larvae and those with brooded larvae. Thus, we were able to test the hypothesis that stronger El Niño events, which are characterized by increased northward water transport along the California coast, led to greater abundance changes in species with long-lived larvae. We found no difference in the patterns of abundance changes between species with non-existent, 1 to 10 day, or longer larval periods. In fact, after analyzing species changes along any of the life-history axes described above we found no significant patterns. Only when we divided species by their geographic ranges was a strong pattern apparent.

As another example, Pounds et al. (1999) found dramatic declines in several species of amphibians in the Monteverde Cloud Forest Preserve in Costa Rica, mirroring worldwide declines in amphibians. Because they saw similar declines in reptiles and birds, however, they were able to discount the effects of fungal diseases known only to infect amphibians in favor of their hypothesis that loss of habitat due to a gradually rising cloud level is ultimately responsible for the declines.

7. Spatial resolution — Biological responses to supposedly global or regional climate changes should be somewhat consistent over a wide spatial area. Unfortunately, few of the historical studies of species' responses to climate change thus far have presented results from more than a single site or limited part of the species' ranges. A notable exception is Parmesan's (1996) study of range shift in Edith's checkerspot butterflies in western North America. Parmesan examined extinction patterns at multiple sites throughout the entire western range of this species and concluded that extinctions were much more likely in the southern part of the range during this century. Interpretations of studies with poor spatial coverage can be bolstered with supporting data from similar studies conducted in other locations. For example, Brown et al.'s (1999) finding of earlier breeding trends in a North American bird corroborates similar findings in spatially limited European phenological records (McCleery and Perrins, 1998). Nevertheless, expected biological responses to climate change, such as geographic range shifts, cannot be determined from data obtained from a single site. Thus, in our study of changes to the intertidal at HMS, we could not confirm range shifts for most species due to lack of spatial coverage.

8. Temporal resolution — By far the most serious and common weakness in historical studies is the lack of high-frequency time-series data tracking the biological changes. Many studies rely on comparisons of two snapshots in time. While this does provide in some cases a dramatic illustration of changes to a system, it tells us little about how a system changed through time. Important questions such as, "Were the changes abrupt and related to an episodic event, or did they occur gradually through time?" must be left unanswered. Denny and Paine (1998) revealed the weakness of our snapshot picture of change in the intertidal in their discussion of an 18.6-year oscillation in the moon's orbital inclination that may be correlated with tidal emersion periods and nearshore sea temperatures. Because we lack regular time-series data on the biological changes at HMS, it is difficult to determine if the species changes we observed are more strongly related to the gradual rise in nearshore sea temperatures over 60 years, the shorter-term sea temperature changes associated with the lunar oscillation, or other changes that have occurred in the last 60 years. Even with the addition of short-term sampling that showed only minor changes over the past 3 years relative to the large changes seen over 60 years (Sagarin et al., 1999), we cannot be certain that interannual changes were always of such a small magnitude. By contrast, scientists with access to long-term time-series data have been able to analyze the signal of change at a range of biologically relevant temporal scales (McGowan et al., 1998) and establish strong correlations between physical and biological variables (Sparks and Yates, 1997).

CONCLUSION

Studies correlating biological changes to climatic changes are a small but rapidly growing area of climate change science. Because global climate change is essentially an uncontrolled experiment in progress, studies of its effects on biological communities will necessarily be observational and historical in nature. As more researchers turn up historical data sets in an attempt to track climate-related changes, they will undoubtedly run into shortfalls in one or several of the areas described above. Likewise, new biological monitoring schemes, which should be a top research priority in climate change science, will always be limited by available funding and effort, and thus will be somewhat incomplete. Nevertheless, particular strengths in one area, such as taxonomic or spatial resolution, have allowed several of the studies cited above to overcome weaknesses in other areas and convincingly show that biotic communities are responding to climate change in the present era.

Weaknesses in individual studies may also be overcome by compiling results of other studies completed on different taxa, in other areas of the world. This is already occurring as evidenced by several meetings sponsored by government and nongovernment agencies in which scientists studying a wide range of biological systems and geographic locales presented their findings (Bird Life International and World Wildlife Fund, 1997; Royal Society for the Protection of Birds et al., 1999). The Southern Hemisphere and Asia are far less represented in these compilations than Europe and North America, but this will change in the future as scientists in those sparsely studied areas are beginning to investigate historical data sets (Broitman, B., personal communication). Essential to this effort is the continued support of existing monitoring programs and the creation of new monitoring programs with the expressed goal of tracking species' responses to climatic changes.

ACKNOWLEDGMENTS

I thank Stephen Schneider and Francisca Saavedra for their invitation to speak at the International Congress on Ecosystem Health. Charles Baxter, Sarah Gilman, and James Barry coauthored the research that is discussed in this chapter. Financial support for this research came from NASA Earth Systems Science Fellowship (Grant NGT 30339); an NSF graduate student fellowship (RTG Grant BIR94-13141 and GRT Grant GER 93-54870); the National Wildlife Federation's Climate Change Fellowship Program; and grants from the Lerner-Gray Fund for Marine Research, Sigma Xi Grants in Aid of Research, the University of California at Santa Barbara Affiliates, and the Myers' Oceanographic and Marine Biology Trust.

REFERENCES

Barry, J.P., Baxter, C.H., Sagarin, R.D., and Gilman, S.E., Climate-related, long-term faunal changes in a California rocky intertidal community, *Science*, 267, 672–675, 1995.

Beebee, T.J.C., Amphibian breeding and climate, *Nature*, 374, 1995.

Beerling, D.J., The impact of temperature on the northern distribution of the introduced species *Fallopia japonica* and *Impatiens glandulifera* in north-west Europe, *J. Biogeogr.*, 20, 45–53, 1993.

Bhaud, M., Cha, J.H., Duchene, J.C., and Nozias, C., Influence of temperature on the marine fauna: what can be expected from a climatic change. *J. Thermal Biol.*, 20, 91–104, 1995.

Bird Life International and World Wildlife Fund, *Climate Change and Wildlife: A Summary of an International Workshop at the National Center for Atmospheric Research*, Briggs, B., Ed., Boulder, CO, 1997.

Breeman, A, Expected effects of changing seawater temperatures on the geographic distribution of seaweed species, in *Expected Effects of Climatic Change on Marine Coastal Ecosystems*, Beukema, J.J., Wolff, W.J., and Brouns, J.J.W.M., Eds., Kluwer Academic Publishers, Dordrecht, 1990, pp. 69–76.

Brown, J.L., Li, S-H., and Bhagabati, N., Long-term trend toward earlier breeding in an American bird: A response to global warming? *Proc. Natl. Acad. Sci.*, 96, 5565–5569, 1999.

Castilla, J.C., Earthquake-caused coastal uplift and its effects on rocky intertidal kelp communities, *Science*, 242, 440–443, 1988.

Chapin, F.S.I., Shaver, G.R., Giblin, A.E., Nadelhoffer, K.J., and Laundre, J.A., Responses of arctic tundra to experimental and observed changes in climate, *Ecology*, 76, 694–711, 1995.

Davis, A.J., Jenkinson, L.S., Lawton, J.H., Shorrocks, B., and Wood, S., Making mistakes when predicting shifts in species range in responses to global warming, *Nature*, 391, 783–786, 1998.

Dayton, P.K., Tegner, M.J., Edwards, P.B., and Riser, K.L., Sliding baselines, ghosts, and reduced expectations in kelp forest communities, *Ecol. Appl.*, 8, 309–322, 1998.

Denny, M.W. and Paine, R.T., Celestial mechanics, sea-level changes, and intertidal ecology, *Biol. Bull.*, 194, 108–115, 1998.

Fields, P.A., Graham, J.B., Rosenblatt, R.H., and Somero, G.N., Effects of expected global climate change on marine faunas, *Trends Ecol. Evol.*, 8, 361–367, 1993.

Francis, R.C. and Hare, S.R., Decadal-scale regime shifts in the large marine ecosystems of the North-east Pacific: a case for historical science, *Fish. Ocean.*, 3, 279–291, 1994.

Frank, K.T., Perry, R.I., and Drinkwater, K.F., Predicted response of Northwest Atlantic invertebrate and fish stocks to CO_2-induced climate change, *Trans. Am. Fish. Soc.*, 119, 353–365, 1990.

Helmuth, B.S.T., Intertidal mussel microclimates: predicting the body temperature of a sessile invertebrate, *Ecol. Monogr.*, 68, 51–74, 1998.

Hewatt, W.G., Ecological studies on selected marine intertidal communities of Monterey Bay, California, *Am. Mid. Nat.*, 18, 161–206, 1937.

Holbrook, S.J., Schmitt, R.J. and Stephens, J.S., Jr., Changes in an assemblage of temperate reef fishes associated with a climatic shift, *Ecol. Appl.*, 7, 1299–1310, 1997.

Intergovernmental Panel on Climate Change, Third Assessment Report, 2001.

Kullman, L., Tree-limits and montane forests in the Swedish Scandes: sensitive biomonitors of climate change and variability, *Ambio*, 27, 312–321, 1998.

Langdon, S.J., Increments, ranges, and thresholds: human population responses to climate change in northern Alaska, in *Human Ecology and Climate Change: People and Resources of the Far North,* Peterson, D.L. and Johnson, D.R., Eds., Taylor & Francis, Washington, D.C, 1995, pp. 139–154.

Lubchenco, J., Navarrete, S., Tissot, B.N., and Castilla, J.C., Possible ecological responses to global climate change: nearshore benthic biota of northeastern Pacific coastal ecosystems, in *Earth System Responses to Global Change: Contrasts between North and South America,* Mooney, H.A., Ed., Academic Press, San Diego, 1993, pp. 147–165.

MacCall, A.D., Patterns of low-frequency variability in fish populations of the California current, *Calif. Coop. Ocean. Fish. Invest. Rep.*, 37, 100–110, 1996.

McCleery, R.H. and Perrins, C.M., Temperature and egg-laying trends, *Nature*, 391, 30–31, 1998.

McGowan, J.A., Cayan, D.R., and Dorman, L.M.,. Climate-ocean variability and ecosystem response in the Northeast Pacific, *Science*, 281, 210–217, 1998.

McGowan, J.A., Chelton, D.B. and Conversi, A., Plankton patterns, climate, and change in the California Current, *Calif. Coop. Ocean. Fish. Invest. Rep.*, 37, 45–68, 1996.

Menzel, A. and P Fabian, P., Growing season extended in Europe, *Nature*, 397, 1999.

Parmesan, C., Climate and species' range, *Nature*, 382, 765–766, 1996.

Parmesan, C. et al., Poleward shifts in geographical ranges of butterfly species associated with regional warming, *Nature*, 399, 579–583, 1999.

Patz J.A., Epstein, P.R., Burke, T.A., and Balbus, J.M., Global climate change and emerging infectious diseases, *JAMA,* 275, 217–223, 1996.

Pounds, J.A., Fogden, M.P.L., and Campbell, J.H., Biological response to climate change on a tropical mountain, *Nature*, 398, 611–615, 1999.

Root, T., Effects of climate change on North American birds and their communities, in *Biotic Interactions and Global Change,* Kareiva, P.M., Kingsolver, J.G., and Huey, R.B., Eds., Sinauer Associates, Inc., Sunderland, MA, 1993, pp. 280–292.

Royal Society for the Protection of Birds et al., *No Place to Go? The Impact of Climate Change on Wildlife,* The Norwich Conference: Climate Change and Wildlife, University of East Anglia, Norwich, U.K., 1999.

Rubenstein, D.I., The greenhouse effect and changes in animal behavior: effects on social structure and life-history strategies, in *Global Warming and Biological Diversity,* Peters, R.L. and Lovejoy, T.E., Eds., Yale University Press, New Haven, 1992, pp. 180–192.

Sagarin, R., Historical studies of species' responses to climate change: promise and pitfalls, in *Wildlife Responses to Climate Change,* Schneider, S. and Root, T., Eds., Island Press, Washington, D.C., 2001, pp. 127–163.

Sagarin, R.D., Barry, J.P., Gilman, S.E., and Baxter, C.H., Climate related changes in an intertidal community over short and long time scales, *Ecol. Monogr.*, 69, 465–490, 1999.

Sanford, E., Regulation of keystone predation by small changes in ocean temperature, *Science*, 283, 2095–2097, 1999.

Scott, D. and Poynter, M., Upper temperature limits for trout in New Zealand and climate change, *Hydrobiologia*, 222, 147–151, 1991.

Southward, A.J., Hawkins, S.J., and Burrows, M.T., Seventy years' observations of changes in distribution and abundance of zooplankton and intertidal organisms in the western English Channel in relation to rising sea temperature, *J. Therm. Biol.*, 20, 127–155, 1995.

Sparks, T.H. and Yates, T.J., The effect of spring temperature on the appearance dates of British butterflies 1883–1993, *Ecography,* 20, 368–374, 1997.

Thomas C.D. and Lennon, J.J., Birds extend their ranges northwards, *Nature*, 399, 213, 1999.

Underwood, A.J., *Experiments in Ecology*, Cambridge University Press, Cambridge, 1997.

Welch, D.W., Ishida, Y., and Nagasawa, K., Thermal limits and ocean migrations of sockeye salmon (*Onco-rhynchus nerka*): long-term consequences of global warming, *Can. J. Fish. Aquat. Sci.*, 55, 937–948, 1998.

Woodward, F.I., A review of the effects of climate on vegetation: ranges, competition, and composition, in *Global Warming and Biological Diversity,* Peters, R.L. and Lovejoy, T.E., Eds., Yale University Press, New Haven, 1992, pp. 105–123.

CHAPTER 52

Use of Long-Term Field Trial Datasets in Forestry to Model Ecosystem Responses to Environmental Change

Csaba Mátyás

INTRODUCTION: RESPONSE LEVELS OF FOREST ECOSYSTEMS TO ENVIRONMENTAL CHANGE

Domestication of forest trees is still in its beginnings. In the majority of cases, forest tree populations still maintain their unimproved, wild condition; this provides a comparatively wide adaptability to the diverse site conditions encountered in forestry. With the widespread use of commercial tree species outside their natural range, however, forest stands became more susceptible to environmental extremes and changing ecological conditions. Mass mortality experienced (e.g., in artificially established Norway spruce forests of Central Europe following a prolonged drought period in the 1990s) directed the interest of ecologists and foresters alike in the ability of forest tree populations to adjust to changing conditions, that is, to their adaptation potential.

The issue of adaptedness appears as a crucial problem for forestry in view of the expected climate instability. The longevity of trees makes a fast adjustment to changed conditions more difficult than in agriculture. Long-term adaptedness and stability should be therefore of higher concern than possible gains in timber or fiber yield. Due to the life span of trees, climate fluctuations, secular changes, and rare events must have shaped the natural adaptive strategy both on the individual and population level. Trees are therefore interesting objects for studying long-term adaptation to the environment.

Adaptation in a genetic sense means the change of the gene pool of a population in response to changes in external conditions. However, genetic consequences of climate or environmental changes in a population cannot be separated from processes in the ecosystem. At least three organizational levels of adjustment that must be considered are studied by different disciplines (ecology, coenology, plant geography, genetics, and physiology):

1. On the ecosystem level, prolonged environmental stress or changing conditions will lead first to the loss of sensitive species of secondary importance, later to the substitution of dominant species. At the same time, other better adapted species will immigrate into the ecosystem. The tolerance limits of the species, and the stability of the species community, will decide to which extent and how fast these changes occur. This phenomenon is described by ecologists as succession or degradation (depending on the direction and quality of species composition changes).
2. On the level of the population (or species), environmental stress triggers a selection process that manifests itself through increased mortality. This process will change the gene frequency of the

population; less adapted individuals will leave less progeny or disappear altogether. The rate of adjustment will be largely influenced by the availability of individuals with suitable traits (i.e., by the genetic variability contained in the population). This is the process of natural selection in the sense of Darwin, and is termed genetic adaptation by geneticists. Genetic adaptation is understood as a change in gene frequency, directed toward a theoretical optimum in a given ecological situation. The genetic transformation is expected to increase the average fitness of the population.

3. Finally, on the individual level the question of survival or death is decided by the resistance or tolerance of the organism to cope with changing conditions. Plasticity (stability or homeostasis) is a physiological trait that is, of course, also determined by heredity. Phenotypic plasticity describes the ability of the organism to change its expressed phenotype as a response to changing environmental conditions. Phenotypic changes are transitory (mostly morphological) and do not involve genetic change of the organism itself.

Therefore, the ability to withstand changes on all three levels depends on the adaptive potential and adaptedness, the determination of which is, however, certainly a question of interpretation. This study examines first of all the genetic adaptation on the level of populations.

Without disregarding conflicting opinions for assessing adaptedness, the suggestion of Ayala (1969) is used: "Adaptedness of a population may be judged on the relative ability (as compared to responses under other conditions) to transform available nutrients and energy into its own living matter." This corresponds to dry-matter production, and in forestry to growth and volume (timber) yield. Adaptational difficulties manifest themselves in deterioration of growth rate, hardiness, and resistance, which affect volume production directly and are therefore of critical importance in forestry. Out of the components of yield of a forest stand, height growth is under strong genetic control. Only this growth trait will be discussed.

Throughout this chapter the term adaptive variation is used to indicate climate-related genetic variation between populations, although it is clear that other forms of variation support adaptation (and fitness) as well. However, climate-related variation is the most prudent interpretation of adaptive genetic variation. This chapter addresses phenomena observed on widespread, outbreeding tree species, mostly conifers. Taxa with divergent reproduction systems or fragmented distribution areas may exhibit different patterns, depending on their evolutionary past.

DISCOVERING LONG-TERM EFFECTS OF CLIMATIC FACTORS ON HEREDITARY TRAITS: A RETROSPECTIVE

The awareness that plant development is affected by climate can be traced back into antiquity. The understanding, however, that species distributed across climatically different regions adapt to local ecological (climatic) conditions through adjusting their genetic variation is relatively recent.

Interestingly, forest trees first attracted attention. In the 18th century, operators of naval shipyards were seriously concerned about the declining supply of good-quality oak and pine timber, and became interested in the question whether seed provenance might influence growth and form of trees in forest plantations. *Provenance* means simply the source of a population sample representing a defined area. The provenances originate from seed collected from identified stands or regions. The inspector general of the French navy, the prominent botanist Duhamel du Monceau, established the first comparative trials in France with Scotch pine seed of Baltic origin between 1745 and 1755. However, no records of his pioneering work survived. Seventy years later, André de Vilmorin endeavored in the 1820s to repeat Duhamel's effort. His results were published after 36 years of observation in 1862, stating differences between provenances and proposing the existence of climate-related genetic variation within the species (Wright and Bull, 1963).

Failures with plantations established from imported seed prompted numerous further experiments in the second half of the 19th century. Outstanding with respect to their extent and design were the provenance experiments in Russia. Similar experiments were established throughout

Central and Northern Europe, Germany, Austria, and Sweden (for details see Langlet, 1971). The interest in provenance trials with forest trees was so great that when the International Union of Forest Research Institutions (IUFRO) was founded in 1892, its first tasks included initiating international provenance experiments. The scope of some of these tests is impressive. For example, the international provenance test with Norway spruce initiated in 1962 includes 1100 provenances from Siberia to Iberia; a total of 20 parallel experiments in 14 countries have been established (Weisgerber et al., 1976).

The importance of ecological factors, especially of climate, in shaping within-species (intraspecific) variability became widely recognized, however, only after experiments with herbaceous plants yielded similar results. The work carried out in California with various perennial species is presumably the best-known investigation of plant populations collected along latitudinal transects. The common garden experiments proved the existence of climate-related genetic differences within the species. The results published in four volumes (Clausen et al., 1940, and following publications) became a citation classic in ecological literature.

AIMS AND SIGNIFICANCE OF PROVENANCE RESEARCH

Provenance research is the expression used in forestry for the analysis of common garden plantations of tree populations originating from geographically different locations.

Provenance tests of trees are time-consuming, require large areas, and are very costly. Nevertheless, provenance testing became a standard procedure because of the economic interest in maintaining vitality and growth vigor of forest stands until maturity. Provenance testing has been applied to practically all major forest tree species of the Northern Hemisphere and also to numerous tropical species of economic interest. Many tests originate from international cooperation coordinated by IUFRO, Food and Agricultural Organization of the United Nations, or other agencies.

The principal goal of provenance tests is to identify stands, populations, or areas that provide the most desired traits and commercially best results at the test location. Beyond strongly practice-oriented intentions serving silviculture, the tests offer an excellent opportunity to analyze intraspecific diversity in areas other than growth traits (molecular genetic, biochemical, physiological, morphological variation) that were also exploited for taxonomic purposes.

The tracing of between-provenance variation probably represents the most powerful available tool for testing hypotheses of climatic adaptation in trees. Instead of analyzing genetic changes in subsequent generation — an unthinkable task in forestry — the observed geographic variation can be interpreted as an adaptive response to changes in climate conditions. The transfer of populations to provenance test sites may be regarded as a simulation of environmental change over time, and the response may be modeled (Mátyás and Yeatman, 1987, 1992; Mátyás, 1994). This approach offers direct applications in forecasting climate change effects on trees and forests.

Compared to common garden experiments with short-lived plants, the long duration of the tests is of great advantage. Effects of annual weather fluctuations and rare anomalies are integrated in the result. Longevity provides for a more reliable response to the given environment than in the case of ephemeral plants. This does not imply a much closer adaptation of trees to local conditions; a perfect adaptation would result in narrowing genetic variation, reducing adaptability, and, in the long run, in the extinction of the species. On the contrary, adaptive strategies must provide for the maintenance of a sufficient level of variability to adjust to continuous and unpredictable fluctuation of conditions during the tree's life span and into an obscure future.

These distinct evolutionary and ecological implications make provenance tests important and interesting objects to study beyond direct silvicultural applications. Although provenance research might be among the most important contributions of forestry to biological sciences, up to now its results have failed to capture much attention outside the forestry community. Even the fact that much of the climatic adaptation research has been initiated and studied on forest trees first is not generally known.

GENETIC VARIATION AND CLIMATE FACTORS

To demonstrate the effects of climate on intraspecific genetic variation, plotting provenance data against an important climate (or geographic) factor of the locations of origin is sufficient (Table 52.1). The various traits show independent patterns depending on the effect the climate has on that specific trait. Correlation between trait variation and climate parameters suggest the selective pressure exerted on the trait, in a sense demonstrating its adaptive significance. Comparing different types of traits, the variation of growth- and phenology-related traits seems to be in closest agreement with climatic factors. Essentially neutral traits, such as morphological and molecular genetic variation, generally do not correlate well with climate factors.

From a forestry point of view, growth and survival traits are the most important. In temperate and boreal forests, the adaptive variation pattern of growth-related traits (including height increment, bud break and formation, and hardiness) are primarily shaped by the thermoperiod. Growth and phenology data of provenances correlate best with thermal parameters, whereas heat sum, temperature average, maxima, and minima yield roughly the same results. These findings are also in agreement with experiments in controlled environments (Mátyás, 1997).

The effect of photoperiod on trees has already been demonstrated by early studies, especially because latitude data of the origins are far easier to obtain than thermal parameters (a classical study is that by Vaartaja, 1959). In most cases, however, a strong confounding effect exists between photoperiod (latitude) and thermal conditions for obvious reasons. Surprisingly, moisture-related intraspecific differentiation is difficult to demonstrate (Table 52.1). Its effect is probably included in the thermal component.

Table 52.1 Correlation of Growth and Morphology Traits of 35 East European Scotch Pine Populations with Climate Factors at the Location of Origin

	Number of Frost-Free Days	Geographic Latitude (photoperiod)	Average July Temperature	Annual Precipitation
11-year height	0.81	−0.69	0.62	−0.14
Height increase, 11th year	0.86	−0.74	0.69	−0.17
Branch number, top whorl	0.44	−0.56	0.41	−0.18
Branch angle	0.70	−0.54	0.35	−0.08
Needle-cast resistance	0.46	−0.15	0.06	0.29
Bud break	−0.84	0.72	−0.62	0.11
Bud set	−0.82	0.75	−0.58	0.17
Winter color of needles	0.80	−0.72	0.75	−0.29

Correlation coefficients indicate the varying selection pressure on the respective trait (measured in Kámon, Hungary.)

From Mátyás, C., *Euphytica*, 92, 45–54, 1996. With permission.

The most striking feature of provenance tests is the generally low sensitivity of populations to changing environments, even with regard to highly adaptive traits such as height growth. Tree populations transferred over large distances into very different environments are able to grow and compete with the native, local populations, even if survival data indicate that only a certain fraction of the population exhibits the necessary plasticity (Table 52.2). Numerous studies verified a high level of individual and population homeostasis (for Norway spruce and Scotch pine: Eriksson, 1982; Southern pines: Wells, 1969; white ash: Kung and Clausen, 1984). As a consequence, between-population genetic differences are not well differentiated within considerably large geographic areas in the absence of steep gradients (e.g., mountain slopes). On contiguous plains the distance between populations with measurable growth differences may exceed 50 to 100 km along ecological gradients.

Table 52.2 Ten-Year Height and Survival of Some Transferred Scotch Pine Provenances in an Experiment near Moscow

	In the Moscow Test		At the Location of Origin	
Provenance	Average Height (cm)	Survival (%)	Annual Mean Temperature	Geographic Latitude
Moscow (local)	400	87	3.4	56
Minsk	409	88	5.0	54
Brest	435	92	7.0	52
Volynsk	397	65	7.0	50
Donetsk	418	68	8.0	48

Note the comparable growth of populations from substantially warmer environments; maladaptation is indicated by the decline of survival.

From Mátyás, C., *Euphytica*, 92, 45–54, 1996. With permission.

PREDICTING EFFECTS OF CLIMATE CHANGE

Numerous approaches can be used to estimate the transfer effect along an ecological gradient (Westfall, 1992) that might be used for modeling the effects of environmental change. The assessment of effects of climate change with the help of multiple regression analysis has been proposed by the author (Mátyás, 1994, 1996). The use of existing provenance tests to predict such effects is based upon the aforementioned concept that spatial (geographic) variation may be interpreted as a simulation of responses to environmental change over time. Accordingly, response regressions of growth of transferred populations on thermal parameters may be interpreted as a model of response to temperature change scenarios.

The investigation of some conifers indicates that a relative growth increase until a certain limit of transfer into cooler environments is a phenomenon regularly encountered in provenance tests, at least into environments not too close to the upper distribution limits (physiological tolerance limits) of the species. (Relative growth means that data are related to the performance of the local populations. In absolute terms, growth decreases naturally under conditions in which the temperature sum is lower.) It is the result of complex hereditary and other biotic effects that demonstrate the repressive character of biological regulation (Mátyás, 1996). On the other hand, raising temperature affects growth negatively through increasing drought stress. As an example, Table 52.3 shows the interpretation of the general response of populations calculated from the dataset of 45 families of ponderosa pine (*Pinus ponderosa*) from the Sierra Nevada in California at four sites. The data were obtained from three-dimensional response regression functions, in which the transfer effect (the ecological distance) was one of the variables, and the ecological conditions at the test site the other (Mátyás, 1997; Morgenstern, 1996; Westfall, 1992).

The growth rate visibly decreases with increasing pace as average annual temperature increases. An increase of only 1°C (or 1.8°F) already leads to a growth loss of roughly 5%, and 3°C (or 5.4°F) corresponds to a loss of 13% of the original height at mid-elevation. The sensitivity difference between high-elevation and mid-elevation conditions, exhibiting a stronger response at higher altitudes, has to be treated with caution and needs further confirmation.

The calculated decline in height growth is only an indication for actual productivity loss, because a similar decline in diameter and basal area has a quadratic effect on stem volume. In addition to growth loss, other consequences of maladaptation increase as a function of growing temperature (or ecological) distances. These include susceptibility to snow break or the loss of resistance to diseases and pests that are otherwise often harmless or of minor importance (e.g., saprophytic fungi may turn into real parasites). A direct link between ecological distance and mortality could be proven for *Pinus banksiana* (Mátyás and Yeatman, 1992).

Temperature increase triggers an upward shift of species distribution bands and, in consequence, of elevational vegetation zones. Within the elevational band of distribution, plant species are clearly

Table 52.3 Estimated Change of Height Growth of California Ponderosa Pine Families at Age 12 as a Result of Different Annual Average Temperature Change Scenarios

Test Site		Annual Mean Temperature Change (°C)					
		4.5	4	3	2	1	0
High elevation	cm	406.9	422.4	449.1	474.8	506.9	537.8
	%	75.6	78.5	83.5	88.3	94.3	100
Mid-elevation	cm			475.1	496.7	520.0	544.0
	%			87.3	91.3	95.6	100
Low elevation	cm					504.6	524.1
	%					96.3	100

Responses are calculated for three elevational positions; no reliable estimates are available for the empty cells.

structured into altitudinal races or clines, as already proven by the classical study of annuals by Clausen. The same is true for tree populations. Because of the altitudinal genetic differences, 300 m of elevation is generally proposed as the critical limit of safe transfer by ecological genetic studies on some western conifers. The temperature lapse rate for the Sierras was established as 0.54°C/100 m or 2.98°F/1000 ft. Elevation of 300 m therefore equals a temperature increase of 1.62°C of yet relatively safe adaptation (Mátyás, 1997). A further increase of temperature will inevitably lead to significant losses in productivity, as shown in Table 52.3.

Considering the long regeneration cycle of trees, the pace of the forecasted climate changes will be too fast to provide enough time for proper genetic adaptation of populations. A temperature shift will in any case pose a threat to most of the populations. At low elevations close to the lower limit of the species, temperature increase certainly will lead to the thinning out and disappearance of the species as it loses its competitive ability against other species. At the upper limits, temperature increase will bring an improvement of site conditions. This offers the possibility of a migration advance for the species in the long run, as often modeled in climate scenarios. However, apart from the fact that natural migration speed of forest trees is relatively low, even in the long run the utilization of climatically improved zones off the present distribution range will be limited by soil conditions. The development of soil profiles takes millennia and the usually shallow, less developed soils of high altitudes and high latitudes will not change for a long time. Thus, contrary to general belief, the site potential will not follow the improvement in climate.

CONCLUDING REMARKS

Common garden tests are carried out in forestry with easy-to-propagate, widely distributed species. The majority of species in large-scale test networks up to now have been conifers that predominantly occupy the initial phases of succession in temperate forest communities. It is therefore not certain that the observed phenomena apply to all widely distributed tree species. The described modeling method has its limitations as well, as the effect of climate variables at a given test location cannot be evaluated in its full complexity. In the described approach the effects of latitudinal (north–south) transfer have been neglected, although changes in the photoperiod might have some effect on light-sensitive species. An additional constraint for the interpretation is the undoubtedly changed environment of a test plantation as compared to a naturally regenerating forest; this cannot be overcome easily.

With the above limitations, the described effects of temperature change on growth and productivity of tree populations seem to have a rather general validity, even though the magnitude and type of response is determined by the actual genetic structure and tolerance of the species. Contrary to the general perception of physiological reaction to environmental change, the response of adapted, local populations may be assumed to be asymmetric due to the manifold constraints of biologic regulation. As a result, temperature increase leads to relatively fast growth and productivity loss and selective mortality. Apart from economic consequences, growth decline also affects the functioning of the forest ecosystem as a carbon sink in a twofold manner. With increasing temperature, the quantity of sequestered carbon decreases and simultaneously the decomposition rate of dead organic matter accelerates, causing additional carbon release into the atmosphere.

An important feature of the modeling of climate change effects, as presented in the example, is the fact that it is based on directly measured data and therefore the conclusions are fairly realistic. Such information is available in provenance tests for many commercially important tree species in many parts of the world. Their evaluation may contribute to a better understanding of natural processes triggered by environmental changes, and may help in formulating protective strategies.

The results indicate that because of the conservative nature of the genetic adaptation process, and of the relative speed of expected changes, even agricultural crops will demand a strategy to facilitate adaptation. Long-lived, immobile organisms such as trees will especially need human interference in order to enhance gene flow and adaptation to altered conditions, in spite of an impressive hereditary capacity. National forest policies have to incorporate this task into the agenda of the next decades (Eriksson et al., 1993). To counteract genetic erosion and extinction, populations along the southern (or low-elevation) limits of species distribution areas will need special attention.

REFERENCES

Ayala, R.J., An evolutionary dilemma: fitness of genotypes versus fitness of population, *Can. J. Genet. Cytol.*, 11, 439–456, 1969.

Clausen, J., Keck, D.D., and Hiesey, W.M., *Experimental Studies on the Nature of Species*, Vol. I–IV, Nos. 520, 564, 581, 615, Carnegie Inst. Publ., Washington, D.C., 1940, 1945, 1948, 1958.

Eriksson, G., Ecological genetics of conifers in Sweden, *Silva Fenn.*, 16, 149–156, 1982.

Eriksson, G., Namkoong, G., and Roberds, J.H. Dynamic gene conservation for uncertain futures, *For. Ecol. Manage.*, 62, 15–37, 1993.

Kung, F.H. and Clausen, K.T., Graphic solution in relating seed sources and planting sites for white ash plantations. *Silva Genet.*, 33, 46–53, 1984.

Langlet, O., Two hundred years of genecology, *Taxon*, 20, 653–722, 1971.

Mátyás, C., Modeling climate change effects with provenance test data, *Tree Physiol.*, 14, 797–804, 1994.

Mátyás, C., Climatic adaptation of trees: rediscovering provenance tests, *Euphytica*, 92, 45–54, 1996.

Mátyás, C., Ed., *Perspectives of Forest Genetics and Tree Breeding in a Changing World*, Vol. 6, IUFRO World Series, Vienna, 1997.

Mátyás, C. and Yeatman, C.W., Adaptive variation of height growth of *Pinus banksiana* populations, *Erd. Faipari. Egy. Közl., Sopron*, 1–2, 191–197, 1987 (in Hungarian).

Mátyás, C. and Yeatman, C.W., Effect of geographical transfer on growth and survival of jack pine populations, *Silva Genet.*, 43, 370–376, 1992.

Morgenstern, E.K., *Geographic Variation in Forest Trees*, UBC Press, Vancouver, 1996.

Vaartaja, O., Evidence of photoperiodic ecotypes in trees, *Ecol. Monogr.*, 29, 91–111, 1959.

Weisgerber H., Dietze, W., Kleinschmidt, J., Racz, J., Dietrich, H., and Dimpflmeier, R., Ergebnisse des internationalen Fichten-Provenienzversuches 1962. Teil 1: Phenologische Beobachtungen und Höhenwachstum bis zur ersten Freilandaufnahme, *Allg. Forst. Jagd.*, 147, 227–235, 1976.

Wells, O.O., Results of the southern pine seed source study through 1968–69, in *Proc. 10th South. Conf. For. Tree Impr.*, Houston, 1969, 117–129.

Westfall, R.D., Developing seed transfer zones, in *Handbook of quantitative forest genetics,* Fins, L., Friedman, S.T., and Brotschol, J.V., Eds., Kluwer Academic Press, Dordrecht, 1992, pp. 313–398.

Wright, J.W. and Bull, W.I. Geographic variation in Scotch pine, *Silva Genet.*, 12, 1–40, 1963.

CHAPTER 53

The Possible Impacts of Climate Change on Pacific Island State Ecosystems

William C.G. Burns

> As we are all aware, mini-ecosystems are amongst the most fragile. Such fragility today places small island states in the frontline of nature's reaction to humanity's overuse and abuse of the environment.
>
> — *Maumoon Abdul Gayoom*
> *President of the Republic of Maldives*

INTRODUCTION

While most nations likely will ultimately suffer adverse consequences from climate change,[1] small island states may face the most dire and immediate consequences. A previous piece by this author focused on the socioeconomic and cultural ramifications of climate change on Pacific Island Developing Countries (PIDCs).[2] This chapter examines the possible impacts of climate change on the ecosystems of PIDCs.

PACIFIC ISLAND DEVELOPING COUNTRIES: AN OVERVIEW

PIDCs consist of 22 political entities, of which 15 are politically independent.[3] The region encompasses well over 1100 islands and islets, of which approximately 500 are inhabited.[4] With the exception of Papua New Guinea and Fiji, all PIDCs fall within the United Nations' definition of "small island states," which are islands with less than 10,000 km^2 in land mass and with less than 500,000 inhabitants.[5] The majority of countries in the region fall well below these thresholds, with populations of less than 200,000 and land areas well below 1000 km^2.

The physical geography of PIDCs can be divided into two broad categories — high and low islands — with a further division into continental and volcanic islands on one hand, and atolls and limestone islands on the other. Most PIDCs, however, do not fall into discrete categories, but rather are a combination of island types.[6]

GLOBAL AND REGIONAL PROJECTIONS OF CLIMATE CHANGE

Incorporating the Intergovernmental Panel on Climate Change's (IPCC) "business as usual" projections for greenhouse gas emissions over the next century into a general circulation model, the Hadley Centre for Climate Prediction and Research recently predicted an increase in global mean surface temperatures of 3°C by 2100,[7] while Wigley's most recent assessment, incorporating the IPCC's latest "marker" scenarios, yields a range of 1.3 to 4.0°C over the next century.[8] Based on varying assumptions of future population and economic growth, land-use patterns, and energy policies, the IPCC projects an increase in sea levels of between 15 and 95 cm by the year 2100, with a best estimate of 49 cm.[9] This is a rate two to four times greater than that experienced in the previous century.[10] Moreover, levels could continue to rise for several centuries after greenhouse gas emissions are stabilized.[11]

In projecting climate trends in the PIDCs over the next century, it must be emphasized at the outset that regional assessments of climate change remain fraught with uncertainty:

> [General circulation models] have difficulty in reproducing regional climate patterns, and large dis-crepancies are found among models. In many regions of the world, the distribution of significant surface variables, such as temperature and rainfall, are often influenced by the local effects of topography and other thermal contrasts, and the coarse spatial resolution of the GCMs can not resolve these effects.[12]

Indeed, given the massive expanse of the Pacific island region and substantial topographical variations, it is likely that climate change will result in markedly different manifestations across the region.[13]

Climate researchers have developed several strategies to conduct regional assessments. Nested models seek to simulate regional climates by the application of limited area models nested in a general circulation model.[14] In recent years, some of these models have yielded high correlations between regional climate predictions and observed climatic phenomena, including precipitation, thermal inertia of water bodies, and temperature.[15] Downscaling by statistical means, deriving statistical relationships between observed local climatic variables and large-scale variables, also have proved successful, *inter alia,* in linking large-scale spatial averages of precipitation and surface temperature to local precipitation and temperature–time series.[16]

However, most efforts to project changes in the region are still conducted with atmosphere-ocean GCMs with insufficient horizontal resolution to simulate island climates.[17] Moreover, future climate projections will most likely continue to be dominated by uncertainties in radiative feedbacks associated with responses of water vapor, cloud cover, and sea ice to warming.[18] At the regional level, these feedbacks may be substantially influenced by systemic nonlinearities that may prove extremely difficult to forecast.[19]

With these caveats in mind, recent research by New Zealand's National Institute of Water and Atmospheric Research (NIWA) in collaboration with France's meteorological service, Meteo France, concludes that a significant change has occurred in the Pacific climate during the past 20 years, including an eastward movement of the South Pacific Convergence Zone (SPCZ), consistent with model predictions. The study reveals that central and western Kiribati, Tokelau, and north-eastern French Polynesia became 0.3°C warmer between 1977 and 1994, with a 30% increase in rainfall compared with pre-1977 averages. New Caledonia, Vanuatu, Fiji, Tonga, Samoa, and the southern Cook Islands experienced a strong warming of 0.4 to 0.6°C from 1900 to 1977, which has since slowed to 0.1°C, with average rainfall declining by 15% after 1977.[20] Based on data from 34 stations in the Pacific, New Zealand's Meteorological Services concluded that surface area temperatures have increased by 0.3 to 0.8°C this century, with the greatest increase in the zone southwest of the SPCZ.[21]

The most recent IPCC assessment predicts that PIDCs will experience "moderate warming."[22] Over the next century, however, a recent experiment with a global coupled ocean-atmosphere GCM projected increases of sea surface temperatures in the Pacific region (under conditions of CO_2 doubling) of 3.49°C in the eastern Pacific and 2.21°C in the western Pacific.[23] These results "resemble not only the climate anomalies associated with present-day El Niño-related events in many areas, but also the decadal timescale climate anomalies observed during the 1980s."[24]

In terms of projected sea level rise in the region, most sea level monitoring sites in the South Pacific are recording accelerated rises of up to 25 mm per year, more than 10 times the trend this century — findings that have been validated by satellite data showing 20 to 30 mm rises from Papua New Guinea southeast to Fiji.[25] While these dramatic increases may be attributable to El Niño/Southern Oscillation (ENSO),[26] recent research indicating that regional climate change in the future may closely track the effects of El Niño[27] leads to the inference that sea levels may increase substantially over the next century in the Pacific. The IPCC projects that thermal expansion alone will raise sea levels in the southwest Pacific by 28 to 32 cm at the time of CO_2 doubling,[28] although regional projections in this context remain highly speculative because dynamic ocean effects have yet to be effectively modeled.[29]

Projected buildups in greenhouse gas emissions will likely raise ocean temperatures and ocean surface water temperatures to above 26°C in the next century.[30] This could result in a greater exchange of energy and add momentum to the vertical exchange processes critical to the development of tropical typhoons and cyclones.[31] Therefore, some researchers estimate that the occurrence of tropical typhoons and cyclones could increase by as much as 50 to 60%,[32] and their intensity by 10 to 20%.[33]

However, by no means do scientists universally agree that climate change will cause an increase in violent weather events on PIDCs. Some researchers believe that the purported linkage between increased ocean temperatures and violent weather events is overly simplistic, citing other factors that influence storm development, including atmospheric buoyancy, instabilities in the wind flow, and vertical wind shear.[34] Moreover, some climate scientists argue that ocean circulation changes associated with climate change may counter the effects of added warmth.[35] In its most recent regional assessment, the IPCC concluded that "model projections suggest no clear trend, so it is not possible to state whether the frequency, intensity, or distribution of tropical storms and cyclones will change."[36]

POSSIBLE ECOSYSTEM IMPACTS OF CLIMATE CHANGE ON PIDCS

Coral Reefs

Temperature Impacts

Coral reefs are rocklike ridges, composed of calcium carbonate in the form of aragonite, formed from the harder outer skeletons, or polyps, of coral animals.[37] Most polyps subdivide as they grow and their skeletons fuse together, creating complex coral colonies.[38] Reefs have been termed the "tropical rainforests of the ocean." While covering less than 0.2% of the ocean's area, they serve as the home for up to one quarter of all marine species.[39] Additionally, reefs protect coastal areas from erosion and storms,[40] and serve as a "sink" or absorber of carbon dioxide, helping to reduce the level of this potent greenhouse gas in the atmosphere.[41]

In the context of small island nations, coral reefs constitute "an extensive and vital"[42] component of the ecosystem. Coral reefs serve as a buffer against coastline erosion,[43] a function that will become even more critical in the future if climate change intensifies storm surges[44] and sea-level rise threatens coastal regions of PIDCs.[45] Reefs provide habitat for fish species that meet 90% of the protein needs of PIDC inhabitants,[46] as well as support the livelihood of small-scale fishers in the region.[47] Moreover, coral reefs are the primary source of carbonate sand that constitutes the majority of beach deposits on PIDCs.[48]

The projected temperature rise over the next century associated with climate change may pose the greatest long-term threat to coral reef ecosystems in PIDCs. Coral reefs have extremely narrow temperature tolerances of 25 to 29°C with some species in PIDCs currently living near their threshold of thermal tolerance.[49] Water temperature increases of 1 to 2°C over an extended period can result in coral "bleaching," whereby dinoflagellates, endosymbiotic algae species that live in coral fish, are expelled or reduced.[50] Because coral derive most of their food from dinoflagellates,[51] they can quickly die in their absence.[52]

In 1998, with global mean surface temperature reaching its highest level in recorded history, and sea surface temperatures increasing 2 to 3°C in many tropical areas where reefs are found, and as much as 4 to 6°C in some areas,[53] the world's reefs "appear to have suffered the most extensive and severe bleaching and subsequent mortality in modern record."[54] Approximately 95% of shallow water corals in the Maldives died, as well as 75% of corals in the Seychelles Marine Park System.[55] Massive bleaching also occurred on the coasts and islands of India, Kenya, Tanzania, and in Australia's Great Barrier Reef.[56]

It is difficult to definitively link reef bleaching and mortality to temperature increases because other stressors also can contribute to the phenomenon, including pollution, sediment loading, a reduction in marine salinity, intense solar radiation, and exposure to the air.[57] However, a U.S. Department of State report on the 1998 events concluded that only warming could have caused such extensive bleaching throughout virtually all of the disparate reef regions of the world, including remote areas.[58] Moreover, the authors of a multifactorial analysis of coral bleaching events in the 1980s recently concluded that "of all stresses which could potentially cause widespread mass bleaching, only excessively high temperature was present in all cases."[59]

Reefs can recover from bleaching if conditions improve, with coral larvae settling on the reef structure to renew the building process. However, if elevated sea surface temperatures and bleaching persists, new building will not occur and the reef frame can gradually erode, resulting in habitat destruction and mortality.[60] For example, 1 year after the 1998 mass-bleaching incident, 80 to 90% of the bleached corals in the more severely affected areas of the Indian Ocean had died, including previously resistant species.[61] Moreover, the synergism between rising sea levels and anthropogenic damage to reefs, such as destructive fishing practices[62] and pollution,[63] may impede reef recovery.[64]

Sea-Level Rise Impacts

Most researchers believe that reefs' maximum sustained vertical accretion rate of 10 mm per year will be adequate to keep up with sea-level rise of 1 m or less,[65] and that even slowly accreting reef flats should be able to cope with projected sea-level rise over the next century.[66] Historical evidence supports this conclusion, with most reefs keeping up with sea level rises of 20 cm per decade between 14,000 and 6000 years ago.[67]

Indeed, some researchers argue that rising sea levels may be beneficial for coral reef ecosystems by inducing vertical growth.[68] However, under the stress of rising sea levels, reefs are likely to develop at deeper average depths in the future, exposing coastlines to greater wave and current effects.[69]

Other Climate Impacts on Reefs

Should climate change in the region result in an increase in violent storm activities,[70] surges could further degrade already weakened coral reef ecosystems.[71] In addition, heightened CO_2 levels in the atmosphere may increase the acidity of surface ocean water, reducing the concentration of calcium carbonate ions, thus inhibiting the ability of coral-forming organisms to create their skeletons.[72] It has been recently projected that a doubling of CO_2 could reduce reef calcification rates by approximately 17% in 2065 and 35% by 2100.[73]

Mangroves

Sea-Level Rise

Mangroves, also known as mangals, constitute a group of 34 tree species that grow in sheltered conditions in shallow tropical and subtropical waters.[74] More than 343,000 hectares of mangroves exist in the Pacific, with the largest stands occurring in Papua New Guinea, the Solomon Islands, Fiji, and New Caledonia.[75]

In addition to providing a range of products for humans, including construction material, firewood, tannin, and herbal medicines,[76] mangroves comprise a critical ecosystem in many PIDCs. Mangroves serve as important nursery and feeding sites for nekton, including many fishery species, with surveys of fish and crustacean assemblages around mangroves recording high levels of diversity and abundance.[77] Also, mangrove trees serve as filters for sediment that threaten coral reefs,[78] and island pests and exotic insects,[79] and help to detoxify contaminants in PIDC waters.[80]

Mangrove communities often can cope with sea-level rise where sedimentation rates are commensurate with or exceed local sea-level rise.[81] However, most small island states are characterized by microtidal, sediment-poor environments. Stratigraphy from low-island mangrove systems in the Pacific, including Tongatapu, Tonga, the Marshall Islands, Kiribati, and Tuvalu, reveal sediment accumulation rates of only 12 cm per 100 years.[82] Thus, low-island mangroves are expected to suffer reductions in geographical distribution from projected sea-level rise over the next century.[83] Should the IPCC middle-range estimates of sea-level rise come to fruition over the next century, high-island mangroves with sediment accumulation rates of 45 cm per century could also be threatened.[84] Increased salinity caused by sea-level rises may also result in decreased net productivity and stunted growth in certain species. Changes in competition between mangrove species also can be anticipated.[85]

Temperature Impacts

Warming in the future should prove beneficial to PIDC mangrove ecosystems. Increased temperatures will increase the diversity of higher latitude marginal mangroves, facilitating expansion into mangrove margins occupied currently only by *Avicennia* species, as well as expansion of mangroves into salt marsh environments. Warming also can be expected to increase mangrove productivity, characterized by increased growth and litter production.[86]

Other Climate Impacts on Mangroves

Mangroves operate in the C_3 pathway of carbon fixation for photosynthesis.[87] Research indicates that increases in atmospheric CO_2 increase the productivity and efficiency of water use by C_3 plants.[88] Thus, the projected increases in CO_2 likely will enhance mangrove tree growth and litter production.[89]

Climate models predicts that islands north of 17°S may experience increases in annual rainfall, while islands south of 17°S may see rainfall decline.[90] Snedaker postulates that increased rainfall may benefit mangroves by reducing salinity and exposure to sulfate, while increasing delivery of terrigenous nutrients.[91] Conversely, he contends that decreased rainfall in some regions, with an attendant increase in evaporation, will reduce the extent of mangrove areas, particularly with the projected loss of the landward zone to unvegetated hypersaline flats.[92]

Seagrasses

Temperature Impacts

Seagrass ecosystems are valuable resources in coastal waters worldwide, including in the shallow, intertidal environments of many island states. Seagrass communities provide habitats for a wide variety of marine organisms, including meiofauna and flora, benthic flora and fauna, epiphytic organisms, plankton, and fish.[93] Additionally seagrass plants provide food for waterfowl, bind and stabilize bottom sediments, and improve water quality by filtering suspended matter.[94]

Seagrass species living near the upper limit of thermal tolerance, as is true in many coastal regions of PIDCs, may experience declines in productivity and distribution as sea surface temperatures rise in the future.[95] Increased temperatures also could alter seagrass distribution and abundance by affecting seed germination.[96]

Sea-Level Rise and Seagrasses

Elevated sea levels will reduce available light, which may limit plant photosynthesis and result in reduced distribution and decreased productivity.[97] The projected 50-cm increase in water depth due to projected sea-level rise over the next century may result in a 30 to 40% reduction in seagrass growth.[98]

Higher sea levels also may increase tidal currents, especially in areas of restricted tidal flow. Increased tidal currents can restrict the depth in which plants grow by exacerbating the stress that accompanies light limitation. This may result in a withdrawal of the deep edge of seagrass beds, which could result in a diminution of total seagrass area.[99]

Rising sea levels will result in increased inland and upstream penetration of salt water in tidal systems. Under high levels of salinity, some seagrass species experience declines in seedling growth, limiting both reproduction and distribution. Seagrasses also may suffer from salinity stress resulting from high external osmotic potentials, ion toxicity from excessive salt intake, and nutrient imbalances resulting from inadequate carrier selectivity during ion uptake.[100]

IMPACTS ON FLORA AND FAUNA SPECIES

Beyond the potential adverse consequences for marine species resulting from damage to coral reef and mangrove ecosystems, other climate change-related ramifications for PIDC flora and fauna species may manifest themselves.

Endemic species on islands are exceptionally vulnerable to extinction or extirpation for several reasons, including the fact that island populations are usually small, isolated, and often endemic to a single island, and typically have evolved with minimal competition, predators, or disease.[101] Moreover, on most islands such species face severe threats from pollution, overexploitation, and poor management.[102] Thus, PIDCs may have the world's greatest rate of endangered species (per capita or per unit of land area),[103] encompassing one of every three threatened plant species,[104] as well as a very high number of avian species.[105]

Climate change could exacerbate threats to island endangered species in several ways. Increased temperatures may result in a diminution of some species, as well as alter species distribution and composition in unpredictable ways.[106] Moreover, clam and sea turtles in coastal areas of PIDCs may be particularly susceptible to temperature change.[107] Warmer waters in bays and lagoons also might stimulate the growth of toxic algae, threatening predator avian species, such as cormorants and pelicans.[108] Inundation of low-lying wetlands also could reduce the abundance and diversity of freshwater species, as well as deltaic species such as turtles and crocodiles.[109]

While nontropical forests probably will be most adversely affected by climate change,[110] PIDC forests also may face threats in the future. In addition to possible forest losses from rising sea levels,[111] PIDC forests may be subject to potentially damaging temperature rises and changes in soil water availability.[112] Moreover, increased temperatures may result in more frequent outbreaks of pathogens and pests deleterious to forests, and may elevate the frequency and intensity of fires.[113] However, increased amounts of carbon dioxide may enable some forest species to use water and nutrients more efficiently, thereby ameliorating or wholly offsetting some of the threatening aspects of climate change.[114]

THE UNITED NATIONS FRAMEWORK CONVENTION ON CLIMATE CHANGE AND SMALL ISLAND STATES

The primary legal instrument to confront the possible ramifications of climate change is the United Nations Framework Convention on Climate Change (UNFCCC),[115] which entered into force in 1994. (The UNFCCC had been ratified by 179 nations as of June 1999.[116]) However, while the parties pledged to "achieve ... stabilization of greenhouse gas concentrations in the atmosphere at a level that would prevent dangerous anthropogenic interference with the climate system,"[117] their record to this point has been disheartening.

Initially, the major greenhouse gas-emitting states agreed to "aim" to reduce their greenhouse gas emissions to 1990 levels by 2000. Yet, all industrialized nations flouted this pledge, leading the Organization for Economic Cooperation and Development to conclude that emissions from industrialized nations would actually rise by 11 to 24% over the next 15 years.[118]

At the Third Conference of the Parties of the UNFCCC, the parties adopted the Kyoto Protocol,[119] under which industrialized nations agree to reduce their collective emissions of six greenhouse gases by at least 5% below 1990 levels by 2008–2012.[120] However, hostility to the protocol by powerful sectors in the U.S., including organized labor, fossil fuel producers, and influential members of the Senate,[121] may thwart its adoption.[122] This would severely undercut the treaty's effectiveness, because the U.S. is responsible for approximately one quarter of greenhouse gas emissions.[123] Moreover, U.S. refusal to adopt the protocol might prompt European nations and Japan to balk.[124] This could doom the agreement, because it requires ratification by Annex I nations producing at least 55% of greenhouse gas emissions before it will come into effect.[125]

Additionally, even full implementation of the Kyoto Protocol would have very little effect over the next century. Because many greenhouse gases persist in the atmosphere for decades, "their radiative forcing — their tendency to warm Earth — persists for periods that are long compared with human life spans."[126] Thus, the legacy of greenhouse gas emissions from the past few decades will persist for many decades to come no matter what policies we adopt.

Additionally, the UNFCCC currently binds only developed countries and economies that are in transition to the reduction of greenhouse gas emissions.[127] However, given the tremendous projected increases in greenhouse gas emissions in developing countries over the next century,[128] the future effectiveness of the UNFCCC is contingent on engaging these nations in the regime's mission.[129] Yet, great trepidation exists among developing countries about possible economic impacts associated with reducing emissions, as well as a sense of unfairness given the tremendous disparity between per capita emissions of industrialized and those of developing nations.[130] Developing countries repulsed an effort at the Third Conference of the Parties in Kyoto to establish emission limitation objectives for wealthier developing states.[131] At the Fourth Conference of the Parties in Buenos Aires, Kazakhstan announced its intention to join Annex I and thus assume commitments under the protocol. However, Argentina's effort to place the issue of voluntary commitments on the agenda was rejected by the G77, most of which declined even to join in informal talks on the topic, and the Argentine proposal was formally excluded from the Fourth Conference agenda.[132]

As a consequence of the factors discussed above, Martin Parry recently projected that full implementation of the Kyoto Protocol will reduce projected warming by a mere 1/20 of 1°C, and will have a minimal effect on elevation of sea levels.[133]

Ultimately, the major greenhouse gas-producing nations may find it cost-beneficial to enact more stringent measures to reduce emissions.[134] As Cline suggests, under a "moderate-central" assumption of the damages caused by climate change, benefits ultimately will begin to exceed abatement costs on a global scale in 2050, with the benefits of damage avoidance rising to about 1% of world gross domestic product (GDP) by 2050, and 5% of GDP by 2275.[135]

Unfortunately, by that time, most small island states may have paid a terrible price, and some may even have ceased to exist. If the UNFCCC mandate to "prevent dangerous anthropogenic interference with the climate system" is to be complied with equitably (i.e., in a manner that seeks to protect all nations from the most serious ramifications of climate change), then major greenhouse–emitting nations should not wait to act until they deem it expedient. However, given the record of industrialized states in confronting climate change over the past few decades, it is difficult to be sanguine about the future. In a world governed more by realpolitik than international law, small island states likely will remain hostages to forces far beyond their control.

ENDNOTES

1. Torn, M.S., Evans, M., Fried, J., Will climate change spark more wildfire damage? *A Summary of Global Change in the Antarctic,* LBNL Rep. 42592, http://www.antcrc.utas.edu.au/scar/newsletter2/2summary.html, SCAR Global Change Programme, Nov., 1998; Blackman, K., Global Warming Worries Indigenous People, *Inter Press Service,* Aug. 13, 1998; Burns, W.C., The Second Session of the Conference of the Parties to the United Nations Framework Convention on Climate Change: More Heat Than Light?, *Colo. J. Int. Environ. Law Policy,* Y.B. 153, 1996.

2. Burns, W.C.G., The impact of climate change on Pacific island developing counties in the 21st century, in *Climate Change in the South Pacific: Impacts and Responses in Australia, New Zealand, and Small Island States,* Gillespie, A. and Burns, W.C.G., Eds., Kluwer, Dordrecht, 2000, p. 113.

3. Campbell, J.R., Contextualizing the effects of climate change in Pacific island countries, in *Climate Change: Developing Southern Hemisphere Perspectives,* Giambelluca, T.W. and Henderson-Sellers, Eds., Wiley Interscience, New York, 1996, p. 354.

4. Campbell, J.R., Contextualizing the effects of climate change in Pacific island countries, in *Climate Change: Developing Southern Hemisphere Perspectives,* Giambelluca, T.W. and Henderson-Sellers, Eds., Wiley Interscience, New York, 1996, p. 354.

5. Pernetta, J.C., Impacts of climate change and sea-level rise on small island states, *Global Environ. Change,* 2, 20, 1992; Bequette, J.F., Small islands: dreams and realities, *UNESCO Courier,* Mar. 1994, 23, 1994.

6. Campbell, J.R., Contextualizing the effects of climate change in Pacific island countries, in *Climate Change: Developing Southern Hemisphere Perspectives,* Giambelluca, T.W. and Henderson-Sellers, Eds., Wiley Interscience, New York, 1996, p. 354.

7. Hadley Centre for Climate Prediction and Research, Climate Change and Its Impacts, http://www.meto.govt.uk/sec5/CR_div/Brochure98/science.html, 1998; Global warming forecast raised one degree, <http://www.gnet.org>, *Global Network Environ. Technol. News,* Dec. 19, 1998.

8. Wigley, T.M.L., The Science of Climate Change, Report of the Pew Center on Climate Change, 1999, p. 21. These results are a shift upwards from the IPCC projections in its Second Assessment of 0.8 to 3.5°C by 2100 and reflect projected lower SO_2 emissions in scenarios being developed for the IPCC's upcoming Special Report on Emissions Scenarios. Hadley Centre for Climate Prediction and Research, Climate Change and Its Impacts, http://www.meto.govt.uk/sec5/CR_div/Brochure98/science.html, 1998; Global warming forecast raised one degree, http://www.gnet.org, *Global Network Environ. Technol. News,* Dec. 19, 1998.

9. Intergovernmental Panel on Climate Change, Contribution of Working Group I to the IPCC Second Assessment Report, IPCC-XI/Doc. 3, 1995, p. SPM.35. As the IPCC notes, its projections are premised on the assumption of minimal melting of the Greenland and Antarctic ice sheet. If this assumption proves incorrect, which the IPCC admits is a possibility, ocean levels could be elevated much more than originally predicted. One researcher at Victoria University in New Zealand recently warned that the Western Antarctic Ice Sheet may be on the point of melting, which could result in a 6-m rise in sea levels in less than a century; Ice sheet "on point of melting," http://www.press.co.nz/04/99012833.htm, *The Press* (New Zealand), Jan. 28, 1999. *See also* Foley, G., The threat of rising seas, *The Ecologist,* 29, 76, 1999; Haimson, L., Concerns grow over west Antarctic ice sheet, *Global Change,* Winter, 5, 1999.

10. Intergovernmental Panel on Climate Change, Third Assessment Report of Working Group 2 (Draft), 1999, p. 4.

11. Intergovernmental Panel on Climate Change, The IPPC Assessment of Knowledge Relevant to the Interpretation of Article 2 of the UN Framework Convention on Climate Change: A System, 1995, sec. 3.16.

12. Solman, S.A. and Nunez, M.N., Local estimates of global climate change: a statistical downscaling approach, *Int. J. Climatol.*, 19, 835, 1999. *See also* Everett, J.T. and Bolton, S., Lessons in Climate Change Projections and Adaptation: From One Living Marine Resource to Another, International Whaling Commission Seminar on Climate Change and Cetaceans, SC/M96/CC11, 1996.

13. Campbell, J.R., Contextualizing the effects of climate change in Pacific island countries, in *Climate Change: Developing Southern Hemisphere Perspectives*, Giambelluca, T.W. and Henderson-Sellers, Eds., Wiley Interscience, New York, 1996, p. 360.

14. Hadley Centre for Climate Prediction and Research, The Greenhouse Effect and Climate Change, 1999, p. 14; Kondratyev, K. Y. and Cracknell, A.P., *Observing Global Climate Change*, Taylor & Francis, London, 1998, p. 381.

15. Kondratyev, K. Y. and Cracknell, A.P. *Observing Global Climate Change*, Taylor & Francis, London, 1998, p. 383; Miller, N., Climatically sensitive California: past, present, and future climate, in *Potential Impacts of Climate Change and Variability for the California Region*, Report to the U.S. Global Change Research Program National Assessment, 1998, p. 25.

16. Solman, S.A. and Nunez, M.N., Local estimates of global climate change: a statistical downscaling approach, *Int. J. Climatol.*, 19, 836, 1999.

17. Watson, R.T. et al., The Regional Impacts of Climate Change, Special Report of IPCC Working Group II, 1998, p. 340.

18. Rind, D., Complexity and climate, *Science*, 284, 105, 1999.

19. Rind, D., Complexity and climate, *Science*, 284, 107, 1999.

20. South Pacific Regional Environmental Program, Research Shows Major Change in Pacific Climate, press release, Aug. 6, 1998.

21. Intergovernmental Panel on Climate Change, Third Assessment Report of Working Group 2 (Draft), 1999, p. 5.

22. Watson, R.T. et al., The Regional Impacts of Climate Change, Special Report of IPCC Working Group II, 1998, p. 340.

23. Meehl, *supra* note 10, at 142–143.

24. *Id.* at 145.

25. Chairpersons of the Third SPREP Meeting on Climate Change and Sea Level Rise, South Pacific climate change, http://www.cru.uea.ac.uk/tiempo/floor0/archive/issue26/t26art2.htm, *Tiempo*, 26, 1997.

26. Chairpersons of the Third SPREP Meeting on Climate Change and Sea Level Rise, South Pacific climate change, http://www.cru.uea.ac.uk/tiempo/floor0/archive/issue26/t26art2.htm, *Tiempo*, 26, 1997.

27. Curtis, S. and Hastenrath, S., Long-term trends and forcing mechanisms of circulation and climate in the equatorial Pacific, *J. Climate*, 12, 1134, 1999; Meehl, supra note 10, at 145-46.

28. Watson, R.T. et al., The Regional Impacts of Climate Change, Special Report of IPCC Working Group II, 1998, p. 341.

29. Klein, R.J.T., Assessment of coastal vulnerability to climate change, *Ambio*, 28, 182, 1999; Rahmstorf, S., Shifting seas in the greenhouse? *Nature*, 399, 523, 1999.

30. NASA Goddard Institute for Space Studies, How Will the Frequency of Hurricanes Be Affected by Climate Change? http://www.giss.nasa.gov/research/intro/druyan.02/, 1999; Karl, T.R., Nicholls, N., and Gregory, J., The coming climate, *Sci. Am.*, http://www.sciam.com/0597issue/0597karl.html, May 1997.

31. Insurers refuse to cover global warming risks, *The Independent*, May 8, 1992, p. 11.

32. NASA Goddard Institute for Space Studies, How Will the Frequency of Hurricanes Be Affected by Climate Change? http://www.giss.nasa.gov/research/intro/druyan.02/, 1999; Schlesinger, M.E., Model projections of CO_2-induced equilibrium climate change, in *Climate and Sea Level Change: Observations, Projections and Implications*, Warrick, R.A., Barrow, E.M., and Wigley, T.M., Eds., Cambridge University Press, Cambridge, 1993, p. 186; Haarsman, R.J., Tropical disturbances in a GCM, *Climate Dyn.*, 8, 247, 1993.

33. Karl, T.R., Nicholls, N., and Gregory, J., The coming climate, *Sci. Am.*, http://www.sciam.com/0597issue/0597karl.html, May 1997; Knutson, T.R., Tuleya, R.E., and Kurihara, Y., Simulated increase of hurricane intensities in a CO_2-warmed climate, *Science*, 279, 1018, 1998. *See also* Druyan, L.M., A GCM investigation of global warming impacts relevant to tropical cyclone genesis, *Int. J. Climatol.*, 19, 607, 1999. (Increased instability of lower troposphere under doubling of CO_2 scenario in GCM "significantly more favorable" for tropical cyclone genesis over tropical oceans.)

34. Karl, T.R., Nicholls, N., and Gregory, J., The coming climate, *Sci. Am.*, http://www.sciam.com/0597issue/0597karl.html, May 1997; Holland, G.J., The maximum intensity of tropical cyclones, *J. Atm. Sci.*, 54, 2519, 1995.

35. Hileman, B., Climate observations substantiate global warming models, http://pubs.acs.org/hotartcl/cenear/951127/pgl.html, *Chem. Eng. News*, Nov. 27, 1995.

36. Intergovernmental Panel on Climate Change, Contribution of Working Group I to the IPCC Second Assessment Report, IPCC-XI/Doc. 3, 1995, p. 341. *See also* Royer, J.F. et al., A GM study of the impact of greenhouse gas increase on the frequency of occurrence of tropical cyclones, *Climatic Change*, 38, 307–322, 1998 (study using high resolution atmospheric model found substantial reduction in number of tropical storms, especially in the Southern Hemisphere, in doubled carbon dioxide simulation); Schneider, D., The rising seas, *Sci. Am.*, 112, Mar. 1997. *But see* Knutson, T.R., Tuleya, R.E., and Kurihara, Y., Simulated increase of hurricane intensities in a CO_2-warmed climate, *Science*, 279, 118, 1998.

37. Wells, S. and Hanna, N., *The Greenpeace Book of Coral Reefs*, Sterling, New York, 1992, p. 14.

38. Highfield, R., Why Coral Reefs Matter, *Daily Telegraph*, Mar. 1, 1995, p. 16. Other constituent elements of reefs include shells, foams, and calcareous algae. McManus, J.W., Coral Growth and Sea-Level Rise, United Nations University Electronic Seminar on Global Warming, Nov. 20, 1995 (Internet document available from author).

39. Roberts, C.M. et al., The Distribution of Coral Reef Fish Biodiversity: The Climate-Biodiversity Connection, Fourth Session of the Conference of the Parties of the United Nations Framework Convention on Climate Change, Buenos Aires, Argentina, Nov. 2–13, 1998. A single reef may contain as many as 3000 different species of marine life. Fact sheet: the coral reef initiative, *Department State Dispatch*, Dec. 26, 1994. Overall, coral reefs support 1 to 9 million species and a far greater number of phyla than rain forests. Sale, P.F., Recruitment in space and time, *Nature*, 397, 25, 1999.

40. Edgerton, L.T., *The Rising Tide*, Island Press, CA, 1991, p. 32.

41. Coral reefs act as sponge for carbon dioxide, *Nikkei Weekly*, Aug. 7, 1995 (LEXIS, World Library). Coral reefs absorb approximately 1.1 billion metric tons of carbon dioxide per year, equal to approximately 2% of annual discharges. Edgerton, L.T., *The Rising Tide*, Island Press, CA, 1991, p. 32.

42. Hinckley, D., Assessing the Condition of Tropical Island Ecosystems and Their Responses to Climatic Change, unpublished report supplied to the author, p. 7.

43. Tangley, L., Will coral reefs be the first victims of global warming? *Earthwatch*, Apr. 1991, p. 27.

44. "Storm surge is the elevation of water generated by strong wind-stress forcing and a drop in atmospheric pressure." Daniel, P., A real time system for forecasting hurricane storm surges over the French Antilles, in Maul, *supra* note 4, at 146. *See also* Hubbert, G.D. and McInnes, K.L., A storm surge inundation model for coastal planning and impact studies, *J. Coastal Res.*, 15, 168, 1999.

45. Coasts at risk, *Global Change*, Oct. 1998, p. 10; Hubbert, G.D. and McInnes, K.L., A storm surge inundation model for coastal planning and impact studies, *J. Coastal Res.*, 15, 184, 1999; Steering Committee of the Climate Change Study, Climate Change Science, Current Understanding and Uncertainties, 1995, p. 46.

46. Serageldin, I., Coral reef conservation: science, economics, and law, in *Coral Reefs: Challenges and Opportunities for Sustainable Management*, Hatziolos, M.E., Hooten, A.J., and Fodor, M., Eds., The World Bank, Washington, D.C., 1998, p. 5; Zurick, D.N., Preserving paradise: environmental degradation in South Pacific island countries, *Geol. Rev.*, 85, 157, 1995. Reefs surrounding Palau support more than 2000 species of fish. Hinrichsen, D., Requiem for reefs? *Int. Wildl.*, Mar. 13, 1997, p. 12.

47. Huber, M.E., An assessment of the status of the coral reefs of Papua New Guinea, *Mar. Pollut. Bull.*, 29, 69, 1994; Lonsdale, S., Hopes rise for coral rainforests of the sea, *The Observer*, Apr. 25, 1993 (LEXIS, World Library). The fishing industry is one of the three primary economic sectors in most PIDCs. Canadian International Development Agency, Environmental change: vulnerability and security in the Pacific, http://www.gechs.org/aviso/January1999.html, *Aviso*, 1, Jan. 1999. In Micronesia, fish products as a percentage of export revenues increased from 43 to 86% during the period of 1988–1993, and in the Marshall Islands this figure increased from 20 to 80%. Acharya, A., Small islands: awash in a sea of troubles, *World Watch*, 8, 24, 1995. Stocks are already declining in many PIDCs as a consequence of overharvesting. Zurick, D.N., Preserving paradise: environmental degradation in South Pacific island countries, *Geol. Rev.*, 85, 160, 1995.; Haydon, S. For Pacific islands, global warming could mean extinction, Reuters, Jun. 6, 1992 (LEXIS, World Library).

48. South Pacific Regional Environmental Programme, International Coral Reef Initiative Pacific Regional Workshop, 1995, p. 92.

49. Intergovernmental Panel on Climate Change, Contribution of Working Group I to the IPCC Second Assessment Report, IPCC-XI/Doc. 3, 1995, p. 342; Legget, J., The AOSIS Summary of Scientific and Policy Issues, Feb. 1991.

50. Wilkinson, C. et al., Ecological and socioeconomic impacts of 1998 coral mortality in the Indian Ocean: an ENSO impact and a warning of future change? *Ambio*, 28, 188, 1999; Carpin, S. Global Warming Killing the Spectacular Seychelles Reefs, http://www.mcbi.org/maritimes/news05.html, *San Antonio News–Express*, Jan. 12, 1999; Chadwick-Furman, A.E., Reef coral diversity and global change, *Global Change Biol.*, 2, 559, 1996.

51. Dinoflagellates, such as zooxanthellae, live within coral fishes and engage in a symbiotic relationship. Coral polyps provide shelter for the zooaxanthellae and their waste provides a source of nutrients; in turn zooxanthellae supply the coral polyps with carbohydrates for food and construction of the limestone skeletons in which they live. Young, L.B., *Islands: Portraits of Miniature Worlds*, W.H. Freeman, San Francisco, 1999, p. 202; Wells, S. and Hanna, N., *The Greenpeace Book of Coral Reefs*, Sterling, New York, 1992, p. 14.

52. Hinckley, D., Assessing the Condition of Tropical Island Ecosystems and Their Responses to Climatic Change, unpublished report supplied to the author, p. 8. *See also* Buddemeier, R.W. and Fautin, D.G., Coral bleaching as an adaptive mechanism: a testable hypothesis, *BioScience*, 43, 320, 1993.

53. Hinckley, D., Assessing the Condition of Tropical Island Ecosystems and Their Responses to Climatic Change, unpublished report supplied to the author, p. 8. *See also* Buddemeier, R.W. and Daphne G. Fautin, D.G., Coral bleaching as an adaptive mechanism: a testable hypothesis, *BioScience*, 43, 320, 1993.

54. Pomerance, R., Coral Bleaching, Coral Mortality, and Global Climate Change, Report to the U.S. Coral Reef Task Force by the Bureau of Oceans and International Environmental and Scientific Affairs, http://www.state.gov/www/global/global_issues/coral_reefs/990305_coralreef_rpt.html, Mar. 5, 1999.

55. Wilkinson, C. et al., Ecological and socioeconomic impacts of 1998 coral mortality in the Indian Ocean: an ENSO impact and a warning of future change? *Ambio*, 28, 191, 1999.

56. Wilkinson, C. et al., Ecological and socioeconomic impacts of 1998 coral mortality in the Indian Ocean: an ENSO impact and a warning of future change? *Ambio*, 28, 188, 1999; Berkelmans, R. and Oliver, J.K., Large-scale bleaching of corals on the Great Barrier Reef, *Coral Reefs*, 18, 55–58, 1999.

57. Parker-Muller, G. and D'Elia, C.F., Interactions between corals and their symbiotic algae, in *Life and Death of Coral Reefs*, Birkeland, C., Ed., Chapman & Hall, New York, 1997, pp. 96–113; Warner, M. and Fitt, W.K., Mechanisms of bleaching of zooxanthellate symbioses, *Am. Zool.*, 31, 28, 1991.

58. Pomerance, R., Coral Bleaching, Coral Mortality, and Global Climate Change, Report to the U.S. Coral Reef Task Force by the Bureau of Oceans and International Environmental and Scientific Affairs, http://www.state.gov/www/global/global_issues/coral_reefs/990305_coralreef_rpt.html, Mar. 5, 1999; Hileman, B., Case grows for climate change, *Chem. Eng. News*, 77, 16, 1999.

59. The Global Coral Reef Alliance, Coral Reef Bleaching and Sea Surface Temperature, http://www.fas.harvard.edu/~goreau/bleach.intro.html, 1998. *See also* Perry, M., South Pacific reefs seen dying as oceans heat up, Reuters World Service, Aug. 1, 1994.

60. Wilkinson, C. et al., Ecological and socioeconomic impacts of 1998 coral bleaching in the Indian Ocean: an ENSO impact and a warning of future change? *Ambio*, in press; U.S. Global Change Research Program, Coral reef bleaching: ecological and economic implications, *Environ. Law*, Feb. 7, 1996.

61. Wilkinson, C. et al., Ecological and socioeconomic impacts of 1998 coral mortality in the Indian Ocean: an ENSO impact and a warning of future change? *Ambio*, 28, 191, 1999.

62. South Pacific Regional Environmental Programme, International Coral Reef Initiative Pacific Regional Workshop, 1995, p. 82, (dynamite fishing); MacKinnon, N., Destructive fishing practices in the Asia-Pacific region, in *Coral Reefs: Challenges and Opportunities for Sustainable Management*, Hatziolos, M.E., Hooten, A.J., and Fodor, M., Eds., The World Bank, Washington, D.C., 1998, p. 32; Cone, M., A toxic solution: the growing use of cyanide to stun and catch tropical fish is killing off coral reefs, researchers say, *Los Angeles Times*, Orange County Edition, Nov. 13, 1995, p. Metro B:2.

63. U.S. Geological Service, Hurricane Effects on Wildlife and Ecosystems, http://biology.usgs.gov/pr/newsrelease/1998/12-8.html, 1998; South Pacific Regional Environmental Programme, International Coral Reef Initiative Pacific Region Strategy, Mar. 1996, p. 3; Flanagan, R. Corals under siege, *Earth*, May 1993, p. 28.

64. Dustan, P., Coral reefs: harbingers of global change? in *Coral Reefs: Challenges and Opportunities for Sustainable Management*, Hatziolos, M.E., Hooten, A.J., and Fodor, M., Eds., The World Bank, Washington, D.C., 1998, p. 140; Wilkinson, C. et al., Ecological and socioeconomic impacts of 1998 coral mortality in the Indian Ocean: an ENSO impact and a warning of future change? *Ambio*, 28, 554, 1999. Moreover, even sublethal stressing of reefs increases their susceptibility to infection by opportunistic pathogens; epizootics can result in significant reef mortality. Pomerance, R., Coral Bleaching, Coral Mortality, and Global Climate Change, Report to the U.S. Coral Reef Task Force by the Bureau of Oceans and International Environmental and Scientific Affairs, http://www.state.gov/www/global/global_issues/coral_reefs/990305_coralreef_rpt.html, Mar. 5, 1999.

65. McManus, J.W., Coral Growth and Sea-Level Rise, United Nations University Electronic Seminar on Global Warming, Nov. 20, 1995 (Internet document available from author). Prepared statement of Fred T. Mackenzie, Global Climate Change and the Pacific Islands, Hearing before the Senate Committee on Energy and Natural Resources, S. Hrg. 102-664, 1992, p. 16.

66. Edwards, A.J., Impact of climate change on coral reefs, mangroves and tropical seagrass ecosystems, in *Climate Change: Impact on Coastal Habitation*, Eisma, D., Ed., Lewis Publishers, Boca Raton, FL, 1994, p. 209.

67. Wilkinson, C.R., Global change and coral reefs: impacts on reefs, economies and human cultures, *Global Change Biol.*, 2, 547, 1996.

68. Intergovernmental Panel on Climate Change, Contribution of Working Group I to the IPCC Second Assessment Report, IPCC-XI/Doc. 3, 1995, p. 342; Roberts, C.M., Coral reefs: health, hazards and history, *Trends Ecol. Evol.*, 8, 425, 1993.

69. Buddemeier, R.W., Coral reef responses to climate change: issues for Pacific Island nations, in Hay and Kaluwin, *supra* note 13, at p. 98.

70. Karl, T.R., Nicholls, N., and Gregory, J., The coming climate, *Sc. Am.*, http://www.sciam.com/0597issue/0597karl.html, May 1997; Holland, G.J., The maximum intensity of tropical cyclones, *J. Atm. Sci.*, 54, 2519, 1995; Hileman, B., Climate observations substantiate global warming models, http://pubs.acs.org/hotartcl/cenear/951127/pgl.html, *Chem. Eng. News*, Nov. 27, 1995; Intergovernmental Panel on Climate Change, Contribution of Working Group I to the IPCC Second Assessment Report, IPCC-XI/Doc. 3, 1995, p. 341. *See also* Royer, J.F. et al., A GM study of the impact of greenhouse gas increase on the frequency of occurrence of tropical cyclones, *Climatic Change*, 38, 307, 322, 1998 (study using high resolution atmospheric model found substantial reduction in number of tropical storms, especially in the Southern Hemisphere, in doubled carbon dioxide simulation); Schneider, D., The rising seas, *Sci. Am.*, 112, Mar. 1997. *But see* Knutson, T.R., Tuleya, R.E., and Kurihara, Y., Simulated increase of hurricane intensities in a CO_2-warmed climate, *Science*, 279, 118, 1998.

71. Weber, P.K., Saving the coral reefs, *Futurist*, July-Aug. 1993, p. 28; Mimura, N., Vulnerability of island countries in the South Pacific to sea level rise and climate change, *Climate Res.*, 12, 137, 1999.

72. Roach, J., Coral reefs threatened by increasing carbon dioxide, nationalgeographic.com, http://www.ngnews.com/news/1999/05/051799/coral_3210.asp, 1999; Global warming a threat to Barrier Reef, *Deutsche Press-Agentur*, Apr. 2, 1999 (LEXIS, World Library).

73. Kleypass, J.A. et al., Geochemical consequences of increased atmospheric carbon dioxide in coral reefs, *Science*, 284, 118, 1999.

74. Hinckley, D., Assessing the Condition of Tropical Island Ecosystems and Their Responses to Climatic Change, unpublished report supplied to the author, p. 8.

75. South Pacific Regional Environmental Programme, International Coral Reef Initiative Pacific Regional Workshop, 1995, p. 118.

76. Wilkinson, C.R. and Buddemeier, R.W., Global Climate Change and Coral Reefs: Implications for People and Reefs, Report of the UNEP-IOC-ASPEI-IUCN Global Task Team on the Implications of Climate Change on Coral Reefs, 1994, p. 72; Ellison, J.C. and Stoddart, D.R., Mangrove ecosystem collapse during predicted sea-level rise: holocene analogues and implications, *J. Coastal Res.*, 7, 159, 1991; Tomlinson, P.B., *The Botany of Mangroves*, Cambridge University Press, New York, 1986, pp. 231–232.

77. Lee, S.Y., Tropical mangrove ecology: physical and biotic factors influencing ecosystem structure and function, *Aust. J. Ecol.*, 24, 355, 1999.

78. Huber, M.E., An assessment of the status of the coral reefs of Papua New Guinea, *Mar. Pollut. Bull.*, 29, 71, 1994; Knight, D., Warmer oceans destroying coral reefs, *Inter Press Service*, Dec. 3, 1998 (LEXIS, World Library).

79. Intergovernmental Panel on Climate Change, Third Assessment Report of Working Group 2 (Draft), 1999, p. 15.

80. Vincente, V.P., Littoral ecological stability and economic development in small island states: the need for an equilibrium, in *Small Island States: Marine Science and Sustainable Development*, Maul, G., Ed., 1996, p. 274.

81. Hendry, M.D. and Digerfelt, G., Palaeogeography and palaeoenvironments of a tropical coastal wetland and adjacent shelf during holocene submergence, Jamaica, *Palaeogeogr., Palaeoclimatol., Palaeoecol.*, 73, 1, 1989.

82. Ellison, J.C., Pollen analysis of mangrove sediments as a sea level indicator: assessment from Tongatapu, Tonga, *Palaeogeogr., Palaeoclimatol., Palaeoecol.*, 74, 327, 1989; Ellison, J.C., Mangrove retreat with rising sea-level, Bermuda, *Estuarine Coastal Shelf Sci.*, 37, 75, 1993.

83. Hinckley, D., Assessing the Condition of Tropical Island Ecosystems and Their Responses to Climatic Change, unpublished report supplied to the author. Ellison, J., How South Pacific mangroves may respond to predicted climate change and sea-level rise, in *Climate Change in the South Pacific: Impacts and Responses in Australia, New Zealand, and Small Island States*, Gillespie, A. and Burns, W.C.G., Eds., Kluwer, Dordrecht, 2000, p. 299.

84. Ellison, J., How South Pacific mangroves may respond to predicted climate change and sea level rise, in *Climate Change in the South Pacific: Impacts and Responses in Australia, New Zealand, and Small Island States*, Gillespie, A. and Burns, W.C.G., Eds., Kluwer, Dordrecht, 2000, p. 294.

85. Ellison, J., How South Pacific mangroves may respond to predicted climate change and sea level rise, in *Climate Change in the South Pacific: Impacts and Responses in Australia, New Zealand, and Small Island States*, Gillespie, A. and Burns, W.C.G., Eds., Kluwer, Dordrecht, 2000, p. 296.

86. Ellison, J., How South Pacific mangroves may respond to predicted climate change and sea level rise, in *Climate Change in the South Pacific: Impacts and Responses in Australia, New Zealand, and Small Island States*, Gillespie, A. and Burns, W.C.G., Eds., Kluwer, Dordrecht, 2000, p. 298.

87. Clough, B.F., Andrews, T.J. and Cowan, I.R., Physiological processes in mangroves, in *Mangrove Ecosystems in Australia*, Clough, B.F., Ed., Australian Institute of Marine Science and Australian National, Australia, 1982, pp. 193–210.

88. Farnsworth, E.J., Ellison, M.A., and Gong, W.K., Elevated CO_2 alters anatomy, physiology, growth and reproduction of red mangrove (*Rhizophora mangle* L.), *Oecologia*, 108, 599, 1996.

89. Ellison, J., How South Pacific mangroves may respond to predicted climate change and sea level rise, in *Climate Change in the South Pacific: Impacts and Responses in Australia, New Zealand, and Small Island States*, Gillespie, A. and Burns, W.C.G., Eds., Kluwer, Dordrecht, 2000, p. 298.

90. Campbell, J.R., Contextualizing the effects of climate change in Pacific island countries, in *Climate Change: Developing Southern Hemisphere Perspectives*, Giambelluca, T.W. and Henderson-Sellers, Eds., Wiley Interscience, New York, 1996, p. 360.

91. Snedaker, S.C., Mangroves and climate change in the Florida and Caribbean region: scenarios and hypotheses, *Hydrobiologia*, 295, 43, 1995.

92. Snedaker, S.C., Mangroves and climate change in the Florida and Caribbean region: scenarios and hypotheses, *Hydrobiologia*, 295, 43, 1995.

93. Murdoch University, The Significance of Seagrass Ecosystems, http://possum.murdoch.edu.au/~seagrass/signif.htm, Western Australia, 1997.

94. Bell, J.D. and Pollard, D.A., Ecology of fish assemblages and fisheries associated with seagrasses, in *Biology of Seagrasses: A Treatise on the Biology of Seagrasses with Special Reference to the Australian Region*, Larkum, A.W.D, McComb, A.J., and Shepherd, S.A., Eds., Elsevier, Amsterdam, 1989, p. 565–609. Short, F.T. and Short, C.A., The seagrass filter: purification of estuarine and coastal water, in *The Estuary as a Filter*, Kennedy, V.S., Ed., Academic Press, New York, 1984, pp. 395–413.

95. Short, F.T. and Neckles, H.A., The effects of global climate change on seagrasses, *Aquat. Bot.*, 63, 169, 1999.

96. Short, F.T. and Neckles, H.A., The effects of global climate change on seagrasses, *Aquat. Bot.*, 63, 169, 1999.

97. Orth, R.J. and Moore, K.A., Distribution of *Zostera marine* L. and *Ruppia maritime* L. *sensu lato* along depth gradients in the Lower Chesapeake Bay, USA, *Aquat. Bot.*, 32, 291, 1988; Drew, E.A., Physiological aspects of primary production in seagrasses, *Aquat. Bot.*, 7, 139, 1979.

98. Short, F.T. and Neckles, H.A., The effects of global climate change on seagrasses, *Aquat. Bot.*, 63, 178, 1999.

99. Koch, E.W. and Beer, S., Tides, light and the distribution of *Zostera marina* in Long Island Sound, USA, *Aquat. Bot.*, 53, 97, 1996.

100. Short, F.T. and Neckles, H.A., The effects of global climate change on seagrasses, *Aquat. Bot.*, 63, 182, 1999; Gorham, J., Wyn Jones, R.G., and McDonnell, E., Some mechanisms of salt tolerance in crop plans, *Plant Soil*, 89, 15, 1985.

101. Lobban, C.S. and Schefter, M., *Tropical Pacific Island Environments*, University of Guam Press, Guam, 1997, p. 257.

102. Intergovernmental Panel on Climate Change, Third Assessment Report of Working Group 2 (Draft), 1999, p. 16.

103. Campbell, J.R., Contextualizing the effects of climate change in Pacific island countries, in *Climate Change: Developing Southern Hemisphere Perspectives*, Giambelluca, T.W. and Henderson-Sellers, Eds., Wiley Interscience, New York, 1996, p. 357.

104. Intergovernmental Panel on Climate Change, Contribution of Working Group I to the IPCC Second Assessment Report, IPCC-XI/Doc. 3, 1995, p. 343.

105. Intergovernmental Panel on Climate Change, Third Assessment Report of Working Group 2 (Draft), 1999, p. 16. "Among birds, approximately 23% of the species found on islands are threatened, compared with only 11% of the global bird population."

106. Intergovernmental Panel on Climate Change, Third Assessment Report of Working Group 2 (Draft), 1999, p. 16.

107. Gomez, E.D. and Belda, C.A., Growth of giant clams in Bolinao, Philippines, in *Giant Clams in Asia and the Pacific*, ACIAR Monograph No. 9, Copland, J.W. and Lucas J.S., Eds., Australian Centre for International Agricultural Research, Canberra, Australia, 1988, p. 178; Lucas, J.S. et al., Selecting optimum conditions for ocean-nursery culture of *Tidacna gigas*, in *Giant Clams in Asia and the Pacific*, ACIAR Monograph 9, Copland, J.W. and Lucas J.S., Eds., Australian Centre for International Agricultural Research, Canberra, Australia, 1988, p. 129.

108. Hinckley, D., Assessing the Condition of Tropical Island Ecosystems and Their Responses to Climatic Change, unpublished report supplied to the author, p. 3.

109. Pernetta, J.C., Impacts of climate change and sea-level rise on small island states, *Global Environ. Change*, 2, 24, 1992.

110. Sedjo, R. and Sohngen, B., Impact of Climate Change on Forests, Resources for the Future Climate Issue Brief 9, 2nd ed., 1998, p. 1420. Existing boreal forest cover could decline by as much as 50 to 90%; Jardine, K., Finger on the carbon pulse, *The Ecologist*, 24, 220, 1994. *See also* Clark, M.E., Consequences of global change for Earth's biosphere, in *Global Warming and the Challenge of International Cooperation: An Interdisciplinary Assessment*, Bryner, G.C., Ed., Kennedy Center Publications, Provo, UT, 1992, p. 51 (shift of deciduous forests in U.S. to Canada); Retallack, S., Wildlife in danger, *The Ecologist*, 29, 102, 1999; Melillo, J.M., Warm, warm on the range, *Science*, 283, 183, 1999.

111. Connell and Lea, *supra* note 13, at 151.

112. Intergovernmental Panel on Climate Change, Contribution of Working Group I to the IPCC Second Assessment Report, IPCC-XI/Doc. 3, 1995, p. 343.

113. Intergovernmental Panel on Climate Change, Climate Change Impacts on Forests, Working Group II, http://www.usgcrp.gov/ipcc/ html/chap01.html, 1998. Heightened temperatures in Siberia, Canada and Alaska have resulted in increased infestation by pests of spruce and pine trees. Cookson, C., UN Conference on Climate Change: Case for Action Strengthens as Signs Point to Global Warming, *Financial Times*, Mar. 28, 1995, p. 8.

114. Intergovernmental Panel on Climate Change, Contribution of Working Group I to the IPCC Second Assessment Report, IPCC-XI/Doc. 3, 1995, p. 343. However, recent research indicates that carbon dioxide's stimulatory effects may decrease as trees age. Report: High Carbon Dioxide Boosts Forest Growth by 25 Percent, http://www.eurekalert.org/releases/dumc-rhc051099.html, 1999.

115. United Nations Framework Convention on Climate Change, 31 ILM, 849, 1992.

116. United Nations Framework Convention on Climate Change had been ratified by 179 nations as of June 1999, web site, http://www.unfccc.int.

117. United Nations Framework Convention on Climate Change, 31 ILM, 849, Art. 2, 1992.

118. U.S. Greenhouse Gas Emissions Continue to Climb, *Global Environ. Change Rep.*, Apr. 26, 1996, p. 2. The UNFCCC Secretariat has recently projected that emissions from developed countries will rise by 18% 1990 levels by 2010 absent effective action. UNFCCC Secretariat, press release, http://www.unfccc.de/text/media/pressre.html, Oct. 1999.

119. Kyoto Protocol to the United Nations Framework Convention on Climate Change, FCCC/CP/ 1997/ L.7/Add. 1, Dec. 10, 1997.

120. Kyoto Protocol to the United Nations Framework Convention on Climate Change, FCCC/CP/ 1997/ L.7/Add. 1, Art. 3 (1), Dec. 10, 1997. For a detailed analysis of the Third Conference of the Parties, *see* Davies, P.G.G., Global warming and the Kyoto Protocol, *Int. Comp. L. Q.*, 47, 446, 1998.

121. Mine workers Gore's treaty will cost 1.7 million U.S. jobs; union bosses should quiz veep on "economic armageddon," *PR Newswire*, Feb. 19, 1999 (LEXIS, World Library); Sultana, R., Campaign against Global Warming: The Differing Positions, *The Independent*, Feb. 13, 1999 (LEXIS, World Library); A New Disinformation Campaign, *Rachel's Environment and Health Weekly*, Apr. 30, 1998; Vogel, G. and Lawler, A., Hot year, but cool response in Congress, *Science*, 280, 1684, Jun, 12, 1998.

122. It is anticipated that the Protocol will not even be submitted to the Senate for consideration until after the 2000 presidential election. International Energy Agency: there is no time to lose, *Petroleum Economist*, May 12, 1999 (LEXIS, World Library). Recently, the Clinton administration and industry have been emphasizing voluntary programs and tax incentives to reduce emissions. GCC industry voluntary actions highlighted; administration focus, legislative efforts in congress emphasize importance, *PR Newswire*, Apr. 22, 1999 (LEXIS, World Library); Clinton calls for clean-air fund to tackle greenhouse gas emissions, *Power Economist*, Mar., 31, 1999; Knight, D., Industry and greens debate climate change bill, *Inter Press Service*, Feb. 11, 1999 (LEXIS, World Library). Congress has recently evinced its contempt for Kyoto by slashing FY 2000 appropriations for renewable energy and energy efficiency programs by 10% from 1999 levels while substantially increasing funding of fossil fuel programs. Renewable energy funding takes a hit, May 28, 1999, http://www.enn.com/news/enn-stories/ 1999/05/ 052899/energy_3460.asp, May 28, 1999. Even if the U.S. ultimately ratifies the Protocol, it faces an imposing task to meet its target of a 7% reduction of greenhouse gas emissions below 1990 levels by 2008–2012. Given a projected 23% increase in emissions between now and 2010, the U.S. would have to slash emissions by 30% or make heavy use of the emissions trading system being developed under the Convention. U.S. economic growth seen forcing emissions cuts, *Reuters*, Sep. 15, 1999.

123. Barnum, A., Can world unite, halt climate threat? *San Francisco Chronicle*, Nov. 28, 1997, p. A21.

124. Reinstein, R.A., In the news, *Federal News Service*, Apr. 29, 1999; Risky business, *Global Change*, Oct. 1998, p. 2. Moreover, recent evidence indicates that the EU has yet to formulate climate policy measures to effectuate the commitments it made at Kyoto. Europe faces climate policy credibility gap, http://ens.lycos.com/ens/may99/1999L-05-19-06.html, *Environ. News Serv.*, May 22, 1999.

125. Kyoto Protocol to the United Nations Framework Convention on Climate Change, FCCC/CP/ 1997/ L.7/Add. 1, Art. 24, Dec. 10, 1997. The Protocol will enter into force 90 days after "not less than 55 parties to the [UNFCCC], incorporating Parties included in Annex 1 which accounted in total for at least 55% of the total carbon dioxide emissions for 1990 of the Parties included in Annex 1" have ratified it. However, as of August 1999, while 84 parties to the UNFCCC have signed the Protocol, only 14 parties, all developing nations, have ratified it. UNFCCC Secretariat, Kyoto Protocol, Status of Ratification, http://www.unfccc.de/resource/kpstats.pdf, 1999.

126. Hileman, B., Climate observations substantiate global warming models, http://pubs.acs.org/hotartcl/ cenear/951127/pgl.html, *Chem. Eng. News*, Nov. 27, 1995. *See also,* Karl, T.R., Nicholls, N., and Gregory, J., The coming climate, *Sci. Am.*, http://www.sciam.com/0597issue/0597karl.html, May 1997, p. 24; ("as much as 40% of [carbon dioxide] tends to remain in the atmosphere for centuries"); Kusidio, S., Climatic changes are no longer preventable, warn experts, *Deutsche Presse-Agentur*, Mar. 22, 1995 (LEXIS, News file) ("Even a worldwide stabilization of the emissions would not a prevent a rise in the greenhouse gases . . . for the next 200 years . . .").

127. UNCED, Framework Convention on Climate Change, opened for signature, Jun. 4, 1992, reprinted in 31 ILM 849 (1992), at Art. 4(2) and Annex I; Kyoto Protocol to the United Nations Framework Convention on Climate Change, FCCC/CP/ 1997/L.7/Add. 1, Art. 3, Dec. 10, 1997.

128. "Though projections vary substantially with assumptions about rates of economic growth, it appears that sometime between 2010 and 2020 China will overtake the United States as the world's largest greenhouse gas emitter, and that sometime thereafter emissions in the developing world will exceed those of the developed world." Thomas, W.L., The Kyoto Protocol: history, facts, figures and projections, *Public Utilities Fortnightly*, 137, Apr. 15, 1999, p. 48. Two thirds of greenhouse gas emissions between 1995 and 2025 may be produced by developing countries. Carbon emissions to rise over next 20 years, *Modern Power System*, Jan. 31, 1999, p. 3 (LEXIS, World Library).

129. Breidenich, C. et al., The Kyoto Protocol to the United Nations framework convention on climate change, *Am. J. Int. Law*, 92, 315, 1998; Kyoto Protocol: the unfinished agenda, *Climatic Change*, 39, 9, 1998.

130. Stevens, W.K., Climate talks enter harder phase of cutting back emissions, *The New York Times*, Apr. 11, 1995, p. 4; Davies, P.G.G., Global warming and the Kyoto Protocol, *Int. Comp. Law Q.*, 47, 457, 1998. "20 percent of the world's population is responsible for 63 percent of carbon dioxide emissions, while another 20 percent is responsible for only 2 percent of these emissions." Engelman, R., Population, consumption and equity, *Tiempo*, Dec. 1998, p. 5. *See also* Grubb, M., Vrolijk, C., and Brack, D., *The Kyoto Protocol. A Guide and Assessment*, Royal Institute of International Affairs, London, 1999, p. 265–269.

131. Stevens, W.K., Climate talks enter harder phase of cutting back emissions, *The New York Times*, Apr. 11, 1995, p. 4; Davies, P.G.G., Global warming and the Kyoto Protocol, *Int. Comp. Law Q.*, 47, 457, 1998; Engelman, R., Population, consumption and equity, *Tiempo*, Dec. 1998, p. 5; Grubb, M., Vrolijk, C., and Brack, D., *The Kyoto Protocol. A Guide and Assessment*, Royal Institute of International Affairs, London, 1999, p. 265–269.

132. Grubb, M., Vrolijk, C., and Brack, D., *The Kyoto Protocol. A Guide and Assessment*, Royal Institute of International Affairs, London, 1999, p. 265–269; The Buenos Aires tango, *Global Change*, Summer 1999, p. 2. Argentina subsequently announced its intention to enter into a binding commitment to reduce greenhouse emissions.

133. Parry, M. et al., Buenos Aires and Kyoto targets do little to reduce climate change impacts, *Global Environ. Change*, 8, 285, 1998.

134. For analysis of how nations' cost–benefit analyses have hindered development of an effective emissions reduction regime, *see* Fox, S., Responding to climate change: the case for unilateral trade measures to protect the global atmosphere, *Georgetown Law J.*, 84, 2499, 1996; LaLonde, M.J., The role of risk analysis in the 1992 framework convention on climate change, *Mich. J. Int. Law*, 15, 215, 1993.

135. Cline, W.R., Socially efficient abatement of carbon emissions, in *Climate Change and the Agenda for Research*, Hanisch, T., Ed., Westview Press, Boulder, CO, 1994, p. 102. However, even if the major emitters of greenhouse gases ultimately decide to make a serious commitment to emissions reductions, the task of stabilizing emissions will be extremely imposing. As Hoffert et al. recently concluded: "Stabilizing atmospheric CO_2 at twice pre-industrial levels while meeting the economic assumptions of 'business as usual' implies a massive transition to carbon-free power, particular [sic] in developing nations. There are no energy systems technologically ready at present to produce the required amounts of carbon-free power." Stabilizing atmospheric CO_2 at double pre-industrial levels will require about 15 terawatts of carbon-free power by 2050, and even more thereafter. By comparison, total world energy consumption is currently only 13.5 terawatts. Hoffert, M.I. et al., Energy implications of future stabilization of atmospheric CO_2 content, *Nature*, 395, 884, 1998. *But see,* Victor, D.G., Strategies for cutting carbon, *Nature*, 395, 837, 1998 ("if policies to improve energy efficiency could accelerate the 1% annual decline in energy use per unit of economic output to 1.5%, then the carbon-free energy required would be cut in half").

CHAPTER 54

Modeling Assessment of the Biological and Economic Impact of Increased UV Radiation on Loblolly Pine in the Middle Atlantic States

John J. Streicher and Keith Endres

INTRODUCTION

Consideration of the economic impact of environmental degradation provides additional justification for stewardship that imposes regulation and operational costs to aid remediation measures. The assessment of the economic impact of an ecological stressor within a geographic region requires knowledge of the distribution of vulnerable receptors within the region, determination of appropriate spatial and temporal resolution of the stressor, and information about the exposure and effect relationship. The increases in the surface flux of solar ultraviolet (UV) radiation due to decreasing stratospheric ozone are a stressor for numerous plant species, including the commercially important loblolly pine (*Pinus taeda*). The extent of species range and biomass of loblolly pine can be estimated for the Middle Atlantic area from U.S. Forest Service Forest Inventory and Analysis (FIA) data. An empirical exposure and effect function may then be applied to approximate biomass reduction and economic impact in a single-stressor perturbation scenario.

Problem Statement

Terrestrial flux of solar ultraviolet-B radiation (UV-B) is a naturally occurring plant stressor. Internal repair and protection mechanisms at the cellular level have evolved to cope with preindustrial levels of UV-B. The intensity of UV-B radiation reaching the earth's surface is mediated by atmospheric gases and aerosols, with ozone being the predominant mediator. With anthropogenic releases of ozone-depleting substances, increases in surface fluxes of UV have been predicted and observed. Loblolly pine is perhaps the most vulnerable of all conifers found near sea level. Typically, the loblolly pine receives relatively less UV-B than high-elevation conifers and adapts less effectively to increased UV-B than conifers indigenous to higher elevations.

Historical Perspective

The Montreal Protocol (1987) and subsequent amendments and adjustments have restricted the atmospheric release of ozone-depleting substances. The abundances of several chlorofluorocarbons (CFCs) have been monitored since 1978. Recent data suggest that tropospheric concentrations of some CFCs are now decreasing. However, a 3- to 5-year lag in transport to the stratosphere implies that stratospheric ozone will continue to decline, with total column ozone decreasing to a minimum near 2010. Biological stress responses to increasing UV-B may be expected during this time (Caldwell et al., 1994). Nearly one half of species evaluated have exhibited physiological or morphological sensitivity to UV-B (Sullivan, 1997).

Definitions

Incident solar radiation is measured as a spectral flux with units of watts-per-square-meter-per-nanometer wavelength, or $W/m^2/nm$. Biologically effective exposure, or explicitly, area-normalized biologically effective exposure, is measured as biologically weighted flux, or W/m^2_{be}. Biologically effective dose, or explicitly area-normalized biologically effective dose, is the time-accumulated biologically effective exposure, and is measured as biologically weighted energy density, with units of biologically weighted joules per square meter, or J/m^2_{be}.

Risk Assessment

Surface fluxes of UV-B at specific locations are determined by geodesic as well as atmospheric variables. Latitude and elevation, in addition to aerosol loading, total column ozone, and cloud cover, all affect the exposure and cumulative dose received by plants. The radiation microclimate of a specific plant is further determined by slope and aspect of the local terrain, and by shading from adjacent plants and geologic structures. The actual UV-B dose received by individual conifer needles will vary greatly within an individual tree. The risk assessment approach taken here is at a spatial resolution of four-digit hydrologic unit code (HUC) watershed, and at a temporal resolution of 3 years. In meteorological terminology, this assessment falls between mesoscale and synoptic scale. This assessment approach makes no statements with respect to effects on finer than a spatial mesoscale, or a 3-year temporal scale.

This study examines the possible effect of increases in UV-B on the loblolly pine within the Middle Atlantic states area. As an economically valuable species, potential economic loss was examined as well. Current atmospheric total column ozone for the latitude and longitude of the watershed centroid were estimated using a regional regression model based on satellite and ground-based ozone measurements. Backcast and forecast estimates of changes in total column ozone derived from the NASA Goddard Institute for Space Studies (GISS) Global Circulation Model (GCM) were used to calculate average daily biologically effective doses for the months of April and July, in the years 1979 and 2010. Average daily biologically effective doses were calculated by month for each watershed in the Middle Atlantic having measurable biomass of loblolly pine. Dose–response relationships developed by Sullivan et al. (1992) were applied to loblolly growth rates for two cases of ozone loss: reasonable worst-case ozone loss and effect-threshold ozone loss.

MODELING METHODS

Overview

The assessment of risk to loblolly pine due to anthropogenically driven increases in surface fluxes of solar UV-B radiation must consider the temporal and spatial profile of the stressor, the spatial distribution of the receptor, the spectral specificity of effects, and the time-dependence of the cause-effect relationship (i.e., the dose metric). Dose–response function then relates a hypothetical exposure scenario to a biological quantity, such as annual growth rate deficit. The economic aspect of the analysis must assign a dollar value to the growth rate deficit, and scale the calculation from a normalized per unit cost, up to a regional economy scale (e.g., the Middle Atlantic states region).

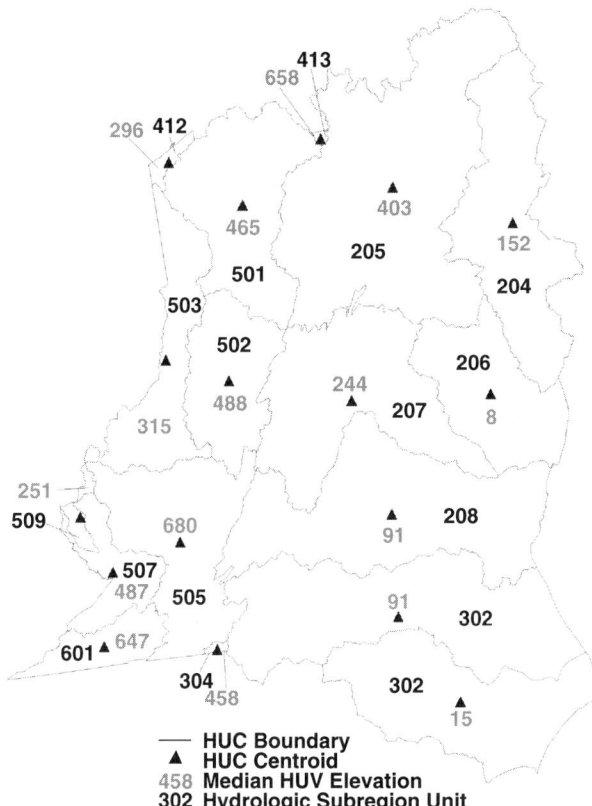

Figure 54.1 Middle Atlantic Integrated Assessment (MAIA) region, showing HUC centroid location and HUC median elevation in meters.

This assessment was completed with spatial subdivision by watershed at the four-digit HUC resolution. This resolution of the Middle Atlantic states region resulted in approximately 20 subregions (watersheds). The UV-B dose profile varies continuously in space across the region. Assignment of a single-dose profile to a subregion required choosing a representative point within the subregion for model calculations. Calculations estimating dose profile were determined for the latitude and longitude of the watershed centroid at the median watershed elevation, as shown in Figure 54.1.

The temporal resolution used in the development of the dose profile varied considerably with model component. A broadband radiative transfer model used to determine solar zenith angle iterated in 1-min time steps. The ozone model iterated in 1-day time steps. A linear dose metric computed a daily biologically effective dose. Daily average biologically effective doses are reported by month, however. Projections of future midlatitude ozone depletion were with 1-year precision. The temporal resolution of the dose–response function, however, was 3 years, and was clearly the limiting factor in the temporal resolution of the assessment. Consequently, dose profiles at 1-month resolution were considered adequate. July is presented as the month of peak biologically effective dose. April is presented as an early growing season month during which emergent growth structures are initially exposed to sunlight.

Model Suite: Description and Usage

No single model existed to perform the entire assessment. Consequently, a series of models and modeling approaches were used in succession to calculate ozone, UV-B surface flux, exposure, dose, growth rate reduction, and cost. A brief description of models/approaches follows.

A regional total column ozone regression model was developed from composite satellite and ground-based ozone measurements. This model was used to calculate current (1999) ozone as a function of Julian day, latitude, and longitude within the Middle Atlantic states region.

A discrete-ordinate spectral radiative transfer model (LibRadtran, version 0.13) was used to calculate the spectral flux W/m^2/nm for selected solar zenith angles, total column ozone levels, and elevations above sea level. Spectral flux was calculated at a resolution of 0.05 nm. Global (beam plus diffuse) radiation was used for subsequent exposure calculation.

The biologically effective exposure W/m$^2_{be}$ was then calculated by convolution of the Caldwell generalized plant spectral weighting function with the spectral flux (Caldwell et al., 1986). Spectral weighting functions, or action spectra, quantify the relative effectiveness by wavelength of incident light. The Caldwell function was normalized at 300 nm in this analysis, following the convention established in previous photobiological work.

Calculation of the biologically effective exposure for several solar zenith angles, ozone levels, and elevations enabled the development of a simple regression model of biologically effective exposure as a function of these input variables. This simple regression model was then incorporated into a sun-tracking model with a 1-min time resolution to calculate a daily biologically effective dose J/m$^2_{be}$ using an assumed linear dose metric.

Ultraviolet radiation dose and response data for loblolly pine (Sullivan et al., 1992) were used to estimate growth rate reductions. Laboratory and field experiments approximated diurnal and seasonal variation of UV with supplemental UV corresponding to various hypothetical ozone depletion scenarios (Sullivan et al., 1988; 1989). A dose–response regression function was developed from the data and used to investigate possible growth rate reductions corresponding to two ozone depletion scenarios.

Data from the FIA were queried by four-digit HUC watershed. Total loblolly biomass by HUC was reported for three size classes, as well as annual loblolly biomass growth rate. Annual growth rate by HUC was determined, and future growth rate was estimated by applying the regressed dose–response function for projected ozone reductions.

The economic assessment of growth rate reduction considered fixed stumpage price by tree size. (The stumpage price refers to unprocessed raw wood; value-added wood products would command a substantial price premium.) While lumber prices fluctuate with market demand and supply, and regional pricing variations can and do exist, the prices used in this assessment were recent fixed quotes for Virginia stumpage, published in *Timber Mart-South*. Tree size categories were saw timber (diameter > 8.9 in.); pole timber (8.9 in. > diameter > 5 in.); and sapling/seedling (5 in. > diameter). The dollar values associated with each size category were as follows: saw timber at $24.16, pole timber at $19.12, sapling and seedling at $8.06. Annual biomass production was similarly assigned a dollar value within each watershed, and projected annual growth deficit due to UV-B stress was then evaluated in dollars.

RESULTS

Dose calculations were completed only for watersheds having recorded loblolly pine biomass. Figure 54.2 presents the total aboveground biomass by hydrologic subregion. Loblolly pine trees are found throughout most of the Middle Atlantic states region, but predominantly in the southernmost areas, and nearer the low elevation coastline than the interior highlands. The Chowan–Roanoke (HUC 0301), Neuse–Pamlico (HUC 0302), Lower Chesapeake (HUC 0208), and Upper Chesapeake (HUC 0206) watersheds account for more than 95% of loblolly biomass within the Middle Atlantic states region.

The 1999 estimated UV-B dose profile by watershed for April and July is shown in Figure 54.3. The April biologically effective doses ranged from 3.2 KJ/m^2 in the Delaware watershed (HUC 0204), to 4.0 KJ/m^2 in the Monongahela (HUC 0502), Chowan–Roanoke, and the Neuse–Pamlico

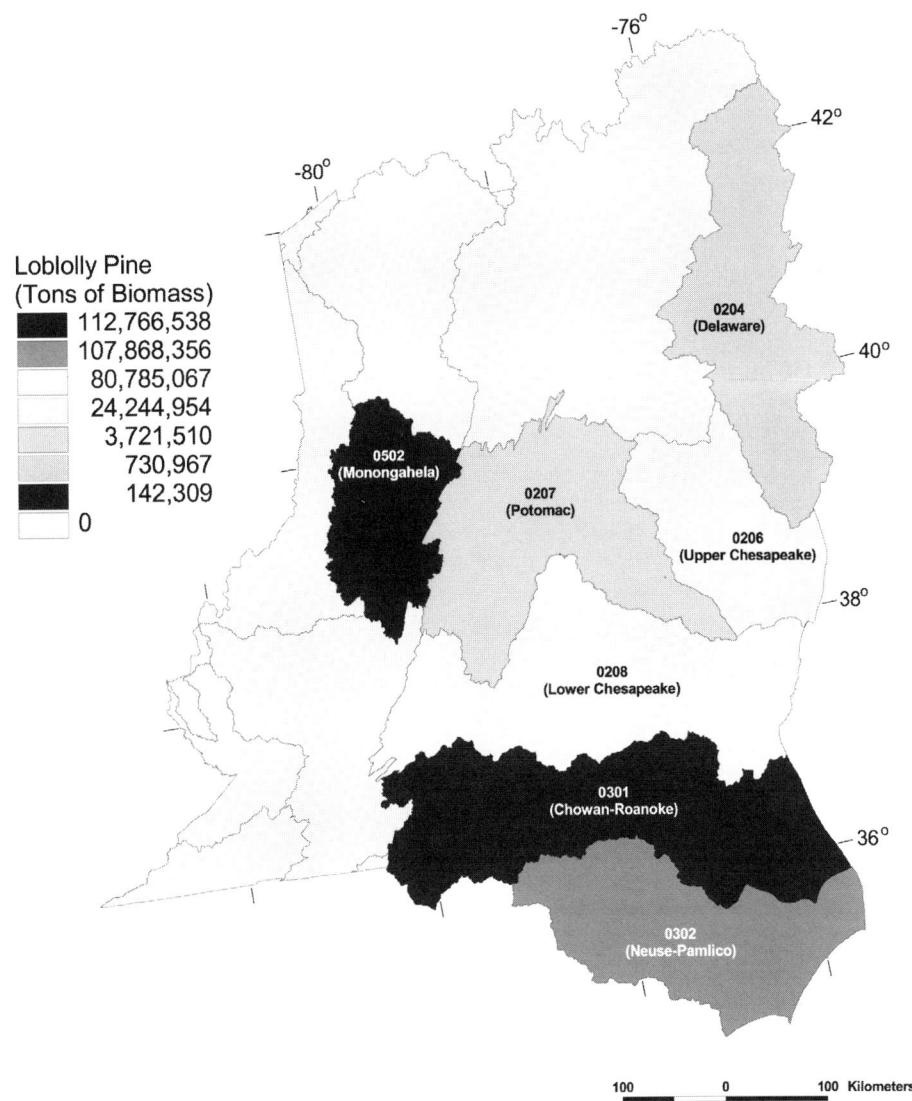

Figure 54.2 Loblolly pine total aboveground biomass (tons) by watershed.

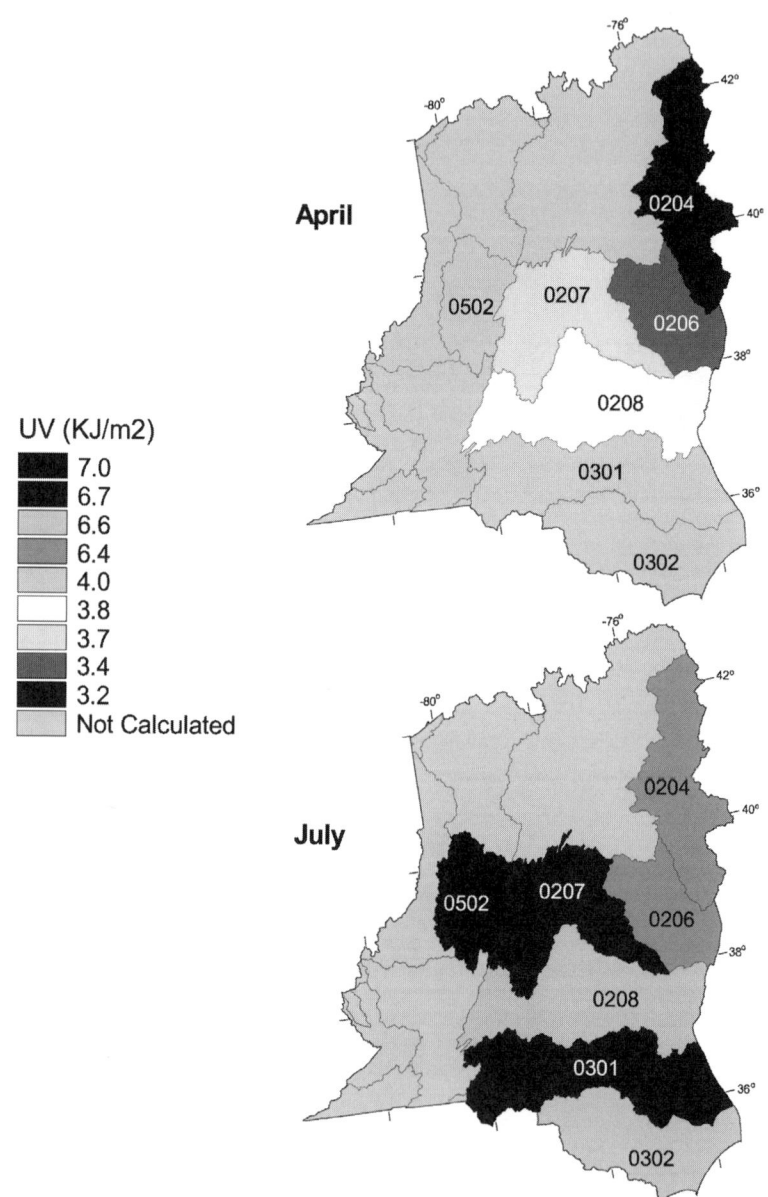

Figure 54.3 Estimated biologically effective UV-B dose profile by watershed for the months of April and July 1990.

watersheds. The July doses ranged from 6.4 KJ/m^2 in the Delaware and Upper Chesapeake watersheds, to 7.0 KJ/m^2 in the Monongahela watershed.

For comparison, a 1979 backcast dose profile is shown in Figure 54.4. Total column ozone has decreased by approximately 5% since 1979, and therefore biologically effective dose would have been less in 1979. Figure 54.3 presents 1979 doses relative to base year 1999, expressed as a percentage of base year doses. April UV-B doses have increased significantly more than July doses in the last 20 years. April 1979 doses were 14 to 17% less than those of April 1999, with the Delaware watershed having experienced the greatest increase in the last 20 years. The April relative increases follow a latitudinal ordination, with the more northerly latitudes showing the greatest increase relative to the lower absolute dose that is expected in the north. The July dose changes ranged from 10 to 11%, with latitude being a less significant factor.

Two hypothetical future cases of midlatitude ozone depletion were examined. A reasonable worst-case scenario of 10% total column ozone decrease from current levels by 2010 is shown in Figure 54.5. April and July dose changes are shown relative to current (1999) doses. Note that the April dose is forecast to increase very significantly relative to 1999. As was seen in the backcast, the greatest relative changes are expected to occur in the northernmost watersheds of the region. The Delaware watershed would experience an April increase of 34% relative to April 1999. Overall, April biologically effective dose would increase by 29 to 34% under an additional 10% total column ozone decrease. The July increases would range from 20 to 22%.

A second hypothetical scenario was examined: an effect threshold scenario with 15% ozone depletion from current levels. This is not a forecast of midlatitude ozone depletion; rather, it is an economic exercise to estimate the magnitude of the cost of lost productivity if UV-B radiation were to exceed the threshold known to cause diminished biomass productivity. Figure 54.6 shows the loblolly average annual growth as biomass accumulation. The ordination of growth rate mirrors the biomass tonnage shown in Figure 54.2, with the ranking from greatest to least being: Chowan–Roanoke, Neuse–Pamlico, Lower Chesapeake, Upper Chesapeake, and Potomac (HUC 0207) watersheds. The FIA database recorded insignificant biomass growth in the Delaware and Monongahela watersheds. Figure 54.7 displays loblolly growth rate reduction for the 15% ozone depletion scenario. Percentage reduction biomass accumulation for a 3-year period is indicated in the legend. The Monongahela and Potomac watersheds would experience a 2% reduction in biomass accumulation; the Neuse–Pamlico, Chowan–Roanoke, Lower Chesapeake, and Delaware watersheds would experience a 1% reduction in biomass accumulation; and the Upper Chesapeake would be unaffected. The growth reduction gradients follow physiography more than latitude, with higher elevation experiencing the greatest potential loss of productivity.

A query of the FIA database by tree size class, and by watershed, is presented as a bar chart in Figure 54.8. The Monongahela and Delaware watersheds (HUC 0502 and 0204, respectively) have virtually no loblolly. The Chowan–Roanoke and Lower Chesapeake watersheds (HUC 0301 and 0208, respectively) have substantial loblolly stands, with the sapling and seedling class predominating. The value of loblolly is a function of size. Figure 54.9 presents a bar chart of the dollar value of loblolly stands by tree size class and by watershed. Since larger trees have greater economic value, this economic weighting scheme gives the pole timber class the greatest value in six of seven watersheds.

Figure 54.10 shows the estimated stumpage dollar value for loblolly stands by watershed. The ordination of these watersheds exactly mirrors their ranking by biomass. Figure 54.11 presents the estimated 3-year economic loss due to growth rate reduction under a 15% additional ozone depletion scenario. The losses are concentrated in the southernmost watersheds, with Chowan–Roanoke experiencing the greatest economic loss. The 3-year loss for the entire Middle Atlantic states region is estimated to be approximately $7.3 million in raw loblolly timber.

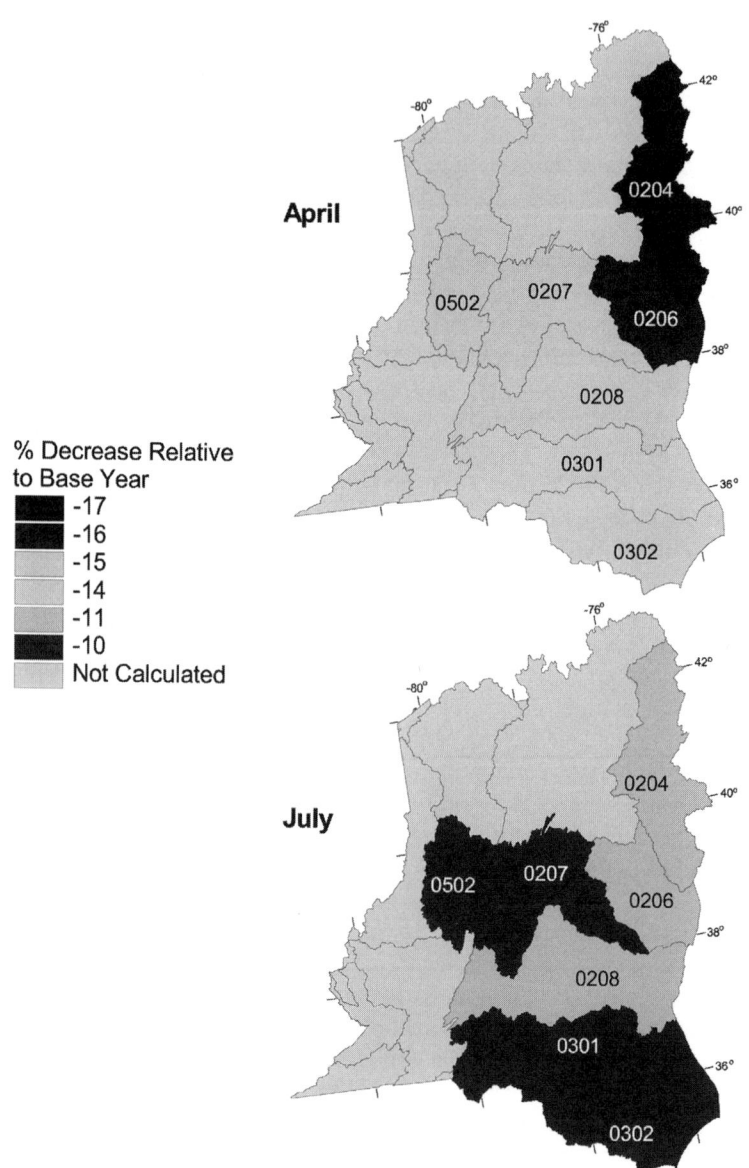

Figure 54.4 Backcast (year 1979) estimated biologically effective UV-B dose profile by watershed for the months of April and July. April 1979 doses are estimated to be 14 to 17% less than April 1999; July 1979 doses are estimated to be 10 to 11% less than July 1999.

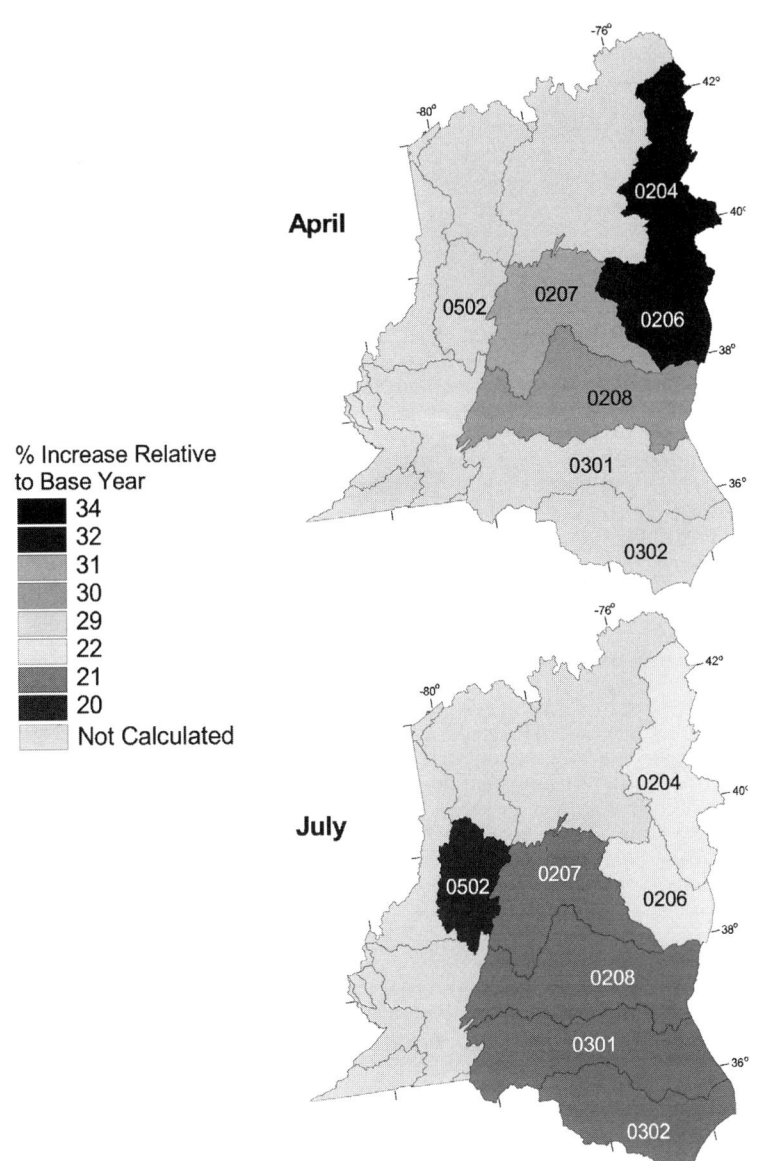

Figure 54.5 Forecast (year 2010) estimated biologically effective UV-B dose profile by watershed for the months of April and July.

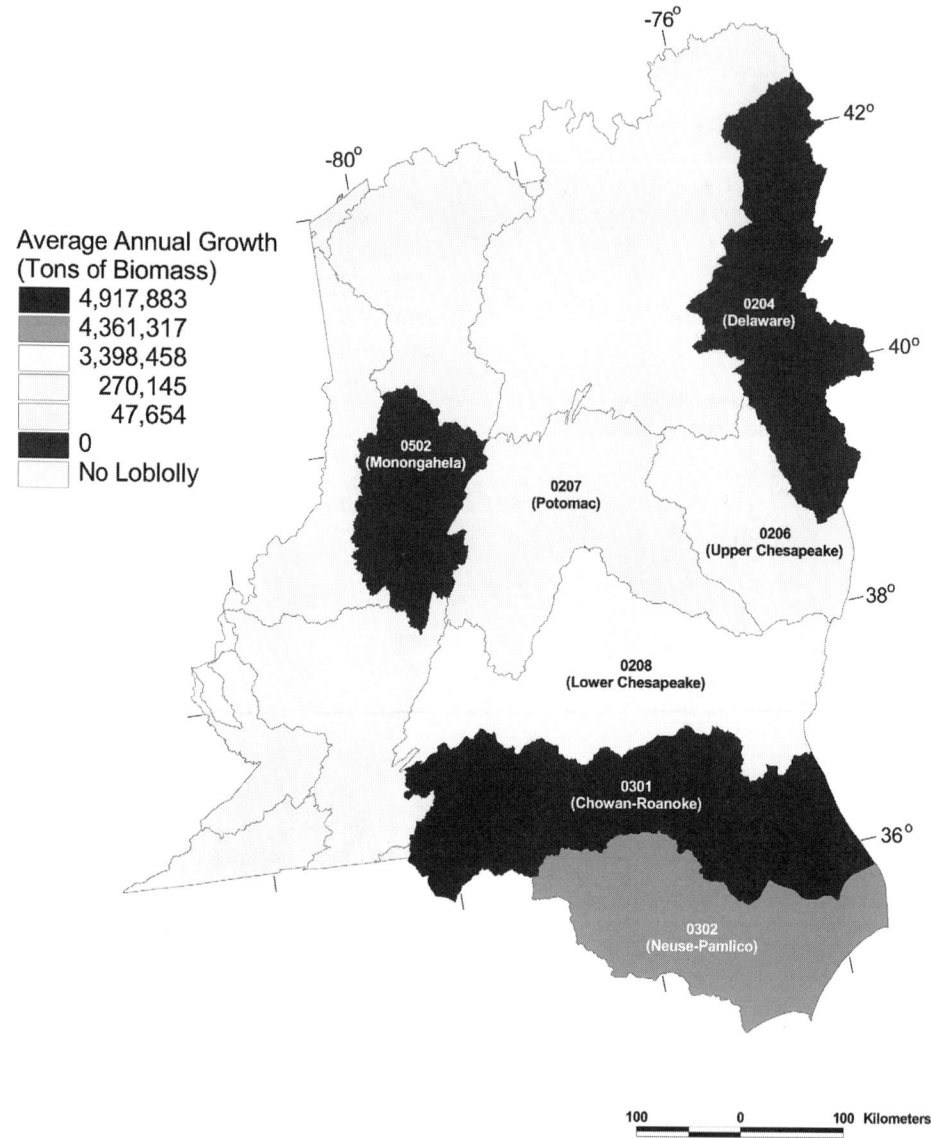

Figure 54.6 Loblolly average annual growth as tons of biomass.

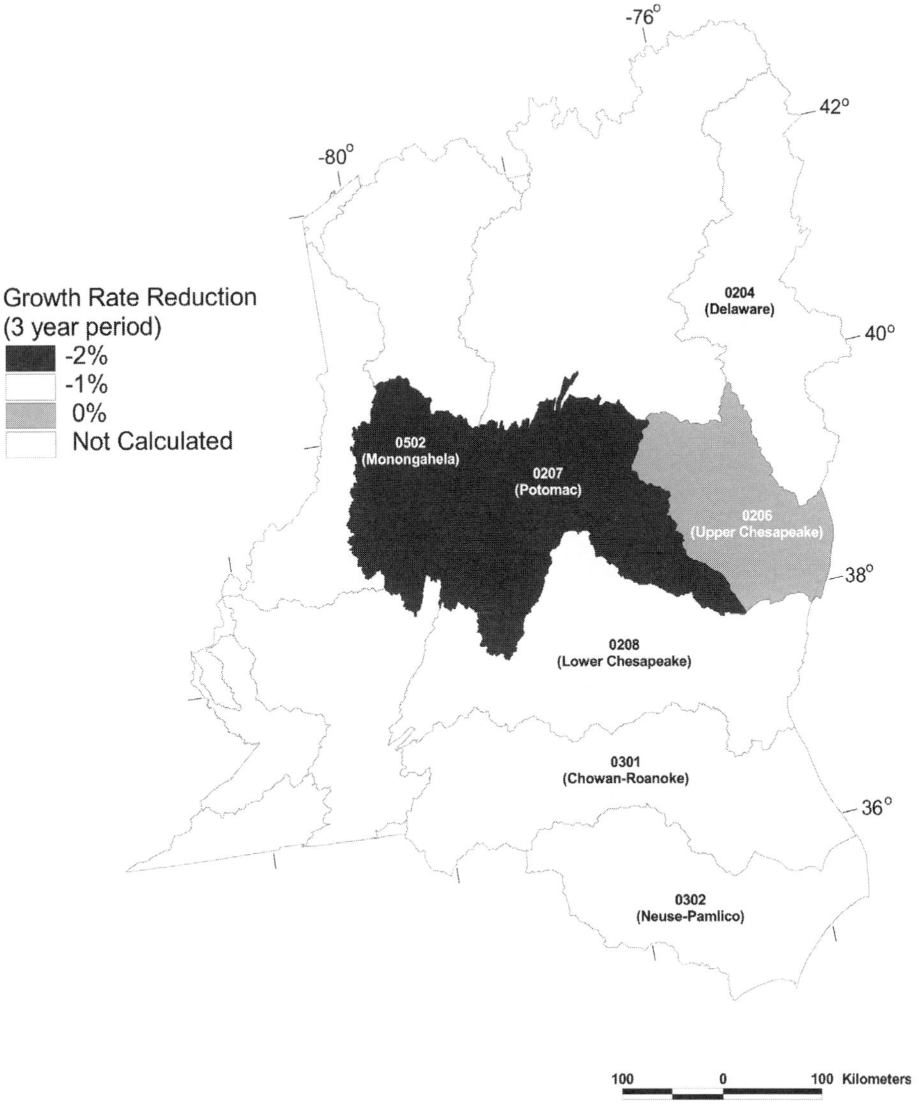

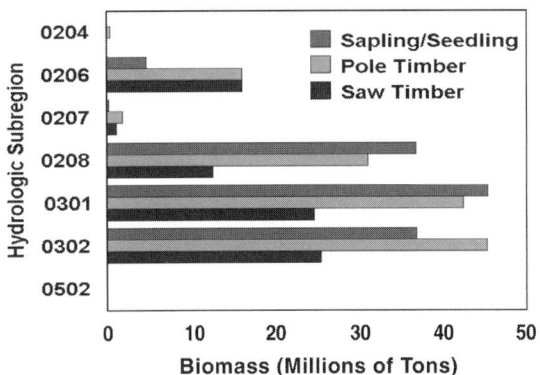

Figure 54.7 Loblolly growth rate reduction for a hypothetical 15% ozone depletion scenario. Percentage reduction in biomass accumulation is indicated in the legend.

Figure 54.8 Biomass by tree size class.

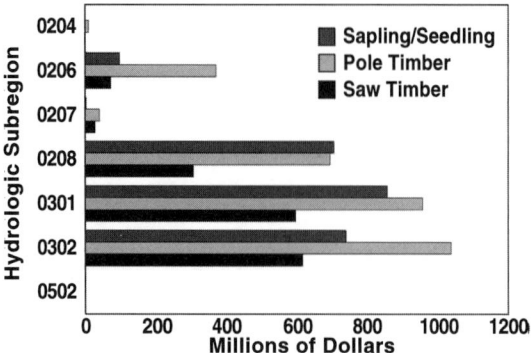

Figure 54.9 Dollar value of loblolly stands by tree size class and by watershed.

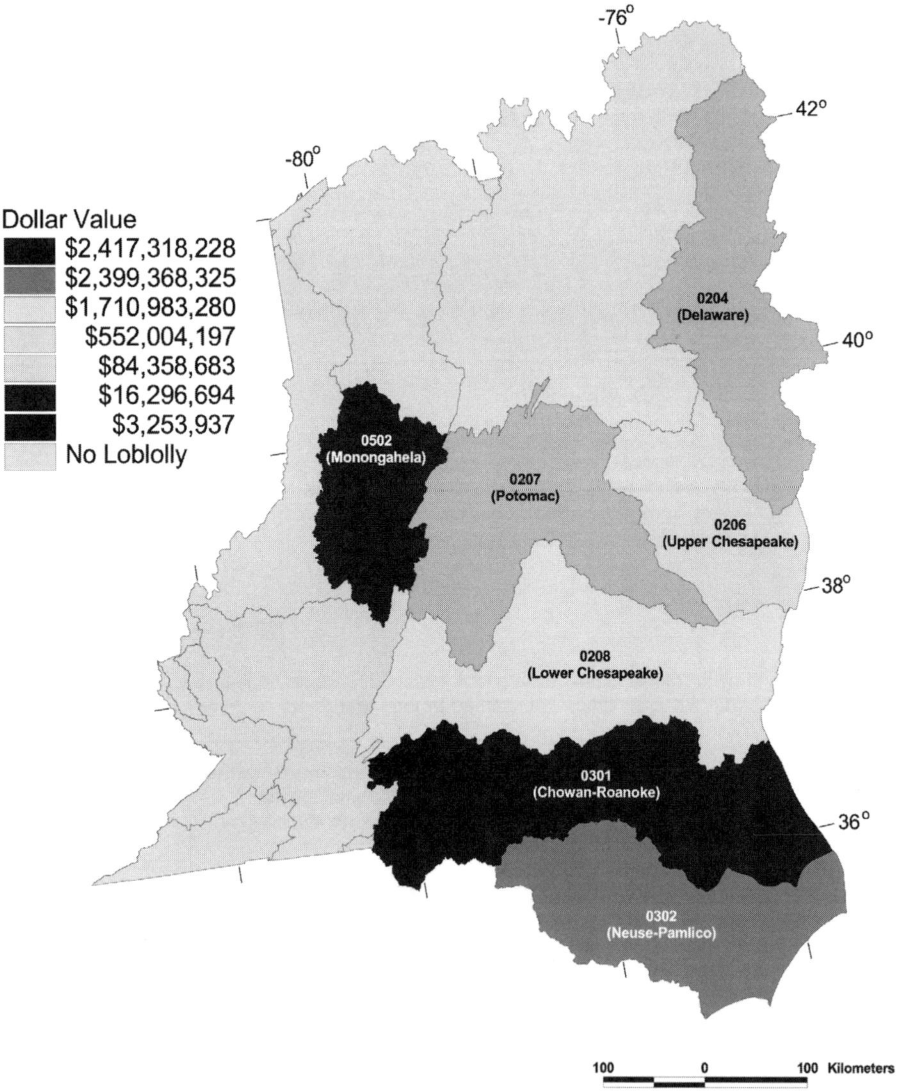

Figure 54.10 Estimated stumpage dollar value for loblolly stands by watershed.

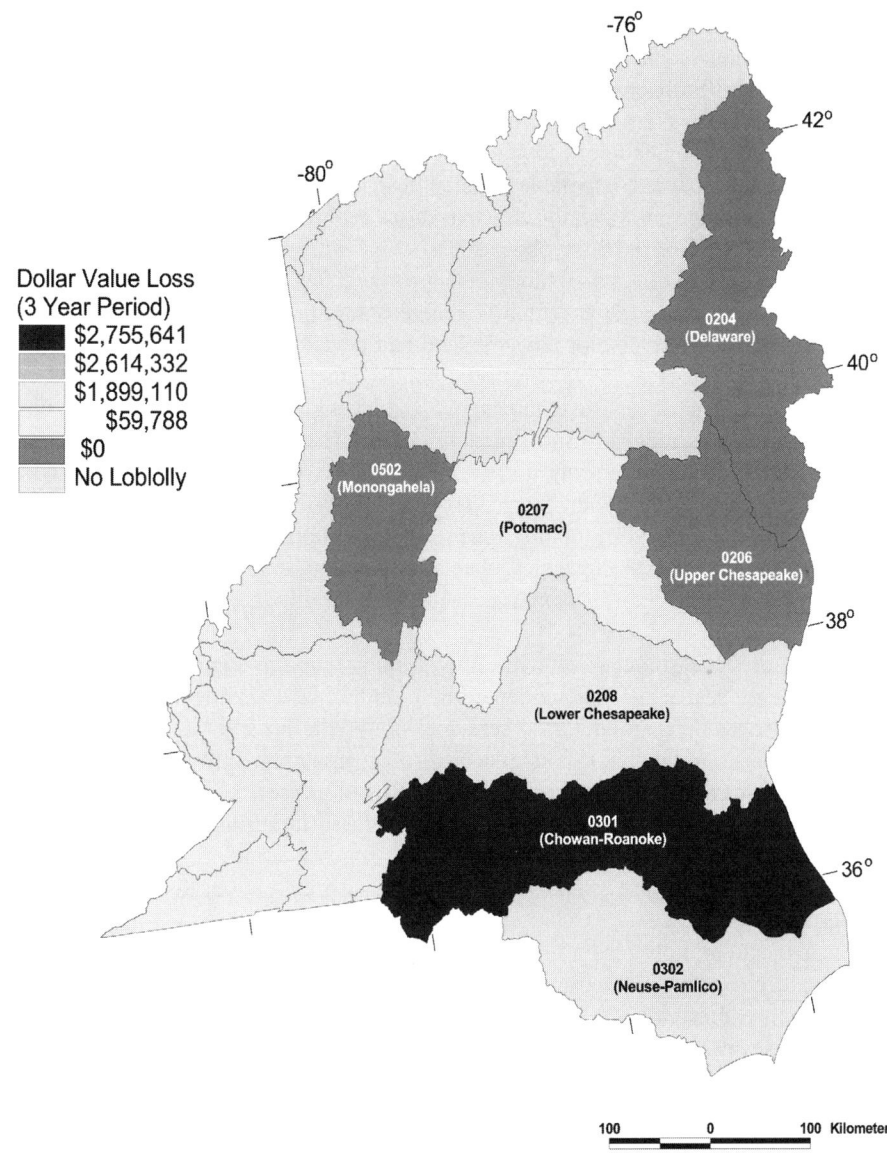

Figure 54.11 Estimated 3-year economic loss due to growth rate reduction under a hypothetical 15% additional ozone depletion scenario (relative to 1999).

DISCUSSION

The sensitivity of the biologically effective dose with ozone, aerosols, and elevation is integral to any consideration of remedial action. The "biological amplification factor," defined here as the biologically effective dose sensitivity to total column ozone, is approximately –2 for ozone (near 300 Dobson units) (i.e., a 1% decrease in total column ozone results in a 2% increase in the biologically effective dose). Background-level aerosols (summer months) decrease the biologically effective dose by approximately 10% (vs. aerosol-free atmosphere). Finally, increasing elevation from sea level to 1000 m will increase the biologically effective dose by approximately 17%.

In field studies conducted on loblolly pine seedlings, ambient light was supplemented with UV-B radiation from lamps (Sullivan et al., 1988, 1989, 1992). The total UV-B radiation applied was intended to replicate clear skies at Beltsville, MD (39°N latitude), as calculated using Green's model. The referenced biologically effective dose for current ozone levels was 8.4 MJ/day$_{be}$ on the summer solstice. This dose was comparable to modeling calculations here only when aerosols were neglected. Monthly average biologically effective doses reported here include background level aerosols that diminish surface doses by about 10%.

The mechanism of UV-B effect on loblolly pine growth reduction is uncertain (Teramura, 1996). Leaf surface reflectance in the UV-B region is generally less than 10%. Absorption of UV-B by chromophores in the cell may disrupt physiological and developmental processes in the needles (Naidu et al., 1993; Staxen et al., 1993). Reduced photosynthetic and evapotranspiration capacity were observed in young needles. Developmentally, a thickening of the epidermis was observed, and needle elongation was reduced (Sullivan et al., 1996). The physiological and developmental deficiencies of young needles maturing under elevated UV-B may begin to explain the overall reduced growth rate of the tree. Production of protective, UV-B–absorbing flavonoid compounds within the cell was not observed in loblolly pine to the extent observed in other conifers (Laakso et al., 1998). A protracted growth deficiency, even after elevated levels of UV-B are removed, may be explained by reduced function of damaged needles for the lifetime of the needles on the tree (Sullivan, 1994; Caldwell et al., 1998).

The threshold of UV-B–induced biomass growth rate reduction appears to be greater than a reasonable worst-case scenario of ozone depletion at midlatitudes. While increases in UV-B are anticipated, the dose–response model used here does not predict loss of biomass production. The relative perturbation in ambient UV-B is expected to be greater in spring than in summer. For the reasonable worst-case scenario, April effective doses will increase by 29 to 34%; July effective doses will increase by 20 to 22%. This substantial increase in spring radiation, coupled with existing increases on the order of 15% relative to 1979, imply a cumulative spring dose increase on the order of 50%, worst case, by 2010. This may have implications for bud structures that emerge in spring.

Economic losses in loblolly pine productivity due to a hypothetical 15% ozone depletion (from 1999 base) would affect the southernmost watersheds of the Middle Atlantic states region. The initial 3-year loss is estimated to be $7.3 million in raw timber; a substantially greater cost would be estimated if losses in value-added processing and finished products were included. Indications of a prolonged, or cumulative, growth rate deficit due to retention of UV-B-damaged needles have appeared; however, extrapolation beyond an initial 3-year period is not possible due to lack of experimental data.

DISCLAIMER

This work has been funded wholly or in part by the U.S. Environmental Protection Agency. It has been subjected to Agency review and approved for publication. Mention of trade names does not constitute endorsement or recommendation for use.

REFERENCES

Caldwell, M.M. and Flint, S.D., Stratospheric ozone reduction, solar UV-B radiation and terrestrial ecosystems, *Climatic Change*, 28, 375–394, 1994.

Caldwell M.M., Bjorn, L.O., Bornman, J.F., Flint, S.D., Kulandaivelu, G., Teramura, A.H., and Tevini, M., Effects of increased solar ultraviolet radiation on terrestrial ecosystems, *J. Photochem. Photobiol. B: Biol.*, 46, 40–52, 1998.

Caldwell M.M., Camp,L.B., Warner, C.W., and Flint, S.D., *Action Spectra and Their Key Role in Assessing Biological Consequences of Solar UV-B Radiation Change*, Vol. G8, NATO ASI Series, Springer-Verlag, Heidelberg, 1986.

Laakso, K. and Huttunen, S. Effects of the ultraviolet-B radiation (UV-B) on conifers: a review, *Environ. Pollut.*, 99, 319–328, 1998.

Naidu, S.L., Sullivan, J.H., Teramura, A.H., and DeLucia, E.H., The effects of ultraviolet-B radiation on photosynthesis of different aged needles in field-grown loblolly pine, *Tree Physiol.*, 12, 151–162, 1993.

Staxen, I., Bergouniux, C., and Bornman, J.F., Effect of ultraviolet radiation on cell division and microtubule organization in peunia hybrida protoplasts, *Protoplasma*, 173, 70–76, 1993.

Sullivan, J.H., *Temporal and Fluence Responses of Tree Foliage to UV-B Radiation*, Vol. I 18, NATO ASI Series, Springer-Verlag, Heidelberg, 1994.

Sullivan, J.H., Effects of increasing UV-B radiation and atmospheric CO_2 on photosynthesis and growth: implications for terrestrial ecosystems, *Plant Ecol.*, 128, 194–205, 1997.

Sullivan, J.H. and Teramura, A.H.,. Effects of ultraviolet-B irradiation on seedling growth in the Pinaceae, *Am. J. Bot.*, 75, 225–230, 1988.

Sullivan, J.H. and Teramura, A.H., The effects of ultraviolet-B radiation on loblolly pine:1. Growth, photosynthesis and pigment production in greenhouse-grown seedlings, *Physiol. Plant.*, 77, 202–207, 1989.

Sullivan, J.H. and Teramura, A.H., The effects of ultraviolet-B radiation on loblolly pine: 2. Growth of field-grown seedlings, *Trees*, 6, 115–120, 1992.

Sullivan, J.H., Howells, B.W., Ruhland, C.T., and Day, T.A., Changes in leaf expansion and epidermal screening effectiveness in *Liquidambar styraciflua* and *Pinus taeda* in response to UV-B radiation, *Physiol. Plant.*, 98, 349–357,1996.

Teramura, A.H., How plants respond to a changing UV-B radiation environment, in *Regulation of Plant Growth and Development by Light*, Briggs, W.R., Heath, R.L., and Tobin, E.M., Eds., American Society of Plant Physiologists, Rockville, MD, 1999, pp. 127–133.

SECTION II.4

Exotic Species: Eradication Revisited

CHAPTER 55

Overview: Exotic Species: Eradication Revisited

James R. Carey

Invasions of nonindigenous species threaten native biodiversity, ecosystem functioning, animal and plant health, and human economies. The best solution is to prevent the introduction of exotic organisms; once introduced, however, eradication might be feasible. Even though the potential ecological and social ramifications of eradication projects make them controversial, these programs provide unique opportunities for experimental ecological studies. Chapters by three leading experts in invasion biology note that deciding whether to attempt eradication is not simple, and alternative approaches might be preferable in some situations.

Judith Myers (University of British Columbia) believes that decisions regarding whether or not to eradicate ultimately must be weighed in terms of costs and benefits, including the possibility of failure. She reviews three eradication programs undertaken in British Columbia, Canada, including programs designed to eliminate the gypsy moth, the codling moth, and the Norway rat. She concludes that such programs should be undertaken only following realistic economic evaluation, and should be carried out only when the benefits are sufficiently large to warrant what may be an extended, expensive effort.

Daniel Simberloff (University of Tennessee) begins his chapter by acknowledging that eradication of nonindigenous species has a bad name among ecologists, largely because many of the highly publicized eradication programs have been unsuccessful. However, he observes that many eradications have been complete successes, identifying in particular three complete eradications: the African malaria vector (*Anopheles gambiae*) from northeastern Brazil, the screwworm fly in the southeastern U.S., and the melon fly and oriental fruit fly in the Ryukyu Archipelago in Japan. After reviewing these and other successes, he outlines five points that appear to demand consideration in successful eradication: sufficient resources, cooperation among parties, biological feasibility (e.g., detectability), prevention of reinvasion, and subsequent problems such as restoration of community. Simberloff's conclusion is similar to that of Dr. Myers: while a useful technique, eradication should be attempted only in certain circumstances.

The chapter by Armand Kuris (University of California, Santa Barbara) focuses on eradication of introduced marine pests. He notes that although prevention is the best strategy under all circumstances, it is particularly important in marine systems because detection is particularly difficult in this environment. Thus, in the past detected marine pests were almost always considered established pests; the detection of introduced marine species was met with defeatism. However, he reviews two propitious eradication programs: a successful one in Cayucos, CA, where several million South African sabellid worms had infested mollusk shells, and another just undertaken in Darwin Harbor, Northwest Territory, Australia, where early results look promising. He concludes that eradication may be a viable alternative to simply accepting the notion of learning to live with the exotics in marine systems.

CHAPTER 56

Eradication: Is It Ecologically, Financially, Environmentally, and Realistically Possible?

Judith H. Myers

INTRODUCTION

Eradication is the removal of every potentially reproducing individual of a species from an area that will not be reinvaded by other members of the species. The undertaking of an eradication program implies that the program organizers have the right, jurisdiction, and technological ability to eliminate the species, and that cost–benefit analyses support the decision. Exotic species are the most frequent targets of eradication programs, and their potential removal from the ecosystem is likely to be acceptable to most people, particularly those who are not directly involved. However, the financial, environmental, and potential health costs resulting from exposure to poisons or insecticides that accompany large-scale eradication programs are usually unacceptable to a portion of society. In addition, individuals often are required to give up certain rights for the sake of the area-wide program; consequently, obtaining the necessary cooperation for eradication programs is sometimes difficult (Collins et al., 1999; Kazmierczac and Smith, 1996).

Complete eradication, which is really the only kind, is likely to be successful only under certain circumstances. Six requirements for successful eradication outlined by Myers et al. (2001) are (1) sufficient funds, (2) clear authority of a lead agency, (3) a target species that is biologically susceptible to eradication procedures, (4) feasible prevention of reinvasion, (5) methods for continued surveillance, and (6) restoration if necessary. Eradication is most likely to succeed when the target population is very small and restricted. For introduced species, by the time the presence of the species is discovered, the number of individuals can be quite large and possibly past the threshold for feasible eradication. Populations in well-defined island habitats to which dispersal does not occur, are the most vulnerable to eradication. For example, it may be possible to eradicate wasps from a building, or plants or animals from small islands or lakes, or a disease organism from a human body. But on a larger scale, eradication is unlikely to succeed except in cases of homogeneous habitats, and in programs in which financial support is plentiful and environmental side effects are of no consideration.

An eradication program may be undertaken when an exotic species is first identified in a new geographic area or it may involve a pest species that is well established. Almost always, programs are initiated by a government agency that has the necessary jurisdiction, but may require further regulations that compromise private property rights. To engender support for an eradication program, the agency involved must inform and convince the public that the program should be undertaken. Public relations are particularly difficult because initially there could be little interest in the project. Furthermore, the agency may already have decided to proceed with the project, and a skeptical public may interpret the agency's promise of success as unrealistic, or doubt the potential damage that the target species can cause. A danger is that the program proponents assume the attitude of the experts: "Just trust us, because we know best." Critical review of proposals may be avoided to reduce controversy, but in the long run doing so is probably not wise.

Justification of eradication programs usually begins with a cost–benefit analysis. The biases involved in cost–benefit analyses have been reviewed in Myers et al. (1998). Estimation of the benefits of eradication is influenced by the length of time over which the costs will be amortized. If a pest species is eliminated, its destructive actions will be removed forever. Therefore, the estimated benefits can be enormous. However, projecting the benefits into the future can be misleading. If, for example, successful eradication increases the supply of a commodity, this could reduce its value and the profit made by the grower. Accurately predicting the influences of a newly established exotic organism can be difficult, thereby biasing the cost-effectiveness estimates. Will it become established? Will it create economic damage? A strong commitment by the lead agency may cause the estimates of benefits to be inflated and those of costs to be conservative. Estimated costs do not always cover all of the contingencies. Unanticipated costs are almost always going to occur. Some factors that may not be adequately estimated include costs for long-term monitoring of:

- Success or failure
- Public relations
- Preventing reintroductions
- Lawsuits
- Human health problems
- Human error
- Compensation for lost crops
- Initial population reduction that may require removal of host trees and extensive spray programs
- Continuing the project until the last individual is eliminated.

Three eradication programs that have been undertaken in British Columbia, Canada, and serve to illustrate prospects for success in estimating cost-effectiveness and in eradicating the target species.

GYPSY MOTH

The gypsy moth (*Lymantria dispar*) is native to Europe and Asia. The European form of gypsy moth was first introduced to North America in Massachusetts in 1869 (Elkinton and Liebhold, 1990). The preferred hosts of the gypsy moth are deciduous trees, particularly oaks, although larvae will feed on conifer foliage. The species has now spread north through Quebec, Ontario, New Brunswick, and Nova Scotia, in eastern Canada, south into Virginia and North Carolina, and west into Michigan. After an efficient pheromone trapping system was developed in the mid-1970s, monitoring for new introductions of the moths was undertaken in western states and provinces. The first captures of male gypsy moths in British Columbia occurred in 1978 (Pacific Forest Centre Web site, August 1999). Since that time, captures have been made at approximately 120 sites (Figure 56.1). Eradication programs have been carried out at 20 of these sites to "prevent" establishment. A microbial spray with the active ingredient *Bacillus thuringiensis* kurstaki (Btk) has been used in these programs. The advantage of this spray is that it is specific to lepidopterans, but the belief of some people that spraying of bacteria threatens human health is disadvantageous. Therefore, proposed aerial spray programs in British Columbia all have been associated with public outcry.

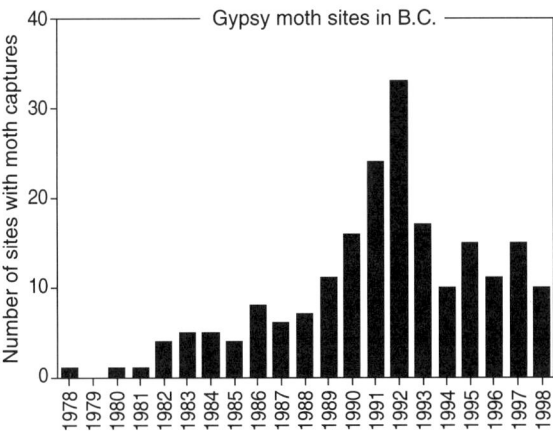

Figure 56.1 Number of sites with captures of male gypsy moths in the pheromone trap monitoring program in British Columbia. (Data from the Canadian Forest Service, Pacific Forestry Centre Web site.)

The largest gypsy moth eradication program in British Columbia was undertaken in 1992. The city of Vancouver (20,000 ha) was sprayed from the air three times at the cost of approximately $6 million. This followed the capture of 23 males of the Asian form of gypsy moth in a relatively restricted area near the grain port in Vancouver in 1991. The Asian gypsy moths are more likely to feed on conifers, particularly larches, and unlike the European gypsy moth, female Asian gypsy moths fly and are attracted to lights. During population outbreaks in the early 1980s and 1990s in eastern Russia, female moths were attracted to the well-lit grain terminals, and many hatching egg masses were found aboard Russian grain ships arriving in Vancouver during spring of those years. Small caterpillars are assumed to have been carried from the ships to trees near the port by the wind.

The proposed eradication program initially aroused very little public interest, but as awareness grew, discussion eventually became heated. Of particular controversy was the use of the microbial spray, its possible effects to human health, and the patent protection that prevented disclosure of all of the ingredients in the spray, Foray B. Experts were imported from the U.S. to describe the potentially devastating result they anticipated if this species should become established. Local officials declared that once the spray program was undertaken, the Asian gypsy moth would be eradicated. That claim was optimistic, and in 1993, 1994, and 1995 several male Asian gypsy moths were captured in the vicinity of Vancouver, although no further spraying was undertaken and no moths have been captured subsequently. The Asian gypsy moth seems to have not become established.

Since 1987, various numbers of male European gypsy moths have been captured at sites in southern Vancouver Island almost yearly. Eradication programs were begun in 1988. Between 1995 and 1998, the number of males collected in traps that were placed to delimit the distribution of the gypsy moth in this vicinity rose from 5 to 280, and egg masses were found (Pacific Forest Centre Web site, 1999). In 1999, Southern Vancouver Island was declared to be a regulated area requiring inspection of Christmas trees, nursery stock, and outdoor household articles before they may be transported from that area into unregulated areas (those without established gypsy moth populations) of the U.S. In 1998, a spray program was proposed, but permits were denied following court action involving health concerns of some citizens. However, in the summer of 1999, a $3 million aerial spray program was undertaken.

Is the gypsy moth a threat to British Columbia? The gypsy moth primarily defoliates deciduous trees and therefore its potential influence is largely as a nuisance in recreational and urban areas. However, Canada regulates gypsy moths to protect export trade with the U.S. Regulated areas are established based on survey results, and certification programs ensure that material shipped from infested areas are inspected for freedom from gypsy moths (Stubbings, G., personal communication). Although expensive, the cost of aerial spray programs is less than that for certification of materials, including lumber, which would result if the whole province were to become a regulated area. Jurisdiction in this case falls to the U.S., which requires eradication of the gypsy moth in British Columbia to prevent trade restrictions.

Gypsy moth introductions to British Columbia from eastern North America will continue, and each aerial spray program is likely to be met with an outcry from concerned citizens. While eradication programs with Btk have reduced gypsy moth populations, moths frequently are caught in these locations at a later date (Myers and Rothman, 1995). Even if not successful as eradication, population outbreaks and further spread of gypsy moths are reduced by spray programs. Slowing the spread is an alternative to eradication for introduced species (Myers et al., 2000). Attempted eradication of the gypsy moth is more about trade and trade barriers than biology.

CODLING MOTH

Codling moth (*Cydia pomonella*) is the most serious pest afflicting apples worldwide. It is native to western Asia and eastern Europe but occurs now in all apple-growing areas. The amount of damage caused by codling moths is partly influenced by the number of generations that occur each year, a figure that can vary from one to three. Soon after hatching from the egg, codling moth larvae enter the fruit where they feed on the seeds. When fully developed they burrow out through the fruit. In British Columbia the crop may be downgraded at the packing house if more than 0.5% of the fruit is damaged by worms. Codling moth is particularly troublesome for organic fruit growers.

In the 1970s, a pilot project was undertaken by the Canadian Department of Agriculture in the Similkameen Valley in the southern Okanagan region of British Columbia to investigate the feasibility of eradicating codling moth through the use of sterile male release (Proverbs et al., 1975, 1982). This project was carried out in a 500-ha apple-growing area over 3 years. By the end of the program most orchards in the area had no codling moths, but 7 of the 476 orchards had apple damage above the acceptable level. The cost of the program was $225/ha/year compared to normal investment of $95/ha/year for insecticide control. The conclusion at the end of the project was that eradication would not be feasible, even considering that insecticide use was reduced for several years. Further expansion of the program was not recommended.

Rather surprisingly, in the late 1980s, renewed interest was expressed by Agriculture Canada in initiating a codling moth eradication program for the whole Okanagan Valley, from Kelowna in the north to Keremeos in the south (Myers et al., 1998, 2000; Winston, 1997). Agriculture Canada and the B.C. Fruit Growers Association became partners in the program that was funded by the federal and provincial governments and by special taxation on both orchardists (approximately $195/ha) and urban and rural landowners ($45/ha). To sell a special tax to the taxpayers required a very positive spin to be put on the project, and the cost–benefit analysis was very optimistic as a result. The aim was initially to eradicate codling moths first from the southern end of the valley and eventually to expand the program to the north. The economic analysis amortized costs over a long period of time and this enhanced the financial benefits side of the equation. The proponents underestimated some costs in the analysis, such as the enormous cost required to initially reduce the moth population with spray programs, fruit culling, and removal of abandoned and urban fruit trees. The costs of monitoring and evaluation were minimized, and other technologies, such as mating disruption and the possible development of viral sprays, were initially ignored. Critical comment on the program was ignored or suppressed. People were easily convinced because everyone wanted the program to succeed. Everyone wanted to reduce insecticide use, and the sterile insect release (SIR) causes no negative environmental side effects.

A state-of-the-art moth rearing facility was built and the program began. Five years into the program the feasibility of eradicating the codling moth was evaluated. One of the major problems influencing the success of the SIR was the failure of the first generation of sterile males to fly actively and to compete for mates with wild males. The initial reduction of wild populations was slow and expensive. Growers began turning to other controls, such as mating disruption or even chemical sprays. Between 1995 and 1998, the percentage of orchards with no codling moth damage increased from 42 to 84%, and the percentage of traps with wild moths declined from 98 to 60% (Anonymous, 1999). But clearly, eradication remained allusive. In the winter of 1998–1999 the program goal was changed from eradication to area-wide suppression with the long-term goal of eventual eradication (Fisher-Fleming et al., 1998). By the summer of 2001, codling moth densities were very low in the southern Okanagan Valley (Edwards, L., personal communication).

Eradication of the codling moth in the Okanagan Valley is an unrealistic goal for several reasons. For one, land use in the area is heterogeneous. In addition to commercial orchards, the region contains urban lots with backyard apple trees and abandoned orchards. Part of the area is in native land reserves under a separate jurisdiction. Reintroduction of moths to treated areas is difficult to stop unless the movement of fruit bins is very carefully controlled. The value of apples has been declining because worldwide production is high and therefore prices are low. Maintaining a program that is more than twice as expensive to the growers as chemical control must show rapid results or support declines. On the other hand, the high capital cost for the rearing facility and the optimistic promises of the program team make the eradication program difficult to stop.

The advantage of the sterile insect release approach to eradication is that it is species-specific and environmentally friendly, although high levels of insecticides are initially required to reduce moth populations. The greatest long-term costs of the failure of this program are the growers' increased mistrust of scientists' advice and the poor value for taxpayers' dollars. Whether it is economically feasible to use SIR as part of an integrated pest management strategy remains to be seen.

THE NORWAY RAT ON LANGARA ISLAND

Langara Island (3100 ha) on the northwest tip of the Queen Charlotte Islands was formerly one of British Columbia's largest seabird colonies with six species of burrow-nesting birds. Between 1971 and 1980, four of these bird species became extinct and populations of the ancient murrelet were greatly reduced. The Norway rat (*Rattus norvegicus*) is known to have a large impact on colonial seabirds and even displaced black rats from Langara Island during the seabird decline (Bertram, 1994; Bertram and Nagorsen, 1995). Diet analysis indicated that Norway rats did attack ancient murrelets (Drever and Harestad, 1998). By 1992, experimentation had shown the effectiveness of a new anticoagulant rat poison, brodifacoum, for eradicating rats from small islands in New Zealand (Taylor and Thomas, 1989). The advantage of this poison is that it kills the rat after one dose, but it is slow acting, which has implications to the social system of rats. In 1993, $2.5 million became available for seabird enhancement as mitigation for the Nestucca oil spill. In hopes of restoring Langara Island as a successful breeding colony, a rat eradication project was undertaken. In 1994, a pilot project was carried out on Lucy Island, a small, nearby island. The use of fixed stations baited with poison pellets was successful in removing rats from Lucy Island, and the next year a large-scale program was launched on Langara Island. In July 1995, the eradication of rats from Langara Island began. Bait stations were placed at 100-m intervals with the aim of having one in each rat home range. In total, 3905 bait stations were used. By August 1995 most rats were gone, and none was captured between 1996 and 2001 (Kaiser et al., 1997; Taylor et al., 2000; Kaiser, G.W., personal communication).

The poisoning program had some impact on other animals on Langara Island (Howald, 1997). Ravens learned how to extract the poison pellets, and approximately 50% of the raven population was killed. Eagles would not feed on dead rats but did eat poisoned ravens and also suffered some mortality. Raven populations have since recovered through immigration from other islands. Ancient murrelets do not breed until the age of 3 or 4 years, so recovery will be slow. The continued success of this program depends on preventing a reintroduction of rats. Because Langara Island is isolated and the sites visited by humans are localized, that may be possible.

The factors that contributed to the success of the program included the effective poison, the territorial behavior of the rats, the isolated nature of the island, and the fact that the rats were not thriving because food was limited. Seasonal laborers and volunteers enabled this program to be carried out on such a large scale. This successful program is also well documented.

CONCLUSION

Eradication programs are difficult to evaluate because little information is available in the published literature. Monitoring is often not carried out in a rigorous manner, and the desire for success is strong. Therefore, undesirable or compromised results are not welcome. In eradication programs of insects, such as those for the gypsy moths, numbers of male moths trapped may be recorded, but the number of traps used is not. Such limited documentation undermines the ability to compare the numbers caught in different areas and in different years. Evaluation of the growth or decline of populations is difficult. Eradication programs are always very expensive, and the large scale of projects sometimes influences the impetus to maintain the project.

Programs can be classified into three categories: (1) those called eradication for the purpose of preventing trade barriers (gypsy moth eradication); (2) those called eradication of well- and widely established native or exotic organisms (usually insects) that should truly be called area-wide management (codling moth eradication); and (3) those that target the removal of a species from a defined area for which eradication is possible (Norway rat eradication). I believe that eradication should be attempted only in very restrictive situations in which it can be complete and permanent. That will occur in situations in which the distribution of the organism is limited, either because it is newly introduced and concentrated in a particular area, or because it is on an island with limited access. Vertebrates are likely to be more susceptible to eradication because they are more readily detected than invertebrates, they may have limited breeding seasons, and baited poisons have been developed that have been shown to be effective.

Programs should be undertaken only following realistic economic evaluation and should be carried out only when the benefits are sufficiently great to warrant what may have to be an extended and expensive program. Having an emergency response plan in place for dealing with introduced species that may become pests is important. A model that could be emulated by other countries is that of Hosking (2001). This document outlines six response phases: (1) detection, (2) evaluation, (3) the response decision, (4) the operation phase, (5) the monitoring stage, and (6) the review stage. In the early stages of a program, communicating honestly with members of the public who will pay the cost of the program is most important. Program directors should realize that in the long run the truth will emerge. If an eradication program is not successful, the credibility of the scientist usually suffers if unrealistic promises have been made. It is best to call area-wide management by its name rather than proposing that a widespread or frequently introduced organism can be successfully eradicated. Slowing the spread might be an alternative to an eradication program (Sharov and Liebhold, 1998) with more achievable goals. And finally, the monitoring and review stages are most important in evaluation of eradication success or reasons for failure. These should not be forgotten.

REFERENCES

Anonymous, Codling moth news, 1999.

Bertram, D.F., The roles of introduced rats and commercial fishing in the decline of ancient murrelets on Langara Island, British Columbia, *Conserv. Biol.*, 9, 865–872, 1994.

Bertram, D.F. and Nagorsen, D.W. Introduced rats, *Rattus* spp. on the Queen Charlotte Islands: implications for seabird conservation, *Can. Field Nat.*, 109, 6–10, 1995.

Collins, R.L. Larson, J.A., Roberts, R.K., and English, B.C., Factors influencing southwestern Tennessee farmers' willingness to participate in the boll weevil eradication program, *J. Cotton Sci.*, 3, 1–10, 1999.

Drever, M.C. and Harestad, A.S., Diets of Norway rats, *Rattus norvegicus*, on Langara Island, Queen Charlotte Islands, British Columbia: implications for conservation of breeding seabirds, *Can. Field Nat.*, 112, 676–683, 1998.

Elkinton, J.S. and Liebhold, A.M., Population dynamics of gypsy moth in North America, *Annu. Rev. Entomol.*, 35, 571–596, 1990.

Fisher-Fleming, B., Judd, G., and Craig, S., Report to the SIR Board, Dec. 4, 1998.

Howald, G.R., The Risk of Non-Target Species Poisoning from Brodifacoum Used to Eradicate Rats from Langara Island, British Columbia, Canada, Thesis, University of British Columbia, Vancouver, Canada, 1997.

Hosking, G., An Emergency Response Guide for Forestry Incursions, Biosecurity, Protection and Risk Management of Forests Programme, Forest Research Institute, Rotorua, New Zealand, 2001.

Kaiser, G.W., Taylor, R.H., Buck, P.D., Elliott, J.E., Howald, G.R., and Drever, M.C., The Langara Island Seabird Habitat Recovery Project: Eradication of Norway Rats (1993–1997), Technical Report Series 304, Canadian Wildlife Service, Pacific and Yukon Region, British Columbia, 1997.

Kazmierczac, R.F., and Smith, B.C., Role of knowledge and opinion in promoting boll weevil (Coleoptera, Curculionidae) eradication, *J. Econ. Entomol.*, 89, 1166–1174, 1996.

Myers, J.H. and Rothman, L., Field experiments to study regulation of fluctuating populations, in *Population Dynamics: New Approaches and Synthesis*, Cappuccino, N. and Price, P.W., Eds., Academic Press, San Diego, 1995, pp. 229–250.

Myers, J.H., Savoie, A., and van Randen, E., Eradication and pest management, *Annu. Rev. Entomol.*, 43, 471–491, 1998.

Myers, J.H., Simberloff, D., Kuris, A., and Carey, J., Eradication revisited: dealing with exotics, *Trends Ecol. Evol.*, 15, 316–320, 2000.

Pacific Forestry Centre Web site, http://www.pfc.cfs.nrcan.gc.ca./biodiversity/gmoth/, Aug. 1999.

Proverbs, M.D., Newton, J.R., Logan, D.M., and Brinton, F.E., Codling moth control by sterile insect release of radiation sterilized moths in a pome fruit orchard and observations of other pests, *J. Econ. Entomol.*, 68, 555–560, 1975.

Proverbs, M.D., Newton, J.R. and Campbell, C.J., Codling moth: a pilot program of control by sterile insect release in British Columbia, *Can. Entomol.*, 114, 363–376, 1982.

Sharov, A. and Liebhold, A., Model of slowing the spread of gypsy moth (Lepidoptera: Lymantriidae) with a barrier zone, *Ecol. Appl.*, 8, 1170–1179, 1998.

Taylor, R.H. and Thomas, B.W., Eradication of Norway rats, *Rattus norvegicus*, from Hawea Island, Fiordland, using brodifacoum, *N. Z. J. Ecol.*, 12, 23–32, 1989.

Taylor, R.H., Kaiser G.W., and Drever, M.C., Eradication of Norway rats for recovery of seabird habitat on Langara Island, British Columbia, *J. Restoration Ecol.*, 8, 151–160, 2000.

Winston, M.L., *Nature Wars: People vs. Pests*, Harvard University Press, Cambridge, MA, 1997.

CHAPTER 57

Why Not Eradication?

Daniel Simberloff

INTRODUCTION

Eradicating a population — removing every last individual — is a seductive but highly controversial idea. Some question the value of attempting eradication of well-established, widespread populations (e.g., Dahlsten, 1986; Rejmanek in Borenstein, 1999). Probably some of the skepticism arises from several highly visible, costly failures (Newsom, 1978). Perhaps the worst was the attempt to eradicate the imported fire ant (*Solenopsis invicta*) from the southeastern U.S. (Davidson and Stone, 1989). This was such a catastrophe in terms of collateral damage (impacts on nontargets, including humans) and expense (over $200 million) that E.O. Wilson has termed it the *Vietnam of entomology* (Brody, 1975).

Part of the controversy surrounding eradication can be traced to the fact that the term is used colloquially and imprecisely to mean partial removal of a pest species and control at some lower, acceptable density. This usage is common among politicians (e.g., Chiles, 1996) but also among scientists (e.g., Langland and Sutton, 1992). Thus, a program not really designed to eliminate every individual of a population can be said to have failed even if it confers substantial control. Of course, if a management program aiming at complete elimination — an eradication campaign — uses the same methods that would have been used had the goal been to lower densities to an acceptable level, even a failed campaign can be useful. No harm need be done even if the species is not eradicated (Simberloff, 1997), and substantial control can be achieved, as in the current campaign to eradicate *Spartina* spp. from New Zealand (Nicholls, 1998).

However, if different means are used in an eradication attempt than would have been used for maintenance control, a real problem can arise (Dahlsten, 1986). Consider, for example, a major premise of biological control: the maintenance of both a pest and its natural enemy at low levels, with a homeostatic relationship between them such that an increase in pest density triggers a rapid increase in that of the natural enemy. An eradication project that attempted to kill every pest individual but failed could nonetheless leave pest densities so low that natural enemy populations are extinguished, and subsequent increase of pest densities would be unimpeded.

One typical approach in insect eradication projects is massive use of chemical pesticides, generally ones that are not highly specific. For example, the chemicals used to attack the imported fire ant are widely believed to have exacerbated the problem, at least partly, by lowering populations of competing ants more than that of the fire ant itself (Davidson and Stone, 1989). The effectiveness of a complex of native natural enemies of the introduced spotted alfalfa aphid (*Therioaphis trifolii*) in California was compromised by broad-spectrum pesticides that attacked native predators and

1-56670-612-2/03/$0.00+$1.50
© 2003 by CRC Press LLC

parasites (Smith and van den Bosch, 1967). And use of massive amounts of chemicals in an eradication attempt can have far-reaching ripple effects, such as secondary pest outbreaks associated with the destruction of natural enemies of other species that had been successfully held in abeyance (Perkins, 1989). Further, of course, broadscale application of generalized pesticides often has generated substantial collateral damage to nontargets, including biological magnification and human health risks. Finally, insects have evolved resistance to many initially promising insecticides.

SUCCESSFUL ERADICATION PROJECTS AND TECHNIQUES

An early remarkable success was the complete eradication of an African malaria vector, the mosquito *Anopheles gambiae*, from a large area of northeastern Brazil (Soper and Wilson, 1943; Davis and Garcia, 1989). Although this insect was recorded in 1930, the eradication campaign did not commence until devastating malaria outbreaks in 1938, by which time it had spread to a much larger geographic area. The project began in earnest in 1939, when the Brazilian government gave full control of the project to the Rockefeller Foundation, with almost dictatorial powers to Fred Soper of the foundation. Separate chemical treatments for adults and larvae achieved the complete eradication of *A. gambiae* by the end of 1940. Several features of the mosquito's life history, including its relatively poor dispersal in Brazil and the fact that its microhabitat requirements were satisfied in Brazil almost exclusively in human habitations, contributed to the success of the campaign. Also, the cost to workers exposed to the chemicals was substantial. Nevertheless, this project yielded potential lessons. First, if really substantial resources are devoted to eradication, even an entrenched, widespread pest may be vulnerable. In this instance, both the Brazilian government and the Rockefeller Foundation provided the resources. Second, the campaign was mobilized quickly and put under a unified command that enforced rapid cooperation among all relevant organizations and agencies. Finally, draconian powers were granted to the campaign, and individuals could not exempt themselves from its mandates.

The concept of eradication of pest insects received a tremendous boost from the work of E.F. Knipling, who developed the sterile-male technique that succeeded in completely ridding the island of Curaçao of the screw-worm fly (*Cochliomyia hominivorax*) in 1954–1955 (Baumhover et al., 1955) and aided in the eradication of the fly from the southeastern United States in 1958–1959 (Perkins, 1989). The method subsequently has been successfully applied in a few instances to other insects. For example, the melon fly (*Bactrocera cucurbitae*) was eradicated from Rota Island by this technique.

Steiner et al. (1955, 1965, 1970) pioneered in developing the male annihilation method to rid islands of fruit flies, such as the Oriental fruit fly (*Dacus dorsalis*) from Rota and Guam. Most recently, the melon fly was eradicated, island by island, from the entire Ryukyu Archipelago, including Okinawa, between 1972 and 1993 (Iwahashi, 1996; Kuba et al., 1996) by the sterile-male technique combined with the male annihilation method.

In the public mind, the sterile-male technique seems the quintessence of surgical removal, as it is autocidal and does not directly affect other species. However, it is often not employed alone. Knipling (1966) suggested that insect eradication programs should use a combination of techniques carefully designed to deal with the entire pest population. In the Ryukyus, male annihilation was used: absorbent materials were impregnated with a lure/toxicant, then aerially distributed by helicopter and manually hung in residential areas (Koyama et al., 1984).

Knipling was particularly enthusiastic about the sequential use of pesticides and the sterile-male technique. He reasoned that insecticides can efficiently kill insects when populations are large but are often inefficient when they are small, whereas the sterile-male technique can be highly efficient against small populations but is inefficient against large ones. This sequence of an insecticide (malathion) followed by sterile males is exactly the one envisioned in the California medfly campaign, although in the 1989 and 1990 program the sterile-male technique had to be abandoned because too few flies were available, and pesticide spraying was substituted (Penrose, 1996). However, even the initial use of pesticides could, in theory, lead to the sorts of problems indicated above: nontarget impacts, secondary pest outbreaks, and evolution of resistance.

Another set of potential problems with the sterile-male technique concerns evolution. Mass-rearing (domestication) of both vertebrates and insects often leads to a deterioration of quality, the exact nature of deterioration depending on the selective regime induced by the culture conditions. For example, low dispersal ability and decreased longevity may result (Boller and Chambers, 1977). In the Ryukyu melon fly eradication program, mass-reared males (which were sterilized) were shown to have evolved lower sexual competitiveness, while wild females evolved discriminatory mating behavior against mass-reared males (Iwahashi, 1996). The existence of these evolutionary processes implies that the initial inundative release of sterile males must be sufficiently large and widespread to prevent an opportunity for evolution to occur.

Several brute-force chemical and manual eradication programs have succeeded against recently introduced species that had not dispersed far. For example, the Asian citrus blackfly (*Aleurocanthus woglumi*) was found in Key West in 1934 and was eradicated over 3 years by spraying with a mixture of paraffin oil, whale oil, soap, and water (Hoelmer and Grace, 1989). It never reached other islands. Eradication of a subsequent invasion by the same species failed for a much larger region centered on Fort Lauderdale.

The giant African snail (*Achatina fulica*) was brought to Miami from Hawaii by a tourist in 1966 and liberated in a backyard. The resulting infestation, encompassing about 42 city blocks, was first discovered in 1969, and another infestation was quickly discovered 40 km away. An eradication campaign employing a granular chemical bait plus handpicking was quickly mounted by the Florida Division of Plant Industry (DPI) (Mead, 1979), entailing quarantining of many properties and a heavy publicity campaign. Despite some setbacks, including the discovery of three other infestations as far as 5.6 km from the original one and perhaps 2 to 3 years old, the DPI persisted in its campaign. At a cost of more than $1 million, success was achieved by 1975, though frequent surveys, baiting, and carbaryl drenches continued for many months afterwards. The snail has not been seen in the region since, and this project served as the inspiration for the only other successful eradication of the giant African snail in Queensland, Australia (Colman, 1978). The fact that these invasions were caught quickly was crucial, as was the fact that *A. fulica* does not self-fertilize.

As noted in the mosquito example above, insects can be eradicated over large areas, if sufficient resources are committed to the attempt, if organization is sufficiently strong, and if rules are universally enforced. The medfly was eradicated from Florida at least once and possibly twice (references in Simberloff, 1997). In 1929, the medfly was found in Orlando and nearby rural areas, eventually spreading to 20 counties. An 18-month, $1.7 million campaign employed 6000 personnel, a strict quarantine with National Guard roadblocks, destruction of produce and plants, trapping, and insecticide sprays. The insect was not seen again until 1956, when 28 counties from Miami to Tampa-St. Petersburg were infested. A $10 million campaign ensued, again with quarantines, roadblocks, insecticide-spraying, and extensive trapping. Victory was declared, although the situation bears some resemblance to that in California. The medfly recurred occasionally in the 1960s (each time it was targeted for a mini-eradication), was not seen in the 1970s, and appeared rarely in the 1980s. Two infestations of 1997 and 1998 may have not even been of the same origin (Prasher, 1999).

Nonindigenous vertebrates, at least when restricted to relatively small areas, are generally easier to eradicate than insects. They are usually more visible and far less numerous. New Zealanders are adept in this area, having extinguished various combinations of 12 mammals that include the house mouse (*Mus musculus*), black rat (*Rattus rattus*), Norway rat (*R. norvegicus*), and European rabbit (*Oryctolagus cuniculus*), and feral pigs and goats from many small islands using a variety of techniques such as chemicals and hunting (Veitch and Bell, 1990; Veitch and Henry, 2002). But they are hardly alone in successful eradication of this sort. For example, black rats were recently eradicated from Great Bird Island, off Antigua, in the northeastern Caribbean by rodenticide baiting (Day and Daltry, 1996a, b), followed by subsequent eradication from four other islands in a *rolling front* operation to prevent recolonization of Great Bird Island (Day and Daltry, 1996b; Varnham et al., 1998) that resembles the sequential eradication of the melon fly from islands in the Ryukyu Archipelago (Kuba et al., 1996). A similar project is under way and apparently successful on White Cay in the Bahamas (Day et al., 1998). Likewise, trapping and poison were used to eradicate the Norway rat from seven islands in the Sept-Îles Archipelago off the coast of Brittany (Pascal, 1996; Pascal et al., 1998), and several small islands in the Mascarenes have been cleared of various combinations of house mice, black and Norway rats, black-napped hares (*Lepus nigricollis*), and feral cats (Bell, 1999). Finally, various combinations of feral cats, Norway and black rats, house mice, European rabbits, goats, sheep, and burros have been eradicated from nine islands in North-west Mexico in an ongoing project (Donlan et al., 2000).

Several small, isolated fish populations have been eradicated, as have small plant populations (references in Simberloff, 1997). Some larger plant infestations have been eradicated, including *Hydrilla verticillata* from 64-ha Lake Murray (California) over 16 years and entailing the use of a quarantine, chemicals, divers with suction dredges, and a massive drawdown (Simberloff, 1997). In Western Australia, a 9-year project to eradicate the agricultural weed *Kochia scoparia* from over 3200 ha is very close to success (Randall, 2001).

The islands cleared of mammals in the West Indies, Northwest Mexico, the Mascarenes, New Zealand, and Brittany are small, as are most sites from which plants have been eradicated. But with sufficient resources and assiduity, much larger sites could possibly be cleared. Inspirational in this regard is the successful 10-year effort to eradicate the South American nutria (*Myocaster coypus*) from England, where it had been established for 50 years and had reached 200,000 individuals (Gosling, 1989). Key to the success, achieved primarily by trapping, was exhaustive scientific study of the natural history and population biology of this mammal in England. A current attempt to replicate this success in Maryland is advised by the director of the English project (Bounds, 1998). This is probably the most ambitious mammal eradication program yet. Since the 1950s nutria have been in Maryland, where they occupy thousands of hectares of wetlands and are remarkably fecund. In one 4000-ha area alone they have expanded from 150 individuals to between 35,000 and 50,000 individuals in 30 years. Nutria are established in at least 15 states, including Virginia and Delaware bordering Maryland, but the Virginia population is isolated from the Maryland one, while the Delaware population is simply an extension of the Maryland one and can easily be incorporated in the project (Colona, 1999). Thus, though the mainland is surely easier to recolonize than is an island, once cleared, this region could quite possibly be kept free of nutria. A partnership of 21 federal and state agencies and private organizations has secured extensive funding and commenced a large study of the biology of the local population, the necessary precursor to eradication. A test project on one 2500-ha site will initiate the eradication attempt.

Possibly the largest plant eradication project is the elimination of witchweed (*Striga asiatica*), an African root parasite of several grass-plant crops, from the Carolinas. Using herbicides, soil fumigation, and stringent regulation of movement of equipment and crops, this project has reduced witchweed from 162,000 ha in the 1950s to approximately 2800 ha now (Cross, 1999).

WHEN TO TRY ERADICATION, AND WHEN TO AVOID IT

The fact that eradication is feasible does not automatically mean a campaign should be mounted, particularly if resources are insufficient to surmount all environmental problems that have been identified. The key factors to consider are likely costs and benefits, as well as probability of success (Simberloff, 2002). A survey of many campaigns suggests some factors that should always be considered (cf. Myers et al. 1998, 2000; Simberloff, 2002).

Perkins (1989) points out that regional eradication efforts, by their nature, go against the grain of American conceptions of the relationship of the individual to the government (especially federal government) and of personal responsibility. The general alternative to insect eradication, integrated pest management (IPM), can usually be used on a field-by-field basis by individual farmers, granted that often its rewards are far greater when accompanied by widespread cooperation. On the other hand, eradication often requires absolute cooperation by all parties on a scale that can be organized and enforced only by the government. Voluntary cooperation is unlikely to suffice. Because costs are very widely shared but benefits often accrue disproportionately to certain segments of society, the matter of perceived equity is a difficult and delicate one. For example, aerial spraying of malathion in medfly eradication projects engendered extensive public hostility in California (Penrose, 1996) and Florida (Anon., 1997). Much of the public was doubtless hostile because they perceived themselves as suffering both the economic costs of the program plus whatever human health consequences might ensue, while the agriculture industry reaped the benefits. Similarly, in Florida, a campaign to eradicate a new infestation of citrus canker has met with armed resistance (Sharp, 2000).

No doubt public education about the costs of a pest to all of society (if such costs really do exist) and ways in which all would benefit from an eradication (if such benefits can be demonstrated) can facilitate the acceptance of a regional eradication plan. When human health is at stake, as in the malaria mosquito example, expense and the heavy and mandatory intervention of government are less controversial than when the target is an agricultural or environmental pest. However, the obligatory nature of eradication activities, the nature of certain techniques, and the key role of government will generally combine to generate a substantial social cost. For instance, animal-rights concerns impeded the successful British nutria eradication, and the nascent Maryland nutria project will surely face similar objections. Nevertheless, if a broad swath of the public can come to see themselves as stakeholders in the operation, a small degree of opposition may not suffice to torpedo a project.

Eco-terrorism may be a problem (Perkins, 1989), as demonstrated by the reintroduction of northern pike (*Esox lucius*) into Lake Davis after its costly but apparently successful eradication (Anon., 1999). This reappearance is probably the result of human activity (Moyle, P., personal communication, 1999). Because eco-terrorists can be single individuals (e.g., Davis, 1990) and such activities do not necessarily require major resources or sophisticated technology, eco-terrorism — or, in some instances, plain carelessness — will be a persistent threat to many eradication programs. On the other hand, eco-terrorism is not as relevant to maintenance control programs, under which low-level ongoing populations will remain.

Even aside from eco-terrorism, reinvasion is an important consideration in whether to undertake an eradication program. Under certain circumstances, particularly where agricultural pests are targeted, an eradication effort may be appropriate even if quick reinvasion is likely (Simberloff, 2002). However, if the probability of reinvasion is very high, a successful eradication could be a wasted effort, and development of a maintenance control method at the outset would have been more productive. In Washington, 130-ha Long Lake was the target of an intensive eradication campaign, mixing mechanical and chemical methods, to eliminate Eurasian water milfoil (*Myriophyllum spicatum*) (Thurston County Department of Water and Waste Management, 1995). This operation achieved success, but reinfestation quickly occurred at a public boat ramp, and the county now controls this plant at an acceptably low level by hand-pulling (Swartout, 1999). Reinvasion is less likely on islands than on mainland, although vagile organisms can often move from mainland to island or island to island surprisingly readily. This is the rationale of the rolling front technique described above.

Unforeseen new impacts may arise after the eradication of a species (references in Towns et al., 1997). For example, eradication of Norway rats from Mokoia Island was followed by greatly increased densities of mice. In many instances, control of top predators at densities far short of eradication has led to proliferation of intermediate predators with ripple effects throughout the community. Herbivores may also rebound with disastrous impacts. The removal of Pacific rats (*Rattus exulans*) from Motuopao Island to protect a native snail led to great increases in an exotic snail, to the detriment of the native one. Paradoxically, great increases in a native species after eradication of a nonindigenous one can also impede restoration. For example, when Pacific rats were eradicated from Burgess Island in the Mokohinau Islands, the existence of rank pastureland permitted the rapid proliferation of an endemic chafer beetle, *Odontria sandageri*, which in turn almost eliminated an endemic shrub, *Hebe aff. bollonsii*. Often, removal of an introduced herbivore leads to proliferation of exotic weeds rather than an increase in the native vegetation. On Motunau Island, eradication of rabbits fostered proliferation of exotic boxthorn (*Lycium ferocissimum*) that apparently had been suppressed by the rabbits. On Santa Cruz Island, removal of large grazers led to dramatic proliferation of fennel (*Foeniculum vulgare*) and other exotic weeds (Dash and Gliessman, 1994).

Finally, even aside from the danger of reinvasion, uncertainty remains about whether restoration is feasible once a species is eradicated, and whether the effort needed for restoration is a serious option. Removal of undesirable species does not by itself automatically constitute restoration (Towns et al., 1997). The original community may be so heavily changed that nothing approximating it can ever be achieved. For example, one or more of the original key species may be extinct. Sometimes related functional equivalents are seen as acceptable substitutes (Atkinson, 1988); other times, no such species exist. Often restorations are unsuccessful and we do not understand why. Forest and prairie restorations sometimes have been impeded by the mysterious absence of natural densities of ants that disperse some species and generate crucial small-scale disturbances (references in Simberloff, 1990). To date, all attempts to reintroduce stitchbirds (*Notiomystis cincta*) to islands following eradication of predators on small New Zealand islands have failed; reasons are unclear (Towns et al., 1997). Perhaps process restoration is the most critical requirement for successful restoration of functional ecosystems (Simberloff et al., 1999), but the technology of reintroducing many ecological processes is new and often rudimentary. In some instances, key processes (such as hydrological cycles) cannot be reestablished without costly, even regional, efforts that are not economically and politically feasible.

CONCLUSION

Eradication is a useful technique, but it should be attempted only in certain circumstances. Islands and other settings in which reinvasion is less likely are propitious. However, with sufficient resources, organization, and mandatory powers granted to an eradication campaign, large areas may be successfully cleared of a pest if its biology renders it vulnerable. Even on islands, the precise goal and feasibility of achieving it must be carefully considered before one embarks on an eradication campaign. Half-hearted efforts may not only waste money, but can also worsen a conservation problem. Particularly worthy of attention are eradication programs in which the techniques employed can either hinder maintenance control of the target pest or, even if successful, cause other species to become more pestiferous.

REFERENCES

Anonymous, Medfly flybys cut back following complaints, *Tallahassee Democrat*, Jul., 9, 1997, p. 10c.
Anonymous, Pike reappear, and a California city is on guard, *New York Times*, Jun. 21, 1999, p. 12.
Atkinson, I.A.E., Presidential address: opportunities for ecological restoration, *N. Z. J. Ecol.*, 11, 1–12, 1998.
Baumhover, A.H., Graham, A.J., Bitter, B.A., Hopkins, D.E., New, W.D., Dudley, F.H., and Buchland, R.C., Screw-worm control through release of sterilized flies, *J. Econ. Entomol.*, 48, 462–466, 1955.
Bell, B.D., The good and bad news from Mauritius, *Aliens*, 9, 6, 1999.

Boller, E.F. and Chambers, D.L., Eds., Quality control: an idea book for fruit fly workers, *IOBC WPRS Bull. 1977/5*, 1977.

Borenstein, S., Government takes aim at alien species, *Corvallis (OR) Gazette-Times*, July 26, 1999, p. B8.

Bounds, D., Marsh Restoration: Nutria Control in Maryland, Marsh Restoration/Nutria Control Partnership, Cambridge, MA, 1998.

Brody, J.E., Agriculture department to abandon campaign against the fire ant, *New York Times*, April 20, 1975, p. 46.

Chiles, L., Proclamation — Invasive Nonnative Plant Eradication Awareness Month, State of Florida Executive Department, Tallahassee, FL, 1996.

Colman, P.H., An invading giant, *Wildl. Aust.*, 15, 46–47, 1978.

Dahlsten, D.L., Control of invaders, in *Ecology of Biological Invasions of North America and Hawaii*, Mooney, H.A. and Drake, J.A., Eds., Springer-Verlag, New York, 1986, pp. 275–302.

Dash, B.A. and Gliessman, S.R., Nonnative species eradication and native species enhancement: fennel on Santa Cruz Island, in *The Fourth California Islands Symposium: Update on the Status of Resources*, Halvorson, W.L. and Maender, G.J., Eds., Santa Barbara Museum of Natural History, Santa Barbara, CA, 1994, pp. 505–512.

Davidson, N.A. and Stone, N.D., Imported fire ants, in *Eradication of Exotic Pests*, Dahlsten, D.L. and Garcia, R., Eds., Yale University Press, New Haven, 1989, pp. 196–217.

Davis, J.R. and Garcia, R., Malaria mosquito in Brazil, in *Eradication of Exotic Pests*, Dahlsten, D.L. and Garcia, R., Eds., Yale University Press, New Haven, 1989, pp. 274–283.

Davis, S., The cape weed caper, *Calif. Waterfront Age*, 6, 22–24, 1990.

Day, M. and Daltry, J., Antiguan racer conservation project, *Flora Fauna News*, April, 1996a.

Day, M. and Daltry, J., Rat eradication to conserve the Antiguan racer, *Aliens*, 3, 14–15, 1996b.

Day, M., Hayes, W., Varnham, K., Ross, T., Carey, E., Ferguson, T., Monestine, J., Smith, S., Armstrong, C., Buckle, A., Alberts, A., and Buckner, S., Rat eradication to protect the White Cay iguana, *Aliens*, 8, 22–24, 1998.

Donlan, C.J. et al., Island conservation action in northwest Mexico, in *Proceedings of the Fifth California Island Symposium*, Browne, D., Mitchell, K., and Chaney, H., Eds., Santa Barbara Museum of Natural History, Santa Barbara, CA, 2000, pp. 330–338.

Gosling, M., Extinction to order, *New Sci.*, 121, 44–49, 1989.

Hoelmer, K.A. and Grace, J.K., Citrus blackfly, in *Eradication of Exotic Pests*, Dahlsten, D.L. and Garcia, R., Eds., Yale University Press, New Haven, CT, 1989, pp. 147–165.

Iwahashi, O., Problems encountered during long-term SIT program in Japan, in *Fruit Fly Pests: A World Assessment of their Biology and Management*, McPheron, B.A. and Steck, G.J., Eds., St. Lucie Press, Delray Beach, FL, 1996, pp. 391–398.

Knipling, E.F., Some basic principles in insect population suppression, *Bull. Entomol. Soc. Am.*, 12, 7–15, 1966.

Koyama, J., Teruya, T., and Tanaka, K., Eradication of the oriental fruit fly (Diptera: Tephritidae) from the Okinawa Islands by a male annihilation method, *J. Econ. Entomol.*, 77, 468–472, 1984.

Kuba, H., Kohama, T., Kakinohana, H., Yamagishi, M., Kinjo, K., Sokei, Y., Nakasone, T., and Nakamoto, Y., The successful eradication programs of the melon fly in Okinawa, in *Fruit Fly Pests: A World Assessment of Their Biology and Management*, McPheron, B.A. and Steck, G.J., Eds., St. Lucie Press, Delray Beach, FL, 1996, pp. 534–550.

Langland, K. and Sutton, D., Assessment of *Mimosa pigra* Eradication in Florida, Florida Department of Natural Resources, Gainesville, FL, 1992.

Mead, A.R., Ecological malacology: with particular reference to *Achatina fulica*, in *Pulmonates*, Vol. 2b, Fretter, V., Fretter, J., and Peake, J., Eds., Academic Press, London, 1979.

Myers, J.H., Savoie, A., and van Randen, E., Eradication and pest management, *Annu. Rev. Entomol.*, 43, 471–491, 1998.

Myers, J.H., Simberloff, D., Kuris, A.M., and Carey, J.R., Eradication revisited: dealing with exotic species, *Trends Ecol. Evol.*, 15, 316–320, 2000.

Newsom, L.D., Eradication of plant pests — con, *Bull. Entomol. Soc. Am.*, 24, 35–40, 1978.

Nicholls, P., Maintaining coastal integrity: eradication of *Spartina* from New Zealand, *Aliens*, 8, 10, 1998.

Pascal, M., Norway rat eradication from Brittany islands, *Aliens*, 3, 15, 1996.

Pascal, M., Siorat, F., and Bernard, F., Norway rat and shrew interactions: Brittany, *Aliens*, 8, 7, 1998.

Penrose, D., California's 1993/1994 Mediterranean fruit fly eradication program, in *Fruit Fly Pests: A World Assessment of their Biology and Management*, McPheron, B.A. and Steck, G.J., Eds., St. Lucie Press, Delray Beach, FL, 1996, pp. 551–554.

Perkins, J.H., Eradication: scientific and social questions, in *Eradication of Exotic Pests*, Dahlsten, D.L. and Garcia, R., Eds., Yale University Press, New Haven, CT, 1989, pp. 16–40.

Prasher, D., Analysis of potential origins of medfly infestations within the continental U.S., paper presented at the annual meeting of the Entomological Society of America, Las Vegas, NV, 1999.

Randall, R., Eradication of a deliberately introduced plant found to be invasive, in *Invasive Alien Species: A Toolkit of Best Prevention and Management Practices*, Wittenberg, R. and Cock, M.J.W., Eds., CAB International, Wallingford, Oxon, U.K., 2001, p. 174.

Sharp, D., Citrus tree inspectors face gunshots in Fla, *USA Today*, Aug. 2, 2000, p. 2A.

Simberloff, D., Reconstructing the ambiguous: can island ecosystems be restored? in *Ecological Restoration of New Zealand Islands*, Towns, D.R., Daugherty, C.H., and Atkinson, I.A.E., Eds., Department of Conservation, Wellington, New Zealand, 1990, pp. 37–51.

Simberloff, D., Eradication, in *Strangers in Paradise: Impact and Management of Nonindigenous Species in Florida*, Simberloff, D., Schmitz, D.C., and Brown, T.C., Eds., Island Press, Washington, D.C., 1997, pp. 221–228.

Simberloff, D., Today Tiritiri Matangi, tomorrow the world! — Are we aiming too low in invasives control?, in press in *Turning the Tide: The Eradication of Invasive Species*, Veitch, C.R. and Clout, M.N., Eds., Invasive Species Specialist Group of the World Conservation Union (IUCN), Auckland, 2002.

Simberloff, D., Doak, D., Groom, M., Trombulak, S., Dobson, A., Gatewood, S., Soulé, M.E., Gilpin, M., Martinez del Rio, C., and Mills, L., Regional and continental restoration, in *Continental Conservation. Scientific Foundations of Regional Reserve Networks*, Soulé, M.E. and Terborgh, J., Eds., Island Press, Washington, D.C., 1999, pp. 65–98.

Smith, R.F. and van den Bosch, R., Integrated control, in *Pest Control*, Kilgore, W.W. and Doutt, R.L., Eds., Academic Press, New York, 1967, pp. 295–340.

Soper, F.L. and Wilson, D.B., *Anopheles gambiae* in Brazil, 1930 to 1940, The Rockefeller Foundation, New York, 1943.

Steiner, L.F. and Lee, R.K.S., Large-area tests of a male-annihilation method for oriental fruit fly control, *Journal of Economic Entomology*, 48, 311–317, 1955.

Steiner, L.F., Hart, W.G., Harris, E.J., Cunningham, R.T., Ohinata, K., and Kamakahi, D.C., Eradication of the oriental fruit fly from the Mariana Islands by the methods of male annihilation and sterile insect release, *J. Econ. Entomol.*, 63, 131–135, 1970.

Steiner, L.F., Mitchell, W.C., Harris, E.J., Kozuma, T.T., and Fujimoto, M.S., Oriental fruit fly eradication by male annihilation, *J. Econ. Entomol.*, 58, 961–964, 1965.

Thurston County Department of Water and Waste Management, Long Lake Eurasian Watermilfoil Eradication Project, Olympia, W.A., 1995.

Towns, D.R., Simberloff, D., and Atkinson, I.A.E., Restoration of New Zealand islands: redressing the effects of introduced species, *Pac. Conserv. Biol.*, 3, 99–124, 1997.

Varnham, K., Ross, T., Daltry, J., and Day, M., Recovery of the Antiguan racer, *Aliens*, 8, 21, 1998.

Veitch, C.R. and Bell, B.D., Eradication of introduced animals from the islands of New Zealand, in *Ecological Restoration of New Zealand Islands*, Towns, D.R., Daugherty, C.H., and Atkinson, I.A.E., Eds., Department of Conservation, Wellington, New Zealand, pp. 137–146.

Veitch, C.R. and Henry, J., Eradication of Pacific rats (*Rattus exulans*) from Tiritiri Matangi Island, in press in *Turning the Tide: The Eradication of Invasive Species*, Veitch, C.R. and Clout, M.N., Eds., Invasive Species Specialist Group of the World Conservation Union (IUCN), Auckland, 2002.

CHAPTER 58

Eradication of Introduced Marine Pests

Armand M. Kuris

INTRODUCTION

The ecological and economic impacts of introduced marine species have attracted considerable recent attention. A substantial research effort is under way to prevent further introductions, study the spread of these invaders, and assess their effects on native organisms (Carlton and Geller, 1993). In some parts of heavily invaded ecosystems, exotics comprise over 90% of the biomass (Cohen and Carlton, 1998). However, unlike terrestrial and freshwater introduced pests, remarkably few attempts, with little evident success, have been made to control of marine pests once they have become established. Typically, news of their establishment engendered responses such as, "We will just have to learn to live with them." Recently, this defeatist fatalism has been pierced by theoretical arguments, indicating that classical biological control may be possible in a marine environment (Lafferty and Kuris, 1996; Meinesz, 1999; Kuris and Lafferty, 2000), and by the development of sustained concerted research efforts to control or eradicate introduced marine pests. Notably, these include the green crab (*Carcinus maenas*) in North America and Tasmania, the sabellid worm pest of abalone (*Terebrasabella heterouncinata*) in California, the north Pacific starfish (*Asterias amurensis*) in Australia, and the alga (*Caulerpa taxifolia*) in the Mediterranean and California (Lafferty and Kuris, 1996; Thresher, 1997; Culver and Kuris, 1999; Kuris et al., 1996; Goggin, 1998; Meinesz and Thibaut, 1998: Meinesz, 1999; Thibaut and Meinesz, 2000). The establishment of the Centre for Research on Introduced Marine Pests (CRIMP) by the Commonwealth Scientific and Industrial Research Organization (CSIRO) Marine Division in Australia, dedicated to the prevention, eradication, and control of non-indigenous marine species in Australia (Thresher and Martin 1995) has brought a sustained and coherent approach to these problems and raised them to a national priority.

In this chapter, I will make a few comparisons of the problems associated with eradication of introduced marine versus terrestrial and freshwater pests. I will then describe two recent apparently successful efforts to eradicate well-established introduced marine pests. I will comparatively analyze these two histories to seek the factors that promoted success in these two cases so that other introduced pests may be better evaluated for possible eradication.

Detection of introduced marine species is generally more difficult than detection of terrestrial or freshwater exotics. The sea is relatively inaccessible (except for species-poor sandy beach habitats). Few observers can actually recognize a species as nonnative, because most importantly, even intertidal species are exposed for only a few hours daily, and subtidal species are only seen by a few divers and by blind sampling (e.g., trawls, fishing lines). Early detection is widely acknowledged to be essential for the development of an eradication campaign. This will only rarely be achieved in the marine environment because very few monitoring programs are capable of detecting a new invader. Hence, detected marine exotics are likely to be well established, with large and increasing populations of reproductive individuals, sustained by further recruitment of young.

TWO RECENT ERADICATION SUCCESSES

Case History 1: The Sabellid Polychaete Worm Pest of Abalone and Other Gastropods

In 1993, a previously unknown and undescribed sabellid polychaete worm was detected in abalone mariculture facilities in California. Investigations demonstrated that it had been accidentally imported with South African abalone for commercial research purposes in the 1980s (Culver and Kuris, 1996; Kuris and Culver, 1999; Ruck and Cook, 1998). Recently described as *Terebrasabella heterouncinata* (Fitzhugh and Rouse, 1999), this worm had a unique and remarkable biology. It preferentially settled in the apertures of abalone and many other species of gastropods, evaded the fouling defenses of the host, and guided the mantle tissue to promote a laminar calcification process that formed a tube for the sabellid within hours of settlement of the worm. The sabellid develops in the tube, feeds with a tentacular crown through the opening of the tube on the outer surface of the host shell, and reproduces. Unlike most sabellids, these worms were hermaphroditic and produced benthic crawling larvae that were soon competent for settlement on a new host (Kuris and Culver, 1999). Heavily infested hosts could not secrete the outer prismatic shell layer and growth slowed or ceased as the laminar secretions were added in thickening layers to form the worm tubes. This soon deformed the aperture and weakened the shell structure so that it became brittle and friable (Kuris and Culver, 1999). The potential to affect growth, survival, and appearance of many abundant native gastropods meant that the sabellid was a sufficiently serious pest to enact regulations to prevent its establishment in nature and to eliminate it from culture facilities (Culver et al., 1997; Kuris et al., 1998).

Some relevant life history features of the sabellid made its establishment, if released in appropriate habitat, seem likely. Its hermaphroditism, competent larvae at or soon after release from the parental tube, wide host range, abundance of susceptible hosts, and short generation time (1 to 2 months) (Fitzhugh and Rouse, 1999; Kuris and Culver, 1999) were all features that promoted its establishment. Fortunately, many infested abalone mariculture facilities did not discharge their outflow in appropriate habitats, and the sabellid's lack of a planktonic stage meant that once established locally it would spread relatively slowly to distant sites.

Ultimately all 18 abalone mariculture facilities became infested. The infested abalone were sent to other states, Mexico and Chile, and the outflow from these facilities posed a significant risk of release and establishment in the natural environment. These fears were realized when, in 1995, we discovered a substantial naturalized population at Cayucos, CA, below the seawater outflow of a mariculture facility that discharged into a rich rocky intertidal zone habitat (Culver and Kuris, 1999). Fortunately, in the 2 years since its discovery in the mariculture facilities, sufficient knowledge of its biology was available to estimate its abundance at this infested site, and to devise an eradication program (Culver and Kuris, 1999).

Sabellid worms were being released from the culture facility in abundance in the shells of live escaped cultured abalone, empty shell debris (cultured abalone shells can be recognized by their light green color), and in the shells of kelp snails (three species), brought into the facility with kelp harvested for abalone food. These infested hosts were readily found over an 80-m-long stretch of rocky intertidal zone habitat in the vicinity of the outfall. Examination of upper and midintertidal zone snails disclosed that the very abundant snail, *Tegula funebralis*, and also *T. brunnea*, had become infested. Since *T. funebralis* was not present in the culture facility, transmission of worms to this host had to have occurred in the wild.

Using marked sentinel host *T. funebralis* from elsewhere, we began a regular monitoring program to estimate infestation rates at Cayucos. In the initial survey, 32% of the sentinel snails were infested in the six-week monitoring period, demonstrating a very high rate of transmission and population increase of the sabellid at this site. The infestation was spreading throughout the *T. funebralis* population at an epidemic outbreak level.

At our recommendation, the facility screened the outflow to prevent further release of live molluscs and shell debris, and a cleanup of cultured abalone, kelp snails, and their debris from the intertidal zone was conducted. A subsequent sentinel snail release indicated the rate of new infestations had decreased, but new infestations were still occurring. This indicated that the sabellid population was established because wild *T. funebralis* were now the primary source of new infestations.

Using conservative estimates of host abundance, prevalence of the sabellids in these hosts, and intensity of the worms on the infested hosts, we estimated that this established population included at least 2.2 million worms.

We rejected an eradication strategy based on detection and removal of infested snails. The worms were too small to be readily detected on the black *T. funebralis* shells without microscopic examination, there were too many snails to examine, and estimates of the prevalence of the worms indicated that the number of infested snails was in the range of 1.5 to 2 million.

Since the sabellid was (1) highly habitat specific (as an epibiont on the shells of gastropods), our previous research (Kuris and Culver, 1999, Culver, 1999) had demonstrated that (2) *Tegula* species were highly susceptible, and (3) large individuals were much more susceptible than were small individuals, we were able to devise an eradication scheme based on the Kermack–McKendrick Threshold Theory from epidemiology (McKendrick, 1940; Bailey, 1957; Stiven, 1964, 1968). The incidence of new infestations increases with the density of susceptible hosts. Hence, the threshold density of new infestations drops lower than the replacement rate necessary to maintain the population of parasites. If this low host density can be maintained for a sufficiently long period of time, the number of infected hosts will dwindle to zero.

With the staff of the Abalone Farm culture facility (Cayucos, CA), California Department of Fish and Game (CDFG) biologists and student volunteers from the University of California, Santa Barbara, and Cuesta College, at least 1.6 million large (> 10 mm) *T. funebralis* and *T. brunnea* were removed. Repeated monitoring using sentinel snails (with a high degree of sensitivity, 1×10^{-9}) has not detected a newly infested snail over the 2 years since the removal of large numbers of susceptible hosts (Culver and Kuris, 1999).

At this point it appeared likely that the once-abundant Cayucos population of the sabellid has been eradicated. This success provided a strategy should other infestations be discovered. In addition to release from other culture facilities, it is also possible that infested abalone stock were outplanted as part of a program to recover overfished abalone populations in the 1980s and early 1990s.

Case History 2: The Brackish Water Black-Striped Mussel

A team of introduced marine species detection specialists from CRIMP conducted a port survey in Darwin Harbor in the Northern Territory in the spring of 1999, as part of a program to survey all of Australia's international ports. During the survey, the team discovered the brackish water black-striped mussel, *Mytilopsis sallei*, in abundance (Bax, 1999). Their previous survey, 6 months earlier, had not detected this species, so it was evidently a recent introduction and was clearly established. Most of the population was in Cullen Bay, a yacht harbor with a narrow seaward entrance protected by a locked gate to cope with the extreme tidal amplitude at Darwin.

This mussel had high pest potential as it has spread to other tropical harbors (Singapore, Suva, Fiji), forming monospecific clumps up to 15 cm thick and weighing as much as 100 kg/m² (Bax, 1999).

At the time of its discovery, the mussel had already reached densities as high as 60,000/m² in Cullen Bay, and its total population was many millions. The scientific and regulatory response was rapid. CRIMP determined lethal dosage levels for chlorine and $CuSO_4$. The Museum of the Northern Territory promptly identified the invader as *M. sallei*. The government of the Northern Territory quarantined and locked the gates of Cullen Bay 5 days after the discovery of the mussels. Just 9 days after discovery of the mussels, Cullen Bay was treated with 160 t of bleach and 54 t of $CuSO_4$ (including applications to localized infestations elsewhere in Darwin harbor). Collateral damage was great, but localized; all living things in this 600-Ml marina were killed. Repeated monitoring over the next few weeks failed to recover any living mussels, and Cullen Bay was reopened to boat traffic. The eradication appears to be a success.

KEY FEATURES OF THE SUCCESSFUL MARINE PEST ERADICATION PROGRAMS

The history of the campaigns against the sabellid pest and the black-striped mussel offer some interesting points of comparison. While they shared important features, they differed in many biological and strategic aspects. This implies that targets for future eradication efforts do not have to meet a very narrow set of criteria, and more introduced marine pests might serve as appropriate targets for eradication campaigns than are under consideration at present. Table 58.1 summarizes these comparative aspects.

Although rapid detection is generally considered a paramount precondition for an eradication effort, the sabellid was likely established for a few years before its discovery. The black-striped mussel was very rapidly detected after its establishment, which is fortunate given its explosive population growth at Cullen Bay.

Both pests were very well established with populations in the millions. Both were also reasonably localized, although other populations of each may be present but not yet detected. The localization of each of these pests was certainly important in the initial decision to develop an eradication plan and likely a key factor in the success of these programs because a relatively small monitoring effort could effectively evaluate the status of the eradication effort.

Both pests were rapidly evaluated as exotics. Since the sabellid was an unknown and undescribed species, California experienced a brief period in which some abalone aquaculturists challenged its nonnative status. If it was an undescribed native species (Culver et al., 1997), no environmental risk could be claimed, and it would have been merely an economic pest in the facilities (as it is in South Africa) (Ruck and Cook, 1998).

Regulatory authorities did not call for a comprehensive risk assessment for the impact of either pest. If such had been required, then the geographic scale and abundance of the pests would have been so great that consideration of eradication would have been precluded, and other forms of

Table 58.1 A Summary of the Features of Successful Marine Pest Eradication Programs

	Sabellid	Black-Striped Mussel
Time to detection	A few years?	<6 months
Estimated population size	Minimum 2.5 million	Millions
Evaluation as an exotic	Rapid; unambiguous	Rapid; unambiguous
Species identification	New species; described after eradication under way	U.S. sea grant non-indigenous species Web site information
Impact assessment	Experimentally evaluated	Assumed
Basis for eradication plan	Host-specificity studies; epidemiological theory	Toxicity studies; brute force
Stakeholder involvement	Decision making: yes implementation: yes	Decision making: yes implementation: no
Government regulation	Minimal; informal	Extensive; formal; swift
Collateral damage	Localized; minimum 1.6 million native snails of two species	Localized; extreme; all taxa
Cost	~ $3000 (U.S.)	$1,200,000 (U.S.)

control would have been much more difficult. This has been the sad history of some other major marine pests such as the alga, *Caulerpa taxifolia,* in the Mediterranean (Meyer et al., 1998; Meinesz, 1999). *Caulerpa taxifolia* was discovered at Monaco in 1984. Its eradication was called for in 1991 when it still was localized and its removal could have probably been achieved. It has since spread from Spain to Croatia, and eradication is no longer an option. The recent arrival of the green crab, *Carcinus maenas,* on the Pacific coast of North America, provides a case in point. As it spread north, each regional Sea Grant program established the effects of the crab as the principal issue to be studied. The director of the University of California program, James Sullivan, mandated a 3-year moratorium on biological control grants to await results of the impact studies (Sullivan, J., personal communication). Over those 3 years, the green crab spread from being localized in South San Francisco Bay to settling in large numbers in Humboldt Bay, 400 km to the north. In the next 3 years, it reached the outer coast of Vancouver Island, Canada.

The potential impact of the sabellid pest in nature had been partially assessed because it was under study for its effect and control within the aquaculture facilities (Culver et al., 1997). Considerable information on its host range, and on the process of deformation of infested shells, indicated the worst-case potential impact would be severe. However, no studies were directly made to assess its effect in the wild before the eradication program was conceived. Likewise, the eradication plan against the black-striped mussel was conceived, planned, and initiated on the general principle that a very abundant, filter-feeding, fouling organism would have a great impact on both benthic communities and plankton, as well as on maritime activities (e.g., boat fouling, pipe clogging).

Cultural differences also play a role in impact assessments of nonindigenous species. Potential economic impact is generally needed to initiate regulatory action in the United States, whereas its ecological impact is fully sufficient to trigger a regulatory response in Australia (Thresher, R., personal communication).

The development of an eradication strategy against the sabellid pest was firmly based in epidemiological theory. The campaign was designed to break the threshold of transmission for a sufficiently long time for sabellid transmission to damp out to the extinction point. The use of this theory required specific knowledge about host species specificity and the much greater susceptibility of large hosts. Armed with this information, it was relatively easy to design, implement, and monitor the eradication effort.

The environmental configuration of the Cullen Bay Marina was probably key to the eradication campaign against the black-striped mussel. Since it could be effectively closed off and quarantined, it was possible to use chemical control, albeit on a rather grand scale. The preliminary toxicology studies to determine lethal dosages were obvious, rapidly conducted, and conservative, as the kill of the mussels appears to have been attained at a lower chlorine concentration than predicted by the laboratory studies (Bax, 1999).

For both these eradication efforts, the private stakeholders were involved in the decision-making processes. In the case of the sabellid, stakeholders provided much of the labor to remove the large, susceptible *T. funebralis* snails.

The Australian and Northern Territory governmental regulatory response was rapid and courageous, as quarantining the major yacht harbor in the territory, with little time to alert and educate the public, certainly carried political risks. The decision to use massive amounts of highly toxic chemicals would have likely met strong public outcry in other countries and could have required extensive deliberation that would have likely been fatal to the successful eradication effort. In contrast, the decision to attempt the sabellid eradication was done by the aquaculture facility and university researchers. The regulatory agency, CDFG, was merely informed and then cooperated in the eradication effort.

Collateral damage in both of these eradication programs was substantial. The estimated 1.6 million *T. funebralis* (and some *T. brunnea*) removed were long-lived snails. Almost all were at least 10 years old, based on their size and growth rates determined elsewhere (Frank, 1985). It is worth emphasizing that these snails were not eliminated at this site. They were still plentiful enough to be routinely collected on the posteradication monitoring transects (Culver and Kuris, 1999).

Although the collateral damage in the chemically based eradication of the black-striped mussels was largely confined to the gated Cullen Bay Marina, it was severe. All living organisms (including vertebrates) in this large marina were killed. In some cultures, this expected prospect would probably have encountered vigorous public opposition.

The cost of the sabellid eradication program was minimal, and no specific funds for its implementation were sought. The eradication of the black-striped mussel was considerable ($1.2 million U.S.). As with the other elements in the decision-making process, the rapid release of substantial funds for this campaign was noteworthy.

CONCLUSION

The eradication of the sabellid can be attributed to a theoretically based plan, local initiative, and private action with government cooperation. In this sense, it was firmly based upon a strong American cultural imperative of individual action and problem solving. In contrast, a nationally coordinated port survey program (Hewitt and Martin, 1996) and a strong organization (CRIMP) with a coherent plan to identify and manage introduced marine pests (Thresher and Martin, 1995) were essential to the scope, cost, and rapidity of the program against the black-striped mussel in Darwin Harbor. The background for this achievement seems clearly derived from an acute Australian cultural awareness of the effect of introduced species on the native biota, and a willingness to take a proactive stand against such pests.

Taken together, these two recent and apparently successful campaigns strongly suggest that other introduced and established marine pests may be vulnerable to eradication. However, these must be appropriately scaled, and a well-conceived plan with a theoretical basis should be developed. In addition, posteradication monitoring must be a key element in the program. Awaiting results of extensive studies of the effects of an exotic marine pest, and the potential for its further geographic spread, will likely foreclose a successful eradication plan because such studies will consume too much time. Instead, an assessment of impact based on knowledge elsewhere, ecological theory, studies of comparable organisms, and common sense should be the basis for eradication campaigns. Eradication programs in any ecosystem may have controversial, social, and ecological ramifications, so it is necessary to evaluate their political and environmental acceptability as well as their ecological feasibility (Myers et al., 2000). In the U.S., the establishment of a coherent scientific program, such as CRIMP in CSIRO of Australia, would aid the prospect for timely and effective responses against introduced marine pests. A possible candidate for urgent consideration of an eradication effort as part of a control program is the Asian veined rapa whelk, *Rapana venosa*, in Chesapeake Bay because it is a large generalist predator that is currently not abundant, but is becoming widespread (Harding and Mann, 1999; Mann and Harding, 2000).

ACKNOWLEDGMENTS

I thank Carrie Culver, Ron Thresher, and Nic Bax for discussion of the two eradication programs. This chapter was funded by a grant from the National Sea Grant College Program, National Oceanic and Atmospheric Administration, U.S. Department of Commerce, under Grant NA36RGO537, project number UCSG-27 through the California Sea Grant College system and by CRIMP of CSIRO Australia. The views expressed herein are my own and do not necessarily reflect the views of NOAA or any of its subagencies. The U.S. government is authorized to reproduce and distribute this for governmental purposes.

REFERENCES

Bailey, N.T.J., *The Mathematical Theory of Epidemics*, Hafner, New York, 1957.

Bax, N., Eradicating a dreissenid from Australia, *Dreissenia*, 10, 1–5, 1999.

Carlton, J.T. and Geller, J.B., Ecological roulette: the global transport of nonindigenous marine organisms, *Science*, 261, 78–82, 1993.

Cohen, A.N. and Carlton, J.T., Accelerating invasion rate in a highly invaded estuary, *Science*, 279, 555–558, 1998.

Culver, C.S. The Aperture of Marine Gastropods: Factors Precluding Settlement of Fouling Organisms, Ph.D. dissertation, University of California, Santa Barbara, 1999.

Culver, C.S. and Kuris, A.M., The apparent eradication of a locally established introduced marine pest, *Biol. Invasions*, 2, 245–253, 1999.

Culver, C.S., Kuris, A.M., and Beede, B. Identification and Management of the Exotic Sabellid Pest in California Cultured Abalone, Publication T-104, California Sea Grant College Program, La Jolla, CA, 1997.

Fitzhugh, K. and Rouse, G.W., A remarkable new genus and species of fan worm (Polychaeta: Sabellidae: Sabellinae) associated with some marine gastropods, *Invertebrate Biol.*, 118, 357–390, 1999.

Frank, P.W., Shell growth in a natural population of the turban snail, *Tegula funebralis*, *Growth*, 29, 395–403, 1985.

Goggin, C.L., Ed., Proceedings of a meeting on the biology and management of the introduced seastar *Asterias amurensis* in Australian waters, Commonwealth Scientific and Industrial Research Organization Centre for Research on Introduced Marine Pests Technical Report, 15, 1–75, 1998.

Harding, J.M. and Mann, R., Observations on the biology of the veined whelk, *Rapana venosa* (Valenciennes, 1846) in the Chesapeake Bay, *J. Shellfish Res.*, 18, 9–17, 1999.

Hewitt, C.L. and Martin, R.B., Port surveys for introduced marine species: background considerations and sampling protocols, Commonwealth Scientific and Industrial Research Organization Centre for Research on Introduced Marine Pests Technical Report, 4, 1–40, 1996.

Kuris, A.M. and Culver, C.S., An introduced sabellid polychaete pest of cultured abalone and its potential spread to other California gastropods, *Invert. Biol.*, 118, 394–403, 1999.

Kuris, A.M. and Lafferty, K.D., Can biological control be developed as a safe and effective mitigation against established introduced marine pests, in *Marine Bioinvasions: Proceedings of a Conference,* January 24–27, 1999, Pederson, J., Ed., MIT Sea Grant College Program, Cambridge, MA, 2000, pp. 102–106.

Kuris, A., Friedman, C., Gordon, R., Trevelyan, G., and Hulbrock, R., Sabellid Worm Report, California Department of Fish & Game: Report to Aquaculture Disease Committee, 1998.

Kuris, A.M., Lafferty, K.D. and Grygier, M.J., Detection and preliminary evaluation of natural enemies for possible biological control of the northern Pacific seastar, *Asterias amurensis*, Commonwealth Scientific and Industrial Research Organization Centre for Research on Introduced Marine Pests Technical Report, 3, 1–17, 1996.

Lafferty, K.D. and Kuris, A.M., Biological control of marine pests, *Ecology*, 77, 1989–2000, 1996.

Mann, R. and Harding, J.M., Invasion of the North American Atlantic coast by a large predatory Asian mollusc, *Biol. Invasions,* 2, 7–22, 2000.

McKendrick, A., The dynamics of crowd infections, *Edinburgh Med. J.* (New Series), 47, 117–136, 1940.

Meinesz, A., *Killer Algae*, University of Chicago Press, Chicago, 1999.

Meinesz, A. and Thibaut, T. The biological control of an invasive species in the open sea: need of an international decision, in *Third International Workshop on Caulerpa taxifolia*, CF Boudouresque, C.F., et al., Eds., Publication GIS Posidonie, Marseille, France, 1998, pp. 113–116.

Meyer, U., Meinesz, A., and de Vaugelas, J., Invasion of the accidentally introduced tropical alga *Caulerpa taxifolia* in the Mediterranean Sea, in *Plant Invasions: Ecological Mechanisms and Human Responses*, Starfinger, U. et al., Eds., Backhuys Publishers, Leiden, 1998, pp. 225–234.

Myers, J.H., Simberloff, D., Kuris, A.M., and Carey, J.R., Eradication revisited: dealing with exotic species, *Tree*, 15, 316–320, 2000.

Ruck, K.R. and Cook, P.A., Sabellid infestations in the shells of South African molluscs: implications for abalone mariculture, *J. Shellfish Res.*, 17, 693–699, 1998.

Stiven, A.E., Experimental studies on the host parasite system hydra and *Hydramoeba hydroxena* (Entz). II. The components of a single epidemic, *Ecol. Monogr.*, 34, 119–142, 1964.

Stiven, A.E., The components of a threshold in experimental epizootics of *Hydramoeba hydroxena* in populations of *Chlorohydra viridissima*, *J. Invert. Pathol.*, 11, 348–357, 1968.
Thibaut, T. and Meinesz, A., Are the Mediterranean ascoglossan mollusks *Oxynoe olivacea* and *Lobiger
   serradifalci* suitable agents for a biological control against the invading tropical alga *Caulerpa
   taxifolia?*, *Compte Rendus de l'Academie des Sciences Paris, Sciences de la Vie*, 323, 477–488, 2000.
Thresher, R.E., Ed., Proceedings of the first international workshop on the demography, impacts and management of introduced populations of the European crab, *Carcinus maenas*, Commonwealth Scientific
   and Industrial Research Organization Centre for Research on Introduced Marine Pests Technical
   Report, 11, 1–101, 1997.
Thresher, R.E. and Martin, R. Controlling marine introductions: CRIMP, *Water,* July/Aug., 24–26, 1995.

SECTION II.5

Landscape Health Assessment

CHAPTER 59

Overview: Landscape Health Assessment

Ganapati P. Patil

LANDSCAPE ECOSYSTEM HEALTH ASSESSMENT USING REMOTE SENSING DATA

Ecosystem health entails both status and trends. Ecosystem processes operate in space and time, so it matters not only what is observed and how it is observed, but also where and when the observations are made. Ecosystems are open systems with various processes functioning in gradients over a range of spatial and temporal scales. Ecological hierarchy theory implies that observations made in a particular spatial and temporal frame of reference will find certain processes more strongly expressed than others.

Ecosystem health has been evolving over the past decade from explorations of the possible relevance of the health concept at the ecosystem and landscape scale (Nielsen, 1999), to the exploration of the concept in the monitoring and assessment of large-scale ecosystems (Rapport et al., 1995), to an ever-widening range of considerations. These include incorporating societal values (Rapport et al., 1998) at ecosystem and landscape scales, economic and social determinants and consequences of ecological conditions (Buckingham, 1998; Rapport et al., 1998), and the increasing dependence of human health on ecosystem conditions (Epstein, 1995; Epstein and Rapport, 1996; Huq and Colwell, 1996; Karr, 1997; McMichael, 1997). It also includes the encouragement of appropriate techniques for rapid, accurate, and economically feasible methods for regional scale assessment of ecological conditions, including developments in geographic information system (GIS) and remote sensing (Rapport, 1999).

We focus on the new generation of techniques and analyses that permit broadscale geographically based assessments of environmental and ecological conditions. These techniques, including recent advances in applications of GIS and remote sensing imagery, will prove invaluable for advances in ecosystem health. For example, these tools may be the basis for assessing ecosystem health parameters (e.g., biotic community structure, biotic cover, primary productivity) in relation to the provision of ecosystem services at regional scales (e.g., nutrient flux in watersheds, sediment loads to drainage basins, biodiversity as expressed in quantities and ranges of habitat, etc.). These techniques will drive the next generation of quantitative assessment of ecosystem health at regional scales (Patil and Myers, 1999; Rapport, 1999).

While considerable effort has been devoted to identification of stressors and development of indicators, protocols for framing observations spatially and temporally are less well established. No one has yet been heroic enough to attempt delineation of ecosystems, but there are several competing strategies for ecological mapping that can help to segregate areas with respect to degree of coupling among biophysical processes. All such mapping strategies lead to hierarchies of zones nested over several levels of spatial detail. Successful application of such ecological mapping strategies requires information on distribution of natural and anthropogenic features across landscapes. Making *in situ* determinations everywhere is infeasible, so remote sensing constitutes the only available recourse by which to establish geographic context and achieve spatial integration. Landscape ecology provides increasingly mature guidance regarding spatial organization among ecosystems, and how remotely sensed data can be used to understand such organization in a particular geographic setting.

Resolution, frequency of coverage, cost, and procurement are not the only concerns with respect to use of remotely sensed data for assessment of ecosystem health. Remotely sensed data seldom constitute a complete information source for ecosystem analysis. Improvements in these respects will not necessarily translate to substantially improved monitoring and assessment unless we learn better ways of incorporating pixelized spectral data into multitiered analysis that integrates intensive site studies, distributed sample plots, and various partial coverages from remote sensors at different resolutions and different times. In the remote sensing community, this is known as the challenge of data fusion. The statistical, environmental, and ecological communities know the problem by a

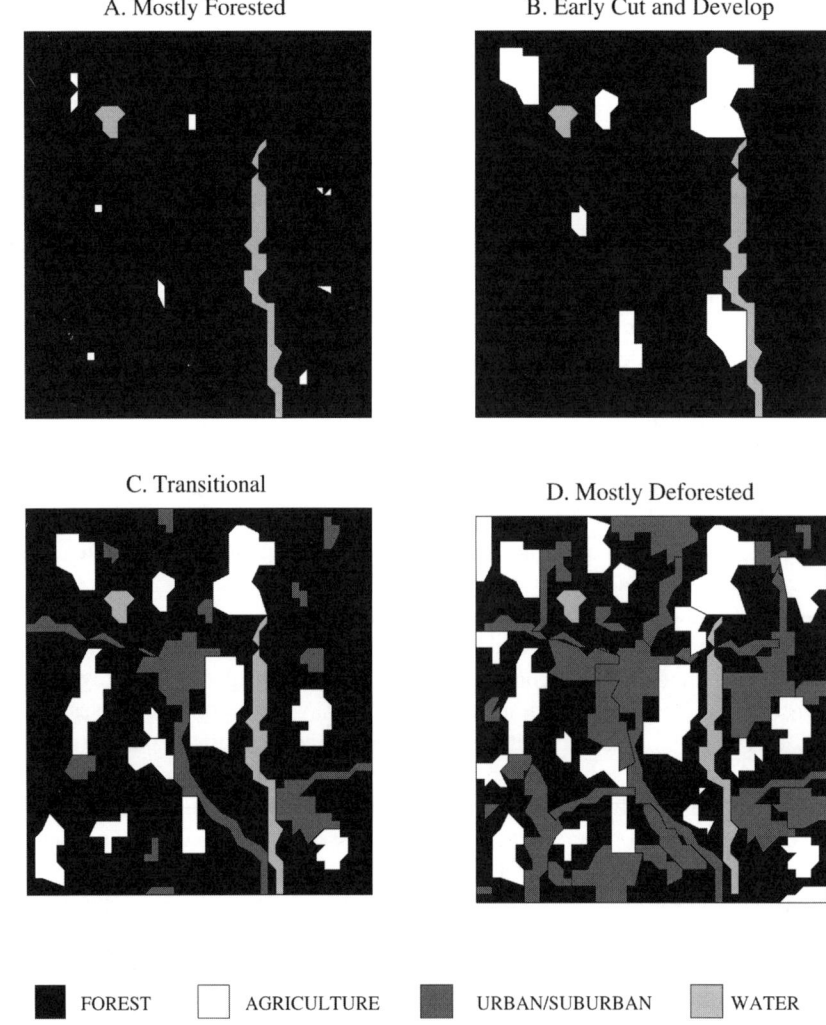

Figure 59.1 Stages in forest defragmentation.

variety of other names. Remotely sensed data is also problematic in the sense that the variates (spectral bands) are not measures that speak to a particular focus of interest like variables conventionally measured in field surveys. They literally constitute a picture of the landscape, and many different sorts of things can be seen from the same picture. Extraneous information is noise to a particular analysis, and segregation of relevant variability from irrelevant variability often requires sophisticated methodology that is the special domain of image analysis. Contemporary improvements in sensors tend to compound such difficulties.

A related concern is for parsimony. Since a massive infusion of high resolution hyperspectral image data could swamp most environmental data analysis facilities, it becomes important to exploit opportunities for compression. Furthermore, the remote sensing community is only beginning to address spatial analysis explicitly in its routine work with image data. Ironically, the spatial structure of image data may ultimately be of most importance for ecosystem analysis.

While compilations of land-use and land-cover maps by remote sensing have a long history, even this basic process is seldom easy or even satisfying with respect to desired detail and accuracy of classification. Determining occurrence and character of landscape change has proven equally problematic. The specific issues are deep, but concrete and finite.

WATERSHEDS, LANDSCAPE PATTERNS, AND THEIR COMPARISON

When a landscape is represented by multiple land-cover types, instead of just forest and nonforest, the challenge is to define a measurement of landscape fragmentation that can be applied to any defined geographic area. Such a measurement would ideally allow quantitative decision making for determining when a landscape pattern has significantly changed, either within the same geographic extent over time, or among different locations within a similar ecoregion. It can also be important to identify ecosystems, such as those that may be delineated by watershed boundaries and are close to the critical point of transition (Figure 59.1C) into a different, possibly degraded, ecosystem where the landscape matrix has become developed land (Figure 59.1D), supporting only small sparsely scattered forest islands that do not provide sufficient forest interior habitat. Along with the collapse of forest-interior species richness, degradation may also be attested to by increasing environmental contamination that is also associated with intensive land development.

Such a measurement of landscape fragmentation can then be a primary component of an ecosystem risk assessment in a manner addressed by Graham et al. (1991). Identifying areas whose landscape level ecosystems are poised for a great reduction in overall species and the elimination of critical functional groups is of utmost concern because such areas may still be salvaged with intervention by proper land-use planning. Meanwhile, other areas that have crossed the line but are not too far beyond the critical point may still be reversible. Indeed, ecosystems that are near critical transition points present both risks and opportunities.

As a Markov transition model, the landscape fragmentation generating model can be fully described by its transition probability matrices. To simulate the null scenario of a self-similar fragmentation process at each resolution, we may invoke a stationary model whereby the same stochastic matrix applies at each transition. Stationary probability transition matrices are based on characteristics of actual watershed-delineated landscapes that are represented by eight land-cover types at a floor resolution of 30-m pixels. The actual landscape maps are reproduced in Figure 59.2. More detail about the data sources can be found through the Web page, http://www.pasda.psu.edu, that includes metadata for the land coverage. The Sinnemahoning Creek watershed is mostly forested, representing a continuum of forest interior wildlife habitat. The Jordan Creek watershed represents a transitional landscape that barely maintains a connected forest matrix that is encroached by agriculture and urban and suburban land use. Meanwhile, the Conestoga Creek watershed represents a landscape that is dominated by open agricultural land and highly aggregated urban and suburban land, with isolated patches of remaining forest. Conditional entropy profiles, measuring landscape fragmentation, appear in Figure 59.3.

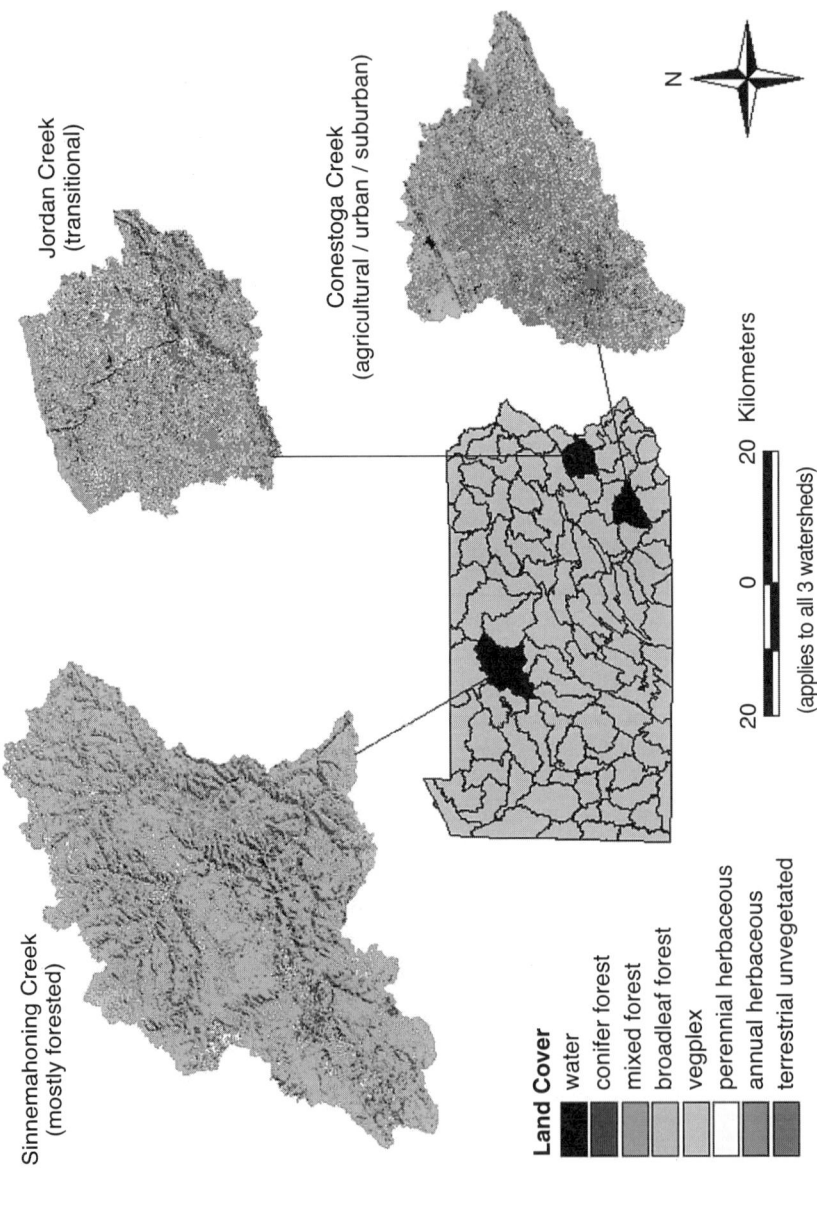

Figure 59.2 Land-cover maps for three watersheds of Pennsylvania.

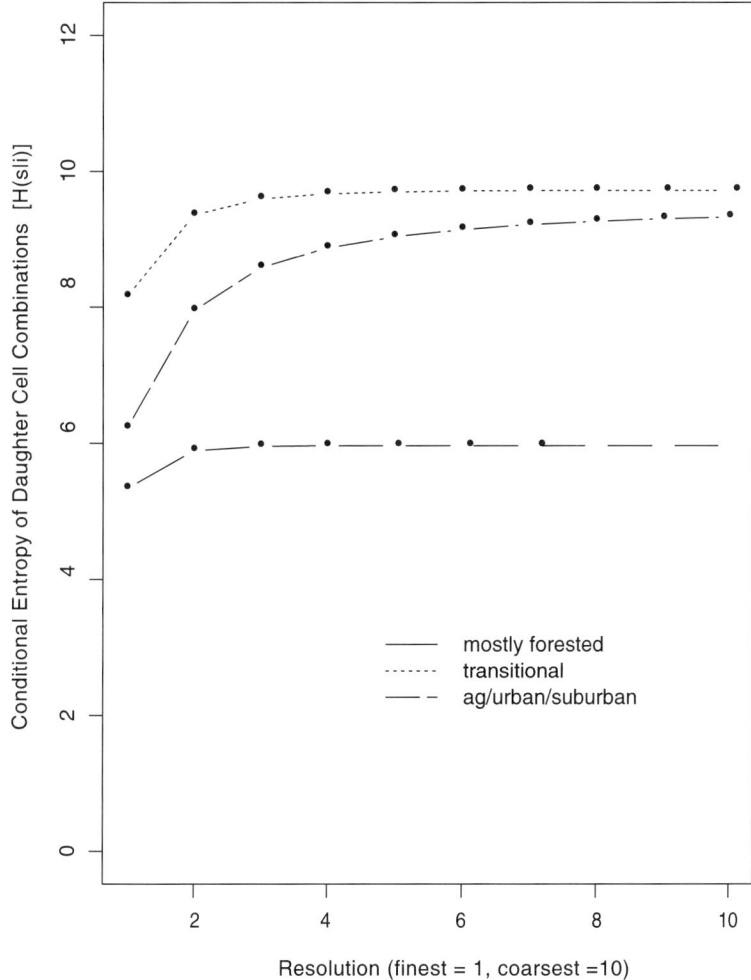

Figure 59.3 Conditional entropy process files as landscape fragmentation profiles for hierarchical Markov transition matrix (HMTM) models whose transition matrices are obtained from watersheds with three distinctly different land-cover patterns.

The stochastic transition matrices can be modeled as appropriate. For example, null landscape models may be obtained by designating a degree of within-patch coherence by the magnitude of diagonal elements (self-preserving probabilities) in a stochastic matrix. Labeling the diagonal value as λ, off-diagonal elements may then be evenly distributed among the remaining probability mass $(1 - \lambda)$ within each row. The conditional entropy profiles for some examples of such models are presented in Figure 59.4.

The shape of a conditional entropy profile appears to be largely governed by two aspects of landcover pattern, as seen in the floor resolution data: the marginal distribution, viewed as the relative frequency of each landcover, and the spatial distribution of landcover types across the given landscape. For more information, see Johnson and Patil (1998), Johnson et al. (1999, 2000), Myers et al. (1997), and Patil (1998a,b).

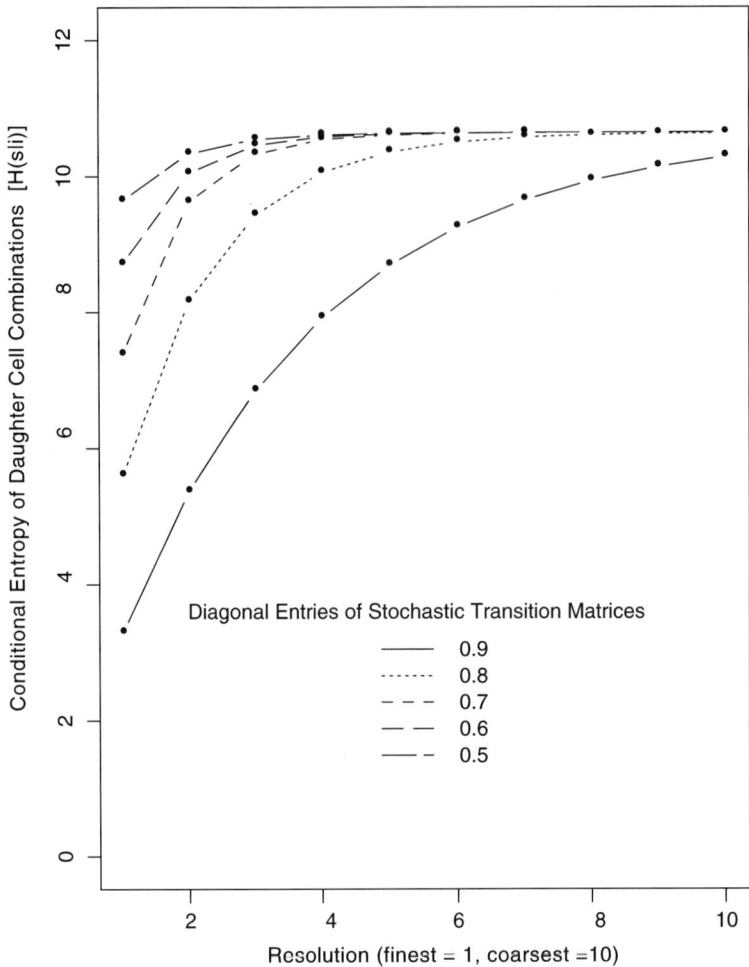

Figure 59.4 Conditional entropy process profiles for HMTM models whose hypothetical $k \times k$ transition matrices have the value λ along the diagonal and the value $(1 - \lambda)/(k - 1)$ off the diagonal. Here $k = 8$ and the values of λ are indicated in the legend. The stationary vector is uniform across the k categories and is used as the initial vector in the model. The floor resolution maps G_L were 1024×1024 (i.e., $L = 10$). Large values of λ result in strong spatial dependence (indicated by large profile relief) that persists at larger distances (indicated by a slowly rising profile). All models have the same stationary vector and therefore the same horizontal asymptote.

CONCLUSION

The chapters in this section were invited to help demonstrate environmental and ecological regional policy research with remote imagery and geospatial information. It is clear that the satellite sensors provide databases of the surface of the Earth. Using various technologies, it is now possible to provide snapshots of landscapes indicative of various features of interest pertaining to human societies, plant and animal communities, aquatic networks, etc. The information is synoptic and we are subsequently able to see various kinds of multicategory maps of different kinds of regions of interest, whether political, natural, or methodological. These multicategory maps provide a basis for comparative assessments of regions within a policy-making and implementation context. It is also possible to extract meaningful profiles of management units, such as watersheds, that can be calibrated and compared in order to assess and manage watersheds of a region. These current developments are in the nature of an environmental and ecological information superhighway, and lead to the cutting-edge frontier of regional policy research with remote imagery and geospatial information with 21st century technology. This frontier initiative can use a focused approach to accomplish the capability and the output within a short time frame.

REFERENCES

Buckingham, D.E., Does the World Trade Organization care about ecosystem health? The case of trade in agricultural products, *Ecosyst. Health*, 4, 92–108, 1998.

Epstein, P.R., Emerging diseases and ecosystem instabilities: new threats to public health, *Am. J. Pub. Health*, 85, 168–172, 1995.

Epstein, P.R. and Rapport, D.J., Changing coastal marine environments and human health, *Ecosyst. Health*, 2, 166–176, 1996.

Graham, R.L., Hunsaker, C.T., O'Neill, R.V., and Jackson, B.L., Ecological risk assessment at the regional scale, *Ecol. Appl.*, 1, 196–206, 1991.

Huq, A. and Colwell, R.R., Vibrios in the marine and estuarine environment: tracking *Vibrio cholerae*. *Ecosyst. Health*, 2, 198–214, 1996.

Johnson, G.D. and Patil, G.P., Quantitative multiresolution characterization of landscape patterns for assessing the status of ecosystem health in watershed management areas, *Ecosyst. Health*, 4, 177–187, 1998.

Johnson, G.D., Myers, W.L., Patil, G.P., and Taillie, C., Multiresolution fragmentation profiles for assessing hierarchically structured landscape patterns, *Ecol. Modelling*, 116, 293–301, 1999.

Johnson, G.D., Myers, W.L., Patil, G.P., and Taillie, C., Characterizing watershed-delineated landscapes in Pennsylvania using conditional entropy profiles, *Landscape Ecol.*, 16, 597–610, 2001.

Karr, J.R., Bridging the gap between human and ecological health, *Ecosyst. Health*, 3, 197–199, 1997.

McMichael, A.J., Global environmental change and human health: Impact assessment, population vulnerability, research priorities, *Ecosyst. Health*, 3, 200–210, 1997.

Myers, W.L., Patil, G.P., and Joly, K., Echelon approach to areas of concern in synoptic regional monitoring, *Environ. Ecol. Stat.*, 4, 131–152, 1997.

Nielsen, N.O., The meaning of health, *Ecosyst. Health*, 5, 65–66, 1999.

Patil, G.P., Environmental and Ecological Regional Policy Research with Remote Imagery and Geospatial Information: Issues, Approaches, and Examples, Technical Report 98-1201, Center for Statistical Ecology and Environmental Statistics, Department of Statistics, Penn State University, University Park, 1998a.

Patil, G.P., Statistical ecology and environmental statistics for cost-effective ecological synthesis and environmental analysis, in *Modern Trends in Ecology and Environment*, Ambasht, R.S., Ed., Backhuys Publishers, the Netherlands, 1998b, pp 5–36.

Patil, G.P. and Myers, W.L., Environmental and ecological health assessment of landscapes and watersheds with remote sensing data, *Ecosyst. Health*, 5, 221–224, 1999.

Rapport, D.J., Gaining respectability: development of quantitative methods in ecosystem health, *Ecosyst. Health*, 5, 1–2, 1999.

Rapport, D.J., Gaudet, C.L., and Calow, P., Eds., *Evaluating and Monitoring the Health of Large-Scale Ecosystems*, Springer-Verlag, Berlin, 1995.

Rapport, D.J., Costanza, R., and McMichael, A.J., Assessing ecosystem health: challenges at the interface of social, natural and health sciences, *Trends Res. Evol. Ecol.*, 13, 397–402, 1998.

CHAPTER 60

Multiscale Advanced Raster Map Analysis System for Measuring Ecosystem Health at Landscape Scale — A Novel Synergistic Consortium Initiative

Ganapati P. Patil, Robert P. Brooks, Wayne L. Myers, and Charles Taillie

SETTING THE STAGE

Geospatial data form the foundation of an information-based society. Remote sensing has been a vastly underutilized resource involving a multimillion dollar investment at the national levels. Even when utilized, its credibility has been at stake, largely because of lack of tools that can assess, visualize, and communicate accuracy and reliability in a timely manner and at desired confidence levels.

Consider an imminent 21st century scenario: What message does a multicategorical map have about the large landscape it represents? And at what scale and what level of detail? Does the spatial pattern of the map reveal any societal, ecological, environmental condition of the landscape? And therefore can it be an indicator of change? How do you automate the assessment of the spatial structure and behavior of change to discover critical areas, hot spots, and their corridors? Is the map accurate? How accurate is it? How do you assess the accuracy of the map? How do we evaluate a temporal change map for change detection? What are the implications of the kind and amount of change and accuracy on what matters, whether climate change, carbon emission, water resources, urban sprawl, biodiversity, indicator species, ecosystem health, human health, or early warning? And with what confidence? The proposed consortium research initiative is expected to find answers to these questions and a few more that involve surface maps and multicategorical raster maps based on remote sensing and other geospatial data in measuring ecosystem health and in managing for healthy ecosystems.

INFERENTIAL GEOSPATIAL INFORMATICS FOR ECOSYSTEM HEALTH

The satellite sensors provide databases of the surface of the Earth. Using various technologies, it is now possible to provide snapshots of landscapes indicative of various features of interest pertaining to human societies, plant and animal communities, and aquatic networks. The information is synoptic and we are subsequently able to see various kinds of maps of different kinds of regions of interest. These maps provide a basis for comparative assessments of regions within a policy-making and implementation context. It is also possible to extract meaningful profiles of management units, such as watersheds, that can be calibrated and compared to assess and manage watersheds of a region.

Space Age and
Stone Age Syndrome

• Data: Space Age/Stone Age
• Analysis: Space Age/Stone Age

Data/ Analysis	Space Age	Stone Age
Space Age	+	+
Stone Age	⊕	+

Figure 60.1 Disappointing report (left) and disappointing software (right).

These multidimensional and multidisciplinary approaches call for linkages and collaborations beyond those traditionally applied in ecosystem health research. Thus, we propose that new consortia be forged to capture, analyze, and model whole ecosystems or subsets of them. The results can be applied to holistic ecosystem management strategies. A purpose of this chapter is to identify the timely need for a consortium initiative on environmental and ecological policy research using remote imagery and geospatial information with appropriate multiscale advanced raster map information science and technology to be accomplished within a short time frame.

The proposed consortium initiative will help advance regional policy research and judge the effectiveness and usefulness of potential or actual policies using satellite technology, computer technology, statistical technology, information technology, multimedia communication technology, and landscape ecology. Initial regional policy research will involve issues that are environmental, ecological, and societal, providing multiscale assessments for management purposes. The proposed initiative will begin with selected prototype case studies involving forests, watersheds, coastal areas, and other regions of interest for issues pertaining to biodiversity, watershed integrity, landscape vulnerability, surface water quality, ecosystem health, human health, global change and disturbance impacts, geographic surveillance, and others.

The urgency of the proposed consortium initiative lies in the disappointing software syndrome depicted in Figure 60.1. The program manager finds the report disappointing and wonders about the software used. It is important that we use space-age analysis for space-age data. Several nonstandard problems arise, and these require nonstandard tools (Patil, 1998; Patil and Myers, 1999) in a timely manner. The proposed consortium initiative is expected to be cost-effective in this emergent need for assessment and management for ecosystem health at landscape scale.

MEASURING ECOSYSTEM HEALTH AT LANDSCAPE SCALE

The Middle Atlantic region studies demonstrate the feasibility and practicality of ecosystem health assessments (Brooks et al., 1998, 2001). This area provides an ideal case study because it is an ecoregion that is rich in synoptic data and contains many of the geographical elements found in the eastern U.S., and other temperate regions. Its natural and human-induced landscapes, gradients, and boundaries provide a wealth of options to explore.

For example, Pennsylvania watersheds have been mapped at scales ranging from 102 units for the state water plan to 9855 units for individually named streams. These watershed units have been studied from diverse perspectives including nonpoint pollution, groundwater pollution potential, land cover, and animal habitats (Johnson, 1999; Johnson et al., 1998, 1999a,b, 2001a,b, Chapter 65, this volume; Johnson and Patil, 1998; Myers et al., 2000; Patil et al., 2000a,b). Pennsylvania watersheds vary in their ecology, geology, hydrology, degree of human influence, and other factors.

Representing this complexity synoptically in a format that enables one to address questions of ecosystem health, integrity, and resilience is our key challenge and goal. Using the collective data from the Middle Atlantic region, we confront the following types of questions in this context:

- What is the health status of a particular watershed, and how does this compare with a similar but less stressed system?
- To what degree is ecosystem degradation associated with cumulative effects from population growth and economic development within the watershed?
- Do changes in spatial biocomplexity of key indicators of ecosystem distress serve as an early warning sign of loss of resilience at regional scales?
- Which watersheds show the greatest degree of fragmentation?
- Do these watersheds also indicate a loss of ecosystem services such as water quality and habitat?
- Is the degree of fragmentation within watersheds correlated with the loss of ecosystems goods and services as measured by synoptic data on water quality, soil erosion, biodiversity, and other factors?

Although spatial landscape analyses have been conducted for years, it has been difficult to compare different locations when using multiple indicators simultaneously. However, with the application of insightful and sophisticated quantitative methods, it is possible to create truly integrative measures that characterize the synergistic relationships among landscape patterns and indicators. In particular, we will explore the techniques described here in the multiscale advanced raster map analysis system for addressing multiple indicators, partial orderings, and multicriterion decision support, along with echelons of spatial variation. This approach will allow us to search for and define consistent and recognizable landscape patterns, while allowing us to define a set of reference conditions to understand the consequences, both favorable and unfavorable, that human actions have on biocomplexity.

There are several national and international projects in progress, such as the Atlantic Slope Consortium, that will benefit with their participation in the proposed geoinformatic methods and tools consortium discussed in this chapter. Several of these projects will find this geospatial and temporal quantitative arm useful in their in-house work in progress from day one of their participation. This chapter constitutes a call for this collaborative synergism in the form of a short course, workshop, project collaboration, or proposal preparation.

MULTISCALE ADVANCED RASTER MAP ANALYSIS SYSTEM FOR MEASURING ECOSYSTEM HEALTH

This section briefly describes applications of the emergent methodologies collectively known as the Multiscale Advanced Raster Map analysis system (MARMAP) for landscape health assessments (Patil, 2000; 2001b,c,d; Chapter 59, this volume; 2002a,b; also see http://www.stat.psu.edu/~gpp/newpage11.htm).

Modeling and Simulation of Thematic Raster Maps

A raster map depicts the landscape as a grid of uniform cells. Modeling and simulation of raster maps are employed for three general purposes. First, model-fitting provides a set of estimated parameter values characterizing the spatial structure of the map (landscape). Second, simulation yields statistical confidence capability, as well as response sensitivity to variation in the fitted parameter values. Third, model validation provides a check on tendencies to overfit the model. Three classes of map models are available. These address issues important to monitoring and diagnostics to determine and discriminate differing status with regard to degradation of habitat integrity across landscapes. From land-cover maps derived by remote sensing, we examine naturalistic vs. more strongly human disturbed situations through an index of conditional entropy to

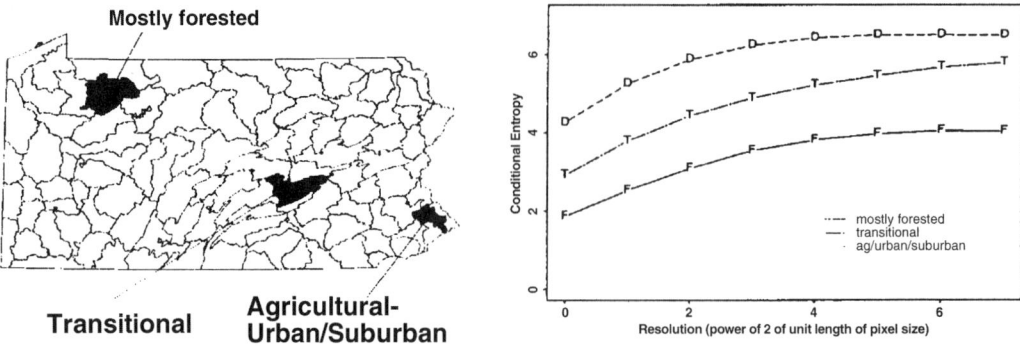

Figure 60.2 Fragmentation profiles for three Pennsylvania watersheds with distinct land cover patterns: mostly forested, transitional and mostly deforested (agricultural – urban and suburban).

obtain profiles of disruption (Figure 60.2). For more information, see Johnson (1999), Johnson and Patil (1998), Johnson et al. (1998, 1999a,b, 2001a), Patil et al. (2000a,b), and Patil and Taillie (1999, 2000a,b,c). The issues of landscape characterization and discrimination, patch structure and patch dynamics, scaling domains, spatial pattern heterogeneity detection, and others arise. The issues of assessment of accuracy and change detection of land-cover and land-use maps also arise for a variety of ecosystems. These issues can be addressed by the methods and tools discussed in the proposed MARMAP System.

Echelons of Spatial Variation, Critical Area Detection, and Delineation

Echelons frame local values of synoptically mapped environmental indicators in regional context for comparative purposes and objective analysis of complex hierarchies in spatial variation across landscapes. The environmental indicators are considered as surface variables in virtual (or real) topographies as depicted in Figure 60.3. Echelons are structural entities consisting of peaks, foundations of peaks, foundations of foundations, and so on in an organizational hierarchy. It is natural to cast the echelon hierarchy as a dendrogram, from which profiles of spatial complexity can be obtained and principal families determined as contiguous areas of criticality from perspectives of either pronounced ecosystem health or pronounced ecosystem distress. Echelons have proven effective for elucidating concentration and connectivity of biodiversity and complexity of landscape change induced by factors such as wildland fire or pattern of propagation for urban sprawl (Myers et al., 1995, 1997, 1999; Kurihara et al., 2000; Smits and Myers, 2000; Patil and Myers, 2002).

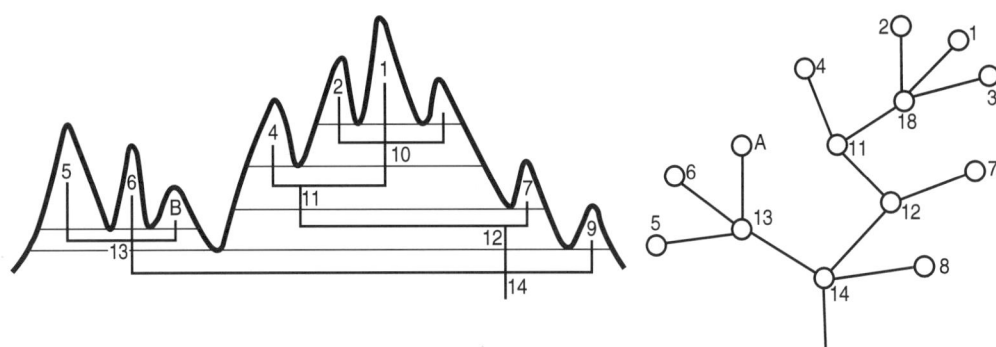

Figure 60.3 Echelon decomposition of a surface (left) and associated echelon tree (right).

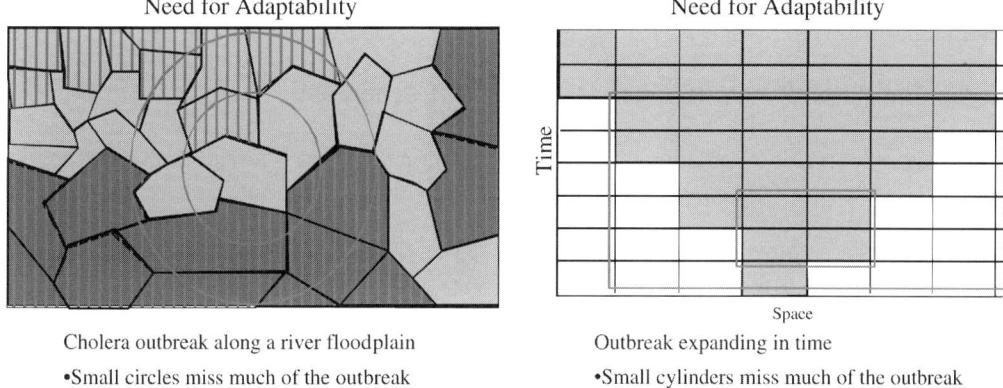

Need for Adaptability — Need for Adaptability

Cholera outbreak along a river floodplain — Outbreak expanding in time

•Small circles miss much of the outbreak — •Small cylinders miss much of the outbreak
•Large circles include many unwanted cells — •Large cylinders include many unwanted cells

Figure 60.4 Scan statistic zonation (left) and space-time scan statistic zonation (right).

Contemporary study of human disease as a component of ecosystem health entails a spatial scan statistic (Kulldorff and Nagarwalla, 1995) for detecting geographic clusters of disease and other responses that are significantly elevated with respect to the regional setting. In conjunction with the spatial scan statistic, echelon analysis is expected to more clearly delineate the cluster boundaries for focus of investigation (Patil and Taillie, 2001d), as depicted in Figure 60.4.

Pattern-Based Landscape Change Analysis

Landscape change analysis is becoming increasingly important for ecosystem monitoring. Deforestation, habitat fragmentation, and land-use conversion are growing concerns for conservation, landscape ecology, and planning. Effective and parsimonious methods are needed to make the combinatorial challenges of comparative analysis manageable. Composite mosaics of multiple images derived by pattern-based segmentation have proven particularly advantageous for extracting and representing change from remotely sensed image sequences, where previously analysis was largely restricted to consideration of image pairs taken at two times with the same sensor (Myers, 2000; Patil et al., 2000b).

Geospatial Data Compression, Segmentation, and Classification

Remote sensing is generating spatial data at a rapidly increasing rate. The increase in data flow has a threefold nature due to increasing spatial resolution, increasing spectral richness, and more frequent acquisition. These data have potential informational utility that often remains unrealized. Facilitating the realization of such potential informational utility is one of the major challenges facing modern information technology, and provides the underlying motivation for emergence of data mining. Data mining is essentially a search for pattern that may have some interpretability, but it is too often an aimless search that is lacking conceptualization of what constitutes pattern. When dealing with image data, however, space and time offer organizing paradigms that have been underexploited.

From both theoretical and practical perspectives, landscapes have a mosaic nature with particular pattern elements emerging at different scales. This compound mosaic nature is fundamental as a basis for landscape ecology. The process of mosaic pattern extraction is one of image segmentation, where the operative partitioning takes place in the spectral domain. The determination of spatial mosaic segments is a direct consequence of partitioning as spectral subspaces. Accomplishment of such segmentation must take into account both distinctiveness and expansiveness of mosaic elements. Distinctiveness is most important for perceptual purposes, and the expansiveness for ana-lytical purposes (Myers, 2000).

Geographic Surveillance, Disease Mapping, and Evaluation

Disease mapping is about the use and interpretation of maps showing the incidence or prevalence of disease. Disease data occur either as individual case events or as groups of case events (count data) within areal units, such as census tracts, zip codes, or counties. Any disease map must be considered with the appropriate background population which gives rise to the incidence. Maps answer the question: Where? The maps in conjunction with the underlying data reveal spatial patterns not easily recognized from lists of statistical data (Kulldorff et al., 1997, 1998; Lawson and Williams, 2001). For example, use of remote sensing data and other relevant geospatial data can help evaluate surrounding landscape characteristics that may be precursors for vector-borne diseases leading to early warning, involving landscape health, ecosystem health, and human health.

Urban Heat Islands, Urban Sprawl, and Environmental Justice

The urban heat island may be visualized as a temperature dome on an urban area. It contributes to the formation of ozone, which is a major urban air pollutant that has serious human health consequences. Analysis of thermal energy characteristics helps us understand how we can modify the city landscape to lessen the impacts of the urban heat island and its subsequent effects on air quality (Quattrochi and Luvall, 1999; Quattrochi et al., 2000; Quattrochi and Gillani, 2001). Three main objectives may be involved:

1. Characterization of thermal landscape in the metropolitan area. This aims at evaluating not only the strength of the urban heat island, but also the spatial variance within the heat island.
2. Evaluation of the relative roles of land cover characteristics and urban structures. This involves the quantification of land-cover characteristics and urban structures, such as percent of impervious surfaces, biomass density, urban canyon geometry, and roadway density.
3. Linking localized thermal characteristics to human health outcome. This attempts to directly and indirectly link illnesses, such as pediatric asthmatic attacks and heat strokes, to localized thermal stress.

Multiple Indicators, Partial Orderings, and Multicriterion Decision Support

To prioritize and to rank means to linearize. Rather than derive a composite index, we will prioritize without having to integrate the indicators. This is now possible, and the approach is relatively novel and innovative. We have developed it for nationwide prioritization for the United Nations Environmental Programme with land, air, and water indicators measuring the human environment interface at the national level (Patil 2001a; Patil and Taillie, 2001b, c). For another example, a landscape atlas published by the U.S. Environmental Protection Agency (Jones et al., 1997) considers 33 indicators of ecological condition on 123 watersheds (7-digit hydrologic unit codes [HUCs]) of the Middle Atlantic region and attempts to rank the watersheds using clustering and quintile-frequency methods. We address the question of ranking such a collection of objects when a suite of indicator values is available for each member of the collection. The objects can be represented as a cloud of points in indicator space, but the different indicators (coordinate axes) typically convey different comparative messages, and there is no unique way to rank the objects. A conventional solution is to assign a composite numerical score to each object by combining the indicator information in some fashion. Every such composite involves judgments (often arbitrary or controversial) about trade-offs or substitutability between indicators.

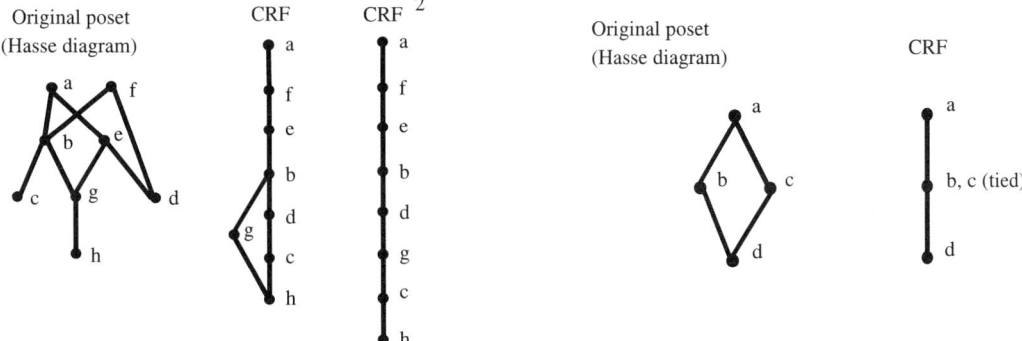

Figure 60.5 The linearizing effect of the CRF operator (left); example of how ties can emerge during linearization (right). A poset is a partially ordered set.

Rather than trying to impose a unique ranking, we take the view that the relative positions in indicator space determine only a partial ordering and that a given pair of objects may not be inherently comparable. Working with Hasse diagrams of the partial order, we study the collection of all rankings that are compatible with the partial order and arrive at the ranking and prioritization as in Figure 60.5, using the cumulative rank frequency (CRF) operator specially developed for the purpose (Patil and Taillie, 2001a).

Hierarchical Structure Analysis

Trees and other nodal graph structures arise in map modeling, echelons, and posets of MARMAP. Also important is the coupling of related structures. For example, when a suite of indicators is partitioned into subgroups (e.g., stressor, integrity, socioeconomic), the Hasse diagrams have a common set of labeled nodes but perhaps different edges (Patil, 2001c,d).

Mining Geospatial Data

Data structures and algorithms are under investigation for exploring associations between ecosystem degradation and spatial patterns employing higher-level models for detecting changes and finding interesting spatiotemporal patterns and trends (Rodriguez, 2001).

Interface Design and Visualization Toolbox

The main goal is to promote the discovery of inherent structures and patterns, enable the study of particular facets and dimensions of data, and provide means to visually assess the utility and accuracy.

PROJECT MARMAP CONSORTIUM: SYNERGISTIC INTEGRATION AND TECHNOLOGY TRANSFER

An essential part of this proposed consortium initiative is to introduce concepts and methods at the core of MARMAP to researchers in ecology, environment, socioeconomics, and the quality of human life. It is timely to think of multidisciplinary groups for ecosystem health measurement at the landscape level. It will be productive and cost-effective to initiate synergistic collaboration, whether in the form of a short course, workshop, project research, or proposal preparation.

The experiences gained from the collaborative studies will feed back into refinements of methods. At the end of the day, the thrust of this enterprise will have quantification of ecosystem health at landscape scales from subwatersheds to major watersheds as an essential, replicable method of assessing our progress toward sustainability (Rapport et al. 1999). You are invited to visit the MARMAP Web site, http://www.stat.psu.edu/~gpp/newpage11.htm, and contact gpp@stat.psu.edu for your participation in this unusual consortium initiative to manage for healthy ecosystems in the spirit of digital governance of the 21st century.

ACKNOWLEDGMENTS

Prepared with partial support from U.S. EPA Star Grant for Atlantic Slope Consortium, Cooperative Agreement Number R82868401, Barbara Levinson, Project Officer. The contents have not been subjected to agency review, and therefore do not reflect the view of the agency. No official endorsement should be inferred. The authors would also like to thank the editors, Calvin Qualset and David Rapport, for their interest and support.

REFERENCES

Brooks, R.P., O'Connell, T.J., Wardrop, D.H., and Jackson, L.E., Towards a regional index of biological integrity: the example of forested riparian ecosystems, *Environ. Monitoring Assess.*, 51, 131–143, 1998.

Brooks, R.P., Brinson, M., Hershner, C., Shortle, J., and Whigham, D., Development, testing, and application of indicators for integrated assessment of ecological and socioeconomic resources of the Atlantic Slope in the Mid-Atlantic, Penn State Cooperative Wetlands Center, University Park, 2001.

Johnson, G.D., Landscape Pattern Analysis for Assessing Ecosystem Condition: Development of a Multi-Resolution Method and Application to Watershed Delineated Landscapes in Pennsylvania, Ph.D. thesis, Penn State University, University Park, 1999.

Johnson, G.D. and Patil, G.P., Quantitative multiresolution characterizations of landscape patterns for assessing the status of ecosystem health in watershed management areas, *Ecosyst. Health*, 4, 177–187, 1998.

Johnson, G.D., Myers, W.L., Patil, G.P., and Walrath, D., Multiscale analysis of the spatial distribution of breeding bird species richness using the echelon approach, in *Assessment of Biodiversity for Improved Forest Planning*, Bachmann, P., Kohl, M., and Paivinen, R., Eds., Kluwer, Dordrecht, 1998, pp. 135–150.

Johnson, G.D., Myers, W.L., Patil, G.P., and Taillie, C., Multiresolution fragmentation profiles for assessing hierarchically structured landscape patterns, *Ecol. Modeling*, 116, 293–301, 1999a.

Johnson, G.D., Myers, W.L., Patil, G.P., and Taillie, C., Characterizing watershed-delineated landscapes in Pennsylvania using conditional entropy profiles, *Landscape Ecol.*, 16, 597–610, 1999b.

Johnson, G.D., Myers, W.L., Patil, G.P., and Taillie, C., Fragmentation profiles for real and simulated landscapes, *Environ. Ecol. Stat.*, 8, 5–20, 2001a.

Johnson, G.D., Myers, W.L., Patil, G.P., and Taillie, C., Predictability of surface water pollution loading in Pennsylvania using watershed-based landscape measurements, *J. Am. Water Resourc. Assoc.*, 37, 821–835, 2001b.

Jones, K.B., Riitters, K.H., Wickham, J.D., Tankersley, R.D., O'Neill, R.V., Chaloud, D.J., Smith, E.R., and Neale, A.C., An Ecological Assessment of the United States Mid-Atlantic Region: A Landscape Atlas, EPA/600/R-97/130, U.S. EPA, Washington, D.C., 1997, pp. 1–105.

Kulldorff, M., and Nagarwalla, N., Spatial disease clusters: detection and inference, *Stat. Med.*, 14, 799–810, 1995.

Kulldorff, M., Feuer, E.J., Miller, B.A., and Freedman, L.S., Breast cancer clusters in the Northeast United States: a geographic analysis, *Am. J. Epidemiol.*, 146, 161–170, 1997.

Kulldorff, M., Athas, W.F., Feuer, E.J., Miller, B.A., and Key, C.R., Evaluating cluster alarms: a space-time scan statistic and brain cancer in Los Alamos, New Mexico, *Am. J. Public Health*, 88, 1377–1380, 1998.

Kurihara, K., Myers, W.L., and Patil, G.P., Echelon analysis of the relationship between population and land cover patterns based on remote sensing data, *Community Ecol.*, 1, 103–122, 2000.

Lawson, A.B. and Williams, F.L.R., *An Introductory Guide to Disease Mapping*, John Wiley, New York, 2001.

Myers, W.L., PHASE-Based Broad-Area Landscape Change Analysis, final report on NASA Research Project NAGS5–6713, Environmental Resources Research Institute Research Report ER2005, Penn State University, University Park, 2000.

Myers, W.L., Patil, G.P., and Taillie, C., Comparative paradigms for biodiversity assessment, invited paper at the IUFRO Symposium in Chiang-mai, Thailand, in *Measuring and Monitoring Biodiversity in Tropical and Temperate Forests*, Boyle, T.J. and Boontawee, B., Eds., Center for International Forestry Research, Bogor, Indonesia, 1995.

Myers, W.L., Patil, G.P., and Joly, K., Echelon approach to areas of concern in synoptic regional monitoring, *Environ. Ecol. Stat.*, 4, 131–152, 1997.

Myers, W.L., Patil, G.P., and Taillie, C., Conceptualizing pattern analysis of spectral change relative to ecosystem health, *Ecosyst. Health*, 5, 285–293, 1999.

Myers, W.L. et al., Pennsylvania Gap Analysis Project: Leading Landscapes for Collaborative Conservation, final report for U.S. Geological Survey, Gap Analysis Program, 2000.

Patil, G.P., Statistical ecology and environmental statistics for cost-effective ecological synthesis and environmental analysis, in *Modern Trends in Ecology and Environment*, Ambasht, R.S., Ed., Backhuys Publishers, Leiden, the Netherlands, 1998, pp. 5–36.

Patil, G.P., Multiscale Advanced Raster Map Analysis for Sustainable Environment and Development: A Research and Outreach Prospectus of Advanced Mathematical, Statistical and Computational Approaches Using Remote Sensing Data, development and implementation of prototype MARMAP remote sensing applications, technology, and education for multiscale advanced raster map analysis program, http://www.stat.psu.edu/~gpp/PDFfiles/prospectus-1.pdf, 2000.

Patil, G.P., Nationwide Indicators and Their Integration, Evaluation, and Visualization Worldwide-UNEP Initiative, invited paper, U.S. EPA Conference on Environmental Statistics and Information, Philadelphia, 2001a.

Patil, G.P., Multiscale Advanced Raster Map Analysis System: Definition, Design, and Development, invited plenary address at the Brazilian Ecological Congress, Porto Allegre, Brazil, 2001b.

Patil, G.P., Multiscale Advanced Raster Map Information Science and Technology: A Research and Outreach Prospectus of Advanced Mathematical, Statistical, and Computational Approaches Using Remote Sensing Data: Development and Implementation of User-Friendly MARMAP System, http://www.stat.psu.edu/~gpp/PDFfiles/Prospectus-2.pdf, 2001c.

Patil, G.P., Multiscale Advanced Raster Map Analysis System: Definition, Design, and Development, invited plenary address at the Portuguese Statistical Congress, Ponte Delgada, Portugal, 2001d.

Patil, G.P., Multiscale Advanced Raster Map Analysis System: Definition, Design, and Development, Joint Statistical Meetings, ASA, invited paper, New York, 2002a.

Patil, G.P., Conditional entropy profiles for multiscale landscape fragmentation and environmental degradation, in *Encyclopedia of Environmetrics*, Vol. 1, El-Shaarawi, A. and Piegorsch, W.W., Eds., John Wiley & Sons, New York, 2002b, pp. 413–417.

Patil, G.P. and Myers, W.L., Echelon analysis. in *Encyclopedia of Environmetrics*, El-Shaarawi, A. and Piegorsch, W.W., Eds., John Wiley & Sons, New York, 2002, 2, 583–586.

Patil, G.P. and Myers, W.L., Environmental and ecological health assessment of landscapes and watersheds with remote sensing data, *Ecosyst. Health*, 5, 221–224, 1999.

Patil, G.P. and Taillie, C., A Markov model for hierarchically scaled landscape patterns, in *Bulletin of the International Statistical Institute*, 58(1), Voorburg, the Netherlands, 1999, pp. 89–92.

Patil, G.P. and Taillie, C., Modeling and Interpreting the Accuracy Assessment Error Matrix for a Doubly Classified Map, Technical Report 99-0502, Center for Statistical Ecology and Environmental Statistics, Department of Statistics, Penn State University, University Park, 2000a.

Patil, G.P. and Taillie, C., A multiscale hierarchical Markov transition matrix model for generating and analyzing thematic raster maps, Technical Report 2000–0603, Center for Statistical Ecology and Environmental Statistics, Department of Statistics, Penn State University, University Park, 2000b.

Patil, G.P. and Taillie, C., Analytic Solution of the Regularized Latent Truth Model for Binary Maps, Technical Report 2000-0601, Center for Statistical Ecology and Environmental Statistics, Department of Statistics, Penn State University, University Park, 2000c.

Patil, G.P. and Taillie, C., On Quantitative Formulation of Nationwide Human Environment Index, final report, Division of Early Warning and Assessment, United Nations Environmental Programme, Nairobi, Kenya, 2001a.

Patil, G.P. and Taillie, C., Environmental Indicators: Comparisons and Rankings without Integration (some statistical and visual tools with application to the proposed UNEP Human Environment Index), invited paper, Plenary Forum on Environmental Indicators and Their Integration for Quality of Life, Index 2001 Congress, Rome, 2001b.

Patil, G.P. and Taillie, C., Multiple Indicators, Partially Ordered Sets, and Linear Extensions: Multi-Criterion Ranking Methods, Technical Report 2001-1204, Center for Statistical Ecology and Environmental Statistics, Department of Statistics, Penn State University, University Park, 2001c.

Patil, G.P. and Taillie, C., Powerpoint presentations, http://www.stat.psu.edu/~gpp, 2001d.

Patil, G.P., Johnson, G.D., Myers, W.L., and Taillie, C., Multiscale statistical approach to critical-area analysis and modeling of watersheds and landscapes, in *Statistics for the 21st Century: Methodologies for Applications of the Future*, Rao, C.R. and Szekely, G.J., Eds., Marcel Dekker, New York, 2000a, pp. 293–310.

Patil, G.P., Myers, W.L., Luo, Z., Johnson, G.D., and Taillie, C., Multiscale assessment of landscapes and watersheds with synoptic multivariate spatial data in environmental and ecological statistics, *Math. Comp. Modeling*, 32, 257–272, 2000b.

Quattrochi, D.A. and Gillani, N.V., Urban heat island and human health effects: a case for using Atlanta, Georgia as a study area, manuscript, 2001.

Quattrochi, D.A. and Luvall, J.C., Thermal infrared remote sensing data for analysis of landscape ecological processes: methods and applications, *Landscape Ecol.*, 14, 577–598, 1999.

Quattrochi, D.A., Luvall, J.C., Rickman, D.L., Estes Jr., M.G., Laymon, C.A., and Howell, B.F., A decision support information system for urban landscape management using thermal infrared data, *Photogrammetric Eng. Remote Sensing*, 66, 1195–1207, 2000.

Rapport, D.J., Christensen, N., Karr, J.R., and Patil, G.P., The centrality of ecosystem health in achieving sustainability in the 21st century: concepts and new approaches to environmental management, in *Human Survivability in the 21st Century: Transactions of the Royal Society of Canada*, University of Toronto Press, Toronto, 1999, pp. 3–40.

Rodriguez, S., Statistical Data Mining of Remote Imagery for Characterization, Classification and Comparison of Landscape and Watersheds of Pennsylvania, Ph.D. thesis, Penn State University, University Park, 2001.

Smits, P.C. and Myers, W.L., Echelon approach to characterize and understand spatial structures of change in multi-temporal remote-sensing imagery, *IEEE Trans. Geosci. Remote Sensing*, 38, 2299–2309, 2000.

CHAPTER 61

Application of Landscape Models to Alternative Futures Analyses

Anne C. Neale, K. Bruce Jones, Maliha S. Nash, Rick D. Van Remortel, James D. Wickham, Kurt H. Riitters, and Robert V. O'Neill

INTRODUCTION

Scientists and environmental managers alike are concerned about broadscale changes in land use and landscape pattern and their cumulative impact on environmental and economic end points, such as water quality and quantity, species habitat, productivity, erosion potential, recreational value, and overall ecological health (Rapport et al., 1998). They also are interested in predicting short- and long-term future impacts on ecological goods and services based on current land management policies and decisions (Steinitz, 1996). Because we have the means to adjust land management policies, it is worthwhile to develop approaches that can predict the consequences (alternative futures) of different land management policies for different environmental end points. This type of analysis can, for example, allow decision makers in resource conservation and restoration programs to estimate how they can get the most ecological benefit for the least cost (Steinitz, 1996).

Modeling alternative futures can be a simple or complicated process depending on the method employed and the environmental end point in question. The questions and end points of interest are paramount in developing a valid model. Very different models, for example, are needed to predict terrestrial wildlife habitat vs. aquatic conditions. Even within a broad category such as aquatic conditions, different models may be appropriate for different aspects of aquatic conditions (e.g., macroinvertebrate health, pesticide toxicity, eutrophication potential). While different models may be required for different end points, common landscape composition and pattern metrics may be employed in a number of models for a suite of environmental end points. For example, percent natural cover or road density in a measurement unit (e.g., watershed) is an important factor for many environmental end points, including water quality and quantity, wildlife habitat, erosion potential, and recreational value (Burns, 1972; Harden, 1992; Saunders et al., 1992; Kattan et al., 1994; Koopowitz et al., 1994; Short and Turner, 1994; Jones et al., 2000).

Different methods have been proposed and used to predict future conditions, but the basic premise is the same; they are predicated on (1) what land managers and the public want based on needs and values, and (2) the biophysical constraints of the environment (Steinitz, 1996). Models may be developed by establishing trends based on past and present conditions and projecting those trends into the future. They also may be developed empirically by assessing how conditions of

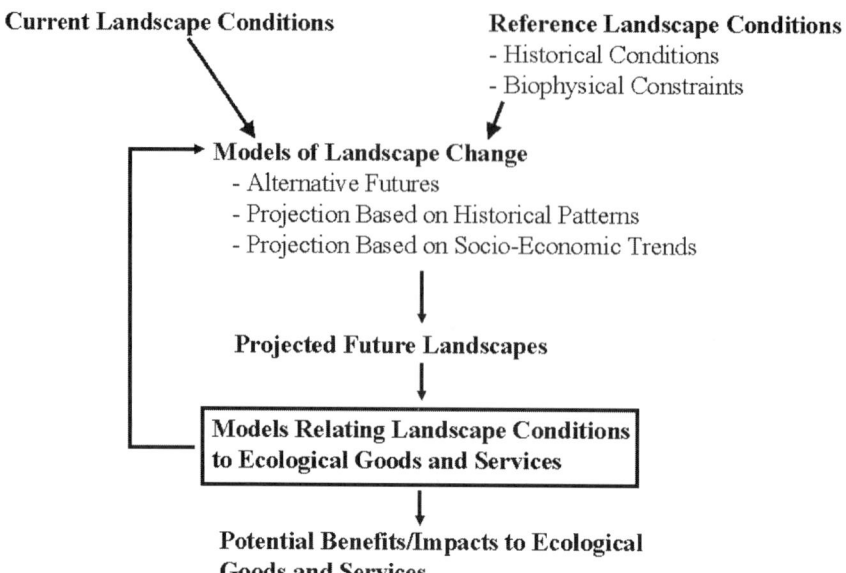

Figure 61.1 The use of landscape models in assessing the consequences of alternative landscape futures.

environmental end points vary with landscape composition and pattern (substituting spatial variability for time) and then manipulating the landscape conditions for different scenarios to project environmental end point conditions based on those different scenarios. Figure 61.1 describes the process used to apply a landscape model to alternative futures analysis.

This chapter will describe a model to predict nitrogen loading, one aspect important to water quality of streams, from a suite of landscape metrics and then will apply this model to a series of alternative future landscapes. This example also will illustrate important issues to consider when developing models for future conditions. Although we will describe only the process for modeling nitrogen loading, the methods presented could easily be applied to other environmental end points. A similar approach, using virtually the same suite of landscape pattern metrics, was used in an assessment relating landscape metrics to breeding bird richness in the Middle Atlantic region (Jones et al., 2000).

BACKGROUND INFORMATION FOR EXAMPLE MODEL

Key to model development was establishing a quantitative relationship between landscape pattern metrics and nitrogen loading to streams. Nitrogen is of particular interest to the U.S. Environmental Protection Agency (EPA) because, as a nutrient, it is essential to the health and continued functioning of natural ecosystems. However, when nutrient inputs exceed the assimilative capacity of a water body system, the system progresses toward hypereutrophic conditions. Excessive nutrient loadings can result in excessive growth of macrophytes or phytoplankton and potentially harmful algal blooms (HAB), leading to oxygen declines, imbalances between prey and predator species, public health concerns, and a general decline of the aquatic resource (U.S. Environmental Protection Agency, 1998).

A number of studies have shown strong relationships between water quality and landscape characteristics. A decrease in natural vegetation indicates a potential for future water quality problems (Likens et al., 1977; Franklin, 1992; Walker et al., 1993; Hunsaker and Levine, 1995; Smith et al., 1997). Many studies have shown that land use within a watershed can account for much of the variability in stream and estuary water quality (Omernik et al., 1981; Omernik, 1987;

Hunsaker et al., 1992; Charbonneau and Kondolf, 1993; Roth et al., 1996; Herlihy et al., 1998; Jones et al., 2001). Changes in landscape conditions in the riparian zone and in areas surrounding water quality sample sites may have a greater influence on water quality than broader-scale watershed conditions (Lowrance et al., 1984). The relationship between intact riparian areas and high water quality is well established, especially in the eastern U.S. (Karr and Schlosser, 1978; Yates and Sheridan, 1983; Lowrance et al., 1984; Cooper et al., 1987). Riparian habitat functions as a sponge, greatly reducing nutrient and sediment runoff into streams (Peterjohn and Correll. 1984; Cooper et al. 1987). Wetlands also play an important role in reducing nutrient loads to surface waters (Weller et al., 1996). High amounts of impervious surface and roads on watersheds also may result in high loadings of nutrients and sediment to streams (Burns, 1972; Harden, 1992; Arnold and Gibbons, 1996), and atmospheric deposition may be a significant source of nitrogen in surface waters (Stensland et al., 1986). Degraded water quality and quantity can, in turn, affect many other environmental end points such as species habitat, productivity, recreational value, and overall ecological health.

In 1996, a regional-scale land-cover database was developed for the five-state area of the U.S. Middle Atlantic region, and this database, along with other regional landscape coverages (e.g., topography, soils, road networks, stream networks, and human population density) were used to assess landscape conditions across the entire region down to a scale of 30 m (Jones et al., 1997). The assessment used a set of landscape metrics (O'Neill et al., 1988, 1997) to evaluate the spatial patterns of human-induced stresses and the spatial arrangement of forest, forest-edge, and riparian habitats. Advances in computer technology and geographic information systems (GISs) have made it possible to calculate landscape metrics over large areas (e.g., regions) at relatively fine scales (e.g., down to 30 m).

Using landscape metric data generated from Jones et al. (1997), and nitrogen loading data provided by the U.S. Geological Survey (USGS) (Langland et al., 1995), we developed a preliminary model predicting nitrogen loading to streams from landscape metrics for a subset of Middle Atlantic watersheds found in the Chesapeake Bay basin (Jones et al., 2001). The analyses presented in this chapter are demonstrative only and should not be construed as an ultimate model for predicting future nitrogen loading. The research to develop predictive nitrogen loading models based on landscape metrics is still ongoing.

METHODS

The obvious first step in developing alternative future models is to collect the appropriate input data for the models. In some cases, the researchers may have the luxury of designing and collecting their own data. In many cases, however, we are limited to existing data, especially if we want to include historical data in our analyses.

In this project, data were compiled from two independent sources. The nitrogen yield data were acquired from the USGS (U.S. Geological Survey, 1995) and the landscape metric data were derived from the data used by Jones et al. (1997).

The USGS calculated annual nutrient and suspended-sediment loads and yields for 148 nontidal streams within the Chesapeake Bay basin. The Chesapeake Bay basin contains more than 150,000 stream miles in the District of Columbia and parts of New York, Pennsylvania, Maryland, Virginia, West Virginia, and Delaware. The basin comprises six major river systems: the Susquehanna, Patuxent, Potomac, Rappahannock, York, and James Rivers (Langland et. al., 1998). The USGS annual nitrogen yield estimates were based on the USGS water year which is October 1 through September 30. We calculated a median annual nitrogen yield based on the yields for the years 1989 through 1996. The inputs for the USGS model were measured concentration of nitrogen in milligrams per liter, measured discharge in cubic feet per second, and time measured in decimal years. USGS model methods and results are described in detail in Langland et al. (1998). The USGS annual loads and yields are based on two different sampling regimes: flow-driven and fixed-interval sampling. In the flow-driven (or total stream flow) sampling regime, samples are collected on the basis of stream flow conditions. Fixed-interval sampling programs collect samples on a regular schedule, usually monthly or quarterly (Langland et al., 1995). Loads are reported by USGS in tons per year. Yields, which have been normalized by watershed area, are reported in pounds per acre. Our analyses were conducted with the yield data only and we combined data for flow-driven and fixed-interval sampling.

Watershed support areas were delineated using Arc/Info GIS software (ESRI, 1996) for each of the USGS water quality monitoring locations so our support areas consisted of only that part of the watershed actually contributing to the water quality monitoring point. For the landscape metrics, we acquired digital coverages of landscape metrics generated by Jones et al. (1997) and then calculated landscape metrics for each of the delineated watersheds. The source of the land cover map from which many of the landscape metrics were derived was the Multi-Resolution Land Characteristics (MRLC) project (Vogelmann et al., 1998). The MRLC data were derived from Landsat Thematic Mapper and had a resolution of 30 m and 15 land cover classes (Vogelmann et al., 1998). However, before calculating the landscape metrics, we used an Arc/Info routine to aggregate the 15 land cover classes into six classes: urban, agriculture, wetland, forest, barren, and water. The landscape metrics used in this analysis are listed in Table 61.1.

Figure 61.2 shows the delineated watersheds used in this analysis in the context of the entire Middle Atlantic region. Although it cannot be detected in Figure 61.2, several of the watersheds are actually nested within larger watersheds. The points on the figure represent water quality data collection sites. Any delineated watershed that did not overlap the Middle Atlantic study area by at least 75%, such as those in the northern portion of the study area, were deleted from the analysis.

Table 61.1. List of Landscape Metrics Compared to Nitrogen Loads[a]

Name of Metric	Explanation
Riparian agriculture	Percentage of watershed with agricultural land cover adjacent to stream edge; 1 pixel wide
Riparian forest	Percentage of watersheds with forest land cover adjacent to stream edge; 1 pixel wide
Forest fragmentation	Forest fragmentation index for watershed; of all edges in the watershed involving at least 1 forested pixel, the percent that joins a forest pixel to a nonforest pixel; higher values indicate higher fragmentation
Road density	Road density for watershed expressed as an average number of kilometers of roads per square kilometer of watershed; normalized to approximate scale of land-cover metrics
Forest land cover	Percentage of watershed with forest land cover
Agricultural land cover	Percentage of watershed with agricultural land cover (pasture/crops)
Agricultural land cover on steep slopes	Percentage of watershed with agriculture occurring on slopes greater than 3%
Nitrate deposition	Estimated average annual wet deposition of nitrate
Potential soil loss	Proportion of watershed with the potential for soil losses greater than 1 ton per acre per year
Roads near streams	Proportion of total stream length that has roads within 30 m; normalized to approximate scale of land-cover metrics
Slope gradient	Average percent slope gradient for watershed
Slope gradient range	Percentage slope gradient range (maximum–minimum) for watershed
Slope gradient variance	Percentage slope gradient variance for watershed; normalized to approximate scale of land-cover metrics
Urban land cover	Percentage of watershed with urban land cover
Wetland land cover	Percentage of watershed with wetland land cover
Barren land cover	Percentage of watershed with barren land cover; includes quarry areas, coal mines, and transitional areas, such as clear-cut areas

[a] Calculation methods and details of each indicator can be found in Jones et al. (1997). Metrics were calculated for each watershed support area.

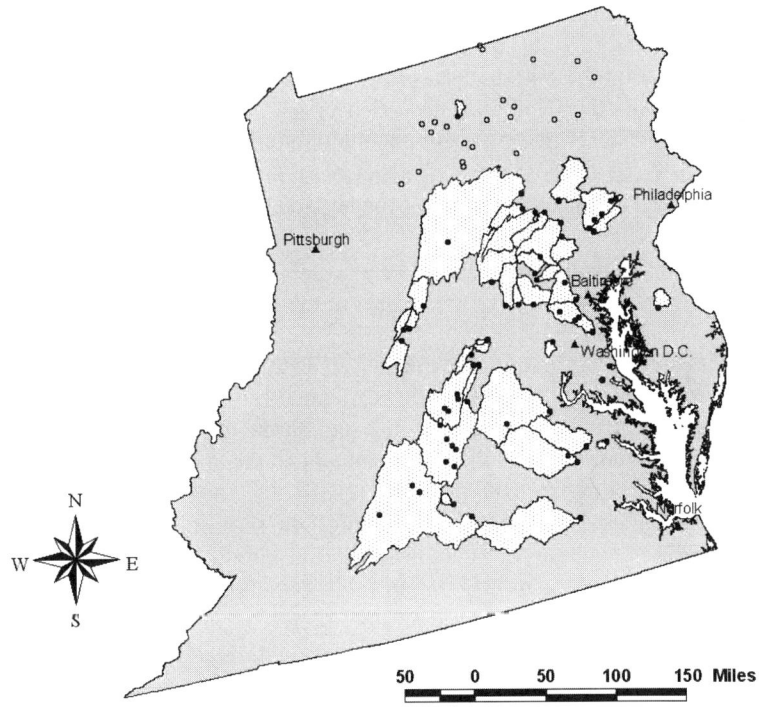

● Location of water monitoring sites (total nitrogen)

Delineated watersheds

○ Additional sites with total nitrate measurements

Figure 61.2 Location of water sampling points and their associated delineated watersheds.

We first examined the individual relationships between nitrogen yield (dependent variable) and landscape metrics (independent variables) using individual scatter plots and preliminary regression analyses (SAS Institute, 1990). From our preliminary analyses, we concluded that a log transformation of the nitrogen yield data was necessary. We also performed a square-root transformation of the nitrate deposition landscape variable to linearize the relationship between it and the dependent variables.

We then ran several different regression analyses (SAS Institute, 1990) for nitrogen with the suite of landscape metrics listed in Table 61.1 to help us understand the importance of each variable. We selected a model based on four requirements:

1. The model had to be valid (i.e., basic principles of regression analysis, such as normality, constancy, and independency of error terms were not violated).
2. The model explained a high proportion of the variance in the nitrogen yield.
3. The model included variables over which we may have more control (e.g., selecting riparian forest as a predictor variable over agricultural land cover; agricultural land cover may have explained more overall variance in the model had it been included, but there is a low likelihood of changing the amounts of agriculture in a watershed significantly).
4. The model did not include variables that were highly collinear, as it is difficult to separate the effects of different variables when multicollinearity is an issue.

After selecting the model, we applied it to three future scenarios for all watersheds (defined by USGS hydrologic accounting units) in the Middle Atlantic region. The variables selected for this regression model, based on the criteria discussed above, were amount of riparian forest cover and air nitrate deposition. Amount of riparian forest cover is defined as the percentage of watershed with forest land cover adjacent to the stream edge with a buffer size of 30 m (i.e., one 30 × 30 m picture element). Air nitrate deposition is a modeled value that represents the estimated average annual wet deposition of nitrate. The resulting model from this analysis was:

$$\text{total nitrogen yield} = \exp[0.9056 - 0.02769(\text{riparian forest cover}) + 0.00168\ (\text{nitrate deposition})]$$

This model, based on 69 observations, explained 87% of the variation in nitrogen loading values with 68% explained by riparian forest and 19% explained by nitrate deposition. We next applied this model to the landscape metrics (riparian forest and nitrate deposition) from Jones et al. (1997) for the USGS hydrologic accounting units and manipulated the amount of riparian forest to create three future scenarios for nitrogen loading.

We subsequently applied this model to all the hydrologic accounting units in the Middle Atlantic region based on current conditions of riparian forest and nitrate deposition. We then projected a future scenario if the current amount of riparian forest were increased by an additional 10% to a maximum value of 100%. We also projected a future scenario if the amount of riparian forest was decreased by 10% to a minimum value of 0%. These predictions assume that the amount of nitrate deposition remains constant. This is a simple model and does not take into account several potentially important variables, such as loadings from point sources, hydrogeologic conditions, groundwater contribution, and other highly correlated but important variables, such as total amount of agriculture on the watershed. It also does not take into account other things that could affect future scenarios (e.g., probable population growth patterns).

RESULTS

Figure 61.3 shows nitrogen loading conditions as predicted by current conditions. Figure 61.4 shows a future scenario if the amount of riparian forest were increased by 10%. Figure 61.5 shows a future scenario if the amount of riparian forest were decreased by 10%. These figures demonstrate that by adjusting the riparian forest percentages, we could have a significant impact on nitrogen loading to streams. The pattern of transition from lower to higher nitrogen loadings to streams is clear in these three figures. The cutoff values for each of the five classes shown in Figures 61.3, 61.4, and 61.5 are based on quantiles and were defined from current condition data (Figure 61.3). Cutoff values perhaps would be more meaningful if they were based on some known water quality criteria for nitrogen such as those developed on an ecoregional basis by the EPA (U.S. Environmental Protection Agency, 2000).

A potentially serious limitation of this analysis is extrapolating the model derived from limited spatial coverage (Figure 61.1) to a larger spatial extent. The farther away we get from those watersheds shown in Figure 61.1 (i.e., to the north and the west), the less reliable our model may be. We may be able to justify extrapolating to the north by examining the relationship between total nitrate and the same landscape metrics. We had total nitrate data for 79 sites, including 20 sampling sites to the north of those shown in Figure 61.1. The resulting model, if we regress total nitrate on the same two landscape variables (riparian forest and nitrate deposition), is as follows:

$$\text{total nitrate load} = \exp[0.9621 - 0.0366(\text{riparian forest cover}) + 0.00163\ (\text{nitrate deposition})]$$

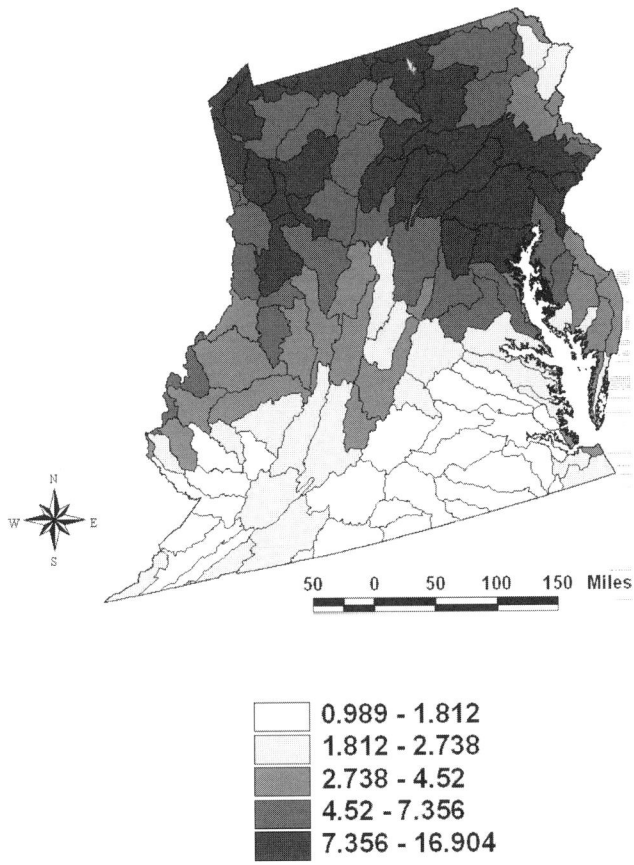

	0.989 - 1.812
	1.812 - 2.738
	2.738 - 4.52
	4.52 - 7.356
	7.356 - 16.904

Figure 61.3 Nitrogen loadings (lbs/acre/year) as predicted by present conditions.

This model, based on 79 observations, explained 78% of the variation in nitrate loading values with 55% explained by riparian forest and 23% explained by nitrate deposition. While this model is somewhat similar to the model for total nitrogen, riparian forest explains less of the variation but has a more severe impact on nitrogen loading (steeper regression slope).

Another potential limitation of the model concerns the important issue of scale. The model was developed based on data from multiple scales ranging from very small to very large watersheds; some of these were larger than the USGS hydrologic accounting units and many of them were smaller. The model parameter estimates could be very different if, for example, all the watersheds were based on either very small or very large watersheds.

DISCUSSION

Wall-to-wall landscape data of relatively fine spatial scale (e.g., 30 m), and field sample measurements of a variety of stream chemistry parameters, permit the development of spatially distributed empirical models that can be used to assess potential loadings to streams across an entire region. As demonstrated in this chapter, the development of these models is critical to assess how future landscape scenarios might affect stream water quality across a region. In our study, riparian extent along streams was a strong predictor of nitrogen yield. Because of this relationship, we were able to assess how two future scenarios of riparian habitat condition would affect nitrogen loadings to streams at a watershed or catchment scale. We used two simple

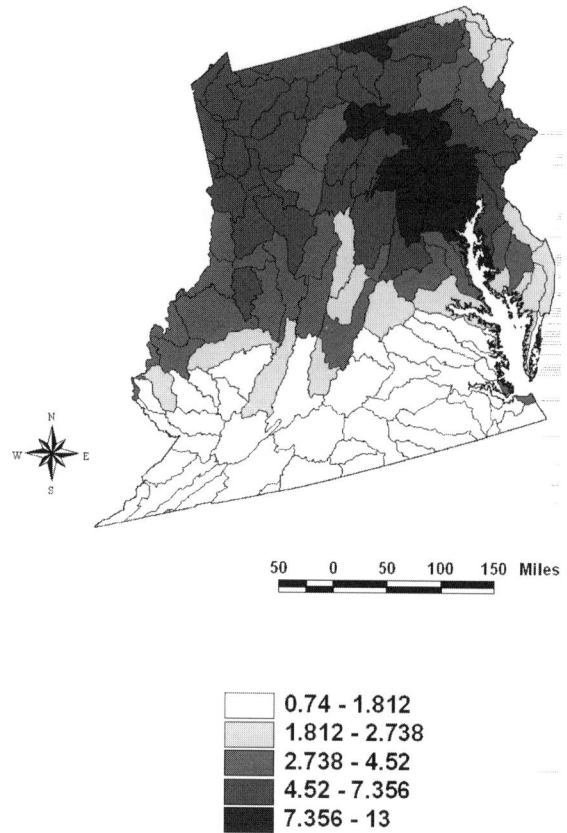

Figure 61.4 Nitrogen loadings (lbs/acre/year) as predicted with a 10% increase of riparian forest.

future scenarios — one with a 10% increase in the amount of riparian habitat at the catchment scale and one with a 10% decrease in riparian habitat — to demonstrate how the approach can be used to target those watersheds that would benefit the most from an increase in riparian habitat (restoration) and those that would be at the greatest risk due to decreases in riparian habitat (catchments needing protection). The decreased riparian habitat scenario is similar to a treadmill stress test in humans (after the ecohealth concept of Rapport et al., 1998). It finds those catchments where streams are most vulnerable to increases in nitrogen load and, therefore, a potential dramatic effect on human (waterborne diseases and drinking water quality) and ecological health (stream biota).

Our modeling approach differs from other watershed or catchment models that predict nutrient loadings in that it considers the spatial pattern and distribution of key landscape features; most existing models only consider the percentage of land cover at the catchment scale (U.S. Department of Agriculture, 1972; Liang et al., 1994). Additionally, most existing models lack riparian and atmospheric deposition parameters. Riparian habitats can be significant filters of nutrient inputs to streams (Lowrance et al., 1984; Peterjohn and Correll, 1984; Cooper et al., 1987) and atmospheric nitrogen deposition can be a significant source of nitrogen input at the catchment scale (Stensland et al., 1986; Appleton, 1995).

Our modeling approach is a significant improvement over the comparative watershed approach used by Jones et al. (1997) and Wickham et al. (1999a) because landscape metrics were quantitatively linked to stream conditions. Jones et al. (1997) and Wickham et al. (1999a) ranked the relative vulnerability of watersheds but did not link their ranking to observed water quality.

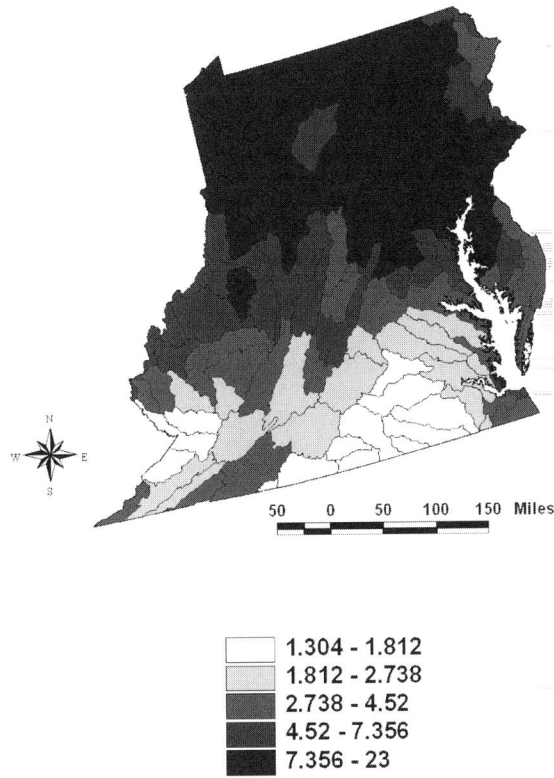

	1.304 - 1.812
	1.812 - 2.738
	2.738 - 4.52
	4.52 - 7.356
	7.356 - 23

Figure 61.5 Nitrogen loadings (lbs/acre/year) as predicted with a 10% loss of riparian forest.

Because our model includes the spatial pattern of key landscape features, it is possible to find those areas that would benefit most from riparian restoration and riparian protection (those catchments at risk under the reduced riparian habitat scenario). Such information can be used by land managers and private landowners to decide where to restore and protect riparian habitats. This is important because budgets for restoration and protection are often limited. Wickham et al. (1999b) used a similar approach to identify those catchments that had the greatest potential to restore forest connectivity. Such targeting approaches are needed to identify restoration and protection opportunities for other ecological goods and services (e.g., flood abatement, wildlife habitat, and forest, rangeland, and agricultural productivity).

The spatially distributed model demonstrated in this chapter can be used to assess the consequences of other landscape scenarios. For example, the model could be applied to assess the potential change in nitrogen loading to streams associated with different land management scenarios generated from public workshops (Steinitz, 1996). It also could be used in combination with socioeconomic, land-cover change models (Wickham et al., 2000) to assess the consequences of socioeconomic futures on nitrogen loadings to streams.

Our modeling approach requires regionally consistent landscape data and a fairly extensive network of water quality samples. Therefore, availability of regional-scale, land-cover data similar to that being developed by the MRLC Consortium in the U.S. (Vogelmann et al., 1998), and stream samples similar to those being collected by the Environmental Monitoring and Assessment Program program in the U.S. and the Waterwatch program in Australia (Waterwatch, 1997), is critical to the use of our modeling approach. Additionally, since quantitative relationships between landscape pattern and stream water quality vary between different biophysical settings (Omernik et al., 1981; Clarke et al., 1991; Jones et al., 2001), water quality sampling must be sufficient to represent different biophysical settings.

ACKNOWLEDGMENTS

We thank Scott Phillips and Judy Denver of the USGS for providing information on nutrient parameters used in this study. The U.S. Environmental Protection Agency (EPA), through its Office of Research and Development, partially funded and collaborated in the research described here. It has not been peer-reviewed by the EPA. Mention of trade names or commercial products does not constitute endorsement or recommendation by the EPA for use.

REFERENCES

Arnold, C.L. and Gibbons, C.J., Impervious surface coverage: the emergence of a key environmental indicator, *J. Am. Plann. Assoc.*, 62, 243–258, 1996.

Burns, J.W., Some effects of logging and associated road construction on Northern Californian streams, *Trans. Am. Fish. Soc.*, 101, 1–17, 1972.

Charbonneau, R. and Kondolf, G.M., Land use change in California, USA: nonpoint source water quality impacts, *Environ. Manage.*, 17, 453–460, 1993.

Clarke, S.E., White, D., and Schaedel, A.L. Oregon, USA, ecological regions and subregions for water quality management, *Environ. Manage.*, 15, 847–856, 1991.

Cooper, J.R., Gilliam, J.W., Daniels, R.D., and Robarge, W.P., Riparian areas as filters for agricultural sediment, *Soil Sci. Soc. Am. J.*, 51, 416–420, 1987.

Environmental Protection Agency, National Strategy for the Development of Regional Nutrient Criteria, EPA/822/R-98/002, Washington, D.C., 1998.

ESRI, Introduction to ArcView GIS, Environmental Systems Research Institute, Redlands, CA, 1996.

Franklin, J.F., Scientific basis for new perspectives in forests and streams, in *Watershed Management*, Naiman, R.J., Ed., Springer-Verlag, New York, 1992, pp. 25–72.

Harden, C.P., Incorporating roads and footpaths in watershed-scale hydrologic and soil erosion models, *Phys. Geog.*, 13, 368–385, 1992.

Herlihy, A.T., Stoddard, J.L., and Johnson, C.B., The relationship between stream chemistry and watershed land cover data in the Mid-Atlantic Region of the U.S., *Water Air Soil Pollut.*, 105, 377–386, 1998.

Hunsaker, C.T. and Levine, D.A., Hierarchical approaches to the study of water quality in rivers, *BioScience*, 45, 193–203, 1995.

Hunsaker, C.T., Levine, D.A., Timmins, S.P., Jackson, B.L., and O'Neill, R.V., Landscape characterization for assessing regional water quality, in *Ecological Indicators*, McKenzie, D.H., Hyatt, D.E., and McDonald, V.J., Eds., Elsevier Applied Science, New York, 1992, pp. 997–1006.

Jones, K.B., Riitters, K.H., Wickham, J.D., Tankersley, R.D., O'Neill, R.V., Chaloud, D.J., Smith, E.R., and Neale, A.C., An Ecological Assessment of the United States Mid-Atlantic Region: A Landscape Atlas, EPA/600/R-97/130, U.S. Environmental Protection Agency, Office of Research and Development, Washington, D.C., 1997.

Jones, K.B., Neale, A.C., Nash, M.S., Riitters, K.H., Wickham, J.D., O'Neill, R.V., and Van Remortel, R.D., Landscape correlates of breeding bird richness across the United States mid-Atlantic region, *J. Environ. Monitoring Assess.*, 63, 159–174, 2000.

Jones, K.B., Neale, A.C., Nash, M.S., Van Remortel, R.D., Wickham, J.D., Riitters, K.H., and O'Neill, R.V., Predicting nutrient and sediment loading from the United States mid-Atlantic region, *Landscape Ecol.*, 16, 301–312, 2001.

Karr, J.R., and Schlosser, I.J., Water resources and the land-water interface, *Science*, 201, 229–233, 1978.

Kattan, G.H., Alvarez-Lopez, H., and Giraldo, M., Forest fragmentation and bird extinctions: San Antonio eighty years later, *Conserv. Biol.*, 8, 138–146, 1994.

Koopowitz, H., Thornhill, A.D., and Andersen, M., A general stochastic model for the Lane prediction of biodiversity losses based on habitat conversion, *Conserv. Biol.*, 8, 425–438, 1994.

Langland, M.J., Leitman, P.L., and Hoffman, S., Synthesis of Nutrient and Sediment Data for Watersheds within the Chesapeake Bay Drainage Basin, Report 95-4233, U.S. Geological Survey Water-Resources Investigations, Baltimore, MD, 1995.

Liang, X., Lettenmaier, D.P., Wood, E.F., and Burges, S.J., A simple hydrologically based model of land surface, water, and energy fluxes for general circulation models, *J. Geophys. Res.*, 99, 14415–14428, 1994.

Likens, G.E., Bormann, F.H., Pierce, R.S., Eaton, J.S., and Johnson, N.M., *Biogeochemistry of a Forested Ecosystem*, Springer-Verlag, New York, 1977.

Lowrance, R.R., Leonard, R., and Sheridan, J., Managing riparian ecosystems to control nonpoint pollution, *J. Soil Water Conserv.*, 40, 87–91, 1984.

Omernik, J.M., Ecoregions of the United States: map at a scale of 1:7,500,000, *Suppl. Ann. Assoc. Am. Geogr.*, **77**, 118–125, 1987.

Omernik, J.M., Abernathy, A.R., and Male, L.M., Stream nutrient levels and proximity of agricultural and forest land to streams: some relationships, *J. Soil Water Conserv.*, 36, 227–231, 1981.

O'Neill, R.V., Krummel, J.R., Gardner, R.H., Sugihara, G., Jackson, B., DeAngelis, D.L., Milne, B.T., Turner, M.G., Zygmunt, B., Christensen, S.W., Dale, V.H., and Graham, R.L., Indices of landscape pattern, *Landscape Ecol.*, 1, 153–162, 1988.

O'Neill, R.V., Hunsaker, C.T., Jones, K.B., Riitters, K.H., Wickham, J.D., Schwarz, P., Goodman, I.A., Jackson, B., and Baillargeon, W.S., Monitoring environmental quality at the landscape scale, *BioScience*, 47, 513–520, 1997.

Peterjohn, W.T. and Correll, D.L, Nutrient dynamics in an agricultural watershed: observations on the role of a riparian forest, *Ecology*, 65, 1466–1475, 1984.

Rapport, D.L., Caudet, C., Karr, J.R., Baron, J.S., Bohlen, C., Jackson, W., Jones, B., Naiman, R.J., Norton, B., and Pollock, M.N., Evaluating landscape health: integrating societal goals and biophysical processes, *J. Environ. Manage.*, 53, 1–15, 1998.

Roth, N.E., Allan, J.D., Erickson, D.L., Landscape influences on stream biotic integrity assessed at multiple scales, *Landscape Ecol.*, 11, 141–156, 1996.

SAS Institute, Inc., *SAS/STAT User's Guide*, Version 6, 4th ed., Vol. 2, Cary, NC, 1990.

Saunders, D.A., Hobbs, R.J., and Margules, C.R., Biological consequences of ecosystem fragmentation: a review, *Conserv. Biol.,* 5, 18–32, 1991.

Short, J., and Turner, B., A test of the vegetation mosaic hypothesis: a hypothesis to explain the decline and extinction of Australian mammals, *Conserv. Biol.,* 8, 439–449, 1994.

Smith, R.A., Schwarz, G.E., and Alexander, R.B., 1997. Regional interpretation of water-quality monitoring data, *Water Resour. Res.*, 33, 2781–2798, 1997.

Steinitz, C., Ed., Landscape Planning for Biodiversity: Alternative Futures for the Region of Camp Pendleton, California, Graduate School of Design, Harvard University, Cambridge, MA, 1996.

Stensland, G.L., Whelpdale, D.M., and Ochlert, G., Precipitation chemistry, in *Acid Deposition: Long-Term Trends*, National Research Council, National Academy Press, Washington, D.C., 1986, pp. 128–199.

U.S. Department of Agriculture, Soil Conservation Service (SCS), National Engineering Handbook, Hydrology Section 4, Chapter 4, Government Printing Office, Washington, D.C., 1972.

Vogelmann, J.E., Sohl, T., Campbell, P.V., and Shaw, D.M., Regional land cover characterization using Landsat Thematic Mapper data and ancillary sources, *Environ. Monitoring Assess.*, 51, 415–428, 1998.

Walker, J., Bullen, F., and Williams, B.G., Ecohydrological changes in the Murray-Darling Basin. I. The number of trees cleared over two centuries, *J. Appl. Ecol.*, 30, 265–273, 1993.

Waterwatch, News from the national Waterwatch conference: getting better at getting wet, *Waterwatch,* Australia Conference Newsletter, Australia, 1997.

Weller, M.C., Watzin, M.C., and Wang, D., Role of wetlands in reducing phosphorus loading to surface water in eight watersheds in the Lake Champlain Basin, *Environ. Manage.*, 20, 731–739, 1996.

Wickham, J.D., Jones, K.B., Riitters, K.H., O'Neill, R.V., Tankersley, R.D., Smith, E.R., Neale, A.C., and Chaloud, D.J., An integrated environmental assessment of the U.S. mid-Atlantic region, *Environ. Manage.*, 24, 553–560, 1999a.

Wickham, J.D., Jones, K.B., Riitters, K.H., Wade, T.G., and O'Neill, R.V., Transitions in forest fragmentation: implications for restoration opportunities at regional scales. *Landscape Ecol.*, 14:137–145, 1999b.

Wickham, J.D., O'Neill, R.V., and Jones, K.B., A geography of ecosystem vulnerability, *Landscape Ecol.*, 15, 495–504, 2000.

Yates, P. and Sheridan, J.M., Estimating the effectiveness of vegetated floodplains/wetlands as nitrate-nitrite and orthophosphorus filters, *Agric. Ecosyst. Environ.*, 9, 303–314, 1983.

CHAPTER 62

Echelon Screening of Remotely Sensed Change Indicators

Wayne L. Myers and Francis R. Beck

INTRODUCTION

Tracking of changes on the landscape is central to assessment of ecosystem health and sustainability, although implications of change phenomena are not always straightforward (Costanza, 1991; Jones et al., 1997; O'Neill et al., 1997; Picket and White, 1985). Determining and recording changes on a parcel basis at local scale is a fairly mechanical matter when modern geographic information systems (GISs) are utilized, but becomes much more problematic for large areas when funds are limited. In the latter case, it becomes a practical necessity to exploit remotely sensed data obtained electronically from instrumentation carried by satellites and transmitted to strategically located receivers. Remotely sensed data are typically recorded in a grid arrangement with a set of spectral reflectance readings for each cell of the grid. The readings correspond to energy of selected wavelengths that is integrated (averaged) over the sector of Earth's surface encompassed by the cell. Remote sensing specialists use the technical term *pixel* (for picture element) instead of the more generic cell terminology (Wilkie and Finn, 1996). The area covered by a pixel (cell) is usually too large to provide high resolution regarding geometric detail of objects on the ground (Lillesand and Kiefer, 1994). Differences in landscape features must be inferred from computed differences among spectral readings, as opposed to detailed visual examination.

Computerized methods of detecting landscape changes involve one of two general modes with numerous variations in either mode (Lunetta and Elvidge, 1998). One mode consists of using multivariate statistics to assign each cell to a feature (thematic) class, such as forest or grassland. Specific types of class changes for a pixel between dates of imaging will constitute landscape changes of interest. The utility of this *classification mode* depends heavily on the accuracy of classification. If classifications for two dates both have substantial inaccuracies that occur independently, then derived inference regarding change will be less accurate than either original classification. The second mode involves spatial comparison of spectral readings without performing classifications as an intermediate step. This produces an *index of change* (change signal) with a value for each pixel that is indicative of where change may have occurred when analyzed comparatively across the extent of the image. The index of change can be displayed directly as an image, but decisions must then be made regarding what constitutes background that indicates lack of change vs. highlights that suggest candidate areas where change may have occurred and further investigation is warranted. The typical approach to making such decisions is by establishing a threshold level in terms of the index. Establishment of the threshold may be done either visually or by a computational heuristic.

Our focus is on the index of change mode, with concern for limitations of thresholds in the context of change detection. Spectral expression of landscape change is given brief consideration in order to provide perspective on the problem. Next, we discuss limitations of thresholds relative to change indicators. We then explain an alternative way of highlighting candidate areas with respect to change. This alternative approach has been called *echelons* (Myers et al., 1997). Echelons overcome many of the limitations for thresholds, and are also applicable to finding areas of prospective interest relative to grid mappings for environmental indexes other than change.

IMAGE EXPRESSION OF LANDSCAPE CHANGE

In an oversimplified manner of statement, a change will be most evident if the conditions before and after have a strongly contrasting appearance. Conversely, the appearance of low contrast changes will be subtle and subdued. *Appearance* is a word that may be somewhat misleading in this context because remotely sensed data usually incorporate infrared wavelengths that are invisible to the unaided human eye. Nevertheless, extending the idea of appearance informally to encompass response of an electronic sensor should convey the intended impression. Building a four-lane highway through a forest thus constitutes a high contrast change, whereas partial regrowth of trees in a brushy forest clear-cut constitutes a change having low contrast.

The task of change detection becomes more complicated because high contrast changes are not all of interest, and not all changes of interest have high contrast. The crop grown in an agricultural field is often changed from year to year for maintenance of fertility and reduction of pest problems. If the successive crops have different tillage and harvest schedules, the resulting appearance of change may exhibit high contrast while being just temporal variants of a consistent and sustainable agricultural cropping regime. On the other hand, exposed earth due to construction may have an appearance similar to former agricultural plowing, which gives rise to a change of low contrast but considerable importance with respect to ecosystem health.

Spatial continuity and extent of change may also be more important than its degree of consistency. A small forest opening due to demise of a couple of trees affecting one pixel may be quite inconsequential, but a pest infestation or fire with varying degree of impact over a larger area becomes something of major concern. High contrast changes are spotty in the latter situation, but these spots connect across bridges of lesser change to form a change complex encompassing substantial area.

LIMITATIONS OF THRESHOLDS FOR CHANGE INDICATORS

In an ideal change detection scenario, a change signal grid would contain a patchy mixture of two populations. One population would have low values of the change index and would represent areas where change had not occurred. The other population would have high values of the change index and would represent areas where change had transpired. It would then be straightforward and appropriate to fix a threshold in the mid-zone between high- and low-change values, choosing an intermediate value that is absent or has the fewest occurrences. Unfortunately, however, such a scenario is practically never encountered for the reasons given above. Consequently, the choice of threshold becomes somewhat arbitrary and inadequate in several respects.

Figure 62.1 poses a hypothetical profile along a transect line through a grid of change index values. The horizontal dashed lines depict possible positions of a threshold to illustrate the attendant problems. The threshold at level A gives the impression that change only occurs in isolated spots, concealing both the substantial extent of change and differing degrees of spatial coherence in the

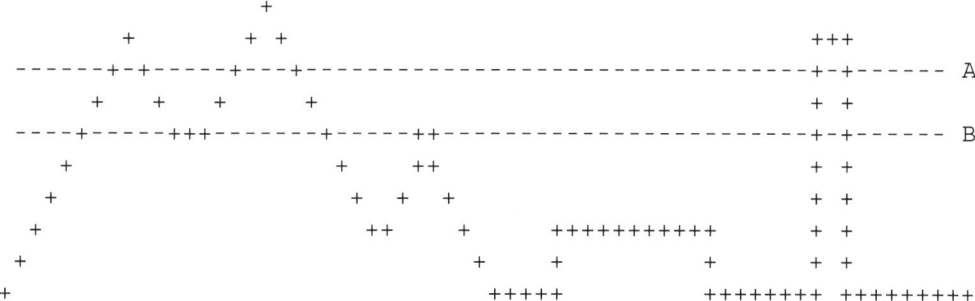

Figure 62.1 Hypothetical profile of a change signal (+) with illustrative thresholds (–) at levels A and B.

change. The threshold at level B begins to catch somewhat more of the spatial extent for change, but still gives an erroneous impression that there is one larger patch and two smaller ones. Both A and B thresholds miss the lower level of more subdued change.

As the example thresholds serve to illustrate, there is an inherent arbitrariness to the setting of thresholds when change occurs in a graded manner. Furthermore, no single threshold can adequately capture spatial pattern of graded change that is not disjunct.

ECHELONS AS AN ALTERNATIVE TO THRESHOLDS

What is needed is to capture emerging components of spatial structure in the change grid as a threshold is lowered progressively from highest peak to lowest point. This is effectively what *echelons* accomplish. Echelons partition a surface variable into peak and foundation components with records of major characteristics for each component.

The peaks are first located as primary components, and the peaks are numbered in order of decreasing summit level. Each peak extends downward and outward from the summit until a saddle is reached on any side of the peak, whereupon the respective peak terminates. The foundation components are then located in order of decreasing upper level as secondary components. As with a peak, a foundation spreads downward and outward until a saddle is encountered on any side. The foundations are numbered in sequence, starting with the next available number after that allocated to the lowest peak. The echelon components of Figure 62.1 are depicted in Figure 62.2, with dashed lines indicating saddles instead of thresholds.

Our usual analogy for echelons is drainage after submersion by floodwaters. The peaks are components that break the water surface to emerge as islands. New foundations are initiated whenever two or more current islands merge to become a larger composite island. There is also

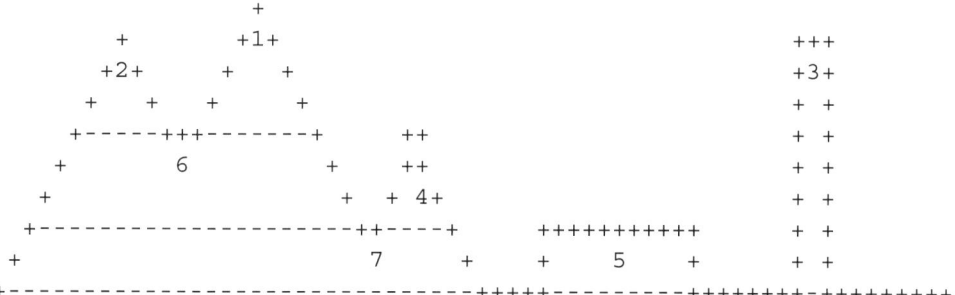

Figure 62.2 Echelon partition of illustrative change signal from Figure 62.1, with dashed lines at saddles.

an ordering scheme for echelons that is useful in the change detection context. Peaks are of first order, as for instance components 1 through 5 in Figure 62.2. A foundation that supports two or more first-order echelons is of second order, with echelon 6 in Figure 62.2 providing an example. A foundation of a given order can also support one other echelon of like order, as does component 7 in Figure 62.2. A foundation becomes third order if it supports two or more second-order components. In general, the order of a foundation depends on the maximum order of components supported and the number of such maximum-order components. If there are two or more maximum-order components, then the foundation is of next order. The order of the grid as a whole is that of its highest-order component.

Echelon organization also entails a family tree of foundation relations. The peaks are *terminals* of the tree. Second-order foundations supporting only first-order components are *twigs* of the tree, as exemplified by component 6 in Figure 62.2. A *stem* is a foundation that is induced only by first-order components, but is more complex than a twig. Component 7 in Figure 62.2 is a stem because its formation is due to presence of the first-order component 4, although it also supports a second-order component 6. Twigs are locally prominent upper structures of a change surface. Terminals supported by stems are bumps on sidehills of change, and are thus somewhat subordinate components of change. An echelon supported by a foundation is called an *ascendant* of the foundation, and a foundation is called the *founder* of a component that it supports.

Software written in C language with a command line interface for computing echelons is available via the Internet at http://www.orser.psu.edu. Computation of echelons proceeds in stages with one program module for each stage. The first two stages serve to determine all first-order echelons. There is a capacity limit of 65,280 echelons for any single grid. If there are more than 65,280 first-order echelons, then the change surface must be simplified (smoothed) in order to proceed. Initial smoothing is accomplished by eliminating single-pixel blips. The most similar immediate neighbor replaces a blip to be eliminated. If there is still insufficient capacity to accommodate first-order echelons, then a median filter is used to further simplify the change grid. When first-order computation is achieved, the results can be displayed to determine whether additional smoothing should be done. If so, then small first-order echelons are shaved off according to size criteria.

Foundation echelons are determined in another stage. Whereas determination of first-order echelons is fairly rapid, foundation echelons are more time-consuming to compute. If computation of foundations becomes unduly protracted, provision is made for controlled termination and subsequent work with partial echelons. Since determination of foundations works progressively downward, the upper echelon components are determined in the beginning. Focal areas of change can be detected by combining first-order echelons with upper foundations. After foundations are determined (whether wholly or partially), a table of echelon properties still must be completed with the appropriate program module.

A fairly sophisticated capability is provided for preparing echelon-based display layers that are compatible with several commercial and no-cost viewers intended for GIS and remote sensing applications. This facility allows one echelon property to serve as a base, with other properties having roles as modifiers. It is recommended that *echelon precedence* be used as a base and rendered as grayscale. Precedence is derived from the order in which echelons have been determined. The first modification is to use the color red for rendering first-order echelons having twigs as foundations. The second modification is to use yellow for rendering first-order echelons having stems as foundations. It may also be desirable to suppress small first-order echelons that lack a foundation by rendering them in the same manner as the (black) background. Modifications involve recoding with a program module, and then editing a textual color lookup file.

Displays using this coding scheme are readily interpreted. Red areas are the strongest indications of change. Red areas situated in whitish or light gray surroundings represent locally strong change that is also spatially coherent but of variable intensity over a vicinity. Yellow colors situated in light gray surroundings are subordinate peaks on sidehills of change. Light gray areas lacking embedded red and yellow highlights are peaks of change having a base below the level to which partial echelons were developed. The lighter ones of these latter areas represent changes of relatively strong contrast, whereas the darker ones represent changes of relatively weak contrast. An example of this presentation scheme is not provided here because color is key to achieving the desired visual effect.

SUMMARY

Capability for detecting spatially coherent areas having changes of varying intensity is a special advantage of echelons. Echelons likewise pick up all sorts of local peaks, many of which would be missed by usual placement of a threshold. Echelons further offer several options for filtering on the basis of multiple criteria in order to suppress certain types of changes that may not be of interest. Finally, the utility of echelons for screening of surface variables is not limited to the change detection context. Potential for characterizing and distinguishing spatial patterns among surface sectors lies in the capacity of echelons to capture spatial organization of a surface in dendrogram mode. In order for this potential to be realized, further research is needed on methods for comparative analysis of dendrograms.

REFERENCES

Costanza, R., Ed., *Ecological Economics: The Science and Management of Sustainability*, Columbia University Press, New York, 1991.
Jones, K.B., Riitters, K.H., Wickham, J.D., Tankersley, Jr., R.D., O'Neill, R.V., Chaloud, D.J., Smith, E.R., and Neale, A.C., An Ecological Assessment of the United States Mid-Atlantic Region: A Landscape Atlas, EPA/600/R-97/130, U.S. Environmental Protection Agency, Office of Research and Development, Washington, D.C., 1997.
Lillesand, T.M. and Kiefer, R.W., *Remote Sensing and Image Interpretation*, John Wiley & Sons, New York, 1994.
Lunetta, R.S. and Elvidge, C.D., Eds., *Remote Sensing Change Detection: Environmental Monitoring Methods and Applications*, Ann Arbor Press, Ann Arbor, MI, 1998.
Myers, W.L., Patil, G.P. and Joly, K., Echelon approach to areas of concern in synoptic regional monitoring, *Environmental and Ecological Statistics*, 4, 131–152, 1997.
O'Neill, R.V., Hunsaker, C.T., Jones, K.B., Riitters, K.H., Wickham, J.D., Schwartz, P.M., Goodman, I.A., Jackson, B.L., and Baillargeon, W.W., Monitoring environmental quality at the landscape scale, *BioScience*, 47, 513–519, 1997.
Picket, S.T.A. and White, P.S., Eds., *The Ecology of Natural Disturbance and Patch Dynamics*, Academic Press, Orlando, FL, 1985.
Wilkie, D.S. and Finn, J.T., *Remote Sensing Imagery for Natural Resources Monitoring: A Guide for First-Time Users*, Columbia University Press, New York, 1996.

CHAPTER 63

Grassland Bird Communities and Environmental Health: The Role of Landscape Features

Douglas H. Johnson, Sharon F. Browder, and Maiken Winter

INTRODUCTION

Ecosystems throughout the world are imperiled by myriad anthropogenic factors. It is critical to develop tools for evaluating and monitoring the health or integrity of various ecosystems. Because of the broad scale of the problem, the urgency to address informational needs, and the limited financial resources committed to the problem, relatively inexpensive tools that encompass large areas are especially desirable. Remote sensing offers possibilities for such large-scale evaluation and monitoring. With aerial or satellite imagery, vast quantities of information from very large areas can be collected periodically.

Before remote sensing can be employed to assess ecosystem health on an operational basis, however, two fundamental questions must be addressed: (1) How do features that can be remotely sensed relate to the health of the ecosystem being evaluated? (2) How can remotely sensed measures be validated with on-the-ground assessments of the ecosystem? Here we address those questions in the case of grasslands and grassland birds in the northern Great Plains of the U.S.

We first describe the importance of grassland systems, notably to bird populations, and why their health or integrity needs to be evaluated. Next, we discuss birds of the grasslands and offer reasons birds are useful as measures of ecosystem health. We then address question 1 above by demonstrating that a variety of features that can be sensed remotely are in fact associated with the health of the ecosystem, as reflected in bird communities and their population dynamics. Much of that material is drawn from Johnson and Winter (1999) and Johnson (2001). We then face question 2 by indicating how a measure of ecosystem health for grasslands relates to a community metric based on breeding-bird populations. Those results are based on Browder (1998) and Browder et al. (submitted).

GRASSLANDS OF THE NORTHERN GREAT PLAINS

Three broad provinces of grassland in the Great Plains are generally recognized, which correspond to a gradient of increasing precipitation from west to east: shortgrass prairie in the west, mixed-grass prairie in the center, and tallgrass prairie in the east (Risser et al., 1978). Prior to settlement by Europeans, the northern plains were a vast grassland; trees were scarce or absent (e.g., Bragg and

Steuter, 1995). Grasslands were maintained by periodic drought, fires, and, especially in the west, grazing by large herds of herbivores such as bison (*Bison bison*). These forces created mosaics of habitat ranging from heavily grazed to undisturbed (England and DeVos, 1969).

Much of the grassland in the northern Great Plains has been cultivated for crops. This conversion is nearly total in the eastern Great Plains; tallgrass prairie is one of the most threatened habitats in the northern plains, with only scattered fragments remaining (Samson and Knopf, 1994; Noss et al., 1995). A smaller proportion of mixed-grass prairie has been cultivated, largely because the terrain is rougher and precipitation is lower and less predictable. The proportion of shortgrass prairie that has been tilled is even lower, although much of it is intensively grazed by domestic livestock. Cultivation, however, has increased in the past few decades in the mixed-grass and shortgrass prairies due to the advent of sprinkler irrigation using water from ancient aquifers. Settlement of the northern plains by Europeans also brought major increases of woodland, with tree claims planted to protect farmsteads from winds, shelterbelts established to reduce soil erosion, and increased woody vegetation resulting from fire suppression by settlers (McNicholl, 1988).

Grasslands of the northeastern Great Plains are dotted with innumerable depressions, called prairie potholes, that formed when glaciers retreated about 10,000 years ago. Individual wetland basins contain water for different lengths of time in most years (Stewart and Kantrud, 1971). This variation among wetlands in hydrological regime induces variation in water chemistry, vegetation, nutrient cycling, and invertebrate communities (van der Valk, 1989). This results in a rich diversity of birds and other vertebrates (Igl and Johnson, 1998).

Like grasslands, prairie wetlands have been altered, notably by drainage intended to increase the area under cultivation. Losses of wetlands from settlement to 1980 were 27% in Montana, 49% in North Dakota, 35% in South Dakota, 42% in Minnesota, and 89% in Iowa (Dahl, 1990). Siltation of wetlands due to agricultural activity has had adverse effects on remaining wetlands.

BIRD POPULATIONS OF THE NORTHERN GREAT PLAINS

Many bird species breed in the northern Great Plains. The avifauna includes species of boreal, eastern, southern, and western origins (Johnsgard, 1979; Stewart, 1975). Despite these varied origins, the regional avifauna is distinctive. Only 29% (56 out of 190) of the breeding species in North Dakota are associated with the north-central, mixed-grass avifauna, but those species make up 80% of the total breeding bird population (Stewart, 1975).

Quantitative information about changes in populations of grassland birds exists only for the period since the mid-1960s, 75 to 100 years after most of the major habitat changes on the northern Great Plains occurred. A systematic survey (the North American Breeding Bird Survey, or BBS) was initiated in 1966 (Robbins et al., 1986). BBS results indicate that grassland-nesting birds as a group have suffered more serious population declines than other groups of birds (Peterjohn and Sauer, 1999). From 1966 to 1994, declines in the Midcontinent Region were reported for common yellowthroat (*Geothlypis trichas*), dickcissel (*Spiza americana*), grasshopper sparrow (*Ammodramus savannarum*), bobolink (*Dolichonyx oryzivorus*), Baird's sparrow (*Ammodramus bairdii*), northern harrier (*Circus cyaneus*), loggerhead shrike (*Lanius ludovicianus*), and lark sparrow (*Chondestes grammacus*). Increasing populations during that time were reported for upland sandpiper (*Bartramia longicauda*), ferruginous hawk (*Buteo regalis*), and sharp-tailed grouse (*Tympanuchus phasianellus*) (Johnson, 1996).

In addition, a statewide survey of North Dakota birds conducted in 1967 (Stewart and Kantrud, 1972) and repeated in 1992 and 1993 (Igl and Johnson, 1997) provides a useful contrast between those two time periods. Numbers of chestnut-collared longspurs (*Calcarius ornatus*), western meadowlarks (*Sturnella neglecta*), Savannah sparrows (*Passerculus sandwichensis*), and Baird's sparrows declined by 39% or more; clay-colored sparrows (*Spizella pallida*) and bobolinks declined at lesser rates (Igl and Johnson, 1997). Of the 12 species with significant decreases during that period, 11 were grassland or wetland dependent. Conversely, 24 of the 34 species that significantly increased were those associated with woody vegetation or human-made structures.

Metrics based on bird communities can provide useful indicators for monitoring ecosystems for several reasons (Browder et al., submitted):

1. Bird species are associated with particular habitats (see, e.g., Johnson and Igl, 1998, for grassland species).
2. Birds occur across a broad gradient of anthropogenic disturbance, from wilderness to urban areas.
3. Groups of bird species can be used to develop associations with cover types that relate to the level of anthropogenic disturbance (Bradford et al., 1998; Croonquist and Brooks, 1991; Szaro, 1986).
4. Most birds live only a few years, so changes in species composition and abundance can be detected in a relatively short time period.
5. Systematic bird surveys, such as the BBS, are currently conducted across the U.S. and southern Canada. These surveys can provide information for use in bird-community metrics.
6. Birds are important to a large segment of the public, which may relate better to concerns about changes in bird communities than to those of other taxa, such as plants or invertebrates.

REMOTELY SENSED FEATURES AND GRASSLAND BIRDS

We now consider how some features of the ecosystem that can be remotely sensed influence the numbers or population dynamics of grassland birds. These features are characteristics of both the grassland patch and the landscape in which the patch is embedded.

The size of a grassland patch and its surrounding landscape markedly influence the use of that patch by grassland birds. Some patches may be too small to be colonized by certain species, or birds using smaller patches may suffer more from competition, predation, or brood parasitism than do birds in larger patches. Also, smaller patches have a relatively greater proportion of their area near an edge, so edge effects can be greater in smaller patches (e.g., Johnson, 2001). Edge effects are phenomena such as avoidance, predation, competition, or brood parasitism, which operate at different levels near a habitat edge than in the interior of a habitat patch (e.g., Faaborg et al., 1993; Winter and Faaborg, 1999). Isolation from other grassland patches is a landscape feature that can affect either the use by or fate of birds in a patch.

Each of these factors, patch size, amount of edge, and isolation, can be assessed remotely and can affect (1) the occurrence or density of birds using a habitat patch; (2) reproductive success, through either predation or brood parasitism; or (3) competition with other species. The following overview of studies that examined these relationships is based on Johnson and Winter (1999) and Johnson (2001).

Patch Size Effects: Occurrence and Density of Birds

Species that are absent from or have lower densities in small habitat patches are referred to as area-sensitive species (Robbins, 1979). Species with large home ranges, such as greater prairie-chicken (*Tympanuchus cupido*), upland sandpiper, and northern harrier, are typical area-sensitive species that are rarely present in small, isolated habitat patches. But even species with territories much smaller than a habitat patch can be area sensitive.

Several grassland passerines, which typically have territories of only a few hectares, have been identified as area sensitive in different geographic areas. Among these are the sedge wren (*Cistothorus platensis*) in North Dakota and Minnesota (Johnson and Igl, 2001), but not in Illinois (Herkert, 1994), and clay-colored sparrow (Johnson and Igl, 2001). The widely distributed grasshopper sparrow (*Ammodramus savannarum*) has been shown to be area sensitive in several regions (Bollinger, 1995; Herkert, 1994; Johnson and Igl, 2001). The Baird's sparrow was suggested to be area sensitive by Johnson and Igl (2001). Also noted as area sensitive were the following:

- Henslow's sparrow (*Ammodramus henslowii*) (Bollinger, 1995; Herkert, 1994; Winter, 1998)
- Savannah sparrow (Bollinger, 1995; Herkert, 1994; Johnson and Igl, 2001; Johnson and Temple, 1990)
- Vesper sparrow (*Pooecetes gramineus*) (Bock et al., 1999)
- Bobolink (Bollinger, 1995; Herkert, 1994; Johnson and Igl, 2001)
- Eastern meadowlark (*Sturnella magna*) (in Illinois [Herkert, 1994] but not New York [Bollinger, 1995]).

These studies demonstrated that several bird species prefer grasslands of much larger areal extent than their territories.

Patch Size Effects: Reproductive Success

In a Missouri study, Burger et al. (1994) found that artificial nests survived at a higher rate in larger than in smaller prairie patches, although artificial nests may not accurately reflect what happens to real nests (Davison and Bollinger, 2000). Results from Johnson and Temple (1990) indicated an effect of patch size on nest success of Savannah sparrows. Winter (1998) showed that dickcissels reproduced more successfully in larger than in smaller prairies in Missouri.

Edge Effects: Occurrence and Density of Birds

Reduced densities of several species near edges have been reported. In Wisconsin, Wiens (1969) indicated that Savannah sparrows and grasshopper sparrows tended to avoid forest edges. Delisle and Savidge (1996) found in their Nebraska study that few grasshopper sparrow nests were located within 60 m of a habitat edge. Helzer (1996) also reported that grasshopper sparrows in Nebraska avoided woody and cornfield edges, and that Bobolinks avoided woody edges. Bock et al. (1999) indicated that Savannah sparrows, grasshopper sparrows, vesper sparrows, bobolinks, and western meadowlarks avoided edges associated with suburban development in Colorado.

Edge Effects: Reproductive Success

Nests of bobolinks and western meadowlarks in Minnesota that were far from a habitat edge were more successful than those that were closer (Johnson and Temple, 1990). Burger et al. (1994) reported a similar finding for artificial nests in Missouri. Winter et al. (2000) reported that predation rates on Henslow's sparrow and dickcissel nests in Missouri were higher near shrubby edges, probably due to greater activity by mid-sized carnivores near edges. In contrast, Delisle and Savidge (1996) found no edge effect for grasshopper sparrows in Nebraska.

Brown-headed cowbirds (*Molothrus ater*) are brood parasites, which lay their eggs in the nests of other species, and thereby reduce the productivity of the hosts. Johnson and Temple (1990) detected higher brood-parasitism rates of nests within 45 m of forest edge for the clay-colored sparrow and western meadowlark in Minnesota. In Missouri prairies, Winter et al. (2000) found higher parasitism rates for nests near shrubby edges.

Isolation Effects

The array of suitable and unsuitable habitats within a landscape can affect the use of individual patches and the reproductive success of birds in those patches. Effects of fragmentation may not occur until the amount of habitat in a landscape is reduced to 10 to 30% of the original (Andrén, 1994), but less vagile species that require larger territories are likely to exhibit a response earlier. Landscape effects have been demonstrated most clearly in forest systems (Donovan et al., 1997; Freemark and Collins, 1992); evidence in grasslands is ambiguous (Winter, unpub. data).

These examples clearly show that landscape features can markedly influence the well-being of grassland birds. We now turn to the question of how grassland bird communities reflect the health of the landscape.

BIRDS AS INDICATORS OF GRASSLAND HEALTH

Concerns about the effects of anthropogenic changes to various ecosystems have encouraged scientists to develop measures to assess biotic integrity over large geographic areas (reviewed by Browder et al., submitted). Such measures were initially proposed for fish assemblages in streams (Karr, 1981, 1991; Karr et al., 1986), and have been developed for birds (Bradford et al., 1998; Canterbury et al., 2000), other vertebrates, and invertebrates. None had been developed for grassland birds until recently.

Browder et al. (submitted) proposed the following measure of grassland integrity (GI):

$$GI = \%Grassland - \%cropland - \%woodland - \%odd\ area$$

This measure is intended to apply at scales ranging from about 1 ha to several square kilometers which encompasses the typical territory areas of grassland birds.

The cover types included in the measure were chosen because of their structure, functions, and influences in grassland ecosystems. Grassland obviously provides habitat structurally most similar to an intact grassland ecosystem. Cropland, woodland, and odd areas (e.g., farmsteads and rock piles) were considered detractors to grassland integrity because they represent effects of anthropogenic changes to the original structure of a grassland ecosystem and have the potential to exert the most negative influence on grassland birds. We summarized these effects for woodland, showing that nearby woody vegetation reduces the value of grassland to breeding birds. Habitats in the odd area category are likely to function similarly, because they may harbor predators and competitors, or offer perch sites for brood parasites. Cropland, with its frequent and intensive agricultural activities, may preclude successful nesting, and pesticide and herbicide applications might affect adjacent habitats.

Browder et al. (submitted) developed four regression models to predict the GI measure described above from either the presence and absence or the abundance of particular bird species. The models were developed for the prairie pothole region of North Dakota. Species presence and absence and abundance data were obtained from roadside surveys encompassing 889 points in 44 study plots in 1995 and 1996. The plots were 4050-ha hexagons, systematically distributed across North Dakota, east and north of the Missouri River, following the sampling scheme designed for the Environmental Monitoring and Assessment Program (U.S. Environmental Protection Agency, 1993).

Browder et al. (submitted) identified cover types using digital aerial photography and quantified them with geographic information systems. Each area of homogeneous habitat was assigned to one of the following classes: cropland, grassland, patch (field borders, rights-of-way, etc.), wetland, woodland, barren land (road surfaces, buildings, etc.), or odd area (farmsteads, rock piles, and other small areas that did not fit any other category). See Browder (1998) for further details about the classification. Areas of patch, wetland, and barren land were not included in the measure of GI because they were expected to exert fairly neutral influences. Two scales were used; one included all habitats within 200 m of the point from which birds were counted, and the other included all habitats within 400 m of that point.

Bird species selected for analysis included grassland species and other species that occurred commonly. Browder et al. (submitted) constructed preliminary models with data from 1 of the 2 years, and then tested the predictive ability of the models by cross-validation with data from the other year. Cross-validation tests indicated that the index consistently predicted GI. The four final models developed involved 11 species that were statistically significant ($p \leq 0.05$) in all preliminary models. Species that had positive coefficients in all models were:

1. Chestnut-collared longspur
2. Baird's sparrow
3. Grasshopper sparrow
4. Western meadowlark
5. Sedge wren
6. Savannah sparrow
7. Clay-colored sparrow (*Spizella pallida*)
8. American coot (*Fulica americana*)
9. American bittern (*Botaurus lentiginosus*)

The first seven are recognized grassland-dependent species (e.g., Johnson and Igl, 1998). The final two are wetland-obligate species. Thus, these nine species are reasonable indicators of relatively intact grassland–wetland ecosystems.

Two species entered in the models with negative coefficients were the vesper sparrow and horned lark (*Eremophila alpestris*). Both species favor bare ground (e.g., Owens and Myres, 1973) such as cropland provides, and indicate the presence of cultivated areas as opposed to intact sod.

Accordingly, Browder et al. (submitted) concluded that the presence of several grassland- and wetland-obligate bird species, combined with the absence of those species associated with cropland, correlates well with their proposed GI measure. Although the development of those indices was labor-intensive, the indices themselves can be applied relatively inexpensively to monitor grassland integrity over a large geographic area. That fortunate situation is due to the annual collection of bird information through the BBS (Robbins et al., 1986). The BBS program includes data collected annually since 1966 on numerous roadside routes (2812 in 1999) in North America. Although the density of routes varies geographically, at least one route occurs within each 1° block of latitude and longitude across the U.S. and southern Canada. Hence, indices could be applied to a specified region using only data that are collected operationally under the BBS program.

SUMMARY

Remote sensing offers opportunities to measure landscape features associated with grasslands (and wetlands), such as size of the habitat patch, amount of edge, extent of bordering by woody vegetation, and isolation from similar habitats. These features have been shown to influence the use of grassland habitats by breeding birds and the reproductive success of those birds. In turn, results of an ongoing operational program, the BBS, can provide a useful measure of the integrity of the grassland–wetland ecosystem in the northern Great Plains. These two sets of relationships offer complementary views of the connections between a landscape and some of its biotic resources. Indices such as those proposed here, in combination with BBS data, could be enlisted to provide a valuable assessment of the health of landscapes, with respect to breeding birds, over large areas and through time. Conversely, remotely sensed data can be applied to determine the potential of grassland habitats in a landscape to support viable populations of breeding birds.

ACKNOWLEDGMENTS

We are grateful to Robert R. Cox, Jr. and Diane L. Larson for comments on this chapter and to our colleagues who have contributed to our understanding of the issues discussed here.

REFERENCES

Andrén, H., Effects of habitat fragmentation on birds and mammals in landscapes with different proportions of suitable habitat: a review, *Oikos,* 71, 355–366, 1994.

Bock, C.E., Bock, J.H., and Bennett, B.C., Songbird abundance in grasslands at a suburban interface on the Colorado high plains, *Stud. Avian Biol.*, 19, 131–136, 1999.

Bollinger, E.K., Successional changes and habitat selection in hayfield bird communities, *Auk*, 112, 720–730, 1995.

Bradford, D.F., Franson, S.E., Neale, A.C., Heggem, D.T., Miller, G.R., and Canterbury, G.E., Bird species assemblages as indicators of biological integrity in Great Basin rangeland, *Environ. Monitoring Assess.*, 49, 1–22, 1998.

Bragg, T.B. and Steuter, A.A., Mixed prairie of the North American Great Plains, *Trans. North Am. Wildl. Nat. Resourc. Conf.*, 60, 335–348, 1995.

Browder, S.F., Assemblages of Grassland Birds as Indicators of Environmental Condition, M.S. thesis, University of Montana, Missoula, MT, 1998.

Browder, S.F., Johnson, D.H., and Ball, I.J., Assemblages of breeding birds as indicators of grassland condition, submitted.

Burger, L.D., Burger, L.W., Jr., and Faaborg, J. Effects of prairie fragmentation on predation on artificial nests, *J. Wildl. Manage.*, 58, 249–254, 1994.

Canterbury, G.E., Martin, T.E., Petit, D.R., Petit, L.J., and Bradford, D.F., Bird communities and habitat as ecological indicators of forest condition in regional monitoring, *Conserv. Biol.*, 14, 544–558, 2000.

Croonquist, M.J. and Brooks, R.P., Use of avian and mammalian guilds as indicators of cumulative impacts in riparian-wetland areas, *Environ. Manage.*, 15, 701–714, 1991.

Dahl, T.E., Wetlands Losses in the United States 1780s to 1980s, U.S. Fish and Wildlife Service, Washington, D.C., 1990.

Davison, W.B. and Bollinger, E., Predation rates on real and artificial nests of grassland birds, *Auk*, 117, 147–153, 2000.

Delisle, J.M. and Savidge, J.A., Reproductive success of grasshopper sparrows in relation to edge, *Prairie Nat.*, 28, 107–113, 1996.

Donovan, T.M., Jones, P.W., Annand, E.M., and Thompson III, F.R., Variation in local-scale edge effects: mechanisms and landscape context, *Ecology*, 78, 2064–2075, 1997.

England, R.E. and DeVos, A., Influence of animals on pristine conditions on the Canadian grasslands, *J. Range Manage.*, 22, 87–94, 1969.

Faaborg, J., Brittingham, M., Donovan, T., and Blake, J., Habitat fragmentation in the temperate zone: a perspective for managers, in Status and Management of Neotropical Migratory Birds, General Technical Report RM-229, Finch, D.M. and Stangel, P.W., Eds., U.S. Department of Agriculture Forest Service, Fort Collins, CO, 1993, pp. 331–338.

Freemark, K. and Collins, B., Landscape ecology of birds breeding in temperate forest fragments, in *Ecology and Conservation of Neotropical Migrant Landbirds*, Hagan, J.M., III and Johnston, D.W., Eds., Smithsonian Institution Press, Washington, D.C., 1992, pp. 443–454.

Helzer, C.J., The Effects of Wet Meadow Fragmentation on Grassland Birds, M.S. thesis, University of Nebraska, Lincoln, 1996.

Herkert, J.R., The effects of habitat fragmentation on midwestern grassland bird communities, *Ecol. Appl.*, 4, 461–471, 1994.

Igl, L.D. and Johnson, D.H., Changes in breeding bird populations in North Dakota: 1967 to 1992–93, *Auk*, 114, 74–92, 1997.

Igl, L.D. and Johnson, D.H., Wetland birds in the northern Great Plains, in Status and Trends of The Nation's Biological Resources, Mac, M.J. et al., Eds., U.S. Department of the Interior, U.S. Geological Survey, Washington, D.C., 1998, pp. 454–455.

Johnsgard, P.A., *Birds of the Great Plains*, University of Nebraska Press, Lincoln, 1979.

Johnson, D.H., Management of northern prairies and wetlands for the conservation of neotropical migratory birds, in Management of Midwestern Landscapes for the Conservation of Neotropical Migratory Birds, General Technical Report NC-187, Thompson, F.R., III, Ed., U.S. Department of Agriculture Forest Service, North Central Forest Experiment Station, St. Paul, MN, 1996, pp. 53–67.

Johnson, D.H., Habitat fragmentation effects on birds in grasslands and wetlands: a critique of our knowledge, *Great Plains Res.*, 11:211–231, 2001.

Johnson, D.H. and Igl, L.D., Coordinators, Effects of Management Practices on Grassland Birds, http://www.npwrc.usgs.gov/resource/literatr/grasbird/grasbird.htm, Northern Prairie Wildlife Research Center, Jamestown, ND, 1998.

Johnson, D.H. and Igl, L.D., Area requirements of grassland birds: a regional perspective, *Auk*, 118, 24–34, 2001.

Johnson, D.H. and Winter, M. Reserve design for grasslands: considerations for bird populations, *Proceedings of the Tenth George Wright Society Biennial Conference,* Hancock, MI, 1999, pp. 391–396.

Johnson, R.G. and Temple, S.A., Nest predation and brood parasitism of tallgrass prairie birds, *J. Wildl. Manage.*, 54, 106–111, 1990.

Karr, J.R., Assessment of biotic integrity using fish communities, *Fisheries*, 6, 21–27, 1981.

Karr, J.R., Biological integrity: a long-neglected aspect of water resource management, *Ecol. Appli.*, 1, 66–84, 1991.

Karr, J.R., Fausch, K.D., Angermeier, P.L., Yant, P.R., and Schlosser, I.J., Assessing biological integrity in running waters: a method and its rationale, *Ill. Nat. Hist. Surv. Spec. Publ.*, 1986, p. 5.

McNicholl, M.K., Ecological and human influences on Canadian populations of grasslands, in *Ecology and Conservation of Grassland Birds*, Technical Publication 7, Goriup, P.D., Ed., International Council on Bird Preservation, Norfolk, England, 1988, pp. 1–25.

Noss, R.F., LaRoe, E.T., III, and Scott, J.M., Endangered ecosystems of the United States: a preliminary assessment of loss and degradation, Biological Report 28, National Biological Service, Washington, D.C., 1995.

Owens, R.A. and Myres, M.T., Effects of agriculture upon populations of native passerine birds of an Alberta fescue grassland, *Can. J. Zoo.*, 51, 697–713, 1973.

Peterjohn, B.G. and Sauer, J.R., Population status of North American grassland birds from the North American Breeding Bird Survey, 1966–1996, *Stud. Avian Biol.*, 19, 27–44, 1999.

Risser, P.G., Birney, E.C., Blocker, H.D., May, S.W., Parton, W.J., and Wiens, J.A., *The True Prairie Ecosystem*, U.S./International Biological Program Synthesis Series 16, Hutchinson Ross, Stroudsburg, PA, 1978.

Robbins, C.S., Effect of forest fragmentation on bird populations, in Management of North Central and Northeastern Forests for Nongame Birds, General Technical Report NC-51, DeGraaf, R.M. and Evans, K.E., Eds., U.S. Department of Agriculture Forest Service, St. Paul, MN, 1979, pp. 198–212.

Robbins, C.S., Bystrak, D., and Geissler, P.H., The breeding bird survey: its first fifteen years, 1965–1979, Resource Publication 157, U.S. Fish and Wildlife Service, Washington, D.C., 1986.

Samson, F. and Knopf, F. Prairie conservation in North America, *BioScience*, 44, 418–421, 1994.

Stewart, R.E., *Breeding Birds of North Dakota*, Tri-College Center for Environmental Studies, Fargo, ND, 1975.

Stewart, R.E. and Kantrud, H.A., Classification of natural ponds and lakes in the glaciated prairie region, Resource Publication 92, U.S. Fish and Wildlife Service, Washington, D.C., 1971.

Stewart, R.E. and Kantrud, H.A., Population estimates of breeding birds in North Dakota, *Auk*, 89, 766–788, 1972.

Szaro, R.C., Guild management: an evaluation of avian guilds as a predictive tool, *Environ. Manage.*, 10, 681–688, 1986.

U.S. Environmental Protection Agency, Draft Program Guide: Environmental Monitoring and Assessment Program, Office of Research and Development, Cincinnati, OH, 1993.

van der Valk A., Ed., *Northern Prairie Wetlands*, Iowa State Press, Ames, IA, 1989.

Wiens, J.A., An approach to the study of ecological relationships among grassland birds, *Ornithol. Monogr.*, 8, 1969.

Winter, M., Effect of Habitat Fragmentation on Grassland-Nesting Birds in Southwestern Missouri, Ph.D. dissertation, University of Missouri, Columbia, 1998.

Winter, M. and Faaborg, J., Patterns of area sensitivity in grassland-nesting birds, *Conserv. Biol.*, 13, 1424–1436, 1999.

Winter, M., Johnson, D.H., and Faaborg, J., Evidence for edge effects on multiple levels: artificial nests, natural nests, and distribution of nest predators in Missouri tallgrass prairie fragments, *Condor*, 102, 256–266, 2000.

CHAPTER 64

An Ecosystem Approach to Human Health

Lada V. Kochtcheeva and Ashbindu Singh

INTRODUCTION

The health and well-being of humans are intricately linked to the natural environment, which provides both sustenance and hazards. Not all risks and threats to the health of the population arising from the degradation of the environment are the result of human activities. Many of the hazards are an intrinsic part of ecosystems themselves. The challenge lies in maintaining people's health while simultaneously improving the health of the ecosystem as a whole.

The ecosystem health approach may be defined as a comprehensive approach to the diagnostic, preemptive, and predictive aspects of ecosystem management, and to the understanding of relationships between ecosystem health and human health. It seeks to understand and optimize the intrinsic capacity of an ecosystem for self-renewal while meeting reasonable human goals. This approach encompasses the role of societal values in shaping our conception of health at human and ecosystem levels.

Health problems resulting from ecosystem degradation vary dramatically from region to region, reflecting geography, climate, and a region's level of economic growth and policy preferences (World Resources Institute, 1998). The total impact of the transformed environment causes considerable harm to natural life-support systems and a threat to the sustainability of human health (Rapport et al., 1998).

Human health problems connected to environmental causes can be understood as coming from (1) lack of development, defined as inability to cope with natural hazards and/or lack of access to essential environmental resources, and (2) unsustainable development, which causes ecosystem degradation (World Resources Institute, 1998). Environmental changes, societal advance, as well as characterization of both biological and chemical environmental hazards are, thus, critical items to be analyzed.

The goal of this study is to establish whether there are strong and direct links between degradation of ecosystems and human health. Therefore, the main objectives are (1) to conceptualize issues of human health and ecosystem health, (2) to review key emerging and reemerging threats to human health on global, regional, and local levels due to ecosystem degradation, and (3) to explore causal links between ecosystem degradation and human health.

METHODOLOGICAL APPROACH

The basic methodology used in this research is a literature search and a critical analysis of the impacts of the transformed ecosystem on the health of population. Environmental conditions that foster the transmission or spread of diseases and exposure to harmful chemicals and hazardous physical conditions are reviewed. By compiling examples and grouping them in accordance with specific regional natural conditions and ecological characteristics, this chapter intends to establish a basis for a global data collection, as well as to communicate the importance of the interrelationships between human health and the state of the ecosystem.

ISSUES RELATED TO ENVIRONMENTAL CHANGE AND HUMAN HEALTH: DEFINING THE PROBLEM

The World Health Organization (WHO) characterizes health as "a state of complete physical, mental, and social well-being and not merely the absence of disease or infirmity" (Purdom, 1980). Major trends of human health in the world today include the following (World Health Organization, 1996):

- Increase in life expectancy
- Decline in infant and child mortality in most developing countries
- Reduction in incidence of certain vaccine-preventable diseases
- Increased incidence of chronic noninfectious diseases and the spread of HIV/AIDS

Still, avoidable illnesses and premature deaths are occurring in large numbers in many regions of the world, and environmental factors contribute to these deaths (National Environmental Health Association, 1998).

Environmental health refers to conditions and characteristics of the environment, which affect the quality of population health. Ecosystem ills are increasingly at the root of suffering within the human community (Rapport et al., 1998). The decline of human health due to the degradation of the ecosystem may be described as an "illness resulting from disrupted internal balances due to external stresses" (Odum, 1995). The consequences of ecosystem collapse include human dimensions such as biological, physical, social, and economic breakdown (Rapport et al., 1998). Discovering original and improved ways to assess ill health and dysfunction in the ecosystem, which represents the basic functional unit of the natural environment, is the emerging goal of the environmental–health interface. Because human activity sometimes results in unpredictable outcomes, a significant element of health is flexibility or adaptability in the face of unexpected transformations and uncertainty.

The status of human health is a reflection of the complex interactions between the internal biological system and external environmental system. However, certain population groups, because of their lifestyle, occupation, location, or consumption patterns, are differentially vulnerable to specific health risks and threats. This, in combination with differential hazard exposure, may put a group at increased risk (Tata Energy Research Institute, 1998). Health effects of one particular change in the environment should be assessed within the context of other coexisting environmental effects and occurrences, such as rapid urbanization, population density, increasing mobility, increasing movement of produce, resources exhaustion, desertification, and pollution.

During the past 20 years, about 30 new diseases (e.g., Legionella, HIV/AIDS, Ebola, Hantavirus pulmonary syndrome, a new strain of cholera, and a host of antibiotic-resistant pathogens) have emerged, possibly as a result of environmental change (Tata Energy Research Institute, 1998). Emerging diseases can be defined as infections that have newly appeared in the population or have existed, but are rapidly increasing in incidence or geographic range. Environmental changes have contributed in one way or another to the appearance of many such diseases. Human activity, resulting in ecosystem degradation, and human behavioral changes also favor the spread of disease (Tata Energy Research Institute, 1998).

GENERAL CONCEPTS OF THE SUSTAINABILITY OF HEALTH
IN THE CHANGING ENVIRONMENT

Sustainability of human health is a priority when the concept of global environmental change is applied to health issues. Indices of the sustainability of health status may be focused on the integrity and stability of the global environment's ecological systems that maintain the life and health of the population. These indices may not directly measure human biology but may assess the degree to which human biophysical needs are being satisfied by the sustainable use of ecosystem services. Possible indicators may include bio-indices predictive of human disease risk, such as vegetation cover and groundwater levels, in relation to infectious disease vectors, or the degree of balance between population size and available resources (McMichael, 1997).

Ecological approach does not necessarily directly link individual exposure to resulting events or to each other, and is subject to multiple confounding variables. In ecological systems everything is connected directly or indirectly to everything else. An approach that takes into consideration medical and individual data demonstrates the causes of illness cases approach, while an approach that takes into account ecological, ecosystemic, and population variables shows the causes of incidence (Ehrenfeld, 1995).

Interdependence between the state of environment and the status of human health is complex and multidisciplinary in nature, and an often wrongly defined issue with uncertain solutions (Schirnding, 1997). Development, environment, and human health concerns require an integrated approach, where development is a major cause of the environmental change; the environment is a mediator; and health, insuring the ability for the development, is an indicator of these changes. The sustainability of human health is a highly important criterion of successful social and economic policy (Figure 64.1).

The key research issue today is an exploration of the fundamental infrastructural significance of the biosphere's natural systems, as related to human health. Potential threats and risks from global environmental change differ from well-recognized, locally found, and direct-acting harmful agents. Disruption of natural ecological systems endangers the health of the population by both direct and indirect ways, and immediate and delayed mechanisms (McMichael, 1997).

CONCEPT OF THE ECOSYSTEM HEALTH

Health of the Earth's Ecosystems

An ecosystem is a functioning unit of nature that combines biotic communities and the abiotic environments with which they interact (LaRoe et al., 1995). The three main features characterizing a healthy ecosystem are vigor, resilience, and organization. A healthy ecosystem is a sustainable component of the biosphere that has the ability to maintain its structure (organization) and function (vigor) in time and in the face of external stresses (resilience) (Mageau et al., 1995). Human community benefits from natural healthy ecosystems, which serve as sources of food, shelter, the capacity to assimilate and recycle wastes, and clean air and water. Ecosystem health is determined not only by the biophysical states of its components, but also by societal values. Still, there is difficulty in characterizing a general index, given the colossal qualitative and quantitative differences among ecosystems. For instance, there are ecosystems in which rich biodiversity is a sure sign of health, and others in which it is a sign of disturbance (Ehrenfeld, 1995).

DEVELOPMENT ⟺ ENVIRONMENT

HEALTH

Figure 64.1 Connection between development, environment, and health.

The concept of the ecosystem distress syndrome was introduced by Rapport and Odum (1985). This syndrome includes such ecosystem disturbances as changes in biotic composition and energy flow, loss of biodiversity and nutrient capital and, in general, the loss of the balance among ecosystem components. Acidified lakes, highly polluted coastal marine systems and estuaries, overgrazed grasslands, loss of valuable fisheries, desertification as a result of overgrazing, and invasions of exotic species are examples of ecosystem pathology that influence human population health and may be readily detected, even by the general public, in all countries (Rapport et al., 1998).

Changes in the distribution and ecological activity of organisms, often resulting from environmental modifications, may give early evidence of environment-related shifts in human health risks. In many cases, the appearance of disease is a symptom of ecosystem dysfunction. Disease, therefore, can be considered a first-recognized impact of an environmental stress (World Health Organization, 1996).

Present State of the Environments and Causes of Ecosystem Degradation

One of the significant aspects of human activity is humanity's total impact on ecosystems, in contrast to the particular contributions that arise from specific pollutants. Recognition of this overall effect is necessary to identify the real problem of ecosystem degradation. From a global perspective, the environment has continued to degrade during the past decade (United Nations Environmental Programme, 1997). According to the United Nations Environmental Programme (UNEP) assessment, different regions in the world experience various environmental concerns. Humans have modified approximately 50% of the land surface, account for more than 20% of the atmospheric carbon dioxide concentration, utilize over 50% of the accessible surface fresh water, and are responsible for about 60% of all nitrogen fixation; the list of such impacts continues (Rapport et al., 1998). The total impact of human activity represents a significant danger to functioning of the natural life-support systems, and to the sustainability of population health. This combination of environmental changes that creates the conditions favorable to disease occurrence is of a significant concern.

Human activities are directly responsible for creating agroecosystems and cultural land-scapes at the expense of many natural communities and the reduction in ecosystem services. Serious loss of forest quality and old-growth habitat in many temperate and boreal forests, due to pollution and other injurious agents, and tropical deforestation with current rates averaging about 0.7% per year are problems of significant concern. Desertification and drought are problems of a global dimension that affect more than 900 million people in 100 countries, some of them among the least developed in the world. Approximately 25% of the Earth's land area is being affected by land degradation. Desertification is occurring on 30% of irrigated areas, 47% of rain-fed agricultural lands, and 73% of rangelands. Hydrological and ecological functions of over one half of all wetlands have been altered due to encroachment. Global freshwater biodiversity is declining significantly. Today, about one third of the world's population is living under moderate to severe water stress, most notably in Middle Asia and North Africa. Coastal waters are being contaminated by land-based sources, particularly by municipal wastes, that cause eutrophication. Many fishery resources are classified as overexploited. Stratospheric ozone has decreased since 1979 by about 5.4% at northern midlatitudes in winter and spring, and about 2.8% in summer and fall. The spatial and temporal patterns of precipitation are changing (Watson et al., 1998). These diverse changes are beginning to have adverse consequences for the human population.

There are some fundamental mechanisms and forms of ecosystem degradation that affect human health. According to Karr (1997), there are three major, multidimensional mechanisms of environmental and human systems alteration:

1. Indirect depletion of ecological systems (soil degradation, water supplies degradation, bio-geochemical cycle alterations, climate changes, ozone layer depletion, and water, air, and soil pollution)
2. Direct depletion of nonhuman living systems (loss of biodiversity, renewable resources exhaustion, pest outbreaks, spread of alien species)
3. Direct depletion of human systems (epidemics, emerging and reemerging diseases, reduced quality of life, failure to thrive in infants and children)

Environmental changes and ecosystem degradation in particular are the result of many different occurrences in natural and human-made systems. Basic causes of the ecosystem degradation can be divided into two major categories: natural and human induced. Natural induced changes in ecosystems include, but are not limited to:

1. Changes in climate (e.g., ocean functioning, cosmic radiation). According to the Intergovernmental Panel on Climate Change (IPCC), the mean surface temperature of the Earth may increase by approximately 1 to 3.5°C in the next hundred years. Sea level rise due to the climate change may lead to increased erosion in coastal zones and loss of natural protective features, such as dunes and mangroves. Potential health impacts are considered to be cumulative and interact synergistically (World Health Organization, 1996). Changes in climatic conditions would enable mosquitoes and other disease-carrying insects to survive and breed at more northern latitudes and higher altitudes. The distribution of species in an ecosystem may vary due to such changes.
2. Natural disasters (e.g., floods, cyclones, droughts, volcanic eruptions, and earthquakes). Natural disasters may have devastating consequences on both natural and human-managed ecosystems. Severe rains and flooding may cause inundation of a floodplain and lowlands, as well as impair top soil layers and wash out nutrients and microelements. Volcanic eruptions can lead to a collapse of an ecosystem by polluting the air and covering the land surface with lava and ash, destroying vegetative cover. Earthquakes may lead to land degradation and droughts may lead to the loss of biodiversity and species migration.

Human-made causes of changes in ecosystems include, but may be not limited to:

1. Development and intensification of agriculture (World Resources Institute, 1998). The direct result of agricultural practices is the conversion of forest and grassland ecosystem into agroecosystems, which are poorer in biodiversity and consequently less stable and resistant to other interventions. Other effects include soil and water contamination with chemicals and pesticides, land degradation, and salinization.
2. Industrialization, increasing energy use, and urbanization (World Resources Institute, 1998). Industrial development and rising energy use lead to direct changes in, and very often destroys, the ecosystem by simply occupying the space and converting natural environments into industrial sites and urban areas. The results of industrialization include habitat fragmentation and loss of biodiversity; alteration and destruction of vegetative cover; removal and disproportional distribution of species; air, water (fish kills and eutrophication), and soil degradation; pollution; and contribution to the climate change (greenhouse gases). Stratospheric ozone depletion is also considered to be the result of industrial development (World Resources Institute, 1998).
3. Other activities (e.g., construction, forestry, hunting, fishing, recreation). Such activities may lead to the loss of biodiversity, habitat fragmentation, river and stream regime alteration, resource extraction, vegetative cover destruction, disproportional distribution of species, and pollution of the environment.

IMPACTS OF THE ECOSYSTEM DEGRADATION ON HUMAN HEALTH

Human health can be a casualty of environmental degradation and change. Ecosystems that are sufficiently stable and biologically diverse tend to maintain the quality of human health.

Degraded or collapsed ecosystems, both aquatic and terrestrial environments, have a significant impact on human health.

Degradation of Aquatic Ecosystems and Human Health

Water pollution continues to degrade freshwater and marine ecosystems, which in turn causes millions of preventable deaths every year, especially among children (United Nations Environmental Programme, 1998). Water affects disease in many ways, (e.g., drinking contaminated water, contact with aquatic invertebrates, lack of water, infection through vectors). Illness due to consumption of contaminated fish and shellfish is an increasing concern. Harmful algal blooms in coastal regions cause poisoning, neurological disorders, gastroenteritis, and other diseases (United Nations Environmental Programme, 1998; HEED, 1999). Eutrophication, a process of water quality degradation caused by excessive nutrient loads, is depriving lakeside residents of good water quality in many densely populated areas of the world (United Nations Environmental Programme, 1994).

Aquatic ecosystems such as ponds and wells, which are affected by climate change, provide breeding grounds for certain parasites and disease vectors. Changes in water flow in these systems could influence the incidence of a number of diseases. Increased flooding, following changes in precipitation, may cause contamination of water supplies, which leads to greater incidence of fecal-oral contamination (World Health Organization, 1996). Natural networks of rivers, lakes, and marshes play a role in the transmission of water-related and vector-borne diseases as well. Table 64.1 demonstrates categorization of water-related diseases.

Table 64.1 Categorization of Water-Related Diseases

Category	Examples
Diseases carried by water or waterborne infections (microbial and chemical)	Typhoid; cholera; fluorosis; infectious hepatitis; leptospirosis; amebiasis
Diseases related to poor sanitation, lack of water, or water-washed infections	Scabies; trachoma; bacillary dysentery; skin and eye infections; diarrheal diseases; bacillary dysentery
Water-based diseases or infections (penetrating skin; ingested)	Schistosomiasis; guinea worm; echinococcosis
Insect vectors-related diseases (vector bite; breeding in water)	Sleeping sickness; yellow fever; malaria; trypanosomiasis

Source: Adapted from Tata Energy Research Institute (1998) and Wolman (1986).

Degradation of Terrestrial Ecosystems and Human Health Impacts

Expanding agriculture, forest clearing, mining activity, dam building, irrigation schemes, and poorly planned urban development and activities, which change the structure and functioning of terrestrial as well as aquatic ecosystems, pose a number of health concerns. These concerns include increased exposure to toxic substances (e.g., pesticides) and increased exposure to infectious agents. As a result, mosquitoes would have new breeding grounds, causing more people to come into contact with them (Tata Energy Research Institute, 1998).

Loss of species and ecosystem diversity erodes genetic diversity. In addition, many of the 20,000 plant species used as traditional medicines around the world are under threat of overexploitation (United National Environmental Programme, 1993). The genetic diversity of species is not only one of the keys to successful agriculture, which prevents malnutrition-related health problems, but also a sphere of promise for medical research. According to an article in *The Wall Street Journal*, deforestation and hunting bring people into contact with animals, which can lead to transmission of various diseases, including the simian virus closely related to HIV and found in African chimpanzees (Waldholz, 1999). At the same time, animals' disease-fighting immune mechanisms are able to control the virus. By studying these mechanisms, scientists may obtain important clues about how to prevent and treat the devastating infection in humans.

Ebola hemorrhagic fever also may serve as an example of a possibly animal-borne disease; however, the exact origin, location, and natural habitat of the virus remain unknown (Centers for Disease Control and Prevention, 1999). An association between forest fragmentation and Ebola cases is still unclear.

Where forest cover is interrupted, nutrients are released into the hydrological cycle. A net nutrient outflow pollutes local river systems and greatly reduces the productive capacity of the cleared land (World Conservation Monitoring Center, 1992). According to the National Institute of Environmental Health Sciences (NIEHS), deforestation changes natural habitats and creates several human health concerns, such as an increase in infectious diseases and depression. Reports from Southeast Asia say that deliberately set fires have increased pollutants in the air, causing respiratory problems and contributing to global climate change. Forest fires have profound impacts on the physical environment including: land cover, biodiversity, and climate change and forest ecosystems. Health impacts are often serious. Estimates suggest that 20 million people are in danger of respiratory problems from fires in southeast Asia. Large fires also occurred in the Caribbean region, Africa, and in some parts of the former Soviet Union (Levine et al., 1999). Deforestation may continue at high rates until more of us see the value of forests for biodiversity, potential medicines, improved environmental quality, and climate mitigation (National Institute of Environmental Health Sciences, 1999).

Consequences of Climate Change in Human Health

Almost all scientists who study climatic dynamics accept the notion that the increase in, and spread of, many diseases is likely to be the single most dangerous threat that climate change poses to human health (Kingsnorth, 1999).

Temperature and weather changes may have direct and indirect health outcomes. Exposure to thermal extremes and altered frequency of weather events may result in altered rates of heat- and cold-related illnesses, psychological disorders, and death. Results indirectly attributable to the effects on range and activity of vectors, altered food productivity, sea level rise, impacts of air pollution and others may include changed incidence of diarrheal and vector-borne diseases, malnutrition, impairment of child growth and development, asthma, allergic and respiratory disorders, and deaths (World Health Organization, 1996).

Global climate changes may create favorable conditions for disease-carrying insects to proliferate at more northern latitudes and higher altitudes. Malaria, dengue, yellow fever, and some other types of viral encephalitis are likely to increase. According to the World Health Organization (1996), approximately one half of the world population is at risk for insect-borne diseases (Tata Energy Research Institute, 1998). Malaria is an important example because it at present accounts for approximately 350 million cases annually, including about 2 million deaths (McMichael, 1997). The combination of climatic factors contributed to appearances of several rodent-borne diseases, such as leptospirosis and viral hemorrhagic fevers (Epstein, 1997). The impacts of extreme weather events on human health may include death, injuries, stress-related disorders, and other adverse health effects that are associated with social disruption and forced migration. Increased frequency of extreme weather events may also result in increased incidences of infectious diseases, mainly due to poor sanitation and lack of clean, fresh water. For instance, floods often cause increases in incidence of diseases associated with fecal-oral transmission (World Health Organization, 1996).

Significant stratospheric ozone losses have occurred, mainly at middle and high latitudes. Ozone depletion is more pronounced in winter and spring than in summer. An increase in ultraviolet exposure may increase the incidence of skin cancers, the severity of sunburn, skin aging, and eye diseases (e.g., cataracts). The negative effects of ozone layer depletion may also include some suppression of immune functioning, increasing the susceptibility to infectious diseases (McMichael, 1997).

Consequences of Ecosystem Degradation and Human Health at Regional and Local Levels

The extent of environmental threats to human health is distributed unevenly between developed and developing countries. The effects of a degraded ecosystem or transformed environment are exacerbated by inadequate sanitation and nutrition, cultural peculiarities, and demographic features. In areas where environmental threats coincide with poor social and economic conditions, risks and threats to population health increase. In general, countries in Africa and parts of Asia face the highest health threats from the degradation and collapse of the ecosystem.

Table 64.2 demonstrates basic information on the linkage between ecosystems, impact on human health, regional and local examples, and consequences and possible occurrences of the threat and risk to human health.

LINKING DEGRADATION OF THE ECOSYSTEM AND HUMAN HEALTH

The Quality and Quantity of Information

An analysis of the literature on environmental health shows that there are many attempts to find a link between human health and the changing environmental conditions. One of the first and most significant contributions was made by the legacy of Rachel Carson's *Silent Spring*. This book brought a new awareness to the public that nature was vulnerable to human intervention. The threats to human health that Carson had outlined — contamination of the food chain, cancer, and genetic damage — appeared to be too important to ignore.

An important contribution to the literature on environmental health was made by a joint publication of the World Resources Institute (WRI), the UNEP, the United Nations Development Programme (UNDP), and the World Bank (World Resources Institute, 1998). This report clearly explains the multiple causes of environmental change and demonstrated humans' dependence on the natural environment, taking into consideration social, economic, and cultural variables. A number of UNEP reports and publications addressing the environmental health issues provide an excellent source of information on ecosystem degradation due to unsustainable development, emphasizing the importance of maintaining the health of the ecosystem in order to sustain and improve the status of human health (United Nations Environmental Programme, 1993, 1994, 1997).

Research articles and reports on different tropical emerging and reemerging diseases, such as the Rift Valley fever in Kenya, attempt to link the changes between the environment and health, showing the naturally occurring causes of the disease, such as seasonal climate conditions changes (Linthicum et al., 1999).

Many of the parameters associated with environmental change and patterns of disease can be sensed remotely by instruments onboard aircraft and satellites, and modeled spatially with specialized computer software. Remote sensing and geographic information system (GIS) technologies can be used to describe local- and landscape-level features that influence the patterns and prevalence of disease and then model their occurrence in space and time (CHAART, 1999). However, the capabilities of remote sensing technology have not been disseminated to the health investigators and agencies that could be using them. It goes without saying that surveillance, modeling, and early warnings are the main goals of a human health system. However, the research needed to create such a system has not been completely done. Good surveillance means that the factors, such as weather and vegetative cover, places, and other geospatial data to be monitored are well known and identified. In addition, methods of relating these components to demographic information to determine the population risk has to be established. The simplistic logic connecting weather patterns, vegetation, and disease incidence usually lack scientific rigor (Guptill, 1998).

Table 64.2 Ecosystem Degradation and Human Health Consequences at Regional and Local Levels

Ecosystem	Driving Forces and Changing Ecological Patterns	Influence on Human Health and Additional Possible Consequences	Regional and Local Examples
Atmospheric	*Climate Change* Extremes in temperature (excessive heat); increased rainfall; frequency and severity of droughts; changes in temperature and humidity; extension of drought areas; deficiency in micronutrients; variations in temperature, precipitation, and humidity	Malaria; meningococcal disease epidemics; starvation; malnutrition-related diseases; dengue; arboviral infections; excess in the rate of heat-related deaths in summer; physiological disturbance; Hantavirus; respiratory diseases are the fourth leading cause of death Further global mean temperature increase may create ecological conditions conductive to malaria in 60% of the world's land area, compared with current 45%; similar outcomes are possible with schistosomiasis; health consequences may include diseases from a breakdown in sanitation	Rwanda; Ethiopia; East African highlands; Madagascar; Benin; Burkina Faso; Chad; Mali; Niger; Nigeria; Northeast Africa; Australia; Oceania; China; U.S.; Mexico; Argentina
	Pollution Pollution from power plants, metallurgy, the coal industry, the chemical industry, and vehicular emissions; burning of bio- and fossil fuels	Respiratory diseases; eye irritation	A large number of developing and developed countries
Aquatic Marine	*Biological Contamination, Pollution* Oil contamination; water contamination with wastewater; further deterioration of marine ecosystems from a severe imbalance due to severe navigation; sewage discharges; harmful (toxic and nontoxic) algal blooms from the rapid reproduction and localized dominance of phytoplankton; shellfish poisoning; wildlife mortalities; sunlight penetration prevention; oxygen shortages; reservoirs for bacteria	Gastroenteritis; eye and skin infections; decrease in life expectancy; typhoid; malaria; diphtheria; poisonings; diarrhea; dehydration; headaches; confusion; dizziness; memory loss; weakness; gastroenteritis; bacterial infections; swimming-related illnesses; neurological diseases; death; cholera	South Africa; Black and Azov Seas; Caspian Sea; former Soviet Union; United Kingdom; France; Southern and Gulf States; a large number of countries in Latin America
Fresh	Pollution; dam construction; degradation; hydrogeological cycle changes; inundating of lands; contaminated river systems; fecal water pollution; fallen water tables; chemical contamination	Epidemic of schistosomiasis; infection rates in the Diama region went from zero before the dams to more than 90% of the population; fecal infections; intestinal diseases; progressive and irreversible kidney damage; diarrhea; hepatitis; cholera outbreaks; infectious diseases; intestinal parasites	Africa (Senegal River, Manantali and Diama Dams, South Africa) Central Asia; former Soviet Union; Bangladesh; India; Palestine; Israel; China

(continued)

Table 64.2 (continued) Ecosystem Degradation and Human Health Consequences at Regional and Local Levels

Ecosystem	Driving Forces and Changing Ecological Patterns	Influence on Human Health and Additional Possible Consequences	Regional and Local Examples
Terrestrial Vegetation	*Deforestation, Natural Disaster, Intensified Agriculture* Soil destabilization; clearing and intervening in the forests; changes in local hydrological cycles; firewood shortage; land degradation; forest and rangeland fires due to high temperature, strong winds, and low soil moisture content; new breeding grounds for insects; agroecosystem development (inundating of lands due to rice growing); conversion of forest into cotton and sugar-cane culture and cattle pasture	Malaria; trauma; allergic reactions; aches; cuts; infections; respiratory diseases; cancer; yellow fever; Ebola epidemics; burns; smoke inhalation; hemorrhagic fever; displacement of population	South Africa, Kenya, Cote d'Ivoire, and some other African countries; Nepal; China; India; Australia; U.S.; Peru; Bolivia; Brazil; Honduras; Central America; Venezuela
Land degradation	*Changed Agricultural Patterns, Desertification, Soil Contamination* Destruction of the agricultural ecosystem; increased availability of breeding sites for insects; fertilizers and organic manure contamination; transformation of agricultural ecosystems; increase in nitrate level leading to the high levels of nutrients in rivers; mining activities leading to land degradation; soil qualities impairment; contamination with pesticides	Malaria; human African trypanosomiasis (sleeping sickness); rising incidence of cancer; birth defects; Lyme disease, lung diseases; possibility of carcinogenesis; infections of respiratory and digestive tracts; population migration; immune suppression	East African highlands and Madagascar; Siberian and Arctic regions; Western Europe; USA; Former Soviet Union; some countries in Latin America

Data from Chen et al. (1997), HEED (1999), Homer-Dixon and Percival (1996), National Institute of Environmental Health Sciences (1999), PAHO (1999), Tata Energy Research Institute (1998), World Health Organization (1998), World Resources Institute (1998).

For most of the potential impacts of the ecosystem degradation on human health, information upon which to build a standard health risk assessment seems to remain inadequate. Most research articles, books, and reports that attempt to connect the deterioration of environment and its impact on human health focus mainly on the social, economic, and demographic consequences of such interactions, rather than on an ecosystem approach. Another problem in connecting human diseases and ecosystem degradation is that these links are not clearly explored. Ecological and medical issues are discussed quite separately, with an emphasis on either, but no connection between the issues.

Reviewed literature shows that there is a continued trend in environmental health publications to connect physical degradation of the Earth with the quality of human health. However, links are generally established through social, economic, and legal spheres (e.g., infirmities in health care, pollution abatement measures, poor sanitation). Declining levels of the world food output is contributing to a reversal in economic progress in many countries, and chemical contamination of agricultural products is often a concern. In addition, many conclusions are built on the inference that the ties between environmental abuse and the population health remain inevitably more presumptive than proven (Feshbach and Friendly, 1992).

Air and water pollution and climate change remain the main spheres, from which the discussion of the human health problems in connection to the ecosystem degradation starts and develops. Moreover, in researching environmental causes of human health, scientists tend to consider the impact of separate substances instead of an aggregate effect.

Many sources provide numerical data on the levels of pollution, health status examples, or medical statistical data, but these sources leave readers with the opportunity to make their own conclusions. Moreover, traditional environmental health risk assessment is based on (1) identifying a discrete health threat, (2) characterizing the resulting health risk, (3) evaluating possible human exposure, and (4) concluding by estimating likely disease outcomes. Such approaches seem to be inappropriate if the health risk is connected to an ecological entity or ecosystem phenomenon, including weather events, quality and quantity of water resources, and vegetative cover (World Health Organization, 1996).

The very concept of an ecosystem is rather broad, taking into account the fact that an ecosystem is a functioning unit of the environment, and can be represented by both a small lake and a watershed. Some sources tend to consider separate parts of the ecosystems. A total reflection of the environmental impact on humans should also consider psychological and mental influences and hazards, even though their precise effects, the magnitude of their influence, or the epidemiology of mental illnesses are not well known (Purdom, 1980).

Environmental health studies do not analyze those causes of ill health that might be attributed to combinations of environmental factors interacting with one another and that are themselves components of complex systems affected by human interventions. Empirical sciences experience difficulties in dealing with the uncertainties arising from such complex systems and the predictive modeling of them. Adequate consideration of the health impacts of ecosystem degradation on human society, and of societal efforts to mitigate or adapt to such change, necessitates multidisciplinary research that draws heavily on the environmental health sciences, epidemiology, anthropology, social psychology, and disciplines (World Health Organization, 1996).

Dealing with the Problem

According to McMichael (1997), better understanding, assessment, and dealing with the problem of the environmental impact on human health requires that three major fundamentals should be accomplished.

1. Practical research into the health impacts of environmental modifications (e.g., dose–response, cause–effect relationships investigations, epidemiological studies, database creation)
2. Environmental modeling for the prediction of the population health outcomes
3. Prediction, forecasting, and estimating of possible future scenarios (e.g., climate change and its influence on human health, geographic range of vector insects)

Public health programs can help to anticipate the health impacts of ecosystem degradation and collapse. Surveillance systems could be improved or installed in sensitive geographic areas. Intersectoral collaboration should be strengthened so that public health considerations are incorporated into the development process through environmental management techniques.

It is worth mentioning that there is no lack of warnings, pronouncements, and declarations on the issue of the human health in connection to the degraded ecosystem from scientists, politicians, humanists, business society, and interested agencies (Rapport et al., 1998). Reports, publications, Internet sources, research articles, and books from UNEP, WRI, the WHO, and other organizations provide the necessary material on the subject of environmental health. The problem is how this information is presented, where the emphasis is placed, and how and to whom the information is disseminated and then used. A proper understanding of the threats from the degraded ecosystem to human health is necessary. Strengthening of collaborative agreements among agencies may help to enhance scientific quality and extend data resources (U.S. Department of Health and Human Services, 1984). Therefore, government agencies that are responsible for health assessment, policy making, regulations, and health quality assurance require an ongoing production, collection, and analysis of information about the impacts of the changed ecosystems on population health (PAHO, 1999).

CONCLUSION

A basic obstacle to the assessment of the human health status, ecosystem health, and the establishment of direct links between ecosystem degradation and human disease lies in coping with the absence of direct, strong, connective mechanisms and scientific uncertainty. For instance, the recognition of the fact that ozone layer depletion, loss of biodiversity, or pesticide accumulation affects human health is based on our understanding that it is the energy or food chain that is affected, often through indirect pathways. Intermediaries connect the change in the ecosystem and human health (Figure 64.2). For example, such environmental changes as climate change, land degradation, pesticide and fertilizer use, and aquifer depletion seriously affect agricultural production. Agricultural production is a major determinant of nutritional status and population health. Hence, human health is affected via processing or consumption of agricultural production and not directly by land degradation or aquifer depletion.

However, there are some environmental changes that directly impact the quality of human health. They include temperature increase, which causes thermal stresses or respiratory problems, and the deterioration of aquatic ecosystems that can subsequently lead to waterborne diseases, while, in general, most of the health impacts of the ecosystem degradation would be mediated by changes in other systems and processes, such as the proliferation of bacteria, distribution of vector organisms, or quality and availability of water supplies. In view of the above, it seems that the problems are relatively well understood. Significant benefits could be achieved if concerns about environmental threats to health are incorporated into development planning at the onset. Changes in health conditions are posing a demand for knowledge and are calling for new solutions in implementation of environmental health policies.

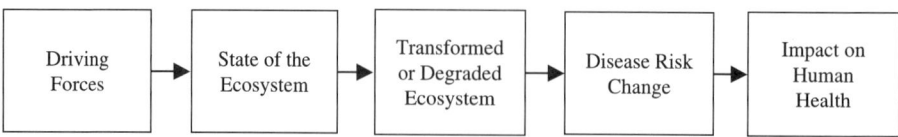

Figure 64.2 Mutually tied changes in ecosystem and human health.

REFERENCES

Center for Disease Control and Prevention, Viral hemorrhagic fevers: ebola hemorrhagic fever, CDC, 1999, http://www.cdc.gov/ncidod/diseases/virlfvr/ebolainf.htm.

CHAART, Sensor evaluation, Center for Health Applications of Aerospace, 1999, http://geo.arc.nasa.gov/esdstaff/health/sensor/sensor.html.

Chen, S.R., Lenhardt, W.C., and Alkire, F.K., Consequences of Environmental Change — Political, Economic, Social. Environmental Flash Points Workshop, Consortium for International Science Information Network, Columbia University, New York, 1997.

Enhrenfeld, D., The marriage of ecology and medicine: are they compatible? *Ecosyst. Health,* 1, 15–21, 1995

Epstein, P.R., Climate, ecology, and human health, *Consequences,* 3, 3–19, 1997.

Feshbach, M. and Friendly, A., *Ecocide in the USSR: Health and Nature under Siege,* Basic Books, Harper Collins, New York, 1992.

HEED, Global change program, marine ecosystems: emerging diseases as indicators of change, Health Ecological and Economic Dimensions, 1999, http://heed.unh.edu/heedreport/exec/exec001.html.

Homer-Dixon, T. and Percival, V., Environmental scarcity and violent conflict: briefing book, *The Project on Environment, Population, and Security,* American Association for the Advancement of Science, University College, University of Toronto, Toronto, 1996.

Karr, J.R., Bridging the gap between human and ecological health, *Ecosyst. Health,* 3, 197–199, 1997.

Kingsnorth, P., Human health on the line, *The Ecologist,* 29, 92–93, 1999.

LaRoe, E.T., Farris, G.S., Puckett, C.E., Doran, E.D., Peter, D., and Mac, M.J., Our Living Resources: A Report to the Nation on the Distribution, Abundance, and Health of U.S. Plants, Animals, and Ecosystems, U.S. Department of Interior–National Biological Service, Washington, D.C., 1995.

Levine, J.S., Bobbe, T., Ray, N., and Singh, A., Wildland Fires and the Environment: A Global Synthesis, UNEP/DEIAEW/TR.99–1, United Nations Environment Programme, Nairobi, Kenya, 1999.

Linthicum, K.J., Anyamba, A., Tucker, C.J., Compton, J., Kelley, P.W., Myers, M.F., and Peters, C.J., Southern Oscillation Index, Sea Surface Temperature, and Satellite Vegetation Index Indicators to Forecast Rift Valley Fever Epizootics/Epidemics in Kenya, NASA Goddard Space Flight Center, Greenbelt, MD, 1999.

Mageau, M.T., Costanza, R., and Ulanowicz, R.E., The development and initial testing of a quantitative assessment of ecosystem health, *Ecosyst. Health,* 1, 201–213, 1995.

McMichael, A.J., Global environmental change and human health: impact assessment, population vulnerability, and research priorities, *Ecosyst. Health,* 3, 200–210, 1997.

National Environmental Health Association, New global health report warns about health risks of environmental degradation, *J. Environ. Health,* 61, 31–33, 1998.

NIEHS, Queensland fever, National Institute of Environmental Health Services, 1999, http://www.niehs.nih.gov/external/a2z/page4.htm.

Odum, E.P., *Ecology and Our Endangered Life-Support Systems,* Sinauer Associates, Inc., Sunderland, MA, 1989.

Odum, E.P., Profile analysis and some thoughts on the development of the interface area of environmental health, *Ecosyst. Health,* 1, 41–45, 1995.

PAHO, Health in America, Scientific publication, 1, 596, PanAmerican Health Organization, 1998.

Purdom, W.P., *Environmental Health,* 2nd ed., Academic Press, New York, 1980.

Rapport, D.J., Christensen, N., Ka, J.R., and Patil, G.P., 1998. Sustainable Health of Humans and Ecosystems, Unpublished report.

Schirding, Y.E.R., Addressing health and environment concerns in sustainable development with special reference to participatory planning initiatives such as healthy cities, *Ecosyst. Health,* 3, 220–228, 1997.

Tata Energy Research Institute, Domestic Environment Associated Health Problems in Women and Children, Project Report No.97EE52, New Delhi, India, 1998.

United Nations Development Programme, Human Development Report 1998, Oxford University Press, New York, 1998.

United Nations Environmental Programme, Global Biodiversity, UNEP/GEMS Environment Library No. 11, Nairobi, Kenya, 1993.

United Nations Environmental Programme, The Pollution of Lakes and Reservoirs, UNEP/GEMS Environment Library No. 12, Nairobi, Kenya, 1994.

United Nations Environmental Programme, *Global Environment Outlook*, 1st ed., Oxford University Press, New York, 1997.

U.S. Department of Health and Human Services, Human Health and the Environment: Some Research Needs, Report of the Third Task Force for Research Planning in Environmental Health Science, U.S. Government Printing Office, Washington, D.C., 1984.

Waldholz, M., Out of Africa, Origins of AIDS emerge, *The Wall Street Journal*, Feb. 1, 1999, pp. B1, B4.

Watson, R.T., Dixon, J.A., Hamburg, S.P., Janetos, A.C., and Moss, R.H., Protecting Our Planet: Securing Our Future, United Nations Environmental Programme, NASA, and the World Bank, 1998.

Williamson, S.J., *Fundamentals of Air Pollution*, Addison-Wesley, Reading, MA, 1973.

Wolman, A., Health and water quality, in *Managing Water Resources*, Cairns, J., Jr. and Patrick, R., Eds., Praeger, New York, 1986, pp. 5–17.

World Conservation Monitoring Center, *Global Biodiversity: Status of the Earth's Living Resources*, Chapman & Hall, London, 1992.

World Health Organization, Climate Change and Human Health: An Assessment Prepared by a Task Group on Behalf of the WHO, WMO, and UNEP, McMichael A.J., Eds., WHO, Geneva, 1996.

World Resources Institute, *A Guide to the Global Environment: The Urban Environment 1996–1997*, Oxford University Press, New York, 1996.

World Resources Institute, *A Guide to the Global Environment: Environmental Change and Human Health. World Resources 1998–1999*, Oxford University Press, New York, 1998.

CHAPTER 65

Predictability of Bird Community-Based Ecological Integrity Using Landscape Measurements

Glen D. Johnson, Wayne L. Myers, Ganapati P. Patil, Timothy J. O'Connell, and Robert P. Brooks

INTRODUCTION

Assessing the ecological condition of large geographic areas is difficult at best. Nothing can replace on-the-ground intensive sampling or surveying by experienced field personnel; however, this is obviously not feasible for areas that are much larger than a site, such as at landscape and regional scales. Remotely sensed data such as land cover maps derived from satellite imagery can provide valuable covariate information. Essentially, local detail is exchanged for spatially synoptic coverage.

The challenge then becomes one of how to analyze remotely sensed data in a way that reflects ecological condition. Many different landscape measurements are available (McGarigal and Marks, 1995) and prior research has been directed toward finding sets of variables that are suitable for distinguishing among different landscape types (Riitters et al., 1995). Meanwhile, Johnson et al. (2001) have assessed the ability to categorize watershed-delineated landscapes in Pennsylvania with respect to landscape patterns.

As part of their study, several conventional landscape measurements were evaluated, along with a new multiresolution characterization of pattern, coined a conditional entropy profile. When applied to actual watershed-delineated landscapes in Pennsylvania, these profiles revealed a valuable graphical tool for presenting a multiscale characterization of landscape fragmentation in a way that is sensitive to pattern changes. Using spatial pattern measurements and land cover proportions, landscapes could be generally categorized according to degree and type of forest fragmentation; however, a more direct characterization of ecological condition was needed for evaluating the performance of these landscape measurements. For this reason, this chapter serves to provide an independent categorization of these watershed-delineated landscapes in Pennsylvania using only bird community data for deciphering categories of ecological integrity.

The ecological data source is a breeding bird atlas (Brauning and Gill, 1983–1989) that is the result of a 5-year field survey performed by trained volunteers. The presence of breeding evidence was recorded for bird species in each of approximately 5000 blocks covering the entire state. Each block is one sixth of a U.S. Geological Survey (USGS) 7.5-min quadrangle. Encountered species were assigned one of four different levels of strength of evidence for breeding. Of these, the three strongest levels were included for this current study: *possible*, *probable*, and *confirmed*. Any species

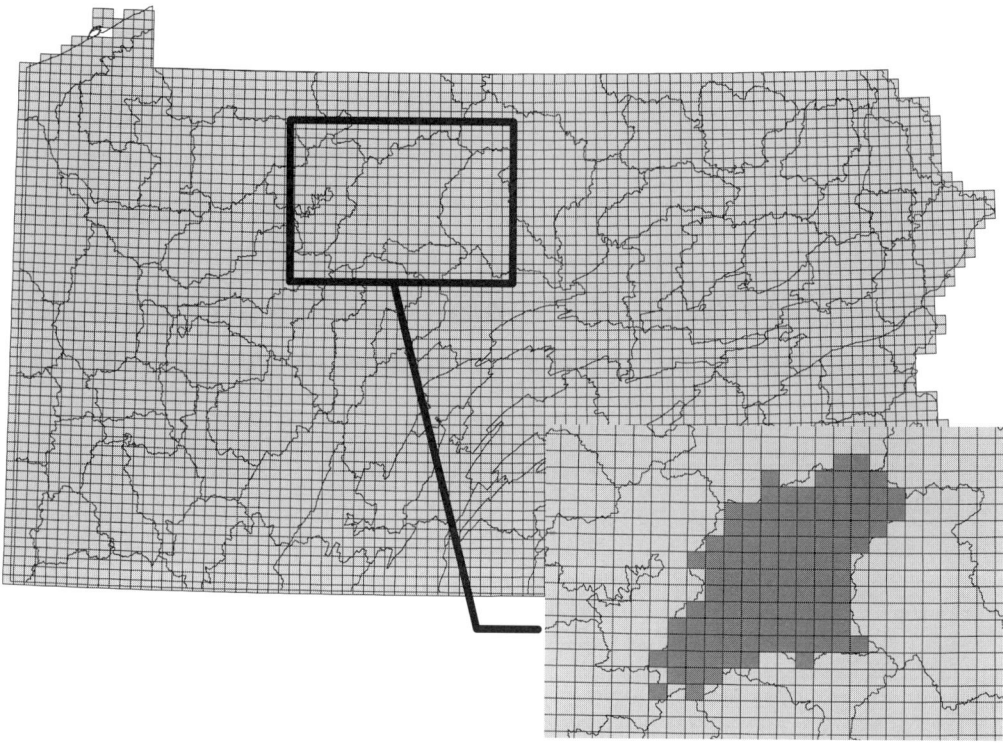

Figure 65.1 Breeding bird atlas blocks (small squares) superimposed on watershed boundaries (large, irregular polygons) in Pennsylvania. An example is shown of how blocks are selected to represent a watershed according to whether the center of a block lies within the watershed.

listed under one of these evidence categories was treated as *present*. Atlas blocks were assigned to respective watersheds by choosing those blocks whose centers were within the watershed boundary, as depicted in Figure 65.1.

Given a set of atlas blocks for each watershed, where each block contains a species list, the objective is to summarize these data in a way that rationally assesses ecological condition for the watershed as a whole. Simple measurements of species richness and diversity are not sensitive to changes in community composition, and yield little ecological information. This is especially true for geographic areas that are the size of these watersheds. However, if a species list is converted to response guilds that represent structural, compositional, and functional components of an ecosystem, then the guild composition may be used to assess ecological condition (Brooks et al., 1998). What follows are approaches to working with guilding schemes that aim to provide a meaningful ecological assessment for each watershed.

ECOLOGICAL INTEGRITY BASED ON A SONGBIRD COMMUNITY INDEX FROM THE MIDDLE ATLANTIC HIGHLANDS ASSESSMENT AREA

Background

A landscape-level indicator of ecological condition was developed by O'Connell, Jackson, and Brooks (1998a,b) for the Middle Atlantic Highlands Assessment (MAHA) area depicted in Figure 65.2. This indicator is based on the breeding songbird community composition, as obtained through representative field sampling. Their motivation was to characterize ecological condition in the same manner as an Index of Biotic Integrity (Karr, 1991; 1993; Bradford et al., 1998). Here,

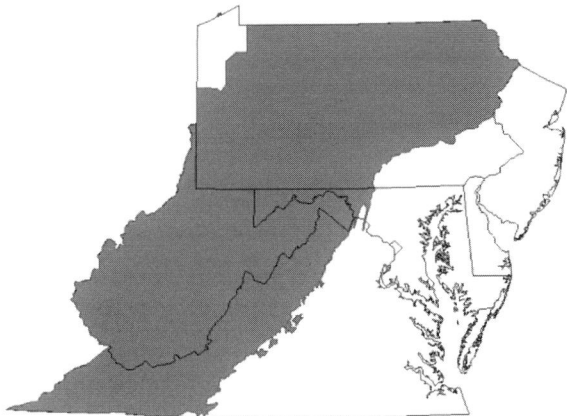

Figure 65.2 The Middle Atlantic Highlands Assessment Area, shaded in gray, within Region 3 of the U.S. EPA.

a bird community that reflects high integrity would be dominated by guilds that depend on native system attributes. The indicator incorporates aspects of ecosystem structure (e.g., vegetation), function (e.g., energy flow), and composition (e.g., demographics). The native vegetation matrix for Pennsylvania on a landscape scale is temperate forest. Therefore, specialist guilds, such as obligate tree-canopy nesters, indicate high integrity, while generalist guilds, such as shrub nesters indicate low integrity.

O'Connell, Jackson, and Brooks (1998a,b) initially evaluated their approach using 34 reference sites. First, these sites were grouped into high, medium, and low integrity categories based on the bird community index (BCI). This grouping was then compared to an *a priori* classification of the same sites as pristine, moderate, and disturbed. The independent *a priori* classification incorporated factors, such as sediment deposition, soil properties, plant and amphibian community, a wildlife community habitat profile, and general landscape context. Upon observing satisfactory correspondence between the two independent ranking approaches, a blocked random sample of 126 sites was taken throughout the MAHA area, based on the national Environmental Monitoring and Assessment Program (EMAP) sampling grid (Overton, 1990). Field sampling details are found in O'Connell et al. (1998a).

Each site was represented by a list of guilds, where each guild was represented by the proportion of species observed at the site that were in the respective guild. Since a species could belong to several guilds, the list does not necessarily sum to one, as with a multinomial distribution. The sites were all clustered in guild space using the complete linkage squared Euclidean distance protocol. Each of the guilds was then subject to analysis of variance, using cluster membership as factor levels and species proportions as the response. Tukey's simultaneous confidence intervals ($\alpha = 0.05$) were then used to determine the statistically separable clusters of sites with respect to each guild. This was performed iteratively, combining clusters that were not significantly different from each other along the way. If a guild did not yield at least two statistically distinguishable clusters of sites, then the guild was considered noninformative and therefore eliminated. In this manner, 16 guilds were retained, where each guild revealed two to five statistically separable clusters.

A ranking score was then assigned to these clusters for each guild, whereby if a guild were a specialist, the ranks increased as the species proportions increased and if a guild was a generalist, the ranks decreased as the species proportions increased. Therefore, a set of ranks was obtained for each guild, where each rank corresponded to a range of species proportions, and is presented in Table 65.1. Fractional ranks occur from averaging ranks of statistically indistinguishable clusters. In cases where clusters failed to be statistically distinguishable, yet only slightly overlapped, then separate groups may have still been identified. The final ranking scheme reported in Table 65.1 is actually the result of interactive statistical analysis and expert judgment.

Table 65.1 Biotic Integrity Ranks for 16 Songbird Guilds in the Mid-Atlantic Highlands Assessment Area

Guild Type	Guild	Proportion	Rank
Structural	Forest birds	0.000–0.280	4.5
		0.281–1.000	2.5
	Interior forest birds	0.000–0.010	1
		0.011–0.080	1.5
		0.081–0.260	3
		0.261–0.430	4
		0.431–1.000	5
	Forest ground-nesters	0	1
		0.001–0.020	1.5
		0.021–0.160	3
		0.161–0.240	4.5
		0.241–1.000	5
	Open ground-nesters	0.000–0.020	1
		0.021–0.110	2.5
		0.111–1.000	5
	Shrub-nesters	0.000–0.210	4
		0.211–0.330	1.5
		0.331–1.000	1
	Tree canopy-nesters	0.000–0.280	1.5
		0.281–0.320	2
		0.321–1.000	4.5
Functional	Bark-probing insectivores	0.000–0.060	1.5
		0.061–0.110	3
		0.111–0.170	4
		0.171–1.000	5
	Ground-gleaming insectivores	0.000–0.050	1.5
		0.051–0.070	2
		0.071–0.140	4.5
		0.141–1.000	5
	Tree canopy insectivores	0.000–0.030	1.5
		0.031–0.050	2
		0.051–0.120	3
		0.121–0.200	4.5
		0.201–1.000	5
	Shrub-gleaning insectivores	0.000–0.140	1.5
		0.141–0.230	2.5
		0.231–1.000	5
	Omnivores	0.000–0.290	5
		0.291–0.410	4
		0.411–0.480	3
		0.481–0.580	1
		0.581–1.000	2
Compositional	Nest predator/brood parasite	0.000–0.100	5
		0.101–0.150	3.5
		0.151–0.180	2
		0.181–1.000	1
	Exotic species	0	5
		0.001–0.020	4.5
		0.021–0.050	3
		0.051–0.110	2
		0.111–1.000	1
	Residents	0.000–0.260	5
		0.261–0.390	3.5
		0.391–0.570	2
		0.571–1.000	1

Table 65.1 (continued)   Biotic Integrity Ranks for 16 Songbird Guilds in the Mid-Atlantic Highlands Assessment Area

Guild Type	Guild	Proportion	Rank
	Temperate migrants	0.000–0.210	4
		0.211–0.300	2
		0.301–1.000	1
	Single-brooded	0.000–0.410	1.5
		0.411–0.450	2
		0.451–0.610	3
		0.611–0.730	4
		0.731–1.000	5

Table 65.2 BCI Scores Corresponding to Qualitative Categories of Ecological Integrity

Highest integrity	60.1–77.0
High integrity	52.1–60.0
Medium integrity	40.1–52.0
Low integrity	20.0–40.0

Each site was then assigned a BCI score by summing the ranks assigned to each guild within the respective site. The logic here is that the BCI reflects biotic integrity because the BCI increases as the ratio of specialists to generalists increases. One can also break down the overall BCI into functional, compositional, and structural scores, a valuable feature of this approach (O'Connell et al., 1998a). The five distinct clusters of sites were ranked according to the relative proportions of specialists and generalist guilds at the sites within each cluster. This ranking scheme allowed the placement of five clusters into four distinct categories of BCI scores, labeled as low, medium, high, and highest. The range of site BCI scores that correspond to each category is presented in Table 65.2. More detailed analysis of individual species proportions in each guild allowed separation of the low-integrity group into low-agricultural and low-urban categories.

Application

Using the same songbird species and the final set of 16 guilds as O'Connell et al. (1998a), data were obtained from the breeding bird atlas (Brauning and Gill, 1983–1989) and applied to each watershed according to the protocol presented below. The cutoff values in Table 65.1 were used to assign species proportions within a guild to a ranking score, and the values in Table 65.2 were used to assign BCI values to an integrity category.

Since the results in Tables 65.1 and 65.2 are based on a representative sample for the MAHA area, we only applied our protocol to those Pennsylvania watersheds that intersected the MAHA region. The native system conditions are expected to be similar at the landscape scale throughout the MAHA area because this area lies within the Appalachian Plateaus and Ridge and Valley physiographic provinces and is affected by common climatic regimes. Note that their sampled sites were 79 ha, whereas an atlas block is approximately 2800 ha, and each was sampled under different protocols. Also, some known sampling bias occurs in both the breeding bird atlas (Brauning, 1992) and the study by O'Connell et al. (1998a). These factors may lead to some anomalous results, but it is anticipated that they will be minimal. Nevertheless, we are presented with a valuable spatially synoptic database that we are analyzing in a logical way to extract ecological integrity assessments for whole watersheds.

Protocol

For each of $w = 1, \ldots, W$ watersheds, containing $i = 1, \ldots, N_w$ breeding bird atlas blocks, let S_{w_i} be the number of species in the ith block and let $S_{w_{ij}}$ be the number of species in the jth guild of the ith block.

For each block, a guild profile is obtained by computing the proportion of overall species in the block that are members of each of $j = 1, \ldots, G$ guilds. Therefore, for the ith block, a species proportion is computed for the jth guild by the equation below.

$$P_{ij} = \frac{S_{w_{ij}}}{S_{w_i}} \qquad (65.1)$$

Note that $\sum_{j=1}^{G} P_{ij} \neq 1$, or if $\sum_{j=1}^{G} P_{ij} = 1$, it is only by chance.

Each watershed is then represented by a matrix

$$\mathbf{P}_w = \begin{bmatrix} P_{11} & \cdots & \cdots & \cdots & P_{1G} \\ \vdots & \ddots & & & \vdots \\ \vdots & & P_{ij} & & \vdots \\ \vdots & & & \ddots & \vdots \\ P_{N_w 1} & \cdots & \cdots & \cdots & P_{N_w G} \end{bmatrix} \qquad (65.2)$$

Each P_{ij} value is then converted to a rank score according to Table 65.1, thus resulting in a matrix of ranks, represented as

$$\mathbf{R}_w = \begin{bmatrix} R_{11} & \cdots & \cdots & \cdots & R_{1G} \\ \vdots & \ddots & & & \vdots \\ \vdots & & R_{ij} & & \vdots \\ \vdots & & & \ddots & \vdots \\ R_{N_w 1} & \cdots & \cdots & \cdots & R_{N_w G} \end{bmatrix} \qquad (65.3)$$

Each row of the $\mathbf{R}_w$ matrix is then summed to yield a BCI for each block (row), whereby the ecological integrity increases with increasing BCI. The resulting BCI value for each block is converted to one of four categories of ecological integrity — low, medium, high, and highest — based on BCI cutoff values in Table 65.2.

Results for all of the Pennsylvania watersheds that are in the MAHA region are presented in Figure 65.3. Each watershed can now be described by the proportion of land (proportion of blocks) that is in each of the four categories of ecological integrity.

Summarizing General Ecological Integrity for Whole Watersheds

The next question is how to summarize the information in Figure 65.3 so that each watershed can be assigned some overall value of general ecological integrity. It seems most informative to characterize each watershed based on the full spatial distribution of ecological integrity; therefore, the following approach was taken.

All of the MAHA watersheds were clustered using the average Euclidean distance clustering protocol, where the response variables were the proportions of land in each of the four integrity categories (low, medium, high, and highest). Average linkage was used for reasons of robustness and consistency. Further, when Manhattan Distance was evaluated, no differences were seen in the resulting clusters. The clustering was done collectively for all the MAHA-region watersheds, as opposed to separate analysis for the physiographic provinces, because the original study that yielded the P_{ij} and BCI cutoffs (O'Connell et al., 1998a) was performed on randomly sampled sites from

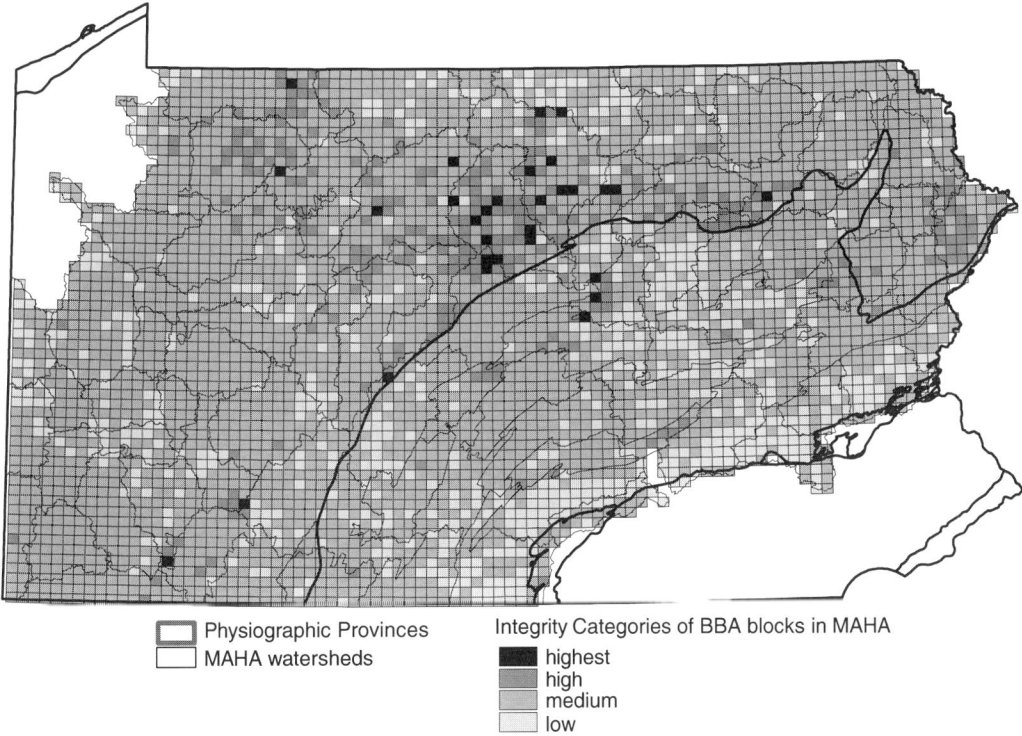

Physiographic Provinces
MAHA watersheds

Integrity Categories of BBA blocks in MAHA
■ highest
■ high
■ medium
□ low

Figure 65.3 Breeding bird atlas blocks coded according to categories of ecological integrity with respect to the songbird community, overlaid with Pennsylvania State Water Plan–based watersheds that are in the MAHA region and the major physiographic provinces of Pennsylvania.

throughout MAHA. The resulting cluster dendrogram is seen in Figure 65.4 and the clustered watersheds are mapped in Figure 65.5. The general category labels for the clusters in Figure 65.4 were based on the distributions of the four integrity categories that are summarized as box plots in Figure 65.6.

All watersheds in cluster 7 have a higher proportion of low-integrity blocks than any watersheds of other clusters. Since cluster 7 also has the lowest distribution of medium- and high-integrity blocks, along with no occurrence of highest-integrity blocks, this cluster is clearly a lowest general integrity group.

At the other extreme, watersheds of clusters 3 and 6 collectively have a higher proportion of high-integrity blocks than any watersheds of other clusters. Since these clusters also yield among the lowest overall proportions of low-integrity blocks, clusters 3 and 6 are clearly in a highest general integrity group.

Cluster 1 is clearly a medium general integrity group, as these watersheds range from having about 85 to 100% medium-integrity blocks, have among the lowest proportions of both low- and high-integrity blocks, and have no blocks that are highest integrity.

Watersheds of clusters 4 and 5 collectively have the second highest proportions of low-integrity blocks. Since cluster 4 watersheds have a relatively high proportion of medium-integrity blocks and also have among the lowest proportions of high-integrity blocks, cluster 4 appears to be in a medium-low general integrity category.

Watersheds of clusters 2 and 5 collectively have the second largest proportions of high-integrity blocks. Since the watersheds of cluster 2 have among the lowest proportions of low-integrity blocks and contain some highest-integrity blocks, yet are predominantly medium-integrity, cluster 2 appears to be a medium-high general integrity category.

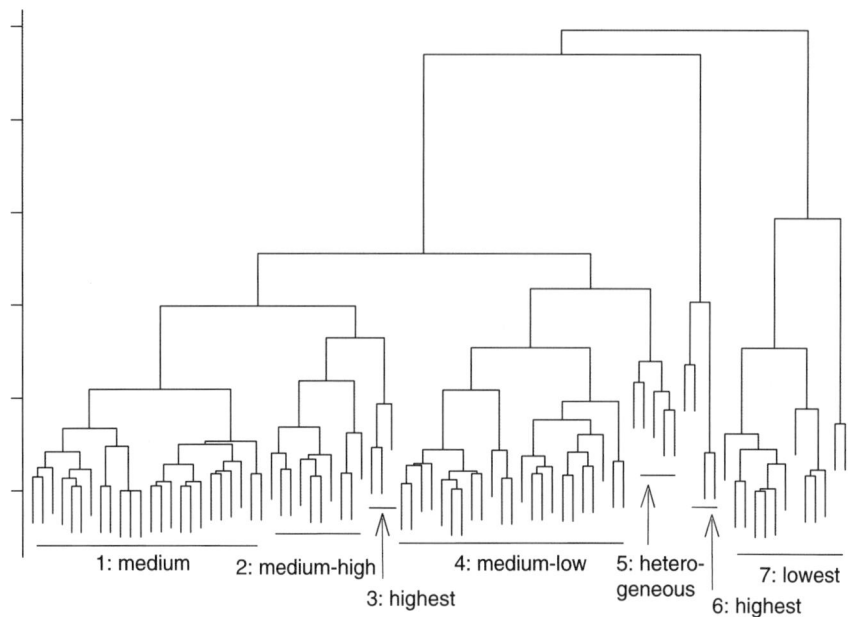

Figure 65.4 Dendrogram of watershed clusters in the Pennsylvania MAHA region obtained from applying the average Euclidean distance agglomerative clustering protocol to the proportions of low, medium, high, and highest integrity atlas blocks in each watershed.

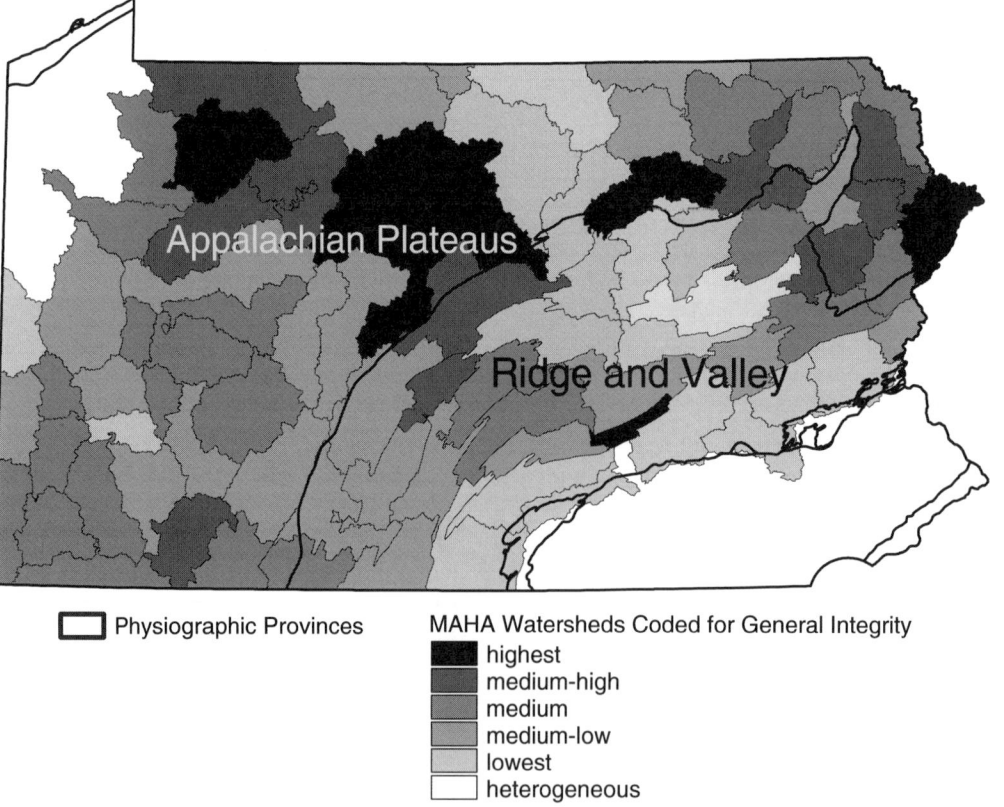

Figure 65.5 Watersheds in the MAHA region coded according to general ecological integrity as determined by cluster membership in Figure 65.4.

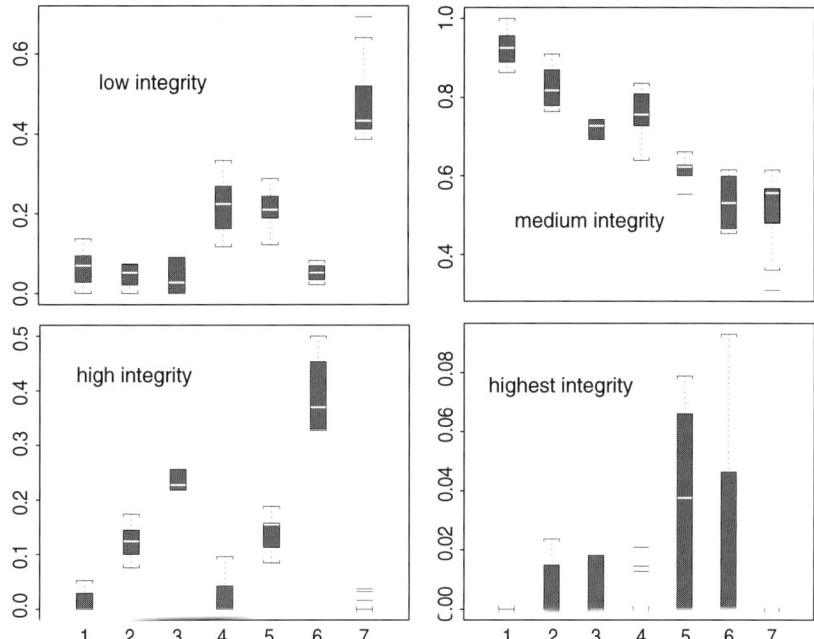

Figure 65.6 Box plots of the proportions of low, medium, high, and highest integrity land in each of the 7 clusters delineated in Figure 65.4. The numbers 1 to 7 appearing along the bottom axes correspond to the labeled clusters in Figure 65.4 as they occur from left to right.

Finally, cluster 5 is labeled heterogeneous because of the simultaneous high proportion of both low- and high-integrity blocks, with some highest integrity. Such heterogeneity can be explained by very different land uses in different areas of the watersheds. Allocating this group into the other categories always resulted in substantially increasing the variability seen in the box plots of Figure 65.6. It was decided to keep these five watersheds in a separate group.

The results mapped in Figure 65.5 generally agree with intuitive expectation based on landscape context, with the only apparent anomaly being the one watershed on the northern border that is labeled as lowest. Based on land cover alone, one may expect a medium classification at worst for this particular watershed (Tioga Creek); however, these results may reflect lower sampling coverage in this area (Brauning, 1992). On the other hand, this protocol did identify one small watershed in the Ridge and Valley province as being in the highest category, and this is indeed expected. This small watershed (Clarke Creek) is completely forested because it is the protected drinking water supply for the city of Harrisburg.

COMPARING LANDSCAPE MEASUREMENTS TO THE SONGBIRD-BASED ASSESSMENTS OF WATERSHED-WIDE ECOLOGICAL INTEGRITY

Now the objective is to determine how well ecological integrity can be predicted by landscape variables on a watershed-wide basis. For this purpose, landscape measurements of both land cover proportions and spatial pattern were evaluated that were previously obtained (Johnson et al., 2001) from an eight-category raster land cover map of Pennsylvania that was in turn derived from LANDSAT™ images (Myers, 1999). Spatial pattern variables were obtained with the FRAGSTATS program (McGarigal and Marks, 1995) for single-resolution maps, and from conditional entropy profiles for multiresolution maps.

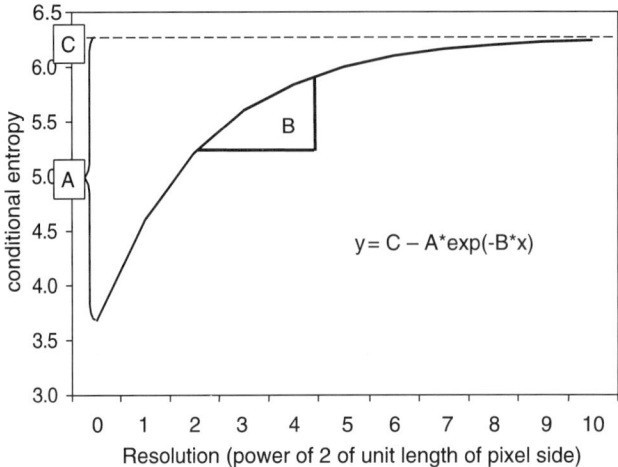

Figure 65.7 Conceptual conditional entropy profile and associated parameters.

The conceptual and methodological development of conditional entropy profiles is found elsewhere (Johnson et al., 1995, 1999, 2001; Johnson and Patil, 1998). Basically, the spatial distribution of 2×2 4-tuples of pixels in a raster map is quantified by its entropy in a way that is conditional on values of parent coarser resolution pixels that the 4-tuples are nested within. When computed for multiple resolutions, ranging from the floor that is provided by the original raster map, up to a resolution beyond which conditional entropy does not change much, a profile is traced out that reflects aspects of the underlying spatial pattern. An example profile and its parameterization is seen in Figure 65.7. Essentially, A is the extent of information that is lost from degrading the map resolution, B is the rate of information loss, and C is the asymptotic conditional entropy that is highly correlated with the entropy of the marginal land cover distribution.

Of the measurements obtained by Johnson et al. (2001), we worked with those reported in Table 65.3, which includes a subset of the spatial pattern measurements (both single and multiresolution) that are uncorrelated with each other, along with key land cover proportions. Initial exploratory data analysis indicated that contagion, total forest and total herbaceous land reveal monotonic trends as one moves from the highest to lowest general integrity.

Table 65.3 Landscape Variables Measured for Pennsylvania Watersheds

Variable Description	Code
Single-resolution variables	
Patch size coefficient or variation	PSCV
Landscape shape index	LSI
Double-log fractal dimension	DLFD
Contagion[a]	CONTAG
Multiresolution variables	
Extent of information loss	A
Rate of information loss	B
Land cover proportions	
Total forest cover	TOT.FOREST
Total herbaceous cover	TOT.HERB
Terrestrial unvegetated	TU

Note: Diagonal pixels were included when determining patches.

[a] Pixel order preserved when measuring contagion.

Linear Regression Modeling

The relationship between ecological integrity and landscape variables was quantified by regression modeling (ordinary least squares), where several response variables were investigated. The proportion of a watershed that was estimated to be in a particular ecological integrity was considered as a response. Since there were many watersheds with no land classified as high integrity and far less classified as highest, the sum of both high and highest integrity proportions were treated as one response. On the other hand, the majority of watersheds had some low-integrity land, and therefore the proportion of low-integrity land was included. Every watershed had some medium-integrity coverage; however, this category is neutral and therefore was not included. Finally, the raw BCI scores were averaged over the watersheds to provide a rational alternative watershed-wide response.

For each of the different responses, a stepwise selection protocol was applied, whereby the criterion for choosing the best set of predictors was a modification of Mallow's Cp statistic (Mallows, 1973), known as the Akaike information criterion (AIC) (Akaike, 1974), which is summarized as follows:

$$AIC = RSS(p) + MSE \times 2 \times p \qquad (65.4)$$

The term $RSS(p)$ is the residual sum of squares from the new model defined by p terms (k predictors plus the intercept), and MSE is the mean squared error from the original model prior to deleting or adding a term.

Using S-Plus (MathSoft, Inc., 1997), the stepwise selection protocol worked by choosing the set of predictors that minimized the AIC statistic. Critical F values for deciding whether or not to include or remove predictor variables were set at 2. Subsequently, the stepwise selection protocol erred in favor of large sets of predictor variables.

Models were checked by the usual diagnostic graphics. In addition, partial residual plots were obtained for each regressor in a model. Following Montgomery and Peck (1982), the ith partial residual for the regressor x_j is

$$
\begin{aligned}
e_{ij}^* &= y_i - \hat{\beta}_1 x_{11} - \cdots - \hat{\beta}_{j-1} x_{i, j-1} - \hat{\beta}_{j+1} x_{i, j+1} - \cdots - \hat{\beta}_k x_{ik} \\
&= MPe_i + \hat{\beta}_j x_{ij} \qquad \text{for } i = 1, \ldots, n
\end{aligned}
\qquad (65.5)
$$

These partial residual plots display the relationship between the response y and the regressor x_j after the effect of the other regressors $x_i (i \neq j)$ have been removed, therefore more clearly showing the influence of x_j, given the other regressors. Along with providing a check for outliers and inequality of variance, these plots also indicate more precisely how to transform the data to achieve linearity than do the usual residual plots.

A dummy variable was forced to be retained by each model in order to indicate the effect of membership in the Appalachian Plateaus, relative to the Ridge and Valley physiographic province. The set of variables in Table 65.3 constituted the initial set of potential predictors from which an optimal subset was chosen. Although the land cover proportions were correlated among themselves and to some degree with the spatial pattern variables, they were all included in order to see which ones may be retained by a stepwise regression protocol as being most informative. Prior to applying the stepwise protocol, a log transform was applied to the proportion of terrestrial unvegetated land in order to stabilize an extreme outlier.

The sum of both high and highest integrity land did not provide a valid response variable for linear modeling, as evidenced by unacceptable partial residual plots, other diagnostic plots and a very low R^2 of 0.37. Since pairwise plots of high plus highest integrity land with the potential

Table 65.4 Coefficients and Corresponding Statistics from Regressing the Proportion of Low-Integrity Atlas Blocks against Quantitative Landscape Variables and an Indicator Variable for Specifying Membership in One of Two Physiographic Groups

Regressor[a]	Value	t Value	Pr (> |t|)
Intercept	6.0968	2.9500	0.0042
Appalachian plateaus	−0.0831	−2.0196	0.0470
DLFD	−2.6258	−2.3951	0.0191
CONTAG	−0.0178	−4.9177	0.0000
A	−0.2812	−3.0365	0.0033
B	−1.1433	−2.9249	0.0046
TOT.HERB	0.6769	3.8952	0.0002

[a] Labels explained in Table 65.3.
Mean squared error (75 d.f.) = 0.97 and multiple R^2 = 0.57.

predictors suggested that a log transform of the response may improve the fit, the response was transformed as log(response + 1); however, an improvement was not seen and the R^2 remained essentially unchanged at 0.38.

When the proportion of low integrity atlas blocks was regressed against the set of potential predictors, the resulting set yielded diagnostic plots that were favorable with respect to a linear fit with no overly influential observations and a residual distribution that was approximately normal. The R^2 was only 0.57; however, besides an overall linear fit, the partial residual plots revealed some fairly strong linear relationships once the other predictors were factored into the model. The computed linear coefficients and summary statistics are reported in Table 65.4, where we see the spatial pattern variables that were retained, including both the A and B parameter estimates of the conditional entropy profiles, show fairly strong influence on the proportion of low-integrity atlas blocks in a watershed. Note that standard errors are not reported because the watersheds constitute a population of units; however, p values are reported to show the relative strength of each predictor.

Finally, the raw BCI score, averaged for all blocks within a watershed, was compared to landscape variables by the stepwise regression procedure, resulting in the model defined in Table 65.5. The corresponding partial residual plots for the quantitative predictors revealed some strong linear relationships. Further, the R^2 for this model is 0.69, which is substantially stronger than the model reported in Table 65.4, especially since it contains two fewer terms.

Clustering

In a separate study (Johnson et al., 2001), the watersheds were clustered using a variety of landscape variables that included those in Table 65.3 along with other spatial pattern measurements. However, qualitative labeling of the watershed clusters was subjective, being based on the relative amount of forest fragmentation in the original land cover maps.

Table 65.5 Coefficients and Corresponding Statistics from Regressing the Raw BCI Score, Averaged for All Blocks within a Watershed, against Quantitative Landscape Variables and an Indicator Variable for Specifying Membership in One of Two Physiographic Groups

Regressor[a]	Value	t Value	Pr (> |t|)
Intercept	37.6907	24.6755	0.0000
Appalachian plateaus	1.1347	2.8318	0.0059
PSCV	−0.0002	−2.0926	0.0397
CONTAG	0.2219	5.8799	0.0000
TOT.HERB	−11.6307	−5.2991	0.0000

[a] Labels explained in Table 65.3.
Mean squared error (77 d.f.) = 13.7 and multiple R^2 = 0.69.

Table 65.6 Comparison of Watershed Clusters Obtained from BCI Proportions to an Independent Clustering Obtained from Landscape Measurements

Landscape Measurements	BCI Proportions						
	B[a]	MB	M	MW	W	H	Total
Appalachian Plateaus							
Best	4	3	3	1	1	1	13
MB	2	5	6	3	0	1	17
MW	0	0	8	7	0	0	15
MW (Pitt.)[b]	0	0	2	2	2	0	6
Total	6	8	19	13	3	2	51
Ridge and Valley							
Best	1	0	1	0	0	0	2
M (lower App. Mtns.)	0	2	2	7	0	2	13
M (upper App. Mtns.)	0	0	2	2	3	1	8
MW (lower Gr. Val)	0	0	0	0	2	0	2
MW (upper Gr. Val)	0	0	0	1	3	0	4
Worst	0	0	0	0	2	0	2
Total	1	2	5	10	10	3	31

[a] Symbols for qualitative cluster labels are B = best, MB = medium-to-best, M = medium, MW = medium-to-worst W = worst, H = heterogeneous.

[b] Symbols for geographic location, where appropriate are Pitt. = Pittsburgh and near vicinity, lower App. Mtns. and upper App. Mtns. = the lower (southwest) and upper (northeast) Appalachian Mountains, respectively, and lower Gr. Val. and upper Gr. Val. = the lower (southwest) and upper (northeast) Great Valley, respectively.

Clustering of the Pennsylvania MAHA watersheds based on the spatial assessment of ecological integrity using the songbird community data (BCI proportions), as presented in Figures 65.4 and 65.5, was compared to clusters obtained from using all of the landscape measurements, as presented by Johnson et al. (2001). Comparisons for each of the physiographic provinces that overlap both Pennsylvania and the MAHA region are tabulated in Table 65.6. The cluster categories that were based solely on ecological integrity are the same for both physiographic provinces because this clustering is based on all watersheds across the two provinces; however, the cluster categories that were based solely on landscape variables are somewhat different for each province because clustering was applied separately within each province due to very different fragmentation patterns.

These results indicate some fairly strong dependence between the two classification schemes. A formal chi-square or likelihood ratio test is not possible for either physiographic province because of the zero entries in their respective contingency tables, but these zero entries are the result of a dependent structure.

For the Appalachian Plateaus, the landscape measurements alone did an excellent job of identifying medium-low watersheds, as evidenced by complete correspondence with clusters that were qualified as having medium to medium-low ecological integrity, and with the two watersheds that were qualified as having the lowest integrity. The landscape measurements also did a good job of identifying the medium-high and highest watersheds, as evidenced by distributions across the ecological integrity clusters that were biased toward the medium-high and highest clusters, respectively. However, one medium-low and one lowest integrity watershed were each misclassified as highest with the landscape variables.

Meanwhile, for the Ridge and Valley, the landscape variables alone identified the single watershed that was qualified as having the highest ecological integrity (Clarke Creek). A medium integrity watershed was also grouped with the highest watershed, but no gross misclassifications into medium-low or lowest general integrity occurred. The medium-low and lowest integrity watersheds of the Great Valley section were clearly identified by the landscape variables alone. Finally, out of the 21 watersheds identified as medium by the landscape variables, all but 4 were qualified as being in one of the medium-integrity categories, with 3 out of these 4 being labeled as lowest integrity.

Conditional Entropy Profiles

Conditional entropy profiles are reported in Figure 65.8 for those watersheds coded as highest, medium, and lowest in Figure 65.5. The medium-high and medium-low watersheds were not included in order to maintain better graphical clarity.

One observation is that the highest-integrity watersheds lie near the bottom of the profiles. This is because the highest-integrity watersheds are mostly forested, and therefore characterized by large coherent forest patches. This highly contagious pattern results in a very low conditional entropy across the resolutions.

The profiles migrate upwards as the underlying landscape becomes increasingly fragmented and land cover becomes more evenly distributed both marginally and spatially. The increasing fragmentation generally corresponds with decreasing ecological integrity. The lowest-integrity watersheds, however, do not yield the top profiles. This is because watersheds of the very lowest integrity are those that have returned to a more contagious condition, but now it is due to large nonforest patches.

Within Figure 65.8, the very lowest watershed was chosen for each province as the one with the highest proportion of low-integrity blocks, the very highest had the highest proportion of high- plus highest-integrity blocks, and the all medium watersheds were covered 100% by medium-integrity blocks.

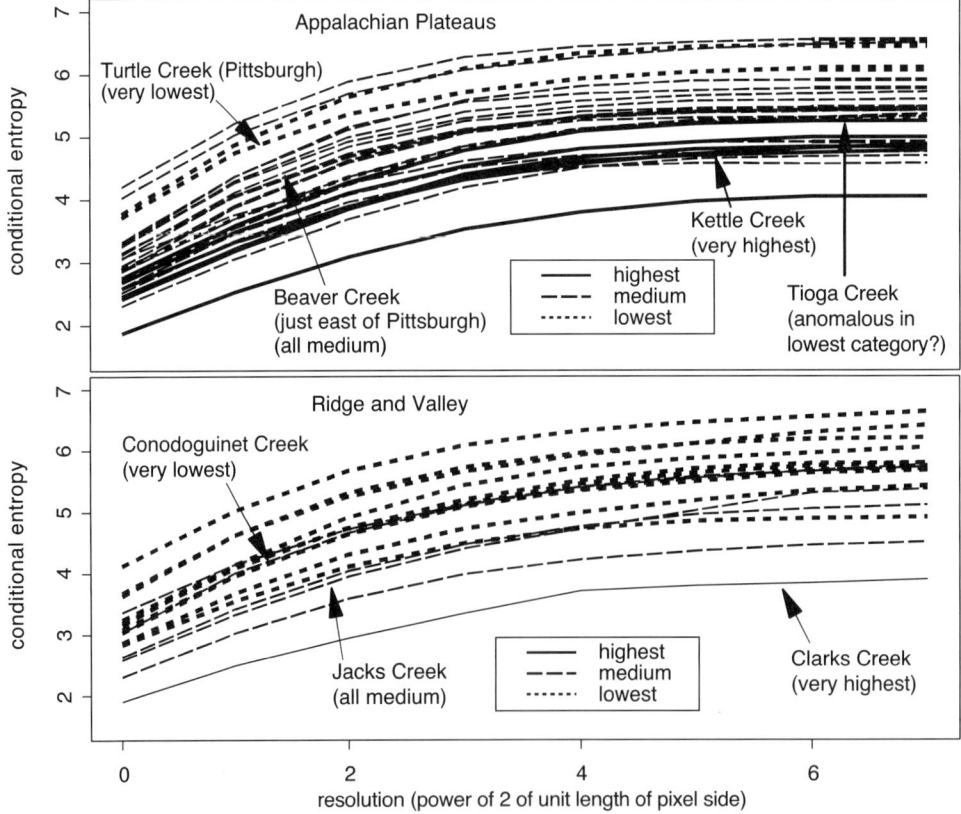

Figure 65.8 Conditional entropy profiles of watersheds in the MAHA section of Pennsylvania that were labeled as having highest, medium, or lowest ecological integrity, as assessed from the songbird community.

This shows that as we go from the very highest to all medium to the very lowest watersheds, the profiles increase for all resolutions; however, when all watersheds are viewed, it appears that as the forest is increasingly fragmented, the maximum conditional entropy profile reflects a sort of critical state, beyond which the landscape matrix is nonforest and ecological integrity gets increasingly lower.

ECOLOGICAL INTEGRITY BASED ON ALL SPECIES IN THE BREEDING BIRD ATLAS

An alternative approach to the one discussed in this chapter for assessing ecological integrity is to use all of the breeding bird atlas blocks (as per Figure 65.1), so that an assessment could be made for every watershed. Since the standards established for songbirds strictly in the MAHA region should not be extended beyond MAHA, nor to other species or guilds, a new basis for categorizing watersheds had to be established. Bishop (2000) defined a set of 34 guilds that incorporate all species from the full breeding bird atlas (Brauning and Gill, 1983–1989). Using these guilds and all the species that were listed in the atlas as possible, probable, or confirmed, two variations on a common protocol were investigated.

Initially, reference watersheds were established to represent both the best and worst overall ecosystem condition, based on results from nonpoint source surface water pollution studies (Johnson et al., 2001) and, where appropriate, results from the MAHA songbird evaluation presented earlier in this chapter. Three watersheds from Pennsylvania's northern tier were chosen for the best condition, and four watersheds from the Piedmont Plateau and Great Valley section were chosen for the worst condition, based on the highest and lowest rankings for Agricultural Pollution Potential, respectively (Johnson et al., 2001). The best watersheds with respect to water pollution potential were also categorized as having the highest general ecological integrity based on the MAHA songbirds. Of the worst watersheds with respect to water pollution potential, the one (in the Great Valley section) that lay within the MAHA region was also categorized as having the lowest ecological integrity based on the MAHA songbirds.

For each of the two reference groups of watersheds, a 34-guild mean vector was obtained by averaging the species proportions in each guild over all atlas blocks in the group of watersheds. This included 274 blocks for the three best watersheds and 166 blocks for the four worst watersheds. Given the reference mean vectors for the best $\mathbf{P}_b$ and worst $\mathbf{P}_w$ watershed groups, a vector of weights was established by subtraction: $\mathbf{D} = \mathbf{P}_b - \mathbf{P}_w$. Therefore, each of $j = 1, \ldots, G$ guilds is represented by a weight $\mathbf{D}_j \in [-1, 1]$, such that for $\mathbf{D}_j > 0$, the guild is representative of increasingly better ecological integrity as $\mathbf{D}_j$ approaches 1, and for $\mathbf{D}_j < 0$ the guild is representative of increasingly lesser integrity as $\mathbf{D}_j$ approaches -1; whereas for $\mathbf{D}_j = 0$, the guild is noninformative.

Now a bird community index can be calculated for each of $i = 1, \ldots, N$ atlas blocks across the state by the following sum:

$$\mathrm{BCI}_i = \sum_{j=1}^{G} P_{ij} D_j \tag{65.6}$$

where, for block i, P_{ij} is the proportion of species in the jth guild.

This protocol was carried out under two different scenarios for calculating the P_{ij} values. First, the P_{ij} was computed as the proportion of species in guild j out of all species in block i. Alternatively, the P_{ij} was computed as the proportion of species in guild j out of all potential species in guild j. The first approach, coined species richness-based, can yield identical multinomial vectors of species proportions for two different blocks, although the overall species richness may be very different between the blocks. Therefore, it has no way of accounting for large differences in species richness that may in turn be due to differing environmental conditions. For this reason, the second approach, coined guild potential-based was evaluated.

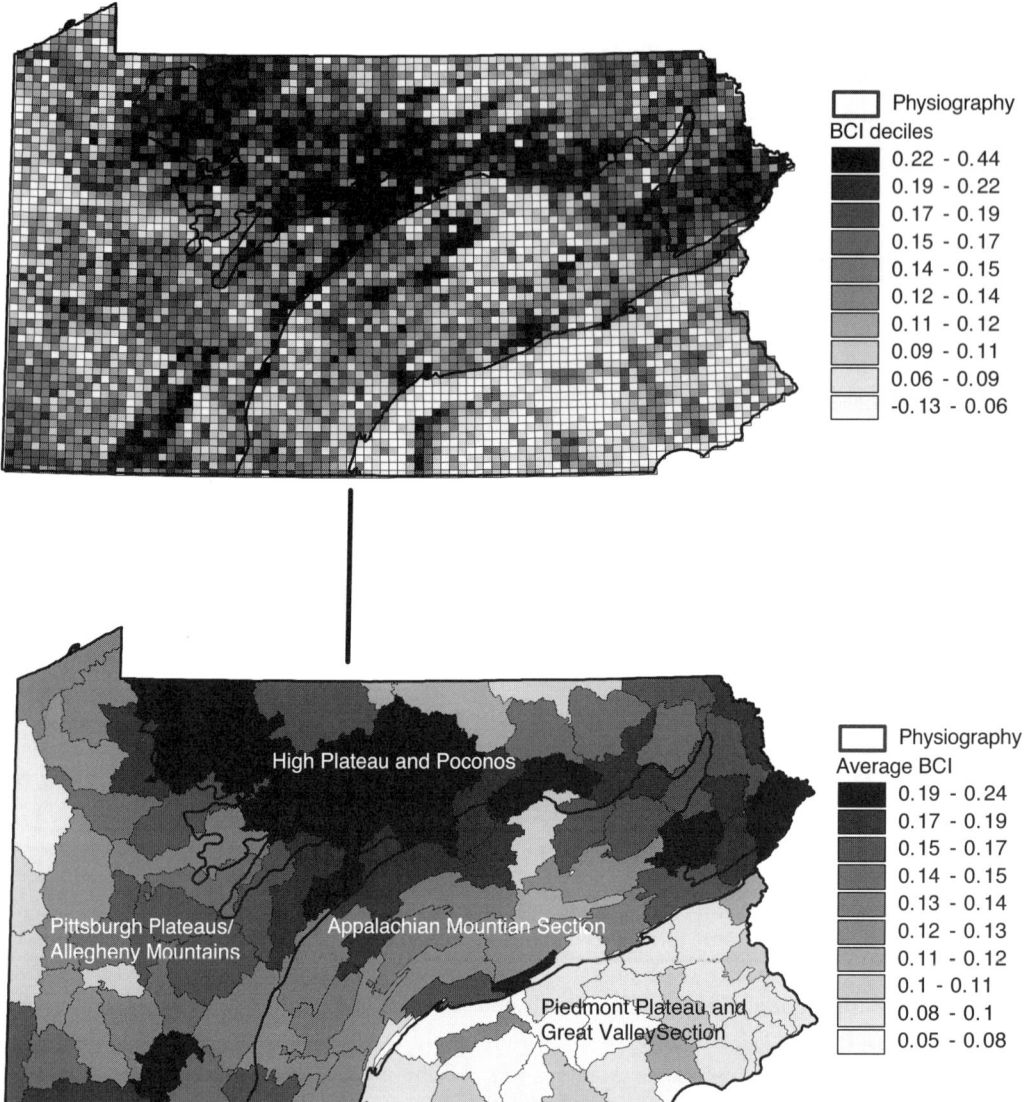

Figure 65.9 Species richness-based bird community index of ecological integrity, presented within each breeding bird atlas block (top) and for blocks averaged within each watershed (bottom). A combination of physiographic provinces and sections have been overlaid to stratify the state into more homogeneous areas of ecological integrity.

The results for both variations on the protocol are presented in Figures 65.9 and 65.10, where the block values of BCI are averaged to obtain a summary BCI value for each watershed. Since there are some gross statewide patterns, the state was stratified by a combination of physiographic provinces and sections to delineate more homogeneous areas using natural boundaries.

Unlike with the MAHA songbirds, there are no preestablished standards for assigning BCI scores to categories of ecological integrity. Conceptually, one may consider clustering the blocks in guild-space; however, with close to 5000 blocks in the state, interpretation would be near impossible at best. Therefore, ecological integrity will be simply represented on the continuous scale of the BCI scores.

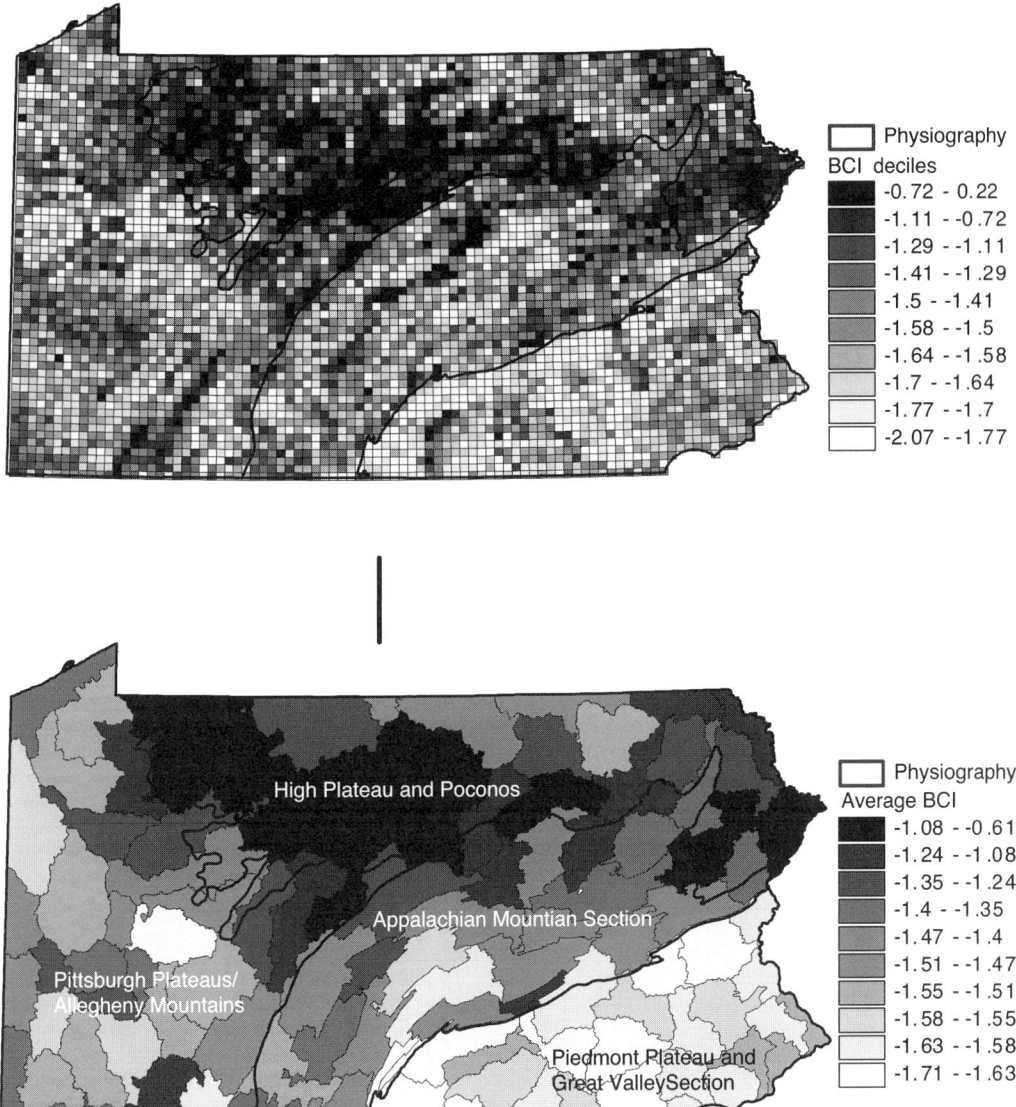

Figure 65.10 Guild potential-based bird community index of ecological integrity, presented within each breeding bird atlas block (top) and for blocks averaged within each watershed (bottom). A combination of physiographic provinces and sections have been overlaid to stratify the state into more homogeneous areas of ecological integrity.

Comparison to Landscape Measurements

Since there is no basis for categorizing the BCI scores, as with the MAHA songbirds approach discussed earlier, a spatial distribution of ecological integrity is not obtainable. Losing this spatial information is a major sacrifice; however, the average BCI scores do lend themselves to regression modeling.

The stepwise selection protocol, as discussed in a previous section, was applied to the same set of potential predictor variables in Table 65.3 in order to assess the strength of relationship between the average BCI value and the landscape variables for the watersheds. Dummy variables indicating physiographic membership of each watershed were forced to be retained by the stepwise

Table 65.7 Coefficients and Corresponding Statistics from Regressing the Species Richness-Based BCI Score, Averaged for All Blocks within a Watershed, against Quantitative Landscape Variables and Indicator Variables for Specifying Membership in a Physiographic Group

Regressor[a]	Value	t Value	Pr (type 1 error)
Intercept	−0.3204	−1.4741	0.1438
App. Mountain	−0.0276	−3.7969	0.0003
High Plat. Pocono	−0.0038	−0.5209	0.6037
Pied. Gr. Valley	−0.0368	−5.1539	0.0000
DLFD	0.2286	1.6553	0.1012
CONTAG	0.0015	3.5774	0.0006
B	−0.0701	−2.2483	0.0269
TOT. FOREST	0.1414	5.9101	0.0000

[a] Labels explained in Table 65.3.
App. Mountain = Appalachian Mountains Section; High Plat. Pocono = High Plateaus Sections and Pocono Area; Pied. Gr. Valley = Piedmont Plateau and Great Valley Section; Applied to all watersheds in Pennsylvania. Mean squared error (94 d.f.) = 0.19 and multiple R^2 = 0.78.

Table 65.8 Coefficients and Corresponding Statistics from Regressing the Guild Potential-Based BCI Score, Averaged for all Blocks within a Watershed, against Quantitative Landscape Variables and Indicator Variables for Specifying Membership in a Physiographic Group

Regressor[a]	Value	t Value	Pr (type 1 error)
Intercept	−1.1653	−7.0529	0.0000
App.Mountain	−0.1277	−2.6240	0.0101
HighPlat.Pocono	0.1063	2.2922	0.0241
Pied.GrValley	−0.1071	−2.4600	0.0157
CONTAG	0.0054	2.1643	0.0329
B	−0.5351	−2.6726	0.0089
TOT. HERB	−0.8786	−5.1467	0.0000

[a] Labels explained in Tables 65.3 and 65.7.
Applied to all watersheds in Pennsylvania. Mean squared error (95 d.f.) = 1.33 and multiple R^2 = 0.66.

protocol. Results for each of the two methods yielded favorable diagnostics, including partial residual plots that indicated strong linear relationships. Corresponding statistics are reported in Tables 65.7 and 65.8. Since these regressions were performed on all watersheds in the Pennsylvania population, the final least squares calculations are for actual parameter values, given this particular set of predictors. Significance levels, represented by p-values, are still reported to reveal the relative strength of each predictor.

The species richness-based method of computing BCI yielded the best diagnostics and an R^2 of 0.78 (8 parameters). The guild potential-based method yielded an R^2 of 0.66 (7 parameters). These results, along with favorable model diagnostics, indicate that a fairly strong relationship is obtained with the species richness-based method of computing BCI.

COMPARISON OF THE DIFFERENT METHODS FOR COMPUTING A BIRD COMMUNITY INDEX

Consistency among the three different approaches to computing a BCI of ecological integrity is evaluated in Table 65.9. For each of the watershed clusters that were determined from landscape

Table 65.9 Area-Weighted Bird Community Indices (BCIs) for Three Different Protocols within Each Cluster Defined by Landscape Measurements

Province	Clusters Based on Landscape Measurements	Species Richness	Guild Potential	MAHA Songbirds
AP	High (13)[a]	0.189	−1.019	47.606
	MH[b]	0.163	−1.243	46.045
	ML1 (6)	0.132	−1.437	44.310
	ML2 (19)	0.115	−1.443	42.302
RV	High (2)	0.167	−1.349	46.223
	M1 (13)	0.136	−1.452	43.767
	M2 (8)	0.134	−1.391	42.924
	ML1 (2)	0.097	−1.606	40.258
	ML2 (4)	0.091	−1.626	40.466
	Very low (2)	0.076	−1.678	39.215
Pied.	MH (1)	0.046	−1.600	N/A
	M (4)	0.084	−1.570	N/A
	ML (4)	0.066	−1.632	N/A
	Very low (7)	0.086	−1.630	N/A

[a] ( ) = number of watersheds in cluster.
[b] Symbols for qualitative cluster labels are: MH = medium-to-high, M = medium, ML = medium-to-low. Within the Appalachian Plateaus (AP), ML2 = Pittsburgh and near vicinity, and ML1 = medium-to-low elsewhere. Within the Ridge and Valley (RV), M1 and M2 refer to medium watersheds in the lower (southwest) and upper (northeast) Appalachian Mountains, respectively, and ML1 and ML2 refer to medium-to-low watersheds in the lower (southwest) and upper (northeast) Great Valley, respectively.

measurements alone, as discussed earlier, the average BCI was obtained by taking an area-weighted average BCI of all watersheds in each respective cluster.

Table 65.9 does indeed show that for the majority of Pennsylvania (all of the Appalachian Plateaus and the Ridge and Valley), the average BCI consistently increased for all three methods of computing the BCI as the watershed clusters went from lowest to highest amount of forest cover.

In the Piedmont, the two methods that could be applied did not reveal consistent results and they also did not coincide with apparent decreasing forest cover. This physiographic region, however, presents several limitations for comparing average BCI values among landscape-based watershed clusters. First, there are no watersheds that are mostly forested to provide a background. The one watershed that is labeled as medium-to-high with respect to forest cover is actually not much different from those in the medium cluster. Second, the Piedmont is a small area of Pennsylvania and therefore the sample of watersheds is much smaller than for the other two provinces.

DISCUSSION

For predicting ecological integrity, landscape variables provided fairly strong predictors for each of three different approaches to obtaining a watershed-wide average BCI of ecological integrity. After factoring out physiographic regimes, an objective stepwise model selection procedure retained landscape variables that explained 66 to 78% of the variation in BCI. As expected, for each method of obtaining the BCI response, either total forest or total herbaceous land served as a very strong predictor. However, several measurements of spatial pattern were also retained. Of these pattern measurements, the conditional entropy term that relates the rate of information loss (parameter B) was retained twice and contagion was retained all three times. Contagion was highly significant in all cases and revealed a fairly strong linear relationship to BCI. Since contagion is very highly linearly correlated with the conditional entropy term C, this indicates that only land cover proportions and terms from the multiresolution conditional entropy profiles (A, B, and C in Figure 65.7) can provide a strong capability to predict general landscape-scale ecological integrity.

For the songbird-based approach to obtaining BCI values within the MAHA region, the watersheds could be clustered based on their respective proportions of breeding bird atlas blocks in low, medium, high, and highest integrity. When these watershed clusters were compared to distinct clusters obtained using only landscape variables, a high degree of correspondence was observed, further supporting the claim that landscape measurements alone can characterize landscape-level ecosystem condition.

Improvements can be made for the two approaches to computing a watershed-wide BCI that were based on reference watersheds within the state and incorporated a large set of guilds. As with the initial approach by O'Connell et al. (1998a), the full set of guilds may be reduced to remove redundancy prior to computing a BCI. Also, one can experiment with different choices of reference watersheds.

We look forward to seeing how well landscape measurements alone can predict the condition of landscape-level ecosystems elsewhere.

ACKNOWLEDGMENTS

Prepared with partial support from the National Science Foundation Cooperative Agreement Number DEB-9524722 and the U.S. Environmental Protection Agency Cooperative Agreement Number CR-825506. The contents have not been subjected to agency review, and therefore do not reflect the views of the agencies. No official endorsement should be inferred.

REFERENCES

Akaike, H., A new look at statistical model identification, *IEEE Trans. Autom. Control*, AC-19, 716–722, 1974.

Bishop, J.A., Associations between Avian Functional Guild Response and Regional Landscape Properties for Conservation Planning, M.S. thesis, Penn State University, University Park, 2000.

Bradford, D.F., Franson, S.E., Miller, G.R., Neale, A.C., Cantebury, G.E., and Heggem, D.T., Bird species assemblages as indicators of biological integrity in Great Basin rangeland, *Environ. Monitoring Assess.*, 49, 1–22, 1998.

Brauning, D.W., *Atlas of Breeding Birds in Pennsylvania*, University of Pittsburgh Press, Pittsburgh, 1992.

Brauning, D.W. and Gill, F.B., *Pennsylvania Breeding Bird Atlas Data*, The Academy of Natural Sciences of Philadelphia, Pennsylvania Game Commission, and Wild Resource Conservation Fund, Harrisburg, PA, 1983–1989.

Brooks, R.P., O'Connell, T.J., Wardrop, D.H., and Jackson, L.E., Towards a regional index of biological integrity: the examples of forested riparian ecosystems, *Environ. Monitoring Assess.*, 51:131–143, 1998.

Johnson, G.D. and Patil, G.P., Quantitative multiresolution characterization of landscape patterns for assessing the status of ecosystem health in watershed management areas, *Ecosyst. Health*, 4, 177–187, 1998.

Johnson, G.D., Tempelman, A.K., and Patil, G.P., Fractal based methods in ecology: a review for analysis at multiple spatial scales, *Coenosis*, 10, 123–131, 1995.

Johnson, G.D., Myers, W.L., Patil, G.P., and Taillie, C., Multiresolution fragmentation profiles for assessing hierarchically structured landscape patterns, *Ecol. Modeling*, 116, 293–301, 1999.

Johnson, G.D., Myers, W.L., Patil, G.P., and Taillie, C., Categorizing watershed-delineated landscapes in Pennsylvania using conditional entropy profiles, *Landscape Ecol.*, 16, 597–610, 2001.

Johnson, G.D., Myers, W.L., and Patil, G.P., Predictability of surface water pollution loading in Pennsylvania using watershed-based landscape measurements, *J. Am. Water Resour. Assoc.*, 37, 821–835, 2001.

Karr, J.R., Biological integrity: a long-neglected aspect of water resource management, *Ecol. Appl.*, 1, 66–84, 1991.

Karr, J.R., Defining and assessing ecological integrity: beyond water quality, *Environ. Toxicol. Chem.*, 12, 1521–1531, 1993.

Mallows, C.L., Some comments on Cp, *Technometrics*, 15, 661–675, 1973.

MathSoft, Inc., *Splus 4 Guide to Statistics*, Data Analysis Products Division, Seattle, 1997.

McGarigal, K., and Marks, B., FRAGSTATS: spatial pattern analysis program for quantifying landscape structure. General Technical Report PNW-GTR-351. U.S. Department of Agriculture, Forest Service, Pacific Northwest Research Station, Portland, OR, 1995.

Montgomery, D.C. and Peck, E.A., *Introduction to Linear Regression Analysis*, John Wiley & Sons, New York, 1982.

Myers, W., Remote Sensing and Quantitative Geogrids in PHASES [Pixel Hyperclusters As Segmented Environmental Signals], release 3.4, Technical Report ER9710, Environmental Resources Research Institute, Penn State University, University Park, 1999.

O'Connell, T.J., Jackson, L.E., and Brooks, R.P., The Bird Community Index: A Tool for Assessing Biotic Integrity in the Mid-Atlantic Highlands, Final Report, Report 98-4, Penn State Cooperative Wetlands Center, Forest Resources Laboratory, Penn State University, University Park, 1998a.

O'Connell, T.J., Jackson, L.E., and Brooks, R.P., A bird community index of biotic integrity for the Mid-Atlantic Highlands, *Environ. Monitoring Assess.*, 51, 145–156, 1998b.

Overton, W.S., White, D., and Stevens, D.L., Design Report for EMAP. EPA/600/3-91/053, U.S. Environmental Protection Agency, Office of Research and Development, Washington, D.C., 1990.

Riitters, K.H., O'Neill, R.V., Hunsaker, C.T., Wickham, J.D., Yankee, D.H., Timmins, S.P., Jones, K.B., and Jackson, B.L., A factor analysis of landscape pattern and structure metrics, *Landscape Ecol.*, 10:23–39, 1995.

CHAPTER 66

Landscape Biodiversity and Biological Health Risk Assessment: The Map of Italian Nature

Giovanni Zurlini, Orazio Rossi, and Vittorio Amadio

INTRODUCTION

Ecosystem health is related to the system's capability to maintain some degree of biophysical integrity while supplying humans with valuable services (Rapport, 1995). Thus, the functional relationship between the diversity of organisms and the set of ecological services on which humanity depends must be addressed (Perrings et al., 1995). Signs of dysfunction, either structural or functional, associated with particular stresses can be identified by suitable indicators in the biophysical domain of ecosystems themselves. In the Mediterranean regions, where ecosystems have been shaped by the millennial historic and evolved interactions between man and nature, many forms of human disturbance are recognized to be important sustaining components of natural systems (Pickett and White, 1985); consequently in the Mediterranean, but not only there, we have often to deal with the question of defining both the extent to which ecosystem services start to be impaired by either natural or anthropogenic perturbation, and the extent up to which the services themselves are unimpaired, or even sustained, by human disturbances.

Studies of natural, managed, and disturbed or impacted ecosystems have shown that ecosystems under stress appear to keep more of their functional performance than their species composition (Holling et al., 1995), as processes are buffered by species redundancy. Thus, species composition appears more sensitive than ecosystem processes since it responds more quickly and recovers more slowly (Angermeir and Karr, 1994). Changes in species composition may also significantly alter ecosystem performances (Naeem et al., 1994), like the resilience and stability of the ecosystem to several sources of disturbance.

However, giving equal emphasis to every piece of biodiversity would be ecologically incorrect and practically unachievable (Walker, 1992). Indeed, one of the fundamental aims of conservation management and reserve design is often not just the preservation of maximum species diversity per se, but the preservation of the diversity of species "typical" (or rare) of the habitat or habitats within a reserve (Usher, 1991), also classified into functional, structural, and compositional groups using a "guilding" concept (O'Connell et al., 1998).

This chapter addresses risk assessment of biological diversity at the landscape level, associated with ecotypes or ecotype mosaics, by a simple multiscale conceptual model incorporating metrics related to current human disturbance, based on native species most threatened with extinction and reduction. We consider all native vertebrates threatened according to the International Union for

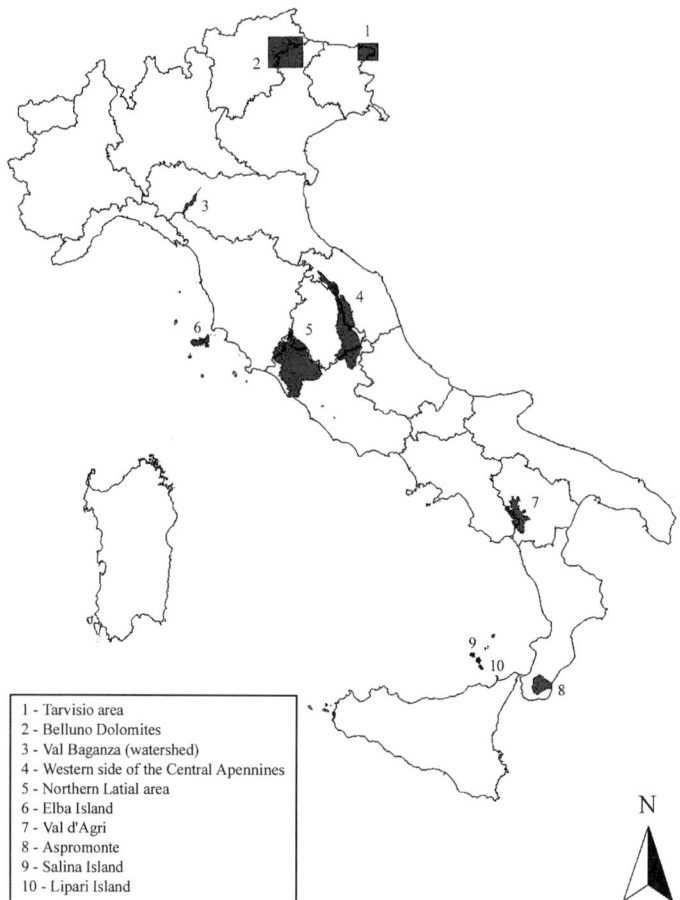

Figure 66.1 Distribution of the current study sites of the Map of Italian Nature, accounting for almost 1,500,000 ha.

Conservation of Nature and Natural Resources–The World Conservation Union (IUCN) (1996), and European Directive 92/43/EU, following Pinchera et al. (1997) updates, overlapping with their distribution range at least 10% of the area considered by the Map of Italian Nature project (Rossi and Zurlini, 1993, 1998; Zurlini et al., 1999). This project accounts for 1,500,000 ha (Figure 66.1), started as a result of the Italian national law 394 (1991) on reserves and protected areas (Ministry of Environment, 1992). Vertebrates considered are mostly at the top of their natural food webs and are end points to a number of ecological processes. They tend to be rare animals with slow rates of population growth, particularly susceptible to overexploitation or habitat loss, and are expected to reduce populations or disappear first, as disturbance intensifies, because of their sensitivity. They are supposed to be reliable witnesses of the ability of ecosystems to maintain their characteristic patterns and rates of process in response to disturbances and to the variability inherent in their climate regimes (Walker, 1992).

While risk and conservation value assessments for a pilot study area already have been presented elsewhere, together with overall data processing procedures (Zurlini et al., 1999), this chapter presents primary results based on real Map of Italian Nature sites with particular regard to the Aspromonte area (Area 8, Figure 66.1). Biological health issues are also addressed, where health stands for health condition or well-being in the native biodiversity structure of ecosystems (Wickert and Rapport, 1998; Zurlini et al., 1999).

In the last centuries, due to anthropic pressure on Mediterranean ecosystems and abandonment of intense agricultural and pastoral practices, Mediterranean plant communities have been shaped into a mosaic-like pattern composed of different human-induced degradation and regeneration stages (Naveh and Liebermann, 1994; Pignatti, 1994). Such mosaics can be physically recognized as composites of CORINE habitats (European Union Directorate General XI, 1991), and risk assessments are made at a nominal scale of 1:50,000, to provide focus and direction for proactive land management activities. Digital thematic maps are generated as geographical information system (GIS) coverages in polygonal (vector) format of (1) existing land cover habitat types and biodiversity component distributions and (2) different kinds of human-induced disturbance.

The project, akin to the Gap Analysis Program (Scott et al., 1996), aims at identifying gaps in the existing reserve network to establish new reserves and protected areas to obtain a more representative network of regional biological diversity, based on (1) their natural values and (2) fragility. The project operates in a multiscale perspective (O'Neill et al., 1986; Allen and Hoekstra, 1992), dealing with scales in which humans generally influence the condition of the landscape. Activities are currently enlarging to cover another 6,000,000 ha, and accounting for almost one fourth of the entire Italian territory (about 30,000,000 ha).

A SEMANTIC RATIONALE FOR THE MAP OF ITALIAN NATURE

While the Italian law deals with values as components of the national natural heritage including different biological, ecological, geological, anthropological, archaeological, and historical issues, the project focuses on biophysical and ecological components. Here values are in reference to biodiversity, biological integrity, and biological and ecosystem health. Biodiversity refers to the variety of components at every level of biotic community organization, whereas biological integrity is reflected in both biotic elements and processes that generate and maintain those elements (Angermeir and Karr, 1994). Biological integrity is "the capability of supporting and maintaining a balanced, integrated, adaptive community of organisms having a species composition, diversity, and functional organization comparable to that of natural habitat of the region" (Karr and Dudley, 1981) prior to industrial settlement by humans. Exotic species also are taken into account by biological integrity. A site can be high in biological diversity, but not have biological integrity if exotics are common, making up a significant percentage of diversity (Karr and Dudley, 1981). Ecosystem health can be deemed as a condition of normality in the linked processes and functions that constitute ecosystems (Rapport, 1995), and defined in terms of vigor, resilience, and organization (Mageau et al., 1995). Ecosystem health and biological integrity are related, since a sure way to maintain ecosystem health is to maintain biological integrity (Karr, 1995), but not vice versa. Along the same line, we might state that biological health and integrity also are related, since a sure way to maintain biological integrity is to maintain biological health, albeit this would not guarantee the maintenance of processes. Thus biodiversity, biological integrity, and biological and ecosystem health, while related, are not identical. It is generally recognized that biological integrity is primary as a conservation value (Angermeir and Karr, 1994).

Fragility relates to the degree of change observed in species abundance and composition following disturbance (displayed fragility, cf. Nilsson and Grelsson, 1995). However, Ratcliffe's (1977) definition seems more useful to provide direction for land management and planning; it states that fragility "reflects the degree of sensitivity of habitats, communities and species to environmental change, and so involves a combination of intrinsic and extrinsic factors." Fragility can be estimated as intrinsic fragility, which relates to intrinsic factors like broad biological differences among taxa, such as the size of the species range, their trophic status, their habitat fidelity, and other aspects related to feeding, breeding, and reproductive strategies. Extrinsic factors are represented by human and naturally induced stresses. The amount of stress, coupled with those intrinsic factors, determines fragility. The type, magnitude, and timing of the stressor; the exposure of the species; and the species' inherent or intrinsic fragility (sensitivity) have important interactive relationships that determine the overall fragility.

CORINE HABITAT MOSAICS

The basic data produced by the Map of Italian Nature project concerns mosaics of contiguous patches (Figure 66.2) corresponding to the different habitat types in the CORINE biotopes classification (European Union Directorate General XI, 1991). The CORINE classification system aims to compile an inventory of biotopes (habitat) of major importance for nature conservation in the European Union, providing a unique basis for evaluation of conservation management issues. The CORINE habitat classification, through satellite, airborne, and ground data, permits the biophysical identification of ecosystems, as defined by Tansley (1935), as patches, based on vegetation covers, physiognomy, soil types and landforms, and an integration of related abiotic and biotic components, like habitats and syntaxa (Usher, 1991). CORINE biotopes, called habitats, are coded at different levels of a hierarchical system of mosaics of patches within patches, that can go from very broad

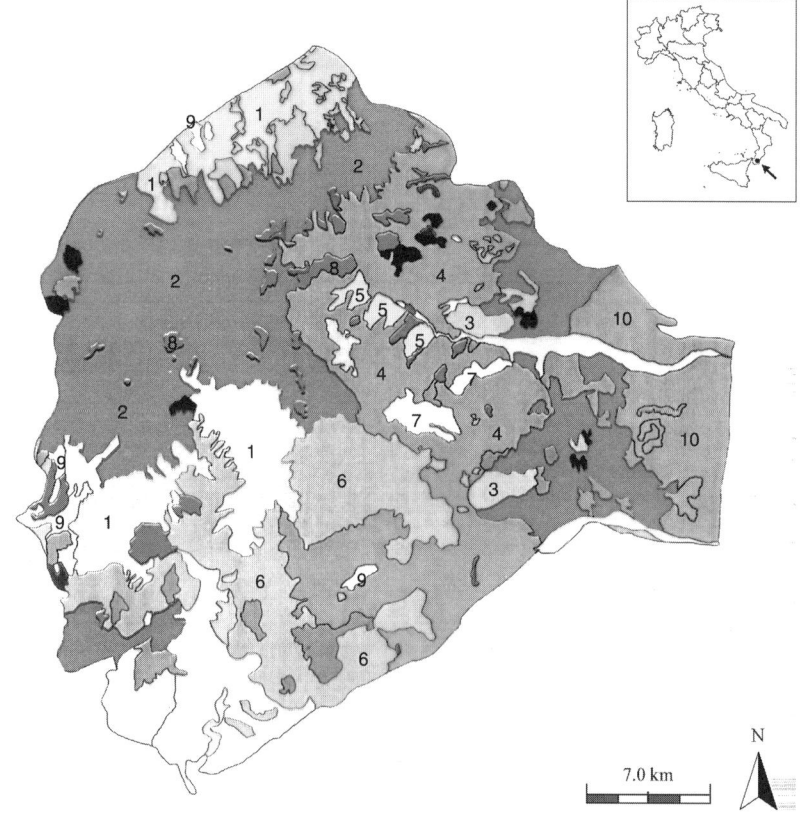

Figure 66.2 From the Map of Italian Nature, distribution of CORINE habitat mosaics (1:50,000) for the Aspromonte area (Area 8 from Figure 66.1) where a total of 168 patches belonging to 30 habitat types were identified. The following primary CORINE habitat types were identified:(1) Sila and Aspromonte Laricio forests (CORINE code 42.651); (2) Aspromonte beech forest (CORINE code 41.185); (3) Meso-Mediterranean holm-oak forests (CORINE code 45.31); (4) Illyrian holm-oak woodland (CORINE code 45.319); (5) Tyrrhenian broom fields (CORINE codes 31.844, 32.21, and 32.23); (6) Thermo-Mediterranean brushes, thickets, and heath garrigues (CORINE codes 32.23, 32.21, and 34.5); (7) Mediterranean xeric grasslands (CORINE codes 34.5 and 32.21); (8) Southern Italian mat-grass swards communities (CORINE code 35.72); (9) Chestnut woods (CORINE code 41.9); (10) Eucalyptus plantations (CORINE code 83.322).

Table 66.1 An Example of the Hierarchical Classification of Habitats within the CORINE Biotopes Classification Scheme

CORINE Habitat Classification	Description	CORINE Code
Sclerophyllous scrub	Mediterranean and sub-Mediterranean evergreen sclerophyllous bush and scrub maquis, garrigue, matorral, recolonization and degradation stages of broad-leaved evergreen forest, supra-Mediterranean garrigues, pseudo-maquis, Macaronesian xerophytic communities	32
Arborescent matorral	*Quercetalia ilicis, Pistacio-Rhamnetalia alaternii*	32.1
Thermo-Mediterranean shrub formations	*Pistacio-Rhamnetalia alaterni, Oleo-Ceratonion, Asparago-Rhamnion oleoidis, Periplocion angustifoliae, Rhamno-Quercion cocciferae, Juniperion lyciae*, etc.	32.2
Meso-Mediterranean silicicolous maquis	*Cisto-Lavanduletea, Pistacio-Rhamnetalia alaternii, Ericion arboreae*; shrubby formations, often tall, on most siliceous soil of the meso-Mediterranean zone of the Iberian peninsula, France, Italy, and the large western Mediterranean islands, degradation stages of evergreen oak forests; very similar formations of the thermo-Mediterranean zone and of the eastern Mediterranean are included	32.3
High maquis	Highest formations, with a tall stratum of *Erica arborea, Arbutus unedo, Quercus* spp., but few or no emergent oaks, in contrast to 32.1	32.31

Note: The high maquis (code 32.31) belongs to the meso-Mediterranean silicicolous maquis (code 32.3), which, in turn, belongs to sclerophyllus scrub (code 32) (modified from European Union Directorate General XI, 1991).

Modified from European Union Directorate General XI, CORINE Biotopes Manual, Habitats of the European Community. A Method to Identify and Describe Consistently Sites of Major Importance for Nature Conservation, EUR 12587/3, Brussels, 1991.

syntaxa at the landscape level down to alliances and associations. This is a provisional classification open to further clarification and precision. An example of hierarchical classification for habitats within the CORINE biotopes classification (European Union Directorate General XI, 1991) is highlighted for the *high maquis* in Table 66.1. In the classification, the brief description of habitats, the list of plants, and the phytosociological terms included are intended first to facilitate identification by data collectors, and second to draw attention to sensitive taxa harbored by the concerned habitat. A habitat is defined as:

1. A patch differing from its surroundings in nature and appearance at the finest resolution of the available spatial data
2. Capable of covering large enough surfaces to be important habitat for animals
3. Physiognomically significant in the landscape
4. Essential to the survival of sensitive or rare plant and animal species
5. A necessary element of a larger ecosystem
6. Remarkable because of either ecological processes or its aesthetic value

Map of Italian Nature uses the Landsat Thematic Mapper (TM) to derive digital maps of habitats (Figure 66.3). First, the images are used to identify and mask out urban areas, villages, and sparse agricultural settlement areas. The TM imagery is then used to focus on the habitat patterns of the remaining more naturalistic areas.

An airborne multispectral infrared and visible imaging spectrometer (MIVIS), available from the Italian National Council of Research (CNR), is also used on some areas of interest to generate images with 102 spectral bands in four spectral ranges (0.43 to 0.83, 1.15 to 1.55, 2.0 to 2.5, 8.2 to 12.7 μm), allowing a spatial resolution up to 3 m for the further identification of structural and functional indicators, subtle gradients, and structures like ecotones and riparian corridors.

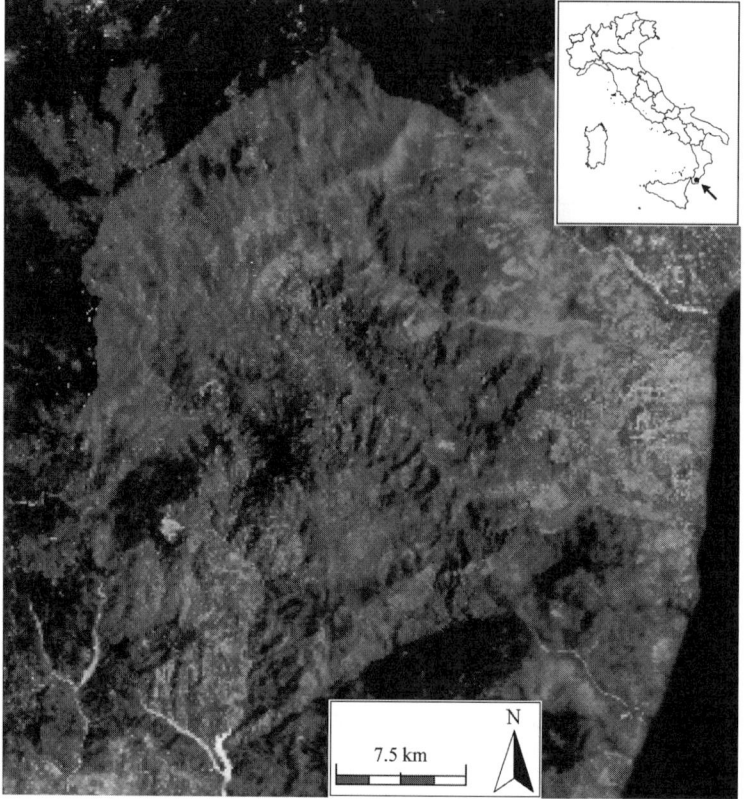

Figure 66.3 Satellite image of the Aspromonte area from the Map of Italian Nature (Landsat TM5, July 1996).

Preprocessed seven-band TM data are used to delineate areas of relative homogeneity and these are then labeled into classes, defined by the CORINE habitat classification. We use true and false color digital TM scenes, together with other available information sources such as MIVIS hyperspectral imagery, air photos, and other existing maps, to create an integrated visual interpretation of habitat patterns. CORINE habitat patterns also are delineated by supervised classification (Colwell, 1983), by selecting known representative CORINE habitats (training areas) and deriving discriminant functions based on their spectral properties. In practice, the final delineation of habitat mosaics on comparable rendering maps is an iterative process based on integrated evidence from processed satellite imagery, true and false color TM scenes, aerial photos, MIVIS imagery, existing vegetation, other coverage maps, and field reconnaissance.

An example of the decision rule for the CORINE classification can be made in reference to olive trees (*Olea europaea*). When sparse, they are evaluated together with other plant species and habitat physical traits and considered within a specific plant syntaxa (e.g., olive-carob forests). When they dominate, accounting for more than 80% of the vegetation cover, they are classified as olive groves if the peculiar habitat traits of this agroecosystem also are met.

About 10,000 CORINE habitats in Italy, representative of habitat and biotic heterogeneity at the national level, have been identified to date at different scales, mainly 1:50,000 (typical project scale), but also at 1:25,000, accounting for 1,500,000 ha.

CORINE habitat mosaics can be regarded as integrators of many of the physical and biological properties of an area. Each habitat type represents a list of typical plant and animal species linked to it (e.g., invertebrates; Council of Europe, 1996), as well as species shared with other habitat types. Since the urgency for conservation action can often be so great that the lack of detailed inventories for most taxa becomes of secondary importance, CORINE habitats could be extended, as a coarse filter approach (Noss, 1983), to threatened and nonthreatened species with no actual recorded distribution data. Of course, no coarse filter can be regarded as a complete substitute for biological health and integrity protection status. In Map of Italian Nature habitats, species passing through the mesh of the coarse filter, such as very narrow endemics or wide-ranging animals, can be recorded in the data bank and are not lost for mapping. They can still be recovered by higher- (or lower-) resolution assessment of potential high-priority biodiversity areas in the embedded hierarchy of organization levels provided by CORINE habitats (Zurlini and Rossi, 1995). For the Map of Italian Nature, the habitat level (coarse filter) protection can be a complement, not a surrogate, for protection of individual species at risk. Thus, the Map of Italian Nature intends to protect landscape mosaics at the broader scale, based on gaps in protection and threats, while including finer scale data on threatened species and habitats to make sure they do not fall through the cracks.

BIODIVERSITY COMPONENTS AND CONSERVATION VALUES

With regard to habitats, digital and alphanumeric data were acquired at a scale of 1:100,000 from the Nature 2000 national network of special areas of conservation (SAC) according to Directive 92/43/EU, listing habitat types whose conservation demands consideration for SAC designation, and whose animal and plant species require similar attention by member states. Nature 2000 information layers were generally enlarged to a scale of 1:50,000 and cross-validated by CORINE habitat mosaics below. Existing national and regional park and reserve areas also were acquired as digital data.

As a rule, priority is given only to officially recognized and geo-referenced data available on threatened species. At this first broad analysis level of the Map of Italian Nature, our analyses were restricted to threatened vertebrates, although we are currently enlarging vertebrate data and collecting distribution data on invertebrates and plants, officially provided by scientific societies.

Of 551 terrestrial vertebrate species present, even if only in Italy seasonally, 398 are native reproducing species while 153 are exotics and not reproducing species (Pinchera et al., 1997). Eighty-eight native species are classified in the threatened categories according to IUCN (1996) and Directive 92/43/EU, following Pinchera et al. (1997) updates. Digital maps of observed distribution for the 88 threatened terrestrial native vertebrates were acquired from the Italian Ministry of Environment data bank (Boitani et al., 1993), based on current agency records and reliable records from research and conservation institutions. Digital maps were selected for 31 vertebrates (Table 66.2) with their distribution range included for at least 10% within the 1,500,000 ha currently investigated. Maps were enlarged to 1:100,000 and intersected with CORINE land use maps (Heymann et al., 1993) and digital elevation model (DEM) coverages. Habitat suitability for each species, either within (expected) or outside the observed distributions (potential), was derived by rules of combination for intersecting areas, based on different land use suitability and the elevation range of species distribution. Suitability was expressed as positive or negative scores (3, 2, 1, –1, –2, –3) according to whether it was high (+) or low so that avoidance might occur (–), with unknown suitability equal to zero (Figure 66.4). These digital maps refer to CORINE land use coverages (1:100,000), and represent to date the best current knowledge, providing predictions of encountering an animal in a particular area. The digital maps were carefully enlarged by experts to CORINE habitat mosaics (1:50,000), so that species could be attributed to patches as small as 5 ha (minimum mapping unit). For some species, such a resolution may be desirable to allow more precise estimation both of habitat area and of the wildlife and habitat relationship model (WHRM). For others, such small patches may be biologically meaningless. Most terrestrial vertebrates considered by the project depend on multiple habitat types and sizes for their life history. In order to establish a WHRM for wide-ranging animals, CORINE habitat mosaics summarized at the landscape or at

Table 66.2 List of 31 Threatened Native Terrestrial Vertebrates and Their Distribution Range from the Map of Italian Nature Project

Genus	Species	Class	Order	Threat Level	Occurrence Sites Corresponding to Figure 66.1
Rhinolophus	*blasii*	Mammalia	Chiroptera	CR	2
Hieraatus	*fasciatus*	Aves	Accipitriformes	CR	8
Neophron	*percnopterus*	Aves	Accipitriformes	CR	7, 8,
Canis	*lupus*	Mammalia	Carnivora	EN	3, 4, 5, 7, 8
Barbastella	*barbastellus*	Mammalia	Chiroptera	EN	2, 3, 4, 5
Myotis	*capaccinii*	Mammalia	Chiroptera	EN	2, 3, 4, 5, 7, 8
Rhinolophus	*hipposideros*	Mammalia	Chiroptera	EN	2, 3, 4, 5, 7, 8
Pyrrhocorax	*pyrrhocorax*	Aves	Passeriformes	EN	4
Charadrius	*morinellus*	Aves	Charadriformes	EN	5
Rana	*latastei*	Amphibia	Salientia	EN	3
Martes	*martes*	Mammalia	Carnivora	VU	2, 3, 4, 5, 7, 8
Felix	*silvestris*	Mammalia	Carnivora	VU	4, 5, 7, 8
Lariius	*minor*	Aves	Passeriformes	VU	5, 4, 7
Sylvia	*hortensis*	Aves	Passeriformes	VU	3, 4, 5, 7, 8
Oenanthe	*hispanica*	Aves	Passeriformes	VU	2, 5, 8,
Melanocorypha	*calandra*	Aves	Passeriformes	VU	5, 7, 8
Picoides	*tridactylus*	Aves	Piciformes	VU	2
Picoides	*leucotos*	Aves	Piciformes	VU	4
Picoides	*medius*	Aves	Piciformes	VU	4, 5, 7, 8
Coracias	*garrulus*	Aves	Coraciiformes	VU	4, 5, 7, 8
Columba	*oenas*	Aves	Columbiformes	VU	4, 7, 8
Bubo	*bubo*	Aves	Strigiformes	VU	2, 3, 4, 5, 7, 8
Scolopax	*rusticola*	Aves	Charadriformes	VU	2, 3, 4
Burhinus	*oedicnemus*	Aves	Charadriformes	VU	5
Tetrao	*urogallus*	Aves	Galliformes	VU	2
Falco	*biarmicus*	Aves	Falconiformes	VU	3, 4, 5, 7, 8
Falco	*subbuteo*	Aves	Falconiformes	VU	2, 3, 4, 5, 7
Circus	*pygargus*	Aves	Accipitriformes	VU	3, 4, 5
Circaetus	*gallicus*	Aves	Accipitriformes	VU	2, 5, 7
Milvus	*milvus*	Aves	Accipitriformes	VU	5, 7
Testudo	*hermanni*	Reptilia	Testudines	VU	3, 4, 5, 7, 8

Note: Vertebrates were selected if their distribution range was included for at least 10% within the 1.5 million ha currently investigated. CR = critically endangered, EN = endangered, VU = vulnerable.

landscape systems level are going to be considered together with corridors (Corsi et al., 1999) and estimates of species home range size.

For freshwater vertebrates in Italy, the status of the native fish (45 species of 71) is seriously compromised; only 17% of the original fauna (mostly endangered diadromous species) have not been involved in human manipulations and trans-introductions (Bianco, 1995). Most native species have been subjected to authorized or clandestine trans-introductions; only 36% have not been contaminated by trans-introductions of cospecific populations or transplanted to localities outside their original range. Today there is no way to trace the original range of any primary freshwater species (Bianco, 1995). Thus, Italian freshwater systems appear, as a whole, to have quite poor biological integrity, according to the definition of Karr and Dudley (1981).

Strategies and planning for conservation at the ecosystem and regional level, rather than the species-by-species approach focusing strictly on target or endangered species, may protect up to 85 to 90% of species by conserving representative natural communities without a separate inventory of individual species (Noss, 1987). We generated integrated conservation value distributions by intersecting several maps based on habitat and threatened species distributions. Intersection of

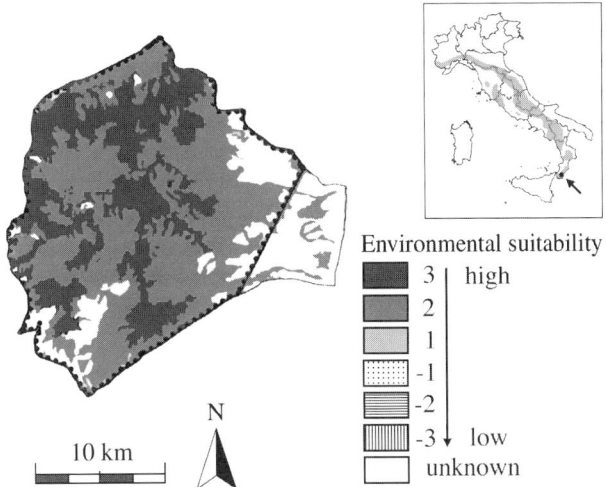

Environmental suitability

3 | high
2
1
-1
-2
-3 ↓ low
unknown

N

10 km

Figure 66.4 Example of vertebrate distribution range map for the wolf (*Canis lupus*) in the Aspromonte area from the Map of Italian Nature. The area within dotted boundaries represents the present observed distribution. The wolf's distribution range for Italy is given in the small map. Habitat suitability of CORINE land use covers is provided through positive (good suitability) and negative scores; unknown suitabilities are equal to zero (see text).

habitat coverages produced a GIS composite coverage (hypercoverage) where each habitat received a score according to the different information layers to which it belonged (Zurlini et al., 1999). Scores were assigned to each CORINE habitat: (a) for its occurrence as habitat per se in the Directive 92/43/EU, (b) for its inclusion within park and reserve areas, or within SAC sites, and (c) for the presence of threatened vertebrates, from observed range distribution maps. An example of map-based conservation value of threatened vertebrates distribution is given in Figure 66.5 for the Aspromonte area. Ranked scores were derived for either absolute species occurrences or species per unit of habitat surface (threatened species concentration), with five ranks based on histogram equalization of data frequencies at the current broadest (1,500,000 ha) analysis level. Multivariate scaling techniques were then applied and, by simple inspection of cluster vector means, it was possible to order habitat clusters according to either absolute conservation values (Figure 66.6) or conservation value concentration (Zurlini et al., 1999).

DISTURBANCE

Since disturbance is currently considered to include every process altering birth and mortality rates of individuals present in a patch (Petraitis et al., 1989), it is difficult to separate its intrinsic and extrinsic factors (Nilsson and Grelsson, 1995). Therefore, it seems more appropriate to consider a combination of those factors as provided by fragility.

We considered two main types of human disturbance: (1) human interference activities and (2) pollution. The former refers to interferences from all human activities that transform, interrupt, and fragment the environment, together with noise and other controlled and uncontrolled presences. The latter refers to pollution activities, which in our case are represented by organic load into the watershed from industry, agricultural practices, human residential impact, and tourism.

Human interference was assessed using GIS to read and quantify the length of networks given by state and regional roads, highways, and railways. Road lengths in kilometers crossing and partially or completely delimiting each CORINE habitat were calculated. In addition, habitat edges with agricultural, urban, and industrial areas were considered. For each habitat, the overall amount of exposure length (roads plus edges) in kilometers per unit of habitat surface were calculated and

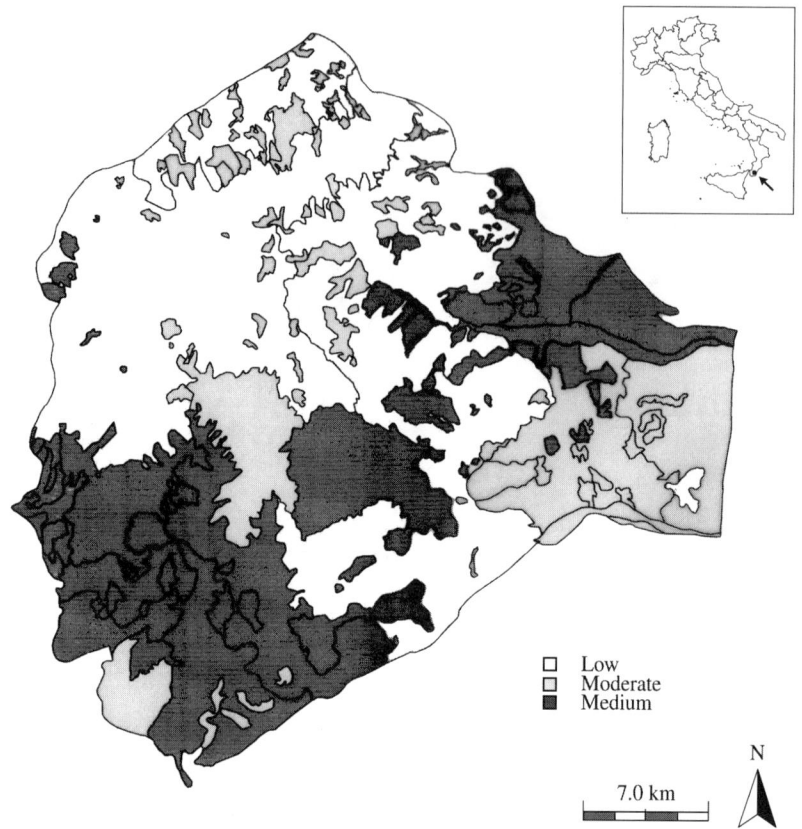

□ Low
□ Moderate
■ Medium

7.0 km

N

Figure 66.5 Distribution map of habitat conservation values according to the occurrences of threatened native vertebrates for the Aspromonte area from the Map of Italian Nature.

ranked to provide a disturbance exposure index (DE) with the first rank expressing slight or very low disturbances (Zurlini et al., 1999). An example of DE distribution is given in Figure 66.7 for the Aspromonte area.

Pollution was based on (a) the number of residents of each municipality, (b) the number of employees in agriculture, (c) the number of employees in industry, based on estimates from the 1991 census of the Italian National Statistical Institute. Numbers were transformed into inhabitant-equivalents by coefficients varying according to the categories (a), (b), and, within (c), to the type of industry, as established by the National Water Research Institute (IRSA) (Marchetti, 1998), to estimate potential pollution for residents (RWL), agriculture (AWL), and industry (IWL), respectively.

An overall estimate of agricultural pollution into bodies of water was obtained by taking the AWL and adding 5% of pollution from animal breeding (BWL). BWL was weighted differently following IRSA procedure based on the number of breeding animals (cattle, sheep, pigs, etc.) present. From these basic metrics, different indices could be derived. A simple metric of pollution stress to bodies of water at the watershed level could be obtained by ranking the metric WL = AWL + BWL + RWL + IWL per unit of surface in five ranks. The remaining 95% of pollution due to animal breeding (BL) is assumed not to go into bodies of water (Marchetti, 1998), but to the portion of agricultural land falling in each watershed. This was done using GIS by intersecting digital watershed maps with municipality boundaries and agricultural land use areas. This resulted, for each municipality, in a ranked estimate of organic load for agricultural use per unit of surface

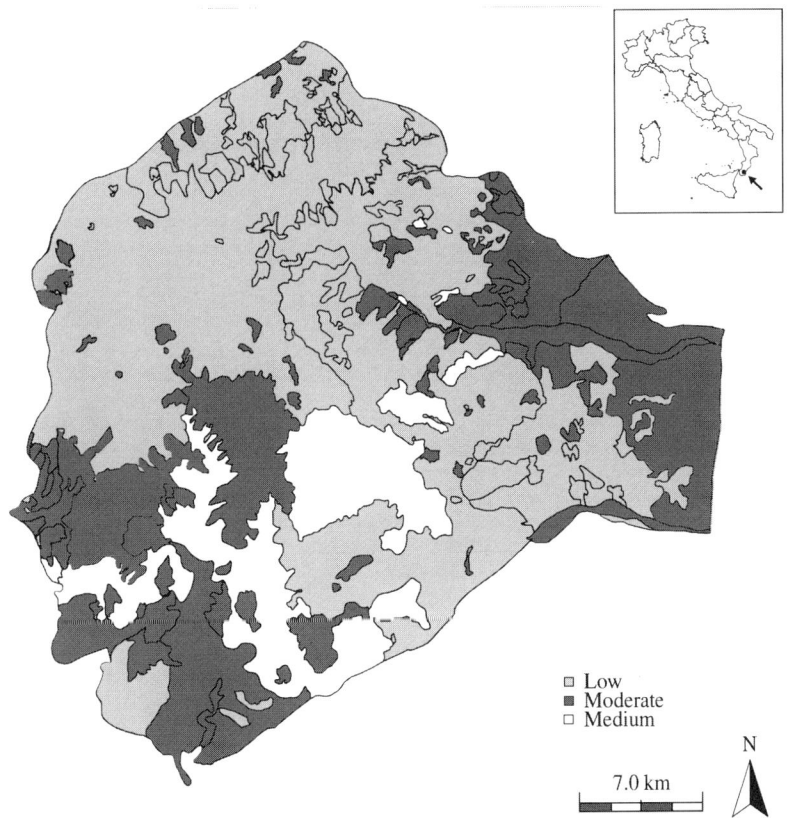

Figure 66.6 Distribution map of overall absolute conservation values of CORINE habitat mosaics in the Aspromonte area from the Map of Italian Nature.

(AL). All anthropogenic disturbance indices were ranked by histogram equalization of all data frequencies at the broadest analysis level (Zurlini et al., 1999).

A SIMPLE MODEL FOR FRAGILITY

At each hierarchical level within the Map of Nature, fragility of an environmental system at time t depends on both the sensitivity of the system and overall disturbance, expressed by natural disturbances plus the intensity and frequency of human activities.

A simple conceptual (linear) model can be expressed by the following equation:

$$Fr = K + \alpha(U) \tag{66.1}$$

where Fr is the degree of fragility on an arbitrary, semiquantitative scale; K is a constant; α is a coefficient of sensitivity; and U represents overall disturbance, radically different from the natural regime, and linked to unexpected events not embodied by the system through homeostatic or homeoretic adaptive mechanisms (Figure 66.8).

Model 66.1 is a specific case of a more general model adopted within the project:

$$Fr = K + \alpha(U)^\beta \tag{66.2}$$

where β takes into account that in a given situation the increase of Fr with U may be different from that expected from a linear relationship. When $\beta = 1$ Model 66.2 reduces to Model 66.1; when β

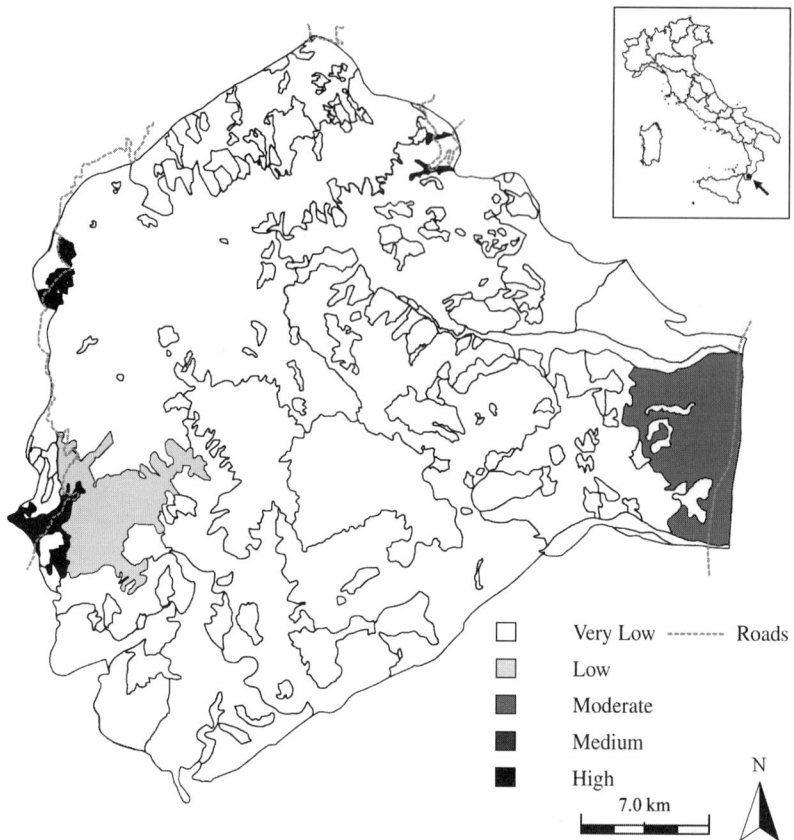

Figure 66.7 Distribution map of the disturbance exposure index (DE) for the Aspromonte area from the Map of Italian Nature.

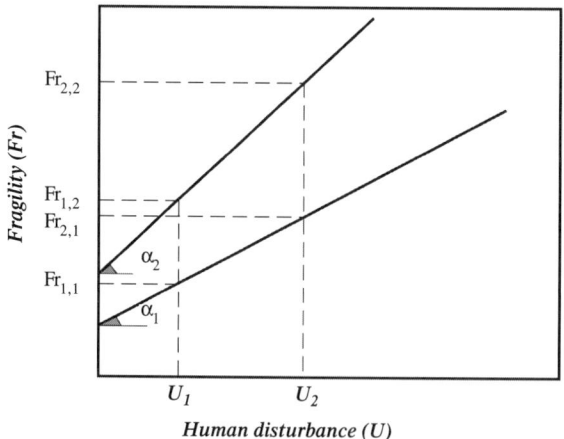

Figure 66.8 Graphic representation of the simple conceptual model for estimating fragility from the Map of Italian Nature. Fr = degree of fragility, α = coefficient of sensitivity, U = overall disturbance.

> 1 Fr increases more than expected according to Model 66.1; for $\beta < 1$ Fr increases less than expected according to Model 1. For $U = 0$, K expresses background fragility typical of the hierarchical level considered. Such fragility is related to specific adaptive mechanisms to expected disturbances, embodied into the memory of the system itself, characterizing the system with relative frequencies in time and space (Norton and Ulanowicz, 1992), and including human disturbances that mimic or simulate natural disturbances. We used Model 66.1 as a first approximate model.

DERIVING SENSITIVITY, FRAGILITY, AND BIOLOGICAL HEALTH IN PRACTICE

Fragility has been often obtained by scores on a semiquantitative scale (Nilsson and Grelsson, 1995), but invariably human disturbances were not explicitly considered. Although empirical, this way of quantifying fragility can be useful since it meets the need to compare habitats, planning intervention, and maintenance measures.

To identify and evaluate fragile habitats, it is important to compare distribution maps of threatened native species. The study area is assumed to be partly or fully covered by species distribution maps as CORINE habitat mosaics. Intersecting areas with one or more threatened species are then considered to be *critical* areas for conservation. However, two critical areas with the same degree (n) of threat, where n is the number of threatened species present, can be different due to the *quality* of threat to the species. Different extinction threat categories for each species, as provided by IUCN (1996) and Pinchera et al. (1997) for vertebrates, must be taken into account. Thus, for threatened vertebrate species in an area, n_1 are critically endangered, n_2 are endangered, and n_3 are vulnerable.

Extinction threats to IUCN threatened species categories are based on the reduction probabilities (p) in the wild within 10 years or three generations, whichever is longer (International Union for Conservation of Nature and Natural Resources–The World Conservation Union, 1996; Walter and Gillet, 1998): critically endangered species (CR) with $p_1 = 0.80$; endangered species (EN) with $p_2 = 0.50$; vulnerable species (VU) with $p_3 = 0.20$. The IUCN criteria which led to extinction threats are given in Table 66.3.

IUCN Red List criteria and categories are applicable to all taxonomic groups, except microorganisms (International Union for Conservation of Nature and Natural Resources–The World Conservation Union, 1996; Walter and Gillet, 1998). The Map of Italian Nature project needs, however, to produce extinction probabilities for a time interval smaller than 10 years. By using extinction curves provided by Akcakaya and Ginzburg (1991), it is possible to estimate extinction probabilities within a 5-year period.

Table 66.3 Criteria Adopted by IUCN (International Union for Conservation of Nature and Natural Resources) for Extinction Threats to Threatened Species Categories

Criteria for Estimating Extinction Threats to IUCN-Threatened Species Categories	Threatened Species Designation		
	Critically Endangered (CR)	Endangered (EN)	Vulnerable (VU)
Observed, estimated, or projected reduction in wild population within 10 years or three generations, whichever is the longer	80%	50%	20%
Extent of occurrence	<100 km²	<5000 km²	<20,000 km²
or			
Area of occupancy	<10 km²	<500 km²	<2000 km²
Population of mature individuals estimated to number less than:	250	2500	10,000

International Union for Conservation of Nature and Natural Resources–The World Conservation Union, *1996 IUCN Red List of Threatened Animals*, Gland, Switzerland, 1996; and Walter, K.S. and Gillet, H.J., Eds., *1997 IUCN Red Lists of Threatened Plants*, compiled by the World Conservation Monitoring Center, International Union for Conservation of Nature and Natural Resources–The World Conservation Union, Gland, Switzerland and Cambridge, 1998.

Risk can be defined as the probability of suffering harm from a hazard. A complete technical analysis of risk describes (1) a hazard that is a substance or an action that can cause harm, (2) the events or conditions that create the possibility of harm, and (3) a statistical estimate of the likelihood that harm will occur. A hazard (risk agent) can be a chemical substance, biological organism, radioactive material or other potentially hazardous substance, or a disturbance activity. Under plausible circumstances, a risk agent can cause harm to human health and ecosystems. A baseline risk assessment estimates current risks to organisms and ecosystems resulting from direct and indirect exposure to risk agents in the absence of any remediation.

The likelihood (I) that a given area will lose one or more threatened native species in a fixed time interval (Δt) tends to increase with $n_1 + n_2 + n_3$, but it also depends on the categories of threat to each species. Therefore:

$$I = (p_1 \cdot n_1) + (p_2 \cdot n_2) + (p_3 \cdot n_3). \qquad (66.3)$$

Thus, I takes threat weights into consideration, using p-values, and gives the minimum number of threatened native species expected to become extinct in a fixed time interval (Δt) for a given area. Differences in I-values between areas can be considered to be proportional to the differences in sensitivity of those areas, because we define sensitivity as a condition due to the presence of threatened species and a trait of the area itself.

I-values do not necessarily correspond to the condition of ecological processes that maintain conservation values or species richness as I-values do not directly address functional traits; neither do they reflect patterns of genetic variation within and among population variability that may illustrate the relative risk of extinction among populations (Dunham et al., 1999). Nevertheless, a large number of the threatened vertebrates are considered at the top of their natural food webs, and they are end points to a number of biological and ecological food web processes. Therefore, when comparing areas, I-values can represent the baseline intrinsic risk to native biodiversity, because I-values incorporate current intrinsic traits of native species (events or situations creating the possibility of harm). Following the IUCN criteria, these intrinsic traits can be given by:

1. Observed, estimated, or projected population reductions, including demographic stochasticity (Caughley, 1994)
2. Extent of occurrence
3. Area occupancy (the ability of a species to occupy a larger or smaller fraction of its potential suitable habitat)
4. Number of mature individuals (representative of population growth potential)

Fragility is not directly measurable, but can be estimated, using Model 66.1 or Model 66.2 through surrogates for U (risk agents, e.g., DE) and α (sensitivity; i.e., the likelihood that the harm will occur together with the exposure conditions creating that possibility). As a surrogate for sensitivity (α), four ranked I-values (I^*) I^*-1 through I^*-4 for threatened native vertebrates were used, based on histogram equalization of data frequencies at the current broadest analysis level.

Two comparable areas can have similar sensitivity but different fragility because of differences in the type, frequency, and extent of the risk agents. Different estimates of current fragility can be made using different kinds of risk agents (disturbances) and their relative impact on the responses of the threatened species in question (receptor). For example, most vertebrates can be assumed to cope immediately with disturbances (e.g., noise) by escaping, whereas for invertebrates with aquatic larval stages, other stressors like water pollution are inescapable and should be addressed. The diverse roles receptor species play in the natural food webs, including producers, decomposers, predators, parasites, herbivores, and pollinators, can also be differently affected by various disturbances.

By comparing habitats within CORINE types and between comparable habitat composites (mosaics) in the landscape according to their sensitivity (**I***-values) at the broadest analysis level, it is possible to perform comparative assessments of biological health by defining well-structured, biologically healthy systems. Habitats that fall outside the current normal bounds and that can become dysfunctional because of different degrees and types of stress also can be identified. In this context, current fragility estimates also can be regarded as baseline risk assessments to biological health.

SOME ILLUSTRATIVE RESULTS

Some examples are given for the Aspromonte area (Area 8 in Figure 66.1, CORINE habitat mosaics in Figure 66.2, and satellite image in Figure 66.3), which is mainly montane, but also contains coastal areas of the Ionian Sea. The montane habitats of the northwest range from 1000 to 1956 m altitude and occupy a metamorphic substrate susceptible to erosion and the typical pedogenesis of acid environments. Climate is Mediterranean-montane type, and morphology is rocky with ridges, slopes, depressions, and streams in sunken valleys; beech woods are prevalent here, with two overlapping associations represented by *Aquifolio-Fagetum* at altitudes up to 1500 m and *Asyneumati-Fagetum* at higher altitudes, belonging to the same CORINE habitat, coded 41.185. *Abies alba* is frequent in the association at intermediate altitudes, but disappears at higher altitudes. *Pinus laricio* woods (CORINE code 42.651) start at lower altitudes on ranker soils. Dependent plant associations are present on rocky outcrops. Coastal landscape habitats, located on the eastern side, are characterized by seasonally torrential streams with very broad beds (*fiumare*), and agricultural practices (citrus orchards) with concentrations of exotic species (*Eucalyptus* plantations). Anthropogenic areas are represented by a few villages and sparse agricultural settlements with negligible industrial activity.

To estimate conservation values, human disturbance indices, and **I***-values, the approach described in previous sections was followed, with rank-values comparable within the current broadest analysis level of the project. Overall, 14 occurrences of threatened native vertebrates were recorded (Table 66.2), including wolves (*Canis lupus*), martens (*Martes martes*), wild cats (*Felis silvestris*), and birds of prey: *Hieraatus fasciatus* (Bonelli's eagle), *Neophron percnopterus*, hawks, and owls (*Bubo bubo*). To find a suitable surrogate for *U* in Model 66.1, we took into account human residential impact as well as agricultural, pastoral, and forest activities, which are quite significant in the area, whereas tourism and industry are negligible. Thus, terrestrial vertebrates were assumed to be affected mainly by DE, RWL, and by inhabitant-equivalents (employees) for AWL per unit of agricultural land use together with those due to animal breeding (AL). These four single metrics (DE, RWL, AWL, AL) were ranked at the broadest analysis level for the 168 habitat units and 30 habitat types identified at a scale of 1:50,000 in the area, then scaling techniques (principal component analysis) were applied to derive two components explaining most of the variance. Those components were subjected to *k*-means cluster analysis to identify five main habitat clusters for easier interpretation. By simple inspection of cluster vector means it was possible to order habitat clusters according to disturbances in five ranks. This was called overall disturbance (Figure 66.9).

For sensitivity, we used four **I***-ranks to index threatened vertebrates (low, moderate, medium, high). Fragility scores were transformed into a series of six ranks from very low (1) to very high (6), based on histogram equalization of habitat fragility frequencies. Sensitivity and current fragility distribution maps for the Aspromonte are presented in Figures 66.10 and 66.11, respectively.

Absolute conservation value distributions are presented in Figure 66.6. However, such distributions are biased since Directive 92/43/EU still neglects beech woods in the priority list.

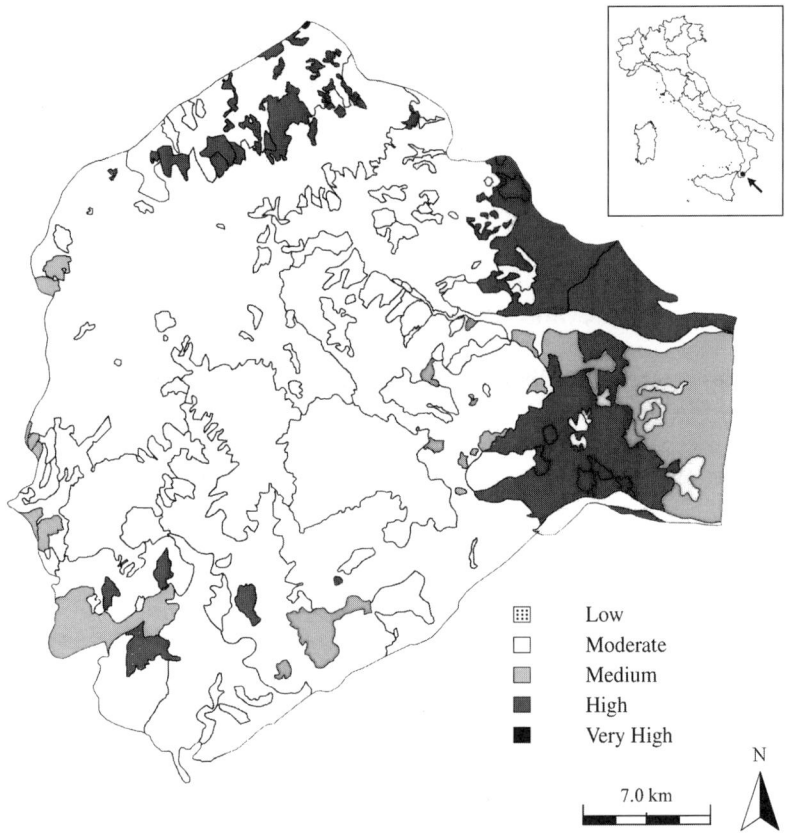

Figure 66.9 Distribution map of the overall disturbance index in the Aspromonte area from the Map of Italian Nature.

At higher elevations, patches with high sensitivities were found embedded within the environmental matrix of beech wood habitats (CORINE code 41.185). *Pinus laricio* woods (CORINE code 42.651) shelter more threatened native vertebrates than beech woods, likely because they are at lower altitudes, and as many as plantations with exotic *Eucalyptus*. By comparing Figures 66.10 and 66.2, we could recognize either similar sensitivity (biological health) levels between different habitat types (e.g., 41.185 and 45.319), or different sensitivity levels within the same habitat type, as illustrated by *Illiryan holm-oak* woods (code 45.319) that at intermediate elevations appear to be a buffer zone between beech woods and anthropogenic coastal areas (Figure 66.2). At the landscape level, sensitivity and fragility distributions were very similar because of relatively low human disturbance in this area compared to other Map of Italian Nature sites. Where sensitivity (biological health) was higher, fragility (risk to biological health) was also higher. The highest fragility and risks to biological health were found for patches of *Castanea sativa-dominated chestnut* woods located in the west (Figures 66.2 and 66.11).

Since landscapes are generally structured according to scaling regions with distinct dimensions connected by transition zones (Holling, 1992), we used fractal geometry to provide a more realistic picture of the geometry of naturally occurring objects. A shift in the fractal boundary dimension, *D*, might indicate a substantial change in the processes generating and in maintaining landscape patches (Krummel et al., 1987; Sugihara and May, 1990). A procedure based on successive, piecewise, log-log perimeter and area linear regressions (Grossi et al., 2001), where regression slopes are related to *D*, was applied to CORINE habitat mosaics and a statistically significant discontinuity was detected. This discontinuity separated two distinct scaling regions of habitats in the Aspromonte

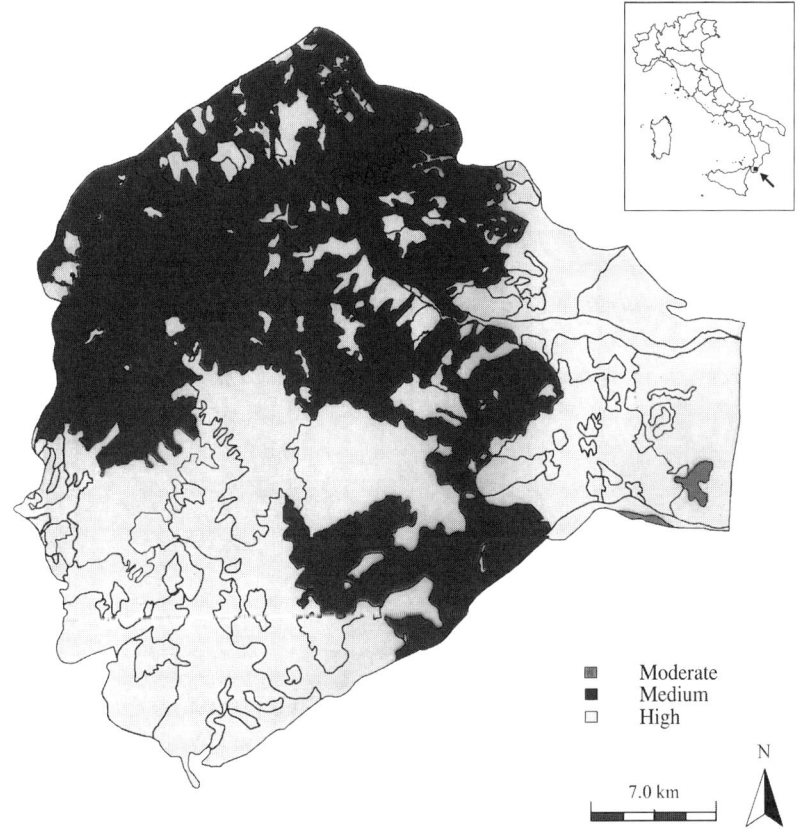

Figure 66.10 Sensitivity distribution map of CORINE habitat mosaics of the Aspromonte area from the Map of Italian Nature. Only ranks present in the area are shown.

area. Habitat patches at higher elevations showed higher fractal D-values, since patch dynamics appear more driven by natural processes, leading to more complex plane-filling fractal edges. In contrast, human and natural (e.g., fiumare) disturbances predominate on smaller spatial scales at lower altitudes, making for smoother patch geometry and lower fractal D-values.

CONCLUDING REMARKS

Integrity applies, at one end of a continuum of human influence, to sites with a balanced, integrated, adaptive biota, that is the product of evolutionary and biogeographic processes, with the full range of elements (genes, species, assemblages) and processes that are expected in the region's natural environment (Karr and Chu, 1999). Integrity as a management goal aims at systems looking like this evolved state as much as possible.

Through the kind of approach presented here (but see also GAP analysis, Scott et al., 1996), we are aware we might not preserve the processes that maintain and sustain diversity. Nevertheless, we recognize that ecosystem elements like species and habitat features are used, from the practical point of view, more frequently than processes as indicators of integrity, since they are more sensitive to degradation, more fully understood, and less expensive to monitor (Angermeier and Karr, 1994). In this sense, biodiversity is an important indicator of biological integrity (Angermeier and Karr, 1994). Diversity of threatened native species, CORINE habitats, and patch size and shape relationships are end points to a number of biological and ecological processes

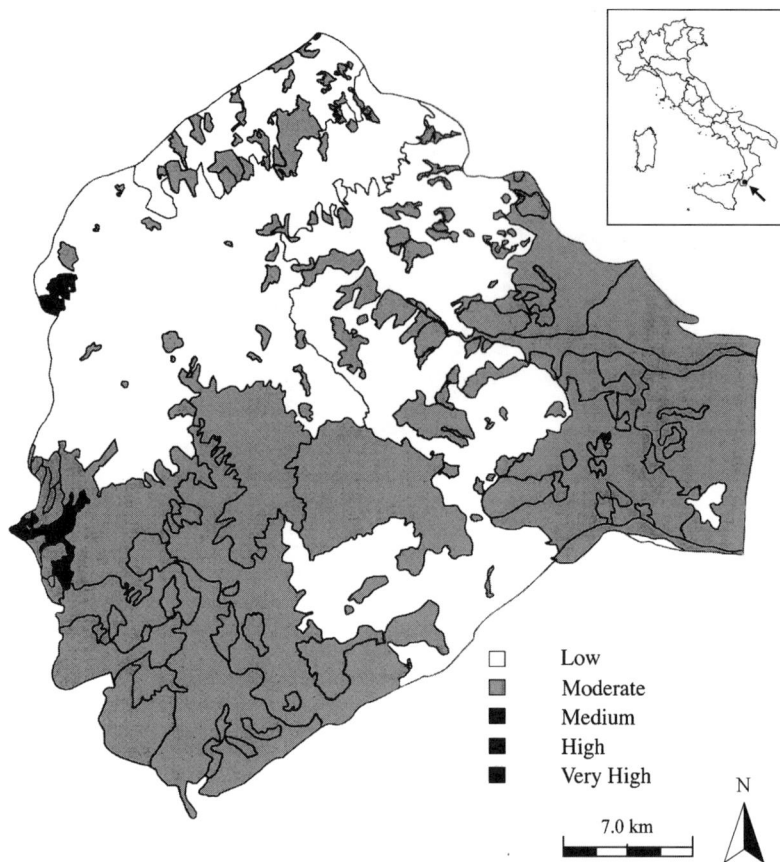

Figure 66.11 Fragility distribution map of CORINE habitat mosaics of the Aspromonte area from the Map of Italian Nature. Only ranks present in the area are shown.

they demonstrate. Indeed, CORINE habitats are not recognized by the Map of Italian Nature as patches themselves, but also for both the ecological processes they stand for, and in the functional perspective of preserving and maintaining habitat-creating processes (Prendergast et al., 1999), including human dynamics.

Albeit we speculatively acknowledge the need and the primacy of integrity as a management goal, we also think that preserving biotic communities, with particular concern for native species populations with their evolved or historic interactions together with their habitat composites, can be still a valuable parallel conservation goal. For example, in regions like the Mediterranean where, due to the millennial synergistic interaction between humans and nature, there might be no easy-to-find historical baseline to judge integrity that predates industrial man. This seems to be the case of freshwater fish communities in Italy (Bianco, 1995). Furthermore, seeking historical benchmarks for evaluating biological integrity addresses elusive issues that must be addressed in planning the restoration of perturbed systems, with regard to the question of resetting the ecological clock, as pointed out by Cairns (1994). Actually, the ecological system, once restored, must be self-maintaining and integrated into the larger landscape where the perturbed patch occurred. When the damage occurred long ago, resetting the ecological clock to predisturbance condition might not always be compatible with the larger landscape where it occurred. Presumably, the larger landscape also has been changing since the patch was perturbed. Therefore, if we want to restore an ecosystem to some approximate earlier condition, at what point should the ecological clock be reset?

Wilderness conservation is dealing with dynamic systems that tend to be more dynamic at interfaces with humans. Under the influence of these human dynamics, conservation has to face an intrinsic paradox (i.e., that we seek to preserve what must change) (Pickett and White, 1985).

The prioritization of areas at risk is a function both of the status of our knowledge and of urgency for conservation management. The Map of Italian Nature is a knowledge system that is evolving and updating due to enlargement, clarification, and increased precision of the data. Ideally, Map of Nature digital maps will never be finished products, since they must continually evolve as new information is gathered at state, regional, and local government levels. Thus, we will prioritize areas at greater risk in a comparative way with approximations based on the best status of knowledge currently available.

Through very simple models and procedures we are generating distribution maps of threatened habitats, plant communities, and animal species. However, our emphasis is on maps of habitat sensitivity and fragility, which can give policy makers and land managers information to foresee the impact their land-use decisions will have on existing risks to ecosystem health.

Assessing the significance of risks to biological health involves judgment and trade-offs to determine what risk level is acceptable. A variety of formal techniques can be used to address this question, including risk–perception analysis, cost–benefit analysis, and decision analysis. However, the use of such techniques to determine an acceptable level of risk is controversial. They provide insight, but also involve uncertainties. Therefore, decision makers, who are going to use the Map of Italian Nature, must be aware that they have also to consider alternatives such as negotiation, consensus building, and other means of broadening involvement in acceptable risk decisions.

ACKNOWLEDGMENTS

The research was conducted under contracts with the Italian Ministry of Environment and the Prime Minister's National Technical Services. We are particularly grateful to G. P. Patil and W. Myers of the Pennsylvania State University for many discussions and helpful comments on this manuscript. Thanks are also given to S. Pignatti, L. Boitani, L. Soliani, L. Grossi, F. Spada, F. Pani, and M. Amadei, as well as a vast number of participants to the Map of Italian Nature project, and an anonymous reviewer. Many thanks are due to N. Zaccarelli for figures and tables.

REFERENCES

Akcakaya, H.R. and Ginzburg, L.R., Ecological risk analysis for single and multiple populations, in *Species Conservation: A Population Biological Approach*, Seitz, L. and Loeschcke, V., Eds., Birkhaeuser Verlag, Basel, Switzerland, 1991, pp. 43–57.

Allen, T.F.H. and Hoekstra, T.W., *Toward a Unified Ecology*, Columbia University Press, New York, 1992.

Angermeir, P.L. and Karr, J.R., Biological integrity versus biological diversity as policy directives: protecting biotic resources, *BioScience*, 44, 690–697, 1994.

Bianco, P.G., Mediterranean endemic freshwater fishes of Italy, *Biol. Conserv.*, 72, 159–170, 1995.

Boitani, L., Corsi, F., Angelici, F.M., Penteriani, V., and Pinchera, F., *Banca Dati della Fauna Terrestre Italiana*, Ministry of the Environment, Rome, 1993.

Cairns, J. Jr., Public policy must respond to the increasing need for ecosystem service production, *Environ.Manage. Health*, 5, 7–15, 1994.

Caughley, G., Directions in conservation biology, *J. Anim. Ecol.*, 63, 215–244, 1994.

Colwell, R.N., Ed., *The Manual of Remote Sensing*, American Society of Photogrammetry, Falls Church, VA, 1983.

Corsi, F., Duprè, E., and Boitani, L. A large-scale model of wolf distribution in Italy for conservation planning, *Conser. Biol.*, 13, 150–159, 1999.

Council of Europe, Ed., *Liste des biotopes d'Europe d'apres leur importance pour les invertebres*, Sauvegarde de la nature N. 77, Strasbourg, 1996.

Dunham, J., Peacock, M., Tracy, C.R., Nielsen, J., and Vinyard, G., Assessing extinction risk: integrating genetic information, http://www.consecol.org/vol3/iss1/art2, *Conserv. Ecol.*, 3,2, 1999.

European Union Directorate General XI, CORINE Biotopes Manual, Habitats of the European Community. A Method to Identify and Describe Consistently Sites of Major Importance for Nature Conservation, EUR 12587/3, Bruxelles, 1991.

Grossi, L., Zurlini, G., and Rossi, O., Statistical detection of multiscale landscape patterns, *Environ. Ecol. Stat.*, 8, 253–267, 2001.

Heymann, Y., Steenmans, C., Croisille, G., and Bossard, M., *CORINE Land Cover — Guide Technique*, Commission of European Community, Jouve, Paris, 1993.

Holling, C.S., Cross-scale morphology, geometry, and dynamics of ecosystems, *Ecol. Monogr.*, 62,,447–502, 1992.

Holling, C.S., Schindler, D.W., Walker, B.H., and Roughgarden, J., Biodiversity in the functioning of ecosystems: an ecological primer and synthesis, in *Biodiversity Loss: Ecological and Economic Issues*, Perrings, C.A. et al., Eds., Cambridge University Press, Cambridge, 1995, pp. 44–83.

International Union for Conservation of Nature and Natural Resources–The World Conservation Union, *1996 IUCN Red List of Threatened Animals*, Gland, Switzerland, 1996.

Karr, J.R., Using biological criteria to protect ecological health, in *Evaluating and Monitoring the Health of Large-Scale Ecosystems*, Rapport, D.J., Gaudet, C., and Calow, P., Eds., Springer-Verlag, Heidelberg, 1995, pp. 137–152.

Karr, J.R. and Chu, E.W., Eds., *Restoring Life in Running Waters: Better Biological Monitoring*, Island Press, Washington, D.C., 1999.

Karr, J.R. and Dudley D.R., Ecological perspectives of water quality goals, *Environ. Manage.*, 5, 55–68, 1981.

Krummel, J.R., Gardner, R.H., Sugihara, G., O'Neill, R.V., and Coleman, P.R., Landscape patterns in a disturbed environment, *Oikos*, 48, 321–324, 1987.

Mageau, M.T, Costanza, R., and Ulanowicz, R.E., The development and initial testing of a quantitative assessment of ecosystem health, *Ecosyst. Health*, 1, 201–213, 1995.

Marchetti, R., Inquinamento delle acque superficiali, in *Ecologia Applicata*, Provini, A., Galassi, S., and Marchetti, R. Eds., Società Italiana di Ecologia e Città Studi Edizioni, Turin, 1998, pp. 237–282.

Ministry of Environment, *Note esplicative alla legge quadro sulle aree protette*, Ministero dell'Ambiente, Servizio Conservazione Natura, Rome, 1992.

Naeem, S., Thompson, L.J., Lawler, S.L., Lawton, J.H., and Woodfin, R.M., Declining biodiversity can alter the performance of ecosystems, *Nature*, 368, 734–737, 1994.

Naveh, Z. and Liebermann, A., *Landscape Ecology: Theory and Application*, Springer-Verlag, New York, 1994.

Nilsson, C.N. and Grelsson, G., The fragility of ecosystems: a review, *J. Appl. Ecol.*, 32, 677–692, 1995.

Norton, B.G. and Ulanowicz, R.E., Scale and biodiversity policy: a hierarchical approach, *Ambio*, 21: 244–249, 1992.

Noss, R.F., A regional landscape approach to maintain diversity, *BioScience*, 33, 700–706, 1983.

Noss, R.F., From plant communities to landscape in conservation inventories: a look at the Nature Conservancy (USA), *Biol. Conserv.*, 41, 11–37, 1987.

O'Connell, T.J., Jackson, L.E. and Brooks, R.P., A bird community index of biotic integrity for the Mid-Atlantic Highlands, *Environmental Monitoring and Assessment*, 51, 145–156, 1988.

O'Neill, R.V., De Angelis, D.L., Waide, J.B., and Allen, T.F.H., *A Hierarchical Concept of Ecosystems*, Princeton University Press, Princeton, NJ, 1986.

Perrings, C.A., Maler, K–G., Folke, C., Holling, C.S., and Jansson, B–O., *Biodiversity Loss: Ecological and Economic Issues*, Cambridge University Press, Cambridge, 1995.

Petraitis, P.S., Latham, R.E. and Niesenbaum, R.A., The maintenance of species diversity by disturbance, *Q. R. Biol.*, 64, 393–418, 1989.

Pickett, S.T.A. and White, P.S., Eds., *The Ecology of Natural Disturbance and Patch Dynamics*, Academic Press, Orlando, 1985.

Pignatti, S., *Ecologia del paesaggio*, UTET, Milan, 1994.

Pinchera, F., Boitani, L., and Corsi, F., Application to the terrestrial vertebrates of Italy of a system proposed by IUCN for a new classification of national Red List categories, *Biodiversity Conserv.*, 6, 959–978, 1997.

Prendergast, J.R., Quinn, R.M., and Lawton, J.H., The gaps between theory and practice in selecting nature reserves, *Conserv. Biol.*, 13, 484–492, 1999.

Rapport, D.J., Ecosystem health: an emerging integrative science, in *Evaluating and Monitoring the Health of Large-Scale Ecosystems*, Rapport, D.J., Gaudet, C., and Calow, P. Eds., Springer-Verlag, Heidelberg, 1995, pp. 5–31.

Ratcliffe, D.A., Ed., *A Nature Conservation Review*, Vol. 1, Cambridge University Press, Cambridge, 1977.

Rossi, O. and Zurlini, G., Primi elementi conoscitivi essenziali per la realizzazione della Carta della Natura (Legge n. 394 del 6/12/1991), *S.It.E. Notizie, Bull. Ital. Ecol. Soc.,* 14, 46–56, 1993.

Rossi, O. and Zurlini, G., Biodiversitá e Carta della Natura, *Proc. Nat. Acad. Lincei*, 145, 91–111, 1998.

Scott, J.M., Tear, T.H. and Davis, F.W., *Gap Analysis – A Landscape Approach to Biodiversity Planning*, American Society for Photogrammetry and Remote Sensing, Bethesda, MD, 1996.

Sugihara, G. and May, R.M., Applications of fractals in ecology, *Trends Ecol. Evol.*, 5, 79–86, 1990.

Tansley, A.G., The use and abuse of vegetational concepts and terms, *Ecology*, 16, 284–307, 1935.

Usher, M.B., Habitat structure and the design of nature reserves, in *Habitat Structure: The Physical Arrangement of Objects in Space*, Bell, S.S., McCoy, E.D., and Mushinsky, H.R., Eds., Chapman & Hall, London, 1991, pp. 373–391.

Walker, H.B., Biodiversity and ecological redundancy, *Conserv. Biol.*, 6, 18–23, 1992.

Walter, K.S. and Gillet, H.J., Eds., *1997 IUCN Red Lists of Threatened Plants*, compiled by the World Conservation Monitoring Center, International Union for Conservation of Nature and Natural Resources–The World Conservation Union, Gland, Switzerland and Cambridge, 1998.

Wickert, G.A. and Rapport, D.J., Fish community structure as a measure of degradation and rehabilitation of riparian systems in an agricultural drainage basin, *Environ. Manage.*, 22, 425–443, 1998.

Zurlini, G. and Rossi, O., L'analisi della complessitá il contributo delle classificazioni ecologiche territoriali, *S.It.E. Notizie, Bull. Ital. Ecol. Soc.*, 16, 40–44, 1995.

Zurlini, G., Amadio, V., and Rossi, O., A landscape approach to biodiversity and biological health planning: the Map of the Italian Nature, *Ecosyst. Health*, 5, 296–311, 1999.

CHAPTER 67

Interior Columbia Basin Forests and Rangelands, 1930s to Present*

Paul F. Hessburg, Bradley G. Smith, R. Brion Salter, Roger D. Ottmar, and Ernesto Alvarado

INTRODUCTION

Forest and rangeland ecosystems of the interior Northwest U.S. are remarkably diverse and productive, owing to great variety in climate, geology, landforms, floral and faunal species, and ecosystem processes (Bailey, 1995; Franklin and Dyrness, 1988). Recurring disturbances, such as those caused by fires, insects, pathogens, and weather are essential to maintaining this diversity (Agee, 1993, 1994; Arno, 1976, 1980; Hall, 1976; Hessburg et al., 1994; Turner, 1987, 1989). Terrestrial plant communities include dry, short-grass prairies and sagebrush hills; dry ponderosa pine and Douglas fir forests; cool and moist midmontane western hemlock and western red cedar forests; high-elevation whitebark pine and subalpine larch forests, krummholz, and heath. Alpine tundra, rock barrens, and glaciers occupy many of the highest elevations.

Here as elsewhere, vegetation patterns at diverse spatial scales are closely related to underlying patterns of biophysical environments and associated disturbance regimes. Discernible patterns of life-forms and physiognomic conditions arise from broadscale differences in topography, surficial geology, geomorphology, annual temperature, precipitation, solar radiation, and large-scale disturbance. Within the mosaic of coarse patterns, mesoscale patterns result from environmental gradients, patch-scale disturbances, stand development, succession, and other processes.

Natural fire regimes of forests range from frequent, nonlethal surface fires typical in dry ponderosa pine and Douglas fir forests to moderately infrequent, mixed-severity fires, characteristic of some moist grand fir, western hemlock, and western red cedar forests, and infrequent, lethal, stand-replacing fires typical in cold subalpine forests (Agee, 1993, 1994). Likewise, native insect and pathogen disturbance regimes vary in their frequency, severity, duration, and spatial extent. Pandemic bark beetle and defoliator outbreaks occur relatively infrequently in any given geographic area, and when outbreaks occur, often are synchronous with climatic extremes or periods of geographically dominant vegetation structure, composition, or condition resulting from preceding events or trends. Insect or pathogen disturbance associated with endemic populations blends seamlessly with other succession and stand-development processes.

The declining health of forest ecosystems in the Interior Northwest has been the subject of much recent study and controversy (e.g., Everett et al., 1994; Harvey et al., 1995; Lehmkuhl et al., 1994; O'Laughlin et al., 1993; Wickman, 1992). Land use practices of the 20th century altered disturbance

* Copublished, with permission, in *Forest Ecology and Management*, 136, 53–83, 2000.

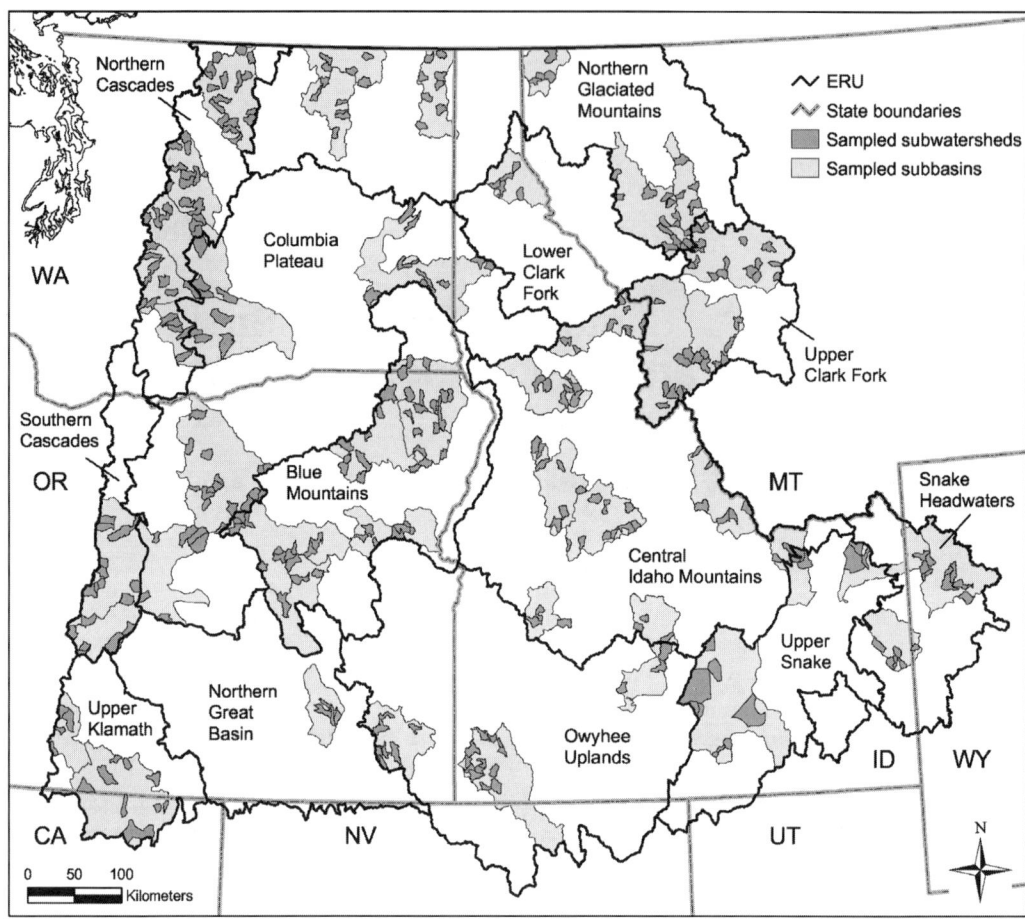

Figure 67.1 Map of Ecological Reporting Units (ERUs), and sampled subbasins and subwatersheds in the midscale assessment of the interior Columbia River basin.

regimes, spatial and temporal patterns of terrestrial habitats and species, and reduced ecosystem resilience to both native and human-caused disturbances (Covington et al., 1994). Fire suppression and exclusion, timber harvest, and domestic livestock grazing have contributed most to increased vulnerability to insect, pathogen, and wildfire disturbance. Concern over declining forest health centers on the perception that management activities have damaged ecosystem structure and function. That perception is founded on a strongly held social value that forest ecosystems should appear natural and function naturally.

This chapter presents results of a study conducted under the aegis of the Interior Columbia Basin Ecosystem Management Project. We report on a mesoscale (map scale = 1:24,000) scientific assessment of change in spatial patterns of forest and rangeland landscapes of the interior Columbia River basin. We also report on associated changes in landscape vulnerability to fire and related PM_{10} (particulate matter <10 µm in diameter) smoke production, and insect and pathogen disturbances, and we discuss management implications of those changes. The study area (58 million ha) included the Columbia River basin east of the crest of the Cascade Range and portions of the Klamath and Great Basins in Oregon ("the basin," Figure 67.1).

Our study had four objectives:

1. To characterize current structure and composition of a representative sample of forest and rangeland landscapes
2. To compare existing vegetation conditions to the oldest historical conditions we could reconstruct at a comparable scale

3. To link historical and current vegetation spatial patterns with patterns of vulnerability to insect and pathogen disturbances
4. To link historical and current landscape vegetation characteristics throughout the basin with fuel conditions, potential fire behavior, and related smoke production.

Linkages in objectives 3 and 4 would enable us to better understand causal connections among historical management activities and current conditions, and assist in evaluating current air quality and human health trade-offs associated with wild and prescribed fires, and trade-offs associated with alternative insect and pathogen vulnerability scenarios.

METHODS

In the midscale assessment (Hessburg et al., 1999a), we quantified change in vegetation patterns and landscape vulnerability to fire, insect, and pathogen disturbances over the most recent 50 to 60 years. Sampling design and change analysis methods were adapted from Lehmkuhl et al. (1994). Our sample of historical conditions corresponded with the start of the period of intensive timber harvest, road construction, fire suppression, and a period of declining intensity in rangeland management. We based our assessment on a two-stage stratified random sample of 337 of 7496 subwatersheds (10,000 ha average area) distributed in 43 of 164 subbasins (400,000 ha average area) across all ownerships. We reported results of change analysis by ecological reporting units (ERUs; Figure 67.1). ERUs were developed during the broadscale assessment of the basin (Quigley and Arbelbide, 1997) as statistical pooling strata for generalizing results of various ecological, social, economic, and integrated assessments. The ERUs represent land areas that are broadly homogeneous in their biophysical and social ecosystem characteristics.

Forest and rangeland composition and structure were derived from raw data developed from aerial photographs taken from 1932 to 1966 (historical), and from 1981 to 1993 (current). Historical conditions of most forested settings were represented by photography from the 1930s and 1940s, while those of rangelands were represented by aerial photography from the 1950s and 1960s. Areas with homogeneous vegetation composition and structure were delineated as patches to a minimum size of 4 ha. Vegetation cover types, structural classes, and potential vegetation types were classified for each patch using the raw attributes, and topographic or biophysical data from other digital sources of comparable scale and image resolution.

Each patch was assigned a rating for 3 to 7 vulnerability factors associated with each of 21 different forest insects and pathogens, including defoliator, bark beetle, dwarf mistletoe, root disease, rust, and stem decay disturbances. Vulnerability factors were unique for each host–pathogen or host–insect interaction modeled, and included items such as site quality, host abundance, canopy layers, host age or host size, stand vigor, stand density, connectivity of host patches, topographic setting, and type of visible logging disturbance. Patch vulnerability factors were taken from the literature or were based on the experience of field pathologists and entomologists with expertise in specific geographic areas (Hessburg et al., 1999b). Resulting models represent a substantial revision of prototypes described in Lehmkuhl et al. (1994).

Similarly, historical and current vegetation patches were matched to 1 of 192 fuel condition classes (Ottmar et al., 1996; Schaaf, 1996) and assigned a fuel loading level. Fuel loads were used to compute fuel consumption, particulate emissions production (PM10), crown fire potential, and fire behavior attributes for an average wildfire scenario using published procedures (Huff et al., 1995; Ottmar et al., in press). We also modeled fuel consumption and related smoke production for a prescribed burn scenario. Algorithms for estimating fuel consumption for both burn scenarios were taken from the CONSUME (Ottmar et al., 1993) and First Order Fire Effects Model (FOFEM; Keane et al., 1994). Fire behavior attributes were rate of spread, flame length, and Byram's fireline intensity. We computed fireline intensity (Rothermel, 1983), rate of spread, and flame length using the published equations of the National Fire Danger Rating System (Cohen and Deeming, 1985; Deeming et al., 1977; Rothermel, 1972).

Vegetation and Landscape Pattern Analysis

This assessment was a map-based characterization of landscape patterns and ecological processes across space and time. We used the ARC/INFO (Environmental Systems Research Institute, 1995) geographical information system (GIS) to manipulate and analyze digital maps, and to develop and run spatially explicit insect and pathogen vulnerability (Hessburg et al., 1999b), and fuel consumption and fire behavior models (Ottmar et al., in press). Spatial and statistical analyses characterized change in patterns and quantified the significance of change. FRAGSTATS (McGarigal and Marks, 1995) was used to compute class and landscape pattern metrics directly from ARC/INFO data tables, and we incorporated three additional metrics (N1, N2, and R21; Table 67.1) into the source code. We used S-PLUS (MathSoft, Inc., 1993) to summarize and analyze ARC/INFO and FRAGSTATS outputs. Vegetation maps and raw and derived patch attributes formed the basic data set for all analysis. For spatial pattern analysis, a variety of unique vegetation maps were derived in a GIS by dissolving on single or combined data items. Patch types of a map submitted to analysis could be defined by any raw attribute, such as canopy layers or total crown cover class, or by any derived attribute, such as cover type or structural class, either singly or in combinations.

To quantify change in individual patch types and patterns of various patch types, we used raster versions of current and historical vegetation maps where patch types were physiognomic conditions, cover types, structural classes, potential vegetation types, or combinations. A raster format was chosen because several useful metrics were only available in FRAGSTATS for raster maps. The appropriate cell (pixel) size was determined by calculating several class metrics in vector and raster form, with cell sizes ranging from 10 to 100 m^2 (1 ha), in 10-m^2 increments, and at 141.4 m^2 (2.0 ha) and 223.2 m^2 (5.0 ha), and plotting each raster-derived metric value against the vector value. When compared with vector values, raster bias was insignificant with 30-m^2 and smaller cell sizes, and we used 30-m^2 raster maps for all pattern analysis.

Sample Statistics

We used percentage of area, patch density, mean patch size, edge density, and mean nearest neighbor class-metrics to describe change in area and connectivity of patch types in subwatersheds of an ERU. We used ten landscape-metrics to describe changes in patch type richness, evenness, diversity, dominance, contagion, interspersion, juxtaposition, and edge contrast (Tables 67.1 and 67.2). For each ERU, means, standard errors, and confidence intervals were estimated using methods for simple random samples with subwatersheds as sample units. Significant ($p \leq 0.2$) change from historical to current conditions was determined by examining the 80% confidence interval (CI) around the mean difference for the ERU. We used a moderately conservative 80% CI because we wanted to be able to detect changes in spatial patterns that were of potential ecological importance. We reasoned that with a more conservative CI, we might increase the probability of Type II error (false positive). When we compared 90 and 95% CI estimates of mean difference with 80% estimates, we noted that important changes went undetected using the more conservative CIs.

To avoid increasing the probability of Type I error (false negative), we supplemented our significance test with two other tests. These enabled us to evaluate the potential ecological importance of change in patch area or connectivity, and the probability of error in rejecting the null hypothesis. First, we estimated a reference variation by calculating for each metric the 75% range around the historical sample median, and then compared the current sample median value with this range. We chose the median 75% range instead of the full range as a measure of reference variation to portray typical variation exclusive of extreme observations. Historical and current data distributions were most frequently right-skewed, the sample median value accurately

Table 67.1 FRAGSTATS Indices Used to Quantify Spatial Patterns of Patch Types in Sampled Subbasins in the Midscale Ecological Assessment of the Interior Columbia River Basin

Abbreviation	Scale	Index Name	Description[a]
%LAND	Class	Percentage of landscape (%)	Percentage of a landscape composed of the corresponding patch type
PD	Class or landscape	Patch density (no. per 10,000 ha)	Number of patches in an area of 10,000 ha
MPS	Class or landscape	Mean patch size (ha)	Average patch size
AWMECI	Class or landscape	Area-weighted mean edge contrast index (%)	Average patch edge contrast as a percentage of maximum contrast with patch edge contrasts weighted by patch area; equals 100 when all edge is maximum contrast; approaches 0 when all edge is minimum contrast
SHDI	Landscape	Shannon's diversity index[b]	Measures proportional abundance of patch types and the equitable distribution of patch type areas; increases with patch richness (PR) and equitability of area
RPR	Landscape	Relative patch richness (%)	Observed number of patch types within a landscape over a realistic potential maximum number of patch types
PR	Landscape	Patch richness	Observed number of patch types within a landscape boundary
MSIEI	Landscape	Modified Simpson's evenness index[c]	Observed distribution of area of patch types within a landscape over evenly distributed area of patch types
IJI	Class or landscape	Interspersion and juxtaposition index (%)	Observed interspersion of edge types over maximum possible interspersion; IJI approaches 0 when patch types are clumped; IJI approaches 100 when all patch types are equally adjacent to all other patch types
CONTAG	Landscape	Contagion index (%)	Observed contagion over the maximum possible contagion for the given number of patch types; approaches 0 when the distribution of adjacencies among unique patch types becomes increasingly uneven; approaches 100 when all patch types are equally adjacent to all other patch types; measures patch type interspersion and patch dispersion
N1	Landscape	Hill's index N1[d]	A transformation of SHDI, computed as e^{SHDI}; rare patch types are weighted less in the calculation than in PR
N2	Landscape	Hill's index N2[d]	A transformation of SIDI, computed as $1/(1 - SIDI)$; rare patch types are weighted less in the calculation than in N1
R21	Landscape	Alatalo's evenness index[e]	Measures evenness of patch types; computed as $(N2 - 1)/(N1 - 1)$, where PR > 1; values approaching 0 indicate uneven distribution of patch type areas; values approaching 1 indicate even distribution of area for the given number of patch types

[a] See McGarigal and Marks (1995) for algorithms and complete descriptions of all indices, except N1, N2, and R21.
[b] Shannon and Weaver (1949).
[c] Simpson (1949).
[d] Hill (1973).
[e] Alatalo (1981).

reflected central tendency, and most observations were clustered within the median 75 to 80% range. Second, we determined the largest changes in absolute area of a patch type within a sample using transition analysis. Transition analysis estimates the percentage of sampled area in each unique historical to current patch type transition. We reasoned that the more extreme variation

Table 67.2 Edge Contrast Weights Used in Calculating the FRAGSTATS Metric Area-Weighted Mean Edge Contrast Index (AWMECI) in Pattern Analyses of Patch Types of Sampled Subwatersheds in the Midscale Ecological Assessment of the Interior Columbia River Basin[a]

Patch Type	Non forest and Non range	Herb land	Shrub land	Wood land	Forest (by structural class[b])				
					si	seoc and secc	ur and yfms	ofss	ofms
Nonforest and nonrange	0[c]	0.2	0.3	0.4	0.5	0.6	0.8	0.9	1
Herbland		0	0.2	0.3	0.4	0.6	0.7	0.8	0.9
Shrubland			0	0.2	0.3	0.5	0.6	0.7	0.8
Woodland				0	0.3	0.4	0.5	0.6	0.7
Forest si					0	0.3	0.4	0.5	0.6
Forest seoc and secc						0	0.3	0.4	0.5
Forest ur and yfms							0	0.3	0.4
Forest ofss								0	0.3
Forest ofms									0

[a] For FRAGSTATS, see McGarigal and Marks (1995).

[b] Forest structural classes are: stand initiation (si); stem exclusion open canopy (seoc); stem exclusion closed canopy (secc); understory reinitiation (ur); young forest multistory (yfms); old forest single-story (ofss); and old forest multistory (ofms). See also Figure 67.8 for graphical illustrations of forest structural classes.

[c] Range of possible values is 0 to 1, with increasing values representing greater edge contrast.

resulted from either unique contexts or environments, or from rare events. By imposing the contrast between current median values and a typical range of historical conditions, we retained the ability to detect conditions resulting from management activities, rare or chance events, or perhaps climate change that were unique in some aspect.

RESULTS

Trends in Physiognomic Conditions

Significant changes in physiognomic types occurred throughout the basin. Cover of forest increased significantly in the Blue Mountains, Columbia Plateau, and Upper Snake ERUs (Figure 67.2, Table 67.3). Transition analysis showed that effective fire prevention, suppression, and exclusion allowed the expansion of forests into areas that were formerly fire-maintained shrublands, herblands, or areas created by early logging. Connectivity (spatial aggregation) of forest increased in the Central Idaho Mountains and Upper Snake ERUs (see Hessburg et al., 1999a). Increased connectivity was the result of expanded forest cover on former shrubland areas (Table 67.3). The Central Idaho Mountains contain large wild and roadless areas. Transition analysis of cover type changes indicated that increased connectivity of forests resulted from effective fire suppression and fire exclusion. We note that in a few subwatersheds increased forest connectivity resulted from regrowth of forest in areas that had experienced large-scale stand replacement fires.

Both area and connectivity of forest cover declined in the Upper Klamath ERU (Figure 67.2). Upper Klamath forests are naturally quite fragmented because broad grassy valley bottoms or grasslands on dry southerly aspects often separate forested slopes. High timber harvest levels and juniper woodland expansion into the ponderosa pine cover type were responsible for the observed reduction, and we suspect that domestic livestock grazing was indirectly involved as well. Observed reductions in area of forest were readily explained via transition analysis of historical and current physiognomic and cover type maps, and through analysis of patch density

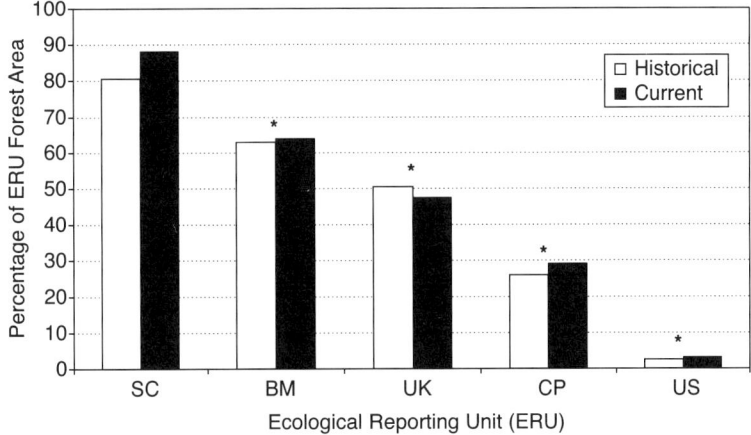

Figure 67.2 Change in percentage of area in the forest physiognomy for selected ERUs. ERU abbreviations are: SC = Southern Cascades; BM = Blue Mountains; UK = Upper Klamath; CP = Columbia Plateau; US = Upper Snake. Asterisk (*) denotes significant difference at $P \leq 0.2$.

and mean patch size trends. Figure 67.3 illustrates altered physiognomic conditions in sampled subwatersheds of the Blue Mountains (Figure 67.3 [top]) and Upper Klamath (Figure 67.3 [bottom]; see color figure following p. 668) ERUs.

Woodland (sparsely wooded rangeland) area increased in 7 of 13 ERUs and declined in none (Figure 67.4). Transition analysis revealed that fire suppression, fire exclusion, and domestic livestock grazing enabled mostly juniper woodland expansion at the expense of herbland and shrubland area. The associated regional decline in area of shrublands was the most dramatic of all changes in physiognomic conditions (Table 67.3, Figure 67.4). Transition analyses showed that losses of native shrublands resulted from forest and woodland expansion (Blue Mountains and Northern Great Basin ERUs), cropland expansion (Northern Great Basin ERU), and conversion to seminative or nonnative herblands (Owyhee Uplands and Snake Headwaters ERUs). Figure 67.5 (see color figure following p. 668) illustrates reduced shrubland and increased woodland area in an Upper John Day subwatershed, Blue Mountains ERU.

Conversely, herbland area increased in the Central Idaho Mountains, Northern Great Basin, Owyhee Uplands, Snake Headwaters, and Southern Cascades ERUs and declined in none. In the Northern Great Basin, herbland increased at the expense of shrubland area, which fell by more than 15%. Transition analysis indicated that half of the lost shrubland area is currently occupied by juniper, and the balance supports montane bunchgrasses or exotic grass and forb cover. Herblands and shrublands followed a similar pattern in the Owyhee Uplands. Across the basin, most increases in herbland area resulted from expanding colline (environments occurring below lower treeline), exotic grass, and forb cover with shrubland conversion. Here, we note that the vast majority of native herblands were converted to agricultural production prior to our historical condition. We describe changes to what are essentially relict herblands.

Forest Cover Type Trends

Shifts from early- to late-seral cover species were evident in most ERUs (Table 67.3), but the most pronounced shifts occurred in the Northern Glaciated Mountains (Figure 67.6). In some ERUs, the shift was partially masked by steep climatic gradients. For example, in the Northern Cascades ERU, Douglas fir is early-seral in several mid- to upper-montane series (e.g., western hemlock, western red cedar), but to the east at lower elevations is climax in the Douglas fir series.

Table 67.3 Historical and Current Percentage of Area of Physiognomic Types, Forest Cover Types, and Structural Classes of Subwatersheds Sampled in Ecological Reporting Units (ERUs) of the Midscale Ecological Assessment of the Interior Columbia River Basin

Patch Types	Blue Mountains H[a]	C	Central Idaho Mountains H	C	Columbia Plateau H	C	Lower Clark Fork H	C	Northern Cascades H	C	Northern Glaciated Mountains H	C	Northern Great Basin H	C	Owyhee Uplands H	C	Snake Head waters H	C	Southern Cascades H	C	Upper Clark Fork H	C	Upper Klamath H	C	Upper Snake H	C
														—percentage of area—												
Physiognomic Types
Forest	**62.8**[b]	**64.1**	73.4	73.5	**26.1**	**29.1**	91.7	94.5	78.8	78.2	81.0	80.8	7.2	7.3	0.2	0.2	74.5	73.8	80.5	88.3	87.2	86.2	**50.5**	**47.5**	2.4	3.2
Woodland	**2.7**	**4.2**	0.1	0.0	**6.7**	**12.2**	—	—	**0.3**	**0.7**	—	—	15.3	22.2	5.5	7.6	**0.2**	**0.3**	0.0	0.4	—	—	**8.4**	**12.8**	3.0	2.9
Shrubland	**14.1**	**10.7**	19.2	17.1	**32.2**	**23.4**	1.9	0.6	4.8	4.1	3.1	2.5	72.8	57.6	88.8	81.0	16.3	13.9	0.5	0.5	2.5	2.1	21.4	18.8	73.8	68.5
Herbland	17.4	18.0	3.2	4.5	12.7	14.0	5.4	3.2	6.7	6.5	7.4	8.1	**3.9**	**12.2**	**1.0**	**7.4**	**6.1**	**8.7**	**0.6**	**2.7**	5.5	5.7	10.6	9.0	10.6	9.9
Other[c]	3.0	2.9	4.2	4.9	22.4	21.4	0.9	1.8	**9.4**	**10.6**	8.5	8.5	0.8	0.8	4.5	3.8	3.0	3.3	**18.4**	**8.1**	4.8	6.0	**9.1**	**12.0**	10.3	15.4
Forest Cover Types
GF/WF[d]	**15.3**	**8.4**	9.6	10.2	1.1	0.4	40.4	42.5	**1.0**	**2.2**	**0.0**	**1.2**	—	—	—	—	—	—	5.9	6.5	0.0	0.1	7.8	8.1	—	—
ES/SAF	**6.3**	**4.4**	22.7	24.1	—	—	2.5	2.2	**16.8**	**13.6**	**11.5**	**13.2**	—	—	—	—	**24.3**	**31.4**	**0.0**	**0.2**	**14.2**	**17.3**	0.1	0.1	—	—
ASP/COT	0.1	0.1	1.1	0.8	0.3	0.3	0.1	0.7	—	—	0.3	1.9	8.4	7.7	0.2	0.2	**8.8**	**5.7**	0.0	0.4	0.3	0.3	0.0	0.1	0.9	1.0
JUN	**2.7**	**4.2**	0.1	0.0	**6.5**	**12.0**	—	—	1.0	1.0	—	—	**14.1**	**21.8**	**5.5**	**7.5**	**0.2**	**0.3**	—	—	—	—	**8.4**	**12.8**	2.6	2.5
WL	2.6	2.2	0.5	0.3	1.0	0.1	0.8	2.6	3.3	4.7	14.8	11.4	—	—	—	—	—	—	0.0	0.8	2.5	3.0	0.0	0.1	—	—
WBP/SAL	**0.0**	**0.7**	5.1	2.5	—	—	—	—	3.3	4.7	0.3	0.2	—	—	—	—	6.9	5.7	—	—	**4.3**	**3.5**	—	—	—	—
LPP	2.4	2.3	9.7	9.5	1.3	0.9	2.1	1.8	5.9	5.2	8.0	8.3	—	—	—	—	**15.6**	**11.3**	19.4	20.6	20.9	19.5	1.4	1.7	0.1	0.2
LP	—	—	0.4	0.4	—	—	—	—	—	—	—	—	—	—	—	—	**0.7**	**1.1**	—	—	0.0	0.4	—	—	—	—
PP	28.4	28.9	6.0	5.9	**19.2**	**21.4**	3.0	5.1	**16.5**	**13.2**	**13.4**	**11.4**	—	—	—	—	—	—	22.7	28.1	**12.3**	**9.5**	**26.7**	**23.5**	—	—
DF	**7.7**	**17.1**	17.6	18.5	**3.0**	**3.9**	26.1	21.1	**23.8**	**25.8**	30.3	30.2	—	—	—	—	18.2	18.6	1.5	1.7	32.7	32.5	2.1	1.2	1.4	2.1
WH/WRC	—	—	0.9	1.3	**0.4**	**2.2**	14.7	17.3	3.0	2.4	0.7	2.8	—	—	—	—	—	—	—	—	**0.0**	**0.1**	4.7	4.2	—	—
MH	—	—	**0.0**	**0.0**	—	—	**1.3**	**0.6**	1.3	1.2	0.1	0.0	—	—	—	—	—	—	30.5	29.7	—	—	—	—	—	—
SP/WWP	—	—	—	—	—	—	0.3	0.6	0.1	0.3	1.5	0.0	—	—	—	—	—	—	0.3	0.3	—	—	—	—	—	—
PSF	—	—	—	—	—	—	—	—	**6.0**	**8.3**	—	—	—	—	—	—	—	—	—	—	—	—	—	—	—	—
OWO	—	—	—	—	—	—	—	—	0.6	0.9	—	—	—	—	—	—	—	—	—	—	—	—	—	—	—	—
SRF	—	—	—	—	—	—	—	—	—	—	—	—	—	—	—	—	—	—	0.2	0.4	—	—	7.8	8.5	0.4	0.5
PJ	—	—	—	—	—	—	—	—	—	—	—	—	—	—	—	—	—	—	—	—	—	—	—	—	—	—

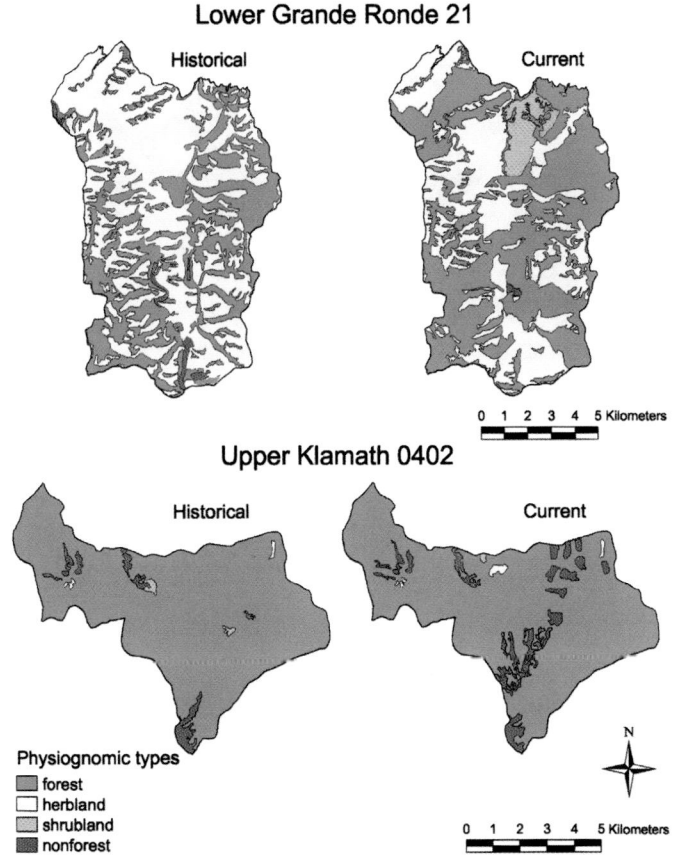

Figure 67.3 Historical and current spatial patterns of physiognomic conditions: subwatershed 21, Lower Grande Ronde subbasin, Blue Mountains ERU (top); and subwatershed 0402, Upper Klamath subbasin, Upper Klamath ERU (bottom).

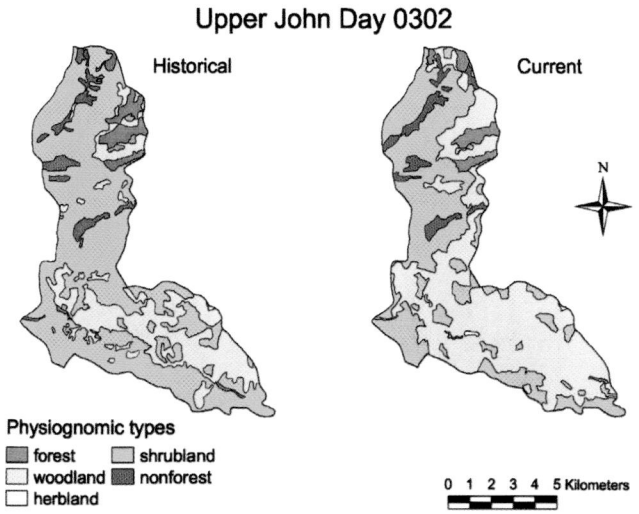

Figure 67.5 Historical and current spatial patterns of physiognomic types in sampled subwatershed 0302, Upper John Day subbasin, Blue Mountains ERU.

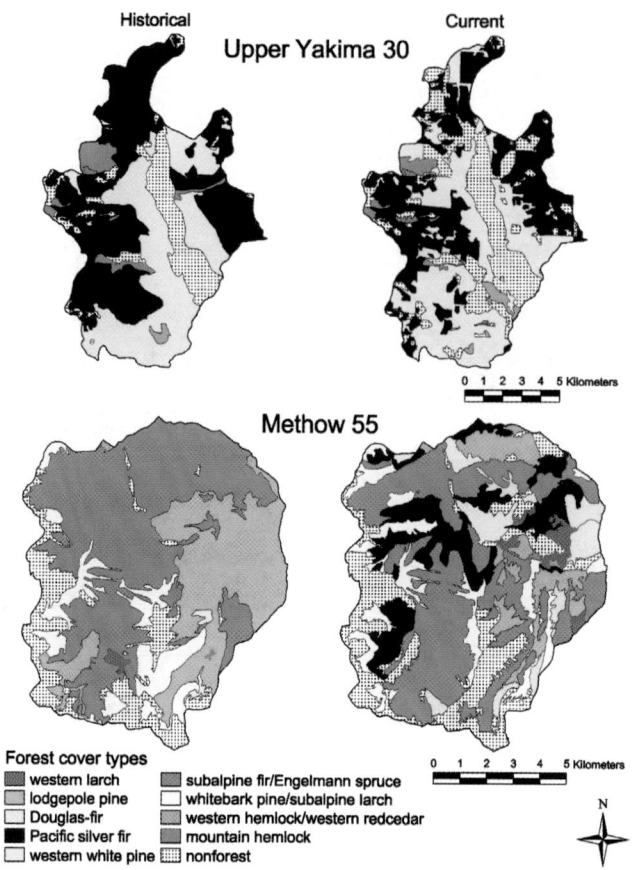

Historical

Current

Upper Yakima 30

0 1 2 3 4 5 Kilometers

Methow 55

Forest cover types

- western larch
- lodgepole pine
- Douglas-fir
- Pacific silver fir
- western white pine
- subalpine fir/Engelmann spruce
- whitebark pine/subalpine larch
- western hemlock/western redcedar
- mountain hemlock
- nonforest

0 1 2 3 4 5 Kilometers

N

Figure 67.7 Historical and current forest cover type area of subwatershed 30, Upper Yakima subbasin (top); and subwatershed 55, Methow subbasin, North Cascades ERU (bottom).

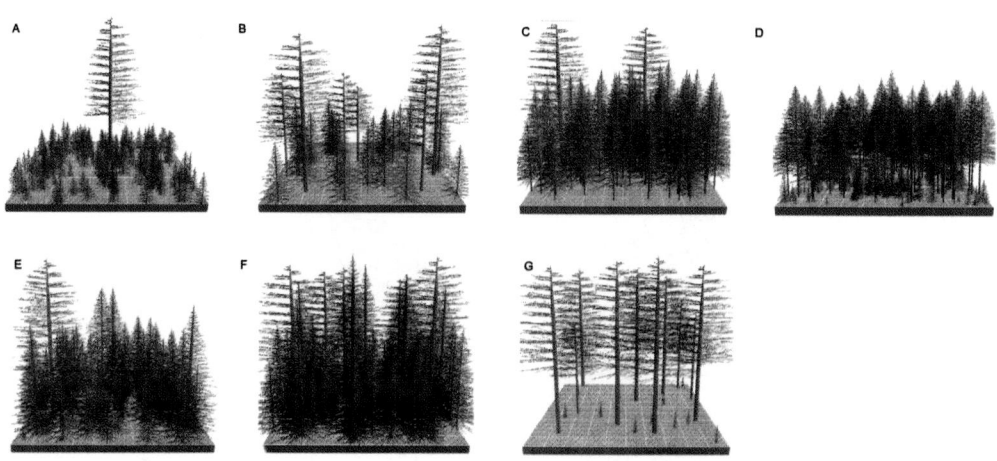

Figure 67.8 Graphical representation of forest structural classes used in the midscale assessment of the interior Columbia River basin: (A) stand-initiation; (B) open stem exclusion; (C) closed stem exclusion; (D) understory reinitiation; (E) young multistory forest; (F) old multistory forest; (G) old single-story forest. Refer to Oliver and Larson (1996) and O'Hara et al. (1996) for expanded descriptions of forest structural classes.

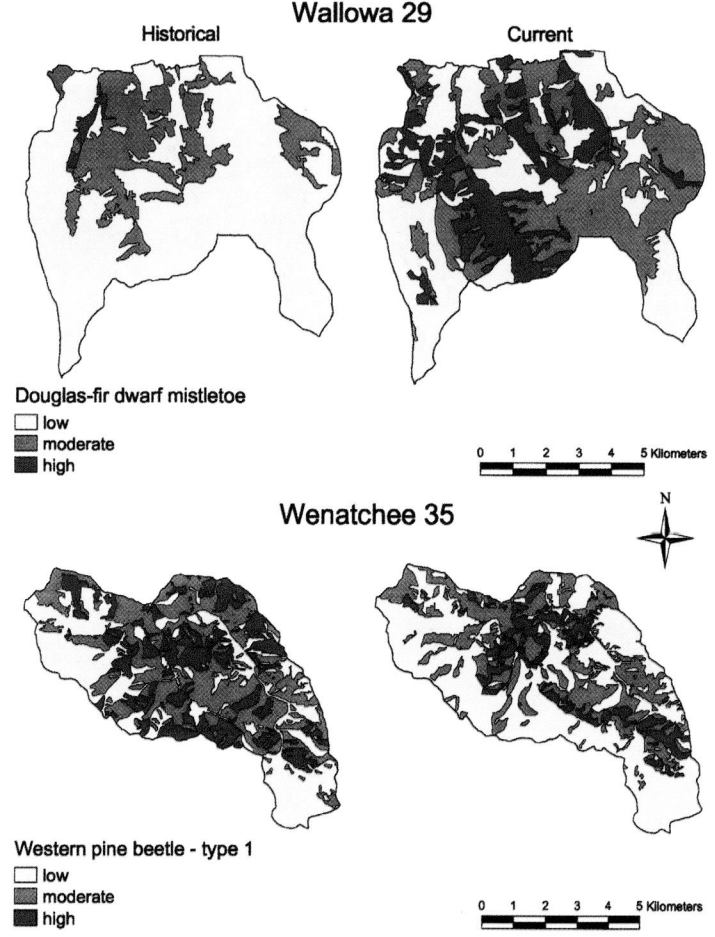

Figure 67.11 Historical and current forest area: (top) vulnerable to Douglas-fir dwarf mistletoe disturbance, subwatershed 29, Wallowa subbasin, Blue Mountains ERU; (bottom) vulnerable to western pine beetle disturbance of medium (40.5 to 63.5 cm d.b.h.) and large (>63.5 cm d.b.h.) ponderosa pine, subwatershed 35, Wenatchee subbasin, Northern Cascades ERU.

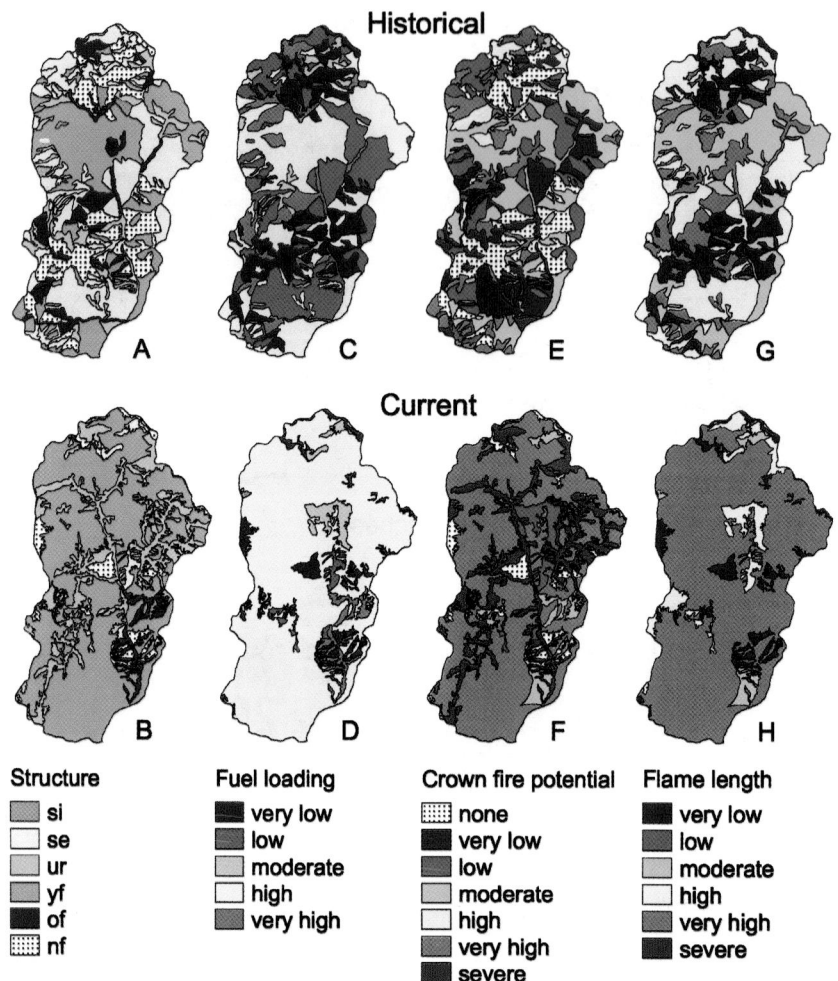

Figure 67.12 Maps of subwatershed 55, Lower Grande Ronde subbasin, Blue Mountains ERU, displaying historical and current structural classes (A and B), fuel loading (C and D), crown fire potential under wildfire conditions (E and F), and flame length under wildfire conditions (G and H), respectively. Structural class abbreviations are si = stand-initiation; se = stem exclusion (both open and closed canopy conditions); ur = understory reinitiation; yfms = young multistory forest; of = old multistory and single-story forest; and nf = nonforest. Fuel loading classes are: very low < 22.5 Mg/ha; low = 22.5 to 44.9 Mg/ha; moderate = 45 to 56.1 Mg/ha; high = 56.2 to 67.3 Mg/ha; and very high > 67.3 Mg/ha. Crown fire potential classes were a relative index. Flame length classes were: very low < 0.6 m; low = 0.7 to 1.2 m; moderate = 1.3 to 1.8 m; high = 1.9 to 2.4 m; very high = 2.5 to 3.4 m; and severe > 3.4 m.

Forest Structural Classes

	1	2	3	4	5	6	7	8	9	10	11	12	13	14	15	16	17	18	19	20	21	22	23	24	25	26
si[e]	3.9	**6.5**	9.7	**5.9**	2.3	**2.8**	32.7	**9.5**	9.2	**10.4**	**16.9**	**9.4**	—	—	—	—	6.4	**7.0**	9.1	9.9	**15.9**	11.1	1.9	3.6	**0.8**	**0.3**
seoc	14.3	**9.6**	18.4	**17.7**	6.7	**7.8**	15.7	**9.2**	13.2	**13.2**	**11.8**	**11.6**	6.5	**6.0**	0.0	0.1	**19.1**	**15.3**	12.3	14.3	**18.5**	18.2	11.3	10.9	**0.4**	**1.0**
secc	5.0	**5.0**	7.7	**8.5**	3.8	**3.6**	10.3	**17.6**	7.6	**7.9**	**7.2**	**12.8**	0.7	**1.3**	—	—	**7.9**	**4.8**	0.5	4.8	**16.7**	21.1	1.2	1.6	0.1	0.1
ur	13.6	**11.2**	16.0	**21.4**	3.1	**3.3**	16.3	**37.7**	17.5	**19.5**	**18.4**	**23.3**	—	—	0.4	1.1	13.8	**12.6**	10.3	8.7	**15.6**	14.0	5.6	8.1	2.5	1.6
yfms	21.3	**29.6**	18.4	**17.1**	7.3	**10.0**	14.3	**17.5**	21.2	**22.0**	**25.5**	**22.8**	—	—	0.1	0.1	**22.0**	**30.9**	46.0	45.6	**19.7**	21.1	21.1	16.4	0.6	1.1
ofms	2.2	**1.0**	1.4	**1.2**	2.3	**1.3**	0.2	**0.5**	5.8	**2.7**	**0.5**	**0.4**	—	—	—	—	**3.2**	**1.8**	0.7	1.4	**0.6**	0.4	4.3	5.5	—	—
ofss	2.7	**0.9**	1.8	**1.7**	1.1	**1.0**	2.2	**2.5**	4.3	**2.4**	**0.7**	**0.6**	—	—	—	—	**2.0**	**1.3**	1.6	3.7	**0.2**	0.3	7.4	4.8	0.1	0.0

[a] H = historical condition; C = current condition.

[b] Mean values shown in bold type are significantly different at $P \leq 0.2$.

[c] "Other" includes anthropogenic cover types and other nonforest and nonrange types.

[d] Forest cover types are grand fir/white fir (GF/WF); Engelmann spruce/subalpine fir (ES/SAF); aspen/cottonwood/willow (ASP/COT); juniper (JUN); western larch (WL); whitebark pine/subalpine larch (WBP/SAL); lodgepole pine (LPP); limber pine (LP); ponderosa pine (PP); Douglas fir (DF); western hemlock/western red cedar (WH/WRC); mountain hemlock (MH); sugar pine/western white pine (SP/WWP); Pacific silver fir (PSF); Oregon white oak (OWO); Shasta red fir (SRF); and pinyon/juniper (PJ).

[e] Forest structural classes are stand initiation (si); stem exclusion open canopy (seoc); stem exclusion closed canopy (secc); understory reinitiation (ur); young forest multistory (yfms); old forest single-story (ofss); and old forest multistory (ofms).

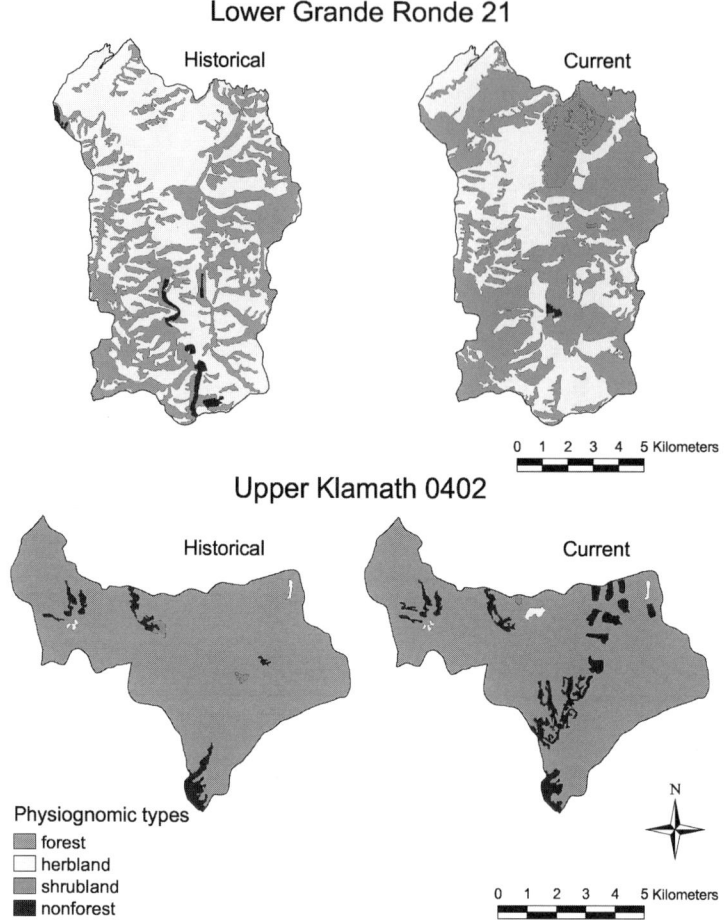

Figure 67.3 Historical and current spatial patterns of physiognomic conditions: subwatershed 21, Lower Grande Ronde subbasin, Blue Mountains ERU (top); and subwatershed 0402, Upper Klamath subbasin, Upper Klamath ERU (bottom). (See color figure following p. 668.)

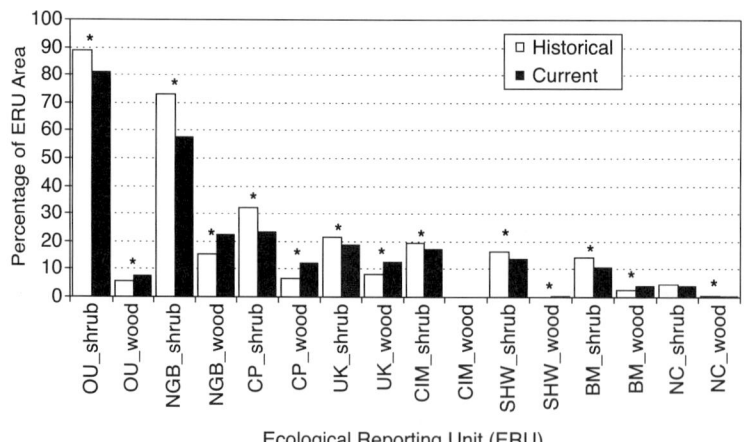

Figure 67.4 Change in percentage of area in the shrubland (shrub) and woodland (wood) physiognomies for selected ERUs. ERU abbreviations are OU = Owyhee Uplands; NGB = Northern Great Basin; CP = Columbia Plateau; UK = Upper Klamath; CIM = Central Idaho Mountains; SHW = Snake Headwaters; BM = Blue Mountains; NC = Northern Cascades. Asterisk (*) denotes significant difference at $P \le 0.2$.

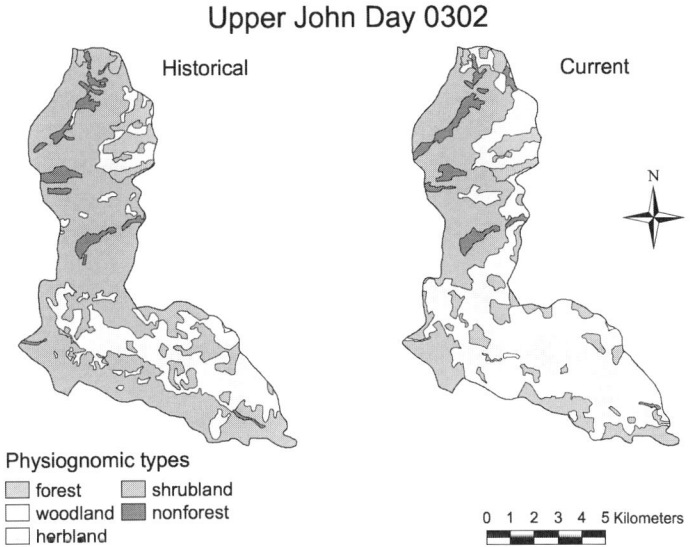

Figure 67.5 Historical and current spatial patterns of physiognomic types in sampled subwatershed 0302, Upper John Day subbasin, Blue Mountains ERU. (See color figure following p. 668.)

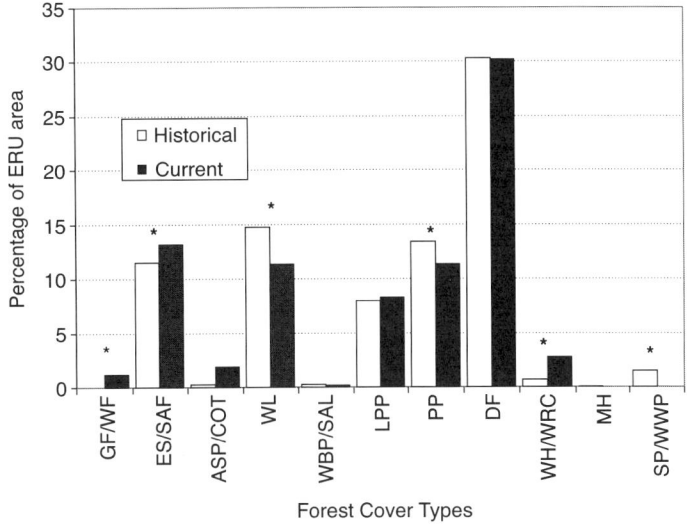

Figure 67.6 Change in percentage of area in forest cover types of the Northern Glaciated Mountains ERU. Asterisk (*) denotes significant difference at $P \leq 0.2$.

Western larch cover declined significantly in the Central Idaho Mountains, Columbia Plateau, and Northern Glaciated Mountains, and ponderosa pine cover decreased in the Northern Cascades, Northern Glaciated Mountains, Upper Clark Fork, and Upper Klamath. Increased ponderosa pine cover in the Southern Cascades resulted from regrowth of forests that were tractor-logged prior to our historical photo coverage. Lodgepole pine cover declined in the Snake Headwaters and in six other ERUs. Western white pine cover decreased in the Northern Glaciated Mountains as a consequence of white pine blister rust and mountain pine beetle mortality, and early selective harvesting. It increased slightly in the Northern Cascades as a result of recent reforestation. Whitebark pine-subalpine larch cover declined in the Central Idaho Mountains, Northern Glaciated Mountains, Snake Headwaters, and Upper Clark Fork ERUs and increased in the Blue Mountains and Northern Cascades. Decline in whitebark pine-subalpine larch cover resulted from ongoing blister rust and mountain pine beetle mortality, and expanded area of subalpine fir and Engelmann spruce.

In contrast, Douglas fir cover increased in the Blue Mountains, Columbia Plateau, and Northern Cascades; grand fir cover increased in the Northern Cascades and Northern Glaciated Mountains; Pacific silver fir cover increased in the Northern Cascades; Engelmann spruce-subalpine fir cover increased in the Northern Glaciated Mountains, Snake Headwaters, Southern Cascades, and Upper Clark Fork; and western hemlock-western red cedar cover increased in the Columbia Plateau and Northern Glaciated Mountains (Table 67.3). Engelmann spruce-subalpine fir cover declined in the Blue Mountains, and Engelmann spruce-subalpine fir and western hemlock-western red cedar cover both decreased in the Northern Cascades. Results of transition analysis showed that noted increases in these shade-tolerant cover types were best explained by fire suppression and exclusion, and selective timber harvest activities.

Added to expanded area of late-seral species, average patch sizes of most forest cover species are smaller and current land cover is more fragmented (Figure 67.7; see color figure following p. 668). In the historical condition, spatial patterns of biophysical environments and disturbance regimes created patches of land cover that were large, and overall patterns were relatively simple. In the current condition, highly fragmented land cover mosaics have replaced simpler patterns. In some heavily roaded subwatersheds (Figure 67.7, top), widely applied patterns of small cutting units were responsible for the change. In roadless subwatersheds, fine- to midscale disturbance processes are responsible for reduced connectivity of land cover. For example, subwatershed 55 of the Methow subbasin (Figure 67.7, bottom) resides in the Pasayten Wilderness. Prior to the era of fire suppression (pre-1930s), large-scale stand replacement fires would naturally occur on an infrequent basis, resulting in simple land cover mosaics consisting of a relatively few large patches. Today, the Methow subwatershed exhibits reduced connectivity of land cover, and causative factors were small fires and bark beetle outbreaks.

In the next section, we discuss changes among forest structural classes. Here, it will be important to keep in mind the scale dependence of observations. Immediately above, we reported that land cover mosaics exhibited increased pattern complexity in the current condition. Below, we show patterns of structural classes changing in a different way.

Trends among Structural Classes

In general, the vertical structure of individual forest patches has become more complex, but the overall landscape pattern of forest structural conditions is simpler when compared with historical forests. Landscape area in old forest structures (multistory and single-story) declined in most forested ERUs (Figure 67.8; see color figure following p. 668), but the most significant declines occurred in the Blue Mountains, Northern Cascades, Snake Headwaters, and Upper Klamath ERUs (Figure 67.9, Table 67.3). Landscape area in stand-initiation structures (patches of young, new forest) declined in five of nine forest-dominated ERUs and increased in only one, the Blue Mountains. Area in stand-initiation structures declined in the Central Idaho Mountains,

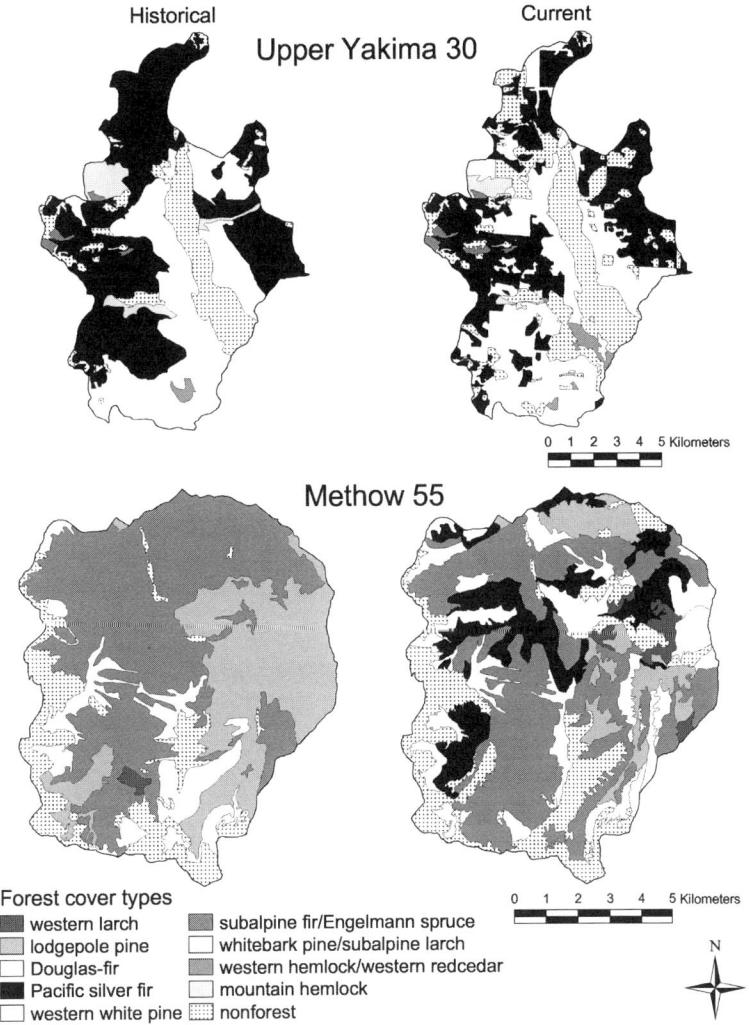

Figure 67.7 Historical and current forest cover type area of subwatershed 30, Upper Yakima subbasin (top); and subwatershed 55, Methow subbasin, North Cascades ERU (bottom). (See color figure following p. 668.)

Lower Clark Fork, Northern Glaciated Mountains, Upper Clark Fork, and Upper Snake ERUs (Figure 67.10). This stands to reason because stand replacement and mixed severity fires were historically dominant in these ERUs across both space and time.

Area and connectivity of intermediate (not new and not old forest) structural classes (stem exclusion, understory reinitiation, and young multistory patches) increased in most forested ERUs. This change toward landscape dominance of intermediate forest structures was the general mechanism of pattern simplification. When viewed simplistically, there is currently less area in stand-initiation and old forest structures, and considerably more area and improved connectivity of intermediate forest structures. The most notable increases in intermediate structures occurred in the Blue Mountains, Central Idaho Mountains, Columbia Plateau, Lower Clark Fork, Northern Glaciated Mountains, Snake Headwaters, Southern Cascades, and Upper Clark Fork ERUs. Area in intermediate structural classes actually declined in the Upper Klamath ERU, where transition analysis implicated extensive past timber harvesting.

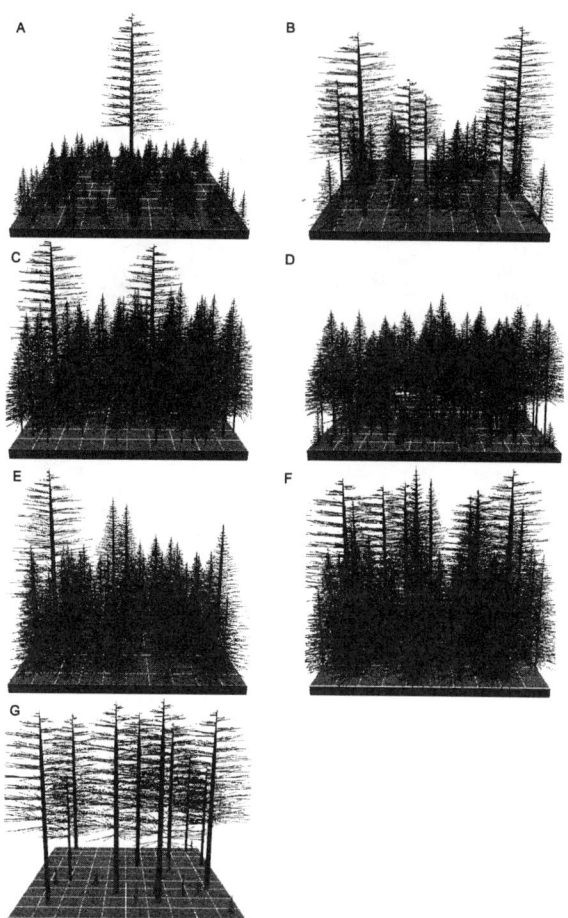

Figure 67.8 Graphical representation of forest structural classes used in the midscale assessment of the interior Columbia River basin: (A) stand-initiation; (B) open stem exclusion; (C) closed stem exclusion; (D) understory reinitiation; (E) young multistory forest; (F) old multistory forest; (G) old single-story forest. Refer to Oliver and Larson (1996) and O'Hara et al. (1996) for expanded descriptions of forest structural classes. (See color figure following p. 668.)

Other Structural Changes

Four additional findings related to forest structural change are worthy of brief mention:

1. In the historical condition, large (>63.5 cm diameter at breast height, d.b.h.) and medium (40.5 to 63.5 cm d.b.h.) trees were once more widely distributed in structures other than old forest as a conspicuous remnant after stand-replacing wildfires. Change analysis indicated that patches with medium and large trees were targeted for timber harvest, regardless of their structural affiliation.
2. Along with other raw attributes interpreted from aerial photos, we estimated dead tree and snag abundance in each forest patch as: none, <10%, 10 to 39%, 40 to 70%, and >70% of trees dead or as snags. Change analysis with these data indicated that dead tree and snag abundance increased significantly in most forested ERUs, but primarily in the pole and small tree (12.7 to 40.4 cm d.b.h.) size classes, because the medium and large trees were depleted by timber harvest.
3. Current forest patches have more canopy layers than were displayed in the historical condition, and understory layers are typically comprised of late-seral species.
4. In the historical condition, forest understories were often absent or they were comprised of shrub and herbaceous species. Current forest understories are less often grass or shrub and mostly coniferous. With livestock grazing and the elimination of surface fires, multilayered conifer understories developed.

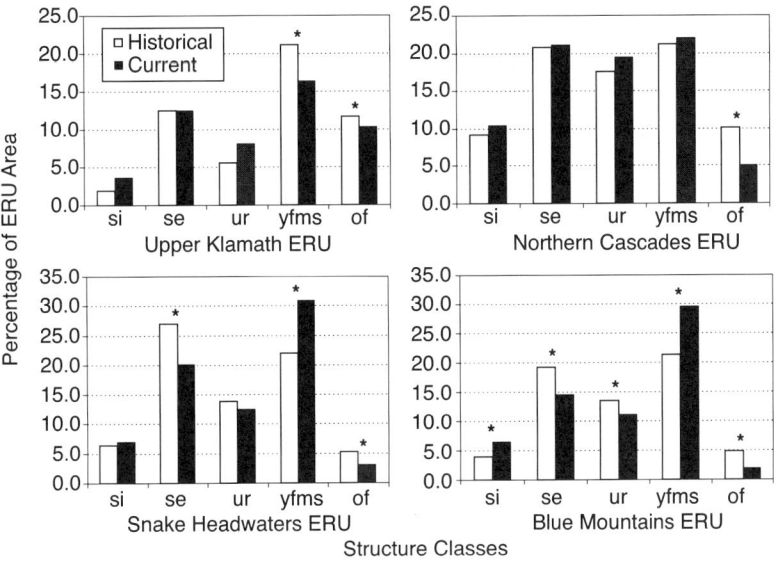

Figure 67.9 Change in percentage of area in old forest and other structures for selected ERUs. Structural class abbreviations are si = stand-initiation; se = stem exclusion (both open and closed canopy conditions); ur = understory reinitiation; yfms = young multistory forest; of = old multistory and single-story forest. Asterisk (*) denotes significant difference at $P \leq 0.2$.

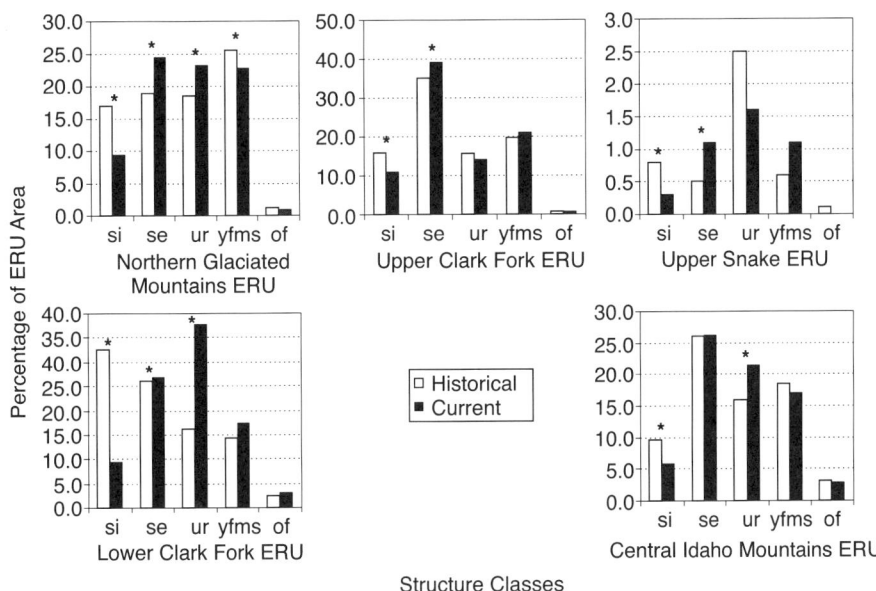

Figure 67.10 Change in percentage of area in stand-initiation and other forest structures for selected ERUs. Structural class abbreviations are si = stand-initiation; se = stem exclusion (both open and closed canopy conditions); ur = understory reinitiation; yfms = young multistory forest; of = old multistory and single-story forest. Asterisk (*) denotes significant difference at $P \leq 0.2$.

Landscape Spatial Patterns

We conducted change analysis with cover type-structural class couplets as patch types. Following, we discuss changes occurring across all ERUs for a given subset of metrics (Table 67.1).

Richness, Diversity, and Evenness

Patch richness (PR), Shannon's (1949) diversity index (SHDI), and the inverse of Simpson's λ (N2; Hill, 1973) provide different views of the diversity of patch types of a landscape. While richness tallies the number of patch types present regardless of their abundance, the SHDI and N2 indices incorporate abundance into the measurement of diversity, but N2 responds to change in the abundance of dominant patch types. Relative patch richness (RPR) relates PR to the total number of patch types (192) present in the basin. The SHDI (or its transformed equivalent, N1; Hill, 1973) is intermediate in responsiveness to abundance changes between RPR and N2.

In general, for the five measures of richness and diversity (RPR and PR, SHDI and N1, and N2), all ERUs displayed a positive mean difference with two notable exceptions (Table 67.4): the Lower Clark Fork and Upper Klamath ERUs exhibited minor declines in PR. Using transition analysis, we could attribute declines in richness to a history of widespread timber harvest activity. Five of 13 ERUs displayed change in PR generally on the order of a 15 to 30% increase, while 8 of the 13 ERUs reflected both increased dominance and diversity (N2). PR values and transition analysis showed that cover-structure patch types not only increased in number, but new patch types (cover type and structural class combinations) occupied considerable landscape area. The ERUs displaying increased dominance and diversity are shown in Table 67.4.

Evenness measures assess how equitably area is distributed among patch types. Both evenness measures (modified Simpson's evenness index, MSIEI, and Alatalo's evenness index, R21) index change in the distribution of area among all patch types. Many ERUs displayed increased diversity, richness, and dominance during the sample period for the diversity measures we used. Such increases typically result in a modest increase in the values of the evenness measures, if any change occurs. Our results confirmed this relation; the MSIEI and R21 indices increased in 6 of 8 ERUs displaying increased diversity and dominance. The Upper Clark Fork and the Central Idaho Mountains were the only 2 ERUs to decline in evenness; the Upper Clark Fork declined in both evenness measures (Table 67.4). To explain changes in evenness, it is helpful to reexamine changes in the area relations of cover types and structural classes. In the Central Idaho Mountains, few cover type changes were significant, but distribution of area in forest structural classes became increasingly uneven (Table 67.3, Figure 67.10). Area in stand-initiation structures declined from 9.7 to 5.9% of the ERU, and area in understory reinitiation structures increased from an average of 16 to 21.4% of the ERU. A similar pattern of change was evident in the Upper Clark Fork; few cover type changes were evident, but distribution of area in stand-initiation, closed canopy stem-exclusion, and young multistory forest structures became uneven (Table 67.3).

Landscape metrics were averaged across sampled subwatersheds, hence, values for all metrics in the historical and current condition reflect the average per subwatershed. But an examination of data in Table 67.4 prompts two additional questions: (1) What is the total richness and diversity of patch types of each ERU, and (2) Have total richness and diversity values for the ERU changed during the sample period? Heltshe and Forrester (1983) describe a jackknife estimator for richness that attempts to estimate total richness for a geographic area of interest. We applied this technique and a related estimator for N2 (Burnham and Overton, 1979) to the historical and current data to estimate difference in total richness and dominance for each ERU (Table 67.5). The jackknife technique results in estimates of the total and standard error. We used these statistics in simple two-way t-tests to evaluate significant change in total

Table 67.4 Landscape Metric Results for 13 Ecological Reporting Units in the Midscale Assessment of the Interior Columbia River Basin Where Patch Types Were Cover Type-Structural Class Doublets

Landscape Metrics	Blue Mountains	Central Idaho Mountains	Columbia Plateau	Lower Clark Fork	Northern Cascades	Northern Glaciated Mountains	Northern Great Basin	Owyhee Uplands	Snake Headwaters	Southern Cascades	Upper Clark Fork	Upper Klamath	Upper Snake
Richness and Diversity													
RPR_h[a,b]	11.4	11.3	7.8	15.8	13.9	11.6	4.8	2.9	11.8	7.7	11.6	9.6	5.1
RPR_c	12.1	12.5	8.2	15.5	16.0	14.3	5.3	2.9	11.9	10.6	13.5	9.5	5.2
RPR_md	0.7	1.2*c	0.3	-0.3	2.0*	2.7*	0.5	0.0	0.1	3.0*	1.9*	-0.1	0.1
PR_h	22.8	22.3	15.2	30.6	27.0	22.6	9.3	5.6	22.9	14.9	22.5	18.6	9.8
PR_c	23.5	24.7	15.8	30.0	31.0	27.9	10.3	5.6	23.0	20.6	26.1	18.5	10.1
PR_md	0.7	2.4*	0.6	-0.6	4.0*	5.3*	1.0	0.0	0.0	5.8*	3.7*	-0.1	0.3
SHDI_h	2.1	2.2	1.5	2.5	2.5	2.2	1.3	0.4	2.3	1.7	2.4	1.8	1.0
SHDI_c	2.2	2.3	1.5	2.6	2.6	2.4	1.5	0.5	2.4	2.1	2.5	1.9	1.1
SHDI_md	0.0	0.1*	0.0	0.1	0.1*	0.2*	0.3*	0.1*	0.1*	0.3*	0.1*	0.0	0.0
N1_h	10.0	10.1	5.2	12.9	13.3	10.3	3.6	1.7	10.7	6.0	11.7	7.4	3.4
N1_c	9.9	10.6	5.5	14.5	15.1	12.7	4.6	2.0	11.6	9.2	12.3	8.2	3.4
N1_md	-0.1	0.4*	0.2	1.6*	1.8*	2.3*	1.1*	0.3*	0.9*	3.1*	0.6	0.8	0.0
N2_h	6.8	7.0	3.8	8.2	9.4	7.2	2.6	1.4	7.7	4.5	8.6	5.2	2.4
N2_c	6.8	7.1	3.8	10.2	10.6	8.9	3.5	1.6	8.5	6.8	8.7	6.1	2.5
N2_md	0.0	0.1	0.1	2.0*	1.3*	1.6*	0.9*	0.2*	0.8*	2.3*	0.0	0.8*	0.0
Evenness													
MSIEI_h[d]	0.58	0.60	0.46	0.61	0.66	0.61	0.44	0.15	0.62	0.55	0.67	0.51	0.34
MSIEI_c	0.58	0.58	0.46	0.67	0.68	0.63	0.54	0.20	0.65	0.60	0.65	0.54	0.35
MSIEI_md	0.00	-0.01	0.00	0.06*	0.01*	0.02	0.11*	0.05*	0.03*	0.04	-0.03*	0.04	0.00
R21_h	0.64	0.64	0.63	0.62	0.67	0.65	0.65	0.44	0.66	0.69	0.71	0.62	0.61
R21_c	0.64	0.62	0.61	0.69	0.68	0.66	0.68	0.46	0.68	0.68	0.67	0.66	0.59
R21_md	0.00	-0.02*	-0.01	0.07*	0.01	0.00	0.03*	0.01	0.02	-0.01	-0.04*	0.04*	-0.02

(continued)

Table 67.4 (continued) Landscape Metric Results for 13 Ecological Reporting Units in the Midscale Assessment of the Interior Columbia River Basin Where Patch Types Were Cover Type-Structural Class Doublets

	Ecological Reporting Units (ERUs)												
Landscape Metrics	Blue Mountains	Central Idaho Mountains	Columbia Plateau	Lower Clark Fork	Northern Cascades	Northern Glaciated Mountains	Northern Great Basin	Owyhee Uplands	Snake Headwaters	Southern Cascades	Upper Clark Fork	Upper Klamath	Upper Snake
Contagion and Interspersion													
CONTAG_h	58.2	57.3	66.3	56.6	55.8	58.1	65.8	86.1	56.2	63.3	55.1	63.3	73.3
CONTAG_c	57.9	57.6	65.8	54.2	54.9	56.4	60.6	74.0	54.8	59.7	55.8	62.1	73.6
CONTAG_md	-0.3	0.3	-0.5	-2.4*	-0.9*	-1.7*	-5.2*	-12.1*	-1.4*	-3.6*	0.7	-1.2	0.3
IJI_h	65.9	67.6	60.2	69.0	68.8	67.6	56.0	42.6	70.1	64.0	70.6	61.5	47.1
IJI_c	65.2	67.1	58.8	71.7	69.7	68.2	56.5	52.4	71.1	65.1	68.7	63.0	56.7
IJI_md	-0.7	-0.6	-1.4	2.7	1.0*	0.6	0.6	9.8*	1.0	1.0	-1.9*	1.5	9.6*
Edge Contrast													
AWMECI_h	37.8	37.3	28.0	33.5	38.8	35.5	24.7	10.5	41.1	37.3	34.7	33.9	17.3
AWMECI_c	38.5	38.3	29.0	38.6	39.1	37.7	24.8	11.4	41.2	40.4	35.3	33.6	18.9
AWMECI_md	0.7	1.1*	1.0	5.1*	0.3	2.2*	0.1	0.9*	0.1	3.1*	0.6	-0.3	1.6*

[a] RPR values represent percentage of relative patch richness where the observed number of patch types (cover type-structural classes) in an ERU is scaled against a realistic maximum number of patch types possible across the entire basin assessment area. PR values simply represent the total number of patch types present within an ERU. N1 is a transformation of SHDI; rare patch types are weighted less than in PR. N2 also counts numbers of patch types like RPR, but N2 gives dominant patch types increased weight and can be considered a count of the average number of dominant patch types in an ERU. With N2, rare patch types are weighted less than in N1.

[b] Suffix h = average historical value among subwatersheds of an ERU; c = average current value among subwatersheds of an ERU; and md = mean difference of pairwise comparisons of sampled subwatersheds within an ERU; RPR = relative patch richness; PR = patch richness; SHDI = Shannon-Weaver diversity index; N1 = Hill's index $N1 = e^{SHDI}$; N2 = Hill's index $N2 = 1/(1/SIDI)$; MSIEI = modified Simpson's evenness index; R21 = Alatalo's evenness index = $(N2 - 1)/(N1 - 1)$; CONTAG = contagion index; IJI = interspersion and juxtaposition index; AWMECI = area-weighted mean edge contrast index (see also Table 67.1 and McGarigal and Marks, 1995).

[c] Asterisk (*) indicates statistical significance at $P \leq 0.2$.

[d] MSIEI is more sensitive to change in abundance among all patch types, whereas R21 is more sensitive to change in abundance of the dominant patch types. Increases indicate that area distributed among patch types is increasingly even. Declines indicate that some patch types are more abundant than others within an ERU. Significant figures are computed to two decimal places.

Table 67.5 Jackknife Estimates of Total Patch-Type Richness and Dominance (N2) for 13 Ecological Reporting Units in the Midscale Assessment of the Interior Columbia River Basin, Where Patch Types Were Cover Type-Structural Class Doublets

Ecological Reporting Unit	No. of Sampled Watersheds	Richness[a]		Dominance (N2)[b]	
		Historical (s.e.)	Current (s.e.)	Historical (s.e.)	Current (s.e.)
Blue Mountains	44	114 (6.0)	123 (4.3)	23 (3.5)	20 (2.4)
Central Idaho Mountains	43	142 (6.4)	135 (5.7)	32 (2.9)	29 (2.7)
Columbia Plateau	38	121 (7.7)	119 (5.5)	10 (1.7)	11 (1.7)
Lower Clark Fork	5	88 (9.5)	73 (2.9)	19 (1.6)	17 (1.8)
Northern Cascade Mountains	47	135 (5.1)	133 (3.8)	36 (4.5)	36 (3.8)
Northern Glaciated Mountains	41	127 (5.1)	136 (5.4)	25 (2.4)	26 (2.7)
Northern Great Basin	4	22 (2.6)	29 (3.6)	4 (0.4)	5 (0.6)
Owyhee Uplands	22	40 (6.0)	41 (4.1)	1 (0.2)	2 (0.3)*c
Snake Headwaters	15	83 (5.4)	92 (5.9)	30 (3.2)	26 (3.1)
Southern Cascades	16	69 (5.7)	80 (8.1)	15 (1.3)	15 (2.6)
Upper Clark Fork	32	113 (5.8)	120 (4.9)	25 (1.9)	23 (2.2)
Upper Klamath	13	107 (7.0)	100 (5.7)	13 (3.9)	16 (3.0)
Upper Snake	15	67 (8.3)	71 (6.5)	3 (0.4)	3 (0.9)

[a] Estimates of total richness and standard error (s.e.) were computed by using the methods of Heltshe and Forrester (1983). Estimates for total richness were rounded to the nearest integer.
[b] Estimates of total dominance (N2) and its standard error were computed by using the methods of Burnham and Overton (1979). Estimates for total dominance were rounded to the nearest integer.
[c] Asterisk (*) indicates significant difference at $P \leq 0.2$.

richness or dominance. All changes but one were insignificant, but 8 ERUs displayed nonsignificant increases in richness. Jackknife estimates of richness are sensitive to differences in sample size and distribution. In this instance, it is best not to make comparisons among ERUs, but comparisons between current and historical values are appropriate. Jackknife estimates for N2 are not restricted in this way. The N2 values across ERUs range from a low of 1 in the Owyhee Uplands to 36 in the Northern Cascades. We expect forest-dominated ERUs to display much larger values of total dominance than rangeland-dominated ERUs because of their greater patch richness and diversity.

Contagion and Interspersion

Contagion (CONTAG) and interspersion-juxtaposition (IJI) metrics quantify the extent to which pixels of differing type intermix; IJI considers length of edge between contrasting patch types, and CONTAG estimates patch dispersion and interspersion for raster maps. Seven of 13 ERUs displayed significant declines in CONTAG, and all significant mean differences values were negative (Table 67.4). A negative mean difference value of CONTAG indicated that across an ERU, cover-structure patches became smaller and more dispersed. Three of 6 ERUs with nonsignificant mean difference values of CONTAG also exhibited a negative sign. These results suggest a basin-wide decrease in cover-structure patch contagion. With the exception of the Northern Great Basin and Owyhee Uplands ERUs, the magnitude of decrease was small relative to initial average historical values.

Only 4 of 13 ERUs displayed significant mean difference values for IJI; 2 were positive and 2 were negative (Table 67.4). The Owyhee Uplands and Upper Snake ERUs were noteworthy because the magnitude of the mean IJI difference for these 2 ERUs was especially large. Unlike CONTAG, there was no consistent pattern across ERUs for this metric, and most changes were small. We conclude that interspersion changes as measured by this metric are minimal at this scale, but that changes in interspersion and juxtaposition may be better observed at smaller pooling scales where environmental variation is controlled.

Edge Contrast

The area-weighted mean edge contrast index (AWMECI) uses a set of user-defined values ranging from 0 to 1 to represent relative edge contrast (Table 67.2) between patch types, weighted by area, to evaluate change in edge contrast of a sample of landscapes. We based edge contrast weights on physiognomic and structural conditions in deference to edge-sensitive and edge-dependent terrestrial species, and their sensitivity to structural edge differences. An increase in area-weighted mean edge contrast was indicated as the percentage of the total edge that was the equivalent of maximum contrast edge. The greater the difference in structure or physiognomic condition (e.g., an old single-story forest patch adjacent to herbland), the greater the edge contrast weights. Significant increase in AWMECI for a sample of landscapes indicated that greater contrast in structural and physiognomic condition was occurring at patch edges. Six of 13 ERUs displayed such an increase. Most increases were relatively modest except in the Lower Clark Fork where increase in maximum contrast edge averaged 5.1% of the total edge (Table 67.4).

Landscape Vulnerability to Disturbances

Insects and Pathogens

Our analysis indicated that forest landscapes have changed significantly in their vulnerability to major insect and pathogen disturbances. Absent frequent fires and influenced by selective harvesting and domestic livestock grazing, overstory and understory cover of shade-tolerant conifers expanded, forest structures became more layered, and grass and shrub understories were replaced by coniferous understories. As a consequence, the vulnerability associated with shade-tolerant forest area increased substantially. Conversely, because patches with medium and large trees of early-seral species were primarily harvested, the vulnerability associated with mature and old early-seral forests declined. A few examples follow (also see Table 67.6).

In many ERUs, the area vulnerable to western spruce budworm increased, but most changes were not significant at the scale of an ERU. To determine whether the lack of observed significant difference was scale dependent, we summarized statistical results for budworm (and other disturbances) by smaller scale pooling strata. For example, we used subregional groupings of sampled subwatersheds based on similarity of climate and environments, and we were able to readily detect significant change in area vulnerable to budworm defoliation. Results of these analyses suggested that ERUs were too large for detecting these and perhaps other changes, and smaller pooling strata would be more informative. In addition, we noted that the conduciveness of both patch- and landscape-scale vegetation conditions to widespread severe budworm defoliation increased over the sample period. This was due to the increased area and connectivity of patches of shade-tolerant conifers, increased abundance of multilayered host patches, increased contagion of host patches reflected both in increased average host patch size, and in some cases, increased host patch density.

Throughout the basin, area vulnerable to Douglas fir beetle and Douglas fir dwarf mistletoe increased in three important ways: (1) forest landscapes in the existing condition exhibit increased cover and connectivity of Douglas fir; (2) current stand densities are elevated well above historical conditions; and (3) patches today tend to be multilayered, often with several layers of understory Douglas fir. Figure 67.11 (top; see color figure following p. 668) provides an illustration of increased forest area vulnerable to Douglas fir dwarf mistletoe in a subwatershed of the Wallowa subbasin of the Blue Mountains ERU. Conversely, area vulnerable to western pine beetle disturbance of mature and old ponderosa pine (WPB1, Table 67.6) fell because medium and large ponderosa pine were selectively harvested from old forest patches and from other structures (Figure 67.11 bottom; see color figure following p. 668).

Table 67.6 Change in Area Highly Vulnerable to Insect and Pathogen Disturbance in 11 Forested ERUs in the Midscale Assessment of the Interior Columbia River Basin

Disturbance Agent[a]	Blue Mountains H[b]	Blue Mountains C	Central Idaho Mountains H	Central Idaho Mountains C	Columbia Plateau H	Columbia Plateau C	Lower Clark Fork H	Lower Clark Fork C	Northern Cascades H	Northern Cascades C	Northern Glaciated Mountains H	Northern Glaciated Mountains C	Snake Headwaters H	Snake Headwaters C	Southern Cascades H	Southern Cascades C	Upper Clark Fork H	Upper Clark Fork C	Upper Klamath H	Upper Klamath C	Upper Snake H	Upper Snake C
											— Percentage of ERU area —											
WSB	38.2	38.9	49.4	51.1	**9.3**	**12.0**	56.8	65.0	51.5	50.4	**44.5**	**47.9**	45.0	51.8	**10.1**	**12.3**	59.1	55.9	14.6	15.9	**1.6**	**2.1**
DFB	**5.2**[c]	**7.8**	4.4	5.0	2.9	2.6	0.2	5.9	8.7	7.4	3.6	5.0	2.1	3.9	**1.8**	**0.1**	**8.0**	**4.8**	0.0	0.0	0.5	1.0
WPB1	2.5	2.5	1.0	1.3	4.6	2.9	0.0	0.6	**3.7**	**1.8**	1.2	0.9	34.6	29.2	5.2	5.1	**2.9**	**0.5**	5.7	4.5	—	—
WPB2	17.8	19.7	3.3	3.3	**14.9**	**17.1**	1.5	3.8	**9.8**	**8.2**	7.9	7.3	—	—	20.5	24.4	**9.9**	**8.1**	19.3	21.3	—	—
MPB1	**6.7**	**5.1**	21.0	22.1	1.8	2.4	**4.0**	**12.9**	5.3	6.8	**15.4**	**18.9**	34.6	29.2	29.0	24.9	36.1	37.6	4.7	4.3	0.6	0.3
MPB2	17.8	19.7	3.3	3.3	**14.9**	**17.1**	1.5	3.8	**9.8**	**8.2**	7.9	7.3	—	—	20.5	24.4	**9.9**	**8.1**	19.3	21.3	—	—
FE	**24.6**	**15.0**	21.3	**26.2**	1.8	2.9	28.3	37.0	20.4	21.5	6.8	8.4	19.3	16.1	9.0	10.2	**7.8**	**9.7**	**17.1**	**18.0**	—	—
SB	**2.6**	**0.7**	3.1	3.6	—	—	0.1	0.5	6.0	5.3	3.0	4.5	8.3	7.6	0.1	0.1	6.9	5.6	0.8	0.2	0.6	1.5
DFDM	**10.1**	**16.5**	10.7	10.5	6.9	6.4	6.0	7.9	18.6	17.9	13.1	14.3	**4.1**	**6.4**	**2.3**	**0.5**	**16.2**	**13.2**	17.8	15.5	—	—
PPDM	**10.4**	**8.1**	2.2	1.8	10.8	7.8	0.0	0.9	**5.6**	**3.9**	**3.8**	**2.5**	—	—	12.9	17.9	**5.0**	**2.3**	—	—	—	—
WLDM	1.3	0.8	0.5	0.1	0.0	0.1	0.2	0.2	0.5	0.4	6.9	4.2	—	—	—	—	**2.8**	**1.3**	—	—	—	—
LPDM	1.5	1.6	13.7	15.1	0.2	0.4	0.2	2.6	2.7	3.1	9.3	9.1	**30.8**	**20.9**	10.2	11.9	22.6	22.5	0.4	0.3	0.3	0.3
AROS	40.7	41.0	37.6	39.2	9.1	10.5	55.0	65.1	48.6	45.2	**37.3**	**40.7**	**20.4**	**31.5**	10.9	12.8	34.2	31.8	13.2	13.6	1.0	1.6
PHWE	34.5	37.0	29.3	27.8	10.4	9.7	59.4	62.0	41.7	39.2	27.8	31.0	**10.9**	**12.8**	31.1	35.4	21.6	20.8	18.5	17.7	1.0	1.6
HEAN_s	**24.3**	**16.9**	36.2	**38.9**	0.8	**5.4**	71.4	77.0	29.6	32.2	**20.0**	**26.8**	**22.0**	**30.6**	37.1	38.3	32.2	34.6	22.8	23.3	1.0	1.7
HEAN_p	11.6	10.4	2.1	1.7	11.8	11.2	0.0	0.2	7.5	5.9	3.1	4.0	—	—	**13.8**	**23.4**	**5.4**	**4.0**	19.0	19.8	—	—
TRBR	**4.4**	**2.5**	9.3	11.0	—	—	1.0	1.5	**11.4**	**9.9**	**7.1**	**9.0**	13.1	15.1	0.8	0.8	9.9	10.3	—	—	—	—
SRBR	**46.7**	**52.1**	57.1	56.2	17.2	15.4	56.2	52.3	61.2	57.2	66.9	65.8	49.9	48.6	25.6	29.4	60.6	59.0	**26.4**	**17.9**	**1.5**	**2.1**
WPBR1	—	—	0.0	0.1	**1.4**	**0.1**	0.8	3.7	0.1	0.2	**1.9**	**0.3**	—	—	0.2	0.3	2.5	1.4	0.0	0.0	—	—
WPBR2	—	—	0.7	0.6	—	—	—	—	0.4	0.9	0.0	0.0	4.0	2.0	—	—	2.9	2.4	—	—	—	—
RRSR	1.1	0.8	0.1	0.3	0.0	0.1	1.0	1.7	0.6	1.1	0.0	0.2	—	—	1.1	1.7	1.2	2.3	4.8	4.1	—	—

[a] WSB = western spruce budworm; DFB = Douglas fir beetle; WPB1 = western pine beetle (type 1 attack of mature and old ponderosa pine); WPB2 = western pine beetle (type 2 attack of immature and overstocked ponderosa pine); MPB1 = mountain pine beetle (type 1 attack of overstocked lodgepole pine); MPB2 = mountain pine beetle (type 2 attack of immature and overstocked ponderosa pine; FE = fir engraver; SB = spruce beetle; DFDM = Douglas fir dwarf mistletoe; PPDM = ponderosa pine dwarf mistletoe; WLDM = western larch dwarf mistletoe; LPDM = lodgepole pine dwarf mistletoe; AROS = *Armillaria* root disease; PHWE = laminated root rot; HEAN_s = S-group annosum root disease; HEAN_p = P-group annosum root disease; TRBR = tomentosus root and butt rot; SRBR = Schweinitzii root and butt rot; WPBR1 = white pine blister rust (type 1 on western white pine/sugar pine); WPBR2 = white pine blister rust (type 2 on whitebark pine); RRSR = rust-red stringy rot. See also Hessburg et al., 1999b.

[b] H = historical condition; C = current condition.

[c] Values shown in bold indicate significant increase or decrease ($P \leq 0.2$) from the historical to the current condition.

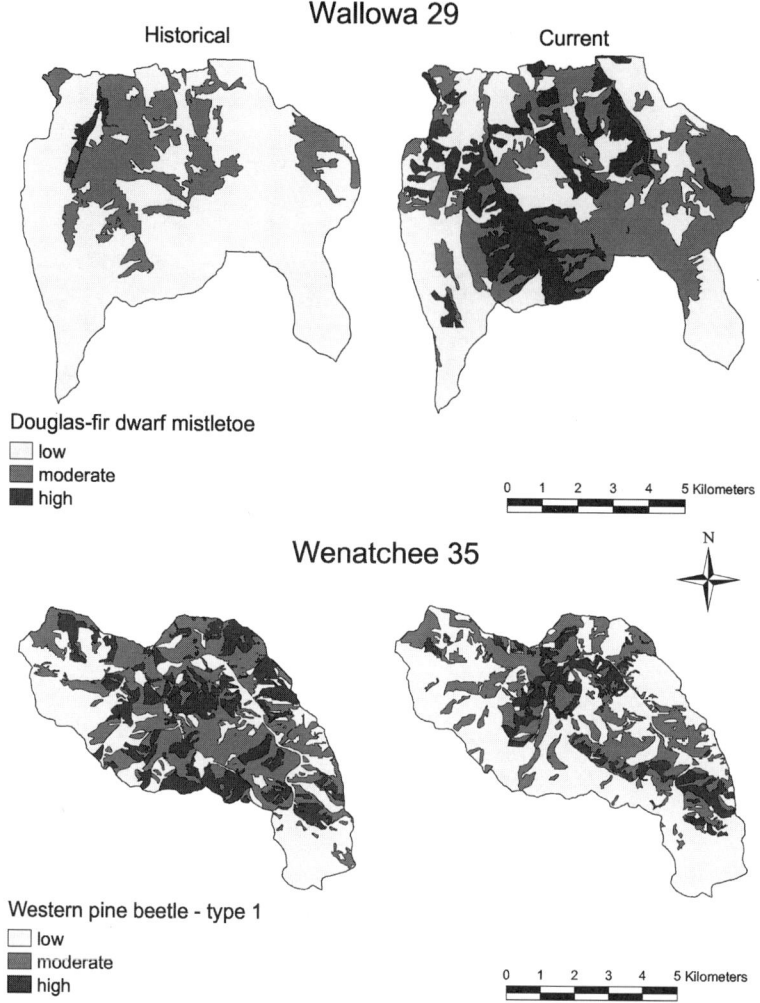

Figure 67.11 Historical and current forest area: (top) vulnerable to Douglas fir dwarf mistletoe disturbance, subwatershed 29, Wallowa subbasin, Blue Mountains ERU; (bottom) vulnerable to western pine beetle disturbance of medium (40.5–63.5 cm d.b.h.) and large (> 63.5 cm d.b.h.) ponderosa pine, subwatershed 35, Wenatchee subbasin, Northern Cascades ERU. (See color figure following p. 668.)

Wildfires, Prescribed Fires, and Smoke Production

Forest landscapes have been significantly altered in their vulnerability to wildfires, increasing the potential for air quality degradation as well. In general, the risk of stand replacement fire has increased throughout the forest-dominated portion of the basin. Elevated risk is indicated by increased ground fuel loads, crown fire potential, flame length, rate of spread, fireline intensity, and smoke production (PM10), each of which are consequences of change in spatial patterns of both living and dead forest cover and structure. Changes in vegetation patterns are the result of effective fire suppression, timber harvest, and fire exclusion. Key factors responsible for fire exclusion were the widespread elimination of flashy fuels through extensive domestic livestock grazing, especially in the first half of the 20th century (Hann et al., 1997, and references therein; Skovlin and Thomas, 1995; Wissmar et al., 1994); reduced connectivity of fire-prone landscapes through placement of extensive road networks; settlement of fire-prone interior valleys and subsequent conversion to irrigated agriculture by European immigrants; and movement of American Indians, who frequently used fire as a management tool on reservations (Robbins and Wolf, 1994).

A number of changes in fuel loading, PM10 smoke production (Table 67.7), crown fire potential, and fire behavior attributes (Table 67.8) were significant at the ERU scale; a few examples follow (see also Ottmar et al. in press). In the Central Idaho Mountains, fuel loads (>45 Mg/ha), smoke production (>448 kg/ha), and flame length (>1.2 m) increased to high levels or above on more than 5% of the area. In the Lower Clark Fork ERU, fuel loads (> 45 Mg/ha), PM10 smoke production (>448 kg/ha) during wildfires, and flame lengths greater than 1.2 m during wildfires increased to moderate levels or above on approximately one third of the ERU area. Crown fire potential increased to high levels or above on 29% of the ERU area. At present, 82% of the area of the Lower Clark Fork ERU exhibits moderate to severe crown fire potential. In the event of a wildfire today, it would be difficult with current technology to suppress expected flames on 94% of the current forest and rangeland area. At present, over 81% of the ERU would exhibit very low smoke production if prescribed fires were implemented in place of wildfires for fuels reduction. Under a wildfire burn scenario, only 14% of the ERU area would exhibit very low smoke production.

Finally, in the Southern Cascades fuel loads (>45 Mg/ha) increased on nearly 5% of the ERU area, rate of spread (>2.4 m/min) during wildfires increased on more than 11% of the area, and extreme fireline intensity (>3459.2 kW/m) increased on nearly 8% of the area. At present, 56% of the ERU area exhibits moderate to extreme crown fire potential, and 82% of the forest and rangeland area would experience flame lengths in excess of 1.2 m during wildfires. In the event of a wildfire, it would be difficult to suppress flames on 8 of every 10 ha. At present, over 75% of the ERU would exhibit very low smoke production if prescribed fires were implemented in place of wildfires for fuels reduction. Under a wildfire burn scenario, 21% of the ERU area would exhibit very low smoke production.

DISCUSSION

The primary utility of landscape assessment is in understanding the characteristics and potential responses of the ecosystems we manage (Morgan et al., 1994). Knowledge of landscape pattern change at regional and subregional scales provides critical context for forest-level planning; watershed analysis and project-level planning; and valuable insight for ecosystem restoration, conservation, and monitoring decisions. Landscape change analysis provides an empirical basis to evaluate the historical and current rarity of landscape pattern features and aids in determining how representative current patterns are compared with historical conditions.

- The basin assessment area is large, and we summarized many changes in vegetation condition and associated vulnerability to disturbance. Here we focus on the most important generalities lest we lose them.
- The most dramatic change in physiognomic conditions that we observed was the widespread decline in shrublands. Losses resulted from forest, woodland, and cropland expansion, and conversion to semi- or nonnative herbland.
- The loss of historical herblands to agriculture was considerably more dramatic, but it had already occurred by the time of our historical sample.
- Shifts from early- to late-seral species were evident in many ERUs. Change in ponderosa pine, western larch, and Douglas fir cover was associated with reduced area with medium and large trees.
- We observed a precipitous decline in area and connectivity of western white pine cover in northern Idaho and northwestern Montana, the heart of the historical range. Losses were attributed to early selective harvesting, an introduced blister rust fungus, and mountain pine beetle mortality.
- Shade-tolerant conifers now dominate interior Columbia Basin forests. Lacking significant pattern restoration, the insects and pathogens that favor shade-tolerant conifers will continue shaping forest patterns by their growth and mortality effects, and by indirect influence on fire regimes.
- Patch area with old forest-structures declined sharply in all ERUs where they historically occupied more than a minor area. The same was true of patches with remnant large trees.

- In several ERUs, area with medium and large trees overshadowed or augmented losses to historical old-forest area. Our results suggested that 20th century timber harvest activities targeted patches with medium- and large-sized trees regardless of their structural affiliation. There are at least two important ramifications: First, it has been broadly assumed that large trees were principally associated with old forests, where they obviously contribute important living and dead structure. In some ERUs, old forest abundance was historically quite minimal (refer to Table 67.3), but medium and large trees were broadly distributed in other forest structures as a remnant after stand-replacing fires. In some cases, large trees comprised as much as 24% of the crown cover of other forest structures, contributing important living and dead late-successional functionality. Second, where old forest area and area with remnant large trees has been depleted, the present and future supply of medium and large dead trees as snags and down logs is substantially diminished. This is especially true of snags and down logs of early-seral species. We propose that terrestrial and aquatic species and processes requiring large dead tree structure may be adversely influenced by this reduction unless the shortfall is remedied through intentional recruitment.
- In several ERUs, we observed a marked reduction in landscape vulnerability to dwarf mistletoes of early-seral species. Comparisons of historical and current subwatersheds showed that timber harvest reduced crown cover of large early-seral trees while one or more shade-tolerant understory strata developed. Wildlife microhabitat ramifications are likely.
- Area and mean patch size of stand-initiation structures dramatically declined where natural stand-replacing fires have been excluded. Such reduction was evident despite widespread timber harvest activity. With recent emphasis on conserving dwindling areas of late-successional and old forests in the Interior Northwest, the role of stand-initiation structures on the landscape may have been underestimated. The immediate effect of stand-replacing fires is to simplify landscape patterns and restore early-seral conditions. After such a fire, forests are regenerated with a new cohort of early-seral seedlings, saplings, grasses, and shrubs. Subsequently, environmental and disturbance gradients interact, recreating some of the lost pattern complexity, but expanses of interior forest emerge as a reminder of prior disturbance. Absent disturbances of this magnitude, what will create the interior forests of the future?
- The absence of wildfires has had profound effects on forest and woodland area and connectivity at subwatershed to regional scales. The history and legacy of fire suppression and prevention programs is well known, but fire exclusion effects have been more difficult to pin down because many interacting factors played a role. As a result, fire prevention and suppression efficacy may have been overstated, and the role of factors responsible for exclusion of fire understated. Basic ecological studies are needed that explore in greater detail effects of road networks and domestic livestock herbivory on historical *and* current exclusion of fires and accretion of trees.
- Patches of current forest cover types and structural classes are more fragmented than before. Patch densities are now higher, mean patch sizes are smaller, the largest patch of any given cover or structural class is smaller, and edge density is greater. These combined outcomes point to reduced landscape contagion as a consequence of timber harvest and road construction. Landscape pattern metrics confirmed the presence of highly fragmented landscapes in the current condition and pointed to increased pattern complexity among patch types in managed landscapes, and decreased complexity in roadless and wilderness-dominated landscapes.

Management Implications

Several important management implications emerge from our assessment of basin conditions, but before we discuss implications, we summarize what was learned from assessment. First, we showed that spatial patterns of historical landscapes were indeed variable, and that those patterns reflected variation resulting from patterns of environments and natural disturbance regimes. Second, we showed that ground fuel and fire behavior conditions can be explicitly linked to vegetation conditions, that fuel and fire behavior conditions changed throughout the basin, and that landscapes varied in the direction and magnitude of changes. Finally, we showed that fire, insect, and pathogen disturbances were commonly associated with historical landscapes, but that the area and pattern of area that was highly vulnerable to disturbance, and the degree of spatial isolation has changed. In

Table 67.7 Historical and Current Percentage of Area of Fuel Loading, Fuel Consumption, and PM10 Smoke Production Attributes of Subwatersheds Sampled in ERUs of the Midscale Ecological Assessment of the Interior Columbia River Basin

Ecological Reporting Units (ERUs)

— Percentage of area —

Attributes	Blue Mountains H[a]	Blue Mountains C	Central Idaho Mountains H	Central Idaho Mountains C	Columbia Plateau H	Columbia Plateau C	Lower Clark Fork H	Lower Clark Fork C	Northern Cascades H	Northern Cascades C	Northern Glaciated Mountains H	Northern Glaciated Mountains C	Northern Great Basin H	Northern Great Basin C	Owyhee Uplands H	Owyhee Uplands C	Snake Headwaters H	Snake Headwaters C	Southern Cascades H	Southern Cascades C	Upper Clark Fork H	Upper Clark Fork C	Upper Klamath H	Upper Klamath C	Upper Snake H	Upper Snake C
Fuel Loading[e]																										
Very low	37.5	36.4	29.0	28.3	74.2	70.8	5.8	4.8	23.4	21.7	21.0	19.4	91.6	92.3	99.8	99.3	28.9	29.6	22.9	12.8	19.5	15.6	51.7	54.9	96.8	96.6
Low	**19.3[b]**	**16.4**	31.5	26.0	8.9	10.1	46.5	17.3	22.6	26.3	28.2	23.3	8.5	7.7	0.3	0.2	**37.6**	28.2	17.5	23.4	35.1	37.4	12.3	11.4	2.1	1.7
Moderate	12.0	13.0	8.1	8.6	8.0	10.0	9.4	14.6	10.4	10.7	7.8	15.1	—	—	—	—	11.4	12.0	21.4	29.5	16.6	17.6	3.6	7.2	0.1	0.2
High	**22.2**	**26.3**	16.4	18.1	6.4	5.8	20.4	44.1	22.9	22.7	33.3	26.7	—	—	—	—	9.5	10.8	35.4	26.9	21.3	16.5	**18.3**	**10.0**	0.9	1.3
Very high	9.0	8.0	15.1	19.0	2.5	3.3	17.8	19.3	20.7	18.7	9.7	15.5	—	—	—	—	**12.7**	**19.4**	2.9	7.4	**7.6**	**13.1**	14.2	16.5	0.2	0.2
Fuel Consumption during Wildfires[e]																										
Very low	38.3	40.4	37.3	33.0	74.7	71.2	35.2	13.9	30.6	31.7	33.9	26.7	91.6	92.3	99.8	99.8	36.7	36.4	24.8	15.6	29.0	24.4	51.9	55.2	96.9	90.0
Low	20.1	**14.4**	23.2	21.4	8.9	10.3	**17.2**	**8.2**	15.4	16.5	15.4	16.6	8.5	7.7	0.3	0.2	**29.8**	21.5	18.0	22.1	25.9	30.3	12.0	11.2	2.0	1.6
Moderate	20.8	21.9	10.7	10.8	13.6	14.7	9.7	19.6	16.4	14.8	15.2	18.6	—	—	—	—	11.4	11.9	24.1	30.9	22.2	18.2	13.8	11.6	0.1	0.2
High	12.0	15.4	13.7	15.9	**0.7**	**1.3**	20.1	39.0	17.3	18.5	25.8	22.7	—	—	—	—	9.5	10.8	30.1	24.9	15.3	14.1	8.7	6.4	0.9	1.3
Very high	8.8	7.8	15.1	19.0	2.2	2.7	17.8	19.3	20.4	18.5	9.7	15.3	—	—	—	—	**12.7**	**19.4**	2.9	6.6	**7.6**	**13.0**	13.6	15.6	0.2	0.2
PM10 Smoke Production during Wildfires[e]																										
Very low	40.1	41.4	39.4	35.2	75.4	72.4	38.5	13.9	32.1	33.3	36.2	27.8	91.6	92.3	99.8	99.8	38.9	37.5	28.5	21.3	32.5	27.9	53.1	56.1	97.5	96.7
Low	25.5	20.9	24.4	22.4	12.3	15.5	14.5	9.9	18.6	20.2	15.8	21.5	8.5	7.7	0.3	0.2	**34.8**	28.1	32.3	41.3	29.0	32.3	13.4	16.5	1.5	1.8
Moderate	14.3	15.5	7.9	8.2	9.8	10.0	15.0	24.6	**12.4**	**11.0**	13.3	14.5	—	—	—	—	7.0	5.8	6.2	7.1	16.2	14.1	13.3	9.0	0.0	0.0
High	17.6	21.5	26.8	32.5	2.4	1.7	30.4	49.5	31.2	31.6	33.9	35.0	—	—	0.0	0.0	**18.9**	**27.5**	32.0	27.2	21.6	25.1	11.6	11.8	0.9	1.4
Very high	2.6	0.6	1.5	1.8	0.2	0.5	1.6	2.2	**5.7**	**3.9**	0.8	1.3	—	—	—	—	0.4	1.2	1.0	3.1	0.7	0.7	8.5	6.6	0.2	0.1
PM10 Smoke Production during Current Wildfires and Prescribed Fires[e]																										
Very low	41.5[c]	87.2[d]	35.2[c]	95.6[d]	72.4[c]	90.7[d]	13.9[c]	81.3[d]	33.3[c]	85.5[d]	27.8[c]	87.7[d]	92.3[c]	100[d]	99.8[d]	100[d]	37.5[c]	97.5[d]	21.3[c]	75.5[d]	27.9[c]	91.1[d]	56.1[c]	79.0[d]	96.7[c]	99.8[d]
Low	20.9[c]	9.3[d]	22.4[c]	2.8[d]	15.5[c]	8.6[d]	9.9[c]	9.0[d]	20.2[c]	8.2[d]	21.5[c]	5.2[d]	7.7[c]	0.0[d]	0.2[c]	0.0[d]	28.1[c]	1.2[d]	41.3[c]	21.6[d]	32.3[c]	4.4[d]	16.5[c]	15.9[d]	1.8[c]	0.1[d]
Moderate	15.5[c]	0.1[d]	8.2[c]	0.1[d]	10.0[c]	0.0[d]	24.6[c]	0.3[d]	11.0[c]	0.4[d]	14.5[c]	0.4[d]	—	—	—	—	5.8[c]	1.0[d]	7.1[c]	0.4[d]	14.1[c]	0.2[d]	9.0[c]	0.1[d]	0.0[c]	0.0[d]
High	21.5[c]	3.5[d]	32.5[c]	1.6[d]	1.7[c]	0.7[d]	49.5[c]	9.4[d]	31.6[c]	5.9[d]	35.0[c]	6.7[d]	—	—	0.0[c]	0.0[d]	27.5[c]	0.3[d]	27.2[c]	2.6[d]	25.1[c]	4.3[d]	11.8[c]	5.0[d]	1.4[c]	0.2[d]
Very high	2.6[c]	0.6[d]	1.8[c]	0.6[d]	0.5[c]	0.0[d]	2.2[c]	0.0[d]	3.9[c]	0.0[d]	1.3[c]	0.0[d]	—	—	—	—	1.2[c]	0.0[d]	3.1[c]	0.0[d]	0.7[c]	0.0[d]	6.6[c]	0.0[d]	0.1[c]	0.0[d]

a H = historical condition; C = current condition.
b Values shown in bold indicate significant increase or decrease (P ≤ 0.2) from the historical to the current condition.
c PM10 smoke production during current wildfires.
d PM10 smoke production during current prescribed fires.
e Fuel loading and fuel consumption classes (Mg/ha) are: very low <22.5; low = 22.5—44.9; moderate = 45—56.1; high = 56.2—67.3; and very high >67.3. PM10 smoke production classes (kg/ha) are: very low = 0–224.2; low = 224.3–448.3; moderate = 448.4–672.5; high = 672.6–896.7; and very high >896.6.

Table 67.8 Historical and Current Percentage Area of Fire Behavior and Crown Fire Potential Attributes of Subwatersheds Sampled in ERUs of the Midscale Ecological Assessment of the Interior Columbia River Basin

Ecological Reporting Units (ERUs) — Percentage of area

Attributes	Blue Mountains		Central Idaho Mountains		Columbia Plateau		Lower Clark Fork		Northern Cascades		Northern Glaciated Mountains		Northern Great Basin		Owyhee Uplands		Snake Headwaters		Southern Cascades		Upper Clark Fork		Upper Klamath		Upper Snake	
	H[a]	C	H	C	H	C	H	C	H	C	H	C	H	C	H	C	H	C	H	C	H	C	H	C	H	C
Fireline Intensity during Wildfires																										
Very low	26.4	27.4	14.9	14.9	50.2	51.1	4.0	3.6	21.5	19.9	19.6	18.3	19.6	35.7	16.2	25.0	18.2	20.1	25.4	15.6	19.4	15.2	32.2	38.3	31.0	33.6
Low	11.3[b]	7.9	15.4	15.3	24.3	19.9	2.4	1.7	2.9	4.0	2.4	2.8	78.7	62.5	83.8	74.9	18.6	13.3	0.1	0.2	3.2	3.9	20.1	17.4	67.0	63.6
Moderate	20.5	19.2	19.8	21.2	12.5	9.8	8.4	7.7	23.4	19.1	20.6	15.5	1.7	1.8	0.0	0.1	22.3	22.6	14.3	18.1	17.0	15.3	16.3	11.4	0.6	0.6
High	15.2	14.0	24.0	20.2	5.0	4.9	49.5	28.4	19.2	23.6	26.1	23.6	—	—	—	—	29.2	29.4	20.3	26.9	35.3	36.7	6.3	7.2	0.5	0.8
Very high	18.0	20.5	23.7	26.0	0.2	5.7	27.4	48.1	24.0	23.4	29.2	29.4	—	—	0.0	0.0	11.4	13.5	33.7	1.9	23.0	24.8	15.6	11.6	0.9	1.4
Severe	1.2	0.9	0.8	0.9	—	0.2	2.3	2.4	3.6	4.2	0.5	0.5	—	—	—	—	0.1	0.4	1.8	12.3	0.4	0.5	3.1	3.2	—	—
Extreme	7.3	7.9	1.4	1.6	3.6	6.2	6.0	8.1	5.3	5.8	1.5	4.0	—	—	—	—	0.2	0.8	4.5	—	1.6	3.5	6.4	10.8	—	—
Rate of Spread during Wildfires																										
Very low	24.4	24.0	26.1	28.9	37.0	36.4	4.8	5.0	30.0	27.1	24.9	24.1	13.0	19.3	6.6	14.0	33.6	35.3	25.4	17.6	24.8	23.2	21.8	24.1	21.7	26.5
Low	45.9	46.7	52.7	50.4	44.7	43.9	56.8	64.3	48.7	48.9	56.0	51.5	86.3	78.9	93.1	85.8	54.5	51.9	59.8	56.5	52.8	52.5	56.5	52.9	76.9	71.5
Moderate	29.7	29.3	21.2	20.6	18.3	19.6	38.4	30.5	21.2	23.6	19.1	24.1	0.7	1.9	0.3	0.2	11.9	12.8	14.8	25.8	22.4	24.2	21.7	22.9	1.5	2.0
High	0.0	0.0	0.0	0.0	0.0	0.0	0.0	0.3	0.0	0.4	0.1	0.3	—	—	—	—	0.1	0.1	0.0	0.1	0.0	0.2	0.0	0.1	—	—
Flame Length during Wildfires																										
Very low	24.0	26.0	12.8	12.6	43.1	47.5	3.8	3.6	20.5	19.1	19.4	17.9	19.6	35.7	13.1	22.6	12.8	15.5	22.9	12.8	17.7	13.5	29.8	35.6	22.4	28.5
Low	14.3	11.2	22.8	23.8	31.5	23.6	3.3	2.2	4.7	5.4	5.1	6.1	80.0	63.9	86.9	77.4	32.1	23.7	2.6	5.3	6.4	8.5	22.6	20.1	75.6	69.0
Moderate	18.5	18.3	13.5	14.2	11.6	9.7	7.4	7.2	20.3	15.9	17.5	11.9	0.4	0.5	—	—	10.7	15.7	14.1	13.0	13.9	11.0	15.2	9.5	0.6	0.3
High	24.6	20.0	33.0	28.7	5.7	6.9	57.2	44.2	25.1	29.7	35.3	37.3	—	—	—	—	35.4	35.1	32.2	34.9	44.0	44.3	9.2	10.9	1.1	1.7
Very high	16.3	21.3	16.6	19.5	7.1	10.2	22.3	35.1	26.7	26.1	21.3	24.8	—	—	—	—	8.9	9.5	27.9	32.9	17.1	19.9	20.0	18.6	0.4	0.5
Severe	2.3	3.2	1.2	1.3	1.0	2.2	6.0	7.7	2.8	3.9	1.2	2.1	—	—	—	—	0.1	0.6	0.4	1.2	0.9	2.8	3.2	5.3	—	—
Crown Fire Potential during Wildfires																										
None	35.3	31.8	30.9	30.4	68.2	60.2	10.1	6.8	24.0	23.6	24.9	22.2	85.9	78.3	94.1	92.6	38.0	33.7	22.5	16.5	19.1	18.9	54.3	56.0	96.9	95.9
Very low	7.1	9.4	4.6	3.9	11.0	16.5	1.4	0.2	5.8	4.9	8.4	3.9	14.1	21.8	5.9	7.4	4.7	3.1	8.0	7.7	8.7	6.7	11.3	13.5	1.7	1.8
Low	12.7	7.2	17.0	14.3	5.2	5.4	24.7	10.8	10.4	11.9	12.0	12.0	—	—	—	—	15.4	18.6	19.7	20.1	15.0	14.3	6.1	6.3	0.3	0.6
Moderate	16.4	18.8	8.9	6.8	7.3	9.5	23.1	11.8	14.6	13.3	11.7	11.3	—	—	—	—	5.5	6.3	15.4	19.7	12.2	11.1	16.6	11.1	0.1	0.1
High	5.4	3.8	9.7	9.7	0.4	0.6	0.7	2.7	6.6	7.6	11.0	13.8	—	—	—	—	13.5	14.2	1.7	3.0	12.3	15.5	1.8	1.6	0.1	0.1
Very high	14.3	17.0	18.6	22.5	7.6	6.9	19.8	28.0	19.4	19.3	18.3	20.6	—	—	—	—	18.4	18.8	23.1	24.3	21.0	23.7	5.3	6.1	0.3	0.4
Severe	8.9	9.9	10.4	12.5	0.4	0.9	20.3	39.7	19.3	19.5	13.7	14.5	—	—	—	—	4.5	5.4	9.6	8.7	11.7	9.9	4.7	5.5	0.6	1.1

[a] H = Historical condition; C = Current condition.

[b] Values shown in bold indicate significant increase or decrease (P = 0.2) from the historical to the current condition.

[c] Fireline intensity classes (kW/m) are: very low = 0.0–172.9; low = 173.0–345.9; moderate = 346.0–1037.8; high = 1037.9–1729.6; very high = 1729.7–2,594.4; severe = 2594.5–3459.2; and extreme >3459.2. Fire rate of spread classes (m/min) are: very low = 0.0–0.6; low = 0.7–1.2; moderate = 1.3–1.8; high = 1.9–2.4; very high = 2.5–3.4; and severe >3.4. Flame length classes (m) are: very low <0.6; low = 0.7–2.4; moderate = 2.5–9.1; and high >9.1. Crown fire potential classes are a relative index.

this section, we highlight several key issues surrounding current public land management of Interior Northwest forests and relevant implications from this assessment.

Fuels and Fire Behavior — Examining and Isolating the Risks

Throughout the basin, there are currently many large, spatially continuous areas that display elevated ground fuel conditions and increased severity in fire behavior attributes. Prior to the 20th century period of active management, areas that were normally influenced by infrequent lethal fires and those of a mixed lethal–nonlethal type also displayed relatively high fuel loads, crown fire potential, rate of spread, flame length, and fireline intensity. Active management has not made the entire landscape wildfire-prone; rather, it has removed the degree of spatial isolation that patches prone to stand replacement once enjoyed (Figures 67.12A to H; see color figure following p. 668). A reasonable target of restoration would be to restore more typical patterns of spatial isolation to affected landscapes.

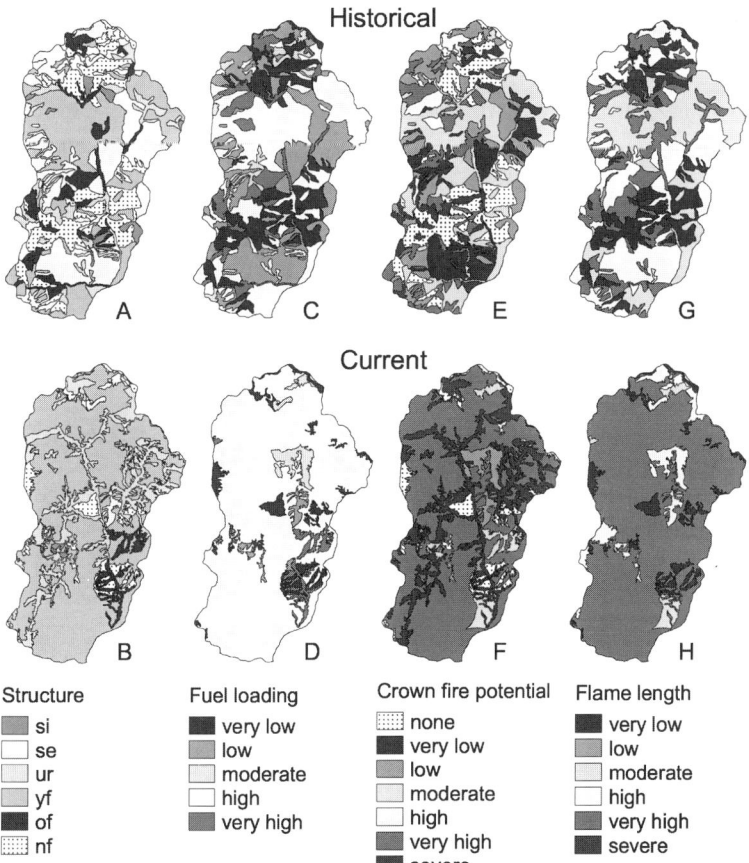

Figure 67.12 Maps of subwatershed 55, Lower Grande Ronde subbasin, Blue Mountains ERU, displaying historical and current structural classes (A and B), fuel loading (C and D), crown fire potential under wildfire conditions (E and F), and flame length under wildfire conditions (G and H), respectively. Structural class abbreviations are si = stand-initiation; se = stem exclusion (both open and closed canopy conditions); ur = understory reinitiation; yfms = young multistory forest; of = old multistory and single-story forest; and nf = nonforest. Fuel loading classes are very low < 22.5 Mg/ha; low = 22.5–44.9 Mg/ha; moderate = 45–56.1 Mg/ha; high = 56.2–67.3 Mg/ha; and very high > 67.3 Mg/ha. Crown fire potential classes were a relative index. Flame length classes were: very low < 0.6 m; low = 0.7–1.2 m; moderate = 1.3–1.8 m; high = 1.9–2.4 m; very high = 2.5–3.4 m; and severe > 3.4 m. (See color figure following p. 668.)

The Wildland–Urban Interface

Most people who live adjacent to National Forest lands of the basin live on or adjacent to dry woodlands, or dry to mesic forests. These specific settings have been most altered by 20th century management activities. As a result, surface fire regimes that once affected lands of the current wildland-urban interface have become lethal or at best mixed regimes, and forest residents are at risk at each outbreak of fire. In the context of declining fire suppression effectiveness, public land managers might consider, as a first priority, restoring vegetation and fuels patterns in the wildland-urban interface. In addition, managers might consider working with local citizens and communities to halt what is currently a rapidly expanding zone of interface. The tacit assumption of citizens who take up private residence in the forest is that the public land manager will come to their aid at the outbreak of fire. At some point, the rescue role of public land managers must be clearly enunciated and the total extent of a wildland-fire interface zone defined. The safety and management implications of not doing so are large.

Fire and Smoke — How Much and When?

Stated simply, the question before public land managers is not whether there will be fire and smoke in their future, but how do citizens want their fire and smoke. As Ottmar et al. (in press) and Huff et al. (1995) have shown, the air quality and smoke production trade-offs between wild and prescribed fires are highly significant; prescribed burning can eliminate 50% or more of the particulate emissions generated by wildfires, and the timing, movement, and disposal of smoke released to the airshed is managed. According to Hann et al. (1997), it is unlikely that fire suppression efficacy will improve given current vegetation and fuels conditions. Provided there is sufficient management latitude, it will be up to resource managers to restore patterns of living and dead vegetation to conditions that are more attuned with natural fire regimes and biophysical environment conditions. In this scenario, citizens would learn to live with some measure of smoke on a regular basis as an alternative to uncertain fire- and smoke-free intervals punctuated by periods with extreme fires and smoke emissions of unchecked magnitude, distribution, and effect.

Insects and Pathogens — Which Ones and How Much?

In the same way that fires visit healthy forests in many areas, some native insects and pathogens also visit healthy forests. Insect and pathogen disturbances motivate forest succession, especially in the fire-free interval. Under natural disturbance regimes, these disturbances produced semi-predictable outcomes, and more importantly, insect and pathogen disturbances were either attuned with biophysical environment conditions and natural fire regimes, or they recalibrated them. Such is not the case today. Owing to effects of past management, current patterns of forest vegetation are out of synchrony with natural fire regimes, and current insect and pathogen disturbance regimes are anomalous. There is no way to eliminate insect and pathogen disturbances from basin landscapes, and we would undoubtedly not wish to do so even if we were able. The by-product of most insect and pathogen disturbances is microhabitat (snags, down wood) of varying quality and residence time essential to a variety of terrestrial and aquatic species. Management, by altering spatial and temporal patterns of vegetation structure, composition, and growing conditions, can influence the suite of agents operating on a landscape, but it cannot, and from a purely ecological point of view, should not attempt to eliminate these disturbances.

Dynamic and Reserve Systems

In the Pacific Northwest, the Northwest Forest Plan (NWFP) defines a network of late-successional reserves throughout the Cascade and Coast Ranges of Oregon, Washington, and

California. While these lands are not explicitly set aside for custodial management, that is how they are managed. It is likely that such an approach may be implemented west of the Cascade crest where most fire regimes are lethal; fire-free intervals may last 250 to 800 years or more; and the most common disturbances, which are caused by native insects and pathogens, result in relatively small canopy gaps. East of the Cascade crest where this assessment has taken place, the picture is different. Wildfires and large-scale insect outbreaks are relatively common, and escaped wildfire frequency is increasing as a consequence of past management (Hann et al., 1997).

Results from the interior Columbia Basin assessments (Hann et al., 1997; Hessburg et al., 1999a; Ottmar et al., in press) suggest that a two-pronged, mixed dynamic and reserve system management approach may be needed to ensure recovery of the northern spotted owl and associated species. In the short term (e.g., 50 to 100 years), it is likely that areas currently functioning as late-successional and old forest habitats will be maintained with only limited success. Risk of disturbance and uncertainty of outcomes will be high. Over that period, some areas will be affected by stand-replacement fires, and will cease to function as late-successional habitat. For example, since 1994, 10 of the 140 to 180 (6 to 7%) northern spotted owl nest stands and neighborhoods were lost to uncontrolled wildfires on the Wenatchee National Forest alone. Patterns of structure and composition within the NWFP reserve network will continue to change as a result of uncontrolled fires, insect outbreaks, and other processes.

What may be needed is an approach that marries a short-term system of reserves with a long-term strategy to convert from a reserve system to a continuous network of landscapes with dynamic properties. In such a system, late-successional elements with semipredictable environmental settings (*sensu* Camp et al., 1997) are continuously recruited, but shifting in landscape position across space and time. Patches of late-successional and old forest structure are ephemeral landscape elements; they have specific contexts in space and across time. Future old forest will grow from some other condition; current old forests will become something else. Taking hold of this notion enables one to identify the dilemma of strategies that rely on a reserve system without backup. Because of the unfortunate legacy of past management actions, late-successional reserves must represent a special case for management. But the special case is an unforeseen consequence of past events, and in the interior, the long-term likelihood of success is low.

Healthy Fish and Healthy Forests

Can there really be an aquatic conservation strategy separate from a landscape pattern restoration strategy? One conclusion of the interior Columbia basin aquatic (Lee et al., 1997) and landscape assessments (Hann et al., 1997) is the notion that disturbance regimes and succession processes associated with terrestrial and aquatic environments are inextricably intertwined. For example, in forested catchments, hydrologic regimes are naturally governed by spatial and temporal patterns of vegetation and disturbance. Wildfires and timber harvest activities have a direct bearing on the timing, purity, temperature, and flow of water through catchments. Disturbance in upland settings can result in either positive or negative effects on aquatic conditions. Disturbances associated with native regime attributes tend to result in the renewal and redistribution of essential aquatic habitats. Disturbances of unprecedented frequency, intensity, distribution, and duration tend to result in habitat damage and losses to fish production. Likewise, by specifying standardized buffer zones and long-term custodial management of riparian zones, aquatic conservation strategies have a direct bearing on spatial and temporal patterns of vegetation. But in neither case are the effects of either strategy jointly considered. While the tug-of-war between healthy aquatic systems and healthy forests is to be expected initially, reason suggests that as the debate comes of age, scientists and managers would pursue strategies that jointly consider long-term spatiotemporal patterns of upland vegetation and disturbance, *and* consequences to hydrologic regimes and aquatic habitats.

ACKNOWLEDGMENTS

The U.S. Department of Agriculture, Forest Service, through the Pacific Northwest and Rocky Mountain Research Stations, and the U.S. Department of the Interior, Bureau of Land Management, provided financial support for this study. We are grateful to John Lehmkuhl, Sandy Boyce, Ann Camp, and two anonymous reviewers for helpful comments and suggestions on an earlier draft.

REFERENCES

Agee, J.K., *Fire Ecology of Pacific Northwest Forests,* Island Press, Washington, D.C., 1993.

Agee, J.K., Fire and Weather Disturbances in Terrestrial Ecosystems of the Eastern Cascades, General Technical Report PNW-GTR-320, USDA Forest Service, Pacific Northwest Research Station, Portland, OR, 1994.

Alatalo, R.V., Problems in the measurement of evenness in ecology, *Oikos,* 37, 199–204, 1981.

Arno, S.F., The Historical Role of Fire on the Bitterroot National Forest, Research Paper INT-187, USDA Forest Service, Intermountain Research Station, Missoula, MT, 1976.

Arno, S.F., Forest fire history in the northern Rockies, *J. For.,* 78, 460–465, 1980.

Bailey, R.G., *Description of Ecoregions of the United States*, 2nd ed., Miscellaneous Publication 1391, USDA Forest Service, Washington, D.C., 1995.

Burnham, K.P. and Overton, W.S., Robust estimation of population size when capture probabilities vary among animals, *Ecology,* 60, 927–936, 1979.

Camp, A.E., Oliver, C.D., Hessburg, P.F., and Everett, R.L., Predicting late-successional fire refugia from physiography and topography, *For. Ecol. Manage.,* 95, 63–77, 1997.

Cohen, J.D, and Deeming, J.E., The National Fire Danger Rating System: Basic Equations, General Technical Report PSW-82, USDA Forest Service, Pacific Southwest Research Station, Berkeley, CA, 1985.

Covington, W.W., Everett, R.L., Steele, R., Irwin, L.L., Daer, T.A., and Auclair, A.N., Historical and anticipated changes in forest ecosystems of the Inland West of the United States, *J. Sustainable For.,* 2, 13–63, 1994.

Deeming, J.E, Burgan R.E., and Cohen, J.D., The National Fire Danger Rating System — 1988, General Technical Report INT-19, USDA Forest Service, Intermountain Research Station, Ogden, UT, 1988.

Environmental Systems Research Institute, ARC/INFO version 7.0 User's Manual, Redlands, CA, 1995.

Everett, R.L., Hessburg, P.F., Jensen, M.E., and Bormann, B., Volume I: Executive Summary, Eastside Forest Ecosystem Health Assessment, General Technical Report PNW-GTR-317, USDA Forest Service, Pacific Northwest Research Station, Portland, OR, 1994.

Franklin, J.F. and Dyrness, C.T., *Natural Vegetation of Oregon and Washington*, OSU Press, Corvallis, OR, 1988.

Hall, F.C., Fire and vegetation in the Blue Mountains: implications for land managers, in *Proceedings of the 15th Annual Tall Timbers Fire Ecology Conference*, 16–17 October 1974, Portland, OR, Tall Timbers Research Station, Tallahassee, FL, 15, 1976, pp. 155–170.

Hann, W.J., Jones, J.L., Karl, M.G., Hessburg, P.F., Keane, R.E., Long, D.G., Menakis, J.P., McNicoll, C.H., Leonard, S.G., Gravenmeier, R.A., and Smith, B.G., Landscape dynamics of the basin, in An Assessment of Ecosystem Components in the Interior Columbia Basin and Portions of the Klamath and Great Basins, General Technical Report PNW-GTR-405, Quigley, T.M. and Arbelbide, S.J, Eds., USDA Forest Service, Pacific Northwest Research Station, Portland, OR, 1997, Chapter 3.

Harvey, A.E., Hessburg, P.F., Byler, J.W., McDonald, G.I., Weatherby, J.C., and Wickman, B.E., Health declines in western interior forests: symptoms and solutions, in *Proceedings of the Symposium: Ecosystem Management in Western Interior Forests*, 3–5 May 1995, Spokane, WA, Everett, R.L. and Baumgartner, D.L., Eds., Washington State University Press, Pullman, WA, 1995.

Heltshe, J.F, and Forrester, N.E., Estimating species richness using the jackknife procedure, *Biometrics,* 39, 1–11, 1983.

Hessburg, P.F., Mitchell, R.G., and Filip, G.M., Historical and Current Roles of Insects and Pathogens in Eastern Oregon and Washington Forested Landscapes, General Technical Report PNW-GTR-327, USDA Forest Service, Pacific Northwest Research Station, Portland, OR, 1994.

Hessburg, P.F., Smith, B.G., Kreiter, S.G., Miller, C.A., Salter, R.B., McNicoll, C.H., and Hann, W.J., Historical and Current Forest and Range Landscapes in the Interior Columbia River Basin and Portions of the Klamath and Great Basins, Part I: Linking Vegetation Patterns and Landscape Vulnerability to Potential Insect and Pathogen Disturbances, General Technical Report PNW-GTR-458, USDA Forest Service, Pacific Northwest Research Station, Portland, OR, 1999a.

Hessburg, P.F., Smith, B.G., Miller, C.A., Kreiter, S.G., and Salter, R.B., Modeling Change in Potential Landscape Vulnerability to Forest Insect and Pathogen Disturbances: Methods for Forested Sub-watersheds Sampled in the Mid-Scale Interior Columbia River basin Assessment, General Technical Report PNW-GTR-454, USDA Forest Service, Pacific Northwest Research Station, Portland, OR, 1999b.

Hill, M.O., Diversity and evenness: a unifying notation and its consequences, *Ecology*, 54, 427–431, 1973.

Huff, M.H., Ottmar, R.D., Alvarado, E., Vihnanek, R.E., Lehmkuhl, J.F., Hessburg, P.F., and Everett, R.L., Historical and Current Forest Landscapes of Eastern Oregon and Washington, Part II: Linking Vegetation Characteristics to Potential Fire Behavior and Related Smoke Production, General Technical Report PNW-GTR-355, USDA Forest Service, Pacific Northwest Research Station, Portland, OR, 1995.

Keane, R.E., Reinhardt, E.D., and Brown, J.K., FOFEM: A First Order Fire Effects Model for Predicting the Immediate Consequences of Wildland Fire in the United States, in *Proceedings of the 12th Conference of Fire and Forest Meteorology*, 26–28 October 1993, Jekyll Island, GA, American Meteorological Society, Boston, 1994, pp. 628–631.

Lee, D.C., Sedell, J.R., Rieman, B.E., Thurow, R.F., and Williams, J.E., Broadscale assessment of aquatic species and habitats, in An Assessment of Ecosystem Components in the Interior Columbia Basin and Portions of the Klamath and Great Basins, General Technical Report PNW-GTR-405, Quigley, T.M. and Arbelbide, S.J., Eds., USDA Forest Service, Pacific Northwest Research Station, Portland, OR, 1997, Chapter 4.

Lehmkuhl, J.F., Hessburg, P.F., Everett, R.L., Huff, M.H., and Ottmar, R.D., Historical and Current Forest Landscapes of Eastern Oregon and Washington, Part I: Vegetation Patterns and Insect and Disease Hazards, General Technical Report PNW-GTR-328, USDA Forest Service, Pacific Northwest Research Station, Portland, OR, 1994.

MathSoft Inc., S-PLUS User's Manual, version 3.2, StatSci, a Division of MathSoft, Inc., Seattle, 1993.

McGarigal, K. and Marks, B.J., FRAGSTATS: Spatial Pattern Analysis Program for Quantifying Landscape Structure, General Technical Report PNW-GTR-351, USDA Forest Service, Pacific Northwest Research Station, Portland, OR, 1995.

Morgan, P., Aplet, G.H., Haufler, J.B., Humphries, H.C., Moore, M.M., and Wilson, W.D., Historical range of variability: a useful tool for evaluating ecosystem change, in *Assessing Forest Ecosystem Health in the Inland Northwest*, Sampson, R.N. and Adams, D.L., Eds., Haworth Press, New York, 1994, pp. 87–111.

O'Hara, K.L., Latham, P.A., Hessburg, P.F., and Smith, B.G., A structural classification of inland Northwest forest vegetation, *West. J. Appl. For.*, 11, 97–102, 1996.

O'Laughlin, J., MacCracken, J.G., Adams, D.L., Bunting, S.C., Blatner, K.A., and Keegan, C.E., Forest health conditions in Idaho, Report 11, University of Idaho, Forest, Wildlife, and Range Policy Analysis Group, Moscow, ID, 1993.

Oliver, C.D. and Larson B.C., *Forest Stand Dynamics*, John Wiley & Sons, New York, 1996.

Ottmar, R.D., Burns, M.F., Hall, J.N., and Hanson, A.D., CONSUME Users Guide, General Technical Report PNW-GTR-304, USDA Forest Service, Pacific Northwest Research Station, Portland, OR, 1993.

Ottmar, R.D., Schaaf, M.D., and Alvarado, E., Smoke considerations for using fire in maintaining healthy forest conditions, in General Technical Report INT-GTR-341, USDA Forest Service, Intermountain Research Station, Ogden, UT, 1996, pp. 2–28.

Ottmar, R.D., Alvarado, E., Hessburg, P.F., Smith, B.G., Kreiter, S.G., Miller, C.A., and Salter, R.B., Historical and Current Forest and Range Landscapes in the Interior Columbia River Basin and Portions of the Klamath and Great Basins, Part II: Linking Vegetation Patterns and Potential Smoke Production and Fire Behavior, General Technical Report PNW-GTR-XXX, USDA Forest Service, Pacific Northwest Research Station, Portland, OR, in press.

Quigley, T.M. and Arbelbide, S.J., Eds., An Assessment of Ecosystem Components in the Interior Columbia Basin and Portions of the Klamath and Great Basins, General Technical Report PNW-GTR-405, USDA Forest Service, Pacific Northwest Research Station, Portland, OR, 1997.

Robbins, W.G. and Wolf, D.W., Landscape and the Intermontane Northwest: An Environmental History, General Technical Report PNW-GTR-319, USDA Forest Service, Pacific Northwest Research Station, Portland, OR, 1994.

Rothermel, R.C., A Mathematical Model for Predicting Fire Spread in Wildland Fuels, Research Paper INT-115, USDA Forest Service, Intermountain Research Station, Ogden, UT, 1972.

Rothermel, R.C., How to Predict the Spread and Intensity of Forest and Range Fires, General Technical Report INT-143, USDA Forest Service, Intermountain Research Station, Ogden, UT, 1983.

Schaaf, M.D., Development of the Fire Emission Tradeoff Model (FETM) and Application to the Grande Ronde River Basin, Oregon, CH_2M Hill Contract Report 53–82FT-03–2, CH_2M Hill Co., Portland, OR, 1996.

Shannon, C. and Weaver, W., *The Mathematical Theory of Communication*, University of Illinois Press, Urbana, 1949.

Simpson, E.H., Measurement of diversity, *Nature*, 163, 688, 1949.

Skovlin, J.M. and Thomas, J.W., Interpreting Long-Term Trends in Blue Mountains Ecosystems from Repeat Photography, General Technical Report PNW-GTR-315, USDA Forest Service, Pacific Northwest Research Station, Portland, OR, 1995.

Turner, M.G., *Landscape Heterogeneity and Disturbance*, Springer-Verlag, New York, 1987.

Turner, M.G., Landscape ecology: the effect of pattern on process, *Annu. Rev. Ecol. Syst.*, 20, 171–197, 1989.

Wickman, B.E., Forest Health in the Blue Mountains: The Influence of Insects and Diseases, General Technical Report PNW-GTR-295, USDA Forest Service, Pacific Northwest Research Station, Portland, OR, 1992.

Wissmar, R.C., Smith, J.E., McIntosh, B.A., Li, H.W., Reeves, G.H., and Sedell, J.R., A history of resource use and disturbance in riverine basins of eastern Oregon and Washington (early 1800s–1990s), *Northwest Sci.*, 68, 1–35, 1994.

SECTION II.6

Communities, Politics, Culture, and Tradition

CHAPTER 68

Overview: Communities, Politics, Culture, and Tradition

Alexander H. Harcourt

The three main themes of this section are the issues of multiple users, means of assessing environmental health, and cooperation to mitigate adverse effects on environmental health. The regions covered extend from the Americas, east through Europe and the Middle East, and then to Bhutan, in the northeast of the Indian subcontinent. The size of the regions ranges from multicountry in North America and the Middle East, to multistate within the U.S., to single countries (Mexico, Netherlands), to a single county in a state in the U.S. (in California), to a single city (Brussels), and to communal pastures in two districts of Bhutan. The features assessed or managed encompass the total environment, both physical and economic (Netherlands); the natural environment (Mexico); water quality in the whole Great Lake catchment basin of Canada and the U.S.; desertification (Middle East); the effects of urbanization on the environment (U.S.); the effects of change in transportation use on air quality (Brussels); agricultural environment and native biota (Yolo County, CA); and pasture and livestock health (Bhutan). The levels of cooperation examined include international treaties, interstate agreements in the U.S., governmental and NGO (nongovernmental organizations associated with the United Nations) coordination in Mexico, cooperation among disciplines in the Netherlands, and communal use of pastures in Bhutan.

To ensure a healthy environment despite increased use, we need to know who the users are, and how they will be affected by measures aimed at the maintenance of environmental health. In all the cases in this section, public land is, effectively, the environment under discussion. Multiple users are therefore involved, with the number of users ranging from a few score people and a few hundred livestock in a communal grazing pasture in Bhutan (Ura) to millions of users in the Great Lakes region of Canada and the U.S. (Bertram et al.), and in the Middle East (Christiansen). With so many users, the types of use are, of course, varied and variably conflicting. Little to no common ground exists between the biologist and the urban developer, especially where most of the natural environment has disappeared already, as in Mexico, leaving set-aside as often the only way to preserve natural environments (Ezcurra). By contrast, cleaner fuel and improved public transport could result in a cleaner, quieter city despite increased urban sprawl and commuting (Safonov et al.).

However complex the multiple use, we obviously have to understand which environmental variables are crucial, and how to measure them. Bertram et al. describe the 80 indicators of environmental health being used by Canada and the U.S. in the Great Lakes region, 19 of which involve water quality. The indicators are biotic (parasite and waterfowl numbers, area of natural habitat) and chemical (phosphorus and acid concentrations), and also include ecosystem function (resilience to disturbance) and health of the users (contaminants in breast milk). The Great Lakes protocol essentially uses the indicators separately, but Zhang et al. describe their attempt to produce a single index of environmental health for Yolo County, CA, that incorporates not only indices appropriate to sustainable agriculture in the county, but also to protection of wildlife. In recognition of the importance of the economics in determining almost all aspects of policy, an assessment protocol developed in the Netherlands involves integration of measures of both environmental and economic health. Furthermore, the protocol includes, most crucially, assessment of progress toward achievement of goals (Verkaar et al.). The sense of the future inherent in such assessment is paramount in the analyses by Matheny and Endres and Safonov et al. of the effects of urbanization on environmental health. Matheny and Endres, for instance, examine the possibility that telecommuting in the eastern U.S. would allow smaller or fewer new offices, resulting in less environmental degradation than traditional commuting would cause. And Safonov et al. start with socioeconomic analyses in order to predict trends in urbanization and consequent trends in transportation, which they then use to predict changes in air quality under a variety of scenarios for urbanization in the region of the European Union capital, Brussels.

Turning to the issue of cooperation, Bertram et al. describe what must be one of the larger attempts at collaborative management in the world, the Great Lakes Water Quality Agreement between Canada and the U.S. They also describe the sort of small-scale cooperation that is a necessary part of the large-scale collaboration. The new Brantford office complex of Union Gas Ltd. in Ontario, Canada, has been designed to be environmentally friendly and, more visibly, the company restored the surrounding lands to nearly their natural prairie state. The local community is using the prairie as an outdoor classroom, and Union Gas itself benefits both from the goodwill so generated, and from the low cost of upkeep of a natural landscape. (Prairies don't need fertilizer.) If only more companies were so farsighted. Perhaps the most ambitious cooperative achievement sought is peace in the Middle East by cooperation among the nations of Egypt, Israel, Jordan, the Palestinian Authority, and Tunisia through concerted research and action on control of desertification (Christiansen). However, the multiorganization involvement in conservation of the natural environment in Mexico (Ezcurra) is the scale to which most of us are probably accustomed.

All themes of the section — multiple users, environmental assessment, and cooperation in maintenance of environmental health — come together in Ura's account of traditional grazing practices in Bhutan. For hundreds of years, common grazing grounds have been sustainably and equitably used through institutionalized assessment of pasturage quality, distribution of grazing rights, and insistence on maintenance duties that accompany the rights. With open and fair assessment, and adherence to rules that work, Ura stresses that a tragedy of the commons can be avoided.

CHAPTER 69

Gambling for Sustainability — Local Institutions for Pasture Management in Bhutan

Karma Ura

INTRODUCTION

Each summer, on the 18th of the 7th lunar month of the Bhutanese calendar, the pastoral communities of the villages of Gechukha, Chumpa, and Chempa in Northwestern Bhutan assemble at their community shrine. The gathering is a scene of day-long prayer sponsored by the community. Supernatural influence is invoked for favorable outcomes. This daylong ritual is just a prelude to the actual business, a brief, statistical affair, that takes place in the evening. Die is cast to allocate the communal pastures to groups of herders for the coming year. The group with the highest score gets first choice, the second highest gets the second choice, and so forth. Until they meet again, 1 year later, herders will be dependent on the allotted pastures to provide their yaks with sufficient feed.

Viewed as rules of the game, a local institution of the type mentioned above is crucial for equitable access to one of the major resources in agropastoral communities because it provides a mechanism for the allocation and division of pastures among herders. But this institution is also important for the many other functions it plays. A tentative outline of these can be made:

- First, an institution is very often eco-specific and therefore it may reflect sustainable uses of resources within that eco-niche. This could also mean that the institution could express accumulated indigenous knowledge about sustainable resource use within that eco-niche.
- Second, an institution, since it is correlated to an eco-niche, is adaptive over eco-niches. Institutional diversity and ecological diversity, therefore, go hand in hand.
- Third, an institution defined as "rules of the game" regulates the behavior of the individual members of an organization, be it formal or informal. In this sense an institution acts as coordinator of collective activities and obligations of the membership of that community or organization.

My intention in this chapter is to explore briefly these three aspects of local institutions with references primarily to pasture distribution rules in Bhutan.

PASTURE DISTRIBUTIVE RULES FOR COMMUNAL YAK PASTURES IN GECHUKHA, CHUMPA, AND CHEMPA IN HAA

Communal pasture is a more usual form of usufruct ownership than private pastures in Haa. The communities of Gechukha, Chumpa, and Chempa in upper Haa have collective usufruct rights over summer and winter pastures for their yak herds. Grazing months are divided equally between these seasons; the period from November to April is spent in winter pastures, and from May to October in summer pastures. These pastures form a contiguous area, but each is divided into five (more or less) equal patches. Boundaries among these patches often follow natural landmarks such as streams, rivers, ridges, watersheds, etc. and these are likely to provide practical and useful perimeters for both herders and yaks.

When herders meet they first reach consensus on the boundaries, then preferential ranking of each of the five patches of summer and the five patches of winter pasture is done. This ranking of the carrying capacity of the patches is based on the herders' knowledge about the quality and area of each. Their understanding of the local ecology, geography, and wildlife is today an exclusive knowledge base about these high mountain and low subtropical pastures. Even indigenous doctors in Bhutan depend upon local yak herders to find rare alpine medicinal plants. It is this experience and knowledge about the local ecology that enables them to rank the 10 patches of pasture.

When consensus has been reached on a ranking, the pairing of the summer and winter patches is done. (See Table 69.1 for the ranking and pairing.) The best patch of summer pasture is paired with the worst patch of winter pasture and so forth. Twinning of winter and summer patches in this manner was aimed to equalize the access to pastures. Pairing the most preferred summer pasture with the least preferred winter pasture is a mechanism that compensates the loss a herder may find in winter months by gains in the summer. Within the competing herders' group, the triumph of getting Jatekha, the best of the summer pastures, is tempered by the prospect of herding in Richey, the worst of winter pastures. Equity and fairness is built into the way herders pair off winter and summer pastures.

Since there are five patches of summer pastures and five patches of winter pastures, the yaks of the Gechukha community are also grouped into five equal herds. No differentiation is made between calves and adult yaks. Every yak, whether cow, oxen, or calf is taken as one unit. Thus, corresponding to the five herds, five groups of herders are formed.

With the formation of five groups of herders and five pairs of winter and summer pastures, the stage is set for the actual allocation of pastures. In turn, a representative from each of the five herders' groups casts three die. The herder who scores highest makes the first choice among the pairs of winter and summer pasture patches. The second choice is given to the herder who scores the second highest and so forth.

Because pasture patches are ranked and paired prior to allocation, all are ignorant about who will get a particular pair of pastures. Anonymity of the recipient at that time ensures that there will be no personal incentives to mismatch the summer and winter pastures in any biased way.

This ritual of casting die randomizes the allocation of pasture in a most ingenious manner, ensuring that the knowledge of all the herders is combined to identify the fairest distribution of communal pasture possible.

Table 69.1 Paired Ranking of Summer and Winter Pastures Communally Owned by Chumpa, Chempa, and Gechukha Villages in Haa, Bhutan

Summer Pastures Ranked by Preference	Winter Pastures Ranked by Preference
1. Jatekha	5. Richey
2. Lunghamekha	4. Shungkhatho
3. Pamling	3. Phodeytshang
4. Jangbana	2. Hingtho
5. Chala	1. Yangathangkha

PASTURE DISTRIBUTIVE RULES FOR COMMUNAL PASTURES IN URA, BUMTHANG

We now move our examination of the rules of dividing pastures to Ura village in Central Bhutan where the design of equitable division of winter pastures shows significant differences from that in Haa. Like the villages in Haa, Ura is a pastoral community. Unlike these villages, Ura depends predominantly on cattle rather than yaks. The cattle of Ura village are grazed in a subtropical belt for over 8 months from October and November to April and May. Almost all the summer and winter pastures are communal, and the division of pastures requires a fair method acceptable to all cattle herders. The rules they use for sharing and dividing pastures have not altered in living memory and are used every year. Surprisingly, cattle herders are of the opinion that neither the quality of the pastures (and their stocking capacity) nor the total population of the cattle grazing these pastures has changed very much. Thus, they find that the validity of the rules has not diminished.

There are 22 winter pastures in subtropical regions, which collectively belong to cattle herders of Ura. The herders and herds descend toward these pastures within weeks of each other. Although some herders may drive their herds a bit earlier than others toward these pastures, unilateral entry is forbidden. They must await the arrival of all the other herds. Finally, all the herders and their herds camp in a spacious meadow for a few days, usually in early November. Here, they convene daylong discussion on allotment of pasture. In practice, this is usually done throughout the night because the day is taken up by chores.

Communal pastures of herders in Ura are, as far as possible, allocated by consensus. But disputes can rarely be resolved through discussion when it comes to good pastures that are vied for by many herders. It is then that the herders resort to allocation by lottery. The names of the pastures are written down on pieces of paper and thrown in a jumble in a bag. Who gets what depends on blind chance. A large number of herders from Bhutan, who own communal pastures, use this particular mode of allocation.

Cattle herders of Ura have a clear idea of how many cows each pasture can and has been supporting in the past. This is a key concept, and also a difficult one, in the allocation of pastures. One would expect that the stocking capacity would generally be decided on the basis of an adult equivalent by taking a number of calves as adult cattle. Oxen in the herd should also be taken into account when calculating carrying capacity. However, among the herders of Ura only cows above 4 years of age are acknowledged for the purpose of allocating pastures. It is as if other animals, not at all productive or at that time unproductive, do not matter in this calculation for finding carrying capacity. Disregard for all other cattle except these cows impart disincentives to keep animals other than adult cows. Oxen, infertile females, and calves are, so to speak, disenfranchised, while the claims to grazing lands are weighed. Denial of suffrage to oxen, infertile females, and calves has environmental justification in subtropical pastures. By contrast, when the pastures are divided among yak herders in Haa, male yaks and calves are not excluded in the head count. This difference will be explained later.

Puzzling though it is, the cattle herders of Ura have a rule of thumb about the carrying capacity of each pasture for cows over 4 years. The convention in Omdaar provides cattle herders with an opportunity to negotiate which of the 22 pastures each would prefer to utilize. These negotiations take place after they have reached agreement about the number of the contenders' cows above 4 years and the carrying capacity of each pasture (defined only with respect to cows above 4 years). The herders know the carrying capacity as a historically and traditionally determined figure, historical and traditional in the sense that the information about the methodology of calculating stocking size is beyond recollection by the present herders. But the number itself is not absolutely sacrosanct. Unexpected changes in the condition of the pastures are surveyed and assessed by the herders before they meet. Possible decline in the quality of pastures could be due to a landslide; encroachment of bamboo forest, which cannot be penetrated by cattle; or lower grass yield due to low rainfall on a slope has been observed by herders. In such pastures, the number of cows would be reduced from the carrying capacity, a fixed number that has been previously determined.

The cattle herders of Ura are expected to occupy the pastures for the winter season after allotment by lottery. During that season, herders have control and utilization of the pasture. They also have obligations and responsibilities to manage the pasture in such a way that its quality is maintained for future herds. A herd's consumption during one season must not deplete the future flow of goods and services, such as grass and fodder and footpaths in the forest. In addition, the herders must use an active form of management termed *brocksel* (i.e., clearing the pasture). It is failure to notice this kind of restorative measure carried out by herders that leads to opinions that herders are environmentally destructive. Migratory livestock farming is part of the forest system since they are both associated in many ways. The species mix and regeneration process is affected by livestock, although not necessarily in the negative way that is widely presupposed in today's antimigratory livestock literature. The replanting of wild fodder trees and plants, the selective thinning of bamboo forest by herders, and the availability of cattle compost as forest plant nutrients are a few of the herders' management activities that have obvious positive effects.

The yield of winter pastures can be reduced because of several factors. As animals graze and fodder is cut, unpalatable plants and shrubs may displace palatable ones. If growth of unpalatable species is not restricted by manual cutting, the species mix is changed in favor of unpalatable ones. If the herder does not carry out restorative measures, especially to reduce the rate of bamboo growth, the situation of the herder who gets that particular pasture the following season will be worsened. Further, the herder may use lops and tops of wild fodder trees and fodder vines, but may not fell them, for this would permanently deprive the succeeding wave of herders who follow him into this pasture. Saplings of wild fodder trees are uprooted from the forest and planted in more convenient places for more intensive care. Herders propagate wild fodder trees and fodder vines that are fed to the cattle only when surface herbage growth is fully consumed. This usually happens by the end of January. From then until the cattle begin their movements toward summer pastures in Bumthang, they subsist heavily on wild fodder trees and fodder vines. Another restorative measure includes repairing the labyrinth of forest tracks that give cattle access to grazing areas in the depth of the jungle.

The obligation to pass on the pastures in a state as productive as when they were received is an ingrained herders' norm, but it is not enforced by explicit sanctions. The propensity to follow the norm is conditioned by the knowledge that mutual cooperation in not overexploiting the pasture will keep the pasture ecology stable. If one herder leaves the pasture allotted to him in a worse state and others are tempted to follow his behavior of negligence toward the pasture every season, it may have a snowball effect and eventually leave the pasture degraded. Such a runaway process could result in environmental disaster where every herder would be worse off.

There are those who tend to predict an almost sure likelihood of an environmental disaster in situations where resources are managed communally, known as the tragedy of the commons. Their apprehension is misplaced in this case. It probably stems from an unquestioned belief that peasants are irrational, a belief that empirical findings could easily disprove, and an assumption in behavioral theories that have long been discarded. That is, individuals will manage resources for their own short-term benefits without taking the future needs of the group, of which they are part, into consideration.

COMPARISON OF THE TWO PASTURE DISTRIBUTIVE RULES

Both systems of pasture distribution show a high regard for an agreed and just principle of transfer of pasture from some herders to others within the same community for a specific period. Both stress establishing broadly equitable access to pastures over both time (winter and summer seasons) and space, in the sense that over the long run the throwing of dice or drawing a lot will assign a herder to any pasture with equal probability. In the long-term perspective, a herder will stay equally long in all the pastures. We find the actual allotment of pastures is determined at random by the throw of dice or the pick of a lot. At the same time, because each herder has an equal chance to get a specific patch of pasture, herders set egalitarian subrules about how many livestock units can be assigned to certain pasture areas and how winter pastures can be combined with summer pastures.

It was mentioned that while determining the occupancy rate in the winter pastures of cattle in Ura, herders take account of only cows above 4 years old. This generates disincentives for these herders to keep oxen, calves, or infertile cows in the herd. This disincentive compels the herders to export their unproductive cattle, mainly oxen, to other parts of Bhutan. The herds of Ura are therefore dominated by cows. The yak herders of Haa, however, do not differentiate yaks either by age or sex during allocation of pastures. Each yak is taken as a livestock unit during the division of communal pastures. To a certain extent, this lack of disincentive to reduce the number of male yaks is economically justified because they are, unlike oxen, profitable meat animals. Male yaks are slaughtered at about 6 years and their carcasses can be sold for about 8000 Nu each ($1U.S. = 47 Nu). In late 1992, the price of yak meat was 65 Nu/kg compared with 15 Nu/ kg for beef in the Thimphu market.

Pasture distributive rules such as the ones described here are profoundly important as rules of the game, yet both policy makers and researchers might pay insufficient attention to them. While we cannot assume that all existing institutions are completely optimal on both distribution and efficiency grounds, the fact that pasture distributive rules have existed for centuries without damaging the forest calls for greater faith and credibility in the system.

CONCLUSION

The migration of native bovine species, both yak and siri cattle, is a particularly skillful adaptation to ecologically varied niches where biomass varies spatially and seasonally. As a sophisticated social and ecological response to varied niches, different bovine species graze on herbage growing and dying at different rates, at different times, and at different altitudes. Traditional herder management systems embody a dynamic relationship among humans, native bovine species, and heterogeneous mountain niches (subtropical forest pasture and alpine grassland pasture).

The decentralization policy of the Royal Government of Bhutan creates an environment for the survival of diverse institutions, including their rules and the organizational forms, as determined by and for the communities. In our current search for sustainability, there is a case for continually feeding enduring local institutions into national policy, though there are also frequent occasions for doing the reverse. I would submit that the pasture distributive rules governing the use of communal pastures is one such valuable institution. It cannot be regenerated easily once destroyed.

ACKNOWLEDGMENTS

I am thankful to Dasho Shingkhar Lam and Tashi Dorji, former gup of Ura for most of the information in the Bumthang section. I would like also to acknowledge encouragement and help received from Dasho Sonam Tobgye, the Chief Justice of Bhutan.

CHAPTER 70

Environmental and Socioeconomic Indicators of Great Lakes Basin Ecosystem Health

Paul Bertram, Harvey Shear, Nancy Stadler-Salt, and Paul Horvatin

INTRODUCTION

The purpose of the Great Lakes Water Quality Agreement (GLWQA) between the U.S. and Canada is to restore and maintain the chemical, physical, and biological integrity of the waters of the Great Lakes basin ecosystem (U.S. and Canada, 1987). The U.S. and Canada have spent billions of dollars and uncounted hours attempting to reverse the effects of cultural eutrophication, toxic chemical pollution, overfishing, habitat destruction, introduced species, and other insults to the ecosystem. Environmental management agencies are now being asked to demonstrate that past remediation programs have been successful and that the results of future or continuing programs will be commensurate with the resources expended (financial and personnel time). The demand for high quality data is forcing environmental and natural resource agencies, which operate with limited resources, to be more selective and more efficient in the collection and analysis of data.

Assessing the health of something as large and complex as the Great Lakes basin ecosystem is a significant challenge. The lakes themselves contain one-fifth of the world's fresh water with over 10,000 miles (17,000 km) of shoreline. The basin consists of over 200,000 square miles (520,000 km²) of land, and about 33.5 million people reside within the basin. The basin is governed by two nations, eight states, two provinces, and hundreds of municipal and local governments. A set of Great Lakes basin ecosystem indicators will enable the Great Lakes community — government and nongovernment organizations, academia, industry, and individual citizens — to work together within a consistent framework to assess and monitor changes in the state of the ecosystem.

In this context, the information collected can be characterized within a series of indicators. An indicator is a parameter or value that reflects the condition of an environmental (or human health) component, usually with a significance that extends beyond the measurement or value itself (Canada and U.S., 1999). An indicator is more than a data point. It consists of both a value (which may be a direct environmental measurement or may be derived from measurements) and a target or reference point. Used alone or in combination, indicators provide the means to assess progress toward one or more objectives: Are conditions improving so that the objective is closer to being met, or are conditions deteriorating? The achievement of these objectives leads toward achievement of higher order goals and vision for the ecosystem (Figure 70.1).

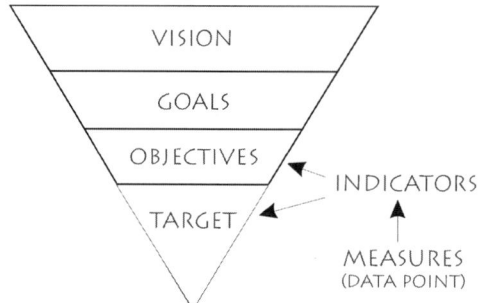

Figure 70.1 Conceptual model of the relationships among indicators, measures, targets, objectives, goals, and visions. (From Canada and the U.S., *State of the Great Lakes 1999*, www.on.ec.gc.ca/solec/ or www.epa.gov/glnpo/solec/98/, Toronto and Chicago, 1999.)

Within the Great Lakes basin, no one organization has the resources or the mandate to examine the state of all the ecosystem components. However, dozens of organizations and thousands of individuals routinely collect data, analyze them, and report on parts of the ecosystem. A consensus by environmental management agencies and other interested stakeholders about what information is necessary and sufficient to characterize the state of Great Lakes ecosystem health, and to measure progress toward ecosystem goals, would facilitate more efficient monitoring and reporting programs. The relative strengths of the agencies could be utilized to improve the timeliness and quality of the data collection and the availability of the information to multiple users.

The State of the Lakes Ecosystem Conferences (SOLEC) were established by the governments of Canada and the U.S. in 1992 in response to reporting requirements of the GLWQA. The conferences are to provide biennial science-based reporting on the state of health of the Great Lakes basin ecosystem.

In the first SOLEC (1994), the governments reported on basin-wide conditions relating to aquatic ecosystem health, human health, aquatic habitat and wetlands, nutrients, contaminants, and the economy. These categories ensured that major components of the ecosystem were assessed, as well as a major component of human activity (the economy). The organizers developed a series of *ad hoc* indicators to provide an assessment of the state of these components and to assess progress toward the goals of the GLWQA. These indicators were based on the best professional judgment of a number of scientists and managers who had prepared background papers on the subject components (U.S. and Canada, 1995). A similar process was followed for SOLEC '96, in which the focus was on the nearshore aquatic and terrestrial environments (Canada and U.S., 1997).

For SOLEC '98, the organizers wanted to support the further development of easily understood indicators that objectively represented the condition of the Great Lakes basin ecosystem, the stresses on the ecosystem, and the human responses to those stresses. These indicators would provide a predictable set of signs of the health of the ecosystem components and the progress being made to remedy existing problems.

INDICATOR SELECTION PROCESS

To guide the development of a suite of indicators for the Great Lakes basin ecosystem, a series of organizing principles was recognized.

- *Build upon the work of others.* There has been a long history of monitoring and of the development of indicators for the Great Lakes. For example, indicators have been proposed in some form by the International Joint Commission, the Great Lakes Fishery Commission, Lakewide Management

Plans, and in the background papers prepared for SOLEC '94 and SOLEC '96. The intent for the SOLEC process was to make use of the work that has gone on before rather than repeat it.

- *Focus on broad spatial scales.* SOLEC indicators are intended to reflect conditions on the scale of the whole Great Lakes basin, whole lake watersheds, and subbasins of the larger lakes. Indicators for local geographic areas are reported through other programs.

- *Select a framework for subdividing the Great Lakes basin ecosystem.* SOLEC indicators were developed within the framework of geographic zones (offshore, nearshore, coastal wetlands, nearshore terrestrial) and nongeographic issues (human health, land use, societal, and unbounded). This framework provides flexibility to incorporate the main elements of other frameworks. For example, physical, chemical, and biological status can be assessed within each geographic zone.

- *Select a system for types of indicators.* There are several classification schemes or models for indicators (for example, GMIED, 1998; IJC, 1991, 1996; Messer, 1992; Regier, 1992), one of which is the State-Pressure-Human Activity (Response) model. This S-P-A model is simple and broadly applicable. *State* indicators address the state of the environment, the quality and quantity of natural resources, and the state of human and ecological health. *Pressure* indicators describe natural processes and the results of human activities that impact, stress, or pose a threat to environmental quality. *Human activity (response)* indicators include individual and collective actions to halt, mitigate, adapt to, or prevent damage to the environment. They also include actions for the preservation and the conservation of the environment and natural resources.

- *Identify criteria for indicator selection.* The primary criteria that were used for the selection of the suite of indicators were: *necessary*, *sufficient*, and *feasible*. For individual indicators, a set of 21 additional criteria was identified. Not all indicators scored high for all criteria. However, for any given indicator, the more criteria that were met, the stronger the case for including that indicator on the proposed SOLEC list (Bertram and Stadler-Salt, 1999).

A group of experts was assembled to review, select, and refine Great Lakes indicators. Some members of this indicators group were also leaders of seven core groups, one for each of the SOLEC indicator categories mentioned above. The core groups were composed of expert panels who participated either directly with hands-on selection and development of indicators, or by reviewing draft products throughout the process. Over 130 people were involved in some way in this project.

The core groups studied documents from multiple sources that identified or discussed ecosystem objectives or environmental indicators. More than 850 indicators were identified through this process. The indicators were then screened according to the criteria, and appropriate indicators were identified. In many cases, indicators from the existing list were modified or combined, or new indicators were developed to create the proposed suite of 80 SOLEC indicators.

For each of the SOLEC indicators, detailed descriptive information was assembled into a meta-database. The information included the following (Bertram and Stadler-Salt, 1999):

- Environmental measurements to be taken
- Environmental components the indicator would be used to assess
- Ecosystem objectives the indicator would be addressing
- Indicator desired end points or reference values
- Descriptions of the indicator's features and limitations
- Illustrations of how the indicator would be presented and interpreted
- Additional comments that would clarify the intent and implementation of the indicator

The list of 80 indicators (Table 70.1) addresses all of the SOLEC categories. Most of the indicators (84%) reflect the state of the ecosystem component or stresses on the ecosystem rather than human activities. The indicators also have applicability to broad environmental compartments (air, water, sediments, land, biota, fish, and humans) and to issues of human concern (toxic substances, nutrients, nonnative species, habitat, climate change, and stewardship). A full description of the indicators can be found in Bertram and Stadler-Salt (1999), which is available online at www.on.ec.gc.ca/solec/solec2000 or www.epa.gov/glnpo/solec/98/.

Table 70.1 Indicators Developed for Each of the SOLEC Groupings, with Cross-Referenced Applicability to Other Groupings of Indicator Type, Environmental Compartments, and Issues

Indicator Name	Indicator Type			Environmental Compartment							Issues					
	State	Pressure	Human Activity	Air	Water	Land	Sediments	Biota[a]	Fish	Humans	Contaminants[b]	Nutrients	Nonnative Species	Habitat	Climate Change	Stewardship
Open and Nearshore Water Indicators																
Lake trout and *Diporeia hoyi*	x							x	x		x	x	x	x		
Toxic chemical concentrations in offshore waters		x			x						x					
Atmospheric deposition of toxic chemicals		x		x	x						x					
Fish habitat	x				x				x					x		
Salmon and trout	x								x		x	x	x	x		
Preyfish populations	x								x		x	x	x	x		
Walleye and *Hexagenia*	x							x	x		x	x	x	x		
Phytoplankton populations	x							x			x	x	x			
Zooplankton populations	x							x			x	x	x			
Benthos diversity and abundance	x							x			x	x		x		
Native unionid mussels	x							x					x			
Sea lamprey		x							x				x			
Contaminants in colonial nesting waterbirds	x							x			x					
Concentrations of contaminants in sediment cores		x					x				x					
Contaminant exchanges between media: air to water and water to sediment		x		x	x		x				x					
Phosphorus concentration and loadings		x			x							x				
Deformities, eroded fins, and lesions and tumors (DELT) in nearshore fish	x								x		x					
Contaminants in young-of-the-year spottail shiners		x							x		x					
Wastewater pollution		x			x						x	x				
Sediment available for coastal nourishment	x				x	x									x	
Coastal Wetland Indicators																
Coastal wetland area by type	x				x	x									x	
Coastal wetland invertebrate community health	x							x							x	
Amphibian diversity and abundance	x							x							x	
Wetland-dependent bird diversity and abundance	x							x							x	
Presence, abundance, and expansion of invasive plants	x							x					x	x		
Coastal wetland fish community health	x								x				x	x		
Deformities, eroded fins, lesions and tumors (DELT) in coastal wetlands fish	x								x		x					
Contaminants in snapping turtle eggs		x						x			x					

Table 70.1 (continued)   Indicators Developed for Each of the SOLEC Groupings, with Cross-Referenced Applicability to Other Groupings of Indicator Type, Environmental Compartments, and Issues

Column groups: *Indicator Type* (State, Pressure, Human Activity); *Environmental Compartment* (Air, Water, Land, Sediments, Biota[a], Fish, Humans); *Issues* (Contaminants[b], Nutrients, Nonnative Species, Habitat, Climate Change, Stewardship)

Indicator Name	State	Pressure	Human Activity	Air	Water	Land	Sediments	Biota[a]	Fish	Humans	Contaminants[b]	Nutrients	Nonnative Species	Habitat	Climate Change	Stewardship
Nitrates and total phosphorus into coastal wetlands		x			x							x				
Sediment flowing into coastal wetlands		x			x	x								x		
Effects of water level fluctuations		x			x									x	x	
Gain in restored coastal wetland area by type			x		x	x								x		x
Nearshore Terrestrial Indicators																
Area, quality, and protection of lakeshore communities	x					x	x							x		x
Extent and quality of nearshore natural land cover	x					x								x		
Nearshore land use	x					x								x		
Nearshore species diversity and stability	x							x					x			
Extent of hardened shoreline		x				x								x		
Artificial coastal structures		x				x								x		
Nearshore plant and animal problem species		x						x					x	x		
Contaminants affecting productivity of bald eagles		x						x			x					
Contaminants affecting the American otter		x						x			x					
Protected nearshore areas			x			x								x		x
Shoreline managed under integrated management plans			x			x										x
Community/species plans			x					x								x
Land Use Indicators																
Habitat fragmentation	x					x								x		
Habitat adjacent to coastal wetlands	x					x								x		
Urban density	x					x										
Land conversion		x				x										
Mass transportation		x		x		x									x	x
Sustainable agricultural practices			x			x										x
Brownfield redevelopment			x			x										x
Green planning process			x		x	x										x
Human Health Indicators																
Geographic patterns and trends in disease incidence	x									x						
Chemical contaminant intake from air, water, soil, and food		x								x	x	x				
Chemical contaminants in human tissue		x								x	x	x				
Contaminants in edible fish tissue		x							x		x					

(continued)

Table 70.1 (continued) Indicators Developed for Each of the SOLEC Groupings, with Cross-Referenced Applicability to Other Groupings of Indicator Type, Environmental Compartments, and Issues

Indicator Name	Indicator Type			Environmental Compartment							Issues					
	State	Pressure	Human Activity	Air	Water	Land	Sediments	Biota[a]	Fish	Humans	Contaminants[b]	Nutrients	Nonnative Species	Habitat	Climate Change	Stewardship
Human Health Indicators (continued)																
Contaminants in recreational fish	x								x		x					
Drinking water quality	x				x			x			x	x				
Escherichia coli and fecal coliform levels in nearshore recreational waters	x				x			x			x					
Radionuclides		x		x	x						x					
Air quality		x		x							x					
Societal Indicators																
Aesthetics	x									x						x
Economic prosperity	x															
Water withdrawal		x			x					x						x
Energy consumption		x		x						x					x	x
Solid waste generation		x		x		x				x	x				x	x
Capacities of sustainable landscape partnerships			x							x						x
Organizational richness of sustainable landscape partnerships			x							x						x
Integration of ecosystem management principles across landscapes			x							x						x
Integration of sustainability principles across landscapes			x							x						x
Citizen/community place-based stewardship activities			x							x						x
Financial resources allocated to Great Lakes programs			x							x						x
Unbounded Indicators																
Breeding bird diversity and abundance	x							x						x		
Threatened species	x							x	x				x	x		
Climate change: number of extreme storms		x													x	
Climate change: first emergence of water lily blossoms in coastal wetlands		x						x							x	
Climate change: ice duration on the Great Lakes		x			x										x	
Acid rain		x		x	x	x					x					
Nonnative species		x						x	x				x			
Count	30	35	13	9	19	19	4	23	13	13	29	11	13	27	7	19

[a] Excluding fish and humans.
[b] Including pathogens.

Because SOLEC conferences are multiagency, multijurisdictional reporting venues, the SOLEC indicators require acceptance by a broad spectrum of stakeholders in the Great Lakes basin. The indicator descriptions were widely circulated for review and revision several times, and most proposed indicators have been accepted by consensus. The indicator list will remain dynamic, however, and it will continue to evolve.

IMPLEMENTING SOLEC INDICATORS

As capacity to monitor and report on the 80 indicators builds over the next several years, a more complete answer to the questions posed by the public about the health of the Great Lakes will emerge. For example, 19 of the 80 indicators are directly concerned with the waters of the Great Lakes (Table 70.1). By analyzing the monitoring data of those indicators and aggregating the results, a picture of the health of the waters of the Great Lakes should emerge. Gaps will no doubt be identified that require both an adjustment in the number of indicators needed and a fine tuning of the indicators in order to more fully describe the state of the waters. This will be true for the other environmental compartments and issues as well.

Currently, however, data are not available for all 80 indicators. The example indicators presented here represent each of the SOLEC organizational groupings. For each indicator, a short overview is followed by a description of the indicator, with examples of the data available for that indicator.

Nearshore and Open Waters Indicators

The nearshore waters of the Great Lakes occupy a band of varying width around the perimeter of each lake between the land and the deeper offshore waters of the lake. Also included as nearshore waters are the Great Lakes connecting channels and the lower reaches of tributaries that are influenced by changes in water levels in the Great Lakes. Human activities have substantially altered the Great Lakes basin landscape and the nearshore waters of the basin ecosystem. Some of the most significant stresses include the following:

- High density patterns of settlement, development, and population growth leading to the release of contaminants and nutrients, as well as habitat loss
- Agricultural development in the southern portion of the basin, which created an abundance of food and fiber leading to increased nutrient and pesticide loading
- High usage of surface water for drinking, manufacturing, power production, and waste disposal into tributaries
- Navigational structures such as dams and canals
- Nonnative species invasions, often facilitated by the construction of canals
- Development of sheltered areas into marinas and deepwater ports

The open waters of the Great Lakes are all of the waters beyond the lakeward edge of the nearshore waters (Edsall and Charlton, 1997). The offshore waters of the Great Lakes are also subject to many of these same stresses plus some unique offshore issues, such as atmospheric deposition of contaminants, the alteration of fish communities, and loss of biodiversity associated with overfishing and fish stocking practices.

Sea Lamprey

This indicator estimates the abundance of sea lampreys in the Great Lakes, which has a direct impact on the structure of the fish community and health of the aquatic ecosystem. Populations of large, native, predatory fishes can be diminished by sea lamprey predation.

The sea lamprey (*Petromyzon marinus*) is a parasitic aquatic vertebrate native to the Atlantic Ocean that is able to spawn and live entirely in fresh water. It has invaded all the Great Lakes in a span of 110 years, 1835 to 1946 (Mills et al., 1993).

The sea lamprey is not selective in its feeding as it preys on all species of large fish including salmon, lake trout, whitefish, walleye, and chubs. During its adult stage, it is possible that an individual sea lamprey can cause the death of more than 40 pounds of fish. Control measures managed by the Great Lakes Fishery Commission and supported by federal, provincial, state, and tribal governments has brought the lamprey population under control in most areas (Klar et al., 1996).

The information presented in Figure 70.2 shows sea lamprey populations (in their spawning phase) throughout the Great Lakes. During the past 20 years, Lake Superior populations have remained at levels less than 10% of peak abundance. For Lake Michigan, populations have been relatively stable, but an increase has been seen in northern sections of the lake due to the large population in Lake Huron moving into Lake Michigan. Lake Huron populations have remained at very high levels since the early 1980s because of large spawning populations in the St. Marys River. An integrated control strategy was initiated in the St. Marys River in 1997 which includes

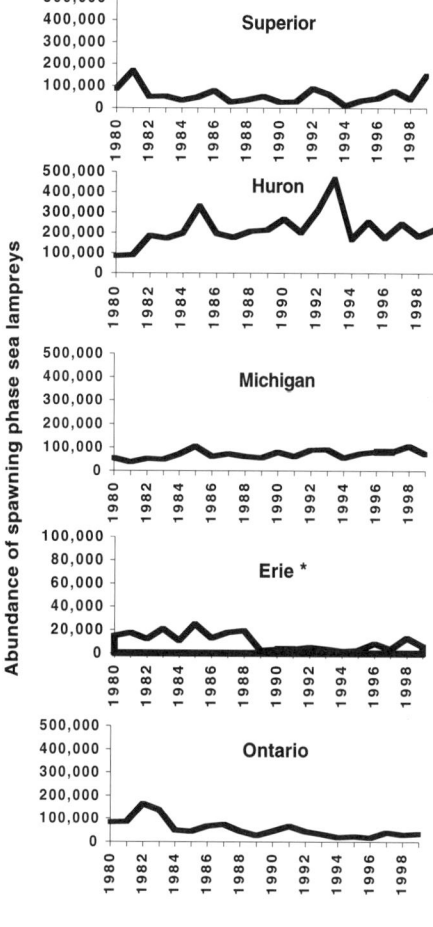

*Note scale change for Lake Erie

Figure 70.2 Total abundance of sea lamprey in each of the five Great Lakes estimated during the spawning phase. (From Canada and the U.S., *State of the Great Lakes 2001*, www.binational.net, Toronto and Chicago, 2001.)

the targeted application of a new bottom-release lampricide, enhanced trapping of spawning animals, and sterile-male release. In Lake Erie, lamprey abundance has increased since the early 1990s. Sources of this increase were several streams to which treatment has been deferred due to low water flows or concerns for nontarget organisms. Lake Ontario populations have remained constant in recent years because of adequate control.

Future pressures include the increased potential for sea lamprey to colonize new locations due to improved water quality in Great Lakes tributaries. Additionally, short lapses in lamprey control can result in quick increases in abundance. Significant additional control efforts, like those on the St. Marys River, may be necessary to maintain reduced lamprey populations.

Phosphorus Concentrations and Loadings

This indicator, which assesses total phosphorus levels in the Great Lakes, is used to support the evaluation of trophic status and food web dynamics in the Great Lakes. Phosphorus is an essential element for all organisms and is often the limiting factor for aquatic plant growth in the Great Lakes.

Sewage treatment plant effluent, agricultural runoff, and industrial processes have released high concentrations of phosphorus into the lakes (IJC, 1996). Strict phosphorus loading targets implemented in the 1980s have been successful in reducing nutrient concentrations in the lakes (Figure 70.3), although high concentrations still occur locally in embayments and harbors (Canada and U.S., 1999).

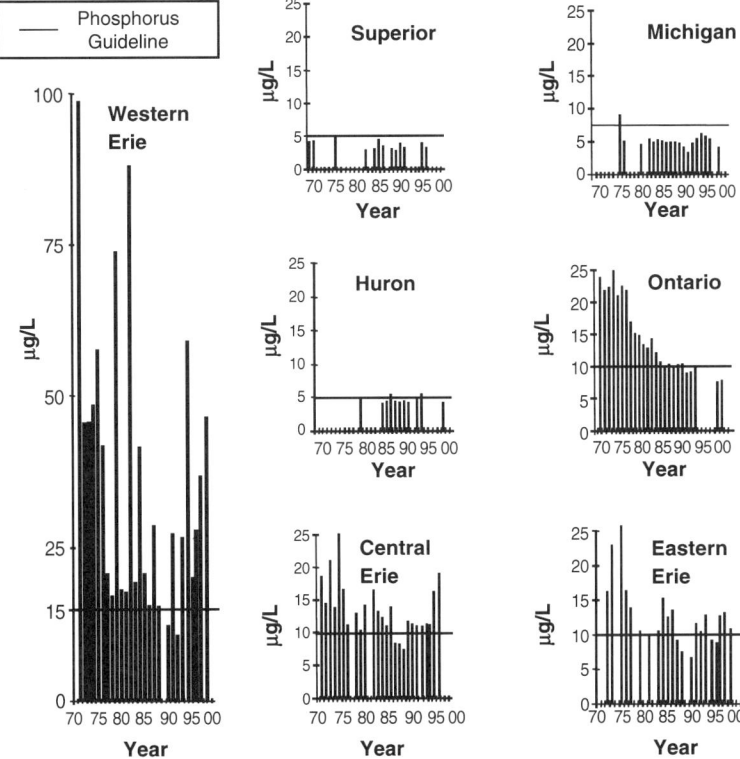

Figure 70.3 Total phosphorus trends in the Great Lakes 1971–2000. Data reflect samples collected at the surface from offshore waters during spring. Blank indicates no sampling (From Canada and the United States, *State of the Great Lakes 2001*, www.binational.net, Toronto and Chicago, 2001.)

Concentrations of total phosphorus in the open waters of the Great Lakes have remained nearly stable since the mid-1980s. Concentrations in the open waters of Lakes Superior, Michigan, Huron, and Ontario are at or below guideline levels. Concentrations in all three basins of Lake Erie exceed phosphorus guidelines, and recent data suggest an increasing trend. This trend may be related to the large populations of nonnative zebra mussels (*Dreissena polymorpha*) and quagga mussels (*Dreissena bugensis*).

With the anticipated human population growth in the basin over the next 25 years (up to 2 million in Canada alone), the nutrient loads to the lakes from this population will be significant and they will require additional treatment, particularly at sewage treatment plants. The phosphorus indicator will be of value in helping governments track trends in nutrients in the lakes and adjust management strategies accordingly.

Coastal Wetland Ecosystem Indicators

Wetlands are important ecologically, socially, and economically and are one of the most productive ecosystems in the world (Maynard and Wilcox, 1997). Despite these values, coastal wetlands are in trouble. Threats include regulation of lake water levels, land-use change, nonnative species, and toxic chemicals.

To select indicators of the health and integrity of coastal wetlands, the following criteria for coastal wetland health were used:

- Capability to self-maintain assemblages of organisms that have a composition and functional organization comparable to natural habitat
- Resiliency to natural disturbances
- Risk factors or human-induced pressures at an acceptable level

There are few existing monitoring programs for Great Lakes coastal wetlands. Efforts are being made to select indicators for which there are existing data and monitoring programs, although many of the indicators will require new or improved monitoring programs.

Wetland-Dependent Bird Diversity and Abundance

Assessments of the diversity and abundance of wetland-dependent birds in the Great Lakes basin, combined with an analysis of habitat characteristics, are used to evaluate the health and function of wetlands.

Birds are among the most visible and diverse groups of wildlife in Great Lakes coastal wetlands. Because breeding wetland birds require an appropriate mix and density of vegetation, sufficient and safe food resources, and freedom from predation and other disturbances, their presence and abundance provides information that integrates the physical, chemical, and biological status of their habitats.

Patterns in the species composition and numbers of breeding wetland birds may reflect changes in the condition of breeding habitats. Five years of Marsh Monitoring Program (MMP, www.bsc-eoc.org) data provided information on numbers of marsh nesting birds. Basin-wide increases were observed for Canada goose, mallard, chimney swift, northern rough-winged swallow, common yellowthroat and common grackle (Table 70.2). Basin-wide declines were observed for pied-billed grebe, blue-winged teal, American coot, and black tern. Each of the declining species depends on wetlands for breeding, but because they use wetland habitats almost exclusively, the pied-billed grebe, American coot, and black tern are particularly dependent on the availability of healthy wetlands.

Continuing and future pressures on these bird species include the continued loss and degradation of wetlands, water level stabilization, sedimentation, contaminant and nutrient inputs, and the invasion of nonnative plants and animals.

Table 70.2 Estimated Annual Percent Changes in Population Indices of Marsh-Nesting and Aerial Foraging Bird Species Detected on Great Lakes Basin Marsh Monitoring Program (MMP) Routes, 1995–1999

Species	% Change	95% Confidence Limits
Canada goose	20.2	4.9, 37.7
Mallard	29.2	17.0, 42.6
Chimney swift	15.8	1.7, 31.8
Northern rough-winged swallow	20.1	0.8, 43.1
Common yellowthroat	6.8	2.0, 11.7
Common grackle	42.5	27.1, 59.7
Pied-billed grebe	−11.8	−19.9, −2.5
Blue-winged teal	−13.2	−24.3, −0.5
American coot	−22.1	−34.7, −7.2
Black tern	−20.2	−28.7, −10.8

Population indices are based on counts of individuals inside the MMP station boundary and are defined relative to 1999 values (see Canada and U.S., 2001).

Nearshore Terrestrial Ecosystem Indicators

The nearshore terrestrial environment is an integral part of the Great Lakes basin ecosystem, the extent of which is defined by the lakes themselves. A description of these areas and major stresses on these natural communities are described in *State of the Great Lakes 1997* (Canada and U.S., 1997).

Thirteen indicators of nearshore terrestrial ecosystem health have been developed to fulfill the need for a cost-effective and easily understood set of measures that will describe how nearshore ecosystems across the basin are changing, identify what is causing the changes, describe the current status of these ecosystems and component parts, and evaluate how effectively humans are responding to the changes.

Area, Quality, and Protection of Alvar Communities

This indicator was designed to measure the area, quality, and protected status of the alvar communities occurring within 1 km of the shoreline. Alvar communities are naturally open habitats occurring on flat limestone bedrock. The information collected to satisfy this measure may also help to identify the sources of threats to these ecologically significant habitats, as well as the success of management activities associated with the protection status.

More than 90% of the original extent of alvar habitats has been destroyed or substantially degraded. Emphasis is focused on protecting the remaining 10%. Approximately 64% of the remaining alvar area exists within Ontario; 16% in New York State; 15% in Michigan; and smaller areas in Ohio, Wisconsin, and Quebec. Less than 20% of the nearshore alvar acreage is currently fully protected, while over 60% is at high risk (Figure 70.4). Michigan has 66% of its nearshore alvar acreage in the fully protected category, while Ontario has only 7%. In part, this is a reflection of the much larger total shoreline acreage in Ontario.

Continuing pressures on alvars include:

- Habitat fragmentation and loss
- Trails
- Off-road vehicles
- Resource extraction uses such as quarrying or logging
- Adjacent land uses such as residential subdivisions
- Grazing or deer browsing,
- Plant collection for bonsai and other hobbies
- Invasion by nonnative plants

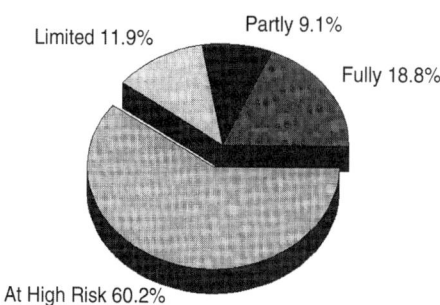

Figure 70.4 Protection status of Great Lakes nearshore alvar communities, 2000 (Canada and the U.S., *State of the Great Lakes 2001*, www.binational.net, Toronto and Chicago, 2001.)

Land-Use Indicators

Changing patterns of land use are a major ecosystem stressor for the Great Lakes basin and its nearshore areas. The many forms of development — including industrial, commercial, residential, agricultural, and transportation-related activities — carry specific, significant, and cumulative impacts for the natural world and particularly for Great Lakes water quality. These activities take place throughout the basin, but their most immediate and direct impact on the Great Lakes appears to be on lands proximate to the lakes themselves and their tributary waters. These nearshore areas suffer from a particular and disproportionate environmental burden because of their unique and sensitive environments and proximity to development.

Sustainable Agricultural Practices

This indicator assesses the number of environmental and conservation farm plans and environmentally friendly practices in place, such as integrated pest management to reduce the unnecessary use of pesticides, conservation tillage and other soil preservation practices to reduce energy consumption, and prevention of ground and surface water contamination.

Agriculture accounts for 35% of the land area of the Great Lakes basin and dominates in the southern portion of the basin. In the past, excessive tillage and intensive crop rotations led to soil erosion and the resulting sedimentation of major tributaries. Agriculture is a major user of pesticides with an annual use of 26,000 tons. These practices led to a decline of soil organic matter. The adoption of more environmentally responsible practices has helped to replenish carbon in the soils to 60% of turn-of-the-century levels. The following are two examples of voluntary programs to encourage the development and implementation of environmental farm plans.

In Ontario, farmers can complete an Environmental Farm Plan identifying environmental areas of concerns on their farms and note activities and specific actions they will take to remediate them. For example, ensuring that farm manure is managed to avoid contaminating surface water courses and groundwater is critical to maintaining safe and clean drinking water for the farmer as well as preventing contamination of downstream water or aquifers. The farm plan will identify the possibility of contamination and identify preventative or remedial solutions and actions. The farm plan then becomes a stewardship guidebook for environmental management by the farmer and a reference document for further remedial or preventative actions. Figure 70.5 depicts the percentage of acreage with approved Environmental Farm Plans in Ontario.

The U.S. Department of Agriculture (USDA) offers landowners financial, technical, and educational assistance to implement conservation practices on privately owned land and promote sustainable agricultural practices. The Conservation Reserve Program (CRP) reduces soil erosion, protects the nation's ability to produce food and fiber, reduces sedimentation in streams and lakes, improves water quality, establishes wildlife habitat, and enhances forest and wetland resources. It

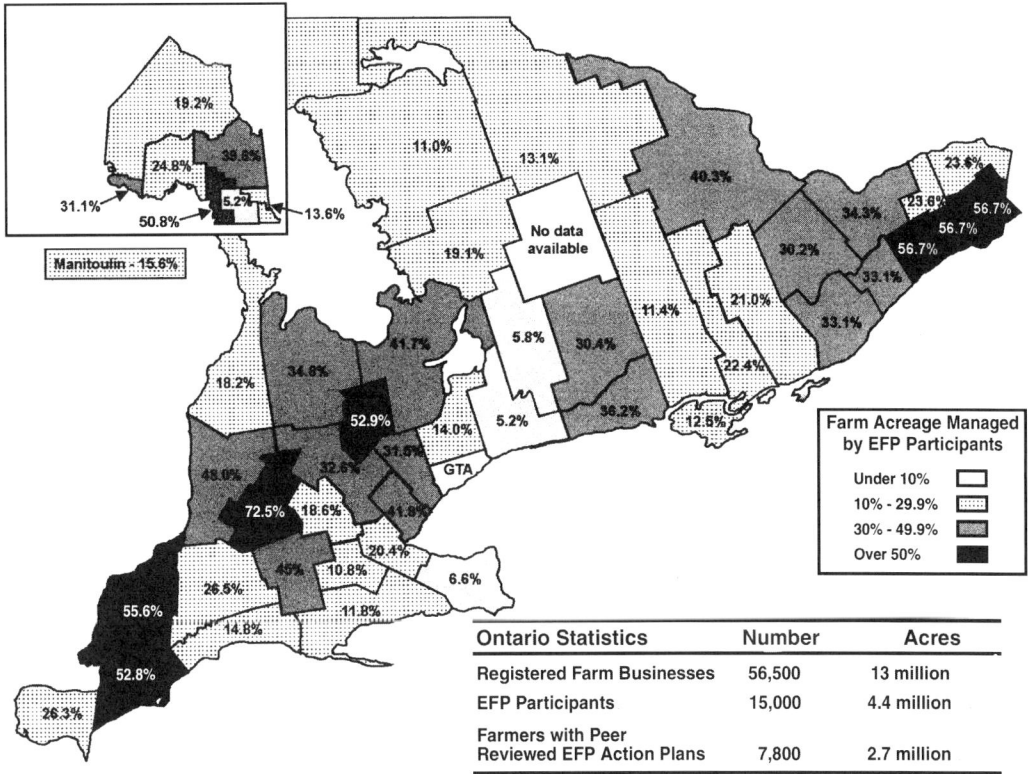

Ontario Statistics	Number	Acres
Registered Farm Businesses	56,500	13 million
EFP Participants	15,000	4.4 million
Farmers with Peer Reviewed EFP Action Plans	7,800	2.7 million

Figure 70.5 Ontario Environmental Farm Plans (EFP). (From Canada and the U.s., *State of the Great Lakes 2001*, www.binational.net, Toronto and Chicago, 2001.)

encourages farmers to convert highly erodible cropland or other environmentally sensitive acreage to vegetative cover, such as tame or native grasses, wildlife plantings, trees, filter strips, or riparian buffers. Figure 70.6 shows the number of acres with conservation planned systems in place.

Human Health Indicators

Many people consider the protection of human health to be one of the more important goals of environmental management. Consequently, there is interest in indicators of changes in human health, or changes in factors that affect health, as they relate to the Great Lakes environment. The premise is that as social, economic, and environmental conditions change in the Great Lakes basin, so could the health of the population. Indicators are also needed to assess the effectiveness of social, economic, health, and environment policies and actions to protect or improve the health of the Great Lakes basin population.

Because the ability to relate the state of human health in the Great Lakes basin to components of the Great Lakes ecosystem per se is difficult at best, the effort to develop human health indicators has focused primarily on exposure to environmental contaminants and pathogens. The indicators include contaminant levels measured in human tissues, such as breast milk or blood; estimates of daily intake of persistent contaminants by the Great Lakes population (e.g., via fish consumption); and contaminant levels in air, drinking water, and recreational water. The contribution of these exposures as causative factors in disease, such as cancer and birth defects, can be difficult to identify. Using an indicator that analyzes geographic patterns and trends in incidence rates, however, can serve to identify potential areas of concern and may lead to testable hypotheses regarding the correlation of environmental exposure with human disease.

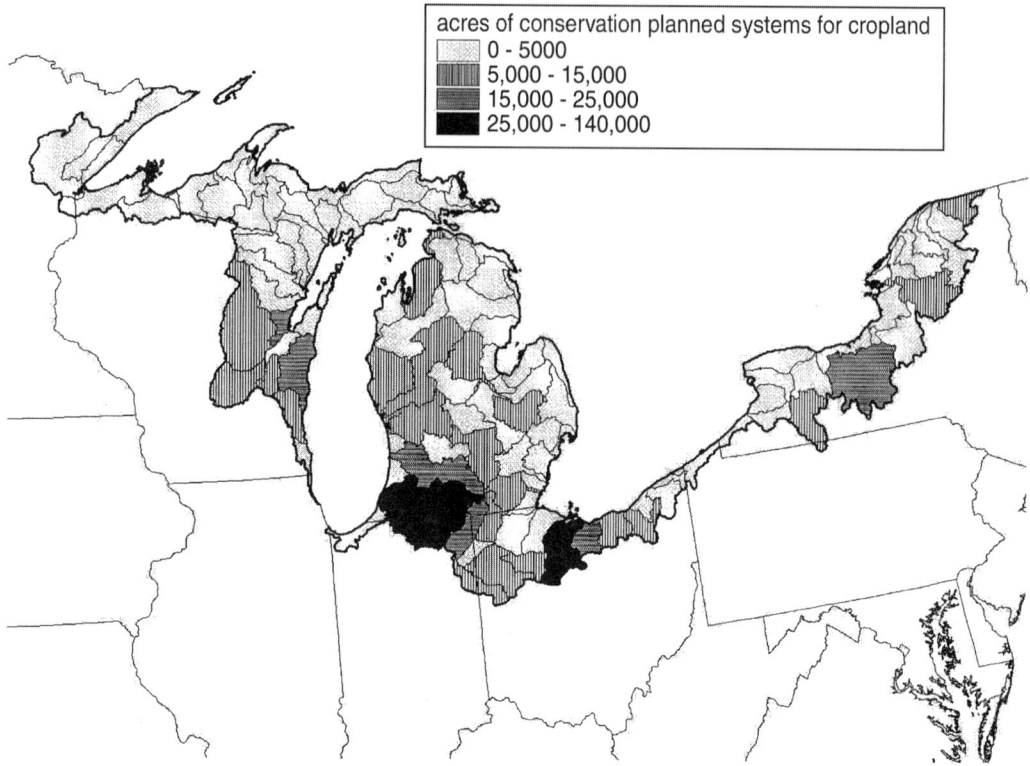

Figure 70.6 Annual U.S. Conservation Planned Systems for 2000. (From Canada and the U.S., *State of the Great Lakes 2001*, www.binational.net, Toronto and Chicago, 2001.)

Escherichia coli *and Fecal Coliform Levels in Nearshore Recreational Waters*

This indicator assesses *E. coli* and fecal coliform abundance in nearshore recreational waters, acting as a surrogate indicator for other pathogen types, and it is used to infer potential harm to human health through body contact with nearshore recreational waters.

Recreational waters may become contaminated with animal and human feces from sources such as combined sewer overflows that occur in certain areas after heavy rains, agricultural runoff, and poorly treated sewage. Gastrointestinal disorders and minor skin, eye, ear, nose, and throat infections have been associated with microbial contamination. Children, the elderly, and people with weakened immune systems are those most likely to develop illnesses or infections after swimming in polluted water.

Survey reports of U.S. beach advisories during the 1998 swimming season (June, July, and August) show that 78% of the reporting beaches were open for the entire 1998 season. Results were similar for Canadian beaches where 79% of the reporting beaches were open the entire season (Figure 70.7). Survey reports of U.S. beach closings or advisories during the 1999 season show that 65% of the reporting beaches were open the entire 1999 season. Several factors may have influenced the apparent increase in percentage of beach closings in 1999 compared with 1998:

- Fewer beach managers responded to survey questionnaires in 1999
- A different mix of beaches was reported in 1999
- More beach managers were using *E. coli* testing in 1999 than in 1998
- An improved system for accounting for beach advisory days was used in 1999

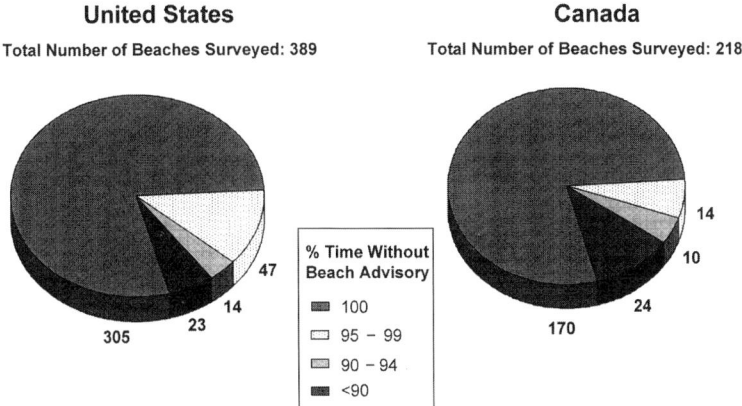

Figure 70.7 Comparison of the frequency of U.S. and Canadian beach advisories for Great Lakes beaches, 1998. The swimming season was defined as June, July and August. Data were obtained from voluntary responses by beach managers to annual requests for information. (From Canada and the U.S., *State of the Great Lakes 2001*, www.binational.net, Toronto and Chicago, 2001.)

While the actual water quality near beaches may not have changed, the increase in beach advisory days is believed to be more protective of human health.

Societal Indicators

Human society is part of the ecosystem, and integrated environmental management requires human activities to be respectful of other ecosystem components. For example, the creation and discharge of waste materials by humans may have an impact on both the habitat and the health of plant and animal species. A responsible society will recognize its collective impact on the surrounding environment, and it will seek to sustain ecosystem integrity indefinitely.

Within the suite of proposed Great Lakes indicators, sustainability is implicit throughout. A set of indicators to assess human activities and other societal issues, however, would reflect our individual and collective actions to halt, mitigate, adapt to, or prevent damage to the environment. Socioeconomics, urban issues, societal responsibility, and other social aspects of Great Lakes communities are not easy to monitor due to the complexity of the relationships between jurisdictions and the lack of coordinated monitoring programs. The SOLEC process is being used to engage a variety of agencies and organizations in the Great Lakes basin to participate in the further selection and implementation of clear, understandable societal indicators.

Economic Prosperity

This indicator assesses the unemployment rates within the Great Lakes basin and, when used in association with other societal indicators, infers the capacity for society in the Great Lakes region to make decisions that will benefit the Great Lakes ecosystem. During periods of low unemployment and economic well-being, public support for environmental initiatives by government agencies may be increased.

By most measures, the binational Great Lakes regional economy is reasonably healthy. The unemployment rate for the Great Lakes states was less than the U.S. national average for most of the 1990s. Canadian and Ontario economic recoveries unfolded later than the U.S., but Ontario unemployment rates in 2000 were at the lowest level since 1990.

Both sides of the border reflect a manufacturing intensity greater than their national economies. The Great Lakes states represent about 27% of national output in manufacturing, whereas Ontario's percentage is twice as large. The manufacturing sector has many cross-border linkages, particularly for the auto industry. About one half of the billion dollar-a-day U.S.–Canada trade is tied to the Great Lakes states, with Ontario as the most prominent province in this relationship.

The overall message is mixed. Good economic times can translate into high levels of consumer spending and home buying, which can lead to increased household waste generation, increased air pollution, and accelerated land use changes.

Unbounded Indicators

Several proposed indicators do not fit neatly into any of the other SOLEC ecological categories. These indicators may have application to more than one of the other organizing categories, or they may reflect issues that affect the Great Lakes but have global origins or implications. For example, indicators related to climate change will affect all the groups yet truly belong in none of them.

Nonnative Species

Currently, this indicator reports introductions of aquatic organisms not naturally occurring in the Great Lakes basin, and it is used to assess the status of biotic communities in the basin. The indicator will expand to terrestrial organisms in the future.

Since the 1830s, there have been 63 nonnative aquatic animal (fauna) species introduced into the Great Lakes that have maintained permanent, breeding populations (Figure 70.8). In almost the same time frame, 83 nonnative plant (flora) species were introduced and established residency.

Some of the main entry mechanisms include ship ballast water, the deliberate release of fish and other faunal species, and releases from hobby aquaria. Some plant species have escaped from cultivation. Even with voluntary and mandatory ballast exchange programs recently implemented in Canada and the U.S., newly introduced species associated with shipping activities have been identified.

Introductions of nonnative species are expected to continue because of increased global trade, new diversions of water into the Great Lakes, aquaculture industries, and changes in water quality and temperature. Even the presence of some key nonnative species could make the Great Lakes more hospitable for other nuisance species.

CONCLUSION

The process for selecting and developing indicators of the health of the Great Lakes basin ecosystem has been open but rigorous, engaging the participation of a wide variety of stakeholders.

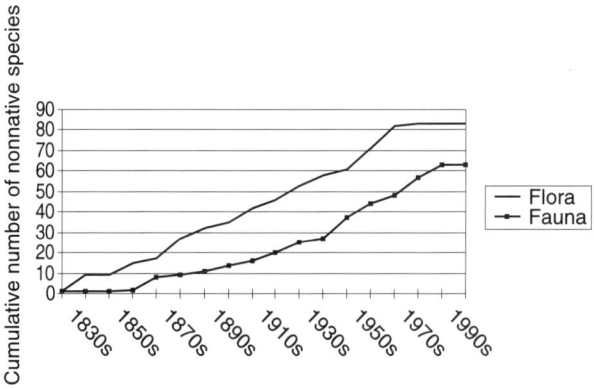

Figure 70.8 Cumulative number of nonnative species introduced into the Great Lakes since the 1830s (From Canada and the U.S., *State of the Great Lakes 2001*, www.binational.net, Toronto and Chicago, 2001.)

Informal consensus on the suite of indicators has been actively sought, and it continues to be important as the indicators are reviewed and refined. The indicators have been extensively reviewed, but the list continues to be dynamic, and individual indicators are subject to testing and further revision.

Based on the overall criteria of *necessary* and *sufficient*, the suite of SOLEC indicators addresses most of the Great Lakes ecosystem components. Some additional indicators may be added as information gaps or other managerial needs for data are identified. Some indicators may also be removed from the list if they no longer provide useful information.

Monitoring and reporting on the state of ecosystem components is not a new concept to Great Lakes programs. The value added by the SOLEC process is the deliberate selection of information requirements and the translation of those requirements into a comprehensive suite of indicators for multiple users. The data collected from collaborative monitoring efforts and the subsequent interpretation of the data should be useful for environmental managers at all levels of government as well as for researchers, industry, and private citizens. The suite of indicators can also be used by the governments of Canada and the U.S. not only as a basis for reporting on progress toward the goals of the GLWQA, but also as a basis for engaging additional monitoring and research.

Several challenges remain to implement fully the SOLEC indicators and the SOLEC reporting venues. They include the following:

- Periodically reviewing and refining the indicator list.
- Gaining acceptance of the list by federal, state, provincial, and municipal partners who have the potential to monitor these indicators.
- Nesting local and lake-wide indicators within basin-wide indicators.
- Building appropriate monitoring and reporting activities into Great Lakes programs at the federal, provincial, state, Tribes/First Nations, and industry levels, including agencies that have not traditionally provided monitoring data.
- Reporting on indicators at future SOLEC conferences and in printed reports or other media in a format that will meet the needs of multiple users.

As more of the underlying data supporting the indicators become available, more audiences can be served, including the general public, local decision makers, and the scientific and engineering community. A well-informed public should facilitate the management decisions and actions that are needed to continue progress toward the goals of the Great Lakes Water Quality Assessment.

REFERENCES

Bertram, P. and Stadler-Salt, N., Eds., *Selection of Indicators for Great Lakes Basin Ecosystem Health*, Version 4, www.on.ec.gc.ca/solec/ or www.epa.gov/glnpo/solec/98/, prepared for the State of the Lakes Ecosystem Conference, U.S. Environmental Protection Agency, Great Lakes National Program Office, and Environment Canada–Ontario Region, Office of the Regional Science Advisor, Burlington, Chicago, and Ontario, 1999. www.on.ec.gc.ca/solec/ or www.epa.gov/glnpo/solec/98/

Canada and the U.S., *State of the Great Lakes 1997*, Toronto and Chicago, 1997.

Canada and the U.S., *State of the Great Lakes 1999*, www.on.ec.gc.ca/solec/ or www.epa.gov/glnpo/solec/98/, Toronto and Chicago, 1999.

Canada and the U.S., *State of the Great Lakes 2001*, www.binational.net, Toronto and Chicago, 2001.

Edsall, T.A. and Charlton, M.N., Nearshore Waters of the Great Lakes, background paper for the State of the Lakes Ecosystem Conference 1996, Windsor, Ontario, www.on.ec.gc.ca/solec/ or www.epa.gov/glnpo/solec/98/, 1997.

Green Mountain Institute for Environmental Democracy (GMIED), *The Resource Guide to Indicators*, 2nd ed., www/gmied.org, Montpelier, VT, 1998.

International Joint Commission (IJC), *Pollution of Lake Ontario and the International Section of the St. Lawrence River*, International Lake Erie Water Pollution Board and the International Lake Ontario-St. Lawrence River Water Pollution Board, Windsor, Ontario, 1969.

International Joint Commission (IJC), *A Proposed Framework for Developing Indicators of Ecosystem Health for the Great Lakes*, Windsor, Ontario, 1991.

International Joint Commission (IJC), *Indicators to Evaluate Progress under the Great Lakes Water Quality Agreement*, prepared by the Indicators for Evaluation Task Force, Windsor, Ontario, 1996.

Klar, G.T., Schleen, L.P., and Young, R.J., Integrated Management of Sea Lampreys in the Great Lakes: Annual Report to the Great Lakes Fishery Commission, www.glfc.org/slar96/slar961.htm, Ann Arbor, MI, 1996.

Maynard, L., and Wilcox, D., Coastal Wetlands of the Great Lakes, background paper for the State of the Lakes Ecosystem Conference 1996, Windsor, Ontario www.on.ec.gc.ca/solec/ or www.epa.gov/glnpo/solec/98/, 1997.

Messer, J.J., Indicators in regional ecological monitoring and risk assessment, in *Ecological Indicators,* Vol. 1, Makenzie, D.H., Hyatt, D.E., and McDonald, V.J., Eds., Elsevier Applied Science, New York, 1992, pp. 135–146.

Mills, E.L., Leach, J.H., Carlton, J.T., Secor, C.L., Exotic species in the Great Lakes: a history of biotic crises and anthropogenic introductions. *J. Great Lakes Res.*, 19, 1–54, 1993.

Regier, H.A., Indicators of ecosystem integrity, in *Ecological Indicators,* Vol. 1, Makenzie, D.H., Hyatt, D.E., and McDonald, V.J., Eds., Elsevier Applied Science, New York, 1992, pp. 183–200.

U.S. and Canada, Great Lakes Water Quality Agreement, as amended by Protocol signed Nov. 18, 1987, 1987.

U.S. and Canada, *State of the Great Lakes 1995*, www.on.ec.gc.ca/solec/ or www.epa.gov/glnpo/solec/98, Chicago and Toronto, 1995.

CHAPTER 71

Control of Natural Resource Degradation to Restore Ecosystem Health and Help Secure Peace in the Middle East

Scott Christiansen

INTRODUCTION

The Multilateral Middle East Peace Process called for, among other things, strengthening regional cooperation to control desertification. The World Bank was instrumental in facilitating the process of preparing and implementing the approved program, the *Initiative for Collaboration to Control Natural Resource Degradation (Desertification) of Arid Lands in the Middle East*, which has been recently shortened to *Regional Initiative for Dryland Management* for Phase II of the project, scheduled to continue through 2003. The initiative involves regional cooperation in technology sharing, application of research findings, provision of training, and preparation of pilot investment priorities. The World Bank, Switzerland, Luxembourg, Japan, the U.S., the Republic of Korea, and Canada provide funding that is complemented by in-kind and financial inputs by the participating countries: Egypt, Israel, Jordan, the Palestinian Authority, and Tunisia. The International Center for Agricultural Research in the Dry Areas (ICARDA) is responsible for program execution through a facilitation unit hosted in the ICARDA-Cairo offices.

This chapter presents a summary of the project setup and achievements in Phase I and outlines how it has been redesigned to better achieve its goals in Phase II. The Phase I program (1995–2000) was subject to an external program review and an external financial audit, both serving to strengthen the project. The Phase I matrix setup (Table 71.1) gave way to a more objective-oriented approach in Phase II, namely: (1) natural resource conservation and watershed development using water harvesting techniques and suitable plant resources for the benefit of the local communities within selected national development projects; (2) treated wastewater and biosolids management, reuse, and guidelines; and (3) identification of policy options for reversing natural resource degradation and reducing poverty.

Administrative Structure and Participating Institutions

A steering committee of donors meets annually and sets policy guidelines; approves work plans; reviews annual reports and budgets; supervises the facilitation unit; approves accounting, procurement, and auditing reports; and evaluates the program overall. It comprises representatives of contributing donor organizations or countries, country counterparts, an international facilitator, an

Table 71.1 The Original Design for the Implementation of Phase I[a]

Regional Support Program (RSP) Theme	National Support Activity (NSA)				
	Egypt	Israel	Jordan	Tunisia	Palestinian Authority
Germ plasm for arid lands	**NSA and RSP leader**	NSA	NSA	NSA	NSA
Economic forestry and orchards	NSA	**NSA and RSP leader**	NSA	NSA	NSA
Rangeland management	NSA	NSA	**NSA and RSP leader**	NSA	NSA
Marginal water and saline soils	NSA	NSA	NSA	**NSA and RSP leader**	NSA
Capacity building	—	—	—	—	**Focal point**

[a] Countries leading regional support programs (RSPs) are indicated in bold. Each of the five countries was required to carry out projects within its territory, known as national support activities (NSAs). In addition, the Palestinian Authority was assisted in capacity building through support for the establishment of the Palestinian Environmental Authority (PEnA), subsequently merged with the Ministry of Environmental Affairs (MEnA). For example, Jordan is to carry out NSAs within its territory and provides regional support to all the other countries in the RSP for the rangeland management theme. NSAs were planned as research and development projects while the RSPs are used for coordination, training, and communication.

ICARDA representative, the World Bank supervisor, and a World Bank steering committee chairperson. The conception and implementation of work programs and budgets are the responsibility of the national agencies in cooperation with the facilitator.

Phase I of the initiative was implemented during 1995–2000, a period during which the political environment and the institutional arrangements were under severe stress. For example, the regional dimensions of the initiative were completely frozen from March 1997 until summer 1999, and the mechanisms for cooperation under the Multilateral Middle East Peace Talks were mostly ineffective. Despite these stresses and some financial constraints, however, the project evolved from its original design into a reasonably effective state, but with limitations on regional cooperation.

PROGRESS UNDER PHASE I

The initiative began in earnest during 1995 with the setup of the facilitation unit and organization of national work plans. The partner countries deployed these technical solutions at specific sites with varying degrees of success (see Boxes 71.1 to 71.3). The original structure elaborated by the project designers (Table 71.1) began to evolve into the current form of the initiative from the 1997/1998 to 1999/2000 agricultural seasons (Table 71.2). When activities failed to occur, or when work plans were weak, team leaders were encouraged to consolidate and refocus their programs. In fact, some activities will not continue into Phase II because of the lack of a supporting national development project, which was a prerequisite for continuation of the work.

THE RATIONALE FOR CONTINUATION

A Vision for Phase II

Phase II of the initiative focuses almost exclusively on regional collaboration, underpinned by national-level activities in the domain of natural resource management. The activities of Phase II build on the achievements of Phase I. Greater attention is given to the role of poverty as a causal factor in environmental degradation. Major efforts are being made to capitalize on strong knowledge bases within the region on runoff farming and agricultural reuse of treated wastewater and biosolids

---

Box 71.1   Accomplishments — Plant Genetic Resources

EGYPT: A systematic plant collection effort in the Sinai characterized and documented the best sites for conservation, providing a useful guide to rangeland scientists, ecologists, botanists, and taxonomists. The work in the northwest coast deserves credit for the manner in which the initiative worked with Matrouh Resource Management Project (MRMP), the Agricultural Research Center, the Desert Research Center, ICARDA, and consultants to review the recommended plant genetic resources considered appropriate for the region. This teamwork resulted in a widescale change in the genetic resource offerings recommended for propagation in more than a dozen private green-houses working with MRMP, each with a capacity of approximately 100,000 seedlings per year.

ISRAEL: Israel carried out research for elucidating drought resistance of forage plants from the desert and nondesert transition zone of Israel. A postdoctoral fellow developed a climatological index to characterize annual fluctuations in aridity of this transition zone, a report that is available in the scientific literature.

JORDAN: Activity focused on promotion, conservation, and utilization of medicinal and herbal plants of Jordan. Work is now expanded for fenugreek, anise, and fennel. It was found that growing thyme under plastic house conditions maximizes yield and return, while for sage open-field condi-tions are recommended. For chamomile, the management target is to develop techniques to extend flowering time to increase yield. Arak, mellisa, and roselle can feasibly be grown under conditions of high heat and salinity, particularly when mulched. The project provided the foundation for a Global Environment Facility (GEF) Project Development Facility Grant of $350,000 (U.S.) that started in 2000.

PALESTINIAN AUTHORITY: A 15-ha botanical garden in Jericho is established. The area is fenced and has plants established around the entire perimeter. The land was cleared of plastic, cultivated, and tested for chemical and physical properties. A drip irrigation system is installed on nearly two thirds of the area. In 1999, eight plastic houses were used for seed production of local vegetable varieties (tomatoes, cucumbers, pumpkins, etc.) in association with a local NGO. A four-room office building (including storerooms) is functional. About two thirds of the area is already planted with trees and shrubs. The vegetable program is based on the technical input from local NGO experts working with about 200 farmers on conservation of 23 local vegetable varieties from 11 main crops. The various units of the garden include herbs and medicinal plants, wild plants, shrubs and forages, citrus, palms, forest species, and cropland races.

TUNISIA: The activities of the germ plasm group were to promote on-farm conservation and propagation of new types of useful arid land germ plasm (shrubs, trees, grass) that can be used to stabilize the fragile ecosystem of the Zammour Valley within the Matmata Mountains. The project has attracted a credible volunteer participation in the program by members of the community. The NGO *Les Jeunes de Zammour* with IRA-Médenine have developed a proposal on the use of medicinal and herbal plants to further enrich the selection of plants with potential economic value that would allow for the development of a project through GEF. A booklet is currently under preparation containing an inventory of plants with pictures and descriptions of the target herbal, aromatic, medicinal, and culinary species with economic potential.

Note: Because of the similarity of the "germ plasm" for arid lands and the "economic forestry and orchards" themes, the accomplishments are consolidated in a single box.

---

because of their potential to add value to land, permitting the intensification of agriculture through the introduction of trees and other higher value crops. These subjects provide excellent areas for cooperation between Arab and Israeli partners owing to their potential to share knowledge and provide outputs that could have significant sociological and economic impacts. Greater attention is devoted to the sharing of knowledge on policy issues common to the region, especially in terms

---

Box 71.2   Accomplishments — Rangeland Management

EGYPT: Committed interactions and a high degree of communication are taking place among the initiative, ICARDA, the International Plant Genetic Resources Institute (IPGRI), Matrouh Resource Management Project (MRMP), and the GTZ/Qasr Rural Development Program (QRDP). A range strategy was written during a meeting held October 17–19, 1998 and a study to investigate the potential of rangeland improvement in the wadi tips has culminated in a successful demonstration in the field. Adapted plant materials, good quality seedlings grown in private nurseries, and well-supervised transplantation techniques can result in successful intensification of rangeland on private land where the farmers have shown willingness to protect the transplanted seedlings. This activity requires follow-up if it is to be expanded to other wadis to assure the success of plantations in the first year (watering when absolutely necessary, monitoring survival, and checking farmer compliance to prohibit grazing). The program covers approximately 35 ha at 10 Wadi Ramla sites using approximate 50,000 plants and representing five adapted species.

ISRAEL: The Israelis have invested a large amount of effort in securing national support to carry out their combined NSA on rangeland and economic forestry. The JNF is now a full partner and a site is identified within the Israeli side of the Yatir Forest to monitor the effect of forests on hydrology, productivity, and biodiversity in the watershed. Bedouin farmers from neighboring countries will visit and share experiences with Bedouin in Israel.

JORDAN: The Ministry of Agriculture (MOA), the Royal Society for the Conservation of Nature (RSCN), the Canadian International Development Agency (CIDA), and the Canadian contractor AGRODEV manage the Sustainable Rangeland Management Project (SRMP), which is established in Faysaliya and Buseira and will now expand to another agroecological zone in Muaggar. The Jordanians agreed to expand a 1-ha spineless cactus nursery in Mujib to a much larger area. The initiative team in Jordan enrolled six of its staff in a master of science rangeland program at the University of Jordan set up through collaboration of the initiative and CIDA/AGRODEV. A successful SRMP "cut and carry" operation was achieved on 40 ha at the 5-year-old Faysaliya range reserve with benefits for the villagers and positive public awareness spin-off for the project. The government has expressed its desire to expand this system of management to four additional rangeland reserves of the MOA in the country.

PALESTINIAN AUTHORITY: The main achievements of the project were to introduce a variety of plants to improve production, minimize land degradation, and diversify feed sources. So far, the team has planted *Atriplex halimus, A. nummularia, Acacia cyanophylla, Cassia sturiti*, and cactus on 25 ha of land at three locations, which currently are perceived by local people as islands of plant conservation that are managed by farmers. The project also assists to establish wells or rehabilitate Roman cisterns.

TUNISIA: The initiative's activities in the activities on rangeland and forestry are complementary activities conducted in an area about 50 to 60 km west of Gabès. The collaboration is synergistic in that the initiative provides operational support in assisting the communities through the provision of added inputs such as 500-l water tanks (20 for 1999) and other supplies as incentives to improve the effectiveness of the Forestry Department in the MOA. In 1998/99, 300 ha of degraded rangeland were planted with adapted trees and shrubs in the Henchir Snoussi community using a participatory approach and 500,000 plants/year produced in the Oued Zayed nursery near Menzel Habib. In addition, 300 ha of olives were planted and irrigated. The farmers appreciate the olives, because even when the trees are planted at a low density, the farmers are convinced that the olives will provide better income than might be achieved with fodder shrubs alone.

---

of improving policy performance when policies are already in place and are not properly implemented. Finally, a concerted effort is being made to create a mechanism to allow national leadership to surface in the various activities of the initiative through delegation of responsibility to the people actually doing the work.

Box 71.3   Accomplishments — Marginal Water and Saline Soils

EGYPT: An integrated farm management system using drainage water, treated wastewater, and fresh water as sources of irrigation was applied to crops grown within a rotation sequence. Since this study began in 1997, the crop sequence was wheat, soybean, sugar beet, sunflower, and canola. The plots were irrigated under both surface and drip irrigation with five sources (fresh, treated, treated alternated with fresh, drainage, and drainage alternated with fresh). In addition to the large plot field trials the same crops were grown concurrently in a set of lysimeters under more controlled conditions. Five field days for farmers and four training courses for extension agents and technicians were hosted.

ISRAEL: No activity to report within the initiative. Israel has many important activities that could be incorporated into Phase II if resources were mobilized for such efforts.

JORDAN: A fodder-growing demonstration is established on a 2-ha site near the Madaba water treatment plant that previously had problems with soil management, irrigation practices, weed control, and agronomy. Cooperating farmers have now established a green cut fodder cooperative for dairy cattle feeding. The project demonstrated proper ground leveling, preparation of soil basins, and management techniques using ryegrass and sudangrass. Plots are irrigated using secondary treated wastewater. Soil moisture and chemical composition are monitored and a similar demonstration is established on a farmer's field. The main aim of the trials at both sites is to provide year-round green forage and to demonstrate better varieties, crops, and management practices to local farmers. The MOA and the Ministry of Water and Irrigation (MOWI) have agreed to cooperate formally through the initiative to fulfill the government of Jordan's policy on reuse of treated wastewater.

PALESTINIAN AUTHORITY: The Ministry of Environmental Affairs has worked diligently during the past 2 years to create a duckweed demonstration at Al-Arroub Agricultural School. The activity was expanded in the past year from a small exploratory trial, and several technical difficulties were overcome. The technical staff diagnosed and solved a problem related to high levels of biological oxygen demand in the influent wastewater that affected the growth of the duckweed. Through discussions with the World Bank modifications were introduced that solved the problem, this teamwork provided the staff with a practical learning experience. A detailed report has been prepared by the team leader summarizing the experience of the past 2 years including a survey of public attitudes concerning agricultural reuse of treated wastewater.

TUNISIA: A continuing program exists to grow tree seedlings (4000 for 1999) at the Gabès wastewater treatment plant for distribution to communities and schools. The partnership with the Association Tunisie Méditerranée Pour le Développement Durable (ATUMED), an NGO, has proved beneficial. The Commissariat Régional de Développement Agricole (CRDA) in Gabès is in the process of training young farmers who will receive land in a perimeter to be irrigated with secondary treated wastewater. A pilot tertiary treatment facility is being developed at the Gabès wastewater treatment plant to reduce pathogens and suspended solids from the treated effluent. This work is undertaken in close cooperation and cofinancing with the Office National de l'Assainissement (ONAS). Important regional workshops were held in Israel, the Palestinian Authority, and Tunisia in 1999 and 2000.

Beyond Phase II

The vision for some 10 to 15 years from now includes:

- Reduced degradation of the natural resources of the region
- Higher per capita incomes for the rural poor
- Fuller integration of all stakeholders in efforts to control desertification
- Fuller economic integration among the countries of the region by harmonization of their economic science, and technology policies
- A more deeply rooted peace process in the region

Table 71.2 The Actual Deployment of National Support Activities (NSAs) at the End of Phase I

RSP Theme	Egypt Activities 1 to 5		Israel 6 to 8	Jordan 9 to 13	Tunisia 14 to 16	Palestinian Authority 17 to 19
			National Support Activities (NSAs)			
Germ plasm for arid lands	1. Germ plasm diversification in the NWC	2. Collection in the Sinai	6. Various semiarid to arid sites north to south in the Negev	9. Medicinal plants countywide with focus at NCARTT's Mushaqqar, Maru, Ghor Safi, and Khaldieh stations	14. Zammour Valley Germ plasm resources Orchards and range	15. Khenchir Snoussi and Menzel Habib Forestry and range	17. Botanical garden near Jericho
Economic forestry and orchards	3. Fruit tree diversification and integrated watershed management in the NW coast		7. Avdat and Mashash sites in the Negev Desert	10. MOA Mastaba and Khushaibah		
				11. Wadi Araba with RSCN, CIDA, and AGRODEV		
Rangeland management	4. Range species for use in the NW Coast: "Intensification of Private Range"			12. MOA Ma'in and Faysaliya		18. Samou'a: Shrubs for rehabilitation of eastern slopes
Marginal water and saline soils	5. Kafr El Sheikh and Abou Rawash		8. Several sites in Israel (closed down in 1999)	13. MOA Madaba and Al-Samra	16. Gabès wastewater treatment (WWT) plant, schools, and nonmunicipal WWT	19. Al-Arroub duckweed pond

The Mission to Rehabilitate or Maintain Ecosystem Health

The mission of Phase II is to promote regional technical collaboration within the framework of national pilot development efforts, as an instrument to arrest natural resource degradation of arid lands in the Middle East. If successful, the pilot program will contribute to the strengthening of the peace process among all countries of the region and have a lasting effect into future generations.

Strategic Objectives

1. To focus regional cooperation more effectively on the application of available technical solutions for poverty reduction in rural areas and refugee camps as a means to reduce natural resource degradation.
2. To make use of pilot projects to demonstrate what can be accomplished through country-to-country collaboration, with the eventual goal of upscaling these projects into larger country programs, focusing the technical themes on runoff water harvesting and reuse of treated wastewater and biosolids.
3. To broaden the impact of the initiative by linking its activities to other relevant ongoing activities to combat desertification in the region, namely, the Sub-Regional Action Plan for West Asia through the United Nations Convention to Combat Desertification (UNCCD).
4. To develop an intercountry dialogue on policy issues shared among the partner countries to reduce poverty and natural resource degradation in the region.
5. To sustain and further the goals of the Multilateral Middle East Peace Process by promoting dialogue and common vision among technical people of the region.

The Structure of Phase II

The scope of activities in Phase II is more focused and institutional arrangements for implementation more clearly defined. Work is primarily focused on regional activities. On-the-ground efforts are anchored in relevant national development projects aimed at reducing rural poverty and arresting natural resource degradation. Efforts are incremental to national efforts, adding regional value to existing work. Activities have the potential for upscaling into larger investment programs, thus increasing the level of national financial resources allocated to the sectors. Phase II has three components, as discussed below.

Regional Support to National Development Programs

Each of the participating countries has a number of development activities under implementation, either ongoing or proposed. Each participating country has at least one development project to serve as the foundation for its regional component.

The regional teams of specialists, each selected by the participating countries, add incrementally to national efforts, filling specific gaps and needs for specialized technical assistance. The designated regional team works jointly with the national team, adds value to the output of the national pilot program, and contributes to its potential for subsequent upscaling. These teams share experiences, transfer state-of-the-art knowledge, and contribute to capacity building and strengthening of private and public institutions. These interactions build bridges for future expansion of collaboration, ensure sustainability, and consolidate the peace process.

Policy Fora

Efforts to counter land degradation processes cannot be achieved through technological improvements alone. Policy changes and effective participation of local communities are equally important. The capacity of national institutions to define their policy agendas and mobilize the

List 71.1 National Management Committees and National Development Projects Identified for Phase II of the *Regional Initiative for Dryland Management*

Egypt
(a) Ministry of Agriculture and Land Reclamation (MALR), including the Agricultural Research Center (ARC) and the Desert Research Center (DRC); Egyptian Environmental Affairs Agency (EEAA); Ministry of Housing (Authority for Sewage Water); Land Reclamation; Private Sector; and the University of Zagazig.
(b) The Matrouh Resource Management Project on the northwest coast of Egypt is a large development project sponsored by the World Bank and the government of Egypt. Additional collaboration exists with the On-Farm Water Management (OWSOM) Project in Kafr El-Sheikh.

Israel
(a) The Blaustein Institute for Desert Research (BIDR); Ministry of Agriculture and Rural Development (MA&RD); Ministry of Environment (MOE); Keren Kayemet Le'Israel, also known as the Jewish National Fund (JNF); and Mashav (Ministry of Foreign Affairs).
(b) The Yatir Forest in Israel, a site of afforestation in the semiarid drylands in the northeast Negev planted by the JNF (~250 mm annual precipitation). The work is complemented by training and demonstration sites at the Evenari experimental farm at Avdat, a resurrected ancient Byzantine farm in a hyperarid zone in the highlands of the central Negev (~90 mm annual precipitation) and Mashash Farm, a runoff farming research site. Wastewater treatment and biosolids reuse takes place at Hamaapil on the coastal plain and Ruchama, Nirim, Nir-Am, and Shuval in the western Negev in association with the MA&RD and MOE.

Jordan
(a) Directorate for Rangeland, in the Ministry of Agriculture; The National Center for Agricultural Research and Technology Transfer (NCARTT); NGO — The Royal Society for the Conservation of Nature (RSCN); The Agricultural Union in Madaba; and the Women's Union in Madaba.
(b) An area east and west of Madaba serves as the base for work in Phase II. The CIDA/AGRODEV/MOA/RSCN Sustainable Rangeland Management Project, supported by the government of Canada, has been central to the achievement attained in Phase I, and it is planned to reinforce the Range Department of the MOA in the years ahead. A Global Environment Facility (GEF) project for herbal and medicinal plants is the national development project located in Wadi Mujib that will be implemented by NCARTT in cooperation with the MOA and RSCN.

Palestinian Authority
(a) Ministry for Environmental Affairs (MENA); Ministry of Agriculture (MOA); The Water Authority (PWA); The Palestinian Hydrology Group (PHG); NGO — Palestinian Agricultural Relief Committee (PARC); and An'Njah and Hebron Universities.
(b) Principal focus will be the Qa'abnah site, an area with customary rights held by the Arab al-Kaabneh tribe, working in collaboration with the MOA/UNDP Rangeland Project; the MOA/UNDP Agro-Biodiversity Project; the MOA/UNDP Land Reclamation Project; and the MOA Green Palestine Project funded by the Dutch. The location of the treated wastewater and biosolids reuse work will be in association with the Gaza Wastewater Treatment Plant in Wadi Gaza and the Al-Bireh Wastewater Treatment Plant in the West Bank.

Tunisia
(a) From the Tunisian Ministry of Environment (Ministere de l'Environnement et de l'Aménagement du Territoire (MEAT)): Centre International des Technologies de l'Environnement de Tunis (CITET) serving as the Focal Point and the Office National de l'Assainissement (ONAS). From the Ministry of Agriculture (MOA): Commissariat Régional de Développement Agricole (CRDA) in Gabès; Direction Générale des Forêts (DFG); Institut National de Recherches Genie Rural Eaux et Fôret (INRGREF). NGO partnerships: Association Tunisie Mediterranée Pour le Développement Durable (ATUMED) in Gabès.
(b) The Menzil Habib project site covers about 300 ha, 50 to 60 km west of Gabès, and is traditionally owned by the Henchir tribe of the Snoussi confederation. A degraded rangeland and agricultural site was protected and planted with adapted trees and shrubs in Phase I. For treated wastewater, the Dissa Project irrigated site covers 400 ha near Gabès. All farmers are identified and participate in the implementation of the project.

ᵃ Institutional composition of national management committee
ᵇ National development projects

necessary means to implement these policies is one of the most critical factors that will affect the future of the region. Key socioeconomic policy issues, such as those bearing upon off-farm determinants of the rural poor's access to resources, markets, and product prices, are key to the management of natural resources. Equally, the organizational and policy practices of relevant institutions, and the nature of the institutions that administer the arid lands, represent an important domain for contributing to the goals of the initiative. Policy dialogues among institutions operating in the region are developing a better understanding of the issues surrounding the management of arid lands. Priority attention is given to issues identified by the partner countries, such as exchange of experiences related to land management by Bedouin, and standardization of legal frameworks for treated wastewater and biosolids reuse.

Capacity Building

Given the young age of its institutions and the importance of the Palestinian Authority in the peace process, the objective of this component is to provide capacity building activities, primarily in support of Palestinian institutions responsible for management of its arid lands. Activities provide financing that leverages contributions by other programs executed by bilateral and multilateral institutions operating within the Palestinian territories.

The approach to be deployed under this component is to identify critical gaps in existing programs that support capacity building in the pertinent institutions. Identification of these gaps is accomplished by the national steering committee, which is composed of ministries, nongovernment organizations, and universities with the objective of identifying the most rational deployment of the initiative's limited resources in areas with the highest payoff. The facilitation unit cooperates with the respective ministries to convene consultation meetings to identify the critical gaps and the modalities for addressing them. With these determined, the facilitation unit establishes contacts with participating countries to identify their potential support to the respective institutions in the Palestinian Authority.

CONCLUSION

The initiative successfully carried out activities during a transitional period from October 1999 to July 2000. Phase II is under way with the involvement of the institutions identified in List 71.1. Regional activities are anchored in selected national development programs that address natural resource management issues within the framework of rural poverty reduction and promote dialogue on common technical, scientific, and economic policy issues that are critical to natural resource management in the region. The facilitation unit assists regional cooperation for capacity building, with particular emphasis on the needs of the relevant institutions of the Palestinian Authority. The facilitation unit also assumes responsibility for monitoring activities and organizing periodic reviews and thematic evaluations. The World Bank assumes leadership for mobilizing the necessary resources from current and potential donors and playing an active role in linking Phase II efforts to investment projects in the partner countries. Emphasis has shifted from research to development, focusing on a strategy that allows for the introduction of rainfall runoff farming and reuse of treated wastewater and biosolids as core elements of the technical package in each of the national development projects. These themes in turn provide the platform in each country upon which incremental regional work can be programmed and financed through the initiative.

CHAPTER 72

Biological Conservation in Mexico: An Overview

Exequiel Ezcurra

INTRODUCTION

The Mexican highlands, and especially the Basin of Mexico, are among the regions of the world that have suffered some of the greatest environmental alterations in history. When the Europeans arrived in Mexico in 1519, the region was already greatly transformed by agriculture, water projects constructed for the development of *chinampa* agriculture, deforestation, and faunal extinction (Ezcurra, 1992). The rate of environmental change, however, has increased continuously since then, and, with the growing ecological deterioration, the awareness of the problem by different social groups has increased accordingly.

The development of the Spanish colony in Mexico hinged on an intentional transformation of the environment to adapt it to the introduced European crops and animals, and to new means of transportation. Toward the end of the 18th century, Alexander von Humboldt, in the *Political Essay on the Kingdom of the New Spain*, expressed serious concern about the future effects of the rapid deforestation that was taking place in the slopes of the Mexican highlands. In the early 20th century, Miguel Ángel de Quevedo campaigned for the planting of trees in Mexico and predicted that large-scale environmental degradation would ensue if deforestation continued at the rapid rate it was undergoing at that time. Unfortunately, Humboldt's and Quevedo's fears proved to be correct. Environmental degradation progressed rapidly during the 20th century, and induced a growing series of responses from both government and nongovernmental organizations (NGOs). Conservation in Mexico is especially important in the context of developing countries, because the country harbors an immense biological diversity, and also because Mexico's industrial growth is typical of many of Latin America's emergent economies (Williams-Linera et al., 1992).

BIOLOGICAL CONSERVATION: A GROWING CONCERN

In spite of the early and dire forecasts of Humboldt, Quevedo, and many other early visionaries, general consciousness about the problem of environmental degradation began to expand only in the late 1970s. The economic development of Mexico during the mid-20th century was conducted essentially with nonsustainable use of natural resources, and with no awareness of the negative environmental consequences that would follow. In the Mexican tropics for example, programs were developed to slash tropical forests and to transform them into grasslands. A national commission promoting deforestation (*Comisión Nacional de Desmontes*) existed within the Secretariat of Agriculture until the late 1970s. Actions were driven by the need to industrialize and develop, with few restrictions and no planning for a sustainable future.

GOVERNMENTAL ACTIONS FOR PROTECTION OF BIOLOGICAL DIVERSITY

By the mid-1970s many Mexicans became aware of the accumulation of environmental problems that needed to be confronted. The conclusions of the 1972 U.N. Environmental Conference and the attention drawn by some books, such as Rachel Carson's *Silent Spring,* confirmed the truth of the growing perception that the environment was quickly deteriorating. During the 1970s both the governmental authorities and Mexican society started to react — first around issues of urban pollution, and later with a growing concern for deforestation and biological extinction. Simultaneously, governmental reports and regulations started to emerge, with the aim of curbing the perceived environmental degradation (e.g., DOF, 1971, 1984, 1988; CONADE, 1992).

In 1992, the United Nations Conference on Environment and Development (UNCED) took place, and the Mexican government participated actively in the preparatory meetings of the previous years and in the final writing of Agenda 21. Following the paradigm of sustainable development that UNCED attempted to implement at a global scale, the Mexican Federal Administration started to regard environmental degradation as a problem related to social and economic development, linked to the problem of poverty and lack of social progress. Also as a result of UNCED, the federal government took an active interest in the protection of the natural heritage of Mexico, and created the National Commission for the Knowledge and Use of Biological Diversity (CONABIO) in early 1992, an organization devoted to the survey of the biological richness of Mexican ecosystems. The federal environmental authority (now known as *Secretaría de Medio Ambiente, Recursos Naturales y Pesca,* or SEMARNAP) also started to implement environmental projects with the World Bank, including a donation from the Global Environmental Facility (GEF) to support the operation of ten priority Mexican Biosphere Reserves (BRs).

In that same year, Article 27 of the Mexican Constitution (see *Constitución Política de los Estados Unidos Mexicanos,* 1995) was modified, ending the process of agrarian reform and land allotment — and opening the possibility of ending the communal ownership of the land under the *ejido* (communal land) system for those *ejidos* and indigenous communities wishing to do so. The overexploitation of natural resources in excessively subdivided communal lands was one of the arguments put forth. Although the new version of Article 27 states that one of its objectives is to preserve the environmental balance, Mexican conservation groups are concerned that this constitutional change has exerted pressures to privatize and sell lands that are potentially valuable and still well preserved, such as tourist areas or pristine forests ecosystems.

The federal environmental authority proposed, among other things, to decentralize decision making in environmental matters. Many attributions that were previously held by the federal government in Mexico City were passed to the state and municipal governments (INAP, 1991). In the early 1990s, many state governments enacted their own environmental laws and regulations, and some started to operate programs that had been previously operated by the federal administration (SEDESOL, 1994b). As an example, during this period the state of Campeche started managing the natural protected areas within its territory, including Calakmul, a federal BR.

The new emphasis led to the publication of 58 official standards dealing with different aspects of the environment, including Mexico's first comprehensive listing of endangered species (SEDESOL, 1994a). The publication of these standards opened for the first time the possibility of governmental control on the trade of endangered species within Mexico, as a complement to the obligations assumed in the international sphere with the signature of the Convention on International Trade in Endangered Species of Wild Flora and Fauna (CITES). These standards also established a formal regulatory framework for environmental impact studies and ecological planning. Additionally, these new regulations brought into the governmental discourse the use of new analytical techniques in decision making and priority setting, such as risk assessment and cost–benefit analysis.

In December 1994, shortly after assuming office, the Zedillo administration created a new secretariat (DOF, 1994) to encompass all federal environmental functions, including those dealing with brown ecology (environmental pollution), and green ecology (natural resource management). The Secretariat of Environment, Natural Resources and Fisheries (SEMARNAP) included functions that were previously dispersed in different federal agencies, including the protection and management of natural resources such as water, forests, and fisheries; waste management and pollution control; the management of national parks and other protected natural areas; and environmental law enforcement. Deconcentrated governmental agencies dealing with environmental matters all shifted to the jurisdiction of this new federal secretariat. The creation of this strong ministry reflects the ever-increasing importance that federal authorities are now giving to the environmental problems of Mexico. It also underscores the perceived need to institutionalize sustainable development initiatives by integrating governmental activities that may affect the environment and that were previously separated, such as forestry, fisheries, and conservation, or energy use, transportation, and urban design (DOF, 1996).

Furthermore, in this new organization, the management of natural protected areas became separated from the management of wildlife, and acquired a higher status as a coordination unit in 1996, and as a national commission in 2000. SEMARNAP faced its great responsibilities and the financial difficulties engendered by the policies of governmental reduction established in 1995 by looking for even stronger programs of decentralization than those implemented by the *Secretaría de Desarrollo Social* (SEDESOL) on the one hand, and by trying to develop alliances with NGOs and the private sector to manage specific aspects of the environmental policies on the other. An ambitious program of decentralization of governmental functions to the states was discussed and established in the early stages of SEMARNAP's creation (SEMARNAP, 1996a, c). More specifically, a program was conceived by SEMARNAP to decentralize the administration of the National Parks, and to cooperate with NGOs and research institutions in the management of BRs (SEMARNAP, 1996b, c). For this purpose, SEMARNAP created an independent Advisory Council on Protected Areas, and worked closely with NGO representatives to convert the GEF $25 million U.S. donation to Mexico into a private trust fund.

NONGOVERNMENTAL ORGANIZATIONS AND THE PROTECTION OF BIOLOGICAL DIVERSITY

NGOs also have changed rapidly and profoundly during the last three decades (Sandoval and Semo, 1985; Ezcurra et al., 1999). Although much of the ecological activism in the 1960s and 1970s hinged on the growing urban problems of Mexico City, some groups started to work from the late 1970s in the protection of biological diversity and opposing the growing deforestation of the Mexican tropics. Under the direction of Gonzalo Halffter, the Instituto de Ecología (Institute of Ecology) — a research center — started promoting the concept of BRs in Mexico. While fairly accepted at present, the idea, which had been developed by a group of ecologists in United Nations Educational, Scientific, and Cultural Organization's (UNESCO) Man and the Biosphere Program (MAB), was radically new in 1975. Biosphere reserves were conceived as natural protected areas where the indigenous populations living inside the area or in the surrounding buffer zones were encouraged to use their natural resources in a sustainable manner. Instead of following the paradigm of the natural park concept, in which protected areas were conceived as pristine areas free of human influence, the BR concept involved the maintenance of the local populations under a sustainable management plan. Instead of opposing conservation against human use of the resources, BRs promoted the idea of sustainable use. The Mexican BRs were among the world's first to become operational, and became integrated into the MAB network of BRs. Many of the concepts discussed in UNCED in 1992 were already operational almost 20 years ago in the MAB concept of BRs:

1. The need to protect biodiversity globally through a network of protected areas
2. The urgency to preserve cultural diversity together with natural diversity
3. The demand to integrate local populations in the protection of biological diversity
4. The promotion of sustainable use of natural resources.

Although the Institute of Ecology is an academic research center and, strictly speaking, not an NGO, its work opened the way for many experiences developed later by NGOs along these basic guidelines.

Similarly, in the 1970s, another research center was created in Xalapa, Veracruz, devoted to research on the use of biological natural resources. The *Instituto de Investigaciones de los Recursos Bióticos* (INIREB; Institute of Research on Biotic Resources) was founded and directed by Arturo Gómez-Pompa, a leading Mexican botanist. Apart from its outstanding research, INIREB played a very important role in drawing the attention of researchers and conservationists to the problem of deforestation in the tropics, and especially toward the logging of the Uxpanapa rain forests. The pioneer work of INIREB opened the way for many NGOs to become conscious of the problem of deforestation in the Mexican tropics, and to organize actions around this problem.

The National University of Mexico (UNAM) also played a very important role in the growth of ecological awareness in Mexico. In the 1970s the Institute of Biology started to manage a tropical rain forest field station at Los Tuxtlas, in Veracruz, followed shortly thereafter by management of a tropical dry-forest field station and nature reserve at Chamela, on the Pacific coast of the state of Jalisco (e.g., González-Soriano et al., 1997). From the Institute of Biology a new Institute of Ecology grew in the late 1980s. Together with the Faculty of Sciences of UNAM these three organizations established a kernel of ecological research and conservation that gave rise to myriad academic organizations in the country interested in ecological conservation.

Toward the end of the 1980s, all political parties in Mexico had included environmental issues in their platforms. By that time, many environmentalist NGOs had merged into a coalition known as *Pacto de Grupos Ecologistas* (Pact of Ecological Groups). Around this time, local chapters of international NGOs were also established in the country, some — such as Greenpeace and Oilwatch — were devoted to the surveillance and monitoring of the attainment of environmental regulations and ecological protection. Other organizations — such as the World Wildlife Fund (WWF) and Conservation International (CI) — were dedicated to the conservation of nature and the protection of biodiversity. These organizations are contributing significantly to link the Mexican environmental groups into the global international perspective.

Environmentally motivated groups have continued to manifest themselves in rural Mexico with an ever-increasing momentum. In 1992, the National Commission on Human Rights (CNDH) published its *Recomendación 100*, calling for federal and state authorities to take urgent action to solve the critical environmental problems confronted by the *campesinos* (peasants) in the Tabasco lowlands, most of them caused by the oil industry. In many parts of Mexico, indigenous groups have started to claim the management of their own natural areas. For this purpose, they have made alliances with urban conservationist organizations, and even with international organizations. Examples of this trend are the *Consejo Regional de X'Pujil* in Campeche, and the *Comité para la Defensa de los Chimalapas* in Oaxaca, among many others. Even the traditional urban environment groups have extended their interests into the conservation of biological diversity in natural ecosystems, bringing the concept of biodiversity into the urban agenda. The *Asociación Ecológica de Coyoacán*, for example, now also hosts a group known as *Comisión de Selvas Mexicanas* (Commission for Mexican Tropical Forests). The *Grupo de los 100* (Group of 100) is a very strong opinion group headed by Homero Aridjis and supported by a hundred leading artists and intellectuals. Although it is still concerned with the environment in general, the organization is now more concerned about conservation issues such as whales, marine turtles, the trade of endangered species, monarch butterflies, and the management of BRs. Additionally, numerous strong grassroots groups working for the conservation of biological diversity have become established in recent years. Some older NGOs based in Mexico City have established regional chapters, working intensely on local problems. In doing so, these NGOs have decentralized much in the same way as the federal government is trying to do, acquiring in the process a renewed grassroots emphasis.

Other initiatives also have prospered vigorously. Supported by international organizations such as the WWF, The Nature Conservancy, and CI, a trust fund was launched in 1993 with the name *Fondo Mexicano para la Conservación*. This fund has received endowments from U.S. Agency for International Development (USAID) Parks in Peril program, from the Mexican federal government, from the GEF, and from Mexican private funds. Together with other NGOs, the *Fondo Mexicano* has given its advice and support in the actions currently taking place to convert the GEF donation to Mexico into a private trust fund. In 1997, the *Fondo Mexicano* was able to establish a specific fund to support the functioning of 10 selected biosphere reserves. This fund, called *Fondo de Áreas Naturales Protegidas*, operates as a private trust fund the GEF's donation discussed in the previous section. For the first time in Mexico, the NGOs started in the late 1990s to play a central role in the financing of conservation efforts and biosphere reserve management.

THE GROWING ROLE OF CONABIO

From the point of view of biological diversity, the creation of CONABIO can be regarded as a true landmark. CONABIO started funding research on biological diversity and promoting the use of biological collections for decision making. Myriad studies were funded to understand the importance of some crucial taxa, and a growing list of publications started to emerge with information and results from CONABIO's impressive portfolio of projects (see CONABIO Bibliography).

Within these projects, large-scale efforts have been done at a national level to map biodiversity for selected taxa of importance to conservation, such as terrestrial mammals or columnar cacti (Figure 72.1a,b). The resulting maps have highlighted the distribution of hot spots for different taxa

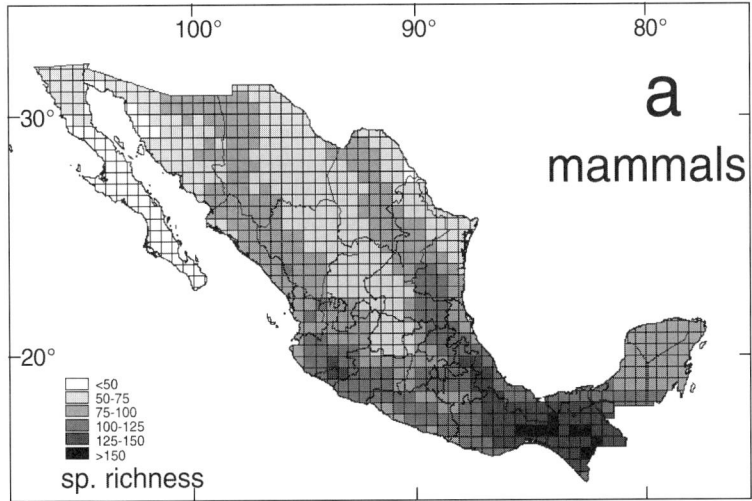

Figure 72.1a Species richness patterns of selected Mexican taxa, developed through CONABIO-funded studies. Mammals (developed by Hector Arita and available at the CONABIO Web site).

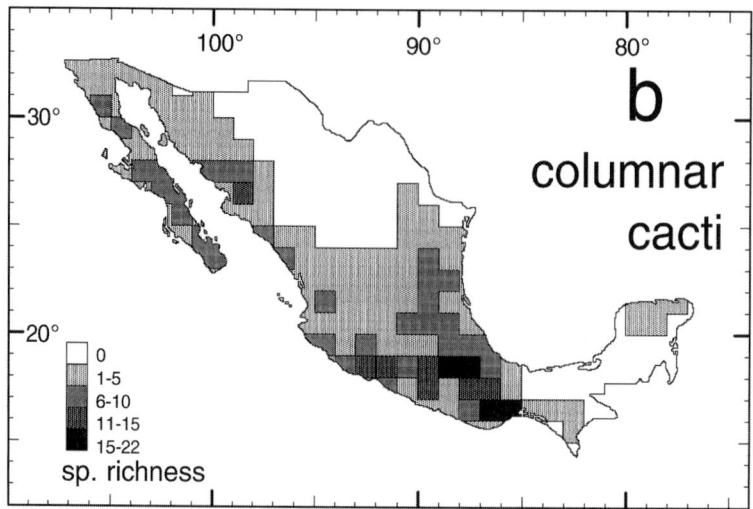

Figure 72.1b Species richness patterns of selected Mexican taxa, developed through CONABIO-funded studies. Columnar cacti (Cactaceae, tribe Pachycereeae [Mourelle, 1997; Mourelle and Ezcurra, 1997]). Note the similarity of the patterns and distribution of the high-diversity corridors in 72.1a and 72.1b.

Figure 72.2 Map of priority areas for conservation in Mexico, developed by CONABIO in 1997, with support from PRONATURA, the World Wildlife Fund, the *Fondo Mexicano para la Conservación de la Naturaleza* (Mexican Fund for the Conservation of Nature), The Nature Conservancy, the USAID program, and the National Institute of Ecology (INE-SEMARNAP). This map has been instrumental in defining new conservation policies and establishing new protected areas.

within Mexico, and have led efforts to protect the natural environment in a more effective manner. Based on the emerging results from these largely academic exercises, and also on the experience of hundreds of field biologists and conservationists, CONABIO has been working to identify priority areas for conservation, both in marine and the terrestrial environments, by organizing expert workshops for the identification of hot spots, for the identification of areas providing critically important ecological services, and for setting conservation priorities (Figure 72.2). These immense undertakings have been tackled with the active participation of conservation groups such as PRONATURA, WWF, The Nature Conservancy, or the *Fondo Mexicano para la Conservación de la Naturaleza* (Mexican Fund for the Conservation of Nature). This latter funding agency, in particular, is directing all its financing priorities toward the areas identified through the CONABIO planning exercises.

CONCLUSION

Environmental destruction is still a major problem in Mexico, and biodiversity loss is likely to remain a major problem in decades to come. However, a rapid response is occurring both at the governmental and at the nongovernmental level. Funding agencies that have emerged help ensure the continuous financing of conservation across the changing federal administrations. Biodiversity protection is now part of the jargon of political parties and governments, and it ranks high among the priorities of NGOs and grassroots organizations.

The problem of biodiversity loss should still rank high among the unresolved problems of Mexico, but mechanisms are being established to approach it in a scientific and well-informed manner. Among these mechanisms, the GEF-endowed *Fondo de Áreas Naturales Protegidas* has ensured continuity of funding for critically important biosphere reserves with independence from potential policy changes in the federal government. Additionally, CONABIO has developed an impressive system of biodiversity information and a large portfolio of funded projects. CONABIO's outstanding work has generated a growing series of online accessible databases, documents, ecoregional maps, and applied studies that have been able to generate an ever-improving theoretical framework to prioritize conservation efforts. The results of these projects are in turn starting to guide and influence public policies, as well as help generate funding for new protected areas.

REFERENCES

CONADE, Informe de la Situación General en Materia de Equilibrio Ecológico y Protección Ambiental 1989–1990, Comisión Nacional de Ecología, México, 1992.

Constitución Política de los Estados Unidos Mexicanos, Editorial Porrúa, México, 1995.

DOF, Ley de Protección al Ambiente, *Diario Oficial de la Federación*, México, Mar. 23, 1971.

DOF, Ley Federal de Protección al Ambiente, *Diario Oficial de la Federación*, México, Jan. 27, 1984.

DOF, Ley General del Equilibrio Ecológico y la Protección al Ambiente, *Diario Oficial de la Federación*, México, Jan. 28, 1988.

DOF, Decreto que reforma, adiciona y deroga diversas disposiciones de la Ley Orgánica de la Administración Pública Federal, *Diario Oficial de la Federación*, México, Dec. 28, 1994.

DOF, Reglamento Interior de la Secretaría de Medio Ambiente, Recursos Naturales y Pesca. *Diario Oficial de la Federación*, México, July 8, 1996.

Ezcurra, E., Crecimiento y colapso en la Cuenca de México, *Ciencias*, 25, 13–27, 1992.

Ezcurra, E. et al., *The Basin of Mexico. Critical Environmental Issues and Sustainability*, The United Nations University Press, Tokyo, 1999.

Gonzàlez-Soriano, E., Dirzo, R., and Vogt, R.C., Eds., 1997. *Historia Natural de Los Tuxtlas*, Instituto Biología, UNAM, Instituto de Ecología, UNAM, and CONABIO, México, 1997.

INAP, *El Municipio y la Ecología*, Gaceta Mexicana de Administración Pìblica Estatal y Municipal Nos. 39, 40, 41, Instituto Nacional de Administración Pública, México, 1991.

Mourelle, C., Biodiversidad de la familia Cactaceae: Un enfoque biogeogràfico, Ph.D. thesis, Centro de Ecología-Colegio de Ciencias y Humanidades, Universidad Nacional Autónoma de México, México, 1997.

Mourelle, C. and Ezcurra, E. Rapoport's rule: a comparative analysis between South and North American columnar cacti, *Am. Nat.*, 150, 131–142, 1997.

Sandoval, J.M. and Semo, I. Los movimientos sociales del ecologismo en México, in *México: Biosociología*, Programa Universitario Justo Sierra, UNAM, México, 1985.

SEDESOL, *Normas Oficiales Mexicanas en materia de protección ambiental*, Instituto Nacional de Ecología, Secretaría de Desarrollo Social, México, 1994a

SEDESOL, Informe de la Situación General en Materia de Equilibrio Ecológico y Protección al Ambiente 1993–94, Instituto Nacional de Ecología, Secretaría de Desarrollo Social, México, 1994b.

SEMARNAP, Estrategia del Proyecto de Descentralización de la SEMARNAP, Primera versión para discusión, Internal unpublished document, 1996a.

SEMARNAP, *Programa de Áreas Naturales Protegidas de México 1995–2000*, Instituto Nacional de Ecología, Secretaría de Medio Ambiente, Recursos Naturales y Pesca, Poder Ejecutivo Federal, México, 1996b.

SEMARNAP, *Programa de Medio Ambiente 1995–2000*, Instituto Nacional de Ecología, Secretaría de Medio Ambiente, Recursos Naturales y Pesca, México, 1996c.

Williams-Linera G., Halffter, G., and Ezcurra, E., El estado de la biodiversidad en México, in *La Diversidad Biológica de Iberoamérica*, Halffter, G., Ed., Instituto de Ecología – CYTED-D, Xalapa, Veracruz, 1992, pp. 285-312.

CONABIO BIBLIOGRAPHY

Publications edited by CONABIO between 1995 and 1998. The list does not include the many publications derived from CONABIO-funded projects and published in other sources.

Allen, G.R. and Robertson, D.R., *Peces del pacífico oriental tropical*, Agrupación Sierra Madre, Cementos Mexicanos, and CONABIO, México, 1998.

Arriaga-Cabrera, L., Vázquez-Domínguez, E., González-Cano, J., Jiménez-Rosenberg, R., Muñoz-López, E., and Aguilar-Sierra, V., Eds., *Regiones prioritarias marinas de México*, CONABIO, México, 1998.

Arriaga-Cabrera, L., Aguilar-Sierra, V., Alcocer-Durán, J., Jiménez-Rosenberg, R., Muñoz-López, E., Vázquez-Domínguez, E., Eds., *Regiones hidrológicas prioritarias: fichas técnicas y mapa*, CONABIO, México, 1998.

Benítez Díaz, H., Vega López, E., Peña Jiménez, A., and Avila Foucat, S., Eds., *Aspectos económicos sobre la biodiversidad de México*, Instituto Nacional de Ecología-SEMARNAP, CONABIO, México, 1998.

Challenger, A., *Utilización y conservación de los ecosistemas terrestres de México: pasado, presente y futuro*, Instituto de Biología, UNAM, CONABIO, and Agrupación Sierra Madre, México, 1998.

CONABIO, *Guía de aves canoras y de ornato*, Instituto Nacional de Ecología, SEMARNAP, and CONABIO, México, 1997.

CONABIO, *Suculentas mexicanas: Cactáceas*, CVS Publicaciones, S.A., and CONABIO, México, 1997.

CONABIO, *La diversidad biológica de México: Estudio de país*, CONABIO, México, 1998.

de la Cruz Agôero, G., *Catálogo de los peces marinos de Baja California Sur*, Centro Interdisciplinario de Ciencias Marinas, IPN, and CONABIO, México, 1997.

Espejo Serna, A. and López Ferrari, A.R., *Las monocotiledóneas mexicanas. Una sinopsis florística*, Universidad Autónoma Metropolitana-Iztapalapa and CONABIO, México, 1998.

Galindo-Leal, C. and Weber, M., *El venado de la Sierra Madre Occidental: Ecología, manejo y conservación*, Ediciones Culturales, S.A. de C.V. and CONABIO, México, 1998.

Glass, C.E., *Guía para la identificación de cactáceas amenazadas de México/Identification guide to threatened cacti of Mexico*, Cante, A.C. and CONABIO, México, 1998.

Gonzàlez-Soriano, E., Dirzo, R., and Vogt, R.C., Eds., *Historia Natural de Los Tuxtlas*, Instituto Biología, UNAM, Instituto de Ecología, UNAM, and CONABIO, México, 1997.

Guzmàn, G., *Los nombres de los hongos y lo relacionado con ellos en América Latina*, Instituto de Ecología, A.C., and CONABIO, México, 1997.

Llorente Bousquets, J.E., García Aldrete, A.N., and González Soriano, E., Eds., *Biodiversidad, taxonomía y biogeografía de artrópodos de México: Hacia una síntesis de su conocimiento*, Instituto de Biología, UNAM, and CONABIO, México, 1996.

Llorente Bousquets, J.E., Oñate-Ocaña, L., Luis-Martínez, A., and Vargas-Fernández, I., (Illustrations by Pál János) Eds., *Papilionidae y pieridae de México: Distribución geográfica e ilustración*, Facultad de Ciencias, UNAM, and CONABIO, México, 1997.

Medellín, R.A., Arita, H.T., and Sànchez, O., *Los murciélagos de México: clave de campo*, Asociación Mexicana de Mastozoología, and CONABIO, México, 1997.

Sada de Hermosillo, M.L., López de Mariscal, B., and Sada de Rosenzweig, L., *Guía de campo para las aves de Chipinque*, CONABIO, México, 1995.

Simonian, L., *La defensa de la tierra del jaguar*, Instituto Nacional de Ecología — SEMARNAP, Instituto Mexicano de Recursos Naturales Renovables, and CONABIO, México, 1999.

Vidal, R.M. and Macías, C., *Aves de los altos de Chiapas: canto, color y tradición*, Pronatura Chiapas, A.C., and CONABIO, México, 1997.

CHAPTER 73

Environmental Impacts of Mobility and Urban Development: A Case Study of the Brussels-Capital Region

Paul Safonov, Vincent Favrel, and Walter Hecq

INTRODUCTION

Aggravation of traffic intensity and air pollution in urban areas requires strategic environmental assessment (SEA) of transportation networks and the integration of land-use planning with transport–environment concerns.

Road traffic in the Brussels-Capital region has increased continuously during the last decade. The reasons for this trend lie mainly in the urban exodus, the growth of employment in the Brussels area and its peripheral region, and the constantly increasing population motorization rate. Recent studies (Plan IRIS, 1989–1995) predict the complete saturation of the road network before the year 2005 in the European capital.

The assessment of the impacts of this road traffic on the environment in general, and on air quality, in particular, helps not only to evaluate the actual situation but also the possible effects of measures toward a more sustainable transport system as an important element of regional sustainable development.

Transport policies alone will not lead to a sustainable level of fuel consumption and emissions. The need to coordinate land-use changes with transport measures in order to achieve an environmentally and economically sustainable transport system is eminent. Transport and land use should be considered as an integrated system. However, as Lobe and Duchâteau (1998) advise, in many urban areas this system resembles a spiral of urban decline, caused by potentially vicious interactions of this system at an urban level.

This chapter presents a study being undertaken in the Center for Economic and Social Studies on the Environment (CEESE) of Université Libre de Bruxelles (University of Brussels, ULB) within a project to analyze the ecological aspects of mobility induced by major regional policy for the case study of Brussels-Capital region (Safonov, 2000). Recognizing the importance of different environmental effects of mobility (such as problems of noise, vibration, and smell, which would stay outside the framework of this study), this project is focused on the assessment of air pollutants emissions [carbon dioxide (CO_2), carbon monoxide (CO), nitrogen oxides (NO_x), sulfur dioxide (SO_2), volatile organic compounds (VOC), particulate matter (PM), as well as others] and the consumption of nonrenewable fuel.

While considering the impacts of air pollution (e.g., in terms of building deterioration, health effects, climate change), the general approach ties to traffic a sequence based on the five steps: human activities, emissions, immissions (pollution concentrations), physical impacts, and external costs. This is the classical approach used in major studies, such as ExternE (European Commission, 1995), for the assessment of externalities in the energy sector, which has recently been updated for its application to the transport sector.

The new model is being synthesized to analyze the influence of recent urban and transport policies on the mobility of people in an urban area. The environmental impacts of such mobility are analyzed on the basis of the information available for the Brussels-Capital region. An integrated modeling approach is proposed, including analysis of socioeconomic factors of mobility, road network assignment model, and estimation of air pollution from traffic. The first version of the model linking urban mobility and emissions from private vehicles is developed using the Professional's Transportation Planning Package (TRIPS, 1999) for network analysis. For possible scenarios of mobility according the policies defined in the Brussels Regional Mobility Plan (Plan IRIS, 1989–1995) emissions of different air pollutants and fuel consumption were projected and evaluated for the year 2005 as a result of road traffic modeling.

DEVELOPMENT OF THE INTEGRATED MOBILITY MODEL FOR THE BRUSSELS-CAPITAL REGION

Conceptual Structure of the Model and Steps of Implementation

Based on existing experience in modeling different socioeconomic issues of mobility, and its environmental impacts (see Favrel and Hecq, 1998, 2000; and Safonov, 2000; for literature survey), an integrated system of models (Figure 73.1) is proposed, which includes several main components:

1. Urban development model. This will include forecasts of population and employment dynamics, in accordance with different economic and urban development scenarios. Such forecasts should be based on indicators of economic development by main sectors of activities and the respective demand for labor resources, as well as trends in population dynamics and labor resources structured with respect to spatial distribution in the region. Office stock dynamics and other urban factors are also considered.

2. Mobility model. The model provides scenarios of road traffic in the region, according to different origin–destination matrices, generated on the basis of the different urban development scenarios. For transportation network analysis the recent version of TRIPS package (TRIPS, 1999) is used, which provides powerful tools for assignment and graphical presentation.

3. A model linking mobility and air pollution. The focus is mainly on local pollution of CO, NO_x, VOC, and PM. Also consumption of nonrenewable fuels with respect to CO_2 and SO_2 emission is analyzed. The Computer Program to Calculate Emissions from Road Transport (COPERT) methodology (Ahlvik et al., 1997) was used for calculation of the emission functions per kilometer driven, taking into consideration climate, composition of the private car fleet with specific speed profiles, as well as new European/Belgian regulations on vehicles, integrated in COPERT Version III. The emissions are calculated spatially on the transportation network as a function of the assigned traffic intensity, average speed on each link, and length of the trip.

4. In the final stage, as part of the Sustainable Mobility in the Brussels-Capital Region project CEESE-ULB is currently developing a methodology for the assessment of the physical effects and external costs caused by air pollution generated by road traffic in an urban area (Favrel and Hecq, 1998, 2000).

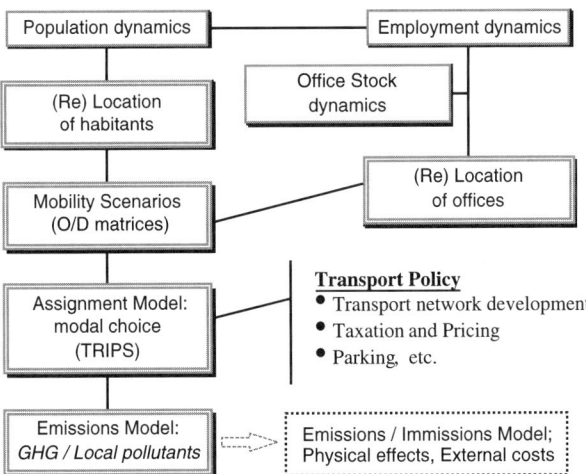

Figure 73.1 General structure of the system of models.

The data necessary for identification of the proposed complex of models are available from various sources, but an in-depth study of the recent literature with a following update of the information base is required along with some additional surveying. Main sources of the data include the National Statistical Institute, administration of the Brussels-Capital region (e.g., Regional Mobility Plan IRIS reports), Brussels office survey (Jones Lang Wootton reports), review of office property (by Brussels-Capital region), and others.

The implementation of the integrated model is planned within several phases. At the initial stage, the mobility demand-related information (socioeconomic, office and dwellings dynamics forecasts) is mainly obtained from the results of previous studies (Plan IRIS, 1989–1995). This enables us to analyze environmental impacts of mobility scenarios within existing urban policy options, along with some additional considerations, such as dynamics of vehicle fleet composition.

As one of the first steps in development of the proposed integrated approach, we have designed a model that allows the assessment of the contribution of mobility by private cars to emission of the air pollutants in the Brussels area. The software was developed to link the mobility model, road traffic assignment, and emissions calculations (Safonov, 2000).

Below we describe the recent situation and the main considerations for the envisaged model design and scenario analysis, and already implemented developments within the proposed modeling approach.

Urban Development

The population of the Brussels metropolitan area has been slowly but regularly decreasing. Households migrating to the close periphery are mainly middle- and upper-class families with a high car ownership level. The decrease in the Brussels-Capital region has been much faster, with an important consequence: the main part of the region's financial resources comes from the taxes levied on its inhabitants' income, and these resources are quickly diminishing as a result of both the decrease of the population and also of the average household income. Development in the suburbs cannot be controlled by the government of the Brussels region, because it is located outside of its borders, nor by the federal government, because the land-use planning jurisdiction has been fully decentralized.

During the last 20 years of the 20th century, the economy of the region has undergone a significant mutation. Initially an industrial center, Brussels progressively became an important administrative center and has recently been designated as the European Union capital. Many

Table 73.1 Share of Brussels in the National Economy (%)

	1980	1990	1997
Share of Brussels in the gross national product	15.5	13.4	12.6
Part of Brussels in the global amount of taxable revenue of individuals of the country	11.6	9.8	9.2

Source: Duchâteau (1998), and authors' calculations, based on the data of Banque Nationale (1999).

industrial activities and wholesale enterprises are also migrating toward the periphery, seeking both better accessibility by road and cheaper land (Table 73.1).

It is difficult to identify the cause of such a change, but this process apparently was accelerated by the dramatic improvement of accessibility to the city by road, which resulted from the building of a high standard motorway network during the 1970s.

The structural changes in the economic activities, especially related to technical change, are important factors influencing the mobility. A model of economic dynamics and labor resources in the region, to be implemented as a part of the discussed integrated approach, operates with the information per sector of economic activity. The sectors of services and communication should be considered in greater detail since these are of particular interest for this study as most Belgian offices in this sector are located in the Brussels area.

Several tendencies determine the office stock dynamics and its spatial distribution. Many companies leave city centers for locations that are considered more accessible by their suppliers, clients, and staff members. At the same time, the comparative advantages of housing in cities are decreasing, which reinforces the tendency of more financially secure inhabitants to look for a place to live in the surrounding area of the city, whereas the poorest inhabitants tend to accept housing in the centers. Because of the narrowness of the territory of the Brussels-Capital region, this phenomenon extends beyond the limits of the region, entailing a loss of its substance and a relative impoverishment (Table 73.1). These global trends, as well as the results of the more precise analyses, show that the Brussels-Capital region is seriously endangered by an evolution similar to that of some North American cities, whose centers have been completely deserted by a lot of companies as well as by the middle-class population or the rich.

Mobility

The number of vehicles on the road in Belgium is constantly increasing. In January 1991, the average number of private cars at the disposal of households per 100 inhabitants in the country amounted to 38. The trend of the change observed this last decade suggests that the growth of the motorization rate continues in the coming years, although possibly at a lower rate because of a progressive saturation effect in the demand for vehicles. The general assumption in Belgium is that this saturation point will be reached at 50 vehicles per 100 inhabitants (Duchâteau, 1998). It is relatively modest compared to the levels already reached in North America (nearly 60 vehicles per 100 inhabitants).

The households also took advantage of the new form of mobility that a private car offers in order to change the way they choose where they live. For many households, the comparison of the advantages and the price of a location in the old urban center with those of a location in the suburbs have led to the choice of the latter. The growth of population in the suburbs of Brussels is a consequence of this choice.

The increase of mobility by car also has had indirect consequences on the suppliers of goods and services by extending their market areas, modifying the conditions of competition that they impose on each other, and forcing them to reconsider their location criteria. For some suppliers, especially traders, the location choice is a question of survival, because accessibility to their sale points is a critical element of competition. The increase of jobs in the periphery of the city indicates that decisions against locating in the city were of large scale in the 1980s.

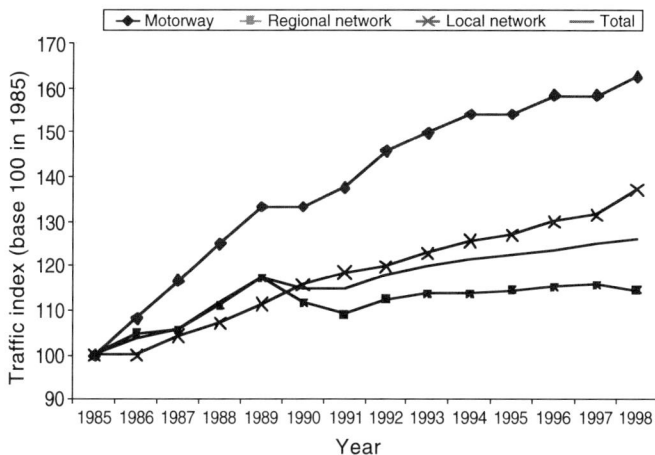

Figure 73.2 Traffic intensity (in vehicle-kilometers) in the Brussels-Capital region, 1985–1998. Calculation based on Recensement de la Circulation, Ministere des Communications et de l'infrastructure, 1998.

The combined effect of the dispersal of housing and employment and the increase in motorization of the population has led to a very important increase in automobile traffic both in and around the city. Figure 73.2 depicts the aggregated dynamics of the mobility in the region for private cars per main categories of roads. If this trend continues, and other variables remain unchanged, especially the offer of public transport, congestion of the road network will develop. The result is a doubling of automobile travel time during the morning peak period.

According to Duchâteau (1998), such a development is not realistic because it would lead to an unbearable deterioration of the functioning conditions of the city:

- Urban economic players cannot accept the worsening of their accessibility; if nothing is done to change the situation, their reaction will be to leave the city for a peripheral location.
- The inhabitants will support neither the impediments to mobility due to congestion nor the increase in pollution that will result from it; those who can afford it will leave the city in huge numbers.

Regional public authorities must react against decline of inhabitants and employers in the city.

Network Model of Private Road Transport

The Brussels-Capital region covers the area of 161 km^2 with 951,580 inhabitants (1997), in the center of Belgium (Figure 73.3). Administratively, the region is divided into 19 municipalities (communes), but the total area under study (Figure 73.4) comprises a wider territory (covering also nearby districts from which the traffic is most intensive), and it is subdivided into 167 smaller districts. The road network (Figure 73.5), due to further disaggregation of the administrative districts in the central part of Brussels, contains in total of 255 zones, from which trips are generated, with 2545 nodes and 8366 links.

As a basic computer tool for spatial analysis of mobility, the latest available version of the TRIPS package is used (TRIPS, 1999). It includes a set of interrelated modules: highway assessment, public transport assignment, demand modeling, matrix estimation, TRIPS graphics, and TRIPS manager–graphical project management tool. Private car transportation has been modeled (Plan IRIS, 1989–1995) according to the origin–destination matrix (Figure 73.5) for the morning peak hour (7:30 to 8:30) on an average day in 1991. The traffic intensity volumes (number of vehicles), average time, and speed per each link are calculated as a result of traffic assignment with TRIPS Highway Assessment module. The model permits different algorithms of traffic assignment: minimal cost paths (so-called "all-or-nothing" assignment), or multiple paths for each origin and

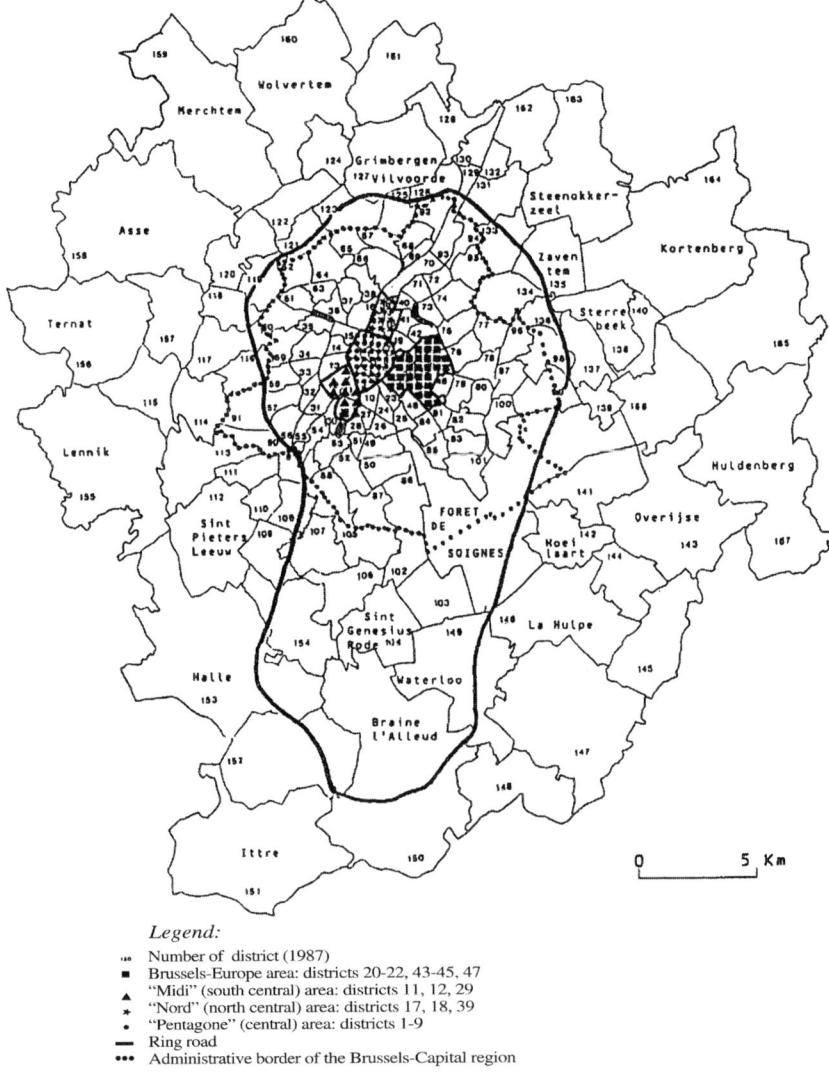

Figure 73.3 Geographical position of the area under study (an enlarged area of Brussels-Capital region). The outer districts within Belgium (from which passengers' mobility is about 20% of the total Brussels area traffic) are numbered 168 to 184.

Legend:

- ₁₄₀   Number of district (1987)
- ■   Brussels-Europe area: districts 20-22, 43-45, 47
- ▲   "Midi" (south central) area: districts 11, 12, 29
- ✳   "Nord" (north central) area: districts 17, 18, 39
- •   "Pentagone" (central) area: districts 1-9
- —   Ring road
- •••   Administrative border of the Brussels-Capital region

Figure 73.4 Area under study.

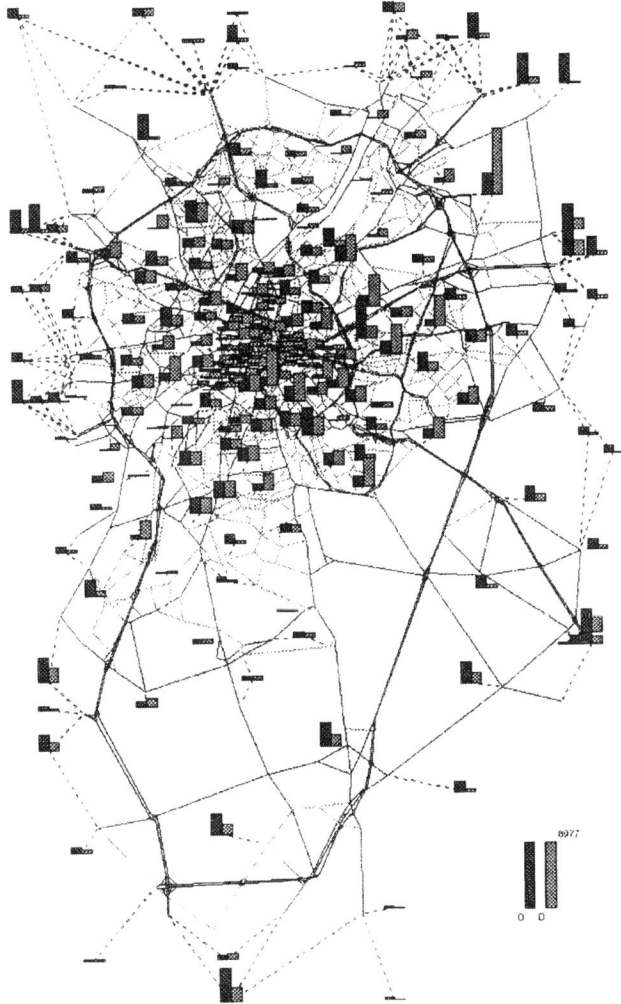

Figure 73.5 Road network and origin–destination matrix for private transport (morning peak hour, 1991).

destination (dial assignment model or Burrell assignment). For this study the Burrell model was used, which provides sufficiently good results from the available data.

Air Pollution

The last step in our modeling scheme is to analyze the impact of mobility on the indicators of air quality. The model considers the following air pollutants: CO_2, CO, SO_2, NO_x, VOC, and PM. In this chapter, given the restricted format, we describe only the emission model. Further analysis of concentrations of air pollutants (emissions–immissions link) can be found in Favrel and Hecq (1998, 2000).

Calculation of the emissions from road traffic was based on two types of data. First, the volume of traffic must be estimated, e.g., in the form of vehicle-kilometers driven by the different vehicle categories within the area considered. We used the results of the traffic assignment within TRIPS package and also regional statistics for model validation. Second, suitable emission factors are required for the different vehicle categories circulating in the Brussels area. Average speed-dependent emission factors, proposed in the COPERT II methodology (Ahlvik et al., 1997), have been used.

Table 73.2 Composition of the Belgian Private Vehicle Fleet in 1991

Vehicles Categories	Number of Cars, 1991	Share (%)
Petrol engines	**2,797,526**	**71.57**
<1.4 Liters	*1,405,917*	*35.97*
PRE ECE [<1971]	31,229	0.80
ECE 15/00-01 [1972–1977]	36,847	0.94
ECE 15/02 [1978–1979]	53,988	1.38
ECE 15/03 [1980–1984]	344,979	8.83
ECE 15/04 [1985–1990]	788,857	20.18
91/441/EEC [1991–1996]	150,016	3.84
94/12/ECE [>1997]	0	0.00
1.4–2.01 Liters	*1,186,085*	*30.34*
PRE ECE [<1971]	26,346	0.67
ECE 15/00-01 [1972–1977]	31,086	0.80
ECE 15/02 [1978–1979]	45,547	1.17
ECE 15/03 [1980–1984]	291,037	7.45
ECE 15/04 [1985–1990]	665,510	17.02
91/441/EEC [1991–1996]	126,559	3.24
94/12/ECE [>1997]	0	0.00
>2.0 Liters	*205,524*	*5.26*
PRE ECE [<1971]	4,565	0.12
ECE 15/00-01 [1972–1977]	5,386	0.14
ECE 15/02 [1978–1979]	7,892	0.20
ECE 15/03 [1980–1984]	50,431	1.29
ECE 15/04 [1985–1990]	93,516	2.39
91/441/EEC [1991–1996]	43,733	1.12
94/12/ECE [>1997]	0	0.00
Diesel engines	**1,089,055**	**27.86**
<2 Liters	*795,512*	*20.35*
Conventional [<1990]	710,628	18.18
91/441/EEC [1991–1996]	84,884	2.17
94/12/ECE [>1997]	0	0.00
>2 Liters	*293,543*	*7.51*
Conventional [<1989]	231,081	5.91
91/441/EEC [1990–1996]	62,462	1.60
94/12/ECE [>1997]	0	0.00
LPG engines	**22,484**	**0.58**
Conventional [<1990]	20,085	0.51
91/441/EEC [1991–1996]	2,399	0.06
94/12/ECE [>1997]	0	0.00
Total fleet	**3,909,065**	**100.00**

COPERT distinguishes for each vehicle category (such as passenger cars, light-duty vehicles, heavy-duty vehicles) different subcategories according to fuel type (petroleum, diesel, and liquid petroleum gases, LPG), cylinder capacity, catalyst type, different legislation and regulations (e.g., Sanger et al., 1997) governing motor vehicle emissions, fuel specifications, and consumption. For the period considered, the Belgian vehicle fleet was distributed according to these subcategories on the basis of the available statistics. An aggregated structure used in the model is presented in the Table 73.2, from which the vehicle fleet of the Brussels-Capital region has been deduced. Using the mileage (number of vehicles multiplied by link length) and the average speed of each vehicle category on each link of the network, the methodology developed provides the spatially distributed emissions, generated by road traffic in the Brussels-Capital region.

Hot and *cold start* emissions are distinguished. Cold start emissions represent the additional emissions resulting from vehicles while they are warming up or with a catalyst below its light-off temperature. The ratio of cold to hot emissions and the fraction of kilometers driven with cold engines are calculated using the yearly average temperature and an estimate of the average trip length following the COPERT methodology.

The overall calculation of emissions on the region's road network can be summarized as follows:

$$ET_i = ET_{i,\text{hot}} + ET_{i,\text{cold}},$$

$$ET_i = \sum_j \sum_k EF_{i,j,\text{hot}}(S_k) \cdot VM_{j,k,\text{hot}} + \sum_j \sum_k EF_{i,j,\text{cold}}(S_k) \cdot VM_{j,k,\text{cold}},$$

where

$i =$ pollutant index

$j =$ vehicle category index

$k =$ link index

$ET_i =$ emission of pollutant i due to road traffic

$ET_{i,\text{hot}} =$ emission of pollutant i due to road traffic with hot engines

$ET_{i,\text{cold}} =$ emission of pollutant i due to road traffic whit cold engines

$EF_{i,j,\text{hot}} =$ emission factor of pollutant i for vehicle category j driven with hot engines

$EF_{i,j,\text{cold}} =$ emission factor of pollutant i for vehicle category j driven with cold engines

$S_k =$ average speed on the link k

$VM_{j,k,\text{hot}} =$ vehicle mileage for vehicle category j driven on link k with hot engines

$VM_{j,k,\text{cold}} =$ vehicle mileage for vehicle category j driven on link k with cold engines

Since each link has its own length, average speed, traffic intensity, and other characteristics, we made an assumption that average emission factors per vehicle can be used based upon COPERT, as functions of the speed on the link. The temperature, the fleet composition (Table 73.2), and the share of cold start and hot emissions were taken as parameters. Given these parameters, the emission functions for an average vehicle can be calculated as functions of the speed only (Figure 73.6).

To calculate the distribution of the emission of the air pollution and the fuel consumption on the road network, an external module was developed (in Visual Basic). This program is linked to the TRIPS project, so that it is possible to spatially visualize the emissions of each pollutant along with the assignment results on the road map of the Brussels-Capital region. For details of the software organization, see Safonov (2000).

SCENARIO SIMULATIONS

The reference calculations with the model have been performed for the morning peak hour on a representative day for the years 1991 and 1996, according the actual statistical information from the Plan IRIS reports. Possible scenarios for sustainable mobility in the region are discussed in this section. The environmental impacts of urban mobility have been estimated using the emissions model described above.

The urban policies for the Brussels-Capital region defined in the Regional Mobility plan (Plan IRIS, 1989–1995) have been used to build the scenarios of prognostic simulations for the year 2005. Global strategy of Plan IRIS includes six groups of actions (Duchâteau, 1998) aimed at urban

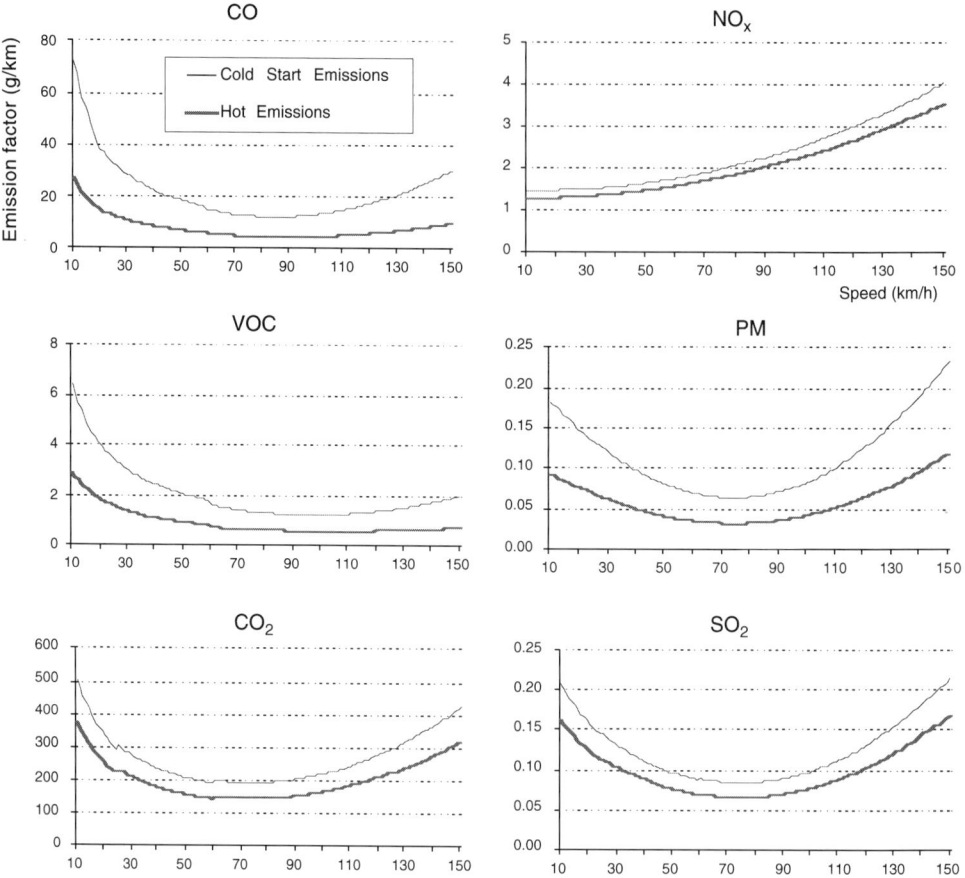

Figure 73.6 Emission functions for an average car in the Brussels-Capital region in 1991 (horizontal axis = speed in km/h, vertical axis = emissions coefficient in g/km).

structure, public transport, car parking, automobile traffic, getting around on foot and by bicycle, and actions on urban road pricing. In particular, the comparison of two scenarios (S1 and S2) that we have developed for the year 2005 are demonstrated below, based on these considerations for possible future actions in order to increase accessibility in the city and reduce the pollution.

Scenario S1

This is the business-as-usual scenario, where the trends are extrapolated from the year 1991 and 1996, from which the latest information is available. The origin–destination matrix for the year 2005 was built upon the forecasts within the Plan IRIS (1989–1995).

The network saturation rate in 2005 increases in comparison to the 1991 situation due to almost 70% growth of total vehicle-kilometers driven on the network. The overcongested links are marked by dark thick lines on Figure 73.7 (saturation rate > 1). This results in a significant decrease of average speed, especially on the ring road and other highways, and brings severe congestion in the center of the city at the peak hours.

It is assumed that the fleet of vehicles develops, proportionally substituting the old cars by the new (Table 73.3), with a growing share (until 1%) of LPG-driven vehicles in the total private car fleet. For this new fleet composition and the climatic conditions (the average temperature is assumed to increase in 2005 approximately to 1.5°C to the level of 1991), the new emission factors were calculated.

Figure 73.7 Network saturation rate (link intensity vs. link capacity), morning peak hour.

Table 73.3 Forecasted Composition of the Belgian Private Vehicle Fleet in 2005

Vehicle Categories	Share (%)
Petrol engines	**60.00**
<1.4 Liters	*25.00*
ECE 15/04 [1985–1990]	0.50
Euro I — 91/441/EEC [1991–1996]	1.50
Euro II — 94/12/ECE [1997–2000]	11.00
Euro III — 98/69/EC Stage 2000 [2001–2005]	12.00
1.4–2.01 Liters	*29.00*
ECE 15/04 [1985–1990]	0.50
Euro I — 91/441/EEC [1991–1996]	1.50
Euro II — 94/12/ECE [1997–2000]	12.00
Euro III — 98/69/EC Stage 2000 [2001–2005]	*15.00*
>2.0 Liters	6.00
Euro I — 91/441/EEC [1991–1996]	1.00
Euro II — 94/12/ECE [1997–2000] and <1997	2.00
Euro III — 98/69/EC Stage 2000 [2001–2005]	3.00
Diesel engines	**37.00**
<2 Liters	*30.00*
Conventional [<1990]	0.50
Euro I — 91/441/EEC [1991–1996]	4.50
Euro II — 94/12/ECE [1997–2000]	15.00
Euro III — 98/69/EC Stage 2000 [2001–2005]	10.00
>2 Liters	*7.00*
Conventional [<1989]	0.10
Euro I — 91/441/EEC [1991–1996]	0.40
Euro II — 94/12/ECE [1997–2000]	3.50
Euro III — 98/69/EC Stage 2000 [2001–2005]	3.00
LPG engines	**3.00**
Euro I — 91/441/EEC [1991–1996]	0.10
Euro II — 94/12/ECE [1997–2000]	1.00
Euro III — 98/69/EC Stage 2000 [2001–2005]	1.90
Total fleet	**100.00**

Table 73.4 Comparative Analysis of Scenarios

Year	Traffic Intensity Vehicle–km	Air Pollutants Emission						Fuel Consumption		
		CO	NO$_x$	VOC	PM	CO$_2$	SO$_2$	Petrol	Diesel	LPG
1991	100	100	100	100	100	100	100	100	100	100
1996	120	86	86	85	99	116	109	103	160	57
2005 (S1)	163	75	80	68	129	179	94	154	255	470
2005 (S2)	124	41	60	40	87	117	69	98	180	346

Due to further expected growth of traffic and respective fuel consumption in 2005, the emissions of CO$_2$ increase significantly and emissions of PM also grow (see Table 73.4). Emissions of NO$_X$ (which are lower at a slow car speed, Figure 73.6), CO, and VOC decrease, as a result of improvement of vehicle characteristics, such as use of advanced autocatalytic converters. This effect on reduction of PM (which has been observed during the period of 1991–1996) is not enough to compensate for the growth in traffic forecasted within business-as-usual scenario S1. Due to the introduction of fuels with low sulfur content (beginning in late 1996) the emissions of SO$_2$ also decrease.

Scenario S2

This scenario is based on so-called voluntarist group of scenarios of Plan IRIS (1989–1995), in which several prospective measures and urban policies were introduced:

- Development of the mixed land-use schemes, favoring the localization of offices and residential areas close to the accessibility points
- Improvement of the links in public transport between the city center and the suburbs (suburban metro/suburban express rail system, RER)
- Introduction of parking control, which would favor the switch from roads to public transportation
- Decrease of automobile traffic in residential areas and limitation of the congestion in the city center by means of traffic-flow control measures
- Improvement of the travel conditions of pedestrians and cyclists

The above measures are aimed at reduction of the need for mobility in general, and in particular, stimulate a modal shift from private cars in favor of public transportation (mainly through introduction of RER). In terms of transport model, these changes were introduced mainly in the origin–destination matrix (reduced demand for private car use), as well as in the road network itself (e.g., penalties on particular segments of the roads, link capacities, and times). This information has been provided by the Equipment and Transport Administration (AED) of the Brussels-Capital region, based on the Plan IRIS (1989–1995) assumptions and calculations.

Comparison of aggregated indicators for the above-discussed scenarios is given in Table 73.4. Significant reduction of traffic intensity, coupled with other urban policy measures in scenario S2, result in dramatic abatement of environmental pressure.

Figures 73.8 through 73.10 depict the spatial distribution of several emissions in 2005 for scenarios S1 and S2. Emission levels are proportional to the bandwidth and darkness of the links on the road map. Substantial decrease of emissions in scenario S2 is observed, especially for CO$_2$, as well as pollution of NO$_X$ and SO$_2$. As already mentioned above, since NO$_X$ emission is proportional to speed, the higher volumes of this pollution are concentrated on the highways, such as the ring road. Other pollutants concentrate on most busy and congested roads. Projected within scenario S2 is better accessibility by public transportation of some important locations in the city, such as Zaventem International Airport (around which many offices are located), which also helps to avoid outrageous levels of pollution in their neighborhood (Figure 73.9).

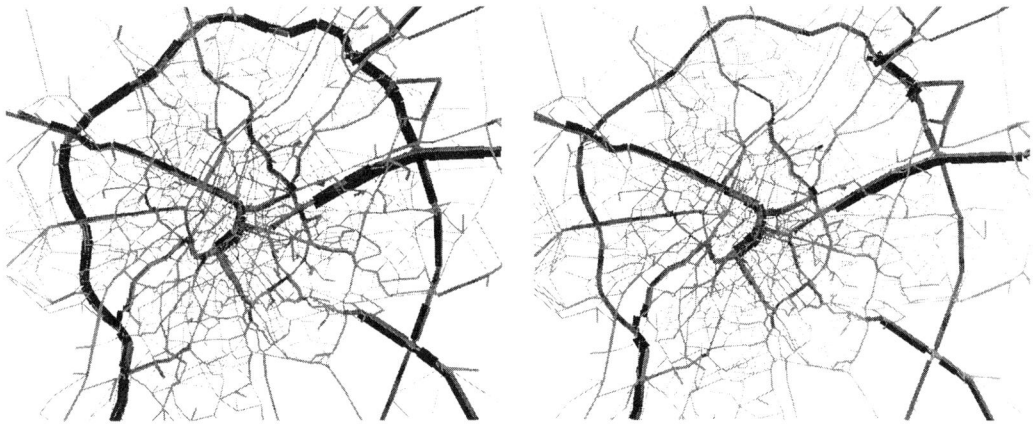

Figure 73.8 NO_x emissions (2005), Scenarios S1 and S2.

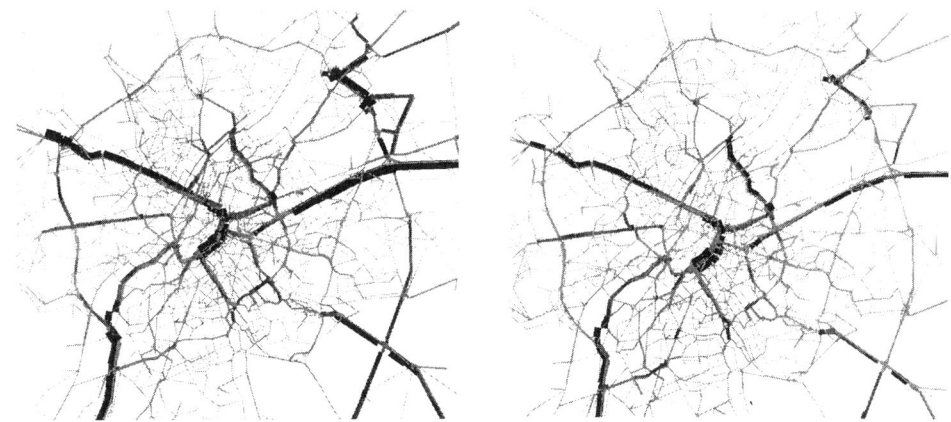

Figure 73.9 SO_2 emissions (2005), Scenarios S1 and S2.

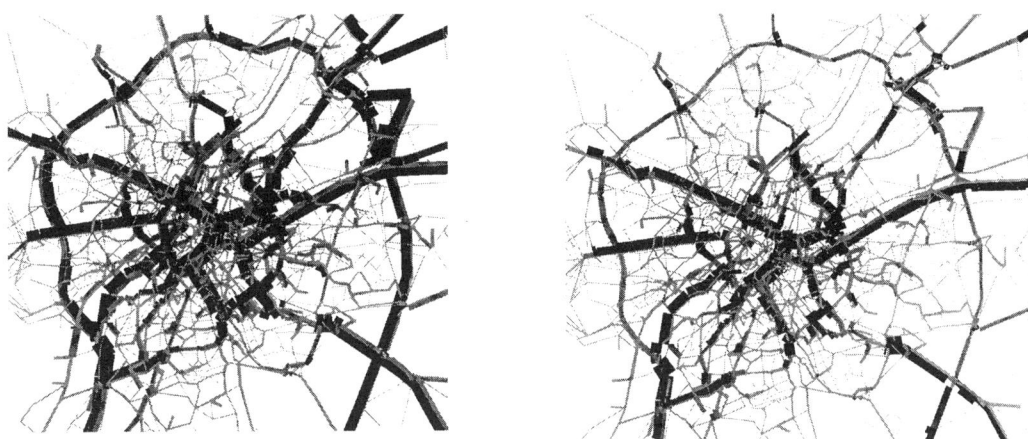

Figure 73.10 CO_2 emissions (2005), Scenarios S1 and S2.

CONCLUSION

The evaluation of the impact of mobility on air pollution is a multicriteria task. When discussing a sustainable mobility, a compromise should be made between simultaneous achievement of several goals. These goals include providing sufficient transportation services and accessibility to match the mobility demand in the urban area, reasonable travel time from origin to destination, traveling comfort and safety for both those in a vehicle and those on the road, and finally, acceptable emission levels from road transportation. The last two issues (safety and emissions) are linked directly when damage to the health of people in urban areas, as a result of local emissions, is discussed.

Restoring mobility by only decreasing the road traffic congestion, that is, trying to suppress the traffic bottlenecks, increasing the capacity of the main road networks, creating new parking lots, or implementing sophisticated techniques of traffic management, is no longer a realistic strategy in the long term. These are the answers that have been applied for years with only one objective: to increase the flow and fluidity of automobile traffic. They are responsible for their own inefficiency because any possibility of an increase in mobility is used by residential and economic players to increase even further their demand of automobile travel.

Experience has shown that in any city, in which measures were taken to increase the fluidity of traffic, the initial problems reappear some years later in a more acute form. Indeed, in cities where road networks are very congested, such measures only result in a slight shift of the thresholds and a postponement of the critical point. It enables a short-term solution, but does not modify the fundamental causes of the problem, nor create the new conditions required for a sustainable development of cities. In the case of Brussels, it would be particularly harmful as it would favor the centrifugal powers that would empty it of its substance because of the region's small area (Duchâteau, 1998).

As a possible way to solve this problem, an integrated approach to modeling urban development, mobility, and its environmental impacts is proposed and discussed in this chapter. At the first step in its implementation, we have designed a mobility model that allows spatial assessment of the road traffic contribution to air pollutant emissions on a network flow model. In particular, a comparison of two scenarios for the year 2005 demonstrated, based on statistical trends, considerable reduction in pollution with implementation of policies that would increase accessibility in the city, reduce demand for mobility, and favor a shift from private cars to public transportation, or to LPG vehicles. However, it is important to realize that implementation of such scenarios would require, above all, serious behavioral changes in mobility patterns and modal choice.

Further development of the described system of models and of methods used for environmental impact assessment is recently under way with ongoing continuation of this study at the CEESE-ULB in cooperation with the Brussels-Capital region administration.

ACKNOWLEDGMENTS

This chapter is based on a research project financed by the Government of the Brussels-Capital region within a Research in Brussels Actions program in 1999. The authors are thankful to the Equipment and Transport Administration (AED) of the Brussels-Capital region, in particular to Mr. Wouters, Mr. Broes, and Mr. Richel, for the information and technical assistance they provided. Our thanks also go to two anonymous reviewers for their remarks.

REFERENCES

Ahlvik, P., Eggleston, S., Gorißen, N., Hassel, D., Hickman, A-J., Joumard, R., Ntziachristos, L., Rijkeboer, R., Samaras, Z., and Zierock, K-H., COPERT II, Computer Program to Calculate Emissions from Road Transport — Methodology and Emission Factors (final draft report), European Environment Agency, European Topic Center on Air Emission, 1997.

Duchâteau, H., Mobility and accessibility – the case of Brussels, in *Operations Research and Decision Aid Methodologies in Traffic and Transportation Management*, Labbé, M. et al., Eds., NATO ASI Series, 1998.

European Commission, *ExternE: Externalities of Energy*, EUR16520, Vol. 1–5, Luxembourg, 1999.

Favrel, V. and Hecq, W., A model for the assessment of the contribution of road traffic to air pollution in the Brussels urban area, *Proceedings of the International Symposium on Technological and Environmental Topics in Transports — Externalities in the Urban Transport: Assessing and Reducing the Impacts*, Oct., Milan, Italy, http://www.feem.it/gnee/libr.html, 1998.

Favrel, V., and Hecq, W., External costs of air pollution generated by road traffic in the Brussels urban area, *Proceedings of the 9th International Symposium "Transports and Air Pollution,"* Avignon, France, June 5–8, Actes INRETS 70, 2000, pp. 199–206.

FIGAZ (Belgian Federation of Gas Industry), *Statistical Yearbook*, 1996.

Lobe, P. and Duchâteau, H., Impacts of transport price on mobility and land use in the Brussels area, *Proceedings of the International Symposium on Technological and Environmental Topics in Transports —Externalities in the Urban Transport: Assessing and Reducing the Impacts*, Oct., Milan, Italy, 1998.

Plan IRIS — Plan Régional de Déplacements (Regional Mobility Plan), Ministère de la Région de Bruxelles-Capitale, Bruxelles, 1989–1995 (in French).

P.R.A.S. — Plan Régional de Affectacion de Sol (Regional Plan of Land Use), Ministère de la Région de Bruxelles-Capitale, Bruxelles, 1995 (in French).

PRD — Plan Régional de Développments (Regional Development Plan), Ministère de la Région de Bruxelles-Capitale, Bruxelles, 1995 (in French).

Safonov, P., Dynamic Ecological-Economic Modeling of Regional Planning: A Case Study of Environmental Impacts of Mobility Induced by Major Policy Options in the Brussels-Capital Region, report on the Research in Brussels project, University of Brussels, Belgium, 2000.

Sanger, R. et al., Motor vehicle emission regulations and fuel specifications, Part 2, Detailed information and historic review (1970–1996), report 6/97, CONCAWE, Brussels, Mar. 1997.

TRIPS – The Professional's Transportation Planning Package, Ver. 7, MVA Systematica, Surrey, U.K., 1999.

CHAPTER 74

Relating Indicators of Ecosystem Health and Ecological Integrity to Assess Risks to Sustainable Agriculture and Native Biota — A Case Study of Yolo County, California

Minghua Zhang, K. Shawn Smallwood, and Erin Anderson

INTRODUCTION

A new holistic, yet quantitative landscape ecology approach to environmental science was born from the availability of spatially referenced data and Geographic Information System (GIS). Scientists have recently begun developing a hierarchically organized, top-down approach for assessing the conditions of, and risks to, ecological resources in large areas (Griffith, 1998). This approach is referred to as the ecological indicators approach. First used in the Netherlands (Rotmans et al., 1994), the ecological indicators approach is now being applied in the U.S. by the U.S. Environmental Protection Agency (EPA) (O'Neill et al., 1994; Schultze and Colby, 1994), U.S. Department of Agriculture (1994), U.S. Geological Survey (Battaglin and Goolsby, 1995), and multiple academic scientists (Rapport et al., 1985; Karr et al., 1986; Bedford and Preston, 1988; Hunsaker et al., 1990; Graham et al., 1991; Cairns and McCormick, 1992). Adopting the approach ourselves, we have gained valuable insight into factors influencing ecological health (Zhang et al., 1997, 1998) and integrity (Smallwood et al., 1998) at the county level. These spatially explicit indicators were relevant to legal compliance issues and useful for focusing planning and regulatory efforts on priority areas in the landscape.

In our study, agricultural chemical application data are integrated and used as an indicator of ecosystem health to examine the risks to ecosystem functions posed by chemical contaminants. Indicators of ecological integrity are developed from habitat quality and landscape contiguity and represent the degree to which functional parts of the landscape are intact from the perspective of biota. The integration of ecosystem health and ecological integrity indicators will become essential for managing agricultural landscapes so that agriculture is sustainable and locally endangered species are conserved.

In this chapter, we present the concept of integrating ecosystem health and ecological integrity indicators to assess agricultural sustainability. We demonstrate the concept by applying the ecological indicators approach to Yolo County, in Sacramento Valley, California, in which we spatially compare the levels of pesticides used to the likely distribution of special-status species. We explore how our indicators of health and integrity may interact and affect each other, and how these most sensitive and vulnerable parts of Yolo County can be monitored effectively for potential impacts. Our study is intended to serve as an example of how to characterize exposures of special-status species to agricultural pesticides and to prioritize further research and regulatory attention toward the areas with the highest exposure on the landscape.

Yolo County (Figure 74.1) is home to numerous special-status species, some of which were selected as target species by Smallwood et al. (1998) for measuring ecological integrity (Table 74.1). Yolo County is an agricultural county. Farmers in the valley portion of the county grow field crops, such as alfalfa, rice, orchard crops, and wine grapes. These growers use many pesticides to control weeds, insects, nematodes, fungi, and rodents. Each ecosystem element has an inherent sensitivity to disturbance (stress), and each has a vulnerability posed by the implementation of management practices (Zhang et al., 1998). The coincidence of sensitivities and vulnerabilities yields a suite of impacts among the ecosystem elements as well as consequences for the societal goals. By predicting these impacts and consequences, or at least understanding them, we can set about making a proper management response.

Pesticide exposure can be assessed by several means:

1. Measuring the active ingredients within environmental media, including within the tissues of biological species
2. Mapping the amounts of pesticides reportedly used by growers or that are typically associated with each crop
3. Mapping the amounts of pesticides used, their likely pathways through the environment, and their likely fates

The U.S. EPA's BASINS software (Best Assessment Science Integrating Point and Nonpoint Sources) has consolidated monitoring programs for environmental contaminants in the waters of the U.S., thus providing the means to assess exposures as described in (1) above. However, the data resulting from point (1) above lack the resolution of geographic coverage that we would prefer for assessing exposure risk. Assessment approaches (2) and (3) have become more practical now that GIS and spatial data have become more available and the ecological indicators approach has matured. The indicators approach complements the more conventional reductionist approaches by offering a top-down perspective on likely exposures. Using the indicators approach, academic scientists and government regulators can focus on selecting locations within the landscape, the crops associated with the most damaging chemical applications, and the species that may be at the greatest risk.

The toxicity of pesticides varies by biological species, exposure pathway, source term, and interactions among active ingredients. For animals, direct toxic effects can cause mortality, reduce foraging efficiencies, or disrupt reproductive and social performance. Indirect effects are suffered through habitat degradation. Pesticide use may therefore result in reduced numbers and spatial extent of animal species. These impacts can be devastating on landscapes where special-status species persist on remnant habitat patches, constrained to the margins of agricultural fields, roadways, and irrigation structures (Smallwood et al., 1998). Mineau et al. (1999) summarized fatalities of raptorial birds due to dermal contact with labeled pesticides, including five of the special-status species occurring in Yolo County.

The Endangered Species Act (ESA) regulates the taking of threatened and endangered species. Often not considered under the regulations are the effects of exposures to hazardous substances in the environment, such as agricultural pesticides. Remnant aggregations of many special-status species exist in or near agricultural landscapes, however, so such species and the biological elements of their habitats are exposed to pesticides. Under the Federal Insecticide, Fungicide, and Rodenticide Act (FIFRA) and the Food Quality Protection Act (FQPA), pesticide registration and use are much more regulated. Exploring spatial patterns of ecological integrity and ecosystem health are essential to ecosystem risk assessment relevant to sustainable agriculture. This is especially true in Yolo County where implementation of a proposed Habitat Conservation Plan (HCP) would also force

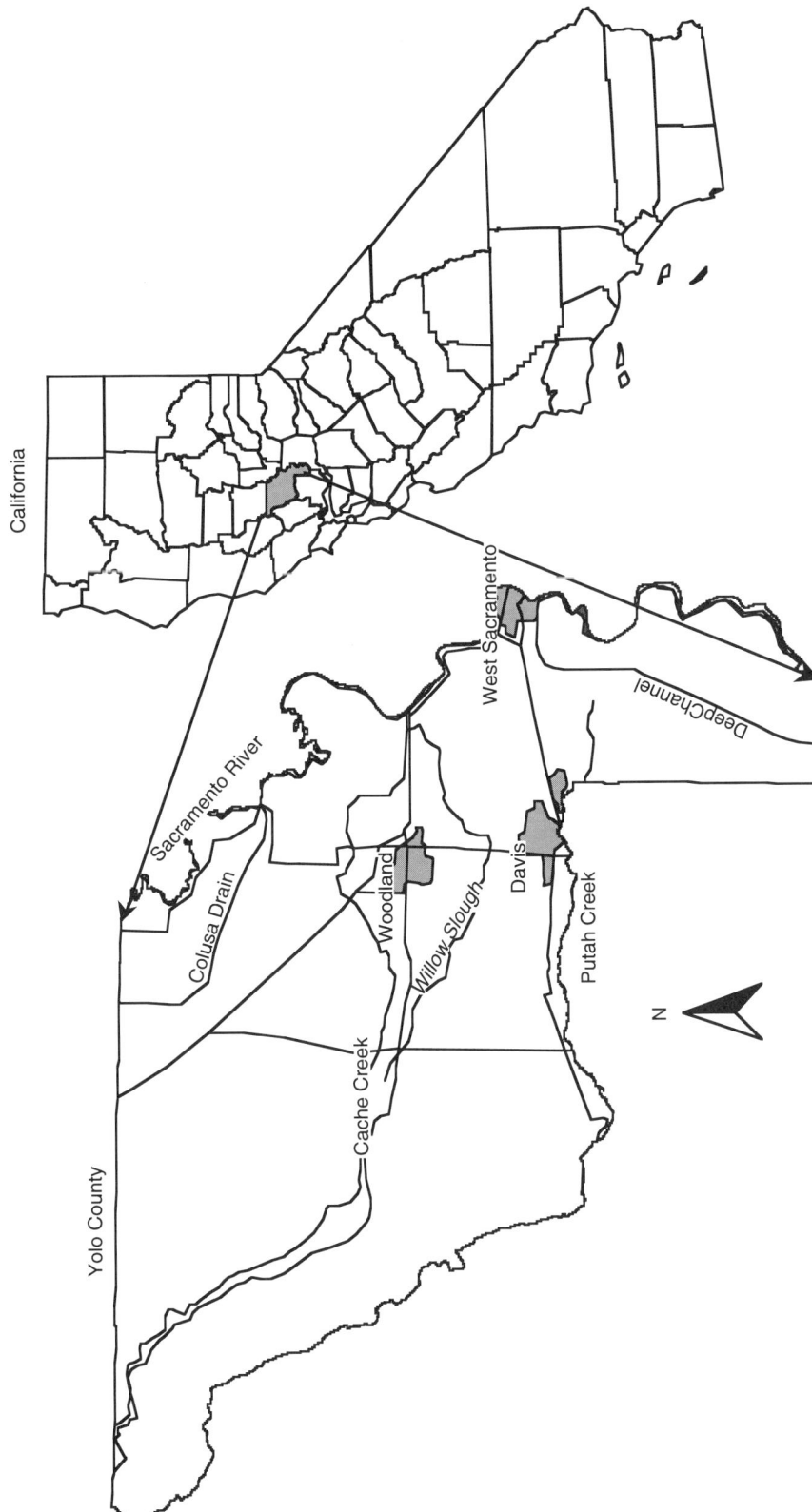

Figure 74.1 Study area in Yolo County, California.

Table 74.1 Target Species Used to Develop the Indicator of Ecological Integrity in Yolo County

Common and Species Names	Legal Status[a]
Heckard's peppergrass (*Lepidium latipes* var. *heckardii*)	— [b]
Brittlescale (*Atriplex depressa*)	— [b]
San Joaquin saltbush (*Atriplex joaquiniana*)	—,[b] C2
Alkali milkvetch (*Astragalus tener* var. *tener*)	— [b]
Palmate bird's-beak (*Cordylanthus palmatus*)	CE,[b] FE
Colusa grass (*Neostrapfia colusana*)	CE,[b] PT
Crampton's tuctoria (*Tuctoria mucronata*)	CE,[b] PT
Conservancy fairy shrimp (*Branchinecta conservatio*)	FE
Longhorn fairy shrimp (*Branchinecta longiantenna*)	FE
Vernal pool fairy shrimp (*Branchinecta lynchi*)	FT
Vernal pool tadpole shrimp (*Lepidurus packardi*)	FE
Valley longhorn elderberry beetle (*Desmocerus californicus dimorphus*)	FT
California tiger salamander (*Ambystoma californiense*)	CSC, C1
Western spadefoot toad (*Scaphiopus hammondi*)	CSC, C2
Northwestern pond turtle (*Clemmys marmorata marmorata*)	CSC, C2
Giant garter snake (*Thamnophis gigas*)	CT, FT
Double-crested cormorant (*Phalacrocorax auritus*)	CSC
White-faced ibis (*Plegadis chihi*)	CSC, C2
Cooper's hawk (*Accipiter cooperii*)	CSC
Swainson's hawk (*Buteo swainsoni*)	CT
Northern harrier (*Circus cyaneus*)	CSC
Greater sandhill crane (*Grus canadensis tabida*)	CT
Western yellow-billed cuckoo (*Coccyzus americanus occidentalis*)	CE
Short-eared owl (*Asio flammeus*)	CSC
Western burrowing owl (*Athene cuniculana hypugea*)	CSC
Bank swallow (*Riparia riparia*)	CT
Loggerhead shrike (*Lanius ludovicianus*)	CSC
Tricolored blackbird (*Agelaius tricolor*)	CSC, C2
California yellow warbler (*Dendroica petechia brewsteri*)	CSC

[a] Listed as CE = California endangered; CT = threatened; CSC = special concern (not threatened with extinction, but rare, very restricted in range, declining throughout range, peripheral portion of species' range, associated with habitat that is declining in extent); FE = federal endangered; FT = threatened; PT = proposed as threatened; C1 = category 1 candidate; C2 = category 2 candidate.
[b] Listed by California Native Plant Society as rare, threatened, or endangered in California.

the county to consider all 29 species listed on the HCP Incidental Take Permit as endangered, even though many are California Species of Special Concern (U.S. Department of the Interior and U.S. Department of Commerce, 1996).

METHODS

Our study area was the valley portion of Yolo County, or the area below the 92 m elevational contour line along the foothills on the western side of the county. A grid of 2.56 km^2 resolution was digitized in the study area at the Agricultural Geographic Information System (AGIS) laboratory of the University of California–Davis. We used GIS overlays of land use, pesticide use, and natural vegetation maps, all at 1:24,000 scale, GIS buffer functions, and simple statistics. The land-use and natural vegetation maps represented conditions in 1989. We obtained the land-use map from the California Department of Water Resources, the pesticide-use data from the California Department of Pesticide Regulation, and the natural vegetation map from Yolo County.

We developed an ecosystem health indicator to represent the cumulative effects of residue concentrations of multiple pesticides. We indicated vulnerability, or exposure risks, of the special-status species to pesticides by adding all the amounts of pesticides used by each farmer in each field for each 2.56 km² grid cell within Yolo County. To account for long-term pesticide use, we used data from 1990 through 1997. The ecosystem health index (EH) was estimated as follows:

$$EH = (100 - (\Sigma \, P_i/P_{hi} \times 100\%)),$$

where P_i was the amount (kg) of the ith pesticide used in a grid cell, and P_{hi} was the maximum amount of pesticides actually used in a grid cell ($P_{hi} \geq 11{,}363$ kg). Setting 11,363 kg as the ceiling for P_{hi} prevented the few extremely high reports of pesticide usage from rendering the majority of EH values as extremely small. It also allowed us to normalize the data for EH. Therefore, EH values ranged from 0 for grid cells using $P_i \geq 11{,}363$ kg of pesticides to 100 for grid cells using no pesticides.

The health indicator therefore measured 7 years of pesticide use, which bears on exposure of the special-status species, so long as these species were likely to occur in the same grid cell. Higher EH values represented higher health conditions in the ecosystem (lower pesticide application).

Ecological integrity (EI) was defined formally as the relative degree to which the functionally significant parts of the ecosystem are intact from the perspective of the 29 target species (Smallwood et al., 1998). More generally, ecological integrity measures the degree to which the endemic species assembly in the cell or region is intact and uninvaded by exotic species. The areas with higher values of ecological integrity should be given higher priority of restoration opportunities. Ecological integrity was an indicator composed of the area index (AI), connectivity index (CI), vegetation variety index (VI), vegetation occurrence index (VO), vegetation use (VU), restoration index (RI), and scarcity index (SI) (modified from Smallwood et al., 1998):

$$\text{Ecological Integrity (EI)} = \left(\frac{AI + CI + VI + VO + VU}{5} + \frac{RI + SI}{2} \right) \div 2 \times 100\%$$

AI indicates the size of the natural vegetation polygons overlapping the grid cell. AI ranges from 1 to 4, 4 indicating the largest polygon; CI measures the distance of the contiguous stretch of riparian vegetation polygon on the grid cell connected with non-riparian vegetation patch; VI is the number of vegetation species presented in each grid cell; VO is the percent of grid cell within the study area boundary consisting of natural vegetation species. The maximum value of VO (VO = 100%) is possible for some grid cells. VU is the sum of the species that are used for the targeted species in each grid cell. RI is a restoration index, which refers to the number of years estimated to restore the slowest growing vegetation species on the grid cell. For example, a grid cell containing valley oak riparian forest and vernal pools would be rated 100 because valley oak riparian forest would take about 100 years to restore to maturity, whereas vernal pools would take about 10 years. SI is a scarcity index which refers to the fraction of the study area covered by each vegetation species divided by the smallest fraction (most scarce type). Vegetation species were then ranked ordinally in ascending order from the most to least scarce and based on expert knowledge of the quality and magnitude of historical reductions in acreage of each species.

EI values ranged from 0 to 100. Higher EI values represented a higher likelihood that a larger subset of the 29 special-status species will be present in the grid cell. We assumed that a positive correlation between EI and EH would indicate a relatively healthy ecosystem, in which pesticide inputs are inversely proportional to ecological integrity.

We interpreted the spatial coincidence of health and integrity as relative levels of risk. The following exposure risk (ER) indicator expressed the coincidence of EH and EI as:

$$ER = (2EI/(EH + 1)).$$

This indicator expressed ER as a function of EH and EI. To emphasize the exposure risk for high-integrity areas, we designed this indicator in such a way that the range of EI values was proportional to EI. There was no point in obtaining high ER values where there was low EI. ER ranges from 0 to 200.

Visual inspection of ecological integrity showed that high integrity areas were clustered around waterways. To adjust for this nonuniform distribution, a 1.6-km buffer zone was established around all major waterways within Yolo County, namely, Putah Creek, Cache Creek, the Sacramento River, the Colusa Drain, the Deep Water Channel, and Willow Slough. All subsequent calculations compared areas within the buffer zone to areas outside the buffer zone. A deviation index was established to quantify the deviation of both ecological integrity and ecosystem health from uniform spatial distributions. Several steps were involved in this process:

1. Integrity and health values were classified according to natural breaks in their respective frequency distributions, and a frequency distribution (F) for each was tabulated inside and outside the buffer zone.

2. A cumulative frequency (CF) for the study area was calculated for the ith of n classes as:

$$CF_i = F_{in,i} + F_{out,I}.$$

3. The observed percent frequency (PF$_o$) for each class was calculated as:

$$\%PF_{o,i} = (F_{in,i}/CF_i) \times 100.$$

If the ecological integrity or ecosystem health distributions were uniform, then we would expect PF$_o$ values to equal the proportion of the total study area which lies within the buffer zone. In other words, if 40% of the study area occurred within the buffer zone, we would have expected 40% of the high values to occur within the buffer zone. The expected percent frequency (PF$_e$) was therefore defined as:

$$PF_e = \Sigma F_{in,i}/(\Sigma F_{in,i} + \Sigma F_{out,i}).$$

4. The deviation index (DI) represented how much the observed integrity or health deviated from a uniform distribution and was calculated as:

$$DI_i = [(PF_{o,i} - PF_e)/PF_e] \times 100.$$

A DI value of 0 indicated a uniform distribution throughout the entire study area. A DI value > 0 indicated that a disproportionately high frequency was observed within the buffer zone, while a DI value < 0 indicated that a disproportionately high frequency was observed outside the buffer zone. DI values for ecological integrity ranged from −32 to 38, while DI values for ecosystem health ranged from −12 to 12.

Similarly, the deviation index for land use was calculated using the above steps. DI values for land use ranged from −0.3 to 1.25. The positive DI values represented high concentrations of land use within the buffer zones while the negative values represented high concentrations of land use outside the buffer zone. Zero values indicated that the land use was uniformly distributed between the waterway buffer zone and the remainder of the study area.

The relative percentage of area (RPA) for each land use was defined as the area of the land use within the buffer zone (LUA) over the total area of the buffer zone (TAB). This index showed the percent allocation of land within, and outside, the buffer zone:

$$\%RPA = (LUA/TAB) \times 100.$$

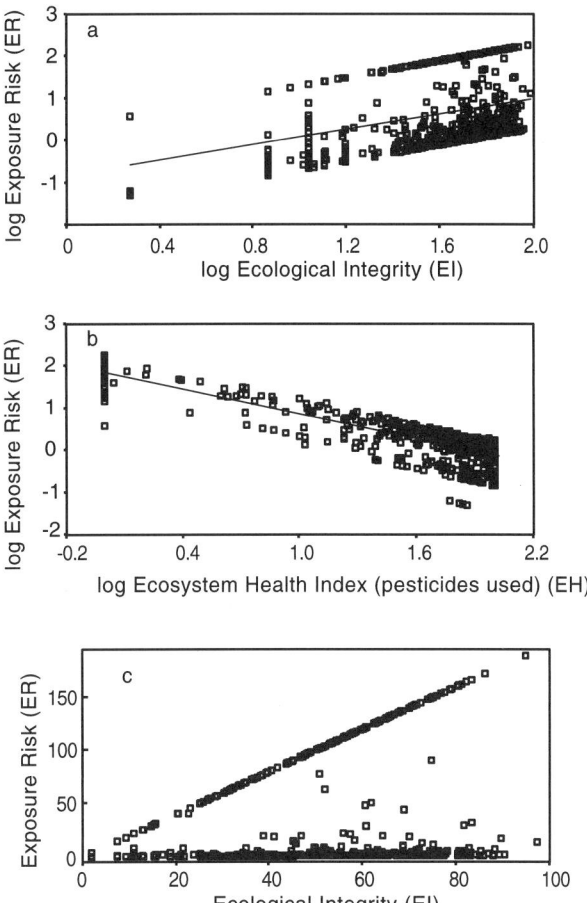

Figure 74.2 Log exposure risk increased linearly with increasing log ecological integrity (a) and decreased with increasing log ecosystem health (b). The value range of exposure risk increased as ecological integrity increased and was forced to 0 where ecological integrity was 0 (c).

RESULTS

As designed, ER values increased with increasing values of EI (Figure 74.2a) and decreased with increasing values of EH (Figure 74.2b). ER was forced to zero when EI = 0, and its maximum and range of values depended on the magnitude of EI (Figure 74.2c).

Higher ecological integrity (EI) values were concentrated within the 1.6-km waterway buffer zone and along the western edge of the study area (Figure 74.3). Patches of lower EI values were found in the northeast corner of Yolo County between the Sacramento River and the Colusa Drain, to the southwest of Willow Slough to the area between Cache Creek and the Colusa Drain, and to the southeast of Willow Slough.

Higher ecosystem health (EH) values were observed between the Colusa Drain and Cache Creek, and along the middle section of the Sacramento River, north of Sacramento (Figure 74.4). Patches of low EH values were found between Davis and West Sacramento (the Yolo County Bypass area), west of Woodland, and along Putah Creek.

According to the frequency deviation index, high EI values occurred at a disproportionately high frequency within the waterway buffer zones, where moderate values of ecosystem health also occurred at a disproportionately high frequency (Figure 74.5, left). High ecosystem health values

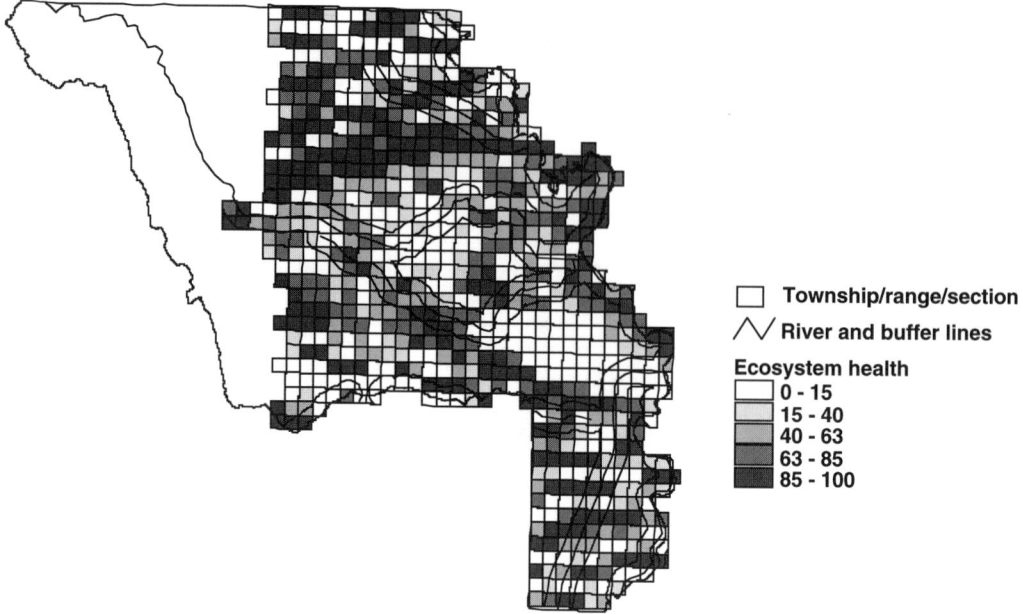

Figure 74.3 The spatial pattern of ecological health in the Yolo County study area.

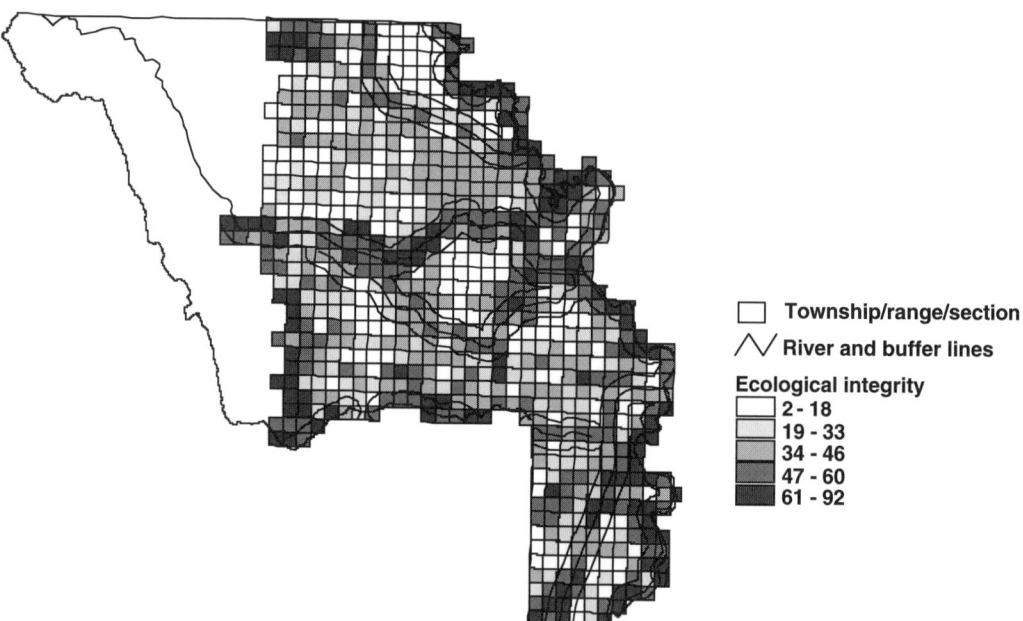

Figure 74.4 The spatial pattern of ecosystem integrity in the Yolo County study area.

occurred outside the waterway buffer zones at a disproportionately high frequency, where low ecological integrity values occurred at a disproportionately high frequency (Figure 74.5, right).

The pesticide exposure risk map (Figure 74.6) depicts distinct bands of high-risk areas oriented from the northwest to the southeast and spanning our study area across Cache Creek, the lower reach of Willow Slough and across the Yolo Bypass to West Sacramento. Another area of high ER occurred along Putah Creek from Winters to Davis. Other high risk areas were the lower reach of

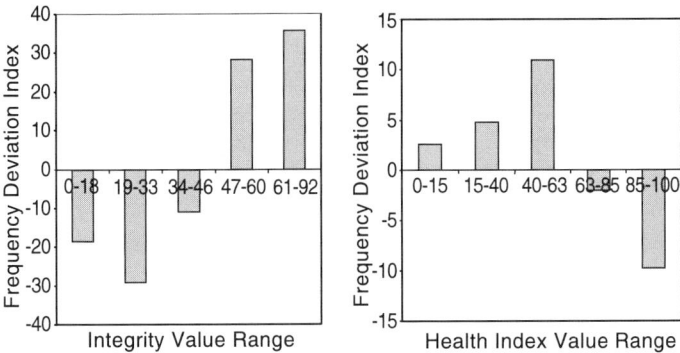

Figure 74.5 The frequency distributions were nonuniform for both ecological integrity (left) and ecosystem health (right) along waterways within 1.6-km buffer zones.

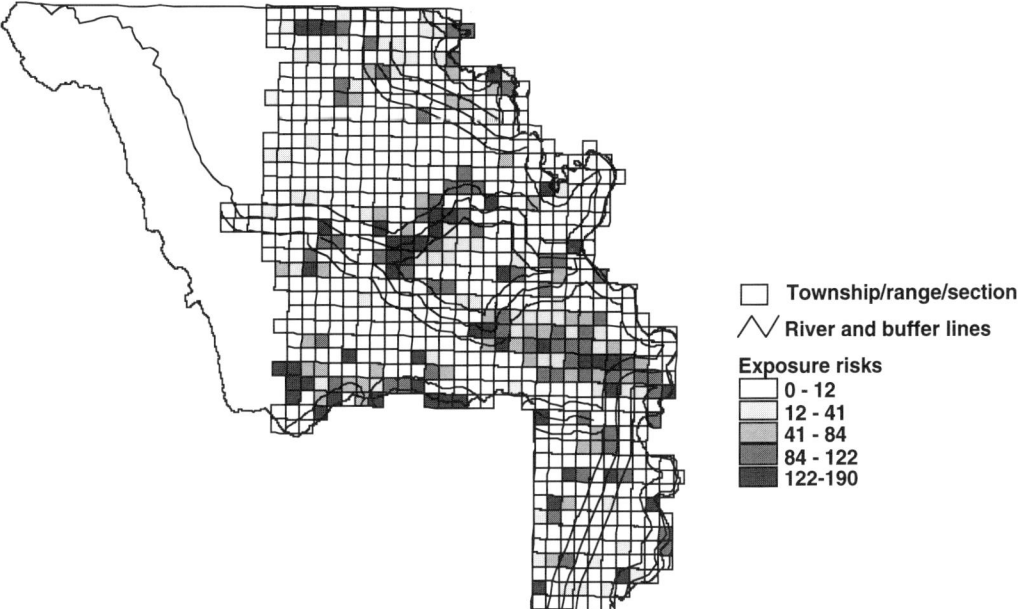

Figure 74.6 The spatial pattern of exposure risk in the Yolo County study area.

Oat Creek (northwest corner of the study area), and the southern portions of Yolo County in the vicinity of the Deep Water Channel. The low-risk areas for pesticide exposure included the upland areas between Putah and Cache Creeks and between Cache Creek and the Colusa Drain, where intensive farming and habitat loss have excluded most of the 29 special-status species used to assess ecological integrity.

Within the study area, EI and EH were not correlated, but the maps showed coincidence of high integrity and low health values that could be explained by considering other factors such as nearness to streams, land use, and possibly soils. The waterway buffer included streams that were artificial (e.g., Colusa Drain, Deep Water Channel) or that were degraded to the point where nearly no riparian vegetation remained (e.g., Willow Slough, segments of Cache Creek near areas of gravel mining, and segments of the Sacramento River).

Land uses associated with high pesticide applications occurred disproportionately within the waterway buffer zone (Figure 74.7). Higher percentages of citrus, orchards, field crops, rice, and

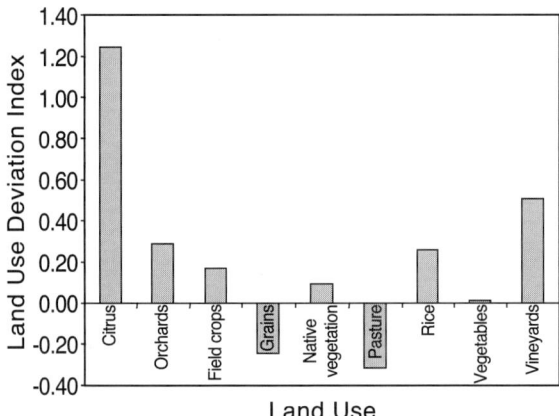

Figure 74.7 Land uses were disproportionately distributed within the buffers. Positive values refer to measures inside the 1.6-km riverine zone, while negative values refer to measures outside the zone.

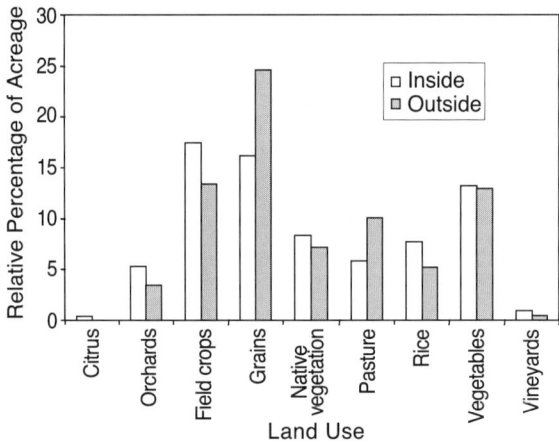

Figure 74.8 Relative percentage of different types of crop acreage within and outside the waterway buffer.

vineyards were grown within 1.6 km of waterways. Figure 74.8 illustrates the relative percentage of crop acreage within and outside the waterway buffer. Orchards, field crops, grains, pastures, rice, and vegetables comprise the majority of agriculture within the buffer zone. Field crops, grains, pastures, rice, and vegetables comprise the majority of agriculture outside the buffer zone.

DISCUSSION

Due to the dominating influences of agriculture in Yolo County, the areas of high ecological integrity were concentrated along the waterways within the 1.6-km buffer zone. The special-status species were likely also concentrated along the waterways within this buffer zone. To better protect the biota along these waterways, exposures to pesticides should be reduced within the 1.6-km buffer zone.

The pesticide use is relatively uniform across the study area in Yolo County compared to the distribution of ecological integrity. Relating a uniform pattern to a clustered pattern forces any combined pattern to match the clustered pattern. The clustered pattern of EI depended on the vegetation cover, connectivity, scarcity, and landscape contiguity, all of which were determined mostly by the spatial extent and practices of agriculture. It is likely that EI would be distributed differently in other counties. We also expect to find more uneven distributions of EH in counties with highly clustered cropping patterns.

Although the direct correlation between EI and EH was not statistically significant, a trend of negative correlation was displayed in Figure 74.5. Nevertheless, we were able to identify parts of Yolo County where exposure risk is likely the greatest, as well as the parts of the county where exposure risk is likely the least. The trends illustrated in Figure 74.5 may mean that the higher EI areas where the county's special-status species likely reside also had lower EH values. This indicated that the areas near waterways within the 1.6-km buffer zone may have received relatively high pesticide exposures.

The map of exposure risk provided us with well-defined patches of high vulnerability. Management efforts can be focused on these areas of high exposure risk to conserve the ecosystem upon which the special-status species depend and to sustain agriculture in Yolo County. This map also helps identify locations where the county can restore habitat that is less likely to be degraded by high exposures to agricultural pesticides.

The disproportionate distribution of land use within and outside the waterway buffer zone illustrated that there may be a soil–crop relationship leading to levels of pesticide use. The best soils for agricultural production tend to be nearby the streams, where sands have accumulated. On these soils, growers are more likely to grow higher-value crops and to protect them with chemical pesticides. Within the areas of high exposure risk, alternative cropping patterns could be considered in an effort to minimize the impact of pesticides on special-status species in these areas.

The concept developed in this study was rather simple, and our study was only a first step in comparing health and integrity indicators to estimate cumulative exposure risk. The environmental fates of many pesticides used need to be considered using pathway analysis, including portions of pesticides leaching into the groundwater, the rates and locations of pesticide residue reappearance at the ground surface, portions drifting with wind to nontarget areas, and portions removed from the environment by harvest or environmental degradation. Our approach is not intended to indicate species-specific or habitat-specific toxicological relationships, but rather cumulative relationships that are useful for top-down assessments in identifying where the greatest risks may occur, where management and conservation decisions may be prioritized, and where the focus of research efforts should be. Future applications of this approach could be improved by considering the toxicological sensitivities of each species to various pesticides, as well as their aggregative effects, and they could be improved by considering the behaviors of the species that modify exposures. Future applications of combining health and integrity indicators will also be improved by using sensitivity analysis and some validation based on sampling for impacts. As more explicit assumptions and exposure pathways are added to our model, we should be able to advance our modeling from one of hazards to risk of exposure dose (Rejesky, 1993; Beyea and Hatch, 1999).

REFERENCES

Battaglin, W.A. and Goolsby, D.A., Spatial Data in Geographic Information System Format on Agricultural Chemical Use, Land Use, and Cropping Practices in the United States, U.S. Geological Survey, Water Resources Investigations Report 94-4176, Denver, CO, 1995.

Bedford, B.L. and Preston, E.M., Developing the scientific basis for assessing cumulative effects of wetland loss and degradation on landscape functions: status, perspectives, and prospects, *Environ. Manage.*, 12, 751–771, 1988.

Beyea, J. and Hatch, M., Geographic exposure modeling: a valuable extension of geographic information systems for use in environmental epidemiology, *Environ. Health Perspect.*, 107, 181–190, 1999.

Cairns, J., Jr. and McCormick, P.V., Developing an ecosystem-based capability for ecological risk assessments, *Environ. Prof.*, 14, 186–196, 1992.

Graham, R.L., Hunsaker, C.T., O'Neill, R.V., and Jackson, B.L., Ecological risk assessment at the regional scale, *Ecol. Appl.*, 1, 196–206, 1991.

Griffith, J.A., Connecting ecological monitoring and ecological indicators: a review of the literature, *J. Environ. Syst.*, 26, 325–363, 1998.

Hunsaker, C.T., Graham, R.L., Suter, II, G.W., O'Neill, R.V., Barnthouse, L.W., and Gardner, R.H., Assessing ecological risk on a regional scale, *Environ. Manage.*, 14, 325–332, 1990.

Karr, J.R., Fausch, K.D., Angermeier, P.L., Yant, P.R., and Schlosser, I.J., Assessing biological integrity in running waters: a method and its rational, *Ill. Nat. Hist. Surv. Spec. Publ.*, 5, 1986.

Mineau, P., Fletcher, M.R., Glaser, L.C., Thomas, N.J., Brassard, C., Wilson, L.K., Elliott, J.E., Lyon, L., Henny, C.J., Bollinger, T., and Porter, S.L., Poisoning of raptors with organophosphorus and carbamate pesticides with emphasis on Canada, U.S., and U.K., *J. Raptor Res.*, 33:1–37, 1999.

O'Neill, R.V., Jones, K.B., Riitters, K.H., Wickham, J.D., and Goodman, I.A., Landscape Monitoring and Assessment Research Plan, U.S. EPA 620/R-94/009, U.S. Environmental Protection Agency, Washington, D.C., 1994.

Rapport, D.J., Reiger, H.A., and Hutchinson, T.C., Ecosystem behavior under stress, *Am. Nat.*, 125, 617–640, 1985.

Rejesky, D., GIS and risk: a three-culture problem, in *Environmental Modeling with GIS*, Goodchild, M.F., Parks, B.O., and Steyaert, L.T., Eds., Oxford University Press, New York, 1993, pp. 318–331.

Rotmans, J., van Asselt, M.B.A., de Bruin, A.J., den Elzen, M.G.J., de Greef, J., Hiderink, H., Hoekstra, A.Y., Janssen, M.A., Koster, H.W., Martens, W.J.M., Niessen, L.W., and de Vries, H.J.M., Global change and sustainable development, *Global Dynamics & Sustainable Development Programme*, GLOBO Report Series 4., RIVM, Bilthoven, the Netherlands, 1994.

Schultze, I. and Colby, M., A Conceptual Framework to Support the Development and Use of Environmental Information, Environmental Indicators Team, U.S. Environmental Protection Agency, Office of Policy, Planning, and Evaluation, Washington, D.C., 1994.

Smallwood, K.S., Wilcox, B., Leidy, R., and Yarris, K., An indicator of ecological integrity across large areas: a case study in Yolo County, California, *Environ. Manage.*, 22, 947–958, 1998.

U.S. Department of Agriculture, Economic Research Service, Natural Resources and Environment Division, Agricultural resources and environmental indicators, Agricultural Handbook No. 705, Washington, D.C., 1994.

U.S. Department of the Interior and U.S. Department of Commerce, Endangered Species Habitat Conservation Planning Handbook, Washington, D.C., 1996.

Zhang, M., Geng, S., Ustin, S.L., and Tanji, K.K., Pesticide occurrence in groundwater in Tulare County, California, *Environ. Monitoring Assess.*, 45, 101–127, 1997.

Zhang, M., Geng, S., and Smallwood, K.S., Nitrate contamination in groundwater of Tulare County, California, *Ambio*, 27, 170–174, 1998.

CHAPTER 75

Nature Policy Assessments: Strategic Surveying and Assessing Progress across Policy Levels

H.J. Verkaar, Hans A.M. de Kruijf, G.W. Lammers, and Rudo Reiling

INTRODUCTION

The need for sustainability in the use of natural and other resources has been widely accepted by decision makers and society as a whole. Following early warnings in Rachel Carson's classic *Silent Spring* (1963) and the Massachusetts Institute of Technology studies (Meadows, 1972), the Brundtland report (United Nations World Commission on Environment and Development, 1987) reflected a real turning point in public opinion. It later began to become clear, however, that it would not be easy to implement sustainability in complicated societal decisions. Apart from the acceptance of sustainability as a guiding principle, bridging the gap between the disciplinary borders of knowledge and political and governmental boundaries through integrated approaches was shown to be a vital prerequisite. In the past decade, an overwhelming number of integrated assessment studies have appeared for policy assessment — often in relation to sustainability — in technology, social welfare, monetary issues, agriculture, physical planning, fisheries, environmental issues, water management, and public health.

In Europe, integrated environmental and nature policy assessment studies are defined by the interdisciplinary process of identification, analysis, and appraisal of all relevant natural and human processes and their interactions that determine both the current and future state of environmental, natural, and landscape quality and resources on appropriate spatial and temporal scales, facilitating the framing and implementation of policies and strategies (European Environment Agency, 1995). In fact, they thereby analyze the degree of effectiveness and efficiency of actual and potential policies and strategies on nature quality and biological diversity for current and next generations. Here, we will demonstrate techniques developed in Europe, particularly in the Netherlands, to assess current nature and landscape policies. Furthermore, we will show the paramount importance of crossing knowledge and political boundaries if the implementation of objectives in nature and landscape policy is to be successful.

ECOSYSTEM HEALTH, CONDITIONING FACTORS, AND SOCIAL DOMAIN

In the Netherlands, values related to natural areas (e.g., forests, heathlands, dunes, and wetlands) have in the past century suffered from habitat destruction and fragmentation, environmental pollution, detrimental site and population management, and desiccation. Since 1900, active bogs, heathlands, and inland drifting sand dunes have decreased by 99, 91, and 86%, respectively. Most of the Dutch

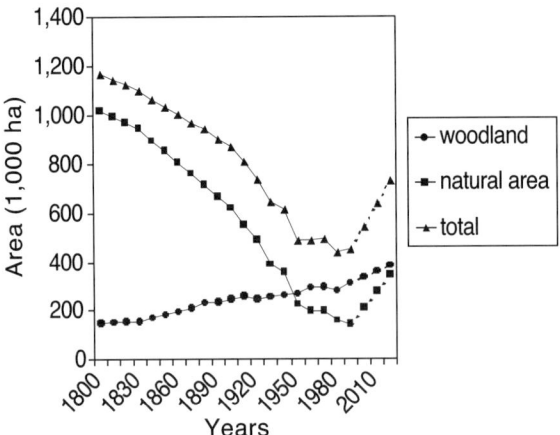

Figure 75.1 Woodlands and natural sites in the Netherlands from 1800 onward. The broken lines indicate the current policy targets. (Adapted from Rijks Instituut voor Volksgezondheid en Milieu, 1997; and Van der Woud, 1998.)

landscape at the beginning of the 19th century was almost treeless (van der Woud, 1998). After 1850, however, many natural areas were reclaimed thanks to new fertilization techniques and other technology. Furthermore, dry and nutrient-poor drifting sands and heathlands were replaced with mainly pine forests, so that during the last two centuries the total area of natural sites has decreased to about 15%, whereas the area of forests, mainly plantations, has increased more than twofold (Figure 75.1). In the meantime, spatial scales in landscape have changed as well. In the western, Holocene, and the relatively open landscapes of reclaimed fenlands and marshes on clayish soil and in the much closer Pleistocene eastern landscapes on wind-borne sand deposits, fragmentation has occurred and is still occurring (Rijks Instituut voor Volksgezondheid en Milieu, 1997).

Apart from the loss of natural areas and landscape integrity, many native species have shown a decline or have even become extinct. A significant number of exotic species have also invaded the Netherlands (Hengeveld, 1989; RIVM, 1998a). These changes in species density and composition can largely be attributed to the habitat change, destruction and fragmentation, environmental pollution, and unsuitable water management. For instance, the decline in vascular plant species of wet and nutrient-poor habitats, and as a consequence, their sensitivity to eutrophication and desiccation was far larger than in species found in other habitats (Figure 75.2). Inappropriate site and population management (e.g., poaching, hunting, and inadequate mowing) has also contributed to the deterioration of natural areas.

There is no doubt that habitat deterioration and destruction and incorrect site management are not isolated factors within an environmental constraint but are embedded in a social and technological structure (i.e., in economic and technological, social and cultural, and institutional and environmental domains; De Kruijf and Van Vuuren, 1998; Van Ierland et al., 1998). Figure 75.3 shows schematically the relationships among these domains. Although the close relationship among them has long been widely accepted, clearly sustainable use of natural resources can be achieved only if the concept of sustainability is also incorporated in all the domains (United Nations World Commission on Environment and Development, 1987; Ul Hacq, 1996; De Kruijf and Van Vuuren, 1998). In this chapter we will confine our remarks to the issue of the relationship between the environmental and institutional domains.

THEORETICAL ASPECTS OF POLICY ASSESSMENT

Acceptance of a societal problem in the political life cycle (Figure 75.4) can cause government authorities to implement policy measures. In general, this acceptance is triggered by information

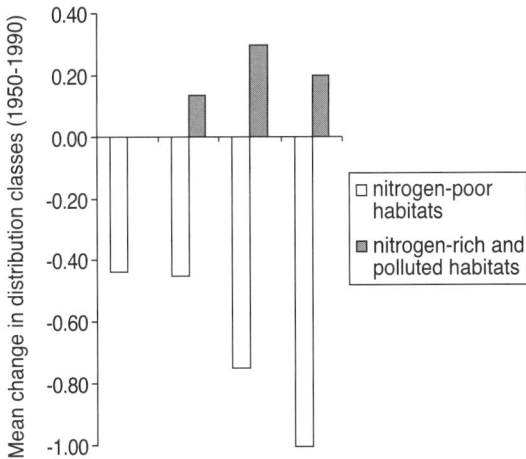

Figure 75.2 The change in distribution of vascular plant species in various habitats in the Netherlands between 1950 and 1990. (Adapted from Rijks Instituut voor Volksgezondheid en Milieu, *Natuurbalans 98*, Samsom, H.D., Willink, T., Alphen aan den Rijn, A., the Netherlands, 1998a.)

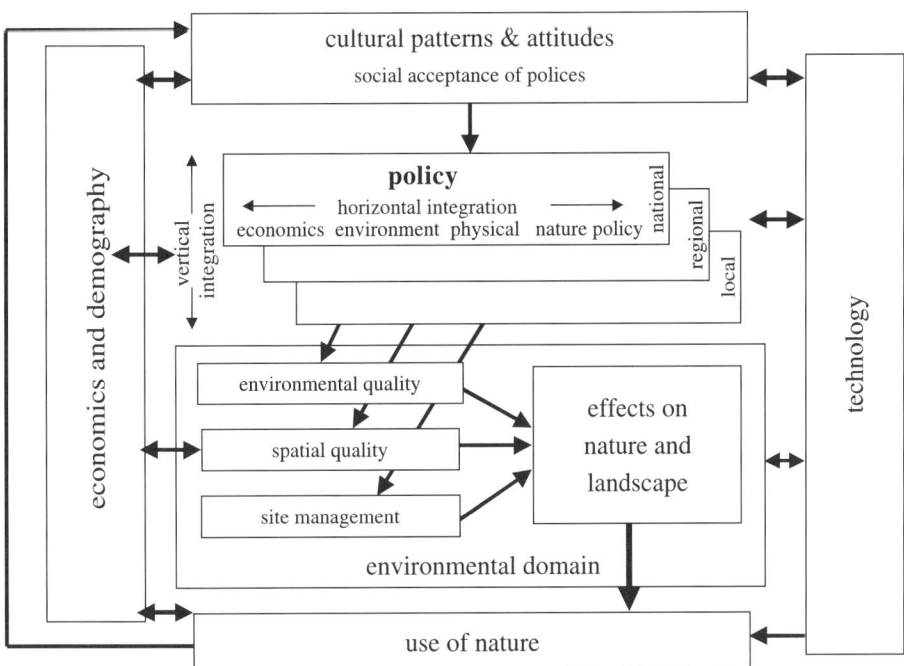

Figure 75.3 The schematic relationship among the economic and technological, social and cultural, and institutional and environmental domains.

on various indicators related to the so-called Driving Forces, Pressures, States, Impacts, and Responses (DPSIR) framework (De Kruijf and Van Vuuren, 1998; European Environment Agency, 1999). These indicators are summarized to four groups:

1. State quality of societally acknowledged values (e.g., nature and its biodiversity)
2. Pressures exerted on this
3. Policy responses to improve this quality aiming driving forces, pressures, and/ the quality directly
4. Impact of these measures on state quality

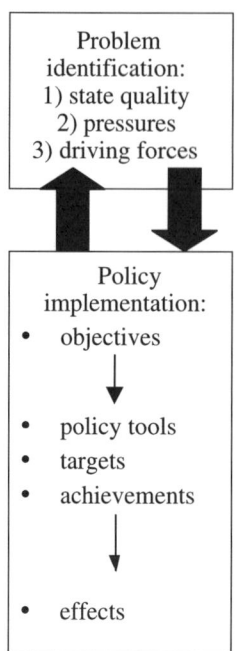

Figure 75.4 Simplified schematic representation of the policy life cycle and evaluation indicators.

All these indicators can be implemented into monitoring systems that judge the efficiency and effectiveness of the applied measures and such an evaluation can then lead to new societal problems. Thus, Figure 75.1 can be regarded as an example of a state indicator, whereas Figure 75.2 shows the impact of a pressure, nitrogen pollution, on natural vegetation. The impact of specific measures to improve natural quality is very difficult to monitor, as nature is generally influenced by a variety of environmental and biotic factors with very complicated dose–response relationships and very often a considerable relaxation time. Therefore, the effectiveness and efficiency of measures in terms of natural quality are very difficult to assess separately for each measure. Hereafter we will focus on indicators of the policy responses.

Within the framework of policy implementation several steps in the policy's life cycle can be distinguished in greater detail:

1. The formulation of objectives
2. The definition of targets and policy tools, i.e., the concrete set of actions to be taken to attain these targets
3. The periodic assessment of achievements
4. The effects of these achievements

Without doubt, monitoring systems to assess the effectiveness and efficiency of policy measures can be successfully developed only if objectives, targets, and tools have been sufficiently defined.

POLICY ASSESSMENTS AND OBSTACLES TO INTEGRATION OF POLICIES

Authorities indeed have gradually come to acknowledge the decline of natural resources taking place since the 1970s as a societal problem. Initially, their efforts were focused mainly on human health and prosperity, as far as these were affected by environmental quality. Since 1950, there have been massive economic growth rates (e.g., an approximate sixfold increase of GNP in the Netherlands and an approximately 2.5-fold increase in Europe on the whole, with an annual expected increase in the GNP for Europe of 2 to 2.5% in the next decade; European Environment Agency, 1999). Despite this, environmental quality has improved in some respects (e.g., water pollution and atmospheric SO_2- and NO_x-emissions), and eco-efficiency has increased by decoupling the resource use and pollutant release from economic activity (OECD, 1998). Conversely, the emissions of CO_2, waste, and other substances are not yet under sufficient control (RIVM, 1998b; European Environment Agency, 1999).

Besides taking measures to improve the quality of the conditioning factors, governmental policy makers see clearly that existing nature reserves should be preserved and new natural areas established to create and enhance ecological networks (Ministry of Agriculture, Nature Management and Fisheries, 1990; Bennett, 1991). In Europe, upwards of 10% of the territory is expected to be designated as nature conservation areas as part of the Natura 2000 Network (European Environment Agency, 1999). Furthermore, some 250,000 ha of new nature areas in the Netherlands will be connected to existing areas to create a National Ecological Network (NEN) in 2018. This has led us to conclude that policy makers will define objectives on a broad base so that we can assess the achievements of current policies on the improvement in the quality of nature areas and the landscape.

A number of policy tools in the European Union have univocal targets (e.g., the Habitat Directive in the framework of the Natura 2000 Network; European Environment Agency, 1999). The member states have defined and assessed such targets (Statistics Finland, 1997; RIVM, 1997, 1998a; Kuijken, 1999). For some policy fields, however, objectives cannot be assessed properly since verifiable targets and tools fail (e.g., in forestry and landscape policies in the Netherlands; RIVM, 1997; Table 75.1]).

Some policy tools are fairly successful (Figure 75.5). Indeed, the planned acreage of national parks, forests, and nature reserves, for instance, has increased; ecological corridors have been

Table 75.1 A Summary Showing General Suitability of Nature, Forestry, and Landscape Policies in the Netherlands

	Nature Policy	Forestry Policy	Landscape Policy
Objectives	+	+/-	+
Policy tools	+	+/-	-
Policy targets	+	+/-	+/-
Achievements	+/-	+/-	-
Effects	?	?	-

Note: + = well defined; +/? = doubtful; ? = not defined or inadequately defined; ? = uncertain.

Data from Rijks Instituut voor Volksgezondheid en Milieu, 1997.

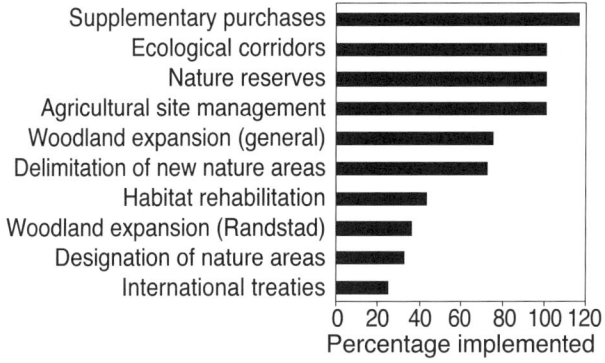

Figure 75.5 An example showing the achievements as a percentage of the targeted figures in nature policy in the Netherlands between 1990 and 1997. (From Rijks Instituut voor Volksgezondheid en Milieu, *Natuurverkenning 97*, Samsom H.D. Willink, T., Alphen aan den Rijn, A., the Netherlands, 1997.)

established and agricultural site management has become popular (RIVM, 1997, 1998a). Without a doubt, natural values will be enhanced in the areas through the implementation of these measures compared to areas where measures are not implemented. It would be intriguing to know, however, why so many targets have not yet been reached. There are some marked examples: until 1997, the Netherlands had not succeeded in implementing international treaties of the European Union due to problems in establishing new forests near the large cities in the western part of the country (Randstad). Furthermore, targets for designating new natural areas and areas for habitat rehabilitation had not been realized. As an example of this problem in implementation we will discuss the creation of the Dutch NEN below in greater detail.

LACK OF INTEGRATION BETWEEN POLICY LEVELS

First of all, both national and regional water policies claim to aim at improving the quality of natural areas. An important feature of the NEN comprises the marshlands, and in particular seepage areas, in which many rare and vulnerable species occur. More than half of all designated marshlands in the NEN are, however, dried up and desiccated. The national water policy aims at the recovery of 150,000 ha desiccated areas before 2000. Unfortunately, the measures will focus on only some 60% of the desiccated areas and on a little more than half of the 115,000 ha desiccated wetlands in the NEN (RIVM, 1998a). This can be considered an example of how horizontal integration between policy levels is being hampered.

The national government has commissioned the regional authorities with the implementation, through land delimitation and acquisition, of the NEN concept by 2018. Although acquisition is on average still taking place on schedule, there are distinct differences among the provinces (Figure 75.6). Provinces state a number of reasons for the delays in implementation (summarized in Figure 75.7),

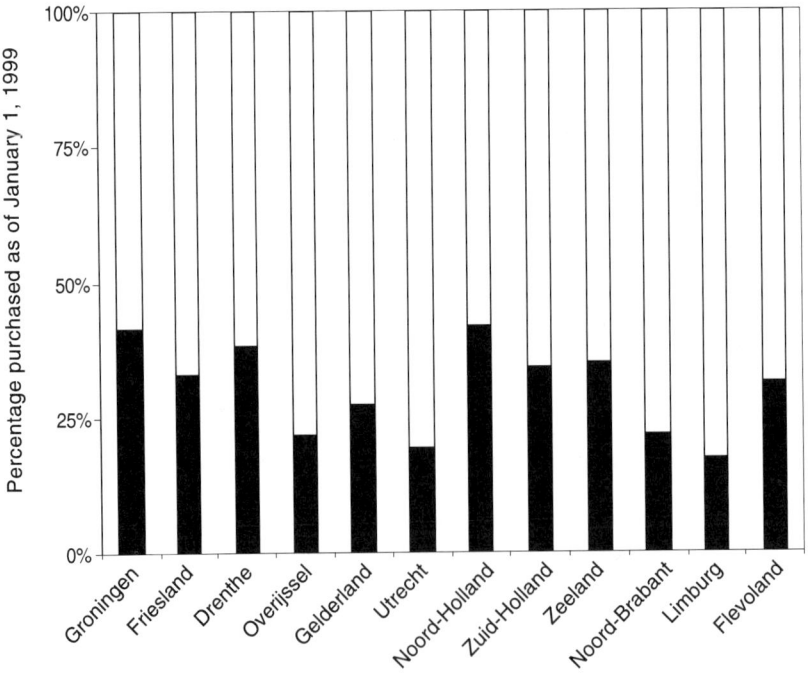

Figure 75.6 Regional achievements after the purchase of land of new nature areas as a percentage of the total targeted area per province. (From Rijks Instituut voor Volksgezondheid en Milieu, *Natuurbalans 99*, Samsom, H.D., Willink, T., Alphen aan den Rijn, A., the Netherlands, 1999.)

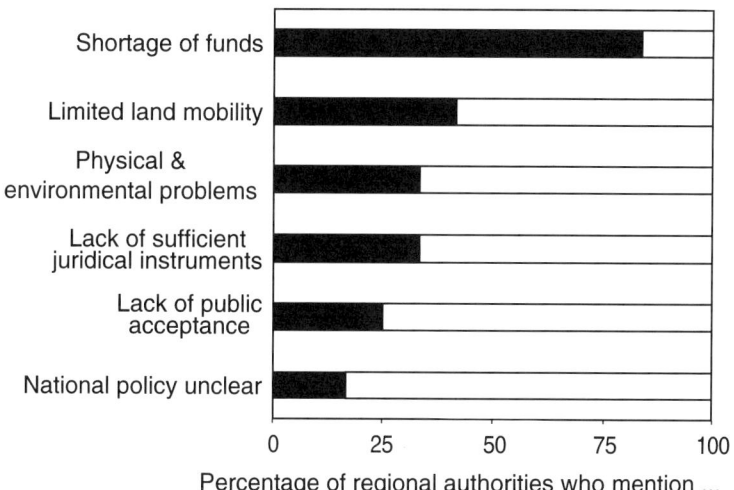

Figure 75.7 Frequency of claims for limitations in land acquisition for new nature areas according to Netherlands' regional authorities. (From Rijks Instituut voor Volksgezondheid en Milieu, *Natuurbalans 99*, Samsom, H.D., Willink, T., Alphen aan den Rijn, A., the Netherlands, 1999.)

including financial problems, as confirmed by an analysis of land prices. Although the national government supplemented some of the major funds for land acquisition in 1998, prices are still increasing, and the current market price in eight provinces now exceeds the fixed standard price per hectare.

Apart from financial problems, there are evidently also communication and attitudinal problems hampering vertical integration of nature policy. These problems include a lack of acceptance of national policy targets by regional and local authorities, and the rural population and the regional authorities' criticism of the lack of a clear national policy (Figure 75.7).

Sustainable use of natural resources requires integrated policies, and the degree of integrating policies in all domains can be properly judged only by integrated assessments in which ecoefficiency and economic development are measured. Integrated assessments need integrated knowledge. We have shown a number of examples of integrated assessments in Europe and confined these to the Netherlands to the assessment of nature policy, which can be considerably upgraded if the ill-defined targets and some of the policy strategies are improved, and if obstacles to political boundaries are overcome.

ACKNOWLEDGMENTS

We thank the Environmental and Nature Policy Assessment Division at Rijks Instituut voor Volksgezondheid en Milieu (RIVM), in which the National Institute of Public Health and the Environment collaborates with the Department of Agricultural Research, for sponsoring the studies. We are also much indebted to N.D. van Egmond, F. Langeweg, R.J.M. Maas, R. van Oostenbrugge, and H.J.W Oosterveld for the stimulating discussions and enthusiastic support given in carrying out this research. We are grateful to Dr. S. Carrizosa for his critical comments on an earlier draft of this chapter. Finally, we acknowledge the editing of Mrs. R.E. de Wijs.

REFERENCES

Bennett, G., Ed., *Vers un réseau écologique européen*, Institute of European Environmental Policy (Institut pour une Politique Européenne de l'Environnement), Arnhem, the Netherlands, 1991.
Carson, R., *Silent Spring*, Hamish Hamilton, London, 1963.

De Kruijf, H.A.M. and Van Vuuren, D.P., Following sustainable development in relation to the north-south dialogue: ecosystem health and sustainability indicators, *Ecotoxicol. Environ. Saf.*, 40, 4–14, 1998.

European Environment Agency, *Strategy for Integrated Environmental Assessment at the European Environmental Agency*, Copenhagen, Denmark, 1995.

European Environment Agency, *Environment in the European Union at the Turn of the Century*, Copenhagen, Denmark, 1999.

Hengeveld, R., *Dynamics of Biological Invasions*, Chapman & Hall, London, 1989.

Kuijken, E., Ed., *Toestand van de natuur in Vlaanderen: Cijfers voor het beleid*, Mededelingen van het Instituut voor Natuurbehoud 6, Brussels, Belgium, 1999.

Meadows, D.L., *The Limits to Growth*, Universe Books, New York, 1972.

Ministry of Agriculture, Nature Management and Fisheries, Nature Policy Plan, SDU, The Hague, the Netherlands, 1990.

Rijks Instituut voor Volksgezondheid en Milieu (RIVM), *Natuurverkenning 97*, Samsom H.D. Willink, T., Alphen aan den Rijn, the Netherlands, 1997.

Rijks Instituut voor Volksgezondheid en Milieu (RIVM), *Natuurbalans 98*, Samsom, H.D., Willink, T., Alphen aan den Rijn, the Netherlands, 1998a.

Rijks Instituut voor Volksgezondheid en Milieu (RIVM), *Milieubalans 98*, Samsom, H.D., Willink, T., Alphen aan den Rijn, the Netherlands, 1998b.

Rijks Instituut voor Volksgezondheid en Milieu (RIVM), *Natuurbalans 99*, Samsom, H.D., Willink, T., Alphen aan den Rijn, the Netherlands, 1999.

Statistics Finland, 1997. *Finland's Natural Resources and the Environment 1997*, Ministry of the Environment, Helsinki, Finland, 1997.

Ul Hacq, M., Sustainable Development for Global Human Safety, paper at the Dutch Conference Action for Environment and Work, Amersfoort, the Netherlands, Mar. 1996.

United Nations World Commission on Environment and Development, *Our Common Future*, Oxford University Press, Oxford, 1987.

Van der Woud, A., *Het lege land, de ruimtelijke ordening van Nederland 1798–1848*, Uitgeverij Contact, Antwerpen, Belgium, 1998.

Van Ierland, E.C., De Kruijf, H.A.M., and Van der Heide, C.M., Eds., *Attitudes and the Value of Biodiversity*, Wageningen Agricultural University/Utrecht University, Wageningen/Utrecht, the Netherlands, 1998.

CHAPTER 76

Land-Use Change Due to Urbanization for the Middle Atlantic Integrated Assessment Region of the Eastern United States

Ronald W. Matheny and Keith Endres

INTRODUCTION

The U.S. Environmental Protection Agency (EPA) Regional Vulnerability Assessment Program (ReVA) is designed to develop and demonstrate approaches to identify the ecosystems at the greatest risk from regional population growth and economic activity (Smith, 1999). A region is a multistate area involving many metropolitan areas, drainage basins, associated ecosystems, and cultural infrastructures. The term *vulnerability* is a variable, ranging from no vulnerability to low and high vulnerability; without a qualifier, the term implies nothing. ReVA is particularly used to compare risks arising from all sources of potential harm, acting alone or in combination, over the entire region. ReVA is beginning with a pilot study conducted as part of the Middle Atlantic Integrated Assessment (MAIA). The study area includes all of Pennsylvania, Maryland, West Virginia, Virginia, the District of Columbia, and parts of North Carolina, Delaware, New York, and New Jersey. This area was selected because it has a wealth of ecological data collected by field surveys, remote sensing, and other ecological monitoring, modeling, and research activities.

The study reported herein addresses two key questions: (1) What will be the land conversion to urban use and nitrogen loading during the next 5 to 25 years? (2) Where are the most vulnerable ecosystems located? Immediate objectives are these:

1. Integrate multiple data sources and existing assessment technologies
2. Expand research to fill critical gaps in our ability to apply existing data at the regional scale
3. Incorporate socioeconomic research to better understand factors driving environmental change and to more accurately assess the true costs of environmental degradation

ReVA will test alternative approaches by applying the technology as it is developed to the MAIA and obtain feedback from decision makers at the regional and local levels and the public.

METHODOLOGY

The Community Growth Model (CGM) is a cellular-based model incorporating land-use categories, slope, roads, railroads, pipelines, transmission lines, streams, federal lands, and population projections. Both the approach and data were selected to take into account the data determined to be significant in the literature. All data used are publicly available. Sewer lines were not used because a complete data set could not be located. Economic data, such as land value, developed through econometric methods will be used in the next version of the model. The model uses a base grid that integrates infrastructure and land use. It then calculates the area to be converted to the three urban uses (low-intensity residential, high-intensity residential, and commercial/industrial/transportation) from population projections. It then allocates new urban growth using a neighborhood function and modifies the base grid to prepare for the next iteration for each of the three urban classes. For this analysis, the model performed these functions at the county level, edge-matched (aligned) all counties together, and clipped the resulting grid to the MAIA region.

Base Grid

The 1992 Multi-Resolution Land Characteristics (MRLC) for the MAIA study area was used as an initial starting grid. This data set is a 30-m^2 grid coverage of the land use and land cover developed for the EPA MAIA project region. The coverage was produced using 1988, 1989, 1991, 1992, and 1993 Earth Resource Observation System (EROS) Landsat Thematic Mapper (TM) data. In addition, we used U.S. census data for population and housing density, U.S. Geological Survey land-use and land-cover data, and National Wetland Inventory (NWI) data as supplemental layers for data classification. The general procedure of the MAIA coverage was as follows:

1. Mosaic multiple summer TM scenes and classify them using an unsupervised classification algorithm (a computer-directed numerical process)
2. Interpret and label classes (land characteristics) into land-cover categories using aerial photographs as reference data
3. Resolve confused classes using the appropriate ancillary data sources
4. Incorporate land-cover information from leaves-off (without foliage) TM data and NWI data to refine and augment the *basic* classification from the National Land Cover Data Classification System. The classification number used in this study corresponds with the classification (see Table 76.1)

Modification of the Initial MRLC Grid

Since this analysis examines urban expansion and land use changes as the result of urbanization (see Table 76.2) the above classification detail was not needed. Eleven categories were examined, collapsed from the 23 categories listed above:

- Three urban classes (low-intensity residential, high-intensity residential, commercial/industrial/transportation)
- Bare rock/sand
- Transitional
- Deciduous forest
- Evergreen forest
- Mixed forest
- Pasture/hay
- Emergent herbaceous wetlands
- Woody wetlands

The urban/recreational grasses were collapsed into the low-intensity residential class, and gravel pits, quarries, and coal mines were merged into the commercial/industrial/transportation class. Digital line graphs (DLGs) were converted into grid cells. Class I roads, railroads, and pipelines were added to the commercial/industrial/transportation class. Class II roads were added to the low-intensity residential class. Assumptions built into the model are that open water, federal

Table 76.1 The National Land Cover Data Classification System and Final Classification System

National Land Cover Data Classification System	Final Classification System
Water	
11–Open water	
12–Perennial ice/snow	
Urban	
21–Low-intensity residential	Low-intensity residential
22–High-intensity residential	High-intensity residential
23– Commercial/industrial/transportation	Commercial/industrial/transportation
Barren	
31–Bare rock/sand/clay	Bare rock/sand/clay
32–Quarries/strip mines/gravel pits	
33–Transitional	Transitional
Vegetated; Natural Forested Upland	
41–Deciduous forest	Deciduous forest
42–Evergreen forest	Evergreen forest
43–Mixed forest	Mixed forest
Vegetated; Natural Shrubland	
51–Deciduous shrubland	
52–Evergreen shrubland	
53–Mixed shrubland	
Vegetated; Non-Natural Woody	
61–Orchards/vineyards/other	
Herbaceous Planted/Cultivated	
81–Pasture/hay	Pasture/hay
82–Row crops	
83–Small grains	
84–Fallow	
85–Urban/recreational grasses	
Wetlands	
91–Woody wetlands	Woody wetlands
92–Emergent herbaceous wetlands	Emergent herbaceous wetlands

lands, and land within 50 m of streams would not be developed. It was also assumed that land would not convert from the urban classes. Thus, any other land use could convert to any of the urban classes.

Calculations of Yearly Land Conversion Pixels

In order to calculate the number of cells that must be added to each year in the three urban uses the modified MRLC grid was used for each county. The grid was examined and the number of cells of each of the urban classes beginning in 1992 was placed in a file and merged with the Woods and Poole (1998) projection data. The merged data for each of the counties were based on the assumptions that the ratio number of households to the number of high-intensity cells, the ratio number of households to the number of low-intensity cells, and the ratio of total earnings (adjusted

Table 76.2 Change in Number of Households in the States within the MAIA Region[a] Households (1000)

	1992	2000	2010	2020
District of Columbia	249	231 (−7.23%)	227 (−1.73%)	218 (−4.00)
Maryland	1756	1954 (11.28%)	2153 (10.18%)	2318 (7.66%)
New Jersey	2799	2947 (5.29%)	3075 (4.34%)	3153 (2.54%)
New York	6645	6828 (2.75%)	7008 (2.64%)	7070 (0.88%)
North Carolina	2530	2960 (16.70%)	3353 (13.28%)	3689 (10.02%)
Pennsylvania	4504	4680 (3.91%)	4863 (3.91%)	4963 (2.06%)
Virginia	2302	2628 (14.16%)	2949 (12.21%)	3214 (9.00%)
West Virginia	690	730 (5.80%)	764 (4.66%)	786 (2.88%)
Total[a]	21,475	22,969 (6.96%)	24,390 (6.19%)	25,411 (4.19%)

[a] The MAIA region contains portions of each of these states. The MAIA region was created by clipping grids at the county level for the MAIA region.

by subtracting out farm earnings and military earnings) to the number of commercial cells to be added each year will remain constant.

The Neighborhood Function

To determine which pixel would be *urbanized* a neighborhood function was used (see Figure 76.1). For each cell, we used a 7×7 area of 30 m^2 cells. Each cell in the grid was assigned a score based on the number of urban classes that were in the neighborhood. Each pixel then had a score for each of the urban classes for a total of three scores for each cell. The model then selected the cells with the highest scores as the most likely to be converted to urban uses and updated the modified MRLC grid for the next iteration. From these results, we calculated the number of cells that would be converted from nonurban uses to one of the three urban uses. This process was repeated for each county in the MAIA region to obtain a total number of cells that would be converted from nonurban to urban uses during the time span of 2000 to 2020.

In more technical terms, neighborhood functions require both a neighborhood (grid) and a combination rule (specifications of what to do to the grid) in order to work (Chrisman, 1997). For this analysis, the focal sum combination rule was used. The FOCALSUM command in ARC/INFO examines each cell location on an input grid, adds the values within a specified neighborhood, and sends the sum to the corresponding cell location on the output grid (ESRI, 1994). It passes a moving window over the input grid. For this model, a 7×7 grid moving window was used for each of the three urban classification. Each cell was assigned a value: urban = 1, nonurban = 0. The sum of the cells was placed in the center of the output grid. This process was repeated for each of the three urban classification 7×7 grids within each county grid.

Those cells with the largest number represent the areas most likely to have an increase in each of the three urban classes, as development will spread near areas that are already similarly developed. The number of cells determined in the calculations of yearly land conversion pixels were then divided between these cells to determine how many and which cells would change from nonurban to each of the three urban uses.

Nitrogen Loadings

Nitrogen loadings for the MAIA region for 1992, 2000, 2010, and 2020 were calculated using nutrient loading coefficients derived from the Chesapeake Bay study (Chesapeake Bay Program, 1995). The model yielded grids with 13 land-use categories. From these 13 categories, 5 categories were derived (see Table 76.3) and used to calculate the total nitrogen loading resulting from increased urbanization. Low-intensity residential, pasture/hay, and row crops remained the same as their output from the model. The high-intensity residential and commercial/industrial/transportation

Table 76.3 Nitrogen Loading Coefficients by Land Use Category

Land Use	Nitrogen Loading Coefficient
Low-intensity residential	11.9821
High-intensity residential/commercial/industrial/transportation	12.62119
Forest	2.62552
Pasture/hay	9.7399742
Row crops	19.74579

Note: It should be noted that the nitrogen loading coefficients for agricultural uses pasture/hay and row crops are very similar to the loading for low-intensity residential and high-intensity residential/and commercial/industrial/transportation classes.

were combined. Forest equaled the sum of deciduous, evergreen, and mixed forests. The area for each category was converted from the 30 m^2 output of the model to hectares. The total hectares for each of the five new categories were summed. The total area for each land-use category was then multiplied by their corresponding nitrogen loading coefficient.

RESULTS

Table 76.4 depicts the land use change in the entire MAIA region using the CGM methodology. The low-intensity residential area, high-intensity residential, and commercial/industrial/transportation area sectors increased by 5.11 and 16.21% in 2020, respectively. Forest area decreased by 0.73%, pasture/hay by 1.0%, and row crop by –1.00.2%. The decreases in the nonurban areas are very small, less than –1.19%, even though the increases in the urban classes seem high. This is because the urban classifications are a minor part of the total land use in the MAIA region, even though they may be close to sensitive ecological areas that might be degraded. The relatively minor decreases may also reflect that projections were made for only 28 years. Many other models have projection time periods of 50 to 100 years.

Also, urban growth occurred where there was already urban growth (sometimes called urban sprawl). There is strong concern over where this growth is occurring; if it is occurring near ecologically sensitive areas it could have severe impacts. This can be seen graphically by comparing the grids for each of the modeled years. Urban sprawl appears to be relatively minor (see Figure 76.1). When examining the growth in urban land use in the MAIA region it becomes clear that there is significant growth in land cover for all three urban sectors (see Figure 76.2).

Nitrogen Loadings Impacts

The significant growth in land cover for all three urban sectors may be applied to nitrogen loadings from 1992 to 2020 for the entire MAIA region. These results are not as insignificant as

Table 76.4 Land Use Change in Hectares (1000)

Year	Low-Intensity Residential Area	High-Intensity Residential/ Commercial/Industrial/ Transportation Area	Forest Area	Pasture/Hay Area	Row Crop Area
1992	13,243 (0.00%)	9780 (0.00%)	248,125 (0.00%)	71,049 (0.00%)	26,726 (0.00%)
2000	14,270 (7.76%)	11,319 (15.74%)	246,607 (–0.61%)	70,436 (–0.86%)	26,400 (–1.22%)
2010	15,348 (7.56%)	13,248 (17.04%)	244,785 (–0.74%)	69,710 (–1.03%)	26,071 (–1.25%)
2020	16,133 (5.11%)	15,395 (16.21%)	242,986 (–0.73%)	69,011 (–1.00%)	25,762 (–1.19%)

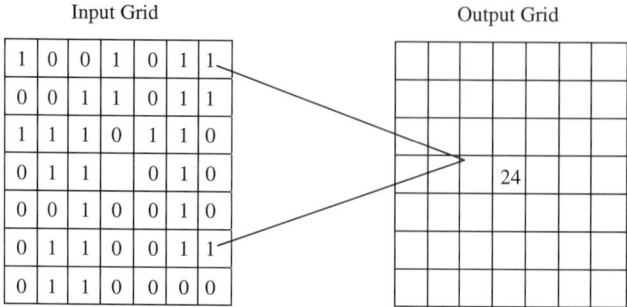

Function: Output grid = FOCALSUM (inputgrid, rectangle, 7, 7, data)

Figure 76.1 The neighborhood function.

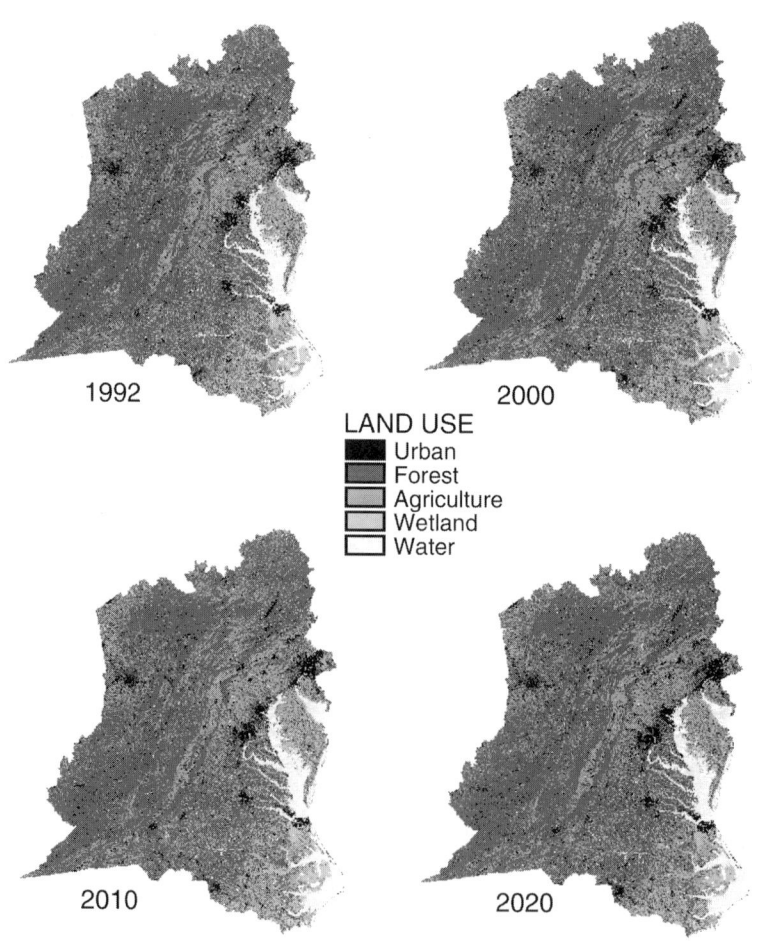

Figure 76.2 Changes in land use from population increases in the MAIA region (1992–2020).

Table 76.5 Nitrogen Loadings by Land Use and Total Nitrogen Loadings for the MAIA Region

Year	Low-Intensity Residential	High-Intensity Residential/Commercial/ Industrial/Transportation	Forest	Pasture/Hay	Row Crops	Total
		Nitrogen Load (millions kg/ha) by Land Use for the MAIA Region				
1992	15.9	12.3	65.1	69.2	52.8	215
2000	17.1	14.3	64.7	68.6	52.1	216
2010	18.4	16.7	64.3	67.9	51.5	219
2020	19.3	19.4	63.8	67.2	50.9	221

Year	Total Nitrogen	Percent Increase
	Total Nitrogen Loadings (millions kg/ha) for the MAIA Region for 1992–2020	
1992	215.33	—
2000	216.87	0.712
2010	218.76	0.872
2020	220.64	0.863

the land-use change. Table 76.5 presents the nitrogen loadings by land-use classification for the entire MAIA region. The urban classes are only 28.2% in 1992, 31.4% in 2000, 35.1% in 2010, and 38.7% in 2020 of the total nitrogen loadings in the region. Still, nitrogen loadings from all urban sources are proportionately small.

Table 76.5 also illustrates that increases in nitrogen loadings for the entire MAIA region for all land-use classes is less than 1%. On a regional scale, nitrogen does not seem to be a problem, but again, it must be cautioned that on a smaller scale, where urbanization is taking place near ecologically sensitive areas, there could be severe problems.

The changes occurring between the urban, forest, and agricultural land-use categories between 1992 and 2020 appear insignificant at the MAIA regional scale. The contribution of each land source to the total nitrogen loading in the MAIA region gives a more accurate assessment of where nitrogen problems could be originating. In relation to urbanization, all years show less than 1% increases over time (+0.712% from 1992 to 2000, +0.872% from 2000 to 2010, and +0.863% from 2010 to 2020).

Another way of showing the small change in nitrogen loadings is by comparing the urban uses with the nonurban uses and related nitrogen loadings (see Table 76.6 and Figure 76.3). Total nitrogen loadings are not increasing significantly. The source of nitrogen loading does change. Total nitrogen

Table 76.6 Comparison of the Nitrogen Loadings for the Three Scenarios

	kg/ha/yr Nitrogen (millions)				Hectares (millions)			
	1992	2000	2010	2020	1992	2000	2010	2020
	Base Case Scenario							
Urban	28.2	31.4	35.1	38.7	2.3	2.5	2.8	3.1
Forest	65.1	64.7	64.3	63.8	24.8	24.7	24.5	24.3
Agriculture	122.0	120.7	119.4	118.1	9.8	9.6	9.6	9.5
	Telecommuting Scenario							
Urban	28.2	NA	33.4	35.0	2.3	NA	2.7	2.8
Forest	65.1	NA	64.5	64.3	24.8	NA	24.6	24.5
Agriculture	122.0	NA	120.0	119.5	9.7	NA	9.6	9.6
	Smart Growth Scenario							
Forest	65.1	NA	64.4	64.0	24.8	NA	24.5	24.4
Agriculture	122.0	NA	119.7	118.7	9.8	NA	9.6	9.5

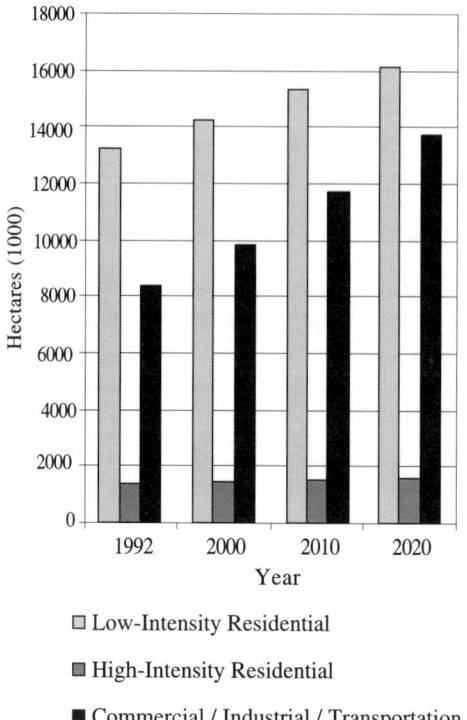

Figure 76.3 Total nitrogen for urban and nonurban land uses.

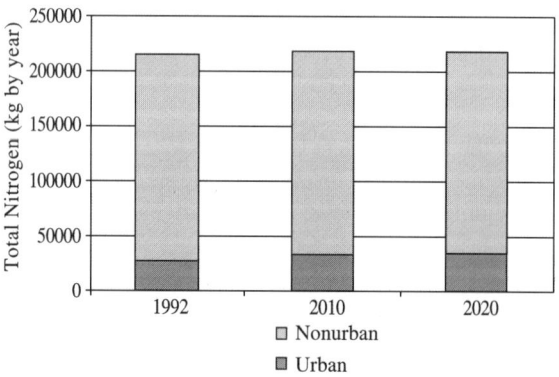

Figure 76.4 Total nitrogen for urban and nonurban land uses.

from urban development increases 11.25% from 1992 to 2000, 11.87% from 2000 to 2010, and 10.4% from 2010 to 2020. Total nitrogen from nonurban areas decreases 1.89% from 1992 to 2000, 2.15% from 2000 to 2010, and 2.11% from 2010 to 2020.

SCENARIO COMPARISON

Besides the base case scenario described above, two other scenarios were analyzed. The *telecommuting* scenario varied the commercial/industrial/transportation category by 25% for each county and each year. It was assumed that now facilities would be smaller to allow for employees to work either at home or at telecommuting centers in order to reduce land demand. The *smart growth* scenario varied household land demand by reducing land demand for the low-intensity residential and high-intensity residential categories by 25% for each county and each year. New households would require less land to be converted from each land-use classification. For both of these scenarios a geographic distribution of land-use change was made and compared with the base scenario. Table 76.7 compares the three scenarios by revised MRLC classification by year. Note that the nitrogen loading and hectare differences are relatively insignificant.

CONCLUSIONS

This approach to land-use change and associated nitrogen loadings has proven to be successful. We need to reconsider the assumption, however, that the land-use proportion in each urban category by county will remain the same since socioeconomic issues may change. To overcome this problem, the next version of CGM will be linked with an econometric model to include socioeconomic elements and to determine the amount of land that will be converted from one land-use category to another. The neighborhood function will be modified slightly to allow the CGM to take the resulting land-use change and allocate it geographically by county.

This version of the model used a single nitrogen loading for each land-use classification. Future model runs will use county or state level data to increase accuracy and, if available, modify the loadings by year to account for efficiencies in agriculture and forest management. Future model runs will also look at other loadings such as phosphorus.

The three scenario runs did not show great differences in nitrogen loadings for the urban categories. This may be a function of the allocation method of land change, poor assumptions on the part of the modeler, or possible poor quality control on the MRLC base grid and other geographic information system data sets. Comparisons with other model output is planned for the near future.

Anthropogenic pressures are often local in scale, and ecologically sensitive areas must be examined at the local scale. Fortunately, the data used in this model can be subdivided into smaller units, such as river basins, cities, or counties. Additional local data could then be added to conduct site-specific analysis. For example, monitoring emission factors on a highway system could produce a possible air quality analysis.

There is a need for integrated modeling of land-use and land-cover change. The methods to integrate the theoretical and simulation models of land use and land cover must be modeled through the interactions of the environment and society. At least two different approaches have been used. One approach is based on the theory of environmental and social interactions and the other approach attempts to join existing models. Neither one is sufficient. An integration of these approaches would provide the most promise for future models and is the direction the CGM is taking.

REFERENCES

Chesapeake Bay Program, The State of the Chesapeake Bay, 1995. Chesapeake Bay Program Report, Annapolis, MD, 1995. http://www.chesapeakebay.net.

Chrisman, N.R., *Exploring Geographic Information Systems*, John Wiley & Sons, New York, 1997.

Clarke, K.C. and Gaydos, L. Long term urban growth prediction using a cellular automaton model and GIS: applications in San Francisco and Washington/Baltimore, *Int. J. Geograph. Inf. Sci.*, 12, 699–714, 1998.

Costanza, R. and Matthias, R. Using dynamic modeling to scope environmental problems and build consensus, *Environ. Manage.*, 22, 183–195, 1998.

Environmental Systems Research Institute, Inc. (ESRI), *Grid Commands*, Redlands, CA, 1994.

Flamm, R.O. and Turner, M.G., Alternative model formulations for a stochastic simulation of landscape change, *Landscape Ecol.*, 9, 37–46, 1994.

Glesson, M.E., Effects of an urban growth management system on land values, *Land Econ.*, 55, 350–365, 1979.

Hazen, B.C. and Berry, M.W., The simulation of land-cover change using a distributed computing environment, *Simulation Pract. Theo.*, 5, 489–514, 1997.

Hunsaker, C.T. and Levine, D.A., Hierarchical approaches to the study of water quality in rivers, *BioScience*, 45, 194–203, 1995.

King, D.A. and Sinden, J.A., Price formation in farm land markets, *Land Econ.*, 70, 38–52, 1994.

Kirtland, D.L., Gaydos, L., Clarke, K., DeCola, L., Acevedo, W., and Johnson, C.B., An Analysis of Human-Induced Land Transformations in the San Francisco Bay/Sacramento Area, http://geo.arc.nasa.gov/usgs/WRR_paper.html, USGS, EROS Data Center, Ames, IA, 1996.

Knapp, G.J., The price effects of urban growth boundaries in metropolitan Portland, Oregon, *Land Econ.*, 61, 26–35, 1985.

La Gro, J.A., Jr. and DeGloria, S.D., Land use dynamics within an urbanizing non-metropolitan county in New York State (USA), *Landscape Ecol.*, 7, 275–289, 1992.

Logsdon, M.G., Bell, E.J., and Westerlund, F.V., Probability mapping of land use changes: a GIS interface for visualizing transition probabilities, *Comput. Environ. Urban Syst.*, 2, 389–398, 1996.

Martinez, F., MUSSA: land use model for Santiago City, *Transp. Res. Rec.*, 1552, 126–134, 1992.

Nelson, A.C., An empirical note on how regional urban containment policy influences an interaction between greenbelt and exurban land markets, *Am. Psychol. Assoc. J.*, 178–184, 1998.

Pond, B. and Yeates, M., Rural/urban land conversion II: identifying land in transition to urban use, *Urban Geogr.* 15, 25–44, 1994.

Smith, E.R., An overview of EPA's Regional Vulnerability Assessment (ReVA) Program, *Environ. Monitoring Assess.*, unpublished manuscript, 1999.

Stopher, P.R. Hartgen, D.T., and Li, Y., SMART: simulation model for activities, resources and travel, *Transportation*, 23, 293–312, 1996.

Turner, M.G., Wear, D.N., and Flamm, R.O. Land ownership and land-cover change in the southern appalachian highlands and the olympic peninsula, *Ecol. Appl.*, 6, 1150–1172, 1996.

Vaillancourt, F. and Monty, L., The effect of agricultural zoning on land prices, Quebec, 1975–1981, *Land Econ.*, 67, 36–42, 1985.

Veldekamp, A. and Fresco, L.O. Exploring land use scenarios, an alternative approach based on actual land use, *Agric. Syst.*, 55, 1–17, 1995.

Vesterby, M. and Heimlich, R.E., Land use and demographic change: results from fast growth counties, *Land Econ.*, 76, 279–291, 1991.

Woods & Poole, Inc., *CEDDS 1998: The Complete Economic and Demographic Data Source*, Washington, D.C., 1998.

Section II.7

Agriculture and Human Health

CHAPTER 77

Overview: Agriculture and Human Health

Marc Schenker

While the impact of ecosystem changes in agriculture on human health have not been accurately characterized, the occupational health hazards of agricultural work have been well described (Schenker, 1996). Agriculture ranks as one of the two or three most hazardous occupations based on fatality rates, with an injury mortality rate approximately three to five times that for all other occupations (Myers, 1990). More than 700 agricultural work-related deaths and 140,000 disabling injuries occur annually in the U.S. (Council, 1999).

Many illnesses (cardiovascular, respiratory, cancer, dermatitis, infection, neuropsychiatric disorders) occur more commonly among agricultural workers than in the general population, but the full burden of illness from acute and chronic occupational exposures in the agricultural workplace is unknown. However, consideration of illnesses from agricultural work illustrates the breadth of hazardous exposures in this environment, and provides some direction for future studies of ecosystem impact on health in the agricultural environment. Consideration of agricultural respiratory disease (Schenker, 1998) helps reveal the scope of the problem. The spectrum of agents causing respiratory disease from agricultural exposures ranges from simple irritant gases, to crystalline silica, to a variety of microorganisms (Table 77.1).

Documented agricultural respiratory diseases from these diverse exposures similarly cover a very broad spectrum. They include:

- Rhinitis (allergic and non-allergic)
- Asthma
- Asthma-like syndrome
- Tracheobronchitis
- Chronic obstructive lung disease
- Organic dust toxic syndrome
- Hypersensitivity pneumonitis
- Interstitial fibrosis
- Respiratory infections (bacterial, viral, mycobacterial, fungal, rickettsial)

While research is beginning to elucidate the risks of respiratory disease from occupational exposures in agriculture, very little is known about the effects of these exposures on agrosystem ecology and the health implications resulting from perturbations to the ecosystem from agriculture. For example, agricultural burning has been shown to increase ambient air pollution significantly, and some studies have shown an association of agricultural burning with acute respiratory disease among area residents (Jacobs et al., 1997). Interstitial fibrosis has been observed among rural

Table 77.1 Agricultural Exposures Affecting Respiratory Health

	Examples
Organic dusts	Pollens, cotton dust, fungal conidia
Microorganisms	Bacteria, fungi, viruses, mycobacterium
Mycotoxins	Aflatoxin, trichothecenes
Endotoxins	Heat-stable lipopolysaccharide protein complexes from Gram-negative bacteria
Decomposition and other gases	NH_3, H_2S, CO_2, CO, NO, NO_2
Pesticides	Methyl bromide, metam-sodium
Fertilizers	Anhydrous ammonia
Feed additives	Vitamins, antibiotics
Inorganic dusts	Crystalline silica, silicates

residents with no history of agricultural work who reside in dry-climate farming regions (Sherwin et al., 1979). These harmful respiratory effects caused by agricultural practices are in contrast to the results of recent studies documenting a *decrease* in asthma prevalence among rural and agricultural residents, suggesting a benefit from environmental exposures while growing up in an agricultural environment (Ernst and Cormier, 2000).

Our knowledge about other health outcomes (e.g., neurologic disease, chronic immune disorders, cancer) is even less than for respiratory disease. For example, some studies have suggested that incidence of Parkinson's disease is elevated among residents in agricultural settings, but other studies have not observed such an association (Goldsmith et al., 1990; Semchuk et al., 1992).

Several agricultural exposures or environments are of particular concern with regard to their effects on the ecosystem and on human health. These include (1) agrochemicals, including pesticides and fertilizers; (2) agricultural wastes and wastewater, and specifically infectious diseases in this environment; (3) air pollutants from a variety of agricultural processes, including chemical spraying, soil preparation, and agricultural burning; and (4) genetic modifications of agricultural products.

Increased efforts at developing sustainable agriculture practices have begun to focus attention of the agricultural community on the broader ecosystem. In addition to managing agrosystems for long-term economic viability, efforts are developing to enhance environmental sustainability and human health. The social and political attention given to health effects of genetically modified foods has certainly increased concern about the social acceptability (and ecologic impacts) of agricultural practices (Longman, 1999).

In summary, scientists and others now increasingly recognize that agricultural policies must be developed with consideration of their ecologic and health implications, and similarly that environmental policies must consider their impact on agriculture. The result of this process will be systems that support both sustainable agriculture and ecologic health. To do less will result in unnecessary injuries, illnesses, and deaths.

The chapters in this section examine several different areas of the ecosystem and human health problems related to agriculture. Gunderson et al. provide an insightful overview of agricultural impacts on ecosystem health, including a historical perspective and a look into future practices that may affect this area. This seminal chapter covers a range of topics from economic and public policy factors to potential ecosystem and health effects of genetically modified foods. Three chapters raise different major issues involved with the potential effects of agriculture on human and ecosystem health — infectious diseases, air pollution, and pesticides.

The topics covered in this section are of profound importance to society. As noted by Gunderson et al., the future of society itself is dependent on preservation of sustainable agricultural production, which is in turn dependent on systems that maintain ecosystem and human health.

REFERENCES

Ernst, P. and Cormier, Y., Relative scarcity of asthma and atopy among rural adolescents raised on a farm, *Am. J. Resp. Crit. Care Med.*, 161, 1563–1566, 2000.

Goldsmith, J.R., Herishanu, Y., Abarbanel, J.M., and Weinbaum, Z., Clustering of Parkinson's disease points to environmental etiology, *Arch. Environ. Health*, 45, 88–94, 1990.

Jacobs, J., Kreutzer, R., and Smith, D., Rice burning and asthma hospitalizations, Butte County, California, 1983–1992, *Environ. Health Perspect.*, 105, 980–985, 1997.

Longman, P.J., The curse of Frankenfood. Genetically modified crops stir up controversy at home and abroad, *US News World Rep.*, Jul. 26, 1999, pp. 38–41.

Myers, J.R., National surveillance of occupational fatalities in agriculture, *Am. J. Ind. Med.*, 18, 163–168, 1990.

National Safety Council, *Accident Facts,* Itasca, IL, 1999.

Schenker, M., Respiratory health hazards in agriculture, *Am. J. Respir. Crit. Care Med.*, 158, S1–S76, 1998.

Schenker, M.B., Preventive medicine and health promotion are overdue in the agriculture workplace, *J. Public Health Policy,* 17, 275–305, 1996.

Semchuk, K.M., Love, E.J., and Lee, R.G., Parkinson's disease and exposure to agricultural work and pesticide chemicals, *Neurology,* 42, 1328–1335, 1992.

Sherwin, R.P., Barman, M.L., and Abraham, J.L., Silicate pneumoconiosis of farm workers, *Lab. Invest.*, 40, 576–582, 1979.

CHAPTER 78

Effects of Agriculture on Ecosystem and Human Health

Paul D. Gunderson, Nancy B. Young, and Dean T. Stueland

INTRODUCTION

For most of the 20th century the American farm has performed at Herculean production levels. From time to time these production levels have been presumed to be at their limits; however, with new discoveries those limits were pushed well beyond what contemporaries thought achievable. En route the American farmer has been characterized as extraordinarily productive and resilient, as an early adapter of technological change, and a master at integrating signals from the world marketplace with individual agricultural enterprises. Described since the inception of our nation in heroic and generous terms (Jefferson, 1785), the American farmer pursues the objective of remaining competitively productive and capable of feeding the world (Pedraza, 1996).

DISCUSSION

Concurrent with the emergence of the American farmer as a key player in rural agricultural politics has been the recognition that sustainable agricultural production cannot remain a luxury. Once an agricultural resource base has been eroded, the larger society it sustains cannot remain viable (Runnel, 1995). Sustainability must be maintained in the face of projections that the world in the year 2060 will contain between 10 and 11.5 billion people and the discomforting evidence that sustainable gains in the yield of agricultural commodities will be difficult to maintain (Faeth, 1993; Brummer, 1998; Mohapatra, 1999; Agnew, 1999; Mann, 1997).

The long-term capability to evaluate the ecological impact of agronomic practice on the larger ecosystem is essential. Fortunately, these evaluations have been under way for some time. The first was initiated in the mid-1800s with the establishment of the Rothamsted experimental station in England in 1843. Other sites in the English-speaking world include Morrow, IL (1876) and Sanborne, MO (1888) in the U.S.; Longerenong (1917) and Waite (1925) in Australia; and Lethbridge (1911) and Breton (1930) in Canada. More recently, a substantial number of long-term agroecosystem experiments have been established in other parts of the world. In general these initiatives have (1) identified management practices capable of either sustaining or foreshortening agricultural productivity across long time frames; and (2) assessed the differential impacts from natural and synthetic crop additives, fertilizers, and pesticides. In addition, these experiments generated insight that has enabled scientists to model biological impact of overall change in global climates and the effect of increased atmospheric CO_2 on carbon sequestration (Houghton et al., 1999). As time passed, scientists involved in these initiatives called for less fragmentation in the agricultural research enterprise and more interaction with other disciplines, including the health sector (National Research Council, 1996; Wilson, 1998).

Unfortunately, these experiments were never designed to assess the human health impact of technological change occurring in agricultural production systems nor the evolutionary impact of these systems on human populations and adjacent forms of biology. Accordingly, we must turn elsewhere in order to marshal the evidence that exists relative to agricultural production on human health. It will be important to differentiate between health effects observable within the larger society vs. those experienced by human populations with the greatest exposure to agricultural practices. The most immediate exposures affect agricultural workers, resident households, and populations immediately adjacent to acreage devoted to agricultural production.

North American agriculture experienced several revolutions during the 20th century. The century began with mechanization of agricultural tasks and the widespread diffusion of information technologies; this was followed by the hybridization of plants, the technological transformation of whole sectors of the industry, and the recent emergence of *accelerated* genetically modified crops. Each revolution has changed the way in which agricultural workers react to these work environments. Wide variations in federal agricultural policy occurred simultaneously with these developments, including the emergence of environmental policy, public investment in agricultural research, response to global warming and other climatic forces, and consumer concern about the safety of food destined for human consumption. Each of these developments has or will have a significant impact upon the work tasks that agricultural populations pursue. Only a few of these impacts can be explored here.

Advances in agricultural technology have reduced the enormous levels of physical drudgery associated with many conventional agricultural tasks. Historically, from basic tillage of the soil, through the handling of agricultural inputs, to the harvesting and postharvest transfer of agricultural production, the human musculoskeletal system has absorbed the load. Now, much of that load is carried by technological invention in technologically advanced societies, resulting in substantially reduced levels of musculoskeletal disease or injury. Today's agricultural workers must concern themselves with the potential for mechanical entanglement; exposure to volatile organic chemicals resulting from use of the internal combustion engine; inhalation of silica, fungi, and spores that are disturbed and transported by mechanical cultivation and harvesting technologies; or exposure to agricultural pesticides. On balance, however, those who treat disease and injury associated with agricultural exposures believe that the incidence of those conditions resulting from the infusion of technology on the North American farm has stabilized.

On another front, until public policy began to encourage landowners, farmers, and ranchers to develop green ecological-filtering masses around fragile stream banks, ponds, and lakes, the potential for transmission of harmful protozoa that can survive in the bovine, human gut, and groundwater was intensifying. This is an important development given the dependence of agricultural populations and their urban neighbors on watersheds. As grazing livestock are moved away from fragile watersheds, the potential for disease among human populations should be substantially reduced.

As livestock operations increased their use of intensive confinement systems, the potential for increased virulence of bacteria and spread of zoonotic disease and the possible development of resistance to antibiotics in conventional human use have emerged (Borchardt et al., 1999; Diez-Gonzalez et al., 1998; New York Academy of Sciences, 1999). These new technologies were developed in an effort to reduce input resources, substantially increase labor efficiency, and improve the health status of livestock through management of wide thermal variances. However, these technologies can opportunistically induce human disease and thwart conventional public health containment measures. Unfortunately, the technologies to store and process the massive accumulations of livestock waste lagged design and development of livestock confinement housing. The result is accelerated environmental risk when these systems do not work or malfunction in use.

Potentially, the most important revolution of the 20th century will result from the manipulation of the DNA of plants and livestock. In spite of current European and Asian objections, the world is likely to derive the majority of its food, natural fiber, renewable fuels, lubricants, cosmetic and pharmaceutical base components, and livestock feed from vegetation or forest preserves in which some genetic component has been changed (Abelson, 1998). Among the numerous impacts that are coupled to this revolution is its potential to reduce human exposure to agricultural pesticides currently in use. Alternative practices for the control of pests will change the nature, duration, and frequency of human exposures to both natural and synthetic crop and livestock pesticides. Genetically modified organisms that permit the use of more environmentally benign pesticides or agronomic practices and that provide alternative approaches to the control of pests (e.g., crop rotation, use of narrow spectrum natural pesticides or beneficial insect populations, or genetically different organisms that can be applied to plants themselves or the soil from which they draw their nutrients) could significantly reduce human exposure. When one couples these developments to technological change in application techniques, it is possible to envision a future where the agronomic tool kit will be radically different from that in use today (Lubchenco, 1998), virtually eliminating human exposure to conventional crop protection products.

The technological transformation of North American agriculture significantly reduced respiratory insult experienced by agricultural workers. New forage and grain-harvesting technologies permitted a concomitant reduction in airborne particulates that caused farmer's lung and other forms of hypersensitivity pneumonitis. Exposures remain, given that the structural integrity of protective work enclosures is often compromised in agricultural settings or workers may refuse to use personal protective equipment due to behavioral or thermal barriers. However, in the absence of major natural calamities, it is reasonable to expect that current levels of respiratory disease among working agricultural populations will continue to decline. Natural calamities, particularly in wet growing cycles, possess the potential of reversing the observed declines in respiratory disease, since microbial particulate found in many soil profiles is washed onto plant stems and grain surfaces, providing a route for aerosolization during harvesting activities. Wide fluctuations in growing climates across the world could dramatically impact on the incidence of respiratory disease of workers and populations adjacent to cultivated areas.

The ecological impact of North American agriculture can also be detected on wider community, regional, or national scales. For example, the change in agronomic practice that has transformed large areas of North America from forested verandas to legume cover has contributed substantially to the increase in the amount of fixed nitrogen that enters the global terrestrial nitrogen cycle (Malakoff, 1998). This development possesses potentially serious consequences for human populations far removed from agricultural production sites. Nitrogen loading is likely to have contributed to the observed declines in aquatic wildlife and the intensification of algal blossoming leading to significant human disease outbreaks (e.g., *Pfeisteria piscicida*). Additionally, since land now devoted to legume production must be productively maintained through the addition of organic and synthetic fertilizers, groundwater and surface water flowing out of these areas into holding ponds and reservoirs, estuaries, and coastal waters could impact human health on a wide scale. As watersheds become overwhelmed by agricultural runoff, public drinking water systems are likely to experience increased concentrations of nitrites and bacterial loading.

A second generalized population effect associated with current agronomic practice can be detected in the nontarget proliferation of selected agricultural pesticides. Although most products in current use such as organophosphates and carbamates are short-lived, they are acutely toxic, and some contain breakdown products that have been demonstrated in laboratory studies to suppress or alter basic immunology as well as mimic hormone activity (Matson et al., 1997). One key control strategy could be the adoption of new technologies which rely on spatial and temporal response to nutrient demand. Advanced farming systems that couple global positioning technologies to on-the-go variable rate capability permit precise metering and response to the nutrient requirements of cereal and other feed grain crops (Moffat, 1996). With advancing biotechnology it is also possible to mate selected geographic or specie-specific characteristics such as perennial durability with maximized uptake of N, K, P, or heavy metals.

The emerging trend to house large numbers of livestock and poultry in concentrated settings holds various dangers. Alarms have been triggered fueled by the discovery that some bacteria that cause human disease have developed specialized resistance to selected antibiotics in common use in animal feeds. This development is particularly unsettling for hospitals where bacterial resistance has been a facet of clinical life for some time. Resistance to antimicrobial pharmaceuticals emerges either as the result of new mutations in the bacterial genome or through acquiring resistant new genes (Witte, 1998). While the empirical verdict is still out (Anon., 1999), prudence would suggest that a rigorous risk assessment must soon be undertaken relative to the use of antibiotic additives in livestock feed (Anon., 1999). This is imperative given that current animal husbandry practice typically calls for background use of these pharmaceuticals to promote disease-free growth of livestock.

Three related potential effects are associated with perpetuation of dominant agronomy practices in existence at the end of the 20th century. The first of these reflects the desire by many sectors of production agriculture, as well as numerous governments, to integrate fully the free enterprise model into most if not all of the world's agricultural sectors. Unfortunately, as Hardin so eloquently stated, "Freedom in a commons brings ruin to all," which from its original state has been extended to, "Individualism is cherished because it produces freedom, but the gift is conditional: the more the population exceeds the carrying capacity of the environment, the more freedoms must be given up" (Hardin, 1968, 1998). Should this postulate hold true, it would appear that unilateral adoption of the free enterprise system in which each productive agricultural entity is locked into a system that causes the entity to maximize production could result in a unsustainable worldwide environment.

Second, an inherent risk associated with manipulation of plant and animal genomic structures is the potential for restriction of biodiversity. While the world's scientists plead for public financing of plant and animal genomic banks, many of those pleas have fallen on deaf ears. Should diversity of genomic structures be reduced, the potential for future food scarcity, given the evolutionary nature of agronomic pests, looms as an ever-present threat.

Third, some evidence exists that migration of modified genetic structures to nontarget species can occur. While much public misunderstanding, indeed specific misinformation, has been transmitted by the popular press, justifiable scientific concern remains about the potential for such movement. Should that occur and the means to regulate and ultimately thwart such movement be restricted, the potential exists for future food shortages, since the ultimate impact could be unplanned retardation and removal of nontarget food species.

The ecological impact of these developments is obviously varied. However, the tools at the disposal of scientific establishments are likely to exist so that the embedded technologies can be harnessed toward the health and well-being of the world's human and animal populations. That we have yet some distance to travel should propel all of us to redouble our effort and focus our scientific zeal.

REFERENCES

Abelson, P.H., A third technological revolution, editorial, *Science*, 279, 2019, 1998.
Agnew, B., Crop scientists seek a new revolution, *Science*, 283, 1038–1043, 1999.
Anon., Experts Call for Sound Science on Antibiotics, *Feedstuffs*, June 7, 1999.
Borchardt, M.A., Spencer, S.K., Moore, F.M., Richard, L., Berg, R.L., and Shukla, S.K., Campylobacter subtype and virulence variability between diarrheic and asymptomatic dairy cows. Marshfield Medical Research Foundation, Marshfield Clinic, Marshfield, WI, 1999.

Brummer, E.C., Editorial, *Agron. J.*, 90, 1, 1998.

Diez-Gonzalez, F., Callaway, T.R., Kizoulis, M.G., and Russell, J.B., Grain feeding and the dissemination of acid-resistant *Escherichia coli* from cattle, *Science*, 281, 1666–1668, 1998.

Faeth, P., Agricultural Policy and Sustainability: Case Studies from India, Chile, and the United States, World Resources Institute, Washington, D.C., 1993.

Hardin, G., Extensions of "the tragedy of the commons," *Science*, 280, 680–682, 1998.

Hardin, G., Tragedy of the "commons," *Science*, 162, 1243–1244, 1968.

Houghton, R.A., Hackler, J.L., and Lawrence, K.T., The U.S. carbon budget: contributions from land-use change, *Science*, 285, 574–578, 1999.

Jefferson, T., Thomas Jefferson to John Jay, 1785.

Lubchenco, J., Entering the century of the environment: a new social contract for science, *Science*, 279, 491–496, 1998.

Malakoff, D., Death by suffocation in the Gulf of Mexico, *Science*, 281, 190–192, 1998.

Mann, C., Reseeding the green revolution, *Science*, 277, 310–314, 322, 1997.

Matson, P.A., Parton, W.J., Power, A.G., and Swift, M.J., Agricultural intensification and ecosystem properties, *Science*, 277, 504–509, 1997.

Moffat, A.S., Higher yielding perennials point the way to new crops, *Science*, 274, 1469–1470, 1996.

Mohapatra, S.C., World hunger, *Resource*, 6, 33, 1999.

National Research Council, *Colleges of Agriculture at the Land Grant Universities: Public Service and Public Policy*, National Academy Press, Washington, D.C., 1996.

New York Academy of Sciences, Infectious diseases and public health, *Annals*, 1999.

Pedraza, J.M., Are we ready to feed the world? *Agric. Week,* 12(1), 28–29.

Runnel, C.N., Environmental degradation in ancient greece, *Sci. Am.*, 272, 96–99, 1995.

Wilson, E.O., Integrated science and the coming century of the environment, *Science*, 279, 2046–2047, 1998.

Witte, W., Medical consequences of antibiotic use in agriculture, *Science*, 279, 996–997, 1998.

CHAPTER 79

Infectious Disease Hazards to Agricultural Workers

Dean O. Cliver

The National Food Safety Initiative of May 1997 (Anonymous, 1997) refers repeatedly to manure management and implies that the use of animal manure in vegetable crop production constitutes a hazard to consumer health. Indeed, raw produce has been the vehicle in a considerable number of foodborne disease outbreaks in the U.S. (Beuchat, 1998), and most of the disease agents involved have been zoonotic infectious agents (agents that cause disease in humans but are usually of animal origin). If this is true, those who work in vegetable-growing fields to which manure is applied should be at greater risk, though little has been done to test this hypothesis (Coia et al., 1998). The zoonotic agents of greatest concern are *Escherichia coli* O157:H7, *Salmonella* spp., and *Cryptosporidium parvum*. These are shed, under certain circumstances, in the manure of food animals. The source animals are generally healthy, although mild diarrhea may sometimes be seen.

The effects of various treatments and storage of manure on the infectivity of zoonotic bacteria and protozoa are under study (Himathongkham et al., 1999); methods that destroy infectivity are known, but not widely used. The most rigorous treatment generally used is composting. It is important to note that composting is not defined in law or regulation, so that some manure handlers regard any stack that heats as undergoing the composting process. In fact, true composting, which entails turning a stack repeatedly to ensure aeration and efficient generation of heat by the microflora of the material, may not be necessary to reduce the named zoonotic pathogens to undetectable levels. The problem is that there are no guidelines or specifications that ensure that, given a certain treatment of manure, the pathogens will no longer be present and transmissible via the produce that may be fertilized with it. Dependence on events in soil to destroy pathogens from manure (California Certified Organic Farmers, 1996) is also inadequately supported by evidence.

People infected with such zoonotic agents may experience mild to severe intestinal illness, will shed the agents in their feces, and are able to transmit them to others (Benenson, 1995). In addition to causing diarrhea that may show frank, red blood in the stool, *E. coli* O157:H7 can cause life-threatening illnesses: hemolytic uremic syndrome (typically in children under 5 years of age) and thrombotic thrombocytopenic purpura (typically in older persons). *Salmonella* spp., after the gastroenteritis has resolved, may cause lifelong arthritis. *Cryptosporidium parvum* typically causes only severe diarrhea, but there is no accepted treatment for the infection, so persons who are immunologically impaired (e.g., AIDS patients) may remain infected for the rest of their lives, with intermittent recurrences of diarrhea.

Because infected persons shed these agents in their feces, surveys — using various methods of fecal sampling — are needed to determine the incidence of such infections among agricultural workers, as related to occupational exposure. For logistical and cultural reasons, sampling arrangements have proven difficult. For example, in its efforts to protect worker health and the environment, the California Occupational Health and Safety Agency has regulations that require provision of portable toilets with effective disinfectants in their sumps and the use of these toilets by agricultural field workers. Our trials indicate that the disinfectants are typically quite effective, which makes the portable toilets of little value as noninvasive fecal sampling facilities. Another serious problem in the establishment of sampling programs is the perceived liability of the grower. If infected workers are identified in a food producing operation, there is a real probability that the product will be embargoed. Therefore, growers would rather trust the judgment of the field foreman as to the health of the workers than have fecal samples being tested in a laboratory. It is certainly true that not all infections with the agents discussed here result in disease, so more workers are likely to be infected than are overtly ill. *Silent infections* are unlikely to be detected other than in the laboratory, yet they could cause serious repercussions to the producer. In surveys of farm environments for infectious agents (outside the context of human feces), it has been necessary to have the sample taker code the samples in such a way that positive test results cannot be associated with the farm of origin. It may be necessary to adopt similar measures in studies of farmworker intestinal infections.

It is also important to recognize that treatments applied to manure to protect the consumer may not protect workers, in that at least some workers may be exposed to the manure before it is treated. This certainly holds true for those working in animal production, who will be in contact with freshly voided cow, chicken, or other animal manure. Therefore, control groups for studies of infections in workers growing produce should include people who work in fields to which no manure is applied and those who work directly with food animals and their manure.

In summary, some disease agents that may be transmitted to humans are shed in animal manure. It is known that these agents may contaminate produce via manure and thus lead to consumer illnesses. The degree to which agricultural workers, who are exposed to contaminated manure in the process of growing the vegetables, are infected needs to be determined. It seems reasonable to suppose that the field workers are more at risk than the consumer.

REFERENCES

Anonymous, Food Safety from Farm to Table: A Report to the President. U.S. Environmental Protection Agency, U.S. Department of Health and Human Services, and U.S. Department of Agriculture, Washington, D.C., 1997.

Benenson, A.S., Ed., *Control of Communicable Diseases Manual*, 16th ed., American Public Health Association, Washington, D.C., 1995.

Beuchat, L.R., Surface Decontamination of Fruits and Vegetables Eaten Raw: A Review, WHO/FSF/FOS/ 98.2, Food Safety Unit, World Health Organization, Geneva, Switzerland, 1998.

California Certified Organic Farmers, *Certification Handbook*, Santa Cruz, CA, 1996.

Coia, J.E., Sharp, J.C.M., Campbell, D.M., Curnow, J., and Ramsay, C.N., Environmental risk factors for sporadic *Escherichia coli* O157 infection in Scotland: results of a descriptive epidemiology study, *J. Infect.*, 36, 317–321, 1998.

Himathongkham, S., Bahari, S., Riemann, H., and Cliver, D., Survival of *Escherichia coli* O157:H7 and *Salmonella typhimurium* in cow manure and cow manure slurry, *FEMS Microbiol. Lett.*, 178, 251–257, 1999.

CHAPTER 80

Size Distribution of PM$_{10}$ Soil Dust Emissions from Harvesting Crops

Robert T. Matsumura, Lowell Ashbaugh, Teresa James, Omar Carvacho, and Robert Flocchini

INTRODUCTION

The majority of prior work on fugitive dust emissions in the U.S. has been related to wind erosion, paved and unpaved roads, agricultural practices, and source sampling of point or area sources such as construction and mining. Throughout this past work, the National Ambient Air Quality Standard for suspended particles has been based primarily on particle size:

- 1971 to 1987 — Total suspended particulate (TSP)
- 1987 to current — PM$_{10}$ or particulate matter less than or equal to 10 µm in aerodynamic diameter

The earlier TSP standard consisted simply of all particulate matter (PM) that could remain suspended in the atmosphere while the current PM$_{10}$ standard consists of suspended particles small enough to be inhaled and pose potential health effects. With the current reevaluation of the PM standard by the U.S. Environmental Protection Agency (EPA), another size cut at 2.5 µm has been recommended for compliance monitoring (Chow, 1995). The inhalation properties of the fine (<2.5 µm) particles are more relevant to health issues since they can be deposited deeper into the human respiratory system than coarse (2.5 to 10 µm) PM$_{10}$ particles. PM$_{2.5}$ particles are also primarily secondary aerosols that form in the atmosphere and are more likely to contain hazardous chemical compounds.

Knowledge of fugitive dust size distributions may be important in determining which chemical compounds are associated with or adhere to suspended soil particles. Each particle size fraction has different inhalation properties that determine how far into the respiratory system the particles are deposited. For these reasons it is important that fugitive dust databases include species and size specific measurements. This is especially important because of the relatively wide particle size distribution of dust emissions.

HARVESTING

The San Joaquin Valley air basin, located in the southern portion of California's Central Valley, has been classified nonattainment for PM$_{10}$ primarily for exceedances that occur in the winter period. Fugitive dust is often the largest fraction of PM$_{10}$ in the summer and fall seasons (Chow

Table 80.1 Inventory of Top Ten Crops by Harvested Acreage in the San Joaquin Valley (SJV) Air Basin for Harvested Acreage and Cash Value in the SJV and California[a]

Crop	Harvested Acreage in SJV[b]	Value (U.S. $) in SJV[b]	Harvested Acreage in California[c]	Value (U.S. $) in California[c]	California Share of U.S.[c] (%)
All cotton — lint	1,134,603	1,110,555,000	1,136,000	991,559,000	19.5
Hay — alfalfa and other	338,578	239,496,000	1,380,000	748,857,000	5.1
Grapes — raisin type	271,204	758,866,000	271,000	613,080,000	89.4
Almonds — all	208,093	577,805,000	402,000	911,430,000	99.9
Corn for silage	190,300	93,644,000	N/A	N/A	N/A
Grapes — wine type	149,340	354,949,000	307,000	854,485,000	89.4
Oranges — navel	107,543	424,250,000	114,000	277,008,000	26.7
Tomatoes — processing	106,900	179,862,000	274,000	529,038,000	92.7
Wheat	90,600	25,834,000	540,000	141,390,000	1.8
Walnuts	81,487	199,711,000	175,000	364,000,000	99

[a] The California share of U.S. production is also included. Note that the dollar values in the SJV and in California are obtained from different sources, so they may not be consistent.
[b] Annual Agricultural Reports (1993). Tabulated from various 1993 county agricultural commissioners reports.
[c] California Agricultural Statistics (1993).

et al., 1990). Many crops are harvested in the later part of this period. Table 80.1 shows the ten most widespread crops by harvested acreage in the San Joaquin Valley in 1993. Of these ten crops, the harvests of almonds, cotton, walnuts, and wheat were identified as possibly having high fugitive dust emissions. The purpose of this study was to evaluate the size of PM emitted during harvesting.

Beginning in 1994, the Air Quality Group of the University of California measured PM_{10} fugitive dust emissions from the harvest of almonds, figs, walnuts, and cotton. The three orchard crops (almonds, figs, walnuts) are almost exclusively mechanically harvested using the same type of harvesting equipment. Their harvest entails three separate operations, although figs and walnuts combine sweeping and pickup into a single operation conducted simultaneously:

- *Shaking* of nuts from the tree onto the ground using a mechanical arm. (Figs naturally fall to the ground as they dry, so this operation does not apply to figs.) Almonds are allowed to dry on the ground for a few days.
- *Sweeping* of nuts or figs into windrows between tree rows using mechanical sweepers and blowers.
- *Pickup* of nuts or figs, using a harvester that gathers the nuts by sweeping, blows away dirt and debris, and unloads the nuts or figs into a trailer or bin container.

Cotton harvesting includes three basic operations. We examined only the first two in 1994.

- *Picking* of cotton from defoliated cotton stalks using revolving spindles to remove the cotton (with seed) from stalks. The cotton is then blown upward through tubes into a cage. Newer harvesters pick four or five rows of cotton during each pass, while older versions pick only two rows.
- *Stalk cutting* by flail shredders that cut and mulch the picked cotton stalks. Covered blades are spaced apart to cut several rows of cotton during each pass.
- *Stalk incorporation* using a disk to turn over the soil and incorporate the cut stalks into the ground.

METHODS

As part of our program to measure fugitive dust emissions from agricultural activities, we measured PM_{10} and $PM_{2.5}$ at locations upwind and downwind of various harvest operations. James et al. (1995), Flocchini et al. (1994), and Matsumura (1992) describe the sample collection and analytical techniques used. Table 80.2 summarizes the measurements. All aerosol samples discussed

Table 80.2 Summary of Aerosol Measurements

Particle Size	Sampler	Size-Cut	Substrate	Analysis (Species)
PM$_{10}$	IMPROVE Module D[a]	Size selective inlet[b]	25 mm Teflon filter	Gravimetric (mass) Laser integrating plate method (optical absorption) Elemental analysis[c] (elements H, Na to Pb)
PM$_{2.5}$	IMPROVE Module A[a]	Cyclone[d]	25 mm Teflon filter	Gravimetric (mass) Laser integrating plate method (optical absorption) Elemental analysis[c] (elements H, Na to Pb)
10 to ~15 µm, 5.0 to 10 µm, 2.5 to 5.0 µm, 1.15 to 2.5 µm, 0.07 to 1.15 µm, and <0.07 µm	DRUM Impactor[e]	Cascade impactor	Greased mylar plus 25 mm Teflon after-filter	PIXE[e] (elements Si to Pb)

[a] Eldred et al. (1988, 1990), Malm et al. (1994).
[b] Sierra-Anderson Model 246b PM$_{10}$ inlet (16.7 lpm).
[c] Elemental analysis includes: (1) Proton Elastic Scattering (PESA) for hydrogen (Cahill et al. 1989); (2) X-Ray Fluorescence (XRF) for S to Pb; and (3) Proton Induced X-Ray Emission (PIXE) (Eldred et al., 1988)
[d] John et al. (1988), Eldred et al. (1988).
[e] Cahill et al. (1987).

here were collected at a height of 3 m above ground at distances ranging from about 3 to 100 m downwind of each harvested field.

Our PM$_{10}$ research focused on total gravimetric mass and geological material (soil dust). Through our suite of particulate measurements, we are often able to account for nearly all of the mass collected on Teflon filters. In this process, measured elemental species are converted to *composite variables* using some basic assumptions. For example, the majority of airborne soil is assumed to be composed of the geological material in their soil oxide forms plus a correction factor for other compounds, such as MgO, Na$_2$O, water, and carbonate. By measuring the soil elements and converting them to their soil oxide forms we are able to reconstruct the soil component of measured aerosols.

We refer to *reconstructed soil* as the sum of normal soil oxides calculated from our elemental measurements of particulate samples:

$$SOIL = (2.20 \times Al) + (2.49 \times Si) + (1.63 \times Ca) + (2.42 \times Fe) + (1.94 \times Ti)$$

Table 80.3 shows how the conversion factors for each soil element were derived. Each conversion factor is divided by 0.86 to account for other nonmeasured compounds since the soil oxides account for 86% of average sediment (Lide, 1992). Potassium (K) is not directly used in the calculation of the soil composite because of a nonsoil fine potassium component from ambient

Table 80.3 Data Table Showing the Calculations Used in Deriving Reconstructed Soil from Elemental Values

Measured Element	Elemental Mass (g)	Soil Oxide	Soil Oxide Mass (g)	Soil Oxide/ Element	Compositional Divisor	Conversion Factor
Al	26.98	Al$_2$O$_3$	101.96	1.89	0.86	2.20
Si	28.09	SiO$_2$	60.08	2.14	0.86	2.49
Ca	40.08	CaO	56.08	1.40	0.86	1.63
Fe	55.85	2*FeO+Fe$_2$O$_3$	303.40	1.36	0.86	1.58
Fe (K)	39.08	K$_2$O * 0.6	56.50	0.72	0.86	0.84
Ti	47.90	TiO$_2$	79.90	1.67	0.86	1.94

smoke. Instead, iron (Fe) is used as a surrogate for soil K assuming that the average K/Fe ratio for coarse particles is 0.6 ± 0.2 (Cahill et al., 1986). Iron is included as the sum of two iron oxides present in equal molar concentration and a potassium component that is derived from the measured Fe.

RESULTS

Table 80.4 compares $PM_{2.5}$ and PM_{10} for mass and reconstructed soil from filter measurements at the downwind field edges where harvesting operations occurred. *Close* sampler locations refer to downwind locations 3 to 10 m from the edge of the field while *far* represents locations 50 to 100 m downwind. To generalize the harvest emissions, all operations are included in the ratios. Upwind measurements were subtracted from the close and far measurements to isolate the harvest emissions.

On average, $18\% \pm 12\%$ of PM_{10} mass was less than 2.5 μm, $74\% \pm 15\%$ PM_{10} mass was soil, and $46\% \pm 22\%$ of $PM_{2.5}$ mass was soil.

DRUM (Davis Rotating Drum Unit for Monitoring) impactor measurements provided reconstructed soil downwind of the operations in four size ranges less than 10 μm: 0.07 to 1.15 μm, 1.15 to 2.5 μm, 2.5 to 5 μm, and 5 to 10 μm. We compared each size range of reconstructed soil against PM_{10} soil or the sum of soils from four DRUM stages (0.07 to 10 μm). Figure 80.1 shows the average ratio of soil to PM_{10} soil in four size ranges for close and far sampling locations. This essentially provides the average size distribution of airborne soil, normalized to PM_{10}, for tested harvest emissions.

As expected, a majority of the PM_{10} soil was in the coarse size range (i.e., greater than 2.5 μm). For the close sampling locations, the average ratio of the coarsest size range (5 to 10 μm) was 0.49 ± 0.12. The average ratio for the size range from 2.5 to 5.0 μm was 0.26 ± 0.08, from 1.15 to 2.5 μm was 0.19 ± 0.09, and from 0.07 to 1.15 μm was 0.08 ± 0.06.

For both size ranges < 2.5 μm, the average ratios of far sampling locations were lower and within ± 0.03 of the close locations. In the coarsest particle size range, the far average ratio was also lower than the close, by 0.08. Only between 2.5 and 5 μm was the far average ratio higher (0.40 ± 0.08) than the close (0.26 ± 0.08).

DISCUSSION

To simplify and generalize the results, our discussion of size distribution is restricted to relative size only and does not account for the actual concentration levels downwind of harvest emissions. For total PM_{10} crop harvest emissions, 80 to 85% were in the coarse size range between 2.5 and 10 μm. Similarly for DRUM measurements, 75 to 81% of the PM_{10} soil emissions were in the coarse size range.

Table 80.4 Summary of Ratios for Mass and Reconstructed Soil by Size from Filter Measurements for Various Harvest Operations

Description	Sampler Location	Average Ratio	Standard Deviation	Variance	Number	Max.	Min.
$PM_{2.5}$ Mass/PM_{10} Mass	Close and far	0.18	0.12	0.01	117	0.72	0.05
$PM_{2.5}$ Mass/PM_{10} Mass	Close	0.15	0.07	0.01	62	0.41	0.05
$PM_{2.5}$ Mass/PM_{10} Mass	Far	0.21	0.15	0.02	52	0.72	0.08
$PM_{2.5}$ Soil/$PM_{2.5}$ Mass	Close and far	0.46	0.22	0.05	118	1.15	0.06
$PM_{2.5}$ Soil/$PM_{2.5}$ Mass	Close	0.50	0.20	0.04	63	0.86	0.13
$PM_{2.5}$ Soil/$PM_{2.5}$ Mass	Far	0.44	0.23	0.06	52	1.15	0.06
PM_{10} Soil/PM_{10} Mass	Close and far	0.74	0.15	0.02	265	1.20	0.33
PM_{10} Soil/PM_{10} Mass	Close	0.78	0.17	0.03	68	1.20	0.41
PM_{10} Soil/PM_{10} Mass	Far	0.71	0.15	0.02	56	1.05	0.33

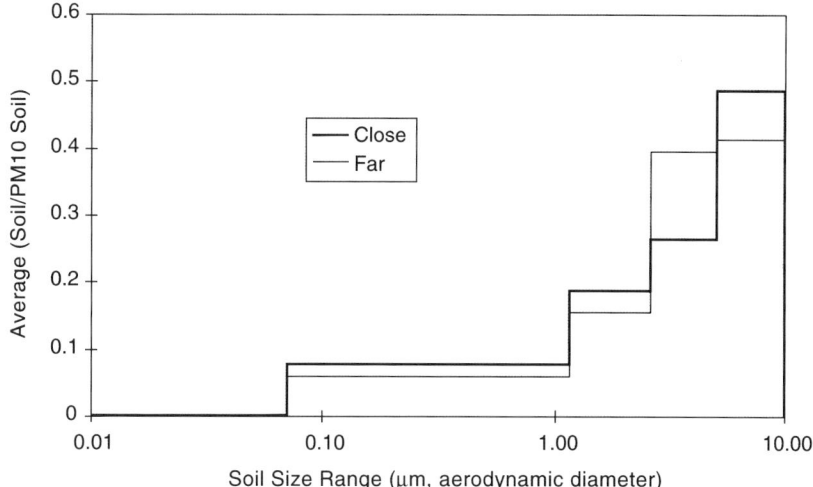

Figure 80.1 Average size distribution of reconstructed soil from harvest emissions, normalized to PM$_{10}$ soil. Soil size ranges include 0.07 to 1.15 µm, 1.15 to 2.5 µm, 2.5 to 5 µm, and 5 to 10 µm.

Our DRUM soil measurements showed an overall increase in soil quantity with increasing particle size. The far sampling locations suggest a natural maximum in soil mass between 2.5 and 10 µm aerodynamic diameter, while the maximum appears to be at 5 µm or greater for the close sampling locations. This difference between sampling locations can be attributed to the atmospheric deposition of coarse particles with transport of the dust plumes.

From filter measurements, soil accounted for 44 to 50% of PM$_{2.5}$ harvest emissions and 71 to 80% of PM$_{10}$ emissions. The nongeological or unaccounted mass, such as organics, sulfates, nitrates, and volatiles, are significant, especially for PM$_{2.5}$. More site-specific composites for reconstructing soil based upon the soil source could increase the accuracy of our measurements.

CONCLUSIONS AND RECOMMENDATIONS

The data presented here provide general information on the size distribution of soil emissions generated by harvest operations. High variations of the calculated averages of PM$_{10}$ harvest emissions can be attributed to significant differences in the emission sources and to plume dispersion.

The concentration and composition of dust from harvest emissions are related to the agricultural practice (e.g., type of operation, implement speed, mechanical energy input, farm management) and the soil source (e.g., silt content, soil moisture, soil texture). On the other hand, the downwind impact of any dust source can be greatly affected by the environmental conditions (e.g., atmospheric stability, wind, humidity) that affect particle transport. Much additional work is needed to isolate these variables so that a generalized emission model can be constructed.

REFERENCES

Annual Agricultural Reports, County Departments of Agriculture for Fresno, Kern, Kings, Madera, Merced, San Joaquin, Stanislaus, and Tulare counties, California, 1993.

Cahill, T.A., Feeney P.J., and Eldred, R.A., Particulate monitoring and data analysis for the National Park Service 1982–1985, Report to National Park Service, University of California, Davis, 1986.

Cahill, T.A., Feeney P.J., and Eldred, R.A., Size-time composition profile of aerosols using the DRUM sampler, *Nuclear Instr. Methods Phys. Res.*, 19, 183, 1987.

Cahill, T.A. Eldred, R.A., Motallebi, N., and Malm, W.C., Indirect measurement of hydrocarbon aerosols across the United States by nonsulfate hydrogen-remaining gravimetric mass correlations, *Aerosol Sci. Technol.*, 10, 421, 1989.

California Agricultural Statistics, California Agricultural Statistics Service, Sacramento, CA, 1993.

Chow, J., Measurement methods to determine compliance with ambient air quality standards for suspended particles, *J. Air Waste Manage. Assoc.*, 45, 320–382.

Chow, J., Watson, J.G., Lowenthal, D.H., Pritchett, L.C., and Richards, L.W., San Joaquin Valley Air Quality Study, Phase 2: PM_{10} Modeling and Analysis, Volume I: Receptor Modeling Source Apportionment, DRI Document Number 8929.1F, report prepared for San Joaquin Valley Air Pollution Agency by Desert Research Institute, Reno, NV, 1990, pp. 7-1–7-4.

Eldred, R.A., Cahill, T.A., Pitchford, M., and Malm, W.C., IMPROVE — A new remote area particulate monitoring system for visibility studies, paper number 88-54.3, Air Pollution Control Association 81st Annual meeting, Dallas, TX, 1988.

Eldred, R.A., Cahill, T.A., Feeney, P.J., Shadoan, D.J., Beveridge, P.J., Wilkinson, L.K., IMPROVE — Standard Operating Procedures, University of California, Davis, 1990.

Flocchini, R.G., Cahill, T.A., Matsumura, R.T., Carvacho, O., and Lu, Z. Study of fugitive PM_{10} emissions from selected agricultural practices on selected agricultural soils, SJV grant file 20960, University of California, Davis, 1994.

James, T. et al., Strategies for measuring fugitive dust emissions, International Conference on Air Pollution from Agricultural Operations, Kansas City, MO, February 7–9, 1996, pp. 161–168,

Lide, D.R,. *CRC Handbook of Chemistry and Physics*, CRC Press, Boca Raton, FL, 1992, pp. 14–17.

Malm, W.C., Sisler, J.F., Hoffman, D., Eldred, R.A., and Cahill, T.A., Spatial and seasonal trends in particle concentration and optical extinction in the United States, *J. Geophys. Res.*, 99, 1347–1370, 1994.

Matsumura, R.T., Measurement of fugitive PM_{10} emissions from selected agricultural practices in the San Joaquin Valley, *PM_{10} Standards and Nontraditional Particulate Source Controls*, Vol. 1, 1992, pp. 417–432.

CHAPTER 81

Evaluating Pesticide Effects on Ecosystems and Human Health: The Rotenone Application at Lake Davis

Marylou Verder-Carlos and Michael A. O'Malley

INTRODUCTION

Lake Davis is located in Plumas County (population 19,739), California, near the town of Portola (population, 2200; Rand, 2000). The lake and its tributaries are popular areas for boating and trout fishing, and also supply drinking water to Portola. In 1994, the California Department of Fish and Game (DFG) discovered illegally planted northern pike in the lake. DFG had successfully used rotenone to eliminate a 1991 infestation of Frenchman's reservoir, 15 miles to the east (Lee, 2001). DFG planned a similar treatment to eradicate the pike from Lake Davis, but quickly became involved in a community controversy that involved both legal challenges and a lengthy environmental assessment process (Roberts, 1996). Threats to DFG employees were even reported (Lee, 2001).

On October 15 and 16, 1997, a total of 49,000 acre-feet in the lake were treated. Deep water was treated with 15,785 gallons of a liquid formulation of rotenone (Nusyn-Noxfish™), containing 2.5% rotenone, 5% resin compounds, 2.5% piperonyl butoxide, and 90% naphthalene range petroleum distillate (a mixture of naphthalene, methyl naphthalene, alkyl benzene, toluene, xylene, and trichloroethylene [TCE]). Approximately 64,000 lb of a dust formulation (Pro-Noxfish™) was used near the shoreline.

The controversy over the eradication continued during the application itself: protesters gathered by the side of the lake and local officials chained themselves to a buoy in the middle of the lake. Other residents witnessed the application out of curiosity and some assisted DFG in the postapplication fish cleanup (Braxton Little, 1997a, b; Nacar, 1997a, b, c; Voet, 1997). After the application (La Ganga, 1998), local government representatives cited DFG officials for violating state water law, citing unexpected low level persistence of the adjuvant compound piperonyl butoxide (Braxton Little, 1997a,b; Martin, 1997).

TOXICOLOGY BACKGROUND

Rotenone is a pesticide of biological origin, extracted from the root, leaves, or seeds of various plants including cubé, derris, barbasco, haiari, nekoe, and timbo. The pure chemical is a colorless crystal with a melting point of 163°C, a vapor pressure of <1 mPa ($<7.6 \times 10^{-6}$ mmHg) at 20°C, and molecular weight of 394.4. It is only slightly soluble in water, but readily soluble in acetone, carbon disulfide, ethyl acetate, and chloroform. It dissolves less readily in diethyl ether, alcohols, petroleum ether, and carbon tetrachloride. It decomposes on exposure to light and air, and is racemized to less insecticidal compounds by alkalis (EXTOXNET, 2000). Racemization occurs more rapidly in organic solvents and oils than in aqueous solutions (Ray, 1991).

Apart from the pike eradication project, reported use in California for 1997 included 386 lb (of active ingredient) in 9338 separate applications. Most of the use occurred on vegetable and nursery crops, with applications to lettuce accounting for slightly more than one third of the total (California Department of Pesticide Regulation, 1998).

Rotenone acts by poisoning electron transport in the mitochondria, a mechanism affecting a wide range of organisms. In the rat, the pure material has an LD_{50} ranging from 25 to 132 mg/kg. Systemic poisoning, usually due to accidental ingestion, is associated with vomiting, respiratory depression or apnea, coma, and renal tubular necrosis.

Commercial rotenone products have presented very minimal hazard to humans (Ray, 1991). Some reports have noted numbness of oral mucous membranes in workers who got dust from the powered derris root in their mouths. Dermatitis and respiratory tract irritation have also been reported in some occupationally exposed individuals (Morgan, 1989).

California agricultural workers have experienced relatively few reported illnesses associated with the use of rotenone products. Between 1982 and 1995, 58 reported cases involved mixed exposure to rotenone and other pesticides, including organophosphates, carbamates, and pyrethroid compounds. These included 22 Ventura County lemon harvesters who reported dermatitis in 1988 after working in an orchard treated previously with rotenone and petroleum oil. The two cases of isolated exposure to rotenone involved both involved applicators, one with eye irritation and the other skin irritation.

Nonpesticidal Ingredients of Formulated Products

The aromatic hydrocarbons in the petroleum distillates contained in Nusyn-Noxfish have well-recognized irritant effects. The prototype compound, naphthalene, occurs commonly as white crystalline flakes, and has a vapor pressure of 0.054 mmHg (7199.4 mPa) at 20°C (American Conference of Governmental Industrial Hygienists, 1997). It also possesses a distinct coal tar, mothball, or menthol odor at concentrations above 40 to 84 ppb.

The threshold limit value (TLV) for naphthalene (10,000 ppb) was set in order to prevent ocular irritation (American Conference of Governmental Industrial Hygienists, 1997). The critical study was an industrial hygiene study reported by Robbins in 1951, who reported that indoor exposure to naphthalene over 15 ppm resulted in eye irritation. The number of subjects on which this observation was based is not described in the paper, but may have been the author's subjective experience of exposures he experienced while taking hygiene samples (Robbins, 1951).

Toluene is a colorless, highly flammable, and volatile liquid with an odor characteristic of aromatic hydrocarbons. Intentional solvent abuse or occupational exposures to 10×10 ppb can cause central nervous system depression and encephalopathy. Symptoms including headache and upper respiratory tract irritation may result from inhalation of 10,000, 75,000, or 100,000 ppb toluene for 4 to 6 h under controlled conditions. One should also note that ethanol has a potentiating effect to toluene. Xylene occurs in three isomeric forms and odor threshold ranges from 70 to 40,000 ppb. In general, various studies have concluded xylene is an irritant to the eyes, mucous membranes, and skin at levels above 200,000 ppb and the TLV has been set at 100,000 ppb. It can also cause narcosis at high concentrations.

TCE is a colorless, nonflammable liquid. It has a distinct sweetish odor similar to chloroform detectable at concentrations ranging from 180 to 5210 ppb. On topical application, it has typical primary irritant effect and pure TCE can cause localized pustular eruptions, pruritus, and erythema. Direct eye contact causes pain but not permanent corneal injury. Nonspecific systemic effects include fatigue, irritability, headache, and gastric disturbances. Ethanol intolerance, known as degreaser's flush, may also occur. Exposure to caffeine and other drugs may also exacerbate the effect of TCE (American Conference of Governmental Industrial Hygienists, 1997).

Results of controlled studies with inhaled TCE in humans are the basis for the current TLV of 50,000 ppb. Although TCE is suspected as a carcinogen, extensive human epidemiological studies have failed to demonstrate development of excess numbers of cancers despite occupational exposure to high levels of TCE (American Conference of Governmental Industrial Hygienists, 1997).

METHODS

Follow-Up of Illness Reports

On October 16, 1997, residents around the lake and those who witnessed the application reported various upper respiratory and irritant symptoms. On October 20, 1997, the Department of Pesticide Regulation (DPR) received media reports of illnesses resulting from the Lake Davis treatment, and on October 23, 1997, DPR received the first Pesticide Illness Report from a physician's office in Portola and began receiving telephone complaints regarding the rotenone application. The illness registry staff initiated a telephone survey of those who consulted a physician, or who made a telephone complaint to DPR, DFG, Air Resources Board (ARB), or the Plumas County Department of Agriculture. The respondents were questioned about their exposure symptoms and medical history. Pesticide Illness Surveillance Program (PISP) staff developed a data file based on the results of the telephone interviews. The file included identifying information, exposure circumstances, and symptoms. Specific questions focused on eye and upper respiratory irritation, systemic complaints (headache, dizziness, nausea, diarrhea, vomiting, abdominal pain, etc.), and lower respiratory symptoms (bronchial pain, wheezing). Background medical history included allergies, cardiorespiratory, reproductive, and other problems of concern for each person interviewed. Similar information was recorded, where possible for subjects identified from illness reports but not available for interview. No medical records were available for review, and no resources were available to survey community members for unreported illnesses.

Air Monitoring

On October 18, 1997, the ARB emergency response team arrived at Lake Davis and established six air monitoring sites near and around the lake for rotenone, trichloroethylene, and aromatic hydrocarbons. Figure 81.1 shows the map of the air monitoring sites and the air monitoring results are displayed in Table 81.4 of the results section. The monitoring site closest to the Portola was the dam spillway, approximately 7 miles from the northern edge of the community.

RESULTS

A total of 71 individuals reportedly experienced symptoms between October 15 and October 29, 1997. Between October 24 and November 15, 1997, PISP staff conducted telephone interviews with 58 citizens of the Portola area who reported illness or injury possibly related to the DFG eradication program. Approximately nine proxy interviews were also conducted for nine children under age 16. Limited information was available for four individuals for whom pesticide illness reports were filed, but who could not be reached by telephone.

The 71 symptomatic residents included 29 males and 42 females (Table 81.1). Occupations varied, but most were retirees, homemakers, or students. The median age of the 66 individuals for whom an age was reported was 47 years, ranging from 2.4 months to 91 years. There were 8 children below 13 years of age (11%) and 11 adults between the ages of 61 and 91 (15%).

Figure 81.1 Map of air monitoring sites at the spillway below the dam for rotenone, TCE, aromatic heavy hydrocarbons, and other hydrocarbons.

Table 81.1 Summary of Age Distribution

Age Range (years)	Number of Subjects
< 1	1
1–12	7
13–18	3
19–30	5
31–40	8
41–50	17
51–60	14
61–70	5
71–100	6
Not specified	5[a]
Total	**71**

[a] Four adults and one student.

Table 81.2 Detection of Odor vs. Distance from the Lake

Distance	Noticed Odor	No Odor	No Information	Total
At lake/creek/dam	37	1	3	41
0.2–1 miles	18	1	2	21
>1 mile–8 miles	5	1	0	6
Not specified	0	2	1	3
Total	**60**	**5**	**6**	**71**

Circumstances of Exposure and Odor Complaints

Of the 71 subjects, 41 (57.7%) reported exposures at the lake, the dam spillway, or the creek immediately downstream during the application or in the week immediately afterward; 27 (38.0%) had exposures at distances ranging from 0.2 to 7 miles (in the town of Portola) but did not have direct contact with the lake, and 3 did not indicate their location at the time of exposure.

For 6 subjects, no information was available regarding the presence or absence of odor from the application. In the remaining 65 subjects, 60 (92.3%) reported smelling an odor after the rotenone application, at distances up to 8 miles from the lake (Table 81.2). Descriptions of the odor ranged from plain chemical smell, very strong odor, or extremely powerful odor, to more specific terms as creosote-like smell. Some described it as mothball-like odor, while others described it as insecticide or Raid-like odor.

Medical Evaluation, Medical History, and Medical Treatment Received

Of the 71 people reported sick (including the 4 subjects not available for interview), 20 were evaluated by either a physician or nurse practitioner, 27 were evaluated by a physician via telephone consultation, and 24 did not seek medical attention. Of the cases, 38 were reported to the illness registry from a single physician's office in the town of Portola, and 3 cases were reported from the community hospital emergency room. The remaining 30 cases were identified from telephone contacts only. These included 6 individuals who sought medical treatment at other primary care physician's offices, but were not reported by their providers to the registry.

The individuals interviewed reported symptoms that included ocular and upper respiratory irritation and nonspecific systemic symptoms. Irritant symptoms included sore throat, eye irritation, tearing, rash, coughing, and runny nose. Nonspecific systemic symptoms included wheezing, diarrhea, dizziness, headache, and nausea. There was no significant difference in the pattern of symptoms between those exposed at the lake and those farther from the site of the application (Table 81.3).

A few subjects reported unrelated, preexisting medical conditions, including six with a history of asthma, five with hypertension, two with a history of sensitivity to chemicals, and two with a history of recurring sinus infections. One individual stated she had bronchitis at the time of the application, and two individuals were recuperating from debilitating illnesses, one from a stroke and another from major surgery. Another individual experienced symptoms of diarrhea and abdominal pain before the treatment of Lake Davis. Other patients had fever in addition to nonspecific complaints of pharyngitis compatible with either chemical irritation or upper respiratory infection.

Table 81.3 Relationship of Illness Pattern to Distance from Lake at Time of Exposure

Illness Pattern	At Lake/ Dam/Creek	% of Total	Other[a]	% of Total	Not Specified	Total
Skin	1		0		0	1
Respiratory	1		2		0	3
Eye/respiratory	4		5		0	9
Topical only	**6**	**14.6**	**7**	**25.9**	**0.0**	**13**
Systemic	5		2		1	8
Eye/systemic	4		0		0	4
Respiratory/systemic	16		4		2	22
Skin/respiratory/systemic	2		1		0	3
Eye/respiratory/systemic	6		10		0	16
Skin/eye/respiratory/systemic	2		3		0	5
Systemic or systemic/topical	**35**	**85.4**	**20**	**74.1**	**3**	**58**
Total subjects	**41**		**27**		**3**	**71**

[a] 0.2–8 miles.

Table 81.4 Summary of Air Monitoring Results at the Spillway below the Dam for Rotenone, TCE, Aromatic Heavy Hydrocarbons, and Other Hydrocarbons

Date of Collection	Rotenone (ppb)	TCE (ppb)	Naphthalene-Range Hydrocarbons (ppb)
October 18	0.02	500	281
October 19	0.02	<3	201
October 20–21	0.03	<3	No report
	0.004		
October 21	0.02	<3	No report
October 22	0.03	<3	No report
October 23	0.007	<3	No report
October 24	Invalid result	<3	No report
October 25	0.005	<1	No report
October 26	0.002	No report	5.34
October 27	< 0.001	No report	2.26
October 28	0.003	<1	0.81
October 29	0.001	<1	3.07
October 30	< 0.001	<1	1.78
October 31	< 0.001	<1	1.13
November 1	< 0.001	<1	1.61

Ten patients were prescribed antibiotics, and one child with gastrointestinal symptoms was sent for laboratory fecal analysis and culture. In one case a streptococcal infection was definitely diagnosed.

Air Monitoring Sites and Results

Among the six air monitoring sites ARB established, the site at the spillway below the dam was the location that generated the highest results and was also the closest to the town of Portola (approximately 7 miles to the south). At the five other sites, all samples confirmed significantly lower or nondetectable levels. Table 81.4 summarizes air monitoring results for the spillway below the dam.

According to the results reported by ARB, rotenone was detected at the spillway at very low levels (0.02 ppb) on October 18 and 19 and declined to nondetectable levels by October 30. TCE was detected only at the spillway on October 18. The levels after the 18 were less than 3 ppb until October 24, and less than 1 ppb on October 25 through November 1. Monitoring for aromatic heavy hydrocarbons showed the most significant levels with 281 ppb on October 18, 201 ppb on October 19 declining to 5.34 ppb on October 26, and dropping to 1.61 ppb on November 1. ARB terminated monitoring at the spillway on November 4 and at all other sites on October 29 because all samples were below the limit of detection.

DISCUSSION

On initial monitoring (October 18, 1997), air concentrations of naphthalene-range aromatics were markedly higher (201 to 281 ppb) around the lake than concentrations of rotenone (0.02 ppb), as would be predicted from the difference in their respective vapor pressures. Concentrations were highest at the dam spillway, with levels at or near the detection limit elsewhere. Although the measured concentrations at the spillway were below the reported 15,000 ppb TLV values, the occupational 8-h time-weighted-average standard (American Conference of Governmental Industrial Hygienists, 1997) is not based upon sufficient information (Robbins, 1951) to be used as an environmental exposure standard. Given the volume of liquid rotenone formulation used in the lake it is conceivable that local concentrations approached or exceeded the concentration necessary to cause eye and upper respiratory irritation and the nonspecific systemic symptoms reported by residents who were at the lake on October

15 to 17. As in previously investigated community environmental exposures to pesticides (Ames and Stratton, 1991; Scarborough et al., 1989) and waste-disposal by products (Shusterman et al., 1991), it is also possible that residents primarily reacted to the odor of naphthalene and related compounds present (Catanach, personal communication, 1997) in the liquid Nusyn-Noxfish formulation.

Given the 7-mile distance between the south end of the lake and the northern end of the community it seems less likely that levels of naphthalene present in Portola approached the existing occupational exposure thresholds. No air monitoring was conducted in the community, so it is not possible to ascertain whether community air concentrations approached the reported odor threshold for naphthalene (40 to 84 ppb). However, two individuals exposed at a distance of greater than 7 miles from the lake also reported noticing the odor from the rotenone application. The symptom complexes reported by those exposed at the lake were similar to those reported by persons exposed at distances of 0.25 to 1 mile, or those exposed at distances more than a mile. Both groups reported symptoms of eye and skin irritation and nonspecific systemic complaints (Table 81.4).

It was not possible, given available staff resources, to survey all 16,000 Plumas County residents or the 2000 residents of the community of Portola. It is therefore probable that symptoms possibly associated with the rotenone application occurred in some community members that were not reported to the illness registry. In this connection, no illness reports were received for six residents interviewed by telephone who stated that they had visited their own physician for symptoms they associated with the rotenone treatment. It could not be ascertained from the available information whether the treating physicians were unaware of the illness reporting requirement or simply considered the reported symptoms to be unrelated.

In addition to underreporting, it is possible that some symptoms actually unrelated to the application were captured in the mix of telephone complaints and physician-reported illnesses. In case 6, for example, a student who spent October 15 picking up dead fish at the lake was reported to have pharyngitis due to a streptococcus infection. It is ambiguous whether any of his symptoms (dyspnea, pharyngitis, abdominal pain) were actually due to petroleum distillate exposure at the lake. This example underscores the possibility that any hypothetical survey to determine the prevalence of eye and upper respiratory irritation and nonspecific systemic symptoms associated with the rotenone application would have been biased by the accompanying community controversy (see above).

Several illness episodes have been reported to the California pesticide illness registry that also involved off-site movement of a pesticide or its inert ingredients to adjacent areas. One 1994 episode involved 49 Pacific Gas and Electric employees who noticed an intense odor and developed symptoms following a chlorpyrifos insecticide formulated with naphthalene pesticide application to a cotton field about a quarter of a mile away. In 1995, the manufacturer changed the formulation to use a less odiferous carrier solvent. Nevertheless, a 1996 Kern County episode, involving chlorpyrifos formulated with the same naphthalene-range aromatic solvent, affected more than 500 workers. In these cases, the odor of the naphthalene solvent was probably mixed with the odor of mercaptan contaminants (O'Malley, 1997) present in the formulation.

Pike Reappear

A final controversy arose when pike were rediscovered in the lake in May 1999 (Lee, 2001). Did some pike survive the treatment or were they replanted in the lake (Lake Davis Steering Committee, 2001)? Was it possible to control the pike by electrofishing, nets, or would it be necessary to attempt retreatment with rotenone (Lassen County News, 2000)?

CONCLUSION

In the pike eradication program, some exposures occurred because of direct contact of community residents with the lake at the time it was treated. Symptoms included eye and respiratory irritation, as well as nonspecific systemic symptoms. In nearly all cases, symptoms were associated with a perceived odor, described as being like menthol or like insecticide. Evaluation of cases was confounded by community controversy about the eradication program, the absence of monitoring data for the days the application took place, and the lack of an appropriate standard for environmental exposure to naphthalene-range aromatic compounds.

REFERENCES

American Conference of Governmental Industrial Hygienists, Documentation of the Threshold Limit Values and Biological Exposure Indices, Cincinnati, OH, 1997.

Ames, R.G., and Stratton, J.W., Acute health effects from community exposure to N-propylmercaptan from an ethoprop (Mocap)-treated potato field in Siskiyou County, California (USA), *Arch. Environ. Health*, 46, 213–217, 1991.

Braxton Little, J., Fish and Game again accused of violations in lake poisoning, *Sacramento Bee*, October 31, 1997a, p. B3.

Braxton Little, J., Portola demands new water supply after lake is poisoned, *Sacramento Bee*, November 1, 1997b, p. B4.

California Department of Pesticide Regulation, 1997 Annual Pesticide Use Report Indexed by Chemical, California Department of Pesticide Regulation, Sacramento, CA, 1998.

EXTOXNET, Pesticide Information Profile, Rotenone, Vol. 2000, Oregon State University, Corvallis, 2000.

La Ganga, M., Effort to rid lake of pike leaves a poisonous legacy, *Los Angeles Times*, April 8, 1998, p. A-1.

Lake Davis Steering Committee, The Pike DNA Report, Vol. 2001, http://www.geocities.com, Portola, CA, 2001.

Lassen County News, Plans in place to revive Lake Davis, Vol. 2001, http://www.Lassennews.com, 2000.

Lee, D., Northern pike control at Lake Davis, California, in *Rotenone in Fisheries: Are the Rewards Worth the Risks?* Cailteux, R.L. et al., Eds., American Fisheries Society, Trends in Fisheries Science and Management 1, Bethesda, MD, 2001.

Martin, G., Showdown over poisoning Lake Davis — County fights state's pike-killing plan to save salmon, *San Francisco Chronicle*, Oct. 13, 1997, p. A1.

Morgan, D., *Recognition and Management of Pesticide Poisonings*, 5th ed., U.S. Environmental Protection Agency, Washington, D.C., 1989.

Nacar, T., Biology students assist in fish sampling, *Feather River Bull.*, Quincy, CA, 1997a, p. 7A.

Nacar, T., Forest services stays out of Davis squabble, *Feather River Bull.*, Quincy, CA, 1997b, p. 7A.

Nacar, T., More than 20 ill from poisoning, *Feather River Bull.*, Quincy, CA, 1997c, p. 7A.

O'Malley, M., Clinical evaluation of pesticide exposure and poisonings, *Lancet*, 349, 1161–1166, 1997.

Rand, Corp., Census and Current Population Survey Statistics, Vol. 2000, The Rand Corporation, Santa Monica, CA, 2000.

Ray, D., Pesticides derived from plants and other organisms, in *Handbook of Pesticide Toxicology*, Hayes, W.J. and Laws, E.R., Eds., Academic Press, New York, 1991, pp. 585–636.

Robbins, M.C., Determination of naphthalene in air, *Arch. Ind. Hyg. Occup. Med.*, 4, 85–87, 1951.

Roberts, J., Cal Fish & Game schedule poison for Portola town's drinking water, *Electric Nevada*, Vol. 2000, 1996.

Scarborough, M.E., Ames, R.G., Lipsett, M.J., and Jackson, R.J., Acute health effects of community exposure to cotton defoliants, *Arch. Environ. Health*, 44, 355–360, 1989.

Shusterman, D., Lipscomb, J., Neutra, R., and Satin, K., Symptom prevalence and odor-worry interaction near hazardous waste sites, *Environ. Health Perspect.*, 94, 25–30, 1991.

Voet, G., The road to Lake Davis goes through Great Lakes, *Sacramento Bee*, October 29, 1997, p. E1.

SECTION II.8

Mining Impacts

CHAPTER **82**

Overview: Mining Impacts

Jean Lebel

The models used for illustrating and conceptualizing the relationship between the environment and human health have advanced considerably over the past century. The modeling of human health from the perspective of interactions between humans and their surroundings was initially colored by the experience of the biomedical world and the spread of infectious diseases. Yet the world is subject to the influences of many complex factors that have the potential to undermine health and that cannot be controlled by a reductionist medical approach alone, no matter how sophisticated such an approach might be.

The ecosystem approach to human health constitutes a bridge between a strategy for integrated management of the environment (healthy ecosystems) and a global, ecological approach to promoting human health. The ecosystem approach to human health offers an unequaled opportunity to promote human health through an enlightened approach to management of the ecosystem. The creation of more sustainable ecosystems relates to the management of natural and environmental resources, of course, but it must also take account of all the many anthropogenic components, by integrating the social, economic, and cultural factors pertaining to the living environment. The ecosystem approach to human health is also dependent on a participatory and transdisciplinary research methodology that will remain sensitive to gender and other issues of social equity.

Within the context of mining, this ecosystem approach to human health represents a challenging and appealing way of trying to better understand the various effects on human health of this important economic activity. This approach also assists in developing intervention strategies that are effective and sustainable.

The chapters presented in this section are some of the keys that will open the door to a new world of knowledge. The particular focus given to the developing nations within this section reflects the importance of this sector in the economy of a growing number of countries. It also reflects the programming of Canada's International Development Research Centre (IDRC), an international aid agency that sponsored the participation of international scientists in the technical sessions related to mining at the 1999 International Congress on Ecosystem Health.

This section begins with a broad overview by Lebel and Burley of the effects of mining operations on the ecosystem, and on communities that are involved directly or indirectly in these activities. The authors point out that the complexity and dynamic nature of many of the relationships require an integrated framework in order to truly understand the damaging as well as the beneficial influences of mining. Following this introduction of the mining world, Shields and Šolar discuss some fundamental issues of scale related to selection and interpretation of mineral indicators. For time as well as space scales, the researchers are always selecting a certain perspective when studying an ecosystem. The importance of knowing the limitations or the appropriateness of the scale in ecosystem approaches to human health research is fundamental.

As mentioned earlier, the human health concept has evolved significantly over the years. Maclean et al. present some of these concepts and make a good case for moving beyond the traditional ill-health conceptual framework, given the broad array of effects of mining activities that are linked not only to the development of diseases. In a related vein, Echavarría Usher invites us to pay particular attention to a population group often affected by mineral development: indigenous peoples. Through her work in Colombia, the meaning of health and well-being takes an entirely different aspect from the mainstream vision — a vision that needs to be included in intercultural research. Mergler integrates many of the previous points on human health in the conceptual frameworks she is presenting. With an overall goal of integrating in a sound manner human health within an ecosystem approach to mining, she demonstrates — through the use of a nested, hierarchal framework — the need to examine different effects of this activity, from the mine sites to distant populations. She also makes an eloquent case for the need to focus more on early changes in biological, psychological, and social parameters if one wants to take a preventive approach to human health.

Researchers are confronted by difficulties in the analyses and in understanding the complex and diverse links of the different issues related to human health that are raised through consultative process by a broad spectrum of stakeholders who are involved directly or indirectly with mining activities. Those complexities and difficulties in understanding can make researchers and policy makers reluctant to use an ecosystem approach to human health. This should not be the case. Noronha presents a brilliant case study in which the use of a multistakeholder consultation process for the identification of issues of the biophysical, economic, and sociopolitical-cultural domains served as the basis for development of tools to measure the contribution of mining over time toward the sustainable development of the iron mining region of Goa, India. The tools are indicators, quality of life instrument, and impact-adjusted income account. The last case study relates to the situation in the Langat Basin, a region of Malaysia. Through the use of different indicators, Pereira and Komoo demonstrate the possibility of evaluating the effects of the exploitation of mineral resources. Through careful use of information available from different sources, they show how measurement over time can reveal progression or failure to achieve a more sustainable use of mincral resources.

We hope that the views provided in this section will help the research community and policy makers to move toward the development of an integrated framework linking findings with actions.

CHAPTER 83

The Ecosystem Approach to Human Health in the Context of Mining in the Developing World

Jean Lebel and Lisa Burley

IDRC AND THE ECOSYSTEM APPROACH TO HUMAN HEALTH: AN INTRODUCTION

The International Development Research Centre (IDRC) is a public corporation created by the Parliament of Canada to help researchers and communities in the developing world find solutions to their social, economic, and environmental problems. IDRC connects people, institutions, and ideas to ensure that the results of the research it supports and the knowledge that research generates are shared equitably among all its partners, north and south.

Program initiatives (PIs) are the IDRC's primary programming unit for funding research in developing countries. Managed by a multidisciplinary team from within the IDRC, PIs are working networks that link southern and northern researchers to address specific research problems and set a research agenda. By linking all the parties involved in the research process, IDRC hopes to add to the likelihood of success. Because of the interdisciplinary focus, PIs will often address issues that fall under several research themes.

The purpose of the PI Ecosystem Approaches to Human Health (ECOHEALTH) is to improve human health while simultaneously maintaining a healthy ecosystem.* We are referring to the concept of health as defined by the World Health Organization in its constitution, "Health is a state of complete physical, mental and social well-being, and not merely the absence of disease or infirmity." To this end, the PI supports research projects, all of which are carried out in developing nations, that explore the structure and function of stressed ecosystems on which people depend for their lives and livelihoods. Research projects have a mix of several characteristics, they are trans-disciplinary,** social and gender sensitive,*** and participatory.****

The Ecosystem Approach at IDRC (1) begins with the ecosystem as the point of departure for understanding the nature of the stressor and for improving human health, (2) embraces a particular

---

* Broadly speaking, an ecosystem is a dynamic complex of plant and animal (including human) communities situated within a given environment. The boundaries of ecosystems often have more to do with the parameters that humans set arbitrarily according to the scientific, management, or policy question they wish to examine rather than with physical dimensions.

** *Multidisciplinary* research involves different disciplines that have parallel input without necessarily consulting one another. *Interdisciplinary* research implies some degree of integration between the different disciplines in relation to the problem at hand (IDRC, 1993). *Transdisciplinary* means transformation, going beyond disciplinary mind-sets into a re-conceptualization of phenomena, problems, goals, and approaches (Peden, unpublished 1999). The need for a transdisci-plinary approach arises from developments in knowledge and culture that are characterized by complexity, hybridity, nonlinearity, and heterogeneity (Klein, 1994).

---

concept of human health, and (3) promotes research methodologies that are derived from a holistic vision and can effectively respond to the various dimensions of the human health concept.

The PI emphasizes improving health using ecosystem management as compared to relying only on health sector interventions. ECOHEALTH encourages and supports research activities intended to further develop methodology, to increase understanding of causal linkages between ecosystem management and human health, to test ecosystem-based interventions, and to disseminate the concept and research results to diverse stakeholders. The knowledge generated from research is applied to the development of effective and sustainable interventions and policies.

The PI focuses on several ecosystem stressors, one of which is mining. Like other stressors, this activity presents risks to human health and the environment. It also creates benefits for communities and individuals in these ecosystems. The PI recognizes that sustainable alternatives that promote the health of both humans and the ecosystem depend on the participation of all members of the community.

This chapter attempts to demonstrate that the sustainability challenges posed by the stress of mining on ecosystems affect the physical, mental, and social well-being of communities through various changes in mediums, such as the quality of air, water and food, and conditions, such as socioeconomics, culture, and politics. The chapter also attempts to illustrate that addressing these challenges requires methodologies that bring together various disciplines, that include and empower those people directly involved in and affected by mining activity, and that incorporate sensitivity to gender and sociocultural issues. In short, the purpose of this introductory chapter is to underline that an ecosystem approach to human health can provide stakeholders with a conceptual framework appropriate to tackling the many concerns and factors involved in mining in the context of developing nations.

MINING IN THE DEVELOPING COUNTRY CONTEXT: WHO ARE THE MAIN STAKEHOLDERS AND WHAT ARE THE MAIN ISSUES?

This section refers to generalities and broad categories in order to identify the principal stakeholders and some of the central challenges they face regarding the ideals of ecosystem and human health.

Governments

A recent change in exploration trends has been geography. With the dual (yet not necessarily causal) phenomena of democratization and the adoption of neoliberal policies for trade and foreign investment (Brown, 1994), many governments of developing nations have created conditions conducive to attracting foreign investment by revising mining law and policy. Walde (1992) points out, however, that the typically centralized structure of many developing nation governments means that authority over mining issues does not rest at the local level. In addition, formal mechanisms usually do not exist for involving local people in decisions made at the central or national government level (Clark, 1998). Many nations have made efforts to decentralize central government

---

*** Socially sensitive research is aware of the implications of age, gender, ethnicity, occupation, class, etc. on various social aspects such as power and decision making. Gender sensitive research recognizes that women and men have different interests, concerns, priorities, and needs. Research is therefore (1) aware of what men's and women's differences and needs are, and addresses the research issue accordingly, and (2) proposes to explore gender differences as part of the research (Kabeer and Subrahmanian, 1996; Singh, 1999).

**** Participatory research is broadly understood, and includes a plethora of tools and methodological approaches, including such commonly used methods as Participatory Rural Appraisal (PRA), Participatory Action Research (PAR), Rapid Rural Appraisal (RRA), and Farmer Participatory Research. Rooted in ideological and radical social movements which mobilized local people to challenge existing power regimes, participatory research has become increasingly popularized as a means of capturing local knowledge and perspectives and for involving local people in research and development activities which affect them (McAllister, 1999).

authority. Labonne (1999) points out that this decentralization has brought to the fore the necessity for good regional and local governance since these authorities will impact on the relationship between investors and the community. Labonne also points out that regional agencies often have weak capacities for management and regulation, and due to rural–urban migration, local governance structures are often in a constant state of change. Tyler (1999) explains that most government reforms involving natural resource management in many developing nations are almost always inadequately supported. Decentralization of authority from central to regional governments has not been accompanied by adequate funding, training, or capacity-building among the officials charged with implementing the policies. Further, Tyler reports that as a result, enlightened policies may either fail to be implemented or be implemented very differently at the local level than intended by policy makers.

Industry: Juniors and Majors

Lured by revised mining laws of many developing nations, and in the case of Canada, discouraged by high tax rates, stringent environmental regulations, expensive labor, and the excessive withdrawal of land from mineral exploration (Natural Resources Canada, 1993), junior* and major companies from this nation, Australia, and the U.S. began investing in exploration. For example, Canadian mining investment in Latin America began to increase in 1991. By 1993, the amount of investment created a veritable boom, which began to reshape the mining industry. Most of the junior firms leading this boom were Canadian and were funded through Canadian capital markets (Natural Resources Canada, 1997).

The studies of Pascó-Font (1998) have outlined that in the absence of proper environmental legislation, junior companies are more likely to behave irresponsibly. McMahon (1998) also observes that demands by communities for consultation are largely accepted by major companies, but less so by the juniors. McMahon also explains that juniors resist formal consultation processes as they wish to avoid a costly procedure given the small likelihood of actually finding an exploitable deposit. Moody (1998) outlines that this situation presents concerns since it is the junior mining companies, supported with venture capital raised on Canadian stock exchanges, that are conducting a large portion of exploration activity not only in Latin America, but also around the world.

The exploration budgets of the larger or major Canadian-based companies for Latin America and the Caribbean have grown at an average annual compound rate of over 50% between 1992 and 1997. In 1997, these companies held the largest share of the larger mining company market in all of Latin America (Canadian Intergovernmental Working Group of the Mineral Industry, 1998).

Although many mining companies now recognize the importance of community participation, many of them struggle with issues regarding *how* to involve communities. Prager (1997) explains that part of this problem is due to corporate culture. She points out that decision-making levels of company management do not often include social scientists, and rarely confront issues with an inter- or transdisciplinary approach. Indeed, the international Advisory Group made up of stakeholder representatives of the IDRC Mining Policy Research Initiative flagged the importance of research responding to corporate concerns of relationship building with civil society (MPRI, unpublished 2000).

---

* Broadly speaking, a junior firm is specialized in the exploration activities needed in order to detect a claim with a sufficient value to develop a mine. Juniors sell their concession rights to larger mining companies to continue with advanced exploration, feasibility studies, and, perhaps, exploitation.

Industry: Small-Scale and Artisan Mining*

This kind of mining is currently experiencing a resurgence around the world. In many cases, it is a response to poverty. It provides employment to unskilled workers and supports local economies. However, through contraband, taxes are evaded thereby precluding economic benefits to the national economy, although direct and collateral benefits are very important. Confrontations often arise with resident communities, and environmental impacts are severe. In addition, working conditions are very precarious. Developing nation governments experience many problems exercising control over small-scale and artisan mining (Wotruba et al., 1998).

Communities and Ecosystems

Another key stakeholder in the context of mining expansion in developing nations are people and ecosystems. Crane-Engel and Schanze (1988) point out that although some developing nations have a mining history, generally speaking, industrial large-scale, privatized mining has traditionally been an activity of wealthier nations of temperate climates. This geographical shift to more southern nations involves several different characteristics including pervasive poverty, considerable wealth disparities, large rural populations, and the conditions of a tropical climate where contaminants may infiltrate food chains more readily than in temperate or permafrost regions (Ripley et al., 1996). In many developing nations, ecosystems are already under stress from the many ways in which they are being used to provide subsistence needs and to support primary industries important to the export sector, such as foodstuffs, flowers, fish and sea products, timber, oil, etc.

In short, the developing nation context involves extremely vulnerable populations where the needs for economic development and the risk of falling to greater depths of poverty are acute. Since many subsistence livelihoods, some national industries and parts of the export sector are often linked to the natural resource base, ecosystems are equally vulnerable to overharvesting that may result from poverty increases and to contamination from misuse.

Clark (1998) observes that although many developing nation governments have emphasized mineral development and the creation of an attractive investment climate, they have not offered guidance to foreign mining companies regarding how to proceed with social and cultural issues. This skill set is precisely what industry is lacking; for junior companies, small-scale and artisan mining, problems also arise with the investment required to appropriately respond to socioenvironmental issues. The development of artisan and small-scale mining face several challenges and the costs and benefits of this activity continue to be the subject of considerable controversy (Wotruba et al., 1998).

Governments are also grappling with either centralized decision-making structures or the challenges posed by the transition to decentralized authority where regional and municipal agencies lack financial support and capacities to implement regulatory policy. Nonetheless, many aspects of revised mining laws have attracted and continue to attract investment in mineral development.

THE ECOSYSTEM AS THE POINT OF DEPARTURE: WHAT IS THE NATURE OF THE STRESSOR AND GENERAL IMPACTS ON THE ENVIRONMENT?

Mining is a complex and diverse activity. This section will speak more to broad categories, than of specific situations with a view to providing the reader with information in order to appreciate the complexity and diversity of the stressor under study.

---

* While there is no universal definition of small-scale and artisan mining, it may generally be characterized as being labor-intensive and utilizing low technology. Workers generally have little or no formal education and often have few options for making a living (Peiter et al., 2000).

The extent of mining impacts on ecosystems and human health are influenced by the several different ways mining can be done. Broadly speaking, four categories of mines can be described according to organizational structure, investment level, production level, and technological capacity: large, intermediate, small, and artisanal scale mines (IENIM, 1996). For the purposes of this chapter, large and intermediate mines will be grouped together, and small-scale and artisan mining will also be referred to as a single category.

The mining of each mineral commodity has its own distinctive set of environmental and health impacts. All of these effects are not equal in their significance, but at every level, there is the possibility of uptake and exchange of residuals by organisms (Ripley et al., 1996). In the case of small-scale and artisan mining, costs and benefits also vary according to the type of mining. For example, alluvial gold mining and the extraction of precious stones have far greater socioenvironmental costs, are more difficult for governments to control, and more likely to become contraband than coal and phosphate mining and quarrying (Wotruba et al., 1998).

Mining development in general involves three broad stages: (1) exploration and construction, (2) operation, and (3) closure and site rehabilitation (Ripley et al., 1996). While considerable randomness characterizes the length of the life cycle of these stages with small-scale and artisan mining, it is an extensive process with large and intermediate mines. Typically, exploration can go on for 5 to 10 years, with mine construction in the order of 1 to 3 years. Operation varies greatly, depending on market conditions but can easily exceed 50 years. Closure can take 2 years, and physical rehabilitation is highly variable, depending on the level of disruption and damage to the environment (Ripley et al., 1996), and the demands and lobbying influence of host communities.

Exploration and Construction

Typically, at the exploration stage of large and intermediate mines, the likelihood of significant biophysical damage to the environment is limited in extent and scope. Generally, exploration involves geochemical or geophysical techniques, followed by the drilling of promising targets and the delineation of orebodies. The early stages of exploration can be carried out using remote-sensing methods from satellites and aircraft. As promising areas are located, low-flying aircraft and ground-based exploration methods are used (Ripley et al., 1996).

In terms of environmental effects, exploration may affect more terrain than subsequent mining stages, but for a shorter duration and with less severity. Ripley et al. (1996) outline some environmental effects:

- Low-flying aircraft can affect people and wildlife.
- Exploration grids may leave visible scars that may also disturb wildlife and lead to considerable erosion.
- Trenching and stripping may have long-term effects on nearby waterways.
- Drilling may contribute to contamination of surface and groundwater.

Overall, however, road construction and transportation are frequently the main source of environmental impacts, depending on who uses them, and how much they are used after they are built (Ripley et al., 1996). Some sources claim that for each operational mine, up to a 1000 exploration programs have been conducted (EMCBC, 1998).

If artisan prospectors and explorers locate a significant deposit exploitable with low technology, often times considerable and immediate migration to the area occurs, resulting in an invasion of artisan miners. Since these areas are usually remote and because the invasion occurs quickly, enforcement of environmental laws is usually ineffective or nonexistent (Peiter et al., 2000). Indiscriminately, trails and roads are opened, timber is extracted from the forest, nearby water sources tapped, and *ad hoc* construction is undertaken. All of this is considerably damaging to the host ecosystem.

Operation

Many of the environmental impacts of small-scale and artisan miners are rooted in their lack of capital to invest in cleaner technologies, their own mentality and occupational culture, and their notorious ignorance of environmental issues and management. In the case of gold mining, host and surrounding ecosystems are seriously affected by (Wotruba et al., 1998):

- The removal of vegetal layers
- Poor air quality
- Dust
- Noise
- Water contamination
- Alteration of water courses
- Mercury contamination
- Effluent mixes of lubricants, oil, sulfur oxides, and heavy metals as part of water and ground contamination
- Erosion

Indeed, all mining is characterized by its awesome generation of gaseous, liquid, and solid waste material, a large portion of which can be classified as hazardous. Wastes are in the forms of overburden, tailings heaps, slag, sludge, mineralized toxic waters, smoke, and gas emissions (Pastizzi-Ferencic, 1992). For example, in the large-scale copper mining industry, the ratio of material handled to units of marketable copper metal is 420:1 with a typical ore grade of 0.6% (Dudka and Adriano, 1997).

Waste materials may be chemically reactive and, consequently, generate soluble acids or alkalis, soluble metal salts, and radionuclides. Liquid wastes can be produced by such reactions or directly by the processes themselves (Ripley et al., 1996). The potential for dispersion of the solid particulate of waste materials in many environments is also enhanced by climatic conditions (Ripley et al., 1996). Many of the mining sites in the developing world are in desertic or semidesertic regions where the eolian factor is important.

The most common and serious problem faced by large and intermediate mines in waste disposal derives from the mining and processing of sulfide ores, which can produce acid drainage. Essentially, this is the result of the chemical reaction that takes place when a metal sulfide combines with oxygen and water, yielding a metal hydroxide precipitate and sulfuric acid (Ripley et al., 1996; Dudka and Adriano, 1997). In Peru, numerous rivers have been declared biologically dead because of the leaching process of toxic materials (Ride, 1998).

Closure

Historically, mining operations were abandoned when they reached the end of their useful lives. Virtually all abandoned mines were operated during a time when environmental standards were less stringent than they are today; disturbed sites, waste dumps, and mine structures were seldom reclaimed. As a result, altered surface topography, surface and subsurface drainage, disturbed vegetation and soil, and abandoned roads and buildings were left behind after the operation ceased (Ripley et al., 1996). Abandoned mines present the risk of environmental problems that include acid drainage and metal contamination.

Most discussions of reclamation have centered on the damage done to land surfaces; however, mining also has environmental effects in other areas that include damage to water bodies, hazards to human health and property, and dangers to fish and other wildlife. An approach to the subject of amelioration requires a comprehensive view that emphasizes the interrelations between the various spheres of influence (Summers and Lewko, 1995).

It is important that developing nation governments often do not have sufficient resources to rehabilitate areas abandoned by large and intermediate mining companies. Small-scale and artisan operations are almost always left abandoned.

WHAT ARE THE IMPLICATIONS FOR HUMAN HEALTH?

McMahon (1998) states that social implications of large mine development have displaced the environmental concerns at the fore of the public debate on mining investment. This section will outline some issues of human health and mining. A full discussion of the topic covering the various types and stages of mining is beyond the scope of this chapter.

Physical Health

In terms of physical health in the occupational context, it has been recognized for a long time that working in a mine is a dirty job. Some of the most ancient cases of occupational disease were documented through the analysis of data coming from the population of miners. Dusty environments have been the basic cause for the development of different respiratory diseases of workers in mining operations (Reger and Morgan, 1993). For example, there is thorough documentation of coal mining and black lung disease (Weeks, 1993; Oxman et al., 1993), quarrying and silicosis (Wagner, 1997), asbestos mining and mesothelioma (McDonald and McDonald, 1996; Wagner, 1997), uranium mining and lung cancer (Samet, 1991). The Pan American Health Organization (PAHO) estimated in its 1994 *Work Illnesses* report that 8.5 million workers in Latin America from the mining, quarrying, ceramics, metallurgical, and construction sectors were at risk of developing silicosis (Joyce, 1998). Noise and vibration are also common problems encountered in the workplace, as are exposure to heat, mine gases, and to chemical vapors (Burgess, 1981).

Mining operations can also potentially affect the health of people who are living at a distance from the mine due to the effects of fallout gases and solids or dispersion of toxic materials to crops, farm animals, and food obtained from undomesticated sources such as fish or game.

For example, a major concern for the physical health of the greater community involves the contamination of groundwater, and thus of wells. Recent studies have shown in region II of Chile, an area of intense copper mining, that bladder and lung cancers were significantly associated to exposure to arsenic present in drinking water. Researchers estimated that arsenic might account for 7% of all deaths among those ages 30 and over (Smith et al., 1998). Similar evidence was found in other regions of Chile where copper is also mined (Ferreccio et al., 1998).

Castro-Larragoitia et al. (1997) provide another good example illustrating the complexity of health issues with the case study from the semiarid region of La Paz/Sao Luis Potosi in Mexico. Two hundred years of mining activities resulted in the significant accumulation of tailings, which had been dispersed by the wind over an area of 100 km^2. There was also leaching of toxic material in the surrounding river system that was used for irrigation purposes. Monitoring results have shown high concentrations of cadmium, lead, and arsenic in the leaves of corn used to feed cattle. The transfer of these contaminants to humans through the consumption of food have yet to be evaluated.

Another missing piece in the evaluation of the impact of mining on physical health involves a lack of clarity regarding the origin of the source of contaminants. To what extent is the source anthropogenic, resulting from mining activity, or naturally occurring? By identifying the origin of the different sources of the contamination, it is then possible to have a more global picture of the situation and to develop intervention measures that are more likely to be effective for the entire community. This information is crucial for the development of mitigation measures that will appropriately reduce human exposure.

For example, research in the Amazon has shown that exposure to low levels of methylmercury originating from the consumption of fish can adversely affect the nervous system of riparian populations at levels lower than what had been reported in other studies (Lebel et al., 1996, 1998; Mergler et al., 1999). Although the original source of mercury was associated with the discharge of the substance from the artisanal mining, transdisciplinary research conducted with an ecosystem approach to human health revealed that agricultural slash-and-burn practices, in addition to artisanal mining, were also responsible for contamination of fish and deleterious human health effects. Erosion and lixiviation of mercury-loaded soils now appear to be the main contributors at the global level of the Amazonian Basin, while mining is the main source for areas surrounding mining sites (Roulet et al., 1999).

The above examples are only an infinitesimal illustration of the literature on physical health and mining. Indeed, there is much ground to cover when it comes to evaluating the health impacts of mining operations ranging from mental health and social well-being of communities as well as physical health.

Mental Health and Social Well-Being

There is a growing body of literature dedicated to the impacts of mining on culture and society; only a few highlights are mentioned here.

Social Well-Being

It cannot be denied that many people have benefited from the products derived from mining. The industry has favorably contributed to economic growth, particularly at the macro level. Yet, debate exists among researchers regarding the extent to which benefits are experienced by host and nearby communities. In the context of developing nations, recent research on large-scale mining questions the potential of this option for rural development. McMahon's (1998) conclusions from various studies state that large, technified mining requiring skilled workers creates little employment in rural areas in developing nations. He also adds that this kind of mining has tended to have a small multiplier effect since local, and sometimes national, economies of developing nations are not sophisticated enough to provide inputs to mining processes.

Pascó-Font (1998), a researcher from Peru, observed that whatever the macroeconomic benefits of a particular mining investment might be, most of the costs it generates, especially those involving the environment, are borne at the local level. In addition, the opening of new roads to rural areas often brings large numbers of migrants, many of whom remain in the area. The work of Sassoon (1998) shows that frequent conflicts arise between original residents and newcomers regarding territory, land use, and resource competition. Pascó-Font (1998) also cautions that his research results do not clarify whether the benefits of large-scale mining outweigh the costs at the community level.

The social impact of large-scale mining expansion also affects the lives of many women. In many Latin American countries, it is against the law for women to be employed in underground mines (Robinson, 1998). In addition, social and cultural values consider women in mines as taboo or bad luck (Robinson, 1998). This means that many new jobs created by mining expansion are not accessible to women. In addition, Robinson (1998) underlines that the low-cost labor of men from developing nations has been sustained and subsidized by unpaid female labor in the household, on farms and at markets.

Based on her work in Indonesia, Robinson (1998) examined several features of the place of women in a large-scale mining community. Some women become contract or second wives of mine workers while workers remain away from their original families. This literally multiplies the effects of migrant labor on family disruption, leaving women left with the burden of raising and providing for children alone. Robinson also points out that male-dominated mining towns create a thriving market for sexual services, in which women, and also some men and children, participate through prostitution. Robinson's observations also included growing incidences of alcoholism, sexual assault, and other forms of violence against women, as well as teenage pregnancy. She concluded her work stating that, for many local people, mining expansion is changing attitudes toward sexuality and the role of women.

In terms of small-scale and artisan mining, the International Labour Organization (ILO) has flagged several important issues related to women. The current type and participation of women in this kind of mining means that they derive fewer benefits than men. Women need improved access to finance and law reforms, and discrimination needs to be removed which prevents women from holding assets and obtaining finance. Women often retain the primary responsibility for child and family care, so their participation in mining is accompanied by a search for balance with reproductive roles. Medical risks are also presented to women in the early stages of pregnancy (ILO, 1999).

Child labor is also found among small-scale and artisan mining operations, and in some countries, there is a gender component where boys are favored over girls. As a result of poverty, lack of education, and other opportunities, children are exposed to many occupational and health risks. Due to the financial crisis in Asia, there is evidence that child labor in small-scale mining is increasing. However, immediate removal of children from hazardous work and small-scale mining might result in further economic hardship through the denial of livelihood and income (ILO, 1999). The ILO also notes that in Ecuador, women and members of mining families were found to be key agents of change on questions of health and the environment.

In some cases, small-scale and artisan mining has become the mainstay of many rural economies (Labonne, 1994). It has created employment and provided workers with the potential to become economically self-sufficient (Davidson, 1993; Traore, 1994). It has also served to curb urban migration, which has been a serious socioenvironmental and health problem in many developing nations (Labonne, 1994).

Nonetheless, this type of mining has led to significant rural-rural migration, bringing instability to resident and indigenous communities. The isolated locations of small and artisan mining centers also gives rise to crime, and confrontations with settled communities who rarely benefit from the value of the extracted mineral. Invasions can bring changes to local systems of values and ethics, and in extreme cases, governments can lose control over the situation, as were the cases in Serra Pelada of Brazil, Muzo in Colombia, Nambija in Ecuador, and Mindanao in the Philippines (Wotruba et al., 1998). It has been pointed out that an extremely important externality of small-scale and artisan mining is the irreversible nature of cultural damage due to entry or invasion of sensitive tribal lands (McMahon et al., 1999).

Culture: Individual and Collective Mental Health

The impacts of mining on culture are better grasped by first defining parameters for the concept. While culture is the subject of intense study and debate within the field of anthropology, James Peacock (1989) provides an explanation useful as a point of reference to discuss impacts of mining. Peacock writes that culture is *partly* a manifestation of continuous negotiation among people regarding the role that established norms and values will play in dealing with routine and complex issues. Peacock argues that culture is dynamic, and can be seen as a subtext in everyday interactions requiring decisions and adaptation. This explanation draws attention to the linkages between culture — a system of beliefs and values as part of the identity of individuals and community — and dealing with practical issues. Peacock's work underscores two key issues to consider when looking at the impacts of mining on culture: (1) the pace of introduced change, and (2) the adaptability of individuals and their community.

Clark (1992, 1997) has argued that mining has its greatest impacts on rural, subsistence-based people. In these contexts the mode of production, often family-based agriculture and husbandry, plays a central role in individual, household, and community identity, and is inextricably linked to cultural values and beliefs. Gender relations, control of resources to nourish children and run households, division of labor, degrees of reciprocity among community neighbors, socialization and nonformal education of children, the relationship to the environment, planting and harvest celebrations, and worship are interwoven to create cultural identity (Keesing, 1981).

Operating mines provide salaried employment to workers, most often male. This can have profound implications for gender relations, and control over resources required to nourish children and run households. In addition, movement away from direct control over one's interaction with natural resources through independent farming to laboring as a mine worker may affect the way in which the natural world is perceived (Simmons, 1993). This can affect the way in which workers, as fathers, socialize and educate their children, bringing another layer of implications for gender relations and control over household resources.

Operating mines in rural areas often result in mining communities made up of newcomers who have migrated to service the needs of the mine and its workers. These communities are typically characterized by high alcohol and drug consumption, prostitution, violence, and crime. These conditions present any community with challenges to maintaining cultural values and beliefs, particularly among youth.

Even the exploration stage of mining, which does not often cause significant environmental impact, can be the source of drastic sociocultural impacts. The presence of mining company personnel can create various expectations among community members, as well as opinions regarding what kinds of impacts might result and how they should be dealt with. These opinions can become set, and differences can affect community cohesion (Clark, 1992). Disruption to community cohesion can affect established reciprocity patterns and divisions of labor, both of which are strongly linked to productive activities. This web of reactions can ultimately influence not only subsistence levels, but also culture (Crehan, 1992).

In summary, the impacts of mineral development demand cultural change. Ideally, a parallel should exist between the pace of introduced change and the adaptive capacity of individuals and the community. In this case everyday interactions and decision making would continue to both draw upon and negotiate the role of the cultural subtext. Conversely, should the pace of change simply overwhelm the ability to adapt, the vital linkage between culture and dealing with practical issues suffers enormous damage.

CONCLUSIONS AND RECOMMENDATIONS

The stress of mining affects the health of ecosystems and, by corollary, the physical and mental health of humans and their social well-being. The exploration, operation, and closure stages of large-scale, small-scale, and artisan mining have deleterious impacts on ecosystem and human health, which are also conditioned by the various operational implications of the particular mineral under exploitation.

An intimate and intricate relationship exists between the environment and human beings. The pace of introduced change and the ability of communities to adapt both have repercussions on the resilience of human culture and ecological integrity. In many developing nations, people and environments are vulnerable to increased poverty and the implications of overharvesting and contamination of the resource base. The degree to which large-scale, small-scale, and artisan mining contribute to national and local economies through the generation of employment and multiplier effects needs to be considered along with various externalities. Some of these include environmental impacts, local and regional stability reflected in the extent of conflict over territory, land-use and resource competition, implications for women and children, and implications for physical health in the community and occupational contexts. In many ways, these externalities reflect many facets of ecosystem and human health.

Mining in developing nations presents challenges to various stakeholders whose concerns tend to be defined in terms of their position and interests. To be effective, research must engage stakeholders in a participatory process and draw on expertise from various disciplines in the social and health sciences to work in an interdisciplinary fashion. Mining, in the context of developing nations, can be characterized by complexity and heterogeneity. Stakeholder needs and ecosystem and human health require transdisciplinary approaches capable of reconceptualizing problems and the goals to be attained. As mentioned earlier in this chapter, research needs to further develop methodology, to increase understanding of causal linkages between ecosystem management and human health, to test ecosystem-based interventions, and to apply results to the development of interventions and policies. The ecosystem approach to human health can provide stakeholders and researchers with a conceptual framework to respond to the many concerns and factors involved in mining in developing nations. This chapter puts forward the following recommendations for research.

Small-Scale Mining

The small-scale and artisan mining sector lacks organization and capacity to deal with the many problems generated by the activity. Participatory research and development on this sector is strongly recommended.

Mining Companies and Skills with Community Consultation and Participation

Community members are very familiar with a variety of conditions within and beyond a mining operation's area of influence, placing them in an excellent position to ensure development of a comprehensive health and socioenvironmental impact assessment. This assessment is one of the key tools for exploring alternatives and mitigation strategies, designing plans for socioenvironmental mine management, and implementing monitoring programs, which can also be led and involve community members (ICGPSIA, 1993).

There are several challenges to achieving community participation, particularly in the context of developing nations. Anderson's work (1998) draws attention to tendencies of community members to respond to companies with political and personal agendas, making it difficult for companies to understand the realities of host communities. Sassoon (1998) notes that many communities that have had a history of paternalistic relations with government are often not accustomed to assuming a central role in their own development. She also emphasizes the lack of information and expertise at the community level regarding the many issues related to how mining may impact upon their lives and environment. Sassoon (1998) underlines that this plays a role in the limited capabilities of communities to evaluate the risks they may be facing.

The importance, however, of community consultation and participation cannot be over emphasized. Consultation may take several forms. Whether an open meeting or among a select few, the consultation event can provoke informal preparation and debriefing among many community members, providing an opportunity to communicate opinions and react to expected change. This communication brings issues to the surface where they are handled within the dynamic parameters of community culture. Ideally, consultation can enable communities to assume elements of control over the pace of introduced change. If equipped and trained, companies can gauge the process of adaptation and avoid overwhelming their host communities. In many ways, community participation is a significant component of the company–community interface, and can play a large role in ensuring that mineral development results in positive sociocultural impacts (Gagnon et al., 1993; ICGPSIA, 1993). Research needs to respond to the needs of companies to develop methodologies for community interaction.

Community Participation and Empowerment

The probability that the results from research will translate into effective action can only be increased if the community is involved at every step of the research development and implementation process. The mining industry, governments, and the research community have encountered difficulties in managing situations in which either their presence or their actions have been questioned and refused by communities. Dealing with the community is not a technique, it is an art. Without having the sensitivity for human relationships and the time required to build trust and confidence through prolonged presence and discussion, community participation will inevitably become a token process leading to dead-end solutions. Community empowerment can only be achieved if the visitors are transparent in their actions and are willing to walk into a field where ultimately they might not have the final word.

The constant in all mining operations is the presence of surrounding communities that also provide the workforce. These men, women, and children are impacted directly or indirectly by mining operations and are in many cases not receiving sufficient benefits. These communities need closer attention.

Government

Authorities of national, regional, and local government may have a significant impact on the relationship between investors and the community. Capacity building is a key issue.

Scaling-Up from the Specific to the Generic

The size of the mine and the type of ores and minerals extracted is also creating a very complex situation where the damage to the environment and to human health can be greatly varied. In this chapter only a few examples were raised in order to illustrate the diversity of the impacts on the environment and on human health. The fact is that basically all mines have their own particularities, and that generalizations, at this point, are difficult to make beyond some very broad and generic issues. One of the most challenging areas for the researcher will be to try to distinguish between what is generic and what is specific in order to extrapolate the results generated from a particular situation.

Long-Term Impacts of Prolonged, Low-Level Exposure

Another area that has received limited attention is the long-term impacts of prolonged exposure to low concentrations of contaminants at a distance from the source. In our view, it is critical to investigate and evaluate the long-term effects of this dynamic process, since it is a situation encountered across the board in all mining regions of the developing world. The situation should be looked at particularly from the point of view of the state of human health and also food resources.

Origins of Contaminants

Continued research aimed at clarifying anthropogenic and natural sources of contaminants will have significant impact on the design, implementation, and effectiveness of mitigation measures.

REFERENCES

Anderson, K., Mining and communities: a discussion paper, in *Mining and the Community: Results of the Quito Conference*, McMahon, G., Ed., The World Bank, Energy, Mining and Telecommunications Department, Washington, D.C., 1998, pp. 57–68.

Brown, B.S., Developing countries in the international trade order, *Northern Ill. Univ. Law Rev.*, 14, 347–406, 1994.

Burgess, W.A., *Recognition of Health Hazards in Industry: A Review of Material and Process*, John Wiley & Sons, New York, 1981.

Canadian Intergovernmental Working Group on the Mineral Industry, *Overview of Trends in Canadian Mineral Exploration*, Natural Resources Canada, Ottawa, Canada, 1998.

Castro-Larragoïtia, J., Kramar, U., and Pulchelt, H. 200 years of mining activities at La Paz/San Luis Potosi/ Mexico: Consequences for Environmental and Geochemical Exploration, *J. Geochem. Explor.*, 58, 81–91, 1997.

Centro de Educación y Capacitación del Campesinado del Azuay (CECCA), *La Vida del Pueblo: Sus Derechos*, Azuay, 1989.

Clark, A., Impact of mining on indigenous peoples' in mining journal books, *Mining and the Environment*, Mining Journal Books, U.K., 1992.

Clark, A., Mining and related social and cultural issues: the east Asian perspective, in *Mining and the Community: Results of the Quito Conference*, McMahon, G., Ed., The World Bank, Energy, Mining and Telecommunications Department, Washington, D.C., 1998, pp. 83–97.

Crane-Engel, M. and Schanze, E., Multilateral exploration assistance: the United Nations programs, in *World Mineral Production*, Tilton, J.E., Eggert, R.G., and Landsberg, H.H., Eds., Resources for the Future, Inc., Washington, D.C., 1988.

Crehan, K., Rural households: survival and change, in *Rural Livelihoods: Crises and Responses*, Bernstein, H., Crow, B., and Johnson, H., Eds., The Open University Press, Walton Hall Milton Keynes, U.K., 1992.

Davidson, J., The transformation and successful development of small-scale mining enterprises in developing countries, *Nat. Resour. Forum*, 17, 4, 1993.

Dudka, S. and Adriano, D.S., Environmental impacts of metal ore mining and processing: a review, *J. Environ. Qual.*, 26, 590–602, 1997.

EMCBC, More Precious than Gold: Mineral Development and the Protection of Biological Diversity in Canada, Environmental Mining Council of British Columbia (EMCBC), 1998.

Ferreccio, C., Gonzalez, P.C., Milosavjlevic, S.V., Marshall, G.G., and Sancha, A.M., Lung cancer and arsenic exposure in drinking water: a case-control study in northern Chile, *Cad. Saude Publ.*, 14 (Suppl. 3), 193–198, 1998.

Gagnon, C., Hirsch, P., and Howitt, R., Can SIA empower communities? *Environ. Impact Assess. Rev.*, 13, 229–253, 1993.

ICGPSIA, Interorganizational Committee on Guidelines and Principles for Social Impact Assessment (ICGP-SIA), Guidelines and principles of social impact assessment, Belhaven, NC, 1993.

International Development Research Centre (IDRC), Research Policy Note, No.2, Beyond Disciplinary Borders, Ottawa, Canada, Aug. 1993.

International Labour Organizations (ILO), Note on the Proceedings: Tripartite Meeting on Social and Labour Issues in Small-Scale Mines, Geneva, Switzerland, May 17–21, 1999.

Interorganizational Committee on Guidelines and Principles for Social Impact Assessment (ICGPSIA), Guidelines and principles for social impact assessment, *Environmental Impact Assessment Review*, 15, 11–41, 1995.

Joyce, S., Major issues in miner health, *Environ. Health Perspect.*, 106, A538–A543, 1998.

Kabeer, N. and Subrahmanian, R., Institutions, Relations and Outcomes: Framework and Tools for Gender-Aware Planning, IDS Discussion Paper 357, IDS, U.K., 1996.

Keesing, R., *Cultural Anthropology: A Contemporary Perspective*, Holt, Rinehart and Winston, Toronto, Canada, 1981.

Klein, J.T., Notes toward a Social Epistemology of Transdisciplinarity Communication au Premier Congrès Mondial de la Transdisciplinarité (Convento de Arrábida, Portugal), 1994.

Labonne, B., Small- and medium-scale mining: the Harare seminar and guidelines, *Nat. Resour. Forum*, 18, 1, 1994.

Labonne, B., The mining industry and the community: joining forces for sustainable social development, *Nat. Resour. Forum*, 23, 315–322, 1999.

Lebel, J., Mergler, D., Lucotte, M., Amorim, M., Dolbec, J., Miranda, D., Arantès, G., Rheault, I., and Pichet, P., Evidence of early nervous system dysfunction in Amazonian population exposed to low levels of methylmercury, *Neurotoxicology*, 17, 157–168, 1996.

Lebel, J., Mergler, D., Branches, F.J.P., Lucotte, M., Amorim, M., Larribe, F., and Dolbec, J., Neurotoxic effects of low-level methylmercury contamination in the Amazon Basin, *Environ. Res.*, 79, 31–44, 1998.

McAllister, K., *Understanding Participation: Monitoring and Evaluating Process, Outputs and Outcomes*, IDRC, Ottawa, Canada, 1999.

McDonald, J.C. and McDonald, A.D., The epidemiology of mesothelioma in historical context, *Euro. Respir. J.,* 9, 1932–1942, 1996.

McMahon, G., Mining and the community: a synthesis, in *Mining and the Community: Results of the Quito Conference*, McMahon, G., Ed., The World Bank, Energy, Mining and Telecommunications Department, Washington, D.C., 1998, pp. 1–10.

McMahon, G., Evia, J.L., Pascó-Font, A., and Sánchez, J.M., *An Environmental Study of Artisanal, Small and Medium Mining in Bolivia, Chile and Peru*, World Bank Technical Report 429, The World Bank, Washington, D.C., 1999.

Mergler, D., Lebel, J., Dolbec, J., Filizolla, L., Gentil, P., Branches, F., Passos, C.J.S., Arantès, G., Roulet, M., and Lucotte, M., Neurotoxic effects of low level methylmercury exposure in the Brazilian Amazon, Fifth International Conference Mercury as a Global Pollutant, Rio de Janeiro, May 23–28, Book of Abstracts, CETEM, Rio de Janeiro, Brazil, 1999, p. 371.

Moody, R., Diamond dogs of war, *N. Internationalist*, 299, 15–17, 1998.

Mining Policy Research Initiative (MPRI), Proceedings from MPRI Advisory Group Meeting, unpublished, Toronto, Canada, Mar. 8, 2000.

Natural Resources Canada, *Canadian Minerals Yearbook 1993*, Ottawa, Canada, 1993.

Natural Resources Canada, *Canadian Minerals Yearbook 1997*, Ottawa, Canada, 1997.

Oxman, A.D., Muir, D.C., Shannon, H.S., Stock, S.R., Hnizdo, E., and Lange, H.J., Occupational dust exposure and chronic obstructive pulmonary disease: a systematic overview of the evidence, *Am. Rev. Resp. Dis.*, 148, 38–48, 1993.

Pascó-Font, A., Economic costs and benefits for a local community: workshop report, in *Mining and the Community: Results of the Quito Conference*, McMahon, G., Ed., The World Bank, Energy, Mining and Telecommunications Department, Washington, D.C., 1998, pp. 29–34.

Pastizzi-Ferencic, D., Introduction to the issues, in *UNDTCD Mining and the Environment: The Berlin Guidelines*, Mining Journal Books Ltd., London, 1992.

Peacock, J., *The Anthropological Lens: Harsh Light, Soft Focus*, Cambridge University Press, New York, 1989.

Peden, D., Mono-, Multi-, Inter-, and Trans-Disciplinarity in IDRC Research Activities, unpublished, IDRC, Ottawa, Canada, Apr. 12, 1999.

Peiter, C., Villas Boas, R.C., and Shinya, W., The stone forum: implementing a consensus building methodology to address impacts associated with small mining and quarry operations, *Nat. Resourc. Forum*, 24, 1–9, 2000.

Prager, S., Changing North America's mind-set about mining, *Energ. Mining J.*, Feb., 36–44, 1997.

Rao, R.K. and Yerpude, R.R., Impact of mining projects on the socio-economics of the region, in *Proceedings from the International Symposium on The Impact of Mining on the Environment: Problems and Solutions*, 11–16 Jan., Nagpur, India, Whitakers, Ed., AA Balkema Publishers, Rotterdam, the Netherlands, 1994.

Reger, R.B. and Morgan, W.K., Respiratory cancers in mining, *Occup. Med.*, 8, 185–204, 1993.

Ride, A., Mining, the facts, *N. Internationalist*, 299, 24–25, 1998.

Ripley, E.A., Redmann, R., and Crowder, A., 1996. *Environmental Effects of Mining*, St. Lucie Press, Boca Raton, FL, 1996.

Robinson, K., A bitter harvest, *New Internationalist*, 299, 24–25, 1998.

Roulet, M., Lucotte, M., Farella, N., Serique, G., Coelho, H., Sousa Passos, C.J., de Jesus da Silva, E., Scavone de Andrade, P., Mergler, D., Guimaraes, J-RD, and Amorim, M., Effects of recent human colonization on the presence of mercury in Amazonian ecosystems, *Water Air Soil Pollut.*, 112, 297–313, 1999.

Samet, J.M., Diseases of uranium miners and other underground miners exposed to radon, *Occup. Med.*, 6, 629–639, 1991.

Sassoon, M., 1998. Social benefits and costs: workshop report, in *Mining and the Community: Results of the Quito Conference*, McMahon, G., Ed., The World Bank, Energy, Mining and Telecommunications Department, Washington, D.C., 1998, pp. 35–43.

Schreckinger, V.I., *Explotación del oro y contaminación por mercurio en el Ecuador*, Ministerio de Energía y Minas, Dirección General de Medio Ambiente, Quito, Ecuador, 1987.

Simmons, I., *Interpreting Nature: Cultural Constructions of the Environment*, Routledge, New York, 1993.

Singh, N., Thinking Gender in Development Research: A Review of IDRC-Funded Projects from a Gender Perspective, IDRC/GSD internal document, Ottawa, Canada, 1999.

Smith, A.H., Goycolea, M., Haque, R., Biggs, M.L., Marked increase in bladder and lung cancer mortality in a region of northern Chile due to arsenic in drinking water, *Am. J. Epidemiol.*, 147, 660–669, 1998.

Summers, C. and Lewko, J., Stakeholder Values in Mine Rehabilitation and Environmental Protection, social science measures paper presented at Sudbury '95, Conference on Mining and the Environment, 28 May–1 June, Laurentian University, Sudbury, U.K., 1995.

Traore, P.A., Constraints on small-scale mining in Africa, *Nat. Resourc. Forum*, 18, 207, 1994.

Tyler, S., Policy implications of natural resource conflict management, in *Cultivating Peace: Conflict and Collaboration in Natural Resource Management*, Buckles, D., Ed., IDRC, Ottawa, Canada, 1999, pp. 263–280.

Wagner, G.R., Asbestosis and silicosis, *Lancet*, 349, 1311–1315, 1997.

Walde, T., Environmental policies towards mining in developing countries, *J. Energ. Nat. Resourc. Law*, 10, 327–357, 1992.

Weeks, J.L., From explosions to black lung: a history of efforts to control coal mine dust, *Occup. Med.*, 8, 1–17, 1993.

World Bank Industry and Mining Division (IENIM), *A Mining Strategy for Latin America and the Caribbean*, Technical Paper 345, Industry and Energy Department, Washington, D.C., 1996.

Wotruba, H., Hruschka, F., Hentschel, T., and Priester, M., *Manejo Ambiental en la Pequena Minería*, Manejo Integrado del Medio Ambiente en la Pequena Mineria MEDMIN and Agencia Suiza para el Desarrollo y la Cooperación COSUDE, La Paz, 1998.

CHAPTER **84**

Issues of Scale in the Selection and Interpretation of Mineral Indicators

Deborah J. Shields and Slavko V. Šolar

INTRODUCTION

In 1992, the United Nations Conference on Environment and Development (UNCED) adopted Agenda 21. Subsequently, many countries embraced the concepts of sustainable development and, by extension, sustainable resource management. Agenda 21 (UNCED, 1992, Chapter 40.4) called for the creation of indicators of sustainable development that could provide a basis for decision making at all levels. In response, academics and governments, as well as nongovernmental and intergovernmental organizations, began the complex processes of first defining and then quantifying sustainability.

To date no single definition of sustainability has been agreed upon. Rather, a variety of different conceptual models have been proposed, including genuine wealth (World Bank, 1998), pressure-state-response (OECD, 1998), and ecological footprint (Wackernagel and Rees, 1997), each of which takes a different philosophical perspective. Whether or not agreement on a definition and modeling approach is reached, making the sustainable development paradigm operational will require research progress in many different areas, one of which is the development of indicators crucial to specific sectors (Billharz and Moldan, 1997).

One such sector is forestry. Several major international efforts, including the Helsinki and Montreal Processes, have developed sets of criteria and indicators (C & I) to assess the sustainability of forest ecosystems and the economic and social systems dependent upon them. The Helsinki Process started in 1990 in Strasbourg, France, with the First Ministerial Conference on the Protection of Forests in Europe. Pan-European C & I for sustainable management of temperate and boreal forests were subsequently developed within the Follow-up of the Second Ministerial Conference, held in Helsinki. Slovenia is a signatory to the Helsinki Process Resolutions (Shields and Šolar, 1999). The Montreal Process resulted in the Santiago Declaration, which contained a consensus list of C & I for sustainable forest management agreed to by the U.S. and nine other countries. Both sets of C & I represent ambitious attempts to provide both a context within which countries can engage in discussions about sustainability and a common framework within which to assess and evaluate progress toward sustainability at the national scale.

As noted above, the governments of both the U.S. and Slovenia have embraced sustainable development, including sustainable resource management. Slovenia has endorsed (or signed, in the case of treaties) all the major international agreements on sustainable development (Agenda 21, Helsinki Resolutions). Sustainability principles are being incorporated into many national documents, such as the Environmental Protection Program (adopted by parliament in 1999) and Long Term Land Use Plan 2000–2020, which is currently being written. The content of the State Mineral Resource Management Programme, which was completed by mid-2001, also reflects sustainability principles. Selected indicators of sustainability, most of which were based on the Organization for Economic Cooperation and Development (OECD) approach, were presented in the 1998 government report, "Environment in Slovenia 1996"; however, no mineral indicators were included.

The U.S. Forest Service (USFS), which manages over 180 million acres of public lands, plans to utilize C & I of ecological, social, and economic sustainability as a major component of its comprehensive monitoring program. As a first step, the USFS has committed to implementing the seven criteria and 67 indicators developed through the Montreal Process on all National Forest System lands by 2003 (Dombeck, 1998).

The concepts of sustainable development and sustainable natural resource management are being extended beyond forests to other ecosystem types and resources, including rangelands and nonrenewable resources. Scientists are currently studying the applicability of the Montreal Process C & I to rangelands and grasslands, and although neither the Montreal nor the Helsinki Processes addressed the role of energy and mineral resources in sustainable development, these resources have not been ignored. The governments in several nations, including Australia, Canada, the U. S., and a number of European countries, have been studying the role of minerals in sustainability. Most of these efforts will result in generic, decontextualized C & I of mineral system sustainability (i.e., in measures defined so as to be applicable at many different locations, rather than being site specific and context dependent).

The Montreal and Helsinki C & I were intended to be applicable at the national scale, but the land management decisions that affect sustainability take place on the ground in specific locations. There is a need for local scale C & I and associated measures that can be used to assess how land management decisions influence ecosystem and social system sustainability. This is the case for all types of indicators, whether the sector of interest is forestry or minerals.

Many of the national and international groups investigating minerals in sustainability have taken a continental or national scale perspective, including the OECD (Australia and mineral/energy indicators) and the U.S. Sustainable Minerals Roundtable. Similarly, the work being done on material flows by the World Resources Institute is national in scope (World Resources Institute, 1997). Mineral indicators applicable at the local scale are also being developed. Berger and Iams (1996) have published a set of geoindicators that measure Earth system processes. Many of the indicators, such as groundwater level, shoreline position, and surface displacement can be used to address local sustainability issues. Scientists participating in the Mining & Energy Research Network (MERN) are studying sustainability as it relates to mine operations. They are designing indicators of process, such as capacity building, participation, and well-being, as well as more conventional input/output indicators of performance and social development (Warhurst, 1998).

The analysis and interpretation of complex phenomena at multiple scales is one of the central issues in the study of ecology. One part of the body of literature addressing these problems focuses on the conceptual structure of interrelationships across scales and methods for dealing with that structure. We apply a subset of that body of knowledge to the question of how mineral indicators of sustainability, which are currently being defined for use at different spatial scales, relate to one another. We begin by introducing concepts from hierarchy theory, applying those concepts to conceptual mineral indicators. We next present a framework within which mineral indicators can be organized and compared across scales, giving a series of examples. We close with a discussion of the policy implications of utilizing scale-specific indicators.

APPLYING CONCEPTS FROM HIERARCHY THEORY TO MINERAL INDICATORS OF SUSTAINABILITY

Every entity is both a part and a whole. It is a part of some larger entity and, at the same time, it is the sum of its own parts. This is the essence of hierarchies. Every level in the hierarchy relates in two directions, upward to the larger whole into which it is integrated, and downward to the parts of which it consists (Allen and Starr, 1982).

Hierarchies can be thought of as partially ordered sets with an asymmetry of relationship among the elements of the set (Sugihara, 1983). An element or entity in the set has a role that is defined by the upper level to which it belongs (i.e., the higher level constrains the context of lower levels). The broad-scale processes that take place at higher levels of the hierarchy constrain the finer-scale phenomena (Allen et al., 1984). Conversely, as fine-scale patterns propagate to larger scales, they can potentially constrain broad-scale patterns (Huston et al., 1988).

Although there is no single correct scale of analysis (Wiens, 1989), this does not mean that all scales serve equally well in all cases. Choice of scale should be an explicit function of the policy and management decisions that need to be made (Rykiel, 1998; Tainter, 1999). The information collected and analyses conducted should be consistent with the scale of the decision context. Moreover, scale must be an explicit function of our knowledge about and understanding of the systems in question rather than being an artifact of anthropogenic perception of phenomenon scale. This is because perception of pattern and process depends upon the scale at which variables are measured. Thus, the scale of investigation can have profound effects on what is found (see, for example, Weaver, 1995; Edmunds and Bruno, 1996).

Detection of pattern and process depends on the grain and extent of the investigation (O'Neill et al., 1986). Grain refers to the individual units of observation; extent refers to the overall area being observed. Grain constrains the inferences that can be drawn; below the grain size, there is no way to observe relevant information. Extent constrains the size of the entity we describe, because it is not possible to observe an entire entity unless the universe of observation is large enough to include all its parts.

A sieve provides a useful example of grain and extent. The size of the mesh in a sieve's screen determines the size of the particles that will be retained in the sieve as well as those that will pass through. Only particles larger than the mesh size can be sampled. It will not be possible to collect samples of particles so small they pass through the sieve. In this way the mesh size defines the grain of the experiment. The size of the sieve itself sets a limit on how much material can be passed through the screen within a specified time period, and hence the volume of material, or size of the universe, that will be examined. This defines the extent of the experiment.

We can apply the concept of grain to open pit mines, such as the one in Jersovec, Slovenia, where clayey to silty chert rubble is being extracted. Silty clay is a product of the weathering of carbonate rocks. Chert has very low Al_2O_3 and Fe_2O_3 content; the concentrations of these undesirable impurities are much higher in the silty clay portion of the sediment. The aluminum and iron oxides are removed by separating the silt/clay from the chert fragments through wet screening. By using a cutoff screen of 3 mm, the larger chert particles (>3 mm) can be separated from the silt/clay and smaller chert particles (<3 mm). Grain size is therefore 3 mm. The extent is volume of material screened within a specified time period.

The grain and extent set the limits of resolution for a study. It is not possible to generalize beyond the extent without making the assumption that pattern and process will not change (i.e., that they are scale-independent) (Wiens, 1989). Since system processes frequently differ across scales, scaling-up with existing information can be problematic. It may result in a biased understanding of broader-scale processes, with the process from the smaller scale being given more importance than is actually warranted in the larger system (MacNally and Quinn, 1998).

Viewed from the resource management perspective, the problem is one of practices that are appropriate for, and contribute to, sustainable resource management at the site level, but which may not contribute to sustainability when applied to a larger geographic area (Fox, 1992). Multiple local optima do not necessarily result in a global optimum. This is why environmental impact statements typically include information on cumulative effects. Nor does it follow that the processes at work at broader scales are the same ones driving a system at smaller scales. Hierarchy theory suggests that using fine-scale data to make predictions about large-scale events is more likely to be accurate than the reverse (Fox, 1992). This is because the grain of the broad-scale data may be too coarse to detect the pattern and process occurring at finer scales. Great care must be taken when predicting from higher to lower levels.

For reasons of cost and tractability, expanding the extent of an experiment often entails collecting data at a coarser grain (i.e., as the study area is enlarged, fewer measurements are taken). In so doing, the observer gains information on the broad-scale pattern, but loses the fine-scale details. This can be a worthwhile trade-off when the policy issues of concern are broad-scale as well. Moreover, just as multivariate statistical methods can be used to summarize large amounts of fine-grain data to reveal broader relationships, moving up in scale can increase the scientist's ability to generalize (Levin, 1992).

Mineral exploration displays this pattern, albeit in reverse. Exploration activities usually begin with studies of very large areas, perhaps as much as several thousand square kilometers. More detailed exploration activities are only undertaken in those areas with "good results," so that as the extent of exploration target area shrinks gradually from the entire landscape to the mine site area (mineral deposit of a few hectares), the sampling grain becomes finer. However, the extent of the area over which sampling takes place must be at least the size of the ore body if the objective of exploration is to define the ore body's boundaries.

Scientists in many fields utilize hierarchies to organize information (Figure 84.1). Those hierarchies may be either spatial (administrative or economic, for example) or temporal (historic, geological). Organizing into hierarchies clarifies information and facilitates the understanding of different aspects of complex systems. The hierarchies chosen for analysis are often interdependent and overlapping. This is because patterns and processes defined within one hierarchy affect the structure and functioning of entities within other hierarchies.

Consider, for example, the interrelationships among three hierarchies, geological (ore body limits within a geologic formation), administrative (mining property boundaries within politically defined boundaries), and economic (a mine within an industry sector). The extent of the ore body is defined within the geological settings hierarchy; however, the political units that make up the administrative hierarchy will define ownership of and access to the ore body. Mine profitability,

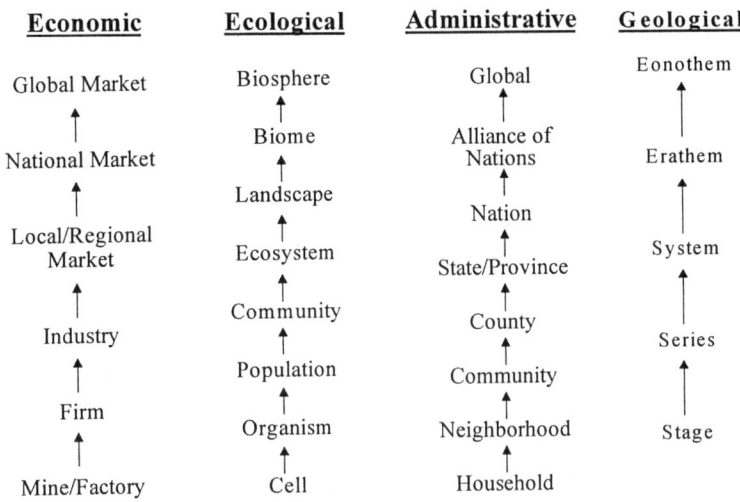

Figure 84.1 Hierarchies such as these are used to organize complex information according to scale.

which resides within the economic hierarchy, is dependent upon the property rights regime, as well as the nature of the ore body. The existence of the ore body can impact decisions made within the administrative hierarchy about access and ownership (i.e., administrative functioning). And these interactions between two hierarchies (geology and administrative) have implications for the way a process within the economic hierarchy, such as mine development, is handled.

The role an entity plays depends on the higher level system to which it is assigned. The higher level system to which it belongs is determined by the phenomenon in question (Allen et al., 1984; Allen and Hoekstra, 1990). This implies that entities can reside in more than one hierarchy, depending upon the phenomenon being investigated.

Returning to our mining example, an ore body can be thought of an as entity. The rock within that ore body could be pure limestone, clay, or a combination thereof (e.g., limestone [less than 5% clay], marly limestone, limy marl, marl, clayey marl, marly clay, or clay [less the 5% limestone]). If the phenomenon of interest is the genesis of the rock, then the hierarchy within which the entity resides is geologic. However, the composition of the rock also has implications for (pit) mine design because the ultimate angle of slope stability ranges from a few degrees in clays to a few tens of degrees in limestone. Thus, if the phenomenon of interest is mining or mine design, then the hierarchy within which the entity resides is economic, or perhaps engineering.

A FRAMEWORK FOR COMPARING INDICATORS AND PHENOMENA ACROSS SCALES

As the foregoing makes clear, inferring pattern and process at one scale based on information collected at another is fraught with potential difficulties. However, linkages can be made when a single phenomenon is significant at different levels of the same hierarchy (Allen et al., 1984). Some phenomena are applicable only at certain scales (e.g., global climate change is meaningful at the global, national, and industry levels, but not at the mine site level). Similarly, underground water balance is a meaningful concept at the mine site scale, but not at the industry-wide scale. Conversely, phenomena such as production and safety are relevant across many different levels of the hierarchies to which they belong (Figure 84.2). In cases where cross-scale linkages exist, it may be possible to predict the status of an entity at one scale in a hierarchy based on information about a shared phenomenon from another scale of the hierarchy.

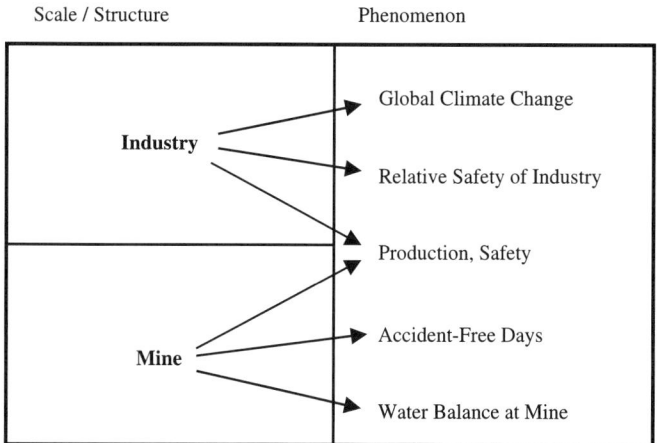

Figure 84.2 Unique and mutually occurring phenomena. Scales or levels within a hierarchy are connected to one another by mutually occurring phenomena.

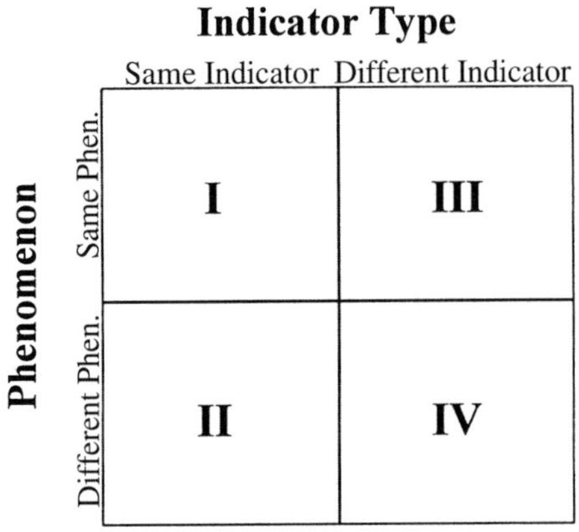

Indicator Type

Same Indicator Different Indicator

Figure 84.3 Two-by-two table of phenomena and indicators. Phenomena and the indicators that measure them may be the same or different across scales of a hierarchy.

Most entities and phenomena can be described by qualitative or quantitative measurements. This is the purpose of indicators: to provide information about the status of entities or functioning of phenomena. Although entities can be linked across scales by shared phenomena, it does not follow that an indicator of that phenomenon that is meaningful at one scale will be equally as informative at a different scale. An indicator can be used to describe a single phenomenon across scales (i.e., aggregated up or disaggregated down) only in those instances when its meaning does not change as grain and extent are changed.

The relationship among phenomena and indicators can be thought of as a 2-by-2 problem (Figure 84.3). A phenomenon may or may not be playing out at two different scales of a hierarchy. Similarly, an indicator that is applicable at one scale may or may not be applicable at a different scale. An additional complication is that an indicator may be useful at more than one scale, but indicative of different processes at the different scales. Each of the alternatives will be considered below.

Case I

Case I represents the straightforward situation in which both the phenomenon itself and the indicator that describes it are meaningful across two scales. Data on production of mineral resources is among the most convenient and understandable of examples. Production can be reported at many different spatial scales: quarry (company), quarrying area, county, state, country, region, continent, and global level. Typically, data are collected at the quarry level and then aggregated (summed) for reporting at broader scales. Aggregation of these data (scaling-up) is acceptable in this case because meaning does not change across the scales. Of course, as noted above, scaling-up results in a loss of detail that is available at the disaggregated lower level. There are also instances in which disaggregation is appropriate. Production data reported at the industry or firm level can be reported as average productivity per mine or per employee.

Case II

Case II represents the situation in which a phenomenon differs from one scale to the next, but the same indicator is descriptive of the phenomena playing out at each scale. This situation is much less common than is Case I, but sometimes does occur. One such situation concerns NO_x emissions. The local scale phenomenon is soil condition. Excess nitrogen makes the soil more fertile than it would normally be, which can lead to changes in ecosystem structure and functioning. Emission of NO_x could be used as a proxy indicator for nitrogen levels in the soil. At the continental or global scale, NO_x has been identified as a greenhouse gas (a contributor to global warming). Thus emissions would act as an indicator of potential global climate change.

Case III

Case III represents the situation in which the phenomenon is the same at difference scales, but the appropriate indicators differ across the scales. Consider the phenomenon of the spread of exotic (nonnative) plant species. An indicator at the site level might be the presence of vehicles and equipment brought in from infected areas. At the landscape scale, an indicator could be fragmentation due to road building. Another example relates to the costs associated with the transport of mineral materials. At the mine site scale, an indicator might be transportation cost per unit of weight, whereas at the regional level, an indicator might be the dollar value of impacts on infrastructures (highways, bridges, etc.) caused by transport vehicles.

Case IV

Finally, Case IV pertains to situations in which the phenomena and related indicators are different across scales of a hierarchy. The concept of material flow, that is, the life cycle for a mineral, is relevant at the national and global scales, but is not meaningful for an individual mine. Conversely, as noted previously, water balance is an extremely important issue at the mine site scale and completely irrelevant at the national or global scale.

CONCLUSION

Commitment to sustainable development necessitates integration of environmental policies and development strategies to satisfy human needs and improve the quality of life today while protecting resources for the future. This will not be an easy task. Humans have developed complex cultural, institutional, and economic structures. The need to maintain these systems is of as great importance as is the need to maintain the health of biological systems. Exclusive focus on any single aspect is both inappropriate and impossible. Exclusive focus on sustainability at only one scale, be it local, national, or regional, is inappropriate as well.

Agenda 21 (UNCED, 1992, Chapter 40.4) calls for the development of indicators of sustainable development (SI) that can provide a basis for decision making at all levels. SI should:

- Be representative of the chosen system (social, economic, environmental, biophysical)
- Have a scientific basis
- Be quantifiable
- Include reference threshold values
- Be without social bias, represent manageable processes
- Have predictive meaning

Indicators of sustainable development may be transformations of underlying information or selected members of a larger set of data. As these indicators are developed, scale will need to be considered. There is no correct scale of analysis, only an appropriate scale for the issue at hand. Rather than trying to determine the correct scale, we must understand how the system description changes across scale (Levin, 1992). Then by moving up in scale we can abstract from unrepeatable and unpredictable individual cases to generalized behaviors. In the case of mineral resource management, site-specific information is needed for local decision making. However, an understanding of the broader phenomena should inform policy making. There is a need for both localized, site-specific indicators, and national-scale indicators of sustainability for mineral resources. The framework presented here is intended as a tool for organizing phenomena and indicators across multiple scales.

REFERENCES

Allen, T.H.F. and Hoekstra, T.W., The confusion between scale-defined levels and conventional levels of organization in ecology, *J. Veg. Sci.,* 1, 5–12, 1990.

Allen, T.H.F. and Starr, T.B., *Hierarchy: Perspectives for Ecological Complexity,* University of Chicago Press, Chicago, 1982.

Allen, T.F.H., O'Neill, R.V., and Hoekstra, T.W., Interlevel Relations in Ecological Research and Management: Some Working Principles from Hierarchy Theory. USDA Forest Service General Technical Report RM-110, U.S. Forest Service, Rocky Mountain Forest and Range Experiment Station, Fort Collins, CO, 1984.

Berger, A.R. and Iams, W.J., *Geoindicators: Assessing Rapid Environmental Changes in Earth Systems,* A.A. Balkema, Rotterdam, 1996.

Billharz, S. and Moldan, B., Elements of a research agenda, in *Sustainability Indicators,* Moldan, B. and Billharz, S., Eds., Wiley Interscience, New York, 1997, pp. 389–395.

Dombeck, M., A Gradual Unfolding of a National Purpose: A Natural Resource Agenda for the 21st Century, Transcription of speech given to USDA Forest Service employees, Mar. 2, 1998.

Edmunds, P. and Bruno, J., The importance of sampling scale in ecology: kilometer wide variation in coral reef communities, *Mar. Ecol. Prog.,* 143, 165–171, 1996.

Huston, M., DeAngelis, D., and Post, W., New computer models unify ecological theory, *BioScience,* 38, 682–691, 1988.

Fox, J., The problem of scale in community resource management, *Environ. Manage.,* 16, 289–297, 1992.

Levin, S.A., The problem of pattern and scale in ecology: the Robert H. MacArthur Award Lecture, *Ecology,* 73, 1943–1967, 1992.

MacNally, R. and Quinn, G., Symposium introduction: the importance of scale in ecology, *Austr. J. Ecol.,* 23, 1–7, 1998.

O'Neill, R.V., DeAngelis, D.I., Waide, J.B., and Allen, T.F.H., *A Hierarchical Concept of Ecosystems,* Princeton University Press, Princeton, NJ, 1986.

Organization for Economic Cooperation and Development (OECD), *Towards Sustainable Development: Environmental Indicators,* Paris, 1998.

Rykiel, E.J., Jr., Relationship of scale to policy and decision making, in *Ecological Scale — Theory and Applications,* Peterson, D.L. and Parker, V.T., Eds., Columbia University Press, New York, 1998, pp. 485–497.

Shields, D. and Šolar, S., Applying indicators of sustainability to surface mining of construction aggregates, in *Mineral Planning in Europe,* Fuchs, P.E.K., Ed., Institute of Quarrying, Nottingham, U.K., 1999, pp. 279–291.

Sugihara, G., Peeling apart nature, *Nature,* 304, 94, 1983.

Tainter, J.A., Rio Grande basin and the modern world: understanding scale and context, in Rio Grande Ecosystems, U.S. Forest Service, Rocky Mountain Research Station, Ogden, UT, 1999, pp. 7–11.

United Nations Conference on Environment and Development (UNCED), Agenda 21, Chapter 40.4, Rio de Janeiro, 1992.

Wackernagel, M. and Rees, W., Perceptual and structural barriers to investing in natural capital: economics from an ecological footprint perspective, *Ecol. Econ.,* 20, 3–24, 1997.

Warhurst, A., Corporate social responsibility and the mining industry, *MERN Res. Bull.,* 13/14, 81–91, 1998.

Weaver, J.C., Indicator species and scale of observation, *Conserv. Biol.,* 9, 939–942, 1995.

Wiens, J.A., Spatial scaling in ecology, *Functional Ecol.,* 3, 385–397, 1989.

World Bank, *Expanding the Measure of Wealth, Indicators of Environmentally Sustainable Development,* Washington, D.C., 1998.

World Resources Institute, *Resource Flows: The Material Basis of Industrialized Economies,* Washington, D.C., 1997.

CHAPTER 85

Conceptual Approaches to Health and Well-Being in Minerals Development: Illustrations with the Case of HIV/AIDS in Southern Africa

Camilla Maclean, Alyson Warhurst, and Philip Milner

BACKGROUND

Introduction

This chapter focuses on conceptual approaches to human health and well-being in the context of the mining industry. It contributes to a multidisciplinary Mining and Environment Research Network (MERN) 3-year research project on environmental and social performance indicators detailed below.* The biomedical model of health and some of the implications that emerge from the predominant use of the biomedical model as a first step toward assessing the extent to which a minerals development project contributes to, or detracts from, the health of local communities is also examined. In addition, this chapter aims to build on, and contribute to, the work by the Tata Energy Research Institute (TERI) team in Goa, led by Dr. Ligia Noronha, and the Instituto de Estudios Regionales (INER) team, University of Antioquia in Colombia, led by Cristina Echavarria, on the effects of mineral development projects on human health and well-being.

Recent research concerning the broader determinants of health supports the supposition that the biomedical concept of health as the absence of disease is a partial and restrictive model. Emerging research consensus advocates a wider conceptualization of health that considers the critical determinants of health as those located within broader societal and environmental contexts, and which includes a subjective sense of well-being as a significant constituent of health. It also considers that the constituents of well-being are best located within a development rights framework (Echavarria, 1999; Warhurst, 1998).

While it is important for a mining company to undertake baseline health impact assessments (HIAs), it is not the intention of this chapter to offer an in-depth assessment of HIAs or community health strategies for mining development projects per se. Rather, we assert that further discussion is still required on the implications emerging from predominantly using the biomedical concept of health that *a priori* shapes the structure of health impact assessments, many health development polices and practices, and community development initiatives.

---

* This work was undertaken with the aid of a grant from the International Development Research Center (IDRC), Canada, and with the support of the Environmental Research Program of the Department for International Development (DfID), U.K.

Thus, the first section reviews several strands of the biomedical model of health. We suggest that the biomedical concept has been internalized by, and applied widely within, the mining industry, through both occupational health strategies and many community development strategies relating to health. Some of the adverse health effects traditionally associated with mining activities are listed to consider if or how they reflect the biomedical model.

The second section of the chapter reviews existing literature critiques of the biomedical model of health and subsequent health determinants and health measurements that are outcomes generated by the logic of this model. The implications of using such a biomedical model of health are considered for individuals and communities associated with mineral development projects. Initial evidence from MERN research in southern Africa suggests that the biomedical model is insufficient to evaluate the extent to which mining projects contribute to, or detract from, community health and well-being. Corollary evidence from an emerging research consensus also suggests that the biomedical model, if it is used in isolation and to the exclusion of community-level health definitions, is an incomplete foundation on which to base health assessments, policies, and strategies (Marmot and Wilkinson, 1999; Milner, 1999; Soskolne and Bertellini, 1998).

The final section sketches out a definition of a more integrated and holistic understanding of health that includes well-being. We conclude that a reconceptualization of health, negotiated and validated at the community level, and including, but not restricted to, the biomedical model, is needed in order to more accurately assess the health and well-being of those host communities affected by mineral development projects. Further, we suggest that new research is necessary to inform and frame the changes that mining companies will need to make in order to ensure that their activities contribute positively toward community health and well-being.

MERN Research Project: Environmental and Social Performance Indicators and Sustainability Markers in Minerals Development

MERN is an international collaborative research program involving centers of excellence in the major mineral-producing countries of the world. It was established in 1991 with the aim of generating analysis to facilitate the improvement of environmental and social performance and competitiveness of mining projects in the context of growing environmental regulation, societal concerns, and technological innovation.

MERN is currently working on a multidisciplinary 3-year research program principally aimed at developing environmental and social performance indicators (ESPIs) and sustainability markers for mining projects. The ESPIs project is core-sponsored by the U.K. Department for International Development (DFID) and Canada's International Development Research Center (IDRC). It includes collaborative support from principal MERN Industry Club members and research users, particularly for the fieldwork and research workshops phase. A key part of this research is collaboration with the INER team and with TERI. Phase two of this research will focus on the testing and verification of well-being indicators and quality of life tracking tools. Thus, one of the essential aims of this research is to provide policy input to enable mineral development projects to contribute positively toward human health and well-being.

MINING AND MEDICAL MODELS OF HEALTH

This section reviews the biomedical model of health and starts from the assumption that a biomedical model of health is a dominant paradigm. This means that the model has been widely internalized within the mining industry. Some of the adverse health effects traditionally associated with mining activities are reviewed to demonstrate how their typical classification as diseases or injuries illustrates this point. Some of the unintended consequences of operationalizing the biomedical model are considered in reflections on the case of HIV/AIDS in southern Africa below.

Models of Health and the Biomedical Paradigm

Debating concepts of health is not new; debates have occurred since the time of the early Greek Empire (Renaud, 1994). Larson (1991) supports this by suggesting that constructing an absolute conceptualization of health is problematic and "dependent upon the historical period in question and the culture in which it is defined." In practice, among and across societies, health is conceptualized and measured in a variety of ways. However, a common thread running through many of the models is that health is understood as efficient human physiological functioning (brain and body). Few models challenge the fundamental principle that health is only the absence of disease or injury. It is this principle that characterizes and defines the biomedical model of health (Green, 1992; Larson, 1991). Within the biomedical model of health it is possible to identify several distinct strands. These are briefly outlined below.

One strand conceptualizes health as a measure of the state of the physical bodily organs. Implicit within this definition is the belief that ill health occurs when one or more parts of the body are malfunctioning. That is, an individual is unhealthy when the body's functioning of one or more constituent parts and organs is impaired. In this view, health is the absence of disease or injury. Thus, the definition is based on an essentially negative concept: health exists when one is free from disease or injury (Bowling, 1991; Evans et al., 1994; Green, 1992).

The biomedical model of health is the dominant paradigm in conceptualizing and defining health status in most industrialized countries, and consequentially in most industries. The biomedical model shapes patterns of health care policy and may account for preoccupations with attention to the absence of disease and disability evident in all aspects of health care services. The biomedical model is established and embedded in industrialized countries, but its influence extends around the globe (Green, 1992).

The biomedical model has its foundation in the ascension of expert scientific medical research that relies upon knowledge gained from reductionist methods of inquiry and investigation. Of course, the benefits of medicine are immense and not to be minimized, as medical science has enabled the cure and prevention of many diseases (Evans and Stoddart, 1994). Indeed, this is a very powerful reason accounting for, in part, this model's success as a dominant paradigm throughout the world. Although other social, political, economic, and cultural factors also contribute to the predominance of the biomedical paradigm, it is not within the scope of this chapter to address these factors. The chapter is more concerned with the implications.

In the biomedical model, the role of health care providers, hospitals, and medical treatments are seen as fundamental in alleviating symptoms, restoring capacity, eradicating disease, and reestablishing health. Disease, or illness, is seen as the result of physiological and/or organic deficiencies and resulting treatment and interventions are aimed at restoring function, usually centered on secondary and tertiary care.

A closely parallel strand found in the literature is a definition that conceptualizes health not just in terms of the mechanics of the different bodily functions, but in the ability of the body as a whole to function (i.e., how the disease or disorder is experienced by, or translated into an impairment or handicap for, the individual; Rapheal, 1996). This model incorporates the element of an individual's subjective perception of their symptoms, but given that disease and injury are still paramount within the model, it nonetheless reflects a predominantly biomedical paradigm.

Last, the behavioral conceptualization of health falls within the biomedical paradigm as well. Here, health is considered to reside within the individual and is determined by individual behaviors. In addition to physiological functioning, health is conceptualized as "energy and physical-functional ability, wellness and the adoption and carrying out of appropriate disease-preventing lifestyles" (Rapheal, 1996). Central to this model is the focus on specific risk factors and the belief that a person's behavior patterns are a critical determinant of health. However, the behavioral model is nested within the biomedical model of health insofar as the essential pattern is unchanged: an individual gets sick, is cared for by health care service providers, and is consequently treated or cured.

The biomedical model of health has been portrayed here as a discrete coherent entity. The reality is that individuals within companies and in other organizations are unlikely to subscribe to a simplistic view of health. Moreover, we are not proposing that there is a right or wrong definition of health, only to suggest that there are different implications associated with using different models. The implications of how the traditional classification of health problems associated with mining, reflecting the internalization of the biomedical conception of health as an absence of disease or injury, need to be explored more fully. These implications are evident in the description of health issues typically associated with mining reviewed in the section below.

Traditional Health Concerns in Mining

The list of adverse occupational health effects associated with mining is relatively well documented: injuries and accidents, respiratory conditions, i.e., tuberculosis or pneumoconiosis, chronic heavy metal exposure, and chemical exposure (Marcus, 1997; Williams and Campbell, 1998).* These may be related to the mining life cycle, the wastes produced, and the hazards potentially generated, but all may be classified within the biomedical paradigm. If health is defined as the absence of disease or injury, then the converse must hold: that an individual is unhealthy if diseased or injured.

Figure 85.1 illustrates some of the possible hazards associated with the mining life cycle and the environmental, social, and, particularly, health effects that may ensue. The authors note that this mining life cycle is indicative of only some of the industrial processes now used to extract base metals. Different flowcharts can be formulated for different processes. Each chart may vary according to the extracted mineral, the processing method, environmental legislation, and degree of effective monitoring, as well as according to the preexisting biogeophysical conditions and social and cultural context of the host country and region.

Another way of looking at broader health effects associated with mining is through decommissioning and how the associated socioeconomic effects of unemployment, changing land-use patterns, and relocation link to health (Warhurst et al., 1999). Some of the socioeconomic effects, including health, of mining are summarized in Figure 85.2. Research by Warhurst et al. (1999) suggests that these relate to the failure of many mining projects to address key issues (e.g., ongoing social impact assessments through the life of the mining project, systematic closure planning, environmental management plans, etc.). Given that further in-depth analysis of these issues is detailed elsewhere, no further comment is offered here, except to note that many of the socioeconomic issues related to mine closure have significant direct and indirect intra- and intergenerational effects on community health and well-being.

The mining industry has historically relied on scientifically determined exposure limits with respect to toxic or heavy metal water pollution, air sulfur dioxide, fluoride, and particulate emission standards, and on technical solutions to mitigate against physical injuries or accidents. These acceptable limits and technocratic environmental solutions are crucial for the mining industry in terms of providing measurements for monitoring in order to provide assurances to their affected communities that mineral development projects are contributing to, and not detracting from, community health and well-being.

There are a number of significant problems with primarily using technocratic environmental solutions and exposure limits. There is the general population's growing inclination to distrust official data or reassurances commensurate with greater technical sophistication in scientific testing that raises doubts about what can be claimed to be known with reasonable certainty (Rodricks, 1994). Furthermore, the growing body of research under way that engenders a greater understanding

---

* This also includes internal MERN working papers: Macfarlane, M., Warhurst, A., and Milner, P., An Overview of Health Issues in Mining, Mining and Environment Research Network Working Paper 144 (1998); Macfarlane, M., Mining and Social Impacts and Their Assessment, Mining and Environment Research Network Working Paper 134 (1998); and Mitchell, P. and Warhurst, A., Mining, Mineral Processing and Extractive Metallurgy: Technologies, Environmental Implications and the Development of ESPIs, Mining and Environment Research Network Working Paper 139 (1998). All are available from the Corporate Citizenship Unit, Warwick Business School, Warwick University, U.K.

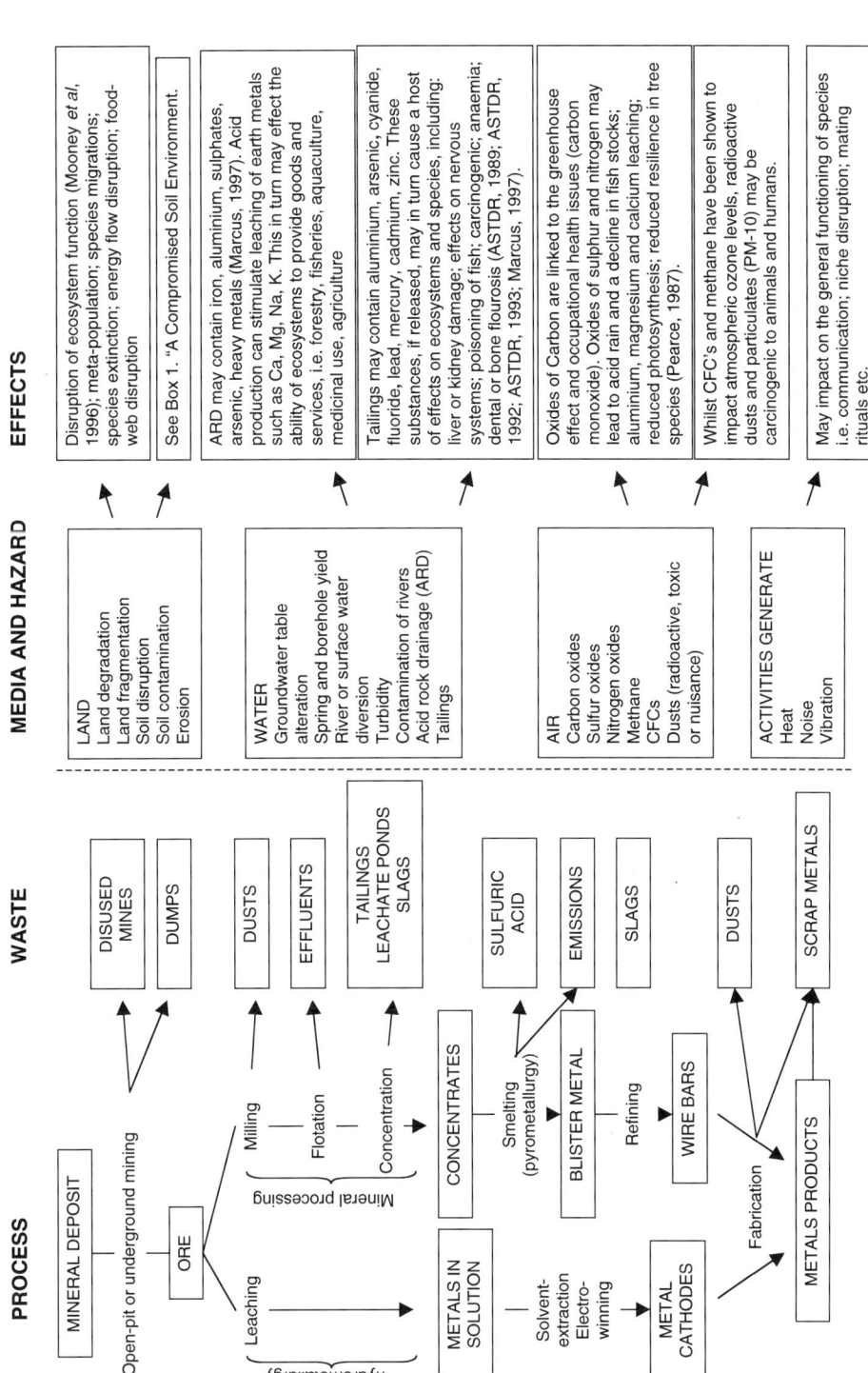

Figure 85.1 Mining life cycle and health-related effects. (From Warhurst, A. and Noronha, L., Eds., *Planning for Closure: Towards Best Practice in Public Policy and Corporate Strategy in Managing the Environmental and Social Effects of Mining*, CRC Press, Boca Raton, FL, 1999. With permission.)

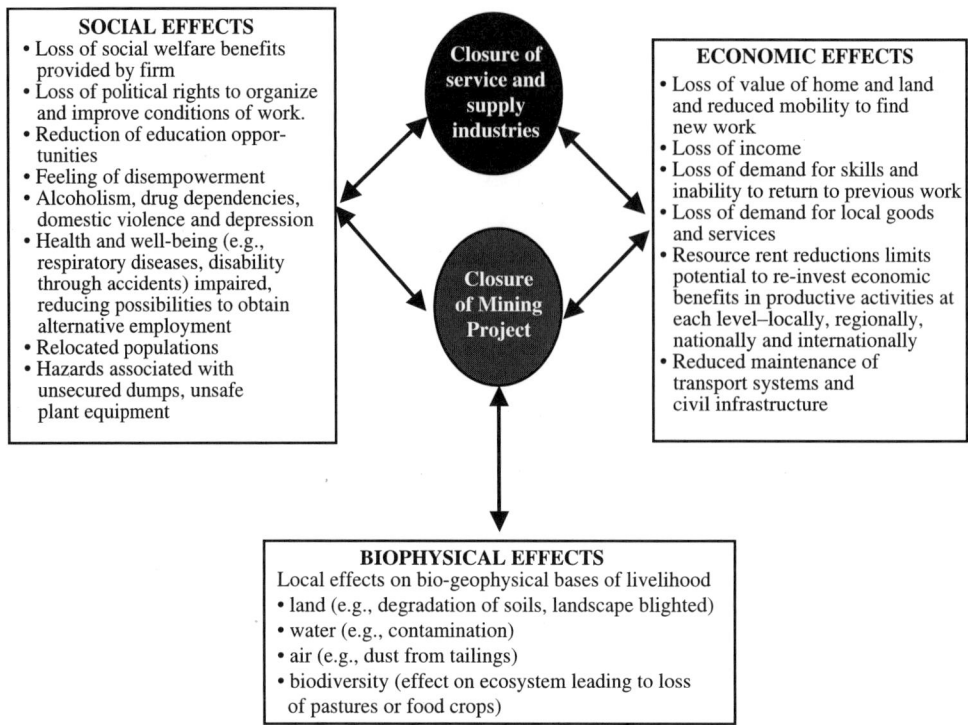

Figure 85.2 Scheme of some of the socioeconomic effects of mine closure. (From Warhurst, A. and Noronha, L., Eds., *Planning for Closure: Towards Best Practice in Public Policy and Corporate Strategy in Managing the Environmental and Social Effects of Mining*, CRC Press, Boca Raton, FL, 1999. With permission.)

of the causal pathways between social and biological effects (Evans, 1994) needs to be considered within minerals development projects and its relevance for individual and community health and well-being. However, the fundamental problem is with how mining's traditional health concerns conflate disease with health (e.g., health concerns are framed within a biomedical model). This is explained further in the following section critiquing the biomedical model and its implications for health determinants and measurements.

PROBLEMS WITH HEALTH PARADIGMS

Following the outline of the biomedical model, this section will now turn to the literature critiquing the model, including its affiliated determinants and measurements of health. The implications of working from a biomedical model of health are considered for individuals and communities associated with mineral development projects. Implications of the biomedical model are detailed to show that, on its own, it is an incomplete model to evaluate the extent to which mining projects contribute to, or detract from, community health and well-being.

Critique of the Biomedical Model

The biomedical concept of health has been criticized as being inadequate for a number of reasons. The high cost of health care in industrialized countries and the persistence of ill health in all countries have prompted more detailed research into exploring health determinants. Noncommunicable diseases have replaced communicable diseases as the main causes of mortality and morbidity in industrialized countries (Birley, 1995; Evans, 1994). Thus, while a specific disease may be eradicated, another disease or illness emerges as a cause of mortality. People in Europe and America now rarely die from tuberculosis. Instead, the World Health Organization reported in 1999 that depression is considered the second highest mortality cause (*The Guardian*, May 12, 1999). Therefore, if high levels of secondary and tertiary care exist in a specific country or location and people are *still* unhealthy, then there must be factors other than health care services, so central to the biomedical model, that impact on health.

While the absence of disease and injury may be a relevant consideration in determining an individual's status as healthy, many people with physiological conditions that can be classified as a disease still feel healthy (e.g., a diabetic whose symptoms are well managed). Individuals also may feel ill without being diagnosed by clinician as having a specific disease or being able to attribute symptoms to an identifiable cause.* If bad health can exist in the absence of disease or injury, and if disease or illness can exist without compromising self-perceived good health or quality of life, then the biomedical model is clearly an incomplete explanation. It is incomplete both for how people define health in its fullest sense, as well as for identifying critical health determinants. Further, an individual's functional capacity and perception of illness will be significant aspects, but not the sum of their health and well-being (Evans et al., 1994; Larson, 1991; Levine, 1987).

The biomedical model of health, defined by clinicians, often misses an important component: the individuals' perceptions of their own state of health. Individuals are ultimately concerned with the impact of illness on their lives, not necessarily with the disease itself. There is a growing awareness that physical health is not the sole purpose in life but more a resource in order to live fully.** This awareness has mobilized policy makers and regulators to consider how public and private activities impact on the elements of well-being or quality of life, as much as on people's physical health.***

With regard to the behavioral conceptualization of health in the biomedical model, it can be argued that insufficient consideration is given to the social and historical processes that shape individual behaviors and perceptions of risk. Research on individual perception of risk, particularly with respect to the nuclear industry (but not irrelevant to mining) suggests that people's own values and concerns represent coherent and reliable bases upon which to explain logical patterns of behavior and responses to potential health hazards (Wynne et al., 1993). Unfortunately, health promotion campaigns aimed at individual behavior may have the dual outcome of diverting attention from broader, perhaps more relevant, determinants of health and placing responsibility for ill health primarily on the shoulders of the individual, who thereby takes the blame for the ill health (Petersen and Lupton, 1996).

For example, with respect to HIV/AIDS health campaigns, in addition to the education and condom distribution programs, there is a movement within South Africa to have HIV/AIDS be classified as a reportable disease, which would in turn add to the risk of further stigmatization of sufferers. Moreover, while health promotion campaigns are often politically inoffensive, comparatively inexpensive, and highly visible, they are also relatively ineffective as the sole mechanism to bring about changes in health outcomes (Campbell and Williams, 1998; Evans, 1994). The Mozambican governmental HIV/AIDS prevention strategy is targeting individual behaviors, thus replicating health promotion campaigns designed in industrialized countries in efforts that seem not to take account of country-specific social and historical processes (Mozambican Deputy Director, personal communication, 1999). Some implications of this approach are discussed in further detail below in Consequences for Communities.

---

* A significant finding in the TERI case study survey is the high percentage of respondents who perceived themselves as being ill but were unable to attribute a reason or cause to their illness.
** World Health Organization definition quoted in The Gottenburg Consensus Paper (1999).
*** There are numerous examples of initiatives that reflect the growing awareness that a person's quality of life or well-being is based on many dimensions and many different aspects of living, such as the U.K. government's "Headline Indicators," (Department of Environment, Transport and the Regions, 1999).

If the biomedical paradigm is only a partial or incomplete conceptualization of health, then greater account needs to be taken of the multidimensional and dynamic aspects of health. This includes aspects of health that are more in accordance with individuals' and society's perceptions and interpretations of what constitutes their definition of health and well-being.

Determinants and Measurements

Part of ongoing MERN research effort is concerned with improving the understanding of the effects of mineral development projects on the determinants of health and well-being for individuals and communities. This is paralleled by medical and social science research's questioning causality of disease in a number of disparate but interrelated ways. Assumptions about genetic and biological determinism, how the social environment influence cellular responses along biological pathways, as well as cultural interpretations of who is considered ill and who is not have all been the theme of recent research (Evans et al., 1994).

It is not the purpose of this chapter to attempt to draw up a definitive list of determinants to health and well-being. Rather, the aim here is to illustrate that there are important health determinants for individuals and communities being missed by mining companies. These significant health determinants are not well understood and, yet, are still considered critical to a more comprehensive definition of health and need to be included in any future effort to measure the health and well-being of communities associated with mineral development projects.

Attempts to advance understanding of health determinants have generally been undertaken by the medical community, with the result that research has tended to focus attention on the narrower concept of health: absence of disease or injury. In part, this is due to the significant utility of being able to measure health through the easily quantifiable events of death (mortality) or the incidence and prevalence of diseases (morbidity).

However, this research has been enriched more recently with contributions from other disciplines (Evans et al., 1994) by including more general dimensions of health and well-being that many people would judge to be important in relation to their own goals and aspirations. For example, indigenous people in Colombia affected by the Cerrajón Norte Coal Project have indicated that well-functioning systems of reciprocity, complementarity, and social justice are critical to their sense of health and well-being (Echavarria, 1999).

The negative and positive impacts of a mineral development project on the physical health of individuals and communities have been the source of extensive MERN research.* However, the negative impacts on the social health of individuals and communities have been well documented across specific projects and regions (Carter, 1999; Macfarlane, 1998). The connections between the two are only now being established, reinforcing the need for improved knowledge and understanding about the significance of social, economic, and cultural factors in the etiology of disease.

According to Evans et al. (1994):

> The factors that determine the health status of a population are not only multiple and complicated but they interact with each other in ways that are much more intricate than previously understood. The link between cause and effect is neither immediate nor direct. They show up in differential susceptibility to threats of illness; and the biological and social mechanisms underlying susceptibility are far from easy to understand, let alone to subject to clear and rigorous testing.

Indeed, the evidence for consideration of a broader set of determinants of health is emerging and accumulating from different disciplines to include not only health care, ecosystem health, nutritional security, and genetic patterns, but also includes identifying the supportiveness of the social environment, group cohesiveness, community resilience, and a sense of personal adequacy

---

* For example, refer to the 1997 World Bank Quito Mining and Community Conference papers.

or control as important determinants. Also increasingly apparent is that what is perceived as a well-being issue may be, or evolve into, a health issue in the narrow sense. New research is investigating the relationship between the two (Evans, 1994; Noronha, 2001). These social determinants affect health at all levels of its definition but also go well beyond traditional models of health care and ecosystem management.

Consequences for Communities: HIV/AIDS and Southern Africa

We argue for the need to use a broader conceptualization of health that evolves beyond the narrow biomedical model, that is, one that can be used as a framework for better evaluating the extent to which the activities and policies of mining projects contribute to, or detract from, community health and well-being *as defined by the community*. The consequences of using a biomedical concept of health for individuals and communities are explored through the following illustrations from the case of HIV/AIDS in southern Africa.

Each conceptualization of health used has critical implications, insofar as what is understood as health problems will differ within each approach. How one views health will affect how health problems are constituted, conceptualized, and solved. Decisions as to disease etiology will clearly structure the type of interventions and planning conceived as logical and possible.

In southern Africa many members of the medical community affiliated with the mining industry have identified one of their leading health problems as HIV/AIDS. While not denying the material dimensions of AIDS as a disease, and the fact that the pain and suffering experienced is very great, the dominant issue characterization has been in terms of the physiological condition of the individual, a reflection of the dominant biomedical model (Campbell and Williams, 1999). A more holistic conception of AIDS might include the social and psychological condition of individuals and those caring for them. The following explains how this can occur.

HIV/AIDS is imposing considerable pressure on those caring for ill or disabled miners, who are typically family members, in the southern African context. The household impacts being felt with HIV/AIDS are multifaceted and immense. As Whiteside and Michael (1998) point out, primary economic providers are dying with little or no provision for their families, a third of which will also be infected. Illness or death results in a household's spending scarce resources, as well as the loss of an adult income (i.e., spending "a full year's income meeting treatment and funeral costs"; Whiteside and Michael, 1998). Moreover, other family members may be left trying to support orphans, further burdening their own household resources. Generally, caring for the sick and dying is left to women in the household, reducing their opportunities to engage in other productive and income-generating activities.

Mining companies have adopted a dual approach to the problem of HIV/AIDS based on their conceptualization of it: first, to reduce the rate of transmission, and, second, to reduce the impact of the disease on the company. Interventions aimed at the former have been based on educational programs such as the Knowledge–Awareness–Prevention program, which is designed to promote changes in individual behavior patterns of condom use and multiple sexual partners (Senior mining industry physician, personal communication, 1999). This is clearly indicative of a behavioral approach within the biomedical paradigm that considers individual lifestyle and lack of personal information as a critical health determinant. This approach is based on the view that with the right information, individual behavior will change and, thus, incidence and prevalence rates will decrease. Interventions aimed at reducing the impact of the disease on the company have included changing employee health benefit packages for cost containment factors (Senior mining industry physician, personal communication, 1999).

However, as Renaud (1994) states:

One's capacity to modify potentially pathogenic behaviors and to "stick with it" is directly related to one's wealth, power and education — in short to the degree of control one has over one's future. The higher up the social hierarchy, the more control one feels capable of exerting over life, the easier it is to change unhealthy habits. In other words, one's "will to change" is largely predetermined by one's social environment.* To be told by an education program or otherwise, that one's lifestyle should change is neither helpful nor effective.

Therefore, these health education campaigns, while laudable in their attempts to protect people's health, may serve to obscure the understanding that lifestyles are themselves determined by the social, economic, and cultural environment. The broader determinants of health are outside the sphere of influence for any one individual, and by focusing on individual risk factors, the status quo is maintained and more fundamental ways of thinking about health are left unchallenged and unchanged (Evans, 1994; Renaud, 1994).

While health promotion programs that focus on lifestyle changes have good intentions, they can also have uncertain or potentially negative outcomes. For example, there exists a twofold risk of failing to meet positive objectives (better health) and a greater risk of the unintended, and not so positive, consequences of individuals being held personally accountable for their ill health. Many health promotion programs take insufficient account of the larger social and cultural processes that are, in themselves, responsible for lifestyle patterns and for health and disease outcomes and health inequities.

The outcome of individuals being blamed for their ill health may arise from health promotion activities that typically, but not solely, target individual behaviors and lifestyles. If individuals are primarily responsible for their health, then the converse must also hold true: that they are also ultimately responsible for their ill health. This logic has important implications. First, it may encourage a disproportionate preoccupation with the physical self and with individual health risks and health hazards, privileging physical health above quality of life or pleasurable activities that contribute to a sense of enhanced well-being. Second, it diverts attention from the fact that people's actions are shaped by the surrounding social and economic context in which they live. As such, there needs to be a collective, rather than solely individual, responsibility for modifying that context if change is required to achieve greater health parity. Typically, health promotion campaigns do not convey the understanding that choice is largely determined by one's social and economic status, which can influence the degree of perceived personal control felt, and by dominant media or societal messages.

For example, while health campaigns in North America encourage people to exercise more and eat less, in reality people are subjected to endless and sophisticated media messages and product and service advertisements designed to encourage people to do the exact opposite. Health promotion campaigns can serve to absolve the larger collective entities, such as corporations and governance structures, of any responsibility for the individual health outcomes and divert attention away from increasing disparities in health, wealth, and power.

Third, and perhaps of most concern, is that health promotion campaigns that focus on individual behavior and chosen lifestyles as the prime determinants of health can create or encourage social and economic discrimination, disadvantage, and exclusion for certain groups and individuals. It allows for the maintenance of the status quo in existing power relations and puts those individuals or groups of people who fail to exhibit appropriate self-control or who fail to take steps to preserve their health in a medically established and accepted manner, at risk for stigmatization.

Health promotion activities in South Africa, with regards to high HIV/AIDS rates among predominantly black miners appear to be at risk for ignoring several factors. These factors include the historical path and logic of multiple partners in Africa and the disparity of economic and political power among different racial groupings in South Africa. Additionally, health promotion activities may ignore the diminished autonomy of women in many African societies. There is also a prevailing cultural belief that many miners articulated in a recent study (Campbell and

* Additionally, the authors would add that one's will to change is also predetermined or influenced by one's biogeophysical, as well as social, environment.

Williams, 1999) that flesh-to-flesh sex, (i.e., as opposed to using condoms) was imperative to maintain their physical health.

Further research would be required for verification, but we suggest that an unintended and unwelcome outcome of some health promotion campaigns in southern Africa might be that, inadvertently, they support paternalistic or, possibly, racist interpretations of patterns of behaviors. For example, one set of actions could be interpreted as being better than another for demonstrating self-control in preserving one's health. At best, such health promotion strategies have failed to stem the tide of HIV/AIDS in South Africa. At worst, they could reinforce the perception that citizens are irresponsible and ignorant through having failed to modify their behavior in light of scientific and rational knowledge (Petersen and Lupton, 1996).

Alternatively, it may be viewed that policy has failed to distinguish between the very different possibilities that men and women possess for exercising control over their own sexuality and the challenges women face. Challenges for women include trying to protect themselves and their children when the main means of control made available are principally in the power of men (i.e., use of condoms, multiple sexual partners, etc.). As Evans (1994) points out, well-intentioned efforts to maintain or improve the health of individuals and the populations they comprise may be wasted if there is poor understanding as to what actually contributes to their health and well-being in the first place. If the principal determinants of health lie elsewhere and are not to be found in individual behavior patterns and biomedical services, then interventions are missing important pieces of the puzzle.

In addition, the underlying reasons for selecting one health problem over another as worthy of attention and receiving targeted resources may obscure better understanding of the health and well-being concerns experienced by the individual. The mining community in southern Africa may be concerned with the alarming rise in HIV/AIDS statistics, and subsequent implications for workers' health, as well as the financial impact on company operations. However, female commercial sex workers working outdoors identified exposure to the flu and colds as their primary daily health concern (Commercial sex workers, personal communication, 1999). Likewise, in another study, mine workers identified concerns about death and injury as their main daily health concern (Campbell and Williams, 1998). In this context there appears to be a disjunction between the identified priorities of the mining and medical community and the priorities of the individuals whose behavior they are trying to influence.

Mining companies may not be optimizing efficient use of resources available for health interventions. Perhaps interventions will not be, in the long term, effectual or sustainable, if they are not cognizant of other, environmental, psychological, social, and spiritual health determinants that are more critical to individuals and communities than their current health and community policies acknowledge. The points outlined above underline some of the complexities and difficulties that need to be considered when assessing a mineral development project's contributions to, or detractions from, the health and well-being of individuals and communities.

National and international mining companies may consider their primary role in society to be wealth creators. However, this is changing. Increasingly, many companies are embracing a triple bottom-line approach, aiming for improved economic, social, and environmental performance. Companies may be making a mistake by focusing on increased individual income through direct employment or trickle-down effects, and contribution to the gross national product (GNP) as the primary benefits of minerals development for communities and countries. Increased income does not necessarily bring greater equity. It may also increase ecosystem disruptions and may disrupt or destroy the determinants of people's health and well-being, and the social fabric of families and communities. Moreover, quantitative goals, such as increased income, longer life expectancies, and higher GNPs, by definition, pay little account of the quality of people's lives.

We are not proposing that a focus on economic goals is exclusively limited to mining companies or mining activities. The effectiveness of a government's policy is measured primarily by that country's GNP growth and many public policy decisions are based on economic cost–benefit analyses. Moreover, financial institutions assess credit ratings of municipalities, corporations, and governments alike on the basis of economic performance and potential rates of return. Although recent research suggests that, increasingly, those financial institutions are placing environmental and social conditions on the provision of finance to protect their returns against future liabilities (Warhurst and Hughes, 1998). However, the focus of our research remains on a mining industry that has been heavily criticized in recent decades over many aspects of their environmental and social performance in an effort to inform new corporate strategy about change and improvement.*

Therefore, we propose that it is within the realm, as well as the responsibility, of mining companies to enhance their social performance with respect to human health. These companies must not only encompass, but also surpass narrow occupational concerns, based on an understanding of health as negotiated with the communities likely to experience the impacts of the mine's operations (Milner, 1999). In turn, the negotiated community-based health definition should then be used in conjunction with mining company personnel and/or with specialist health expertise to develop health impact assessment methodologies and subsequent monitoring programs for community health during the operation of the mine and in planning for decommissioning and closure.

These community health investment activities can be considered to fall under the rubric of corporate social responsibility. Some of the drivers influencing mining companies to act in a socially responsible manner and expand the broader determinants to health are considered in the section below. A brief discussion about the implications for emerging strategies of corporate social responsibility is outlined in the chapter's final section.

Drivers

Over the last decade, multinational corporation (MNC) activity has expanded. In 1970, there were only 7000 MNCs and now there are around 40,000, with over 200,000 globally spread affiliates. In the minerals sector they are particularly active in developing countries and potentially major conduits for technology transfer and economic benefit.

Since 1989, over 75 countries have liberalized their investment regimes for mining to promote further investment and have privatized the large old state dinosaur mining companies, such as COMIBOL in Bolivia and CENTROMIN in Peru, leading to downsizing and drastic direct and indirect employment effects, which brings serious social and health consequences, as mentioned earlier. In turn, this has led to a reduction of the welfare-providing role of the state, which previously supplied a social wage to those large workforces and their families through the provision of subsidized food stuffs, health services, schools, and a salary. This constitutes a form of corporate social responsibility, a paternalistic one, as distinct from the proactive and more deeply embedded concept of corporate social responsibility being discussed now. These benefits were not sustained and were often taken away from one day to the next if metal prices fell, leading to increased poverty and inequity, particularly at old mining regions.

As a result, a growing voice of society is demanding that regardless of formal requirements MNCs, particularly those that have benefited from privatization, adopt a longer-term, forward-looking approach. This should be an approach that proactively anticipates these drivers and implications and takes responsibility for addressing some of the past inefficiencies and inequalities that have, in this new market context, provided them with their opportunities. In this context, it should not be a surprise to the industry that this voice of society is controversially demanding that MNCs apply their capabilities more broadly to address some of the more far-reaching and indirect effects of their activities on the health and quality of life of local communities affected by minerals development in general. This includes addressing the sins of the past (i.e., environmental damage and potential health liabilities resulting from past pollution, generated in the absence of regulation).

---

* The numerous criticisms and examples of poor performance have been well documented by MERN researchers and research associates. For example, see Echavarria (1999), Macfarlane (1998), Noronha (1999), Carter (1999), Warhurst and Mitchell (1998).

Societal concerns are increasingly expressed as demands for information, accountability, and, particularly, community participation. These concerns are also reflected in Agenda 21, the action plan resulting from the Rio Earth Summit in 1992, which, among other imperatives, obliges industry, morally not legally, to contribute to local capacity building in developing countries and to transfer clean technology.

At a sector level there are industry codes of conduct. For example, in mining there is the International Council of Metals and Environment Code of Practice, and within the quarry industry in the U.K. there is the Quarry Products Association Environmental Code of Practice. The Amnesty International Human Rights Principles for companies, in particular, urges companies to adopt an explicit commitment toward respecting and protecting human rights, including human health and well-being (Article 25 of the UN Declaration of Human Rights; Frankental and House, 2000).

Action groups, specifically nongovernmental organizations (NGOs), are increasingly important drivers of change. Friends of the Earth and Greenpeace have launched high profile exposé campaigns and, more recently, have sought to develop a solutions agenda with the business community as NGOs themselves have recognized the creative and shaping role that business can play in the global economy. The role of special interest NGOs has recently grown in importance, parallel with the retreat of government and the diminution of scientific authority, as seen with the scares surrounding bovine spongiform encephalitis, salmonella, and more recently, genetically modified foods. Those special interest groups that are especially active with respect to mining include Minewatch, Third World Network, and Survival International.

Regulation has always been considered a key driver of environmental and social performance. We are not underemphasizing the role that public policy has to play in framing the route toward the sustainable development goal. However, in a developing country context, environmental law is often weakly developed and poorly enforced; it plays its principal role in defining the conditions attached to the permitting of industrial activity (the license to operate) rather than the performance of an ongoing operation or its closure. Notwithstanding, it is important to note that there has been a fundamental shift from the regulation paradigm of command and control, single medium, pollution clean-up incrementally enforced through inspectorates, the courts, and penalties, to a paradigm of integrated pollution control and pollution prevention, promoted through market incentives and innovative rehabilitation bonds.

The environmental and social conditions attached to the provision of mine project finance (as noted above) are proving to be an even more significant driver of improved environmental and social performance on the part of business. In mining, the investment costs are so high (most projects are financed, as a rule of thumb, one third equity, two thirds debt) that equity investment or credit has environmental or, increasingly now, social conditions attached to it to promote environmental and social performance that reduces potential future liabilities.* A major bottom-line concern of financial institutions, as well as companies, is to protect themselves specifically against health damage compensation claims and class-action suits.

Finally, there are also growing internal pressures from shareholders and mining company employees to be more environmentally and socially responsible. Some NGO groups are even buying shares so as to be able to ask questions at company annual general meetings. These questions frequently concern the health and well-being of local communities.

PROACTIVE APPROACH

Mineral development projects have the capacity to contribute to and detract from the health and well-being of individuals and communities. We argue here that adopting a well-defined and

---

* For example, one senior mining physician in South Africa commented on the likelihood that in the future international, if not national, finance institutions would require a company HIV/AIDS prevention programs as part of their loan conditionality (personal communication, 1999).

monitored strategy of corporate social responsibility is key to ensuring that community health and well-being improves and does not decline. Public policy may provide the enabling framework, but detailed below is the argument that a proactive, anticipative strategy of corporate social responsibility can make the difference. For that reason, it is paramount that from the outset of mineral exploration, an integrated definition of health and well-being be employed. This definition will inform the constituents of that strategy of social responsibility, and show that there is consultation and agreement with regard to whose definition of health and well-being matters. This section briefly considers what might constitute an integrated definition of health and the implications of using a different model for assessing the effects of mineral development projects on community health and well-being.

Corporate Social Responsibility

The development of the concept of corporate social responsibility has fast expanded since the days when it was considered that, "the social responsibility of business is to increase profits" (Friedman, 1970). For example, as Andrews (1988) argues:

> Corporate strategy is the pattern of decisions in a company that determines and reveals its objectives, purposes, or goals, produces the principal policies and plans for achieving those goals, and defines the range of business the company is to pursue, the kind of economic and human organization it is or intends to be, and the nature of the economic and non-economic contribution it intends to make to its shareholders, employees, customers and communities.

More recently, Drucker (1993) stated, "corporate citizenship means active commitment. It means responsibility. It means making a difference in one's community, one's society, and one's country."

Corporate social responsibility is defined here as the internalization by the company of the social and environmental effects of its operations through proactive pollution prevention and social impact assessment so that harm is anticipated and avoided and benefits are optimized (Warhurst, 2000). The concept is about companies seizing opportunities and targeting capabilities that they have built up for competitive advantage to contribute to sustainable development goals in ways that go beyond traditional responsibilities to shareholders, employees, and the law; they focus on sustained progress toward improvements in the health, well-being, and quality of life of mining projects' host communities.

There is a need for consultation and agreement with regard to whose responsibility it is to protect and promote health and well-being on the part of individuals, community, government, and companies in different situations and at different times, particularly where political power is not equal. An evolving area of MERN research is, therefore, the concept of good governance and corporate social responsibility with respect to human health and well-being. We suggest that adopting an integrated view of health with regard to how the mining company's activities directly and indirectly affect population health outcomes could usefully form one component of a company's corporate social responsibility strategy.

Integrated Conception of Health

In 1946, the World Health Organization defined health as "a state of physical, mental and social well-being and not merely the absence of disease or infirmity" (Bowling, 1991). Putting this holistic and positive view of health into operation has been slow in developing. However, over the last few decades an emerging health consensus advocating a more integrated view of health has been gaining momentum. An integrated definition views health as a positive state with the ability to do things that are important or have meaning, and which result in a sense of enhanced well-being (Bowling, 1991).

This positive definition conceptualizes health as the integration of the body, mind, and spirit within the individual and the individual's integration within wider social and cultural spheres. In this view, the individual is understood as already embodying, and being embedded in, dynamic and interactive ecosystems and social systems. In contrast to other concepts of health outlined earlier, a key feature of this positive model is that constituents of health are determined by individuals in a dialectic relationship with their other psychological, social, spiritual, and biogeophysical environments.

This positive definition of health is reinforced by evidence of a strong correlation between mortality and social support networks (Evans, 1994; Levine, 1987). Social support networks provide integral mechanisms for coping with stress and adverse life circumstances. Recent research in Colombia suggests that the links between health and well-being and social support networks may be even more pronounced among indigenous peoples (Echavarria, 1999). This research, undertaken participatively with the Wayuu indigenous peoples of Colombia, demonstrates clearly the urgent need for those within the mining industry to understand the broader view of health and well-being and encourages respect for multicultural definitions.

Additionally, we argue that an understanding of a broader view of health should not be limited to health specialists hired by a mining company, given that many health specialist are themselves agents and proponents of the biomedical model, but should also reside within the company's human resources. This includes those responsible within the company for managing community development and environmental protection programs. The recognition that people are not only individuals, but also members of a family, community, and society* should have important implications for the way mining companies view and plan interventions to community health and well-being.

Analysis and Research Implications

We have argued in this chapter that a clearer definition of health is needed before progress can be made in measuring health. The emergent understanding within the medical community that health is more than the absence of disease or injury is reflected in the growth of health measurement scales that go beyond physical functioning. Efforts today are focused on measuring more than just physical aspects of health; they include efforts to measure subjective well-being or an individual's perception of their quality of life. The World Health Organization's research project to develop a quality of life measurement tool is one such example. Indeed, Bowling (1991) calls quality of life the missing measurement. This must be, therefore, taken into account in any attempt to "measure" a mineral development project's contribution to, or detraction from, community health and well-being.

One of the difficulties with most conceptualizations of health and well-being is that listing the constituents implicitly assumes equal importance of each constituent; however, not all people value these components similarly. Therefore, the question of weighting and explicitly clarifying whose values are determining the weighting toward one particular aspect over another becomes paramount in any attempt to measure mining's contributions to or detractions from community health and well-being. Individual and community-level values must also be factored in any measurement attempt.

An integrated health concept suggests that much broader interventions, including community empowerment and antipoverty measures, are necessary to promote health effectively. Evans, citing Wilkinson (1992), suggests that the health of a population depends on the *equality* within a group not the *average* income distribution. Averaging incomes may conceal the reality that the income is unevenly skewed or restricted to a small portion of the population and, therefore, rising incomes can still be

---

* "The protective sense of self-esteem or coping ability may well be a collective as well as an individual possession" (Evans and Stoddart, 1994, p. 52).

associated with declining health.* Evolving research on the full connections between socioeconomic standing and health status goes well beyond the initial analysis than higher incomes bring better health.

A mining company's efforts at being a good neighbor are often, like health promotion schemes, aimed at individual lifestyles. But they are likely to be ineffective if they are missing these several critical considerations listed above. Again, the conclusion is that greater regard must be given to the social and cultural weighting of what each society and its members considers to be important determinants for them and their health and well-being.

Moreover, the very act of beneficence could further undermine the very determinants that appear to be critical to our health, the degree of control people feel capable of exerting over their lives (Evans and Stoddart, 1994). Whereas a company may believe it is contributing to a community by building a school, unless this has been done in careful consultation and as a priority identified by the community itself, the company may undermine the community's sense of control and may foster dependency.

Additionally, the high incidence and prevalence of communicable diseases in developing countries should not obscure the reality that social support networks are also critical determinants to health and well-being. A mine-built health clinic may provide services that eradicate certain types of diseases and illnesses, but that does not automatically rule out the possible, equally negative, physical health impacts inadvertently caused or associated with the adverse social impacts associated with mineral development projects if they are poorly managed. The potential significance of social processes operating on health of group and population clearly needs greater attention and research. A mineral development project's impact on its community** health and well-being is wide-ranging and is not yet as well understood by those "coming in" to the community as it is experienced by those "already in" the community.

More work must be done to facilitate an integrated and multidisciplinary approach — within companies and between organizations — to ensure that health interventions are not tackled in isolation from their cultural and social contexts. Our research suggests that health information alone is not sufficient to change behavior, as exemplified by the growing HIV/AIDS epidemic in southern Africa.

Another line of research inquiry is to explore if differences in stakeholder perceptions as to what comprises their constituents of health and well-being, and determinants, are a source of conflict between stakeholders. Communities must be able "to determine priorities for improving the quality of their environment, and thus their own health and well-being" (U.K. Environmental Health Commission, 1997). Thus, to respond appropriately to the health needs of a particular community, those needs must be identified by, through, and with the people who live in that community, and priorities and targets set through a collaborative negotiation process.

In summary, as outlined above, preliminary research in southern Africa suggests that mining companies conceptualize health in the physical sense (disease and injury) based on the biomedical model. This model misses many health issues and may act as a conceptual, thus in turn, organizational, barrier to understanding the full impact of their activities on an individual's and a community's health and well-being. The links and interrelatedness between and among physical, social, spiritual, psychological, human, and ecosystem health are so far only being made in a very limited sense. This area needs a greater focus of research work.

* This is supported by resource-rent research undertaken by Markandya, A. and Warhurst, A., Analysis of the Distribution of Rents from Mining Operations, Mining and Environment, working paper, Corporate Citizenship Unit, Warwick Business School, University of Warwick, U.K. (1999). They have been advocating for greater distribution of economic benefits derived from mining.
** However the company chooses to define and understand the "community" associated with its mineral development projects. Most companies are likely to understand community spatially within a set geographical definition. However, due to the migrant labor system in South Africa several mining companies also define the communities associated with particular mineral development projects as those communities from which the migrant labor are drawn.

CONCLUSION

We have argued for the need of a broader, more integrated conceptualization of health beyond physiological functioning to include social, psychological, and spiritual aspects sometimes referred to as well-being or quality of life (Bowling, 1991; Bruton, 1997; Fallowfield, 1990; Nordenfelt, 1994). The inclusion of these terms in recent health literature and research efforts reflects a move toward the reconceptualization of health as being more than the absence of disease or illness. This argument is based on two interrelated lines of argument.

The first argument is that research into the determinants of health, even in the biomedical sense, suggests that they are still not well known. However, recent evidence indicates that social conditions are more critical determinants to our health than higher incomes alone and the existence of tertiary health services. Evidence is supporting the observation that social support networks and a personal sense of control have direct positive effects on human health in terms of physiological functioning (Evans et al., 1994).

The second line of thought is that any definition of health structures and limits our understanding of what health problems are and what the solutions might be. A biomedical definition of health typically characterizes problems as diseases or individual behaviors and as a result solutions are targeted at medical solutions or changing individual risk factors, behaviors, and hazards. This, therefore, has implications for the need to change business strategy in mining and policy mechanisms with regard to the implementation of strategies of corporate social responsibility, particularly in the areas of community health and well-being. Health policy based on the narrow biomedical model of health and health care will continue to have limited results. A narrow set of assumptions about what constitutes health will result in a limited view of which are the key health problems and lead to a limited view of what possible solutions exist.

It should be added that, although the implications of a broader conceptualization of health are daunting, we are not proposing that it is the sole responsibility of companies to address these socioeconomic and cultural contexts. Rather, we suggest that this very broadening of the concept of health broadens responsibilities too, and that the key lies in collaboratively addressing the problem and negotiating the solutions together across business, government, and community sectors.

While disease relating to the physical environment may in some instances be resolved, to a certain extent, with technical solutions, the structural causes of ill health associated with a mineral development project will be more difficult to act upon as solutions relate to social and spatial structures, as well as to intergenerational inequities. Although the links between mineral development projects and physical health of individuals have received some degree of attention and research, the understanding of mining's impact on the broader health and well-being of host communities, in its totality, is in its infancy and remains critically underresearched. It is the goal of the joint IDRC-funded research to advance knowledge in this key area of the sustainable development and development policy debates.

ACKNOWLEDGMENTS

The authors would like to acknowledge the input and helpful comments of Cristina Echavarria, Dr. Jean Lebel, Professor Donna Mergler, Dr. Eamonn Molloy, Dr. Ligia Noronha, Dr. Assheton Stewart-Carter, the MERN research team, and two anonymous reviewers.

REFERENCES

Andrews, K.R., The concept of corporate strategy, in *The Strategy Process: Concepts, Contexts and Cases*, Quinn, J.B., Mintzberg, H., and James, R.M. Eds., Prentice Hall, Englewood Cliffs, NJ, 1988.

Birley, M., *The Health Impact Assessment of Development Projects*, HMSO, London, 1995.

Bowling, A., *Measuring Health a Review of Quality of Life Measurement Scales*, Open University Press, Buckingham, 1991.

Bruton, H., *On the Search for Well-Being*, University of Michigan Press, Ann Arbor, 1997.

Campbell, C. and Williams, B., Managing disease on the goldmines, *S. Afr. Med. J.*, 88, 789–795, 1998.

Campbell, C. and Williams, B., Evaluating HIV-prevention programs: conceptual challenges, *Psychol. Soc.*, 24, 57–68, 1998.

Campbell, C. and Williams, B., Beyond the biomedical and behavioral: towards an integrated approach to HIV-prevention in the southern African mining industry, *Soc. Sci. Med.*, 48, 1625–1639, 1999.

Carter, A.S., Community Participation: An Indicator of Social Performance in Minerals Development, Ph.D. thesis, Corporate Citizenship Unit, Center for Creativity, Strategy and Change, Warwick Business School, Warwick University, U.K., 1999.

Corin, E., The social and cultural matrix of health and disease, in *Why Are Some People Healthy and Others Not? The Determinants of Health of Populations*, Evans, R.G., Barer, M., and Marmor, T., Eds., Walter de Gruyter, New York, 1994, pp. 93–132.

Drucker, P.F., *Post Capitalist Society*, Butterworth-Heinemann, Oxford, 1993.

Echavarria, C., Human Health and Well-Being Issues of Significance to Indigenous Peoples in the Context of Mining, Mining and Environment Research Network Working Paper 155, Instituto de Estudios Regionales, Universidad de Antioquia, Colombia, 1999.

Evans, R., Introduction, in *Why Are Some People Healthy and Others Not? The Determinants of Health of Populations*, Evans, R.G., Barer, M., and Marmor, T., Eds., Walter de Gruyter, New York, 1994, pp. 3–26.

Evans, R. and Stoddart, G., Producing health, consuming health care, in *Why Are Some People Healthy and Others Not? The Determinants of Health of Populations*, Evans, R.G., Barer, M., and Marmor, T., Eds., Walter de Gruyter, New York, 1994, pp. 27–64.

Evans, R., Barer, M., and Marmor, T., Eds., *Why are Some People Healthy and Others Not? The Determinants of Health Populations*, Walter de Gruyter, New York, 1994.

Fallowfield, L., *The Quality of Life: The Missing Measurement in Health Care*, Souvenir Press, London, 1990.

Frankental, P., and House, F., Human Rights — Is It Any of Your Business?, The Prince of Wales Business Leaders Forum, London, 2000.

Friedman, M., The social responsibility of business is to increase its profits, *New York Sunday Times Magazine*, Sep. 13, 1970, pp. 32–33.

Green, A., *An Introduction to Health Planning in Developing Countries*, Oxford University Press, Oxford, 1992.

Larson, J., *The Measurement of Health: Concepts and Indicators*, Greenwood Press, CT, 1991.

Levine, S., The changing terrains in medical sociology: emergent concern with quality of life, *J. Health Soc. Behav.*, 28, 1–6, 1987.

Macfarlane, M., Social Impact Assessment in Minerals Development, Ph.D. thesis, Corporate Citizenship Unit, Center for Creativity, Strategy and Change, Warwick Business School, Warwick University, U.K., 1998.

Marcus, M.J., Ed., *Mining Environmental Handbook*, Imperial College Press, London, 1997.

Marmot, M. and Wilkinson, R., *Social Determinants of Health*, Oxford University Press, Oxford, 1999.

Milner, S., The Health Impact Assessment of Non-Health Public Policy in Gothenburg Consensus Paper, World Health Organization, Regional Office for Europe, European Center for Health Policy, 1999.

Nordenfelt, L., Ed., *Concepts and Measurements of Quality of Life in Health Care*, Kluwer Academic Publishers, Dordrecht, 1994.

Noronha, L., Environmental and social performance indicators and sustainability markers in minerals development: reporting progress towards improved health and human well-being: Indian case study, *Nat. Resour. Forum*, 25, 53–65, 2001.

Petersen, A. and Lupton, D., *The New Public Health: Health and Self in the Age of Risk*, Sage Publications, London, 1996.

Rapheal, D., Quality of Life and Public Health, North York Community Health Promotion Research Unit, University of Toronto, Canada, 1996.

Renaud, M., The future: Hygeia versus Panakeia.? in *Why Are Some People Healthy and Others Not? The Determinants of Health of Populations*, Evans, R.G., Barer, M., and Marmor, T., Eds., Walter de Gruyter, New York, 1994, pp. 317–334.

Rodricks, J., *Calculated Risks: The Toxicity and Human Health Risks of Chemicals in Our Environment*, Cambridge University Press, Cambridge, 1994.

Soskolne, C. and Bertellini, R., 1998. Global Ecological Integrity and "Sustainable Development": Cornerstones of Public Health, World Health Organization, European Center for Environment and Health, Rome Division, 1998.

U.K. Environmental Health Commission, Agendas for Change, Chadwick House Group Ltd., U.K., 1997, p. 3.

Warhurst, A., Corporate Social Responsibility: A Pro-Active Approach, Mining and Environment Research Network Working Paper 126, Corporate Citizenship Unit, Warwick Business School, Warwick University, U.K., 1998.

Warhurst, A., and Hughes, N., Financial Drivers of Environmental and Social Performance Indicators Mining and Environment Research Network Working Paper 142, Corporate Citizenship Unit, Warwick Business School, Warwick University, U.K., 1998.

Warhurst, A. and Noronha, L., Eds., *Planning for Closure: Towards Best Practice in Public Policy and Corporate Strategy in Managing the Environmental and Social Effects of Mining*, CRC Press, Boca Raton, FL, 1999.

Warhurst, A., Macfarlane, M., and Wood, G., Issues in the management of the socioeconomic impacts of mine closure: a review of challenges and constraints, in *Planning for Closure: Towards Best Practice in Public Policy and Corporate Strategy in Managing the Environmental and Social Effects of Mining*, Warhurst, A. and Noronha, L., Eds., CRC Press, 1999, pp. 81–99.

Whiteside, A. and Michael, K., AIDS and development, *Indicator S. Afri.*, 15, 55–60, 1998.

Williams, B. and Campbell, C., Creating alliances for disease management in industrial settings, *Int. J. Occup. Environ. Health*, 4, 257–264, 1998.

Wynne, B., Waterton, C., and Grove-White, R., Public Perceptions and the Nuclear Industry in West Cumbria. working paper available from the CSEC, Lancaster University, U.K., 1993.

CHAPTER 86

Mining and Indigenous Peoples: Contributions to an Intercultural and Ecosystem Understanding of Health and Well-Being*

Cristina Echavarría Usher

INTRODUCTION

Economic globalization and the concurrent liberalization of investment regimes have promoted the rapid expansion of mining activities to areas previously isolated from mainstream development processes in developing countries (Warhurst, 1999). Many of these areas coincide with territories traditionally occupied and used by indigenous peoples (IIPP), African Americans, and traditional *campesino* communities (IITC, 1996; World Bank, 1997).

Although demographic data on indigenous populations in many countries are often incomplete, it is estimated that there are some 300 million individuals who identify themselves as IIPP in the world today (Coordinadora de Organizaciones Indígenas de la Cuenca Amazònica, 1999). These IIPP inhabit 70 countries and are characterized by an enormous cultural diversity. In Colombia alone there are 81 ethnically and culturally distinct IIPP, although they total only some 600,000 individuals (Dirección General de Asuntos Indígenas, 1998).

On a global scale, indigenous peoples are increasing their levels of empowerment, while their issues of concern with regard to the manner in which large development projects in their territories affect their ethnic integrity,** their livelihoods, health, and well-being, all of which have reached greater visibility in both academic literature and mass media. Important international institutions such as the World Bank, the European Union, the International Labor Organization (ILO), and the United Nations (UN), as well as international nongovernment organizations (NGOs), are increasingly addressing these same concerns.

National legislation varies widely between countries in its degree of protection and recognition of the rights of IIPP. The general tendency is toward a growing recognition of IIPP rights, in a context characterized by cultural and territorial conflict resulting from the dominance of resource intensive development models and national fiscal objectives over local and regional social and economic development priorities.

---

* This chapter is based on the technical report, Social and Environmental Performance Indicators: Evaluating Progress Towards Improved Health and Well Being, Colombian Component: Indigenous Peoples, Vol. 1–3, Echavarría, C., Correa, H.D., and Puerta, C., Instituto de Estudios Regionales (INER), final phase 1 technical report to Mining and Energy Research Network, Warwick University (MERN), International Development Research Center (IDRC) (1999). This report is available from INER, University of Antioquia, Medellín, Colombia; and IDRC, Ottawa, Canada. The author made all the translations from Spanish.

** Ethnic integrity is understood as the maintenance of the essential factors of livelihood and of the recreation of the ethnic, economic, political, social, spiritual, and cultural systems of indigenous peoples (DGAI, 1998, p. 57).

The physical disruption that characterizes most mining activities often leads to conflict over natural resource use and development and to rapid changes in local livelihoods (Hoon et al., 1997). The pollution and degradation of the resource base (water, air, soil, flora, fauna, etc.) has a greater impact on the more vulnerable sectors of society that depend directly on the land for subsistence. Nevertheless, the problem goes beyond contamination to direct impacts on nutritional security through the destruction of livelihoods. This is often without proper compensation or implementation of livelihood alternatives that are culturally compatible or that respect cultural diversity.

The social impact caused by the immigration of foreign workers into marginal or remote regions also has multiple effects on the social fabric, health, and well-being of traditional communities. The construction of large infrastructure developments required by mining may fragment the traditional territories of these communities, disarticulating their use of space and disrupting food-gathering and herding circuits. Often land and resources required by miners constitute important assets of the typically diversified livelihoods of indigenous peoples, many of whom rely on a low-intensity, but extensive and diverse, use of natural resources. This itself may reduce the ability of IIPP to sustain their traditional ways of life and place severe stress on their ethnic integrity and well-being, generating intraethnic and interethnic conflict over both local natural resources and the economic benefits generated by mining activity.

In many of these regions and localities of the developing world, local communities still have high indices of unresolved basic needs, and great expectations are likely to be placed on mining companies to deliver tangible social benefits. In many cases, companies want to develop appropriate social policy, but local and regional state agencies are corrupt or inefficient. In addition, citizens do not have the capacity or the power to implement public auditing systems to ensure greater transparency and distributive justice in the management of royalties at the local level.

In this context of inefficient state institutions and poor governance, communities fear that pollution will go unreported and unnoticed until it is too late and the damage has been done. They are painfully aware that a diseased ecosystem will mean hunger and illness for them, and that political corruption is likely to exclude them from economic benefits that may be produced by the minerals development.

All these factors are fuel for social and political conflict. They increase the risk of operations being closed down, and in consequence, they tend to reduce opportunities for the attraction of the foreign direct investment that has become crucial for the sustainability of macroeconomic goals of many developing countries.

For many of the above reasons, and on the basis of their negative historical experience with development, it is common to find that indigenous peoples who oppose natural resource development projects (in their territories or close to them) are labeled by the press and governments as the obstacles of development* without sufficient analysis of the trade-offs involved. In countries like Colombia, this may easily lead to the physical extermination of indigenous leaders by extremist groups.

If improved well-being is a necessary outcome of minerals developments, then all of the above aspects need to be addressed by the main stakeholders, in particular, the mining companies and the national and local governments. This points to the need to develop conceptualizations and methodologies that incorporate intercultural criteria and decision-making mechanisms throughout the life cycle of a mineral's development. These should facilitate authentic indigenous participation, in both the planning and proper consultation of development projects that may affect them, and in the prediction, evaluation, compensation, and mitigation of the social and environmental impacts generated by mining activity.** Mining, in a multicultural world, and specifically in local contexts

---

* For instance, with regard to the negative of the U'wa peoples of Colombia to allow oil exploration and development in what they consider their traditional territory, the Minister of Mines in 1997 was reported as saying "one cannot compare the interests of 38 million Colombians with those of an indigenous community" (CCEU,1997 cited by Project Underground 1998, available from project_underground@moles.org).

** It is not the object of this chapter to analyze present practice in environmental (EIA) and social impact assessment (SIA). For indigenous guidelines for use of traditional knowledge in EIA, see Emery et al. (1997).

characterized by the presence of several worldviews and cultures, calls for the development and application of intercultural forms of communication and understanding.

The object of this chapter is to report the advances in research to this end made by the University of Antioquia (Colombia) in the context of the Environmental Social Performance Indicator (ESPI) research program. It aims to contribute toward an integral understanding of health and well-being, to analyze some of the elements that have been identified as key issues for health and well-being by indigenous peoples, and to propose a working integration of these into the main ESPI framework.

KEY APPROACHES TO UNDERSTANDING HEALTH AND WELL-BEING

This section draws on key elements that have been produced by number of paradigms of health and well-being in order to develop an ecosystem and intercultural conceptual framework useful for developing tools for tracking health and well-being in multicultural mining regions. Not intending to make an in-depth analysis of each of these approaches, the aim is to make explicit the approaches that are being incorporated by the ESPI conceptual framework for indicators of health and well-being.

ESPIs: Beyond the Biomedical Model of Health

Due to the complexity of the social and environmental impacts involved, recent research supports the idea that the traditional biomedical concept of health (understood as the absence of disease or injury) is a limiting model for dealing with the impacts of mining on health and well-being (MacLean and Warhurst, 1999).

ESPI research team discussions and papers advocate that health concepts need to take into account the social, political, cultural, environmental, and economic conditions and processes that determine the well-being of communities and individuals. Evidence indicates "social conditions are more critical determinants of our health than higher incomes alone and tertiary health services" (MacLean and Warhurst, 1999).

Working definitions of health, well-being, and quality of life are used in ESPI research in an almost interchangeable manner, with subtle differences in emphasis. This results in the reconceptualization of health beyond that expounded by the predominant biomedical paradigm (MacLean and Warhurst, 1999), to a conception more akin to well-being (Max-Neef et al., 1996; Nussbaum and Sen, 1996) and quality of life. The World Health Organization's (WHO) definition of health directly emphasizes the interdependency between health and well-being: "*Health* is a dynamic state of physical, mental, social and spiritual *well-being*, not merely the absence of disease or infirmity."

The identification between well-being and quality of life, so evident in the writings of Nussbaum and Sen (1996), is also reflected in policy documents, such as the WHO definition of quality of life: "An individual's perception of their positions in life in the context of the culture and value systems in which they live and in relation with their goals, expectations, standards and concerns" (WHO, 1993).

This definition recognizes another key characteristic: that conceptions of health, well-being, and quality of life are culturally determined and constructed. This is particularly important to understand in scenarios where macrodevelopment projects establish themselves in traditional indigenous and rural communities. Here, the values and cultures of various stakeholders may vary significantly with regard to what is meant by well-being and quality of life, as an expected result of the proposed development. In order for these projects to be sustainable in social and environmental terms, both the mining companies and governments need to ensure that they deliver culturally compatible community programs that generate improvements in well-being, health, and quality of life that are recognized as such by the local communities.

Well-Being and Quality of Life*

Despite cultural diversity, authors like Max-Neef et al. (1996) and Sen (1996), among others, have argued that some categories of well-being are universal.

According to Sen (1996), the core aspects of both well-being and quality of life are the capability and ability of individuals to achieve valuable functionings in terms of being and doing. "*Functionings* represent parts of the state of a person — in particular the various things that he or she manages to do or be in leading a life." While "the capability of a person reflects the alternative combinations of functionings the person can achieve, and from which he or she can choose." In this sense, "capability reflects the freedom of an individual to choose between different ways of living" (Sen, 1996).

Sen and others (Cohen, 1996; Corredor, 1998; Korsgaard, 1996; Max-Neef et al., 1996; Nussbaum and Sen, 1996) have offered extensive critiques of conventional measures of well-being and quality of life arguing that they overestimate the importance of primary goods. According to Sen (1985): "Although goods and services are valuable, they are not so by themselves. Their value is found in what they can do for people, or better, in what people can do with these goods and services."

According to these authors, a theory of well-being must be based on a new approach that clearly incorporates ethical principles and rights. This would not give so much importance to the quantity of production of goods and services, as it would to the effectiveness of the economic and institutional systems to ensure the rights, valuable functionings, and capabilities of individuals.

What determines the absence of individual well-being is not only a budgetary restriction, but also the absence of capacity for freedom, in particular, the individual moral freedom of deciding and choosing between different options. Nevertheless, these options and the capacity and will to put them into action are directly dependent on the fulfillment of fundamental human needs. If basic needs, now recognized as social, cultural, economic, and developmental rights, are not resolved, there is little possibility that individuals and social groups will be able to develop the capabilities required to actually exercise these rights (Corredor, 1998).

For instance, if local indigenous communities in developing countries have not solved meeting their basic needs, such as in the case of the Wayúu peoples in Colombia, they are not free to exercise freedom of choice. When consulted by a mining company for their approval to have access to their territory in 1998, the Wayúu decision was heavily influenced by the fact that they did not have access to drinking water or sufficient territory for their livelihoods. They agreed to allow access for exploration on the part of mining companies in hope that these needs would be at least partially fulfilled, but with the risk of negatively affecting their ethnic integrity because the eventual development of the mine would inevitably change their way of life (Echavarría et al., 1999).

Most indigenous peoples want to improve their health and well-being without losing their identity, traditional territory, and autonomy. Often, they do not want to change their way of life or their own model of development or self-determination, but they do want to participate in the decisions that affect them. The promise of access to clean water, formal intercultural education and health care, more land, or compensation money is likely to put short-term trade-offs in the forefront, risking long-term ethnic integrity and well-being.

Freedom of choice in the absence of social, economic, cultural, and development rights is clearly difficult to exercise for many communities in developing countries. From this perspective, the insistence of indigenous peoples on the rights issues becomes clearly pertinent in understanding conceptions of well-being that give greater emphasis to freedom of choice as a crucial category.

Social Processes, Cultural Representations, Ecosystem Health, and Rights

The construction of an intercultural and ecosystem conceptual approach for tracking health and well-being in mining regions poses the challenge of articulating, in a meaningful manner, concepts

* The author acknowledges the contributions in this section of Betancur, O.L., *Enfoques sobre las Teorías del Bienestar* (Theoretical approaches to well being), Instituto de Estudios Regionales (INER) working paper, 2000.

from a number of other theoretical approaches, while at the same time responding to the actual diversity of conceptions of health and well-being. Abundant literature has been produced on these approaches and has been analyzed elsewhere (Puerta and Echavarría, 1999). The main contributions of these models to our approach are summarized below.

Social medicine proposes that health and disease need to be understood as social processes, both in terms of the historical relationships between human individuals and groups, and between people and the environment in specific contexts. In particular, the historical materialist view of social medicine gives importance to the relationship between type of work, living conditions, and health profiles (Laurell, 1982; Quevedo, 1992, 1993; Pinzón et al., 1993). The use of this last approach is particularly pertinent to occupational health in mining, and has produced important results in Colombia, such as in analyzing the health problems of the coal miners in the Guajira coal-mining region (CENSAT et al., 1995). More generally, the social medicine approach states that:

> [D]iseases are of a historical and social character ... the social nature of disease is not verified in the clinical case but in the characteristic manner in which different human groups become ill and die ... it is not in the study of ill persons that we are going to better understand the social nature of disease, but in the pathological profiles of different social groups (Laurell, quoted by Quevedo, 1993).

The underlying assumption of this approach is that health promotion must link with the regulation of social processes — governance, participation, rights — and not only with the identification and management of specific vectors of disease or of particular biophysical pressures.

Researchers in medical anthropology such as Bibeau (1993), Chaumeil (1993), Langdon (1991), Pederssen (1993), Pinzón et al. (1993), and many others, especially in Canada and Latin America, emphasize, "the role of the social and personal life of the diseased in the genesis, configuration and meaning of illness" (Quevedo, 1992). Here the understanding of health–illness as a process also implies the need to recognize the existence of diverse cultural representations and interpretations, that is, the particular conceptions of health and disease that a cultural group has developed based on its historical experience and everyday living. "All illness is social illness ... it is the expression of the forms of social organization and symbolic representation of the group" (Quevedo, 1993).

Medical anthropology in ecological perspective is relevant in this case since it addresses precisely this problem:

> Anthropology has usually emphasized that health and healing are best understood in terms of a given society's system of ethno medicine, and that the "insider's view" is necessary to understand how a society defines and diagnoses disease. Western medicine, on the other hand, usually considers disease as a clinical entity that can be diagnosed and treated while ignoring the cultural context (McElroy and Townsend, 1989).

The ecosystem approach to human health promoted by International Development Research Centre (IDRC) in its research initiative (1997) was initially based on the ecosystem health paradigm (Costanza, 1992; Holland, 1995; Interagency Ecosystem Management Task Force, 1995; Rapport, 1999). In complementing the ecosystem health paradigm with socially based approaches, this paradigm incorporates key elements that will make it more useful as a conceptual model for the purpose of developing intercultural and ecosystem tools that track health and well-being in mining regions, urban centers, and agroindustrial regions. As McElroy and Townsend (1989) state, "When the unit of analysis is a total ecosystem, rather than simply the individual or the society, health and disease become indicators of the group's effectiveness in dealing with the environment."

This recognizes the health implications of the interactions between human groups and their environment. The biophysical environment may in itself pose natural threats to health, for instance, in environments with naturally occurring acid rock drainage of heavy metals or mangrove environments where malaria and yellow fever vectors have optimal breeding grounds. These naturally occurring risks, which we call restrictions or conditioning factors, are bound to become severe health problems for incoming human population attracted by large-scale development projects, such as mining. These risks and restrictions may be understood as vulnerabilities that need to be identified and managed from the outset of mineral development in order to prevent disease and conflict.

Our case study of the Cerrejón coal development serves to illustrate this issue.* The Guajira desert ecosystem in north Colombia has naturally occurring high levels of atmospheric particulate matter that increases during the dry season with the incoming northeastern trade winds. The opening of one of the largest open-cast coal mines in the world, El Cerrejón, in an environment with critical levels of dust, tends to exacerbate the incidence of respiratory illness in the rapidly growing local population. An ecosystem approach would indicate the need to implement an epidemiological monitoring system that takes into account this risk in order to monitor and evaluate the different types of respiratory illnesses reported at local medical centers and hospitals. The present system only identifies acute respiratory disease, which does not allow medical authorities to establish causal relationships between mining activities and respiratory health in the region and, as a result, fails to implement more stringent dust mitigation measures on the mining companies operating in the region.

World Heath Organization and Pan American Health Organization Global Perspectives on Health and Well-Being

The WHO Health for All 2000 is a policy framework for health development adopted by all its member states. At a guideline level this policy has incorporated elements of many of these approaches (Ochoa, 1994). In particular, it has recognized the influence of factors, such as access to education, nutritional security, drinking water, basic sanitation, attention to mothers and infants, immunizations, prevention and attention to endemic diseases, adequate treatment of disease, and access to basic medicines, in the improvement of health. In practice, full implementation has not become a reality for many communities in the developing world. Nevertheless, it offers a framework that reinforces the political aspects of health and, as such, complements the ecosystem approach to human health paradigm (International Development Research Center [IDRC], 1997).

According to Alderslade (1998), "Health has multiple determinants, and is rooted in the political, economic, social and institutional circumstances of individual countries." Health for All 2000, aims at 38 targets that broadly encompass better health, lifestyles conductive to health, healthy environments, and appropriate health care services. It has also promoted the principles for sustainable development in health, as follows (Alderslade, 1998):

- Equity in health
- Solidarity
- Health as an essential component of quality of life
- Sustainability
- Human rights, dignity and integrity, and patient's rights
- Cost effectiveness
- Basic health care for all
- The systematic, rational application of current knowledge, using agreed health outcome measurements
- Clear, appropriately directed public information campaigns helping to promote healthy lifestyles
- A focus on health promotion, disease prevention, therapy, and rehabilitation
- Multisector health impact assessments
- The strengthening of primary health care, and family care
- An appropriate hospital system

---

* For further information on health in La Guajira, see Puerta and Echavarría (1999).

This understanding needs to be complemented if it is to be useful in the analysis of health–disease processes in many indigenous populations where illness is interpreted and addressed as an expression of conflict, disharmony, and imbalance of the different dimensions of individual and collective social and symbolic life.

Responding to these concerns, the Pan American Health Organization (PAHO) proposed at the Winnipeg Conference (1993) the SAPIA* initiative for addressing indigenous peoples' health programs in the Americas (Stout and Coloma, 1993; Pan American Health Organization, 1996; Rossi, 1996). Its principles incorporate key elements of indigenous conceptions of health and well-being:

- The need for an integral approach to health, "since for indigenous peoples the interrelations between health, environment, territory, work, nutrition, social relations, etc., constitute a holistic reality without the fragmentation characteristic of 'modern' societies" (Rossi, 1996).
- The right to self-determination of indigenous peoples in that they must have the possibility of benefiting from technologic and scientific advances, while maintaining control over decisions taken about their health.
- The right to participation and empowerment.
- Respect and revitalization of indigenous cultures.
- Reciprocity in relationships between medical personnel and indigenous patients and with traditional healers.

Clearly there are many coincidences and complementarities between these approaches. Figure 86.1 illustrates the models from which contributions are being taken to produce a working paradigm of well-being that is meaningful to indigenous peoples, as well as other local communities in different cultures. These approaches will be complemented in the following section with the analysis of those elements that are fundamental to most indigenous peoples.

INDIGENOUS PEOPLE'S CONCEPTIONS OF HEALTH AND WELL-BEING: KEY COMMON ELEMENTS

The general objective of international environmental and sustainable development law is the formation of a global partnership of all peoples and nations to ensure for present and future generations *the well-being of humanity and the larger community of life* [author's emphasis] by promoting equitable and sustainable development and by protecting and restoring the health and integrity of the Earth's biosphere, of which all life is a part and apart from which humanity cannot survive or realize its creative potential. This global alliance should be founded on commitment to an integrated framework of shared ethical principles and practical guidelines (World Council on Indigenous Peoples, 1993).

The pertinence and ongoing validity of core common elements in the interpretation of health and well-being on the part of indigenous peoples of very diverse origin is evidenced in their recurrence in numerous international agreements,** in the rights accorded to them by some national constitutions (such as, those of Colombia, Canada, Venezuela, Australia, among others) and in declarations and consultations*** made by IIPP in different fora and processes.

---

* SAPIA is Spanish for "Iniciativa de salud de los pueblos indígenas de América" (Health initiative for indigenous peoples of the Americas).
** See for example, UN Draft Declaration of the Rights of Indigenous and Tribal Peoples, World Bank O.D. 4.20, ILO Convention No. 169, The Biodiversity Convention, among others.
*** See Earth Council on Indigenous Peoples, www.ecocouncil.ac.cr/indig; Indigenous Peoples and Global Governance Special report for Rio +5, www.ecouncil.ac.cr/rio/focus/report/wcip.htm; Center for World Indigenous Studies, www.halcyon.com/FWDP/cwisinfo.html; Global Initiative for Traditional Systems of Health (1995); Indigenous health cultural exchange group, http://members.tripod.com/~Whitney/INDIGENO.HTM; Indigenous health network, http://yg.cchs.usyd.edu.au/IHN/; International Human Rights Conference, Vienna (1993); International Indian Treaty Council (1996); DGAI et al. (1997); etc.

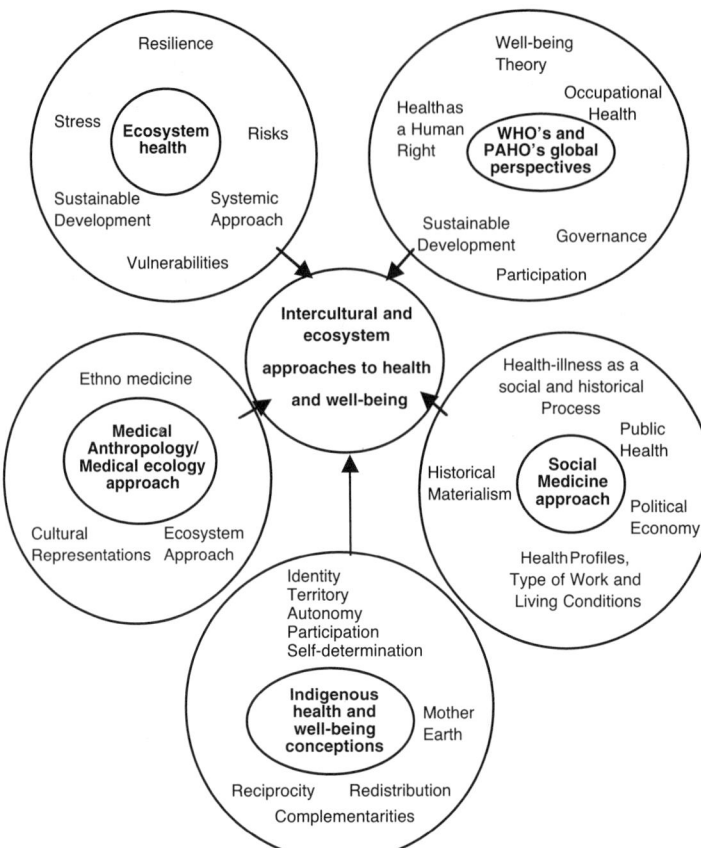

Figure 86.1 Conceptual sources for an intercultural and ecosystem approach to health and well-being.

The following common elements were systematically revised in our research and provide the basis for the identification of what we have described as fundamental elements of indigenous conceptions of health and well-being. It is important to indicate that these are not rigid and exclusive concepts and values that apply to *all* indigenous peoples, who are themselves very diverse, and have suffered different historical processes. They are true for many Latin American and Caribbean IIPP, and for some IIPP in the developed world (i.e., those who consider themselves as stewards of nature). The degree of present-day validity of these principles also varies depending on the ethnic integrity of each peoples, on their history of relationship with other regional and national social groups, and on their forms and degree of linkage to the market economy. Nevertheless, in the representations that IIPP make of themselves and of their understanding of health and well-being, the overall tendency is to emphasize the aspects proposed below.

Integral and Collective Conceptions of Well-Being

Although well-being is experienced on the individual level, many indigenous peoples conceptualize it as a collective state, where the individuals are usually seen in their relational dimension to other humans and to other beings of the world (nonhumans and superhumans).* For indigenous peoples, disease and conflict are not usually individual issues. Their treatment involves both the

---

* Understood as the humanization of natural phenomenon. For the Wayúu of Colombia and Venezuela, rain is the spirit of dead male ancestors.

immediate social network of the ill person, as well as other beings, all of which calls for a diversity of prophylactics.

> Perhaps the most salient common characteristic of indigenous medicine is that both humans and animals, plants, rivers, mountains, stars participate in a totality. The elements that constitute this totality are all alive, they are closely related and interdependent. Therefore, the survival of the totality is a phenomenon which depends on the balance of the relationships between its various components (Arbeláez, 1990).

Well-being and health are seen to depend on the dynamic equilibrium* between humans and other beings in nature based on the ongoing validity and vitality of systems of reciprocity, redistribution, and complementarities (see below). This is not to ignore the processes of individuation that exist among IIPP. Many IIPP in the developing world emphasize the collective nature of well-being in the sense that due to historical disadvantage and the need to ensure ethnic integrity in the long term, it is necessary to elevate their collective rights as a people over the rights of their individual members. In Colombia, for example, IIPP have been recognized as collective social subjects by the Constitutional Court,** with five fundamental collective rights: identity, territory, autonomy, participation, and self-determination.*** If this position can be understood as a defensive strategy for maintaining ethic integrity in a context of enormous external pressures, then the importance of maintaining the vitality of these systems reinforcing the collective nature of well-being becomes evident.

The integral nature of most indigenous worldviews also applies to their insistence on the integrity and interdependence of the five fundamental collective rights mentioned above. In Colombian jurisprudence, these rights, as components of the Constitution and its juridical system of human rights, are conceived of as an indivisible and integrated set.

Preventive and Participatory Health Management Systems

> Elements such as wind, rain, thunder and other living beings of nature, like animals, plants, water … are U'wa people of other times, with whom they still relate to, according to their own rigorous systems of reciprocity, redistribution and complementarities. [This is sustained by] a logic of harmony in the recreation of the totality of life and the survival of Mother Earth (Dirección General de Asuntos Indígenas de Colombia [DGAI], Letter to a Magistrate of the Constitutional Court of Colombia, 1997) (Perafán and Azcárate, 1995).

The values, beliefs, knowledge, traditions, and environment of a given social group are the source from which common law and ritual actions aimed at preventing illness and managing potential social or natural conflict are developed. In this way, the risk of becoming ill is controlled by actions framed within a specific social and cultural context (McElroy and Townsend, 1989).

An ideal state of well-being for many IIPP today may be represented in terms of dynamic stability in the context of a changing world. In other words, the ideal state of well-being depends on the ongoing capacity to regulate natural and social life through the exercise of conflict resolution mechanisms and health prevention measures, the essence of which is based on the principles of reciprocity, redistribution, and complementarities (see below).

This is not to imply that IIPP live in a permanent state of harmony. In many segmented indigenous societies, such as the Wayúu, conflict is the matrix that mobilizes alliances. It is

---

* Equilibrium (and harmony) is understood, not as absence of conflict, but as the existence of mechanisms for the effective regulation of social and environmental conflict. The Kaggaba peoples of Colombia use the word *yuluka*, which may be translated as "to be in agreement" (Reichel-Dolmatoff, 1985), to describe collective health.

** Colombian Constitutional Court Sentence T380/93 (DGAI, 1998, p. 70).

*** This does not mean that these are the only rights of IIPP in Colombia, but that under these fundamental collective rights can be grouped many other rights that they have acquired through historical resistance to integration, extensive lobbying, and organization.

important to understand that creating and recreating well-being in society is an ongoing process, the rules of which are established in reference to the particular worldviews and traditional laws of each IIPP. The challenge is to learn to understand these views and incorporate them into the actions and procedures of all public and private actors in their relations with IIPP.

Trust building between corporations, governments, and local communities around mining and other large-scale developments in IIPP territories is a key component for sustainable and equitable development. The real intercultural challenge lies in understanding how these traditional systems of reciprocity, redistribution, and complementarities work in each case, and how this understanding can be applied to the process of consultation and negotiation, impact evaluation, and compensation, mitigation, and management, in order to prevent and reduce conflict from the outset. This will improve the opportunities for delivering culturally compatible benefits when natural resource development is accepted as an option by indigenous peoples in a particular territory.

A Spiritual Dimension at the Base of an Ethic of Reciprocity

There will be cases where intensive natural resource development is not an option. This will be due to the lack of trust in the will and capabilities of the responsible stakeholders to deliver promised benefits, or because of ethical principles in relation to the nature of these resources.

Most IIPP consider that they belong to broad universes inhabited by interrelated beings of diverse nature. These usually include their ancestors, spirits of elements of the cosmos, and other nonhumans and superhumans. *Every element, event or phenomenon is provided with a spirit.* This knowledge alone has ethical and moral implications for human/other relationships, that make for all-encompassing conceptions of well-being and respect for the "diversity of spirits" (Achito, 1997).

Some indigenous peoples of Australia, Canada, the U.S., and Latin America see themselves as stewards of life on Earth, despite their present situations. Territorial reduction has forced many groups to overexploit the little they have left. Ethnic disintegration associated with natural resource development has been a common experience for many communities during the past century (CECOIN et al., 1995). The adaptation difficulties produced by rapid cultural change (such as that generated by mining projects in remote areas) has usually brought with it loss of traditional knowledge about the local territory-ecosystem (identity-territory); the generation or deepening of intraethnic conflict through the disruption of traditional systems of reciprocity, redistribution, and complementarities (participation and self-determination); and the breakdown or invisibilization of indigenous systems of common law (autonomy). Often changes in resource availability and access, caused by an activity such as mining, will directly affect livelihoods, while forced relocation will cause the breakdown of social networks built into the territory.

On the other hand, full economic and management participation of IIPP in mining development, even with the best of intentions, consultation, transparency, and equitable negotiations, has also been the cause of rapid cultural change bringing undesirable effects, as in the case of the Golden Bear Mine in Canada (Asp, J., personal communication). Unprecedented economic benefits rapidly transformed the values of the younger generation, many of whom adopted the consumer patterns of the dominant society. This was sometimes in excess of what would seem reasonable from the standpoint of indigenous cultural values (as understood by more traditional IIPP of Latin America and the Caribbean), especially because these changes have emphasized individual profit rather than collective well-being.

Yet, the past decades have seen the empowerment of indigenous organizations all over the world and a growing concern over indigenous issues, particularly in relation to natural resource development, such as mining, fossil fuel extraction and transport, and dam building. New realities are accentuating the political trend of indigenous peoples the world over to present themselves as stewards of nature.* This reflects the *ethic of reciprocity* that opposes that of extraction and fragmentation of nature, based on the predominant instrumentalization of relationships between humans and nature driven by globalization and the expansion of the market economy to all corners of the planet.

DETERMINANTS OF WELL-BEING FOR INDIGENOUS PEOPLES

The ESPI conceptual framework for health and well-being proposes that satisfactory levels of well-being require that ecological, social, and economic systems function efficiently (Noronha et al., 1999). These "well-functioning systems" may be regarded as determinants of well-being (MERN et al., 1999), i.e., *What is instrumental to our flourishing?* (Dasgupta, 1993; O'Neill, 1997).

Building on the previous points, this last section explores the categories that constitute the determinants of well-being for indigenous peoples. In the case of indigenous peoples, the proper functioning these systems is equally pertinent, but the indigenous systems need to be further understood in terms of well-functioning systems of reciprocity, redistribution, and complementarities. These systems form the basis of traditional livelihoods, even though they increasingly include linkages to local, regional, national, and international economies and societies.

Our research indicates that these systems will only function adequately** in an international, national, and regional context where it is possible for indigenous peoples to exercise their five fundamental collective rights of identity, territory, autonomy, participation, and self-determination.

Exercise of the Five Fundamental Collective Rights of Indigenous Peoples as a Determinant of Well-Being

Five generic issues are of great concern to indigenous peoples in terms of their ethnic integrity, and especially in the context of resource development in or near their territories: identity, territory, autonomy, participation, and self-development. These core issues are at the heart of most declarations by indigenous peoples. They appear in international agreements, such as the International Labor Organization—ILO 169), in international charters (the United Nations Draft Declaration of the Rights of Indigenous and Tribal Peoples), and many other documents, and to a greater or lesser degree, as rights in the constitutions of some countries, such as Colombia, where:

> These ... form the integral context of the fundamental collective rights that cannot be separated, isolated or negated without increasing the vulnerability of the ethnic and cultural integrity of these peoples (DGAI et al., 1998).***

In the Colombian Constitution these rights are generically defined as follows (DGAI, 1997):

- *The right to identity* as indigenous peoples is related with the right to be different and the right not to be discriminated against in their relations with the state and with society in general.
- *The right to a territory* is understood as habitat and as enough space and resources for the cultural reproduction as a peoples.

---

* Some people may not agree with this, but stewardship is a common representation that indigenous peoples and organizations make of themselves and their self-determination models.
** In terms of contributing to progress toward increased health and well-being, as they are locally understood.
*** In Spanish: "Estos ... forman un contexto integral de los derechos prioritarios y fundamentales colectivos que no pueden ser separados, aislados ni negados sin vulnerar la integridad étnica y cultural de estos pueblos."

- *The right to autonomy* in the various spheres of their life as peoples: government, justice, education, health, and territorial management, social and economic reproduction.
- *The right to participation* in the different spheres of national life. Participation includes the right to *prior consultation** with regards to laws, plans, programs, and projects that may affect their ethnic integrity, their territories, or the resources found in them.
- *The right to self-determination*, meaning the ability to participate in determining the future evolution of the social groups and their culture. Also in the improvement of their own quality of life according to their conceptions of well-being, as well as in the life plans that they build as a peoples and in their intercultural relationships with other social groups.

Systems of Reciprocity, Redistribution, and Complementarities as Determinants of Well-Being

On another level, key determinants of well-being also need to incorporate the systems of reciprocity, redistribution, and complementarities that have been mentioned above. Insofar as these systems act as regulators of social conflict and, at the same time, as territorial resource management systems, their proper functioning and adaptation to changing circumstances facilitate improved well-being among indigenous peoples.

Systems of Reciprocity

"Systems of reciprocity" refers to the historical links woven by indigenous peoples among their members (kinship systems, work organization systems, authorities of different jurisdictions of same ethnic territory, etc.), with other human groups (members of government, private agencies, other indigenous peoples, and NGOs), and with other beings in nature (totemic kin, spirits, ancestors, etc.). This is the social fabric that sustains livelihood systems. The aim of livelihood systems may be understood as the maintenance of *stability* required for the prevention of disease and the proactive management of social and environmental conflict.

Systems of reciprocity with other beings of the world,** are usually managed through the religious sphere. Usually it is the responsibility of shamans,*** spiritual leaders, or healers to monitor and control the fulfillment and maintenance of these reciprocities. This may be done through the interpretation of dreams, visions, or through divination and the use of diverse curing practices. It also may be done through established forms of payment or compensation by humans with objects of symbolic value to these other spirits, in reciprocity for the use of resources or the curing of disease. Illness and conflict are thus interpreted as early warning indicators of the breakdown of stability in existing systems of reciprocity. Prophylactics call for, among other actions, the reestablishment of the social and spiritual fabric with the reaffirmation of reciprocities through ritual symbolic interchange, usually associated with collective ceremonies.

Among humans, the system of reciprocity may be understood as:

> [A] distribution system characterized by the transfer of goods and services from one person to another without immediate payment or exchange for other goods, but where it is expected that the receiver will repay the favor at some future date, when the giver needs it. Long term relationships between partners along social networks are thus established and maintained (Perafán and Azcárate, 1995).

Reciprocity is particularly important in relation to proper compensation in justice arrangements (also known as systems of social regulation and control), both for the prevention of conflict and as a mechanism for managing conflict once it has arisen. In segmented societies that have no unified,

---

* See ILO Convention 169.
** There are also superhumans (natural elements with human attributes) and dead ancestors, perceived as humans (Puerta and Echavarría, 1999).
*** Shamanism is the predominant religion among traditional indigenous peoples.

central representation, such as the Wayúu peoples of Colombia and Venezuela, the smooth functioning of traditional social control and regulation systems is crucial for the maintenance of social stability and harmony. Although conflict arises easily, its presence and the existence of conflict resolution mechanisms is a powerful driving force for alliance building in segmented societies, and a crucial aspect of well-being.

Likewise, in the context of compensations related to negative impacts of mining activity, the development of intercultural compensation systems needs to build on the traditional local legal practice (common law) of the particular indigenous peoples affected. This will facilitate improved well-being and ensure that agreements are both compatible and sustainable with the local culture.

Systems of Redistribution

Reciprocity is directly linked with the maintenance of systems of redistribution among humans. Although there are everyday reciprocities within the close family, the main collective instances of redistribution are rite of passage celebrations (birth, puberty, matrimony, and wakes) and other traditional gatherings related to religious and economic calendars. When a centralized authority exists, communal produce is usually managed by this authority and redistributed among the community during celebrations. Otherwise, it is the head of the family who is in charge of redistribution and who ensures its equity (understood in the sense that all give according to their capacity).

The determination of how economic or social benefits are distributed by a mining company requires proper consultation with the indigenous peoples in each case. The articulation of social policy through an understanding of how local redistribution systems flow may optimize social benefits in the short term, and prevent many conflicts between mining companies and local communities.

In addition, the integrity of the ceremonial cycle is crucial for the exercise of systems of redistribution, and its continued observance is an indicator of ethnic integrity and well-being. In this perspective, consultation processes and meetings related to mineral development and labor agreements need to take into account the ceremonial cycle, respecting the time they require and understanding their importance for social stability and ethnic integrity.

System of Complementarities: Territory, Identity, and Well-Being

La tierra para el indígena es como la salud. Un indígena sin tierra vive enfermo

— **Jacanamijoy**
quoted in: Cárdenas et al. (1998)*

The maintenance of complementarities in the use of territory and resources refers to the diversified and extensive subsistence strategies that characterize many indigenous peoples and many traditional *campesino* societies. This means that livelihoods based on combinations of agriculture, horticulture, herding, hunting, fishing and gathering, etc., often depend on the extensive use of resources either from niches at different altitudes (as in Andean societies) or from several, more or less adjacent ecosystems.

An extensive family may have different segments living in different niches, and would be then regarded as having a polyresidential settlement pattern.** In this way, specific social groups (such as extensive families) maintain permanent obligations of product interchange and resource use in different econiches, both as a strategy to diversify livelihoods and to strengthen interdependence between segments and as a strategy to prevent conflict. This strategy reflects both recent historical and long-term adaptation strategies aimed at the reduction of vulnerability through the development

---

* In English: "Land for IIPP is the same as health. An Indian without land lives in disease."
** We use polyresidential as opposed to nomadic since the second has the connotation of weak rooting to the land, while the first clearly emphasizes the complementary nature of subsistence strategies and settlement patterns.

of social systems and territorial management systems adapted to ensure availability and access to diverse and complementary resources, which today usually include paid labor and commerce.

By affecting one particular place, station, or resource, an industrial development may be affecting, not only the livelihood of the local family, but also that of other families that form part of a particular network as well as others who may only use it seasonally (or eventually through established cultural practices embedded in local common law).

The impact of territorial reduction due to mineral development would then need to be analyzed from the perspective of how livelihoods are affected in the light of complementarities in resource use on a particular territory. An indigenous territory must be regarded, not only as a particular parcel of land, but also as a network of resources located in different places, to be used in different seasons, on the basis of the existence of social networks that sustain complex livelihood systems.

CONCLUSION

Each social and biogeophysical context presents specific vulnerabilities and opportunities for mineral development. Different cultural groups interpret, represent, and cure illness in different ways. If mineral development is going to deliver improved well-being and health, all of these variables must be identified and managed from the outset.

We have argued that the construction of an intercultural and ecosystem approach, as a basis for developing tools for tracking health and well-being in multicultural mining regions, needs to incorporate concepts from a number of theoretical models and the key common elements and determinants of indigenous conceptions of health and well-being.

In addition to the proper functioning of natural and social systems in general, an intercultural approach to the identification of the determinants of well-being for indigenous peoples must incorporate two key elements: an understanding of the actual systems of reciprocity, redistribution, and complementarities that characterize each particular indigenous peoples and clarity in regard to how the five fundamental collective rights unfold in each context.

Although a global, generic, and broad definition of well-being is possible, further research is required to develop tools to aid in the understanding of how determinants of well-being unfold in each cultural context. To this end, the INER research team, within the ESPI program, has developed a cultural matrix that will aid researchers in identifying the constituents of well-being. This matrix is a descriptive tool that may orient companies and governments to the challenges of intercultural communication and management. It will allow the recognition of the message of "the other," and an understanding the internal significance and implications of general and specific impacts of mining on the total ecosystem and the issues raised by indigenous peoples themselves, from the outset of mineral development.

The matrix in mineral development is structured on the basis of the five fundamental collective rights of IIPP, identified above as determinants of well-being. It does not intend to be a final and complete description of any particular culture. In other words, it cannot be taken as officializing or freezing a culture, but it can be an important aid in understanding its strengths and vulnerabilities in the face of specific pressures, such as mining activities.

The elements described above provide a new understanding of the ecosystem approach to human health. They bring intercultural perspectives and identify synergies and complementarities of a number of approaches that contribute new elements to the ecosystem health paradigm. Further research expects to put evaluation tools in operation that will respond to the actual cultural diversity in which mining develops in the world today.

ACKNOWLEDGMENTS

Thanks to Carolina Quintana for the final editing of this chapter, and to Claudia Puerta for her valuable input.

REFERENCES

Achito, A., *Autonomía Territorial, Jurisdicción Especial Indígena y Conflictos Interétnicos en el Pacífico*, in Dirección General de Asuntos Indígenas (DGAI)–Ministerio del Interior, Consejo Regional Indígena del Cauca (CRIC), y Ministerio de Justicia y del Derecho, *Del Olvido Surgimos para Traer Nuevas Esperanzas. La Jurisdicción Especial Indígena*, Imprenta Nacional, Santafé de Bogotá, Colombia, 1997.

Alderslade, R., Health and Human Rights in Europe, Conference University of Bath, April 21, Bath U.K., http://www.warwick.com/, Mining and Environment Research Network, 1998.

Arbeláez, C., *Medicinas Tradicionales en Colombia*, Gaceta, Mar. 6–Apr., Santafé de Bogotá, Colombia, 1990.

Bibeau, G,. ¿Hay una Enfermedad en las Américas? Otro camino de la antropología médica para nuestro tiempo, in *Cultura y salud en la Construcción de las Américas. Reflexiones sobre el sujeto social*, Vol. I and II, Pinzón, C., Suárez, R., and Garay, G., Eds., Memorias Primer Simposio internacional de cultura y salud, VI Congreso de Antropología en Colombia. Instituto colombiano de Antropología (ICAN)–Comitato Internazionale per lo sviluppo del popoli, Santafé de Bogotá, Colombia, 1993.

Cárdenas, M., Mesa, C., and Riascos, J.C., Eds., *Planificación ambiental y ordenamiento territorial. Enfoques, conceptos y experiencias*, Fundación Friedrich Ebert de Colombia–FESCOL, Departamento Nacional de Planeación y Centro de Estudios de la Realidad Colombiana–CEREC, Santafé de Bogotá, Colombia, 1998.

CECOIN, GHK, and Organización Nacional Indígena de Colombia (ONIC), *Tierra Profanada. Grandes Proyectos en Territorios Indígenas de Colombia*, Disloque Editores, Santafé de Bogotá, Colombia, 1995.

CENSAT Agua Viva, Instituto de los Seguros Sociales y Ministerio de Salud, *Riesgos de la Minería,* Módulo 4, Estudio Epidemiológico Social y Ambiental de la Minería de Carbón en el Cerrejón, Zona Centro y la Jagua de Ibirico, Santafé de Bogotá, Colombia, 1995.

Chaumeil, J.P., Del Proyectil al Virus. El Complejo de Dardos – Mágicos en el Chamanismo del Oeste Amazónico, in *Cultura y salud en la Construcción de las Américas. Reflexiones sobre el sujeto social*, Vol. I and II, Pinzón, C., Suárez, R., and Garay, G., Eds., Memorias Primer Simposio internacional de cultura y salud, VI Congreso de Antropología en Colombia. Instituto colombiano de Antropología (ICAN)–Comitato Internazionale per lo sviluppo del popoli, Santafé de Bogotá, Colombia, 1993.

Cohen, G.A., Igualdad de qué? Sobre el bienestar, los bienes y las capacidades, in *La Calidad de Vida*, Nussbaum, M.C. and Sen, A.K., Eds., Fondo de Cultura Económica, México, 1996.

Coordinadora de Organizaciones Indígenas de la Cuenca Amazònica (COICA), Consideraciones sociales y ambientales de las actividades hidrocarburíferas en áreas sensibles de la cuenca Sub-Andina, in Banco Mundial and OLADE, Tercera Reunión del Grupo Energía, Población y Ambiente, Cartagena, Colombia, 1999.

Costanza, R., Toward an operational definition of ecosystem health, in *Ecosystem Health: New Goals for Environmental Management*, Constanza, R., Norton, B., and Haskell, B.D., Eds., Island Press, Washington, D.C., 1992, pp. 239–256.

Costanza, R., Norton, B., and Haskell, B.D., Eds., *Ecosystem Health: New Goals for Environmental Management*, Island Press, Washington, D.C., 1992.

Corredor, C., Ética, Desarrollo y Pobreza, Presentation to Universidad de Antioquia, Medellín, Colombia, Sept. 23, 1998.

Dasgupta, P., *An Inquiry into Well-Being and Destitution*, Clarendon Press, Oxford, 1993.

Dirección General de Asuntos Indígenas (DGAI)–Ministerio del Interior, Letter to a Magistrate of the Constitutional Court of Colombia, Mario Germán Iguarán Arana, re the expedient T-84771, of *Defensoría del Pueblo y comunidad U'wa contra Occidental de Colombia y Otros*, photocopy, Santafé de Bogotá, Colombia.

Dirección General de Asuntos Indígenas (DGAI)–Ministerio del Interior, *Palabras de Mama 3. Salud y Enfermedad*, Proyecto Gonawindua, Santafé de Bogotá–Santa Marta, Colombia, 1997a.

Dirección General de Asuntos Indígenas (DGAI)–Ministerio del Interior, Consejo Regional Indígena del Cauca (CRIC), y Ministerio de Justicia y del Derecho, *Del Olvido Surgimos para Traer Nuevas Esperanzas. La Jurisdicción Especial Indígena*, Imprenta Nacional, Jimeno G., Constitución Política, Jurisdicción EspecialIndígena, y Autonomía Territorial, Santafé de Bogotá, Colombia, 1997b, p. 182.

Dirección General de Asuntos Indígenas (DGAI)–Ministerio del Interior, *Los Pueblos Indígenas en el país y en América: Elementos de política nacional e internacional*, Serie Retos de la Nación Diversa No. 1, Santafé de Bogotá, Colombia, 1998.

Echavarría, C., Indigenous Peoples and Mining: Core Issues — The Case of the Wayúu People and El Cerrejón Coal Development, MERN Working Paper 184, University of Warwick, U.K., 1999.

Emery, A.R. and Associates Center for Traditional Knowledge, *Guidelines for Environmental Assessments and Traditional Knowledge*, Canada, 1997.

Holland, A., Ed., Special theme issue on ecosystem health, *Review: Environmental Values*, Vol. 4, No. 4, White Horse Press, Cambridge, 1995.

Hoon, P., Singh, N., and Wanmali, S.S., Sustainable Livelihoods: Concepts, Principles and Approaches to Indicator Development, draft discussion paper, prepared for the workshop on Sustainable Livelihoods Indicators, United Nations Development Programme, New York, 1997.

Interagency Ecosystem Management Taskforce, The Ecosystem Approach: Healthy Ecosystems and Sustainable Economics, National Technical Information Service (NTIS) U.S. Department of Commerce, Washington, D.C., 1995.

International Development Research Centre, *Ecosystem Approaches to Human Health*, http://www.idrc.ca/ecohealth/, Ottawa, Canada, 1997.

International Indian Treaty Council, Declaration: Mining and Indigenous Peoples Consultation, London, U.K., 1996. http://www.hookele.com/iite/mining-96.htm.

Korsgaard, C.M., Comentario a Igualdad de qué ? y a Capacidad y bienestar. Nussbaum, M.C. and Sen, A., *La calidad de vida*. Fondo de Cultura Económica de México, México, 1993.

Langdon, J., Interethnic processes affecting the survival of shamans: a comparative analysis, *Otra América en Construcción. Medicinas Tradicionales Religiones Populares*, Pinzón, C. and Suárez, R., Eds., Instituto Colombiano de Antropología, 46 Congreso de Americanistas Universidad de Amsterdam, Santafé de Bogotá, Colombia, 1991.

Laurell, A.C., La salud-enfermedad como proceso social, *Cuadernos Médico Sociales*, 19, Rosario, Argentina, 1982.

Maclean, C. and Warhurst, A. Conceptual Approaches to Human Health and Well-Being in Minerals Development and a Case Study in Southern Africa, MERN Working Paper 168, http://users.wbs.warwick.ac.uk/ccu/mern/working_papers.pdf, University of Warwick, U.K., 1999.

Max-Neef, M., Elizalde, A., and Hopenhayn, M., *Desarrollo a Escala Humana. Una opción para el futuro*, CEPAU –Fundación Dag Hammarskjöld, Proyecto 20 Editores, Medellín, Colombia, 1996.

McElroy, A. and Townsend, P.K., *Medical Anthropology in Ecological Perspective*, 2nd ed., Westview Press, New York, 1989.

Mining and Energy Research Network (MERN), Tata Energy Research Institute (TATA), Instituto de Estudios Regionales (INER), Environmental and Social Performance Indicators in Minerals Development: Reporting Progress Towards Improved Ecosystem Health and Human Well Being, technical report, Ottawa, Canada and Bath, U.K., 1999.

Noronha, L. et al., Environmental and Social Performance Indicators in Minerals Development Reporting Progress Towards Improved Ecosystem Health and Human Well Being, Tata Energy Research Institute (TERI) Project Report 97WR52, 1999.

Nussbaum, M.C. and Sen, A.K., Eds., Virtudes no relativas: un enfoque aristotélico, *La Calidad de Vida*, Fondo de Cultura Económica, México, 1996.

Ochoa, G.L., Antropología Ecológica, Salud y Hábitat, El encanto real de una utopía, Tésis de grado, Universidad Nacional, Medellín, Colombia, 1994.

O'Neill, J., 1997. Nature, intrinsic value and human well-being, in *Human Well-Being and Economic Goals*, Ackerman, F.D. et al., Eds., Island Press, Washington, D.C., 1997, pp. 40–44, 273–275.

PAHO/OPS, Organización Panamericana de la salud, Américas en armonia: la salud y el ambiente en el desarrollo humano sostenible ilna oportunidad para el cambio y un llamado a la nación, Washington, D.C., 1996.

Pederssen, D., La Construcción Cultural de la Salud y Enfermedad en la América Latina, in *Cultura y salud en la Construcción de las Américas. Reflexiones sobre el sujeto social*, Vol. I. Pinzón, C., Suárez, R., and Garay, G., Eds., Instituto colombiano de Antropología (ICAN)–Comitato Internazionale per lo sviluppo del popoli, Santafé de Bogotá, Colombia, 1993.

Pinzon, C., Suarez, R., and Garay, G., Eds., Cultura y Salud en la Construcción de las Américas. Vols. 1 and 2, Instituto Colombiano de Antropologia (ICAN), Comitato Internazionale per lo suiluppo del popoli. Santafé de Bogotá, Colombia, 1993.

Perafán, C.C. and Azcárate, L.J., *Caracterización Comunidades Indígenas: Tipologías. Organización Social, Nivel de Vida*, Fondo DRI, Consultoría DRI/IICA/94, Feb., Santafé de Bogota, Colombia, 1995.

Puerta, C. and Echavarría, C., INER final phase 1 technical report to MERN-IDRC, Vol. 2, *Aporte de las Concepciones Indígenas a los Enfoques Ecosistémicos Sobre Salud y Bienestar Humano*, Available from Instituto de Estudios Regionales (INER), University of Antioquia, Medellín, Colombia, and IDRC, Ottawa, Canada, 1999.

Quevedo, E., El proceso salud-enfermedad: hacia una clínica y una epidemiología no positivistas, in *Sociedad y Salud*, Cardona, A., Ed.,. Zeus Asesores Limitada, Medellín, Colombia, 1992.

Quevedo, E., La Cultura desde la Medicina Social, in *Cultura y salud en la Construcción de las Américas. Reflexiones sobre el sujeto social*, Vol. I and II, Pinzón, C., Suárez, R., and Garay, G., Eds., Memorias Primer Simposio internacional de cultura y salud, VI Congreso de Antropología en Colombia. Instituto colombiano de Antropología (ICAN)–Comitato Internazionalc per lo sviluppo del popoli, Santafé de Bogotá, Colombia, 1993.

Rapport, D.J., Ecosystem Health: An Emerging Integrative Science, discussion paper, Ecoresearch Chair in Ecosystem Health, University of Guelph Canada, 1999.

Reichel-Dolmatoff, A., Los Kogi: Una tribu de la Sierra Nevada de Santa Maria, Colombia, 2nd ed., 2 Vols., Procultura, S.A., Santafé de Bogotá, Colombia, 1985.

Rossi, F., *Iniciativa Salud de los Pueblos Indígenas de América S.A.P.I.A.*, in, Carrillo, J.C., Plan de salud Etnia Wayuu. Memorias Foro, Cuadernos de Auxología 6, TEA Fundación Auxológica, Santafé de Bogotá, Colombia, 1996.

Sen, A.K., *Commodities and Capabilities*, Oxford University Press, Oxford, 1985.

Sen, A.K., Funcionamientos y capacidad, in *Nuevo Examen de la Desigualdad*, Alianza Editorial, Madrid, Spain, 1995.

Sen, A.K., Capacidad y Bienestar, in *La Calidad de Vida*, Nussbaum, M.C. and Sen, A.K., Eds., Fondo de Cultura Económica, México, 1996.

Stout, M. and Coloma, C., Los Pueblos Indígenas y la Salud, Base Document, Seminar-workshop- OPS–OMS, Winnipeg, Canada, 1993.

Tata Energy Research Institute Team and University of Bath, Environmental/Social Performance Indicator (ESPIs) and Sustainability Markers in Minerals Development: Reporting Progress Towards Improved Ecosystem Health and Human Well-being, The Indian Case study, New Delhi, India, 1998.

United Nations Conference on Human Rights, Vienna, Austria, June 1993, Review of the International Commission of Jurists, No. 50, International Commission of Jurists, Geneva, 1993.

Warhurst, A., Conceptual Approaches to Researching Human Health, Quality of Life and Ecosystem Health in Minerals Development, Draft presentation, First IDRC Working Group Meeting: Ecological Framework for Mining and Health in Latin America, Montevideo, Uruguay, 1999.

World Bank, Expanding the measure of wealth: indicator of environmentally sustainable development, in *Environmentally Sustainable Development Studies and Monographs*, Series 17., Washington, D.C., 1997.

World Council on Indigenous Peoples, Participation of Indigenous Peoples in the Consultation Process of the Earth Charter, 1993. www.ecocouncil.ac.cr/indig/ec/mcf.htm.

World Health Organization (WHO), Division of Mental Health, *Quality of Life Assessment: International Perspectives*, Proceedings of the Joint Meeting Organized by the World Health Organization and the Foundation IPSEN in Paris, July 1993, Orley, J. and Kuyken, W., Eds., Springer-Verlag, Berlin, 1994.

CHAPTER 87

Integrating Human Health into an Ecosystem Approach to Mining

Donna Mergler

INTRODUCTION

The ecosystem approach to human health is an innovative concept that incorporates human health into the dynamic interrelations of ecosystem analyses. In this approach, humans are not solely the drivers and disrupters of the ecosystem or the passive victims of environmental destruction, but are in continuous interaction with changing environmental, social, and economic conditions. In this dynamic process, ecosystem changes, resulting from human interventions (mining, agriculture, urbanization, etc.), driven by the need for subsistence or the forces of economic development, affect current and future well-being, not only of the present, but also for coming generations. Human health and well-being are, in turn, determinants of social, economic, and cultural development. Appropriately applied to environmental and health impact studies, the ecosystem approach can be a useful tool toward attaining sustainable and equitable development.

Human health deterioration and its consequences for the quality of life are most often omitted from environmental and cost–benefit impact studies. Here, we propose a means of integrating human health into the ecosystem analysis, using mining activities as an example. Four major issues, keys to the understanding of the relation between mining and human health, are examined: (1) miners' health; (2) a framework for focusing on the impact of mining development and activities on human health; (3) a model for examining human health deterioration; and (4) the notion of fragile ecosystem, using the example of gold mining in the Brazilian Amazon.

MINING ACTIVITIES

The extraction and transformation of minerals and quarrying are almost as old as civilization, and so are the accidents and diseases that have afflicted those who are engaged in these activities. In ancient Greece, in the 4th century B.C., Hippocrates described illness among lead miners (Landrigan et al., 1990). Five hundred years later, Pliny the Elder called lead poisoning "one of the diseases of slaves" and described a bladder-derived protective mask to be used by laborers subjected to large amounts of dust or lead fumes (Hamilton, 1943).

The lure of easy riches brought the Spanish and Portuguese to South America, in the search for gold- and silver-bearing ores, which native populations had been mining for centuries. Anxious for large and quick profits, the *conquistadores* set up mining operations in areas such as the mountain of Potosi in Bolivia, one of the first to be exploited; even today, the Spanish expression *vale un potosi* means that something has much worth. Indians were brought in as miners within a harsh system called *la mita*. In 1585, Fray Rodrigo de Laoyza, a predecessor of the modern Latin American activist priests, describing the conditions and lives in this silver mine, stated that the Indians were being consumed like sardines (Gumucio, 1988). The recent photographs by Sebastão Salgado of Brazilian gold mining (Salgado, 1997) evoke the sardines of Fray Rodrigo, and show that harsh working conditions and child labor are not things of the past.

Workers' health is most often completely ignored when considering the environmental impact of mining. The workforce is usually incorporated into the production process and considered in terms of labor and liability costs. Although increased income certainly contributes positively to the health of workers and their families, the history of mining is fraught with occupational diseases and accidents. In countries with mining activities, mining is almost always at the head of the list of sectors with the highest incidence of accidents and occupational diseases.

Although mining is not a major employer, accounting for about 1% of the world's workforce (some 30 million people), it is responsible for about 7% of fatal accidents at work, which is estimated by the International Labor Office (ILO) to be around 15,000 per year. No reliable data exist for accidental injuries, but they are significant, as is the number of workers affected by occupational diseases (such as pneumoconioses, hearing loss, vibration disease, metal poisoning, cancers), whose premature disability and even death can be directly attributed to their work (International Labor Office, 1998). Small-scale mining is expanding rapidly, and often uncontrollably, in many developing countries, employing large numbers of women and children in dangerous conditions and generating a workplace fatality rate up to 90 times higher than mines in industrialized countries (International Labor Office, 1999).

Mining takes its toll not only on humans, but also on the environment in its yearly production of approximately 23 billion tons of minerals, including coal. For high-value minerals, the quantity of waste produced is many times that of the final product. For example, each ounce of gold is the result of dealing with about 12 tons of ore, and each ton of copper comes from about 30 tons of ore. For lower value materials, such as sand, gravel, and clay — which account for the bulk of the material mined — the amount of waste material that can be tolerated is clearly minimal. According to a recent report of the ILO (1998), the world's mines must process at least twice the final amount of earth required (excluding the removal of surface overburden that is subsequently replaced and therefore handled twice); some 50 billion tons of ore are mined each year, which is the equivalent of digging a 1-m-deep hole the size of Switzerland.

SPHERES OF IMPACT: A FRAMEWORK FOR EXAMINING HUMAN HEALTH

This disruption of the ecosystem has effects not only on the workers, who are in direct contact with the mining process, but also on other sectors of the community. The framework presented in Figure 87.1 is generic and can be applied to many situations of environmental disturbance. Here, it is used as a framework analysis of the human health consequences of mining activities whereby the positive and negative effects of economics, social, and cultural reorganization, geophysical alterations, and chemical changes in the environment can be examined within a nested hierarchy of four spheres: workers, the local community, remote communities, and the general population. The arrows are bidirectional. For example, toward the center: the economic situation in a country can determine whether mining activities are permitted as well as the control on its environmental impact; workers' economic condition can bring them to mining and are an important determinant for their health. On the other hand, mining activities can provide wealth for the different sectors

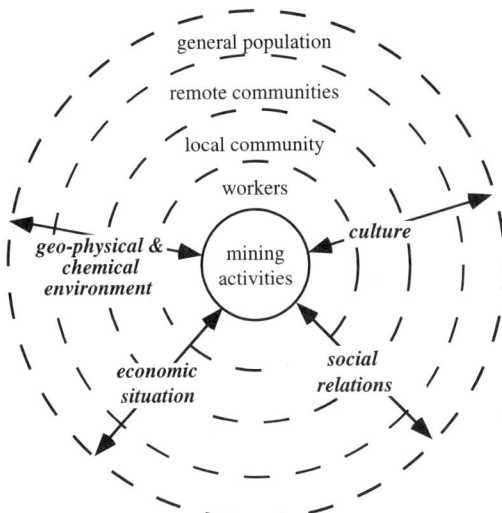

Figure 87.1 A concentric framework for studying the positive and negative impacts of mining activities on human health in an ecosystem perspective.

of the population, which may or may not be equally distributed; this wealth can contribute to improving the health of the four levels of the population. The spheres are not completely closed since the activities of one can affect the others, influencing their respective health status.

Mine Workers

Mining activities are situated in the center of the sphere and the workers constitute the first group whose health and well-being is affected by the working conditions. The statistics from the ILO on accidents and illnesses constitute only the tip of the iceberg. Since reporting these events is dependent on the capacity and the will of different countries, they are often underreported for various reasons. Among these reasons is the question of liability and compensation. Moreover, illegal small mining operations are rarely included in official reports. In addition, there is the healthy worker effect, which occurs when workers' health deteriorates to the point where they are unable to continue and abandon the workplace, leaving behind only those who are physically and emotionally capable of carrying out the work. The workers in poor health leave the mines and their illnesses and disabilities are not attributed to mining, while those who are healthy enough to continue to work constitute the healthy workforce. A high turnover of workers in mining activities can be an indicator of poor health; the statistics on accidents and illnesses often do not reflect this reality.

Working conditions in mines have improved over the centuries, and these improvements have accelerated during the last century, with resulting improvement in miners' health. Much of this improvement has been the result of major battles waged by workers, their wives, and their unions to obtain better working conditions and health services for themselves and their families. In 1975, International Women's Year, Domitila Barrios de Chungara told the world of struggles of the tin miners and their families in Bolivia, high in the Andes (Viezzer, 1976). Films, such as *Harlan County USA*, *Miner's Daughter*, *The Salt of the Earth*, and *Germinal*, have brought vividly to our screens stories of mining families' battles to improve their working and living conditions.

Threatened with the possibility of losing their jobs if the mine closes down, miners may often oppose environmental measures. Their voice must be heard and included in all multistakeholder endeavors and research for good ecosystem management.

Local Communities

Mining activities have a major direct and indirect impact on the health of local communities. People from outside the region, looking for jobs in mining or associated areas, can upset the structure of the existing society. Gender relations are affected since mining jobs are mainly available to men, with women in satellite roles (Wasserman, 1999). Hierarchical relations are modified through the introduction of a new power structure in the community. The movement of populations, particularly in areas with a potential for dissemination of infectious diseases, creates a further health hazard. Sexually transmitted diseases are known to increase in such situations (Campbell and Williams, 1999), and, in the case of gold-mining in areas with endemic tropical diseases, miners also constitute a vector for the transmission of malaria and other infectious diseases into local communities (Rambajan, 1994; Carvalho et al., 1999). The well-being and health of the local community can also be affected by the increased burden to families of workers who have been injured in mining accidents or are suffering occupational diseases.

There are also economic benefits to the local communities, which can have a positive impact on health, the most important being job opportunities. Many communities, living near mines or proposed mining areas, in different Latin American countries have used organized protests to obtain the setting up of clinics, sewage systems, and other collective installations, which positively affect health. After undertaking a cost–benefit analysis, a Peruvian private mining company set up an integrated health care and family planning program, resulting in improved children's health and reduced cost for pharmaceuticals (Foreit et al., 1991).

The greatest biophysical environmental destruction is most often, but not always, in the direct vicinity of the mining activities. Deforestation can have devastating and long-lasting effects, changing ecosystem dynamics. Many mining activities require great quantities of water, affecting the aquifers that are essential to human health and well-being. Pollution from the particular metal being mined or the waste products of the production process enter the surrounding ecosystem, potentially affecting the health of plant, animal, and human populations.

Remote Communities

Even communities living far from mining activities may be affected through water and airborne pollution (Figure 87.2). Rivers can carry pollutants over hundreds of kilometers into the sea, affecting the wildlife, fish, and seafood, which are then unknowingly consumed by the communities living far from the source. Ores are often transported by trucks or rail over large distances, polluting the areas through which they pass, and thus affecting the health of people living near these routes. The effects of low-level, long-term exposure to environmental pollutants are a subject of current concern, particularly in relation to their effects on the reproductive, neurological, immune, and endocrine systems.

General Population

The general population within a country can benefit from the accrued income from mining activities if regulations and policies so permit. However, while different segments of the population may reap benefits, others may become more impoverished. In the 1999 Pan-Canadian Ecohealth Lecture Tour, in which the theme was mining, Labonne (1999) pointed out that the activities of the extractive industry are now increasingly located in developing countries. These are countries where, paradoxically, mineral resources endowment, including oil, has had a corruptive effect on many governments, causing both impoverishment and restlessness of the population living in the mineral and oil producing regions.

Finally, since to date there has been little effort to plan for closure on the part of mining companies, the social and economic costs of cleanup are often taken on by the country, particularly

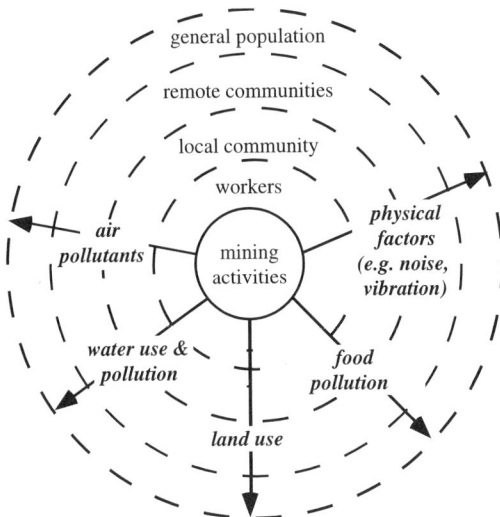

Figure 87.2 A concentric framework for studying the health of populations affected by geophysical and chemical environmental changes from mining activities.

when a mine has closed down, leaving behind waste and devastation and a population whose health has been affected by poor working and living conditions.

HUMAN HEALTH DETERIORATION ON A CONTINUUM

Human health is much more than the absence of disease. Ecosystem disruption associated with mining (with the exception of major disasters or spills) usually does not produce dramatic immediate effects on human health. Rather, there are slow, insidious changes in biological functions, reflecting diminishing well-being and affecting the quality of life and the collective capacity of the community to intervene and improve the situation. Schematically, the deterioration of human health can be represented on a continuum (Figure 87.3). The first stage corresponds to an increase in the prevalence of nonspecific symptoms of dysfunction that reflect early biological impacts among human populations resulting from ecosystem stress. The second stage includes early functional changes that are also not apparent in individuals, but can be measured, on a group basis, in exposed populations (Mergler, 1998). By the third stage, individuals are manifesting subclinical signs of

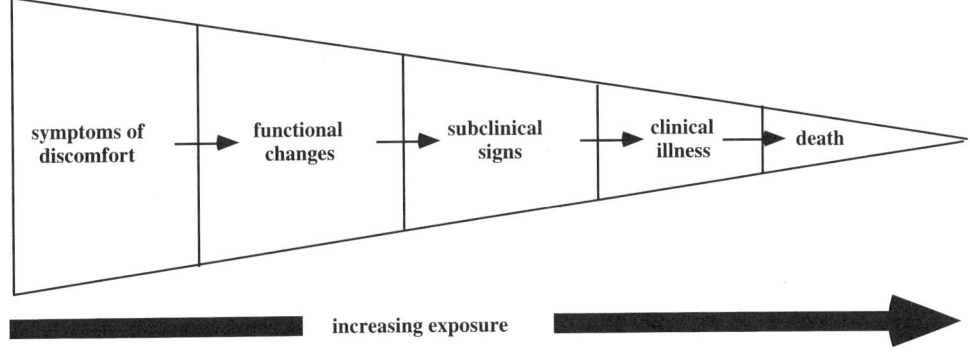

Figure 87.3 Stages of the health deterioration continuum. Exposure dose contributes to increasing deterioration. The areas within the triangle correspond to the relative frequency of the population at risk at different stages.

illness, while at the fourth stage frank illness is present. For the most part, environment-related disease processes are irreversible, or even relentlessly progressive. The final stage is death.

The rapidity with which persons progress along the continuum depicted in Figure 87.3 depends not only on the degree of ecosystem disruption, but also on individual factors (genetic makeup, age, lifestyle, other illness) and social factors (poverty, malnutrition, discrimination, lack of autonomy). These factors often interact with each other.

In the early stages of health deterioration, it is not possible to make diagnoses in individuals, but diagnosis of dysfunction within populations (community health) can be linked to ecosystem stressors. Early biological alterations constitute not only an early warning for increased risk of eventual illness, but, more importantly, reflect diminished well-being and impaired functional capacities. It is at this stage that intervention can be effective.

Diminished well-being, associated with an environmental pollutant, is well illustrated by the numerous studies on the relation between lead exposure from gasoline and children's IQ. While the levels of lead exposure from gasoline may not be sufficient to provoke clinical illness in the major cities where it is still used, the levels are sufficient to affect children's development, reducing the mean IQ of the exposed populations by about 4 to 10 points. Individually, this effect is not necessarily detectable; it is, however, measurable on a group basis, using sensitive, quantitative tests. A reduction of IQ of 5 points means that approximately twice as many children will have learning problems (<70 IQ points) and half as many children will have very high IQs (>130 IQ points) (Rice, 1998). Higher blood levels of lead are also associated with hyperactivity in children and reduced ability to concentrate. It is society that suffers, along with its capacity to develop fully. As a result, an increasing number of children become aggressive, have learning disabilities, drop out of school, and miss out on higher education (Needleman et al., 1996). Even very low levels of lead exposure can affect well-being and become dramatic when coupled to poverty, malnutrition, poor living conditions, inadequate health care, and poor schooling. From the perspective of an ecosystem approach to human health, these factors should be considered and included in analyses, rather than controlled or adjusted, when assessing the situation and working with communities to propose mitigating measures.

A Triangle of Risk

The proportion of the population at different stages along the continuum of health deterioration can be represented by a triangle with illness and death at the top. At similar levels of environmental destruction, those who are in poorest conditions and the most vulnerable may become very ill or die. At the base of the triangle, there are the many more persons who are, to some extent, capable of resisting and adapting to the loss of adequate life support systems and increasing toxic exposure, but who present the subtle, subclinical effects indicative of loss of well-being. As ecosystem stress increases, breaking down the human organism's adaptive capacity, more and more persons suffer from severe illness and die. Consequently, there is the need for more efforts in occupational and environmental health research to develop and apply quantitative and qualitative indicators of early health deterioration and loss of well-being, with a view to preventive intervention (Mergler, 1999).

Methodological considerations differ with respect to where one studies the impact of mining activities on population health along this continuum. Traditionally, we have examined the health impacts from mining at the tip of the triangle, where very large study populations are required, since fewer persons will be suffering from specific illness or die of specific causes. The outcomes are dichotomous (ill/not ill, dead/not dead). Moreover, the most vulnerable persons reach this level more quickly, making it difficult to distinguish between the health impact of ecosystem disruption resulting from mining and the effect of individual factors. As one goes farther down the continuum, more people are affected. Toward the base of the triangle, performance measurements can be made on everyone within the designated population, as was the case with studies of the effects of lead on IQ. With this approach, smaller populations can provide statistically relevant information.

An example of a study of early alterations associated with mining is provided by a study of 36 children in the Njamba gold-mining area of Ecuador, where mercury is used in the process of extracting gold (Counter et al., 1998). In this study, a significant relation was observed between blood mercury and hearing threshold. Although the children were not deaf, their hearing was increasingly impaired with increasing mercury exposure. In a comprehensive approach to health, the consequences of these changes are important in terms of learning capacity and future capabilities. Moreover, they should be further examined within the context of the other stresses on humans arising from the ecosystem disruption associated with these mining activities.

FRAGILE ECOSYSTEMS

Certain ecosystems are more fragile than others with respect to resistance to stress. The ecosystem approach requires a sound knowledge of the biophysical, chemical, and social parameters in management for ecosystem sustainability. For example, in a study of the mercury cycle in the soil, river sediment, and waters of the Amazonian Basin, my colleagues found that there were high levels of natural mercury in the soil. They also found that mercury used in gold mining was not the sole contributor to the high mercury levels observed in carnivorous fish and in the human populations who consumed these fish (Lebel et al., 1997; Roulet et al., 1998, 1999). Slash-and-burn agricultural practices throughout the Amazonian basin, resulting in soil erosion and lixiviation, were also responsible for releasing mercury into the extensive water system of this region.

The implication of this observation for ecosystem management of mining activities is that mercury released through mining activities cannot be isolated from the other activities that were likewise increasing the mercury burden in this ecosystem. The argument that mining activities are only adding a small proportion to the total mercury load is inappropriate. Because the fragile ecosystem already has high levels of naturally occurring mercury, the addition of even very small amounts from any source can have very important effects on ecosystem disruption and human health. Adequate mitigation measures depend on a comprehensive assessment of the ecosystem, as opposed to a random grab bag of measurements.

Because mines are established in areas where there are mineral ore deposits, in many cases the populations living on these deposits may already have relatively high levels of exposure to the substance through water, soil, and possibly food. For example, congenital disorders, identified in the Groote Eylandt population, that has been living since World War I in a manganese-bearing ecosystem, were subsequently exacerbated by manganese-mining activities (Kilburn, 1987).

The interaction among exposure to harmful metals, mining activities, ecosystem dynamics, and human health was recently demonstrated in studies of malaria in gold-mining areas. The inordinately high prevalence of malaria observed in gold-mining areas has been traditionally attributed solely to environmental destruction. In Brazil, the Evandro Chagas Institute, which specializes in tropical diseases, has traced most cases of malaria in Brazil to the gold-mining areas (Strickland et al., 1999). Recent studies suggest that exposure to mercury vapors, released during gold extraction processes that involve mercury amalgamation, can reduce resistance to malaria (Silbergeld et al., 1999). An ecosystem approach examining malaria in gold-mining areas should consider the many factors that potentially contribute to increased malaria: breeding conditions for mosquitoes, influx of miners from remote areas with minimal resistance to the disease, exposure to mercury vapors, poor health services, etc. Indeed, exposure to toxic substances may not only have direct effects, but also indirect effects, through modulation of different physiological functions. In regions where tropical diseases are endemic, the use of chemicals that possibly affect people's capacity to resist these diseases must be considered within a larger ecosystem context for adequate mitigation. Ecosystem management must strive to integrate rather than juxtapose information from various disciplines.

Within the context of ecosystem management, human health needs should be pursued in a holistic manner, by focusing on early changes in biological, psychological, and social parameters. In this way, emphasis is placed on characterizing the health and well-being of the population, disease prevention, and ecosystem equilibrium, rather than on treatment, disease, the assignation of respective responsibility and liability, and perhaps costly ecosystem remediation. Integrating human health considerations into environmental impact studies will help contribute to more equitable mining development.

REFERENCES

Campbell, C. and Williams, B., Beyond the biomedical and behavioural: towards an integrated approach to HIV protection in the southern African mining industry, *Soc. Sci. Med.*, 48, 1625–1639, 1999.

Carvalho, L.H., Fontes C.J., and Krettli, A.U., Cellular responses to *Plasmodium falciparum* major surface antigens and their relationship to human activities associated with malaria transmission, *Am. J. Trop. Med. Hyg.*, 60, 674–679, 1999.

Counter, S.A., Buchanan, L.H., Laurell, G., and Ortega, F., Blood mercury and auditory neuro-sensory responses in children and adults in the Njamba gold mining area of Ecuador, *Neurotoxicology*, 19, 185–196, 1998.

Foreit, K.G., Haustein, D., Winterhalter, M., and La Mata, E., Costs and benefits of implementing child survival services at a private mining company in Peru, *Am. J. Public Health*, 81, 1055–1057, 1991.

Gumucio, M.B., Part 1 in *Potosí, Patrimonio Cultural de la Humanidad*, Compañia Minera del Sur, La Paz, Brazil, 1988.

Hamilton, A., *Exploring the Dangerous Trades*, Little, Brown, Boston, 1943.

International Labor Office, Sectorial Activities: Mining (coal, other mining), ILO, Geneva, 1998.

International Labor Office, Social and Labour Issues in Small Scale Mines, Report for discussion at tripartite meeting on social and labour issues in small scale mining, ILO, Geneva, 1999.

Kilburn, C.J., Manganese, malformation and motor disorders: findings in a manganese-exposed population, *Neurotoxicology*, 8, 421–430, 1987.

Labonne, B., Cleaning-Up Our Mining Act: A North-South Dialogue, Keynote address in the Pan-Canadian 1999 EcoHealth Lecture Tour organized by the International Development Research Centre (IDRC) Academic Fellowship, 1999.

Landrigan, P.J., Silbergeld, E.K., Froines, J., and Pfeffer, R.M., Lead in the modern workplace (editorial), *Am. J. Public Health*, 80, 907–908, 1990.

Lebel, J., Roulet, M., Mergler, D., Lucotte, M., and Larribe, F., Fish diet and mercury exposure in a riparian Amazonian population, *Water Air Soil Pollut.*, 97, 31–44, 1997.

Lebel, J., Mergler, D., Branches, F., Lucotte, M., Amorim, A., Dolbec, J., and Larribe, F., Neurotoxic effects of low-level methylmercury contamination in the Amazonian basin, *Environ. Res.*, 79, 20–32, 1998.

Mergler, D., Manifestations of acute and early chronic poisoning, in *Encyclopaedia of Occupational Health and Safety*, 4th ed., Stellman, J.M., Ed., International Labour Office, Geneva, Vol. 1, 1998, pp. 7.13–7.14.

Mergler, D., Combining quantitative and qualitative approaches in occupational health: towards a better understanding of the impact of work-related disorders, *Scand. J. Work Environ. Health*, 25 (Suppl. 4), 54–60, 1999.

Needleman, H.L., Riess, J.A., Tobin, M.J., Biesecker, G.E., and Greenhouse, J.B., Bone lead levels and delinquent behavior, *JAMA*, 275, 363–369, 1996.

Raizenne, M., Dales, R., and Burnett, R., Air pollution exposures and children's health, *Can. J. Public Health*, 89 (Suppl. 1), S43–S48, 1998.

Rambajan, I., Highly prevalent falciparum malaria in north west Guyana: its development, history and control patterns, *Bull. Pan Am. Health Organ.*, 28, 193–201, 1994.

Rice, D., Issues in developmental neurotoxicology: interpretation and implications of the data, *Can. J. Public Health*, 89 (Suppl. 1), S31–S36, 1998.

Roulet, M., Lucotte, M., Canuel, R., Rheault, I., Tran, S., De Frietos Gogh, Y.G., Farella, N., Souza do Valle, R., Sousa Passos, C.J., De Jesus da Silva, E., Mergler, D., and Amorim, M., Distribution and partition of total mercury in waters of the Tapajós River Basin, Brazilian Amazon, *Sci. Total Environ.*, 213, 203–211, 1998.

Roulet, M., Lucotte, M., Farella, N., Serique, G., Coelho, H., Sousa Passos, C.J., de Jesus da Silva, E., Scavone de Andrade, P., Mergler, D., and Amorim, M., Effects of recent human colonization on the presence of mercury in Amazonian ecosystems, *Water Air Soil Pollut.*, 112, 297–313, 1999.

Salgado, S., *Trabalhadores: uma arquelogia da era industrial*, Companhia das Letras, Sao Paulo, Brazil, 1997.

Silbergeld, E.K., Sacci, J., Azad, A., McKenna, K., Woodruff, S., Strickland, G.T., Trevant, C., and Liggans, G., Mercury impairs host resistance to malaria, *Mercury as a Global Pollutant*, 1999, p. 458 (abstract).

Strickland, G.T., Nash, D., Trevant, C., Silbergeld, E.K., Souza, J.M., and Silva, R.S.U., Mercury exposure and disease prevalence among *garimpeiros* in Para, Brazil, *Mercury as a Global Pollutant*, 1999, p. 387 (abstract).

Viezzer, M., *Domitila. Si on me donne la parole...*, Petite Collection Maspero, Paris, 1976.

Wasserman, E., Environment, health and gender in Latin America: trends and research issues, *Environ. Res.*, 80, 253–273, 1999.

CHAPTER 88

A Conceptual Framework for the Development of Tools to Track Health and Well-Being in a Mining Region: Report from an Indian Study*

Ligia Noronha

INTRODUCTION

The purpose of this chapter is to present a framework for the development of a set of tools to track human health and well-being in a mining region.** More specifically, it will help identify, link, and guide the collection, organization, and analysis of data concerning the implications of mining activity to human health and well-being using a multi-stakeholder perspective. The chapter is divided into two main sections: In the first, existing frameworks are discussed that assess links between economic activity and human well-being, and in the second, the proposed framework is presented with illustrations from the case study in a mining region in Goa, India.

EXISTING FRAMEWORKS THAT ASSESS LINKS BETWEEN ECONOMIC ACTIVITY AND HUMAN WELFARE

The literature on frameworks that seek to assess the links between economic activity and human welfare can be classified as follows: (1) accounting, (2) reporting, and (3) sustainability frameworks. This classification is based on the type of function the frameworks serve.

Accounting frameworks are those that explicitly quantify and value economic activity and the use of resources in such activity, either through physical or monetary accounts. Earlier accounting frameworks only included economic activities. More recently, environmental and natural resources utilized in such activities are included in the accounting. The best known of these are natural and environmental resource accounting frameworks that require that gross domestic product (GDP), or net domestic product (NDP), be adjusted for the cost of using up natural resources and causing

---

* This chapter forms part of the project on Environmental/Social Performance Indicators (ESPIs) and Sustainability Markers in Minerals Development: Reporting Progress towards Improved Health and Human Wellbeing: Phase I sponsored by the International Development Research Centre (IDRC), Canada. A revised (modified) version of this chapter has appeared in *Natural Resources Forum*, 25, 53–65, 2001.
** The conceptual map that informs this framework only clearly emerged at the end of the empirical study of the mining region in Goa, India. This empirical study was initiated on the basis of certain preanalytical insights, which served to organize the data, obtained in part from the TERI 1997 study, and was, in turn, refined by the data. This process of formulating hypotheses and testing them has finally yielded the conceptual map mentioned earlier.

environmental degradation. Over the last few years, there has been a growing literature on natural resource accounting. Ahmad et al. (1989) provide a compilation of developments in this subject area. In 1993, the United Nations (UN) revised the system of national accounts (SNA) for the first time in 25 years, and recommended that member countries implement it. As part of this revision, integrated economic and environmental accounting was introduced. The concept and structure of such accounting are shown in the revised SNA manual, as well as in a separately published handbook for integrated environmental and economic accounting (UN, 1993). More recent developments in theory and practice include the work of Peskin (1996), Chopra and Kadekodi (1997), TERI (1997a), and Uno and Bartlemus (1998).

Reporting frameworks are those that report on environmental, economic, and socioeconomic conditions. The better-known economic indicators at the country level are the gross natural product (GNP) and the net natural product (NNP) or national income (see various World Development Reports). The Human Development Report (HDR) has improved upon these indicators by including a social component in its Human Development Index (HDI). This is the sum of normalized indices of life expectancy at birth, adult literacy, and per capita national income (UNDP, 1990). The conceptual framework that guides the construction of the HDI index sees development as an enlargement of the choices of people, indicators of which are the expansion of education, health, and employment opportunities. This framework's main shortcoming, from the point of view of our objective of assessing health and well-being in the mining region, is the lack of attention to the biophysical domain and its implications to health. There is a conceptual distance between longevity and good health, a distance that life expectancy at birth does not capture. In later versions of this index, separate HDIs for different population groups were constructed. The disaggregated HDI is said to be useful in detecting societal stress and potential conflicts (UNDP, 1994). The HDI offers an alternative to GNP for measuring the relative socioeconomic progress at national and subnational levels. Comparing HDI and per capita income ranks of countries, regions, or ethnic groups within countries highlights the relationship between their material wealth and income, on the one hand, and their human development, on the other. The HDR 1997 introduced a human poverty index (HPI) in an attempt to formulate in a composite index for the different features of deprivation in the quality of life to arrive at an aggregate judgment on the extent of poverty in a community. The HDR (1996) attempted this through a particular version of the capability measure (Doraid, 1997).

Over the last decade, reporting on environmental issues has become the fashion led by organizations such as the UN Environmental Programme (UNEP), the Food and Agricultural Organization of the United Nations (FAO), the Organization for Economic Cooperation and Development (OECD), the World Bank, and the World Resources Institute (WRI) (UNEP, 1994; OECD, 1993; FAO, 1997; World Bank, 1996; Hammond, 1995) Environmental issues are discussed in terms of a stress–response framework (UNEP, 1994). The FAO has also been involved in building indicators for tracking state and trends of natural resources (FAO, 1997).

Frameworks to assess sustainable development seek to assess the long-term viability of systems, activities, or sectors. They include (FAO, 1996):

1. Ecosystem approaches that are based on assessing the performance of a sector according to ecological, economic, and social dimensions and use four criteria for sustainability: productivity, resilience, stability, and equity.
2. Total factor productivity where systems are deemed to be sustainable when total factor productivity (the ratio of the value of all outputs divided by all inputs, economic and environmental, normalized to remove change in prices) show a nondeclining trend.

Prescott-Allen (1996) developed the "sustainability barometer" in which one can have a static representation of sustainability. In this tool, human and ecological well-being are given equal status and are taken to be the fundamental dimensions of sustainability. Rennings and Wiggering (1997) propose the use of ecological and economic sustainability indicators in a complementary fashion by targeting indicators, such as critical loads and levels, that should build the core of indicator sets for sustainable development. Socioecological indicators for sustainability are discussed in Azar et al. (1996).

THE PROPOSED CONCEPTUAL FRAMEWORK

The framework proposed in this chapter draws on these frameworks and has the following as its main features:

1. An ecosystem approach for assessing the impacts of minerals development to the economic, biophysical, social domains, and issues of relevance to stakeholders
2. Highlighting of the relations between mining and well-being
3. A determinants and constituents approach to understanding the concept of well-being
4. A multistakeholder perspective in the validation of issues of concern for indicator development
5. Tools for tracking well-being, which involve:
    a. The development of environmental and social performance indicators based on these issues
    b. The development of a quality of life tool to measure the satisfaction of individuals/communities with the conditions found in the region
    c. An income accounting exercise to account more broadly for the environmental, economic, and social impacts of minerals development in the region

An Ecosystem Approach to Assessing the Impacts of Minerals Development

The approach used in this study to identify impacts is consistent with an emerging ecosystem approach, in which data are spatially referenced and organized on the basis of ecologically defined geographic units as well as administrative units. The point of entry for the study is mining activity, which is seen as the source of change in the ecosystem. Since mining is the driving force, the study area is defined with reference to clusters of mining villages. Using a cluster of mining villages as the ecosystem also allows the communication of what is involved in this definition because the context is known and clear.* We call these clusters of mining villages mining ecosystems, a man-made ecosystem created or developed due to mining activity, which has over the years developed certain specific characteristics. Thus, in the Indian study,** the clusters consist of villages under direct or indirect mining activity and as well as those under direct or indirect impacts of mining-related activities and those bounded by a natural activity; in this case, it is the river. In strict terms, these clusters are not geomorphic units, even though at places they are bounded by rivers.

---

* We follow Shugart (1998) here, and define ecosystems relative to the objectives of our study. This restricts the "definition to the important processes of the particular case of interest. The definition of which aspects of a system are important (or unimportant) to a particular problem is stated explicitly. The ecosystem, as an abstract term, is closer to systems concepts in other sciences" (p. 49).

** The Indian study was in the state of Goa, located on the west coast of India, bounded on the north by Maharashtra, on the east and south by Karnataka, and on the west by the Arabian Sea. It has a coastline 105 km long and a total area of about 3701 km². Goa can be perceived to have four main ecozones: the Sahaydrian watershed located in the Western Ghat region; the plateau area, which is in the midland region; and the alluvial flats and coastal zone in the coastal region (Alvares, 1993). Most of the mining activity is in the Western Ghat region and occurs amid dense settlements, agricultural fields, and forests. The state is drained by a number of rivers, namely, Mandovi, Zuari, Tiracol, and Chapora, flowing westward into the Arabian sea, of which the Mandovi has the largest drainage basin in Goa. The river basins of Mandovi and Zuari cover about 69% of the total area of the state and are together called the lifeline of Goa. These waterways are extensively used to transport the iron ore from the beneficiation plants to the main port from where they are exported. For details on the conditions in the mining belt in Goa, see TERI (1997b). This was a study to arrive an environmental management plan for the mining belt of Goa. It employed a multidisciplinary approach. The study combined spatial data obtained from aerial photographs and remotely sensed products, environmental data, and statistical data obtained from miner level and socio-economic surveys. The various data sets were integrated using geographical information systems tools.

Mining and Human Well-Being

Rees (1985) lists five major criteria that are commonly used to judge the impacts of resource development on human welfare. These impacts are (1) economic efficiency, (2) distributive justice, (3) economic growth and employment generation, (4) supply security, and (5) maintenance of environmental quality. This list indicates that mining has implications for human welfare mainly through three domains: economic, social, and environmental.

Goodwin (1997) discusses three world views on the relationship between economic growth (mining is a subset of this) and human well-being (HWB). These worldviews can be discussed in the context of minerals development as it directly contributes to economic growth. The dominant view in neoclassical economics, before the environmental and social imperative began to assume importance, was that growth would always lead to increased well-being. In developing countries, this is often the implicit view because it is generally believed that jobs are the bottom line for any improvement in well-being. Hence, any job-creating activity, such as minerals development, will, *ipso facto*, contribute to improved well-being. Since mining creates jobs, it is seen as contributing positively to HWB. The negative impacts are not recognized.

Another view holds that there are some aspects of well-being to which growth cannot contribute at all; growth cannot actually reduce well-being, but is neutral to it. In the context of mining, this translates to saying that the income, job, and revenue creation aspects of minerals development contribute to HWB, but that HWB involves many other aspects that minerals development has nothing to offer, positively or negatively.

More recently, there is a growing understanding that some aspects of minerals development do not contribute to well-being but are actually detrimental to well-being. This refers, for example, to the externalities from mining activity which are detrimental to well-being because of effects, such as disruption in community life, growth of alcoholism, destruction of indigenous communities and cultural values, growth of dust pollution, water quality deterioration, and land degradation. Excessive direct or indirect dependency on the mine in the absence of alternative opportunities can lead to considerable stress on a community when a mine closes down as well.

These perspectives on the relationship between minerals development and induced growth and human well-being suggest an investigation along two axes: (1) the different conceptions of well-being that are used, and (2) the different views of what minerals development actually involves in terms of contributions to society. Those who see human well-being in a materialistic sense of jobs and income, and who see minerals development essentially as contributing to these objectives, will subscribe to the first view of the relationship of well-being to minerals development. In contrast those who see human well-being as being more than just material prosperity and who have a wider definition of the good life will see minerals development as contributing to improved well-being, but only to an extent. Some aspects of minerals development do not, in fact, contribute to it, but remain neutral. A third view sees human well-being as including not only material prosperity, but something much wider.

Travers and Richardson (1993) discuss the concept of human well-being as having four components: material well-being, happiness, health, and social participation. They point to the argument made by some economists that because material well-being expands options and so contributes to well-being, it can adequately serve as a proxy indicator of individual and national welfare. However, they refer to evidence that suggests that while the association between subjective evaluation of happiness and wealth is positive, the relationship is weak. This is because the causes of unhappiness go beyond the realm of material well-being and the fact that there can be trade-offs between material well-being and overall human well-being. They further discuss the relation between material well-being and health, and suggest that the former is important at the national scale insofar as they determine the level of public investment in health services. At the individual level, health is far more related to the interaction between material and cultural inputs into health and of the social circumstances that make an egalitarian distribution of these inputs possible.

Some thinkers approach a definition of well-being by composing lists of the elements that go into it (Goodwin, 1997). Others discuss the concept with reference to two social goals: individual happiness and economic justice. These goals are then decomposed into lists of social values of which each is composed and the socioeconomic outputs required to achieve these values (Wilson, 1991).

This conception is similar to that discussed in Dasgupta (1993). He suggests two ways of assessing human well-being. One is to measure the constituents of well-being (e.g., indices of health), and the other is to value the commodity determinants of well-being that refer to the goods and services that are inputs in the production of well-being (e.g., real national income). The concept of well-being can be seen more broadly in terms of constituents (i.e., what constitutes our flourishing) and the determinants (i.e., what is instrumental to our flourishing) (O'Neil, 1993). This goes beyond commodity determinants.

Understanding Human Well-Being in the Context of This Study

From the literature, it is evident that there is no consensus on a single definition of well-being. However, since our objective is to track changes in well-being, we proposed to examine the concept of well-being in terms of constituents and determinants (O'Neil, 1993; Dasgupta, 1993). Nussbaum (1990) (quoted in Ackerman et al., 1997) suggests that what constitutes our flourishing can be approximated by a listing of various capabilities. These capabilities can be grouped under the following: activity* and health, freedom to make choices, and diversity.** While these constituents may not vary between cultures, the weight of each in the scale of well-being may vary.***

What is instrumental to our flourishing? The literature suggests that the ingredients of flourishing lie in well-functioning social, economic, political, and ecological systems. These systems contain the determinants of our flourishing. An ecosystem is considered to be well-functioning and healthy if it is active; maintains its organization, connectivity, and autonomy over time; and is resilient to stress (Costanza, 1992). Karr et al. (1986) state that a biological system can be considered healthy when its inherent potential is realized, its condition is stable, its capacity for self-repair when perturbed is preserved, and when minimal external support for management is needed (quoted in Rapport, 1995).

The physical and psychological relationships between individuals and the ecosystems and the social and political relationships they have with the rest of the community in terms of participation and civic engagement are central to well-being. It is important to recognize here that while the centrality of the individual as the repository of rights is accepted, some variations in the processes by which the individual is constituted has to be conceded. The social and cultural context through which an individual acquires a sense of the cognitive and the moral matrix (i.e., a sense of self and a sense of rights and duties) needs to be given due attention. By doing so, the framework will be achieving both the centrality of the individual and conceding that this individual can be embodied differently in different locations. These differences can vary from minor influences to more radical differentiation. This is important because the framework hopes to be relevant to countries in which the emphasis on community is different (e.g., U.S., India, Colombia), and it is different for an economic activity, such as mining, which generates communities of its own wherever the location. The framework needs to be read along two independent axes: (1) the region and the culture that goes with it, and (2) mining as an activity that shapes the individual and the community. Figure 88.1 illustrates our conception of well-being as being dependent and determined by well-functioning ecological and social systems.

---

* By activity, we mean that agents are in a position to engage in development, productivity, growth, and are in a position to recover from stress.
** The idea of diversity refers to the existence of many identities.
*** For example, health, as an end objective, may have a higher weight in an individual's welfare function in the U.S., whereas freedom may be more important in a country such as Colombia. For the latter, any improvements in levels of freedom may have a greater impact on the overall well-being than improvements in health.

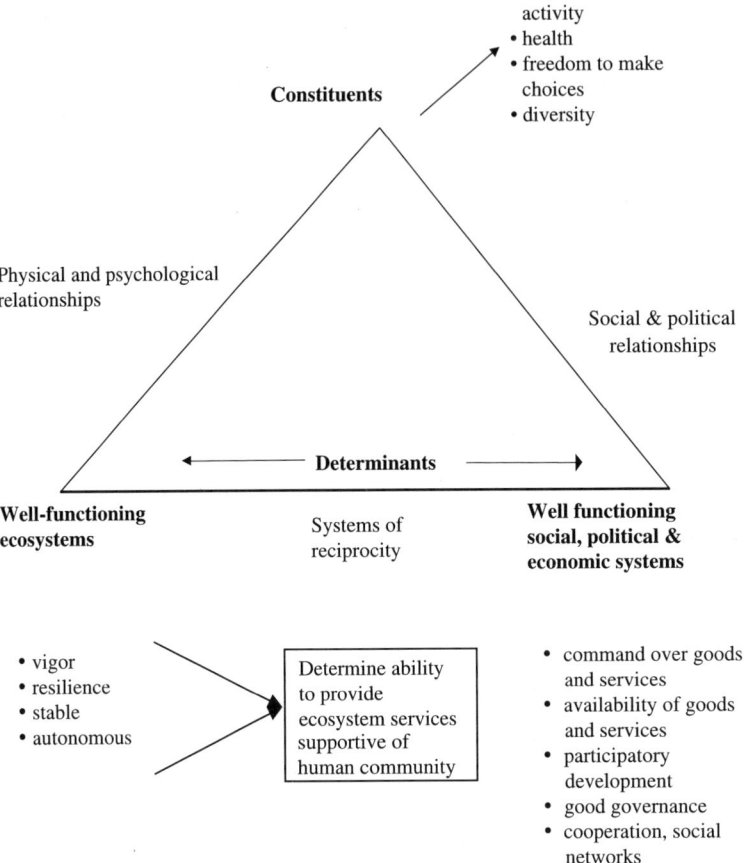

Figure 88.1 Constituents and determinants of well-being.

It can be argued that the following are important determinants of human health and well-being in a mining region context:

- *Command over goods and services* as made possible through increased employment and income. Purchasing power is crucial to a material well-being.
- *Availability of goods and services*, both natural and produced. This includes the availability of good quality and quantity air, land, and water as basic determinants of health, or the production of minerals and the ability to exchange these for other goods and services. Goods and services are important to support and maintain capabilities.
- *Participatory development*, a process by which stakeholders voluntarily come together to share, negotiate, and control decision-making in development issues. Stakeholders participate in development for two sets of benefits: development benefits and negative benefits. Participatory development in mining regions can enhance well-being through its impacts on the constituents as well as the other determinants of well-being.
- *Good governance* by which is meant the ability of governments and other institutions in civil society to create an enabling environment with which goals can be achieved. Three sets of rules are of relevance to good governance:
  - Rules of the game such as laws, rights (positive and negative), regulations, policies, customary laws, and court decisions.
  - Rules that constrain all law making at the level of the constitution.

- Rules of law and the standard of legality that informs the judiciary in any given society (IRIS, 1996).
- *Community and social cohesiveness and systems of reciprocity*, such as the building of networks, community relations, etc., which build up social capital are particularly valuable where the efficiency of legal rule enforcement is in short supply. Cooperation in terms of playing by unwritten rules and social norms then become important in progressing toward well-being goals.

Minerals development can affect all of these determinants and, through them, the constituents of human well-being. Such impacts can be positive or negative. In our case study in India, we identify how these determinants are impacted through an identification of issues of relevance: (1) from the social and environmental assessment and, (2) by stakeholders in minerals development. These issues are aspects of the determinants. For example, a concern with air quality is connected with the need for good quality natural goods and services that are important to our flourishing. A concern with poor rule enforcement reflects on governance and has implications for well-being. A concern with investment of a portion of mining income in developing nonmining skills reflects a concern with command over goods and services in a future period, after mine closure. Issues reflect various aspects of the determinants of our flourishing. In order for these issues to be representative, we need to adopt a multistakeholder perspective to issue development.

Multistakeholder Perspective to Issue Development

At least three sets of stakeholders and user groups are of importance here:

1. The mining company, both corporate management and plant level staff
2. The government, both in its capacity as regulator and as a development agent
3. The public, both local communities and therefore immediate stakeholders, and the wider public

The mapping of stakeholders and the identification of issues and their validation is illustrated with examples from the Indian case study below.

Stakeholder Mapping

In the Indian study, the following stakeholders have been mapped at the following levels:

1. Industry
2. Community and the public groups
3. Government

Mining companies in Goa are all in the private sector, and except for one large company are all family owned and managed. Five companies who represent 80% of all production and export of ore from the state dominate the group. Many of the companies are bound together by familial or kinship relationships. The companies do come together formally to present a joint front to the government and the public in the Goa Mineral Ore Exporters' Association, which is the main iron ore-exporting lobby group. The exporters operate in an international buyers' market that is mostly Japanese. There is no domestic market for Goan ore (Noronha, 1999).

The consumer's product needs and specifications rather than the price of iron ore are important factors in the supply responses of the mining companies. Decisions on how much to produce are based on contracts arrived at through negotiations with the buyers at international prices. The consumers have so far been unconcerned with the externalities created by the product. Their sole concern is that the end product come up to their specifications. In fact, the stringent specifications demanded by the consumers impose an additional burden on the environment since the ore needs to be beneficiated, giving rise to a large number of residuals. Because prices have not changed in real terms since 1981, this has increased costs to the producer and, more importantly, increased indirect costs to the region through the degradation created by effluent discharges and improper tailings dams. Rising costs of production and inelastic prices faced by the industry are often used as an argument by the companies to justify not spending more on environmental protection.* This lack of spending protects the profit margins of producers, but degrades the environment. Except for one mining company that gets part of its capital through shares, the rest of the companies are all financed either through their own funds or bank funds. Investors and creditors are still not important drivers of environmental performance in India.

Transport operators and especially truck drivers are a very important feature of the mining industry in Goa. Ore is carried from the mine pits to the beneficiation plants by trucks and from there by trucks to loading points. From here the ore is loaded onto barges and then these carry ore through Goa's main rivers to the port where the ore is loaded on to the ships. The existence of the waterways is a key factor explaining the competitiveness of the Goan ore relative to ore from Brazil and Australia. Truck and barge operators and their associations are important factors in this cycle. These trucks and barges do not belong to the mining companies, but are contracted from outside the industry.

Informal public groups, such as environmental nongovernmental organizations (NGOs), and the local community provide the most important drivers for improved environmental performance in the mining areas. However, at the local community level, lack of organizational, technical, and financial resources prompt individuals affected by mining to approach companies directly. The solutions arrived at are short term, arbitrary, and *ad hoc,* and the problem surfaces again. The NGOs are better equipped to negotiate with the companies. They, however, prefer to approach the judicial courts for redress, since they find that the government agencies responsible for monitoring the activities of mining do not provide them with satisfactory answers.

Finally, the third category of stakeholders is the government with a dual stake: to promote the development of the industry and to protect the environment. Government has a positive stake in the industry because it creates jobs, income, foreign currency, and revenues for the economy and, hence, needs to ensure a climate that promotes the growth of the industry. But since it has the mandate to protect the environment, government has to constrain the options available to the industry through various rules and statutes. These rules are at three levels: the center, the state, and the village.** Because mining creates impacts on sectors other than itself, there is a need for action on the part of departments. For example, the irrigation department is needed for the pollution of waterways and canals, and the agriculture department is needed because mining reduces productivity and causes loss of farmlands. The transport department is needed because of the dust problem created by truck operations, while the town and country planning department is necessary because of the changes in land-use patterns that are occurring in the mining region due to creeping expropriation of present land uses. Figure 88.2 provides a map of the concerns of the main stakeholders as analyzed from the study.

The issues identified from the case study were then validated with the three major groups: the government, the community, and the companies. The predominant issues from individual and common stakeholder perspectives are organized using a Venn diagram (Figure 88.3.). This tool helps us to see which issues are common to all stakeholders and which to groups of two or just to individual stakeholders.

This multistakeholder perspective to issue identification and validation is a central methodological feature of this framework. It is favored here for many reasons.

---

* This is also the reasoning provided sometimes by government officials to explain the deteriorating environmental quality in the region.
** In Goa, there is on an average one panchayat for every two to three villages. A panchayat is made up of representatives elected from these villages.

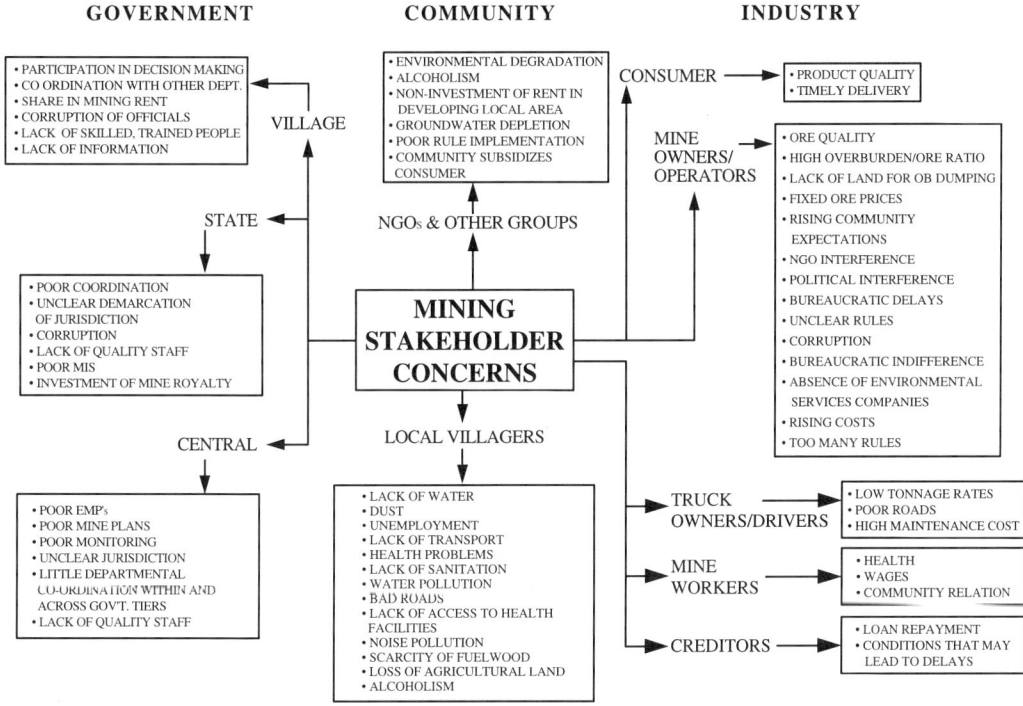

Figure 88.2 Stakeholder concern map. (From Noronha, L., *Nat. Res. Forum*, 25, 53–65, 2001. With permission.)

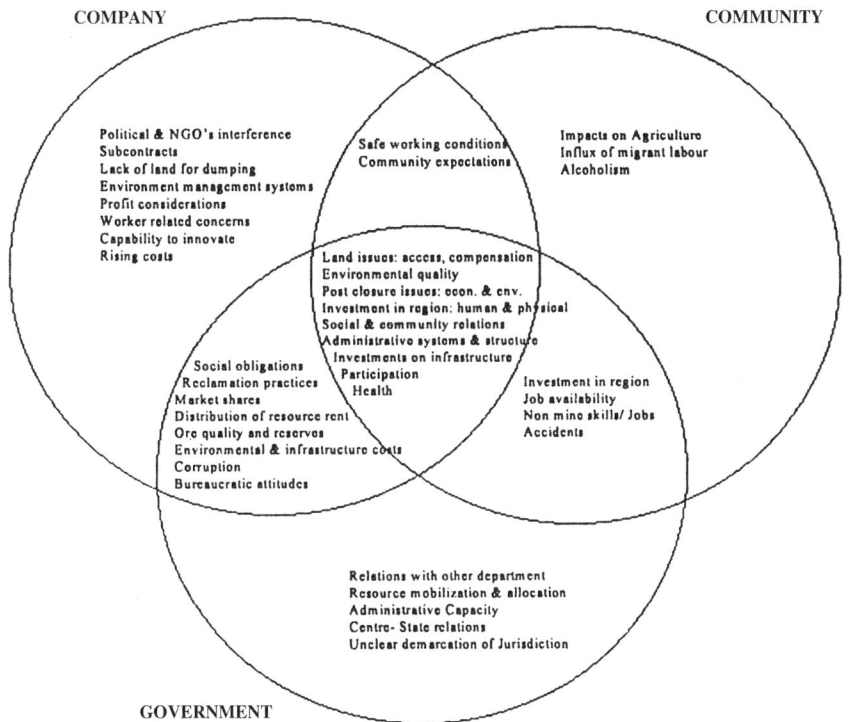

Figure 88.3 Issues of relevance arising from the Indian study. (From Noronha, L., *Nat. Res. Forum*, 25, 53–65, 2001. With permission.)

1. Tools that are based on a multistakeholder perspective will address the concerns and needs of the stakeholders. Such a perspective is more participative since the issues are those that have emerged from direct interaction with stakeholders.
2. A core set of issues common to all stakeholders can be identified from these lists. Tool development can then be based only on this smaller set of common issues. This makes the task more manageable.
3. Such a core common set, rather than an excessive focus on issues common to just one stakeholder, will also have acceptability from a wider audience. Indicators based on such issues will therefore have a wider appeal.

Tracking Well-Being

In order to be able to track how we are doing and to measure the movement toward improved health and human well-being, there is need to have a way of tracking well-being through measuring its constituents and determinants of well-being. Three sets of tools are suggested:

1. Determinants (using issues as aspects of determinants) can be measured using environmental, economic, and social performance indicators (ESPIs). Sets of environmental and social performance indicators are developed for the issues that have been identified by the various stakeholders. ESPIs provide a measure of the conditions that affect human health and well-being. Changes in these indicators represent improvement or deterioration of conditions in the mining region and provide relevant domain specific information to the main stakeholders.
2. A quality of life (QOL) instrument provides a measure of satisfaction with life of the individual and the community in a mining region. This instrument measures an individual's responses with regard to his level of satisfaction with the conditions in the social, cultural, biological, economic, and political environment that he faces.* The QOL instrument tries to capture the relationship between conditions or determinants found in the mining regions, and the subjective responses of the individuals and the community living in the region.
3. An impact adjusted income account of the mining region. This tool seeks to assess the long-term viability of the activity and of the region in terms of the kind of impacts that the activity is creating. Such impacts may be consuming the very resources — natural, human, and social — that make the activity possible and may, over the long run, cause the region to go into a decline.

In Figure 88.4 the main elements of the framework are summarized.

How do these tools track well-being? This question is examined in the next section.

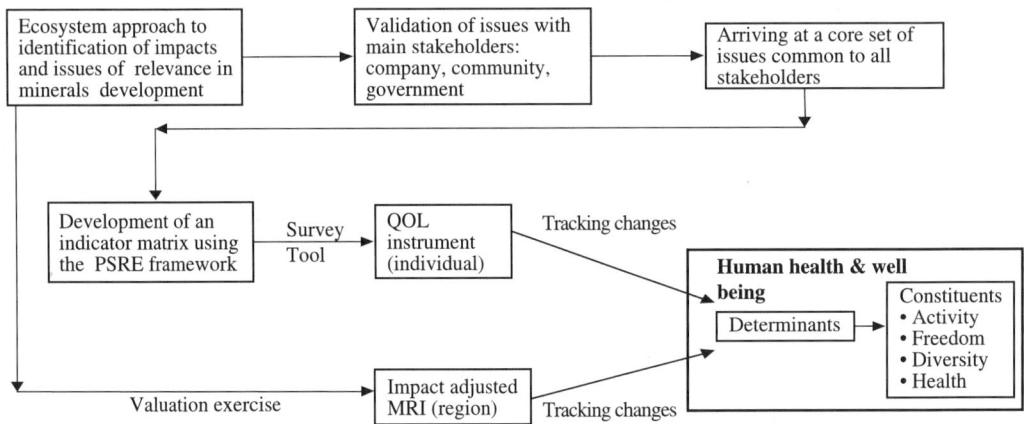

Figure 88.4 Main elements of this conceptual framework. (From Noronha, L., *Nat. Res. Forum*, 25, 53–65, 2001. With permission.)

---

* The literature on QOL is now vast. For a taste of the issues, see Nussbaum and Sen (1993).

Environmental and Social Performance Indicators (ESPIs)

An indicator has been defined as "a statistic or measure which facilitates interpretation and judgments about the condition or an element of the world or society in relation to a standard or goal." Further, an indicator is a parameter (i.e., a measured or observed property), or some value derived from parameters (e.g., via an index or model), which provides managerially useful information about patterns or trends (changes) in the state of the environment, human activities that affect or are affected by the environment, or about relationships among such variables (EPA, 1995).

Indicators can be of use to various audiences. The type of audience in mind when working with developing indicators will influence the components of the framework that is developed. ESPIs are based on issues identified and validated by the stakeholders in minerals development. For purposes of this discussion, just the set common to all stakeholders is concentrated upon as this is the minimum set of issues that need to have indicators developed in order to be able to track well-being in the region.*

Quality of Life Instrument

The QOL instrument is connected to the indicator framework. Here the state indicators can be used as descriptors of the conditions faced by local communities. The survey instrument is then used to try to arrive at the sense of personal impact and satisfaction with these conditions. If it is desired that the two be related (i.e., that the indicator framework be related to the subjective well-being) this can be done by seeing one as related to the other (i.e., for each objective condition faced by respondents, what is their level of satisfaction with it).

For example, we have the following as suspended particulate matter standards for air quality: 200 µg/m^3 to indicate good air quality and 500 µg/m^3 to indicate poor air quality. These are then transformed to a normal scale varying from 0 to 1 (value function cure) in which 0 corresponds to poor air quality and 1 to good air quality. The level of satisfaction of individuals with this air quality is obtained through the use of this QOL instrument. The indicators track the conditions, and this instrument tracks the subjective dimension (e.g., the level of satisfaction with the conditions). The interconnections are illustrated in the matrix in Table 88.1

Issues in each domain can be given equal weights or an expert meeting can be had where weights are assigned for each issue area based on observations and data available. Alternatively, each country can have its own weighting since different cultures may have different relative

Table 88.1 Developing a Quality of Life Instrument

Domains	Multistakeholder Set of Issues Identified	Environmental and Social Performance Indicators	QOL Instrument
	Common	State indicators and exposure indicators	Level of individual/ community satisfaction and perception of impacts on constituents of WB: health, freedom, diversity, and activity
Social Political Economic Biophysical			

* In practice, issues relevant to the community should be given added weight in the choice of issues that are used for indicator development, in order to concede the point that decisions are made within an asymmetric power matrix where state and company collude to disadvantage community.

priorities. These can then be combined into an overall index of quality of the biosphere, quality of the sociocultural political sphere, or quality of the economic sphere.

An Impact-Adjusted Income Account of the Mining Region

A tool to track sustainability in the mining region is one that accounts for impacts caused by mining activity, values these monetarily, and then adjusts the income obtained from mining to account for such impacts. For each unit of analysis, for example, a village or cluster, we estimate the net income using appropriate approaches. We then adjust this net income of the village or the cluster for the impacts created by the economic activities using the following formula:

$$\text{IMRI}_c = \text{MRI}_c \pm V_{\text{MD/ME}} \pm V_{\text{ED/EI}} \pm V_{\text{SI}}$$

where IMRI_c = impact adjusted income of cluster/village, MRI_c = income of the cluster/village, $V_{\text{MD/ME}}$ = value of mineral resource depletion/enhancement, $V_{\text{ED/EI}}$ = value of environmental degradation/improvement attributed to mining, and V_{SI} = value of social impacts attributed to mining.

Different methodologies can be used to calculate resource change, environmental change (Markandya, 1998), and social impacts. In valuing social impacts of mining, care has to be taken to avoid double accounting. It is evident that jobs, income, and fiscal gains for example, are all accounted for in the traditional definition of income from the mining activity. The social impacts, such as those on health that are related to a declining environmental quality, are already captured when measuring environmental costs. Those aspects that are not reflected in those calculations need to be explicitly included. More difficult to estimate are sociocultural impacts such as change in traditional values, improved social mobility, etc.

The difference between the environmental and social impact adjusted income for the region and the MRI of the region is an indicator of the resource, social, and environmental changes in the region. The IMRI over time can serve as a sustainability marker. Accounting for impacts enables us to keep track of what minerals development is doing to mineral and environmental resources, to local communities, and provides information that is essential for corrective action.

CONCLUSION

In this chapter, a framework for the development of tools to track health and well-being in minerals development is presented. The main features of this framework are summarized below.

1. All domains that affect or are affected by mining activity need to be studied, such as the biophysical, the sociopolitical-cultural, and the economic domain.
2. Human well-being is understood as being made up of constituents and determinants, where the former refers to what contributes to our flourishing and the latter to what is instrumental to this flourishing.
3. A multistakeholder perspective consisting of the government, the mining company, and the community is adopted to identify issues of relevance in each domain.
4. Economic and social performance indicators for the issues are developed. These are sets of measures of the conditions found in a mining region of mining practices and activities that affect health and well-being, of responses to such conditions and pressures, and of the relationships among such variables.
5. An instrument can be developed to measure the satisfaction of the individuals and community in the mining region with their quality of life.
6. The use of an accounting framework helps develop an estimate of income from a mining region from which values of the impacts, both positive and negative, caused by mining are adjusted. This account, seen over time, will tell us if the mining region is able to continue functioning into the

indefinite future without being forced into a decline through the exhaustion and deterioration of its key natural, social, and human resources.

If these tools are used regularly, an information system will emerge that will provide markers of what mining is doing to the region and the communities. This set of tools, if taken together, provide the users with the following information:

1. Economic and social performance indicators that will provide trends of what companies are doing to prevent or mitigate the environmental and social impacts of mining
2. The QOL tool that will provide information and feedback on the levels of satisfaction among local communities with existing conditions
3. The adjusted income accounts that will provide a macroperspective of the impacts, both positive and negative

The trends revealed by these tools will enable appropriate and timely action by the users and the stakeholders to protect resources and human well-being.

ACKNOWLEDGMENT

The author would like to thank IDRC for this sponsorship and would like to acknowledge the considerable intellectual inputs provided by Professor Alyson Warhurst and her Mining Energy Research Network (MERN) colleagues at the University of Bath; Cristina Echavarría of Instituto de Estudios Regionales (INER) at the University of Antioquia, Medellín, Colombia; Professor Peter R. de Souza of Goa University; and several colleagues at TERI. This framework has benefited greatly from this interaction. Special thanks to Sudhir Khalwadekar of Tata Energy Research Institute (TERI) for help with figures and with formatting.

REFERENCES

Ackerman, F.D., Kiron, N.R., Goodwin, N.R., Harris, J.M., and Gallagher, K., Eds., *Human Well-Being and Economic Goals*, Island Press, Washington, D.C., 1997.

Ahmed, Y.J., Serafy, S., and Lutz, E., *Environmental Accounting for Sustainable Development*, World Bank, Washington, D.C., 1989.

Azar, C., Holmberg, J., and Lindgren, K., Socio-ecological indicators for sustainability, *Ecological Economics*, 18, 89–112, 1996.

Center for Institutional Reform and the Informal Sector (IRIS), Governance and the Economy in Africa: Tools for Analysis and Reform of Corruption, http://www.geog.umd.eduIRIS/toolkit.html, College Park, MD, 1996.

Chopra, K. and Kadekodi, G.K., *Natural Resource Accounting in the Yamuna Basin: Accounting for Forest Resources*, Institute of Economic Growth, New Delhi, India, 1997.

Costanza, R.B., Toward an operational definition of ecosystem health, in *Ecosystem Health: New Goals for Environmental Management*, Costanza, R., Norton, B.G., and Haskell, B.D., Eds., Island Press, Washington, D.C., 1992, pp. 239–256.

Dasgupta, M., Chen, L.C., Krishnan, T.N., Eds., *Health, Poverty & Development in India*, Oxford University Press, Mumbai, 1998.

Dasgupta, P., *An Inquiry into Well-Being and Destitution*, Clarendon Press, Oxford, 1993.

Doraid, M., Analytical tools for Human Development, Human Development Report Office, Aug. 1997.

Environmental Protection Agency (EPA), Conceptual Framework to Support Development and Use of Environmental Information in Decision-Making. Document 239-R-95–012, Washington, D.C., Apr. 1995.

Food and Agricultural Organization of the United Nations, Land quality indicators and their use in sustainable agriculture and rural development, *Land Water Bull.*, 5, 1997.

Goodwin, N., Overview essay, in *Human Well-Being and Economic Goals*, Ackerman, F.D. et al., Eds., Island Press, Washington, D.C., 1997, pp. 1–14.

Hammond, A., *Environmental Indicators: A Systematic Approach to Measuring and Reporting on Environmental Policy Performance in the Context of Sustainable Development*, World Resource Institute, Washington, D.C., 1995.

Karr, J.R., Fausch, K.D., Angermeir, P.L., Yant, P.R., and Schlosser, L.R., Assessing biological integrity in running waters: a method and its rationale, *Ill. Nat. Hist. Surv. Spec. Publ.* 5, Champagne, IL, 1986.

Markandya, A., *Methodology for Evaluating Environmental Damages over Time and the Progressive Benefits of Prevention and Mitigation: Application to the Mining, Mineral Processing and Metallurgy Industries*, Department of Economics and International Development, University of Bath, Bath, U.K., 1998.

Noronha, L., Designing tools to track health and well-being in mining regions of India, *Nat. Resour. Forum*, 25, 53–65, 2001.

Noronha, L., Mining in Goa: the need to integrate local, regional and national interests, in *Development Policies in Natural Resource Economies*, Mayer, J., Chambers, B., and Farooq, A., Eds., Edward Elgar Ltd., Cheltenham, U.K., 1999, pp. 155–171.

Nussbaum, M., Aristotelian social democracy: a summary, in *Human Well-Being and Economic Goals*, Ackerman, F.D. et al., Eds., Island Press, Washington D.C., 1997, pp. 273–275.

Nussbaum, M. and Sen, A., Eds., *The Quality of Life*, Oxford University Press, Oxford, 1993.

O'Neill, J., Nature, intrinsic value and human well-being, in *Human Well-Being and Economic Goals*, Ackerman, F.D. et al., Eds., Island Press, Washington D.C., 1997, pp. 40–44.

Organization for Economic Cooperation and Development (OECD), *OECD Core Set of Indicators for Environmental Performance Reviews*, Paris, 1993.

Peskin, H.M., Alternative Resource and Environmental Accounting Approaches and Their Contribution to Policy, draft paper, 1996.

Prescott-Allen, R., *Barometer of Sustainability: What It's for and How to Use it?* The World Conservation Union (IUCN), Gland, Switzerland, 1996.

Rapport, D.J., Ecosystem health: an emerging integrative science, in *Evaluating and Monitoring the Health of Large scale Ecosystems*, Rapport, D.J., Gaudet, C.L. and Calow, P., Eds., Springer-Verlag, Berlin, 1995, pp. 5–31.

Rees, J., *Natural Resources: Allocation, Economics and Policy*, Metheun, London, 1985.

Rennings, K. and Wiggering, H., Steps towards indicators of sustainable development: Linking economic and ecological concepts, *Ecol.l Econo.*, 20, 25–36, 1997.

Shugart, H., *Terrestrial Ecosystems in Changing Environments*, Cambridge University Press, Cambridge, 1998.

Tata Energy Research Institute, National Resource Accounting in the Yamuna Sub-basin for the Ministry of Environment and Forest, Government of India, Nov. 1997a.

Tata Energy Research Institute, Areawide Environmental Quality Management (AEQM) Plan for the Mining Belt of Goa State, Directorate of Planning and Statistics, Government of Goa, Goa, India, Nov. 1997b.

Travers, P. and Richardson, S., Material well-being and human well-being, in *Human Well-Being and Economic Goals*, Ackerman, F.D. et al., Eds., Island Press, Washington, D.C., 1997, pp. 29–29.

United Nations, *Integrated Environmental and Economic Accounting: Handbook of National Accounting*, New York, 1993.

United Nations Development Programme (UNDP), *Human Development Report*, 1990, 1994, 1996, 1997.

United Nations Environmental Programme (UNEP), An Overview of Environmental Indicators: State of the Amount and Perspectives, UNEP Environmental Assessment Technical Report, 1994–2001, Nairobi, 1994.

Uno, K., and Bartlemus, P., *Environmental Accounting in Theory and Practice*, Kluwer Academic Publishers, Dordrecht, 1998.

Warhurst, A. and Noronha, L., Eds., *Environmental Policy in Mining: Corporate Strategy and Planning for Closure*, CRC/Lewis Publishers, London, 1999.

Wilson, J.O., Human values and economic behavior: a model of moral economy, in *Socio-economics: Towards a New Synthesis*, Etiozioni, A., Ed., M.E. Sharpe, New York, 1991, pp. 233–263.

World Bank, *Performance Monitoring Indicators*, Washington, D.C., 1996.

World Bank, *World Development Report*, several issues.

CHAPTER 89

Addressing Gaps in Ecosystem Health Assessment: The Case of Mineral Resources

Joy Jacqueline Pereira and Ibrahim Komoo

INTRODUCTION

Mineral resources have been concentrated by a variety of geological processes, which operate so slowly by human standards that the rates of replenishment are infinitesimally small relative to its consumption. Once a mineral is extracted and used, it is gone forever. As mineral deposits are finite and exhaustible, they are considered a nonrenewable resource. Modern civilization is very dependent on mineral resources. Minerals are used in the construction industry, the electronics industry, the chemical industry, and many major manufacturing industries. Infrastructure development projects, in particular, require adequate supplies of minerals, especially aggregates, sand, and gravel. One of the most significant impacts of land development in certain parts of Malaysia is restriction on the availability of minerals to sustain the economic growth, due to lack of proper planning (Selangor State Government, 1999). The expansion of urban and industrial areas that encroach upon existing mines and quarries prevent the exploitation of and access to undeveloped mineral resources. If such resources become sterilized, minerals have to be transported into the area concerned, resulting in increased costs to the community.

The issue of ecosystem health is closely linked to sustainability. At the most basic level, a healthy ecosystem requires conditions that are sustainable. Sustainability has been described as a system's ability to maintain structure and function indefinitely (Costanza, 1992). This description is applicable to a variety of complex systems ranging from cells to ecosystems as well as economic systems. Indicators facilitate the assessment of ecosystem health. The selection of indicators is based on an integrated approach involving a combination of biophysical, socioeconomic, and human health parameters (Bertollo, 1998). The concept of ecosystem health is deeply rooted in the fields of biological and ecological sciences. The parameters at present utilized to assess ecosystem health are predominantly based on ecology and biology (Table 89.1).

Parameters developed to assess ecosystem health in the context of minerals are limited. One parameter that has been proposed as an indicator of the threat to ecosystem integrity is the depletion rate of mineral reserves (Munn, 1993, cited in Bertollo, 1998). Indicators of reduced threats to ecosystem integrity include increasing output of production per unit of nonrenewable resources, increasing recycling efforts, as well as decreasing economic and energy subsidies provided for the natural resource sectors. Several indicators have been proposed for minerals as part of the efforts to monitor the consumption of nonrenewable resources (Table 89.2). These indicators are not specifically designed to

Table 89.1 Types of Indicators to Assess Ecosystem Health

Ecosystem process rates and storage	Abiotic zones
Community structure	Biotic composition
Population structure	Disease prevalence
Diversity	Bioaccumulation of contaminants
Fish abundance and condition	Grazing and decomposer food chains
Exotic rare species	Lake size
Plant biomass	Stream/river width
Nutrient cycling	Contaminant levels in fish
Productivity	Algae bloom in water bodies

Adapted from Bertollo, P., *J. Ecosystem Health*, 4, 33–50, 1998.

Table 89.2 Types of Indicators Proposed for Mineral Resources

Per capita consumption	Waste generation
Shortage	Reclamation expenses
Intensity of use	Soil quality
Available reserves	Groundwater quality
Price of commodity	Reduction/recycling
Mining activity	Waste management expenditure
Ex-mining land	Waste disposal charges
Erosion	Total material requirement
Recycling and substitution rates	Contribution to GDP and employment
Emission	Loss of biodiversity
Metal losses in waste	Consumption of energy and water

Adapted from Peterson, P., in Vol. 1: —*Management Response Strategies*, Penerbit UKM, Bangi, Malaysia, 1997; Adriaanse, A. et al., in *Resource Flows: The Material Basis of Industrial Economies*, World Resources Institute, Washington, D.C., 1997; Natural Resources Canada, in *Proceedings of an International Experts Workshop*, Quebec, May 7, 1998.

assess ecosystem health; however, they can be adapted to serve that purpose. One such example is the total material requirement (TMR) proposed by Adriaanse et al. (1997). The TMR is a summary indicator that integrates mineral resource utilization directly to economic activity at the national level (Adriaanse et al., 1997). The focus is on broad categories, such as metals, industrial minerals, construction minerals, infrastructure excavation, fossil fuels, and renewable resources. The TMR includes both renewable and nonrenewable resources, and its methodology can also be applied to assess nonrenewable resource flows within the ecosystem framework. The Canadian government has also embarked on an initiative to develop sustainable development indicators for minerals and metals (Natural Resources Canada, 1998). A preliminary menu has been compiled, based on the Organization for Economic Cooperation and Development pressure–state–response framework for national reporting purposes. The parameters proposed are the most comprehensive selection for the minerals sector to date.

This chapter provides an overview of mineral utilization trends in the Langat Basin, which is located in Peninsular Malaysia. All the minerals exploited in the basin to support economic activities are accounted for, including nonmineral materials removed during extraction and processing. The issues of mineral sterilization and information gaps that threaten the management of mineral resources within the basin are also highlighted. Finally, potential indicators are proposed for assessing the health of the Langat Basin, within the context of the minerals sector.

METHODOLOGY

The main sources of data for mineral production in Malaysia are the minerals and geoscience department and the statistics department of the Institute for Environment and Development (LESTARI). Data on mineral production at the state level is estimated based on surveys conducted by these two departments. The mineral production data for the Langat Basin are based on that of the state production.

One method to chart material movement supporting economic activities has been developed by Adriaanse et al. (1997). The method accounts for the major flows of both renewable and nonrenewable resources in an industrialized nation's economic activities. The boundary between nature and the economy is defined at the point when humans first extract or move materials from natural sites.

The method developed to chart material movement in the Langat Basin is loosely based on the methodology of Adriaanse et al. (1997). However, the flow is calculated only for nonrenewable resources that are extracted in the Langat Basin. Due to lack of data, it has been difficult to estimate the amount of minerals imported into the basin. The total amount of minerals consumed in the Langat Basin has not been quantified.

The extraction of mineral resources is often accompanied by the removal of material that is associated with it. Aggregate extraction requires the removal of large quantities of overburden that can modify or damage the environment even though they are not of economic value. The washing of tin mine tailings to obtain sand results in the removal of fine material, which is eventually disposed along the waterways. All such flows are part of the economic activity in the Langat Basin, but never enter the monetary economy as commodities. The flows that do not enter the monetary economy are referred to as hidden flows. Hidden flows are estimates of overburden or gangue removed during mineral extraction. These estimates are obtained from the literature and modified based on field visits to extraction sites in the Langat Basin. In charting the mineral movement within the Langat Basin, hidden flows are separated from those that enter the economy.

MINERAL EXTRACTION IN THE LANGAT BASIN

The minerals produced in the Langat Basin include both metallic and industrial minerals. The industrial minerals extracted are aggregates, kaolin, clay, and earth materials, as well as sand and gravel. The main metallic mineral is tin. At present, there are 21 granite quarries, 3 clay pit sites, and 1 kaolin extraction site reported to be operating in the Langat Basin (Figure 89.1). In addition, 43 earth material extraction sites and 86 sand mining permits have been approved in the Langat Basin since 1997 (GSD-IM, 1997, 1998). In terms of tonnage, over 95% of the minerals extracted are consumed within the Langat Basin. Additional minerals required for the manufacturing industries located within the basin are imported from other Malaysian and foreign sources.

Aggregates

Aggregates refer to crushed rock materials obtained from quarries that are produced in various sizes. The main rock type exploited as aggregates in the Langat Basin is granite. The aggregates are utilized in the construction industry. Between 1980 and 1989, the annual production rate of aggregates in the Langat Basin increased by about 5% (Table 89.3). Between 1990 and 1996, the annual rate of aggregate production increased by nearly 30%. This increase coincides with the opening of the southern corridor for development, which includes part of the Langat Basin. In 1997, aggregate production dropped by about 60% compared to the previous year. This is due to the economic slowdown that has affected Malaysia and other Southeast Asian economies.

Kaolin

Kaolin, or china clay, is a special kind of high quality clay material suitable for a variety of industrial applications. Premium grade kaolin is used as paper coating and the manufacture of fine tableware, porcelain bone china, and electric insulators. Normal grade kaolin is best suited as fillers and extenders in the ceramic, paper, paint, and rubber industries. The production of kaolin in the Langat Basin has increased significantly (about 50 times) since 1980 (Table 89.3). The fourfold

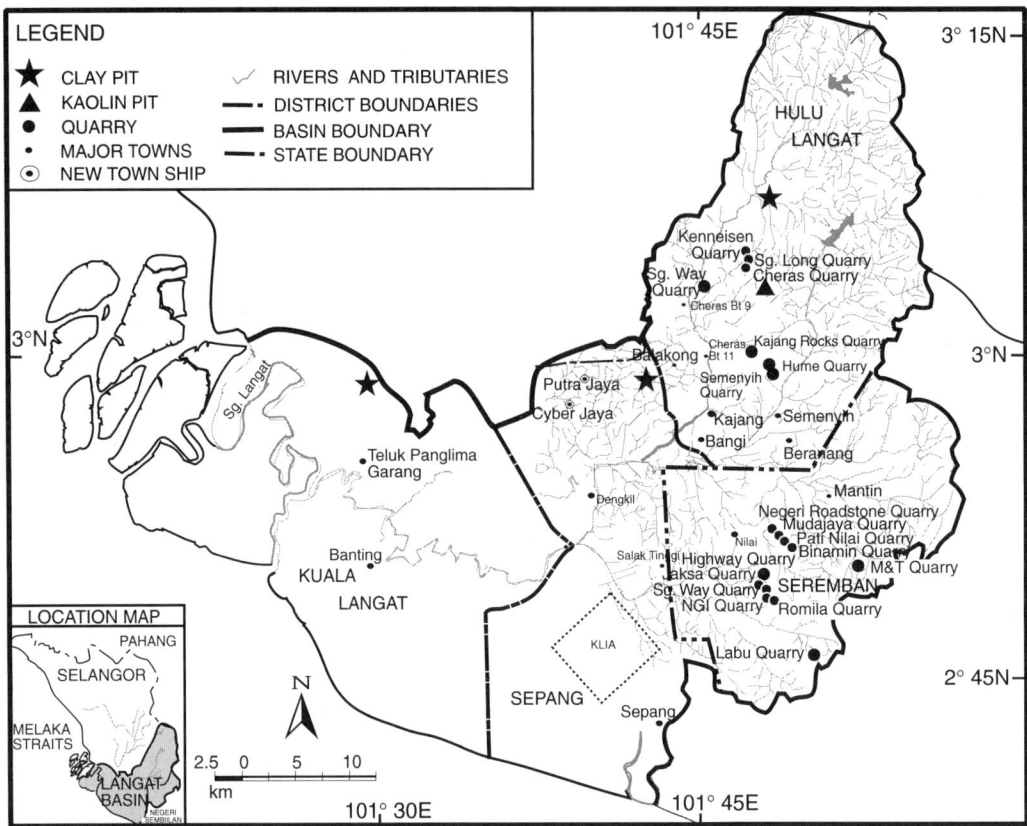

Figure 89.1 Location of selected extraction sites in the Langat Basin.

increase of kaolin production in the 1995–1996 period is attributed to the upgrading of facilities in the processing plants to make them integrated and capable of producing kaolin suitable for export to Taiwan and Korea (GSD-IM, 1996).

Clay and Earth Materials

Clay and earth materials include material used for landfill, common clay, brick clay, plastic clay, and fire clay. It excludes special clays such as kaolin described earlier. The various types of clay are used for different purposes depending on their properties and the extent to which they can withstand high temperatures (GSD-IM, 1996a,b). Ball clay and plastic clay are used in the manufacture of high-quality whiteware, pottery, clay pipes and tiles, ceramics, and other structural clay products. Common and brick clays are suitable for making building bricks while fireclay is essential for the manufacture of refractory products such as firebricks, refractory mortars and mixes, crucibles, and castable materials. Since 1980, the production of clay and earth materials in the Langat Basin has been stable, ranging from between 0.5 million and 1.6 million tons annually (Table 89.3).

Sand and Gravel

The basic materials used in the construction industry are sand and gravel. About 40% of the mines in the Langat Basin exploit sand from mine tailings while the rest extract sand from rivers. In line with the development of the Langat Basin, the production of sand has increased steadily since 1980. The highest production of sand and gravel in the Langat Basin was in the 1993–1994

Table 89.3 Mineral Production in the Langat Basin[a]

	Aggregates	Kaolin	C & E Material[b]	S & G[c]	Tin Conc.[d]	Total
1980	3,183,430	926	1,172,400	142,310	28,860	4,527,926
1981	3,757,050	882	1,416,400	180,460	28,170	5,382,962
1982	4,731,560	887	1,240,000	314,930	24,600	6,311,977
1983	4,556,530	1148	1,348,800	337,120	19,442	6,263,040
1984	5,017,680	1449	1,354,800	588,700	19,414	6,982,043
1985	4,424,050	1651	1,242,400	331,590	15,852	6,015,543
1986	5,128,310	1701	958,000	552,020	9166	6,649,197
1987	4,362,410	1938	595,600	655,833	14,982	5,630,763
1988	4,066,170	2337	746,800	283,850	13,699	5,112,856
1989	5,074,720	2167	1,163,600	333,340	14,016	6,587,843
1990	6,310,055	3059	1,140,800	561,540	11,666	8,027,120
1991	7,545,390	4026	1,309,600	493,780	8787	9,361,583
1992	14,616,840	2215	614,642	681,700	4824	15,920,221
1993	16,113,900	—	549,060	1,024,200	3556	17,690,716
1994	17,965,000	—	1,661,056	1,134,400	2318	20,762,774
1995	28,287,140	8965	697,086	762,500	2066	29,757,757
1996	27,682,610	10,159	1,074,734	848,100	1643	29,617,246
1997	10,136,735	—	963,244	1,169,600	—	12,269,579
Total	172,959,580	43,510	19,249,022	10,395,973	22,3061	202,871,146

[a] Units are in metric tons.
[b] C&E Material = clay and earth material.
[c] S & G = sand and gravel.
[d] Tin Conc. = tin concentrates.
— Data not available.

Extrapolated from state production data.

period (Table 89.5). Subsequent to this, sand and gravel production dropped slightly. In 1997, production of sand and gravel increased by about 38% compared to the previous year. There were 86 extraction sites for sand and gravel within the Langat Basin in 1997, compared to 115 the year before.

Tin Concentrates

The production of tin concentrates has declined substantially since 1980 (Table 89.5). The tin-mining industry used to be the major economic activity in the Langat Basin prior to the 1980s. The tin industry is now insignificant, with fewer than three mines operating at any one time, reworking tailings from dredges and old mines. The by-product of such operations includes gold, ilmenite, monazite, zircon, pyrite, and silica. Tin concentrate is at present being imported into Malaysia from Australia, China, and Bolivia for smelting. The refined tin is then exported to countries, including the Netherlands, Japan, the U.S., and South Korea. Since 1996, the consumption of tin metal in the country has been higher than the country's tin production. The metal is being consumed in the solder, tinplate, and pewter industries in Malaysia (GSD, 1996).

MINERAL UTILIZATION TRENDS AND ECOSYSTEM INTEGRITY

Total Mineral Extraction

Total mineral extraction (TME) is the sum of locally extracted minerals per capita that supports economic activity within the basin. The contributors to the TME for the Langat Basin for the period of 1980–1997 are aggregates, kaolin, clay and earth materials, sand and gravel, and tin concentrates.

Table 89.4 The TME per Capita in the Langat Basin, 1980–1997[a]

Year	TME/ Capita	Aggregates	C & E Material[b]	Kaolin	S & G[c]	Tin Conc.[d]	Population
1980	11.0	7.7	2.8	0.0023	0.35	0.070	411,495
1981	12.5	8.7	3.3	0.0020	0.42	0.065	431,247
1982	14.0	10.5	2.7	0.0020	0.70	0.054	451,947
1983	13.2	9.6	2.8	0.0024	0.71	0.041	473,640
1984	14.1	10.1	2.7	0.0029	1.19	0.039	496,375
1985	11.6	8.5	2.4	0.0032	0.64	0.030	520,201
1986	12.2	9.4	1.8	0.0031	1.01	0.017	545,170
1987	9.9	7.6	1.0	0.0034	1.15	0.026	571,339
1988	8.5	6.8	1.2	0.0039	0.47	0.023	598,763
1989	10.5	8.1	1.9	0.0035	0.53	0.022	627,503
1990	12.2	9.6	1.7	0.0047	0.85	0.018	657,624
1991	14.1	11.3	2.0	0.0060	0.74	0.013	666,129
1992	22.8	20.9	0.9	0.0032	0.98	0.007	698,103
1993	24.2	22.0	0.8	—	1.40	0.005	731,612
1994	27.1	23.4	2.2	—	1.48	0.003	766,730
1995	37.0	35.2	0.9	0.0112	0.95	0.003	803,533
1996	35.2	32.9	1.3	0.0121	1.01	0.002	842,102
1997	13.9	11.5	1.1	—	1.3	—	882,523

[a] Units are in metric tons/capita. An annual average growth of 5% was used for the population estimate.
[b] C & E Material = clay and earth material.
[c] S & G = sand and gravel.
[d] Tin Conc. = tin concentrates.
— Data not available.

The TME per capita rose steadily in the Langat Basin by about 6% annually, from 1980 up to 1984 (Table 89.4). Between 1985 and 1988, the TME registered a drop of about 10%. This drop coincides with the economic recession in the mid-1980s, where the building and extractive industries were affected. From 1989, the TME per capita of the Langat Basin increased by about 20% annually up to 1996. In 1997, the TME dropped by about 60% due to the economic slowdown.

Aggregate is the largest contributor to the TME in the Langat Basin (Table 89.4). In the early 1980s, aggregates contributed to about 70% of the TME. This increased to about 90 to 95% in the mid-1990s. The national aggregate consumption for 1995 is 4.7 tons/capita, while in the Langat Basin the consumption for that year was nearly eight times higher, at about 37 tons/capita. This high level of aggregate production reflects the intense physical development in the Langat Basin, with the opening of the southern development corridor in Selangor. Aggregate is a basic resources that is required to support physical development to ensure societal well-being. However, the levels of aggregate consumption in the Langat Basin are unhealthy for the ecosystem. The aggregate depletion rates could not be established due to lack of data on the reserves. Based on the available data for 1980 to 1997, it is proposed that aggregate consumption be maintained between 11 and 14 tons/capita.

In contrast, the contribution of tin to the TME/capita has declined significantly since 1980 (Table 89.4). Tin extraction was one of the main economic activities in the Langat Basin where it contributed nearly 1% of the TME. In recent years, the contribution of tin to the TME is negligible. The contribution of clay and earth materials to the TME has reduced by about 5% annually between 1980 and 1996, while that of kaolin by 10% annually during the same period. On the other hand, the contribution of sand and gravel does not show any particular trend, ranging from 3 to 5% of the TME/capita.

Hidden Flows

Hidden flows (HFs) are associated with the extraction and processing of minerals. Hidden flows in the Langat Basin are associated with overburden and gangue (unused material) (Table 89.5). The

Table 89.5 Hidden Flows Associated with the Production of 1 Metric Tonne of Mineral

Aggregate Overburden: 5% Gangue: Minimal	Modified from Adriaanse et al. (1997); the high level of weathering in tropical climate generally results in very thick overburden; there is very little gangue associated with quarrying of rock aggregates
Sand and Gravel Overburden: Minimal Gangue: 2%	Modified from Adriaanse et al. (1997); there is very little overburden associated with the extraction of sand and gravel from ex-mining land; the washing of sand releases clay (gangue) into the drainage system
Clay and Earth Material Overburden: 1 tons/ton Gangue: Minimal	Modified from Adriaanse et al. (1997); the existence of rich residual soil enables the direct excavation of earth material; the amount of overburden removed is less than that recommended for clay in Adriaanse et al. (1997).
Kaolin Overburden: 3 tons/ton Gangue: 25%	Obtained from Adriaanse et al. (1997)
Tin Concentrate Overburden: 2 tons/per ton ore Gangue: 10%	Obtained from Adriaanse et al. (1997)

Adapted from Adriaanse, A. et al., in *Resource Flows: The Material Basis of Industrial Economies*, World Resources Institute, Washington, D.C., 1997; and field observation in the Langat Basin.

material removed is transported by stormwaters and cause siltation of river waters. In most long-term extraction sites, such as aggregate quarries and kaolin mines, silt traps are installed to mitigate this problem. Sand and gravel mines, which are generally small, short-term operations, normally do not install silt traps.

An examination of trends since 1980 shows that HFs associated with aggregate production are generally between 0.3 and 0.6 ton/capita except for the period between 1992 and 1996 where this number was almost doubled (Table 89.6). HFs associated with the extraction of clay and earth materials as well as tin has declined, while the reverse is observed for kaolin and sand and gravel. HFs associated with the total extraction of minerals in the Langat Basin have declined from about

Table 89.6 Hidden Flows (HFs) Associated with the Production of Minerals in the Langat Basin[a]

Year	Aggregate	C & E Material[b]	Kaolin	S & G[c]	Tin Conc.[d]	Total HF	% TME
1980	0.4	2.8	0.007	0.007	0.147	3.4	31
1981	0.4	3.3	0.007	0.008	0.137	3.9	31
1982	0.5	2.7	0.006	0.014	0.114	3.4	24
1983	0.5	2.8	0.008	0.014	0.086	3.4	26
1984	0.5	2.7	0.009	0.024	0.082	3.4	24
1985	0.4	2.4	0.010	0.013	0.064	2.9	25
1986	0.5	1.8	0.010	0.020	0.035	2.3	19
1987	0.4	1.0	0.011	0.023	0.055	1.5	15
1988	0.3	1.2	0.013	0.009	0.048	1.7	20
1989	0.4	1.9	0.011	0.011	0.047	2.3	22
1990	0.5	1.7	0.015	0.017	0.037	2.3	19
1991	0.6	2.0	0.020	0.015	0.028	2.6	18
1992	1.0	0.9	0.010	0.020	0.015	2.0	9
1993	1.1	0.8	—	0.028	0.010	1.9	8
1994	1.2	2.2	—	0.030	0.006	3.4	12
1995	1.8	0.9	0.036	0.019	0.005	2.7	7
1996	1.6	1.3	0.039	0.020	0.004	3.0	8
1997	0.6	—	—	0.027	—	0.6	4

[a] Units are in metric tons/capita.
[b] C & E Material = clay and earth material.
[c] S & G = sand and gravel.
[d] Tin Conc. = tin concentrates.
— Data not available.

30% of the TME in 1980, to about 20% in 1990, and less than 10% since 1995, while production has increased in proportion (Table 89.6). It appears that the overall efficiency of mineral resource extraction in the Langat Basin has improved.

STERILIZATION OF MINERAL RESOURCES

Aggregate Resources

In terms of area, about 35% of the Langat Basin has high potential for aggregate resources. The major high potential area is located in the upper eastern section of the Langat Basin (Figure 89.2). Four other high potential areas of limited extent have also been identified within the basin. In terms of land cover, it was revealed that 15% of the high potential area has been built-up, and its aggregate resources are effectively sterilized. About 54% of the high potential area is under water bodies and covered by forests, such as the Langat, Semenyih, and Galla Forest Reserves, and the resources therein are inaccessible. Only 31% of the high potential area in the Langat Basin is available for future exploration and exploitation. All the other areas with high potential within the basin are either sterilized or unavailable for exploration and exploitation.

Prior to 1981, the annual rate of aggregate sterilization by built-up areas in the Langat Basin was not significant (Table 89.7). The annual rate of aggregate sterilization between 1981 and 1998 was about 10%. Assuming that the forest cover and water bodies are retained, and that the focus of development remains on the high potential areas, it is estimated that total sterilization of aggregate

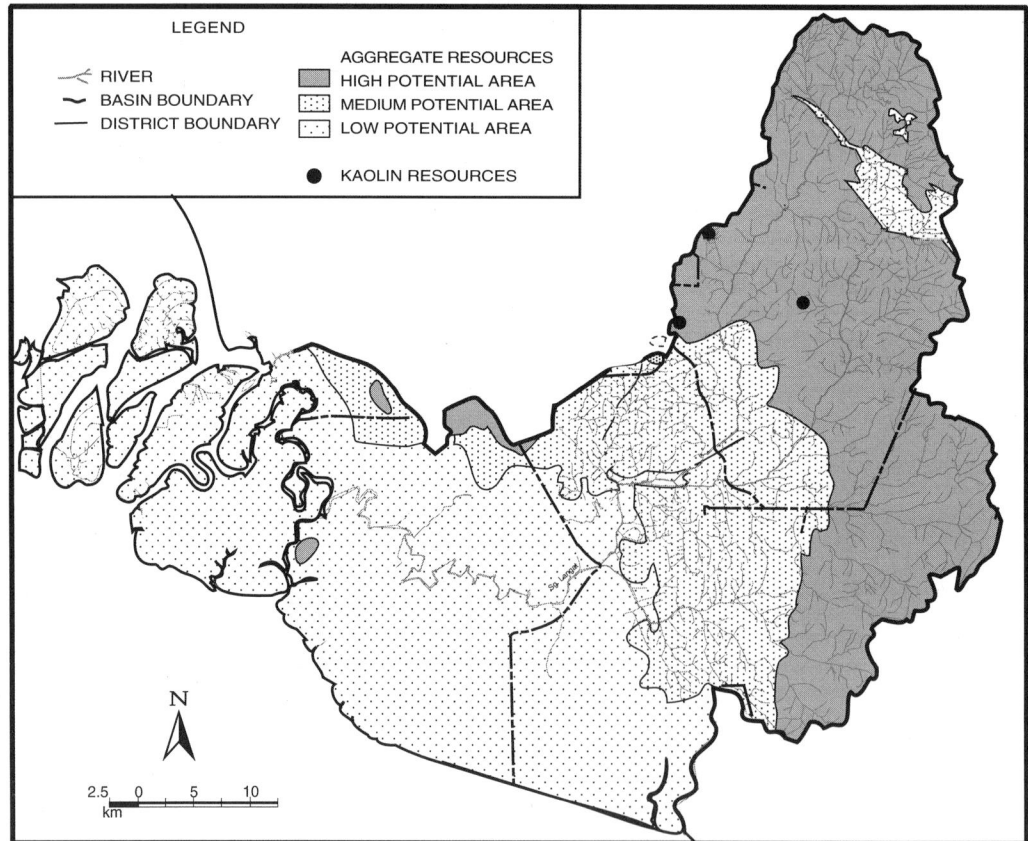

Figure 89.2 Potential areas for aggregate and kaolin resources in the Langat Basin.

Table 89.7 Land-Use Percentage of the High Potential Area for Aggregates, which Represents 35% of the Langat Basin (which has an area of 2938 km²)

Land Use	1974	1981	1995	1998
Built-up area	3	3	13	15
Forest	53	52	50	50
Water body	4	4	4	4
Agriculture-based	40	41	33	31
Total	100	100	100	100

The 1998 data are for land cover based on the TM Landsat image. All other data are derived from land-use maps produced by the Agriculture Department of Malaysia.

resources within the Langat Basin would occur within the next 15 years. As a result of this sterilization, the pressure to exploit high potential areas that are currently overlain by forest reserves will increase to support development in other parts of the basin. The removal of forest reserves in the Langat Basin for quarrying, particularly in the highlands, will adversely impact the overall health of the ecosystem.

Compounding the sterilization of potential aggregate resources, existing quarries are increasingly being encroached upon by housing development and industrial parks. This limits the expansion of such quarries to meet future demand in the Langat Basin. In addition, fly-rocks, dust pollution, vibrations, and sound pollution from quarry operations endanger humans when new development projects are located near existing quarry operations.

There is a pressing need to establish mineral land banks for aggregate resources. Mineral land banks refer to sites with substantial reserves of specific minerals or rock materials that can be extracted for future use. With the expected urbanization and development in the Langat Basin, the requirement for aggregates is expected to increase to meet the demand of the building industry. Mineral land banks should be delineated for aggregates to ensure that their supply is not threatened in the future. These land banks should be maintained for mineral development to sustain future physical development in the Langat Basin. In addition, buffer zones should also be identified around existing quarries. This is to prevent the encroachment of development that would limit the expansion of existing quarries.

Other Resources

About 12% of the national reserves of clay and earth materials (72 Mt), as well as 0.1% of kaolin (0.1 Mt), have been identified in the state of Selangor, where a major part of the Langat Basin is located (Loh and Hamadi, 1997). There are three areas containing kaolin resources in the Langat Basin (Figure 89.2). However, the reserves in the Langat Basin have not been inventoried. The availability of kaolin and clay and earth material resources is necessary to meet the demands of the expanding ceramics, advanced ceramics, and composite products industries in Selangor. Long-term planning should be considered to prevent the sterilization of existing kaolin and clay and earth material reserves in the Langat Basin.

A total of 60% of the sand and gravel in the Langat Basin is derived from rivers and about 40% are from ex-mining areas. There are no available data on the sand and gravel reserves in the Langat Basin. No study has been conducted on the requirement and availability of sand and gravel to support development within the basin.

Exploration for new tin deposits has declined substantially in the Langat Basin. The Kuala Langat area, located near the coast, is believed to contain significant tin resources. Unfortunately, with the depressed tin prices and the increasing competition for land, the area is under increasing pressure to be developed. Permanent nonmineral development in such areas will sterilize the existing tin resources and prevent its future extraction. The trade-off between using the land for the present population and denying the next generation an opportunity for wealth creation should be carefully studied and understood before the area is developed.

Potential Indicators to Assess Ecosystem Health

The health of an ecosystem is assessed using indicators, and a healthy ecosystem requires conditions that are sustainable. In the context of minerals, a healthy ecosystem would have sufficiently accessible basic resource stocks to support economic growth and societal well-being in an effective and equitable manner, and would experience minimal adverse environmental impacts from its extraction. In the case of the Langat Basin, aggregate and other building materials (e.g., sand and gravel and clay and earth materials) are considered basic minerals. These are essential to support physical development within the basin and to meet the needs of society. The lack of these minerals may threaten future infrastructure development or make it more costly if these minerals have to be imported into the Langat Basin. Minerals such as kaolin and tin are also important, but more so for the creation of wealth. Indicators that are useful for the assessment of ecosystem health in the context of minerals can be divided into three major categories: mineral availability and consumption, mineral extraction, and minerals management.

Availability and Consumption of Minerals

Indicators that are useful to gauge the availability of minerals in the Langat Basin are the amount of reserves. Currently, data are not available for all the minerals extracted in the Langat Basin. Kaolin is important for the resource-based sector and the advanced materials sector under the Second Industrial Master Plan for Malaysia. The Langat Basin has been delineated to host small- and medium-sized industries that will form the backbone of economic development over the next two decades. The availability of data on kaolin reserves will encourage investment in the advanced materials sector. Data on mineral reserves will facilitate long-term planning, with respect to the economic sustainability of the Langat Basin. In the case of aggregates, data will assist in determining the amount of mineral landbanks that should be established.

Another useful indicator related to the availability of mineral resources is rates of sterilization. The problem of mineral sterilization is not being seriously addressed in the Langat Basin, particularly for aggregates and other building materials. The high rate of mineral sterilization of high potential aggregate areas threatens the health of the ecosystem as the high potential areas currently locked under forest reserves may be exploited to meet the demand for aggregates in about 15 years into the future.

The Langat Basin is a net importer of minerals, but levels of actual mineral consumption could not be established. Only the minerals that have been extracted from the basin have been examined, with respect to their consumption in the various economic sectors. The consumption level of aggregates within the basin has been extremely high between 1992 and 1996, at levels between 22 and 37 tons/capita. This indicator on aggregate consumption reflects the intense physical development within the basin due to rapid economic growth during that period. Reasonable levels of aggregate consumption for the basin would be between 11 and 14 tons/capita.

There are other indicators useful in relation to ecosystem health but these have not been examined in detail. These include recycling and substitution rates. At the ecosystem level, the only way to estimate the quantity of the minerals that are recycled is by extrapolating its use from industries already existing in the basin.

Extraction of Minerals

Indicators relating to the extraction of minerals serve to assess environmental impacts that are associated with the extraction of minerals and the direct effect of extraction on the health of the ecosystem. The indicator identified from this study is the amount of hidden flows associated with the overburden and gangue during the extraction and processing of the minerals. Other hidden flows not considered in this study include:

- Loss of biodiversity
- Emissions into the air
- Discharge into waterways
- Accidental spillage during transfer of substances (e.g., explosives and diesel during mineral extraction)

Indicators directly related to the well-being of humans are the number of quarry-related occupational accidents, the number of and types of complaints related to quarrying activities, such as fly rock and subsidence from residents, and the number of violations of conditions set in operating licenses, as well as environmental infractions.

Management of Minerals

The maintenance of ecosystem health requires effective management of minerals with respect to their availability, consumption, and extraction. The indicators required for this purpose should measure the effectiveness of actions taken rather than the number of or type of action taken. For instance, reduced frequency of encroachment of housing and industrial development onto areas adjacent to quarries would indicate the effectiveness of the planning process within the basin. Other indicators of increased effectiveness in the planning process would be mineral sterilization rates, increased area for mineral land banks and buffer zones, and reduced number and amount of ex-mining areas. Examples of indicators showing improved management on the part of the industry would include:

- Reduced number of violations of conditions set in operating licenses, as well as environmental infractions
- Reduced consumption of energy and water
- Reduced emissions
- Occupational accidents and land used for waste disposal
- Increased tendency to self-regulate through certification to ISO 14000 and other standards

CONCLUSION

In terms of the balance of mineral flows, the Langat Basin is a net importer of nonrenewable resources. The minerals extracted from the Langat Basin for the period of 1980 to 1997 are aggregates, kaolin, clay and earth materials, sand and gravel, and tin concentrates. Aggregate is the largest contributor to TME/capita, reflecting the intense physical development in the Langat Basin. The high rate of aggregate sterilization threatens the health of the ecosystem, as the high potential areas now locked under forest reserves may be exploited in the near future. An examination of trends since 1980 shows that hidden flows in the Langat Basin have reduced from about 30% of the TME to less than 10%. It appears that the overall efficiency of mineral resource use in the Langat Basin has improved. Parameters such as estimates of reserves, rates of sterilization, rates of mineral consumption, TME, HFs, as well as total areas of mineral land bank, buffer zones, and ex-mining land were found to be useful indicators for ecosystem health assessment in the Langat Basin.

ACKNOWLEDGMENTS

This chapter is part of the research funded by MATREM CP/5220–97–02, a project of the United Nations Environment Programme's Network for Environmental Training at Tertiary Level in Asia and the Pacific (NETLAPP). The encouragement and support of Professor Mohd Nordin Hj. Hasan, the project leader, is gratefully acknowledged.

REFERENCES

Adriaanse, A., Bringezu, S., Hammond, A., Moriguchi, Y., Rodenburg, E., Rogich, D., and Schutz, H., *Resource Flows: The Material Basis of Industrial Economies*, World Resources Institute, Washington, D.C., 1997.

Bertollo, P., Assessing ecosystem health in governed landscapes: a framework for developing core indicators, *J. Ecosyst. Health*, 4, 33–50, 1998.

Costanza, R., Toward an operational definition of health, in *Ecosystem Health: New Goals for Environmental Management*, Costanza, R., Norton, B., and Haskell, B., Eds., Island Press, Washington, D.C., 1992, pp. 181–204.

Department of Statistics, Annual census of mining and stone quarrying, Kuala Lumpur, Malaysia, 1996.

Geological Survey Department Malaysia (GSD), *Malaysian Minerals Yearbook 1995*, Kuala Lumpur, Malaysia, 1995.

Geological Survey Department Malaysia (GSD), *Malaysian Minerals Yearbook 1996*, Kuala Lumpur, Malaysia, 1996a.

Geological Survey Department Malaysia (GSD-IM), Industrial Mineral Production Statistics and Directory of Producers in Malaysia 1996, Kuala Lumpur, Malaysia, 1996b.

Geological Survey Department Malaysia (GSD-IM), Industrial Mineral Production Statistics and Directory of Producers in Malaysia 1997, Kuala Lumpur, Malaysia, 1997.

Geological Survey Department Malaysia (GSD-IM), Industrial Mineral Production Statistics and Directory of Producers in Malaysia 1998, Kuala Lumpur, Malaysia, 1998.

Hindon, A., Policies, incentives and opportunities for investment in the ceramic industry, in *Opportunities in the Clay-Based Industries*, Hasnida, Z. and Khoo, K.K., Eds., Geological Survey Department, Malaysia, Kuala Lumpur, Malaysia, 1997, pp. 1–8.

Loh, C.H. and Hamadi, C.H., Clay resources of Malaysia: an overview of Geological Survey Department's findings to date, in *Opportunities in the Clay-Based Industries*, Hasnida, Z., and Khoo, K.K., Eds., Geological Survey Department, Malaysia, Kuala Lumpur, Malaysia, 1997, pp. 9–18.

Natural Resources Canada, Sustainable development criteria and indicators for minerals and metals: moving from words to action, *Proceedings of an International Experts Workshop*, Quebec, May 7, 1998.

Peterson, P., Indicators of sustainable development in industrializing countries, Vol. 1: *Management Response Strategies*, Penerbit UKM, Bangi, Malaysia, 1997.

Schaffalitzky, C., Recent developments in the classification of mineral reserves and resources, in *Case Histories and Methods in Mineral Resource Evaluation*, Annels, A.E., Ed., Geological Society, London, 1992.

Selangor State Government, *Strategi Pembangunan Mampan dan Agenda 21 Selangor — Kawasan Sensitif Alam Sekitar*, Institute for Environment and Development (LESTARI), Bangi, Malaysia, 1999.

Forest Health Monitoring and Restoration

CHAPTER 90

Overview: Forest Health Monitoring and Restoration

Bruce A. Wilcox

Among all ecosystems, forests are simultaneously the most globally extensive, studied, historically exploited, and the most subject to management. It follows that forest health monitoring should be the most advanced of any ecosystem health monitoring efforts. Generally speaking, that is the case. As described elsewhere in this volume (see Wilcox et al., Chapter 32) more than 150 counties are involved in initiatives to define national-level criteria and indicators for forest management, including forest health and vitality. However, as also suggested, monitoring programs actually employing these indicators generally have not yet been widely implemented. Further, as the reader will discover from the chapters in this section, forest health monitoring is a highly complex specialty requiring not only a transdisciplinary approach, but one that crosses ecological and institutional scales, ranging from the local management unit or forest patch to an entire forest biome. Of the four chapters in this section, two based on work in Australia and Switzerland provide restoration and management policy perspectives bearing on monitoring at the forest or stand level. The other two describe national forest monitoring programs at the country (Canada) and regional levels (West Coast of the U.S.). As a group, these chapters provide a wide breadth of coverage on the science and policy issues bearing on forest health monitoring.

Brang, in his chapter, gets to the heart of the concept of forest health from a forest stand perspective, drawing on current understanding of forest dynamics in Swiss coniferous forestland. The chapter argues that determining the components, structures, and processes necessary for ecological stability essential to the long-term delivery of products and services lies at the core of the challenge of forest health monitoring. In the context of Switzerland's forests, whose intensive utilization for centuries has diminished the flow of human benefits, forest monitoring necessarily focuses on restoration of health. This situation forces Switzerland to explicitly confront questions such as which products and services are desired of a given forest, thus defining health according to social values, and, in turn, what ecosystem characteristics are required for their long-term delivery.

Calver picks up where Brang leaves off by dealing with how to determine whether forest health is significantly impaired once stakeholders agree on these characteristics. Focusing on management issues within jarrah forest in western Australia, Calver demonstrates how the Precautionary Principle can be used to establish quantitative standards for determining impairment of health, given a particular level of impacting activity (e.g., logging). The approach, along with others under development, represents an important area of progress in the field. It demonstrates that, despite the subjectivity inherent in deciding when an ecosystem is healthy and despite the imperfect state of ecological knowledge, establishing objective standards of measurement is nonetheless possible.

1-56670-612-2/03/$0.00+$1.50
© 2003 by CRC Press LLC

Hirvonen describes Canada's national program for forest ecosystem health indicators. Given Canada's immense size and extent of forested land, and its historic role in leading international ecosystem health and sustainability initiatives, the experience of Canada in developing a monitoring framework provides valuable lessons about the scientific and policy challenges of monitoring forest health. Paramount among the challenges is how to design a system for gathering data that can be meaningfully aggregated, accounting for both spatial and temporal scale differences, workable at the country level, as well as meeting international reporting requirements.

And finally, Campbell et al. focus on the U.S. Forest Service's present monitoring efforts on the West Coast. By contrast to Hirvonen's panoramic discussion of Canada's program and the monitoring system design challenges in general, Campbell et al. bring the focus of section back down to the forest or stand level. They describe the U.S. West Coast program in terms of the specific indicators being measured to indicate ecosystem change and stress, for example, as well as those pertinent to societal values.

Overall, the chapters in this section reflect how far we have come toward incorporating the concept of ecosystem health in forest management science and policy. They also show that, perhaps more so than in other areas dealing with the conservation of global biodiversity, the attempt to distinguish and reconcile natural attributes and social values of ecosystems has been carried further for forests. Yet we also see from these chapters that forest health monitoring is still in its infancy.

CHAPTER 91

Resistance and Elasticity: A Conceptual Framework for Managing Secondary Forest Ecosystems in Switzerland

Peter Brang

INTRODUCTION

The purpose of this chapter is to propose a novel approach to the design of **human disturbance*** regimes in forests. Human[§] disturbances are sudden changes in forest composition and structure caused by forest management, while **natural disturbances** are those caused by natural agents such as storm, snow load, and bark beetles. The dynamics of managed forest ecosystems need to be restricted to those system states where desired products and services are continuously provided. These restrictions need to be both sufficiently narrow to ensure that human demands are met, and sufficiently large to allow for **ecosystem dynamics** to occur. Implementing a strict regime to control ecosystem dynamics has led, in some examples, to long-term management failures (Holling, 1987). To what extent such failures generally occur is an open question.

In order to ensure the supply of products and services, managers should strive to create forests that are both resistant to changes that lead to unacceptable states and resilient (Grimm and Wissel, 1997) (i.e., able to return from unacceptable to acceptable states after disturbances). The speed of this return has been termed "**elasticity**" (Grimm and Wissel, 1997).

In this chapter, I will show that in Switzerland and elsewhere in Europe the **resistance** and elasticity of forest ecosystems have been degraded by inadequate management practices. I will demonstrate this by comparing these man-made forests with remnants of **old-growth forests** on similar sites, most of which are located in Slovakia (Korpel, 1995, 1997). These comparisons reveal important differences in **ecosystem composition, ecosystem structure**, and **processes**, which are causing management problems today. Solutions to these problems may lie in partially restoring old-growth composition and structure to **secondary forests**. I explore how the concepts of resistance and elasticity can be used to evaluate the extent of such restoration.

---

* Terms in bold type are defined in the glossary at the end of the chapter.

CURRENT USES OF FORESTS IN SWITZERLAND

In Switzerland, 68% of the forest is publicly owned (Brassel and Brändli, 1999), and most forests are managed for multiple products and services. The main forest product is timber, and the main services are protection against natural hazards (e.g., snow avalanches, rockfall, torrents, and soil erosion), recreation, and conservation. Although most forests serve multiple purposes, providing protection against natural hazards is often the most important. It is the main service in 8 to 45% of the forest area in the Swiss mountains, depending on the region and the definition of a protection forest (Brassel and Brändli, 1999).

HUMAN IMPACT HAS SHAPED SWISS FOREST ECOSYSTEMS

For centuries, human activities have strongly shaped almost all forests in Switzerland. People have used the forests for timber harvesting (including clear-cutting and permanent deforestation), grazing, litter-raking, and hunting. Only 20 ha of old-growth forest remain. Air pollution and anthropogenic climate change have emerged as new influences during the past few decades. As a result, Swiss forests are secondary forests. Their compositions, structures, and processes differ markedly from those of old-growth forests (Figure 91.1).

Important examples of changes in ecosystem composition are reductions in forest area, growing stock, and volume of coarse woody debris, and an altered species composition, which has shifted toward *Picea abies* (L.) Karst. The average growing stock of all Swiss forests is currently around 366.5 ± 3.5 m³/ha (mean $\pm$ SE; Brassel and Brändli, 1999). This is much less than the 500+ m³/ha reported for Slovakian old-growth forests (Korpel, 1997) on the majority of those sites that are similar to those found in Switzerland. The low-growing stock in Swiss forests is due to the underrepresentation of late seral stages in managed forests (Burschel and Huss, 1987).

The volume of coarse woody debris in Swiss forests amounted to 11.9 ± 0.5 m³/ha (mean $\pm$ SE; Brassel and Brändli, 1999) around 1994. It has certainly increased since, especially as a result of Hurricane Lothar in 1999, but is still much smaller than the volumes of 50 to 280 m³/ha reported for Slovakian old-growth forests (Korpel, 1997), boreal old-growth forests (27 to 201 m³/ha; Linder et al., 1997), and for temperate old-growth forests (60 to 1189 m³/ha; Harmon et al., 1986).

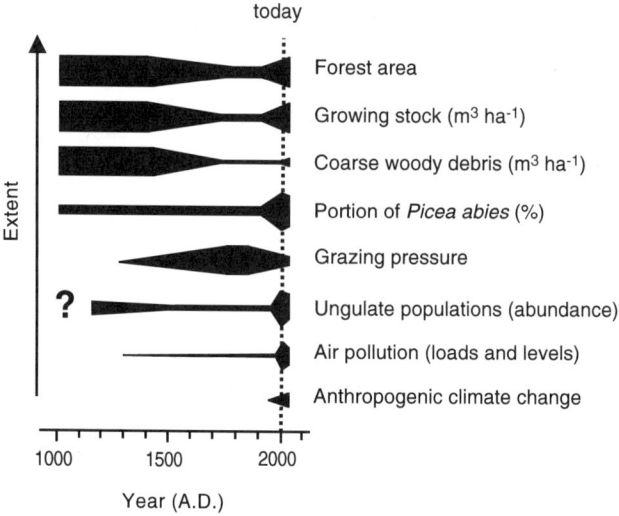

Figure 91.1 Qualitative evaluation of the magnitude of human-mediated disturbances and their effect on Swiss forest ecosystems during the past 1000 years, and projected through the year 2040.

In comparison with what is assumed to be the species composition in old-growth forests, *P. abies* is overrepresented in Swiss forests. Using a site-classification model, Kienast et al. (1994) classified 6793 forest inventory plots out of a total of 10,400 plots as sites where deciduous trees should be dominant. However, inventory data showed that one fifth of these plots had more than 75% basal area of *P. abies*, and two fifths had more than 75% basal area of *P. abies* or other coniferous species. In contrast, *Abies alba* Mill. is underrepresented. This shade-tolerant species is not competitive in large open areas and is much more heavily browsed by ungulates than most other tree species (Liss, 1988). Its distribution was greatly reduced by large clear-cuts and grazing in the 19th century, and by the intense browsing pressure of recent decades. In eastern Bavaria, a region more heavily impacted than most parts of Switzerland, the proportion of *A. alba* was reduced by 75% over 200 years (Eichenseer, 1997). The two examples of *P. abies* and *A. alba* show that the current species composition deviates considerably from the species composition of old-growth forests.

The stand structures are much more uniform in the current man-made forests than in old-growth forests. On most sites, the remaining old-growth stands in central Europe are structurally diverse, with patch sizes rarely exceeding 1 ha (Zukrigl, 1991). In the Swiss Forest Inventory, 68% of the plots were classified as having uniform stand structure (Brassel and Brändli, 1999). There are no representative data from old-growth forests on similar sites. However, many case studies in Slovakian old-growth forests are available (Korpel, 1995). Given those studies, I conclude that without human interference, over three quarters of Switzerland's forests would be structurally diverse. This structural diversity would be found in *Fagus sylvatica* L. forests, mixed *A. alba–F. sylvatica* and *P. abies–A. alba* forests, and subalpine *P. abies* forests (Brassel and Brändli, 1999). The increase in uniform stand structure is the result of clear-cutting, afforestation, and repeated thinning (Ott et al., 1997).

Ecosystem processes also differ between secondary and old-growth forests. Tree growth was probably adversely affected by centuries of litter-raking and other nutrient-exporting practices (Spiecker et al., 1996; Küchli, 1997), which decreased site productivity. Recently, an increase in tree growth was reported for many parts of Europe (Spiecker et al., 1996). The abandonment of practices that depleted nutrient stocks coupled with increased nitrogen deposition, elevated CO_2 concentrations, and higher temperatures is likely to account for the increased tree growth (Spiecker et al., 1996), although definite proof is lacking. Other examples of ecosystem processes that have been interfered with include tree death and decay, which have been short-circuited by timber harvesting, and only partly been allowed to take place again in recent years. Even if no timber were harvested from now on, which is unacceptable, it would take many decades, perhaps centuries, to accumulate the amount of coarse woody debris characteristic of old-growth forests.

The lack of seedbeds created by decaying logs has also inhibited tree regeneration, another ecosystem process, on wet sites, particularly for those species that preferentially establish on nurse logs (Korpel, 1995). Grazing and browsing have also had a negative impact on regeneration. Although grazing pressure from domestic livestock has decreased in recent decades, browsing by wild ungulates has increased, mainly by *Cervus elaphus* L., *Capreolus capreolus* L., and *Rupicapra rupicapra* L. The absence of large predators, such as the wolf, lynx, and bear, and decreasing poaching have allowed ungulate populations to increase (e.g., Righetti and Huber, 1983; Brückner, 1993). Further exacerbating the situation, agricultural development has reduced the area of suitable ungulate habitat, thereby forcing animals into forested areas. These factors are serious obstacles to forest renewal.

Many of the reported changes in forest ecosystem composition, structure, and processes have been substantial. The current man-made forests clearly differ from old-growth forests, and must be considered secondary forests.

RELEVANT ECOSYSTEM CHARACTERISTICS HAVE BEEN IMPAIRED

Differences in the ability of old-growth and secondary forests to meet human demands cannot be inferred from differences in ecosystem characteristics. Indeed, many forests have been deliberately and successfully altered to enhance their ability to provide specific products or services. This is the case with thinnings, for instance, that are aimed at enhancing wood quality and stand resistance to natural disturbances such as windfall and snow break (Slodicák, 1995).

Nevertheless, many of the changes in Swiss forest ecosystems must be seen as the unintended side effects of other activities, which have impaired, rather than enhanced, the forests' ability to meet human needs in the long-term. A striking example of this is the reduction in *A. alba*. Previous loss of *A. alba*, and its continuing elimination by browsing, have greatly contributed to the current and future regeneration difficulties in many forests (Ott et al., 1997). This species has several silvicultural properties that facilitate its management, and make it an important component of forests managed for protection against natural hazards. It is rarely attacked by bark beetles, in contrast to the highly susceptible *P. abies*. Unlike most other tree species, *A. alba* is able to establish in dense shade (Mayer, 1984). It can avoid strong competition from herbs, grasses, and shrubs that need more light (Ott et al., 1997). It is also much less dependent on the availability of coarse woody debris as a seedbed than *P. abies* (Mayer-Wegelin and Schulz-Brüggemann, 1949; Hunziker, 1999; but see Mayer et al., 1972).

The regeneration problems for *P. abies*, the most important tree species in Switzerland, which have been caused by the loss of coarse woody debris, are another example of unintended side effects of past management practices. As a consequence, natural regeneration is often sparse, and costly investments in planting are necessary. Moreover, management has often promoted uniform stand structures, which are more susceptible to large-scale disturbances, such as windfall and bark beetle damage (Mayer, 1984), and therefore less able to protect against natural hazards and to produce timber. This has happened where pure stands of *P. abies* have been established on sites that, without human intervention, would have supported mixed stands of *P. abies–A. alba* or *A. alba–F. sylvatica*, which usually exhibit a small-scale natural disturbance regime (Korpel, 1995). The negative effect of the lack of old trees on species diversity (Perry, 1994; Samuelsson et al., 1994) is another unwanted side effect of past forest management. Reduced species diversity means that the forest is less valuable as a habitat.

RESTORATION — WHAT SHOULD BE RESTORED AND TO WHAT EXTENT?

As we have seen in the last section, some of the side effects of past forest management are undesirable. Restoration efforts are needed, and such efforts are about to be made in Switzerland and elsewhere in Europe. However, it remains unclear which ecosystem characteristics should be restored, and how far restoration should proceed toward a more natural condition. What is the goal of ecosystem restoration in this case?

I propose that, for the managed forests in Switzerland, the restored ecosystem characteristics should be components, structures, and processes that are crucial for the long-term delivery of products and services. I elaborate on these characteristics in the section, "Restoration Should Enhance the Resistance and Elasticity of Forest Ecosystems."

Full restoration of old-growth forest dynamics would entail undesirable fluctuations in the delivery of products and services (De Leo and Levin, 1997). Like some human disturbances (see above), some natural disturbances can impair management goals. Large insect outbreaks, for instance, cause considerable economic damage, and affected stands may provide ineffective protection against natural hazards. The same applies to disturbances, such as storms and fires, that have the potential to affect forest ecosystems on a large scale. Forest management aims to constrain forest dynamics within specific limits. These limits include only a portion of the possible states of a forest, as expressed by composition and structure. Forests should be kept in a stable condition — although the inherently dynamic nature of ecosystems (Holling, 1996) has to be taken into account. This means that the goal is not to restore old-growth forest dynamics entirely, but only to such an extent that human demands can be better met than they are today. In contrast to past management, the long-term side effects of forest management on the ecosystem itself (De Leo and Levin, 1997) and on adjacent systems need to be considered. Such managed ecosystems can be called healthy, according to the utilitarian definition of ecosystem health (Kolb et al., 1994; Kimmins, 1996), or be said to have integrity (De Leo and Levin, 1997).

CURRENT APPROACHES TO DESCRIBING FOREST DYNAMICS FOR MANAGEMENT PURPOSES ARE OVERLY SIMPLISTIC

To cope with the dynamic nature of forest ecosystems, forest managers need to understand their dynamics. They also need to know which ecosystem characteristics are most important in influencing management goals, and the thresholds beyond which these goals could be impaired. This understanding takes as its starting point a description of the patterns at the stand and landscape scales and of the factors causing them (Brang et al., 2001). From this, the forest's future dynamics as a result of natural disturbances can be predicted. Finally, the management actions that are most likely to retain the system within the desired limits, or to bring it to such a state, can be designed.

Intuitively, forest managers have followed these steps for decades in Switzerland. However, there is still much guesswork involved. Long-term forest dynamics are often poorly understood so that predictions about future dynamics tend to be highly uncertain, and management tends to follow the lines of tradition and belief. Moreover, managers often delineate the system too narrowly, and fail to consider the side effects described above. In contrast, those limits where management goals are impaired seem to be better known. For instance, site- and hazard-specific target values for forest condition have been established for forests that provide protection against natural hazards (Wasser and Frehner, 1996).

A general framework for describing those aspects of long-term forest dynamics that are relevant for forest management has yet to be developed. Yield tables have been used to predict stand growth (Assmann, 1970). The coefficient of slenderness of the stem, crown length, stand age, species of tree, stem rot, and variables characterizing the site conditions have been used to explain and to predict a tree's susceptibility to windthrow and snow break (Rottmann, 1985; Schmid-Haas and Bachofen, 1991; Slodicák, 1995; Nykanen et al., 1997; Herold and Ulmer, 2001). Predictions from these studies have, however, usually been very rough, with the outcomes predicted in a few classes, such as high, intermediate, and low risk. Several ecologists have developed models to help understand and predict forest dynamics better (Shugart, 1984; Kräuchi and Kienast, 1993; Bugmann and Cramer, 1998), but most of these models have not been aimed at solving management questions (but see Erni and Lemm, 1995). Risk-based assessments (Mitchell, 1998; Pukkala, 1998) have come closer to getting to the root of the problem by providing predictions of how likely specific outcomes of *leaving* and *doing* are. *Leaving* means simply accepting natural disturbances as they occur, whereas *doing* means altering the natural disturbance regime by introducing human disturbances.

ECOLOGICAL STABILITY PROPERTIES: A USEFUL APPROACH FOR DESCRIBING FOREST DYNAMICS

The concept of *ecological stability* (e.g., Holling, 1973) has received great attention in ecology, and it may also be useful for describing forest dynamics. However, the term itself has caused so much confusion that Grimm and Wissel (1997) wrote a paper titled, "Babel, or the Ecological Stability Discussions: An Inventory and Analysis of Terminology and a Guide for Avoiding Confusion." They advocate avoiding the term ecological stability and referring to "ecological stability properties" instead, which include the basic concepts of (1) **constancy**, (2) **resilience**, and (3) **persistence**. Constancy includes the property resistance, and resilience the property elasticity.

Rather than deal with the broad properties of constancy, resilience, and persistence, I would like to focus on resistance and elasticity for the following reasons. Constancy occurs only on short timescales or for the most inactive system components such as the bedrock chemistry. Resilience as a general concept is also not very relevant in central Europe since the ability to return to predisturbance forest dynamics seems to be high for most disturbances (see below) and most forest ecosystems. Evidence for this is the fast recovery after the mid-19th century of many forests that had been heavily exploited prior to 1850. In contrast, resilience is important in ecosystems that are vulnerable to long-term degradation, such as many Mediterranean forests. In central Europe, however, only a specific aspect of resilience is relevant, namely, the *speed* of return to referential states or dynamics (i.e., elasticity). Persistence is a less important issue since the resilience of the forests seems to be high, and the existing forest area is protected by strict legislation.

High resistance means that the *probability* that the desired service or product of a forest will not be delivered is reduced, and vice versa. An estimation of how resistant a particular forest is to various disturbances is part of an assessment of the risk that products and services will be impaired. A resistance assessment contributes to an ecological risk assessment (Cairns, 1998). High elasticity, on the other hand, means that the *period* during which the desired service or product of a forest may be unavailable is short, and vice versa. An elasticity assessment is needed since disturbances will sometimes impair the delivery of the desired effects or products of a forest, and managers will want to know how long the gap in service will be.

RESISTANCE AND ELASTICITY APPLIED TO FOREST ECOSYSTEMS

A stability statement, such as "resistance and elasticity are important for forest management in central Europe," is not helpful unless I specify the ecological situation or context (Grimm and Wissel, 1997). Do I mean the resistance of a landscape to changes in forest patch patterns? Am I addressing the resistance of an ant to a leaf falling to the ground? Grimm and Wissel stress the importance of defining the ecological situation to which a stability statement applies and provide a checklist with six characteristics (Table 91.1). In the context of forests managed to provide specific products and services, the answer to many of the questions about resistance and elasticity is given by the scale of these products and services.

To apply the concepts of resistance and elasticity usefully to the management of forest ecosystems in Switzerland, those disturbances that can affect management aims should be known, as should how and to what extent they affect these aims. The most important **disturbing agents**, the paths through which they disturb the tree layer, their occurrence and effect on the ecosystem, and their impact on two management aims (protection against natural hazards and timber production) are listed in Table 91.2. The disturbing agents are interconnected. The occurrence of the disturbances varies greatly in space and time and is species and site specific. Some impacts on the delivery of productions and services are immediate, while others become noticeable only after a time lag.

RESTORATION SHOULD ENHANCE THE RESISTANCE AND ELASTICITY OF FOREST ECOSYSTEMS

I proposed earlier that, for managed forests, the ecosystem characteristics to be restored are the components, structures, and processes that are crucial for the long-term delivery of products and services. Having seen the importance of resistance and elasticity, we can narrow our search for the ecosystem characteristics to be restored in central European forests to those that confer resistance and elasticity on a forest ecosystem. For the most widespread disturbing agents, a preliminary list of these characteristics is given in Table 91.3. Some of these characteristics can be altered by management and others not. The most important means of influencing resistance to disturbances

Table 91.1 Checklist for Ecological Stability Properties Applied to Forest Management

Characteristics of the Ecological Situation	Question to Be Answered	Possible Answers in the Context of Forest Management[a]
Level of description	On which hierarchical level of an ecosystem are we describing stability?	Individual to landscape, depending on the products and services to be delivered; for products and services that depend directly on trees, the population level is used to describe ecological stability properties
Variable of interest	Which (measurable) state variables are we referring to?	Indices of the spatial distribution of trees, dbh (diameter of breast height) distribution, distribution of gap sizes and orientations, volume increment, estimated nutrient cycling rates
Reference state or dynamics of the variable of interest	What is the state variables *acceptable*[b] behavior?	*Acceptable* values of state variables depend on management aims that are defined in order to maintain a permanent delivery of products and services
Disturbing agents and disturbances	Which disturbing agents are affecting the ecosystem's behavior, and to what extent?[c]	Disturbing agents that can cause the unwanted death of trees (storm, snow load, bark beetles), or that have the potential to alter tree species composition, and growth and mortality patterns (pathogenic fungi, extreme climatic events, bad management practices, browsing ungulates), frequency and magnitude of the disturbances
Spatial scale	To which spatial scale does the stability statement refer?	Tree to stand level (10–1000 m)
Temporal scale	To which temporal scale does the stability statement refer?	Years to several centuries

The table describes how the ecological situation can be defined to make valid ecological stability statements for managed forest ecosystems.

[a] All answers depend on the goals of the particular forest management.

[b] Grimm and Wissel (1997) focus on *normal* reference states or dynamics without external influences. This seems inappropriate in the context of an ecosystem managed for specific purposes. *Acceptable* behavior seems to be more precise.

[c] Grimm and Wissel (1997) define disturbance as the result of dynamics exceeding reference dynamics. In the context of forest management, reference dynamics are dynamics where acceptable levels of disturbance are not exceeded. Management usually aims at replacing part of the highly variable natural disturbances by more predictable human disturbances.

Adapted from Grimm, V. and Wissel, C., *Oecologia*, 109, 323, 1997. With permission.

that directly affect the tree layer involves changing the composition of the tree species since the susceptibility of trees to disturbances varies among species. Another less influential option lies in changing the stand structure. Tree regeneration is the key to the elasticity of the tree layer. Resistance to soil acidification is a special case since it does not directly affect the above-ground part of the trees, but rather affects their roots through changes in soil chemistry.

Interestingly, the majority of the characteristics outlined in Table 91.3 are also characteristics of old-growth forests. This supports the hypothesis that, at least in central Europe, **natural disturbance** regimes lead to forests that have a fairly high resistance and elasticity to disturbances, and justifies restoring more **naturalness** in secondary forests. There is also some empirical evidence for this: old-growth forests are often less affected by disturbances, such as windfall, than adjacent secondary forests (Zukrigl, 1991). However, some old-growth forests show characteristics that make them less resistant and elastic than well-managed secondary forests. One example is the occurrence of pure *P. abies* old-growth forests on productive sites that are susceptible to large-scale disturbances (Korpel, 1995), in particular windfall and bark beetle mortality. This shows that even if our aim is to maximize resistance and elasticity, managed forests should not be made to resemble old-growth forests absolutely. Naturalness is not a goal in itself in managed forests. A certain naturalness, however, is a means of enhancing the probability that the long-term delivery of products and services is ensured.

Table 91.2 Most Important Disturbing Agents, Their Path of Effect, and the Effect of Disturbances Caused by These Agents, for a Forest Providing Protection against Natural Hazards, and a Forest Used for Timber Production (both in Switzerland)

Disturbing Agent	Path of Effect	Effect of Disturbance on Tree Layer	Impact on Management for Protection against Natural Hazards	Impact on Management for Timber Production
Storm	Direct impact on trees	From single treefalls to gaps > 10 ha; lower resistance to bark beetles	Protection impaired, often increased management costs	Lower wood quality, economic losses
Snow load	Direct impact on trees	From single treefalls to gaps < 1.0 ha; lower resistance to bark beetles	Protection impaired, increased management costs	Lower wood quality, economic losses
Bark beetles	Direct impact on trees, triggered by presence of weakened trees, often after storm or snow break	From single trees killed to gaps >1.0 ha; lower resistance to other disturbances	Protection impaired, increased management costs	Lower wood quality, economic losses
Ungulates	Browsing and fraying on tree regeneration	Long-term changes in tree species composition, slower juvenile stand development; lower elasticity in the face of other disturbances	Protection impaired in the long term, increased management costs	Economic losses as a result of slower juvenile stand development
Pathogenic fungi	Root rot, stem rot	Mortality of single trees, lower resistance of single trees to storm, snow load, and bark beetles	In most cases no direct short-term consequence	Economic losses
Acid deposition	Decrease of ratio base cations/Al, soil nutrient leaching, in advanced stages damage to roots and mycorrhizae	Shallower roots, lower resistance to storm and snow load, nutrient deficiencies; growth reduction	No direct short-term consequences, in the long term increased management costs and impairment of protection	No direct short-term consequences, in the long term economic losses
Drought	Reduced tree vigor, making trees susceptible to bark beetle attack, seedling mortality	Lower resistance to storm, snow load, and bark beetles, growth reduction, lower elasticity in the face of other disturbances	No direct short-term consequences	Economic loss
Inappropriate harvesting practices	Damage to remaining stand, infection by wood-decaying fungi, root and stem rot, soil compaction	Lower resistance to storm and snow load, growth reduction	No direct short-term consequences, in the long term increased management costs and impairment of protection	Lower wood quality, economic losses

All disturbances vary highly in time and space and among sites and tree species.

Table 91.3 Characteristics of Swiss Forest Ecosystems and of Their Management That Make the Tree Layer More Resistant and More Elastic in the Face of Widespread Disturbing Agents

Disturbing Agent	Characteristics Increasing Resistance	Characteristics Increasing Elasticity
Storm and snow load	Highly structured stands with long internal edges, small-sized trees, tree species not susceptible to stem and root rot, low gravity point, low coefficient of slenderness of the stems, long crown, no silvicultural treatments creating edges exposed to storm, no thinnings that lower the individual or group stability of trees, *low site productivity (unless soils are waterlogged), aspect not exposed to prevailing winds (applies only to storm), deep soil*	Presence of tree seedlings waiting for release; if tree seedlings are absent, presence of seed-producing trees, presence of safe sites for seedling establishment, low competition from vegetation, low browsing pressure, *no drought during germination and early establishment, high genetic variability of the regeneration*
Bark beetles	Tree species not susceptible to bark beetles, mixed-species stands, low susceptibility to other disturbances (e.g., storm, snow load, and drought), high tree vigor (low defoliation, long crown, fast growth)	See storm and snow load
Ungulates	Tree species that are not or rarely browsed, low population density of ungulates, high forage value of habitat, defense measures (e.g., fencing)	Seedlings able to sprout, *large-scale disturbances that increase the forage value of the habitat, heavy winters reducing ungulate population density*
Fungi causing stem and root rot	Tree species not susceptible to stem and root rot, low damage to roots and to the bark by harvesting, tree fall, and rockfall	See storm and snow load
Acid deposition	Mycorrhizal fungi, *high buffering capacity of the soil,* tree species able to pump nutrients from deep soil layers, ground vegetation able to retain nutrients, liming, absence of nutrient-exporting treatments (e.g., whole-tree harvesting)	*High input of nutrients from weathering of the bedrock,* tree species able to pump nutrients from deep soil layers
Drought	Tree species able to cope with water stress (low evaporative demand, strong stomatal aperture control, extensive and deep-reaching root system)	Tree species able to grow a second set of leaves or to sprout
Inappropriate harvesting practices	Reward and punishment system depending on harvesting quality, high educational standards, low stand density	Tree species able to grow over bark damage fast

Italics indicate characteristics that cannot be altered by management, or only to a small extent, within <500 years, such as aspect, soil moisture conditions, and site productivity.
Note: A complete description of the ecological situation (Table 91.1) is not given since only relative stability statements are made.

OUTLINE OF A PROCEDURE FOR REACHING DECISIONS ABOUT RESTORATION

Most of the ecosystem characteristics given in Table 91.3 have already been considered by forest management so that experience and a body of scientific knowledge about how to influence these characteristics are available. However, resistance and elasticity with respect to acceptable ranges of ecosystem states have rarely been explicitly included in management decisions.

When deciding on the best management strategies for each particular forest, which often involves restoring impaired forest ecosystem characteristics, the following questions need to be answered:

1. Which products and services should a particular forest deliver continuously?
2. Which ecosystem characteristics are most important for the current and long-term delivery of specific products and services?

3. Which disturbances have the potential to cause ecosystem states that do not comply with management aims?
4. Which ecosystem characteristics are most relevant in enhancing resistance and elasticity in the face of disturbances?
5. Which variables and which assessment procedures should be considered in monitoring characteristics related to products and services, disturbances, resistance, and elasticity?
6. Which value ranges of the variables identified in question 5 are acceptable for specific sites and management aims?
7. How resistant and how elastic are the ecosystems in question at present, in relation to disturbances that have the potential to drive ecosystems to an unacceptable state?
8. Which ecosystem characteristics, if any, should be restored in order to make the ecosystems in question more resistant and more elastic?
9. Which management practices are most appropriate to achieve this restoration?
10. How much investment required for restoration is justified given the potential result?

CONCLUSION

It is commonly accepted among Swiss forest managers that many of the problems they are facing are due to failures of past management, and that more naturalness needs to be restored (Ott et al., 1997). For instance, the role of natural disturbances in shaping the forest has been underestimated, and the potential for replacing them with specific management practices overestimated. Ecosystem characteristics have often been considerably altered in order to enhance the forest's output of desired products and services, but without taking into account the long-term consequences for the resistance of the forest to disturbances and for its elasticity. Restoration measures in secondary forests in Switzerland to enhance their resistance and elasticity in the face of disturbances are therefore required.

The concepts of resistance and elasticity are not yet operational for use in describing forest dynamics, although a qualitative use of some of their aspects has been incorporated into management guidelines for forests that provide protection against natural hazards (Wasser and Frehner, 1996). Using a list of questions, I have developed a procedure that could make resistance and elasticity more generally applicable in forest management. I have tried to answer, in a preliminary way and for Swiss conditions, some of these questions for both forests providing protection against natural hazards and forests producing timber. The challenge is now to describe resistance and elasticity quantitatively, to develop suitable indicators for monitoring, and to apply these concepts to forests that serve primarily other purposes such as recreation and conservation. Achieving this will take us a step closer to designing forest management systems based on a rigorous assessment of the chances and risks that the forest will continue to provide the desired products and services.

ACKNOWLEDGMENTS

I would like to thank two reviewers for their helpful comments on the manuscript, and S. Dingwall for improving the English.

DEFINITIONS OF TERMS AS USED IN THIS CHAPTER*

Constancy   "Staying essentially unchanged" (Grimm and Wissel, 1997).

Disturbance   "Any relatively discrete event in time that disrupts ecosystem, community, or population structure and changes resources, substrate availability, or the physical environment" (White and Pickett, 1985). I would like to extend this definition of disturbance to include "events" caused by continuously acting agents, such as acid deposition, that lead to slow changes in the ecosystem.

Disturbance regime   Type, frequency, spatial distribution, and magnitude of the disturbances acting on an ecosystem, which may interact with each other.

Disturbing agent   Factor that has the potential to change ecosystem characteristics substantially.

Ecosystem composition   Sum of the components in an ecosystem.

Ecosystem dynamics   Temporal development of a whole ecosystem.

Ecosystem processes   Temporal development of ecosystem components.

Ecosystem structure   Spatial arrangement of ecosystem components

Elasticity   "Speed of return to the reference state or dynamics after a disturbance" (Grimm and Wissel, 1997).

Human disturbance   Disturbance caused by people.

Natural disturbance   Disturbance caused by disturbing agents other than people.

Naturalness   Degree to which an ecosystem, at a given site, resembles an old-growth forest in composition, structure, and processes.

Old-growth forest   Forest that has been subject to natural disturbances only.

Persistence   "Permanent existence of an ecosystem" (Grimm and Wissel, 1997).

Resilience   The ability of an ecosystem to return to the referential state or dynamics after a disturbance.

Resistance   The ability of an ecosystem to remain unchanged in spite of the presence of a disturbing agent.

Secondary forest   Forest that has been subject to substantial human disturbances.

REFERENCES

Assmann, E., *The Principles of Forest Yield Study: Studies in Organic Production, Structure, Increment and Yield of Forest Stands*, Pergamon Press, Oxford, 1970.

Brang, P., Schönenberger, W., Ott, E., and Gardner, B., Forests as protection from natural hazards, in *The Forests Handbook, Vol. 2: Applying Forest Science for Sustainable Management*, Evans, J., Ed., Blackwell Science, Oxford, 2001, pp. 53–81.

Brassel, P. and Brändli, U.-B., *Schweizerisches Landesforstinventar: Ergebnisse der Zweitaufnahme 1993–1995*, Paul Haupt, Bern, Switzerland, 1999.

Brückner, E., Die Entwicklung des Rotwildbestandes und der Waldbiotope im Westerzgebirge-Vogtland von 1591 bis 1990, *Z. Jagdwiss.*, 39, 46–59, 1993.

Bugmann, H.K.M. and Cramer, W., Improving the behaviour of forest gap models along drought gradients, *For. Ecol. Manage.*, 103, 247–263, 1998.

Burschel, P. and Huss, J., *Grundriss des Waldbaus. Ein Leitfaden für Studium und Praxis*, Paul Parey, Hamburg, Germany, 1987.

Cairns, J.J., Ecological risk assessment: a predictive approach to assessing ecosystem health, in *Ecosystem Health*, Rapport, D. et al., Eds., Blackwell Science, Oxford, 1998, pp. 216–228.

De Leo, G.A. and Levin, S.A., The multifaceted aspects of ecosystem integrity, *Conserv. Ecol.*, [online] 1, 1997.

Eichenseer, F., Entwicklung der Tannenanteile in Ostbayern im 19 und 20. Jahrhundert, *Forst Holz*, 52, 498–501, 1997.

Erni, V. and Lemm, R., Ein Simulationsmodell für den Forstbetrieb — Entwurf, Realisierung und Anwendung, *Ber., Eidg. Forschungsanstalt Wald Schnee Landschaft*, 341, 1995.

Grimm, V. and Wissel, C., Babel, or the ecological stability discussions: an inventory and analysis of terminology and a guide for avoiding confusion, *Oecologia*, 109, 323–334, 1997.

* Adapted from Grimm, V. and Wissel, C., *Oecologia*, 109, 323, 1997. With permission.

Harmon, M.E., Franklin, J.F., Swanson, F.J., Sollins, P., Gregory, S.V., Lattin, J.D., Anderson, N.H., Cline, S.P., Aumen, N.G., Sedell, J.R., Lienkaemper, G.W., Cromack, K.J., and Cummins, K.W., Ecology of coarse woody debris in temperate ecosystems, *Adv. Ecol. Res.*, 15, 133–302, 1986.

Herold, A. and Ulmer, U., Stand stability in the Swiss National Forest Inventory: assessment technique, reproducibility and relevance, *For. Ecol. Manage.*, 145, 29–42, 2001.

Holling, C.S., Resilience and stability in ecological systems, *Ann. Rev. Ecol. Syst.*, 4, 1–23, 1973.

Holling, C.S., Simplifying the complex: the paradigms of ecological function and structure, *Eur. J. Operational Res.*, 30, 139–146, 1987.

Holling, C.S., Surprise for science, resilience for ecosystems, and incentives for people, *Ecol. Appl.*, 6, 733–735, 1996.

Hunziker, U., Zusammenhänge zwischen Kleinstandort und Nadelholzansamung in einem *Abieti-Piceetum* bei Chironico TI, diploma thesis, Department of Forest Sciences, ETH, Zurich, Switzerland, 1999.

Kienast, F., Brzeziecki, B., and Wildi, O., Computergestützte Simulation der räumlichen Verteilung naturnaher Waldgesellschaften in der Schweiz, *Schweiz. Z. Forstwes.*, 145, 293–309, 1994.

Kimmins, J.P., The health and integrity of forest ecosystems: are they threatened by forestry? *Ecosyst. Health*, 2, 5–18, 1996.

Kolb, T.E., Wagner, M.R., and Covington, W.W., Utilitarian and ecosystem perspectives: concepts of forest health, *J. For.*, 92, 10–15, 1994.

Korpel, S., *Die Urwälder der Westkarpaten*, Gustav Fischer, Stuttgart, Germany, 1995.

Korpel, S., Totholz in Naturwäldern und Konsequenzen für Naturschutz und Forstwirtschaft, *Forst Holz*, 52, 619–624, 1997.

Kräuchi, N. and Kienast, F., Modelling subalpine forest dynamics as influenced by a changing environment, *Water Air Soil Pollut.*, 68, 185–197, 1993.

Küchli, C., *Forests of Hope: Stories of Regeneration*, Earthscan, London, 1997.

Linder, P., Elfving, B., and Zackrisson, O., Stand structure and successional trends in virgin boreal forest reserves in Sweden, *For. Ecol. Manage.*, 98, 17–33, 1997.

Liss, B.M., Der Einfluss von Weidevieh und Wild auf die natürliche und künstliche Verjüngung im Berg mischwald der ostbayerischen Alpen, *Forstwiss. Centralbl.*, 107, 14–25, 1988.

Mayer, H., *Waldbau auf soziologisch-ökologischer Grundlage*, Gustav Fischer, Stuttgart, Germany, 1984.

Mayer, H., Schenker, S., and Zukrigl, K., Der Urwaldrest Neuwald beim Lahnsattel, *Centralbl. Gesamte Forstwes.*, 89, 147–190, 1972.

Mayer-Wegelin, H. and Schulz-Brüggemann, M., Untersuchungen über den Bestandesaufbau im kleinen Urwald des Rothwaldes, *Zentralbl. Gesamte Holz- Forstwirtsch.*, 71, 303–331, 1949.

Mitchell, S.J., A diagnostic framework for windthrow risk estimation, *For. Chron.*, 74, 100–105, 1998.

Nykanen, M-L., Peltola, H., Quine, C., Kellomäki, S., and Broadgate, M., Factors affecting snow damage of trees with particular reference to European conditions, *Silva Fenn.*, 31, 193–213, 1997.

Ott, E., Frehner, M., Frey, H-U., and Lüsche, P., *Gebirgsnadelwälder: Ein praxisorientierter Leitfaden für eine standortgerechte Waldbehandlung*, Paul Haupt, Bern, Switzerland, 1997.

Perry, D.A., *Forest Ecosystems*, Johns Hopkins University Press, Baltimore, 1994.

Pukkala, T., Multiple risks in multi-objective forest planning integration and importance, *For. Ecol. Manage.*, 111, 265–284, 1998.

Righetti, A. and Huber, W., Ausrottung und Wiedereinwanderung des Rothirsches (*Cervus elaphus* L.) im Kanton Bern (Schweiz), *Rev. Suisse Zool.*, 90, 863–870, 1983.

Rottmann, M., Waldbauliche Konsequenzen aus Schneebruchkatastrophen, *Schweiz. Z. Forstwes.*, 136, 167–184, 1985.

Samuelsson, J., Gustafsson, L., and Ingelög, T., *Dying and Dead Trees: A Review of Their Importance for Biodiversity*, Swedish Threatened Species Unit, Uppsala, Sweden, 1994.

Schmid-Haas, P. and Bachofen, H., Die Sturmgefährdung von Einzelbäumen und Beständen, *Schweiz. Z. Forstwes.*, 142, 477–504, 1991.

Shugart, H.H., *A Theory of Forest Dynamics: The Ecological Consequences of Forest Succession Models*, Springer-Verlag, New York, 1984.

Slodicák, M., Thinning regime in stands of Norway spruce subjected to snow and wind damage, in *Wind and Trees*, Coutts, M.P. and Grace, J., Eds., Cambridge University Press, Cambridge, 1995, pp. 436–448.

Spiecker, H., Mielikäinen, K., Köhl, M., and Skovsgaard, J.P., Discussion, in *Growth Trends in European Forests: Studies from 12 Countries*, Spiecker, H., Mielikäinen, K., Köhl, M., and Skovsgaard, J.P., Eds., European Forest Institute Research Report 5, Springer-Verlag, Heidelberg, 1996, pp. 355–367.

Wasser, B. and Frehner, M., *Wegleitung Minimale Pflegemassnahmen für Wälder mit Schutzfunktion*, Bundesamt für Umwelt, Wald und Landschaft (BUWAL), Bern, Switzerland, 1996.

White, P.S. and Pickett, S.T.A., Natural disturbance and patch dynamics: an introduction, in *The Ecology of Natural Disturbance and Patch Dynamics*, Pickett, S.T.A. and White, P.S., Eds., Academic Press, New York, 1985, pp. 3–16.

Zukrigl, K., Succession and regeneration in the natural forests in central Europe, *Geobios (Jodhpur)*, 18, 202–208, 1991.

CHAPTER 92

The Precautionary Principle and Ecosystem Health: A Case Study from the Jarrah Forest of Southwestern Australia

Michael C. Calver

INTRODUCTION

Forest health, sustainability, and ecosystem management are of interest to many people. The ambiguity of these terms encourages people to engage in forest policy discussions because their own ideas have not been predefined out of the debate. If foresters can't define forest health, or if their definition is unacceptably narrow, people will likely seek opinions about forest management elsewhere. Thus the forest health analogy is a powerful, but imperfect, metaphor for communicating with the public. (O'Laughlin, 1996).

[W]hat is a sufficient precautionary approach to forest use given the uncertainties and risks arising from imperfect knowledge, predicted climate change and unforeseen events? Norton (1996).

In January 1849, members of surveyor-general John Roe's expedition became the first Europeans to pass through the heart of the jarrah *Eucalyptus marginata* hardwood forests of southwestern Western Australia. In his journal, Roe commented with awe at the sight of "the splendid straight mahogany or jarrah trees, growing within 3 or 6 feet of each other, reaching to the height of 50 and 80 feet without a branch or blemish, and apparently quite sound" (Roe, 1852). However, he was a practical man and moved by considerations other than aesthetics. He further reflected:

It is also to be hoped that, as one of the most valuable and most readily available sources of wealth in this colony, the superb naval timber which I observed in such inexhaustible quantity in the forests behind Bunbury, will not much longer be suffered to remain there idle, but that, on the formation of practicable roads, the axe and saw will shortly resound amongst it, to the mutual advantage of the colony and of its parent country (Roe, 1852).

Roe's wish was soon fulfilled. By 1920, only 71 years after his expedition, "nearly one million acres of the jarrah forest were cut over for the removal of 750 million cubic feet of logs, causing a reduction of almost 50% in the forest canopy"* (Wallace, 1965). This period of almost unrestrained exploitation saw intense land-use conflicts between agricultural clearing and long-term forestry,

---

* One acre = ~0.404 hectares and one cubic foot = ~0.028 cubic meters. The original extent of the jarrah forest is estimated as 13,000,000 acres (Wallace, 1965).

bitter complaints of wastage in the timber industry, and large-scale government inquiries. Furthermore, in 1894 the Australasian Association for the Advancement of Science lobbied successfully to create a 65,000 ha conservation reserve in the heart of the jarrah forest, only to have it overturned in 1911 following pressure from the timber industry (Rundle, 1996). The timbers that Roe saw in "inexhaustible quantity" proved finite and, as always, people argued bitterly over a finite resource.

In 1918, the Western Australian state parliament passed a Forests Act, designating large areas of jarrah forest as State Forest for long-term sustainable timber production, and forest management has evolved since then. However, the basic land-use conflicts involving agriculture, forestry, and conservation in the jarrah forest persist today with unabated intensity (e.g., Abbott and Christensen, 1994, 1996; Calver and Dell, 1998a, b; Calver et al., 1999). In the late 1990s, the State Government of Western Australia and the Commonwealth Government of Australia attempted to resolve these issues through a Regional Forests Agreement (RFA), which balanced all competing interests for the next 20 years. However, both the process and its outcome have been criticized and disavowed so the RFA cannot claim to have achieved consensus (Horwitz and Calver, 1998).

Despite these difficulties, the debate over jarrah forest management during the 1990s highlighted strong community interests in the values inherent in the Precautionary Principle and in the concept of ecosystem health. The Precautionary Principle argues:

> Where there are threats of serious or irreversible damage, lack of full scientific certainty should not be used as a reason for postponing measures to prevent environmental degradation. In the application of the precautionary principle, public and private decisions should be guided by: (i) careful evaluation to avoid, wherever practicable, serious or irreversible damage to the environment; and, (ii) an assessment of the risk-weighted consequences of various options (The Intergovernmental Agreement on the Environment, May 1992, quoted in Deville and Harding, 1997).

The ecosystem health concept argues that ecosystems may become dysfunctional under disturbance and that such dysfunction can be detected and corrected (Rapport, 1998a).

The Precautionary Principle is explicitly invoked in a ministerial condition governing the management of the jarrah forest (Calver et al., 1999). It was also the subject of an unsuccessful legal action against the Western Australian Department of Conservation and Land Management (CALM), which manages logging and associated practices in the region (Calver et al., 1999). The concept of ecosystem health was introduced into management issues in the jarrah forest by Calver and Dell (1998b) and Calver et al. (1998), after earlier authors had raised the related concept of ecological condition, a measure of the continuance of fundamental ecological processes (Abbott and Christensen, 1994; see also Rapport, 1998a). Ecosystem health caught the popular imagination recently when local physicians ran large advertisements in daily newspapers claiming that the consequences of some current forest management practices were unhealthy.

The widespread interest aroused by the Precautionary Principle and ecosystem health make these concepts ideal facilitators to bring together the diverse stakeholders involved in the forest debate to achieve consensus on management goals and practices. Preliminary frameworks for applying the Precautionary Principle are already available and there is a growing literature on forest ecosystem health. However, application of both concepts is vague and subjective unless measurable, quantitative standards are chosen for selected response variables that reflect possible changes in an ecosystem. In this chapter I illustrate the value of the Precautionary Principle in setting such quantitative standards for maintaining the health of the jarrah forest ecosystem in conjunction with a viable timber industry contributing to the economic well-being of local communities. Brief histories of the concept of forest ecosystem health and the Precautionary Principle follow, together with a background to management issues in the jarrah forest and an outline of a framework for applying both the concept of ecosystem health and the Precautionary Principle to forest management. The discussion reveals that ecosystem health and the Precautionary Principle can be used as quantitative bases for forest management in this case, but that this has not yet been achieved.

BACKGROUND

History of the Ecosystem Health Concept and the Precautionary Principle

Ecosystem health is a new concept scarcely a decade old (Rapport, 1998a,b). It centers on ecosystem resilience and sustainability, absence of significant environmental stresses, and the capacity of the ecosystem to sustain healthy human communities, making it ideally suited to forest management issues (e.g., O'Laughlin, 1994; Vora, 1997). Several authors have proposed quantitative criteria to assess forest ecosystem health (e.g., O'Laughlin, 1994, 1996; Yazvenko and Rapport, 1996), although value judgments are inherent in most measures (Kimmins, 1996). However, this is not necessarily a weakness if there is extensive consultation among all stakeholders in deciding the criteria for evaluation. Importantly, the concept of ecosystem health need not proscribe logging or other uses of natural resources, providing that such use does not compromise other ecosystem services, such as water catchments, erosion prevention, or wildlife conservation. Indeed, such resource use may be essential in sustaining local human communities. With regard to forest management, ecosystem health emphasizes reducing the risk of forest change undesirable to the wider community and, where forests are degraded, restoration of forest conditions that the community wishes (O'Laughlin, 1996). Criteria for ecosystem health include vigor (sustaining normal levels of energy throughput and primary productivity), resilience (capacity to recover from stress), and organization (maintenance of species diversity and a full spectrum of ecological interrelationships) (Rapport 1998a). The emphasis on risk reduction establishes a clear link with the Precautionary Principle.

The origins of the Precautionary Principle are unclear, but lie in politics rather than science. Most commentators accept that it was first explicitly included in an international agreement at the 1987 Second International North Sea Conference in London. The new approach focused on scientific uncertainty; required industry to demonstrate that its activities were low-risk; and gave priority to preventive management to avert possible, but not certain, damage (Gray and Bewers, 1996). Statements of the Precautionary Principle in national legislation and international declarations are widespread and no longer restricted to the marine environment (Deville and Harding, 1997). However, there is frustration in application, not least because there has been no consistent, widely held definition, and key terms within existing definitions may be undefined (Gray and Bewers, 1996). The formulation of legal and scientific definitions of the Precautionary Principle and guidelines for their application are important topics in the current literature (Gray and Bewers, 1996; Deville and Harding, 1997; Rogers et al., 1997). However, some authors take the view that the Precautionary Principle should remain a political or ethical issue distinct from science. It has been parodied as meaning "do nothing until you know everything" (Abbott and Christensen, 1996), or claimed to function best when understood clearly not to be a scientific concept (Santillo et al., 1998).

Ecosystem Health Issues in the Jarrah Forest

The jarrah forest bioregion of southwestern Australia is unique in existing on some of the most highly leached and impoverished forest soils in the world (Abbott and Christensen, 1996). It lies mainly within the Dale and Menzies subdistricts of the Darling Botanical District (Figure 92.1A) and has been fragmented greatly by agricultural clearing, leaving the main forest estate in State Forest managed by government agencies (Figure 92.1B). This discussion focuses on management issues within State Forest, which is managed for multiple uses including timber production, water catchments, and flora and fauna conservation and recreation.

There is considerable debate whether past management and, more particularly, the current multiple-use practices have impacted on values of the forests other than timber production such as tourism and species conservation (e.g., Abbott and Christensen, 1994, 1996; Calver et al., 1996,

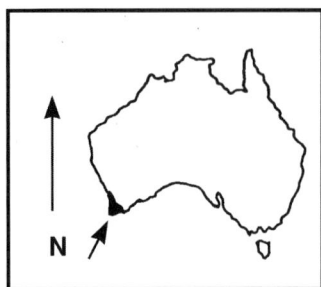

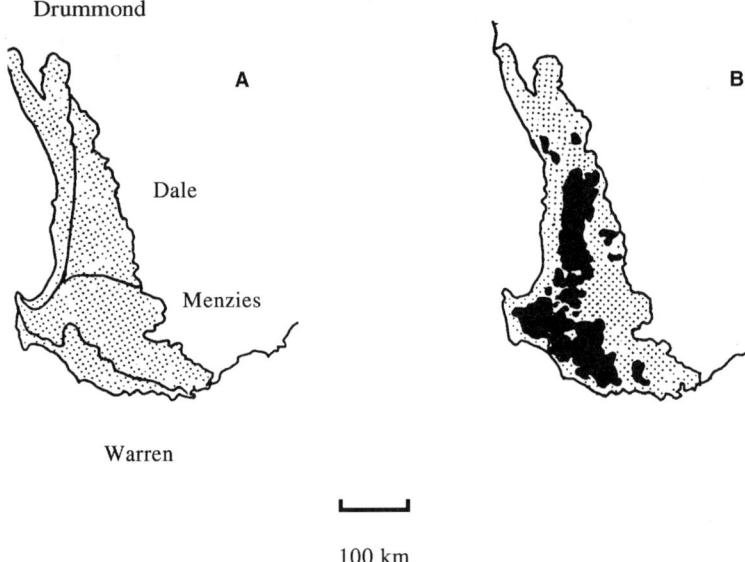

100 km

Figure 92.1 The location of the Darling Botanical District in southwestern Australia. (Adapted from Wardell-Johnson and Nichols, 1991). The jarrah forest lies predominantly within the Dale and Menzies subdistricts, with Drummond and Warren being dominated by other species (A). The boundaries of State Forest, which is managed by government agencies, are shown in B.

1998). The region is rich in endangered and threatened marsupials and relictual plant and invertebrate taxa (Abbott and Christensen, 1994). Therefore, the conservation value of the area is unquestionable, as is its value for tourism, recreation and protection of water catchments (RAC, 1993). This is recognized by CALM, which, until recently, had a dual role as principal conservation authority for State Forest as well as director of forestry operations (Stoneman et al., 1997; see Conservation Commission of Western Australia, http://www.conservation.wa.gov.au/forest_plan.html for recent reforms).

Perceived threats to ecosystem values include introduced predators, extensive agricultural clearing, and the spread of the exotic plant pathogen, *Phytophthora cinnamomi*, which infects plant roots, invades the vascular system, and ultimately kills susceptible hosts. Many plants in southwestern Australia are susceptible so *P. cinnamomi* may alter the structure and composition of plant communities and through this change hydrological regimes (Wardell-Johnson and Nichols, 1991). There is also debate on applying appropriate fire regimes to reduce the risk of wildfire while maintaining other ecosystem values (Russell and Rowley, 1998, and included references). Logging impacts on ecosystem values are controversial. Abbott and Christensen (1994, 1996) claim that adequate knowledge has been achieved and that the impact is negligible and not cumulative, while

Nichols and Muir (1989) admit that data are lacking but doubt that there are serious impacts. By contrast, Calver et al. (1996, 1998) and Calver and Dell (1998a, b) consider the question open for lack of evidence, while Mawson and Long (1994) believe that deleterious impacts are already occurring. In this chapter I focus on the logging issue, because it is the most contentious and the need for new approaches to resolve disputes is greatest.

APPLYING THE PRECAUTIONARY PRINCIPLE TO PROTECT ECOSYSTEM HEALTH

An important first step is to propose and defend criteria for ecosystem health and precaution. Procedures for applying the criteria can then be considered. The following outline is for illustration only and is intended to stimulate discussion, because final choice of criteria should involve all stakeholders. Such open, consultative processes are essential to the successful implementation of the approach proposed.

Setting Criteria for Ecosystem Health and Precaution

A useful starting point is the existing literature on assessing the ecological condition of forests. Definitions of ecological condition exclude the value judgments inherent in an assessment of "health" (Kimmins, 1996; O'Laughlin, 1996), so they are a less emotive beginning. Indices of condition have already been proposed for the jarrah forest (Abbott and Christensen, 1994). A valuable initiative was made in the 1995 Santiago Declaration on Sustainability of Boreal and Temperate Forests, in which Australia participated (Turner and Lambert, 1997). This declaration proposed criteria and indicators for quantifiable assessment of forest condition. Criteria included conservation of biological diversity, maintenance of the productive capacity of forest ecosystems, maintenance of forest ecosystem vitality, conservation and maintenance of soil and water resources, and maintenance of forest contribution to global carbon cycles. There was also reference to maintenance and enhancement of long-term multiple socioeconomic benefits to meet the needs of society and legal, institutional, and economic frameworks for forest conservation and sustainable management. However, these social and legal categories remained vague and poorly defined.

Extensive indicators accompanied each criterion. For example, in the case of the conservation of biological diversity, indicators included classification of ecosystem types and their conservation within reserves, determination of the conservation status of forest-dependent species, and estimates of the genetic diversity of species on the basis of any reduction in range they might have incurred. Updates on these criteria for implementation in Australia are given on the Australian Department of Primary Industries and Energy Web site, http://www.dpie.gov.au/agfor/forests/montreal/international.html. The combination of ecological processes and the needs of local human communities is an excellent basis from which to build locally applicable definitions of the health of the jarrah forest and to approach recent attempts to define forest ecosystem health (e.g., Kimmins, 1996; O'Laughlin, 1996; Yazvenko and Rapport, 1996).

Once criteria for forest ecosystem health have been agreed to the point where all participating stakeholders accept them as legitimate, the next step is to decide on a precautionary approach to activities that may compromise particular criteria of health. Several authors have considered the thorny problem of developing quantitative approaches to the Precautionary Principle for those cases where the principle is called for in legislation or a policy directive, but there is no agreed mechanism for its implementation. Suggested means for quantifying precaution include Bayesian statistics (Varis and Kuikka, 1997), risk analysis (Rogers et al., 1997), and statistical power analysis (Buhl-Mortensen and Welin, 1998; Calver et al., 1999, and included references). The following discussion considers only the power analysis approach because it is the dominant one in the literature, but this does not imply that the alternatives lack merit or might not prove superior.

Detailed accounts of the concept of statistical power can be found in basic biostatistics textbooks and specialist texts such as Cohen (1988). It is extremely relevant to environmental issues, since it deals with cases where an experimental test of a putative environmental impact is not significant. These cases are of special concern as they may arise either because there is no effect or because the experiment/monitoring lacked the sensitivity to detect an impact that actually occurred (Buhl-Mortensen and Welin, 1998). Power analysis helps in these situations by determining the probability that an impact of given magnitude would have been detected if it had actually occurred. This is known as the power of the test. If power is low (commonly defined as less than 0.85 or 0.9, Thomas and Juanes, 1996), then the test is insensitive and a possible impact has not been discounted. Power is directly related to sample size/replication in the study, the level of significance (α) chosen for the test, and the effect size (ES), which is the minimum difference/change that one wishes to detect. High power in a test can be achieved by increasing sample size, relaxing α above the traditional level of 0.05, or seeking to detect only a large ES. Given the likelihood of low power in many ecological studies and environmental monitoring, several authors have argued for a reversal of the burden of proof in impact studies (e.g., Underwood, 1997, and included references) or for explicit quantification of the power of studies (e.g., Di Stefano, 2001; see Craig, 2000, and Williams et al., 2001, for practical applications). This requires industry to demonstrate with high power that an unacceptable impact is unlikely, just as manufacturers of drugs must take reasonable steps to identify possible hazards or side effects (see reviews in Calver et al., 1999; Calver, 2000). Such a reversal is entirely consistent with the Precautionary Principle.

Applying the Criteria

Deville and Harding (1997) developed a four-step approach to applying the Precautionary Principle. Discussion of each step with reference to ecosystem health issues in the jarrah forest follows. The headings stated for each step are taken directly from Deville and Harding (1997).

Step 1. Are Precautionary Measures Needed?

Logging impacts on many proposed indices of ecosystem health in the Northern Hemisphere and in eucalypt forests in eastern Australia are confirmed or strongly suspected (e.g., see many detailed examples in the book edited by De Graaf and Miller, 1996). They are a sound scientific basis for believing that damage to fundamental ecosystem processes in jarrah forest might follow logging. Impacts could include local extinction of populations of endangered wildlife during logging operations, increased activity of introduced predators in disturbed areas, weed introductions, changes in hydrological regimes or soil properties, pest outbreaks, and possible spread of plant pathogens. These possible costs need to be balanced against the benefits of a timber industry to local communities, as well as the possibility that other industries such as tourism might provide the benefits with fewer environmental costs.

There is a long history of debate on such concerns with regard to the jarrah forest (see reviews in Calver and Dell, 1998a, b; Calver et al., 1999). Significant reduction in the forest canopy as a result of logging is documented (Wallace, 1965), introduced predators are widespread (Abbott and Christensen, 1994), and *Phytophthora cinnamomi* is altering vegetation patterns widely in the jarrah forest bioregion (Davison and Shearer, 1989). However, it has also been argued strongly that properly managed logging operations are not a threat to the long-term conservation of biodiversity or viability of ecosystem processes in the jarrah forest (Abbott and Christensen, 1994, 1996; Stoneman et al., 1997). On balance, I believe that most parties would agree that at present there is at least the potential of significant impact on ecosystem health values, but that disagreement arises on the question of whether these threats are ameliorated by current management practices.

Resolution is impeded by gaps in our knowledge of putative environmental impacts in the jarrah forest. The Australian Commonwealth Government's Resource Assessment Commission (RAC, 1993) completed the most recent detailed review of ecological impacts of forest use in Australia.

The RAC considered not only the number of impact studies (defined as work measuring impact of human activity on one or more ecological variables and reported in one or more publications) done in each Australian state, but also classified them according to other factors. These included:

- The major forest uses investigated
- The specific activities involved (of which 33 environmental variables were studied)
- The study design adopted
- The geographic scale of the studies

An analysis of the 51 Western Australian studies was summarized by Calver et al. (1998). This indicated a considerable emphasis on wood production and mining, which comprised 96% of the broad forest uses studied and 56% of the specific forest activity studies. By contrast, conservation management studies comprised only 2% of the forest uses studied, while recreation and tourism had not been studied as either a major forest use or a specific activity. Calver et al. (1998) also doubted the quality of the studies. Nearly half (49%) reported only a single post-disturbance observation, while just 16% used the powerful option of a before/after design. Finally, 92% of all studies were restricted in scale to a single site or local area, while 74% were short term, lasting only 2 years or less. This situation contrasts strongly to that in eastern Australia, where a larger number of more substantive studies have been completed (e.g., chapters in the edited volume by Lunney, 1991).

By the early 1990s many conservation or environmental impact issues in Western Australia's forests had not been researched intensively and there were limitations to some of the work completed. Although the research pace has accelerated and important work has been published recently, significant gaps remain and coverage is poor compared with eastern Australia (Calver et al., 1998). Overall, I believe that the important conservation values of the jarrah forest, the scientific plausibility of the hypothesis that deleterious impacts could occur, and the gaps in the relevant studies satisfy the criteria for a precautionary approach.

Step 2. How Precautious Should We Be?

According to Deville and Harding's (1997) framework, precaution should increase if either the threat or the uncertainty is great. If the threat is minor, little precaution is required, while if the likelihood of the risk is well known and grave, then the appropriate strategy is prevention, not precaution. The risk of significant declines in endemic organisms and decline of fundamental ecosystem services such as water catchments are substantial threats. When coupled with the gaps in published studies of putative impacts, these observations argue for very strong levels of precaution.

Step 3. What Precautionary Measures Can Be Applied?

The multiple-use policies for the jarrah forest appear to qualitatively follow the Precautionary Principle through their use of reserves, buffer zones, habitat tree requirements, and so on. However, quantitatively they may not detect serious environmental impacts before an irreversible consequence occurs. This is because the policies include no direct measurements of response variables, such as population trends of indicator species, which relate directly to chosen criteria of ecosystem health. It would be a daunting task to prepare monitoring plans or experimental tests for each component of ecosystem health, but priorities for evaluation could be set through the same consultation process proposed for setting the original ecosystem health criteria. If the power analysis approach to quantifying precaution were adopted, it would also be important to agree on the levels of power and effect size that would be required for precaution in the context of each criterion of ecosystem health. In turn, these would determine the sample sizes and level of α required in individual studies. Two examples from published studies of environmental impacts in the jarrah forest illustrate the

Table 92.1 Median Numbers of Counts of Individual Birds Noted per 100 m of Transect per Hour in Jarrah Forest

Species	Untreated Forest			Forest Logged between 1982 and 1983		
	1981	1982	1983	1981	1982	1983
Port Lincoln ringneck	0.7	0.1	0.8	0.5	0.2	—
Red-capped parrot	0.3	0.4	0.3	0.7	0.5	—
Fan-tailed cuckoo	0.2	—	—	—	—	0.2
Shining bronze cuckoo	0.4	0.3	—	0.5	—	—
Laughing kookaburra	—	—	0.3	—	0.2	—
Scarlet robin	0.6	0.3	0.1	—	0.5	0.7
Western yellow robin	0.7	1.0	0.4	0.5	0.5	0.2
Golden whistler	2.4	1.1	2.7	1.9	4.2	2.2
Grey shrike thrush	—	0.4	0.5	0.7	—	0.4
Grey fantail	1.1	0.9	0.8	2.8	1.8	4.2
Western gerygone	0.8	3.0	1.8	1.3	1.5	2.2
Inland thornbill	0.2	0.3	1.8	2.2	2.1	1.7
Western thornbill	0.6	1.2	0.9	0.5	2.1	0.4
Varied sittella	—	0.8	—	1.2	0.5	—
Rufous treecreeper	—	—	—	0.5	0.2	0.7
Spotted pardalote	0.1	0.4	0.6	0.7	0.7	0.4
Striated pardalote	4.8	2.8	2.8	4.4	13.7	4.9
Silvereye	—	—	0.1	—	0.5	0.4
Brown honeyeater	—	0.2	—	0.5	0.2	0.2
White-naped honeyeater	—	0.6	1.1	0.5	0.7	0.4
Western spinebill	1.2	1.9	0.5	1.6	0.7	0.2
Little wattlebird	0.4	0.3	—	1.3	0.5	—
Dusky woodswallow	—	—	—	—	—	0.4
Pied currawong	—	0.1	0.3	—	0.5	—
Australian raven	0.1	—	0.1	—	0.2	—

Data are a subset of Appendix 1 from Abbott and van Heurck (1985), showing only those birds that occurred at some stage in the treated forest.

power analysis approach. They are both applicable to a criterion of maintenance of endemic species diversity, which could be used as an indicator of ecosystem health.

The first example is a study of logging impacts on bird counts in jarrah forest by Abbott and van Heurck (1985). The authors surveyed 32 bird species along set transects in two jarrah forest sites during 1981–1983. In early 1983, one of the sites was thinned, reducing the canopy by approximately 50%. Nonparametric Mann–Whitney tests were used to compare the median number of individual birds in each species/100 m of transect/h between sites and between years (see summary data in Table 92.1). At the logged site, no species was recorded as significantly less frequent after treatment and only two species were recorded as significantly more frequent after treatment. The authors were cautious in interpreting their results, concluding: "These findings suggest that most bird species in the jarrah forest may be adaptable to gross disturbance of the forest" (Abbott and van Heurck, 1985).

Power analysis can aid in interpreting the data by determining the probability of rejecting the null hypothesis of no significant impact of the thinning operation if it is truly false. The results of such an analysis for large, medium, and small effect sizes at α levels of 0.05, 0.1, and 0.2 are shown in Table 92.2. The effect sizes were based on the recommendations of Cohen (1988) and, assuming an approximately normal distribution, correspond to the difference between the means of samples divided by the common standard deviation (($\mu_1 - \mu_2$)/σ) giving 0.8, 0.5, and 0.2, respectively. The analysis used the nonparametric test option in the nQuery Advisor software (Elashoff, 1997).

Under most conditions in this study, power is very low. At the conventional level of $\alpha = 0.05$ used in the original study, there was only a 6% chance of rejecting a false null hypothesis at a small effect size and this only improved to 44% at a large effect size. If α is relaxed to 0.2, power

Table 92.2 Results of a Power Analysis for the Mann–Whitney Tests Used by Abbott and van Heurck (1985) to Detect Differences in Bird Counts before and after Logging in Jarrah Forest (%)

	$\alpha = 0.05$	$\alpha = 0.10$	$\alpha = 0.20$
Small effect size $(\mu_1 - \mu_2)/\sigma = 0.2$	6	12	21
Medium effect size $(\mu_1 - \mu_2)/\sigma = 0.5$	21	31	45
Large effect size $(\mu_1 - \mu_2)/\sigma = 0.8$	44	56	70

Power is shown as a percentage for three different levels of α and large, medium, and small effect sizes (based on Cohen, 1988). μ_1 and μ_2 are the means of the two groups being compared in the test, and σ is the common standard deviation of the groups.

improves to only 21% for a small effect size and a more respectable 70% for a large effect size. Unfortunately, the exact probabilities for individual tests were not presented in Abbott and van Heurck (1985) so the number of significant results following relaxation of α cannot be determined. However, the authors' original caution is vindicated, because the power analysis shows that their study had only a very low probability of detecting significant impacts. To follow a precautionary approach, quantitative standards would be agreed in advance to determine a standard of proof to be met before such logging operations could proceed on a wider scale. Stakeholders might also raise concerns about the cumulative impact of successive disturbance (see, for example, Recher, 1996). The results of the power analysis could be compared with the *a priori* criteria to determine objectively if permission should be granted.

A second example concerns a test of the hypothesis that fuel-reduction burning practiced in the jarrah forest could influence the fate of egg clutches of the rare frog, *Geocrinia lutea* (Driscoll and Roberts, 1997). As part of a larger study, the authors monitored the survival of 60 *G. lutea* egg clutches at six pairs of burned and unburned sites in mixed stands of jarrah and karri *Eucalyptus diversicolor* forest near the southern extremity of the jarrah forest. They found no significant association between burning and egg clutch survivorship ($\chi_1^2 = 1.72$, $p = 0.19$), with survivorship of clutches in the absence of fire being ~0.5 (18/32) and in the presence of fire being ~0.4 (11/28). Assuming a constant sample size of 60 clutches overall, half of them exposed to fire and half unexposed, with a survivorship of 0.5 in the clutches from unburned areas, the power of detecting a range of survivorships less than 0.5 in clutches in burned areas was determined using Elashoff (1997) (Table 92.3).

The power of detecting a survivorship of 0.4 at $\alpha = 0.05$ was only 11%. Relaxing α to 0.1 improved power to 19%, while at $\alpha = 0.2$ power rose to 30%. A power of greater than 80% at $\alpha = 0.05$ could be achieved by seeking to detect only a survivorship of 0.1 in clutches in burned areas, considerably less than the survivorship of 0.5 established in the unburned controls. Assuming that the overall sample size is a fixed constraint, the only option for testing for smaller changes in survivorship (assuming that this is desirable) with moderate power is to relax α above 0.05.

However, relaxing α to improve power increases the probability of concluding falsely that fire had an impact on egg clutch survival (a Type I error: Di Stefano, 2001). A recent review (Calver

Table 92.3 Results of a Power Analysis for the Egg Clutch Data of Driscoll and Roberts (1997) to Detect Differences in Survivorship in Burned and Unburned Forest

Significance Level (α)	Proportion Egg Clutches Surviving in Unburned/Proportion Egg Clutches Surviving in Unburned Areas (%)			
	0.5/0.4	0.5/0.3	0.5/0.2	0.5/0.1
0.05	11	34	69	94
0.10	19	47	79	97
0.20	30	62	88	99

Sample size is assumed fixed at 60 clutches and power is shown as a percentage for three different levels of α and four different combinations of survivorship in burned and unburned areas.

et al., 1999, and included references) argued that this is acceptable when the cost of a Type I error (falsely concluding an impact) is less than that of a Type II error (falsely concluding no impact). Ideally, these errors should be costed with, perhaps, some special monetary value built in for noneconomic costs, such as intergenerational equity or aesthetic values. Then it is possible to express the ratio of cost of Type II error/cost of Type I error. Cohen (1988) suggested using this ratio to judge the acceptable ratio of Type II/Type I errors in a study and the concept was applied specifically to forests by Di Stefano (2001). Ideally, the ratio of Type I/Type II errors acceptable to all parties would be decided as part of the initial consultative process preceding implementation of a quantitative, precautionary approach to monitoring forest ecosystem health. This approach would also lead to better assessment of research needs and justification of the necessary level of funding to meet them.

Step 4. What Precautionary Measures Should Be Applied?

The significant strength of a quantitative monitoring process is that it measures the direct response variables to putative impacts, rather than indirect variables, such as width of buffer zones which it is hoped will correlate with ecosystem health. However, it is also important to specify what steps should be taken if significant impacts are evident.

Most authors have concluded, albeit with important qualifications, that sustainable logging in eucalypt forests is possible without significant detriment to environmental values (see reviews in Calver and Dell, 1998a, b). Under this view, sustainable logging requires a correct balance of a range of management practices and incorporation of environmental values into projected timber yields (e.g., Calver et al., 1998). Dissenting views such as that proposed by Mawson and Long (1994) argue that a timber industry is incompatible with conservation interests in the jarrah forest, which by extension suggests that it would be incompatible with ecosystem health. Therefore, the options being argued range from different perspectives on multiple-use policies to a shift in timber production from native forest to plantations. Under the Deville and Harding (1997) framework and the Intergovernmental Agreement on the Environment (1992) that they cite, choice of precautionary measures would be sensitive to social and economic costs and integrated with other principles of ecologically sustainable development. It is appropriate to consider these factors at the stage of deciding the criteria to be used to monitor forest ecosystem health and to quantify the level of precaution required.

DISCUSSION

The Precautionary Principle and the concept of ecosystem health both share popular appeal with laypersons (Earll, 1992; O'Laughlin, 1996). When combined, they encourage a range of stakeholders to participate in drafting management guidelines for the jarrah forest. Several authors have already identified important, quantifiable components of forest ecosystem health (e.g., O'Laughlin, 1994, 1996). Others have provided quantitative definitions of precaution with a basis in statistical power analysis (e.g., Gray and Bewers, 1996; Buhl-Mortensen and Welin, 1998) and a framework for applying the Precautionary Principle (Deville and Harding, 1997). This application of the Precautionary Principle to assessing the ecosystem health of the jarrah forest is unoriginal in its components, but rather is a case study using tools that others have developed. Nevertheless, it demonstrates that despite much effort and good intentions, current management of the jarrah forest falls short of a quantitative, precautionary assessment of the possible significant impacts of forestry practices on ecosystem health because it does not set quantitative standards for response variables. As a result, this discussion provides little confidence in the ability of government regulatory agencies in Western Australia to regulate an economically viable timber industry, which does not compromise ecosystem health.

Two important weaknesses of the approach are that (1) decisions based on a small set of indicators would not anticipate or preclude deleterious impacts in other variables of ecosystem health, and (2) that the case study does not endorse one action over others when evaluating precautionary measures. The solution to both problems lies in full involvement of stakeholders in determining initial definitions and indicators and complete availability of relevant information for decision making. Many authors have called for these measures (e.g., Norton, 1996; Calver and Dell, 1998b; Calver et al., 1998) and they are implicit in mediation processes for forest management developed in eastern Australia (Pressey, 1998, and included references). The acknowledged popular acceptance of both ecosystem health (O'Laughlin, 1996) and the Precautionary Principle (Earll, 1992) should contribute to wide involvement of stakeholders. Criteria for monitoring ecosystem health can then be agreed on, and steps can be taken if any criterion is compromised.

Significant successes in applying the Precautionary Principle as a political process have occurred in an atmosphere of accord regarding goals (e.g., Kruger et al., 1997; Santillo et al., 1998). In other circumstances, such as the case discussed by Calver et al. (1999), applications of the Precautionary Principle have been contested politically or in the courts and judgment has been complicated by the lack of clear, quantitative definitions. This is a driving force in the development of both legal and scientific definitions of the principle (e.g., O'Riordan and Jordan 1995) and is a strong basis for attempting the participatory, quantitative approach proposed in this chapter.

Perhaps the true role of quantitative approaches will come in monitoring activities permitted under the Precautionary Principle, subject to the condition that they will be monitored and halted before any serious or irreversible damage occurs (Deville and Harding, 1997). Such a position gives primacy to the political process in initial applications of the Precautionary Principle and uses science and statistics effectively in placing unambiguous quantitative guidelines on the standards expected of any activity permitted. The jarrah forest case is a clear example of the value of developing such quantitative guidelines. If stakeholders can agree on quantitative criteria to maintain forest ecosystem health in the face of economic use and clear standards to ensure that each criterion is not violated, then progress will be made in resolving current disputes over forest management.

ACKNOWLEDGMENTS

The ideas presented in this chapter have benefited greatly from discussions with Stuart Bradley, Barbara Jones, Harry Recher, and Ian Wright and from a review by L.D. Ford. However, these colleagues do not necessarily endorse all the author's opinions.

REFERENCES

Abbott, I. and Christensen, P., Application of ecological and evolutionary principles to forest management in Western Australia, *Austr. For.*, 57, 109–122, 1994.

Abbott, I. and Christensen, P., Objective knowledge, ideology and the forests of Western Australia, *Austr. For.*, 59, 206–212, 1996.

Abbott, I. and van Heurck, P., Response of bird populations in jarrah and yarri forest in Western Australia following removal of half the canopy of the jarrah forest, *Aust. For.*, 48, 227–234, 1985.

Buhl-Mortensen, L. and Welin, S., The ethics of doing policy relevant science: the precautionary principle and the significance of non-significant results, *Sci. Eng. Ethics*, 4, 401–412, 1998.

Calver, M.C., Lessons from preventive medicine for the precautionary principle and ecosystem health, *Ecosyst. Health*, 6, 99–107, 2000.

Calver, M.C. and Dell, J., Conservation status of mammals and birds in south-western Australian forests, I. Is there evidence of direct links between forestry practices and species decline and extinction? *Pac. Conserv. Biol.*, 4:296–314, 1998a.

Calver, M.C. and Dell, J., Conservation status of mammals and birds in south-western Australian forests, II. Are there unstudied, indirect or long-term links between forestry practices and species decline and extinction? *Pac. Conserv. Biol.*, 4, 315–325, 1998b.

Calver, M.C., Hobbs, R.J., Horwitz, P., and Main, A.R., Science, principles and forest management: a response to Abbott and Christensen, *Austr. For.*, 59, 1–6, 1996.

Calver, M.C., Dickman, C.R., Feller, M.C., Hobbs, R.J., Horwitz, P., Recher, H.F., and Wardell-Johnson, G., Towards resolving conflict between forestry and conservation in Western Australia, *Austr. For.*, 61, 258–266, 1998.

Calver, M.C., Bradley, J.S., and Wright, I.W., Towards scientific contributions in applying the precautionary principle: an example from Western Australia, *Pac. Conserv. Biol.*, 5, 1–10, 1999.

Cohen, J., *Statistical Power Analysis for the Behavioural Sciences*, Lawrence Erlbaum Associates, Hillsdale, NJ, 1988.

Craig, M.D., The Short-Term Impacts of Timber Harvesting on the Jarrah Forest Avifauna, Ph.D. thesis, University of Western Australia, Crawley, 2000.

Davison, E.M. and Shearer, B.L., Phytophthora species in indigenous forests in Australia, *N.Z. J.For. Sci.*, 19, 277–289, 1989.

DeGraaf, R.M. and Miller, R.I., Eds., *Conservation of Faunal Diversity in Forested Landscapes*, Chapman & Hall, London, 1996.

Deville, A. and Harding, R., *Applying the Precautionary Principle*, Federation Press, Sydney, 1997.

Di Stefano, J., Power analysis and sustainable forest management, *Forest Ecology and Management*, 154, 141–153, 2001.

Driscoll, D.A. and Roberts, J.D., Impact of fuel-reduction burning on the frog *Geocrinia lutea* in southwest Western Australia, *Austr. J. Ecol.*, 22, 334–339, 1997.

Earll, R.C., Commonsense and the precautionary principle — an environmentalist's perspective, *Mar. Pollut. Bull.*, 24, 182–186, 1992.

Elashoff, J.D., *nQuery Advisor Version 2.0 Users' Guide*, Dixon Associates, Los Angeles, 1997.

Gray, J.S. and Bewers, J.M., Towards a scientific definition of the precautionary principle, *Mar. Pollut. Bull.*, 32, 768–771, 1996.

Horwitz, P. and Calver, M.C., Credible science? Evaluating the Regional Forest Agreement process in western Australia, *Austr. J. Environ. Manage.*, 5, 213–225, 1998.

Kimmins, J.P., The health and integrity of forest ecosystems: are they threatened by forestry? *Ecosyst. Health*, 2, 5–18, 1996.

Kruger, F.J., van Wilgen, B.W., Weaver, A.B., and Greyling, T., Sustainable development and the environment: lessons from the St. Lucia environmental impact assessment, *S. Afr. J. Sci.*, 93, 23–33, 1997.

Lunney, D., Ed., *Conservation of Australia's Forest Fauna*, The Royal Zoological Society of New South Wales, Mosman, New South Wales, 1991.

Mawson, P.R. and Long, J.L., Size and age parameters of nest trees used by four species of parrot and one species of cockatoo in south-west Australia, *Emu*, 94, 149–155, 1994.

Nichols, O.G. and Muir, B., Vertebrates of the jarrah forest, in *The Jarrah Forest: A Complex Mediterranean Ecosystem*, Dell, B., Havel, J.J., and Malajczuk, N., Eds., Kluwer Academic Publishers, Dordrecht, 1989, pp. 133–153.

Norton, T.W., Conservation of biological diversity in temperate and boreal forest ecosystems, *For. Ecol. Manage.*, 85, 1–7, 1996.

O'Laughlin, J., Defining and measuring forest health, *J. Sustainable For.*, 2, 65–85, 1994.

O'Laughlin, J., Forest ecosystem health assessment issues: definition, measurement and management implications, *Ecosyst. Health*, 2, 19–39, 1996.

O'Riordan, T. and Jordan, A., The precautionary principle in contemporary environmental politics, *Environ. Values*, 4, 191–212, 1995.

Pressey, R., Algorithms, politics and timber: an example of the role of science in a public, political negotiation process over new conservation areas in production forests, in *Ecology for Everyone*, Wills, R. and Hobbs, R., Eds., Surrey Beatty & Sons, Chipping Norton, New South Wales, 1998, pp. 73–87.

Rapport, D., Defining ecosystem health, in *Ecosystem Health*, Rapport, D. et al., Eds., Blackwell Science, Malden, MA, 1998b, pp. 18–33.

Rapport, D., Answering the critics, in *Ecosystem Health*, Rapport, D. et al., Eds., Blackwell Science, Malden, MA, 1998a, pp. 41–50.

Recher, H.F., Conservation and management of eucalypt forest vertebrates, in *Conservation of Faunal Diversity in Forested Landscapes*, DeGraaf, R.M. and Miller, R.I., Eds., Chapman & Hall, London, 1996, pp. 339–388.

Resource Assessment Commission (RAC), Ecological Impacts of Forest Use: A Survey of Completed Research, research paper 9, Canberra, Australia, 1993.

Roe, J.S., Report of an expedition under the surveyor-general, Mr. J S Roe, to the south-eastward of Perth, in Western Australia, between the months of September, 1848, and February, 1849, to the Hon. the colonial secretary, *J. R. Geogr. Soc. London*, 22, 1–57, 1852.

Rogers, M.F., Sinden, J.A., and De Lacy, T., The precautionary principle for environmental management: a defensive-expenditure application, *J. Environ. Manage.*, 51, 343–360, 1997.

Rundle, G.E., History of conservation reserves in the south-west of Western Australia, *J. R. Soc. West. Austr.*, 79, 225–240, 1996.

Russell, E. and Rowley, I., The effects of fire on a population of red-winged fairy-wrens *Malurus elegans* in Karri forest in southwestern Australia, *Pac. Conserv. Biol.*, 4, 197–208, 1998.

Santillo, D., Stringer, R.L., Johnson, P.A., and Tickner, J., The precautionary principle: protecting against failures of scientific method and risk assessment, *Mar. Pollut. Bull.*, 36:939–950, 1998.

Stoneman, G.L., Rayner, M.E., and Bradshaw, F.J., Size and age parameters of nest trees used by four species of parrot and one species of cockatoo in south-western Australia: critique, *Emu*, 97, 94–96, 1997.

Thomas, L. and Juanes, F., The importance of statistical power analysis: an example from *Animal Behaviour*, *Anim. Behav.*, 52, 856–859, 1996.

Turner, J. and Lambert, M., Development of indicators of sustainable development in Australia's forests, in *Conservation Outside Nature Reserves*, Hale, P. and Lamb, D., Eds., Centre for Conservation Biology, The University of Queensland, Brisbane, 1997, pp. 471–475.

Underwood, A.J., Environmental decision-making and the precautionary principle: what does this principle mean in environmental sampling practice? *Landscape Urban Plann.*, 37, 137–146, 1997.

Varis, O. and Kuikka, S., Joint use of multiple environmental assessment models by a Bayesian meta-model — the Baltic salmon case, *Ecological Modelling*, 102, 341–351, 1997.

Vora, R.S., Developing programs to monitor ecosystem health and effectiveness of management practices on lakes states national forests USA, *Biol. Conserv.*, 80, 289–302, 1997.

Wallace, W.R., Fire in the jarrah forest environment, *J. R. Soc. West. Austr.*, 49, 33–44, 1965.

Wardell-Johnson, G. and Nichols, O., Forest wildlife and habitat management in southwestern Australia: knowledge, research and direction, in *Conservation of Australia's Forest Fauna*, Lunney, D., Ed., The Royal Zoological Society of New South Wales, Mosman, New South Wales, 1991, pp. 161–192.

Williams, M.R., Abbott, I., Liddelow, G.L., Vellios, C., Wheeler, I.B., and Mellican, A.E., Recovery of bird populations after clearfelling of tall open eucalypt forest in Western Australia, *J. Appl. Ecol.*, 38, 910–920, 2001.

Yazvenko, S.B. and Rapport, D.J., A framework for assessing forest ecosystem health, *Ecosyst. Health*, 2, 40–51, 1996.

CHAPTER 93

National Indicators of Forest Ecosystem Health: A Science Perspective on the Canadian Initiative

Harry Hirvonen

INTRODUCTION

In 1993, the Canadian Council of Forest Ministers (CCFM) embarked on an initiative to define, measure, and report on the forest values Canadians want to sustain and enhance. An extensive consultative process was carried out involving scientists and others from the federal, provincial, and territorial governments, as well as experts from the academic community, industry, and non-governmental organizations. Two years later, as an outcome of this initiative, the report "Defining Sustainable Forest Management: A Canadian Approach to Criteria and Indicators" was published (Canadian Council of Forest Ministers, 1995). This report represented the best-available scientific knowledge on indicators related to sustainable forest management in the Canadian context. Appendix 93.I lists the six criteria and elements associated with each criterion.

Soon after the release of the national framework for criteria and indicators for sustainable forest management in 1995, the CCFM created a task force to report on Canada's ability to measure the various indicators. Those findings are the basis of the report "Criteria and Indicators of Sustainable Forest Management in Canada: Technical Report 1997" (Canadian Council of Forest Ministers, 1997), which provides information on Canada's capacity to measure forest sustainability.

This CCFM report, in terms of the ecological criteria and indicators, was more of a status report on the various indicators than an analytical assessment. For example, in discussion of one of the indicators, "rates of pollutant deposition," the amount of wet-sulfate deposition was depicted. However, discussion was minimal on the impact of this deposition on forest ecosystems. The report was good at describing what was happening, and sometimes on why it was happening, but not as good at describing its impact or significance. The report's value lay in introducing the framework in a concrete sense and as a foundation upon which to build. The 2000 CCFM report aims to look more closely at this significance aspect. It strives to better address questions of sustainability. A specific objective is to discuss cause-and-effect relationships.

Ecosystem health is inherent in the determination of sustainable forest management. Several of the ecological indicators in the Canadian criteria and indicators framework relate directly to assessment of the status and change in forest ecosystem health. This chapter focuses on the practical considerations associated with reporting on progress toward sustainability, touching on aspects such as definition, scaling, targets, and cumulative effects. Although the CCFM criteria and indicators include economic and social components, concentration here is on the ecological criteria and indicators pertinent to forest ecosystem health. Assessment of progress toward sustainability, however, must include the complete package of criteria and associated indicators.

THE POLICY CONNECTION

Reporting on progress toward sustainability is complicated for many reasons. One of the more complicated factors is the definition of sustainability itself. CCFM defines sustainable forest management as "the development of forests to meet current needs without prejudice to their future productivity, ecological diversity, or capacity to regenerate" (Canadian Council of Forest Ministers, 1997). The term *needs* encompasses ecological, social, and economic considerations. This definition leaves much latitude for interpretation. Some may view it as a way to limit development; others may see it as a means to ensure continued and enhanced development through minimal interference. These basic philosophical approaches to a common definition are hard to reconcile.

Nonetheless, these political spins on sustainability do not diminish or alter the task of the scientist who strives to understand and measure changes in ecosystem health. The objective is to provide the best science-based information as input to policy questions. Specifically, policy questions that guide forest health research and monitoring and reporting within the Canadian Forest Service (CFS) include:

How is the health of Canada's forest ecosystems changing, and why?

What trends and changes can be predicted regarding the health of major forest regions in Canada?

How do conditions and predictions for Canada's forests compare with the forest conditions reported or predicted in other countries?

What methods are required to improve the accuracy and effectiveness of forest health assessments and provide comprehensive data for the national initiative on criteria and indicators of sustainable forest management?

These questions drive the forest health program; still, answers to them are difficult. Several factors serve as obstacles to this national reporting: the current inability within Canada to measure ecological change over time, the limited understanding of cumulative impacts, and a myriad of perceptions of what a healthy forest really looks like. These science concerns are discussed in more detail later.

SUSTAINABILITY AND FOREST ECOSYSTEM HEALTH

Sustainable forest management and forest ecosystem health go hand in hand. A sustainably managed forest is a healthy forest. Inherent in sustainability is the obligation to think at regional scales of planning (Hirvonen, 1990; Marshall et al., 1993; King, 1993). Sustainable forest management must consider ecological dynamics over time, ecological cycles, and interactions with adjoining ecosystems. One cannot sustain a tree or even a stand; each is part of a bigger whole. If the regional landscape is not sustainable, certainly the components therein are not. More to the point, what impacts the regional landscape will inevitably impact the components within. This is not necessarily the case the other way around. For example, one may harvest a stand of trees and the sustainability of the regional landscape may not be affected.

An analogy, depicting the interdependence of the site and the regional ecosystem, may be of an individual farming operation. No matter how ecologically friendly and forward-thinking a farmer is with regard to his or her own operation, that farm will not be sustainable if the region within which it exists is characterized by unsustainable activities. Pollution may impact on the farm through the air, the waterways, or the ground. Regional activities may alter insect or disease cycles, climate, or wildlife habitat, all of which could be detrimental to the isolated farmer in one way or another.

Such ecological isolation from the surrounding area has been termed *islandization* in the context of national parks and other protected areas (Panel on the Ecological Integrity of Canada's National Parks, 2000; Loo and Hirvonen, 1999). Protected areas surrounded by incompatible land-use activities cannot maintain their ecological integrity, regardless of plausible efforts within the boundaries. Ecological islandization is not compatible with sustainability.

In this context of sustainability, ecosystem health must also be considered as a regional or landscape concept (Hirvonen, 1990; Forest Health Network, 1999). Certainly, identification of ecosystems is dependent on scale. A pond, wetland, or forest plantation may be considered an ecosystem, as may components of these ecosystems. Much depends on the perspective of the user and the scale of application (King, 1993). Regardless, in terms of health all these ecosystems, at whatever scale of conceptualization, are linked. An adverse impact on one ecosystem will certainly reflect on the health of adjacent ones. To understand ecosystem structure and function, one must understand the broad picture. This broad ecological picture can be encapsulated by the pressure, exposure, and response model of stressors on ecosystems (Hammond et al., 1995; Environment Canada, 1991; Hunsacker and Carpenter, 1990). Managers must understand the origin of the stress, its pathways, its impacts, and finally its significance to the health of desirable forest ecosystems. Such an analysis necessarily takes a regional and sometimes national, continental, or global view of ecological interactions and relationships.

SCIENCE ISSUES AND INDICATORS OF FOREST ECOSYSTEM HEALTH

As ecology is not a science with black-and-white answers, many factors come to play in understanding interrelationships between external pressures and ecological cycles. A brief discussion ensues on key scientific considerations for sustainability and indicators of forest ecosystem health.

Scale of Application

Scale, both spatial and temporal, defines the indicator. As mentioned earlier, sustainability is a regional concept and progress toward it must be measured by regional-scale indicators. It is important to understand the relationship between spatial scales and ecological processes and functions to develop indicators to track and report on ecosystem change. Each spatial scale is associated with natural and anthropogenic influences or stressors that impact ecosystem components. Broadscale influences impact broad-scale ecological cycles. For example, at global or continental scales, stressors such as global change or long-range transport of toxic chemicals may be an issue. At regional scales, perhaps the issue becomes one associated with acid rain. At local levels, a myriad of factors including soils, vegetation, and aspect come into play to determine an appropriate indicator (Hirvonen, 1990; Marshall et al., 1993). Local indicators do not report on sustainability directly. They may serve to track specific forest management practices which, in total, may be looked at to determine progress toward sustainability.

The linking of indicators to appropriate scales of application is paramount. The ecological context of the influence must be understood. The mixing of broadscale indicators with local conditions or site-specific indicators to track broad-scale influences is not good science. The Canadian framework of criteria and indicators for measuring progress toward sustainable forest management is for national reporting purposes. The intent is to report on changes to forest ecosystems at regional and national scales. The ecological indicators must reflect this. The current review process is looking at this question of scale among its list of issues. In part, the original 83 indicators have been pared to 49 for the CCFM 2000 report because of this consideration. For example, an original indicator, "water quality as measured by water chemistry, turbidity, etc.," is too specific to provide a regional context, much less a national one.

Monitoring Status and Change of Forest Ecosystems

A basic need for measuring progress toward sustainability is the ability to measure status and change over time in forest ecosystems. On a national basis in Canada, we can measure status but we are woefully inadequate in tracking change over time. The largest national ecological database for the country is Canada's Forest Inventory (CanFI), a compilation of individual provincial and territorial forest inventories for national reporting purposes (Lowe et al., 1996). A major weakness of this integrated database is that the information contained therein cannot be used to measure ecological change over time. The data are the most-recent available but they may be a few months old or over 30 years old, depending on the inventory cycles of the individual provinces (Lowe et al., 1996; Gray and Power, 1997).

In addition, the provincial inventories are not random, plot-based inventories. Data are not compiled from the same piece of land during the inventory cycles. Without permanent plots inventoried in comparable ways, measures of forest ecosystem change are not possible. CanFI has served its historical purpose very well: provision of information on the commercial forest useful for timber management purposes. In recent years, provinces and territories, along with the federal government, have recognized the need for forest inventories to provide information on biodiversity and other nontimber values, as well as timber values. These jurisdictions also recognize the need to measure ecological change over time in order to meet reporting obligations associated with the CCFM criteria and indicators of sustainable forest management. With this in mind, through the efforts of the Canadian Forest Inventory Committee, these jurisdictions are embarking jointly on a new random, plot-based National Forest Inventory (NFI) that will allow reporting on change over time in a statistically defensible manner. Recognizing the need to report on a regional basis, a minimum of 50 ground plots per ecological zone has been determined as necessary to provide minimal acceptable information for extrapolation to this ecozone level.

Initial reporting, based on NFI, is not expected until 2005. Until then, there does not exist a valid scientific basis for determination of changes in the health of Canada's forests. This lack of ecological monitoring data severely tests the interpretive application of the ecological indicators associated with the CCFM initiative.

Targets for Sustainability

Measurement of ecosystem change is one prerequisite in determining progress toward sustainability. Another prerequisite is the understanding in measurable terms of what the end points are for sustainability. There is a need for targets of sustainability. Ecological indicators are most useful if they can be compared with an acceptable target. A measure of change in itself is not helpful in answering the question of whether things are getting better or worse. The current problem is that no such generally acceptable targets exist for ecological indicators.

Targets are available for some pollution concerns. In Canada, for example, target loads of 8 to 20 kg/ha/year of acid rain have been suggested for receiving waters, depending on the sensitivity of these waters (Research and Monitoring Coordinating Committee, 1990; Federal/Provincial/ Territorial Ministers of Energy and Environment, 1998). Levels at or below these loads are not considered harmful to these lakes. No such targets exist for air pollution or other stressors on forest ecosystems (Hall et al., 1998).

For acidic deposition, critical loads for forest ecosystems are being developed. Some preliminary results carried out through the University of New Brunswick in conjunction with CFS forest health data are available (Moayeri and Arp, 1996). The Eastern Premiers/New England Governors Secretariat is carrying out a major forest-mapping project for eastern Canada and the northeastern U.S. to establish critical loads for major forest ecosystems that characterize this part of the continent.

Ecological Baseline

In the CCFM process, there is mention of a historical baseline as a reference point for change, the inference being that such a baseline is a surrogate for a healthy forest. The term *historical,* in the Canadian context, is really a red herring. There is no argument that a baseline is needed for any determination of change. Otherwise, no reference exists for assessing whether forest health is improving, degrading, or remaining constant. However, all baselines are *de facto* historical, whether they are pre-European settlement, post-World War II, or present-day; the question is whether current forest ecosystems are considered healthy.

Holling (1986) states that ecosystems normally undergo both internal and external perturbations and have adjusted to these disturbances over long periods of time. These perturbations are in fact essential to the sustainability of the ecosystems. In other words, they are part and parcel of the ecological integrity of the ecosystem. Forests that maintain their ecological integrity can be considered healthy (Rowe, 1961; Holling, 1973; Woodley, 1993; Forest Health Network, 1999).

In Canada, large-scale forest harvesting and other intensive forest uses are barely 100 years old. For much of central, northern, and western Canada, current harvesting occurs within previously nonharvested areas. Second-rotation crops cover a relatively small portion of the commercial forest. With some regional exceptions, the present forest ecosystems have remained much the same since pre-European settlement. These forest ecosystems have evolved over time, having adjusted to natural perturbations such as ice storms, fire, and insect and disease attacks. Their ecological integrity has not been compromised and, as such, may serve as a baseline for change. This fact is reinforced through a recent national forest health report by CFS which states that Canada's forested ecozones are generally healthy. Some forested ecozones are under sustained stress from anthropogenic influences and continue to serve as priority cases for regional monitoring of forest health (Forest Health Network, 1999).

Inherent in the notion of an ecological baseline for healthy forests must be variability. Ecosystems are not static; they encompass spatial and temporal aspects. Succession, natural influences, and other ecological processes all come into play and must be recognized. Protected areas serve to illustrate this principle. The term *protected* is, in many ways, a misnomer. No one stage of ecosystem evolution can be protected in perpetuity. Ecosystems, by their nature, are dynamic. The key to conserving desirable ecosystems is to conserve the ecological functions and processes of these ecosystems. Succession, changing species composition and age classes, death and renewal, and associated nutrient, hydrological, and geochemical cycles must be included in the understanding of baseline.

Conservation of ecological function and process also has a spatial consideration. For example, protected areas have to be large enough to maintain ecological integrity. Size and fragmentation are key measures of this integrity. Also, ecosystems cannot be isolated from surrounding ecosystems. Some of Canada's highly prized national parks are threatened because of the islandization of these parks over the years. One national park within the Prairie Provinces has essentially turned into a forest refuge. Farming operations surround the park boundary. There are other examples in the mountain parks of British Columbia where clear-cutting operations abut the parks. These forested parks fall into the category of significant exceptions to healthy forests mentioned earlier. If they do not currently show signs of deterioration, the constant pressure from adjacent land-use activities certainly makes them prone to deterioration. To reiterate, sustainability is a regional concept. A bog, a stand of trees, or a national park cannot be sustained if not linked in a sound, ecological manner with surrounding ecosystems.

Human-Induced vs. Natural Variation

A recently released national forest health report for Canada states that a "healthy forest is one that maintains and sustains desirable ecosystem functions and processes" (Forest Health Network,

1999). Healthy forests are essential to sustainability. The determination of change from "desirable ecosystem functions and processes" is not straightforward. Certainly, *desirable* is a value judgment and can cause confusion. Yet, even if desirable is acceptable, the challenge to measure change is one thing; the challenge to measure unacceptable change is daunting. This challenge stems from the fact that both healthy and unhealthy ecosystems are characterized by temporal and inherent natural variability in their ecological cycles. The presence of dead and dying trees alone does not reflect a deteriorating ecosystem. To reiterate: life, death, and rejuvenation are all dynamic parts of healthy ecosystems. So are natural influences of ecological change such as insect and disease infestations and fire. In fact, without cyclical occurrence of these influences, our forested ecosystems will not sustain themselves. The absence of these influences, in fact, may be signs of ecological deterioration of desired ecosystems.

In essence, assessment of forest ecosystem health involves both an understanding of successional change and delineation of human-induced variation on ecological function and processes (Freedman, 1995). Certain anthropogenic influences are easy to quantify, such as air pollution. If levels of pollutants such as smog, heavy metals, and acidic deposition increase, the assumption is that this is undesirable and not sustainable. We are very good at tracking emission and deposition levels, and we can model trajectories for long-range transport of these pollutants; cause-and-effect relationships on receiving environments have had mixed results. As an example, good correlation exists regarding acidic deposition and aquatic impacts on fish populations (Jeffries, 1997). There is still uncertainty, however, concerning impacts of acidic deposition to forest productivity or site deterioration (Hall et al., 1998). Effects of land-use activities such as harvesting on cycles of insects and diseases, for example, are difficult to determine without long-term monitoring because the natural variability of cycles of infestation is great.

Ecological Reporting Framework

Canada is a huge country comprising close to 1 billion hectares. Its forest land comprises some 417 million hectares. A myriad of forest ecosystems characterize the country. These forests have both national and regional significance to the country's health and well-being. Because of the diverse nature of the country's forests, national averaging of the health of these forests is meaningless.

The use of Canada's national ecological classification system as a reporting framework for the Canadian Forest Service allows for a consistent national reporting framework, while at the same time there is flexibility to focus on major regional forest ecosystems. This framework divides Canada into 15 broad terrestrial ecozones. These ecozones form the most general level of a nested hierarchical system that divides them into 214 ecoregions, which are further subdivided into over 1000 ecodistricts (Ecological Stratification Working Group, 1996). Nine of the ecozones have a substantial forestland component and form the basis for national reporting on the country's forests (Forest Health Network, 1999). The CCFM ecological indicators are reported, to the extent possible, by ecozone.

The use of ecozones as reporting frameworks has advantages and disadvantages to reporting on forest ecosystem health. A major advantage lies in the fact that reporting boundaries are based on ecology rather than jurisdiction. Ecozones transcend provincial and other jurisdictional boundaries. They also have been linked with ecological units of the adjacent states of the U.S., allowing for a common international reporting framework. Ecozones and their subdivisions are used as the ecological reporting framework by several provinces and other federal departments. This fact aids in communicating information to the public, and in exchange of information.

Compilation of data is a concern. Historically, any ecological data gathered were normally aggregated by jurisdictional boundaries such as county, state, or provincial level. Conversion of these data to conform with ecological boundaries is difficult. Another characteristic of the ecological framework is its holistic nature. Ecological units were developed through consideration of major ecological influences at the given scale of application. It is not a specific resource sector survey. Compromises had to be made by each resource sector to allow for development and use of a common ecological framework that most sectors can apply for reporting purposes. Naturally, a specific forest-based ecosystem classification would better delineate forest ecosystems at various scales. However, these kinds of specific resource classifications would have minimal use by other resource sectors. A result may be several maps covering the same region, a nightmare for the general public to understand, or to allow for spatial comparisons. A common acceptable spatial framework facilitates reporting and data exchange while at the same time facilitating communication with the general public (Carpenter et al., 1998).

Cumulative Effects

Cumulative impacts and complex interrelationships among ecosystem components serve to complicate the measurement of ecosystem health. To gain some understanding of cause-and-effect relationships, cumulative knowledge gained from several research and monitoring initiatives is needed more often than not. Such analyses must be carried out at the scientific level and then put into a format understandable by decision makers. Not only must data be translated to appropriately address the criteria and indicators of sustainable forest management, but the indicators themselves must be analyzed together to provide the desired scientific advice to the policy process. The CCFM criteria and indicators process emphasizes that all criteria and related indicators must be considered together in the assessment of sustainable forest management. Such an assessment is meaningless if based on some but not all criteria or indicators.

NATIONAL VS. INTERNATIONAL REPORTING REQUIREMENTS

Not only does Canada have a responsibility to report on progress toward sustainable forest management nationally, but it has international obligations as well. Canada has signed on to the Montreal Process Criteria and Indicators for Conservation and the Sustainable Management of Temperate and Boreal Forests. One country report has already been produced (Montreal Process Liaison Office, 1997). In terms of science, the choice between the two processes becomes one of compatibility. Are these two processes mutually compatible or are we comparing apples with oranges?

The Montreal Process is an international response to the call of the 1992 United Nations Conference on Environment and Development to improve the quality and management of the global forest estate (Canadian Forest Service, 1995). It is informally named after a seminar of experts held in Montreal, November 1993, looking at Sustainable Development of Temperate and Boreal Forests. The process brings together in common purpose 12 countries from the Northern and Southern Hemispheres. At its sixth meeting (Santiago, Chile, February 1995), the Montreal Process Working Group, which now includes Argentina, Australia, Canada, Chile, China, Japan, Republic of Korea, Mexico, New Zealand, the Russian Federation, the U.S., and Uruguay, endorsed the Santiago Declaration, a statement of political commitment with an associated, comprehensive set of criteria and indicators for the conservation and sustainable management of temperate and boreal forests outside Europe. European countries are part of a similar process to identify and measure indicators of sustainable forest management, known as the Helsinki Process. Canada serves as the Secretariat for the Montreal Process and has "Observer" status within the Helsinki Process.

Canada is an exporting nation as far as its forest sector is concerned. How customers in Europe and other international markets interact and react with the Canadian forest-sector forest practices can and does affect exports. More and more overseas customers are insisting that Canadian forest products and services be derived from sustainable forest management practices. Indicators of progress toward such sustainability are key in this process.

Table 93.1   Criteria of Sustainable Forest Management Frameworks: Comparison of the CCFM Initiative and the Montreal Process

CCFM Initiative	The Montreal Process
Conservation of biological diversity	Conservation of biological diversity
Ecosystem condition and productivity	Maintenance of productive capacity of forest ecosystems
	Maintenance of forest ecosystem health and vitality
Conservation of soil and water resources	Conservation and maintenance of soil and water resources
Global ecological cycles	Maintenance of forest contribution to global carbon cycles
Multiple benefits	Maintenance and enhancement of long-term, multiple socioeconomic benefits to meet the needs of societies
Society's responsibility	Legal, institutional, and economic framework for forest conservation and sustainable management

Thus, it is essential that the CCFM criteria and indicators initiative be compatible and in general agreement with the Montreal Process and other international processes. Domestic measures of progress toward sustainability must have international credibility, as well as serve to guide domestic forest management practices. The Montreal Process comprises seven criteria and 67 associated indicators; the CCFM process has six criteria and 83 indicators. Discrepancies do exist in the wording and context of some of the indicators (see Appendices 93.I and 93.II).

Still, there is substantial agreement at the criterion level (Table 93.1). The Montreal Process has an additional criterion not evident within the Canadian context. This is related to the legal and institutional framework in place to carry out the process, which may be an issue for certain countries that are part of the Montreal Process.

In terms of ecological criteria, the CCFM process has four such criteria and the Montreal Process has five. The latter essentially breaks down the CCFM criterion of "ecosystem condition and productivity" into the "productivity" and "condition" components. With this breakdown, the Montreal Process relates the criterion "maintenance of productive capacity of forest ecosystems" to the commercial forest. The CCFM indicators reflect a broader ecological interpretation.

All the science issues mentioned earlier apply to the international Montreal Process, as they do for the CCFM process. A close look at the ecological indicators of both processes reveals concerns in terms of appropriate scales of application. For example, the criterion relating to soil conservation and water resources, found in both processes, has indicators with doubtful application at national scales. Indicators relating to erosion, water chemistry, and aquatic biodiversity are local and watershed scale indicators within the Canadian context. The depiction of these indicators nationally would be inappropriate for a large and ecologically diverse country such as Canada.

Slight differences in the wording of indicators under a given criterion may cause concern when reporting is required under the national and international processes. The criterion related to conservation of biological diversity has consistent indicators for both processes and should not cause any interpretive concern in reporting on these indicators. Problems arise with some of the indicators associated with the criterion related to maintenance and enhancement of forest condition and productivity. For example, the CCFM process emphasizes area and severity of disturbances such as fire, insects, and diseases. The Montreal Process emphasizes area and percent of forest affected by disturbances beyond the range of historic variation. Undoubtedly, the same data, where available, would be used to address both processes but the spin would be very different. In this situation, the author would support modifying the CCFM process indicator to conform with the Montreal Process because annual area and severity of disturbances are meaningless in an ecological sense without understanding historical variation. Of course, data are not available to compile measures of severity or historical variation on a national basis. This lack of long-term monitoring data is prevalent for most ecological indicators.

The criterion "conservation and maintenance of soil and water resources" offers another example of similar yet different indicators that may provide confusing messages to a general audience. In this criterion, the CCFM Process emphasizes indicators related to soil erosion and loss of organic matter to harvested forestland. The Montreal Process applies these indicators to all forestland. The public concern in Canada is focused on forest management practices and impacts on the environment; thus the Canadian emphasis makes sense. There is a direct connection to cause and effect missing in the Montreal Process.

These are only two examples that illustrate the need to bring together these two processes on reporting progress toward sustainable forest management. In doing this, analyses and interpretations required by Canadian forest scientists to meet domestic and international reporting requirements in relation to criteria and indicators of forest ecosystem health and more broadly of sustainable forest management would be less onerous. Certainly, the perspectives and associated constraints of the two processes serve to justify, to some degree, the slight differences between the two processes: one national in scope and focused on Canadian views, the other international in scope and, by necessity, having to consider the views of many nations. Still, in the end, the objective of both processes is to communicate to a broad audience progress toward sustainability. For Canada, this objective may not be met because the differences due to nuances in definition may confuse rather than enlighten the intended audience.

REFERENCES

Canadian Council of Forest Ministers, Defining sustainable forest management: a Canadian approach to criteria and indicators, Natural Resources Canada, Canadian Forest Service, Ottawa, Ontario, 1995.

Canadian Council of Forest Ministers, Criteria and indicators of sustainable forest management in Canada: technical report, Natural Resources Canada, Canadian Forest Service, Ottawa, Ontario, 1997.

Canadian Forest Service, Criteria and indicators for the conservation and sustainable management of temperate and boreal forests: the Montreal Process, Natural Resources Canada, Canadian Forest Service, Ottawa, Ontario, 1995.

Carpenter, C., Busch, W-D., Cleland, D., Gallegos, J., Harris, R., Holm, R., Topik, C., and Williamson, A., The use of ecological classification in management, in *Ecological Stewardship: A Common Reference for Ecosystem Management,* Szaro, R.C. et al., Eds., Elsevier Science, the Hague, the Netherlands, 1998, pp. 395–430.

Ecological Stratification Working Group, A national ecological framework for Canada. Agriculture and Agri-Food Canada, Centre for Land and Biological Resources Research, Research Branch; and Environment Canada, Environment Conservation Service, State of the Environment Directorate, Ottawa, Ontario, 1996.

Environment Canada, A report on Canada's progress towards a national set of ecological indicators, SOE Report 90–1, Indicators Task Force, State of the Environment Reporting, Environment Canada, Ottawa, Ontario, 1991.

Federal/Provincial/Territorial Ministers of Energy and Environment, The Canada-wide acid rain strategy for post-2000, Strategy and Supporting Document, 1998.

Forest Health Network, Forest health in Canada: an overview 1998, Natural Resources Canada, Canadian Forest Service, Fredericton, New Brunswick, 1999.

Freedman, B., *Environmental Ecology: The Ecological Effects of Pollution, Disturbance, and Other Stresses,* 2nd ed., Academic Press, San Diego, 1995.

Gray, S. and Power, K., Canada's forest inventory 1991: the 1994 Version, Information Report BC-X-362E, Natural Resources Canada, Canadian Forest Service, Pacific Forestry Centre, Victoria, British Columbia, 1997.

Hall, P., Bowers, W., Hirvonen, H., Hogan, G., Foster, N., Morrison, I., Percy, K., Cox, R., and Arp, P., Effects of acidic deposition on Canada's forests, Natural Resources Canada, Canadian Forest Service, Ottawa, Ontario, 1998.

Hammond, A., Adriaanse, A., Rodenburg, E., Bryant, D., and Woodward, R., Environmental indicators: a systematic approach to measuring and reporting on environmental policy performance in the context of sustainable development, World Resources Institute, Washington, D.C., 1995.

Hirvonen, H., The development of regional scale indicators: the Canadian experience, in *Ecological Indicators,* Vol. II, McKenzie, D., Hyatt, E., and McDonald, V., Eds., Elsevier Applied Science, London, 1990, pp. 901–915.

Holling, C.S., Resilience and stability of ecological systems, *Annu. Rev. Ecol. Syst.,* 4, 1–23, 1973.

Holling, C.S., The resilience of terrestrial ecosystems: local surprise and global change, in *Sustainable Development in the Biosphere,* Clark, W. and Munn, R., Eds., Oxford University Press, Oxford, 1986, pp. 292–320.

Hunsacker, C. and Carpenter, D., Eds., Ecological indicators for the environmental monitoring and assessment program, EPA 600/3–90/060, Environmental Protection Agency, Office of Research and Development, Research Triangle Park, NC, 1990.

Jeffries, D.S., Ed., 1997 Canadian Acid Rain Assessment, Vol. III, The Effects on Canada's Lakes, Rivers and Wetlands, Environment Canada, Ottawa, Ontario, 1997.

King, A.W., Considerations of scale and hierarchy, in *Ecological Integrity and the Management of Ecosystems,* Woodley, S., Kay, J., and Francis, G., Eds., St. Lucie Press, Boca Raton, FL, 1993, pp. 19–46.

Loo, J. and Hirvonen, H., Reporting and indicators: mechanisms in ecologically sustaining the forested sector, The George Wright Forum, 16(2), 64–74, 1999.

Lowe, J.J., Power, K., Gray, S., Canada's forest inventory 1991: the 1994 version, Information Report BC-X-362E, Natural Resources Canada, Canadian Forest Service, Pacific Forestry Centre, Victoria, British Columbia, 1996a.

Lowe, J., Power, K., and Marsan, M., Canada's forest inventory 1991: summary by terrestrial ecozones and ecoregions, Information Report BC-X-364E, Natural Resources Canada, Canadian Forest Service, Pacific Forestry Centre, Victoria, British Columbia, 1996b.

Marshall, I., Hirvonen, H., and Wiken, E., National and regional scale measures of Canada's ecosystem health, in *Ecological Integrity and the Management of Ecosystems,* Woodley, S., Kay, J., and Francis, G., St. Lucie Press, Boca Raton, FL, 1993, pp. 117–130.

Moayeri, M. and Arp, P., Assessing critical soil acidification load effects for ARNEWS sites: preliminary results, draft paper, Faculty of Forestry and Environmental Management, University of New Brunswick, Fredericton, New Brunswick, 1996.

Montreal Process Liaison Office, Canada's report on the Montreal Process criteria and indicators for the conservation and sustainable management of temperate and boreal forests, Natural Resources Canada, Canadian Forest Service, Ottawa, Ontario, 1997.

Panel on the Ecological Integrity of Canada's National Parks, Unimpaired for future generations: conserving ecological integrity within Canada's national parks, Report of the Panel on Ecological Integrity of Canada's National Parks, Ottawa, Ontario, 2000.

Research and Monitoring Coordinating Committee, The 1990 long range transport of air pollutants and acid precipitation assessment report: Part One, Executive Summary, Environment Canada, Ottawa, Ontario, 1990.

Rowe, J.S., The level-of-integration concept and ecology, *Ecology,* 42, 231–244, 1961.

Woodley, S., Monitoring and measuring ecosystem integrity in Canadian national parks, in *Ecological Integrity and the Management of Ecosystems,* Woodley, S., Kay, J., and Francis, G., Eds., St. Lucie Press, Boca Raton, FL, 1993, pp. 155–176.

Canadian Environmental Criteria and Indicators of Sustainable Forest Management

Six criteria define sustainable forest management in the Canadian context. Each of these six criteria has elements associated with it and progress within each element is measured through several indicators. In total, the six criteria comprise 22 elements and 83 indicators.

Elements and indicators are presented for the three criteria with direct relevance to forest ecosystem health.

Criterion 1: Conserving Biological Diversity
    Element 1.1: Ecosystem diversity
    Indicators:
        Percent and extent in area of forest types relative to historical condition and total forest area
        Percent and extent of area by forest type and age class
        Area percent and representativeness of forest types in protected areas
    Element 1.2: Species diversity
    Indicators:
        Number of known forest-dependent species classified as extinct, threatened, endangered, rare, or vulnerable relative to the total number of known forest-dependent species
        Population levels and changes over time for selected species and species guilds
        Number of known forest-dependent species that occupy only a small portion of their former range
    Element 1.3: Genetic diversity
    Indicators:
        Implementation of an *in situ/ex situ* genetic conservation strategy for commercial and endangered forest vegetation species
Criterion 2: Ecosystem Condition and Productivity
    Element 2.1: Disturbance and stress
    Indicators:
    Area and severity of insect attack
    Area and severity of disease infestations
    Area and severity of fire damage
    Rates of pollutant deposition
    Ozone concentration in forested regions
    Crown transparency in percentage by class
    Area and severity of occurrence of exotic species detrimental to forest condition
    Climate change as measured by temperature sums

Element 2.2: Ecosystem resilience

Indicators:

Percentage and extent of area by forest type and age class

Percentage of area successfully naturally regenerated and artificially regenerated

Element 2.3: Extant biomass

Indicators:

Mean annual increment by forest type and age class

Frequency of occurrence within selected indicator species

Criterion 3: Conserving Soil and Water Resources

Element 3.1: Physical environmental factors

Indicators:

Percentage of harvested area having significant soil compaction, displacement, erosion, puddling, loss of organic matter, etc.

Area of forest converted to nonforest land use

Water quality as measured by water chemistry, turbidity, etc.

Trends and timing of events in stream flows from forest catchments

Changes in the distribution and abundance of aquatic fauna

Element 3.2: Policy and protection

Indicators:

Percentage of forest managed primarily for soil and water protection

Percentage of forested area having road construction and stream crossing guidelines in place

Area, percentage, and representativeness of forest types in protected areas

Criterion 4: Forest Ecosystem Contributions to Global Ecological Cycles

Criterion 5: Multiple Benefits of Forests to Society

Criterion 6: Accepting Society's Responsibility for Sustainable Development

Criteria and Indicators for the Conservation and Sustainable Management of Temperate and Boreal Forests: The Montreal Process

Only the indicators associated with the first four ecological criteria of the Montreal Process are outlined.

Criterion 1: Conservation of Biological Diversity
    Indicators:
        Extent and area by forest type relative to total forest area
        Extent and area by forest type and by age class or successional stage
        Extent of area by forest type in protected area categories as defined by IUCN or other classification system
        Extent of area by forest type in protected area defined by age class or successional stage
        Fragmentation of forest types
        Number of forest-dependent species
        Status of forest-dependent species at risk of not maintaining viable breeding populations, as determined by legislation or scientific assessment
        Number of forest-dependent species that occupy a small portion of their former range
        Population levels of representative species from diverse habitats monitored across their range
Criterion 2: Maintenance of Productive Capacity of Forest Ecosystems
    Indicators:
        Area of forestland and net area of forestland available for timber production
        Total growing stock of both merchantable and nonmerchantable tree species on forestland available for timber production
        Area and growing stock of plantations of native and exotic species
        Annual removal of wood products compared to the volume determined sustainable
        Annual removal of nontimber forest products compared to the level determined to be sustainable
Criterion 3: Maintenance of Forest Ecosystem Health and Vitality
    Indicators:
        Area and percent of forest affected by processes or agents beyond the range of historic variation
        Area and percent of forestland subjected to levels of specific air pollutants
        Area and percent of forestland with diminished biological components indicative of changes in fundamental ecological processes and/or ecological continuity
Criterion 4: Conservation and Maintenance of Soil and Water Resources
    Indicators:
        Area and percent of forestland with significant soil erosion
        Area and percent of forestland managed primarily for protective functions

Percent of stream kilometers in forested catchments in which stream flow and timing have significantly deviated from the historic range of variation

Area and percent of forestland with significantly diminished soil organic matter and/or changes in other soil chemical properties

Area and percent of forestland with significant compaction or change in soil physical properties resulting from human activities

Percent of water bodies in forest areas with significant variance of biological diversity from the historical range of variability

Percent of water bodies in forest areas with significant variation from the historic range of variability in pH, dissolved oxygen, levels of chemicals, sedimentation, or temperature change

Criterion 5: Maintenance of Forest Contribution to Global Carbon Cycles

Criterion 6: Maintenance and Enhancement of Long-Term Multiple Socioeconomic Benefits to Meet the Needs of Societies

Criterion 7: Legal, Institutional, and Economic Framework for Forest Conservation and Sustainable Management

CHAPTER 94

Forest Health Monitoring in the U.S.: A West Coast Perspective

Sally Campbell, Barbara Conkling, John Coulston, Peter Neitlich, Iral Ragenovich, and William Smith

INTRODUCTION*

Forest Health Monitoring (FHM) is a multiagency national program to monitor, assess, and report on the long-term status, changes, and trends in forest ecosystem health in the U.S. The program was initiated in 1990 in response to a legislative mandate and public concern over the nation's forests (Heggem et al., 1993). The Forest and Rangeland Renewable Resources Research Act of 1978 (92 Stat. 353, as amended; 16 U.S.C. 1600, 1641–1648), per the 1988 amendment, directed the Forest Service to establish a 10-year program to monitor long-term trends in the health and productivity of domestic forests (USDA [U.S. Department of Agriculture] Forest Service, 1997). The program initially was designed and implemented jointly with the U.S. Environmental Protection Agency (USEPA) as part of its Environmental Monitoring and Assessment Program (EMAP) (Palmer et al., 1991). FHM is managed by the USDA Forest Service in cooperation with other agencies and organizations. The primary partner is State Forestry or Agriculture in each state; other partners include the USDA Natural Resource Conservation Service; U.S. Department of the Interior Bureau of Land Management and the National Parks Service; the Department of Defense; the USEPA; and various universities. The National Association of State Foresters provides essential program support, guidance, and assistance.

PURPOSE AND GOALS OF FHM

Forest health information is available from a variety of sources, but frequently it has been for one particular geographic area or for addressing a single issue. FHM is designed to provide more complete, accurate, and unbiased information on which to base forestland-management action decisions. To achieve this, FHM has the following goals:

1. Developing environmental indicators and better ways for assessing forest health.
2. Developing national standards and guidelines for both plot and survey components of detection monitoring (discussed later).

* All general information about the FHM program is adapted from the Forest Health Monitoring Fact Sheet Series (USDA Forest Service, 1996–1999).

3. Integrating FHM, the Forest Service's Forest Inventory and Analysis (FIA), National Forest Inventories, and Forest Health Protection surveys to provide the basic data on forest health status and trends.
4. Bringing other key environmental information such as data on weather, soils, and air pollution into forest health analyses.
5. Increasing or reducing the base sampling grid intensity to answer national, regional, state, and local forest health questions.
6. Providing quality assured and quality controlled data.
7. Providing systematic follow-up via evaluation monitoring (discussed later) to forest health problems found in detection monitoring.
8. Providing a research link for understanding ecosystem processes in intensive site ecosystem monitoring (discussed later).

PROGRAM STRUCTURE

The FHM program is composed of the following four interrelated activities: detection monitoring, evaluation monitoring, intensive site ecosystem monitoring, and research on monitoring techniques.

Detection Monitoring

Detection monitoring is the most extensive of the FHM program's three monitoring activities. It is designed to provide data to determine baseline or current conditions of forest ecosystems, and to detect changes and trends over time. This information is analyzed to determine if detected changes are within normal bounds or indicate improving or deteriorating conditions. Detection monitoring covers all forested lands and has two components, which together describe forest condition: (1) the plot component, measurements made on the FHM network of permanent plots (in 2000, responsibility for these plots shifted to FIA); and (2) the survey component, forest surveys of insect, disease, and other stressor effects, conducted independently of the FHM plot network. Detection monitoring data are used to make national, regional, and state assessments of forest condition.

As of 2001, the survey component of the FHM program was funded in all 50 states; the plot component, now administered by FIA, was funded in 46 states. FHM is managed regionally by Forest Health Protection programs; FIA is managed regionally by Research Stations in the Northeast, North Central, South, Interior West, and the Pacific Northwest.

Evaluation Monitoring

Evaluation monitoring examines the extent, severity, and probable causes of undesirable changes or improvements in forest health beyond that provided in detection monitoring. Results of evaluation monitoring address likely cause-and-effect relationships; identify associations between forest health indicators and particular stressors such as insects, disease, or pollutants; identify management consequences; and identify follow-up research needs that can be met experimentally or by information from intensive site ecosystem monitoring.

Intensive Site Ecosystem Monitoring

Intensive site ecosystem monitoring (ISEM) occurs at selected sites throughout the U.S. At these sites, the FHM indicators are measured intensively along with other site attributes such as climate variables to obtain detailed information on key components and processes of that forest ecosystem. ISEM allows for a rigorous evaluation of cause-and-effect relationships, identifies indicators of key processes that shape forest ecosystems, and provides biological and abiotic databases that support monitoring and forest-health-related research at these sites.

Research on Monitoring Techniques

Monitoring techniques research is work specifically directed to improve all three of the previously discussed monitoring activities. Its purpose is to identify new indicators, improve current indicators, evaluate sampling designs and impacts from repeated sampling, and improve analytical and reporting approaches.

PLOT COMPONENT AND INDICATORS

The FHM plot sampling design is a spatially and temporally systematic sample on a national hexagonal base grid, with one plot per 38,364 ha; these FHM plots are now a subsample of the more intensive FIA plot network (one plot per 2428 ha). One fifth of the FHM plots is visited each year in each state, so that all plots are remeasured every 5 years. Analysis techniques allow for estimates of change from year to year.

FHM refers to the measurements associated with these plots as indicators. It defines an indicator as any biological or nonbiological component of the environment that quantitatively estimates the condition or change in condition of ecological resources, the magnitude of stress, or the exposure of a biological component to stress. The basic indicators used include:

- Lichen community diversity and abundance
- Ozone injury on bioindicator plants
- Tree crown condition
- Soil morphology, chemistry, and erosion
- Plant diversity
- Vegetative structure and diversity
- Down woody debris

Collection of data for the above indicators aids in the assessment of environmental or societal values (forest productivity, biodiversity and vitality, carbon cycles, soil and water conservation) identified in the Criteria and Indicators for the Conservation and Sustainable Management of Temperate and Boreal Forests, endorsed by 12 countries in 1995 in Santiago, Chile (known as the Santiago Agreement or Declaration) (Anon., 1995).

Selection of indicators in FHM follows a set process:

- Identify relevant environmental or societal values of concern
- Formulate key questions relating to those values
- Review the scientific literature and available databases
- Test indicators in pilot and demonstration studies
- Formulate plot-level indices
- Review indicators by partners and scientific peers

Part of the review process is the evaluation of indicators relative to six criteria to increase scientific objectivity, consistency, and depth of evaluation:

1. Low environmental impact
2. Simple field quantification
3. Unambiguous interpretation
4. Stable during sampling window
5. Meaningful when applied across various strata of interest (e.g., ecosections or forest types)
6. High signal-to-noise ratio regarding data error

Methods for measuring indicators on FHM plots are documented and updated yearly in field manuals (e.g., USDA Forest Service, 1999b). Some regional differences exist and regional manuals are published as well.

SURVEY COMPONENT

The purpose of the survey component of the FHM program is to detect broadscale disturbance events from forest insects, diseases, and other disturbance agents that the plot component may not be able to adequately detect. This information provides a context for interpreting plot data and for identifying likely factors that contribute to forest health changes. Both aerial and ground surveys are employed. The aerial surveys focus on tree mortality, defoliation, discoloration, dieback, tree uprooting, branch breakage, and branch mortality. Ground surveys focus on damage by specific insects, diseases, or other disturbance agents.

Implementation of the survey component differs widely from state to state. Unlike the plot component that was designed specifically for national consistency, survey programs were already established in most states prior to the FHM program. Some survey programs, such as the aerial survey program in Oregon and Washington, have been functioning for many decades. Each was initially established to provide for state and local needs such as locating dead timber for salvage or tracking defoliation to determine insecticide spray boundaries. By establishing reasonable survey standards, FHM aims to achieve survey conformance among previously existing programs, and thereby provide comparable data as a basis for regional and national analysis and reporting. Standards for aerial survey data collection, quality control, and data processing have been developed and are now being used by participating states and Forest Service Regions (USDA Forest Service, 1999a). General information and guidelines for aerial survey programs and associated geographic information systems (GIS) have also been written recently (McConnell et al., 2000; USDA Forest Service, 2001). In addition, standardized methods for particular ground surveys are available for the eastern U.S. (USDA Forest Service, 1995).

INTEGRATION WITH OTHER PROGRAMS

To achieve the goal of integrating FHM, FIA, and National Forest System Inventories, FIA began administering FHM plot data collection in 2000. In fact, FHM plots are now a subset of the FIA plot network and share the same plot design, core set of measurements, and data collection standards. In 2001, FIA also administered collection of plot data on national forests in Washington, Oregon, and California (previously, the national forest inventory was administered separately by Forest Service regions in these states rather than FIA). Progress is also under way to develop more uniform standards for surveys for damage by insects, diseases, and other disturbance agents conducted by Forest Health Protection programs in the Forest Service and the states.

ANALYSIS AND REPORTING

Results from plots and survey data have been reported in various publications, both nationally and regionally (Burkman et al., 1998; Campbell and Liegel, 1996; Dale, 1996; Stolte, 1997; Stoyenoff et al., 1998; USDA Forest Service, 1994; Wittwer, 1998). FHM data analysis methods, including procedures for estimating change, are presented by Smith and Conkling (2002).

WEST COAST PROGRAM

The West Coast Region in the FHM program consists of Alaska, California, Oregon, Washington, Hawaii, and other U.S.-affiliated Pacific islands. As of 2001, only California, Oregon, and Washington had both plot and survey components (detection monitoring) fully funded and operational. FHM plots are planned for Hawaii, the Pacific islands, and Alaska and are expected to be in place within a few years. Aerial surveys are being conducted in all four West Coast continental states. A variety of ground surveys or monitoring networks for insect populations or damage by insects, diseases, and other disturbance agents are also carried out as needed in all states.

Over the past 4 years (1998–2001), 22 West Coast evaluation monitoring projects have been funded by the FHM program. Projects have included investigating impacts from native and exotic insects or diseases, determining accuracy of aerial surveys, evaluating impacts of ozone on pine growth, determining fuel loads and fire risks from native and exotic vegetation, and conducting special aerial surveys.

EXAMPLES OF FHM FINDINGS

Changes in Carbon Sequestration

Maintaining forest contributions to global carbon cycles is Criterion 5 of the Santiago Agreement. It is widely suggested that the increased concentration of greenhouse gases, including carbon dioxide, will result in climate change in most regions of the world. There are many ways of mitigating this effect that relate to trees and forests. Among them is increasing forest growth, tree planting, minimizing loss of carbon to the atmosphere through catastrophic mortality, efficient use of harvested wood products, and salvage of mortality.

Carbon is removed from the atmosphere in the process of growth. Carbon is lost to the atmosphere when dead trees or wood products decay. Approximately one half of the carbon harvested as biomass is stored for long periods as durable wood products. FHM data (specifically tree growth, size, and density, and tree harvest) have been used to estimate carbon storage or loss for variable time periods between 1990 and 1996. Other published data and relationships were used to estimate carbon in belowground portions of trees and in down woody debris (Birdsey, 1996).

Carbon storage was estimated by determining the biomass of the living boles and roots of all trees and saplings, and subtracting the biomass of the trees that have died and approximately one half the biomass of the trees that were harvested over the same time period. This one half represents the proportion of harvested biomass utilized in a durable form, e.g., bound books, wooden structures, etc. (Birdsey, 1996). A net gain can result from an increased stand growth and the efficient utilization of harvest trees and salvage of mortality, or some combination thereof. The estimated net carbon loss or gain for FHM plots in 10 ecosections in California is presented in Table 94.1. Most of the losses are in the Sierra Nevada ecosection where there was substantial mortality due to drought and insect-caused mortality in the early 1990s (USDA Forest Service, 1994).

Lichen Monitoring on the West Coast

Lichens are a component of most temperate North American forest communities, contributing to nutrient cycling and providing food and habitat for other organisms (Pike, 1978; Rominger and Oldemeyer, 1989; Sharnoff and Rosentreter, 1998). They are also indicators of air quality (Nash and Wirth, 1988; Seaward, 1993). Although trees may respond to moderate, chronic levels of air pollution, all of the other influences on tree growth, such as variation in soils, make the responses of trees to pollutants difficult to measure in the field. Therefore, lichen communities provide not

Table 94.1   Tree Carbon (kg/ha) of Softwoods and Hardwoods by Ecosection in California

Ecosection	N	Estimate of Change	Standard Error of Estimate	R^2	Degrees of Freedom	Value of t	Pr > t
263A Northern California Coast	21	4539.35	1327.90	0.9974	5	3.42	0.019
300 Desert and Semidesert	10	201.83	190.96	0.9999	1	1.06	0.482
M261A Klamath Mountains	34	−1370.21	1887.23	0.9901	8	−0.73	0.488
M261B Northern Coast Range	17	2649.00	668.52	0.9995	3	3.96	0.029
M261C Northern Interior Coast Range	26	283.65	404.38	0.9968	5	0.70	0.514
M261D Southern Cascades	24	231.22	482.80	0.9987	6	0.48	0.649
M261E Sierra Nevada	74	−1991.37	1715.26	0.9825	16	−1.16	0.263
M261G Modoc Plateau	17	183.88	169.77	0.9998	3	1.08	0.358
M262A Central Coast and Range	13	2709.54	1282.19	0.9908	2	2.11	0.169
M262B Southern Coast and Range	9	979.05	760.35	0.9998	1	1.29	0.420

only a measure of air pollution impacts on lichens, but also suggest air pollution impacts on trees and other aspects of forest health that are difficult to measure directly. Some small-scale studies have also shown that lichen communities respond to certain forest management practices. These studies found that young managed stands host relatively few species compared with mature forest, and, likewise, densely stocked, dark stands host far fewer species than stands with open structure (Neitlich, 1993; Neitlich and McCune, 1997; Peck and McCune, 1997). The FHM program includes lichen community diversity and abundance as one of the indicators of forest health. The results of FHM lichen monitoring help address Santiago Criteria 1 (Conservation of biological diversity), 2 (Maintenance of productive capacity of forest ecosystems), and 3 (Maintenance of forest ecosystem health and vitality).

Lichen communities were first sampled in West Coast FHM plots in 1998. The conclusions drawn from the data gathered in 1998 and subsequent years are limited until gradient models that describe climatic and air quality gradients for West Coast regions are constructed (McCune et al., 1997a,b). At that time, the model can be applied to the data to make more accurate interpretations of the results. However, the results from the first year of sampling are still useful and are presented here briefly and in more detail by Neitlich and co-workers (1999).

Field sampling was designed to be carried out by nonlichenologist field crews according to the procedures described in the FHM Manual (USDA Forest Service, 1999b). Professional lichenologists conducted the lichen training, quality assurance, sample identification, and data analysis.

In Oregon and Washington, the mean lichen species richness (number of species) was 17 species per plot (Figure 94.1), approximately 75% higher than species richness in California and 85% higher than Idaho. The region west of the Cascades hosted approximately 60% higher total number of species than the region east of the Cascades, with species richness in western Washington approximately 13 species per plot and western Oregon with approximately 20 species per plot.

Western Oregon and Washington also had a larger number of factors depressing species richness, including air quality (although determination of the extent of the problem awaits the regional gradient model). Willamette Valley sites and those near or downwind of Seattle and Portland had fairly low to intermediate species richness for this zone and are probably hosting lower diversity than they would in a clean air regime. Forest management probably plays a role in the low diversity in some sites in the Oregon Cascades and on the immediate coast (especially in Washington). It is likely (though yet untested) that species richness, where poor in these areas, was negatively influenced by the current forest age or structure.

In California, lichen species richness was predictably low given the relatively dry climate, with a mean of 9.6 species per plot (Figure 94.2). The steep natural gradients in California lichen community composition combined with small sample size over a large area made inferences

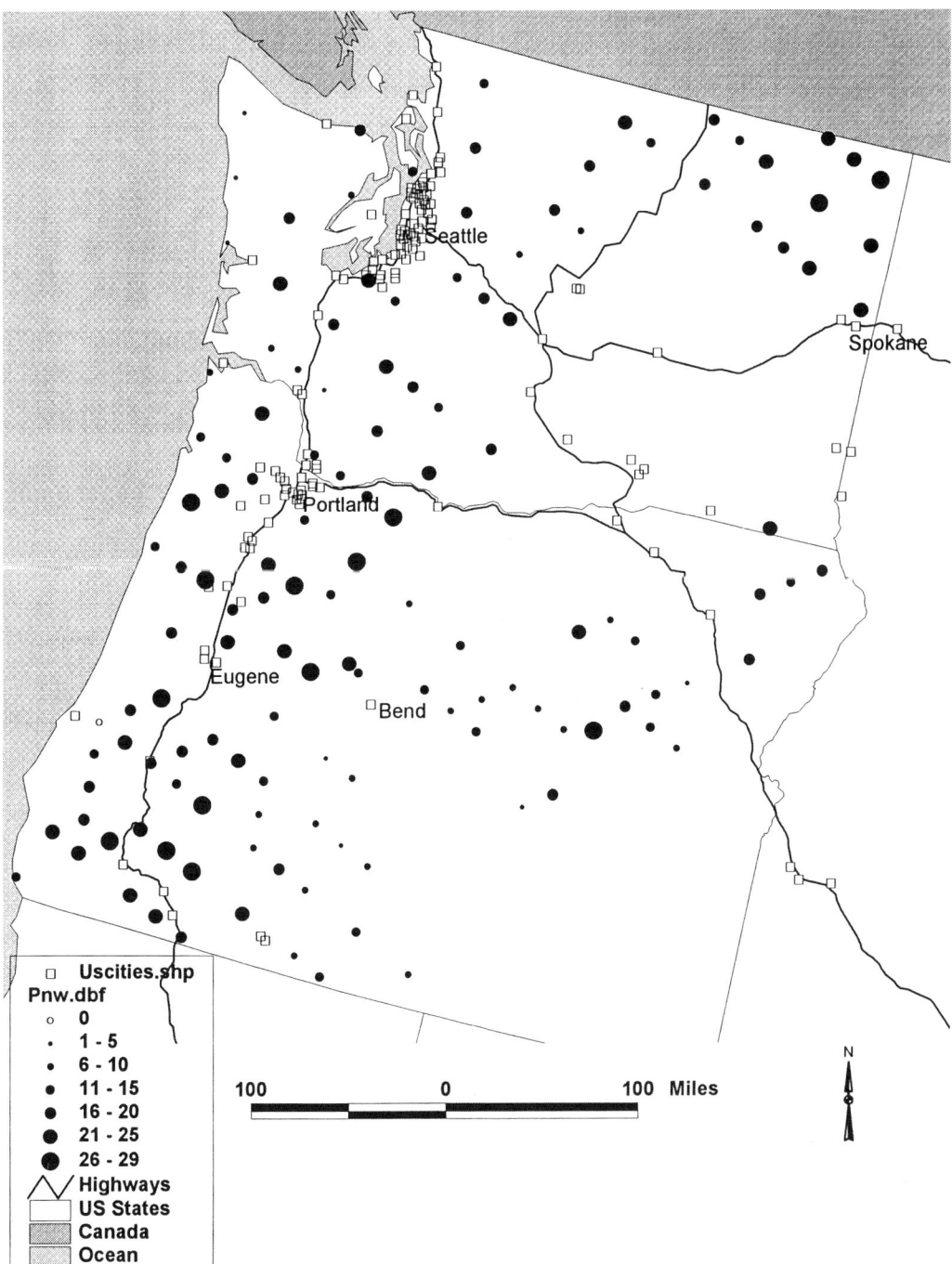

Figure 94.1 Lichen species richness on 170 Forest Health Monitoring plots in Oregon and Washington in 1998. Circular dot size is graded to indicate increasing species richness, with number of species per plot shown under "Pnw.dbf" in the legend.

about anthropogenic impacts difficult prior to developing regional gradient models. Pollution-related declines around large urban areas were expected, but the number of plots was not sufficient to address the question. However, anecdotal evidence is suggestive of air-quality concerns. First, observation and studies from other areas in the U.S. (e.g., McCune et al., 1997a) suggest that

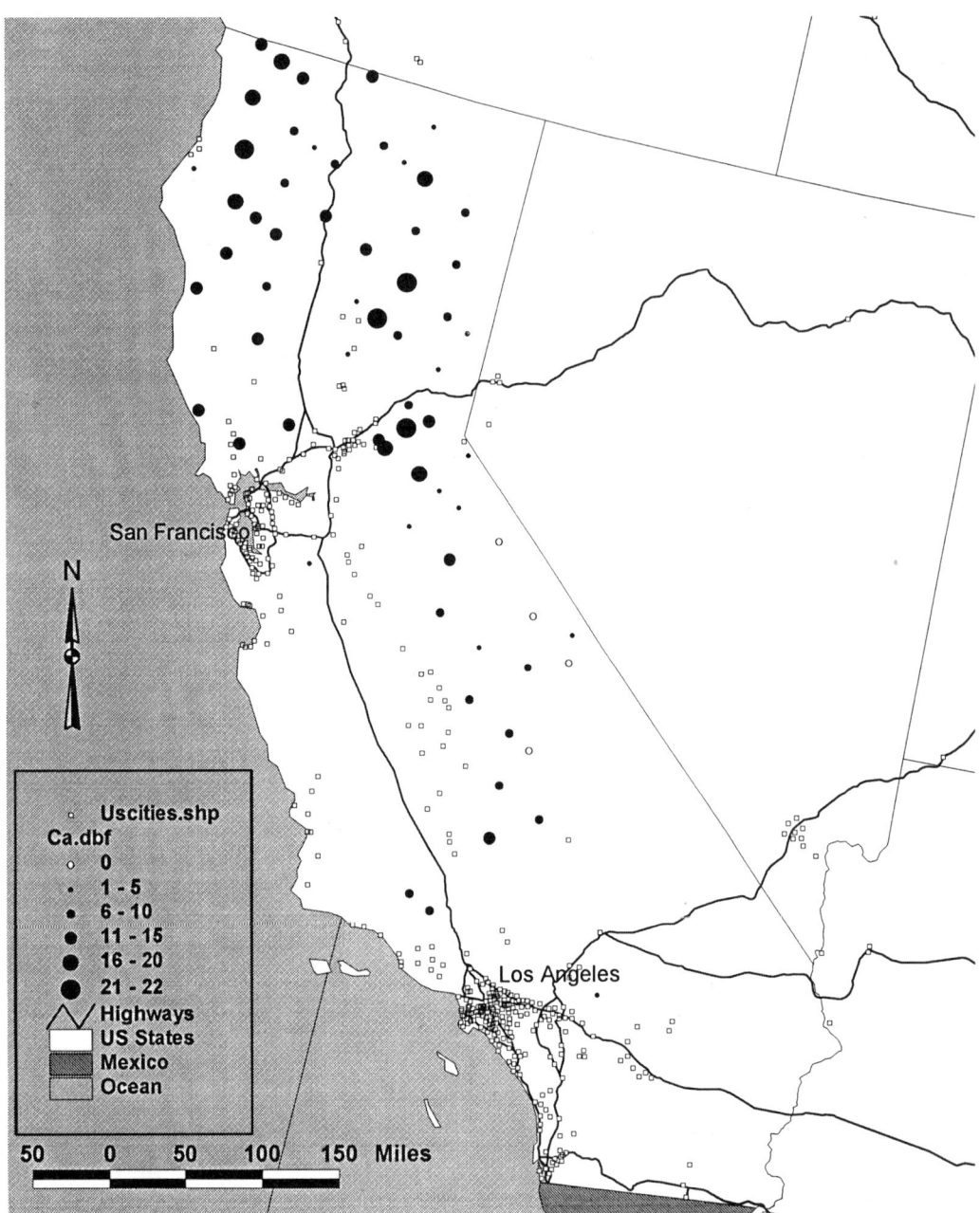

Figure 94.2 Lichen species richness on 76 Forest Health Monitoring plots in California in 1998. Circular dot size is graded to indicate increasing species richness, with number of species per plot shown under "Ca.dbf" in the legend.

the areas surrounding large- and medium-sized urban centers are likely to be most heavily impacted. Second, nitrification from agriculture in the Central Valley is likely to have fostered a bloom of nitrophilous species in areas adjacent to the Valley. Third, the air-pollution plumes from both Orange County (Los Angeles) and the Bay Area (San Francisco, Oakland) are likely to have an effect on downwind oak woodland (*Quercus* spp.) communities. In clean air conditions, these would normally host a high diversity of lichens, including some pollution-sensitive species that currently are not present.

The 1998 lichen data enhanced the base of knowledge of lichen flora in all three states. Many new point locations were found for several species, ranges of some species were expanded or changed, and knowledge was gained regarding ecological relationships.

Survey for Balsam Woolly Adelgid in Oregon and Washington

Balsam woolly adelgid (*Adelges piceae*) is an introduced insect that has had significant impact on true fir species (*Abies* spp.) in the Pacific Northwest. During the late 1950s and 1960s, balsam woolly adelgid caused extensive mortality in grand fir (*A. grandis*), subalpine fir (*A. lasiocarpa*), and Pacific silver fir (*A. amabilis*), primarily in the Cascade Mountains in Oregon and Washington. It also affects tree growth and reproduction, ultimately resulting in a significant decrease of the host species in many areas. In some cases, the host species may be the major tree species for that system.

The insect occurs in other states, such as Idaho, and in the eastern U.S. During the initial outbreak, a substantial amount of information was collected and outbreaks were characterized, based on the information at the time. Some permanent research plots were established to determine the effects of the insect on the host trees under various site conditions. Owing to suspected changes in balsam woolly adelgid distribution and impact on various hosts, a survey to gather up-to-date information was proposed and funded as a 3-year (1998–2000) FHM Evaluation Monitoring project. The objectives of the survey were to:

1. Conduct a ground survey of host type (subalpine fir, Pacific silver fir, grand fir) throughout Washington and Oregon to confirm occurrence and distribution of balsam woolly adelgid and supplement the forest health monitoring information collected during annual aerial surveys.
2. Determine effects of balsam woolly adelgid on the host species and changes in the local ecosystem.
3. Determine if the existing parameters for occurrence and risk described the situation at the time.
4. Explore opportunities for adapting this system for use in monitoring effects of other introduced species.

In order to accomplish the first objective, a map of risk and known occurrence was developed using historical aerial survey information and risk parameters of host type and elevation developed by Mitchell in the early 1960s (Mitchell, 1966). Also, areas were identified where balsam woolly adelgid was suspected to occur but had not been reported, or where there was personal but possibly undocumented on-the-ground knowledge. Ground surveys within those areas identified on the hazard/risk map would then be undertaken to verify and document the occurrence and impact of the balsam woolly adelgid.

From 1998 to 2000, approximately 1038 plots were visited, primarily in Oregon and Washington. At each plot, information was taken on location, elevation, tree species, tree size (seedling, sapling, or tree), balsam woolly adelgid symptoms (gouting, stem infection, crown abnormality, or adelgid-caused mortality), land use, tree structure, species contribution, and site description.

Also, seven permanent plots established in the early 1960s were revisited and data collected. Initial results from the permanent plots show that:

1. Tree damage was most severe in the first decade of the balsam woolly adelgid outbreak.
2. Environment played a significant role; higher mortality occurred on wet sites and at lower elevations.
3. Host species had unique responses. Shasta red fir (*A. magnifica*) and red fir (*A. procera*) showed considerable resistances at high elevations; subalpine fir was susceptible at all locations.
4. Grand fir was being eliminated from low-elevation landscapes and coastal streams in several western Oregon and Washington locations.
5. Subalpine fir was being removed as a pioneer species in important mountain environments such as alpine meadows, avalanche tracks, and old lava beds.

More detailed information can be found in Mitchell and Buffam (2001). Analysis of the ground survey plot data is currently under way.

Analytic Questions and Methods for Combining Aerial Survey Data and Plot Data

One of the goals of the FHM program is to develop methods for determining relationships between diverse data sets to create a clearer picture of insect and disease impacts on a landscape level. In particular, co-analysis of aerial survey and plot data presents some unique challenges. Aerial survey data are relatively inaccurate, owing to the method of collection (sketching areas of damage on a paper map from a small plane, 160 to 900 m aboveground, traveling at approximately 193 km/h). The data's inaccuracy, however, does not negate its usefulness in detecting overall trends in broadscale disturbance events or aiding local forest managers in locating or assessing damage. In contrast, plot data are more precise and geographically accurate but are a relatively small sample.

A method for analyzing aerial survey and plot data was tested by the FHM program. Washington was selected as the study area, using data from 1996–1998 aerial surveys and 1997–1998 plot data (Figure 94.3). Much of the state is fairly continuously forested, and has several national forests. Both FHM plot data and aerial survey data are available for the state, and aerial survey data show a relatively large area in south central Washington where defoliation damage increased over several years. Just a few insects and diseases account for 91% of the damage (mortality and defoliation) detected in aerial surveys in 1996–1998: western spruce budworm, balsam woolly adelgid, larch needlecast, fir engraver, western pine beetle, mountain pine beetle, and Douglas fir beetle.

The primary objective of this analysis was to predict damage or probability of damage at plot locations and unmeasured locations. Several questions must be answered to make these predictions:

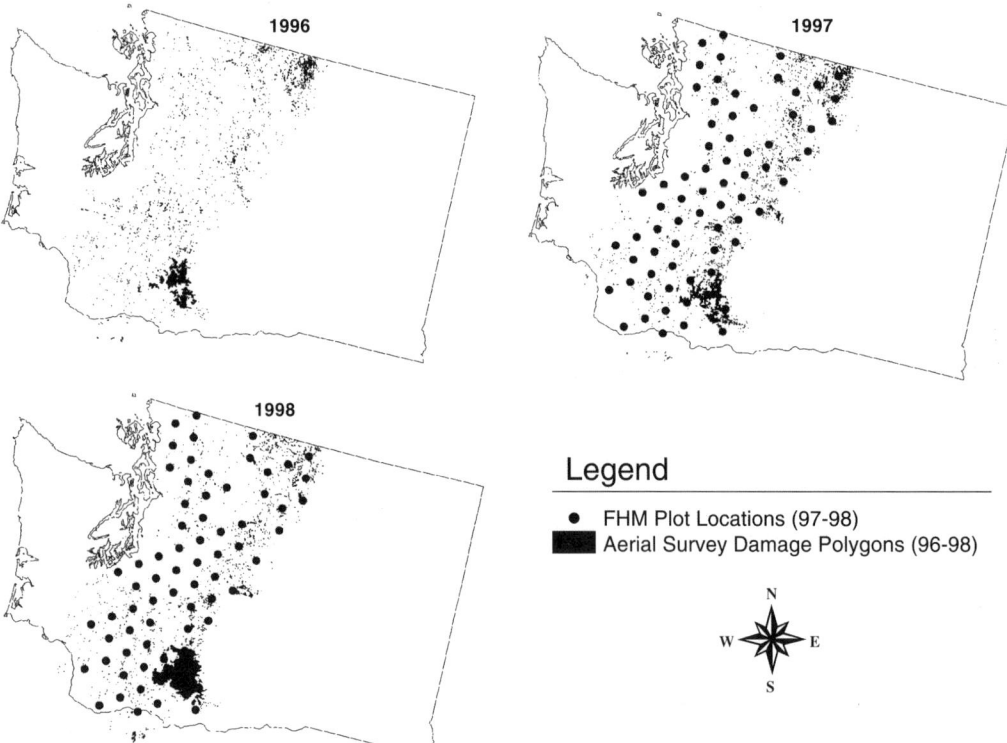

Figure 94.3 Maps of Washington state showing aerial survey damage polygons (irregular areas) and FHM plots (circles) for 1996, 1997, and 1998.

1. What is the relationship of damage polygons from one year to the next? Are they contiguous across years? Does damage expand from an epicenter or move linearly? Or is there no relationship across years?
2. Are there higher levels of crown damage (e.g., defoliation) or mortality on FHM plots when the plots are closer to aerial survey damage polygons?
3. What is the time lag between survey and plot data? What symptoms show up first on the plots and what symptoms show up first from the air? Is it dependent on the damaging agent?
4. What is the spatial relationship regarding damage between FHM plots, between aerial survey polygons, and between plots and polygons?

Cokriging and universal kriging are two predictive analytical tools that were used. Cokriging is spatial prediction based on covariables in space; universal kriging is a combination of regression and kriging where spatial predictions are based on one variable and spatial trend or drift from other variables (Cressie, 1993).

CONCLUSION

In less than a decade since it was initiated, FHM had made significant strides toward development of a national program to monitor, assess, and report on the long-term status, changes, and trends in the health of the nation's forested ecosystems. Each of its four interrelated activities, including detection, evaluation, intensive site ecosystem monitoring, and research on monitoring techniques, are being implemented at some level. The detection activity covers all forested lands. Its two components, a network of permanent plots on which measurements are made each year, and a survey component that includes monitoring the effects of insects, diseases, and other disturbance agents, are in place in most states. A set of forest health indicators, including tree crown condition, ozone injury to bioindicator plants, soil condition, plant diversity and structure, down woody debris, and lichen communities, continues to be measured on the permanent plots. In the West Coast region (Alaska, Washington, Oregon, California, and U.S.-affiliated Pacific islands), baseline FHM plot data were collected in the 1990s in California, Oregon, and Washington; remeasurement of one fifth of the plots is made annually. Plots are planned for the Pacific islands and Alaska. Examples of national and West Coast monitoring activities and analyses were described, including analyzing plot data to assess forest health issues such as species diversity or air quality (via lichen data) or carbon sequestration (via harvest, mortality, and tree measurement data); implementing FHM evaluation monitoring projects such as the balsam woolly adelgid survey in Oregon and Washington; and investigating methods to analyze diverse forest health data sets, such as the co-analysis of aerial survey data and FHM plot data.

REFERENCES

Anon., Sustaining the world's forests: the Santiago Agreement, *J. For.*, 93(4), 18–21, 1995.

Birdsey, R.A., Carbon storage for major forest types and regions in the conterminous United States, in *Forests and Global Change,* Vol. II, *American Forests*, Sampson, N.R. and Hair, D., Eds., Washington, D.C., 1996, pp. 1–25.

Burkman, W.G., Vissage, J.S., Hoffard, W.H., Starkey, D.A., and Bechtold, W.A., Summary report: forest health monitoring in the South, 1993 and 1994, USDA Forest Service, Southern Research Station, Resource Bulletin SRS-32, Asheville, NC, 1998.

Campbell, S. and Liegel, L., tech. coords., Disturbance and forest health in Oregon and Washington, USDA Forest Service, Pacific Northwest Research Station and Pacific Northwest Region, Oregon Department of Forestry, Washington Department of Natural Resources, Gen. Tech. Rep. PNW-GTR-381, Portland, OR, 1996.

Cressie, N., *Statistics for Spatial Data*, John Wiley & Sons, New York, 1993.

Dale, J., tech. coord., California forest health in 1994 and 1995, USDA Forest Service, Pacific Southwest Region, Report R5-FPM-PR-002, San Francisco, 1996.

Heggem, D., Alexander, S.A., and Barnard, J.E., Forest Health Monitoring: 1992 Activities Plan, U.S. Environmental Protection Agency, EPA/620/R-3/002, Washington, D.C., 1993.

McConnell, T.J., Johnson, E.W., and Burns, B., A guide to conducting aerial sketch mapping surveys, USDA Forest Service, Rocky Mountain Research Station, Forest Health Technology Enterprise Team Publication, FHTET 00–01, Ft. Collins, CO, 2000.

McCune, B., Dey, J., Peck, J., Cassell, D., Heiman, K., and Will-Wolf, S., Regional gradients in lichen communities of the southeast United States, *Bryologist,* 100(2), 145–158, 1997a.

McCune, B., Dey, J., Peck, J., Cassell, D., Heiman, K., Will-Wolf, S., and Neitlich, P., Repeatability of community data: species richness versus gradient scores in large-scale lichen studies, *Bryologist,* 100(1), 40–46, 1997b.

Mitchell, R.G., Infestation characteristics of the balsam woolly aphid in the Pacific Northwest, USDA Forest Service, Pacific Northwest Forest and Range Experiment Station, Research paper PNW-35, Portland, OR, 1966.

Mitchell, R.G. and Buffam, P., Patterns of long-term balsam woolly adelgid infestation and effects in Oregon and Washington, *J. Appl. For.,* 16(3), 121–126, 2001.

Nash, T.H. and Wirth, V., Eds., Lichens, bryophytes and air quality, *Bibl. Lichen.,* 30, 1–297, 1988.

Neitlich, P., Lichen Abundance and Biodiversity Along a Chronosequence from Young Managed Stands to Ancient Forest, Master's thesis, University of Vermont, Burlington, 1993.

Neitlich, P. and McCune, B., Hotspots of epiphytic lichen diversity in two young managed forests, *Conserv. Biol.,* 11(1), 172–182, 1997.

Neitlich, P., Hasselbach, L., and Szewczak, S., Forest Health Monitoring lichen community data from Washington, Oregon, and California, Pre-gradient model interim summary, White Mountain Research Station Report, Bishop, CA, 1999.

Palmer, C.J., Riitters, K.H., Strickland, T., Cassell, D.L., Byers, G.E., Papp, M.L., and Liff, C.I., Monitoring and Research Strategy for Forests: Environmental Monitoring and Assessment Program (EMAP), U.S. Environmental Protection Agency, EPA/600/4–91/012, Washington, D.C., 1991.

Peck, J.E. and McCune, B., Effects of green tree retention on epiphytic lichen communities: a retrospective approach, *Ecolog. Appl.,* 7, 1181–1187, 1997.

Pike, P.L., The importance of epiphytic lichens in mineral cycling, *Bryologist,* 81, 247–257, 1978.

Rominger, E.M. and Oldemeyer, J.L., Early-winter habitat of woodland caribou, Selkirk Mountains, British Columbia, *J. Wildl. Manage.,* 53, 238–243, 1989.

Seaward, M.R.D., Lichens and sulphur dioxide air pollution: field studies, *Environ. Rev.,* 1, 73–91, 1993.

Sharnoff, S. and Rosentreter, R., Wildlife use of lichens, http://www.lichen.com/fauna.html, 1998.

Smith, W.D. and Conkling, B.L., Analyzing forest health data, USDA Forest Service, Southern Research Station, National Forest Health Monitoring Program Report, Research Triangle Park, NC, 2002.

Stolte, K., 1996 National technical report on forest health, USDA Forest Service, Southern Research Station, National Forest Health Monitoring Program Report, Research Triangle Park, NC, 1997.

Stoyenoff, J., Witter, J., and Leutscher, B., Forest health in the North Central States, University of Michigan, School of Natural Resources and Environment, Unnumbered publication, Ann Arbor, 1998.

USDA Forest Service, California Forest Health: Past and Present, USDA Forest Service, Pacific Southwest Region, Forest Pest Management Report R5-FPM-PR-001, San Francisco, 1994.

USDA Forest Service, Working Plan: integration of forest health and forest health protection activities with Forest Health Monitoring for the Eastern United States, USDA Forest Service, Northeastern Area Forest Health Protection and Region 8 Forest Health Publication, St. Paul, MN, 1995.

USDA Forest Service, Forest Health Monitoring Fact Sheet Series, USDA Forest Service, Southern Research Station, National Forest Health Monitoring Program Publication, Research Triangle Park, NC, 1996–1999.

USDA Forest Service, Forest Health Monitoring Program Overview Fact Sheet, USDA Forest Service, Southern Research Station, National Forest Health Monitoring Program Publication, Research Triangle Park, NC, 1997.

USDA Forest Service, Forest Health Monitoring 1999, Aerial Survey Standards, Forest Health Protection Web site (http://www.fs.fed.us/foresthealth/id/detect.html), Washington, D.C., 1999a.

USDA Forest Service, Forest Health Monitoring 1999, Field Methods Guide, Forest Health Protection Web site (http://www.fs.fed.us/foresthealth/id/detect.html), Washington, D.C., 1999b.

USDA Forest Service, Aerial survey geographic information system (GIS) handbook, Rocky Mountain Research Station, National Forest Health Monitoring Program Publication, Ft. Collins, CO, 2001.

Wittwer, D., Forest insect and disease conditions in Alaska, USDA Forest Service, Alaska Region, and State of Alaska, Department of Natural Resources, General Technical Report R10-TP-74, Anchorage, 1998.

SECTION II.10

Agroecosystems

CHAPTER 95

Overview: Integrating Agricultural Production with Ecosystem Health

Terrell P. Salmon

In the 21st century, ecosystem health, regardless of scale, will depend almost entirely on how humans manage the environment. Many examples of negative land-use patterns and practices exist, including some for agricultural production. Despite cogent examples of agricultural practices that degrade the environment, agriculturists throughout the world are more concerned with ecosystem health than most realize. Clearly, without a healthy environment, agricultural production will decline.

The scale of agriculture amplifies its effects on the ecosystem because much of the environment is under agricultural production. In California's Central Valley, for example, more than 40% (6.2 million acres) of the ecosystem is devoted to agricultural production. The health of those agricultural acres and their influences on the remaining land affect the overall health of the Central Valley. While some people argue that agriculture should not dominate the landscape, most recognize it as a necessary part of managing the ecosystem. How we manage these lands is more important than the overall amount of land devoted to this resource use.

This section is devoted to understanding agriculture and how it relates to and alters the environment. This information is necessary to better define agricultural practices that are compatible with, and potentially supportive of, environmental health. The chapters cover four important aspects of agricultural production.

First, we deal with the overall impact of agricultural production on the ecosystem. Debates and discussions continue about large-scale agriculture and how it modifies the biodiversity of a region. John Hopkins' chapter, "Fallow Land Patches and Ecosystem Health in California's Central Valley Agroecosystem," examines habitat conservation and restoration programs and their potential for improving wildlife habitat. He also discusses the ways in which farming practices affect the ecosystem. He argues that greater availability and use of technical assistance and demonstration programs could significantly improve the overall health of the Central Valley ecosystem. By implication, his work has significance to other areas throughout the world.

While the spatial relationships of agricultural and other resources with the ecosystem are critically important, the specific agricultural practices used in producing food and fiber also compromise ecosystem health. One area of interest is soil disturbance resulting from normal farming practices. Mitchell et al. discuss their long-term research in "Reduced-Disturbance Agroecosystems in California." Conservation tillage is more widely accepted outside California, although the beneficial aspects of this farming practice on farm operations and the overall health of the environment probably apply to California as well. They investigated soil health in relation to conservation tillage practices. Their studies are essential in demonstrating the importance of these techniques and in convincing farmers to adopt new farming methods. Often, adoption is the most difficult part of change because farmers resist uncertainty with its potential for devastating impact on a farming enterprise. Research on conservation tillage showing the long-term benefits to soil quality and productivity is essential to bring about necessary change that will lead to improved ecosystem health.

Pest management is another area of agricultural operations that can have huge impacts on ecosystem health. Pest management includes synthetic and natural inputs to the ecosystem, as well as techniques for altering the interactions of various organisms. Each of these has implications for environmental health. Altieri and Nicholls discuss pest management and ecosystem health in "Ecologically Based Pest Management: A Key Pathway to Achieving Agroecosystem Health." They argue that proper pest management techniques can restore and enhance the resiliency, sustainability, and health of agroecosystems. The key component to ecosystem health and sustainability is preservation of the cultural diversity of local agricultural ecosystems. They show how managing agricultural areas to improve diversity will lead to less reliance on artificial inputs such as pesticides. They also recognize that the biological system must exist in the context of social organization that encourages the harmonious interaction of humans, the agroecosystem, and the overall environment.

Farming practices have effects on the environment outside the agricultural system. Water is a vital resource for agriculture and is also a major transporter of nutrients and other materials through the environment. The impact of nutrients can have far-reaching implications on ecosystem health and must be understood for better management of agricultural operations to occur. Kaffka and coworkers have researched the effects of nutrients in northern California and southern Oregon. Their chapter, "Irrigation, Agricultural Drainage, and Nutrient Loading in the Upper Klamath Basin," evaluates nutrient loading related to irrigation practices. Understanding the effects of irrigation on water quality is important and will lead to improved ecosystem health.

Ecosystem health and agricultural production are inextricably linked. A clear and thorough understanding of the effects of agriculture on the ecosystem, and the changes that the ecosystem and all its components exert on agricultural systems, is key to maintaining the sustainability and health of agriculture and the environment.

CHAPTER 96

Fallow Land Patches and Ecosystem Health in California's Central Valley Agroecosystem

John D. Hopkins

INTRODUCTION

California's Central Valley is 450 miles long and encompasses over 15 million acres (Great Valley Center, 1998). It includes the Sacramento Valley as the northern portion, the delta where the Sacramento and San Joaquin Rivers join, and the San Joaquin Valley as the southern portion.

The Central Valley was an extremely rich wildlife region prior to European settlement (Thelander and Crabtree, 1994). Estimates of presettlement vegetation suggest that there were over 800,000 acres of riparian forest along many of the rivers and streams, forming a band of trees up to several miles wide (Dawdy, 1989). There were additional groves of oaks scattered across grassy plains. Tule marsh encompassed 600,000 to 700,000 acres, much of it in the delta and around large lakes at the southern end of the San Joaquin Valley (Dawdy, 1989). Most of the remaining valley floor was a grassy plain, although there were large areas of alkali-tolerant scrub and alkali saltbush scrub in the southwest portion of the San Joaquin Valley. These different plant communities provided habitat for a rich array of animals, from huge flocks of wintering waterfowl; to large numbers of antelope, elk, and grizzly bears; to forest-dwelling song birds (Barry, 1972; Thelander and Crabtree, 1994).

The valley has been transformed into the most productive agricultural region on the planet (Platzek, 1995). With $13.5 billion worth of agricultural products (U.S. Department of Agriculture, 1999), the industry produces about 25% of the U.S. supply of vegetables, fruits, etc. (Platzek, 1995), as well as major amounts of food and fiber for export to other countries. Natural areas are limited to small remnants for the most part (U.S. Fish and Wildlife Service, 1998). The only vertebrate groups that remain in large numbers are wintering waterfowl and shorebirds that benefit from wildlife refuges and duck clubs developed as waterfowl habitat and from winter flooding of rice fields (Heitmeyer et al., 1989; Shuford et al., 1998).

Farming is now very intensive, with economic pressures on individual farmers increasing that intensity. Big farmers, generally speaking, rely on large acreages, massive machines, and planting of every available acre to remain profitable. *Clean farming,* where field edges and stream banks are kept devoid of vegetation by disking and herbicide application, is common in the valley (Anderson, 1995). As of 1997, there were 6.2 million acres of harvested cropland in the Central Valley (field crops, orchards, and vineyards), forming the core of intensive agricultural production (Table 96.1) (U.S. Department of Agriculture, 1999). Also, since 1970, pasturelands and grain crops

Table 96.1   Changes in Acres of Harvested Cropland in Central Valley Counties, 1992–1997[a]

| County | Acres Harvested | | Change, 1992–1997 | |
	1992	1997	Acres	Percent
Butte	171,901	222,209	50,308	29.3
Colusa	222,244	287,630	65,386	12.9
Fresno	1,013,424	1,157,367	143,943	14.2
Glenn	181,258	212,848	31,590	31.6
Kern	760,159	893,221	133,062	17.5
Kings	431,212	445,537	14,325	3.3
Madera	271,352	294,706	23,354	8.6
Merced	417,346	434,074	16,728	4.0
Sacramento	135,170	120,220	−14,950	−11.1
San Joaquin	468,572	498,985	30,413	6.5
Stanislaus	301,310	315,978	14,668	4.9
Sutter	232,014	266,399	34,385	14.8
Tulare	607,361	639,578	32,217	5.3
Yolo	286,085	324,291	38,206	13.4
Yuba	70,006	79,586	9,580	13.7
Total	5,569,414	6,192,629	623,215	11.2

[a] Harvested croplands include field crops, orchards and vineyards but not pasture or rangeland.

Based on data from U.S. Department of Agriculture, 1997 Census of Agriculture, California State and County Data, Vol. I, Geographic Area series, Part 5, Washington, D.C., 1999.

have been giving way to increasing acreages of intensive vegetable fields, orchards, and vineyards (Kroll et al., 1991; U.S. Department of Agriculture, 1999), further depleting available wildlife-foraging habitat.

In contrast, less-intensive farming practices can provide a great deal of wildlife habitat through hedgerows, woodlots, strips of riparian woodland, vegetated field edges, and keeping areas of poor soils out of production (Biadi, 1998; Clark and Rollins, 1996; Yolo County Resource Conservation District, 1999). Studies by several authors highlight the tremendous loss of biodiversity in Europe and the central portion of the U.S. as this wildlife-friendly form of farming has given way to extremely intensive agriculture in recent decades (Biadi, 1998; Freemark and Boutin, 1995; Fuller et al., 1995; Ribic et al., 1998).

While the habitats that the wildlife-friendly practices could provide in the Central Valley are often different in composition and form from presettlement plant communities, they can provide habitat for a wide variety of native animals, including several that are imperiled or undergoing significant decline. This chapter examines some of the obstacles to creation of more wildlife habitats in Central Valley cropland areas, solutions to these obstacles, and the types of strategic thinking necessary to provide farmland habitat patches of maximum benefit to overall wildlife populations in the Central Valley.

CONSERVATION RESERVE PROGRAM DEMONSTRATES LACK OF HABITAT CONSERVATION ON CENTRAL VALLEY FARMLAND

The minimal enrollment of Central Valley farmland in the U.S. Department of Agriculture's Conservation Reserve Program (CRP) is one indicator of the lack of habitat conservation. Farmers enroll croplands in the CRP under 10-year contracts, agreeing to take them out of crop production and carry out one or more specified conservation practices. In return they receive an annual per-acre rent and 50% of the cost of establishing any long-term cover such as perennial grasses or trees.

Congress authorized the CRP under Title XII of the 1985 Food Security Act. The initial purpose was to reduce soil erosion by taking highly erodible land out of production. The vast majority of acres enrolled were in midwestern and Plains states, and nearly all were converted to grasslands. Various studies have shown how the CRP has aided grassland birds (Dunn et al., 1993; Ryan et al., 1998), even though the original program did not target areas important for wildlife (Szentandrasi et al., 1995).

Subsequent legislation, most recently the 1996 Federal Agriculture Improvement and Reform Act (FAIR), expanded the CRP to include a strong wildlife habitat component (Farm Service Agency, 1997a). There is a special effort to enroll riparian areas. Enrolled lands must be planted with permanent vegetation, from grass to trees for the 10-year contract lifetime. Riparian acres should provide a vegetated buffer of 50 to 150 ft or more in width along the waterway.

Applications for annual CRP enrollments are analyzed with an Environmental Benefits Index (EBI), which awards points for land attributes and proposed practices. EBI includes a major wildlife component, with items such as establishment of wildlife habitat cover, provision of cover for endangered species, and proximity to protected areas. In addition, a water-quality component includes attributes met by planting riparian vegetation (Farm Service Agency, 1997b).

Furthermore, farmers may enroll lands in the CRP at any time (continuous sign-up) if they will carry out high-priority practices, most of which provide wildlife habitat on fallowed land. These practices include filter strips along waterways, riparian buffers, shelterbelts, and shallow water areas for wildlife.

Conservation Reserve Program Enrollment in the Central Valley

In October 1999, 31,261,014 acres were enrolled nationwide. Just 131,758 acres (4.2%) were in California, and only 25,994 acres (0.41%) were in the Central Valley. The 25,994 acres enrolled in the Central Valley contrasts to 6.2 million acres of harvested cropland in the region. Nearly all of the Central Valley acreage was located in Yolo and Merced Counties (Table 96.2). None of the Yolo County acreage (15,243 acres) was enrolled through the CRP continuous sign-up program (Weatherford, 1999).

Examination of the 18th sign-up (1998) shows how, nationwide, new CRP lands provided for wildlife almost as much as reduction in soil erosion, with 18.6% of the enrolled acres being permanent wildlife habitat, compared with 21.7% having highly erodible soils. None of the California acres enrolled in the 18th sign-up was wildlife habitat (Table 96.3). For the 10 new contracts in the Central Valley, 2070 out of the 2138 acres were existing grassland (Table 96.4).

The CRP is only one of a number of programs that provide funding for wildlife habitat, albeit the one with the broadest potential for habitat conservation. The U.S. Department of Agriculture's Wetlands Reserve Program has a larger number of enrolled farms in the Central Valley, but is

Table 96.2   Acres of Land in Central Valley Counties Enrolled in the U.S. Department of Agriculture's Conservation Reserve Program

County	1997 Acres Harvested Cropland[a]	1999 Conservation Reserve Program Acreage[b]
Colusa	287,630	876
Glenn	212,848	1231
Kern	893,221	1504
Kings	445,537	357
Merced	434,074	6783
Yolo	324,291	15,243
Other Central Valley counties	3,595,028	0
Totals	6,192,629	25,994

[a] 1997 Harvested cropland excludes irrigated pasture and non-irrigated grasslands. Data source: U.S. Department of Agriculture.
[b] Data from Farm Service Agency, 1999b.

Table 96.3   Conservation Reserve Program 18th Sign-Up: Nationwide vs. California

Item	U.S.	California
Acres	4,987,061	8776
Average rent/acre	$45.50	$23.45
Acres of rare and declining habitat	216,883	0
Acres of permanent wildlife habitat	930,402	0
Acres of highly erodible soils	1,082,700	2766

From Farm Service Agency, Conservation Reserve Program 18th sign-up booklet, U.S. Department of Agriculture, Washington, D.C., 1999.

Table 96.4   Conservation Reserve Program in the Central Valley — New Contracts under the 18th Sign-Up (1998)

County	Number of Contracts	Average Rent/Acre ($)	Acres	Acres as Existing Grass Planting
Colusa	1	23.73	373	373
Merced	7	19.00	1443	1375
Yolo	2	33.29	322	322

From Farm Service Agency, County CRP signup 18 information, U.S. Department of Agriculture, Washington, D.C., 1999.

focused almost exclusively on seasonal wetlands for wildlife. Most other programs provide cost sharing for wildlife habitat enhancements but not direct funding for setting aside cropland.

OBSTACLES AND SOLUTIONS

Financial

Agriculture is a business and farmers are businesspeople, focused on making a profit under often-difficult circumstances. While Central Valley croplands have high-yield potential, there are many uncertainties that make farming a precarious venture. Many growers are just surviving (Rollins, 1999; Weatherford, 1999). Profit is determined by crop yield, selling price, production costs, and subsidies. Crop yield per acre is affected by weather, disease, and other factors. Furthermore, farmers in some places have to reduce the number of acres they plant when dry winters result in limited irrigation water. The purchase price for crops varies considerably from year to year but the trend has not kept up with inflation. Costs continue to climb each year.

While yields and income per acre for individual crops are available from individual counties, costs per acre are not. Thus, it is difficult to get an accurate figure of net profit or loss per acre for individual crops in different areas. The University of California Cooperative Extension Service developed very detailed cost estimates for growing a variety of crops on hypothetical farms in different counties (University of California Cooperative Extension, 1994–1999). They show that in many cases higher yields and higher purchase prices are needed to make better profits and that it is easy to realize a loss on individual crops.

These fiscal uncertainties focus farmers on maximizing their net income. Activities that seem superfluous to the business of farming and provide little or no economic return are unlikely to get attention. As a result, one of the biggest obstacles to habitat conservation on Central Valley farmland is the low payments from government programs such as the CRP. The CRP payments vary with county or conservation practice and average $45.50 per acre across the U.S. (Table 96.3). For the 18th annual sign-up contracts in the Central Valley, the Colusa County contract had an annual rent of $23.73 per acre; the seven Merced County contracts averaged $19; and the Yolo County contracts averaged $33.29 (Table 96.4).

Payment for land taken out of production is just one fiscal need. A second need is financial assistance with the often considerable costs of establishing native vegetation. The CRP pays half the cost for improvements such as permanent vegetation. Recent estimates for riparian enhancement along a slough amounted to $8433.75 for a 20-ft strip along a quarter-mile stretch of slough, plus $1722.12 for annual maintenance required over the first 3 years (Yolo County Resource Conservation District, 1999). A 50% cost sharing still leaves the farmer with significant out-of-pocket expenses. It is possible for a farmer to obtain additional cost sharing from a variety of other federal and state programs (Yolo County Resource Conservation District, 1999), but this will require development of grant applications.

The essential solution to this fundamental obstacle is to provide more significant funding for provision of wildlife habitat on farmland, and to make it as easy as possible to apply for and receive funds. A state-funded program providing annual rental payments should augment the CRP, for example. In addition, the different agencies and programs need a coordinated approach, focused on individual counties, watersheds, or other geographic regions, to make it simple for landowners to apply for additive funding from multiple sources.

Coupling of fallow land conservation practices with projects such as flood control that provide direct and obvious benefits to farmers is a helpful approach (Jones and Stokes Associates, 1996). Local flooding from creeks and streams occurs along the length of the Central Valley, making field croplands inaccessible for weeks at a time, often during planting season. Orchards can suffer from direct flooding impacts and to a lesser extent from damage by floating debris.

A focus on flood control can also provide additional sources of funding. In Fresno County, the California Department of Fish and Game and the Fresno Metropolitan Flood Control District are cooperating on a 20-year program to improve flood control, and to preserve and restore riparian habitat along 100 miles of streams on the east side of the Valley floor (Rush, 1998). This approach brings in funding sources not normally available for habitat projects.

Practices and Awareness

Farm practices in the Central Valley center around intensive agricultural production using modern equipment. In the San Joaquin Valley, in particular, land is laid out in large rectangular fields, often a half mile long. Cultivation with large tractors and farm equipment utilizes entire fields, with no patches of poor soils or field corners left fallow (Pinter, 1999). Areas with winding streams or sloughs, on the other hand, have fields with odd corners that cannot be cultivated as easily. Turning those areas into patches of native vegetation, such as perennial grasses, can solve the farmer's weed problem over time as well as provide wildlife cover.

Land ownership is another significant issue, as tenant farmers are less interested in conservation practices that, at best, provide long-term economic benefits (Soule et al., 1999). A growing share of U.S. farmland is rented out to operators, exceeding 40% by 1992.

In California's Central Valley, generally an area of rapid urban development, there is an aging farm-ownership population and nonfarming interests buying land for its speculative value. Anecdotal information suggests increasing rental acreage. Data are available to distinguish all rented and owner-operated acres in California. However, we can distinguish between farms operated by landowners who farm the land they own, and farms operated by tenant farmers. Comparison of 1992 and 1997 data for Central Valley counties (U.S. Department of Agriculture, 1999) suggests that there is not an overall shift from owner-operated to tenant-operated acres occurring in the Central Valley. Most counties show large increases in the owner-operated acres of harvested crop land between 1992 and 1997. One county, Sacramento, shows decreases in owner- and tenant-operated acres. Sacramento County has particularly severe pressure from spreading urbanization.

Some landowners who lease their property for agricultural operations still take an interest in providing wildlife habitat, e.g., one large corporate owner in Yolo County with over 16,000 acres. This farm provides significant acreage of wetland and saline soil upland habitat, and also provides a great deal of vegetation along sloughs and ditches and in odd field corners. While this is often not native vegetation, it provides habitat for a variety of birds, including a number of nongame species, such as the rapidly declining northern harrier, and one species of concern, the tri-colored blackbird. In this instance, the public relations value of providing habitat is a major factor for the corporate landowner, and there are several tours each year for local residents (Hall, 1999).

Farmer and landowner awareness of the values of fallow land and techniques for providing wildlife habitat is another issue. For example, a farmer may spend considerable time and money keeping field and waterway edges bare of vegetation in order to control weed growth. He may not know that planting of perennial vegetation such as native grasses will solve this weed problem, as well as provide wildlife habitat. A strip of native perennial grasses requires minimal attention after 2 or 3 years and avoids repeated disking and spraying (Yolo County Resource Conservation District, 1999). Furthermore, the U.S. Department of Agriculture's Natural Resource Conservation Service (formerly Soil Conservation Service) promoted clean farming in earlier decades and successfully developed a culture that "only clean farming is good farming" (Saffel, 1998).

The Yolo County Resource Conservation District has promoted a variety of wildlife-friendly farming techniques for a number of years, many developed by local farmer John Anderson. It has enrolled a growing number of local farmers in conservation practices, held many workshops, and obtained cost-sharing grants to fund conservation practices (Pye, 1998). It is now taking a "whole watershed" approach, in a collaborative project with Audubon California (Anderson, 1999; Boshoven, 1999) (discussed later).

There are also individual farm showcases in other counties. For example, Glenn County farmer Allen Garcia has shown that rice farming on marginal land in the central Sacramento Valley can blend well with wildlife-friendly farming, using techniques that include providing vegetation along creeks and runoff channels and allowing some fields to lie fallow for a year (California State Coastal Conservancy, 1994).

An important valley-wide solution is to increase the education and outreach to local farmers. There are many resource conservation districts (RCDs) in the Central Valley, with varying degrees of activity but all with little money. RCD directors are all local farmers, so these organizations can effectively reach out to the agricultural community. Funding for a central or southern San Joaquin Valley RCD to develop a multiyear program that promotes wildlife-friendly farming techniques and tests the conservation practices in the more-arid climate of this southern valley is a great need. In addition, each RCD and the umbrella organization, the California Association of Resource Conservation Districts, needs funding to develop individual projects around the entire Central Valley to provide farmers with local success stories. Finally, more research is needed on the medium- and long-term economic benefits of wildlife habitat patches because a solid economic justification will provide the best approach to enroll more farmers.

Regulation

Endangered Species Act

Farmers and other landowners often fear Endangered Species Act restrictions on how they use their land if a listed species occurs on their property. As a result, they are unwilling to carry out habitat enhancements or leave land fallow in case these steps attract listed species (Environmental Defense Fund, 1999). The Central Valley, and especially the southern half of the San Joaquin Valley, is particularly rich in endangered species; therefore, this is a major issue.

The U.S. Fish and Wildlife Service has developed the Safe Harbor Policy to address some of these concerns (U.S. Fish and Wildlife Service, 1999). Under this policy, landowners can enter into an agreement with the service to carry out and maintain specific habitat enhancements on some of their property for a defined time period. If this habitat enhancement results in listed species moving on to the property, the landowner is not subject to any additional restrictions under the Federal

Endangered Species Act regarding species covered by the agreement. The landowner is free to remove the habitat enhancement after the agreement period. There is no equivalent policy for the California Endangered Species Act.

Establishment of Safe Harbor agreements covering regions within the Central Valley, and certain habitat enhancements such as establishment of riparian vegetation, would be a great help in promoting conservation practices. In addition, it must be simple for landowners to obtain permits or agreements. Currently, if a landowner obtains a Conservation Reserve Program or Wetlands Reserve Program contract with the U.S. Department of Agriculture, he or she does not have to deal directly with the U.S. Fish and Wildlife Service. However, there is still a review process for the individual contract (Lowrie, 1999). The Department of Agriculture's Natural Resources Conservation Service (NRCS) makes an initial determination whether the conservation practices that the landowner will use will have any effect on listed species. If NRCS determines that there will be no effect, it issues documentation that provides some assurance to landowners. If it determines that there will be some effect, it carries out a consultation with the U.S. Fish and Wildlife Service under Section 7 of the Federal Endangered Species Act. The result is a Biological Opinion that provides for the practices that the landowner will carry out. The landowner receives a letter from NRCS saying that it has obtained a Biological Opinion from the service. It does not appear that this procedure is as effective as a Safe Harbor agreement in alleviating concerns in the agricultural community.

Need for a One-Stop Permit Process

Many projects to provide fallow land wildlife habitat require various permits and a farmer will have to deal with a multitude of federal, state, and local agencies. Concern that obtaining these permits will be time-consuming and complex, and could result in restrictions of farm operations, limits landowners' interest in carrying out conservation projects. Cost is another consideration as there is a fee for each permit (Jones and Stokes Associates, 1996; Mountjoy, 1999).

A 2-year-old pilot project on the Elkhorn Slough watershed in coastal California's Monterey County provides an example of how one-stop-permitting aids conservation projects (Mountjoy, 1999). The local NRCS office can provide an overall permit to farmers carrying out any of ten conservation practices and using NRCS technical or cost-share assistance. This permit encompasses six separate permitting processes under federal and state Endangered Species Acts, the federal Clean Water Act, Monterey County ordinances, and other legal provisions. Because the conservation project serves a public purpose, reduction in the level of sediment and pollution entering the slough, there is no fee (Johnson, 1999).

Establishment of this one-stop permit had a major impact on local farmers' attitudes. In the first year, twice as many farms participated as were expected. Individuals that would normally refuse to participate in such a project participated because they obtained an easy way to deal with erosion and flooding on their land, without having to obtain their own permits from multiple agencies (Mountjoy, 1999). In addition, farmers accepted higher standards for the practices that they used.

Strategic Approaches

Provision of wildlife habitat in fallow patches and along waterways occurs in a random manner at present, driven by individual landowner decisions. Scattered patches of habitat, however, often have limited biological value for most wildlife, particularly sensitive and declining species. These small patches may provide for small populations of some vertebrates, but are unlikely to provide for long-term, viable populations. For example, small, very narrow riparian strips may attract a variety of riparian-dependent songbirds to breed. But these areas will provide for too few nesting pairs and are likely to have severe cowbird parasitism, as well as predation by additional birds or small carnivores. Songbirds need at least a 20% nesting success for a population to survive (Robinson et al., 1995). The result is that small patches of trees could become population sinks, reducing the viability of overall populations (Donovan et al., 1995a,b). Unfortunately, research has not been done on this issue for Central Valley conditions and species.

For optimum effect, it is necessary to have strategic approaches at the regional scale. Following are some of the strategies that are pertinent to the Central Valley.

Focus on Watersheds Important for Declining Riparian Birds

There is a need for the creation of many, or more continuous, riparian strips in watersheds deemed important for particular riparian species, such as rare and declining birds (e.g., the yellow warbler, *Dendroica petechia*; the yellow-breasted chat, *Icteria virens*; and the song sparrow, *Melospiza melodia*), in a manner that is beneficial to those species. A Riparian Bird Conservation Plan (Riparian Habitat Joint Venture, 2000) provides guidance as to the most important locations for restoration of additional riparian habitat, and the types and extents of riparian vegetation required for different species. In addition, the CALFED project (see Chapter 14, this volume), while focused on restoration of delta waterways and aquatic species, also addresses needs in the larger Sacramento and San Joaquin River watersheds.

Focus on Entire Smaller Watersheds

Focusing on riparian vegetative strips and other habitat patches in small watersheds, with a long-term goal of involving many of the landowners, will provide for greater continuity of riparian vegetation and usefulness of upland habitat patches. The new Audubon California/Yolo RCD project for Union School Slough in Yolo County is taking this approach, reaching out to landowners, and developing a number of specific projects (Boshoven, 1999). Several years ago, the Napa County RCD in the San Francisco Bay area enrolled 63 Huichica Creek watershed landowners in a land stewardship project that involved a variety of conservation practices, including planting native vegetation along the creek. This project was so successful that landowners shifted from hostility toward the endangered freshwater shrimp in the creek to distress if it was absent from their stretch of creek, because the shrimp is an indicator of good water quality (Bowker, 1996).

Fallow Land Habitat Adjacent to or Near Permanently Protected Lands or Other Lands Managed for Wildlife

The Central Valley has a number of wildlife areas, most of which are managed by the U.S. Fish and Wildlife Service and the California Department of Fish and Game. In addition, there is the Grasslands Ecological Area, with over 100,000 acres of public land, duck clubs, and other lands. Most of these areas have waterfowl as their primary focus, in particular providing wintering habitat for Pacific Flyway birds. But they also provide for a variety of rare vertebrates, vernal pool invertebrates, and a number of rare plant species (Silveira, 1998). Additional and adjacent patches of fallow land, especially riparian and wetland habitat, will add to usable habitat for birds and other animals, and thus help increase population sizes. Winter flooding of rice fields now provides major wintering grounds for waterfowl, as well as nearly all the winter habitat for over 300,000 shorebirds (Shuford et al., 1998; Elphick and Oring, 1998).

In the future, there will be more protected areas, including lands managed for a variety of rare and endangered species. In addition, large blocks of land will be taken out of agricultural production to reduce groundwater and runoff water salinity on the west side of the San Joaquin Valley. Decades of cropland irrigation have raised the water table to just a few feet below ground level and have caused a buildup of salts and particularly toxic levels of selenium in that groundwater. The saline groundwater causes significant crop damage and the area with shallow groundwater is expanding rapidly. For example, in the Westlands Water District in the west-central portion of the San Joaquin Valley, the area where the water table is within 5 ft of the surface increased from 38,000 to 328,000 acres between 1991 and 1997. The high levels of selenium cause major biological problems in the San Joaquin River system and associated wetlands, particularly birth defects in birds. Retiring sufficiently large blocks of land from irrigated agriculture will both lower the groundwater tables and reduce the transport of selenium into waterways (Land Retirement Technical Committee, 1999; U.S. Army Corps of Engineers, 1999).

Also, there is increased emphasis on protection of critical lands for endangered and rare species in the Central Valley. For example, several endangered vertebrates are now found only in limited patches of native habitat in the southern San Joaquin Valley (U.S. Fish and Wildlife Service, 1998). Fallow land providing wildlife habitat adjacent to one of these blocks would provide a temporary expansion of occupied habitat for key upland species and help build up their populations (Kelly, 1999). Additional patches of wildlife habitat on temporary fallow land adjacent or near to permanently protected habitat areas are likely to provide greater wildlife benefits (Kelly, 1999).

Creating Patches of Habitat in Lands Identified as Critical Wildlife Corridors

The recovery plan for upland species of the San Joaquin Valley identifies specific agricultural areas where wildlife corridors are necessary (U.S. Fish and Wildlife Service, 1998). These are mainly for the San Joaquin kit fox, an umbrella species that will travel significant distances and will make some use of agricultural lands. Kit foxes are subject to predation by coyotes and red foxes, and they construct numerous dens for year-round use in order to avoid predation. An agricultural kit fox corridor will need periodic dens on fallow land patches to provide for effective movement of the species and to avoid becoming a population sink.

Conservation of Habitat in Small Specialized Areas Such as Rare Plant Communities

There are small patches of natural habitat or poor-quality agricultural land scattered across the Central Valley that provide important biological values. For example, Yolo County has a small area of alkali soils supporting a population of the palmate-bracted bird's beak (*Cordylanthus palmatus*), an endangered plant (U.S. Fish and Wildlife Service, 1998). Patches such as this need to be permanently protected natural habitat.

CONCLUSION

Adoption of programs that overcome the obstacles to creation of fallow land wildlife habitat, and projects that focus on key areas such as those mentioned previously are essential for the long-term ecological health of California's Central Valley. With widespread implementation, they will provide for a tremendous increase in wildlife habitat and in the populations of declining and commoner species that can utilize agricultural lands or habitat patches located in a farmland matrix. Without these programs, most of the valley will continue to be biologically impoverished and society's ability to recover declining wildlife species will be severely diminished.

REFERENCES

Anderson, J., Wildlife habitat and clean farming can be compatible, *Linkages,* 1, 2–3, 1995.
Anderson, J., Yolo County farmer, personal communication, July 1999.
Barry, W.J., The Central California Prairie, California Department of Parks and Recreation, Sacramento, CA, 1972.

Biadi, F., Wildlife as an indicator of the changes in agriculture: impacts of farming systems and practices on wildlife, *C. R. Acad. Agric. France,* 84, 125–138, 1998.

Boshoven, J., Watershed Coordinator, Audubon California/Yolo County RCD, personal communication, August 1999.

Bowker, D., Napa County Resource Conservation District, personal communication, May 1996.

California State Coastal Conservancy, Options for wetland conservation: A guide for California landowners, California State Coastal Conservancy, Oakland, 1994.

Clark, J. and Rollins, G., Farming for wildlife: voluntary practices for attracting wildlife to your farm, California Department of Fish and Game, Sacramento, 1996.

Dawdy, D.R., Feasibility of mapping riparian forests under natural conditions in California, in *Proceedings of the California Riparian Conference: Protection, Management, and Restoration for the 1990s,* Abell, D.L., Ed., Gen. Tech. Rep. PSW-110, Pacific Southwest Forest and Range Experimental Station, Forest Service, U.S. Department of Agriculture, Berkeley, CA, 1989, pp. 63–68.

Donovan, T.M., Thompson, F.R., Faaborg, J., and Probst, J.R., Reproductive success of migratory birds in habitat sources and sinks, *Conserv. Biol.,* 9, 1380–1395, 1995a.

Donovan, T.M., Lamberson, R.H., Kimber, A., Thompson, F.R., and Faaborg, J., Modeling the effects of habitat fragmentation on source and sink demography of neotropical migrant birds, *Conserv. Biol.,* 9, 1396–1407, 1995b.

Dunn, P. et al., Ecological benefits of the Conservation Reserve Program, *Conserv. Biol.,* 7, 132–139, 1993.

Elphick, C.S. and Oring, L.W., Winter management of California rice fields for waterbirds, *J. Appl. Ecol.,* 35, 95–108, 1998.

Environmental Defense Fund, Safe harbor: helping landowners help endangered species, Washington, D.C., 1999.

Farm Service Agency, The Conservation Reserve Program: innovation in environmental improvement, U.S. Department of Agriculture, Washington, D.C., 1997a.

Farm Service Agency, Conservation Reserve Program Sign-Up 16: Environmental Benefits Index, U.S. Department of Agriculture, Washington, D.C., 1997b.

Farm Service Agency, Conservation Reserve Program 18th sign-up booklet, U.S. Department of Agriculture, Washington, D.C., 1999a.

Farm Service Agency, County CRP signup 18 information, U.S. Department of Agriculture, Washington, D.C., 1999b.

Freemark, K. and Boutin, C., Impacts of agricultural herbicide use on terrestrial wildlife in temperate landscapes: a review with special reference to North America, *Agric. Ecosyst. Environ.,* 52, 67–91, 1995.

Fuller, R.J., Gregory, R.D., Gibbons, D.W., Marchant, J.H., Wilson, J.D., Baillie, S.R., and Carter, N., Population declines and range contractions among lowland farmland birds in Britain, *Conserv. Biol.,* 6, 1425–1441, 1995.

Great Valley Center, Agricultural land conservation in the Great Central Valley, Great Valley Center, Modesto, CA, 1998.

Hall, M., Conaway Ranch manager, personal communication, May 1999.

Heitmeyer, M.E., Connelly, D.P., and Pederson, R.L., The Central, Imperial and Coachella Valleys of California, in *Habitat Management for Migrating and Wintering Waterfowl in North America,* Smith, L.M., Pederson, R.L., and Kiminski, R.M., Eds., Texas Technical University Press, Lubbock, 1989, pp. 475–505.

Johnson, B., Quality control: a coalition of six Central Coast Farm Bureaus develops a unique program to control erosion and reduce sediment in area waterways, *Calif. Farmer,* March 8–10, 1999.

Jones and Stokes Associates, Inc., Willow Slough watershed integrated resources management plan, Report JSA 95–232, Sacramento, 1996.

Kelly, P., San Joaquin Valley Endangered Species Recovery Program, personal communication, June 1999.

Kroll, C., Goldman, G., and Phelan, M., The Central Valley: changing economic structure and the costs of doing business, in *People Pressures: California's Central Valley, Agricultural Issues Center,* Carter, H.O. and J. Spezia, Eds., University of California, Davis, 1991, pp. 13–36.

Land Retirement Technical Committee, Land retirement, Final report, The San Joaquin Valley Drainage Implementation program and the University of California Salinity/Drainage Program, California Department of Water Resources, Sacramento, 1999.

Lowrie, J., Natural Resource Conservation Service, U.S. Department of Agriculture, personal communication, August 1999.

Mountjoy, D., One Stop Regulatory Shopping for Resource Conservation on Farms, U.S. Department of Agriculture, Natural Resources Conservation Service, Salinas, CA, 1999.

Pinter, J., Yolo County farmer, personal communication, March 1999.

Platzek, R., California's most important choice: an agricultural or an urban Central Valley? *Calif. Planner,* 6(6), 8–9, 1995.

Pye, K., Yolo County Resource Conservation District Executive Director, personal communication, March 1998.

Ribic, C.A., Warner, R.E., and Mankin, P.C., Changes in wildlife habitat on farmland in Illinois: 1920–1987, *Environ. Manage.,* 22, 303–313, 1998.

Riparian Habitat Joint Venture, The riparian bird conservation plan: a strategy for reversing the decline of riparian associated birds in California, Point Reyes Bird Observatory, Stinson Beach, CA, 2000.

Robinson, S.K., Thompson, F.R., Donovan, T.M., Whitehead, D.R., and Faaborg, J., Regional forest fragmentation and the nesting success of migratory birds, *Science,* 267, 1987–1990, 1995.

Rollins, G., California Department of Fish and Game, personal communication, June 1999.

Rush, A.V., Two agencies embark on a 20-year program to restore Fresno County waterways, *Fresno Bee,* December 8, 1998.

Ryan, M.R., Burger, L.W., and Kurzejeski, E.W., The impact of CRP on avian wildlife: a review, *J. Prod. Agric.,* 11, 61–66, 1998.

Saffel, J., California Association of Resource Conservation Districts, Region 9, personal communication, October 1998.

Shuford, W.D., Page, G.W., and Kjelmyr, J.E., Patterns and dynamics of shorebird use of California's Central Valley, *The Condor,* 100, 227–244, 1998.

Silveira, J., U.S. Fish and Wildlife Service, personal communication, May 1998.

Soule, M., Adebayehu, T., and Wiebe, K., Conservation on rented farmland: a focus on U.S. corn production, in *Agricultural Outlook: January–February 1999,* Economic Research Service, U.S. Department of Agriculture, Washington, D.C., 1999, pp. 15–19.

Szentandrasi, S., Polansky, S., Berrens, R., and Leonard, J., Conserving biological diversity and the Conservation Reserve Program, *Growth Change,* 26, 383–404, 1995.

Thelander, C.G. and Crabtree, M., *Life on the Edge: A Guide to California's Endangered Natural Resources: Wildlife,* Biosystems Books, Santa Cruz, CA, 1994.

University of California Cooperative Extension, Sample costs to produce crops series, Davis, 1994–1999.

U.S. Army Corps of Engineers, Draft environmental assessment and draft finding of no significant impact, Central Valley Project Improvement Act, Land Retirement Program Demonstration Project, Mid-Pacific Regional office, U.S. Army Corps of Engineers, Sacramento, 1999.

U.S. Department of Agriculture,. 1997 Census of Agriculture, California State and County Data, Vol. I, Geographic Area series, Part 5, Washington, D.C., 1999.

U.S. Fish and Wildlife Service, Recovery plan for upland species of the San Joaquin Valley, California, Region 1, Portland, OR, 1998.

U.S. Fish and Wildlife Service, Announcement of final Safe Harbor policy, *Fed. Regis.,* 64, 32717–32726, 1999.

Weatherford, J., Natural Resource Conservation Service, U.S. Department of Agriculture, personal communication, June 1999.

Yolo County Resource Conservation District, *Bring Farm Edges Back to Life!,* 3rd ed., P. Robins, Ed., Yolo County Resource Conservation District, Woodland, CA, 1999.

CHAPTER 97

Reduced-Disturbance Agroecosystems in California

Jeff P. Mitchell, W. Thomas Lanini, Steve R. Temple, Peter N. Brostrom, E.V. Herrero, E.M. Miyao, Timothy S. Prather, and Kurt J. Hembree

INTRODUCTION

Over the last several decades, growers in California's Central Valley, as in many other agricultural areas throughout the world, have become increasingly aware of the importance of soil organic matter (SOM) in relation to sustained soil quality or function (West Side On-Farm Demonstration Project Participant Survey, 1999; Romig et al., 1995; Magdoff, 1992). Soil carbon (C), which typically constitutes about one half of SOM, is closely linked to many desirable soil physical, chemical, and biological properties that are associated with enhanced soil productivity and quality (Reicosky, 1996b; Ismail et al., 1994). More recently, awareness has also increased concerning the negative impact that tillage has on SOM storage (Reicosky, 1995; Jackson, 1998). While moderate tillage may provide more favorable soil conditions for crop growth and development and weed control over the short term (Carter, 1998), intensive tillage of agricultural soils has historically led to substantial losses of soil C that range from 30 to 50%. Conventional tillage practices disrupt soil aggregates exposing more organic matter to microbial degradation and oxidation (Reicosky, 1996a), and are one of the primary causes of tilth deterioration over the long term (Karlen et al., 1990). Micro- and macrochannels within the soil, created by natural processes, such as decaying roots and worms, may also be destroyed by tillage (Carter, 1998). Deep tillage, as is customarily done as a routine soil preparation operation, is also costly and requires high energy and increased subsequent effort to prepare seedbeds. A recent survey by Jackson (1998), documenting a 40% decline in SOM since intensive tillage practices began in the Salinas Valley, confirms the conclusion drawn from other long-term crop rotation studies (e.g., the Morrow Plots at the University of Illinois; the Sanborn Field Plots in Columbia, MO; and the Columbia Plateau Plots near Pendleton, OR) that intensive tillage typically leads to decreased soil C via gaseous CO_2 emissions in virtually all crop production systems (reviewed by Reicosky, 1996b). There is mounting evidence as well as concern that this C source has been a significant component of the historic increase in atmospheric CO_2 (Wilson, 1978; Post et al., 1990), and the potentially associated greenhouse effect (Lal et al., 1998) that is attracting intense attention worldwide.

Recent pioneering studies by Reicosky and Lindstrom (1993), involving a variety of tillage methods, indicate major gaseous losses of C immediately following tillage but point to the potential for reducing soil C loss and enhancing soil C management through the use of conservation tillage (CT) crop production practices. Though these practices have been developed over several decades primarily for erosion control in other parts of the U.S., recent concerns have reemphasized the

Table 97.1   Typical Tillage/Land Preparation Practices before Planting Processing Tomatoes and Cotton in the San Joaquin Valley's West Side Region

Practice	No. of Times
Stubble disk (shredding required following cotton, broccoli, and peppers)	2
Chisel to average 27-in. depth	1
Disk/"float"	1–2
Landplane	1
Preplant fertilize	1
List beds	1
"Till and pack"	1
Bed shape	1
Preirrigate	
Shank apply fumigant	1
Plant	1
In-season cultivation	2–3

importance of CT and how it might be implemented on a broader scale. These concerns include the need to sustain soil quality (Abdul-Baki, 1998) and profitability (Mitchell and Goodell, 1999) in areas such as California, where CT is virtually nonexistent, and potential global climate change. CT implementation will help reduce soil C losses (Carter, 1998) and thereby sustain soil quality and agricultural productivity.

Most current San Joaquin Valley (SJV) annual crop production systems are very tillage intensive. Typically 9 to 11 tillage operations are made in most fields following harvest in preparation for a succeeding crop at about 18 to 24% of a farmer's seasonal budget (West Side On-Farm Demonstration Project Participant Survey, 1999; Table 97.1). Tillage in these production systems is typically done in a "broadcast" manner throughout a field, without deliberate regard to preserving dedicated crop growth or traffic zones. Studies by Carter (1985, 1998) over the last several decades, however, have confirmed the potential to eliminate deep tillage, decrease the number of soil preparation operations by as much as 60%, reduce unit production costs, lower soil impedance, and maintain productivity in a number of SJV cropping contexts using reduced, precision, or zone tillage practices that limit traffic to permanent paths throughout a field. The result is the reduction of soil compaction and preservation of an optimum soil volume for root exploration and growth (Carter et al., 1991; Rechel et al., 1991). There have been relatively few other studies during the past several decades that have investigated aspects of CT (Table 97.2). Recent work by Herrero et al. (2001a,b), also in the SJV, evaluating a mulch-transplanted processing tomato production system has also demonstrated the potential to sustain productivity under certain mulches and increase soil C and earthworms using a CT approach. Given the successes of these component research studies and in anticipation of increasing energy (Carter, 1998) and production costs (Goodell, P., personal communication), and out of a growing desire among producers to improve soil quality via C sequestration, there is interest among many Central Valley farmers to develop CT practices that may be feasible in this region.

Table 97.2   Conservation Tillage Evaluation Sites in California

Year	Crop	Location	Evaluation Specifics
1961	Vegetables	Santa Maria	W.C. Snyder, personal communication
1987–1991	Barley/wheat	Yolo, San Luis Obispo, Tehama, and Tulare Counties	Pettygrove, 1995
1997	Tomatoes	Five Points	Herrero et al., 2001a,b
1998	Tomatoes	Parlier	Mitchell, unpublished
1998	Mixed vegetables	Indio	Hutchison and McGiffen, 2000
1998	Cotton	Five Points	S. Goyal, personal communication

CURRENT CENTRAL VALLEY ANNUAL CROP PRODUCTION SYSTEMS

California's Central Valley is one of the world's most productive agricultural regions and leads the U.S. in the production of many crops (California Department of Food and Agriculture, 1998). A wide variety of annual crops is produced throughout the Central Valley. Corn, processing tomatoes, wheat, safflower, and edible dried beans are common annual rotation crops in the southern Sacramento Valley (Clark et al., 1998). Dominant cropping patterns of the central SJV include rotations of cotton, processing tomatoes, onions, garlic, cantaloupes, wheat, sugar beets, and lettuce (Mitchell et al., 2001). The productive capacity of these valleys is made possible in large part by intensive irrigation practices; agrichemical inputs, including fertilizers and pesticides; and frequent tillage of large, monoculture production fields that predominate throughout the area (Mitchell et al., 2001). While there are instances in which more than one crop is grown in a given field during one year, as in the case of wheat double-cropped with beans in the Sacramento Valley (Clark et al., 1998), or back-to-back spring and fall lettuce in the central SJV (Mitchell et al., 1999b), in general, most row crop acreage throughout the Central Valley produces one crop per year. It is not uncommon, therefore, for large acreages to be bare during periods between crops, particularly during the winter months from October through March.

Recently, there is evidence of growing interest in the use of cover crops to improve the productivity and sustainability in a number of agroecosystems in California (Ingels et al., 1994), including row crop regions in the central SJV (Mitchell and Goodell, 1999). Potential benefits as well as management issues associated with cover cropping in the Central Valley have been widely reported (Lanini et al., 1989; Groody, 1990; Stivers and Shennan, 1991). Growing off-season cover crops may also contribute to the goal of carbon sequestration, and may facilitate capture of winter precipitation if higher infiltration rates result from their use (Colla et al., 2000). Legume-based cover crop systems that have low carbon-to-nitrogen organic residue inputs have recently been shown to significantly increase soil carbon retention (Drinkwater et al., 1998). *In situ* production and retention of high levels of crop residues, part of which may derive from cover crops, in conjunction with no-till practices, may be the most cost- and time-efficient way to increase soil organic matter (Crovetto, 1996). Despite these potential benefits, adopting cover crops into many cropping systems has been slow, largely because innovative adjustments in traditional cropping practices have not been made.

CHANGES IN TRADITIONAL CENTRAL VALLEY AGROECOSYSTEM MANAGEMENT

To continue the Central Valley's phenomenal productive capacity, many farmers in the region are increasingly becoming interested in management approaches that reduce costs, increase grower revenues, and sustain the region's resource base (Mitchell et al., 2001). Recent on-farm demonstration work conducted through the Biologically Integrated Farming Systems (BIFS) Project in the West Side area of the SJV has resulted in 60% of farmer participants intending to increase their use of cover crops in this region during coming years, in contrast to the established practice of not using cover crops (Mitchell et al., 2001). The use of cover crops in such intensively managed ecosystems is not without potential costs, and ongoing research throughout Central Valley annual cropping systems is attempting to evaluate the long-term feasibility of this practice by answering such questions as whether cover cropping increases infiltration and reduces surface runoff enough to offset the soil moisture it depletes, and whether expected increases in infiltration that might result from cover cropping are high enough to increase groundwater recharge (Joyce et al., 2002).

There is also interest in evaluating and developing management systems that reduce tillage in the Central Valley, as evidenced by recent surveys and farmer–scientist focus sessions that have

Table 97.3   Central Valley Annual Crop Rotations in Which
              Conservation Tillage Practices Are Currently Being
              Evaluated

Region	Crops	Production System
San Joaquin Valley	Processing tomatoes	No-till
		Strip-till
	Cotton	Strip-till
		No-till
	Fresh market tomatoes	No-till
	Cantaloupe	Strip-till
Sacramento Valley	Processing tomatoes	No-till
	Field corn	Ridge-till
	Beans	Ridge-till
	Wheat	No-till
	Safflower	No-till

been conducted (Benefield, C., personal communication). Participant feedback from conferences in Five Points and Davis, CA in 1998 that focused on relationships between soil organic matter, tillage, and soil quality indicates that conservation tillage may become more widely adopted throughout the region if successful examples are demonstrated (Table 97.3). A variety of factors or motivations may drive the development of CT systems in this highly intensive agricultural region in which they have not been used widely before.

Unlike other regions of the U.S., where the need to reduce soil erosion has spawned the conversion of considerable farm acreage to CT practices over the last several decades (CTIC, 1999), erosion has not yet been identified as a research and management priority. There is, however, increasing attention being paid to potential benefits of reduced tillage production systems in terms of mitigating airborne respirable dusts that are known to result from heavily tilled agricultural production systems (Clausnitzer and Singer, 1997). Currently, however, reducing production costs and improving soil quality surface are primary drivers of the growing interest in developing CT or reduced disturbance production systems (Robertson, H., personal communication).

Many issues will need to be addressed and resolved, however, before CT practices are adopted in the intensive agricultural row crop regions of California's Central Valley. These issues include the complete array of decisions faced by producers throughout the area: profitability, weed and other pest management, irrigation systems that permit uniform water application in surface residue fields, appropriate CT equipment, crop rotation and bed configuration schemes that are compatible with CT, as well as the degree to which soil and water conservation is achieved through reduced disturbance systems. Several pilot studies have recently been initiated throughout the Central Valley (Table 97.4) and based on the interest that these have generated so far, there is a strong likelihood that more will be started in the near future.

Table 97.4   Response to Conference Evaluation Question by Participants of Emerging
              Soil Management Options for California (Soil Quality/Soil Organic Matter/
              Tillage) Conferences April 22 and 23, 1998, Five Points and Davis, CA

What do you think about the future adoption of the various conservation tillage (CT) practices in California that were presented at this Conference? (Number indicated is the number of respondents selecting a given option.)

0	CT practices will not be widely adopted in California.
11	Adoption of CT will likely be on a very limited scale.
34	CT may become more widely adopted if successful examples are demonstrated.
7	It is inevitable that CT will have a far wider role in California.

ACKNOWLEDGMENTS

The broad-based, ongoing research that is described here has been funded by the U.S. Department of Agriculture National Research Initiative, the University of California's Sustainable Agriculture Research and Education Program, the California Tomato Research Institute, the California Tomato Commission, the California Melon Board, the California Department of Pesticide Regulation, and the U.S. Environmental Protection Agency.

REFERENCES

Abdul-Baki, A., Soil management: the key to sustainability, in *Proceedings: Emerging Soil Management Options for California,* Mitchell, J.P., Ed., Davis, CA, 1998, pp. 32–34.

California Department of Food and Agriculture, *Agric. Stat.,* Sacramento, 1998.

Carter, L.M., Wheel traffic is costly, *Trans. ASAE,* 28, 430–434, 1985.

Carter, L.M., Tillage, in *Cotton Production,* University of California Division of Agriculture and Natural Resources Publication, 1998, pp. 1–14.

Carter, L.M., Meek, B.D., and Rechel, E.A., Zone production system for cotton: soil response, *Trans. ASAE,* 34, 354–360, 1991.

Clark, M.S., Horwath, W.R., Shennan, C., and Scow, K.M., Changes in soil chemical properties resulting form organic and low-input farming practices, *Agron. J.,* 90, 662–671, 1998.

Clausnitzer, H. and Singer, M.J., Intensive land preparation emits respirable dust, *Calif. Agric.,* 51(2), 27–30, 1997.

Colla, G., Mitchell, J.P., Joyce, B.A., Huyck, L.M., Wallender, W.W., Temple, S.R., Hsiao, T.C., and Poudel, D.D., Soil physical properties and tomato yield and quality in alternative cropping systems, *Agron. J.,* 92, 924–932, 2000.

Crovetto, C.L., *Stubble Over the Soil: The Vital Role of Plant Residue in Soil Management to Improve Soil Quality,* American Society of Agronomy, Madison, WI, 1996.

CTIC, National survey of conservation tillage practices, Conservation Tillage Information Center, West Lafayette, IN, 1999.

Drinkwater, L.E., Wagoner, P., and Sarrantonio, M., Legume-based cropping systems have reduced carbon and nitrogen losses, *Nature,* 396, 262–265, 1998.

Groody, K., Implications for Cover Crop Residue Incorporation and Mineral Fertilizer Applications Upon Crust Strength and Seedling Emergence, Master's thesis, University of California, Davis, 1990.

Herrero, E.V., Mitchell, J.P., Lanini, W.T., Temple, S.R., Miyao, E.M., Morse, R.D., and Campiglia, E., Soil properties change in no-till tomato production, *Calif. Agric.,* 55(1), 30–34, 2001a.

Herrero, E.V., Mitchell, J.P., Lanini, W.T., Temple, S.R., Miyao, E.M., Morse, R.D., and Campiglia, E., Use of cover crop mulches in a no-till furrow-irrigated processing tomato production system, *HortTechnology,* 11(1), 43–48, 2001b.

Hutchison, C.M. and McGiffen, M.E., Jr., Cowpea cover crop mulch for weed control in desert pepper production, *HortScience,* 35(2), 196–198, 2000.

Ingels, C., Van Horn, M., Bugg, R.L., and Miller, P.R., Selecting the right cover crop gives multiple benefits, *Calif. Agric.,* 48(5), 43–48, 1994.

Ismail, I., Blevins, R.L., and Frye, W.W., Long-term no-tillage effects on soil properties and continuous corn yields, *Soil. Sci. Soc. Am. J.,* 58, 193–198, 1994.

Jackson, L., Carbon and nitrogen dynamics after tillage in California soils, in *Proceedings: Emerging Soil Management Options for California,* April 22–23, 1998, Five Points and Davis, CA, pp. 29–31.

Joyce, B.B., Wallender, W.W., Mitchell, J.P., Huyck, L.M., Temple, S.R., Brostrom, P.N., and Hsiao, T.C., Infiltration and soil water storage under winter cover cropping in California's Sacramento Valley, *Trans. ASAE,* 45: 315–326, 2002.

Karlen, D.L., Erbach, D.C., Kiaspar, T.C., Colvin, T.S., Berry, E.C., and Timmons, D.R., Soil tilth: a review of past perceptions and future needs, *Soil Sci. Soc. Am. J.,* 54, 153–161, 1990.

Lal, R., Kimble, J.M., Follett, R.F., and Cole, C.V., *The Potential of U.S. Cropland to Sequester Carbon and Mitigate the Greenhouse Effect,* Ann Arbor Press, Chelsea, MI, 1998.

Lanini, T., Pittenger, D.R., Graves, W.L., Munoz, F., and Agamalian, H.S., Subclovers as living mulches for managing weeds in vegetables, *Calif. Agric.,* 43:6, 25–27, 1989.

Magdoff, F., Building soils for better crops, in *Organic Matter Management,* University of Nebraska Press, Lincoln, 1992.

Mitchell, J.P. and Goodell, P.B., Biologically integrated farming systems, Annual Report 1999, University of California Sustainable Agriculture Research and Education Program, Davis, 1999.

Mitchell, J.P., Thomsen, C.D., Graves, W.L., and Shennan, C., Cover crops for saline soils, *J. Agron. Crop Sci.,* 183, 167–178, 1999a.

Mitchell, J.P., Hartz, T.K., Pettygrove, G.S., May, D.M., Munk, D.S., Menezes, F., O'Neill, T., and Diener, J., Organic matter recycling varies with crops grown, *Calif. Agric.,* 53(4), 37–40, 1999b.

Mitchell, J.P., Goodell, P.B., Krebill-Prather, R., Prather, T.S., Coviello, R.L., May, D.M., Munk, D.S., and Menezes, F., Innovative agricultural extension partnerships in California's Central San Joaquin Valley, *J. Extension,* 39(6), 1–12, 2001.

Pettygrove, G.S., No-till wheat and barley production in California, University of California Cooperative Extension/California Energy Commission, 1995.

Post, W.M., Peng, T.H., Enamuel, W.R., King, A.W., Dale, V.H., and DeAngelis, D.L., The global carbon cycle, *Am. Sci.,* 78, 310–326, 1990.

Rechel, E.A., DeTar, W.R., Meek, B.D., and Carter, L.M., Alfalfa (*Medicago sativa* L.) water use efficiency as affected by harvest traffic and soil compaction in a sandy loam soil, *Irrig. Sci.,* 12:61–65, 1991.

Reicosky, D.C., Impact of tillage methods on water and carbon dioxide dynamics, in *Proceedings of the Second Asian Crop Science Conference,* Fukui, Japan, August 21–23, 1995, pp. 1–2.

Reicosky, D.C., Tillage-induced CO_2 emission from soil, *Nutrient Cycling in Agrosystems,* 49:273–285, 1996a.

Reicosky, D.C., Tillage, residue management and soil organic matter, *National Conservation Tillage Digest,* 3(8): 21–22, December, 1996b.

Reicosky, D.C. and Lindstrom, M.J., Fall tillage method: effect on short-term carbon dioxide flux from soil, *Agron. J.,* 85, 1237–1243, 1993.

Reicosky, D.C., Dugas, W.A., and Torbert, H.A., Fall tillage method: effect on short-term carbon dioxide flux from soil, *Agron. J.,* 85, 1237–1248, 1996.

Romig, D.E., Garlynd, M.J., Harris, R.F., and McSweeney, K., How farmers assess soil health and quality, *J. Soil Water Conserv.,* 50, 229–236, 1995.

Stivers, L.J. and Shennan, C., Meeting the nitrogen needs of processing tomatoes through winter cover cropping, *J. Prod. Agric.,* 4, 330–335, 1991.

West Side On-Farm Demonstration Project Participant Survey, Survey results are being prepared for publication as part of the West Side Biologically Integrated Farming Systems Project's Final Report, March 30, 1999.

Wilson, E.T., Pioneer agricultural explosion in CO_2 levels in the atmosphere, *Nature,* 273, 40–41, 1978.

CHAPTER 98

Ecologically Based Pest Management: A Key Pathway to Achieving Agroecosystem Health

Miguel A. Altieri and Clara Ines Nicholls

INTRODUCTION

The ultimate goal of all farmers should be to grow healthy and productive plants and animals, while maintaining the ecological integrity of the resource base of their farms. For agroecologists, such goals would translate into healthy agroecosystems exhibiting a high degree of integrity; a strong capacity to respond and adapt; and high levels of efficiency, productivity, stability, and self-dependence (Altieri and Nicholls, 1999).

Sustainable yield in the agroecosystem derives from the proper balance of crops, soils, nutrients, sunlight, moisture, and coexisting organisms. The agroecosystem is productive and healthy when this balance of rich growing conditions prevail, and when crop plants remain resilient to tolerate stress and adversity. Occasional disturbances can be overcome by vigorous agroecosystems, which are adaptable, and diverse enough to recover once the stress has passed (Altieri and Rosset, 1995). If the cause of disease, pests, soil degradation, etc. is understood as imbalance, then the goal of agroecological treatment is to recover balance, setting in motion the agroecosystem's natural tendency toward repairing itself. This tendency is known in ecology as homeostasis, the maintenance of the system's internal functions and defense mechanisms to compensate for external stress factors. But achieving and maintaining homeostasis requires a deep understanding of the nature of agro-ecosystems and the principles by which they function. Fortunately, there are new integrative scientific approaches that allow for such understanding. Among them is agroecology, which has emerged as a discipline that provides basic ecological principles on how to study, design, and manage agroecosystems that are productive and natural resource conserving (Altieri, 1995).

Agroecology goes beyond a one-dimensional view of agroecosystems — their genetics, agronomy, edaphology, etc. — to embrace an understanding of ecological and social levels of coevolution, structure, and function. Instead of focusing on one particular component of the agroecosystem, agroecology emphasizes the interrelatedness of all agroecosystem components and the complex dynamics of ecological processes such as nutrient cycling and pest regulation (Gliessman, 1998).

From a management perspective, the agroecological objective is to provide a balanced environment, sustained yields, biologically mediated soil fertility, and natural pest regulation through the design of diversified agroecosystems and the use of low-input technologies (Altieri, 1994). The strategy is based on ecological principles that lead management to optimal recycling of nutrients and organic matter turnover, closed energy flows, water and soil conservation, and balanced pest–natural enemy populations. The strategy exploits the complementation that results from the various combinations of crops, trees, and animals in spatial and temporal arrangements (Altieri and Nicholls, 1999). These combinations determine the establishment of a planned and associated functional biodiversity that, when correctly assembled, delivers key ecological services that subsidize agroecosystem processes that underlie agroecosystem health.

In other words, ecological concepts are utilized to favor natural processes and biological interactions that optimize synergies so that diversified farms are able to sponsor their own soil fertility, crop protection, and productivity through the activation of soil biology, the recycling of nutrients, the enhancement of beneficial arthropods and antagonists, etc.

In this chapter, we provide an agroecological framework to achieve crop health through agroecosystem diversification and soil quality enhancement, key pillars of agroecosystem health. The main goal is to enhance the *immunity* of the agroecosystem (i.e., natural pest control mechanisms) and regulatory processes (i.e., nutrient cycling and population regulation) through management practices and agroecological designs that enhance plant species and genetic diversity in time and space, and the organic matter content and biological activity of the soil (Altieri, 1999).

MONOCULTURES AND THE FAILURE OF CONVENTIONAL
PEST MANAGEMENT APPROACHES

Many scientists today agree that conventional modern agriculture faces an environmental crisis. Land degradation, salinization, pesticide pollution of soil, water and food chains, depletion of groundwater, genetic homogeneity, and associated vulnerability raise serious questions regarding the sustainability and health of modern agroecosystems. The loss of yields due to pests in many crops, despite the substantial increase in the use of pesticides, is a symptom of the environmental crisis affecting agriculture (Altieri and Rosset, 1995). It is well known that cultivated plants grown in genetically homogeneous monocultures do not possess the necessary ecological defense mechanisms to tolerate outbreaking pest populations. Modern agriculturalists have selected crops for high yields and high palatability, making them more susceptible to pests by sacrificing natural resistance for productivity (Robinson, 1996). On the other hand, modern agricultural practices negatively affect natural enemies (predators and parasites), which do not find the necessary environmental resources and opportunities in monocultures to effectively suppress pests (Altieri, 1994). While monocultures are maintained as the structural base of modern agricultural systems, pest problems will continue to be the result of a negative treadmill that reinforces itself, as vegetational simplification, nutrient imbalances caused by excess fertilizers and pesticide residues compound pest invasions (Figure 98.1). Thus the major challenge for those advocating ecologically based pest management (EBPM) is to find strategies to overcome the ecological limits imposed by monocultures.

Integrated Pest Management (IPM) approaches have not addressed the ecological causes of the environmental problems in modern agriculture. There still prevails a narrow view that specific causes affect productivity, and overcoming the limiting factor (i.e., insect pest) via new technologies continues to be the main goal. In many IPM projects, the main focus has been to substitute less-noxious inputs for the agrochemicals blamed for so many of the problems associated with conventional agriculture. Emphasis is now placed on purchased biological inputs such as *Bacillus thuringiensis,* a microbial pesticide that is now widely applied in place of chemical insecticides. This type of technology pertains to a dominant technical approach called *input substitution.* The thrust is highly technological, characterized by a limiting-factor mentality that has driven conventional agricultural research in the past. Agronomists and other agricultural scientists have for generations been taught the *law of minimum* as a central dogma. According to this dogma, at any given moment there is a single factor limiting yield increases, and that factor can be overcome with an appropriate external input. Once the hurdle of the first limiting factor has been surpassed — nitrogen deficiency,

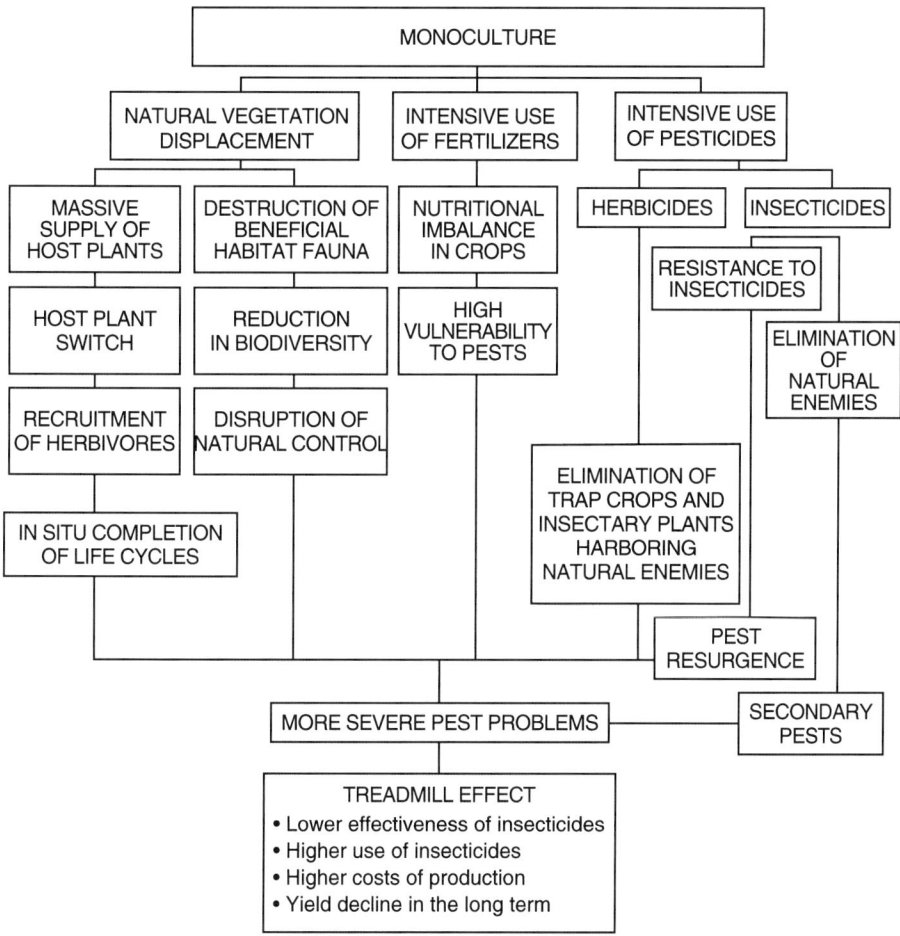

Figure 98.1 The ecological consequences of monoculture with special reference to pest problems and the agrochemical treadmill.

for example, with urea as the correct input — then yields may rise until another factor, e.g., insect pests, becomes limiting, in turn owing to increased levels of free nitrogen in the crop foliage. The factor then requires another input — insecticide in this case — and so on, perpetuating a process of treating symptoms rather than dealing with the real causes that evoked ecological imbalance (Altieri and Rosset, 1995).

Emerging biotechnological approaches do not differ as they are being pursued to patch up problems (e.g., pesticide resistance, pollution, soil degradation, etc.) caused by previous agrochemical technologies promoted by the same companies now leading the biorevolution. Transgenic crops developed for pest control closely follow the paradigm of using a single control mechanism (a pesticide) that has proven to fail over and over again with insects, pathogens, and weeds. Transgenic crops are likely to increase the use of pesticides and to accelerate the evolution of *super weeds* and resistant insect pests (Rissler and Mellon, 1996).

The *one-gene–one-pest* approach can be easily overcome by pests that are continuously adapting to new situations and evolving detoxification mechanisms. There are many unanswered ecological questions regarding the impact of the release of transgenic plants and microorganisms into the environment (Snow and Moran, 1997). Among the major environmental risks associated with genetically engineered plants are the unintended transfer to plant relatives of the *transgenes* and the unpredictable ecological effects on nontarget organisms. In fact, the effects of transgenic crops are not as localized or as transient as advocated by biotechnology supporters. Rather, many studies suggest that the ecological effects of transgenic crops might spread via wind, move through food webs, and accumulate in the soil (Marvier, 2001).

Given the above considerations, agroecological theory predicts that biotechnology will exacerbate the problems of conventional agriculture (Altieri, 2000). By promoting monocultures, it will also undermine ecological methods of farming, such as rotations and polycultures, key strategies to break the homogeneous nature of monocultures. As currently conceived, biotechnology does not fit into the broad ideals of a truly sustainable agriculture.

ACHIEVING CROP HEALTH THROUGH ECOLOGICALLY BASED PEST MANAGEMENT (EBPM)

Nowhere are the consequences of biodiversity reduction more evident than in the realm of agricultural pest management. The instability of agroecosystems, which is manifested as the worsening of most insect pest problems, is increasingly linked to the expansion of crop monocultures at the expense of the natural vegetation, thereby decreasing local habitat diversity (Altieri and Letourneau, 1982). Plant communities that are modified to meet the special needs of humans become subject to heavy pest damage, and generally the more intensely such communities are modified, the more abundant and serious the pests (Andow, 1991). The inherent self-regulation characteristics of natural communities are lost when humans modify such communities through the shattering of the fragile thread of community interactions. Agroecologists maintain that this breakdown can be repaired by restoring the shattered elements of community homeostasis through the addition or enhancement of biodiversity (Altieri and Nicholls, 1999).

The key is to identify the type of biodiversity that is desirable to maintain, enhance, and carry out ecological services, and then to determine the best practices that will encourage the desired biodiversity components. Figure 98.2 shows that there are many agricultural practices and designs that have the potential to enhance functional biodiversity, and others that negatively affect it. The idea is to apply the best management practices to enhance or regenerate the kind of biodiversity that subsidizes the health and sustainability of agroecosystems by providing ecological services such as biological pest control, nutrient cycling, water and soil conservation, etc. (Gliessman, 1998). As depicted in Figure 98.3, crop health can be achieved by regulating insect pests through two routes: a healthy soil, and a rich natural-enemy biodiversity harbored by a diversified agroecosystem.

Healthy Soils, Healthy Plants

A key feature of modern cropping systems is the frequency of soil disturbance regimes, including periodic tillage and pesticide applications, which reduce soil biotic activity and species diversity in agroecosystems. Such soil biodiversity reductions are negative because the recycling of nutrients and proper balance between organic matter, soil organisms, and plant diversity are necessary components of a productive and ecologically balanced soil environment (Hendrix et al., 1990). The ability of a crop plant to resist or tolerate pests is tied to optimal physical, chemical, and biological properties of soils. Adequate moisture, good soil tilth, moderate pH, adequate amounts of organic matter and nutrients, and a diverse and active community of soil organisms all contribute to plant health. Organic-rich soils generally exhibit good soil fertility as well as complex food webs and beneficial organisms that prevent infection by disease-causing organisms such as *Pythium* and *Rhizoctonia* (Giller et al., 1997). On the other hand, farming practices that cause nutrition imbalances can lower crop resistance. High nitrogen fertilizer levels can enhance the incidence of diseases, such as *Phytophtora* and *Fusarium,* and stimulate outbreaks of homopteran insects, such as aphids and leafhoppers

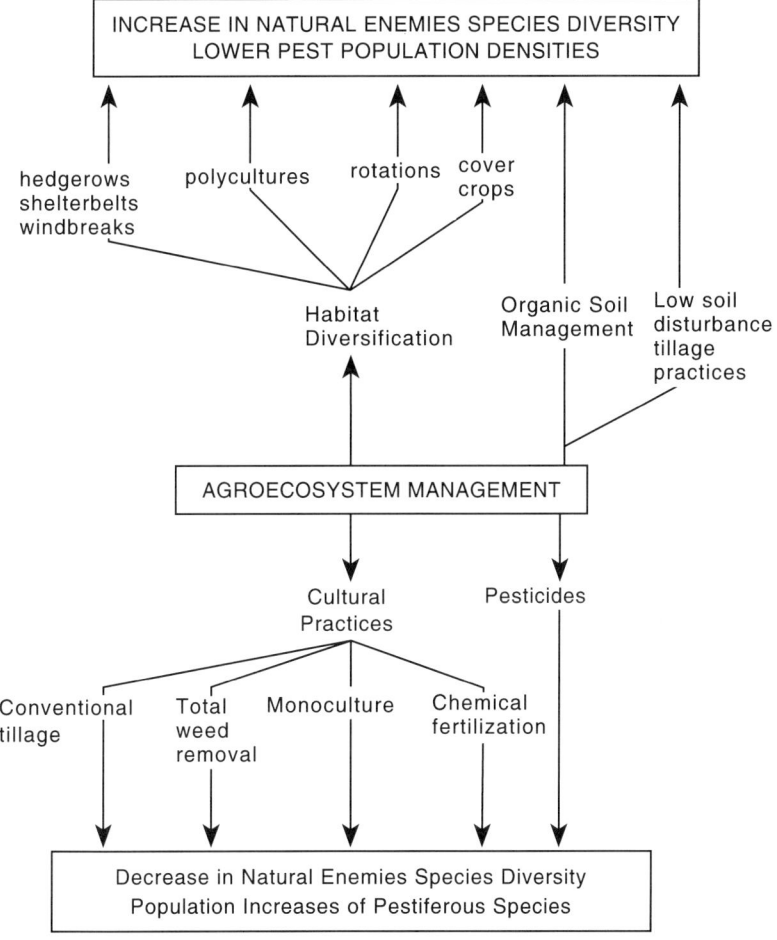

Figure 98.2 The effects of agroecosystem management and associated cultural practices on the biodiversity of natural enemies and the abundance of insect pests.

(Campbell, 1989). In fact, there is increasing evidence that crops grown in organic-rich and biologically active soils are less susceptible to pest attacks. Many studies suggest that the physiological susceptibility of crops and pathogens may be affected by the form of fertilizer used (organic vs. chemical fertilizer). Studies documenting lower diversity of several insect herbivores in low-input systems have partly attributed such reduction to a low nitrogen content in organically farmed crops (Magdoff, 1992). In California, a series of comparative experiments conducted on various growing seasons between 1989 and 1996 where broccoli was subjected to varying fertilization regimes (conventional vs. organic) showed that agroecological techniques can reduce the abundance of key insect pests, cabbage aphid (*Brevicoryne brassicae*) and flea beetle (*Phyllotreta cruciferae*), while sustaining yields. Lower herbivore numbers in organically managed plots were attributed to low foliage nitrogen content in compost-fed broccoli plants (Altieri, 1994).

In Japan, density of immigrants of the planthopper *Sogatella furcifera* was significantly lower while the settling rate of female adults and survival rate of immature stages of ensuing generations were lower in organic rice fields. The number of eggs laid by a female of the invading and following generations was smaller, and the percentage of brachypterous females in the next generation was also lower. Consequently, the density of nymphs and adults in the ensuing generations decreased in organically farmed fields (Kajimura, 1995).

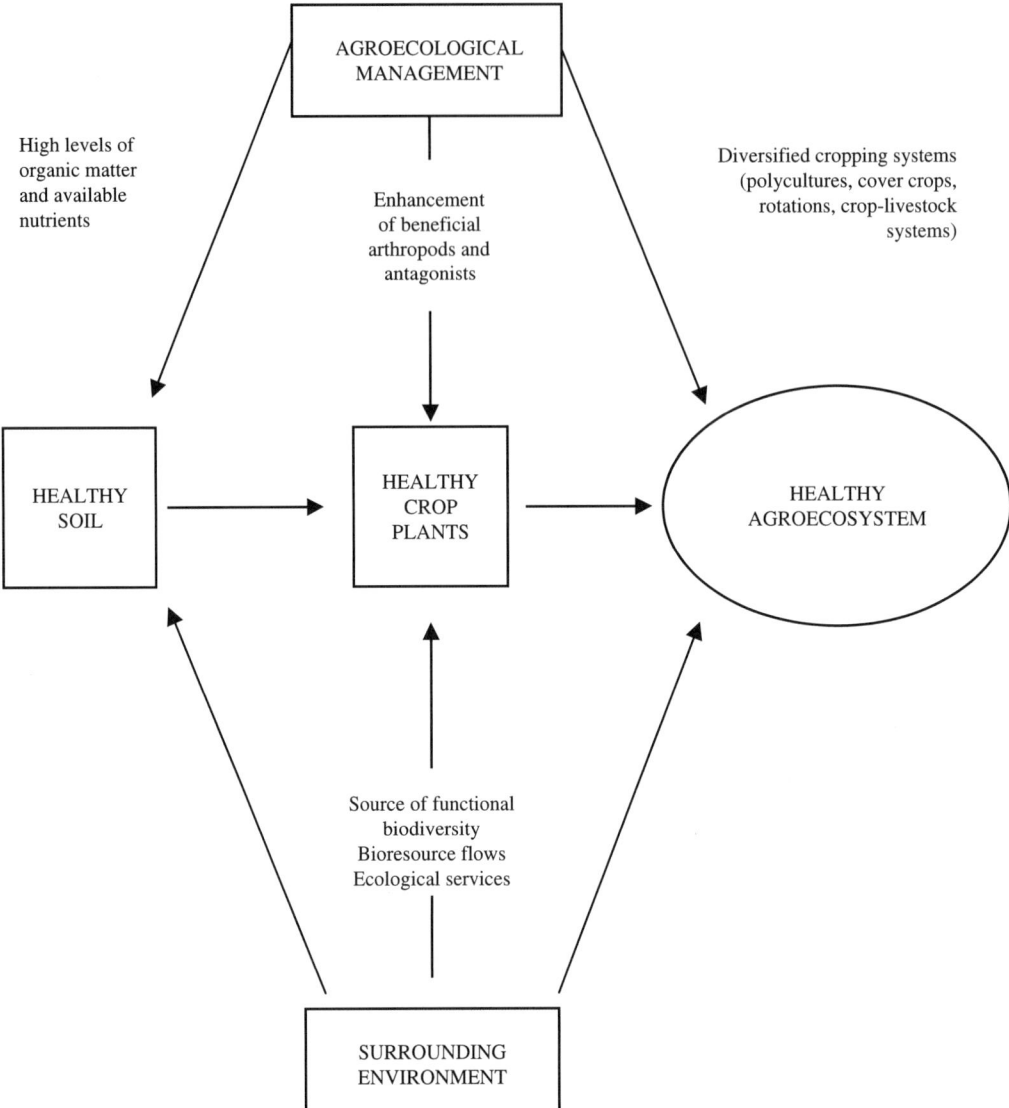

Figure 98.3 The pillars of agroecosystem health.

In England, conventional winter wheat fields developed a larger infestation of the aphid *Metopolophium dirhodum* than its organic counterpart. During June, this crop also had higher levels of free protein amino acids in its leaves, which were believed to have resulted from a nitrogen top dressing of the crop in early April. However, the difference in the aphid infestations between crops was attributed to the aphid's response to relative proportions of certain nonprotein to protein amino acids in the leaves at the time of aphid settling in the crops (Kowalski and Visser, 1979). In greenhouse experiments when given a choice of maize grown on organic vs. chemically fertilized soils, European corn borer females preferred to lay significantly more eggs in chemically fertilized plants (Phelan et al., 1995).

Nutrition is also important in determining susceptibility or resistance of plants to pathogens. Mineral nutrients are essential metabolic regulators of plant growth, and most studies indicate that nutrition affects pathogens and diseases indirectly. Biological activity in the soil becomes very intense in response to organic amendments and increases fungistasis as well as populations of existing microbial antagonists. Composts of diverse organic materials have proven effective in controlling diseases. Many types of compost host beneficial organisms that feed directly on pathogens, compete with them for nutrients, or produce antibiotics (Tjamos et al., 1992)

Balanced soil nutrition helps plants stay more vigorous, increases the growth rate, makes better use of soil water, and improves anatomical or histological characteristics. These factors enable plants to produce greater numbers of roots, allowing more surface area for root absorption and nutrient uptake, and possibly shortening susceptible stages of plant growth. This allows plants to function more efficiently even when some roots are infected. Histological changes strengthen plants and possibly create barriers more difficult for pathogens to breach. A healthier plant may also increase quality of exudates, which can stimulate increases in populations of antagonistic microorganisms. These, in turn, may compete with pathogens for nutrients or possibly produce toxins that directly affect pathogen development and survival (Palti, 1981).

Application of nutrients (especially N) may increase competitive suppression of crops by weeds, as most weeds (especially C_4 species) are often more responsive to application of nitrogen than other crops. A study reported that application of chemical N fertilizer to wild oat–spring wheat mixtures increased wild oat growth and decreased wheat yields. On the other hand, other studies have shown that N fertilizer can improve the competitive status of crops. What seems critical in determining the outcome of competitive interactions is the timing of nutrient availability relative to crop and weed demands upon nutrient supplies (Liebman and Gallandt, 1997). Emerging research shows that the species composition and general ecology of weeds is radically different in organic vs. conventionally fertilized systems. Yield reductions of wheat due to interference from Italian ryegrass (*Lolium multiflorum*) were greater under conventionally fertilized conditions than under organic-fertilized conditions. In many cases, weed suppression is related to delayed N release from the organic N source compared to nitrate fertilizer. In other cases, soil-incorporated organic residues increase phytotoxicity or pathogen activity, which suppresses weed seed and seedlings (Liebman and Ohno, 1998).

Diversified Agroecosystems and Pest Management

Diversified cropping systems, such as those based on intercropping and agroforestry or cover cropping of orchards, have been the target of much research recently. This interest is largely based on the emerging evidence that these systems are more stable and more resource conserving (Vandermeer, 1995). Much of these attributes are connected to the higher levels of functional biodiversity associated with complex farming systems. In fact, an increasing amount of data reported in the literature documents the effects that plant diversity has on the regulation of insect herbivore populations by favoring the abundance and efficacy of associated natural enemies (Altieri and Letourneau, 1984). One of the key elements in diversified agroecosystems is the presence of flowering plants that provide pollen and nectar, which serve as alternative food for natural enemies. A number of studies have shown that many parasitoids require nectar for normal fecundity and longevity, and thus spectacular parasitism increases have been observed in annual crops and orchards with rich floral undergrowth (Leius, 1967).

Flowering plants can also increase populations of nonpestiferous herbivores (neutral insects) in crop fields. Such insects serve as alternative hosts or prey to entomophagous insects, thus improving the survival and reproduction of these beneficial insects in the agroecosystem. Many researchers have reported that the presence of neutral insects on flowering plants within or near crop fields increase predation and parasitism of specific crop pests (Altieri and Whitcomb, 1979).

Our research in northern California vineyards confirms the importance of maintaining full season floral diversity to enhance natural-enemy abundance and diversity in the grape agroecosystem. Growing summer cover crops of buckwheat and sunflower had a substantial impact on the abundance of western grape leafhoppers and associated natural enemies. During two consecutive years (1996–1997), vineyard systems with flowering summer cover crops were characterized by lower densities of leafhoppers (Figure 98.4). Such reduced pest numbers in diversified vineyards were due to the impacts of generalist predators. With increased plant diversity, insect pests remained at lower levels than in clean-cultivated vineyards, partly because the summer cover crop vegetation

1996

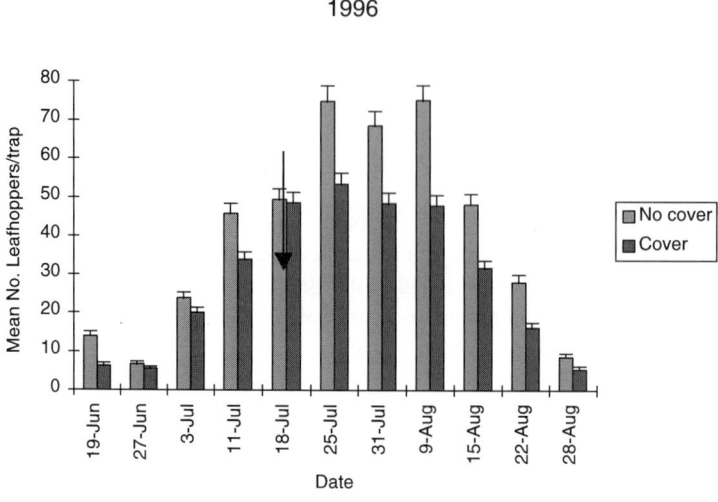

1997

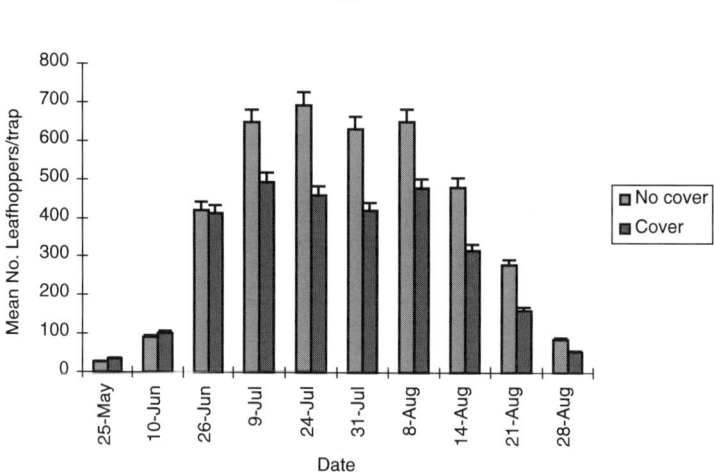

Figure 98.4 Densities of adult leafhoppers, *E. elegantula*, in cover-cropped and monoculture vineyards in Hopland, CA, in two growing seasons. Mean densities (number of adults per yellow sticky trap) and standard errors of two replicate means are indicated. In some cases, error bars were too small to appear in the figure.

harbored pollen, nectar, and neutral insects that served as alternative food and hosts for important predators, along with the parasitic wasp *Anagrus* (Nicholls et al., 2000).

Plant diversity around crop fields is also important in determining the diversity and abundance of natural enemies within agroecosystems. Emerging data demonstrate that there is enhancement of natural enemies and more effective biological control where wild vegetation remains at field edges in close association with crops (Altieri, 1994). These habitats are important as overwintering sites for predators, or they may provide increased resources such as pollen and nectar for parasitoids from flowering plants (Thies and Tscharntke, 1999). The presence of plant-rich habitats enhances predator colonization and abundance of adjacent crop fields but this influence is limited to the distance to which natural enemies disperse into the vineyard (Corbett and Plant, 1993). A corridor, however, could amplify this influence by allowing enhanced and timely circulation and dispersal movement of predators from edges into the center of crop fields.

In northern California, taking advantage of an existing 600-m corridor connected to a riparian forest that cut across a monoculture vineyard, we tested whether such a corridor served as a biological highway for the movement and dispersal of natural enemies from the forest into the center of the vineyard. The goal was to determine if the corridor acted as a consistent, abundant, and well-dispersed source of alternative food (pollen, nectar, and neutral insects) for a diverse community of generalist predators and parasitoids (Nicholls et al., 2001).

During 1996 and 1997, a 2.5-ha organic vineyard dissected by a corridor composed of 65 flowering plants species, which was connected to the surrounding riparian habitat, was monitored to assess the distributional and abundance patterns of the western grape leafhopper, its parasitoid *Anagrus* spp., and generalist predators. In both years, leafhopper adults and nymphs tended to be more numerous in middle rows of the vineyard and less abundant in border rows close to the forest and corridor where predators were more abundant (Figure 98.5). The complex of predators sup-

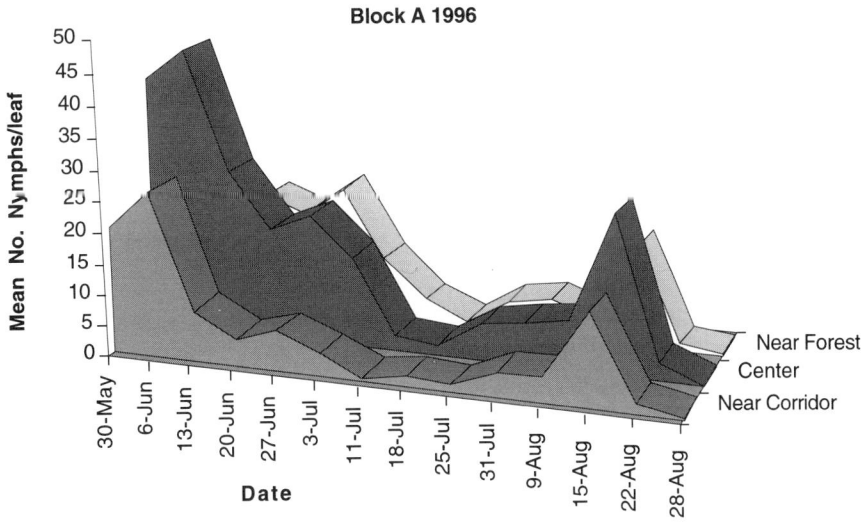

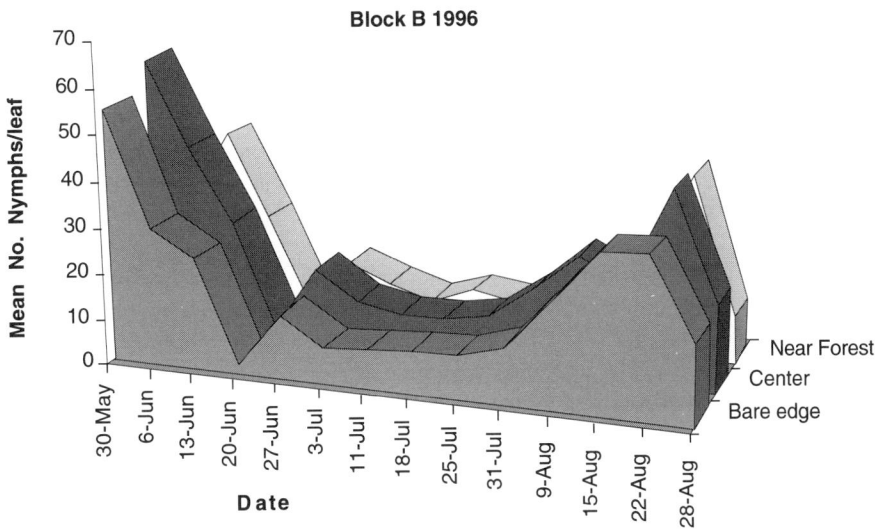

Figure 98.5 Seasonal patterns of leafhopper nymphs in both vineyard blocks, as influenced by the presence or absence of forest edges and the corridor (Hopland, CA, 1996).

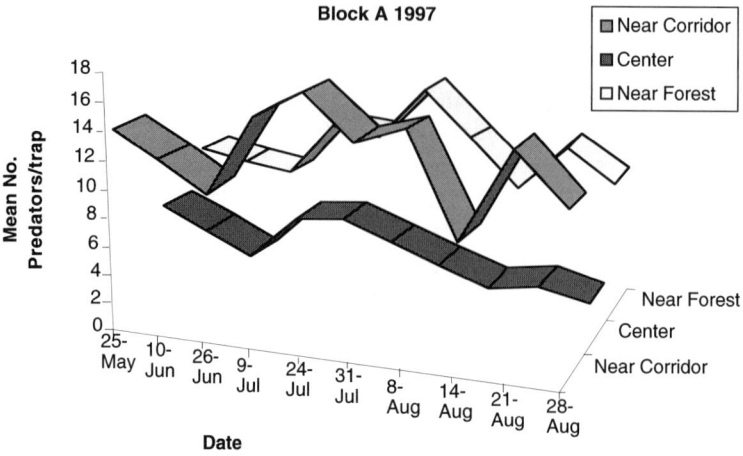

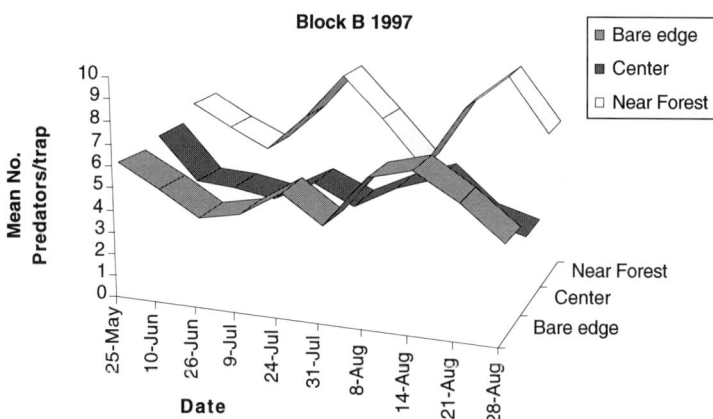

Figure 98.6 Seasonal patterns of predator catches (numbers per yellow sticky trap) in both vineyard blocks, as influenced by the presence or absence of forest edge and the corridor (Hopland, CA, 1997).

ported by the corridor moved to the adjacent vine rows and exerted a regulatory impact on herbivores present in such rows. Although it is suspected that the leafhopper parasitic wasp *Anagrus* depended on food resources of the corridor, it did not display a gradient from this rich flowering area into the middle of the field. Likewise, no differences in rates of egg parasitism of leafhoppers could be detected in vines near the corridor or in the vineyard center. The presence of riparian habitats enhanced predator colonization and abundance on adjacent vineyards, although this influence was limited by the distance (about 25 m) to which natural enemies dispersed into the vineyard (Figure 98.6). However, the corridor amplified this influence by enhancing timely circulation and dispersal movement of predators into the center of the field.

CONCLUSION

Achieving health in agroecosystems requires that management be directed at improving soil and plant quality, as the link between healthy soils and healthy plants is fundamental to EBPM.

Of key importance is also the realization that the level of internal regulation of function in agroecosystems is largely dependent on the level of plant and animal biodiversity present. In fact, there is overwhelming consensus that agroecosystem function may be significantly impaired by loss of biodiversity. In agroecosystems, biodiversity performs a variety of ecological services beyond the production of food, including recycling of nutrients and regulation of pest populations. For this reason, agroecologists promote multifunctional technologies that conserve or enhance biodiversity, as their adoption usually means favorable changes in various components of the farming systems at the same time. Many of these techniques do not demand a radical change in the farming system, but rather the establishment of the right type of ecological infrastructure that promotes key regulatory processes. An example is legume-based crop rotations, one of the simplest forms of biodiversification, that can simultaneously optimize soil fertility and pest regulation. It is well known that rotations improve yields by the known action of interrupting weed, disease, and insect life cycles. However, they can also have subtle effects, such as enhancing the growth and activity of soil biology, including vesicular arbuscular mycorrhizae, which allow crops to more efficiently use soil nutrients and water and thus better resist pest attack.

Many scientists are concerned that, with accelerating rates of habitat simplification through monoculture expansion, the contribution to pest suppression by biocontrol agents using these habitats will decline (Fry, 1995). This is why many researchers have proposed options to rectify this decline by increasing vegetational diversity within agroecosystems and throughout agricultural landscapes. Options include diversifying cropping systems with polycultural designs and maintaining wild vegetation adjacent to crop fields (Thomas et al., 1991).

The ultimate goal of agroecological design is to integrate components so that overall biological efficiency is improved, biodiversity is preserved, and the agroecosystem productivity and its self-sustaining capacity is maintained. The goal is to design a quilt of agroecosystems within a landscape unit, each mimicking the structure and function of natural ecosystems, i.e., systems that include:

Vegetative cover as an effective soil- and water-conserving measure, met through the use of no-till practices, mulch farming, and use of cover crops and other appropriate methods.

A regular supply of organic matter through the regular addition of organic matter (manure, compost, and promotion of soil biotic activity).

Nutrient recycling mechanisms through the use of crop rotations, crop livestock systems based on legumes, etc.

Pest regulation assured through enhanced activity of biological control agents achieved by introducing and conserving natural enemies and antagonists.

REFERENCES

Altieri, M.A., *Biodiversity and Pest Management in Agroecosystems,* Haworth Press, New York, 1994.

Altieri, M.A., *Agroecology: The Science of Sustainable Agriculture,* Westview Press, Boulder, CO, 1995.

Altieri, M.A., The ecological role of biodiversity in agroecosystems, *Agric. Ecosyst. Environ.,* 74, 19–31, 1999.

Altieri, M.A., The ecological impacts of transgenic crops on agroecosystem health, *Ecosyst. Health,* 6, 13–23, 2000.

Altieri, M.A. and Letourneau, D.K., Vegetation management and biological control in agroecosystems, *Crop Prot.,* 1, 405–430, 1982.

Altieri, M.A. and Letourneau, D.K., Vegetation diversity and insect pest outbreaks, *CRC Crit. Rev. Plant Sci.,* 2, 131–169, 1984.

Altieri, M.A. and Nicholls, C.I., Biodiversity, ecosystem function and insect pest management in agricultural systems, in *Biodiversity in Agroecosystems,* Collins, W.W. and Qualset, C.O., Eds., CRC Press, Boca Raton, FL, 1999, pp. 69–99.

Altieri, M.A. and Rosset, P., Agroecology and the conversion of large-scale conventional systems to sustainable management, *Int. J. Environ. Stud.,* 5, 1–21, 1995.

Altieri, M.A. and Whitcomb, W.H., The potential use of weeds in the manipulation of beneficial insects, *HortScience,* 14, 12–18, 1979.

Andow, D.A., Vegetational diversity and arthropod population response, *Annu. Rev. Entomol.,* 36, 561–586, 1991.

Campbell, R., *Biological Control of Microbial Plant Pathogens,* Cambridge University Press, Cambridge, U.K., 1989.

Corbett, A. and Plant, R.E., Role of movement in response of natural enemies to agroecosystem diversification: a theoretical evolution, *Environ. Entomol.,* 22, 519–531, 1993.

Fry, G., Landscape ecology of insect movement in arable ecosystems, in *Ecology and Integrated Farming Systems,* Glen, D.M., Greaves, M.P., and Anderson, H.M., Eds., John Wiley & Sons, Bristol, 1995, pp. 236–242.

Giller, K.E., Beare, M.H., Lavelle, P., Izac, A.M.N., and Swift, M.J., Agricultural intensification, soil biodiversity and agroecosystem function, *Appl. Soil Ecol.,* 6, 3–16, 1997.

Gliessman, S.R., *Agroecology: Ecological Processes in Sustainable Agriculture,* Ann Arbor Press, Chelsea, MI, 1998.

Hendrix, P.H., Crossley, D.A., Jr., and Coleman, D.C., Soil biota as components of sustainable agroecosystems, in *Sustainable Agricultural Systems,* Edwards, C.A. et al., Eds., Soil and Water Conservation Society, Ankeny, IA, 1990, pp. 637–654.

Kajimura, T., Effect of organic rice farming on planthoppers. 4: Reproduction of the white backed planthopper, *Sogatella furcifera* (Homoptera: Delphacidae), *Res. Popul. Ecol.,* 37, 219–224, 1995.

Kowalski, R. and Visser, P.E., Nitrogen in a crop-pest interaction: cereal aphids, in *Nitrogen as an Ecological Parameter,* Lee, J.A., Ed., Blackwell Scientific, Oxford, 1979, pp. 67–74.

Leius, K., Influence of wild flowers on parasitism of tent caterpillar and codling moth, *Can. Entomol.,* 99, 444–446, 1967.

Liebman, M. and Ohno, T., Crop rotation and legume residue effects on weed emergence and growth: Implications for weed management, in *Integrated Weed and Soil Management,* Hotfield, J.L. and Stewert, B.A., Eds., Ann Arbor Press, Chelsea, MI, 1998, pp. 181–122.

Liebman, M. and Gallandt, E.R., Many little hammers: ecological management of crop-weed interactions, in *Ecology in Agriculture,* Jackson, L.E., Ed., Academic Press, San Diego, 1997, pp. 291–343.

Magdoff, F.R., *Building Soils for Better Crops: Organic Matter Management,* University of Nebraska Press, Lincoln, 1992.

Marvier, M., Ecology of transgenic crops, *Am. Sci.,* 89, 160–167, 2001.

Nicholls, C.I., Parrella, M.P., and Altieri, M.A., Reducing the abundance of leafhoppers and thrips in a northern California organic vineyard through maintenance of full season floral diversity with summer cover crops, *Agric. For. Entomol.,* 2, 107–113, 2000.

Nicholls, C.I., Parrella, M.P., and Altieri, M.A., The effects of a vegetational corridor on the abundance and dispersal of insect biodiversity within a northern California organic vineyard, *Landscape Ecol.,* 16, 133–146, 2001.

Palti, J., *Cultural Practices and Infectious Crop Diseases,* Springer, New York, 1981.

Phelan, P.L., Mason, J.F., and Stinner, B.R., Soil fertility management and host preference by European corn borer, *Ostrinia nubilalis,* on *Zea mays*: a comparison of organic and conventional chemical farming, *Agric. Ecosyst. Environ.,* 56, 1–8, 1995.

Rissler, J. and Mellon, M., *The Ecological Risks of Engineered Crops,* MIT Press, Cambridge, MA, 1996.

Robinson, R.A., *Return to Resistance: Breeding Crops to Reduce Pesticide Resistance,* AgAcess, Davis, CA, 1996.

Snow, A.A. and Moran, P., Commercialization of transgenic plants: potential ecological risks, *BioScience,* 47, 86–96, 1997.

Thies, C. and Tscharntke, T., Landscape structure and biological control in agroecosystems, *Science,* 285, 893–895, 1999.

Thomas, M.B., Wratten, S.D., and Sotherton, N.W., Creation of "islands" habitats in farmland to manipulate populations of beneficial arthropods: predator densities and emigration, *J. Appl. Ecol.,* 28, 906–917, 1991.

Tjamos, E.C., Papavizas, G.C., and Cook, R.J., *Biological Control of Plant Diseases: Progress and Challenges for the Future,* NATO ASI Series, Plenum Press, New York, 1992.

Vandermeer, J., The ecological basis of alternative agriculture, *Annu. Rev. Ecol. Syst.,* 26, 201–224, 1995.

CHAPTER 99

Irrigation, Agricultural Drainage, and Nutrient Loading in the Upper Klamath Basin

Stephen R. Kaffka, Ashk Dhawan, and Donald W. Kirby

INTRODUCTION

The Upper Klamath Basin (UKB) is a high desert region straddling the California–Oregon border east of the Cascade Range. Because precipitation averages approximately 25% of potential evapotranspiration, irrigation is necessary for crop production. Irrigated agriculture is the basis of the local economy. The UKB is the location of wildlife refuges that are critical for migrating waterfowl using the Pacific Flyway, and one of the sources of the Klamath River, which is important for its scenery, trout, and anadramous fisheries. Two native fish species in Upper Klamath Lake are listed as endangered, and steelhead trout and coho salmon also are listed or are likely to be listed as threatened or endangered in the Klamath River in the near future.

Irrigation and other agricultural practices in the U.S. Bureau of Reclamation (USBR) Klamath Project may result in impaired surface water quality, reducing its use for wildlife and fish in the Tule Lake National Wildlife Refuge (TLNWR), the Lower Klamath National Wildlife Refuge (LKNWR), and the Klamath River. Surface water within the basin can have high pH levels (> 9.0 in summer), high levels of nonionized ammonia (NH_3) (>1.0 mg l^{-1}), and low dissolved oxygen (DO) during warm months (MacCoy, 1994). These quality characteristics are related to the growth, death, and decomposition of algae and other aquatic organisms.

The relationship between agricultural drainage and surface water quality in the region is not well understood. Shallow waters in the UKB have always been eutrophic, but likely have been further enriched by human activity, particularly farming, grazing, and logging. Farming may enrich surface water with nitrogen and phosphorus from fertilizer applications or from the mineralization of the region's organic-matter-rich soils resulting from cultivation. Any effects on surface water quality resulting from agricultural drainage likely occur through stimulation of aquatic plant growth at higher than background levels by nutrients, especially phosphorus, lost from farm fields.

Nonpoint source (NPS) pollution of surface waters has been recognized as a significant problem nationwide for many years (Novotny and Chesters, 1981). The loss of nutrients from agricultural landscapes can lead to impairment of surface waters for other uses, including wildlife conservation. In the UKB, important conservation objectives for waterfowl and fish are linked directly to the use of land and water for irrigated agriculture because much of the water for the wildlife refuges is derived directly from agricultural drainage, and return flows from the project enter the Klamath River. To protect surface waters from NPS pollution, the current federal Clean Water Act provides for the establishment of standards or limits, called total maximum daily loads (TMDL). By court order, TMDL standards must be set for the Klamath River by 2004.

In addition to water quality concerns, local farmers are faced with the loss of part or all of their irrigation supplies owing to restrictions related to the Endangered Species Act. The loss of irrigation water or narrow restrictions on the quality of drainage water will undermine the economic viability of the local community. The resolution of regulatory issues in the UKB provides a test of whether more ambitious conservation objectives related to species preservation or recovery can be adopted while maintaining the economic use of natural resources in rural, irrigation-dependent areas of the western U.S.

The USBR has measured surface water transfers since the project began in 1907, and water quality over a 60-year period, but sampling and analysis have not been consistent throughout this period. The U.S. Fish and Wildlife Service and the U.S. Geological Service (USFWS/USGS) carried out a multiyear sampling and analysis program for surface waters in the region (Sorenson and Schwarzbach, 1991; MacCoy, 1994). Snyder and Morace (1997) analyzed and modeled the rates of mineralization of organic soils surrounding Upper Klamath Lake in Oregon. Their model predicted large losses of N and P from mineralizing organic soils, but the concentrations they measured in agricultural drainage water were less than they predicted. Kaffka and co-workers (1995) reviewed all available historical data, and collected limited surface and subsurface tile drain data with the objective of formulating initial hypotheses about the relationship between agricultural nutrient losses and surface water quality. Their analyses included estimations of water and nutrient application rates by farmers and an assessment of long-term trends in surface water quality in the region. On the basis of USBR data, they reported a decline in the salinity of surface waters (measured as electrical conductivity) draining the Tule Lake Sumps and the Tule Lake Irrigation District (TID) from 1945 to 1990 (Figure 99.1). A salt balance calculated for the irrigated areas of the TID by USBR supports the hypothesis that by the 1980s, salt inputs and outputs were in approximate balance (Kaffka et al., 1995). Fewer data were available for determining trends for total P concentrations in surface waters. Sample collections and analyses made by the USBR and reported in Kaffka and co-workers (1995), revealed no apparent trend over the 10-year period from 1980 to 1990 (Figure 99.2). Salinity and P concentrations in water samples were not correlated during the period (Figure 99.3), suggesting that P concentrations in surface waters are not related linearly to the leaching and drainage processes resulting in the loss of salts from soils. They also collected a limited number of agricultural tile drain samples to estimate salt and nutrient losses but the number collected was insufficient to evaluate season-long nutrient concentrations and possible transfers from soils to surface waters.

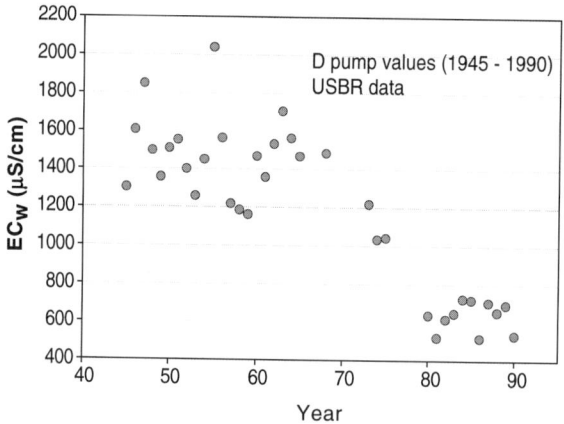

Figure 99.1 EC_w (μS cm^{-1}) in the Tule Lake Sumps, measured at the D pump outlet. (From Kaffka, S.R. et al., Res. Prog. Rep. 108, University of California Intermountain Research and Extension Center, Tulelake, 1995.)

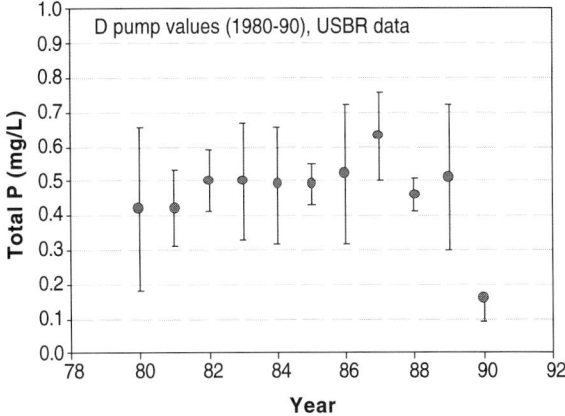

Figure 99.2 Total P measurements (mg l⁻¹) at the D pump outlet reported in Kaffka, S.R. et al., Res. Prog. Rep. 108, University of California Intermountain Research and Extension Center, Tulelake, 1995.

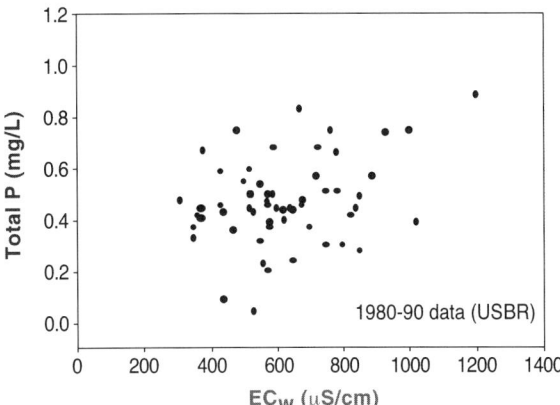

Figure 99.3 Total P and EC_w data over the 1980 to 1990 period. (From Kaffka, S.R. et al., Res. Prog. Rep. 108, University of California Intermountain Research and Extension Center, Tulelake, 1995.)

To investigate further the relationships between agricultural practices and surface water quality in the UKB, we have undertaken a multiyear reconnaissance survey of surface water and agricultural tile drain locations. The objectives of this effort are:

- To collect and analyze subsurface tile drain and surface water quality samples throughout the TID and surrounding locations.
- To integrate results with previous analyses into a hypothesis linking surface water quality with agricultural drainage.

This chapter presents results to date from the ongoing investigation.

MATERIALS AND METHODS

Starting in 1995, a reconnaissance sampling program for surface water and agricultural subsurface tile drains was carried out in the southern portion of the USBR Klamath Project, focusing

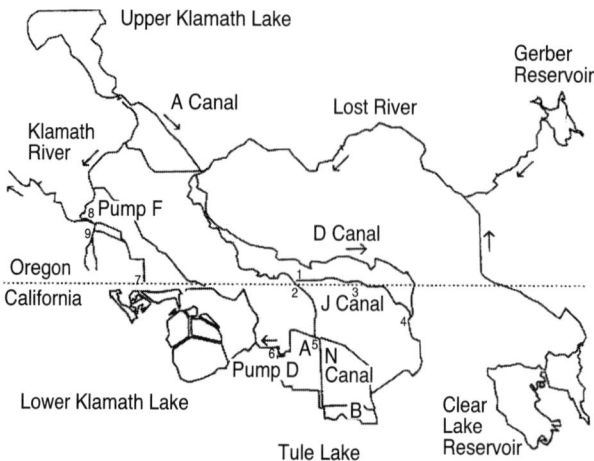

Figure 99.4 The Klamath Project in southern Oregon and northern California. The Tule Lake Irrigation District at the bottom of the map is the primary focus of monitoring and mass balance calculations. For an explanation of the numbers next to the sample points, see Table 99.1.

particularly on the TID (Figure 99.4; Table 99.1). Triplicate samples were collected at approximately 10-day intervals throughout the growing season (April through October) from tile drains and surface water sampling locations; in winter, surface water samples were collected twice a month or as often as possible. In 1995, five tile drain locations were sampled; in 1996, eight additional subsurface tile drains and additional surface sample sites were included. The 13 tile drain sample points are located throughout the irrigation district, with eight in the Lease Lands area, part of the TLNWR. Samples were collected directly from tile drain outlets, or if tile outlets were submerged, directly from tile lines near the drain outlets fitted with backflow check valves, using a small portable pump. Some samples were collected from drainage sumps fitted with lift pumps, integrating drainage water from several tile lines. Surface water samples were collected from inflow locations (J canal diversion [1], the Lost River as it enters California from Oregon [2], pump 7 [3], D canal

Table 99.1   Description of Surface Water Sample Locations

Location No.	Name	Description
1	J canal diversion	Anderson-Rose (A-R) Dam on the Lost River where water is diverted for irrigation use in the TID.
2	Lost River	The Lost River at the California–Oregon border. Quantities are estimated upstream at the A-R Dam, but there are additional, unaccounted return flows from farm fields added to the river before it reaches the state line.
3	Pump 7	Drainage canal from the Klamath Irrigation District to the north of TID. Water volume is poorly quantified.
4	D canal	Irrigation diversion-drainage canal from the KID adding to water supplies in the TID. Water volume is poorly quantified.
5	N canal	Irrigation return flow-Irrigation diversion canal.
6	D pump	Outlet for surplus water from the Tule Lake Sumps to the Lower Klamath Refuge and eventually the Klamath River.
7	Lower Klamath Refuge outlet (LKL)	Water leaving the Lower Klamath Refuge.
8	Klamath Straits Drain outlet (KSD)	Pumps F and FF at highway 97 in Oregon. Water leaving the straits drain enters marshlands and the Klamath River at this point.
9	ADY canal	Water is re-diverted from the Klamath River to the Lower Klamath Refuge and to provide irrigation water for the Klamath Drainage District (KDD).

[4], and selected distribution canals [5]) and outflows (D pump [6]; and the Klamath Straits Drain at its origin, leaving the LKNWR [7], and at its end by the lift pump intake near highway 97 [8]). These surface water locations trace water quality as it enters the TID, as it leaves, and as it returns to the Klamath River. They allow for an estimation of the amounts of nutrients transferred within and outside of the southern Klamath Project. One of the possible measures of ecosystem health, a subjective term, might be the retention or loss of ecologically important nutrients such N and P.

At collection, water temperature, pH, and electrical conductivity (EC_w: a measure of salinity) were determined. Samples were stabilized with toluene and refrigerated until analysis. Sample aliquots were filtered through 5-μm filters and analyzed for filterable, reactive (largely inorganic) P by direct colorimetric analysis (AOCS 973.55; Watanabe and Olsen, 1965) and for soluble N (NO_3-N) using HPLC methods (Thayer and Huffaker, 1980). Typically, 0.45-μm filters are used for this procedure (Haygarth and Jarvis, 1999). In the UKB, most of the surface water samples are so laden with algae, aquatic plants, and other materials that filtration with smaller-diameter filters has proved impractical. A second aliquot was digested using a procedure modified from Johnes and Heathwaite (1992), using persulfate digestion. Digested samples were then analyzed for N and P using HPLC and colorimetric methods as above. Starting in 1999, filtered samples were redigested and analyzed for P as above. A schematic diagram of the analysis protocol is presented in Figure 99.5. Ammonia is collected separately in acidified sample bottles and refrigerated until analysis. HPLC and spectrophotometry methods are used for determination (Goyal et al., 1988).

Farmers were surveyed about fertilizer use, and fertilizer suppliers were interviewed about typical fertilizer application rates in the TID. Crop water use was estimated by Kaffka and co-workers (1995) based on data from California's statewide CIMIS weather data collection system combined with crop water use estimates developed over several years for a variety of crops in the region. A modified Penman–Montieth equation was used to predict ET_c (Snyder et al., 1987).

The proportion of irrigation water resulting in drainage was derived from Kaffka and co-workers' (1995) estimate of irrigation efficiency and TID data. Water use efficiency (WUE) was calculated as:

$$\text{WUE (\%)} = [\Sigma\ ET_c/(\Sigma\ ID \pm P)] * 100 \tag{99.1}$$

where ET_c is crop evapotranspiration; ID are net irrigation diversions derived from TID records; and P is yearly precipitation derived from weather station records (CIMIS) in Tulelake. Irrigation diversions include measurements at the J canal and net diversions from the Tule Lake Sumps. Water is transferred back and forth between the sumps and irrigation canals during the growing season.

Phosphorus analysis

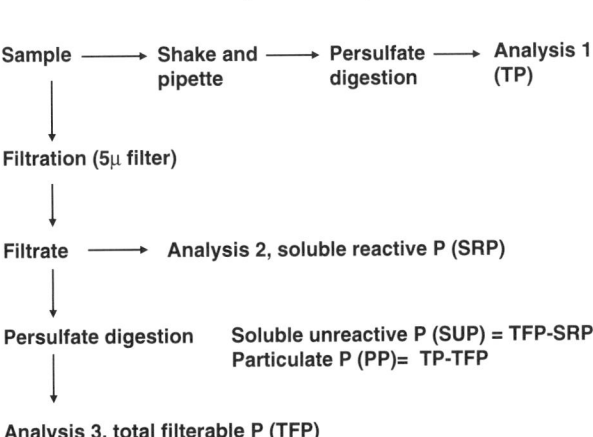

Figure 99.5 Protocol for the analysis of P in water samples.

Net transfers can be calculated from TID pump records (Kaffka et al., 1995). The water diverted but not used as ET_c is assumed to be agricultural drainage returned to the system. This value is 67% and corresponds to best estimates from agronomists working in the area. If precipitation is included in the equation, then the final value for crop water use is approximately 60%.

These assumptions about drainage water volume are conservative with respect to the potential contribution of agricultural drainage to surface water P and N levels. By conservative we mean that it is a maximum possible amount, and we may overestimate net drainage losses but are unlikely to underestimate them. This is because in practice less water is specifically diverted for irrigation than is applied to crops as irrigation in the TID. Diversions at the J canal roughly match ET_c estimates. Additional irrigation water is derived from other sources. During the irrigation season, water returning to the system as drainage is reused as an irrigation source by farmers downstream (roughly south) within the district. This water becomes enriched upon reuse with salts and possibly with nutrients. Also, there are poorly quantified drainage inflows, which serve as inputs into the TID from the irrigation district to its north, the Klamath Irrigation District (KID). Water is also diverted from the Tule Lake Sumps. Thus, the pathways and disposition of water within the district are extremely complex and poorly understood.

A recent study provided a higher estimate for WUE for the TID subregion of the Klamath Project. Davids Engineering (1998) estimated that water diverted for irrigation was used with very high efficiency by farmers in the project, with an estimate of 74% for WUE, and 93% for effective efficiency, a term attempting to account for reuse of drainage water by farmers on a project-wide basis, accounting for inputs and drainage among the sequentially related irrigation districts within the Klamath Project. The reuse process within the TID, however, is just now being modeled specifically and results are not yet available; thus the larger estimate for drainage is used instead.

Deep percolation below tile line depth is considered to be a small loss, and is discounted in this analysis because arable soils are underlain by deep, largely lucustrine clay deposits with extremely low rates of infiltration and hydraulic conductivity that are saturated most of the year (NRCS, 1994). Artificial drainage is required for farming. Additional justification for discounting deep soil drainage is indirect and is provided by the observation that overall water balances account for water inputs and outputs without large drainage losses to deeper soil profiles. Water balances for the area including the TID are calculated as:

$$I_P \pm I_J \pm I_{LR} \pm I_{KID} = O_D \pm O_{ETc} \pm O_{ETL} \qquad (99.2)$$

where I_P is a water input from precipitation; I_J is the water diversion at the J canal; I_{LR} is the water entering the Tule Lake Sumps from the Lost River; I_{KID} is water entering the TID as drainage from KID to the north. O_D is an output, water pumped form the Tule Lake Sumps at the D pump plant; O_{ETc} is crop evapotranspiration; and O_{ETL} is evapotranspiraton from the Tule Lake Sumps, considered to be equivalent to losses from an open water body. The water equation is balanced by adjusting I_{KID}, but the values determined this way are similar to those estimated by TID personnel and in the Davids Engineering study.

Data for the amount of water transferred from place to place within the Klamath Project are available from the USBR. Long-term data (1961 to 1995) were used to develop yearly and monthly averages using statistical software. For this report, an average year is evaluated using average total P and salinity concentrations and yearly water volumes. Using this approach, mass balances for salts (as TDS) and total P were calculated. P is emphasized because it is generally the limiting element for algal and aquatic plant growth in most freshwater systems and is the subject of heightened concern in the analysis of agricultural systems worldwide (Haygarth and Jarvis, 1999; Heckrath et al., 1995; Novotny and Chesters, 1981). Also, N is not thought to limit aquatic plant life in the surface waters of the UKB. Large amounts of blue-green algae are commonly present and some of the species are capable of N fixation, providing that sufficient P is available. Phosphorus transfers in aquatic systems can be complex and the form of P present can be important in regulating aquatic productivity

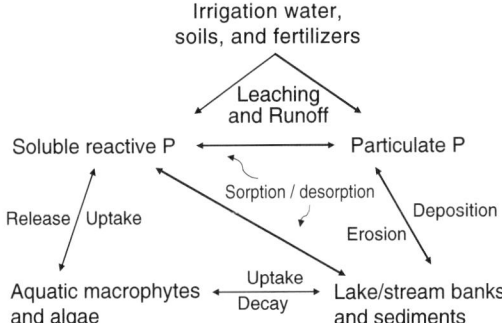

Figure 99.6 Phosphorus transformations during transfer from agricultural to aquatic systems. (Adapted from Sharpley, A.N. et al., in *Phosphorus in the Global Environment,* Tiessen, H., Ed., John Wiley & Sons, New York, 1995, pp. 171–199.)

(Figure 99.6; Sharpley et al., 1995). This analysis ignores unknown transfers within the aquatic portion of the ecosystem. There has been no study of these processes to our knowledge, and studying the mechanisms regulating them in the UKB is beyond the scope of our analysis.

RESULTS

Fertilizer Use and Crop Uptake

Responses to a questionnaire about fertilizer application rates from farmers ranged from less than 15% of all planted acres for sugar beets to greater than 25% of all planted acres for small grains. Nevertheless, average farmer responses agreed closely with estimates provided independently by the primary fertilizer supplier in the region for all the important crops. On average, crops take up and remove more N but less P than is applied as fertilizer. In particular, potato crops are heavily fertilized, thus most of the P surplus results from applications to potatoes (Table 99.2).

Table 99.2 Estimated Annual Average Fertilizer Application and Crop Nutrient Removal in the TID (mg yr⁻¹)

Crops and Land Area	N			P		
	Fertilizer N Application	Crop N Removal	Difference	Fertilizer P Application	Crop P Removal	Difference
Small grains (13,200 ha)	1300	1400	−100	170	330	−160
Potatoes and onions (4970 ha)	1110	780	330	530	210	320
Sugar beets (2390 ha)	240	530	−290	110	60	50
Alfalfa (4,660 ha)		(2200)		130	175	−45
Total (25,220 ha)	2640	2710	−60	940	775	165
Mg ha⁻¹ yr⁻¹			−0.0024			0.0065

Fertilizer inputs are based on recommended rates developed by the university, a survey of farmers and fertilizer dealers, supplemented by reports from individual farmers about their fertilizer use and the amount of land they cultivate. Crop removal is calculated based on literature values, or in the case of small grain crops, average crude protein values for the barley, wheat, and oats produced in the region. Crop yields and land areas are derived from annual reports by the TID, and are averages. All figures are rounded. Per unit area, potatoes and onions receive the most fertilizer, and for P, applications result in a district-wide surplus. N is in approximate balance if the assumption is made that all N removed by alfalfa is fixed from the atmosphere. If some N is recovered from soil by alfalfa, then N removed by crops is greater than N applied.

Water Balance and Use by Crops

Water transfers within the southern portion of the Klamath Project are presented in Figure 99.7 and Table 99.3. Approximately 40% of the water entering the TID from all sources including precipitation is diverted to the Lower Klamath National Wildlife Refuge. This water is supplemented and mixed with water from the Klamath River via the ADY and North Canals primarily and then returned to the Klamath River, acquiring additional agricultural drainage along the way as it passes through the Klamath Drainage District (Figure 99.4).

A water balance for the TID, the primary agricultural area in the southern Klamath Project is shown in Figure 99.8, and a water balance for irrigated cropland is shown in the TID in Figure 99.9. A quantity of water approximately equal to estimated ET_c in the TID is diverted into the J canal for irrigation. Other inputs include flows in the Lost River, and drainage from the KID to the north of TID or into the Lost River below the Anderson-Rose Dam but above the Tule Lake Sump. KID drainage and return flows to the Lost River are not well accounted for, and are assumed to be equal

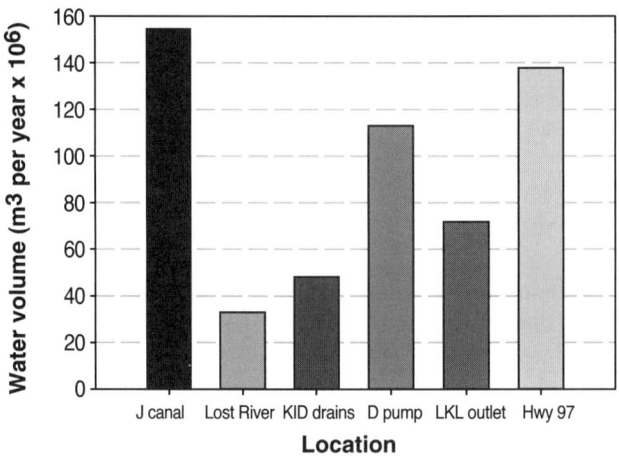

Figure 99.7 Long-term average water flows in the Klamath Project at the most important surface water sample points (USBR data, 1962 to 1995).

Table 99.3   Water Balance Data

Location or Source	Amount (m³ yr⁻¹ × 10⁶)	Source of Data
J canal	154.5	TID and USBR records
Lost River (LR)	33.0	TID and USBR records
Precipitation	64.6	CIMIS data for Tule Lake
Klamath Irrigation District (KID)	48.1	Estimated (see text)
D pump	112.7	TID and USBR records
Crop evapotranspiration (ET_c)	148.0	Estimated based on crop acreage records from TID, CIMIS weather data, and the use of a modified Penman–Montieth equation (see text)
Sump evaporation (ET_{TL})	39.5	Estimated based on CIMIS data and corrected for the effects of open water bodies on lake evaporation
Cropland drainage	107.8	Estimated: Equal to average J canal diversions + precipitation + diversions from the Tule Lake Sumps to cropland during the growing season - ETc - evaporation from irrigation canals;[a] data for diversions from the Tule Lake Sumps from Kaffka et al. (1995)

[a] Losses from irrigation canals and drains are assumed to be equal to 80% of the year-long panevaporation amount derived from the CIMIS weather data system (880 mm) and a surface area of 4.86 km².

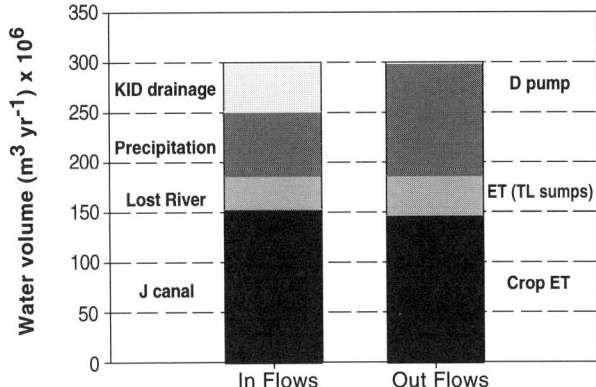

Figure 99.8 A water balance for the TID. All values are long-term averages and are based on USBR data for surface water flows, estimated crop evapotranspiration, calculated amounts of evaporation from the Tule Lake Sumps based on surface area and meteorological values. KID values are calculated by difference (see text).

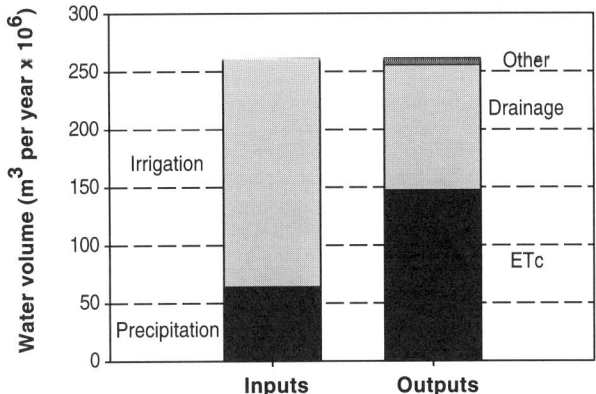

Figure 99.9 A water balance for irrigated crop land in the TID (see Table 99.3 for details).

to the amount of water needed to balance water use and transfers among the regions within the study area. Unaccounted outputs (Figure 99.9, Other) are largely due to canal evaporation losses.

Mass Balances for Salts and P in the TID

The most common salts in surface waters in the UKB are calcium and sodium sulfates (Sorenson and Schwarzbach, 1991; Kaffka et al., 1995). The quantity of salt leaving the TID is approximately equal to the amount introduced in surface waters from all sources combined (Figure 99.10). This result supports the conclusions USBR project managers reported a decade ago (Kaffka et al., 1995). Salts also increase as water is reused, passing from the northeast to the southwest of the TID. Values for tile drains in the northeastern region of the TID average approximately 600 μS cm^{-1}, while in the southwest portion of the district they increase to 2000 μS cm^{-1}. This increase is a function of the reuse of water and higher levels of salinity in soils in some of the southwest areas of the district (Sorenson and Schwarzbach, 1991). The EC$_w$ measured in surface waters over the 1995 to 1999 period also increase as water passes through the region. Water entering the district averages somewhat less than 300 μS cm^{-1}, water leaving the Tule Lake Sumps at the D pump increases to 600 μS cm^{-1}, and water reentering the Klamath River increases further to approximately 900 μS cm^{-1}.

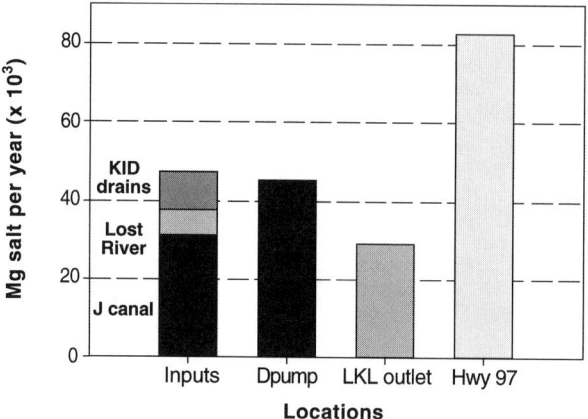

Figure 99.10 Calculated mass balance for salt entering and leaving the TID, based on EC_w values for surface water samples. The amount of salt entering and leaving the TID (columns 1 and 2) is in approximate balance. EC_w was converted to TDS using the equation TDS (mg l^{-1}) = EC_w (dS m^{-1}) × 640 (Mg = megagrams = metric ton).

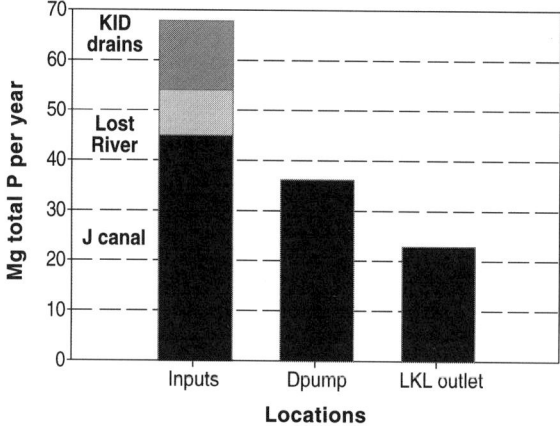

Figure 99.11 Calculated mass balance for P entering and leaving the southern portion of the Klamath Project. Less P leaves the TID in surface water at the D pump, then enters with irrigation water.

Phosphorus behaves differently in soils and surface waters. At the scale of the watershed, the concentration of P increases as water travels from its entry into the TID to its return to the Klamath River. But the quantity of total P returned to the Klamath River is estimated to be lower than the amount of P entering the TID in surface waters because the amount of water returned is less than that diverted (Figure 99.11). The diversion of water through the southern portion of the Klamath Project, exclusive of use by the Klamath Drainage District north of LKNWR, including its use for agriculture and wetlands, results in an apparent storage or removal of P from the surface waters transferred through the system.

At the irrigation district scale, mean yearly total P concentrations in representative subsurface tile drains and drainage sumps from 1995 through mid-1999 are reported in Figure 99.12. These values on average do not reflect a large increase over the average total P concentration of waters diverted for irrigation at the J canal (Figure 99.13). If this increase (adjusted for drainage water volume) is assumed to equal the contribution to surface waters from agricultural soils, the net amount of P derived from agricultural tile drains is less than the amount of P imported in irrigation water, and is small with respect to the amount of P applied as fertilizer. This includes other types of drains in the district as well.

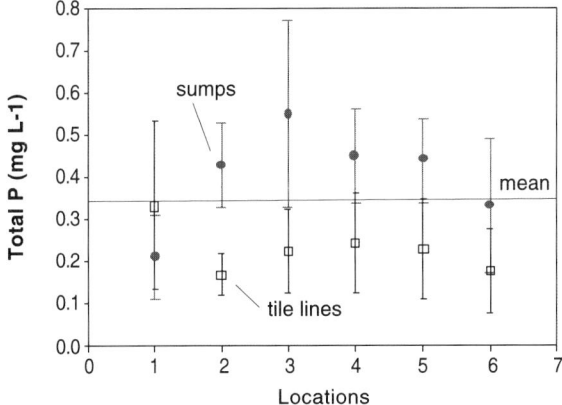

Figure 99.12 The average P concentrations (mg l⁻¹) observed in tile drain samples from 1995 to mid-1999. P losses in samples collected in the northeastern area of the TID are larger on average than those collected in the southwest portion of the district.

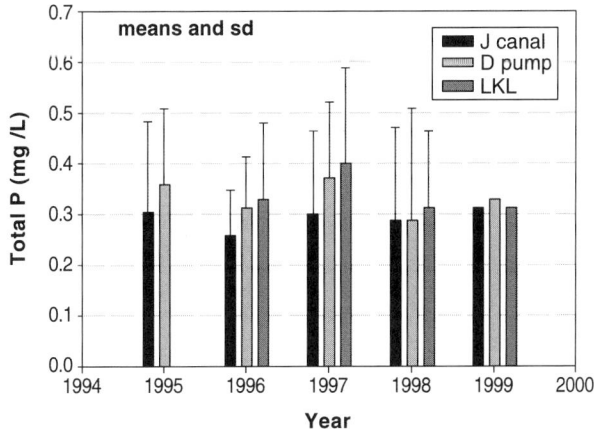

Figure 99.13 The average P concentrations (mg l⁻¹) observed in surface water samples collected from the J canal input location, compared with the D pump and Lower Klamath Refuge (LKL) outflows.

At the farming system scale, the amount of P introduced as fertilizer in the TID is approximately an order-of-magnitude greater than the amount of P entering the TID in surface waters. The amount of fertilizer P applied in excess of crop removal equals an amount approximately five times larger than the amount of P leaving as drainage water (Figure 99.14). Even if the highest average values for tile drainage concentrations are used instead of the district average, tile drainage losses remain small compared with other sources. If all inputs of P are added, and all outputs including crop removal and water transferred at the D pump are subtracted, then there appears to be a net accumulation of P of approximately 8 kg/ha per year. This P, derived both from imported water and from surplus fertilization, appears to be accumulating in soils and lake sediments in the region, but primarily in soils (Table 99.2).

DISCUSSION

Sources of Uncertainty in the Estimates

There are several sources of uncertainty associated with the estimates for salt and total P provided. The measurement of water volumes transferred within the basin is based on the field-

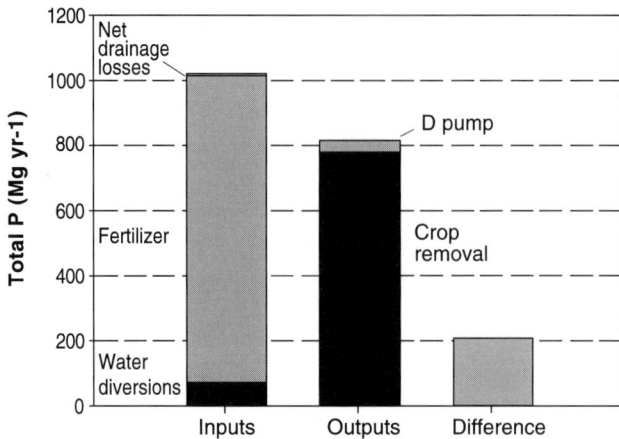

Figure 99.14 A phosphorus balance for the Tule Lake Irrigation District, including the TLNWR. All known inputs are included. Atmospheric deposition is unknown. Tile drain outflows are estimated using an irrigation efficiency of 67% and an average total P concentration of 0.34 mg l⁻¹.

rated capacity of pumps and measurements at weirs and gauging stations by USBR. All of these measurements are subject to error. In particular, Davids Engineering estimated recently that water balances for the southern- and westernmost portion of the project, comprising LKNWR and the farming areas of the Klamath Drainage District, were poorly accounted for. They suggested that pump ratings for the Klamath Straits Drain outlet (Figure 99.4, no. 8) may underreport the amount of water transferred at that location, while an apparently larger volume of water enters that part of the watershed than can be accounted for by the measured flows at the facilities maintained by USBR, particularly the ADY canal. If more water is being imported into this portion of the watershed than is accounted for, then the amount of nutrients imported also is underestimated. Alternatively, if export amounts are underreported, then the amount of nutrients leaving the Klamath Project and entering the Klamath River at this location also are underreported.

Agricultural fields are drained using three different types of systems. One is a combined series of collector tiles feeding a common sump and also usually intercepting water passing along one side of the field in an irrigation supply canal. Five of our sample locations are of this type. Another type of tile system consists of individual lines draining directly into surface drains at the ends of fields. Eight of our sample locations are of this type. The most common system is a deep ditch along one end of a field, emptied periodically into a master drain with a pump. Identifying the source of nutrients in the water derived from this last source is uncertain, because nutrients may originate from neighboring fields, from sediments, or from mobilization within the ditch or canal due to the biological and physical processes occurring in the ditches themselves, or from both. Because of this uncertainty, we believe that the water samples collected from simple tile drains best represent the subsurface soil water nutrient concentrations that have the potential to appear in surface waters. Nevertheless, we average all subsurface tile drain values together in estimating a value to use for drainage transfers.

The reuse of drainage water for irrigation throughout the Klamath Project further complicates the estimation of nutrient contributions from agriculture to tile drains. Each time water passes through the soil profile, it dissolves soluble salts and nutrients. Salt concentrations increase in the TID from north to south, reflecting increasing water reuse in that direction. The tile lines located in the southwestern portion of the district, however, average 0.24 mg total P l⁻¹, while those in the northeastern area average 0.42 mg total P l⁻¹. The latter are primarily drainage sumps. The data suggest that total P in tile drains declines as water is reused. There are several possible reasons for this trend. One reason is that more row crops are grown in the northeastern portion of the district on private lands than in the southwestern portion on public or refuge lands, where grains are more common. Thus greater nutrient concentrations in these locations may reflect higher rates of fertilization, especially over time (Table 99.2). Another reason is a difference in the behavior of the drainage systems sampled. In the north portion of the district, sumps that integrate several field tile drains with water intercepted at field edges from agricultural drains are sampled. These values have been consistently higher than the simple tile line values. Differences between the two systems may reflect different hydrologic behaviors of the drainage system and the interception of P from other sources in the more-complex sump drainage systems. Additionally, N and P applied with irrigation water serve as inputs to the cropping system, contributing ultimately to the nutrient supply used by crops. In contrast, salts are not taken up except by sugar beets, and then only in small quantities. Thus, salt concentrations increase while N and P concentrations do not, suggesting that soils and crops act as nutrient sinks.

Our sampling of tile drainage is not linked directly to irrigation cycles. Upon initial soil saturation, drainage volumes increase, and concentrations of salts and some nutrients will be greater than later in the cycle. Over the period of sampling, tile drainage samples were collected at various times during the irrigation–drainage cycle and reflect the range in concentrations possible. Averaging them may underestimate the loading rates from waters derived from tile drains, particularly for salts. In any case, the behavior of soluble constituents in the soil water system of these unique organic soils is poorly understood. In particular, salt leaching is more complex in bicarbonate-dominated soil water systems such as those of the UKB than otherwise (Oster and Rhoades, 1975). Simple leaching ratios may not predict salt-loading rates because bicarbonate can increase salt solubility. Salt solubility may also be increased by oxidizing organic matter in the organic-matter-rich soils of the region. The behavior of N and P is more complex still because of their role in plant growth and microbiological activity and the complex solubility chemistry of P in soils.

We have adopted a simplified approach to determining mass balances in the region. By using an average year, we have simplified within-year variation. To date, we have detected no consistent pattern of variation in surface water nutrient concentrations, other than a tendency for concentrations to be higher during the early winter and spring periods when no water is being transferred from farm fields (data not shown). Fewer samples have been collected during this period of the year, however, and a seasonally adjusted mass balance will be attempted when sufficient data have been collected after the current cycle of sampling is completed.

Finally, we have not attempted to interpret the differences in the relative contribution of agricultural drainage to differences in bioavailable P or other types of P transferred from location to location in the UKB (see Figure 99.4). This may underestimate the importance of agriculturally derived P in nutrient cycles in the UKB. Sharpley et al. (1995) note that the form of P entering surface waters is important for its effects on productivity. However, in the UKB, surface waters are all shallow (typically less than 2 m), and subject to constant agitation from wind and from turbulent flow of water due to diversions and transfers. This suggests that sedimentary P is constantly resuspended in the shallow surface waters, and internal cycles between sediments and aquatic plants have the largest potential to influence levels of productivity (Figure 99.6). Phosphorus concentrations measured in the outflows of the two refuges (TLNWR/LKNWR) have been fairly consistent over the years (data not shown). This suggests that these values reflect the dynamics of the wetland systems rather than external loading rates.

Agriculture and Ecosystem Health in the Southern, Upper Klamath Basin

All older reports about water quality in the UKB, including simple observations by early explorers, report eutrophic and saline conditions (Kaffka et al., 1995). With the construction of the Klamath Project starting in 1905, the natural landscape was altered substantially, particularly by the draining of wetlands and the removal of surplus water as drainage. This alteration doubtless affected a number of native species of fish and waterfowl in the UKB, and may have caused some species extinctions in recent history that are not recognized. The hydrology of the UKB also was altered, particularly by linking the Klamath and Lost River watersheds, which were previously separate, and by conserving more water in the upper basin late in the season than would have been conserved naturally. Organic soils developed in the region's shallow wetlands under anaerobic conditions may have released large amounts of organically bound nutrients when first exposed to air and cultivation (Snyder and Morace, 1997), enriching sediments in the region's lakes and streams. These sediments currently support abundant populations of algae and other aquatic plant species.

Total P concentrations are somewhat higher in the Tule Lake Sumps and in the surface waters draining LKNWR than they are in water in the Lost River used for irrigation. But waters tend to increase in nutrient content in all watersheds as they pass from headwater sources to the rivers delivering them to the sea. It is not likely that the small amount of nutrients derived from agricultural drainage, estimated we believe as a maximum, is ecologically significant when transferred to the already nutrient-rich surface waters of the UKB.

On a mass-balance basis, fertilizer is the primary source of accumulating P in the soils and sediments of the TID. Irrigation district or watershed scale calculations have only recently begun to be made and there are few references to provide perspective. A recent summary reported seven different calculated rates of P accumulation in arable farming systems, from largely temperate areas ranging from 2 to 94 kg/ha/yr^{-1} (Haygarth and Jarvis, 1999). Five of the estimates were greater than 11 kg/ha/yr^{-1}, placing the TID loading rate in the lower end of the range reported. Nevertheless, accumulating P in soils above crop needs has the potential to result in P release to the environment over time. Heckrath et al. (1995) found in an analysis of the long-running Broadbalk experiment at Rothamsted, England, that P losses in drain tiles increased to undesirable levels only after soil P analyses rose to >60 mg kg^{-1} Olsen P (NaHCO$_3$ extraction; see Olsen et al., 1954). There has been no systematic survey of soil P levels in the TID. Kaffka and co-workers (1998) reported Olsen P levels from a sugarbeet irrigation cutoff experiment that ranged between 10 and 20 mg kg^{-1} on average over the soil profile to tile drain depth (1.1 m). It is not known if these values are typical for the region, nor can it be surmised that lower values in the soils in the UKB result in low values of P loss. Additional systematic surveys are required. In any event, reduction of P fertilization rates to potato crops would reduce loading rates, should be agronomically achievable over time, and would lessen further the potential for agricultural contributions of P to surface waters in the region. Improved fertilization practices for the region's potato crops, where possible, would reduce the amount of surplus P applied as fertilizer, and reduce this rate of accumulation to low or background levels. This strategy would be a positive step for agricultural producers in further reducing any possible contribution of P by farming to the surface waters (Sharpley et al., 1995).

Health is a subjective term, so it has no rightful place in science. The health of the UKB is a function of the observer's preferred reference point, and hence, more a matter of politics than science. The UKB's landscape was substantially altered early in the 20th century by the creation of USBR's Klamath Project. Conditions for many species of waterfowl, fish, and other wildlife were changed. If, for example, the health of the ecosystem is judged from the point of view of waterfowl use and the benchmark is presettlement conditions, then the health of the ecosystem has undoubtedly declined. A more reasonable reference point may be whether the current mixed landscape use results in ongoing or increasing impairment of surface waters for other purposes, particularly wildlife conservation. The capacity to retain or lose ecologically active nutrients like N and P is one possible measurement of ecosystem function that might be useful to decision makers in evaluating the health of the UKB ecosystem. By this standard, P losses do not appear to be quantitatively significant, and the system as a whole appears to conserve P on a mass basis.

Since the alteration of the landscape starting at the turn of the 20th century, there appears to be no compelling evidence that farming in the southern portion of the project, including TID, has significantly increased the rate at which nutrients are transferred to surface waters, or worsened ecosystem function. This conclusion may not be true for the northern portion of the UKB basin, however (Snyder and Morace, 1997). The use of the refuges by waterfowl has remained steady or increased in recent years, as waterfowl populations in the Pacific Flyway overall have recovered from low numbers in previous decades, though this increase has not been uniform in both refuges (Mauser, D., USFWS, personal communication). The USFWS carries out its own farming operations on the refuges to increase waterfowl-carrying capacity by producing grain crops, and there is abundant evidence that residues from agricultural production elsewhere in the region also increase waterfowl-carrying capacity.

SUMMARY

The salt and nutrient content of surface waters increases as water moves through the watershed from the Lost River and J canal diversion to the Klamath Straits Drain. EC_w levels in input waters at the J canal diversion were approximately 300 μS cm^{-1}, while water sampled at the D pump increased to 600 μS cm^{-1} on average over the sample period, and by the time water reenters the Klamath River, salt concentrations have increased further to approximately 900 μS cm^{-1}. The salt content of surface waters increased approximately threefold as water moved through the watershed.

The EC_w values observed in tile drains were higher on average than in input waters and surface waters elsewhere in the region. EC_w values averaged approximately 2000 μS cm^{-1}. Recycling irrigation water through soils in the TID increases the salinity of the water, especially by the time it reaches and is reused in the lease lands area in the southern portion of the district.

For total P, input waters averaged approximately 0.29 mg l^{-1}. Water leaving the Tule Lake Sumps at the D pump increases to 0.34 mg l^{-1}. Water leaving the LKNWR, sampled at the start of the Klamath Straits Drain, averaged 0.34 mg l^{-1}, similar to those at the D pump. The overall increase in P concentration in surface waters was much less than for salt, suggesting that processes other than simple enrichment are occurring, particularly those associated with the exchange of sedimentary P and aquatic plant species.

The average seasonal total P value in tile drains beneath farm fields calculated to date is approximately 0.34 mg l^{-1}. While average total P values in subsurface tile drains were not different from those found at the D pump and the LKNWR outlet, the range in values was great (0.1 to 0.8 mg l^{-1}). Similarly, high NO_3-N values were observed at times in tile drains (data not reported). Very high values in tile drains lead to the inference that some fertilizer N and P is lost in drainage water. The amount estimated as lost, however, is much less than the amount of surplus fertilizer P applied and the amount of P surmised to be mineralized from decaying soil organic matter. Snyder and Morace (1997) reported lower tile drain N and P concentrations than their estimated organic matter decay rates would suggest. P from fertilizer and decaying organic matter appears to be accumulating in soils and lake sediments in the region.

Some leaching of soluble soil nutrients is unavoidable when crops are irrigated. Improved irrigation efficiency and reduced fertilizer used together can help bring P inputs and outputs into balance and may reduce further any avoidable losses of nutrients. These objectives should be the subject of an agronomic research program in the region.

Surface waters entering the TID, the TLNWR, and the LKNWR are already enriched with P. It is not clear that further reducing avoidable P losses from farming in the TID can reduce the amount of P present in surface waters sufficiently to make them significantly less eutrophic. The threshold for algae growth in fresh waters is variously reported as between 0.05 and 0.01 mg l^{-1} (Grobbelaar and House, 1995), far below the values of waters entering the TID.

REFERENCES

Davids Engineering, Klamath Project Historical Water Use Analysis, Review Draft, USBR, Klamath Falls, OR, 1998.
Goyal, S.S., Rains, D.W., and Huffaker, R.C., Determination of ammonium ions by fluorometry or spectrophotometry after on-line derivatization of O-phthalaldehyde, *Anal. Chem.*, 60, 175–179, 1988.

Grobbelaar, J.H. and House, W.A., Phosphorus as a limiting resource in inland waters: interactions with nitrogen, in *Phosphorus in the Global Environment,* Tiessen, H., Ed., John Wiley & Sons, New York, 1995, pp. 225–273.

Hanson, B., Grattan, S.R., and Fulton, A., *Agricultural Salinity and Drainage,* University of California Irrigation Program, Dept. of Land, Air, and Water Resources, University of California, Davis, 1993.

Haygarth, P.M. and Jarvis, S.C., Transfer of phosphorus from agricultural soils, *Adv. Agron.,* 66, 195–249, 1999.

Heckrath, G., Brookes, P.C., Poulton, P.R., and Goulding, K.W.T., Phosphorus leaching from soils continuing different phosphorus concentrations in the Broadbalk experiment, *J. Environ. Qual.,* 24, 904–910, 1995.

Johnes, P.J. and Heathwaite, A.L., A procedure for the simultaneous determination of total nitrogen and total phosphorus in freshwater samples using persulphate microwave digestion, *Water Res.,* 26(10), 1281–1287, 1992.

Kaffka, S.R., Lu, T., and Carlson, H., An assessment of the effects of agriculture on water quality in the Tulelake region of California, Res. Prog. Rep. 108, University of California Intermountain Research and Extension Center, Tulelake, 1995.

Kaffka, S.R., Peterson, G.P., and Kirby, D., Earlier irrigation cutoff for sugarbeets conserves water, *Calif. Agric.,* 52(1): 21–241998.

MacCoy, D.E., Physical, Chemical, and Biological Data for Detailed Study of Irrigation Drainage in the Klamath Basin, California and Oregon, 1990–1992, Open file report 93-497, U.S. Geological Survey, Sacramento, 1994.

Novotny, V. and Chesters, G., *Handbook of Nonpoint Pollution, Sources and Management,* Van Nostrand Rheinhold, New York, 1981.

NRCS, Soil Survey of Butte Valley–Tule Lake Area, Parts of Siskyou and Modoc Counties, USDA, 1994.

Olsen, S.R., Cole, C.V., Watanabe, F.S., and Dean, L.A., Estimation of available phosphorus in soils by extraction with sodium bicarbonate, USDA Circ. 939, USDA, Washington, D.C., 1954.

Oster, J. and Rhoades, J.R., Calculated drainage water compositions and salt burdens resulting from irrigation with river waters in the western United States, *J. Environ. Qual.,* 4, 33–40, 1975.

Sharpley, A.N., Hedley, M.J., Sibbesen, E., Hilbriecht-Ilkowska, A., House, W.A., and Ryzkowski, L., Phosphorus transfers from terrestrial to aquatic systems, in *Phosphorus in the Global Environment,* Tiessen, H., Ed., John Wiley & Sons, New York, 1995, pp. 171–199.

Snyder, D.T. and Morace, J.L., Nitrogen and phosphorus loading from drained wetlands adjacent to Upper Klamath and Agency Lakes, OR, Water Resources Investigation Rep. 97-4059, USGS/USBR, 1997.

Snyder, R.L., Lanini, D.A., Shaw, D.A., and Pruitt, W.O., Using reference evapotranspiration (ET_o) and crop coefficients to estimate crop evapotranspiration for agronomic crops, grasses, and vegetable crops, Leaflet 21427, Cooperative Extension of the University of California, DANR, Oakland, 1987.

Sorenson, S.K. and Schwarzbach, S.E., Reconnaissance Investigation of Water Quality, Bottom Sediment, and Biota Associated with Irrigation Drainage in the Klamath Basin, California and Oregon, 1988–89, U.S. Geological Survey Water-Resources Investigations Report 90–42304, Sacramento, 1991.

Thayer, J.R. and Huffaker, R.C., Determination of nitrate and nitrite by high pressure liquid chromatography: comparison with other methods for nitrate determination, *Anal. Biochem.,* 102, 110–119, 1980.

Watanabe, F.S. and Olsen, S.R., Test of an ascorbic acid method for determining phosphorus in water and $NaHCO_3$ extracts from soil, *Soil Sci. Soc. Am. Proc.,* 677–678, 1965.

Wilson, L.G., Luthien, J.N., and Biggar, J.W., Drainage-salinity investigation of the Tulelake lease lands, Bulletin 779, California Agricultural Experiment Station, 1961.

Grazing Animals and Rangelands

CHAPTER 100

Overview: Grazing Animals and Rangelands

Albert C. Medvitz

California is known for the diversity and importance of its environmental resources. It is a large state with a landmass of about 100 million acres, about the same as that of Japan. California also has a current population of approximately 30 million people, which is growing rapidly and which will double in size in about 30 years. The state also has an important agricultural sector that includes substantial livestock production.

Roughly 40% of the state's land is used for grazing domestic animals. These grazing lands are composed of forest, desert, grasslands, and oak woodlands. Of these grazing lands, about 18 million acres of mostly grasslands and oak woodlands are in private ownership. Somewhat more than 20 million acres of mostly forestlands and desert are in public ownership.

Rangelands in California serve important environmental functions. They maintain the quality and quantity of the state's water supply, support diverse wildlife and native plants, and provide a valuable aesthetic resource for the state's growing urban population. But rangelands are also an important part of California's agricultural production. The animals they support contribute well over $1 billion per year to the California economy. As a result, conflicts erupt over the use of grazing animals, creating problems in the management of both public and private rangelands. Grazing animals and the ranchers who maintain them are often seen by urban environmentalists as purveyors of ill health to range ecosystems.

Overall, however, properly managed grazing animals have constituted an important mechanism for maintaining the health of California rangelands. This series of chapters describes the extent, diversity, and importance of California's grazing land, summarizes available information on the beneficial impacts of grazing on ecosystem health, describes problems in the protection and beneficial management of range ecosystems, discusses federal and state policy about these lands, and presents the perspectives of scientific managers and sheep and cattle ranchers on the cooperation necessary to achieve scientific management of grazing lands in California.

1-56670-612-2/03/$0.00+$1.50

CHAPTER **101**

Changing Public Perceptions of the Ranch: A Preliminary Review of Ancient and Contemporary Claims and Processes

Sally K. Fairfax

INTRODUCTION

The American people in general, and specifically those with environmental/conservation interests, have long had a love/hate relationship with ranching. On the one hand, ranches and ranchers are frequently discussed among our unique contributions to world popular culture. The cowboy image evokes a yearning, if not precisely for cows and sheep, for wide open spaces, simple virtues, masculine self-sufficiency, and healthy living. This image, though widely at variance with the actual working conditions of the real "cowboy," nevertheless drives movie stars and Los Angeles dentists to purchase Montana ranches. The less well-heeled among us are inspired by the marketing of the ranch mystique to buy all sorts of products including some that are not good for our health, such as high-heeled, pointy-toed shoes and cigarettes. On the other hand, followers of John Muir* have also reviled ranching, ranchers, and livestock as land grabbers, cattle barons, hooved locusts, and whatever else the rhetoric of the day required.

In this chapter, I will discuss these conflicted and — I will suggest, rapidly changing — configurations of the ranch in terms of changing claims on ranch land. Different groups in society have different expectations about ranches, what academics frequently call different claims to ranch land. The claims are manifest in expectations about how the land ought to be managed and the goods and services that the lands ought to provide. These general public claims exist on land that ranchers typically consider their own, and frequently own outright. And, claimants have some pretty effective tools and processes for making their priorities felt. Not surprisingly, therefore, accompanying what I will describe as changing claims and claimants, and their expectations about ranch management, are changing processes for making decisions about how the land will be manipulated to meet the changing expectations.

The claims and processes discussed here present different challenges to the rancher and ranching community. These challenges include theft and trespass. If I think cows are inherently

---

* John Muir (1838–1914) is widely acknowledged as the dominant figure in American preservationism. He wrote numerous books and magazine articles that are fundamental to the canon of American nature writing. One of his best, *My First Summer in the Sierra* (1911), talks about his adventures and observations working as a shepherd in the Sierra in 1869. He grew to despise the sheep he tended, coining the now famous epithet, "hooved locusts."

evil, following the familiar rhetoric of Edward Abbey,* "I might feel justified in shooting one. If I think your ranch is my open space, I might feel free to trespass. If I think that your cows are negatively impacting my drinking water or fishing stream, I might lobby for the adoption of federal, state, or local regulation of stream use, on either public or private land. But on the other hand, if I think ranching is a healthy antidote to city life, I might seek a way to come to a ranch and "help out," so to speak, and I might even be willing to pay for the privilege. So, in and among these challenges are also many opportunities if the rancher — as an individual or as a group — is willing and able to capture them."

These claims and processes have a long history, one that for many decades focused primarily on the challenges and opportunities inherent in national-level political activity, and on regulatory activity more generally. More recently, the process has become increasingly responsive to individual and market-based decision making.

In this chapter, I am going to do two things. First, I am going to run through a brief history of claims and processes. I am not going to dwell overly on current events. Once I get to the present — the new claims, claimants, and so on — I think things start to look pretty familiar. I will conclude with my opinion of the changing claims, claimants, and processes.

CHANGING CLAIMS AND CLAIMANTS

Ranching's long history of wearing a black hat, in spite of its rarefied place as a cultural icon, emanates initially, I believe, from the view of ranchers as stealers of land — not just any land, but land belonging to the public and appropriately destined, had it not been stolen, for settlement by hardy yeomen, the "little guy" or "actual settler" who would build a home and become a part of a western community.

Cattle kings is a term of art — it goes along with timber barons — and it arose in a period of our history when "little guys" were increasingly frozen out of settlement and homesteading by the huge holdings which ranchers and timber interests assembled — claimed I should say — in rather flagrant violation or fraudulent manipulation of the 19th century settlement laws.

The legal structure of the disposition era was not inspired. As Robert Nelson has argued in *Public Lands and Private Rights* (1995), almost everything that Congress tried to do failed. Policies pertaining to the arid intermountain west were a particular flop, based as they were on the assumption that family farms would settle the arid lands even as they had succeeded in the Shenandoah Valley or the "old North West." There was, in fact, no *legal* way for ranchers to take title to the amount of land needed to support a family ranch, as opposed to a family farm in much of the rangelands of the western U.S. Congress refused to heed John Wesley Powell's** insights and accept the fact that 160 acres, or 320, or even 640 acres, just would not cut it in regions where 1000 acres were routinely needed to support one cow. Faced with these Congressional blind spots, a dominant but not exclusive strategy was for ranchers to simply occupy the land without benefit of title, and control it by force, through the cooperation of the western congressional delegations.

---

* Edward Abbey (1927–1989) is perhaps most famous for his advocacy of environmental sabotage in his book, *The Monkey Wrench Gang* (1975). A variety of environmental groups, taking a cue from this text, or not, have engaged in sabotage (referred to as *ecotage*, or even *monkey-wrenching*) against the ranching industry by destroying livestock facilities, cutting fences, and so on.

** John Wesley Powell (1834–1902) was a famous explorer, scientist, and ethnographer who wrote the prescient *Report on the Lands of the Arid Region of the United States* for Congress in 1879. Powell advocated distributing the lands of the arid western half of North America differently from the eastern half, where rain-fed agriculture was possible. He argued that land should be classified according to its productive capacity and characteristics, and that settlement should be based on what he called *hydrogeographic basins*. Instead of allocating land in small farm lots of a few hundred acres, Powell advocated community management of pasturelands, timber, and water, coupled to individual ownership of irrigated agricultural plots within a basin. In doing so, he drew on Hispano and Mormon models of settlement, and his extensive knowledge of the native peoples and geography of the West. Powell founded the Bureau of American Ethnography, and was the second director of the U.S. Geological Survey.

And although this method of claiming their place on the public domain worked extremely well for the ranchers — who enjoyed most of the benefits of control and relatively few of the disadvantages of formal title — the strategy was not unchallenged. Those who would homestead were successful in making a number of rival claims to both the unreserved, unentered public domain, and later, to the reserved public domain in national forests. An interesting book written by Helene Zahler (1941) recounts the efforts of eastern labor groups to secure homesteading legislation in order to assure a fair wage to factory workers. They claimed that the public domain land belonged first and foremost to those laborers who were trapped in slavery without the safety valve that the option of western settlement provided. Partially in response to that pressure, Congress enacted the 1862 Homestead Act, and in 1885, a law that made it illegal for ranchers to fence the public domain. More of that later.

Throughout the late 19th century, wrapping themselves in the mantle of the little guy struggling against the Cattle King, homesteaders and potential homesteaders were successful in procuring more and more advantageous terms and conditions for claiming and taking actual title to land ranchers had illegally claimed as their own. The Enlarged Homestead Act of 1909 and the Stock-raising Homestead Act of 1916 encouraged hapless would-be farmers to go west — abetted by promotional brochures from the railroads proselytizing the theory of *dry farming* — and fence in what the ranchers had tried to claim and control. Jonathan Raban's *Bad Land* (1996) is a wonderful recitation of how those homesteaders or *honyokers* made successful legal claims but could not survive economically. Gradually, the railroad towns died and the ranchers reclaimed the lands. Those who followed the morphing of the Montana railroad town of Ismay into Joe, Montana, will find Raban's volume particularly poignant.

The Taylor Grazing Act: The Zenith

With the natural endowments of the arid west on their side, ranchers prevailed over the farmers. The lack of water severely constrained row-crop agriculture. A high point in the rancher's strategy for making claims was the passage of the Taylor Grazing Act (TGA) in 1934. The entire unreserved and unentered public domain in the continental U.S. was turned over to the ranching claimants to eventually become the domain of the Department of Interior's Bureau of Land Management (BLM). The fences that were illegal in 1885 were accepted as evidence of historical use for the purposes of adjudicating rival rancher's claims to portions of the range. The land was semipermanently allocated to ranchers for grazing during specified periods of the year on a permit basis as grazing allotments. Although the TGA specifically states that it confers no title, a series of ambiguous court decisions makes clear that although formal title did not pass to the ranchers, something of value in the permit did. We are still trying to figure out exactly how control is divided between the feds and the ranchers. The claims of ranchers, illegal though they clearly were, have, with the passage of time, ripened into something that apparently cannot be altered without considerable due process and perhaps compensation.

New Claimants, Old Processes

Thus, the informal claim strategy of ranchers worked fairly well until after World War II. At that point, a different and ultimately more powerful group of claimants emerged as a factor in the ranch political economy. Recreation and environmental claimants became active for just about the first time — ranch land and BLM grazing land did not attract much attention as an environmental resource until the mid-1970s — it was, for example, completely ignored in the Wilderness Act of 1965. In the Earth Day era, the arid lands previously more or less defaulted to the ranchers were resurrected in the public view — they had many more values than previously suspected and burgeoning interest groups were forming to claim them. Wildlife habitat, recreation (not merely quiet contemplation but off-road vehicles and dirt bikes), water quality, and wilderness values were soon extensively embraced.

Early effective interests were manifest, just like those of the ranchers, in the federal congressional and regulatory arena. First, new claimants emerged in the form of advocates for the protection of wild horses and burros. At just about the same time, the Endangered Species Act passed in 1973, creating a whole group of new claimants and advocates. Slightly later, wilderness advocates were successful, as part of the Federal Lands Planning and Management Act of 1976, in asserting claims to wilderness areas managed by the BLM. Determining what specifically these claims are is a matter of some dispute, especially in Utah, but the direction of the discussion is unmistakable. Similarly, water pollution control regulations have allowed still other advocates to make claims concerning the management necessary for achieving high standards of water quality on public lands. All of these advocates can be seen as claimants to land and resources that ranchers once considered their own.

Most of these claims have been pursued in familiar, quite well-defined processes: congressional advocacy and the courts.

New Processes: The Emerging Market and the Federal Collapse

More interesting are the changes that have taken place in the claims and processes in the last 20 years. The federal regulatory and political processes have declined in importance, leaving the stage clear for a number of different arenas, actors, and approaches. For the range livestock industry, the major news has been in the decline of funding for federal agencies and the growing legitimacy of local processes and market mechanisms of decision making.

Much of this erosion of the federal process is familiar. I shall highlight one element — recently in the news — which is less discussed and arguably more important for the long-term configuration of the ranch, what I call the Land and Water Conservation Fund Bust.

The tremendous growth of government participation in public recreation that followed World War II was manifest not just in the Wilderness Act but also in the Land and Water Conservation Fund Act of 1965 (LWCF). Under the LWCF, federal funds were made available to federal agencies, states, and localities to acquire private land for recreational purposes. Ultimately, a small portion of revenues from outer continental shelf oil leasing was allocated to the fund. This constituted a direct effort to change claims and claimants by giving one set of them — recreation interests — federal funds specifically for making claims, that is, for acquisition of recreation lands by federal, state, and local governments.

This program collapsed in the early 1980s, a victim, it is frequently argued, of James Watt* and his hostility to environmental values. Funding since that time has been severely limited. I think a different story can be told: that the LWCF died not merely of Watt's hostility — which was real — but of multiple causes. Two other factors contributed: (1) growing hostility to federal ownership of formerly private land and (2) the lethal embrace of the environmentalists. Failing to see that about 30 to 40% of the funds were allocated to state and local claimants, and arguing instead that the funds were intended exclusively for the acquisition of endangered species habitat and wetlands, one wing of the environmental groups so dominated the allocation of funds that the coalition supporting it simply fell apart.

That this vacuum was filled by private groups, nongovernmental organizations able to raise money and leverage it to acquire land for open space, is well known. Growing from almost nowhere, we now have a plethora of organizations designed exclusively or primarily to acquire title to environmentally valuable land. Much of what we see appears to be designed to provide ranchers with sufficient income to stay on the land in real estate markets where the escalating value of the land vastly outstrips the value of the crops that can be produced. Ranchers are reasonable trying to develop new and more-valuable products that will raise their income. A lot of creativity and energy is consumed in both individual and group efforts along these lines. All kinds of tourism

---

* James Watt (1934–) was Secretary of the Interior under Ronald Reagan from 1981 to 1983. He was generally disliked by the environmental community for taking a market-oriented approach to natural resources and federal lands.

have developed to take advantage of the ranch's open space and environment. In addition, diverse marketing efforts have evolved to develop new products and new market niches to justify a higher price for those products. Green labeling of beef as hormone-free, organic, humanely produced, or produced on operations that are fair to workers; farmstead and specialty cheeses; and a variety of local and regional appellations have become a part of the supermarket and farmers' market scene.

OPINIONS OFFERED

In the face of the federal collapse, new, market-driven approaches to claiming ranches have emerged. I have two thoughts that may be worth considering.

First, most of the niche markets, farmstead cheese, and similar efforts to boost rancher return seem to be band-aids, stop-gap measures at best, which may work for some adaptable folks in some places long enough for us to think of something better. I do not see them as adequately responsive to the economic and political forces threatening continued ranching.

The second point is that the strategy with more staying power may not satisfy ultimately, either. I am fearful of the conservation easement that separates the bundle of sticks associated with property owning into two or more parcels. Perhaps as a public domain specialist, I am overly impressed by the fragmenting of the landscape that went on in the 19th century. However, one way to understand what happened is in terms of everybody going out, grabbing what they could get with very little thought about the resulting pattern of land ownership. Railroads, miners, farmers, federal agencies, and so on all grabbed what they could. The result is a mess that we decry and that makes us yearn for landscape-level management. The uniformly embraced solution seems to be to fragment the landscape further, to sever the development rights and other elements of ownership from what now appear as the more salubrious agricultural elements. It is not clear to me that the agricultural uses are rendered sustainable thereby. If the basic problem is inadequate income from ranch products, I am not sure a one-time payment will solve it, and I am fairly convinced that a further fragmenting of the landscape will not in the end prove a good idea. If all you are trying to do is stop development, then in some places and circumstances, there seems to be a temporary gain. I am concerned, however, about the long term.

REFERENCES

Abbey, E., *Desert Solitaire,* University of Arizona Press, Tucson, 1988.

Abbey, E., *The Monkey Wrench Gang,* Lippincott, Philadelphia, 1975.

Muir, J., *My First Summer in the Sierra,* Houghton Mifflin, Boston, 1911.

Nelson, R.H., *Public Lands and Private Rights: The Failure of Scientific Management,* Rowman & Littlefield, Lanham, MD, 1995.

Powell, J.W., *Report on the Lands of the Arid Region of the United States,* 2nd ed., U.S. Government Printing Office, Washington, D.C., 1879

Raban, J., *Bad Land: An American Romance,* Pantheon Books, New York, 1996.

Zahler, H., *Eastern Working Men and National Land Policy, 1829–1862,* Columbia University Press, New York, 1941.

CHAPTER 102

A Rancher's Eye View of Grazing Native Grasslands in California

Richard R. Hamilton

I am a fourth-generation farmer/rancher from Rio Vista in Solano County. My family has a diversified farming and ranching operation, which is located in the Sacramento Delta, Montezuma Hills, and the Dixon Range area (also known as the Jepson Prairie). Our livestock operation includes both sheep and cattle, which feed off the crop residues and native grasslands that encompass our ranches.

My family's involvement in the Dixon Range of California goes back approximately 80 years. In 1920, my great-grandmother's brother, Alex Cook, purchased acreage in the Dixon Range area from the W.Q. Wright Corporation. The corporation had tried to make an irrigation district out of the Dixon Range and, in the process, created Calhoun Cut. Calhoun Cut runs through the southern end of the Solano County Open Space Preserve and our property. Alex Cook also leased ground in the Dixon Range, and he grew hay on the property, which is now Field 19 in the Solano County Open Space Preserve. This is an area where some rare vegetation can be found.

In 1956, Champion Paper Company (CPC) purchased 1500 acres in the Dixon Range; 418 acres of the purchase was from the land owned by my family. CPC was planning to build a paper factory on the property. My family leased back from CPC the property it had sold. When CPC withdrew its plans to build a paper factory, it sold the land to Southern Pacific Railroad Company (SPRC) and we continued to lease the same portion of the property from the new owners. In 1981, SPRC sold its entire property to The Nature Conservancy (TNC) and the property became known as the Jepson Prairie. Once again my family continued to lease its original acreage from TNC, and in 1983 leased the entire acreage from TNC.

Our lease agreements with CPC and SPRC were cash rent agreements with no type of grazing restriction. We manage the property much the same as the way we manage the property we continue to own. It was not until we entered into a lease arrangement with TNC that we experienced any type of grazing restrictions or plans that were derived by a landowner. For my family, this was a new experience — to be in a position where we were being questioned about our grazing practices and their threat to the sensitive plants and the environment. In his first meeting with TNC, my father's response was that we could not be bad land managers, if these sensitive plants were still around after his family had been pasturing this ground for more than 60 years. With our family history, our love for the land, and our openness to listen to new ideas, we were very committed to making our relationship with TNC work. It is true that there was a faction within TNC that wanted to lock up the property and not allow grazing (which would create a blight condition), but most of the people from TNC realized that some form of grazing needed to occur. We appreciate the attitudes of TNC's Barbara Malloch, Owen Pollack, and Rich Reiner, with whom we have worked. They listen, and respect our opinions (see Chapter 106, this volume).

In the grazing operation on the Jepson Prairie, we incorporate TNC's grazing needs into the grazing rotation of our own properties. Following are some of the grazing restrictions on the Jepson Prairie:

1. Sheep are the only livestock allowed to graze.
2. There is no use of supplemental feed (hay or corn).
3. There is no control to stop the predation by coyotes.
4. The grazing rates of the pasture are controlled by AUMs (animal unit per month, the amount of forage needed for 1 month by an animal unit, usually a 1000-lb cow) per acre, and there are periods of no grazing. During the wildflower tour season, March to April, the sheep are moved on and off the preserve before and after weekend tours.

For our operation, our adjoining property provides the flexibility to operate with these restrictions; operators who do not have adjoining property have a harder time complying.

Each year, when we know the grazing priority, we incorporate the preserve pastures with our own pastures. The grazing pattern depends on the needs and type of vegetation control TNC desires. For example, to control grasses in the brome family and to utilize these grasses for the sheep, it is necessary to graze these grasses early in their development because that is when they are most palatable to the sheep. Once brome gets past a certain stage, sheep will not eat it. To manage these grasses, which are low in nutritional value, we use sheep that are nonproductive and able to survive without supplementation of feed. To control all grasses during the grass-growing period (January to April) thereby promoting a good propagation of wildflowers, we use high-stocking and short-duration grazing practices, which usually include ewes and lambs. To control yellow-star thistle, mustard, wild lettuce, and wild radish, we use a late-spring and early-summer grazing practice of high stocking rate and longer duration per field. To protect the desirable plants, such as bunchgrasses, we monitor closely what the sheep are eating on a daily basis; that is not too different from the way we manage our own property. The grazing practices in terms of stocking rate, time period of grazing, and frequency and duration of grazing are dependent on other factors, such as the weather. The timing of rain and temperature conditions varies each year; thus, each year's grazing schedule and types of grazing practices can never be permanently set. There has to be some flexibility in the grazing plan for the unexpected conditions, like early or late rainfall. As part of our lease arrangement with TNC we periodically get together and talk about grazing plans and we keep monthly grazing reports for each field. We assist with the maintenance of the preserve and with the conducting of grazing experiments in test plots at the University of California, Davis.

Our philosophy in the management of the grazing of native grasses on our ranches involves many factors. Making improvements to native grasslands with grazing can take up to 10 years, and there must be a way to monitor progress. The ability to make the right decisions is based on 70% practical knowledge and 30% technical knowledge. The practical knowledge comes from two sources: (1) experiencing the various climatic conditions that can affect the property, and (2) the experience of running livestock through production cycles on the property. Both the technical and practical knowledge provide the basis to help with the management of the land.

Some of the factors to consider in managing livestock in native grassland are:

1. Survey the grasses to determine the palatability to livestock.
2. Be familiar with the behavior of the livestock you are managing.
3. Have a knowledge of livestock food preferences (their priority of grasses and forbs).
4. Have a knowledge of livestock grazing routines (so you can distinguish stress from contentment).

5. Know the nutritional requirements of livestock during its life and production cycle (nonlactating vs. lactating animal, offspring vs. mature animal).
6. Multispecies grazing is often better in terms of utilization and management of native grasslands (cattle and sheep utilize different grasses on the same acreage).

The livestock segment is often blamed unfairly for the demise of native grasslands and open range. The real culprit is the ownership of the properties. The abuse of native grasslands occurs in two ways:

1. The purchase of grasslands by both public and private entities which do not develop a plan of utilization for the property; thus the property becomes a blight area.
2. The practice of establishing undesirable leases by public and private entities. Leases of short duration and high cost per acre or AUM tend to lead to high turnover of lessees, which in turn leads to mismanagement of the resources of the property. A lessee has a hard time making physical improvements and recovering his investment in short-term leases, then loses interest in the well-being and future of the land.

Finally, as private property owners of native grassland and farmland, we face a constant pressure of trespassing from both the private and government sectors on our property. The courtesy of asking for permission to gain access is hardly ever considered, and the lack of respect of fences and gates is common. It seems we are always dealing with illegal hunting during the season; out of season, we are dealing with vehicles driving through our crops and pastures, and governmental agencies and private companies not asking for permission to gain access nor stating their intentions. Over time, we have been burdened with tremendous expense in damage done to our crops, livestock, and assets such as buildings, fences, windmills, and water troughs. These problems are never caused by our neighbors. All our neighbors, including TNC, have been very respectful of our rights as property owners. Being good neighbors is very important to them.

To have success in the management of native grasslands in both the public and private sectors, long-term relationships must be formed. Our relationship with TNC and the Solano County Open Space Foundation has been a good example of how long-term relationships can be managed for the betterment of the land. As ranchers, we feel we are in the best position as keepers and historians of the environment. With our factors of production being tied to nature, and for ranchers to be successful in business, we must respect and nurture the environment.

CHAPTER 103

California Grazing Lands:
Science Policy and the Rancher

Jack Hanson

I am a cattle rancher from Susanville, CA, a small rural community in the remote northeast corner of our state. I would like first to briefly touch on rangelands in California, and what ranchers are doing to better manage the animals that are grazing these lands. Then I would like to broaden the subject and share with you my observation on the changing social fabric that defines the rural west, and how this change may affect the environment and the structure of the ranching industry.

I would like to share with you a glimpse of my personal history and my current situation, as I believe this will give you clearer insight into my observations. I was born and raised in the San Francisco Bay area and received my college education at a small, liberal university in the Berkeley Hills, which was, at the time, known best for its extracurricular activities and its poor football teams. After a stint as an officer in the U.S. Army, I settled into an occupation for which I had been groomed since birth: working as a stockbroker for Dean Witter, wearing a three-piece suit. I was, however, disillusioned with this calling and wanted to find a more honorable and useful way to contribute to society. To this end in 1975, I left behind my home in Tiburon, the Golden Gate Bridge, the famous three-martini lunches and the weekend golf games, and went into ranching. After 23 years in the business, and divided between three locations, I find myself, along with my wife, Darcy, and our two young sons, owning and operating a cow/calf and hay ranch in the open expanses of the Great Basin. The irony of all this, and as a demonstration of the keen insight I gained through my degree in economics from Berkeley, is that I left the brokerage business the year the market bottomed at about 490 on the Dow, and when the NYSE average daily trading volume was running around 13 million shares. I totally immersed myself in a business where the actual value of the unit of production is the same now as in 1975. However, our costs are substantially higher and our relative position in the economy is far worse.

Let me make a few quick comments about rangelands in California and their management. The U.S. has about 770 million acres of rangeland, more than 50% of which is privately owned, 43% owned and managed by the federal government, and the remainder owned by state and local governments. In California, there are over 40 million acres of land utilized for grazing; approximately one half of these acres is privately owned. Cattle grazing is the most benign, productive, economic activity that can occur on these valuable rangelands.

In addition to being the foundation for the state's livestock industry, these rangelands contribute property tax revenue, fuel load management, groundwater recharge areas, wildlife habitat and open spaces to the public. I strongly believe that the vast majority of ranchers who use these public and private rangelands are keenly aware of the importance of maintaining the lands in a healthy condition so these landscapes will provide high-quality forage on a sustainable basis as well as clean water and an appealing appearance. Healthy, diverse ecosystems and clean water translate into healthy, more-productive cows and calves. Healthier soil and grass and greater control over livestock use patterns increase agricultural production. Good grazing practices promote plant diversity, protect waterways, reduce erosion, reduce fuel loads, and are key to the long-term health of our watersheds. There is a belief that persists among ranching and farming families that we do not inherit this land from our parents, we borrow it from our children. If you subscribe to this, you will over time do your best to treat the land with respect and use it gently.

To me, it is too bad that the poor cow is such a maligned animal in some circles today. A rancher's greatest advantage is that he raises ruminants. Cows are residue gleaners, by-product users, and an efficient way to productively use these marginal lands. On many of these acres of rangelands, it is only through the use of the cow that we can convert energy from the sun into high-quality, digestible protein for human consumption.

The key ingredients of successful management of rangelands are simple: you have to develop a vision, then design a management plan to achieve that vision. Ranchers have for years attempted to utilize informal plans, mostly developed over time and based on trial and error, and evaluated only by common sense. For the last 20 years or so, these plans have become more formal with detailed strategies applying new science and techniques. Animal behavior, plant responses, hydrology, and practicality are the basic considerations when developing a grazing strategy. I would just add the following points about grazing systems or plans and how we must view them:

1. Plans must, by their nature, be dynamic. Implement them as best you can, monitor your results, and make adjustments.
2. Keep in mind that no matter how scientific, precise, and well thought out a strategy is, you are still dealing with unpredictable variables that can humble even the best manager or most experienced scientist — the cow and mother nature — drought, flood, freezes, early grass, late grass, etc. may all necessitate adjustments. The potential need for adjustments dictates that we have alternatives and flexibility built into the plans. These are expensive. While this flexibility is necessary, it does have a cost associated with it.
3. Plans or strategies must be site specific. What works in the coastal ranges north of us will not necessarily work in the Great Basin or the high meadows of the Sierras.

Grazing systems, in their simplest form, are designed to control intensity, season of use, and frequency of grazing. It should be noted that they must consider the nuisances of specific watersheds, in specific settings, during specific weather patterns, with specific livestock and big game herds. Especially in California, it is likely that the most important and significant result of a grazing management plan is water quality. If regulators and the public are going to become more involved in the management of our private land and our grazing livestock, it will be through the concern over nonpoint-source pollution in relation to water quality. The California Cattlemen's Association has implemented a proactive approach to deal with water quality issues on rangelands. It is a voluntary program in which agencies assist ranchers in preparing a grazing plan, which ensures protection of the quality of the water. It was approved by California's Water Quality Control Board in response to the requirements of the federal Clean Water Act.

One item I mentioned earlier that is an important ingredient in a grazing strategy is practicality. If a strategy or management plan strays too far from being practical, it will not succeed. In grazing livestock, from a business sense a rancher is essentially trading time for money. A grazing strategy that takes excessive time is just as impractical as one that takes excessive money to implement. Consequently, in the final analysis, management strategies have to be accomplished within the financial and skill limitations of ranchers who ultimately have to make grazing plans work.

The previous comment, I believe, adequately transitions into discussion of the economics of the ranching business, the dynamic changes occurring in the livestock business and agriculture in general, and how these changes are affecting the social fabric of the rural west and the fate of the rangelands. Some of these changes are not new at all. They have been occurring since the 1870s when the phrases *cattle kings* and *cattle barons* became popular. But the changes are now moving at an accelerated pace and affecting a segment of society that, by its nature, does not easily and quickly adjust to change. The changes are not necessarily good or bad but surely inevitable. They are not unique to agriculture either: consolidation, mega units, globalization, vertical integration, alliances, consumer orientation, biodiversity, ecosystem management, and nonpoint-source pollution. But what makes these changes important to everybody is the fact that they can affect the supply of safe food and the management and preservation of the resources that produce our food supply as well as the environment.

When discussing the current economic situation in the livestock industry, a little history may increase our understanding. Returns to agricultural producers have never been great. An average 3% return on investment over time was typical in the 1970s and early 1980s but it has been in a steady state of decline since. Today, only the top 20% of agricultural producers have returns on investment of 3% or more. In 1951, a beef producer received 71% of the beef dollars; in 1996, it was 45%. At the same time, retail margins have increased from 19 to 46% of the beef dollar. Along with increased retail margins, we have seen an overwhelming concentration in the beef-packing industry. In 1981, we had the four largest packers slaughtering 36% of the fed cattle. Today IBP, ConAgra, and Cargill slaughter 82% of all the fed cattle and are quickly moving into the cattle feed business. What is really going on? For years producers have been told that our cattle were cheap because of market forces: oversupply, slack demand, and competition from other meats. However, looking at 1998 as an example, we moved a near-record supply of beef at pretty good prices at the retail level. The catch is that these high consumer prices have occurred while cattle prices have been low. As Jerry Siebert, agricultural economist at the University of California, Berkeley, stated, "There is really a major and significant difference between retail and farm value. What we see is that retail prices go up, the processor or the packer price goes up, and the producer prices are relatively stable or declining." Robert Young, co-director of the Food and Agricultural Policy Institute, University of Missouri, stated the producers had always been under financial stress and going out of business, and that the trend toward larger farms will continue. In a hand-out, Young projected commodity prices from 1997 through 2007. Returns for the cow/calf unit annually over this 10-year period were all negative, except for the year 2000.

As an aside, hog prices in late 1998 hit a 50-year low, not inflation-adjusted price but an actual 50-year low. For the first time since the 1940s, live hog prices fell below $10 per 100 lbs. Jim Winder, a rancher from New Mexico who was on a panel at the 1999 International Congress on Ecosystem Health, has stated, "These are tough times for ranchers … although we win some of the battles, we are losing the war. We are not losing to environmentalists, we are losing to economics and demographics. If the American public ever sweeps us off the land, it will be because we are not producing enough value for society."

This brings me to what I believe is a critical point in discussing agricultural economics: that is, the erosion of the farmer's and rancher's relative position in the country's economy. What kind of value should society place on its food at the farm gate and the human and natural resources needed to produce it? Think about it. Our society has valued a 4-year-old company, with estimated sales of $90 million and profit of less than $50 million, at about $25 billion. This is just below the $26.8 billion value placed on the 1996 agricultural production in the state of California, about 12 to 13% of the national total. An even more glaring example of this anomaly is another company, which has been referred to as "an online garage sale," with earnings of less than $6 million in 1998 on $36 million in sales. This company is worth $12 billion. Obviously, those astute investors in the audience who bought stock in these companies on initial public offerings know that I am referring to Yahoo and eBay. In fact, Morgan Stanley's index of 73 Web companies collectively lost $1.5 billion last year, yet has a combined market value of $115 billion. The online traders are an amazing group. In their world, you throw out everything you learned about rational pricing —

earnings, price–earning ratio, book values, even revenues. They are all meaningless. Do not worry about the long haul. Invest for the moment. Make a killing. With this attitude becoming more commonplace in society, many ranchers feel alienated. They find themselves with their more stable, traditional approach further from the mainstream.

Wendell Berry offers another interesting fact, which gives us insight into where farmers and ranchers fit in the big picture, in his essay "Conserving Communities." In October 1993, the *New York Times* announced that the U.S. Census Bureau "would no longer count the number of Americans who live on farms." California, being the trendsetter, had suspended this survey as far back as 1970. By 1950, the number of farmers had declined to 23 million, and to 4.6 million in 1991. In addition, in 1991 32% of our farm managers and 86% of our farm workers did *not* live on the land they farmed. These figures, at least on the national level, indicate we no longer have an agricultural class that can demand to be recognized by government; we no longer have a "farm vote." From a demographic standpoint, farmers and ranchers have become statistically insignificant.

How does this play out for the environment and the social structure of rural communities? It might be dangerous, but for the sake of discussion, let us divide the farmers and ranchers into three groups: part-time farmers and ranchers, full-time family farmers and ranchers, and the so-called mega-family farms and ranchers including the corporate farms. Note that I am going to be generalizing when referring to ranchers and their situations. I am sure you realize that there are many exceptions. My comments do not refer to everybody all the time, but do reflect what I believe to be the trends in the industry.

The first group obviously is relatively small producers, who have full-time jobs off the farm. This group will survive, but does not necessarily define agriculture nor does it provide strong leadership for agriculture's industry and infrastructure. The second group is, in my opinion, the most important group to the rural communities and the ecosystems in which its members live and work and to the infrastructure of the industry. Sadly, this is the group that is under the most financial stress. In order to survive with the current low margins, this group must bring in outside income or grow. Growth is difficult without outside capital. In fact, many in this group are eating into existing capital, just to survive. The third group has the capital to take advantage of opportunities and new technology, and its size offers the members a market advantage. They have economies of scale, which lower their production costs and allow them to continue to expand at the same time.

With this division in mind, if the current economics continue, family ranchers are in danger and, therefore, the rangelands that they own are in danger of either being subdivided or gobbled up by larger, impersonal operators for better or worse. Obviously, smaller units are not as environmentally healthy as larger ones. Larger mega or corporate units are not as desirable for the rural communities as are the more-numerous family operations where the owners live on the property. This trend in subdividing or consolidation will surely change the character of the rural West, and not necessarily for the better. People may think they would like to move to the country, but that is not what they really want. What they want is to move everything they have, the shopping malls, sewer lines, mail delivery service, and so on, into what was once the country.

What do we lose over and above the important production value when a working ranching landscape is carved up into 40-acre ranchettes? For the rancher, it is often a personal tragedy. For society in general, we lose the viability of the agricultural economy that is the lifeblood of many rural communities and an essential element of the West's sense of place. Piecemeal subdivision of working ranch landscape threatens the viability of the dynamic ecosystem, and fragments the most productive wildlife habitat, but more importantly and sadly, it alters the relationship between humans and the land. On subdivided land humans no longer enter into an active relationship with the natural process. They are no longer stewards of the land but rather temporary tenants. Obviously, as land is divided and removed from serious production (and economics remain difficult here), we will push production offshore. Is this good? Production initially will be cheaper, but it will not necessarily be better for the planet. Dennis Avery, of the Hudson Institute, has observed that the worst use of land through slash-and-burn techniques in underdeveloped countries is driven at least by the restrictions and reduction of more-efficient production on American farmlands themselves.

It is obviously not all bad; many of us will survive. We will be willing to work with neighbors and adapt to the changing economics. Tomorrow's ranchers must be prepared to use all the private land tools at their disposal: agricultural conservation easement, estate planning, voluntary zoning district, collaborative planning, and others. Fortunately, the rancher will not be alone in this effort. Despite a long history of conflict over management practices, ranchers, environmentalists, range managers, wildlife biologists, and others are forging new ground to facilitate change toward the common goal of adapting and sustaining ranching as a way to maintain healthy ecosystems. We must, through this new alliance, develop a mechanism to somehow recognize the nonproduction value which, in fact, working ranches give to society.

I would like to bring to your attention one initiative to do this. A new effort to ensure the preservation of rangelands and cattle grazing and to take a proactive approach in addressing environmental and resource use issues took shape recently. A group of California ranchers, including myself, started down the path to form the cattlemen's own land trust. This trip culminated with the formation of the California Ranchland Trust (CRT). While the trust was conceived within the ranks of the California Cattleman's Association, it is a stand-alone organization run by a board composed of cattlemen. Even though the number of land trusts in California doubled to 120 over the past 10 years, I believe CRT can fill a unique niche in the preservation of open space in California and keep ranchers on the land. Since this organization will have a good understanding of the needs of ranchers and also a sensitivity of what society gains from healthy rangelands and open spaces, it can provide a vehicle by which society can recognize those nonproduction values of ranches, and in some fashion apply that recognition to keeping families on ranches and cattle grazing on the rangelands. During its formation, we worked diligently to gain support from groups outside the livestock industry that share our concern for the fate of the rangelands and ranchers. In this I think we have been successful, but we have a long way to go. On the other side, many ranchers are skeptical of land trusts that do not understand the livestock and that may not be interested in perpetuating livestock grazing as a practice on their rangelands. We hope that CRT can bring these parties together in an atmosphere of trust, mutual understanding, and common goals and ideals, in other words a win–win situation.

In summary, we do not want government programs or hand-outs. These are almost always abused and unfairly applied. But during this era of globalization of our economy and huge multinational food companies with near-monopolistic control over certain commodities, what we want is not only a free and fair trade, but also a level playing field when dealing with imports and large food-processing corporations, and realistic and practical environmental requirements that are fairly and evenly applied. Most importantly, through the market mechanism, society will put a higher valuation on our products at the farm gate and somehow recognize the nonproduction value of having us as the stewards of these rangelands and as a crucial ingredient in the fabric of the rural West.

One more thing. It is still an honorable profession. I am proud to be a rancher and proud to represent my family and my industry. Obviously, I have faith that we will survive and prosper.

CHAPTER 104

Reinventing the Range: To Graze or Not to Graze Is Not the Question

Lynn Huntsinger

INTRODUCTION

Domestic livestock can be used to maintain and improve ecosystem health. Grazing is a complex interaction between herbivore and vegetation that can be done to "prescription" in order to meet ecosystem management needs. Herein is discussed the ecological dimensions of what can be called "grazing to prescription" in Mediterranean California. The chapter is organized loosely by scale, starting at the landscape level, then going to the community level, then to the individual plant level, and, finally, returning to the landscape level to close. This chapter is not an attempt to argue the desirability or appropriateness of grazing for any particular site or situation, but only to explain some of the possibilities of livestock grazing management and what it can offer in stewarding diverse and changing ecosystems.

California's most productive rangelands, the Mediterranean-type grasslands and woodlands, are in the majority privately owned. They are increasingly at risk of conversion to residential housing as the population grows, commuting and telecommuting become commonplace, and wealthy retirees seek retreats in the countryside. California ranchers own most of these lands. Of late, open space and biodiversity advocates have become interested in finding out how they can work with these landowners to protect California's landscapes. In addition, conservation easements and land trusts are becoming more common on ranchlands. California has more acres in conservation easements than any other state, some 1.25 million acres. For these conservation efforts to work, it is important to understand how these "working landscapes" can be managed for the objectives of landowners and investors in easements.

The purposes of grazing have been reinvented many times to fit the political, economic, or social ends of the times. For the Spanish who settled California starting in 1769, livestock were grazed to produce hides and tallow for export, and to support colonization and missionary efforts. Throughout the 19th century, Spain and the U.S. used livestock as a visible means of occupying and claiming territory. For New World speculators toward the close of the century, grazing and the land grabs that went with it was a way of producing returns for distant investors. In the progressive era after 1900, grazing livestock to produce food and fiber for a flourishing and ever-more-populous young nation was widely accepted as its highest purpose; as a corollary to that, the highest and best use of rangelands was to graze livestock.

The costs and benefits of livestock grazing are weighted with the purposes of the past, but today must include the role of grazing in conserving and supporting cultural heritage, wildlife habitat, open space, water quality, scenic amenities, recreation areas, fire resistant landscapes, and community well-being. Too many times, although the various demands on rangelands are recognized as complicated, grazing itself is approached as simply a question of "to graze or not to graze." This is a disservice to what we understand about ecosystem dynamics and herbivory. Grazing is not a black box, but is a complex set of interactions with ecosystem processes that creates multiple effects. We need to know what kind of grazing, where, when, to whose benefit, and for what purpose, in order to make decisions about what the role of grazing in an ecosystem management scenario should be. The quality of the answers depends on the quality of information about the ecosystem and its response to the various impacts of grazing, as well as the quality of the decision-making process.

THE CHANGING RANGELAND LANDSCAPE

California's rangelands renew themselves annually, with a unique mix of herbaceous species in response to patterns of rainfall and temperature. California's oak woodlands, savannas, grasslands, and shrublands make up the state's Mediterranean-type rangelands. Forage for livestock is generally the annual grasses, broad-leaved plants, and shrubs that grow during the wet, mild winter and spring. Rainfall is highly variable, from year to year, north to south, and with elevation. San Diego's average precipitation is about 250 mm per year; Eureka, in the north, averages 1000 mm per year. Forage production can vary from almost nothing to 7000 lbs per acre, and among the annuals, can be vastly different in species composition from year to year and place to place. Attempts to predict rainfall and hence forage production and composition have generally been unsuccessful, regardless of their sophistication. About 70% of the rainfall comes in the winter, and summers are characteristically hot with a drought that lasts from 2 to 11 months.

The rangelands of California are also undergoing change over a longer time frame. About 2% of the Earth's surface has a Mediterranean climate, including the region around the Mediterranean Sea and parts of Chile, Australia, and South Africa. Today, California shares much of its flora and fauna with these parts of the world. Walk out on a rangeland in southern Spain and you are likely to see a familiar grassland. Wild boar, several kinds of deer, pheasants, and other fauna have immigrated to California and are flourishing. Commercial hunting, particularly during the Gold Rush, severely reduced the numbers of native tule elk and pronghorn antelope, the major native range herbivores.

Burcham (1981) diagrammed what he thought were likely dates of invasion of several important rangeland plant species into California, beginning with Spanish colonization in 1769 (modified as Figure 104.1). Many of these species, particularly annual grasses and legumes, though competitors

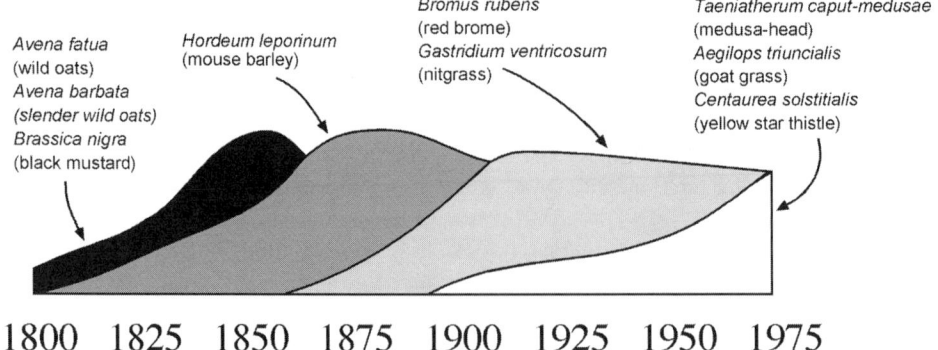

Figure 104.1 Plant species invasions into California. (Modified from Burcham, L.T., *California Range Land: A Historico-Ecological Study of the Range Resources of California*, Center for Archaeological Research 7, University of California, Davis, 1981 [1957].

with the native flora, have at least proved valuable as forage. Some of them, such as yellow star thistle (*Centaurea solsticialis*), are considered major pests without economic or ecological benefits. In the 1960s, there was an estimated 1.2 million acres of star thistle in California; in the 1980s, 7.9 million acres. Today, the species is thought to extend over more than 20 million acres. The nonnative flora that has been successful here is characterized by competitive aggressiveness and an exceptional capacity to cope with disturbance, perhaps most notably, periods of drought.

There are no comprehensive vegetation surveys available for the period prior to Spanish settlement, and as a result, there is considerable debate about what the native herbaceous flora was like and what happened to it. The perceived wisdom has often been that California was once carpeted by an evergreen bunchgrass flora, predominantly *Nassella pulchra* or purple needlegrass. Recent work suggests a more varied grassland, with a high annual component in some areas and bunchgrasses concentrated in more mesic environments (Hamilton, 1997; Webster, 1981). What we do know with certainty is that today a rich mixture of natives and nonnatives makes up the California rangeland flora. If number of species per unit area is used as the measure, it is among the most diverse floras in the world. In most places, native grasses are present, though often difficult to spot in a sea of nonnative annual grasses. Where endemic soils restrict the nonnatives, the native species flourish. The heterogeneous soils, microsites, aspects, and elevations characteristic of California rangelands contribute to their diversity.

The California rangeland landscape is a new and renewing landscape. It is unprecedented, and continuing to change.

COMMUNITY LEVEL GRAZING EFFECTS

Grazing affects vegetation at the plant community level in a number of ways, but perhaps most notably by influencing the relationships among plants growing together. Around the world, drought, fire, and grazing by livestock or wildlife are major disturbance factors for Mediterranean rangeland plant communities. Because of the high productivity and long dry period on California rangelands, fire has been common for millennia, ignited by lightning or by people. Like burning, grazing by native and nonnative herbivores removes vegetation and redistributes nutrients. But grazing is also quite different, in part because it is a selective process.

Selection

Grazing animals *select* the plants and the plant parts they eat. This selection is influenced by the species, age, and physiological status of the animal; the time of year; the availability and physiological status of the plant; and so forth. While fire may not burn everything, or burn evenly, it is not as selective by species as grazing. Much of what we believe about the influence of grazing on plant communities is predicated on the idea that this selection influences the competitive interactions among plants: to put it simply, plants that get eaten are at a disadvantage compared with plants that grazers avoid. This can be used to advantage. For example, cattle can be used to control grass growth around young trees, as long as the grasses are more attractive to the livestock than the trees. Goats are becoming popular for controlling brush, because they preferentially consume shrubs.

In some cases, and in perhaps much of California's rangelands, climatic and other abiotic effects outweigh the effects of grazing on plant-to-plant relationships (Westoby et al., 1989). The disturbance (most notably drought) and variability characteristic of California rangelands may overwhelm the influence of selection by grazers. This is no doubt true when considering relationships among annual plants.

An alternative model widely used in California considers only that grazing removes vegetative matter, and what influences the composition and structure of the returning plants is the amount of biomass left on the soil through the summer, termed *residual dry matter*. This dry matter protects soil and creates microsites that encourage the growth of particular annual species in the coming growing season. A large amount of dry matter encourages plants that can come up through a thick layer of dead vegetation, often creeping wild rye (*Leymus triticoides*) in much of California. Grazing until there is little dry matter left results in a very different species composition as seeds germinate in drier, lighter, conditions. All this in turn, of course, can have very local effects on wildlife populations. Small rodents flourish under a thick thatch of dry grass. Raptors like the red-tailed hawk are seen waiting at the fence line or overlooking a patchily grazed rangeland, snagging the rodents when they venture out from beneath the protective cover into grazed areas.

Perevolotsky and Seligman (1998) argued that intermediate grazing intensities result in the greatest biodiversity on Mediterranean rangelands. This follows from the intermediate disturbance hypothesis, which states that intermediate levels of disturbance foster diversity. Intermediate disturbance prevents any one plant or group of plants from dominating the ecosystem everywhere or all the time, maximizing the diversity of the vegetation as a broader variety of plant species and structural types can find niches.

For each grazing application, the model or combination of models of ecosystem dynamics upon which management prescriptions are based needs to be carefully considered and subject to ongoing evaluation.

Animal Distribution

Variable impacts on plant communities are also caused by the behavior of animals in distributing themselves across the landscape, referred to as *animal distribution*. Grazing animals use plant communities and plants to different degrees depending on slope, aspect, proximity of fences, size of pasture, location of water, predator presence, and so forth. Where animals spend the most time, more grazing tends to occur. Cattle like low-lying areas, horses and goats the uplands, all of them like to spend time under the tree canopy during hot weather, and so forth. This can have consequences for soils, if animals concentrate in particular areas, and for the redistribution of nutrients.

As an example, sometimes when much of the grassland is brown and fairly dry, the area under an oak tree is still quite green. These areas often green up earlier in the season compared to areas outside the canopy. *Fog drip* and species differences due to the canopy have been used as explanations. Other explanations that have been put forth are that concentrations of grazing animals seeking shade have left more nutrients under the trees, or that intensive grazing under the trees results in a different species composition or different levels of residual biomass, affecting the timing of when the herbaceous understory turns green. Some believe that the tree itself acts as a "nutrient pump," bringing nutrients up from the depths to the surface. The explanation or combination of explanations probably varies from locale to locale.

A recent article by Fuhlendorf and Engle (2001) suggests that grazing can be managed to enhance biodiversity, by managing for patchy effects. Though the usual paradigms of range management emphasize even livestock distribution and use of vegetation, managing for biodiversity may instead mean managing for heterogeneity, including allowing an uneven distribution of grazing use. They argue that this is more in line with the grazing patterns with which plant communities and ecosystems evolved.

THE PLANT LEVEL

Though California has only a 230-year history of livestock grazing, it has a history of fire and herbivory that goes back thousands of years. California native plants, as well as introduced Mediterranean species, are generally well equipped to cope with fire. Native herbivores, such as elk and deer, once roamed in huge herds (McCullough, 1971), so the native grass and shrub species also have reason to be adapted to some kinds of herbivory. Many of the adaptations and coping mechanisms that help plants respond to fire also help plants respond to grazing.

Annual species simply grow back from seed after a disturbance, such as fire. They mature quickly and can produce large volumes of seed in a short period of time. The perennial native bunchgrasses resprout after fires that are not intense enough to damage the deep root crown. One of them, purple needlegrass, today is often observed on what appear to be highly disturbed sites — on road cuts, for example.

Many shrub species resprout from the base or from stems. With herbivory, they are able to develop shapes and structures that control grazing impacts and allow them eventually to grow to a height where they are safe from grazing. Rhizomatous growth forms and tough, fibrous stems and leaves are found in some species. Many native shrubs are full of toxins and oils that give them a bad taste. Some species are sensitive to grazing at certain times of the year, but are relatively unaffected most other times. For example, research has found that purple needlegrass is most sensitive in the spring when it is putting out its flowers (Huntsinger et al., 1996). Other studies show that grazing influences native species in different ways at different times (Dennis, 1989). Knowledge of the ecology of the plants of interest is needed to develop grazing prescriptions.

PRESCRIPTIONS

Traditionally, range managers have developed four principles that should be considered in planning a grazing program:

1. The kind and class of animal
2. The spatial distribution of animals
3. The temporal distribution of animals
4. The number of animals

The kind and class of animal is important, because as discussed earlier, different kinds of animals prefer different kinds and amounts of forage. Goats and deer tend to browse shrubs; cattle and elk graze grasses. Pronghorn antelope often seek broad-leaved herbs. Animals of different ages and in different physiological states also have dietary differences.

The spatial distribution of animals is important. The more spread out animals are, the more selective they can be. A lot of animals confined to a small pasture do not have much opportunity to be picky, either in choice of plants or in where they congregate. Some grazing systems use this to reduce selective effects, keeping a large number of animals in small pastures for short periods. Placement of water, fences, and salt blocks can be used to influence animal distribution and keep animals away from sensitive species or soils.

The timing of grazing can also be important. It can affect the types of plants present and their edibility, hence the choices of the grazing animals. Some plants are more sensitive to grazing during certain times of year, during flowering or in the growing season. During wet season, some soils are more vulnerable to trampling. In the dry season, livestock are more likely to concentrate on the moist riparian areas. The length of time animals are in an area also affects the grazing choices they make.

The number of animals per unit area affects the intensity of grazing. For example, when managing to leave residual dry matter, particular intensities of grazing will result in different amounts of biomass being left behind at the end of the season. A large number of animals on a unit area for a long period will result in more removal of plant parts than will a few animals. Removal of a lot of the vegetation results in a homogeneously grazed look, somewhat like a golf course. Under this scenario the grazing animal cannot be particularly selective. As the grazing intensity gets lighter, the grassland will look less smooth, more heterogeneous, as animals pick and choose what they eat, or happen to concentrate in one place or another. Typically, range managers manage for an intermediate grazing intensity, removal of about half of the forage, but that varies a great deal depending on management goals. This can achieve a patchy effect. Light grazing can also make the vegetation look patchy, or it may not even be easily detectable. Animals will select the plants and places they prefer, leaving many places and plants untouched.

Example: Forest Grazing

An example of how grazing can be used to achieve particular goals can be drawn from studies of using cattle as a vegetation management tool to reduce shrub competition with young trees in conifer plantation (Huntsinger, 1996; Allen and Bartolome, 1989). For the system to work, of course, the shrubs and the grasses needed to be selectively grazed by the cattle. To test this, cattle were fenced in to areas and forced to graze at a high intensity. They still avoided the less-palatable pines and firs, and ate only the shrubs and grasses that grew among them. By studying the response of the shrubs to the timing, intensity, and frequency of grazing, it was found that the timing of grazing was not so important within the late spring through early fall season, but the frequency and the intensity (proportion of edible twigs removed) was. On the basis of the results, it was possible to develop prescriptions for grazing for at least three different kinds of objectives at this particular site:

1. To suppress competing shrubs as much as possible requires repeated, intense grazing during the grazing season. If started when shrubs are seedlings, this could eliminate most shrubs. This would also reduce fire hazard.
2. To maintain balanced shrub and tree growth, providing lots of wildlife and cattle forage while accepting slower tree growth, the area can be grazed heavily, removing all succulent stems, but only once or for a very limited period during the season. This would keep the shrubs low, producing a high proportion of edible and reachable forage, but would allow them to continue to grow. This would reduce fire hazard, but not as much as Prescription 1.
3. To allow shrubs to grow to their full size, light grazing, removing half or less of the succulent stems, still results in rapid growth of the shrubs. These shrubs would not produce as much succulent or reachable forage as the ones in Prescription 2, but they would produce a thick understory cover. Tree growth would be slow.

There are a wide variety of management goals for which grazing prescriptions might be developed, including controlling poison oak, reducing grass competition with oaks, encouraging flowers or native plant species, and reducing fire hazard. Some of California's native species benefit when the heavy growth of introduced annuals is opened up by grazing.

GRAZING AND CHANGE

At the landscape, community, and plant levels, livestock grazing interacts with ecological features and processes and can be manipulated to achieve particular effects. Knowledge of the dynamics of landscapes, plant communities, and plant species, as well as an understanding of herbivores and their activities, is needed to create grazing prescriptions for particular goals. Even so, unpredicted changes will occur year to year and over the long term. The effects of prescriptions or management activities need to be monitored and must be adjusted over time, to reflect the inevitable changes in the ecosystem, and in goals.

An understanding of the complex set of grazing effects in ecosystems is also important to the conservation of wild herbivores and their habitat. The grazing of wild herbivores takes place within ecosystems that have profoundly changed in the last 200 years, and will continue to change. A dispassionate look at the effects of wildlife grazing is needed, rather than operating on the blind assumption that any grazing by wild herbivores is "natural," and hence always benign in its effects on ecosystem characteristics. The factors outlined in this chapter can help sort through the likelihood of certain types of ecosystem change in response to changes in the behavior, habitat, or numbers of wild herbivores.

The activity of management can minimize conflicts and help achieve goals. This chapter does not attempt to grapple with the social complexities of setting goals for rangelands; that is left for others. The salient fact is that when rangeland changes to housing development, vineyards, or freeways, it is, for all intents and purposes, the final change. The wide spectrum of environmental goods and services that can come from an extensive range are lost, as it is reinvented for more singular purposes.

REFERENCES

Allen, B.H. and Bartolome, J.W., Cattle grazing effects on understory cover and tree growth in mixed conifer clearcuts, *Northwest Sci.,* 63, 214–20, 1989.

Burcham, L.T., *California Range Land: A Historico-Ecological Study of the Range Resources of California,* Center for Archaeological Research 7, University of California, Davis, 1981 [1957].

Dennis, A., Effects of Defoliation on Three Native Perennial Grasses in the California Annual Grassland, Ph.D. dissertation, University of California, Berkeley, 1989.

Fuhlendorf, S.D. and Engle, D.M., Restoring heterogeneity on rangelands: ecosystem management based on evolutionary grazing patterns, *BioScience,* 51, 625–632, 2001.

Hamilton, J.G., Changing perceptions of pre-European grasslands in California, *Madroño,* 44, 311–333, 1997.

Huntsinger, L., Grazing in a California silvopastoral system: effects of season, intensity, and frequency of defoliation on deerbrush, *Ceanothus integerrimus* (Hook and Arn.), *Agrofor. Syst.,* 34, 67–82, 1996.

Huntsinger, L., McClaran, M.P., Dennis, A., and Bartolome, J.W., Defoliation response and growth of *Nassella pulchra* (A. Hitchc.) Barkworth from serpentine and non-serpentine populations, *Madroño,* 43, 46–57, 1996.

McCullough, D.R., *The Tule Elk: Its History, Behavior, and Ecology,* University of California Press, Berkeley, 1971.

Perevolotsky, A. and Seligman, N.G., Role of grazing in Mediterranean rangeland ecosystem: inversion of a paradigm, *BioScience,* 48, 1007–1017, 1998.

Webster, L., Composition of native grasslands in the San Joaquin Valley, California, *Madroño,* 28, 231–241, 1981.

Westoby, M., Walker, B., and Noymeir, I., Opportunistic management for rangelands not at equilibrium, *J. Range Manage.,* 42, 266–274, 1989.

CHAPTER 105

California Grazing Lands: Wither They Go*

Albert G. Medvitz

INTRODUCTION: ISSUES IN THE MANAGEMENT OF GRAZING LAND IN CALIFORNIA

Forty years ago, L.T. Burcham provided "An Historico-Ecological Study of the Range Resource of California" (1957), in which he discussed the history of livestock production in California. In that work, he described range degradation and growing threats to the continuing economic development of livestock production in a state with a rapidly growing population. Thirty years later, the California Department of Forestry and Fire Protection (Ewing et al., 1988) published "California's Forest and Rangelands: Growing Conflicts over Changing Uses," a comprehensive assessment of the state's forests and rangelands in which it laid out not only the economic and social issues, but also the new (post–World War II) prominent concerns of environmental quality and habitat maintenance. Since then, the problems and conflicts over rangeland in California have continued, as have the plethora of research and policy documents. Examples of the conflicts include debates about:

- The appropriate strategies for addressing the continuing expansion of yellow star thistle
- The appropriateness of rules to protect watershed management
- Public protests and political actions when oak trees are removed to install vineyards in Santa Barbara
- Efforts to demand expensive environmental impact reports (EIRs) from East Bay ranchers who wish to continue to graze lands they have customarily grazed for years
- The efficacy and safety of applying urban sewage sludge to grazing land

In this conflicted environment, developing effective policies to promote the management of grazing lands in California to achieve ecosystem health is difficult. This chapter discusses three factors important to the problem of developing and implementing public policy which would lead to the management of California rangelands to achieve environmental goals. These factors are the economic conditions of livestock production in California, patterns of rangeland ownership, and rapid population growth.

---

* This is a revision of an unpublished paper presented at the 1999 Annual Meeting of the American Association for the Advancement of Science, Anaheim, CA, January 24, 1999.

DOMESTIC GRAZING ANIMALS AND THE LAND

Grazing animals* in California are a substantial economic component of the state's agriculture. In 1999, grazing meat animals — primarily beef cattle and sheep — accounted for $1.3 billion in California farm-gate sales, that is, about 4% of the state's total agricultural sector output (CDFA [California Department of Food and Agriculture], 2000). Cattle and calves alone are the state's fourth-ranked commodity in dollar terms, following milk and cream, grapes, and nursery products (CDFA, 2000). With an inventory of 2.3 million beef cattle and calves at the beginning of 2001, California ranked second nationally in beef cattle and calf inventory, and second in sheep and lamb inventory with 840,000 head (USDA [U.S. Department of Agriculture], 2001).

The most recent (1997) Census of Agriculture (USDA, 1999) more readily allows us to see patterns of beef cattle ownership. It reports that in 1997 roughly 12,000 farms in California reported owning at least one beef cow, calf, or steer, and about 3000 reported owning sheep (USDA, 1999). Relatively few operations, however, own medium to large herds or flocks of grazing animals. Only 3000, or just under 25% of California cattle owners, own more than 100 head (USDA, 1999). Only 445, or 11%, of 3014 California sheep operations reported having an inventory of 100 or more sheep or lambs (USDA, 1999). Large flocks or herds are managed by an even smaller minority of operators.

Consistent with the large number of domestic grazing animals in California are the large amounts of land devoted to them. Ownership of this land is divided between privately owned farms and ranches, and publicly owned forests, grasslands, scrubland, and desert. The distributions of ownership described here suggest that grazing animals in California occupy private lands ranging from small fenced-in plots to larger fields of irrigated and nonirrigated pastures, to stubble remaining after crop harvests, to large and extensive ranches, as well as publicly owned forests, rangelands and deserts.

Because of the diversity of California's topography and climate and the size and dispersion of livestock operations in the state, describing grazing animals and rangeland in California is complex. California is a large state with about 100 million acres (156,000 mi^2) of land within its borders. Few Californians realize that roughly half of all land in the state is owned by their governments; 44.1 million acres are owned by the federal government alone (Bureau of Land Management, 1998). Ownership patterns are shown in Table 105.1 and Figure 105.1A (Ewing et al., 1988; California Department of Forestry and Fire Protection–FRAP, 2001).

Table 105.1 California Land Ownership by Vegetation Type (in thousands of acres)

Type	Private	Public[a]	Private/Public (%)	Total
Conifer	7645	15,368	33.2	23,013
Hardwood	6785	2762	71.1	9547
Shrub	8359	10,792	43.7	19,151
Grass	8391	1166	87.8	9557
Desert	5058	16,220	23.8	21,278
Alpine barren and rock	288	1832	13.6	2120
Total	36,526	48,140	43.1	84,666
Agriculture and urban				15,000

Note: Hardwood and grassland ecosystems in California are overwhelmingly privately owned. These systems are largely used for the production of grazing animals.

[a] Lands administered by the U.S. Forest Service, Bureau of Land Management, U.S. National Park Service, other federal, state, and local agencies.

From Ewing. R.A. et al., California's forests and rangelands: growing conflict over changing uses (an assessment prepared by Forest and Rangeland Resources Assessment Program [FRRAP]), California Department of Forestry and Fire Protection, July 1988.

---

* For purposes of this chapter, consideration of grazing livestock is limited to beef cattle and sheep. Dairy cows are by and large confined, and unconfined horses, burros, goats, and other species are small in number. Inclusion of these nonbovine species would not substantially change the description or conclusions of this analysis.

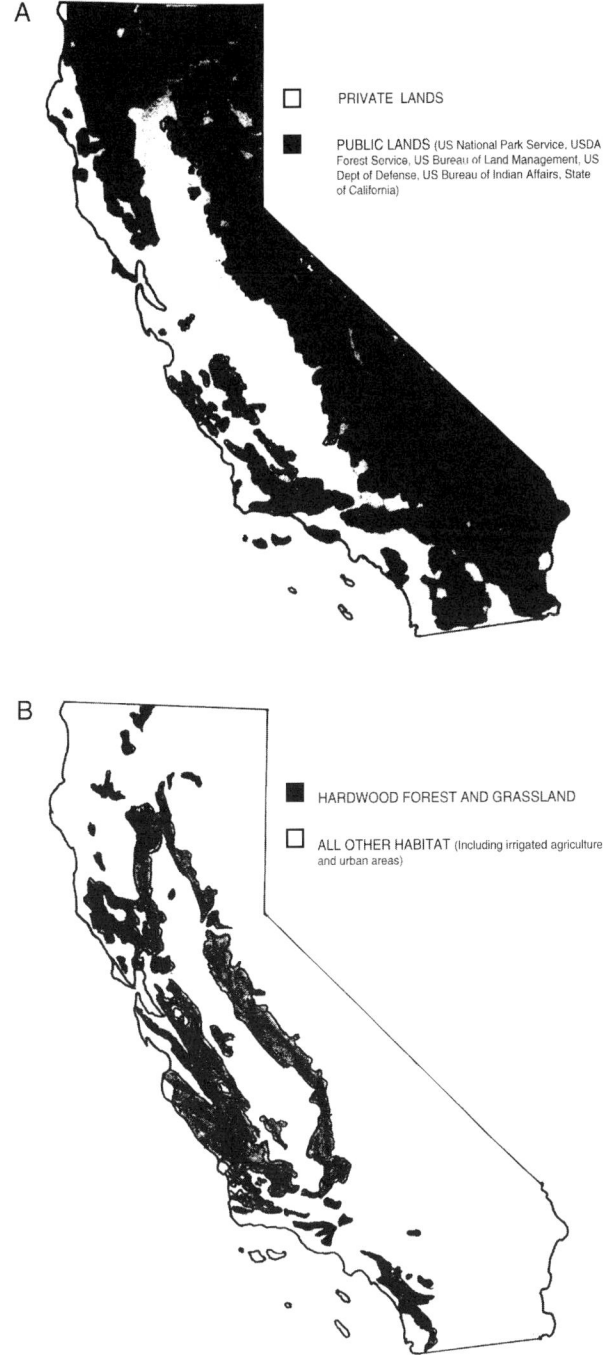

Figure 105.1 Private lands and grazing land in California. Private land in California makes up about 50% of the state. Most private land is in the agriculturally rich central and coastal valleys and in coastal plains (A). Grassland and hardwood forests (B) provide 78% of grazing forage for livestock, are largely privately owned, and are located mostly in the hilly regions of the central and coastal valleys, regions undergoing rapid population growth. (From Ewing, R.A. et al., California's forests and rangelands: growing conflict over changing uses (an assessment prepared by Forest and Rangeland Resources Assessment Program [FRRAP]), California Department of Forestry and Fire Protection, July 1988; and California Department of Forestry and Fire Protection–FRAP, California Major Land Cover Classes, map (http://frap.cdf.ca.gov/data/browsegraphic/frapwhr10_map.gif), and Government Ownership, map (http://frap.cdf.ca.gov/data/browsegraphic/govtownmap), 2001.)

As a place of great tectonic activity on the western edge of the continent and spanning roughly 10° of latitude, California contains enormous diversity in climate and habitat. Eighty-five percent of the land can be classified as forest and rangeland, and the state also can be characterized as being part of at least five major vegetation regimes plus those lands which have been transformed by agriculture and urbanization (Ewing et al., 1988). Overall, roughly 40 million acres of these lands are grazed by domestic livestock. About 18 million of these acres are private lands and 23 million acres are publicly owned. Although more public lands are used by grazing animals, the vast majority of the forage consumed by the grazing animals, 91%, is on private lands (Ewing et al., 1988). Grasslands and hardwood forests are the most prolific sources of forage, generating about 78% of all the forage consumed. About 71% of hardwood forests (6.8 out of 9.5 million acres) and 88% of grasslands (8.4 out of 9.6 million acres) are privately owned (Ewing et al., 1988) (Figure 105.1B). Thus, 13.2 million acres of hardwood forests and grassland, or nearly two thirds of these specific habitats, are under private management for animal production.

These ownership patterns are of profound significance in developing strategies for ensuring ecosystem health for California's hardwood and grassland habitat. A comparison of Figures 105.1A and B shows that California's habitats under private grazing management are proximate to areas of rapid population growth in the central and coastal valleys, as well as Southern California.

THE DYNAMICS OF RANGELAND DEVELOPMENT IN CALIFORNIA

The dynamics of the evolution of ranching and the human use of rangelands in California is intimately tied to the rapidly growing population and to a rapidly evolving, highly technological agricultural system. California's population growth over the last century and a half has been very rapid and largely urban based (Medvitz, 1999; Olmstead and Rhode, 1997; Rothstein, 1999). Figure 105.2 shows California's population growth from 1850 to the present. For 150 years of California's history, its population growth has exceeded an exponential growth rate of 3% per year. Although in the past 20 years the rate has declined to about 2% per year, it is still high by international standards, and some of the county growth rates are very high indeed, as shown in Table 105.2. There is little doubt that this growing population will continue to increase well into

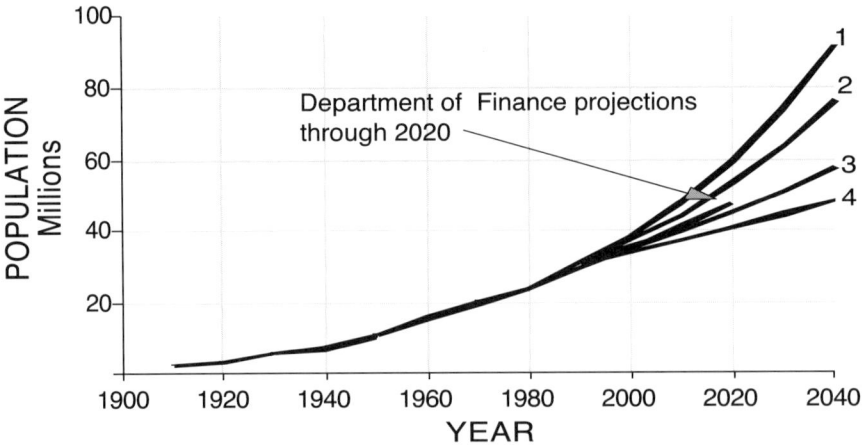

Figure 105.2 California population 1850–2000, extrapolated to 2040 under varying assumptions of future growth. Projection assumptions: (1) 2.29% per year = average from 1980 to 1990; (2) 1.8% per year as continuation of 1992 rate; (3) 1.24% as continuation of 1993–1994 rate; (4) 0.9% per year scenario with no foreign or domestic immigration and slight outmigration as a continuation of 1994–1995 rate.

Table 105.2 Five-Year Average Yearly Population Growth Rates[a] of Selected Countries Compared with California and Selected Counties in California (in bold)

	%		%
Imperial	4.4	Orange	2.0
Israel	3.8	Bangladesh	2.0
Madera	3.5	India	1.9
Peru	3.3	**San Bernadino**	1.9
Saudi Arabia	3.0	Haiti	1.8
Nigeria	2.9	Brazil	1.6
Kenya	2.8	**California**	1.4
Afghanistan	2.8	**Santa Clara**	1.3
Riverside	2.8	Argentina	1.2
Zimbabwe	2.6	China	1.1
Kern	2.3	**Los Angeles**	1.0
South Africa	2.3	United States	1.0
Fresno	2.3	Switzerland	1.0
Vietnam	2.2	France	0.5
Tulare	2.2	Japan	0.3
Ecuador	2.1	Great Britain	0.3
Stanislaus	2.0	Russia	0.1
Mexico	2.0	Italy	0.1

Note: California and its counties are growing rapidly in population. Historical population growth rates of California as a whole have exceeded those of many developing countries. Many counties in California have population growth rates that are very high in world terms. Over the past few years, California's population growth rate has varied between that of Mexico and China.

[a] California and county rates calculated for the period 1990 to 1995 are mean yearly rates. They are somewhat less than 1980 to 1990 averages and are based on Department of Finance estimates rather than U.S. census counts. Country estimates are 1990 to 1995 averages from the World Bank's *World Population Projections: 1994–95.*

the 21st century; the population will almost certainly double by 2050. Even a historically low rate of 1.09% will increase California's population to 50 million people in about 50 years.

This rapid population growth, the character of the immigrants who made it up, and the unique conditions of California's climate have generated a highly technological, large-scale agricultural system with strong links to international trade (Olmstead and Rhode, 1997; Rothstein, 1999). After initially exploiting the unfenced wildlands of the state to produce small grains (wheat and barley) and livestock, by the late 19th century, California farmers had mastered reclamation of delta lands and irrigation and had begun the continuing evolution toward increasing crop variety and the transformation of drylands to irrigated lands. This trend continues today. Figure 105.3 uses existing census bureau data to show the trends in agricultural land use during the 20th century: Figure 105.3A shows the dramatic changes in grazing land since 1950; Figure 105.3B correlates these changes to population growth. It is worthwhile noting that the changes in farmland and numbers of farms have varied over time in relation to population growth.

From the period of statehood through the 1960s, California's growing population was associated with an expansion of agricultural lands. Agriculturalists, with the aid of massive irrigation projects, continued to expand not only irrigated lands, but also grazing lands. It was not until the 1960s that the supply of farmland began to diminish in the face of continuing population pressures and a growing desire to maintain large segments of the environment in more or less wild states.

Despite a declining land base, California agriculture continued to increase production in the latter half of the 20th century. The effects of the land loss have been compensated for by enhanced technologies, new crops, and new management strategies. In terms of production, the effects of the continuing loss of land and changing management are shown in Figure 105.4. Clearly, the trends

A

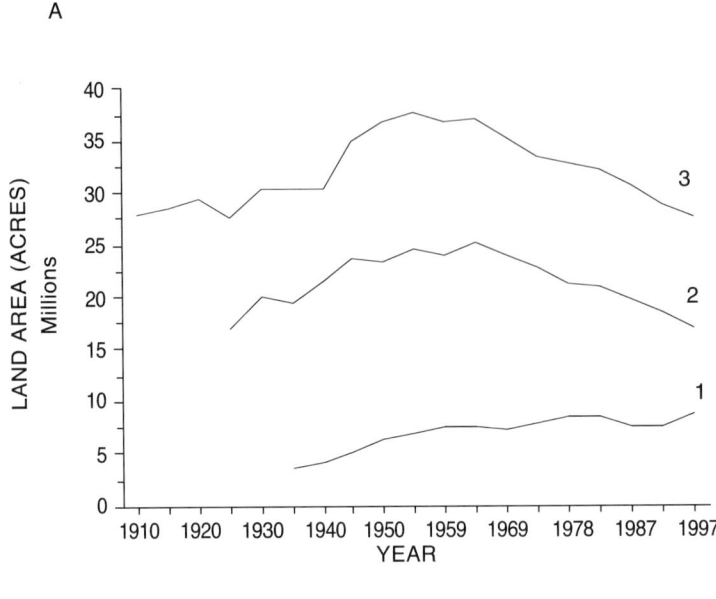

1 IRRIGATED LAND   2 GRAZING LAND   3 LAND IN FARMS

B

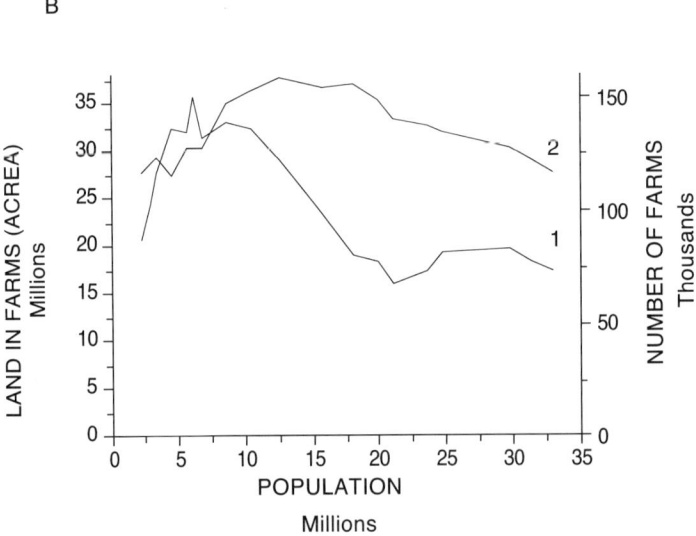

1 NUMBER OF FARMS   2 LAND IN FARMS

Figure 105.3 California farmland changes and population growth. California's population was approximately 15 million in 1960, and reached approximately 30 million in 1990. After an extended period of increasing numbers of farms and land in farms (acres) with growing population, farmland and numbers of farms have been decreasing at the expense of grazing land. (A) Changes in farmland, 1910–1997. (B) Farmland changes related to population. (From Bureau of the Census, Censuses of agriculture, Department of Commerce, Washington, D.C., 1925–1994.)

A

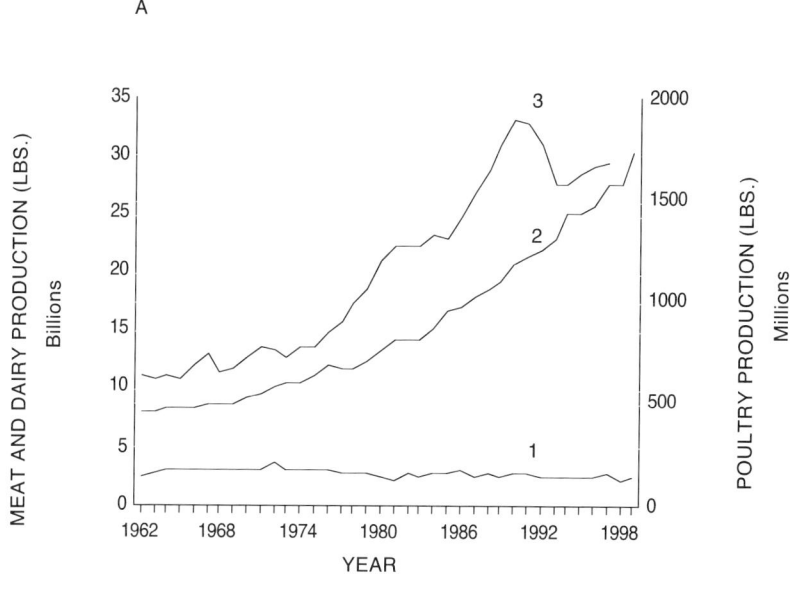

1 BEEF, SHEEP, HOGS   2 DAIRY   3 POULTRY

B

Increasing

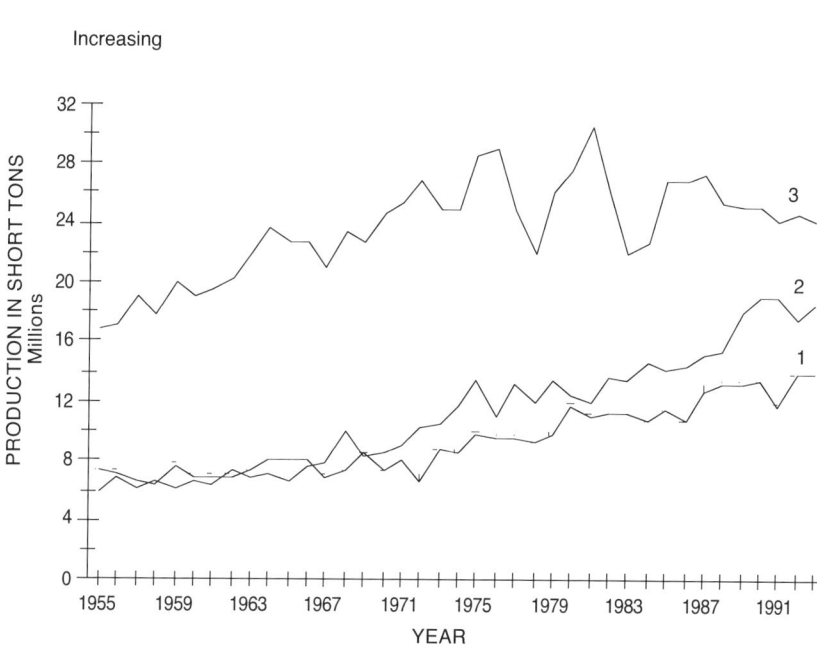

1 FRUIT AND NUTS   2 VEGETABLES AND MELONS   3 FIELD CROPS

Figure 105.4 California agricultural production trends, 1962–1999. Increasing production of intensive irrigated crops and intensive livestock production are correlated with decreased production from extensive rangeland field crops. (A) Annual total production of dairy, red meat, and poultry, 1962–1999. (B) Annual production of field crops, fruits and nuts, and vegetables, 1955–1993.

are for increased production in intensive crops and livestock (poultry and dairy) production, and steady and declining production in extensive crops and grazing livestock. Privately owned grazing lands that remain in agriculture continue to give way to orchard and tree crops, intensive livestock facilities, or to housing developments. Obviously, the trends in production are the results of many causal factors besides population pressures. The changing nature of international markets, new transportation, food processing and distribution technologies, and price volatility in livestock markets related to institutional structure of production and processing and distribution are all powerful forces in pushing change in agriculture.

A simple extrapolation of Figure 105.3A shows that if population growth continues at high rates (1 to 2%) and technology and markets continue to evolve as they have in the past, then we can expect privately owned rangelands in the state to decline by another 5 million acres, or 46% or so in the next 30 years, thus increasing pressures for environmental maintenance on remaining rangelands lands, both public and private.

A major concern in this scenario is that urban pressures along with environmental demands and industrially concentrated, highly volatile livestock markets will ultimately push livestock agriculture to operations of confined animals nourished by feed produced through intensive irrigation. It would also speed the conversion of important grasslands and forestlands to urban or suburban development. It is difficult to imagine this scenario as contributing to the maintenance of the health of California's rangeland ecosystems.

RECONCEPTUALIZING THE FARM AND RANCH

The traditional social contract of the farmer in Western society has been that society allows ownership of land and associated environmental resources for the farmer or rancher to manage. The farmer or rancher then purchases inputs, provides labor, and acquires knowledge and experience to produce tangible products of food and fiber, which are sold in open markets. Some ranchers also sell entertainment by providing "dude" experiences. The contract allows the farmer to keep the market earnings and to pay for inputs and social costs and services, taxes and rent, and to support herself or himself. There is also a strong tradition that owners of farmland may sell it for nonfarm purposes, for whatever reason they choose. The currency and operation of commodity and land markets under this social contract have come to be widely understood, even to the extent that when they operate unfairly, as in the case with concentrated markets, both farm and nonfarm populations can agree to intervene in markets to make them fair.

In recent years, a new and very undefined market of California rangeland values has developed: the market for environmental benefits. The new market has arisen from a new set of social values demanding the designation and protection of special wild species,* the maintenance and protection of clean air and water, and the preservation of aesthetically pleasing landscape. Not only are good husbandry and sustainable production desired outcomes of the new social contract with farmers, but also detailed and specific environmental outcomes. But the transaction system for this market is not clearly defined: methods of pricing, of conveying value to producers, or, indeed, even of identifying production are not all precisely known. In the absence of a clearly defined market mechanisms, other than expensive fee purchase (and more recently experiments in conservation easements) on land, society has developed a set of regulatory burdens on farmers to obtain newly defined environmental value from farmlands. These burdens increase social benefits but at increased costs of operation for the farmer. The regulatory system does not allow farmers to benefit from the newly defined goods. Because the nature of agricultural commodities and the institutional structure of agricultural livestock markets prohibits passing additional costs on to consumers, some of these new environmental market interventions can dramatically reduce the asset value of rangeland. For

---

* These species include not only endangered species, but other species that, for whatever reason, have acquired political champions: wild horses and burros, predators such as coyotes, feral cats, etc.

example, by defining a parcel rangeland as habitat for an endangered species, the parcel can lose both urban development and agricultural value, depriving operators of wealth and hence potential loan collateral with which to obtain working capital. In any case, the regulatory burden often dramatically increases management and labor costs in addition to opportunity costs. Such burdens push livestock operators and landowners to change agricultural land use to higher-value crops or to convert land for residential or recreational purposes. In so doing, they undermine the integrity of the grazing land ecosystem.

This analysis suggests that adequate protection of ecosystem health of California's grasslands and hardwood forests demands more than continued learning of the technical management of grazing animals to achieve environmental health. It demands deep study and understanding of the social history and the economics of the interactions of commodity and land markets in the context of a rapidly growing population. With this understanding and a political will to proceed, explicit market structures need to be developed to reframe ranching from an enterprise of simple commodity production to one of environmental management where income from environmental outcomes can be made part of the income stream of livestock agriculture.

REFERENCES

Burcham, L.T., California Rangeland: An Historico-Ecological Study of the Range Resource of California, Center for Archaeological Research at Davis, Publication 7, Davis, CA, 1957 [1982].

Bureau of the Census, Censuses of agriculture, Department of Commerce, Washington, D.C., 1925 to 1994.

Bureau of Land Management, Department of the Interior, *Public Land Statistics 1997*, Vol. 182, BLM/BC/ST-98/001+1165, March 1998, http://www.blm.gov/natacq/pls97/, Table 1–3.

California Department of Forestry and Fire Protection–FRAP, California Major Land Cover Classes, map (http://frap.cdf.ca.gov/data/browsegraphic/frapwhr10_map.gif), and Government Ownership, map (http://frap.cdf.ca.gov/data/browsegraphic/govtownmap), 2001.

CDFA (California Department of Food and Agriculture), *California Agricultural Resource Directory* 2000, CDFA, Sacramento, 2001.

Ewing. R.A. et al., California's forests and rangelands: growing conflict over changing uses (an assessment prepared by Forest and Rangeland Resources Assessment Program [FRRAP]), California Department of Forestry and Fire Protection, July 1988.

Medvitz, A.G., Population growth and its impacts on agricultural land in California: 1850 to 1998, in *California Farmland and Urban Pressures: Statewide and Regional Perspectives,* Medvitz, A.G., Sokolow, A.D., and Lemp, C., Eds., Agricultural Issues Center, University of California, Davis, 1999, pp. 11–32.

Olmstead, A.L. and P.W. Rhode, An overview of the history of California agriculture, in *California Agriculture: Issues and Challenges,* University of California, Giannini Foundation, Division of Agriculture and Natural Resources, August 1997.

Rothstein, M., California agriculture over time, in *California Farmland and Urban Pressures: Statewide and Regional Perspectives,* Medvitz, A.G., Sokolow, A.D., and Lemp, C., Eds., Agricultural Issues Center, University of California, Davis, 1999, pp. 33–50.

USDA, The 1997 census of agriculture: California state and county data, Vol. I, Geographic Area Series Part 5, USDA, Washington, D.C., 1999.

USDA, National Agricultural Statistics Service, Agricultural Statistics: 2001, http://www.usda.gov/nass/pubs/agr98/98_ch7.pdf, 2001.

CHAPTER 106

Protecting the Biodiversity of Grasslands Grazed by Livestock in California

Richard J. Reiner

INTRODUCTION

California annual grasslands are found in wide-open expanses, as well as interspersed in many of California's vegetation types. They are best represented in portions of the Great Central Valley, the Sierra and Cascade foothills, the Coast Range, and the Transverse Range. Annual grasslands represent a huge portion of the undeveloped California landscape, supplying tens of millions of California residents with open space, clean water, recreation, beef, and wool. They are also important contributors to California's biodiversity, providing an ecosystem composed of plants and animals adapted to frequent fires, summer drought, little shade, and intense herbivory. Kuchler's (1964) map of the potential natural vegetation of the U.S. shows 5.35 million ha of grassland in the Central Valley and surrounding foothill ranges, and an additional 3.87 million ha with an oak overstory.

Early references to the botanical composition of California grasslands are sketchy at best. The first accounts simply describe it as excellent pasture (Heady, 1988). Most scientists concur, however, that a large percentage of the original upland valley grassland was composed of perennial grass species, especially purple needlegrass (*Nassella pulchra*). Early writings also indicate that the original grasslands were rich in native wildflowers that most likely occurred in the spaces between bunchgrasses. In April 1868, John Muir wrote of the Santa Clara Valley, "the hills were so covered with flowers that they seem to be painted" (Adams and Muir, 1948). On the same trip, Muir referred to the Central Valley as a "garden of yellow Compositae" (Heady, 1988).

Grasslands provide society many benefits. The most obvious products, such as meat, are easily valued in familiar currency. Other products, such as clean water, open space, and biodiversity, are more difficult to value monetarily. Yet it is these environmental services that are becoming important in the discussion regarding the future of our remaining grasslands.

THREATS TO GRASSLANDS

To date, the greatest threat to grasslands is outright conversion to other land uses. Nearly all of the grasslands in the Central Valley were lost to cultivation near the turn of the 20th century and, as of 1987, only 1% of the valley grassland remained. The current large expanses of grassland almost exclusively reside in large private ranches in foothill regions and many of these ranches are now in the process of being subdivided. The reasons for the demise of these ranches include housing development, poor returns in the cattle market, and subsequent conversion to other land uses such as vineyards. Some ranches are being broken up because California tax laws make intergenerational transfer prohibitively expensive.

The spread of nonnative species also threatens the diversity of California grasslands as the abundance of these weeds has increased dramatically. The grasslands observed today are only a shadow of their former selves, having been invaded by annual grasses from Europe and other Mediterranean climates. Some of these species, such as wild oats and bromes, are so well established in today's grassland that they are considered naturalized. With careful management, the percentage of these species can be decreased in favor of natives, but it is unlikely that they can ever be eradicated. Grasslands with the highest native species composition are now found in areas where annual grasses are least successful, such as vernal pools, serpentine soils, and rocky landscapes.

Most troublesome to biodiversity are newer weed arrivals such as star thistle (*Centaurea solstitialis*), medusa head (*Taeniatherum caputmedusae*), and goatgrass (*Aegilops triuncialis*). These species threaten to eliminate the remaining native species and, because they are not desirable forage, livestock production is decreased. It is critical that we begin to make significant progress on stabilizing grassland weeds. If current trends continue, it is entirely possible that conservationists could expend huge sums of money protecting ranches from development and still lose their biodiversity to introduced weeds.

BIODIVERSITY AND LIVESTOCK GRAZING

After hearing such a bleak view of the condition of California grasslands, many people are surprised that the remaining grasslands are so extensive and biologically diverse. To be inspired, one need only to drive through the coastal and Sierran foothills during a wet spring to view the spectacular wildflower displays. In addition to wildflowers, there are many animals that are specialized for life in grasslands. Some of those animals, such as black-tailed hare (*Lepus californicus*), Beechey ground squirrels (*Otospermophilus beecheyi*), and burrowing owls (*Athene cunicularia*) are visible to the casual viewer. Many others, such as San Joaquin kit foxes (*Vulpes macrotis*) and California tiger salamanders (*Ambystoma tigrinum* subsp. *califoriense*), are best viewed with a spotlight at night.

Research from grassland sites as varied as South Africa and the American Midwest indicate that grasslands are dynamic systems influenced by natural disturbance regimes (Collins and Barber, 1985; McNaughton, 1994). Among the best-understood disturbance regimes affecting grasslands are fire and grazing. Both of these natural disturbances are patchy, reduce the standing biomass of vegetation, and occur on varying scales. Each temporarily affects the balance between organisms competing for space, light, water, and nutrients. Research in mixed-grass prairie (Collins and Barber, 1985) shows that if grasslands are subjected to moderate levels of multiple disturbances, they contain greater biodiversity (Figure 106.1). In other words, grasslands that are protected from fire and grazing eventually lose species.

Examples of grasslands being degraded by the removal of grazing and fire are numerous both in the literature and in practice. In a study of a rare population of the Sonoma spine flower (*Chorizananthe valida*), Davis and Sherman (1992) found that the population dramatically declined when protected from livestock. On the other hand, it appears clear that grasslands that are periodically grazed and burned have fewer weed species and greater abundance of native species (Kan, 1998).

The fact that properly managed livestock grazing and biodiversity can be mutually compatible is significant. Livestock grazing not only becomes another tool that can be used against the threat of weed species, it adds an important strategy for conserving large grassland landscapes.

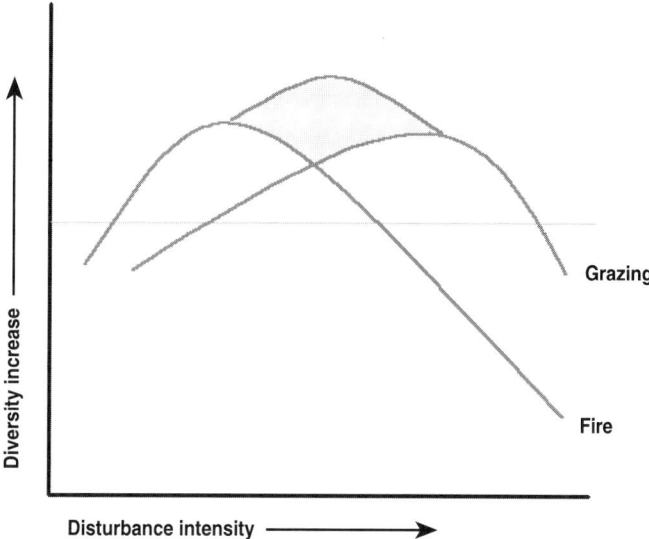

Figure 106.1 A generalized model showing the relationship between disturbance and diversity in grasslands. The area that is shaded represents the increase in diversity when both grazing and fire are represented on the landscape.

THE JEPSON PRAIRIE EXPERIENCE

Jepson Prairie, located 12 miles south of Dixon, CA, is an outstanding example of Central Valley grassland. It is an especially diverse landscape with sizable remnants of purple needlegrass (*Nassella pulchra*) grassland and high-quality vernal pool communities. At last count, the preserve had 212 species of plants belonging to 45 families. The preserve is habitat for many extremely rare plants and is one of only two known locations for Solano grass (*Tuctoria mucronata*). It is also home to many rare and interesting animals, such as the California tiger salamander, and the only known population of delta green ground beetle (*Elaphrus viridis*). The Nature Conservancy (TNC) recognized the importance of Jepson Prairie and purchased it in December 1980.

Jepson Prairie had been grazed by livestock since before the turn of the 20th century and the question of how to manage it has been controversial from the time of its initial purchase by TNC. The Hamilton family had operated a sheep operation there for over 100 years and when TNC contemplated removing sheep, the family quickly pointed out that they saw no need for change since the native species we were most interested in protecting still existed under their long-term livestock management. Yet, some people in the environmentalist community argued that TNC was advocating livestock grazing because of political pressure from the locals and a desire for rent income. Tempers flared, emotional letters were written, and the agricultural and conservation communities were further distanced.

TNC temporarily eased the friction caused by livestock on the prairie by excluding a portion of the preserve from sheep. This area, known as the Docent Pasture, became the focus for public visitation and included a self-guided nature trail. TNC also began to engage the University of California in grazing research, and in 1987 funded a study to investigate further the effect of grazing and fire on the prairie grasslands. Over a 5-year period, it became clear to all involved that without fire and grazing the Docent Pasture was being invaded by weeds, and that biodiversity was indeed decreasing. The nature trail became infested with yellow star thistle, prickly lettuce (*Lactuca serriola*), and medusa head. The beautiful spring wildflower displays that still occurred on the sheep-grazed pastures were no longer found on the "protected" land.

Research at the University of California, Davis, lead by John Menke and later by Kevin Rice and their students, began to shed light on the mechanisms that caused a decline in biodiversity in the Docent Pasture. A major factor was competition with the introduced annual grasses and especially the effect of the buildup of annual grass mulch. Without grazing and fire, the previous years' grass production was left matted on the ground, leaving the native California species in an environment where germination conditions were poor. Even worse, weed species were encouraged by these conditions.

Weeds quickly became the common focus of both biodiversity and livestock interests at the prairie, and therefore formed the foundation of a management plan. The plan was developed in a team atmosphere with input from a broad group of stakeholders including TNC; the University of California, Davis; Solano County Farmlands and Open Space Foundation; Hamilton Brothers Livestock; U.S. Forestry Service Pacific Southwest Research Station; California Department of Fish and Game; and the Jepson Prairie Docents. Such a broad base of involvement led to a very holistic set of planning goals and a greater appreciation from all involved as to the complexities of managing the prairie.

The plan focuses on biodiversity objectives while recognizing the need to keep the sheep operation as profitable as possible. It lays out a clear set of objectives, a grazing and burning schedule for each pasture, and a monitoring protocol. The plan recognizes the need for flexible grazing schedules as weather patterns change and different weeds wax and wane. One of the first actions under the plan was to reduce grazing to increase the fuel load in a pasture infested by medusa head, and then to burn the pasture in the early spring. Oren Pollak and Tamara Kahn of TNC studied the effect of the fire and found a 100% reduction of medusa head in the first year. By the third year, the weed had regained some ground, but was still reduced to below the pre-burn levels (Kan, 1998).

One point that should not be overlooked from our experience managing Jepson Prairie is the importance of working with local knowledge. From the Hamiltons we learned management tricks it would have taken years to learn on our own; specifically their local knowledge regarding the diets and behavior of sheep proved invaluable. They knew from years of experience at the prairie which season sheep would select specific weeds. Rip-gut brome (*Bromus diandrus*), for example, would have to be targeted early in the year because sheep rarely eat it after its three-leaf stage. The Hamiltons also understood sheep behavior. When we became concerned that sheep would impact rare plants in vernal pools, we learned that sheep really do not like to get their feet wet. They would avoid vernal pools if we waited until the pools were full of water before putting the sheep out.

A workable plan also needed to consider sheep–public interactions. Having sheep in the pastures during high visitation periods proved to be problematic. We had weathered several wildflower seasons of fiery letters, condemning both the sheep and TNC's management of the preserve. A solution was found by considering the need to concentrate and rotate livestock during the spring season. Working with the Hamiltons, we developed a spring schedule that assigned wildflower-viewing pastures to be grazed from Monday through Thursday. Sheep were then concentrated on a back pasture during the high visitation weekends.

The Jepson Prairie experience convinced TNC that working closely with livestock operators could be to the advantage of both. Our view of the conscientious livestock producer thus changed from that of resource consumer to resource steward.

LANDSCAPE-SCALE CONSERVATION STRATEGIES ON GRAZED GRASSLANDS

TNC recently completed an analysis that indicates that the remaining grasslands in California's coast ranges, Central Valley, and Sierra foothill regions are mostly on large, privately owned ranches. It is becoming increasingly clear that saving grassland biodiversity will require protecting large private ranches from being broken up and sold off to subdivisions or other land uses. The need to work on a very large scale is especially evident when one considers the conservation of large native mammals such as elk and mountain lions.

TNC has developed a systematic approach to identifying and protecting threatened natural communities. In the case of grasslands, the first step is a recognizance-level study that identifies portfolios of examples of each native grassland community that, if protected, could preserve the natural community into the future. At least three examples of sufficient size of each community type are selected for the portfolio. Once the properties are identified, a team of scientists and planners identify the major ecological systems, land tenure, and connectivity to other large protected areas. They then determine a minimum project size to assure the protection, maintenance, or restoration of large-scale ecological processes such as fire and grazing.

After identifying the threats to a portfolio site, the team attempts to isolate the sources of those threats. Most threats have more than one source. For example, if the threat is subdivision, one source might be due to a landowner's financial debt with no ability to reap income except from the highly volatile and cyclic livestock market. A second threat might be related to intergenerational transfer taxes. An effective conservation solution must then be a strategy that helps landowners reduce debt and one that allows transfer of the land to their children.

One of the most promising conservation tools being used today by TNC and others is the purchase of conservation real-estate easements from private landowners. The purchase of easements enables a conservation group to identify, value, and purchase certain rights to a property. These rights are permanently removed from the property deed. For example, the right to subdivide a property or the right to convert the property to intensive cropping can be stripped from a property deed. Conservation easements are promising in that they can be written to address specific threats unique to a property, while allowing the owner to continue income-producing activities such as ranching and hunting.

CONCLUSION

It is evident that the long-term interests of organizations concerned with protecting biodiversity and those making a living raising livestock are not mutually exclusive. The Nature Conservancy has successfully owned and held conservation easements on livestock-grazed grasslands for nearly 20 years. In 1999, TNC purchased conservation easements on over 100 mi^2 of grazed grasslands.

TNC is working to help new land trusts become established. In 1997, TNC donated Jepson Prairie Preserve to the Solano County Farmlands and Open Space Foundation along with a stewardship endowment to fund ongoing management. A promising new organization is California Cattleman's Association Rangeland Trust. Ranchers are realizing that the purchase of conservation easements is a way of placing monetary value on the environmental services their properties provide to society. Perhaps most important, the recognition of the value of these environmental services can form the catalyst for environmentalists and ranchers to creatively work together to protect large ranching landscapes in perpetuity. Long live the great wide open.

ACKNOWLEDGMENTS

I would like to thank Kevin Rice, John Menke, and their students, Oren Pollak, Tamara Kan, Mark Homrighausen, Robin Cox, Ann Dennis, and Pam Muik for their input and ideas.

REFERENCES

Adams, A. and Muir, J., *Yosemite and the Sierra Nevada,* Houghton Mifflin, Boston, 1948, 130 pp.
Collins, L.C. and Barber, S.C., Effect of disturbance on diversity in mixed-grass prairie, *Vegetatio,* 64, 87–94, 1985.
Davis, L.H. and Sherman, R.J., Ecological study of the rare *Chorizizanthe valida* (Polygonaceae) at Point Reyes National Seashore, California, *Madroño,* 39, 271–280, 1992.

Heady, H.F., Valley grassland, in *Terrestrial Vegetation of California,* Barbour, M.G. and Major, J., Eds., California Native Plant Society, Sacramento, 1988, pp. 491–514.

Kan, T., The Nature Conservancy's approach to weed control, California Native Plant Society, *Fremontia,* 26, 44–48, 1998.

Kuchler, A.W., Potential natural vegetation of the conterminous United States, American Geographical Society, Special Publication 36, New York, 1964.

McNaughton, S.J., Biodiversity and function of grazing ecosystems, in *Biodiversity and Ecosystem Function,* Schulze, E.D. and Mooney, H.A., Eds., Springer-Verlag, New York, 1994, pp. 361–405.

PART III

Case Studies

SECTION III.1

The Colorado River Delta Ecosystem, U.S.–Mexico

Overview: The Colorado River Delta Ecosystem: Ecological Issues at the U.S.–Mexico Border

Daniel W. Anderson

The continuing loss of natural habitats to human settlement is the principal cause of declining biodiversity in North America and the largest impediment to conservation. Consequently, the key to effective conservation and long-term maintenance of biodiversity is habitat conservation and restoration, but in a context of sustainability and wise use (Bildstein et al., 1991; Brady, 1988; Howe, 1987; Zedler and Powell, 1993; and others). This means that many habitats, although forever altered from their natural states, are still capable of supporting rich natural biodiversity. Wetlands, because of high primary and secondary productivity, have long been attractive sites for human settlement or intense human activities, and consequently are some of the most disturbed natural habitats in North America, and the world. Howe (1987) and Dahl (1990) documented the extensive wetland losses in North America. In Mexico, habitat changes are occurring rapidly, and many of Mexico's coastal wetlands are now degraded and severely threatened by human activities. The Río Colorado delta region (CRDR), which spans the U.S.–Mexico border, was noted by early biological explorers to have the greatest biological diversity and natural productivity in the southwestern United States and northwestern Mexico (Sykes, 1937). But today, the region may be one of the most ecologically degraded in North America.

This section contains chapters based on research papers identifying some of the current problems threatening the ecological health of this region, as well as potential conflicts and problems likely to emerge as enhancement occurs or expands.

Several interesting contrasts emerge from these chapters. First, the CRDR is extremely changed and ecologically degraded from its original condition; but in contrast to this, the region still represents critical and important habitat for fish and wildlife. Second, many ecological threats (past and present) exist in the CRDR; some of these seem sufficiently immense and complex to discourage any kind of response. Generations of husbandry and environmentally oriented management will be required to rectify problems, and no simple solutions are apparent. But in contrast to this discouraging scenario, abundant opportunity remains in the CRDR for sound and sustainable natural resource management and restoration.

But perhaps most apparent in analysis of this ecosystem and its threats is the physical–biological connectedness of various parts of the CRDR. Because many of the problems in the CRDR ecosystem are entwined in international political policies and practices, the management and restoration of the region must be considered in an international context, and the CRDR must be considered as a singular whole in future natural resource management activities. Nowhere else is a need for close international collaboration and coordination for successful future improvements more apparent than in the Colorado Delta ecosystem.

ACKNOWLEDGMENTS

I acknowledge the following colleagues for providing critical manuscript review: S. Alvarez-Borrego, C.J. Bahre, D.E. Blockstein, W.M. Boyce, M. Friend, E.P. Glenn, O. Hinojosa-Huerta, S.N. Hurlbert, J.O. Keith, M.A. Mora, R.D. Ohmart, and T.E. Rocke. The University of California Institute for Mexico and the United States, Riverside, CA, was the unifying and driving force in bringing people and organizations together; the institute also kindly provided funds supporting this series of research papers. We especially thank Kathryn Vincent, Juan-Vicente Palerm, and all the others at UC MEXUS for continuing support and encouragement.

REFERENCES

Brady, N.C., International development and the protection of biological diversity, in *Biodiversity,* Wilson, E.O., Ed., National Academy of Sciences, Washington, D.C., 1988, pp. 409–418.

Bildstein, K.L., Bancroft, G.T., Dugan, P.J., Gordon, D.H., Erwin, R.M., Nol, E., Payne, L.X., and Senner, S.E., Approaches to the conservation of coastal wetlands in the Western Hemisphere, *Wilson Bull.,* 103, 218–254, 1991.

Dahl, T.T., Wetlands of the United States 1780s to 1980s, Unpublished report, U.S. Fish and Wildlife Service, Washington, D.C., 1990.

Howe, M.A., Wetlands and waterbird conservation, *Am. Birds,* 41, 204–209, 1987.

Sykes, G., *The Colorado Delta,* Lord Baltimore Press, Washington, D.C., 1937.

Zedler, J.B. and Powell, A.N., Managing coastal wetlands: complexities, compromises, and concerns, *Oceanus,* 36, 19–28, 1993.

CHAPTER 108

Physical and Biological Linkages between the Upper and Lower Colorado Delta

Saúl Alvarez-Borrego

INTRODUCTION

In 1539, Hernán Cortés sent Francisco De Ulloa with three ships to explore north from Acapulco. De Ulloa reported that the island of Calafia (Baja California) was in reality a peninsula. De Ulloa circumnavigated the Gulf of California and named it *Mar Bermejo*, the Vermilion Sea, because of the reddish color of the muddy waters of the Colorado Estuary (León-Portilla, 1972; van Andel and Shor, 1964). The Gulf of California is very fertile, mainly due to upwelling and tidal mixing (Alvarez-Borrego et al., 1978).

The only river in the northern Gulf of California is the Colorado. The record of flow of Colorado River water across the Mexico–U.S. border shows that before 1935, when the filling of the Hoover Dam (Lake Mead) began, the large amounts of fresh water that were discharged into the Upper Gulf had a seasonal modulation with peak discharges in June. After construction of the Glen Canyon Dam, regular input of Colorado River water to the Upper Gulf stopped. During 1979–1987, water releases became necessary due to abnormal snowmelts in the upper basin of the river (Lavín and Sánchez, 1999). Water releases also occurred in 1993 and 1997–2002. Due to the construction of dams, the delta is a negative estuary (Alvarez-Borrego et al., 1973), with the exception of years with very high precipitation or abnormal snowmelts in the upper river basin. It behaves as a very fertile coastal lagoon (Hernández-Ayón et al., 1993).

The modern investigation of the gulf began with the E.W. Scripps cruise in 1939, which made a series of 53 hydrographic stations throughout the gulf (Sverdrup, 1941). In March 1939, salinity of the northernmost hydrographic station, some 70 km from the Colorado River mouth, was 35 ppt. Thus, only a very weak dilution effect of the Colorado River fresh water was evident in the Upper Gulf during the period Lake Mead was being filled (February 1935 to July 1941).

The Upper Gulf of California is the shallow, northernmost part of the gulf (Figure 108.1). It has unique oceanographic characteristics. Maximum registered spring tidal range at San Felipe is 6.95 m (Gutiérrez and González, 1989), with even larger amplitudes at the entrance to the delta. Depth is less than 30 m, with shallower waters at the Baja California side than at the Sonora side. Tidal current speeds exceed 1 m s^{-1} (Alvarez-Sánchez et al., 1993). It has a vertically mixed coastal regime due to tidal stirring. Tidal energy dissipation rates are very high, more than 0.5 W m^{-2} (Argote et al., 1995).

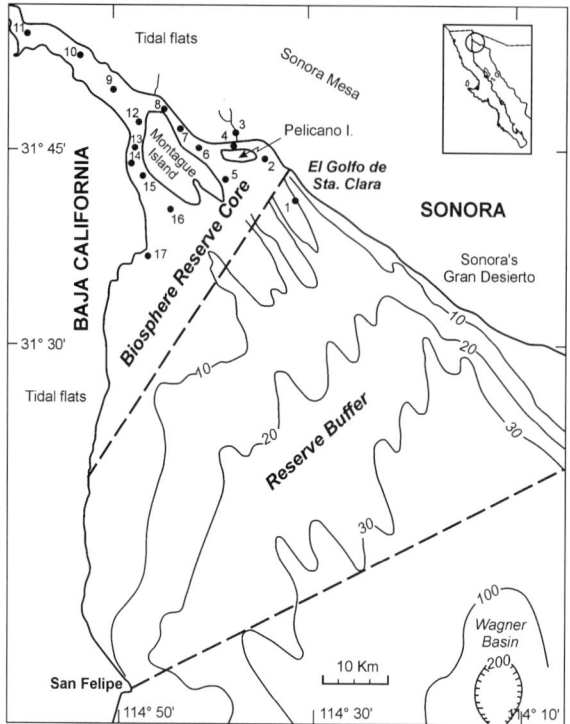

Figure 108.1 The Upper Gulf of California and Colorado River delta. Circles are the sampling locations of Hernández-Ayón et al. (1993). Bathymetry in m.

There are no towns immediately at the shores of the Colorado River estuary, and there is only a small city and a small town at the shores of the Upper Gulf. San Felipe is on the Baja California side, with a population of ~15,000; and El Golfo de Santa Clara is on the Sonora side, with a population of ~600. The economy of both towns is sustained with fishing and recreational activities. The Upper Gulf and the delta comprise an area of reproduction and nursery for many fish species, some of which are considered in danger of extinction, such as the endemic totoaba (*Totoaba macdonaldi*) (Cisneros-Mata et al., 1995). Marine mammals and some commercial shrimp species are also abundant. This area is part of the habitat of the endangered small dolphin, *Phocoena sinus,* commonly known as *la vaquita,* another endemic species. Due to this, the Mexican government decreed this area as a Reserve of the Biosphere in June 1993 (Reserva de la Biósfera del Alto Golfo de California y Delta del Río Colorado). Part of the Reserve of the Biosphere is comprised of important delta wetlands supported by agricultural drainage water, with Cienega de Santa Clara in the core zone. Cohen et al. (Chapter 112, this volume) present a description of these wetlands and some restoration alternatives. The present chapter is mostly concerned with the estuary and the Upper Gulf.

HYDROLOGY OF THE ESTUARY

Development and evolution of the Colorado River delta has been controlled by the interaction of two opposing geologic processes: the Colorado River with a sediment supply calculated to have been on the order of 160×10^6 ton year^{-1}, and a tidal regime considered to be among the largest in the world. The interplay of these constructive and destructive processes has yielded a delta dominated by tidal processes (Carriquiry and Sánchez, 1999). Human intervention has led to almost total elimination of water and sediment discharge into the delta (Meckel, 1975).

In the absence of new sediment supply from the river, the delta has become subject to destructive processes such as strong tidal currents and wind waves. A suspended sediment budget at the mouth of the Colorado River reveals that the system is now exporting sediments to the northern Gulf of California at rates similar to those when the river was active. The exported sediments are derived from the delta itself, which is eroding. The entire northern Gulf is the "Colorado River sedimentary province" (Carriquiry and Sánchez, 1999; Cupul, 1994; van Andel and Shor, 1964).

The fluvial channel of the Colorado River rapidly widens, forming a 50-km-long estuarine basin. It is 16 km wide at the mouth and constricts to 2 to 8 km wide for most of its length. The semidiurnal tidal regime produces tidal currents with maximal velocities of 3 m s^{-1} inside the estuarine basin and 1.5 m s^{-1} in the shallow marine platform adjacent to the delta (Cupul, 1994). Although the Colorado River is responsible for having excavated one of the deepest canyons of the world, it is now tamed. The construction of the major dams in the Colorado River, the Hoover Dam and the Glen Canyon Dam, had the most drastic effect upon the amount and timing of the fresh water that reached the Gulf of California. In spite of the great ecological impact caused by the construction of dams, life is abundant in the inverse-estuary. Due to sediment resuspension, the water in the estuary looks brownish. Microalgae are abundant in the estuary, possibly because of turbulence and low tides that allow enough light for a healthy growth. Crustaceans and mollusks are abundant there. Dolphins are seen often inside the estuary.

Hernández-Ayón et al. (1993) reported salinities and nutrients measured in the estuary during 1989–1990 (sampling locations are shown in Figure 108.1). During winter, salinity increased from a minimum of about 36 outside the estuary to about 39 off the northern end of Montague Island, and then decreased to about 34 at the internal extreme of the inverse-estuary. During summer, salinity increased from about 38 outside the estuary to about 40 off the northern end of Montague Island, and then decreased to about 39 at the northernmost few kilometers of the inverse-estuary.

Nutrient concentrations are high in the estuary waters throughout the year, with no clear seasonal pattern and lower values toward the oceanic region. Maximum nutrient values were 15, 53, 12, and 92 μM, for NO_2, NO_3, PO_4, and SiO_2, respectively. Most values were under 2, 40, 5, and 60 μM, respectively. NO_3 vs. salinity diagrams often showed a "U" shape with both variables increasing from the oceanic locations to those near the northern end of Montague Island; then salinity decreased while NO_3 increased toward the internal extreme (Figures 108.1 and 108.2) (Hernández-Ayón et al., 1993). Agricultural drainage water is carried from Mexicali Valley to the estuary through the Hardy and Colorado Rivers (Glenn et al., 1999). Relatively low salinity and high NO_3 concentrations at the internal extreme of the inverse-estuary are an effect of this flux.

HYDROGRAPHIC CONDITIONS IN THE UPPER GULF

The northern half of the gulf is dry and desertlike. The number of rainy days per year is about 5 from the central Baja California coast to the Upper Gulf. The estimated mean evaporation rate for the northern Gulf is 1.1 m year^{-1}, while the yearly mean precipitation is almost nil (Lavín et al., 1998). The bathymetry of the Upper Gulf of California is irregular with channels and shoals oriented northwest–southeast. Close to Montague Island, navigation is difficult for boats with draft greater than 1 m, especially for those not familiar with the area and its tidal currents. Montague and Pelícano Islands are low and flat and covered by water at high spring tides. Large areas of lowlands are flooded periodically with the tides, on both the western and eastern shores.

The surface horizontal temperature gradient reverses at the beginning of spring and fall due to the annual cycle of atmospheric temperature. Surface temperature increases from the southeast to the northwest in summer, while the opposite occurs in winter (Figure 108.3a,b). Minimum and maximum surface temperatures have been recorded west of Montague Island: 8.25°C in

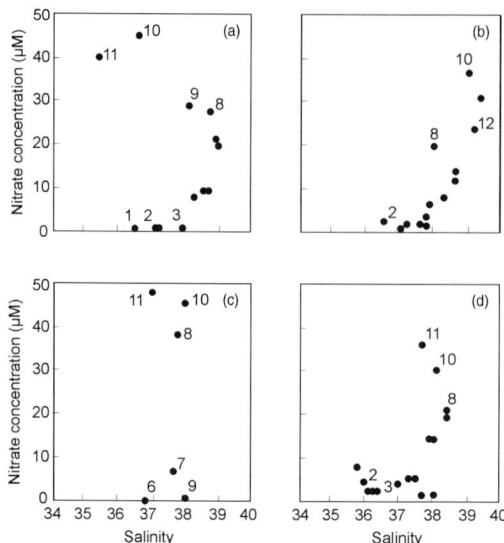

Figure 108.2 Nitrate vs. salinity for (a) June, (b) August, (c) November, and (d) December 1989. The numbers next to the symbols correspond to the locations shown in Figure 108.1. (From Hernández-Ayón, M. et al., Nutrient concentrations are high in the turbid waters of the Colorado River delta, *Estuarine Coastal Shelf Sci.*, 37, 593–602, 1993. With permission.)

December and 32.58°C in August. In the deeper waters of the southeast region, the seasonal range is 17 to 30.75°C. A monthly hydrographic sampling carried during 1972–1973 (years of no Colorado River water release) showed that salinity in general maintained a surface gradient with values increasing northwestward (Figure 108.3c,d). It ranged from a minimum of 35.28 in October to a maximum of more than 38.50 in July (Alvarez-Borrego et al., 1973). With data obtained between December 1993 and June 1996, this hydrographic behavior has been shown to persist essentially the same at present (Lavín et al., 1998), when there is no Colorado River freshwater release. Lavín et al. (1998) reported that density maintains the same horizontal gradient throughout the year with values increasing from the southeast to the northwest. In spite of higher temperatures during summer off Baja California, density is higher there than off Sonora due to the higher salinity values. Density is highest in winter and lowest in summer. Surface dissolved oxygen varies from a minimum of 1.33 ml l^{-1}, north of Montague Island in October, to supersaturating values higher than 130% in the more southern and central regions, during different times of the year (Alvarez-Borrego et al., 1973). This type of thermohaline behavior clearly indicates that any kind of input of high saline and warm water into the Upper Gulf could have a drastic ecological impact, with extreme effects during summer. Recently, there have been discussions about the possibility of carrying Salton Sea water into the Upper Gulf (the exchange solution for the Salton Sea). Concern has been expressed about potential negative effects on the ecology of the Reserve of the Biosphere through this type of thermal pollution (Alvarez-Borrego et al., 1999).

Lavín and Sánchez (1999) used the opportunity provided by the Colorado River freshwater release of March to April 1993 to observe the effect of the positive-estuary behavior on the Upper Gulf. In spite of the 1993 discharge being only ~25% of the mean discharge before 1935 (in m^3 s^{-1}), and of very short duration, a clear hydrographic impact was recorded. In opposition to the "normal" inverse estuarine situation, salinity and density decreased toward the head. Dilution was detectable in a coastal band flowing to the right-hand side of the river discharge, up to San Felipe, some 70 km from the river mouth (Figure 108.4). Stratification was established during neap tides, in an estuarine salt-wedge structure. Surface salinity decreased from 35 near San Felipe to 32 at ~10 km south of Montague Island, their northernmost sampling location. Surface temperature was

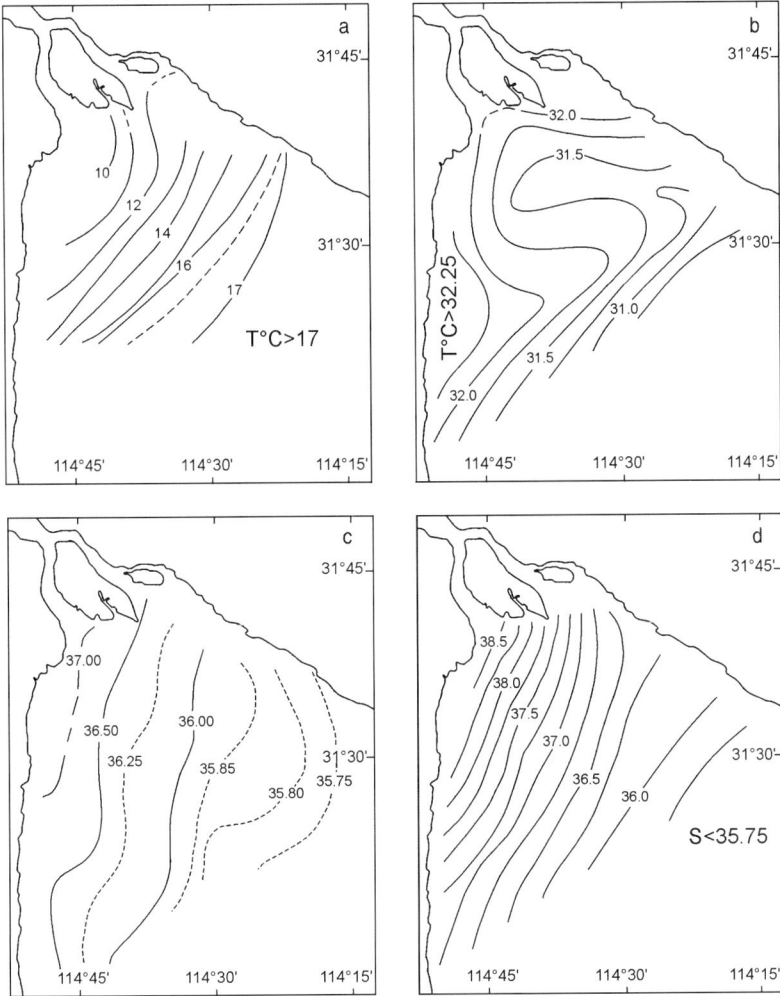

Figure 108.3 Upper Gulf surface temperature and salinity distributions: (a and b) temperature for December and August, respectively; (c and d) salinity for January and August, respectively.

normal for the time of year. This type of hydrographic behavior was relatively common until the early 1960s, but since filling of the Glen Canyon Dam (Lake Powell) began in March 1963, the inverse estuarine behavior predominates.

Rodriguez et al. (2001) used faunal and isotopic evidence to reconstruct the zone of freshwater influence caused by Colorado River flow prior to the construction of dams. The bivalve mollusk *Mulinia coloradoensis* is a brackish-water species that was very abundant when the river flowed into the Gulf. Shells of *M. coloradoensis* dating from before the construction of the dams have $\delta^{18}O$ values significantly more negative than those from species living in the delta at present. Radiocarbon- and amino acid–dated shells range in age from 215 to 650 radiocarbon years old (Kowalewski et al., 1998). Macrofaunal evidence indicates a mixing zone extending as far as 65 km south of the river's mouth along the western shore of the Upper Gulf. Average $\delta^{18}O$ values in shells of *M. coloradoensis* become more positive with increasing distance from the river's mouth, reflecting the greater dilution of river water with gulf water. The zone of influence of the Colorado River fresh water extended beyond 65 km during times of spring snowmelt (Rodriguez et al., 2001). This agrees very well with Lavín and Sánchez's (1999) report on the effect of the freshwater release of March to April 1993.

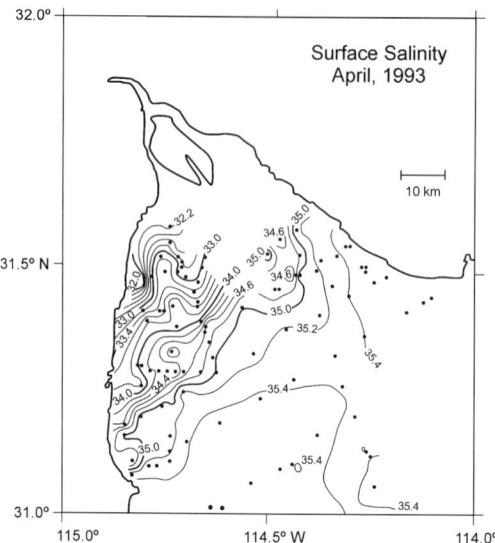

Figure 108.4 Upper Gulf salinity distribution in April 1993. (From Lavín, M.F. and S. Sánchez, On how the Colorado River affected the hydrography of the Upper Gulf of California, *Continent. Shelf Res.*, 19, 1545–1560, 1999. With permission of Pergamon Press.)

In the northern gulf, there is a net heat flux from the water to the atmosphere during winter. However, the annual mean of the net surface heat flux was found to be into the sea ($\sim$100 W m^{-2}) (Castro et al., 1994). These results require an oceanic export of heat and salt out of the northern gulf to achieve a balance, and imply that the annual mean thermohaline circulation must have an inflowing component at depth. This has a profound ecological implication, because inflowing deep water has higher inorganic nutrient content than outflowing surface water. Thus, thermohaline circulation in the northern gulf is a natural fertilization mechanism. Very high tidal energy dissipation rates in the Upper Gulf are another fertilization mechanism. Tidal turbulence brings sediments into suspension and mixes the nutrient-rich pore waters with surface waters. Surface nutrient concentrations are usually high in the central and northern Gulf of California. Alvarez-Borrego et al. (1978) reported PO$_4$, NO$_3$, and SiO$_2$ surface concentrations in the ranges 0.7 to 1.0 μM, 0.0 to 4.0 μM, and 6.1 to 18 μM, respectively, for the northern Gulf of California. Their northernmost station was a little south of 31°N. In the area of the Upper Gulf adjacent to Montague Island, values range from 0.3 to 3.1 μM and 3.3 to 18.3 μM for PO$_4$ and NO$_3$, respectively (Hernández-Ayón et al., 1993).

Nieto-García (1998) compared nutrient concentrations in the Upper Gulf for the springs of 1993 (wet year) and 1996 (dry year). She found that NO$_2$, NO$_3$, and PO$_4$ were lower in spring of 1993, with freshwater input, than during 1996. Only silicate was significantly higher during 1993, indicating a clear influence of river water. A possible explanation for the relatively low NO$_2$, NO$_3$, and PO$_4$ values in 1993 is that the river flow diluted the agricultural drainage water carried from Mexicali Valley and also "cleaned" the surface sediments of the estuary.

MARSH VEGETATION AND PLANKTON

The intertidal wetlands have a marsh vegetation restricted to a dozen halophyte species that can survive the hypersaline seawater of the Upper Gulf. Endemic saltgrass, *Distichlis palmeri*, predominates and produces an edible grain that was harvested by the indigenous Cucapá people up to the 20th century. There are some 3×10^4 ha of *D. palmeri* marshes in the delta. They are feeding stations for birds that prey on crabs and small fish in the marshes, and are nurseries for juvenile stages of fish and invertebrates (Glenn et al., 1999; Kniffen, 1931).

Using coastal zone color scanner (CZCS) imagery, Santamaría-Del-Angel, Alvarez-Borrego, and Muller-Karger (1994) generated time series of phytoplankton pigment concentrations for a series of locations throughout the gulf, from November 1978 to June 1986. The time series for the Upper Gulf showed that pigment concentrations are, in general, greater than 4 mg m^{-3} and a seasonal cycle is not evident. The effect of strong tidal mixing is superimposed on seasonal variation in the Upper Gulf. Eight days chlorophyll a concentration (Chl) time series were generated by direct water sampling, in September 1990 and February and May 1991, for locations near El Golfo de Santa Clara, Sonora, and San Felipe, Baja California. In spite of high turbidity, Chl may reach values higher than 5 mg m^{-3}, and on occasion, higher than 20 mg m^{-3} (Figure 108.5). These high values are typical of very rich coastal waters. In both locations Chl varied irregularly with time, sometimes changing by a factor of two within a few hours. This is the result of a patchy phytoplankton distribution.

Zooplankton biomass (B dry organic weight per m^3) in the Upper Gulf of California had no clear seasonal cycle during October 1972 to October 1973. The highest B values (as high as 154 mg m^{-3}) were always found in the channels around Montague Island. In general, B values increased

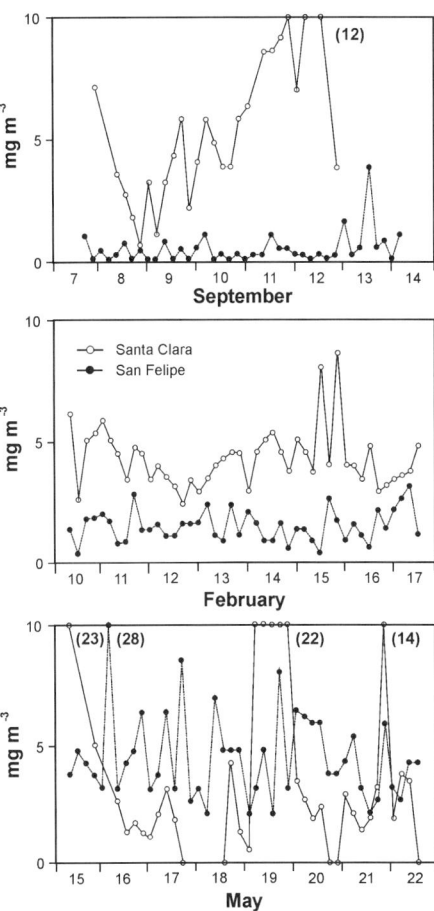

Figure 108.5 Chlorophyll a concentration time series from locations off El Golfo de Santa Clara and San Felipe. Upper graph: September 1990; middle graph: February 1991; and lower graph: May 1991. Numbers on the horizontal axis are the dates; the vertical lines between numbers mark midnight. Sampling was done every 4 h. Closed circles represent data from San Felipe. Numbers in parentheses are maximum values of the peaks off scale.

westward or northwestward. In general, calanoid copepods were the most abundant taxonomic group, numerically and in biomass. Summer B values were similar to those of the rich central Gulf of California; values for the channels around Montague Island were as high as those of estuaries and coastal lagoons (Farfán and Alvarez-Borrego, 1992).

FISH AND INVERTEBRATES

The Mexican government recognized the importance of this part of the gulf as an area of reproduction and nursery for fish species adjacent to the Colorado River mouth, and decreed a permanently closed season for fishing in 1955. This was done based on a few direct observations without any systematic study. Berdegué (1955) reported on the totoaba fishery of San Felipe, Baja California, and indicated that this area was almost totally unknown biologically.

Guevara-Escamilla et al. (1973) concluded that there is a clear spatial distribution pattern of juveniles of different fish species, with most present in the area north of a line connecting El Golfo de Santa Clara, Sonora, with a point 5 nautical miles north from San Felipe, Baja California (Figure 108.1). The outstanding characteristic of this area is the high turbidity that may provide fish juveniles with protection and food. Félix-Pico and Mathews (1975) also concluded that this area is very important for juvenile blue shrimp (*Litopenaeus stylirostris*).

Using commercial shrimp boats, Guevara-Escamilla et al. (1973) sampled the Upper Gulf in the summer of 1973. They reported a total of 73 fish species identified with the help of R.H. Rosenblatt by comparison with specimens of the Scripps Institution of Oceanography collection. All of these species were collected from soft bottoms. They reported that 13 of these species had not been collected outside the gulf and therefore they might be endemic; 31 species had been collected in California waters; the other 29 species were categorized as panamic fauna.

Since Berdegué's (1955) report, the totoaba has been treated with special interest in the scientific literature because it supported an important fishery in the Upper Gulf. This fishery started in 1929 and declined drastically at the end of the 1960s. Annual total catch was 2261 metric tons in 1942 and only 58 metric tons in 1975 (Flanagan and Hendrickson, 1976). It was this decline that made people aware of the possibility of extinction of the species. The Mexican government declared an indefinite fishing moratorium for totoaba in 1975. The totoaba is the largest member of the family Sciaenidae; maximum reported lengths are almost 2 m and weights exceed 135 kg. Totoaba fishing was done mainly for the swimming bladder, to prepare a gourmet soup. Guevara-Escamilla et al. (1973) reported a total of 247 totoaba juveniles caught, with total length between 7 and 27 cm, and this was a clear indication of damage the shrimp fishing nets were doing to the totoaba population. Greatest totoaba abundance was found in the channels around Montague Island and off Baja California near the island.

In April to June 1994, the School of Marine Sciences of the Autonomous University of Baja California developed a field technique that permitted successful capture and transport of totoaba broodstock from the Upper Gulf to the laboratory at Ensenada (True et al., 1997). They were able to keep these specimens of totoaba alive and made them reproduce successfully. In October 1997, they set 250 juveniles back into the Upper Gulf. These specimens were 4 months of age and 20 to 25 cm long.

Félix-Pico and Mathews (1975) studied the invertebrates sampled in the collections with the shrimp nets in 1973. They identified 17 of the most abundant species around Montague Island and 48 in the oceanic region. Their list includes species of Cnidaria, Mollusca, Arthropoda, and Echinodermata. Together with that of Guevara-Escamilla et al. (1973) for fish species, it gives an approximation of the variety of species that were being devastated by shrimp fishery (trawlers) before the decree of the Reserve of the Biosphere.

The problem of ecological alteration of the Upper Gulf of California due to changes in the Colorado River flow has been ignored by government agencies. Galindo-Bect et al. (2000) explored the relationship between the shrimp catch landed in San Felipe (>90% *L. stylirostris*) during 1975–1997, and the Colorado River flow at the Mexico–U.S. border. They found a significant statistical correlation ($p = 0.003$) between shrimp catch and the logarithm of the previous year's river flow. The log of river flow explained 45% of the shrimp catch variance. Their analysis showed that the base catch in the absence of flow into the Upper Gulf is ~200 tons year^{-1}, and it increases to ~600 tons year^{-1} or greater following years of freshwater input.

MARINE MAMMALS AND BIRDS

Few studies of marine mammals have been conducted in the Upper Gulf. Silber et al. (1994) carried a total of 1715 km of boat-based surveys and 1521 km of aircraft-based surveys from 1986–1989 to assess the distribution, relative abundance, and ecological relationships of cetaceans in the Upper Gulf and adjacent areas (1986–1987 was the end of a freshwater release period). On four occasions, surveys were conducted 4 to 12 km into the Colorado estuary. Their primary objective was to characterize habitat use and distribution of the vaquita. Seven cetacean species were seen: common dolphins, *Delphinus delphis* (total individuals seen 14,239); bottlenose dolphins, *Tursiops truncatus* (1416); fin whales, *Balaenoptera physalus* (215); vaquitas, *Phocoena sinus* (110); killer whales, *Orcinus orca* (17); Bryde's whales, *B. edeni* (7); and gray whales, *Eschrichtius robustus* (3). Common dolphins were numerically dominant and bottlenose dolphins were seen most often. Bryde's whales and vaquitas had the smallest group sizes. Bottlenose dolphins were the only marine mammals seen in the Colorado estuary and in waters as shallow as 2.5 m. Common sea lions, *Zalophus californiensis,* are abundant in the Upper Gulf. An important colony is found on Roca Consaga off San Felipe. Fishermen report that they often find sea lions eating from their gill net catches.

Before experiencing changes in freshwater fluxes, the Colorado River delta was important for waterbirds. Now the area is considered to be of moderate to poor quality, although still offering important habitat for winter waterfowl. Because of the importance of the delta for shorebirds, it was one of the first two Mexican sites incorporated into the Western Hemisphere Shorebird Reserve Network. A January 1992 aerial survey detected 163,744 shorebirds in the area (Morrison et al., 1992). The vast majority were sandpipers (80%) (mainly *Calidris mauri*), but included over 9000 American avocets (*Recurvirostra americana*) and nearly 8000 willets (*Catoptrophorus semipalmatus*). The small shorebirds were found mainly on mudflats at the southern end of Montague and Pelícano Islands. The medium and large shorebirds were found mainly on the western margin of the delta on somewhat harder mudflats. A seasonal study showed a clear pattern of maximum waterbird counts in winter (>100,000 with 36 species) and minimum in summer (~500 with 15 species) (Mellink et al., 1997). The avifauna of the Colorado River delta is dominated by Charadriiformes that congregate there to rest and feed during the winter, but the area is also important to seabirds. Brown pelicans (*Pelecanus occidentalis*) were by far the most common seabirds observed during early winter. Anderson et al. (this volume, Chapter 109) give a detailed regional perspective for migratory wildlife resources of this region.

POLLUTANTS

One of the major threats of the Colorado River delta wetlands is selenium, which can be bioaccumulated to levels toxic for wildlife and causes high rates of embryonic mortality and deformity. Selenium is a naturally occurring element originating from cretaceous formations in the upper Colorado River and, due to its high solubility, is distributed along the Colorado River waters (Presser et al., 1994).

Low concentrations of mercury in fish and clams have been detected in the Rio Hardy wetlands. However, high levels of selenium, boron, and arsenic were found in birds from the same area. Dissolved selenium in water samples collected in 1997–1998 from the Hardy/Colorado wetlands (range 9 to 71 µg l^{-1}) was up to 14 times the Environmental Protection Agency's criterion for the protection of freshwater aquatic life. Bioaccumulation factors for several fish species ranged from 350 to 900 times the average waterborne concentration, depending on the tissue and species (Valdés-Casillas et al., 1998, and others cited therein). Samples of fish, mollusks, amphibians, and aquatic insects were collected from the Cienega de Santa Clara, the El Doctor, and the Rio Hardy wetlands and from the mainstream of the Colorado River from spring 1998 to summer 1999. Selenium concentrations in these samples did not exceed significantly the 4 ppm toxic threshold. The highest selenium concentration was detected in samples from the Rio Hardy wetlands (Garcia-Hernandez et al., 2000). Further sampling of biota in the area is needed to determine levels of this element that constitute risk for fish, birds, and humans.

One study analyzed the concentrations of Cu, Mn, Zn, Al, Cd, As, and Se in specimens of the mussel *Modiolus capax* sampled from a coastal site near San Felipe (Punta Estrella) and from other places around the Baja peninsula (Gutierrez-Galindo et al., 1999). Manganese, zinc, aluminum, and selenium concentrations were higher in specimens from Punta Estrella than in samples from the other sites, and this was attributed to the effect of sedimentary material from the Colorado River delta. Selenium concentrations were <4 ppm. Concentrations of all these metals were considered to be low, reflecting natural levels at the sites studied.

Since the early 1970s, there has been concern about the possibility of pesticide transport from the Mexicali Valley into the Upper Gulf. The first study to assess pesticide levels in organisms of the Mexicali Valley irrigation canals and the Upper Gulf of California was conducted in 1972–1973. Guardado-Puentes et al. (1973) reported on DDT metabolites in specimens of the clam *Chione californiensis* from a location at the southern end of Montague Island and another one 30 km north from San Felipe, at shore. They reported that, on average, total DDT (the sum of all detected metabolites of DDT) in *C. californiensis* was 0.067 µg g^{-1} dry weight for specimens from Montague Island and 0.145 for specimens from the other location, with opDDD the most abundant metabolite. These concentrations were between one and two orders of magnitude lower than those in specimens of the freshwater clam *Corbicula fluminea* from the irrigation canals of the Mexicali Valley sampled the same year by these authors.

In 1985–1986, more than 10 years later, Gutierrez-Galindo and co-workers (1988) sampled specimens of mollusks for pesticide analysis. They sampled *C. fluminea* from the irrigation canals of the Mexicali Valley, *Chione californiensis* from the beaches of El Golfo de Santa Clara, and specimens of the mussel, *M. capax,* from a location some 20 km south from San Felipe. Again, the specimens from the Mexicali Valley had total DDT concentrations one order of magnitude greater than the marine specimens. However, total DDT concentrations were much lower compared to those reported for the specimens sampled in 1973. Gutierrez-Galindo and co-workers (1988) analyzed a wide range of organochlorinated pesticides and polychlorinated biphenyls. In the marine environment, only ppDDE was found. Its concentrations in *C. californiensis* were 0.005 to 0.011 µg g^{-1}, and in *M. capax* were 0.008 to 0.009. Given the persistence of the organochlorinated pesticides in the environment, their presence in the Mexicali Valley and the Upper Gulf of California in 1985–1986 was due to their use in the past. Mora et al. (this volume, Chapter 111) more thoroughly evaluate contaminants in the Colorado Delta region.

CONCLUSION

Due to almost total elimination of water discharge into the Colorado River delta, the delta behaves as an inverse-estuary most of the time. The semidiurnal tidal regime produces tidal currents with maximal velocities of 3 m s^{-1} inside the estuarine basin. This causes resuspension of sediments and mixing of porewaters with the water column, which in turn increases nutrient concentrations in the water column. Nutrient concentrations are high in the estuary waters throughout the whole year. Agricultural drainage water is carried from the Mexicali Valley to the estuary through the Hardy and Colorado Rivers. Relatively low salinity and high NO_3 concentrations at the internal extreme of the inverse-estuary are a result of this flux. Phytoplankton and zooplankton abundances are high in the estuary and the adjacent Upper Gulf of California. These support a rich marine ecosystem with abundant and diverse fauna, including birds and marine mammals. These waters appear to be relatively clean of pollutants, with low levels of pesticides and heavy metals. However, more detailed studies are needed concerning metals in the estuary and coastal waters and their effect on biota. Charismatic endemic species, such as the totoaba and the vaquita, are now protected under the regime of the Reserve of the Biosphere. However, studies are needed to find and describe other endemic invertebrate and fish species and characterize the situation of their populations. Due to very saline and warm waters during summer, this ecosystem is very much under stress and any other anthropogenic factor, such as the input of hot brines, could be catastrophic.

ACKNOWLEDGMENTS

J.M. Dominguez and F. Ponce produced the art work for this chapter.

REFERENCES

Alvarez-Borrego, S., Anderson, D.W., Fernández-de-la-Garza, G., Letey, J., Matsumoto, M., Orlob, G.T., and Palerm, J.V., Alternative futures for the Salton Sea, The UC MEXUS Border Water Project Issue Paper 1, Available from the University of California Institute for Mexico and the United States, Riverside, CA, 1999.

Alvarez-Borrego, S., Galindo-Bect, L.A., and Flores-Baez, B.P., Hidrología, in *Estudio Químico sobre la Contaminación por Insecticidas en la Desembocadura del Río Colorado,* Tomo I, Reporte a la Dirección de Acuacultura de la Secretaría de Recursos Hidráulicos, Available from Universidad Autónoma de Baja California, Ensenada, B.C., Mexico, 1973, pp. 6–177.

Alvarez-Borrego, S., Rivera, J.A., Gaxiola-Castro, G., Acosta-Ruiz, M.J., and Schwartzlose, R.A., Nutrientes en el Golfo de California, *Cienc. Marinas,* 5, 53–71, 1978.

Alvarez-Sánchez, L.G., Godínez, V., Lavín, M.F., and Sánchez, S., Patrones de turbidez y corrientes en la Bahía de San Felipe al NW del Golfo de California. Comunicaciones Académicas CICESE CTOFT-9304, Available from CICESE, Ensenada, B.C., Mexico, 1993.

Argote, M.L., Amador, A., and Lavín, M.F., Tidal dissipation and stratification in the Gulf of California, *J. Geophys. Res.,* 100, 16103–16118, 1995.

Berdegué, A.J., La pesquería de la totoaba (*Cynoscion macdonaldi,* Gilbert) en San Felipe, Baja California, *Rev. Soc. Mex. Hist. Nat.,* 16, 45–78, 1955.

Carriquiry, J.D. and Sánchez, A., Sedimentation in the Colorado River Delta and Upper Gulf of California after nearly a century of discharge loss, *Mar. Geol.,* 158, 125–145, 1999.

Castro, R., Lavín, M.F., and Ripa, P., Seasonal heat balance in the Gulf of California, *J. Geophys. Res.,* 99, 3249–3261, 1994.

Cisneros-Mata, M., Montemayor-Lopez, G., and Roman-Rodriguez, M., Life history and conservation of *Totoaba macdonaldi, Conserv. Biol.,* 9, 806–814, 1995.

Cupul, A.L., Flujos de sedimentos en suspensión y nutrientes en la cuenca estuarina del Río Colorado, Master's thesis, available from Universidad Autónoma de Baja California, Ensenada, B.C., Mexico, 1994.

Farfán, C. and Alvarez-Borrego, S., Zooplankton biomass of the northernmost Gulf of California, *Cienc. Marinas,* 18, 17–36, 1992.

Félix-Pico, E. and Mathews, C.P., Estudios preliminares sobre la ecología del camarón en la zona cercana a la desembocadura del Río Colorado, *Cienc. Marinas,* 2, 68–85, 1975.

Flanagan, C.A. and Hendrickson, J.R., Observations on the commercial fishery and reproductive biology of the totoaba *Cynoscion macdonaldi* in the northern Gulf of California, *Fish. Bull.,* 74, 531–544, 1976.

Galindo-Bect, S., Glenn, E.P., Page, H.M., Fitzsimmons, K., Galindo-Bect, L.A., Hernández-Ayón, J.M., Petty, R.L., Garcia-Hernandez, J., and Moore, D., Analysis of penaeid shrimp landings in the Upper Gulf of California in relation to Colorado River discharge, *Fish. Bull.,* 98, 222–225, 2000.

Garcia-Hernandez, J., Glenn, E.P., Artiola, J., and Baumgartner, D.J., Bioaccumulation of selenium (Se) in the Cienega de Santa Clara Wetland, Sonora, Mexico, *Ecotoxicol. Environ. Saf.,* 45, 298–304, 2000.

Glenn, E.P., Garcia, J., Tanner, R., Congdon, C., and Luecke, D., Status of wetlands supported by agricultural drainage water in the Colorado River Delta, Mexico, *HortScience,* 34, 16–21, 1999.

Guardado-Puentes, J., Núñez-Esquer, O., Flores-Muñoz, G., and Nishikawa-Kinomura, K.A., Contaminación por pesticidas organoclorados, in *Estudio Químico sobre la Contaminación por Insecticidas en la Desembocadura del Río Colorado,* Tomo I, Reporte Final a la Dirección de Acuacultura de la Secretaría de Recursos Hidráulicos, available from Universidad Autónoma de Baja California, Ensenada, B.C., Mexico, 1973, pp. 179–200.

Guevara-Escamilla, S., Huerta-Díaz, M.A., Félix-Pico, E., Farfán, C., and Mathews, C., Biología, in *Estudio Químico sobre la Contaminación por Insecticidas en la Desembocadura del Río Colorado,* Tomo II, Reporte Final a la Dirección de Acuacultura de la Secretaría de Recursos Hidráulicos, available from Universidad Autónoma de Baja California, Ensenada, B.C., Mexico, 1973, pp. 236–364.

Gutierrez, G. and González, J.I., Predicciones de mareas de 1990: estaciones mareográficas del CICESE. Informe Técnico OC-89-01, available from CICESE, Ensenada, B.C., Mexico, 1999.

Gutierrez-Galindo, E.A., Flores-Muñoz, G., and Villaescusa-Celaya, J., Chlorinated hydrocarbons in mollusks of the Mexicali Valley and Upper Gulf of California, *Cienc. Marinas,* 14, 91–113, 1998.

Gutierrez-Galindo, E.A., Villaescusa-Celaya, J.A., and Arreola-Chimal, A., Bioaccumulation of metals in mussels from four sites of the coastal region of Baja California, *Cienc. Marinas,* 25, 557–578, 1999.

Hernández-Ayón, M., Galindo-Bect, S., Flores-Báez, B.P., and Alvarez-Borrego, S., Nutrient concentrations are high in the turbid waters of the Colorado River delta, *Estuarine Coastal Shelf Sci.,* 37, 593–602, 1993.

Kniffen, F., The primitive culture landscape of the Colorado Delta, *Univ. Calif. Publ. Geol.,* 5, 43–66, 1931.

Kowalewski, M., Goodfriend, G.A., and Flessa, K.W., High resolution estimates of temporal mixing within shell beds: the evils and virtues of time averaging, *Paleobiology,* 24, 287–304, 1998.

Lavín, M.F. and Sánchez, S., On how the Colorado River affected the hydrography of the Upper Gulf of California, *Continent. Shelf Res.,* 19, 1545–1560, 1999.

Lavín, M.F., Godínez, V.M., and Alvarez, L.G., Inverse-estuarine features of the Upper Gulf of California, *Estuarine Coastal Shelf Sci.,* 47, 769–795, 1998.

León-Portilla, M., Descubrimiento en 1540 y primeras noticias de la Isla de Cedros, *Calafia,* 2, 8–10, 1972.

Meckel, L.D., Holocene sand bodies in the Colorado delta area, northern Gulf of California, in *Deltas, Models for Exploration,* Broussard, M.C., Ed., Houston Geological Society, 1975, pp. 239–265.

Mellink, E., Palacios, E., and Gonzalez, S., Non-breeding waterbirds of the delta of the Rio Colorado, Mexico, *J. Field Ornithol.,* 68, 113–123, 1997.

Morrison, R.I.G., Ross, R.K., and Torres, M.M., Aerial surveys of Neartic shorebirds wintering in Mexico: some preliminary results, Progress Note, Canadian Wildlife Service, Available from Canadian Ministry of the Environment, 1998.

Nieto-García, E., Nutrientes en el Norte del Golfo de California durante condiciones estuarinas y antiestuarinas, Master's thesis, Available from CICESE, Ensenada, B.C., Mexico, 1998.

Presser, T.S., Sylvester, M.A., and Low, W.H., Bioaccumulation of selenium from natural geologic sources in western states and its potential consequences, *Environ. Manage.,* 183, 423–436, 1994.

Rodriguez, C.A., Flessa, K.W., Téllez-Duarte, M.A., and Dettman, D.L., Macrofaunal and isotopic estimates of the former extent of the Colorado River estuary, Upper Gulf of California, Mexico, *J. Arid Environ.,* 49, 183–193, 2001.

Santamaría-Del-Angel, E., Alvarez-Borrego, S., and Muller-Karger, F.E., Gulf of California biogeographic regions based on coastal zone color scanner imagery, *J. Geophys. Res.,* 99, 7411–7421, 1994.

Silber, G.K., Newcomer, M.W., Silber, P.C., Pérez-Cortéz, H., and Ellis, G.M., Cetaceans of the northern Gulf of California: distribution, occurrence, and relative abundance, *Mar. Mammal Sci.,* 10, 283–298, 1994.

Sverdrup, H.U., The Gulf of California: preliminary discussion on the cruise of the E.W. Scripps in February and March 1939, *Sixth Pacific Sci. Cong. Proc.,* 3, 161–166, 1941.

True, C.D., Silva-Loera, A., and Castro-Castro, N., Acquisition of broodstock of *Totoaba macdonaldi*: field handling, decompression, and prophylaxis of an endangered species, *Progr. Fish Cult.*, 59, 246–248, 1997.

Valdés-Casillas, C., Hinojosa-Huerta, O., Muñoz-Viveros, M., Zamora-Arroyo, F., Carrillo-Guerrero, Y., Delgado-García, S., López-Camacho, M., Glenn, E.P., García, J., Riley, J., Baumgartner, D., Briggs, M., Lee, C.T., Chavarría-Correa, E., Congdom, C., and Luecke, D., Information database and local outreach program for the restoration of the Hardy River wetlands, lower Colorado River Delta, Baja California and Sonora, Mexico, North American Wetland Conservation Council and Instituto Nacional de Ecología, Guaymas, available from http://uib.campus.gym.itesm.mx/hardy/hardy/index.htm, 1998.

van Andel, T.J.H. and Shor, G.G., Jr., Eds., *Marine Geology of the Gulf of California: A Symposium*, American Association of Petroleum Geologists Memoir 3, 1964.

CHAPTER 109

Migratory Bird Conservation and Ecological Health in the Colorado River Delta Region

Daniel W. Anderson, Eduardo Palacios, Eric Mellink, and Carlos Valdés-Casillas

Ecological integrity can be defined as the maintenance of the structural and functional attributes of a particular ecosystem.... Our present human society globally has reached its level of affluence and density because of two life support systems — one technological and the other ecological. However, the events since the beginning of the Industrial Revolution clearly indicate that most human societies place far more value on the technological life support system than on the ecological one. In the last few decades, the shift has been toward a recognition of the duality of the life support system and an attempt to give more value ... to the ecological life support system.... Ecological restoration is a means of buying more time for human society to develop life styles less threatening to natural systems and for ecologists to develop more robust methods for restoring the earth's damaged ecosystems.

John Cairns, Jr. (1994)

THE COLORADO RIVER DELTA REGION: A HABITAT MOSAIC OF RICH PAST AND PRESENT BIODIVERSITY

The Colorado River Delta region (CRDR) was historically one of the richest areas for biodiversity in western North America; it included desert, riparian, estuarine, and marine environments. As stated by Felger (1976) in a little-known but important volume regarding conservation in the CRDR and the Gulf of California:

This was the life-center of the Gulf, and probably the delta was biotically the richest place in the southwestern part of the North American continent (citing Sykes, 1937).

Originally there were well over 500 species of plants (aquatic and terrestrial), and currently the best representation of this former biodiversity is near Cienega Santa Clara, where about 400 species are known (see also Ezcurra et al., 1988). California condors (*Gymnogyps californicus*) once flew over the CRDR (Murphy, 1917); Leopold (1966) told of jaguars (*Panthera onca*) in the delta; and Funcke (1919; Murphy, 1917) collected pronghorns (*Antilocapra americana*) nearby on the periphery. Endemic vegetation includes the saltgrass, *Distichlis palmeri*. Yuma clapper rail (*Rallus longirostris yumanensis*), now classified an endangered species, is also endemic in the CRDR. It has been estimated that Cienega Santa Clara alone supports a significant

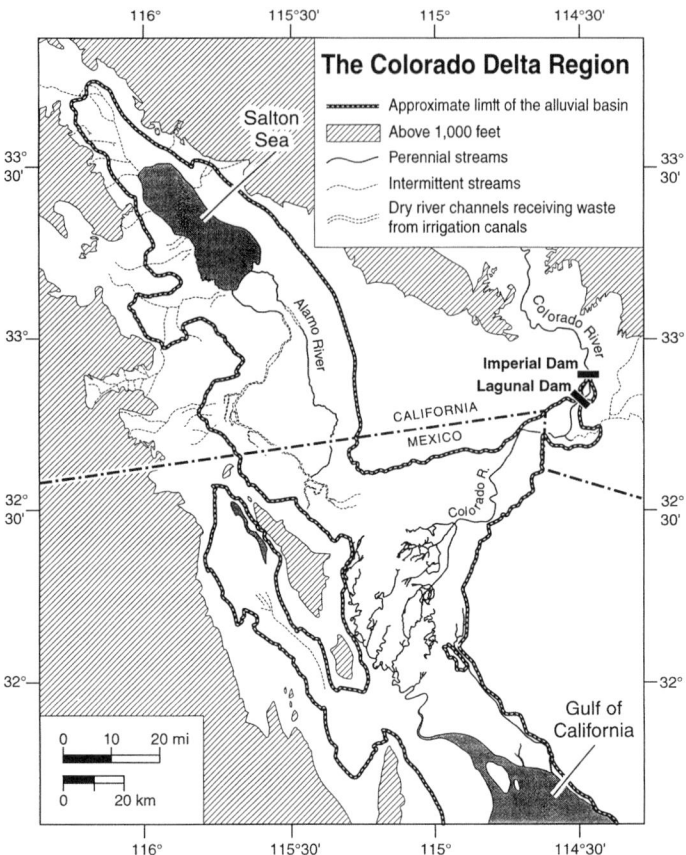

Figure 109.1 Map of the Rio Colorado Delta region, redrawn from Sykes (1937) to show the extensive wetlands of the area in the early 1900s, The map is simplified; consultation of the original map for specific details is recommended.

percentage of the total Yuma clapper rail population (Hinojosa-Huerta, O., personal communication, 2000). Hundreds of thousands of shorebirds and other waterfowl still pass through and utilize this area as a wintering or stopover area in migration (Shuford et al., 2000; Patton et al., 2000; Alvarez-Borrego, Chapter 108, this volume).

Yet this high productivity and richness has not gone unnoticed to land developers, especially agricultural interests (Bonillas and Urbina, 1912, 1913; Lumholz, 1912; Fradkin, 1981; Sykes, 1937). Ironically, even agriculture made a small contribution to the biodiversity of the region (Rosenberg et al., 1991; Mellink, 1995). Present habitat diversity in the delta hints at what was there in the past and it includes marine habitats, marine coastal habitats, deltaic islands, intertidal and upland wetlands, riparian zones, vernal pools, coastal deserts, and modified habitats, such as aquaculture ponds, salt works, agroecosystems, tourism facilities, and urban/industrial areas.

Explorers and naturalists recognized early that the lower CRDR was an integrated and diverse wetland ecosystem, comprised of interacting and connected parts (Grinnell, 1914; Murphy, 1917; Sykes, 1937; Goldman, 1951; Leopold, 1966), including the local separation of many of its hydric and mesic parts by expanses of more xeric habitats. All this formed a local mosaic of physical diversity that provided unparalleled, seasonally variable biodiversity to the region. This region (Figure 109.1) included:

1. The entire lower Colorado River and associated wetlands, today defined as riparian habitats and wetlands that extend as far as the first major dam north on the Colorado (Rosenberg et al., 1991).
2. The diverse delta area and associated wetlands (Glenn et al., 1996).

3. The northern part of the Gulf of California itself (Hendrickson, 1974; Maluff, 1983; Farfan and Alvarez-Borrego, 1992; Hernandez-Ayon et al., 1993; Santamaria-del-Angel et al., 1994).
4. The more recently formed Salton Sea (Sykes, 1937).
5. The incidentally formed Cienega de Santa Clara, the largest remaining wetland in the CRDR along with several other important wetlands nearby, but with an unknown future (Glenn et al., 1992).

Sykes (1937) delimited the CRDR an area of some 860,000 ha (Figure 109.1), not including the related marine zone of the upper Gulf of California, which is even larger in area, depending on where one delineates strictly oceanic influence (Figure 109.2). The present-day core area of the lower CRDR (comprised mostly of wetlands in Mexico, but not including the Salton Sea and associated wetlands, the Colorado River and associated wetlands, or the expansive terrestrial areas to the north) has about 250,000 ha. Thus, the CRDR is mainly comprised of the present mouth of the Colorado River with its intertidal flats and regularly flooded islands (Islas Montague-Gore and Pelícano), Esteros Santa Clara and La Ramada, Cienegas de Santa Clara and El Doctor, Laguna del Indio, Pescaderos and Hardy Rivers, the remaining riparian corridor along the Colorado River, and various drainage canals such as the Wellton-Mohawk Canal and Riito-Santa Clara canal (Glenn et al., 1996; Briggs and Cornelius, 1998; Valdés-Casillas et al., 1998a). The lower CRDR also includes an important portion of native riparian vegetation, today estimated at about 1650 ha (Valdés-Casillas et al., 1998a, 1999), which recently has been identified as an important corridor for migratory neotropical birds, such as the willow flycatcher (*Empidonax trailii*).

Alvarez-Borrego (this volume) terms the northern Gulf of California the *sedimentary province,* and his analysis places the northern gulf within the direct ecological influence of the CRDR and its uplands. The core aquatic components in this system are interconnected by riparian corridors

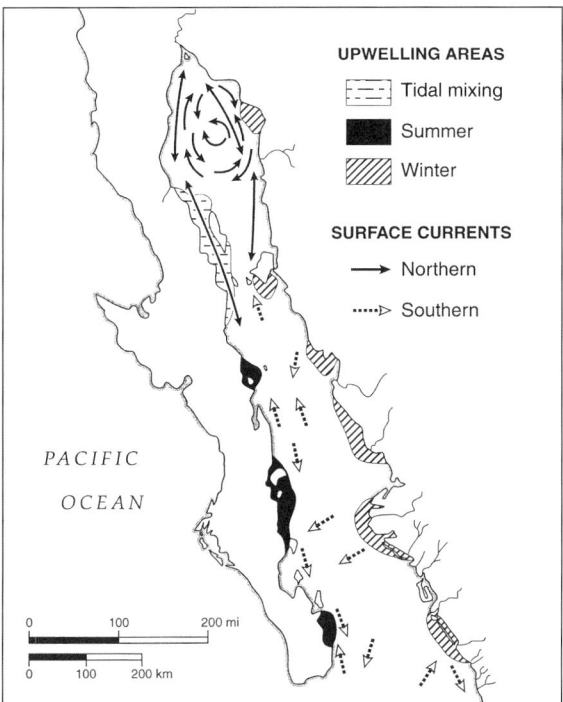

Figure 109.2 Surface water characteristics of the Gulf of California, showing the distinctiveness of the northern region. The northern gyre and likely more of this oceanic region are closely linked ecologically to the CRDR. (See references in text; redrawn from Maluff, L.Y., Physical oceanography, in *Island Biogeography in the Sea of Cortez,* Case, T.J. and M.L. Cody, Eds., University of California Press, Berkeley, 1983, pp. 26–45. With permission.).

(but much more extensively in the past); other aquatic components are separated by expanses of desert; and in the northern Gulf of California, zones of primary productivity are delineated by sediment and nutrient deposition patterns and current patterns. Yet most potential barriers to dispersal are transcended by the mobile avifauna that occupy the CRDR during various times of their annual cycles.

There is also an inter- and intraseasonal nature to the less-mobile biota of this region. CRDR habitat mosaics supported, to a much greater degree than currently, a rich biota of less-mobile species, whose viability was largely dependent on size, diversity, and local connectivity of habitat patches. These local aspects of habitat structure, based largely on the dynamics of patch size, patch quality, and connectivity, are currently of major interest in modern conservation biology (Harrison and Bruna, 1999) and are likely to be key issues in conservation strategies for restoration in the CRDR.

Changes in the Colorado River basin involve extreme ecological modifications in one of the largest and most spectacular past watersheds in western North America (Findley, 1973; Fradkin, 1981; Reisner, 1986). Most of these changes were brought about by activities that had little or no regard for the natural and sensitive floral and faunal elements of this large system. The lower components of the Colorado River basin have been modified most in the watershed, partly because of their low or terminal position, but also because the CRDR spans the U.S.–Mexico border and for many years was not considered a U.S. issue.

Saunders and Saunders (1981) conducted long-term U.S. Fish and Wildlife Service (USFWS) waterfowl surveys in Mexico from the late 1930s through the early 1960s, mostly because of interests in the status of hunted migratory waterfowl populations. Their surveys were essentially the only data available from the coastal ecosystems south of the U.S.–Mexico border through that period. Regarding the CRDR, they remarked:

> The Rio Colorado Delta is very disappointing as a waterfowl wintering ground . . . it has few ducks in comparison to its great flights of earlier years.

Today, wetlands in the CRDR vary considerably in extent, depending almost entirely on unpredictable floodwaters, agricultural drain waters, and municipal wastewaters; based on figures given by Glenn et al. (1996) we estimate that less than 3% of the original wetlands remain today. However, more recent studies using remote sensing techniques have identified more than 60,000 ha of active wetlands present in years of adequate water, representing about 20% of the original extent; however, this figure includes modified habitats and areas with salt cedar (*Tamarix* spp.) intrusion (Valdés-Casillas et al., 1999).

> ...the Delta has probably been made safe for cows, and forever dull for adventuring hunters. Freedom from fear has arrived, but a glory has departed from the green lagoons (Leopold, 1966).

The picture that is so very different today in the CRDR is characterized by widely scattered and reduced wetlands that are changed and degraded, dominated by exotic plants and animals, challenged by many human-induced ecological problems, some newly created by wastewater disposal (Glenn et al., 1996; USFWS, 1997; Mellink and Ferreira-Bartrina, 2000). Conflicts over water are, in fact, an ongoing *worldwide* issue in conservation biology today. *World Conservation* (volume 30, 1999) is devoted entirely to the issue of water needs in wetland conservation.

A remarkably similar scenario has been described in Australia (Kingsford and Johnson, 1998), where large water diversions have resulted in universal population declines of aquatic birdlife and widespread degraded natural habitats. Referring to the Aral Sea, Kindler (1998) aptly stated, "The time to address freshwater challenges is running out." This applies everywhere. Views on conflicting uses of water in the CRDR are discussed by Rosen (1992), Morrison et al. (1996), Stephens (1997), and others. The messages are all abundantly clear: which will be the uses of scarce water supplies?

Today, another integral part of the CRDR, the Salton Sea, having unique problems of its own as well as problems common to all similar wetlands in western North America (USFWS, 1997; Cohen et al., 1999; UC MEXUS, 1999), remains one of the few dominant wetlands of this once vast and rich ecosystem (Ornithological Council, 1998), along with those remnant wetlands (mentioned previously) south of the U.S.–Mexico border. Anywhere in the CRDR, ample opportunity still exists for significant restoration (highlighted by Glenn et al., 1996; Briggs and Cornelius, 1998; Valdés-Casillas, 1999; Luecke, 1999).

THE PACIFIC FLYWAY: A BROAD MOSAIC UNIFIED BY MIGRATORY BIRDS

The most far-reaching biogeographical connectedness of the CRDR, however, involves migratory birds that come from nearly every part of North America, including elements from the Paleoarctic (Bellrose, 1980; Linduska, 1964; Shuford et al., in press), although that connectedness might be severely challenged by many of the factors mentioned. For want of a better definition, the classic representation of the Pacific Flyway is depicted in Figure 109.3 (Linduska, 1964; Hawkins et al., 1984). Although the original flyway concept was based largely on band recoveries and distributional data from hunted migratory ducks and geese (Lincoln, 1935; Hawkins et al., 1984), in reality (and from an ecological viewpoint representing many more migratory bird species), a flyway is more realistically and perhaps more theoretically a dynamic, ever-changing mosaic of avian movement patterns within and among habitat patches over space and time. For migratory birds and other mobile fauna, total connectivity in such a system is based largely on their life-history requirements: *availability of suitable habitat* plus its distribution or *accessibility*) (see Haig and co-workers, 1998, and Plissner and co-workers, 2000, for specific discussions of these concepts in the conservation of specific migratory birds). It is a system of connected and accessible habitats: wetlands, riparian corridors, or terrestrial habitats (connected by movements in the case of migratory or mobile birds). Each species of bird responds to the presence and accessibility of suitable habitat patches (managed refuges and natural areas, remnants of natural habitat not yet destroyed, patches of restored habitat, artificial and unnatural habitats, agricultural habitats, and sometimes even by other man-modified habitats*). Thus, habitat patterns in such a system must be of sufficient diversity and variety to enable these patches and edges to support a wide variety of bird (and other wildlife) species over the portions of their annual cycles that overlap any given geographical area, and they must be accessible. On a geographic landscape, these patterns merge into a conceptual picture (Figure 109.3) and, therefore, still serve as useful management units for resource managers, and as conceptual units in conservation biology (e.g., combining concepts put forth by Hawkins et al., 1984, and Haig et al., 1999).

For purposes of resource management and administration, the North American continent has been divided into four flyway-based units (administrative flyways, as Hawkins and co-workers, 1984, call them), separated on the basis of state legal jurisdictions (still based, however, mostly on economically valuable bird species, such as hunted migratory waterfowl). Bartonek (1984) described the administrative structure and history of the administrative Pacific Flyway. The definition of their habitats was based largely on hunted migratory waterfowl, but this does not seriously detract from conservation needs of other nonhunted migratory bird species, such as passerines, herons, grebes, pelicans, cormorants, and others, because these species utilize different aspects of the same general habitat mosaics and *all* are protected by the Migratory Bird Treaty Act. Each flyway is managed on a separate basis (recommendations come from knowledgeable expert biologists appointed to each Flyway Council) and the entire system is managed by appropriate federal agencies (Canadian Wildlife Service, U.S. Fish and Wildlife Service [USFWS], and Secretaria de Medio Ambiente y Recursos Naturales [SEMARNAT] in Mexico), working together primarily

---

* For example, waterfowl communities comprised of wintering species on the Lower Colorado River have been influenced and enhanced by downstream habitats associated with dams (Anderson and Ohmart, 1988).

Figure 109.3 The most recent depiction of the Pacific Flyway. Key central and southern subregions within this broad mosaic include (in rough order of current importance to migratory birds): Sacramento Valley, Klamath Basin and associated areas, San Joaquin Valley, marshes of mainland Western Mexico, Rio Colorado Delta region, San Francisco Bay Delta region, inland lake system in western Nevada, and marshes of western Baja California. (From the U.S. Department of the Interior; taken from Linduska, J.P., Ed., *Waterfowl Tomorrow,* U.S. Fish and Wildlife Service, Washington, D.C., 1964.)

under the Migratory Bird Treaty Act, the Endangered Species Act, and the North American Wetlands Conservation Act.

Recently, significant advances in wetland conservation in the flyways have been made through the North American Waterfowl Management Plan, which includes the CRDR in its Mexican Priority Wetlands designation (Williams et al., 1999) as well as the Wetlands for the Americas programs (Patterson, 1995). North American management plans are now being developed for shorebirds and colonial waterbirds (Kushlan, J., U.S. Department of the Interior, personal communication, 2000). Schmidt and co-workers (1999) have summarized these approaches and philosophy for current migratory bird management in North America.

Regardless of the precise definition of a Pacific Flyway (taxonomic, administrative, ecological, or some artificial combination of the three), it is abundantly clear that the CRDR, despite its current state of ecological degradation, remains a critical stopover habitat for migratory birds in the southern Pacific Flyway (Ornithological Council, 1998). In fact, most other areas of current or previously suitable migratory bird habitat in the Pacific Flyway to the north of the CRDR are

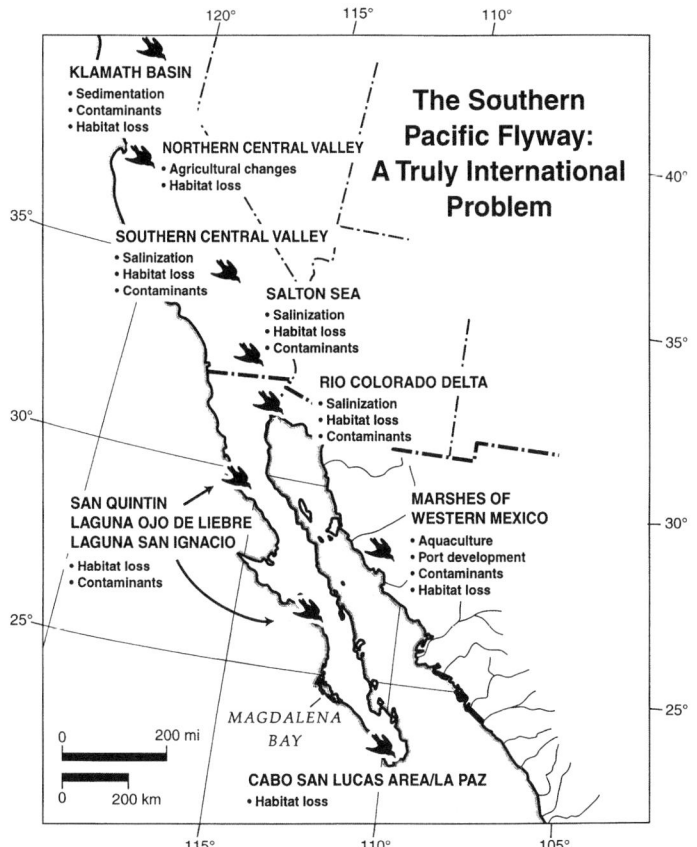

Figure 109.4 Map of the central and southern Pacific Flyway region (western U.S. and Mexico), showing the major general, human-related ecological factors that have impacted habitats for migratory birds. This map was summarized from the many references cited in the text and personal experiences of the authors.

also challenged by many factors related to their degradation, water development projects and related habitat losses being one of the major concerns throughout the western U.S., plus numerous factors in areas to the north (Figure 109.4) and south (Figure 109.5) of the CRDR. In fact, wetlands throughout western North America have been considerably reduced and changed (Gilmer et al., 1982, 1995; Nichols et al., 1986; Glenn et al., 1996). Most of the remnants of such habitats remain as state or National Wildlife Refuges, or in some cases as private land (e.g., hunting clubs in California; Gilmer et al., 1982). It is also clear that the remnants of these habitats are highly modified, human-influenced systems (Anderson, 1995; Mellink and Ferreira-Bartrina, 2000). The agriculture–wildland interface is not likely to change much in the future; these habitats are no longer free of the influences of human development. In fact, an agriculture and wildland interface is probably the dominant characteristic in the central and southern portions of the entire Pacific Flyway.

The far-reaching ecological connections of this region are increasingly important to all migratory birds of the Pacific Flyway, which is still a vast and ever-changing system of migratory corridors and critical wintering and stop-over habitats (both wetland and terrestrial) for birds from all parts of this continent. Native riparian vegetation corridors have largely disappeared or are disappearing along the lower reaches of the Colorado River system (Rosenberg et al., 1991) due to outright destruction resulting from factors mostly related to water use and management, the invasion of

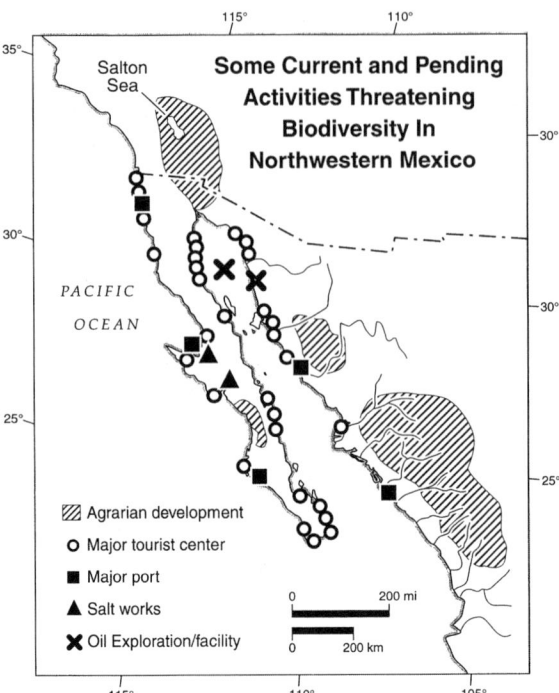

Figure 109.5 Map of the Gulf of California, summarizing the range of human-related activities that threaten ecological connectedness and integrity in this part of the Pacific Flyway, This map was summarized from the many references cited in the text and personal experiences of the authors.

exotic plants and animals, and other factors (Glenn et al., 1996; Valdés-Casillas, 1999; Luecke et al., 1999). Although most highly migratory bird species do not require continuous habitat corridors, less migratory species do; and small, isolated oases are important for migrating birds as well as for local breeding populations (Hinojosa-Huerta, O., personal communication, 2000; Skagan et al., 1998).

Harrison and Bruna (1999), in reviewing habitat fragmentation theory, have concluded that neither corridors nor configuration can substitute for overall habitat loss. If improvements are to be made, no more native or even altered habitat that is suitable for wildlife need be further degraded; instead, wetland habitats need to be enhanced and restored (Zedler and Powell, 1993). If migratory birds are to be protected and enhanced, it is clear that, for the most part, the southern Pacific Flyway must enter an era of restoration and *increased* protection (Glenn et al., 1996; Briggs and Cornelius, 1997; Briggs and Cornelius, 1998).

In the northern Gulf of California, still a part of the CRDR (Figure 109.2), native avian biodiversity is rich and abundant, comparatively unchanged (Anderson, 1983; Velarde and Anderson, 1994; Everett and Anderson, 1991; Mellink et al., 1997; Ruiz-Campos and Rodríguez-Meraz, 1997; Patten et al., 2000). Yet, here one of the world's most endangered marine mammals, the vaquita or *little cow* (*Phocoena sinus*), still resides in very low and dwindling numbers (Vidal, 1995) and a formerly important commercial fish species, the totoaba (*Totoaba macdonaldi*), is now also threatened by human activities and habitat loss in the CRDR (Cisneros-Mata et al., 1995). In fact, the general region of the CRDR and northern Gulf of California faces continuing threats of degradation and ecological decline (Anderson et al., 1976; Tershy et al., 1977; Glenn et al., 1996; Galindo-Bect et al., 1999).*

---

* Some interesting newspaper accounts, mostly relating to the Gulf of California, can be found in a series of articles by Tom Knudson, *Sacramento Bee,* 1995 and 1999; http://www.sacbee.com/news/projects/dyingsea.

THREATS TO BIODIVERSITY AND ECOLOGICAL HEALTH IN THE COLORADO RIVER DELTA REGION

What happens in one part of the CRDR potentially affects its other parts, even outside this geographic area. What happens to migratory birds in this region has potential to affect their populations in distant places; linked to that are the dominant and modifying influences of past and present human activities.

Ohmart and co-workers (1975) clearly recognized the potential for properly managed riparian wetlands of the Lower Colorado River, for both land bird and waterbird conservation. Briggs and Cornelius (1997) have stated, "Despite the tremendous changes that have occurred in the Colorado River Delta, it is important to emphasize that the delta is not beyond recovery." Saunders and Saunders (1981) similarly stated, "Undoubtedly many thousands more ducks and geese would winter in this area if conditions were more favorable."

Ohmart and co-workers (1975) and their many colleagues recognized the ecological values of this region some 30 years ago; they conducted intensive management and restoration research activities along the lower Colorado River (Rosenberg et al., 1992) that will serve as a model for future restoration.

Understanding the mechanisms whereby human-induced and natural variables affect a system and its components, how negative effects can be alleviated and positive effects enhanced, is the major goal of conservation biologists, resource managers, governmental leaders, and the people who utilize or otherwise interact with these resources. Whether and how these systems ultimately become managed and conserved rest in the uneasy realm between knowledge and reality, policy and politics; necessarily, it includes the aspects of use (conservation is most commonly defined as "wise use") (Bildstein et al., 1991; Cairns, 1994), as well as some major readjustments in the way things are currently done.

Numerous problems concerning the conservation and population viability of CRDR migratory birds and other wildlife have been identified by various authors (Anderson et al., 1976; Velarde and Anderson, 1994; Glenn et al., 1996; USFWS, 1997; Valdés-Casillas, 1998b; Cohen et al., 1999; and others). Based on our experiences in the region over the past several decades, we have attempted to summarize threats that might in various combinations apply to most categories of wildlife and wildlife habitat (including migratory birds) in the CRDR and Gulf of California region (Figure 109.6). Importantly, they encompass the following general factors, ranked by us in order of importance:

1. Outright habitat destruction and loss of connectivity through water diversions; flooding by poor quality wastewaters, and other water development activities; development of shoreline facilities; drainage of coastal wetlands; agricultural and urban developments.
2. Undesirable habitat modification and degradation through the introduction and subsequent competition from exotic plants and animals, eutrophication, pollution from agriculture and aquaculture (see Naylor et al., 1998; Páez-Osuna et al., 1998; Mellink and Ferreira-Bartrina, 2000), salinization, introduction of exotic diseases, sedimentation, and habitat destruction (Watling and Norse, 1998).
3. Direct mortality of wildlife (Rocke and Friend, Chapter 110, this volume) through unmanaged exploitation, further disease or contaminant interactions, incidental or accidental "take"; or through pollution per se (Mora et al., Chapter 111, this volume): direct toxic effects of biocides, salts, toxic levels of nutrients, oil, and other chemicals.
4. Interferences and disturbances, sublethal effects of contaminants, simple disturbances in various wildlife concentration areas, feeding areas, loafing areas, and breeding areas (such as recreational and vehicular traffic on beaches and areas adjacent to wetlands), competition with fisheries, aquaculture, and agriculture (also note that some of these interactions occasionally have a positive effect on some species of wildlife).
5. Shortsighted planning and lack of understanding of ecosystem functions.

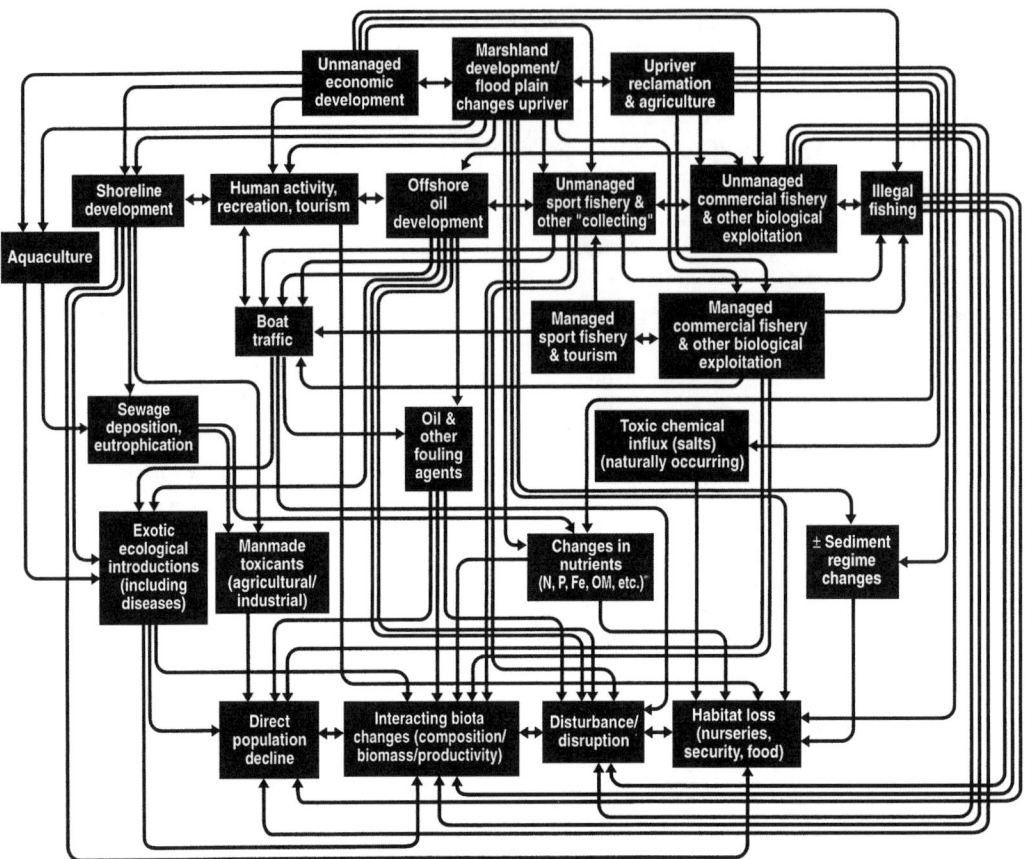

Figure 109.6 A conceptual flow diagram summarizing the interrelatedness of the various threats to biodiversity in the CRDR. Different biotic elements in the system are affected through various pathways that lead to ultimate population decline and biodiversity loss.

Interestingly, this condensed list, although specifically drawn from the CRDR, reads almost like a catalog of world migratory-bird conservation problems. We cannot go into specific detail here, outlining each and every situation (if even all were known). But in a general sense, it can be seen that different categories of wildlife (migratory birds in this discussion) would follow different and multiple pathways to their degradation. A purely single-species approach would in most cases be "too-little, too-late." We need to ask questions such as "what happens when we are confronted with an extinction of habitats as has been shown to be associated with many endangered freshwater, aquatic communities in interior North America" (Moyle and Williams, 1990). Whole-system or watershed approaches and philosophy are ultimately needed (National Research Council, 1992). This has been a new trend in conservation programs, such as the World Wildlife Fund and Nature Conservancy ecoregional approach. It can and should be done, also considering the "small pieces," such as very specific management plans for local habitats (marshes, blocks of riparian vegetation, offshore islands, or even single species in certain instances).

Cairns (1994) advocates that only a "top-down" approach, starting with the entire system and working down to components, will ultimately succeed. Cairns states, "The component-by-component approach is not as effective when a dense population with high per capita energy consumption is competing for an ever-diminishing natural resource base." And that is just what the CRDR represents. Yet CRDR complexity suggests that in order to solve large problems, step-by-step solutions and approaches need to be identified soon. This will require strong commitments from governments and local communities. Some situations, such as conservation for the vaquita, seem simple but may in reality be highly complex and difficult for strictly nonbiological reasons, such as illegal fishing activities and net-entrapments, open access, and lack of economic alternatives (Vidal, 1995; Rojas-Bracho and Taylor, 1999) (Figure 109.6). This critically endangered marine mammal requires a "bottom-up" emergency approach as a prelude to more encompassing approaches in the overall conservation and management of the CRDR system. Other situations indicate that a large suite of regional problems might eventually contribute to a degradation of some specific aspect of the system; this seems to be the case for migratory birds (Figure 109.6; a large proportion of the illustrated factors contribute to overall population declines). Once a societal commitment to conservation of biodiversity is secured for the CRDR, individual problems can be attacked one by one. Emergency situations, such as high gill-net mortality in the vaquita, still need immediate attention.

When dealing with the potential population effects of habitat deterioration, developers and politicians often argue that migratory birds will merely move elsewhere (Anderson, 1999). This question can be addressed for the CRDR by simply stating that habitat deterioration is characteristic of the entire CRDR region and for most of the Pacific Flyway (Figures 109.4 and 109.5). Sutherland (1998) has proposed a quantitative framework for evaluating habitat deterioration effects on global populations of migratory birds and other species. Regarding widespread regional deterioration, he states, "as the population is displaced from the site of local deterioration, there will be increases in density in other sites resulting in density-dependent mortality"; this will extend to breeding success elsewhere with the ultimate degradation of species' populations. Sutherland's (1998) framework requires much knowledge and information (basic research) to derive various terms in his equations, but it can be extended to any species. Yet in many cases, there may not be enough time to gather all the ecological data necessary to adequately quantify a complex model.

Worse, however, are occasions where there is no "elsewhere" for displaced birds, and this may well be the case for some species of CRDR birds. For much of the CRDR, the idea can be simply stated as "more protections and more restorations are needed, in as many places as possible" (our paraphrase). We know of no other situation so similar in an ecological, political, and social milieu of the CRDR as the Aral Sea, where river deltas have also historically been the most productive and richest parts of the ecosystem and where "linking environmental and economic objectives is not an easy task if one takes a short time perspective" (Kindler, 1998). Kindler proposes, as a first step in mitigation, the creation of "wetland belts (wetland buffer zones)" in the ecologically important delta areas, but difficulties arise because upland water uses and delta ecological needs are separated by political boundaries. Yet wetland belts and riparian areas function in one importantly defined "ecological service": they significantly improve water quality in flow-through waters (Gilliam, 1994).

BIODIVERSITY CONSERVATION IN THE CRDR

The major foundation for migratory bird conservation in North America is the Migratory Bird Treaty Act of 1936 (with modifications in 1972). Specifically in the CRDR and the Gulf of California, efforts toward conservation of wildlife resources and associated habitats are active on both sides of the U.S.–Mexico border, in some cases utilizing a whole-system or "top-down" approach. These efforts are being supported by strong bottom-up approaches through species- or local-area-specific management plans. Conservation approach in Mexico has been mostly a matter of top-down, whole-ecosystem conservation and management rather than by single species or groups of species, such as migratory waterfowl. Yet, interspecies aggregations, categories of birds or mammals, and single species have recently come to represent immediate and short-term conservation problems that have remained unsolved. Much conservation-oriented activity is occurring in the CRDR, although much of it depends on opportunistic supplies of water that come south more by accident or stochasticity than by design (Glenn et al., 1996; Briggs and

Cornelius, 1998). The USFWS has participated with SEMARNAT and several institutions through the North American Free Trade Agreement (NAFTA) for more effective international wildlife management. More recently, the Department of the Interior has entered into a cooperative agreement with SEMARNAT to enhance U.S.–Mexico collaboration (Babbitt and Carabias, 1997). One of these collaborations is through the "sister reserves," which coordinate activities between the Biosphere Reserve of the Upper Gulf of California and the Imperial Valley National Wildlife Refuges. Other efforts include assessments of abundance, distribution, and habitat use of the Yuma clapper rail in the CRDR, as well as monitoring of waterbirds in the Biosphere Reserves. The North American Wetland Conservation Council is also supporting projects for conservation and restoration of wetlands in the CRDR.

In Mexico in the last decade, several bioregional plans to manage and conserve that nation's resources have grown through the Biosphere Reserve concept (Halfter, 1980; Dyer and Holland, 1991). In 1993, Mexico established the Reserva de la Biosfera Alto Golfo de California y Delta del Río Colorado (together with an adjacent reserve, the two form one of the largest protected areas in the world) (see figure in Alvarez-Borrego, this volume).* In 1995, Biosphere Reserve plans for this area were recognized as a biodiversity management plan by the Mexican federal agency, SEMARNAT.

In 1985, Mexico became a contracting party to the Convention on Wetlands of International Importance (RAMSAR Convention) and, in 1996, Mexico designated these same marshes (Colorado Delta) as one of two new worldwide RAMSAR sites called "Humedales del Delta del Río Colorado" (Hubert and Frazier, 1997). The RAMSAR Convention is a treaty among various governments to conserve the world's wetlands (Frazier, 1996). During 1998–1999, in support of the program, "Partners in Flight," a series of workshops throughout Mexico was carried out for the definition of important bird areas (IBAs). The riparian corridor and wetlands of the lower RCD were put on the list of IBAs for northwestern Mexico.

In 1992, the lower CRDR wetlands (along with the Marismas Nacionales, 1200 km south) were the first sites in Mexico to be designated "international reserves" in the Western Hemisphere Shorebird Reserve Network. Ducks Unlimited, Inc., Memphis, TN, and the Canadian Wildlife Service are currently assisting Mexican ornithologists to produce a Mexican Shorebird Conservation Plan that includes the CRDR as one of the most important shorebird habitats in all of Mexico (McKnight, K., personal communication, 2000). And in 1994, Mexico also became a cooperator in the North American Waterfowl Management Plan. The many restoration opportunities, consequences of water use, and feasibility of enhanced water acquisition for wetland use are outlined by Glenn et al. (1996) and Briggs and Cornelius (1997). There is yet another active program involving cooperation among many U.S. and Mexican institutions aiming to restore the Rio Hardy wetlands (Figure 109.1; Valdés-Casillas et al., 1998b, 1999; Payne et al., 1991).

Another biosphere reserve, the Islas del Golfo de California, Zona de Reserva y Refugio de Aves Migratorias y Fauna Silvestre, whose regional offices are in Guaymas, Ensenada, Loreto, and La Paz, operates mainly regarding management, protection, and research on the many islands in the Gulf of California (a management plan is currently in draft) (SEMARNAT, 1999; Zavala, A. and Figueroa, A.L., personal communication, 1999; http://uib.gym.itesm.mx/). A third reserve, the Reserva de la Biosfera, El Vizcaino operates on the Pacific Coast of Baja California in the San Ignacio Lagoon area (Sanchez-Sotomayor, V., Director, personal communication, 2000).

In the U.S., additional programs are emerging that combine the international aspects of migratory bird conservation and management. In 1994, the USFWS initiated an ecosystem-approach policy to address national fish and wildlife issues. The Salton Sea Science Subcommittee is considering logical steps for ecological restoration of the Salton Sea, mostly under

---

* There are also two issues of *CEDO News* (summer 5(2)1993 and fall/winter 6(2) 1994) that describe this significant event (Ramirez, 1994), *CEDO News* is a bilingual, educational newsletter published by the Intercultural Center for the Study of Deserts and Oceans, P.O. Box 249, Lukeville, AZ 85341.

a recently passed House of Representatives bill, the Sonny Bono Memorial Salton Sea Rec-lamation Act. It is also developing plans that include wetland restoration projects and fisheries enhancements (Szijj, L., California State Polytechnic University, Pomona, personal commu-nication, 2001; Salton Sea Authority and U.S. Bureau of Reclamation, 1998). The Environ-mental Protection Agency (USEPA) is executing an "Environmental Plan for the Mexican–U.S. Border Area" that involves environmental protection and pollution reduction under a 1990 agreement between Mexico and the U.S. (USEPA, 1992; Mumme, 1992), also called "Programa Ambiental Frontera 21." The UC MEXUS (University of California Institute for Mexico and the United States) Committee and Pacific Institute, nongovernmental organizations (NGOs)* involved in Rio Colorado Delta problems (Cohen et al., 1999; UC MEXUS, 1999), have provided additional guidelines for management and restoration of the region's resources, including the proposal for new partnerships between agricultural and natural resource interests (such as proposed in the San Francisco Bay Delta, CALFED Program; Meadows and Goetz, 1999). Another major recommendation of these two NGOs is to expand the view of problems at Salton Sea to the entire CRDR.

SOME COMPARISONS WITH OTHER PROGRAMS

We do not have space here to discuss this issue adequately, but as a beginning, we have briefly compared the Colorado Delta programs with two other well-known conservation and management programs, the Great Lakes Water Quality Agreement (Gilbertson, 1997, 1999; Gilbertson et al., 1999) and the CALFED San Francisco Bay Delta Program (Table 109.1). Some outstanding features of effective international and regional programs like these include (1) long history of interest in the region on both sides of the border, (2) long- and short-term international agreements (that include international working bodies) with continual evolution, (3) heavily shared resources in the system, (4) shared and equal responsibility for the resources and their uses, and (5) a strong funding commitment. Apparently, successful programs involve many additional attributes (Table 109.1).

Most importantly for the Colorado River Delta, strong international cooperation and aid are needed to bolster the advancing but sometimes uncoordinated conservation programs of the region. Working in mostly restored and forever-altered ecosystems is a major challenge for conservationists in the CRDR wetland areas (Glenn et al., 1996). It will be necessary for agricultural and natural resource interests to work together more closely than in the past (UC MEXUS, 1999). We believe it would be better to substitute large engineering projects with large-scale, international habitat-restoration projects and land stewardships, guided by an international coordinating commission. Techniques for the restoration of native vegetation in these habitats were pioneered by Rosenberg et al. (1991), Zedler (1995), and others; activities such as these must be an active element in the future conservation biology of the area (Environmental Defense Fund, 1996; Briggs and Cornelius, 1997, 1998; Valdés-Casillas et al., 1998a).

Ingram and co-workers (1995) have aptly stated: "Environmental problems have become so large that they can be intelligently grasped and dealt with only on a scale that transcends national boundaries." Szekely (1992) has studied and proposed a trilateral convention and treaty among the U.S., Canada, and Mexico to deal with current and future environmental problems. For wildlife and ecosystem health, we believe an international commission is needed to coordinate these efforts.

Yet, in Mexico there has been little long-term planning for enforcement activities. The creation of reserves and management plans is still in its infancy; funds are generally insufficient or lacking;

---

* UCMEXUS is the University of California Institute for Mexico and the United States (University of California, 3324 Olmsted Hall, Riverside, CA 92521; www.ucr.edu/ucmexus/). The Pacific Institute for Studies in Development, Environment, and Security (654 13th Street, Preservation Park, Oakland, CA 94612; www.pacinst.org) is an NGO dedicated to sustainable development and international security.

Table 109.1 Some Subjective Comparisons between Large-Scale Ecosystem-Oriented Restoration and Management Programs in North America and Developing Ecosystem Restoration Programs in the Rio Colorado Delta Region

Characteristic	Great Lakes	CALFED Bay Delta	Rio Colorado
Ecological crisis defined	X	X	X
Long-term treaties or agreements negotiated, continually refined	X	(X)	0
Realistic, attainable goals, objectives[a]	X	X	(X)
Associated programs integrated, allowed to continually evolve	X	(X)	0
High economic or social value of resources involved	X	X	X
Program involves "top-down" approach	X	X	X
Program involves "bottom-up" approach	X	X	(X)
Multiple stressors approached	X	X	0
Resources heavily shared by many parties, with mutual benefits	X	X	(X)
Lack of control by either party; lack of or weak differential access to resources	X	X	0
Outstanding, extensive database; continuing scientific study; "hard science" (basic and applied)	X	X	0
Healthy debate on issues	X	X	0
Results (e.g., population recoveries) have been and continue to be seen	X	X	0
Generous, enviable, funding; strong societal commitment	X	X	0
Long-term commitment	X	X	(0)
Protected zones part of solution	X	X	X
Relatively less demand for water itself	X	0	0
Extensive local involvement	X	X	(X)
"Inspired" team of lawyers and biologists	X	X	(0)

Notes: Characteristics of each regional situation are subjectively compared in their major defining characteristics and will hopefully serve as a basis for further discussions. References for the Great Lakes and CALFED (San Francisco Bay Delta) are mainly from this volume and Gilbertson (1997, 1999). Symbols under each category are strictly subjective and generally represent: X = we think that this characteristic is strongly present in the program, (X) = we think this characteristic is weakly or potentially present in the program, 0 = we think this characteristic is absent from the program, and (0) = we think this characteristic is potentially absent from the program.

[a] This is, of course, dependent on who defines these objectives; we are using "ecological health" objectives discussed in the text.

implementation of specific management plans is slow or lacking; protection and monitoring of endangered species are almost nonexistent; and the ecology and avifauna of many of the wetlands remain little studied and cataloged. The problem of wetland and biodiversity conservation in the region is complex and any conservation strategy must consider the following aspects:

1. Governments must ensure the protection of these ecosystems through enforcement, using data from academia for better decision making.
2. Support for essential research, such as biodiversity monitoring and basic ecological studies, will be essential.

3. Direct conservation of natural habitat must continue to be an important approach, and it should be followed by completion and implementation of management plans.
4. Cooperative programs among the U.S., Canada, and Mexico should expand, but solutions to the problems must be site-specific and operational, taking into account not only the biological importance of the resources but also the uses made of them by local inhabitants and their importance to local economies.
5. Local residents must not feel deprived of their rights and resources.
6. Enforcement should be complemented with a nonregulatory management program, which can also be adaptive with a participatory approach.

Regarding the Salton Sea but applicable to the entire CRDR, Hurlbert (personal communication, 1999) mentioned four natural-resource management objectives that appear achievable for the CRDR:

1. The protection and restoration of viable and functional wetlands and associated habitats
2. A continuing but environmentally sensitive agriculture
3. Commercial fishing ventures
4. Sportfishing and outdoor recreation activities

The Department of the Interior (USDI) has begun a process that more fairly allocates water supplies from the Colorado River for bioconservation purposes (*ENN News*, 5 August 1999, "California deal cuts use of Colorado River water"). The USDI also recently signed an agreement with Mexico for a binational watershed plan for the nearby San Pedro River Basin (*ENN News*, 23 June 1999, "U.S., Mexico team up to protect watershed"). Yet, we have to ask whether the powerful and influential water-use interests of the large cities of southern California, the extensive monotypic and polluting agricultural uses of Colorado River water, golf courses, resort and casino development with high-impact recreational uses on the Salton Sea and Lower Colorado River, a growing aquaculture industry, and burgeoning human populations will prevail at the cost of biodiversity, ecosystem health, and local human populations. Furthermore, will both the U.S. and Mexico, now being aware of the unique ecological and economic importance of the CRDR, allow for its further degradation and disappearance?

ACKNOWLEDGMENTS

This chapter originated in 1995 as a lecture by DWA on Gulf of California seabird conservation at the Cooper Ornithological Society and Mexican Section of Birdlife International (CIPAMEX) joint meeting in La Paz, Baja California, and then became the basis of a briefing to southern California congressional representatives in Washington, D.C. in 1996. It was further developed at a UC MEXUS workshop, a USFWS workshop, U.S. Geological Survey (BRD) workshops, a Pacific Institute workshop on the Salton Sea and Rio Colorado Delta area in 1997–1998, various workshops and meetings facilitated by the Salton Sea Science Subcommittee and then into its finalized form here. This contribution also represents a project of the New World Pelican Working Group of the World Conservation Union and Wetlands International (DWA, co-chair with A. Crivelli). Support for CV-C was provided by the North American Wetland Conservation Council, PRONATURA Sonora, the Sonora Institute, and the Summit and Compton Foundations. The UC MEXUS Program was a unifying and driving force for all of this. Thanks also go to Juan Guzman-Po and Lloyd Kiff for setting up the original paper, and to Milton Friend and others of the Salton Sea Science Subcommittee for encouragement and support along the way. Claudia Graham produced the illustrations.

NOTE ADDED IN PROOF

At the initial writing of this review, conservation activities in the CRDR were moving so rapidly that it was difficult to keep up and we feared this chapter would be outdated by the time it was published. Ironically, since 2000, little has happened and our chapter remains more current than we had imagined. One recent book adds to the CRDR literature:

Pattern, M.A. and McCaskie, R.G., *Birds of the Salton Sea: Status, Biogeography, and Ecology*, University of California Press, Berkeley, 2002.

Several scientific journals have produced or will be producing issues devoted wholly or in part to the CRDR: *Hydrobiologia*, 466, 2001: *J. Arid Environments*, 49, 2001; *Hydrobiologia*, 473, 2002; *Studies in Avian Biology*, in press, 2003.

REFERENCES

Anderson, B.W. and Ohmart, R.D., Structure of the winter duck community on the Lower Colorado River: patterns and process, in *Waterfowl in Winter,* Weller, M.W., Ed., University of Minnesota Press, Minneapolis, 1988, pp. 191–236.

Anderson, D.W., The seabirds, in *Island Biogeography in the Sea of Cortez,* Case, T.J. and Cody, M.L., Eds., University of California Press, Berkeley, 1983, pp. 246–264.

Anderson, D.W., Changes in pest control practices reduce toll on wildlife, *Calif. Agric.,* 49, 65–72, 1995.

Anderson, D.W., Saving more than the Salton Sea, *UC MEXUS News,* Summer, 2–3, 1999.

Anderson, D.W., Mendoza, J.E., and Keith, J.O., Seabirds in the Gulf of California: a vulnerable, international resource, *Nat. Resourc. J.,* 16, 483–505, 1976.

Babbitt, B. and Carabias, J., Letter of intent between the Department of the Interior and the Secretariat of Environment, Natural Resources and Fisheries of the United Mexican States for joint work in natural protected areas on the U.S.–Mexican border, May 5, 1997.

Bartonek, J., Pacific Flyway, in *Flyways: Pioneering Waterfowl Management in North America,* Hawkins, A.S. et al., Eds., U.S. Fish and Wildlife Service, Washington, D.C., 1984, pp. 395–403.

Bellrose, F., *Ducks, Geese, and Swans of North America,* Stackpole Books, Harrisburg, PA, 1980.

Bildstein, K.L., Bancroft, G.T., Dugan, P.J., Gordon, D.H., Erwin, R.M., Nol, E., Payne, L.X., and Senner, S.E., Approaches to the conservation of coastal wetlands in the Western Hemisphere, *Wilson Bull.,* 103, 218–254, 1991.

Bonillas, Y.S. and Urbina, F., Informe acerce de los recursos naturales de la parte norte de la Baja California, especialamente del Delta del Río Colorado, *Paragones del Institutó Geologio de México,* 4, 161–235, 1912–1913.

Briggs, M.K. and Cornelius, S., Opportunities for ecological improvement along the lower Colorado River and delta, Defenders of Wildlife, Washington, D.C., 1997.

Briggs, M.K. and Cornelius, S., Opportunities for ecological improvement along the lower Colorado River and delta, *Wetlands,* 18, 513–529, 1998.

Cairns, J., The current state of watersheds in the United States: ecological and institutional concerns, in *Watershed '93: A National Conference on Watershed Management,* Environmental Protection Agency, EPA 840-R-94–002, Washington, D.C, 1994, pp. 11–17.

Cisneros-Mata, M.A., Montemayor-Lopez, G., and Roman-Rodriguez, M.J., Life history and conservation of *Totoaba macdonaldi, Conserv. Biol.,* 9, 806–814, 1995.

Cohen, M.J., Morrison, J.I., and Glenn, E.P., *Haven or Hazard: The Ecology and Future of the Salton Sea,* Pacific Institute for Studies in Development, Environment, and Security, Oakland, CA, 1999.

Dyer, M.I. and Holland, M.M., The biosphere-reserve concept: needs for a network design, *BioScience,* 41, 319–336, 1991.

Environmental Defense Fund, Concept paper on the Colorado River delta restoration, Environmental Defense Fund (www.edf.org), New York, 1996.

Everett, W.T. and Anderson, D.W., Status and conservation of the breeding seabirds on offshore Pacific islands of Baja California and the Gulf of California, *Int. Counc. Bird Prot. Tech. Publ.,* 11, 115–139, 1991.

Ezcurra, E., Felger, R.S., Russell, A.D., and Equihua, M., Freshland islands in a desert sand sea: the hydrology, flora, and phytogeography of the Gran Desierto oases of northwestern Mexico, *Desert Plants,* 9, 35–63, 1988.

Farfan, C. and Alvarez-Borrego, S., Zooplankton biomass of the northernmost Gulf of California, *Cienc. Marinas,* 18, 17–36, 1992.

Felger, R.S., The Gulf of California: an ethno-ecological perspective, *Nat. Resourc. J.,* 16, 451–464, 1976.

Findley, R., The bittersweet waters of the lower Colorado, *Nat. Geogr.,* Oct., 540–569, 1973.

Fradkin, P.L., *A River No More: The Colorado River and the West,* Alfred A. Knopf, New York, 1981.

Frazier, S., *An Overview of the World's RAMSAR Sites,* Information Press, Oxford, U.K., 1996.

Funcke, E.W., Hunting antelope for museum specimens, *Field Stream,* March, 834–836, 1919.

Gilbertson, M., Are causes knowable? Some consequences of successional versus toxicological interpretations of the Great Lakes Water Quality Agreement, *Can. J. Fish. Aquat. Sci.,* 54, 483–495, 1997.

Gilbertson, M., Water quality objectives: yardsticks of the Great Lakes Water Quality Agreement, *Environ. Health Perspect.,* 107, 239–241, 1999.

Gilbertson, M. et al., Is it time for a Great Lakes ecosystem management agreement separate from the Great Lakes Water Quality Agreement? *J. Great Lakes Res.,* 25, 237–238, 1999.

Gilliam, J.W., Riparian wetlands and water quality, *J. Environ. Qual.,* 23, 896–900, 1994.

Gilmer, D.S., Miller, M.R., Bauer, R.D., and LeDonne, J.R., California's Central Valley wintering waterfowl: concerns and challenges, *Trans. North Am.Wildl. Nat. Resourc. Conf.,* 47, 441–452, 1982.

Glenn, E.P., Felger, R.G., and Burquez-Montijo, J.A., Oasis in the Colorado Delta, Cienega de Santa Clara: a remnant wetland, *CEDO News* (Centro Intercultural de Estudios de Desiertos y Oceanos, Tucson, AZ), 4, 14, 19, 31–32, 1992.

Glenn, E.P., Lee, C., Felger, R., and Zengel, S., Effects of water management on the wetlands of the Colorado River Delta, Mexico, *Conserv. Biol.,* 10, 1175–1186, 1996.

Goldman, E.A., Biological investigations in Mexico, *Smithson.Misc. Collect.,* 115, 1–476, 1951.

Grinnell, J., Birds of a voyage on Salton Sea, *Condor,* 10, 185–191, 1908.

Grinnell, J., An account of the mammals and birds of the Lower Colorado Valley, with special reference to the distributional problems presented, *Univ. Calif. Publ. Zool.,* 12, 51–294, 1914.

Haig, S.M., Mehlman, D.W., and Oring, L.W., Avian movements and wetland connectivity in landscape conservation, *Conserv. Biol.,* 12, 749–758, 1998.

Halfter, G., Biosphere reserves and national parks: complementary systems for national protection, *Impact Sci. Soc.,* 30, 269–277, 1980.

Harrison, S. and Bruna, E., Habitat fragmentation and large-scale conservation: what do we know for sure?, *Ecography,* 22, 225–232, 1999.

Hawkins, A.S., Hanson, R.C., Nelson, H.K., and Reeves, H.M., *Flyways: Pioneering Waterfowl Management in North America,* U.S. Fish and Wildlife Service, Washington, D.C., 1984.

Hendrickson, J.R., Study of the marine environment of the Northern Gulf of California, Final report, National Technical Information Service Publication N74-16008, 1974, 1–95.

Hernandez-Ayon, J.M., Galindo-Bect, M.S., Flores-Baez, B.P., and Alvarez-Borrego, S., Nutrient concentrations are high in the turbid waters of the Colorado River Delta, *Estuarine Coastal Shelf Sci.,* 37, 593–602, 1993.

Hubert, E. and Frazier, S., Update on new RAMSAR contracting parties and sites, *Wetlands,* 3, 10–11, 1997.

Ingram, H., Laney, N.K., and Gillilan, D.M., *Divided Waters: Bridging the U.S.-Mexico Border,* University of Arizona Press, Tucson, 1995.

Kindler, J., Linking ecological and development objectives: trade-offs and imperatives, *Ecolog. Appl.,* 8, 591–600, 1998.

Kingsford, R.T. and Johnson, W., Impact of water diversions on colonially-nesting waterbirds in the Macquarie Marshes of arid Australia, *Colonial Waterbirds,* 21, 159–170, 1998.

Leopold, A., *A Sand County Almanac, with Other Essays on Conservation from Round River,* Oxford University Press, New York, 1966.

Lincoln. F.C., *The Waterfowl Flyways of North America,* U.S. Fish and Wildlife Service, U.S. Department of Agriculture Circular 342, Washington, D.C., 1935.

Linduska, J.P., Ed., Waterfowl Tomorrow, U.S. Fish and Wildlife Service, Washington, D.C., 1964.

Luecke, D.F., Pitt, J., Congdon, C., Glenn, E.P., Valdes-Casillas, C., and Briggs, M., *A Delta Once More: Restoring Riparian and Wetland Habitat in the Colorado River Delta,* Environmental Defense Fund Publications, Washington, D.C., 1999.

Lumholz, C., *New Trails in Mexico: An Account of One Year's Exploration in Northwestern Sonora, Mexico, and Southwestern Arizona 1909–1910,* Charles Scribner's Sons, New York, 1912.

Maluff, L.Y., Physical oceanography, in *Island Biogeography in the Sea of Cortez,* Case, T.J. and Cody, M.L., Eds., University of California Press, Berkeley, 1983, pp. 26–45.

Meadows, R. and Goetz, J., UC shores up research for CALFED: how do you slice the water pie? *Calif. Agric.,* 53(1): 6–10, 1999.

Mellink, E., Status of the muskrat in the Valle de Mexicali and Delta del Río Colorado, Mexico, *Calif. Fish Game,* 81, 33–38, 1995.

Mellink, E. and Ferreira-Bartrina, V., On the wildlife of wetlands of the Mexican portion of the Río Colorado Delta, *Bull. South. Calif. Acad. Sci.,* 99, 2000.

Mellink, E., Palacios, E., and Gonzalez, S., Non-breeding waterbirds of the Delta of the Río Colorado, Mexico, *J. Field Ornithol.,* 68, 113–123, 1997.

Morrison, J.I., Postel, S.L., and Gleick, P.H., *The Sustainable Use of Water in the Lower Colorado River Basin,* Pacific Institute for Studies of Development, Environment, and Security, Oakland, CA, 1996.

Moyle, P.B. and Williams, J.E., Biodiversity loss in the Temperate Zone: decline of the native fish fauna of California, *Conserv. Biol.,* 4, 275–284, 1990.

Mumme, S.P., New directions in United States–Mexican transboundary environmental management: a critique of current proposals, *Nat. Resourc. J.,* 32, 539–562, 1992.

Murphy, R.C., Natural history observations from the Mexican portion of the Colorado Desert, *Proc. Linnean Soc. N.Y.,* 28–29, 43–101, 1917.

National Research Council, *Restoration of Aquatic Ecosystems: Science, Technology, and Public Policy,* National Academy Press, Washington, D.C., 1992.

Naylor, R.L., Goldburg, R.J., Mooney, H., Beverage, M., Clay, J., Folke, C., Kautsky, N., Lubchenco, J., Primavera, J., and Williams, M., Nature's subsidies to shrimp and salmon farming, *Science,* 282, 883–884, 1998.

Nichols, F.H., Cloern, J.E., Luoma, S.N., and Peterson, D.H., The modification of an estuary, *Science,* 231, 567–573, 1986.

Ohmart, R.D., Deason, W.O., and Freeland, S.J., Dynamics of marsh land formation and succession along the lower Colorado River and their importance and management problems as related to wildlife in the arid Southwest, *Trans. North Am. Wildl. Nat. Resourc. Conf.,* 40, 240–251, 1975.

Ornithological Council, The Salton Sea: a bird's eye view, *Bird Issue Brief,* 1(6), http://www.nmnh.si.edu/ BIRDNET, 1998.

Páez-Osuna, F., Guerro-Galván, S.R., and Ruiz-Fernández, A.C., The environmental impact of shrimp aquaculture and the coastal pollution in Mexico, *Mar. Pollut. Bull.,* 36, 65–75, 1998.

Patten, M.A., Mellink, E., and Gomez de Silva, H., Status and taxonomy of the Colorado Desert avifauna of Baja California, *Monogr. Field Ornithol.,* 3, 2000.

Patterson, J.H., The North American Waterfowl Management Plan and Wetlands for the Americas programmes: a summary, *Ibis,* 137, S215-S218, 1995.

Payne, J.P., Reid, F.A., and Carrera-Gonzalez, E., Feasibility study for the possible enhancement of the Colorado Delta wetlands of Baja California Norte, Mexico, unpublished report to Ducks Unlimited and Ducks Unlimited de Mexico, 1991.

Plissner, J.H., Haig, S.M., and Oring, L.W., Postbreeding movements of American avocets and implications for wetland connectivity in the western Great Basin, *Auk,* 117, 290–298, 2000.

Ramirez, R.E., Commitment in the upper gulf: the challenge of the upper Gulf of California and Colorado River Delta Biosphere Reserve, *CEDO News,* 6, 3–4, 17, 1994.

Reisner, M., *Cadillac Desert: The American West And Its Disappearing Water,* Penguin Books, New York, 1986.

Rojas-Bracho, L. and Taylor, B.L., Risk factors affecting the vaquita (*Phocoena sinus*), *Mar. Mammal Sci.,* 15, 974–989, 1999.

Rosen, M.D., Conflict within irrigation districts may limit water transfer gains, *Calif. Agric.,* 46(6): 4–7, 1992.

Rosenberg, K.V., Ohmart, R.D., Hunter, W.C., and Anderson, B.W., *Birds of the Lower Colorado River Valley,* University of Arizona Press, Tucson, 1991.

Ruiz-Campos, G. and Rodríguez-Meraz, M., Composición taxonómica y ecológia de la avifauna de los ríos El Mayor y Hardy, y areas adyacentes en el Valle de Mexícali, Mexíco, *An. Inst. Biol., Ser. Zool.,* 68, 291–315, 1997

Salton Sea Authority and U.S. Bureau of Reclamation, Salton Sea Project Work Plan, U.S. Bureau of Reclamation, Boulder City, NV, http://www.lc.usbr.gov/~g2000/sswp.html, 1998.

Santamaria-del-Angel, E., Alvarez-Borrego, S., and Muller-Karger, F.E., Gulf of California bio-geographic regions based on coastal zone color scanner imagery, *J. Geophys. Res.,* 99, 7411–7421, 1994.

Saunders, G.B. and Saunders, D.C., Waterfowl and their wintering grounds in Mexico, 1937–64, *USFWS Res. Publ.,* 138, 1–151, 1981.

Schmidt, P., Butler, D., and Petit, D., Moving migratory bird management to the next level in North America, *Wetlands Int. Publ.,* 55, 13–16, 1999.

SEMARNAT, *Programa de Manejo,* Instituto Nacional de Ecologia, Mexico City, 1999.

Shuford, W.D., Warnock, N., and Molina, K.C., The importance of the Salton Sea, California, to the Pacific Flyway waterbirds, *Stud. Avian Biol.,* in press.

Shuford, W.D., Warnock, N., Molina, K.C., Mulroony, B., and Black, A.E., Avifauna of the Salton Sea: abundance, distribution, and annual phenology, Contribution, 931, Point Reyes Bird Observatory, Final report for EPA Contract R826552-01-0 to the Salton Sea Authority, La Quinta, CA, 2000.

Skagan, S.K., Melcher, C.P., Howe, W.H., and Knopf, F.L., Comparative use of riparian corridors and oases by migrating birds in southeast Arizona, *Conserv. Biol.,* 12, 896–909, 1998.

Stephens, T., A river runs through desert agriculture, *Calif. Agric.,* 51(3): 6–10, 1997.

Sutherland, W.J., The effect of local change in habitat quality on populations of migratory species, *J. Appl. Ecol.,* 35, 418–421, 1998.

Sykes, G., *The Colorado Delta,* Lord Baltimore Press, Washington, D.C., 1937.

Szekely, A., Establishing a region for ecological cooperation in North America, *Nat. Res. J.,* 32, 563–622, 1992.

Tershy, B.R., Breese, D., and Croll, D.A., Human perturbations and conservation strategies for San Pedro Martir Island, Islas del Golfo de California Reserve, Mexico, *Environ. Conserv.,* 24, 261–270, 1997.

UC MEXUS (University of California Institute for Mexico and the United States), Alternative futures for the Salton Sea, Border Water Project Issue Paper 1, 1999, 1–23.

USEPA, EPA Summary: Environmental Plan for the Mexican-U.S. Border Area, Environmental Protection Agency (Office of International Activities), Washington, D.C., 1992.

USFWS (U.S. Fish and Wildlife Service), *Saving the Salton Sea: A Research Needs Assessment,* U.S. Department of the Interior, Washington, D.C., 1997.

Valdés-Casillas, C. et al., Information database and local outreach program for the restoration of the Hardy River wetlands, lower Colorado River delta, Baja California and Sonora, Mexico, Final report to the North American Wetland Conservation Council and others, 1998a.

Valdés-Casillas, C., Glenn, E.P., Hinojosa-Huerta, O., Carillo-Guerrero, Y., García-Hernandez, J., Zamora-Arroyo, F., Muñoz-Viveros, M., Briggs, M., Lee, C., Chavarría-Correa, E., Riley, J., Baumgardner, D., and Congdon, C., Wetland Management and Restoration in the Colorado River Delta: The First Steps, special publication of CECARENA-ITESM Guaymas and NAWCC, 1998b, pp. 1–32.

Velarde, E. and Anderson, D.W., Conservation and management of seabird islands in the Gulf of California: setbacks and successes, in *Seabirds on Islands: Threats, Case Studies and Action Plans,* Birdlife International (Birdlife Conservation Series 1), Cambridge, 1994, pp. 229–243.

Vidal, O., Population biology and incidental mortality of the vaquita, *Phocoena sinus, Rep. Int. Whaling Comm.,* 16 (special issue), 247–272, 1995.

Watling, L. and Norse, E.A., Disturbance of the seabed by mobile fishing gear: a comparison to forest clearcutting, *Conserv. Biol.,* 12, 1180–1197, 1998.

Williams, B.K., Koneff, M.D., and Smith, D.A., Evaluation of waterfowl conservation under the North American Waterfowl Management Plan, *J. Wildl. Manage.,* 63, 417–440, 1999.

Zedler, J.B., Salt marsh restoration: lessons from California, in *Rehabilitating Damaged Ecosystems,* Cairns, J., Jr., Ed., CRC Press, London, 1995, pp. 75–95.

Zedler, J.B. and Powell, A.N., Managing coastal wetlands: complexities, compromises, and concerns, *Oceanus,* 36, 19–28, 1993.

CHAPTER 110

Wildlife Disease in the Colorado Delta as an Indicator of Ecosystem Health

Tonie E. Rocke and Milton Friend

INTRODUCTION

Ecosystem health and management are relatively new paradigms for environmental management (Costanza et al., 1992) and will be continually redefined and reevaluated relative to their purpose and utility in serving human society (Costanza et al., 1992; Sutter, 1993; Grumbine, 1994; Lackey, 1998, 1999). The stimulus for ecosystem management is often associated with sustaining or restoring ecosystem health to provide values sought by humans (Costanza et al., 1992). We accept the concepts of ecosystem health and management as useful precepts. However, we argue that the orientation in serving human values often has resulted in inadequate attention being given to nonhuman species. The need to address ecosystem health for the benefit of all species that share Planet Earth is noted by Merchant (1997). This humanistic viewpoint also has economic benefits that are part of the wealth that we will pass on to future generations (Pulliam, 1995). Despite the viewpoints just noted, disease occurrence in wildlife receives inadequate consideration with regard to perspectives of ecosystem health and as a focus for ecosystem management. Evidence for this view is the limited resources allocated for wildlife disease programs and prevention efforts, and persistent attitudes within the wildlife conservation community that disease is a "natural event," not influenced by environmental conditions or anthropogenic insults (May, 1988; Friend, 1995).

Wildlife species are frequently viewed to serve as the "canary in the coal mine" for protecting and predicting human health problems, particularly in association with monitoring zoonotic diseases (Korch et al., 1989; Reisen et al., 1997) and chemical contamination of the environment (University of Connecticut, 1979; Beyer et al., 1996). However, the complexity of interacting factors that result in health impacts from contaminants requires more than measuring the presence of chemical residues in animal tissues or the environment (Anderson, 1998). Likewise, detecting antibody for infectious diseases in animals or virus in insect vectors does not directly translate to cases of human disease. Clinical cases of animal disease are generally of greater importance as an index for human health risks.

The recent resurgence of infectious diseases within human populations and concerns related to changing patterns of disease associated with predicted global climate change (Colwell, 1996; McMichael et al., 1996; Patz et al., 1996) have also created renewed interest in wildlife diseases as indices for human health (Epstein et al., 1998). However, disease emergence during recent years among free-ranging wildlife populations (Murphy, 1994; Friend, 1995, 1999) has not been seriously considered as an index of ecosystem health for wildlife. In general, the response to wildlife disease events has been typical of the response to other catastrophic events, focusing on cleanup and documentation of the event, rather than stimulating a collective focus on the environmental conditions that are influencing disease outcomes. We contend that greater attention to the health of free-ranging wildlife will provide substantial benefits for wildlife and humans alike. This philosophy goes beyond the common perspective of wildlife serving as the canary in the coal mine or as an index to impending doom for humans. We argue that wildlife disease occurrence, in addition to being a component of an environmental distress syndrome for human values (Rapport and Whitford, 1999), has important considerations for wildlife conservation (sustainability) and species biodiversity (Friend et al., 2001).

We use the Salton Sea in southern California as a case study of an ecosystem where an increase in documented disease occurrence in wild birds and fish is an obvious indicator of declining ecosystem health. In addition, we contend that, for migratory birds, the concept of ecosystem health should be expanded from individual units to that of nested but disparate units (wetlands and flyways) that provide for species well-being during the entire annual migration cycle. We also describe recent findings on wetland characteristics associated with avian botulism in waterbirds to illustrate the point that environmental indices can be used to assess and perhaps reduce the risk of disease in specific ecosystems. Our viewpoints are based on our personal opinions of what constitutes ecosystem health and are influenced by our experiences as practitioners dealing with disease in free-ranging wildlife populations.

DISEASE AS AN OUTCOME

Traditionally, population and evolutionary biologists have generally viewed infectious diseases as agents of natural selection in higher organisms (Levin et al., 1999) and, in some cases, we concur that disease agents may act in this capacity. However, this concept of disease as solely a natural event fails to recognize that disease is an outcome rather than a cause (Friend, 1995). In a very simple model (Figure 110.1), disease is a product of interactions

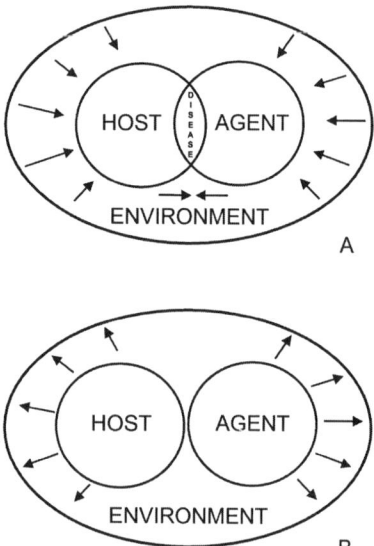

Figure 110.1 (A) Disease may be generically thought of as an outcome of interactions (area of overlap) between susceptible hosts, agents capable of causing disease, and environmental factors. In many instances, environmental conditions are the dominant factor. (B) Understanding and managing the environmental component of this triad in a manner that minimizes the potential for major disease outbreaks is a basic ecosystem management need for the conservation of wildlife populations.

between a host population, one or more agents, and environmental factors, which encompass insect vectors for transmission of some diseases, nutritional factors that affect susceptibility, and other considerations too numerous to list. Because environmental conditions play such a major role in the outcome of host–agent interactions and can influence the outcome for better or worse (from the perspective of the host population), it is reasonable to consider disease occurrence and ecology as parameters for evaluating ecosystem health, regardless of the etiologic agent involved.

Human health has benefited greatly from application of this concept. Environmental standards for sanitation are testimony to the recognized relation between environmental quality and disease outcomes that periodically reappear in the scientific literature (Dubos, 1958, and Van Cleave, 1945, as notable examples). These lessons are continually relearned as a result of infrastructure destabilization and breakdown due to war, political turmoil, and catastrophic natural disasters (Kohn, 1995; Maurice, 1995). Human-induced changes in environmental conditions are the major reason that environmental standards are needed to protect human health. Nevertheless, the relation between disease occurrence and environmental conditions has largely been ignored in the application of ecosystem health concepts to nonhuman species. An exception is the common focus on chemical contaminants. Thus, it is noteworthy that the first annual meeting of the Society for Conservation Biology included a symposium titled, "Conservation and Disease." During the meeting, it was stated that:

> Given the conspicuous role that diseases have played, and in many parts of the world continue to play, in human demography, it is surprising that ecologists have given so little attention to the way diseases may affect the distribution and abundance of other animals and plants. Until recently, for example, ecology textbooks had chapters discussing how vertebrate and invertebrate predators may influence prey abundance, but in most cases you will search the index in vain for mention of infectious diseases (May, 1988).

We contend that the conservation (sustainability) of wildlife is a valued function and product of ecosystems. Therefore, we view some wildlife mortality events as a reflection of declining ecosystem health that may negatively impact wildlife conservation, biodiversity, and human benefits derived from those ecosystems. However, the mere presence of disease in a population, even an event causing catastrophic losses, is not necessarily an indicator of poor ecosystem health. Rather, evaluations should be based on changes in the frequency of disease occurrence, the etiologic agents involved, and the magnitude of disease losses in relation to the capacity of an ecosystem to sustain its wildlife species and populations. Further, we do not advocate the absence of disease as necessarily being a desirable aspect of ecosystems, especially for diseases caused by endemic infectious agents. Actually, species naivety to some infectious organisms can result in enhanced vulnerability to disease outbreaks. Also, some host–parasite interactions may truly function to regulate populations (Jellison et al., 1958; Bell and Reilly, 1981; Hudson, 1986; Dobson, 1988), as hypothesized for other causes of mortality, such as predator–prey interactions (Crawley, 1992; Greenwood and Sovada, 1996). Depending on the functional relations of the affected species within an ecosystem, disease impacts (or lack of disease) may result in broader ecosystem perturbations.

We have emphasized two key points: (1) the outcome of disease in human and nonhuman species alike has a strong linkage to environmental quality, supporting the need for disease evaluations as an index for ecosystem health; and (2) disease impacts on wildlife populations can, in some cases, reduce their recreational, economic, and aesthetic value to human society. Therefore, evaluations of ecosystem health and ecosystem management should consider the effects of disease on the sustainability of wildlife populations at levels consistent with human values (consumptive and nonconsumptive uses).

AVIAN DISEASE AND THE SALTON SEA

The Salton Sea, located in the Imperial Valley of southern California, is a unique ecosystem. A brief review of its history will provide the necessary context for our discussion. This largest inland water body in California is of recent origin. In 1901, water was diverted from the Colorado River to irrigate the Imperial Valley and was quickly followed by settlers who cultivated the land. Flooding by the Colorado River resulted in a breach of an irrigation structure in March 1905 and extensive water flow into the Salton Trough. Unprecedented multiple flood events resulted in further diversion, eventually causing the entire flow of the river to discharge into the Salton Trough. Following Herculean efforts to close the breach, the Colorado River was finally returned to its riverbed on February 10, 1907 (Kennan, 1917). Left behind was a freshwater lake with a surface elevation 195 feet below sea level. By 1925, evaporation exceeded inflows, and the sea subsided to 250 feet below sea level (Laflin, 1995). The sea is currently stabilized at 227 feet below sea level, augmented primarily by increased irrigation plus waste-water from industry and sanitation. Evaporation and inflows have equilibrated at about 1.363 million acre-feet (Tetra Tech, Inc., 1999), resulting in a current sea that is 35 miles long, 9 to 15 miles wide, with a surface area of 380 square miles, and a volume of approximately 7.3 million acre-feet. The sea has a maximum depth of 51 ft and an average depth of about 31 ft (Cohen et al., 1999).

Since its formation, the salinity of the sea has risen to approximately 44 ppt due to precipitation of salts upon evaporation of surface water. This level of salinity exceeds ocean water by 26% and is increasing at a rate of approximately 0.5% per year (Figure 110.2) due to an annual input of approximately 5.2 million tons of dissolved salts (Tetra Tech, Inc., 1999). In addition, nutrient loading from the inflows has created a highly eutrophic water body. Approximately 15,000 tons of nutrients such as nitrogen and phosphorus annually flow into the sea (Cohen et al., 1999). Despite the hypersaline, eutrophic conditions, the Salton Sea supports the most productive sportfishery in California inland water bodies based on angler success (Black, 1985) and is habitat for an extraordinary number of bird species. With more than 400 species of birds reported at the sea and

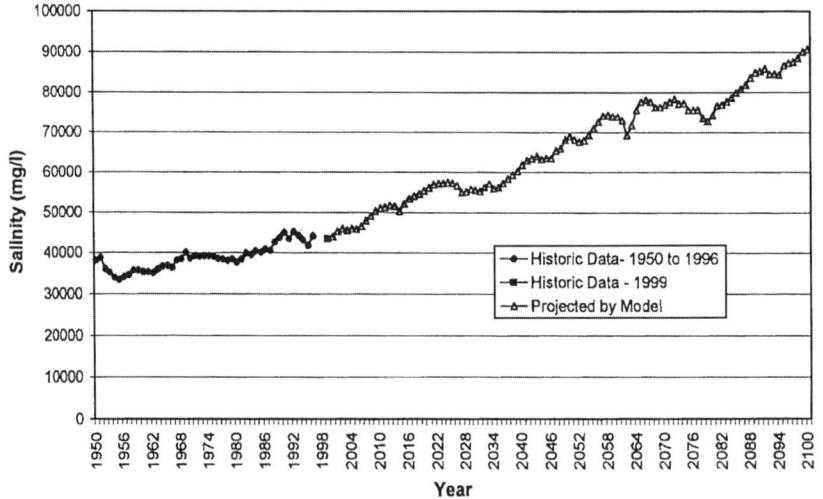

Figure 110.2 Projected increases over time in the level of Salton Sea salinity based on continuation of the current inflow of 1.363 million acre-feet of water. The annual rate of increase is approximately 0.5%. (Salinity graph provided by P.A. Weghorst, Principal Hydrologic Engineer, Salton Sea Reclamation Project, Bureau of Reclamation.)

surrounding environs (Shuford et al., 1999), the avian biodiversity of this ecosystem is greater than nearly all other wetlands in the contiguous U.S.

Avian disease is not new to the Salton Sea. An outbreak of "duck sickness" was reported at the Salton Sea as early as 1917 (Kalmbach and Gunderson, 1934), and correspondence of the Bureau of Biological Survey in the National Archives indicated major die-offs of waterfowl and other waterbirds in the mid-1920s (Holmes, 1933). The focal points for these events were the deltas of the New and Alamo Rivers at the southern end of the sea. Initial descriptions of sick birds and field signs strongly suggest type C avian botulism as the causative agent, a disease that was diagnosed as the cause of major mortality events at the sea during 1933 and 1934 (Kalmbach, 1938). A review of the Salton Sea National Wildlife Refuge (now the Sonny Bono National Wildlife Refuge) narrative reports from 1939 to the present (no reports could be located for the period from 1930 [year of establishment] to 1939, nor during the war years from 1942 through 1946) disclosed periodic occurrences of avian botulism and little else until the 1980s (Friend, 2002).

Avian cholera (*Pasteurella multocida*) first appeared within the Salton Sea ecosystem during 1978. The following year, the first major epizootic from avian cholera killed several thousand waterbirds (Brand and Duncan, 1983). The next appearance of avian cholera did not occur until February 1983, when a small event resulted in the collection of 170 carcasses. In 1984, 312 carcasses were collected during an October die-off that the California Department of Fish and Game, Wildlife Disease Investigations Laboratory diagnosed as avian botulism. Salton Sea National Wildlife Refuge records indicated an estimated loss of 500 birds from avian cholera in 1987. The following year, an estimated 2000 birds died from avian botulism and an additional 600 from avian cholera. Three different diseases were reported during 1989: 215 birds were found dead from avian botulism, approximately 50 died from avian cholera, and more than 4500 birds died from salmonellosis at a wetland near the sea.

The 1990s stand in marked contrast to the losses reported during most of the previous decade. Disease outbreaks have been reported at the Salton Sea during every year of the 1990s, and the frequency of events, magnitude of losses, and variety of causes is unprecedented for any previous decade (Tables 110.1 and 110.2). The most noteworthy of these mortality events, in 1992, is the estimated loss of 155,000 eared grebes (*Podiceps nigricollis*), approximately 7% of the North American population of this species. A similar number may have perished during a 1989 die-off that was not investigated (Jehl, 1996). The causative agent for most of these deaths remains unknown, although algal toxins are suspected of playing a role in those die-offs and in losses of thousands of grebes since then. During 1996, more than 15,000 waterbirds died from avian botulism. Approximately 9000 of those that died were white pelicans (*Pelecanus erythrorhynchos*), a loss that was estimated at between 10 to 15% of the western population of this species (U.S. Fish and Wildlife Service, 1997), and more than 1200 that died were endangered California brown pelicans (*P. occidentalis californicus*), making this the largest recorded loss from disease of an endangered species. Virtually the entire production of nestling double-crested cormorants on Mullet Island in the Salton Sea died from Newcastle disease during 1997. A similar event occurred the following year killing an estimated 6000 birds and, although virus was not isolated from dead birds, the clinical and field signs (Figure 110.3) were strongly suggestive of Newcastle disease as the cause of mortality.

Frequent large-scale fish kills that range from tens of thousands to a million or more fish (Figure 110.4) have also occurred during the 1990s, some of which have also been linked to parasites and other microbial pathogens, such as *Vibrio* spp. (National Wildlife Health Center, unpublished data), *Amyloodinium* infestations (Kuperman and Matey, 1999), and algal toxicoses. An estimated 7.6 million fish succumbed during the first week of August 1999 (Sonny Bono National Wildlife Refuge, unpublished data). These massive fish kills have been linked to disease mortality in birds, at least for the occurrence of botulism outbreaks in fish-eating birds (Rocke, unpublished data). We believe this drastic increase in mortality in fish and birds at the Salton Sea is reflective of an

Table 110.1 Reported Avian Disease Events at the Salton Sea and Nearby Wetlands

Time Period	Years with Total Losses Reported to Be					Data Source/Comments
	1–100	101–500	501–1000	1001–2000	2000+	
1925–1935[a]					Annual for entire period except 1932	Holmes, 1933;[b] Kalmbach, 1938
1936–1945[c]	1940				1939	Salton Sea National Wildlife Refuge (SSNWR) narrative reports
1946–1955[d]	1949, 1953, 1954	1947, 1955	1951		1952	SSNWR; bird losses of approx. 4000–5000 in 1952
1956–1965	1956–1959	1965	1964		1962, 1963	SSNWR; bird losses of approx. 10,000 in 1963 and 7000 in 1962
1966–1975	1966–1969, 1973	1971	1970, 1975	1974	1972	SSNWR; bird losses of approx. 5500
1976–1985[e]	1978	1983–1984			1979	SSNWR; bird losses of approx. 10,000 in 1979; National Wildlife Health Center (NWHC)
1986–1995	1990	1987, 1993, 1995			1988–1992, 1994	SSNWR; NWHC; more than 150,000 grebes died in 1992; more than 20,000 birds died in 1994
1996–1999[f]					1996–1999	SSNWR; NWHC; more than 20,000 birds died, including approx. 10,000 pelicans in 1996; more than 10,000 birds died in 1997 and approx. 20,000 birds died in 1998

a First year information is available by extrapolation from Holmes.
b Holmes, S.W., Letter to H.P. Sheldon, Bureau of Biological Survey, National Archives, Washington, D.C., November 17, 1933.
c Information only available for 1939–1941.
d No data for 1946 or 1950.
e No SSNWR data for 1977, 1978, 1981, and 1982.
f July 31, 1999.

Table 110.2 Causes of Avian Disease Reported for the Salton Sea and Nearby Wetlands[a]

Time Period	Avian Botulism	Avian Cholera	Newcastle Disease	Cause of Avian Mortality		Lead Poisoning	Pesticides	Unknown	Data Source/Comments
				Salmonellosis	Algal Toxins				
1936–1945[f]	1939, 1940								Salton Sea National Wildlife Refuge narrative reports (SSNWR)
1925[d,e]–1935	Annual except 1932					1933			Holmes, 1933;[c] Kalmbach, 1938; Tonkin, 1933[c]
1946–1955[g]	1947, 1949, 1951–1955				1955[h]			1955	SSNWR
1956–1965	1956–1959, 1962–1965					1956	1962–1963[h], 1964[j]	1956	SSNWR
1966–1975	Annual				1975[h]		1974[j]		SSNWR
1976–1985[e]	1979, 1980, 1984	1979, 1983					1978[i]		SSNWR; National Wildlife Health Center (NWHC) confirmed first cases of avian cholera in Imperial Valley in 1979
1986–1995	1988, 1989, 1991, 1994, 1995	1987–1989, 1990–1994		1989, 1991, 1995	1989[h], 1992[h], 1994[h], 1995[h]	1989, 1991	1986[j], 1989[j], 1995[j]	1992	SSNWR, NWHC; 1997 Newcastle disease outbreak was first in wild birds west of Rocky Mountains; aspergillosis 1994–1995
1996–1999	1996, 1997, 1999	1996, 1998	1997, 1998[f]	1996					SSNWR; NWHC

a Diagnosis of causes of mortalities prior to 1980 are based on laboratory analysis in some instances and evaluations of field and clinical signs in others. Where there are major questions regarding cause, we have considered the event to be of unknown cause.
b Holmes, S.W., Letter to H.P. Sheldon, Bureau of Biological Survey, National Archives, Washington, D.C., November 17, 1933.
c Tonkin, G., 1933. Letter to Chief, Biological Survey, National Archives, Washington, D.C., November 26, 1933.
d First year data are available based on extrapolation from Holmes.[b]
e No laboratory evaluations prior to 1933.
f Information only available for 1939–1941.
g No data for 1946 or 1950.
h Reported as suspected cause of death; no laboratory diagnosis.
i No SSNWR data for 1977–1978, 1981–1982.
j Die-offs on agricultural fields or in agriculture drains rather than the sea.

Figure110.3 A double-crested cormorant at the Salton Sea with unilateral paralysis, a sign of previous infection with Newcastle disease virus. (Photo by M. Friend.)

Figure 110.4 A massive die-off of gulf croaker at the Salton Sea. (Photo by M. Friend.)

ecosystem in severe stress (Rapport and Whitford, 1999). Certainly, the magnitude of bird losses during the past decade can only be viewed as a decline in the functional ability of this ecosystem to provide for the sustainability of some of its species. Studies are in progress to define the factors contributing to those disease events and to provide a foundation for management and restoration of the sea.

ECOSYSTEM HEALTH AND MIGRATORY BIRDS

We have used the Salton Sea as an example where changes in wildlife disease patterns reflect declining ecosystem health and have noted the challenges imposed by those diseases for the affected species. Another important consideration for migratory species is that their sustainability is dependent upon more than one geographic location. Therefore, since the scale of an ecosystem is determined by the management problem at hand (Lackey, 1999), an area much greater than the Salton Sea must be considered for migratory birds, including all of the geographic areas that provide for critical stages in the species' annual migratory cycle. Since these geographic areas are not contiguous, nested ecological units at disparate locations (e.g., wetlands within flyways) should be considered the functional management units (ecosystems) for the various species involved. The functional processes and products to be provided from each of these ecosystems are related to the life stage needs of those species. The collective contributions from each of these disparate ecosystems provide for species sustainability. For migratory waterbirds using the Salton Sea, this would necessitate consideration of wetlands along both the Pacific and Central Flyways, as well as wetlands and coastal areas to the south in Mexico and along the Gulf.

ENVIRONMENTAL INDICES FOR EVALUATING DISEASE RISK IN ECOSYSTEMS

Because environmental conditions play such a key role in defining the occurrence and outcome of host–disease agent interactions, in some cases environmental indices other than disease mortality or morbidity can be used to assess the potential risk of disease outbreaks in specific ecosystems. To illustrate this concept, we considered recent findings regarding wetland characteristics associated with type C avian botulism outbreaks in interior wetlands (Rocke and Samuel, 1999). Avian botulism, a paralytic disease caused by ingestion of a neurotoxin, often kills thousands of waterbirds in a single outbreak, with annual continental losses in North America approaching the hundreds of thousands to millions (Rocke and Friend, 1999). In recent years (1994 to 1997), outbreaks of avian botulism killed more than 4 million waterfowl in Canada and the U.S. (National Wildlife Health Center, unpublished data), and focused renewed attention on the potential effects of this disease in North American waterfowl populations. This is particularly relevant for species like the northern pintail (*Anas acuta*), which has accounted for a large portion of the recorded botulism mortality (15 to 20%) and has failed to show increased population levels despite generally favorable breeding conditions (U.S. Fish and Wildlife Service, 1998).

The disease is caused by a neurotoxin produced by *Clostridium botulinum* type C, an anaerobic bacterium that forms dormant spores in aerobic or other environmental conditions adverse to the bacterium (Smith and Sugiyama, 1988). These spores are highly resilient (Hofer and Davis, 1972) and are widely distributed in wetland sediments (Smith and Sugiyama, 1988). In addition, spores are highly prevalent in the tissues of aquatic invertebrates (Jensen and Allen, 1960) and vertebrates (Reed and Rocke, 1992). However, despite the ubiquitous distribution of botulinum spores, outbreaks of avian botulism are unpredictable, often occurring annually in certain wetlands but not in adjacent ones. In one study conducted on a northern California wetland, no relationship was found between botulinum cell and spore density and the occurrence of botulism outbreaks (Sandler et al., 1993). This finding suggests that density of the agent is probably not a limiting factor in the occurrence of outbreaks. Instead, results of studies conducted in wetlands with ongoing botulism outbreaks and paired control wetlands throughout the U.S. suggested that the relative risk of a botulism outbreak was predicted by complex relations among redox potential, pH, and temperature measured at the sediment–water interface (Rocke and Samuel, 1999).

To illustrate these findings, a series of predictive models was developed that relate environmental conditions to the relative risk of botulism outbreaks in wetlands (Figure 110.5). A similar risk assessment for avian botulism is currently being conducted for both fresh and brackish wetlands associated with the Salton Sea. These risk models can potentially provide a new method to identify wetlands with high risk of botulism outbreaks, to develop and evaluate alternative strategies for reducing the risk of botulism in high-risk areas, and to evaluate the effect of current wetland management practices on the risk of botulism outbreaks. A substantial advantage provided by these models is that evaluation of different management strategies can be based on the predicted relative

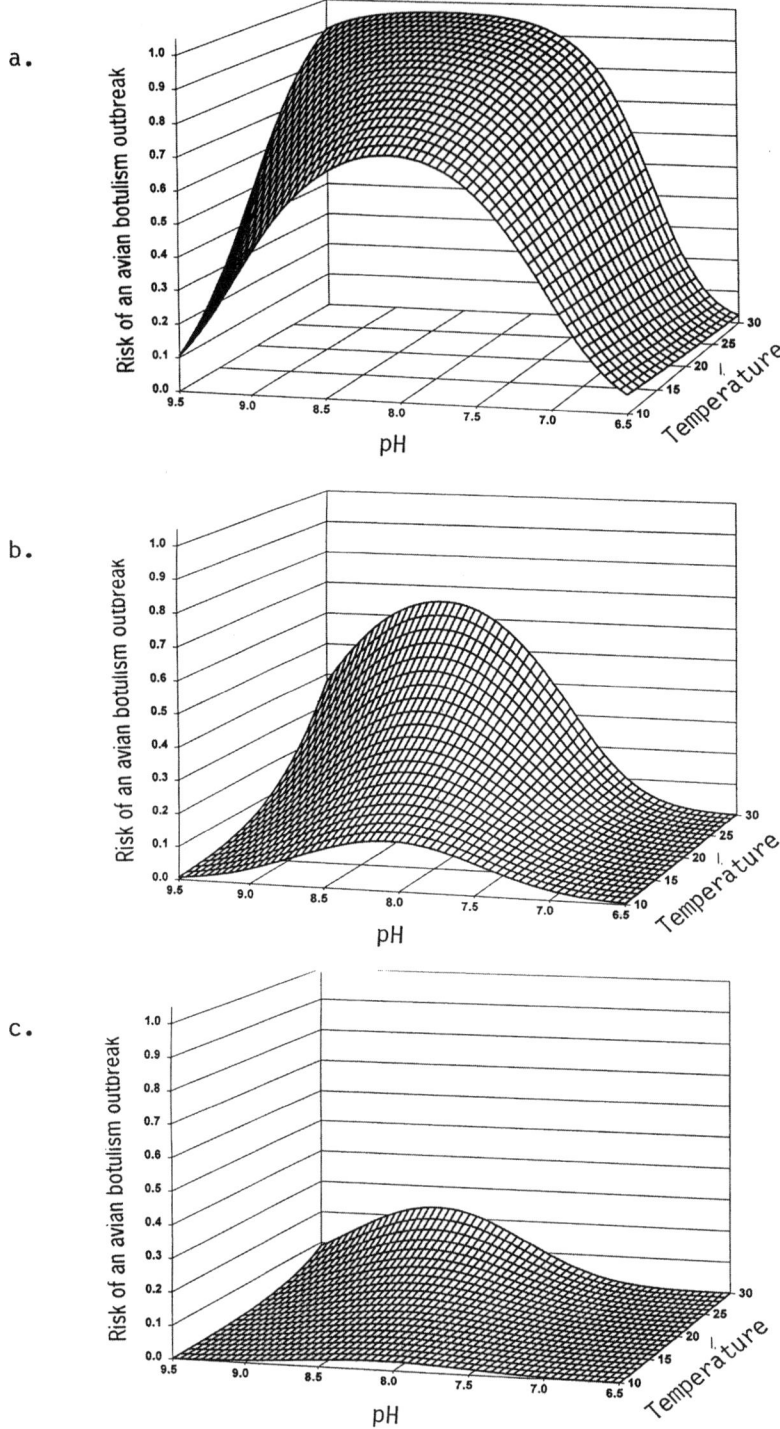

Figure 110.5 Predictive models showing the relationship between water pH and temperature, and the relative risk of botulism outbreaks in U.S. wetlands with water redox potentials of (a) −200 mv, (b) 100 mv, and (c) 300 mv. We predicted relative risk of botulism outbreaks using average water values (pH = 8.17, redox potential = −65.38 mv, standardized redox potential = 2.61 mv) for outbreak wetlands as the baseline. (From Rocke, T.E. and M.D. Samuel, Water and sediment characteristics associated with avian botulism outbreaks in wetlands, *J. Wildl. Manage.*, 63, 1249–1260, 1999. With permission.)

risk of an outbreak and would not be dependent on the occurrence of avian mortality from botulism. With additional research, these risk models also may be useful in evaluating the suitability of specific wetland acquisition and management actions in providing for the conservation of wetland-dependent avifauna.

CONCLUSION

The contemporary paradigms of ecosystem health and management are conceptually debated and perused on the basis of perspectives that address values for human society; both are useful but arbitrary concepts in the manner in which they are considered and applied. We offer perspectives that expand the considerations and application of these concepts to evaluations of wildlife health as an end point in itself; expand the concepts from evaluation and management of a single ecosystem to the need for considering nested, but disparate ecosystems when addressing the sustainability of migratory wildlife populations, and suggest that the occurrence of some diseases and environmental indices for the risk of disease can be used for evaluating ecosystem health. We challenge the notion that all wildlife disease events are "natural" regulators within ecosystems. We consider that disease is an outcome rather than a cause and note two important associations: (1) environmental conditions are a primary factor underlying disease occurrence, and (2) ecosystems in which disease occurrence is of sufficient prevalence to be of concern are rarely "natural systems." These systems are often altered by human actions or are human-created systems. Human interventions that created the alterations frequently change the relationship between hosts, disease agents, and the environment. Disease outbreaks within altered environments tend to have more-lasting negative biological impacts and can seriously challenge the ability of the ecosystem to support the wildlife values sought by human society.

Unfortunately, much of the world's wildlife habitat has undergone significant human-induced alterations and an increasing amount is human-created habitat, such as the Salton Sea. This trend will continue for the foreseeable future due to landscape changes and other actions associated with the continual growth and changing demography of human populations (Vitousek et al., 1997; Ayensu et al., 1999). Greater attention must be given to defining ecosystem health and the application of ecosystem management to these altered and created systems as habitat for the sustainability of wildlife populations and biodiversity. In many ways, these "ecosystems" are different from natural ecosystems and must be evaluated and managed in their own image and purpose, not against standards established for natural systems. In both instances, human values are the motivation for outcomes to be derived.

If contemporary society is to serve humankind and the other species of our world, we must view all natural, altered, and human-created ecosystems as important components of the mosaic that is our environment and not pit one against another as superior and inferior components. To do less would express arrogance regarding our level of understanding of ecosystem health and our ability to manage it. To place this statement in context for wildlife and ecosystems, we quote Aldo Leopold (1933). For wildlife:

> ... doctoring is of recessive importance in health control ... the real determinants of disease mortality are the environment and the population, both of which are being "doctored" daily, for better or for worse, by gun and axe, and by fire and plow (Leopold, 1933).

For ecosystems:

> The effort to control the health of the land has not been very successful.... In general, the trend of the evidence indicates that in the land, just as in the human body, the symptom may be in one organ and the cause in another. The practices we now call conservation are, to a large extent, local alleviations

to biotic pain. They are necessary, but they must not be confused with cures. The art of land-doctoring is being practiced with vigor, but the science of land-health is a job for the future (Leopold, 1941).

The underlying thesis for Leopold's land-health evaluation is society's failure to understand and respond to cause rather than symptom. We close by asking the question: how far have we come relative to Leopold's statements of more than a half-century ago regarding our understanding and ability to address wildlife and ecosystem health? Our answer: not nearly far enough; the science of land health still remains a challenge for the future.

REFERENCES

Anderson, D.W., Evaluation and impact of multiple-stressors on ecosystems: three classic case histories, in *Multiple Stresses in Ecosystems,* Cech, J.J., Jr., Wilson, B.W., and Crosby, D.G., Eds., Lewis Publishers, New York, 1998, pp. 3–8.

Ayensu, E., van R. Claasen, D., Collins, M., Dearing, A., Fresco, L., Gadgil, M., Gitay, H., Glaser, G., Juma, C., Krebs, J., Lenton, R., Lubchenco, J., McNeely, J.A., Mooney, H.A., Pinstrup-Anderson, P., Ramos, M., Raven, P., Reid, W.V., Samper, C., Sarukhan, J., Schei, P., Tundisi, J.G., Watson, R.T., Guanhua, X., and Zakri, A.H., International ecosystem assessment, *Science,* 286, 685–686, 1999.

Bell, J.F. and Reilly, J.R., Tularemia, in *Infectious Diseases of Wild Mammals,* 2nd ed., Davis, J.W., Karstad, L.H., and Trainer, D.O., Eds., Iowa State University Press, Ames, 1981, pp. 213–231.

Beyer, W.N., Heinz, G.H., and Redmon-Norwood, A.W., *Environmental Contaminants in Wildlife: Interpreting Tissue Concentrations,* CRC Press, Boca Raton, FL, 1996.

Black, G., The Salton Sea sportfisheries, California Department of Fish and Game Region V, Information Bulletin 0010, Sacramento, 1985.

Brand, C.J. and Duncan, R.M., Avian cholera in the American flamingo, *Phoenicopterus ruber*: a new host record, *Calif. Fish Game,* 69, 190–191, 1983.

Cohen, M.J., Morrison, J.I., and Glenn, E.P., *Haven or Hazard: The Ecology and Future of the Salton Sea,* Pacific Institute, Oakland, CA, 1999.

Colwell, R.R., Global climate and infectious disease: the cholera paradigm, *Science,* 274, 2025–2031, 1996.

Costanza, R., Norton, B.G., and Haskell, B.D., Eds., *Ecosystem Health: New Goals for Environmental Management,* Island Press, Washington, D.C., 1992.

Crawley, M.J., Ed., *Natural Enemies: The Population Biology of Predators, Parasites and Diseases,* Blackwell Scientific, Boston, 1992.

Dobson, A.P., Restoring island ecosystems: the potential of parasites to control introduced mammals, *Conserv. Biol.,* 2, 31–39, 1988.

Dubos, R.J., The evolution of infectious diseases in the course of history, *Can. Med. Assoc. J.,* 79, 445–451, 1958.

Epstein, P.R., Sherman, B.H., Spanger-Siegfried, E., Langston, A.N., Prasad, S., and McKay, B., Marine Ecosystems: Emerging Diseases as Indicators of Change, Technical report, Center for Health and the Global Environment. Harvard Medical School, Boston, 1998.

Friend, M., Increased avian diseases with habitat change, in *Our Living Resources,* LaRoe, E.T. et al., Eds., National Biological Service, Washington, D.C., 1995, pp. 401–405.

Friend, M., Duck plague: emergence of a new cause of waterfowl mortality, in *Status and Trends of the Nation's Biological Resources,* Mac, M.J. et al., Eds., U.S. Department of the Interior, U.S. Geological Survey, Washington, D.C., 1999, pp. 458–460.

Friend, M., Avian disease at the Salton Sea, *Hydrobiologia,* 473, 293–306, 2002.

Friend, M. and Franson, J.C., Newcastle disease, in *Field Manual of Wildlife Diseases*, 1999, pp. 175–190.

Friend, M., McLean, R.G., and Dein, F.J., Disease emergence in birds: Challenges for the twenty first century, *Auk,* 118, 290–303, 2001.

Greenwood, R.J. and M.A. Sovada, Prairie duck populations and predation management, *Trans. North Am. Wildl. Nat. Resourc. Conf.,* 61, 31–42, 1996.

Grumbine, R.E., What is ecosystem management? *Conserv. Biol.,* 8, 27–38, 1994.

Hofer. J.W. and Davis, J., Survival and dormancy, *Tex. Med.,* 68, 80–81, 1972.

Holmes, S.W., Letter to H.P. Sheldon, Bureau of Biological Survey, National Archives, Washington, D.C., November 17, 1933.

Hudson, P.J., The effect of a parasitic nematode on the breeding production of red grouse, *J. Anim. Ecol.,* 55, 85–92, 1986.

Jehl, J.R., Jr., Mass mortality events of eared grebes in North America, *J. Field Ornithol.,* 67, 471–476, 1996.

Jellison, W.L., Bell, J.F., Vertrees, J.D., Holmes, M.A., Larson, C.L., and Owen, C.R., Preliminary observations on disease in the 1957–58 outbreak of *Microtus* in western United States, *Trans. North Am. Wildl. Conf.,* 23, 137–145, 1958.

Jensen, W.I. and Allen, J.P., A possible relationship between aquatic invertebrates and avian botulism, *Trans. North Am. Wildl. Nat. Resourc. Conf.,* 25, 171–180, 1960.

Kalmbach, E.R., Botulism: A Recurring Hazard to Waterfowl, Wildlife Research and Management leaflet BS-120, U.S. Department of Agriculture, Bureau of Biological Survey, Washington, D.C., 1938.

Kalmbach, E.R. and Gunderson, M.F., Western Duck Sickness: A Form of Botulism, U.S. Department of Agriculture Technical Bulletin 411, Washington, D.C., 1934.

Kennan, G., *The Salton Sea: An Account of Harriman's Fight with the Colorado River,* Macmillan, New York, 1917.

Kohn, G.C., Ed., *Encyclopedia of Plague and Pestilence,* Facts on File, New York, 1995.

Korch, G.W., Childs, J.E., Glass, G.E., Rossi, C.A., and LeDuc, J.W., Serologic evidence of hantaviral infections within small mammal communities of Baltimore, Maryland: spatial and temporal patterns and host range, *Am. J. Trop. Med. Hyg.,* 41, 230–240, 1989.

Kuperman, G.I. and Matey, V.E., Massive infestation by *Amyloodinium ocellatum* (Dinoflagellida) of fish in a highly saline lake, Salton Sea, California, USA, *Dis. Aquat. Organ.,* 39, 65–73, 1999.

Lackey, R.T., Seven pillars of ecosystem management, *Landscape Urban Plann.,* 40, 21–30, 1998.

Lackey, R.T., Radically contested assertions in ecosystem management, *J. Sustainable For.,* 9, 21–34, 1999.

Laflin, P., *Salton Sea: California's Overlooked Treasure,* Coachella Valley Historical Society, Indio, CA, 1995.

Leopold, A., *Game Management,* Charles Scribner's Sons, New York, 1933.

Leopold, A., Wilderness as a land laboratory, *Living Wilderness,* 6, 3, 1941.

Levin, B.R., Lipsitch, M., and Bonhoeffer, S., Population biology, evolution, and infectious disease: convergence and synthesis, *Science,* 283, 806–809, 1999.

Maurice, J., Russian chaos breeds diphtheria outbreak, *Science,* 267, 1416–1417, 1995.

May, R.M., Conservation and disease, *Conserv. Biol.,* 2, 28–30, 1988.

McMichael, A.J., Haines, A., Slooff, R., and Kovats, S., Eds., *Climate Change and Human Health,* World Health Organization, Geneva, Switzerland, 1996.

Merchant, C., First first! The changing ethics of ecosystem management, *Hum. Ecol. Rev.,* 4, 25–30, 1997.

Murphy, F.A., New, emerging, and re-emerging infectious diseases, *Adv. Virus Res.,* 43, 1–52, 1994.

Patz, J.A., Epstein, P.R., Burke, T.A., and Balbus, J.M., Global climate change and emerging infectious diseases, *JAMA,* 275, 217–223, 1996.

Pulliam, H.R., Foreword, in *Our Living Resources,* LaRoe, E.T. et al., Eds., National Biological Survey, Washington, D.C., 1995, p. v.

Rapport, D.J. and Whitford, W.G., How ecosystems respond to stress, *BioScience,* 49, 193–203, 1999.

Reed, T.M. and Rocke, T.E., The role of avian carcasses in botulism outbreaks, *Wildl. Soc. Bull.,* 20, 175–182.

Reisen, W.K., Lathrop, H.D., Presser, S.B., Hardy, J.L., and Gordon, E.W., Landscape ecology of arboviruses in southeastern California: temporal and spatial patterns of enzootic activity in Imperial Valley, 1991–1994, *J. Med. Entomol.,* 34, 179–188, 1997.

Rocke, T.E. and Friend, M., Avian botulism, in *Field Manual of Wildlife Diseases,* Friend, M. and Franson, J.C., Eds., U.S. Geological Survey, Washington, D.C., 1999, pp. 271–281.

Rocke, T.E. and Samuel, M.D., Water and sediment characteristics associated with avian botulism outbreaks in wetlands, *J. Wildl. Manage.,* 63, 1249–1260, 1999.

Sandler, R.J., Rocke, T.E., Samuel, M.D., and Yuill, T.M., Seasonal prevalence of *Clostridium botulinum* type C in sediments of a northern California wetland, *J. Wildl. Dis.,* 29, 533–539, 1993.

Shuford, W.D., Warnock, N., and Molina, K.C., The Avifauna of the Salton Sea: A Synthesis, Report to Salton Sea Science Subcommittee, Point Reyes Bird Observatory, Stinson Beach, CA, 1999.

Smith, L. and Sugiyama, H., *Botulism: The Organism, Its Toxins, the Disease,* Charles C Thomas, Springfield, IL, 1988.

Sutter, G.W. II, A critique of ecosystem health concepts and indexes, *Environ. Toxicol. Chem.,* 12, 1533–1539, 1993.

Tetra Tech, Inc., Salton Sea Restoration Project, Administrative draft EIS/EIR, San Francisco, 1999.

University of Connecticut, *Animals as Monitors of Environmental Pollutants,* National Academy of Science, Washington, D.C., 1979.

U.S. Fish and Wildlife Service, Saving the Salton Sea: A Research Needs Assessment, U.S. Fish and Wildlife Service, Washington, D.C., 1997.

U.S. Fish and Wildlife Service. Waterfowl Population Status, U.S. Fish and Wildlife Service, Washington, D.C., 1998.

Van Cleave, H.J., Some influences of global war upon problems of disease, *J. Am. Diet. Assoc.,* 21, 513–515, 1945.

Vitousek, P.M., Human domination of earth's ecosystems, *Science*, 277, 494, 1997.

CHAPTER 111

Contaminants without Borders: A Regional Assessment of the Colorado River Delta Ecosystem

Miguel A. Mora, Jaqueline García, Maria Carpio-Obeso, and Kirke A. King

INTRODUCTION

The Colorado River Delta Region (CRDR) is comprised of large agricultural areas in the U.S. (Imperial and Yuma districts) and Mexico (Mexicali and San Luis districts), the Salton Sea, and the Lower Colorado River from southern Arizona to the Gulf of California (Valdés-Casillas et al., 1998). Before agricultural development, the lower delta comprised large riparian and wetland areas (Sykes, 1937; Glenn et al., 1999). Currently, most water from the Colorado River below Imperial Dam is used for irrigation in California and Mexico, and only remnants of the former CRDR wetlands remain. However, the Cienega de Santa Clara and the Rio Hardy wetlands (Figure 111.1) provide important habitat for many resident and migratory birds, including endangered species such as the desert pupfish (scientific names are provided in Table 111.1) and the Yuma clapper rail (Abarca et al., 1993; Mellink et al., 1997). Detailed descriptions of the CRDR and its wetlands are provided elsewhere (Sykes, 1937; Glenn et al., 1995; 1999; Friederici, 1998; Valdés-Casillas et al., 1998).

The CRDR is affected by multiple contaminant stressors, including pesticides and heavy metals from agriculture and other anthropogenic activities. The Mexicali Valley in Mexico is located just south of the Imperial Valley in California. Approximately 70% of the cultivated land in Mexicali (182,000 ha) is irrigated by gravity with water from the Colorado River (Secretaria de Agricultura y Recursos Hidraulicos, 1984). Because most water is used first for agriculture in Mexico and the U.S., it is likely that organisms using wetlands in the CRDR are exposed to considerable amounts of pesticides and other contaminants. Runoff from the Mexicali Valley is discharged directly to drainages and to the Rio Hardy. The Rio Hardy also receives brine waste, with potentially high concentrations of arsenic (As) and boron (B), from a geothermal energy plant located at Cerro Prieto, approximately 30 km south of the city of Mexicali (Comisión Federal de Electricidad, 1987, unpublished report); however, the plant stopped releasing this brine to the river in 1989 (Campoy, J., SEMARNAP, personal communication, 2000). The Cienega de Santa Clara also receives runoff from the Wellton-Mohawk irrigation and drainage district located near the confluence of the Gila and Colorado Rivers in the U.S. (Figure 111.1; Burnett et al., 1993).

This chapter reviews studies of contaminants conducted from 1970 to the present in the environment and biota of the CRDR, including published papers, unpublished reports, and unpublished data of the coauthors. The earliest studies were conducted in the mid-1970s on

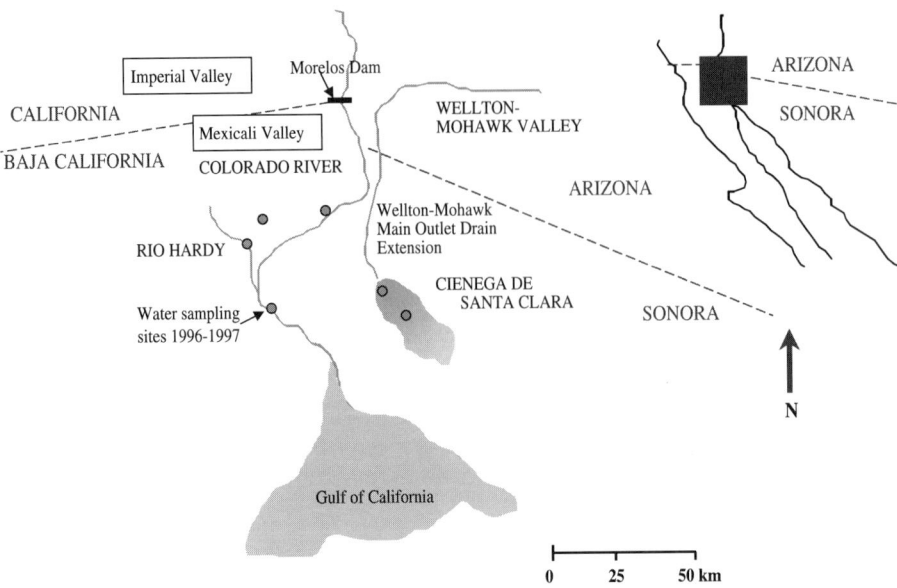

Figure 111.1 Map of the Colorado River Delta Region (Mexican side) showing locations of Cienega de Santa Clara, Rio Hardy, and water sampling locations.

Table 111.1 Scientific Names of Species Mentioned in Text and Tables

Group	Common Name	Scientific Name
Birds	Pied-billed grebe	*Podylimbus podiceps*
	Double-crested cormorant	*Phalacrocorax auritus*
	Black-crowned night heron	*Nycticorax nycticorax*
	Least bittern	*Ixobrychus exilis*
	Green heron	*Butorides virescens*
	Cattle egret	*Bubulcus ibis*
	Great egret	*Ardea alba*
	Mallard	*Anas platyrhyncos*
	Ruddy duck	*Oxyura jamaicensis*
	Yuma clapper rail	*Rallus longirostris yumanensis*
	Sora	*Porzana carolina*
	Common moorhen	*Gallinula chloropus*
	American coot	*Fulica americana*
	Mourning dove	*Zenaida macroura*
	Southwestern willow flycatcher	*Empidonax traillii extimus*
	Great-tailed grackle	*Quiscalus mexicanus*
	Red-winged blackbird	*Agelaius phoeniceus*
Fish	Desert pupfish	*Cyprinodon macularius*
	Tilapia	*Tilapia* spp.
	Mullet	*Mugil* spp.
	Carp	*Cyprinus carpio*
	Channel catfish	*Ictalurus punctatus*
	Largemouth bass	*Micropterus salmoides*
Invertebrates	Clams	*Corbicula fluminea*

aquatic invertebrates and fish from irrigation canals and drainages in the Mexicali Valley (Guardado-Puentes, 1976; Gutierrez-Galindo et al., 1988a,b). However, contaminant-oriented studies in northwest Mexico were conducted until the mid-1980s (Mora and Anderson, 1991; Mora et al., 1987). These studies assessed hazards of environmental pollution in northwest Mexico on migratory and local wildlife, and investigated variations in seasonal dynamics of pollutants in coastal and inland wetland ecosystems. Initial studies with waterfowl suggested the occurrence of some locally contaminated areas in northwest Mexico (Mora et al., 1987). Subsequently, the levels of contaminants in CRDR wildlife were evaluated from 1985 to 1989 (Mora, 1991; Mora and Anderson, 1991, 1995).

In the 1990s, residues of selenium (Se) and other trace elements in water, sediments, and biota of the Cienega de Santa Clara were measured (García-Hernández, 1998). Those efforts continue and are focused on evaluating bioaccumulation of Se and mercury (Hg) and their potential threats to wildlife of the Colorado delta in general. Other contaminant monitoring studies have been conducted over the years along the lower Colorado River and in the Salton Sea in the U.S. (White et al., 1987; Rasmussen, 1988; Setmire et al., 1990; King et al., 1993). Additionally, water, sediment, and fish samples from eight stations in the lower Colorado River and New River were analyzed for organic and inorganic contaminants during 1995–1996 as part of a binational study (Crane et al., 1996). In this chapter, we emphasize studies conducted during the past 30 years in the CRDR on the Mexican side, and we make some comparisons with data collected in the U.S.

CONTAMINANT STUDIES IN THE COLORADO RIVER DELTA

Pesticides

The use of the organochlorine insecticide DDT (1,1,1-trichloro-2,2-bis[p-chlorophenyl)ethane) in Mexico declined considerably during the 1970s but continued in agriculture at least until 1978 (Fertilizantes Mexicanos, 1981). In 1972, approximately 220,000 kg of DDT were used on cotton in the Mexicali Valley (Alvarez-Borrego et al., 1973). Another 275,000 kg of DDT were used in all of northwest Mexico during 1978 (Fertilizantes Mexicanos, 1981). During the 1980s, production of DDT in Mexico averaged 5000 tons per year, with its use restricted for malaria control (Lopez-Carrillo et al., 1996).

The earliest organochlorine (OC) contaminant data from the Mexicali Valley were obtained from analyses of sediments and clams in the 1970s and 1980s (Table 111.2; Alvarez-Borrego et al., 1973; Guardado-Puentes, 1976; Gutierrez-Galindo et al., 1988a,b). In the early 1970s, sediments from the upper Gulf of California had p,p'-DDT and p,p'-DDE concentrations of 4.4 and 7 ng/g dry weight (dw), respectively (Alvarez-Borrego et al., 1973), whereas clams from the Mexicali Valley had mean levels of DDE, the most persistent metabolite of DDT, as high as 7.7 µg/g wet weight (ww) (Table 111.2; Guardado-Puentes, 1976). By the mid-1980s, mean DDE in clams and fish collected from drains and canals in the Mexicali Valley were all <0.2 µg/g ww (Table 111.2;

Table 111.2 Mean DDE Concentrations in Sediments (µg/g dw) from the Gulf of California, and Clams and Fish (µg/g ww) from the Mexicali Valley, Mexico

Species	N	DDE	Year	Reference
Marine sediments		0.007	1972	Alvarez-Borrego et al., 1973
Clams	9	7.7	1973	Guardado Puentes, 1976
Tilapia	11	0.036	1985	Gutierrez-Galindo et al., 1988b
	14	0.056	1986	Gutierrez-Galindo et al., 1988b
Carp	12	0.183	1985	Gutierrez-Galindo et al., 1988b
	11	0.106	1986	Gutierrez-Galindo et al., 1988a
Clams	14	0.134	1985	Gutierrez-Galindo et al., 1988a
	14	0.100	1986	Gutierrez-Galindo et al., 1988a
	2	0.146	1998	King, unpublished data

Table 111.3 Geometric Mean DDE Concentrations (µg/g ww, ranges in parentheses) in Carcasses of Birds from the Mexicali Valley, Mexico, 1986

Species	N	Summer	N	Winter
Pied-billed grebe	2	1.2		
Double-crested cormorant	1	11.5	8	1.8 (0.8–4.2)
Cattle egret	7	2.0 (1.0–1.4)	8	1.0 (0.5–2.0)
Great-tailed grackle	6	3.1 (1.3–7.0)	8	4.1 (1.8–9.4)
Red-winged blackbird	4	1.7 (0.5–6.1)	4	1.0 (0.4–2.6)
Mourning dove	6	0.04 (0.02–0.09)	9	0.03 (0.02–0.07)

Compiled from Mora and Anderson, 1991.

Gutierrez-Galindo et al., 1988; Gutierrez-Galindo et al., 1988a,b). In 1998, levels of DDE in clams from three sites in the CRDR were also <0.2 µg/g ww (Table 111.2; King, unpublished data).

In a detailed study of the accumulation of OCs in wildlife, 63 specimens from 6 bird species (pied-billed grebes, double-crested cormorants, cattle egrets, mourning doves, great-tailed grackles, and red-winged blackbirds) were collected from agricultural fields in the Mexicali Valley and from the Rio Hardy during July–August (summer) and December (winter) 1986 (Mora and Anderson, 1991). Results showed that many OCs were present in carcasses in low amounts, but DDE was present at the highest concentrations with mean levels ranging from 0.04 to about 12 µg/g ww (Table 111.3). Mean DDE concentrations were not different between summer and winter. Hexachlorocyclohexane (HCH) concentrations (not shown in Table 111.3) were higher in mourning doves (up to 1.4 µg/g) and grackles (up to 0.27 µg/g) during the winter, indicating some winter accumulation of HCH at Mexicali. The mean DDE concentrations reported by Mora and Anderson (1991) were similar to those reported in carcasses of birds from the southwestern U.S. in general (Clark and Krynitsky, 1983; Henny et al., 1985; White and Krynitsky, 1986). Mean DDE concentrations in eggs of cattle egrets from Mexicali were lower than those in egrets and herons from the Salton Sea; however, these differences might be due to differences in diets rather than differences between sites (Table 111.4; Mora, 1991; Ohlendorf and Marois, 1990).

Mora and Anderson (1991) also sampled in the Yaqui and Culiacan Valleys, and found exposure of birds to DDE and other OCs to be lower than in the Mexicali Valley. The Yaqui and Culiacan Valleys are located at approximately 850 and 1400 km south, respectively, from the Mexicali Valley. On average, 52,070 ha per year were cultivated with cotton at Mexicali from 1978 to 1988 (Covarrubias, D., SARH, personal communication, 1989). Consequently, application of OC insecticides in this valley may have been more intense than in other areas of northwest Mexico which produce less cotton. One conclusion from this study was that some wetland habitats and agricultural ecosystems of northwest Mexico represented potential contamination sites for wintering and resident birds.

Table 111.4 Geometric Mean DDE Concentrations (µg/g ww, ranges in parentheses) in Bird Eggs from the Colorado River Delta Region

Species	Location	N	DDE	Year	Ref.
Cattle egret	Mexicali Valley	40	3.2 (2.3–4.5)	1987	Mora, 1991
		10	3.5 (2.1–5.7)	1988	Mora, 1991
Great egret	Salton Sea	10	24 (16–48)	1985	Ohlendorf and Marois, 1990
Black-crowned night heron	Salton Sea	10	8.6 (2.5–20)	1985	Ohlendorf and Marois, 1990

In 1995 to 1996, a binational study measured organic and inorganic contaminants in water, sediments, and fish from the lower Colorado River and New River (Crane et al., 1996). The lower Colorado River monitoring stations were located at the All American Canal, Northern International Border, Yuma Main Drain, Gila River, and Wellton-Mohawk Drain. The New River monitoring stations were located at Mexicali, Calexico, and Westmoreland. In all, 59 fish (carp, channel catfish, and largemouth bass) were collected and composited into 17 samples that were analyzed for trace and organic chemicals. Fish were not collected from Wellton-Mohawk, Gila River, and Mexicali. Some of the OC concentrations in fish exceeded selected criteria (set concentration values for the protection of aquatic life, wildlife, and human health) at some monitoring stations. Particularly, carp and channel catfish from the Westmoreland site in the New River and the Yuma main drain exceeded the National Academy of Sciences (NAS) recommended guidelines of 0.1 µg/g ww for toxaphene and endosulfan and 1 µg/g ww for total DDT (Crane et al., 1996). Of 88 pesticides analyzed in water, 6 (chlorpyrifos, EPTC, simazine, diazinon, pronamide, and DCPA) were detected in the lower Colorado River at concentrations ≤0.1 µg/l; and 13 (fonofos, malathion, atrazine, metribuzinsencor, trifluralin, linuron, disulfoton, and penidmethalin, and those detected in the lower Colorado River) in the New River had concentrations ≤0.21 µg/l (M. Carpio-Obeso, unpublished data).

Heavy Metals and Metalloids

The accumulation of heavy metals and metalloids on biota of the Colorado River delta has been addressed more frequently than the accumulation and effects of organochlorine contaminants and other organic chemicals. Heavy metals and metalloids have been detected in irrigation drain water in some areas of the southwestern U.S. at concentrations that are associated with reproductive, teratogenic, and behavioral anomalies in birds. For example, increased bird mortalities were linked with high concentrations of Se at the Kesterson Reservoir in California (Ohlendorf et al., 1986a). Agricultural ecosystems in northwest Mexico (i.e., Mexicali and San Luis Rio Colorado valleys) are intensively irrigated and have great potential for accumulation of elevated concentrations of trace elements in drainwater. Consequently, high concentrations of some elements in these areas could also result in detrimental effects on resident and migratory birds and other wildlife of the Colorado River delta.

Until the mid-1980s, there had only been a few studies on fish and clams that evaluated the environmental hazards and concentrations of trace elements on wildlife of the Mexicali Valley and Colorado delta. Concentrations of Hg in fish and clams from irrigation canals and drainages of the Mexicali Valley were measured to evaluate the extent of Hg contamination by the geothermal energy plant at Cerro Prieto (Gutierrez-Galindo et al., 1988). Mercury concentrations in fish and clams (0.05 and 0.11 µg/g dw, respectively) (Gutierrez-Galindo et al., 1988a,b), were low and not of public health concern.

During 1985 and 1986, in order to determine accumulation of trace elements in birds, a portion of liver was taken from 61 carcasses collected from the Mexicali Valley (Mora and Anderson, 1995). Fish samples from tilapia and mullet were also collected during the same period. Selenium concentrations were significantly higher ($p < 0.05$) in double-crested cormorants and lower in mourning doves than in other species (Table 111.5). This is not surprising because cormorants are the most aquatic of the species sampled and Se accumulates primarily in aquatic food chains (Ohlendorf et al., 1986a,b; Saiki et al., 1993). Selenium concentrations were similar among cattle egrets, great-tailed grackles, and red-winged blackbirds. These species are all intermediate in type of habitat used (aquatic–terrestrial) and probably share some common food sources. The concentrations of Se in double-crested cormorants from the Mexicali Valley were near or above the threshold at which reproductive effects were apparent in American coots at the Kesterson Reservoir during 1984 (Ohlendorf et al., 1987), and were 1.3 to 2.5 times less than those reported in birds from the Salton Sea in 1986 (Table 111.6; Setmire et al., 1990; White

Table 111.5 Geometric Mean Concentrations (µg/g dw, ranges in parentheses) of Selenium and Boron in Livers of Birds from the Mexicali Valley, Mexico, 1985–1986[a]

Species	N	Se	B
Double-crested cormorant	9	16.7 A (9.5–23.9)	13.8 AB (9.5–26.9)
Cattle egret	15	4.6 B (1.8–7.8)	18.9 B (9.3–27.8)
Red-winged blackbird	8	5.1 B (3.8–15)	13.3 AB (8.9–27)
Great-tailed grackle	14	5.3 B (3.3–10.6)	7.6 A (4–25.2)
Mourning dove	15	2.3 C (0.7–3.3)	32.7 C (14.1–93.2)

[a] Compiled from Mora and Anderson, 1995. Means not sharing the same letter (among species) are significantly different.

Table 111.6 Mean and Individual Selenium Concentrations (µg/g dw, ranges in parentheses) in Livers of Cormorants from the Colorado River Delta Region, 1986–1987

Location	N	Se	Ref.
Mexicali Valley	9	16.7 (9.5–23.9)	Mora and Anderson, 1995
Salton Sea	9	38.8 (20.8–72)	White et al., 1987
New River	1	21.2	Setmire et al., 1990
Alamo River	1	18.0	Setmire et al., 1990
Alamo River	1	42.0	Setmire et al., 1990

Table 111.7 Mean Selenium Concentrations (µg/g dw) in Liver of Waterbirds from Mittry Lake and Imperial National Wildlife Refuge, Lower Colorado River[a]

Species	Mittry Lake 1991		Imperial NWR[b] 1993	
	N	Se	N	Se
American coot	2	8.1	15	6.4
Sora	2	7.7		
Common moorhen	2	4.2	17	9.2
Least bittern			20	13.6
Green heron	1	5.6	10	16.0
Pied-billed grebe			3	26.0
Black-crowned night heron	2	8.5		
Ruddy duck	1	2.6		

[a] Taken from Martinez, 1994, and Rusk, 1991.
[b] Geometric means.

et al., 1987). Selenium levels in livers of cormorants from the CRDR were on average 27.3 µg/ g dw, similar to those reported in pied-billed grebes from the Imperial National Wildlife Refuge (NWR). However, they were 1.5 to 10 times higher than residues in most aquatic birds from Mittry Lake and Imperial NWR in the lower Colorado River (Tables 111.6 and 111.7; Rusk, 1991; Martinez, 1994). These differences among species may be related more to differences in diet than location.

Boron was another element of concern because of the possibility of elevated discharges of brine from the geothermal energy plant at Cerro Prieto and because agricultural drain waters are considered major sources of B (Eisler, 1990; Saiki et al., 1993). Mora and Anderson found that B was significantly greater in livers of mourning doves (mean = 32.7 μg/g dw; $p < 0.05$) than in any other species (Table 111.5). Since mourning doves are mostly seed-eaters, it is likely that the main source of B resulted from high concentrations in agricultural seed crops, but no documentation of B in seeds was possible. Adult mallards fed B in the diet at 300 μg/g dw had mean residues in liver of 17 μg/g dw (Smith and Anders, 1989). Therefore, it is reasonable to assume that mourning doves at the Mexicali Valley were probably ingesting B at mean levels greater than 300 μg/g dw. The toxicological significance of B concentrations observed in mourning doves from the Mexicali Valley has not been addressed. As indicated earlier, high concentrations of B in the Mexicali Valley also could have resulted from brine discharges from the Cerro Prieto geothermal energy plant during its operation until 1989 when discharges stopped. Concentrations of B in water from 14 wells near the plant averaged 12.09 μg/g in 1972 (Alvarez-Borrego et al., 1973). Levels of B in Rio Hardy sediments reached maximum concentrations of 26 μg/g ww in 1986 at the site closest to the brine discharge from the geothermal plant (Comisión Federal de Electricidad, 1987, unpublished report). Concentrations of B diminished with increasing distance from the discharge point, but remained high for most of the river throughout the valley.

Heavy metal data for fish also suggest elevated exposure to Se in the CRDR (Table 111.8). Concentrations of 6.4 and 6.8 μg/g dw Se in fish from the Mexicali Valley were above the U.S. national mean of 0.42 μg/g ww (approximately 1.7 μg/g dw) in whole fish during 1984 (Schmitt and Brumbaugh, 1990), but were lower than the maximum concentration observed in common carp (2.3 μg/g ww, approximately 9 μg/g dw) from the lower Colorado River, Arizona (Schmitt and Brumbaugh, 1990), and the average (10.8 μg/g dw) measured during the same year in tilapia from the New River (Setmire et al., 1990; Table 111.8). Selenium residues in tilapia from the Mexicali Valley were about two times higher (3.8 μg/g dw) than levels found in tilapia from the Cienega de Santa Clara 10 years later (García-Hernández, 1998). Selenium residues in other fish species from the Cienega de Santa Clara were similar to those in tilapia from Mexicali, and below the dietary levels which are known to cause mortality or negative reproductive effects on birds (Skorupa and Ohlendorf, 1991).

Clams collected in the Mexicali Valley in 1998 showed concentrations of Se of 1.6 μg/g dw (King and Garcia, unpublished data). Water, sediment, invertebrates, plants, and fish were collected during 1996 and 1997 in the Cienega de Santa Clara, Colorado River, and Rio Hardy wetlands for a study of Se bioaccumulation in the Colorado River delta (García-Hernández, 1998). Selenium concentrations in water ranged from 5 to 19 μg/l in the Santa Clara Cienega and from 7 to 28 μg/l at six stations in the Santa Clara, Rio Hardy, and Colorado River wetlands (Figure 111.1). These concentrations in water exceeded the EPA (Environmental Protection Agency) criterion of

Table 111.8 Mean Selenium Concentrations (μg/g dw) in Fishes from the Colorado River Delta Region

Species	Location	N	Se	Year	Ref.
Striped mullet	Colorado Yuma, AZ	2	4.1	1984	Schmitt and Brumbaugh, 1990
Mullet	Mexicali Valley, MX	4	6.4	1986	Mora and Anderson, 1995
Tilapia	New River, CA	2	10.8	1986	Setmire et al., 1990
	Mexicali Valley, MX	6	6.8	1986	Mora and Anderson, 1995
	Cibola Lake, AZ	2	8.6	1989	King et al., 1993
	Santa Clara, MX	2	3.8	1997	García-Hernández, 1998
Largemouth bass	Colorado Yuma, AZ	1	3.7	1984	Schmitt and Brumbaugh, 1990
	Martinez Lake, AZ	1	8.3	1984	Schmitt and Brumbaugh, 1990
	Martinez Lake, AZ	3	10.4	1988	King et al., 1993
	Cibola Lake, AZ	2	7.5	1989	King et al., 1993
	Santa Clara, MX	2	5.3	1997	García-Hernández, 1998
Common carp	Martinez Lake, AZ	2	9.0	1984	Schmitt and Brumbaugh, 1990
	Martinez Lake, AZ	3	10.0	1988	King et al., 1993
	Cibola Lake, AZ	2	5.3	1989	King et al., 1993
	Santa Clara, MX	3	6.4	1997	García-Hernández, 1998

5.0 µg/l for the protection of aquatic life (U.S. EPA, 1987). However, Se levels in sediment (0.8 to 1.8 µg/g dw), aquatic plants (0.03 to 0.17 µg/g dw), and fish (1.5 to 6.6 µg/g dw) did not exceed threshold levels (García-Hernández, 1998).

In the binational study of 1995–1996, the following trace elements were detected in the water column at stations in the lower Colorado River: aluminum, arsenic, barium, boron, chromium, copper, iron, manganese, mercury, molybdenum, nickel, selenium, uranium, vanadium, and zinc (M. Carpio-Obeso, unpublished data). Only manganese at the Gila River station exceeded the California Safe Drinking Water Act standards. Arsenic was detected in all samples although at low concentrations. Arsenic concentrations exceeded the maximum tissue residue levels established by the California Water Control Board in fish tissues from the Northern International Border station, and in two samples from the All American Canal (Crane et al., 1996). The highest concentrations of B were detected in water samples from the Wellton-Mohawk station that collects the agriculture discharges from some areas in Arizona.

CONCLUSION

Monitoring studies of organochlorine pesticides in biota of the Colorado River delta have been few. The only major studies in wildlife in the mid-1980s suggested some concern for elevated concentrations of DDE and possible accumulation of other OCs during the winter. The most persistent OC, p,p'-DDE, was detected in wildlife at levels that could be of concern for top predators, particularly raptors and fish-eating birds. Data from clams in the CRDR (Alvarez-Borrego, this volume) suggest that DDE levels, at least, have declined in recent years. More current assessments of OCs in wildlife are needed to determine if DDE residues have generally declined in the Colorado delta, as has been observed in other areas of the southwestern U.S. (Mora and Wainwright, 1998). Also, new-generation and more-toxic organophosphorus and carbamate pesticides are currently in use in the Mexicali and San Luis Rio Colorado Valleys. However, the potential hazards of new-generation pesticides on wildlife of the Colorado have not been addressed. Addressing the impacts of a broader range of contaminants on endangered species such as the Yuma clapper rail, south-western willow flycatcher, and the desert pupfish in the Colorado River delta, should be an international priority for the near future.

In contrast to the fewer studies of chlorinated pesticides in the Colorado River delta, monitoring studies of trace elements, particularly Se, have been more abundant. Selenium concentrations were elevated during the 1980s in samples collected in the Mexicali Valley, Salton Sea, and a few other areas in the CRDR. Most Se data seemed to indicate a general pattern of Se contamination in the CRDR. King et al. (1993) noted that mean Se concentrations remained high in fish from the lower Colorado River from 1976 to 1989 and pointed out that there was widespread Se contamination of lower Colorado River backwater habitats. A similar observation applies to fish and other biota samples from the Colorado River delta in Mexico. Data from 1986 and 1996 also support the assumption that Se concentrations in biota of the Colorado River delta remain high. Overall, Se concentrations in biota of the CRDR have not decreased in the past 30 years. In some cases, Se levels in biota were near the threshold for effects on the species themselves or in predators at the top of the food chain that may feed on such species.

King et al. (1993) proposed a "single unified contaminant monitoring plan" for the Colorado River wildlife refuges in the U.S. We propose that an integrated approach for contaminant research should also be followed in the Colorado River delta, so that a single unified contaminant monitoring plan is implemented for the whole Colorado River delta ecosystem. This monitoring should be uniform and include samples that represent the whole ecosystem, from water to sediments, and species at the top of the food chain. Selected locations in the U.S. could include portions of the lower Colorado, Alamo, and New Rivers, as well as the Salton Sea. Sites in Mexico could include the Rio Hardy and Santa Clara wetlands, Rio Nuevo, and selected drainages and canals in the Mexicali Valley. The implementation of this plan would require the participation of federal and state agencies in both the U.S. and Mexico, as well as nongovernmental organizations from both countries.

ACKNOWLEDGMENTS

This chapter has been greatly improved by comments from J.O. Keith, D. Buckler, D. Papoulias, and an anonymous reviewer.

REFERENCES

Abarca, F.J., Ingraldi, M.F., and Varela-Romero, A., Observations on the desert pupfish (*Cyprinodon macularius*), Yuma clapper rail (*Rallus longirostris yumanensis*) and shorebird communities in the Cienega de Santa Clara, Sonora, Mexico, Nongame and Endangered Wildlife Program Technical Report, Arizona Game and Fish Department, Phoenix, 1993.

Alvarez-Borrego, S., Nishikawa-Kinomura, K.A., and Flores-Muñoz, G., Reporte Preliminar a la Dirección de Acuacultura de la Secretaría de Recursos Hidráulicos, de la Segunda Etapa del Estudio Químico sobre la Contaminación por Insecticidas en la Desembocadura del Río Colorado, Contrato SRH, No. E1–49 Clave LL-11, Unidad de Ciencias Marinas, Instituto de Investigaciones Oceanológicas, Ensenada, Baja, CA, Mexico, 1973.

Burnett, E., Kandl, E., and Croxen, F., Cienega de Santa Clara Geologic and Hydrologic comments, U.S. Bureau of Reclamation, Yuma, AZ, 1993.

Clark, D.R. and Krynitsky, A.J., DDT: recent contamination in New Mexico and Arizona? *Environment*, 25, 27–31, 1983.

Crane, D., Regalado, K., Linn, J., Munoz, G., and Smith, L., Toxic Substances Bioaccumulation Monitoring 1995–1996 Data Report, California Department of Fish and Game, Rancho Cordova, 1996.

Eisler, R., Boron hazards to fish, wildlife, and invertebrates: a synoptic review, U.S. Fish and Wildlife Service Biological Report 85 (1.20), Washington, D.C., 1990.

Fertilizantes Mexicanos, Plan de desarrollo de Fertimex en la producción, formulación y comercialización de insecticidas, Vol. II, Gerencia general de Programación y Desarrollo, Mexico, 1981.

Friederici, P., Stolen River, The Colorado and its delta are losing out, *Defenders*, Spring, 11–33, 1998.

García-Hernández, J., Bioaccumulation of Selenium in the Cienega de Santa Clara, Colorado River Delta, Sonora, Mexico, Master's thesis, University of Arizona, Tucson, 1998.

Glenn, E.P., Lee, C., Fleger, R., and Zengel, S., Effects of water management on the wetlands of the Colorado River Delta, Mexico, *Conserv. Biol.*, 10, 1175–1186, 1995.

Glenn, E.P., Garcia, J., Tanner, R., Congdon, C., and Luecke, D., Status of wetlands supported by agricultural drainage water in the Colorado River Delta, Mexico, *HortScience*, 34, 16–21, 1999.

Guardado-Puentes, J., Concentración de DDT y sus metabolitos en especies filtroalimentadoras y sedimentos en el Valle de Mexicali y Alto Golfo de California, *Calif. Coop. Oceanic Fish. Invest. Rep.*, 18, 73–80, 1976.

Gutierrez-Galindo, E.A., Flores-Muñoz, G., and Villaescusa-Celaya, J.A., Chlorinated hyrocarbons in molluscs of the Mexicali Valley and upper Gulf of California, *Cienc. Marinas*, 14, 91–113, 1988a.

Gutierrez-Galindo, E.A. et al., Organochlorine insecticides in fishes from the Mexicali Valley, Baja California, Mexico, *Ciencias Marinas*, 14, 1–22, 1988b.

Gutierrez-Galindo, E.A., Flores-Muñoz, G., and Aguilar-Flores, A., Mercury in freshwater fish and clams from the Cerro Prieto geothermal field of Baja California, Mexico, *Bul. Environ. Contam. Toxicol.*, 41, 201–207, 1988.

Henny, C.J., Blus, L.J., and Hulse, C.S., Trends and effects of organochlorine residues on Oregon and Nevada wading birds, 1979–1983, *Colonial Waterbirds*, 8, 117–128, 1985.

King, K.A., Baker, D.L., Kepner, W.G., and Martinez, C.T., Contaminants in sediment and fish from national wildlife refuges on the Colorado River, Arizona, U.S. Fish and Wildlife Service, Phoenix, 1993.

Lopez-Carrillo, L., Torres-Arreola, L., Torres-Sanchez, L., Espinosa-Torres, F., Jimenez, C., Cebrian, M., Waliszewski, S., and Saldate, O., Is DDT use a public health problem in Mexico? *Environ. Health Perspect.*, 104, 584–588, 1996.

Martinez, C.T., Selenium Levels in Selected Species of Aquatic Birds on Imperial National Wildlife Refuge, Master's thesis, University of Arizona, Tucson, 1994.

Mellink, E., Palacios, E., and Gonzalez, S., Non-breeding waterbirds of the delta of the Rio Colorado, Mexico, *J. Field Ornithol.*, 68, 113–123, 1997.

Mora, M.A., Organochlorines and breeding success in cattle egrets from the Mexicali Valley, Baja California, Mexico, *Colonial Waterbirds*, 14, 127–132, 1991.

Mora, M.A. and Anderson, D.W., Seasonal and geographical variation of organochlorine residues in birds from northwest Mexico, *Arch. Environ. Contam. Toxicol.*, 21, 541–548, 1991.

Mora, M.A. and Anderson, D.W., Selenium, boron, and heavy metals in birds from the Mexicali Valley, Baja California, Mexico, *Bull. Environ. Contam. Toxicol.*, 54, 198–206, 1995.

Mora, M.A. and Wainwright, S.E., DDE, mercury, and selenium in biota, sediments, and water of the Rio Grande-Rio Bravo Basin, 1965–1995, *Rev. Environ. Contam. Toxicol.*, 158, 1–52, 1998.

Mora, M.A., Anderson, D.W., and Mount, M.E., Seasonal variation of body condition and organochlorines in wild ducks from California and Mexico, *J. Wildl. Manage.*, 51, 132–141, 1987.

Ohlendorf, H.M. and Marois, K.C., Organochlorines and selenium in California night-heron and egret eggs, *Environ. Monitor. Assess.*, 15, 91–104, 1990.

Ohlendorf, H.M., Hothem, R.L., Aldrich, T.W., and Moore, J.F., Relationships between selenium concentrations and avian reproduction, *Trans. North Am. Wildl. Nat. Resourc. Conf.*, 51, 330–342, 1986a.

Ohlendorf, H.M., Hoffman, D.J., Saiki, M.K., and Aldrich, T.W., Embryonic mortality and abnormalities of aquatic birds: apparent impacts of selenium from irrigation drainwater, *Sci. Total Environ.*, 52, 49–63, 1986b.

Ohlendorf, H.M., Hothem, R.L., Aldrich, T.W., and Krynitsky, A.J., Selenium contamination of the grasslands, a major California waterfowl area, *Sci. Total Environ.*, 66, 169–183, 1987.

Rasmussen, D., Toxic Substances Monitoring Program 1986, State Water Resources Control Board, Water Quality Monitoring Report 88-2, Sacramento, 1988.

Rusk, M.K., Selenium Risk to Yuma Clapper Rails and other Marsh Birds of the Lower Colorado River, Master's thesis, University of Arizona, Tucson, 1991.

Saiki, M.K., Jennings, M.R., and Brumbaugh, W.G., Boron, molybdenum, and selenium in aquatic food chains from the lower San Joaquin River and its tributaries, California, *Arch. Environ. Contam. Toxicol.*, 24, 307–319, 1993.

Schmitt, C.J. and Brumbaugh, W.G., National contaminant biomonitoring program: concentrations of arsenic, cadmium, copper, lead, mercury, selenium, and zinc in U.S. freshwater fish, 1976–1984, *Arch. Environ. Contam. Toxicol.*, 19, 731–747, 1990.

Secretaria de Agricultura y Recursos Hidraulicos, Guía para la asistencia técnica agrícola, area de influencia del campo agrícola experimental Valle de Mexicali, Mexicali, BC, Mexico, 1984.

Setmire, J.G., Wolfe, J.C., and Stroud, R.K., Reconnaissance investigation of water quality, bottom sediment, and biota associated with irrigation drainage in the Salton Sea area, California, 1986–87, U.S. Geological Survey, Water Resources Investigations Report 89-4102, Sacramento, 1990.

Skorupa, J. and Ohlendorf, H.M., Contaminants in drainage water and avian risk thresholds, in *The Economics and Management of Water and Drainage in Agriculture*, Dinar, A. and Zilberman, D., Eds., Kluwer Academic, Norwell, MA, 1991, pp. 345–368.

Smith, G.J. and Anders, V.P., Toxic effects of boron on mallard reproduction, *Environ. Toxicol. Chem.*, 8, 943–950, 1989.

Sykes, G., *The Colorado Delta*, Carnegie Institution of Washington, Special Publication 19, New York, 1937.

U.S. Environmental Protection Agency, Ambient water quality criteria for selenium, Office of Water Regulations and Standards, Washington, D.C., 1987.

Valdés-Casillas, C., Glenn, E.P., Hinojosa-Huerta, O., Carrillo-Guerrero, Y., Garcia-Hernandez, J., Zamora-Arroyo, F., Munoz-Viveros, M., Briggs, M., Lee, C., Chavarria-Correa, E., Riley, J., Baumgartner, D., and Congdon, C., Wetland Management and Restoration in the Colorado River Delta: The First Steps, Special publication of CECARENA-ITESM Campus Guaymas and NAWCC, Guaymas, SO, Mexico, 1998.

White, D.H. and Krynitsky, A.J., Wildlife in some areas of New Mexico and Texas accumulate elevated DDE residues, 1983, *Arch. Environ. Contam. Toxicol.*, 15, 149–157, 1986.

White, J.R., Hofmann, P.S., Hammond, D., and Baumgartner, S., Selenium verification study 1986, A report to the California State Water Resources Control Board from the California Department of Fish and Game, Agreement 5-096-300-2, Sacramento, 1987.

CHAPTER 112

Conservation Value and Water Management Issues of the Wetland and Riparian Habitats in the Colorado River Delta in Mexico

Michael Cohen, Edward P. Glenn, Jason Morrison, and Robert J. Glennon

INTRODUCTION

Human activity has disrupted most of the world's large river ecosystems. Arid rivers are especially at risk due to their modest base flows and the many human demands on their waters. An example is the Colorado River, which is a major source of water for the desert region of the southwest U.S., northwestern Mexico, and the southern California coastal plain. Approximately 23 million people now use Colorado River water (Morrison et al., 1996). The delta of a regulated river system, below the last dams and diversions, is the natural collecting place for wastewaters and surplus flows exiting the watershed. The delta is the depositional section of the river, where the reduction in grade broadens the floodplain and permits the expansion of wetland and riparian habitats. It is also the mixing zone with the ocean, which is enriched by brackish and freshwater sources from land. River deltas may retain natural functions that have disappeared from upstream stretches. That is certainly true of the Colorado River delta, which supports a large riparian habitat (Glenn et al., 1996, 1999; Valdes-Casillas et al., 1998, 1999). The Colorado River delta is a binational resource, located mainly in Mexico, yet supplied primarily by water from the U.S.

On the lower Colorado River from Grand Canyon to the border with Mexico, dams built since 1935 have regulated the river flow and prevented overbank flooding which previously germinated seeds and washed excess salts from the river banks. Other disturbances have included channelization and dredging of the river and clearing of floodplain land for farming (Szaro, 1989; Stromberg and Patten, 1991; Bush and Smith, 1995; Briggs, 1996). As a consequence, the quality of the riparian habitat has declined dramatically. Along most of the river, the native riparian forest of *Populus fremontii* (cottonwood) and *Salix gooddingii* (willow) trees has been replaced by drought- and salt-tolerant shrub vegetation. By 1997, only 100 ha of the riparian zone in the U.S. were still classified as gallery forests. The most common plants along the river are now *Tamarix ramosissima* (salt cedar), an introduced species, and *Pluchea sericea* (arrowweed) (Ohmart et al., 1988). The decline of the native forest vegetation has reduced the habitat value of the riparian zone for many of the native fauna (Ohmart et al., 1988).

Human use has also had a severe impact on the delta of the Colorado River, from Morelos Dam (site of the last major diversion in Mexico) to the mouth of the river in the Gulf of California. Before the construction of upstream dams, the delta contained several hundred thousand hectares of riparian and wetland habitat that supported numerous species of plants and animals and provided a critical ecological interface with the Gulf of California (Sykes, 1937; Leopold, 1949; Glenn et al., 1996). Most of this habitat disappeared after the construction of the Hoover and Glen Canyon Dams and the diversion of water for human uses (Fradkin, 1981; Richardson and Carrier, 1992). In the past 20 years, however, an interesting phenomenon has taken place in the delta. Surplus flows of river water plus agricultural drainage water from the U.S. have reestablished a vegetated floodplain from Morelos Dam south to the intertidal zone in the Gulf of California (Glenn et al., 1996, 1999; Valdes-Casillas et al., 1998, 1999). We describe below the source of the surplus flows, the wetland and riparian habitats they support, and the research and management tasks needed if these resources are to be conserved for future generations.

DELTA WATER SOURCES AND HABITATS

Extent of the Delta Floodplain and Sources of Water

A series of earthen, flood control levees were constructed along the channel of the Colorado River in Mexico in the 1980s to contain flood releases from the U.S. following the filling of Glen Canyon Dam (Glenn et al., 1996). Inside the levees, agriculture and settlement are for the most part prohibited, and riparian and wetland vegetation are the dominant plant associations. The floodplain enclosed between the levees is narrow as it passes through the agricultural zone, but widens to greater than 30 km as it approaches the Gulf of California (Figure 112.1). This area encompasses approximately 170,000 ha, of which 60,000 ha is at least partially vegetated while the remainder consists of extensive tide flats (Valdes-Casillas et al., 1998).

Water discharged into the floodplain comes from two main sources: surplus river flows released from the Hoover Dam by the U.S. Bureau of Reclamation, and agricultural drainage water from the U.S. and Mexico. The Bureau of Reclamation releases Colorado River water based on a three-tiered system of priorities: (1) flood control and river regulation, (2) off-stream diversions, and (3) hydroelectric power generation (U.S. Bureau of Reclamation, 1996). With the exception of 1993, flows generated by flooding along the Gila River, Colorado River instream flows in the delta are a direct result of flood control releases from the Hoover Dam (Figure 112.2). In addition, the influence of seawater extends 35 km up the mouth of the river due to the high tidal range that characterizes the upper Gulf of California. An additional source of water is upwellings of freshwater springs (*pozos*) onto the mudflats from artesian sources. These water sources support a variety of riparian and wetland habitats, which are briefly described later.

Cienega de Santa Clara

Cienega de Santa Clara is perhaps the largest example of emergent wetland habitat in the Sonoran Desert. It has been created by the discharge of agricultural drain water into the eastern part of the delta (Glenn et al., 1992, 1995; Zengel et al., 1995). Most (85%) of the water originates from the Wellton–Mohawk Irrigation and Drainage District in the U.S. Drain water from Wellton-Mohawk originates from a series of shallow wells which are pumped to control the depth and salinity of the water table in this irrigation district. Since 1977, this water has been conveyed in a 80-km lined canal (the Main Outlet Drain Extension) to the southeast part of the delta in Mexico where it is deposited in a former small-marsh system that drains into the intertidal zone of the Gulf of California. The mean annual discharge is approximately 1.8×10^8 m^3 per year at a salinity of approximately 2940 ppm, but there has been a trend toward decreased volume and salinity over the years, perhaps due to more-efficient irrigation practices in the irrigation district. This discharge is augmented by 3.2×10^7 m^3 per year of local irrigation return water from the Riito Drain, with a mean salinity of approximately 4000 ppm.

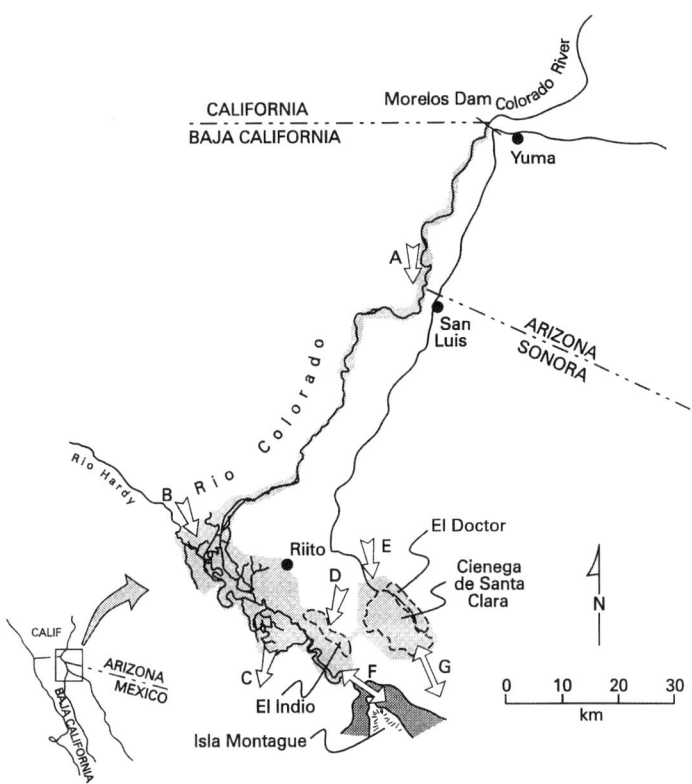

Figure 112.1 Vegetated portion (gray area) of the Colorado River delta floodplain in Mexico. Arrows show the floodplain's main entry and exit points of water. (A) Colorado River flood release flows from the U.S.; (B) Mexicali agricultural drainage water entering Rio Hardy; (C) Colorado River flows exiting to intertidal zone and Laguna Salada (not shown); (D) San Luis Valley agricultural drainage water entering El Indio wetlands; (E) Wellton-Mohawk water entering Cienega de Santa Clara wetland; (F) river water and tide water mixing at the mouth of the Colorado River; and (G) water from Cienega de Santa Clara mixing with tidewater in the intertidal zone.

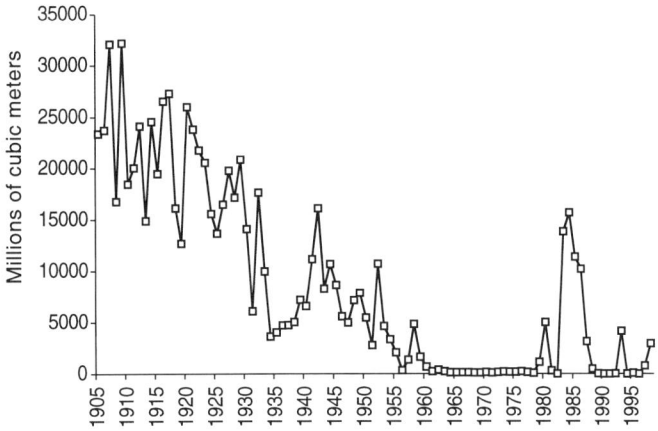

Figure 112.2 Annual flows in the Colorado River at the southerly international boundary, below Morelos Dam, the last diversion point of water on the Colorado River.

The drain waters have created a densely vegetated *Typha domengensis* marsh covering approximately 4200 ha. Besides *Typha,* the Cienega contains 23 other species of emergent and submerged wetland plants. Water flowing through the marsh increases in salt content due to water lost to evapotranspiration; at the southern end the salinity exceeds the tolerance limit for *Typha.* Ultimately, the discharge exits onto the mudflats below the Cienega, where it mixes with tide water from the Gulf of California.

The Cienega supports a rich fauna, including crustaceans and fish, upon which numerous migratory and resident waterfowl feed. A recent study (Hinojosa and DeStephano, unpublished data) shows that as many as 5000 individuals of the endangered Yuma clapper rail (*Rallus longirostris yumanensis*) are resident in the Cienega, many more birds than are found anywhere else in the Lower Colorado River ecoregion. The backwater areas of the Cienega also support large populations of the endangered desert pupfish (*Cyprinodon macularius*) (Zengel and Glenn, 1996).

The El Doctor Wetlands

These pocket wetlands are supported by artesian springs along the eastern escarpment of the delta, where it joins Gran Desierto. The water originates as rainfall in the Pinacate Mountains and migrates under the Gran Desierto dune fields to the delta, where it bubbles to the surface in isolated pozos, which support a unique assemblage of wetland and halophytic plant species (Felger et al., 1997). Although these wetlands cover only 500 to 750 ha, they are important because they are among the few natural wetlands remaining in the delta. Because their source water is less saline than the drain water entering Cienega de Santa Clara, the pozos support a wider variety of plant life. The El Doctor wetlands contain 29 plant species, including *Cyperus, Eleocharis,* and *Hydrocotyle* species not found in the Cienega. Some of the same birds as are found in the Cienega are found in the El Doctor wetlands. In June 1999, the southwest subspecies of willow flycatchers (*Empidonax traillii extimus*), an endangered species, were documented in these wetlands, but they appeared to be migrant, as they were no longer present during the July nesting season (Vanda Gerhart, University of Arizona Environmental Research Laboratory, unpublished data).

Rio Hardy Wetlands and the Intertidal Marshes

The Rio Hardy drains the northwest corner of the Mexicali Valley and joins the Colorado River approximately 50 km from the mouth of the river. The source water for the Rio Hardy is agricultural drain water from the Mexicali Valley. About half of the valley's agricultural discharge enters the Rio Hardy, while the remainder flows north in the New River to the Salton Sea. The flows in the Rio Hardy are not well documented, but were reported to be 1.3×10^8 m^3 per year with a mean salinity of 4200 ppm in the 1970s (Valdes-Casillas et al., 1998). The Rio Hardy has a perennial water supply, and it supports bird habitat as well as recreational fishing and aquaculture. It is basically a brackish marsh system, with the emergent reed, *Phragmites australis,* growing along the water line and with *Tamarix ramosissima* thickets dominating the exposed banks.

Downstream from the junction with the Rio Hardy, the Colorado River becomes increasingly saline due to tidal influence. Along the lower reach of the river near the mouth and on Montague Island there are approximately 33,000 ha of *Distichlis palmeri* flats. This saltgrass is the only endemic grass species in the Sonoran Desert. It produces a large seed that was harvested as a grain by the indigenous Cocopa people (Felger et al., 1997) and it supports nesting shorebirds (Mellink et al., 1997).

Riparian Corridor of the Colorado River

The perennial portion of the Colorado River ends at Morelos Dam, which diverts water into Mexico's Alamo Canal, the last major diversion on the river, located 1.8 km downstream from the northerly international border between the U.S. and Mexico. Below Morelos Dam, the river flows only when excess water must be released from upstream impoundments. From 1964 (when Glen Canyon Dam was completed) to 1981, virtually no excess water was released because Lake Powell behind the dam was not filled to capacity. Any excess water in the river system could simply be stored behind the dam. After 1981, however, there have been several major releases of water during wet, El Niño cycle years (Figure 112.2). In fact, during the period 1981–1998, mean annual flows past Morelos Dam have been approximately 20% of the total river flows.

As a result of these flood release flows, a 1997 preliminary study (Valdes-Casillas et al., 1998, 1999) revealed that the delta now supports more riparian habitat than the perennial stretch of river from Grand Canyon to the international border, even though that stretch of river is five times longer. Although the riparian corridor is dominated by *T. ramosissima,* several thousand hectares of native gallery forest have also reestablished in the delta.

Wildlife Habitat Value of the Delta

Only a few surveys have been conducted on wildlife in the delta. However, it is clear that the delta and its estuary now provide habitats for several U.S. and internationally listed endangered species, including desert pupfish (Eddleman, 1989; Abarca et al., 1993; Zengel and Glenn, 1996), vaquita porpoise (*Phocoena sinus*), totoaba fish (*Cynoscion macdonaldi*) (Glenn et al., 1996), Yuma clapper rails (Eddleman, 1989; Abarca et al., 1993), southwestern willow flycatchers and other endangered species of birds (J. Campoy, SEMARNAP, Mexico, unpublished data). It is a major stop on the Pacific Flyway for migratory waterfowl (Mellink et al., 1997) and nesting birds (Mellink et al., 1996). Due to its importance, much of the delta is included in the core zone of the Biosphere Reserve of the Upper Gulf of California and Colorado River delta and in the Western Hemisphere Shorebird Reserve Network. More recently, it was designated a RAMSAR site (Valdes-Casillas et al., 1998). These wild areas coexist with a local population of nearly 1.5 million people in Mexicali and San Luis, Mexico.

MANAGEMENT NEEDS FOR DELTA HABITATS

Although the delta riparian and wetland ecosystems are the largest, most important examples of these habitat types left on the Colorado River, and despite their location in a Biosphere Reserve, they have received little study and limited binational management attention. Indeed, these critical ecosystems are the inadvertent creation of water management decisions made in the U.S. and Mexico. In the future, these habitats face severe threats; yet they present remarkable opportunities. Threats include population increases along both sides of the border and, in the Mexicali Valley, a conversion from a predominantly agricultural to an industrial economy, which removes the rural character of the region. In the U.S., competition for Colorado River water is fierce. Cities, farmers, and Native American tribes seek larger pieces of an already over-allocated pie, while environmental organizations and federal wildlife agencies push for policies that would protect and enhance the environment. It is unclear how these pressures will impact management of the river and, ultimately, the flows that reach the delta. Water that now flows to the delta habitats is the unintended result of upstream management decisions and is not guaranteed for the future. Active management is necessary in order to conserve and enhance these habitats.

Although the threats are substantial, so too are the opportunities for binational management of the river to protect the ecosystems of the delta. Mexico has an interest in augmenting flows to the delta and Gulf of California in order to improve the health of the shrimp fishery (Galindo-Bect et al., 2000). Although located in Mexico, the delta wetlands and associated fauna are a major environmental asset for the U.S. as a reservoir of species diversity for the entire Southern Basin and Range Ecoregion (Felger et al., 1997). The delta is a major stop for birds on the Pacific Flyway and for resident shorebirds. Given the number of threatened and endangered species in this region, the U.S. Fish and Wildlife Service has an opportunity to enforce the Endangered Species Act by approving a Habitat Conservation Plan that includes assurances of flood flow releases and continued agricultural drain flows to the delta.

Despite the fact that some of the delta habitats are in a Biosphere Reserve, for the most part these natural areas are not well protected. The riparian corridor in Mexico is controlled by the Comisión Nacional de Agua (CNA), whose primary responsibility is to distribute irrigation water and control property damage when water is released below Morelos Dam. CNA has engaged in clearing large areas of native trees and channeling sections of the riverbed to mitigate damage to roads and bridges during flood flows. It has been suggested (Valdes-Casillas et al., 1998) that the riparian corridor could be designated as a wildlife management zone, or incorporated into the buffer zone of the Biosphere Reserve, so that environmental considerations are taken into account before the habitats are altered. The Cienega de Santa Clara and the El Doctor wetlands, though inside the core zone of the Biosphere Reserve, are also highly affected by human use. Cattle grazing in the El Doctor wetlands reduces the vegetation to stubble height wherever it is accessible. Cattle also graze along the edges of the Cienega, and in winter the *Typha* is sometimes deliberately burned off to increase penetration by cattle the following spring. Burning appears to improve the quality of the marsh when it regrows but, unfortunately, burning often takes place during the Yuma clapper rail nesting season, which must impact the population of these birds (Campoy, J., Biosphere Reserve manager, personal communication).

The most serious external threat to the delta wetlands is from water management decisions made in the U.S. that affect the flows of water to these areas. The lower Colorado River delta has no explicit water entitlement. Water that has reached the riparian corridor in recent years has done so solely because flood release flows have exceeded the diversion capacity of upstream users. Nor does the Cienega enjoy a legal entitlement to Wellton-Mohawk drainage. The Yuma Desalting Plant was constructed to treat this drainage, to bring the U.S. into compliance with a 1973 treaty with Mexico addressing the salinity of water delivered to Mexico (Getches, 1985). Because the U.S. has been able to satisfy these salinity requirements without treating Wellton-Mohawk drainage, this water has been routed to the Cienega. Similarly, the flood release flows in the Colorado River are threatened by U.S. water management decisions. The Bureau of Reclamation is currently reviewing several proposed changes to Colorado River management, including the development of surplus (and shortage) criteria, off-stream storage, and diverting a portion of flood release flows into the Salton Sea to reduce its salinity. Each of these proposals would likely reduce the quantity and frequency of flood release flows reaching the delta.

The delta wetlands need to receive binational management attention. In Mexico, the riparian corridor which is outside the Biosphere Reserve would benefit from protected status, since this area contains the largest remaining stands of native riparian forest, and appears to serve as a migration corridor for migratory songbirds. The Cienega de Santa Clara and The El Doctor wetlands within the Biosphere Reserve would benefit from a more-active management program than is now in place, one which would restrict grazing and other human impacts (Campoy, J., personal communication). The potential for wildlife poisoning from heavy metals and pesticides must be monitored and addressed throughout the delta habitats. In particular, the Rio Hardy wetlands, supported by agricultural drainage water from Mexico, may present a selenium hazard to wildlife (Mora and Garcia, this volume). This portion of the floodplain receives restricted flushing during floods because it is above the juncture with the Colorado River. However, flushing to reduce loads of selenium and other toxins could conceivably be accomplished using control gates and pump stations that were installed during the floods of the early 1980s to move backed-up floodwaters from the Rio Hardy system into the Rio Colorado system to prevent flooding of fields (Valdes-Casillas et al., 1998).

The most urgent need is to develop a binational water management plan for flood release flows and agricultural wastewaters that cross the international border. Even without any explicit management attention, these flows now support the most valuable examples of wetland and riparian habitat left in the Lower Colorado River ecoregion. A management plan might not involve providing *more* water to the delta; it might only require that the timing of flows be adjusted to provide maximum ecosystem benefit, in a way that is still consistent with operation of the river system for irrigation, power generation, recreation, and flood control. However, such a management plan will require much more knowledge of the vegetation and hydrology of the delta ecosystems than is currently available, so research must precede action.

IMPEDIMENTS TO BINATIONAL MANAGEMENT

A host of institutional impediments challenges the implementation of a binational conservation plan. The present system of water appropriation for the Colorado River is the result of a complex history of conflict and negotiation among U.S. water users, and between Mexico and the U.S. (Hundley, 1966; Getches, 1985; Getches and Myers, 1986). Flows of the Colorado River are controlled and regulated based on a complex legal framework of treaties, interstate compacts, state and federal laws, regulations, Supreme Court decisions, and contracts, known as the Law of the River (Hundley, 1975; Getches, 1985; Glennon, 1995). The degree of institutional control of the river cannot be overstated; the 1983 flood has been the only instance since the construction of the Hoover Dam in 1935 when discharge from the dams was not completely controlled. Furthermore, environmental considerations have only recently been included in water management decisions (Bolin, 1993).

Legal Impediments

Controversy over rights to Colorado River water has been a major theme in the 20th century history of the American West and of the relations between the U.S. and Mexico. Fights between and among states have led to legendary court battles. The State of Arizona first sued California in the 1930s and the dispute is ongoing. As the population of the West has mushroomed, competition for the Colorado's water has intensified. This will increase because demographers predict continued rapid population growth in the West (Case and Alward, 1997). The fight over water is not just over high-quality supplies, but extends even to poor-quality water such as the saline, agricultural drainage water from Wellton-Mohawk that currently sustains Cienega de Santa Clara (Glenn et al., 1992).

The Law of the River has created legal rights in the Colorado River basin states and in Mexico to over 18 million acre-feet ($22,200 \times 10^6 \text{ m}^3$ [mcm]) per year of water from the Colorado, even though the mean flow of the river over the past century has been only 15 million acre-feet (18,500 mcm) per year (Pontius, 1997). The first task in attempting to develop a water management plan is to determine how the various existing water rights may impact management of the river and to secure sufficient wastewater for the Cienega de Santa Clara and flood release flows to maintain the ecosystem in the delta.

To date, Upper Basin states have yet to divert their full entitlement, enabling California to divert as much as 18% more than its legal entitlement. Increases in Upper Basin consumption rates and Lower Basin population, and proposed changes in river management intended to address these changes, will reduce the availability of flood release flows for the delta region in the future. Under the Law of the River, the U.S. is required to deliver 1850 mcm per year of Colorado River water to Mexico (2009 mcm in years when a surplus is declared). Demand for water in Mexico's cities of Mexicali and San Luis, and for irrigation in the Mexicali Valley, already exceeds supply; the shortfall is partially met by unsustainable pumping of groundwater (Morrison et al., 1996). Projected population and industrial growth in the region will place additional strain on limited water resources (Ganster, 1996). In the absence of binational agreement, these factors reduce the likelihood that Mexico would allow future flood release flows to reach the delta.

While the struggle over the Colorado has largely involved issues of water *quantity,* the rise of the environmental movement in the past 30 years has produced new legal obligations and constraints. In the U.S., the Clean Water Act, the Endangered Species Act, the National Environmental Policy Act, and the Colorado River Basins Salinity Control Act focus on issues of water *quality* and environmental protection. In Mexico, a large portion of the delta and upper Gulf of California are now part of a protected Biosphere Reserve. As noted earlier, human-induced changes in the river in terms of dams and diversions have wreaked havoc with the natural riparian system, and have destroyed or substantially degraded the habitat of various species of animals and plants. Degradation of habitat is one of the main factors that has contributed to the listing of species as threatened or endangered under the Endangered Species Act. Such listing triggers obligations on the part of federal agencies to attempt to protect the habitat of the particular species and to take other measures to ensure its continued existence. In the Colorado River basin, these requirements have resulted in the Upper Basin Recovery Implementation Program and, in the Lower Basin, the Lower Colorado River Multispecies Conservation Program (LCR MSCP).

In 1997, after consultation between the U.S. Bureau of Reclamation and the U.S. Fish and Wildlife Service, a Biological Opinion was released that concluded that existing operations of the Lower Colorado River dams were likely to jeopardize the continuing existence of the bonytail chub, razorback sucker, and southwestern willow flycatcher. The LCR MSCP currently applies only to the section of the river within the borders of the U.S.; yet the environmental issues, particularly with respect to endangered species, do not stop at the border. There is considerable interest among federal agencies and environmental groups in developing information and exploring alternatives to requirements of the Endangered Species Act. In particular, an unanswered legal question is the extent to which the Endangered Species Act would permit considering the ecosystems of the delta as part of the solution for promulgating a Habitat Conservation Plan for the lower Colorado River.

Even assuming that the Endangered Species Act would allow for the consideration of the ecosystem beyond the territorial boundaries of the U.S., there remains the problem of assuring that the wastewater and surplus flows that cross the border into Mexico ultimately reach the delta region. The U.S. government and the Colorado River basin states will want assurance from Mexico that such flows will not be diverted for agricultural irrigation or to supply the burgeoning municipal demands of Mexicali and San Luis.

As a matter of Mexican law, issues of water allocation are determined by federal law, not by the laws of the states of Sonora or Baja California Norte. The relevant binational agreements are the Mexican Water Treaty of 1944; Minutes 218, 241, and 242 of the International Boundary and Water Commission; the North American Free Trade Agreement (NAFTA); and the Colorado River Basin Salinity Control Act of 1974. There is some support for an argument that the 1944 Water Treaty would allow for water releases for environmental purposes and not require changes to the treaty. However, our working hypothesis is that the best solution for ensuring that such flows reach the delta would be an explicit agreement between Mexico and the U.S. Such an agreement is a realistic option because both countries have an interest in maintaining the ecosystem of the delta.

The complexity and specificity of the institutions governing the management of the Colorado River frustrate efforts to implement binational conservation strategies. The Law of the River is flawed by its arbitrary division of the Colorado River based on political boundaries rather than watersheds. Distinct federal and state agencies separately regulate surface and groundwater resources, while yet other agencies monitor water quality, an arrangement that complicates comprehensive water management efforts (Morrison et al., 1996). Efforts to manage the basin in a coordinated comprehensive manner have been frustrated by institutional inertia and political opposition. In 1978, the U.S. Congress rejected a call for a comprehensive environmental assessment of the basin, due to fears of a centralization of power in federal hands and opposition from agriculture (Fradkin, 1981).

OPPORTUNITIES FOR BINATIONAL MANAGEMENT

Although the institutional barriers described above hinder efforts to implement binational conservation efforts for the Colorado delta, several recent developments suggest opportunities for change. Public recognition and support for water left instream, as demonstrated by the growing body of environmental and instream flow protections (Howe, 1996) provide a foundation for binational conservation efforts. Comments by the U.S. administration indicate a new willingness to consider binational management of the watershed. Upon Mexico's declaration of the Colorado River Delta and Upper Gulf of California Biosphere, former U.S. Secretary of the Interior Babbitt remarked that "[b]oth nations have an interest in setting aside some of these lands and in acknowledging the environmental linkages that connect us" (Wilson, 1994). In June 1999, during a speech at the Natural Resource Law Center in Boulder, CO, former Secretary Babbitt said, "It is time to acknowledge that the natural values of river systems can no longer be treated as table scraps, left over after every conceivable consumptive appetite has been fully satisfied."

A series of legal precedents supports the establishment of dedicated flows for the delta. Mexico and the U.S. have entered into several formal and informal agreements regarding management of binational resources. Agenda 21, a binational environmental initiative signed in 1992, calls for management of water resources at the basin level (Wilson, 1994). The Border XXI Framework Document of 1996 establishes a foundation for dedicated flows to the delta. Further precedent is established by the 1996 agreement between the U.S. and the United Mexican States on Cooperation for the Protection and Improvement of the Environment in the Border Area; the 1992 Rio Declaration on Environment and Development, Convention on Biological Diversity; the 1994 North American Agreement on Environmental Cooperation; and the 1997 Letter of Intent between the U.S. Department of the Interior and the Secretariat of Environment, Natural Resources, and Fisheries of the United Mexican States for joint work in natural protected areas on the U.S.–Mexico border. These precedents bolster recent calls for the establishment of instream flows for the delta (Wilson, 1994) through such mechanisms as a Minute (a subsidiary agreement) to the 1944 Treaty Respecting the Utilization of Waters of the Colorado and Tijuana Rivers, and of the Rio Grande (Snape, 1998).

The recently concluded negotiations between the Metropolitan Water District of Southern California, the Imperial Irrigation District, and the Coachella Valley Water District, purportedly facilitating water transfers from agricultural to urban users, offer another useful precedent for the delta. Although significant institutional obstacles remain, the potential exists for negotiating long-term lease agreements with U.S. water-rights holders to guarantee instream flows for the delta. Other parties to such long-term lease agreements are the potentially water-rich yet cash-poor sovereign Indian tribes, whose water rights continue to be the source of contention and litigation.

Finally, recent commentary has suggested the need for changes in the management of the Colorado River. As noted above, Colorado River management is fragmented and often arbitrarily divided among a variety of different federal and state agencies, each with its own focus. The limited binational cooperation further impedes efforts to manage the region sustainably. Pontius (1997) has proposed creating a Colorado River Basin Coordinating Council, while Getches (1997) has advocated a new type of management of the river that would require an interstate compact between and among the Colorado River basin states. Meeting this potential will require broad changes in the institutions governing the Colorado River.

REFERENCES

Abarca, F., Ingraldi, M., and Varela-Romero, A., Observations on the desert pupfish (*Cyprinodon macularius*), Yuma clapper rail (*Rallus longirostris yumanensis*), and shorebird communities in the Cienega de Santa Clara, Sonora, Mexico, Nongame and Endangered Wildlife Program Technical Report, Arizona Game and Fish Department, Phoenix, 1993.

Anderson, B.W. and Ohmart, R.D., Vegetation, in *Inventory and Monitoring of Wildlife Habitat*, Cooperrider, A.M. et al., Eds., U.S. Department of the Interior, Bureau of Land Management, Washington, D.C., 1986, pp. 639–660.

Anon., Programmatic environmental assessment for proposed rule making for offstream storage of Colorado River water and interstate redemption of storage credits in the lower division states, U.S. Department of the Interior, Bureau of Reclamation, Boulder City, NV, 1997.

Bolin, J., Of razorbacks and reservoirs: the Endangered Species Act's protection of endangered Colorado River Basin fish, *Pace Environ. Law Rev.*, 11, 35–87, 1993.

Briggs, M., *Riparian Ecosystem Recovery in Arid Lands*, Strategies and References, University of Arizona Press, Tucson, 1996.

Case, P. and Alward, G., Patterns of Demographic, Economic and Value Change in the Western United States: Implications for Water Use and Management, Report to the Western Water Policy Review Advisory Commission, Phoenix, AZ, 1997.

Eddleman, W.R., Biology of the Yuma Clapper Rail in the Southwestern U.S. and Northwestern Mexico (Special Publication), U.S. Bureau of Reclamation and U.S. Fish and Wildlife Service, Yuma, AZ, 1989.

Felger, R.B., Broyles, Wilson, M., and Nabhan, G., The binational Sonoran Desert biosphere network and its plant life, *J. Southwest*, 39, 411–560, 1997.

Fradkin, P., *A River No More: The Colorado River and the West*, Alfred A. Knopf, New York, 1981.

Galindo-Bect, M., Glenn, E.P., Page, H., Fitzsimmons, K., Galindo-Bect, L., Hernandez-Ayon, J., Petty, R., Garcia-Hernandez, J., and Moore, D., Penaeid shrimp landings in the upper Gulf of California in relation to Colorado River freshwater discharge, *Fish. Bull.*, 98, 222–225, 2000.

Ganster, P., Environmental issues of the California-Baja Border Region, Border Environmental Research Reports 1:1, Institute for Regional Studies of the Californias, San Diego, 1996.

Getches, D., Colorado River governance: sharing federal authority as an incentive to create a new institution, *Colo. Law Rev.*, 68, 574–658, 1997.

Getches, D., Competing demands for the Colorado River, *Univ. Colo. Law Rev.*, 56, 413–479, 1985.

Getches, D. and Myers, C., The River of Controversy, in *New Courses for the Colorado River: Major Issues for the Next Century*, Weatherford, G. and Brown, F., Eds., University of New Mexico Press, Albuquerque, 1986, pp. 51–86.

Glenn, E., Felger, R., Burquez, A., and Turner, D., Cienega de Santa Clara: endangered wetland in the Colorado River delta, Sonora, Mexico, *Nat. Resourc. J.*, 32, 817–824, 1992.

Glenn, E., Garcia, J., Congdon, C., and Luecke, D., Status of wetlands supported by agricultural drainage water in the Colorado River delta, Mexico, *HortScience*, 34, 16–21, 1999.

Glenn, E., Lee, C., Felger, R., and Zengel, S., Effects of water management on the wetlands of the Colorado River delta, Mexico, *Conserv. Biol.*, 10, 1175–1186, 1996.

Glenn, E., Thompson, T., Frye, R., Riley, J., and Baumgartner, D., Effects of salinity on growth and evapotranspiration of *Typha domengensis* Pers., *Aquatic Bot.*, 52, 75–91, 1995.

Glennon, R., Coattails of the past: using and financing the Central Arizona Project, *Ariz. State Law J.*, 27, 677–756, 1995.

Howe, C., Water resources planning in a federation of states: equity versus efficiency, *Nat. Resourc. J.*, 36, 29–36, 1996.

Hundley, N., *Water and the West: The Colorado River Compact and the Politics of Water in the American West*, University of California Press, Los Angeles, 1975.

Hundley, N., *Dividing the Waters: A Century of Controversy between the United States and Mexico*, University of California Press, Los Angeles, 1966.

Leopold, A., The green lagoons, in *A Sand County Almanac*, Leopold, A., Ed., Oxford University Press, New York, 1949, pp. 150–158.

Mellink, E., Palacios, E., and Gonzalez, S., Notes on nesting birds of the Cienega de Santa Clara saltflat, northwestern Sonora, Mexico, *West. Birds*, 27, 202–203, 1996.

Mellink, E., Palacios, E., and Gonzalez, S., Non-breeding waterbirds of the delta of the Rio Colorado, Mexico, *J. Field Ornithol.*, 68, 113–123, 1997.

Morrison, J., Postel, S., and Gleick, P., *The Sustainable Use of Water in the Lower Colorado River Basin*, Pacific Institute for Studies in Development, Environment, and Security, Oakland, CA, 1996.

Ohmart, R., Anderson, B., and Hunter, W., Ecology of the Lower Colorado River from Davis Dam to the Mexico–United States Boundary: A Community Profile, National Technical Information Service, Alexandria, VA, 1988.

Owen-Joyce, J. and Raymond, L., An Accounting System for Water and Consumptive Use along the Colorado River, Hoover Dam to Mexico, United States Geological Survey Water-Supply Paper 2407, Washington, D.C., 1996.

Pontius, D., Colorado River Basin Study, Final report to the Western Water Policy Review Advisory Commission, Phoenix, AZ, 1997.

Richardson, J. and Carrier, C., *The Colorado: A River at Risk,* Westcliffe Publishers, Engelwood, CA, 1992.

Snape, W., Adding an environmental minute to the 1994 Water Treaty, in Workshop Proceedings: Water and Environmental Issues of the Colorado River Border Region: A Roundtable Workshop, Sponsored by the Pacific Institute for Studies in Development, Environment, and Security and Defenders of Wildlife, Cohen, M. and Morrison, J., Eds., Pacific Institute, San Diego, 1998.

Stromberg, J. and Patten, D., Flood flows and dynamics of Sonoran riparian forests, *Rivers,* 2, 221–235, 1991.

Sykes, G., *The Colorado Delta,* Publication 460, Carnegie Institution of Washington, Washington, D.C., 1937.

Szaro, R., Riparian forest and scrubland community types of Arizona and New Mexico, *Desert Plants*, 9, 1–138, 1989.

U.S. Bureau of Reclamation, Description and Assessment of Operations, Maintenance, and Sensitive Species of the Lower Colorado River, U.S. Bureau of Reclamation, Lower Colorado Region, Boulder City, NV, 1996.

Valdes-Casillas, C., Hinojosa-Huerta, O., Munoz-Viveros, M., Zamora-Arroyo, F., Carrillo-Guerrero, Y., Delgado-Garcia, S., Lopez-Camacho, M., Glenn, E.P., Garcia, J., Riley, J., Baumgartner, D., Briggs, M., Lee, C.T., Chavarria-Correa, E., Congdon, C., and Luecke, D., Information Database and Local Outreach Program for the Restoration of the Hardy River Wetlands, Lower Colorado River Delta, Baja California and Sonora, Mexico, Instituto Tecnológico y de Estudios Superiores de Monterrey (ITESM), Campus Guaymas, Sonora, Mexico, 1998.

Wilson, F., A fish out of water: a proposal for international stream flow rights in the lower Colorado River, *Colo. J. Int. Environ. Law Pol.,* 5, 249–272, 1994.

Zengel, S. and Glenn, E., Presence of the endangered desert pupfish (*Cyprinodon macularius,* Cyprinidontidae) in Cienega de Santa Clara, Mexico, following an extensive marsh dry-down, *Southwest. Nat.,* 41, 73–78, 1996.

Zengel, S., Mertetsky, V., Glenn, E.P., Felger, R., and Ortiz, D., Cienega de Santa Clara, a remnant wetland in the Rio Colorado delta (Mexico): vegetation distribution and the effects of water flow reduction, *Ecol. Eng.,* 4, 19–36, 1995.

SECTION III.2

Canadian Prairie Ecosystem

Overview: Sustainability of the Semiarid Prairie Ecosystem, Canadian Prairie Ecosystem Study (PECOS)

Darwin W. Anderson

The Prairie Ecosystem Study (PECOS) involved the University of Saskatchewan and the University of Regina in an interdisciplinary study that included the natural, agricultural, and health sciences. PECOS was funded by the Tri-Council Ecosystem Research Program (TRICERP), a granting council to administer Green Plan funding from Environment Canada. TRICERP projects were available only to universities, and had to include research in at least two of the three areas of the national granting councils: the Natural Sciences and Engineering Research Council (NSERC), the Social Sciences and Humanities Research Council (SSHRC), and the Medical Research Council (MRC, now the Canadian Institute for Health Research). PECOS involved more than 40 professors and research scientists, guiding the graduate study of 29 students, as well as 5 postdoctoral fellows, several undergraduate research assistants, research associates, health professionals, and technicians. PECOS was community based in that residents of the study area made suggestions about the overall design of the study, helped to connect PECOS with the community, participated in a community canvass to encourage completion of a questionnaire discussed in the chapter by McDuffie et al., and served as subjects in several health-related studies.

Because of the breadth of the PECOS study, investigators decided to manage the large and complex research effort by forming focus groups. Focus 1 concentrated on social, economic, and cultural aspects, with an emphasis on the attitudes of the PECOS-area residents on conservation and environmental issues, and the health of rural communities. Focus 2 concentrated on human health; specifically, the environmental effects of pesticide exposure on health. Focus 3 dealt with the health of the land and the biota. To avoid the potential for creation of divisions within PECOS along lines paralleling focus group content, investigators created study-wide programs through publication of a newsletter and scheduling of inclusive activities such as seminars and field excursions.

The objective of PECOS was to assess the sustainability of the semiarid prairie ecosystem, particularly the Palliser Triangle — the semiarid grassland region of southwestern Saskatchewan and southeastern Alberta. The Palliser Triangle, or mixed grassland ecoregion, is the northernmost extension of the Great Plains — the huge region just to the east and in the rain shadow of the western mountains. The Palliser Triangle is an area of more than 500,000 km^2; PECOS elected to study a smaller and representative area, Agricultural Census District 3BN. The PECOS study included an area of 15,700 km^2 in southwestern Saskatchewan, with 14 rural municipalities

(RMs), and a human population of 30,000. About one half of the people live in the one urban center, Swift Current.

The PECOS area included 2775 farms in 1996, a decline from 2970 in 1991, which reflected a continuing trend to larger farms that began soon after settlement, and that has intensified since the 1970s, particularly in the 1990s. The average farm is just over 500 ha (1 ha = 2.47 acres) in size; 795 farms are larger than 900 ha, and 146 are six sections (1550 ha) or larger. About 70% of the land is cultivated in wheat (hard red spring and durum), barley, oil seeds (flax, canola, and mustard), pulse crops (lentil, pea, and chickpea), and canary seed. Cattle are raised on 1200 farms or ranches, using mainly native grasslands for grazing. Several large, modern hog-producing operations were established in the area in the 1990s, with farmers as major shareholders in arrangements that permit sale of feed grains such as barley to the hog operation, as well as providing local opportunities for employment.

Despite its appearance as a reasonably well maintained agricultural landscape with generally large, prosperous-looking farms, the area is in crisis. The health of the ecosystem is compromised, which is particularly evident when it is evaluated within the broad context of biophysical, economic, and social dimensions (Nielsen, 1999).

When the PECOS proposal to TRICERP was being prepared in 1992, the area had just experienced several years of drought. Prices for cereal grains at that time were extremely low, a consequence of depressed world prices for wheat and other cereals. Government support programs and emergency (drought) aid contributed more to the agricultural incomes than sales from crop production. Proponents justified reduction of subsidies on the basis of elimination of budgetary deficits, both provincially and federally. A consistent movement favoring fewer government subsidies for agriculture — at least directly to farmers — began in the 1990s. In Saskatchewan, for example, direct government support for agriculture has been reduced by 71% since 1991–1992, according to the Provincial Auditor's report for 1999.

Several years of above-average rainfall in the middle-to-late 1990s resulted in better yields of cereal grains. Crops including canola and pea that have been grown generally in more-humid regions of the prairies replaced wheat on a substantial part of the land. Yields of the new crops were good, a consequence of the moist growing seasons and of reduced disease pressure. The changes appeared to mark a new period of prosperity, but the effect was short-lived. Declining prices for almost all crops, and steadily increasing costs of production inputs, became common by the late 1990s. The elimination of the rail transportation subsidy known as the Crowsnest Pass rate, which had been in effect since the 1880s, resulted in considerably higher freight charges for shipping grain to ocean ports. Profit margins for export grains became even smaller or disappeared, particularly in parts of Saskatchewan that are distant from Atlantic and Pacific ocean ports. Government officials have predicted that farm incomes will remain low for the near future.

Many farms expanded, with considerable investment in large machinery, particularly new technologies such as zero tillage or minimal tillage equipment. The new technologies offer the promise of improved soil conservation, greater efficiency (at least in terms of labor), and lower costs of production. Calculations by farm management specialists suggest that many of the large tractors, conservation tillage equipment, and combine harvesters should be used on farms of no less than 2000 to 3000 ha in order to reconcile their high initial cost.

The increase in farm size has resulted in fewer farmers, and a recent trend among farmers to rent rather than own a substantial proportion of their cropland. In fact, many farmers who retire or stop farming are not able to sell their land for what they consider a fair price, and instead rent the land to neighbors. The steadily declining farm population, and the increasingly common practice of consolidating farm supply outlets in only a few large centers, as well as a markedly reduced number of grain delivery points, have resulted in a marked decline in the on-farm population and smaller farming communities.

The changes have been described by proponents as the industrialization of agriculture, the progressive process of moving agricultural production from dependence on biological systems and the attendant vagaries of nature to a more-predictable production system modeled on the factory. The industrial system depends on high yields obtained from greater use of purchased inputs such as fertilizers and pesticides, in comparison to earlier practices that were lower yielding and more dependent on the nutrients provided free (and generally without natural replacement) by the soil and nature. In 1999, these changes coupled with low commodity prices resulted in a renewed crisis in farming, and instigated the call for substantial government aid for farmers in serious financial difficulty.

Four chapters are presented in the case study. In the chapter "Ecosystem Level Functional Changes in Breeding Bird Guilds in the Mixed Grassland since Agricultural Settlement," Todd Radenbaugh, a Ph.D. candidate at the University of Regina, describes the ecoregion and examines the change in bird populations using mainly historical data. Cultivation and related changes have altered the character of the prairie landscape and the birds that utilize the prairie as habitat. The research demonstrates that ecosystems change over time, and that the health of present systems cannot always be evaluated using a standard of a past, supposedly pristine condition.

MacDonald and Remus' chapter, "Health and Well-Being in a Changing Environment: Perceptions of a Rural Senior Saskatchewan Population," presents a broad assessment of the largest single group of residents, those over 55 years of age. Generally, the survey found that most older people consider their health as good or excellent, particularly those with higher incomes. Most residents were satisfied with their lives — in spite of declining conditions in their communities and related commercial, health, and educational services — pointing to the great resiliency of those who have lived and prospered in this uncertain environment. Despite the present uncertainty, most seniors remain optimistic about the future, a marked contrast to the high school students who see little or no economic or cultural prospects within the rural areas (Butler, 1998).

The chapter by McDuffie et al., "Prairie Ecosystem Study (PECOS): From Community to Chemical Elements, the Essential Role of Questionnaires," discusses the involvement of the local community and health-science researchers to advance our understanding of potential human health risks associated with pesticide exposure. The essential role of questionnaires is stressed, as are the impediments and benefits of interdisciplinary research. The data gathered by questionnaires are conceptual links between different researchers but it is the interaction and discussion of the data that challenges the conventional wisdom inherent in each specialty, and leads to new insights concerning the broader picture.

In their chapter, "Evaluating Agroecosystem Sustainability Using an Integrated Model," Belcher and Boehm discuss a research approach for making the dynamic linkages between system compo-nents. A computer simulation model has been developed to dynamically integrate the ecological (a soils submodel that simulates soil quality and crop growth parameters) and economic components. The chapter examines biophysical constraints to agroecosystem sustainability, as well as the impli-cations of the constraints for policy development.

The chapters are representative of the research done in PECOS. The Belcher and Boehm and Remus and MacDonald papers are from Focus 1, in which emphasis was placed on the social, economic, and cultural dimensions of evaluating the health of ecosystems. McDuffie et al., of Focus 2, deal with human health aspects, and discuss the process of working in an interdisciplinary way, especially by learning the language of related disciplines. Radenbaugh's chapter is part of his Ph.D. study, based in Focus 3. All studies in Focus 3 were graduate research projects, making the integration and discussion that promote interdisciplinarity even more important. Interestingly, perhaps not encumbered by the many years of discipline-based research, the students were more completely interdisciplinary than their professors, not only taking ownership of their individual projects, but also appreciating those of fellow students. PECOS and TRICERP shared a common objective: for the interdisciplinary perspective developed by students to follow them as they develop their own careers, helping them to remain conscious of the many interrelated dimensions of assessing the health of ecosystems.

REFERENCES

Butler, J.L., The Transition from School to Work in the Age of Globalization: Southwest Saskatchewan, Master's thesis, University of Saskatchewan, Saskatoon, Saskatchewan, Canada, 1998.

Nielsen, N.O., The meaning of health, *Ecosyst. Health,* 5, 65–66, 1999.

CHAPTER **114**

Historical Land Use and Ecosystem Health of the Canadian Semiarid Prairie Ecosystem

Lisa L. Dale-Burnett and Darwin W. Anderson

INTRODUCTION

The Palliser Triangle is an area of about 500,000 km^2 in southwestern Saskatchewan and southeastern Alberta coinciding with the Mixe Grassland Ecoregion of the Prairie Ecozone and the Brown Soil Zone (Acton et al., 1998). The southern boundary of the Palliser Triangle corresponds with the international boundary at 49°N latitude; however, ecological conditions of the land to the south in Montana and western North Dakota are similar. The Palliser Triangle was the focus of a major interdisciplinary study that examined the sustainability of the semiarid prairie ecosystem from the perspective of the health of the land and the well-being of the people, the Prairie Ecosystem Study (PECOS). The subject of this chapter is the history of land use and agriculture in the PECOS area, and the larger Palliser Triangle from the late 1800s until the present. It emphasizes the history of agricultural development, explores the difficulties that have periodically resulted in unhealthy ecosystems, and, finally, discusses today's ecosystem health in relation to the past.

EVALUATION OF THE PECOS AREA BY EARLY EXPLORERS

The Palliser Triangle was explored by an expedition financed by the British government and led by Captain John Palliser in 1857–1859 (Spry, 1968). The objective of the expedition was to explore and evaluate the settlement potential of the lands in western Canada that were then under the control of the Hudson Bay Company. Palliser's party traveled through the area during a decade that was warmer and drier than average. In his final report to the British government, Palliser described a triangular area as "desert, or semidesert in character, which can never be expected to become occupied by settlers" (Spry, 1968).

Palliser's view of the region was echoed in the report from H.Y. Hind, who led a Canadian expedition that explored western Canada in 1857–1859. Hind described "that part of the Canadian Plains west of the Missouri Coteau from the character of its soil and aridity of its climate ... as permanently sterile and unfit for the abode of civilized man" (cited in Mackintosh, 1934). Hind considered the region as the northern extension of what was described as the Great American Desert.

Two decades later, John Macoun, a botanist accompanying an 1879–1881 railway survey team, traveled through the southern part of the Palliser Triangle, probably during a moist period. He described the area quite differently from Palliser and Hind:

> The appearance of the country passed through was altogether different from what I expected, having been led to believe that much of it was little else but desert.... [F]uture settlers will prefer the prairie as there is less broken land, less marsh and swamp and less labour required to make a home. The open treeless prairie generally condemned to sterility is by far the best farming land (Macoun, cited in Waiser, 1989).

The transcontinental Canadian Pacific Railway (CPR), which was originally routed through the northern fertile belt, was rerouted through the Palliser Triangle at about 50°N latitude. Although the influence of Macoun's reassessment of the agricultural potential of the semiarid region in this relocation has been questioned (Waiser, 1985), his opinions certainly strengthened the CPR's decision. The relocation of the CPR mainline was the first step in the agricultural development of the semiarid prairie region that is the economic basis of the PECOS area.

EARLY HISTORY

Archaeological evidence indicates that the Prairie Ecozone has been inhabited for several thousand years (Potyondi, 1995). Aboriginal hunting pressure probably contributed to the extinction of species, such as a larger type of bison, thousands of years ago. Fire was used to manage the grasslands by the aboriginal peoples, who periodically burned the prairie to encourage new vegetative growth, which in turn attracted the bison. Many recognize today that it was fire combined with a very dry climate that kept trees from growing on the prairie.

Potyondi (1995) summarized the influence of the aboriginal people on the prairie ecosystem as follows:

> As many as seven different native groups frequented the area until the 1870s. Typically, they used the oases on its borders more than they did its interior, which was poorer in natural resources and therefore less hospitable to people. As they had for millennia, the Native Peoples depended primarily on the herds of buffalo that ranged through southwestern Saskatchewan seasonally. The abiding abundance of animals suggest that the native population did not exceed the natural carrying capacity of the land, and their semi-nomadic way of life, linked intimately to the seasonal availability of sustenance, helped to maintain the ecological balance.

Between 1830 and 1880, major change came to the area in the form of buffalo hunts. Metis hunters and traders from the Red River settlement in southern Manitoba traveled by ox cart to southwestern Saskatchewan to systematically harvest the buffalo. Hides were collected and buffalo meat was used to make pemmican. For the first time, the buffalo were seen to have economic value rather than as a source of meat and clothing for the local Indian tribes (Potyondi, 1995). This commodification resulted in the hunting of the buffalo to virtual extinction by 1880.

THE RANCHING PERIOD

In the 1880s, cattle replaced bison in the Palliser Triangle. The first ranchers were the Metis and French Canadians who, with the decline in the bison herd and the related activities of freighting and trading, began to graze cattle on the now-unused prairie (Potyondi, 1995). Markets were limited, however, with main sales being to the government. About the same time, some American ranchers moved their cattle north to graze because of increased pressure on the Montana range. Some held Canadian government grazing leases, but others simply took advantage of the open range and herded their cattle north. By the 1890s, thousands of American cattle, ranging as far north as the CPR, grazed Canadian grass.

In the late 1880s and 1890s, the Canadian government encouraged ranching in the Palliser Triangle by granting inexpensive, long-term grazing leases. By 1888, Sir John Lister-Kaye had started to establish three large ranch-farms in the vicinity of the PECOS area. Lister-Kaye's company was known as the Canadian Agricultural, Coal and Colonization Company, or the "76" for the brand displayed on many of the cattle that had been purchased in the U.S. The farms were initially set up as mixed operations with grain, cattle, sheep, and hogs, but after 2 years of crop failure it was decided to specialize in cattle and sheep. The 76 was only the first of several large ranches that were established in the PECOS area. In 1902, the Cresswell Cattle Company, an American outfit, trailed 25,000 cattle and 600 horses from South Dakota into what was to become southwestern Saskatchewan and southeastern Alberta. The ranch was known as the Turkey Track, for the shape of its brand.

The harsh winters of 1903–1904 and 1906–1907, combined with government support for homesteading, spelled the end of the 76 and Turkey Track ranches. The cattle had grazed year round on the unfenced range through the mild winters of the 1890s. But during the long, cold winter of 1906–1907, 60 to 70% of the range cattle were lost (Potyondi, 1995). The era of the open range ended as settlers arrived to homestead quarter- or half-section entitlements.

An exception to the demise of the large ranches was the Matador Land and Cattle Company, a Scottish-owned ranching company which was part of a larger U.S.-based ranching company of the same name. In 1905, the Matador obtained a 21-year lease to 50,000 acres (20,000 ha) of grazing lands north of the South Saskatchewan River. The Matador's Canadian range was essentially a giant finishing operation for 2- and 3-year-old steers. Shipped into the area by rail, they grazed on the rich pasture for 2 or 3 years, and were then sold as 4- or 5-year-olds, either back to the U.S. or to Great Britain, depending on the price and tariffs of the day. High American tariffs that were imposed after World War I, coupled with generally depressed prices, resulted in the close of the Matador's Canadian operation in 1922.

Today, a provincially owned community pasture takes up much of the Matador range. An ecological reserve owned by the University of Saskatchewan is also within the original leasehold area. As part of the International Biosphere Program, the university established a major research project on grassland ecology, known as the Matador Project, on this reserve in the 1970s.

THE HOMESTEADERS

Although there were a few farmers in the PECOS area before, general agricultural settlement did not take off until the first decade of the 1900s. The CPR mainline went through the PECOS area in 1882–1883, but construction did not resume until 1911 when a number of branch lines were started. Railway construction continued until the early 1930s. In the PECOS area, much settlement preceded branch line construction. Settlers lived far from established towns, transportation, and postal services. The arrival of a railway meant not only the development of grain elevators for the more convenient marketing of grain and other agricultural products, but also regular mail service, delivery of commercial goods, and easier travel to larger centers.

Often the last branch rail lines constructed in the PECOS area were the first to be abandoned as better road transportation and larger farms obviated the need for many small communities and grain delivery points. One example is the Mawer to Main Centre line, which served the east-central part of the PECOS area just south of the South Saskatchewan River. Built in 1930–1931, it was taken up in the late 1960s, and most of the villages along this route have disappeared.

The rural population* of the PECOS area was 3395 in 1906, and rapidly increased to just over 30,000 in 1931. At this time, there were 19 incorporated towns and villages and one city (Swift Current) in the study area. After a peak in 1931, the rural population of the PECOS area began a decline that continues to this day.

The agricultural methods and implements brought by settlers from more humid regions were poorly suited to the semiarid climate of the PECOS area. A classic example was the use of the plow for tilling the soil. Appropriate in more-humid regions, moldboard plows buried crop residues and left the soil in a dry, pulverized condition, easily eroded by wind. Within two decades of agricultural settlement, marked deterioration of the soil was becoming apparent. Inappropriate tools and techniques led to soil erosion, loss of organic matter, and lowered fertility (Shutt, 1925). In fact, many homesteaders had tried and given up farming within a few years, especially on poorly suited lands such as the Seward Sand Hills, west of Swift Current.

The droughts and hardships of the 1917–1926 period were particularly harsh. Concerns arising from soil-drifting and loss in soil fertility resulted in the Better Farming Conference held by professional agriculturists and farmers at Swift Current in 1920 (Anstey, 1986). A month after the conference, a provincial government royal commission was established to investigate the conditions surrounding the drought and farm abandonment in southwestern Saskatchewan. Also in 1920, the federal government established an agricultural research station based in Swift Current to research and develop cropping and range-management strategies suitable for semiarid regions (Campbell, 1971). In 1921, the Better Farming Commission made several recommendations, including the establishment of a soil survey for Saskatchewan and more comprehensive collection of meteorological records (Province of Saskatchewan, 1921). These actions enhanced the scientific side of farming, and gradually dryland farming methods evolved that slowed the process of soil deterioration and stabilized production to a considerable degree.

In the early years of agricultural settlement, the main approach to dryland farming was based on the better application of summer fallow. Summer fallowing is the practice of letting a field lay idle once every 2 or 3 years and tilling it to keep the weeds down on the fallow year. It was believed that loosening the soil allowed rain to penetrate to the subsurface, while a fine dust-mulch protected this subsoil moisture from evapotranspiration. Working the soil also killed moisture-loving weeds. In this way, a portion of rainfall in the fallow year was retained for the following growing season. From the beginning, summer fallow was recognized as a major contributor to the loss of soil organic matter or fiber, which increased susceptibility of the soil to wind erosion, but it was still recommended because alternatives were not yet available (Shutt, 1925).

The comparative success of summer fallow as a risk-averse cropping system that exploited reserves of plant nutrients in soil organic matter was undoubtedly a reason for its general adoption in the region. Historically, about 40% of the cultivated land in the PECOS area was in summer fallow, but in recent decades this proportion has been smaller. Decomposition leading to decreased organic matter also released large amounts of nitrogen as nitrate. Because of these high levels of nitrate in the soil, wheat grown after summer fallow had high protein content even without applying fertilizer. In the first years of cultivation, the amount of nitrate mineralized from the reserve stores was generally adequate to supply crops. Overall, the practice of summer fallow led to a loss of nitrate, however. A large proportion of the nitrogen removed from the prairie soils since cultivation cannot be accounted for by crop uptake. A great deal of soil nitrate has been lost either through leaching beyond the root zone, or by loss back to the atmosphere by denitrification. In the past few decades, the use of nitrogen fertilizer has become common in the PECOS area.

The 1930s was a period of drought and economic depression when all three components of the prairie ecosystem, the biophysical, socioeconomic, and human well-being were consistently unhealthy. An estimated 10,000 farm families left the semiarid region of Saskatchewan in the 1930s, some leaving farming, but others moving to the forest areas in the north to homestead again under very

---

* Rural population is defined as people living outside of the boundaries of any city, town, or village and including those living in communities of fewer than 100 people.

difficult conditions. The farmers and ranchers who remained in the semiarid region endured considerable hardship. Year after year, events conspired to deprive farmers of an income. If there was enough moisture for a crop, then the grasshoppers invaded; if there was a good harvest, then prices were low. Records of low wheat yields, low prices, and high unemployment were set during the 1930s.

In 1935, the Canadian government established the Prairie Farm Rehabilitation Administration (PFRA) by an act of parliament. PFRA's mandate was to provide for the rehabilitation of the drought and drifting areas, and to develop and promote systems of farm practice, tree culture, water supply, land utilization, and land settlement that would lead to greater economic security for the agricultural region of western Canada (PFRA, 2000). One early PFRA program was the setting up of community pastures on eroded, low-quality lands that had been inappropriately used for farming.

GRADUAL CHANGE IN AGRICULTURE

The generally more benign climate and more favorable economic prospects of the 1940s through the 1960s was a period of recovery for farming in the PECOS area. Part of the change was the gradual increase in farm size and the related reduction in number of farms. The number of farms decreased from a maximum of 6960 in 1931, to 5911 in 1946, and to 4052 in 1961. The trend continues, with 2942 farms in 1991, and 2775 in 1996.

The increase in area of land farmed by individual farmers was made possible by the mechanization of farming, generally declining real prices for grain, and the comparatively low productivity of land in the Brown soil zone. PECOS area farms were consistently larger and more mechanized than the Saskatchewan average. The transition from horse to tractor power was almost complete in the PECOS area by 1945, unlike the rest of Saskatchewan; combine harvesters replaced threshing machines in the PECOS area before the province as a whole.

Although some government programs have supported smaller, more-diversified farms, there has been a consistent philosophy (whether intended or simply expedient) that the farming sector will be better off as smaller farms become absorbed by larger, more technically refined producers.

Today, the model of prairie agriculture most often recommended by agricultural scientists and many farmers is the large, highly mechanized, capital-intensive farm. These farms operate on a narrow margin (small profit per acre or per animal) in an intensely competitive global market. Survival depends on the efficiency of the farmer in managing thousands of hectares or hundreds of animals. This shift in farming from a nature-dependent activity, susceptible to natural variations, to a highly managed, technologically dependent activity has been referred to as the industrialization of agriculture. For example, in industrial agriculture the nitrogen requirements of a crop are provided by fertilizers, not animal or green manure; and pests are controlled by pesticides, not natural predators. In essence, agriculture becomes more like a factory, with greater stability but, interestingly, less resiliency. Stability is considered to be the ability of a system to resist or withstand change or disturbance, whereas resilience is the capacity of a system to return to an optimum state following a stress, as when a willow bends in the wind but does not break.

RECENT TRENDS IN PECOS AREA AGRICULTURE

Ranching and farming have substantially altered the prairie ecosystem. In the PECOS area, 61% of the land had been broken by 1961. Today, about 70% of the prairie has been converted to cropland, resulting in the loss and fragmentation of habitat. Habitat loss is linked to the extirpation (loss of a species from a particular area) of animal species, such as the swift fox and prairie dog, and reductions in numbers of many others, such as the sharp-tailed grouse and burrowing owl.

Certain plant communities are almost completely gone. The wheatgrass-junegrass type that was characteristic of the clayey lacustrine soils in the Brown and Dark Brown soil zones (Coupland

and Rowe, 1969) remains in only a few small protected areas such as the Matador Grassland. These clay soils and landscapes are particularly well suited to dryland farming because of their fertility and high moisture-storage capacity, and so are very desirable as farmland.

Evaluations of soils under different cropping systems have consistently shown that erosion and the loss of soil organic matter are more severe in rotations where tilled summer fallow is used frequently. Summer fallowing every second year, which was a common practice in the Palliser Triangle for many decades is considered to be the worst-case scenario. Organic matter contents of surface horizons are only about one half of the level found in native grassland soils, with related decreases in nutrient supplying power or fertility, soil tilth, and resistance to soil erosion. Intensively farmed soils have fewer and less stable soil aggregates than soils under grass. Poorly aggregated soils are more difficult to till, more easily eroded, and take in water more slowly than well-aggregated soils. Cultivated soils are less porous than soils under prairie, with fewer large pores or macrovoids that are so important to the intake of water. Water from rain or snowmelt that does not enter the soil cannot be stored for future use by crops, and runs off, thereby increasing the potential for water erosion. Crop rotations that avoid tilled summer fallow and increase the amount of crop residues or manures added to the soils are likely to result in increased soil organic matter, better soil aggregation, and more-efficient use of scarce moisture supplies.

Many farmers, particularly in the driest parts of the Palliser Triangle, continue to summer fallow about one half of their land each year as a moisture conservation strategy. Increasingly in recent years, herbicides are used to control weeds on summer fallow rather than tillage. This practice is known as chem-fallow. When combined with seeding methods that disturb the soil minimally, chem-fallow maintains a cover of crop residues that protects the soil from wind erosion.

The present-day increase in land under conservation tillage reflects a transition to a more efficient farming system (one farmer can farm more land) that conserves soil and is generally more productive. Conservation tillage is a general term used to describe farming methods that protect the soil by maintaining a protective cover of crop residues or stubble from previous crops, yet use little or no tilling. Zero tillage is a tillage system that completely avoids tillage, except for the minimal disturbance made when fields are seeded. The proportion of land in the Rosetown area (generally typical of the PECOS area) seeded using conventional tillage methods decreased from 88 to 60% between 1994 and 1998, as reported in a recent field survey completed by PFRA (Haak, D., PFRA, Rosetown, Saskatchewan, Canada, personal communication, 2000). There was a related increase in zero tillage from 3.3 to 12.2%, and in minimal tillage from 6 to 18%. Both zero-till and minimum till are described as direct seeding methods, in that there is no cultivation prior to seeding the land. Broad-spectrum herbicides such as glyphosate are used to control weeds. These recommended methods are increasingly dependent on the more intensive use of agrichemicals, particularly herbicides, pesticides, and fertilizers. More fertilizers, especially nitrogen fertilizers, are needed as the supply of nitrogen present in the soil organic matter is used up, and as farmers move to continuous cropping with conservation tillage.

The general reduction in area of land in summer fallow and the adoption of conservation tillage are considered to be indicators of a more-sustainable agriculture, an agriculture that maintains or enhances soil and ecosystem health (Acton and Gregorich, 1995). In a recent report on the environmental sustainability of Canadian agriculture, McRae and co-workers (2000) write:

> The risk of erosion on the Prairies declined by about 30% from 1981 until 1996, with two-thirds of the decline occurring between 1981 and 1991. About three-quarters of the reduction can be attributed to a change in tillage practice. The remainder is mainly the result of changes in cropping practices, specifically less summer fallow.

However, farming methods that require greater use of pesticides raise concerns about both the health of the environment and the people applying or otherwise exposed to the agrichemicals. Studies of surface water bodies, particularly farm dugouts or reservoirs that collect surface runoff from cultivated land, found herbicides in the water months after their application to crops (Cessna and Elliott, 2000).

Technologies for conservation tillage are expensive and have resulted in a marked increase in the size of "economically viable" farms, a further reduction in the number of farms, and declines in communities supported by farms. Livestock production has become more intensive, concentrated in large mechanized operations that themselves present environmental and social problems.

Recent decades have seen a welcome trend to farming systems less dependent on wheat. Although spring and durum wheats are still the main crops in terms of area planted, there has been a substantial and continuing increase in land under pulse crops, especially lentil, pea, and chickpea. The increases are most evident on the agriculturally superior clay soils (PFRA, 2000). Pulse crops generally result in better economic returns and require less nitrogen fertilizer, an important factor as energy prices and the cost of inputs dependent on fossil fuels continue to increase.

THE PECOS AREA IN 2000

Despite large, prosperous-looking farms, an agricultural landscape that seems to be well cared for, and uses improved soil conservation practices, there is little doubt that the semiarid prairie region is a community in crisis. This crisis is particularly clear when ecosystem health is evaluated within the broad context of biophysical, economic, and social dimensions (Nielsen, 1999). In the early 1990s, when commodity prices were low and the area had just experienced several years of drought, government support programs and emergency aid contributed more to agricultural incomes than did production. Beginning in the 1990s, there has been a consistent trend to fewer agricultural subsidies at both the national and provincial levels. These cutbacks have been justified on the basis of eliminating budgetary deficits. In Saskatchewan, for example, direct government support for agriculture has been reduced by 71% since 1991–1992, according to the provincial auditor's report for 1999.

A period of above-average rainfall in the mid- to late-1990s resulted in the expansion of crops, such as canola and pea, from more-humid regions of the prairies into the more-arid areas, and in better yields overall. These changes, plus the introduction of new crops such as chickpeas appeared to mark a new period of prosperity, but the effect was short-lived. By the late 1990s, declining crop prices and steadily increasing costs of production canceled most gains.

In 1983, the elimination of the Crowsnest Pass Agreement, a transportation subsidy that had been in effect since the late 1880s, resulted in income margins for export grains becoming even smaller or disappearing, particularly in parts of Saskatchewan that are distant from both Atlantic and Pacific ocean ports. The wheat economy, the mainstay for most Saskatchewan farmers, relied on two subsidies. First was the Crow Rate, which moved grain to ocean ports at considerably less than the actual cost; second was the natural subsidy of nitrogen reserves stored in the soil organic matter, which allowed farmers to grow high-quality wheat without fertilizer.

With considerable investment in large machinery to support new practices such as zero or minimal tillage, many farms became even larger. The new technologies offer the promise of improved soil conservation, greater efficiency (at least in terms of labor), and lower costs of production. Calculations by farm management specialists suggest that many of the large tractors, conservation tillage equipment, and combine harvesters should be used on at least 5000 or 6000 acres each year, in order to reconcile their high initial cost.

The increase in size of farms has resulted in fewer farmers and the growing importance of farmland rental as opposed to ownership. In fact, many farmers who retire or stop farming are not able to sell their land for what they consider a fair price, so the land is rented to neighbors. This change in land tenure is of concern to those who deal with soil conservation. Studies have shown that farmers with a long-term stake in land (e.g., owners) are more likely than renters to adopt soil-conserving management strategies.

There are also concerns about persons being interested in farming in the future. Most farmers in the area today, as in most of western Canada, are older and thinking of retirement. There are few young people who anticipate a career in farming, or even a career in the rural areas where employment prospects are limited (Butler, 1998).

As described in this chapter, agriculture in the PECOS area has moved from a way of producing food that was dependent on biological systems and the attendant vagaries of nature to a more-predictable, industrial production system modeled on the factory. The latter industrial system depends on high yields obtained from greater use of purchased inputs, such as fertilizers and pesticides, as opposed to the former practices that were more dependent on the preexisting soil nutrients that were free, but generally without natural replacement. These changes, coupled with a return to low commodity prices in 1999–2000, have resulted in a renewed crisis in farming and the call for substantial government aid for farmers in serious financial difficulty.

REFERENCES

Acton, D.F. and Gregorich, L.J., The health of our soils, toward sustainable agriculture in Canada, Publication 1906/E Centre for Land and Biological Resources Research, Research Branch, Agriculture and Agri-Food Canada, Ottawa, Ontario, 1995.

Acton, D.F., Padbury, G.A., and Stushnoff, C.T., *The Ecoregions of Saskatchewan,* Canadian Plains Research Center, University of Regina, Saskatchewan, 1998.

Anstey, T.H., *One Hundred Harvests, Research Branch, Agriculture Canada 1886–1986,* Canadian Government Publishing Centre, Ottawa, Ontario, 1986.

Butler, J.L., The transition from school to work in the age of globalization: a southwest Saskatchewan perspective, Master's. thesis, University of Saskatchewan, Saskatoon, Saskatchewan, 1998.

Campbell, J.B., *Swift Current Research Station 1920–70,* Canada Department of Agriculture, Ottawa, Ontario, 1971.

Cessna, A.J. and Elliott, J.A., Seasonal variation of concentrations of herbicides and inorganic ions in prairie farm dugouts, Final report of the Sustainable Water Management Technical Committee and the Water Quality Technical Committee, available from PFRA, Agriculture and Agri-Food Canada, Regina, Saskatchewan, 2000.

Coupland, R.T. and Rowe, J.S., Natural vegetation of Saskatchewan, in *Atlas of Saskatchewan,* Richards, J.H. and Fung, K.I., Eds., University of Saskatchewan, Saskatoon, Saskatchewan, 1969, pp. 73–77.

Hind, H.Y., Narrative of the Canadian Red River exploring expeditions of 1857 and of the Assiniboine and Saskatchewan exploring expeditions of 1858, M.G. Hurtig Ltd., Edmonton, Alberta, 1971 [1860].

Mackintosh, W.A., *Prairie Settlement: The Geographic Setting*, Vol. I, Canadian Frontiers of Settlement Series, Macmillan, Toronto, Ontario, 1934.

McRae, T.S., Smith, C.A.S., and Gregorich, L.J., Eds., Environmental Sustainability of Canadian Agriculture: Report of the Agri-environmental Indicator Project, Agriculture and Agri-Food Canada, Ottawa, Ontario, 2000.

Nielsen, N.O., The meaning of health, *Ecosyst. Health,* 5, 65–66, 1999.

Potyondi, B., *In Palliser's Triangle: Living in the Grasslands 1850–1930,* Purich Publishing, Saskatoon, Saskatchewan, 1995.

Prairie Farm Rehabilitation Administration (PFRA), Prairie agricultural landscape: a land resource review, available from PFRA, Agriculture and Agri-Food Canada, Regina, Saskatchewan, 2000.

Province of Saskatchewan, *Report of the Royal Commission Inquiry into Farming Conditions,* J.W. Reid, King's Printer, Regina, Saskatchewan, 1921.

Shutt, F.T., Influence of grain growing on the nitrogen and organic matter content of western prairie soils in Canada, Canadian Department of Agriculture Bulletin 44, 1925.

Spry, I.M., Ed., *The Papers of the Palliser Expedition 1857–1869*, The Champlain Society, Toronto, Ontario, 1968.

Waiser, W.A., A Willing Scapegoat: John Macoun and the Route of the CPR, *Prairie Forum,* 10, 65–81, 1985.

Waiser, W.A.,*The Field Naturalist: John Macoun, the Geological Survey, and Natural Science*, University of Toronto Press, Toronto, Ontario, 1989.

CHAPTER 115

Health and Well-Being in a Changing Environment: Perceptions of a Rural Senior Saskatchewan Population

Mary B. MacDonald and Gail Remus

INTRODUCTION

Environmental sustainability has become a major concern throughout the world. The purpose of the Prairie Ecosystem (PECOS) interdisciplinary research project was to gain a better understanding of the interactions between humans and the environment. A description of the PECOS project, Sustainability of the Semiarid Prairie Ecosystem: Imperatives for Agriculture, Environment, and Rural Communities, is covered in the overview.

Concern has been expressed about the lack of services and supports for rural seniors. Commonly reported stressors for rural elderly people include loss, isolation, financial concerns, and decreased ability to manage (Johnson et al., 1993).

Better understanding of the needs of older rural women is also a challenge. The proportion of older women living in rural areas is increasing. Previous research has shown that older rural women have more chronic illnesses than do their urban counterparts (Palmore, 1983–1984). "Their increasing age and susceptibility to disability, along with limited access to acute and chronic care facilities, present a challenge for health care providers" (Dietz, 1991).

Self-reliance and independence have been demonstrated as strong characteristics of rural dwellers. Rural residents often manage on their own or turn to family and friends for assistance rather than to formal agencies for health care (Bushy, 2000; Stewart, 1997). As more of the younger generation migrate from the rural area, the question arises whether older residents can continue to have their health care needs met in rural areas.

The stressors experienced by rural elderly people are related not only to personal characteristics, but also by the physical and social environment in which they live (Bushy, 2000). Thus, there is a need to better understand this population.

The purpose of the study was to determine seniors' perception of their health, satisfaction with life, and optimism regarding the future for residents living within the study area.

1-56670-612-2/03/$0.00+$1.50

METHODOLOGY

Data were obtained through a telephone survey done by a marketing firm using a stratified random sampling technique. In all, 1000 residents within the study area were surveyed: 508 (51%) men and 492 (49%) women. Respondents had to qualify on the basis of gender and rural munici-pality residence to assure representativeness and parallel "the actual distribution of people as given in the 1991 Statistics Canada data" (Dickinson, 1998/1999).

The questionnaire used in the telephone survey was developed by a multidisciplinary team of researchers, including faculty and graduate students in the areas of health, sociology, and economics. Issues related to health and well-being, health care reform, crime and security, education and work, agricultural practices, and environmental ethics were explored.

This chapter addresses those aspects of the study related to health and well-being of seniors (ages 55 and over) within the study area.

RESULTS

Demographic Data

Of the seniors surveyed, 195 (51%) were men and 187 (49%) were women, aged 55 to 95. Most of the respondents (80%) lived in town, defined as being other than living on a farm. Of a total of 72 widowed seniors, 13 were men and 59 were women. When analyzed for marital status, more widowed residents lived in town (94%) than did single persons (80.5%) or married persons (77%) ($p = 0.004$) (Table 115.1). Further analysis of the data according to gender showed that place of residence was significant for women ($p = 0.002$), but not for men. For women, 97% of the widows as compared to 81% of those who were single and 75% of those who were married lived in town (Table 115.1); 75% of the women and 70% of the men had retired; 27% of those who had retired had changed their place of residence, while 28% of those who had not yet retired planned to change their place of residence upon retirement. More men than women had changed their place of residence since retiring (Table 115.1).

Gross income was significantly different for those living in town vs. on the farm ($p < 0.0001$) (Table 115.2). Most town residents (58%) earned $15,000 to less than $40,000 per year, whereas

Table 115.1 Place of Residence of Seniors, Stratified Based on Selected Attributes

	Married	Single	Widowed	Male	Female
Percent living in town	77	80.5	94		
Percent senior females living in town	75	81	97		
Percent retired				70	75
Percent since retirement who changed place of residence				33	20

Table 115.2 Gross Income (Can. $) for Seniors According to Place of Residence and Gender

Factor	<15k	15 ≤ 40K	40 ≤ 80K	≥80K
Place of Residence				
Farm (%)	11	27	36	26
Town (%)	16	58	22	3
Gender				
Male (%)	9	58	24	9
Female (%)	22	46	26	5

most of the farm residents earned $40,000 to less than $80,000 per year (35.5%). More farm residents (26%) than town residents (3%) had a gross income of $80,000 or more per year. There was also a significant difference in gross income according to gender ($p = 0.004$). Generally, there was a higher or similar portion of men than women for each income category except for that of $15,000 or less per year; 22% of the women and only 9% of the men earned less than $15,000 per year. There was also a significant difference between gross income and education level ($p = 0.006$). Regardless of education level, most seniors had a gross income between $15,000 and under $40,000 per year. More seniors with at least some elementary school education earned less than $15,000 per year (30%), compared with those with more education. More of the seniors earning $40,000 to less than $80,000 per year had some college or university undergraduate education (31.5%) or had completed at least a university undergraduate degree (35.5%).

Perception of Health

Overall, the residents' perception of their health was positive. Of the total senior population surveyed, 71% rated their health as good or excellent. This remained unchanged regardless of gender, marital status, or whether one lived alone, with a spouse, or with extended family.

Perception of health was definitely affected by gross income ($p = 0.0002$) (Table 115.3). As gross income level improved, so did the percentage of persons who perceived their health to be excellent or good. Perception of health according to gross income was significant for both genders, but particularly so for men ($p = 0.002$). This was most evident in the income category of less than $15,000 where 71% of the men rated their health as fair or poor while most of those earning

Table 115.3 Percentage of Senior Residents Who Perceive Themselves to Be in Excellent/ Good Health and Fair/Poor Health, in Relation to Income, Education, and Satisfaction with Life

Related Factor	Excellent/Good Health	Fair/Poor Health
Income		
< 15,000	48	52
15 to < 40K	71	29
40 to < 80K	81	19
≥ 80K	88	12
Income (males)		
< 15,000	29	71
15 to < 40K	69	31
40 to < 80K	77	23
≥ 80K	82	18
Education Level		
Some/completed elementary school	48	52
Some/completed high school	73.5	26.5
Some college/university	79	21
Completed university	72	28
Satisfaction with Life (all)		
Satisfied	73	27
Dissatisfied	54	46
Satisfaction with Life (males)		
Satisfied	72	28
Dissatisfied	36	64

$15,000 or more per year rated their health as excellent or good. Most women in each income level rated their health as excellent or good ($p = 0.02$), although for the lowest income level (less than $15,000 per year) only 58% rated their health as excellent or good.

Other than for the lowest educational level, most residents rated their health as excellent or good. As expected, more of those with the lowest education (e.g., not having attended high school) perceived their health to be only fair or poor, compared with those who had at least some high school ($p = 0.0005$) (Table 115.3). There was a similar pattern for men ($p = 0.03$) and for women ($p = 0.04$).

There was a significant difference in perception of health according to satisfaction or dissatisfaction with life ($p = 0.05$). Of those who were satisfied with their lives, 73% rated their health as excellent or good, while only 54% of those who were dissatisfied with their lives rated their health as excellent or good (Table 115.3). When analyzed for gender, there was no significant difference in perception of health between women who were satisfied with their lives and those who were dissatisfied. However, for men there was a significant difference in perception of health according to satisfaction with life ($p = 0.014$). Most men who were satisfied with their lives (72%) rated their health as excellent or good, while only 36% of those who were dissatisfied with their lives rated their health as excellent or good.

There was no significant difference in perception of health whether one was neutral, agreed, or disagreed with the statements that (1) availability of a local physician was an important factor in where they lived and (2) availability of the hospital was an important factor in where they lived. Most of the residents rated their health as excellent or good regardless of their opinions about the importance of the availability of a physician or hospital in determining where they lived. This view remained unchanged regardless of gender. As well, most of the residents agreed with the statements, "The availability of a doctor in my community is an important factor in determining where I live" (75%) and "The availability of a hospital in my community is an important factor in determining where I live" (78%).

When analyzed according to residents' optimism or pessimism in regard to the future, perception of health was not significant. There was no significant difference in perception of health whether residents had changed their place of residence since retirement or planned to change their place of residence after retirement.

Satisfaction with Life

Most of the residents were satisfied with their lives. There was no significant difference in satisfaction with life when analyzed in terms of gender, marital status, or agreement that availability of a doctor or hospital was important in determining where they lived. Regardless of whether children, or other family, lived within 50 miles, most residents were satisfied with their lives. Those living alone or with spouses were significantly more satisfied with life than those who lived with extended family (98, 94, and 77%, respectively; $p = 0.01$) (Table 115.4).

Level of education was not significant in terms of satisfaction with life for the total senior population or when analyzed by gender. The same was found for gross income level and satisfaction or dissatisfaction with life.

Agreement with the statement "upcoming retirement causes me stress or worry" and satisfaction with life was significant ($p < 0.00001$). More of the residents who were not worried about their upcoming retirement were satisfied with their lives (97%) than those who were worried about their upcoming retirement (79%). This was significant for men ($p = 0.0002$) and for women ($p = 0.0001$) (Table 115.4).

Optimism Regarding the Future

Table 115.4 indicates a significant difference ($p = 0.009$) in satisfaction with life depending on whether or not one was optimistic regarding the future. While most residents were satisfied with

Table 115.4 Percentage of Seniors Who Are Satisfied/Dissatisfied with Life and Optimistic/Pessimistic about the Future

Related Factor	Satisfied with Life	Dissatisfied with Life
Living Arrangements		
Lives alone	98	2
Lives with spouse	94	6
Lives with extended family	77	23
Worry about upcoming retirement	79	21
Not worried about upcoming retirement	97	3
Optimism about the Future		
Optimistic	96	4
Pessimistic	88	12

their lives regardless of their optimism or lack of optimism regarding the future, 96% who were optimistic about the future were satisfied with their lives, compared with 88% who were pessimistic about the future yet satisfied with their lives.

Most people were optimistic about the future regardless of their level of education ($p = 0.02$). However, more residents with at least an undergraduate degree were optimistic about the future (85%) as compared with those who were pessimistic about the future (15%) (Figure 115.1). A similar pattern existed for each gender but was significant only for men ($p = 0.02$). Those men with the least amount of education were more pessimistic regarding the future (47%) than those with more education. Although not significant, most people were also optimistic about the future regardless of income level.

More seniors who had not changed their place of residence since retirement were optimistic regarding the future (83%), compared with those who had changed their place of residence (69%; $p = 0.03$). However, when analyzed by gender there was no significant difference. There was no significant difference in optimism or pessimism regarding the future for those who planned or did not plan to change their place of residence after retirement.

DISCUSSION

Income status and education have been identified as key determinants of health (Epp, 1986; World Health Organization et al., 1986; Federal, Provincial and Territorial Advisory Committee

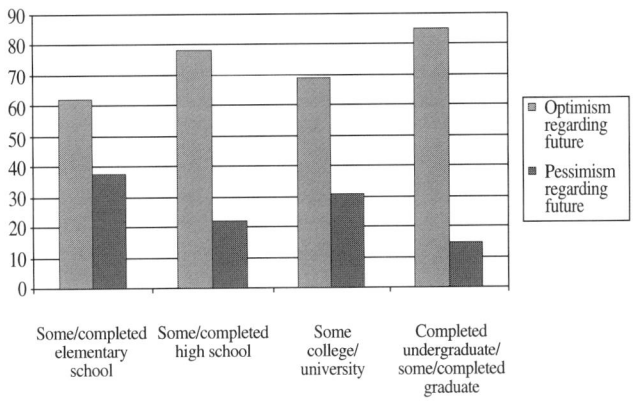

Figure 115.1 Seniors' optimism regarding the future according to education level ($p = 0.02$).

on Population Health, 1994; Saskatchewan Provincial Health Council, 1996). The results of this study support the importance of these two health determinants. Overall, the perception of health by rural seniors was positive. However, more of those in the lowest income category (52%) rated their health as fair or poor. Generally, as income level increased, so did the proportion who rated their health as excellent or good. Men in the lowest income bracket were more likely to perceive their health as fair or poor than were women, although a high percentage of each gender in this income category perceived their health as only fair or poor. The reason for this could not be explored because the data were collected by questionnaire. Future studies, however, need to determine whether individuals are in the lowest income categories because of poor health or if poor health is a result of having a lower income. Level of education was also linked to perception of health. Those with less education perceived their health as being poorer than those with a higher level of education. Mookherjee (1992) also found that those with less education perceived their health as only fair or poor.

Most of the residents perceived their health as excellent or good regardless of whether they were satisfied or dissatisfied with life. However, when men and women were looked at separately, satisfaction with life was significant in terms of perception of health for men. Although 28% of men viewed their health as fair or poor, they still indicated that they were satisfied with life. For those dissatisfied with life, 36% perceived their health as excellent or good, while 64% perceived their health as fair or poor. Again, future studies need to explore whether poorer health leads to feeling dissatisfied with life, or vice versa.

Most of the residents were satisfied with their lives and were optimistic about the future. Interestingly, more residents were satisfied with life if they lived alone or with a spouse than if they lived with extended family. This finding might be linked to seniors' desire to be independent, although the study did not explore this aspect.

As might be expected, residents who were not worried about their upcoming retirement were most satisfied with their lives. Of those who had already retired, those who had not changed their place of residence were more likely to be optimistic about the future. The reason for change in residence was not explored, but if it was because of failing health or increasing dependence, it would stand to reason that seniors would be less optimistic about the future. Residents were asked their perceptions of their health, but were not asked to indicate specific illnesses or chronic conditions.

It would appear from the results of this study that the majority of seniors in the study area perceived their health as good to excellent, were satisfied with their lives, and were optimistic about the future. Some of the literature, generally from the U.S., suggests that elderly people in rural areas have lower incomes and are in relatively poorer health (Eggebeen and Lichter, 1993; Johnson et al., 1993; Mainous and Kohrs, 1995; Birchfield, 1996; Henson et al., 1997). While comparisons cannot be made because of differences between the two health care systems, this did not seem to be true for our study population. A recent Canadian study (Johnson et al., 1995) also found a high percentage of the sample perceived their health as excellent or good (81%) and determined that there was not a significant difference between rural and urban settings in terms of perception of health. A qualitative study by Stewart (1997) looked at senior women in the PECOS study area. She found that social and family support systems were key determinants for perceived health and well-being for these women and for their sustainability in a rural environment.

In the study, 72% of the seniors were retired. Of these, only 27% had changed their place of residence and only 28% of those who had yet to retire planned to change their place of residence. This would suggest that most of the seniors were satisfied with their present place of residence and plan to remain in their home community. In order to provide appropriate health services to seniors in rural areas, it is important for health care workers and administrators to have an understanding of the perceptions of residents living in that area. Policy makers need to be aware of the relationship between education and income on perception of health. The factors that have been reported in this and other studies need also to be considered when making decisions regarding the provision of health services and improving the health of the population.

The role of nurses in rural health and the skills needed by nurses in our changing environment, as well as the health needs of seniors in rural environments, were not explored in this study. However, more research is required in these areas. As nurses, we need to assess each individual community for assets and barriers to health promotion, as each rural population varies in age of residents, income level, occupation, culture, and health status (Long, 1993). As the population decreases and young people continue to move away from rural communities, we must assess the needs of the remaining population, and be creative and innovative in determining the services essential to maintaining a healthy and satisfying lifestyle.

REFERENCES

Birchfield, P.C., Elder health, in *Community Health Nursing: Promoting Health of Aggregates, Families, and Individuals,* 4th ed., Stanhope, M. and Lancaster, J., Eds., Mosby, St. Louis, 1996, pp. 581–600.

Bushy, A., *Orientation to Nursing in the Rural Community,* Sage Publications, London, 2000.

Dickinson, H., Sustainability and health, The Prairie Ecosystem Study (PECOS), *Health Can. Soc.,* 5, 177–187, 1998/1999.

Dietz, M., Stressors and coping mechanisms of older rural women, in *Rural Nursing,* Bushy, A., Ed., Sage Publications, Thousand Oaks, CA, 1991, 267–280.

Eggebeen, D. and Lichter, D., Health and well-being among rural Americans: variations across the life course, *J. Rural Health,* 9, 86–98, 1993.

Epp, J., *Achieving Health for All: A Framework for Health Promotion,* Health and Welfare Canada, Ottawa, Ontario, 1986.

Federal, Provincial and Territorial Advisory Committee on Population Health, *Strategies for Population Health: Investing in the Health of Canadians,* Health Canada, Ottawa, Ontario, 1994.

Henson, D., Chafey, K., and Butterfield, P.G., Rural health, in *Community Health Nursing: Promoting the Health of Aggregates,* 2nd ed., Swanson, J.M. and Nies, M.A., W.B. Saunders, Philadelphia, 1997, pp. 407–431.

Johnson, J.E., Waldo, M., and Johnson, R.G., Research considerations: stress and perceived health status in the rural elderly, *J. Gerontol. Nurs.,* 19, 24–29, 1993.

Johnson, J.L., Ratner, P.A., and Bottoroff, J.L., Urban-rural differences in the health-promoting behaviours of Albertans, *Can. J. Public Health,* 86, 103–108, 1995.

Long, K.A., The concept of health: rural perspectives, *Nurs. Clin. North Am.,* 28, 123–130, 1993.

Mainous, A.G., III, and Kohrs, F.P., A comparison of health status between rural and urban adults, *J. Commun. Health,* 20, 423–431, 1995.

Mookherjee, H., Perceptions of well-being by metropolitan and nonmetropolitan populations in the United States, *J. Soc. Psychol.,* 132, 513–524, 1992.

Palmore, E., Health care needs of the rural elderly, *Int. J. Aging Hum. Dev.,* 18, 39–45, 1983–1984.

Saskatchewan Provincial Health Council, Your health, my health, our health: Our individual and collective responsibilities: a discussion paper on the determinants of health, Saskatchewan Provincial Health Council, Regina, Saskatchewan, 1996.

Stewart, H.C., The determinants of health and well-being as perceived by senior rural women in southwestern Saskatchewan, Master's thesis, University of Saskatchewan, available from College of Nursing, University of Saskatchewan, Saskatoon, Saskatchewan, 1997.

World Health Organization, Health and Welfare Canada, and Canadian Public Health Association, *Ottawa Charter for Health Promotion,* Canadian Public Health Association, Ottawa, Ontario, 1986.

CHAPTER 116

Prairie Ecosystem Study (PECOS): From Community to Chemical Elements, the Essential Role of Questionnaires

Helen H. McDuffie, K.M. Semchuk, Margaret M. Crossley, A. Senthilselvan,
Alan M. Rosenberg, Louise Hagel, Allan J. Cessna, D.G. Irvine, and D.L. Ledingham

INTRODUCTION

An evaluation of the sustainability of the semiarid prairie ecosystem, including the land, the biota, and the people, was the primary objective of the Prairie Ecosystem Study (PECOS). To further our understanding of potential human health risks associated with pesticide exposure, we selected a defined, rural geographical area and, in cooperation with the residents, we designed and undertook a study to:

1. Characterize human environmental and occupational exposure to pesticides by analyzing data from questionnaires and by analysis of blood samples
2. Assess pulmonary function using spirometry
3. Examine neuropsychological function using a battery of psychometric tests
4. Evaluate the frequency of autoimmunity as reflected by the presence of antinuclear antibodies
5. Assess the determinants of unintentional injury in this rural population
6. Determine the concentrations of 44 chemical elements in drinking water samples

In design, execution, statistical analyses, and interpretation of our results, all were completed with the consensus of a multi-, trans-, and interdisciplinary study group which included community representation. The objective of this chapter is to provide an overview of the complexity of the study design and methodology. Limited results are included for illustrative purposes.

THE PROCESS AND THE PROJECT

Community Involvement

Details of the organizational aspects and recruitment of community involvement in this project have been previously published (Semchuk et al., 1998a; Rosenberg et al., 1999). Briefly, the investigators engaged the cooperation of Saskatchewan rural residents in defining priorities for research into the sustainability of the semiarid prairie ecosystem by holding meetings in their communities. We encouraged a diversity of opinions by inviting farmers, their families, professionals in the agricultural and health disciplines, members of local government, and interested citizens to express their views. The ideas generated during these encounters were incorporated into our research protocol.

Community members agreed to support the research, provided that (1) they had continuing representative involvement in design, conduct, and interpretation of the study and (2) they would not be "guinea pigs." In all, 93 community members served as representatives on various study committees, publicists, organizers, canvassers, and research and phlebotomist technicians. The local health board provided space to conduct the study within its centrally located facilities.

Questionnaires

For the purposes of this study, pesticides are defined as chemicals specifically formulated to kill or control pests. Our definition was further restricted to herbicides, insecticides, and fungicides. To evaluate environmental and occupational exposure to pesticides, with permission, we used modified versions of a questionnaire previously utilized and validated by Hoar-Zahm et al. (1986) in a study of the association of rare tumors and pesticide exposure in Kansas.

We used two versions of the questionnaire: baseline (February and March 1996) and retest (June and July 1996). The modifications from Hoar-Zahm's telephone-interview questionnaire maintained the structure of the conditional questions and included conversion to a self-report, structured, written version (baseline) and to a personal interview (retest) format, with the subject volunteering the names of chemicals to which he or she was exposed between February and March, and June and July 1996. In addition, the Hoar-Zahm questionnaire had previously been modified, with permission, by McDuffie et al. (1994) to incorporate names of pesticides licensed for use in Canada, deleting those that were not licensed in Canada, and adding names of pesticides that were classified by the IARC (International Agency for Research on Cancer, 1980, 1986, 1987, 1991) as possible or probable human carcinogens.

The baseline questionnaire solicited information on occupational and environmental exposure to 52 herbicides, 38 insecticides, and 14 fungicides during an individual's lifetime. If an individual indicated exposure to a specific pesticide, supplemental questions designed to permit dose–response characterizations (quantity, time frame, days per year, which crops or pest, indoor vs. outdoor application) were asked. Accidental spills on skin or clothing, inhalation, and acute illness were assessed by general questions, with supplemental questions concerning which pesticide was involved and when the incident occurred. The use of personal protective equipment and hygiene practices were assessed without reference to specific pesticides. The methods most often used to apply herbicides, insecticides, and fungicides, without further categorization by chemical group, were also determined.

In addition to detailed characterization of pesticide exposure, the baseline questionnaire requested demographic and health information, smoking history, source of drinking water, and occupational history. The requested demographic information included date of birth, sex, handedness, farm or nonfarm residence, and level of education. Health variables collected were related to vision, hearing, sense of touch, and hand coordination, responses (yes or no) to a list of respiratory, cardiovascular, and neurological chronic conditions, and consumption of cigarettes, alcohol, and caffeine-containing beverages, as well as prescription and nonprescription medication, and dietary supplements. The baseline and retest questionnaires also incorporated two measures of neuropsychological functioning: the Brief Symptoms Inventory (BSI) (Derogatis et al., 1993) and the Cognitive Failures Questionnaire (CFQ) (Broadbent et al., 1982). The BSI is a 53-item self-report questionnaire that includes the Global Severity Index (GSI). Evaluation of the GSI combines the number of symptoms reported and the intensity of reported distress. Scores on the GSI are a sensitive indicator of psychological distress. The Cognitive Failures

Questionnaire measures self-reported frequency of commonly experienced errors in perception, memory, and motor functions. Scores of the BSI and CFQ are interpreted by comparison to well-established normative values. Table 116.1 outlines the similarities and differences between the baseline and retest questionnaires (McDuffie et al., 1997).

Table 116.1 PECOS: From Community to Chemical Elements Comparison of Baseline and Retest Questionnaires

Baseline Questionnaire	Retest Questionnaire
Winter Assessment February and March 1996	**Summer Assessment June and July 1996**
Mailed questionnaire	Personal interview
Structured and open-ended questions	Structured and open-ended questions
Demographics	Demographics
Birth month, day, year	Birth month, day, year
Sex	Sex
Residence (farm and nonfarm)	Farmer (yes and no)
Education	
Medical history	
Specific conditions	
Medications	
Psychological	Psychological
Brief Symptom Inventory	Brief Symptom Inventory
Global Severity Index	Global Severity Index
Cognitive Failures Questionnaire	Cognitive Failures Questionnaire
Occupation	
Current occupation	
Previous occupation	
Description of duties	
If farmer, main and any type of farming operation	

Pesticide Exposure

Work, home, garden	Work, home, garden
Pesticide use (lifetime)	Pesticide use between February and March 1996 and June and July 1996
List of commonly used pesticides	Subject volunteers name of pesticide
Major groupings (Herbicides, Insecticides, Fungicides):	
Method of application	
Any pesticide:	
Accidental inhalation	
Accidental spills on hands or clothing	
Protective clothing and respirator use	
Safety practices	
Individual products	Individual products
Acres sprayed	Acres sprayed
Bushels treated	Days applied/mixed
Days per year exposed	Hours per day applied
Indoor/outdoor exposure	Indoor/outdoor exposure
Personal mixing/applying	Accidental spills on hands or clothing
Year of first application	Accidental inhalation
	Ill within 24 hours of exposure
	Application rate
	Application method
Crop treated	Crop treated
	Protective gloves or clothing
Pest controlled	Safety practices
	Pest controlled

Study subjects who participated in pulmonary function screening at baseline and at retest completed an additional short questionnaire that focused on respiratory and cardiovascular symptoms and conditions.

Recruitment of Participants

Participants were recruited (Semchuk et al., 1998a) by 29 community residents who personally visited each town or farm household within three towns and three rural municipalities, and delivered a letter explaining the study and a postage-paid reply card. Residents of long-term care facilities, boarders, and students who were studying away from home were excluded. Residents of 259 households (62% farm, 38% town) returned the reply card and expressed interest in obtaining more information about the proposed study. Each of these households was contacted by telephone, a census of household members was completed, and additional information concerning the study design and objectives was provided. The respondent households included 490 adults (farm 321, town 169) and 94 children (farm 80, town 14) between 12 and 17 years of age. All of these individuals were invited to participate by completing questionnaires, by providing blood samples for antinuclear antibody testing and pesticide concentration, and by having objective assessments of their pulmonary and neuropsychological functioning. Those individuals who agreed to participate were mailed baseline questionnaires.

After completion of their retest assessments, study subjects were asked if they would consent to participate in a study of unintentional injuries. When the response was positive, most individuals preferred to participate immediately and the interviewer arranged appointments for the remainder. The injury questionnaire was interviewer administered. It was designed to capture details of farm work exposure as well as any unintentional injury which necessitated medical attention, or which resulted in 4 hours or more away from regular activities within the 2 years prior to the interview. It also was designed to utilize information collected in the baseline questionnaire to minimize duplication.

The sections of the injury questionnaire that elicit descriptions of the major factors of interest, the injury and farm work exposure questions, as well as medication use and lifetime serious injury, were taken verbatim from a questionnaire published in the doctoral dissertation of Dr. Jane Elkington (1993) with permission of her supervisor, Dr. Susan Gerberich. Additional questions concerning exposure to motor vehicles, recreational vehicles, and sports were developed, tested, and validated by Hagel (1998). Factors of interest in the injury study stratified by injury questionnaire and by the baseline questionnaire are outlined in Table 116.2.

Biological Testing

The detailed methodology related to the biological testing aspects of our study have been published or presented elsewhere (Rosenberg et al., 1999; Semchuk et al., 1998b; Senthilselvan et al., 1998, 2000; Crossley et al., 1998, 1999a,b). Only a short summary of each aspect is included in this chapter.

Pulmonary Function Measurements

Participants were offered pulmonary function testing at both baseline and retest sessions (Senthilselvan et al., 1998, 2000). The pulmonary function tests were conducted using Sensor Medics volume displacement spirometers and following American Thoracic Society guidelines (1987). The variables measured were forced expiratory volume in one second (FEV1), forced vital capacity (FVC) and the mid-maximal flow rate (MMFR). The ratio of FEV1/FVC × 100 was calculated. In addition, height, weight, blood pressure, and age of each subject were recorded. The study excluded 17 subjects because of pregnancy, high blood pressure, recent chest or abdominal surgery, or chest

Table 116.2 PECOS: From Community to Chemical Elements[a]

Injury Study Questionnaire[b]	PECOS Environmental Pesticide Exposure and Human Health Baseline Questionnaire[c]
Characteristics of Injury	Demographics
Body part injured	Age
Type of injury	Sex
Treatment of injury	Education
Characteristics of Injury Event	Location of residence
Activity at time of injury	Occupation
Cause of injury	Handedness
Agent of injury	Health Characteristics
Time of day	Preexisting medical conditions
Month and year	Medication use
Contributing factors (weather, etc.)	Sleep difficulties
Health Characteristics	Use of alcohol, tobacco, caffeine, and exercise
Previous lifetime serious injury	Brief Symptoms Inventory (BSI)
Medication use	Global Severity Index (GSI)
Sleep difficulties	Cognitive Failures Questionnaire (CFQ)
Exposure to Farm Work	Exposure to Pesticides
Hours per week	Type of pesticide
Exposure to Motor Vehicles	Type of use
Kilometers traveled per week	Years of exposure
Travel on gravel roads	Number of acres treated
Travel at slow speeds	Personal protective equipment used
Use of seat belt	Pesticide related spills and illness
Exposure to Recreational Vehicles	
Kilometers traveled per year	
Use of helmet	
Exposure to Sports	
Type and number of sports	
Hours per year for each sport	

[a] Factors of interest in the injury study stratified by instrument used for measurement.
[b] Administered by interviewer, June and July 1996.
[c] Administered by self-report, February and March 1996.

Source: Hagel, L.M., A Descriptive Study of Non-fatal, Unintentional Injury in a Rural Population, Master's dissertation, University of Saskatchewan, Saskatoon, Saskatchewan, 1998. With permission.

injuries. Participants received the results of their pulmonary function tests and an evaluation of their respiratory health immediately following completion of the tests.

Neuropsychological Functioning

Participants completed objective tests of higher brain function such as memory, attention, motor strength and speed, somatosensation, problem solving, and general intellectual ability at baseline and retest sessions (Crossley et al., 1997, 1998, 1999a,b). The function measured, the names of the tests, and references are outlined in Table 116.3. The tests were administered by six specially trained psychometricians according to the standardized procedures described in published manuals for the respective tests. Two psychologists provided professional supervision and support to the psychometricians. At the Centre for Agricultural Medicine, the test data were hand or computer scored and coded, and subsequently double-entered into a computer database.

Antinuclear Antibody (ANA) Assay

Participants were requested to donate blood samples for antinuclear antibody assays on two occasions, baseline and retest. Sera from consenting persons were assayed for ANA by indirect

Table 116.3 PECOS: From Community to Chemical Elements. The Neuropsychological Test Battery

Brain Function	Test	Ref.
Somatosensation	Two-Point Discrimination	Spreen et al., 1998
Motor speed and dexterity	Grooved Pegboard	Heaton et al., 1991
Motor strength	Grip Strength	Spreen et al., 1998
Attention and speeded visuo-motor scanning	Concentration Endurance (d2 Test)	Brickenkamp, 1981
		Heaton et al., 1991
	Trail Making Test	Smith, 1991
	Symbol Digit Modalities	
Learning and memory	California Verbal Learning Test	Delis et al., 1987, 1991
	Visual Reproductions	Wechsler, 1945
		Stone et al., 1946
Working memory	Digits Backwards	Wechsler, 1987
Verbal fluency	Animal Naming	Goodglass et al., 1983
Constructional skill and problem solving	Block Design	Wechsler, 1981
Intellectual ability	Wide Range Achievement Test-3 (Reading)	Wilkinson, 1993
Self-rating of cognitive functions	Cognitive Failures Questionnaire	Broadbent et al., 1982
Mood	Brief Symptoms Inventory	Derogatis, 1993

immunofluorescence on a Hep-2 cell substrate (Antibodies Incorporated, Davis, CA) (Rosenberg, 1988) using polyclonal antihuman immunoglobulin conjugated to fluorescein isothiocyanate. Unequivocal immunofluorescence at a dilution of 1:40 as observed independently by two readers was considered positive. Complete details of the methodology are provided in Rosenberg et al. (1999).

Environmental Testing: Chemical Elements in Drinking Water

One member of each household provided a sample of drinking water drawn from their kitchen faucet into containers provided by us. We collected one sample per household in midwinter. The water sources were community based (groundwater and surface water) and private wells. We conducted chemical elemental analyses using inductively coupled plasma mass spectrometry (ICP MS) (Jarvis et al., 1992). Quantification of the concentrations of 44 chemical elements in ppb was included in the assay. We grouped the elements (McDuffie et al., 1998) into (1) those nine tested for which there are Saskatchewan municipal drinking water objectives (boron, chromium, arsenic, selenium, cadmium, barium, mercury, lead, and uranium); (2) those with known or suspected human toxicity, teratogenicity, or carcinogenicity when exposure occurs in sufficient dose (chromium, nickel, copper, zinc, arsenic, selenium, strontium, molybdenum, cadmium, antimony, barium, lead, thorium, and uranium); and (3) those present in high concentration or with wide variation in concentration in samples of four herbicide formulations supplied by area farmers (Song, 1997). The herbicides were Fortress®, Achieve®, Sencor®, and Edge®. The concentrations of each of sodium, magnesium, phosphorus, sulfur, calcium, chromium, manganese, iron, copper, strontium, silicon, titanium, and arsenic differed by an order of magnitude or more among the herbicides. The concentrations of barium, aluminum, cadmium, and lead also varied substantially. Figure 116.1 shows an example of the graphic information provided to each participant describing the household water quality with respect to elements for which there are municipal drinking water objectives. In addition, a written report was provided that described the municipal drinking water objectives and provided information related to further testing if the sample had exceeded any objective.

Quality Control and Ethical Approval

Prior to commencement of the study, each potential participant was provided with detailed written information concerning the study design, the proposed use of the information obtained,

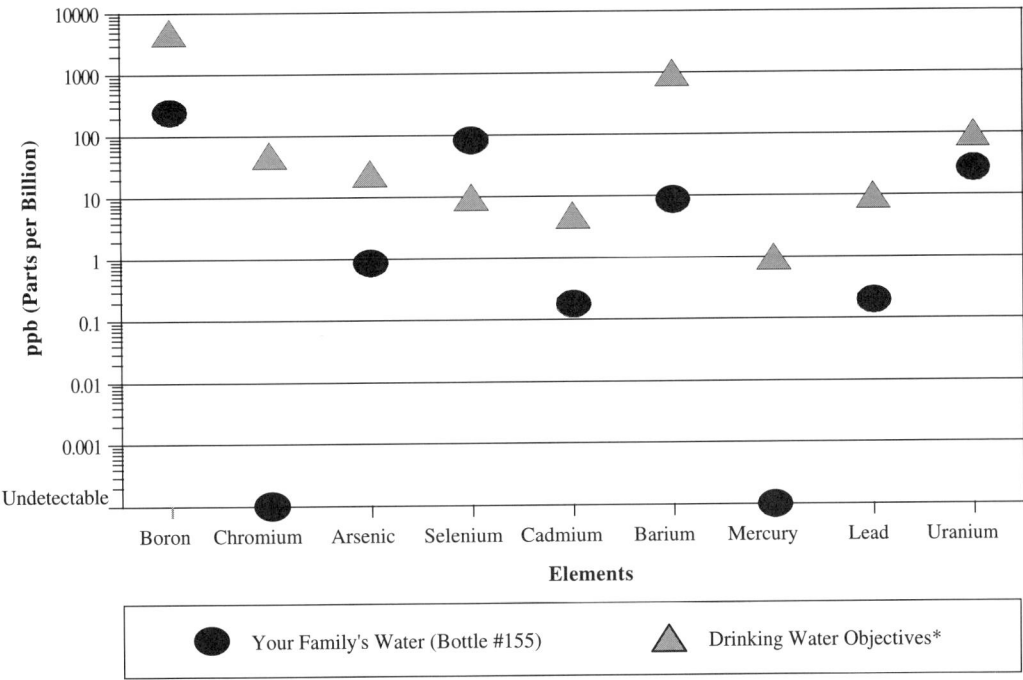

Figure 116.1 An example of the graphic information provided to each participant describing household water quality with respect to elements for which there are municipal drinking water objectives.

and the assurance of confidentiality of personal data acquired by the researchers. When individuals presented at the health centers in their community with a completed baseline questionnaire, they were registered by a postdoctoral fellow or an occupational health nurse, and given a code number that identified the individual's information. Stickers were provided to each individual with his or her code to be presented to each interviewer or technician as the individual moved from station to station.

As quality control measures, all data obtained by questionnaire or through biological testing were reviewed for internal consistency before being computer-coded using double-entry by two independent coders (Ledingham et al., 1998). Discrepancies were resolved by review by the appropriate researchers. For questionnaire data, we used a custom-designed data entry program, SPSS-DE® (1998). We present limited descriptive results relating to the evaluation of lifetime and recent (interval between February and March, and June and July 1996) pesticide exposure. Figure 116.2 demonstrates the connections among various aspects of our project.

The study received ethical approval from the University of Saskatchewan, Royal University Hospital, and the regional Health Board.

ILLUSTRATIVE RESULTS

Information obtained by questionnaire permitted us to classify exposure to pesticides on several different levels. The range included:

1. Minimal environmental exposure
2. Exposure to broad classes such as herbicides, insecticides, fungicides

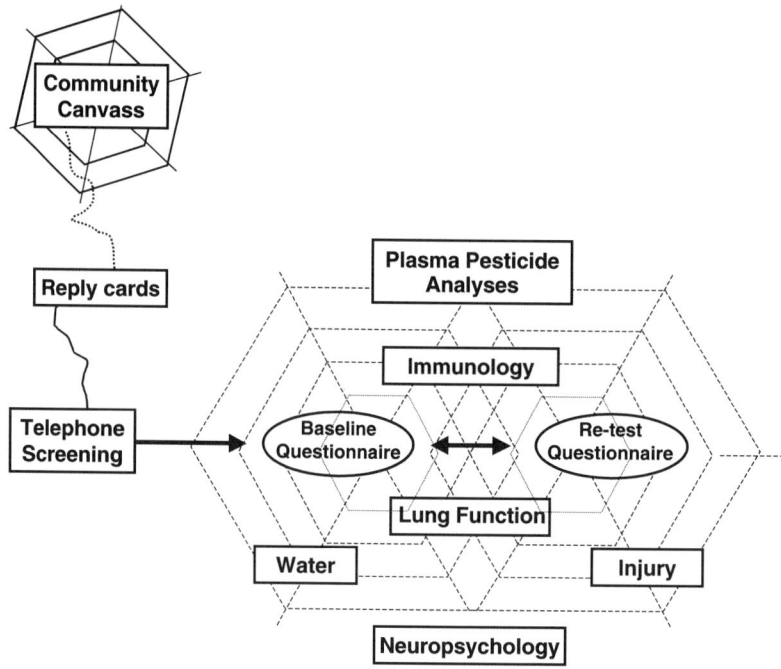

Figure 116.2 Connections between various aspects of the project.

3. Exposure to specific chemical classes such as phenoxyherbicides, carbamates, chlorinated hydro-carbons
4. Exposure to specific chemicals such as 2,4-D
5. Use of protective equipment or clothing to minimize exposure
6. Accidental dermal contact or inhalation which would maximize exposure

Table 116.4 describes the exposure to pesticides by applicators classified broadly during two time periods (lifetime and recent). An applicator was defined as an individual who personally applies or mixes pesticides as part of an occupation. This definition excluded individuals who mix or apply pesticides for use in their home, garden, or personal greenhouse. Table 116.5 describes data obtained from the baseline questionnaire concerning occupational use of pesticides organized into chemical classes and individual compounds. Table 116.6 describes the use of personal protective equipment and clothing while handling pesticides, and the frequency of accidental inhalation and dermal exposure to pesticides by applicators.

The study design facilitated exploring relationships among the biological tests (pulmonary function, autoimmunity, neuropsychological measurements, plasma concentrations of selected pesticides), environmental assessments (concentrations of chemical elements in drinking water),

Table 116.4 PECOS: From Community to Chemical Elements[a]

| | Ever in Lifetime (Baseline Questionnaire) | | Recent: February to July 1996 (Retest Questionnaire) | |
	n	%	*n*	%
Applicators	177	100.0	98	100.0
Herbicides	150	84.7	79	80.6
Insecticides	139	78.5	30	30.6
Fungicides	105	59.3	65	66.3
"Other pesticides": algicides	27	15.3	0	0.0

[a] Occupational pesticide use stratified by reports of lifetime or recent exposure.

Table 116.5 PECOS: From Community to Chemical Elements[a]

Chemical Groups and Individual Compounds	Number of Applicators $n = 177$	Percent of Applicators
Herbicides		
Phenoxyacetic acids	129	72.9
2,4-D	125	70.6
2,4,5-T	0	0.0
MCPA	69	39.0
Phenoxy-2-propionic acids	20	11.3
Phenoxybutanoic acids	5	2.8
Other phenoxyalkanoic acids	53	29.9
Carbamates	3	1.7
Thiocarbamates	108	61.0
Dinitroanilines	72	40.7
Phosphonic acids (glyphosate)	94	53.1
Amides	24	13.6
Phenols	107	60.5
Sulfonylureas	55	31.1
Sulfonylureas: Group A[b]	41	23.2
Sulfonylureas: Group C	30	16.9
Triazines	7	4.0
Chlorinated alkanoic acids	3	1.7
Bipyridiniums	58	32.8
Ureas	4	2.3
Pyridine carboxylic (picolinic) acids	8	4.5
Cyclohexenones	41	23.2
Dicamba	107	60.5
Metribuzin	43	24.3
Other herbicides	30	16.9
Insecticides		
Carbamates	88	49.7
Organochlorines	64	36.2
Organochlorines: Group A[c]	54	30.5
Organochlorines: Group B	20	11.3
Organophosphorus compounds	87	49.2
Pyrethroids	77	43.5
Biologicals	1	0.6
Natural products	0	0.0
Metal compounds	2	1.1
Other insecticides	14	7.9
Fungicides		
Carbamates	1	0.6
Amides	90	50.8
Aldehydes	10	5.6
Dithiocarbamates	90	50.8
Manganese-containing compounds	1	0.6
Mercury compounds	13	7.3
Other fungicides	13	7.3

[a] Occupational use of pesticides organized into chemical groups and individual compounds: Ever in lifetime (baseline questionnaire).
[b] Sulfonylureas, Group A: chlorsulfuron; Group C: tribenuron-methyl mixtures.
[c] Organochlorines, Group A: aldrin, chlordane, dieldrin, endrin, heptachlor, lindane; Group B: DDT, methoxychlor.

Table 116.6 PECOS: From Community to Chemical Elements

Personal Protective Equipment Used When Handling Pesticides	Number of Applicators n = 177	Percent of Applicators
Use of Personal Protective Clothing by Pesticide Applicators Ever in Lifetime (Baseline Questionnaire)		
Gloves	89	50.3
Cloth gloves	8	4.5
Rubber gloves	124	70.1
Rubber apron	11	6.2
Goggles	26	14.7
Rubber boots	55	31.1
Waterproof hat	2	1.1
Coveralls without sleeves	7	4.0
Coveralls with sleeves	86	48.6
Respirator/mask	43	24.3
Accidental Dermal or Inhalation Contact with Pesticides by Applicators Ever in Lifetime (Baseline Questionnaire)		
Inhalation Directly into Lungs		
Yes	53	29.9
No	116	65.5
Missing	8	4.5
Spilled Pesticide on Skin/Clothing		
Yes	89	50.3
No	84	47.5
Missing	4	2.3

and exposure to pesticides as determined by questionnaire. For example, Rosenberg et al. (1999) examined the possible association of environmental and occupational exposure to pesticides using the hierarchy described above and the presence of antinuclear antibodies. In univariate statistical analyses, antinuclear antibody detection at sample dilutions of 1:40 or greater (positive) was significantly associated with lifetime exposure to insecticides, specifically carbamates, organochlorine group A (chlordane, dieldrin, endrin, heptachlor, and lindane, excluding DDT and methoxychlor), pyrethroids, and with phenoxyacetic acid herbicides. In multivariate statistical analyses, including adjustment for age, sex, and other insecticide exposures, antinuclear antibody positivity was associated with a history of oilseed production and with exposure to pyrethroid insecticides.

DISCUSSION

The overall study design and methodology benefited from the expertise of dedicated and committed community members who have been and continue to be active on PECOS management and human health committees. Their input led to sharpening of the research questions and to an enhanced sensitivity by the researchers to the community's real-life concerns. Additional questions were formulated to accommodate community members, and the format of other questions was altered to respect their knowledge of local conditions. We highly recommend inviting community representatives to be partners in conducting complicated research and in investigating issues as complex as sustainability of an ecosystem.

Our team of researchers decided to operate by consensus. This was an important early decision, as many of us had not previously been involved in multidisciplinary research. At first, the process of achieving consensus was prolonged and difficult. Gradually, an interdisciplinary approach evolved that respected the integrity and standards of each discipline. We began to understand the traditions and language of other disciplines. In the beginning, each of us used words with definitions explicit to our own discipline without realizing that others used the same words to describe entirely different phenomena. Eventually, we developed mutual trust.

Two years of planning were required before commencing field studies due to the complexity of our study design and methodology, and the researcher transportation and housing logistics associated with collecting data in very small communities 2 hours' drive from the university. Three intensive and coordinated data-collection periods within 8 months resulted in equally intensive and coordinated data scoring, coding, and computer entry using custom-designed data-entry programs, which facilitated the integration of the various components of the study. Subcommittees with specialized knowledge in consultation with the entire research team guided the statistical analyses of these data.

There are many studies that focus on the occupational exposures to pesticides of adult males and health indicators (Blair et al., 1985; Hoar et al., 1986; Hoar-Zahm et al., 1990). Fewer have included female farmers (McDuffie, 1994; Zahm, 1994; Fincham et al., 1994). Occupational studies usually define exposure by industry (agriculture), by occupation (farming), by broad categories of pesticides (herbicides), by use of specific chemical classes of compounds (phenoxy-herbicides), or by reported use of specific compounds (2,4-D). We have incorporated all these categories in our research, while also including home and garden exposures to pesticides and variables to measure extent of exposure, including accidental spills or inhalation incidents For specific statistical analyses, we further subdivided participants by their recent and lifetime exposure to pesticides (1) at work, (2) as a family member of a pesticide applicator, and (3) other study participants. The majority of the occupational pesticide applicators were men, each of whom handled a variety of different chemical pesticides. The most commonly used chemical classes of herbicides (lifetime) were phenoxyacetic acids, thiocarbamates, phosphonic acid (glyphosate) and dicamba; of insecticides, carbamates, organochlorines, and organophosphorus compounds; and of fungicides, amides and dithiocarbamates. Although the primary focus of our research was acute and not chronic health effects of exposure to pesticides, several of the most commonly used pesticides have been declared possible or probable human carcinogens by the IARC (1980, 1986, 1987, 1991).

The use of personal protective clothing or equipment by occupational pesticide applicators while mixing or applying pesticides was inconsistent. Few reported using safety goggles or self-contained respirators. Accidental spills of pesticides on skin or clothing and accidental inhalation into the lung were relatively frequent occurrences. The lack of use of protective equipment and the accidental spills and inhalation incidents potentially increase the absorbed dose of pesticides.

Explicit and detailed characterization of exposure to pesticides by questionnaire was the conceptual link between baseline and retest measurements of neuropsychological functioning, of assessment of pulmonary health, of measurement of autoimmunity, and the detection and quantification of herbicide levels in human peripheral blood samples.

ACKNOWLEDGMENTS

This study was funded by the Tri-Council Secretariat of Canada through the Ecoresearch Programme and through a National Health Canada Research Scholar Award to Dr. K.M. Semchuk, the lead investigator. The substantial contributions of our fellow researchers, Dr. D. Anderson, Dr. J.A. Dosman, Dr. K. Arbuthnott, Ms. M. Masley, Dr. R. Kerrich, Dr. V. Laxdal, Mr. Juorio, and our community representative, Mr. P. Hanke, are gratefully acknowledged. We thank Mrs. S. de Freitas for formatting the questionnaires and for secretarial support. We reserve our greatest appreciation and thankfulness to members of the community who helped us design, conduct, and interpret this study, as well as provided us with information and biological samples.

REFERENCES

American Thoracic Society, Standardization of spirometry, an update, *Am. Rev. Respir. Dis.,* 136, 1285–1296, 1987.

Blair, A., Malker, H., Cantor, K.P., Burmeister, L., and Wiklund, K., Cancer among farmers: a review, *Scand. J. Work Environ. Health,* 11, 397–407, 1985.

Brickenkamp, R., Test d2: Aufmerksamkeits-Belastungs-Test [Test d2; Concentration-Endurance-Test: Manual], 7th ed., C.J. Hogrefe, Toronto, 1981.

Broadbent, D.E., Cooper, P.F., FitzGerald, P., and Parkes, K.R., The cognitive failures questionnaire (CFQ) and its correlates, *Br. J. Clin. Psychol.,* 21, 1–16, 1982.

Crossley, M., Arbuthnott, K., Semchuk, K.M., McDuffie, H.H., Senthilselvan, S., Cessna, A.J., Dosman, J.A., Hanke, P., Irvine, D.G., Rosenberg, A.M., Laxdal, V., Ledingham, D.L., and Snodgrass, P., The development of a neuropsychological test battery for a Prairie Ecosystem Study of pesticide exposure and health, International Neuropsychological Society Annual Conference, Orlando, *J. Int. Neuropsychol. Soc.,* 3 (abstr.), 56, 1997.

Crossley, M., Arbuthnott, K., Semchuk, K.M., McDuffie, H.H., Senthilselvan, A., Cessna, A.J., Dosman, J.A., Irvine, D.G., Rosenberg, A.M., Ledingham, D.L., and Snodgrass, P., Rural norms for a neuropsychological battery designed to assess pesticide exposure effects, *Can. Psychol.,* 39 (abstr.), 80, 1998.

Crossley, M., McDuffie, H.H., Semchuk, K.M., Arbuthnott, K., Cessna, A.J., Ledingham, D.L., Irvine, D.G., Hanke, P., Senthilselvan, A., Rosenberg, A.M., Dosman, J.A., Hagel, L.M., and Masley, M.L., Neuropsychological effects of occupational and environmental exposure to herbicides in a Prairie Ecosystem Study (PECOS), International Neuropsychological Society Annual Conference, Boston, *J. Int. Neuropsychol. Soc.,* 5 (abstr.), 129, 1999a.

Crossley, M., Semchuk, K.M., McDuffie, H.H., Ledingham, D.L., Hagel, L.M., Cessna, A.J., Irvine, D.G., Senthilselvan, A., and Dosman, J.A., Neuropsychological functioning and health-related symptoms in a commercial pesticide applicator during high- and low-level exposure seasons, *J. Agric. Safety Health,* 5, 279–287, 1999b.

Delis, D.C. et al., *California Verbal Learning Test: Adult Version Manual,* The Psychological Corporation, San Antonio, 1987.

Delis, D.C. et al., Alternate form of the California Verbal Learning Test: development and reliability, *Clin. Neuropsychol.,* 5, 154–162, 1991.

Derogatis, L.R., *Brief Symptom Inventory (BSI): Administration, Scoring, and Procedures Manual,* 3rd ed., NCS Inc., Minneapolis, 1993.

Elkington, J.M., A Case-Control Study of Farmwork Related Injuries in Olmstead County, Minnesota, Ph.D. dissertation, University of Minnesota, Minneapolis, 1993.

Fincham, S., Macmillan, A., and Berkel, J., Cancer patterns among female farmers in Alberta, in *Agricultural Health and Safety: Workplace, Environment, Sustainability,* Lewis Publishers, Boca Raton, FL, 1994, pp. 169–174.

Goodglass, H. and Kaplan, E., *The Assessment of Aphasia and Related Disorders,* 2nd ed., Lea & Febiger, Philadelphia, 1983.

Hagel, L.M., A Descriptive Study of Non-fatal, Unintentional Injury in a Rural Population, Master's thesis, University of Saskatchewan, Saskatoon, 1998.

Heaton, R.K., Grant, I., and Matthews, C.G., *Comprehensive Norms for an Expanded Halstead-Reitan Battery: Demographic Correction, Research Findings, and Clinical Applications,* Psychological Assessment Resources, Odessa, FL, 1991.

Hoar, S.K., Blair, A., Holmes, F., Boysenc, C., Rohel, R.J., Hoover, R., and Fraumeni, J.F., Agricultural herbicide use and risk of lymphoma and soft tissue sarcoma, *JAMA,* 256, 1141–1147, 1986.

Hoar-Zahm, S., Weisenburger, D.D., Babbitt, P.A., Saal, R.C., Vaught, J.B., Cantor, K.P., and Blair, A., A case-control study of non-Hodgkin's lymphoma and the herbicide 2,4-dichlorophenoxyacetic acid (2,4-D) in eastern Nebraska, *Epidemiology,* 1, 349–356, 1990.

International Agency for Research on Cancer, Working Group, an evaluation of chemicals and industrial processes associated with cancer in humans based on human and animal data, *Cancer Res.,* 40, 1–12, 1980.

International Agency for Research on Cancer, Monographs on the Evaluation of the Carcinogenic Risk of Chemicals to Humans, Vol. 41, *Some Halogenated Hydrocarbons and Pesticide Exposures,* Lyon, France, 1986.

International Agency for Research on Cancer, Overall Evaluation of Carcinogenicity: An Updating of the International Agency for Research on Cancer Monographs, Vol. 1–42, Suppl. 7, Lyon, France, 1987.

International Agency for Research on Cancer, Monographs on the Evaluation of Carcinogenic Risks to Humans, Vol. 53, *Occupational Exposures in Insecticide Application and Some Pesticides,* Lyon, France, 1991.

Jarvis, I., Gray, A.L., and Hauk, R.S., *Handbook of ICP–MS,* Chapman & Hall, New York, 1992.

Ledingham, D.L., Semchuk, K.M., McDuffie, H.H., Cessna, A.J., Hanke, P., Senthilselvan, A., Crossley, M., Irvine, D.G., Rosenberg, A.M., Hagel, L.M., Masley, M.L., Dosman, J.A., and Laxdal, V.A., PECOS: Integration and Analysis of Data from an Interdisciplinary Health Study, presented at International Society for Environmental Epidemiology/International Society for Exposure Assessment Joint Meeting, Boston, *Epidemiology,* 9, suppl., 1998.

McDuffie, H.H., Women at work: agriculture and pesticides, *J. Occup. Med.,* 36, 1240–1246, 1994.

McDuffie, H.H., Dosman, J.A., McLaughlin, J.R., Theriault, G., Pahwa, P., Choi, N.W., Fincham, S., Robson, D., Spinelli, J., Skinnider, L.F., and White, D., Non-Hodgkin's lymphoma (NHL) and pesticide exposure: Canada, American Association of Cancer Research, San Francisco, *Proc. Am. Assoc. Cancer Res.,* 35 (abstr.), 1721, 1994.

McDuffie, H.H., Semchuk, K.M., Cessna, A.J., Irvine, D.G., Senthilselvan, A., Rosenberg, A.M., Crossley, M., Dosman, J.A., Hanke, P., Laxdal, V.A., Ledingham, D.L., Holfeld, L., Hagel, L., and Masley, M.L., PECOS: Environmental Pesticide Exposure and Human Health: Characterization of Exposure by Questionnaires, presented at Canadian Society of Epidemiology and Biostatistics, London, Ontario, 1997.

McDuffie, H.H., Semchuk, K.M., Kerrich, R., Cessna, A.J., Irvine, D.G., Senthilselvan, A., Ledingham, D.L., Juorio, V., Hanke, P., Hagel, L.M., Masley, M.L., Dosman, J.A., and Crossley, M., The Prairie Ecosystem Study (PECOS): Drinking Water Quality, presented at International Society for Environmental Epidemiology/International Society for Exposure Assessment Joint Meeting, Boston, *Epidemiology,* 9, suppl., 1998.

Rosenberg, A.M., The clinical associations of antinuclear antibodies in juvenile rheumatoid arthritis, *Clin. Immunol. Immunopathol.,* 49, 19–27, 1988.

Rosenberg, A.M., Semchuk, K.M., McDuffie, H.H., Ledingham, D.L., Cordeiro, D.M., Cessna, A.J., Irvine, D.G., Senthilselvan, A., and Dosman, J.A., Prevalence of anti-nuclear antibodies in a rural population, *J. Toxicol. Environ. Health,* Part A, 57(4), 225–236, 1999.

Saskatchewan Municipal Drinking Water Objectives, Regina, Saskatchewan, 1994.

Semchuk, K.M., McDuffie, H.H., Hanke, P., Dosman, J.A., Senthilselvan, A., Cessna, A.J., Crossley, M.F.O., Irvine, D.G., Rosenberg, A.M., Hagel, L.M., Masley, M.L., Ledingham, D.L., and Laxdal, V.A., Partnering with the community — The Prairie Ecosystem Study: environmental pesticide exposure and human health, presented at International Society for Environmental Epidemiology/International Society for Exposure Assessment Joint Meeting, Boston, *Epidemiology,* 9, suppl., 420, 1998a.

Semchuk, K.M., Rosenberg, A.M., McDuffie, H.H., Ledingham, D.L., Senthilselvan, A., Cessna, A.J., Irvine, D.G., Hanke, P., Dosman, J.A., Crossley, M., Hagel, L.M., Masley, M.L., and Laxdal, V.A., Antinuclear antibodies: association with farming and occupational use of pesticides, presented at International Society for Environmental Epidemiology/International Society for Exposure Assessment Joint Meeting, Boston, *Epidemiology,* 9, suppl., 60, 1998b

Senthilselvan, A., Dosman, J.A., Semchuk, K.M., McDuffie, H.H., Cessna, A.J., Irvine, D.G., Rosenberg, A.M., Crossley, M., Laxdal, V.A., Holfeld, L., Masley, M.L., Hagel, L., Ledingham, D.L., and Hanke, P., Changes in Pulmonary Function Measurements between Winter and Summer Seasons in a Rural Population in Saskatchewan: Prairie Eco-System Study, presented at American Thoracic Society International Conference, Chicago, 1998.

Senthilselvan, A., Dosman, J.A., Semchuk, K.M., McDuffie, H.H., Cessna, A.J., Irvine, D.G., Crossley, M.F.O., and Rosenberg, A., Seasonal changes in lung function in a farming population, *Can. Resp. J.,* 7, 320–325, 2000.

Smith, A., *Symbol Digit Modalities Test,* Western Psychological Services, Los Angeles, 1991.

Spreen, O. and Strauss, E., *A Compendium of Neuropsychological Tests: Administration Norms, and Commentary,* 2nd ed., Oxford University Press, New York, 1998.

SPSS, Data Entry II® Statistical Packages for the Social Sciences: Statistical Data Analysis, SPSS Inc., Chicago, 1998.

Song, L., Micronutrients, Toxic and Other Trace Elements in the Soil-Plant-Water Systems in the PECOS Study Area: Implications for Natural and Anthrogenic Inputs or Outputs to Agricultural Land, Master's thesis, University of Saskatchewan, Saskatoon, 1997.

Stone, C. and Wechsler, D., *Wechsler Memory Scale Form II,* Psychological Corporation, San Antonio, TX, 1946.

Wechsler, D., A standardized memory scale for clinical use, *J. Psychol.,* 19, 87–95, 1945.

Wechsler, D., *Wechsler Adult Intelligence Scale — Revised,* Psychological Corporation, New York, 1981.

Wechsler, D., *Wechsler Memory Scale — Revised,* Psychological Corporation, San Antonio, TX, 1987.

Wilkinson, G.S., *WRAT 3 Administration Manual,* Wide Range Inc., Wilmington, DE, 1993.

Zahm, S.H., Weisenburger, D.D., Saal, R.C., Vaught, J.B., Babbit, P.A., and Blair, A., Pesticide use, genetic susceptibility and non-Hodgkin's lymphoma in women, in *Agricultural Health and Safety: Workplace, Environment, Sustainability,* Lewis Publishers, Boca Raton, FL, 1994, pp. 127–133.

CHAPTER 117

Ecosystem-Level Functional Changes in Breeding Bird Guilds in the Mixed Grassland since Agricultural Settlement

Todd A. Radenbaugh

INTRODUCTION

There is considerable concern that human activities are having detrimental impacts on Earth's biological diversity and the function of biophysical systems. These human processes may influence, interact with, alter, and even control the operations of biophysical functions at many levels of biological organization. Studies have shown that within many ecosystems, human activities have roles directly controlling entire biophysical systems (such as land-use practices that alter native habitat and expatriate species) (Kaiser and Gallagher, 1997; Tilman et al., 1997; Vitousek et al., 1997). Examples of direct human alteration of ecosystems include desertification of marginal lands (Schlesinger et al., 1990; Mainguet, 1994), eutrophication of lakes (Schindler et al., 1990) and coastal areas (Lapointe and O'Connell, 1989), loss of coral reefs in the Caribbean (Hughes, 1994), loss of soil organic matter due to agriculture in the Great Plains (Seastedt, 1995) and collapse of marine fisheries (Botsford et al., 1997). Many times, however, impacts are indirect, such as changes in regional species composition and the addition of pollutants to the soil, surface and groundwater, and atmosphere. Indirect effects are a secondary result of direct species interactions and may alter species composition and community structure and as such may have strong effects through natural selection (Miller and Travis, 1996; Rosemond, 1996). Recent studies have shown that both direct and indirect effects have large impacts on the functioning of ecosystems, thus affecting human health (Botsford et al., 1997; McMichael, 1997).

At lower ecological levels, human-caused direct and indirect influences have altered species abundance and distribution of local species assemblages and have allowed invasion of species into a region (Naeem et al., 1994; Nee and May, 1997). Further, ecological function of local species assemblages has been shown to change with declining diversity (Tilman, 1996). More specifically, function depends on the role of the species in the particular system of which it was a member (Valentine and Jabonski, 1983; Jackson, 1994; Symstad et al., 1998). Not all species in an assemblage, however, are equal in terms of how they function in an ecological system; some species have disproportionate roles, e.g., keystone species, dominant herbivores, and system engineers (Lawton, 1987). In species assemblages, studies suggest that it is not simply declining species diversity but the functional diversity within the species pool that best explains variations in system function (e.g., plant biomass and productivity) (Grime, 1997; Hooper and Vitousek, 1997; Tilman et al., 1997). Further, human-controlled processes act on biotic systems at many levels but it is the broadscale influences that affect ecosystem integrity the most. Thus, measurements of regional functional diversity may be better indicators of ecosystem health than the traditionally used measures of local species richness or diversity.

This chapter explores the broadscale influences of human interactions within a complex hierarchical system in North America's northern Great Plains (specifically the Mixed Grassland Ecoregion of Saskatchewan) by examining breeding bird guilds. The baseline for this study is the introduction of European agricultural methods early in the 1900s. Although there has been an aboriginal presence recorded in the region for the past 10,000 years, this earlier system is assumed to have been in a relatively steady state (compared to changes in the past 100 years) and serves as the template to measure modern change. From the early 1900s to the mid-1990s, approximately 80% of the northern Great Plains has been converted from native habitats to other land cover types (Samson and Knopf, 1996; Selby and Santry, 1996). Moreover, since the late 1880s the two dominant structuring forces, fire (Rowe, 1969; Wright and Bailey, 1980; Collins and Wallace, 1990; Collins, 1992) and large grazing ungulates (Frank et al., 1998; Knapp et al., 1999), have been eliminated or heavily managed. The magnitude of these and other impacts raises concerns about changing functional roles in this ecosystem.

Spatial Scales

Because biotic processes operate at many spatial and temporal scales, a hierarchical approach is taken here where multiple ecological scales are identified. This study then examines only the broader ecosystem level processes or the "big picture," thus eliminating some of the confusing "ecological noise" caused by local fluctuations and stochastic processes that operate at species assemblage and population levels. This approach allows for the investigation of the broad-scale changes that agricultural settlement has imposed on both vegetation and animals. The focus of this study is a physiologically and ecologically similar region, the Mixed Grassland Ecoregion of Saskatchewan (defined in ESWG, 1995; Acton et al., 1998) and was investigated as a whole unit in terms of system attributes. Although data were collected at lower levels (e.g., species assemblage and bird populations), no attempt was made to interpret data at these levels. Because processes at different levels occur at different rates (Allen and Starr, 1982; Salthe, 1985), processes at lower levels appear to occur more rapidly than those occurring at the focal level (and vice versa for higher levels). Thus, in order to investigate focal-level processes and functions, the inputs of lower levels must be scaled up. This scaling up is achieved through the averaging of inputs so that cycles and fluctuations are taken into account. This is done by grouping together species into vegetation classes and guilds that occur within fixed time periods. In this way, the detailed and dynamic nature of lower-level processes may be bypassed.

Breeding Bird Guilds

Breeding birds were used as indicators of broad regional changes for many reasons. First, this group has historical abundance data available. Second, as a group they use the majority of the habitats in the ecosystem for nesting and foraging. Most birds in this region are highly mobile, often moving when conditions are severe or resources become low. Moreover, approximately 80% of the bird species in the northern Great Plains are migratory, and winter in the southern U.S., Central America, or South America. Thus, bird populations are spatially dynamic on an annual or even a monthly basis. For all these reasons, averages over multiple years were taken when estimating bird abundance. Last, breeding birds were used to study regional functional roles because they generally spend the 3-month growing season in limited areas defending territories, choosing mates, and raising fledglings. During the summer, breeding bird populations have a significant impact in terms of resource, space, and food utilization.

Because birds are highly mobile and use many of a region's resources, as a group they can be used as a broad indicator of ecological changes. Further, due to cascading effects within the regional interactions web, broad changes in one group will usually show up in others that are linked spatially. To analyze ecological functions and measure change on a broad scale, guild theory (Root, 1967) was used to group species into functional categories. Guild is defined as Root (1967) used it: a group of species that exploits the same class of environmental resources in a similar way without regard to taxonomic position. Thus, a guild is composed of those species that use similar ecological resources or employ similar strategies for access to resources such as nesting sites and feeding strategies. Thus, changes in guilds in terms of their relative abundance may be used to indicate broad changes in functional roles in an ecosystem.

METHODS

Study Site

The study site encompasses the entire Mixed Grassland Ecoregion of Saskatchewan (Figure 117.1). This ecoregion was chosen as the geographic area of study because historically

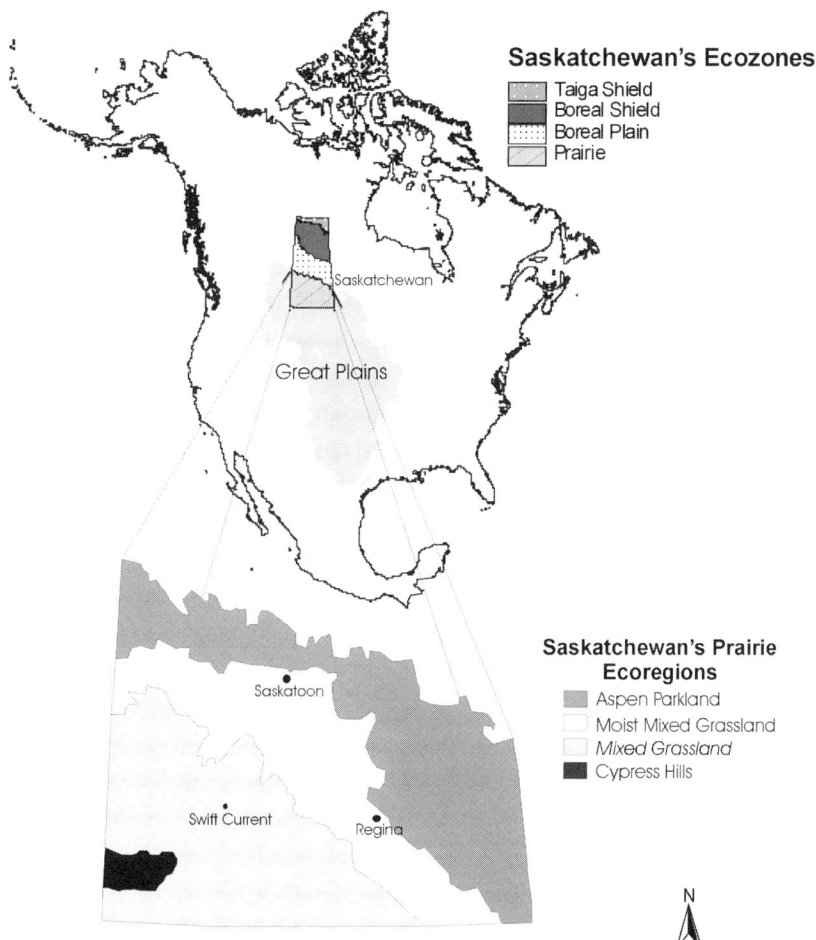

Figure 117.1 Study site: the position of Saskatchewan's Mixed Grassland Ecoregion within the Great Plains of North America and the Prairie Ecozone.

(past 5000 years) it has had limited species assemblage diversity (Axelrod, 1985). Prior to agricultural settlement, the primary vegetation types were limited to grasslands, wetlands, and brushlands, with no extensive forests or other major vegetation types (Watts, 1960; Archibold and Wilson, 1980). The dynamics controlling the ecosystem structure are also suggested to be relatively simple (i.e., climate, fire, and grazing) (Coupland, 1961; Joern and Keeler, 1995; Knapp et al., 1999). In the past 100 years, vegetation patterns have increasingly become influenced by human interactions. The majority of this influence in this study region is due to agriculture rather than multiple land uses such as forestry, urban development, or large-scale strip mining, although human habitation (villages, towns, and one city) and road construction also have significant impacts. The limited types of human activities allow a focus on the effects of agroecosystem development. Since social institutions and land-use practices can vary in different political systems, the political borders of Saskatchewan define the extent of the study area in the Mixed Grassland.

Vegetative Habitats

To examine how society has modified this region, four major vegetative habitats were studied between ca. 1900 and the 1990s. Vegetation habitats were used due to the sessile nature of plants, which makes them relatively easy to identify and assess in geographic terms. Moreover, many bird populations are dependent on these specific assemblage types for mating and foraging; thus, changes in major vegetation habitats also could influence bird populations.

Unfortunately, few accurate historical documents give an account of vegetation distribution. The best data sources are the Dominion Lands Survey and Statistics Canada's Agricultural Censuses. Statistics Canada, however, provides farm-related vegetation categories, which are defined using Census Canada definitions. When Saskatchewan became a province in 1905, the prairie region had been divided into townships and ranges, local improvement districts (LID), and later rural municipalities (RM). These are the smallest units reported in Agricultural Censuses and are the units used to report vegetation categories.

Four vegetation habitats were recorded for each LID or RM in Saskatchewan's Mixed Grassland Ecoregion including:

1. Native and reestablished grassland or native pasture
2. Woodland/brushland
3. Wetland, "waste" areas, and open water
4. Improved land (includes cultivated, summerfallow, and improved pasture)

Although these categories are not individual plant assemblages, they were used because they are the reported categories in both the Dominion Lands Survey and most of the Canada Agricultural Censuses.

The following general definitions are used to classify vegetation types:

- *Grassland or native pasture:* Areas of vegetation dominated by native herbaceous grasses and forbs, covered by less than 10% trees or shrub.
- *Woodland/brushland:* Areas of vegetation dominated by short trees and shrubs that comprise greater than 70% of the total vegetation, and in a relatively closed stand.
- *Wetland/waste:* Areas of low-lying land that are submerged or inundated periodically by fresh or saline water, or land areas recognized as poor farming areas such as salt flats and open water.
- *Improved land:* Areas where existing vegetation has been removed for the cultivation of target species mainly used for agricultural production, including land in summer fallow; also land that has been drained and seeded into pasture to produce improved pasture.
- *Unimproved land:* The sum of native pasture, woodland/brushland, and wetland/waste.

To measure vegetation changes within the mixed grassland, the major vegetation categories were estimated as a time series from pre-European settlement to the 1990s; ca. 1880 was chosen as a baseline because it is just prior to the massive influx of European settlers. Archibold and Wilson (1980) reconstructed the native vegetation ca. 1880 by using the Dominion Lands Survey's township and range maps. These maps document major landscapes and vegetation of southern Saskatchewan using vegetation types similar to those used by Statistics Canada. To compare these data to Statistics Canada data, the township and range data of Archibold and Wilson (1980) were scaled up into LIDs and RMs. This is possible since each LID and RM is defined in terms of townships and ranges. To complete the time series, the vegetation types in later periods were based on reported condition of occupied land using eight Agricultural Censuses (1881, 1886, 1921, 1936, 1941, 1971, 1991, and 1996).

Breeding Bird Guilds

The goal of the guild analysis was to construct a reasonable ecological grouping of species that "exploit the same class of environmental resources in the similar way" (Root, 1967, p. 335). A list of breeding birds for Saskatchewan's Mixed Grassland Ecoregion was compiled using Smith's *Atlas of Saskatchewan Birds* (Smith, 1996). This list is provided in Appendix 117.I and references to bird species are listed as common names but use the standard by Godfrey (1986) for taxonomic classifications. Only bird species that breed in significant numbers annually (i.e., not "hypothetical," "incidental," or "sporadic") were included. The breeding bird lists reported all confirmed and probable breeders, but excluded any species listed as "possible." Additional species were added to the list when published accounts from the 1990s back through the 1860s showed that a species historically bred in the study region in significant numbers.

Guild Characterization

Four broad breeding bird guild types were identified because they indicate resources or characteristics important to the survival of breeding bird populations: foraging method, foraging habitat, nesting habitat, and diet. Within each of these guild types, four to eight guilds were then identified based on recurring habitat use, food items, or bird foraging behaviors (Table 117.1) and are described later. These guild types provide an overall measure of resource use in the Mixed Grassland Ecoregion. Any significant change in these guild types may also indicate a functional change in the ecology of the entire region.

Foraging Method Guilds

An exploratory data analysis was conducted to identify the type and number of foraging method guilds. First, as species accounts were searched for the development of the bird database, all methods that authors used to describe foraging methods and behaviors were recorded. The terms and definitions of how species acquired food grew to more than 20 potential guilds. This list was reduced to eight guilds after careful investigation of bird behavior and expansion of categories to include similar groups. The resulting eight guilds were dabbler, diver, gleaner, picker, prober, sallier, stalker, and swooper, as defined in Table 117.3.

Table 117.1 Guild Types and Guilds Used for Studying Breeding Birds in the Mixed Grassland Ecoregion of Saskatchewan

Guild Type	Guild
Foraging method	Dabbler, diver, gleaner, picker, prober, sallier, stalker, swooper
Foraging habitat	Woodland/brushland, grassland, wetland, other
Nesting habitat	Cavity, cliff, ground, floating mat, vegetation
Diet	Aquatic invertebrates, aquatic vertebrates, aquatic vegetation, omnivore, seeds, terrestrial invertebrates, terrestrial vertebrates, terrestrial vegetation

Table 117.2 Breeding Bird Guilds and Their Assignment Based on Specialty

Guild Type	Specialist	Opportunist	Generalist
Foraging Method	One primary method and rarely switches to another	One primary method but may use up to three	Uses three or more methods based on local food abundance
Foraging Habitat	Forages in only one habitat	One primary habitat but able to feed in others	Multiple habitats used
Nesting Habitat	Only one nesting type	One primary nesting type but able to switch to another	Nesting type not limited
Diet	One primary food category	Ability to switch, may have two primary food categories any many secondary	Large diet range with no primary food category

Foraging Habitat Guilds

Foraging habitat guild types were divided into guilds based on units used by the U.S. and Canadian Wildlife services (Sauer et al., 1999). Table 117.2 defines the four broad vegetation types identified as foraging habitats and includes grassland, woodland/brushland, wetland, and other. The other category includes cliff and urban foraging species but also includes species without any clear habitat associations.

Nesting Habitat Guilds

Nesting habitat guilds were defined using an exploratory data analysis similar to the one described for foraging method guilds. Out of a potential list of ten guilds that included both human-constructed and native habitats, five guilds were defined because half the terms were similar and, thus, able to be grouped together (e.g., nest boxes and tree cavities or building and cliff habitats). Five nesting guilds were used: cavity, cliff, ground, floating mat, and vegetation (Table 117.3).

Diet Guilds

To find the important breeding bird diet guilds, 15 food categories were identified that recurred in the literature (seven animal, seven plant, and one refuse categories). Next, for each species, the food categories eaten by a species were ranked qualitatively from 1 to 4 (1 = primary food). To develop diet guilds from food categories, the food category rank for each species was counted. Following this, the percentage and total number of times a food category was the primary source for a species (i.e., a score of 1) was calculated. Any food category that received over 15 hits and was a primary food source for more than 30% of the species was considered for guild status. Six food categories were identified as important: seeds, grain, terrestrial vertebrates, aquatic vertebrates, terrestrial insects, and aquatic insects. There were, however, species that had more than one primary food source; thus, an omnivore guild was created that encompassed most of the species that utilized multiple primary-food categories. Food categories that were not earmarked as significant were added to a similar food category (such as seeds and grain) in an identified diet guild. The results of this method created eight guilds: aquatic invertebrates, aquatic vertebrates, aquatic vegetation, omnivore, seeds, terrestrial invertebrates, terrestrial vertebrates, and terrestrial vegetation (Table 117.3).

Migration Classes

Concern over possible declines in many North American long-distance migrant species or neotropical migrants (Sauer and Droege, 1992; Sherry and Holmes, 1996; Latta and Baltz, 1997) prompted

Table 117.3 Definitions of Bird Guilds and Migratory Classes Used

Guild	Interface	Description
	Foraging Method	
Dabbler	Water surface to water surface or just below	Moves bill side to side or back and forth on mudflats or in shallow edges of wetlands probing mud for plant or animal material including tipping for submerged vegetation
Diver	Water surface to depth	Feeding submerged at depths of 0.5 to 10 m on plants or animals
Gleaner	Vegetation to vegetation	Forages in vegetation by picking seeds, berries, or insects from branches, leaves, bark, or other vegetation
Picker	Ground to ground	Forages about using bill to dig or pick at ground, branches, carrion, or plants
Prober	Probed cavity or mud	Uses bill to search cavities for plant or animal material
Sallier	Ground or air to air	Chases and catches insects in flight; also hovering to glean berries or insects (flycatcher)
Stalker	Water	Stands motionless at water's edge, darts bill downward to seize prey
Swooper	Air to air or ground	Dives from a height to catch prey in bill or talons, prey on land, water, or wing

Guild	Description
	Foraging Habitat
Brushland/ woodland	Primary diet is found in areas dominated by short trees and shrubs that comprise greater than 70% of the total vegetation
Grassland	Primary diet is found in herbaceous grasses and forbs, covered by less than 20% trees or shrub
Wetland	Primary diet is found in aquatic areas and low-lying land that is submerged or inundated periodically by fresh or saline water
Other	Primary diet is found in all other vegetation types

Guild	Description
	Nesting Habitat
Cavity	Nests found in cavities (i.e., trees and nest boxes)
Cliff	Nests made on cliff faces, ledges, and bridges and sides of buildings
Ground	Nests made on ground
Floating mat	Nests constructed of vegetation that floats on water or in vegetation just above water surface
Vegetation	Nests constructed in vegetation above the ground in trees, shrubs, or forbs

Diet

Diet guilds: Aquatic invertebrates, aquatic vertebrates, aquatic vegetation, omnivore, seeds, terrestrial invertebrates, terrestrial vertebrates, terrestrial vegetation

Class	Description
	Migration Class
Resident	Lives in Mixed Grassland ecoregion year round
Short-distant	Winters in southern U.S.
Long-distant	Winters south of the continental U.S. (neotropical migrants)

analysis of trends in three migration classes (Table 117.3). The three migrant classes investigated were resident birds (live in the study region throughout the year), short-distance migrators (winter in the southern U.S.), and long-distance migrators (winter in Central and South America).

Guild Specialty Assignment

Each bird species was then assigned to one guild in each of the four guild types that were based on resource use and natural history accounts. Within each guild type, a particular species was associated with one primary guild. However, in nature many species exhibit opportunistic behavior and participate in more than one guild. To quantify this, each species was assigned to one of three specialization ranks (specialist, opportunist, and generalist) using the rules outlined in Table 117.2.

Guild Trends

To test whether society has altered breeding bird guilds in the Mixed Grassland Ecoregion of Saskatchewan, guild trends over the past 100 years were examined. Trends for each guild were explored using relative abundance ranks derived from records dating ca. 1900 and the 1990s. To standardize bird abundance, many ornithological groups (e.g., American Ornithologist's Union and the Federation of New York State Bird Clubs) have developed strict guidelines for estimating bird numbers in a region based on sightings (see Godfrey, 1986; DeSante and Pyle, 1986; Smith, 1996). These guidelines are used to define abundance ranks, so that bird numbers for large geographic regions may be qualitatively estimated. Regardless of the organization that uses them, most abundance rank standards are similar in that they are an estimate of abundance for a given species based on its detectability by a "competent observer" observing at the "appropriate time." Because of these similarities, it is possible to make conversions and comparisons between standards. They also take into account yearly and monthly variations in local distribution due to habitat and food source availability. However, these abundance ranks are designed to be relative and are not defined using mathematical precision.

Breeding bird trends were calculated using historical accounts of relative bird abundance ranks, but when conflicts were found, expert opinion was used to identify the appropriate rank (e.g., consulted Canadian Wildlife Service or estimated abundance on amount of primary habitat available at the time). The abundance rank standards from the different sources were converted into DeSante and Pyle's (1986) four categories (common, fairly common, uncommon, and rare). The abundance ranks were collected for two periods: ca. 1900 (pre-European settlement) and the 1990s.

Circa 1900 (Native Prairie) Abundance Ranks

By the early 1900s, ornithological experts had compiled extensive breeding bird lists for Saskatchewan including records of their relative abundance (Richardson and Swainson, 1831; Hind, 1860; Selwyn, 1874; Coues, 1878; Raine, 1892; Houseman, 1895; Macoun, 1900; Bent 1907, 1908). Agricultural change had begun by the time of Bent's 1906–1907 surveys (Bent, 1908), but he generally recorded bird abundances prior to the major settlement period. Therefore, only studies conducted prior to 1908 were used to develop the native bird species accounts. Any breeding occurrence recorded prior to this date was assumed to be a "native" mixed grassland species. New species listed in literature after this date were recorded as "not present" in the early ca. 1900 native ecosystem.

1990s Abundance Ranks

The 1990s abundance ranks were based on Price and co-workers (1995), Roy (1996), and Smith (1996).

Quantifying Abundance Ranks

A relative abundance score (RAS) was used to quantify species abundance rank data between each period. The RAS was based on the "percentage of days that a few to many birds are expected

Table 117.4 Relative Abundance Score (RAS) Assigned to DeSante and Pyle's (1986) Abundance Rank Standards[a]

Abundance Rank	Description	RAS
Common	Always or almost always encountered	0.70
Fairly common	Usually or often encountered	0.50
Uncommon	Not usually encountered	0.20
Rare	Generally not expected	0.05
Not present	Not present or does not breed in area	0.00

[a] Based on the "percentage of days that a few to many birds are expected to be seen" at the appropriate time and in their appropriate habitat.

to be seen" as defined by DeSante and Pyle's (1986) abundance rank standard. Table 117.4 gives the RAS assigned to each of the relative abundance ranks.

As seen in Equation 117.1, the relative change for each guild ($T_{\text{Abundance}}$) may then be calculated for the past 100 years.

$$T_{\text{Abundance}} = [(\Sigma A_I) - (\Sigma A_j)]/\Sigma A_i \qquad (117.1)$$

where A_i denotes the RAS in the ca. 1900 period of the paired combination; A_j denotes the RAS for the later 1990s; and the summation extends over all species within a guild. These calculations were conducted for each of the 25 guilds in Table 117.2.

RESULTS

Vegetation Habitats

Vegetation changes within the Mixed Grassland Ecoregion of Saskatchewan were estimated as a time series from pre-European settlement (ca. 1880) to the present. Using the original percentages for each township estimated by Archibold and Wilson (1980), the percent coverage was calculated for five vegetation classes within the study area. Figure 117.2 shows that 95.8% of the region was dominated by grasslands covering 8,276,764 ha. The remaining 4.2% of the area comprised the four other vegetation classes: woodland (0.3%), brushland (0.5%), wetland (1.1%), and open water (2.3%) (inset of Figure 117.2). The grassland figure, however, may be a slight overestimate because

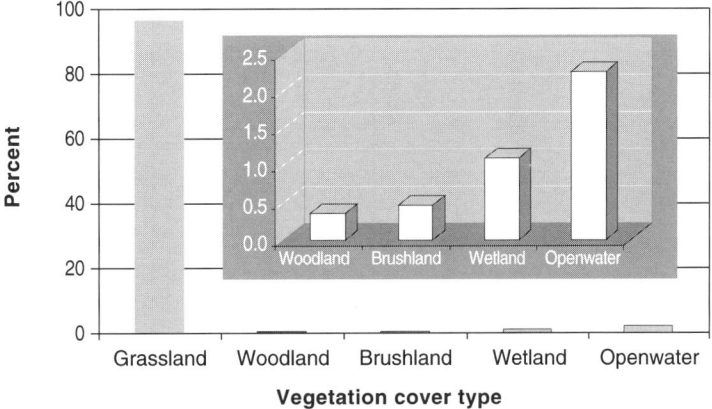

Figure 117.2 Vegetation cover types ca. 1880 based on Dominion Lands Survey township and range maps as compiled by Archibold and Wilson (1980). Inset graph details the percent cover for vegetation classes other than grassland.

Table 117.5 Area (Acres) and Percent Cover of Vegetation Classes in the Mixed Grassland Ecoregion of Saskatchewan[a]

| | Total Area Improved | | Farm Area | | Estimated Land Use (as reported on farms) | | | | | | | | | | |
| | | | | | Improved Area | | Native Pasture | | Wood/ Brushland | | Wetland/ Waste | | Unimproved | |
Year	Acres	%	Acres	%	Acres	%	Acres	%	Acres	%	Acres	%	Acres	%
1885	1141	0.0	11,360	0.1	1141	10.0								
1891	3799	0.0	57,860	0.3	3799	6.6	53,975	93.3	86	0.1			54,061	93.4
1921	7,450,881	34.9	13,160,610	61.6	7,450,881	56.6	5,355,232	40.7	28,799	0.2	325,640	2.5	5,709,671	43.4
1936	10,602,453	61.8	17,167,457	80.4	10,602,453	61.8	6,111,821	35.6	37,822	0.2	415,467	2.4	6,565,119	38.2
1941	10,687,302	50.0	17,921,847	84.1	10,687,302	59.6	6,808,604	38.0	115,443	0.6	355,551	2.0	7,279,598	40.6
1971	12,904,876	60.4	19,899,283	93.2	12,904,876	64.9							6,994,407	35.1
1991	13,895,029	65.1	20,190,397	94.5	13,895,029	68.8	5,787,325	28.7					6,295,368	31.2
1996	13,816,050	64.7	19,873,456	93.0	13,816,050	69.5	5,567,599	28.0					6,045,557	30.4

[a] As reported on farms from the Canadian Agricultural Census.

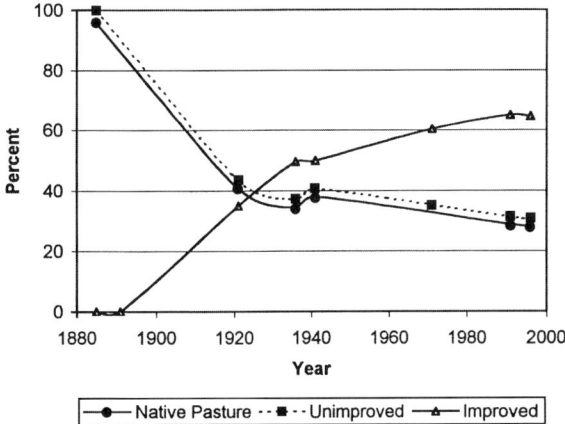

Figure 117.3 Percentage of total land area classified as improved land, unimproved land, and native pasture in the Mixed Grassland Ecoregion of Saskatchewan, as reported by Canadian Agricultural Censuses from 1885 to 1996.

the line transects used were separated by 1 mile, and thus underrepresented many of the small sloughs and wooded coulees (see Archibold and Wilson, 1980).

Eight Agricultural Censuses (1885, 1881, 1921, 1936, 1941, 1971, 1991, and 1996) were used also to estimate the condition of occupied land. For each census, up to four vegetative habitats were recorded in each LID or RM in the study region, including (1) native and reestablished grassland (native pasture), (2) woodland/brushland, (3) wetland, "waste" areas, and open water, (4) cultivated and improved pasture and summer fallow. Because of changes in both the survey questionnaires used and how terms were defined, not all of the vegetation data are available in each census. Table 117.5 shows the areas and the percent coverage for vegetation classes on farms. Improved lands in the study region dramatically increased from 0.5% of the study region in 1885 to 50% in 1936. During the Depression in the 1930s, the breaking of sod leveled off but by the 1970s, it was on the rise again; by 1996, 65% of the region was classified as improved land. Figure 117.3 illustrates the large increase in improved lands in the study region over the past 111 years, from 0.5% in 1885 to more than 65% in 1996.

Since the reported vegetative cover for each agricultural census includes only land classified as farmland (areas organized for farm activity), areas not listed as farms, such as parks (national and regional), large water bodies, road rights-of-way, and urban areas are not reported. Because of this, the area of unreported land changed over time. The percentage of farm-related activity grew rapidly until 1936 when it reached 80.4% of the ecoregion. After this, area converted into farms slowed but continued. By 1991, more than 94% of the study area was reported as farmland by the census.

Figure 117.3 also graphs the native pasture (grassland) against the total unimproved land in the study region as reported on farms from agricultural censuses. These curves illustrate the dominance that grassland areas had in the region until 1921. At this time, improved lands bypassed native grassland as the most-abundant vegetation cover.

One enigma in using census data deals with how vegetation categories are reported in each census. Because the many rules for defining categories were "redefined" between census years, direct comparisons between vegetation classes could not always be made (note the missing data in Table 117.5). To investigate these classes, Figure 117.4 subtracts both the improved land and native pasture (curve with triangle data points) from the total farm area. This remaining vegetation cover is a small percentage of the total and shows a slightly decreasing curve from 2.7% in 1921 to 2.4% in 1996. This indicates that not only does most of the census data account for land as either improved land or native pasture, but these land uses have remained generally constant since the 1920s. However, the wetland and woodland/brushland curves in Figure 117.4 are moving in opposite directions for the three censuses (1921, 1936, and 1941) in which both wetland and woodland/

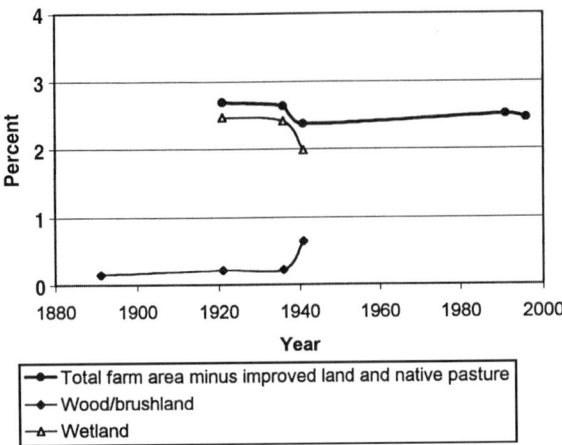

Figure 117.4 Percentage of farmland not classified as improved land or native pasture, with lands classified as wetland and wood/brushland, in the Mixed Grassland Ecoregion of Saskatchewan, as reported by Canadian Agricultural Censuses from 1885 to 1996.

brushland data are available, with the largest decline in wetland/waste area correlating with the largest increase in woodland/brushland. This trend has continued into the 1990s, based on 1994 LANDSAT data that estimated woodlands covered 1.9% and wetland 1.6% of the ecoregion (Prairie Farm Rehabilitation Administration, GIS division, personal communication).

Breeding Bird Guilds

Table 117.6 lists the richness in bird species by season visited for those species that breed (confirmed and probable) and migrate through the study area (transient). A total of 135 bird species (51%) was recorded as confirmed or probable breeders. Of these, 81.5% are summer visitants (110 species), and the remaining 18.5% are permanent residents (25 species). Appendix 117.I lists each of the 135 breeding bird species found in the Mixed Grassland Ecoregion of Saskatchewan, along with their migration patterns and relative abundance ranks.

Table 117.7 lists the number of bird species found in each guild for the four guild types (foraging method, foraging habitat, nesting habitat, and diet). It also divides each guild into specialization classes. In each guild type, generalists had the lowest membership percentage with 3.7% by foraging method, 20.7% by foraging habitat, 4.5% by nesting habitat, and 14.8% by diet guilds. Thus, to keep analysis simple, guilds were analyzed as a single unit since guild-switching generalists occupied the minority of all the guild types, and analysis by specialty is not examined further in this chapter.

The number of bird species in the three migratory classes is shown in Table 117.8. By far the largest group was short-distance migrants, comprising 54%, or 73 species. Resident birds make up the smallest class, comprising only 13%, with long-distance migrants at 33%.

Table 117.6 Visitation and Species Richness for Breeding and Transient Birds on the Mixed Grassland Ecoregion of Saskatchewan

	All Birds Records		Breeding Birds		Transient Birds	
Visitation	No.	%	No.	%	No.	%
Summer	136	51.3	110	81.5	26	20.0
Spring/Fall	69	26.0	0	0.0	69	53.1
Winter	32	12.1	0	0.0	32	24.6
All Year	28	10.6	25	18.5	3	2.3
Total	265	100.0	135	100.0	130	100.0

Table 117.7 Number and Percent of Species in Specialty Classes by Guilds in Each Guild Type (Foraging Method, Foraging Habitat, Nesting Habitat, and Diet)

Guild	Specialty			Sum	%
	Specialist	Opportunist	Generalist		
Foraging Method Guilds					
Dabbler	2	6	1	9	6.7
Diver	10	2	0	12	8.9
Gleaner	6	13	1	20	14.8
Picker	21	23	1	45	33.3
Prober	5	3	1	9	6.7
Sallier	11	5	0	16	11.9
Stalker	2	1	1	4	3.0
Swooper	15	5	0	20	14.8
Sum	72	58	5	135	
Percentage	53.3	43.0	3.7		
Foraging Habitat Guilds					
Grassland	17	8	5	30	31.1
Wetland	28	11	3	42	22.2
Woodland	15	16	9	40	29.6
Other	4	8	11	23	17.0
Sum	64	43	28	135	
Percentage	47.4	31.9	20.7		
Nesting Habitat Guilds					
Floating mat	8	5	0	13	9.7
Ground	40	13	0	53	39.6
Vegetation	22	17	6	45	33.6
Cavity	7	3	0	10	7.5
Cliff	6	7	0	13	9.7
Sum	83	45	6	134[a]	
Percentage	61.9	33.6	4.5		
Diet Guilds					
Aquatic invertebrates	2	9	1	12	8.9
Aquatic vegetation	1	4	1	6	4.4
Aquatic vertebrates	3	7	0	10	7.4
Omnivore	0	13	12	25	18.5
Seeds	2	10	3	15	11.1
Terrestrial invertebrates	15	27	3	45	33.3
Terrestrial vegetation	1	4	0	5	3.7
Terrestrial vertebrates	10	7	0	17	12.6
Sum	34	81	20	135	
Percentage	25.2	60.0	14.8		

[a] The cowbird is a parasitic breeder and is not included.

Table 117.8 Number of Breeding Bird Species in Migratory Classes for the Mixed Grassland Ecoregion of Saskatchewan

Migratory Class	Total
Resident	17
Short-distance	73
Long-distance	45

Table 117.9 Relative Abundance Ranks and Values for Breeding Birds in the Mixed Grassland Ecoregion of Saskatchewan.

Abundance Rank	Circa 1900		1990s	
	No.	%	No.	%
Common	41	30.4	27	20.0
Fairly common	13	9.6	47	34.8
Uncommon	24	17.8	34	25.2
Rare	32	23.7	25	18.5
Not present	25	18.5	2	1.5
Sum	135	100.0	135	100.0

Note: Values are based on relative abundance ranks of historical accounts (see text for sources).

Relative Abundance Ranks

Table 117.9 lists the five relative abundance ranks for the two periods, the number of bird species assigned to each, and the percentage of bird species in each rank. From ca. 1900 to the 1990s, species classified as common and rare declined from 30% (41 species) and 24% (32 species) to 20% (27 species) and 19% (25 species), respectively. During this period, however, the species ranked as fairly common and uncommon substantially increased by 34 to 35% and 10 to 25%, respectively. Moreover, since the early 1900s, 25 species were introduced or expanded their ranges into the study region.

Figure 117.5 summarizes relative abundance trends (RAT) by guild. Overall, of the 25 guilds, 10 had positive trends, 11 had negative, and 4 showed no trend (RAT between 0.1 and –0.1). Of the foraging method guilds, two (gleaner and sallier) showed large increasing trends, while three primarily associated with wetland habitats decreased (foraging method guilds: divers, probers, and stalkers) (Figure 117.5A). Figure 117.5B shows that during this period, two of the four foraging habitat guilds increased (woodland and other) and two decreased (grassland and wetland). The five nesting habitat guilds (Figure 117.5C) had three increasing guilds (cavity, vegetation, and cliff) and two decreasing guilds (ground and floating mat). Finally, with regard to diet (Figure 117.5D), four of the guilds (aquatic invertebrates, aquatic vegetation, aquatic vertebrates, and terrestrial vegetation) showed large decreasing trends and only two increasing (seeds and terrestrial invertebrates). Interestingly, for each guild type those associated with aquatic habitats or wetland diets showed the largest declines.

Figure 117.6 plots the relative changes in three migration classes (resident, short-distance, and long-distance), showing that the three migration classes increased in the study region over the past 100 years. The largest RAT increase was in resident birds (0.62), followed by long-distance migrants (0.16). Short-distance migrants increased only slightly, with a relative change score of 0.03.

DISCUSSION

Vegetation Cover

The vegetative cover of the Mixed Grassland Ecoregion of Saskatchewan has changed significantly over the past 100 years. According to Archibold and Wilson's (1980) original estimation of late-19th century survey maps, over 90% of the study region was covered in grassland. By the mid-1990s, native grassland had been reduced to less than 28%. Further, woodland and brushland have increased since the late 1800s (due to the loss of grazing pressure, control of fire, and the promotion of growing shelterbelts), while wetland area has decreased (due to increasing cultivation).

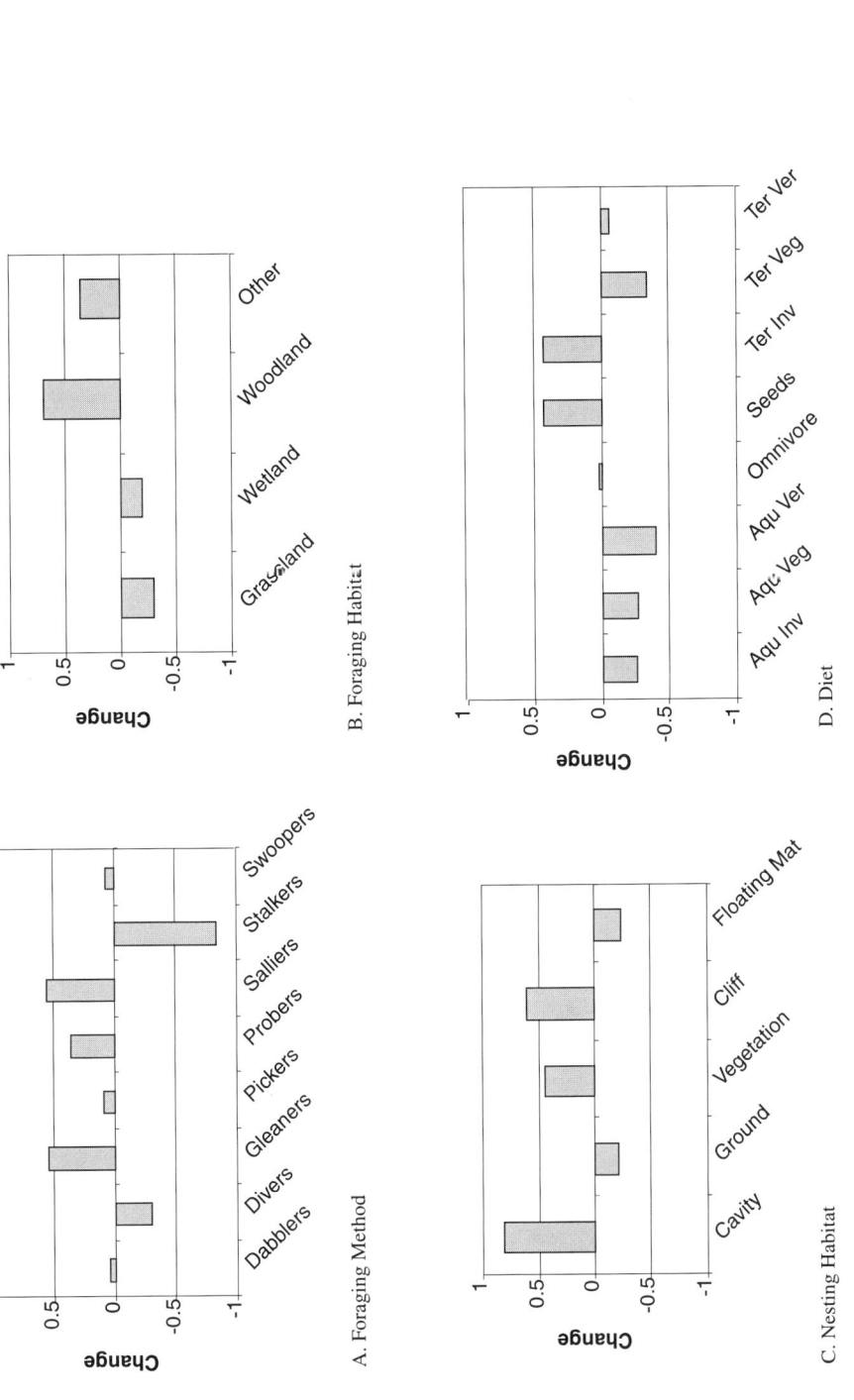

Figure 117.5 Relative abundance trends in breeding bird guilds from ca. 1900 to the 1990s for the four guild types in the Mixed Grassland Ecoregion of Saskatchewan.

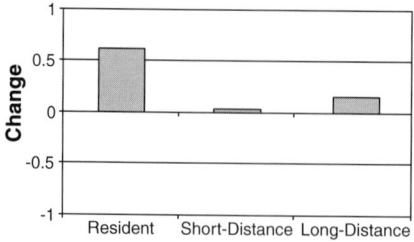

Figure 117.6 Migration class relative changes in the Mixed Grassland Ecoregion of Saskatchewan from ca. 1900 to the 1990s.

The dominant land cover in the 1990s was improved farmland, which comprises nearly 70% of the land area of the study region (Table 117.5).

Guilds and Relative Change Ranks

Since agricultural settlement, there have been many changes in breeding bird guilds. The early 1900s began a period of rapid change in bird populations on the northern prairies. This was observed as early as 1907, when Bent (1908, p. 407) observed that the opening up of new lands for agricultural settlement was "making such rapid and marked changes in the great wild-fowl breeding grounds of northwest Canada." He continued, "Even during the one year intervening between my two visits to this region, the change was so striking as to indicate the passing away within the near future of nearly all the great breeding resorts of the interesting region."

This change, however, has not been due to losses in bird diversity. Table 117.10 shows that between ca. 1900 and the 1990s there has been an overall increase of approximately 21% in species richness and 12% in diversity. Therefore, if there were no new selection pressures, each guild trend should have increased. However, along with the RAT increases there were decreases. For example, in foraging method guilds, there were large RAT increases in gleaners, probers, and salliers, with moderate increases in the traditional picker, dabbler, and swooper guilds, while there were large decreases in the diver and stalker guilds. For the foraging habitat, the grassland and wetland guilds had declined RATs, while the trends for both woodland and other guilds increased.

Ten of the 25 total guilds exhibited decreasing trends (RAT > 0.1). Of these, all but three (grassland foraging, ground nesting by habitat guild types, and terrestrial vegetation by diet) were associated with aquatic environments (i.e., divers, stalkers by foraging method, wetland by foraging habitat, aquatic invertebrate, aquatic vegetation, and aquatic vertebrate by diet) (Figure 117.5). This indicates the importance of wetland and lacustrine environments to regional breeding bird populations, confirming that their numbers have drastically declined since the early 1900s (Smith, 1996; Igl and Johnson, 1997).

In bird diet guilds, RATs increased for the seeds and terrestrial invertebrates guilds (Figure 117.5D). The omnivore and terrestrial vertebrate guilds had minor increases and the remaining four diet guilds (aquatic invertebrates, aquatic vegetation, aquatic vertebrates, and terrestrial vegetation) had large negative trends. Of the four diet guilds exhibiting negative trends, three were associated with aquatic habitats, indicating a decline in aquatic food sources.

Table 117.10 Breeding Bird Species Richness and Diversity (ca. 1900 and the 1990s) in the Mixed Grassland Ecoregion of Saskatchewan

Index	Circa 1900	1990s
Species Richness	110	133
Species Diversity[a]	1.89	2.11

[a] Calculated using the Shannon–Weaver index; relative abundance values were substituted for number of individuals.

Each breeding migratory class (resident, short-distance, and long-distance) showed a positive RAT. Thus, migratory classes have little influence on negative guild RATs observed. Further, although the resident bird class comprises the fewest bird species, it had the largest relative change. In the 1990s data, there were seven more resident species living in the study region, an increase of 41% from ca. 1900. The increase in resident birds is primarily due to seven introduced species, two of which (gray partridge and ring-necked pheasant) were directly introduced into the study region. The remainder expanded their range from other parts of North America, i.e., greater prairie chicken from the southern prairies, and house sparrow, northern mockingbird, and rock dove from eastern North America. Moreover, in the ca. 1900 period, resident species that were important on the open grasslands had by the 1990s dropped sharply in abundance, such as the sage grouse (fairly common to rare), and the sharp-tailed grouse (common to fairly common). During this time, however, the introduced house sparrow and rock dove became common species.

Long-distance migrants had a RAT of 0.16; this increase was mostly due to woodland native songbirds, such as house wren, veery, least flycatcher, and the western kingbird, becoming more abundant. The least-affected class was the short-distance migrants (RAT = 0.04), but it also contained 54% of all bird species (73 members). The slight increase in short-distance migrants is due primarily to the range expansion of eastern deciduous forest species, such as the American robin and cedar waxwing and the introduced European starling and mourning dove, all of which have experienced large population increases.

Changing Habitat

Most of the changes observed in breeding bird guilds have been due to changing habitats since agricultural settlement. In terms of nesting habitat, breeding bird guilds that have benefited from human activity include vegetation, cavity, and cliff nesters. Human-constructed nest boxes mimic tree cavities and the development of this resource may result in the proliferation of cavity-nesting species in a given area. For example, Bittner (1988) (for Saskatchewan) and Pinel (1980) (for Alberta) reported that the introduction of nesting boxes in a region significantly increased the breeding populations of mountain bluebirds. This increase also has been shown for house sparrows, tree swallows, and flycatchers (Smith, 1996). Similarly, buildings, bridges, and overpasses are surrogates for cliffs by swallows (Knapton, 1988; Belanger, 1989), doves, vultures (Beaulieu, 1985; Wapple, 1985), and phoebes (Friesen, 1983; Desment, 1987).

Since the early 1900s, the control of fire and a massive reduction in unmanaged ungulate grazing, coupled with the planting of farm shelterbelts, have all increased the area of woodland and shrub habitats (Figure 117.4). As the woodland habitats increased, the species that use them have become more abundant, including northern flicker, least flycatcher, western kingbird, tree swallow, rock wren, veery, American robin, brown thrasher, European starling, lark sparrow, and Baltimore oriole. Further, the woodland and shrubland predators such as red-tailed hawk, American kestrel, merlin, and great horned owl have increased while the once-common prairie raptors such as the ferruginous and Swainson's hawks have experienced large population declines.

Human-made structures such as utility lines, communication towers, and fence posts are used by birds as they would use tall trees or snags, providing additional nesting and perching habitat that has been historically limited (Houston, 1996; Curtis, 1997). Further, breeding birds are increasingly incorporating human-made materials in nest construction (Bancroft, 1978), perhaps showing adaptable behavior in response to a changing resource base.

The relative changes illustrated in habitat guilds (foraging habitat in Figure 117.5B and nesting habitat in Figure 117.5C) closely parallel the overall habitat trends observed in vegetation (Table 117.5) and nesting sites as described previously. Table 117.11 matches these habitat trends with the habitat guilds, with each habitat having the same negative or positive trend as the habitat guild's relative change value.

Table 117.11   Correlation between Bird Habitat Guilds and Overall Habitat Trend in the Mixed Grassland Ecoregion of Saskatchewan from Approximately 1900 to the 1990s

Guild Type	Guild	Habitat Trend	Guild Trend
Foraging Habitat	Brush/woodland	+	0.58
	Grassland	−	−0.28
	Wetland	−	−0.18
	Other	+	1.57
Nesting Habitat	Cavity	+	0.73
	Cliff	+	0.45
	Ground	−	−0.19
	Floating mat	−	−0.18
	Vegetation	+	0.38

Note: + = increasing habitat trend; − = decreasing habitat trend.

CONCLUSION

For the abundance of breeding bird populations in the Mixed Grassland Ecoregion of Saskatchewan, agricultural development has been both a positive and negative force by both created and reduced specific habitats and food supplies. For example, there has been a consistent decrease in native grassland and aquatic related guilds, while those related to woodland vegetation have increased. The areas converted to croplands not only reduced native species diversity but also the energy available to birds, because much of the biomass is harvested and transported from the field (Matson et al., 1997). Therefore, agricultural settlement has altered not only local plant communities but also the regional structure and energy flow of the ecoregion.

Studies on northern prairies at spatial scales of species assemblages have shown losses in species diversity (Leach and Givnish, 1996; Lesica and DeLuca, 1996; Bakker and Berendse, 1999; Davis and Duncan, 1999; Pepper, 1999). Thus, at the level of species associations, the prairie is becoming more homogeneous with many species becoming locally extinct. This is especially evident in agricultural fields and tame pastures (Peltzer, 2000) that together, by the mid-1990s, have expanded to cover over 70% of the prairie landscape. However, the relationship between habitat loss and diversity is not so clear because local habitats do not necessarily constrain regional diversity (Simberloff and Gotelli, 1984). In the Mixed Grassland Ecoregion since 1900, an increase in regional breeding bird diversity has occurred despite major changes in vegetation cover. Therefore, at the broader ecosystem levels (i.e., Mixed Grassland Ecoregion), agricultural practices and the development of human living spaces have added landscape heterogeneity and species diversity to the region. The changes in breeding bird guilds suggest that human land-use practices have modified the available regional habitats, altering the broad functional diversity and species composition at both the species assemblage and ecosystem levels.

The importance of society in shaping the present northern prairie environment has been primarily explored in terms of human disturbance to individual species living in conspicuous habitats (e.g., wetlands). Perhaps this is because the most visible agricultural impact is the replacement of native prairie cover with other vegetation types, and the resulting decline of individual "endangered" or "threatened" species that once lived there. However, human interactions are influencing more than a few fragmented districts or declining species. Society has carved out an ecological role within the prairie ecosystem, the effects of which may be measured by changes and responses in functional roles of key species.

Persson and co-workers (1996) examined food web structure in species associations and identified four factors which, when modified, alter the dynamics of ecological systems:

1. Habitat structure
2. Disturbance fluctuations

3. Population size and structure
4. Animal behavior and defenses

Within species assemblages in the Mixed Grassland Ecoregion of Saskatchewan, agricultural development can be linked directly to the alteration of each of these factors. Over 60% of the native plant assemblages has been significantly modified through tilling, drainage, and pesticide use. Further, it may be argued that much of the remaining 40% has been altered by secondary effects such as chemical and fertilizer runoff, air pollution, or introduced species. These impacts reduce functional diversity by excluding all but a few species in these areas. On the other hand, the introduction of woody vegetation through the expansion of brushland and aspen groves, along with the construction of shelterbelts, parks, and urban areas have added species and perhaps functional diversity. Disturbance fluctuations and population size and structure have been changed through the introduction of annual tillage, pest control, fencing, game management, and irrigation.

Since agricultural systems impact the Mixed Grassland Ecoregion at both fine and broad hierarchical levels, there are major ecological repercussions. The greatest immediate repercussion may be that modern agriculture represents an innovation that is constantly affecting prairie biota by altering food web interactions and abiotic resource distribution. Because the temporal pace of ecosystem level processes is very slow, it generally takes hundreds of years to measure the full response to a perturbation. Thus, if modern society is significantly altering the functional components of the prairie ecosystem, then it is changing the ways in which resources are partitioned for other species and diversifying selection pressures. The mechanisms that are altering prairie landscapes include human innovations that reallocate energy, space, and resources away from native to new, human-dominated systems that partition resources in new and innovative ways. Therefore, by changing the ways resources are used and divided among species, society has created new ecological space that is being exploited by many species. This new ecological space may be allowing more species and individuals to inhabit a landscape (as seen by the invasion breeding birds and the increases in regional bird richness and diversity). This suggests that measurements of an ecosystem's integrity should not rely heavily on biodiversity measurements alone.

The important changes to the Mixed Grassland Ecoregion of Saskatchewan that have occurred over the past 100 years are primarily due to recent human interactions. Thus, society has converted this prairie system into an agroecosystem. If we are to understand fully the implications of this, an integrated approach must be taken that investigates ecosystems as broad dynamic units in which society is a dominant player. Thus, in addition to lower-level mechanistic studies of species interactions, future studies must investigate ecosystems in terms of spatiotemporal scales and include societal components. In doing this, we should enhance our understanding of all ecosystem functions and how they are structured.

ACKNOWLEDGMENTS

I wish to thank Dave Gauthier and Brian Mlazgar at the Canadian Plains Research Center for their helpful input and suggestions. This manuscript was greatly improved by suggestions of three anonymous reviewers. Also thanks to Alan R. Smith of the Canadian Wildlife Service (CWS) for his suggestions in developing the bird database. This study was supported by grants from the Prairie Ecosystem Study (PECOS), an interdisciplinary research endeavor funded by the Tri-Council Natural Sciences and Engineering Research Council (NSERC), the Social Science and Humanities Research Council (SSHRC), the Medical Research Council (MRC), and the University of Regina.

REFERENCES

Acton, D.A., Padbury, G.A., Stuchniff, C.T., Gallagher, L., Gauthier, D.A., Kelly, L., Radenbaugh, T.A., and Thorpe, J., *Ecoregions of Saskatchewan,* Canadian Plains Research Center and Saskatchewan Environment and Resource Management, Regina, Saskatchewan, 1998.

Allen, T.F.H. and Starr, T.B., *Hierarchy: Perspectives of Ecological Complexity,* University of Chicago Press, Chicago, 1982.

Archibold, O.W. and Wilson, M.R., The natural vegetation of Saskatchewan prior to agricultural settlement, *Can. J. Bot.,* 58, 2032–2042, 1980.

Axelrod, D.I., Rise of the grassland biome, central North America, *Bot. Rev.,* 51, 163–201, 1985.

Bakker, J.B. and Berendse, F., Constraints in the restoration of ecological diversity in grassland and heathland communities, *Trends Ecol. Evol.,* 14, 63–68, 1999.

Bancroft, J., Variations on bird nesting habitats, *Blue Jay,* 36, 120, 1978.

Beaulieu, R., Abandoned house nest site for turkey vulture, *Blue Jay,* 43, 46–48, 1985.

Belanger, R., Cliff swallows invade farmhouse, *Blue Jay,* 47, 215, 1989.

Bent, A.C., Summer birds of southwestern Saskatchewan, *Auk,* 24, 407–431, 1907.

Bent, A.C., Summer birds of southwestern Saskatchewan, *Auk,* 25, 25–35, 1908.

Bittner, R.A., Bluebirds at Abernethy: history and 1988 results, *Blue Jay,* 46, 215–218, 1988.

Botsford, L.W., Castilla, J.C., and Peterson, C.H., The management of fisheries and marine ecosystems, *Science,* 277, 509–515, 1997.

Collins, S.L., Fire frequency and community heterogeneity in tallgrass prairie vegetation, *Ecology,* 73, 2001–2006, 1992.

Collins, S.L. and Wallace, L.L., *Fire in North American Tallgrass Prairies,* University of Oklahoma Press, Norman, 1990.

Coues, E., Field-notes on birds observed in Dakota and Montana along the 49th parallel during the seasons of 1873 and 1874, *Bull. U.S. Geol. Geogr. Surv. Terr.,* 4, 545–661, 1878.

Coupland, R.T., A reconsideration of grassland classification on the northern great plains of North America, *J. Ecol.,* 49, 135–67, 1961.

Curtis, C., Birds and transmission lines, *Blue Jay,* 55, 43–47, 1997.

Davis, S.D. and Duncan. D.C., Grassland songbird occurrences in native and crested wheatgrass of southern Saskatchewan, *Stud. Avian Biol.,* 19, 211–218, 1999.

DeSante, D. and Pyle, P., Distributional Checklist of North American Birds to Neotropical Forest Habitats, Latin American Program, Canadian Wildlife Service, Ottawa, 1986.

Desment, K.D., Eastern range extension of the Say's phoebe into south-central Manitoba, *Blue Jay,* 45, 108–109, 1987.

ESWG (Ecological Stratification Working Group), A National Ecological Framework for Canada, Agricultural and Agri-food Canada, Centre for Land and Biological Resources Research and Environment Canada, State of the Environment Directorate, Ecozone Analysis Branch, Ottawa/Hull, 1995.

Frank, D.A., McNaughton, S.J., and Tracy, B.F., The ecology of the earth's grazing ecosystems, *BioScience,* 48, 513–521, 1998.

Friesen, V.C., (Un)Usual nesting site of the eastern phoebe, *Blue Jay,* 41, 166–167, 1983.

Godfrey, W.E., *The Birds of Canada,* rev. ed., National Museums of Canada, Ottawa, 1986.

Grime, J.P., Biodiversity and ecosystem function: the debate deepens, *Science,* 277, 1260–1261, 1997.

Hind, H.Y., *Narrative of the Canadian Red River Exploring Expedition of 1857 and of the Assinniboine and Saskatchewan Exploring Expedition of 1858,* Vol. I, Longman, Green, Longman and Roberts, London, 1860.

Hooper, D.P. and P.M. Vitousek, The effects of plant composition and diversity on ecosystem processes, *Science,* 277, 1302–1305, 1997.

Houseman, J.E., North west notes for 1894, *Oologist,* 12, 8–11, 1895.

Houston, C.S., Great horned owl nest sites in Saskatchewan, *Blue Jay,* 54, 125–33, 1996.

Hughes, T.P., Catastrophes, phase shifts, and large-scale degradation of a Caribbean coral reef, *Science,* 265, 1547–1551, 1994.

Igl, L.D. and Johnson, D.H., Changes in breeding bird populations in North Dakota: 1967 to 1992–1993, *Auk,* 114, 74–92, 1997

Jackson, J.B.C., Community unity? *Science,* 264, 1412–1413, 1994.

Joern, A. and Keeler, K.H., *The Changing Prairie: North America's Grasslands,* Oxford University Press, New York, 1995.

Kaiser, J. and Gallagher, R., How humans and nature influence ecosystems, *Science,* 277, 1204–1205, 1997.

Knapp, A.K., Blair, J.M., Briggs, J.M., Collins, S.L., Hartnett, D.C., Johnson, L.C., and Towe, E.G., Keystone role of bison in North American tallgrass prairie, *BioScience,* 48, 39–50, 1999.

Knapton, R.W., Unusual bank and barn swallow nesting on Lake Winnipegosis, Manitoba, *Blue Jay,* 46, 37, 1988.

Lapointe, B.E. and O'Connell, J., Nutrient-enhanced growth of *Cladophora prolifera* in Harrington Sound, Bermuda: eutrophication of a confined, phosphorus-limiting marine system, *Estuarine Coastal Shelf Sci.,* 28, 347–360, 1989.

Latta, S.C. and Baltz, M.E., Population limitation in neotropical migratory birds: commentary on Rappole and McDonald (1994), *Auk,* 114, 754–762, 1997.

Lawton, J.H., Are there assembly rules for successional communities? in *Colonization, Succession, Stability,* Gray, A.J., Crawely, M.J., and Edwards, P.J., Eds., Blackwell Scientific Press, Oxford, 1987, pp. 225–245.

Leach, M.K. and Givnish, T.J., Ecological determinants of species loss in remnant prairies, *Science,* 273, 1555–1558, 1996.

Lesica, P. and DeLuca, T.H., Long-term harmful effects of crested wheatgrass on Great Plains grassland ecosystem, *J. Soil Water Conserv.,* 51, 408–409, 1996.

Macoun, J., *Catalogue of Canadian Birds,* S.E. Dawson, Queen's Printer, Ottawa, 1900.

Mainguet, M., *Desertification: Natural Background and Human Mismanagement,* Springer-Verlag, New York, 1994.

Matson, P.A., Parton, W.J., Power, A.G., and Swift, M.J., Agricultural intensification and ecosystem processes, *Science,* 277, 504–509, 1997.

McMichael, A.J., Global environmental change and human health: impact assessment, population vulnerability and research priorities, *Ecosyst. Health,* 3, 200–210, 1997.

Miller, T.E., and Travis, J., The evolutionary role of indirect effects in communities, *Ecology,* 77, 1329–1335, 1996.

Naeem, S.L., Thompson, J., Lawler, S.P., Lawton, J.H., and Woodfin, R.M., Declining biodiversity can alter the performance of ecosystems, *Nature,* 368, 734–737, 1994.

Nee, S. and May, R.M., Extinction and the loss of evolutionary history, *Science,* 278, 692–693, 1997.

Peltzer, D.A., Ecology and ecosystem functions of native prairie and tame grasslands in the northern Great Plains, *Prairie Forum,* 25, 65–82, 2000.

Pepper, J., Diversity and Community Assemblage of Ground-dwelling Beetles and Spiders on Fragmented Grasslands of Southern Saskatchewan, Master's thesis, University of Regina, Saskatchewan, 1999.

Persson, L., Bengtsson, J., Menge, B.A., and Power, M.E., Productivity and consumer regulation: concepts, patterns, and mechanisms, in *Food Webs: Integration of Pattern and Dynamics,* Polis, G.A. and Winemiller, K.O., Eds., Chapman & Hall, Toronto, 1996, pp. 396–434.

Pinel, H.W., Reproductive efficiency and site attachment of tree swallows and mountain bluebirds, *Blue Jay,* 38, 177–183, 1980.

Polis, G.A. and Winemiller, K.O., Eds., *Food Webs: Integration of Pattern and Dynamics,* Chapman & Hall, New York, 1996.

Price, J., Droege, S., and Price, A., *The Summer Atlas of North American Birds,* Academic Press, San Diego, 1995.

Raine, W., *Bird-Nesting in North-West Canada,* Hunter, Rose, Toronto, 1892.

Richardson, J. and Swainson, W., *Fauna Boreali-Americana, Part Second: Birds,* John Murry, London, 1831.

Root, R.B., The niche exploitation pattern of the blue-grey gnatcatcher, *Ecol. Monogr.,* 37, 317–350, 1967.

Rosemond, A.D., Indirect effects of herbivores modify predicted effects of resources and consumption of plant biomass, in *Food Webs: Integration of Pattern and Dynamics,* Polis, G.A. and Winemiller, K.O., Eds., Chapman & Hall, New York, 1996, pp. 149–159.

Rowe, J.S., Lighting fires in Saskatchewan grasslands, *Can. Field Nat.,* 83, 317–324, 1969.

Roy, J.F., *Birds of the Elbow,* Special Publication 21, Saskatchewan Natural History Society, Regina, Saskatchewan, 1996.

Salthe, S.N., *Evolving Hierarchical Systems: Their Structure and Representation,* Columbia University Press, New York, 1985.

Samson, F.B. and Knopf, F.L., Eds., *Prairie Conservation: Preserving North America's Most Endangered Ecosystem*, Island Press, Washington, D.C., 1996.

Sauer, J.R. and Droege, S., Geographic patterns in populations trends of Neotropical migrants in North America, in *Ecology and Conservation of Neotropical Migrant Landbirds*, Hagen, J.M., III and Johnston, D.W., Eds., Smithsonian Institution Press, Washington, D.C., 1992, pp. 26–42.

Sauer, J.R., Hines, J.E., Thomas, I., Fallon, J., and Goush, G., The North American Breeding Bird Survey, Results and Analysis 1996–1998, Version 98.1, U.S.G.S. Patuxent Wildlife Research Center, Laurel, MD, 1999.

Schindler, D.W., Beaty, K.W., Fee, E.J., Cruikshank, D.R., DeBruyn, E.R., Findlay, D.L., Linsey, G.A., Shearer, J.A., Stainton, M.P., and Turner, M.A., Effects of climatic warming on lakes of the central boreal forest, *Science*, 250, 1043–1048, 1990.

Schlesinger, W.H., Reynolds, J.F., Cunningham, G.L., Huenneke, L.F., Jarrell, W.M., Virginia, R.A., and Whitford, W.G., Biological feedbacks in global desertification, *Science*, 247, 1043–1048, 1990.

Seastedt, T.R., Soil systems and nutrient cycles of the North American Prairie, in *The Changing Prairie: North American Grasslands*, Joern, A. and Keller, K.H., Eds., Oxford University Press, New York, 1995, pp. 157–174.

Selby, C.J. and Santry, M.J., A National Ecological Framework for Canada: Data Model, Database and Programs, Agriculture and Agri-Food Canada, Research Branch, Center for Land and Biological Resources Research and Environment Canada, State of the Environment Directorate, Ecozone Analysis Branch, Ottawa/Hull, 1996.

Selwyn, A.R.C., Notes on a journey through the North-West Territory, from Manitoba to Rocky Mountain House, *Can. Nat. Q. J. Sci.*, 7, 193–215, 1874.

Sherry, T.W. and Holmes, R.T., Winter habitat quality, population limitation, and conservation of Neotropical-Neoarctic migrant birds, *Ecology*, 77, 36–48, 1996.

Simberloff, D. and Gotelli, N., Effects of insularisation on plant species richness in the prairie-forest ecotone, *Biol. Conserv.*, 29, 27–46, 1984.

Smith, A.R. *Atlas of Saskatchewan Birds*, Special Publication 22, Saskatchewan Natural History Society, Regina, Saskatchewan, 1996.

Symstad, A., Tilman, J.D., Willson, J., and Knops, J.M.H., Species loss and ecosystem functioning: effects of species identity and community composition, *Oikos*, 81, 389–397, 1998.

Tilman, D., Biodiversity: populations versus ecosystem stability, *Ecology*, 77, 350–363, 1996.

Tilman, D., Knops, J., Wedin, D., Reich, P., Ritchie, M., and Siemann, E., The influence of functional diversity and composition on ecosystem processes, *Science*, 277, 1300–1302, 1997.

Valentine, J.W. and D. Jablonski, Fossil communities: compositional variation at many scales, in *Species Diversity in Ecological Communities: Historical and Geographical Perspectives*, Ricklefs, R.E. and Schluter, D., Eds., University of Chicago Press, Chicago, 1983, pp. 341–349.

Vitousek, P.M., Mooney, H.A., Lubchenco, J., and Melillo, J.M., Human domination of the Earth's ecosystems, *Science*, 277, 494–499, 1997.

Wapple, G.J., Nesting turkey vulture near Biggar, Saskatchewan, *Blue Jay*, 43, 44–46, 1985.

Watts, F.G., The natural vegetation of the southern Great Plains of Canada, *Geog. Bull.*, 14, 25–43, 1960.

Wright, H.A. and Bailey, A.W., Fire Ecology and Prescribed Burning in the Great Plains, USDA Forest Service General Technical Report INT-77, Intermountain Forest and Range Experiment Station, U.S. Department of Agriculture, Washington, D.C., 1980.

Breeding Bird Species List

Breeding bird species list showing migration pattern and abundance ranks for the Mixed Grassland Ecoregion (Migration Pattern: SD = Short Distance, LD = Long Distance, and RS = Resident; Abundance Ranks: C = Common, F = Fairly Common, U = Uncommon, R = Rare, and n/p = not present).

Common Name	Migration Class	Abundance Rank	
		Circa 1900	1990
Pied-billed grebe	SD	R	F
Horned grebe	SD	U	F
Eared grebe	SD	C	C
Western grebe	SD	C	F
American white pelican	SD	F	U
Double-crested cormorant	SD	U	U
American bittern	LD	C	R
Great blue heron	SD	U	U
Cattle egret	SD	n/p	R
Black-crowned night heron	SD	n/p	U
Canada goose	SD	C	F
Green-winged teal	SD	U	U
Mallard	SD	C	C
Northern pintail	SD	C	C
Blue-winged teal	LD	C	C
Cinnamon teal	LD	R	R
Northern shoveler	SD	C	C
Gadwall	SD	C	C
American wigeon	SD	F	C
Canvasback	SD	C	F
Redhead	SD	F	U
Lesser scaup	SD	C	F
White-winged scoter	SD	U	R
Ruddy duck	SD	C	U
Turkey vulture	SD	F	U
Northern harrier	SD	F	F
Sharp-shinned hawk	SD	U	U
Cooper's hawk	SD	R	R
Swainson's hawk	LD	C	F
Red-tailed hawk	SD	n/p	U
Ferruginous hawk	SD	C	U
Golden eagle	SD	R	R
American kestrel	SD	U	F

Common Name	Migration Class	Abundance Rank	
		Circa 1900	1990
Merlin	RS	R	F
Prairie falcon	SD	U	R
Peregrine falcon	RS	R	n/p
Gray partridge	RS	n/p	C
Ring-necked pheasant	RS	n/p	R
Sage grouse	RS	F	R
Greater prairie chicken	RS	n/p	R
Sharp-tailed grouse	RS	C	F
Sora	SD	C	F
American coot	SD	C	C
Snowy plover	SD	n/p	R
Piping plover	SD	F	R
Killdeer	SD	C	F
American avocet	SD	C	C
Willet	LD	C	F
Spotted sandpiper	LD	U	U
Upland sandpiper	LD	C	U
Long-billed curlew	LD	C	U
Marbled godwit	SD	C	F
Common snipe	SD	R	U
Wilson's phalarope	LD	C	F
Franklin's gull	LD	C	C
Ring-billed gull	SD	C	F
California gull	SD	F	F
Common tern	LD	F	U
Forster's tern	SD	R	U
Black tern	LD	F	U
Rock dove	RS	n/p	C
Mourning dove	SD	n/p	C
Black-billed cuckoo	LD	R	U
Great horned owl	RS	R	F
Burrowing owl	LD	U	U
Long-eared owl	RS	U	R
Short-eared owl	SD	U	R
Northern saw-whet owl	RS	n/p	R
Common nighthawk	LD	U	U
Common poorwill	SD	n/p	R
Ruby-throated hummingbird	LD	n/p	R
Belted kingfisher	SD	U	U
Red-headed woodpecker	SD	R	R
Northern flicker	RS	R	F
Willow flycatcher	LD	R	U
Least flycatcher	LD	U	F
Say's phoebe	SD	R	U
Western kingbird	LD	R	C
Eastern kingbird	LD	F	C
Horned lark	SD	C	F
Tree swallow	SD	R	F
Violet-green swallow	LD	n/p	R
Northern rough-winged swallow	LD	n/p	R
Bank swallow	LD	U	F
Cliff swallow	LD	F	F
Barn swallow	LD	R	C
Black-billed magpie	RS	U	C
Common crow	SD	U	F
Common raven	RS	U	n/p
Black-capped chickadee	RS	R	U
Rock wren	SD	R	F

Common Name	Migration Class	Abundance Rank	
		Circa 1900	1990
House wren	LD	U	C
Marsh wren	SD	n/p	F
Mountain bluebird	SD	R	U
Veery	LD	U	F
American robin	SD	U	F
Gray catbird	LD	C	F
Northern mockingbird	RS	n/p	R
Sage thrasher	SD	n/p	R
Brown thrasher	SD	R	F
Sprague's pipit	SD	C	F
Cedar waxwing	SD	R	F
Loggerhead shrike	SD	R	F
European starling	SD	n/p	F
Warbling vireo	LD	R	F
Red-eyed vireo	LD	R	U
Yellow warbler	LD	F	C
Common yellowthroat	LD	U	F
Yellow-breasted chat	LD	n/p	U
Lazuli bunting	LD	n/p	U
Spotted towhee	SD	R	U
Chipping sparrow	LD	R	F
Clay-colored sparrow	LD	C	C
Brewer's sparrow	LD	n/p	U
Vesper sparrow	SD	C	C
Lark sparrow	LD	R	F
Lark bunting	LD	C	C
Savannah sparrow	SD	C	C
Baird's sparrow	LD	C	F
Grasshopper sparrow	LD	n/p	R
Le Conte's sparrow	SD	R	R
Song sparrow	SD	U	U
McCown's longspur	SD	C	F
Chestnut-collared longspur	SD	C	F
Bobolink	LD	R	U
Red-winged blackbird	SD	C	C
Western meadowlark	SD	C	F
Yellow-headed blackbird	LD	C	F
Brewer's blackbird	SD	C	F
Common grackle	SD	R	U
Brown-headed cowbird	SD	C	C
Orchard oriole	LD	n/p	R
Baltimore oriole/Bullock's	LD	n/p	F
American goldfinch	SD	R	C
House sparrow	RS	n/p	C

CHAPTER 118

Evaluating Agroecosystem Sustainability Using an Integrated Model

Ken W. Belcher and M. Boehm

INTRODUCTION

The increasing intensity of agricultural production in recent years has resulted in a degradation of natural ecosystems at a range of scales. Soil degradation, groundwater and surface water pollution, destruction of wildlife habitat and biodiversity loss have prompted questions about the sustainability of agroecosystems in recent years (Brown, 1997; World Commission on Environment and Development, 1987). The importance of agroecosystems to social and economic health has driven a demand for policy development and analysis focusing on the long-term sustainability of these systems. Agricultural ecosystems (agroecosystems), distinguished from natural ecosystems by the dominant role of human management for specific marketable products, are an important component of global land use and resource appropriation. Agroecosystems, like natural ecosystems, depend on the fundamental processes, primary production, consumption, and decomposition in combination with abiotic energy flows and nutrient cycling. Within an agroecosystem these processes are inextricably linked with economic and social components through the role of human management. These natural and social processes can be characterized as the flow of goods and services coming from the capital stock of the agroecosystem. The capital stock of an agroecosystem includes man-made, human, and natural capital corresponding to the factors of production, capital, labor, and land (Costanza and Daly, 1992; Prugh, 1995). Man-made capital is the manufactured means of production including machinery, equipment, buildings, technology, etc. (Prugh, 1995). Human capital is the collective knowledge, skills, and culture of the human population. Natural capital is comprised of components that are natural in origin. The maintenance of these capital stocks over time is an important condition for the preservation of the function of these agroecosystems.

While there has been ongoing debate about the characterization of sustainable development, there is some consensus that the concept is based on maintaining capital stocks for future generations (Batie, 1989). Costanza (1996) argues that sustainability is not a definitional problem, but a prediction problem and that it should be characterized as a long-term goal whereby research focuses on policies and conditions that will lead to this goal. The present study evaluates agroecosystem sustainability based on changes in components of systems' capital stock. It is assumed that natural and man-made capital are largely complements in the target system and, as such, sustainability requires that each component of the capital stock is maintained separately. This is consistent with strong sustainability as described by Daly (1994). Based on the strong sustainability framework, the sustainability of an agroecosystem can be evaluated by monitoring changes in the capital stock to determine if the system is on a sustainable path.

To evaluate how a system develops with respect to a long-term goal of sustainability, researchers usually adopt some form of predictive model. To better understand system dynamics, models have traditionally been developed using spatial and temporal boundaries that minimize the interactions between the target system and the rest of the universe. Although this simplification facilitates the development of tractable models, an assumption of insignificant or linear interactions between the target system and its surroundings is necessary. Complex systems are characterized by strong, often nonlinear, interactions between the components, complex feedback loops that obscure the distinction between cause and effect, and significant time and space lags, discontinuities, thresholds, and limits (Costanza et al., 1993). An effective modeling framework for complex systems, therefore, cannot isolate the target system from its context.

The present study integrates economic and environmental components of an agroecosystem within a simulation modeling framework to assess the sustainability of the complex system. The Sustainable Agroecosystem Model (SAM) identifies an agroecosystem as a system that can be represented by economic and environmental subsystems. By integrating these component models, SAM can make explicit the effect that economic development has on the natural capital and, in turn, how changes in the natural capital stock affect man-made capital and future economic decisions. The integrated structure of the SAM model facilitates the simulation of system dynamics whereby land-use decisions in period t will influence land-use decisions in period $t + n$ through changes in components of the production systems' environmental context.

This chapter concentrates on describing the structure of SAM. The following sections provide details of the relationships used to construct the economic and soils and crop components of the modeling framework. The later sections of the chapter present results from SAM simulation runs to illustrate how the model can be used to evaluate agroecosystem sustainability.

THE SUSTAINABLE AGROECOSYSTEM MODEL

SAM is comprised of an economic model that simulates constrained profit maximizing production decisions, and a soils-and-crop-growth model that simulates changes in soil quality and the ability of the soil to provide nutrients and water for crop growth. The dynamic interdependence of the economic and environmental functions of an agroecosystem is made explicit in SAM by linking these component models. An important goal in developing SAM was to create a relatively simple model that would provide information on general system dynamics. Generality comes at the expense of precision (Costanza et al., 1993) but it was felt that models already exist that can provide high levels of precision simulating economic and environmental systems. The structure of the component models and the linkages used to integrate the models are represented in Figure 118.1.

An important characteristic of SAM is the treatment of temporal linkages in the economic and soils submodels. Although production decisions in one time period affect future soil quality – and thus future land-use decisions – the model assumes that farmers are myopic with respect to the effect of their land-use decisions. Although this unequal treatment of the temporal linkages in the soil and economic submodels appears to be inconsistent, there are good reasons to believe that this inconsistency is empirically valid.

From an economic perspective, the assumption that farmers are myopic is introduced by setting the user cost to zero. User cost is defined as the change in the future stream of profits that could be expected due to the changes in soil quality and soil function. Excluding user cost obviously simplifies the modeling process and makes solutions easier to simulate. However, the main reason for leaving out the user cost is because the requirements for its inclusion are quite stringent and unlikely to be fulfilled. For user cost to be included in land-allocation decisions, it must be assumed that (1) farmers are fully aware of the effect of soil quality on current and future production, (2)

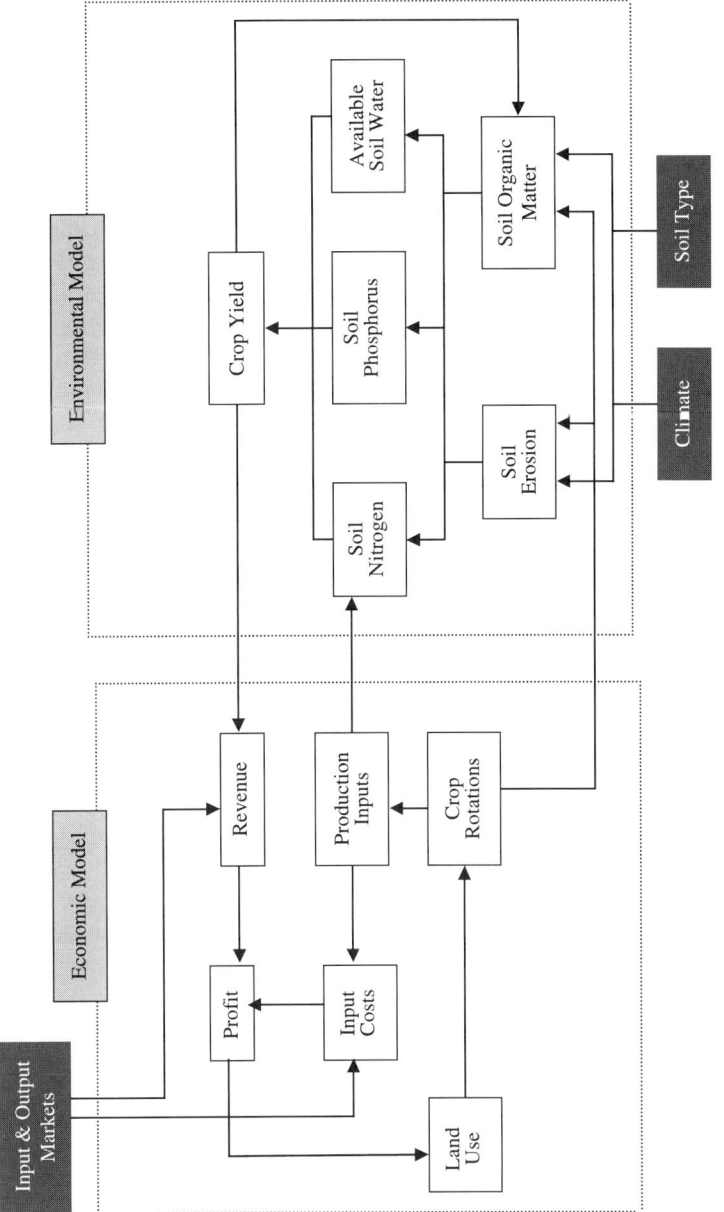

Figure 118.1 Schematic of agroecosystem modeling structure highlighting the important system linkages. Black shaded cells identify exogenous parameters.

land prices effectively reflect the capitalization of future rents, which are dependent on soil quality, and (3) land tenure enables farmers to realize the benefits of maintaining soil quality.

Research has shown that the assumptions required for user cost to be included in land-use decisions are often violated. Van Kooten and Furtan (1987) reported that farmers do not or are not often able to recognize the effect of soil quality on current and future production due to such factors as:

- A lack of knowledge regarding the relationship between soil quality and management practices.
- Technological advances that compensate for soil degradation such that yields are nondecreasing with lower soil quality.
- Fluctuations in interest rates and output prices causing an increase in the producer's discount rate such that the soil resource is allocated to the current generation.
- A nonlinear relationship between soil quality and crop yield.

With respect to the length of planning horizons and land tenure, McConnell (1983) suggests that family-owned land, and particularly rented land, are land tenure arrangements that generally do not have infinite planning horizons. Family-owned farms and rented family farms are the dominant tenure arrangements in Saskatchewan agriculture. In the absence of infinite planning horizons, the recognition of user cost requires that land prices reflect land quality. Clark and co-workers (1993) showed that land price and land rent are not highly correlated, implying that land prices do not reflect the discounted sum of the expected value of future rents. Based on these findings, the assumption of myopic decision making is more likely to reflect the actual behavior of farmers.

Many farm-scale economic models assume that farmers make production decisions using simple profit-maximizing criteria. The economic component of SAM assumes that short-run profit maximizing management decisions are constrained, with inadequate soil water conditions or low previous year profits leading to suboptimal provision of nitrogen fertilizer (see soils-and-crop-growth model, discussed later).

The spatial scale of SAM is an ecodistrict, an area defined by landform, vegetation, and soil that is relatively homogeneous in these characteristics as compared to the surrounding landscape (Acton et al., 1998). It is assumed that the farms within an ecodistrict are also relatively homogeneous in terms of their economic characteristics. An agroecosystem is the agricultural production system and its ecosystem context within a single ecodistrict.

SAM was constructed using the STELLA® modeling software, which facilitates the construction of models using stocks and flows. As such, the model was comprised of stocks such as organic matter, plant residue, and topsoil depth in the soils, and crop growth component and land allocation stocks in the economic component. The simulations are based on a 1-year time step; each simulation comprises 50 years. The results of the simulation are presented as mean and standard deviation values for 50 separate runs of the model. The framework used to construct the economic and soils-and-crop-growth models are now described in detail.

THE ECONOMIC MODEL

The economic component of SAM simulates the allocation of the land input by farms to four annual crop rotations and a perennial tame forage crop. The modeled rotations include summer fallow, a common management practice in southwestern Saskatchewan. In a summer fallow year, the land is idled from crop production to store water and control weeds with tillage or chemicals. The four annual crop rotations are (1) wheat–fallow (WF), (2) wheat–wheat–fallow (WWF), (3) wheat–canola–peas (WCP), and (4) wheat–fallow–peas (WFP). Native land (native grass, shrub, bush, and wetlands) is included in the model as a separate stock that provides no economic returns to the farm. However, native land may be converted to cultivated land when it is profitable to do so.

The central part of the economic model focuses on the allocation of land to the four annual crop rotations. Mathematically, the economic model incorporates the following production function:

$$x_j^t = f(g_j^t, l_j^t) \tag{118.1}$$

where x_j^t represents the output quantity from rotation j in time period t; g_j^t denotes a vector of variable inputs (nitrogen fertilizer, soil nutrients, soil water, etc.) for rotation j in time period t; and l_j^t denotes the land input for rotation j in time period t. In each time period, the stock of land available for cultivation is allocated to the rotations, based on their relative profitability.

The land input, in turn, is assumed to be used in fixed proportions with chemical inputs, phosphorus (P) fertilizer input, and "other inputs." The model is calibrated such that one unit of land is 1 ha, one unit of output is the quantity produced on 1 ha, and one unit of the chemical, phosphorus, and other inputs is the quantity required to produce one unit of output on 1 ha of land. The per-hectare cost of the chemical, phosphorus, and other inputs are distinct for each rotation.

Farmers are assumed to be price takers in both the output and input markets, and output and input prices are fixed throughout the simulations. Farmers are also assumed to allocate land among rotations so as to maximize constrained short-run profits. The profit for each rotation is the difference between rotation revenue and rotation costs. Rotation revenue is calculated at each time step using fixed output prices and crop yield. Crop yields are calculated by the soils-and-crop-growth model and is discussed later. Rotation revenue is represented by average revenue per hectare. For example, gross revenue for the WFP rotation at time $t(R_{wfp}^t)$ is calculated:

$$R_{wfp}^t = \frac{(p_w \times y_{wwfp}^t) + (p_p \times y_{pwfp}^t) + 0}{3} \tag{118.2}$$

where p_w and p_p are the output prices for wheat and peas, respectively; and y_{wwfp}^t and y_{pwfp}^t represent the yield at time t for wheat and peas, respectively. Note that revenue during the fallow year is zero.

Rotation costs include the cost of chemicals, fertilizer, and other inputs. The chemical input includes herbicides, insecticides, and fungicides. Chemical costs in the model reflect the average cost per hectare in each rotation with chemical quantities based on recommended rates for the specific crops in the specific region. For example, the chemical input cost for the WFP rotation (C_{chwfp}) is calculated:

$$C_{chwfp} = \frac{c_{chsw} + c_{chf} + c_{chfp}}{3} \tag{118.3}$$

where c_{chsw}, c_{chf}, c_{chfp} are, respectively, the per hectare chemical costs for wheat seeded into land that was cropped the previous year (stubble wheat), fallow, and peas seeded into land that was fallow the previous year (fallow peas).

The cost of fertilizer for an average hectare in each rotation (C_{fj}^t) is calculated as the sum of phosphorus and nitrogen costs. Phosphorus input cost represents the cost for an average hectare in the rotation and is calculated using the same format as described for the chemical input (Equation 118.3). Phosphorus cost is fixed in the simulation based on recommended application rates for the modeled crops. The quantity of nitrogen fertilizer used in each time step is determined endogenously in the model and is a function of soil water and soil nitrogen stocks and the previous year's profit. Therefore, the quantity of nitrogen fertilizer applied is not necessarily the profit maximizing rate. This procedure is described in detail in a later section.

The other inputs category captures those inputs not represented by the land, fertilizer, and chemical input markets. Other inputs is an index for machinery, fuel, buildings, and management skill. The price of other inputs is determined endogenously using a separate supply curve for each rotation. The development of a separate supply curve for each rotation means that other inputs are assumed to be rotation specific. These supply curves are upward sloping to reflect the increasing marginal costs associated with these inputs, particularly management.

The supply relationships for other inputs were assumed to be linear:

$$Q^t_{oj} = \alpha_{oj} + \beta_{oj}P^t_{oj} \tag{118.4}$$

where P^t_{oj} denotes the per hectare price of other inputs for rotation j at time t; Q^t_{oj} is the quantity of other inputs for output j at time t; $\beta_{oj} = E_{oj}(Q^t_{oj}/P^t_{oj})$, where E_{oj} is the elasticity of supply of input j; and $\alpha_{oj} = Q^t_{oj} - \beta_{oj}P^t_{oj}$.

To derive the other input supply functions, the initial other-input price, initial quantity, and elasticity values are used to derive the parameters β_{oj} and α_{oj}. Some important assumptions were made about the target agroecosystem to derive these values. The initial ($t = 0$) quantity of land assigned to each of the rotations (Q^0_{lj}) in the target agroecosystem is based on recent land-use statistics. Assuming that other inputs are used in fixed proportions with the land input, the initial quantity of other inputs is derived ($Q^0_{oj} = Q^0_{lj}$).

It is further assumed that the initial land allocation represents an economic equilibrium so that total costs equal total revenue:

$$P^0_{oj} = R^0_j - (P^0_l + C_{chj} + C^0_{fl}) \tag{118.5}$$

where P^0_{oj} represents the initial other-input price; P^0_l represents the initial input cost of land or land rent; and C^t_{fl} represents the cost of fertilizer. The assumption of zero profits in Equation 118.5 implies that there are no excess profits in farming. The return to land is determined by land rent, while the return to management is included in the price of other inputs.

The other input supply elasticities were derived by noting that the own-price and cross-price area response elasticities for wheat, peas, and canola can be written as functions of the other, input supply elasticities, which were selected to provide a wheat own-price supply elasticity of 0.5, a value consistent with literature value estimates (see Clark and Klein, 1992; Schmitz, 1968; for estimates of supply elasticities). Given the absence of own-price elasticities estimates for peas and canola in the literature, these elasticities were simply constrained to be positive. The cross-price elasticities were not constrained to be negative (as would normally be the case), because individual crops in the model are tied to specific rotations. For example, an increase in the price of peas may not result in a decrease in the area of canola because canola and pea areas are linked within the WCP rotation. Using this procedure the other input supply elasticities were derived (E_{oj}) (Table 118.1).

The economic model requires the inverse form of the other input supply function. The inverse function is written as follows:

$$P^t_{oj} = \delta_{oj} + \omega_{oj}Q^t_{oj} \tag{118.6}$$

where $\delta_{oj} = -a_o/\beta_{oj}$ and $\omega_{oj} = 1/\beta_{oj}$.

The revenues and costs outlined previously can be used to derive the inverse demand for land for each rotation. The demand for land for each rotation is a derived demand that reflects the amount a farmer is willing to pay for a hectare of land, given the revenues received and the total costs of nonland inputs required to produce the output on that hectare of land. The assumption is that if

Table 118.1 Simulated "Other Input" Supply Elasticities for Modeled Rotations in the Brown Soil Zone

Crop Rotation	Other Input Supply Elasticity
Wheat–fallow	1.5
Wheat–wheat– fallow	3.0
Wheat–canola–peas	5.0
Wheat–fallow–peas	17.0

revenues exceed costs for a particular rotation, farmers will shift land into that rotation. Similarly, if costs exceed revenues in a rotation, farmers will shift land out of that rotation. This shifting land use captures farmers' desire to increase their profits. Land use will be in economic equilibrium when gross revenues for a rotation equal the costs associated with that rotation.

The inverse demand for land is determined by equating gross revenues in period $t(R_j^t)$ with total costs in period $t(P_l^t + P_{oj}^t + C_{chj}^t + C_{fl}^t)$, substituting Equation 118.6 for P_{oj}^t and solving for the common land rent (P_l^t) as a function of the quantity of land (recall that with fixed proportions, the quantity of land equals the quantity of other inputs, $Q_{oj}^t = Q_{lj}^t$). The linear inverse demand curve for land for rotation j is

$$P_l^t - \eta_{lj}^t - \omega_{lj} Q_{lj}^t \qquad (118.7)$$

where $\eta_{lj}^t = R_j^t - (\delta_{oj} + C_{chj}^t + C_{fl}^t)$ and $\omega_{lj} = \omega_{oj}$.

Inverting Equation 118.7 gives the standard demand curve for land for each rotation:

$$Q_{lj}^t = \rho_{lj}^t - \theta_{lj} P_l^t \qquad (118.8)$$

where $\rho_{lj}^t = \eta_{lj}^t / \omega_{lj}$ and $\theta_{lj}^t = 1/\lambda_{lj}$.

The total demand for all land dedicated to annual crop production is obtained by summing the demand for land for each rotation:

$$Q_l^t = \sum_j Q_{lj}^t = \rho_l^t - \theta_l P_l^t \qquad (118.9)$$

where $\rho_l^t = \Sigma_j \rho_{lj}^t$ and $\theta_t = \Sigma_j \theta_{lj}$.

Within SAM, it is assumed that all land in annual cultivation is allocated to the four annual crop rotations represented in the model. This assumption implies that the supply of annually cultivated land ($\overline{Q}_l^t$) at any given point in time is fixed. The quantity parameter is expressed with a time superscript to indicate that the stock of annually cultivated land will increase with conversion of perennially vegetated land (tame forage and native land) to annual cultivation, or decrease with conversion of annual cultivation to perennial tame forage.

The market clearing land rent, is derived by setting Equation 118.9 equal to $\overline{Q}_l^t$ and solving for P_l^t:

$$P_l^t = \frac{\rho_l^t - \overline{Q}_l^t}{\theta_l} \qquad (118.10)$$

Substituting P_l^t into the derived land demand functions (Equation 118.7) provides the constrained profit-maximizing equilibrium quantity of land allocated to each of the rotations at a point in time.

Agroecosystems are generally comprised of a variety of land uses including annually cultivated land, land dedicated to some form of tame perennial vegetation, and native land (including native range, native grassland, shrubs, bush, and wetlands). The stock of annually cultivated land ($\overline{Q}_l^t$) is altered by the conversion of land from perennial cover to annual cultivation, or from annual cultivation to perennial forage.

Within SAM, land is allocated to the use (annual cultivation or perennial forage) that provides the highest potential economic returns to that land at each point in time. Land rent (P_l^t), as calculated in Equation 118.6, represents the opportunity cost of allocating land to perennial crops. Tame forage net revenue (NR_f^t) is calculated by the model, at each time step, using a fixed hay price, forage conversion costs, forage production costs, and hay yield. Hay yield is a function of precipitation based on the following relationship:

$$y_h^t = \phi + \mu(gs^t) + \tau(ws^t) \tag{118.11}$$

where y_h^t denotes tame hay yield; gs^t, ws^t denote growing season and nongrowing season precipitation, respectively; and ϕ, μ, and τ denote empirically derived coefficients. The coefficients in Equation 118.11 were statistically derived using the GRASSGRO® pasture growth model (Cohen et al., 1995) (Table 118.2). Coefficients were derived for different climatic and soil conditions to reflect the specific biophysical conditions of the target landscapes.

The process used by SAM to simulate the dynamics of land conversion between tame forage and annual crop is depicted in Figure 118.2. The economic threshold for converting land from forage to annual crop or breakup threshold, and the threshold for converting annual crop to perennial forage or set-aside threshold, are labeled, BUT and SAT, respectively. The line labeled D^t represents the difference between land rent (P_l^t) and forage net revenue in a given year (NR_f^t), i.e., $D^t = P_l^t - NR_f^t$. The zero line is where annual crop and forage net revenues are equal. If it is assumed that initially ($t = 0$) land is allocated to forage production, when D^t crosses the BUT ($t = X$), land is converted to annual crop production. This land remains in annual crop production until D^t crosses the SAT ($t = Y$), at which time the land is converted to forage production. Figure 118.2 shows how conversion costs buffer land-use shifts. Land remains in either annual crop production or forage production for a number of periods despite significant decreases in relative profitability.

The BUT and SAT values discussed above were calculated using Saskatchewan hay and grain net income data, and provincial grain and tame forage areas for the period 1971 to 1996 (Saskatchewan Agriculture and Food, 1996) and the conversion threshold calculation procedure described by Gray and co-workers (1993). The BUT and SAT are distinct due to the large costs associated with converting land from annual crops to forage.

Within SAM, it is assumed that cultivated land cannot be converted to native land. Thus, the stock of native land is nonincreasing during the simulations. Native land is converted to annual crop production when the opportunity cost of land, represented by land rent in the model, is greater than the cost of converting native land. The procedure used to describe the native land supply function

Table 118.2 Empirically Derived Hay Yield Coefficients for Brown Clay (Eston Plain), Loam (Gull Lake Plain), and Sandy Loam (Beechy Hills and Antelope Creek Plain) Soil Textures in the Study Area

Coefficient	Brown Clay	Brown Loam	Brown Sandy Loam
ϕ	1.16 (2.71)[a]	1.31 (2.38)	0.48 (1.20)
μ	0.04 (2.50)	0.03 (1.53)	0.08 (5.05)
τ	0.05 (2.94)	0.06 (3.01)	0.04 (2.40)
R-squared	0.29	0.24	0.46

[a] Values in parentheses are the *t* statistics for the coefficients.

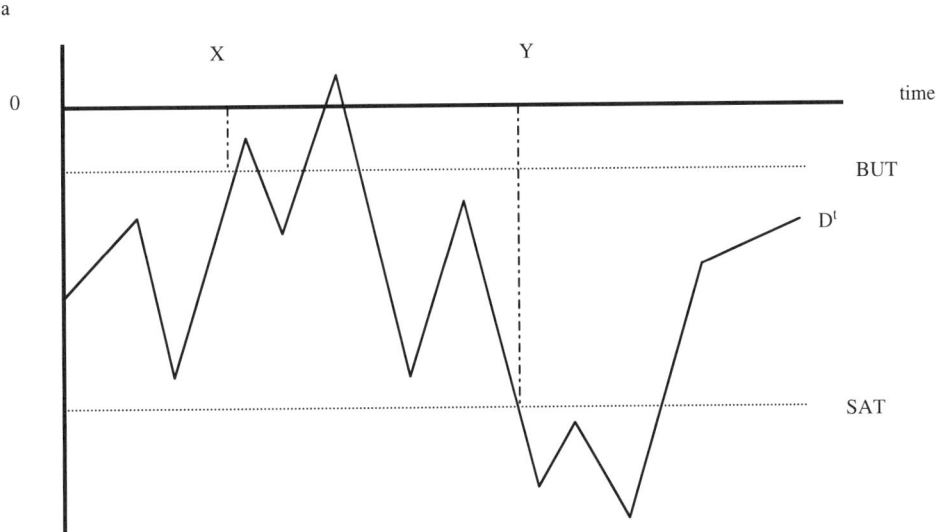

Figure 118.2 Economic thresholds for conversion of land between perennial forage and annual crop production. D^t represents the difference between annual crop and forage net revenue at a point in time (annual crop net revenue^t – forage net revenue^t = D^t).

was informed by a framework developed by Van Kooten (1993). It was assumed that native land conversion costs are a function of the stock of native land. The native land supply function reflects the condition that as native land is converted to cultivation — e.g., the stock of native land decreases — the average quality of the native land stock decreases and marginal conversion cost increases.

SOILS-AND-CROP-GROWTH MODEL

The soils-and-crop-growth component of SAM is based on the SIMPle model developed by Greer and co-workers (1992). This part of SAM simulates crop production as a function of climate and soil quality, with soil quality being influenced in turn by management in previous years. The model simulates wheat, canola, peas, and summer fallow management in context of the WF, WWF, WFP, and WCP rotations.

The model was parameterized using ecodistrict-specific biophysical data: (1) soil type — texture, thickness, initial organic matter content; (2) climate — growing season precipitation, nongrowing season precipitation, growing degree days, mean daily temperature; and (3) initial soil stocks — surface residue, organic matter carbon, nitrogen, phosphorus. The sectors that comprise the soils-and-crop-growth model are now presented in detail.

Crop yields are simulated in the model based on the law of the minimum, whereby the maximum potential yield is achieved only if all plant growth requirements are optimal. It was assumed that annual crop yields are limited by available water, nitrogen, and phosphorus, such that one of these factors will be the binding constraint:

$$y^t_{crj} = y_{cr\,max} \times ys^t_{rcrj} \times ys^t_{ncrj} \times ys^t_{pcrj} \qquad (118.12)$$

where y^t_{crj} denotes the yield for crop cr in rotation j at time t; $y_{cr\,max}$ denotes the maximum potential yield for crop cr; and ys^t_{rcrj}, ys^t_{ncrj}, ys^t_{pcrj} represent crop input sufficiencies, or the proportion of maximum yield $(y_{j\,max})$ possible, given available water, nitrogen, and phosphorus, respectively. These sufficiency values fall between zero and one, with a value of one reflecting conditions where available water, nitrogen, and phosphorus are nonconstraining. The yield equation was parameterized using empirically derived boundary conditions of scatter plot relationships between annual crop yield and water, nitrogen, and phosphorus availability. The remainder of this section discusses how SAM simulates the provision of the crop growth parameters by the soil.

Growing season precipitation and snow melt equivalent precipitation (nongrowing season precipitation) provide the soil water that will potentially be available for crop growth. The proportion of total precipitation that is available to the crop in each time step is determined by the infiltration rate, recharge rate, and the water storage capacity of the soil. It is assumed that approximately 50% of the overwinter water enters cropland soil. Soil water storage capacity determines the water available for crop growth. The water storage capacity of the soil is a function of soil texture (percent clay) and the organic carbon content of the soil and is consistent with the universal soil loss equation (Wischmeier and Smith, 1978).

Soil organic matter carbon (SOMC) dynamics reflect the difference between the rate of organic matter addition to the soil as crop residues and the rate of organic matter loss as CO_2 through microbial decomposition. Additions of crop residue are modeled as a function of grain yield and the type of residue (Douglas and Rickman, 1992). More specifically, crop residues are determined as follows:

$$CR_{cr}^t = y_{cr}^t \times CA_{cr} \times HI_{cr} \tag{118.13}$$

where CR_{cr}^t denotes the quantity of crop residue added by crop cr in time period t; y_{cr}^t is the yield of crop cr; CA_{cr} is the carbon content of the residue for crop cr (%); and HI_{cr} is the harvest index ratio of grain to total biomass for crop cr.

The rate of crop residue decomposition (Decomp^t) is a function of the nitrogen content of the residue, soil water, and growing degree days (Douglas and Rickman, 1992). Decomposition is calculated using the following relationship:

$$\text{Decomp}^t = Rd^t \times (\exp(NC_{cr} \times W \times k \times \text{GDD})) \tag{118.14}$$

where Rd^t is the stock of crop residue at time t; NC_{cr} represents the nitrogen content of plant residue from crop cr; W is a scalar tillage factor (summer fallow = 0.3, crop = 0.2); k is a constant (-0.0004); and GDD is growing degree days, a measure of solar energy input. Increasing the frequency of tillage increases the rate of decomposition by increasing soil temperature, soil aeration, and available nitrogen (Hendrix et al., 1988). Soil organic matter carbon stocks will also decrease under summerfallow management primarily due to the lack of crop residues being added during the fallow year (Tiessen et al., 1982).

The amount of nitrogen available to a crop is soil N plus fertilizer N. Soil N is mineralized from soil organic matter through microbial decomposition. It is assumed that SOMC has a carbon to nitrogen ratio of ~10:1. Based on this assumption, available nitrogen (Avail.N) is calculated:

$$\text{Avail.N}^t = N_{\min} \times \left(\frac{\text{SOMC}^t}{10}\right) \tag{118.15}$$

where $N_{\min}$ is the rate of nitrogen mineralization; and SOMC^t is the stock of soil organic matter carbon at time t. The nitrogen mineralization rate varies with soil thickness, soil temperature, and soil moisture. For example, processes that decrease soil thickness (e.g., erosion) cause a loss of soil organic matter and a decrease in soil nitrogen. In contrast, production practices that increase the SOMC stock, soil temperature, or soil moisture will increase the rate of nitrogen release from the soil organic matter. In SAM, soil nitrogen is made available through the production of nitrogen fixing crops such as peas.

It is assumed that nitrogen fertilizer is available to the crop only in the year of application. An upper limit on the quantity of nitrogen fertilizer permitted at each time step is imposed by assuming that soil N plus fertilizer N cannot exceed 100% sufficiency for each crop. Nitrogen fertilizer rates are further adjusted based on previous year profits for the particular rotation, and available soil water at the beginning of the growing season. For example, the quantity of nitrogen applied would be scaled back from optimum rates if spring soil moisture conditions were less than 50% of average or the previous year revenues less than average. These relationships were developed based on the assumption that low net revenue or poor soil moisture conditions will cause farmers to be less willing to invest in fertilizer inputs.

The quantity of available soil phosphorus (Pa_j^t) is determined by the rate of phosphorus release from mineral, organic, and fertilizer sources. Mathematically, available soil phosphorus is calculated:

$$Pa_j^t = (Po^t + Pi^t + (Pf_{cr} \times 0.25)) \times Pu_{cr} \qquad (118.16)$$

where Po^t represents the mineralization rate of organic phosphorus; Pi^t is the mineralization rate of inorganic phosphorus; Pf_{cr} represents the quantity of phosphorus fertilizer provided for crop cr in the economic model; and Pu_{cr} represents the rate of phosphorus uptake by the crop. Organic P occurs in soil organic matter with a C:N:P ratio of 100:10:1. Inorganic phosphorus is determined by soil clay content and, as such, is influenced in the model by the soil texture. The proportion of fertilizer phosphorus that becomes available is based on the assumption that only 25% of fertilizer phosphorus is available for plant growth (Greer et al., 1992).

Erosion of soil by wind, water, and tillage is assumed to remove a fixed quantity of top soil in each time step: (1) WF, 10 t/ha/year; (2) WWF, 5 t/ha/year; (3) WCP, 0 t/ha/year; (4) WFP, 5 t/ha/year (Monreal et al., 1995). These erosion rates provide an estimate of the level of background erosion associated with management regimes (tillage intensity and frequency) and biophysical properties (soil type and climate).

SUSTAINABILITY INDICATORS

The sustainability of a complex system can be evaluated effectively only by monitoring changes across the system, including the economic and biophysical components. SAM produces a range of output parameters that reflect the condition of the agroecosystem under examination. The following parameters are the sustainability indicators used in this chapter:

1. SOMC: SOMC is an important indicator of soil quality and soil function due to its role in soil nutrient availability, soil structure, water-holding capacity, and soil microbial ecology (Karlen et al., 1997). SOMC is responsive to tillage, residue management, cropping intensity, and nitrogen and phosphorus fertilization.
2. Carbon dioxide emissions: CO_2 is an important greenhouse gas. CO_2 emissions in SAM originate primarily from decomposition of soil organic matter, with smaller contributions from direct and indirect fossil fuel use (e.g., input embodied carbon).
3. Crop yield: Annual crop yield is an integrative indicator reflecting the ability of the soil to provide water and nutrients for crop growth. The ability of the economic component of the agroecosystem to maintain or improve soil function through nitrogen fertilizer application, soil management, and rotation choice is revealed through crop yield.
4. Land rent: Economic returns to land indicate whether the farming system is economically viable. Lower land rent values generally imply a lower probability of economic sustainability. The standard deviation of land rent provides insight into the economic risk of the farming systems.
5. Land use: The allocation of land to annual and perennial cropping strategies provides a landscape-scale sustainability indicator. An agroecosystem that is dominated by a single crop is assumed to be less economically and environmentally sustainable than one that has a more diverse landscape.

SIMULATING AGROECOSYSTEMS

The target area of the present research lies within the 15,700 km² crop district 3BN in the Mixed Grassland Ecoregion of southwestern Saskatchewan. Crop district 3BN was the primary study area of the Prairie Ecosystem Study (PECOS). This semiarid region has a severe moisture limitation to annual crop production with evapotranspiration rates double the total annual precipitation received.

Four ecodistricts within crop district 3BN are the focus of this research: Eston Plain, Beechy Hills, Antelope Creek Plain, and Gull Lake. Soils in the study area range from predominantly Brown or Vertisol soils of clay texture in the Eston Plain ecodistrict, to a mixture of Brown loam and sandy loam in the Gull Lake Plain ecodistrict. Other biophysical characteristics of the four target ecodistricts are presented in Table 118.3.

The wheat–summer fallow production system is dominant in this area, with wheat being produced on approximately 45% of the annual cropland, and summer fallow comprising another 40% of this land (Statistics Canada, 1997) (Table 118.4). The data presented in Tables 118.3 and 118.4 are used to parameterize the soils and crop growth and economic component models. Growing season and nongrowing season precipitation at each time step is randomly selected from gamma distributions that were developed based on the statistical characteristics of the appropriate precipitation data. It should be noted that it was assumed that precipitation patterns were unaffected by climate change.

Table 118.3 Biophysical and Climatic Characteristics of Target Ecodistricts

Parameter	Eston Plain	Beechy Hills	Antelope Creek Plain	Gull Lake Plain
Soil texture	Brown clay	Brown loam	Brown loam	Brown loam/ sandy loam
Growing season precipitation (cm)[a]	19	18	19	19
Nongrowing season precipitation (cm)[b]	15	14	17	18
Growing degree days	1,479	1,507	1,459	1,397
Initial solum (cm)[c]	60	60	60	60
Initial SOMC (t/ha)[d]	50,000	50,000	50,000	50,000
Initial surface trash (kg/ha)[e]	4,200	4,200	4,200	4,200

[a] Total precipitation received between May and July.
[b] Total snowmelt equivalent precipitation received between August and April.
[c] Total depth of the A and B horizons of the soil, specific to soil zone.
[d] Organic matter carbon content of the soil to depth of solum, specific to soil zone.
[e] Quantity of residue carbon on the surface of the soil, specific to soil zone.

Table 118.4 Initial Land-Use Statistics (ha) for the Four Target Ecodistricts

Land Use	Eston Plain	Beechy Hills	Antelope Creek Plain	Gull Lake Plain
Annual crop	370,604	91,104	101,481	42,291
Summerfallow	272,442	46,039	70,168	26,095
Tame hay	4,379	4,295	6,496	5,005
Improved pasture	11,714	13,383	11,998	13,339
Native pasture	108,822	58,453	35,444	41,181
Other land[a]	15,614	4,726	4,973	4,002
Total farmland	783,575	218,000	230,560	131,913

[a] Other land includes farm yards including buildings, shelterbelts, fencerows, ditches, wetlands, shrubland, and woodland.

Source: Statistics Canada, 1996.

Production cost data for an average 660-ha farm in the Brown soil zone of southwestern Saskatchewan (Saskatchewan Agriculture and Food, 1998) and 1998 output prices were used to parameterize the supply-and-demand relationships of the economic component of SAM. An important assumption required to set up these relationships, as discussed earlier, is that the most-recent ecodistrict land-use statistics (Statistics Canada, 1996) represent an equilibrium position with respect to land allocation.

AGROECOSYSTEM SIMULATIONS

The results discussed in this chapter represent two categories of simulations of the target agroecosystems by SAM: (1) baseline scenario, 50-year simulations for the four target ecodistricts to highlight the responsiveness of the model to a range of climatic and physical characteristics; and (2) carbon emissions policy scenarios, 50-year simulations of a single ecodistrict under a carbon credit policy.

BASELINE SIMULATIONS

The baseline simulations provide insight into changes in the sustainability of the target agroecosystems, based on the selected indicators, in two dimensions: (1) within the ecodistrict, calibrated with current biological, physical, and social parameters; and (2) between ecodistricts, each with distinctly different biological, physical, and social parameters.

As discussed earlier, the allocation of land to different crops and management schemes is an important driver of agroecosystem change at a range of scales. The baseline simulations indicated that these management decisions are set in context of the biophysical constraints specific to the production system. Annual crop yields were higher, with lower variability, in the Eston Plain compared to the other ecodistricts. The Eston Plain has clay-textured soils, adequate precipitation, and a less-severe temperature constraint. The simulated yields of the Eston Plain result in higher and less-variable revenues for the annual crop rotations, and therefore, higher and less-variable economic returns to land (Table 118.5). In contrast, the less clayey textured soils, lower rainfall or greater temperature stress of the remaining ecodistricts resulted in lower and more-variable yields, which cause lower revenues and less-favorable economic conditions.

The economic signals simulated by SAM drove the land-allocation decisions over the course of the simulations. In the Eston Plain, the area of land allocated to each of the land-use categories was relatively stable over the course of the simulations (Figure 118.3). In contrast, land in the Beechy Hills and Gull Lake Plain generally shifted to WF and WWF with virtually no land being allocated to the WCP and WFP rotations. In addition, the area of native and tame forage declined. It should be noted that because the Eston Plain ecodistrict had the smallest initial stock of native land of the four ecodistricts, a relatively inelastic segment of the native land supply curve influenced land-allocation decisions. This resulted in smaller decreases in the native land stock than may be expected given the high opportunity cost of land in this landscape.

Table 118.5 Simulated Mean and Standard Deviation of Land Rent ($/ha), Averaged over 50-Year Simulation

| | Ecodistrict | | | |
	Eston Plain	Beechy Hills	Antelope Creek	Gull Lake
Mean	30.96	24.13	30.60	23.39
Standard deviation	51.15	65.73	61.38	56.01

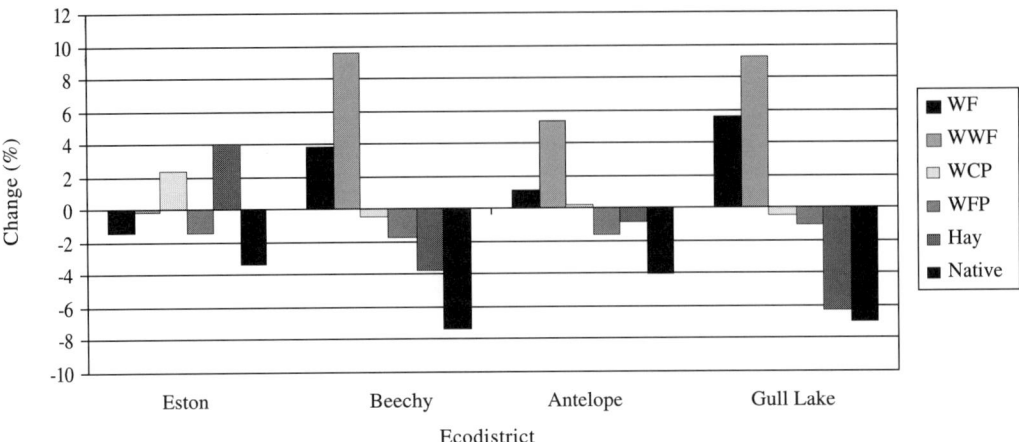

Figure 118.3 Simulated change in land allocation in the four ecodistricts over the 50-year simulations.

Within SAM, annual crop yields are a function of soil quality, as determined by physical and climatic conditions, and soil management practices. Therefore, soil quality changes, as influenced by yield and revenues, are an important factor determining land use. An important positive feedback captured by the model is as follows: higher yields result in larger additions of plant residues to the soil, which are allocated to the SOMC stock or released to the atmosphere as CO_2 through decomposition. Greater stocks of SOMC result in increases in available soil water, soil nitrogen, and soil phosphorus, thereby providing conditions for greater yields. In addition, the greater availability of organic nitrogen decreases the requirement for nitrogen fertilizer, thereby decreasing input costs.

The SOMC–yield feedback relationship resulted in a "rich get richer, poor get poorer" scenario with increasing yields and improved economic indicators in the ecodistricts with minor biophysical constraints (Eston Plain), and decreasing yields and decreasing economic parameters in the eco-districts with greater biophysical constraints (Beechy Hills, Gull Lake Plain) (Table 118.6). In the Beechy Hills and Gull Lake Plain ecodistricts, the lower inherent crop yields resulted in smaller additions to the SOMC stock, thereby decreasing soil quality and future yield. In addition, the combination of low revenues and limited soil water, as determined by inherent soil texture and soil quality attributes, provided little incentive to increase nitrogen fertilizer inputs, further limiting future yields. The land-use effect in these ecodistricts was to allocate more land to rotations that provide more-consistent yields such as WF and WWF. These rotations facilitate higher yields under poor water conditions due to the presence of wheat, a relatively drought-tolerant crop, and summer fallow, which increases soil water stocks. This land allocation limited the increases in SOMC stocks due to greater decomposition rates and greater erosion rates associated with the more-frequent fallow management on lighter textured soils.

Based on the strong sustainability criteria, a synthesis of the indicators discussed facilitates a system-based view of the relative sustainability of the target ecodistricts. The economic indicators simulated for the Eston Plain suggest that components of the man-made capital stock of this system

Table 118.6 Simulated Changes in Soil Indicators over Initial Value for an Average, Annually Cropped Hectare

	Ecodistrict			
	Eston Plain	**Beechy Hills**	**Antelope Creek**	**Gull Lake**
SOMC (kg/ha)	4947.18	2799.67	3968.79	3383.16
N fertilizer (kg/ha)	1503.72	1559.46	1567.59	1569.83
Soil CO_2 (kg/ha)	75,697.59	52,150.06	58,214.48	51,326.74

Table 118.7 Simulated Carbon Balance over the 50-Year Simulations per Average Hectare of Annual Crop Plus Tame Forage Land (Mt/50-year)

Carbon Balance	Eston	Beechy	Antelope	Gull Lake
Input-embodied C	10.27	9.41	9.08	8.61
Soil CO_2 emissions	70.63	49.15	52.57	44.86
Total CO_2 emissions	80.90	58.56	61.65	53.47
Total C stored	6.44	4.98	6.73	7.01
Net CO_2/ha emissions	74.46	53.58	54.92	46.46

would be maintained. In addition, the soil natural capital would be maintained or improved. However, simulation results also indicate that relatively large levels of CO_2 are being emitted from the Eston Plain agroecosystem, primarily due to decomposition of large SOMC stocks in annually cropped land, and greater levels of input-embodied carbon. This result may be interpreted as indicating an unsustainable path with respect to climate change contributions (Table 118.7).

In contrast to the Eston Plain results, the biophysical characteristics of the Beechy Hills and Gull Lake Plain restricted the viable annual crop production options to WF and WWF. As a result, SOMC stocks were smaller and land rent values were approximately 25% lower than the other two ecodistricts. These indicators suggest a lower probability of maintaining the man-made and soil natural capital stocks. The Antelope Creek Plain ecodistrict produced indicators that fell between the extremes presented in the other ecodistricts.

In general, the simulations indicated that the WF and WWF rotations were the only economically viable rotations in all of the ecodistricts, with 61 to 74% of the landscapes being allocated to these rotations. These low diversity landscapes, created by the dominance of wheat and summer fallow, will likely have lower levels of economic and ecological resilience, and, therefore, a lower probability of agroecosystem sustainability. One of the important policy implications of these simulation results is that any menu of incentives or disincentives aimed at moving these ecosystems toward sustainability must take into account the production options available to the farmers, as influenced by the specific biophysical constraints. For example, a set of policies forcing farmers to adopt management practices that disturb the soil less (i.e., continuous cropping, perennial crops) may result in a very different set of incentives to farmers in Eston Plain compared to the Gull Lake Plain.

CARBON POLICY SCENARIOS

Increases in atmospheric greenhouse gas concentrations and the associated concerns about global climate change have stimulated such global initiatives as the United Nations Framework Convention on Climate Change in 1992, and the Kyoto Protocol in 1997. These agreements focus on "stabilizing atmospheric concentrations of greenhouse gases at a level that will prevent dangerous anthropogenic interference with the climate system" (Bruce et al., 1998). Agriculture has been identified as both an important source of greenhouse gases and potential sink for atmospheric carbon. As a result, policy tools, such as government or industry paying landowners for soil carbon sequestration services, may be an important strategy in initiatives addressing climate change.

The relative sustainability of the Antelope Creek ecodistrict under the influence of a carbon credit policy is evaluated using a series of simulations. A carbon credit serves as a subsidy to those land-use practices that sequester carbon in the soil. In this analysis, it was assumed that a carbon credit applies only to the incremental carbon sequestered in annual cropland and tame forage land, while native vegetated soils are assumed to be in equilibrium with respect to organic carbon. Simulations were carried out using carbon credit levels of 25, 75, and $125 per tonne (Mg) of carbon to discern the effect of varying incentive levels on the capital stock indicators.

The SAM simulation results indicated that a carbon credit increased the quantity of land allocated to those land uses that sequester more carbon. Tame forage has a greater rate of carbon sequestration

Table 118.8 Simulated Percentage Change in Land Use with Different Levels of Carbon Credit

Scenario	Land-Use Category (% change)					
	WCP	WFP	WF	WWF	Hay	Native
Baseline (ha)	3479.00	1083.00	93,161.00	78,429.00	15,875.00	39,139.00
$25 credit	53.39	−62.10	−8.52	−7.60	87.91	−2.33
$75 credit	−32.82	−99.77	−30.35	−22.00	340.50	−5.07
$125 credit	−47.02	−100.00	−49.24	−33.48	525.38	−8.73

than any of the modeled annual crop rotations. As a result, the implementation of a carbon credit resulted in land being allocated from annual crop to tame forage (Table 118.8). In addition, there was a relative shift in annual crop land allocation toward the WWF rotation as carbon credit levels increased. At higher carbon credit levels, the marginal productivity of the nitrogen fertilizer input, in terms of soil carbon increments, was greater with WF and WWF, which are less constrained by water than the WCP or WFP rotations. As a result, revenues for WF and WWF increased at a relatively greater rate than the other annual crop rotations with carbon credit. Nitrogen fertilizer inputs in these rotations increased the marginal change in soil carbon which translates into a revenue-optimizing strategy in this scenario. In general, the WWF rotation was more positively influenced by the SOMC yield feedback process, with larger SOMC stocks providing greater levels of organic soil nitrogen, less demand for nitrogen fertilizer, and more-stable yields leading to larger decreases in economic risk. The economic conditions created by a carbon credit resulted in WWF being the most economically attractive annual crop rotation in this ecodistrict.

Increased revenues, coupled with constant nonland input costs under a carbon credit, translated into an upward shift in the overall demand for annual cropland. This increased demand for annual cropland put upward pressure on land rent. This upward pressure on land rent was complemented by an increase in tame forage land, which decreased the stock of annual cropland. The relatively large increases in land rent (19% increase at $25/Mg carbon credit; 70% increase at $75/Mg carbon credit; and 93% increase at $125/Mg carbon credit) increased the opportunity cost of native land, resulting in an increase in the rate of conversion of native lands to annual crop and tame forage production (Table 118.8). The decrease in native land has negative consequences for the ecological sustainability (i.e., biodiversity) of the agroecosystem.

Simulated carbon dioxide emissions decreased with increasing carbon credit levels primarily due to the shift in land allocation toward tame forage production (Table 118.9). The cost of decreased carbon dioxide emissions was calculated as the total payments made for sequestered carbon divided by the change in net emissions over the simulations. The $25/Mg credit was the most economically efficient due to the moderate decreases in total input embodied carbon emissions (with decreased annual cropland) and a moderate increase in sequestration simulated at this credit level. At credit rates above $25/t, the shift in annual crop land to WF and WWF, which increased organic matter decomposition and greater levels of input-embodied carbon associated with greater nitrogen fertilizer use, increased CO_2 emissions from annual cropland. At these higher credit levels, decreases in net emissions were realized primarily through greater sequestration in tame forage land which must be paid for under a carbon credit policy, thereby increasing its cost. It should be noted that total emission calculations did not include the mineralization of soil carbon associated with the

Table 118.9 Simulated Carbon Dioxide Balance with Carbon Credit and Carbon Tax Policies (Mt/50-year)

Policy Scenario	Total Emission	Total Soil Sequestered C	Net Emission	Emission Change	Emission Decrease Cost ($/t)
Baseline	11.81	1.29	10.51	—	—
$25 credit	11.13	1.65	9.48	−1.04	$39.77
$75 credit	9.98	2.43	7.55	−2.96	$61.43
$125 credit	8.91	2.95	5.96	−4.55	$80.85

conversion of native land to annual cultivation. This may be an important source of CO_2 considering the increased rate of native land conversion caused by a carbon credit.

These results reveal an interesting implication for policy development. A carbon credit increases the economic returns to land and therefore implies maintenance of the man-made capital stock of the agroecosystem. The carbon credit also provides an incentive to adopt management practices that have a higher marginal productivity of nitrogen fertilizer in terms of incremental soil carbon. However, in this moisture-constrained ecodistrict, those rotations that include fallow management are more productive. Therefore, the carbon credit policy encourages greater frequency of fallow, causing an increase in soil CO_2 emissions on annually cultivated land. The carbon credit also encourages greater additions of nitrogen fertilizer, which contains high levels of embodied carbon. In addition, the higher opportunity cost of land increases the incentive to convert native land to annual crop production that decreases the environmental sustainability of the system. As a result, the carbon credit policy results in a land-use pattern that may be less sustainable.

CONCLUSION

Agroecosystems, as complex systems, are characterized by multiple feedback and linkages between environmental and social components. These characteristics make it difficult to meaningfully assess the sustainability of these systems. The integrated, interdisciplinary model used in this study facilitates the simulation of dynamic system effects within a biophysical context, and provides insight into the agroecosystem changes over time. While the results obtained should be evaluated considering the assumptions and structure of the SAM, insight into the sustainability trends are revealed using appropriate indicators within the strong sustainability framework.

The baseline and carbon policy simulations presented in this chapter reveal the importance of feedback processes in the dynamics of agroecosystem structure and function over time. In addition, the biophysical constraints had a strong influence on the management options available to farmers and, therefore, on the sustainability of the agroecosystem. These results highlight the importance of recognizing the biophysical characteristics of a target system when designing policy.

None of the simulated landscapes showed systemwide improvement or degradation of all components of the capital stock. This result suggests the difficulty in developing a policy, or set of policies, that will engender a sustainable system without assigning relative societal or environmental system values to the capital stock components. Identifying those capital stock components in which degradation will not be tolerated or accepted is a process requiring knowledge of societal preferences, ethical responsibilities, degradation thresholds, and system coevolution.

ACKNOWLEDGMENTS

K.W.B. has been supported during his Ph.D. research by the Prairie Ecosystem Study (PECOS), an interdisciplinary research program examining the sustainability of prairie agriculture, based at the Universities of Saskatchewan and Regina.

REFERENCES

Acton, D.F., Padbury, G.A., and Stushnoff, C.T., *The Ecoregions of Saskatchewan,* Saskatchewan Environment and Resource Management, Canadian Plains Research Center, Regina, Saskatchewan, 1998.

Batie, S.S., Sustainable development: challenges to the profession of agricultural economics, *Am. J. Agric. Econ.,* 71, 1083–1101, 1989.

Brown, L.R., Facing the prospect of food scarcity, in *State of the World 1997: A Worldwatch Institute Report on Progress toward a Sustainable Society,* Starke, L., Ed., W.W. Norton, New York, 1997, pp. 22–41.

Bruce, J.P., Frome, M., Haites, E., Janzin, H., Lal, R., and Paustian, K., Carbon Sequestration in Soils, A report presented to the Soil and Water Conservation Society's Carbon Sequestration in Soils Workshop, Calgary, Alberta, May 21–22, 1998.

Clark, J.S. and Klein, K.K., Restricted estimation of crop and summerfallow acreage response in Saskatchewan, *Can. J. Agric. Econ.*, 75, 147–155, 1992.

Clark, J.S., Fulton, M., and Scott, J.T., Jr., The inconsistency of land values, land rents, and capitalization formulas, *Am. J. Agric. Econ.*, 75, 147–155, 1993.

Cohen, R.D.H., Donnelly, J.R., Moore, A.D., Leech, F., Bruynooghe, J.D., and Lardner, H.A., GRASSGRO®: a computer decision support system for pasture and livestock management, *J. Range Manage.*, 46, 376–379, 1995.

Costanza, R., Designing sustainable ecological economic systems, in *Engineering within Ecological Constraints,* Schulze, P.C., Ed., National Academy Press, Washington, D.C., 1996, pp. 79–95.

Costanza, R. and Daly, H., Natural capital and sustainable development, *Conserv. Biol.,* 6(March), 37–46, 1992.

Costanza, R., Wainger, L., Folke, C., and Maler, K.-G., Modeling complex ecological economic systems, *BioScience,* 43(8), 545–555, 1993.

Daly, H.E., Operationalizing sustainable development by investing in natural capital, in *Investing in Natural Capital: The Ecological Economics Approach to Sustainability,* Jansson, A.M., Eds., Island Press, Washington, D.C., 1994, pp. 22–37.

Douglas, D.L. and Rickman, R.W., Estimating drop residue decomposition from air temperature, initial nitrogen content, and residue placement, *Soil Sci. Soc. Am. J.,* 56, 272–278, 1992.

Gray, R., Furtan, H., Conacher, G., and Manaloor, V., *Set Aside Options for Western Canada,* Agricultural Economics, University of Saskatchewan, Saskatoon, 1993.

Greer, K.J., Hilliard, C.R., Schoenau, J.J., and Anderson, D.W., Developing simplified synergistic relationships to model topsoil erosion and crop yield, in *Proceedings of the Soil and Crop Workshop,* University of Saskatchewan, Saskatoon, 1992, pp. 108–125.

Hendrix, P.F., Chun-Ru, H., and Groffman, P.M., Soil respiration in conventional and no-tillage agroecosystems under different winter cover crop rotations, *Soil Tillage Res.,* 12, 135–148, 1988.

Karlen, D.L., Mausbach, M.J., Doran, J.W., Cline, R.G., Harris, R.F., and Schuman, G.E., Soil quality: a concept, definition, and framework for evaluation, *Soil Sci. Soc. Am. J.,* 61, 4–10, 1997.

McConnell, K.E., An economic model of soil conservation, *Am. J. Agric. Econ.,* February, 83–89, 1983.

Monreal, C.M., Zentner, R.P., and Robertson, J.A., The influence of management on soil loss and yield of wheat in Chernozemic and Luvisolic soils, *Can. J. Soil Sci.,* 75, 567–574, 1995.

Prugh, T., *Natural Capital and Human Economic Survival,* International Society for Ecological Economics, Solomons, MD, 1995.

Saskatchewan Agriculture and Food, *Agricultural Statistics 1996,* Regina, Saskatchewan, 1997.

Saskatchewan Agriculture and Food, *Crop Planning Guide 1998: Brown Soil Zone,* Regina, Saskatchewan, 1998.

Schmitz, A., Canadian wheat acreage response, *Can. J. Agric. Econ.,* 16, 79–86, 1968.

Statistics Canada, *Ecodistrict Agricultural Statistics Database,* Agriculture and Agri-Food Canada, Ottawa. Ontario, 1996.

Tiessen, H., Stewart, J.W.B., and Bettany, J.R., Cultivation effects on the amounts and concentration of carbon, nitrogen and phosphorus in grassland soils, *Agron. J.,* 64, 831–835, 1982.

Van Kooten, G.C. and Furtan, H., A review of issues pertaining to soil deterioration in Canada, *Can. J. Agric. Econ.,* 35, 33–54, 1987.

Van Kooten, G.C., Bioeconomic evaluation of government agricultural programs on wetlands conversion, *Land Econ.,* 69, 27–38, 1993.

Wischmeier, W.H. and Smith, D.D., *Predicting Rainfall Erosion Losses*, Agricultural Handbook 537, U.S. Department of Agriculture, Agricultural Research Service, Washington, D.C., 1978.

World Commission on Environment and Development, *Our Common Future,* The World Commission on Environment and Development, Oxford University Press, Oxford, 1987.

CHAPTER 119

Prediction of Soil Salinity Risk by Digital Terrain Modeling in the Canadian Prairies*

Igor V. Florinsky, Robert G. Eilers, and Glenn W. Lelyk

INTRODUCTION

Soil salinization is a typical process for steppe, semiarid, and desert landscapes. Saline soils have increased content (more than 0.25% of dry weight of a soil sample) of soluble salts, such as chlorides and sulfates of sodium, magnesium, and calcium. These soils are observed commonly in flat, closed, and poorly drained areas with nonpercolative water regimes, in territories with saline surface deposits and discharges of saline groundwater. There are primary and secondary salinizations. Primary salinization is caused by integration of natural phenomena, such as climatic, geological, and geomorphic factors; secondary salinization is a result of human activities, such as land-use and water management, which may lead to vertical and lateral redistribution of salts in the landscape. Salinization is one of the main factors of soil cover degradation (Kovda, 1946, 1971, 1973; Mordkovich, 1982; Lobova and Khabarov, 1983).

Topography is one of the major natural factors determining spatial distribution and dynamics of soil salinization. Different manifestations of topographic control of soil salinization can be seen at different scales. At macro-topographic scale (typical sizes reach several hundreds and thousands of meters), build up of salts occurs in closed and partly drained macro-depressions, which are natural accumulators of substances moved by gravity from macro-crests and macro-slopes. At meso-topographic scale (typical sizes range between several tens and hundreds of meters), meso-depressions have additional moisture due to overland runoff, and so they are marked by more leached soils than meso-slopes and meso-crests. Bands of solonetzs and solonchaks are typical for meso-slopes around meso-depressions through redistribution and secondary accumulation of leached substances. At micro-topographic scale (typical sizes are several meters), build up of salts is typical for micro-crests marked by low moisture and high evaporation (Kovda, 1946, 1971, 1973). So, topography determines a spatial mosaic of soil salinization, all other factors being equal.

One of the essential steps in studies of soil salinity is compilation of maps of soil salinization (Khakimov et al., 1983). A map allows one to conveniently represent regularities of spatial distribution of an object, a phenomenon or a process studied corresponding to the given scale (Aslanikashvili, 1974). Among these scale-dependent regularities are relationships between the relief and soil salinization indicated earlier. In other words, small- and middle-scale maps should represent relations

* This chapter appeared in the *Canadian Journal of Soil Science*, Volume 80, No. 3, pp. 455–463, 2000. It is reprinted here with permission.

Table 119.1 Definitions and Physical Interpretations of Some Topographic Variables

Variable	Definition	Interpretation
G, °	An angle between a tangent plane and a horizontal one at a given point on the land surface (Shary, 1991)	Velocity of substance flows
k_v, m^{-1}	A curvature of a normal section of the land surface by a plane, including gravity acceleration vector at a given point (Shary, 1991)	Relative deceleration of substance flows
k_h, m^{-1}	A curvature of a normal section of the land surface; this section is orthogonal to the section of vertical curvature at a given point on the land surface (Shary, 1991)	Convergence of substance flows
H, m^{-1}	A half-sum of k_h and k_v (Shary, 1991)	Flow convergence and relative deceleration with equal weights
K_a, m^{-2}	A product of k_h by k_v (Shary, 1995)	Degree of flow accumulation

Data from Florinsky and Kuryakova, 1996; Florinsky, 1998a.

between soil salinity and macro-depressions; salinization of meso-slopes should be shown on large-scale maps, while salinization of micro-crests should be indicated on detailed maps.

However, existing approaches for compilation of soil salinity maps, as a rule, cannot allow one to display these regularities (Khakimov et al., 1983). On these maps, polygons characterizing type and extent of soil salinization include depressions, slopes, and crests (Kondorskaya, 1967; Novikova, 1975; Eilers et al., 1995, 1997). As this takes place, authors of these maps mean that actual salinization can be observed within landforms corresponding to the given scale rather than within all polygons. However, since these relationships are not graphically displayed, users of such types of maps insert corresponding corrections to the best of their ability. This is because (1) there are no rigorous quantitative definitions of depressions, slopes, and crests and (2) there are no methods for reproducible delineation of these landform types using topographic maps and digital terrain models (DTMs).

To compile small-scale maps of soil salinization of great areas of Middle Asia, Stepanov (1979) and Khakimov and co-workers (1983) tried to solve this problem using data on horizontal land-surface curvature (k_h) (Table 119.1). For depressions, they considered areas of flow convergence described by negative values of k_h. However, k_h data are not sufficient to recognize depressions because, from the geomorphic point of view, values $k_h < 0$ correspond to valley spurs rather than depressions (Evans, 1980; Shary, 1995).

Metternicht and Zinck (1997) used a qualitative geomorphic map to compile small-scale predictive maps of secondary soil salinization for a mountainous valley in Bolivia. Lack of this technique is connected to a subjectivity of landform presentation on a traditional geomorphic map transferred to a salinization map.

To improve small-scale predictive maps of secondary soil salinization for the Crimean steppe, Florinsky (2000) used DTMs and a concept of accumulation, transit, and dissipation zones of the land surface (Shary et al., 1991; Shary, 1995). Applying this approach, one can partition a landscape into polygons marked by relative accumulation, transit, and dissipation of flow. These quantitative terms have the qualitative geomorphic analogues of depression, midslope, and crest, respectively.

In this chapter, we present some results of compilation of a small-scale map of salinity risk index for the Canadian prairies and adjacent territories using DTMs and the concept of accumulation, transit, and dissipation zones.

STUDY AREA

Soil salinization is a typical process for the Prairie Ecozone within the Canadian provinces of Manitoba, Saskatchewan, and Alberta, and some adjacent territories of the Boreal Plains ecozone (Figure 119.1) (Eilers et al., 1995, 1997).

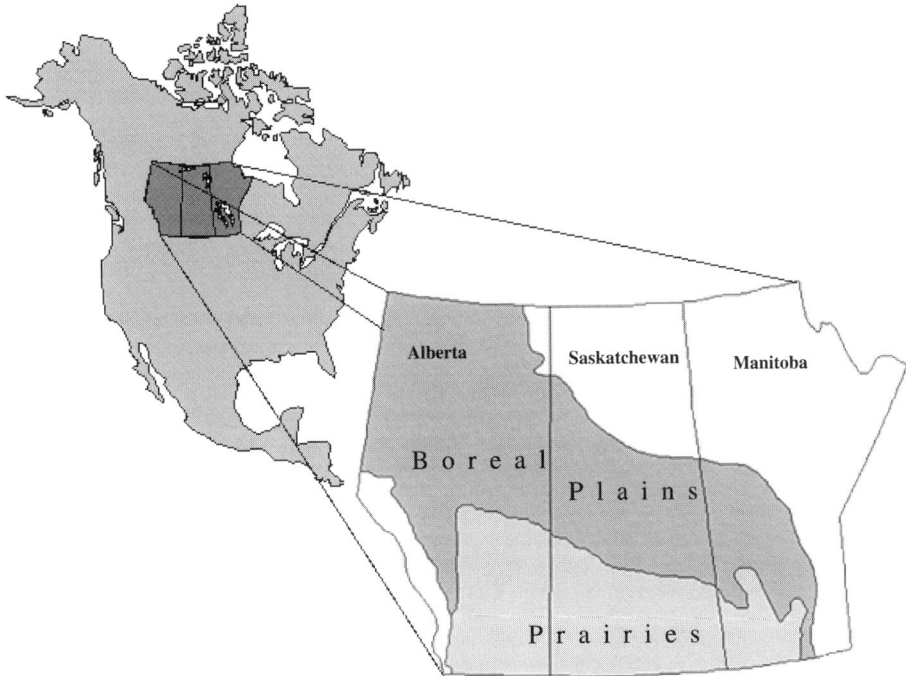

Figure 119.1 Geographical position of the study area.

The Canadian prairies are the northern extension of open grasslands in the Great Plains of North America. The area has a continental climate; subhumid to semiarid with long, cold winter and short, hot summer; low levels of precipitation; and high evaporation. Mean winter temperature is −10°C, and mean summer temperature is 15°C. Mean annual precipitation ranges from 250 mm in the arid grassland areas in southwest Saskatchewan and southeast Alberta, to 700 mm in Manitoba. The boreal plains are situated to the north of the prairies. Mean winter temperature is −14°C, and mean summer temperature is 14°C. Mean annual precipitation ranges from 300 mm in Alberta to 625 mm in Manitoba (Ecological Stratification Working Group, 1995).

The prairies and the boreal plains are underlain mostly by Cretaceous shales. These are nearly level to rolling landscapes consisting of hummocky glacial moraine and undulating to nearly level lacustrine deposits. Most rivers flow in an easterly direction across these territories. The soils are Black Chernozems with groves of trembling aspen, balsam poplar, and intermittent grassland in the northern edge of the prairies. The driest shortgrass prairie section with Brown Chernozems occurs in southwestern Saskatchewan and southeastern Alberta. The moist mixed grasslands and Dark Brown Chernozems can be observed in other parts of the prairies as a transition zone between these two zones. Dark Grey Chernozems Luvisols occur within the boreal plains. White birch, trembling aspen, and balsam poplar are typical in areas of the boreal plains adjacent to the prairies, while white and black spruce, jack pine, and tamarack are typical in the north of the boreal plains (Ecological Stratification Working Group, 1995).

Agriculture is the dominant land use in the prairie ecozone and in the southern and northwestern parts of the boreal plains. Saline soils are present in many areas throughout these regions (Figure 119.2). Sulfate salinization dominates there; 62% of the farmland in the prairies has a low extent of salinity (less than 1% of lands affected). Low-salinity areas usually have well-drained soils and are found generally on major uplands, over rapidly drained sandy soils, next to deep river channels that help to drain the soil, and in northern areas with soils formed under forest vegetation in moist conditions. About 36% of agricultural soils on the prairies are moderately affected by salinity (1 to 15% of lands affected). These are mainly medium-textured soils that occur next to small wetlands

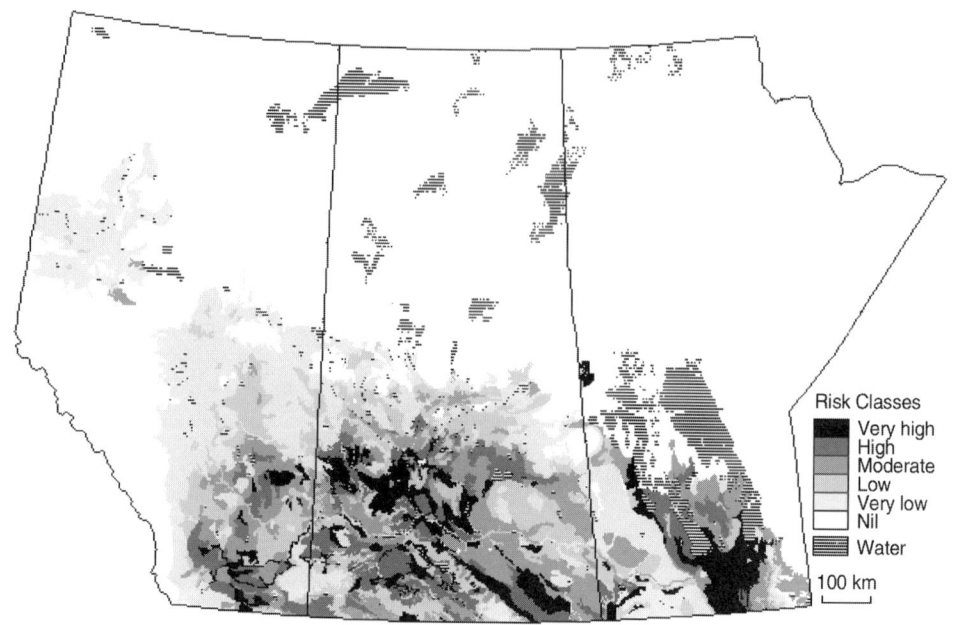

Figure 119.2 Map of salinity risk.

and in depressions. Areas with a greater extent of saline soils are fairly small and scattered throughout southern regions. These areas often receive regional groundwater discharge, such as those found on nearly level plains at the base of prominent uplands, nearly level plains where the discharge is from major aquifers, and where drainage is limited (Figure 119.2) (Eilers et al., 1995).

MATERIALS AND METHODS

The following initial data were used in this analysis:

- A map of salinity risk for Manitoba, Saskatchewan, and Alberta (Eilers et al., 1995, 1997) (Figure 119.2).
- A digital elevation model (DEM) of an area between 48°55′ and 60°05′ N, and 88°55′ and 120°05′ W. The DEM was given by a spheroidal trapezoidal grid with the grid size of 5′ (NOAA, 1988). This DEM included 50,625 points.

Salinity risk index (Figure 119.2) is used to rank individual land areas according to the chance that the salinity level will change with changing conditions. Evaluation of salinity risk is based on the concept that salinity is a dynamic condition of the soil cover, and on the premise that functional relationships exist among components affecting the salinization, that these components can be given a relative numerical weighting for their influences, and that each of these weightings can be combined resulting in a dimensionless index. The following factors are used to assess salinity risk:

1. Current extent of salinity derived from existing salinity maps at a scale of 1:1,000,000 indicating presence, extent, and position of moderate or greater salinity levels in each soil landscape (the electrical conductivity of a saturated-paste extract of a soil sample is more than 8 dS/m)
2. Slope gradient (G) (Table 119.1)
3. Soil drainage class
4. Aridity (the difference between precipitation and potential evapotranspiration)
5. Surface cover representing land-use and land management practices (Eilers et al., 1995, 1997)

The notion that soil salinization is typical for macro-depressions, meso-slopes, and micro-crests was not used in estimation of salinity risk index.

The concept of topographically expressed accumulation, transit, and dissipation zones is based on the following assumptions. Gravity-driven overland and intrasoil transport can be interpreted in terms of divergence or convergence, and deceleration or acceleration of flows (Shary, 1995). Flow tends to accelerate when vertical land-surface curvature (k_v) is positive, and to decelerate when k_v is negative (Table 119.1) (Speight, 1974; Shary, 1991). Flow diverges when $k_h > 0$, and converges when $k_h < 0$ (Table 119.1) (Kirkby and Chorley, 1967; Shary, 1991). Flow convergence and deceleration result in accumulation of substances in soils. At different scales, the spatial distribution of accumulated substances can depend on the distribution of the following landforms (Shary et al., 1991):

1. Landforms marked both by convergence and deceleration of flow; i.e., both by $k_h < 0$ and by $k_v < 0$ (accumulation zones)
2. Landforms offering both divergence and acceleration of flow; i.e., both $k_h > 0$ and $k_v > 0$ (dissipation zones)
3. Landforms that are free of a concurrent action of flow convergence and deceleration as well as divergence and acceleration; i.e., values of k_h and k_v have different signs or are zero (transit zones)

Recognition of accumulation, transit, and dissipation zones can be realized by simple registration of k_h and k_v maps (Koshkarev, 1982; Lanyon and Hall, 1983). However, in this case one can visualize only spatial distribution of these zones without quantitative estimation of a probable degree of flow accumulation. To solve this problem, Shary (1995) proposed use of data on accumulation (K_a) and mean (H) land surface curvatures (Table 119.1). Negative values of K_a correspond to transit zones, while positive values of K_a correspond to both accumulation and dissipation zones. Accumulation and dissipation zones can be distinguished using values of H. Positive values of K_a with negative values of H correspond to accumulation zones, whereas positive values of K_a with positive values of H correspond to dissipation zones (Shary, 1995).

We derived digital models of H and K_a from the DEM by the method of Florinsky (1998b). These were calculated using a grid spacing of 5′ because it is best matched to the regional scale and readability of maps to be obtained. H and K_a digital models include 49,609 points. Derivation of H and K_a digital models was carried out with LandLord software (Florinsky et al., 1995). A map of accumulation zones (Figure 119.3) was obtained by combination of H and K_a data.

To produce an improved map of salinity risk index taking into consideration typical occurrence of soil salinization in macro-depressions, we linked the map of salinity risk index for Manitoba, Saskatchewan, and Alberta (Figure 119.2) and the map of accumulation zones for this area (Figure 119.3). The resulting map (Figure 119.4) represents only those landscape polygons marked by different rates of soil salinization that are located within accumulation zones.

RESULTS AND DISCUSSION

A comparative analysis of the old and the improved maps of salinity risk index (Figures 119.2 and 119.4) demonstrates that once data on macro-depressions have been taken into account, areas marked by different levels of salinity risk decrease significantly. Therefore, the method proposed can prevent an overestimation in predictions of soil cover degradation due to salinization. This is not to say that soil salinization cannot occur within areas marked by nil salinity risk. However, this possible salinization relates to other scale levels and should be indicated on more-detailed maps.

The method applied can be used to compile maps of soil salinization at other scales. To produce middle-scale maps of soil salinization, one must consider data on accumulation zones; to compile large-scale maps, one should use information on transit zones; and to produce detailed maps, one

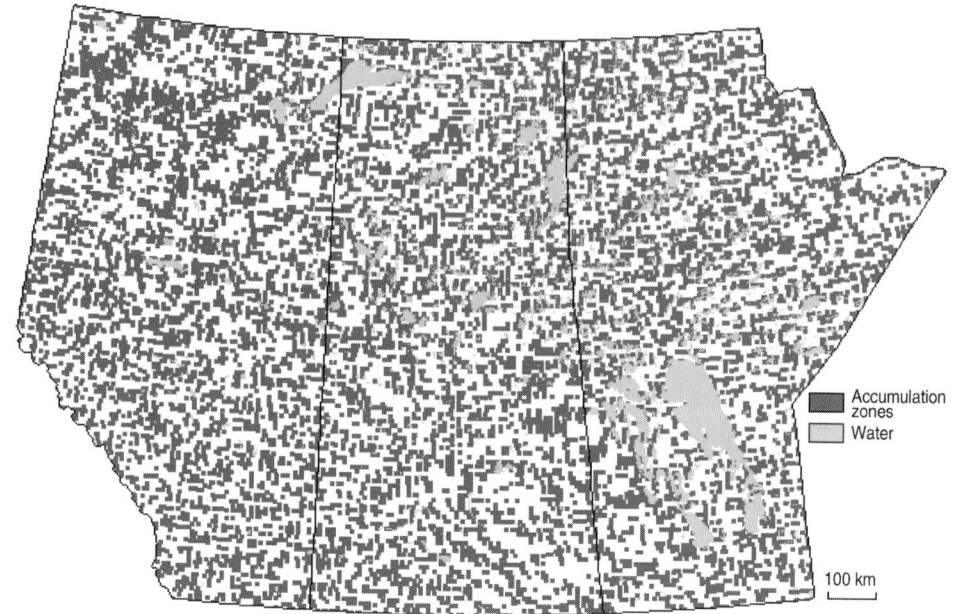

Figure 119.3 Map of accumulation zones.

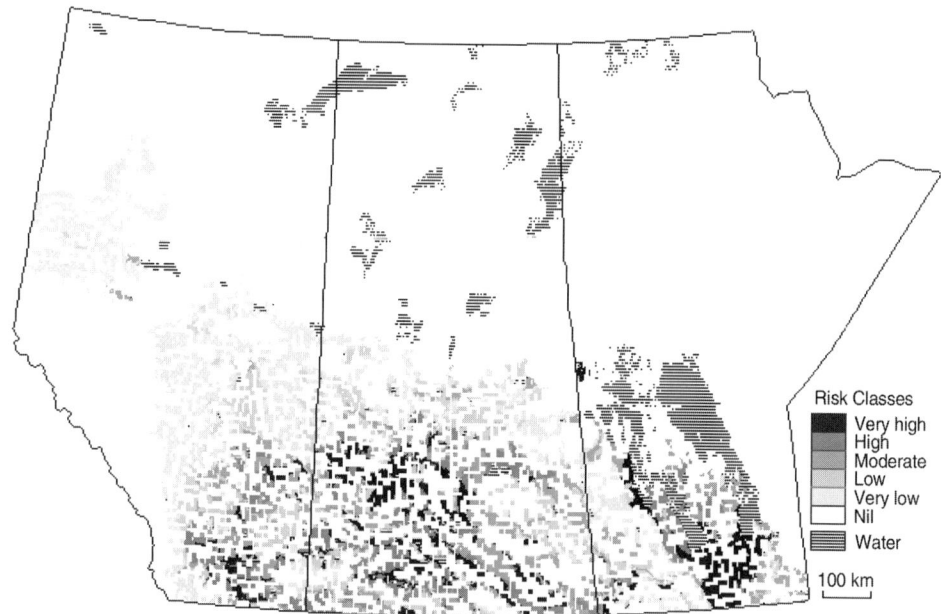

Figure 119.4 Improved map of salinity risk.

should apply data on dissipation zones. In all these cases, data on spatial distribution of accumulation, transit, and dissipation zones of the land surface can be derived from DEMs characterized by an appropriate resolution.

The method proposed can extend the use of DTMs in landscape investigations (Moore et al., 1991; Shary et al., 1991; Florinsky, 1998).

CONCLUSION

1. The following main principles should be taken into consideration in compilation of soil salinization maps: small- and middle-scale maps should represent relations between soil salinity and macro-depressions; salinization of meso-slopes should be shown on large-scale maps; salinization of micro-crests should be indicated on detailed maps.

2. The concept of accumulation, transit, and dissipation zones of the land surface can be applied to recognize depression, slope, and crest of different scales. Maps of accumulation, transit, and dissipation zones can be produced by linking K_a and H digital models derived from DEMs.

3. The small-scale map of the salinity risk index for the Canadian prairies and adjacent territories was compiled using existing data on soil salinity, DTMs, and the concept of accumulation, transit, and dissipation zones.

ACKNOWLEDGMENTS

The authors are grateful to P.V. Kozlov (ZAO "IC Protek," Moscow, Russia) for development of the Gecon Module of the LandLord software.

REFERENCES

Aslanikashvili, A.F., *Metacartography: Main Problems,* Metsniereba, Tbilisi, Georgia, U.S.S.R. (in Russian), 1974.

Ecological Stratification Working Group, *A National Ecological Framework for Canada,* Centre for Land and Biological Resources Research, Agriculture and Agri-Food Canada, State of the Environment Directorate, Environment Canada, Ottawa Ontario, 1995.

Eilers, R.G., Eilers, W.D., and Fitzgerald, M.M., A salinity risk index for soils of the Canadian prairies, *Hydrogeol. J.,* 5, 68–79, 1997.

Eilers, R.G., Eilers, W.D., Pettapiece, W.W., and Lelyk, G., Salinization of soil, in *The Health of Our Soils: Towards Sustainable Agriculture in Canada,* Acton, D.F. and Gregorich, L.J., Eds., Centre for Land and Biological Resources Research, Agriculture and Agri-Food Canada, Ottawa, Ontario, 1995, pp. 77–86.

Florinsky, I.V., Combined analysis of digital terrain models and remotely sensed data in landscape investigations, *Prog. Phys. Geog.,* 22, 33–60, 1998a.

Florinsky, I.V., Derivation of topographic variables from a digital elevation model given by a spheroidal trapezoidal grid, *Int. J. Geogr. Inf. Sci.,* 12, 829–852, 1998b.

Florinsky, I.V., Relationships between topographically expressed zones of flow accumulation and sites of fault intersection: analysis by means of digital terrain modeling, *Environ. Modeling Software,* 15, 87–100, 2000.

Florinsky, I.V. and Kuryakova, G.A., Influence of topography on some vegetation cover properties, *Catena,* 27, 123–141, 1996.

Florinsky, I.V., Grokhlina, T.I., and Mikhailova, N.L., LANDLORD 2.0: the software for analysis and mapping of geometrical characteristics of relief, *Geod. Cartogr.,* 1995–5, 46–51 (in Russian), 1995.

Khakimov, F.I., Deeva, N.F., and Brynskikh, M.N., *Modern Requirements for Soil Salinization Maps And Methods of Their Compilation,* Biological Research Centre Press, Pushchino Moscow Region, U.S.S.R. (in Russian), 1983.

Kirkby, M.J. and Chorley, R.J., Throughflow, overland flow and erosion, *Bull. Int. Assoc. Sci. Hydrol.,* 12, 5–21, 1967.

Kondorskaya, N.I., Areals of present-day salt accumulations in USSR soils, *Sov. Soil Sci.,* 1967–4, 462–473, 1967.

Koshkarev, A.V., Topography as an input parameter for mathematical and cartographic models of geosystems, in *Geographical Cartography in Scientific Research and National Economic Practices,* Moscow Branch of the Soviet Geographical Society, Moscow, U.S.S.R. (in Russian), 1982, pp. 117–131.

Kovda, V.A., *Origin and Regime of Saline Soils,* Vol. I, Soviet Academic Press, Moscow, U.S.S.R. (in Russian), 1946.

Kovda, V.A., *Origin of Saline Soils and Their Regime,* Vol. I, Israel Program for Scientific Translations, Jerusalem, Israel, 1971.

Kovda, V.A., *The Principles of Pedology, General Theory of Soil Formation,* Vols. I–II, Nauka, Moscow, U.S.S.R. (in Russian), 1973.

Lanyon, L.E. and Hall, G.F., Land surface morphology: 2, Predicting potential landscape instability in eastern Ohio, *Soil Sci.,* 136, 382–386, 1983.

Lobova, E.V. and Khabarov, A.V., *Soils,* Mysl, Moscow, U.S.S.R. (in Russian), 1983.

Metternicht, G.I. and Zinck, J.A., Spatial discrimination of salt- and sodium-affected soil surfaces, *Int. J. Remote Sensing,* 18, 2571–2586, 1997.

Moore, I.D., Grayson, R.B., and Ladson, A.R., Digital terrain modelling: a review of hydrological, geomorphological and biological applications, *Hydrol. Proc.,* 5, 3–30, 1991.

Mordkovich, V.G., *Steppe Ecosystems,* Nauka, Novosibirsk, U.S.S.R. (in Russian), 1982.

NOAA Data announcement 88-MGG-02, digital relief of the surface of the Earth, NOAA, National Geophysical Data Center, Boulder, CO, 1988.

Novikova, A.V., Prediction of secondary salinization of soils produced by irrigation, Estimation of terrain suitability for irrigation exemplified by the south part of the Ukrainian Soviet Socialist Republic, Urozhai, Kiev, Ukraine, U.S.S.R. (in Russian), 1975.

Shary, P.A., The second derivative topographic method, in *Geometry of Landsurface Structures,* Stepanov, I.N., Ed., Pushchino Research Centre Press, Pushchino Moscow Region, U.S.S.R. (in Russian), 1991, pp. 30–60.

Shary, P.A., Land surface in gravity points classification by complete system of curvatures, *Math. Geol.,* 27, 373–390, 1995.

Shary, P.A., Kuryakova, G.A., and Florinsky, I.V., On the international experience of topographic methods employment in landscape researches (a concise review), in *Geometry of Landsurface Structures,* Stepanov, I.N., Ed., Pushchino Research Centre Press, Pushchino Moscow Region, U.S.S.R. (in Russian), 1991, pp. 15–29.

Speight, J.G., A parametric approach to landform regions, in *Progress in Geomorphology: Papers in Honour of D.L. Linton,* Brown, E.H. and Waters, R.S., Eds., Institute of British Geographers, London, 1974, pp. 213–230.

Stepanov, I.N., Ed., Map of quantitative and qualitative estimation of salinization within soil strata of 1 m for Turkmen SSR, scale 1:600,000, Central Board of Geodesy and Cartography, Moscow, U.S.S.R. (in Russian), 1979.

Aquatic Ecosystems:
Lake Tahoe and Clear Lake, California

CHAPTER 120

Overview: Aquatic Ecosystems: Lake Tahoe and Clear Lake, California

Dennis E. Rolston

Lake Tahoe and Clear Lake are two very important lakes in California, and each have very different characteristics from which lessons can be learned to aid in understanding and managing complex, multiple-stressed ecosystems. Clear Lake is an ancient, shallow, highly productive lake at an elevation of 395 m in an oak woodland–dominated northern Coast Range mountain watershed with rich habitats for breeding and migrating waterfowl, fish-eating birds, and mammals. The lake has a mean depth of about 7 m. Lake Tahoe lies in the crest of the Sierra Nevada mountains at an elevation of 1898 m and is located in a montane-subalpine watershed dominated by coniferous vegetation and nutrient-poor soils. It is a deep lake with a mean depth of 313 m. Each of these lakes has been studied extensively for a number of years.

This section consists of three chapters. The chapter by Suchanek et al. gives the history of various natural and anthropogenic stressors to the Clear Lake ecosystem and describes a holistic approach to understanding and managing the ecological health of this lake. It is one of the most comprehensive studies of the effects of multiple stressors on a well-defined ecosystem, and has contributed greatly to fundamental understanding of ecosystem function and dynamics. It represents an example of an interdisciplinary approach to studying ecosystem health.

The chapter by Cech, Choi, and Houck describes a specific study of the uptake of methyl mercury by one species of Clear Lake fish. Since mercury is one of the major contaminants in the lake, this study is important in understanding possible exposure pathways to birds and humans.

The chapter by Reuter et al. summarizes the history of various natural and anthropogenic stressors to the Lake Tahoe ecosystem and describes an integrated approach to studying the watershed. The decreasing transparency of the lake has been well documented by more than 30 years of data, and the chapter points out the need for long-term data sets to study complex systems. It also describes the necessary role that monitoring and fundamental research play in adaptive watershed management.

CHAPTER 121

Evaluating and Managing a Multiply Stressed Ecosystem at Clear Lake, California: A Holistic Ecosystem Approach

Thomas H. Suchanek, P.J. Richerson, D.C. Nelson, C.A. Eagles-Smith, D.W. Anderson, Joseph J. Cech, Jr., R. Zierenberg, G. Schladow, J.F. Mount, S.C. McHatton, D.G. Slotton, L.B. Webber, Brian J. Swisher, A.L. Bern, and M. Sexton

INTRODUCTION

Clear Lake (Figure 121.1) and its watershed in Northern California constitute a serene and beautiful environment that has been used extensively by inhabitants of the surrounding basin for millennia. It has one of the oldest documented North American "early man" sites, with paleo-indian occupation of the Clear Lake Basin about 10,000 years before present (ybp) at Borax Lake, immediately adjacent to Clear Lake (Heizer, 1963). Native American settlement was relatively dense during European contact in the early 1800s, with about 3000 people scattered among 30 or so villages within the basin (Baumhoff, 1963). These people utilized the lake's abundant fish, as well as tens of thousands of waterfowl and runs of native fishes in adjacent streams (Simoons, 1952) to supplement their staple acorn diet. Variously named *Lupiyoma, Hok-has-ha,* or *Ka-ba-tin* by early Native Americans (Mauldin, 1960); *Big Waters* by some of the early European pioneers in the 1820s; *Laguna* for a short time by Spanish Californians in the 1830s; and finally, Clear Lake in the 1840s, this lake has had a long and fascinating history. European or American trappers first started visiting the lake seasonally in 1833, but more permanent agricultural settlers did not arrive until the 1850s (Simoons, 1952).

Volcanic activity in the area of the lake provided heat to drive hydrothermal systems that created rich mineral deposits. Almost immediately, settlers began to extract these minerals from the landscape. The first commercial mines were small-scale operations that exploited borax in 1864 and sulfur in 1865. Mercury mining became a significant industry with the development of the Sulphur Bank Mercury Mine in 1872. Beginning in the mid-1870s, abundant mineral springs attracted thousands of health-conscious citizens to the region (Simoons, 1952).

The population of Lake County has grown from approximately 3000 in 1860 to more than 55,000 in 1999 (Table 121.1). Associated with that population growth have come dramatic land-use changes, altering the watershed and limnological and ecological dynamics of Clear Lake. Today, the Clear Lake watershed basin includes a high proportion of forested land, oak woodlands, orchards and vineyards, other croplands, and little remaining original wetlands (Figure 121.2). Clear Lake is used for storage of agricultural irrigation water for downstream Yolo County. The Yolo County

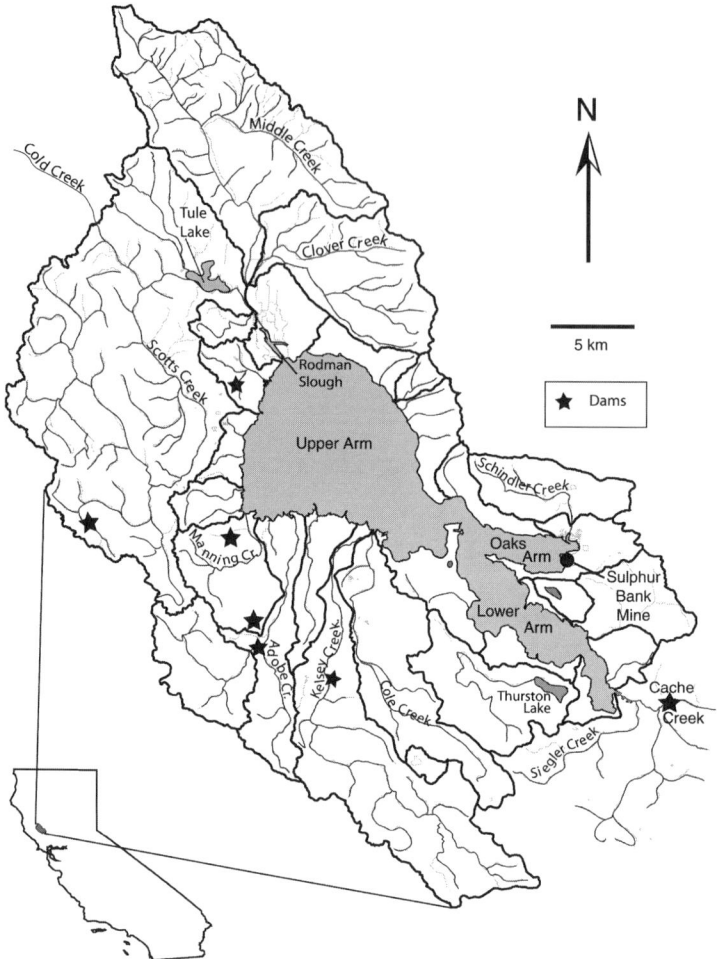

Figure 121.1 Map of Clear Lake and surrounding watershed, with locations of dams.

Table 121.1 Lake County Population since European Settlement[a]

Year	Lake County Population
1850	2210[b]
1870	2900[b]
1890	7100[b]
1900	6017
1910	5526
1920	5402
1930	7166
1940	8069
1950	11,481
1960	13,786
1970	19,548
1980	36,366
1990	50,631
2000	55,000

[a] Note especially the exponential growth beginning in the 1960s.
[b] Simoons, 1952.

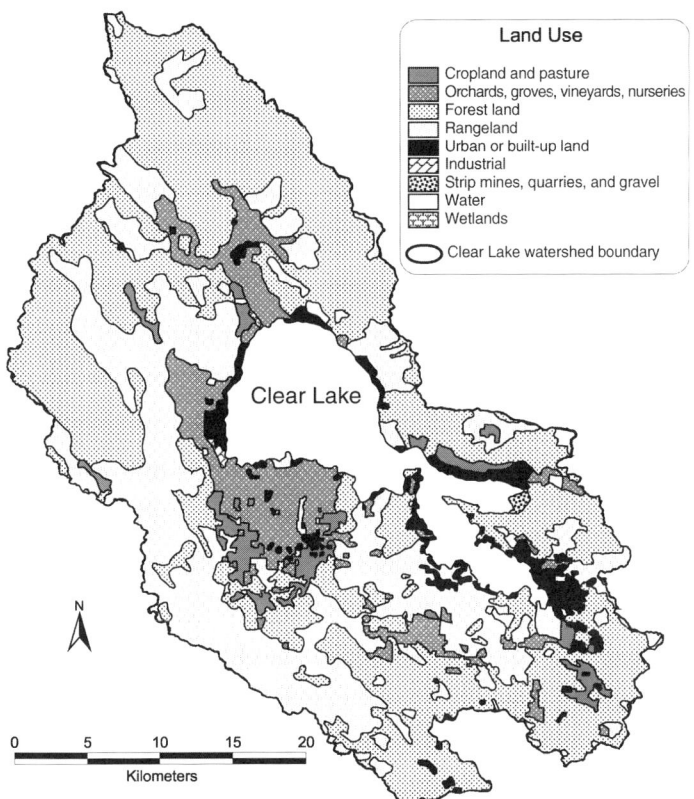

Figure 121.2 Land use within the Clear Lake watershed as of 1998.

Flood and Water Conservation District owns the rights to use the water in the lake and regulates the flow of releases from the single outlet dam to Cache Creek. The lake's rimlands and surrounding watershed also support extensive agricultural production, including pears, walnuts, grapes, and wild rice. The area supports breeding populations of several important bird species associated with the lake itself, including the western grebe (*Aechmophorus occidentalis*), Clark's grebe (*A. clarkii*), double-crested cormorant (*Phalacrocorax auritus*), great blue heron (*Ardea herodias*), osprey (*Pandion haliaetus*), and bald eagle (*Haliaeetus leucocephalus*). In addition, this aquatic setting supports many other species of birds associated with surrounding wetlands and uplands. It supports the only commercial fishery on a lake in California and extensive sport fisheries, especially for bass and catfish. Clear Lake is used heavily for water sports (swimming, boating, water skiing, jet skiing); thus, it attracts significant recreational tourism, especially during summer months.

   This aquatic ecosystem has also been subject to multiple stresses, both natural and anthropogenic. It has experienced periodic flooding and fires and has lost over 85% of its original natural wetlands habitat. However, dam construction in Lake County has added approximately 6500 acres of impoundment water. The lake experienced increased nutrient loading and decreased water clarity between 1925 and 1938, likely due to the introduction of more-efficient heavy earth-moving equipment in the basin and the loss of original wetlands on the major tributaries (Richerson et al., 2000). By 1938, the lake had become too turbid for rooted aquatic vegetation to flourish and noxious cyanobacterial scums became a perennial problem (Lindquist and Deonier, 1943; Murphy, 1951). The lake also has elevated concentrations of mercury, among other contaminants; a U.S. EPA Superfund site is located on the eastern shore of Clear Lake at the former Sulphur Bank Mercury Mine. Clear Lake was also the first site at which the deleterious effects of large concentrations of organochlorine pesticides on bird populations were documented (original research reported by Hunt and Bischoff [1960] and Rudd [1964], and popularized in Rachel Carson's *Silent Spring* [1962]). The lake contains a fish fauna comprised of over 75% introduced species. Clear Lake has been designated by the State of California as an "impaired water body" under Section 303d of the Clean Water Act, which mandates the state to identify water bodies or stream segments that currently are not meeting, or are not expected to meet, designated beneficial uses.

Managing multiple, frequently conflicting, uses of the lake and watershed is challenging; it involves ecological, economic, and political balances. Before rational management can be successful, however, managers must understand, to a useful approximation, the complex processes that drive the ecosystem's behavior. Our goal is to document the historical and modern-day multiple uses and multiple stresses associated with the Clear Lake aquatic ecosystem, including the lake proper and the surrounding watershed, in an attempt to demonstrate what rational management of this and similar valuable, multiply stressed ecosystems demands of applied science programs.

This chapter provides a prehistoric and historic background from which to evaluate the importance and impacts of a multitude of natural and anthropogenic stresses imposed upon Clear Lake and its watershed. Specifically, we identify major geological, climatological, ecological, political, and economic factors that influence the outcome of management and policy-level decisions on the health and well-being of the entire ecosystem, especially the lake proper. Natural stressors include geologic and tectonic events, regional and global climate change, fires, droughts, and floods. Anthropogenic stressors include:

1. Modifications to the landscape, including fires, logging and deforestation, dam construction, and other creek modifications
2. Contaminants, including pesticides, mining, sewage and septic overflows
3. Land-use changes, including original wetland losses (and some wet habitat increases, see previous) dredging, filling, and creek bed alterations, water table and shoreline modifications, road building, agriculture, soil exposure and transport, and livestock grazing
4. Intentional and accidental species introductions

Finally, we illustrate how scientific research has contributed significantly to addressing multiple management objectives in this extremely complex ecosystem.

NATURAL SETTING

Clear Lake (Figure 121.1) is located at 39°00′ N, 122°45′ W within Lake County in the Coast Range of California at an elevation of 402 m with surrounding ridges rising up to 1500 m. Through evidence from a series of cores to ~177 m maximum sediment depth, collected by the U.S. Geological Survey in the 1970s and 1980s, Clear Lake is believed to be the oldest lake in North America with continuous lake sediments of about 480,000 years (Sims, 1988; Sims et al., 1988). It also is possible that Clear Lake is a remnant of an ancestral lake represented by deposits of the Cache Formation, dating back to the early Pleistocene, making it 1.8 to 3.0 million years old (Casteel and Rymer, 1981; Hearn et al., 1988). The Clear Lake Basin represents a fault-bounded subsiding graben related to movement along the San Andreas Fault. The lake is set within an active volcanic region. Abundant geothermal springs release both fluids and gases (primarily CO_2, H_2S, and methane) from the lake bottom. Clear Lake is polymictic (typically well mixed), alkaline (pH 8), shallow (average depth about 6.5 m), and highly productive (eutrophic). Subsidence (caused by block faulting beneath the lake) has kept up with natural sedimentation. Thus, Clear Lake has been a relatively shallow lake since its initial formation, but at least parts of the current basin have been an open water system since the middle Pleistocene. Although early reports claim that "it derives its name from the clearness of its waters" (Menefee, 1873), and it may have been clearer in the 19th century than it is today, it was almost certainly a eutrophic system during the course of its entire existence (Bradbury, 1988). However, many recent land-use changes (described later) likely have exacerbated sediment and nutrient loading to the lake, enhancing noxious cyanobacterial (blue-green "algae") blooms (Richerson et al., 1994). In modern times, the lake supports populations of ~100 species of green and yellow-green algae and cyanobacteria, about 115 species of diatoms, 23 species of aquatic macrophytes, 94 species of invertebrates, and 34 species of fishes, plus numerous lake-associated or lake-dependent mammals such as otter, mink, raccoon, and numerous species of birds such as osprey, bald eagle, and grebes (Horne, 1975; Macedo, 1991; Richerson et al., 1994; Meillier et al., 1997; Moyle, 2002; Lake County Vector Control, unpublished).

NATURAL STRESSORS

Geologic and Tectonic Events

Two different hypotheses have been proposed for the origin of Clear Lake's drainage route. Davis (1933) proposed that Clear Lake was originally two lakes separated by an isthmus at the Narrows: one lake essentially comprising the Upper Arm (which drained westward into the Russian River system via Cold Creek), and another lake comprising the present-day Oaks Arm and Lower Arm (which drained eastward via Cache Creek into California's Central Valley and the Sacramento River) (Figure 121.1). A lava flow in the eastern drainage was believed responsible for forming a dike high enough to build up the water level in the eastern lake allowing water to overspill and cut through the narrow isthmus, thereby connecting the two lakes. The drainage flowed from the eastern lake to the Upper Arm and out Cold Creek into the Russian River system (Mauldin, 1968). Later, perhaps 10,000 ybp, an earthquake likely initiated a landslide in the Cold Creek drainage which blocked the westward flow, forcing the entire lake (both halves) to drain again into the Central Valley through Cache Creek. Alternatively, Becker (1888) and Brice (1953) argue that Clear Lake occupies the lowest part of a shallow fault depression. Hodges (1966) later concurred with this hypothesis and proposed that Scotts Creek formed a fan delta and clogged the western outlet valley of Cold Creek with fluvial debris; thus Cache Creek remained the only outlet. In either case, fish from the Central Valley colonized the Russian River system and there was some speculation that the reverse might have occurred as well (Hopkirk, 1973, 1988).

Regional and Global Climate Change

California's climate has been undergoing dramatic and continuous change over the past 100 years. This change is coincident with documented global climate change over the same time period (Mann et al., 1999; Hileman, 1999); yet it is uncertain how changes in regional climate are linked to more global processes. In California, the coefficient of variation in rainfall and the frequency of 1000-year storm events has been increasing dramatically over the past century (Goodridge, 1998; Figure 121.3). In addition, the frequency of El Niño events has become unusually common in recent years (Trenberth and Hoar, 1997) and this likely is linked with much of the fluctuation in regional and global climatological events at Clear Lake. Furthermore, if global warming enhances El Niño events, for which there is growing evidence (Meehl and Washington, 1996), then California's climate over the next several decades may look like an amplified version of the extreme events that have taken place over the past several decades (Field et al., 1999).

Fires

The Mediterranean climate and vegetation of this region are naturally conducive to fire. The change in regional weather described above also may contribute to an increase in large fires. Lightning is a common cause of fires in this region and although we do not have documentation of fire origins, especially before European contact, we do have moderately accurate records on the

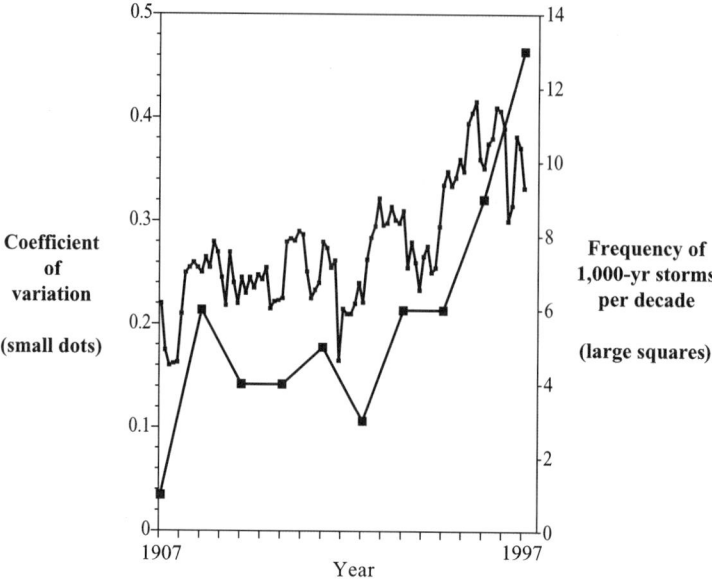

Figure 121.3 Regional California climate changes over the past century showing dramatic increases in the coefficient of variation in rainfall and increase in the frequency of 1000-year storms. (Adapted from Goodridge, J., The impact of climate change on drainage engineering in California, Report to the Alert Users Group Conference in Palm Springs, CA, 26–29 May, 1998.)

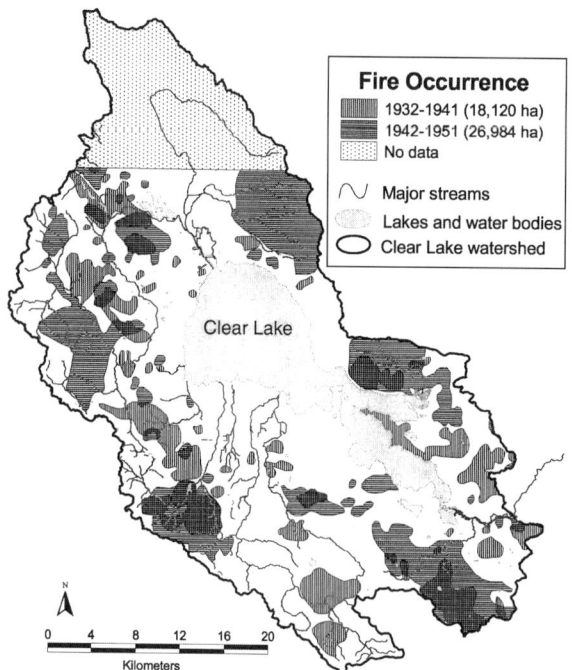

Figure 121.4 Location of fires in the Clear Lake Basin from 1932 to 1951. (Adapted from Simoons, F.J., The Settlement of the Clear Lake Upland of California, Master's thesis, University of California, Berkeley, 1952.)

extent of fires in these watersheds since the turn of the century (see Figure 121.4). It has been estimated that in more modern times (1960s to 1980s) lightning accounted for only about one third of all forest fires (West, 1989).

Droughts and Floods

Short- to medium-length droughts have been documented in the historical record of California. A significant 3-year drought (1975 to 1977), and 6- to 7-year droughts during recent (1987 to 1992) and historical (1928 to 1934) periods lowered the lake level substantially (see Figure 121.5). Tree ring data from California back to the 1500s suggest decade-scale droughts (Earle, 1993). More substantial 46- to 140-year-long droughts (before A.D. ~1350) and >200-year-long droughts (before A.D. ~1112) also have been documented in California by dating submerged tree stumps (Stine, 1994). The most comprehensive understanding we have on how such droughts can impact ecological and limnological processes in and around Clear Lake comes from a 24-year Department of Water Resources water quality data set (1969 to 1992) that includes the 1975 to 1977 and 1987 to 1992 droughts. An analysis of those data by Richerson and co-workers (1994) shows that the drought period correlated strongly with some moderate increases in pH and Secchi disk readings (water clarity) during the longer drought in the Oaks Arm and Lower Arm of Clear Lake (but not necessarily the Upper Arm or during the shorter drought) (Figure 121.6A). In addition, dramatic increases in water column phosphorus (especially dissolved phosphorus) and electrical conductivity have been observed in all three arms (Figure 121.6B), especially during the longer 1985 to 1992 drought. Interestingly, during the past 3 years (1998 to 2000) Clear Lake has exhibited exceptional water clarity (high Secchi disk readings) during nondrought conditions. Clear Lake also has experienced one of the most severe seasons of cyanobacteria (blue-green "algae") blooms in recorded history toward the end of these documented drought periods (Figure 121.7), likely because of increased sediment phosphorus releases.

With increasing frequency of 1000-year storm events resulting from regional climate change (see previous), there also is increased risk from natural flooding to lakeside real estate and public utilities, and additional risk of increased sedimentation, nutrient inputs, and acid mine drainage from the local Sulphur Bank Mercury Mine (discussed later). Figure 121.5 provides documented water levels (high and low water) for Clear Lake relative to a baseline depth of "0 ft Rumsey" (which is equivalent to 1318.256′ 1929 National Geodetic Vertical Datum, NGVD) originally established in 1872. Flooding also has increased recently, and the winter of 1997 to 1998 produced the highest recorded lake levels since the emplacement of the Cache Creek dam in 1914 (Figure 121.5).

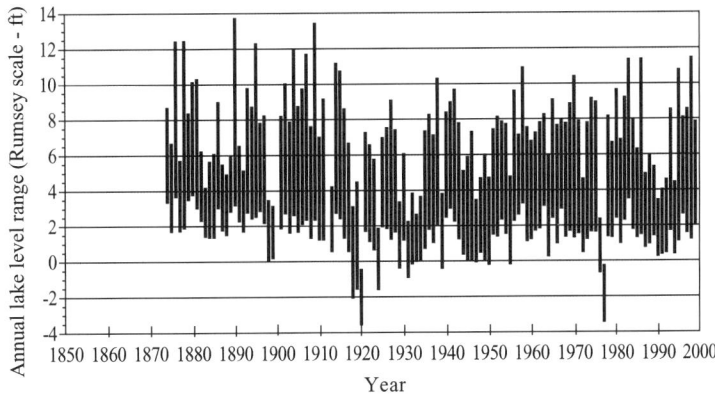

Figure 121.5 Clear Lake water levels (annual minimum and maximum) relative to the Rumsey scale.

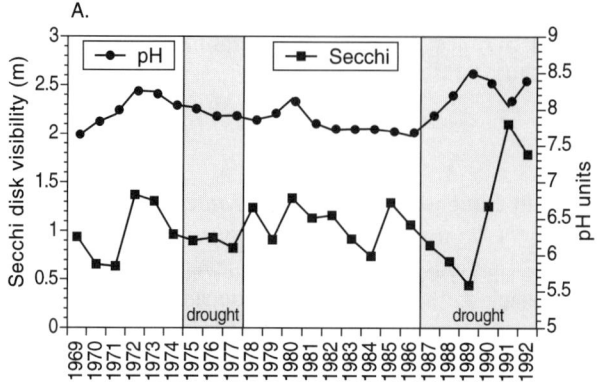

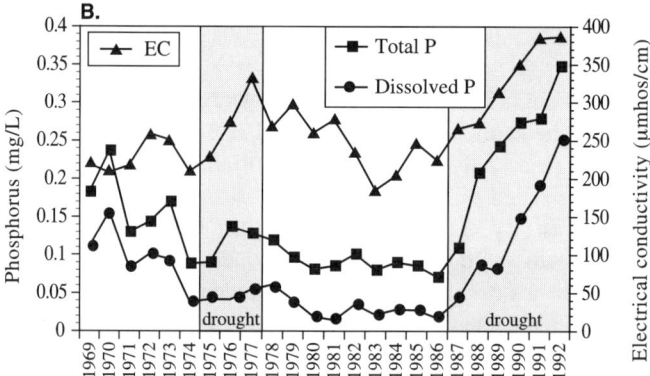

Figure 121.6 A 24-year data set (1969 to 1992) of limnological conditions within the Lower Arm of Clear Lake showing dramatic changes associated with the drought from 1987 to 1992 (A) Water quality as quantified by Secchi disk values and pH readings (B) Phosphorus and electrical conductivity values. (Adapted from Richerson, P.J., T.H. Suchanek, and S.J. Why, The Causes and Control of Algal Blooms in Clear Lake, Clean Lakes Diagnostic/Feasibility Study for Clear Lake, California, prepared for EPA Region IX, 1994.)

ANTHROPOGENIC STRESSORS

Modifications of the Landscape

Fires

Often it is difficult to discriminate between natural and anthropogenic causality for fires. Humans set fires, but also attempt to control them; yet, both processes impact natural ecosystem dynamics. It is likely that pre-Spanish fires in the Coast Range were relatively frequent (10 years or less) (West, 1989). When Europeans first settled the basin during the mid-1800s, widespread intentional burning was common. Herdsmen intentionally burned brush on such a large scale to graze sheep, goats, and cattle that California state legislation was passed in 1872 making fire-setting in wooded or forested lands illegal; yet, this had little impact on burning (Simoons, 1952). These activities were many more times as destructive as lumbering (discussed later). Early European settlers set fires primarily for three reasons: (1) to encourage grass growth for spring feed, (2) to burn needles in coniferous forests to prevent accumulations that would cause unplanned catastrophic fires, and (3) to improve deer feed. While lightning is often the cause of modern fires, anthropogenically

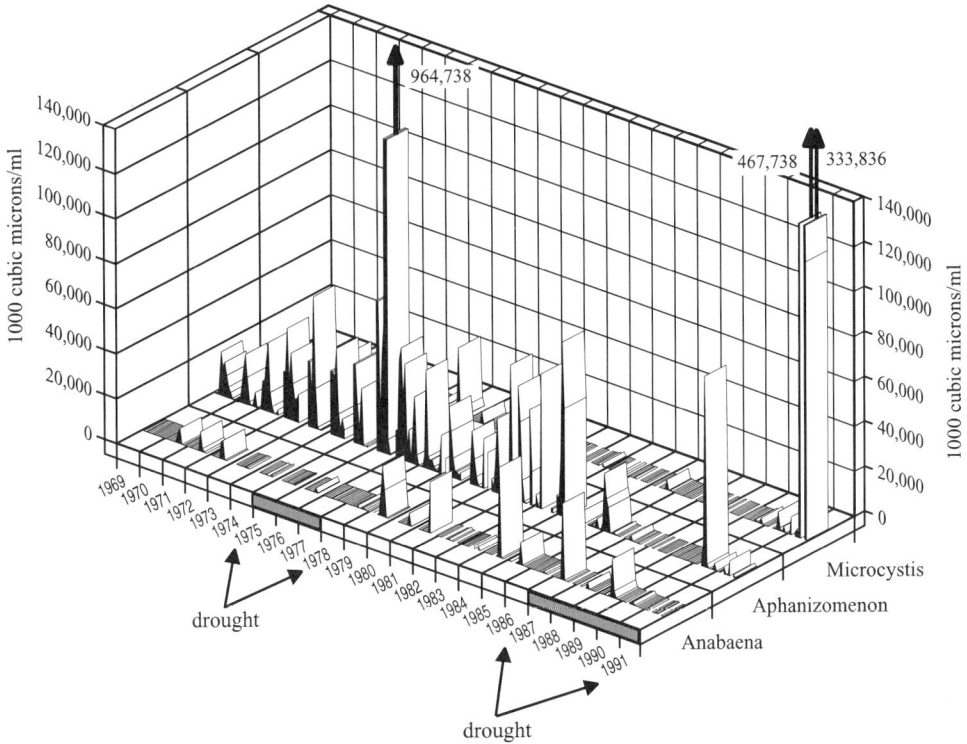

Figure 121.7 Volume of three species of scum-forming cyanobacteria (blue-green "algae") over a 24-year period (1969 to 1992) in the Lower Arm of Clear Lake. Note especially the abundance of *Microcystis* at the end of the drought period, a particularly buoyant species that drifts into shore and accumulates in large windrows, then dies, lyses, and creates noxious odors which affect shoreline aesthetics. (Adapted from Richerson, P.J., T.H. Suchanek, and S.J. Why, The Causes and Control of Algal Blooms in Clear Lake, Clean Lakes Diagnostic/Feasibility Study for Clear Lake, California, prepared for U.S. EPA Region IX, 1994.)

mediated fires are now more likely caused by accident or by arson rather than as a deliberate attempt to manage the landscape using fire. The two main watersheds that provide over 50% of the inflow to Clear Lake (Scotts Creek and Middle Creek) have experienced numerous natural and anthropogenically influenced fires. An example of fires documented during the periods from 1932 to 1941 and from 1942 to 1951 (adapted from Simoons, 1952), as well as more-recent fires within the Scotts Creek and Middle Creek watershed, are shown in Figure 121.4 and Table 121.2. As elsewhere in California, control practices probably make fires fewer and less frequent, but more devastating when they do occur because of increased fuel accumulation. For example, the "Forks Fire" decimated over 86,000 acres just north of Clear Lake in 1996, including about 12,600 acres in the Middle Creek watershed that drains into Clear Lake.

Logging and Deforestation

Major modifications of the forested landscape by Mexican and American ranchers began as early as the 1840s. In the 1850s some of the landscape was cleared for orchards, vineyards, and especially for timber harvest. The most significant timber harvest was Douglas fir (used for mine timbering), sugar pine, and Ponderosa pine (used mostly for lumber) and oak for fuel (Simoons, 1952). Much of this wood was used as fuel for the borax and mercury mining operations (see below). By 1870, no fewer than five commercial sawmills were operating on the lake; by 1905, there were 11 mills that processed over 1.5×10^6 board feet of lumber annually, and in 1946 more

Table 121.2 Acres Burned by Fires in the
Middle Creek Watershed during
the 20th Century

Year of Fire	Acres Burned
1911	400
1912	400
1913	400
1914	400
1915	400
1916	1364
1917	4743
1922	2995
1923	19,077
1928	6394
1929	2339
1930	105
1931	1994
1932	419
1933	7362
1934	1021
1936	184
1939	3304
1941	563
1946	1889
1947	11,648
1951	619
1958	117
1959	593
1960	0
1964	1361
1971	1500
1975	0
1980	3000
1981	0
1991	224
1996	12,685

Note: Although the 1996 "Forks Fire" burned
more than 86,000 acres, only a portion of
this fire affected the Clear Lake Basin.

than 11×10^6 board feet was processed (Simoons, 1952). As a result of forest removal and soil exposure, sediment transport probably increased into Clear Lake, depositing additional entrained nutrients into the system. Nevertheless, this exploitation was not severe enough to be apparent in the pollen record of Clear Lake as deduced from sediment cores (Richerson et al., 2000).

Dam Construction

The Clear Lake dam along Cache Creek was completed in 1914 for the purpose of regulating agricultural irrigation waters to downstream Yolo County, which owns the rights to Clear Lake water (discussed in Richerson et al., 1994). An earlier dam at Clear Lake's outlet (about 2.5 km upstream of the present dam) was first constructed in 1867 to increase water levels to operate a mill at that end of the lake, but was destroyed in 1868 by about 300 angry rimlander property owners who were getting flooded during periods of heavy rain. This is one of the most colorful accounts of Clear Lake history, when a vigilante group took into custody several individuals (including the sheriff, his deputies, the county judge, the superintendent of the mill works, and other prominent citizens) who appeared sympathetic to the dam's owners and operators (Anon.,

1881). A pumping station was installed in this same location around 1910 to increase water flow to irrigate rice crops in downstream Capay Valley, but that station also was intentionally destroyed about 1912. Numerous other dams have been erected within the Clear Lake watershed from 1955 to 1980, primarily for irrigation or recreation purposes (see Figure 121.1 for locations). All of these dams have slowed and altered natural flow of waters from the watershed to Clear Lake and may have prevented some species of Clear Lake fishes (such as the Clear Lake splittail, *Pogonichthys ciscoides*) from migrating upstream to spawn (Moyle, 2002). While there was the potential for increased flooding as the result of the construction of the Clear Lake dam, ironically lake levels have been significantly lower (see Figure 121.5) since the construction of the Cache Creek dam in 1914. Other dams higher up in the watershed have water-holding capacity that lowers flooding risks, and they likely have contributed to the lowered lake levels mentioned previously. They also retain sediments and nutrients that have the potential to become deposited into Clear Lake and thus help to limit eutrophication in the lake.

Other Creek Modifications

Lowering of Cache Creek

The Clear Lake outlet through Cache Creek was deepened in 1938 to accommodate more efficient agricultural irrigation, although further work was halted by a suit filed by a citizen's group.

Kelsey Creek Downcut

In 1965, the delta of Kelsey Creek was dredged to accommodate the installation of a marina. This caused destabilization of the creek bed, providing increased erosional products to be transported into Clear Lake. This increased deposition appears to be recorded in sediment cores from the region of the lake immediately offshore from Kelsey Creek and is represented as apparent increased sedimentation rate during the period starting ca. 1965 (see Suchanek et al., 1997; Richerson et al., 2000).

Contaminants

Aquatic Pesticides

Pesticide applications in Clear Lake have been used primarily to eradicate insects (the Clear Lake gnat *Chaoborus astictopus*) and emergent macrophytes (the exotic and prolific aquatic weed *Hydrilla verticillata*).

In the 1940s, the Clear Lake gnat (at larval densities of 640 larvae per square foot of lake bottom) was creating a serious aesthetic nuisance; while the adult gnats are nonbiting, they would aggregate in huge clouds around the lake's rimlands (Dolphin, 1959). University of California researchers had studied the gnat biology and possible control from 1916 to 1936, and in 1938 Congress appropriated funds to the Department of Agriculture (USDA) for further study of the gnat and its control. The USDA and the California Department of Fish and Game began lab treatment studies in 1945 to 1946 and field trials in 1947. The Lake County Mosquito Abatement District (which became the Lake County Vector Control District in 1995) was created in 1948 to deal with this problem. To reduce populations of the gnat, the California Department of Health Services contributed to funding three large applications of dichloro diphenyl dichloroethane (DDD), about 4×10^4 gal of 30% DDD per treatment in Clear Lake: in 1949, 1954, and 1957 (Dolphin, 1959; Hunt and Bischoff, 1960; Rudd, 1964; Cooke, 1981). DDD was added in extremely high concentrations (Clear Lake water contained ~0.02 ppm DDD) and resulted initially in 99% kill of the gnat larvae in sediments, although the effectiveness of this kill rate declined in future years (Hunt and Bischoff, 1960). Additional treatments of DDD also were applied to 20 other small lakes and reservoirs within about 25 km of Clear Lake.

Unfortunately, it also killed many other benthic invertebrates and had a devastating impact on the resident breeding populations of the western grebe (*Aechmophorus occidentalis*) (Herman et al., 1969). Five years after the initial DDD applications (in 1954), over 100 grebes were found dead in one survey season; in 1957, another 75 dead grebes were documented. No disease was identified, but in 1957 analyses of grebe fat tissues revealed extremely high concentrations of DDD (1600 ppm). In 1958, Clear Lake fishes finally were collected and also found to have excessively high DDD concentrations (40 to 2500 ppm); the largest concentrations were found in brown bullhead (*Ameiurus nebulosus*), largemouth bass (*Micropterus salmoides*), and black crappie (*Pomoxis nigromaculatus*). In addition, there was a substantial crash of the breeding grebe populations. Prior to 1949, over 1000 nesting pairs were estimated at Clear Lake. In the period 1958 to 1959, there were fewer than 25 pairs observed, but none was nesting; and at the end of that season no fledglings were found (Hunt and Bischoff, 1960). Concentrations of DDD in plankton were about 265-fold higher than the water in which they were found, about 500-fold higher in small fishes and ~80,000- to 85,000-fold higher in predaceous birds (grebes) (Lindquist and Roth, 1950; Rudd, 1964). This was the first identification of the process of pesticide bioaccumulation in food webs and the phenomenon of delayed expression of toxic symptoms among biological concentrators of pesticides (Carson, 1962; Rudd, 1964). Interestingly, the period when the western grebe populations were declining (1950s) was the same period during which the last major mercury mining was taking place at the Sulphur Bank Mercury Mine. Yet no one has investigated the possibility that mercury also played a significant role (additively or synergistically) in the decline of the western grebe populations. We (DWA and THS) are in the process of evaluating museum specimens of western grebes collected from Clear Lake during this era to determine whether mercury levels also were elevated. The signal of residual DDD within Clear Lake sediments from this period is so prominent that it also has been used as a dating tool (Chamberlin et al., 1990; Suchanek et al., 1993, 1997).

Within a few years, DDD applications were ineffective in reducing gnat populations, likely because the insects became resistant to the pesticide (Apperson et al., 1978). As a result, a series of other insecticide compounds (including EPN, parathion, dicapthon, methyl trithion, methyl parathion, 2,4-dimethyl menzyl ester of chrysanthemumic acid, barthrin, trithion, Co-ral, DDVP, delnav, diazinon, dylox, ethion, guthion, korlan, malathion, phosphamidon, phostex, and sevin) were used in laboratory tests for their efficacy in killing the *Chaoborus* gnat, but most were never used in Clear Lake proper (Dolphin and Peterson, 1960). Because of the strong resurgence of the gnat populations, two stop-gap measures were implemented in 1959 (Dolphin and Peterson, 1960). The first involved spraying a petroleum product (Richfield Larvicide™) on gnat eggs located on 5500 acres of rafts of drifting debris material along the shorelines of Clear Lake. The other stop-gap measure involved the spraying of malathion to tree and shrub resting areas of adult gnats. In 1962, three applications of methyl parathion were made to Clear Lake (at a concentration of 3.3 ppb), which effectively controlled *C. astictopus*. Methyl parathion was subsequently applied each summer from 1962 to 1975, at which point the treatments failed to control the gnat which had, once again, developed a resistance.

Hydrilla verticillata (see more documentation under introduced species) has become an aggressively growing nuisance macrophyte since about 1994. Beginning in the summer of 1996, annual efforts to control and eliminate this weed have been undertaken primarily using two aquatic herbicides, Komeen™ (copper sulfate), which acts on the emergent vegetation, and SONAR™ (fluridone), which acts on the tubers and is intended to stop production of propagules. Because copper is applied in relatively high concentrations locally, and because mercury is an ongoing contaminant in Clear Lake, it is valuable to understand the interaction of copper and mercury on aquatic biota. Only one formal study has undertaken the task of evaluating the impacts of the multiple contaminants mercury and copper on Clear Lake zooplankton (Gilmartin, 1998; Gilmartin et al., 1998). These preliminary results indicate that copper and mercury act independently (i.e., additively) with respect to their effects on zooplankton behavior and reproduction.

Finally, there are many private (i.e., homeowner) uses of pesticides that are not subject to pesticide use reporting. Several over-the-counter pesticide products are available without a permit (defined as an operator identification number in pesticide use law) and are not regulated by California Pesticide Use Reporting requirements. These pesticides have the potential to enter Clear Lake waters, but no studies have quantified their importance, either from a loading or impacts perspective.

Terrestrial Pesticides

While the human population in the Clear Lake Basin has increased over the past century, so too has the conversion of the landscape to agricultural production (discussed later). The widespread use of highly toxic organophosphate and organochlorine pesticides (OPs), mostly in the lake (e.g., DDD, see previous) has been reduced dramatically over the past 50 years. However, intimately associated with high agricultural production, and particularly with some of the high value crops, has been an increase in the use of pesticides to increase crop yield and ensure crop quality. Since many of the orchards and vineyards surround the lake or eventually drain into the lake, it is important to evaluate whether the use of such xenobiotics may affect the ecology of the watershed. Figure 121.8 illustrates trends of the ten highest use categories of pesticide application, and Figure 121.9 provides trends on the application of the top 10 of the approximately 165 most-heavily used pesticides in Lake County between 1990 and 1998 (California Department of Pesticide Regulation, 1990 to 1999).

Nearly 1 million pounds annually of various petroleum and mineral oils are used primarily on pears (to delay or discourage egg laying by psylla or to smother mite eggs and in some cases codling moth eggs). On rare occasions, oils are used on grapes. During the application of dormant and emergent sprays, these oils often are used simultaneously with some organophosphate pesticides (usually during summer). Currently, alternative approaches to control codling moth involve the use of pheromone mating disrupters, especially when the density of moths is relatively low (Elkins and Shorey, 1998; Bentley et al., 1999). The second most heavily used product, which is not acutely toxic to humans or wildlife at low doses, is sulfur and lime-sulfur. The primary use of these compounds has been as

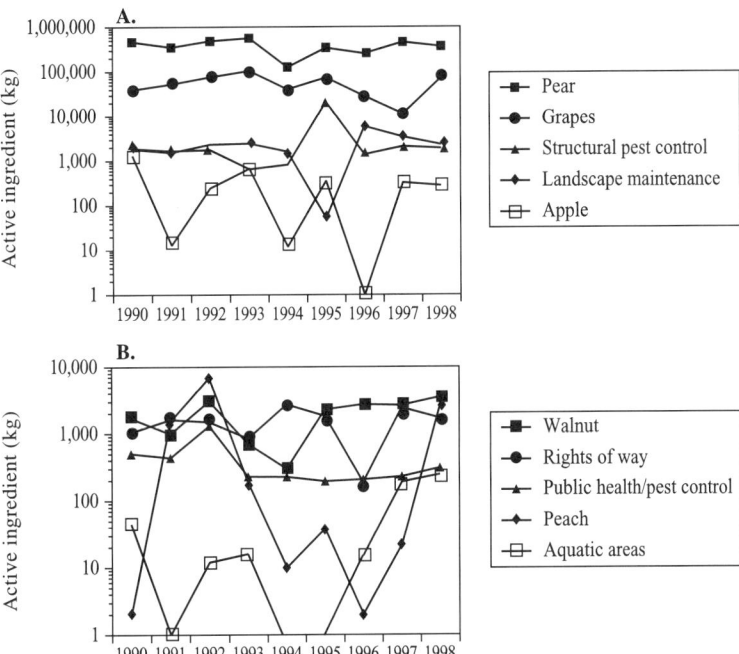

Figure 121.8 Annual cumulative pesticide applications for the top ten crops and uses in Lake County from 1990 to 1998 (by weight). (A) Total pesticide applications for the top five crops (by weight). (B) Total pesticide applications for the second five crops with the heaviest pesticide applications (by weight).

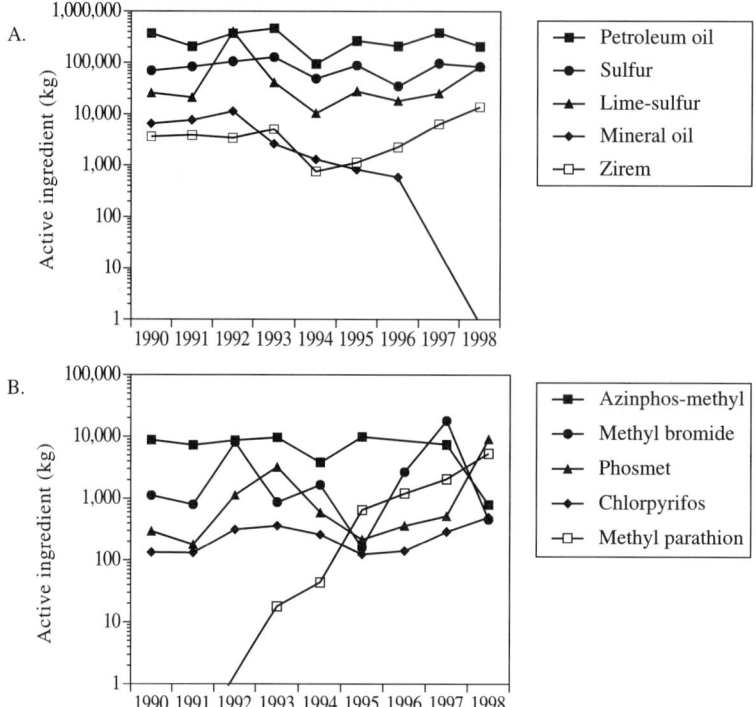

Figure 121.9 Annual pesticide applications for the ten most heavily used pesticides in Lake County from 1990 to 1998 (by weight). (A) Total applications for the top five pesticides. (B) Total applications for the second five most heavily applied pesticides.

fungicides on vineyards for protection against powdery mildew. While not highly toxic, the high application rates have the potential to increase sulfur loading into the lake, which also may interact with sulfate-reducing bacteria in the conversion of inorganic mercury to methyl mercury. However, no studies have been conducted to determine the origin of sulfur compounds in Clear Lake.

Although the four most widely used pesticides have a low environmental risk (both acute and chronic), the use of more-toxic compounds is still widespread. Examples of this have been the two most heavily applied broad-spectrum OPs, methyl parathion and azinphos-methyl (Guthion) (see Figure 121.9). Methyl parathion increased dramatically from approximately 30 lbs in 1993 to nearly 12,000 lbs in 1998, mostly for use against codling moth. Guthion, on the other hand, has seen nearly constant usage from 1990 to 1998, when usage dropped significantly, mostly because codling moth began to show signs of resistance. Due to the U.S. EPA's recent restrictions on methyl parathion residues, and a ban against its use after 1999, we expect to see a rise in Guthion rates in response, although recent restrictions on Guthion will likely keep its use low as well. Although the use of these insecticides is high, their potential ecological impact is somewhat offset by the fact that they have relatively short environmental persistence. Both compounds have terrestrial half-lives from 3 to 16 days, and aquatic half-lives of 1 to 6 days. This significantly reduces their risk to wildlife, unless large amounts are used near the lake as fish and aquatic invertebrates are particularly sensitive to these compounds. Nevertheless, additional research is needed to determine whether terrestrial pesticides are being transported into Clear Lake and, if so, what their potential impacts to the aquatic community might be.

MTBE

Methyl tertiary butyl ether (MTBE), a synthetic chemical *oxygenate* added to gasoline to improve air quality as required by the Clean Water Act, had been used in gasoline in limited quantities since the 1970s. It is considered a possible human carcinogen. It is highly soluble in water and does not readily degrade in the environment; most public water systems are not equipped to remove it completely from drinking water. It has a turpentine-like taste and smell, and initial studies show that some people can detect it in drinking water at concentrations as low as 2.5 ppb. In 1992, oil companies began using it extensively, and in recent years it has been detected in Clear Lake water. A 1998 University of California study indicated that there are significant risks and costs associated with water contamination due to the use of MTBE. In addition, it found that the use of gasoline containing MTBE in motor boats and crafts, in particular those with older two-stroke engines, results in the contamination of surface water reservoirs. In January 1999, the California Department of Health Services established a secondary (taste and odor-based) maximum contaminant level (MCL) allowable standard of 5 ppb, and a primary human health-based MCL of 13 ppb in May 2000. In March 1999, Governor Gray Davis issued an Executive Order for California to phase out the use of MTBE by 2003, and legislation to ratify the order (Senate Bill 989) was passed during the 1999 session. In March 2000, the federal government announced it would ban the use of MTBE under the Toxic Substances Control Act.

Some water samples collected from various locations within Clear Lake have detectable levels of MTBE, but those levels were always lower than the MCL for taste and odor. Nine of 53 Lake County water samples collected from January 1984 to December 1999 showed detectable MTBE concentrations. Of those 53 samples, 23 were collected within Clear Lake proper and 8 of those 23 showed a range of MTBE concentrations of 1.1 to 4.5 ppb. These values are below the taste and odor MCL standard of 5 ppb and significantly below the human health-based standard of 13 ppb, now in effect.

Mining

Clear Lake has contributed significantly to mining operations within the U.S. In 1864, this region was the first location to be mined for borax within the U.S. (Bailey, 1902), and a year later California's first sulfur was extracted from a surface deposit of elemental sulfur called the Sulphur Bank Mine (California Division of Mines, 1950). The deeper deposits of sulfur from this site were contaminated with cinnabar (mercury sulfide), and in 1872 the site was converted to mercury mining as the Sulphur Bank Mercury Mine. At that time California accounted for 89% of the nation's mercury production and the Sulphur Bank Mercury Mine produced about 10% of California's total mercury output (Simoons, 1952). Because of elevated concentrations of mercury in fishes first documented during the 1970s (Curtis, 1977), the Sulphur Bank Mercury Mine was placed on the U.S. EPA's National Priority List as a Superfund site in 1990 (Suchanek et al., 1993).

Mining at the Sulphur Bank Mercury Mine is believed to have contaminated the lake with both mercury and arsenic (Sims and White, 1981; Chamberlin et al., 1990; Suchanek et al., 1993, 1997). Inorganic mercury concentrations in lake bed sediments are significantly elevated (over 400 ppm) close to the mine and decline exponentially with distance from the mine (Suchanek et al., 1997, 1998a, 2000b). Arsenic is only slightly elevated close to the mine, but also exhibits a recognizable background signal throughout the lake, likely as a result of outflow from numerous lakebed springs (Suchanek et al., 1993). Since 1992 detailed studies have been conducted on mercury contamination from the Sulphur Bank Mercury Mine (Suchanek et al., 1993, 1995, 1997, 1998a,b, 2000a,b; Anderson et al., 1997; Mack et al., 1997; Slotton et al., 1997; Webber and Suchanek, 1998; Wolfe and Norman, 1998). Research at this site has identified acid mine drainage, which is low in pH and high in sulfate, as the most likely point source for methyl mercury contamination in Clear Lake (Suchanek et al., 1993, 1997, 2000b; Nelson et al., unpublished). Remediation of the mine is expected to begin around 2003.

Other mining operations have been widespread throughout the Clear Lake Basin. Figure 121.10 provides a location map for the diverse array of mining operations that have existed within this watershed over the past 150 years. With the advent of home and especially road

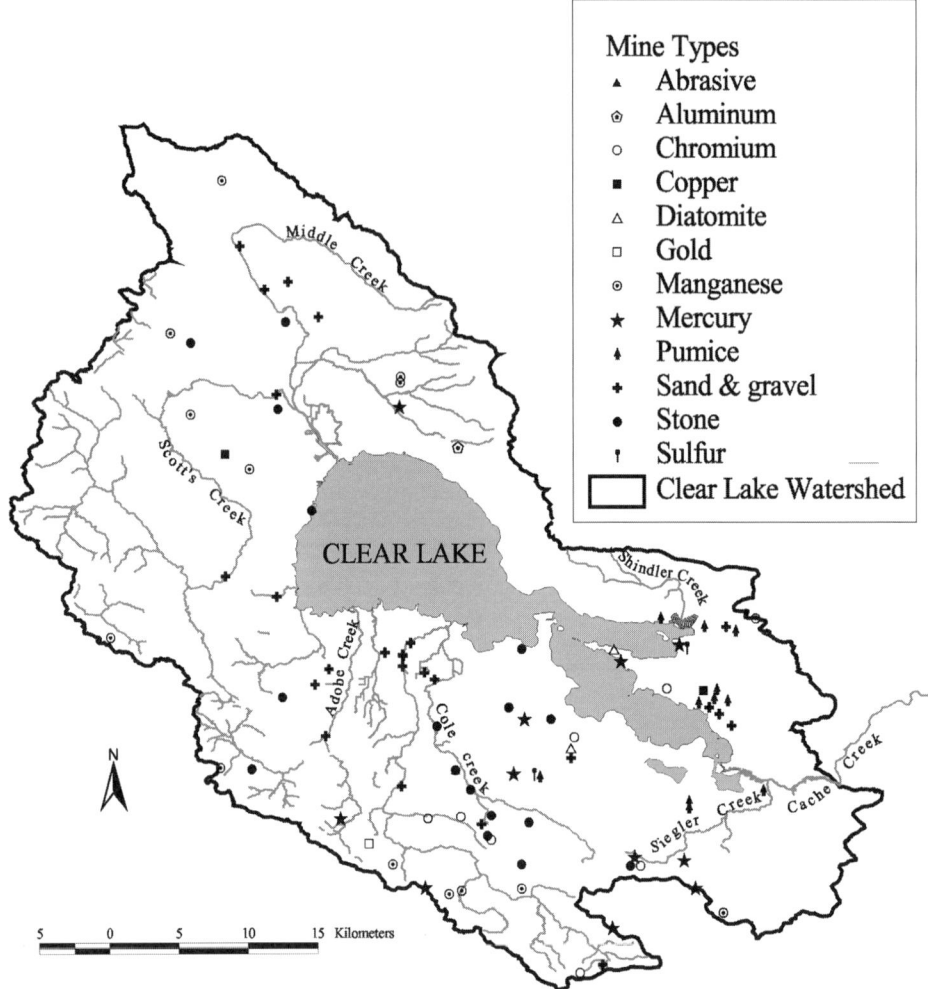

Figure 121.10 Map of mining operations within the Clear Lake watershed over the past 150 years.

construction in the early 1970s, Lake County needed an easily accessible source of gravel as road base. Creek beds provided this resource; thus large volumes of gravel were extracted from Scotts Creek, Middle Creek, Kelsey Creek, Adobe Creek, Forbes Creek, Cole Creek, and Burns Valley Creek (see Figure 121.10). Gravel mining, which was common until 1987 (Zalusky, 1992), changed the level of stream beds as much as 15 ft and caused destabilization and increased erosion during the next flooding season, carrying higher loads of sediments and associated nutrients into the lake. Additionally, roads now block upstream areas that once were used for spawning (e.g., hitch). Volcanic cinder cones also have been mined as a source of aggregate and decorative rock, but the environmental impacts have been relatively small compared with mining in active stream beds. Mining operations that are still active include primarily sand and gravel, cinders, and decorative stone and rock.

Sewage and Septic Overflows

Although there have been many sewage overflows, especially during major flooding periods, this probably does not represent a significant nutrient loading to Clear Lake (see Richerson et al., 1994, for discussion).

Land-Use Changes

Wetland Losses and Gains

Original natural wetland acreage around the rimland of Clear Lake declined by nearly 85% from about 9000 acres in 1840 to about 1500 acres in 1977 (Table 121.3). One of the most dramatic land-use changes experienced within the Clear Lake watershed has been the process of wetland conversion to agriculture. Beginning in the 1890s, a large area of natural wetlands from the Robinson Lake and Tule Lake region in the northwest side of Clear Lake was reclaimed primarily for agricultural production of string beans and lima beans (Simoons, 1952). This was followed by another reclamation project in 1927 in the Middle Creek watershed for agricultural production, and the 1959 Rodman wetland reclamation by the Army Corps of Engineers. These projects collectively increased agricultural production, but also altered the transport of sediments and associated nutrients into Clear Lake, with the likely result that noxious cyanobacteria (blue-green "algae") blooms significantly increased beyond previous levels (Richerson et al., 1994). Wetland reclamation is not unique to Clear Lake; it is a process that was similarly initiated in the late 1800s throughout the country, and especially in the San Francisco Bay–Delta system where 79% of the marshlands have been lost in the last 200 years, and over 538,000 acres were converted to agricultural land (Monroe et al., 1999).

There are several categories of newly created water bodies or wetland habitats, including sewage treatment ponds, agricultural ponds, irrigated crops (such as rice), or pastures and reservoirs, which can have both positive or negative feedbacks to the Clear Lake ecosystem. For example, several tracts of original wetland habitat, such as Tule Lake, are now in rice production. These habitats are managed by flooded irrigation practices, so for much of the year they are functional wetlands and enhance populations of some invertebrates, which in turn provide food resources for some wildlife. These practices can both enhance wildlife and produce potential problems for disease vector agents such as mosquitoes. Some of the more recently created reservoirs can also be beneficial in trapping sediments with associated nutrients before they enter Clear Lake, which would also help to limit nutrient loading. However, we are not aware of any studies that have compared the functionality of original vs. newly created wetlands for any of these categories.

The intrinsic ecological values, in terms of biodiversity and natural filtering capacity of wetlands, have now become more recognized, and a trend to restore and rehabilitate previously reclaimed wetlands has been popular in recent years. At present, reclaimed former wetlands in both the San Francisco Bay–Delta system and the Clear Lake system are being considered for wetland restoration and rehabilitation. In the case of Clear Lake, the Army Corps of Engineers has completed a Reconnaissance Phase and is currently in the Feasibility Phase of a plan to restore up to 1200 acres of wetlands in the Robinson Lake/Middle Creek region of Clear Lake (Jones and Stokes, 1997; Smythe, 1997; Van Nieuwenhuyse, 1997).

Table 121.3 Wetland Loss in the Region Surrounding Clear Lake

Year	Acres	% of 1840 Acreage
1840	9000	100
1920	5400	60
1952	5300	59
1958	5200	58
1966	3400	38
1968	1900	21
1977	1500	17

Dredging and Filling

Over the past 150 years, vast areas around the rimland of Clear Lake that previously were wetlands have been filled or converted to either private, commercial, or public use (for homes, businesses, or roads), especially in response to the population increase after about 1925. These projects also tended to deliver large volumes of sediment and nutrients to Clear Lake (Richerson et al., 1994). The residential subdivision of Clearlake Oaks (known as the Keys) resulted in the extensive dredging of wetlands during the 1960s, and in the generation of about 6.5 miles of navigable channels. Other similar developments include Corinthian Bay and Lands End. For other smaller projects, these sometimes subtle yet continuous changes have been the most difficult to document, for there are few records that quantify conversion of small- to medium-sized areas of wetlands, especially before the turn of the 20th century.

Within the past 5 years there has been a movement by one citizens' group to dredge Clear Lake to a depth of about 60 ft (about 18 m) or more, with the stated purpose of generally "improving water quality," but without any documented specifics. Because the average depth of Clear Lake in most locations is 8 to 10 m, this would involve dredging another 9 m deeper. This concept is fraught with problems, not only economic, but ecological as well. Because relatively high concentrations of mercury and pesticides (DDD) are buried in the sediments of Clear Lake (especially within the top 60 to 70 cm) (Suchanek et al., 1993, 1997; Richerson et al., 2000), these materials would have a high probability of becoming remobilized into the water column and bioaccumulated into the trophic web in the course of dredging. Another problem with increasing the depth of Clear Lake to 60 ft is related to the issue of summer water column stratification. Most deep lake systems undergo temperature and oxygen stratification during the summer, whereby a temperature discontinuity (thermocline) is formed in the water column with an associated water mass at the bottom (hypolimnion) that has dramatically different physical and chemical characteristics from the surface water mass. This hypolimnion is typically cooler and is much more likely to be anoxic than surface waters. Clear Lake's present depth creates a situation in which bottom waters (near the sediment–water interface) occasionally become anoxic, but this condition does not typically last long because of wind-driven mixing. If the lake were significantly deeper, this anoxic condition would persist for long periods of time, perhaps months. This is exactly the condition under which sulfate-reducing bacteria flourish. As this microbial group has been implicated in the conversion of inorganic mercury into toxic methyl mercury, a deeper lake would most likely promote a higher production of toxic methyl mercury in the bottom waters. Winter turnover of the water column would allow this methyl mercury to be remobilized and potentially bioaccumulated by organisms in the lake. Because mercury is already an issue of great concern in Clear Lake, any changes that would promote further methyl mercury production would be undesirable. Nevertheless, such proposals point to the need for ongoing studies and monitoring to understand how natural and anthropogenically mediated processes may influence the complex dynamics imposed by multiple stresses on this ecosystem.

Creek Bed, Water Table, and Shoreline Modifications

The lower reaches of Clear Lake hitch (*Lavinia exilicauda*) spawning streams have usually dried up naturally, but due to groundwater extraction, these areas dry up earlier, resulting in spawning failures; additionally, the loss of marshy wetland areas surrounding the lake limits the habitat available to larval hitch.

Agriculture

Private fruit and nut orchards were planted early (around 1860) within the Clear Lake Basin with the production of pears, plums, prunes, apples, almonds, peaches, nectarines, and grapes, and commercial orchards started in the 1880s (Simoons, 1952). Figure 121.11 provides estimates

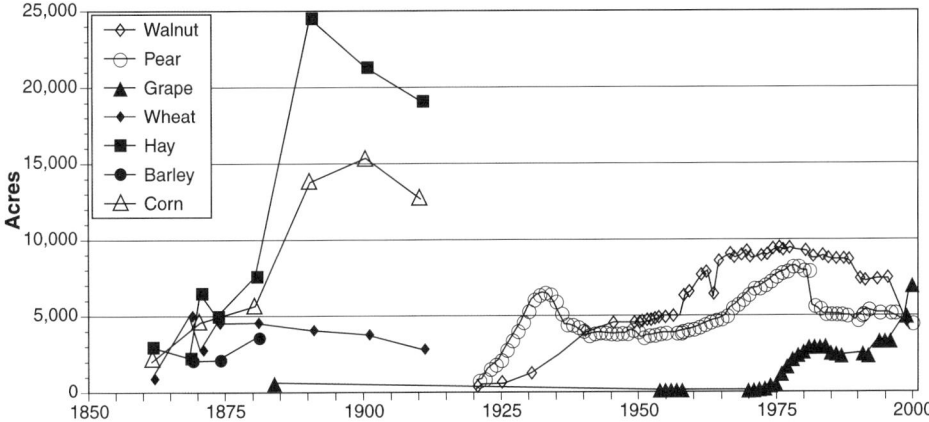

Figure 121.11 Trends in conversion of landscape to agriculture in the Clear Lake Basin. *Note:* No data available for hay, corn, wheat, and barley after 1910; no data on walnuts and grapes before 1920; no data on grapes before 1880.

of acreage planted in the top five crops and orchards since the early 1900s, when reasonably accurate records were first kept. Much of the increase in wetland reclamation was driven by a desire to convert rich, easily irrigatable soils into profit-making tracts of land. The types of crops grown were driven, to a large degree, by market value. At present, wine has made a significant resurgence in the marketplace and a large movement to convert existing orchards (especially walnut orchards) into grape production is under way. Over the past 10 years, the land area committed to growing grapes has increased dramatically from about 2500 acres in 1989 to 7000 in 1999, with no immediate end of the surge in sight (see Figure 121.11). During at least the first 2 years of the establishment of a vineyard (before any ground cover can be established), there is the potential for sheet wash erosion to transport large volumes of sediment and associated nutrients into Clear Lake. This potential problem can be reduced by the early establishment of a ground cover crop that will help stabilize the soil. While this is a topic of considerable debate within the agricultural and environmental community in Lake County, to date we are unaware of any specific studies that have quantitatively addressed the impacts of vineyards or orchards on erosion within the Clear Lake Basin.

Cattle and Sheep Grazing

The first cattle (initially Longhorn, then many other breeds including Shorthorn, Hereford, Polled Hereford, Angus, and Scotch Highland) were brought to the Clear Lake area in 1839, and their grazing impacts on the landscape increased thereafter (Mauldin, 1968). Richerson and co-workers (2000) discuss some of the early farming and ranching practices. Sheep herds were abundant and peaked sharply in the late 1870s at about 50,000 head; declined dramatically thereafter due to declining pasture quality; increased again to nearly 40,000 head around the 1930s; and then steadily declined to today's low levels. Goats and hogs peaked around the turn of the 20th century at about 10,000 each, yet cattle numbers were more stable with about 10,000 head from 1861 through the 1980s. Even in early days, there was a problem with overgrazing, which resulted in the eventual closing of most of the local forest to cattle ranchers in 1947 (Mauldin, 1968). Overgrazing causes decreased soil stability and increased erosion, and results in greater nutrient input to the lake. Recently, grazing has become a problem in regions of emergent vegetation, especially in low water years when cattle have access to tule beds (as happened in 1999), because it can impact grebe nesting colonies.

Soil Exposure and Transport

Construction of residential and commercial facilities along the rimlands and on the upper slopes of the Clear Lake watershed have the continual effect of transporting sediments and associated nutrients into Clear Lake during the rainy season. Nonpaved roads also contribute significant loading of erosional materials to Clear Lake. During the summer, large numbers of off-road vehicles traverse back country roads (e.g., about 20 miles along Cow Mountain), loosening soil from the roadbed. At the end of the summer season, there is often a 5- to 10-cm layer of fine-grained soil (and associated nutrients) on the surface of the roadbed, which is flushed into drainage pathways and the lake with the first major rainstorm.

Species Introductions

Due in large part to its high productivity, Clear Lake has continually supported a very rich fish fauna. The structure of the fish community has changed a great deal over the past 100 years in response to numerous alien fish introductions. Prior to European settlement, there were 13 naturally occurring fish species in the lake, 4 of which were endemic (Hopkirk, 1973). The current fish community consists of 21 species, only 5 (24%) of which are native to the lake (see Table 121.4). The introductions began in the late 1800s with the establishment of three species, the white catfish (*Ameiurus catus*), brown bullhead (*A. nebulosus*), and common carp (*Cyprinus carpio*) (Murphy, 1951) and all three species continue to flourish in the lake.

The introduction with perhaps the greatest impact on the Clear Lake fish community in recent years was the inland silverside in 1967 (Li et al., 1976). Like the western mosquitofish (*Gambusia affinis*), the inland silverside was introduced primarily for control of the Clear Lake gnat. Within 1 year, they established themselves as one of the most abundant fish in the lake and now are the dominant planktivore of the littoral zone (Moyle, 2002). This niche dominance likely acted as a catalyst for sweeping changes throughout the lake. Upon its introduction, the inland silverside reduced zooplankton populations in the nearshore regions of Clear Lake, outcompeting other planktivorous fishes in that region (Moyle, 2002). Moyle also credits the introduction of the inland silverside as a major factor in the final demise of the planktivorous Clear Lake splittail (*Pogonichthyes ciscoides*), a species already in serious decline. One example of the inland silverside's effects on other fish is provided by the black crappie (*Pomoxis nigromaculatus*) and white crappie (*P. annularis*). After the introduction of the inland silverside, the mean standard length of young crappie (also planktivores) decreased and that of adult crappie increased from pre-introduction levels (Li et al., 1976). This suggests that the inland silverside may have competed with young crappie, and that adult crappie were likely preying on silversides. The silversides also have become the primary forage fish for other predatory species and undoubtedly have helped the already booming sportfishing industry in Clear Lake.

Other introductions that have had dramatic impacts on the structure of the aquatic community are the myriad sportfishes such as several species of catfish (e.g., *Ameiurus catus*, *Ictalurus punctatus*), the largemouth bass (*Micropterus salmoides*), and other centrarchids. Moyle (2002) notes that the success of the white catfish (*A. catus*) was associated with a decline of the native cyprinids. He additionally credits the elimination of the pikeminnow (*Ptychocheilus grandis*, previously called squawfish) and the extinction of the thicktail chub (*Gila crassicauda*) to the presence of the largemouth bass and other exotic predators.

The most-recent fish introduction is that of the threadfin shad (*Dorosoma petenense*) in 1985 (Anderson et al., 1986). The impacts of the shad introduction have not been thoroughly quantified; however, it is generally assumed that they compete heavily with the silversides for food, leading to greatly fluctuating zooplankton populations. It also has been observed that the shad themselves go through very large population fluctuations, as evidenced by numerous die-offs that have occurred since their introduction. These generally occur in winter (as occurred during the winters of 1990

Table 121.4   Past and Present Fishes in Clear Lake, including Dates of Introduction for Alien Species

Species	Family	Native/ Introduced	Trophic Position (Juvenile/Adult)	Current Status	Reason for Introduction
Thicktail chub (Gila crassicauda)	Cyprinidae	Native	Omnivorous piscivore and invertivore assumed trophic position[a]	Extinct (1941–1950)	—
California roach (Lavinia symmetricus)	Cyprinidae	Native[b]	Benthic browser	Extinct (<1963)	—
Clear Lake splittail (Pogonichthys ciscoides)	Cyprinidae	Native	Littoral planktivore/pelagic planktivore	Extinct (1972)	—
Hitch (Lavinia exilicauda)	Cyprinidae	Native	Littoral planktivore/pelagic planktivore	Abundant	—
Sacramento blackfish (Orthodon microlepidotus)	Cyprinidae	Native	Pelagic planktivore/benthic detritivore	Abundant	—
Sacramento pikeminnow (Ptychocheilus grandis)[c]	Cyprinidae	Native[b]	Pelagic invertivore/pelagic piscivore	Unknown	—
Threespine stickleback (Gasterosteus aculeatus)	Gasterosteidae	Native	Benthivore	Extinct (<1894)	—
Rainbow trout (Oncorhynchus mykiss)	Salmonidae	Native	Pelagic and benthic invertivore	Extinct (<1963)	—
Pacific lamprey (Lempetra tridentata)	Petromyzontidae	Native[b]	Parasite	Unknown	—
Sacramento sucker (Catostomus occidentalis)	Catostomidae	Native[b]	Benthic detritivore and invertivore	Unknown	—
Sacramento perch (Archoplites interruptus)	Centrarchidae	Native	Littoral planktivore/benthic invertivore and piscivore	Rare	—
Tule perch (Hysterocarpus traski)	Embiotochidae	Native	Planktivore and invertivore	Common	—
Prickly sculpin (Leptocottus armatus)	Cottidae	Native	Benthic invertivore	Abundant	—
Lake whitefish (Coregonus clupeaformis)	Salmonidae	Introduced (1873)[d]	Planktivore/benthic invertivore	Unsuccessful introduction	Sportfishery
Common carp (Cyprinus carpio)	Cyprinidae	Introduced (1880)[d]	Benthic omnivore	Abundant	Accidental
Brown bullhead (Ameiurus nebulosus)	Ictaluridae	Introduced (1880)[d]	Benthic littoral omnivore	Abundant	Sportfishery
Largemouth bass (Micropterus salmoides)	Centrarchidae	Introduced (1888)[d]	Top predator	Abundant	Sportfishery
Golden shiner (Notemigonus crysoleucas)	Cyprinidae	Introduced (1896)[d]	Littoral and pelagic planktivore	Rare	Forage fish
Bluegill (Lepomis macrochirus)	Centrarchidae	Introduced (1910)[d]	Planktivore/omnivore	Abundant	Sportfishery
Black crappie (Pomoxis nigromaculatus)	Centrarchidae	Introduced (1910)[d]	Pelagic planktivore/omnivore	Common	Sportfishery

(continued)

Table 121.4 (continued) Past and Present Fishes in Clear Lake, including Dates of Introduction for Alien Species

Species	Family	Native/ Introduced	Trophic Position (Juvenile/Adult)	Current Status	Reason for Introduction
Yellow perch (*Perca flavescens*)	Percidae	Introduced (1910)[d]	Littoral omnivore	Unsuccessful introduction	Sportfishery
Goldfish (*Carassius auratus*)	Cyprinidae	Introduced (1920)[e]	Littoral herbivore and detritivore	Common	Forage fish
Channel catfish (*Ictalurus punctatus*)	Ictaluridae	Introduced (1920)[e]	Omnivore, mostly invertebrates and fish	Common	Sportfishery
White catfish (*Ameiurus catus*)	Ictaluridae	Introduced (1923)[e]	Benthic invertivore/benthic omnivore	Abundant	Sportfishery
Brown trout (*Salmo trutta*)	Salmonidae	Introduced (1924)	Invertivores/predators	Extinct (<1963)	Sportfishery
Western mosquitofish (*Gambusia affinis*)	Poeciliidae	Introduced (1925)[e]	Invertivore	Common	Pest control
Green sunfish (*Lepomis cyanellus*)	Centrarchidae	Introduced (1935)[e]	Littoral omnivore	Common	Sportfishery
Fathead minnow (*Pimephales promelas*)	Cyprinidae	Introduced (1955)[e]	Benthic omnivore	Rare	Forage fish
White crappie (*Pomoxis annularis*)	Centrarchidae	Introduced (1955)[e]	Pelagic omnivore	Common	Sportfishery
Redear sunfish (*Lepomis microlophus*)	Centrarchidae	Introduced (1965)[e]	Molluscivore	Rare	Sportfishery
Inland silverside (*Menidia beryllina*)	Atherinidae	Introduced (1967)[e]	Littoral planktivore	Abundant	Pest control
Threadfin shad (*Dorosoma petenense*)	Clupeidae	Introduced (1985)	Epipelagic planktivore	Abundant	Forage fish

a From Moyle, 2002.
b Found in watershed streams.
c Formerly known as squawfish.
d From Murphy, 1951.
e From Li and Moyle, 1979.

to 1991 and 1998 to 1999 when tens of thousands of threadfin shad died and floated to shore), and likely is due to their poor tolerance of cold water, because they have great difficulty feeding below 9°C (Griffith, 1978). The introduction of the threadfin shad resulted in a significant decline in *Daphnia* populations that serve as the primary food source for young-of-year largemouth bass. Thus, a cascade effect that reduced the numbers of largemouth bass was correlated with the introduction of shad. Furthermore, when the shad population crashed in 1990, the largemouth bass population made a significant rebound, which most likely was due to the resurgence in the number of *Daphnia* found in the lake (Colwell et al., 1997). One study suggests that *Daphnia* consume primarily planktonic diatoms, but typically leave the less-palatable blue-green "algae" (Elser et al., 1990). Additionally, Colwell and co-workers (1997) have shown a strong positive correlation between threadfin shad abundance and that of various piscivorous birds such as Clark's grebes (*Aechmophorus clarkii*) and western grebes (*A. occidentalis*), double-crested cormorants (*Phalacrocorax auritus*), and California gulls (*Larus californicus*). This may be because the shad prefer well-lighted surface waters, making them easy prey for the birds. After the 1990 population crash, the populations of the aforementioned birds returned to pre-introduction levels. This combination of top-down and bottom-up strong interactions suggests that if this species ever establishes permanently high densities, its introduction may result in dramatic changes throughout the ecosystem, perhaps greater than the changes caused by any previous introduction.

Over the past decade, there typically have been three to four commercial fishing licenses on Clear Lake, mostly focused on the Sacramento blackfish (*Orthodon microlepidotus*) and common carp (*Cyprinus carpio*), which typically are sold live to Asian markets in the San Francisco Bay region. In the 1960s and 1970s, there were years when more than 317,000 kg of carp were taken from the lake and sold mostly to processing plants for cat food. A single haul of carp in about 1990, toward the end of the 6-year drought, yielded about 45,400 kg; yet in recent years carp and blackfish have been in relatively low abundance (Meadows, personal communication, 2000).

Today, there still exist numerous threats to the survival of many Clear Lake fishes. For example, threats to hitch (*Lavinia exilicauda*) involve loss of spawning habitat and nursery areas. The establishment of the threadfin shad (*Dorosoma petenense*) tended to greatly reduce populations of the zooplankton *Daphnia*, a principal food of hitch.

Hydrilla (*Hydrilla verticillata*), indigenous to Southeast Asia (Langeland, 1996), is a noxious, nonnative, submerged aquatic weed that is spreading rapidly throughout the U.S. It was first discovered in Florida in the 1960s and for the first time in Clear Lake in August 1994 (O'Connell and Dechoretz, 1997; Dechoretz, 1998). Where it occurs it causes substantial economic hardships, interferes with various water uses, displaces native aquatic plant communities, and adversely affects freshwater habitats (Langeland, 1996). In 1994, about 175 to 200 surface acres in Clear Lake experienced infestation. This area increased to 648 acres by 1998 (Dechoretz, 1998), and in 1999 the affected area expanded to 845 acres* (Lockhart, personal communication, 2000). This weed is adapted to grow at low light levels, competes effectively for sunlight, and has a rapid growth rate (up to 1 in. per day). It has four different modes of reproduction and the potential to spread with enormous speed to clog enclosed and open waterways, including water bodies as large as Clear Lake (Langeland, 1996).

Ecosystem Complexity

The aquatic ecosystem of Clear Lake and its supporting basin and watershed are extremely productive and complex. The trophic structure of the lake's biota is complicated and dynamic. Figure 121.12 represents a simplified version of an elaborate, multitiered trophic network for some of the more common species in Clear Lake. Richerson and co-workers (1994; Table 121.3.3) also provide documentation of a preliminary food web for Clear Lake, including known predator–prey

---

* In addition to regions actively infested, this figure for 1999 also includes all areas that still are being treated in which *Hydrilla* has ever been found, and may not represent current infestation.

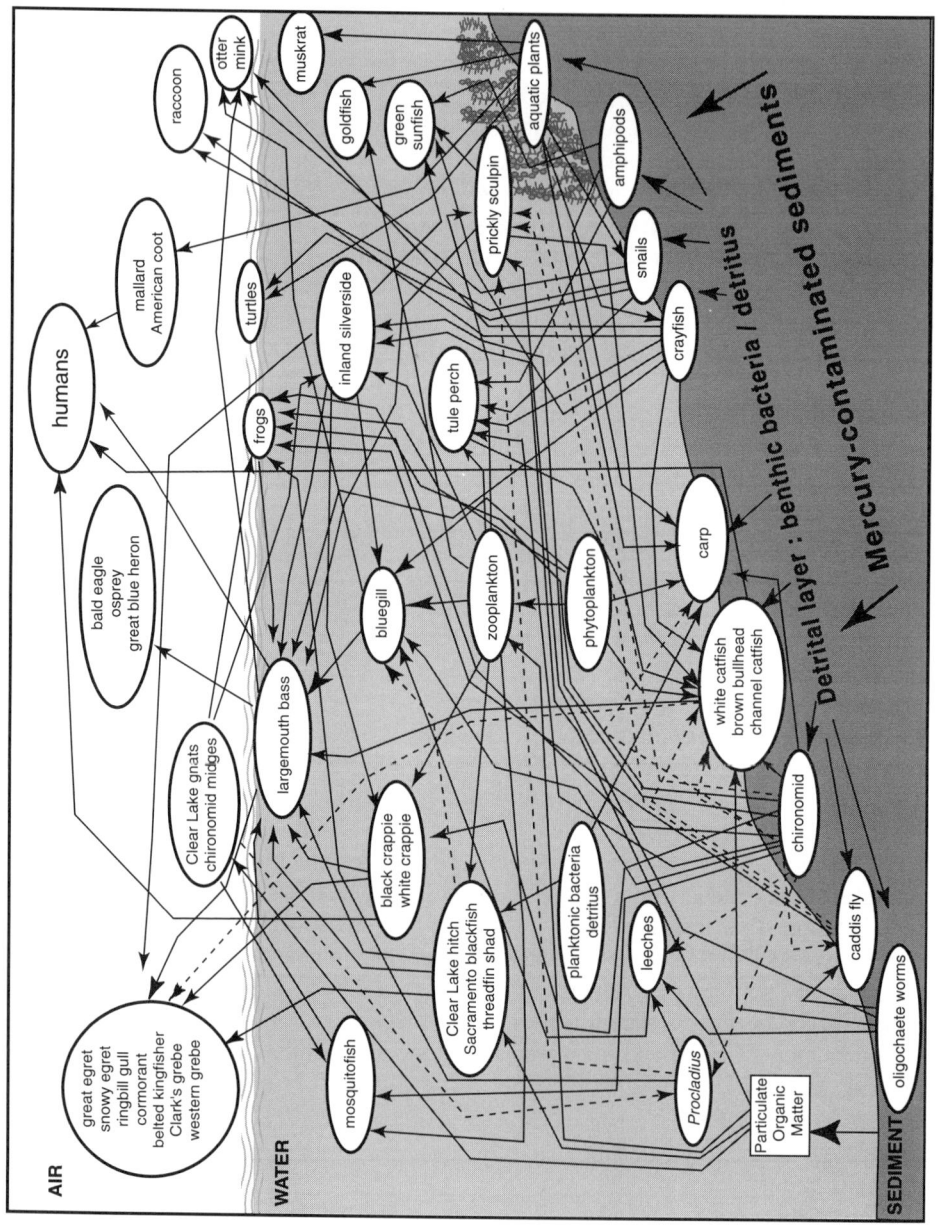

Figure 121.12 Simplified food web for Clear Lake.

relationships. Many of the factors identified above are dynamic, affecting the ecological relationships between species, and present significant challenges to resolving systemwide ecological problems. While considerable research has been conducted on the Clear Lake aquatic ecosystem, there exists a large degree of uncertainty about our knowledge of the impacts from the multiple stresses outlined previously.

Multiple Management Objectives

The previous discussion highlights many multiple stresses that affect the aquatic ecosystem of Clear Lake proper and the surrounding watershed and basin. These impacts affect not only the natural ecology of the system, but interact with aesthetic and economic dimensions as well. The Clear Lake Basin is an aesthetically beautiful setting and, as it has for over a century, attracts many thousands of tourists each year. Tourist visitation revenues for Lake County were estimated at $162 million in 1992 and increased on average about 7.9% per year to $237.5 million in 1998 (Dean Runyan Associates, 1999). It is estimated that $7 to 10 million of tourist revenue is lost annually due to the influence of poor water quality associated with cyanobacterial blooms alone (Goldstein and Tolsdorf, 1994). Given the complexities of managing a productive and multiuse watershed, many conflicts exist. For example, what is the best solution for maintaining the lake level? Downstream farmers in Yolo County would like to accumulate as much water as possible in Clear Lake during periods of winter precipitation for use the following summer; yet, the higher the lake levels rise, the greater risk of flooding to citizens living along the rimlands, as occurred when the first Clear Lake dam was erected in 1867 (see above). Another example relates to the use of pesticides and fertilizers for agricultural crops. Farmers typically use pesticides and fertilizers to maximize production; yet, the health of the aquatic ecosystem (especially its resources for sport and commercial fisheries) may be negatively affected by these practices. Some Lake County residents strive to enact regulations or change land-use patterns to reduce the free-floating cyanobacterial blooms and the noxious odors associated with their stranding on shore. During years when the blooms are minimal, water clarity is improved but shallow-rooted aquatic macrophytes proliferate and clog boat propellers. Yet, macrophytes are desirable as refuges and habitats for increased biodiversity and often act as nursery grounds for young fishes. Trade-offs will always exist, but it is the responsibility of science to provide informed decision-making options (and their inherent consequences) before exploiting multiple-use resources found within Clear Lake and its watershed.

Lake County has been one of the most progressive counties in the state in developing a Coordinated Resource Management and Planning (CRMP) process (Follansbee, 1996). Simply put, using a consensus model, the CRMP process brings together citizen interest groups and representatives from local, state, and federal agencies to develop integrated approaches to complex resource management issues. The first use of the CRMP process in 1949 to address grazing for an allotment in Oregon was documented by Anderson and Baum (1988). Over 200 such CRMPs are in existence within California and several have been started within Lake County. One of the first Clear Lake CRMPs dealt with the problems associated with the cyanobacterial blooms. A committee initially established as the Algae Abatement Committee in 1984 (and now operating under a broader mandate as the Clear Lake Advisory Subcommittee [CLAS]) has met regularly since its inception to address problems and potential solutions to eutrophication and other issues related to water use and quality within Clear Lake and its surrounding watershed. It has since evolved into an entity that addresses county-wide resource issues in an integrated manner while attempting to maintain and enhance the ecosystems and economy of Lake County (Follansbee, 1996). This group has been folded into a more-structured Lake County Resource Management Committee (RMC), which began in 1989 and now includes representatives from county-level resource agencies, agricultural interests, Native Americans, fishing and hunting groups, environmental special interest groups, realtors, Chambers of Commerce, and the University of California. The structure of the RMC includes four subcommittees: Clear Lake Advisory, Biological Resources, Land and Water Resources, and Database and

Information Outreach. These subcommittees seek out information from each of the representative participants and draw heavily upon past and ongoing scientific research that is conducted on Clear Lake and its surrounding watershed. In general, the RMC seeks to improve coordination of planning, research, and land and resource management by obtaining input from all interested parties, sharing information, collecting data, conducting research, and developing policies and regulations that will maximize benefits for the citizenry of the county (Follansbee, 1996). In addition to the Clear Lake (Upper Cache Creek) Watershed CRMP, there are several other local and regional CRMPs that interact closely to resolve environmental issues. These include High Valley and Schindler Creek CRMP, Lake Pillsbury Watershed CRMP, Middle Creek CRMP, and the Scott's Creek CRMP.

Recommendations from the RMC, which receives input from the various CRMPs, are passed to the county Board of Supervisors, who then vote on whether to adopt or modify them. One example of a process initiated and mediated by the Clear Lake CRMP has been an investigation of "The Causes and Control of Algal Blooms in Clear Lake" (see Richerson et al., 1994). University of California, Davis (UCD), scientists conducted a study (funded by the U.S. EPA Clean Lakes Program) to evaluate possible remediation options to reduce blue-green "algal" blooms in Clear Lake. A series of recommendations from this study were provided to the RMC and voted on by the Board of Supervisors. As a result, Lake County has been engaged in several remediation strategies to reduce nutrient inputs to Clear Lake. Two examples include (1) stabilization of creek bed channels (disrupted by gravel mining) by replanting willows and cottonwoods, and (2) initiation of a process to restore and rehabilitate up to 1200 acres of wetlands in the northwest region of Clear Lake (formerly Robinson Lake) that were converted to agricultural production over the past 100 years. The CRMP process is holistic and appears to be working well, yet is always dependent on accurate scientific data for proper strategic planning.

Research

Some of the earliest environmental and ecological research at Clear Lake was a series of observations dealing with fish biology that were initiated shortly after Europeans arrived in the basin (Stone, 1874), and a more broadly based biological survey of the lake in the 1920s (Coleman, 1930). Unfortunately, we have found no documented history of the ecology of the lake before 1870, which would have provided a true aboriginal baseline, especially regarding the level of eutrophication in Clear Lake. The most productive periods of scientific research in Clear Lake have been associated with:

1. Application of DDD to control the Clear Lake gnat (Lindquist and Deonier, 1943; Dolphin, 1959; Hunt and Bischoff, 1960).
2. Early water quality investigations (Goldman and Wetzel, 1963; Lallatin, 1966; Kaiser Engineers, 1968).
3. Cyanobacterial bloom research undertaken by the Clear Lake Algal Research Unit (CLARU) during the period from 1969 to 1972 (Wrigley and Horne, 1974; Horne and Commins, 1987; Horne and Goldman, 1972, 1974; Horne, Javornicky, and Goldman, 1971; Horne et al., 1972; see Richerson, Suchanek, and Why, 1994, for a more complete list).
4. A multidisciplinary investigation of the tectonic, stratigraphic, and paleoclimatic history of the Clear Lake Basin, especially as evidenced by deep (168-m depth) sediment cores collected in the 1970s and early 1980s (Sims, 1988; Sims and White, 1981; Sims, Adam, and Rymer, 1981; Sims, Rimer, and Perkins, 1988; West, 1988).
5. Studies on mercury and arsenic contamination from the Sulphur Bank Mercury Mine that are still ongoing (Columbia Geosciences, 1988; Chamberlin et al., 1990; Suchanek et al., 1993, 1995, 1997, 1998a,b, 2000a,b).
6. An ongoing monitoring program by the Lake County Vector Control District that encompasses water quality, plankton, benthic invertebrates and fishes.
7. A series of ongoing integrated studies by UCD researchers since the early 1990s on the impacts of multiple stresses on the Clear Lake aquatic ecosystem.

The earliest studies (e.g., water quality work, Clear Lake gnat research, cyanobacterial bloom studies), especially in the 1940s and 1950s, were typically "special purpose" programs designed to address specific problems. There was little integration and little continuity; the science was relatively unsophisticated, even though the scientific teams were utilizing the most modern standards of the day. Initially, DDD treatments of the lake were effective in cutting back the populations of the Clear Lake gnat, but some deleterious side effects did occur. Later observations, however, suggested failure of the large pesticide applications and fish predator introductions to reduce and eliminate populations of the Clear Lake gnat, and indicated secondary consequences of population crashes of the western grebe, presumably from DDD contamination. However, western grebes have rebounded in recent years, and exhibited the largest population size ever recorded at Clear Lake (about 210,000 individuals) in the year 2000. The 1970s spawned an era of more-consistent data-collection efforts, with the California Department of Water Resources (DWR) launching a water-quality monitoring program begun in 1969 and continuing to the present day. These data collection efforts provided continuity, but still lacked directed investigation or integration with lake-wide environmental issues. As addressed previously, results of Goldstein and Tolsdorf's (1994) study suggested that compromised water quality in Clear Lake likely is diminishing tourism by $7 to 10M per year. The economic impact suggests that there has been far too little investment in acquiring management-relevant data. Unfortunately, Lake County's responsibilities to support these efforts far outstrips its financial resources to do so.

The U.S. EPA has supported much of the recent work (since 1990) that has been conducted at Clear Lake, both on hypereutrophication and associated cyanobacterial blooms (Clean Lakes Program, about $100,000) and studies on multiple stresses affecting Clear Lake and its watershed (U.S. EPA Office of Research and Development, UCD Center for Ecological Health Research, about $100,000 per year from 1992 to 2001). The U.S. EPA also has funded studies on the impacts of mercury from the Sulphur Bank Mercury Mine on the lake's aquatic ecosystem (U.S. EPA Superfund Program, about $500,000 per year from 1994 to 1998). As a cumulative effort since 1990, funding has supported about five master's theses and seven Ph.D. dissertations (the result of about 42 graduate student study-years) to date.

The U.S. EPA-funded Center for Ecological Health Research (CEHR) at UCD, operating at the watershed scale, is utilizing Clear Lake as one of three model ecosystems (the other two are the Sierra Nevada Ecosystem, including Lake Tahoe, and the Sacramento River Watershed) to evaluate the impacts of multiple stresses on ecosystem health. Another U.S. EPA watershed project (funded by the National Center for Environmental Research Quality Assurance [NCERQA] program) involves policy and management issues associated with assessing water management options. The CEHR program at Clear Lake involves numerous subprojects that are focused on many interrelated questions associated with the dynamics of this system subjected to diverse natural and anthropogenic stresses and how these relate to more regionally based water-quality issues. The intimate relationship between the various disciplines and projects being undertaken by this program is diagrammed in Figure 121.13. Virtually every project has its own set of ongoing investigations (with a rich publication record too extensive to list here), but is well coordinated and integrated with the other projects. The CEHR work is interdisciplinary, reviewed semiannually through scoping and evaluation meetings with a CEHR national Scientific Advisory Committee, and is in the process of producing a synthetic publication summarizing nearly 10 years of research on this ecosystem.

To date, we have made significant progress in understanding the complex dynamics of natural and anthropogenic stressors on the Clear Lake aquatic ecosystem from first principles. For example, the ecological assessment of the impacts of the Sulphur Bank Mercury Mine on Clear Lake was nearing completion in 1995 when we discovered a significant amount of ongoing acid mine drainage (AMD) from the mine site entering Clear Lake through underground seepage, not just as surface runoff. This shifted the entire emphasis of the investigations from the lake bed sediments (which previously were believed to be the point source for the lake's mercury contamination) to ongoing

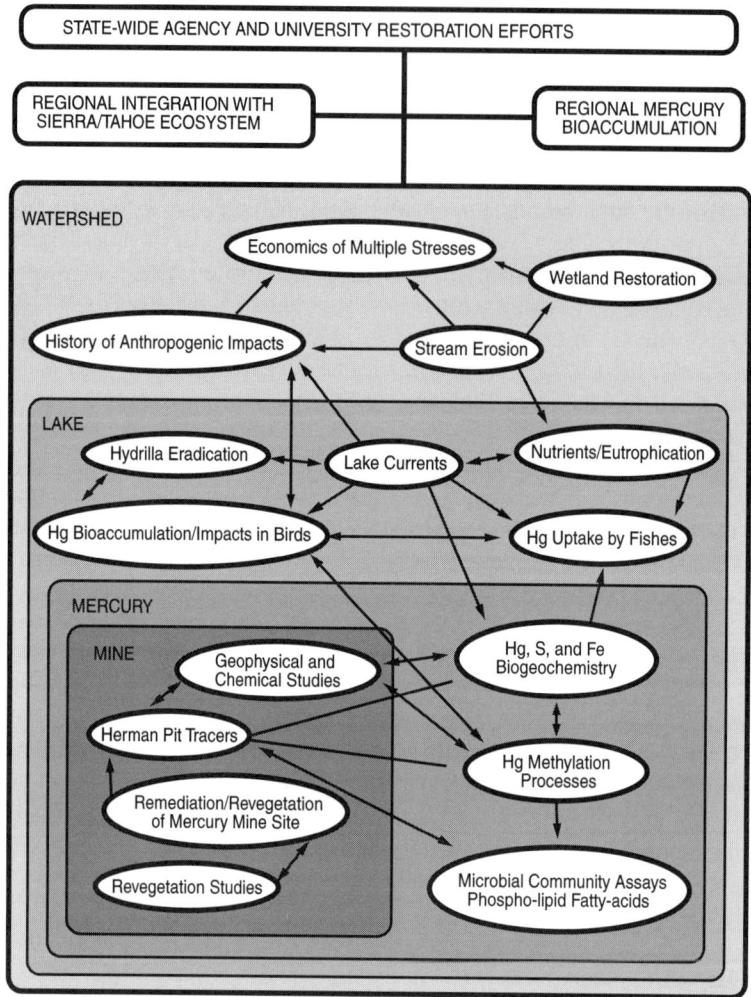

Figure 121.13 Relationship between Clear Lake research projects within the EPA-funded Center for Ecological Health Research.

investigations of AMD. This discovery was made possible only through an active and ongoing monitoring program (Suchanek et al., 2000b). It is possible that our current understanding of the dynamics and influence of sediment and nutrient loading on Clear Lake's productivity also may be incomplete. No "silver bullets" have been found to resolve the complexities of the many natural and anthropogenic stresses imposed on Clear Lake and its surrounding watershed, nor are they likely to be. However, as our work demonstrates, although uncertainties about ecosystem processes remain, we continue to improve our abilities to predict the effects of anthropogenic impacts and manage this productive resource. With unpredictable changes associated with global and regional climate change, new species introductions (purposeful or accidental), and other as-yet-unidentified stressors, this system will need continued and sophisticated adaptive management. In a complex ecosystem such as Clear Lake, research can (1) make steady progress in understanding the complexities of the system, (2) guard against indefensible plans such as the deep dredging of Clear Lake, and (3) offer recommendations for future actions based in the spirit of adaptive management.

If society wants to manage complex ecosystems like Clear Lake at a state-of-the-art level, it must be prepared to make substantial investments in monitoring, research, and modeling on an ongoing basis. Because these investments involve contributions to understanding ecosystems at a fairly fundamental level that will be applicable to many other similarly stressed systems, a substantial federal contribution is warranted. However, a healthy and meaningful program requires more state and local investment than has been forthcoming thus far. A strong commitment from federal, state, and local funding agencies will be needed to provide a reliable technical basis for science-based decision making and management. Lake Tahoe, with its much higher profile nationally, has a relatively well-funded research and monitoring program. However, even this program is weakly institutionalized, has periodic funding crises, and has a funding base that is none too generous considering the scope of its problems. Multiply stressed ecosystems require a sustained and rather costly investment of scientific and technical resources. Recent requests by the National Science Foundation for a major increase in environmental science funding are a step in the right direction, but the more applied and management-oriented part of the investment portfolio still will be neglected unless other agencies step forward.

ACKNOWLEDGMENTS

We would like to thank the County of Lake and specifically Supervisor Karan Mackey for continued support throughout the past decade. In addition, Art Colwell and Norm Anderson of the Lake County Vector Control District, Bob Reynolds of the Lake County Air Quality Management District, Bob Lossius and Tom Smythe of the Lake County Public Works Department, and Marty Winston of the Lake County Department of Public Health have given generously of their time and data to help evaluate and understand many of the stresses affecting Clear Lake. Art Colwell significantly improved the manuscript by providing historical accuracy. Peter Moyle helped us correct the simplified food web diagram. Many other people too numerous to mention have contributed in many ways to this effort. This work has been supported by the following grants or contracts: U.S. EPA Clean Lakes Program (contract 0-166-250-0 to the California State Water Resources Control Board), U.S. EPA Center for Ecological Health Research (R819658 and R825433), U.S. EPA-NSF Watershed Grant (R-825285-01-0), U.S. EPA Region IX SuperFund Program (68-S2-9005), and a UCD Jastro Shields grant to E.J. Gilmartin. Although the information in this document has been funded in part by the United States Environmental Protection Agency, it may not necessarily reflect the views of the agency and no official endorsement should be inferred.

REFERENCES

Anderson, D.W., Cahill, Jr., T.M., Suchanek, T.H., and Elbert, R.A., Relationships between mercury and yearly trends in osprey production and reproductive status at Clear Lake, in *First Annual Clear Lake Science and Management Symposium*, September 13, 1997, Proceedings Volume, 1997, pp. 66–70.

Anderson, E.W. and Baum, R.C., How to do coordinated resource management planning, *J. Soil Water Conserv.*, 43, 216–220, 1988.

Anderson, N.L., Woodward, D.L., and Colwell, A.E., Pestiferous dipterans and two recently introduced aquatic species at Clear Lake, *Proceedings of the California Mosquito Vector Control Association*, 54, 163–167, 1986.

Anon., *History of Napa and Lake Counties, California*, Slocum, Bowen & Co., 1881.

Apperson, C.S., Yows, D., and Madison, C., Resistance to methyl parathion in *Chaoborus astictopus* from Clear Lake, California, *J. Econ. Entomol.*, 71, 772–773, 1978.

Bailey, G.E., The saline deposits of California, *Calif. State Min. Bur. Bull.*, 24, 33, 1902.

Baumhoff, M.A., Ecological determinants of aboriginal California populations, *Univ. Calif. Publ. Am. Archaeol. Ethnol.*, 49, 153–236, 1963.

Becker, C.F., Geology of the quicksilver deposits of the Pacific slope, *USGS Monogr.*, 13, 1888.

Bentley, W.J., Caprile, J.L., Coates, W.W., Elkins, R.B., Pickel, C., Walker, K., Weddle, P.W., and Zalom, F.G., Insects and mites, in *Integrated Pest Management for Apples and Pears*, 2nd ed., Ohlendorf, B.L. et al., Eds., Univ. Calif. Div. Agric. Nat. Res. Publ. 3340, Oakland, CA, 1999, pp. 70–153.

Bradbury, J.P., Diatom biostratigraphy and the paleolimnology of Clear Lake, Lake County, California, in *Late Quaternary Climate, Tectonism, and Sedimentation in Clear Lake*, Sims, J.D., Ed., USGS Special Paper 214, 1988 pp. 97–130.

Brice, J.C., Geology of Lower Lake Quadrangle, California, *Calif. Div. Mines Geol. Bull.*, 166, 1953.

California Department of Pesticide Regulation, Summary of Pesticide Use Report Data, Indexed by Commodity, Years 1990–1999, State of California, Environmental Protection Agency, Department of Pesticide Regulation, 1990–1999.

California Division of Mines, Mineral Commodities of California, Bulletin 156, 247, 1950.

Carson, R., *Silent Spring*, Houghton Mifflin, Boston, 1962.

Casteel, R.W. and Rymer, M.J., Pliocene and Pleistocene fishes from the Clear Lake area, *USGS Prof. Paper*, 1141, 231–235, 1981.

Chamberlin, C.E., Chaney, R., Finney, B., Hood, M., Lehman, P., McKee, M., and Willis, R., Abatement and control study: Sulphur Bank Mine and Clear Lake, Prepared for California Regional Water Quality Control Board, Environmental Resources Engineering Department, Humboldt State University, Arcata, CA, 1990.

Coleman, G.A., A biological survey of Clear Lake, Lake County, *Calif. Fish Game*, 16, 221–227, 1930.

Columbia Geosciences, Hydrological Assessment Report and Appendices, Columbia Geoscience for Bradley Mining Company, July 1988.

Colwell, A.E., Anderson, N.L., and Woodward, D.L., Monitoring of Dipteran Pests and Associated Organisms in Clear Lake (California), in First Annual Clear Lake Science and Management Symposium, September 13, *Proceedings Volume*, 1997, pp. 15–32.

Cooke, S.F., The Clear Lake example: an ecological approach to pest management, *Environment*, 23, 25–30, 1981.

Curtis, T.C., Pesticide laboratory report, California Department of Fish and Game, E.P. No. P-133, 1977.

Davis, W.M., The Lakes of California, *Calif. J. Mines Geol.*, 29, 175–236, 1933.

Dean Runyan Associates, California Travel Impacts by County, 1992–1997, Report prepared for California Trade and Commerce Agency, Division of Tourism, 1999.

Dechoretz, N., *Hydrilla* eradication in Clear Lake, in *Second Annual Clear Lake Science and Management Symposium*, October 24, 1998, *Proceedings Volume*, Webber, L.B. and Suchanek, T.H., Eds., 1998, pp. 17–21.

Dolphin, R., Lake County Mosquito Abatement District Gnat Research Program, *Proceedings and Papers of the 27th Annual Conference*, California Mosquito Control Association, 1959, pp. 47–48.

Dolphin, D.E. and Peterson, R.N., Developments in the research and control program of the Clear Lake gnat *Chaoborus astictopus*, *Proceedings and Papers of the 28th Annual Conference*, California Mosquito Control Association, 90–94, 1960.

Earle, C.J., Asynchronous droughts in California streamflow as reconstructed from tree rings, *Quat. Res.*, 39, 290–299, 1993.

Elkins, R. and Shorey, H.H., Mating disruption of codling moth (*Cydia pomonella*) using "puffers," in *Proceedings of the 7th International Symposium on Pear Growing*, Talca, Chile, Acta Horticulturae No. 475, Retamales, J.B., Ed., International Society of Horticultural Science, 1998, pp. 503–510.

Elser, J.J., Carney, H.J., and Goldman, C.R., The zooplankton-phytoplankton interface in lakes of contrasting trophic status: an experimental comparison, *Hydrobiologia*, 200/201, 69–82, 1990.

Field, C.B., Daily, G.C., Davis, F.W., Gainess, S., Matson, P.A., Melack, J., and Miller, N.L., *Confronting Climate Change in California*, Union of Concerned Scientists and the Ecological Society of America, Washington, D.C., 1999.

Follansbee, B.A., Wetland Restoration at Clear Lake, California: Species Selection, Phosphorus Monitoring, and Coordinated Resource Management Planning, Ph.D. dissertation, University of California, Davis, 1996.

Gilmartin, E.J., Multiple Stresses on Zooplankton in Clear Lake, California: Copper and Methyl Mercury Toxicity to Cladocerans, Master's dissertation, University of California, Davis, 1998.

Gilmartin, E.J., Suchanek, T.H., and Richerson, P.J., Effects of multiple stressors (mercury and copper) on zooplankton: studies on *Ceriodaphnia* and *Daphnia*, in *Second Annual Clear Lake Science and Management Symposium*, October 24, 1998, *Proceedings Volume*, Webber, L.B. and Suchanek, T.H., Eds., 1998, pp. 70–76.

Goldman, C.R. and Wetzel, G.R., A study of the primary productivity of Clear Lake, Lake County, California, *Ecology*, 44, 283–294, 1963.

Goldstein, J.J. and Tolsdorf, T.N., An economic analysis of potential water quality improvement in Clear Lake: benefits and costs of sediment control, including a geological assessment of potential sediment control levels, USDA Soil Conservation Service, Davis and Lakeport Offices, 1994.

Goodridge, J., The impact of climate change on drainage engineering in California, Report to the Alert Users Group Conference in Palm Springs, CA, 26–29 May, 1998.

Griffith, J.S., Effects of low temperature on the behavior and survival of threadfin shad, *Dorosoma petenense*, *Trans. Am. Fish. Soc.*, 107, 63–70, 1978.

Hearn, B.C., Jr., McLaughlin, R.J., and Donnely-Nolan, J.M., Tectonic framework of the Clear Lake Basin, California, in *Late Quaternary Climate, Tectonism, and Sedimentation in Clear Lake*, Sims, J.D., Ed., USGS Special Paper 214, 1988, pp. 9–20.

Heizer, R.F., The West Coast of North America, in *Prehistoric Man in the New World, A.D.*, Jennings and Norbeck, Eds., University of Chicago Press, Chicago, 1963, pp. 117–148.

Herman, S.G., Garrett, R.L., and Rudd, R.L., Pesticides and the western grebe: a study of pesticide survival and trophic concentration at Clear Lake, Lake County, California, in *Chemical Fallout: Current Research on Persistent Pesticides*, Miller, M.W. and Berg, G.G., Eds., Charles C Thomas, Springfield, IL, 1969, pp. 24–53.

Hileman, B., Case grows for climate change, *Chem. Eng. News*, August 9, 1999, pp 16–23.

Hodges, C.A., Geomorphic History of Clear Lake, California, Ph.D. dissertation, Stanford University, Palo Alto, 1966.

Hopkirk, J.D., Endemism in the fishes of the Clear Lake region of Central California, *Univ. Calif. Publ. Zool.*, 96, 1973.

Hopkirk, J.D., Fish evolution and the late Pleistocene and Holocene history of Clear Lake, California, in Late Quaternary Climate, Tectonism, and Sedimentation in Clear Lake, USGS Special Paper 214, Sims, J.D., Ed., 1988, pp. 183–193.

Horne, A.J., The ecology of Clear Lake phytoplankton, Clear Lake Algal Research Unit Special Report, Lakeport, CA, 1975.

Horne, A.J. and Commins, M.L., Macronutrient controls on nitrogen fixation in planktonic cyanobacterial populations, *N.Z. J. Mar. Freshwater Res.*, 21, 413–423, 1987.

Horne, A.J. and Goldman, C.R., Nitrogen fixation in Clear Lake, California, I, Seasonal variation and the role of the heterocysts, *Limnol. Oceanogr.*, 17, 678–692, 1972.

Horne, A.J. and Goldman, C.R., Suppression of nitrogen fixation by blue-green algae in a eutrophic lake with trace additions of copper, *Science*, 83, 409–411, 1974.

Horne, A.J., Javornicky, P., and Goldman, C.R., A freshwater red tide on Clear Lake, California, *Limnol. Oceanogr.*, 16, 684–689, 1971.

Horne, A.J., Dillard, J.E., Fujita, D.K., and Goldman, C.R., Nitrogen fixation in Clear Lake, California: II. Synoptic studies on the Anabaena bloom, *Limnol. Oceanogr.*, 17, 693–703, 1972.

Hunt, E.G. and Bischoff, A.I., Inimical effects on wildlife or periodic DDD applications to Clear Lake, *Calif. Fish Game*, 46, 91–106, 1960.

Jones & Stokes, Middle Creek ecosystem restoration reconnaissance study, Final report, May 1997 (JSA96–239), prepared for U.S. Army Corps of Engineers, Sacramento, 1997.

Kaiser Engineers, Task Force Report on Upper Eel River routing studies prepared for Lake County Flood Control and Water Conservation District, Chapter V, Clear Lake algae and water quality, Report 68-2-RE, 1968.

Lallatin, R.D., Clear Lake water quality investigation, Sacramento Department of Water Resources Bulletin 143-2, 1966.

Langeland, K.A., *Hydrilla verticillata*: "the perfect aquatic weed," *Castanea*, 61, 293–304, 1996.

Li, H.W., Moyle, P.B., and Garrett, R.L., Effect of the introduction of the Mississippi silverside (*Menidia audens*) on the growth of black crappie (*Pomoxis nigromaculatus*) and white crappie (*P. annularis*) in Clear Lake, California, *Trans. Am. Fish. Soc.*, 105, 404–408, 1976.

Lindquist, A.W. and Deonier, C.C., Seasonal abundance and distribution of larvae of the Clear Lake gnat, *J. Kan. Entomol. Soc.,* 16, 143–149, 1943.

Lindquist, A.W. and Roth, A.R., Effect of dichlorodiphenyl dichloroethane on larva of the Clear Lake gnat in California, *J. Econ. Entomol.,* 43, 328–332, 1950.

Macedo, R.A., Creel survey at Clear Lake, California, March-June, 1988, California Department of Fish and Game, Inland Fisheries Administrative Report 91-3, 1991.

Mack, E.E., Nelson, D.C., Brister, L.L., and Suchanek, T.H., Methyl mercury production from unamended sediment cores (core tube microcosms), in First Annual Clear Lake Science and Management Symposium, September 13, 1997, *Proceedings Volume,* 1997, pp. 94–104.

Mann, M.E., Bradley, R.S., and Hughes, M.K., Northern Hemisphere temperatures during the past millennium: inferences, uncertainties and limitations, *Geophy. Res. Lett.,* 26, 759–762, 1999.

Mauldin, H.K., *History of Lake County, Vol. I, Clear Lake and Mt. Konocti,* East Wind Printers, San Francisco, 1960.

Mauldin, H.K., *History of Clear Lake, Mt. Konocti and the Lake County Cattle Industry,* Anderson Printing, Kelseyville, CA, 1968.

Meehl, M.J. and Washington, W.M., El Niño-like climate change in a model with increased atmospheric CO_2 concentrations, *Nature,* 382, 56–60, 1996.

Meillier, L.M., Becker, J.C., Richerson, P.J., and Suchanek, T.H., Late Holocene diatom biostratigraphy of three short cores retrieved from Clear Lake, Lake County, California, in First Annual Clear Lake Science and Management Symposium, September 13, 1997, *Proceedings Volume,* 1997, pp. 155–163.

Menefee, C.A., *Historical and Descriptive Sketchbook of Napa, Sonoma, Lake and Mendocino,* Reporter Publishing House, Fairfield, CA, 1873.

Monroe, M., Olofson, P.R., Collins, J.N., Grossinger, R., Haltiner, J., and Wilcox, C., Baylands Ecosystem Habitat Goals, San Francisco Bay Area Wetlands Ecosystem Goals Project, 1999.

Moyle, P.B., *Inland Fishes of California: Revised and Expanded,* University of California Press, Berkeley, 2002.

Murphy, G.I., The fishery of Clear Lake, Lake County, California, *Calif. Fish Game,* 37, 439–484, 1951.

O'Connell, R.A. and Dechoretz, N., *Hydrilla* eradication in Clear Lake, in *First Annual Clear Lake Science and Management Symposium,* September 13, 1997, *Proceedings Volume,* 1997, pp. 33–36.

Richerson, P.J., Suchanek, T.H., and Why, S.J., The Causes and Control of Algal Blooms in Clear Lake, Clean Lakes Diagnostic/Feasibility Study for Clear Lake, California, prepared for EPA Region IX, 1994.

Richerson, P.J., Suchanek, T.H., Becker, J.C., Heyvaert, A.C., Slotton, D.G., Kim, J.G., Li, X., Meillier, L.M., Nelson, D.C., and Vaughn, C.E., The history of human impacts in the Clear Lake watershed (California) as deduced from lake sediment cores, in *The Integrated Assessment of Ecosystem Health,* Fogg, G. et al., Eds., Ann Arbor Press, Chelsea, MI, 2002, pp. 119–145.

Rudd, R.L., *Pesticides and the Living Landscape,* University of Wisconsin Press, Madison, 1964.

Simoons, F.J., The Settlement of the Clear Lake Upland of California, Master's thesis, University of California, Berkeley, 1952.

Sims, J.D., Ed., Late Quaternary climate, tectonism and sedimentation rate in Clear lake, northern California coast ranges, Geological Society of America, Special Paper 214, 1988.

Sims, J.D. and White, D.E., Mercury in the sediments of Clear Lake, in *Research in the Geysers — Clear Lake Geothermal Area, Northern California,* McLaughlin, R.J. and Donnelly-Nolan, J.M., Eds., Geological Survey Professional Paper, Washington, D.C., 1141, 1981, pp. 237–242.

Sims, J.D., Adam, D.P., and Rymer, M.J., Late Pleistocene stratigraphy and palynology in Clear Lake, in *Research in the Geysers — Clear Lake Geothermal Area, Northern California,* McLaughlin, R.J. and Donnelly-Nolan, J.M., Eds., Geological Survey Professional Paper, Washington, D.C., 1141, 1981, pp. 219–230.

Sims, J.D., Rymer, M.J., and Perkins, J.A., Late Quaternary deposits beneath Clear Lake, California: physical stratigraphy, age, and paleogeographic implications, in *Late Quaternary Climate, Tectonism, and Sedimentation in Clear Lake,* Sims, J.D., Ed., USGS Special Paper 214, 1988, pp. 21–44.

Slotton, D.G., Suchanek, T.H., Mullen, L.H., and Richerson, P.J., Mercury partitioning trends in fish from Clear Lake, a mine-impacted EPA Superfund Site in California, in *First Annual Clear Lake Science and Management Symposium,* September 13, 1997, *Proceedings Volume,* 1997, pp. 115–121.

Smythe, T.R., Overview of the Middle Creek Marsh restoration project, in *First Annual Clear Lake Science and Management Symposium,* September 13, 1997, *Proceedings Volume,* 1997, pp. 164–169.

Stine, S., Extreme and persistent drought in California and Patagonia during mediaeval time, *Nature,* 369, 546–549, 1994.

Stone, L. XX, Report on operations in California in 1873, A, Clear Lake, 1, Field work in the winter of 1872–3. U.S. Commission of Fish and Fisheries, Part II, Report of the Commissioner for 1872 and 1873, 1874, pp. 377–381.

Suchanek, T.H., Richerson, P.J., Woodward, L.A., Slotton, D.G., Holts, L.J., and Woodmansee, C.E., Ecological Assessment of the Sulphur Bank Mercury Mine Superfund Site, Clear Lake, California: A survey and evaluation of mercury, in sediment, water, plankton, periphyton, benthic invertebrates and fishes within the aquatic ecosystem of Clear Lake, California, Phase 1, Preliminary Lake Study report prepared for EPA Region IX Superfund Program, 1993.

Suchanek, T.H., Richerson, P.J., Holts, L.J., Lamphere, B.A., Woodmansee, C.E., Slotton, D.G., Harner, E.J., and Woodward, L.A., Impacts of mercury on benthic invertebrate populations and communities within the aquatic ecosystem of Clear Lake, California, *Water Air Soil Pollut.,* 80, 951–960, 1995.

Suchanek, T.H., Richerson, P.J., Mullen, L.J., Brister, L.L., Becker, J.C., Maxson, A., and Slotton, D.G., The role of the Sulphur Bank Mercury Mine site (and associated hydrogeological processes) in the dynamics of mercury transport and bioaccumulation within the Clear Lake aquatic ecosystem, report prepared for the EPA Region IX Superfund Program, 1997.

Suchanek, T.H., Mullen, L.H., Lamphere, B.A., Richerson, P.J., Woodmansee, C.E., Slotton, D.G., Harner, E.J., and Woodward, L.A., Redistribution of mercury from contaminated lake sediments of Clear Lake, California, *Water Air Soil Pollut.,* 104, 77–102, 1998a.

Suchanek, T.H., Nelson, D.E., Richerson, P.J., Slotton, D.G., and McHatton, S.C., Methyl mercury production at Clear Lake is decoupled from bulk inorganic mercury loading: biotic contamination is lower than expected, in Webber, L.B. and Suchanek, T.H., Eds., *Proceedings: Second Annual Clear Lake Science and Management Symposium*, 1998b, pp. 95–103.

Suchanek, T.H., Lamphere, B.A., Mullen, L.H., Woodmansee, C.E., Richerson, P.J., Slotton, D.G., Woodward, L.A., and Harner, E.J., Mercury in lower trophic levels of the Clear Lake aquatic ecosystem, California, in *The Integrated Assessment of Ecosystem Health,* Fogg, G., Ed., Ann Arbor Press, Chelsea, MI, 2000a, pp. 249–268.

Suchanek, T.H., Richerson, P.J., Flanders, J.R., Nelson, D.C., Mullen, L.H., Brister, L.L., Becker, J.C., and McHatton, S.C., Monitoring inter-annual variability reveals sources of mercury contamination in Clear Lake, California, *Environ. Monitoring Assess.,* 64(1), 299–310, 2000b.

Trenberth, K.E. and Hoar, T.J., El Niño and climate change, *Geophys. Res. Lett.,* 24, 3057–3060, 1997.

Van Nieuwenhuyse, E.E., Effects of Middle Creek Marsh restoration on Clear Lake water quality, in First Annual Clear Lake Science and Management Symposium, September 13, 1997, *Proceedings Volume,* 1997, pp. 170–174.

Webber, L.B. and Suchanek, T.H., Eds., Second Annual Clear Lake Science and Management Symposium, October 24, 1998, *Proceedings Volume,* 1998.

West, G.J., Early historic vegetation change in Alta California: the fossil evidence, in *Columbian Sequences: Vol. 1, Archaeological and Historical Perspectives on the Spanish Borderlands West,* Thomas, D.H., Ed., Smithsonian Institution Press, Washington, D.C., 1989, pp. 333–348.

Wolfe, M.F. and Norman, D., Effects of waterborne mercury on terrestrial wildlife at Clear Lake: evaluation and testing of a predictive model, *Environ. Toxicol. Chem.,* 17(2), 214–227, 1998.

Wrigley, R.C. and Horne, A.J., Remote sensing and lake eutrophication, *Nature,* 250, 213–214, 1974.

Zalusky, S., Ed., Lake County Aggregate Resource Management Plan, Lake County Planning Department, 1992.

CHAPTER 122

Trans-Gill and Dietary Uptake of Methyl Mercury by the Sacramento Blackfish, a Planktivorous Freshwater Fish

Joseph J. Cech, Jr., Monica Heekyoung Choi, and Ann G. Houck

INTRODUCTION

Clear Lake, CA, is a focus ecosystem of the EPA-supported University of California, Davis (UCD), Center for Ecological Health Research. Several Clear Lake fish species carry high levels of methyl mercury (MeHg) in their flesh (Stratton et al., 1987). Mercury (Hg) leaches into the lake from naturally occurring cinnabar ores and from anthropogenic sources such as the Sulphur Bank Mercury Mine, an EPA Superfund site. Hg-contaminated sediments near the abandoned mine from the lake's Oaks Arm are classified as hazardous waste by the California Department of Health Services. By contact or ingestion, Hg is taken up by aquatic organisms and, presumably, biomagnifies up the lake food web (Watras et al., 1994; Suchanek et al., 1997). The abundant Clear Lake fishes are important components of the food web (Moyle, 1976), but little is known about the fishes' Hg uptake routes and rates in the ecosystem.

Heavy metals in the environment and their effects on living organisms have received increasing attention in recent years (Heath, 1995). Mercury, especially in the methyl (MeHg) form, has received much of this attention (Watras and Huckabee, 1994). Inorganic mercury has low water solubility, is readily complexed or adsorbed to particulates, and precipitates to sediments (Benes and Havlik, 1979). Jensen and Jernolöv (1969) showed that inorganic mercury could be methylated in the aquatic environment to the highly toxic MeHg. In freshwater systems, the majority of MeHg appears to be generated by bacteria in the surface sediments and transferred by biota (Beijer and Jernelöv, 1979). Environmental pH, temperature, redox potential, alkalinity, and concentrations of dissolved organic carbon (DOC), oxygen, sulfate, and calcium can affect the Hg methylation rate and MeHg's bioavailability (Benes and Havlik, 1979; Carty and Malone, 1979; Fitzgerald and Clarkson, 1991; Matilianen and Verta, 1995; Choi et al., 1998). When methylated, mercury becomes less water soluble and more lipid soluble. MeHg is extremely mobile, very stable, and can easily penetrate membranes in living organisms (Beijer and Jernelöv, 1979; Hodson, 1988). The high mobility facilitates rapid MeHg uptake and penetration of sensitive tissues, particularly the lipid membranes of neurons, where its toxicity is well known (Wiener and Spry, 1996). Importantly, MeHg then preferentially forms bonds with thiol-containing molecules, such as proteins and enzymes, moving it into the water-soluble cytosolic fraction, slowing depuration.

Although the Clean Water Act of 1972 effectively ended intentional discharges into most U.S. waters, anthropogenic sources of mercury such as old mine tailings and new impoundments continue to contaminate aquatic systems. Much data on human health and MeHg intoxication have accumulated since the outbreak of MeHg poisoning in Minamata, Japan (1956), but less is known about the actual effects of intoxication on wildlife. The effects of mercury toxicity can occur at all levels of the food chain from reduced photosynthesis of phytoplankton to neurotoxic effects on vertebrates including subtle effects on the rates of population reproduction, growth, and mortality (Peterle, 1991). However, the greatest hazards remain in the aquatic food chain where bioaccumulation and bioconcentration occur. Fish species that are shorter lived, those that feed low on the food chain, or those that grow very fast accumulate less mercury than long-lived piscivorous fish (Armstrong, 1979).

In fish, free MeHg is rapidly and efficiently absorbed through the gut and across the gills (Armstrong, 1979). Its high affinity for sulfhydral groups in proteins dramatically increases its biological half-life (Grieb et al., 1990; Bloom, 1992). These factors account for its very high bioaccumulation and biomagnification factors in aquatic food chains (e.g., concentrated in fish muscle tissue (Huckabee et al., 1974) and in terrestrial animals feeding on aquatic organisms. Methyl mercury is one of the most toxic forms of Hg, causing irreversible and likely cumulative damage to the central nervous system (Choi, 1990). Several studies have demonstrated that DOC is an important factor affecting fish Hg bioaccumulation that also includes pH, temperature, sulfide concentration, redox potential, fish age, and growth rate (Driscoll et al., 1995). The ubiquitous DOC in lakes and other freshwater systems originates from the decomposition and degradation of organic detritus, and generally ranges from 0.5 to 4.0 mg C/l in lakes and rivers to 10 to 50 mg C/l in wetlands and marshes (MacCarthy, 1989). It has been well established that because of their polyelectrolytic nature, DOC complexes with or chelates heavy metal ions in natural waters (Mantoura et al., 1978). This metal complexation capacity of DOC therefore plays an important role in natural systems, influencing the environmental fate, bioavailability, toxicity, and mobility of heavy metals by controlling their speciation in natural waters (MacCarthy, 1989) and their uptake by living organisms. For example, complexation of inorganic Hg with organic carbon has been shown to reduce bioavailability to fish (Ramamoorthy and Blumhagen, 1984).

Clear Lake, in the Coastal Range of northern California, is the largest natural lake wholly within California's borders. This shallow, naturally eutrophic lake is located 100 km north of San Francisco and is surrounded by the many communities and parks that dot its shores. While the lake supports a productive commercial and sportfishery, it is also a highly disturbed ecosystem subject to many environmental stressors: chlorinated hydrocarbon residues from an early 1960s attempt to control a nuisance midge, arsenic from natural sources, and mercury from the abandoned Sulfur Bank Mine (Suchanek et al., 1997). Fish are additionally stressed by seasonal extremes in temperature, low dissolved oxygen, and heavy parasite loads (Murphy, 1950; Goldman and Wetzel, 1963). Clear Lake presents a unique opportunity to model the routes and rates of Hg movement through an ecosystem, including its complex food web. Unfortunately, the available data on the fishes' roles in moving Hg through the ecosystem, especially through the food web, are simply inadequate. Our objective was to quantify trans-gill and dietary MeHg dietary uptake rates in Sacramento blackfish (*Orthodon microtepidotus*), a planktivorous, commercially fished species.

MATERIALS AND METHODS

Trans-Gill MeHg Uptake Studies

Sacramento blackfish (900 to 1200 g wet weight) were seined from Clear Lake and held in 600-l fiberglass tanks at the UCD under natural photoperiod for about 1 year before experiments. Tanks were aerated and received a continuous flow of air-equilibrated well water (mean total hardness: 456 mg/l as $CaCO_3$, total alkalinity: 485.1 mg/l as $CaCl_3$, and pH: 8.1), and blackfish were fed Silver Cup trout pellets (Nelson and Nelson, Murray, UT) up to 48 h before surgery. Fish were anesthetized (100 mg/l, 3-aminobenzoic acid, ethyl ester [MS-222]; Sigma Chemical, St. Louis, MO, buffered with $NaHCO_3$ and NaCl), weighed (electronic balance), measured for total length (linear rule), cannulated (PE-50, dorsal aorta) for blood sampling, immobilized by spinal transection (McKim and Goeden, 1982; Schmieder and Weber, 1992), fitted with an oral latex membrane to separate inspired and expired water, and recovered in the respirometer in aerated UCD well water at least 20 h before MeHg exposure.

The water-jacketed respirometer ($20 \times 20 \times 66$ cm, 1-cm thick acrylic plastic) incorporated three compartments: A (pre-gill), B (post-gill), and C (posterior body) were separated by latex membranes (Choi et al., 1998). The water containing $Me^{203}Hg$ was pumped from reservoir 1 by a metering pump (Alidos, Marietta, GA) to compartment A, and the water in compartment A was pumped to compartment B by the fish's gill ventilation. Ventilation volume (Vg) was the volume of water displaced from the compartment B drain by the fish's gill ventilation (Choi et al., 1998). Water was simultaneously sampled from compartments A and B by a peristaltic pump (Minipuls, Gilson, Middleton, WI) for dissolved O_2 samples (calibrated O_2 meters, Model 113, Instrumentation Laboratories, Lexington, MA) and for MeHg samples (discussed later). Oxygen consumption rate (MO_2, mg O_2/h/kg) was calculated by multiplying Vg by the inspired–expired O_2 content difference. Whereas O_2 extraction efficiency (UO_2) was calculated from $100 \times$ the inspired–expired O_2 content difference/inspired O_2 content, Hg extraction efficiency (UHg) was calculated from $100 \times$ the inspired–expired MeHg content difference/inspired MeHg content. Finally, trans-gill Hg uptake rate (RHg) was calculated by multiplying the Hg concentration in the water by the product of Vg and UHg/100. Equal water levels in all compartments eliminated pressure head–induced leakages (checked after each experiment with food dye) between compartments. The whole system was closed, except for the vents connected to the In-Line Gas Purifier (400 ml, Alltech, Deerfield, IL) in the fume hood, to prevent the release of volatilized MeHg to the atmosphere. Volatilized MeHg levels in the Gas Purifier were measured by Geiger–Mueller counter, and water temperature was maintained by a thermoregulator (CFT-75, Neslab, Portsmouth, NH) circulating water through the water jackets of the chamber and reservoirs. The supply water in reservoir 1 was carefully aerated (to minimize volatilization) by using chemically inert porous PTFE tubing (IMPRA, Tempe, AZ).

Blackfish were exposed to aqueous solutions of the MeHg (1.4 ± 0.25 ng/l, reference treatments) or to MeHg (1.4 ± 0.25 ng/l) with 2 and 5 mg DOC/l treatments in the respirometer at 20°C for 1 h (Choi et al., 1998). The $Me^{203}Hg$ or DOC-equilibrated $Me^{203}Hg$ stock solution was added to reservoir 1 and allowed to mix with water via aeration for 1 h before starting $Me^{203}Hg$ exposure. Radiolabeled MeHg was chosen for this experiment due to the simple and sensitive analysis at low concentrations. During the experiment, three, 1-ml water samples from reservoir 1, compartment A, and compartment B were taken and placed in 20-ml liquid scintillation (LS) vials for $Me^{203}Hg$ concentration measurements. Heavy-walled, Teflon (PFA) tubing was used (except viton tubing in peristaltic pump head) for the water samples to minimize losses of $Me^{203}Hg$ and O_2; 1 ml of water containing $Me^{203}Hg$ was mixed with 10 ml of LS cocktail (Universol, ICN Biomedicals, Costa Mesa, CA) in a 20-ml borosilicate LS vial. The ^{203}Hg activity was measured using the LS counter, and quenching effects were corrected. MeHg concentrations were calculated from the specific activities of the stock solution and corrected for radioactive decay. To minimize stress, the chamber was partially covered with black plastic sheet during experiments. After each experiment, wastewater was decontaminated by pumping through a series of water purification filters (Adsorber/Universal System, Cole-Parmer, Niles, IL), and its radioactivity was checked using a Wallac 1410 LS counter (Pharmacia, Turim, Finland) before disposal.

Humic acid (Aldrich, Milwaukee, WI) was dissolved in deionized water and filtered through binder-free glass fiber filters (TCLP, 0.7 μm; Gelman Sciences, Ann Arbor, MI) to remove particulate material to make our DOC stock solution (1100 mg C/l; Choi et al., 1998). The carbon content of the DOC stock solution was measured using a total carbon analyzer (Shimadzu TOC 500, Tokyo, Japan). Initial trace metal concentrations (mg/l) of the DOC stock solution were determined by inductively coupled plasma-mass spectrometry (ICP-MS): As (<0.2), Cd (<0.03), Cu (0.02), Fe (28.2), Pb (<0.1); Mn (0.59), Mo (<0.04), and Zn (0.25). The DOC stock solution was refrigerated (4°C) until diluted with UCD well water to 2 or 5 mg C/l for the experiments. The Hg (<0.001 ng/l) and DOC (<1 μg C/l) concentrations of UCD well water were not detectable. The DOC stock solution was equilibrated with our $Me^{203}HgCl$ stock solution for 48 h before DOC and $Me^{203}Hg$ exposure experiments started (Hintelmann et al., 1995). Immediately after the 1-h exposure and sampling period, each fish was over-anesthetized with MS-222 and dissected for tissue $Me^{203}Hg$ concentrations (Choi et al., 1998).

SuperANOVA (Abacus Concepts, Berkeley, CA) was used for one-factor analyses of variance to compare the DOC effects on the different treatment means. Multiple comparisons for proportions were used to compare the DOC effects on the ratio between blood and water (Zar, 1984). The statistical significance of the differences in variables among the three treatments were determined by Student–Newman–Keuls test.

Dietary MeHg Uptake Studies

Sacramento blackfish broodstock were purchased from commercial fishers at Clear Lake, and spawned in a UCD campus pond (Cech et al., 1982). Juvenile fish were distributed among 24 aquaria (six replicate aquaria of each of four MeHg-exposure treatment groups) for two, sequential 35-day MeHg dietary exposure periods (70-day total exposure). Weighed (Mettler electronic balance) fish epaxial muscle samples were taken for total Hg analysis. Because previous work has shown that >95% of the Hg in fish is in the form of MeHg (Bloom, 1992), the less expensive (compared with MeHg) total Hg assays (cold vapor atomic absorption spectrometry, developed by Slotton et al., 1995) were used. Similarly, we measured fish food samples (total of 8) from all treatment groups and aquarial water samples (total of 8) using cold vapor atomic fluorescence. Dietary Hg uptake rates were calculated following Rodgers and Beamish's (1982) methods. Concentration of Hg (in tissues, food, water) and uptake rates treatment means were compared using ANOVA, Kruskal–Wallis, and *post-hoc* tests (Sokal and Rohlf, 1981), using Systat and SigmaStat software.

Food consumption rates were measured simultaneously in the dietary MeHg exposure blackfish. Experimental diets were prepared from a complete, commercial trout chow (Silver Cup Trout Chow) mash and mixed with an aqueous methyl mercuric chloride solution to reach the target concentration. Wet diets were extruded to the correct diameter, dried at 37°C, cut into pellets of the correct size, and frozen until use. Calculated treatment concentrations of MeHg were 0 mg/kg dry weight of feed (control), 0.45 mg/kg (mean concentration of these species' prey in Clear Lake; Suchanek et al., 1997), 20 mg/kg, and 50 mg/kg (two "threshold" concentrations for comparisons with other studies; Phillips and Buhler, 1978; Rodgers and Beamish, 1982). Fish were situated in groups of 35 in each of six replicate 38-l aquaria per treatment. Each aquarium was considered an experimental unit to avoid pseudoreplication. At the start of the experiment, fish were weighed in water (calibrated, self-taring electronic balance), measured (standard length: tip of snout to caudal peduncle; fork length: tip of snout to end of fin rays in the center of the caudal fin; and total length: tip of snout to end of rays in the longest lobe of the caudal fin), and distributed among the aquaria. Each aquarium was partially immersed in a temperature-controlled (24°C) fiberglass water bath and supplied with a continuous flow of (nonchlorinated) air-equilibrated well water. Pilot experiments showed that water flows of 0.5 l/min were adequate to maintain high dissolved oxygen concentrations (7 to 8 mg/l, Corning O_2 meter) and low dissolved ammonia (<0.2 mg/l, Hach colorimetric kit) with aquarial aeration. Fish were fed *ad libitum* (indicated by excess food on bottom of aquaria after the feeding bout has ceased) twice daily, and tanks were siphoned twice daily to remove uneaten pellets and feces.

Siphoned pellets were filtered (individual, tared paper filters) from siphon water and dried at <37°C to a constant weight immediately prior to weighing. Food consumption rates (g dry weight feed/tank/day) were calculated by subtracting the amounts not eaten from those provided. Treatment means were compared using ANOVA (Sokal and Rohlf, 1981).

RESULTS

Adult Sacramento blackfish take up about 36% of waterborne MeHg (at Clear Lake concentrations of 1.4 ng/l) with one ventilatory pass over their gills. Interestingly, the addition of 2 to 5 mg DOC/l decreased the MeHg uptake rate by about 80% (Table 122.1). In contrast, Sacramento blackfish respiratory variables were not affected by DOC addition. Mean Vg, MO_2, and UO_2 of blackfish exposed to $Me^{203}Hg$ were not influenced by the presence of DOC ($p > 0.05$; Table 122.1). In contrast, mean UHg and RHg significantly ($p < 0.05$) decreased by 78 and 74%, respectively, with 2 mg C/l of DOC, and 85 and 82% with 5 mg C/l of DOC (Table 122.1). Decreased RHg indicates slower $Me^{203}Hg$ bioaccumulation in organs as $Me^{203}Hg$ become less available to fish, presumably due to binding with DOC. The whole blood to inspired water ratio of mean $Me^{203}Hg$ concentrations significantly ($p < 0.05$) decreased by 63 and 70%, respectively, with 2 mg C/l and 5 mg C/l of DOC from the MeHg reference treatment. There were no significant differences ($p < 0.05$) in this ratio, UHg, or RHg means between the 2 mg C/l and the 5 mg C/l DOC treatments.

Although body burdens of dietary-exposed blackfish have not yet been completed, preliminary food consumption rates (overall mean: 24.0 ± 0.47 SE g dry weight of food consumed per tank of 35 fish, after 35 day) showed no effect of dietary MeHg concentrations (Table 122.2).

Table 122.1 Mean (± SE) Values of Respiratory and Methyl Mercury (1.4 ± 0.3 ng/l in inspired water) Uptake Variables in Sacramento Blackfish as a Function of DOC Additions[a]

	Treatment Group		
Variable	Reference Group (no DOC)	With 2 mg DOC/L	With 5 mg DOC/L
Ventilation volume (l/h/kg)	21.53 ± 2.87	22.38 ± 3.07	18.34 ± 1.27
O_2 consumption rate (mg O_2/h/kg)	55.85 ± 6.64	71.93 ± 7.25	70.05 ± 5.53
O_2 extraction efficiency (%)	33.34 ± 5.92	40.30 ± 3.62	43.05 ± 4.13
MeHg extraction efficiency (%)	35.65 ± 3.51	7.69* ± 0.73	5.33* ± 0.60
MeHg uptake rate (ng/h/kg)	8.96 ± 1.56	2.35* ± 0.22	1.57* ± 0.23

Note: Asterisk indicates significantly different ($p < 0.05$) from reference group.

[a] $N = 5$ in each treatment group.

Adapted from Choi et al., 1998.

Table 122.2 Mean (± SE) Preliminary (35-day Exposure) Food Consumption Rates of Juvenile Sacramento Blackfish Exposed to Dietary MeHg[a]

Dietary MeHg Concentration (mg MeHg/kg dry weight of feed)	Food Consumption Rate (g dry weight of feed eaten/aquarium)
0 (control)	24.46 ± 0.27
0.45	23.83 ± 0.48
20	24.13 ± 0.43
50	23.58 ± 0.71

[a] $N = 35$ fish per aquarium.

DISCUSSION

Clear Lake, located in Lake County, is one of California's outstanding recreational areas, and Hg pollution is a very important problem (EPA, 1997). Lake County relies on recreational fishing to support a large segment of the economy. The area surrounding the lake has a sizable population of fixed-income retirees who depend heavily on fish from the lake as a dietary supplement; there is, therefore, a potentially important human health benefit from the Clear Lake studies on fishes. The Sulphur Bank Mine, an EPA Superfund site, is a primary source of Hg pollution, especially in the Oaks Arm of the lake. Hg levels in edible fish (including largemouth bass) muscle tissue have exceeded the maximum limits set for safe human consumption (Suchanek et al., 1997). Little is known about how the mercury accumulates to these high levels in fish and what their physiological ramifications are for the fish. Our study will make a significant contribution to a more-complete understanding of the movement of mercury through the Clear Lake food web and its physiological effects on Clear Lake fishes.

To best manage a mercury-polluted site, we must first have a sufficient understanding of how mercury moves through the system, including the aquatic and terrestrial food webs. The roles of fishes in the pathways of mercury movement in Clear Lake, from mine-contaminated sediments through the water and through the Clear Lake aquatic and surrounding terrestrial food webs are unknown. Several fishes (19 species), including Sacramento blackfish, are preyed upon by a variety of birds and terrestrial aquatic mammals.

To quantify trans-gill MeHg uptake, we directly measured MeHg uptake across the gills of Sacramento blackfish, emphasizing DOC effects on the MeHg bioavailability, using the fish metabolic chamber (McKim and Goeden, 1982; Choi et al., 1998). Direct measurements of Hg uptake rate eliminate potential kinetic analyses errors of bioaccumulation due to rapid biotransformation of the compound, pharmacodynamics within the organism, or statistical interdependence (co-correlation) of fitted parameters in the iterative program used to estimate uptake rate coefficients (Black and McCarthy, 1988),

The presence of DOC significantly decreased Me^{203}Hg uptake by blackfish, but did not influence the fish respiratory variables (Table 122.1). DOC additions also generally decreased Me^{203}Hg accumulations in Sacramento blackfish organs depending on tissue-specific blood flow (Choi et al., 1998). Decreased Me^{203}Hg uptake can be accounted for by Me^{203}Hg binding to DOC in solution, thereby inhibiting its ability to pass across gill membranes due to both the DOC's polar nature and the large size of the macromolecule (Kerndorff and Schnitzer, 1980). These results support the contention that in an alkaline aquatic environment DOC could be an important factor to reduce Me^{203}Hg bioaccumulation in aquatic organisms (Choi et al., 1998). Mantoura and co-workers (1978) reported that in pH 8.0 freshwater 90% of the Hg is complexed by humic substances, especially the strong complexation between inorganic Hg and fulvic acids. Methyl mercury and inorganic Hg complexes of humic substances are very stable due to binding to sulfhydryl functional groups (Hintelmann et al., 1995). Ramamoorthy and Blumhagen (1984) demonstrated that chelation of inorganic Hg to DOC reduces uptake by rainbow trout, *Oncorhynchus mykiss* (pH 8.5 water). Thus, it is crucial to include the physicochemical interactions of Hg when predicting toxic effects on organisms in the natural environments due to DOC-associated decreases in MeHg uptake resulting in reduced bioavailability and toxicity. For example, it is important to look at environmental factors such as eutrophy that would influence DOC and MeHg complexation. Clear Lake (pH 8.3) is naturally eutrophic and algal blooms are frequent during summer (Goldman and Wetzel, 1963). Lake eutrophication and nuisance growth of blue-green algae may be linked with the presence of aquatic humus. In spite of high Hg-contaminated sediments from the mine tailings, Clear Lake fish may have lower than expected MeHg uptake from the waters due to decreased DOC-linked bioavailability. Mercury concentrations are typically lower in fish from eutrophic lakes due to the DOC complexation (Huckabee et al., 1974; McKnight et al., 1983).

Concentrations of MeHg in water are very low. Partly because of this, Spry and Wiener (1991) concluded that direct uptake of aqueous Hg probably accounted for little (perhaps <10%) of the Hg accumulated by fish. More recently, Post and co-workers (1996) modeled Hg uptake in juvenile yellow perch and concluded that uptake from water was much more important, and that the relative contributions of water and diet varied seasonally (due to temperature, food availability, and metabolism) and with ontogenic diet changes. Post and co-workers (1996) suggested that the rate of Hg uptake and the relative contribution of the two uptake pathways are seasonally dynamic because temperature, body size, and diet often vary substantially with season. They also argued that the pathways depend on the food habits and their relative intake of algae or aquatic plants with low Hg concentrations or of animal prey already highly contaminated with Hg (Post et al., 1996).

DOC plays complicated roles in bioavailability and speciation of Hg in lake ecosystems. It may enhance or retard methylation, serve as a carrier to transport Hg into waters, or compete with other binding sites (e.g., gills or sediments) for inorganic and organic Hg species. Driscoll and co-workers (1995) reported that in acidic lakes, DOC controlled the solubility and watershed export of total Hg and MeHg deposited in precipitation or bound to MeHg, limiting its availability to aquatic organisms. Results from Choi and co-workers (1998) showed that gill MeHg uptake is inhibited in the presence of DOC, yet a number of other studies show that in natural systems, fish from colored waters have higher Hg concentrations. This might imply that dietary uptake is more important in high DOC waters. Therefore, DOC effects on MeHg bioavailability at fish gills are only important if branchial uptake represents a significant route of uptake. Because nonoligotrophic lakes typically contain at least 2 mg/l DOC (McKnight et al., 1983), MeHg dietary uptake may be the principal route for the resident fishes in these systems.

These MeHg uptake studies contributed significantly to our understanding of the Clear Lake ecosystem and associated modeling effort. They also address important questions of applied science that have implications beyond the specific problems of the Clear Lake system, toward ecosystem health improvements (Cech et al., 1998).

ACKNOWLEDGMENTS

We appreciate the technical assistance of S. Ayers, S. Bennett, B. Bentley, A. Chan, K. English, T. Essert, R. Kaufman, P. Lutes, B. Nathaniel, A. Robb, J. Schmidt, T. Siddiqui, M. van de Water, and J. Watters; discussions with D. Conklin, M. Lagunas-Solar, J. McKim, J. Nichols, G. Gill, A. Heath, P. Schmieder, D. Sijm, D. Slotton, T. Suchanek, D. Jones, and R. Higashi; and reviews of preliminary drafts of parts of this manuscript by D. Anderson and J. Zinkl. This study was supported in part by grants from the Ecotoxicology Program of the University of California Toxic Substances Research and Teaching Program, the EPA (R819658) Center for Ecological Health Research at the University of California, Davis, and an EPA RARE Grant (CR824192–01–0). Although this research was funded by the EPA, it may not necessarily reflect the views of the agency and no official endorsement should be inferred.

REFERENCES

Armstrong, F.A.J., Effects of mercury compounds on fish., in *The Biogeochemistry of Mercury in the Environment,* Nriagu, J.O., Ed., Elsevier-North Holland, Amsterdam, 1979, pp. 657–670.

Beijer, K. and Jernelöv, A., Methylation of mercury in aquatic environments, in *The Biogeochemistry of Mercury in the Environment,* Nriagu, J.O., Ed., Elsevier-North Holland, Amsterdam, 1979, pp. 205–210.

Benes, P. and Havlík, B., Speciation of mercury in natural waters, in *The Biogeochemistry of Mercury in the Environment,* Nriagu, J.O., Ed., Elsevier-North Holland, Amsterdam, 1979, pp. 175–202.

Black, M.C. and McCarthy, J.F., Dissolved organic macromolecules reduce the uptake of hydrophobic organic contaminants by the gills of rainbow trout (*Salmo gairdneri*), *Environ. Toxicol. Contam.*, 7, 593–600, 1988.

Bloom N.S., On the chemical form of mercury in edible fish and marine invertebrate tissue, *Can. J. Fish. Aquat. Sci.*, 49, 1010–1017, 1992.

Carty, A.J. and Malone, S.F., The chemistry of mercury in biological systems, in *The Biogeochemistry of Mercury in the Environment*, Nriagu, J.O., Ed., Elsevier-North Holland, Amsterdam, 1979, pp. 433–479.

Cech, J.J., Jr., Massingill, M.J., and Stern, H., Growth of juvenile Sacramento blackfish, *Orthodon microlepidotus* (Ayres), *Hydrobiologia*, 97, 75–80, 1982.

Cech, J.J., Jr., Wilson, B.W., and Crosby, D.G., *Multiple Stresses in Ecosystems*, CRC/Lewis, Boca Raton, FL, 1998.

Choi, B.H., Effects of methyl mercury on the developing brain, in *Advances in Mercury Toxicology*, Suzuki, T., Imura, N., and Clarkson, T.W., Eds., Plenum Press, New York, 1990, pp. 315–337.

Choi, M.H., Cech, J.J., Jr., and Lagunas-Solar, M.C., Bioavailability of methyl mercury to Sacramento blackfish (*Orthodon microlepidotus*): dissolved organic carbon (DOC) effects, *Environ. Toxicol. Chem.*, 17, 695–701, 1998.

Driscoll, C.T., Blette, V., Yan, C., Schofield, C.L., Munson, R., and Holsapple, J., The role of dissolved organic carbon in the chemistry and bioavailability of mercury in the remote Adirondack lakes, *Water Air Soil Pollut.*, 80, 499–508, 1995.

EPA (Environmental Protection Agency), Mercury study report to Congress, Office of Air Quality Planning and Standards and Office of Research and Development, Washington, D.C., 1997.

Fitzgerald, W.F. and Clarkson, T.W., Mercury and monomethyl mercury: present and future concerns, *Environ. Health Perspect.*, 96, 159–166, 1991.

Goldman, C.R. and Wetzel, R.G., A study of the primary productivity of Clear Lake, Lake County, California, *Ecology*, 44, 283–294, 1963.

Grieb, T.M., Driscoll, C.T., Gloss, S.P., Schofield, C.L., Bowie, G.L., and Porcella, D.B., Factors affecting mercury accumulation in fish in the Upper Michigan Peninsula, *Environ. Toxicol. Chem.*, 9, 919–930, 1990.

Heath, A.G., *Water Pollution and Fish Physiology*, CRC/Lewis, Boca Raton, FL, 1995.

Hintelmann, H., Welbourn, P.M., and Evans, R.D., Binding of methyl mercury compounds by humic and fulvic acids, *Water Air Soil Pollut.*, 80, 1031–1034, 1995.

Hodson, P.V., The effect of metal metabolism on uptake, disposition and toxicity in fish, *Aquat. Toxicol.*, 11, 3–18, 1988.

Huckabee, J.W., Feldman, C., and Talmi, Y., Mercury concentrations in fish from the Great Smoky Mountains National Park, *An. Chim. Acta*, 70, 41–47, 1974.

Jensen, S. and Jernelöv, A., Biological methylation of mercury in aquatic organisms, *Nature*, 223, 753–754, 1969.

Kerndorff, H. and Schnitzer, M., Sorption of metals on humic acids, *Geochim. Cosmochim. Acta*, 44, 1701–1708, 1980.

MacCarthy, J.F., Bioavailability and toxicity of metals and hydrophobic organic contaminants, in *Aquatic Humic Substances*, Suffet, I.H. and MacCarthy, P., Eds., American Chemical Society, Washington, D.C., 1989, pp. 263–277.

Mantoura, R.F.C., Dickson, A., and Riley, J.P., The complexation of metals with humic materials, in natural waters, *Estuarine Coastal Mar. Sci.*, 6, 387–408, 1978.

Matilainen, T. and Verta, M., Mercury methylation and demethylation in aerobic surface waters, *Can. J. Fish. Aquat. Sci.*, 53, 1597–1608, 1995.

McKim, J.M. and Goeden, H.L., A direct measure of the uptake efficiency of a xenobiotic chemical across the gills of brook trout (*Salvelinus fontinalis*) under normoxic and hypoxic conditions, *Comp. Biochem. Physiol.*, 72, 65–74, 1982.

McKnight, D.M., Feder, G.L., Thurman, E.M., and Wershaw, R.L., Complexation of copper by aquatic humic substances from different environments, *Sci. Total Environ.*, 28, 65–76, 1983.

Moyle, P.B., *Inland Fishes of California*, University of California Press, Berkeley, 1976.

Murphy, G.I., The life history of the greaser blackfish (*Orthodon microlepidotus*) of Clear Lake, Lake County, California, *Calif. Fish Game*, 36, 119–133, 1950.

Peterle, T., *Wildlife Toxicology,* Van Nostrand Reinhold, New York, 1991.

Phillips, G.R. and Buhler, D.R., The relative contributions of methyl mercury from food or water to rainbow trout (*Salmo gairdneri*) in a controlled laboratory environment, *Trans. Am. Fish. Soc.,* 107, 853–861, 1978.

Post, J.R., Vandenbos, R., and McQueen, D.J., Uptake of food-chain and waterborne mercury by fish: field measurements, a mechanistic model, and an assessment of uncertainties, *Can. J. Fish. Aquat. Sci.,*51, 2482–2492, 1996.

Ramamoorthy, S. and Blumhagen, K., Uptake of Zn, Cd, and Hg by fish in the presence of competing compartments, *Can. J. Fish. Aquat. Sci.,* 41, 750–756, 1984.

Rodgers, D.W. and Beamish, F.W.H., Dynamics of dietary methyl mercury in rainbow trout *Salmo gairdneri, Aquat. Toxicol.,* 2, 271–290, 1982.

Schmieder, P.K. and Weber, L.J., Blood and water flow limitations on gill uptake of organic chemicals in the rainbow trout (*Oncorhynchus mykiss*), *Aquat. Toxicol.,* 24, 103–122, 1992.

Slotton, D.G., Reuter, S.E., and Goldman, C.R., Mercury uptake patterns of biota in a seasonally anoxic northern California reservoir, *Water Air Soil Pollut.,* 80, 951–960, 1995.

Sokal, R.R. and Rohlf, F.J., *Biometry,* 2nd ed., W.H. Freeman, New York, 1981.

Spry, D.J. and Wiener, J.G., Metal bioavailability and toxicity to fish in low-alkalinity lakes: a critical review, *Environ. Pollut.,* 71, 243–304, 1991.

Stratton, J.W., Smith, D.F., Fan, A.M., and Book, S.A., Methyl mercury in northern coastal mountain lakes: guidelines for sport fish consumption for Clear Lake (Lake County), Lake Berryessa (Napa County), and Lake Herman (Solano County), Administrative report, Hazard Evaluation Section and Epidemiological Studies and Surveillance Section, Office of Health Services, 1987.

Suchanek, T.H., Richerson, P.J., Mullen, L.I., Brister, L.L., Becker, J.C., Maxson, A.E., and Slotten, D.G., Sulphur Bank Mercury Mine Superfund Site, Clear Lake, California, Interim final report, EPA Region IX, Superfund Program, Davis, 1997.

Watras, C.J. and Huckabee, J.H., *Mercury Pollution: Integration and Synthesis,* Lewis Publishers, Boca Raton, FL, 1994.

Watras, C.J. et al., Sources and fates of mercury and methyl mercury in Wisconsin lakes, in *Mercury Pollution: Integration and Synthesis,* Watras, C.J. and Huckabee, J.H., Eds., Lewis Publishers, Boca Raton, FL, 1994, pp. 153–177.

Wiener, J.G. and D.J. Spry, Toxicological significance of mercury in freshwater fish, in *Environmental Contaminants in Wildlife, Interpreting Tissue Concentrations,* Redmon-Norwood, A.W., Ed., CRC/Lewis, Boca Raton, FL, 1996, pp. 297–339.

Zar, J.H., *Biostatistical Analysis,* Prentice-Hall, Englewood Cliffs, NJ, 1984.

CHAPTER 123

An Integrated Watershed Approach to Studying Ecosystem Health at Lake Tahoe, CA–NV

John E. Reuter, Thomas A. Cahill, Steven S. Cliff, Charles R. Goldman, Alan C. Heyvaert, Alan D. Jassby, Susan Lindstrom, and David M. Rizzo

INTRODUCTION: CAUSE FOR CONCERN

Lake Tahoe lies in the crest of the Sierra Nevada mountains at an elevation of 1898 m within both California and Nevada. The drainage area is 812 km^2 with a lake surface of 501 km^2, producing a ratio of only 1.6:1 (Figure 123.1). This ratio is much smaller than the 10:1 ratio found for typical watersheds. The lake is located in a montane-subalpine watershed dominated by coniferous vegetation and nutrient-poor soils. It is the world's 11th deepest lake at approximately 505 m with a mean depth of 313 m. Its volume is 156 km^3, with a hydraulic residence time of about 650 to 700 years, and is ice-free.

Continuous, long-term evaluation of water quality in Lake Tahoe since the early 1960s has shown that algal growth is increasing at a rate > 5% per year (Goldman, 1988). Correspondingly, there has been a decline of clarity at an alarming rate of 0.25 m per year (Jassby et al., 1999) (Figure 123.2). This long-term trend is both statistically significant and visually noticeable. If the rate of clarity loss continues, the lake could lose a total of 15 m of transparency by the year 2030. The resulting Secchi depth will most likely be accompanied by a change of color and a permanent change in trophic status. Lake Tahoe was once classified as ultraoligotrophic (Goldman, 1974), i.e., low nutrient content, low plant productivity, and high transparency. However, because of the ongoing decline in clarity and rise in algal growth rate, its trophic status (level of fertility) has been moving away from this extraordinary condition.

Research on the spatial distribution of phytoplankton (free-floating algae) indicates a marked correspondence between the highest algal growth rates and the most extensive shoreline development. Synoptic studies have shown that the central portion of the lake has historically been characterized by relatively less algae, with areas near shoreline development exhibiting enhanced production (Goldman, 1974). Similar studies of the attached algae or periphyton also demonstrate this pattern. The dramatic differences in algal growth on rocks at various shoreline locations is linked to nearby development (Loeb and Reuter, 1984).

Ironically, some of the same features that maintained the exceptional historical water quality in Lake Tahoe now threaten its health under current conditions of increased nutrient and sediment loading. Tahoe's large depth and volume once acted to dilute pollutants to a level of no significant effect; however, this is no longer the case. We now know that after nutrients enter the lake, they

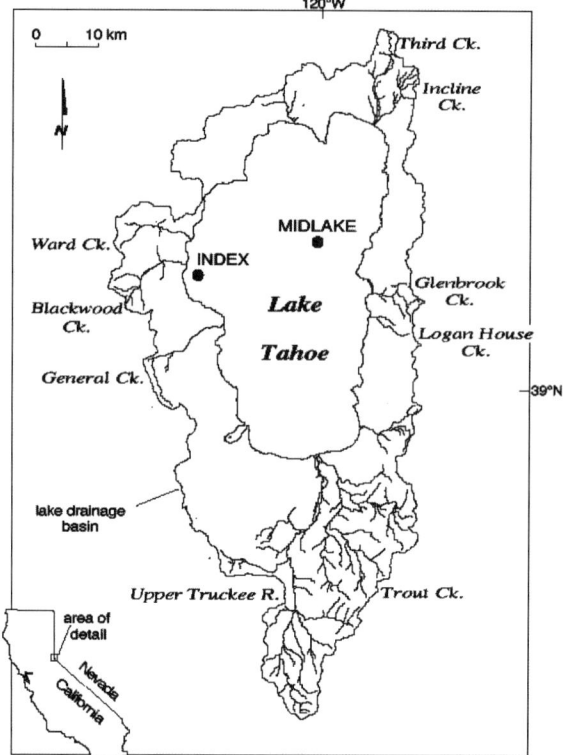

Figure 123.1 Location of Lake Tahoe and its watershed, showing some of the most important tributaries.

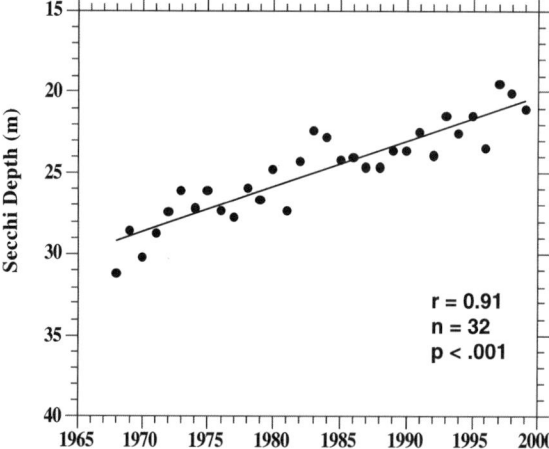

Figure 123.2 Mean annual Secchi depth based on long-term data collected by the Tahoe Research Group, showing the significant decline in lake clarity as measured by Secchi depth transparency. Each annual point represents the mean of 30 to 35 individual observations taken on a regular 7- to 10-day schedule.

remain in the water and can be recycled for periods of more than 10 years (Jassby et al., 1995). As a consequence, these pollutants accumulate from year to year.

Research has shown a fundamental shift from frequent stimulation of algal growth by nitrogen additions to primarily phosphorus stimulation (Goldman et al., 1993). The response of Lake Tahoe algae to nutrient enrichment has been measured since the 1960s, with the observed shift occurring in the early 1980s. Since phosphorus is typically transported along with sediment, these findings underscore the importance of sediment control and erosion mitigation. Atmospheric deposition of nitrogen from both in-basin and out-of-basin sources is now considered a significant factor contributing to the observed shift in nutrient stimulation (Jassby et al., 1994). Because much of the phosphorus input to the lake is still derived from the watershed, erosion control, acquisition of sensitive lands, and other watershed restoration practices remain appropriate courses of action.

LINKAGE BETWEEN SCIENCE AND POLICY FOR THE BENEFIT OF ECOSYSTEM MANAGEMENT

One of the cornerstones of integrated watershed management is a comprehensive understanding of (1) hydrologic, atmospheric, and ecological processes and their interactions; (2) real-time assessment of environmental conditions (e.g., air quality, water quality, forest health, etc.); (3) response to anthropogenic and natural disturbance; and (4) the ability to predict environmental improvement based on various management strategies. Indeed, serious concerns regarding ecological condition and long-term environmental protection underscore the need to provide the highest quality science to aid in problem resolution. Ecosystem health, sustainable environment, and watershed management are interrelated and part of the growing view that the fabric of the natural landscape is a complex weave of interacting influences that include physical, chemical, and biological factors. Without a sound scientific foundation, critical discussions are too easily misdirected toward narrowly focused interests.

The watershed approach taken at Lake Tahoe for many decades recognizes that lake water quality is linked to upland watershed processes and air quality. Disruption of natural ecosystem processes that naturally treat runoff (e.g., wetlands, groundwater infiltration, vegetation) and a changed landscape, which alters hydrology and promotes the accelerated loading of nutrients and sediment (e.g., impervious cover, road network, habitat disruption, land disturbance), have affected natural watershed processes. Successful implementation of land, air, and water quality restoration projects is considered the only likely avenue to arrest further decline in lake clarity. Scientific efforts must be focused toward restoration objectives and coordinated so that information needed for adaptive management can be obtained.

OBJECTIVES

In this chapter, we provide an initial look at integrating a number of disciplines, all important to ecosystem health and management at Lake Tahoe. These include air quality, water quality, and forest health, all of which are affected by human development. The following discussions come from individual presentations given at the International Congress on Ecosystem Health held in Sacramento, California, in August 1999. The theme of the Congress was "Managing for Ecosystem Health," and the session was entitled, "Moving Beyond the Conflict at Lake Tahoe." Participants agreed that a multidisciplinary approach for developing effective water management strategies was critical in the Tahoe Basin, and that a marriage between policy and science was essential. This type of approach was the focus of the recently completed Lake Tahoe Watershed Assessment (USFS, 2000), which was funded through the U.S. Forest Service (Lake Tahoe Basin Management Unit) and included contributions by academic and agency scientists.

PAST HUMAN LAND-USE AND ENVIRONMENTAL CONDITIONS

The past is part of the living present. The Lake Tahoe basin embodies the consequences of a long legacy of human and environmental history. Humans have been a component of the Lake Tahoe ecosystem for at least 8000 years, and land disturbances were initiated here with low-intensity land management by the Washoe Indians and their prehistoric predecessors. Within a century's time, indigenous practices were replaced by profound resource exploitation by incoming Euroamerican populations (beginning in the late 1800s). Within the last few decades, agency regulation has struggled to control explosive community growth induced by millions of people who visit the Tahoe Basin each year.

Contemporary land management attempts to restore the Lake Tahoe ecosystem benefit from an understanding of the long-term ecological role of aboriginal people and historic Euroamericans in the dynamics of wild plant and animal populations and their physical environments. Past land management practices engendered environmental impacts that varied in space, time, scale, intensity, and consequence. At Lake Tahoe, human disturbances range widely in scale, from pruning a patch of native shrubs to clear-cutting thousands of acres of timberland. Some resources were targeted in a single brief event, while others were affected for decades or generations. Furthermore, prehistoric and historic impacts may not have extended basin-wide, and some areas may have been relatively unaffected.

Changing attitudes and assumptions about Tahoe's environment have both enshrined and desecrated its landscape. Native Americans considered themselves stewards of the land and sustained a balanced relationship between human society and the environment. The 19th century arrivals viewed the Tahoe Basin as a natural setting for capital investment and profit associated with the silver mining activities of Nevada's Comstock era. Devastating practices, such as clear-cutting forests, were acclaimed by a society that celebrated human conquest of nature as progress. During the 20th century, the emerging dominance of the tourism industry was accompanied by a growing awareness of resource protection rather than resource extraction. However, the environmental pressure of large numbers of tourists and the growing residential population that exists to serve them has transformed the Tahoe Basin into a landscape that is, paradoxically, increasingly imperiled by its own attractiveness (Raymond, 1992). Tahoe's future well-being directly depends on a healthy physical and socioeconomic environment, not only the public's perception of one. With this latest shift in land-use paradigms, the direction of progress for 21st century users of the Tahoe Basin is less clear. In this fragile context, the Lake Tahoe environment is becoming intensively managed. This action has been partly prompted by research findings developed over the last few decades, which show that the ecosystem is a victim of multiple stresses and is now degrading at a pace observable to the largely shore-bound populace.

To shape an approach to ecosystem planning and resource management in the Tahoe Basin, we pose a number of key questions regarding the physical and cultural conditions that existed in the past, and the scope and scale of anthropogenic or human-induced disturbances that have altered conditions.

- How have climate changes affected changes in Tahoe's overall physical and cultural environment? Under varying climatic regimes, what sustainable environmental conditions are possible in the future and what type and scale of management treatments would be required to achieve them?
- What did presettlement terrestrial and aquatic ecosystems look like, and how have prehistoric and historic anthropogenic disturbances affected changes in plant and animal communities?
- What were the prehistoric and historic fire regimes in the Tahoe Basin?
- What are the past and present sources of sediment and nutrients in Tahoe's watershed that affect water quality?
- How has air quality and atmospheric visibility changed from prehistoric to present times?
- What are the historic underpinnings of causal relationships among the many socioeconomic and environmental factors in the Tahoe Basin?
- What are the culturally important locales and biotic species in the basin, and what threatens these resources?

To answer these and other questions, historical data sources that are uniquely tied to the human dimension of the ecosystem must be explored, in concert with past, present, and future biological, hydrological, and atmospheric sciences studies. In this way, we hope to establish standards by which environmental and socioeconomic health can be measured. History directs future decision making by setting a baseline of reference conditions to determine (1) how present conditions differ from past conditions, (2) the reasons for those differences, and (3) what sustainable conditions may be possible in the future. With a knowledge of how peoples have interacted with their landscape in the past, scientists and land managers not only reconstruct and interpret how people shaped past ecosystems, but link this information to the restoration and maintenance of future ecosystems. Archaeological, ethnographic, and historic documentation offer great time depth and are used as independent and corroborative tools to achieve a nexus between historic conditions and contemporary research, monitoring, and adaptive management (see discussion of paleolimnological reconstruction). Within this context of interplay among multiple sources of information, it may be possible to achieve an understanding that would not be possible using any single source of information.

THE STATE OF AIR QUALITY AND MODELING, AND ITS ROLE IN ECOSYSTEM MANAGEMENT

Lake Tahoe's unique geographical setting is marked by a cold, deep lake surrounded by an air basin defined by high mountains with dramatic vertical relief. The presence of the cold lake at the bottom of this basin defines an atmospheric regime that, in the absence of strong synoptic weather systems, develops very strong (to 10°C), shallow (30 m) subsidence and radiation inversions at all times throughout the year. In addition, the rapid radiation cooling at night generates gentle (1 m/s) but predictable downslope winds each evening, moving from the ridge tops over the developed areas at the edge of the lake and out over the lake itself. Local pollutant sources within this bowl are trapped by inversions that greatly limit the volume of air into which they can be mixed; as a result, pollutants build up to elevated concentrations. Further, the downslope winds each night move local pollutants from developed areas around the periphery of the lake out over the lake, thereby increasing the opportunity for these pollutants to deposit into the lake itself. This meteorological regime, weak or calm winds and a strong inversion, is the most common pattern throughout the year (Cahill et al., 1996).

The location of Lake Tahoe in the crest of the Sierra Nevada creates the second most common meteorological regime, that of atmospheric transport from the Sacramento Valley into the basin by mountain upslope winds. This patterns develops when the western slopes of the Sierra Nevada are heated, causing the air to rise in a chimney effect and move upslope and over the Sierra crest. The strength of this pattern depends on the amount of heating, and thus is strongest in summer, beginning in April and essentially ceasing in late October (Cahill et al., 1997). This upslope transport pattern is strengthened and made even more frequent by the alignment of the Sierra Nevada range across the prevailing westerlies common at this latitude, which combine with the terrain winds to force air up and over the Sierra Nevada from upwind sources in the Sacramento Valley.

Both the proximity of the basin to upwind urban areas and the preponderance of automobiles, which provide access to Tahoe's tourism-based economy, contribute to air pollution in the basin. Due to topography and location, the simplified Lake Tahoe air basin is defined by three major meteorological regimes: the summer daytime westerly winds, the summer nighttime inversion, and a persistent wintertime stagnant inversion. These meteorological regimes transport and then trap pollutants near the lake surface. Recently, measurements at Lake Tahoe indicate a loss of lake clarity which is significantly coupled to an atmospheric source of both nutrients and fine particles (Jassby et al., 1994, 1999; Cliff and Cahill, 2000). An air-quality model specific to this region was developed as part of the Lake Tahoe Watershed Assessment program to aid in management decisions with respect to air quality and ecosystem health.

The Lake Tahoe Airshed Model (LTAM) is a heuristic eulerian model designed to provide predictive capabilities for management of the basin. It also allows us to gather the disparate sources of air quality data at Lake Tahoe into a consistent framework. Pollution sources including automobiles and forest fires (both wildfires and prescribed fires) are put into the model; transport and a factor for deposition across the basin are predicted. The LTAM is an array of 1248 individual 2.56 km^2 cells across the basin encoded on a Microsoft® Excel spreadsheet. The domain is 72 km north to south (Truckee to Echo Summit) and 42 km west to east (Ward Peak to Spooner Summit). All but the southernmost end of the watershed is taken into account by the model. The LTAM is semiempirical in design, and incorporates all available air quality measurements at Lake Tahoe, plus aspects of meteorological and aerometric theory. Free variables (traffic flow, acres burned in the forest, population density, etc.) are assumed to be linear with pollutant emissions. Additional research is proposed to test these and other model assumptions.

LTAM is designed to provide information on the role of the atmosphere in the health and welfare concerns of the Lake Tahoe basin. It has two major immediate goals: (1) to predict the concentration of air quality pollutants in the Tahoe Basin at spatially diverse locations where no data exist and (2) to predict the potential for atmospheric deposition of nutrients and fine particles to the watershed and lake by determining spatial concentration of pollutants within the basin. A thorough description of the LTAM, inputs to the model, and several output scenarios is given in Cliff and Cahill (2000).

Scenarios used as examples for the Lake Tahoe Watershed Assessment modeling integration were developed to present the type of results that are possible with single medium (in this case air quality) models designed specifically for the Tahoe Basin. We evaluated prescribed burns within the Ward Creek watershed, in October, on 40-year (50 ha) and 20-year (100 ha) return cycles, as well as a wildfire (occurring in August) that burned approximately 75% of the forested part of the Ward Creek watershed (1500 ha). From the calculated emission parameters for the wildfire and prescribed fire scenarios, the LTAM calculates the falloff in smoke PM2.5 (particulate matter captured on a 2.5-μm filter) across the basin. The resultant values from the LTAM are graphed as a concentration vs. location and then plotted over a map of the area represented by the model. A comparison of the 20- and 40-year return cycle scenarios is shown in Figure 123.3.

Contrasted to the historical wildfire comparison (Figure 123.3A), it is seen that the LTAM predicts massive violations of federal and state PM$_{2.5}$ standards for the 20-year return scenario (Figure 123.3C). The historical wildfire is an analysis of past conditions based on a 40-year return in the basin divided among the total burn season, equaling about 12 ha burned per day in three small (4 ha) wildfires (Figure 123.3A). The 40-year return scenario predicts localized violations for the October period (Figure 123.3B). A model run for the same scenario for a typical summer period (not shown) predicts lesser violation, mostly due to the increased ventilation of the basin during that period. The hypothetical wildfire during August (1500 ha) is predicted to completely fill the basin and beyond with smoke (Cliff and Cahill, 2000). Although the wildfire burns an order of magnitude more land than the prescribed fires, the number of resultant violation days is predicted by LTAM to be roughly equivalent. That is about 2 to 3 violation days for the 40-year fire, 3 days for the 20-year fire, and 4 to 5 days for the wildfire. This apparent discrepancy is due mostly to the increased ventilation of the basin during the late spring and summer months. Furthermore, increased lofting of smoke in a wildfire results in impact at greater downwind distance than in a prescribed fire situation. It should be noted that the LTAM is capable of only limited predictions at this time. Further study of the impact of fire on the Lake Tahoe ecosystem, especially the impact of smoke on lake clarity, visibility, and human health, is necessary to better define parameters for integrated management models in general, and the LTAM in particular.

Air quality modeling at Lake Tahoe, although at the nadir of development, is capable of aiding in ecosystem health management. The most significant finding from the construction and use of

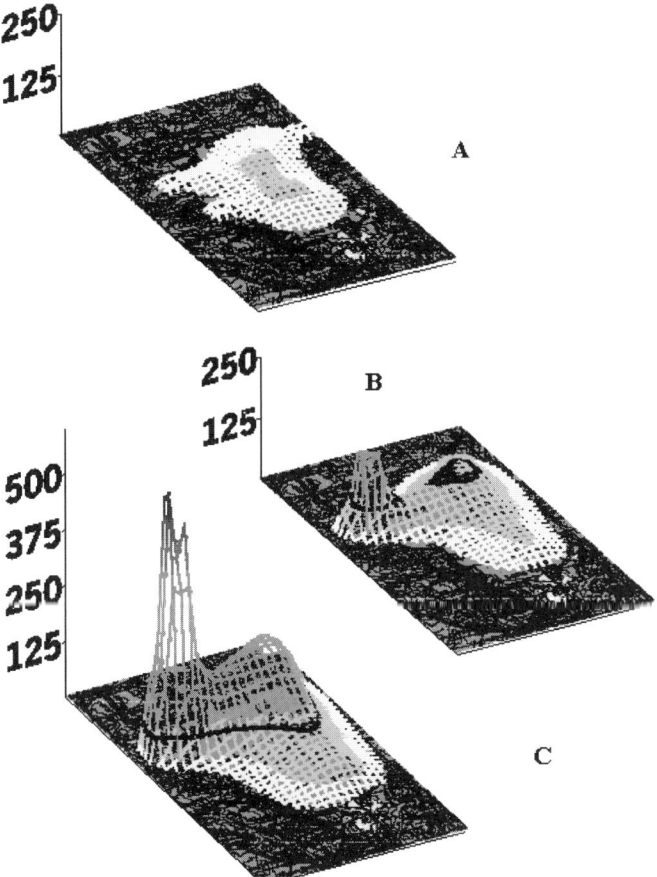

Figure 123.3 PM$_{2.5}$ concentration (mg m^{-3}) predictions from LTAM, based on a 24-h average superimposed on the basin map. Comparison of three fire scenarios in the Lake Tahoe basin. (A) The historical wildfire (12 to 16 ha). (B and C) Hypothetical prescribed fire scenarios located in the Ward Creek watershed; (B) a 50-ha prescribed fire; and (C) a 100-ha prescribed fire representing a 40- and 20-year total fire return time to the basin, respectively. The black isoplith is set at 65 mg m^{-3}, which is the (proposed) federal 24-h standard for PM$_{2.5}$. The peak concentrations for A, B, and C are 29, 165, and 500 mg m^{-3}, respectively.

LTAM is that pollutants associated with dry deposition are most likely to deposit to the lake surface and hence potentially degrade lake clarity at times of intense inversion; this provides support to direct measurements of deposition (Jassby et al., 1994). Atmospheric inversion at Lake Tahoe is the most predominant meteorological condition in the evening during the summer months, and all day throughout the interstorm winter period. An evaluation of PM$_{2.5}$ for hypothetical wildfire and prescribed fires in the Ward Creek watershed has been performed. The result of this evaluation indicates that, for the prescribed fires, a fall burn is particularly troublesome from the standpoint of air quality. The atmospheric inversions that tend to be present during the fall inhibit ventilation and hence allow buildup of the concentration of PM$_{2.5}$. Currently, not enough is known about the chemical composition and speciation of prescribed fire smoke to evaluate the impact of these prescribed fires on the clarity of Lake Tahoe. It is known, however, that PM$_{2.5}$ is a pollutant from the perspective of human health. The potential for violations of federal, state, and basin air quality standards based on visibility is also expected from prescribed fires. In general, to better evaluate the impact of air quality and ecosystem health at Lake Tahoe requires further study. It is imperative to understand the link between emission, transport, and deposition of air constituents throughout the basin to better design integrated modeling tools for management use.

LAKE TAHOE NUTRIENT BUDGETS

For decades, planning, regulatory, and implementation actions in the Tahoe Basin have focused on controlling nutrient and sediment inputs to the lake. Examples include, but are not limited to, acquisition of environmentally sensitive lands, building restrictions, BMP retrofitting, erosion control, installation of BMPs for treatment of surface runoff, permits, and education. However, these have largely been done on a project-by-project basis rather than as part of a unified, comprehensive plan. Now that the public and private sectors are working in cooperation to achieve environmental restoration activities in the basin, it is more important than ever that budgets quantify the critical sources of nutrients and sediment. Setting priorities for restoration projects depends on identifying the critical sources of these materials.

Five major categories of nutrient loading to Lake Tahoe have been identified: (1) direct atmospheric deposition and precipitation, (2) stream discharge, (3) overland runoff directly to the lake, (4) groundwater, and (5) shoreline erosion. The major losses include settling of material from the water column to the bottom and discharge to the Truckee River, the sole tributary outflow. We provide preliminary estimates for total phosphorus (TP), dissolved P, and total nitrogen (TN) loading (Table 123.1).

Atmospheric Deposition

In 1994, the first estimate was made of the contribution by atmospheric deposition of N and P to the annual nutrient load of Lake Tahoe (Jassby et al., 1994). This study analyzed both wet (rain and snow) and dry fallout. It was concluded that atmospheric deposition provides most of the dissolved inorganic nitrogen and total nitrogen to the loading budget of Lake Tahoe. It was further concluded that deposition also contributes significant amounts of soluble reactive P and total P loading, but to a lesser extent than for nitrogen. Comparisons of atmospheric loading at the Ward Lake level, located along the west shore, showed that (1) deposition of nitrate and ammonium were similar, regardless of whether the wet or dry forms were being considered; (2) wet deposition of nitrate and ammonium in the form of snow and rain had approximately twice the loading rate as deposition from dry fallout; (3) conversely, the loading of dry soluble reactive P was 2.4 times that for wet; (4) the ratio of total N to dissolved inorganic N (i.e., nitrate plus ammonium) was 2:1 with dry fallout comprising 64% of TN deposition; and (5) the ratio of total P to soluble reactive P was also just over 2:1, with dry fallout comprising 70% of TP deposition.

Using data contained in Jassby et al. (1994), combined with other portions of the historic deposition monitoring database, and the existing isohyetal map for Lake Tahoe, which shows the spatial distribution of precipitation over the entire lake and watershed, loading values for N and P, which fall directly on the lake surface were calculated. Nutrients deposited on the watershed and

Table 123.1 Budget of Total Nitrogen, Total Phosphorus, and Dissolved Phosphorus Inputs to Lake Tahoe

	Total N	Total P	Soluble P
Atmospheric deposition	234	12.4	5.6
Stream loading	82	13.3	2.4
Direct runoff	23	12.3	2.4
Groundwater	60	4	4
Shoreline erosion	1	1.6	—
Total	400	43.6	14.4

Note: Values were obtained from direct measurements of concentration and discharge, and scaled to the whole lake. They represent an average during the period 1980–1993. Data are presented as metric tons (MT) per year.

which are subsequently transported to the lake are included in the calculations of stream discharge, direct runoff to the lake, and groundwater. For the entire lake surface area, the contribution of P was 12.4 metric tons (MT; where 1 MT = 1000 kg or 2205 lbs). Direct N loading to the lake surface was estimated at 234 MT. This accounts for 28 and 59% of the annual TP and TN budgets, respectively. At this time, we hypothesize that P present in precipitation and dry fallout results from wood smoke, road dust, and aeolian (wind) transport from disturbed land. Sources of N come from within and outside the Tahoe Basin, and include automobile emissions, agrochemical residues, wood smoke, etc.

Stream Loading

A total of 63 streams drain into Lake Tahoe. These streams are characterized by different levels of urban development and disturbance. The Lake Tahoe Interagency Monitoring Program (LTIMP) has been sampling up to ten of these streams since 1980. Because of variation in watershed characteristics around the basin and the significant "rain shadow" effect along the west-to-east direction across the lake (i.e., decline on total precipitation from west to east), no single location is representative of all watersheds. Flow from these monitored streams comprises 50 to 55% of the total discharge from all tributaries. Each stream is monitored on at least 30 to 35 dates each year. N and P loading calculations were done using the LTIMP flow and nutrient concentration database.

Using data from the early 1980s to 1993, the Tahoe Research Group has calculated stream loads for N and P as part of two separate studies. The results for annual N loading were 81 and 55 MT for the beginning and end of this period, respectively. Comparable loading values for total P were 13 and 11 MT (Marjanovic, 1989; Jassby et al., 1994). Differences from period to period reflect the variation in precipitation and runoff. Contributions for N and P in our budget were taken as the mean, or 68 and 12 MT, respectively. Thodal (1997) also provided a very preliminary estimate of nutrient loading to Lake Tahoe during the period 1990 to 1993. Annual nutrient load associated with stream flow was estimated by multiplying the mean annual volume of surface water runoff by the mean annual nutrient concentration. Using this simple approach, loading was reported as 70 MT for total N and 20 MT for total P (Thodal, 1997). A very early estimate of stream flow nutrient load by Dugan and McGauhey (1974) estimated 120 MT of total N and 9 MT for total P. Taking the mean of these four values, which represent different time periods and consequently different precipitation conditions, loading estimates of 82 and 13.3 MT were obtained for total N and total P, respectively. These account for 21 and 31% of the N and P budgets.

Direct Runoff

The Tahoe Basin has 52 intervening zones which drain directly into the lake without first entering the streams. These intervening zones are generally found between the individual watersheds and as such are distributed around the entire lake. These zones range in size from 0.1 km^2 to 10.5 km^2. The range for covered or otherwise disturbed land within these intervening zones ranges from 0 to 63%. The overall ratio of disturbed to total area is 27% with runoff from the intervening zones accounting for 10% of the entire drainage.

Calculations of loading from direct runoff requires quantification of flow and concentration. Flow from each of the intervening areas was estimated by Marjanovic (1989). Data on N and P concentrations are not extensive since this type of study has not received priority funding in the basin. However, based on a limited database presented and reviewed by Reuter et al. (2001), we estimated concentration data for this preliminary nutrient budget. For the purpose of calculation, an intervening area was considered urban if 25% or more of its areas was classified as covered or disturbed. Concentrations representative of urban and rural conditions were taken from the field studies cited previously and used in the quantification of loads (for urban conditions, total Kjeldahl N [TKN] = 0.48 mg l^{-1}, nitrate = 0.08 mg N l^{-1}, total P = 0.31 mg l^{-1}, and soluble reactive P = 0.05 mg l^{-1}, for rural conditions, TKN = 0.19 mg l^{-1}, nitrate = 0.01 mg N l^{-1}, total P = 0.10 mg l^{-1}, and soluble reactive P = 0.01 mg l^{-1}).

N loading was calculated at 23 MT or 6% of the total N budget while P loading was 12.3 MT or 28% of the total P budget. The observation regarding the high contribution of P loading from direct runoff is particularly important because a significant portion of the urbanization at Tahoe is found in the intervening zones. It provides project planners with the understanding that control and treatment of urban and direct runoff to the lake is critical and should be a high priority.

Groundwater

The most comprehensive, basin-wide effort to date comes from Thodal (1997) as part of a hydrogeology study of the Tahoe Basin. Data on the results of a groundwater quality monitoring study done from 1990 through 1992 are presented. By multiplying mean nutrient concentrations from their groundwater survey (N = 1.0 mg l^{-1}, P = 0.074 mg l^{-1}) and estimates of total annual groundwater discharge to the lake (5.15 × 107 m^3), Thodal calculated "rounded estimates" of 60 MT for N loading and a 4 MT for P loading. This accounted for 15% of the TN budget and 9% of the TP budget.

Shoreline Erosion

The process of shoreline erosion and its quantitative importance to the nutrient and sediment budgets of Lake Tahoe have received very little attention. However, the importance of shoreline erosion has been highlighted in recent years when the combination of high lake levels and strong and sustained winds altered some of the west shoreline by many feet. Recently, Adams and Minor (2000) estimated shoreline changes in Lake Tahoe by analyzing historical photographs and combined this evaluation with measurements of the nutrient content of shoreline sediments. They estimated an annual loading of only about 1 MT for nitrogen and 1.6 MT for phosphorus.

Budget Review

This budget clearly suggests the importance of direct runoff as an important P source from urban areas and highlights the need for additional study of loading from this source. At the same time, the contribution of atmospheric deposition to the N budget clearly dominates other sources. Heyvaert (unpublished data) has found that nutrient sedimentation losses to the bottom of Lake Tahoe are 402 MT for total N and 53 MT for total P. These data agree remarkably well with the independent loading estimates given previously. This close agreement give us increased confidence that the loading rates are representative.

Phosphorus reduction strategies at Lake Tahoe will have to address multiple sources including direct runoff, atmospheric deposition, and stream loading. Using the estimated loading of dissolved P as a first approximation of biologically available P (BAP), this budget further shows that BAP is on the order of 33% of TP. This value is not uncommon (Rechow and Chapra, 1983); however, the 14.4 MT value may underestimate true BAP to the extent that bioavailable P is released from particulate P entering the lake's water column. Research to investigate this further has been proposed. Looking at dissolved P alone, the relative importance of the groundwater contribution increases.

As restoration projects are being targeted and adaptive management proceeds, it will be very helpful to have more detailed data on the specific sources of nutrients within each of the major categories discussed previously. Restoration should give priority to those areas that make the greatest contribution to the nutrient loading budget.

PALEOLIMNOLOGICAL RECONSTRUCTION OF BASELINE CONDITIONS AND ECOSYSTEM RESPONSE TO ANTHROPOGENIC DISTURBANCE

Lake sediments constantly accumulate material derived from the watershed and from the overlying water column and, over time, a physical record accrues. Biogeochemical analysis of this record can provide useful information about lake response to changing conditions, whether from natural causes or from anthropogenic disturbance in the watershed. The analysis of lake sediment cores allows us to examine ecosystem processes at a longer and more-relevant time scale than is usually attained from available monitoring data. When used in conjunction with process-oriented research that includes both modeling and the analysis of long-term data, this approach can also significantly improve our efforts to forecast ecosystem response to contemporary watershed disturbance.

In the Tahoe Basin, there have been two major episodes of watershed disturbance since it was located and described by Lt. John Fremont in 1844. The first event was extensive clear-cut logging associated with the Comstock era that began in the 1860s and continued into the 1890s. Timber from the Tahoe Basin was exported to adjacent communities for construction of railroads, mines, and towns. It is estimated that over 60% of the watershed was clear-cut during this period, with much of the remaining land either alpine, barren, or inaccessible. The second period of disturbance began with rapid urbanization in the late 1950s. To this day, development continues within the Lake Tahoe watershed. Among other significant disturbances, an extensive road network now surrounds the lake, located mainly in lower reaches of individual watersheds, but extending a considerable distance up accessible slopes in many areas. Of particular interest are the effects on lake function from logging into the late 1800s, and the system response after that disturbance ended. Such information could be instructive for evaluating modern environmental impacts from urbanization and the potential response from mitigation and restoration.

Over the years, several sediment cores have been extracted from various points within the lake to determine spatial and long-term patterns of sediment composition and accumulation (Heyvaert, 1998). These cores have been analyzed for many chemical and biological constituents and for characterization of their physical attributes. The specific goals of this project were to (1) identify biogeochemical markers that indicate lake and watershed responses to ecological stress; (2) establish the baseline predisturbance condition of these markers and its natural background variability; (3) assess watershed response to historical periods of fire, drought, and timber harvest; (4) determine lake responses to urbanization since the late 1950s; and (5) establish a database for the calibration and verification of watershed–lake models in the Tahoe Basin.

The most important preliminary step in this study was to establish an accurate sediment chronology. We now have a relatively reliable geochronology for the Tahoe sediments, constructed from ^{210}Pb and ^{14}C data. These data indicate significant basin-wide changes in mass sedimentation rates over the last 150 years. Specifically, high sedimentation rates were associated with clear-cut logging in the Tahoe Basin from 1860 through 1900. That period was followed by a three- to five-fold decrease in mass sedimentation rates during the early 20th century. These lower sedimentation rates persisted until urbanization began in the Tahoe Basin after World War II.

From ^{210}Pb data, the average mass sedimentation rate (90% confidence interval) during the Comstock logging era from 1860 to 1900 was 0.043 ($\pm$ 0.011) g cm^{-2} per year. By comparison, the average mass sedimentation rate for the recent period from 1970 to 1990 was 0.027 ($\pm$ 0.006) g cm^{-2} per year. Notably, both of these rates are significantly higher than the average mass sedimentation rate of 0.009 ($\pm$ 0.004) g cm^{-2} per year determined for the intervening period from 1900 to 1970.

Predisturbance sedimentation rates were estimated from ^{14}C measurements in several deep sections of two cores. The long-term average rate was 0.006 ($\pm$ 0.002) g cm^{-2} per year, which is just slightly less than the sedimentation rate that was estimated for the period intervening between Comstock logging and urbanization. Since these rates are comparable, it would suggest that landscape recovery was rapid after clear-cut logging ended and that lake sedimentation rates dropped back almost to predisturbance levels.

Diagenesis and organic decomposition preclude a quantitative reconstruction of historical algal primary productivity (PPr) from the carbon record. However, diatoms represent greater than 80% of the phytoplankton biomass in Lake Tahoe, so biogenic silica accumulation rates were used as a proxy for historical algal productivity. The PPr reconstructed for that interval from 1900 to 1970 was on average 28 g cm^{-2} per year, which is about 25% less than the earliest ^{14}C PPr measurements conducted at Lake Tahoe in 1959 (39 g cm^{-2} per year). Primary productivity reconstructed for the historic period of Comstock logging gave an average annual rate of about 176 g cm^{-2} per year, which is comparable to annual average PPr measured in 1993 (183 g cm^{-2} per year). Our estimate of baseline, predisturbance PPr in Lake Tahoe before 1850, was 27 g cm^{-2} per year. Apparently the lake nearly returned to its baseline PPr during the intervening period, soon after logging ended, despite a 650- to 700-year hydraulic retention.

The fact that mass sedimentation rates and biogenic silica flux decreased shortly after the logging disturbance ended is testimony to rapid landscape stabilization and recovery with second-growth forest. It also indicates that Comstock logging produced a pulse disturbance. By contrast, the disturbance from urbanization could persist as a chronic perturbation for a considerable time. These data also suggest, however, that effective mitigation of the watershed erosion caused by urbanization could directly improve water quality on a decadal timescale. This corresponds to the 50% response times calculated independently from sediment trap data for nitrogen and phosphorus settling velocities.

At this time we continue to refine the Tahoe Basin sediment chronology and our estimates of mass and nutrient sedimentation rates. Work has also begun on the interpretation of additional sediment markers for understanding other disturbance patterns, including drought and forest fires. Recently, piston cores representing several thousand years of lake and watershed history were recovered from the lake. These data will yield, among other factors, long-term profiles of natural variation in sedimentation rates and historical PPr at Lake Tahoe.

FOREST HEALTH IN THE TAHOE BASIN

As in many areas of the western U.S., periodic fire in the coniferous forests of the Tahoe Basin is considered to be a defining influence on ecosystem function and health, altering fundamental plant, soil, and microclimate conditions. Historically, frequent low-intensity fire has been the main disturbance driving ecosystem structure, function, and composition. Within the last century, however, fire suppression and logging at Lake Tahoe (both clear-cutting and selective harvests) have significantly altered stand conditions by increasing stem densities of shade-tolerant species, soil litter depth and understory shading. Both understory and overstory tree densities have increased dramatically throughout basin forests; in particular, there has been a doubling in the importance of white fir and incense cedar and a decline in the importance of Jeffrey pine and sugar pine by 50% (Barbour et al., 2002). Sites that have been previously logged currently exhibit the highest stem densities for both overstory and understory trees. In association with a recent drought, highly visible levels of tree mortality in the Tahoe Basin began in the mid-1980s (Smith et al., 1994; Elliot-Fisk et al., 1996). Based on 1997 and 1998 ground surveys of 31 stands in the basin, overall cumulative tree mortality in the lower montane forests ranges from 9 to 33% in 31 sampled stands with mean mortality of 25% for previously logged stands (seral stands) and 21% for old growth stands that have never been logged (Maloney and Rizzo, unpublished data).

In many stands these conditions are outside the forest's historic range of variability, fundamentally altering ecosystem processes. It has become apparent that in fire-dependent ecosystems, the absence of fire will lead to pests (often in conjunction with drought) and mechanical thinning, which become the primary drivers of both mortality and recruitment. Epidemic levels of bark beetles are clearly the most important cause of tree death in the past 10 years in the Tahoe Basin. A number of reports show increased mortality of Jeffrey pine due to the Jeffrey pine beetle (*Dendroctonus jeffreyi*) and white fir due to the fir beetle (*Scolytus ventralis*) during the drought years of 1987 to 1993 (Ferrell et al., 1994; Smith et al., 1994; Dale, 1996). Bark beetles are known to build to epidemic levels by attacking trees under stress from strong competition for available resources (Ferrell et al., 1994; Smith et al., 1994). Such conditions were readily apparent in the Tahoe Basin during the drought. In addition, a number of pathogens (e.g., root disease and dwarf mistletoe) play a role in tree mortality in the basin, with diseased trees often serving as susceptible hosts for resident populations of bark beetles during interdrought periods (Maloney and Rizzo, unpublished data).

The condition of Lake Tahoe forests has become an important concern for residents and forest managers of the basin. The presence of increased fuel loads, due to many years of fire suppression and the recent tree mortality event, and current high densities of living trees has led to fears of a catastrophic fire. Such a fire can threaten human life and property, as well as potentially upset historical ecological processes. Current fire models suggest that the probability of a large-scale fire (i.e., >4 ha) in the Tahoe Basin is relatively low because of the high elevation climate, basin topography, and quick response times of local fire officials (Manley et al., 2000). However, when such a fire does occur, it most likely will be a crown fire that will cause significant tree mortality.

In light of the current state of Lake Tahoe forests, how should these lands be managed? The most critical need for Lake Tahoe forests is to set the goals that will be used to determine the direction of forest management. Scientific studies of the ecology of coniferous forests are an important part of this process. Research to determine reference conditions to serve as points of comparison and experiments into forest ecosystem processes are ongoing in the basin (e.g., Manley et al., 2000; Barbour et al., 2002). While such information is important, science cannot set societal goals; the major role scientists will have is to offer a series of alternative management schemes and potential outcomes from an initial set of objectives. A number of citizen groups (e.g., Tahoe Forest Health Consensus Group) and government agencies have taken the initiative of setting these goals. In California, Lake Tahoe forests, along with the National Parks, are unique in that there is little pressure for a sustained timber harvest to maintain traditional logging communities. In an urbanized environment dependent on tourism, prevention of catastrophic fire and aesthetics are likely be the main concerns. However, goals such as prevention of catastrophic fire to protect property may not necessarily be compatible with goals that call for old growth restoration. There are many definitions of forest health; these range from commodity-based utilitarian views to ecosystem-based ideas that do not include humans as part of the definition (Kolb et al., 1994). The ability to partition the needs of an urbanized population from needs of natural ecosystem functioning will require much discussion among interested parties.

A number of proposals have been put forth as ways to improve forest health at Lake Tahoe. Restoration of basin forests could potentially focus on both stand thinning and prescribed fire. In theory, restoring forests to their natural densities should mitigate most serious pest outbreaks (e.g., reduced bark beetle epidemics during drought periods) (Ferrell, 1996). As we gain a better understanding of precontact forest, the temptation will be to use these data to return forests to previous stand densities. Moving seral stands toward conditions found in old-growth stands may prevent catastrophic fire, as well as maintain or increase biological diversity. But restoring ecosystem process is more than just restoring patterns and historical tree densities. The long-term impacts of procedures such as thinning dense understories and prescribed burns on ecosystem processes are not explicitly known. For example, thinning may potentially reduce pest problems by lowering stand densities and relieving the potential for drought and competition-induced stress, but such treatments also have the potential to increase pest incidence in some instances. Logging operations often damage residual trees and can increase the buildup of root disease inoculum due to saprobic survival in stumps. Scorched trees, due to hot-burning prescribed fires, may be more susceptible to bark beetle attack. Studies to elucidate the relative roles of mechanical thinning and fire on various organisms and ecosystem processes are currently under way at several locations in California. However, none of these studies is currently being conducted in the Tahoe Basin.

Because of urbanization and population growth in the basin, letting "nature run its course" is clearly not an option in managing Lake Tahoe forests. It is thought that 3500 ha per year historically were burned in the basin; currently less than 400 ha are underburned (Manley et al., 2000). Current air quality regulations will continue to influence the amount of forestland that can be burned. Even if large-scale prescribed burns can be implemented, it is not clear what intensities of prescribed burns should be used to effectively mimic natural wildfires (Manley et al., 2000). Due to uncertainties with the long-term consequences of current management options, scientifically based monitoring will be critical for evaluating the efficacy of these large-scale experiments. If true ecosystem-based management is to take place in Lake Tahoe forests, managers must design any forest treatment with the collection of scientific data in mind. Otherwise, the ability of managers in the future to adapt to changing conditions will be compromised.

CONCLUSION

By its special status as an Outstanding National Resource Water under the federal Clean Water Act (within the state of California), Lake Tahoe is recognized as unique and must be protected. As presented in this chapter, water quality, air quality, forest condition, along with past, current, and future socioeconomics conditions are all integrated under the heading of ecosystem health. They must be considered in concert and not as individual issues. The agency approach for decades at Lake Tahoe has focused on watershed management for the purpose of water quality protection. This is clearly stated in the many planning documents produced over the years for the Lake Tahoe Basin, and is central to the mission of nearly all the participating local, state, and federal resource agencies. Now a wide array of public and private institutions form an extensive foundation for linking policy to scientific research and monitoring.

For effective lake management, we need to know:

1. The specific sources of sediment and nutrients to the lake (within each of the major categories) and their respective contributions
2. The amount of reduction in loading is necessary to achieve the desired water quality standards and total daily maximum loads for Lake Tahoe (i.e., lake response)
3. How the necessary reduction can be achieved

Continued scientific efforts at Lake Tahoe must address these questions.

ACKNOWLEDGMENTS

A special thank you is given to Dennis Rolston (Director) and Cheryl Smith (Manager) of the Center for Ecological Health Research (CEHR), located at the University of California, Davis. For many years, the CEHR has promoted and encouraged interdisciplinary environmental research, and has provided the intellectual climate for productive, integrated studies. The Tahoe story, as we know it today, represents the combined effort of literally hundreds of faculty colleagues, staff, graduate students, undergraduates, and volunteers. We thank Charles Soderquist and Dennis Rolston for their helpful reviews. Major sources of funding for the research described in this chapter come from the EPA through the CEHR (R819658), an EPA/NSF-sponsored Water and Watersheds grant through the National Center for Environmental Research and Quality Assurance (R826282), and support from the U.S. Forest Service, Lake Tahoe Watershed Assessment. The views in this document do not necessarily reflect those of the federal funding agencies, and no official endorsement should be inferred.

REFERENCES

Adams, K.D. and Minor, T.B., Historic shoreline change at Lake Tahoe from 1938 to 1998: Implications for water clarity, Desert Research Institute, University and Community College System of Nevada, Reno, 2000.

Barbour, M. et al., Present and past old-growth forests of the Lake Tahoe Basin, Sierra Nevada, *J. Veg. Sci.,* in press, 2002.

Cahill, T.A. et al., Air Quality, in Sierra Nevada Ecosystem Project, Final Report to Congress, Vol. II, Assessments and Scientific Basis for Management Options, University of California, Centers for Water and Wildland Resources, Davis, 1996, pp. 1227–1261.

Cliff, S.S. and Cahill, T.A., Air quality, in *The Lake Tahoe Watershed Assessment,* Murphy, D.D. and Knopp, C.M., Eds., U.S. Department of Agriculture Forest Service, General technical report, PSW-GTR-175, Pacific Southwest Research Station, Albany, CA, 2000, pp. 131–211.

Dale, J.W., California Forest Health in 1994 and 1995, Rep. R5-FPM-PR-002, U.S. Department of Agriculture Forest Service, Pacific Southwest Research Station, Albany, CA, 1996.

Dugan, G.L. and McGauhey, P.H., Enrichment of surface waters, *J. Water Pollut. Control Fed.,* 46, 2261–2280, 1974.

Elliott-Fisk, D. et al., Lake Tahoe case study, in Sierra Nevada Ecosystem Project, Final Report to Congress, Addendum, Assessments and Scientific Basis for Management Options, University of California, Centers for Water and Wildland Resources, Davis, 1997, pp. 217–268.

Ferrell, G.T., The influence of insect pests and pathogens on Sierra forests, in Sierra Nevada Ecosystem Project, Final Report to Congress, Vol. II, Assessments and Scientific Basis for Management Options, University of California, Centers for Water and Wildland Resources, Davis, 1996, pp. 1177–1192.

Ferrell, G.T., Otrosina, W.J., and Demars, C.J., Jr., Predicting susceptibility of white fir during a drought-associated outbreak of the fir engraver, *Scolytus ventralis,* in California, *Can. J. For. Res.,* 24, 302–305, 1994.

Goldman, C.R., Eutrophication of Lake Tahoe Emphasizing Water Quality, EPA Ecological Research Series, EPA-660/3–74–034, Washington, D.C., 1974.

Goldman, C.R., Primary productivity, nutrients, and transparency during the early onset of eutrophication in ultra-oligotrophic Lake Tahoe, California–Nevada, *Limnol. Oceanogr.,* 33, 1321–1333, 1988.

Goldman, C.R., Jassby, A.D., and Hackley, S.H., Decadal, interannual, and seasonal variability in enrichment bioassays at Lake Tahoe, California–Nevada, USA, *Can. J. Fish. Aquat. Sci.,* 50, 1489–1496, 1993.

Heyvaert, A.C., Biogeochemistry and Paleolimnology of Sediments from Lake Tahoe, California–Nevada, Ph.D. dissertation, University of California, Davis, 1998.

Jassby, A.D., Reuter, J.E., Axler, R.P., Goldman, C.R., and Hackley, S.H., Atmospheric deposition of nitrogen and phosphorus in the annual nutrient load of Lake Tahoe (California–Nevada), *Water Resourc. Res.,* 30, 2207–2216, 1994.

Jassby, A.D., Goldman, C.R., and Reuter, J.E., Long-term change in Lake Tahoe (California–Nevada, U.S.A.) and its relation to atmospheric deposition of algal nutrients, *Arch. Hydrobiol.,* 135, 1–21, 1995.

Jassby, A.D., Goldman, C.R., Reuter, J.E., and Richards, R.C., Origins and scale dependence of temporal variability in the transparency of Lake Tahoe, California–Nevada, *Limnol. Oceanogr.,* 44, 282–294, 1999.

Kolb, T.E., Wagner, M.R., and Covington, W.W., Concepts of forest health, *J. For.,* 92: 10–15, 1994.

Loeb, S. and Reuter, J.E., Littoral Zone Investigations, Lake Tahoe 1982 — Periphyton, Tahoe Research Group, Institute of Ecology, University of California, Davis, 1984.

Manley, P.N., Fites-Kaufmann, J.A., Barbour, M.G., and Schlesinger, M.D., Biological integrity, in The Lake Tahoe Watershed Assessment, Murphy, D.D. and Knopp, C.M., Eds., U.S. Department of Agriculture Forest Service, General Technical Report, PSW-GTR-175, Pacific Southwest Research Station, Albany, CA, 2000, pp. 403–598.

Marjanovic, P., Mathematical modeling of eutrophication processes in Lake Tahoe: water budget, nutrient budget, and model development, Ph.D. dissertation, University of California, Davis, 1989.

Raymond, C.E., A place one never tires of: changing landscape and image at Lake Tahoe, in *Stopping Time: A Rephotographic Survey of Lake Tahoe,* Goin, P., Ed., University of New Mexico Press, Albuquerque, 1992, pp. 11–23.

Reckhow, K.H. and Chapra, S.C., *Engineering Approaches for Lake Management,* Volume I: *Data Analysis and Empirical Modeling,* Butterworth, Woburn, MA, 1983.

Reuter, J.E., Heyvaert, A.C., Luck, M., Hackley, S.H., Dogrul, E.C., Kavvas, M.L., and Askoy, H., Investigations of stormwater monitoring, modeling and BMP effectiveness in the Lake Tahoe Basin, John Muir Institute for the Environment, University of California, Davis, 2001.

Smith, S.L., Dale, J., DeNitto, G., Marshall, J., and Owen, D., California forest health: past and present, Rep. R5-FPM-PR-001, Department of Agriculture Forest Service, Pacific Southwest Research Station, Albany, CA, 1994.

Thodal, C.E., Hydrogeology of Lake Tahoe Basin, California and Nevada, and results of a ground-water quality monitoring network, water years 1990–92. Water-Resources Investigations Report 97-4072, U.S. Geological Survey, Carson City, NV, 1997.

U.S. Forest Service, Lake Tahoe Watershed Assessment: Volume I, Murphy, D.D. and Knopp, C.M., Eds., Gen. Tech Rep. PSW-GTR-195, Albany, CA, Pacific Southwest Station, Forest Service, U.S. Department of Agriculture, 2000.

Aquatic Ecosystems: New York, Maryland, and Florida

CHAPTER 124

Overview: Aquatic Ecosystems: New York, Maryland, and Florida

Dennis E. Rolston

This section contains chapters describing three case studies dealing with aquatic ecosystems in three states. Each of the three studies used a different approach to evaluate the effects of various human activities or stresses on the aquatic ecosystems of the study area.

Weisskoff used economic analysis to measure the implications of different restoration scenarios for the Florida Everglades within affected counties of South Florida. Two commonly used regional economic models were compared to the predicted impacts. The investigator advocates use of an economic analysis because it may help citizens and policy makers to gain a deeper appreciation of the restoration challenge and the need for imaginative collaboration among the stakeholders.

The objective of the study by Mehaffey and co-workers was to evaluate the relationships between land use and surface-water quality in New York watersheds, and to evaluate the risk for New York City's drinking-water supply system. A traditional stepwise regression analysis was used to relate human use of the land to water-quality parameters of the streams in the watersheds.

The chapter by Yuan describes the results of a pilot study in which a knowledge base (expert system) was used to relate human activities to stream ecological conditions. The indicator of stream ecological condition was benthic macroinvertebrates in a Maryland watershed. Fuzzy logic was used to analyze situations for which absolute rules in the expert system that weighs and chooses different options are difficult or impossible to define.

CHAPTER 125

A Tale of Two Models: IMPLAN and REMI on the Economics of Everglades Restoration

Richard Weisskoff

INTRODUCTION AND BACKGROUND

The restoration of the Everglades ecosystem in South Florida is the largest of its type in the world today. The Central and South Florida (C&SF) Comprehensive Review Study (known as the Restudy) was sent to the Congress on July 1, 1999, and passed as the Comprehensive Everglades Restoration Plan (CERP) in December 2000. The total project is expected to cost $8 billion over the next decades, with funding split half from the federal government and half from the state of Florida. If fully implemented, the Everglades Plan is expected to revitalize the ecosystem by improving the quantity, quality, distribution, and timing of the natural water flows, while assuring more-reliable flood protection for the cities, and a more-regular supply of water for agriculture. The Everglades Plan thus promises the best of all worlds to the major stakeholders of the region. This is ironic indeed, given the long-term litigation concerning the finding that upstream agriculture has been responsible for polluting the Everglades in the first place.*

The hypothesis of this chapter is that the major investments required to save the Everglades may inadvertently accelerate its destruction for the following three reasons. First, by assuring an adequate, low-cost supply of water and more extensive flood protection, the plan will allow the high levels of economic growth of the region to continue. Second, the very magnitude of the construction and operation of the rescue mission itself, together with its economic linkages, might cause a significant increase in jobs and income. Third, enhanced amenities that restoration will bring to the region will encourage new activities, such as ecotourism, further home construction, and greater levels of in-migration. All these will create even greater pressures on the ecosystem, especially higher demands for water and land.

The effects of all of these factors will be evaluated in this chapter through the application of two different regional economic models which will measure the impacts of alternative economies. First, the models will be used to project the current trends without restoration (the baseline projection). Then, the regional economy will be altered to include the restoration spending plus other "missing pieces" of economic activity omitted from the baseline projections, namely, large-

---

* Litigation began with the seminal 1988 case, *U.S. v. South Florida Water Management District, et al.*, in the U.S. District, So. Dist. of FL, Case No. 88-1886-CIV-HOEVELER, in which the federal government sued the state of Florida for failing to protect its waters and named the sugar growers as the primary polluters of the Everglades. A complete collection of Everglades litigation can be viewed at www.law.miami.edu/library/everglades. For background to the history of land and water in the region, see McCluney, 1969; Derr, 1989; Douglas, 1988; Light and Dineen, 1994; and Watercourse, 1996.

---

scale restructuring of agricultural activity, plus higher levels of tourism, new construction, and in-migration.

An approach similar to the present study was taken by Duchin and Lange (1994) for the world economy. Using the world model developed by Leontief and co-workers (1977), Duchin and Lange found no quantitative support for the optimistic vision of the 1987 Brundtland Report that an expanding economy was compatible with an improved environment unless a major change in consumption and production patterns were to occur. Our analysis carries a similar urgency, because plans regarding the future of the Everglades are now being formalized and implemented.

There may also be a parallel between the circumstances in South Florida today and the conditions that led to the Latin American debt crisis of the 1980s. Prior to that crisis, international banks lent money to many autonomous government entities throughout the continent, and these loans were guaranteed by each respective national government in the event of default.* As the individual entities began to fail in the worldwide stagflation of the early 1980s, national governments were apparently caught off-guard by the magnitude of the accumulated debts for which they had become responsible. Beginning with Mexico in 1982, nation after nation defaulted, initiating an era that was to undo much of the economic progress of the previous decades.

So, too, in South Florida, the ecosystem is under enormous pressure to accept the ambitious projects of many donors. But there is a growing awareness that environmental degradation has resulted from rapid economic growth, which, until recently, had been viewed almost universally as the solution to, rather than the cause of urban problems.

Unfortunately, the analysis of the economic impact of Everglades restoration comes as an after-thought, rather than as an integral part of the plan.** The Restudy, after all, was conceived as an attempt to solve a huge and complex hydrological–ecological problem. What could possibly have been gained by connecting computer models that simulate the water flows to computer models that simulate the economic growth of the region?

The answer, of course, is simple: Everglades restoration is primarily an economic problem, as well as a scientific one, for it is the very success of the regional economy that has doomed the natural system. Furthermore, if the goal of policy makers is to prepare the ecological setting for "what will be" rather than for "what is," then we must know how the economy is growing and how the economy and the environment are interrelated.

There is perhaps another factor that makes models of economic growth even more alien to the ecological modelers, and that is the manner in which the two professions view the influence of time. Models of the environment rely on the historical records of the past: how are we to restore the natural system, given the historic rainfall patterns and the current needs for flood protection of the farmlands and cities. Models of the economy are just the opposite: what was will never again be repeated; historical trends and interrelationships are used only to forecast the new economic levels which may be expected in future years. The economy of the year 2002 will never be seen again, and the economy of the year 2030 will surely differ, quantitatively and qualitatively, from the present one. To the extent that the demands of this new economy can be anticipated, we are forced to repeat the basic question of this chapter: Is a restored Everglades of the future compatible with the expected economy of the future?

This chapter offers a broader framework for researchers facing this question by extending my earlier study, "Missing Pieces in Ecosystem Restoration: The Case of the Florida Ever-glades" (Weisskoff, 2000). In that paper, I attempted to measure the impact of restoration spending plus other simultaneous investments on the present and future regional economy. The present work improves on the earlier work by comparing the results of two very different regional economic models.

In the first section of this chapter, I review the characteristics of the South Florida economy, the historical record of its growth, and the overall projections for the region. I then apply the

---

* See Lissakers, 1991.
** See U.S. Army Corps of Engineers, 1998, Appendix E, "Socio-Economics."

IMPLAN and REMI models, which have been tailored to the region by the author in several innovative ways, in order to trace out the impacts of the new projects. Then the projects are combined into different superscenarios and the results of the two models compared.

It is hoped that the conclusions will encourage economists to work hand in hand with ecologists to interpret and anticipate the consequences of the full array of activities occurring in the region. Just as the original 1948 C&SF Comprehensive Project opened the region to massive new investments, which had the effect of transforming the environment and setting the stage for the present crisis, so, too, the current Everglades restoration may trigger another prolonged spiral of spectacular growth by enhancing the region's amenities.*

Is there an alternative? The failure to restore will simply permit the present trend to continue, a trend in which economic growth has gone hand in hand with the demise of the natural wetlands, saltwater intrusion into the coastal wells, destruction of estuarine breeding grounds, and the disintegration of the coral reefs. Perhaps by looking closely at the economic picture, all citizens of the region, as well as the Everglades' supporters throughout the nation, may gain a broader picture of the restoration challenge and the need for more imaginative collaboration and compromise among all the stakeholders.

THE REGION AND ITS GROWTH

The South Florida region, which comprises the Everglades and its headwaters, covers almost 40,000 km^2 (15,000 mi^2) and 13 full counties, accounting for 28.1% of Florida's land area, 39.2% of the state's residents, and almost half the state's consumption of fresh water (Table 125.1, lines A.1, A.2, A.4, cols. 1, 4). The region is home to nearly 6 million people, 6.5 million motor vehicles, and over 600,000 cattle and calves (Table 125.1, lines A.2, A.5, B.6). If we consider the time spent by the 18.8 million visitors in the region, the full-year population would be increased by 7.1% on an annual basis (lines A.6–A.8). The region accounted for 3.6% of all new housing permits in the U.S. in 1997, although it contains but 2.2% of the total population. These figures underscore the relative magnitude of the regional housing boom (Table 125.1, lines A.2, A.3, col. 6).

Forty-one percent of the region's land is in agriculture. These farms represent 62% of the state's irrigated farmland, 46% of the nation's sugarcane land, and 38% of the nation's citrus trees. Included also are extensive livestock, sod, and vegetable plantations, as well as many small farms and nurseries (Table 125.1, lines B.3, B.6–B.9, cols. 1, 4, 6).

In terms of environmental heritage, nearly two thirds of the threatened and endangered species in Florida can be found in these counties (Table 125.1, lines C.1, C.2, col. 4). The region is characterized by low elevation and a semitropical climate. Its average temperatures and rainfall are higher than the rest of Florida and much higher than the rest of the U.S. (Table 125.1, lines C.3, C.4, cols. 4–6).

South Florida is one of North America's last frontiers. Large-scale settlement began only in the first decades of the 20th century once the Florida coastal railroad was extended to Miami and drainage of interior swampland for farms began. A series of violent hurricanes in the late 1920s and the Great Depression suspended the frenzy of land speculation until the 1950s, when the improvements in infrastructure brought by the C&SF Project provided better flood protection for the cities and superior drainage for new farmland.

From 1970 to 1990, regional population nearly doubled, while employment and real income have grown much faster (Table 125.2, lines 1–3). Although growth rates have slowed in the most recent decade and are expected to fall further, population surpassed 6 million by the year 2000 and is expected to reach almost 10 million by 2045. The population boom is reflected in the remarkably

* Compare the two Central and South Florida Project proposals, U.S. Army Corps of Engineers, 1948 and 1998. See also www.evergladesplan.org.

Table 125.1 Some Economic Characteristics of the Greater Everglades (South Florida) Ecosystem

	Region [1]	State [2]	USA [3]	Region/State (%) [4]	State/U.S. (%) [5]	Region/U.S. (%) [6]
A. General						
1. Size (13 full counties) (thousand square miles)	15.1	53.9	3718	28.1	1.5	0.41
2. Resident population (1997) (millions)	5.9	15.0	268	39.2	5.6	2.20
3. New housing permits (millions)	52	134	1441	39.1	9.3	3.63
4. Freshwater use (1995) (million gallons per day)	3571	7215	341,000	49.5	2.1	1.05
5. Motor vehicle tags sold (millions)	6.5	18.1	206	35.7	8.8	3.15
6. Visitors (air and auto) (millions)	18.8	47.0	n.c.	40.0	n.c.	n.c.
7. Visitors per year (thousands)	411	1536	n.c.	26.8	n.c.	n.c.
8. Visitors per year as % of residents	7.1	10.4	n.c.	n.a.	n.a.	n.a.
B. Agricultural						
1. Number of farms (thousands)	6.7	34.8	1925	19.1	1.8	0.35
2. Average farm size (acres)	604	300	487	201.3	61.6	124.02
3. Share of land in farms (%)	41.2	30.1	39.2	38.5	1.1	0.43[a]
4. Irrigated farmland (thousand acres)	1149.2	1862.4	55,058	61.7	3.4	2.09
5. Value of all farm produce sold ($M)	2861.5	6004.6	196,865	47.7	3.1	1.45
6. Cattle and calves inventory (thousands)	626.2	1808.9	98,989	34.6	1.8	0.63
7. Sugarcane (thousand acres)	421.0	421.4	919	99.9	45.8	45.81
8. Vegetables (thousand acres)	133.1	250.6	3733	53.1	6.7	3.57
9. Citrus trees (millions)	56,826	107,219	151,530	53.0	70.8	37.50
C. Environment and Climate						
1. Threatened and endangered animals, federal. list (number)	30	46	480	65.2	9.6	6.25
2. Threatened and endangered plants, federal. list (number)	34	54	703	63.0	7.7	4.84
3. Highest elevation (feet)	50	345	20,320	n.a	n.a	n.a.
4. Climate						
Normal temperature January average	67.2	52.4	37.6	128.2	139.4	178.72
Normal annual precipitation (inches)	55.9	51.3	36.5	108.9	140.5	153.18

Note: All data refer to 1997 unless otherwise noted; n.a. = not applicable; n.c. = data not comparable.

[a] In columns 4–6, refers to share of farmland in the smaller unit to farmland in the larger unit of column heading.

Methods and Sources: All county, state, and some national-level data are from Bureau of Economic and Business Research 1998b, cited below as "sa" (Florida Statistical Abstract), unless otherwise noted. Other national data are from U.S. Census Bureau, 1998. County data were summed for the region and compared to state and national totals.

Section A

Line 1   fsa, t. 8.01 and 8.03.

Line 2   Smith and Nogle, 1999. Florida ranks fourth after California, New York, and Texas.

Line 3   fsa, t. 11.05.

Line 4   fsa, t. 8.39 (refers to all freshwater withdrawals, including irrigation).

Line 5   fsa, t. 13.21 (includes out of state registrations).

Line 6–8 Office of Tourist Research, 1998, table, p.i, gives numbers and destinations; table, p. 39 gives number and length of stay of auto and air visitors.

Lines 1–2 National list of species is available for each Florida county from EnviroTools, Inc.,1998. These species were compared to those appearing on the lists for the state and the nation, which are posted on the Web site of the U.S. Fish and Wildlife Service, www.fws.gov/, dated April 30, 1999.

Lines 3   fsa, t. 8.01.

Line 4   fsa, t. 8.74. Data for selected cities are taken to represent the regions: column 1 (Miami); column 2 (Jacksonville); column 3 (average of Los Angeles, Atlanta, Chicago, New York).

Section B

All data from U.S. Department of Agriculture, 1999. The region is summed from county data given in the state files.

Line 8 includes land in sugarcane for both seed and sugar production.

Line 9 refers to land in vegetables harvested for sale.

Section C

The 13 full counties included in the Greater Everglades region are Broward, Collier, Miami-Dade, Glades, Hendry, Highlands, Lee, Martin, Monroe, Okeechobee, Osceola, Palm Beach, and St. Lucie.

Table 125.2 South Florida Growth, Performance, and Projections, 1970–2045

	Historical			Projections			Annual Growth Rate for Period				
	1970	1980	1990	2000	2010	2045	1970–1980	1980–1990	1990–2000	2000–2010	2010–2045
1. Population (thousands)	2593	3868	5112	6097	7011	9867	4.08	2.83	1.78	1.41	0.98
2. Employment (thousands)	1026	1672	2439	2810	3337	4124	5.01	3.85	1.43	1.73	0.61
3. Real income (billion dollars, 1992)	40.9	74.5	118.9	155.5	205.5	385.6	6.18	4.79	2.72	2.83	1.81
Real income per capita (dollars, 1992)	15,773	19,261	23,259	25,506	29,311	39,080	2.02	1.90	0.93	1.40	0.83
4. Private housing starts (thousands) (average annual per decade)	1970–1979	1980–1989	1990–1996								
a. Multifamily	40.3	33.4	14.8	18.2	25.5	n.a.	−1.86	−7.82	2.09	3.43	n.a.
b. Single-family	24.8	32.6	31.1	32.1	31.6	n.a.	2.77	−0.47	0.32	−0.16	n.a.
c. All	65.1	66.0	45.9	50.2	57.0	n.a.	0.14	−3.57	0.90	1.28	n.a.
5. Population distribution (%)											
a. 0–19 years	29.1	25.3	24.8	25.2	23.7	20.4					
b. 20–65 years	53.3	55.6	56.3	56.2	57.4	51.3					
c. 65+ years	17.5	19.1	19.0	18.6	18.9	28.3					
6. Employment distribution (%)[a]											
a. Manufacturing	14.1	10.9	8.0	6.4	5.2	4.8					
b. Services	24.2	26.8	32.5	35.3	38.0	39.8					
7. South Florida in state (%)											
a. Population	38.2	39.8	39.5	39.1	39.1	39.1					
b. Employment	40.1	41.6	40.1	39.2	39.3	39.3					
c. Income	44.3	44.8	44.4	43.7	43.6	43.6					
d. Housing starts:											
1. Multifamily	53.5	49.0	53.6	53.6	54.0	n.a.					
2. Single family	36.8	34.3	36.2	36.9	36.6	n.a.					
3. All	45.6	40.4	40.4	41.6	42.7	n.a.					
8. Farm employment (%)[b]	2.8	2.2	1.5	1.6	1.3	0.8					

Notes: Data for 1970–2010 are derived from Bureau of Economic and Business Research (1998a) and for 2045 from U.S. Department of Commerce (1999). Lines 1–5 are aggregated from the county estimates given Bureau of Economic and Business Research (1998a, Vol. 1), and figures for 2000–2020 from Smith and Nogle (1999) for "medium" projections for the 13 counties which fall entirely or almost entirely within the South Florida Water Management District. Excluded are three "partial" counties. Line 6 is constructed from Bureau of Economic and Business Research (1998a, Vol. 2), by summing and averaging data for the six metropolitan statistical areas (MSAs) that fall in the region, plus Monroe County and the South Central nonmetropolitan area (which includes four of the rural counties in the Everglades region), for the years 1970–1996. For lines 4–6, average shares were computed for 1970–1979, 1980–1989, and 1990–1996.

Projections for 2045 for South Florida were made for lines 1–3 by applying the BEA annual growth rates computed for the state to the latest available BEBR region projection for each corresponding concept. The BEA benchmark years are 2010, 2015, 2020, 2025, and 2045. BEA growth rates for total personal income in constant 1987 dollars were applied to the BEBR real income projections. The age distribution of the population (line 5) for 2045 was estimated by applying BEBR's 2010 shares for South Florida in the state for each age cohort to the BEA forecasts for the absolute number of people in each of the statewide age cohorts for 2045. Shares were then recomputed. The employment distribution (line 6) for 2045 was estimated by applying BEBR's 2010 sector shares of South Florida in the state to BEA's estimates of statewide sectoral employment for 2045 and then converting these numbers into shares.

Line 8, farm employment for South Florida for 1970–2045 was computed by applying the region's share of farmland in the state to BEA's estimates of statewide agricultural employment. Farm area for the 13 counties was computed from the U.S. Census of Agriculture for the years 1974, 1982, 1992, and 1997, corresponding to the years in columns 1–4 above, and the 1997 shares were applied to the years 2010 and 2045. The region's agricultural employment was then added to BEBR's total nonagricultural employment (line 2 above) and the share of A to total employment computed for the corresponding years.

[a] Refers to nonagricultural jobs.
[b] Share agricultural (A) to total A plus non-A employment (BEA).

high number of multifamily housing starts during the 1970s and the subsequent rise in the number of single-family units in later decades* (Table 125.2, lines 4a, 4b). The state's forecast of a strong upward trend in both single- and multifamily housing will have great implications for continuing land pressures in these counties.

The age distribution of the residents (Table 125.2, lines 5a to 5c) indicates an inverting of the age pyramid by 2045. The share of young (≤19 years), which was 29% of the population in 1970, is expected to fall to 20% by 2045. The oldest cohort (>65 years), which accounted for 17.5% of the total in 1970, is expected to rise to 28%. Even these trends understate the importance of the older population due to the exclusion of seasonal residents ("snow birds") from the permanent population.

A summary of the distribution of nonfarm employment (Table 125.2, line 6a) highlights the decline in the share of the combined construction–manufacturing–utilities sectors from 30.6% of the labor force in 1970 to 15.1% by 2045. It is the rise of employment in the services sector (Table 125.2, line 6b) from 69.4 to 80.5% between 1970 and 1990, and its expected increase to nearly 85% by 2045, that is determining the general character of the region.

In summary, South Florida's strong economy compounds its environmental problems: an expanding agricultural and commercial society serving a large tourist, retiree, and transient population. All this surrounds North America's only semitropical and largest remaining wetland, the Florida Everglades.

The struggle between the economic growth and Everglades restoration is today occurring on all four fronts of South Florida (see right-hand map in Figure 125.1). In the northern and central stretches of the ecosystem, the growth of Orlando's suburbs, cattle and dairy lands of the Kissimmee River Valley, diking of Lake Okeechobee, and drainage of the Everglades Agricultural Area (EAA) have altered the downstream flow to the remaining Everglades. To the southwest of Lake

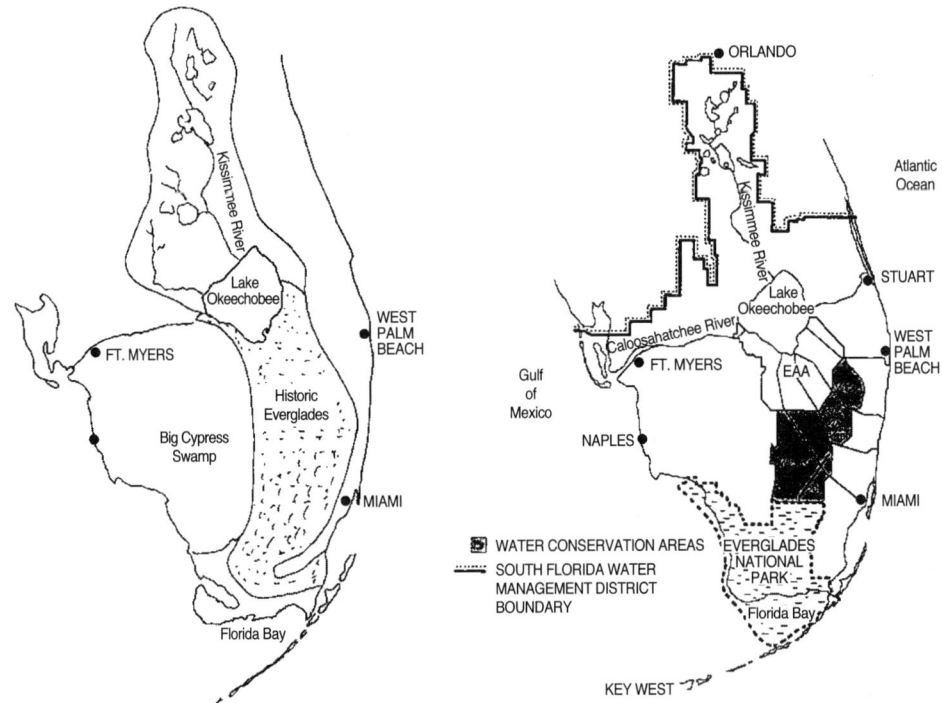

Figure 125.1 Comparison of the Greater Everglades Region in 1890 and the geographical features today.

---

* The number of housing starts has been averaged for each decade due to the wide annual fluctuations. South Florida has accounted for more than half of the state's multifamily starts since the 1970s, a share considerably greater than the region's population share.

Okeechobee, new areas are drained for expanded cattle, citrus, and vegetable production. To the east, only the 100-mile-long protective levee restrains the expansion of the Miami–Ft. Lauderdale metropolitan areas into the Everglades, and three large water conservation areas, actually vast holding tanks, compartmentalize what once was the "River of Grass." To the south, Florida Bay and the Keys receive nutrient-rich runoff from canals that drain the southeastern counties. The cities sprawl and agriculture thrives as the Everglades die, despite the $1.1 billion already spent.*

MODELS

A model of the regional economy that would be useful to policy makers, stakeholders, and taxpayers in general must be detailed enough to pinpoint specific activities, flexible enough to replicate a wide range of alternatives, and clear enough to be comprehensible to layperson and expert alike. But the design and calibration of a new regional economic model requires major investments of time and money. Fortunately, the economic researcher may choose from among three ready-made, off-the-shelf models, designed for any "generic" region in the U.S. Each of these models can then be "fitted" with actual data gathered by different federal and state agencies for any configuration of Florida counties.

The Regional Industrial Multiplier System (RIMS-II)

RIMS-II was first developed in the early 1970s by the Bureau of Economic Analysis of the Department of Commerce. It currently adapts the national BEA input–output table to local conditions on the basis of location quotients (LQs), adjusts for regional household leakages, and estimates regional multipliers for detailed industries (U.S. Department of Commerce, 1992).

Impact Analysis for Planning (IMPLAN)

In contrast to RIMS-II, IMPLAN is a package of flexible software and county-level databases that permits researchers to estimate their own regional model for any configuration of counties and for any level of aggregation of sectors. IMPLAN was developed by the Forest Service of the Department of Agriculture in the late 1970s and has been improved and maintained since 1993 by the Minnesota IMPLAN Group (MIG). The IMPLAN system regionalizes the national BEA input–output table by applying regional purchase coefficients (RPCs) after adding more detail to agriculture. The IMPLAN software then constructs a complete social accounting matrix (SAM) for 528 sectors and 21 economic and demographic variables, as outlined in Figure 125.2, and also computes the regional impact multipliers (see Minnesota IMPLAN Group, 1997).**

While the IMPLAN coefficients provide an up-to-date x-ray of a regional economy in great detail, they measure the impact of new spending on output, employment, and several types of income for only a single year. IMPLAN thus assumes that all the economic effects of the spending are played out in the single time period and that there are no dynamic effects or "carry-overs" of this year's spending into the future. It is up to the user to adapt the model to trace out patterns if the initial spending is spread over many years.

Regional Economic Models, Inc. (REMI)

REMI provides a dynamic modeling capability that allows the user to simulate scenarios involving a much larger number of economic and demographic variables for a set region or group

---

* See South Florida Ecosystem Task Force (1998) for detailed expenditures.
** IMPLAN was extended to Puerto Rico by Ruiz and co-workers (1994). For a detailed discussion of the IMPLAN model, see Weisskoff, 2000.

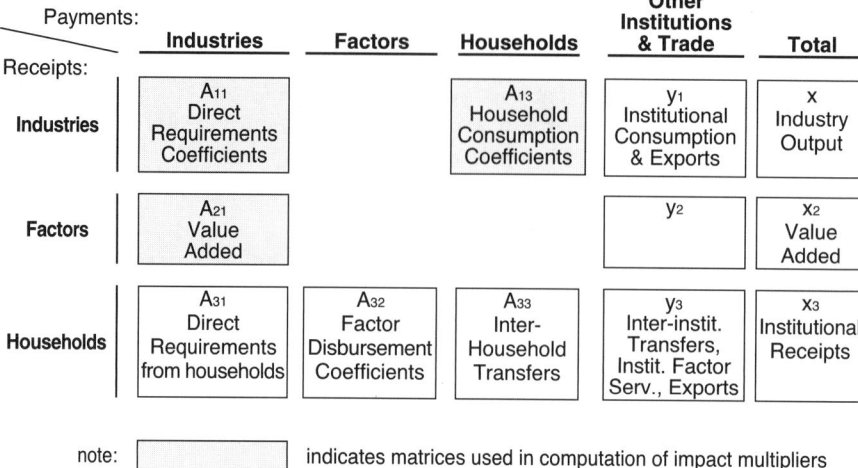

Figure 125.2 Social accounting framework, IMPLAN Type II formulation.

of regions (Treyz, 1993). The historical coefficients are based on time series from the previous 15 years. The modeling horizon extends 35 years into the future, and the blocks of variables include output, population, labor, market shares, and prices.

REMI is actually a massive system of demographic and economic equations set up automatically in a dynamic, multiperiod framework. Unlike IMPLAN in which Spending A causes Impact A in the same year, REMI allows Spending A to be spread over many years and its impacts to be played out on all the dynamic features and feedbacks in the coming decades. It is thus better suited for a long-term project such as Everglades restoration, in which construction occurs in the early years and maintenance expenses are incurred in the later years.

Each model has its own strengths. RIMS-II is a convenient tool for regional agencies that require computed multipliers to assist in project evaluation for a preset region at a modest price. IMPLAN gives the researcher the software and data, and hence the ability to change the model and counties comprising the region. Both RIMS and IMPLAN are relatively low-cost tools and provide detail for 471 and 528 sectors, respectively. REMI, on the other hand, uses historical data as well as input–output data for fewer sectors (14, 53, or 172, depending on cost), allowing the model to play out over a future up to 35 years. The researcher, therefore, is also able to simulate different policy interventions or new projects over time. The effects of these projects and their trajectories are then compared to the "base" or "control" forecast for the same time period.

In one sense, REMI is an agile or fleet-footed runner for any future course, allowing the researcher access to a great many variables applied to fewer, but still an adequate number of sectors. IMPLAN, by contrast, is an old-fashioned clunker, a disaggregated input–output model, augmented with only a few other variables and made, in this chapter, to move slowly over time, like an awkward robot climbing steps.

Several recent articles have compared the performance of the three models by collapsing the dynamic framework of REMI and imposing a uniform way of "closing" the consumption–income link.* Our approach is to use both IMPLAN and REMI to examine the hypothesis that the spending on Everglades restoration could contribute to greater growth of the region. First, we compare both models in terms of their reactions to impacts on single sectors and on simple combinations of sectors. Then we scale up the combined sectors into more realistic scenarios to replicate ongoing and future projects, for example, Everglades restoration and tourist promotion. We apply annual projections of these expenditures to an annual IMPLAN model and to the REMI model. The scenarios are then

---

* See Rickman and Schwer (1995) and Crihfield and Campbell (1991).

further combined into Summary Scenarios and the increases in output and employment compared to the growth of the "base region" in the absence of the new projects. Both models are applied to the South Florida region constructed of 13 South Florida counties. For data, see Appendix 125.I.

RESULTS: SECTORAL ANALYSIS

Understanding the sectoral characteristics of the model is essential to evaluating the complex scenarios. By adding an additional $10 million of final demand to each sector, we compare the total output and total number of jobs generated within each of the models (Table 125.3). Of the agricultural sectors in Panel A, rice (No. 11), vegetables (No. 18), and sugar crops (No. 19) compete for the organic soils of the EAA, but each has its own drainage needs and creates different runoff characteristics. Citrus (i.e., fruit trees, No. 16) requires good drainage and is grown on a wider range of sandy soils. Greenhouses (No. 23) are concentrated closer to the cities and compete for land needed by new residential housing (No. 48). The future of commercial fishing (No. 25) on both Atlantic and Gulf Coasts is threatened by the periodic discharges of fresh water drained from the interior farmlands and coastal cities. The selected manufacturing sectors (Nos. 70–97) process the region's agricultural products. The selected transport and service sectors (Nos. 437–488) are all components of tourist spending.

In the IMPLAN model, the sectors generate between $13.9 (fish processing) and $19.0 million (fruit trees, agricultural services) in total output (Table 125.3, col. 1), but new employment ranges from 90 jobs in rice milling to 867 jobs for agricultural services (col. 2). In general, the service and agricultural sectors generate the most employment, while manufacturing generates the least. The profiles of car rentals and airlines are more similar to manufacturing than to services, and commercial fishing is more similar to services than to agriculture.

In the REMI model, new output ranges from $10 million in greenhouses to $23 million in recreation services. New jobs range from 109 in rice milling and 463 in recreational services. REMI's agricultural sectors generate less output but more jobs than IMPLAN, and REMI's service sectors generate more output and less employment than IMPLAN's. The similarity in the averages of new output and employment between the two models for the 16 comparable sectors is striking: $16.8 million and 284 jobs for IMPLAN vs. $15.6 million and 274 jobs for REMI.

The impacts of $10 million of new demand on the composite scenarios (Table 125.3, Panel B) are based on actual engineering data and on projects in progress which affect combinations of sectors. For IMPLAN, the scenarios yield about the same output (Table 125.3, col. 1, lines B1 to B4), except for the Operation and Maintenance Program of the current Everglades Protection Project (EPP), due to the higher proportion of direct wages in this program. The tourist package (Table 125.3, col. 2, line B.3), which is based on several different statewide and regional visitor spending surveys, yields the highest employment of the four scenarios. For REMI, the EPP construction generates the most output and employment, and other scenarios yield remarkably similar employment totals.

The consumption patterns of the three income classes available with IMPLAN but not REMI (Panel C) suggest the contradiction between the slightly higher output generated by the $10 million spending by the lowest class (col. 1) and the slightly higher employment generated by the same level of spending of the highest class (col. 2). A growth pattern that results in higher incomes to the top classes, therefore, stimulates more jobs, due to the service-intensive component of that expenditure bundle, while the same increase in spending of the lower class generates greater output.

RESULTS: COMPONENTS OF THE SCENARIOS

In the following sections, we review the components and their combination into scenarios (Table 125.4).

Table 125.3 IMPLAN and REMI: Economic Impacts of $10 Million New Demand for Sectors, Programs, and Income Classes, South Florida

Sector Name	IMPLAN		REMI		REMI Sector No.
IMPLAN Sector No.	Output Total ($M)ᵃ [1]	Jobs Total (No.)ᵃ [2]	Output Total ($M) [3]	Jobs Total (No.) [4]	
A. Single sector patterns					
Agriculture					
11 rice	15.9	218	11.1	314	T6
16 fruit trees	19.0	388	12.5	457	T11
18 vegetables	15.9	261	12.4	450	T12
19 sugar crops	15.3	241	12.0	398	T13
23 greenhouses	17.9	319	10.0	330	T17
25 commercial fishing	16.5	666	n.c.	n.c.	
Average: 5 sectors	16.8	285	11.6	390	
Manufacturing					
70 frozen juice	16.9	163	n.c.	n.c.	
74 rice milling	14.7	90	16.3	109	T119
81 sugar mills	18.3	148	16.7	132	T121
97 fish processing	13.9	141	19.0	236	T103
393 boat building	15.6	167	17.3	191	T104
Average: 4 sectors	15.6	137	17.3	167	
Construction					
48 new residence	17.1	219	18.1	231	T23
Transportation					
437 airlines	17.1	172	13.2	132	T264
Services					
26 ag services	19.0	867	16.1	164	T250
454 eat/drink	17.1	415	15.2	301	T233
463 hotels	17.0	356	16.2	169	169
477 car rental	18.4	163	21.1	281	T183

Average: 5 sectors	17.7	436	18.3	276	
Average: 16 comparable sectors	16.8	284	15.6	272	
B. Composite scenarios					
1. Everglades Protection Project (EPP)	17.0	220	21.1	293	T39
2. Operation and Maintenance EPP	13.1	266	16.9	209	T41
3. Tourist Package	17.1	313	13.4	206	pts. 8[b]
4. New Devel Package	17.6	216	15.8	205	604
C. Inc. Patterns standardized					
Low	17.13	264	n.c.	n.c.	
Middle	17.11	267	n.c.	n.c.	
High	17.03	274	n.c.	n.c.	

[a] Indicates sum of direct, indirect, and induced (Type II) effects.
[b] Indicates parts of eight different sectors.

n.c. = Indicates "not comparable."

Table 125.4 Summary of Economic and Employment Impacts on South Florida Scenarios: IMPLAN and REMI

Scenarios		New Output ($M)				New Employment (No.)			
		2000 [1]	2010 [2]	2020 [3]	2030 [4]	2000 [5]	2010 [6]	2020 [7]	2030 [8]
1. Agricultural restructuring									
a. Sugar/rice rotation	IMPLAN	−158	−158	−158	−158	−2794	−2794	−2794	−2794
	REMI	−402	−421	−450	−462	−5167	−5097	−5073	−4797
b. Organic rice replaces sugar in EAA	IMPLAN	−632	−632	−632	−632	−11,176	−11,176	−11,176	−11,176
	REMI	−257	−341	−429	−471	−12,240	−12,730	−12,960	−12,510
c. Water storage only in EAA	IMPLAN	−2428	−2428	−2428	−2428	−24,960	−24,960	−24,960	−24,960
	REMI	−1981	−2103	−2273	−2342	−28,270	−28,260	−28,330	−26,950
2. Everglades restoration									
a. Current EPP ($50 million/yr, 10 yrs then $5.6 mill. O&M)	IMPLAN	92	0	0	0	1249	0	0	0
	REMI	95	11	21	18	1337	233	320	251
b. Total Restudy ($$300 mill./yr, 20 yrs plus $164 mill./yr O&M)	IMPLAN	0	576	576	199	0	7015	7015	3062
	REMI	0	498	585	273	0	6771	7900	3772
3. Tourist promotion package									
a. Auto and air tourism (4.53% incr. on base of $5.02 bill.)	IMPLAN	486	757	1179	1681	8888	14,007	21,579	33,623
	REMI	340	562	917	1433	5268	8213	12,460	17,940
b. Regional nature tourism 6.8% ann. incr. on base of $390.2 mill./yr	IMPLAN	54	74	105	136	982	1384	1916	2677
	REMI	43	88	175	336	671	1275	2349	4148
4. New construction over present levels 3.7% ann. increase on base of $7.4 billion	IMPLAN	482	693	997	1333	5914	8505	12,230	17,381
	REMI	457	672	1020	1493	5967	8835	13,000	18,080
5. Consumption of new residents (2x current rate: 86 thou. hh., $107.5 mill./yr)	IMPLAN	184	184	184	184	3362	3362	3362	3362
	REMI	154	169	191	203	2389	2545	2697	2677

Agricultural Restructuring

- Sugar/rice rotation (3 years sugar/1 year rice) replaces the uninterrupted monoculture of sugarcane and may lead to higher sugar yields and reduced soil subsidence and pesticide use.* In the fourth year, two rice crops are harvested under this plan prior to the replanting of sugarcane in the fifth year. In any single year, one quarter of the EAA sugar lands would be growing rice.
- Organic rice could become the most profitable crop for the EAA as soil subsidence lowers sugar yields, environmental standards regarding downstream runoff are tightened, and low-priced foreign sugar is given a larger share of the U.S. market.
- Water storage only in the EAA removes all agricultural activity and converts the sugarcane lands into reservoirs and marshes. The loss is calculated as the reduction of all current sugarcane production and milling.

Everglades Restoration

- EPP. The ongoing Everglades Protection Project, which is comprised of 14 projects, will be completed by 2005, with Operation and Maintenance (O&M) costs continuing into the future.
- Restudy, completed by the Army Corps of Engineers and now known as the Everglades Plan, will be under construction by 2005 and may end by 2025, with full annual O&M costs continuing thereafter. The project includes 287 conventional water control structures (pumping stations, spillways, culvert, canals, and levees) as well as several untried technologies, such as 345 aquifer storage-and-recovery wells (ASRs) and 47 miles of underground seepage barriers, 10 to 28 ft in depth. The impacts measured here result from construction expenditure only and exclude land acquisition costs.

Tourist Promotion

- Auto and air tourism is based on continuing the 4-year trend in expenditure of visitors who report South Florida as their primary destination.
- Everglades Region nature tourism assumes a doubling of the historical rate of increase in the number of visitors to the major national parks, wildlife preserves, and waterways in the region, but for a fixed expenditure per visitor (U.S. Army Corps of Engineers, 1998, Appendix E, pp. 209–213).**
- New construction. The historical statewide rate for all new construction activity was extended to the region for the entire period.
- Consumption of new residents reflects the expenditure of new retirees and in-migrants at twice their historical numbers in the region.

Summary Scenarios

The sectors and their combinations shown in Table 125.3 comprise the scenarios of Table 125.4. All represent additional pieces to the base projections of output and employment and assume steady growth of the existing economy. In a regional economy where tourism is a key sector, the state's growth model (BEBR in Table 125.2) excludes a tourists sector per se. Rather, elements of tourism, such as spending on food, hotels, and amusements, are absorbed into the spending patterns of local residents, making it virtually an endogenous sector. In our model, however, new spending on tourism is treated as an array of exogenous demands, despite the fact that these demands themselves are promoted by the state through extensive advertising. A restored Everglades, the flagship of the U.S. environmentalist movement, will also add to the region as an important destination for ecotourism and recreation.

* On the compatibility of profitable agriculture and improved environment, see Glaz (1995). On subsidence, see Shih and co-workers (1997).
** The historical rate of increase in nature tourism (3.4%) is also similar to the average annual growth rate of the number of visitors (3.61%) to all the state parks in the region from 1993–1997.

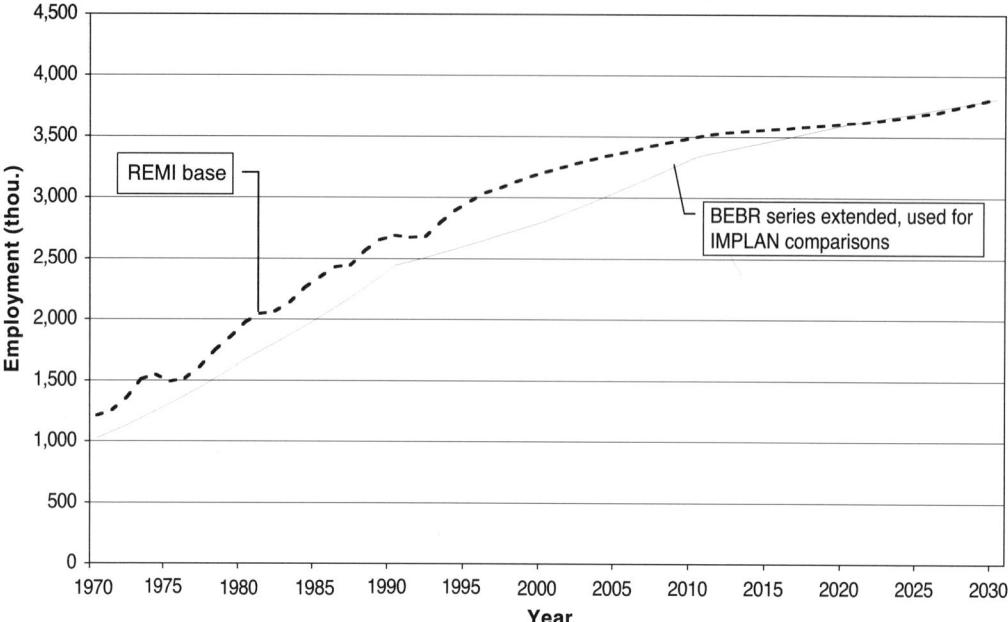

Figure 125.3 South Florida employment, 1970–2030.

In each of the following summary scenarios, the output and employment generated by each model is compared to its respective base, which is the future of the region if present trends continue and no Everglades restoration occurs. We use county and statewide forecasts from Florida's Bureau of Economic and Business Research (BEBR) and the Department of Commerce to estimate the base for the IMPLAN projections. The REMI model generates its own base or regional control on the basis of the regional model with no intervention and the national macro-model which it also generates. The basic employment series from 1970 through the forecast year 2030 for BEBR and REMI are shown in Figure 125.3.

- The "Growth as Usual" Scenario (Table 125.5, line A.1) sums up the projects already underway and ongoing trends: finishing the EPP projects; continuing the greater shares of air visitors but a reduced share of auto visitors to the region; continuing the rate of overall construction activity and the rate of in-migration.* During the projected years, increases due to tourism and construction dominate the output and job impacts. The "Growth as Usual" scenario, according to IMPLAN, results in 0.65% increase in output and 1.43% increase in employment over the total output and total employment, respectively, for 2030 (Table 125.5, line B1, cols. 4 and 8). The REMI model gives the increase in output as 0.95% and the increase in employment as 1.02%.
- But when compared to the annual increase in output and employment of the 2030 economy, the output and employment newly generated by the "Growth as Usual" scenario is quite significant, i.e., 36.0 and 233.6% over the respective annual increases for the basic economy in 2030 (Table 125.5, line C1, cols. 4 and 8), according to IMPLAN, and 60.2 and 129.5%, respectively, according to REMI. These scenarios highlight the degree to which the official projections understate the effects summed here.
- The "New Projects" Scenario adds the rice rotation, the full Everglades Restoration, and additional regional nature tourism (Table 125.4, lines 1a, 2b, and 3b). The sum of the elements in the "New

---

* This is a conservative scenario, as it excludes the growth of other leading sectors, such as greenhouses and commercial fishing. The recent introduction of best management practices (BMPs) in sugar growing has led to dramatic declines in nutrient runoff and a surprising rise in sugarcane output, perhaps a hint of the potential impact that further changes in cropping patterns and growing practices might have on the environment.

Table 125.5 Summary of Scenarios

		New Output ($mil.)				New Employment (no.)			
		2000 [1]	2010 [2]	2020 [3]	2030 [4]	2000 [5]	2010 [6]	2020 [7]	2030 [8]
A. Annual Missing Pieces Impacts	IMPLAN / REMI								
1. Growth as Usual (add 2a, 3a, 4, 5) Cur. restoration; cont. tourism trends; new const; new res.	I	1244	1634	2360	3198	19,413	25,874	37,171	54,366
	R	1046	1414	2149	3144	14,960	19,820	28,460	38,920
2. New Projects (add 1a, 2b, 3b) Rice rotation; full Restudy; nature tourism	I	−104	492	523	177	−1812	5605	6137	2945
	R	37	578	749	597	365	7726	9925	7607
3. Full Speed Ahead! (add 1, 2) Growth as Usual plus New Projects	I	1139	2126	2882	3375	17,600	31,479	43,308	57,311
	R	1083	1992	2899	3743	15,320	27,550	38,400	46,550
4. Naturalist Moderation (add 1b, 2a; half 2b, 3a,b, 4, 5) Rice in EAA; modify other trends	I	63	510	888	1134	−354	5961	11,875	18,877
	R	487	917	1453	2062	−1961	4472	11,720	18,770
5. Naturalist Moderation minus EAA (Same as 4, but without rice or cane)	I	−1289	−842	−464	−218	−14,026	−7712	−1797	5205
	R	−1178	−739	−230	429	−17,240	−9679	−1598	7219
Base 1: Normal Forecast Totals: Output in bil 93$, Empl. in thou. Output in Bil. 92$, Emply. in thou.	I	259.3	342.6	410.0	490.9	2810	3337	3604	3815
	R	137.8	165.5	186.5	214.9	3207	3497	3610	3822
B. Missing Pieces as % of Base 1 Totals									
1. Growth as Usual	I	0.48	0.48	0.58	0.65	0.69	0.78	1.03	1.43
	R	0.49	0.56	0.75	0.95	0.47	0.57	0.79	1.02
2. New Projects	I	−0.04	0.14	0.13	0.04	−0.06	0.17	0.17	0.08
	R	0.02	0.23	0.26	0.18	0.01	0.22	0.28	0.20
3. Full Speed Ahead!	I	0.44	0.62	0.70	0.69	0.63	0.94	1.20	1.50
	R	0.51	0.64	0.84	0.95	0.48	0.79	1.06	1.22
4. Naturalist Moderation	I	0.02	0.15	0.22	0.23	−0.01	0.18	0.33	0.49
	R	0.23	0.36	0.51	0.63	−0.06	0.13	0.33	0.49
5. Nat. Moderation minus EAA	I	−0.50	−0.25	−0.11	−0.04	−0.50	−0.23	−0.05	0.14
	R	−0.56	−0.29	−0.08	0.13	−0.54	−0.28	−0.04	0.19

(continued)

Table 125.5 (continued) Summary of Scenarios

		New Output ($mil.)				New Employment (no.)			
		2000 [1]	2010 [2]	2020 [3]	2030 [4]	2000 [5]	2010 [6]	2020 [7]	2030 [8]
Base 2: Normal Forecast (Annual Changes)									
Change in output in mil. 93$ and no. of jobs	I	7338	6201	7421	8885	48,613	20,356	21,984	23,272
Change in output in mil. 92$ and no. of jobs	R	4537	4037	3257	5224	36,742	24,468	10,208	30,047
C. Missing Pieces as % of Base 2									
1. Growth as Usual	I	17.0	16.9	31.8	36.0	48.3	44.8	169.9	233.6
	R	23.1	35.0	66.0	60.2	40.7	81.0	278.8	129.5
2. New Projects	I	-1.4	7.9	7.0	2.0	-3.7	27.5	27.9	32.7
	R	0.8	14.3	23.0	11.4	1.0	31.6	97.2	25.3
3. Full Speed Ahead!	I	15.5	34.3	38.8	38.0	36.2	154.6	197.0	246.3
	R	23.9	49.3	89.0	71.7	41.7	112.6	376.2	154.9
4. Naturalist Moderation	I	0.9	8.2	12.0	12.8	-0.7	29.3	54.0	80.1
	R	10.7	22.7	44.6	39.5	-5.3	18.3	114.8	62.5
5. Nat. Moderation minus EAA	I	-17.6	-13.6	-6.2	-2.4	-28.9	-37.9	-8.2	22.4
	R	26.0	-18.3	-7.1	8.2	-46.9	-39.6	-15.7	24.0

Panel A: Annual IMPLAN impacts refer to sum of direct, indirect, and induced impacts of scenarios in each of the benchmark years at col. head. Annual REMI impacts refer to the absolute differences in output and employment between the scenario generated by the complete dynamic model and the REMI base control.

Base 1:   The first line (I) is the base line used for IMPLAN comparisons. It refers to the joint BEBR-BEA forecast for each year estimated by applying growth rates to the most recent published base year estimate. Base year output is for 1995 ($225 billion in 1995 dollars) from IMPLAN data for the counties in the study area. Base year total employment is from Bureau of Economic and Business Research (1998a) for 2010. Different growth rates for output and employment were then derived from Bureau of Economic and Business Research (1998a) for 1995–2000, 2000–2010, and from U.S. Department of Commerce (1999) for 2010–2015, 2015–2025, 2025–2045, and applied to forecast "normal" output and employment from the base to the benchmark years. The second line (R) is the base for the REMI comparisons and refers to output and employment given in the REMI database.

Panel B: Compares the IMPLAN and REMI figures for annual increases in output and employment to their respective total bases.

Base 2:   The annual changes in output and employment in each of the bases between the preceding year and the benchmark year in the heading. These figures are also used in Figures 125.4A and 125.4B, where annual changes are also shown for 1969–1970, 1980–1981, and 1989–1990, for both IMPLAN and REMI base series.

Panel C: Compares the IMPLAN and REMI figures for annual increases in output and employment to their respective annual bases.

Projects" adds 27.9% to the annual increases in new jobs according to IMPLAN in the peak years 2010 to 2020 and 97.2% more jobs in 2020 according to REMI (Table 125.5, line C.2, col. 7).

- The "Full Speed Ahead!" Scenario sums "Growth as Usual" and "New Projects." By 2030, annual increases in output and employment will be 38.0 and 246.3%, respectively, above IMPLAN's forecast annual increase and 71.7 and 154.9%, respectively, above REMI's annual increase in 2030 (Table 125.5, line C.3, cols. 4, 8).

- The "Naturalist Moderation" Scenario would substitute rice for sugarcane, the latter a more labor-intensive crop, and reduce the levels of the activities that make up "Growth as Usual" and "New Projects." This scenario generates an annual increase in output and employment of 13 and 81%, respectively, above the annual increases for 2030, according to IMPLAN, and 40 and 63%, respectively, according to REMI (Table 125.5, line C.4).

- Naturalist Moderation minus the EAA cuts out all sugarcane growing and milling in the EAA, further reducing output and employment below the levels of "Naturalist Moderation." This allows some of all the new projects to be undertaken, but replaces the agricultural zone with water storage. The results for both IMPLAN and REMI are remarkably similar: by 2030, the increase in jobs would be 22 to 24% of the annual increase, a fraction of the "Full Speed Ahead!" scenario discussed previously.

The annual number of jobs generated for each of the summary scenarios is compared to the annual increase in jobs in the control base for the IMPLAN in Figure 125.4A and for REMI in Figure 125.4B. Both sketches show similar trends. The annual number of new jobs generated in both bases continues to decline from its peak growth in the 1980 period; the expected increase in new jobs due to Full Speed Ahead! and the reduction in these annual numbers through Naturalist Moderation.

CONCLUSIONS AND CAVEATS

The apparent coincidence of IMPLAN and REMI results for the scenarios presented here is somewhat surprising. IMPLAN is an extremely disaggregated, static input–output model. In this exercise, it has been made to leap into the future by providing it with forecasts of exogenous demands, and its impacts have then been compared to the state–federal forecasts of future employment without the new projects. REMI, by comparison, is a quick-footed, dynamic econometric model with a more aggregated input–output model within it, as well as the mechanism to grow on its own, once the future exogenous demands are provided. REMI also generates its own control base as the model runs into the future with no exogenous demands.

Both models clearly warn us of the high and unanticipated impacts of continued growth. The path of "Natural Moderation" and its variant suggests one way of modifying the magnitude of those impacts.

While the magnitudes of the exogenous demands reflect many judgments by the author, the general procedure developed here is intended to capture the forces that are currently at work in the region. The current trends may change, but if they do not, the results here reflect the probable impact of the current course on the economy of the region.

We have found that the annual increases in output and employment due to the major new trends "missing" from the basic forecast are relatively small compared to the large base. Even in the "Full Speed Ahead!" scenario, the IMPLAN and REMI models suggest annual increases of $3.4 to 3.8 billion in output and from 47,000 to 57,000 new jobs by 2030. The former would add 0.69 to 0.95% to the region's total output and the latter would add from 1.2 to 1.5% to total employment (base 1) for 2030.

The annual increases in output and employment, however, are very large numbers, when compared to the annual increase in the base, especially in the later years. The two models indicate that the Full Speed Ahead! scenario generates from 38 to 72% more output and from 155 to 246%

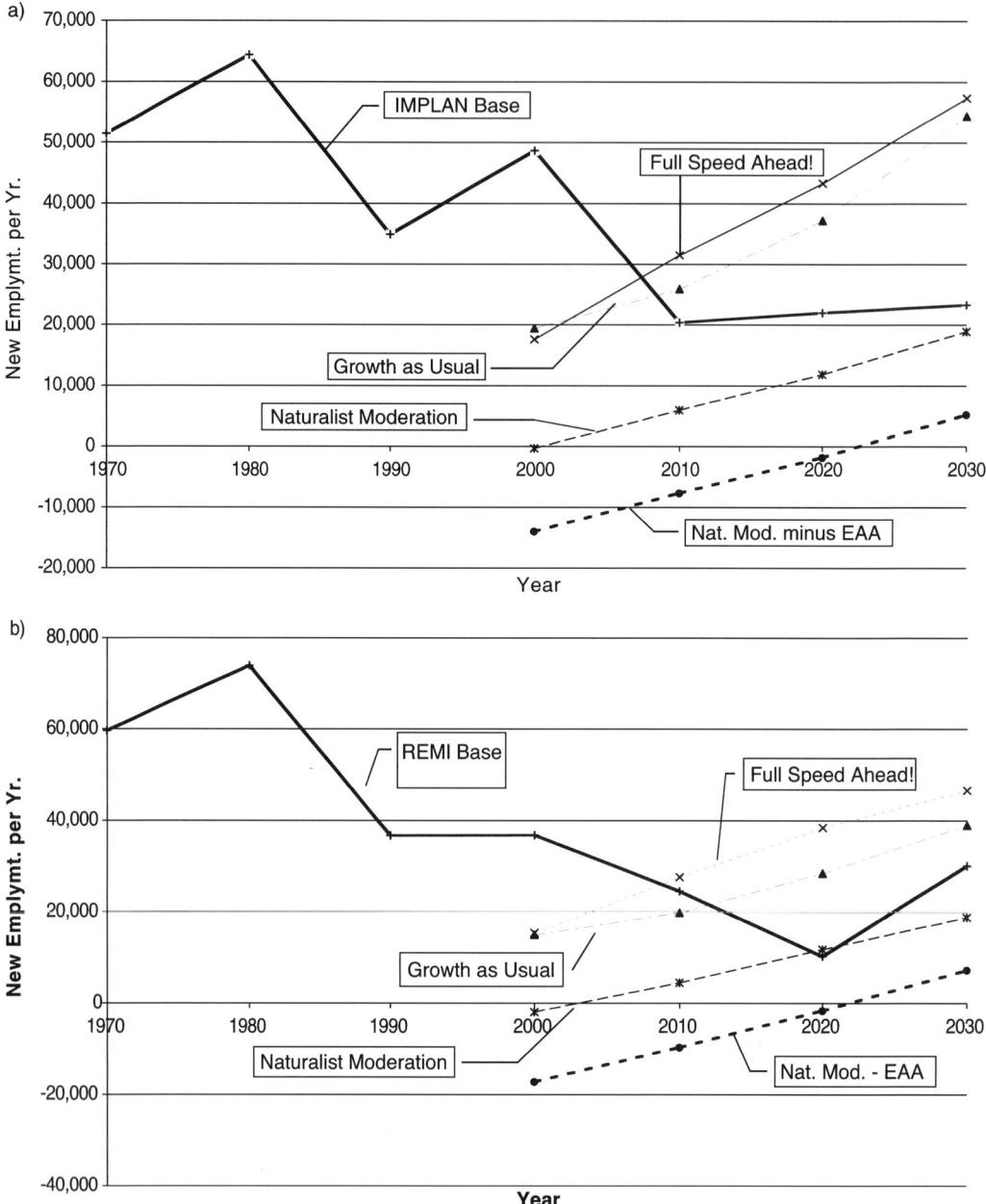

Figure 125.4 (a) IMPLAN: Annual jobs generated by base and scenario; (b) REMI: Annual jobs generated by base and scenario.

more employment than the annual increase generated by the base in 2030. This finding has important implications for those lumpy public investments, such as water and sewage treatment plants for which the annual increase in demand is critical for determining the need for new capacity. The notion of relative annual increases is also important in determining spending on education, police, and other services.

Our findings take issue with the findings of the Restudy's regional economic impact study (Appendix E of the Restudy, pp. E-304 to E-308) which the present author had also computed. I compared the impacts of new construction expenditures, as modeled by IMPLAN, to the total number of jobs and output in the region. All other growth trends were deliberately ignored, because they did not technically fall within the domain of the project. But when the task of the economist is expanded to include the evaluation of other major growth trends that are, in fact, facilitated by the Everglades restoration and are progressing in the region, then we are compelled to reach quite opposite conclusions. It is, after all, not only the contribution that Everglades restoration will make to new output or to new employment that should be of concern, but rather the contribution that all new projects make to the growth of the basic economy. The $8 billion for Everglades restoration may be indeed crucial in that it will permit the basic economy to reach, or as we are suggesting, to overshoot its forecast levels.

Our application of two independent growth models suggests that, in the absence of any intervening natural or social calamities, large numbers of jobs are being created in the region on an annual basis. It is urgent that the implications of this development on land use and water consumption now be evaluated.

Everglades restoration will allow the region to continue its current expansion by providing more water and, in effect, more usable land through better flood protection of the outlying areas. But it is not the only force that contributes to the supercharged economy. Other state and local policies also work to encourage growth, urban sprawl, and agricultural expansion. Tax exemptions, wage subsidies, low impact fees, and low utility rates all attract new manufacturing, commercial, and tourist facilities and promote new housing. Generally low energy rates encourage the construction of larger air-conditioned homes, and low water rates promote green lawns all year round.

Is all this growth desirable? Slower growth itself need not lead to reduced consumption. In the absence of extremely aggressive policies of conservation, the present trend is to encourage greater, not lesser, resource use in urban areas. In the countryside, the water-intensive and land-extensive technologies borrowed from temperate agriculture for the region's crops, dairies, and fishponds are also proving highly inappropriate for the semitropical Everglades ecosystem. Continuing on the current path will surely accelerate ecological pressures. But the total costs of these pressures cannot alone be computed by the complicated but narrow economic models applied here. The restoration effort, designed by engineers from outside the region, needs to be complemented by efforts at smart production and consumption, designed and imposed by taxpayers inside the region, as the broad population comes to recognize the true social and environmental costs of living within a fragile ecosystem.

Hurricanes and rising sea levels may be nature's way of imposing limits on growth. Hurricane Andrew caused $425 billion in damage in just a few days in August 1992, and two powerful hurricanes in the late 1920s helped shut down South Florida land speculation for two decades. New hurricanes could put a stop to the current building frenzy, or simply shift it northward to less densely populated counties. Rising sea levels may also be part of a violent rather than a gradual process, with major storms resulting in changes that irrevocably alter the shoreline.* If the present trend continues, much of South Florida could be under water by the middle of the 21st century.

The economy may exert brakes of its own, for the South Florida engines of growth, such as tourism and construction, are sensitive to the national business cycle. The opening of Cuba and access to its sugarcane lands could result in the reduction of Everglades agriculture, especially as the peat soils of the EAA subside and productivity falls in much of the area. But putting this type of speculation aside, it may be only the intentional redesign of South Florida's urban sprawl and changes in farm practices that together will enable the Everglades to survive the 21st century.

---

* See Wanless and co-workers (1994).

ACKNOWLEDGMENTS

The author wishes to acknowledge the helpful comments of J. Alvarez, J. Englehardt, B. Glaz, D. Lenze, B. Peacock, G. Snyder, and J. Yeasted. Early versions of this chapter were presented as papers at research seminars at the campuses of the University of Florida's Institute for Food and Agricultural Sciences (IFAS) at Gainesville, Belle Glade, and Ft. Pierce.

REFERENCES

Crihfield, J.B. and Campbell, H.S., Jr., Evaluating alternative regional planning models, *Growth Change,* 22, 1–16, 1991.

Derr, M., *Some Kind of Paradise: A Chronicle of Man and the Land in Florida,* William Morrow, New York, 1989.

Douglas, M.S., *The Everglades, River of Grass,* Pineapple Press, Sarasota, FL, 1988.

Duchin, F. and Lange, G.M., *The Future of the Environment: Ecological Economics and Technological Change,* Oxford University Press, New York, 1994.

Glaz, B., Research seeking agricultural and ecological benefits in the Everglades, *J. Soil Water Conserv.,* 50, 609–612, 1995.

Governor's Commission for a Sustainable South Florida, Initial report, Coral Gables, FL, 1995.

Leontief, W., Carter, A.P., and Petri, P.A., *The Future of the World Economy,* Oxford University Press, New York, 1977.

Light, S.S. and Dineen, J.W., Water control in the Everglades: a historical perspective, in *Everglades: The Ecosystem and Its Restoration,* Davis, S.M. and Ogden, J.C., Eds., St. Lucie Press, Boca Raton, FL, 1994, pp. 47–84.

Lissakers, K., *Banks, Borrowers, and the Establishment: A Revisionist Account of the International Debt Crisis,* Basic Books, New York, 1991.

McCluney, W.R., Ed., *The Environmental Destruction of South Florida,* University of Miami Press, Coral Gables, FL, 1969.

Minnesota IMPLAN Group, Inc., IMPLAN system (1995 data and vers. 1.1 software) (www.implan.com), Stillwater, MN, 1997.

Rickman, D.S. and Schwer, R.K., A comparison of the multipliers of IMPLAN, REMI, and RIMS II: benchmarking ready-made models for comparison, *Ann. Reg. Sci.,* 29, 363–374, 1995.

Ruiz, A.L., Weisskoff, R., Alward, G., Siverts, E., Hussain, A., and Maki, W., Puerto Rico IMPLAN System: Model and Database Construction and Application, U.S. Forest Service, Atlanta, 1994.

Shih, S.F., Glaz, B., and Barnes, R.E., Jr., Subsidence lines revisited in the Everglades agricultural area, Technical bulletin 902, Agricultural Experiment Station, University of Florida, Gainesville, 1997.

South Florida Ecosystem Restoration Task Force Working Group, Integrated financial plan 1998: South Florida ecosystem project activities, Miami, 1998.

Treyz, G.I., *Regional Economic Modeling: A Systematic Approach to Economic Forecasting and Policy Analysis,* Kluwer Academic Publishers, Boston, 1993.

U.S. Army Corps of Engineers, Comprehensive report on central and southern Florida for flood control and other purposes, reprinted from 80th Congress, 2nd session, House Doc. 643, Government Printing Office, Washington, D.C., 1948.

U.S. Army Corps of Engineers, Central and South Florida Project, comprehensive review study, draft feasibility report and environmental impact statement, Jacksonville, 1998.

U.S. Department of Commerce, Bureau of Economic Analysis, Regional multipliers: a user handbook for the regional input-output modeling system (RIMS II), U.S. Government Printing Office, Washington, D.C., 1992.

Wanless, H.R., Parkinson, R.W., and Tedesco, L.P., Sea level control on stability of Everglades wetlands, in *Everglades: The Ecosystem and Its Restoration,* Davis, S.M. and Ogden, J.C., Eds., St. Lucie Press, Boca Raton, FL, 1994, pp. 199–224.

Watercourse and South Florida Water Management District, Discover a Watershed: The Everglades, Montana State University, Bozeman, 1996.

Weisskoff, R., Missing pieces in ecosystem restoration: the case of the Florida Everglades, *Econ. Syst. Res.,* 12, 271–303, 2000 (http://exchange.law.miami.edu/everglades/science/weisskoff/missingpieces.pdf).

Data Sources and the Scenarios

SCENARIOS AND INCOME PATTERNS (FOLLOWING THE ROWS OF TABLE 125.4)

1. Agricultural data: Details on sugarcane and rice are found in Snyder et al., 1978; Alvarez et al., 1994; U.S. Department of Agriculture, 1997, 1998; and Izuno et al., 1999.
2. Everglades restoration: Line items for the detailed construction costs and also operation and maintenance costs for each of the projects within the Everglades Protection Project are specified in the appendices of Burns and McDonnell (1994). Details on the Restudy are in U.S. Army Corps (1998), Appendix E.11, computed by the present author on the basis of seven prototypical structures.
3. Tourist expenditures are constructed from the Office of Tourist Research (1996 to 1998) for statewide spending patterns; Strategy Research Corps (1997 to 1999) for urban Miami; English and Kriesel (1996) for detailed regional recreation in the Keys; and Visitor Services Project (1989, 1996) for Everglades National Park visitor spending. The growth of spending for auto and air visitors were averaged for 1994 to 1998 for each mode and then deflated by the travel price index for the state. Total spending level was computed from per-day expenditures and total number of person-nights reported for each mode and for the regional destinations by quarter. Almost half the air visitors but only 13 to 16% of auto visitors reported South Florida as their regional destination.
4. The "new development" package was estimated on the basis of the 1996 sector profiles for new construction, given in Bureau of Economic and Business Research (1998b).
5. Expenditure patterns are given by IMPLAN for three levels of spending over the full range of BLS commodities. New residents are assumed to be attracted at twice the present rate of 43 thousand middle-income households per year.

DATA SOURCES (FOR TABLES 125.1 AND 125.2 AND APPENDIX 125.I)

Alvarez, J., Lynne, G.D., Spreen, T.H., and Solove, R.A., The economic importance of the EAA and water quality management, in *Everglades Agricultural Area (EAA): Water, Soil, Crop, and Environmental Management,* Bottcher, A.B. and Izuno, F.T., Eds., University Press of Florida, Gainesville, 1994, pp. 194–223.

Bureau of Economic and Business Research (BEBR), *The Florida Long-Term Economic Forecast 1998,* University of Florida, Gainesville, 1998a.

Bureau of Economic and Business Research (BEBR), *1998 Florida Statistical Abstract,* University of Florida, Gainesville, 1998b.

Burns and McDonnell, Inc., Everglades protection project, Palm Beach County, Florida and Kansas City, MO, 1994.

English, D. and Kriesel, W., Linking the Economy and Environment of Florida Keys/Florida Bay: Economic Contribution of Recreating Visitors to the Florida Keys/Key West, National Oceanic and Atmospheric Administration, Washington, D.C., 1996.

EnviroTools, Inc., Threatened and endangered species software (TESS), (CD-ROM) Gainesville, FL, 1998.

Izuno, F.T., Rice, R.W., and Capone, L.T., Best management practices to enable the coexistence of agriculture and the Everglades environment, *HortScience,* 34, 26–33, 1999.

Office of Tourism Research, Florida visitor study (various years), Florida Commerce Department, Tallahassee, FL, 1996–1998.

Smith, S.K. and Nogle, J., Projections of Florida population by county, 1998–2020, *Fla. Pop. Stud.,* 32, 1–8, 1999.

Snyder, G.H., Burdine, H.W., Crockett, J.R., Gascho, G.J., Harrison, D.S., Kidder, G., Mishoe, J.W., Myhre, D.L., Pate, F.M., and Shih, S.F., Water table management for organic soil conservation and crop production in the Florida Everglades, Bulletin 801, Agricultural Experimental Station, University of Florida, Gainesville, 1978.

Strategy Research Corporation, Visitor profile and tourism impact: Greater Miami and the beaches annual report, Miami, FL, 1997–99.

U.S. Census Bureau, *Statistical Abstract of the United States: 1998,* U.S. Government Printing Office, Washington, D.C., 1998.

U.S. Department of Agriculture, *Census of Agriculture 1997,* Web site release, www.nass.usda.gov/census/census97/volume 1/1999.

U.S. Department of Agriculture, Economic Research Service, Rice: Situation and Outlook Yearbook, U.S. Government Printing Office, Washington, D.C., 1997.

U.S. Department of Agriculture, Economic Research Service, Sugar and Sweetener, Situation and Outlook Report, U.S. Government Printing Office, Washington, D.C., 1998.

U.S. Department of Commerce, Bureau of Economic Analysis (BEA), State Projections of Employment, Earnings, and Product, Web site release, www.bea.doc.gov/gsp/gspdata/.

Visitor Services Project, *Everglades National Park Visitor Study,* Cooperative Park Studies Unit, University of Idaho, Moscow, ID, 1996.

CHAPTER 126

Analysis of Land Cover and Water Quality in the New York Catskill–Delaware Basins

Megan H. Mehaffey, Timothy G. Wade, Maliha S. Nash, and Curtis M. Edmonds

INTRODUCTION

Prior to 1776, New York City's drinking water was supplied by public wells, but demands for water eventually resulted in construction of the city's first reservoirs. The Croton Reservoir (1842) supplied over 90 million gal/day to the city, but eventually even this amount could not meet the city's needs. To ensure the water supply into the future, the Catskill and Delaware systems were developed (Hecker, 1991; Weidner, 1974). The construction of the impoundments, aqueducts, and tunnels needed to supply water to New York City spanned a total of six decades (1905 to 1965). The six reservoirs are the Cannonsville, Pepacton, Schoharie, Neversink, Rondout, and Ashokan (Figure 126.1). Together, these reservoirs provide 90% of the 1.4 billion gallons of drinking water consumed by New York City residents. With the development of the Catskill–Delaware systems, New York City's drinking water remains one of the cleanest in the nation. Currently, New York City is trying to maintain the high quality of the drinking water supply through the development and implementation of a long-range watershed protection program. The watershed protection plan was signed in 1997 and unites efforts by local communities, the City and State of New York, environmental groups, and the EPA (Ashendorff et al., 1997; Okun et al., 1997). The plan includes upgrading current sewage treatment plants, implementing new watershed regulations, designing and constructing a filtration system for the 10% of water supplied by the Croton system, and acquiring land deemed critical to the preservation of high water quality in the Catskill–Delaware system. Land acquisition was selected, partly to protect the environment, but also to save the taxpayers of New York State billions of dollars. Installing a filtration system for the city's water supply would cost an estimated $2 to 8 billion, while land purchase is estimated to cost $250 to 300 million (Featherstone, 1996; MOA, 1997). The majority (90%) of money spent for land acquisition will be used in the Catskill–Delaware basins. The city has set as its goal the solicitation of 350,000 acres of land (MOA, 1997). Under the agreement, the New York Department of Environmental Conservation issued a permit to acquire land through outright purchase and conservation easement. Priority is given to the purchase of nonresidential land around reservoirs, streams, and wetland areas. Secondary land acquisition will focus on the purchase of sensitive lands surrounding streams and rivers. The expected result of land acquisition and conservation practices is the protection of more than 165 stream miles, the preservation of thousands of acres of natural areas, and continued high water quality without the cost of a multibillion dollar filtration system (Murphy et al., 1995; Featherstone, 1996; MOA, 1997).

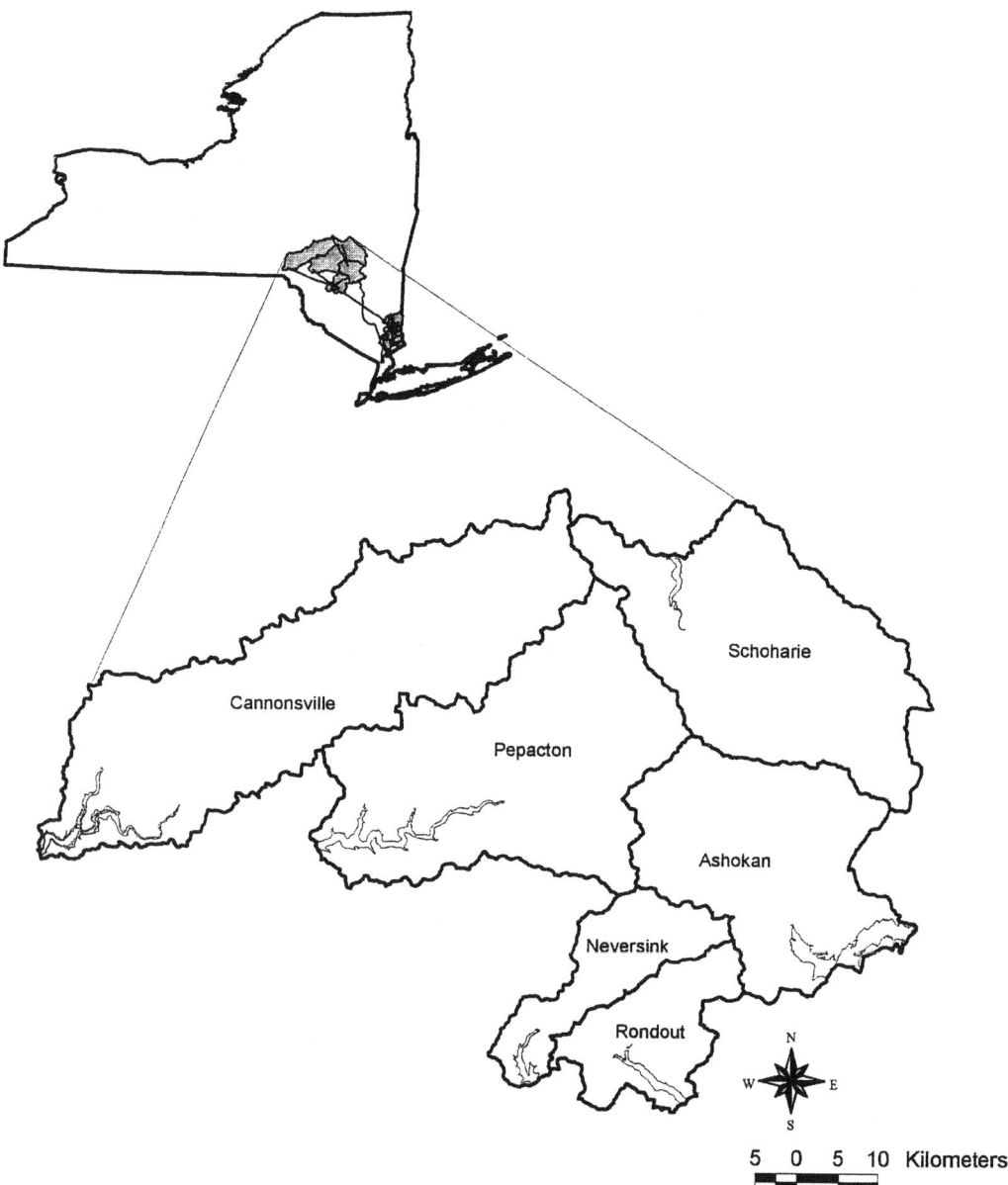

Figure 126.1 The Catskill–Delaware basins and location within New York State.

As a result of topographic constraints, both agricultural and residential land use is concentrated close (250 m) to river and stream riparian zones. Forested riparian zones act as a buffer, and diminish excessive runoff of water and pollutants such as fertilizer, pesticides, and other chemicals (Finlayson, 1991; Manlanson, 1993). When forests are removed, the potential for flow of pollutants into streams is greatly increased. Fertilizers, pesticides, and other pollutants can be transported into streams flowing through or very close to agricultural or residential land more easily than streams that flow through forest (Lowrance et al., 1986; Osborne and Wiley, 1988). Agriculture has the potential to increase soil erosion. When cropping takes place on soils with a 3% or greater slope there is increased likelihood of erosion (Lowrance, 1984). The human population in the Catskill–Delaware basins has increased only 17% in the past 30 years, resulting in approximately 12 people/km^2 (U.S. Census, 1990; FGEIS, 1993). However, the continued effluent inputs from waste treatment plants, nonpoint agricultural sources, and residential runoff increase the potential risk of contamination above acceptable water quality standards for drinking, and survival of freshwater fish and macroinvertebrates. The economy of the area depends largely on seasonal tourism and dairy farming (Stave, 1995), the latter making fecal coliforms, nutrients, and sediment load a particular concern (Hansan, 1996; Wall et al., 1998).

The objective of this study was to explore the relationship between land-cover metrics and surface water quality measurements at the scale of the basin, subbasin, and riparian zone in the Catskill–Delaware basins. Results will be used to evaluate potential risk to New York City's drinking water supply system. Data generated from this study can be used to target basin, subbasin, and riparian zones in need of protection within the basins.

METHODS

Site Characteristics

The 4104-km^2 Catskill–Delaware basins are located in the southeast corner of New York State, 160 km northwest of New York City (Figure 126.1). Average monthly precipitation ranges from 50 to 150 mm. The highest precipitation events occur during the spring and fall months. Temperature ranges from –10 to 27°C; July and August are the hottest months. The topography of the area is diverse and, except for the Adirondacks to the north, has the highest elevation in the state. The bedrock geology of the area is sedimentary and consists of sandstone, shale, and conglomerates formed during the Devonian time period (Miller, 1970). The bedrock is exposed or within 1 m of the surface in about 30% of the basin area. Of the remaining surface geology, almost 60% is glacial till. The basins are bisected by numerous streams including the Delaware, Schoharie, and the Neversink. The alluvial deposition and outwash from these streams and their tributaries account for the remaining 10% of the basin surficial geology.

Historically, the Catskill–Delaware basins were dominated by northern hardwood trees, including maple, birch, and beech. Other species present in smaller numbers were white and red pine, hemlock, elm, and ash (van Valkenburg, 1996). Much of the area was logged prior to the mid-1800s. However, due to the need to eliminate a substantial debt owed by Ulster County, lands owned by the county were conveyed back to New York State in 1884. With the transfer of ownership, 13,759 ha of forest were brought under state protection and the Catskill Park and Forest Preserve was born. In the decades since the Park and Forest Preserve's inception, the forest has rebounded from its previous losses and is once again made up of a mixture of hardwoods and other deciduous and evergreen trees, encompassing over 250,000 ha of land (van Valkenburg, 1996).

Land-Cover Data

Land-cover data are required to calculate most land cover metrics, and for regional and larger areas is generally derived from satellite imagery. The New York City Department of Environmental Protection (NYCDEP) supplied the following Arc View coverages for the study area: basin and subbasin boundaries, topography, soils, roads, streams (including lake shorelines), and water-sampling site locations. Land-cover data from the National Land Cover Database (NLCD) were acquired from the EROS Data Center, a consortium of four federal agencies. The land cover data were derived from a composite of two Landsat thematic mapper (TM) images, an early spring (leaf off) and summer scenes (leaf on) (April 28, 1989 and June 21, 1991). The land-cover data set was generated using an unsupervised classification technique. Clusters of spectrally similar areas were grouped and assigned a land-cover class using National High Aerial Photography program (NHAP) aerial photos. Areas that contained more than one land-cover class were labeled with the aid of ancillary data such as population or elevation (EROS Data Center, 1998). For the purposes of this study, the NLCD classes were aggregated into six categories: water, forest, barren, row crop, pasture, and residential (Table 126.1).

Table 126.1 Aggregation of the National Land-Cover Database (NLCD) Land-Use Classes

NLCD Class	Aggregated Class
Open water	Water
Low intensity residential	
High intensity residential	
High intensity commercial	Residential
Corn, alfalfa, specialty crops	Row crop
Pasture/hay	
Other grasses (e.g., parks, lawns, golf courses)	Pasture
Evergreen forest	
Deciduous forest	
Mixed forest	
Woody wetlands	
Emergent herbaceous wetlands	Forest
Quarries/strip mines/gravel pits	
Transitional	Barren

Land-Cover Metrics

Land-cover metrics of characteristics believed to have a relationship to drinking-water quality were calculated for six basins (average area 700 km^2) and 79 subbasins (average area 50 km^2) in the Catskill–Delaware basins. Four metrics were chosen based on results of a previous study in the Mid-Atlantic region (Jones et al., 1997). These metrics are (1) percentage of the six land-cover categories listed above, (2) human use (U-index), (3) percentage of land cover in and near (100-m buffer) stream riparian zones, and (4) percentage of agriculture on steep slopes.

Because forest is the dominant vegetation cover of the Catskill–Delaware basins (Kuchler, 1964), percentages of other land cover were used as indicators of disturbance. The land-cover metrics were generated by overlaying the land cover with basin and subbasin. Water was excluded from the total area when calculating percentages. Percentages of the six land-cover categories were calculated by basin and subbasin. Total human use, or U-index (O'Neill, 1988), was calculated by summing the percentages of agriculture, residential, and nonnatural barren (e.g., mines or gravel pits) land use.

The MOA (1997) limits development within 100 m of the reservoirs and 50 m from streams. We selected the largest buffer distance (100 m) for use in our analysis. Two metrics were generated for the riparian zone: the percentage of each land cover in a 100-m distance from streams by basin and subbasin, and the percentage of each land cover 100 m upstream of 25 water quality sampling sites. The 25 sampling sites were selected based on the quantity and quality of available water quality data. The two metrics were calculated by overlaying land cover with streams and buffered area polygons. Water was again excluded when calculating percentages.

Erosion potential is related to the steepness of slopes in agricultural use and soil type (Comeleo et al., 1996; Lowrance, 1986). The threshold for increased erosion is a 3% slope (USDA, 1951). Percent slope was generated from 30-m Digital Elevation Model (DEM) data. The soil erosion potential metric was calculated by overlaying the distribution of slope with land cover. The metric was reported by subbasin as the percentage of agriculture in the subbasin on slopes greater than 3%.

Surface Water Quality

An 8-year window (1986 to 1994), enclosing the classified NLCD image of years 1989 and 1991, was used to select water quality data. Data from 243 sampling sites across the Catskill–Delaware basins were captured for this period (Figure 126.2). The main source for the chemical and physical water quality parameters was the EPA storage and retrieval database called STORET.

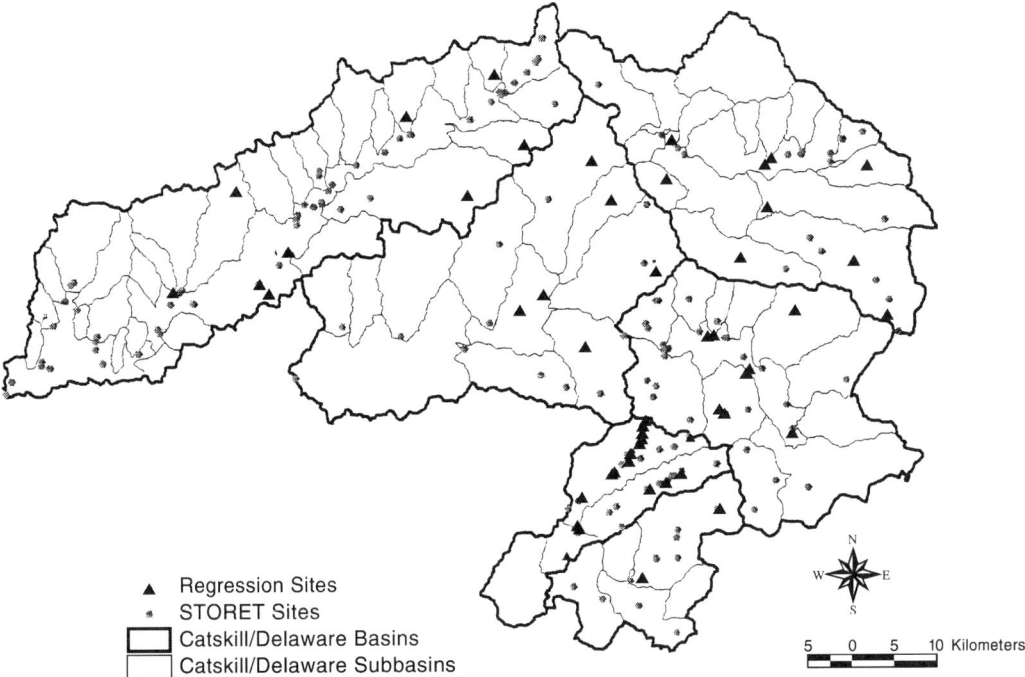

Figure 126.2 STORET sample locations in the Catskill–Delaware basins and those used in the regression analysis.

STORET is a repository for water-parameter data for the contiguous U.S. Many organizations contribute to the database including federal, state, interstate, universities, contractors, individuals, and water laboratories.

Compiling water quality data from a number of different sources gave rise to problems of inconsistency and lack of comparability due to differences in measurement and analytical methods. In the course of our data compilation we encountered the following:

1. Differences in methodologies between agencies and with time.
2. Changes in detection limits of chemicals and metals over time, making comparison with earlier samples difficult.
3. Sample sites monitored for short time spans (1 to 3 years) rather than continuously.
4. Some sites sampled more frequently than others.
5. Discontinuation of monitoring and initiation of new sites in different locations.

Data were not temporally and spatially adequate to confidently characterize the changes in many water quality parameters with time and their relationship to land-cover metrics. Therefore, attention was focused on water parameters with data collected during most of the 1986 to 1994 time period, which have a known relationship to land use. Because of differences in vegetative productivity, nutrient cycling, and climate, the influence of land cover may vary with season. Therefore, we used seasonal average for water quality data. The chemical water quality parameters included nitrate-nitrogen (NO_3–N), chloride, silica, conductance, and dissolved oxygen (DO); physical measures included temperature, suspended sediment, and flow rate.

Statistical Analysis

An analysis of variance (ANOVA) and Tukey's studentized range test of means were conducted to determine significant differences between basin land cover metrics (SAS, 1990). In order to relate land use to the water quality data, we assumed minimal land-cover changes between 1986 and 1994. To examine the relationship of water quality and land cover, data from all sampling sites were used in the correlation analyses at three scales, basin ($n = 6$), subbasin ($n = 79$), and riparian zone ($n = 79$). For these analyses, independence between sampling sites was assumed. For each scale, we used a seasonal average for water quality data.

A stepwise regression ($n = 25$) was run for a subset of water sampling sites to determine the effect of nearby land cover (100-m upstream buffer) on seasonal water quality. The locations of the subset sampling sites were chosen for comprehensive spatial coverage of the basins. From the previous correlations, seasonal water quality data of four parameters (total chloride, silica, pH, and NO_3–N) that had a significant correlation ($P \leq 0.05$) with land cover were used in regression analysis. A new variable was introduced to the regression that represents a multiplication of two land-cover metrics. This new variable, statistically known as an interaction term, is represented in the tables by the symbol "/" set between two land-use cover metrics and in the text with "and" between the two variables combined in the interaction.

RESULTS

The Catskill–Delaware basins were dominated by evergreen and deciduous forest with an average forest cover of 92% across the basins (Table 126.2, Figure 126.3). Three of the six major basins have more than 95% forest cover. The greatest human use of the basins was agriculture (8%), of which 6% was pasture and 2% row crops. However, the majority of the agriculture was concentrated in the Cannonsville basin. Using ANOVA ($P \leq 0.05$), Cannonsville basin was found to have significantly greater percentages of crops (3%) and pasture (17%) than the other five basins. The Schoharie and Pepacton basins had the next highest concentration of agricultural land use, averaging around 10%, of which 3% was row crop, and 7% pasture. The remaining basins had 3% or less total agricultural use. All agriculture (8%) in the basins was located on slopes greater than 3%. Residential land use was minimal and averaged less than 1% of the total area. The Ashokan basin had the highest concentration of residential land use (1.12%). Cannonsville and Schoharie basins were second and third with 0.70 and 0.73% cover, respectively. Basins with the least residential cover were the Neversink, Rondout, and Pepacton, all with less than 0.40%.

The higher percentage of agriculture in the Cannonsville, Pepacton, and Schoharie basins resulted in a greater U-index in these basins (Table 126.2). Although the Ashokan basin had the highest percentage of residential use, its U-index was similar to that of both Neversink and Rondout basins. The U-index in the Neversink and Rondout basins was related to agricultural use. Within the basin riparian areas (100-m stream buffer), the percentage of forest was lower, and agriculture and residential use was higher. Higher percentages of both agricultural and residential land use were found in subbasins containing a major stream (i.e., Esopus in the Ashokan, Batavia Kill and Schoharie in Schoharie, East Branch Delaware in Pepacton, and West Branch Delaware in Cannonsville).

The Cannonsville basin had the highest mean total chloride, silica, and NO_3–N (6.88, 3.88, and 0.52 mg l^{-1}, respectively) (Table 126.3). The Ashokan basin had the second highest total chloride and silica values. The Pepacton basin had similar NO_3–N value as the Cannonsville basin (Figure 126.4). The lowest median total chloride, silica, and NO_3–N values were found in the Neversink basin. The Neversink basin also had a relatively low pH, which ranged between 3 and 7.88 (Table 126.3). Conductance (112.16 µmol) and sediment (7 mg l^{-1}) mean values were greatest in the Cannonsville basin. However, the range of sediment values was greatest in the Ashokan (0 to 27 mg l^{-1}). Dissolved oxygen mean values were similar among all basins (8.99 to 10.95 mg l^{-1}).

Correlation analysis suggested several significant relationships ($P \leq 0.05$) between basin and subbasin land-cover percentages and seasonal water quality parameter values. Of the eight water quality parameters, total chloride, silica, pH, and NO_3–N were significantly related to one or more of the land-cover metrics ($r = 0.35$ to 0.65). During all seasons, a greater percentage of forest cover

Table 126.2 Land-Cover and Use Metric Areas and Percentages for the Catskill–Delaware Basins and Riparian Buffer Areas

	Area (km²)	Forest (%)	Residential (%)	Pasture (%)	Crop (%)	Barren (%)	U-index (%)	Crop on 3% Slope (%)	Pasture 3% Slope (%)
Total	4103	91.63	0.57	5.80	1.79	0.02	8.18	1.79	5.80
Basin Metrics									
Ashokan	662	97.11	1.12	0.66	0.70	0.08	2.57	0.70	0.66
Cannonsville	1178	78.92	0.70	17.41	2.89	0.02	21.02	2.88	17.42
Neversink	238	97.83	0.19	0.98	0.87	0.00	2.03	0.86	0.97
Pepacton	961	90.16	0.36	6.78	2.64	0.00	9.78	2.64	6.78
Rondout	247	96.07	0.34	2.45	0.82	0.00	3.61	0.82	2.45
Schoharie	18	89.68	0.73	6.49	2.83	0.03	10.09	2.83	6.49
Riparian Metrics (100-m buffer)									
Total		89.51	1.00	6.77	2.33	0.01	10.10	—	—
Ashokan		96.21	2.01	0.91	0.59	0.00	3.51	—	—
Cannonsville		72.54	1.15	21.69	4.26	0.02	27.12	—	—
Neversink		98.92	0.12	0.37	0.36	0.00	0.86	—	—
Pepacton		86.33	0.74	8.42	4.20	0.00	13.36	—	—
Rondout		96.14	0.72	1.89	0.74	0.00	3.34	—	—
Schoharie		86.91	1.22	7.36	3.85	0.01	12.44	—	—

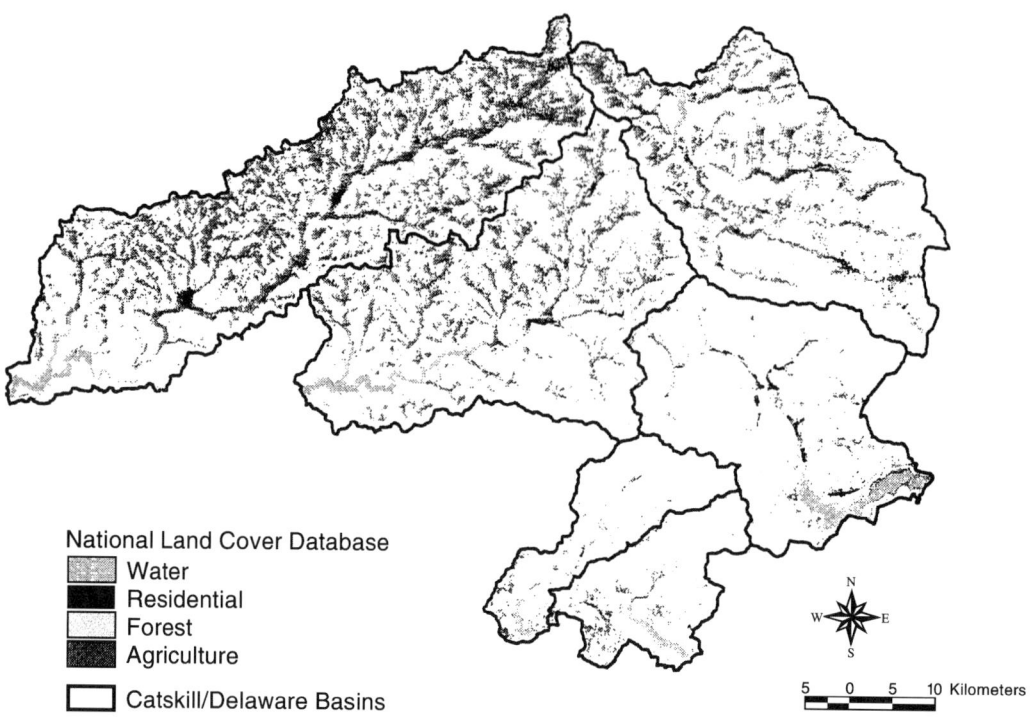

Figure 126.3 Land cover and use in the Catskill–Delaware basins. The land-cover and use categories were aggregated from the National Land Cover Database (NLCD) classification.

was related to lower values of all (total chloride, silica, pH, and NO_3–N) water quality parameters. In winter and fall NO_3–N was higher in basins and subbasins having greater amounts of pasture, while in spring and summer NO_3–N was lower. However, greater percentages of pasture were related to higher total chloride, silica, and pH values in all seasons. Greater residential land-use percentages were associated with higher total chloride in the fall and NO_3–N in the winter. Greater percentages of row crops were associated with higher silica values in the fall and spring.

The inclusion of interaction variables in the regressions generally resulted in more-complex models with higher R^2 (Table 126.4). The land cover influencing total chloride values varied with season. In winter, the interaction of crop and residential land use was related to an increase in total chloride, while in spring, greater percentages of forest cover were related to decreased total chloride. In summer, approximately one half of the variation in total chloride was explained by the percentage of pasture. Greater percentages of pasture were related to higher total chloride levels. The remaining main and interaction variables of forest, crop and pasture, forest and pasture, and residential and pasture were related to lower total chloride levels, each variable accounting for 5% of the remaining variation. The interaction of residential and pasture land use was associated with higher total chloride and accounted for 40% of the variation (Table 126.4). Greater percentages of forest were related to lower silica in winter, spring, and summer and accounted for close to half the model variation. In winter, forest cover was the only variable related to silica. During spring, summer, and fall forest and pasture and crop and pasture were related to higher silica levels. Interaction between crop and pasture accounted for 40% of the variation in the summer and 16% in the fall, while forest/pasture accounted for 6% in the spring. The overriding land cover affecting pH was percentage of forest. Greater forest cover was related with lower pH and accounted for 20 to 40% of the variation. However, in spring, the interaction term of forest and pasture was included in the model accounting for 5% of the variation associated with higher pH. The NO_3–N was significantly related to pasture,

Table 126.3 Water Quality Minimum, Maximum, and Median Values for the Catskill Basins

	N	Min	Max	Median	Mean
ASHOKAN	129	0.50	11.00	4.00	3.19
Total chloride	127	1.40	3.80	2.40	2.49
Silica	36	0.00	27.00	3.00	5.78
Sediment	52	305.70	133,489.00	7948.20	17,254.73
Stream flow	0	—	—	—	—
Conductance	107	6.08	8.37	7.02	7.11
pH	55	0.09	0.87	0.56	0.52
NO_3–N	84	5.00	14.00	9.30	9.97
Total DO					
NEVERSINK	647	0.10	5.00	0.50	0.63
Total chloride	585	0.80	3.20	2.10	2.01
Silica	4	1.00	3.00	1.00	1.50
Sediment	222	81.52	262,902.00	917.10	5495.47
Stream flow	0	—	—	—	—
Conductance	305	3.00	7.88	5.98	5.82
pH	640	0.02	1.10	0.37	0.41
NO_3–N	278	10.30	12.00	10.30	10.36
Total DO					
RONDOUT	57	0.40	30.00	2.00	2.58
Total chloride	25	1.80	2.80	2.20	2.25
Silica	19	1.00	9.00	1.00	2.11
Sediment	17	815.20	33,627.00	4789.30	9488.93
Stream flow	17	32.00	62.00	48.00	49.82
Conductance	52	0.00	7.90	6.33	6.30
pH	29	0.18	0.99	0.49	0.50
NO_3–N	44	9.10	13.40	9.70	10.28
Total DO					
SCHOHARIE	20	0.40	11.00	1.00	2.80
Total chloride	26	0.01	4.40	2.80	2.70
Silica	0	—	—	—	—
Sediment	2	24,557.90	216,028.00	120,292.95	120,292.95
Stream flow	0	—	—	—	—
Conductance	44	5.50	8.39	6.95	6.98
pH	21	0.00	0.88	0.54	0.54
NO_3–N	11	7.30	10.20	9.20	8.99
Total DO					
CANNONSVILLE	72	0.90	17.00	7.00	6.88
Total chloride	22	2.60	4.50	3.40	3.38
Silica	13	2.00	12.00	8.00	7.00
Sediment	9	7642.50	347,479.00	23,946.50	80,354.26
Stream flow	25	57.00	160.00	112.00	112.16
Conductance	67	6.10	8.80	7.48	7.44
pH	16	0.23	1.35	0.73	0.74
NO_3–N	50	7.30	15.80	10.50	10.95
Total DO					
PEPACTON	42	0.50	14.00	3.00	3.93
Total chloride	23	1.60	3.70	2.40	2.40
Silica	0	—	—	—	—
Sediment	3	16,100.20	276,149.00	157,945.00	150,065.47
Stream flow	21	10.20	132.00	80.00	72.20
Conductance	38	6.54	8.50	7.37	7.26
pH	17	0.22	1.41	0.73	0.76
NO_3–N	23	7.60	13.30	10.50	10.33
Total DO					

Note: Units for total chloride, silica, sediment, total DO, and NO_3–N are mg l^{-1}. Units for conductance, pH, and stream flow are micromhos, standard units and $m^3\ h^{-1}$. Parameters in bold type were used in regression analysis.

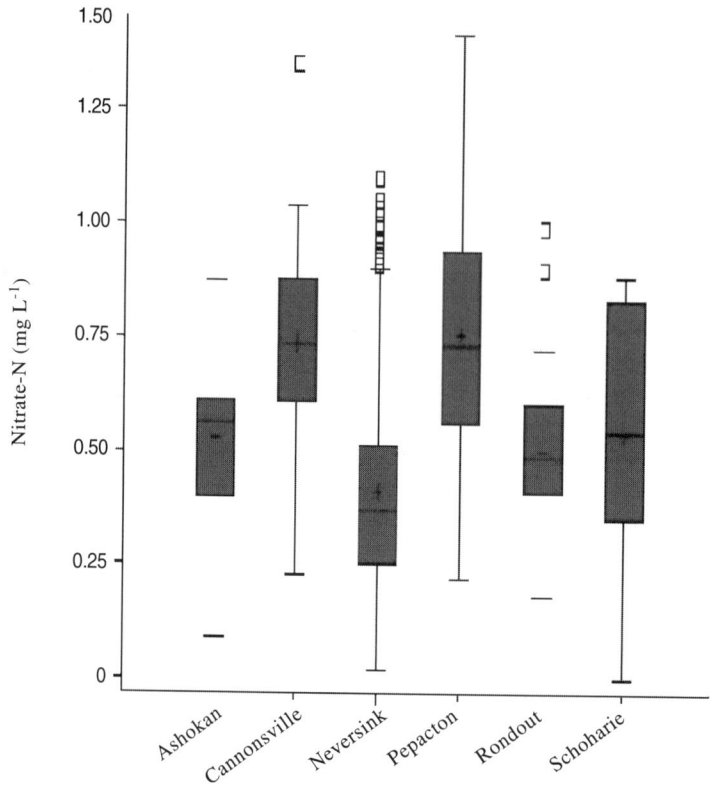

Figure 126.4 Box plot of surface water nitrate nitrogen measurements in the Catskill–Delaware basins.

forest, crop and forest, crop and pasture, and crop and residential. None of the variables accounted for more than 20% of the variation in NO_3–N. Crops were related to NO_3–N as part of an interaction variable during all seasons, except summer. However, summer NO_3–N values were not significantly related to any land-cover variable. Greater percentages of forest, pasture, and crop and forest were associated with lower NO_3–N during the winter and spring months. The interaction terms of crop and pasture and crop and residential were related to higher NO_3–N values during winter and fall.

DISCUSSION

With the exception of a maximum pH in the Cannonsville basin and a minimum pH in the Neversink basin, the water quality parameters during the 1986 to 1994 period of this study never exceeded the maximum contaminant levels (MCLs) set by the New York State Department of Health or the Environmental Protection Agency (NYCDEP, 1997). Considering the fairly limited human use in most of the Catskill–Delaware basins, water quality was expected to be fairly high. The low residential and agricultural land use in the basins is related not only to presence of the Park and Forest Preserve, but also to low population density. However, the concentration of human use in the riparian zones of rivers and streams remains a potential source of pollution. New York City has set as a first priority land acquisition around the reservoirs to ensure that potential sources of pollution are not located near the water supply (Murphy et al., 1995; MOA, 1997; Ashendorff et al., 1997). Second priority has been given to lands around the Esopus River that feeds the Ashokan Reservoir. Results from this study support the priority area. The Ashokan basin and subbasins intersecting Esopus Creek

Table 126.4 Stepwise Regression for Surface Water Quality and Land Cover and Use Metrics

Water Parameter	Season	Variable(s)	Direction of Relation	Model P > F	Model R²	Partial R²
Total chloride	Winter	Crop/Residential	+	0.0001	0.11	
	Spring	Forest	−	0.002	0.22	
	Summer	Pasture	+	0.03	0.71	0.51
		Crop/Pasture	−	0.006		0.06
		Residential/Pasture	−	0.01		0.05
		Forest	−	0.0005		0.04
		Forest/Pasture	−	0.03		0.04
	Fall	Residential/Pasture	+	0.0001	0.57	0.41
		Forest/Pasture	+	0.027		0.09
		Crop/Forest	+	0.0001		0.06
		Residential	−	0.03		0.05
Silica	Winter	Forest	−	0.0001	0.48	
	Spring	Forest	−	0.0002	0.49	0.38
		Forest/Pasture	+	0.04		0.06
		Residential/Pasture	−	0.05		0.06
	Summer	Crop/Pasture	+	0.001	0.44	
	Fall	Forest	−	0.001	0.62	0.46
		Crop/Pasture	+	0.001		0.16
pH	Winter	Forest	−	0.001	0.20	
	Spring	Forest	−	0.001	0.37	0.32
		Forest/Pasture	+	0.09		0.05
	Summer	Forest	−	0.02	0.21	
	Fall	Pasture	+	0.04	0.23	
NO₃–N	Winter	Crop/Residential	+	0.006	0.39	0.18
		Crop/Pasture	+	0.008		0.14
		Crop/Forest	−	0.09		0.06
	Spring	Crop/Forest	−	0.03	0.28	0.15
		Pasture	−	0.04		0.14
	Summer	–		ns		—
	Fall	Forest	−	0.01	0.28	0.20
		Crop/Pasture	+	0.06		0.08

Note: Symbol (/) denotes an interaction of two land cover/use metrics. Dashes and ns refer to nonsignificant results at $P \leq 0.10$ ($n = 25$).

were found to have the highest percentage of residential land use (Figure 126.5a). Regression analyses suggest that residential land use was associated with higher NO_3–N and total chloride values (most likely due to contributions from sewage treatment plants and leach field effluent release into the river). Third priority is land purchase or protection around the West Branch of the Delaware River. The riparian zones around the West Branch have the highest agricultural usage and the second highest residential use (Figure 126.5a,b). In our analysis, greater pasture and row crops were related to higher NO_3–N values in the fall and winter months. In spring, pasture and the interaction between crop and forest had a negative relationship with NO_3–N (most likely due to uptake during spring greenup). The relationship between human use and water quality became stronger as the scale decreased from basin to subbasin and riparian zones. Other third-priority areas for land acquisition include subbasins in the Pepacton and Schoharie basins. These selected areas coincide with our findings of subbasins having high agricultural and residential land use. Land in the Neversink and Rondout basins fall in the fourth-priority level for acquisition. These two basins had the highest forest cover, lowest human use, and, except for low pH levels, better overall water quality values. The link between greater forest cover and lower pH was evident in results from the correlation and regression analyses (Table 126.4). The pH values observed in the Neversink have been noted in other surface water studies and linked to the effects of acidic deposition and a decrease in soil and stream acid-neutralizing capacity (Stoddard, 1991; Murdoch and Stoddard, 1992).

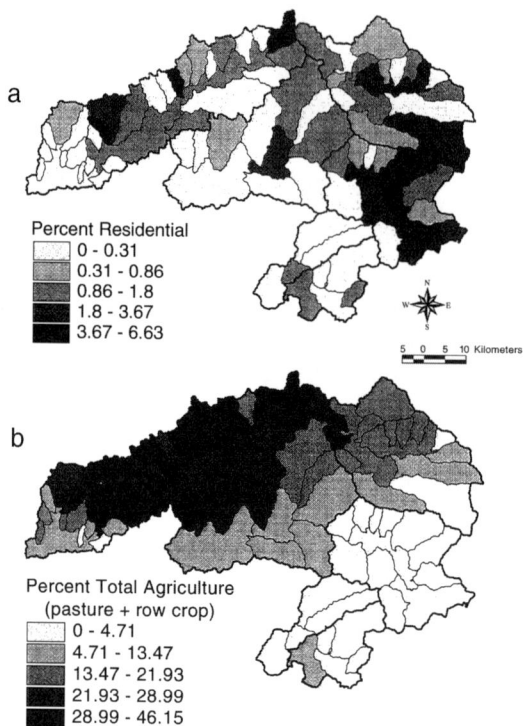

Figure 126.5 Percent of (a) residential and (b) total agriculture in a 100-m riparian buffer by subbasin in the Catskill–Delaware basins. Increased percentages of land use correspond to darker colors.

In regression analysis, land-cover metrics are often analyzed separately to determine their relationship to water quality. Results from this study suggest that combinations of land covers are important in explaining specific water quality parameter measurements. Furthermore, because of differences in vegetative productivity, nutrient cycling, and climate, the influence of upstream (100 m) land cover may vary with season. Besides spatial and seasonal influences on the relationship between land cover and water quality, other potential synergistic components are past land cover history, acid deposition, soils, and geology. In the Catskill–Delaware basins, acid deposition has been shown to contribute to low pH in the Rondout and Neversink basins. In the Schoharie and Ashokan basins, streams like Stoney Clove and Batavia Kill have a history of high sediment and turbidity. However, sediment contributions more likely come from eroding glacial clay lakes than from current agricultural practices (Rich, 1934; Miller, 1970). Analysis of different seasons and spatial scales provided a more-detailed description of the association between land cover and changes in water quality. This study was a first step in examining land-cover influences on water quality in the basins of the Catskill–Delaware basins. Future research could better defined these relationships by conducting a multiyear study.

ACKNOWLEDGMENTS

The authors would like to thank all the people who have participated in the creation of this study. Special thanks goes to the staff of the New York City Department of Environmental Protection, particularly Barbara Dibler, who willingly shared her knowledge of GIS, the Catskill–Delaware basins, and its community whenever we visited. Thanks also goes to members of the Landscape Ecology Branch, many of whom provided invaluable assistance in the production, assessment, processing, analyzing, and editing of this chapter. Without these people this study could not have been completed.

The Environmental Protection Agency, through its Office of Research and Development (ORD), funded and performed the research described. This manuscript has been subjected to the EPA's peer and administrative review and has been approved for publication. Mention of trade names or commercial products does not constitute endorsement or recommendation by the EPA for use.

REFERENCES

Ashendorff, A., Principe, M.A., and Mantus, J., Watershed protection for New York City's supply, *J. Am. Water Works Assoc.,* 89, 75–75, 1997.

Comello, R.L., Paul, J.F., August, P.V., Copeland, J., Baker, C., Hale, S.S., and Latimer, R.W., Relationships between watershed stressors and sediment contamination in Chesapeake Bay estuaries, *Landscape Ecol.,* 11, 307–319, 1996.

EROS Data Center, New York Land Cover Data Set, available at ftp://edcftp.cr.usgs.gov/pub/edcuser/vogel/ states/newyork_readme_ver9807.txt (11/1/1998) ref, July 1998.

Featherstone, J., Conservation in the Delaware River Basin, *J. Am. Wat. Works Assoc.,* 88, 42, 1996.

FGEIS, Final Generic Environmental Impact Statement for the Proposed Watershed Regulations for the Protection from Contamination, Degradation, and Pollution of the New York City Water Supply and Its Sources, prepared for the New York City Department of Environmental Protection, Valhalla, NY, 1993.

Finlayson, C.M., Production and major nutrient composition of three grass species on the Magela floodplain, Northern Territory, Australia, *Aquat. Bot.,* 38, 163–176, 1991.

Hansen, G., New York State Water Quality 1996, NYS Department of Environmental Conservation, Division of Water, New York, 1996.

Hecker, G., The New York City water supply: past, present and future, Civil Engineering Practice, *J. Bos,* 6, 7, 1991.

Jones, K.B., Riitters, K.H., Wickham, J.D., Tankersly, R.D., O'Neill, R.V., Chaloud, D.J., Smith, E.R., and Neal, A.C., An Ecological Assessment of the United States Mid-Atlantic Region, EPA/600/R-97/130, U.S. Environmental Protection Agency, Office of Research and Development, Washington, D.C., 1997.

Kuchler, A.W., *Potential Natural Vegetation of the Conterminous United States, Map and Manual,* American Geographical Society, Special Publication 36, New York, 1964.

Lowrance, R., Todd, R., Fail, Jr., J., Hendrickson, Jr., O., Leonard, R., and Asmussen, L., Riparian forest as nutrient filters in agricultural watersheds, *BioScience,* 34, 374–377, 1984.

Lowrance, R., Sharpe, J.K., and Sheridan, J.M., Long-term sediment deposition in the riparian zone of a coastal plain watershed, *J. Soil Water Conserv.,* 41, 266–271, 1986.

Malanson, G.P., *Riparian Landscapes,* Cambridge University Press, Cambridge, U.K., 1993.

Miller, W.J., *The Geological History of New York State,* Kennilcat Press, Port Washington, NY, 1970.

MOA, Watershed Memorandum of Agreement, NYCDEP Office of Watershed Communications, Kingston, NY, 1997.

Murdoch, P.S. and Stoddard, J.L., The role of nitrate in the acidification of streams in the Catskill Mountains of New York, *Water Resourc. Res.,* 28, 2707–2720, 1992.

Murphy, S., Land acquisition for water quality protection: New York City and the Catskills Watershed System, *Water Resourc. Update,* Summer, 60, 1995.

New York City Department of Environmental Protection (NYSDEP), 1997 New York City Drinking Water Supply and Quality Statement, New York, 1997.

Okun, D.A., Craun, G.F., and Rose, J.B., New York City: to filter or not to filter?, *J. Am. Water Works Assoc.,* 89, 62, 1997.

O'Neill, R.V., Krummel, J.R., Gardner, R.H., Sugihara, G., Jackson, B., DeAngelis, D.L., Milne, B.T., Turner, M.G., Zygmunt, B., Christensen, S.W., Dale, V.H., and Graham, R.L., Indices of landscape pattern, *Landscape Ecol.,* 1, 153–162, 1988.

Osborne, L.L. and Wiley, M.J., Empirical relationships between land use/cover and stream water quality in an agricultural watershed, *J. Environ. Manage.,* 26, 9–27, 1988.

Rich, J.L., *Glacial Geology of the Catskills,* New York State Museum Bulletin, New York, 1934.

SAS. *SAS/SAT User's Guide, Version 6,* 4th ed., Vol. II, SAS Institute, Inc., Cary, NC, 1990.

Stave, K.A., Resource conflict in New York City's Catskill watersheds: a case for expanding the scope of water resource management, *Am. Water Resourc. Assoc.*, 95, 61–67, 1995.

Stoddard, J.L., Trends in Catskill stream water quality: evidence from historical data, *Wat. Resourc. Res.*, 27, 2855–2864, 1991.

United States Bureau of the Census, 1990 U.S. Census of Population and Housing: New York State, Washington, D.C., 1990.

United States Department of Agriculture, Soil Survey Manual, Agricultural Handbook 18, U.S. Department of Agriculture, Washington, D.C., 1951.

van Valkenburg, N.J., *The Forest Preserve of New York State in the Adirondack and Catskill Mountains: A Short History,* Purple Mountain Press, New York, 1996.

Wall, G.R., Riva-Murray, K., and Phillips, P.J., Water quality in the Hudson River basin, New York and adjacent states, 1992–95, U.S. Geological Survey Circular 1165, 1998.

Weidner, C.H., *Water for a City,* Rutgers University, Quinn & Boden, Newark, NJ, 1974.

CHAPTER 127

Using a Knowledge Base to Assess the Effects of Stream Stressors

Lester Yuan

INTRODUCTION

Watershed management is rapidly becoming the preferred method for controlling the damage done to natural systems by human activities. This approach focuses on managing the activities of people residing and working in a given catchment through stakeholder involvement and incentive-based methods, instead of through traditional command-and-control regulations (U.S. EPA [Environmental Protection Agency], 1996; NAS, 1999). Watershed management has arisen partly in response to changes in the dominant stressors upon stream ecological integrity, as attention has shifted in recent years from controlling individual point source releases to addressing nonpoint sources of degradation. Nonpoint source stressors, such as habitat loss, invasive species, and nutrient runoff, are much more difficult to control than point sources, as they are dispersed across large areas and usually involve many different stakeholders.

Computational models that link human activities to specific stressors and to specific ecological effects have great potential as tools for facilitating watershed management. First, a robust, predictive model would allow managers to rapidly assess various management alternatives for their effects upon the ecosystem. Second, models could enhance monitoring efforts, as they could be used to extrapolate measurements from a finite number of field sites to all areas of the watershed. Managers would then be able to use these results to identify regions in a watershed that are either minimally impacted and therefore candidates for preservation, or that are highly degraded and to be targeted for restoration work. Finally, models could facilitate discussions at stakeholder meetings and focus groups by illuminating the pathways along which certain human activities eventually cause degradation in aquatic systems. To fulfill these potential applications, models must be reasonably reliable and must include most major watershed processes.

Many models for watershed processes exist, ranging from those predicting precipitation runoff to those predicting fish population dynamics. Unfortunately, current models fall short of their promise as management tools, primarily because scientists still lack a quantitative understanding of many of the processes by which human activities affect stream ecology. Much of the existing knowledge regarding the response of stream biota to different stressors remains largely qualitative. A second factor that inhibits greater adoption of models for management purposes involves a lack of holistic models that include most watershed processes, particularly those that relate physical stressors and human activities to biological end points. For example, the modeling package BASINS includes modules for hydrology and the in-stream evolution of toxics, but lacks a component linking the hydrological and toxic calculations to biological responses (Lahlou et al., 1998). Because many states and regions are gradually moving toward using biological indicators of water quality (Karr and Chu, 1999), developing holistic models that combine physical and ecological processes seems particularly important. Intuitively though, the prospect of building a model that encompasses all human activities and all ecological end points in a given watershed is a daunting task.

In this chapter, I propose that a knowledge base may be a viable, alternative modeling approach. A knowledge base is a means of representing expert knowledge using linguistic rules. In the application discussed in this chapter, the knowledge base provides a coarse level of modeling that allows one to represent qualitative knowledge and to adopt a more holistic approach to the problem. To assess its potential for application to the field of watershed management, I have designed a demonstration knowledge base that relates some of the human activities in a watershed to stream stressors. Many relationships between human activities and stream ecological effects have been greatly simplified for this knowledge base prototype, as the main goal of this work is to demonstrate the potential of this approach.

KNOWLEDGE BASES

Knowledge bases, also commonly known as expert systems, have been used for many years in the computer science community wherein researchers have sought methods for representing expert knowledge in computer programs. Recently, they have received increasing attention as a possible alternative to traditional modeling approaches for both physical and biological systems (Silvert, 1997; Wenger et al., 1990; Mitra et al., 1998). A knowledge base has two basic components: a set of linguistic rules that defines relationships between variables in the system, and an inference engine that uses the rules to draw conclusions using input data. Each rule is structured as follows:

IF condition THEN conclusion.

Condition and *conclusion* are defined by simple logical clauses in the following form:

$$\text{IF } X \text{ is } A \text{ THEN } Y \text{ is } B \tag{127.1}$$

where X and Y are variables, and A and B are words that quantify X and Y. For example, in the clause, "urban land use is high," the phrase "urban land use" is the variable, and the word high quantifies urban land use. A complete rule using this clause might be "IF urban land use is high, THEN pollutant level is high." The word high only quantifies urban land use approximately, so to define explicitly the meaning of the word high, one uses fuzzy logic.

Fuzzy logic was developed to address situations in which defining absolute rules that decide between different options is difficult or impossible. For example, if one were to define the meaning of the word tall, a reasonable definition could be that anyone over 6′ in height would be considered "tall." However, with this definition someone who is 5′11″ would never be classified as tall, which is clearly a limiting situation. Fuzzy logic addresses this difficulty by capturing the approximate logic people employ in making decisions. Instead of clearly etched values of true or false, fuzzy logic permits one to specify gradations of truth. That is, if true is defined numerically as 1, and false is defined as 0, then fuzzy logic allows values that lie between.

These gradations of truth are represented by relationships known as membership functions. The term *membership function* refers to the idea that we must explicitly define the degree to which objects belong in a group. In the previous example, a person who is 5′11″ is a partial member of the group of tall people. Returning now to the urban land-use example, the notion that urban land use is high is expressed as a function that relates the fraction of urban land use in a catchment to

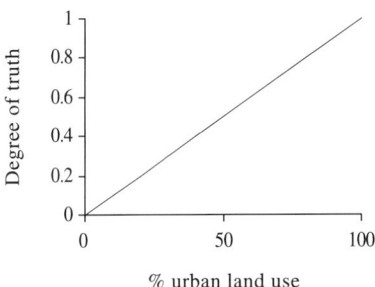

Figure 127.1 Membership function for "urban land use is high."

the degree of truth (Figure 127.1). In this case, a catchment with no urban land use would imply that the degree of truth for the statement "urban land is high" is 0, while a catchment entirely composed of urban land use would generate a degree of truth of 1. Intermediate values of urban land use would produce intermediate degrees of truth. In general, any functional form can be used to specify the membership function; however, in the current work membership functions have been restricted to trapezoidal shapes for simplicity.

The second component of the knowledge base, the inference engine, uses the rule base to draw conclusions from an input database. A significant amount of literature on various inference techniques exists, and the reader is referred to Zimmermann (1991) for complete coverage. Details of the inference methods employed for this work are provided in the example in Zimmermann's Section 4.

METHODS AND MATERIALS

Study Area

The Patapsco River Basin (Figure 127.2) spans two distinct physiographic provinces. The headwaters of the Patapsco lie in the ridges of the Piedmont Province, where the soils are derived from granite rock and consist of loams and clays with rock fragments and gravel. As it flows south

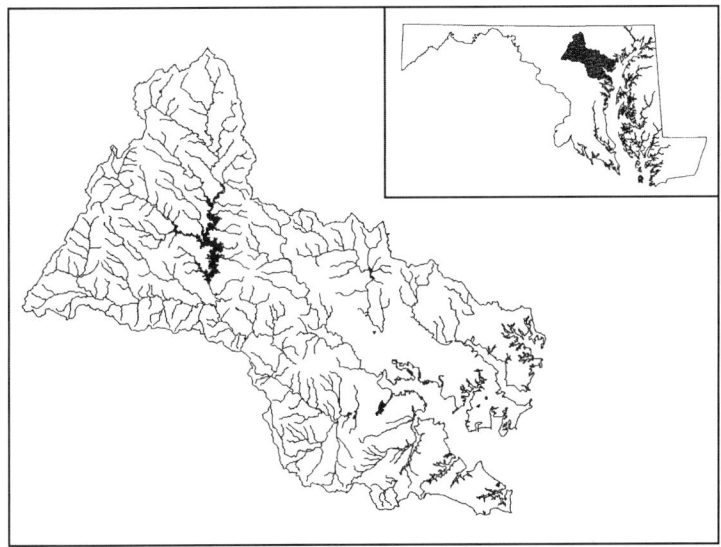

Figure 127.2 Patapsco River Basin, Maryland (inset).

and east, the river enters the Northern Coastal Plain, which is characterized by sands, sandy loams, and silt loams resulting from sea deposits. Land use in the watershed is predominantly agricultural in the northwest, but changes to urban development as one approaches metropolitan Baltimore in the southeast.

Data

Because of the holistic approach employed in this study, the data requirements are large and diverse. For land-cover information, the Multi-Resolution Land Characteristics (MRLC) digital data set is used. This data set was developed by a consortium of federal agencies and is based on LandSat 1991 to 1993 Thematic Mapper images. The data classify land cover into 15 different categories and have a resolution of 30 m. Access to the data was provided by EPA Region 3 (Delaware, Maryland, Pennsylvania, Virginia, West Virginia, and the District of Columbia) and the EPA Office of Water. Delineation of the boundaries of the Patapsco River Basin was provided by the Maryland Natural Resource Conservation Service, while the delineations of catchments corresponding to individual sampling sites were computed by the author from 30-m digital elevation data obtained from the U.S. Geological Survey (USGS). Point source data were obtained from the EPA Permit Compliance System, which is now housed in the Envirofacts warehouse (http://www.epa.gov/enviro/html/ef_overview.html).

Local observations of stream biota and physical habitat were obtained from the Maryland Biological Stream Survey (Roth et al., 1997, 1998). This statewide survey is conducted by the Maryland Department of Natural Resources and employs a stratified random sampling design to assess the biological condition of the nontidal small (first through third order) streams of Maryland. Sampling teams visited the Patapsco River basin in two different years, measuring 61 sites in 1995 and 68 sites in 1996. At each site, teams sampled both fish and benthic macroinvertebrates, and conducted systematic surveys of the quality of the physical habitat (Roth et al., 1997). Indices of biotic integrity based on fish (Roth et al., 1997) and benthic communities (Stribling et al., 1998) were developed using accepted multimetric index procedures. These indices have been shown to consistently distinguish between least impacted *reference* sites and sites that show substantial human influence. Although the benthic index of biological integrity (B-IBI) measures ecological condition relative to a predetermined reference condition, in this chapter it will be used as a surrogate for ecological integrity. I adopt the definitions of Karr and Chu (1999), who define ecosystems with good ecological integrity as those systems that are pristine and whose biota are the result of natural evolutionary and biogeographic processes.

Knowledge Base Development

Because a knowledge base consists only of a set of semantic rules, developing the rules requires one to elicit the qualitative knowledge of experts in the appropriate field. In many cases the knowledge-elicitation process consists of extensive interviews with experts, whose responses are then combined to form a collective set of rules (Chang et al., 1998). Because the primary focus of this current project is only to illustrate the usefulness of knowledge bases in a new arena, it was decided that multiple expert opinions would not be solicited. Instead, the rule base was determined primarily by the qualitative knowledge of the author and by informal consultation with Versar, Inc. and EPA staff familiar with the Maryland data. The simplicity of this approach substantially reduced the time required to develop the knowledge base.

Conceptual Model and Rule Base

The conceptual model illustrated in Figure 127.3 provides a qualitative representation of the linkages between human activities and stream ecology, based on an extensive literature review (see,

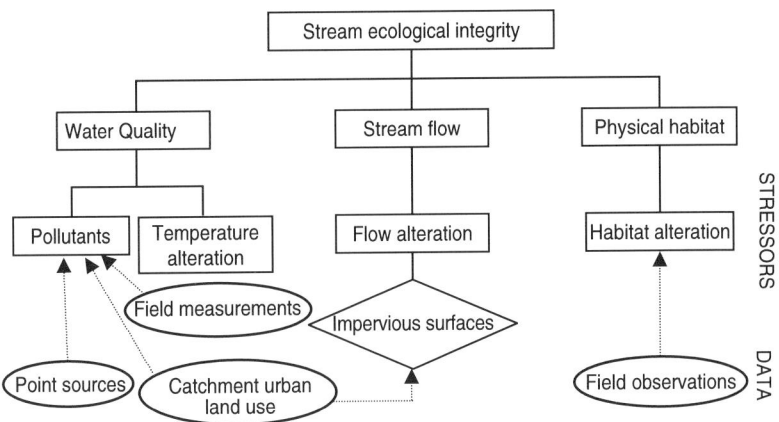

Figure 127.3 Conceptual model.

for example, Schlosser, 1991; Richards and Host, 1994; Richter et al., 1996; Allan et al., 1997). The first level of the conceptual model broadly asserts that the ecological integrity of a stream depends on three factors: water quality, physical habitat, and stream flow. Many researchers include a fourth category that includes ecological effects such as predation and migration (Karr, 1991), but this level of ecological detail is beyond the scope of this work. At the next level in the figure, different stressors are listed that fall within the main groups. This list is not exhaustive, as the stressors listed are only those that are observed to be dominant in the Patapsco basin. The list has also been shortened to reflect the coarse modeling approach of this pilot study. So, instead of listing different stressors for nutrients, toxics, and dissolved oxygen, these factors are all grouped together as *pollutants*. At the final level of the figure, data sources are shown enclosed within ovals. Arrows link these data to the appropriate stressor. One final entry in the conceptual model is *impervious surfaces*. It is enclosed in a diamond to indicate that it is an intermediate value, distinguishing it from directly measured data in ovals and final stressor levels in boxes. Again, the list is not a complete accounting of all possible sources of data, but merely represents those data that are currently available.

The qualitative rule base (listed in Table 127.1) follows directly from the conceptual model, further clarifying the linkages between data sources and stressors and between stressors and ecological condition. Each rule is composed of a condition and a conclusion. Intermediate levels in the conceptual model are represented by intermediate variables. For example, setting pollutant level is an action at the lowest level, but pollutant level then becomes one of the variables used in the conditions for determining water quality.

The knowledge base software used for this work is the Ecological Management Decision Support System (Reynolds et al., 2000). This system integrates a knowledge base inference engine with a geographical information system; thus, it allows one to use geospatial data as inputs to the knowledge base.

Calibration

Once the qualitative linguistic rules were established, the knowledge base was calibrated by adjusting the bounds of individual membership functions, using data from the 1995 Maryland Biological Stream Survey (MBSS) in the Patapsco. Calibration procedures differed depending on whether a given membership function was used in the condition clause or in the conclusion clause in the rule base. For those that were used in the condition clause, calibration procedures depended on the spatial scale at which the variable is relevant, i.e., catchment scale stressors affect all sites, while local stressors affect only a small fraction of the sites, so the calibration techniques applied

Table 127.1 Rule Base

Stream Ecological Integrity

If water quality is good and flow flashiness is low, then ecological integrity is good.
If physical habitat is poor or water quality is poor or flow flashiness is high, then ecological integrity is poor.

Physical Habitat

If minimum score is poor and mean score is poor, then habitat is poor.
If minimum score is good and mean score is good, then habitat is good.
If minimum score is fair or mean score is fair, then habitat is fair.

Stream flow

If impervious surfaces are high, then flow flashiness is high.
If impervious surfaces are low, then flow flashiness is low.

Water Quality

If pollutant level is low, then water quality is good.
If pollutant level is high or measurements are poor, then water quality is poor.

Pollutants

If point source is near and urban land use is high, then pollutant level is very high.
If point source is near or urban land use is high, then pollutant level is high.
If urban land use is low, then pollutant level is low.

Measurements

If DO is low or NO_3 is high, then measurements are poor.
If DO is high and NO_3 is low, then measurements are good.

---

to each were quite different. For the membership functions used in conclusion clauses, additional constraints existed which limited the types of functions that could be used.

Catchment Scale Stressors

Urbanization is the main human activity affecting stream ecological integrity in the Patapsco. Its effects can be seen in Figure 127.4, in which B-IBI decreases with increased catchment urban land fraction. The amount of urbanization in a catchment can affect stream biota in many ways, and two main mechanisms are represented in the conceptual model. First, urban catchments are

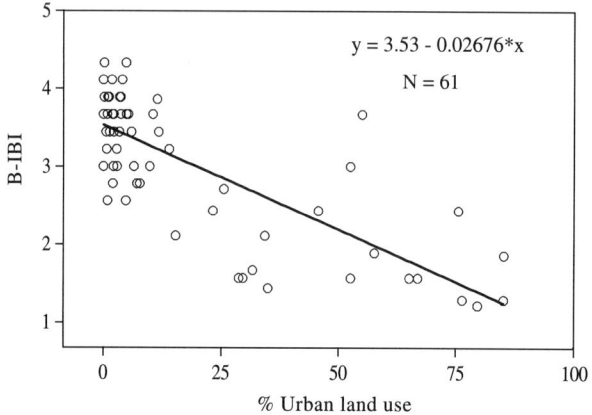

$$y = 3.53 - 0.02676 \cdot x$$

$$N = 61$$

Figure 127.4 Fraction of urban land use vs. B-IBI. Regression relation shown on graph.

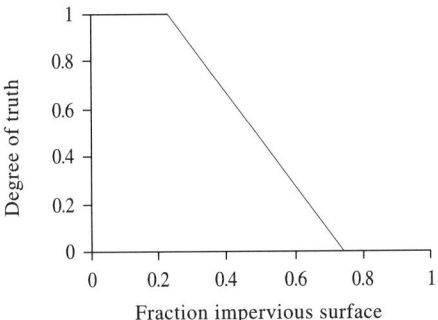

Figure 127.5   Membership function for "fraction impervious surface is low."

covered by more impervious surfaces; therefore, they alter runoff volumes and change flow patterns in the stream. Second, the precipitation runoff from urban areas can carry significantly higher concentrations of pollutants than runoff from forested catchments. So, urban land cover links directly to stream flow alteration and to pollutant level in the rule base and in the conceptual model. To determine appropriate bounds for the membership function expressing "high urban land use," the regression line relating urban land use to B-IBI was used, but only to identify the general trend in the data, as the membership function does not translate directly to a regression line. The final version of the membership function for urban land use takes the form of a straight line, varying smoothly from no urban land use to 100% land use (Figure 127.1). This illustrates the basic approach employed in establishing the membership functions, in which the simplest relation was initially assumed. Substantial evidence of deviations from the simple linear relationship had to be observed before the basic form was altered.

The fraction of impervious surface can be computed from land-cover data using estimates for impervious surfaces (Roth et al., 1998). As mentioned previously, urban land use correlates strongly with the amount of impervious surfaces in the catchment, which impacts runoff patterns and can lead to substantial flow alteration. Here again, regression lines relating the fraction of impervious surface to B-IBI were examined and used as guidance in determining membership functions. The range of values for the fraction of impervious surface in a catchment was 0.20 to 0.65 and the membership function (Figure 127.5) reflects the reduced range.

Local Stressors

Local scale stressors presented difficulties for calibration, as they generally affected only a small fraction of sites within the study area. Only 4 sites of the 61 sampled in 1995, for example, have at least one point source within 1000 m upstream. A similarly small number have low habitat assessment ratings or poor water quality measurements. Because of these small sample sizes, calibrating the membership functions was, by necessity, somewhat *ad hoc*. However, the model itself retains its validity because the rule base remains grounded in a qualitative understanding of the relationships. The three local stressors considered in this work were habitat quality, local water quality measurements, and the presence of point source releases.

Habitat quality was assessed by the MBSS in nine different categories (in-stream habitat; epifaunal substrate; velocity and depth diversity; quality of pools, glides, and eddies; riffle quality; channel alteration; bank stability; aesthetic rating; and remoteness), each on a 20-point scale (Roth et al., 1997). To aggregate the habitat scores into a smaller number of ecologically meaningful statistics, the minimum and mean value for the nine observations were calculated for each site. This combination of statistics captures situations in which one category is rated particularly poorly, as well as situations in which the overall habitat quality of the site is poor. To determine membership functions, the aggregated habitat scores were plotted vs.

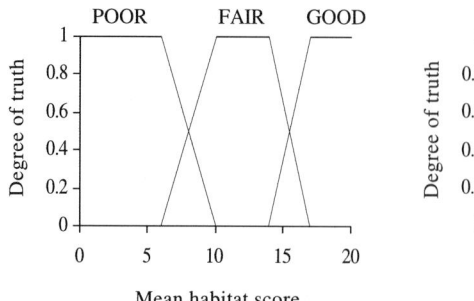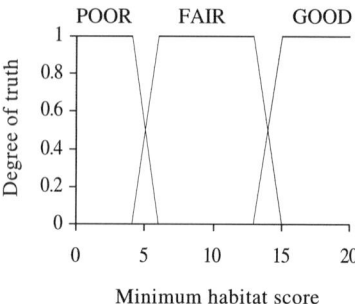

Figure 127.6 Membership functions for mean and minimum habitat scores.

B-IBI, and the trends noted. In contrast to the catchment scale stressors, ecological integrity does not increase consistently with increasing values of habitat quality. Instead, threshold values exist, below which B-IBI values are depressed. These threshold values are reflected in the steep changes in the membership functions shown in Figure 127.6. Other habitat variables, which are not rated on a 20-point scale (e.g., percent embeddedness), were also measured in the MBSS, but these observations have not been included in the knowledge base.

Limited water chemistry measurements were also available from the MBSS data, including dissolved oxygen (DO) and nitrate measurements. Membership functions were initially specified for concentrations of these compounds based upon published water quality criteria (U.S. EPA, 1986, 1999). These values were then slightly adjusted to reflect observations from the Patapsco data. The threshold value for DO was set at 6 ppm, below which a significant drop in B-IBI values was observed. Similarly, the threshold value for nitrate was set at 10 mg/l, above which B-IBI scores were depressed. Membership functions that use these threshold values, representing agreement with the statements "nitrates are high" and "DO is low" are depicted in Figure 127.7.

Point sources, as mentioned earlier, were close to only a small number of sites within the study area. As with other factors, the treatment of point sources is very simplistic for this prototype knowledge base. The knowledge base tests only for the presence or absence of a single point source within a fixed distance upstream from the sampling site. Other relevant factors, such as the composition and quantity of released effluents and the status of compliance with release permits, are not included. To define the membership function for "point source is near" B-IBI scores were again examined at all sites for which any point sources lay upstream. Of these sites, only those with point sources within 1000 m showed significant degradation. This threshold distance was then used to define a membership function, in which the degree of truth decreases linearly with increasing distance (Figure 127.8).

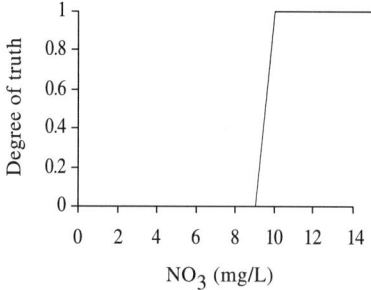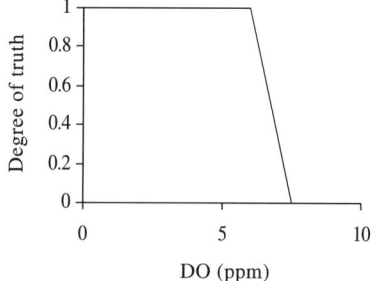

Figure 127.7 Membership functions for "nitrates are high" and "dissolved oxygen is low."

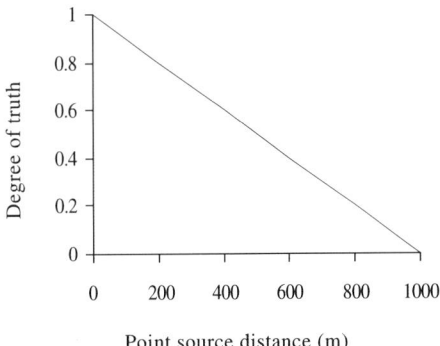

Figure 127.8 Membership function for "point source is near."

Intermediate Variable Membership Functions

As alluded to earlier, intermediate membership functions are constrained by two factors. First, limitations in the capabilities of the current knowledge base software restrict the range of possibilities for the membership functions located in the conclusion part of the rule; all membership functions must be the same rectangular shape (as shown in Figure 127.9). The only variation allowed is in the location of the function along the horizontal axis. Second, all intermediate level variables represent abstract concepts, rather than measurable quantities. In theory, one could devise a way of measuring a variable such as pollutant level as some aggregated index of individual toxics and nutrients. However, in the present knowledge base, pollutant level refers more to an assessment of the various human activities that give rise to waterborne pollutants, rather than an actual value. Because of these two constraints, all membership functions for conclusion variables are defined as ordinal numbers. For example, pollutant levels can be either low or high, so these are assigned ordinal values of 1 and 5, respectively. Evaluating the pollutant level portion of the rule base then leads to values that lie between 1 and 5 and that characterize the relative amount of pollutants produced by activities in the catchment. Because the final result of the knowledge base is itself an ordinal value, using ordinal values in the intermediate stages does not reduce the effectiveness of the model.

EXAMPLE EVALUATION OF A FUZZY RULE

The max-min compositional rule (Zimmermann, 1991) has been chosen as the method for evaluating fuzzy rules in this study. To illustrate this method consider the following example. Suppose we would like to evaluate the rule "IF urban land use is high OR point source is near

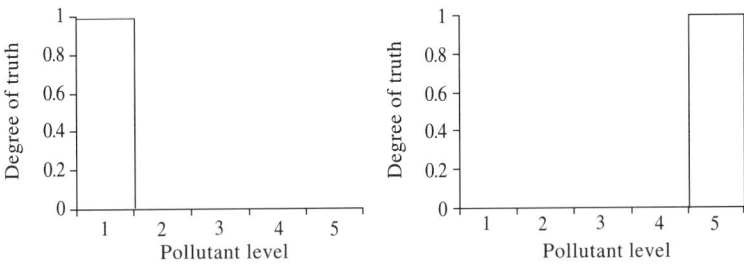

Figure 127.9 Membership functions for "pollutant level is low" and "pollutant level is high."

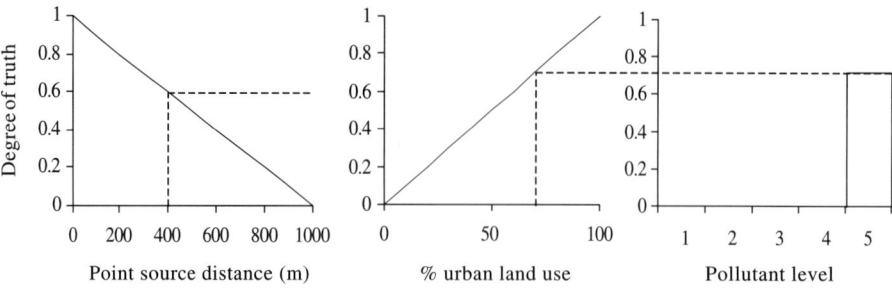

Figure 127.10   Graphical illustration of rule evaluation.

THEN pollutant level is high." Data sources provide numerical values for the fraction of urban land use in the catchment and for the distance to the nearest point source. For this example, take these values to be 72% and 400 m, respectively. To evaluate the rule, we use the appropriate membership functions to determine the degree of truth (DT) for each of the two antecedent conditions. To account for the OR conjunction, we then take the maximum degree of truth of all antecedent conditions (Zimmermann, 1991). In this case, "urban land use is high" has a DT equal to 0.72, while "point source is near" has a DT equal to 0.6, so the DT of the antecedent conditions is 0.72. The rule therefore produces a conclusion that the DT of the statement "pollutant level is high" is also 0.72. We represent this conclusion in terms of membership functions by truncating the membership function for "pollutant level is high" at DT = 0.72. This entire process is shown graphically in Figure 127.10.

Overall pollutant level is represented by a group of different rules, each with a different truncated membership function. In the rule base listed in Table 127.1, three different rules predict pollutant level. With the max-min compositional technique, the union of the results of the three rules determines the membership function for overall pollutant level. Thus, the resulting membership function for pollutant level can be a collection of irregular shapes (Figure 127.11). A single value for pollutant level is required for subsequent computations in the knowledge base, and so we must reduce the irregularly shaped membership function to one number. A variety of techniques exist for this process, more details of which can be found in Zimmermann (1991). For this study, the center of mass method is used, which reduces the membership function to a single value by computing the location of the center of mass. In Figure 127.11, two distinct parts of the membership function are evident, a rectangle with pollutant level of 1 and DT = 0.30, and a rectangle with pollutant level of 5 and DT = 0.72. The center of mass (1 * 0.30 + 5 * 0.72) of this membership function is then 3.9.

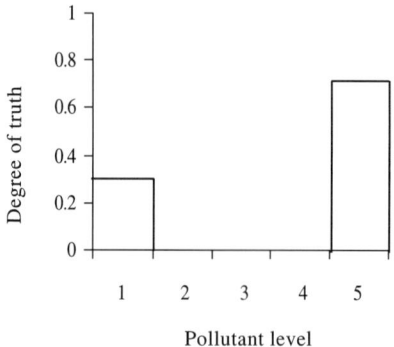

Figure 127.11   Union of all rules predicting pollutant level.

Table 127.2 Narrative Descriptions for Knowledge Output

KB Output	Narrative Description
1–2	Poor
2–3	Marginal
3–4	Suboptimal
4–5	Optimal

Because pollutant level is an intermediate value, the numerical value for pollutant level is then used as an antecedent condition in subsequent rules. In this way, the effects of widely disparate data sources can be combined to draw a single conclusion. The final result from the knowledge base provides a value ranging from 1 to 5 that characterizes the degree to which human activities stress the stream biota at that site. Narrative ratings are assigned to whole number intervals (Table 127.2). The assessment values span the same range as B-IBI scores to facilitate comparison.

RESULTS

Calibration

The ability of the knowledge base to distinguish between four different levels of human influence is illustrated in Figure 127.12 for the data collected in 1995. Ideally, one would want a perfect separation, in which sites classified by the knowledge base as poor would have B-IBI ranging only between 1.0 and 2.0; sites classified as optimal would have B-IBI between 4.0 and 5.0; and sites classified as marginal or suboptimal would also only have B-IBI between 2 and 4. However, classification by the knowledge base is not perfect, despite the fact that the knowledge base was calibrated using this data set. Nevertheless, several aspects of the results are encouraging; at the extreme ends of the spectrum (poor and optimal), the distribution of B-IBI values is quite narrow, so it appears that the knowledge base is correctly identifying sites that also lie at the extreme upper and lower ends of the range of possible B-IBI. Classification at the intermediate ranges (marginal

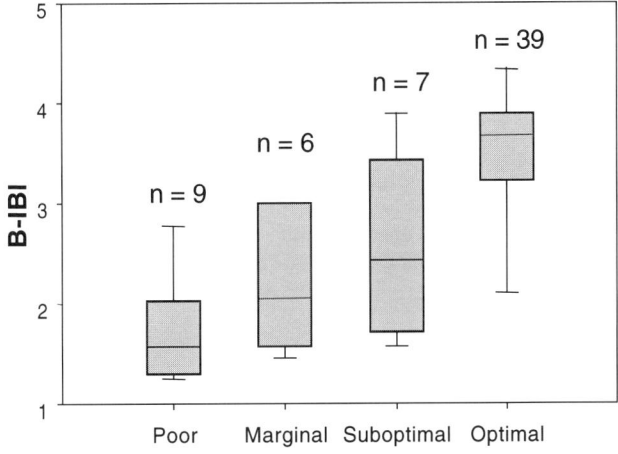

Figure 127.12 Box plot showing knowledge base results compared to B-IBI values for 1995. The horizontal boundaries of each box indicate the locations of the 25th percentile, the median, and 75th percentile, respectively. The bars below and above each box indicate the locations of the minimum and maximum values.

and suboptimal) are encouraging as well. Here, the spread of the data is noticeably higher, but the median values in the intermediate range increase monotonically and lie between median values for poor and optimal sites. This monotonic increase in median B-IBI through all four classifications indicates that, on the average, the knowledge base is identifying a gradient of human influence that is reflected by the biological indicator.

One can also evaluate the knowledge base performance at a coarser level by classifying sites into only two categories: less impacted sites with knowledge base ratings of 3.0 or greater, and degraded sites with ratings below 3.0. Under this coarse classification scheme, 78% of those sites the knowledge base classified as less impacted had B-IBI greater than 3.0, while 87% of sites classified as degraded had B-IBI less than 3.0. The difference in classification efficiencies between less impacted and degraded sites reflects the conservative approach taken in the design of the knowledge base, as it was thought that if misclassifications were to occur, it would be preferable to err on the side of classifying bad sites as good. Therefore, the knowledge base is incorrect more often in misclassifying degraded sites.

Incorrectly classified sites are present at all levels. These discrepancies can be attributed to two possible sources. First, natural systems are inherently temporally variable, so incorrectly classified sites may have had particularly high or low measurements on a given day. The magnitude of the temporal variability in the B-IBI has not yet been adequately quantified. Second, all possible human activities or stressors are not included in the knowledge base. As with any model, the knowledge base is limited by the knowledge of the modeler, so this preliminary version of the knowledge base is not a complete representation.

Validation

A validation test for the knowledge base was performed by running it with 1996 data from the Patapsco. Because of the random sampling monitoring design of the MBSS, sites sampled in 1996 were all different than 1995. In other respects (variables measured, sampling procedure, etc.), 1996 data were identical to 1995 data. A comparison between knowledge base predictions and B-IBI values computed from 1996 data is shown in Figure 127.13. Overall, the knowledge base classifications were higher than measured B-IBI. This overestimation was particularly evident in the sites classified as optimal. Of the 32 sites classified by the knowledge base as optimal, only two had B-IBI greater than

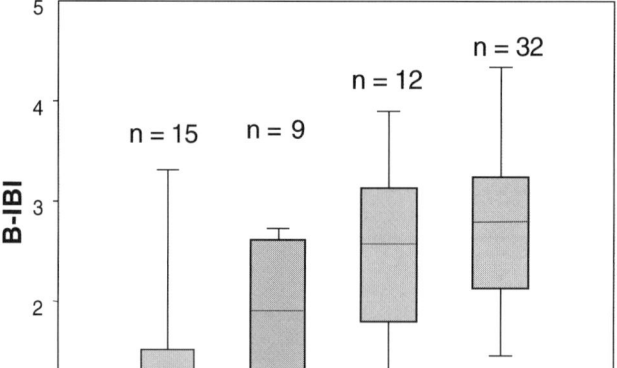

Figure 127.13 Box plot showing knowledge base results compared to B-IBI values for 1996.

4.0. The same discrepancies were present when results are aggregated into only two categories. 43% of sites classified by the knowledge base as least impacted had B-IBI less than 3.0. Conversely, 96% of sites classified as degraded by the knowledge base also had B-IBI less than 3.0.

Despite the general trend toward overestimation, several positive features in the validation results could still be observed. First, median B-IBI values again showed a monotonic increase as one moved from poor to optimal, so the knowledge base was still predicting a gradient of human influence which was mirrored in the measured biotic integrity. Second, the classification efficiencies were heavily skewed toward incorrectly classifying degraded sites as least impacted. Thus, the knowledge base predictions were conservative, as desired.

Comparison to Simple Regression Approaches

Many models relating human activity to ecological end points are based on linear regressions, so one can ask whether the knowledge base approach improves on linear regression models. For this case, in particular, much of the variance in B-IBI could be attributed to urban land use, so a regression model using urban land cover would seem to produce results that were comparable to the knowledge base. To address this question, a linear regression model was computed for 1995 data, and the model tested with 1996 data. The regression with the 1995 data yielded the following predictive equation,

$$\text{BIBI} = 3.53 - 0.268U \qquad (127.2)$$

where U is the fraction of urban land use in the catchment ($r^2 = 0.58$, $p < 0.0001$). This same predictive equation was then applied to the 1996 data. To facilitate comparison to the knowledge base results, the predicted results were compared to field measurements using the same format as in Figures 127.12 and 127.13; i.e., the data were binned into four categories based on predicted B-IBI scores, and box and whisker plots presented for the distribution of actual measurements within each bin (Figure 127.14). In both years the linear regression model produced results that were less accurate than those from the knowledge base. First, a monotonic behavior was not observed in validation results from 1996. Clearly, other factors in addition to urban land use had a strong effect on ecological integrity. Second, the linear regression model classified many more sites in the optimal group than the knowledge base, and so the range of values observed in this group was quite large. This behavior was expected because a large number of sample sites had small urban land use fractions. This example, however, illuminates the limitations of using a regression approach. Both of these shortcomings may be mitigated by regressing against additional variables, but a multiple linear regression analysis was beyond the scope of this current work.

DISCUSSION

In this pilot study, the knowledge base has been shown as a viable alternative to other modeling approaches for predicting the effects of human activities on stream ecological integrity. However, much further work remains before knowledge bases can be considered for general use in watershed management. One immediate concern is the failure of the knowledge base to accurately predict ecological integrity as measured by B-IBI in both the calibration and the validation data sets used for this work. Several points are relevant to this issue. First, scientific explanations for much of the variance in the B-IBI in the Patapsco do not at present exist. The knowledge base used for this work is, admittedly, simple in design, but additional stressors that could account for the unexplained variance were not obvious from the literature. The current lack of complete scientific explanations does highlight an advantage of the knowledge base approach, which is that as new information becomes available, it can be easily incorporated into the existing knowledge base.

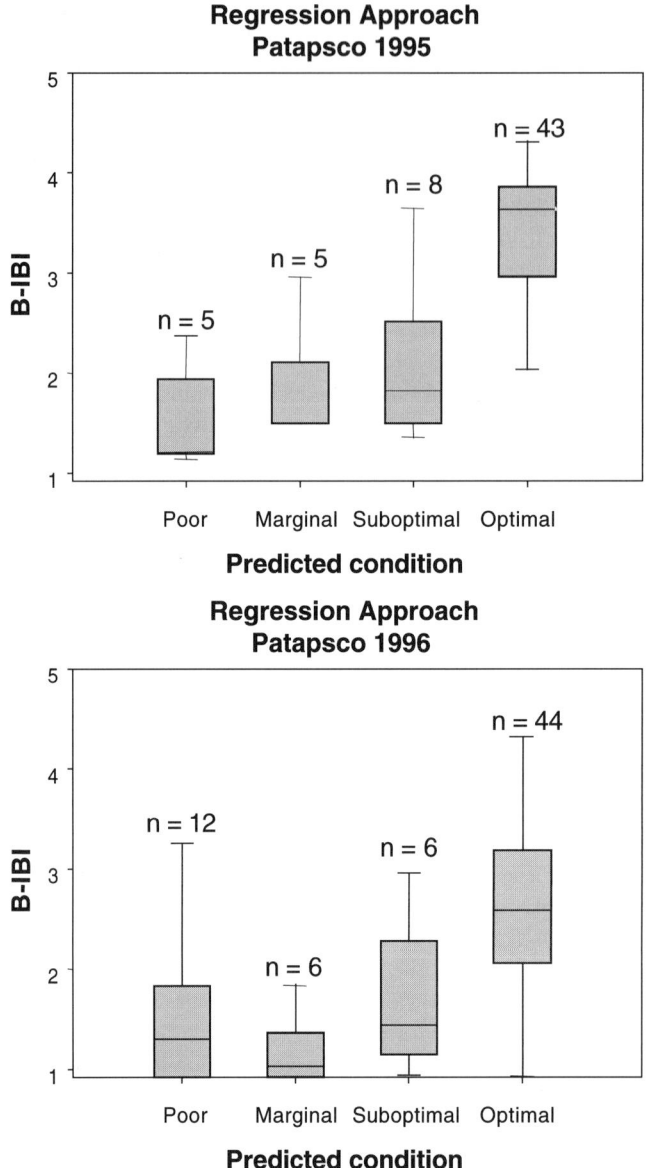

Figure 127.14 Box plots showing results of simple linear regression model.

A second point is that B-IBI is an indicator of a biological system; therefore, a certain amount of temporal variance in the measurements is unavoidable. A means of assessing the performance of the knowledge base relative to a background of temporal variance would be useful. The MBSS is conceived as an ongoing monitoring effort, so as additional samples are taken, an estimate of natural temporal variability may be possible.

Third, natural stressors may be responsible for some of the discrepancies between predictions and measurements in the validation data. The main discrepancy one notices in the 1996 validation plots is the overestimation of ecological condition by the knowledge base. However, part of the overestimation may be a result of lower B-IBI scores. Indeed, if we statistically compare scores in 1995 with 1996 (using only those sites with urban land uses less than 10% to reduce variance), we find that the mean B-IBI score in 1995 was 3.53, while the mean in 1996 was significantly lower, at 2.72

($p < 0.001$). The reason for this general decrease in B-IBI may be attributable to harsher weather conditions in 1996. Weather records from the National Climatic Data Center (http://www.ncdc.noaa.gov/) reveal that 1996 had heavier and more frequent rainfall than 1995. The effects of more frequent and energetic floods upon benthic macroinvertebrates are uncertain, but in certain situations, a decrease in B-IBI might be expected. Also, the records show that the Baltimore area was hit by a late winter storm, March 8–9, 1996, in which temperature dropped to lows of approximately 10°F (–26°C) on two consecutive days. Because most of the samples in the Patapsco were taken within 10 days after the storm, this late freeze may have contributed to a general reduction in B-IBI. The knowledge base in this study focused on anthropogenic sources of stress, so these natural stressors might explain some of the lack of agreement between knowledge base results and field measurements.

In the future, the ability of a single knowledge base to successfully predict ecological integrity across different watersheds and between different ecoregions must be demonstrated. The current version of the knowledge base has been calibrated to match data from the Patapsco River basin, so sets of stressors and human activities included in its rule base are not only abbreviated but are unique to the study area. To apply the same knowledge base to watersheds in other ecoregions, the rule base must be expanded to encompass a broader set of stressors. Expanding the rule base in this way should not affect previously established rules, so the generality of the knowledge base can be increased in a natural way. Ultimately, a knowledge base would be most useful if it could accurately predict stream ecological integrity in an arbitrary ecoregion.

The limitations imposed by the software used for this study have been quite severe. Other inference engines that allow more general definitions of membership functions in the conclusion clauses may enhance the performance of the knowledge base. Future work will include an evaluation of software alternatives for the knowledge base.

This pilot study has demonstrated that knowledge bases can be a promising approach for watershed modeling. They provide a very flexible modeling environment that can represent qualitative scientific knowledge and integrate disparate sources of data. The easily interpreted linguistic rules on which the model is based can facilitate model development, and as new information emerges, it can be readily incorporated into existing knowledge bases.

ACKNOWLEDGMENTS

This chapter was prepared while the author was an American Association for the Advancement of Science Fellow at the National Center for Environmental Assessment of the U.S. Environmental Protection Agency. The author gratefully acknowledges many useful discussions with Jeff Frithsen (U.S. EPA); also, Nancy Roth, Mark Southerland, and Ginny Mercurio (Versar, Inc.) provided valuable assistance in acquiring MBSS data and interpreting results. The Maryland Biological Stream Survey is a statewide stream sampling program conducted by Paul Kazyak and the Maryland Department of National Resources, Monitoring and Non-Tidal Assessment Division, and the author is grateful for their efforts in collecting this comprehensive set of data. The author also acknowledges Keith Reynolds (Forest Service, USDA) and Jim Andreasen (U.S. EPA) for providing the EMDS software. The views expressed herein are those of the author and do not necessarily reflect the views or policies of the U.S. Environmental Protection Agency. Mention of trade names or commercial products does not constitute endorsement or recommendation for use.

REFERENCES

Allan, J.D., Erickson, D.L., and Fay, J., The influence of catchment land use on stream integrity across multiple spatial scales, *Freshwater Biol.*, 37, 149–161, 1997.
Chang, Y.-H., Yeh, C.-H., and Cheng, J.-H., Decision support for bus operations under uncertainty: a fuzzy expert system approach, *Omega Int. J. Manage. Sci.*, 26, 367–380, 1998.

Karr, J.R., Biological integrity: a long-neglected aspect of water resource management, *Ecol. Appl.*, 1, 66–84, 1991.

Karr, J.R., and Chu, E.W., *Restoring Life in Running Waters: Better Biological Monitoring,* Island Press, Washington, D.C., 1999.

Lahlou, M., Shoemaker, L., Choudhury, S., Elmer, R., Hu, A., Manguerra, H., and Parker, A., BASINS 2.0 Users Manual, EPA-923-B-98–006, Washington, D.C., 1998.

Mitra, B., Scott, H.D., Dixon, J.C., and McKimmey, J.M., Applications of fuzzy logic to the prediction of soil erosion in a large watershed, *Geoderma,* 86, 183–209, 1998.

National Academy of Science, New strategies for America's watersheds, Committee on Watershed Management, National Research Council, Washington, D.C., 1999.

Reynolds, K.M., Jensen, M., Andreasen, J., and Goodman, I., Knowledge-based assessment of watershed condition, *Comput. Elec. Agric.*, 27, 315–333, 2000.

Richards, C. and Host, G., Examining land use influences on stream habitats and macroinvertebrates: A GIS approach, *Water Resourc. Bull.,* 30, 729–738, 1994.

Richter, B.D., Baumgartner, J.V., Powell, J., and Braun, D.P., A method for assessing hydrologic alteration within ecosystems, *Conserv. Biol.,* 10, 1163–1174, 1996.

Roth, N.E., Southerland, M.T., Chaillou, J.C., Volstad, J.H., Weisberg, S.B., Wilson, H.T., Heimbuch, D.G., and Seibel, J.C., Maryland biological stream survey: ecological status of non-tidal streams in six basins sampled in 1995, Report CBWP-MANTA-EA-97-2, Maryland Department of Natural Resources, Annapolis, 1997a.

Roth, N.E., Chaillou, J.C., and Gaughan, M., Guide to using 1995 Maryland biological stream survey data, Report no. CBWP-MANTA-EA-97-3, Maryland Department of Natural Resources, Annapolis, 1997b.

Roth, N.E., Southerland, M.T., Chaillou, J.C., Wilson, H.T., Heimbuch, D.G., and Seibel, J.C., Maryland biological stream survey: ecological status of non-tidal streams sampled in 1996, Report CBWP-MANTA-EA-98-1, Maryland Department of Natural Resources, Annapolis, 1998.

Schlosser, I.J., Stream fish ecology: a landscape perspective, *BioScience,* 41, 704–712, 1991.

Silvert, W., Ecological impact classification with fuzzy sets, *Ecol. Model.,* 96, 1–10, 1997.

Stribling, J.B., Jessup, B.K., and White, J.S., Development of a benthic index of biotic integrity for Maryland streams, Report CBWP-EA-98-3, Maryland Department of Natural Resources, Annapolis, 1998.

U.S. EPA, Ambient water quality criteria for dissolved oxygen, Report 440/5–86–003, National Technical Information Service, Springfield, VA, 1986.

U.S. EPA, Watershed '96: Moving ahead together: Plenary proceedings, Report EPA84–R-97–002, National Technical Information Service, Springfield, VA, 1996.

U.S. EPA, National recommended water quality criteria: Correction, Report 822-Z-99–001, National Technical Information Service, Springfield, VA, 1999.

Wenger, R.B., Rong, Y., and Harris, H.J., A framework for incorporating stream use in the determination of priority watersheds, *J. Environ. Manage.,* 31, 335–350, 1990.

Zimmermann, H.-J., *Fuzzy Set Theory and Its Applications,* 2nd ed., Kluwer Academic Publishers, Boston, 1991.

SECTION III.5

The Langat Basin of Malaysia

CHAPTER 128

Overview: The Langat Basin of Malaysia

Nicholas W. Lerche

Ecosystem health is an evolving, integrative field that explores the interrelations among human activity, social organization, natural systems, and human health (Rapport et al., 1998). One product of such integration is the development of a holistic approach to ecosystem health assessment and environmental management (Rapport et al., 1998). This collection of five chapters focuses on several key aspects of ecosystem health assessment and environmental management of the Langat Basin, a rapidly developing region in the state of Selangor, Malaysia. The principal authors, all of whom are affiliated with the Institute of Environment and Development (LESTARI) of the University of Kebangsaan, Malaysia, present a body of research documenting historical trends and the current status of critical components of the basin's ecosystem, including land cover and biodiversity, industrial development and economic growth, and water quality, as well as a novel consideration of geohazards. The results of this research provide baseline data essential not only for assessing the current state of the ecosystem, but also for identifying useful indicators of ecosystem health or dysfunction, and for evaluating the efficacy of any preventive, palliative, or restorative interventions.

LESTARI's long-term project in the Langat Basin is a valuable case study of the application of ecosystem health concepts on a regional basis to the evaluation and management of a large-scale tropical ecosystem (Nordin and Azrina, 1998). The Langat Basin is located adjacent to the Klang Valley, the most highly developed urban center in Malaysia, and site of the nation's capital, Kuala Lumpur. Proposed development projects in the Langat Basin — including a new administrative capital and a new international airport — will undoubtedly induce a large and rapid influx of commercial and industrial development in the basin, with resultant changes in landscape, water and air quality, demography, and demand on regional resources and ecosystem services (Nordin and Azrina, 1998). To the extent that the Langat Basin is representative of other rapidly developing regions throughout Asia that are facing similar ecosystem stressors and perturbations, the research methodologies presented in this collection of chapters may contribute to a conceptual framework for ecosystem health assessment and management of other large-scale tropical ecosystems.

The main goal of LESTARI's long-term project in the Langat Basin is to establish and monitor the health of the basin's ecosystem through research, training, and policy influence, ultimately contributing to a sustainable pattern of development in the basin (Nordin and Azrina, 1998). This research represents a significant contribution toward achievement of that objective.

REFERENCES

Nordin, M. and Azrina, L.A., Training and research for measuring and monitoring ecosystem health of a large-scale ecosystem: The Langat Basin, Selangor, Malaysia, *Ecosyst. Health,* 4, 188–190, 1998.

Rapport, D.J., Ecosystem health: An integrative science, in *Ecosystem Health,* Rapport, D.J. et al., Eds., Blackwell Science, London, 1998, pp. 3–50.

Rapport, D.J., Costanza, R., and McMichael, A.J., Assessing ecosystem health, *Trends Ecol. Evol.,* 13, 397–402, 1998.

CHAPTER 129

Ecosystem Health in Malaysia:
A Case Study of the Langat Basin

**M. Nordin, Joy Jacqueline Pereira, Ahmad Fariz Mohamed, Saiful Arif Abdullah,
Ibrahim Komoo, Shahruddin Idrus, Pauzi Abdullah, Abdul Hadi Harman Shah,
Abdullah Samat, Rospidah Ghazali, W. Mohd Muhiyuddin, and Muhammad Abu Yusuf**

INTRODUCTION

Tropical ecosystems are among the most complex in the world. Defining their health is a difficult task but concepts, approaches, and methods emerging from the literature on ecosystem health increasingly provide insights on how this complexity can be properly examined and understood. This large-scale ecosystem health study in the Langat Basin of the state of Selangor, Malaysia, aims to find and provide a rational framework within which the health of a tropical ecosystem may be assessed and monitored. The work is funded by the United Nations Environment Program, Regional Office for Asia and the Pacific (UNEP/ROAP) with funds from the Danish Cooperation on Environment and Development (DANCED). It is part of UNEP/ROAP's initiative on training and research on environmental management (TREM) in Thailand and Malaysia.

The Langat Basin is situated south and adjacent to the Klang Valley, Malaysia's highly developed urban conurbation where the nation's capital, Kuala Lumpur, is located (Figure 129.1). It has an area of about 2930 km^2 and a population of about 962,649 in 1998. It comprises a watershed catchment area to the northeast along the ridge of Banjaran Titiwangsa (Main Range) of Peninsular Malaysia and a coastal fringe along the Straits of Malacca to the southwest. Being adjacent to Kuala Lumpur, the basin has seen rapid expansion in areas developed for housing, industrial estates, and business centers over the last decade (Nordin and Azrina, 1998). This has largely been in response to the need for urban housing as well as industrial, business, and commercial property development in the basin.

Several large federal government projects, such as the Kuala Lumpur International Airport in Sepang; the new federal government administrative capital at Putra Jaya; and Malaysia's "cyber city," Cyber Jaya, are located within the basin. These development projects have spurred an influx of commerce and industry into the basin, affecting not only the price of land, but also the pattern of human settlements, landscapes, and ecology.

With the exception of the Putra Jaya and Cyber Jaya areas, development in the Langat Basin typifies, in general, the pattern of urban and industrial development in other parts of Malaysia and Southeast Asia. Urban sprawl extends from existing urban centers, in this case Kuala Lumpur, Petaling Jaya, Shah Alam, and Puchong to the north of the basin, into areas that are less developed,

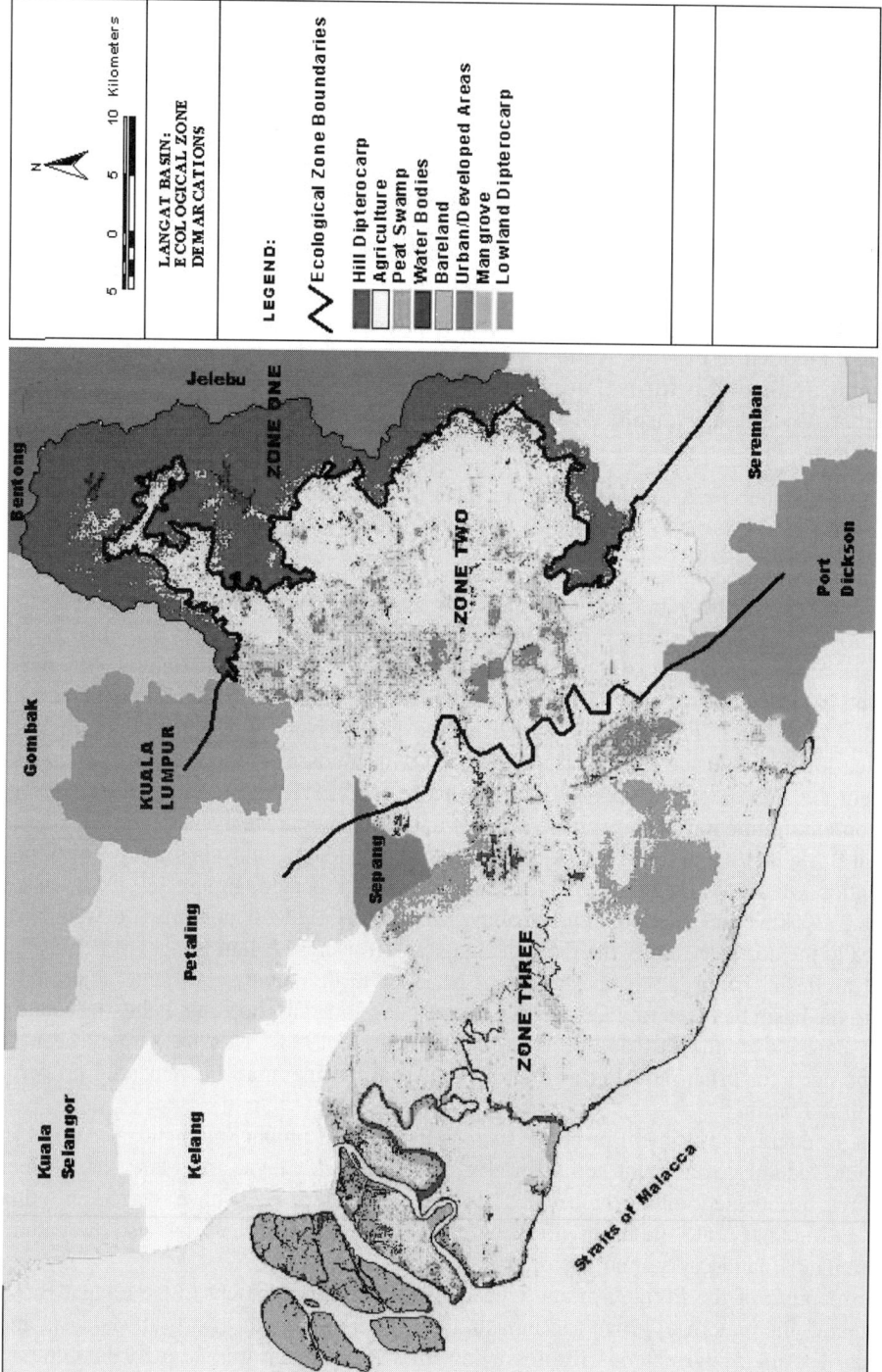

Figure 129.1 The Langat Basin and its ecological zones.

when new infrastructure (the most important being roads, highways, and bridges) is put in place. This leads to the conversion of agricultural land and, in some instances, forest and other natural areas, into housing and industrial estates, as well as business and commercial centers. Such development leads to the establishment of new townships, which then coalesce with new growth from existing towns and villages. These areas form new urban centers that replace natural and agricultural landscapes.

The Langat Basin ecosystem is undergoing rapid change. In the 1960s and 1970s, much of its original lowland dipterocarp and peat swamp forests were lost through conversion to agricultural plantations, tin-mining areas, and human settlements (Wong, 1974). More recently, its natural areas have come under increasing pressure from urbanization and industrialization. Its rivers have become increasingly polluted from industrial, agricultural, and domestic wastes. Rivers have also been realigned and concrete lined to enable more rapid flow rates. Increases in areas of bare and developed land with increasing urban growth cause increased surface runoff. River alignment is also undertaken to minimize riverbank erosion. Large areas of its coastal zone have also been converted to ports and sites for development of industrial parks and aquaculture farms.

The purpose of this chapter is to provide an overview of the features of the Langat Basin and describe a proposed framework for assessing and monitoring the health of the basin ecosystem. Patterns of land cover, ecological zones, population, industrial sites, river water quality, agriculture, water supply and demand, geohazards, mining activities, forests, and wildlife including fish are briefly described. Criteria and indicators of ecosystem health within the methodological framework needed for assessing and monitoring the health of the Langat Basin ecosystem are still under construction. Some of these criteria and indicators are presented in this chapter.

GENERAL LAND COVER

A 1996 LANDSAT thematic mapper, seven-band satellite image was used to prepare a digital base map of the basin with land-cover calculations. Satellite data of the basin were geocoded and projected to a RSO grid using Kertau 1948 datum. The image was initially classified using unsupervised classification. In this way, 30 classes were generated and used as a guide in selecting sites for the supervised classification process. From this, eight initial land cover categories were identified using the maximum likelihood method followed by post-classification processing (Lillesand and Kiefer, 1987). Recent land-cover and land-use maps, as well as simple ground-truthing, were used to aid classification. The results of the original classification were modified during the recoding stage. The final results of the classification scheme were achieved after obtaining the consensus viewpoints from researchers conducting studies in the Langat Basin and remote-sensing experts familiar with the study area. Statistics for each land-cover category were generated from the satellite image.

Land cover in the basin is predominantly agricultural (Table 129.1) with almost 64% of the basin under this category. Large-scale rubber and oil palm plantations are found especially on the coastal plain. Small-scale farms with individual ownership and rural areas and villages where the home-garden and orchard (*kebun*) typically surround each house make up the rest of the agricultural landscape.

Only about 25% of the total area of land in the basin is covered by forests, the majority (15.7% of basin area) is hill dipterocarp forest in the eastern and northeastern parts (Figure 129.1). Mangrove forests cover 5.5% of the land area and peat-swamp forests about 4%. Lowland dipterocarp forests have been almost eliminated from the basin. Only 13.2 km^2 (0.5% of area) of lowland dipterocarp forest remained in 1996, mostly as small remnant fragments of what used to be extensive stretches of forests.

Built-up areas occupy 6.8% of the basin but affect the health of the ecosystem in many and disproportionate ways. Solid and industrial waste generated by industrial and urban centers cause river pollution and increased surface water runoffs that extend beyond their immediate areas.

Table 129.1 Land-Cover Categories and Their Area in the Langat Basin

Category	km²	Total Area (%)
All Forests	739.2	25.2
Hill dipterocarp forest	461.4	15.7
Lowland dipterocarp forest	13.2	0.5
Peat-swamp forest	103.0	3.5
Mangrove forest	161.5	5.5
Agricultural	1865.6	63.6
Bareland	47.3	1.6
Developed areas	200.2	6.8
Water bodies	79.4	2.7
Total	2931.7	99.9

Bare land constituted just about 1.6% of the total land area of the basin in 1996. However, its impact on the ecosystem, especially on water quality of the rivers of the basin, is extensive and severe. Total suspended solids (TSS) values from river water in the basin frequently exceed the Department of Environment's recommended threshold value of 150 mg/l required for the maintenance of aquatic life, and can reach more than 1700 mg/l. High sediment loads are especially prevalent immediately after rain.

The total area of the basin made up of water bodies is 79.4 km², or just 2.7% of the land area. These consist of the surface area of rivers, two major reservoirs (Langat and Semenyih reservoirs) and ex-mining ponds created by open-cast tin mines.

ECOLOGICAL ZONES

The Langat Basin may be divided into three main ecological zones based on its geomorphology and physiognomy (Figure 129.1). The northeastern portion of the basin is hilly and forested, and population densities are low to nonexistent. The demarcation for this zone is the hill-foot boundary that distinguishes the forested hills from the cultivated lowlands. By definition, hill dipterocarp forests are found above the 300-m contour (Whitmore, 1984) but the hill-foot boundary starts at a lower elevation where the land rises sharply above the river floodplain. Vegetational and faunal characteristics, especially of birds, differ markedly across this boundary (Medway and Wells, 1976). The rivers in this section of the basin are relatively unpolluted and all fall within Class II (slightly polluted) of the Malaysian River Water Quality Criteria and Standards (Department of Environment, 1986, 1990, 1994). This section of the basin is designated as Ecological Zone 1 in this study.

The central section of the basin is undulating, with hills and knolls interspersed with relatively flat land. Very little natural forest cover is left and it is the most urbanized section of the basin. Population density in this section is the highest in the basin. The rivers in this section of the basin are badly polluted with chemicals and sediments (Yusuf et al., 2000). They fall within the category of Class III (polluted) and Class IV (highly polluted). It is designated as Ecological Zone 2.

The western and coastal section of the basin is almost entirely flat with the exception of Jugra Hill (232 m). Population density is relatively low and urban centers are widely scattered. Agricultural plantations occupy most of the intervening space. Two large patches of peat-swamp forests (Kuala Langat North Permanent Forest Reserve and Kuala Langat South Permanent Forest Reserve) occur inland, while the coastline is intermittently vegetated by mangrove swamp and strand forests. The Langat River in this section of the basin meanders through the alluvial plain. It falls under the river category of Class IV with high sediment and pollutant loads. It is designated as Ecological Zone 3.

POPULATION DENSITY

The estimated overall population density in the Langat Basin in 1999 was about 289 persons/ km^2 but population distribution is uneven. Densities are highest (562 persons/km^2) in the central section of the basin designated as Ecological Zone 2, especially in areas adjacent to the Federal Territory of Kuala Lumpur and the southern parts of the Klang Valley. Population density in Ecological Zone 1 is 217 persons/km^2, while in Zone 3 it is 143 persons/km^2.

MANUFACTURING INDUSTRIES

Manufacturing industries are located primarily in the central section of the basin, coinciding with the area designated as Ecological Zone 2. A total of 24 of 30 industrial estates found in the Langat Basin are located in this zone in 1997. This zone also has the highest population densities and some of the most polluted sections of the rivers and tributaries within the basin. In 1996, fewer than 10% of all factories located in the basin had installed effluent treatment systems (Fariz and Nordin, 1999). Most industrial estates are located adjacent or close to major rivers and many small- and medium-scale enterprises (SMEs) are located close to the banks of rivers. The discharge of effluent from SMEs is difficult to quantify accurately.

AGRICULTURAL ACTIVITIES

Agricultural land stretches from the coastal alluvial plain in Ecological Zone 3 to the hill-foot boundary in Ecological Zone 1. Oil palm and rubber plantations occupy the majority of the agricultural land. Agricultural smallholdings, as well as orchards and home gardens, make up the rest. Use of pesticides and inorganic fertilizers in agriculture can contribute pollution of rivers and the level of phosphates in some sections of the river in Ecological Zone 2 exceeds the minimum allowable threshold set in the Interim Water Quality Standards of Malaysia (INWQS) for the well-being of aquatic life. Livestock farms are also a major source of river pollution. Discharge of untreated animal waste from farms causes high biological oxygen demand (BOD), ammoniacal nitrogen, and TSS levels in river water. There are more than 400 livestock farms in the basin, most of which are found in Ecological Zones 2 and 3.

MINING

Tin was mined on a wide scale in the Langat Basin, and tin mining used to be a major economic activity prior to the 1980s. Production has since declined substantially and only three mines using dredges are currently in operation.

There are at present 21 granite quarries, 3 clay pits, 43 earth materials, and 1 kaolin extraction site in the basin. The majority of these are located in Ecological Zone 2. Quarrying for earth materials, clay, and granite creates bare land and accelerates soil loss. There are also 86 sand and gravel extraction sites. Sand mining from rivers increases total suspended solids and turbidity of rivers (Rozali et al., 1999).

There are no quarries in Ecological Zone 1 and only one clay pit is found in Ecological Zone 3.

RIVER WATER QUALITY

The quality of water in the rivers of the Langat Basin is variable. Estimation of overall quality is problematical as water quality monitoring stations have fixed locations and several factors determine water quality at any one time. These include river flow, pollutant discharge rates and frequencies, type of land use in surrounding areas, and prevailing weather conditions. Thus, location and river characteristics influence detectable pollution loads and a river's ability to assimilate pollution. Based on the results of 12 monthly samplings, five stretches of the Langat River may be classified as Class II, another five Class III, and the remaining four stretches as Class IV. Only rivers categorized as Class II and less are considered healthy rivers. Although water from Class III rivers can continue to supply drinking water by treatment using conventional water treatment plants, such rivers are ecologically stressed and do not support healthy populations of aquatic life.

Rivers in Ecological Zone 1 are healthy. They are all categorized as belonging to Class II in the Malaysian river classification scheme. Rivers in Ecological Zone 2 are most unhealthy because of high inputs of chemical, biological, and other pollutants from densely populated urban and industrialized areas. Fecal coliform, *Giardia* cysts, and *Cryptosporidium* oocysts were detected at higher levels in rivers in this Ecological Zone compared to other zones (Rohani, 1999). Total suspended solids in the rivers of this zone are also high because of the prevalence of bare land from extensive ongoing infrastructure and other development projects. The river in Ecological Zone 3 are also unhealthy because of pollutants brought downstream from Ecological Zone 2, as well as from within the zone.

WATER SUPPLY AND DEMAND

Water from Langat Basin rivers is used to supply the needs of the Klang Valley as well as the Langat Basin. With further increases in population and industrial activities, water supply in the basin will go into deficit if steps are not taken to further develop its water resources and to make efficient use of water for domestic purposes, as well as for industrial and agricultural production. Loss in the form of nonrevenue water exceeds 40% at present. With further urbanization and industrialization in the basin, there will also be increasing pressure on the availability of raw water of adequate quality for extraction and processing, especially during drought months.

The estimated total demand for water in the Langat River Basin area for the year 2000 was about 468 mld (million liters per day), and this is anticipated to increase to 2141 mld by 2050 (Economic Planning Unit, 2000). The basin currently produces 1052 mld, but 65% is transferred to supply the Klang Valley areas. The estimated population in the Langat River Basin in 2005 is anticipated to be in excess of one million persons. This will double to two million by 2035. Plans are being made to ensure adequate supply to meet the demand for more water in the Langat Basin in the future through an inter-basin transfer of water resources from the adjacent state of Pahang. The Kelau Dam in Pahang, anticipated to be the main reservoir for this interstate water transfer, is planned to be operational in 2007.

GEOHAZARDS

The intensity of landslides is higher in Ecological Zone 2 of the basin, the most developed and urbanized area. The size and volume of material displaced owing to landslides are also the highest in this zone. Estimates of average annual soil loss due to erosion on a basin-wide scale is 4.2 t/km^2/year. In areas being cleared for development soil loss can reach 152 t/km^2/year.

WILDLIFE

Wildlife diversity in Langat Basin remains relatively high (Jasmi, 1998; Abdullah, 1999) because of the varied landscapes and habitat types found in the basin. Major natural habitats include the coastal mangroves along the coast and the peat-swamp forests in Ecological Zone 3, remnant lowland dipterocarp forests in Ecological Zone 2, and the dipterocarp forests of the basin's catchment areas in Ecological Zone 1. Additionally, wildlife adapted to agricultural landscapes continue to survive and urban-areas continue to support viable populations of a limited number of urban adapted species.

Although the mosaic of habitats and ecotypes in the basin is conducive to the perpetuation of a wide variety of wildlife, it will only do so if the existing natural areas are preserved. Threatened species such as tiger (*Panthera tigris corbetti*), elephants (*Elephas maximus*), and Malay tapir (*Tapirus indicus*) are no longer found in Ecological Zone 2. In Ecological Zone 1, wildlife diversity remains relatively high, although logging is ongoing.

FISH DIVERSITY

A total of 41 species of fish belonging to 16 families have been recorded (Samat, 1999). The estimated number of species in Peninsular Malaysia is about 250 (Mohsin and Ambak, 1991). The basin is relatively small compared to other river basins in Malaysia, but fish diversity is moderately high for a basin of this size. Work is in progress to determine species distribution and factors affecting it. In a badly polluted section of the river in Ecological Zone 2, only nine species have so far been recorded (Samat, 1999).

CRITERIA AND INDICATORS OF ECOSYSTEM HEALTH IN LANGAT BASIN

Operationalizing the ecosystem health concept in environmental management is a major challenge. Not many examples exist where ecosystem health has been put to practical use and made the basis of environmental management, especially in tropical countries. Quantitative methods of assessment have yet to be formalized into a framework that will enable replicability across a range of environments. The assessment of ecosystem health is contextual (Rapport et al., 1999). The work undertaken by LESTARI in Langat Basin is of vital importance, as it would provide a basis for testing the applicability and functionality of the concept of ecosystem health in a tropical ecosystem.

A healthy ecosystem is said to be free from ecosystem stress syndrome, resilient, self-sustaining, does not stress neighboring areas, is relatively free of environmental risk factors, economically viable, and supports healthy communities (Rapport et al., 1995). These criteria were explored as the basis for the assessment of ecosystem health of Langat Basin. Ecosystem stress can originate from many causes. With regard to the terrestrial ecosystem in the basin, stress occurs primarily when natural areas are converted to agricultural or built-up areas causing loss of natural landscapes, biological diversity, and soil erosion. With the aquatic ecosystem, stress relates closely to high inputs of sediment, chemicals, and other pollutants into waterways and water bodies.

Two other interdependent criteria were also explored, as they relate closely to ecological well-being. They are ecological integrity and signs of ecosystem dysfunction. Ecological integrity was assessed by determining the degree of diversity of biotic communities in an area using mammal, bird, and fish species as indicators. Ecosystem dysfunction was assessed by noting the frequency of occurrence of extreme events and failure of the environment to adequately provide ecosystem services, such as providing water of adequate quality for extraction and processing as potable water. In the context of the Langat Basin, this includes the frequency with which critical chemical pollutants in the rivers of the basin exceeded IWQS threshold levels, shut down of water treatment plants, water rationing, and incidence of mudflows and slope failures.

Based on these criteria, Langat Basin ecosystems that closely approximate pristine conditions were considered to be of high ecological integrity and thus most healthy. Highly modified nonnatural areas on the other hand would be of low ecological integrity and thus of poor ecological health. They are also usually low in biological diversity and exhibit signs of dysfunction, especially under stress. To maintain a high level of environmental quality in highly modified ecosystems, such as in urban and industrial areas, requires high inputs of energy and resources.

Pristine ecological conditions can be found mostly in Ecological Zone 1 of the basin, where most of the land area is covered with hill dipterocarp forest. This area has high ecological integrity, but is environmentally highly sensitive because of steep terrain. The area around two major water supply impounds, the Langat and Semenyih reservoirs, is legally protected as Catchment Forest to preserve the health and integrity of the forest ecosystem around the reservoirs. Additionally, some forest compartments in the area are designated as Virgin Jungle Reserves (VJR) that serve as a conservatory of the species representative of the forests in the area. These reserves are never logged and remain pristine. Rivers in this area remain relatively healthy and are categorized as only slightly polluted (Class II), mainly by sediments from soil erosion, as no industrial manufacturing or housing are located here. The area is thus ecologically healthy, as it is relatively free from ecosystem stress, does not stress neighboring areas, is relatively free of environmental risk factors, and supports healthy biotic communities. Human population density is relatively low.

An area designated as a VJR is also found within the Kuala Langat South Permanent Forest Reserve in Ecological Zone 3, a large Permanent Forest Reserve (PFR) of peat-swamp forest. However, the peat-swamp forest reserves of Ecological Zone 3 are fundamentally isolated forest islands surrounded at present mainly by oil palm plantations. In the future they are likely to be surrounded by urban and industrial areas and would come under increasing pressure to be converted to other forms of land use.

Some of the in-shore islands of the coast of Langat Basin also remain under pristine mangrove forests. These are ecologically rich environments and contain elements of terrestrial and aquatic diversity characteristic of coastal areas. They provide habitats for coastal biodiversity and function as nursery grounds for coastal fisheries. The coastal mangroves on the mainland of Ecological Zone 3 used to be sustainably harvested, as timber from them is widely used in the construction industry and the production of high-grade charcoal. They have increasingly been replaced by aquaculture farms, or reclaimed for development of industrial warehousing to serve nearby ports, and for other uses including housing estates to serve needs for housing.

The area of the basin designated as Ecological Zone 3 is not free from ecosystem stresses. Much of it has been converted to agricultural plantations growing single crops, and coastal reclamation has transformed natural areas into new housing and industrial estates. Natural resilience has been lost and the area can no longer be regarded as naturally self-sustaining. The draining of large areas of the peat swamp forest causes the peat to dry, and during drought periods it has been known to catch fire, releasing smoke that severely pollutes the air of neighboring areas. The river in this zone is badly polluted (Class IV) from materials originating upstream and from within the zone. However, the area is economically viable and supports healthy human communities. Biotic communities have been compromised, but some diversity remains, especially in the peat-swamp forest reserves and coastal mangroves.

The health of the Langat Basin ecosystem in Ecological Zone 2 is the most compromised as it is furthest from pristine ecological conditions, has low biological diversity, and high output of domestic and industrial pollution. The ecosystem in this zone is highly stressed, has lost most of its biological integrity and resilience, and is no longer naturally self-sustaining. Maintaining environmental quality and ecosystem health in this area requires high inputs of energy and resources, and it needs the most management intervention. At more than 500 persons/km^2, its population density is the highest in the basin. It is, at present, a major source of environmental pollution and generates the most domestic and industrial wastes compared with the other ecological zones. High outputs of sediments originating from soil erosion from bare land under development further stresses neighboring areas. Many environmental risk factors exist and the ecosystem no longer supports healthy biotic communities.

Work is in progress on the selection of indicators of ecosystem health in Langat Basin, vital for monitoring trends and managing the environment. An initial framework of indicators is outlined in Table 129.2. The indicators and benchmarks are suggested for the entire Basin rather than for the individual ecological zones. The air pollution index (API) is based on the index developed under the

Table 129.2 Initial Framework of Criteria and Indicators for Monitoring Ecosystem Health in the Langat Basin

Criteria	Indicator	Benchmarks
Ecosystem stress (pressure indicators)	1. Population density 2. Soil erosion rate 3. TME/capita	500 persons/km^2 1.6 t/km^2/yr 15 t/yr/capita
Resilience	Natural areas (%)	30%
Self-sustainability	Natural areas (%)	30%
Stress to neighboring areas	1. Air Pollution Index (API) 2. Total suspended solids in river water	80 150 mg/l
Economic viability	This indicator has not been selected; economic statistics are not available at the district level	
Health of communities 1. Human	1. Perinatal mortality rate 2. Toddler mortality rate 3. Infant mortality rate	3/1000 0/1000 1/1000
2. Biotic	Threatened species of mammals (%)	20%
Ecological integrity	Natural areas (%)	30%
Ecological dysfunction 1. Freshwater aquatic ecosystem	Presence of *Lyposarcus pardalis* Absence of *Macrobrachium rosenberghii*	
2. Terrestrial ecosystem	Landslides per 10 km^2 per 3 years	Not yet determined

national air monitoring program. It is a composite index based on a suite of levels of air pollutants including SOx, NOx, and suspended particulates (PM$_{10}$). The presence of introduced fish species (e.g., *Lyposarcus pardalis*) and the absence of certain indigenous species (e.g., *Macrobrachium rosenberghii*) are indicative of river ecosystem dysfunction resulting from pollution and high sediment loads.

These indicators and benchmark values are tentative and much work still needs to be done to properly define and describe their construction and use. Additionally, they have to be evaluated within the context of the rapid development that the basin is undergoing and the prevailing social and cultural context of development in Malaysia.

ACKNOWLEDGMENTS

This work was supported by funding from UNEP/ROAP (Project Number CP/5220-97-02/Rev.1) with funds from DANCED. We thank ISEH for enabling our participation in the 1999 International Congress on Ecosystem Health, thus providing the opportunity to discuss and refine our ideas on operationalizing the concept of ecosystem health in environmental management. We thank Universiti Kebangsaan Malaysia for permission to attend the Congress. Numerous friends and colleagues shared ideas and provided valuable inputs that contributed to the success of this challenging initiative. We thank them all.

REFERENCES

Abdullah, S.A., Changes of forested areas and wildlife diversity in the Langat Basin, Langat Basin Research Symposium, 5 and 6 June 1999, Glenmarie Resort, Shah Alam, 1999.
Department of Environment, WQS phase 1 study: development of water quality criteria and standards for Malaysia, Ministry of Science, Technology and Environment, Kuala Lumpur, Malaysia, 1986.
Department of Environment, WQS phase 2 study: development of water quality criteria and standards for Malaysia, Ministry of Science, Technology and Environment, Kuala Lumpur, Malaysia, 1990.

Department of Environment, Classifications of Malaysian rivers, Ministry of Science, Technology and Environment, Kuala Lumpur, Malaysia, 1994.

Economic Planning Unit, *National Water Resource Study (Peninsular Malaysia 2000–2050), State Report Volume 12 Selangor.* Public Works Department Malaysia and Selangor State Water Supply Department, 2000.

Fariz, A. and Nordin, M., Manufacturing industries and ecosystem health: the case of the Langat Basin, Langat Basin Research Symposium, 5 and 6 June 1999, Glenmarie Resort, Shah Alam, 1999.

Jasmi, A., Wildlife Conservation Issues in the Langat Basin, Master's thesis, Universiti Kebangsaan Malaysia, 1998.

Lillesand, T.M. and Kiefer, R.W., *Remote Sensing and Image Interpretation,* John Wiley, New York, 1987.

Medway, Lord and Wells, D.R., *The Birds of the Malay Peninsula: Vol. V, Conclusion and Survey of Every Species,* Witherby, London, 1976.

Mohsin, A.K.M. and Ambak, A., Ikan Air Tawar Semenanjung Malaysia [Freshwater Fish of Peninsular Malaysia], Dewan Bahasa dan Pustaka, Malaysia, 1991.

Nordin, M. and Azrina, L.A., Training and research for measuring and monitoring ecosystem health of a large-scale ecosystem: the Langat Basin, Selangor, Malaysia, *Ecosyst. Health,* 4, 188–190, 1998.

Rapport, D.J., Gaudet, C.L., and Callow, P., Eds., *Evaluating and Monitoring the Health of Large-Scale Ecosystems,* Springer-Verlag, Berlin, 1995.

Rapport, D.J., Bohm, G., Buckingham, D., Cairns, J., Costanza, R., Karr, J.R., de Kruijf, H.A.M., McMichael, A.J., Nielsen, N.O., and Whitford, W.G., Ecosystem health: the concept, the ISEH and the important tasks ahead, *Ecosyst. Health,* 5, 82–90, 1999.

Rohani, A., Pencemaran mikrob dalam air permukaan di Lembangan Langat, Langat Basin Research Symposium, 5 and 6 June 1999, Glenmarie Resort, Shah Alam, 1999.

Rozali, M.O., Samat, A., Suah, P.C., Lee, S.B., and Rohaida, E., Kualiti air Sungai Semenyih dan anak sungainya, Langat Basin Research Symposium, 5 and 6 June 1999, Glenmarie Resort, Shah Alam, 1999.

Samat, A., Senario komuniti ikan dalam anak sungai yang mengalami pelbagai perubahan akibat aktiviti-aktiviti antropogenik, Langat Basin Research Symposium, 5 and 6 June 1999, Glenmarie Resort, Shah Alam, 1999.

Whitmore, T.C., *Tropical Rain Forests of the Far East,* 2nd ed., Clarendon Press, Oxford, U.K., 1984.

Wong, K.H., Landuse in Malaysia, Ministry of Agriculture, Malaysia, 1974.

Yusuf, M.A., Nordin, M., and Pauzi, A., River water quality assessment and ecosystem health: Langat River Basin, Selangor, *Proceedings of Langat Basin Research Symposium,* Nordin, M., Ed., Institute for Environment and Development, Universiti Kebangsaan, Malaysia, Bangi, 2000, pp. 171–188.

CHAPTER 130

Manufacturing Industries and Ecosystem Health: The Case of the Langat Basin

Ahmad Fariz Mohamed and M. Nordin

INTRODUCTION

The manufacturing industry has become the main engine for Malaysia's economic growth. It plays a vital role in sustaining Malaysia's development. In 1996, it contributed RM 45.2 billion to the gross domestic product (GDP), 34.6% of overall GDP, a 13.3% increase over the previous year's value (Ministry of Finance, 1996).

The Malaysian government has identified a strategy and plan to accelerate manufacturing industries development and to assure their sustainability. According to Ministry of International Trade and Industry (MITI, 1996), during the seventh Malaysia Plan period (1996 to 2000), Malaysian manufacturers were encouraged by the federal government to adopt new growth strategies and to venture into large-scale production for the global market. New industrial estates continue to be developed as state governments respond to national plans and the need to create additional socio-economic opportunities through further development of manufacturing industries. Industrial estates' development grew at its highest rate up to 1996. Supported by well-planned strategy and policy set by federal and state governments, more investors invested in manufacturing industries.

The economic downturn in Malaysia which started in 1997 has not affected the government's long-term measures for economic growth through industrial development. New industrial parks established by the public and private sectors provide much-needed physical infrastructure for enhancing the capacity of states to increase manufacturing output. This is because manufacturing industries play a very important role in making Malaysia's economic bounce-back.

Although industrial development has accelerated the growth of the Malaysian economy, there has been a negative impact on the environment. In Malaysia, industrial development and activities contributed to water pollution, air pollution, toxic and hazardous waste, soil erosion and degradation, and affected human health (Abdullah, 1997). Current water and air pollution is directly related to industrial activity. Jamaluddin (1999) also stated that in 1982, 42 rivers were classified as heavily polluted, 16 rivers polluted, and 7 rivers as potentially polluted. In 1996, 30% of rivers monitored had been found heavily polluted with various pollutants.

Industrial activity also produced a vast amount of solid, toxic, and hazardous waste. The industries have problems in managing and disposing of this waste, thereby increasing the pollution problem. Yong (1998) stated that wastes from industrial activities are generally more hazardous and polluting, and require intensive management and treatment before they can be safely discharged.

The development of vast industrial estates would have a significant negative impact on the environment and resources if it is not planned carefully and strategically. To reduce the impact of such development and subsequent operational activities on the environment and natural resources, measures must be incorporated at the planning stage of industrial development. Therefore, encouraging the development of factories in industrial estates provides opportunities for cleaner production which will enhance future industrial environmental performances (Bateman and Tan, 1998).

OBJECTIVES

The overall aim of this substudy of the MATREM Project *Research and Training for Measuring and Monitoring Ecosystem Health in a Large-Scale Ecosystem* is to identify and create an inventory of the types of industries and their location in the Langat Basin, and examine their impact on the health of its ecosystem. Manufacturing industries in each zone within the basin were classified by type, products manufactured, number of factories, and distribution.

Available information on pollution was used to estimate the impact of manufacturing industries on every major river within the Langat Basin, especially on its water quality. Enforcement and legal issues related to the industrial activities were analyzed from case charges made by the Department of Environment between 1989 and 1995. Data obtained from other government departments were used to determine the overall level of enforcement by agencies responsible for ensuring environmental quality in the basin.

The immediate objectives of this substudy are to:

- Identify trends in the development of manufacturing industries in the Langat Basin.
- Determine the current status of industrial development and activity in the basin.
- Assess the implications of industrial development on the health of the Langat Basin ecosystem.

OVERVIEW OF THE DEVELOPMENT OF THE MANUFACTURING INDUSTRY IN MALAYSIA

The development of a modern manufacturing industry in Malaysia commenced in earnest during the early 1960s. With the introduction of the Investment Incentives Act (1968) and the establishment of Free Trade Zones in 1971, growth in the manufacturing sector accelerated. In 1970, the manufacturing sector contributed 13.4% to GDP in Malaysia;. by 1990, it had doubled to 26.6%. The share of manufactured product in export by value was 11.1% in 1970, but by 1990 it had soared to 58.9% (Taylor and Ward, 1994).

The Industrial Master Plan (IMP) and its revision, the Second Industrial Master Plan (IMP2) commissioned in March 1995, guide the development of the manufacturing industry in Malaysia. During the IMP period from 1985 to 1995 manufacturing output increased significantly, surpassing most of the targets set (MITI, 1996). Export of manufactured goods increased by 28.6% (target 9.4%), and the share of manufacturing exports to total merchandise export increased from 32.8% in 1985 to 79.6% in 1995. Value added in the sector registered an average growth rate of 13.5% per annum (target 8.8%), and the share of manufacturing value added to GDP increased to 33.1% (target 23.9%). The share of manufactured exports increased considerably from 11.1% in 1970 to 79.6% in 1995 (MITI, 1996).

MANUFACTURING INDUSTRY DEVELOPMENT IN THE LANGAT BASIN

The Langat Basin Ecosystem encompasses areas in two states: Selangor and Negeri Sembilan (Figure 130.1). Policies and decisions made by both states influence the development of manufac-

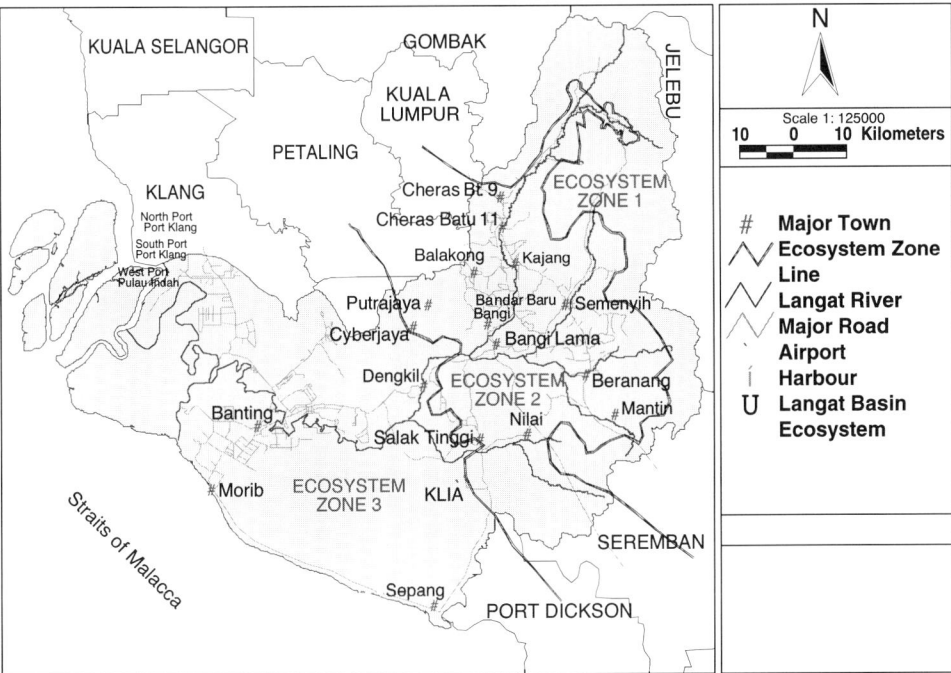

Figure 130.1 The Langat Basin ecosystem.

turing in the basin. Strategies and policies expounded in the IMP and IMP2 allowed both states to achieve considerable growth in manufacturing over the past two decades.

With the development of infrastructure in the Langat Basin and increasing congestion in the adjacent Klang Valley, new manufacturing industrial estates have increasingly moved into the Langat Basin. The federal government's decision to develop the new Kuala Lumpur International Airport (KLIA), the new administrative capital Putra Jaya, and the Multimedia Super-Corridor (MSC) in the Langat Basin provided additional impetus for industries to congregate in the basin. Emphasis will be given to the development of high-technology information and telecommunications industries, as well as their support industries. With good infrastructure, availability of labor, and proximity to sea- and airports, industrial development in the Langat Basin will attract more local and foreign investors to the basin.

INVESTMENT PATTERNS FOR MANUFACTURING INDUSTRY IN THE LANGAT BASIN

From 1991 to September 1998, Selangor and Negeri Sembilan attracted more than RM 33 billion and RM 11 billion in local and foreign capital investments, respectively (Table 130.1). In

Table 130.1 Investment (RM) of Manufacturing Industry Projects for Selangor and Negeri Sembilan from 1991 to September 1998

State	Number of Projects	Local Investment	Foreign Investment	Total Investment
Selangor	1736	19,858,206,600	13,708,730,767	33,566,937,367
N. Sembilan	298	4,079,052,301	7.087,967,844	11,167,020,145

From Selangor DoE, Investment Centre, Industrial Project and Capital Investment in Selangor, Office of the State Secretary of Selangor, Shah Alam, 1998.

Table 130.2 Industrial Project and Capital Investment (in RM) for Langat Basin Industry in Selangor 1991 to September 1998

Area	Number of Projects	Local Investment	Foreign Investment	Total Investment
Teluk Panglima Garang IE	18	11,4361,600	75,326,607	189,688,207
Teluk Panglima Garang FTZ	5	107,000,000	0	107,000,000
Bangi IE	87	611,948,007	498,071,936	1,110,019,943
Cheras Jaya IE/Inch Kenneth/Balakong	18	68,178,747	49,696,946	117,875,693
Beranang	13	83,059,000	44,907,000	127,966,000
Banting	11	25,048,043	13,427,600	38,475,643
Balakong IE	16	81,400,977	7,941,000	89,341,977
Banting IE	2	500,000	3,050,000	3,550,000
Beranang IE	14	93,516,890	16,440,290	109,957,180
Balakong	23	87,425,957	299,215,862	386,641,819
Bangi	11	172,931,753	75,272,124	248,203,877
Cheras	14	127,582,157	30,871,825	158,453,982
Dengkil	1	0	0	0
Hulu Langat	7	53,042,830	15,510,170	68,553,000
Kajang	22	124,289,570	18,779,770	143,069,340
Kuala Langat	14	2,682,947,685	535,355,305	3,218,302,991
Olak Lempit	4	31,712,000	14,400,000	46,112,000
Semenyih	8	58,167,654	5,000,000	63,167,654
Sepang	3	25,170,000	7,530,000	32,700,000
Telok Panglima Garang	9	54,109,685	12,312,540	66,422,225
Total	300	4,602,392,555	1,723,108,975	6,325,501,531

From Selangor DoE, Investment Centre, Industrial Project and Capital Investment in Selangor, Office of the State Secretary of Selangor, Shah Alam, 1998.

Selangor, local capital exceeded that from foreign sources, but in Negeri Sembilan, the opposite was true. However, total capital investments in manufacturing in the Langat Basin during the period was RM 6.3 billion, more than one fifth of all investment in manufacturing for the state of Selangor. Local capital exceeded foreign by more than twice the amount (Table 130.2).

Well-built infrastructure in the Langat Basin enabled the basin to attract investors. Ports, services, production, and consumption centers are linked by excellent highways such as the Kuala Lumpur–Seremban Highway, North–South Link Highway, Shah Alam Highway, and Central Link Highway (connecting North–South Link with Kuala Lumpur–Seremban Highway). The newly completed South Klang Valley Highway and the Kuala Lumpur–Kajang Highway provide excellent transportation networks for manufacturers.

In addition to transport infrastructure, both states have well-developed energy and water supplies. Two power-generating stations located in Serdang and Bukit Changgang are now supplying power to the basin and some parts of the Klang Valley. In addition, each industrial estate built by the state government agency will have its own back-up power generators to ensure uninterrupted power supply.

New water processing plants and additional water intake points have been built. Eight water treatment plants are located along the Langat River. These have a combined capacity to supply more than 220 million gallon per day (mgd) to the basin and some parts of the Klang Valley. In the southern end of the basin (industrial estates around Nilai, Pajam, and Mantin) energy is generated at the Port Dickson power station on the coast, while its water supply is obtained from the Linggi River (MIDA, 1997).

Extensive investment in the Langat Basin is also influenced by the federal government's decision to locate the new administrative capital, Putra Jaya, the MSC, and KLIA in the district of Sepang, in the middle of the Langat Basin. Highways and railways linking Putra Jaya, MSC, and KLIA with Kuala Lumpur and other major cities in the Klang Valley and the existing North–South

Highway would make all industrial areas in Langat Basin highly accessible. With the increased availability of support services and transportation corridors, industries in the Langat Basin would have easy access to Port Klang and to other major towns and cities.

To ensure the development of Putra Jaya, MSC, and KLIA, as well as to facilitate their activities, plans also have been made to increase energy and water supply to these areas (PBT Kuala Langat, Sepang, Putra Jaya, JPBD S. Malaysia, 1996). These plans will also benefit industrial estates around these facilities.

The federal and state government policies and strategies, as well as the availability of well-built infrastructure and better supply of energy and water, played major role in influencing the investment pattern for manufacturing industry development in the Langat Basin.

LOCATION OF INDUSTRIAL ESTATES IN THE LANGAT BASIN

Industrial estates in Langat Basin concentrated in zone 2 of the basin are shown in Figure 130.2. In 1997, 24 industrial estates were located in zone 2, and 6 in zone 3 (Table 130.3).

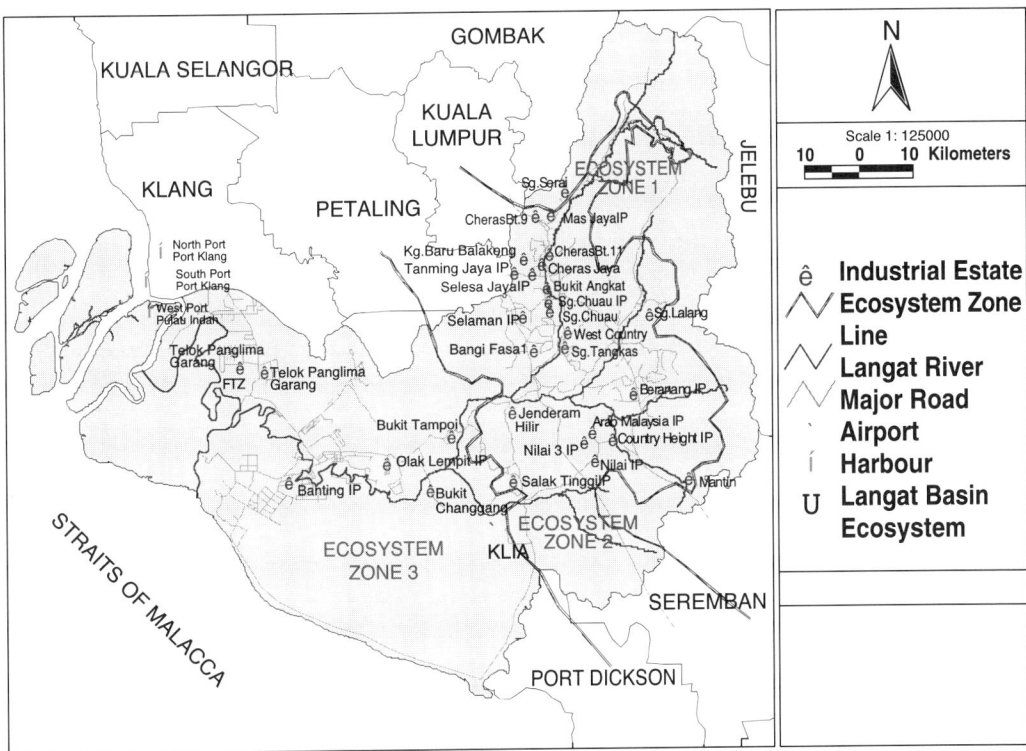

Figure 130.2 Industrial estates in the Langat Basin.

Table 130.3 Number of Industrial Estates in Each Langat Basin Ecosystem Zone

Zone 1	Zone 2	Zone 3
—	24	6

Modified from Malaysian Industry Development Authority (MIDA), Position of industrial estates developed by the State Economic Development Corporations, Regional Development and Port Authorities and Municipalities, Malaysian Industry Development Authority, Kuala Lumpur, 1997b.

Table 130.4 Number of Industries According to Category in Langat Basin Ecosystem

Industry Category	Zone 1	Zone 2	Zone 3	Total
Light A	—	—	—	0
Light B	—	46	1	47
Medium	—	103	15	118
Heavy	—	147	21	168
Special industry	—	4	—	4

Modified from MIDA (1997a) and DoE (1994).

All are located within easy reach of major highways and trunk roads linking the major towns in and outside the basin.

Besides easy access to the transportation network, the ready availability of local labor, proximity to KLIA and the seaport in Klang, business centers and go-down space in the Langat Basin have influenced the location of the industrial estates. Thus, the Langat Basin has grown rapidly as the most important hub of industrial production in Selangor after the Klang Valley.

In all, 30 industrial estates housing more than 330 factories are located in the Langat Basin; 20 were developed by the private sector and the rest by state government. These engage in 22 categories of industrial production ranging from the manufacture of agricultural end products to high-tech products such as consumer electronics. Heavy industries predominate with 168 establishments, medium industry follows (Table 130.4)

Table 130.5 shows the number of factories by product category in each zone of the Langat Basin ecosystem. Electrical and electronic industries predominate with 53 factories, followed by industrial and engineering products, furniture, textiles and fabrics, building materials, and chemicals.

Most industries are concentrated in and around the urban areas such as Kajang, Nilai, Bangi, Cheras, Balakong, Banting, and Beranang (Figure 130.2). The siting of individual factories within

Table 130.5 Number of Factories According to Product in Each Langat Basin Ecosystem Zone

Product	Zone 1	Zone 2	Zone 3
Agricultural machinery	—	—	1
Automotive parts and accessories	—	23	1
Bicycle	—	—	1
Chemicals and adhesive	—	16	2
Building materials and related products	—	24	—
Computer and telecommunications	—	3	—
Electrical and electronics	—	48	5
Food and beverage	—	18	—
Furniture and related products	—	25	11
Household	—	9	—
Industrial and engineering products	—	38	5
Leather and footwear products	—	5	—
Nonmetallic minerals	—	4	—
Paper, printing, packaging, and labeling	—	15	2
Pharmaceuticals, medical equipment, cosmetics, and toiletries	—	4	1
Plastic and PVC	—	21	—
Rubber and rubber products	—	12	3
Stationery	—	4	—
Textiles, fabrics, and button	—	24	—
Tobacco	—	1	—
Waste treatment	—	1	—
Wood and wood products	—	5	—
Total	0	300	37

Modified from MIDA (1997a).

these industrial estates does not follow specific criteria. Most have mixed establishments, with the exception of the Olak Lempit industrial estate, which is restricted to factories processing wood and wood products. In mixed establishments, multinational companies (MNCs) and small and medium industries (SMIs) are located in the same estate; most of the SMIs engage in support industries for the MNC, such as in the production of parts and components. However, there are SMIs that produce their own products for the Malaysian market, for MNC outside the Langat Basin, and for export. As a whole, SMIs predominate, with many engaged in the production of industrial and engineering products.

The absence of guidelines on the possible types and scale of manufacturing industries allowed in each industrial estate means that each estate can host many types of factories, which vary in capital investment, production, and capacity to pollute.

Plans to zone industries by product type have not been successful, and to date only one industrial park for the wood and wood products industry has been established at Olak Lempit. The development of Cyberjaya in Sepang will eventually be the hub for the development of industries related to information technology and high-tech products.

In the future, state and local governments plan to identify zones for industrial estates and relocate industrial estates that have been established in unsuitable areas. This will include an unknown number of illegal factories located in the basin. In 1998, the Selangor state government estimated that there were more than 163 illegal factories along Langat and Semenyih Rivers. Measures have been taken to close these illegal factories and relocate them to industrial estates in the Langat Basin and elsewhere (*The Star,* 1998).

IMPACTS OF THE DEVELOPMENT OF MANUFACTURING INDUSTRY ON THE HEALTH OF THE LANGAT BASIN ECOSYSTEM

The health of an ecosystem is reflected in its ability to sustain a high-quality life for all its residents. Healthy ecosystems provide resources and services vital for a productive life as well as the milieu for human development. A degraded ecosystem is costly to restore, threatens human well-being, and constrains further development.

Rapid industrial development in Langat Basin has had some impacts that are detrimental to its ecosystem. Much of this has originated from factories located along the Langat and Semenyih Rivers and affects the river water quality. The following subsections examine the extent to which pollution by the manufacturing industries has affected river water quality as well as the enforcement efforts by the relevant agencies to ensure the water quality of the rivers.

RIVER WATER QUALITY

It is generally well known that the quality of the water in the Langat River has been gradually deteriorating. According to records of the Department of Environment (DoE) the river classification fell from "polluted" to the worst category, "highly polluted." Actual values of the Water Quality Index for the river from 1985 to 1997 are given in Figure 130.3. The period of rapid decline of the river's water quality coincided with the rapid industrial development in the basin beginning in 1991. Through spatial analysis it has been found that most of the industrial estates were located more than 500 m away from the main Langat River. Table 130.6 gives the number of industrial estates and their distance from the river system. This analysis examines only the distance of each industrial estate from the main river system and does not include the small river system; industrial estates are too close. Most of the drainage systems in the estates discharge industrial wastewater straight to the river system.

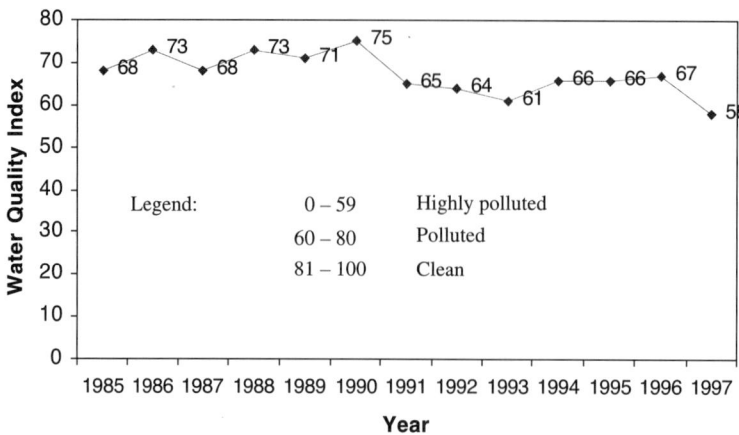

Figure 130.3 Trends of Langat River water quality, 1985 to 1997. (From Selangor DoE, Monitoring Annual Report, 1997, Department of Environment Selangor, Shah Alam, 1997a.)

Table 130.6 Distance of Industrial Estate to Main River System in Each Zone

Distance from Rivers (m)	Zone 1	Zone 2	Zone 3
0–100	—	—	—
100–200	—	1	—
200–300	—	2	—
300–400	—	1	—
400–500	—	1	1
More than 500	—	19	5

From ArcView Spatial Analysis 1.0 and MIDA (1997b).

Records also show many manufacturing facilities along the Langat River are not equipped with effluent treatment facilities. However, not all sources of pollution are from the manufacturing industry. Sources of pollution identified by the DoE are shown in Table 130.7. Oil storage facilities and sewage treatment plants predominate as potential sources of pollution. Nevertheless, 131 of the total 231 potential pollution sources along the Langat River are industrial plants. Furthermore, of 337 legal industries in Langat Basin, only 38 are equipped with an effluent processing system (EPS). Table 130.8 shows industries with EPS and their current status.

Degradation of river water quality in the Langat Basin was exacerbated by the long drought of June to August 1998. During this period, rivers were unable to "wash out" pollution produced from domestic, commercial, and industrial establishments that discharged wastes into the river. The ability of the Langat Basin Rivers to supply water for domestic and industrial use was severely

Table 130.7 Potential Pollution Sources along Langat River

Source	Number
Oil storage premise	85
Sewage treatment plant	69
Industry discharged effluent and stored oil	46
Water treatment plant	5
Landfills	2
Pig farm	12

From Selangor DoE, Inventory Report of Pollution Sources in Langat River Water Catchment, Department of Environment Selangor, Shah Alam, 1997b.

Table 130.8 Industries Equipped with Effluent Processing System (EPS) in the Langat Basin

Industry	Number of Premises	EQA Compliance	EQA Noncompliance	Compliance Percentage
Food	4	3	1	75
Electrical/metals-based	12	10	2	83
Paper based	2	1	1	50
Textiles	3	2	1	67
Rubber-based	7	7	—	100
Concrete/soil-based	3	3	—	100
Chemical-based	6	6	—	100
Palm oil	1	1	—	100

From Selangor DoE, Inventory Report of Pollution Sources in Langat River Water Catchment, Department of Environment Selangor, Shah Alam, 1997b.

Table 130.9 List of Water Treatment Plants in the Langat River Basin, 1997

Treatment Plant	Zone	Production Rate Million Gallons per Day (mgd)	Source	Area of Distribution
Langat	1	85.00	Langat River	Cheras, Kajang, Shah Alam, Wadieburn Camp, Wangsa Maju, and Klang Lama
Semenyih	1	120.00	Semenyih River	South West K.L, Petaling Jaya, Shah Alam, and Klang
Bukit Tampoi	2	6.90	Langat River	Dengkil
Salak Tinggi	2	2.40	Labu River	Salak Tinggi and Sepang
11th Miles Cheras	2	6.00	Langat River	Cheras and Balakong
Lolo River	1	0.90	Lolo River	Lolo River
Serai River	2	2.00	Serai River	Hulu Langat 18th Mile
Pangsoon River	1	4.00	Pangsoon River	Pangsoon

From Selangor DoE, Monitoring Annual Report, 1997, Department of Environment Selangor, Shah Alam, 1997a.

hampered and more than RM 50 million was spent to supply water from alternative sources (old mining pools, underground water) to residents in the basin and elsewhere. Water rationing was instituted in the Klang Valley and the Langat Basin for more than 2 months.

On more than three occasions during the drought, water extracted from the Langat and Semenyih Rivers could not be treated. There are eight water treatment plants with varying capacities located along the Langat and Semenyih Rivers that supply a wide range of areas (Table 130.9). Four of the treatment plants were located in zone 1 and four in zone 2.

LEGISLATION AND ENFORCEMENT ACTIVITY

Enforcement of environmental regulations to curb pollution by manufacturing industries in the Langat Basin is an ongoing process. The number of legal proceedings instituted against industrial concerns in the Langat Basin for breaches of environmental regulations from 1989 to 1996 totaled 80 cases, with the highest number of offenses recorded in 1991, 1992, and 1996 (Figure 130.4). From 1989 to 1996, 80 charges were made to industry under various environmental regulations (details are shown in Table 130.10).

The DoE has adopted strategies to enhance its efficiency in monitoring and enforcing environmental regulations. Issues of lack of manpower and technology know-how have been taken into account. Training staff to equip them with the latest information and technology is an ongoing process. Therefore, enforcement activity in the future is expected to improve.

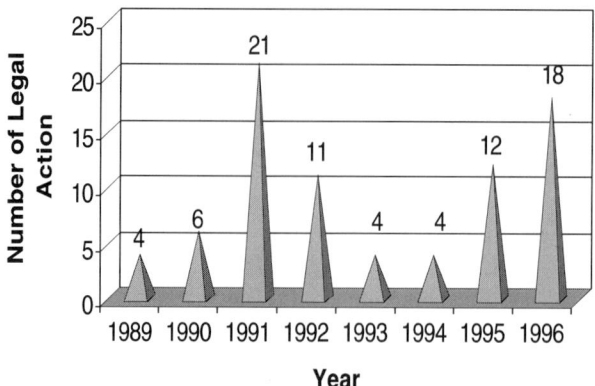

Figure 130.4 Number of legal actions taken against the manufacturing industry in the Langat Basin. (From Selangor DoE, Inventory Report of Pollution Sources in Langat River Water Catchment, Department of Environment Selangor, Shah Alam, 1997b.)

Table 130.10   Charges to Industry in the Langat Basin 1989 to 1996, According to EQA 1974

Charge	Number
Regulation 11, 49(1) Clean Air Regulation 1978	3
Regulation 15(1) Clean Air Regulation 1978	5
Regulation 16(1) Clean Air Regulation 1978	13
Regulation 40 Clean Air Regulation 1978	4
Regulation 4 Sewage and Industrial Effluent 1979	5
Regulation 8 Sewage and Industrial Effluent	9
Section 16(1) EQA 1974	6
Section 18(1) EQA 1974	13
Section 22(1) EQA 1974	3
Section 24(1) EQA 1974	4
Section 25(1) EQA 1974	15
Total	80

From Selangor DoE, Inventory Report of Pollution Sources in Langat River Water Catchment, Department of Environment Selangor, Shah Alam, 1997b.

GENERAL DISCUSSION: DEVELOPMENT OF MANUFACTURING INDUSTRIES AND ECOSYSTEM HEALTH IN THE LANGAT BASIN

The development of manufacturing industries in the Langat Basin has had many positive impacts on the economic well-being of the residents through the creation of increased job opportunities for local communities, improved infrastructure, and the increased availability of modern housing. With policy and incentives by the federal and state governments, manufacturing industries in the future will become the main engine for economic growth in the basin. With MSC development, high-tech and information technology would be the focus of industries.

Nevertheless, industrial development has had undesirable impacts related to the health of the ecosystem and the quality of the environment. Since industrial development requires changes to the physical environment, it can create ecosystem stress. Jamaluddin (1999) stated that industrial development causes habitat loss and creates disturbance and stress to natural and human-made ecosystems. This disturbance includes degradation of environmental quality through loss of aesthetic value, air and water pollution, waste, hazardous toxic waste, and other environmental problems.

Moreover, development of manufacturing industries has resulted in the clearance of large tracts of land originally under plantation crops, secondary forest, and other vegetation. This has created exposed land areas from which soil has eroded and entered river systems as sediment. High sediment loads are prevalent in areas that are rapidly developing and are the predominant cause of river pollution in Malaysia. Suspended sediments change the ecological characteristics of the river and limit its value as habitat for aquatic life. Land cleared for industrial estates is located close to river systems more often than not.

In addition to increased sediment load, the rivers in the Langat Basin have been affected by the discharge of polluting effluent from industries such that their pollution status has been upgraded from "polluted" (Class II rivers) to "highly polluted" (Class III rivers). These highly polluted rivers were found in zones 2 and 3 of the basin ecosystem. The frequent closure of water treatment plants along rivers in the basin in 1998 was a reflection of the "highly polluted" status of the river. Two plants in zone 2 were shut down due to high ammonia content. River water quality fell so much that water was no longer suitable for extraction and processing to supply potable water. The inability of the river to supply water for processing for potable water reflects poor ecosystem health.

The success of manufacturing industry development in the basin has also meant that waste generation by industries increased in type and quantity. Poor compliance by the food, textile, paper, electrical, and metal-based industries indicates that much of their wastes are discharged into the environment and threaten ecosystem health.

It is also possible that industrial estate areas may become contaminated lands. DoE of Selangor has reported that one factory located in zone 3 was found disposing of a scheduled waste into the ground behind the factories. Groundwater testing is being conducted and this case has been taken to court. Although establishing proof will take time, this example indicates that industrial estates in the basin are likely to contaminate the land and may contaminate the groundwater, too.

Inventories of individual manufacturing plants are not available in Malaysia; therefore, data on resources, chemicals used, and pollution types and loads by industries can only be estimated. With more plants becoming operational in the Langat Basin in the next few years, the health of the ecosystem and the quality of the living environment in the Langat Basin are anticipated to decline further, and we are unable to know which factory or industrial estate will be the next contaminated area. Investments in clean production by these plants are unlikely to be significant in the near future because of cost constraints and lack of market incentives. Therefore, data on individual manufacturing plants are needed to establish better management and monitoring, to reduce pollution loads, and to avoid any hazardous contamination or accidents in the future.

Local authorities are entirely dependent on DoE for enforcement of environmental regulations. Also manufacturers focus narrowly on manufacturing processes and constantly work toward optimizing profits by reducing costs of inputs and environmental care. Data on industry location, employees, chemicals and resources used, products, process and technology used are not easily available. Small- and medium-scale industries are rarely monitored for environmental performance.

Many factories are being charged in court each year for polluting the environment. There is a need to tighten enforcement, and the DoE has adopted strategies to enhance its efficiency in monitoring and enforcing environmental regulations. Issues of lack of manpower and technological know-how have been addressed, as this is an ongoing process. However, DoE cannot tackle the pollution issues alone, as some cases are very complex. There is a need to get other relevant agencies and local governments involved, especially in monitoring activities. Integrating legislative tools would enhance enforcement efficiency. Nevertheless, enforcement has not been sufficient to reduce pollution. Industries should also be proactive in reducing their waste and pollution, and local communities should participate in monitoring activities. Therefore, the command-and-control approach should include education and other incentives to minimize industrial pollution in the future.

Planners for industrial development lack adequate appreciation of ecosystem concepts. This has contributed to the unsuitable location of factories and industrial estates in the basin. Moreover, with the absence of guidelines on the type of industry allowed, pollution has increased. Since 1976, guidelines have been developed by DoE on siting and zoning of industry; these guidelines were amended in 1994. They can be used to assist the decision maker, local governments, and developers in deciding the siting and zoning of factory and industrial estates as well as in determining buffer zones. However, DoE does not have the full power to determine location, because decision on land matters falls under the jurisdiction of the state government.

Field observation shows that existing siting and zoning of industries in the basin have created problems to ecosystem health and humans. Jamaluddin (1999) stated that 64% of heavy industry, 59% of medium industry, and 50% of light industry in zone 2 do not comply with the buffer zone guidelines. In zone 3, 100% of heavy and light industry were also found not complying with these guidelines. Buffer distance in these estates only takes into account the distance between factory zone and the nearest house, open land, or roads. Industrial estates in Cheras Batu 9, Cheras Jaya, Balakong, Bukit Angkat in zone 2, and Telok Panglima Garang in zone 3 have been found too near to river drainage systems and housing areas. There are situations where the distance between factories and housing areas is less than 20 m, while the minimum buffer distance suggested is 250 m. However, there are industrial estates that comply with the guidelines, such as Beranang, Bangi Phase I and II, and Selaman Industrial estates.

The issues raised here reflect the failure of developers and decision makers to look at the possibilities of pollution and threats to local communities by industry. At the same time, DoE industrial development guidelines, especially on the siting and zoning of factories and industrial estates, were ignored. To address this issue, state and local governments must ensure that developers comply with guidelines, and must strictly enforce the existing guidelines and standards for industrial development. The siting and zoning of industrial estates are very important. Development should ensure that pollution discharges can be easily monitored and managed, that the health of the basin ecosystem will not deteriorate, and that the safety and health of the people are protected.

On the positive side, all environmentally sensitive areas in the basin have been identified and mapped. Future planning for the development of manufacturing industries in the basin must take into account the sensitivity of these areas and their importance in maintaining ecosystem health.

Growth of the manufacturing industry is anticipated to continue, as it is the main engine of economic growth in Malaysia and contributes toward its vision as a developed country by year 2020. To ensure sustainability of industrial development while maintaining the health of the ecosystem, there is a need to develop a strategic policy and planning environment that aims to harmonize both needs. Specifically, there is the need to:

- Promote the use of clean technologies in industrial processes.
- Encourage the participation of industries, businesses, and the public in planning and decision making about the location of industries.
- Improve management and monitoring system capabilities to reduce pollution.
- Enhance legislative framework for better management and enforcement.
- Ensure SMIs use clean technologies and processes as well as optimize their use of resources by providing adequate and appropriate incentives.
- Encourage R&D by industries, local universities, and research institutions to develop clean technologies and processes as well as develop management and monitoring systems to achieve a greening of industry in the future.
- Establish data on individual industry by the state and local governments and relevant agencies for management and monitoring purposes.

Future planning would also benefit from a decision-support tool that taps into a comprehensive database on the location of industries, chemicals, and resources used, by-products produced, and processes and technologies used by all industries in the Langat Basin. Without strategic planning to incorporate an ecosystem approach for managing and monitoring by state and local governments as well as the relevant agencies, zone 2 (which is expecting a high density of urban and industrial development) would be highly polluted. At the same time, environmental stress in zone 3 would be increasing, as pollution discharges from zone 2 will accumulate in this zone. Therefore, future planning for industrial development must balance the needs of development and economic growth with the effort to sustain the health of the basin ecosystems as well as the nation's environment. To address these issues, databases and ecosystem scenario tools are being developed at the Institute for Environment and Development (LESTARI), Universiti Kebangsaan Malaysia.

ACKNOWLEDGMENTS

The work reported in this chapter funded by the MATREM Project of UNEP/ROAP CP/ 5220–97–02/Rev.1 Research and Training for Measuring and Monitoring Ecosystem Health in a Large-Scale Ecosystem.

REFERENCES

Abdullah, M.A., Pembangunan Perindustrian Di Malaysia: Perkembangan dan Permasalahan, Fajar Bakti Sdn Bhd, Shah Alam, Malaysia, 1997.
Bateman, B.O. and Tan, J., Developing industrial estates in the Asia-Pacific Region: is there a room for the environment, *J. ENSEARCH,* 11, 35–39, 1998.
Department of Environment (DOE), Guidelines on Siting and Zoning of Industries, Ministry of Science, Technology and the Environment, Kuala Lumpur, 1994.
Economic Planning Unit, Negeri Sembilan (EPUNS), Negeri Sembilan gross domestic product 1990–2000, Office of the State Secretary of Negeri Sembilan, Seremban, Malaysia, 1996.
Jamaluddin, J., Letakan dan pengezonan industri di lembangan Langat: Mengundang masalah, paper presented in Research Symposium on Langat Basin Ecosystem (SPELL), 5–6 June 1999.
Malaysian Industry Development Authority (MIDA), Malaysia statistic on the manufacturing sector 1992–1996, Malaysian Industry Development Authority, Kuala Lumpur, 1997a.
Malaysian Industry Development Authority (MIDA), Position of industrial estates developed by the State Economic Development Corporations, Regional Development and Port Authorities and Municipalities, Malaysian Industry Development Authority, Kuala Lumpur, 1997b.
Ministry of Finance, Malaysia, Economic Report 1995/96, Percetakan Nasional Malaysia, Kuala Lumpur, 1996.
Ministry of International Trade and Industry (MITI), Second Industrial Master Plan 1996–2005, Ministry of International Trade and Industry, Kuala Lumpur, 1996.
PBT Kuala Langat, Sepang, Putra Jaya dan JPBD Semenanjung Malaysia, Laporan Pemeriksaan: Rancangan Struktur Kuala Langat Sepang dan Putra Jaya, Town Country and Planning Department of Peninsular Malaysia, Kuala Lumpur, 1996.
Selangor DoE, Investment Centre, Industrial Project and Capital Investment in Selangor, Office of the State Secretary of Selangor, Shah Alam, 1998.
Selangor DoE, Court Case Status 1989–1995, Department of Environment Selangor, Shah Alam, 1996.
Selangor DoE, Monitoring Annual Report, 1997, Department of Environment Selangor, Shah Alam, 1997a.
Selangor DoE, Inventory Report of Pollution Sources in Langat River Water Catchment, Department of Environment Selangor, Shah Alam, 1997b.
Selangor State Secretary Office, Selangor Darul Ehsan in Figures 1995, Office of the State Secretary of Selangor, Shah Alam, 1995.
The Star, 163 factories to move, *The Star,* Kuala Lumpur, March 14, 1998
Taylor, M. and Ward, M., The regional dimension of industrialization, in *Transformation with Industrialization in Peninsular Malaysia,* Brookfield, H., Ed., Oxford University Press, New York, 1994, pp. 95–121.
Unit Perancang dan Pembangunan Negeri Selangor (UPPNS), Rancangan Malaysia ke 7 Negeri Selangor D.E. 1996–2000, Office of the State Secretary of Selangor, Shah Alam, 1995.
Yong, K.L., Waste minimisation for small and medium industries, *J. ENSEARCH,* 11, 1–6, 1998.

CHAPTER 131

Diagnosing Ecosystem Health of the Langat Basin in the Context of Geohazards

Ibrahim Komoo, Joy Jacqueline Pereira, and W. Mohd Muhiyuddin

INTRODUCTION

The concept of ecosystem health dates back to 1788 in the writings of James Hutton, a geologist, who perceived the Earth as an integrated system (Rapport et al., 1995). It is also recorded in the 1940s composition by Aldo Leopold, a naturalist, who described unhealthy ecosystems and referred to "land sickness" (Callicott, 1992). The concept was revitalized as "ecosystem medicine" in the late 1970s and early 1980s (Rapport et al., 1979, 1981, 1985). This work formed the basis for subsequent principles and concepts of ecosystem health (Rapport, 1989; Rapport et al., 1995, 1998; Schaeffer et al., 1988). The process of assessing ecosystem health is complex and requires the establishment of indicators to differentiate healthy ecosystems from unhealthy ones, postulate the causes of sickness, and develop methods for subsequent preventive and rehabilitative measures (Rapport et al., 1995).

The concept of ecosystem is closely tied to the evolution of ecology as a discipline. Thus, it is hardly surprising that attempts to develop the methods of measuring ecosystem health are predominantly ecology based (Fausch et al., 1990; Karr and Chu, 1995; Karr et al., 1986). Indicators identified thus far emphasize the ecological aspects of ecosystem health such as loss of productivity and species diversity, nutrient loadings, acid depositions, etc. It has been argued, however, that societal values should underpin the assessment of ecosystem health, and the criteria for such evaluation require indicators for socioeconomic activity, biophysical conditions, and human health (Rapport et al., 1995).

In recent times, several topics have been studied in the context of ecosystem health, including ocean and water policy; international trade law; biological, economic, and cultural impacts of environmental contamination; integrated water catchment management;, human health; vegetarianism; and gender differences (Boetzkes, 1998; Buckingham, 1998; Fox, 1999; Pellerin and Grondin, 1998; King and Hood, 1999; Yassi et al., 1999). Topics specific to physical aspects, such as mineral utilization, conservation of physical heritage, and prevention of geohazards have not been discussed within the ecosystem health framework, although aspects such as erosion and characterization of landscape patterns have been discussed (Aguilar, 1999; Johnson and Patil, 1998).

This chapter describes two major geological hazards that occur within the Langat Basin in Peninsular Malaysia, i.e., landslides and land erosion. The approach taken for their assessment, and the critical issues related to their occurrence are identified and discussed within the context of ecosystem health. Ultimately, the status of ecosystem health is determined for the Langat Basin using a preliminary set of indicators.

GEOHAZARDS

A geological hazard or geohazard is a geological condition, process, or potential event that poses a threat to the health, safety, or welfare of human beings, or the functions or economy of a community. All processes that involve the Earth, oceans, and atmosphere, including geophysics, meteorology, climatology, and glaciology, come under the scope of geohazards (McCall, 1992). The range of geohazards is quite extensive and includes volcanic eruptions, earthquakes, tsunamis, cyclones, floods, coastal and fluvial erosion, landslides, ground subsidence, and dam failures.

Geohazards may occur due to natural or technological causes, or a combination of both (Aswathanarayana, 1995). Examples of natural hazards are earthquakes and volcanic eruptions, which cannot be controlled by humans. Technological hazards generated by human actions include subsidence near underground mines, burial of toxic waste, or a dam failure. Naturally occurring hazardous land conditions may culminate in the occurrence of a catastrophic event as a result of human intervention. Examples of these are landslides, erosion, floods, and flash floods in developed areas, and subsidence in karstic terrain. The occurrence of geohazards generally results in damage to property, disruption of normal human activities, some form of environmental degradation, or, in extreme events, loss of lives.

Currently, natural hazards are the major cause of loss of life and damage to property (McCall, 1992). Geohazards due to other causes are increasing rapidly and may eventually outnumber natural causes worldwide. In the urban areas of some rapidly developing countries, geohazards occur predominantly due to a combination of technological and natural causes (Komoo et al., 1996; Pereira et al., 1996). The main factors that contribute to the occurrence of geohazards are excessive emphasis on economic growth resulting in the forbearance of shortcuts in the development of infrastructure and other construction projects, inherent weaknesses in the planning process, as well as poor prior geological investigation in land-use planning and construction (Pereira and Komoo, 1998).

GEOHAZARDS IN THE LANGAT BASIN

The Setting

The Langat Basin, with an area of about 2938 km², is located in the central western portion of peninsular Malaysia, between the latitudes of 2°35' and 31°07' and the longitudes of 101°12' and 102°00'. The basin, defined based on the geographical concept, is drained by the Langat, Labu, and Semenyih Rivers, which flow into the Straits of Melaka. The highest peak in the basin is Nuang Mountain at a height of 1493 m.

In terms of land use, the eastern highlands in the Langat Basin are covered with primary forests and serves as the catchment for two dams, i.e., the Langat Dam and Semenyih Dam. The central portion is highly urbanized; most built-up areas in the basin are located here. The western basin is agriculture based, with tracts of oil palm and rubber plantations. Mangrove swamps are restricted to a narrow zone along the coast. The geology of the Langat Basin is relatively simple, where meta-sediments and granite, with a surficial cover of alluvium (Figure 131.1), essentially underlie the basin. The meta-sediments consist of quartzite, schist, phyllite, and shale, and are located in the central and western parts of the basin. A broad alluvial plain overlies the meta-sediments near the coast. Granitic terrain is mainly confined to the upper reaches of the basin and represents an extension of the Main Range Granite.

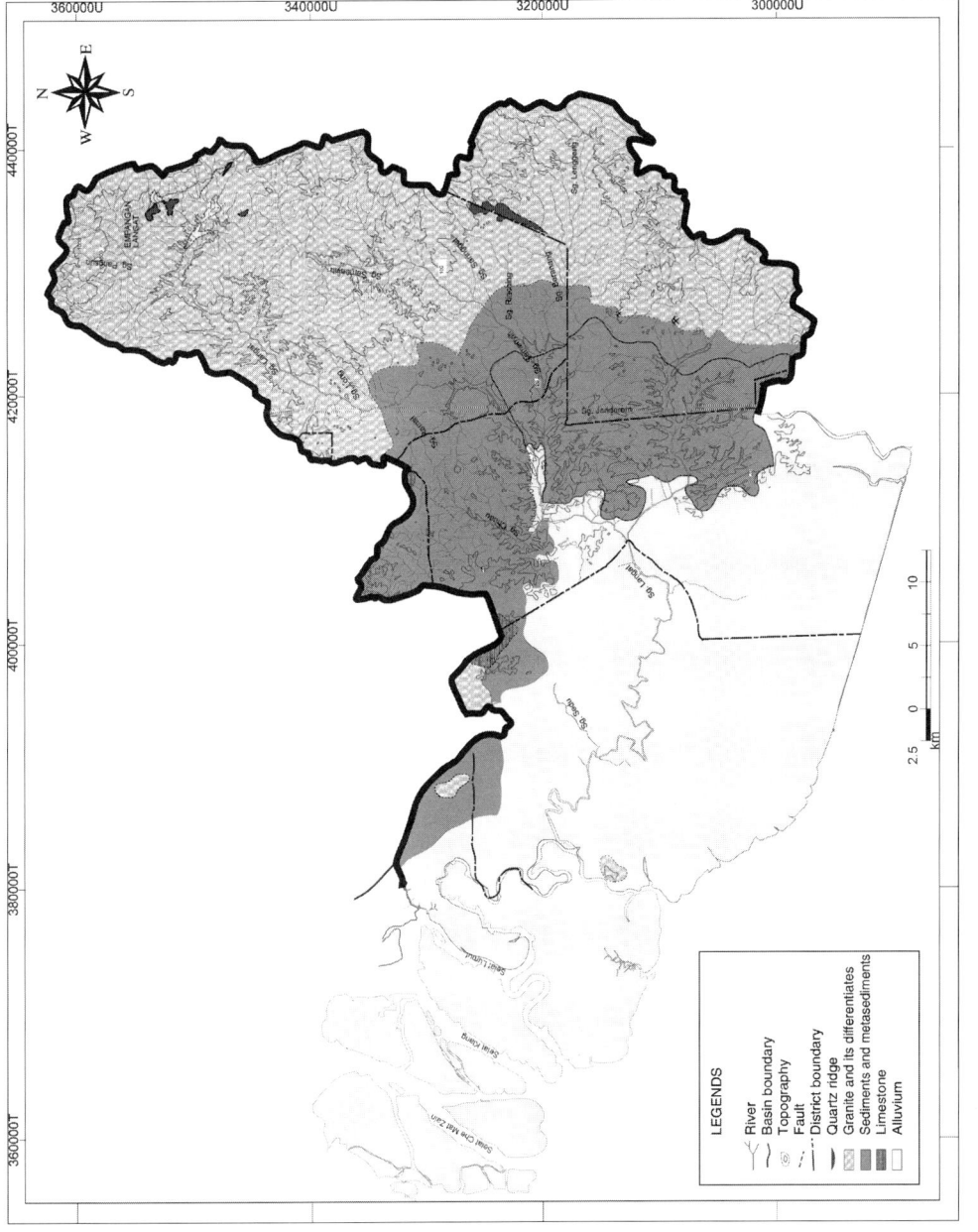

Figure 131.1 Geology of the Langat Basin.

Previous Work

Geohazard studies have been conducted in the Langat Basin only since the mid-1990s. Prior to this, geohazard investigations were conducted mainly in the adjacent northern Klang Basin, which is the focus of development activities (Komoo et al., 1996; Shu, 1986; Pereira et al., 1996; Shu and Chow, 1979). The capital city, Kuala Lumpur, is located within the Klang Basin. The Minerals and Geoscience Department of Malaysia has prepared terrain classification maps for selected parts of the Langat Basin that have been earmarked for national development projects. A macroscale study of the entire basin was initiated only recently as part of the effort to determine the overall state of ecosystem health in the Langat Basin (Wan et al., 1998, 1999).

Assessment of Geohazards

The geohazards that have been reported in the Langat Basin are landslides, land and coastal erosion, floods and flash floods, and ground settlement. Landslides and land erosion are the predominant geohazards in the Langat Basin; thus, geohazard assessment focuses on them.

Landslide is a general term employed to cover a variety of gravity-dominated processes that transport material down to lower ground, and result in slope failures with different form and behavior (Jones, 1992). There are three basic approaches to the assessment of landslides. These are direct mapping, land systems mapping, and indirect mapping (Hansen, 1984; Jones, 1992). The purpose of all three approaches is to produce landslide susceptibility maps that can be used for regional development and hazard minimization. The landslide susceptibility maps provide spatial divisions of the Earth's surface into areas of different levels of threat. It is these divisions that provide the essential framework for land-use planning, building regulation, and engineering practices.

The assessment of landslides in the Langat Basin was based on the direct mapping approach. This involved the preparation of a spatial inventory of existing landslides based on information from air-photo interpretation, field survey, and literature review. During the field survey, information was collected on the locality, type, and size of landslides; the amount of displaced material; and the geology, topography, and land use of the site. It was not possible to establish the age of the landslide from the field survey. The lack of annual systematic mapping of landslides and the limited temporal air-photo coverage made it difficult to differentiate the number of landslides that occurred annually between 1995 and 1998. The information on landslides was then used in conjunction with land cover and landform analysis to evaluate the ecosystem health of the Langat Basin. The study focused on the upper subbasin of the Langat Basin because it was representative with respect to geology and land use.

Land erosion in the tropics refers to the detachment and removal of soil and rock by the action of running water. The dominant processes are action by raindrops, running water, subsurface water, and mass wasting (Selby, 1993). This natural process transforms the surface of the Earth and contributes to the geomorphologic expression of the basin. However, human intervention has resulted in extreme soil loss and subsequent siltation of the rivers. The conventional approach to assess levels of soil loss involves mapping the factors that contribute toward soil erosion. The advent of geographical information systems (GIS) and remote sensing capabilities have led to the development of systematic and simpler mapping techniques. Assessment of soil loss can now be conducted using a combination of GIS and remote sensing techniques, based on the universal soil-loss equation model (Omakupt, 1987; Ravishankar, 1994; Saha and Pande, 1993).

The state of soil loss in the Langat Basin was established based on the TM Landsat image for three time domains — 1993, 1996, and 1998. Land cover for these images was classified into five categories: forests, agriculture, water bodies, built-up areas, and cleared land. These images were then overlaid with the erosion risk map produced by the Agriculture Department in 1996. Availability of sequential images allowed the construction process to be classified into before, during, and after construction phases. Change of phases could be clearly identified in the images for the three time domains. Soil loss for the Langat Basin was then calculated based on appropriately weighted factors, such as land cover, level of risk, and ruggedness of the topography (Wan et al., 1999).

Landslides In The Langat Basin

The Langat Basin consists of four major subbasins: Upper Langat, Labu, Semenyih, and Lower Langat (Figure 131.2). The study on landslides focused on the Upper Langat subbasin. Ecological Zones 1, 2, and 3, as defined by Nordin and co-workers (this volume), occur within the Upper Langat subbasin; only Ecological Zones 1 and 2 occur within the Semenyih subbasin; the Labu and Lower Langat subbasins correspond to Ecological Zones 2 and 3, respectively.

The Upper Langat subbasin can be divided into four topographical categories: mountainous terrain, undulating lowlands, hilly terrain, and floodplains. Ecological Zone 1 is generally represented by mountainous terrain, Ecological Zone 2 by undulating lowlands and hilly areas, and Ecological Zone 3 by floodplains.

About 40% of the Upper Langat subbasin is covered by undulating lowlands and 30% by hilly areas, while floodplains and mountainous terrain each cover 10% of the area. The subbasin lies on granites and meta-sediments, which encompass 93 and 7% of the area, respectively. In terms of land cover, 40% of the basin is agriculture based, 28% forest, 23% built-up areas, 7% cleared area, and the balance is water bodies.

The combination of heavy rainfall and removal of vegetation has caused landslides in 70 localities within the subbasin, between the period of 1995 and 1998. So far, only two deaths from

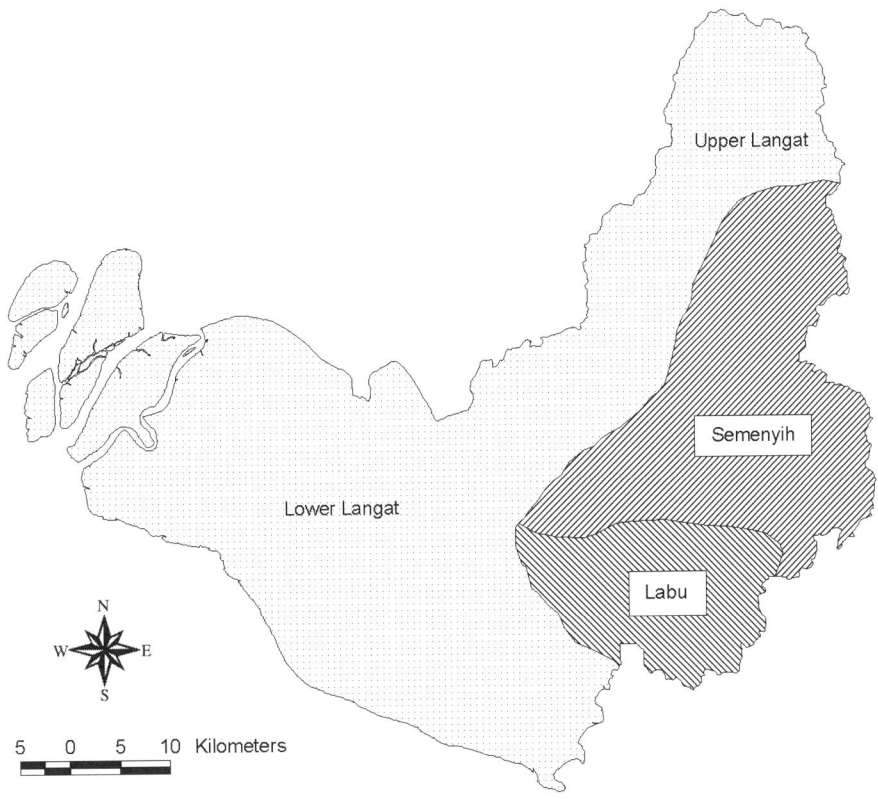

Figure 131.2 Subbasins within the Langat Basin.

Table 131.1 Landslide Intensity and Displaced Material in the Langat Basin

Landslides	No.	Volume (m³)	Area (ha)	Displaced Material per Landslide (m³)	No. per 1000 ha	Displaced Material per ha (m³/ha)
Topography						
Floodplain (<20 m)	12	1534	7255	128	1.6	0.21
Lowland (20–100 m)	49	17,315	30,000	353	1.6	0.58
Hilly (100–500 m)	9	4261	25,454	473	0.35	0.16
Mountainous (>500 m)						
Geology						
Meta-sediment	56	7062	4802	126	11.7	1.47
Granite	14	16,048	64,290	1146	0.2	0.25
Land Cover						
Forest	6	553	19,504	92	0.3	0.03
Agriculture	28	13,391	28,919	478	1.0	0.46
Built-up area	26	7921	15,835	305	1.6	0.50
Cleared area	10	1245	4590	124	2.2	0.27

landslides have been reported in the urban center of the Langat Basin. Economic losses associated with damage to infrastructure such as roads and buildings are unknown. In Malaysia, the government rarely compiles information pertaining to social costs and economic losses from geohazards.

Landslides in the Langat Basin have occurred due to a combination of natural and technological causes. Six types of slope failure are observed: earth slump, earth fall, rock fall, earth slide, complex landslide, and debris flow. The two most common types of slope failure, earth fall and earth slump, have been recorded at 26 and 25 localities, respectively, from a total of 70 incidents. Earth fall often occurs in slopes with steepness more than 60°, while earth slump is common in slopes with steepness less than 45°. The other types of failures have occurred in fewer than five localities within the subbasin.

The parameter obtained by dividing the total volume of displaced material with the number of landslides in a given area provides an indication of the average size of the slope failure. This parameter was calculated for each category of topography, geology, and land cover within the subbasin (Table 131.1). For the category of topography, the volume of displaced material per landslide is relatively smaller in the floodplains compared with the hilly areas. This indicates that the size of the slope failure increases with the ruggedness of the topography. In the case of geology, it was observed that the slope failures over granitic terrain are larger than those within meta-sediments. As the granitic terrain forms a significant proportion of the rugged areas, this result is consistent with the earlier observation for the various categories of topography. In the context of land cover, larger slope failures occur within agriculture and built-up areas compared with cleared areas and forests.

The intensity of landslides is determined from the number of landslide occurrences per unit area (per 1000 ha). The displaced material is obtained by dividing the total volume of material associated with all slope failures in a given area (Table 131.1). Generally, landslide intensity and displaced material are higher in the floodplains and undulating lowland terrain than in hilly areas. The intensity of landslides is nearly 60 times greater and displaced material six times higher within meta-sediments compared with granitic terrain. In terms of land cover, the lowest intensity of landslides and displaced material occurs in forests. The intensity of landslides is the highest in cleared area, followed by built-up and agricultural areas. However, more material is displaced per unit area within the agriculture and built-up areas compared with cleared areas. This means that cleared areas have a high number of small slope failures.

Land Erosion in the Langat Basin

The study on land erosion focused on the entire basin. The overall percentage of land cover within the Langat Basin for the time domains of 1993, 1996, and 1998 is shown in Table 131.2. Between 1993 and 1998, the forest coverage in the Langat Basin has been reduced annually by about 10% to accommodate increases in other types of land use. During the same period, cleared, built-up, and agriculture areas increased annually by 15, 10, and 3%, respectively.

Table 131.3 shows the amount of soil loss within the different types of terrain and land cover in the Langat Basin for 1993, 1996, and 1998. Generally, the highest amount of soil loss occurs in undulating lowlands and hilly terrain, followed by mountainous areas and floodplains. The average soil loss is generally very high for cleared areas (152 t/ha/year) within the Langat Basin. In comparison, soil losses in the forest and agriculture areas are not significant (less than 4 t/ha/year). Between 1993 and 1998, soil loss from cleared areas fluctuated between 4 and 13% annually, while no significant changes were observed for soil loss in forest and agriculture areas. Changes in soil loss could be attributed to the location of the cleared land, i.e., floodplains, undulating lowlands, hilly or mountainous areas. The clearing of land, particularly in hilly areas, makes it susceptible to gully, sheet, and splash erosions as a result of the heavy tropical rainfall. The eroded material generally finds its way down to nearby channels and causes siltation of rivers.

The increase in soil loss during the period between 1993 and 1998 has been correlated to increased levels of suspended sediment in the automatic monitoring stations of rivers draining the Langat Basin (Wan et al., 1999). It was also noted that the highest levels of suspended sediments were recorded during the phase construction at monitoring stations located near areas being cleared. The levels observed normally exceeded the standards for aquatic life, drinking water, and irrigation set by the Department of Environment Malaysia. Concentrations of suspended sediments usually dropped to their original levels after the construction phase.

Table 131.2 Percentage of Land Cover in the Langat Basin for 1993, 1996, and 1998

Type of Land Cover	Coverage (%)		
	1993	1996	1998
Forest	38	32	23
Agriculture	48	52	55
Water body	1	1	1
Built-up area	11	13	17
Cleared area	2	2	4
Total coverage	100	100	100

Table 131.3 Soil Loss in Different Terrain and Land Cover within the Langat Basin for the Years 1993, 1996, and 1998

Soil Erosion	1993	1996	1998
Topography			
Floodplain (<20 m)	0.74	0.99	1.36
Lowland (20–100 m)	6.52	6.36	10.04
Hilly (100–500 m)	2.87	5.74	12.85
Mountainous (>500 m)	3.27	5.23	5.57
Land Cover			
Forest	0.189	0.197	0.21
Agriculture	3.7	3.62	3.6
Cleared area	114	165	152
Average basin soil loss	1.65	2.2	4.2

DIAGNOSIS OF ECOSYSTEM HEALTH

The ecosystem health of the Langat Basin, with respect to landslides and land erosion, can be assessed based on spatial and temporal perspectives. For landslides, it was not possible to establish trends in annual distribution and intensity in the context of different land uses. This is because the current assessment could not differentiate the number of annual landslide occurrences between the period of 1995 and 1998. Continuous annual assessment for at least 3 to 4 years into the future is required before such trends can be established. Trends for soil erosion can be established more readily from the manipulation of remote sensing and GIS modeling.

Based on the spatial perspective, the Langat Basin can be divided into three areas according to ecosystem health: forested mountain terrain, undulating lowlands and hilly areas, and coastal and floodplains. The best state of ecosystem health is within the forested areas in the mountainous terrain, located in the eastern part of the Langat Basin. Soil loss here is below 2 t/ha/year and landslides do not exceed natural levels in size, intensity, and material displacement. The worst state of ecosystem health occurs in undulating lowland and hilly areas, in the central part of the basin where development is focused. Soil loss in this area is generally between 12 to 13 t/ha/year. In areas with land-clearing activities for major construction projects, such as the Kuala Lumpur International Airport and Putra Jaya, soil loss is about 152 t/ha/year. The highest intensity and displaced material associated with landslides occur within this area. The ecosystem health of the coastal and floodplains in the eastern part of the Langat Basin is relatively moderate. Soil losses in the mainly agriculture-based areas here generally do not exceed 3.6 t/ha/year. The intensity of landslide occurrences is quite high, but the size of slope failures and volume of displaced material are relatively low.

Overall soil loss in the Langat Basin can be attributed to increased land-clearing activity to accommodate infrastructure and other forms of physical development. The average soil loss for the basin in 1993 was 1.6 t/ha/year. In the absence of allowable limits for soil erosion within the Langat Basin, the soil erosion value for 1993 was taken as the baseline level. Compared with 1993, soil loss has increased annually by about 20% to 4.2 t/ha/year in 1998. This situation is very unhealthy for the ecosystem. Soil loss results in increased amounts of suspended sediments in rivers, particularly during the construction phase, and further deterioration of the health of the ecosystem. Thus, land-clearing should not exceed 60 km² (or 2% of the total basin) in any one year in order to retain basin soil-loss conditions similar to 1993 and maintain the overall health of the Langat Basin ecosystem. This limit of land clearing would also ensure that landslide occurrences do not exceed natural levels.

Stringent management measures should be enforced to arrest the problem of excessive soil erosion during the construction phase of development projects in the Langat Basin. Soil loss in forest and agriculture areas of the Langat Basin have not changed significantly (less than 0.2 and 3.7 t/ha/year, respectively) between 1993 and 1998. In terms of soil management, it appears that the efficiency of soil management in these sectors, particularly for agriculture, has neither improved nor deteriorated. Thus, these two areas have not contributed significantly to the deterioration of ecosystem health with respect to soil loss.

CONCLUSION

The best state of ecosystem health in the Langat Basin is within the forested areas in the mountainous terrain, located in the upper eastern part of the basin. Soil loss here is below 2 t/ha/year and landslides do not exceed natural levels in term of size, intensity, and material displacement. The worst state of ecosystem health is recorded in the undulating lowland and hilly areas, within the central part of the basin, which is the focus of development. Soil loss in this area is generally between 12 to 13 t/ha/year, attaining levels of 152 t/ha/year where there are land-clearing activities. The highest intensity and displaced material associated with landslides occur within this area. The ecosystem health of the coastal and floodplains located in the eastern part of the Langat Basin is relatively moderate, with soil losses generally not exceeding 3.6 t/ha/year. The intensity of landslide occurrences is quite high, but the size of slope failures and volume of displaced material are relatively low.

Overall soil loss in the Langat Basin can be attributed to increased land-clearing activity to accommodate infrastructure and other forms of physical development. Since 1993, soil loss increased by 20% annually to 4.2 t/ha/year in 1998. The soil loss resulted in increased amounts of suspended sediments in rivers, particularly during the construction phase, and further deterioration of the health of the ecosystem. It is estimated that in order to improve the Langat Basin ecosystem, land clearing should not exceed 60 km^2 (or 2% of the total basin) in any one year. Furthermore, stringent measures should be enforced to arrest the problem of excessive soil erosion during the construction phase.

ACKNOWLEDGMENTS

This chapter is part of the research funded by MATREM CP/5220-97-02, a project of the United Nations Environment Programme's Network for Environmental Training at Tertiary Level in Asia and the Pacific (NETLAPP). The encouragement and support of Prof. Mohd Nordin Hj. Hasan, the project leader, is gratefully acknowledged.

REFERENCES

Aguilar, B.J., Application of ecosystem health for the sustainability of managed systems in Costa Rica, *Ecosyst. Health,* 5, 36–48, 1999.

Aswathanarayana, U., *Geoenvironment: An Introduction,* A.A. Balkema, Rotterdam, the Netherlands, 1995.

Boetzkes, E., Gender, risk and scientific proceduralism, *Ecosyst. Health,* 4, 170–176, 1998.

Buckingham, D., Does the World Trade Organization care about ecosystem health? *Ecosyst. Health,* 4, 92–108, 1998.

Callicott, J.B., Aldo Leopold's metaphor, in *Ecosystem Health: New Goals for Environmental Management,* Costanza, R., Norton, B.G., and Haskell, B.D., Eds., Island Press, Washington, D.C., 1992.

Department of Agriculture, Erosion Risk Map, Department of Agriculture, Kuala Lumpur, Malaysia, 1996.

Fausch, K.D., Lyons, J., Karr, J.R., and Angermeir, P.L., Fish communities as indicators of environmental degradation, *Am. Fish. Soc. Symp.,* 8, 123–144, 1990.

Fox, M.A., The contribution of vegetarianism to ecosystem health, *Ecosyst. Health,* 5, 70–74, 1999.

Hansen, A., Landslide hazard analysis, in *Slope Stability,* Brunsen, D. and Prior, D.B., Eds., John Wiley & Sons, New York, 1984, pp. 523–602.

Johnson, G.D. and Patil, G.P., Quantitative multiresolution characterization of landscape patterns for assessing the status of ecosystem health in watershed management areas, *Ecosyst. Health,* 4, 177–187, 1998.

Jones, D.K.C., Landslide hazard assessment in the context of development, in *Geohazards: Natural and Man-Made,* McCall, G.J.H., Laming, D.J.C., and Scott, S.C., Eds., Chapman & Hall, London, 1992.

Karr, J.R., and Chu, E.W., Ecological integrity: reclaiming lost connection, in *Perspective on Ecological Integrity,* Westra, L. and Lemons, J., Eds., Kluwer, Dordrecht, the Netherlands, 1995, pp. 34–48.

Karr, J.R., Fausch, K.D., Angermeier, P.L., Yant, P.R., and Schlosser., I.J., Assessing Biological Integrity in Running Waters: A Method and Its Rationale, Illinois Natural History Survey Special Publication 5, Champaign, 1986.

King, L.A. and Hood, V.L., Ecosystem health and sustainable communities: north and south, *Ecosyst. Health,* 5, 49–57, 1999.

Komoo, I., Pereira, J.J., and Maziah, S., Geobencana di Lembangan Klang, paper presented at the Seminar Geologi Sekitaran, 7–8 Dis., Malaysia, 1996.

McCall, G.J.H., Natural and man-made hazards: their increasing importance in the end twentieth century world, in *Geohazards: Natural and Man-Made,* McCall, G.J.H. et al., Eds., Chapman & Hall, London, 1992, pp. 1–4.

Omakupt, M., Soil erosion mapping using remote sensing data and GIS, Ministry of Agriculture and Cooperative, Thailand, 1987.

Pellerin, J. and Grondin, J., Assessing the state of Arctic ecosystem health: bridging Inuit viewpoints and biological endpoints on fish health, *Ecosyst. Health,* 4, 236–247, 1998.

Pereira, J.J. and Komoo, I., Using geological information as planning tools for urban centers: the case of Klang Valley, Malaysia, in *Engineering Geology: A Global View from the Pacific Rim,* Moore, D.P. and Hungr, O., Eds., Balkema, Rotterdam, 1998, pp. 2573–2578.

Pereira, J.J., Komoo, I., and Maziah, S., Geohazards and the urban ecosystem, in *Geohazards: Landslide and Subsidence,* Kuala Lumpur, Geological Society of Malaysia, 1996.

Rapport, D.J., Thorpe, C., and Regier, H.A., Ecosystem medicine, *Bull. Ecol. Soc. Am.,* 60, 180–182, 1979.

Rapport, D.J., Regier, H.A., and Thorpe, C., Diagnosis, prognosis and treatment of ecosystem under stress, in Stress *Effects on Natural Ecosystems,* Barrett G.W. and Rosenberg, R., Eds., John Wiley & Sons, New York, 1981, pp. 269–280.

Rapport, D.J., Regier, H.A., and Hutchinson, T.C., Ecosystem behaviour under stress, *Am. Nat.,* 125, 617–640, 1985.

Rapport, D.J., What constitutes ecosystem health? *Perspect. Biol. Med.,* 33, 120–132, 1989.

Rapport, D.J., Gaudet, C., and Calow, P., *Evaluating and Monitoring the Health of Large-Scale Ecosystems,* Springer-Verlag, Berlin, 1995.

Rapport, D.J., Epstein, P.R., Levins, R., Costanza, R., and Gaudet, C., *Ecosystem Health,* Blackwell Science, Malden, MA, 1998.

Ravishankar, H.M., Watershed prioritization through the universal soil loss equation using digital satellite data and an integrated approach, *Asian-Pac. Remote Sensing J.,* 6, 56–60, 1994.

Saha, S.K. and Pande, L.M., An integrated inventory for environment conservation using satellite and agro-meteorological data, *Asian-Pac. Remote Sensing J.,* 5, 21–28, 1993.

Schaeffer, D.J., Henricks, E.E., and Kerster, H.W., Ecosystem health: measuring ecosystem health, *Environ. Manage.,* 12, 445–455, 1988.

Selby, M.J., *Hillslope Materials and Processes,* Oxford University Press, Oxford, 1993.

Shu, Y.K., Investigation of land subsidence and sinkhole occurrences in the Klang Valley and Kinta Valley, Peninsular Malaysia, in *Landplan II Proceedings — Role of Geology in Planning and Development of Urban Centres in Southeast Asia,* Tan, B.K. and Rau, J.L., Eds., AIT, Bangkok, 1986, pp. 15–28.

Shu, Y.K. and Chow, W.S., Land Subsidence at Serdang Lama, Selangor, Report E(F)4/1979, Geological Survey of Malaysia, Malaysia, 1979.

Wan, M.M., Komoo, I., and Pereira, J.J., Degradasi tanah Lembangan Langat: Kajian Kes di sub-lembangan Ulu Sungai Langat, paper presented at Dialog Geosains: Geologi Alam Sekitar dan Bencana Geologi di Lembah Kelang-Langat, Geological Survey Department, 25 November 1997, Kuala Lumpur, Malaysia, 1998.

Wan, M.M., Komoo, I., and Pereira, J.J., Land degradation in Langat Basin: a case study of the Ulu Sungai Langat subbasin, paper presented at the Symposium of Langat Basin Ecosystem Health, 5–6 June 1999, Shah Alam, Malaysia, 1999.

Yassi, A., Mas, P., Bonet, M., Tate, R.B., Fernandez, N., Spiegel, J., and Perez, M., Applying an ecosystem approach to determinants of health in Centro Habana, *Ecosyst. Health,* 5, 3–19, 1999.

CHAPTER 132

River Water Quality Assessment and Ecosystem Health: Langat River Basin, Selangor, Malaysia

Muhammad Abu Yusuf, M. Nordin, and Pauzi Abdullah

INTRODUCTION

The Langat River Basin, which is adjacent to the Klang Valley, has an area of about 2938 km^2 and an estimated population of about 725,000 in 1997. Administratively it is under three whole districts — Hulu Langat, Sepang, and Kuala Langat of Selangor — as well as the southern part of the Seremban District in the state of Negeri Sembilan. The basin has been the focus of major recent development that included the new federal government administrative capital of Putrajaya, the multimedia supercorridor linking Putra Jaya with KLIA and Kuala Lumpur, Cyber Jaya, and the Kuala Lumpur International Airport (KLIA) at Sepang (Nordin and Azrina, 1998). Additionally, the development of new townships such as Bandar Baru Nilai and expansion of existing towns and villages such as Dengkil and Semenyih contribute to the growth of an increasingly dense urban sprawl with concomitant urbanization and industrialization that can potentially cause complex environmental problems.

The rivers of the Langat Basin play a significant role in the ecology of the basin, providing potable water to residents within the basin and the highly urbanized Klang Valley, and supplying industries and agricultural areas with water for manufacturing and irrigation. At present, a total of eight water treatment plants are located in the basin. Recent development activities in the basin can threaten the water quality in its rivers and disrupt its ability to meet water requirements in the basin and elsewhere. This chapter presents the findings of a 12-month study of water quality in the rivers of the Langat Basin and relates them to ecosystem health issues highly relevant to the future of the Langat Basin ecosystem. The study is a meso-scale study that examines river water quality in the entire basin.

Malaysian rivers are classified in the Interim Water Quality Standards for Malaysia (INWQS) based on water quality criteria and standards for several beneficial uses (DOE, 1986, 1990). The six river-use classifications as defined by INWQS are shown in Appendix 132.I.

MATERIALS AND METHODS

The Langat River Basin is divided into three ecological zones based on forest and soil types. Ecological Zone 1 comprises upland forest, and is located in the northern part of the basin;

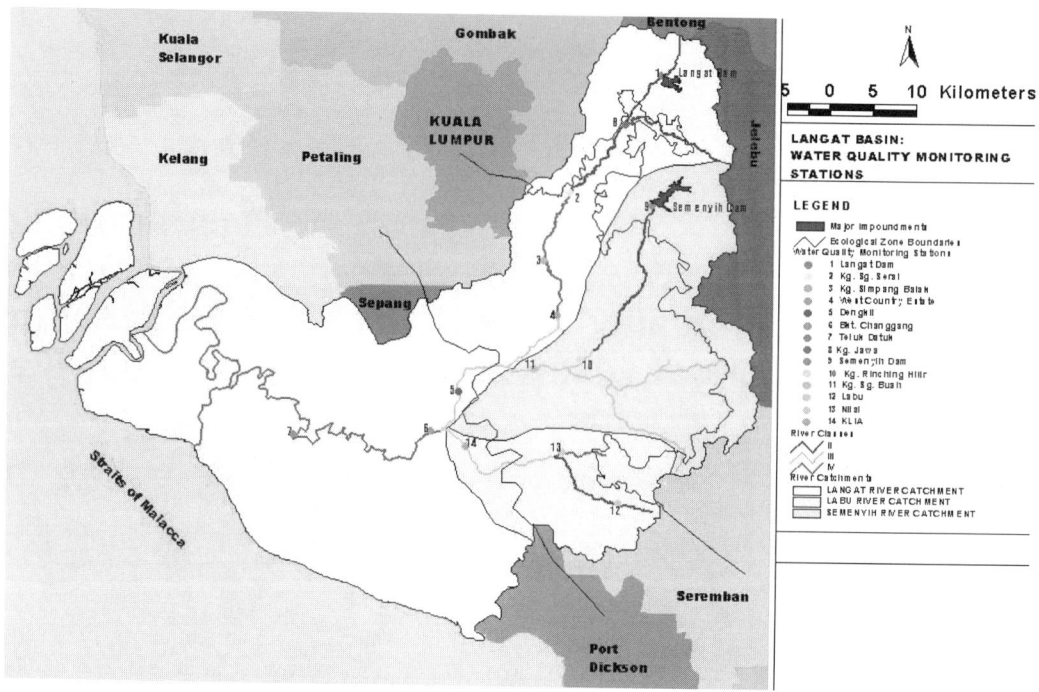

Figure 132.1 Map of the Langat River Basin and water quality monitoring stations along the rivers of the Langat Basin.

Ecological Zone 2 is comprised of lowland forest and highly urbanized areas, and is located in the middle part of the basin. Ecological Zone 3 is comprised of peat-swamp mangrove forest and agricultural land, and is located in the southern part of the basin near the Straits of Malacca.

A total of 14 monitoring stations was established along six rivers of the Langat Basin (Figure 132.1). On-site surveys were conducted prior to selection of sampling sites. Monitoring stations of the Department of Environment (DOE), the Department of Irrigation and Drainage (DID), and other study groups in the Langat Basin were reviewed and visited before selecting the monitoring stations specific to this study. Monitoring stations were selected based on accessibility and ability to sample under all weather conditions; homogeneity of the water column, i.e., the water must be fully mixed at the point of sampling so that representative samples of river water can be easily obtained at the place and time of sampling; stream characteristics so that at least one station was selected to represent subcatchment areas; water supply intake points; towns in the basin; and location of industrial sites. Stations were located a few kilometers downstream of any point source of pollution, not immediately below it.

Grab samples were collected monthly for 1 year from May 1998 to April 1999. Water sampling and analytical methods followed GEMS and Water Operations Guide (Anon., 1978), APHA Standard Methods (APHA, 1989, 1992), and HACH Quick Method (HACH, 1989). During the first 6 months, sampling commenced from the upper reaches of the river. The sampling order was reversed on the seventh month when sampling began from downstream sampling sites and proceeded upstream. All samples were collected and sent to the laboratory on the same day.

The following parameters were studied: water temperature, pH, salinity, conductivity, turbidity, total suspended solids (TSS), volatile solids, fixed solids, dissolved oxygen (DO), biochemical oxygen demand (BOD), chemical oxygen demand (COD), ammoniacal-nitrogen, nitrate-nitrogen, nitrite-nitrogen, phosphate-phosphorus, hexane extractable materials (oil and grease), total phenol, pesticides, and heavy metals (arsenic [As], aluminum [Al], cadmium [Cd], chromium [Cr], copper [Cu], cobalt [Co], iron [Fe], lead [Pb], mercury [Hg], nickel [Ni], and zinc [Zn]). Six physical parameters (temperature, pH, DO, salinity, turbidity, and conductivity) were measured *in situ* using portable instruments (DO meter, pH meter, turbidimeter, and conductivity meter). Three other physical parameters were determined in the laboratory along with 19 chemical parameters.

APHA method 2540D (1989) and 2540E (1989) were employed for the analysis of total suspended solids, volatile solids, and fixed solids. The HACH quick method based on DR/2000 was used to analyze inorganic nutrients. Program 385 (salicylate method), program 351 (cadmium reduction method), program 371 (diazotization method), and program 490 (ascorbic acid method) were used for the analyses of ammoniacal–nitrogen, nitrate–nitrogen, nitrite–nitrogen, and phosphate–phosphorus, respectively.

APHA method 5210 (1989) was used for the analysis of BOD. HACH colorimetric method (program 430) was used for COD analysis. Oil and grease (O&G) were extracted with n-hexane according to EPA method 1664 (liquid–liquid extraction). APHA method 5530 (1989) was employed for total phenol determination.

The trace metals Cu, Cd, Cr, Co, Ni, Fe, Zn, and Pb were analyzed using atomic absorption spectrophotometry (AAS) according to APHA method 3111B (direct air–acetylene method). Arsenic was analyzed using APHA method 3114B (1989) (hydride generation method), and mercury was analyzed using APHA method 3112B (1989) (cold-vapor generation method). Aluminum was determined using HACH method 10 (aluminum method).

Strict quality assurance procedures were used for laboratory analysis. These included field and laboratory blank analysis, replicate sample spiked with known concentrations of parameters to be analyzed, and duplicate sample analysis. Analytical determinations were checked using reference materials, while instruments were routinely calibrated.

Water quality data were stored in Microsoft Excel 7.0. The data were converted to MS-DOS for statistical analysis with the aid of an SAS statistical package. Data were also converted to Microsoft Access 7.0 to prepare the map employing a geographical information system (GIS) using Arc-View 3.0 and Arc-Info 7.1.2 software.

Water quality assessment results were compared to present INWQS with respect to aquatic life, water supply, and irrigation (DOE, 1994).

Statistical tests used for examining water quality data included analysis of variance (ANOVA) and Duncan's multiple-range test with the aid of an SAS statistical package. The two-way ANOVA was conducted for each parameter to test for significant differences in means between stations, time of sampling (morning and afternoon), and interactions (Station * Time). Duncan's multiple-range test was performed to determine which ranges of means were significantly different.

The DOE-WQI index was used to identify the status of river water quality and classify stretches of the Langat Basin rivers into river classes, according to the system adopted by the DOE (DOE, 1994, 1998). Six water quality parameters (DO, pH, BOD, COD, TSS, and ammoniacal-nitrogen) were used to calculate this index (Norhayati, 1981). The index was established based on the results of an opinion poll of a panel of experts who determined the choice of parameters and weights assigned to each chosen water quality parameter (DOE, 1994). Calculations were performed not on the parameters themselves but on their subindices whose value are obtained from a series of equations. These are best-fit equations obtained from rating curves (Norhayati, 1981). The subindices for these parameters are named SIDO, SIBOD, SICOD, SIAN, SISS, and SIPH. The formula used to calculate WQI is as follows:

$$WQI = 0.22 * SIDO \pm 0.19 * SIBOD \pm 0.16$$
$$* SICOD \pm 0.15 * SIAN \pm 0.16 * SISS \pm 0.12 * SIPH$$

(* indicates multiplication), where the multipliers are the weights for the corresponding parameters with a total value of 1.

The formula used to calculate subindices are shown in Appendix 132.III.

The health of the aquatic ecosystem of the Langat Basin was assessed using the Schaeffer and Cox (1992) definition of ecosystem health and by the assessment criteria for physicochemical components of ecosystems suggested by Ramade (1995).

RESULTS AND DISCUSSION

Sampling during the duration of the study proceeded successfully and no major problems were encountered. Duplicate samples were analyzed and average results are presented below for each parameter studied. All values are averages of 12 samples collected monthly throughout the duration of the study. It should be stated that the period of sampling spanned the extended period of drought that occurred in 1998.

Means of each parameter for all monitoring stations are shown in Appendix 132.III and displayed in bar charts (Figures 132.2 to 132.14). The results of ANOVA tests are shown in Tables 132.1 and 132.2.

Table 132.1 Significant and Nonsignificant Differences for Physical Parameters Determined in This Study

Parameters	Significant/Nonsignificant		
	Between Stations	Sampling Time (AM and PM)	Interaction (Station * Time)
Temperature	b	b	b
pH	NS	a	NS
Dissolved oxygen	b	b	NS
Turbidity	b	NS	NS
Conductivity	b	a	NS
Total suspended solids	b	NS	NS
Volatile solids	b	NS	NS
Fixed solids	b	NS	NS

a = significant; b = highly significant; NS = nonsignificant.

Table 132.2 Significant and Nonsignificant Differences for Chemical Parameters Determined in This Study

Parameters	Significant/Nonsignificant		
	Between Stations	Sampling Time (AM and PM)	Interaction (Station * Time)
Ammoniacal-nitrogen	b	NS	a
Nitrate-nitrogen	b	NS	NS
Nitrite-nitrogen	b	NS	NS
Phosphate-phosphorus	b	NS	NS
Biochemical oxygen demand	b	NS	NS
Chemical oxygen demand	b	NS	NS
Oil and grease	b	NS	NS
Phenol	NS	NS	b
Lead	NS	NS	b
Cadmium	NS	NS	NS
Iron	b	NS	b
Aluminum	b	NS	b
Copper	NS	NS	NS
Cobalt	NS	NS	NS
Nickel	NS	NS	a
Zinc	b	NS	NS

a = significant; b = highly significant; NS = nonsignificant.

PHYSICAL PARAMETERS

Temperature

The average water temperature measured along the rivers of the Langat Basin ranged from 25.9 to 29.4°C (Figure 132.2). This is within the normal range of water temperature for Malaysian rivers (Ainon and Sukiman, 1987; Anhar et al., 1988; DOE, 1997). Water temperature in the upper reaches near the Langat and Semenyih dams averaged 25.9 and 28.4°C, respectively. Temperature gradient between the upper and lower reaches of the rivers was monitored. Water temperatures increased from Kg. Sg. Serai to Teluk Datuk along the Langat River and from Semenyih Dam to Sg. Buah along the Semenyih River. The highest average temperature measured (29.4°C) was along the Batang Nilai River.

Differences of mean temperature levels between stations, sampling time (AM and PM) and interaction (Station * Time) were highly significant ($p \leq 0.01$) (Table 132.1). Duncan's multiple-range test results grouped the range of means at different stations into four. The difference of mean temperatures at Nilai, Teluk Datuk, Dengkil, West Country Estate, and Kg. Simpang Balak were significantly higher than at other stations; Langat Dam was significantly lower ($p \leq 0.05$) than all other stations. Significant variations of mean temperatures ($p \leq 0.01$) were observed between sampling times and the interaction between sampling stations and sampling time was highly significant ($p \leq 0.01$).

Several factors influence river water temperature including location and time of sampling. Downstream areas of the river are less shaded by trees compared with the upper reaches and tend to show higher temperatures because of direct exposure to sunlight.

Salinity

Zero salinity was observed *in situ* along the rivers of the Langat Basin. Teluk Datuk and Kg. Bukit Changgang monitoring stations are near the coast, and zero salinity was also measured there as all the samples were collected during low tide. Salinity in the Langat can reach 0.15 ppt in Kg. Bukit Changgang (DOE, 1997) and 0.37 ppt in Teluk Datuk (DOE, 1998). The INWQS maximum of salinity for Malaysian rivers is 1 ppt.

pH

The average pH for measurements made at all stations ranged from 6.4 to 7.2 (Figure 132.3). This is within the INWQS range of pH for Malaysian rivers to support aquatic life. In Teluk Datuk (Station 7), low pH values of 4.9, 4.4, 4.1, and 5.0 were recorded for the months of August 1998, September 1998, January 1999, and February 1999, respectively. During the other months, pH

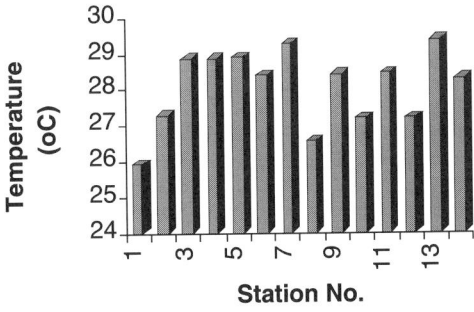

Figure 132.2 Temperature at different stations along the rivers of the Langat Basin.

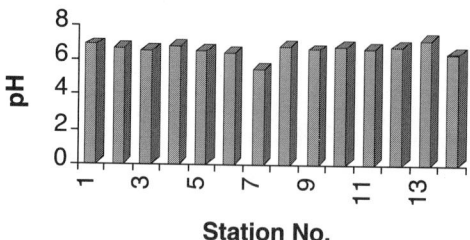

Figure 132.3 The pH value at different stations along the rivers of the Langat Basin.

values ranged between 5.3 and 6.4. The average pH range along the rivers of the Langat Basin was within range for Malaysian rivers.

Differences in mean pH levels between stations and sampling times were highly significant ($p \leq 0.01$) and significant ($p < 0.05$), respectively. Interactions between stations and sampling times were not significant ($p < 0.05$) (Table 132.1). Duncan's multiple-range test results grouped the range of means at different stations into five. The differences in mean pH at Nilai, Langat Dam, Kg. Jawa, and Kg. Rinching Hilir were significantly higher than those of all other stations, while at Teluk Datuk the difference was significantly lower ($p \leq 0.05$) than all other stations.

Previous research records a pH range of 6.3 to 7.6 along the rivers of the Langat Basin (Ainon and Sukiman, 1987; Anhar et al., 1988; DOE, 1997, 1998). Extreme values of pH 5.6 to 8.9 have been recorded in the past (Sukiman, 1985). The pH of most natural waters ranged between 6.0 and 8.5, although lower values can occur in dilute waters high in organic content, and high values in eutrophic waters (Chapman, 1992). A pH range of 6.0 to 9.0 is suitable for supporting freshwater fish and bottom-dwelling invertebrates. The INWQS range for pH in Malaysian rivers to support aquatic life is 5 to 9.

Turbidity

The average turbidity along the rivers of the Langat Basin ranged from 20.67 to 544.17 NTU. INWQS does not propose any threshold value for turbidity of fresh waters for the support of aquatic life. The Ministry of Health has set a threshold level of raw water turbidity at 1000 NTU. The turbidity of river water in the Langat Basin is below this threshold level. Previous turbidity measurements of rivers in the Langat Basin reported values ranging from 17 to 98 NTU (DOE, 1997).

Differences in mean turbidity between stations were highly significant ($p < 0.01$), but differences between sampling time and interactions (Station * Time) were not significant ($p > 0.05$) (Table 132.1). Duncan's multiple-range test results grouped the means at different stations into four. Mean turbidities at Teluk Datuk were significantly higher ($p < 0.05$) than all other stations and at Langat Dam, Semenyih Dam, Labu, Nilai, Kg. Jawa, and Kg. Sg. Serai they were significantly lower ($p < 0.05$).

The highest turbidity and TSS values were obtained at Teluk Datuk (Station 7). Waters containing suspended solids and organic matter generally show high turbidity. Turbid water is not suitable for industrial use, can harbor pathogenic microorganisms and thus increase the risk of waterborne diseases. The rivers of the Langat Basin can still support aquatic life, as turbidity was within the acceptable range for Malaysian rivers.

Conductivity

Conductivity reflects the concentration of electrolytes in water. Average conductivity in the rivers of the Langat Basin ranged from 39.16 to 406.08 µmhos. The highest level of conductivity was measured at Nilai (406.08 μmhos) along Batang Nilai River and the lowest at the Langat Dam (39.16 μmhos) along the Langat River. Previously reported measurements of conductivity ranged between 42.2 to 18 μmhos (DOE, 1997), 30.84 to 523.60 μmhos (DOE, 1998), and 50 to 150 μmhos (Anhar et al., 1988). These values were below the INWQS maximum threshold level (1000 NTU) for Malaysian rivers to support aquatic life.

Differences in mean conductivity between stations and sampling times were highly significant ($p < 0.01$ and $p < 0.05$, respectively); interaction between stations and time (Station * Time) was not significant ($p > 0.05$) (Table 132.1). Duncan's multiple-range test results grouped means at different stations into four. The difference of mean conductivity at stations Nilai and Teluk Datuk were significantly higher than at all other stations. The conductivity at Langat Dam, Kg. Sg. Serai, Kg. Jawa, Labu Semenyih Dam, Kg. Sg. Buah, and Kg. Rinching Hilir were significantly lower than those at all other stations.

Differences in conductivity between sampling times indicated that at some locations values were higher in the afternoon than in the morning. High temperature in the afternoon increasingly dissolves some salts in river water and can explain the higher conductivity.

Total Suspended Solids

Differences in average TSS content of the Langat Basin rivers were considerable, ranging from 6.33 to 528.75 mg/l (Figure 132.4). The lowest TSS levels were obtained in the Langat Dam and highest in Teluk Datuk (Station 7) along the Langat River.

Differences in means between stations were highly significant ($p < 0.01$); differences in sampling times and interactions between stations and time (Station * Time) were not significant ($p \leq 0.05$) (Table 132.1). Duncan's multiple-range test grouped mean TSS at the different stations into five. The difference in mean TSS at Teluk Datuk was significantly higher ($p < 0.05$) than at all other stations. Total suspended solids at the Langat Dam, the Semenyih Dam, Labu, Nilai, Kg. Jawa, and Kg. Sg. Serai were significantly lower ($p < 0.05$) than all other stations.

The INWQS level of TSS for supporting healthy aquatic life in a freshwater ecosystem is 150 mg/l. Particular sections of the rivers of the Langat Basin are badly polluted by suspended sediments. A very high TSS was obtained at Kg. Rinching Hilir, with a value of 1704 mg/l for September 1998. However, the sample was collected during storm flow after heavy rain. Furthermore, a rubber plantation was being clear-felled, and soil from bare land was being transported directly to the river with surface runoff. These events explain the high TSS readings from this station. High levels of TSS were also obtained along the Langat River from Kg. Simpang Balak to Teluk Datuk. Upstream of Kg. Simpang Balak (Station 3), a highway leading to Cheras was being constructed. In March 1999, a high level of TSS was obtained from the station at West Country Estate (1549 mg/l) when samples were collected after rain. TSS were repeatedly high at stations Kg. Bukit Changgang and Teluk Datuk. During the study period, development of KLIA was in progress and large tracts of land were not fully protected by vegetation. The highest level of TSS obtained in Teluk Datuk

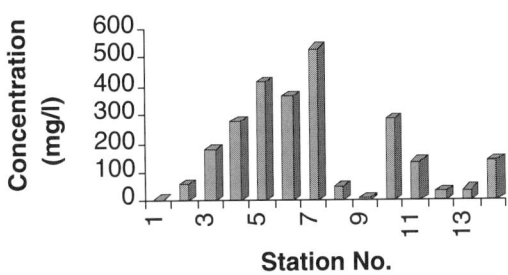

Figure 132.4 Total suspended solids levels at different stations along the rivers of the Langat Basin.

station was 1030 mg/l in March 1999. Previous studies had reported TSS values ranging from 6.5 to 1059.2 mg/l (DOE, 1997, 1998).

Waters high in suspended solids are aesthetically unsatisfactory for bathing, can harbor harmful microorganisms, deter the hatching of fish eggs, and are lethal to some aquatic organisms. Because sections of the rivers of the Langat Basin are already high in TSS, further deterioration will increase the risk of ecological collapse and make the river dead, without aquatic life.

Volatile Solids and Fixed Solids

The average results of volatile solids and fixed solids along the rivers of the Langat Basin ranged from 6 to 88.86 mg/l and 0 to 428.77 mg/l, respectively. The highest level of volatile solids was obtained in Teluk Datuk (Station 7) and the lowest near Langat Dam (Station 1). On the other hand, high levels of fixed solids were obtained also in Teluk Datuk (Station 7) and low levels in the Semenyih River (Station 9).

Differences in means of volatile and fixed solids between stations were highly significant ($p \leq 0.01$). The means between sampling time and interaction of stations and time (Station * Time) were not significant ($p > 0.05$) (Table 132.1). Duncan's multiple-range test results grouped mean volatile and fixed solids at different stations into five. The difference in mean volatile solids at Teluk Datuk was significantly higher ($p \leq 0.05$) than all other stations. Volatile solids at the Langat Dam, the Semenyih Dam, Labu, Kg. Sg. Serai, and Kg. Jawa were significantly lower ($p \leq 0.05$) than all other stations. On the other hand, the differences in mean fixed solids at Teluk Datuk and Dengkil were significantly higher ($p \leq 0.05$) while fixed solids at the Semenyih Dam, the Langat Dam, Nilai, and Labu were significantly lower ($p \leq 0.05$) than all other stations.

Volatile and fixed solids reflect the organic and inorganic metal oxide contents of suspended solids in the river water. These correlate with DO concentrations in water and organic content in suspended solids. Generally, suspended solids containing highly volatile solids result in low DO and high BOD. In Teluk Datuk, the DO level was persistently low, while BOD levels were persistently high during the study period. INWQS did not propose any threshold level of volatile solids and fixed solids for Malaysian rivers.

Dissolved Oxygen (DO)

Average DO levels in the rivers of the Langat Basin ranged from 1.38 to 7.30 mg/l (Figure 132.5). The acceptable range of DO for Malaysian rivers by INWQS standards is 3 to 5 mg/l. The DO level along the Langat River at West Country Estate (2.72 mg/l), Teluk Datuk (1.38 mg/l), and in Nilai (2.69 mg/l) along the Batang Nilai river were below the INWQS acceptable range. At other stations, DO ranged between 3.78 to 7.3 mg/l and were within acceptable the range for Malaysian rivers. The lowest DO was measured in Teluk Datuk (Station 7). DO ranges between

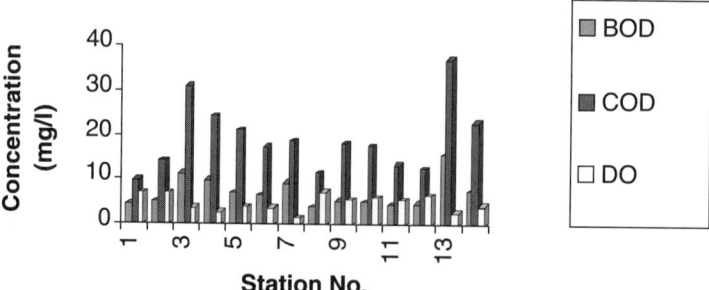

Figure 132.5 Dissolved oxygen, biochemical oxygen demand, and chemical oxygen demand levels at different stations along the rivers of the Langat Basin.

2.73 to 8.3 mg/l have been reported previously for rivers in the Langat Basin (Ainon and Sukiman, 1987; Anhar et al., 1988; DOE, 1997).

Differences in mean DO between stations and between sampling times were highly significant ($p < 0.01$); interactions between station and time (Station * Time) were not significant ($p > 0.05$) (Table 132.1). Duncan's multiple-range test grouped mean DO at the different stations into six. Differences in means at Kg. Jawa, the Langat Dam, and Kg. Sg. Serai were significantly higher ($p < 0.05$) than at other stations. DO at Teluk Datuk was significantly lower ($p < 0.05$) than all other stations.

The differences between DO and sampling time indicated that at some locations DO values were lower in the afternoon than in the morning. High temperature in the afternoon releases DO from river water and this could be the reason for the low DO value. On the other hand, at some stations DO value were low even in the morning. Organics from anthropogenic sources can cause low DO in river water even in the morning. Stations with low DO in the morning and afternoon are generally located downstream from urban and industrial areas. Effluents from industries, sewage treatment plants, urban runoff, and pig farm waste containing high levels of oxygen-demanding substances could all contribute to low DO levels in river water.

Generally, 3 mg/l DO or less is stressful to aquatic vertebrates and other aquatic life. DO levels below 2 mg/l kills most fish species. However, a few tropical fish species, especially air-breathing species, can survive at low DO levels of between 0.8 to 1.1 mg/l (Lim, 1974). Water quality criteria for acceptable DO levels that support aquatic life vary from country to country (Anonymous 1976; Hart, 1974; Pescod, 1973).

Ammoniacal-Nitrogen

The average ammoniacal-nitrogen concentration obtained in rivers of the Langat Basin ranged from 0.59 to 4.51 mg/l (Figure 132.6). INWQS recommends a maximum level of ammoniacal-nitrogen for the support of aquatic life in Malaysian rivers of 0.9 mg/l. The lowest level of ammoniacal-nitrogen was obtained in Kg. Sg. Serai (0.59 mg/l) and the highest in Kg. Simpang Balak (4.51 mg/l). Except for stations at Kg. Sg. Serai (0.59 mg/l), the Semenyih Dam (0.67 mg/l), Kg. Jawa (0.75 mg/l), and Labu (0.77 mg/l), ammoniacal-nitrogen levels in all other stations exceeded the recommended INWQS threshold. Many parts of the rivers of the Langat Basin are badly polluted by ammoniacal-nitrogen.

Differences in means between stations were highly significant ($p < 0.01$); interactions of stations and time of sampling (Station * Time) were significantly different ($p > 0.05$) (Table 132.2). Duncan's multiple-range test results grouped the mean ammoniacal-nitrogen at different stations into six. Mean differences at Kg. Simpang Balak and West Country Estate were significantly higher

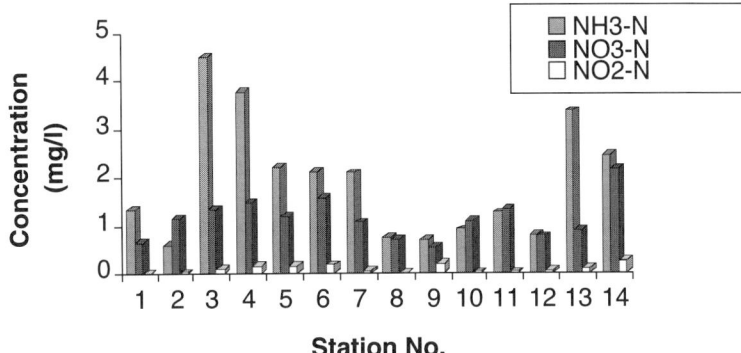

Figure 132.6 Ammoniacal-nitrogen, nitrate-nitrogen, and nitrite-nitrogen levels at different stations along the rivers of the Langat Basin.

($p < 0.05$) than all other stations, while ammoniacal-nitrogen at Kg. Sg. Serai was significantly lower ($p < 0.05$) than all other stations.

Simpang Balak is located downstream of the second biggest industrial park in the Balakong industrial area and industrial effluents contribute to the high level of ammoniacal-nitrogen at this station. West Country Estate is located downstream of Kajang, the largest town in the basin. A significant number of industries are located near Kajang town and industrial effluents are likely to contribute to obtain high ammoniacal-nitrogen at West Country Estate.

Previous reports suggested that the ammoniacal-nitrogen levels in the rivers of the Langat Basin ranged from 0 to 5.69 mg/l (DOE, 1997, 1998). In 1997 to 1998, the water quality of the Langat Basin rivers occasionally deteriorated to such an extent that water treatment plants had to cease operation. Conventional water treatment plants cannot treat raw water when ammoniacal-nitrogen concentration is higher than 3 mg/l (DOE, 1998). The 11th-mile Cheras water treatment plant was closed eight times between October 1997 and March 1998 for a total of 96 h, as the ammoniacal-nitrogen and O&G levels in raw river water were higher than this maximum (DOE, 1998). The Bukit Tampoi water treatment plant was closed three times in March 1998 for a total of 27 h for the same reasons (DOE, 1998). Salak Tinggi and Langat water treatment plants were closed once in 1998 due to high O&G and ammoniacal-nitrogen levels (NST, 1998). The aquatic ecosystem of the Langat River Basin is severely polluted with ammoniacal-nitrogen; further deterioration will aggravate matters, prevent rivers from supporting aquatic life, and make them unfit for processing into drinking water.

Nitrate- and Nitrite-Nitrogen

The INWQS levels of nitrate- and nitrite-nitrogen for the support of aquatic life in Malaysian rivers are 10 and 0.4 mg/l, respectively. The average nitrate- and nitrite-nitrogen concentrations in samples of river water from the Langat Basin ranged from 0.56 to 2.14 mg/l and 0.004 to 0.243 mg/l, respectively (Figure 132.6), below INWQS levels. This suggests that the rivers of the Langat Basin are generally not badly polluted by runoff of nitrate- and nitrite-nitrogen.

Differences in means of nitrate- and nitrite-nitrogen between stations were highly significant ($p < 0.01$), while differences between sampling times and interactions between stations and sampling times (Station * Time) were not significant ($p > 0.05$) (Table 132.2). Duncan's multiple-range test grouped mean nitrate- and nitrite-nitrogen at the different stations into four. The difference of mean nitrate-nitrogen at KLIA was significantly higher ($p < 0.05$) than all other stations. The nitrate-nitrogen levels at Semenyih Dam, Langat Dam, and Kg. Jawa were significantly lower ($p < 0.05$) than at all other stations. On the other hand, the difference in means of nitrite-nitrogen at Nilai was significantly higher ($p < 0.05$) than all other stations. The nitrite-nitrogen levels at Kg. Bukit Changgang, the Semenyih Dam, Dengkil, Teluk Datuk, KLIA, the Langat Dam, Kg. Jawa, and Kg. Sg. Serai were significantly lower ($p < 0.05$) than all other stations.

Surface waters normally contain nitrate concentrations up to 5 mg/l nitrate-nitrogen, but often lower than 1 mg/l nitrate-nitrogen. On the other hand, nitrite concentrations in fresh waters are usually very low, about 0.001 mg/l nitrite-nitrogen and rarely higher than 1 mg/l nitrite nitrogen.

Phosphate-Phosphorus

The average phosphate-phosphorus concentration in rivers of the Langat Basin ranged from 0.16 to 0.44 mg/l (Figure 132.7). The highest average concentration was in Nilai (0.44 mg/l), the lowest were in Dengkil (0.16 mg/l), Kg. Bukit Changgang (0.16 mg/l), and Teluk Datuk (0.16 mg/l) along the Langat River, and from the Semenyih Dam (0.16 mg/l) along the Semenyih River. The INWQS level for phosphate-phosphorus in Malaysian rivers is 0.10 mg/l. Phosphate-phosphorus levels exceeded the INWQS level for Malaysian rivers in all the stations studied in the Langat Basin.

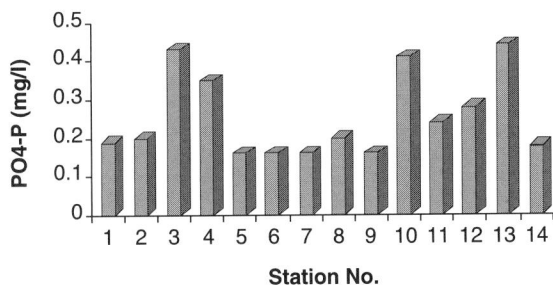

Figure 132.7 Phosphate-phosphorus level at different stations along the rivers of the Langat Basin.

Differences in means between stations were highly significant ($p < 0.01$); sampling times and interactions between stations and time of sampling (Station * Time) were not significant ($p > 0.05$) (Table 132.2). Duncan's multiple-range test grouped the mean phosphate-phosphorus at different stations into four. The difference in means at station Nilai was significantly higher ($p < 0.05$) than all other stations. The phosphate-phosphorus means at Kg. Bukit Changgang, Semenyih Dam, Dengkil, Teluk Datuk, KLIA, the Langat Dam, Kg. Jawa, and Kg. Sg. Serai were significantly lower ($p < 0.05$) than all other stations.

With respect to phosphorus, parts of the rivers in the Langat Basin are unsuitable for supporting aquatic life, as the excess phosphorus causes eutrophication and reduced DO. Further deterioration of water quality with respect to phosphate-phosphorus will have adverse effects on the distribution of aquatic life in the rivers of the Langat Basin.

Biochemical Oxygen Demand

The average value of BOD in rivers of the Langat Basin ranged from 4.08 to 15.68 mg/l (Figure 132.5). BOD was lowest in Kg. Jawa (3.4 mg/l) along the Lui River and highest in Nilai (16.4 mg/l) along the Batang Nilai River. The INWQS level for BOD in Malaysian rivers is 6 mg/l. Five stations along Langat River in Kg. Simpang Balak (11.28 mg/l), West Country Estate (9.79 mg/l), Dengkil (7.16 mg/l), Kg. Bukit Changgang (6.64 mg/l), and Teluk Datuk (9.28 mg/l), as well as the Labu River in KLIA (7.67 mg/l) and the Batang Nilai River in Nilai (15.68 mg/l), exceeded the INWQS level for supporting aquatic life.

Differences in means between stations were highly significant ($p < 0.01$), while differences between sampling times and interactions between stations and time of sampling (Station * Time) were not significant ($p > 0.05$) (Table 132.2). Duncan's multiple-range test grouped the mean BOD levels at different stations into five. The difference in means at Nilai and Kg. Simpang Balak was significantly higher ($p < 0.05$) than all other stations. The BOD at Kg. Jawa, Kg. Sg. Buah, and the Langat Dam were significantly lower ($p < 0.05$) than all other stations.

BOD levels in most parts of the Langat River are high. Pollution is attributed to organics discharged by industries, sewage treatment plants, and animal farms as the major causes of high BODs (DOE, 1997). Anaerobic conditions occur in some parts of the Langat Basin rivers and such areas will be unable to support aquatic life. Further deterioration of water quality will increase the adverse impacts on the health of the Langat Basin aquatic ecosystem.

Chemical Oxygen Demand

The average levels of COD obtained in the rivers of the Langat Basin ranged from 9.75 to 37.33 mg/l (Figure 132.5). The INWQS level for COD in Malaysian rivers is 50 mg/l. COD levels in rivers of the Langat Basin were within INWQS recommended levels for the support of aquatic life. Concentrations of COD in surface waters can range from 20 mg/l or less in unpolluted waters to >200 mg/l in waters receiving effluents.

Differences in means between stations were highly significant ($p < 0.01$). Differences in means between sampling time and interactions between stations and time of sampling (Station * Time) were not significant ($p > 0.05$) (Table 132.2). Duncan's multiple-range test grouped mean CODs at different stations into three. The mean difference at station Nilai was significantly higher ($p < 0.05$) than all other stations and at Langat Dam, Kg. Jawa, Labu, Kg. Sg.Buah, Kg. Sg. Serai, and Kg. Bukit Changgang mean CODs were significantly lower ($p < 0.05$) than all other stations.

Oil and Grease

The average level of O&G in the rivers of the Langat Basin ranged from 1.97 to 4.43 mg/l (Figure 132.8). The recommended INWQS level to support aquatic life for oil and grease in Malaysian rivers is no more than 0.04 mg/l. O&G levels in the Langat Basin rivers were all above the INWQS recommended level. A visible film of oil was often observed in some sections of the Langat River.

Differences in means between stations were highly significant ($p < 0.01$); differences between sampling times and interactions between stations and sampling times (Station * Time) were not significant ($p > 0.05$) (Table 132.2). Duncan's multiple-range test grouped means at different stations into three. The means at Nilai station were significantly higher ($p < 0.05$) than all other stations and those at Kg. Rinching Hilir and the Semenyih Dam were significantly lower ($p < 0.05$) than all other stations.

In 1998, accidental spillage of diesel fuel into the river caused the closure of the 11th-mile Cheras water treatment plant (DOE, 1998; NST, 1998) as the level of O&G in raw water was higher than the maximum possible level for treatment.

The permissible level of O&G in water depends on the intended uses for the water. The recommended concentrations for drinking water supplies and fisheries protection are generally between 0.01 to 0.1 mg/l. Concentrations of crude oil greater than 0.3 mg/l can cause toxic effects in freshwater fish (Chapman, 1992). By these criteria the rivers of the Langat Basin are mostly unsuitable for the support of healthy aquatic life.

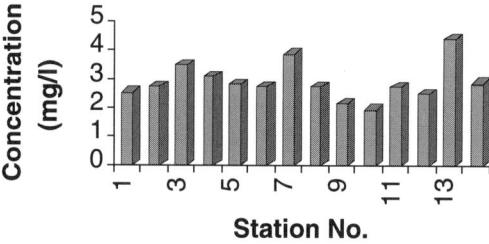

Figure 132.8 Oil and grease level at different stations along the rivers of the Langat Basin.

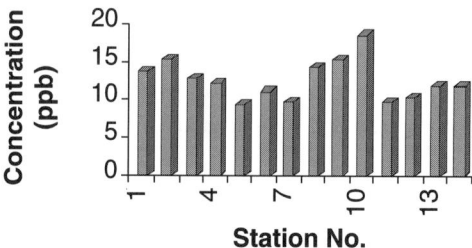

Figure 132.9 Total phenol level at different stations along the rivers of the Langat Basin.

Total Phenol

The average level of phenol in rivers of the Langat Basin ranged from 9.36 to 18.59 µg/l (Figure 132.9). The INWQS maximum level of phenol for Malaysian rivers to support aquatic life is 10 µg/l. Samples from three stations (Dengkil, Teluk Datuk, and Kg. Sg. Buah) were below this level, samples from all other stations exceeded the recommended INWQS level. Md. Pauzi Abdullah and Naiggolan (1990) also reported phenol levels in the Linggi River that exceeded the maximum level for raw river water recommended by the Ministry of Health, Malaysia.

Differences in means of phenol between stations and between sampling times were not significant ($p > 0.05$); interactions between stations and sampling times (Station * Time) were highly significant ($p < 0.01$) (Table 132.2).

The presence of phenol causes a marked deterioration in the organoleptic characteristics of water. Concentrations of phenols in unpolluted waters are usually less than 0.02 mg/l. However, toxic effects on fish can be observed at concentrations of 0.01 mg/l and higher.

Trace Metals

The average concentration of Pb, Cd, Fe, and Cu in water samples obtained from the Langat Basin rivers ranged from 0.074 to 0.100 mg/l, 0.016 to 0.032 mg/l, 0.82 to 4.87 mg/l, and 0.034 to 0.201 mg/l, respectively (Figures 132.10 through 132.12). The recommended INWQS maximum levels for Pb, Cd, Fe, and Cu for Malaysian rivers are 0.05, 0.01, 1, and 0.20 mg/l, respectively. The concentrations obtained for these elements from samples collected from all sampling stations exceeded the INWQS recommended levels.

The average level of Ni, Zn, As, Hg, and Al obtained from rivers in the Langat Basin ranged from 16.42 to 31.83 µg/l, 14.63 to 91.56 µg/l, 0.75 to 4.75 µg/l, 0 to 0.67 µg/l, and 0.14 to 0.24 mg/l,

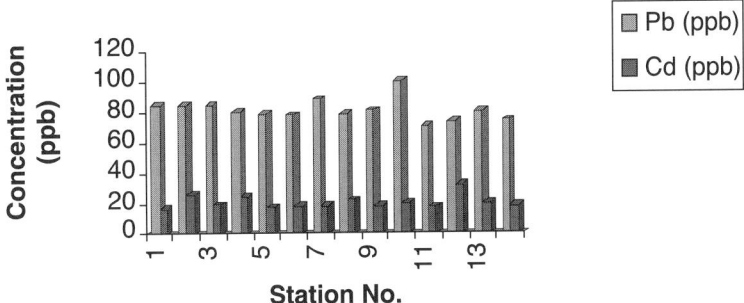

Figure 132.10 Lead and cadmium levels at different stations along the rivers of the Langat Basin.

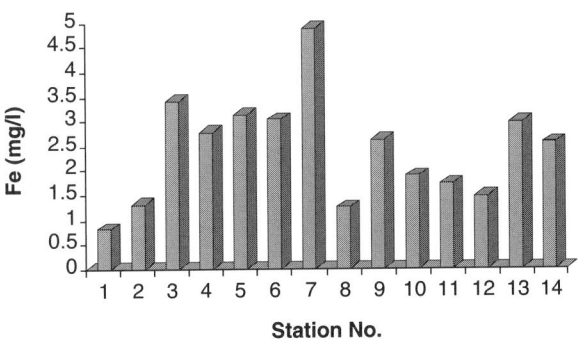

Figure 132.11 Iron level at different stations along the rivers of the Langat Basin.

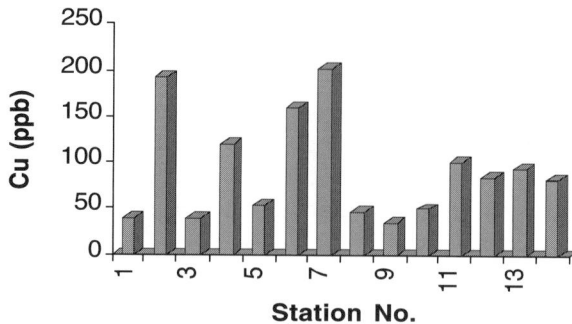

Figure 132.12 Copper level at different stations along the rivers of the Langat Basin.

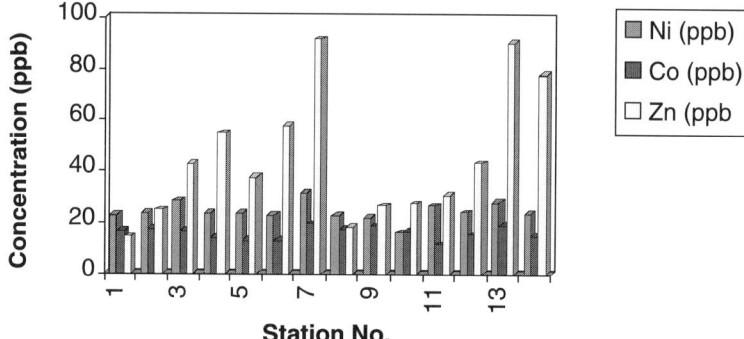

Figure 132.13 Nickel, cobalt, and zinc levels at different stations along the rivers of the Langat Basin.

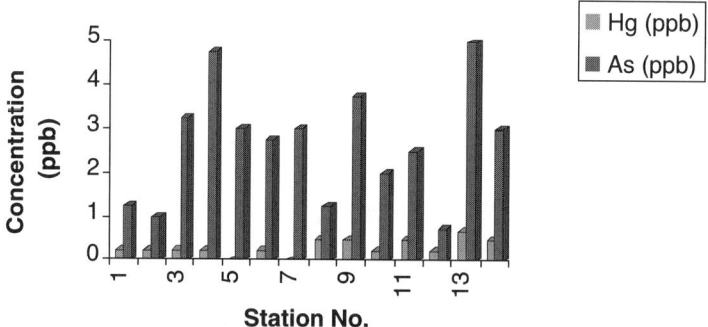

Figure 132.14 Arsenic and mercury levels at different stations along the rivers of the Langat Basin.

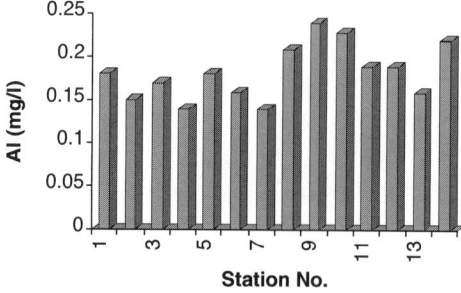

Figure 132.15 Aluminum level at different stations along the rivers of the Langat Basin.

respectively (Figures 132.13 through 132.15). The INWQS maximum levels for Ni, Zn, As, Hg, and Al are 0.90, 5, 0.4, 0.004, and 0.50 mg/l, respectively. Thus, levels of all trace minerals in rivers of the Langat Basin were below the INWQS limits that support aquatic life in Malaysian rivers.

In addition to these findings, Cr was always below detection limit (0.07 ppm); the average Co level ranged from 12.08 to 19.75 ppb. The INWQS has not proposed any maximum levels for these elements. Co was monitored to detect potential discharge from high-tech industries.

Mean differences in Fe between stations and interactions between stations and sampling times (Station * Time) were highly significant ($p < 0.01$); sampling times were not significant ($p > 0.05$) (Table 132.2). Duncan's multiple-range test grouped the mean Fe at different locations into six. Mean iron levels at Teluk Datuk, Kg. Simpang Balak, Dengkil, Kg. Bukit Changgang, and Nilai were significantly higher ($p < 0.05$) than all other stations. The level of iron at the Langat Dam was significantly lower ($p < 0.05$) than all other stations.

Differences in mean of Zn between stations were highly significant ($p < 0.01$); sampling time interactions (Station * Time) were not significant ($p > 0.05$) (Table 132.2). Duncan's multiple-range test grouped mean Zn levels at different stations into three. The mean Zn levels at Teluk Datuk and Nilai stations were significantly higher ($p < 0.05$) than at all other stations; levels at the Langat Dam, Kg. Jawa, Kg. Sg. Serai, the Semenyih Dam, Kg. Rinching Hilir, and Kg. Sg. Buah stations were significantly lower ($p < 0.05$) than all other stations.

The interaction in Al between stations and sampling times (Station * Time) was highly significant ($p < 0.01$); differences between sampling times were not significant ($p > 0.05$) (Table 132.2). Duncan's multiple-range test grouped mean Al levels between stations into five. Duncan's test results indicated that the level of Al between the Semenyih Dam and other stations was significantly higher ($p < 0.05$). Aluminum levels at Teluk Datuk, West Country Estate, and Kg. Sg. Serai were significantly lower ($p < 0.05$) than all other stations.

The differences in means of Pb, Cd, Cu, Co, and Ni between stations and sampling times were not significant ($p > 0.05$) (Table 132.2). While the interactions (Station * Time) for Cd, Cu, and Co were not significant ($p > 0.05$), they were significant for Pb and Ni ($p < 0.01$ and $p < 0.05$, respectively). Some of the stations received industrial effluents in the morning and evening. If industries discharge effluent in the morning, this can lead to high Pb and Ni levels and in the afternoon, due to dilution, levels of Pb and Ni would be lower than in the morning. On the other hand, the high level of Pb and Ni measured in the afternoon at some stations may result from industries discharging their effluents in the afternoon. In addition, high afternoon temperature might raise the amount of Pb dissolved in the river and account for the high afternoon values.

Pollution by Pb, Cd, Fe, and Cu pose considerable health risks to aquatic organisms and organisms higher up the food chain, including humans. The Langat Basin must be considered badly polluted by these trace metals and steps must be taken to determine and eliminate the sources of this pollution.

Water Quality Index Values (WQI)

The calculated DOE-WQI values (Table 132.3 and Appendix 132.IV) for the main rivers in the Langat Basin suggest they would all be classified as polluted and belong to Class II to IV rivers. Of all sampling stations, four gave values that would place them in Class IV: Kg. Simpang Balak, West Country Estate, Nilai, and Teluk Datuk; five qualify as Class III: Kg Rinching Hilir, Kg Sg Buah, KLIA, Dengkil, and Bkt Changgang; the rest would be Class II.

Two water quality monitoring stations are located in Ecological Zone 1, the Langat Dam and Semenyih. Both stations are classified as Class II. Ecological Zone 2 has eight monitoring stations. Among them, three stations are classified as Class IV: Kg. Simpang Balak, West Country Estate, and Nilai; four stations are classified as Class II: Kg. Jawa, Kg. Sg. Serai, Kg. Rinching Hilir, and Labu; one station, Kg. Sg. Buah, is classified as Class III. Ecological Zone 3 has four monitoring

Table 132.3 Water Quality Classification of the Langat River Basin Based on DOE-WQI

Monitoring Station No.	Monitoring Station	Calculated WQI Value	Class
1	Langat Dam	83.45	II
2	Kg. Sg. Serai	80.80	II
3	Kg. Simpang Balak	40.03	IV
4	West Country Estate	48.78	IV
5	Dengkil	56.94	III
6	Kg. Bukit Changgang	57.64	III
7	Teluk Datuk	47.64	IV
8	Kg. Jawa	80.52	II
9	Semenyih Dam	80.33	II
10	Kg. Rinching Hilir	72.14	III
11	Kg. Sg. Buah	67.29	III
12	Labu	81.70	II
13	Nilai	48.91	IV
14	KLIA	60.98	III

Note: DOE-WQI value 91.76 = Class I; 75.37 = Class II; 51.68 = Class III, and 29.61 = Class IV.

stations with three classified as Class III: Dengkil, Bukit Changgang, and KLIA; one station, Teluk Datuk, is classified as Class IV.

HEALTH OF THE LANGAT BASIN RIVER ECOSYSTEM

There are many ways to assess the health of the Langat Basin river system and interpret what that means to the overall health of the basin ecosystem. The approach used in this study relied on the assessment of risk to aquatic life from pollutants in river water. The INWQS gives maximum values for a wide range of chemicals and parameters considered essential for the maintenance of healthy populations of aquatic life, suitability for drinking water, and water for irrigation. As the most-stringent standards are set for the maintenance of aquatic life, maximum levels for this functional category have been used as the basis for assessing the health of the rivers in the basin. Fundamentally, the health of the river ecosystem depends on the absence of pollution, as suggested by Ramade (1995). The concept of functional ecosystem thresholds was first proposed by Schaeffer and Cox (1992). Additionally, the rivers may be exposed to other degradations, such as the spilling of suspended matters or other disturbances impacting both the water column and sediments in the riverbed. Accordingly, a "healthy" aquatic ecosystem has to be devoid of such alterations (Ramade, 1995).

Minns et al. (1990) also used the risk factor approach in assessing aquatic ecosystem health. However, he argued for the incorporation of integrated modeling of ecosystem distress syndromes and the ecosystem's counteractive capacity in assessing risks in order to obtain a better perspective of the health of aquatic ecosystems. Thus far, studies in the Langat Basin rivers have not reached a stage to enable adoption of this latter approach.

In general, the rivers of the Langat Basin are badly polluted and would not support the range of aquatic life expected if the river were pristine. High organic load, siltation, and the presence of ammoniacal-nitrogen, phosphate-phosphorus, oil and grease, phenol, iron, cadmium, lead, and copper at levels exceeding recommended values contribute to the poor state of many sections of the river.

Highly polluted (Class IV by the DOE river classification scheme) stretches occur along the Langat River in the mukim of Cheras between Kg. Sg. Serai, south to Kg. Simpang Balak, and on to the areas around West Country Estate. At these locations the river passes areas where most manufacturing industries and urban centers are located, and where population densities are high. This stretch of river between Bkt. Changgang and Teluk Datuk also qualifies as highly polluted but the main contributory factors are the high levels of suspended sediments and low dissolved oxygen. The sediments originate from land that has been laid bare for future development, while low dissolved oxygen reflects a high level of decaying organic matter and possibly iron in the water column. Iron is rapidly oxidized in water, a process that consumes dissolved oxygen. The Batang Nilai River that flows into the Labu River is also highly polluted, owing to a high input of organic waste from domestic and urban sources, as well as waste from upstream industries. The largest industrial park in the Langat Basin is located around Nilai.

The particular sections of the rivers of the Langat Basin classified as Class IV rivers have probably lost their capacity to self-purify, as the level of pollutants they receive exceeds their assimilative capacity. Rivers polluted beyond their counteractive capacity will take a long time to recover naturally. River sections with a lower level of pollution must be prevented from degrading to a lower quality status.

Overall, the Langat Basin river ecosystem is not as "healthy" as it ought to be. Any further deterioration of the quality of its river water will adversely affect the health of the basin as a whole, as these rivers provide clean water for domestic, industrial, agricultural, and other uses, and habitats for the variety of life required to sustain natural processes, such as nutrient and biogeochemical cycles. Approaching basin-wide management of the environment and development from the framework of ecosystem health will enable a better grasp of the essential ecological processes that must be preserved to ensure both sustainability of the environment and development in the Langat Basin.

ACKNOWLEDGMENT

The work reported in this chapter was funded by the MATREM Project of UNEP/ROAP CP/ 5220-97-02/Rev.1 Research and Training for Measuring and Monitoring Ecosystem Health in a Large-Scale Ecosystem.

REFERENCES

Ainon, H. and Sukiman, S., Water quality survey of Langat River, Selangor, *Malaysia Appl. Biol.,* 16, 369–377, 1987.

Anhar, S., Yusoff, M.K., and Poe, M.T., Water quality profile of Sg. Langat, *Pertanika,* 11, 273–281, 1988.

Anon., Water quality criteria for water, EPA-44/976-023, U.S. EPA, Washington, D.C., 1976.

Anon., GEMS/Water operation guide, WHO, Geneva, Switzerland, 1978.

Anon., National guidelines for drinking water quality, Ministry of Health, Kuala Lumpur, Malaysia, 1983.

Anon., Water quality monitoring programme 1978–84, Department of Environment, Kuala Lumpur, Malaysia, 1985.

APHA, Standard Methods for the Examination of the Water and Waste Water, 17th ed., American Public Health Association, Washington, D.C., 1989.

Chapman, D., *Water Quality Assessments: A Guide to the Use of Biota, Sediments and Water in Environmental Monitoring,* Chapman & Hall, London, 1992, p. 588.

DOE, WQS Phase I Study: Development of Water Criteria and Standards for Malaysia, Department of Environment, Ministry of Science, Technology and the Environment, Kuala Lumpur, Malaysia, 1986.

DOE, WQS Phase II Study: Development of Water Criteria and Standards for Malaysia, Department of Environment, Ministry of Science, Technology and the Environment, Kuala Lumpur, Malaysia, 1990.

DOE, Classifications of Malaysian Rivers, Department of Environment, Ministry of Science, Technology and the Environment, Kuala Lumpur, Malaysia, 1994.

DOE, Environment Management: Water Quality Monitoring Biomonitoring, Department of Environment, Ministry of Science, Technology and the Environment, Malaysia, 1997.

DOE, Water Pollution Control in the Upper Langat River Basin, Department of Environment, Ministry of Science, Technology and the Environment, Kuala Lumpur, Malaysia, 1998.

EPA, Water quality criteria for water, EPA-44/976–023, U.S. EPA, Washington, D.C., 1986.

HACH, *HACH Water Analysis Handbook,* HACH Company, Loveland, CO, 1989.

Lim, S.T., Tropical freshwater fish tolerance and toxicity study: special study, Asian Institute of Technology, Bangkok, 1974.

Md. Pauzi, A. and Nainggolan, H., Phenolic water pollutants in Malaysian river basins, *Environ. Monitor. Assess.,* 19, 423–431, 2000.

Minns, C.K., Moore, J.E., Schindler, D.W., and Jones, N.L., Assessing the potential extent of damage to inland lakes in eastern Canada due to acidic deposition IV. Predicting impacts on species richness in seven groups of aquatic biota, *Can. J. Fish Aquatic Sci.,* 47: 821–830, 1990.

Nordin, M. and Azrina, L.A., Training and research for measuring and monitoring ecosystem health of a large-scale ecosystem: the Langat Basin, Selangor, Malaysia, *Ecosyst. Health,* 4, 188–190, 1998.

Norhayati, L., Indices for water quality in a river, Master's thesis, Environmental Engineering Department, Asian Institute of Technology, Bangkok, 1981.

NST, High ammonia levels: two water treatment plants shut down, *New Straits Times,* 19 October 1997.

Pescod, M.B., Investigation of rational effluent and stream standards for tropical countries, Asian Institute of Technology, Bangkok, 1974.

Schaeffer, D.J. and Cox, D.K., Establishing ecosystem threshold criteria, in *Ecosystem Health: New Goals for Environmental Management,* Costanza, R., Norton, B., and Haskell, B., Eds., Island Press, Washington, D.C., 1992.

Sukiman, S., Kajian pemonitoran kualiti air lembangan Sungai Langat, Selangor, *Sains Malaysiana,* 14, 245–255, 1985.

Ramade, F., Qualitative and quantitative criteria defining a "Healthy" Ecosystem, in *Evaluating and Monitoring the Health of Large-Scale Ecosystems,* Rapport, D.J., Gaudet, C.L., and Calow, P., Eds., Springer Verlag, New York, 1995.

Turk, A.J., Wittes, J.T., and Wittes, R.E., *Environmental Science,* 2nd ed., W.B. Saunders, Philadelphia, 1978.

The Six River-Use Classifications as Defined by INWQS

River Class	Water Uses
I[a]	Conservation of natural environment Water Supply I: Practically no treatment necessary Fishery I: Very sensitive aquatic species
IIA[b]	Water Supply II: Conventional treatment required Fishery II: Sensitive aquatic species
IIB[c]	Recreational use with body contact Fishery II: Sensitive aquatic species
III[d]	Water Supply III: Extensive treatment required Fishery III: Common and tolerant species Livestock drinking
IV[e]	Irrigation
V[f]	None of the above

[a] Class I represents water bodies of excellent quality. Standards are set for the conservation of natural environment in its undisturbed states. Water bodies in this category meet the most stringent requirements of human health and aquatic protection.

[b] Class IIA represents water bodies of good water quality. Most existing raw water supply sources come under this category. Class IIA standards are set on the basis of the criteria developed for the protection of human health and sensitive aquatic species known to exist in these waters. In practice, no body-contact activity is allowed in these waters to prevent infection by pathogens.

[c] Class IIB standard is based on criteria for recreational use and protection of sensitive aquatic species.

[d] Class III is defined with the primary objective of protecting common and moderately tolerant aquatic species with economic value. Water under this classification may be used for water supply with extensive/advanced treatment. This class of water is also suitable for livestock drinking needs.

[e] Class IV defines water quality required for major agricultural irrigation activities, which may not cover minor applications to sensitive crops.

[f] Class V represents other waters which do not meet any of the above uses.

Best-Fit Equations for the Estimation of the Various Subindex Values of WQI

Subindex for DO (in % Saturation)

SIDO = 0	for $x < 8$
SIDO = 100	for $x > 92$
SIDO = $-0.395 + 0.030x^2 - 0.00020x^3$	for $8 < x < 92$

Subindex for BOD

SIBOD = $100.4 - 4.23x$	for $x < 5$
SIBOD = $108 * e^{-0.055x} - 0.1x$	for $x > 5$

Subindex for COD

SICOD = $-1.33x + 99.1$	for $x < 20$
SICOD = $103 * e^{-0.0157x} - 0.04x$	for $x > 20$

Subindex for AN

SIAN = $100.5 - 105x$	for $x < 0.3$
SIAN = $94 * e^{-0.573x} - 5 * \lvert x - 2 \rvert$	for $0.3 < x < 4$
SIAN = 0	for $x > 4$

Subindex for SS

SISS = $97.5 * e^{-0.0016x} + 0.05x$	for $x < 100$
SISS = $71 * e^{-0.0016x} - 0.015x$	for $100 < x < 1000$
SISS = 0	for $x > 1000$

Subindex for pH

SIPH = $17.7 - 17.2x + 5.02x^2$	for $x < 5.5$
SIPH = $-242 + 95.5x - 6.67x^2$	for $5.5 < x < 7$
SIPH = $-181 + 82.4x - 6.67x^2$	for $7 < x < 8.75$
SIPH = $536 - 77.0x + 2.76x^2$	for $x > 8.75$

Note: x = Concentration in mg/l for all parameters except pH.

Source: Data from DOE (1994) and Norhayati (1981).

APPENDIX 132.III

Mean Values of All Parameters Monitored for 14 Stations from the 6 Rivers of the Langat Basin

Monitoring Stations	Temperature (°C)	pH	Turbidity (mg/l)	Conductivity (mg/l)	DO[a] (mg/l)	COD[b] (mg/l)	BOD[c] (mg/l)	NH$_3$-N[d] (mg/l)	NO$_3$-N[e] (mg/l)	NO$_2$-N[f] (mg/l)	PO$_4$-P[g] (mg/l)	TSS[h] (mg/l)	Volatile Solids (mg/l)	Fixed Solid (mg/l)
1. Langat Dam	25.92	6.89	18.67	39.16	7.18	9.75	4.36	1.31	0.65	0.004	0.19	6.33	6.00	0.55
2. Kg. Sg. Serai	27.29	6.66	84.33	40.75	6.88	14.00	4.99	0.59	1.15	0.027	0.20	59.67	20.00	39.09
3. Kg. Simpang Balak	28.87	6.55	222.25	210.00	3.78	31.17	11.28	4.51	1.31	0.095	0.43	181.17	59.55	140.55
4. West Country Estate	28.88	6.78	364.08	183.33	2.72	24.17	9.79	3.76	1.47	0.163	0.35	275.00	55.82	241.18
5. Dengkil	28.92	6.56	499.25	149.42	4.03	21.21	7.16	2.22	1.19	0.149	0.16	414.67	68.91	363.36
6. Kg. Bukit Changgang	28.38	6.38	400.17	157.75	3.75	17.50	6.64	2.09	1.59	0.172	0.16	366.25	65.23	319.86
7. Talok Datok	29.29	5.49	672.33	230.42	1.38	18.83	9.28	2.08	1.06	0.036	0.16	528.75	88.86	428.77
8. Kg. Jawa	26.54	6.82	67.00	42.08	7.30	11.71	4.08	0.75	0.67	0.004	0.20	51.25	21.72	33.55
9. Semenyih Dam	28.42	6.60	27.16	51.25	5.54	18.20	5.28	0.67	0.56	0.195	0.16	11.42	11.64	0.00
10. Kg. Rinching Hilir	27.21	6.80	314.42	69.81	5.89	17.58	5.15	0.91	1.09	0.028	0.41	285.58	57.90	246.727
11. Kg. Sg. Buah	28.46	6.66	222.17	67.75	5.65	13.46	4.34	1.25	1.13	0.027	0.24	134.25	31.18	109.10
12. Labu	27.17	6.77	51.17	46.67	6.61	12.71	4.67	0.77	0.76	0.048	0.28	34.25	19.96	16.50
13. Nilai	29.38	7.23	53.08	406.08	2.69	37.33	15.68	3.36	0.88	0.089	0.36	38.33	32.73	4.73
14. KLIA	28.29	6.43	194.75	167.08	4.29	23.13	7.67	2.44	2.14	0.243	0.18	138.42	27.95	112.41

[a] Dissolved oxygen; [b] chemical oxygen demand; [c] biological oxygen demand; [d] ammoniacal-nitrogen; [e] nitrate-nitrogen; [f] nitrite-nitrogen; [g] phosphate-phosphorus; [h] total suspended solids; [i] oil and grease; nd = no data.

Monitoring Stations	O&G[i] (mg/l)	Total Phenol (ppb)	Lead (ppb)	Cadmium (ppb)	Copper (ppb)	Iron (mg/l)	Zinc (ppb)	Chromium (ppb)	Aluminum (ppb)	Cobalt (ppb)	Nickel (ppb)	Arsenic (mg/l)	Mercury (mg/l)
1. Langat Dam	2.55	13.85	84.42	16.42	39.17	0.82	14.67	nd	0.18	16.92	27.5	1.25	0.25
2. Kg. Sg. Serai	2.76	15.31	85.33	25.33	192.75	1.31	25.00	nd	0.15	17.67	28.2	1.00	0.25
3. Kg. Simpang Balak	3.50	12.71	85.25	18.83	38.50	3.41	42.78	nd	0.17	16.5	34.2	3.25	0.25
4. West Country Estate	3.09	12.10	79.67	23.73	118.17	2.79	54.78	nd	0.14	14.33	28.6	4.75	0.25
5. Dengkil	2.83	9.36	79.17	17.17	52.92	3.12	37.89	nd	0.18	13.42	28.2	3.00	0.00
6. Kg. Bukit Changgang	2.76	11.09	77.83	18.25	158.42	3.03	58.11	nd	0.16	13.5	27.5	2.75	0.25
7. Talok Datok	3.88	9.77	88.75	17.75	200.92	4.87	91.56	nd	0.14	19.75	38.2	3.00	0.00
8. Kg. Jawa	2.73	14.43	78.33	21.83	46.25	1.25	18.56	nd	0.21	17.5	27.5	3.75	0.50
9. Semenyih Dam	2.15	15.31	81.17	17.50	33.75	2.61	26.89	nd	0.24	18.83	26.1	2.00	0.25
10. Kg. Rinching Hilir	1.97	18.59	100.41	19.58	49.67	1.89	27.56	nd	0.23	16.92	19.7	2.50	0.50
11. Kg. Sg. Buah	2.75	9.77	71.33	17.17	100.42	1.73	30.67	nd	0.19	12.08	32.2	1.25	0.50
12. Labu	2.51	10.27	73.50	31.63	84.08	1.47	43.44	nd	0.19	15.42	28.9	3.00	0.50
13. Nilai	4.43	11.91	80.17	19.33	93.75	2.97	90.00	nd	0.16	19.42	34.0	5.00	0.67
14. KLIA	2.87	11.84	74.67	17.92	44.50	2.56	78.22	nd	0.22	15.25	28.5	0.75	0.25

Water Quality Classification of the Rivers of the Langat Basin Based on DOE-WQI[a]

	Water DO[b]	Quality BOD	Criteria				DOE-WQI	Class
			COD	SS	pH	NH₃-N		
	7	1	10	25	7	0.1	91.76	I
	5	3	25	50	6	0.3	75.37	II
	3	6	50	150	5	0.9	51.68	III
	1	12	100	300	5	2.7	29.61	IV
Station			**Existing Water Quality**					
1	7.18	4.36	9.75	6.33	6.89	1.31	83.45	II
2	6.88	4.99	14	59.67	6.66	0.59	80.80	II
3	3.78	11.28	31.17	181.17	6.55	4.51	40.03	IV
4	2.72	9.79	24.17	275	6.78	3.76	48.78	IV
5	4.03	7.16	21.21	414.67	6.56	2.22	56.94	III
6	3.75	6.64	17.5	366.25	6.38	2.09	57.64	III
7	1.38	9.28	18.83	528.75	5.49	2.08	47.64	IV
8	7.30	4.08	11.71	51.25	6.80	0.75	80.52	II
9	5.54	5.28	18.2	11.42	6.60	0.67	80.33	II
10	5.89	5.15	17.58	285.58	6.80	0.91	72.14	III
11	5.65	4.34	13.46	134.25	6.69	1.25	67.29	III
12	6.61	4.67	12.71	34.25	6.77	0.77	81.70	II
13	2.69	15.68	37.33	38.33	7.23	3.36	48.91	IV
14	4.29	7.67	23.13	130.08	6.43	2.44	60.98	III

[a] Class I = not polluted; Class II = slightly polluted; Class III = polluted; and Class IV = highly polluted.
[b] For abbreviations, see Appendix 132.III.

CHAPTER 133

Forest Areas and Wildlife Diversity in the Langat Basin: Indicators for Assessing Langat Basin Ecosystem Health

Saiful Arif Abdullah

INTRODUCTION

The complex ecosystem of tropical rain forest, which is basically through trophic levels and food web, has contributed to extremely rich fauna and flora. However, the complexity of tropical rain forest around the world has been degraded, especially in the 20th century. In the early stage of the century the major contribution of forest loss in tropical areas was due to human population growth that increased demand on forest resources, especially for fuelwood. Since then, development activities have increased gradually and many forested areas in tropical regions are cleared and converted to agricultural areas, industrial estates, urbanization, and human settlements.

In Malaysia, loss of biodiversity is one of the major issues related to forest loss or degradation. This issue is among a main agenda that has been debated from local to international levels. It is notable that Malaysian tropical rain forest harbors a great diversity of species, including 300 species of mammals, 700 to 750 species of birds, 350 species of reptiles, 165 species of amphibians, and more than 300 species of freshwater fish (Anonymous, 1998).

The Langat Basin, which is situated adjacent to Klang Valley (the most urbanized region in Malaysia), is the most developing area in Malaysia. The basin has experienced intensification of forest conversion to agriculture, which began in the late 1960s (Wong, 1974). The situation began rapidly worsening in the late 1980s as more forested areas were converted into housing and industrial estates and were fragmented by infrastructure projects such as the development of highways. The basin was once totally covered by pristine forests, but with growing human population (Rospidah and Nordin, 2000) and development since the 19th century only 737 km² of forested areas remain (Nordin, this volume). Due to this, there has been a considerable reduction of wildlife habitat and diversity in the basin (Lim, 1997). Saiful and Nordin (2001) reported that the reduction of mammalian diversity was concomitant with depletion of forest cover in the basin. As a large-scale ecosystem, any further disruption or persistent forest clearance and fragmentation will adversely affect the ecosystem health of the basin.

The ecosystem health concept is a new paradigm in environmental management that integrates dimensions of social, economics, biophysical, and human health (Rapport, 1998). In Malaysia, a basic understanding of the concept of ecosystem health that may be applied to environmental management is still in infancy and being developed in the Langat Basin (Nordin and Azrina, 1998). In the ecosystem health approach, a set of indicators is required as a tool for assessing the health of an ecosystem (Whitford, 1998). The changes of forest areas and wildlife diversity have been identified as indicators for assessing the health of the Langat Basin ecosystem. Those indicators are characteristics of spatial changes that are directly related to ecological services upon which the human community depends (Rapport, 1998). The purpose of this chapter is to present the changes of forested areas in the Langat Basin and the basin's relation to the number of wildlife diversity (with emphasis on mammalian species) and their conservation status. The importance of forested areas and wildlife diversity as indicators in the contexts of assessing ecosystem health of the Langat Basin are discussed.

TREND OF FOREST COVER CHANGES IN THE LANGAT BASIN

Over three decades, the total forested areas in the basin declined by approximately 27%. The reduction of forested areas in the basin occurred especially in Zones 2 and 3. The largest forested area in the basin is in Zone 1 (Figures 133.1 through 133.3). For each decade, secondary forest represented 50 to 70% of forested areas in the basin. The proportion of primary and secondary forests reduced gradually (Figure 133.4). In the period of 1971 to 1972, the proportion of primary forest cover in the basin was approximately 16% (467.6 km^2) of the total land area, which consequently declined to 11% (324.2 km^2) in the period of 1981 to 1982. By 1991 to 1992, only 8% (243.5 km^2) of primary forest remained in the Langat Basin. The proportion of secondary forest in 1981 to 1982 increased by about 6% (676.9 km^2) of the total area of the basin compared with 22% (636.9 km^2) in 1971 to 1972. The proportion in 1991 to 1992, however, declined nearly 20% (560.3 km^2).

The dipterocarp and peat-swamp and mangrove forests are the major types of forest that can be found in the basin. Dipterocarp forests (lowland and hill dipterocarps) are found only in Zone 1, whereas peat-swamp forest are mostly found in Zones 2 and 3 and mangrove are distributed on the coastal islands and margins of the mainland (Zone 3) (Figures 133.5, through 133.7). In three decades, dipterocarp forests remained the main type of forest cover in the basin, making up 40 to 55% of the total forest cover. The proportion of peat-swamp forest declined rapidly and by 1991 to 1992 constituted only 19% (149.4 km^2) of the total forest type. The proportion of mangrove forests remained about the same from 1971 to 1982 (207.0 km^2), but because of the rapid loss of peat swamp forests had increased to 27% (215.5 km^2) by 1992.

DIVERSITY OF MAMMALIAN SPECIES IN THE LANGAT BASIN

Mammalian species are vital to maintaining the quality, stability, and resilience of forest ecosystems. This group of animal can also provide an indication of the state of health of the ecosystem. Species richness and the number of threatened species are among the most common indicators for the status of biodiversity (Reid et al., 1993) and can provide insight into the health of particular ecosystems.

Based on surveys conducted in the 1950s to 1990s, a total of 140 mammal species from 28 families and 11 orders had been reported to exist in the Langat Basin (DWNP, 1996, and unpublished reports; Fletcher, 1999; Lim, 1997; Ruslan, 1984; Shabrina, 1987; Siti, 1983; Zubaid, 1993; Zubaid and Khairul, 1997). The total represents about 47% of mammal species that have been reported to exist in Malaysian tropical rain forest. The number of families and species for each order of mammals is shown in Figure 133.8.

Threatened Mammals in the Langat Basin

In Malaysia, mammals under threat are known by three sources, as perceived by various researchers, conservation status in Malaysia by the Department of Wildlife and National Parks (DWNP) Malaysia, and their conservation status by the International Union for the Conservation of Nature (IUCN).

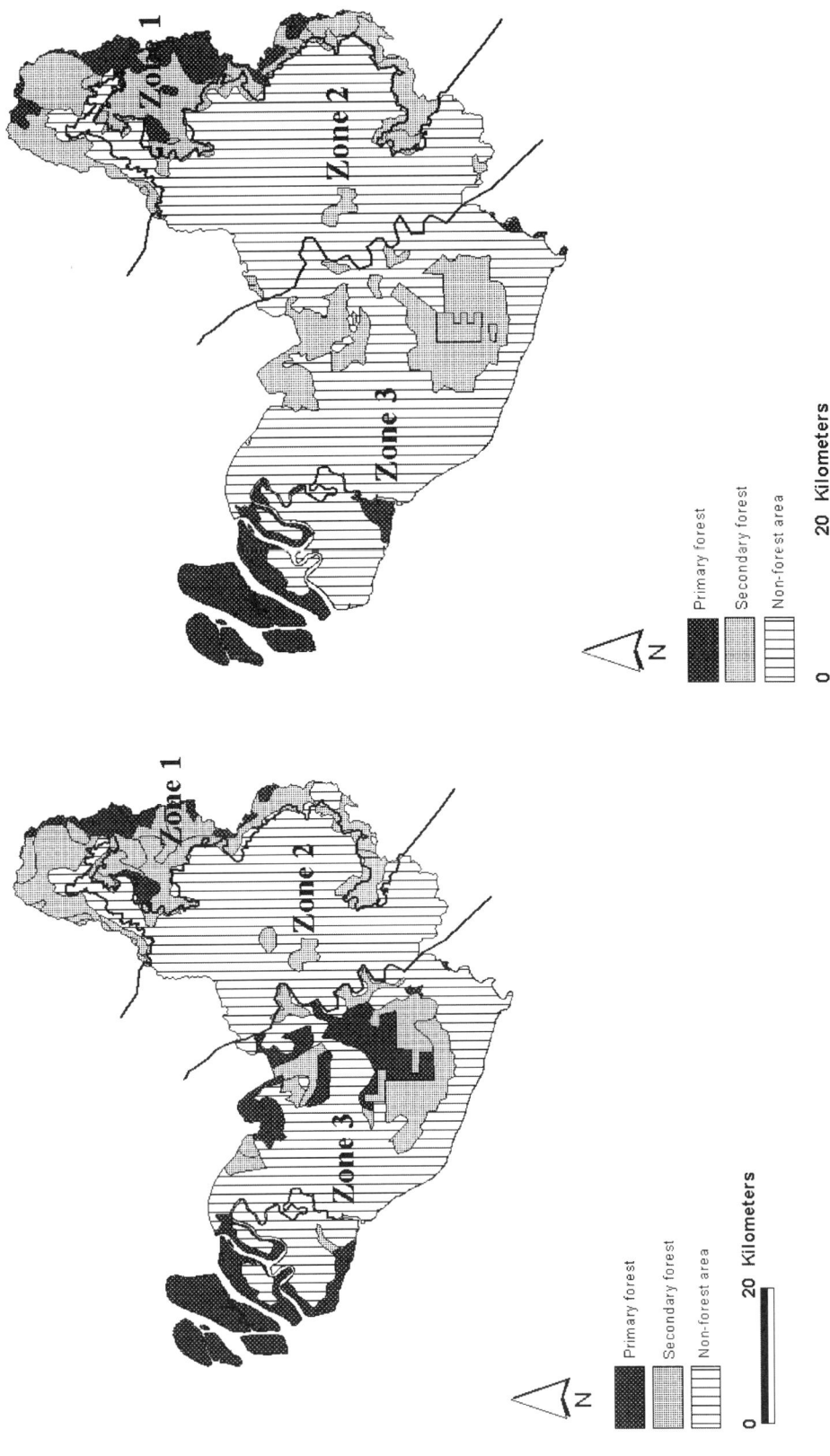

Figure 133.1 Distribution of primary and secondary forests of the Langat Basin in 1971 to 1972.

Figure 133.2 Distribution of primary and secondary forests of the Langat Basin in 1981 to 1982.

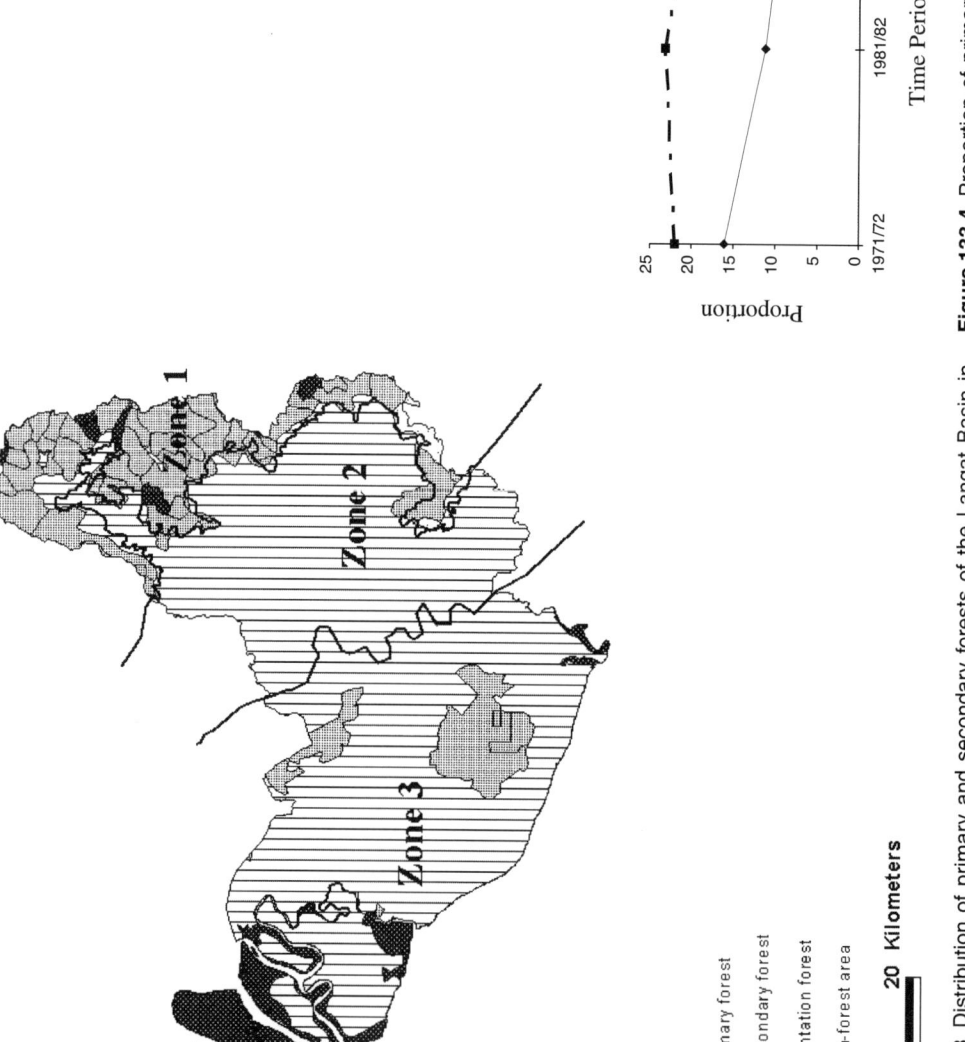

Figure 133.3 Distribution of primary and secondary forests of the Langat Basin in 1991 to 1992.

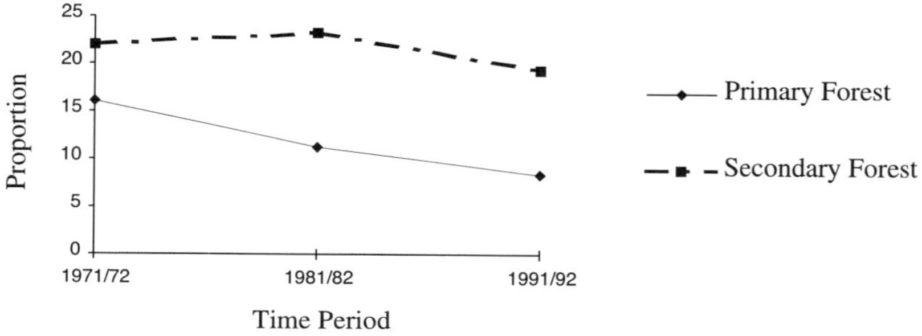

Figure 133.4 Proportion of primary and secondary forests in the Langat Basin for three decades.

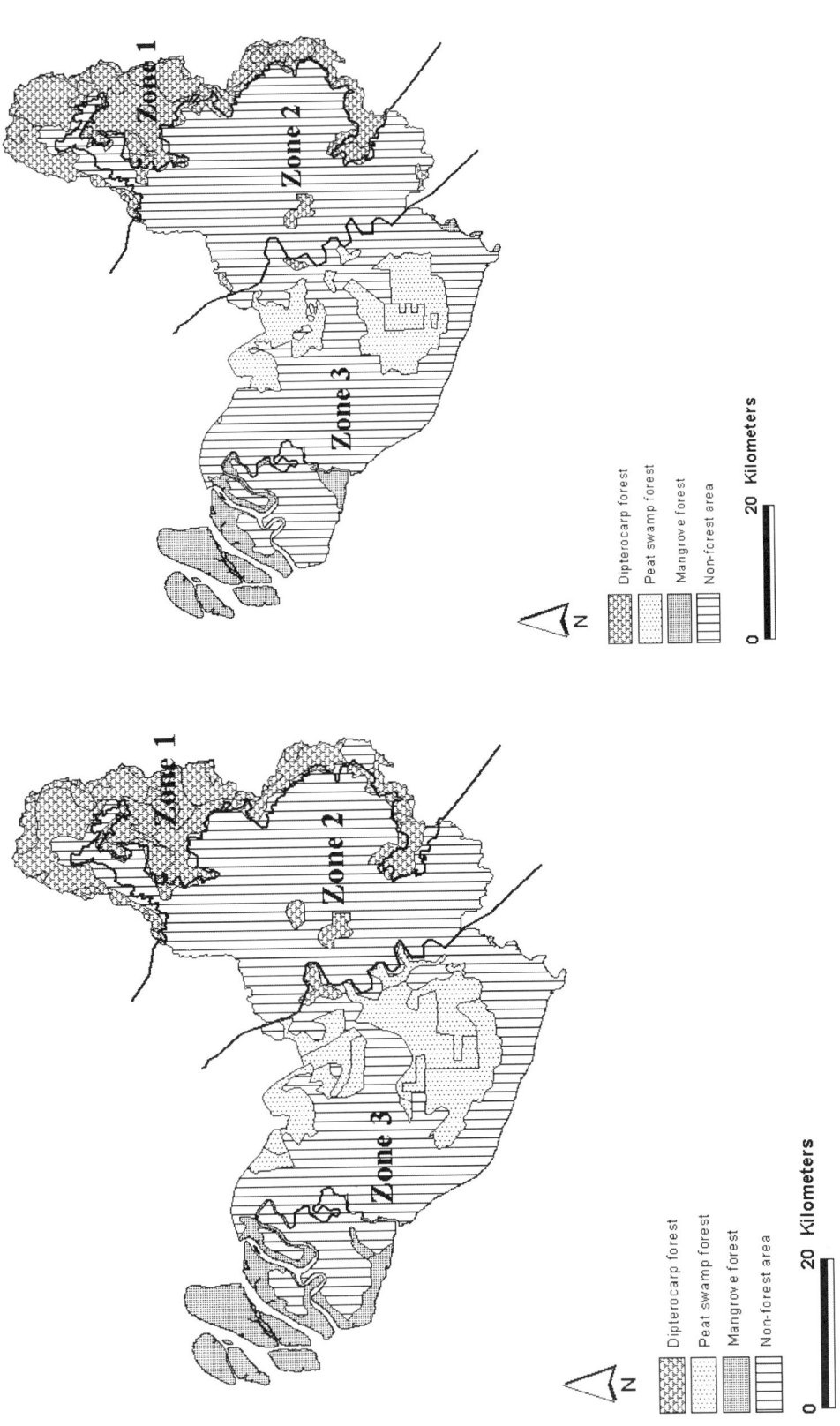

Figure 133.6 Distribution of dipterocarp, peat-swamp, and mangrove forests of the Langat Basin in 1981 to 1982.

Figure 133.5 Distribution of dipterocarp, peat-swamp, and mangrove forests of the Langat Basin in 1971 to 1972.

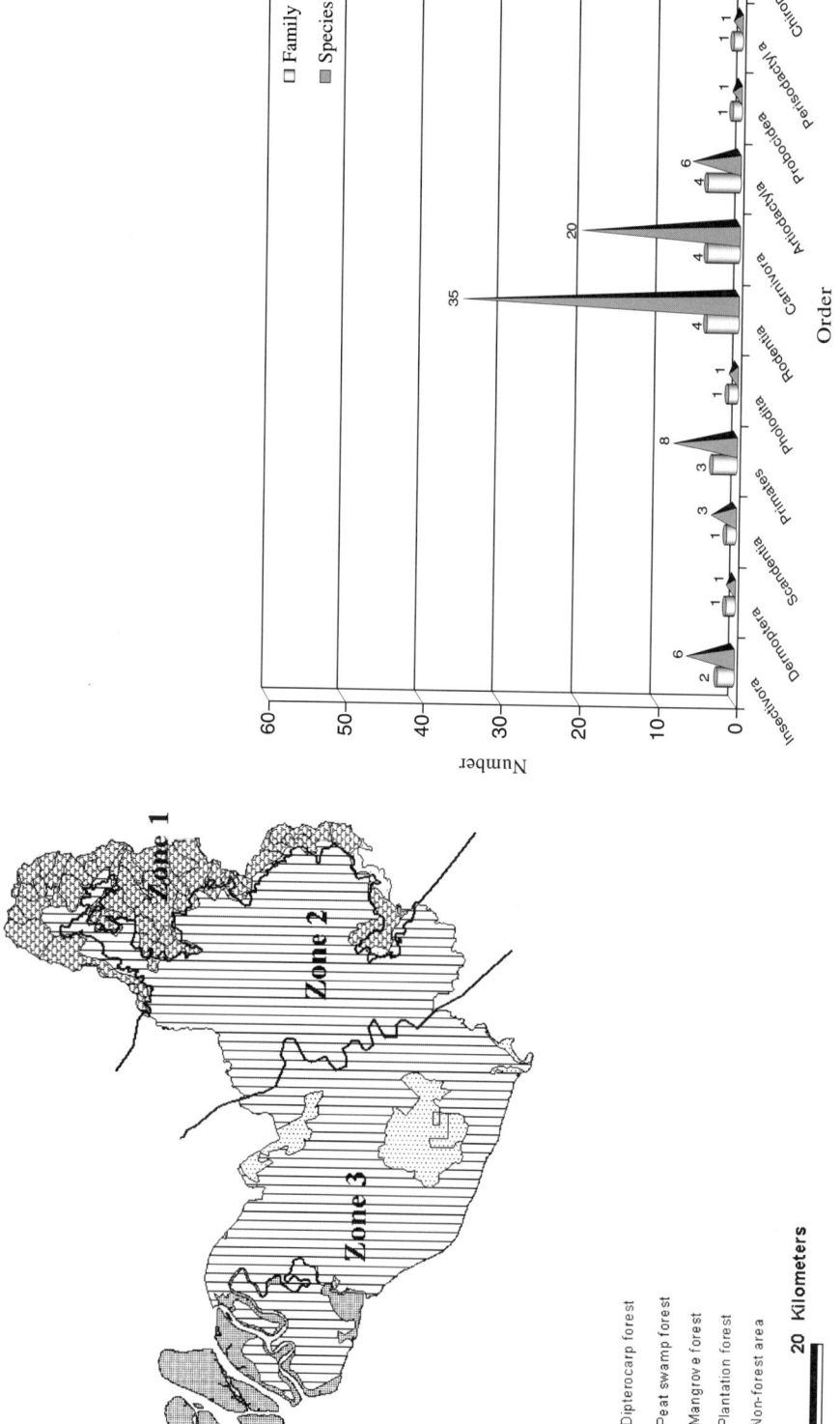

Figure 133.7 Distribution of dipterocarp, peat-swamp, and mangrove forests of the Langat Basin in 1991 to 1992.

Figure 133.8 Number of family and species for each order of mammal in the Langat Basin.

Perceived by Various Researchers

Threatened species of mammals in the Langat Basin under this category were based from sources first compiled by Ratnam and co-workers (1991), and a second list by Bennet (1991). However, data by later sources were focused only on primate species.

Based on a list by Ratnam and co-workers (1991), 23 species of mammals (6 large and 17 small mammals) from 13 families in the Langat Basin are classified as threatened. Of these, 12 species, or 52%, are classified as likely to become threatened, 9 species are threatened, and 2 species are seriously threatened, whereas no species are classified as endangered. A total of eight species of primates from three families had been reported in the Langat. Based on the conservation status by Bennett (1991), only two species, *Trachypithecus cristatus* (silver-leafed monkey) and *Hylobates syndactylus*, are threatened. The former species is classified as an endangered species, whereas the latter falls into vulnerable species.

Conservation Status in Malaysia

Threatened species of mammals by DWNP Malaysia are divided into five categories: endangered, vulnerable, rare, insufficiently known, and indeterminate. The classification and definition for each category follows the IUCN. Of the total number species recorded, approximately 19%, or 27 species, of mammals in the Langat Basin are classified as threatened. Of these, 20 are small and 7 are large mammals. Table 133.1 shows that most threatened species are categorized as insufficiently known and rare while the rest are vulnerable. No species were classed endangered or indeterminate.

Conservation Status by the IUCN

Based on the IUCN 1994 Red List of threatened animals, a total of 14 species of mammals in the Langat Basin have been identified as threatened. Compared with conservation perceived by various researchers and conservation status in Malaysia, the number of large mammals (nine species) in the Langat Basin that fall into the list of threatened species is greater than small mammals species (four species). Of large mammals, two species (*Panthera tigris* and *Elephus maximus*) are classified as endangered, four are vulnerable (*Macaca nemestrina*, *Neofelis nebulosa*, *Capricornis sumatrensis,* and *Tapirus indicus*), and another three species (*Trachypithecus cristatus*, *Hylobates lar*, and *H. syndactylus*) are near threatened. Two (*Lutra perspicillata* and *Hystrix brachyura*) of four species of small mammals are vulnerable, whereas the other two species (*Macaca fascicularis* and *Manis javanicus*) are classified as near threatened.

IMPACT OF FOREST LOSS ON MAMMAL DIVERSITY IN THE LANGAT BASIN

Depletion of dipterocarp forests in Zone 1 has caused a drastic reduction of animal diversity and abundance in the Langat Basin (Lim, 1997). Several large mammals, such as tiger (*Panthera tigris*), Malayan sun bear (*Helarctos malayanus*), and serow (*Capricornis sumatrensis*) were encountered in a 1980s survey; however, they were not encountered in survey conducted in the 1990s by DWNP Malaysia (Lim, 1997). The Asian elephant (*Elephus maximus*) has not been encountered in the forest since surveys in the 1960s. These observations suggest that the quality of forest habitat in Zone 1 is being degraded and cannot maintain or support a viable population of these mammals.

The reduction of forest areas in Zones 2 and 3 was more intense than in Zone 1, making them more stressed than Zone 1. The forests in Zones 2 and 3 are being fragmented and, as isolated forests, will become too small to support those animals with low population densities that require a large home range (Terborgh 1992). Large animals usually occur in small numbers and require

Table 133.1 Number of Large and Small Mammals under Threat in the Langat Basin

Conservation Status[a]	Large Mammals	Small Mammals
Endangered	—	—
Vulnerable	2	1
Rare	5	6
Indeterminate	—	—
Insufficiently known	1	12
Total	8	19

[a] Conservation status is based on categories by the International Union for the Conservation of Nature.

large areas for roaming (Payne, 1990). Thus, these isolated forests will also become too small to support viable populations of larger animals. Forest fragmentation may also affect the diversity and abundance of small mammals in the basin. Studies by Zubaid (1993) on bats in the basin suggest that continuous primary forest had significantly more species and individuals compared with fragmented secondary forest. Fragmentation will also increase risk of exposure to predators and hunting. According to Musa (1983), hunting is one of the major factors contributing to the decline of the wildlife population in Malaysia.

FOREST AREAS AND WILDLIFE DIVERSITY: INDICATORS FOR ASSESSING LANGAT BASIN ECOSYSTEM HEALTH

It is vital to evaluate the health of modified ecosystems such as the Langat Basin (Haila, 1998). One of the tools used to assess the health of an ecosystem is sets of indicators (Whitford, 1998). In the context of Langat Basin ecosystem, the changes of forest areas and the number of species and their conservation status are indicators that can be used.

Natural forests are the finest habitat not only for wildlife to maintain their populations and survive but they also provide ecological services vital for human existence. Destruction and fragmentation of forested areas will reduce the complexity of forest structure and habitat diversity, and affect the interaction between animals and plants. Several studies revealed that fragmentation has some effect on small mammals (Mortberg, 2001) and birds (Lindenmayer and Lacy, 2002). Destruction of forest areas will disrupt ecological processes such as nutrient cycling, energy flow, sequestration of toxic substances, and maintenance of the hydrological cycle. Dysfunction of these features is a characterization of the ecosystem distress syndrome (Rapport, 1998), which affects the biodiversity of forest ecosystem.

Forest areas in the Langat Basin are a habitat for various species of mammalian vertebrates. Several species of mammalian vertebrates are categorized as threatened species; thus they are potential indicators for assessing the Langat Basin ecosystem health.

However, it is also important to evaluate the risk of forest and biodiversity changes in determining the health of an ecosystem (Whitford, 1998). Forest clearance or destruction is mainly due to human activities. In a large-scale ecosystem such as the Langat Basin, we can say that forest is the most integral part in maintaining stability or equilibrium of the ecosystem. Forest degradation will increase the vulnerability of the basin to various geohazard incidents such as soil erosion, flash flood, and mud flood.

Destruction of forested areas will change the landscape composition (e.g., amount of habitat) and configuration (e.g., fragmentation) of the ecosystem. Forest is also a reservoir for various types of bacteria and virus, and several mammalian species are carriers of or hosts for some of them. Studies by Langlois and co-workers (2001) suggest that, when the forest landscape was changed, the transmission of disease (which is harmful to humans and domestic or farm animals) caused by virus accelerated. The bacteria or virus that transmits through air will easily spread; some will find another host to maintain their life cycle, usually among insects or farm animals such as pigs. Due to this, human and domestic or farm animals will be more exposed to various diseases. When an outbreak occurs, it may cause loss of human life, and ultimately the social, economic, and environment of the Langat Basin large-scale ecosystem is affected as a whole.

CONCLUSION

Clearance of forest areas increases the heterogeneity of the forest landscape and the entire landscape of the Langat Basin. Ecologically, forest is a main element to maintain the equilibrium of some ecological process, such as hydrological cycle of the Langat Basin ecosystem. Thus, any disturbances or changes of forest landscape will affect the equilibrium of ecological processes and, in the context of large-scale ecosystem, the environmental health of the Langat Basin.

On a broader scale, healthy environment not only refers to environmental pollution (air or water pollution) but also to social, economic, biological, and human dimensions. Geohazard events such as siltation, flash flood, mud flood, and soil erosions, as well as disease outbreak, are mainly caused by forest loss or destruction. If measures are not taken, these will ultimately affect the social, economic, and human health of the basin. The direct effect of forest loss in the basin is reduction of wildlife (terrestrial and aquatic) habitat, which will reduce population and diversity. This is another indication of distress syndrome that has occurred in the basin. In conclusion, changes of forest areas and wildlife diversity are vital components of indicators for assessing Langat Basin ecosystem health.

ACKNOWLEDGMENT

This chapter is based on my study, which was funded by UNEP/ROAP under the MATREM project CP/5220-97-02/Rev. 1, Research and Training for Measuring and Monitoring Ecosystem Health in a Large-Scale Ecosystem. I would like to thank Prof. Mohd Nordin Hj. Hasan, Senior Fellow of LESTARI, for his valuable comments.

REFERENCES

Anon., National Policy on Biological Diversity, Ministry of Science, Technology and the Environment, Malaysia, 1998.

Bennett, E.L., Diurnal primates, in *The State of Nature Conservation in Malaysia*, Kiew, R., Ed., Malayan Nature Society, Selangor, Malaysia, 1991, pp. 150–172.

DWMP, State of Selangor, Data inventori hidupan liar di Daerah Hulu Langat, Sepang dan Kuala, Langat, Selangor, 1996.

Fletcher, C.D., Perbandingan struktur komuniti kelawar insektivor pada pelbagai altitud di Gunung Nuang, Selangor, Tesis Sm. Sn. (UKM), 1999.

Haila, Y., Assessing ecosystem health across spatial scale, in *Ecosystem Health,* Rapport, D. et al., Eds., Blackwell Science, Malden, MA, 1998, pp. 81–102.

Langlois, J.P., Fahrig, L., Merriam, G., and Artsob, H., Landscape structure influences continental distribution of hantavirus in deer mice, *Landscape Ecol.,* 16, 255–266, 2001.

Lekagul, B. and McNeely, J.A., *Mammals of Thailand,* Kurusapha Ladprao Press, Bangkok, Thailand, 1977.

Lim, B.L., Small mammals studies (1950–1988) in relation to environmental changes in the Langat Basin, Report submitted to the UNDP/ISIS programme of research grants for small-scale projects on the environment and development, 1997.

Lindenmayer, D.B. and Lacy, R.C., Small mammals, habitat patches and PVA models: a field test of model predictive ability, *Biol. Conserv.,* 103, 247–265, 2002.

Medway, L., *The Wild Mammals of Malaya (Peninsular Malaysia) and Singapore,* 2nd ed., Oxford University Press, Kuala Lumpur, Malaysia, 1983.

Mortberg, U.M., Resident bird species in urban forest remnants: landscape and habitat perspective, *Landscape Ecol.,* 16, 193–203, 2001.

Musa, N., Management of wildlife reserve in peninsular Malaysia, *J. Wildl. Parks,* 2, 106–118, 1983.

Nordin, M. and Azrina, L.A., Training and research for measuring and monitoring ecosystem health of a large-scale ecosystem: the Langat Basin, Selangor, Malaysia, *Ecosyst. Health,* 4, 188–190, 1998.

Ondoi, G., Kajian perbandingan kepelbagaian dan ketumpatan mamalia kecil pada 3 kawasan hutan di Semenanjung Malaysia, Tesis Sm. Sn (UKM), 1996.

Payne, J., Rarity and extinction of large mammals in Malaysian rainforests, paper presented at the International Conference on Tropical Biodiversity, 12–16 June 1990, Kuala Lumpur, Malaysia, 1990.

Rapport, D., Need for a new paradigm, in *Ecosystem Health,* Rapport, D. et al., Eds., Blackwell Science, Malden, MA, 1998, pp. 3–17, 1998a.

Rapport, D., Dimension of ecosystem health, in *Ecosystem Health,* Rapport, D. et al., Eds., Blackwell Science, Malden, MA, pp. 34–40, 1998b.

Ratnam, L., Hussein, N.A., and Lim, B.L., Small mammals in Peninsular Malaysia, in *The State of Natural Conservation in Malaysia,* Kiew, R., Ed., Malayan Nature Society, Selangor, Malaysia, 1971, pp. 143–149.

Reid, W.V., McNeely, J.A., Tunstoll, D.B., Bryant, D.A., and Winograd, M., Biodiversity indicators for policy-makers, World Resource Institute, Washington, D.C., 1993.

Rospidah, G. and Nordin, H.H., Population growth and ecosystem health: the case of the Langat Basin, in *Proceedings of the 1999 Langat Basin Research Symposium,* Nordin, H.H., Ed., Institute for Environment and Development, Universiti Kebangsaan Malaysia, 2000, pp. 47–54.

Ruslan, C.P., Kajian kepelbagaian haiwan vertebrata di hutan contoh dan hutan simpan, UKM, Tesis Sm. Sn, 1984.

Saiful, A.A. and Nordin, H.H., Forest cover changes and its implication on wildlife diversity in Langat Basin, Malaysia, in *Proceedings of the International Conference on Tropical Ecosystems,* Ganeshaiah, K.N, Shaanker, R.U., and Bawa, K.S., Eds., Oxford-IBH, New Delhi, India, 2001, pp. 155–158

Shabrina, M.S., The small mammals in the lowland habitats of Pahang and Selangor, *J. Wildl. Parks,* 6, 10–28, 1987.

Siti, H.Y., A preliminary survey on inventory habitat and wildlife in the Ulu Langat Forest Reserve and Sungai Dusun Game Reserve, *J. Wildl. Parks,* 2, 119–140, 1983.

Steven, W.E., Habitat requirement on Malayan mammals, *Malayan Nat. J.,* 22, 3–9, 1968.

Terborgh, J., Maintenance of diversity in tropical forests, *Biotropica,* 24, 283–292, 1992.

Wan, M.M.W.I, Komoo, I., and Pereira, J.J., Hubungan di antara degradasi fizikal dan perubahan litupan tanah di Lembangan Langat, in *Proceedings of the 1999 Langat Basin Research Symposium,* Nordin, H.H., Ed., Institute for Environment and Development, Universiti Kebangsaan Malaysia, 2000, pp. 126–134.

Whitford, W.G., Validation of indicators, in *Ecosystem Health,* Rapport, D. et al., Eds., Blackwell Science, Malden, MA, 1998, pp. 205–209.

Wong, K.H., Landuse in Malaysia, Ministry of Agriculture, Kuala Lumpur, Malaysia, 1974.

Zubaid, A., A comparison of the bat fauna between a primary and fragmented secondary forest in Peninsular Malaysia, *Mammalia,* 57, 201–206, 1993.

Zubaid, A. and Khairul, E.A., A comparison of small mammal abundance between a primary and disturbed lowland rain forest in Peninsular Malaysia, *Malayan Nat. J.,* 50, 201–206, 1997.

SECTION III.6

Pollution: MTBE

Overview: Environmental Impacts of a Motor Fuel Additive: Methyl Tertiary Butyl Ether (MTBE)

Daniel P.Y. Chang

The introduction of a chemical into widely used consumer products such as gasoline can lead to the rapid dispersal of the chemical into the environment. The data provided by the U.S. Geological Survey's National Water Quality Assessment (USGS–NAQWA) program, as presented in the work by Moran and co-workers, illustrate how widespread the contaminant methyl tertiary butyl ether (MTBE) has become in U.S. groundwaters. Although we have benefited from the historical lessons of DDT, leaded gasoline, and the appearance of halogenated dioxins and PCBs throughout the global environment, the relatively recent introduction of MTBE into gasoline and its rapid appearance in the environment caught U.S. regulatory agencies by surprise.

At the time MTBE was adopted for use in reformulated gasoline (RFG), it contributed to the reduction of air pollution and resulted in substantial air quality and health benefits — e.g., reduction of benzene in gasoline, reduction of CO emissions, and to a lesser extent, reduction of reactive volatile organic carbon (VOCs), which are ozone precursors. A review of its potential health effects was undertaken, but as a result of its relatively benign acute toxicity in various environmental media, its introduction into gasoline as an additive was not prevented. A reexamination of the means by which regulatory agencies approve the introduction of chemicals into commerce seemed warranted in light of MTBE's now-apparent harmful effects. One purpose of the session was to examine MTBE as a case study of the potential for adverse impacts on ecosystem health, within the context of the relatively comprehensive review process applied in the U.S.

During the companion panel discussion that followed the session, participants achieved consensus in determining the continuing need for screening methods for chemicals that take into account multimedia fate and transport. However, when applied, *those screening procedures must be supplemented by multidisciplinary "expert" review.* As an example, Morgenroth and Arvin discuss recent German experience in their review of MTBE use in Europe. The German review process would find use of MTBE acceptable, even today, because the procedures applied show that it eventually returns to the atmosphere where it is destroyed, while when it is in the aquatic environment its expected levels did not lead to significant environmental impacts. Nevertheless, panelists and investigators were uncertain whether the German people will regard the benefits of continued use of MTBE as sufficient to justify acceptance of low levels of contaminants in their water supplies.

When significant health or environmental toxicity is not demonstrated or does not arouse public concern, chemicals in the current regulatory environment can fall through the cracks. MTBE, associated largely with aesthetic, "taste and odor" objections despite its economic importance, poses uncertain chronic risks but is not believed to be related to acute health problems. When it was introduced, MTBE was known as a compound that would compromise water quality in the interest of improving air quality. Because the degree of water quality impacts is better understood now, other states and nations considering that trade-off should benefit from the lessons learned in the U.S. As a toxicant, MTBE pales compared to the ecosystem health concerns discussed at this Congress. Nevertheless, study of its regulations provides valuable lessons to improve the process. *Nonhealth-based standards must be considered when reviewing new chemicals; if they are not, the process of evaluation may be much less comprehensive.*

Many of our environmental regulatory agencies have been established along single media lines and — to confuse jurisdictional issues even more — defined by the type of activity, such as solid waste or pesticide regulation. These agencies have been sensitized to the need for multimedia review and improved communication among themselves. Other consumer product reformulations in paints and coatings are being enacted, along with significant changes in the biotechnology and electronic industries. *Ideally, environmental protection agencies would one day become truly multidisciplinary in their thought processes at both the scientific and policy levels.*

In the short term, gasoline will continue to be used worldwide as a motor vehicle fuel. Its constituents and the by-products of its combustion will be present for some time to come. Other countries will be faced with determining the point at which MTBE's drawbacks outweigh its benefits as discussed by Keller and Fernandez in weighing the costs and benefits in California compared to alternative reformulated gasolines. The short-term answer to the Congress forum question "Environmental Pollutants: Must We Live with Them?" is yes. Nevertheless, the crux of the problem for society boils down to doing a better job of managing chemical introduction and use, particularly in transportation fuels as pervasive as gasoline and in other widely used consumer products, e.g., in electronics and pharmaceuticals.

In retrospect, we can take some consolation that a monitoring network not originally targeted for MTBE, i.e., the USGS NAQWA program, alerted us to its widespread appearance. The current regulatory system in the U.S. responded to the discovery of MTBE in drinking water wells, albeit belatedly. *Monitoring capabilities and means to act upon those data effectively are important if we are to minimize long-term damage to ecosystem health.* We hope that these insights and recommendations will continue to improve our ability to manage chemical introduction and use in our environment, and will serve as a valuable case study for others.

CHAPTER 135

MTBE in Groundwater of the U.S.: Occurrence, Potential Sources, and Long-Range Transport

Michael J. Moran, John S. Zogorski, and Paul J. Squillace

INTRODUCTION

Groundwater is an important and valuable resource in the U.S. To some extent, groundwater is used in all 50 states to provide drinking water. In 1995, more than 50% of the population, more than 130 million people, relied on groundwater for drinking water supplies (Solley et al., 1998). About 30%, or 42 million, of these people obtained their water from privately owned wells (U.S. Geological Survey [USGS], 1999). The remaining 90 million people relied on groundwater from public water suppliers for their drinking water.

The gasoline additive methyl *tert*-butyl ether (MTBE) has been frequently detected in ambient groundwater in areas of the U.S. that use substantial amounts of the compound as a fuel oxygenate. Some of this water is used for domestic and public water supplies. MTBE in groundwater has been associated with taste and odor concerns and potential human-health effects.

The primary purpose of this chapter is to summarize current information on the occurrence and distribution of MTBE in groundwater. Information is also presented on the sources and transport of MTBE to groundwater. Data from the USGS's National Water-Quality Assessment (NAWQA) Program (Gilliom et al., 1995), as well as data from other sources, are summarized to describe: (1) the occurrence and distribution of low concentrations of MTBE in ambient groundwater, (2) the effects of high concentrations of MTBE in groundwater used for domestic and public supplies, (3) the potential sources of MTBE, and (4) the long-range transport of MTBE in groundwater.

BACKGROUND ON MTBE

Compounds that contain oxygen, referred to as oxygenates, are added to gasoline in some areas of the U.S. to reduce ozone formation and carbon monoxide emissions and to enhance the octane level in gasoline. Octane enhancement began in the late 1970s with the phase-out of tetraethyl lead from gasoline. The use of oxygenates was expanded due to enactment of the Clean Air Act (CAA) Amendments of 1990, which required that oxygen be added to gasoline in areas where concentrations of ozone in the summer are most severe or where concentrations of carbon monoxide in the winter exceed air-quality standards. The CAA Amendments mandate that gasoline must contain at least 2% oxygen by weight in ozone nonattainment areas and at least 2.7% oxygen by weight in carbon monoxide nonattainment areas (USEPA [U.S. Environmental Protection Agency], 1990).

A number of ether and alcohol compounds have been used as oxygenates in commercial applications or in pilot programs (Zogorski et al., 1997). Although the CAA Amendments do not specify which oxygenate must be added to gasoline, the one used most commonly today is MTBE. The second most frequently used oxygenate is ethanol. Under the mandates of the CAA Amendments, two programs of oxygenate use were established: (1) the Oxygenated Fuels Program (OXY) in which 15% MTBE by volume is added for wintertime use in carbon monoxide nonattainment areas, and (2) the Reformulated Gasoline Program (RFG) in which 11% MTBE by volume is added for year-round use in severe ozone nonattainment areas. Most RFG areas use MTBE whereas most OXY areas do not use MTBE in favor of another oxygenate (typically ethanol).

In general, MTBE is preferred over other oxygenates because of its low cost, ease of production, high octane level, lower evaporative emissions in gasoline, and favorable transfer and blending characteristics (Zogorski et al., 1997). Annual production of MTBE has increased from 0.26 billion l/year in 1980 (Chemical Manufacturers Association, 1997) to 11.9 billion l in 1998 (Department of Energy, 1998) (Figure 135.1). The production of MTBE was the fourth largest of all chemicals produced in 1996 (Chemical Manufacturers Association, 1997).

The large-scale production and use of MTBE and its physical and chemical properties have resulted in its detection in many groundwater systems (Squillace et al., 1996; Zogorski et al., 1998). MTBE is highly soluble in water, does not substantially partition to organic carbon in aquifer material, and biodegrades slowly (Squillace et al., 1997; Zogorski et al., 1998). In water at 25°C, MTBE has a solubility of about 5000 mg/l for a gasoline that is 10% MTBE by weight (Squillace et al., 1997). In a sand aquifer with moderate organic carbon content (0.1%), only 8% of the total mass of MTBE will be sorbed to the organic matter whereas 92% remains in solution. By contrast, 39% of the total mass of benzene and 72% of the total mass of ethylbenzene will be sorbed to the organic matter. The high water solubility and low sorption of MTBE allow dissolved plumes of the compound to migrate farther and faster than other common gasoline components such as benzene, toluene, and ethylbenzene in aquifers with moderate to high organic carbon content. The longer travel distances for MTBE can result in an increased probability that it will be detected in a drinking water well.

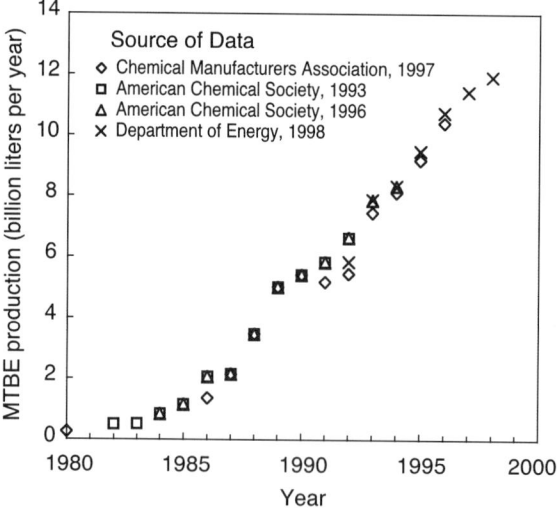

Figure 135.1 Estimated annual production of MTBE in the U.S. from four sources, 1980 to 1998.

Although rates of biodegradation of MTBE may be high in aerobic aquifer systems (Borden et al., 1997; Schirmer and Butler, 1998; Schirmer and Barker, 1998) or in microbial cultures or biofilters (Eweis et al., 1998; Salanitro et al., 1994; Steffan et al., 1997), most studies indicate relatively slow rates of MTBE removal under the anaerobic conditions that are common to most gasoline releases (Suflita and Mormille, 1993; Yeh and Novak, 1994). Biodegradation rate constants for MTBE are estimated to be several orders of magnitude lower than for other gasoline components such as benzene and toluene (Baehr et al., 1999). Lower biodegradation rates mean that MTBE can persist longer in groundwater and can be transported farther and deeper in an aquifer. In addition, a potential biodegradation product for MTBE is *tert*-butyl alcohol (TBA) (Hyman et al., 1998; Landmeyer et al., 1998), which itself has a relatively long removal time under anaerobic conditions (Suflita and Mormille, 1993; Yeh and Novak, 1994).

MTBE is a concern in drinking water because of its low taste and odor thresholds and its potential human-health effects. The USEPA has issued a drinking water advisory to address taste and odor concerns and recommends a range of 20 to 40 µg/l as a limit to help prevent these organoleptic effects (USEPA, 1997). The USEPA advisory concentration range for MTBE is also thought to provide a safety margin for potential carcinogenic effects and to provide a large margin of safety for noncancer effects (USEPA, 1997). The USEPA has tentatively classified MTBE as a Group C possible human carcinogen (USEPA, 1996). The USEPA defines the carcinogenic potential of a Group C carcinogen as "limited evidence from animal studies and inadequate or no data for humans" (USEPA, 1996).

OCCURRENCE AND DISTRIBUTION OF LOW CONCENTRATIONS OF MTBE IN AMBIENT GROUNDWATER

Groundwater samples were collected from wells that are part of NAWQA Program monitoring networks designed to define ambient water quality in aquifers or parts of aquifers. The wells sampled provide water for a wide range of uses including domestic, monitoring, public supply, irrigation, livestock, and commercial. Ambient refers to groundwater that is not associated with known point-source releases. Table 135.1 lists the frequency of MTBE occurrence at concentrations ≥0.2 µg/l in water from wells sampled by the NAWQA Program from 1993 to 1998. A minimum concentration of 0.2 µg/l was selected for reporting concentrations in this table because most samples were analyzed at a laboratory with a censoring level for MTBE of 0.2 µg/l; some samples, however, had detectable concentrations less than 0.2 µg/l.

For all wells sampled, about 5% had detectable concentrations of MTBE greater than 0.2 µg/l. The concentrations of most detections of MTBE were in the range of 0.2 to 20 µg/l with a frequency of detection of these low concentrations of about 5%. Only 0.4% of the detected concentrations were higher than 20 µg/l. Of the samples in which high concentrations were detected, only one was from a well used for drinking.

Table 135.1 MTBE Detections in Ambient Groundwater[a]

Concentration (µg/l)	Frequency (%)	Number of Wells
<0.2[b]	94.7	2598
0.2–20	4.9	133
>20	0.4	12
Total	100	2743

[a] USGS NAWQA data, 1993–1998; all water-use categories; concentrations ≥ 0.2 µg/l. Some information is based on provisional USGS data.
[b] These generally represent nondetections; however, data from some samples had detectable concentrations less than 0.2 µg/l.

Table 135.2 Detection of MTBE in Ambient Groundwater in Areas for MTBE Use and Population Density[a]

MTBE Use Areas	MTBE Detected in Groundwater? (No. of Wells)		Frequency of Detection (%)
	Yes	No	
Substantial Use Area[b]			
Urban areas[c]	49	135	27
Rural areas[d]	50	246	17
All areas	99	381	21
Nonsubstantial Use Area[e]			
Urban areas[c]	14	284	5
Rural areas[d]	32	1933	2
All areas	46	2217	2

[a] USGS NAWQA data, 1993 to 1998; all water-use categories; concentration ≥ 0.2 µg/l. Some information is based on provisional USGS data.
[b] Substantial use areas are defined as RFG/OXY areas where the median annual MTBE content in gasoline is >5% by volume.
[c] Urban areas are defined as areas with a population density of ≥ 386 persons per km^2.
[d] Rural areas are defined as areas with a population density of < 386 persons per km^2.
[e] Nonsubstantial use areas are defined as any areas where the median annual MTBE content in gasoline is $\leq 5\%$ by volume.

Low concentrations of MTBE were detected frequently in ambient groundwater in areas that use substantial amounts of MTBE. These include RFG or OXY areas where the median annual MTBE content in gasoline is >5% by volume. Areas that do not use substantial amounts of MTBE are defined here as any areas in which the median annual MTBE content in gasoline is $\leq 5\%$ by volume. Most areas in this designation use considerably less than 5% MTBE in gasoline. The differentiation of these MTBE use areas is based on currently available information and does not indicate a definitive knowledge of the precise amount of MTBE usage in all designated areas. Table 135.2 lists MTBE detection frequencies in groundwater for areas of MTBE use and population density. This table shows that the distribution of MTBE detections is primarily a function of the use of MTBE but is also related to population density. Urban areas are defined as areas with a population density of ≥ 1000 persons per mi^2 (386 persons per km^2) and rural areas are defined as areas with a population density of < 1000 persons per mi^2 (386 persons per km^2) (Hitt, 1994).

For urban areas, the frequency of detection of MTBE in groundwater in areas of substantial MTBE use was about 27%, whereas the frequency of detection in other areas was about 5%. In rural areas, the frequency of detection of MTBE in areas of substantial MTBE use was about 17%, whereas the frequency of detection in other areas was about 2%. Regardless of population density, the frequency of detection of MTBE in areas of substantial MTBE use was about 21% compared to a frequency of detection of about 2% in other areas. Therefore, being located within an area of substantial MTBE use gives an increased probability of detecting MTBE in groundwater without controlling for any other variables. The relative differences in MTBE detection between areas of substantial MTBE use and other areas are greater than the relative differences in MTBE detection between urban and rural areas (Figure 135.2). In other words, the use of MTBE is a more important factor in determining the likelihood of detecting MTBE in groundwater than is population density.

In addition to differences in frequency of detection between areas of substantial MTBE use and other areas, the question arises as to whether MTBE concentrations differ between the two areas. Figure 135.3 illustrates the frequency of detection of MTBE for selected concentrations for both types of usage. As seen in Figure 135.3, the differences between the frequency of detection of MTBE in different use areas are quite pronounced regardless of the concentration. For each concentration, the frequency of occurrence of MTBE is higher in substantial use areas

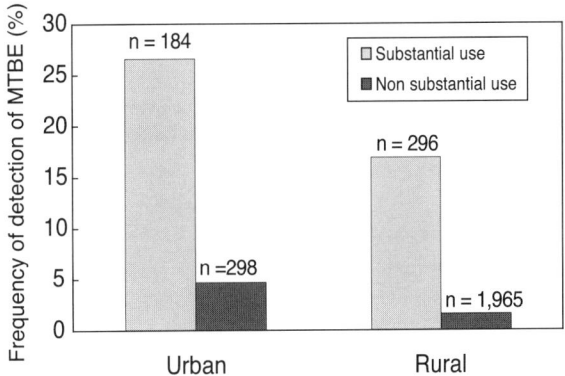

Figure 135.2 Frequency of detection of MTBE in ambient groundwater for MTBE use areas and population density, 1993 to 1998.

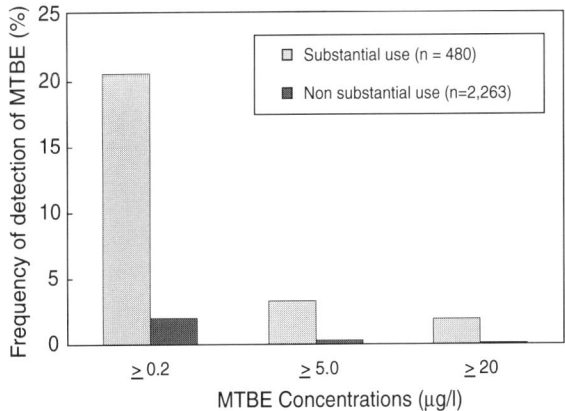

Figure 135.3 Frequency of detection of MTBE in ambient groundwater in for selected MTBE concentrations, 1993 to 1998.

compared to other areas. The differences in frequency of detection are pronounced at higher MTBE concentrations.

MTBE has been detected at low concentrations in groundwater from public water supply wells throughout the U.S. Of 216 public water supply wells that the NAWQA Program sampled from 1993 to 1998, 15 (6.9%) had detectable concentrations of MTBE without regard to a minimum concentration (based on some provisional USGS data). All of the concentrations were less than 2.0 μg/l and the median was 0.14 μg/l. Table 135.3 lists examples of other instances in which MTBE was detected in public water supplies. The data referred to in Table 135.3 are from a variety of sources outside the USGS and point out the widespread detection of low concentrations of MTBE in groundwater used for drinking water.

HIGH CONCENTRATIONS OF MTBE IN SOME DOMESTIC AND PUBLIC WATER SUPPLIES

High concentrations of MTBE have been reported in groundwater used for some domestic and public water supplies. A number of public water supply systems in Illinois, Texas, and California

Table 135.3 MTBE in Public Water Supplies in the U.S.

Source of Data	Number of Public Water Supply Systems/Wells/ Sources with MTBE
Hitzig et al., 1998	251 to 422 wells in 19 states[a]
Zogorski et al., 1997	51 public water systems in 6 of 7 states that provided information[a]
State of Maine, 1998	125 public water-supply systems (16% of tested supplies)[a,b]
California Department of Health Services, 1999[c]	48 sources (0.9% of sources)[a]

[a] Most concentrations were less than 20 µg/l.
[b] All concentrations were less than 35 µg/l.
[c] Monitoring is continuing and number of sources with MTBE may change.

have reported groundwater sources with concentrations of MTBE greater than 20 µg/l (California Department of Health Services, 1999; Zogorski et al., 1997). In most of these cases, treatment of the water was necessary or the wells had to be removed from service. Figure 135.4 illustrates the number of domestic wells, by state, that may have been affected by releases from leaking underground storage tank (LUST) sites. The data in this figure come from a survey completed by the University of Massachusetts in 1998 (Hitzig et al., 1998). The survey queried state LUST programs on the effect of MTBE releases from LUST sites. The concentrations of MTBE in water from these domestic wells are not known; in the vicinity of a LUST site release, however, concentrations can easily exceed 20 µg/l and can be in excess of 1000 µg/l (Buxton et al., 1997; Happel et al., 1998).

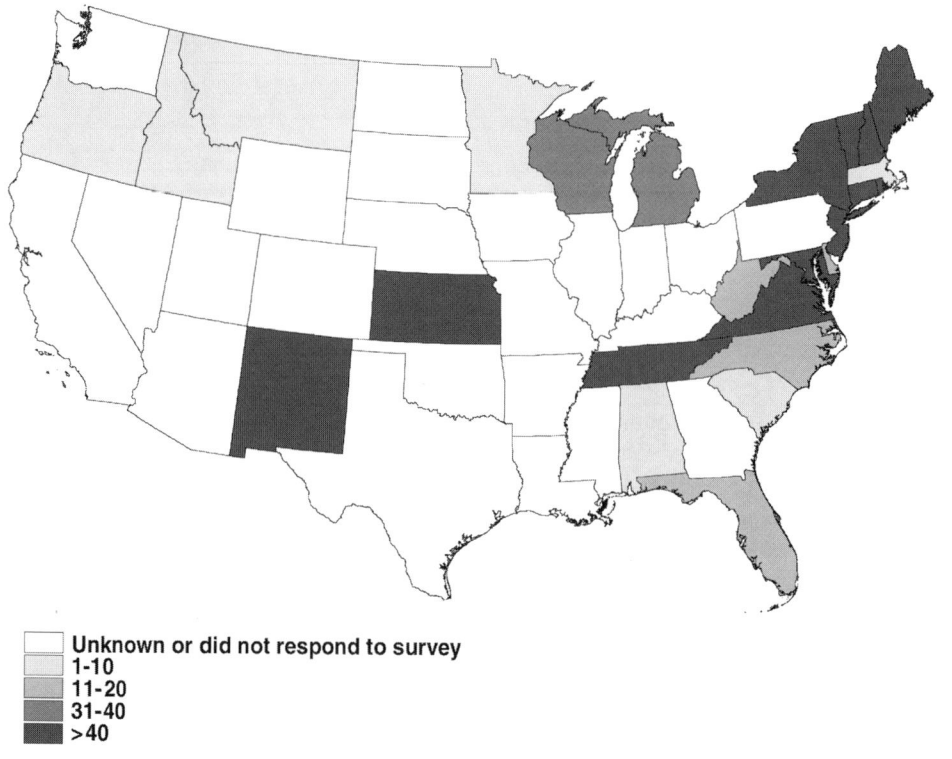

Figure 135.4 Number of domestic wells with detectable MTBE concentrations possibly from LUST site sources. (From Hitzig, R. et al., *Soil Groundwater Cleanup*, August/September, 15–19, 1998. With permission.)

Nationwide data on the occurrence of MTBE in groundwater used for public water supplies are currently not available. The USGS, the Metropolitan Water District of Southern California, and the Oregon Graduate Institute, in conjunction with the American Water Works Association Research Foundation, have initiated a study of MTBE and other volatile organic compounds in raw, untreated ground and surface waters that are sources of drinking water for public supplies. The survey covers the entire U.S. and will focus on groundwater sources serving systems of all sizes (Koch, B., personal communication, 1998).

Additional investigations on the quality of water used for public water supplies are being conducted by the USGS and the USEPA, and include an analysis of MTBE data from wells in 12 states of the northeastern U.S. This additional work consists of two phases: (1) an assessment of existing data on MTBE in finished (treated) water for more than 2000 public water systems, most of which use groundwater as a source; and (2) an assessment of existing data on MTBE in ambient groundwater of the region. The data will be related to land-use patterns, population density, locations of known point-source MTBE releases, and other anthropogenic factors to determine possible sources and transport mechanisms of MTBE in groundwater (Grady, S., personal communication, 1999).

Public water supplies in most states are currently not required to be monitored for MTBE because it is not a regulated chemical; however, regulations by the USEPA place MTBE under the Unregulated Contaminant Monitoring Regulation (UCMR). The regulation requires all large public supplies and some small public supplies to analyze finished (treated) water for MTBE and report the results to the USEPA. These analyses will provide additional information on the occurrence of MTBE in public water systems.

Because domestic wells are not regulated by the USEPA, owners are not required to analyze for MTBE. On the basis of a domestic well survey, the State of Maine estimated that water from 2200 to 6900 domestic wells in Maine may have MTBE concentrations greater than the lower limit of USEPA's drinking water advisory (State of Maine, 1998).

POTENTIAL SOURCES OF MTBE

A variety of point and nonpoint sources may be responsible for the occurrence of MTBE in groundwater (Table 135.4). For point sources, the 1995 Toxic Release Inventory estimated that 1.7 million kg of MTBE were released into the environment from production sources (USEPA, 1998). Of these releases, 97% went to air and <3% went to surface water. Mobile sources, such as automobiles, are estimated to have released 40 million kg of MTBE to the air of the U.S. in 1996 (USEPA, 1998).

No estimates were available for total releases of MTBE to groundwater from underground storage tanks. Since 1988, however, 330,000 point-source releases from LUST sites have been reported to the USEPA Office of Underground Storage Tanks (USEPA, 1998). It is estimated that 100,000 additional releases will be reported in the future as existing systems are upgraded, closed, or replaced (USEPA, 1998). Some of these future releases may be gasoline containing MTBE.

Table 135.4 Potential Point and Nonpoint Sources of MTBE in Groundwater

Point Sources	Nonpoint Sources
Refineries	Vehicle emissions
Pipelines	Vehicle evaporative losses
Storage tanks (above or belowground)	Atmospheric deposition
Accidental spillage	Urban runoff
Homeowner disposal	
Spillage during fueling	
Fueling evaporative losses	

Although it is difficult to estimate the mass of MTBE released from LUST sites, it is believed that these sources contribute significantly more of the total MTBE mass to the environment than either production or mobile sources.

Releases from LUST sites have affected public water supply wells in different areas of the U.S. (Hitzig et al., 1998). According to a national LUST survey, it was estimated that 65 public supply wells in New Jersey have been affected by MTBE released from LUST sites. In New Hampshire and Vermont, between 31 and 40 public supply wells were estimated to have been affected by MTBE released from LUST sites, and between 21 and 30 public supply wells have been affected in Kansas, Massachusetts, and New Mexico (Hitzig, R., personal communication, 1998).

In some instances, high concentrations of MTBE from point sources have even caused the closure of public supply wells. One well-known example is the city of Santa Monica, CA. In 1997, the city ceased pumping groundwater from two wellfields due to persistent and increasing concentrations of MTBE (Brown et al., 1997). This loss accounted for 50% of the Santa Monica's total drinking water supply. Concentrations of MTBE in water from wells in the two wellfields ranged from 8.2 to 610 µg/l. A number of LUST sites and underground petroleum pipelines were identified as potential point sources of the MTBE (Brown et al., 1997).

Concentrations of MTBE may have a hierarchy associated with different potential sources. Figure 135.5 illustrates three ranges of MTBE concentrations and the associated potential sources. The y-axis represents the detection frequency of MTBE in groundwater sampled by the NAWQA Program for the selected concentration ranges.

Concentrations greater than 20 µg/l are generally thought to be associated with point-source releases because it would be unusual for nonpoint sources to cause such high concentrations in groundwater. However, concentrations of MTBE higher than 20 µg/l could be associated with certain high-concentration nonpoint sources. For example, on the basis of partitioning theory and air-quality monitoring, if atmospheric concentrations of MTBE are systematically higher in some local areas such as a parking garage, atmospheric deposition could result in concentrations of MTBE in precipitation and groundwater that are >20 µg/l (Zogorski et al., 1997). Likewise, low concentrations (≤20 µg/l) could be associated with point sources, such as at the leading edge of a groundwater plume.

In some cases, a potential nonpoint source for MTBE has been identified. Using colocated shallow groundwater and atmospheric concentration data, it has been shown that deposition of MTBE from the atmosphere is one of several possible sources for low concentrations of MTBE in shallow groundwater. Under conditions in which biodegradation is very slow, the source of low concentrations of MTBE in groundwater could be atmospheric deposition (Baehr et al., 1999).

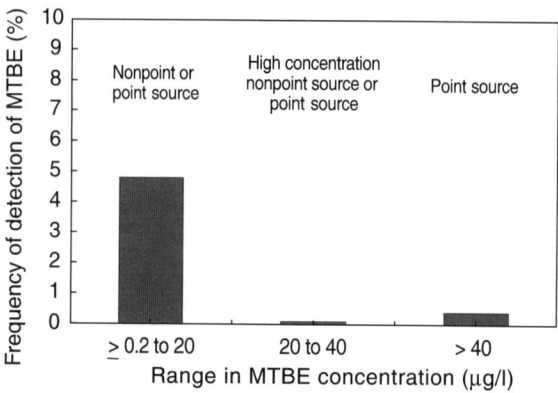

Figure 135.5 Frequency of detection of MTBE in ambient groundwater for selected concentration ranges and associated potential sources, 1993 to 1998.

LONG-RANGE TRANSPORT OF MTBE

At some gasoline release sites, MTBE has migrated much farther than other common gasoline components. In these situations, treatment may be required if the long-range transport of MTBE threatens or affects a drinking water well. Table 135.5 lists some examples of instances in which MTBE has migrated farther than the common gasoline components benzene, toluene, ethylbenzene, and xylenes (BTEX). These observations are consistent with the physical and chemical properties of MTBE that allow it, under certain conditions, to be transported farther than other common gasoline components.

An example of long-range transport of MTBE is the release at the Port Hueneme naval facility in California. Between September 1984 and March 1985, approximately 15,000 l of regular gasoline and 26,000 l of premium unleaded gasoline were released from fuel-delivery lines at the Naval Exchange gasoline station (Kram and Lory, 1998). The exact percentage of MTBE in the released gasoline is not known but is estimated to have ranged between 6 and 11% by volume. The plume of dissolved MTBE occurs in a semiperched aquifer at a depth of about 3 to 6 m below the ground surface, has traveled more than 1300 m, and covers 0.157 km^2. In addition, the MTBE plume has extended far beyond the BTEX plume that traveled approximately 390 m (Kram and Lory, 1998). It is unknown if the MTBE plume at Port Hueneme has affected any domestic or public water supply wells such that treatment would be required; however, this site illustrates the occurrence of long-range MTBE transport.

In some instances, treatment of MTBE in drinking water supplies has been required even though only a small volume of MTBE was released and it did not migrate a great distance. In December 1997, an example of this occurred in Standish, ME, when a vehicle overturned and lost most of the gasoline from its tank. The gasoline was probably reformulated gasoline containing about 11% MTBE. It was estimated that 26 to 57 l spilled onto the soil (Hunter, B., personal communication, 1998). In May 1998, a Standish resident close to the accident site called the Maine Department of Environmental Protection and complained of a bad odor in his water. Subsequent groundwater sampling in the area revealed that more than 20 domestic wells had detectable concentrations of MTBE. Although the exact area of groundwater that was affected by this spill is unknown, it is estimated to be less than 1 km^2.

At one point, water samples from 11 wells had concentrations of MTBE greater than the Maine action level of 35 µg/l, and some of the concentrations exceeded 100 µg/l. No common gasoline hydrocarbons such as BTEX were detected in any of the samples analyzed. Point-of-use treatment of water from wells with concentrations of MTBE exceeding 100 µg/l was performed until affected homes were connected to a public water supply. A granular activated carbon system was used at a cost of about $1600 per year per home (Hunter, B., personal communication, 1998).

SUMMARY

The gasoline additive methyl *tert*-butyl ether (MTBE) has been detected in low concentrations (≤20 µg/l) in ambient groundwater in some areas of the U.S. Some of these occurrences are in

Table 135.5 Examples of MTBE Migrating Farther Than BTEX

Laurel Bay gasoline station, Beaufort, SC (Landmeyer et al., 1998)
Borden Air Force Base, Canada (Schirmer et al., 1998)
North Windam[a] and North Berwick,[a,b] ME (Lawyui and Fingas, 1997)
Port Hueneme, CA[a] (Kram and Lory, 1998)
East Patchogue, NY[a] (Weaver et al., 1996)
Spring Green, WI[a,c]

[a] Migrated >300 m.
[b] A. Smith, personal communication, Maine Department of Human Services, Augusta, 1999.
[c] S. Ales, personal communication, Wisconsin Department of Natural Resources, Fitchburg, 1999.

water from wells that are used for domestic and public water supplies. High concentrations (>20 μg/l) of MTBE also have been detected in ambient groundwater, some of which is used for domestic and public water supplies. In some cases, treatment of the water has been necessary for MTBE removal or the wells had to be taken out of service.

The occurrence of MTBE in groundwater is strongly related to the use of MTBE in gasoline. The probability that MTBE will be detected in ambient groundwater is much higher in areas that use substantial amounts of the compound (>5% by volume in gasoline) compared to other areas (≤5% by volume in gasoline). Urban areas are also more likely than rural areas to have detectable concentrations of MTBE in groundwater.

Low concentrations of MTBE may have point or nonpoint sources. The origin of high concentrations is likely some type of point source; however, some nonpoint sources can result in relatively high concentrations of MTBE in groundwater. At some sites, MTBE has migrated much farther than other common gasoline components and those long distances increase the probability that MTBE will be detected in a drinking water well and that treatment may be required.

Data from the NAWQA Program provide insight into the quality of ambient groundwater resources with respect to the occurrence of MTBE. Information from other sources has also helped to reveal the extent and scope of this water-quality issue and to document various sources of MTBE in groundwater. Additional work by the USGS, the USEPA, and others will provide a more detailed and complete understanding of the occurrence of this compound in groundwater resources.

REFERENCES

Baehr, A.L., Stackelberg, P.E., and Baker, R.L., Evaluation of the atmosphere as a source of volatile organic compounds in shallow groundwater, *Water Resourc. Res.,* 35, 127–136, 1999.

Borden, R.C., Daniel, R.A., LeBrun, IV, L.E., and Davis, C.W., Intrinsic bioremediation of MTBE and BTEX in a gasoline-contaminated aquifer, *Water Resourc. Res.,* 33, 1105–1115, 1997.

Brown, A.F., Rodriquez, R.A., Johnson, B.J., and Bellomo, A.J., Methyl tertiary butyl ether (MTBE) contamination of the city of Santa Monica drinking water supply, in *Proceedings of the NWWA/ API Petroleum Hydrocarbons and Organic Chemicals in Groundwater: Prevention, Detection and Remediation Conference,* 12–14 Nov. 1996, Houston, Stanley, A., Ed., National Water Well Association and American Petroleum Institute, Houston, TX, 1997, pp. 35–39.

Buxton, H.T., Landmeyer, J.E., Baehr, A.L., Church, C.D., and Tratnyek, P.G., Interdisciplinary investigation of subsurface contaminant transport and fate at point-source releases of gasoline containing MTBE, in *Proceedings of the NWWA/API Petroleum Hydrocarbons and Organic Chemicals in Groundwater: Prevention, Detection and Remediation Conference,* 12–14 Nov. 1996, Houston, Stanley, A., Ed., National Water Well Association and American Petroleum Institute, Houston, TX, 1997, pp. 2–18.

California Department of Health Services, Summary of sampling of public water systems for MTBE: California Department of Health Services, accessed at http://www.dhs.cahwnet.gov/ps/ddwem/ chemicals/MTBE/mtbe_summary, 1999.

Chemical Manufacturers Association, *U.S. Chemical Industry Statistical Handbook: 1997,* Chemical Manufacturers Association, Arlington, VA, 1997.

Department of Energy, Petroleum supply monthly, table(s) B1-B4 — digital data files accessed April 14, 1998 at http://ftp.eia.doe.gov/pub/petroleum/data/monthly/oxydata.txt, 1998.

Eweis, J.B., Schroeder, E.D., Chang, D.P.Y., and Scow, K.M., Biodegradation of MTBE in a pilot-scale biofilter, in *Natural Attenuation of Chlorinated and Recalcitrant Compounds,* First International Conference on Remediation of Chlorinated and Recalcitrant Compounds, 18–21 May 1998, Monterey, CA, Wickramanayake, G.B. and Hinchee, R.E., Eds., Battelle Press, Columbus, OH, 1998, pp. 341–346.

Gilliom, R.J., Alley, W.M., and Gurtz, M.E., Design of the National Water-Quality Assessment Program — Occurrence and distribution of water-quality conditions, U.S.G.S. Circular 1112, U.S. Government Printing Office, Washington, D.C., 1995.

Happel, A., Beckenbach, E.H., and Halden, R.U., An evaluation of MTBE impacts to California ground-water resources, report submitted to the California State Water Resources Control Board, Underground Storage Tank Program, UCRL-AR-130897, Lawrence Livermore National Laboratory, Livermore, CA, 1998.

Hitt, K.J., Refining 1970's land-use data with 1990 population data to indicate new residential development, U.S.G.S. WRIR 94–4250, U.S. Government Printing Office, Washington, D.C., 1994.

Hitzig, R., Kostecki, P., and Leonard, D., Study reports LUST programs are feeling effects of MTBE releases, *Soil Groundwater Cleanup,* August/September, 15–19, 1998.

Hyman, M., Kwon, P., Williamson, K., and O'Reilly, K., Cometabolism of MTBE by alkane-utilizing microorganisms, in *Natural Attenuation of Chlorinated and Recalcitrant Compounds*, First International Conference on Remediation of Chlorinated and Recalcitrant Compounds, 18–21 May 1998, Monterey, CA, Wickramanayake, G.B. and Hinchee, R.E., Eds., Battelle Press, Columbus, OH, 1998, pp. 321–326.

Kram, M. and Lory, E., Use of SCAPs suite of tools to rapidly delineate a large MTBE plume, in *Proceedings of the Symposium on the Application of Geophysics to Environmental and Engineering Problems*, 22–26 March 1998, Chicago, Bell, R.S., Powers, M.H., and Larson, T., Eds., 1998, pp. 85–99.

Landmeyer, J.E., Chapelle, F.H., Bradley, P.M., Pankow, J.F., Church, C.D., and Tratnyek, P.G., Fate of MTBE relative to benzene in a gasoline-contaminated aquifer (1993–98), *Ground Water Monitoring Remed.,* 18, 93–102, 1998.

Salanitro, J.P., Diaz, L.A., Williams, M.P., and Wisniewski, H.L., Isolation of a bacterial culture that degrades methyl *t*-butyl ether, *Appl. Environ. Microbiol.,* 60, 2593–2596, 1994.

Schirmer, M. and Barker, J.F., A study of long-term MTBE attenuation in the Borden Aquifer, Ontario, Canada, *Ground Water Monitor. Remed.,* Spring 113–122, 1998.

Schirmer, J.F. and Butler, B.J., Natural attenuation of MTBE at the Borden field site, in *Natural Attenuation of Chlorinated and Recalcitrant Compounds,* First International Conference on Remediation of Chlorinated and Recalcitrant Compounds, 18–21 May 1998; Monterey, CA., Wickramanayake, G.B. and Hinchee, R.E., Eds., Battelle Press, Columbus, OH, 1998, pp. 327–331.

Solley, W.B., Pierce, R.R., and Perlman, H.A., Estimated use of water in the U.S. in 1995, U.S.G.S. Circular 1200, U.S. Government Printing Office, Washington, D.C., 1998.

Squillace, P.J., Zogorski, J.S., Wilber, W.G., and Price, C.V., Preliminary assessment of the occurrence and possible sources of MTBE in groundwater in the U.S., 1993–1994, *Environ. Sci. Technol.,* 30, 1721–1730, 1996.

Squillace, P.J., Pankow, J.F., Korte, N.E., and Zogorski, J.S., Review of the environmental behavior and fate of methyl *tert*-butyl ether, *Environ. Toxicol. Chem.,* 16, 1836–1844, 1997.

State of Maine, The presence of MTBE and other gasoline compounds in Maine's drinking water: a preliminary report, October 13, 1998, Maine Department of Health Services, Augusta, 1998.

Steffan, R.J., McClay, K., Vainberg, S., Condee, C.W., and Zhang, D., Biodegradation of the gasoline oxygenates methyl *tert*-butyl ether, ethyl *tert*-butyl ether, and *tert*-amyl methyl ether by propane-oxidizing bacteria, *Appl. Environ. Microbiol.,* 63, 4216–4222, 1997.

Suflita, J.M. and Mormille, M.R., Anaerobic biodegradation of potential gasoline oxygenates in the terrestrial subsurface, *Environ. Sci. Technol.,* 27, 976–978, 1993.

USEPA, The Clean Air Act Amendments, Washington, D.C., 101st Congress of the U.S., Sec. 219, p. S.1630–1938, accessed at http://www.epa.gov/oar/caa/caaa.txt, 1990.

USEPA, Drinking water regulations and health advisories, EPA 822-B-96-002, USEPA, Office of Water, Washington, D.C., 1996.

USEPA, Drinking water advisory — consumer acceptability advice and health effects analysis on methyl tertiary-butyl ether, EPA-822-F-97-009, USEPA, Office of Water, Washington, D.C., 1997.

USEPA, Oxygenates in water — critical information and research needs, EPA/600/R-98/048, Environmental Protection Agency, Office of Water, Washington, D.C., 1998.

U.S. Geological Survey, Budget request for fiscal year 2000, 1999.

Yeh, C.T. and Novak, J.T., Anaerobic biodegradation of gasoline oxygenates in soils, *Water Environ. Res.,* 66, 744–752, 1994.

Zogorski, J.S., Morduchowitz, A., Baehr, A.L., Bauman, B.J., Conrad, D.L., Drew, R.T., Korte, N.E., Lapham, W.W., Pankow, J.F., and Washington, E.R., Fuel oxygenates and water quality, in *Interagency Assessment of Oxygenated Fuels*, Office of Science Technology Policy, Executive Office of the President, U.S. Government Printing Office, Washington, D.C., 1997.

Zogorski, J.S., Delzer, G.C., Bender, D.A., Baehr, A.L., Stackelberg, P.E., Landmeyer, J.E., Boughton, C.J., Lico, M.S., Pankow, J.F., Johnson, R.L., and Thomson, N.R., MTBE — summary of findings and research by the U.S. Geological Survey, in *Proceedings of the American Water Works Association Annual Conference*, 1998, Denver, American Water Works Association, Washington, D.C., 1998, pp. 287–309.

CHAPTER 136

The European Perspective of MTBE as an Oxygenate in Fuels

Eberhard Morgenroth and Erik Arvin

INTRODUCTION

In Europe, methyl *tert*-butyl ether (MTBE) has developed into an important organic, high-production-volume chemical since its introduction into gasolines in the late 1970s and mid-1980s. MTBE was introduced as an octane booster to replace leaded compounds that had to be reduced in gasolines at that time. Unlike in Europe, MTBE is used to a greater extent in the U.S. because the 1990 Clean Air Act Amendments require fuel oxygenates, such as MTBE or ethanol, to be added to gasoline used in some metropolitan areas to reduce atmospheric concentrations of carbon monoxide (CO) or ozone (O_3). However, recent reports on drinking water contaminations in the U.S. (Squillace et al., 1996; Reuter et al., 1998) have raised concerns whether those compounds that help reduce air pollution may, on the other hand, result in severe problems as a groundwater pollutant. Alarmed by the discussion in the U.S., European environmental protection agencies are reviewing their position on MTBE, and new measurement campaigns are currently being initiated. In this chapter, environmental data and policies toward MTBE are reviewed for a number of European countries and are compared to the situation in the U.S.

MTBE IN THE ENVIRONMENT

MTBE is a volatile, flammable, colorless ether, which is miscible in gasoline. It has a relatively high vapor pressure (33,000 Pa at 25°C), a high water solubility (23 to 54 g/l at 25°C), and a low octanol/water partition coefficient (log K_{OW} = 0.9 to 1.2) (Squillace et al., 1996). Reported dimensionless Henry's law constants vary from 0.0069 to 0.0242 for temperatures ranging from 10 to 25°C, respectively (Rippen, 1999). In gasolines, MTBE is the preferred fuel oxygenate because of its low cost, ease of production, and favorable transfer and blending characteristics. MTBE released into the environment can be found in air, soils, and water, but major environmental concerns are related to the contamination of groundwater. For different media, accumulation and degradation of MTBE in the environment are significantly different:

- *Atmosphere.* MTBE is released into the atmosphere from incomplete combustion in cars and during refueling. In addition, MTBE will evaporate into the atmosphere from accidental spills and during the manufacturing process.
- *Surface water.* In surface waters that are used for recreational purposes, watercraft exhausts have been identified as the major source of MTBE. Two-stroke engine, personal watercraft such as jet skis release approximately 25% of their fuel directly into the water, unburned (McCord and Schladow, 1998). In addition, MTBE can be transported into surface waters through direct diffusion from the atmosphere, through rain, stormwater runoff, and with groundwater.
- *Soil and groundwater.* Major sources of MTBE in groundwater, when it occurs at relatively high concentrations (μg/l to mg/l), are leaking gasoline storage tanks with connecting piping (Squillace et al., 1996). Exchange of MTBE between atmosphere and groundwater through direct diffusion or rain is only a minor source of MTBE that can be expected solely for shallow groundwaters (Pankow et al., 1997). However, those widespread occurrences of MTBE in groundwater resources in low concentrations (around the detection limit of 0.1 μg/l; e.g., Miljøstyrelsen, 1998) can raise concerns of decision makers and the public.

Degradation. Different mechanisms lead to a removal of MTBE from the environmental compartments. In the atmosphere, MTBE is removed through photochemical degradation at rates of 3 d^{-1} (Squillace et al., 1997). In soil and groundwater environments, reports on biological degradation of MTBE range from very slow degradation to no detectable biological degradation at all (Landmeyer et al., 1998; Squillace et al., 1997). In one field study, higher degradation rates were observed (0 to 0.001 d^{-1} first-order degradation rate; Borden et al., 1997). Bradley and co-workers (1999) showed in a microcosm study that 30 to 70% of MTBE was oxidized to CO_2 over a period of 100 days. In engineered reactors, a number of reports have shown that MTBE can be degraded biologically, which can make biological processes in soil remediation a feasible option (Eweis et al., 1998).

It can be concluded that, for groundwater protection, control of underground storage tanks and the rapid detection of leaks and spills followed by effective soil and groundwater remediation are of prime importance. Emissions into the atmosphere based on normal use in car engines are not expected to result in a significant contamination of groundwater. However, deposition from the atmosphere can be the cause of widespread occurrence of MTBE traces in groundwater.

Benefits of using MTBE in gasolines include more complete combustion leading to a reduction of CO emissions, a reduced emission of precursors for ozone formation, and also a reduction of benzene and other toxic compounds. The California Air Resources Board predicted that the addition of oxygenates to fuels would reduce ozone precursors by 15%, reduce benzene emissions by 50%, and reduce CO emissions by 11%; these reductions are equivalent to removing emissions from 3.5 million vehicles. Expected air quality benefits were the basis for introducing a minimum oxygen content for gasolines with the 1990 Clean Air Act Amendments. However, benefits and risks of adding high amounts of MTBE to gasolines have recently come under discussion.

USE OF MTBE IN EUROPE

In the late 1970s, Italy was the first country in Europe to introduce MTBE into gasolines, and was followed by other countries where introduction of MTBE took place mainly in the mid-1980s. MTBE was introduced as a replacement for leaded compounds to increase the octane level to prevent engine knocking. Before regulations required the decrease of the lead content, leaded compounds constituted a major contribution to the octane level in gasolines. To prevent technical problems in automobile engines and fueling system components, gasoline producers initially added very high amounts of MTBE and also increased the aromatics content. With experience, refiners decreased the MTBE content of gasoline, as MTBE is, next to alcylate, the most expensive gasoline-blending component (Pahlke, G., personal communication, 1999). A major reason for a decreased overall consumption of MTBE has, however, been the introduction of many new cars with improved

Table 136.1   Maximum Concentrations of Gasoline Compounds in E.U. and U.S. Regulations

Starting	E.U.				U.S.	
	1985	1998	2000	2005	RFG[a]	CA2RFG[b]
Aromatics, by volume (%)		—	42	35	25	30
Benzene, by volume (%)		5	1	1	1	1.2
Oxygen content, by mass (%)	2.5 or 3.7	2.5 or 3.7	2.7	2.7	min: 2.7 max: 3.5	min: 1.8[c] max: 2.7
Ethers containing 5 or more carbon atoms per molecule, by volume (%)	10 or 15	10 or 15	15	15		
Sulfur content, mg/kg		500	150	50		80
Lead content, g/l			0.005	0.005		0.013
Reference	Anon., 1985	Anon., 1999b	Anon., 1998	Anon., 1998	Anon., 1999a	Anon., 1999a

[a] Absolute limits (cap) for U.S. federal reformulated gasoline for all states except for California.
[b] Absolute limits (cap) California Phase 2 reformulated gasoline.
[c] No minimum oxygen required in summer.

engine technology, which run on 95 octane gasoline. In many countries, a downward trend in the MTBE content of gasolines has been observed over the past 10 years. However, with new EU regulations for gasoline this downward trend is currently being reversed. Other compounds that added to the octane rating of the gasoline, such as benzene and overall aromatics, have to be reduced in steps by the years 2000 and 2005 (Table 136.1), which in turn is expected to result in an increase in MTBE consumption.

European guidelines for gasoline compositions include a maximum MTBE concentration of 15% by volume (Table 136.1). However, in Europe there is no minimum requirement for MTBE. In the U.S., high amounts of MTBE are added to gasolines due to minimum requirements for oxygen in oxygenated and in reformulated gasolines of 2.7 and 2.0% by mass, respectively (Squillace et al., 1996). An overview of gasoline use and MTBE content of gasolines is provided for individual European countries in Appendix 136.I. The MTBE content calculated as an average for all grades of gasolines is below 4% by volume, except for Finland where the overall MTBE content was estimated to be 8.8% by volume based on the high MTBE content of 12.4% by volume in their 98 grade gasoline (Figure 136.1). Based on an estimate from Fortum Oil (Kekalainen, J.P., personal communication, 1999), in 1998 roughly 3 million tons of MTBE were produced in the EU, 0.8 million tons were exported, 0.15 million tons were imported. Overall 2.4 million tons of MTBE were consumed within the EU, compared to an estimated consumption of 8 to 10 million tons of MTBE per year in the U.S. Thus, even with uncertainties for individual estimates for individual countries (Appendix 136.I), the overall order of magnitude appears reasonable (Vahervuori, H., personal communication, 1999). It can be concluded that concerns and problems with respect to MTBE come with a time lag in Europe where MTBE was introduced at a later time and in smaller concentrations compared to the U.S.

MONITORING OF MTBE IN THE ENVIRONMENT

Very limited monitoring data on MTBE in air, soil, or water are available and have been published for European countries because MTBE has not been routinely monitored. National environmental protection agencies are only slowly starting to realize the extent of problems connected to drinking water contamination with MTBE. However, water suppliers in Europe are becoming concerned about possible MTBE contaminations leading to a growing demand for measuring campaigns and a clear policy toward MTBE. Only now are those measuring campaigns being initiated in some countries. In the following discussion, some data for selected European

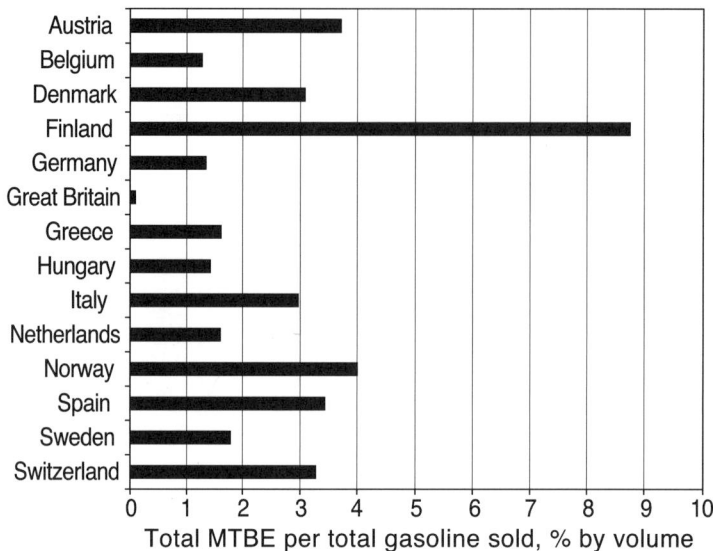

Figure 136.1 Use of MTBE in gasoline calculated as total volume of MTBE per total volume of gasoline consumed (data for 1996 from Appendix 136.I). Note that, according to Fortum Oil and Gas (Vahervuori, H., personal communication, 1999), the use of MTBE is underestimated for Belgium and Great Britain and overestimated for Denmark. In Finland, *tert*-amyl methyl ether (TAME) is added in the same order of magnitude as MTBE (Kekalainen, J.P., personal communication, 1999).

countries are presented that reveal that MTBE is ubiquitously present in very small amounts but it is also found in the mg/l range in connection to gasoline spills.

Denmark

Recently a report from the Danish Environmental Protection Agency on MTBE was published (Miljøstyrelsen, 1998). Analysis of MTBE in groundwater started in the early 1990s. From 1997, the Danish EPA has required that MTBE must be included in the monitoring programs for all wellfields that might be influenced by gasoline storage tanks. MTBE has been found in some counties in very low concentrations, i.e., μg/l range. In one major water supply, some of the drinking water contains –0.1 μg/l (Table 136.2). At several locations downstream of leaking underground storage tanks, concentrations are in the mg/l range and some contain up to several hundred mg/l (Table 136.3). The concentrations of MTBE were not correlated to other gasoline hydrocarbons (e.g., BTEX compounds) for those few simultaneously measured data sets available in Denmark.

Table 136.2 MTBE Concentrations in Groundwater in the Vicinity of Gasoline Stations

Measuring Campaign by Local Authorities		Measuring Campaign by Oil Companies	
Location	Concentration, μg/l	Location	Max. Concentration, μg/l
Fyn	3–6000	Copenhagen 1	480
Ribe	1000–3700	Copenhagen 2	<1
Århus	22,000–550,000	Frederiksborg	30,000
Copenhagen	1–42	Fyn 1	1700
Frederiksberg	0	Fyn 2	5
		Fyn 3	240
		Ringkøbing 1	<10
		Ringkøbing 2	400
		Ringkøbing 3	45
		Viborg	<10

Data from Miljøstyrelsen, Handlingsplan for MTBE (Action plan for MTBE), Miljø-og Energiministeriet Miljøstyrelsen (Danish Environmental Protection Agency), 1998.

Table 136.3 MTBE in Danish Drinking Water Supplies, in µg/l

Location	First Sampling Campaign		Second Sampling Campaign	
	Laboratory 1	Laboratory 2	Laboratory 1	Laboratory 2
Thorsbro trykpumpe	0.1	n.d.[a]	0.11	n.d.
Thorsbro trykpumpe			0.11	n.d.[a]
Islevbro trykpumpe	n.d.	n.d.	n.d.	n.d.
Regnemark trykpumpe	n.d.	n.d.	n.d.	n.d.
Lejre trykpumpe	n.d.	n.d.	n.d.	n.d.
Slangerup trykpumpe	n.d.	n.d.	n.d.	n.d.
Søndersø trykpumpe	n.d.	n.d.	n.d.	n.d.
Marbjerg trykpumpe	n.d.	n.d.		
Thorsbro kildeplads			n.d.	n.d.
Ishøj kildeplads			n.d.	n.d.
Tåstrup-Valby Vest kildeplads	0.15	n.d.[a]	0.12	n.d.[a]
Solhøj kildeplads	0.14	n.d.[a]	0.12	n.d.[a]
Solhøj kildeplads			0.13	n.d.[a]
Kilde XIII + XIV	n.d.[a]	n.d.	n.d.	n.d.
Kilde XIII + XIV	n.d.[a]	n.d.[a]		
Vardegård kildeplads			n.d.	n.d.
Haraldsted Sø			0.59	n.d.[a]
Gyrstinge Sø			n.d.	n.d.
Milli-Q-vand	n d	n d	n d	n.d.
Hvidovre vandforsyning	n.d.			
Frederiksberg vandforsyning	n.d.			
Fåborg vandværk	30			

Note: n.d.= not detected (detection limit 0.1 µg/l).

[a] Laboratory found traces (<0.04 µg/l).

Data from Miljøstyrelsen, Handlingsplan for MTBE (Action plan for MTBE), Miljø-og Energiministeriet Miljøstyrelsen (Danish Environmental Protection Agency), 1998.

This discrepancy has also been observed in a very comprehensive investigation for Texas (Mace and Choi, 1998).

Germany

According to the German Environmental Protection Agency (Umweltbundesamt), no coordinated screening for MTBE has been done (Pahlke, G., personal communication, 1999). In Bavaria, samples from drinking water supplies were tested for MTBE and concentrations were below the detection limit. However, from the vicinity of gasoline stations no data are available. High concentrations of MTBE were discovered in the groundwater at two contaminated industrial sites (Pahlke, G., personal communication, 1999). One site is a former Russian gasoline storage and experimental station in Thüringen that was operated during World War II. Even today after 50 years, large amounts of MTBE can be found in the groundwater at that site. The other contaminated site where MTBE has been identified in the groundwater is the refinery and fuel storage at Leuna, located in the former German Democratic Republic. At the industrial site at Leuna, a large refinery was put into operation in 1965 where unleaded gasoline with an MTBE content of 5 to 6% by volume was produced for the transit highway connection to West Berlin. In the groundwater, BTEX and MTBE are today found in the mg/l range.

Switzerland

Since 1996, representative groundwater and springwater supplies used for drinking water in the district of Zürich were analyzed for MTBE and other volatile contaminants (Schmidt, T.C., personal communication, 1999). In 1996 and 1997, concentrations of MTBE were below the detection limit of 0.05 µg/l for all 62 samples analyzed (EPA method 524.2). However, in

1998 MTBE was found in 18 of 100 samples with concentrations ranging from 0.07 to 0.81 µg/l. No benzene was found in any of the samples for the entire period. It turned out that MTBE was, after trichloroethylene and perchlorethylene, the third most abundant volatile compound before 1,1,1-trichlorethane and chloroform. It should be noted that samples containing MTBE were found in very different groundwater supplies and some of the samples containing MTBE were drawn from very deep groundwaters. No explanation can be given why MTBE concurrently appeared in deep and shallow groundwater supplies after the first sampling campaign in 1996.

United Kingdom

According to a survey of the U.K. Environment Agency in 1995, only 6% of water companies are routinely analyzed for MTBE, although 15% intended to start or increase monitoring. For 9% of all water companies, instances of contamination with MTBE have been reported (Harris, 1999).

REGULATIONS AND POLITICAL DIRECTION

In a recent report prepared for the World Health Organization (IPCS, 1998), it was stated that under common exposure conditions it appears unlikely that MTBE alone induces adverse acute health effects in the general population. However, low threshold values for odor and taste make water inappropriate for drinking water supplies before toxic levels have been reached. So far, there are no coordinated regulations for MTBE in water, air, and soil in Europe, and there also is not a common policy toward the development of MTBE as a fuel oxygenate.

Denmark

To reduce the amount of MTBE released into the environment, the Danish EPA plans to improve the safety of existing and newly built gasoline storage tanks. However, alternatives for MTBE in gasoline are also being investigated, and the use of high-octane fuels may be discouraged in the future through increased taxes. In addition, an initiative to encourage the use of new engine technology that does not require 95 octane fuels is being discussed. With respect to guideline concentrations for MTBE in drinking water (Table 136.4), maximum MTBE concentrations in the future may not be determined by toxicity or taste considerations, but by the contribution of MTBE to the available organic carbon in the water. The available organic carbon determines the potential for microbiological growth in the water distribution network. Virtually 100% of the Danish drinking water is produced from groundwater and is not chlorinated prior to distribution. If an MTBE degrading culture establishes in a water pipeline, its growth will lead to increased bacterial numbers in the water. With a guideline value of 30 µg/l MTBE, this could lead to unacceptable secondary bacterial contaminations. A guideline value of 10 µg/l of assimilable organic carbon (AOC) is considered reasonable for drinking water (van der Kooij, 1992); therefore, in many cases there is only a small margin, probably a few µg/l, left for MTBE from that point of view. Political consideration may also lead to much lower guideline values for MTBE in the future. In Denmark,

Table 136.4 Comparison of Suggested Maximum Target Concentrations for MTBE in Different Environmental Media for Denmark and the U.S.

	Air	Soil	Water	Ref.
Denmark[a]	0.26 mg/m^3 (toxicity) 0.03 mg/m^3 (odor)	500 mg/kg (toxicity) 0.3 mg/kg (odor)	350 µg/l (toxicity) 30 µg/l (odor)	Larsen, 1998
U.S.[b]	n.a.	0.2–3300 mg/kg	20–510 µg/l	Leonard, 1998

[a] Always two concentrations are given based on toxicological concern and on odor threshold.
[b] Overall range for different states. For soil: range of cleanup levels; for water: drinking water standards.

guideline values for chemicals in groundwater are in the order of 1 µg/l, for pesticides in the order of 0.1 µg/l. Political considerations play a significant role for pesticides, and according to the political will they should not be present in the groundwater at all. The same argument may be applied for MTBE in the future. In October 2001, Denmark established a new maximum concentration for MTBE in drinking water of 5 µg/l.

Remediation

As in other European countries, there are no specific measures on groundwater remediation of MTBE. MTBE is removed to some extent when the groundwater is remediated for gasoline compounds by air-sparging, soil-ventilation, by pump-and-treat, etc., but it is not well documented how efficiently MTBE can be removed by those conventional methods. Investigations are in the planning phase in Denmark.

Substitution of MTBE with Ethanol

The discovery of significant groundwater contamination with MTBE in Denmark has initiated a discussion of alternatives to MTBE. The most obvious substitute is ethanol. There is a strong interest from agricultural organizations to produce ethanol from various plant materials in the search for new markets. However, there are several reservations:

1. The oil companies prefer MTBE to ethanol because MTBE is less costly, ethanol causes problems in relation to compliance with vapor-pressure regulations during the summertime, and the use of ethanol becomes complicated when water is present in pipelines and storage tanks.
2. From an environmental point of view, it is poorly documented how ethanol affects the natural degradation of gasoline compounds in the subsurface when they leak from underground storage tanks. In otherwise aerobic aquifers, the ethanol will consume the oxygen and change the redox condition to anaerobic, which will slow the natural degradation of the gasoline hydrocarbons. If the leak happens in an anaerobic aquifer, it is not known whether the ethanol will retard or stimulate the natural anaerobic attenuation of gasoline compounds.
3. Another environmental concern is the production of the plant material, which is the raw material for ethanol production. The production may require significant use of pesticides, which may convert the MTBE problem into a pesticide problem.
4. Moreover, since the price of ethanol from agricultural production is significantly higher than the price of MTBE, the taxes on gasoline may have to be lowered in case of ethanol use, which leads to a reduced revenue to the state.

Germany

The German Environmental Protection Agency (Umweltbundesamt) has the position that contamination of drinking water supplies can be effectively prevented if direct contamination into the ground can be prevented. For MTBE that is spilled and is in contact with the atmosphere, a Mackay distribution (Mackay, 1991) was calculated, resulting in a distribution with 94% in air, 5.9% in water, 0.1% in soil, and 0.1% in sediments (Pahlke, G., personal communication, 1999). Based on the photochemical degradation of MTBE in the atmosphere with a half-life of 3 to 7 days, it was concluded that no significant contamination of ground or surface waters can be expected from the atmosphere. Thus, the threat of MTBE to drinking water supplies can be controlled if a direct contamination of the groundwater can be prevented. In the U.S., leaking underground storage tanks have been identified as a major source of MTBE in the groundwater. In recent years, all German gasoline stations had to be restructured according to stricter standards with a waterproof ground cover and with double-wall storage tanks. With the higher standards of leak prevention at gasoline stations, it is stated that MTBE contamination of water is of lesser concern than in the U.S. where the standard of gasoline stations is believed to be inferior. As carcinogenic effects of MTBE remain unclear, in 1995 MTBE was added to the German MAK-List (list of maximum workplace concentrations) as a potential carcinogen that should be further evaluated. In conclusion, it can be stated that the threat of MTBE to groundwater is evaluated to be not as severe as in the U.S. as effective measures have been taken to prevent direct contamination of the groundwater. Also the extent of MTBE used in Germany is only approximately 1.3% by volume MTBE per total gasoline sold (Figure 136.1). The German Environmental Protection Agency does not encourage an increased use of MTBE but also does not take steps against the current use patterns.

Finland

In a reaction to the increasing concern about MTBE, the Finnish Ministry of the Environment stated that pollution of groundwater must be prevented independent of the type of contaminant (Kekalainen, J.P., personal communication, 1999). Biodegradability and transport properties of the component in groundwater are not taken into consideration and it is stated:

> [I]t is not necessary to distinguish the question of the MTBE from the overall groundwater protection in Finland. The release of gasoline to groundwater is prohibited even if it contains no additional substances. When the pollution of groundwater by gasoline is prevented, it is not relevant whether or not the gasoline contains MTBE. On the basis of the facts the Finnish Ministry of the Environment does not, for the time being, consider that the use of MTBE in gasoline should be restricted because of groundwater protection in Finland.

Strict safety regulations are enforced and "as a rule, new liquid-fuel distribution stations are located outside groundwater areas. If this is not possible, risks to groundwater are eliminated by structural protection measures."

United Kingdom

Air quality is a major concern for the U.K. Environment Agency and the use of improved fuel technology is part of its strategy. Thus, the currently relatively small MTBE content in gasoline (Figure 136.1) is likely to increase in the future. At the moment, the threat to groundwater supplies from MTBE is of comparatively low profile in the U.K. In the last few years, a number of pollution incidents have occurred in the U.K. where gasoline containing MTBE has contaminated groundwater. These pollution incidents may represent only a fraction of the total groundwater polluted with MTBE. The U.K. environment agency is stating that the scale of MTBE pollution may be grossly underestimated in the U.K. for a number of reasons (Anonymous, 1999c). One of the most significant is the comparatively recent appearance of MTBE. To date, regulators and water supply companies have not responded to reports suggesting a risk to groundwater from MTBE and hence many laboratories do not routinely screen for this pollutant. The major focus to decrease groundwater pollution by MTBE is on the improvement of gasoline storage safety. One obstacle in establishing a high safety standard for underground storage tanks at gasoline stations is that, in the U.K., 30% of the gasoline is sold at supermarkets where the environmental concern is less than compared to oil companies (Harris, B., personal communication, 1999).

There is currently no control on MTBE in drinking water in the U.K. This is probably because the available toxicity data derived from both inhalation and oral routes of exposure suggest that MTBE is of low acute and subacute toxicity. It does not appear to be either a reproductive toxicant or a mutagenic. There are equivocal data relating to the carcinogenicity of MTBE. Despite the current scientific debate, it appears that the threshold taste and odor concentration of MTBE (around 15 µg/l) is likely to be lower than a future health-based guideline value (Anonymous, 1999c).

CONCLUSION

- Introduction of MTBE into gasolines in Europe lagged behind the U.S., as does current use. As a result, concerns and possible problems with MTBE develop in Europe with a time lag.
- Very little monitoring data for MTBE in groundwater, surface waters, soil, or air are available for Europe. However, recent discussions in the U.S. have increased public attention on fuel additives, and many European environmental protection agencies are currently initiating monitoring programs for MTBE.
- While there is no minimum requirement for MTBE, its future use is expected to increase in Europe due to its properties as an octane booster. Upcoming EU regulations require a decrease of benzene and overall aromatics content, which contribute to the octane rating of the gasoline. For decision makers, this situation poses a political dilemma in which the reduction of one unwanted compound in gasoline can be the cause for an increased use of another unwanted compound.
- In many European countries, improved gasoline technology is seen as an important contribution to the reduction of air pollution. The position is that fuel oxygenates such as MTBE should be used because they help increase combustion and reduce emissions of carbon monoxide and organic compounds. The increased risk associated with leaking underground storage tanks is generally seen as a technical problem that can be managed by implementing improved safety regulations.
- In a recent study in Denmark, it was shown that MTBE concentration in groundwater used as a source for drinking water were usually below the detection limit (0.1 μm). However, in water samples from the vicinity of gasoline stations, concentrations up to a few hundred mg/l were measured. Thus, it may be only a question of time until MTBE contaminations reach drinking water resources. Groundwater problems are largely unknown and therefore groundwater remediation of MTBE has not yet started.

REFERENCES

Anonymous, Richtlinie zur Einsparung von Rohöl durch die Verwendung von Ersatz-Kraftstoffkomponenten im Benzin (85/536/EWG), 1985.

Anonymous, Worldwide gasoline and diesel fuel survey, OCTEL, 1997.

Anonymous, Directive 98/70/EC of the European parliament and of the council relating to the quality of petrol and diesel fuels and amending council directive 93/12/EEC, 1998.

Anonymous, Characterization of California Phase 2 RFG, Chevron company information, http://www.chevron.com/prodserv/bulletin/phase2rfg/index.html, 1999a.

Anonymous, Kraftstoff-Spezifikationen heute und morgen, Shell – Deutschland, http://www.deutsche-shell.de/sug/kraspz.htm, 1999b.

Anonymous, MTBE — Air quality vs. groundwater quality, http://www.environment-agency.gov.uk/modules/MOD43.16.html, 1999c.

Anonymous, MTBE and Oxygenates 1999 Annual, DeWitt & Company, Houston, January, 2000.

Borden, R.C., Daniel, R.A., LeBrun, IV, L.E., and Davis, C.W., Intrinsic biodegradation of MTBE and BTEX in a gasoline-contaminated aquifer, *Water Resourc. Res.*, 33(5), 1105–1115, 1997.

Bradley, P.M., Landmeyer, J.E., and Chapelle, F.H., Aerobic mineralization of MTBE and *tert*-butyl alcohol by stream-bed sediment microorganisms, *Environ. Sci. Technol.*, 33 (11), 1877–1879, 1999.

CONCAWE, A survey of European gasoline qualities — summer 1996, 1998.

Eweis, J.B., Schroeder, E.D., Chang, D.P.Y., Scow, K.M., Morton, R., and Caballero, R.C., Degradation of MTBE in a pilot-scale biofilter, Conference on Remediation of Chlorinated and Recalcitrant Compounds, Monterey, CA, 1998.

Harris, B., MTBE — A benefit to air quality or a threat to groundwater? A UK Perspective, Conference on MTBE as a Pollutant: Implications for Groundwater Resources, 8 June 1999, Imperial College, London, 1999.

IPCS, Environmental health criteria No. 206: Methyl tertiary-butyl ether, International Programme on Chemical Safety (IPCS), 1998.

Landmeyer, J.E., Chapelle, F.H., Bradley, P.M., Pankow, J.F., Church, C.D., and Tratnyek, P.G., Fate of MTBE relative to benzene in a gasoline-contaminated aquifer (1993–98), *Ground Water Monitoring Remed.,* 18(4), 93–102, 1998.

Larsen, P.B., Evaluation of health hazards by exposure to methyl tertiary-butyl ether (MTBE) and estimation of limit values in ambient air, soil and drinking water, The Institute of Food Safety and Toxicology, Danish Veterinary and Food Administration, 1998.

Leonard, D., State MTBE standards preliminary data based on UMass survey of states, March 1998, http://www.epa.gov/swerust1/mtbe/sumtable.htm, 1998.

Mace, R.E. and Choi, W.-J., The size and behavior of MTBE plumes in Texas, Petroleum Hydrocarbons and Organic Chemicals in Ground Water — Prevention, Detection, and Remediation Conference, Houston, Nov. 11–13, National Ground Water Association and American Petroleum Institute, 1998.

Mackay, D., *Multimedia Environmental Models: The Fugacity Approach,* Lewis Publishers, Chelsea, MI, 1991.

McCord, S.A. and Schladow, S.G., Transport and fate modeling of MTBE in lakes and reservoirs, Report from the University of California, Davis, 1998.

Miljøstyrelsen, Handlingsplan for MTBE (Action plan for MTBE), Miljø-og Energiministeriet Miljøstyrelsen (Danish Environmental Protection Agency), 1998.

Pankow, J.F., Thomson, N.R., Johnson, R.L., Baehr, A.L., and Zogorski, J.S., The urban atmosphere as a non-point source for the transport of MTBE and other volatile organic compounds (VOCs) to shallow groundwater, *Environ. Sci. Technol.,* 31(10), 2821–2828, 1997.

Reuter, J.E., Allen, B.C., Richards, R.C., Pankow, J.F., Goldman, C.R., Scholl, R.L., and Seyfried, J.S., Concentrations, sources, and fate of the gasoline oxygenate methyl *tert*-butyl ether (MTBE) in a multiple use lake, *Environ. Sci. Technol.,* 32(23), 3666–3672, 1998.

Rippen, G., Methyl-*tert*-butylether, in *Handbuch Umweltchemikalien, Stoffdaten,* Rippen, G., Ed., Ecomed, Landsberg/Lech, Germany, 1999.

Squillace, P.J., Pankow, J.F., Korte, N.E., and Zogorski, J.S., Review of the environmental behavior and fate of methyl *tert*-butyl ether, *Environ. Toxicol. Chem.,* 16(9), 1836–1844, 1997.

Squillace, P.J., Zogorski, J.S., Wilber, W.G., and Price, C.V., Preliminary assessment of the occurrence and possible sources of MTBE in groundwater in the U.S., 1993–1994, *Environ. Sci. Technol.,* 30(5), 1721–1730, 1996.

van der Kooij, D., Assimilable organic carbon as an indicator of bacterial regrowth, *J. Am. Water Works Assoc.,* 84(2), 57–65, 1992.

Approximation of the Use of MTBE and ETBE as Fuel Oxygenates in Europe[a,b]

Country	Grade (Ron)	N	Estimated Gasoline Sales (mill. tons/a)	MTBE Content (vol. %)	ETBE Content (vol. %)	Amount MTBE (1000 T/A)
Austria	91	10	0.718	0.9	0	6.462
	95	10	1.046	3.7	0.2	30.702
	98	10	0.452	8.3	0.1	37.516
Belgium	95	9	1.005	0.1	0	1.005
	98	9	1.022	2.5	0	25.55
	98L	8	0.712	1.2	0	8.544
Denmark	92	4	0.228	0	n.d.	0
	95	18	0.794	0.4	n.d.	3.176
	98	14	0.888	6.3	n.d.	55.944
Finland[c]	95	7	1.291	8	n.d.	103.28
	98	5	0.27	12.4	n.d.	33.48
	99	0	0.27	n.d.	n.d.	
France	95	20	2.159	n.d.	n.d.[d]	[d]
	98	20	6.23	n.d.	n.d.[d]	[d]
	SL	0	6.609	n.d.	n.d.[d]	[d]
Germany	91	96	11.291	0.3	0	33.873
	95	96	16.342	1.6	0	261.472
	98	80	1.6	6.2	0	99.2
Great Britain	95	11	14.522	0.1	0	14.522
	98	8	0.709	0.21	0	1.4889
	SL	11	7.178	0.1	0	7.178
Greece	95	27	1.104	3.5	n.d.	38.64
	98	3	0.085	8	n.d.	6.8
	96L	37	1.833	0.2	0	3.666
Hungary	91	0	0.215	n.d.	n.d.	
	95	15	0.59	2	n.d.	11.8
	98	4	0.03	9.5	n.d.	2.85
	92L	0	0.114	n.d.	n.d.	
	98L	15	0.322	1.1	n.d.	3.542
Italy	95	62	7.932	4.3	1.4 ($n=3$)	341.076
	97L	62	9.742	1.9	8.1 ($n=1$)	185.098
Netherlands	95	8	3.145	0.8	0	25.16
	98	8	0.711	5.4	0	38.394
	SL	8	0.335	1.2	0	4.02

Country	Grade (Ron)	N	Estimated Gasoline Sales (mill. tons/a)	MTBE Content (vol. %)	ETBE Content (vol. %)	Amount MTBE (1000 T/A)
Norway	95	22	1.135	3.8	n.d.	43.13
	98	20	0.226	5.1	n.d.	11.526
	SL	0	0.362	n.d.	n.d.	
Portugal	95	0	0.447	n.d.	n.d.	
	98	0	0.355	n.d.	n.d.	
	SL	0	1.143	n.d.	n.d.	
Spain	95	8	1.809	4.7	n.d.	85.023
	98	5	1.266	8.5	n.d.	107.61
	97L	8	5.97	2	n.d.	119.4
Sweden	95	47	3.146	0.3	n.d.	9.438
	98	34	1.116	6	n.d.	66.96
Switzerland[e]	95	20	3.119	2.5	n.d.	77.975
	98	17	0.347	10.4	n.d.	36.088
	SL	0	0.36	n.d.	n.d.	
Sum			122.3			1949.6

Note: n.d. = not determined; RON: Research Octane Number.

[a] Note that the total amount of MTBE used in Europe varies with time and that amount may be higher than the given figure because in some countries no data is available on the average MTBE content of gasoline (France, Portugal) and a number of countries are missing in the statistics (eastern European countries, Ireland, Luxembourg).

[b] According to Fortum Oil and Gas (Kekalainen, J.P., Fortum Oil and Gas, Finland, personal communication, 1999), data compiled in this table are in the right order of magnitude even though, according to their sources, the use of MTBE is underestimated for Belgium and Great Britain and overestimated for Denmark.

[c] In addition *tert*-amyl methyl ether (TAME) is added in the same order of magnitude as MTBE in Finland (Vahervuori, H., Fortum Oil and Gas, Finland, personal communication, 1999). It was estimated that Finland uses 95,000 tons of TAME per year (Anonymous, 2000).

[d] It was estimated that France uses 151,000 tons of MTBE and 157,000 tons of ETBE per year (Anonymous, 2000).

[e] The share of premium gasoline (98 RON) was estimated to be 10%.

Compiled by Torsten C. Schmidt (EAWAG, Switzerland) based on CONCAWE (1998) and Anonymous (1997).

CHAPTER 137

Cost–Benefit Considerations for the Introduction of Gasoline Additives Such as MTBE

Arturo A. Keller and Linda Fernandez

INTRODUCTION

The search for solutions to air quality problems has led to the development of gasoline additives that can have a positive impact on the combustion efficiency, significantly reducing emissions of carbon monoxide, ozone precursors, and hazardous air pollutants, such as benzene. However, a careful consideration of the impact of these gasoline additives on other media must be made for each circumstance. The case of methyl *tert*-butyl ether (MTBE) has served to highlight the potential for cross-media contamination when the environmental impact assessment is incomplete.

A recent study by Keller and co-workers (1998a), as part of a wider evaluation of the health and environmental impacts of MTBE (Keller et al., 1998b), concluded that the air quality benefits for California derived from the use of MTBE as a gasoline additive at 11 to 15% by volume, are relatively small and are decreasing with time. The decrease is based on the fact that other policies and technologies implemented to reduce air emissions are becoming more important as the vehicle fleet modernizes. Older vehicles (pre-1990) do not have many of the emissions control devices (e.g., advanced catalytic converters, oxygen sensor feedback, fuel injection), and thus may emit large amounts of carbon monoxide and other air pollutants. The addition of MTBE (or other oxygenated compounds) can reduce the emissions of carbon monoxide from older vehicles. In addition, MTBE is used to replace some of the high-octane ratings that benzene and other aromatics provide; the substitution reduces the aromatic fraction of the gasoline and thus lowers the emission of air toxics.

The cost–benefit analysis conducted for California (Keller et al., 1998a) examined the human-health benefits derived from controlling air pollution, and systematically analyzed the costs associated with the use of MTBE across the following categories:

- Human health costs due to air pollution from MTBE and its combustion by-products
- Human health costs due to water pollution derived from MTBE
- Water treatment or alternative water supply costs
- Fuel price increase costs
- Costs due to increased fuel consumption
- Monitoring costs
- Recreational costs
- Ecosystem damages

Table 137.1 Composition of Gasoline Formulations

Property	Typical Conventional Gasoline	Oxygenated CaRFG2	Typical CaRFG2 with MTBE	Nonoxygenated CaRFG2[a]
Aromatics, (vol.%)	32	max. 25	25	22.7
Olefins, (vol.%)	9.2	max. 10	4.1	4.6
Benzene, (vol.%)	1.53	max. 1	0.93	0.94
Oxygen content (%)	0	1.8 to 2.7	2.1	0
Sulfur (ppm by weight)	339	max. 40	31	38
Reid vapor pressure (psi)	8.7	max. 7	6.8	6.9
T90, °F	330	max. 300	293	297
T50, °F	218	max. 210	202	208

[a] Based on AQIRP (1997).

Although cost–benefit analysis has been used extensively to evaluate alternative policies, there have been no studies to date that examine the cross-media implications of different gasoline (or other fuel) formulations. For example, a recent study focused on the costs and benefits of the Clean Air Act of 1970 (USEPA, 1997), but considered only the impact of reducing criteria air pollutants on human health. The study did not address specific policies to achieve the reductions, such as modifications to fuel formulations, new vehicle technologies, or emission control technologies on stationary sources. Schwing and co-workers (1980) used cost–benefit analysis to compare leaded and unleaded fuels, but did not consider the impact of leaded fuels on water supplies. In developing countries where MTBE is being evaluated as a substitute for tetraethyl lead, the water quality damages associated with the use of leaded fuels must also be considered. Krupnick and Walls (1992) compared methanol to conventional gasoline in terms of reducing motor vehicle emissions and urban ozone, yet avoided discussing health effects or potential impacts on water quality.

For policy makers charged with deciding between gasoline formulations to achieve improved air quality, a cost–benefit analysis can provide answers to the following questions:

Do the costs of using MTBE outweigh the benefits it produces?

What are the policy options that are available to reduce the costs of using MTBE and what are the trade-offs?

How do alternative formulations compare in terms of costs and benefits?

METHODOLOGY

The first step in the analysis is to identify the benefit and cost categories. In addition to the categories indicated in the introduction, it is possible to add other costs, such as litigation, or replacement land for sites contaminated with MTBE, or the cost of drilling new wells for public drinking water supply. The number of categories considered depends to a large extent on the availability of adequate data; some cost categories have too much uncertainty associated with them, or may be deemed not as significant as the principal cost categories. Although there may be cultural differences that place a different value on some of the cost categories, we believe that for most studies the categories we have identified will serve to make adequate policy decisions.

The next step is to identify alternative policies, or in our case, alternative formulations. In the case of MTBE, there are gasoline formulations that use ethanol, toluene, or iso-octane, which can provide essentially the same air quality benefits with different costs (AQIRP, 1997; Koshland et al., 1998). In addition, since we were interested in determining the cost of adding MTBE, we used as our baseline conventional unleaded gasoline, which was the gasoline formulation sold prior to the introduction of the reformulated oxygenated gasolines in California and other parts of the U.S. Table 137.1 presents the typical composition of these gasoline formulations.

Complete details of the methods used to value each benefit or cost category are presented in Fernandez and Keller (2000) or Keller and co-workers (1998a). Briefly, health benefits were valued using cost of illness (Abdalla et al., 1992) or the value of a statistical life (Fisher et al., 1989), depending on whether the pollutant causes morbidity, mortality, or both. The health effects associated with MTBE and other pollutants were derived from the study by Froines (1998). Water treatment costs were based on experimental and field studies conducted by Keller and co-workers (1998c), plus data from Reuter and co-workers (1998) on the number of water reservoirs that are contaminated, and Fogg and co-workers (1998) on the number of groundwater sites requiring treatment. With this information, we were able to integrate an estimate of the cost of water treatment for the State of California.

Market prices were used to estimate the direct cost paid by consumers at the pump due to the mandated use of oxygenated fuels. The increase in gasoline consumption was based on engineering estimates of the decrease in fuel energy content (NSTC, 1997), which result in an increased cost of operation. Market prices were also used to calculate monitoring costs incurred to track the extent of contamination in surface and groundwaters, or in ambient air concentrations. We use the travel cost method to value recreational costs from possible restrictions of boats and jet-skis on bodies of water which also serve as drinking water sources. The factor income and restoration cost methods are used to value environmental health costs, which account for damage to important environmental goods, such as fish and other sensitive fauna and flora (Anderson and Rockel, 1991; Bell, 1989; Shabman and Batie, 1987).

Conceptually, the valuation of environmental impacts of each alternative is straightforward. In practice, there are significant difficulties because important elements of the valuation process are not measured or have large uncertainties associated with them. For example, although MTBE may be associated with asthma, conclusive epidemiological studies have not been conducted. Similarly, there is a large uncertainty in the valuation of the effects on human health of reducing air pollutant levels when they are below the air quality standards, or even slightly above, because the toxicological data are derived from much higher concentrations.

RESULTS

Details of the assumptions used for constructing the figures can be found in Keller and co-workers (1998a). We present here the main features of our results. Figures 137.1 through 137.3 present the

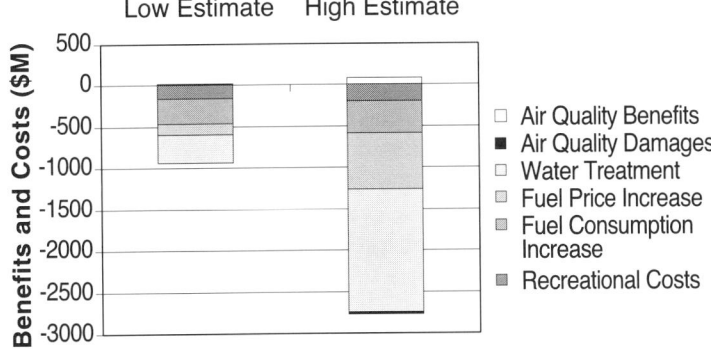

Figure 137.1 Cost and benefits of reformulated gasoline with MTBE. A low and a high estimate were made based on the uncertainties associated with our assumptions. The air quality benefits are quite small, while the various cost categories add up significantly, in particular the water treatment costs.

Gasoline with Ethanol

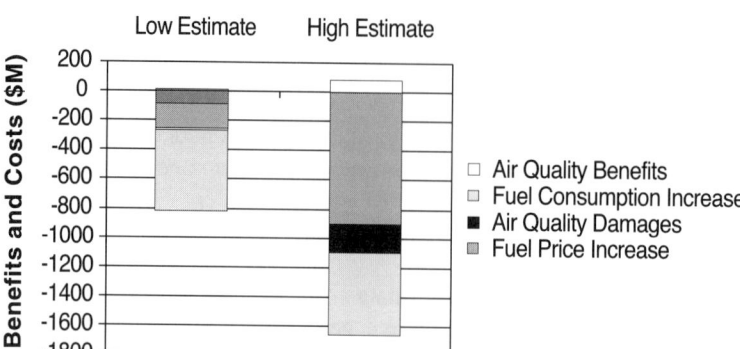

Figure 137.2 Cost and benefits of reformulated gasoline with ethanol. The same air quality benefits are assumed for this formulation. The major cost categories are the fuel price increase and the increased fuel consumption. Air quality damages may be large if significant amounts of acetaldehyde are produced as a result of incomplete combustion.

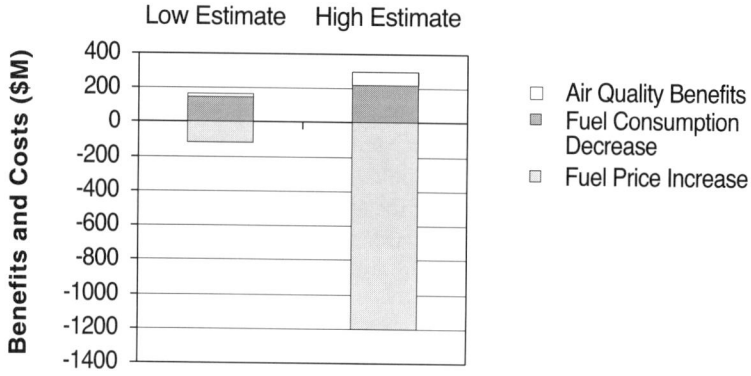

Figure 137.3 Cost and benefits of nonoxygenated reformulated gasoline. Air quality benefits are the same as for the other reformulated gasolines, but in this case the fuel consumption decreases and thus is an added benefit. This may be offset if the fuel price increases significantly.

costs and benefits for the three formulations studied (gasoline with MTBE, gasoline with ethanol, and nonoxygenated gasoline, with either toluene or iso-octane). The health benefits are essentially the same across all three formulations, because studies have shown that all these formulations achieve essentially the same carbon monoxide and ozone precursors emissions reductions, within statistical significance. An additional benefit achieved in the California Phase II Reformulated Gasoline formulations (CARB, 1991), is the reduction of benzene content. Benzene is a known human carcinogen (ATSDR, 1991; IARC, 1985), and the reduced content (less than 1%) in reformulated gasoline results in decreased volatile emissions. We have estimated that this alone reduces 33 to 920 cancer cases per year in California, which are valued at $165 million to $4.6 billion over a lifetime (70 years) using the value of a statistical life of $5 million. On an annual basis, this represents $2 to $66 million in savings. The uncertainty in this estimate is derived largely from the uncertainty in the cancer potency factor. In addition, we estimate that the reduction in carbon monoxide may have decreased the number of hospital admissions due to congestive heart failure by up to 840 cases per year. Similarly, the reductions in ozone concentrations may lead to a reduction in hospital admissions due to acute respiratory illnesses of up to 300 cases per year. However, it is difficult to determine the fraction of the air pollutant's reduction due strictly to the fuel formulation, or due to other factors such as vehicle technology, meteorology, driving patterns, economic activity, etc. Thus, these estimates should be seen as an upper bound. Significant as they are with respect to human health, we estimate that these benefits only represent around $14 to $78 million per year.

There are several reasons these benefits are small in California.

- There have been many programs successfully implemented by the California Air Resources Board to reduce the emissions of air pollutants, and thus the overall concentrations of carbon monoxide and ozone are much lower than in previous decades. For example, carbon monoxide levels prior to the introduction of the current version of oxygenated, reformulated gasoline were less than 12 ppm in practically all air basins in California. The National Ambient Air Quality Standard (NAAQS) for carbon monoxide is 9 ppm for an 8-h average. Only three monitoring stations in California exceeded the NAAQS in 1996, and the downward trend in concentrations since the late 1970s continues (USEPA, 1999). At present, there is only one nonattainment area in California for carbon monoxide, namely, the South Coast air basin. The maximum carbon monoxide concentrations registered in this region since 1994 are around 11 ppm; these levels are only observed a few days during each month, and are concentrated in episodes when adverse meteorological conditions intersect with high levels of emissions. Thus, the benefit of oxygenating the gasoline is small, since it is generally considered that the NAAQS are set such that, if they are attained, there is no health impact even to sensitive populations. A decrease of 2 ppm in peak carbon monoxide concentrations would only benefit people with ischemic heart disease, who represent around 3% of the population. Only a fraction of this population would be affected enough to require hospitalization (Graves and Owings, 1998; Morris et al., 1995).
- Similarly, ozone concentrations have been declining even in the South Coast region over the past two decades. However, in this case, although the number of days the California standard is exceeded has decreased noticeably, from a high of 261 days of exceedance in 1981 to around 135 to 180 days of exceedance in recent years (CARB, 1999), the levels are significantly above the California standard of 0.09 ppm (1-h average) during the summer months, peaking at 0.18 to 0.20 ppm. The decrease in ozone concentrations due to the oxygenate could be up to 0.015 ppm, which would benefit sensitive populations with reduced respiratory function (Burnett et al., 1997).
- The average age of the vehicle fleet has decreased in recent years, due to the economic recovery in southern California since 1993. This means that newer vehicles with improved emissions controls are replacing older vehicles. In addition, programs implemented to remove older vehicles are reducing the number of high emitters.
- The emissions from stationary sources and other mobile sources (trucks, airplanes, ships, trains, etc.) are becoming more important relative to emissions from vehicles that use gasoline with MTBE. Programs to control emissions from these other sources lag the successful control of emissions from vehicles.

These conditions may be quite different in other cities and air quality basins, especially in developing countries. For example, Mexico City has seen increasing ozone and carbon monoxide concentrations in the past few years. A program targeted at these air pollutants would have a major impact on human health. In addition, the vehicle fleet is much older, and instead of a somewhat steady-state vehicular population, it is increasing rapidly. Transportation bottlenecks and adverse meteorological conditions compound the problem. Thus, a full evaluation of the health benefits for these situations should be conducted prior to making any decision to use or ban MTBE. Alternative formulations (e.g., nonoxygenated) should also be considered.

There are air quality costs associated with the new gasoline formulations. For example, the use of MTBE results in volatilization of MTBE (from fuel lines, fueling stations, pipelines, etc.). In addition, the combustion of MTBE increases the emissions of formaldehyde (AQIRP, 1997; Koshland et al., 1998) in controlled combustion studies. Formaldehyde is a known human carcinogen (IARC, 1985; USEPA, 1988, 1993a). Although the observed MTBE concentrations are below the cancer risk level (Froines, 1998), the potential increase in formaldehyde concentrations could result in up to 380 cancer cases per year, negating some of the benefits of reducing benzene emissions. It should be noted that, to date, the concentration of formaldehyde in various California cities has shown no upward (or downward) trend since the introduction of MTBE.

Similarly, adding ethanol to gasoline (as a replacement for MTBE) increases the emissions of acetaldehyde (AQIRP, 1997; Gaffney et al., 1997; Koshland et al., 1998). Acetaldehyde is also a probable human carcinogen (CARB, 1993; IARC, 1985; USEPA, 1987, 1993b). If the concentrations were to increase as they did in New Mexico during the introduction of ethanol-gasoline mixtures, by 1 to 2 ppb, the number of cancer cases could increase to 2800 per year. However, it should be mentioned that ethanol has been used in the midwestern U.S. with no noticeable increase in acetaldehyde concentrations. A more complete study is required to determine whether there is really a concern with the use of ethanol.

If either toluene or iso-octane is used to replace MTBE in nonoxygenated formulations, the concentrations of either chemical would increase in ambient air. However, the levels of toluene would probably be below the reference concentration (RfC) in air of 0.4 mg/m^3 or 400 μg/m^3 (ATSDR, 1992; USEPA, 1993c). In California, the mean concentration in air is 8.5 μg/m^3. This concentration could increase significantly and still not be close to the RfC at which adverse effects would be measurable. None of the data suggests that toluene is carcinogenic. Iso-octane is not classified as a hazardous air pollutant by USEPA, and there is no toxicological information from the Agency for Toxic Substances and Disease Registry (ATSDR). It is a normal component of gasoline, and can produce acute effects on the central nervous system when inhaled at high concentrations, but the risk is similar to conventional gasoline. There have been limited studies of the combustion by-products of these formulations, so it is highly advisable to make a full assessment before proceeding to their widescale introduction.

The introduction of reformulated gasoline with MTBE resulted in an estimated price increase of 1 to 5 cents per gallon. This translates into a cost to the economy of $135 to $675 million. In addition, due to the lower energy content of MTBE (NSTC, 1997), there is an additional cost to the California economy of $300 to $380 million due to the increase in fuel consumption to maintain the same driving pattern. A recent study by the California Energy Commission (CEC, 1999) estimated that the cost of using ethanol as a substitute for MTBE would be around 1.9 to 6.7 cents per gallon, or an annual cost of $260 to $900 million. The increase in fuel consumption due to the use of ethanol would cost an additional $560 million (a 3% increase in consumption). For non-oxygenated gasoline, CEC (1999) estimates a price increase from 0.9 to 8.8 cents per gallon, or $121 million to $1.3 billion per year. The CEC estimates that in the short term (1 to 3 years), the price increase would be at the high end of the range (4.3 to 8.8 cents per gallon), whereas once refiners have been able to install the necessary equipment or long-term import contracts are established (3 to 6 years), the price increase should be only around 0.9 to 3.7 cents per gallon. However, in this case, the use of either toluene or iso-octane would result in decreased fuel consumption, due to the higher energy content of these chemicals, a savings of $150 to $220 million.

Fogg and co-workers (1998) and Reuter and co-workers (1998) estimated the number of groundwater supplies, leaking fuel tanks, and surface water reservoirs that are currently contaminated with MTBE. Based on their information, and the study by Keller and co-workers (1998c), we made an estimate of the aggregate cost of water treatment in California of around $340 to $1480 million per year (Figure 137.1). These costs are based on the premise that contaminated water must be treated to a concentration below the 5 μg/l level set by California's EPA as the Secondary Water Quality Standard, based on taste and odor considerations.

A literature review indicates that the cost of using ethanol, in terms of risk to the water supplies, is low. Ethanol plumes biodegrade fairly rapidly. In the event that water supplies become contaminated with ethanol, the available toxicological information does not support treating the water to the low levels required by MTBE, and filtration in biologically active granular activated carbon (GAC) would probably be a cost-effective option. We consider the incremental costs of water treatment to be negligible relative to conventional gasoline, since BTEX compounds in the gasoline fraction would determine the treatment design, rather than ethanol.

For nonoxygenated gasoline, the differential cost of remediation and water treatment relative to conventional gasoline is small. The increased volumetric fraction of toluene in nonoxygenated

California Reformulated Gas Phase II (CaRFG2) will result in higher initial toluene concentrations, but toluene is easily biodegraded by the intrinsic microbial communities. If iso-octane is used instead of MTBE, it has a very low solubility in water, and it is readily biodegraded along with other components of conventional gasoline. It is likely that natural attenuation will be applicable at the same rates as for conventional gasoline. Aboveground treatment costs may increase at most 10% relative to treating water contaminated by conventional gasoline.

Some utilities may be forced to purchase water from other supplies, at least in the short to intermediate term. For example, the city of Santa Monica has been purchasing water from the Los Angeles Metropolitan Water District due to the contamination of most of its drinking water wells with MTBE. The cost per year for alternative water supply, assuming that 20% of the contaminated water has to be replaced at a cost of $1.65 per 1000 gallons that Santa Monica pays for water from the Metropolitan Water District (Rodriguez, 1997), is around $1 to $30 million. These costs would not be significant for other gasoline formulations, relative to conventional gasoline.

There are some incremental monitoring costs, because water utilities are required to sample more frequently, in particular in surface water reservoirs where boating is allowed. Statewide, this cost is expected to amount to $1 to $4 million. For groundwater sources, the current costs could increase to $1 and $2 million annually. Monitoring air quality is done by collecting samples on a regular basis and running a standardized analysis, which provides information on a number of air toxics. We do not expect any additional costs will be incurred to monitor ambient air concentrations of MTBE, formaldehyde, acetaldehyde, benzene, or combustion by-products. We consider that this cost would not be significant for ethanol-based gasoline formulations or nonoxygenated gasoline, relative to conventional gasoline.

One alternative that can be considered for minimizing the impact of MTBE is banning motorcraft from surface water reservoirs. If boating was completely eliminated from all the reservoirs in California, we estimate that the cost, in terms of recreational value lost, would be on the order of $160 to $200 million. It is likely that only a partial ban would be implemented, and probably not a year-round ban. MTBE is volatile enough to quickly escape to the atmosphere from a contaminated reservoir, in the order of weeks. We consider that this cost would not be significant for ethanol-based gasoline formulations or nonoxygenated gasoline, relative to conventional gasoline.

Ecological risk assessment studies by Werner and Hinton (1998) indicate that the concentrations of MTBE that have been detected in lakes and water reservoirs should not result in significant damages to biota in aquatic ecosystem. Localized spills may have an impact, but there are insufficient data to estimate the ecosystem damages, and they are likely to be small relative to other MTBE costs. Note that all damages and costs are estimated relative to the use of conventional gasoline. For example, local ecosystem damages due to a pipeline rupture would be similar whether the gasoline contained MTBE or not. We consider that this cost would not be significant for ethanol-based gasoline formulations or nonoxygenated gasoline, relative to conventional gasoline.

DISCUSSION

Comparing the bottom line for each formulation (Figure 137.4), it is clear that, for California, using MTBE in reformulated gasoline is a very expensive option. The costs far outweigh the benefits. More significantly, the benefits are small and decreasing with time, as the vehicle fleet modernizes, incorporating emissions control technologies. The main cost driver is water treatment, based on the very tight standards set for MTBE in California. Ethanol-based formulations do not fare much better, although the midpoint of the costs range is smaller than the midpoint of the MTBE formulation. The biggest uncertainty is the impact of a much larger demand for ethanol, which could drive prices up significantly, at least in the short term. One hidden advantage of ethanol is the fact that it can be produced from agricultural wastes such as rice straw, reducing greenhouse

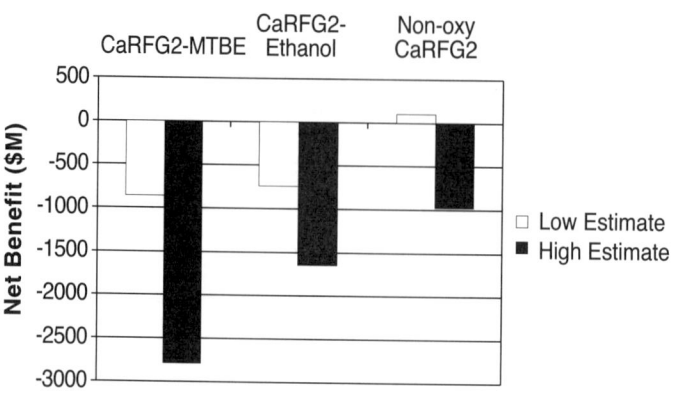

Figure 137.4 Comparison of net benefits or costs for the various gasoline formulations.

emissions. Many developing countries have the potential of producing ethanol from these sources, thus reducing their dependence on imported oil, while improving their air quality.

Nonoxygenated gasoline formulations are apparently the best option for California. There is currently one refiner commercializing a nonoxygenated formulation that meets the strict California Phase II Reformulated Gasoline specifications, except for the oxygen content, presumably at a profitable price. However, the technologies may not be available to the entire industry. Capital expenditures are needed to convert to these formulations. In the meantime, imported toluene or iso-octane may drive the costs toward the high end of the range. If air quality must be maintained, this appears to be the lowest-cost strategy for California. An important consideration is the need to evaluate the toxicology (human and ecological) and fate and transport of any gasoline additives before making drastic changes. The mistakes made with MTBE should be avoided at all costs.

For developing countries, there are a number of factors that must be considered when choosing gasoline formulations. First, an assessment of the air quality benefits must be made to evaluate whether the change in formulations is warranted. This requires information on trends of ambient air concentrations of pollutants, average vehicle age and technology, as well as number of hospitalizations from congestive heart failure and acute respiratory illnesses, and their cost. Next, an assessment of the vulnerability of water sources must be made. The use of geographical information systems, which can overlay well locations with the location of underground storage tanks, can serve to make an assessment of vulnerability. An inventory of leaking tanks is also needed. California was able to reduce the impact on water resources due to the decade-long program to upgrade underground storage tanks. If the tanks do not have double-containment and leak detectors, the probability of failure is around 2% per year (Couch and Young, 1998), and the water treatment costs will certainly overwhelm the air quality benefits.

REFERENCES

Abdalla, C., Roach, B., and Epp, D., Valuing environmental quality changes using averting expenditures: an application to groundwater contamination, *Land Econ.*, 68(2), 163–169, 1992.

Anderson, R. and Rockel, M., Economic valuation of wetlands, Discussion paper 065, American Petroleum Institute, Washington, D.C., 1991.

AQIRP, Auto/Oil air quality improvement research program, Program final report, January 1997.

ATSDR, Toxicological profile for benzene, Agency for Toxic Substances and Disease Registry, U.S. Public Health Service, U.S. Department of Health and Human Services, Atlanta, 1991.

ATSDR, Toxicological profile for toluene, Agency for Toxic Substances and Disease Registry, U.S. Public Health Service, U.S. Department of Health and Human Services, Atlanta, 1992.

Bell, F., Application of wetland valuation theory to Florida fisheries, SGR-95, Sea Grant Publication, Florida State University, Tallahassee, FL, 1989.

Burnett, R.T., Brook, J.R., Yung, W.T., Dales, R.E., and Krewski, D., Association between ozone and hospitalization for respiratory diseases in 16 Canadian cities, *Environ. Res.,* 72(1), 24–31, 1997.

CARB, Proposed regulations for California Phase 2 reformulated gasoline, Staff report, California Air Resources Board, Stationary Source Division, Sacramento, 1991.

CARB, Acetaldehyde as a toxic air contaminant: health assessment for the stationary source division, California Air Resources Board, Sacramento, 1993.

CARB, California air quality data, California Air Resources Board, Sacramento (http://arbis.arb.ca.gov/aqd/aqd.htm), 1999.

CEC, Staff Findings: Timetable for the Phaseout of MTBE from California's Gasoline Supply, Publication 300-99-003, California Energy Commission, Staff Findings, June 1999.

Couch, A. and Young, T., Failure rate of underground storage tanks, in *Health and Environmental Assessment of MTBE,* Vol. III, Last, J., Ed., Toxics Research and Teaching Program, Davis, CA, 1998, pp. 30–51.

Fernandez, L.F. and Keller, A.A., Cost benefit analysis of MTBE and alternative gasoline formulations, *Environ. Sci. Pol.,* 3, 173–188, 2000.

Fisher, A., Chestnut, L., and Violette, D., The value of reducing risks to death: a note on new evidence, *J. Pol. Anal. Manage.,* 8(1), 88–100, 1989.

Fogg, G.E., Meays, M.E., Trask, J.C., Green, C.T., LaBolle, E.M., Shenk, T.W., and Rolston, D.E., Impacts of MTBE on California groundwater, in *Health and Environmental Assessment of MTBE,* Vol.III, Last, J., Ed., Toxics Research and Teaching Program, Davis, CA, 1998, pp. 75–93.

Froines, J., An evaluation of the scientific peer-reviewed research and literature on the human health effects of MTBE, in *Health and Environmental Assessment of MTBE,* Vol. II, Last, J., Ed., Toxics Research and Teaching Program, Davis, CA, 1998, pp. 1–157.

Gaffney, J.S., Marley, N., Martin, R.S., Dixon, R.W., Reyes, L.G., and Popp, C.J., Potential air quality effects of using ethanol-gasoline fuel blends: a field study in Albuquerque, New Mexico, *Environ. Sci. Technol.,* 31, 3053–3061, 1997.

Graves, E.J. and Owings, M.F., 1996 Summary: National hospital discharge survey, Advance data from vital and health statistics, National Center for Health Statistics, Hyattsville, MD, 1998.

IARC, IARC Monographs on the evaluation of the carcinogenic risk of chemicals to humans: allyl compounds, aldehydes, epoxides and peroxides, Vol. XXXVI, International Agency for Research on Cancer, World Health Organization, Lyon, France, 1985.

Keller, A.A., Fernandez, L.F., Hitz, S., Kun, H., Peterson, A., Smith, B., and Yoshioka, M., An integral cost–benefit analysis of gasoline formulations meeting California Phase II reformulated gasoline requirements, in *Health and Environmental Assessment of MTBE,* Vol. V, Last, J., Ed., Toxics Research and Teaching Program, Davis, CA, 1998a, pp. 45–89.

Keller, A.A., Froines, J., Koshland, C.P., Reuter, J., Suffet, I., and Last, J., Health and environmental assessment of MTBE: summary and recommendations, UC TSR&TP Report to the Governor of California, Toxics Research and Teaching Program, University of California, Davis, CA, 1998b.

Keller, A.A., Sandall, O.C., Rinker, R.G., Mitani, M.M., Bierwagen, B.G., and Michael, M.J., Cost and performance evaluation of treatment technologies for MTBE-contaminated water, in *Health and Environmental Assessment of MTBE,* Vol. IV, Last, J., Ed., Toxics Research and Teaching Program, Davis, CA, 1998c, pp. 58–92.

Koshland, C.P., Sawyer, R.F., Lucas, D., and Franklin, P., Evaluation of automotive MTBE combustion byproducts, in *Health and Environmental Assessment of MTBE,* Vol. 2, Last, J., Ed., Toxics Research and Teaching Program, Davis, CA, 1998, pp. 22–35.

Krupnick, A. and Walls, M., The cost effectiveness of methanol for reducing motor vehicle emissions and urban ozone, *J. Pol. Anal. Manage.,* 11(3), 373–396, 1992.

Morris, R.D., Naumova, E.N., and Munasinghe, R.L., Ambient air pollution and hospitalization for congestive heart failure among elderly people in seven large US cities, *Am. J. Public Health,* 85(10), 1361–1365, 1995.

NSTC, Interagency assessment of oxygenated fuels, Executive Office of the President of the U.S., National Science and Technology Council, Committee on Environment and Natural Resources, 1997.

Reuter, J.E., Allen, B.C., and Goldman, C.R., Methyl *tert*-butyl ether in surface drinking water supplies, in *Health and Environmental Assessment of MTBE,* Vol. 3, Last, J., Ed., UC Toxics Research and Teaching Program, Davis, CA, 1998, pp. 102–125.

Rodriguez, R., MTBE in groundwater and the impact on the City of Santa Monica drinking water supply, in *Technical Papers of the 13th Annual Environmental Management and Technology Conference West*, Nov. 4–6, 1997.

Schwing, R., Southwark, B., Von Buseck, C., and Jackson, C., Benefit-cost analysis of automotive emission reductions, *J. Environ. Econ. Manage.*, 7(1), 44–64, 1980.

Shabman, L. and Batie, S., Mitigating damages from coastal wetlands development: policy, economics and financing, *Mar. Resource Econ.*, 4(3), 227–248, 1987.

USEPA, Health assessment document for acetaldehyde, EPA/600/8-86-015A, U.S. Environmental Protection Agency, Environmental Criteria and Assessment Office, Office of Health and Environmental Assessment, Office of Research and Development, Cincinnati, OH, 1987.

USEPA, Health and environmental effects profile for formaldehyde, EPA/600/x-85/362, U.S. Environmental Protection Agency, Environmental Criteria and Assessment Office, Office of Health and Environmental Assessment, Office of Research and Development, Cincinnati, OH, 1988.

USEPA, Integrated Risk Information System (IRIS) on formaldehyde, U.S. Environmental Protection Agency, Environmental Criteria and Assessment Office, Office of Health and Environmental Assessment, Office of Research and Development, Cincinnati, OH, 1993a.

USEPA, Integrated Risk Information System (IRIS) on acetaldehyde, U.S. Environmental Protection Agency, Environmental Criteria and Assessment Office, Office of Health and Environmental Assessment, Office of Research and Development, Cincinnati, OH, 1993b.

USEPA, Integrated Risk Information System (IRIS) on toluene, U.S. Environmental Protection Agency, Environmental Criteria and Assessment Office, Office of Health and Environmental Assessment, Office of Research and Development, Cincinnati, OH, 1993c.

USEPA, The benefits and costs of the Clean Air Act, 1970 to 1990, Environmental Protection Agency, Washington, D.C., 1997.

USEPA, Air quality criteria for carbon monoxide, EPA Report EPA-600/P-99/001, Office of Research and Development, U.S. Environmental Protection Agency, Washington, D.C., 1999.

Werner, I. and Hinton, D.E., Toxicity of MTBE to freshwater organisms, in *Health and Environmental Assessment of MTBE*, Vol. 4, Last, J., Ed., Toxics Research and Teaching Program, Davis, CA, 1998, pp. 10–23.

Index

A

Abalone, 549–554
Abbey, Edward, 1032
Abdullah, Pauzi, 1361–1370, 1395–1419
Abortions, 414
Abu Yusuf, Muhammad, 1361–1370, 1395–1419
Acetaldehyde, 1464–1465
Acevedo, Heiner, 449–458
Acid mine drainage, 1253, 1265
Acid rain, 23, 952
Adaptive restoration, 167–174, 243
Addo Elephant National Park, 359–374
Adelgid, balsam woolly, 972
Adger, W. Neil, 389–402
Advocacy science, 124
Afghanistan, 13
Africa, *see also* specific countries
    agriculture in, 318, 338–339
    biodiversity decline, 321
    cattle breeds, 322
    extinctions in, 45
    life expectancy level, 70
    meningococcal meningitis epidemics, 78
    Rift Valley fever, 81, 610
    river water in, 336
    rural development in, 308–309
    schistosomiasis infection rate, 611
Agency for Toxic Substances and Disease Registry, 1464
Agenda 21 initiatives, *see also* The Rio Declaration on
        Environment and Development
    ecological footprint analysis in, 102
    for ecological health, 8
    indicators for, 835
    industry obligations, 854
    in Mexico, 732, 1143
    water management, 1143
Agriculture
    agrobiodiversity, 317–329, 336, 1002–1003, 1009
    agroecology, 999–1009, 1209–1225
    agroforestry, 326, 1005
    biologically integrated farming systems, 995
    biotechnology for, 1000–1002
    in California, 1059–1062
    in Catskill–Delaware watershed, 1332
    in Central Valley watershed, 981–989, 993–996

    chem-fallow, 1158
    clean farming, 981, 986
    in Clear Lake watershed, 1256–1257
    climate sensitivity, 81
    conservation tillage practices, 979–980, 993–996,
        1158–1159
    cover crops, 995, 1005–1006
    damage to ecosystem from, 50
    desertification after, 606
    direct seeding methods, 1158
    dry farming, 1033
    eradication programs for, *see* Eradication
    experimental stations, 793–794
    field drainage systems, 1022–1023
    forest impact on, 923
    in the Great Lakes basin, 714–715
    high-yield varieties, 320
    human health impact, 612, 789–790, 793–797
    industrialization of, 1157
    input substitution approach to, 1000–1001
    insect control, 474; *see also* Pest management;
        Pesticides
    intercropping, 326, 1005–1006
    land management, 1062–1063
    in Langat Basin, 1363, 1365
    law of minimum, 1000, 1217
    modeling of pollution from, 648
    modeling of South Florida, 1317
    monoculture consequences, 1000–1002
    nitrate migration to watershed, 426
    oil seed crops, 1150
    in the Palliser Triangle, 1155–1160
    in PECOS area, 1150, 1157, 1169–1179
    pulse crops, 1150, 1159
    vs. ranching, 1033
    release of sequestered carbon by, 468–469
    "slash-and-burn," 13–14, 46, 468, 825–826, 887,
        1044–1045
    summer fallow, *see* Summer fallow
    surface area, global, 467
    sustainability of, 340–342, 999–1009
    tillage/land preparation for, 993–995, 1158–1159, 1218
    transgenic crops, 1001–1002
    water quality impact, 1332–1338
Air pollution
    acid rain, 23, 952
    Air Quality Objective, 57

G

Health, human, *see* Human health
Health impact assessments, 843
Hecq, Walter, 741–755
Hedonic pricing method, 132
Heggem, Daniel, 403–411
Helsinki Process, 835, 955
Hembree, K.J., 993–998
Hemlock, western, 672
Hemolytic uremic syndrome, 799
Hemorrhagic fevers, 609
Henderson Island, 45
Hepatitis, 18, 608
Heptachlor, 1177–1178
Herbicides, *see also* Pesticides
    in chem-fallow agriculture, 1158
    for *Hydrilla verticillata* infestations, 1250
    PECOS exposure study, 1170–1179
    trade names, 1174
Heron
    black-crowned night (*Nycticorax nycticorax*), 1126,
        1128, 1130
    great blue (*Ardea herodias*), 1241
    green (*Butorides virescens*), 1126, 1130
Herrero, E.V., 993–998
Hessburg, Paul, 661–692
Hexachlorocyclohexane, 1128
Heyvaert, Alan, 1283–1298
Hidden flows, 907, 910–915; *see also* Pollution
Hierarchy, biological organization, 43
High-yielding varieties, 320
Hippopotamus, 45
Hirvonen, Harry, 949–958
Hitch, Clear Lake (*Lavinia exilicauda*), 1256, 1261
HIV/AIDS
    Cyptosporidium parvum and, 799
    prevention strategies, 849, 851–854
    simian virus related to, 608
Holdridge World Life Zone Classification System, 449–450
Holland, *see* The Netherlands
Homeostasis, 999
Homestead Acts, 1033
Honey, 394
Hoover Dam, 1077
Hopkins, John, 981–991
Hopkins Marine Station, California, 478–483
Horses
    extinction of species of, 45, 46
    genetic uniformity of, 322
    grazing activities, 1050–1051
    western equine encephalomyelitis host, 87
    wildlife immunocontraception program, 148
Horvatin, Paul, 703–720
Hotelling model, 209
Houck, Ann, 1273–1281
Human Development Index, 99, 892
Human Development Report, 892
The Humane Society of the United States, 145–149
Human health
    antinuclear antibody assay, 1173–1174
    biomedical model, 843–854, 865
    capabilities approach, 895

components of, 894
definition of, 843, 845, 856–857, 868, 895
deterioration continuum, 885–886
economic growth and, worldviews, 894–895
an ecosystem approach to, 603–614, 867–868, 870, 881
impact assessment criteria, 894
indicators for, 715
integrated approach to, 857–859
mercury contaminated fish, 1278
mining issues, 825–827, 896–897
neuropsychological function tests, 1173–1174
pulmonary function tests, 1172–1173
risk assessment, 886
senior care, 1161–1167
social goals, 895
tools for assessing, 900, 902–903
wildlife disease as indicators for, 1111–1113
Human Poverty Index, 892
Humboldt, Alexander von, 731
Hungary, 49
Hunter-gatherer lifestyle, 45–46
Hunting, 147, 1039
Huntsinger, Lynn, 1047–1053
Hurricanes, *see also* Cyclones; Typhoons
    factor, climate, 81
    Florida growth impact, 1323
    wetland loss impact, 158
Hydrilla (*Hydrilla verticillata*), 544, 1249–1250, 1261
Hydrologic Unit Code, 404–405
Hyena, 360–361

I

Ibis, white-faced (*Plegadis chihi*), 760
Idaho
    American Fisheries Society chapter, 125
    Columbia basin study, 666–689
    lichen richness, 968
Idrus, Shahruddin, 1361–1370
Illinois, 78, 1439–1440
Illness, 15
Ilmenite, 909
Impact Analysis for Planning, 1311–1323
Improved land, 1186, 1192–1194
Index of Biological Integrity, 416–427
Index of Biotic Integrity, 271–274, 618–621
Index of change, 589–591
Index of Hydrologic Alteration, 295
Index of Sustainable Economic Welfare, 99
India
    agriculture case study, 339
    biodiversity on large farms, 327
    coral reefs, 498
    ecological footprint analysis of, 49
    Goa case study, 893, 897–900
    high-yield varieties impact, 320
    natural capital intervention programs, 221
    rice famine, 322

X

Y